GW00472464

Complete Study Pack For

Engineering Entrances

Objective

PHYSICS

Volume 1

DC Pandey

arihant

ARIHANT PRAKASHAN (SERIES), MEERUT

arihant

Arihant Prakashan (Series), Meerut

Administrative & Production Offices

Regd. Office

'Ramchhaya' 4577/15, Agarwal Road, Darya Ganj, New Delhi -110002
Tele: 011- 47630600, 43518550

Head Office

Kalindi, TP Nagar, Meerut (UP) - 250002
Tel: 0121-7156203, 7156204

Sales & Support Offices

Agra, Ahmedabad, Bengaluru, Bareilly, Chennai, Delhi, Guwahati, Hyderabad, Jaipur, Jhansi, Kolkata, Lucknow, Nagpur & Pune.

ISBN 978-93-26193-34-4

PO No : TXT-XX-XXXXXXX-X-XX

Published by Arihant Publications (India) Ltd.

For further information about the books published by Arihant, log on to www.arihantbooks.com or e-mail at info@arihantbooks.com

Follow us on

PREFACE

Engineering offers the most exciting and fulfilling of careers. As a Engineer you can find satisfaction by serving the society through your knowledge of technology. Although the number of Engineering colleges imparting quality education and training has significantly increased after independence in the country, but simultaneous increase in the number of serious aspirants has made the competition difficult, it is no longer easy to get a seat in a prestigious Engineering college today.

For success, you require an objective approach of the study. This does not mean you 'prepare' yourself for just 'objective questions'. Objective Approach means more than that. It could be defined as that approach through which a student is able to master the concepts of the subject and also the skills required to tackle the questions asked in different entrances such as JEE Main & Advanced, as well other regional Engineering entrances. These two-volume books on Physics 'Objective Physics (Vol.1 & 2)' fill the needs of such books in the market in Physics and are borne out of my experience of teaching Physics to Engineering aspirants.

The plan of the presentation of the subject matter in the books is as follows

- The whole chapter has been divided under logical topic heads to cover the syllabi of JEE Main & Advanced and various Engineering entrances in India.
- The Text develops the concepts in an easy going manner, taking the help of the examples from the day-to-day life.
- Important points of the topics have been highlighted in the text. Under Notes, some extra points regarding the topics have been given to enrich the students.
- The Solved Examples make the students learn the basic problem solving skills in Physics. In some of the example problems How to Proceed has been provided to make the students skilled in systematically tackling the problems.
- An Extra Knowledge Points frequently follows the discussion of few topics which includes the important theorems, results and formulae to increase the grasp of the subject matter.
- Additional Solved Examples given at the end of text part to make the students practice of the complete chapter as a whole.
- Assignments at the end of the chapters have been divided into two parts Objective Questions and Entrance Corner.
- The answers / solutions to all the assignment questions have been provided.
- The Objective Questions have been divided into two levels. Level 1 contains elementary MCQs while Level 2 has MCQs which are relatively tougher and take the students to a level required for various Engineering Entrances require in the present scenario.
- Entrance Gallery and includes the previous years' questions asked in various Engineering entrances. At the end of the book, JEE Main & Advanced & Other Regional Entrances Solved Papers have been given.

I am extremely thankful to Mrs. Sarita Pandey, Mr. Anoop Dhyani & Mr. Shubham Sharma of their endless effort during the Project. I would open-heartedly welcome the suggestions for the further improvements of this book (Vol.1) from the students and teachers.

DC Pandey

CONTENTS

DEDICATION

This book is dedicated to my honourable grandfather

(LATE) SH. PITAMBER PANDEY

a Kumaoni poet; resident of village Dhaura (Almora) Uttarakhand

1

Units, Dimensions and Error Analysis

1.1 Introduction

To measure a physical quantity, we need some standard units of that quantity. The measurement of the quantity is mentioned in two parts, the first part gives how many times of the standard unit and the second part gives the name of the unit. Thus, suppose I say that the length of this wire is 5 m. The numeric part 5 says that it is 5 times of the unit of length and the second part metre says that unit chosen here is metre.

Chapter Snapshot

- Introduction
- The International System of Units
- Error Analysis
- Significant Figures
- Dimensional Analysis

Fundamental and Derived Units

There are a large number of physical quantities and every quantity needs a unit. However, not all the quantities are independent. e.g. if a unit of length is defined, a unit of volume is automatically obtained. Thus we can define a set of fundamental quantities and all other quantities may be expressed in terms of the fundamental quantities. Fundamental quantities are only seven in numbers. Units of all other quantities can be expressed in terms of the units of these seven quantities by multiplication or division.

Many different choices can be made for the fundamental quantities. For example, if we take length and time as the fundamental quantities then speed is a derived quantity and if we take speed and time as fundamental quantities then length is a derived quantity.

Several systems of units are in use over the world. The units defined for the fundamental quantities are called fundamental units and those obtained for derived quantities are called the derived units.

Extra Edge

- Fundamental units are also called the base units.
- Both fundamental units (or base units) and derived units are jointly called the system of units.

1.2 The International System of Units

Three most commonly used systems of units used from the earlier times are the CGS system, the FPS (or British) system and the MKS system. The fundamental units for length, mass and time in these systems are as follows:

Table 1.1 System of units

System	Fundamental unit of length	Fundamental unit of mass	Fundamental unit of time
CGS	centimetre	gram	second
FPS	foot	pound	second
MKS	metre	kilogram	second

SI Units

In 1971, General Conference on Weights and Measures held its meeting and decided a system of units which is known as the International System of Units. It is abbreviated as SI from the French name Le Systeme International d' unites. This system is widely used throughout the world. Table below gives the seven fundamental quantities and their SI units.

Table 1.2 Fundamental quantities and their SI units

S.No.	Quantity	SI unit	Symbol
1.	Length	metre	m
2.	Mass	kilogram	kg
3.	Time	second	s
4.	Electric current	ampere	A
5.	Thermodynamic temperature	kelvin	K
6.	Amount of substance	mole	mol
7.	Luminous intensity	candela	cd

Definitions of Seven SI Units

(i) **Metre** 1 m = length of the path travelled by light in vacuum during a time interval of $\frac{1}{299792458}$ of a second.

(ii) **Second** $1\,s = 9192631770$ time periods of a particular radiation from cesium-133 atom.

(iii) **Kilogram** 1 kg = mass of the international prototype of the kilogram (a platinum-iridium alloy cylinder) kept at International Bureau of Weights and Measures, at Sevres, near Paris, France.

(iv) **Ampere** It is the current which when flows through two infinitely long straight conductors of negligible cross-section placed at a distance of one metre in vacuum produces a force of 2×10^{-7} N/m between them.

(v) **Kelvin** 1 K = 1/273.16 part of the thermodynamic temperature of triple point of water.

(vi) **Mole** It is the amount of substance of a system which contains as many elementary particles (atoms, molecules, ions, etc.) as there are atoms in 12 g of carbon-12.

(vii) **Candela** The candela is the luminous intensity, in a given direction, of a source that emits monochromatic radiation of frequency 540×10^{12} hertz and that has a radiant intensity in that direction of 1/683 watt per steradian.

Extra Edge

- **Two Supplementary Units** Besides the seven base units, there are two supplementary units namely plane angle and solid angle. Their units are radian (rad) and steradian (sr).
  (i) **Plane Angle** ($d\theta$) Plane angle ($d\theta$) is the ratio of arc length ds to the radius r.

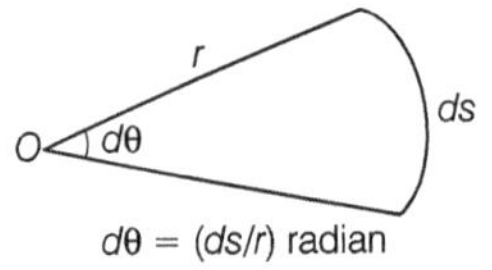

Fig. 1.1

  (ii) **Solid Angle** ($d\Omega$) Solid angle $d\Omega$ is the ratio of the intercepted area dA of the spherical surface described about the apex O as the centre to the square of its radius r.

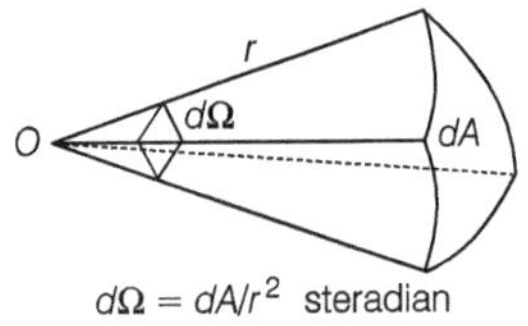

Fig. 1.2

Table 1.3

Power of 10	Prefix	Symbol
6	mega	M
3	kilo	k
–2	centi	c
–3	milli	m
–6	micro	μ
–9	nano	n
–12	pico	p

Table 1.4

Size of object or distance	Length (m)
Distance of the boundary of observable universe	10^{26}
Size of our galaxy	10^{21}
Distance of the sun from the earth	10^{11}
Distance of the moon from the earth	10^{8}
Radius of the earth	10^{7}
Wavelength of light	10^{-7}
Size of hydrogen atom	10^{-10}
Size of atomic nucleus	10^{-14}
Size of a proton	10^{-15}

- 1 parsec = 3.08×10^{16} m (Parsec is the distance at which average radius of earth's orbit subtends an angle of 1 arc second)
- 1 light year = 1 ly = 9.46×10^{15} m (Distance that light travels with velocity of 3×10^8 ms^{-1} in 1 year)
- 1 astronomical unit = 1 AU (Average distance of the sun from the earth) = 1.496×10^{11} m
- 1 angstrom = 1 Å = 10^{-10} m
- 1 fermi = 1 f = 10^{-15} m

Table 1.5 Measurement of mass

Object	**Mass** (kg)
Sun	10^{30}
Earth	10^{25}
Moon	10^{23}
Proton	10^{-27}
Electron	10^{-30}

Table 1.6 Measurement of time

Event	**Time interval** (s)
Age of the universe	10^{17}
Revolution period of the earth	10^7
Rotation period of the earth	10^5
Travel time for light from the sun to the earth	10^2
Travel time for light from the moon to the earth	10^0
Time between successive human heart beats	10^0
Life time of an excited state of an atom	10^{-8}
Period of light wave	10^{-15}
Life span of most unstable particle	10^{-24}

1.3 Error Analysis

No measurement is perfect, as the errors involved in a measurement cannot be removed completely. Measured value is always somewhat different from the true value. This difference is called an error.

Errors can be classified in two ways. First classification is based on the cause of error. Systematic and random errors fall in this group. Second classification is based on the magnitude of error. Absolute error, mean absolute error and relative (or fractional) error lie in this group.
Now, let us discuss them separately.

Systematic Error

These are the errors whose causes are known to us. Such errors can therefore be minimised.
Following are few causes of these errors:

(a) Instrumental errors may be due to erroneous instruments. These errors can be reduced by using more accurate instruments and applying zero correction, when required.

(b) Sometimes errors arise on account of ignoring certain facts. For example, in measuring time period of simple pendulum error may creep because no consideration is taken of air resistance. These errors can be reduced by applying proper corrections to the formula used.

(c) Change in temperature, pressure, humidity, etc., may also sometimes cause errors in the result. Relevant corrections can be made to minimise their effects.

Random Error

The causes of random errors are not known. Hence, it is not possible to remove them completely. These errors may arise due to a variety of reasons.

For example, the reading of a sensitive beam balance may change by the vibrations caused in the building due to persons moving in the laboratory or vehicles running nearby. The random errors can be minimised by repeating the observation a large number of times and taking the arithmetic mean of all the observations. The mean value would be very close to the most accurate reading.

Thus, $$a_{\text{mean}} = \frac{a_1 + a_2 + \ldots + a_n}{n}$$

Absolute Error

The difference between the true value and the measured value of a quantity is called an absolute error. Usually, the mean value a_m is taken as the true value. So, if

$$a_m = \frac{a_1 + a_2 + \ldots + a_n}{n},$$

then by definition, absolute errors in the measured values of the quantity are,

$$\Delta a_1 = a_m - a_1$$
$$\Delta a_2 = a_m - a_2$$
$$\ldots \quad \ldots \quad \ldots$$
$$\Delta a_n = a_m - a_n$$

Absolute error may be positive or negative.

Mean Absolute Error

Arithmetic mean of the magnitudes of absolute errors in all the measurements is called the mean absolute error. Thus,

$$\Delta a_{\text{mean}} = \frac{|\Delta a_1| + |\Delta a_2| + \ldots + |\Delta a_n|}{n}$$

The final result of measurement can be written as

$$a = a_m \pm \Delta a_{\text{mean}}$$

This implies that value of a is likely to lie between $a_m + \Delta a_{\text{mean}}$ and $a_m - \Delta a_{\text{mean}}$.

Relative or Fractional Error

The ratio of mean absolute error to the mean value of the quantity measured is called relative or fractional error. Thus,

$$\text{Relative error} = \frac{\Delta a_{\text{mean}}}{a_m}$$

Relative error expressed in percentage is called as the percentage error, i.e.

$$\text{Percentage error} = \frac{\Delta a_{\text{mean}}}{a_m} \times 100$$

Example 1.1 *The diameters of a wire as measured by screw gauge were found to be 2.620, 2.625, 2.630, 2.628 and 2.626 cm. Calculate:*

(a) mean value of diameter.
(b) absolute error in each measurement.
(c) mean absolute error.
(d) fractional error.
(e) percentage error.
(f) express the result in terms of percentage error.

Sol. (a) Mean value of diameter

$$a_m = \frac{2.620 + 2.625 + 2.630 + 2.628 + 2.626}{5}$$
$$= 2.6258 \text{ cm}$$
$$= 2.626 \text{ cm (rounding off to three decimal places)}$$

(b) Taking a_m as the true value, the absolute errors in different observations are,

$$\Delta a_1 = 2.626 - 2.620 = +0.006 \text{ cm}$$
$$\Delta a_2 = 2.626 - 2.625 = +0.001 \text{ cm}$$
$$\Delta a_3 = 2.626 - 2.630 = -0.004 \text{ cm}$$
$$\Delta a_4 = 2.626 - 2.628 = -0.002 \text{ cm}$$
$$\Delta a_5 = 2.626 - 2.626 = 0.000 \text{ cm}$$

(c) Mean absolute error,

$$\Delta a_{\text{mean}} = \frac{|\Delta a_1| + |\Delta a_2| + |\Delta a_3| + |\Delta a_4| + |\Delta a_5|}{5}$$
$$= \frac{0.006 + 0.001 + 0.004 + 0.002 + 0.000}{5}$$
$$= 0.0026$$
$$= 0.003 \quad \text{(rounding off to three decimal places)}$$

(d) Fractional error $= \pm \dfrac{\Delta a_{\text{mean}}}{a_m}$

$$= \frac{\pm 0.003}{2.626} = \pm 0.001$$

(e) Percentage error $= \pm 0.001 \times 100 = \pm 0.1\%$

(f) Diameter of wire can be written as $d = 2.626 \pm 0.1\%$

Combination of Errors

(i) **Errors in sum or difference** Let $x = a \pm b$.

Further, let Δa is the absolute error in the measurement of a, Δb is the absolute error in the measurement of b and Δx is the absolute error in the measurement of x.

Then, $x + \Delta x = (a \pm \Delta a) \pm (b \pm \Delta b)$

$$= (a \pm b) \pm (\pm \Delta a \pm \Delta b)$$
$$= x \pm (\pm \Delta a \pm \Delta b)$$

or $\Delta x = \pm \Delta a \pm \Delta b$

The four possible values of Δx are $(\Delta a - \Delta b)$, $(\Delta a + \Delta b)$, $(-\Delta a - \Delta b)$ and $(-\Delta a + \Delta b)$.

Therefore, the maximum absolute error in x is,

$$\Delta x = \pm (\Delta a + \Delta b)$$

i.e. the maximum absolute error in sum and difference of two quantities is equal to sum of the absolute errors in the individual quantities.

Example 1.2 *The volumes of two bodies are measured to be* $V_1 = (10.2 \pm 0.02)\, cm^3$ *and* $V_2 = (6.4 \pm 0.01)\, cm^3$. *Calculate sum and difference in volumes with error limits.*

Sol. $V_1 = (10.2 \pm 0.02)\ \text{cm}^3$

and $V_2 = (6.4 \pm 0.01)\ \text{cm}^3$

$$\Delta V = \pm (\Delta V_1 + \Delta V_2)$$
$$= \pm (0.02 + 0.01)\ \text{cm}^3 = \pm 0.03\ \text{cm}^3$$
$$V_1 + V_2 = (10.2 + 6.4)\ \text{cm}^3 = 16.6\ \text{cm}^3$$

and $V_1 - V_2 = (102 - 6.4)\ \text{cm}^3 = 3.8\ \text{cm}^3$

Hence, sum of volumes $= (16.6 \pm 0.03)\ \text{cm}^3$

and difference of volumes $= (3.8 \pm 0.03)\ \text{cm}^3$

(ii) **Errors in a product** Let $x = ab$

Then, $(x \pm \Delta x) = (a \pm \Delta a)(b \pm \Delta b)$

or $$x\left(1 \pm \frac{\Delta x}{x}\right) = ab\left(1 \pm \frac{\Delta a}{a}\right)\left(1 \pm \frac{\Delta b}{b}\right)$$

or $$1 \pm \frac{\Delta x}{x} = 1 \pm \frac{\Delta b}{b} \pm \frac{\Delta a}{a} \pm \frac{\Delta a}{a} \cdot \frac{\Delta b}{b} \quad (\text{as } x = ab)$$

or $$\pm \frac{\Delta x}{x} = \pm \frac{\Delta a}{a} \pm \frac{\Delta b}{b} \pm \frac{\Delta a}{a} \cdot \frac{\Delta b}{b}$$

Here, $\dfrac{\Delta a}{a} \cdot \dfrac{\Delta b}{b}$ is a small quantity, so can be neglected.

Hence, $$\pm \frac{\Delta x}{x} = \pm \frac{\Delta a}{a} \pm \frac{\Delta b}{b}$$

Possible values of $\dfrac{\Delta x}{x}$ are $\left(\dfrac{\Delta a}{a} + \dfrac{\Delta b}{b}\right)$, $\left(\dfrac{\Delta a}{a} - \dfrac{\Delta b}{b}\right)$, $\left(-\dfrac{\Delta a}{a} + \dfrac{\Delta b}{b}\right)$ and $\left(-\dfrac{\Delta a}{a} - \dfrac{\Delta b}{b}\right)$.

Hence, maximum possible value of

$$\frac{\Delta x}{x} = \pm\left(\frac{\Delta a}{a} + \frac{\Delta b}{b}\right)$$

Therefore, maximum fractional error in the product of two (or more) quantities is equal to the sum of fractional errors in the individual quantities.

(iii) **Errors in division** Let $x = \frac{a}{b}$

Then, $$x \pm \Delta x = \frac{a \pm \Delta a}{b \pm \Delta b}$$

or $$x\left(1 \pm \frac{\Delta x}{x}\right) = \frac{a\left(1 \pm \frac{\Delta a}{a}\right)}{b\left(1 \pm \frac{\Delta b}{b}\right)}$$

or $$\left(1 \pm \frac{\Delta x}{x}\right) = \left(1 \pm \frac{\Delta a}{a}\right)\left(1 \pm \frac{\Delta b}{b}\right)^{-1} \quad \left(\text{as } x = \frac{a}{b}\right)$$

As $\frac{\Delta b}{b} << 1$, so expanding binomially, we get

$$\left(1 \pm \frac{\Delta x}{x}\right) = \left(1 \pm \frac{\Delta a}{a}\right)\left(1 \mp \frac{\Delta b}{b}\right)$$

or $$1 \pm \frac{\Delta x}{x} = 1 \pm \frac{\Delta a}{a} \mp \frac{\Delta b}{b} - \frac{\Delta a}{a} \cdot \frac{\Delta b}{b}$$

Here, $\frac{\Delta a}{a} \cdot \frac{\Delta b}{b}$ is small quantity, so it can be neglected. Therefore,

$$\pm \frac{\Delta x}{x} = \pm \frac{\Delta a}{a} \mp \frac{\Delta b}{b}$$

Possible values of $\frac{\Delta x}{x}$ are $\left(\frac{\Delta a}{a} - \frac{\Delta b}{b}\right)$, $\left(\frac{\Delta a}{a} + \frac{\Delta b}{b}\right)$, $\left(-\frac{\Delta a}{a} - \frac{\Delta b}{b}\right)$ and $\left(-\frac{\Delta a}{a} + \frac{\Delta b}{b}\right)$.

Therefore, the maximum value of $\frac{\Delta x}{x}$

$$= \pm\left(\frac{\Delta a}{a} + \frac{\Delta b}{b}\right)$$

or the maximum value of fractional error in division of two quantities is equal to the sum of fractional errors in the individual quantities.

(iv) **Errors in quantity raised to some power**

Let $$x = \frac{a^n}{b^m}$$

Then, $\ln(x) = n \ln(a) - m \ln(b)$

Differentiating both sides, we get

$$\frac{dx}{x} = n \cdot \frac{da}{a} - m\frac{db}{b}$$

In terms of fractional error, we may write

$$\pm \frac{\Delta x}{x} = \pm n\frac{\Delta a}{a} \mp m\frac{\Delta b}{b}$$

Therefore, maximum value of

$$\frac{\Delta x}{x} = \pm\left(n\frac{\Delta a}{a} + m\frac{\Delta b}{b}\right)$$

Example 1.3 *The mass and density of a solid sphere are measured to be* (12.4 ± 0.1) *kg and* (4.6 ± 0.2) *kg/m*^{3}. *Calculate the volume of the sphere with error limits.*

Sol. Here, $m \pm \Delta m = (12.4 \pm 0.1)$ kg

and $\rho \pm \Delta\rho = (4.6 \pm 0.2)\text{kg/m}^3$

Volume, $V = \frac{m}{\rho} = \frac{12.4}{4.6} = 2.69 \text{ m}^3 = 2.7 \text{ m}^3$

(rounding off to one decimal place)

Now, $$\frac{\Delta V}{V} = \pm\left(\frac{\Delta m}{m} + \frac{\Delta\rho}{\rho}\right)$$

or $$\Delta V = \pm\left(\frac{\Delta m}{m} + \frac{\Delta\rho}{\rho}\right) \times V$$

$$= \pm\left(\frac{0.1}{12.4} + \frac{0.2}{4.6}\right) \times 2.7$$

$$= \pm 0.14$$

$\therefore$ $V \pm \Delta V = (2.7 \pm 0.14)\text{ m}^3$

Example 1.4 *Calculate percentage error in determination of time period of a pendulum,*

$$T = 2\pi\sqrt{\frac{l}{g}}$$

where, l and g are measured with ±1% *and* ±2% *errors.*

Sol. $$\frac{\Delta T}{T} \times 100 = \pm\left(\frac{1}{2} \times \frac{\Delta \ell}{\ell} \times 100 + \frac{1}{2} \times \frac{\Delta g}{g} \times 100\right)$$

$$= \pm\left(\frac{1}{2} \times 1 + \frac{1}{2} \times 2\right) = \pm 1.5\%$$

Example 1.5 *The period of oscillation of a simple pendulum is* $T = 2\pi\sqrt{L/g}$. *Measured value of L is* 20.0 *known to* 1mm *accuracy and time for* 100 *oscillations of the pendulum is found to be* 90 *s using a wrist watch of* 1s *resolution. What is the accuracy in the determination of g?*

Sol. Time period of oscillation,

$$T = \frac{t}{n} \Rightarrow \Delta T = \frac{\Delta t}{n}$$

Therefore, $$\frac{\Delta T}{T} = \frac{\Delta t/n}{t/n} = \frac{\Delta t}{t}$$

Now, from the equation $T = 2\pi\sqrt{\frac{L}{g}}$

We can see that $$g = (4\pi^2)\left(\frac{L}{T^2}\right)$$

$\therefore$ Percentage error in $g = 100\left(\frac{\Delta g}{g}\right)$

$$= 100\left(\frac{\Delta L}{L}\right) + 2 \times 100\left(\frac{\Delta T}{T}\right)$$

$$= \frac{0.1}{20.0} \times 100 + 2 \times 100 \times \left(\frac{1}{90}\right)$$

$$= 2.72\ \%$$

Example 1.6 *Calculate focal length of a spherical mirror from the following observations. Object distance $u = (50.1 \pm 0.5)$ cm and image distance $v = (20.1 \pm 0.2)$ cm.*

Sol. Mirror formula is given by $\frac{1}{f} = \frac{1}{v} + \frac{1}{u}$

or $$f = \frac{uv}{u+v} = \frac{(50.1)(20.1)}{(50.1+20.1)} = 14.3 \text{ cm}$$

Also, $$\frac{\Delta f}{f} = \pm\left[\frac{\Delta u}{u} + \frac{\Delta v}{v} + \frac{\Delta u + \Delta v}{u+v}\right]$$

$$= \pm\left[\frac{0.5}{50.1} + \frac{0.2}{20.1} + \frac{0.5+0.2}{50.1+20.1}\right]$$

$$= \pm[0.00998 + 0.00995 + 0.00997]$$

$$= \pm(0.0299)$$

$$\therefore \quad \Delta f = 0.0299 \times 14.3 = 0.428 = 0.4 \text{ cm}$$

$$\therefore \quad f = (14.3 \pm 0.4) \text{ cm}$$

Extra Edge

- **Least Count** The smallest value that can be measured by a measuring instrument is called its least count. For example, least count of a metre scale is 0.1 cm of a vernier calliper is 0.01 cm and of a screw gauge (or a spherometer) is 0.001 cm.
- **Least Count Error** Least count error is associated with the resolution of the error. Using instruments of higher precision, we can reduce the least count error. For example, least count error by a vernier calliper is definitely less than the least count error by a metre scale.

1.4 Significant Figures

Significant figures in the measured value of a physical quantity tell the number of digits in which we have confidence. Larger the number of significant figures obtained in a measurement, greater is the accuracy of the measurement.

"*All accurately known digits in a measurement plus the first uncertain digit together form significant figures.*"

For example, when we measure the length of a straight line using a metre scale and it lies between 7.4 cm and 7.5 cm, we may estimate it as $l = 7.43$ cm. This expression has three significant figures out of these 7 and 4 are precisely known but the last digit 3 is only approximately known.

Note Points

- As the significant figures indicate the precision of measurement, it depends on the least count of the measuring instrument.
- In different units, the number of significant figures are same. For example, an area 2.4 m^2 has two significant figures. In CGS system, this area can be written as 24000 cm^2 or 2.4×10^4 cm^2. In this system of unit also, significant figures are two (digits 2 and 4).

Rules for Counting Significant Figures

For counting significant figures, we use the following rules:

Rule 1 All non-zero digits are significant. For example, $x = 2567$ has four significant figures.

Rule 2 The zeros appearing between two non-zero digits are counted in significant figures, no matter where the decimal point is, if any.
For example, 6.028 has 4 significant figures.

Rule 3 If the number is less than 1, the zero(s) on the right of decimal point but to the left of first non-zero digit are not significant.

For example, 0.0042 has two significant digits.

Rule 4 The terminal or trailing zero(s) in a number without a decimal point are not significant. Thus, 426 m = 42600 cm = 426000 mm has three significant figures.

Rule 5 In a number with decimal, zeros to the right of last non-zero digit are significant.

For example, 4.600 or 0.002300 have four significant figures each.

Rule 6 A choice of change of different units does not change the number of significant digits or figures in a measurement.

For example, the length 7.03 cm has three significant figures. But in different units, the same value can be written as 0.0703 m or 70.3 mm. All these measurements have the same number of significant figures (digits 7, 0 and 3) namely three.

This shows that location of decimal point is of no consequence in determining the number of significant figures.

Point of Confusion and its Remedy

Suppose we change the units, then we will write

$$2.30 \text{ m} = 23.0 \text{ cm} = 2300 \text{ mm} = 0.00230 \text{ km}$$

When we are writing 2300 mm, then from Rule -4, we would conclude erroneously that the number has two significant figures, while in fact it has three significant figures and a mere change of units cannot change the number of significant figures.

To remove such ambiguities in determining the number of significant figures, apply following rule:

Rule 7 The power of 10 is irrelevant in determination of significant figures.
For example, in the measurements

$$230 \text{ m} = 2.30 \times 10^2 \text{ cm}$$
$$= 2.30 \times 10^3 \text{ mm}$$
$$= 2.30 \times 10^{-3} \text{ km}$$

Table 1.7 Measurement of significant figures

Measured value	Number of significant figures	Rule number
7285	4	1
7.0025	5	2
0.00078	2	3
78200 cm	3	4
7.20	3	5
0.006300	4	5
2.30×10^{-3} km	3	7

Rounding Off a Digit

Following are the rules for rounding off a measurement

***Rule* 1** If the number lying to the right of cut-off digit is less than 5, then the cut-off digit is retained as such. However, if it is more than 5, then the cut-off digit is increased by 1.

For example, $x = 6.24$ is rounded off to 6.2 to two significant digits and $x = 5.328$ is rounded off to 5.33 to three significant digits.

***Rule* 2** If the insignificant digit to be dropped is 5, then the rule is

(i) if the preceding digit is even, the insignificant digit is simply dropped.

(ii) if the preceding digit is odd, the preceding digit is raised by 1.

For example, $x = 6.265$ is rounded off to $x = 6.26$ to three significant digits and, $x = 6.275$ is rounded off to $x = 6.28$ to three significant digits.

Algebraic Operations with Significant Figures

In addition, subtraction, multiplication or division the final result should not have more significant figures than any of the original data from which it was obtained. To understand this, let us consider a chain, of which all links are strong except the one. The chain will obviously break at the weakest link. Thus, the strength of the chain cannot be more than the strength of the weakest link in the chain.

(i) **Addition and Subtraction** Suppose in the measured values to be added or subtracted, the least number of significant digits after the decimal is n. Then, in the sum or difference also, the number of significant digits after the decimal should be n.

Example $1.2 + 3.45 + 6.789 = 11.439 \approx 11.4$

Here, the least number of significant digits after the decimal is one. Hence, the result will be 11.4 (when rounded off to smallest number of decimal places).

e.g. $12.63 - 10.2 = 2.43 \approx 2.4$

(ii) **Multiplication or Division** Suppose in the measured values to be multiplied or divided, the least number of significant digits be n, then in the product or quotient, the number of significant digits should also be n. e.g. $1.2 \times 36.72 = 44.064 \approx 44$

The least number of significant digits in the measured values are two. Hence, the result when rounded off to two significant digits become 44. Therefore, the answer is 44.

e.g. $\dfrac{1100}{10.2} = 107.8431373 \approx 110$

As 1100 has minimum number of significant figures (i.e. 2), therefore the result should also contain only two significant digits. Hence, the result when rounded off to two significant digits becomes 110.

e.g. $\dfrac{1100 \text{ m/s}}{10.2 \text{ m/s}} = 107.8431373 \approx 108$

- In this case, answer becomes 108. Think why?

Extra Edge

- Exact numbers that appear in formulae, like 2π in $T = 2\pi\sqrt{\dfrac{l}{g}}$ or $S = 2\pi r$ have a large (infinite) number of significant figures. The value of $\pi = 3.1415926...$ can be written as 3.142 or 3.14 with limited number of significant figure as required in specific cases.
- Similarly in $r = \dfrac{d}{2}$, the factor is an exact number, which can be written as 2.0, 2.00 or 2.000 as required.
- **Order of Magnitude** In scientific notation, a measurement can be written as $(a \times 10^b)$. Here a, which is called the base number lies between 1 and 10. b is any positive or negative exponent (or power) of 10.
- For $a \le 5$, the order of magnitude is called 10^b.
  For $5 < a \le 10$, the order of magnitude is called 10^{b+1}.
- For example; diameter of the earth (1.28×10^7 m) is of the order of 10^7 m. Similarly, mass of an electron (9.11×10^{-31} kg) is of the order of 10^{-30} kg.

Example 1.7 *Write down the number of significant figures in the following:*

(a) 6428 *(b)* 62.00 *m* *(c)* 0.00628 *cm*

Sol. (a) 6428 has four significant figures.

(b) 62.00 m has four significant figures.

(c) 0.00628 cm has three significant figures.

Example 1.8 *Round off to four significant figures*

(a) 45.689 *(b)* 2.0082

Sol. (a) 45.69 (b) 2.008

Example 1.9 *A cube has a side of length* 1.2×10^{-2} *m. Calculate its volume with regard to significant figures.*

Sol. $V = l^3 = (1.2 \times 10^{-2}\ m)^3 = 1.728 \times 10^{-6}\ m^3$

$\because$ Length (l) has two significant figures, the volume (V) will also have two significant figures. Therefore, the correct answer is

$$V = 1.7 \times 10^{-6}\ m^3$$

Example 1.10 *Three masses, $m_1 = 226.2$ g, $m_2 = 127.36$ g and $m_3 = 10.214$ g are kept together. What is the total mass with regard to significant figures?*

Sol. Total mass $= m_1 + m_2 + m_3$

$= (226.2 + 127.36 + 10.214)\ g = 363.774\ g$

But the least precise measurement (226.2 g) is correct to only one decimal place.

$\therefore$ Total mass = 363.8 g

1.5 Dimensional Analysis

Dimensions of a physical quantity are the powers to which the fundamental quantities must be raised to represent the given physical quantity.

For example, $$\text{density} = \frac{\text{mass}}{\text{volume}} = \frac{\text{mass}}{(\text{length})^3}$$

or $$\text{density} = (\text{mass})\,(\text{length})^{-3} \quad \ldots\text{(i)}$$

Thus, the dimensions of density are 1 in mass and −3 in length. The dimensions of all other fundamental quantities are zero. For convenience, the fundamental quantities are represented by one letter symbols. Generally, mass is denoted by M, length by L, time by T and electric current by A.

The thermodynamic temperature, the amount of substance and the luminous intensity are denoted by the symbols of their units K, mol and cd, respectively. The physical quantity that is expressed in terms of the base quantities is enclosed in square brackets.

Thus, Eq. (i) can be written as

$$[\text{density}] = [ML^{-3}]$$

Such an expression for a physical quantity in terms of the fundamental quantities is called the dimensional formula. Here, it is worthnoting that constants such as 5, π or trigonometrical functions such as $\sin\theta$, $\cos\theta$, etc., have no units and dimensions.

$$[\sin\theta] = [\cos\theta] = [\tan\theta] = [\log x] = [e^x] = [M^0L^0T^0]$$

Example 1.11 *Find the dimensions of the physical quantities*
(a) area (b) force (c) velocity (d) work and (e) power

Sol. (a) Area = length × breadth = length × length

$\therefore$ $[\text{Area}] = [L] \times [L] = [L^2]$

(b) Force = mass × acceleration = mass × (length) / (time)^2

$\therefore$ $[\text{Force}] = [M]\,[L]/[T^2] = [MLT^{-2}]$

(c) Velocity = displacement / time = length / time

$\therefore$ $[\text{Velocity}] = [L]/[T] = [LT^{-1}]$

(d) Work = force × displacement = Force × length

$\therefore$ $[\text{Work}] = [\text{Force}] \times [\text{Length}]$

$= [MLT^{-2}]\,[L] = [ML^2T^{-2}]$

(e) Power = work / time

$\therefore$ $[\text{Power}] = [\text{Work}]/[\text{Time}]$

$= [ML^2T^{-2}]/[T] = [ML^2T^{-3}]$

Table 1.8 Dimensional formulae and SI units of some physical quantities frequently used in physics

S.No.	Physical quantity	SI units	Dimensional formula
1.	Velocity = displacement/time	m/s	$[M^0LT^{-1}]$
2.	Acceleration = velocity/time	m/s^2	$[M^0LT^{-2}]$
3.	Force = mass × acceleration	kg-m/s^2 = newton or N	$[MLT^{-2}]$
4.	Work = force × displacement	kg-m^2/s^2 = N-m = joule or J	$[ML^2T^{-2}]$
5.	Energy	J	$[ML^2T^{-2}]$
6.	Torque = force × perpendicular distance	N-m	$[ML^2T^{-2}]$
7.	Power = work/time	J/s or watt	$[ML^2T^{-3}]$
8.	Momentum = mass × velocity	kg-m/s	$[MLT^{-1}]$
9.	Impulse = force × time	N-s	$[MLT^{-1}]$
10.	Angle = arc/radius	radian or rad	$[M^0L^0T^0]$
11.	Strain = $\frac{\Delta L}{L}$ or $\frac{\Delta V}{V}$	no units	$[M^0L^0T^0]$
12.	Stress = force/area	N/m^2	$[ML^{-1}T^{-2}]$
13.	Pressure = force/area	N/m^2	$[ML^{-1}T^{-2}]$
14.	Modulus of elasticity = stress/strain	N/m^2	$[ML^{-1}T^{-2}]$

S.No.	Physical quantity	SI units	Dimensional formula
15.	Frequency = 1/time period	per sec or hertz (Hz)	$[M^0L^0T^{-1}]$
16.	Angular velocity = angle/time	rad/s	$[M^0L^0T^{-1}]$
17.	Moment of inertia = (mass) × (distance)^2	kg-m^2	$[ML^2T^0]$
18.	Surface tension = force/length	N/m	$[ML^0T^{-2}]$
19.	Gravitational constant = $\frac{\text{force} \times (\text{distance})^2}{(\text{mass})^2}$	N-m^2/kg^2	$[M^{-1}L^3T^{-2}]$
20.	Angular momentum	kg-m^2/s	$[ML^2T^{-1}]$
21.	Coefficient of viscosity	N-s/m^2	$[ML^{-1}T^{-1}]$
22.	Planck's constant	J-s	$[ML^2T^{-1}]$
23.	Specific heat (*s*)	J/kg-K	$[L^2T^{-2}K^{-1}]$
24.	Coefficient of thermal conductivity (*K*)	watt/m-K	$[MLT^{-3}K^{-1}]$
25.	Gas constant (*R*)	J/mol-K	$[ML^2T^{-2}K^{-1}mol^{-1}]$
26.	Boltzmann constant (*k*)	J/K	$[ML^2T^{-2}K^{-1}]$
27.	Wien's constant (*b*)	m-K	$[L^0K]$
28.	Stefan's constant (σ)	watt/m^2-K^4	$[ML^{-3}K^{-4}]$
29.	Electric charge	C	[AT]
30.	Electric intensity	N/C	$[MLT^{-3}A^{-1}]$
31.	Electric potential	volt	$[ML^2T^{-3}A^{-1}]$
32.	Capacitance	farad	$[M^{-1}L^{-2}T^4A^2]$
33.	Permittivity of free space	$C^2N^{-1}m^{-2}$	$[M^{-1}L^{-3}T^4A^2]$
34.	Electric dipole moment	C-m	[LTA]
35.	Resistance	ohm	$[ML^2T^{-3}A^{-2}]$
36.	Magnetic field	tesla (T) or weber/m^2 (Wb/m^2)	$[MT^{-2}A^{-1}]$
37.	Coefficient of self-induction	henry	$[ML^2T^{-2}A^{-2}]$

Uses of Dimensions

Theory of dimensions have following main uses:

(i) **Conversion of units** This is based on the fact that the product of the numerical value (n) and its corresponding unit (u) is a constant, i.e.

$$n[u] = \text{constant}$$

or $$n_1[u_1] = n_2[u_2]$$

Suppose, the dimensions of a physical quantity are a in mass, b in length and c in time. If the fundamental units in one system are M_1, L_1 and T_1 and in the other system are M_2, L_2 and T_2, respectively. Then, we can write

$$n_1[M_1^a L_1^b T_1^c] = n_2[M_2^a L_2^b T_2^c] \quad \text{...(i)}$$

Here, n_1 and n_2 are the numerical values in two system of units, respectively. Using Eq. (i), we can convert the numerical value of a physical quantity from one system of units into the other system.

Example 1.12 *The value of gravitation constant is $G = 6.67 \times 10^{-11}$ N-m^2/kg^2 in SI units. Convert it into CGS system of units.*

Sol. The dimensional formula of G is $[M^{-1}L^3T^{-2}]$

Using Eq. (i), i.e.

$$n_1[M_1^{-1}L_1^3T_1^{-2}] = n_2[M_2^{-1}L_2^3T_2^{-2}]$$

$$n_2 = n_1\left[\frac{M_1}{M_2}\right]^{-1}\left[\frac{L_1}{L_2}\right]^3\left[\frac{T_1}{T_2}\right]^{-2}$$

Here, $n_1 = 6.67 \times 10^{-11}$

$M_1 = 1$ kg, $M_2 = 1$ g $= 10^{-3}$ kg, $L_1 = 1$ m,

$L_2 = 1$ cm $= 10^{-2}$ m, $T_1 = T_2 = 1$ s

Substituting these values in the above equation, we get

$$n_2 = 6.67 \times 10^{-11}\left[\frac{1\text{ kg}}{10^{-3}\text{ kg}}\right]^{-1}\left[\frac{1\text{ m}}{10^{-2}\text{ m}}\right]^3\left[\frac{1\text{ s}}{1\text{ s}}\right]^{-2}$$

or $n_2 = 6.67 \times 10^{-8}$

Thus, value of G in CGS system of units is

6.67×10^{-8} dyne-cm^2/g^2.

(ii) **To check the dimensional correctness of a given physical equation** Every physical equation should be dimensionally balanced. This is called the 'Principle of Homogeneity'. The dimensions of each term on both sides of an equation must be the same. On this basis, we can judge whether a given equation is correct or not. But, a dimensionally correct equation may or may not be physically correct.

Example 1.13 *Show that the expression of the time period T of a simple pendulum of length l given by* $T = 2\pi\sqrt{\frac{l}{g}}$ *is dimensionally correct.*

Sol. $$T = 2\pi\sqrt{\frac{l}{g}}$$

Dimensionally, $[T] = \sqrt{\frac{[L]}{[LT^{-2}]}} = [T]$

As in the above equation, the dimensions of both sides are same. The given formula is dimensionally correct.

Principle of Homogeneity of Dimensions

This principle states that the dimensions of all the terms in a physical expression should be same.

For example, in the physical expression $s = ut + \frac{1}{2}at^2$, the dimensions of s, ut and $\frac{1}{2}at^2$ all are same.

- The physical quantities separated by the symbols +, −, =, >, <, etc. have the same dimensions.

Example 1.14 *The velocity v of a particle depends upon the time t according to the equation* $v = a + bt + \frac{c}{d+t}$ *· Write the dimensions of a, b, c and d.*

Sol. From principle of homogeneity,

$$[a] = [v] \quad \text{or} \quad [a] = [LT^{-1}]$$

$$[bt] = [v] \quad \text{or} \quad [b] = \frac{[v]}{[t]} = \frac{[LT^{-1}]}{[T]}$$

or $$[b] = [LT^{-2}]$$

Similarly, $[d] = [t] = [T]$

Further, $\frac{[c]}{[d+t]} = [v]$ or $[c] = [v][d+t]$

or $$[c] = [LT^{-1}][T] \quad \text{or} \quad [c] = [L]$$

(iii) To establish the relation among various physical quantities, if we know the factors on which a given physical quantity may depend, we can find a formula relating the quantity with those factors. Let us take an example.

Example 1.15 *The frequency (f) of a stretched string depends upon the tension F (dimensions of force), length l of the string and the mass per unit length μ of string. Derive the formula for frequency.*

Sol. Suppose that the frequency f depends on the tension raised to the power a, length raised to the power b and mass per unit length raised to the power c. Then,

$$f \propto [F]^a [l]^b [\mu]^c$$

or $$f = k[F]^a [l]^b [\mu]^c \quad \ldots(i)$$

Here, k is a dimensionless constant. Thus,

$$[f] = [F]^a [l]^b [\mu]^c$$

or $$[M^0 L^0 T^{-1}] = [MLT^{-2}]^a [L]^b [ML^{-1}]^c$$

or $$[M^0 L^0 T^{-1}] = [M^{a+c} L^{a+b-c} T^{-2a}]$$

For dimensional balance, the dimensions on both sides should be same.

Thus, $$a + c = 0 \quad \ldots(ii)$$
$$a + b - c = 0 \quad \ldots(iii)$$
and $$-2a = -1 \quad \ldots(iv)$$

Solving these three equations, we get

$$a = \frac{1}{2}, \quad c = -\frac{1}{2} \text{ and } b = -1$$

Substituting these values in Eq. (i), we get

$$f = k(F)^{1/2}(l)^{-1}(\mu)^{-1/2}$$

or $$f = \frac{k}{l}\sqrt{\frac{F}{\mu}}$$

Experimentally, the value of k is found to be $\frac{1}{2}$.

Hence, $$f = \frac{1}{2l}\sqrt{\frac{F}{\mu}}$$

Limitations of Dimensional Analysis

The method of dimensions has the following limitations:

(i) By this method, the value of dimensionless constant cannot be calculated.

(ii) By this method, the equation containing trigonometrical, exponential and logarithmic terms cannot be analysed.

(iii) If a physical quantity depends on more than three factors, then relation among them cannot be established because we can have only three equations by equalising the powers of M, L and T.

Extra Edge

- **Dimensional Constants** The quantities like gravitational constant, Planck's constant, etc., possess dimensions and also have a constant value. They are called dimensional constants.
- **Non-dimensional Constants** The constant quantities having no dimensions are called non-dimensional constants. These include pure numbers (1, 2, 3...), π, e (= 2.718...) and all trigonometric functions.
- In some cases, an equation seems to be dimensionally incorrect although it is correct. For example, in the equation $F = ma$, if we substitute $m = 1$ kg, then this equation becomes,

  $$F = a$$

  Now, this equation seems dimensionally incorrect, because F and a have different dimensions. But actually, this equation is correct because a is multiplied with 1 kg mass, whose dimension is [M]. So, we will have to multiply this dimension with the dimensions of a to find the dimension of F.

Extra Knowledge Points

Least Count

The minimum measurement that can be measured accurately by an instrument is called the least count.

The least count of a metre scale graduated in millimetre mark is 1 mm. The least count of a watch having second's hand is 1 second.

Least count of vernier calliper = {Value of 1 part of main scale(s)} − {Value of one part of vernier scale (V)}

or Least count of vernier calliper = 1 MSD − 1 VSD

where, MSD = Main scale division

VSD = Vernier scale division

$$\text{Least count of vernier callipers} = \frac{\text{Value of 1 part of main scale } (s)}{\text{Number of parts on vernier scale } (n)}$$

In the ordinary vernier callipers, value of 1 part of main scale division is 1 mm and number of parts on vernier scale are 10.

$$\therefore \quad LC = \frac{1 \text{ mm}}{10} = 0.1 \text{ mm}$$

$$\text{Least count of screw gauge} = \frac{\text{Pitch } (p)}{\text{Number of parts on circular scale } (n)}$$

In the ordinary screw gauge, pitch is 1mm and number of parts on circular scale are 100.

$$\therefore \quad LC = \frac{1 \text{ mm}}{100} = 0.01 \text{ mm}$$

Additional Examples

Example 1. *Check the correctness of the relation $s = ut + \frac{1}{2}at^2$, where u is initial velocity, a the acceleration, t the time and s the displacement.*

Sol. Writing the dimensions of either side of the given equation.

$$\text{LHS} = s = \text{displacement} = [M^0LT^0]$$

$$\text{RHS} = ut = \text{velocity} \times \text{time} = [M^0LT^{-1}][T] = [M^0LT^0]$$

and $$\frac{1}{2}at^2 = (\text{acceleration}) \times (\text{time})^2 = [M^0LT^{-2}][T]^2 = [M^0LT^0]$$

As LHS = RHS, formula is dimensionally correct.

Example 2. *Write the dimensions of a and b in the relation,*

$$P = \frac{b - x^2}{at}$$

where, P is power, x the distance and t the time.

Sol. The given equation can be written as

$$Pat = b - x^2$$

Now, $$[Pat] = [b] = [x^2] \text{ or } [b] = [x^2] = [M^0L^2T^0]$$

and $$[a] = \frac{[x^2]}{[Pt]} = \frac{[L^2]}{[ML^2T^{-3}][T]} = [M^{-1}L^0T^2]$$

Example 3. *The centripetal force F acting on a particle moving uniformly in a circle may depend upon mass (m), velocity (v) and radius (r) of the circle. Derive the formula for F using the method of dimensions.*

Sol. Let $$F = k(m)^x (v)^y (r)^z \quad ...(i)$$

Here, k is a dimensionless constant of proportionality.

Writing the dimensions of RHS and LHS in Eq. (i), we have

$$[MLT^{-2}] = [M]^x [LT^{-1}]^y [L]^z = [M^x L^{y+z} T^{-y}]$$

Equating the powers of M, L and T of both sides, we have $x = 1$, $y = 2$ and $y + z = 1$

or $z = 1 - y = -1$

Putting the values in Eq. (i), we get

$$F = kmv^2r^{-1} = k\frac{mv^2}{r} = \frac{mv^2}{r} \quad (\text{where, } k = 1)$$

Example 4. *Write down the number of significant figures in the following:*

(a) 6428 *(b)* 62.00 *m*

(c) 0.00628 *cm* *(d)* 1200 *N*

Sol. (a) 6428 has four significant figures.

(b) 62.00 m has four significant figures.

(c) 0.00628 cm has three significant figures.

(d) 1200 N has four significant figures.

Example 5. *Round off to four significant figures :*

(a) 45.689 *(b)* 2.0082

Sol. (a) 45.69 (b) 2.008

Example 6. *Add 6.75×10^3 cm to 4.52×10^2 cm with regard to significant figures.*

Sol. $a = 6.75 \times 10^3$ cm

$b = 4.52 \times 10^2 \text{ cm} = 0.452 \times 10^3 \text{ cm}$

$= 0.45 \times 10^3$ cm (up to 2 places of decimal)

$\therefore \; a + b = (6.75 \times 10^3 + 0.45 \times 10^3)$ cm

$= 7.20 \times 10^3$ cm

Example 7. *A thin wire has a length of 21.7 cm and radius 0.46 cm. Calculate the volume of the wire to correct significant figures.*

Sol. Given, $l = 21.7$ cm, $r = 0.46 \text{ mm} = 0.046$ cm

$$\text{Volume of wire, } V = \pi r^2 l = \frac{22}{7}(0.046)^2 (21.7) = 0.1443 \text{ cm}^3 = 0.14 \text{ cm}^3$$

- The result is rounded off to least number of significant figures in the given measurements, i.e. 2 (in 0.46 mm).

Example 8. *The refractive index (n) of glass is found to have the values* 1.49, 1.50, 1.52, 1.54 *and* 1.48. *Calculate*
(a) the mean value of refractive index,
(b) absolute error in each measurement,
(c) mean absolute error,
(d) fractional error, and
(e) percentage error.

Sol. (a) Mean value of refractive index,

$$n_m = \frac{1.49 + 1.50 + 1.52 + 1.54 + 1.48}{5}$$

$$= 1.505 = 1.51$$

(rounded off to two decimal places)

(b) Taking n_m as the true value, the absolute errors in different observations are,

$$\Delta n_1 = 1.51 - 1.49 = +\ 0.02$$
$$\Delta n_2 = 1.51 - 1.50 = +\ 0.01$$
$$\Delta n_3 = 1.51 - 1.52 = -\ 0.01$$
$$\Delta n_4 = 1.51 - 1.54 = -\ 0.03$$
$$\Delta n_5 = 1.51 - 1.48 = +\ 0.03$$

(c) Mean absolute error,

$$\Delta n_{\text{mean}} = \frac{|\Delta n_1| + |\Delta n_2| + |\Delta n_3| + |\Delta n_4| + |\Delta n_5|}{5}$$

$$= \frac{0.02 + 0.01 + 0.01 + 0.03 + 0.03}{5}$$

$$= 0.02$$

(d) Fractional error $= \dfrac{\pm \Delta n_{\text{mean}}}{n_m} = \dfrac{\pm 0.02}{1.51}$

$$= \pm 0.0132$$

(e) Percentage error $= (\pm 0.0132 \times 100)$

$$= \pm 1.32\%$$

Example 9. *The radius of sphere is measured to be* (2.1 ± 0.5) *cm. Calculate its surface area with error limits.*

Sol. Surface area, $S = 4\pi r^2$

$$= (4)\left(\frac{22}{7}\right)(2.1)^2$$

$$= 55.44 = 55.4 \text{ cm}^2$$

Further, $\dfrac{\Delta S}{S} = 2\dfrac{\Delta r}{r}$

or $\Delta S = 2\left(\dfrac{\Delta r}{r}\right)(S)$

$$= \frac{2 \times 0.5 \times 55.4}{2.1}$$

$$= 26.38$$

$$= 26.4 \text{ cm}^2$$

$\therefore$ $S = (55.4 \pm 26.4) \text{ cm}^2$

Example 10. *Calculate focal length of a spherical mirror from the following observations. Object distance* $u = (50.1 \pm 0.5)$ *cm and image distance* $v = (20.1 \pm 0.2)$ *cm.*

Sol. Since, $\dfrac{1}{f} = \dfrac{1}{v} + \dfrac{1}{u}$

or $f = \dfrac{uv}{u + v} = \dfrac{(50.1)(20.1)}{(50.1 + 20.1)}$

$$= 14.3 \text{ cm}$$

Also, $\dfrac{\Delta f}{f} = \pm\left[\dfrac{\Delta u}{u} + \dfrac{\Delta v}{v} + \dfrac{\Delta u + \Delta v}{u + v}\right]$

$$= \pm\left[\frac{0.5}{50.1} + \frac{0.2}{20.1} + \frac{0.5 + 0.2}{50.1 + 20.1}\right]$$

$$= [0.00998 + 0.00995 + 0.00997]$$

$$= \pm(0.0299)$$

$\therefore$ $\Delta f = 0.0299 \times 14.3$

$$= 0.428 = 0.4 \text{ cm}$$

$\therefore$ $f = (14.3 \pm 0.4)$ cm

Example 11. *Can a quantity have units but still be dimensionless. Give an example.*

Sol. Yes, the example is angle. Unit of angle is radian. But, it is a dimensionless quantity.

Example 12. *Can a quantity have neither units nor dimensions.*

Sol. Yes, strain has neither units nor dimensions.

Example 13. *Write four pairs of physical quantities, which have the same dimensional formula.*

Sol. 1. Work and energy
2. Pressure and stress
3. Velocity gradient and frequency
4. Angular momentum and Planck's constant.

Example 14. *Give two examples which are not constants but they are dimensionless.*

Sol. Strain and angular displacement are the two examples.

Example 15. *We measure the period of oscillation of a simple pendulum. In successive measurements, the readings turn out to be* 2.63*s*, 2.56*s*, 2.42*s*, 2.71*s and* 2.80*s*. *Calculate the absolute error, relative error or percentage error.*

Sol. Mean period of oscillation of the pendulum

$$T = \frac{(2.63 + 2.56 + 2.42 + 2.71 + 2.80) \text{ s}}{5}$$

$$= 2.624 \text{ s} = 2.62 \text{ s}$$

As the periods are measured to a resolution of 0.01 s, all times are to the second decimal; it is proper to put this mean period also to the second decimal.

The absolute errors in the measurements are

$$2.63\text{ s} - 2.62 = 0.01\text{ s}$$
$$2.56\text{ s} - 2.62\text{ s} = -0.06\text{ s}$$
$$2.42\text{ s} - 2.62\text{ s} = -0.20\text{ s}$$
$$2.71\text{ s} - 2.62\text{ s} = 0.09\text{ s}$$
$$2.80\text{ s} - 2.62\text{ s} = 0.18\text{ s}$$

$$\Delta T_{\text{mean}} = [(0.01 + 0.06 + 0.20 + 0.09 + 0.18)\text{ s}]/5 = 0.11\text{ s}$$

That means, the period of oscillation of the simple pendulum is (2.62 ± 0.11) s, i.e. it lies between $(2.62 + 0.11)$ s and $(2.62 - 0.11)$ s or between 2.73 s and 2.51 s. A more correct way will be to write

$$T = 2.6 \pm 0.1\text{ s}$$

For this reading, the percentage error is

$$\frac{\Delta a}{a} \times 100 = \frac{0.1}{2.6} \times 100 = 4\%$$

Example 16. *The radius of sphere is measured to be (2.1 ± 0.5) cm. Calculate its surface area with error limits.*

Sol. Surface area, $S = 4\pi r^2 = (4)\left(\frac{22}{7}\right)(2.1)^2$

$$= 55.44 = 55.4\text{ cm}^2$$

Further, $\frac{\Delta S}{S} = 2 \cdot \frac{\Delta r}{r}$

or $\Delta S = 2\left(\frac{\Delta r}{r}\right)(S) = \frac{2 \times 0.5 \times 55.4}{2.1}$

$$= 26.38\text{ cm}^2$$
$$= 26.4\text{ cm}^2$$

$\therefore$ $S = (55.4 \pm 26.4)\text{ cm}^2$

NCERT Selected Questions

***Q* 1.** The SI unit of energy is $J = \text{kg m}^2\text{s}^{-2}$; that of speed v is ms^{-1} and of acceleration a is ms^{-2}. Which of the formulae for kinetic energy (K) given below can you rule out on the basis of dimensional arguments (m stands for the mass of the body)

(a) $K = m^2v^3$ (b) $K = (1/2)\,mv^2$
(c) $K = ma$ (d) $K = (3/16)\,mv^2$
(e) $K = (1/2)\,mv^2 + ma$

Sol. In a correct equation, both sides should have same dimensions. Kinetic energy which has the SI unit kg-m^2s^{-2} should have the dimensions,

$$[K] = [\text{M}]\,[\text{L}^2]\,[\text{T}^{-2}] = [\text{ML}^2\,\text{T}^{-2}]$$

Right hand side of option

(a) has the dimensions,

$$[m^2v^3] = [\text{M}^2]\,[\text{LT}^{-1}]^3 = [\text{M}^2\,\text{L}^3\,\text{T}^{-3}]$$

Since, these dimensions do not match with the dimensions of kinetic energy. So, this option is ruled out.

(b) has the dimensions,

$$\left[\frac{1}{2}mv^2\right] = [\text{M}]\,[\text{LT}^{-1}]^2 = [\text{ML}^2\,\text{T}^{-2}]$$

These dimensions match with the dimensions of kinetic energy.

(c) has the dimensions,

$$[ma] = [\text{M}]\,[\text{LT}^{-2}] = [\text{MLT}^{-2}]$$

These dimensions do not match with the dimensions of kinetic energy. So this option is ruled out.

(d) has also the dimensions of option (b). So, these dimensions match with the dimensions of kinetic energy.

(e) has two different dimensions of two different terms. Hence, this option can also be ruled out.

Note that dimensional correctness on two sides cannot tell which of the two, (b) or (d) is the correct formula. But we know the correct formula is (b).

***Q* 2.** A student measures the thickness of a human hair by looking at it through a microscope of magnification 100. He makes 20 observations and finds that the average width of the hair in the field of view of the microscope is 3.5 mm. What is the estimate on the thickness of hair?

Sol. Width of hair after 100 times magnification is given 3.5 mm.

$\therefore$ Thickness of hair $= \frac{3.5\text{ mm}}{100} = 0.035\text{ mm}$

***Q* 3.** Fill in the blanks by suitable conversion of units.

(a) $1\text{ kg m}^2\text{ s}^{-2} = \ldots\ldots\text{ g cm}^2\text{ s}^{-2}$
(b) $1\text{ m} = \ldots\ldots$ ly (light year)
(c) $3.0\text{ ms}^{-2} = \ldots\ldots\text{ kmh}^{-2}$
(d) $G = 6.67 \times 10^{11}\text{ Nm}^2\text{kg}^{-2} = \ldots\ldots\text{ cm}^3\text{ s}^{-2}\text{ g}^{-1}$

Sol. (a) $1\text{ kg m}^2\text{s}^{-2} = 1\,(10^3\text{g})\,(10^2\text{cm})^2\text{s}^{-2}$

$$= 10^3 \times 10^4\text{ g cm}^2\text{s}^{-2} = 10^7\text{ g cm}^2\text{ s}^{-2}$$

(b) 1 light year (ly) $= 9.46 \times 10^{15}$m

$\therefore$ $1\text{ m} = \frac{1}{9.46 \times 10^{15}}\text{ ly} = 1.057 \times^{-16}\text{ ly}$

(c) $3.0\text{ ms}^{-2} = 3 \times 10^{-3}\text{ km} \times \left(\frac{1}{60 \times 60}\text{ h}^{-1}\right)^{-2}$

$$= 3 \times 10^{-3} \times (3600)^2\text{ kmh}^{-2}$$
$$= 3.888 \times 10^4\text{ kmh}^{-2} = 3.9 \times 10^4\text{ kmh}^{-2}$$

(d) $G = 6.67 \times 10^{-11}\ \text{Nm}^2\ \text{kg}^{-2}$

$= 6.67 \times 10^{-11}\ (10^5 \text{dyne})\ (10^2\ \text{cm})^2\ (10^3\ \text{g})^{-2}$

$= 6.67 \times 10^{-11} \times 10^5 \times 10^4 \times 10^{-6} \text{dyne cm}^2 \text{g}^{-2}$

$= 6.67 \times 10^{-8}\ (\text{g cm s}^{-2})\ \text{cm}^2\ \text{g}^{-2}$

$= 6.67 \times 10^{-8}\ \text{cm}^3\ \text{g}^{-1}\ \text{s}^{-2}$

***Q* 4.** Fill in the blanks

(a) The volume of a cube of side 1 cm is equal to ... m^3.

(b) The surface area of a solid cylinder of radius 2.0 cm and height 10.0 cm is equal to...... mm^2.

(c) A vehicle moving with a speed of 18 kmh^{-1} covers...... m in 1s.

(d) The relative density of lead is 11.3. Its density is ...... gcm^{-3} or ...... kgm^{-3}.

Sol. (a) The volume of a cube of side 1 cm is given by

$$V = (10^{-2}\ \text{m})^3 = 10^{-6}\ \text{m}^3$$

(b) The surface area of a solid cylinder of radius r and height h is given by

A = Area of two circles + curved surface area

$= 2\pi r^2 + 2\pi rh = 2\pi r\,(r + h)$

Here, $r = 2\ \text{cm} = 20\ \text{mm}$, $h = 10\ \text{cm} = 100\ \text{mm}$

$$\therefore \quad A = 2 \times \frac{22}{7} \times 20\ (20 + 100)\ (\text{mm})^2$$

$= 15099\ \text{mm}^2 = 1.5099 \times 10^4\ \text{mm}^2$

$= 1.5 \times 10^4\ \text{mm}^2$

(c) Here, $v = 18\ \text{km h}^{-1} = \dfrac{18 \times 1000\ \text{m}}{3600\text{s}} = 5\ \text{ms}^{-1}$

$t = 1\ \text{s}$

$\therefore \quad x = vt = 5 \times 1 = 5\ \text{m}$

(d) Relative density of lead = 11.3

Density of water = 1 g cm^{-3}

$$\text{Relative density of lead} = \frac{\text{density of lead}}{\text{density of water}}$$

$\therefore$ Density of lead

= relative density of lead × density of water

$= 11.3 \times 1\ \text{g cm}^{-3}$

$= 11.3\ \text{g cm}^{-3}$

In SI system, density of water $= 10^3\ \text{kg m}^{-3}$

$\therefore$ Density of lead $= 11.3 \times 10^3\ \text{kg m}^{-3}$

$= 1.13 \times 10^4\ \text{kg m}^{-3}$

***Q* 5.** A calorie is a unit of heat or energy and it equals about 4.2 J, where $1\ \text{J} = 1\ \text{kgm}^2\text{s}^{-2}$. Suppose we employ a system of units in which the unit of mass equals α kg, the unit of length β m and the unit of time is γ s. Show that a calorie has a magnitude $4.2\,\alpha^{-1}\,\beta^{-2}\,\gamma^2$ in terms of new units.

Sol. $n_1 u_1 = n_2 u_2$

or $\quad n_2 = n_1 \dfrac{u_1}{u_2} = n_1 \dfrac{[\text{M}_1^a\,\text{L}_1^b\,\text{T}_1^c]}{[\text{M}_2^a\,\text{L}_2^b\,\text{T}_2^c]}$

$$= n_1 \left[\frac{\text{M}_1}{\text{M}_2}\right]^a \left[\frac{\text{L}_1}{\text{L}_2}\right]^b \left[\frac{\text{T}_1}{\text{T}_2}\right]^c$$

$1\ \text{cal} = 4.2\ \text{J} = 4.2\ \text{kg m}^2\ \text{s}^{-2}$,

$\therefore \quad a = 1,\ b = 2,\ c = -2$

$$\therefore \quad n_2 = 4.2 \left[\frac{1\ \text{kg}}{a\ \text{kg}}\right]^1 \left[\frac{1\ \text{m}}{b\ \text{m}}\right]^2 \left[\frac{1\ \text{s}}{g\ \text{s}}\right]^{-2}$$

$n_2 = 4.2\,\alpha^{-1}\,\beta^{-2}\,\gamma^2$

$\therefore \quad 1\ \text{cal} = 4.2\,\alpha^{-1}\,\beta^{-2}\,\gamma^2$ in new system.

***Q* 6.** A new unit of length is chosen such that the speed of light in vacuum is unity. What is the distance between the sun and the earth in terms of the new unit if light takes 8 min and 20 s to cover this distance?

Sol. We are given that velocity of light in vacuum, $c = 1$ new unit of length s^{-1}.

Time taken by light of sun of reach the earth, t = 8 min 20 s.

$= 8 \times 60 + 20 = 500\ \text{s}$

$\therefore$ Distance between the sun and the earth,

$x = c \times t = 1$ new unit of length $\text{s}^{-1} \times 500\ \text{s}$

= 500 new units of length

***Q* 7.** Which of the following is the most precise device for measuring length?

(a) A vernier callipers with 20 divisions on the sliding scale.

(b) A screw gauge of pitch 1 mm and 100 divisions on the circular scale.

(c) An optical instrument that can measure length to within a wavelength of light.

Sol. The most precise device is that whose least count is minimum.

(a) Least count of vernier callipers = 1 MSD − 1 VSD

$$= 1\ \text{MSD} - \frac{19}{20}\ \text{MSD} = \frac{1}{20}\ \text{MSD}$$

$$= \frac{1}{20}\ \text{mm} = \frac{1}{200}\ \text{cm} = 0.005\ \text{cm}$$

(b) Least count of screw gauge

$$= \frac{\text{Pitch}}{\text{Number of divisions of circular scale}}$$

$$= \frac{1}{100}\ \text{mm} = \frac{1}{1000}\ \text{cm}$$

$= 0.001$ cm

(c) Wavelength of light, $\lambda \approx 10^{-5}\ \text{cm} = 0.00001\ \text{cm}$

$\therefore$ Least count of optical instrument = 0.00001 cm

Thus, clearly the optical instrument is the most precise.

Q 8. State the number of significant figures in the following.

(a) 0.007 m^2 (b) 2.64×10^{24} kg

(c) 0.2370 g cm^{-3} (d) 6.320 J

(e) 6.032 Nm^{-2} (f) 0.0006032 m^2

Sol. The number of significant figures is given as below:

(a) 1 (b) 3 (c) 4

(d) 4 (e) 4 (f) 4

Q 9. The length, breadth and thickness of a rectangular sheet of metal are 4.234 m, 1.005 m and 2.01 cm respectively. Give the area and volume of the sheet to correct significant figures.

Sol. Here, length, $l = 4.234$ m, breadth, $b = 1.005$ m

Thickness, $h = 0.0201 \text{ m} = 2.01$ cm

Area of the sheet $= 2(lb + bh + hl)$

$= 2(4.234 \times 1.005 + 1.005 \times 0.0201 + 0.0201 \times 4.234)$

$= 8.7209468 \text{ m}^2$

As the least number of significant figures in thickness is 3,

$\therefore$ Area $= 8.72 \text{ m}^2$

Volume $= l \times b \times h$

$= 4.234 \times 1.005 \times 0.0201 \text{ m}^3$

$= 0.0855 \text{ m}^3$

Q 10. The mass of a box measured by a grocer's balance is 2.3 kg. Two gold pieces of masses 20.15 g and 20.17 g are added to the box. What is (a) the total mass of the box, (b) the difference in the mass of the pieces to correct significant figures?

Sol. (a) Total mass $= (2.300 + 0.02015 + 0.02017)$ kg

$= 2.34032$ kg

As the least number of significant figures in the mass of box is 2, so maximum number of significant figures in the result can be 2.

$\therefore$ Total mass $= 2.3$ kg

(b) Difference in masses $= 20.17 - 20.15 = 0.02$ g

Since, there are two significant figures, so the difference in masses to the correct significant figures is 0.02 g.

Q 11. A physical quantity P is related to four observables a, b, c and d as follows $P = \frac{a^3 b^2}{\sqrt{cd}}$. The percentage errors of measurement in a, b, c and d are 1%, 3%, 4% and 2%, respectively. What is the percentage error in the quantity P? If the value of P calculated using the above relation turns out to be 3.763, to what value should you round off the result?

Sol. Percentage error in P is given by

$$\frac{\Delta P}{P} \times 100 = 3\left(\frac{\Delta a}{a} \times 100\right) + 2\left(\frac{\Delta b}{b} \times 100\right) + \frac{1}{2}\left(\frac{\Delta c}{c} \times 100\right) + \left(\frac{\Delta d}{d} \times 100\right) \quad \ldots\text{(i)}$$

$$\left.\begin{array}{l} \frac{\Delta a}{a} \times 100 = 1\%, \frac{\Delta c}{c} \times 100 = 4\% \\ \frac{\Delta b}{b} \times 100 = 3\%, \frac{\Delta d}{d} \times 100 = 2\% \end{array}\right] \quad \ldots\text{(ii)}$$

$\therefore$ From Eqs. (i) and (ii), we get

$$\frac{\Delta P}{P} \times 100 = 3 \times 1\% + 2 \times 3\% + \frac{1}{2} \times 4\% + 2\%$$

$$= 3 + 6 + 2 + 2 = 13\%$$

The calculation of error clearly shows that the number of significant figures is 2, so the result of P may be rounded off to two significant digits, i.e. $P = 3.763 = 3.8$.

Q 12. A book with many printing errors contains four different formulae for the displacement y of a particle under going a certain periodic motion:

(a) $y = a \sin \frac{2\pi t}{T}$

(b) $y = a \sin vt$

(c) $y = \frac{a}{T} \sin (t/a)$

(d) $y = \left(\frac{a}{\sqrt{2}}\right)\left(\sin \frac{2\pi t}{T} + \cos \frac{2\pi t}{T}\right)$

(where, a = maximum displacement of the particle, v = speed of the particle, T = time period of motion). Rule out the wrong formulae on dimensional grounds.

Sol. The argument of a trigonometrical function, i.e. angle is dimensionless. Now, here in each case dimensions of LHS is [L] and dimensions of RHS in

(a) $= [\text{L}]$ $\left(\text{angle } \frac{2\pi t}{T} \text{ is dimensionless}\right)$

(b) $= [\text{L}] \sin [\text{LT}^{-1} (\text{T})] = [\text{L}] \sin [\text{L}]$

(angle is not dimensionless here)

(c) $= \frac{[\text{L}]}{[\text{T}]} \sin \frac{[\text{T}]}{[\text{L}]} = [\text{LT}^{-1}] \sin [\text{TL}^{-1}]$

(angle is not dimensionless here)

(d) $= [\text{L}]\left[\sin \frac{T}{T} + \cos \frac{T}{T}\right] = [\text{L}]$

$\therefore$ Formulae (b) and (c) are wrong.

Q 13. The unit of length convenient on the atomic scale is known as an angstrom and is denoted by Å. $1 \text{ Å} = 10^{-10}$ m. The size of the hydrogen atom is about 0.5 Å. What is the total atomic volume in m^3 of a mole of hydrogen atoms?

Sol. $r = 0.5 \text{ Å} = 0.5 \times 10^{-10}$ m

V_1 = volume of each hydrogen atom $= \frac{4}{3} \pi r^3$

$= \frac{4}{3} \times 3.14 \times (0.5 \times 10^{-10})^3$

$= 5.236 \times 10^{-31} \text{ m}^3$

According to Avogadro's hypothesis, one mole of hydrogen contains

$$N = 6.023 \times 10^{23} \text{ atoms}$$

$\therefore$ Atomic volume of 1 mole of hydrogen atoms,

$$V = NV_1$$

or $$V = 6.023 \times 10^{23} \times 5.236 \times 10^{-31}$$

$$= 3.154 \times 10^{-7} \text{ m}^3$$

$$\cong 3 \times 10^{-7} \text{ m}^3$$

Q 14. One mole of an ideal gas at standard temperature and pressure occupies 22.4 L (molar volume). What is the ratio of molar volume to the atomic volume of a mole of hydrogen? (Take the size of hydrogen molecule to be about 1 Å). Why is this ratio so large?

Sol. d = diameter of hydrogen molecule = 1 Å

Molar volume of one mole of hydrogen

$$= 22.4 \text{ L}$$

$$= 22.4 \times 10^{-3} \text{ m}^3$$

r = radius of one molecule of hydrogen

$$= \frac{d}{2} = 0.5 \text{ Å} = 0.5 \times 10^{-10} \text{ m}$$

Volume of one molecule of hydrogen

$$= \frac{4}{3}\pi r^3 = \frac{4}{3}\pi (0.5 \times 10^{-10})^3 = 5.236 \times 10^{-31} \text{ m}^3$$

1 mole has 6.023×10^{23} atoms or molecules of H_2.

$\therefore$ Atomic volume of one mole of hydrogen

$$= 6.023 \times 10^{23} \times 5.236 \times 10^{-31} \text{ m}^3 = 3.154 \times 10^{-7} \text{ m}^3$$

$$\therefore \frac{\text{Molar volume}}{\text{Atomic volume}} = \frac{22.4 \times 10^{-3} \text{ m}^3}{3.154 \times 10^{-7} \text{ m}^3}$$

$$= 7.1 \times 10^4 = 7 \times 10^4$$

The large value of the ratio shows that the inter molecular separation in a gas is much larger than the size of a mulecule.

Q 15. The nearest star to our solar system is 4.29 ly away. How much is this distance in terms of parsecs?

Sol. Distance = 4.29 ly = $4.29 \times 9.46 \times 10^{15}$ m

$$(\because 1 \text{ ly} = 9.46 \times 10^{15} \text{ m})$$

$$= \frac{4.29 \times 9.46 \times 10^{15}}{3.08 \times 10^{16}} \text{ parsec}$$

$$(\because 1 \text{ parsec} = 3.08 \times 10^{16} \text{ m})$$

$$= 1.318 \text{ parsec} = 1.32 \text{ parsec}$$

Q 16. It is claimed that the two cesium clocks if allowed of run for 100 yr free from any disturbance, may differ by only about 0.02 s. What does this imply for the accuracy of the standard cesium clock in measuring a time interval of 1s?

Sol. Time interval = 100 yr

$$= 100 \times 365 \times 24 \times 60 \times 60 \text{ s}$$

$$= 3.155 \times 10^9 \text{ s}$$

Difference in time = 0.2 s

$$\therefore \text{Fractional error} = \frac{\text{Difference in time(s)}}{\text{Time interval(s)}}$$

$$= \frac{0.2}{3.155 \times 10^9} = 6.34 \times 10^{-12}$$

$$= 10 \times 10^{-12} \approx 10^{-11}$$

$\therefore$ In 1s, the difference is 10^{-11} to 6.34×10^{12}.

Hence, degree of accuracy shown by the atomic clock in 1s is 1 part in $\frac{1}{10^{-11}}$ to $\frac{1}{6.34 \times 10^{-12}}$

or 10^{11} to 10^{12}.

Q 17. Estimate the average mass density of sodium atom assuming, its size to be about 2.5 Å (Use the known values of Avogadro's number and the atomic mass of sodium). Compare it with the density of sodium in its crystalline phase 970 kg m^{-3}. Are the two densities of the same order of magnitude? If so, why?

Sol. Average radius of sodium atom,

$$r = 2.5 \text{ Å} = 2.5 \times 10^{-10} \text{ m}$$

$$\therefore \text{Volume of sodium atom} = \frac{4}{3}\pi r^3$$

$$= \frac{4}{3} \times 3.14 \times (2.5 \times 10^{-10})^3$$

$$= 65.42 \times 10^{-30} \text{ m}^3$$

Mass of a mole of sodium = 23 g = 23×10^{-3} kg

One mole contains 6.023×10^{23} atoms, hence the mass of sodium atom,

$$M = \frac{23 \times 10^{-3}}{6.023 \times 10^{23}} \text{ kg}$$

$$= 3.82 \times 10^{-26} \text{ kg}$$

$\therefore$ Average mass density of sodium atom,

$$\rho = \frac{M}{V} = \frac{3.82 \times 10^{-26}}{65.42 \times 10^{-30}} \text{ kgm}^{-3}$$

$$= 0.64 \times 10^3 \text{ kgm}^{-3}$$

Density of sodium in crystalline phase = 970 kgm^{-3}

$$= 0.970 \times 10^3 \text{ kgm}^{-3}$$

$$\therefore \frac{\text{Average mass density of sodium atom}}{\text{Density of sodium of crystalline phase}} = \frac{0.64 \times 10^3}{0.970 \times 10^3}$$

$$= 0.66$$

Both densities are of the same order, i.e. of the order of 10^3.

This is because in the solid phase atoms are tightly packed, so the atomic mass density is close to the mass density of the solid.

***Q* 18.** A SONAR (sound navigation and ranging) uses ultrasonic waves to detect and locate objects under water. In a submarine equipped with a SONAR, the time delay between generation of a probe wave and the reception of its echo after reflection from an enemy submarine is found to be 77.0 s. What is the distance of the enemy submarine? (speed of sound in water=1450 ms^{-1}).

Sol. Time taken by the wave to go from submarine to enemy submarine is

$$t = \frac{77}{2} = 38.5\,s$$

Speed of sound, $v = 1450\ ms^{-1}$

Distance of enemy submarine,

$$\therefore \quad s = vt = 1450 \times 38.50$$
$$= 55825\ m$$
$$= 55.825\ km$$

***Q* 19.** The farthest objects in our universe discovered by modern astronomers are so distant that light emitted by them takes billions of years to reach the earth. These objects (known as quasars) have many puzzling features which have not yet been satisfactorily explained. What is the distance in km of a quasar from which light takes 3.0 billion years to reach us?

Sol. Time taken, $t = 3 \times 10^9$ yr

$$= 3 \times 10^9 \times 365 \times 24 \times 60 \times 60\,s$$

Velocity of light, $c = 3 \times 10^8\ ms^{-1}$

$\therefore$ Distance of quasar from earth $= ct$

$$= 3 \times 10^8 \times 3 \times 10^9 \times 365 \times 24 \times 3600\ m$$
$$= 2.8 \times 10^{25}\ m$$
$$= 2.8 \times 10^{22}\ km$$

Objective Problems

[Level 1]

Measurement of Length and Time

1. Which one is not a unit of time?
(a) Leap year (b) Year (c) Shake (d) Light year

2. "Parsec" is the unit of
(a) time (b) distance
(c) frequency (d) angular acceleration

3. 1 light year distance is equal to
(a) 9.46×10^{10} km (b) 9.46×10^{12} km
(c) 9.46×10^{9} km (d) 9.46×10^{15} km

4. Parallactic second is equal to
(a) 9.4605×10^{15} m (b) 3.07×10^{16} m
(c) 1.496×10^{11}m (d) 3×10^{8} m

5. A new unit of length is chosen such that the speed of light in vacuum is unity. What is the distance between the sun and the earth in terms of the new unit, if light takes 8 min and 20 s to cover this distance?
(a) 300 (b) 400 (c) 500 (d) 600

6. A student measures the thickness of a human hair by looking at it through a microscope of magnification 100. He makes 20 observations and finds that the average width of the hair is 3.5 mm. What is the estimate on the thickness of the hair?
(a) 0.0035 mm (b) 0.035 mm
(c) 0.01 m (d) 0.7 mm

7. Which of the following is the most precise device for measuring length?
(a) A vernier calliper with 20 divisions on the sliding scale
(b) An optical instrument that can measure length to within a wavelength of light
(c) A screw gauge of pitch 1 mm and 100 divisions on the circular scale
(d) All the above are equally precise

Units and Dimensions

8. The dimensions of impulse are equal to that of
(a) force (b) linear momentum
(c) pressure (d) angular momentum

9. In the SI system, the unit of temperature is
(a) degree centigrade (b) kelvin
(c) degree celsius (d) degree fahrenheit

10. Which one of the following have same dimensions?
(a) Torque and force
(b) Potential energy and force
(c) Torque and potential energy
(d) Planck's constant and linear momentum

11. Which of the following does not posses the same dimensions as that of pressure?
(a) Stress (b) Bulk modulus
(c) Thrust (d) Energy density

12. Which of the following is a dimensional constant?
(a) Poission's ratio (b) Refractive index
(c) Relative density (d) Gravitational constant

13. Which one of the following is not the dimensionless quantity?
(a) Planck's constant (b) Dielectric constant
(c) Solid angle (d) Strain

14. Joule $\times$ second is the unit of
(a) energy (b) momentum
(c) angular momentum (d) power

15. Which of the following is not equal to watt?
(a) joule/second (b) ampere $\times$ volt
(c) $(\text{ampere})^2 \times$ ohm (d) ampere/volt

16. Which of the following is not the units of surface tension?
(a) N/m (b) J/m^2
(c) kg/s^2 (d) None of these

17. Wb/m^2 is equal to
(a) dyne (b) tesla (c) watt (d) henry

18. Dimensional formula for electromotive force is same as that for
(a) potential (b) current
(c) force (d) energy

19. Which of the following has the dimensions of pressure?
(a) $[ML^{-2}T^{-2}]$ (b) $[M^{-1}L^{-1}]$
(c) $[MLT^{-2}]$ (d) $[ML^{-1}T^{-2}]$

20. Dimensions of torque are
(a) $[M^2L^2T^2]$ (b) $[ML^2T^{-2}]$
(c) $[ML^0T^{-1}]$ (d) $[ML^2T^{-1}]$

21. Dimensions of impulse are
(a) $[ML^{-2}T^{-3}]$ (b) $[ML^{-2}]$
(c) $[MLT^{-1}]$ (d) $[MLT^{-2}]$

22. What is the dimensional formula of gravitational constant?
(a) $[ML^2T^{-2}]$
(b) $[ML^{-1}T^{-1}]$
(c) $[M^{-1}L^3T^{-2}]$
(d) None of the above

23. Dimensions of surface tension are
(a) $[M^2L^2T^{-2}]$ (b) $[M^2LT^{-2}]$
(c) $[MT^{-2}]$ (d) $[MLT^{-2}]$

24. The dimensional formula for Young's modulus is
(a) $[ML^{-1}T^{-2}]$ (b) $[M^0LT^{-2}]$
(c) $[MLT^{-2}]$ (d) $[ML^2T^{-2}]$

25. Which of the following is the dimensions of the coefficient of friction?
(a) $[M^2L^2T]$ (b) $[M^0L^0T^0]$
(c) $[ML^2T^{-2}]$ (d) $[M^2L^2T^{-2}]$

26. The dimensional formula for the action will be
(a) $[MLT^{-2}]$ (b) $[M^2LT^{-2}]$
(c) $[ML^2T^{-1}]$ (d) $[M^2L^2T^{-2}]$

27. $[ML^{-1}T^{-1}]$ stand for dimensions of
(a) work (b) torque
(c) linear momentum (d) coefficient of viscosity

28. Dimensions of relative density is
(a) $[ML^{-2}]$ (b) $[ML^{-3}]$
(c) dimensionless (d) $[M^2L^{-6}]$

29. The dimensions of the ratio of angular to linear momentum are
(a) $[M^0LT^0]$ (b) $[MLT^{-1}]$
(c) $[ML^2T^{-1}]$ (d) $[M^{-1}L^{-1}T^{-1}]$

30. The dimensional formula for thermal resistance is
(a) $[ML^2T^{-3}K^{-1}]$ (b) $[ML^2T^{-2}A^{-1}]$
(c) $[ML^2T^{-3}K^{-2}]$ (d) $[M^{-1}L^{-2}T^3K]$

31. $[ML^2T^{-3}A^{-1}]$ is the dimensional formula for
(a) capacitance (b) resistance
(c) resistivity (d) potential difference

32. Temperature can be expressed as a derived quantity in terms of any of the following.
(a) length and mass (b) mass and time
(c) length, mass and time (d) None of these

33. Given that $y = a\cos\left(\frac{t}{p} - qx\right)$, where t represents time. In The following statements which is true?
(a) The unit of x is same as that of q
(b) The unit of x is same as that of p
(c) The unit of t is same as that of q
(d) The unit of t is same as that of p

34. The dimensional formula $[ML^0T^{-3}]$ is more closely associated with
(a) power
(b) energy
(c) intensity
(d) velocity gradient

35. Which of the following is dimensionally correct?
(a) Pressure = energy per unit area
(b) Pressure = energy per unit volume
(c) Pressure = force per unit volume
(d) Pressure = momentum per unit volume per unit time

36. Assuming that the mass m of the largest stone that can be moved by a flowing river depends upon the velocity v of the water, its density ρ and the acceleration due to gravity g. Then, m is directly proportional to
(a) v^3 (b) v^4
(c) v^5 (d) v^6

37. If p represents radiation pressure, c represents speed of light and Q represents radiation energy striking a unit area per second, then non-zero integers x, y and z such that $p^xQ^yc^z$ is dimensionless are
(a) $x = 1,\ y = 1, z = -1$
(b) $x = 1,\ y = -1, z = 1$
(c) $x = -1,\ y = 1, z = 1$
(d) $x = 1,\ y = 1, z = 1$

38. The units of length, velocity and force are doubled. Which of the following is the correct change in the other units?
(a) Unit of time is doubled
(b) Unit of mass is doubled
(c) Unit of momentum is doubled
(d) Unit of energy is doubled

39. Which of the following pairs has the same units?
(a) Wavelength and Rydberg constant
(b) Relative velocity and relative density
(c) Thermal capacity and Boltzmann constant
(d) Time period and acceleration gradient

40. The dimensional representation of specific resistance in terms of charge Q is
(a) $[ML^3T^{-1}Q^{-2}]$ (b) $[ML^2T^{-2}Q^2]$
(c) $[MLT^{-2}Q^{-1}]$ (d) $[ML^2T^{-2}Q^{-1}]$

41. Which of the following will have the dimensions of time?
(a) LC (b) $\frac{R}{L}$ (c) $\frac{L}{R}$ (d) $\frac{C}{L}$

42. If C and R denote capacity and resistance, the dimensions of CR are
(a) $[M^0L^0T]$
(b) $[ML^0T]$
(c) $[M^0L^0T^2]$
(d) not expressible in terms of M, L and T

43. The force F on a sphere of radius a moving in a medium with velocity v is given by $F = 6\pi\eta a v$. The dimensions of η are
(a) $[ML^{-3}]$ (b) $[MLT^{-2}]$
(c) $[MT^{-1}]$ (d) $[ML^{-1}T^{-1}]$

44. The equation of a wave is given by $y = a\sin\omega\left(\frac{x}{v} - k\right)$ where ω is angular velocity and v is the linear velocity. The dimension of k will be

(a) $[T^{-2}]$ (b) $[T^{1}]$ (c) $[T]$ (d) $[LT]$

45. A force is given by $F = at + bt^2$, where t is the time. The dimensions of a and b are

(a) $[MLT^{-4}]$ and $[MLT]$ (b) $[MLT^{-1}]$ and $[MLT^{0}]$
(c) $[MLT^{-3}]$ and $[MLT^{-4}]$ (d) $[MLT^{-3}]$ and $[MLT^{0}]$

46. The dimensional formula for Planck's constant and angular momentum is

(a) $[ML^2T^{-2}]$ and $[MLT^{-1}]$ (b) $[ML^2T^{-1}]$ and $[ML^2T^{-1}]$
(c) $[ML^3T^{-1}]$ and $[ML^2T^{-2}]$ (d) $[MLT^{-1}]$ and $[MLT^{-2}]$

47. The dimensions of $\frac{1}{2}\varepsilon_0 E^2$ (ε_0 is the permittivity of the space and E is electric field), are

(a) $[ML^2T^{-1}]$ (b) $[ML^{-1}T^{-2}]$ (c) $[ML^2T^{-2}]$ (d) $[MLT^{-1}]$

48. The dimensions of $\frac{a}{b}$ in the equation $p = \frac{a - t^2}{bx}$, where p is pressure, x is distance and t is time, are

(a) $[M^2LT^{-3}]$ (b) $[MT^{-2}]$
(c) $[LT^{-3}]$ (d) $[ML^3T^{-1}]$

49. Dimensions of velocity gradient are

(a) $[M^0L^0T^{-1}]$ (b) $[ML^{-1}T^{-1}]$
(c) $[M^0LT^{-1}]$ (d) $[ML^0T^{-1}]$

50. The dimensional formula for emf e in MKS system will be

(a) $[ML^2T^{-2}Q^{-1}]$ (b) $[ML^2T^{-1}]$
(c) $[ML^{-2}Q^{-1}]$ (d) $[MLT^{-2}Q^{-2}]$

51. The velocity v of a particle at time t is given by $v = at + \frac{b}{t + c}$, where a, b and c are constants. The dimensions of a, b and c are, respectively

(a) $[LT^{-2}]$, $[L]$ and $[T]$ (b) $[L^2]$, $[T]$ and $[LT^2]$
(c) $[LT^2]$, $[LT]$ and $[L]$ (d) $[L]$, $[LT]$ and $[T^2]$

52. What is the units of $k = \frac{1}{4\pi\varepsilon_0}$?

(a) $C^2N^{-1}m^{-2}$ (b) Nm^2C^{-2}
(c) Nm^2C^2 (d) Unitless

53. Pressure gradient has the same dimensions as that of

(a) velocity gradient (b) potential gradient
(c) energy gradient (d) None of these

54. The unit of permittivity of free space, ε_0 is

(a) coulomb/newton-metre
(b) newton-metre^2/coulomb^2
(c) coulomb^2/newton-metre^2
(d) coulomb^2/(newton-metre)^2

55. Dimensions of electrical resistance are

(a) $[ML^2T^{-3}A^{-1}]$ (b) $[ML^2T^{-3}A^{-2}]$
(c) $[ML^3T^{-3}A^{-2}]$ (d) $[ML^{-1}L^3T^3A^3]$

56. The magnetic moment has dimensions of

(a) $[LA]$ (b) $[L^2A]$ (c) $[LT^{-1}A]$ (d) $[L^2T^{-1}A]$

57. The dimensional representation of specific resistance in terms of charge Q is

(a) $[ML^3T^{-1}Q^{-2}]$ (b) $[ML^2T^{-2}Q^2]$
(c) $[MLT^{-2}Q^{-1}]$ (d) $[ML^2T^{-2}Q^{-1}]$

Significant Figures

58. The significant figures of the number 6.0023 is

(a) 2 (b) 5
(c) 4 (d) 1

59. What is the number of significant figures in 0.0310×10^3?

(a) 2 (b) 3
(c) 4 (d) 6

60. The number of significant figures in 11.118×10^{-6} V is

(a) 3 (b) 4
(c) 5 (d) 6

61. In which of the following numerical values, all zeros are significant?

(a) 0.2020 (b) 20.2
(c) 2020 (d) None of these

62. A student measured the diameter of a wire using a screw gauge with least count 0.001 cm and listed the measurements. The correct measurement is

(a) 8.320 cm (b) 5.3 cm
(c) 5.32 cm (d) 5.3200 cm

63. The length, breadth and thickness of rectangular sheet of metal are 4.234 m, 1.005 m and 2.01 cm, respectively. The volume of the sheet to correct significant figures is

(a) 0.0855 m^3 (b) 0.086 m^3
(c) 0.08556 m^3 (d) 0.08 m^3

64. Three measurements are made as 18.425 cm, 7.21 cm and 5.0 cm. The addition should be written as

(a) 30.635 cm (b) 30.64 cm (c) 30.63 cm (d) 30.6 cm

65. Subtract 0.2 J from 7.26 J and express the result with correct number of significant figures.

(a) 7.1 (b) 7.05
(c) 7 (d) None of these

66. Multiply 107.88 by 0.610 and express the result with correct number of significant figures.

(a) 65.8068 (b) 64.807 (c) 65.81 (d) 65.8

67. When 97.52 is divided by 2.54, the correct result is

(a) 38.3937 (b) 38.394
(c) 65.81 (d) 38.4

68. The radius of a thin wire is 0.16 mm. The area of cross-section of the wire in mm^2 with correct number of significant figures is
(a) 0.08 (b) 0.080
(c) 0.0804 (d) 0.080384

69. What is the number of significant figure in $(3.20 + 4.80) \times 10^5$?
(a) 5 (b) 4
(c) 3 (d) 2

70. What is the value of $[(5.0 \times 10^{-6})(5.0 \times 10^{-8})]$ with due regards to significant digits?
(a) 25×10^{-14} (b) 25.0×10^{-14}
(c) 2.50×10^{-13} (d) 250×10^{-15}

71. The mass of a box is 2.3 kg. Two gold pieces of masses 20.15 g and 20.17 g are added to the box. The total mass of the box to correct significant figures is
(a) 2.3 kg (b) 2.34 kg
(c) 2.3432 kg (d) 2.31 kg

72. Subtract 0.2 kg from 34 kg. The result in terms of proper significant figure is
(a) 33.8 kg (b) 33.80 kg
(c) 34 kg (d) 34.0 kg

73. The length, breadth and thickness of a block are given by $l = 12$ cm, $b = 6$ cm and $t = 2.45$ cm. The volume of the block according to the idea of significant figures should be
(a) 1×10^2 cm^3 (b) 2×10^2 cm^3
(c) 1.763×10^2 cm^3 (d) None of these

Error Analysis

74. The length of a rod is (11.05 ± 0.2) cm. What is the length of the two rods?
(a) (22.1 ± 0.05) cm
(b) (22.1 ± 0.1) cm
(c) (22.10 ± 0.05) cm
(d) (22.10 ± 0.2) cm

75. The radius of a ball is (5.2 ± 0.2) cm. The percentage error in the volume of the ball is approximately
(a) 11% (b) 4%
(c) 7% (d) 9%

76. A physical quantity Q is calculated according to the expression

$$Q = \frac{A^3 B^3}{C\sqrt{D}}$$

If percentage errors in A, B, C, D are 2%, 1%, 3% and 4%, respectively. What is the percentage error in Q?
(a) ± 8% (b) ± 10%
(c) ± 14% (d) ± 12%

77. A body travels uniformly a distance of (13.8 ± 0.2) m in a time (4.0 ± 0.3) s. The velocity of the body within error limit is
(a) (3.45 ± 0.2) ms^{-1}
(b) (3.45 ± 0.3) ms^{-1}
(c) (3.45 ± 0.4) ms^{-1}
(d) (3.45 ± 0.5) ms^{-1}

78. If the error in the measurement of momentum of a particle is (+ 100%), then the error in the measurement of kinetic energy is
(a) 100% (b) 200%
(c) 300% (d) 400%

79. If the error in measuring diameter of a circle is 4%, the error in measuring radius of the circle would be
(a) 2% (b) 8%
(c) 4% (d) 1%

80. The values of two resistors are (5.0 ± 0.2) kΩ and (10.0 ± 0.1) kΩ. What is the percentage error in the equivalent resistance when they are connected in parallel?
(a) 2% (b) 5%
(c) 7% (d) 10%

81. The heat generated in a wire depends on the resistance, current and time. If the error in measuring the above are 1%, 2% and 1%, respectively. The maximum error in measuring the heat is
(a) 8% (b) 6%
(c) 18% (d) 12%

82. A force F is applied on a square plate of side L. If the percentage error in the determination of L is 2% and that in F is 4%. What is the permissible error in pressure?
(a) 8% (b) 6%
(c) 4% (d) 2%

83. A cuboid has volume $V = l \times 2l \times 3l$, where l is the length of one side. If the relative percentage error in the measurement of l is 1%, then the relative percentage error in measurement of V is
(a) 18% (b) 6%
(c) 3% (d) 1%

Miscellaneous Problems

84. The ratio of the SI unit to the CGS unit of modulus of rigidity is
(a) 10^2 (b) 10^{-2}
(c) 10^{-1} (d) 10

85. Imagine a system of unit in which the unit of mass is 10 kg, length is 1 km and time is 1 min. Then, 1 J in this system is equal to
(a) 360 (b) 3.6 (c) 36×10^5 (d) 36×10^{-5}

86. The dimensional formula for molar thermal capacity is same as that of

(a) gas constant (b) specific heat
(c) Boltzmann's constant (d) Stefan's constant

87. In measuring electric energy, 1kWh is equal to

(a) 3.6×10^4 J (b) 3.6×10^6 J
(c) 7.3×10^6 J (d) None of these

88. Out of the following four dimensional quantities, which one qualifies to be called a dimensional constant?

(a) Acceleration due to gravity
(b) Surface tension of water
(c) Weight of a standard kilogram mass
(d) The velocity of light in vacuum

89. The square root of the product of inductance and capacitance has the dimensions of

(a) length
(b) time
(c) mass
(d) no dimension

90. With usual notation, the following equation, said to give the distance covered in the nth second, i.e. $s_n = u + a\dfrac{(2n-1)}{2}$ is

(a) numerically correct only
(b) dimensionally correct only
(c) both dimensionally and numerically only
(d) neither numerically nor dimensionally correct

[Level 2]

Only One Correct Option

1. A quantity is given by $X = \dfrac{\varepsilon_0 lV}{t}$, where V is the potential difference and l is the length. Then, X has dimensional formula same as that of

(a) resistance (b) charge
(c) voltage (d) current

2. The length of a strip measured with a metre rod is 10.0 cm. Its width measured with a vernier callipers is 1.00 cm. The least count of the metre rod is 0.1 cm and that of vernier callipers 0.01 cm. What will be error in its area?

(a) ± 13% (b) ± 7%
(c) ± 4% (d) ± 2%

3. The length of cylinder is measured with a metre rod having least count 0.1 cm. Its diameter is measured with vernier calipers having least count 0.01 cm. Given that length is 5.0 cm and radius is 2.0 cm. The percentage error in the calculated value of the volume will be

(a) 1.5% (b) 2.5%
(c) 3.5% (d) 4%

4. The random error in the arithmetic means of 100 observations is x, then random error in the arithmetic mean of 400 observations would be

(a) $4x$ (b) $\dfrac{1}{4}x$
(c) $2x$ (d) $\dfrac{1}{2}x$

5. Dimensions of ohm are same as

(a) $\dfrac{h}{e}$ (b) $\dfrac{h^2}{e}$
(c) $\dfrac{h}{e^2}$ (d) $\dfrac{h^2}{e^2}$

(where h is Planck's constant and e is charge)

6. Given that $\displaystyle\int \frac{dx}{\sqrt{2ax - x^2}} = a^n \sin^{-1}\left[\frac{x-a}{a}\right]$

where, a = constant. Using dimensional analysis, the value of n is

(a) 1 (b) zero
(c) − 1 (d) None of these

7. If E = energy, G = gravitational constant, I = impulse and M = mass, then dimensions of $\dfrac{GIM^2}{E^2}$ are same as that of

(a) time (b) mass
(c) length (d) force

8. The dimensional formula for magnetic flux is

(a) $[ML^2T^{-2}A^{-1}]$
(b) $[ML^3T^{-2}A^{-2}]$
(c) $[M^0L^{-2}T^{-2}A^{-2}]$
(d) $[ML^2T^{-1}A^2]$

9. Using mass (M), length (L), time (T) and current (A) as fundamental quantities, the dimensions of permeability are

(a) $[M^{-1}LT^{-2}A]$
(b) $[ML^{-2}T^{-2}A^{-1}]$
(c) $[MLT^{-2}A^{-2}]$
(d) $[MLT^{-1}A^{-1}]$

10. Let g be the acceleration due to gravity at earth's surface and K the rotational kinetic energy of the earth. Suppose the earth's radius decreases by 2%. Keeping mass to be constant, then

(a) g increases by 2% and K increases by 2%
(b) g increases by 4% and K increases by 4%
(c) g increases by 4% and K increases by 2%
(d) g increases by 2% and K increases by 4%

11. If the energy (E), velocity (v) and force (F) be taken as fundamental quantities, then the dimension of mass will be

(a) $[Fv^{-2}]$ (b) $[Fv^{-1}]$
(c) $[Ev^{-2}]$ (d) $[Ev^2]$

12. In a system of units, the units of mass, length and time are 1 quintal, 1 km and 1 h, respectively. In this system, 1 N force will be equal to

(a) 1 new unit (b) 129.6 new unit
(c) 427.6 new unit (d) 60 new unit

13. If force F, length L and time T are taken as fundamental units, the dimensional formula for mass will be

(a) $[FL^{-1}T^2]$ (b) $[FLT^{-2}]$
(c) $[FL^{-1}T^{-1}]$ (d) $[FL^5T^2]$

14. Given that $y = A\sin\left[\left(\frac{2\pi}{\lambda}(ct - x)\right)\right]$, where y and x are measured in metre. Which of the following statements is true?

(a) The unit of λ is same as that of x and A
(b) The unit of λ is same as that of x but not of A
(c) The unit of c is same as that of $\frac{2\pi}{\lambda}$
(d) The unit of $(ct - x)$ is same as that of $\frac{2\pi}{\lambda}$

15. The frequency of vibration of string is given by $f = \frac{p}{2l}\left[\frac{F}{m}\right]^{1/2}$. Here, p is number of segments in the string and l is the length. The dimensional formula for m will be

(a) $[M^0LT^{-1}]$ (b) $[ML^0T^{-1}]$
(c) $[ML^{-1}T^0]$ (d) $[M^0L^0T^0]$

16. You measure two quantities as $A = 1.0\text{ m} \pm 0.2\text{ m}$, $B = 2.0\text{ m} \pm 0.2\text{ m}$. We should report correct value for $\sqrt{AB}$ as

(a) 1.4 m± 0.4 m (b) 1.41 m± 0.15 m
(c) 1.4 m± 0.3 m (d) 1.4 m± 0.2 m

17. Which of the following measurement is most precise?

(a) 5.00 mm (b) 5.00 cm
(c) 5.00 m (d) 5.00 km

18. The mean length of an object is 5 cm. Which of the following measurements is most accurate?

(a) 4.9 cm (b) 4.805 cm
(c) 5.25 cm (d) 5.4 cm

19. The vernier scale of a travelling microscope has 50 divisions which coincide with 49 main scale divisions. If each main scale division is 0.5 mm, then what is the minimum inaccuracy in the measurement of distance?

(a) 0.02 mm (b) 0.05 mm
(c) 0.01 mm (d) 0.1 mm

20. Time for 20 oscillations of a pendulum is measured as $t_1 = 39.6$ s, $t_2 = 39.9$ s and $t_3 = 39.5$ s.
What is the accuracy of the measurement?

(a) ± 0.1 s (b) ± 0.2 s (c) ± 0.01 s (d) ± 0.5 s

More than One Correct Options

1. Given, $x = \frac{ab^2}{c^3}$. If the percentage errors in a, b and c are $\pm 1\%, \pm 3\%$ and $\pm 2\%$, respectively.

(a) The percentage error in x can be ±3%
(b) The percentage error in x can be ±7%
(c) The percentage error in x can be ±18%
(d) The percentage error in x can be ±19%

2. If P, Q, R are physical quantities, having different dimensions, which of the following combinations can never be a meaningful quantity?

(a) $(P - Q)/R$ (b) $PQ - R$ (c) PQ/R (d) $(R + Q)/P$

3. If Planck's constant (h) and speed of light in vacuum (c) are taken as two fundamental quantities, which one of the following can, in addition, be taken to express length, mass and time in terms of the three chosen fundamental quantities?

(a) Mass of electron (m_e)
(b) Universal gravitational constant (G)
(c) Charge of electron (e)
(d) Mass of proton (m_p)

Assertion and Reason

Directions (Q. Nos. 1-17) *These questions consists of two statements each printed as Assertion and Reason. While answering these questions, you are required to choose any one of the following five responses.*

(a) If both Assertion and Reason are correct and Reason is the correct explanation of Assertion
(b) If both Assertion and Reason are correct but Reason is not the correct explanation of Assertion
(c) If Assertion is true but Reason is false
(d) If Assertion is false but Reason is true
(e) If both Assertion and Reason are false

1. Assertion Pressure has the dimensions of energy density.

Reason Energy density $= \frac{\text{energy}}{\text{volume}} = \frac{[ML^2T^{-2}]}{[L^3]}$

$= [ML^{-1}T^{-2}] = \text{pressure}.$

2. Assertion Method of dimension cannot be used for deriving formulae containing trigonometrical ratios.

Reason This is because trigonometrical ratios have no dimensions.

3. **Assertion** When percentage errors in the measurement of mass and velocity are 1% and 2% respectively, the percentage error in KE is 5%.

Reason KE or $E = \frac{1}{2} mv^2, \frac{\Delta E}{E} = \frac{\Delta m}{m} + \frac{2\Delta v}{v}$

4. **Assertion** The error in the measurement of radius of the sphere is 0.3%. The permissible error in its surface area is 0.6%.

Reason The permissible error is calculated by the formula

$$\frac{\Delta A}{A} = 4\frac{\Delta r}{r}.$$

5. **Assertion** The light year and wavelength consist of dimensions of length.

Reason Both light year and wavelength represent time.

6. **Assertion** Number of significant figures in 0.005 is one and that in 0.500 are three.

Reason This is because zeros before decimal are non-significant.

7. **Assertion** Out of two measurements $l = 0.7$ m and $l = 0.70$ m, the second one is more accurate.

Reason In every measurement, more the last digit is not accurately known.

8. **Assertion** When we change the unit of measurement of a quantity, its numerical value changes.

Reason Smaller the unit of measurement smaller is its numerical value.

9. **Assertion** L/R and CR both have same dimensions.

Reason L/R and CR both have dimensions of time.

10. **Assertion** $\sqrt{\frac{\text{Magnetic dipole moment} \times \text{moment induction}}{\text{Moment of inertia}}}$

Dimensional formula = $[M^0L^0T]$

Reason The given dimension is that of frequency.

11. **Assertion** $\sqrt{\frac{\text{Modulus of elasticity}}{\text{Density}}}$ has the unit ms^{-1}.

Reason Acceleration has the dimensions of $\frac{1}{(\sqrt{\varepsilon_0 \mu_0})t}$.

12. **Assertion** If $x = \frac{a^n}{b^m}$, then $\frac{\Delta x}{x} = n\left(\frac{\pm \Delta a}{a}\right) - m\left(\frac{\pm \Delta b}{b}\right)$

The change in a or b, i.e. Δa or Δb may be comparable to a and b.

Reason The above relation is valid when $\Delta a << a$ and $\Delta b << b$.

13. **Assertion** Systematic errors and random errors fall in the same group of errors.

Reason Both systematic and random errors are based on the cause of error.

14. **Assertion** Absolute error may be negative or positive.

Reason Absolute error is the difference between the real value and the measured value of a physical quantity.

15. **Assertion** The watches having hour hand, minute hand and seconds hand have least count as 1 s.

Reason Least count is the maximum measurement that can be measured accurately by an instrument.

16. **Assertion** Pendulum bob is preferred to be spherical.

Reason Sphere has minimum surface area.

17. **Assertion** A screw gauge having a smaller value of pitch has greater accuracy.

Reason The least count of screw gauge is directly proportional to the number of divisions on circular scale.

Match the Columns

1. Match the following columns.

Column I	Column II
(A) R/L	(p) Time
(B) C/R	(q) Frequency
(C) E/B	(r) Speed
(D) $\sqrt{\varepsilon_0 \mu_0}$	(s) None of the above

2. Match the following columns.

Column I	Column II
(A) Stress	(p) Pressure
(B) Strain	(q) Energy density
(C) Modulus of elasticity	(r) Angle
(D) Torque	(s) Energy

3. Suppose force (F), area (A) and time (T) are the fundamental units, then the match the following columns.

Column I	Column II
(A) Work	(p) $[A^{1/2}T^{-1}]$
(B) Moment of inertia	(q) $[FA^{1/2}]$
(C) Velocity	(r) $[FA^{1/2}T^2]$

4. Match the following columns.

Column I	Column II
(A) Electrical resistance	(p) $[M^{-1}L^{-2}T^4A^2]$
(B) Capacitance	(q) $[ML^2T^{-2}A^{-2}]$
(C) Magnetic field	(r) $[ML^2T^{-3}A^{-2}]$
(D) Inductance	(s) $[MT^{-2}A^{-1}]$

5. Match the following two columns.

	Column I		Column II
(A)	GM_eM_s	(p)	$[M^2L^2T^{-3}]$
(B)	$\frac{3RT}{M}$	(q)	$[ML^3T^{-2}]$
(C)	$\frac{F^2}{q^2B^2}$	(r)	$[L^2T^{-2}]$
(D)	$\frac{GM_e}{R_e}$	(s)	None of the above

Entrance Gallery

2014

1. To find the distance d over which a signal can be seen clearly in foggy conditions, a railway engineer uses dimensional analysis and assumes that the distance depends on the mass density ρ of the fog, intensity (power/area) S of the light from the signal and its frequency f. The engineer finds that d is proportional to $S^{1/n}$. The value of n **[JEE Advanced]**

(a) 2 (b) 3
(c) 1 (d) 4

2. During Searle's experiment, zero of the vernier scale lies between 3.20×10^{-2} m and 3.25×10^{-2} m of the main scale. The 20th division of the vernier scale exactly coincides with one of the main scale divisions. When an additional load of 2 kg is applied to the wire, the zero of the vernier scale still lies between 3.20×10^{-2} m and 3.25×10^{-2} m of the main scale but now the 45th division of vernier scale coincides with one of the main scale divisions. The length of the thin metallic wire is 2 m and its cross-sectional area is 8×10^{-7} m^2. The least count of the vernier scale is 1.0×10^{-5} m. The maximum percentage error in the Young's modulus of the wire is **[JEE Advanced]**

(a) 1 (b) 2
(c) 4 (d) 8

3. A student measured the length of a rod and wrote it as 3.50 cm. Which instrument did he use to measure it? **[JEE Main]**

(a) A metre scale
(b) A vernier calliper where the 10 divisions in vernier scale match with 9 divisions in main scale and main scale has 10 divisions in 1 cm
(c) A screw gauge having 100 divisions in the circular scale and pitch as 1 mm
(d) A screw gauge having 50 divisions in the circular scale and pitch 1 mm

4. The current voltage relation of diode is given by $I=(e^{1000V/T}-1)$ mA, where the applied V is in volts and the temperature T is in degree kelvin. If a student makes an error measuring $\pm$ 0.01 V while measuring the current of 5 mA at 300 K, what will be the error in the value of current in mA? **[JEE Main]**

(a) 0.2 mA (b) 0.02 mA (c) 0.5 mA (d) 0.05 mA

5. In a simple pendulum experiment, the maximum percentage error in the measurement of length is 2% and that in the observation of the time period is 3%. Then, the maximum percentage error in determination of the acceleration due to gravity g is **[Kerala CEE]**

(a) 5% (b) 6% (c) 7% (d) 8%
(e) 10%

6. The pitch and the number of circular scale divisions in a screw gauge with least count 0.02 mm are respectively. **[Kerala CEE]**

(a) 1 mm and 100 (b) 0.5 mm and 50
(c) 1 mm and 100 (d) 0.5 mm and 100
(e) 1 mm and 200

7. A physical quantity Q is found to depend on observables x, y and z obeying relation $Q=\frac{x^3y^2}{z}$. The percentage error in the measurements of x, y and z are 1%, 2% and 4% respectively. What is percentage error in the quantity Q? **[Karnataka CET]**

(a) 4% (b) 3%
(c) 11% (d) 1%

2013

8. Match List I with List II and select the correct answer using the codes given below the lists. **[JEE Advanced]**

List I		List II
P. Boltzmann constant	1.	$[ML^2T^{-1}]$
Q. Coefficient of viscosity	2.	$[ML^{-1}T^{-1}]$
R. Planck's constant	3.	$[MLT^{-3}K^{-1}]$
S. Thermal conductivity	4.	$[ML^2T^{-2}K^{-1}]$

Codes

	P	Q	R	S
(a)	3	1	2	4
(b)	3	2	1	4
(c)	4	2	1	3
(d)	4	1	2	3

9. The diameter of a cylinder is measured using a vernier callipers with no zero error. It is found that the zero of the vernier scale lies between 5.10 cm and 5.15 cm of the main scale. The vernier scale has 50 divisions equivalent to 2.45 cm. The 24th division of the vernier scale exactly coincides with one of the main scale divisions. The diameter of the cylinder is **[JEE Advanced]**

(a) 5.112 cm (b) 5.124 cm
(c) 5.136 cm (d) 5.148 cm

10. Let $[\varepsilon_0]$ denotes the dimensional formula of the permittivity of vacuum. If M = mass, L = length, T = time and A = electric current, then **[JEE Main]**

(a) $[\varepsilon_0]=[M^{-1}L^{-3}T^2A]$ (b) $[\varepsilon_0]=[M^{-1}L^{-3}T^4A^2]$
(c) $[\varepsilon_0]=[M^{-2}L^2T^{-1}A^{-2}]$ (d) $[\varepsilon_0]=[M^{-1}L^2T^{-1}A^2]$

11. The number of significant figures in the numbers 4.8000×10^4 and 48000.50 are, respectively. [JEE Main]

(a) 5 and 6 (b) 5 and 7
(c) 2 and 7 (d) 2 and 6

12. The dimensional formula for inductance is [Karnataka CET]

(a) $[ML^2T^{-2}A^{-2}]$ (b) $[ML^2TA^{-2}]$
(c) $[ML^2T^{-1}A^{-2}]$ (d) $[ML^2T^{-2}A^{-1}]$

2012

13. In the determination of Young's modulus $\left(Y = \frac{4MLg}{\pi l d^2}\right)$ by using Searle's method, a wire of length $L = 2$ m and diameter $d = 0.5$ mm is used. For a load $M = 2.5$ kg, an extension $l = 0.25$ mm in the length of the wire is observed. Quantities d and l are measured using a screw gauge and a micrometer, respectively. They have the same pitch of 0.5 mm. The number of divisions on their circular scale is 100. The contributions to the maximum probable error of the Y measurement is [IIT JEE]

(a) due to the errors in the measurement of d and l are the same
(b) due to the error in the measurement of d is twice that due to the error in the measurement of l
(c) due to the error in the measurement of l is twice that due to the error in the measurement of d
(d) due to the error in the measurement of d is four times that due to the error in the measurement of l

14. Resistance of a given wire is obtained by measuring the current flowing in it and the voltage difference applied across it. If the percentage errors in the measurement of the current and the voltage difference are 3% each, then error in the value of resistance of the wire is [AIEEE]

(a) 6% (b) zero
(c) 1% (d) 3%

15. A spectrometer gives the following reading when used to measure the angle of a prism.

Main scale reading 58.5 degree.

Vernier scale reading 9 divisions.

Given that, 1 division on main scale corresponds to 0.5 degree. Total division on the vernier scale is 30 and match with 29 divisions of the main scale. The angle of the prism from the above data is [AIEEE]

(a) 58.59° (b) 59.77°
(c) 58.65° (d) 59°

16. The dimensional formula of physical quantity is $[M^a L^b T^c]$. Then, that physical quantity is [Karnataka CET]

(a) spring constant, if $a = 1, b = 1$, and $c = -2$
(b) surface tension, if $a = 1, b = 1$, and $c = -2$
(c) force, if $a = 1, b = 1, c = 2$
(d) angular frequency, if $a = 0, b = 0, c = -1$

17. In a slide calliper, $(m + 1)$ number of vernier division is equal to m number of smallest main scale divisions. If d unit is the magnitude of the smallest main scale division, then the magnitude of the vernier constant is [WB JEE]

(a) $d/(m+1)$ unit
(b) d/m unit
(c) $md/(m+1)$ unit
(d) $(m+1)d/m$ unit

2011

18. The density of a solid ball is to be determined in an experiment. The diameter of the ball is measured with a screw gauge, whose pitch is 0.5 mm and there are 50 divisions on the circular scale. The reading on the main scale is 2.5 mm and that on the circular scale is 20 divisions. If the measured mass of the ball has a relative error of 2%, the relative percentage error in the density is [IIT JEE]

(a) 0.9% (b) 2.4%
(c) 3.1% (d) 4.2%

19. A screw gauge gives the following reading when used to measure the diameter of a wire.

Main scale reading : 0 mm

Circular scale reading : 52 divisions

Given that, 1 mm on main scale corresponds to 100 divisions of the circular scale.

The diameter of wire from the above data is [AIEEE]

(a) 0.052 cm
(b) 0.026 cm
(c) 0.005 cm
(d) 0.52 cm

20. The mass and volume of a body are found to be 5.00 ± 0.05 kg and 1.00 ± 0.05 m^3 respectively. Then, the maximum possible percentage error in its density is [Kerala CEE]

(a) 6% (b) 3%
(c) 10% .(d) 5%
(e) 7%

Answers

Level 1

Objective Problems

1. (d)	**2.** (b)	**3.** (b)	**4.** (b)	**5.** (c)	**6.** (b)	**7.** (b)	**8.** (b)	**9.** (b)	**10.** (c)
11. (c)	**12.** (d)	**13.** (a)	**14.** (c)	**15.** (d)	**16.** (d)	**17.** (b)	**18.** (a)	**19.** (d)	**20.** (b)
21. (c)	**22.** (c)	**23.** (c)	**24.** (a)	**25.** (b)	**26.** (a)	**27.** (d)	**28.** (c)	**29.** (a)	**30.** (d)
31. (d)	**32.** (d)	**33.** (d)	**34.** (c)	**35.** (b)	**36.** (d)	**37.** (b)	**38.** (c)	**39.** (c)	**40.** (a)
41. (c)	**42.** (a)	**43.** (d)	**44.** (c)	**45.** (c)	**46.** (b)	**47.** (b)	**48.** (b)	**49.** (a)	**50.** (a)
51. (a)	**52.** (b)	**53.** (d)	**54.** (c)	**55.** (b)	**56.** (b)	**57.** (a)	**58.** (b)	**59.** (b)	**60.** (c)
61. (b)	**62.** (a)	**63.** (a)	**64.** (d)	**65.** (a)	**66.** (d)	**67.** (d)	**68.** (b)	**69.** (c)	**70.** (a)
71. (a)	**72.** (c)	**73.** (b)	**74.** (d)	**75.** (a)	**76.** (d)	**77.** (b)	**78.** (c)	**79.** (c)	**80.** (c)
81. (b)	**82.** (a)	**83.** (c)	**84.** (d)	**85.** (d)	**86.** (c)	**87.** (b)	**88.** (d)	**89.** (b)	**90.** (c)

Level 2

Only One Correct Option

1. (d)	**2.** (d)	**3.** (b)	**4.** (b)	**5.** (c)	**6.** (b)	**7.** (a)	**8.** (a)	**9.** (c)	**10.** (b)
11. (c)	**12.** (b)	**13.** (a)	**14.** (a)	**15.** (c)	**16.** (d)	**17.** (a)	**18.** (a)	**19.** (c)	**20.** (b)

More than One Correct Option

1. (a,b) **2.** (a,d) **3.** (a,b,d)

Assertion and Reason

1. (a)	**2.** (a)	**3.** (a)	**4.** (c)	**5.** (a)	**6.** (c)	**7.** (b)	**8.** (c)	**9.** (a)	**10.** (d)
11. (b)	**12.** (d)	**13.** (a)	**14.** (a)	**15.** (d)	**16.** (a)	**17.** (c)			

Match the Columns

1. $A \to q$; $B \to p$; $C \to r$; $D \to s$

2. $A \to r$; $B \to p$; $C \to s$; $D \to q$

3. $A \to q$; $B \to r$; $C \to p$

4. $A \to s$; $B \to p$; $C \to r$; $D \to q$

5. $A \to q$; $B \to r$; $C \to r$; $D \to s$

Entrance Gallery

1. (b)	**2.** (c)	**3.** (b)	**4.** (a)	**5.** (d)	**6.** (c)	**7.** (c)	**8.** (c)	**9.** (b)	**10.** (b)
11. (b)	**12.** (a)	**13.** (a)	**14.** (a)	**15.** (c)	**16.** (d)	**17.** (a)	**18.** (c)	**19.** (a)	**20.** (a)

Solutions

Level 1 : Objective Problems

1. Leap year, year and shake are the units of time.
3. 1 light year $= (3\times10^5)(365)(24)(3600) = 9.416\times10^{12}$ km
8. Impulse = change in linear momentum.
13. Solid angle, strain and dielectric constant are dimensionless quantities.
14. Since, $mvr = n\cdot\frac{h}{2\pi}$ and $E = hv$

    So, unit of h = joule-second = angular momentum
17. Wb/m^2 and tesla are the units of magnetic field.
21. Impulse = Force × time
24. Young's modulus and pressure have the same dimensions.
26. Action is a force.
28. Relative density $= \frac{\text{Density of substance}}{\text{Density of water at 4°C temperature}}$

    = Dimensionless
36. $m \propto v^a \rho^b g^c$. Writing the dimensions on both sides

    $[M] = [LT^{-1}]^a\,[ML^{-2}]^b\,[LT^{-2}]^c$

    $[M] = [M^b L^{a-3b+c} T^{-a-2c}]$

    $\therefore\quad b = 1$

    $a - 3b + c = 0;\ -a - 2c = 0$

    Solving these, we get $a = 6$

    Hence, $m \propto v^6$
37. Since, $p^x Q^y c^z$ is dimensionless. Therefore,

    $[ML^{-1}T^{-2}]^x\,[MT^{-3}]^y\,[LT^{-1}]^z = [M^0L^0T^0]$

    Only option (b) satisfies this expression.

    So, $x = 1, y = -1, z = 1$
38. Since, units of length, velocity and force are doubled,

    Hence, $[m] = \frac{[\text{force}]\,[\text{time}]}{[\text{velocity}]}$, $[\text{time}] = \frac{[\text{length}]}{[\text{velocity}]}$

    Hence, unit of mass and time remains same.

    Momentum is doubled.
40. Since, $R = \frac{\rho l}{A}$, where ρ is specific resistance.

    $\therefore\quad [\rho] = \left[\frac{RA}{l}\right], R = \frac{V}{i}, V = \frac{W}{Q}, [\rho] = [ML^3T^{-1}Q^{-2}]$
41. $i = i_0\{1 - e^{-t/(L/R)}\}$, where $\frac{L}{R}$ is time constant and its dimension is same as for time.
42. CR is time constant.
44. ωk is dimensionless.
45. $[a] = \left[\frac{F}{t}\right]$ and $[b] = \left[\frac{F}{t^2}\right]$
47. $\frac{1}{2}\varepsilon_0 E^2$ is energy density or energy per unit volume.
48. $p = \frac{a - t^2}{bx}$, where p = pressure, t = time

    $[pbx] = [a] = [t^2]$

    Hence, $[b] = \frac{[t^2]}{[px]}$

    Dimensions of $\frac{a}{b} = [px] = [MT^{-2}]$
49. Velocity gradient is the change in velocity per unit length.
50. Unit of emf e is volt.
51. $[a] = \left[\frac{v}{t}\right] : [b] = [vt] : [c] = [t]$
54. $F = \frac{1}{4\pi\varepsilon_0}\times\frac{q_1q_2}{r^2}$

    $\Rightarrow\quad \varepsilon_0 = \frac{1}{4\pi}\times\frac{q_1q_2}{Fr^2} \Rightarrow \varepsilon_0 = \frac{(\text{coulomb})^2}{\text{newton - metre}^2}$
55. From definition of time constant $t = RC$, where R is resistance and C is capacitance.

    $\therefore\quad R = \frac{t}{C} = \frac{[T]}{[M^{-1}L^{-2}T^4A^2]} = [ML^2T^{-3}A^{-2}]$
56. $M = NIA$
57. Since, $R = \frac{\rho l}{A}$, where ρ is specific resistance

    $[\rho] = \left[\frac{RA}{l}\right], R = \frac{V}{i}, V = \frac{W}{Q} = [ML^3T^{-1}Q^{-2}]$
68. $R = 0.16$ mm

    Hence, $A = \pi R^2 = \frac{22}{7}\times(0.16)^2 = 0.080384$

    Since, radius has two significant figures so answer also will have two significant figures.

    $\therefore\quad A = 0.080$
73. Minimum number of significant figure should be 1.
75. Radius of ball = 5.2 cm

    $V = \frac{4}{3}\pi R^3 \Rightarrow \frac{\Delta V}{V} = 3\left(\frac{\Delta R}{R}\right)$

    $\left(\frac{\Delta V}{V}\right)\times100 = 3\left(\frac{0.2}{5.2}\right)\times100 = 11\%$
78. Since, error in measurement of momentum is + 100%

    $\therefore\quad p_1 = p, p_2 = 2p$

    $K_1 = \frac{p^2}{2m}, K_2 = \frac{(2p)^2}{2m}$

    % in $K = \left(\frac{K_2 - K_1}{K_1}\right)\times100 = \left(\frac{4-1}{1}\right)\times100 = 300\%$
81. $H = i^2 Rt$

    $\therefore$ % error in H = 2 (% error in i) + (% error in R) + (% error in t)
82. $p = \frac{F}{A} = \frac{F}{L^2} = FL^{-2}$

    % error in pressure = (% error in F) + 2 (error in L)

    = (4%) + 2 (2%) = 8%
89. $f = \frac{1}{2\pi\sqrt{LC}}$ or $\sqrt{LC} = \frac{1}{2\pi f} = \frac{T}{2\pi}$

    Thus, $\sqrt{LC}$ has the dimensions of time.

Level 2 : Only One Correct Option

3. Volume of cylinder, $V = \pi r^2 L, r = \left(\frac{D}{2}\right)$

$$\therefore \quad \left(\frac{\Delta V}{V}\right)\times 100 = 2\left(\frac{\Delta D}{D}\right)\times 100 + \left(\frac{\Delta L}{L}\right)\times 100$$

$$= 2\left(\frac{0.01}{4.0}\right)\times 100 + \left(\frac{0.1}{0.5}\right)\times 100 = 2.5\%$$

4. Since, error is measured for 400 observations instead of 100 observations. So, error will reduce by 1/4 factor.

Hence, $= \frac{x}{4}$

5. Dimensions of ohm, $R = \frac{h}{e^2}$ (e = charge = current × time)

$$= \frac{[Et]}{[it]^2} = \frac{P}{i^2} = (R) \text{ as } P = \left(\frac{E}{t}\right)$$

8. $[\phi] = [BS] = [MT^{-2}A^{-1}]\,[L^2] = [ML^2T^{-2}A^{-1}]$

10. $g = \frac{GM}{R^2} : K = \frac{1}{2} I\omega^2 = \frac{L^2}{2I}$

Further, L will remain constant.

$$\therefore \quad K \propto \frac{1}{I} \quad \text{or} \quad K \propto \frac{1}{\frac{2}{5}MR^2}$$

or $\quad K \propto R^{-2}$

and $\quad g \propto R^{-2}$

11. Energy $= \frac{1}{2} mv^2$

$$[m] = \frac{[E]}{[v^2]} = [Ev^{-2}]$$

12. [Froce] = $[MLT^2]$

$$\therefore \quad 1\text{ N} = \left(\frac{1}{100}\right)\left(\frac{1}{1000}\right)(3600)^2$$

$$= 129.6 \text{ units}$$

13. $[FL^{-1}T^2] = [MLT^{-2}]\,[L^{-1}]\,[T^2] = [M]$

14. Here, $\frac{\pi}{\lambda}(ct - x)$ is dimensionless.

Hence, $\frac{ct}{\lambda}$ is also dimensionless and unit of ct is same as that of x.

Therefore, unit of λ is same as that of x. Also, unit of y is same as that of A, which is also that unit of x.

15. m is mass per unit length.

16. Given, $A = 1.0\text{ m} \pm 0.2\text{ m}$, $B = 2.0\text{ m} \pm 0.2\text{ m}$

Let, $\quad Y = \sqrt{AB} = \sqrt{(1.0)(2.0)} = 1.414\text{ m}$

Rounding off to two significant digits $Y = 1.4\text{ m}$

$$\frac{\Delta Y}{Y} = \frac{1}{2}\left[\frac{\Delta A}{A} + \frac{\Delta B}{B}\right] = \frac{1}{2}\left[\frac{0.2}{1.0} + \frac{0.2}{2.0}\right] = \frac{0.6}{2\times 2.0}$$

$$\Rightarrow \quad \Delta Y = \frac{0.6Y}{2\times 2.0} = \frac{0.6\times 1.4}{2\times 2.0}$$

$$= 0.212$$

Rounding off to one significant digit, $\Delta Y = 0.2\text{ m}$

Thus, correct value for $\sqrt{AB} = Y + \Delta Yr = 1.4 \pm 0.2\text{ m}$

17. All given measurements are correct up to two decimal places. As here 5.00 mm has the smallest unit and the error in 5.00 mm is least (commonly taken as 0.01 mm if not specified), hence 5.00 mm is most precise.

Note In solving these types of questions, we should be careful about units although their magnitude is same.

18. Given length, $l = 5$ cm

Now, checking the errors with each options one by one, we get

$$\Delta l_1 = 5 - 4.9 = 0.1\text{ cm}$$

$$\Delta l_2 = 5 - 4.805 = 0.195\text{ cm}$$

$$\Delta l_3 = 5.25 - 5 = 0.25\text{ cm}$$

$$\Delta l_4 = 5.4 - 5 = 0.4\text{ cm}$$

Error Δl_1 is least.

Hence, 4.9 cm is most precise.

19. By question, it is given that 50 VSD = 49 MSD

$$1\text{ VSD} = \frac{49}{50}\text{ MSD}$$

Minimum inaccuracy = 1 MSD − 1VSD

$$= 1\text{ MSD} - \frac{49}{50}\text{ MSD} = \frac{1}{50}\text{ MSD}$$

Given, $\quad 1\text{ MSD} = 0.5\text{ mm}$

Hence, minimum inaccuracy $= \frac{1}{50} \times 0.5\text{ mm}$

$$= \frac{1}{100} = 0.01\text{ mm}$$

20. Given, $t_1 = 39.6\text{ s}$, $t_2 = 39.9\text{ s}$ and $t_3 = 39.5\text{ s}$

Least count of measuring instrument = 0.1 s

(As measurements have only one decimal place)

Precision in the measurement = Least count of the measuring instrument = 0.1 s

Mean value of time for 20 oscillations is given by

$$t = \frac{t_1 + t_2 + t_3}{3} = \frac{39.6 + 39.9 + 39.5}{3} = 39.7\text{ s}$$

Absolute errors in the measurements

$$\Delta t_1 = t - t_1 = 39.7 - 39.6 = 0.1\text{ s}$$

$$\Delta t_2 = t - t_2 = 39.7 - 39.9 = -0.2\text{ s}$$

$$\Delta t_3 = t - t_3 = 39.7 - 39.5 = 0.2\text{ s}$$

$$\text{Mean absolute error} = \frac{|\Delta t_1| + |\Delta t_2| + |\Delta t_3|}{3}$$

$$= \frac{0.1 + 0.2 + 0.2}{3}$$

$$= \frac{0.5}{3} = 0.17 \approx 0.2$$

(rounding off up to one decimal place)

$\therefore$ Accuracy of measurement = ± 0.2 s

More than One Correct Option

1. Maximum percentage error in

$$x = (\%\text{ error in } a) + 2(\%\text{ error in } b) + 3(\%\text{ error in } c)$$

$$= 1\% + 6\% + 6\% = 13\%$$

Therefore, the correct options are (a) and (b).

2. In this question, it is given that P, Q and R are having different dimensions, hence they cannot be added or subtracted, so we can say that (a) and (d) are not meaningful.

3. We know that dimensions of $h = [h] = [ML^2T^{-1}]$

$$[c] = [LT^{-1}], [m_e] = M$$

$$[G] = [M^{-1}L^3T^{-2}]$$

$$[e] = [AT], [m_p] = [M]$$

$$\left[\frac{hc}{G}\right] = \frac{[ML^2T^{-1}][LT^{-1}]}{[M^{-1}L^3T^{-2}]} = [M^2]$$

$$M = \sqrt{\frac{hc}{G}}$$

Similarly, $$\frac{h}{c} = \frac{[ML^2T^{-1}]}{[LT^{-1}]} = [ML]$$

$$L = \frac{h}{cM} = \frac{h}{c}\sqrt{\frac{G}{hc}} = \frac{\sqrt{Gh}}{c^{3/2}}$$

As, $c = LT^{-1}$

$\Rightarrow$ $$[T] = \frac{[L]}{[c]} = \frac{\sqrt{Gh}}{c^{3/2} \cdot c} = \frac{\sqrt{Gh}}{c^{5/2}}$$

Hence, (a), (b) or (d) any can be used to express L, M and T in terms of three chosen fundamental quantities.

Match the Columns

3. $[A] = [L^2]$

$\therefore$ $$[L] = [A^{1/2}]$$

$$[T] = [T]$$

$$[F] = [MLT^{-2}]$$

$\therefore$ $$[M] = [FL^{-1}T^2] = [FA^{1/2}T^2]$$

Now, $$[W] = [FL] = [FA^{1/2}]$$

$$[I] = [ML^2] = [FA^{-1/2}T^2A] = [FA^{1/2}\,T^2]$$

$$[v] = [LT^{-1}] = [A^{1/2}T^{-1}]$$

Entrance Gallery

1. According to analysis, $d \propto \rho^x s^y f^z$

$\Rightarrow$ $$d = k\rho^x s^y f^z$$

Equating dimensions,

$$[L] = [ML^{-3}]^x \left[\frac{ML^2T^{-3}}{L^2}\right]^y [T^{-1}]^z$$

$$[L] = [M^{x+y}\,L^{-3x}\,T^{-3y-z}]$$

Equating powers on both sides, we get

$$x + y = 0,\quad 1 = -3x,\quad 0 = -3y - z$$

$$x = \frac{-1}{3},\quad y = \frac{1}{3},\quad z = -1$$

$\Rightarrow$ $$d = k\rho^{-1/3}\,s^{1/3}\,f^{-1}$$

So, according to question, $\frac{1}{n} = \frac{1}{3} \Rightarrow n = 3$

2. As, $Y = \frac{F/A}{l/L} \Rightarrow \frac{\Delta Y}{Y} = \frac{\Delta l}{l}$ as, F, A, L are exact quantities.

Here, Δl = least count of vernier scale

= × $^{-5}$.

Measurement of l is along vernier scale only

$$l = (45 - 20) \times 1.0 \times 10^{-5}$$

$$= 2.5 \times 10^{-4}\text{ m}$$

Here, one movement of vernier scale implies one least count measurement.

$$\frac{\Delta Y}{Y} = \frac{\Delta l}{l} = \frac{1.0 \times 10^{-5}}{2.5 \times 10^{-4}} = \frac{1}{25}$$

$$\frac{\Delta Y}{Y} \times 100 = \frac{1}{25} \times 100 = 4\%$$

3. As, 3.50 cm = 35 mm

A metre scale or foot scale gives measurement 35 mm than 36 mm. It does not give more precision.

For vernier scale with 1 MSD = 1 mm and 9 MSD = 10 VSD

Least count = 1 MSD − 1 VSD

$$= 1\text{ mm} - \frac{9}{10}\text{ mm} = \frac{1}{10}\text{ mm} = 0.1\text{ mm}$$

So, 35.0 mm can be measured by vernier calliper.

(c) Screw gauge will give 35.00 mm to 35.09 mm.

(d) Screw gauge will give 35.00 or 35.02 or 34.98 mm.

4. According to question, $I = (e^{1000V/T} - 1)$

$$I = 5\text{ mA gives, } 5 = e^{1000V/T} - 1$$

$\Rightarrow$ $$e^{1000V/T} = 6 \quad \text{...(i)}$$

Differentiating given equation,

$$\frac{dI}{dV} = e^{1000V/T} \times \frac{1000}{T} \Rightarrow \frac{dI}{dV} = \frac{6 \times 1000}{300} = 20$$

$$dI = 20\,dV = 20\,(\pm 0.01) = \pm\ 0.2\text{ mA}$$

Here, voltage V is in volt, current I in mA.

5. As, $T = 2\pi\sqrt{\frac{l}{g}}$

$\Rightarrow$ $$T^2 = 4\pi^2\frac{l}{g} \Rightarrow g = 4\pi^2\frac{l}{T^2}$$

$$\frac{\Delta g}{g} = \frac{\Delta l}{l} + 2\frac{\Delta T}{T}$$

$$= 2\% + 2(3\%) = 8\%$$

6. (a) $\frac{1\text{ mm}}{100} = 0.01$ mm least count wrong.

(b) $\frac{0.5\text{ mm}}{50} = 0.01$ mm least count wrong.

(c) $\frac{1\text{ mm}}{100} = 0.02$ mm least count right.

(d) $\frac{0.5\text{ mm}}{100} = 0.005$ mm least count wrong.

(e) $\frac{1\text{ mm}}{200} = 0.005$ mm least count wrong.

7. Given, $Q = \frac{x^3y^2}{z}$

$\Rightarrow$ $$\frac{\Delta Q}{Q} = 3\frac{\Delta x}{x} + 2\frac{\Delta y}{y} + \frac{\Delta z}{z}$$

$$= 3(1\%) + 2(2\%) + (4\%)$$

$$= (3 + 4 + 4)\% = 11\%$$

8. As, $pV = Nk_BT$

Boltzmann constant, $[k_B] = \left[\frac{pV}{NT}\right]$

$$= [ML^2\,T^{-2}\,K^{-1}] \Rightarrow P \to 4$$

Coefficient of viscosity, $\eta = \frac{\text{Stress}}{\text{Velocity change per unit length}}$

$\Rightarrow$ $$[\eta] = \left[\frac{MLT^{-2}/L^2}{T^{-1}}\right] = [ML^{-1}\,T^{-1}]$$

$\Rightarrow$ $$Q \to 2$$

$E = h\nu \Rightarrow$ Planck's constant, $[h] = \left[\frac{E}{\nu}\right] = \left[\frac{ML^2T^{-2}}{T^{-1}}\right]$

$= [ML^2T^{-1}]$

$\Rightarrow \quad R \to 1$

$\frac{\Delta Q}{\Delta t} = \frac{KA(T_2 - T_1)}{l} \Rightarrow [K] = \left[\frac{ML^2T^{-3}L}{L^2 K}\right]$

So, thermal conductivity = [K] = $[MLT^{-3}K^{-1}]$

$\Rightarrow \quad S \to 3$

9. Length of one Main Scale Division = 1 MSD

$= 5.15 - 5.10 = 0.05$ cm

Length of one Vernier Scale Division = 1 VSD

$= \frac{2.45 \text{ cm}}{50} = 0.049$ cm

Least count = 1 MSD − 1 VSD = 0.05 cm − 0.049 cm

= 0.001 cm

Vernier scale reading = (0.001 cm) (24) = 0.024 cm

Main scale reading = 5.10 cm

Total reading = 5.10 + 0.024 = 5.124 cm

Note The least count is also one division on main scale divided by number of division on vernier scale

$\left(\frac{(5.15 - 5.10)\text{ cm}}{50} = \frac{0.05}{50}\right) = 0.001$ cm

10. As, force $F = \frac{1}{4\pi\varepsilon_0}\frac{q_1q_2}{r^2} \Rightarrow \varepsilon_0 = \frac{q_1q_2}{4\pi Fr^2}$

$[\varepsilon_0] = \left[\frac{AT\,AT}{MLT^{-2}L^2}\right]$

$= [M^{-1}L^{-3}T^4A^2]$

11. 4.8000×10^4 has 4,8, 0, 0, 0 $\Rightarrow$ 5 significant digits.

48000.50 has 4, 8, 0, 0, 0, 5, 0 $\Rightarrow$ 7 significant digits.

For number more than 1 digits, after decimal place are significant.

Power of 10 does not affect number of significant figures or digits.

12. emf, $\varepsilon = -L\frac{di}{dt}$

$\Rightarrow \varepsilon i = -L\frac{di}{dt}i$

$\Rightarrow$ [Power] = [L] $[A^2T^{-1}]$

$\Rightarrow$ [L] = [inductance] $= \left[\frac{ML^2T^{-3}}{A^2T^{-1}}\right]$

$= [ML^2T^{-2}A^{-2}]$

13. Given, $Y = \frac{4MLg}{\pi ld^2}$,

here M, L, g are exact (assumed)

$\Rightarrow \frac{\Delta Y}{Y} = \frac{\Delta l}{l} + \frac{2\Delta d}{d}$

Contribution to maximum error, due to l measurement

$\frac{\Delta l}{l} = \frac{0.05/100}{0.25} = \frac{1}{50}$

Contribution to maximum error, due to d measurement

$= 2\frac{\Delta d}{d} = 2\left(\frac{0.05/100}{0.05}\right) = \frac{1}{50}$

So, contribution due errors in measurement of d and l are same.

14. Resistance, $R = \frac{\text{Voltage difference}}{\text{Current}} = \frac{V}{I}$

$\frac{\Delta R}{R}\% = \frac{\Delta V}{V}\% + \frac{\Delta I}{I}\%$

$= 3\% + 3\% = 6\%$

15. Least count $= \frac{0.05}{30}$ degree

Total reading = Main scale reading + (Vernier scale reading) × (Least count)

$= 58.5$ degrees $+ (09)\left(\frac{0.5}{30}\right)$ degrees

$= \left(58.5 + \frac{1.5}{10}\right) = 58.5 + 0.15$

$= 58.65$ degree $= 58.65°$

16. (a) $[MLT^{-2}]$ = Force $\Rightarrow$ Wrong

(b) $[MLT^{-2}]$ = Force $\Rightarrow$ Wrong

(c) $[MLT^2]$ = Not force $\Rightarrow$ Wrong

(d) $[T^{-1}]$ = Angular frequency (rad s^{-1}) $\Rightarrow$ Right

17. 1 Main scale division = 1 MSD = d unit

1 Vernier scale division = 1 VSD $= \frac{md}{m+1}$ unit

Least count = Smallest measurement by instrument

$= d - \frac{m}{m+1}d = d\left(\frac{m+1-m}{m+1}\right)$

$= \frac{d}{m+1}$ unit

18. Least count of screw gauge $= \frac{0.5}{50}$

$= 0.01 \text{ mm} = \Delta r$

Diameter, $r = 2.5 \text{ mm} + 20 \times \frac{0.5}{50}$

$= 2.70$ mm

$\frac{\Delta r}{r} = \frac{0.01}{2.70}$ or $\frac{\Delta r}{r} \times 100 = \frac{1}{2.7}$

Now, density $d = \frac{m}{V} = \frac{m}{\frac{4}{3}\pi\left(\frac{r}{2}\right)^3}$

Here, r is the diameter.

$\therefore \frac{\Delta d}{d} \times 100 = \left\{\frac{\Delta m}{m} + 3\left(\frac{\Delta r}{r}\right)\right\} \times 100$

$= \frac{\Delta m}{m} \times 100 + 3 \times \left(\frac{\Delta r}{r}\right) \times 100$

$= 2\% + 3 \times \frac{1}{2.7} = 3.11\%$

19. Diameter of wire, $d = \text{MSR} + \text{CSR} \times \text{LC}$

$= 0 + 52 \times \frac{1}{100} = 0.52$ mm

$= 0.052$ cm

20. Density, $d = \frac{m}{V} \Rightarrow \frac{\Delta d}{d} = \frac{\Delta m}{m} + \frac{\Delta V}{V}$

$= \frac{0.05}{5} \times 100 + \frac{0.05}{1} \times 100 = 6$

$\Rightarrow \frac{\Delta d}{d} = 6\%$

2

Vectors

2.1 Scalars and Vectors

Any physical quantity is either a scalar or a vector. A scalar quantity can be described completely by its magnitude only. Addition, subtraction, division or multiplication of scalar quantities can be done according to the ordinary rules of algebra. Mass, volume, density, etc., are few examples of scalar quantities. If a physical quantity in addition to magnitude has a specified direction as well as obeys the law of parallelogram of addition, then and then only it is said to be a vector quantity. Displacement, velocity, acceleration, etc., are few examples of vectors.

Chapter Snapshot

- Scalars and Vectors
- General Points Regarding Vectors
- Addition and Subtraction of Two Vectors
- Components of a Vector
- Product of Two Vectors

Extra Edge

- **Some Scalar Quantities** Distance, speed, mass, volume, density, energy, power, specific heat, time, frequency, work, pressure, surface tension, current, resistance, latent heat, charge, etc.
- **Some Vector Quantities** Displacement, velocity, acceleration, force, weight, momentum, torque, electric field, magnetic field, etc.

2.2 General Points Regarding Vectors

Vector Notation

Usually a vector is represented by a bold capital letter, as $\mathbf{A}, \mathbf{B}, \mathbf{C}$, etc.

The magnitude of a vector $\mathbf{A}$ is represented by A or $|\mathbf{A}|$.

Graphical Representation of a Vector

Graphically, a vector is represented by an arrow drawn to a chosen scale, parallel to the direction of the vector. The length and the direction of the arrow thus represent the magnitude and the direction of the vector respectively.

Thus, the arrow in Fig. 2.1 represents a vector **A** in xy-plane making an angle θ with x-axis.

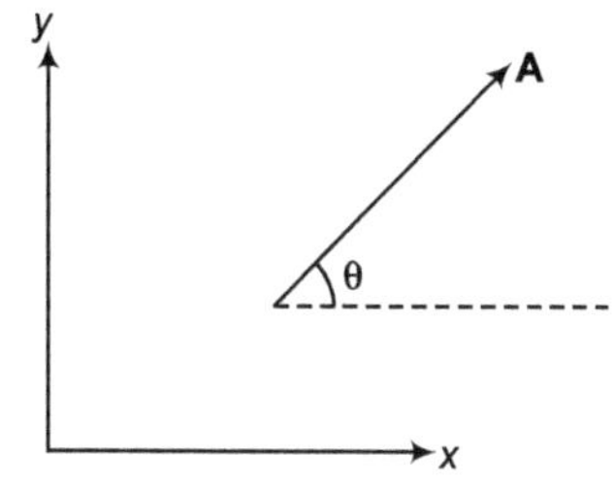

Fig. 2.1

Equality of Vectors

All vectors with the same magnitude and direction are equal despite their different locations in space. Thus, if a vector is displaced parallel to itself, it does not change.

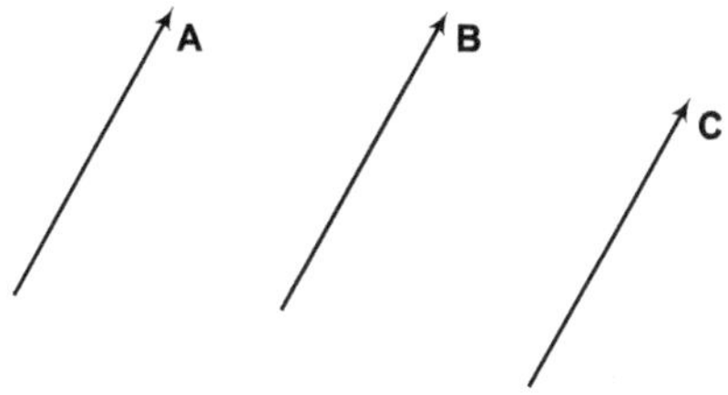

Fig. 2.2

In Fig. 2.2, **A**, **B** and **C** are all equal, since they have the same magnitude and direction even though they are differently located in space.

Angle between Two Vectors (θ)

To find angle between two vectors both the vectors are drawn from a point in such a manner that arrows of both the vectors are outwards from that point. Now, the smaller angle is called the angle between two vectors.

For example in Fig. 2.3, angle between **A** and **B** is 60° not 120°. Because in Fig. (a) they are wrongly drawn while in Fig. (b) they are drawn as we desire.

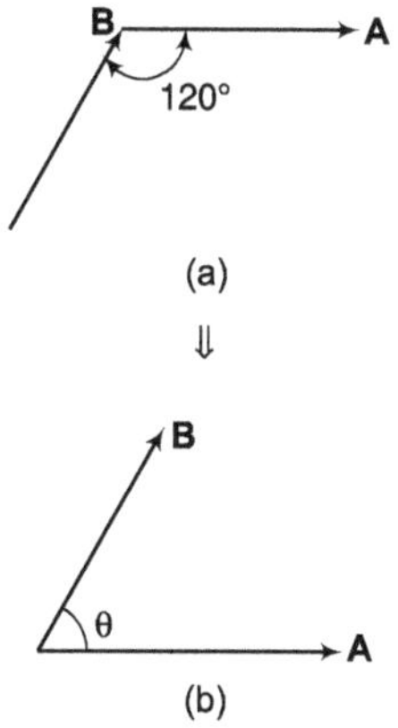

Fig. 2.3

Unit and Zero Vector

A vector of unit magnitude is called a unit vector and the notation for it in the direction of **A** is $\hat{\mathbf{A}}$ read as '*A* hat or *A* caret'.

Thus, $\mathbf{A} = A\hat{\mathbf{A}}$

A unit vector merely indicates a direction. Unit vectors along x, y and z-directions are $\hat{\mathbf{i}}$, $\hat{\mathbf{j}}$ and $\hat{\mathbf{k}}$.

A vector of zero magnitude is called a zero or a **null vector**. Its direction is arbitrary.

Negative of a Vector

It means a vector of same magnitude but opposite in direction.

Fig. 2.4

Multiplication and Division of Vectors by Scalars

The product of a vector **A** and a scalar m is a vector $m\mathbf{A}$ whose magnitude is m times the magnitude of **A** and which is in the direction or opposite to **A** according as the scalar m is positive or negative. Thus,

$$|m\mathbf{A}| = mA$$

Further, if m and n are two scalars, then

$$(m+n)\mathbf{A} = m\mathbf{A} + n\mathbf{A} \text{ and } m(n\mathbf{A}) = n(m\mathbf{A}) = (mn)\mathbf{A}$$

The division of vector **A** by a non-zero scalar m is defined as the multiplication of **A** by $\frac{1}{m}$.

Example 2.1 *What is the angle between the two vectors* **A** *and* 2**A**?

Sol. Angle between **A** and 2**A** is 0°, since 2**A** is in the direction of **A** and its magnitude is double.

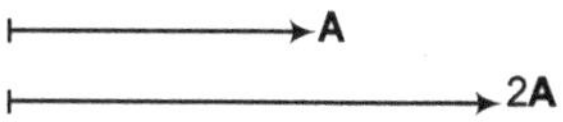

Fig. 2.5

2.3 Addition and Subtraction of Two Vectors

Addition

(i) **The parallelogram law** Let **R** be the resultant of two vectors **A** and **B**. According to parallelogram law of vector addition, the resultant **R** is the diagonal of the parallelogram of which **A** and **B** are the adjacent sides as shown in figure. Magnitude of **R** is given by

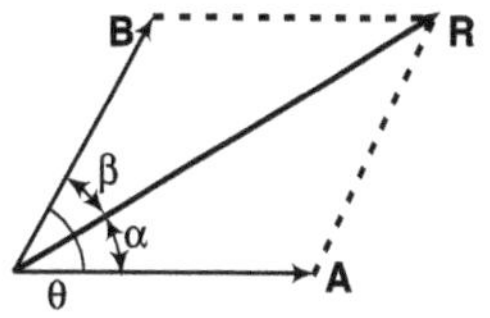

Fig. 2.6

$$R = \sqrt{A^2 + B^2 + 2AB\cos\theta} \qquad \text{...(i)}$$

Here, θ = angle between **A** and **B**. The direction of **R** can be found by angle α or β of **R** with **A** or **B**.

Here, $\tan\alpha = \dfrac{B\sin\theta}{A + B\cos\theta}$

and $$\tan \beta = \frac{A \sin \theta}{B + A \cos \theta} \qquad \dots \text{(ii)}$$

Special Cases

If $\theta = 0°$, R = maximum $= A + B$

$\theta = 180°$, R = minimum $= A \sim B$

and if $\theta = 90°$, $R = \sqrt{A^2 + B^2}$

In all other cases, magnitude and direction of **R** can be calculated by using Eqs. (i) and (ii).

(ii) **The triangle law** According to this law, if the tail of one vector be placed at the head of the other, their sum or resultant **R** is drawn from the tail end of the first to the head end of the other. As it is evident from the figure that the resultant **R** is the same irrespective of the order in which the vectors **A** and **B** are taken, thus, **R = A + B = B + A**

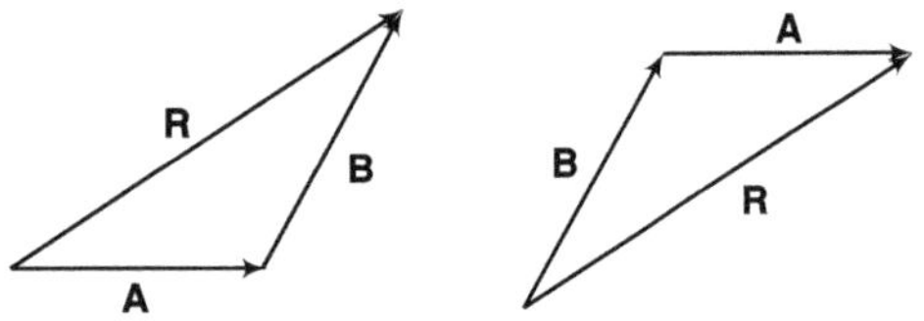

Fig. 2.7

Subtraction

Negative of a vector say **–A** is a vector of the same magnitude as vector **A** but pointing in a direction opposite to that of **A**.

Fig. 2.8

Thus, **A – B** can be written as **A** + (**–B**) or **A – B** is really the vector addition of **A** and **–B**.

Suppose angle between two vectors **A** and **B** is θ. Then, angle between **A** and **–B** will be $180° - \theta$ as shown in Fig. 2.9(b).

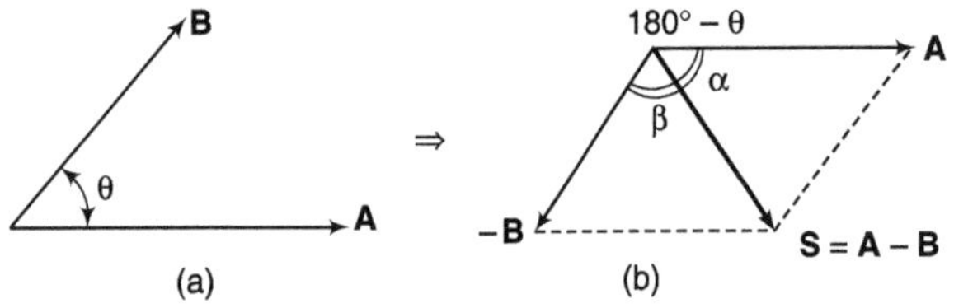

Fig. 2.9

Magnitude of **S = A – B** will be thus given by

$$S = |\mathbf{A} - \mathbf{B}| = \sqrt{A^2 + B^2 + 2AB \cos (180° - \theta)}$$

or $$S = \sqrt{A^2 + B^2 - 2AB \cos \theta} \qquad \dots \text{(i)}$$

For direction of **S**, we will either calculate angle α or β, where,

$$\tan \alpha = \frac{B \sin (180° - \theta)}{A + B \cos (180° - \theta)} = \frac{B \sin \theta}{A - B \cos \theta} \qquad \dots \text{(ii)}$$

or $$\tan \beta = \frac{A \sin (180 - \theta)}{B + A \cos (180 - \theta)} = \frac{A \sin \theta}{B - A \cos \theta} \qquad \dots \text{(iii)}$$

Extra Edge

- **A**–**B** or **B** – **A** can also be found by making triangles as shown in Fig. 2.10 (a) and (b).

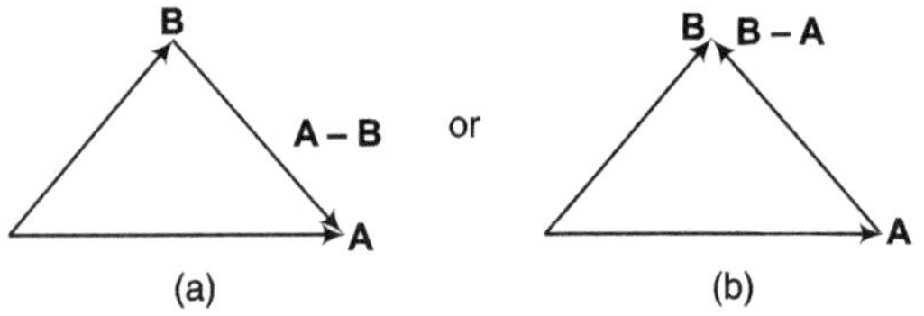

Fig. 2.10

Example 2.2 *Prove the results*

$$R = \sqrt{A^2 + B^2 + 2AB \cos \theta}$$

and $\tan \alpha = \dfrac{B \sin \theta}{A + B \cos \theta}$ *in vector addition method.*

Sol. Let **OP** and **OQ** represent the two vectors **A** and **B** making an angle θ. Then, using the parallelogram method of vector addition, **OS** represents the resultant vector **R**.

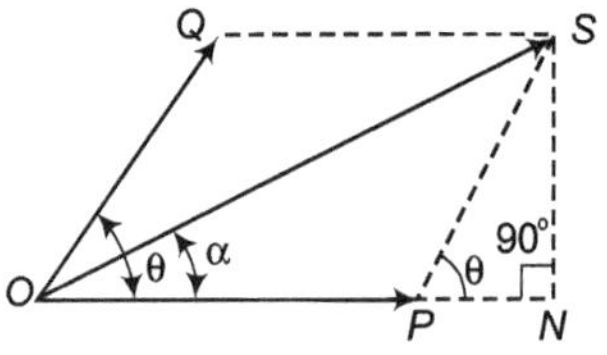

Fig. 2.11

$$\mathbf{R} = \mathbf{A} + \mathbf{B}$$

$$OP = A,\ OQ = PS = B,\ OS = R$$

SN is normal to OP.

From the geometry of the figure,

$$OS^2 = ON^2 + SN^2$$

but $$ON = OP + PN = A + B\cos\theta$$

$$SN = B \sin\theta$$

$$OS^2 = (A + B\cos\theta)^2 + (B\sin\theta)^2$$

or $$R^2 = A^2 + B^2 + 2AB\cos\theta$$

$$R = \sqrt{A^2 + B^2 + 2AB\cos\theta}$$

$$\tan \alpha = \frac{SN}{OP + PN} = \frac{B \sin \theta}{A + B \cos \theta}$$

Example 2.3 *Find* **A + B** *and* **A – B** *in the diagram shown in figure. Given* $A = 4$ *units and* $B = 3$ *units.*

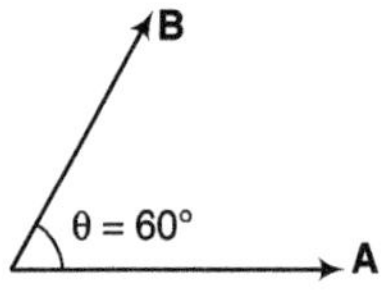

Fig. 2.12

Sol. **Addition**

$$R = \sqrt{A^2 + B^2 + 2AB \cos \theta}$$

$= \sqrt{16 + 9 + 2 \times 4 \times 3 \cos 60°} = \sqrt{37}$ units

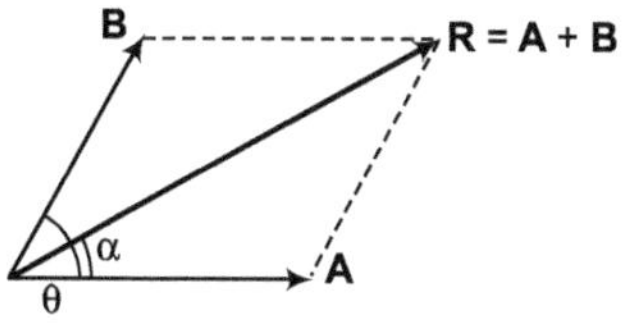

Fig. 2.13

$$\tan \alpha = \frac{B \sin \theta}{A + B \cos \theta} = \frac{3 \sin 60°}{4 + 3 \cos 60°} = 0.472$$

$\therefore \quad \alpha = \tan^{-1} (0.472) = 25.3°$

Thus, resultant of **A** and **B** is $\sqrt{37}$ units at angle 25.3° from **A** in the direction shown in figure.

Subtraction $S = \sqrt{A^2 + B^2 - 2AB \cos \theta}$

$= \sqrt{16 + 9 - 2 \times 4 \times 3 \cos 60°}$

$= \sqrt{13}$ units

and $\tan \alpha = \dfrac{B \sin \theta}{A - B \cos \theta}$

$= \dfrac{3 \sin 60°}{4 - 3 \cos 60°} = 1.04$

$\therefore \quad \alpha = \tan^{-1} (1.04) = 46.1°$

Thus, **A** − **B** is $\sqrt{13}$ units at 46.1° from **A** in the direction shown in figure below.

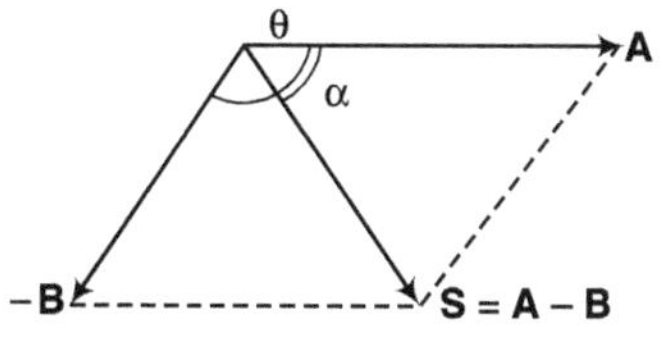

Fig. 2.14

2.4 Components of a Vector

Two or more vectors which, when compounded in accordance with the parallelogram law of vector **R** are said to be components of vector **R**. The most important components with which we are concerned are mutually perpendicular or rectangular ones along the three co-ordinate axes *ox, oy* and *oz*, respectively. Thus, a vector **R** can be written as $\mathbf{R} = R_x \hat{\mathbf{i}} + R_y \hat{\mathbf{j}} + R_z \hat{\mathbf{k}}$.

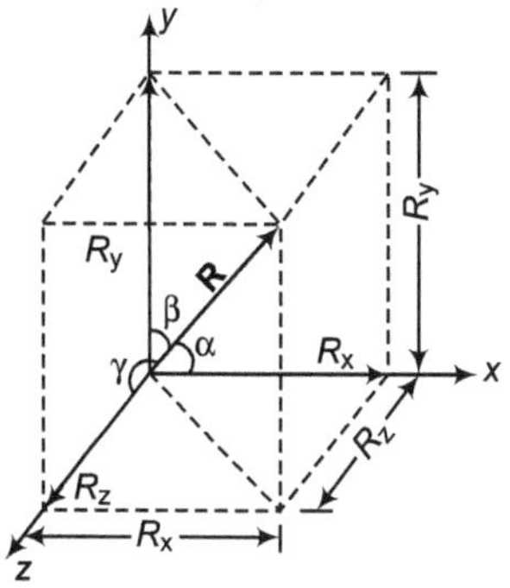

A vector **R** resolved into components along *x,y* and *z*-axes.

Fig. 2.15

Here, R_x, R_y and R_z are the components of **R** in *x, y* and *z*-axes, respectively and $\hat{\mathbf{i}}, \hat{\mathbf{j}}$ and $\hat{\mathbf{k}}$ are unit vectors along these directions. The magnitude of **R** is given by

$$R = \sqrt{R_x^2 + R_y^2 + R_z^2}$$

This vector **R** makes an angle of $\alpha = \cos^{-1} \left(\dfrac{R_x}{R} \right)$ with *x*-axis or $\cos \alpha = \dfrac{R_x}{R}$

$$\beta = \cos^{-1} \left(\frac{R_y}{R} \right) \text{ with } y\text{-axis or } \cos \beta = \frac{R_y}{R}$$

and $$\gamma = \cos^{-1} \left(\frac{R_z}{R} \right) \text{ with } z\text{-axis or } \cos \gamma = \frac{R_z}{R}$$

- cos α, cos β and cos γ are called direction cosines of **R**.

Exercise *Prove that*

(*a*) $\cos^2 \alpha + \cos^2 \beta + \cos^2 \gamma = 1$

(*b*) $\sin^2 \alpha + \sin^2 \beta + \sin^2 \gamma = 2$

Refer Fig. 2.16 (a)

We have resolved a two-dimensional vector **R** in mutually perpendicular directions *x* and *y*.

Component along *x*-axis $= R_x = R \cos \alpha$ or $R \sin \beta$ and component along *y*-axis $= R_y = R \cos \beta$ or $R \sin \alpha$.

If $\hat{\mathbf{i}}$ and $\hat{\mathbf{j}}$ be the unit vectors along *x* and *y*-axis respectively, we can write

$$\mathbf{R} = R_x \hat{\mathbf{i}} + R_y \hat{\mathbf{j}}$$

- Components of a vector in mutually perpendicular directions are called the rectangular components.

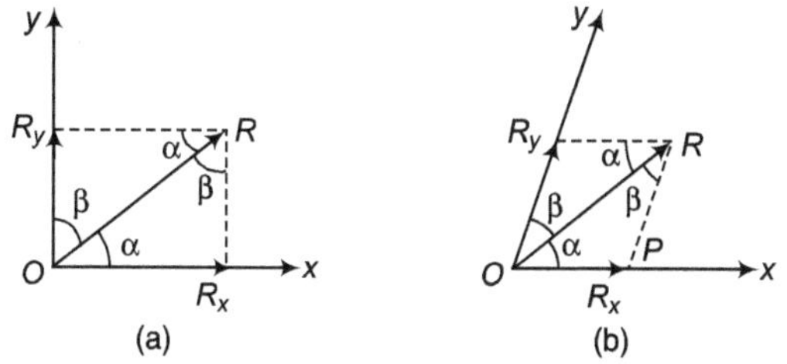

Fig. 2.16

Refer Fig. 2.16 (b)

Vector **R** has been resolved in two axes *x* and *y* not perpendicular to each other. Applying sine law in the triangle *OPR* shown above, we have

$$\frac{R}{\sin [180° - (\alpha + \beta)]} = \frac{R_x}{\sin \beta} = \frac{R_y}{\sin \alpha}$$

or $$R_x = \frac{R \sin \beta}{\sin (\alpha + \beta)}$$

and $$R_y = \frac{R \sin \alpha}{\sin (\alpha + \beta)}$$

If $\alpha + \beta = 90°$, $R_x = R \sin \beta$ and $R_y = R \sin \alpha$

Extra Edge

- Representation of a vector in terms of $\hat{\mathbf{i}}, \hat{\mathbf{j}}$ and $\hat{\mathbf{k}}$

  A force **F** is of 5 N as shown in figure. Let us represent this vector in terms of $\hat{\mathbf{i}}$ and $\hat{\mathbf{j}}$.

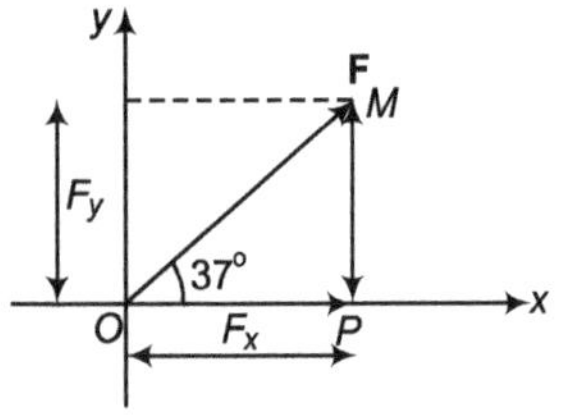

Fig. 2.17

$$F_x = \text{component of } \mathbf{F} \text{ along } x\text{-axis}$$
$$= F\cos 37° = (5)\left(\frac{4}{5}\right) = 4\text{ N}$$
$$F_y = \text{component of } \mathbf{F} \text{ along } y\text{-axis}$$
$$= F\sin 37° = (5)\left(\frac{3}{5}\right) = 3\text{ N}$$

Now, $\mathbf{F} = \mathbf{OP} + \mathbf{PM}$ (From polygon law of vector addition)

or $\mathbf{F} = (4\hat{\mathbf{i}} + 3\hat{\mathbf{j}})$ N

- If $\mathbf{F} = (-6\hat{\mathbf{i}} + 8\hat{\mathbf{j}})$ N, then its magnitude and direction can be found as under

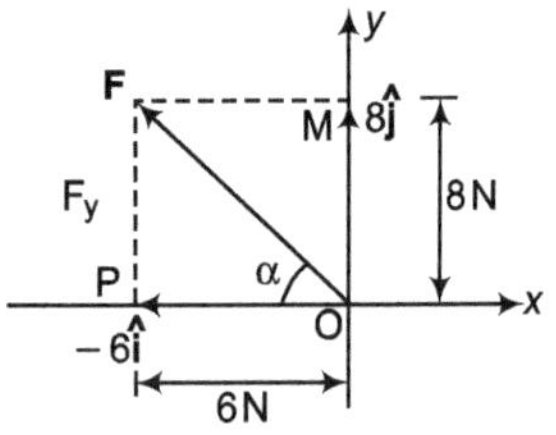

Fig. 2.18

From parallelogram law,

$$\mathbf{F} = \mathbf{OP} + \mathbf{OM} = (-6\hat{\mathbf{i}} + 8\hat{\mathbf{j}})\text{ N}$$

$\therefore$ $$|\mathbf{F}| = F = \sqrt{(6)^2 + (8)^2} = 10\text{ N}$$

$$\tan\alpha = \frac{8}{6} = \frac{4}{3}$$

$\Rightarrow$ $$\alpha = \tan^{-1}\left(\frac{4}{3}\right) = 53°$$

Hence, the given **F** has a magnitude of 10 N and it makes an angle of 53° from negative x-axis towards positive y-axis.

- **Position Vector** In the figure shown, suppose co-ordinates of a point P are (x, y).

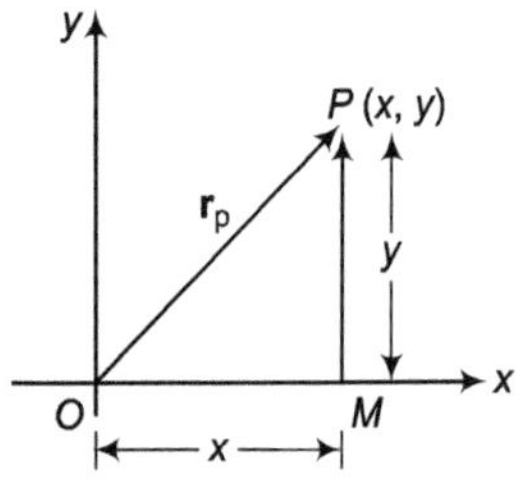

Fig. 2.19

Then,

OP is called position vector of point P with respect to the origin O. This is also denoted by $\mathbf{r}_P$. From polygon law of vector addition, in the triangle OMP we can also see that,

$$\mathbf{OP} = \mathbf{OM} + \mathbf{MP}$$

where, $\mathbf{OM} = x\hat{\mathbf{i}}$ and $\mathbf{MP} = y\hat{\mathbf{j}}$

$\therefore$ $$\mathbf{OP} = \mathbf{r}_P = (x\hat{\mathbf{i}} + y\hat{\mathbf{j}})$$

In the similar manner, we can show that if co-ordinates of point P are (x, y, z) then

Position vector of $P = \mathbf{OP} = \mathbf{r}_P = (x\hat{\mathbf{i}} + y\hat{\mathbf{j}} + z\hat{\mathbf{k}})$

- **Displacement Vector** In the figure shown, let :

$$A = (x_A, y_A) \text{ and } B = (x_B, y_B)$$

Then, $\mathbf{r}_A$ = position vector of A
$= (x_A\hat{\mathbf{i}} + y_A\hat{\mathbf{j}})$ and
$\mathbf{r}_B$ = position vector of B
$= (x_B\hat{\mathbf{i}} + y_B\hat{\mathbf{j}})$

If a particle is displaced from point A to point B, then **AB** is called its displacement vector **S**.

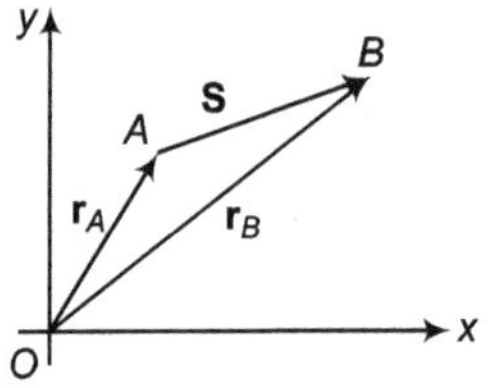

Fig. 2.20

In triangle OAB, from subtraction of vectors we can say that,

$$\mathbf{AB} \text{ or } \mathbf{S} = \mathbf{r}_B - \mathbf{r}_A$$
$$= (x_B\hat{\mathbf{i}} + y_B\hat{\mathbf{j}}) - (x_A\hat{\mathbf{i}} + y_A\hat{\mathbf{j}})$$
$$= (x_B - x_A)\,\hat{\mathbf{i}} + (y_B - y_A)\,\hat{\mathbf{i}}$$

In the similar manner, we can show that, if a particle is displaced from point $A\ (x_1, y_1, z_1)$ to point $B\ (x_2, y_2, z_2)$ then, displacement of particle

$$\mathbf{S} = \mathbf{r}_B - \mathbf{r}_A$$
$$= (x_2\hat{\mathbf{i}} + y_2\hat{\mathbf{j}} + z_2\hat{\mathbf{k}}) - (x_1\hat{\mathbf{i}} + y_1\hat{\mathbf{j}} + z_1\hat{\mathbf{k}})$$
$$= (x_2 - x_1)\,\hat{\mathbf{i}} + (y_2 - y_1)\,\hat{\mathbf{j}} + (z_2 - z_1)\,\hat{\mathbf{k}}$$

Example 2.4 *Resolve a weight of 10 N in two directions which are parallel and perpendicular to a slope inclined at 30° to the horizontal.*

Sol. Component perpendicular to the plane

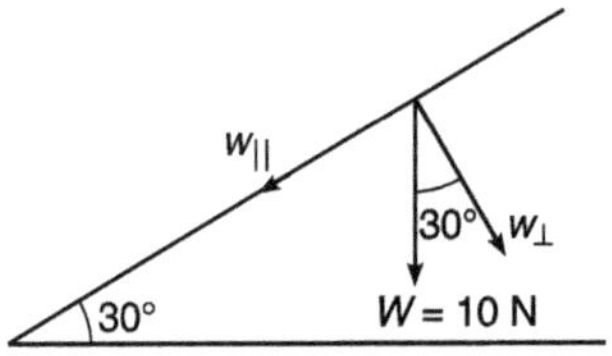

Fig. 2.21

$$W_{\perp} = W\cos 30° = (10)\frac{\sqrt{3}}{2} = 5\sqrt{3}\text{ N}$$

and component parallel to the plane

$$W_{||} = W\sin 30° = (10)\left(\frac{1}{2}\right) = 5\text{ N}$$

Example 2.5 *Resolve horizontally and vertically a force $F = 8$ N which makes an angle of 45° with the horizontal.*

Sol. Horizontal component of **F** is

$F_H = F\cos 45° = (8)\left(\frac{1}{\sqrt{2}}\right) = 4\sqrt{2}$ N

and vertical component of **F** is

$F_V = F\sin 45° = (8)\left(\frac{1}{\sqrt{2}}\right) = 4\sqrt{2}$ N

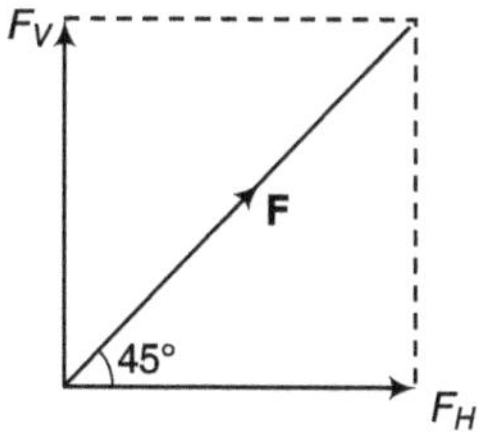

Fig. 2.22

Two vectors in the form of $\hat{\mathbf{i}}$, $\hat{\mathbf{j}}$ and $\hat{\mathbf{k}}$ can be added, subtracted on multiplied by a scalar directly as done in the following example.

Example 2.6 *Obtain the magnitude of* $2\mathbf{A} - 3\mathbf{B}$ *if* $\mathbf{A} = \hat{\mathbf{i}} + \hat{\mathbf{j}} - 2\hat{\mathbf{k}}$ *and* $\mathbf{B} = 2\hat{\mathbf{i}} - \hat{\mathbf{j}} + \hat{\mathbf{k}}$

Sol. $2\mathbf{A} - 3\mathbf{B} = 2(\hat{\mathbf{i}} + \hat{\mathbf{j}} - 2\hat{\mathbf{k}}) - 3(2\hat{\mathbf{i}} - \hat{\mathbf{j}} + \hat{\mathbf{k}})$

$= -4\hat{\mathbf{i}} + 5\hat{\mathbf{j}} - 7\hat{\mathbf{k}}$

$\therefore$ Magnitude of

$2\mathbf{A} - 3\mathbf{B} = \sqrt{(-4)^2 + (5)^2 + (-7)^2}$

$= \sqrt{16 + 25 + 49} = \sqrt{90}$

$= -4\hat{\mathbf{i}} + 5\hat{\mathbf{j}} - 7\hat{\mathbf{k}}$

2.5 Product of Two Vectors

The product of two vectors is of two kinds.

1. A scalar or dot product
2. A vector or a cross product

Scalar or Dot Product

The scalar or dot product of two vectors **A** and **B** is denoted by $\mathbf{A}\cdot\mathbf{B}$ and is read as **A** dot **B**.

It is defined as the product of the magnitudes of the two vectors **A** and **B** and the cosine of their included angle θ.

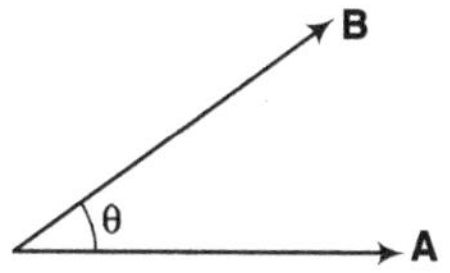

Fig. 2.23

Thus, $\mathbf{A}\cdot\mathbf{B} = AB\cos\theta$ (a scalar quantity)

Important Points Regarding Dot Product

The following points should be remembered regarding the dot product :

(i) $\mathbf{A}\cdot\mathbf{B} = \mathbf{B}\cdot\mathbf{A}$

(ii) $\mathbf{A}\cdot(\mathbf{B}+\mathbf{C}) = \mathbf{A}\cdot\mathbf{B} + \mathbf{A}\cdot\mathbf{C}$

(iii) $\mathbf{A}\cdot\mathbf{A} = AA\cos 0° = A^2$

(iv) $\mathbf{A}\cdot\mathbf{B} = A(B\cos\theta) = A$ (Component of **B** along **A**)

or $\mathbf{A}\cdot\mathbf{B} = B(A\cos\theta) = B$ (Component of **A** along **B**)

(v) $\hat{\mathbf{i}}\cdot\hat{\mathbf{i}} = \hat{\mathbf{j}}\cdot\hat{\mathbf{j}} = \hat{\mathbf{k}}\cdot\hat{\mathbf{k}} = 1, 1\cos 0° = 1$

(vi) $\hat{\mathbf{i}}\cdot\hat{\mathbf{j}} = \hat{\mathbf{j}}\cdot\hat{\mathbf{k}} = \hat{\mathbf{i}}\cdot\hat{\mathbf{k}} = 1, 1\cos 90° = 0$

(vii) $(a_1\hat{\mathbf{i}} + b_1\hat{\mathbf{j}} + c_1\hat{\mathbf{k}})\cdot(a_2\hat{\mathbf{i}} + b_2\hat{\mathbf{j}} + c_2\hat{\mathbf{k}}) = a_1a_2 + b_1b_2 + c_1c_2$

(viii) $\cos\theta = \frac{\mathbf{A}\cdot\mathbf{B}}{AB}$ = (cosine of angle between **A** and **B**)

(ix) Two vectors are perpendicular if their dot product is zero. $(\theta = 90°)$

(x) Component of **A** along $\mathbf{B} = A\cos\theta = \frac{\mathbf{A}\cdot\mathbf{B}}{B}$

Similarly, component of **B** along $\mathbf{A} = B\cos\theta = \frac{\mathbf{A}\cdot\mathbf{B}}{A}$

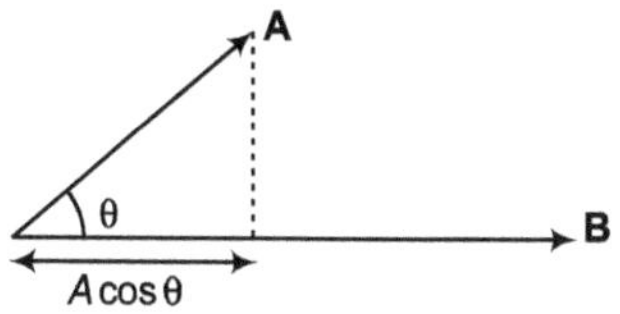

Fig. 2.24

Example 2.7 *Work done by a force* **F** *on a body is* $W = \mathbf{F}\cdot\mathbf{s}$, *where* **s** *is the displacement of body. Given that under a force* $\mathbf{F} = (2\hat{\mathbf{i}} + 3\hat{\mathbf{j}} + 4\hat{\mathbf{k}})$ *N a body is displaced from position vector* $\mathbf{r}_1 = (2\hat{\mathbf{i}} + 3\hat{\mathbf{j}} + \hat{\mathbf{k}})$ *m to the position vector* $\mathbf{r}_2 = (\hat{\mathbf{i}} + \hat{\mathbf{j}} + \hat{\mathbf{k}})$ *m. Find the work done by this force.*

- Displacement vector is given by $\mathbf{s} = \mathbf{r}_f - \mathbf{r}_i$

Sol. The body is displaced from $\mathbf{r}_1$ to $\mathbf{r}_2$. Therefore, displacement of the body is

$\mathbf{s} = \mathbf{r}_2 - \mathbf{r}_1 = (\hat{\mathbf{i}} + \hat{\mathbf{j}} + \hat{\mathbf{k}}) - (2\hat{\mathbf{i}} + 3\hat{\mathbf{j}} + \hat{\mathbf{k}}) = (-\hat{\mathbf{i}} - 2\hat{\mathbf{j}})$ m

Now, work done by the force is $W = \mathbf{F}\cdot\mathbf{s}$

$= (2\hat{\mathbf{i}} + 3\hat{\mathbf{j}} + 4\hat{\mathbf{k}})\cdot(-\hat{\mathbf{i}} - 2\hat{\mathbf{j}})$

$= (2)(-1) + (3)(-2) + (4)(01) = -8$ J

Example 2.8 *Find the angle between two vectors* $\mathbf{A} = 2\hat{\mathbf{i}} + \hat{\mathbf{j}} - \hat{\mathbf{k}}$ *and* $\mathbf{B} = \hat{\mathbf{i}} - \hat{\mathbf{k}}$.

Sol. $A = |\mathbf{A}| = \sqrt{(2)^2 + (1)^2 (-1)^2} = \sqrt{6}$

$B = |\mathbf{B}| = \sqrt{(1)^2 + (-1)^2} = \sqrt{2}$

$\mathbf{A}\cdot\mathbf{B} = (2\hat{\mathbf{i}} + \hat{\mathbf{j}} - \hat{\mathbf{k}})\cdot(\hat{\mathbf{i}} - \hat{\mathbf{k}})$

$= (2)(1) + (1)(0) + (-1)(-1) = 3$

Now, $\cos\theta = \frac{\mathbf{A}\cdot\mathbf{B}}{AB} = \frac{3}{\sqrt{6}\cdot\sqrt{2}} = \frac{3}{\sqrt{12}} = \frac{\sqrt{3}}{2}$

$\therefore$ $\theta = 30°$

Example 2.9 *Prove that the vectors* $\mathbf{A} = 2\hat{\mathbf{i}} - 3\hat{\mathbf{j}} + \hat{\mathbf{k}}$ *and* $\mathbf{B} = \hat{\mathbf{i}} + \hat{\mathbf{j}} + \hat{\mathbf{k}}$ *are mutually perpendicular.*

Sol. $\mathbf{A}\cdot\mathbf{B} = (2\hat{\mathbf{i}} - 3\hat{\mathbf{j}} + \hat{\mathbf{k}})\cdot(\hat{\mathbf{i}} + \hat{\mathbf{j}} + \hat{\mathbf{k}})$

$$= (2)(1) + (-3)(1) + (1)(1) = 0 = AB\cos\theta$$

$\therefore$ $\cos\theta = 0$ (as $A \neq 0, B \neq 0$)

or $\theta = 90^\circ$

or the vectors **A** and **B** are mutually perpendicular.

Vector or Cross Product

The cross product of two vectors **A** and **B** is denoted by $\mathbf{A}\times\mathbf{B}$ and read as **A** cross **B**. It is defined as a third vector **C** whose magnitude is equal to the product of the magnitudes of the two vectors **A** and **B** and the sine of the angle between **A** and **B**, θ.

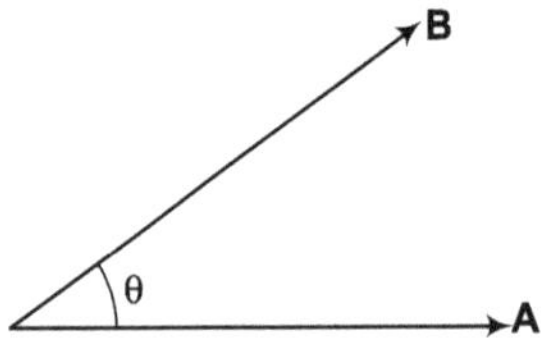

Fig. 2.25

Thus, if $\mathbf{C} = \mathbf{A}\times\mathbf{B}$, then $C = AB\sin\theta$.

The vector **C** is normal to the plane of **A** and **B** and points in the direction in which a right handed screw would advance when rotated about an axis perpendicular to the plane of the two vectors in the direction from **A** to **B** through the smaller angle θ between them or alternatively, we might state the rule as :

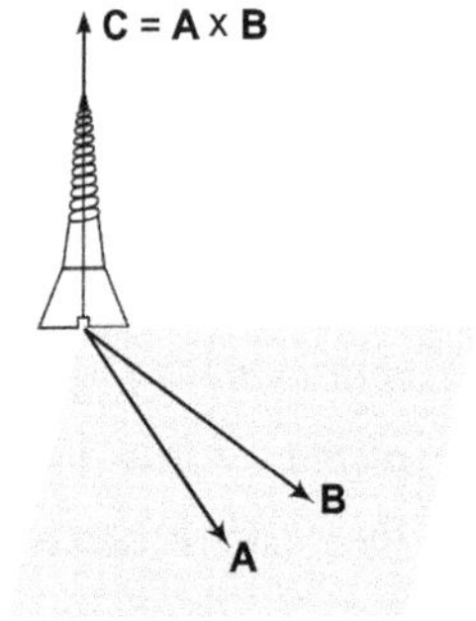

Plane of **A** and **B**
Fig. 2.26

If the fingers of the right hand be curled in the direction in which vector **A** must be turned through the smaller included angle θ to coincide with the direction of vector **B**, the thumb points in the direction of **C** as shown in Fig. 2.26.

Either of these rules is referred to as the right handed screw rule. Thus, if $\hat{\mathbf{n}}$ be the unit vector in the direction of **C**, we have

$$\mathbf{C} = \mathbf{A}\times\mathbf{B} = AB\sin\theta\,\hat{\mathbf{n}},$$

where $0 \le \theta \le \pi$

Important Points About Vector Product

(i) $\mathbf{A}\times\mathbf{B} = -\mathbf{B}\times\mathbf{A}$

(ii) The cross product of two parallel or anti-parallel vectors is a null vector, as $|\mathbf{A}\times\mathbf{B}| = AB\sin\theta$ and $\sin\theta = 0$ for two parallel or anti-parallel vectors (as $\theta = 0^\circ$ or 180°).

Thus, $\hat{\mathbf{i}}\times\hat{\mathbf{i}} = \hat{\mathbf{j}}\times\hat{\mathbf{j}} = \hat{\mathbf{k}}\times\hat{\mathbf{k}} =$ a null vector

(iii) If two vectors are perpendicular to each other, we have $\theta = 90^\circ$ and therefore, $\sin\theta = 1$. So that $\mathbf{A}\times\mathbf{B} = AB\,\hat{\mathbf{n}}$. The vectors **A**, **B** and $\mathbf{A}\times\mathbf{B}$ thus form a right handed system of mutually perpendicular vectors. It follows at once from the above that in case of the orthogonal triad of unit vectors $\hat{\mathbf{i}}, \hat{\mathbf{j}}$ and $\hat{\mathbf{k}}$ (each perpendicular to each other)

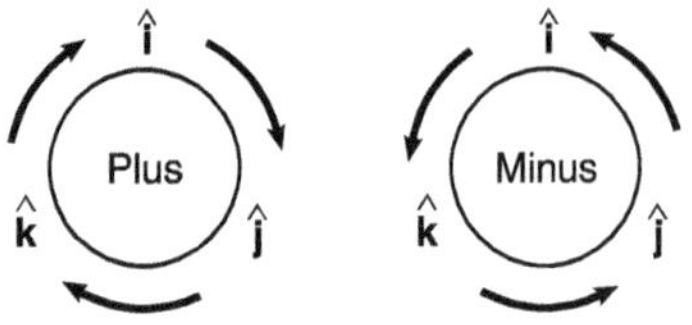

Fig. 2.27

$\hat{\mathbf{i}}\times\hat{\mathbf{j}} = -\hat{\mathbf{j}}\times\hat{\mathbf{i}} = \hat{\mathbf{k}}$

$\hat{\mathbf{j}}\times\hat{\mathbf{k}}$

$= -\hat{\mathbf{k}}\times\hat{\mathbf{j}} = \hat{\mathbf{i}}$ and $\hat{\mathbf{k}}\times\hat{\mathbf{i}} = -\hat{\mathbf{i}}\times\hat{\mathbf{k}} = \hat{\mathbf{j}}$

(iv) $\mathbf{A}\times(\mathbf{B}+\mathbf{C}) = \mathbf{A}\times\mathbf{B} + \mathbf{A}\times\mathbf{C}$

(v) A vector product can be expressed in terms of rectangular components of the two vectors and put in the determinant form as may be seen from the following:

Let $\mathbf{A} = a_1\hat{\mathbf{i}} + b_1\hat{\mathbf{j}} + c_1\hat{\mathbf{k}}$ and $\mathbf{B} = a_2\hat{\mathbf{i}} + b_2\hat{\mathbf{j}} + c_2\hat{\mathbf{k}}$

Then, $\mathbf{A}\times\mathbf{B} = (a_1\hat{\mathbf{i}} + b_1\hat{\mathbf{j}} + c_1\hat{\mathbf{k}})\times(a_2\hat{\mathbf{i}} + b_2\hat{\mathbf{j}} + c_2\hat{\mathbf{k}})$

$$= a_1a_2(\hat{\mathbf{i}}\times\hat{\mathbf{i}}) + a_1b_2(\hat{\mathbf{i}}\times\hat{\mathbf{j}}) + a_1c_2(\hat{\mathbf{i}}\times\hat{\mathbf{k}})$$
$$+ b_1a_2(\hat{\mathbf{j}}\times\hat{\mathbf{i}}) + b_1b_2(\hat{\mathbf{j}}\times\hat{\mathbf{j}})$$
$$+ b_1c_2(\hat{\mathbf{j}}\times\hat{\mathbf{k}}) + c_1a_2(\hat{\mathbf{k}}\times\hat{\mathbf{i}})$$
$$+ c_1b_2(\hat{\mathbf{k}}\times\hat{\mathbf{j}}) + c_1c_2(\hat{\mathbf{k}}\times\hat{\mathbf{k}})$$

Since, $\hat{\mathbf{i}}\times\hat{\mathbf{i}} = \hat{\mathbf{j}}\times\hat{\mathbf{j}} = \hat{\mathbf{k}}\times\hat{\mathbf{k}} =$ a null vector and $\hat{\mathbf{i}}\times\hat{\mathbf{j}} = \hat{\mathbf{k}}$, etc., we have

$$\mathbf{A}\times\mathbf{B} = (b_1c_2 - c_1b_2)\,\hat{\mathbf{i}} + (c_1a_2 - a_1c_2)\,\hat{\mathbf{j}} + (a_1b_2 - b_1a_2)\,\hat{\mathbf{k}}$$

or putting it in determinant form, we have

$$\mathbf{A}\times\mathbf{B} = \begin{vmatrix} \hat{\mathbf{i}} & \hat{\mathbf{j}} & \hat{\mathbf{k}} \\ a_1 & b_1 & c_1 \\ a_2 & b_2 & c_2 \end{vmatrix}$$

It may be noted that the scalar components of the first vector **A** occupy the middle row of the determinant.

Example 2.10 *Find one unit vector perpendicular to* $\mathbf{A} = 2\hat{\mathbf{i}} + 3\hat{\mathbf{j}} + \hat{\mathbf{k}}$ *and* $\mathbf{B} = \hat{\mathbf{i}} - \hat{\mathbf{j}} + \hat{\mathbf{k}}$ *both.*

Sol. As we have read, $\mathbf{C} = \mathbf{A}\times\mathbf{B}$ is a vector perpendicular to both **A** and **B**. Hence, a unit vector $\hat{\mathbf{n}}$ perpendicular to **A** and **B** can be written as

$$\hat{\mathbf{n}} = \frac{\mathbf{C}}{C} = \frac{\mathbf{A}\times\mathbf{B}}{|\mathbf{A}\times\mathbf{B}|}$$

Here, $\mathbf{A}\times\mathbf{B} = \begin{vmatrix} \hat{\mathbf{i}} & \hat{\mathbf{j}} & \hat{\mathbf{k}} \\ 2 & 3 & 1 \\ 1 & -1 & 1 \end{vmatrix}$

$$= \hat{\mathbf{i}}(3+1) + \hat{\mathbf{j}}(1-2) + \hat{\mathbf{k}}(-2-3)$$
$$= 4\hat{\mathbf{i}} - \hat{\mathbf{j}} - 5\hat{\mathbf{k}}$$

Further, $|\mathbf{A}\times\mathbf{B}| = \sqrt{(4)^2 + (-1)^2 + (-5)^2}$

$$= \sqrt{42}$$

$\therefore$ The desired unit vector is

$$\hat{\mathbf{n}} = \frac{\mathbf{A} \times \mathbf{B}}{|\mathbf{A} \times \mathbf{B}|}$$

or $$\hat{\mathbf{n}} = \frac{1}{\sqrt{42}}(4\hat{\mathbf{i}} - \hat{\mathbf{j}} - 5\hat{\mathbf{k}})$$

Example 2.11 *Show that the vector* $\mathbf{A} = \hat{\mathbf{i}} - \hat{\mathbf{j}} + 2\hat{\mathbf{k}}$ *is parallel to a vector* $\mathbf{B} = 3\hat{\mathbf{i}} - 3\hat{\mathbf{j}} + 6\hat{\mathbf{k}}$.

Sol. A vector **A** is parallel to an another vector **B** if it can be written as

$$\mathbf{A} = m\mathbf{B}$$

Here, $$\mathbf{A} = (\hat{\mathbf{i}} - \hat{\mathbf{j}} + 2\hat{\mathbf{k}})$$
$$= \frac{1}{3}(3\hat{\mathbf{i}} - 3\hat{\mathbf{j}} + 6\hat{\mathbf{k}})$$

or $$\mathbf{A} = \frac{1}{3}\mathbf{B}$$

This implies that **A** is parallel to **B** and magnitude of **A** is $\frac{1}{3}$ times the magnitude of **B**.

Note Points

- Two vectors can be shown parallel (or anti-parallel) to one another if
  (i) the coefficients of $\hat{\mathbf{i}}$, $\hat{\mathbf{j}}$ and $\hat{\mathbf{k}}$ of both the vectors bear a positive constant ratio. For example, a vector $\mathbf{A} = a_1\hat{\mathbf{i}} + b_1\hat{\mathbf{j}} + c_1\hat{\mathbf{k}}$ is parallel to an another vector $\mathbf{B} = a_2\hat{\mathbf{i}} + b_2\hat{\mathbf{j}} + c_2\hat{\mathbf{k}}$ if : $\frac{a_1}{a_2} = \frac{b_1}{b_2} = \frac{c_1}{c_2}$ = a positive constant. If the vectors are anti-parallel, then this constant ratio is negative.
  (ii) the magnitude of cross product of both the vectors is zero. For instance, **A** and **B** are parallel (or antiparallel) to each other if

$$\mathbf{A} \times \mathbf{B} = \begin{vmatrix} \hat{\mathbf{i}} & \hat{\mathbf{j}} & \hat{\mathbf{k}} \\ a_1 & b_1 & c_1 \\ a_2 & b_2 & c_2 \end{vmatrix} = 0$$

or $$A \times B = 0\hat{\mathbf{i}} + 0\hat{\mathbf{j}} + 0\hat{\mathbf{k}}$$

Example 2.12 *Let a force* **F** *be acting on a body free to rotate about a point O and let* **r** *be the position vector of any point P on the line of action of the force. Then, torque* (τ) *of this force about point O is defined as*

$$\tau = \mathbf{r} \times \mathbf{F}$$

Given, $$\mathbf{F} = (2\hat{\mathbf{i}} + 3\hat{\mathbf{j}} - \hat{\mathbf{k}})\ N$$

and $$\mathbf{r} = (\hat{\mathbf{i}} - \hat{\mathbf{j}} + 6\hat{\mathbf{k}})\ m$$

Find the torque of this force.

Sol. $$\tau = \mathbf{r} \times \mathbf{F} = \begin{vmatrix} \hat{\mathbf{i}} & \hat{\mathbf{j}} & \hat{\mathbf{k}} \\ 1 & -1 & 6 \\ 2 & 3 & -1 \end{vmatrix}$$

$$= \hat{\mathbf{i}}(1 - 18) + \hat{\mathbf{j}}(12 + 1) + \hat{\mathbf{k}}(3 + 2)$$

or $$\tau = (-17\hat{\mathbf{i}} + 13\hat{\mathbf{j}} + 5\hat{\mathbf{k}})\ \text{N-m}$$

Extra Knowledge Points

- Pressure, surface tension and current are not vectors.
- To qualify as a vector, a physical quantity must not only possess magnitude and direction but must also satisfy the parallelogram law of vector addition. For instance, the finite rotation of a rigid body about a given axis has magnitude (the angle of rotation) and also direction (the direction of the axis) but it is not a vector quantity. This is so for the simple reason that the two finite rotations of the body do not add up in accordance with the law of vector addition. However, if the rotation be small or infinitesimal, it may be regarded as a vector quantity.
- Area can behave either as a scalar or a vector and how it behaves depends on circumstances.
- Moment of inertia is neither a vector nor a scalar as it has different values about different axes. It is tensor. Although tensor is a generalized term which is characterized by its rank. For example, scalars are tensor of rank zero. Vectors are tensor of rank one.
- Area (vector), dipole moment and current density are defined as vectors with specific direction.
- Vectors associated with a linear or directional effect are called polar vectors or usually, simply as vectors and those associated with rotation about an axis are referred to as axial vectors. Thus, force, linear velocity and acceleration are polar vectors and angular velocity, angular acceleration are axial vectors.
- Students are often confused over the direction of cross product. Let us discuss a simple method. To find direction of $\mathbf{A} \times \mathbf{B}$ curl your fingers from **A** to **B** through smaller angle. If it is clockwise, then $\mathbf{A} \times \mathbf{B}$ is perpendicular to the plane of **A** and **B** and away from you and if it is anti-clockwise then, $\mathbf{A} \times \mathbf{B}$ is towards you perpendicular to the plane of **A** and **B**.
- The area of triangle bounded by vectors **A** and **B** is $\frac{1}{2}|\mathbf{A} \times \mathbf{B}|$.

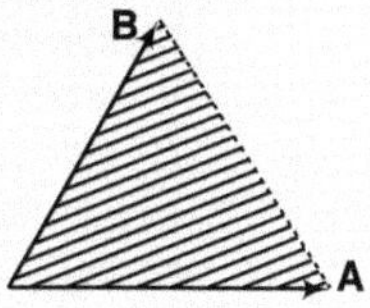

- **Exercise :** Prove the above result.

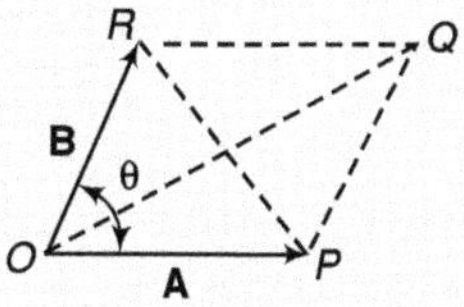

- Suppose **A** and **B** represent the two adjacent sides of a parallelogram *OPQR*, then,

Diagonal $OQ = |\mathbf{A} + \mathbf{B}| = \sqrt{A^2 + B^2 + 2AB\cos\theta}$

while diagonal $RP = |\mathbf{A} - \mathbf{B}| = \sqrt{A^2 + B^2 - 2AB\cos\theta}$

We can see that $OQ = RP$ when $\theta = 90°$.

- $\mathbf{a} \cdot \mathbf{b} = ab\cos\theta$. Here, a and b are always positive as these are the magnitudes of **a** and **b**. Hence,

$0° \le \theta < 90°$ if $\mathbf{a} \cdot \mathbf{b}$ is positive

$90° \le \theta < 180°$ if $\mathbf{a} \cdot \mathbf{b}$ is negative.

and $\theta = 90°$ if $\mathbf{a} \cdot \mathbf{b}$ is zero.

Chapter Summary with Formulae

(i) **Vector Addition**

(a) Law of parallelogram of vector addition

$$\mathbf{R} = \mathbf{A} + \mathbf{B}$$

$$R = |\mathbf{R}|$$

$$R = \sqrt{A^2 + B^2 + 2AB\cos\theta}$$

$$\tan\alpha = \frac{B\sin\theta}{A + B\cos\theta}$$

and $\tan\beta = \dfrac{A\sin\theta}{B + A\cos\theta}$

(b) Vector addition of more than two vectors

Polygon law of vector addition can be applied for addition of two or more than two vectors.

(ii) **Vector Subtraction**

If $\mathbf{S} = \mathbf{A} - \mathbf{B}$ and $S = |\mathbf{S}|$ then

$$S = \sqrt{A^2 + B^2 - 2AB\cos\theta}$$

(iii) Dot Product of Two Vectors

$$\mathbf{A}\cdot\mathbf{B} = AB\cos\theta$$

- **Important Points in Dot Product**

(a) $\mathbf{A}\cdot\mathbf{B} = \mathbf{B}\cdot\mathbf{A}$

(b) $\mathbf{A}\cdot(-\mathbf{B}) = -\mathbf{A}\cdot\mathbf{B}$

(c) $\mathbf{A}\cdot(\mathbf{B} + \mathbf{C}) = \mathbf{A}\cdot\mathbf{B} + \mathbf{A}\cdot\mathbf{C}$

(d) If θ is acute, dot product is positive. If θ is obtuse dot product is negative and if θ is $90°$ dot product is zero. Hence dot product of two perpendicular vectors is zero.

(e) $\mathbf{A}\cdot\mathbf{A} = A^2$

(f) Dot product is a scalar quantity.

(g) Work done $W = \mathbf{F}\cdot\mathbf{S} = \mathbf{F}\cdot(\mathbf{r}_f - \mathbf{r}_i)$

(h) $\hat{\mathbf{i}}\cdot\hat{\mathbf{i}} = \hat{\mathbf{j}}\cdot\hat{\mathbf{j}} = \hat{\mathbf{k}}\cdot\hat{\mathbf{k}} = 1$, $\hat{\mathbf{i}}\cdot\hat{\mathbf{j}} = \hat{\mathbf{j}}\cdot\hat{\mathbf{k}} = \hat{\mathbf{i}}\cdot\hat{\mathbf{k}} = 0$

(iv) **Position Vector and Displacement Vector.**

If coordinates of point A are (x_1, y_1, z_1) and coordinates of point B are (x_2, y_2, z_2). Then,

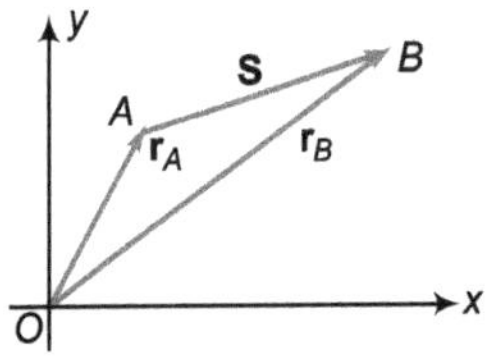

$\mathbf{r}_A$ = position vector of A

$= x_1\hat{\mathbf{i}} + y_1\hat{\mathbf{j}} + z_1\hat{\mathbf{k}}$

$\mathbf{r}_B = x_2\,\hat{\mathbf{i}} + y_2\,\hat{\mathbf{j}} + z_2\,\hat{\mathbf{k}}$

$$\mathbf{S} = \mathbf{r}_B - \mathbf{r}_A = (x_2 - x_1)\hat{\mathbf{i}} + (y_2 - y_1)\,\hat{\mathbf{j}} + (z_2 - z_1)\,\hat{\mathbf{k}}$$

= displacement vector from A to B

(v) **Cross Product of Two Vectors**

Let $\mathbf{C} = \mathbf{A}\times\mathbf{B}$, then $C = |\mathbf{C}| = AB\sin\theta$

Direction of $\mathbf{C}$ is perpendicular to both $\mathbf{A}$ and $\mathbf{B}$ given by right hand screw law. We can also say that $\mathbf{C}$ is perpendicular to the plane formed by $\mathbf{A}$ and $\mathbf{B}$.

- **Important points in cross product**

(a) $\mathbf{A}\times\mathbf{B} = -\mathbf{B}\times\mathbf{A}$

(b) Cross product of two parallel or antiparallel vectors is a null vector having zero magnitude and arbitrary direction.

(c) Cross product of two vectors of given magnitudes has maximum value when they act at $90°$.

(d) $\hat{\mathbf{i}}\times\hat{\mathbf{j}} = \hat{\mathbf{k}}$, $\hat{\mathbf{j}}\times\hat{\mathbf{k}} = \hat{\mathbf{i}}$, $\hat{\mathbf{k}}\times\hat{\mathbf{i}} = \hat{\mathbf{j}}$,

$\hat{\mathbf{i}}\times\hat{\mathbf{k}} = -\,\hat{\mathbf{j}}$, $\hat{\mathbf{k}}\times\hat{\mathbf{j}} = -\,\hat{\mathbf{i}}$

$\hat{\mathbf{j}}\times\hat{\mathbf{i}} = -\,\hat{\mathbf{k}}$

$\hat{\mathbf{i}}\times\hat{\mathbf{i}} = \hat{\mathbf{j}}\times\hat{\mathbf{j}} = \hat{\mathbf{k}}\times\hat{\mathbf{k}} =$ a null vector

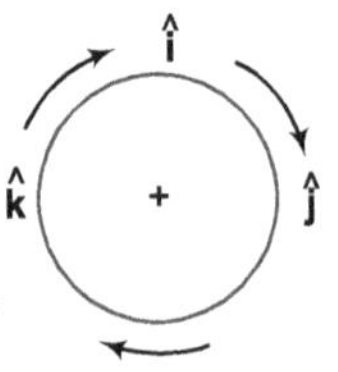

(vi) **A Unit Vector in the Direction of A**

$$\hat{\mathbf{A}} = \frac{\mathbf{A}}{A} \quad \text{or} \quad \mathbf{A} = A\hat{\mathbf{A}}$$

(vii) **A Unit Vector Perpendicular to both A and B**

Let us call it $\hat{\mathbf{C}}$.

Then $\hat{\mathbf{C}} = \pm\dfrac{\mathbf{A}\times\mathbf{B}}{|\mathbf{A}\times\mathbf{B}|}$

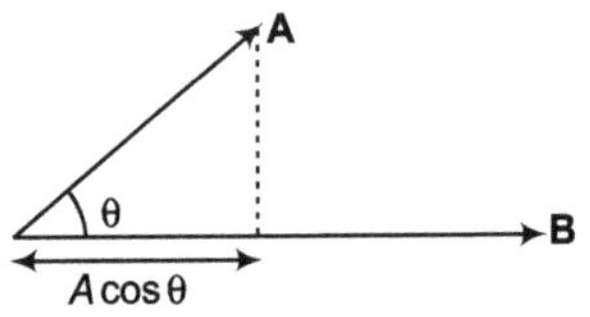

(viii) **Component of A along B** $= A\cos\theta = \dfrac{\mathbf{A}\cdot\mathbf{B}}{B}$

Similarly, component of $\mathbf{B}$ along $\mathbf{A} = B\cos\theta = \dfrac{\mathbf{A}\cdot\mathbf{B}}{A}$

(ix) Angle between Two Vectors

$$\theta = \cos^{-1}\left(\frac{\mathbf{A}\cdot\mathbf{B}}{AB}\right)$$

(x) If Vectors are given in Terms of $\hat{\mathbf{i}}$, $\hat{\mathbf{j}}$ and $\hat{\mathbf{k}}$

Let $\mathbf{A} = a_1\hat{\mathbf{i}} + a_2\hat{\mathbf{j}} + a_3\hat{\mathbf{k}}$ and $\mathbf{B} = b_1\hat{\mathbf{i}} + b_2\hat{\mathbf{j}} + b_3\hat{\mathbf{k}}$, then

(a) $|\mathbf{A}| = A = \sqrt{a_1^2 + a_2^2 + a_3^2}$ and

$|\mathbf{B}| = B = \sqrt{b_1^2 + b_2^2 + b_3^2}$

(b) $\mathbf{A} + \mathbf{B} = (a_1 + b_1)\hat{\mathbf{i}} + (a_2 + b_2)\,\hat{\mathbf{j}} + (a_3 + b_3)\,\hat{\mathbf{k}}$

(c) $\mathbf{A} - \mathbf{B} = (a_1 - b_1)\,\hat{\mathbf{i}} + (a_2 - b_2)\hat{\mathbf{j}} + (a_3 - b_3)\,\hat{\mathbf{k}}$

(d) $\mathbf{A}\cdot\mathbf{B} = a_1b_1 + a_2b_2 + a_3b_3$

(e) $|\mathbf{A}\times\mathbf{B}| = \begin{vmatrix} \hat{\mathbf{i}} & \hat{\mathbf{j}} & \hat{\mathbf{k}} \\ a_1 & a_2 & a_3 \\ b_1 & b_2 & b_3 \end{vmatrix}$

$$= (a_2b_3 - b_2a_3)\,\hat{\mathbf{i}} + (b_1a_3 - b_3a_1)\,\hat{\mathbf{j}} + (a_1b_2 - b_1a_2)\,\hat{\mathbf{k}}$$

(f) Component of $\mathbf{A}$ along $\mathbf{B}$

$$= A\cos\theta = \frac{\mathbf{A}\cdot\mathbf{B}}{B} = \frac{a_1b_1 + a_2b_2 + a_3b_3}{\sqrt{b_1^2 + b_2^2 + b_3^2}}$$

(g) Unit vector parallel to $\mathbf{A} = \hat{\mathbf{A}} = \dfrac{\mathbf{A}}{A} = \dfrac{a_1\,\hat{\mathbf{i}} + a_2\,\hat{\mathbf{j}} + a_3\,\hat{\mathbf{k}}}{\sqrt{a_1^2 + a_2^2 + a_3^2}}$

(h) Angle between $\mathbf{A}$ and $\mathbf{B}$,

$$\theta = \cos^{-1}\left(\frac{\mathbf{A}\cdot\mathbf{B}}{AB}\right)$$

$$\therefore\ \theta = \cos^{-1}\left(\frac{a_1b_1 + a_2b_2 + a_3b_3}{\sqrt{a_1^2 + a_2^2 + a_3^2}\times\sqrt{b_1^2 + b_2^2 + b_3^2}}\right)$$

Additional Examples

Example 1. *State, for each of the following physical quantities, if it is a scalar or a vector : volume, mass, speed, acceleration, density, number of moles, velocity, angular frequency, displacement, angular velocity.*

Sol. Volume, mass, speed, density, number of moles and angular frequency are scalars. The rest are vectors.

Example 2. *Explain why pressure and surface tension are not vectors?*

Sol. Pressure (force per unit area normal to it) or surface tension (force per unit length) are scalars. They have direction which is unique so need not to be specified.

Example 3. *Discuss why an infinitesimal displacement is regarded as a vector quantity.*

Sol. Because two or more infinitesimal displacements are added by vector method.

Example 4. *Is a quantity which has a magnitude and direction always a vector? Give examples.*

Sol. No. A quantity which possesses both magnitude and direction may not be a vector. For example, electric current.

Example 5. *If* $\mathbf{a}$ *and* $\mathbf{b}$ *are the vectors* $\mathbf{AB}$ *and* $\mathbf{BC}$ *determined by the adjacent sides of a regular hexagon. What are the vectors determined by the other sides taken in order?*

Sol. Given, $\mathbf{AB}=\mathbf{a}$ and $\mathbf{BC}=\mathbf{b}$

From the method of vector addition (or subtraction) we can show that,

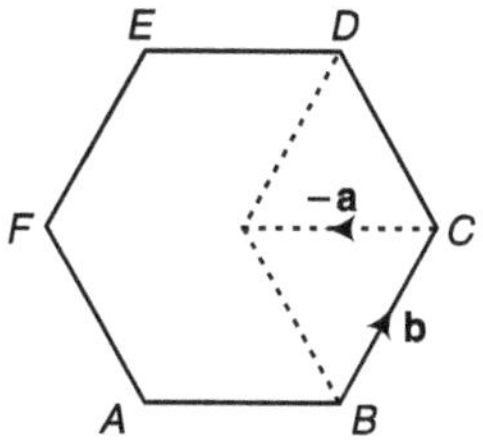

$$\mathbf{CD}=\mathbf{b}-\mathbf{a}$$

Then, $\mathbf{DE}=-\mathbf{AB}=-\mathbf{a}$

$$\mathbf{EF}=-\mathbf{BC}=-\mathbf{b}$$

and $\mathbf{FA}=-\mathbf{CD}=\mathbf{a}-\mathbf{b}$

Example 6. *If the mid-points of the consecutive sides of any quadrilateral are connected by straight line segments, prove that the resulting quadrilateral is a parallelogram.*

Sol. Let $\mathbf{a},\mathbf{b},\mathbf{c}$ and $\mathbf{d}$ be the position vectors of A, B, C and D.

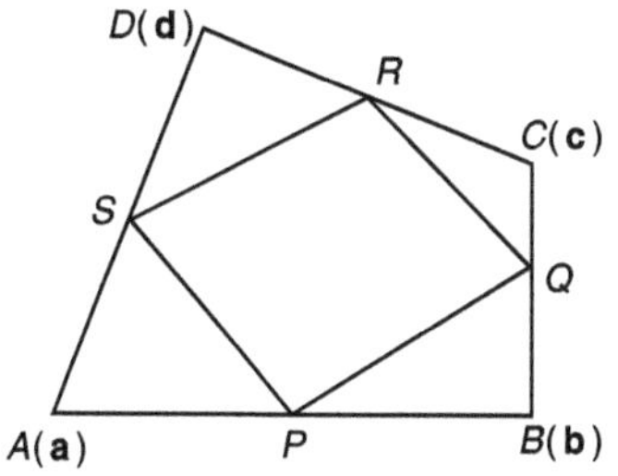

Hence, position vectors of mid-points P, Q, R and S will be

$$\frac{\mathbf{a}+\mathbf{b}}{2}, \frac{\mathbf{b}+\mathbf{c}}{2}, \frac{\mathbf{c}+\mathbf{d}}{2} \text{ and } \frac{\mathbf{d}+\mathbf{a}}{2}$$

We have, $\mathbf{PQ}=$ position vector of Q − position vector of P

$$=\frac{\mathbf{b}+\mathbf{c}}{2}-\frac{\mathbf{a}+\mathbf{b}}{2}=\frac{\mathbf{c}-\mathbf{a}}{2}$$

and $\mathbf{SR}=$ position vector of R − position vector of S

$$=\frac{\mathbf{c}+\mathbf{d}}{2}-\frac{\mathbf{d}+\mathbf{a}}{2}$$

$$=\frac{\mathbf{c}-\mathbf{a}}{2}$$

$\therefore$ $\mathbf{PQ}=\mathbf{SR}$

Similarly, we can show that

$$\mathbf{QR}=\mathbf{PS}$$

$\therefore$ $PQRS$ is a parallelogram.

Example 7. *If* $\mathbf{a}\times\mathbf{b}=\mathbf{b}\times\mathbf{c}\neq\mathbf{0}$ *with* $\mathbf{a}\neq-\mathbf{c}$, *then show that* $\mathbf{a}+\mathbf{c}=k\,\mathbf{b}$, *where k is a scalar.*

Sol. $\mathbf{a}\times\mathbf{b}=\mathbf{b}\times\mathbf{c}$

$$\mathbf{a}\times\mathbf{b}=-\mathbf{c}\times\mathbf{b}$$

$\therefore$ $\mathbf{a}\times\mathbf{b}+\mathbf{c}\times\mathbf{b}=0$

$$(\mathbf{a}+\mathbf{c})\times\mathbf{b}=0$$

$\therefore$ $\mathbf{a}\times\mathbf{b}\neq 0$, $\mathbf{b}\times\mathbf{c}\neq 0$, $\mathbf{a},\mathbf{b},\mathbf{c},\mathbf{d}$ are non-zero vectors.

$$(\mathbf{a}+\mathbf{c})\neq\mathbf{0}$$

Hence, $\mathbf{a}+\mathbf{c}$ is parallel to $\mathbf{b}$.

$\therefore$ $\mathbf{a}+\mathbf{c}=k\,\mathbf{b}$ $\quad$ (k = scalar)

Example 8. *If* $\mathbf{A}=2\hat{\mathbf{i}}-3\hat{\mathbf{j}}+7\hat{\mathbf{k}}, \mathbf{B}=\hat{\mathbf{i}}+2\hat{\mathbf{j}}$ *and* $\mathbf{C}=\hat{\mathbf{j}}-\hat{\mathbf{k}}$. *Find* $\mathbf{A}\cdot(\mathbf{B}\times\mathbf{C})$.

Sol. $\mathbf{A}\cdot(\mathbf{B}\times\mathbf{C})=[\mathbf{A}\,\mathbf{B}\,\mathbf{C}]$,

$$\text{Volume of parallelopiped}=\begin{vmatrix} 2 & -3 & 7 \\ 1 & 2 & 0 \\ 0 & 1 & -1 \end{vmatrix}$$

$$=2(-2-0)+3(-1-0)+7(1-0)$$

$$=-4-3+7$$

$$=0$$

Example 9. *Find the resultant of three vectors* **OA**, **OB** *and* **OC** *shown in figure. Radius of circle* is *R*.

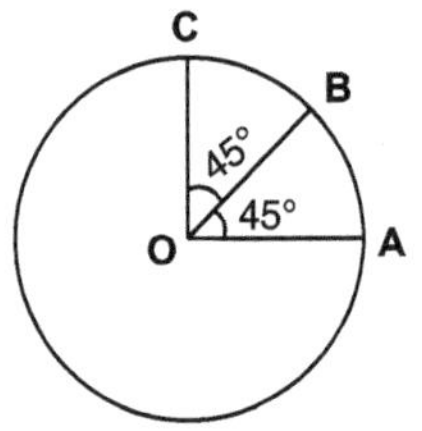

Sol. **OA** = **OC**

OA + **OC** is along **OB**, (bisector) and its magnitude is

$$2R\cos 45° = R\sqrt{2}$$

(**OA** + **OC**) + **OB** is along **OB** and its magnitude is

$$R\sqrt{2} + R = R(1+\sqrt{2})$$

Example 10. *Prove that*

$$|\mathbf{a}\times\mathbf{b}|^2 = a^2b^2 - (\mathbf{a}\cdot\mathbf{b})^2$$

Sol. Let $|\mathbf{a}| = a$, $|\mathbf{b}| = b$

and θ be the angle between them.

$$|\mathbf{a}\times\mathbf{b}|^2 = (ab\sin\theta)^2$$
$$= a^2b^2\sin^2\theta$$
$$= a^2b^2(1-\cos^2\theta)$$
$$= a^2b^2 - (a\cdot b\cos\theta)^2$$
$$= a^2b^2 - (\mathbf{a}\cdot\mathbf{b})^2$$

Example 11. *Show that the vectors* $\mathbf{a} = 3\hat{\mathbf{i}} - 2\hat{\mathbf{j}} + \hat{\mathbf{k}}$, $\mathbf{b} = \hat{\mathbf{i}} - 3\hat{\mathbf{j}} + 5\hat{\mathbf{k}}$ *and* $\mathbf{c} = 2\hat{\mathbf{i}} + \hat{\mathbf{j}} - 4\hat{\mathbf{k}}$ *form a right angled triangle.*

Sol. We have

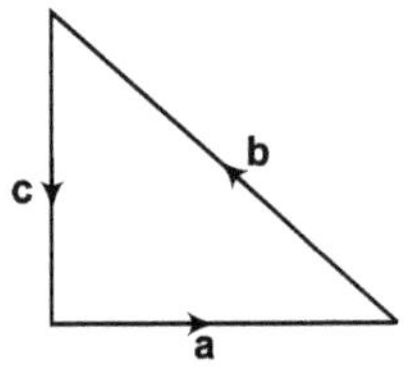

$$\mathbf{b}+\mathbf{c} = (\hat{\mathbf{i}} - 3\hat{\mathbf{j}} + 5\hat{\mathbf{k}}) + (2\hat{\mathbf{i}} + \hat{\mathbf{j}} - 4\hat{\mathbf{k}})$$
$$= 3\hat{\mathbf{i}} - 2\hat{\mathbf{j}} + \hat{\mathbf{k}} = \mathbf{a}$$

Hence, **a**, **b** and **c** are coplanar.

Also, we observe that no two of these vectors are parallel, therefore the given vectors form a triangle. Further,

$$\mathbf{a}\cdot\mathbf{c} = (3\hat{\mathbf{i}} - 2\hat{\mathbf{j}} + \hat{\mathbf{k}})\cdot(2\hat{\mathbf{i}} + \hat{\mathbf{j}} - 4\hat{\mathbf{k}}) = 0$$

Dot product of two non-zero vectors is zero. Hence, they are perpendicular so they form a right angled triangle.

$$|\mathbf{a}| = \sqrt{9+4+1} = \sqrt{14},$$
$$|\mathbf{b}| = \sqrt{1+9+25} = \sqrt{35}$$

and $$|\mathbf{c}| = \sqrt{4+1+16} = \sqrt{21}$$

Example 12. *Let* **A**, **B** *and* **C** *be unit vectors. Suppose that* $\mathbf{A}\cdot\mathbf{B} = \mathbf{A}\cdot\mathbf{C} = 0$ *and that the angle between* **B** *and* **C** *is* $\frac{\pi}{6}$, *then prove that*

$$\mathbf{A} = \pm 2(\mathbf{B}\times\mathbf{C})$$

Sol. Since, $\mathbf{A}\cdot\mathbf{B} = 0$, $\mathbf{A}\cdot\mathbf{C} = 0$

Hence, $(\mathbf{B}+\mathbf{C})\cdot\mathbf{A} = 0$

So, **A** is perpendicular to (**B** + **C**) and **A** is a unit vector perpendicular to the plane of vectors **B** and **C**.

$$\mathbf{A} = \frac{\mathbf{B}\times\mathbf{C}}{|\mathbf{B}\times\mathbf{C}|}$$

$$|\mathbf{B}\times\mathbf{C}| = |\mathbf{B}||\mathbf{C}|\sin\frac{\pi}{6} = 1\times 1\times\frac{1}{2} = \frac{1}{2}$$

$\therefore$ $$\mathbf{A} = \frac{\mathbf{B}\times\mathbf{C}}{|\mathbf{B}\times\mathbf{C}|} = \pm 2(\mathbf{B}\times\mathbf{C})$$

Example 13. *A particle moves on a given line with a constant speed v. At a certain time, it is at a point P on its straight line path. O is fixed point. Show that* (**OP** × **v**) *is independent of the position P.*

Sol.

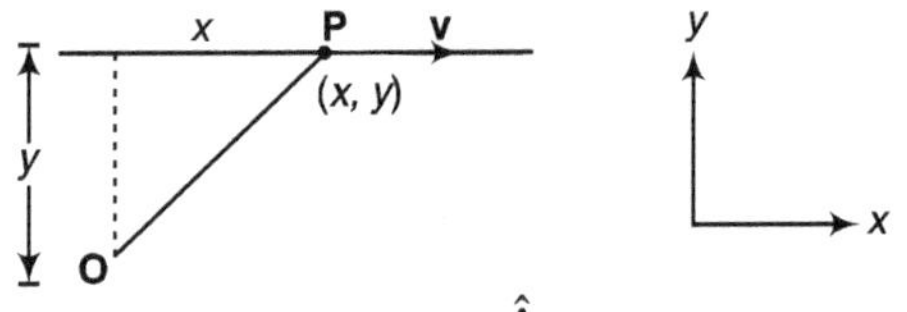

Let $$\mathbf{v} = v\hat{\mathbf{i}}$$
$$\mathbf{OP} = x\hat{\mathbf{i}} + y\hat{\mathbf{j}}$$

Take $$\mathbf{OP}\times\mathbf{v} = (x\hat{\mathbf{i}} + y\hat{\mathbf{j}})\times v\hat{\mathbf{i}}$$
$$= -yv\hat{\mathbf{k}} \quad (\because y \text{ is constant})$$

which is independent of position.

Example 14. *Prove that the mid-point of the hypotenuse of right angled triangle is equidistant from its vertices.*

Sol. Here, $\angle CAB = 90°$, let D be the mid-point of hypotenuse, we have

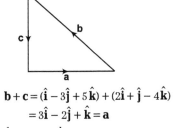

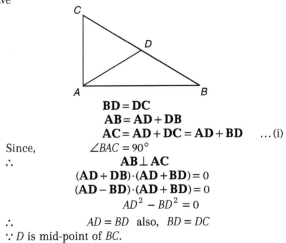

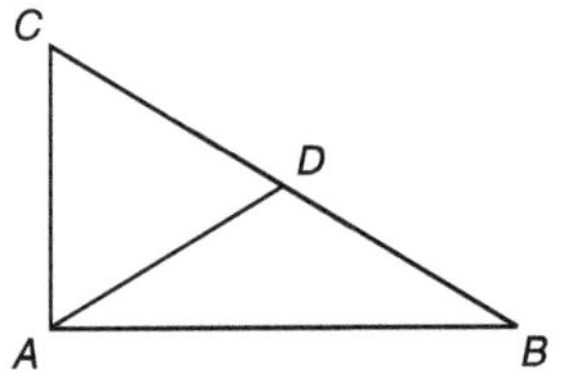

$$\mathbf{BD} = \mathbf{DC}$$
$$\mathbf{AB} = \mathbf{AD} + \mathbf{DB}$$
$$\mathbf{AC} = \mathbf{AD} + \mathbf{DC} = \mathbf{AD} + \mathbf{BD} \quad \ldots\text{(i)}$$

Since, $\angle BAC = 90°$

$\therefore$ $\mathbf{AB} \perp \mathbf{AC}$

$$(\mathbf{AD}+\mathbf{DB})\cdot(\mathbf{AD}+\mathbf{BD}) = 0$$
$$(\mathbf{AD}-\mathbf{BD})\cdot(\mathbf{AD}+\mathbf{BD}) = 0$$
$$AD^2 - BD^2 = 0$$

$\therefore$ $AD = BD$ also, $BD = DC$

$\because$ D is mid-point of BC.

Thus, $|AD| = |BD| = |DC|$. Hence proved.

NCERT Selected Questions

Q 1. State for each of the following physical quantities, if it is a scalar or a vector : Volume, mass, speed, acceleration, density, number of moles, velocity, angular frequency, displacement, angular velocity.

Sol. **Scalar quantities** Volume, mass, speed, density, number of moles, angular frequency.

Vector quantities Acceleration, velocity, displacement, angular velocity.

Q 2. Pick out the two scalar quantities in the following list: Force, angular momentum, work, current, linear momentum, electric field, average velocity, magnetic moment, reaction as per Newton's third law, relative velocity.

Sol. Work, current.

Q 3. Pick out the only vector quantity in the following list: Temperature, pressure, impulse, time, power, total path, length, energy, gravitational potential, coefficient of friction, charge.

Sol. Impulse.

Q 4. State, with reasons, whether the following algebraic operations with scalar and vector physical quantities are meaningful

(a) adding any two scalars,
(b) adding a scalar to a vector of the same dimensions,
(c) multiplying any vector by any scalar,
(d) multiplying any two scalars,
(e) adding any two vectors,
(f) adding a component of a vector to the same vector.

Sol. (a) No, adding any two scalars is not meaningful, because only the scalars of same dimensions can be added.

(b) No, adding a scalar to a vector of the same dimensions is not meaningful because a scalar cannot be added to a vector.

(c) Yes, multiplying any vector by any scalar is meaningful algebraic operation. It is because when any vector is multiplied by and scalar, then we get a vector having magnitude equal to scalar number times the magnitude of the given vector, e.g. when acceleration a is multiplied by mass m, we get force $F = m\mathrm{a}$, which is a meaningful operation.

(d) Yes, the product of two scalars gives a meaningful result e.g. when power P is multiplied by time t, then we get work done (W) i.e. $W = Pt$, which is a useful algebraic operation.

(e) No, as the two vectors of same dimensions can only be added, so addition of any two vectors is not a meaningful algebraic operation.

(f) Yes, a component of a vector can be added to the same vector as both of them are the vectors of same nature.

Q 5. Read each statement below carefully and state with reasons, if it is true or false.

(a) The magnitude of a vector is always a scalar.
(b) Each component of a vector is always a scalar.
(c) The total path length is always equal to the magnitude of the displacement vector of a particle.
(d) Three vectors not lying in a plane can never add up to give a null vector.

Sol. (a) True,

(b) False. As each component of a given vector is always a vector.

(c) False. It is true only if the particle moves along a straight line in the same direction.

(d) True.

Q 6. Establish the following vector inequalities geometrically or otherwise

(a) $|\mathbf{a}+\mathbf{b}| \leq |\mathbf{a}| + |\mathbf{b}|$
(b) $|\mathbf{a}+\mathbf{b}| \geq |\mathbf{a}| \sim |\mathbf{b}|$
(c) $|\mathbf{a}-\mathbf{b}| \leq |\mathbf{a}| + |\mathbf{b}|$
(d) $|\mathbf{a}-\mathbf{b}| \geq |\mathbf{a}| \sim |\mathbf{b}|$

When does the equality sign above apply?

Sol. Let the two vectors $\mathbf{a}$ and $\mathbf{b}$ be represented by the sides OP and PQ of the ΔOPQ taken in the same order, then their resultant is represented by the side OQ of the triangle such that

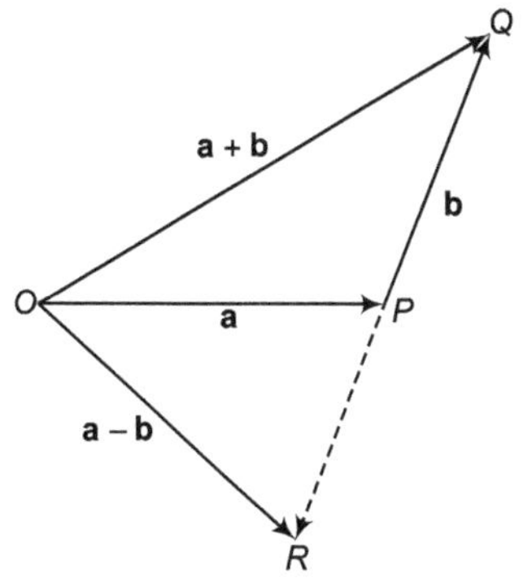

$$\mathbf{OQ} = \mathbf{a} + \mathbf{b}$$

$$\mathbf{OP} = \mathbf{a}, \ \mathbf{PQ} = \mathbf{b}$$

$\therefore$ $$|\mathbf{OP}| = |\mathbf{a}|, |\mathbf{PQ}| = |\mathbf{b}|$$

and $$|\mathbf{OQ}| = |\mathbf{a} + \mathbf{b}|$$

(a) $|\mathbf{a} + \mathbf{b}| \le |\mathbf{a}| + |\mathbf{b}|$

Proof We know from the property of a triangle that its one side is always less than the sum of the lengths of its two other sides.

$\therefore \quad OQ < OP + PQ \quad$ or $\quad |\mathbf{a} + \mathbf{b}| < |\mathbf{a}| + |\mathbf{b}|$

The equality sign applies if **a** and **b** are collinear and act in same direction.

(b) $|\mathbf{a} + \mathbf{b}| \ge |\mathbf{a}| - |\mathbf{b}|$

Proof One side is more than the difference of two other sides

$\therefore \quad OQ > OP \sim PQ \quad$ or $\quad |\mathbf{a} + \mathbf{b}| > |\mathbf{a}| \sim |\mathbf{b}|$

The equality sign applies if **a** and **b** are collinear and act in opposite directions.

(c) $|\mathbf{a} - \mathbf{b}| \le |\mathbf{a}| + |\mathbf{b}|$

Proof Again in ΔOPR

$$\therefore \quad OR < OP + PR$$

$$\text{or} \quad |\mathbf{a} - \mathbf{b}| < |\mathbf{a}| + |-\mathbf{b}|$$

$$< |\mathbf{a}| + |\mathbf{b}|$$

The equality sign applies if **a** and **b** are collinear and act in opposite directions.

(d) $|\mathbf{a} - \mathbf{b}| \ge |\mathbf{OP} - \mathbf{PR}|$

As one side of a triangle is more than the difference of its two other sides, thus is ΔOPR

$$OR > |OP - OR|$$

$$\therefore \quad |\mathbf{a} - \mathbf{b}| > ||\mathbf{a}| - |\mathbf{b}||$$

$$\text{or} \quad |\mathbf{a} - \mathbf{b}| > ||\mathbf{a}| - |\mathbf{b}||$$

The equality sign applies if **a** and **b** are collinear and act in same direction.

Q 7. Given $\mathbf{a} + \mathbf{b} + \mathbf{c} + \mathbf{d} = 0$, which of the following statements are correct

(a) **a**, **b**, **c** and **d** each must be a null vector.

(b) The magnitude of $(\mathbf{a} + \mathbf{c})$ equals the magnitude of $(\mathbf{b} + \mathbf{d})$.

(c) The magnitude of **a** can never be greater than the sum of the magnitudes of **b**, **c** and **d**.

Sol. (a) Statement is wrong because $\mathbf{a} + \mathbf{b} + \mathbf{c} + \mathbf{d}$ can be zero in many ways other then **a**, **b**, **c** and **d** must each be a null vector.

(b) Statement is correct, as $\mathbf{a} + \mathbf{b} + \mathbf{c} + \mathbf{d} = 0$

$$\therefore \quad \mathbf{a} + \mathbf{c} = -(\mathbf{b} + \mathbf{d})$$

$$\text{or} \quad |\mathbf{a} + \mathbf{c}| = |\mathbf{b} + \mathbf{d}|$$

(c) Statement is correct, as $\mathbf{a} + \mathbf{b} + \mathbf{c} + \mathbf{d} = 0$

$$\therefore \quad \mathbf{a} = -(\mathbf{b} + \mathbf{c} + \mathbf{d})$$

$$\text{or} \quad |\mathbf{a}| = |\mathbf{b} + \mathbf{c} + \mathbf{d}| \quad \text{...(i)}$$

From Eq. (i), we see that the magnitude of **a** is equal to the magnitude of vector $(\mathbf{b} + \mathbf{c} + \mathbf{d})$. Since, the sum of the magnitudes of **b**, **c**, **d** may be equal or greater than the magnitude of **a**, hence the magnitude of **a** can never be greater than the sum of the magnitudes of **b**, **c** and **d**.

Objective Problems

[Level 1]

Scalar and Vector Quantities

1. Which is not a vector quantity?
(a) Current (b) Displacement
(c) Velocity (d) Acceleration

2. Which one is a vector quantity?
(a) Temperature (b) Momentum
(c) Work (d) Speed

3. Which is a vector quantity?
(a) Work (b) Power
(c) Torque (d) Gravitational constant

4. Out of the following quantities, which is scalar?
(a) Displacement (b) Momentum
(c) Potential energy (d) Torque

5. Which is a vector quantity?
(a) Angular momentum (b) Work
(c) Potential energy (d) Electric current

6. Which of the following is a vector?
(a) Pressure (b) Surface tension
(c) Moment of inertia (d) None of these

7. Pressure is
(a) scalar (b) vector
(c) Both (a) and (b) (d) None of these

8. Surface area is
(a) scalar (b) vector
(c) Neither scalar nor vector (d) Both (a) and (b)

9. Which of the following is not the vector quantity?
(a) Torque (b) Displacement
(c) Dipole moment (d) Electric flux

Position Vector and Displacement Vector, Addition and Subtraction of two Vectors, Equilibrium of Vectors

10. If the resultant of two unequal vectors is equal to sum of their magnitudes, the angle between the vectors is
(a) $90°$ (b) $180°$
(c) $0°$ (d) None of these

11. Resultant of two vectors $\mathbf{A}$ and $\mathbf{B}$ is given by $|\mathbf{R}| = \{|\mathbf{A}| - |\mathbf{B}|\}$, angle between $\mathbf{A}$ and $\mathbf{B}$ will be
(a) $90°$
(b) $180°$
(c) $0°$
(d) None of the above

12. If $\mathbf{A} + \mathbf{B} = \mathbf{A} - \mathbf{B}$, then magnitude of $\mathbf{B}$ is
(a) $|\mathbf{A}|$ (b) 0
(c) 1 (d) None of these

13. If $|\mathbf{A}| = 2$ and $|\mathbf{B}| = 4$ and angle between them is $60°$, then $|\mathbf{A} - \mathbf{B}|$ is
(a) $\sqrt{13}$ (b) $3\sqrt{3}$
(c) $\sqrt{3}$ (d) $2\sqrt{3}$

14. Two vectors having magnitudes 8 and 10 can have maximum and minimum value of magnitude of their resultant as
(a) 12, 6 (b) 10, 3
(c) 18, 2 (d) None of these

15. The resultant of $\mathbf{A}$ and $\mathbf{B}$ makes an angle α with $\mathbf{A}$ and β with $\mathbf{B}$, then
(a) $\alpha < \beta$ (b) $\alpha > \beta$ if $A < B$
(c) $\alpha < \beta$ if $A = B$ (d) $\alpha < \beta$ if $A < B$

16. At what angle should the two forces $2P$ and $\sqrt{2}P$ act so that the resultant force is $P\sqrt{10}$?
(a) $45°$ (b) $60°$
(c) $90°$ (d) $120°$

17. Two vectors are such that $|\mathbf{A} + \mathbf{B}| = |\mathbf{A} - \mathbf{B}|$. The angle between the vector is
(a) $0°$ (b) $30°$
(c) $60°$ (d) $90°$

18. If two vectors are equal and their resultant is also equal to one of them, then the angle between the two vectors is
(a) $60°$ (b) $120°$
(c) $90°$ (d) $0°$

19. If $\mathbf{A} = \mathbf{B} + \mathbf{C}$ and the magnitudes of $\mathbf{A}$, $\mathbf{B}$ and $\mathbf{C}$ are 5, 4, 3 units, the angle between $\mathbf{A}$ and $\mathbf{C}$ is
(a) $\cos^{-1}(3/5)$ (b) $\cos^{-1}(4/5)$
(c) $\frac{\pi}{2}$ (d) $\sin^{-1}(3/4)$

20. If $\mathbf{A}$ and $\mathbf{B}$ are two non-zero vectors having equal magnitude, the angle between the vectors $\mathbf{A}$ and $\mathbf{A} - \mathbf{B}$ is
(a) $0°$
(b) $90°$
(c) $180°$
(d) dependent on the orientation of $\mathbf{A}$ and $\mathbf{B}$

21. For the resultant of two vectors to be maximum, what must be the angle between them?
(a) $0°$ (b) $60°$ (c) $90°$ (d) $180°$

22. Minimum number of forces of unequal magnitudes whose vector sum can equal to zero is
(a) two (b) three
(c) four (d) None of these

23. For the figure,

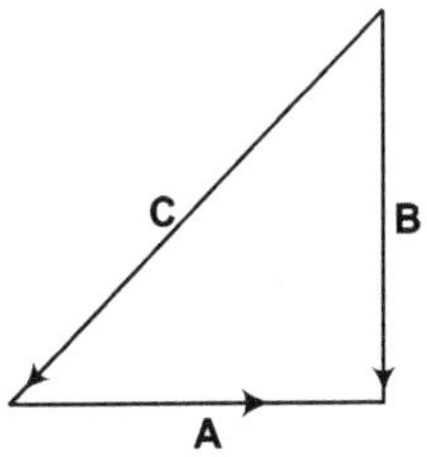

(a) $\mathbf{A}+\mathbf{B}=\mathbf{C}$ (b) $\mathbf{B}+\mathbf{C}=\mathbf{A}$
(c) $\mathbf{C}+\mathbf{A}=\mathbf{B}$ (d) $\mathbf{C}+\mathbf{A}+\mathbf{B}=0$

24. A force of 6 N and another of 8 N can be applied together to produce the effect of a single force of
(a) 1 N (b) 11 N
(c) 15 N (d) 20 N

25. Position of a particle in a rectangular coordinate system is (3, 2, 5). Then, its position vector will be
(a) $3\hat{\mathbf{i}}+5\hat{\mathbf{j}}+2\hat{\mathbf{k}}$ (b) $3\hat{\mathbf{i}}+2\hat{\mathbf{j}}+5\hat{\mathbf{k}}$
(c) $5\hat{\mathbf{i}}+3\hat{\mathbf{j}}+2\hat{\mathbf{k}}$ (d) None of these

26. If a particle moves from point $P(2,3,5)$ to point $Q(3,4,5)$, then its displacement vector will be
(a) $\hat{\mathbf{i}}+\hat{\mathbf{j}}+10\hat{\mathbf{k}}$ (b) $\hat{\mathbf{i}}+\hat{\mathbf{j}}+5\hat{\mathbf{k}}$
(c) $\hat{\mathbf{i}}+\hat{\mathbf{j}}$ (d) $2\hat{\mathbf{i}}+4\hat{\mathbf{j}}+6\hat{\mathbf{k}}$

27. Five equal forces of 10 N each are applied at one point and all are lying in one plane. If the angles between them are equal, the resultant force will be
(a) zero (b) 10 N
(c) 20 N (d) $10\sqrt{2}$ N

28. Two vectors having equal magnitudes A make an angle θ with each other. The magnitude and direction of the resultant are respectively
(a) $2A\cos\frac{\theta}{2}$, along bisector
(b) $A\cos\frac{\theta}{2}$, at 45° from one vector
(c) $2A\sin\frac{\theta}{2}$, along bisector
(d) $A\cos\frac{\theta}{2}$, along bisector

29. Given that $\mathbf{P}+\mathbf{Q}+\mathbf{R}=0$. Which of the following statement is true?
(a) $|\mathbf{P}|+|\mathbf{Q}|=|\mathbf{R}|$
(b) $|\mathbf{P}+\mathbf{Q}|=|\mathbf{R}|$
(c) $|\mathbf{P}|-|\mathbf{Q}|=|\mathbf{R}|$
(d) $|\mathbf{P}-\mathbf{Q}|=|\mathbf{R}|$

30. Two vectors **A** and **B** inclined at angle θ have a resultant **R** which makes an angle ϕ with **A**. If the directions of **A** and **B** are interchanged and resultant will have the same
(a) magnitude
(b) direction
(c) magnitude as well as direction
(d) neither

31. What is correct?
(a) $|\mathbf{a}-\mathbf{b}|=|\mathbf{a}|-|\mathbf{b}|$ (b) $|\mathbf{a}-\mathbf{b}|\le|\mathbf{a}|-|\mathbf{b}|$
(c) $|\mathbf{a}-\mathbf{b}|\ge|\mathbf{a}|-|\mathbf{b}|$ (d) $|\mathbf{a}-\mathbf{b}|>|\mathbf{a}|-|\mathbf{b}|$

32. $\mathbf{A}=2\hat{\mathbf{i}}+\hat{\mathbf{j}}$, $\mathbf{B}=3\hat{\mathbf{j}}-\hat{\mathbf{k}}$ and $\mathbf{C}=6\hat{\mathbf{i}}-2\hat{\mathbf{k}}$.
Value of $\mathbf{A}-2\mathbf{B}+3\mathbf{C}$ would be
(a) $20\hat{\mathbf{i}}+5\hat{\mathbf{j}}+4\hat{\mathbf{k}}$ (b) $20\hat{\mathbf{i}}-5\hat{\mathbf{j}}-4\hat{\mathbf{k}}$
(c) $4\hat{\mathbf{i}}+5\hat{\mathbf{j}}+20\hat{\mathbf{k}}$ (d) $5\hat{\mathbf{i}}+4\hat{\mathbf{j}}+10\hat{\mathbf{k}}$

33. If $\mathbf{A}=4\hat{\mathbf{i}}-3\hat{\mathbf{j}}$ and $\mathbf{B}=6\hat{\mathbf{i}}+8\hat{\mathbf{j}}$, then magnitude and direction of $\mathbf{A}+\mathbf{B}$ with x-axis will be
(a) $5, \tan^{-1}(3/4)$ (b) $5\sqrt{5}, \tan^{-1}(1/2)$
(c) $10, \tan^{-1}(5)$ (d) $25, \tan^{-1}(3/4)$

34. The position vector of a moving particle at time t is $\mathbf{r}=3\hat{\mathbf{i}}+4t\,\hat{\mathbf{j}}-t\,\hat{\mathbf{k}}$. Its displacement during the time interval $t=1$ s to $t=3$ s is
(a) $\hat{\mathbf{j}}-\hat{\mathbf{k}}$ (b) $3\hat{\mathbf{i}}+4\hat{\mathbf{j}}-\hat{\mathbf{k}}$
(c) $9\hat{\mathbf{i}}+36\,\hat{\mathbf{j}}-27\,\hat{\mathbf{k}}$ (d) None of these

35. Resultant of which of a following may be equal to zero?
(a) 10 N, 10 N, 10 N (b) 10 N, 10 N, 25 N
(c) 10 N, 10 N, 35 N (d) None of these

36. A man first moves 3 m due east, then 6 m due north and finally 7 m due west, then the magnitude of the resultant displacement is (in metre)
(a) $\sqrt{16}$ (b) $\sqrt{24}$ (c) $\sqrt{52}$ (d) $\sqrt{94}$

37. The resultant of two forces $3P$ and $2P$ is R. If the first force is doubled, then the resultant is also doubled. The angle between the two forces is
(a) 60° (b) 120° (c) 90° (d) 180°

38. Two forces 8 N and 12 N act at 120°. The third force required to keep the body in equilibrium is
(a) 4 N (b) $4\sqrt{7}$ N
(c) 20 N (d) None of these

39. If **a** and **b** are two units vectors inclined at an angle of 60° to each other, then
(a) $|\mathbf{a}+\mathbf{b}|>1$ (b) $|\mathbf{a}+\mathbf{b}|<1$
(c) $|\mathbf{a}-\mathbf{b}|>1$ (d) $|\mathbf{a}-\mathbf{b}|<1$

40. If 12 different coplanar forces (all of equal magnitudes) maintain a body in equilibrium, then the angle between any two adjacent forces is
(a) 15° (b) 30°
(c) 45° (d) 60°

41. If $\mathbf{P}+\mathbf{Q}=\mathbf{R}$ and $|\mathbf{P}|=|\mathbf{Q}|=\sqrt{3}$ and $|\mathbf{R}|=3$, then the angle between **P** and **Q** is

(a) $\pi/4$ (b) $\pi/6$
(c) $\pi/3$ (d) $\pi/2$

42. Given that $\mathbf{A}+\mathbf{B}+\mathbf{C}=0$, out of three vectors two are equal in magnitude and the magnitude of third vector is $\sqrt{2}$ times that of either of two having equal magnitudes. Then, angle between vectors are given by

(a) $30°, 60°, 90°$ (b) $45°, 45°, 90°$
(c) $90°, 135°, 45°$ (d) $90°, 135°, 135°$

43. Out of the following set of forces, the resultant of which cannot be zero

(a) 10, 10, 10 (b) 10, 10, 20
(c) 10, 20, 20 (d) 10, 20, 40

44. Two vectors **A** and **B** are such that $\mathbf{A}+\mathbf{B}=\mathbf{C}$ and $\mathbf{A}^2+\mathbf{B}^2=\mathbf{C}^2$. If θ is the angle between **A** and **B**, then the correct statement is

(a) $\theta=\pi$ (b) $\theta=\frac{2\pi}{3}$
(c) $\theta=0$ (d) $\theta=\frac{\pi}{2}$

45. If **A** and **B** are two vectors such that $|\mathbf{A}+\mathbf{B}|=2|\mathbf{A}-\mathbf{B}|$, the angle between vectors **A** and **B** is

(a) 45° (b) 60°
(c) 30° (d) data insufficient

Dot Product and Cross Product of Vectors

46. Given $\mathbf{A}=3\hat{\mathbf{i}}+4\hat{\mathbf{j}}$ and $\mathbf{B}=6\hat{\mathbf{i}}+8\hat{\mathbf{j}}$, which of the following statement is correct?

(a) $\mathbf{A}\times\mathbf{B}=0$ (b) $\frac{|\mathbf{A}|}{|\mathbf{B}|}=\frac{1}{2}$
(c) $|\mathbf{A}|=15$ (d) $\mathbf{A}\cdot|\mathbf{B}|=48$

47. In a clockwise system

(a) $\hat{\mathbf{j}}\times\hat{\mathbf{k}}=\hat{\mathbf{i}}$ (b) $\hat{\mathbf{k}}\cdot\hat{\mathbf{i}}=1$
(c) $\hat{\mathbf{i}}\cdot\hat{\mathbf{i}}=0$ (d) $\hat{\mathbf{j}}\times\hat{\mathbf{j}}=1$

48. A vector **A** points vertically upwards and **B** points towards north. The vector product $\mathbf{A}\times\mathbf{B}$ is

(a) zero
(b) along west
(c) along east
(d) vertically downward

49. $|\mathbf{A}\times\mathbf{B}|=\sqrt{3}\,\mathbf{A}\cdot\mathbf{B}$, then the value of $|\mathbf{A}+\mathbf{B}|$ is

(a) $\left(A^2+B^2+\frac{AB}{\sqrt{3}}\right)^{1/2}$
(b) $A+B$
(c) $(A^2+B^2+\sqrt{3}\,AB)^{1/2}$
(d) $(A^2+B^2+AB)^{1/2}$

50. The work done by a force $\mathbf{F}=(\hat{\mathbf{i}}+2\hat{\mathbf{j}}+3\hat{\mathbf{k}})$ N, to displace a body from position A to position B is [The position vector of A is $\mathbf{r}_1=(\hat{\mathbf{i}}+3\hat{\mathbf{j}}+\hat{\mathbf{k}})$ m and the position vector of B is $\mathbf{r}_2=(2\hat{\mathbf{i}}+2\hat{\mathbf{j}}+3\hat{\mathbf{k}})$ m]

(a) 5 J (b) 3 J (c) 2 J (d) 10 J

51. The condition $(\mathbf{a}\cdot\mathbf{b})^2=a^2b^2$ is satisfied when

(a) **a** is parallel to **b** (b) $\mathbf{a}\neq\mathbf{b}$
(c) $\mathbf{a}\cdot\mathbf{b}=1$ (d) $\mathbf{a}\perp\mathbf{b}$

52. If $|\mathbf{A}|=2, |\mathbf{B}|=5$ and $|\mathbf{A}\times\mathbf{B}|=8$. Angle between **A** and **B** is acute, then $\mathbf{A}\cdot\mathbf{B}$ is

(a) 6 (b) 3 (c) 4 (d) 7

53. The modulus of the vector product of two vectors is $\frac{1}{\sqrt{3}}$ times their scalar product. The angle between vectors is

(a) $\frac{\pi}{6}$ (b) $\frac{\pi}{2}$
(c) $\frac{\pi}{4}$ (d) $\frac{\pi}{3}$

54. The area of the parallelogram determined by $\mathbf{A}=2\hat{\mathbf{i}}+\hat{\mathbf{j}}-3\hat{\mathbf{k}}$ and $\mathbf{B}=12\hat{\mathbf{j}}-2\hat{\mathbf{k}}$ is approximately

(a) 43 (b) 56
(c) 38 (d) 74

55. Three vectors **A**, **B** and **C** satisfy the relation $\mathbf{A}\cdot\mathbf{B}=0$ and $\mathbf{A}\cdot\mathbf{C}=0$. Then, the vector **A** is parallel to

(a) **B** (b) **C**
(c) $\mathbf{B}\cdot\mathbf{C}$ (d) $\mathbf{B}\times\mathbf{C}$

56. What is the dot product of two vectors of magnitude 3 and 5, if angle between them is 60°?

(a) 5.2 (b) 7.5
(c) 8.4 (d) 8.6

Unit Vector, Angle between two Vectors, Component of Vector, Conditions of two Vectors

57. Unit vector perpendicular to vector $\mathbf{A}=-3\hat{\mathbf{i}}-2\hat{\mathbf{j}}-3\hat{\mathbf{k}}$ and $\mathbf{B}=2\hat{\mathbf{i}}+4\hat{\mathbf{j}}+6\hat{\mathbf{k}}$ both is

(a) $\frac{3\hat{\mathbf{j}}-2\hat{\mathbf{k}}}{\sqrt{13}}$ (b) $\frac{3\hat{\mathbf{k}}-2\hat{\mathbf{j}}}{\sqrt{13}}$
(c) $\frac{-\hat{\mathbf{j}}+2\hat{\mathbf{k}}}{\sqrt{13}}$ (d) $\frac{\hat{\mathbf{i}}+3\hat{\mathbf{j}}-\hat{\mathbf{k}}}{\sqrt{13}}$

58. Vector $\mathbf{P}=6\hat{\mathbf{i}}+4\sqrt{2}\hat{\mathbf{j}}+4\sqrt{2}\,\hat{\mathbf{k}}$ makes angle from z-axis equal to

(a) $\cos^{-1}\left(\frac{\sqrt{2}}{5}\right)$ (b) $\cos^{-1}(2\sqrt{2})$
(c) $\cos^{-1}\left(\frac{2\sqrt{2}}{5}\right)$ (d) None of these

59. Given $\mathbf{A} = \hat{\mathbf{i}} + \hat{\mathbf{j}} + \hat{\mathbf{k}}$ and $\mathbf{B} = -\hat{\mathbf{i}} - \hat{\mathbf{j}} - \hat{\mathbf{k}}$, then $(\mathbf{A} - \mathbf{B})$ will make angle with **A**

(a) 0° (b) 180°
(c) 90° (d) 60°

60. Component of the vector $\mathbf{A} = 2\hat{\mathbf{i}} + 3\hat{\mathbf{j}}$ along the vector $\mathbf{B} = (\hat{\mathbf{i}} + \hat{\mathbf{j}})$ is

(a) $\frac{5}{\sqrt{2}}$ (b) $4\sqrt{2}$
(c) $\frac{\sqrt{2}}{3}$ (d) None of these

61. Which of the following is a unit vector?

(a) $\hat{\mathbf{i}} + \hat{\mathbf{j}}$ (b) $\cos\theta\, \hat{\mathbf{i}} - \sin\theta\, \hat{\mathbf{j}}$
(c) $\sin\theta\, \hat{\mathbf{i}} + 2\cos\theta\, \hat{\mathbf{j}}$ (d) $\frac{1}{\sqrt{3}}(\hat{\mathbf{i}} + \hat{\mathbf{j}})$

62. The component of a vector along any other direction is

(a) always less than its magnitude
(b) always greater than its magnitude
(c) always equal to its magnitude
(d) None of the above

63. A vector $\mathbf{P} = 3\hat{\mathbf{i}} - 2\hat{\mathbf{j}} + a\hat{\mathbf{k}}$ is perpendicular to the vector $\mathbf{Q} = 2\hat{\mathbf{i}} + \hat{\mathbf{j}} - \hat{\mathbf{k}}$. The value of a is

(a) 2 (b) 1
(c) 4 (d) 3

64. If three vectors along coordinate axis represent the adjacent sides of a cube of length b, then the unit vector along its diagonal passing through the origin will be

(a) $\frac{\hat{\mathbf{i}} + \hat{\mathbf{j}} + \hat{\mathbf{k}}}{\sqrt{2}}$ (b) $\frac{\hat{\mathbf{i}} + \hat{\mathbf{j}} + \hat{\mathbf{k}}}{\sqrt{36}}$
(c) $\hat{\mathbf{i}} + \hat{\mathbf{j}} + \hat{\mathbf{k}}$ (d) $\frac{\hat{\mathbf{i}} + \hat{\mathbf{j}} + \hat{\mathbf{k}}}{\sqrt{3}}$

65. Consider a vector $\mathbf{F} = 4\hat{\mathbf{i}} - 3\hat{\mathbf{j}}$. Another vector perpendicular of **F** is

(a) $4\hat{\mathbf{i}} + 3\hat{\mathbf{j}}$ (b) $6\hat{\mathbf{i}}$
(c) $7\hat{\mathbf{k}}$ (d) $3\hat{\mathbf{i}} - 4\hat{\mathbf{j}}$

66. The angle between vectors $(\mathbf{A} \times \mathbf{B})$ and $(\mathbf{B} \times \mathbf{A})$ is

(a) zero (b) π
(c) $\pi/4$ (d) $\pi/2$

67. A vector perpendicular to both the vectors $2\hat{\mathbf{i}} - \hat{\mathbf{j}} + 5\hat{\mathbf{k}}$ and x-axis is

(a) $\hat{\mathbf{j}} + 5\hat{\mathbf{j}}$ (b) $\hat{\mathbf{j}} - 5\hat{\mathbf{k}}$ (c) $5\hat{\mathbf{j}} + \hat{\mathbf{k}}$ (d) $\hat{\mathbf{i}} + \hat{\mathbf{j}} + \hat{\mathbf{k}}$

68. The angle between **A** + **B** and **A** − **B** will be

(a) 90° (b) between 0° and 180°
(c) 180° only (d) None of these

69. Unit vector parallel to the resultant of vector $8\hat{\mathbf{i}} + 8\hat{\mathbf{j}}$ will be

(a) $(24\hat{\mathbf{i}} + 5\hat{\mathbf{j}})/13$ (b) $(12\hat{\mathbf{i}} + 5\hat{\mathbf{j}})/13$
(c) $(6\hat{\mathbf{i}} + 5\hat{\mathbf{j}})/13$ (d) None of these

70. The angle between the two vectors $-2\hat{\mathbf{i}} + 3\hat{\mathbf{j}} + \hat{\mathbf{k}}$ and $\hat{\mathbf{i}} + 2\hat{\mathbf{j}} - 4\hat{\mathbf{k}}$ is

(a) 45° (b) 90°
(c) 30° (d) 60°

71. The angle between the vector $2\hat{\mathbf{i}} + 4\hat{\mathbf{k}}$ and the y-axis is

(a) 0° (b) 90°
(c) 45° (d) 180°

72. A force of 5 N acts on a particle along a direction making an angle of 60° with vertical. Its vertical component will be

(a) 10 N (b) 3 N
(c) 4 N (d) 2.5 N

73. If $\mathbf{A} = 3\hat{\mathbf{i}} + 4\hat{\mathbf{j}}$ and $\mathbf{B} = 7\hat{\mathbf{i}} + 24\hat{\mathbf{j}}$, the vector having the same magnitude as **B** and parallel to **A** is

(a) $5\hat{\mathbf{i}} + 20\hat{\mathbf{j}}$ (b) $15\hat{\mathbf{i}} + 10\hat{\mathbf{j}}$
(c) $20\hat{\mathbf{i}} + 15\hat{\mathbf{j}}$ (d) $15\hat{\mathbf{i}} + 20\hat{\mathbf{j}}$

74. The vector that must be added to the vector $\hat{\mathbf{i}} - 3\hat{\mathbf{j}} + 2\hat{\mathbf{k}}$ and $3\hat{\mathbf{i}} + 6\hat{\mathbf{j}} - 7\hat{\mathbf{k}}$ so that the resultant vector is a unit vector along the y-axis is

(a) $4\hat{\mathbf{i}} + 2\hat{\mathbf{j}} + 5\hat{\mathbf{k}}$ (b) $-4\hat{\mathbf{i}} - 2\hat{\mathbf{j}} + 5\hat{\mathbf{k}}$
(c) $3\hat{\mathbf{i}} + 4\hat{\mathbf{j}} + 5\hat{\mathbf{k}}$ (d) null vector

75. The expression $\left(\frac{1}{\sqrt{2}}\hat{\mathbf{i}} + \frac{1}{\sqrt{2}}\hat{\mathbf{j}}\right)$ is a

(a) unit vector (b) null vector
(c) vector of magnitude $\sqrt{2}$ (d) scalar

76. A vector is represented by $3\hat{\mathbf{i}} + \hat{\mathbf{j}} + 2\hat{\mathbf{k}}$. Its length in X-Y plane is

(a) 2 (b) $\sqrt{14}$ (c) $\sqrt{10}$ (d) $\sqrt{5}$

77. What is the angle between $\hat{\mathbf{i}} + \hat{\mathbf{j}} + \hat{\mathbf{k}}$ and $\hat{\mathbf{j}}$?

(a) $\frac{\pi}{6}$ (b) $\frac{\pi}{4}$
(c) $\frac{\pi}{3}$ (d) None of these

78. Let $\mathbf{A} = \hat{\mathbf{i}}\, A\cos\theta + \hat{\mathbf{j}}\, A\sin\theta$ be any vector. Another vector **B** which is perpendicular to **A** can be expressed as

(a) $\hat{\mathbf{i}}\, B\cos\theta - \hat{\mathbf{j}}\, B\sin\theta$
(b) $\hat{\mathbf{i}}\, B\sin\theta - \hat{\mathbf{j}}\, B\cos\theta$
(c) $\hat{\mathbf{i}}\, B\cos\theta + \hat{\mathbf{j}}\, B\sin\theta$
(d) $\hat{\mathbf{i}}\, B\sin\theta + \hat{\mathbf{j}}\, B\cos\theta$

79. Which of the following is the unit vector perpendicular to **A** and **B**?

(a) $\frac{\hat{\mathbf{A}} \times \hat{\mathbf{B}}}{AB\sin\theta}$ (b) $\frac{\hat{\mathbf{A}} \times \hat{\mathbf{B}}}{AB\cos\theta}$
(c) $\frac{\mathbf{A} \times \mathbf{B}}{AB\sin\theta}$ (d) $\frac{\mathbf{A} \times \mathbf{B}}{AB\cos\theta}$

80. If $\frac{|\mathbf{a}+\mathbf{b}|}{|\mathbf{a}-\mathbf{b}|}=1$, then angle between **a** and **b** is
(a) 0° (b) 45°
(c) 90° (d) 60°

81. A man walks 20 m at an angle 60° north of east. How far towards east has he travalled?
(a) 10 m (b) 20 m
(c) $20\sqrt{3}$ m (d) $10\sqrt{3}$ m

82. If two vectors $2\hat{\mathbf{i}}+3\hat{\mathbf{j}}+\hat{\mathbf{k}}$ and $-4\hat{\mathbf{i}}-6\hat{\mathbf{j}}-\lambda\hat{\mathbf{k}}$ are parallel to each other, then value of λ is
(a) 0 (b) 2
(c) 3 (d) 4

83. What is the angle between $(\mathbf{P}+\mathbf{Q})$ and $(\mathbf{P}\times\mathbf{Q})$?
(a) Zero (b) $\frac{\pi}{2}$
(c) $\frac{\pi}{4}$ (d) π

84. If a unit vector is represent by $0.5\hat{\mathbf{i}}+0.8\hat{\mathbf{j}}+c\hat{\mathbf{k}}$, then the value of c is
(a) 1 (b) $\sqrt{0.11}$
(c) $\sqrt{0.01}$ (d) $\sqrt{0.39}$

85. If $\mathbf{A}=a_1\hat{\mathbf{i}}+b_1\hat{\mathbf{j}}$ and $\mathbf{B}=a_2\hat{\mathbf{i}}+b_2\hat{\mathbf{j}}$, the condition that they are perpendicular to each other is
(a) $\frac{a_1}{b_1}=-\frac{b_2}{a_2}$ (b) $a_1b_1=a_2b_2$
(c) $\frac{a_1}{a_2}=-\frac{b_1}{b_2}$ (d) None of these

86. For what value of x, will the two vectors $\mathbf{A}=2\hat{\mathbf{i}}+2\hat{\mathbf{j}}-x\hat{\mathbf{k}}$ and $\mathbf{B}=2\hat{\mathbf{i}}-\hat{\mathbf{j}}-3\hat{\mathbf{k}}$ are perpendicular to each other?
(a) $x=-2/3$
(b) $x=3/2$
(c) $x=-4/3$
(d) $x=2/3$

87. **A** and **B** are two vectors given by $\mathbf{A}=2\hat{\mathbf{i}}+3\hat{\mathbf{j}}$ and $\mathbf{B}=\hat{\mathbf{i}}+\hat{\mathbf{j}}$. The magnitude of the component of **A** along **B** is
(a) $\frac{5}{\sqrt{2}}$ (b) $\frac{3}{\sqrt{2}}$
(c) $\frac{7}{\sqrt{2}}$ (d) $\frac{5}{\sqrt{13}}$

88. The values of x and y for which vectors $\mathbf{A}=(6\hat{\mathbf{i}}+x\hat{\mathbf{j}}-2\hat{\mathbf{k}})$ and $\mathbf{B}=(5\hat{\mathbf{i}}+6\hat{\mathbf{j}}-y\hat{\mathbf{k}})$ may be parallel are
(a) $x=0,\ y=\frac{2}{3}$ (b) $x=\frac{36}{5},\ y=\frac{5}{3}$
(c) $x=-\frac{15}{3},\ y=\frac{23}{5}$ (d) $x=-\frac{36}{5},\ y=\frac{15}{4}$

89. The angles which the vector $\mathbf{A}=3\hat{\mathbf{i}}+6\hat{\mathbf{j}}+2\hat{\mathbf{k}}$ makes with the coordinate axes are
(a) $\cos^{-1}\frac{3}{7}$, $\cos^{-1}\frac{6}{7}$ and $\cos^{-1}\frac{2}{7}$
(b) $\cos^{-1}\frac{4}{7}$, $\cos^{-1}\frac{5}{7}$ and $\cos^{-1}\frac{3}{7}$
(c) $\cos^{-1}\frac{3}{7}$, $\cos^{-1}\frac{4}{7}$ and $\cos^{-1}\frac{1}{7}$
(d) None of the above

Miscellaneous Problems

90. If $\mathbf{A}=\mathbf{B}$, then which of the following is not correct?
(a) $\hat{\mathbf{A}}=\hat{\mathbf{B}}$ (b) $|\mathbf{A}|=|\mathbf{B}|$
(c) $A\hat{\mathbf{B}}=B\hat{\mathbf{A}}$ (d) $\mathbf{A}+\mathbf{B}=\hat{\mathbf{A}}+\hat{\mathbf{B}}$

91. The area of the parallelogram represented by the vectors $\mathbf{A}=2\hat{\mathbf{i}}+3\hat{\mathbf{j}}$ and $\mathbf{B}=\hat{\mathbf{i}}+4\hat{\mathbf{j}}$ is
(a) 14 units (b) 7.5 units (c) 10 units (d) 5 units

92. The forces, which meet at one point but their lines of action do not lie on one plane, are called
(a) non-coplanar non-concurrent forces
(b) non-coplanar concurrent forces
(c) coplanar concurrent forces
(d) coplanar non-concurrent forces

93. If the angle between two vectors **A** and **B** is 120°, its resultant C will be
(a) $C=|A-B|$ (b) $C<|A-B|$
(c) $C>|A-B|$ (d) $C=|A+B|$

94. The condition under which the vectors $(\mathbf{a}+\mathbf{b})$ and $(\mathbf{a}-\mathbf{b})$ are parallel is?
(a) $\mathbf{a}\perp\mathbf{b}$ (b) $|\mathbf{a}|=|\mathbf{b}|$
(c) $\mathbf{a}\neq\mathbf{b}$ (d) **a** is parallel to **b**

95. A cyclist is moving on a circular path with constant speed v. What is the change in its velocity after it has described an angle of 30°?
(a) $v\sqrt{2}$ (b) $\frac{v}{2}$
(c) $v\sqrt{3}$ (d) None of these

96. If **A** and **B** are two non-zero vectors having equal magnitudes, the angle between the vectors **A** and the resultant of **A** and **B** is
(a) 0°
(b) 90°
(c) 180°
(d) dependent on the orientation of **A** and **B**

97. Resultant of two vectors of equal magnitude A is
(a) $\sqrt{3}A$ at 60° (b) $\sqrt{2}A$ at 90°
(c) $2A$ at 120° (d) A at 180°

98. If $a\hat{\mathbf{i}}+b\hat{\mathbf{j}}$ is a unit vector and it is perpendicular to $\hat{\mathbf{i}}+\hat{\mathbf{j}}$, then value of a and b is
(a) 1, 0 (b) −2, 0
(c) 0.5, −0.5 (d) None of these

99. If $\mathbf{a}+\mathbf{b}+\mathbf{c}=0$ then $\mathbf{a}\times\mathbf{b}$ is

(a) $\mathbf{b}\times\mathbf{c}$ (b) $\mathbf{c}\times\mathbf{b}$
(c) $\mathbf{a}\times\mathbf{c}$ (d) None of these

100. Consider a force vector $\mathbf{F}=\hat{\mathbf{i}}+\hat{\mathbf{j}}+\hat{\mathbf{k}}$. Another vector perpendicular of $\mathbf{F}$ is

(a) $4\hat{\mathbf{i}}+3\hat{\mathbf{j}}$ (b) $6\hat{\mathbf{i}}$
(c) $2\hat{\mathbf{i}}-\hat{\mathbf{j}}-\hat{\mathbf{k}}$ (d) $3\hat{\mathbf{i}}-4\hat{\mathbf{j}}$

101. A vector perpendicular to both the vector $2\hat{\mathbf{i}}-3\hat{\mathbf{j}}$ and $3\hat{\mathbf{i}}-2\hat{\mathbf{j}}$ is

(a) $\hat{\mathbf{j}}+5\hat{\mathbf{k}}$ (b) $\hat{\mathbf{j}}-5\hat{\mathbf{k}}$
(c) $6\hat{\mathbf{k}}$ (d) $\hat{\mathbf{i}}+\hat{\mathbf{j}}+\hat{\mathbf{k}}$

102. A force $\mathbf{F}=(6\hat{\mathbf{i}}-8\hat{\mathbf{j}}+10\hat{\mathbf{k}})$N produces an acceleration of $1\ \text{m/s}^2$ in a body. The mass of body would be

(a) 200 kg (b) 20 kg
(c) $10\sqrt{2}$ kg (d) $6\sqrt{2}$ kg

103. If $\mathbf{A}=4\hat{\mathbf{i}}-4\hat{\mathbf{j}}-3\hat{\mathbf{k}}$ and $\mathbf{B}=6\hat{\mathbf{i}}+8\hat{\mathbf{j}}$, then magnitude and direction of $\mathbf{A}+\mathbf{B}$ with x-axis will be

(a) 5, $\tan^{-1}(3/4)$
(b) $5\sqrt{5}$, $\tan^{-1}(1/2)$
(c) 10, $\tan^{-1}(5)$
(d) 25, $\tan^{-1}(3/4)$

104. When a force F acts on a body of mass m, then the acceleration produced in the body is a. If three equal forces $F_1=F_2=F_3=F$ act on the same body as shown in figure, then the acceleration produced is

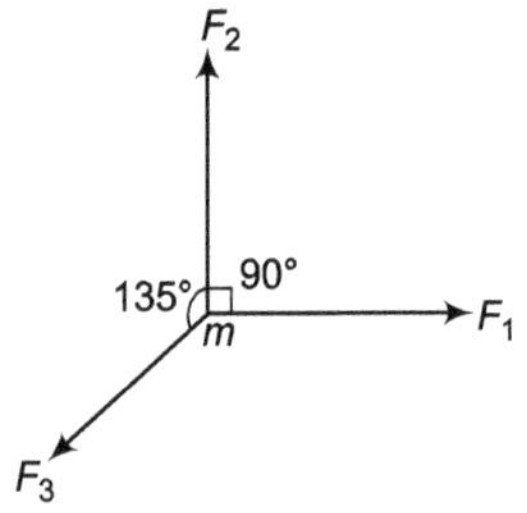

(a) $(\sqrt{2}-1)\,a$ (b) $(\sqrt{2}+1)\,a$
(c) $\sqrt{2}a$ (d) a

105. The direction cosines of vector $(\mathbf{A}-\mathbf{B})$, if $\mathbf{A}=2\hat{\mathbf{i}}+3\hat{\mathbf{j}}+\hat{\mathbf{k}}$, $\mathbf{B}=2\hat{\mathbf{i}}+2\hat{\mathbf{j}}+3\hat{\mathbf{k}}$ are

(a) $0, \frac{1}{\sqrt{5}}, \frac{-2}{\sqrt{5}}$
(b) $0, \frac{2}{\sqrt{5}}, \frac{1}{\sqrt{5}}$
(c) $0, 0, \frac{1}{\sqrt{5}}$
(d) None of the above

106. The vector sum of two forces is perpendicular to their vector differences. In that case, the forces

(a) are not equal to each other in magnitude
(b) are parallel
(c) are perpendicular
(d) are equal to each other in magnitude

[Level 2]

Only One Correct Option

1. If vectors $\mathbf{A}$ and $\mathbf{B}$ have an angle θ between them, then value of $|\hat{\mathbf{A}}-\hat{\mathbf{B}}|$ will be

(a) $2\cos\frac{\theta}{2}$ (b) $2\tan\frac{\theta}{2}$
(c) $2\sin\frac{\theta}{2}$ (d) None of these

2. A particle undergoes three successive displacements given by $\mathbf{s}_1=\sqrt{2}$ m north-east, $\mathbf{s}_2=2$ m due south and $\mathbf{s}_3=4$ m, 30° north of west, then magnitude of net displacement is

(a) $\sqrt{14+4\sqrt{3}}$
(b) $\sqrt{14-4\sqrt{3}}$
(c) $\sqrt{4}$
(d) None of the above

3. The resultant of $\mathbf{A}$ and $\mathbf{B}$ is $\mathbf{R}_1$. On reversing the vector $\mathbf{B}$, the resultant becomes $\mathbf{R}_2$. What is the value of $R_1^2+R_2^2$?

(a) A^2+B^2 (b) A^2-B^2
(c) $2(A^2+B^2)$ (d) $2(A^2-B^2)$

4. If the sum of two unit vectors is a unit vector, then magnitude of difference in two unit vectors is

(a) $\sqrt{2}$ (b) $\sqrt{3}$
(c) $1/\sqrt{2}$ (d) $\sqrt{5}$

5. The sum of two vectors $\mathbf{A}$ and $\mathbf{B}$ is at right angles to their difference. Then,

(a) $A=B$
(b) $A=2B$
(c) $B=2A$
(d) None of the above

6. A vector having magnitude 30 unit makes equal angles with each of X, Y and Z axes. The components of vector along each of X, Y and Z axes are

(a) $10\sqrt{3}$ unit (b) $10\sqrt{3}$ unit
(c) $15\sqrt{3}$ unit (d) 10 unit

7. Figure shown $ABCDEF$ as a regular hexagon. What is the value of

$$\mathbf{AB} + \mathbf{AC} + \mathbf{AD} + \mathbf{AE} + \mathbf{AF}$$

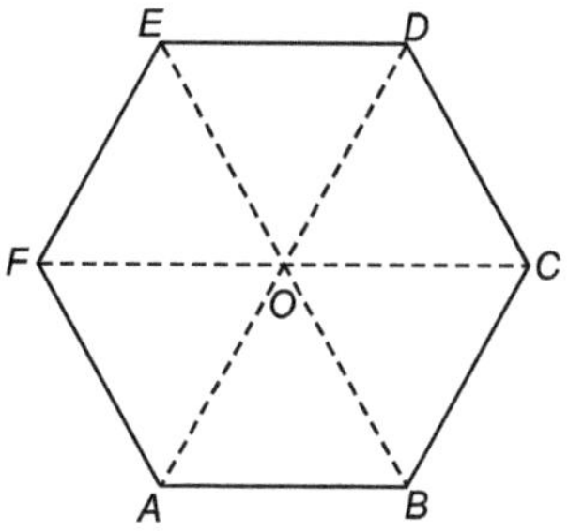

(a) $\mathbf{AO}$ (b) $2\mathbf{AO}$ (c) $4\mathbf{AO}$ (d) $6\mathbf{AO}$

8. If $\mathbf{A}$ is a unit vector in a given direction, then the value of $\hat{\mathbf{A}} \cdot \frac{d\hat{\mathbf{A}}}{dt}$ is

(a) 0 (b) 1 (c) $\frac{1}{2}$ (d) 2

9. If $\mathbf{F}_1$ and $\mathbf{F}_2$ are two vectors of equal magnitudes F such that $|\mathbf{F}_1 \cdot \mathbf{F}_2| = |\mathbf{F}_1 \times \mathbf{F}_2|$, then $|\mathbf{F}_1 + \mathbf{F}_2|$ equals to

(a) $\sqrt{(2+\sqrt{2})}\,F$ (b) $2F$
(c) $F\sqrt{2}$ (d) None of these

10. Figure shows three vectors $\mathbf{p}$, $\mathbf{q}$ and $\mathbf{r}$, where C is the mid point of AB. Then, which of the following relation is correct?

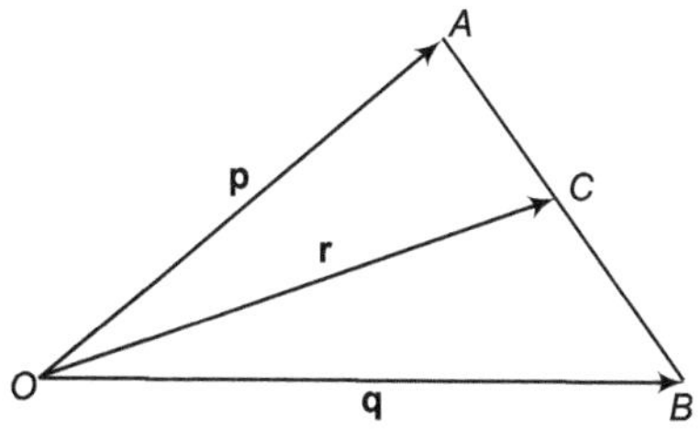

(a) $\mathbf{p} + \mathbf{q} = 2\mathbf{r}$ (b) $\mathbf{p} + \mathbf{q} = \mathbf{r}$
(c) $\mathbf{p} - \mathbf{q} = 2\mathbf{r}$ (d) $\mathbf{p} - \mathbf{q} = \mathbf{r}$

11. If a vector $\mathbf{p}$ is making angles α, β and γ respectively with the X, Y and Z axes, then $\sin^2\alpha + \sin^2\beta + \sin^2\gamma$ is equal to

(a) 0 (b) 1 (c) 2 (d) 3

12. What is the angle between $\mathbf{P}$ and the cross product of $(\mathbf{P} + \mathbf{Q})$ and $(\mathbf{P} - \mathbf{Q})$?

(a) $90°$ (b) $\tan^{-1}(P/Q)$
(c) $\tan^{-1}(Q/P)$ (d) $0°$

13. The resultant of two vectors $\mathbf{P}$ and $\mathbf{Q}$ is $\mathbf{R}$. If $\mathbf{Q}$ is doubled, the new resultant is perpendicular to $\mathbf{P}$. Then, R equals to

(a) P (b) $(P+Q)$
(c) Q (d) $(P-Q)$

14. The angle between the vector $\mathbf{A}$ and $\mathbf{B}$ is θ. The value of the triple product $\mathbf{A} \cdot (\mathbf{B} \times \mathbf{A})$ is

(a) A^2B (b) zero
(c) $A^2B\sin\theta$ (d) $A^2B\cos\theta$

15. The sum of the magnitudes of two forces acting at a point is 18 and the magnitude of their resultant is 12. If the resultant is at 90° with the force of smaller magnitude, what are the magnitudes of forces?

(a) 12, 6 (b) 14, 4
(c) 5, 13 (d) 10, 8

16. A vector $\mathbf{a}$ is turned without a change in its length through a small angles $d\theta$. The value of $|\Delta\mathbf{a}|$ and Δa are respectively

(a) $0, a\,d\theta$ (b) $a \cdot d\theta, 0$
(c) 0, 0 (d) None of these

17. Find the resultant of three vectors $\mathbf{OA}, \mathbf{OB}$, and $\mathbf{OC}$ shown in the figure. Radius of the circle is R.

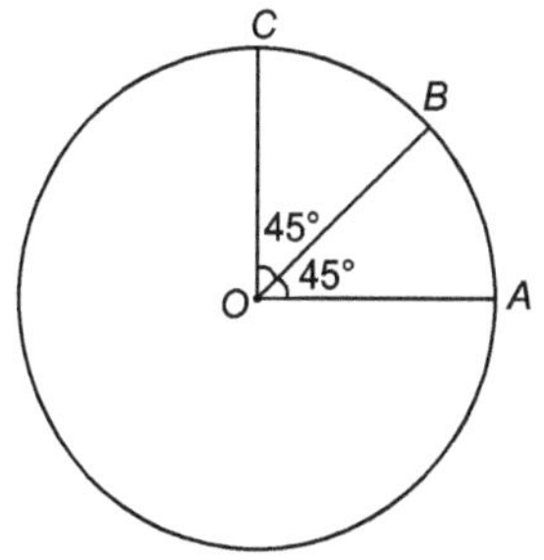

(a) $2R$ (b) $R(1+\sqrt{2})$
(c) $R\sqrt{2}$ (d) $R(\sqrt{2}-1)$

18. In the diagram shown in figure

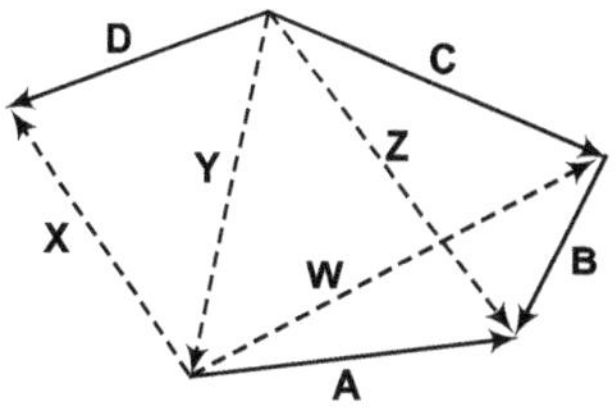

(a) $\mathbf{X} = \mathbf{A} + \mathbf{B} - \mathbf{C} + \mathbf{D}$ (b) $\mathbf{Y} = \mathbf{B} + \mathbf{C} - \mathbf{A}$
(c) $\mathbf{Z} = \mathbf{B} + \mathbf{C}$ (d) $\mathbf{W} = \mathbf{A} + \mathbf{B}$

19. If $\mathbf{a}$ and $\mathbf{b}$ are two vectors, then the value of $(\mathbf{a}+\mathbf{b}) \times (\mathbf{a}-\mathbf{b})$ is

(a) $2(\mathbf{b} \times \mathbf{a})$
(b) $-2(\mathbf{b} \times \mathbf{a})$
(c) $\mathbf{b} \times \mathbf{a}$
(d) $\mathbf{b} \times \mathbf{a}$

20. Given that $\mathbf{A}+\mathbf{B}=\mathbf{C}$ and that $\mathbf{C}$ is perpendicular to $\mathbf{A}$ further if $|\mathbf{A}|=|\mathbf{C}|$, then what is the angle between $\mathbf{A}$ and $\mathbf{B}$?

(a) $\frac{\pi}{4}$ (b) $\frac{\pi}{2}$ (c) $\frac{3\pi}{4}$ (d) π

21. The resultant of two forces, one double and the other in magnitude, is perpendicular to the smaller of the two forces. The angle between the two forces is

(a) 120° (b) 135°
(c) 90° (d) 150°

22. If $\mathbf{A}+\mathbf{B}=\mathbf{C}$, $|\mathbf{A}|=2|\mathbf{B}|$ and $\mathbf{B}\cdot\mathbf{C}=0$, then

(a) $|\mathbf{A}+\mathbf{C}|=|\mathbf{A}+\mathbf{B}|$ (b) $|\mathbf{A}+\mathbf{C}|=\mathbf{B}$
(c) $\mathbf{A}\cdot\mathbf{B}<0$ (d) $\mathbf{A}\cdot\mathbf{C}$ may be zero

23. Two unit vectors when added give a unit vector. Then, choose the correct statement.

(a) Magnitude of their difference is $\sqrt{3}$
(b) Magnitude of their difference is 1
(c) Angle between the vectors is 90°
(d) Angle between the sum and the difference of the two vectors is 90°

24. The velocity of a particle is $\mathbf{v}=6\hat{\mathbf{i}}+2\hat{\mathbf{j}}-2\hat{\mathbf{k}}$. The component of the velocity parallel to vector $\mathbf{a}=\hat{\mathbf{i}}+\hat{\mathbf{j}}-\hat{\mathbf{k}}$ in vector form is

(a) $6\hat{\mathbf{i}}+2\hat{\mathbf{j}}+2\hat{\mathbf{k}}$ (b) $2\hat{\mathbf{i}}+2\hat{\mathbf{j}}+2\hat{\mathbf{k}}$
(c) $\hat{\mathbf{i}}+\hat{\mathbf{j}}+\hat{\mathbf{k}}$ (d) $6\hat{\mathbf{i}}+2\hat{\mathbf{j}}-2\hat{\mathbf{k}}$

25. The resultant $\mathbf{R}$ of vectors $\mathbf{P}$ and $\mathbf{Q}$ is perpendicular to $|\mathbf{P}|$ and $|\mathbf{Q}|$ both, then angle between $|\mathbf{P}|$ and $|\mathbf{Q}|$ is

(a) 45° (b) 135°
(c) 120° (d) All of these

26. The value of $\hat{\mathbf{i}}\times(\hat{\mathbf{i}}\times\mathbf{a})+\hat{\mathbf{j}}\times(\hat{\mathbf{j}}\times\mathbf{a})+\hat{\mathbf{k}}\times(\hat{\mathbf{k}}\times\mathbf{a})$ is

(a) $\mathbf{a}$ (b) $\mathbf{a}\times\hat{\mathbf{k}}$
(c) $-2\mathbf{a}$ (d) $-\mathbf{a}$

27. What is the angle between $\mathbf{P}$ and the resultant of $(\mathbf{P}+\mathbf{Q})$ and $(\mathbf{P}-\mathbf{Q})$?

(a) Zero (b) $\tan^{-1}(P/Q)$
(c) $\tan^{-1}(Q/P)$ (d) $\tan^{-1}(P-Q)/(P+Q)$

Assertion and Reason

Directions (Q. Nos. 1-13) *These questions consists of two statements each printed as Assertion and Reason. While answering these questions you are required to choose any one of the following five responses.*

(a) If both Assertion and Reason are correct and Reason is the correct explanation of Assertion
(b) If both Assertion and Reason are true but Reason is not the correct explanation of Assertion
(c) If Assertion is true but Reason is false
(d) If Assertion is false but Reason is true
(e) If both Assertion and Reason are false

1. Assertion $\mathbf{A}\times\mathbf{B}$ is perpendicular to both $\mathbf{A}+\mathbf{B}$ as well as $\mathbf{A}-\mathbf{B}$.

Reason $\mathbf{A}+\mathbf{B}$ as well as $\mathbf{A}-\mathbf{B}$ lie in the plane containing $\mathbf{A}$ and $\mathbf{B}$ while $\mathbf{A}\times\mathbf{B}$ lies perpendicular to the plane containing $\mathbf{A}$ and $\mathbf{B}$.

2. Assertion Angle between $\hat{\mathbf{i}}+\hat{\mathbf{j}}$ and $\hat{\mathbf{i}}$ is 45°.

Reason $\hat{\mathbf{i}}+\hat{\mathbf{j}}$ is equally inclined to both $\hat{\mathbf{i}}$ and $\hat{\mathbf{j}}$ and the angle between $\hat{\mathbf{i}}$ and $\hat{\mathbf{j}}$ is 90°.

3. Assertion Finite angular displacement is not a vector quantity.

Reason If does not obey the vector laws.

4. Assertion Vector product of two vectors may be greater than, equal to or less than the scalar product.

Reason At $\theta=45°$, two are equal.

5. Assertion Vector addition of two vectors is always greater than their vector subtraction.

Reason At $\theta=90°$, two are equal.

6. Assertion Component of $\mathbf{A}$ along $\mathbf{B}$ is equal to component of $\mathbf{B}$ along $\mathbf{A}$.

Reason Compound of $\mathbf{A}$ along $\mathbf{B}$ is $A\cos\theta$. Where, θ is the angle between two vectors.

7. Assertion Small displacement is a vector quantity.

Reason Pressure and surface tension are also vector quantities.

8. Assertion Component of $\mathbf{A}$ along $\mathbf{B}$ is equal to component of $\mathbf{B}$ along $\mathbf{A}$.

Reason Value of component is always less than the magnitude of vector.

9. Assertion $(\mathbf{A}+\mathbf{B})\cdot(\mathbf{A}-\mathbf{B})$ is always positive.

Reason This is positive if $|\mathbf{A}|>|\mathbf{B}|$.

10. Assertion We can find angle between two vectors by using the relation,

$$\theta=\sin^{-1}\left(\frac{|\mathbf{A}\times\mathbf{B}|}{AB}\right)$$

Reason $\frac{|\mathbf{A}\times\mathbf{B}|}{AB}$ is always positive.

11. Assertion $(\mathbf{A}\times\mathbf{B})\cdot(\mathbf{B}\times\mathbf{A})$ is $-A^2B^2\sin^2\theta$. Here θ is the angle between $\mathbf{A}$ and $\mathbf{B}$.

Reason $(\mathbf{A}\times\mathbf{B})$ and $(\mathbf{B}\times\mathbf{A})$ are two anti-parallel vectors provided $\mathbf{A}$ and $\mathbf{B}$ are neither parallel nor anti-parallel.

12. Assertion If angle between $\mathbf{a}$ and $\mathbf{b}$ is 30°. Then, angle between $2\mathbf{a}$ and $-\frac{\mathbf{b}}{2}$ will be 150°.

Reason Sign of dot product of two vectors tells you whether angle between two vectors is acute or obtuse.

13. Assertion If $|\mathbf{A}|=|\mathbf{B}|$, then $(\mathbf{A}+\mathbf{B})$, $(\mathbf{A}-\mathbf{B})$ and $(\mathbf{A}\times\mathbf{B})$ are three mutually perpendicular vectors.

Reason Dot product of a null vector with any other vector is always zero.

Match the Columns

1. For component of a vector $\mathbf{A}=(3\hat{\mathbf{i}}+4\hat{\mathbf{j}}-5\hat{\mathbf{k}})$, match the following columns.

Column I	Column II
(A) x-axis	(p) 5 unit
(B) Along another vector $(2\hat{\mathbf{i}}+\hat{\mathbf{j}}+2\hat{\mathbf{k}})$	(q) 4 unit
(C) Along $(6\hat{\mathbf{i}}+8\hat{\mathbf{j}}-10\hat{\mathbf{k}})$	(r) Zero
(D) Along another vector $(-3\hat{\mathbf{i}}-4\hat{\mathbf{j}}+5\hat{\mathbf{k}})$	(s) None

2. If θ is the angle between two vectors **A** and **B**, then match the following two columns.

Column I	Column II
(A) $\mathbf{A}\cdot\mathbf{B}=\|\mathbf{A}\times\mathbf{B}\|$	(p) $\theta=90°$
(B) $\mathbf{A}\cdot\mathbf{B}=B^2$	(q) $\theta=0°$ or $180°$
(C) $\|\mathbf{A}+\mathbf{B}\|=\|\mathbf{A}-\mathbf{B}\|$	(r) $\mathbf{A}=\mathbf{B}$
(D) $\|\mathbf{A}\times\mathbf{B}\|=AB$	(s) None

3. Two vectors **A** and **B** have equal magnitude x. Angle between them is 60°. Then, match the following two columns.

Column I	Column II
(A) $\|\mathbf{A}+\mathbf{B}\|$	(p) $\frac{\sqrt{3}}{2}x$
(B) $\|\mathbf{A}-\mathbf{B}\|$	(q) x
(C) $\mathbf{A}\cdot\mathbf{B}$	(r) $\sqrt{3}x$
(D) $\|\mathbf{A}\times\mathbf{B}\|$	(s) None

4. Vector **A** is pointing eastwards and vector **B** northwards. Then, match the following two columns.

Column I	Column II
(A) $(\mathbf{A}+\mathbf{B})$	(p) North-east
(B) $(\mathbf{A}-\mathbf{B})$	(q) Vertically upwards
(C) $(\mathbf{A}\times\mathbf{B})$	(r) Vertically downwards
(D) $(\mathbf{A}\times\mathbf{B})\times(\mathbf{A}\times\mathbf{B})$	(s) None

5. A vector has a magnitude x. If it is rotated by an angle θ, then magnitude of change in vector is nx. Match the following two columns.

Column I	Column II
(A) $\theta=60°$	(p) $n=\sqrt{3}$
(B) $\theta=90°$	(q) $n=1$
(C) $\theta=120°$	(r) $n=\sqrt{2}$
(D) $\theta=180°$	(s) $n=2$

Entrance Gallery

2014

1. Two airplanes A and B are flying with constant velocity in the same vertical plane at an angles 30° and 60° w.r.t. the horizontal, respectively as shown in the figure given below. The speed of A is $100\sqrt{3}\ \text{ms}^{-1}$. At time $t=0$ s, an observer in A finds B at a distance of 500 m. This observer sees B moving with a constant velocity perpendicular to the line of motion of A. If at $t=t_0$, A just escapes being hit by B, t_0 (in second) is **[JEE Advanced]**

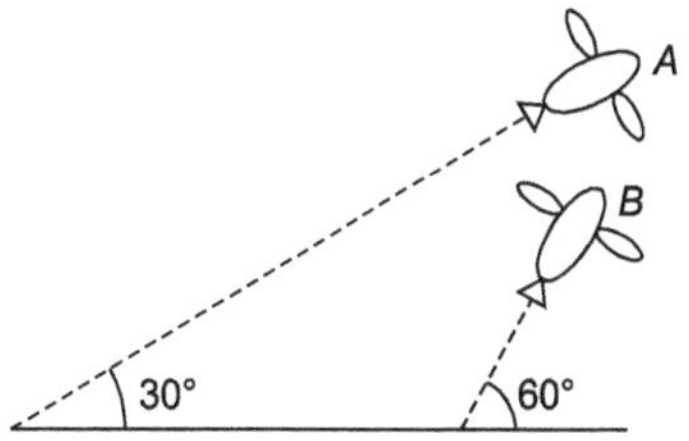

(a) 3 (b) 5
(c) 7 (d) 9

2. The coordinates of a particle moving in XY-plane at any instant of time t are $x=4t^2$, $y=3t^2$. The speed of the particle at that instant is **[Kerala CEE]**

(a) $10\,t$ (b) $5\,t$
(c) $3\,t$ (d) $2\,t$
(e) $\sqrt{13}\,t$

3. A particle has the position vector $\mathbf{r}=\hat{\mathbf{i}}-2\hat{\mathbf{j}}+\hat{\mathbf{k}}$ and the linear momentum $\mathbf{p}=2\hat{\mathbf{i}}-\hat{\mathbf{j}}+\hat{\mathbf{k}}$. Its angular momentum about the origin is **[Kerala CEE]**

(a) $-\hat{\mathbf{i}}+\hat{\mathbf{j}}-3\hat{\mathbf{k}}$ (b) $-\hat{\mathbf{i}}+\hat{\mathbf{j}}+3\hat{\mathbf{k}}$
(c) $\hat{\mathbf{i}}-\hat{\mathbf{j}}+3\hat{\mathbf{k}}$ (d) $\hat{\mathbf{i}}-\hat{\mathbf{j}}-5\hat{\mathbf{k}}$
(e) $\hat{\mathbf{i}}-\hat{\mathbf{j}}+5\hat{\mathbf{k}}$

4. Which of the following is not a vector quantity? **[Karnataka CET]**

(a) Weight (b) Nuclear spin
(c) Momentum (d) Potential energy

5. A force $\mathbf{F}=5\hat{\mathbf{i}}+2\hat{\mathbf{j}}-5\hat{\mathbf{k}}$ acts on a particle whose position vector is $\mathbf{r}=\hat{\mathbf{i}}-2\hat{\mathbf{j}}+\hat{\mathbf{k}}$. What is the torque about the origin? **[Karnataka CET]**

(a) $8\hat{\mathbf{i}}+10\hat{\mathbf{j}}+12\hat{\mathbf{k}}$ (b) $8\hat{\mathbf{i}}+10\hat{\mathbf{j}}-12\hat{\mathbf{k}}$
(c) $8\hat{\mathbf{i}}-10\hat{\mathbf{j}}-8\hat{\mathbf{k}}$ (d) $10\hat{\mathbf{i}}-10\hat{\mathbf{j}}-\hat{\mathbf{k}}$

6. Consider three vectors $\mathbf{A}=\hat{\mathbf{i}}+\hat{\mathbf{j}}-2\hat{\mathbf{k}}$, $\mathbf{B}=\hat{\mathbf{i}}-\hat{\mathbf{j}}+\hat{\mathbf{k}}$ and $\mathbf{C}=2\hat{\mathbf{i}}-3\hat{\mathbf{j}}+4\hat{\mathbf{k}}$. A vector **X** of the form $\alpha\mathbf{A}+\beta\mathbf{B}$ (α and β are numbers) is perpendicular to **C**.

The ratio of α and β is **[WB JEE]**

(a) 1 : 1 (b) 2 : 1 (c) −1 : 1 (d) 3 : 1

7. What is the torque of a force $3\hat{i}+7\hat{j}+4\hat{k}$ about the origin, if the force acts on a particle whose position vector is $2\hat{i}+2\hat{j}+1\hat{k}$? **[J&K CET]**

(a) $\hat{i}-5\hat{j}+8\hat{k}$
(b) $2\hat{i}+2\hat{j}+2\hat{k}$
(c) $\hat{i}+\hat{j}+\hat{k}$
(d) $3\hat{i}+2\hat{j}+3\hat{k}$

2012

8. If $\mathbf{a}\cdot\mathbf{b}=|\mathbf{a}\times\mathbf{b}|$, then θ will be **[O JEE]**

(a) 45° (b) 60°
(c) 75° (d) 30°

9. If $\mathbf{A}=\mathbf{B}+\mathbf{C}$ and have A, B, C scalar magnitudes of 5, 4, 3 units respectively, then the angle between $\mathbf{A}$ and $\mathbf{C}$ is **[WB JEE]**

(a) $\cos^{-1}(3/5)$ (b) $\cos^{-1}(4/5)$
(c) $\pi/2$ (d) $\sin^{-1}(3/4)$

2011

10. Given, $\mathbf{A}=2\hat{i}+3\hat{j}$ and $\mathbf{B}=\hat{i}+\hat{j}$. The component of vector $\mathbf{A}$ along vector $\mathbf{B}$ is **[WB JEE]**

(a) $\frac{1}{\sqrt{2}}$ (b) $\frac{3}{\sqrt{2}}$
(c) $\frac{5}{\sqrt{2}}$ (d) $\frac{7}{\sqrt{2}}$

Answers

Level 1

Objective Problems

1. (a)	**2.** (b)	**3.** (c)	**4.** (c)	**5.** (a)	**6.** (d)	**7.** (a)	**8.** (d)	**9.** (d)	**10.** (c)
11. (b)	**12.** (b)	**13.** (d)	**14.** (c)	**15.** (b)	**16.** (a)	**17.** (d)	**18.** (b)	**19.** (a)	**20.** (d)
21. (a)	**22.** (b)	**23.** (c)	**24.** (b)	**25.** (b)	**26.** (c)	**27.** (a)	**28.** (a)	**29.** (b)	**30.** (a)
31. (c)	**32.** (b)	**33.** (b)	**34.** (d)	**35.** (a)	**36.** (c)	**37.** (b)	**38.** (b)	**39.** (a)	**40.** (b)
41. (c)	**42.** (d)	**43.** (d)	**44.** (d)	**45.** (d)	**46.** (a,b)	**47.** (a)	**48.** (b)	**49.** (d)	**50.** (a)
51. (a)	**52.** (a)	**53.** (a)	**54.** (a)	**55.** (d)	**56.** (b)	**57.** (a)	**58.** (c)	**59.** (a)	**60.** (a)
61. (b)	**62.** (a)	**63.** (c)	**64.** (d)	**65.** (c)	**66.** (b)	**67.** (c)	**68.** (b)	**69.** (b)	**70.** (b)
71. (b)	**72.** (d)	**73.** (d)	**74.** (b)	**75.** (a)	**76.** (c)	**77.** (d)	**78.** (b)	**79.** (c)	**80.** (c)
81. (a)	**82.** (b)	**83.** (b)	**84.** (b)	**85.** (a)	**86.** (a)	**87.** (a)	**88.** (b)	**89.** (a)	**90.** (d)
91. (d)	**92.** (b)	**93.** (c)	**94.** (d)	**95.** (d)	**96.** (d)	**97.** (a,b)	**98.** (d)	**99.** (a)	**100.** (c)
101. (c)	**102.** (c)	**103.** (b)	**104.** (a)	**105.** (a)	**106.** (d)				

Level 2

Only One Correct Option

1. (c)	**2.** (b)	**3.** (c)	**4.** (b)	**5.** (a)	**6.** (a)	**7.** (d)	**8.** (a)	**9.** (a)	**10.** (a)
11. (c)	**12.** (a)	**13.** (c)	**14.** (b)	**15.** (c)	**16.** (b)	**17.** (b)	**18.** (b,c)	**19.** (a)	**20.** (c)
21. (a)	**22.** (c)	**23.** (a,d)	**24.** (b)	**25.** (b)	**26.** (c)	**27.** (a)			

Assertion and Reason

1. (a)	**2.** (b)	**3.** (a)	**4.** (b)	**5.** (d)	**6.** (d)	**7.** (c)	**8.** (e)	**9.** (d)	**10.** (d)
11. (a,b)	**12.** (b)	**13.** (b)							

Match the Columns

1. (A → q, B → r, C → s, D → s) **2.** (A → s, B → q,r, C → p, D → p) **3.** (A → r, B → q , C → s, D → p)
4. (A → p,s, B → s, C → q, D → s) **5.** (A → q, B → r, C → p, D → s)

Entrance Gallery

1. (b)	**2.** (a)	**3.** (b)	**4.** (d)	**5.** (a)	**6.** (a)	**7.** (a)	**8.** (a)	**9.** (a)	**10.** (c)

Solutions

Level 1 : Objective Problems

8. In $\phi = \mathbf{B} \cdot \mathbf{S}$, area is a vector. So area is sometimes as scalar and sometimes a vector.

13. $|\mathbf{A}-\mathbf{B}| = \sqrt{A^2 + B^2 - 2AB\cos\theta}$

$$= \sqrt{4 + 16 - 2\times 2\times 4\times \frac{1}{2}}$$

$$= \sqrt{12} = 2\sqrt{3}$$

14. Maximum $= A + B$ and minimum $= A - B$

15. Resultant is inclined towards a vector having greater magnitude.

16. $P\sqrt{10} = \sqrt{4P^2 + 2P^2 + 4\sqrt{2}\,P^2\cos\theta}$

$\therefore \quad \theta = 45°$

17. $\sqrt{A^2 + B^2 + 2AB\cos\theta}$

$$= \sqrt{A^2 + B^2 - 2AB\cos\theta}$$

$\therefore \quad \cos\theta = 0 \quad \text{or} \quad \theta = 90°$

18. $R = A = B$

$\therefore \quad R^2 = R^2 + R^2 + 2RR\cos\theta$

or $\quad \cos\theta = -\frac{1}{2}$

$\therefore \quad \theta = 120°$

19. $\cos\theta = \frac{C}{A} = \frac{3}{5}$

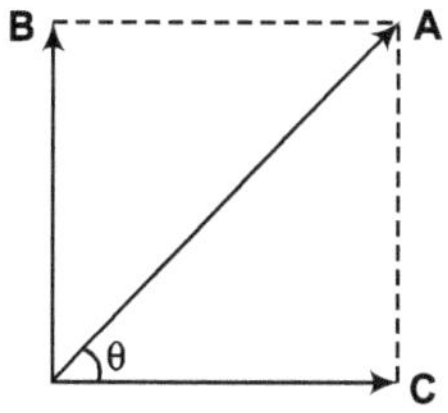

$\therefore \quad \theta = \cos^{-1}(3/5)$

20. Suppose angle between two vectors **A** and **B** of equal magnitude is θ. Then, angle between **A** and **A** − **B** will be $\frac{180° - \theta}{2}$ or $90° - \frac{\theta}{2}$. Hence, this angle will depend on the angle between **A** and **B** or θ.

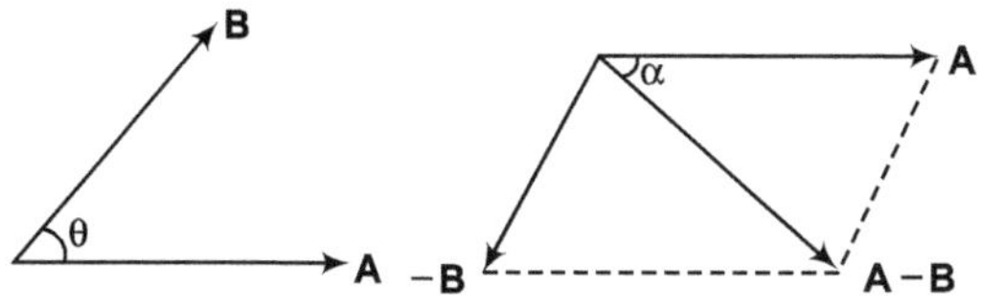

23. From polygon law of vector addition, **C** + **A** = **B**.

24. Resultant should lie between $F_1 + F_2$ and $F_1 - F_2$ or it should lie between 14 N and 2 N.

26. $\mathbf{S} = \mathbf{r}_Q - \mathbf{r}_P = (3\hat{\mathbf{i}} + 4\hat{\mathbf{j}} + 5\hat{\mathbf{k}}) - (2\hat{\mathbf{i}} + 3\hat{\mathbf{j}} + 5\hat{\mathbf{k}})$

$$= \hat{\mathbf{i}} + \hat{\mathbf{j}}$$

27. When drawn as per polygon law of vector addition, they will form a closed regular polygon. Hence, resultant will be zero.

28. Resultant will be $2A\cos\frac{\theta}{2}$ along bisector.

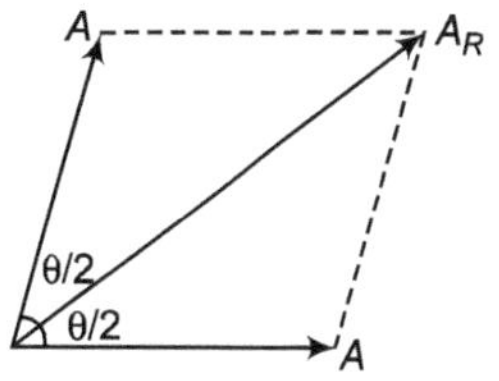

31. **a** − **b** is nothing but addition of **a** and −**b**.

So, the magnitude of **a** − **b** will lie between $|\mathbf{a}| + |\mathbf{b}|$ and $|\mathbf{a}| - |\mathbf{b}|$.

32. $\mathbf{A} - 2\mathbf{B} + 3\mathbf{C} = (2\hat{\mathbf{i}} + \hat{\mathbf{j}}) - 2(3\hat{\mathbf{i}} - \hat{\mathbf{k}}) + 3(6\hat{\mathbf{i}} - 2\hat{\mathbf{k}})$

$$= 20\hat{\mathbf{i}} - 5\hat{\mathbf{j}} - 4\hat{\mathbf{k}}$$

33. $\mathbf{A} + \mathbf{B} = 10\hat{\mathbf{i}} + 5\hat{\mathbf{j}}$

$\therefore \quad |\mathbf{A} + \mathbf{B}| = \sqrt{100 + 25}$

$$= 5\sqrt{5}$$

Angle of **A** + **B** with x-axis,

$$\theta = \tan^{-1}\left(\frac{5}{10}\right) = \tan^{-1}\left(\frac{1}{2}\right)$$

34. $\mathbf{S} = \mathbf{r}_f - \mathbf{r}_i = \mathbf{r}_3 - \mathbf{r}_1$

$$= (3\hat{\mathbf{i}} + 12\hat{\mathbf{j}} - 3\hat{\mathbf{k}}) - (3\hat{\mathbf{i}} + 4\hat{\mathbf{j}} - \hat{\mathbf{k}})$$

$$= 8\hat{\mathbf{j}} - 2\hat{\mathbf{k}}$$

35. $|\mathbf{A} + \mathbf{B}|_{max} = A + B$

and $|\mathbf{A} + \mathbf{B}|_{min} = A - B$

36. $\mathbf{S} = 3\hat{\mathbf{i}} + 6\hat{\mathbf{j}} - 7\hat{\mathbf{i}} = -4\hat{\mathbf{i}} + 6\hat{\mathbf{j}}$

$\therefore \quad |\mathbf{S}| = \sqrt{16 + 36} = \sqrt{52}$ m

37. $R = \sqrt{9P^2 + 4P^2 + 12P\cos\theta}$ ...(i)

$2R = \sqrt{36P^2 + 4P^2 + 24P\cos\theta}$ ...(ii)

Solving them, we get $\cos\theta = -\frac{1}{2}$

$\therefore \quad \theta = 120°$

38. Third force should be equal and opposite to the resultant of the given vectors.

$$|\mathbf{a} + \mathbf{b}| = 2A\cos\frac{\theta}{2}$$

$$= (2)(1)\cos 30°$$

$$= \sqrt{3}$$

$$|\mathbf{a} - \mathbf{b}| = 2A\cos\frac{\theta}{2}$$

$$= (2)(1)\cos 30°$$

$$= 1$$

40. For given condition, $\theta = \frac{360°}{n} = 30°$

41. $3 = \sqrt{(\sqrt{3})^2 + (\sqrt{3})^2 + 2(\sqrt{3})(\sqrt{3})\cos\theta}$

$\Rightarrow \quad \theta = 60°$

42. Angle between **A** and **B** is 90°

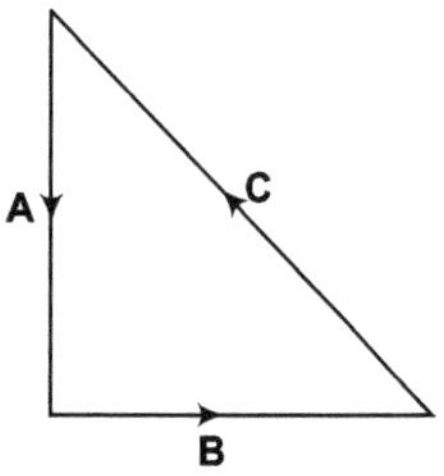

Between **B** and **C** is 135° and that between **A** and **C** is 135°.

43. Resultant of two vectors **A** and **B** lies between $A+B$ and $A-B$.

44. $R=\sqrt{A^2+B^2+2AB\cos\theta}$ ...(i)

From Eq. (i), we get

$$R^2=C^2=A^2+B^2+2AB\cos\theta$$
$$=A^2+B^2$$
$$\Rightarrow \quad \cos\theta=0$$
$$\Rightarrow \quad \theta=\frac{\pi}{2}$$

45. $\sqrt{A^2+B^2+2AB\cos\theta}=2\sqrt{A^2+B^2-2AB\cos\theta}$

From this equation, we cannot find θ.

46. $|\mathbf{A}|=\sqrt{9+16}=5$ and $|\mathbf{B}|=\sqrt{36+64}=10$. $\mathbf{B}=2\mathbf{A}$ or **B** is parallel to **A**. Ratio of their coefficients are equal, so they are parallel or their cross product is zero.

49. $AB\sin\theta=\sqrt{3}\,AB\cos\theta$ or $\tan\theta=\sqrt{3}$

$$\therefore \quad \theta=60°$$

Now, $$|\mathbf{A}+\mathbf{B}|=\sqrt{A^2+B^2+2AB\cos 60°}$$
$$=\sqrt{A^2+B^2+AB}$$

50. $W=\mathbf{F}\cdot(\mathbf{r}_2-\mathbf{r}_1)$

51. $(\mathbf{a}\cdot\mathbf{b})^2=a^2b^2\cos^2\theta=a^2b^2$ (given)

$$\therefore \quad \theta=0°$$

52. $$8=(5)(2)\sin\theta$$
$$\therefore \quad \sin\theta=4/5$$
$$\text{or} \quad \cos\theta=3/5$$
$$\therefore \quad \mathbf{A}\cdot\mathbf{B}=AB\cos\theta=(5)(2)(3/5)=6$$

53. $AB\sin\theta=\frac{1}{\sqrt{3}}AB\cos\theta$

$\therefore \tan\theta=\frac{1}{\sqrt{3}}$ or $\theta=30°$

54. Area of parallelogram $=|\mathbf{A}\times\mathbf{B}|$

55. $\mathbf{A}\cdot\mathbf{B}=0 \Rightarrow \therefore \mathbf{A}\perp\mathbf{B}$

$$\mathbf{A}\cdot\mathbf{C}=0$$
$$\therefore \quad \mathbf{A}\perp\mathbf{C}$$

i.e. **A** is perpendicular to both **B** and **C**.

56. $\mathbf{A}\cdot\mathbf{B}=(3)(5)\cos 60°=7.5$

57. $\hat{\mathbf{n}}=\frac{\mathbf{A}\times\mathbf{B}}{|\mathbf{A}\times\mathbf{B}|}$

58. $\gamma=\cos^{-1}\left(\frac{P_z}{P}\right)$

Here, $P_z=4\sqrt{2}$

and $P=\sqrt{(6)^2+(4\sqrt{2})^2+(4\sqrt{2})^2}=10$

59. $\mathbf{A}-\mathbf{B}=2\hat{\mathbf{i}}+2\hat{\mathbf{j}}+2\hat{\mathbf{k}}=2\mathbf{A}$

i.e. $\mathbf{A}-\mathbf{B}$ and **B** are parallel.

60. Component of **A** along **B** $=A\cos\theta=\frac{\mathbf{A}\cdot\mathbf{B}}{B}=\frac{2+3}{\sqrt{1+1}}=\frac{5}{\sqrt{2}}$

61. Magnitude of which is 1.

63. $\mathbf{P}\cdot\mathbf{Q}=0$

64. Diagonal vector $\mathbf{A}=b\hat{\mathbf{i}}+b\hat{\mathbf{j}}+b\hat{\mathbf{k}}$

or $$A=\sqrt{b^2+b^2+b^2}=\sqrt{3}b$$

$$\therefore \quad \hat{\mathbf{A}}=\frac{\mathbf{A}}{A}=\frac{\hat{\mathbf{i}}+\hat{\mathbf{j}}+\hat{\mathbf{k}}}{\sqrt{3}}$$

65. **F** lies in xy-plane. Hence, $7\hat{\mathbf{k}}$ which is perpendicular to xy-plane is also perpendicular to **F**.

67. Dot product of two perpendicular vectors should be zero.

69. Resultant of two given vector is $12\hat{\mathbf{i}}+5\hat{\mathbf{j}}$. Magnitude of this resultant will be $\sqrt{144+25}=13$. Hence, a unit vector parallel to this resultant would be $\frac{12\hat{\mathbf{i}}+5\hat{\mathbf{j}}}{13}$.

70. Dot product of two given vectors is zero. Hence, they are mutually perpendicular.

71. The given vector is in xz-plane. Hence, this is perpendicular to y-axis.

72. $F_V=5\cos 60°=2.5$ N

73. A vector parallel to **A** will be $n\mathbf{A}$ or $(3n\hat{\mathbf{i}}+4n\hat{\mathbf{j}})$

Now, $|n\mathbf{A}|=|\mathbf{B}|$ is given.

Hence, $$n=\sqrt{9+16}$$
$$=\sqrt{49+576}$$
$$\text{or} \quad n=5$$
$$\therefore \quad n\mathbf{A}=15\hat{\mathbf{i}}+20\hat{\mathbf{j}}$$

74. $\mathbf{A}=(\hat{\mathbf{i}}-3\hat{\mathbf{j}}+3\hat{\mathbf{k}})+(3\hat{\mathbf{i}}+6\hat{\mathbf{j}}-7\hat{\mathbf{k}})=\hat{\mathbf{j}}$ (given)

Hence, $$\mathbf{A}=-4\hat{\mathbf{i}}-2\hat{\mathbf{j}}+5\hat{\mathbf{k}}$$

75. Magnitude of given vector is 1.

76. In xy-plane vector is $3\hat{\mathbf{i}}+\hat{\mathbf{j}}$

$\therefore$ Length in xy-plane $=\sqrt{9+1}=\sqrt{10}$

77. $\theta=\cos^{-1}\left(\frac{\mathbf{A}\cdot\mathbf{B}}{AB}\right)=\cos^{-1}\left(\frac{1}{\sqrt{3}}\right)$

78. Dot product of two perpendicular vectors should be zero.

80. $\frac{|\mathbf{a}+\mathbf{b}|}{|\mathbf{a}-\mathbf{b}|}=1$

i.e. $$|\mathbf{a}+\mathbf{b}|=|\mathbf{a}-\mathbf{b}|$$
$$\text{or} \quad a^2+b^2+2ab\cos\theta=a^2+b^2-2ab\cos\theta$$
$$\therefore \quad \cos\theta=0° \text{ or } \theta=90°$$

81. $20\cos 60°=10$ m

82. The coefficients of $\hat{\mathbf{i}}$, $\hat{\mathbf{j}}$ and $\hat{\mathbf{k}}$ should bear a constant ratio.

$$\text{or} \quad \frac{2}{-4}=\frac{3}{-6}=\frac{1}{-\lambda}$$
$$\text{or} \quad \lambda=2$$

83. $\mathbf{P}\times\mathbf{Q}$ is perpendicular to the plane formed by **P** and **Q**.

$\mathbf{P}+\mathbf{Q}$ lies in this plane. Hence, $\mathbf{P}+\mathbf{Q}$ is perpendicular to $\mathbf{P}\times\mathbf{Q}$.

84. $\sqrt{(0.5)^2 + (0.8)^2 + (c)^2} = 1$

$\therefore \quad c^2 = 0.11 \quad \text{or} \quad c = \sqrt{0.11}$

85. Their dot product should be zero.

86. Dot product should be zero.

87. Component of **A** along **B** $= \dfrac{\mathbf{A} \cdot \mathbf{B}}{B}$

88. For vectors to be parallel, ratio of coefficients should be same.

$\therefore \quad \dfrac{6}{5} = \dfrac{x}{-6} = \dfrac{-2}{-y}$

89. $A = \sqrt{9 + 36 + 4} = 7$

$\therefore \quad \alpha = \cos^{-1}\left(\dfrac{3}{7}\right)$ etc.

91. Area $= |\mathbf{A} \times \mathbf{B}|$

Here, $\mathbf{A} \times \mathbf{B} = (2\hat{\mathbf{i}} + 3\hat{\mathbf{j}}) \times (\hat{\mathbf{i}} + 4\hat{\mathbf{j}})$

$= 8\hat{\mathbf{k}} - 3\hat{\mathbf{k}}$

$= 5\hat{\mathbf{k}}$

$\therefore \quad \text{Area} = |\mathbf{A} \times \mathbf{B}| = 5 \text{ Units}$

93. Resultant of two vectors lies between $\mathbf{A} + \mathbf{B}$ and $|\mathbf{A} + \mathbf{B}|$.

94. Their cross product should be zero.

$(\mathbf{a} + \mathbf{b}) \times (\mathbf{a} - \mathbf{b}) = 0$

$\therefore \quad 2(\mathbf{a} \times \mathbf{b}) = 0$

$\mathbf{a} = \mathbf{b}$

97. $A_R = 2A\cos\dfrac{\theta}{2}$

98. $(a\hat{\mathbf{i}} + b\hat{\mathbf{j}}) \cdot (\hat{\mathbf{i}} + \hat{\mathbf{j}}) = 0$ or $a + b = 0$ ...(i)

Further, $\sqrt{a^2 + b^2} = 1$

or $a^2 + b^2 = 1$ ...(ii)

99. $\mathbf{a} + \mathbf{b} + \mathbf{c} = 0$

$\therefore \quad \mathbf{a} + \mathbf{c} = -\mathbf{b}$

or $(\mathbf{a} + \mathbf{c}) \times \mathbf{b} = -\mathbf{b} \times \mathbf{b} = 0$

or $(\mathbf{a} \times \mathbf{b}) + (\mathbf{c} \times \mathbf{b}) = 0$

or $\mathbf{a} \times \mathbf{b} = \mathbf{b} \times \mathbf{c}$

100. $\mathbf{F} \cdot \mathbf{F}' = 0$. Here, $\mathbf{F}'$ = unknown force.

101. Dot product should be zero.

102. $a = \dfrac{F}{m}$

$\therefore \quad m = \dfrac{F}{a}$

103. $\mathbf{A} + \mathbf{B} = 10\hat{\mathbf{i}} + 5\hat{\mathbf{j}}$

$\therefore \quad |\mathbf{A} + \mathbf{B}| = \sqrt{100 + 25}$

$= 5\sqrt{5}$

Angle of $\mathbf{A} + \mathbf{B}$ with x-axis,

$\theta = \tan^{-1}\left(\dfrac{5}{10}\right)$

$= \tan^{-1}\left(\dfrac{1}{2}\right)$

104. $a = \dfrac{F}{m}$

Resultant of F_1 and F_2 will be

$F_{12} = \sqrt{2}\,F$ (in opposite direction of F_3)

Now, resultant of F_{12} and F_3 will be,

$F_{\text{net}} = (\sqrt{2} - 1)\,F$

$\therefore \quad a' = \dfrac{F_{\text{net}}}{m} = (\sqrt{2} - 1)\dfrac{F}{m}$

$= (\sqrt{2} - 1)\,a$

105. $\mathbf{A} - \mathbf{B} = \hat{\mathbf{j}} - 2\hat{\mathbf{k}} = \mathbf{C}$ (say)

$C = \sqrt{1 + 4} = \sqrt{5}$

$\cos\alpha = \dfrac{0}{\sqrt{5}} = 0, \cos\beta = \dfrac{1}{\sqrt{5}}$

and $\cos\gamma = \dfrac{-2}{\sqrt{5}}$

106. $(\mathbf{A} + \mathbf{B}) \cdot (\mathbf{A} - \mathbf{B}) = 0$

$\therefore \quad \mathbf{A} \cdot \mathbf{A} + \mathbf{B} \cdot \mathbf{A} - \mathbf{A} \cdot \mathbf{B} - \mathbf{B} \cdot \mathbf{B} = 0$

$\therefore \quad A^2 + AB\cos\theta - AB\cos\theta - B^2 = 0$

$\therefore \quad A = B$

Level 2 : Only One Correct Option

1. $|\mathbf{A} - \mathbf{B}| = \sqrt{1 + 1 - 2 \times 1 \times 1 \times \cos\theta} = 2\sin\dfrac{\theta}{2}$

Angle between **A** and **B** is also θ, but their magnitudes are 1.

2. $\mathbf{s}_1 = (\sqrt{2}\cos 45°)\,\hat{\mathbf{i}} + (\sqrt{2}\sin 45°)\,\hat{\mathbf{j}} = \hat{\mathbf{i}} + \hat{\mathbf{j}}$

$\mathbf{s}_2 = 2\hat{\mathbf{j}}$ and $\mathbf{s} = (-4\cos 30°)\,\hat{\mathbf{i}} + (4\sin 30°)\,\hat{\mathbf{j}}$

$= -2\sqrt{3}\,\hat{\mathbf{i}} + 2\hat{\mathbf{j}}$

Now $\mathbf{s} = \mathbf{s}_1 + \mathbf{s}_2 + \mathbf{s}_3 = (1 - 2\sqrt{3})\,\hat{\mathbf{i}} + \hat{\mathbf{j}}$

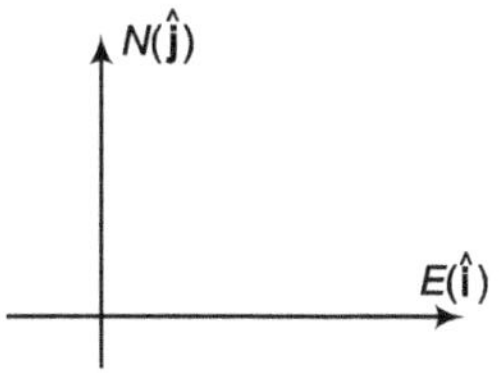

$\therefore \quad s = \sqrt{1 - 2\sqrt{3})^2 + (1)^2}$

$= \sqrt{1 + 12 + 1 - 4\sqrt{3}}$

$= \sqrt{14 - 4\sqrt{3}}$ m

3. $R_1^2 = A^2 + B^2 + 2AB\cos\theta$

$R_2^2 = A^2 + B^2 - 2AB\cos\theta$

$\therefore \quad R_1^2 + R_2^2 = 2(A^2 + B^2)$

4. Sum of two unit vectors is a unit vector, means angle between those two unit vectors is $120°$.

$\therefore \quad |\mathbf{S}| = \sqrt{1 + 1 - 2 \times 1 \times 1 \times \cos 120°} = \sqrt{3}$

5. $\mathbf{R} \cdot \mathbf{S} = (\mathbf{A} + \mathbf{B}) \cdot (\mathbf{A} - \mathbf{B})$

$0 = \mathbf{A} \cdot \mathbf{A} - \mathbf{A} \cdot \mathbf{B} + \mathbf{B} \cdot \mathbf{A} - \mathbf{B} \cdot \mathbf{B}$

or $0 = A^2 - B^2$ (as $\mathbf{A} \cdot \mathbf{B} = \mathbf{B} \cdot \mathbf{A}$)

$\therefore \quad A^2 = B^2 \quad \text{or} \quad A = B$

6. $A_x = A_y = A_z$

Now, $A = \sqrt{A_x^2 + A_y^2 + A_z^2} = \sqrt{3}\,A_x$

$\therefore \quad A_x = \dfrac{A}{\sqrt{3}} = \dfrac{30}{\sqrt{3}} = 10\sqrt{3}$

7. $\mathbf{AB} + \mathbf{AC} + \mathbf{AD} + \mathbf{AE} + \mathbf{AF}$

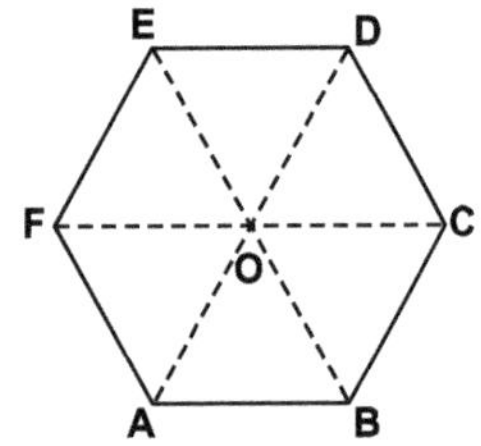

$= \mathbf{AB} + (\mathbf{AB} + \mathbf{BC}) + (\mathbf{AB} + \mathbf{BC} + \mathbf{CD}) + (\mathbf{AF} + \mathbf{FE}) + \mathbf{AF}$
$= (\mathbf{AB} + \mathbf{CD}) + (\mathbf{BC}) + (\mathbf{BC}) + 2(\mathbf{AB} + \mathbf{AF}) + (\mathbf{FE})$
$= \mathbf{AO} + \mathbf{AO} + \mathbf{AO} + 2\,\mathbf{AO} + \mathbf{AO}$
$= 6\mathbf{AO}$

8. Since, **A** is a unit vector in a given direction. It should be a constant unit vector.

or $$\frac{dA}{dt} = 0$$

9. $F \cdot F \cdot \cos\theta = F \cdot F \cdot \sin\theta$ or $\tan\theta = 1$ or $\theta = 45°$

$$|\mathbf{F}_1 \cdot \mathbf{F}_2| = \sqrt{F^2 + F^2 + 2 \cdot F \cdot F \cdot \cos 45°}$$
$$= (\sqrt{2 + \sqrt{2}})\,F$$

10. $\mathbf{OB} + \mathbf{BC} = \mathbf{r}$ or $\mathbf{q} + \mathbf{BC} = \mathbf{r}$ ...(i)

$\mathbf{OA} + \mathbf{AC} = \mathbf{r}$

or $\mathbf{p} + \mathbf{AC} = \mathbf{r}$ ... (ii)

Adding these two equations, we get, $\mathbf{p} + \mathbf{q} = 2\mathbf{r}$, as **AC** and **BC** are equal and opposite vectors.

11. $\cos^2\alpha + \cos^2\beta + \cos^2\gamma = 1$ (a standard result)

or $(1 - \sin^2\alpha) + (1 - \sin^2\beta) + (1 - \sin^2\gamma) = 1$

$\therefore$ $\sin^2\alpha + \sin^2\beta + \sin^2\gamma = 2$

12. Cross product of $(\mathbf{P} + \mathbf{Q})$ and $(\mathbf{P} - \mathbf{Q})$ is perpendicular to the plane formed by $(\mathbf{P} + \mathbf{Q})$ and $(\mathbf{P} - \mathbf{Q})$.

13. $2Q\sin\theta = P$

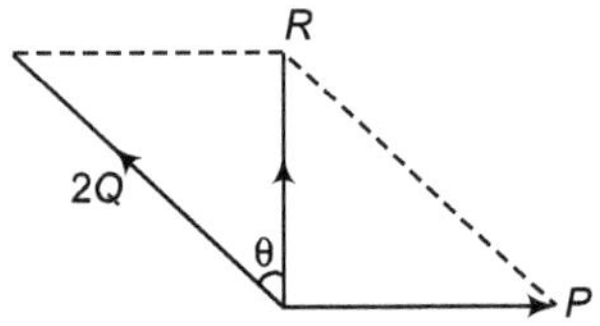

Now, $$R = \sqrt{P^2 + Q^2 + 2PQ\cos(90° + \theta)}$$
$$= \sqrt{P^2 + Q^2 - 2PQ\sin\theta}$$
$$= Q$$

14. $\mathbf{B} \times \mathbf{A}$ is perpendicular to **A**. Hence $\mathbf{A} \cdot (\mathbf{B} \times \mathbf{A})$ will be zero.

15. $P + Q = 18$ ...(i)

$R = 12$ ...(ii)

$Q\sin\theta = P$ and $Q\cos\theta = R$

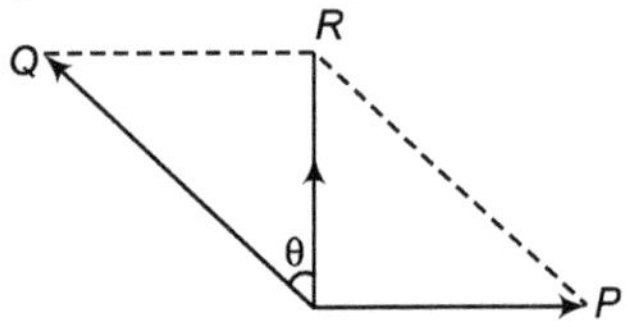

Squaring and adding, we get

$$P^2 + R^2 = Q^2$$

or $Q^2 - P^2 = R^2 = 144$ ...(iii)

Solving the above equations, we get $P = 5$ and $Q = 13$.

16. $|\Delta \mathbf{a}| = \sqrt{a^2 + a^2 - 2 \cdot a \cdot a \cdot \cos(d\theta)} = 2a\sin\left(\frac{d\theta}{2}\right)$

For small angles, $\sin\frac{d\theta}{2} \approx \frac{d\theta}{2}$

$\therefore$ $$|\Delta \mathbf{a}| = 2a \times \frac{d\theta}{2} = a \cdot d\theta$$

17. Resultant of **OA** and **OC** is $\sqrt{2}\,R$ along OB.

Hence, $R_{net} = R + \sqrt{2}R = (1 + \sqrt{2})\,R$

18. Apply polygon law.

19. $(\mathbf{a} + \mathbf{b}) \times (\mathbf{a} - \mathbf{b}) = \mathbf{a} \times \mathbf{a} - \mathbf{a} \times \mathbf{b} + \mathbf{b} \times \mathbf{a} - \mathbf{b} \times \mathbf{b}$
$= 2(\mathbf{b} \times \mathbf{a})$

20. $B\cos\theta = C$ and $B\sin\theta = A$

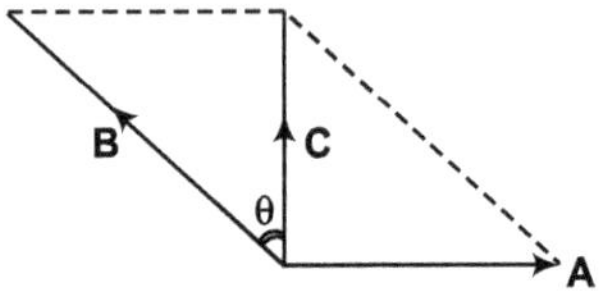

But $A = C$

$\therefore$ $B\cos\theta = B\sin\theta$

or $\theta = 45°$

$\therefore$ Angle between **A** and **B** is $135°$.

21. $\tan 90° = \dfrac{2A\sin\theta}{A + 2A\cos\theta} = \infty$

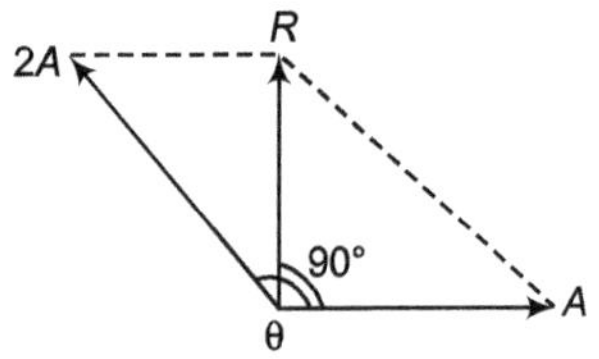

$\therefore$ $A + 2A\cos\theta = 0$

or $\cos\theta = -\frac{1}{2}$ or $\theta = 120°$

22. $\mathbf{B} \cdot \mathbf{C} = 0$

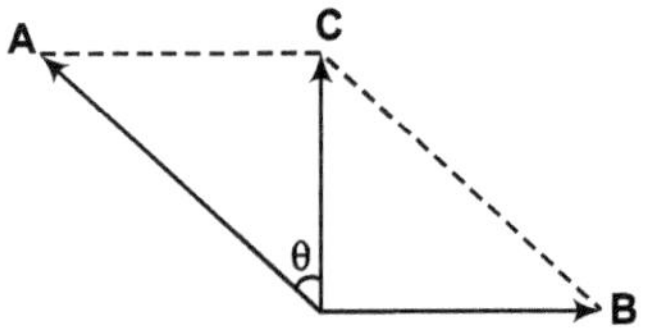

$\therefore$ $\mathbf{B} \perp \mathbf{C}$

$A\sin\theta = B$ or $\sin\theta = \frac{B}{A} = \frac{1}{2}$

$\therefore$ $\theta = 30°$

or $\mathbf{A} \cdot \mathbf{B} < 0$

23. From $R = \sqrt{A^2 + B^2 + 2AB\cos\theta}$

We have,

$$1 = \sqrt{1 + 1 + 2\cos\theta} \quad \text{or} \quad \cos\theta = -\frac{1}{2}$$

$\therefore$ $\theta = 120°$

$$S = \sqrt{A^2 + B^2 - 2AB\cos\theta}$$
$$= \sqrt{1+1-2\times1\times1\times\left(-\frac{1}{2}\right)}$$
$$= \sqrt{3}$$

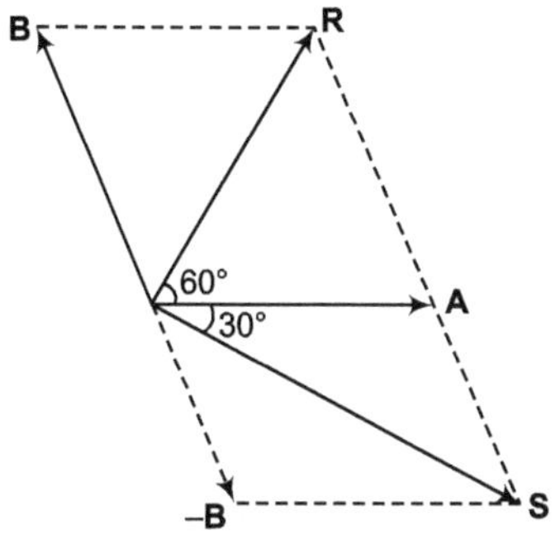

24. Magnitude of component of **v** along **a**

$$= \frac{\mathbf{v}\cdot\mathbf{a}}{a} = \frac{6+2-2}{\sqrt{3}}$$
$$= 2\sqrt{3}$$

Now, $\hat{\mathbf{a}} = \dfrac{\hat{\mathbf{i}} + \hat{\mathbf{j}} + \hat{\mathbf{k}}}{\sqrt{3}}$

∴ Component in vector form

$$= 2\sqrt{3}\,\hat{\mathbf{a}} = (2\hat{\mathbf{i}} + 2\hat{\mathbf{j}} + 2\hat{\mathbf{k}})$$

25. $Q\sin\theta = P$

$$Q\cos\theta = R = P$$

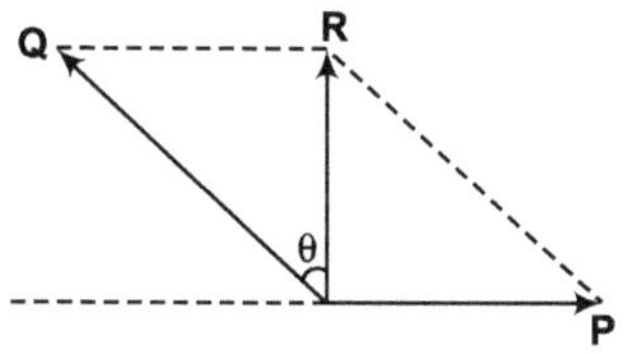

Dividing these two equations, we get

$$\tan\theta = 1$$

∴ $\theta = 45°$

∴ Angle between **P** and **Q** is 135°

26. Suppose $\mathbf{a} = a_1\hat{\mathbf{i}} + a_2\hat{\mathbf{j}} + a_3\hat{\mathbf{k}}$

Now, $(\hat{\mathbf{i}} \times \mathbf{a}) = a_2\hat{\mathbf{k}} - a_3\hat{\mathbf{j}}$

Now, $\hat{\mathbf{i}} \times (\hat{\mathbf{i}} \times \mathbf{a}) = -a_2\hat{\mathbf{j}} - a_3\hat{\mathbf{k}}$

Similarly, $\hat{\mathbf{j}} \times (\hat{\mathbf{j}} \times \mathbf{a}) = -a_1\hat{\mathbf{i}} - a_3\hat{\mathbf{k}}$

and $\hat{\mathbf{k}} \times (\hat{\mathbf{j}} \times \mathbf{a}) = -a_1\hat{\mathbf{i}} - a_2\hat{\mathbf{j}}$

∴ $\hat{\mathbf{i}}\,(\hat{\mathbf{i}} \times \mathbf{a}) + \hat{\mathbf{j}}\,(\hat{\mathbf{j}} \times \mathbf{a}) + \hat{\mathbf{k}}\,(\hat{\mathbf{k}} \times \mathbf{a}) = -2\mathbf{a}$

27. Resultant of (**P** + **Q**) and (**P** – **Q**) is **P** + **Q** + **P** – **Q** or 2**P** which is parallel to **P**.

Assertion and Reason

4. $|\mathbf{A}\times\mathbf{B}| = AB\sin\theta$ and $\mathbf{A}\cdot\mathbf{B} = AB\cos\theta$

5. $|\mathbf{A}+\mathbf{B}| = \sqrt{A^2 + B^2 + 2AB\cos\theta}$

and $|\mathbf{A}-\mathbf{B}| = \sqrt{A^2 + B^2 - 2AB\cos\theta}$

10. $\dfrac{|\mathbf{A}\times\mathbf{B}|}{AB}$ is $\dfrac{AB\sin\theta}{AB} = \sin\theta$

And the angle between two vectors.

$0° \le 180°$. Therefore, $\dfrac{|\mathbf{A}\times\mathbf{B}|}{AB}$ may be positive or zero.

Further, $\theta = \sin^{-1}\left[\dfrac{|\mathbf{A}\times\mathbf{B}|}{AB}\right]$ will give two values of θ. Those are θ and 180° – θ.

11. Angle between **A** × **B** and **B** × **A** is 180°.

∴ $(\mathbf{A}\times\mathbf{B})\cdot(\mathbf{B}\times\mathbf{A}) = (AB\sin\theta)(AB\sin\theta)(\cos 180°)$
$$= -A^2 - B^2\sin^2\theta$$

12. Further, $\mathbf{A}\cdot\mathbf{B} = AB\cos\theta$. Here, AB is always positive value of cosθ determines the sign of $\mathbf{A}\cdot\mathbf{B}$.

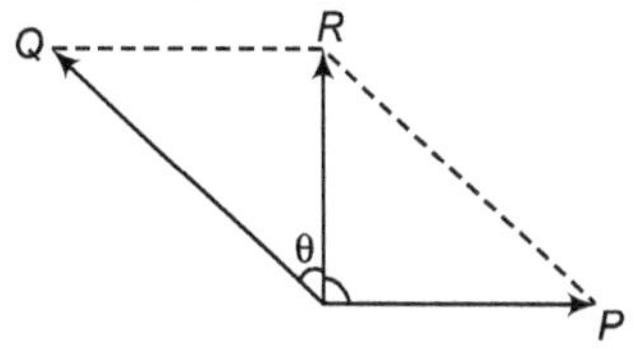

13. $(\mathbf{A}\times\mathbf{B})\cdot(\mathbf{A}-\mathbf{B}) = A^2 - B^2$

This dot product is zero only if $A = B$.

Therefore, they are perpendicular. Further, $(\mathbf{A}\times\mathbf{B})$ is perpendicular to both $(\mathbf{A}+\mathbf{B})$ and $(\mathbf{A}-\mathbf{B})$.

Match the Columns

1. $(2\hat{\mathbf{i}} + \hat{\mathbf{j}} + 4\hat{\mathbf{k}})$ is perpendicular to **A**, because the dot product of these two vectors is zero.

Further $(6\hat{\mathbf{i}} + 4\hat{\mathbf{j}} - 10\hat{\mathbf{k}})$ vectors is parallel to **A**. So, component of **A** along this vector is magnitude of **A** which is $5\sqrt{2}$ unit. The last vector i.e. $(-3\hat{\mathbf{i}} - 4\hat{\mathbf{j}} + 5\hat{\mathbf{k}})$ is anti-parallel to **A** along this vector is negative of magnitude of **A** or $-5\sqrt{2}$ unit.

Entrance Gallery

1. At $t = 0$, airplane B is at location Q, airplane A is at location P.

Relative velocity of B w.r.t. to A

i.e. $\mathbf{v}_{BA} = \mathbf{v}_B - \mathbf{v}_A = \mathbf{v}_B + (-\mathbf{v}_A)$ has direction from Q to P.

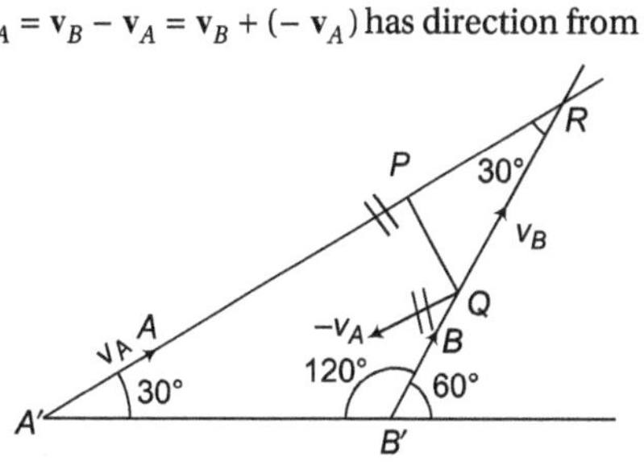

Given, $PQ = 500\text{ m}$

From right angled triangle QPR

$$\frac{PQ}{PR} = \tan 30° \Rightarrow \frac{500}{PR} = \frac{1}{\sqrt{3}}$$

$\Rightarrow$ $PR = 500\sqrt{3}\text{ m}$

Given, speed of $A = 100\sqrt{3}\text{ ms}^{-1}$

At $t = t_0$, airplane A reaches at location R

$$t_0 = \frac{PR}{\text{Speed of plane } A} = \frac{500\sqrt{3}}{100\sqrt{3}\text{ ms}^{-1}} = 5\text{ s}$$

2. As, position vector, $\mathbf{r} = x\hat{\mathbf{i}} + y\hat{\mathbf{j}}$

$$\mathbf{r} = 4t^2\,\hat{\mathbf{i}} + 3t^2\,\hat{\mathbf{j}}$$

Velocity $$\mathbf{v} = \frac{d\mathbf{r}}{dt} = 4(2t)\,\hat{\mathbf{i}} + 3(2t)\,\hat{\mathbf{j}}$$

or $$\mathbf{v} = 8t\,\hat{\mathbf{i}} + 6t\,\hat{\mathbf{j}}$$

$\therefore$ Speed $= |\mathbf{v}| = \sqrt{64t^2 + 36t^2} = \sqrt{100t^2} = 10t$

3. Angular momentum, $\mathbf{L} = \mathbf{r} \times \mathbf{p}$

$$= \begin{vmatrix} \hat{\mathbf{i}} & \hat{\mathbf{j}} & \hat{\mathbf{k}} \\ 1 & -2 & 1 \\ 2 & -1 & 1 \end{vmatrix}$$

$$= \hat{\mathbf{i}}\,(-2+1) - \hat{\mathbf{j}}\,(1-2) + \hat{\mathbf{k}}(-1+4) = -\hat{\mathbf{i}} + \hat{\mathbf{j}} + 3\hat{\mathbf{k}}$$

4. Weight, nuclear spin, momentum are vector quantities, because they have magnitude and direction also.

Potential energy has magnitude only, it does not have direction. So, it is not a vector.

5. Torque $\tau = \mathbf{r} \times \mathbf{F}$

$$= \begin{vmatrix} \hat{\mathbf{i}} & \hat{\mathbf{j}} & \hat{\mathbf{k}} \\ 1 & -2 & 1 \\ 5 & 2 & -5 \end{vmatrix} = \hat{\mathbf{i}}(10-2) - \hat{\mathbf{j}}(-5-5) + \hat{\mathbf{k}}(2+10)$$

$$= 8\hat{\mathbf{i}} + 10\hat{\mathbf{j}} + 12\hat{\mathbf{k}}$$

6. The given vector, $\mathbf{X} = \alpha\mathbf{A} + \beta\mathbf{B} = \alpha(\hat{\mathbf{i}} + \hat{\mathbf{j}} - 2\hat{\mathbf{k}}) + \beta(\hat{\mathbf{i}} - \hat{\mathbf{j}} + \hat{\mathbf{k}})$

$$= (\alpha + \beta)\,\hat{\mathbf{i}} + (\alpha - \beta)\,\hat{\mathbf{j}} + (-2\alpha + \beta)\,\hat{\mathbf{k}}$$

and $$\mathbf{C} = 2\hat{\mathbf{i}} - 3\hat{\mathbf{j}} + 4\hat{\mathbf{k}}$$

Now, $\alpha\mathbf{A} + \beta\mathbf{B}$ is perpendicular to $\mathbf{C}$ when

$$(\alpha\mathbf{A} + \beta\mathbf{B}) \cdot \mathbf{C} = 0$$

$\Rightarrow$ $(\alpha + \beta)(2) + (\alpha - \beta)(-3) + (-2\alpha + \beta)(4) = 0$

$\Rightarrow$ $2\alpha + 2\beta - 3\alpha + 3\beta - 8\alpha + 4\beta = 0$

$\Rightarrow$ $-9\alpha + 9\beta = 0$

$\Rightarrow$ $-9\alpha = -9\beta \Rightarrow \dfrac{\alpha}{\beta} = 1$

$\therefore$ $\alpha : \beta = 1:1$

7. Torque τ is perpendicular to the plane of $\mathbf{r}$ and $\mathbf{F}$ and is given by $\tau = \mathbf{r} \times \mathbf{F}$

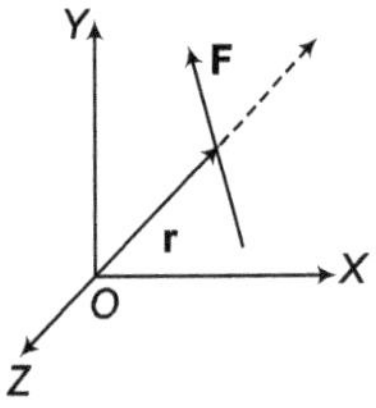

i.e. $$\tau = (2\hat{\mathbf{i}} + 2\hat{\mathbf{j}} + 1\hat{\mathbf{k}}) \times (3\hat{\mathbf{i}} + 7\hat{\mathbf{j}} + 4\hat{\mathbf{k}})$$

$$= \begin{vmatrix} \hat{\mathbf{i}} & \hat{\mathbf{j}} & \hat{\mathbf{k}} \\ 2 & 2 & 1 \\ 3 & 7 & 4 \end{vmatrix} = \hat{\mathbf{i}}(2\times4 - 1\times7) - \hat{\mathbf{j}}(2\times4 - 3\times1)$$

$$+ \hat{\mathbf{k}}(2\times7 - 2\times3) = \hat{\mathbf{i}} - 5\hat{\mathbf{j}} + 8\hat{\mathbf{k}}$$

8. As, $\mathbf{a} \cdot \mathbf{b} = |\mathbf{a} \times \mathbf{b}|$

$$|\mathbf{a}||\mathbf{b}|\cos\theta = |\mathbf{a}||\mathbf{b}|\sin\theta$$

$$1 = \tan\theta$$

$\therefore$ $$\tan 45^\circ = \tan\theta$$

$$\theta = 45^\circ$$

9. See the vector diagram given below $OR = 5$ units

$$OP = 4 \text{ units}$$

$$OQ = 3 \text{ units}$$

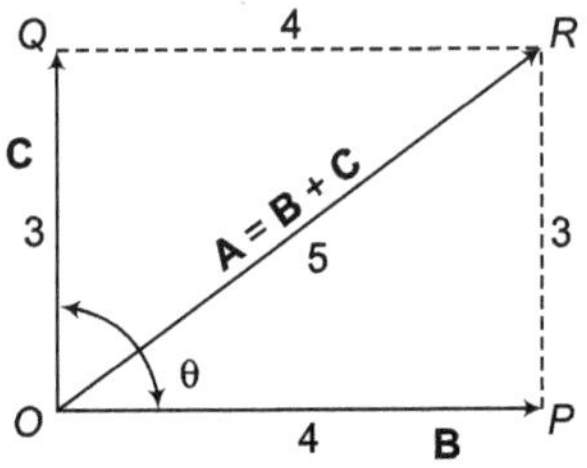

As, $$5 = \sqrt{4^2 + 3^2 + 2\times4\times3\times\cos\theta}$$

$\Rightarrow$ $$25 = 16 + 9 + 24\cos\theta$$

$\Rightarrow$ $$\cos\theta = 0$$

$\Rightarrow$ $$90^\circ = \frac{\pi}{2}$$

Angle between $\mathbf{A}$ and $\mathbf{C} = \angle QOR$

$$\sin(\angle QOR) = \frac{4}{5}$$

$\Rightarrow$ $$\angle QOR = \sin^{-1}\left(\frac{4}{5}\right)$$

Also, $$\cos(\angle QOR) = \frac{3}{5}$$

$\Rightarrow$ $$\angle QOR = \cos^{-1}\left(\frac{3}{5}\right)$$

10. Component of $\mathbf{A}$ along $\mathbf{B} = \dfrac{\mathbf{A}\cdot\mathbf{B}}{|B|}$

Component of $\mathbf{A}$ along $\mathbf{B} = \dfrac{2+3}{\sqrt{2}} = \dfrac{5}{\sqrt{2}}$

3

Motion in One Dimension

Chapter Snapshot

- Introduction
- Position, Position vector, Distance and Displacement
- Average speed, Average velocity, Instantaneous speed, Instantaneous velocity
- Uniform Motion
- Acceleration
- Uniformly accelerated motion
- Graphs
- Relative Motion

3.1 Introduction

Motion is change in position of an object with time. In the present chapter, we shall confine ourselves to the study of motion along a straight line, also called as **rectilinear motion**. If size of the object is much smaller than the distance it moves in a reasonable duration of time, then objects can be treated as point objects.

Kinematics is the branch of physics which deals with the motion regardless of the causes producing this motion. The study of causes of motion is called **dynamics.**

In the chapter of kinematics, the physical quantities involved are : distance (or path length), displacement, speed, velocity, acceleration and time. Of these, displacement, velocity and acceleration are vector quantities. So, they can be added or subtracted like vectors. But we can take liberty with vectors and they can be added/subtracted like scalars if all vectors are one dimensional (for example in rectilinear motion). In this method, one direction is taken as the positive direction and the other as the negative direction. Now, all vector quantities are represented with sign. Then, they can be added/subtracted like scalars.

The origin and the positive direction of an axis are a matter of choice. You should first specify this choice before you assign, signs to quantities like displacement, velocity and acceleration.

For example, if rectilinear motion is taking place along horizontal direction, then sign convention may be taken as shown in Fig. 3.1.

Fig. 3.1

Similarly, if motion is taking place in vertical direction, then sign convention may be taken as shown in Fig. 3.2

Fig. 3.2

3.2 Position, Position Vector, Distance and Displacement

Position and Position Vector

In cartesian coordinate system, position of any point (say A) is represented by its coordinates (x_A, y_A, z_A) with respect to an origin O.

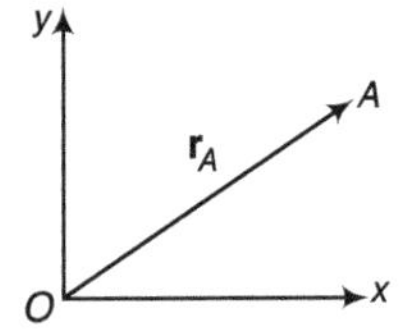

Fig. 3.3

Position vector of point A with respect to O will now be

$$\mathbf{r}_A = \mathbf{OA} = x_A\hat{\mathbf{i}} + y_A\hat{\mathbf{j}} + z_A\hat{\mathbf{k}}$$

Now, suppose coordinates of two points A and B are known to us and we want to find position vector of B with respect to A, then

$$\mathbf{AB} = \mathbf{r}_B - \mathbf{r}_A$$

$$\mathbf{AB} = (x_B - x_A)\hat{\mathbf{i}} + (y_B - y_A)\hat{\mathbf{j}} + (z_B - z_A)\hat{\mathbf{k}}$$

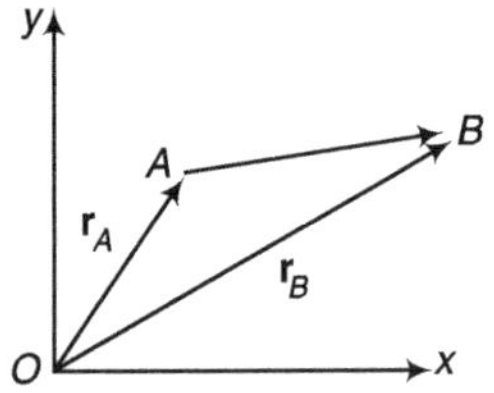

Fig. 3.4

Distance and Displacement

Distance is the actual path length covered by a moving particle or body in a given time interval, while displacement is the change in position vector, i.e. a vector joining initial to final positions. If a particle moves from A to C (Fig. 3.5) through a path ABC, then distance (Δs) travelled is the actual path length ABC, while the displacement is

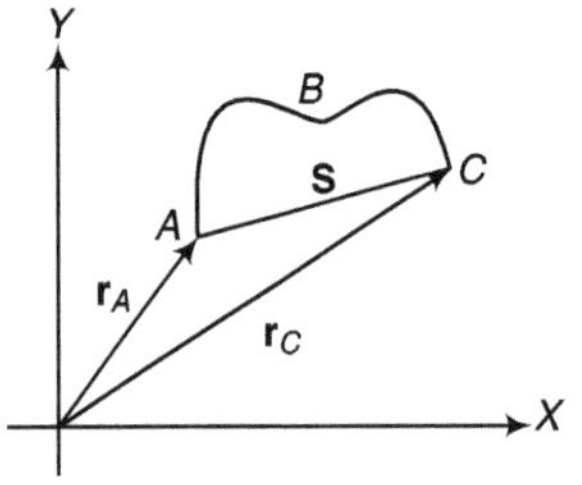

Fig. 3.5

$$\mathbf{s} = \mathbf{r}_C - \mathbf{r}_A$$

If a particle moves in a straight line without change in direction, the magnitude of displacement is equal to the distance travelled, otherwise, it is always less than it. Thus,

$$|\text{displacement}| \le \text{distance}$$

Other Important Points

(i) Distance travelled by a particle is also called the path length.

(ii) In rectilinear motion (or one-dimensional or 1-D motion), we can choose an axis (say x-axis). Now, position of the particle at any point can be represented by its x-coordinate.

(iii) In this case, all vector quantities (displacement, velocity and acceleration) can be represented simply as scalars with proper signs (positive or negative).

(iv) Displacement in this case is equal to the final x-coordinate minus the initial x-coordinate or

$$s = x_f - x_i$$

(v) Distance travelled is always positive but position of the particle (or the x-coordinate) and displacement may be positive or negative.

Example 3.1 *A particle is moving along x-axis. At some given time x-coordinate of the particle is $x = -4$m. First it moves towards positive x-axis and reaches upto $x = +6$m. Then it moves along negative x-axis and reaches upto $x = -10$m. In the complete journey, find*

(a) distance travelled by the particle

(b) displacement of the particle.

Sol.

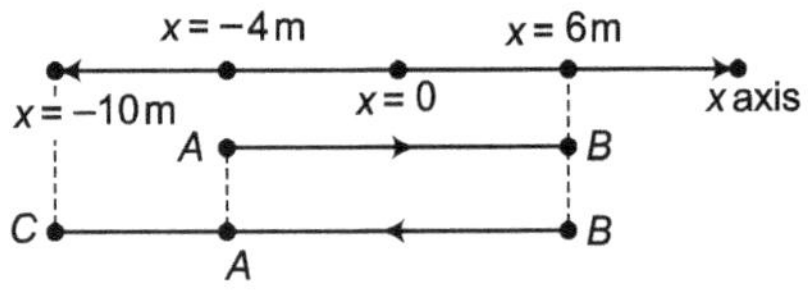

Fig. 3.6

(a) Distance traversed = $AB + BC$
$= 10 + 16 = 26$ m

(b) Displacement $= S = AC = X_f - X_i$
$= (-10 \text{ m}) - (-4 \text{ m}) = -6$m

- Displacement is a vector quantity. Negative sign with any vector quantity represents its direction. For example, $S = -6$m, means 6m towards negative x-direction.

3.3 Average Speed, Average Velocity, Instantaneous Speed, Instantaneous Velocity

Average Speed and Velocity

The average speed of a particle in a given interval of time is defined as the ratio of distance travelled to the time taken while average velocity is defined as the ratio of displacement to time taken.

Thus, if the distance travelled is Δs and displacement of a particle is $\Delta \mathbf{r}$ in a given time interval Δt, then

$$v_{av} = \text{Average speed} = \frac{\Delta s}{\Delta t} \quad \text{and}$$

$$\mathbf{v}_{av} = \text{Average velocity} = \frac{\Delta \mathbf{r}}{\Delta t}$$

Instantaneous Speed and Velocity

Instantaneous speed and velocity are defined at a particular instant and are given by

$$v = \lim_{\Delta t \to 0} \frac{\Delta s}{\Delta t} = \frac{ds}{dt}$$

and $$\mathbf{v} = \lim_{\Delta t \to 0} \frac{\Delta \mathbf{r}}{\Delta t} = \frac{d\mathbf{r}}{dt}$$

Other Important Points

(i) In rectilinear motion (or 1-D motion), when position of any particle can be represented by its x-coordinate only the above formulae can be represented as

(a) $\text{Average speed} = \dfrac{\text{Total distance}}{\text{Total time}}$

(b) $\text{Average velocity} = \dfrac{\text{Displacement}}{\text{Time}} = \dfrac{s}{t} = \dfrac{x_f - x_i}{t}$

(c) $\text{Instantaneous velocity} = \dfrac{dx}{dt}$

(d) Instantaneous speed = magnitude of instantaneous velocity.

(ii) Instantaneous speed is always equal to the magnitude of instantaneous velocity.

(iii) Average and instantaneous values can also be represented as

(a) $\text{Average velocity} = \dfrac{\Delta \mathbf{r}}{\Delta t} = \dfrac{\Delta \mathbf{s}}{\Delta t}$ or $\dfrac{\mathbf{s}}{t}$

$= \dfrac{\Delta s}{\Delta t}$ or $\dfrac{s}{t} = \dfrac{X_f - X_i}{t}$ (in 1-D motion)

(b) $\text{Instantaneous velocity} = \dfrac{d\mathbf{r}}{dt}$ or $\dfrac{d\mathbf{s}}{dt}$

$= \dfrac{ds}{dt}$ or $\dfrac{dx}{dt}$ (in 1-D motion)

Example 3.2 *In Example 3.1, if total time of journey is 4 s, then find*
(a) average speed
(b) average velocity

Sol. (a) $\text{Average speed} = \dfrac{\text{total distance}}{\text{total time}} = \dfrac{26}{4}$

$= 6.5 \text{ m/s}$

(b) $\text{Average velocity} = \dfrac{\text{total displacement}}{\text{total time}} = -\dfrac{6}{4}$

$= -1.5 \text{ m/s}$

Note Points

- Average velocity is a vector quantity. So, negative sign in it merely represents its direction. For example, -1.5 m/s means average velocity is 1.5 m/s along negative x-direction.
- Since, distance $\geq$ |displacement|

$\therefore$ Average speed $\geq$ |Average velocity|

Example 3.3 *Instantaneous velocity (or we can simply say velocity) of a particle moving in a three dimensional space is*

$$\mathbf{v} = (2\hat{\mathbf{i}} - 3\hat{\mathbf{j}} + 4\hat{\mathbf{k}})\ m/s$$

Find instantaneous speed of the particle at this instant.

Sol. Instantaneous speed

= magnitude of instantaneous velocity

$= \sqrt{(2)^2 + (-3)^2 + (4)^2} = \sqrt{29}$ m/s

Example 3.4 *A particle is moving along x-axis. Its x-coordinate varies with time as*

$$x = 2t^2 - 4$$

Here, x is in metre and t in second. Find
(a) average velocity of the particle in a time interval from $t = 0$ to $t = 4$ s.
(b) instantaneous velocity of the particle at $t = 1$ s.

Sol. (a) $\text{Average velocity} = \dfrac{s}{t} = \dfrac{x_f - x_i}{t}$

$x_f = x_{4s} = (2)(4)^2 - 4 = 32$ m

$x_i = x_{0s} = (2)(0)^2 - 4 = -4$ m

$\therefore \text{Average velocity} = \dfrac{(32) - (-4)}{4} = 9$ m/s

(b) Expression of instantaneous velocity at a general time t is given by

$v = \dfrac{dx}{dt} = \dfrac{d}{dt}(2t^2 - 4)$

$= (4t)$

$\therefore$ At $t = 1$ s,

$v = (4)(1) = 4$ m/s

3.4 Uniform Motion

If velocity of a particle is constant, then it is called a uniform motion. Velocity is a vector quantity. So, velocity is constant means its direction and magnitude (or speed) both are constant. Since direction is constant, hence the motion is rectilinear (or 1-D) in the same direction. Displacement and distance both are equal in this case. Particle covers equal distances (or equal displacements) in equal intervals of time. We can apply the formula,

$$\text{Speed or velocity} = \frac{\text{distance or displacement}}{\text{time}}$$

or $$v = \frac{s}{t}$$

$\therefore$ $$s = vt$$

Extra Edge

- In case of uniform motion, instantaneous speed, instantaneous velocity average speed (in any time interval) and average velocity (again in any time interval) all are same.

Example 3.5 *A particle is moving along negative x-direction with uniform velocity of 4 m/s. At time $t = 0$, particle is at $x = 6$m. Find x-coordinate of the particle at $t = 2$ s.*

Sol. Velocity is a vector quantity and motion is one dimensional. So, we can write velocity vector (and other vector quantities also) with sign. Hence,

$$v = -4 \text{ m/s} \quad \text{(Given)}$$

In uniform motion, we can apply the equation

$$s = vt$$

but $$s = x_f - x_i$$

$\therefore$ $$x_f - x_i = vt$$

$\therefore$ $$x_f = x_i + vt$$

or $$x_{2s} = x_{0s} + vt$$

Substituting the values, we get

$$x_{2s} = 6 + (-4)(2) = -2 \text{ m}$$

Example 3.6 *A particle travels half of the total distance with constant velocity v_1 and rest half distance with constant velocity v_2. Find average velocity (or average speed) during the complete journey.*

Sol. In uniform motion, we can apply the equation

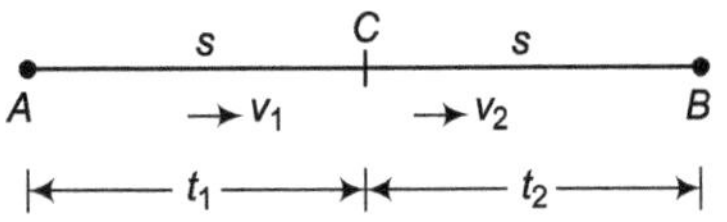

Fig. 3.7

$$s = vt \quad \text{or} \quad t = \frac{s}{v}$$

In the figure, $t_1 = \frac{s}{v_1}$ and $t_2 = \frac{s}{v_2}$

$$\text{Now, average velocity} = \frac{\text{total displacement}}{\text{total time}}$$

$$= \frac{s + s}{t_1 + t_2} = \frac{2s}{(s/v_1) + (s/v_2)}$$

$$= \frac{2 v_1 v_2}{v_1 + v_2}$$

3.5 Acceleration

Velocity is a vector quantity consisting of both magnitude and direction. If velocity of particle changes (either in magnitude or in direction or both) it is said to be accelerated. If complete velocity vector is constant (uniform motion), then acceleration of the particle is zero.

Average acceleration is defined as the ratio of change in velocity, i.e. $\Delta \mathbf{v}$ to the time interval Δt in which this change occurs. Hence,

$$\mathbf{a}_{av} = \frac{\Delta \mathbf{v}}{\Delta t} = \frac{\mathbf{v}_f - \mathbf{v}_i}{\Delta t}$$

The instantaneous acceleration is defined at a particular instant and is given by

$$\mathbf{a} = \lim_{\Delta t \to 0} \frac{\Delta \mathbf{v}}{\Delta t} = \frac{d\mathbf{v}}{dt}$$

- In rectilinear motion (say along x-axis), the formulae of average and instantaneous acceleration can be written as

$$a_{av} = \frac{\Delta v}{\Delta t} = \frac{v_f - v_i}{\Delta t} \quad \text{and} \quad a = \frac{dv}{dt}$$

Here, acceleration and velocity are vector quantities and motion is one-dimensional. So, they can be treated like scalars by assigning them proper signs.

Example 3.7 *At time $t = 0$ s, velocity of a particle is $\mathbf{v}_1 = (2\hat{\mathbf{i}} - 3\hat{\mathbf{j}})$ m/s and at $t = 10$ s velocity is $\mathbf{v}_2 = (4\hat{\mathbf{i}} - 2\hat{\mathbf{j}})$ m/s. Find average acceleration in the given time interval.*

Sol. Average acceleration is given by

$$\mathbf{a}_{av} = \frac{\Delta \mathbf{v}}{\Delta t} = \frac{\mathbf{v}_f - \mathbf{v}_i}{\Delta t} \quad \text{...(i)}$$

Here, $\mathbf{v}_f = \mathbf{v}_{2s} = (4\hat{\mathbf{i}} - 2\hat{\mathbf{j}})$ m/s

and $\mathbf{v}_i = \mathbf{v}_{0s} = (2\hat{\mathbf{i}} - 3\hat{\mathbf{j}})$ m/s

and $\Delta t = 10$ s.

Substituting the values in Eq. (i) we have,

$$\mathbf{a}_{av} = \frac{(4\hat{\mathbf{i}} - 2\hat{\mathbf{j}}) - (2\hat{\mathbf{i}} - 3\hat{\mathbf{j}})}{10} = (0.2\hat{\mathbf{i}} + 0.1\hat{\mathbf{j}}) \text{ m/s}^2$$

Example 3.8 *A particle is moving along the X-axis. The X-coordinate of the particle varies with time as*

$$X = (3t^3 + 4t^2 - 6)$$

Here, X is in metre and t in second. Find

(a) instantaneous acceleration of the particle at $t = 1$ s.

(b) average value of acceleration in a time interval from $t = 0$ to $t = 2$ s.

Sol. (a) Expression of velocity as a function of time (called the instantaneous velocity) is given by

$$v = \frac{dX}{dt} = \frac{d}{dt}(3t^3 + 4t^2 - 6)$$

or $$v = (9t^2 + 8t) \qquad \text{...(i)}$$

By further differentiating Eq. (i) with respect to time we will get expression of acceleration as a function of time (called the instantaneous acceleration)

$$\therefore \quad a = \frac{dv}{dt} = \frac{d}{dt}(9t^2 + 8t)$$

or $$a = 18t + 8 \qquad \text{...(ii)}$$

Now, at $t = 1$ s,

$$a = 18(1) + 8 = 26 \text{ m/s}^2$$

(b) Average acceleration is given by

$$a_{av} = \frac{\Delta v}{\Delta t} = \frac{v_f - v_i}{\Delta t} \qquad \text{...(iii)}$$

From Eq. (i),

$$v_f = v_{2s} = (9)(2)^2 + (8)(2) = 52 \text{ m/s}$$

$$v_i = v_{0s} = (9)(0)^2 + (8)(0) = 0$$

Substituting the values in Eq. (iii) we have,

$$a_{av} = \frac{52 - 0}{2} = 26 \text{ m/s}^2$$

3.6 Uniformly Accelerated Motion

Equations of motion for uniformly accelerated motion (**a** = constant) are as under

$$\mathbf{v} = \mathbf{u} + \mathbf{a}t, \ \mathbf{s} = \mathbf{u}t + \frac{1}{2}\mathbf{a}t^2, \ \mathbf{v} \cdot \mathbf{v} = \mathbf{u} \cdot \mathbf{u} + 2\,\mathbf{a} \cdot \mathbf{s}$$

Here, **u** = initial velocity of particle, **v** = velocity of particle at time t and **s** = displacement of particle in time t

- If initial position vector of a particle is $\mathbf{r}_0$, then position vector at time t can be written as

$$\mathbf{r} = \mathbf{r}_0 + \mathbf{s} = \mathbf{r}_0 + \mathbf{u}t + \frac{1}{2}\mathbf{a}t^2$$

One-dimensional Uniformly Accelerated Motion

If the motion of a particle is taking place in a straight line, there is no need of using vector addition (or subtraction) in equations of motion. We can directly use the equations,

$$v = u + at, \quad s = ut + \frac{1}{2}at^2$$

and $$v^2 = u^2 + 2as$$

Just by taking one direction as the positive (and opposite to it as negative) and then substituting u, a, etc., with sign. Normally we take vertically upward direction positive (and downward negative) and horizontally rightwards positive (or leftwards negative).

Sign convention for (a) motion in vertical direction (b) motion in horizontal direction is shown in Fig. 3.8.

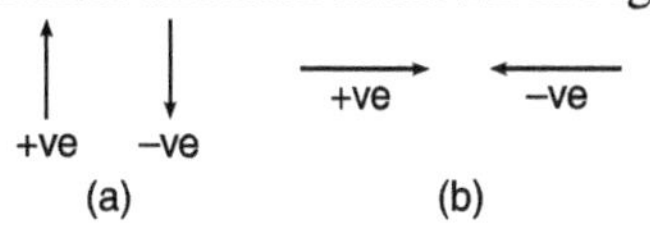

Fig. 3.8

Note Points

- Value of acceleration due to gravity g is 9.8 m/s^2 always in downward direction. So, according to our sign convention (upward positive and downward negative)

$$g = -9.8 \text{ m/s}^2$$

For making the calculations simple we can also take $g = -10 \text{ m/s}^2$ in some problems.

This value of g is independent of the direction of velocity

For example

A particle is projected upwards and it moves from A to B and then returns back from B to A.

Now, in upward journey from A to B velocity is upwards (means positive) and acceleration is g ($= -9.8$ m/s^2). At highest point B, velocity is zero and acceleration is g ($= -9.8$ m/s^2). Finally in downward journey from B to A, velocity is downwards (means negative) but acceleration is still g ($= -9.8$ m/s^2)

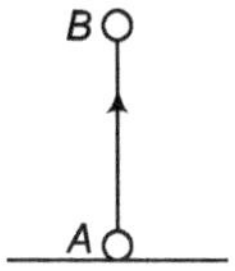

Fig. 3.9

- In rectilinear motion, if velocity and acceleration are in the same direction (both positive or both negative) then speed of the particle increases. For example, in downward journey from B to A both velocity and acceleration are negative. Hence speed is increasing. If velocity and acceleration are in opposite directions (one positive and the other negative), speed of the particle decreases. For example, in upward journey from A to B, velocity is upwards (positive) and acceleration is downwards (negative). So, speed of the particle is decreasing.
- For small heights acceleration due to gravity g is constant. So, equations of constant acceleration ($v = u + at$ *etc.*) can be applied.
- A particle is projected from the top of a tower AB in upward direction. It first moves from B to C, then from C to D and finally hits the ground at time t. Then, there are two methods of findin t.

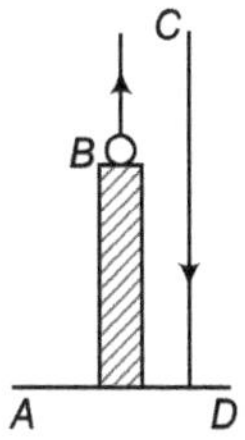

Fig. 3.10

***Method* 1** We directly apply $s = ut + \frac{1}{2}at^2$ and substitute all quantities (s, u and a) with sign. But in this equation, displacement s is measured from the starting point. Here the starting point is B.

***Method* 2** Find $t_{B \to C}$ and $t_{C \to D}$ and then $t = t_{B \to C} + t_{C \to D}$

Example 3.9 *A jet plane starts from rest with an acceleration of 3 ms^{-2} and makes a run for 35 s before taking off. What is the minimum length of the runway and what is the velocity of the jet at take off ?*

Sol. Here $u = 0$, $a = 3\,ms^{-2}$, $t = 35\,s$

Minimum length of the runway required is

$$s = ut + \frac{1}{2}at^2 = 0 + \frac{1}{2} \times 3 \times (35)^2$$

$$= 1837.5\text{ m}$$

Velocity of the jet at take off is

$$v = u + at = 0 + 3 \times 35 = 105\,ms^{-1}$$

Example 3.10 *An electron travelling with a speed of 5×10^3 ms^{-1} passes through an electric field with an acceleration of 10^{12} ms^{-2}.*

(a) How long will it take for this electron to double its speed?

(b) What will be the distance covered by the electron in this time?

Sol. Here, $u = 5 \times 10^3\ ms^{-1}$,

$v = 2 \times 5 \times 10^3\ ms^{-1}$, $a = 10^{12}\ ms^{-2}$

(a) $v = u + at$

$$\therefore \quad t = \frac{v-u}{a} = \frac{2 \times 5 \times 10^3 - 5 \times 10^3}{10^{12}}$$

$$= \frac{5 \times 10^3}{10^{12}} = 5 \times 10^{-9}\text{ s}$$

(b) $s = ut + \frac{1}{2}at^2$

$$= 5 \times 10^3 \times 5 \times 10^{-9} + \frac{1}{2} \times 10^{12} \times (5 \times 10^{-9})^2$$

$$= 3.75 \times 10^{-5}\text{ m}$$

Example 3.11 *A ball is thrown upwards from the top of a tower 40 m high with a velocity of 10 m/s. Find the time when it strikes the ground. Take $g = 10\ m/s^2$.*

Sol. In this problem,

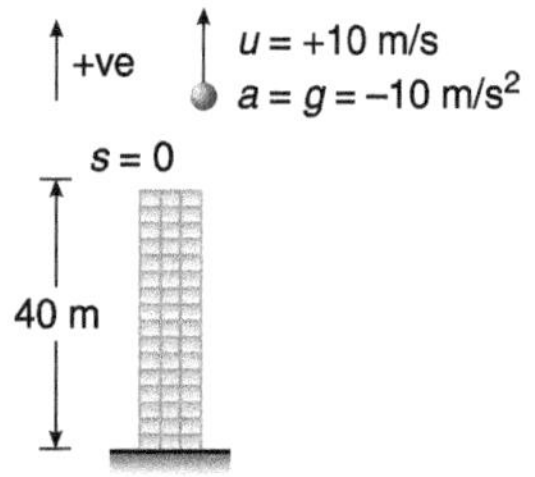

Fig. 3.11

$u = +10\text{ m/s}$, $a = -10\text{ m/s}^2$

and $s = -40$ m (at the point where stone strikes the ground)

Substituting the above values in $s = ut + \frac{1}{2}at^2$, we have

$$-40 = 10t - 5t^2$$

or $\quad 5t^2 - 10t - 40 = 0$

or $\quad t^2 - 2t - 8 = 0$

Solving this, we have

$$t = 4\text{ s} \quad \text{and} \ -2\text{ s.}$$

Taking the positive value, answer is $t = 4$ s.

- The significance of $t = -2$ s can be understood by following figure :

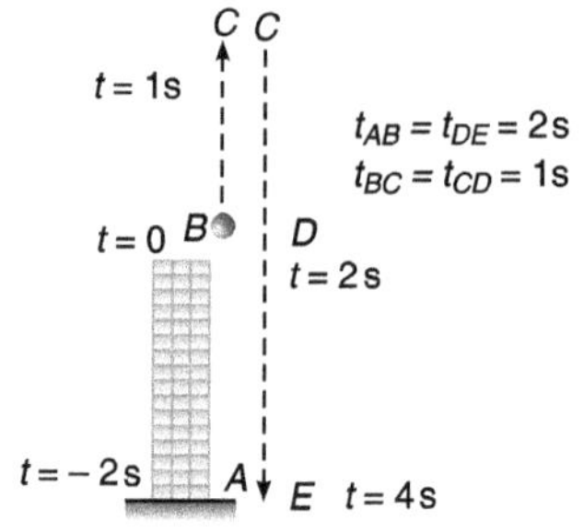

Fig. 3.12

Alternate Method

From *B* to *C* Using $v = u + at$

or $\quad 0 = +10 + (-10)(t)$

or $\quad t = 1\text{s} = t_{BC}$

$$S = ut + \frac{1}{2}at^2$$

$$= (+10)(1) + \frac{1}{2}(-10)(1)^2$$

$$= 5\text{m} = BC$$

From *C* to *A* $CA = CD + DA = 5 + 40 = 45$ m

Using the equation, $S = ut + \frac{1}{2}at^2$ we have

$$(-45) = (0)t + \frac{1}{2}(-10)t^2$$

Solving this equation, the positive value of t comes out to be,

$$t = 3\text{s} = t_{CA}$$

$\therefore$ Total time, $\quad t = t_{BC} + t_{CA}$

$$= 1 + 3$$

$$= 4\text{ s}$$

Example 3.12 *In rectilinear motion with constant acceleration derive the expression of displacement in t^{th} second S_t.*

Sol. S_t = displacement in t^{th} second

= [Total displacement in t seconds]

− [Total displacement in $(t-1)$ seconds]

$$= \left[ut + \frac{1}{2}at^2\right] - \left[u(t-1) + \frac{1}{2}a(t-1)^2\right]$$

Simplifying this equation, we get

$$S_t = u + at - \frac{1}{2}a$$

- In the above equation S_t is the displacement of tth second, not the distance.

3.7 Graphs

Before studying this article, students are advised to go through the theory of graphs. A brief summary is given below.

(i) A linear equation between x and y represents a straight line, e.g. $y = 4x - 2$, $y = 5x + 3$, $3x = y - 2$ equations represent straight line on x-y graph.

(ii) $x \propto y$ or $y = kx$ represents a straight line passing through origin.

(iii) $x \propto \frac{1}{y}$ represents a rectangular hyperbola on x-y graph.

Shape of rectangular hyperbola is as shown in Fig. 3.13.

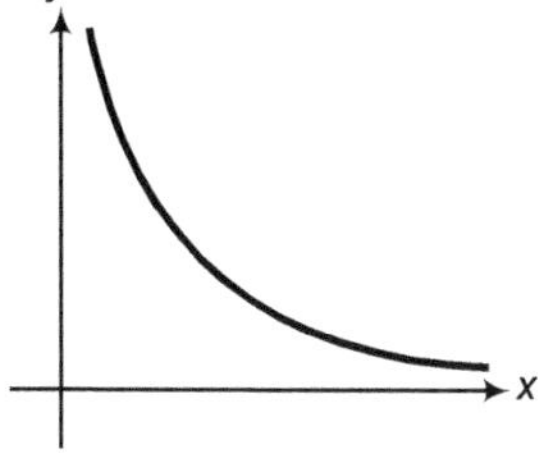

Fig. 3.13

(iv) A quadratic equation in x and y represents a parabola on x-y graph, e.g. $y = 3x^2 + 2$, $y^2 = 4x$, $x^2 = y - 2$ equations represent parabola on x-y graph.

(v) If $z = yx$ or $y\,(dx)$ or $x\,(dy)$, then value of z between x_1 and x_2 or between y_1 and y_2 can be obtained by the area of graph between x_1 and x_2 or y_1 and y_2.

(vi) If $z = \frac{dy}{dx}$, then value of z is given by the slope of x-y graph.

From the above six points, we may conclude that in case of a one dimensional motion :

(a) slope of displacement-time graph gives velocity $\left(\text{as } v = \frac{ds}{dt}\right)$.

(b) slope of velocity-time graph gives acceleration $\left(\text{as } a = \frac{dv}{dt}\right)$.

(c) area under velocity-time graph gives displacement (as $ds = v\,dt$).

(d) area under acceleration-time graph gives change in velocity (as $dv = a\,dt$).

Table 3.1

Graph	Area	Slope
S-t		v
v-t	S	a
a-t	$v_f - v_i$	

(e) displacement-time graph in uniform motion is a straight line passing through origin, if displacement is zero at time $t = 0$ (as $s = vt$).

(f) velocity-time graph is a straight line passing through origin in a uniform accelerated motion, if initial velocity $u = 0$ and a straight line not passing through origin if initial velocity $u \neq 0$ (as $v = u + at$).

(g) displacement-time graph in uniformly accelerated or retarded motion is a parabola $\left(\text{as } s = ut \pm \frac{1}{2}at^2\right)$.

Now, we can plot v-t and s-t graphs of some standard results in tabular form as under. But note that all the following graphs are drawn for one-dimensional motion with uniform velocity or with constant acceleration.

Table 3.2

S.No.	Different cases	v-t graph	s-t graph	Important points
1.	Uniform motion	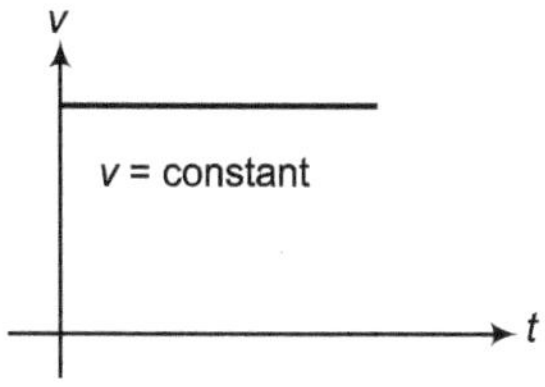	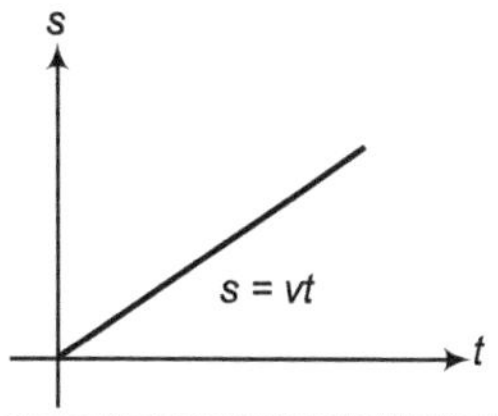	(i) Slope of s-t graph = v = constant (ii) In s-t graph $s = 0$ at $t = 0$

Table 3.2 Contd.

S.No.	Different cases	v-t graph	s-t graph	Important points
2.	Uniformly accelerated motion with $u = 0$ and $s = 0$ at $t = 0$	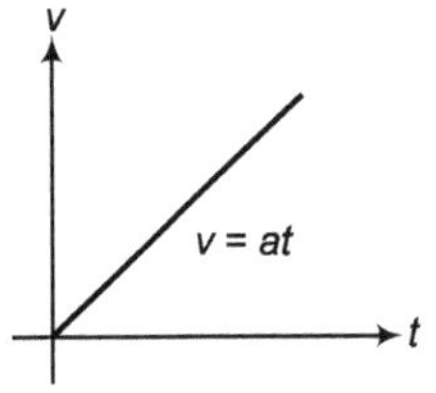	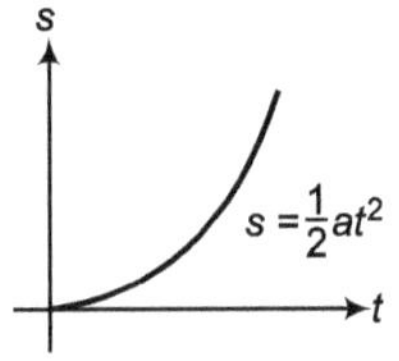	(i) $u = 0$, i.e. $v = 0$ at $t = 0$ (ii) a or slope of v-t graph is constant (iii) $u = 0$, i.e. slope of s-t graph at $t = 0$, should be zero
3.	Uniformly accelerated motion with $u \neq 0$ but $s = 0$ at $t = 0$	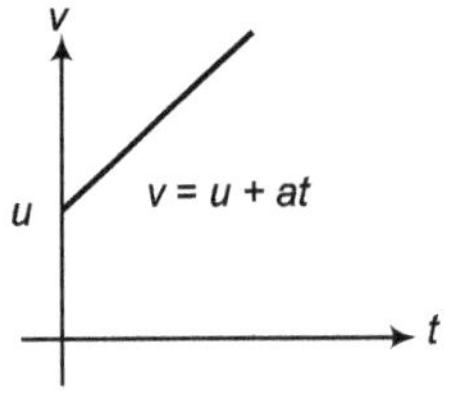	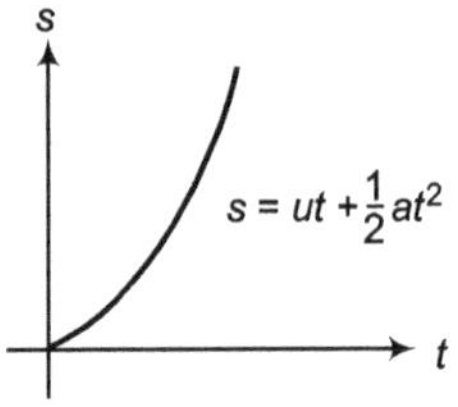	(i) $u \neq 0$, i.e. v or slope of s-t graph at $t = 0$ is not zero (ii) v or slope of s-t graph gradually goes on increasing
4.	Uniformly accelerated motion with $u \neq 0$ and $s = s_0$ at $t = 0$	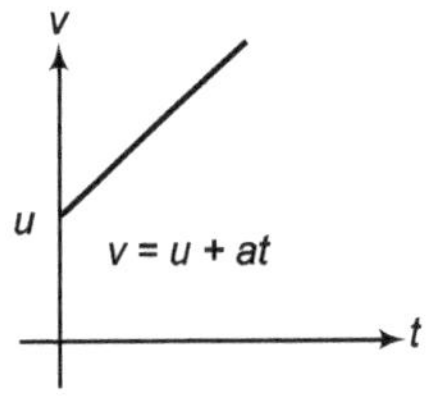	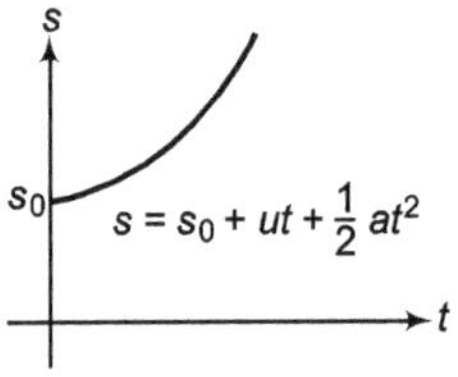	(i) $v = u$ at $t = 0$ (ii) $s = s_0$ at $t = 0$
5.	Uniformly retarded motion till velocity becomes zero	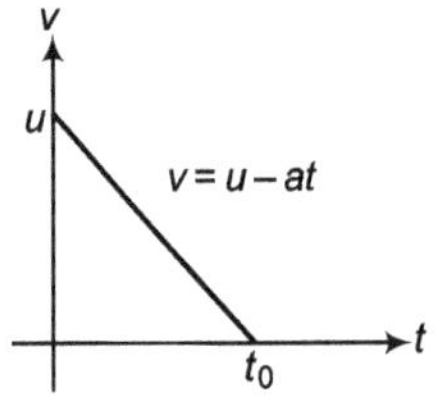	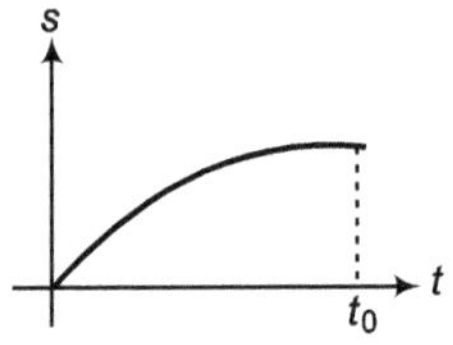	(i) Slope of s-t graph at $t = 0$ gives u (ii) Slope of s-t graph at $t = t_0$ becomes zero (iii) In this case u can't be zero
6.	Uniformly retarded and then accelerated in opposite direction	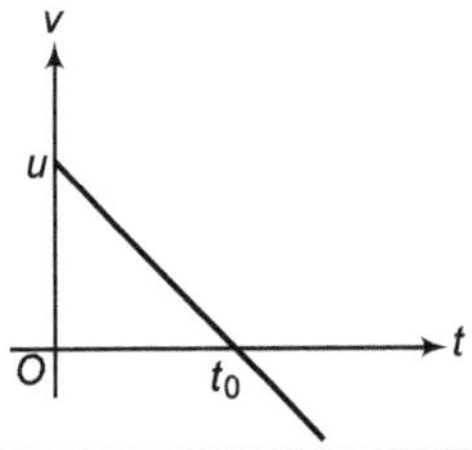	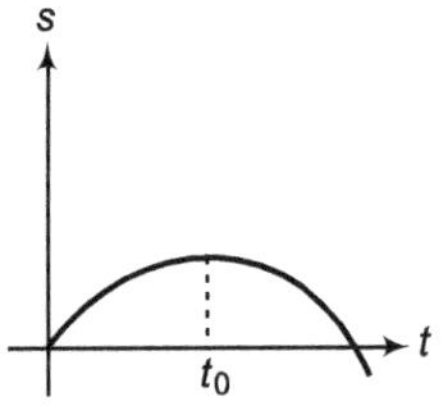	(i) At time $t = t_0$, $v = 0$ or slope of s-t graph is zero (ii) In s-t graph slope or velocity first decreases then increases with opposite sign.

Important Points in Graphs

- Slopes of v-t or s-t graphs can never be infinite at any point, because infinite slope of v-t graph means infinite acceleration. Similarly, infinite slope of s-t graph means infinite velocity. Hence, the following graphs are not possible :

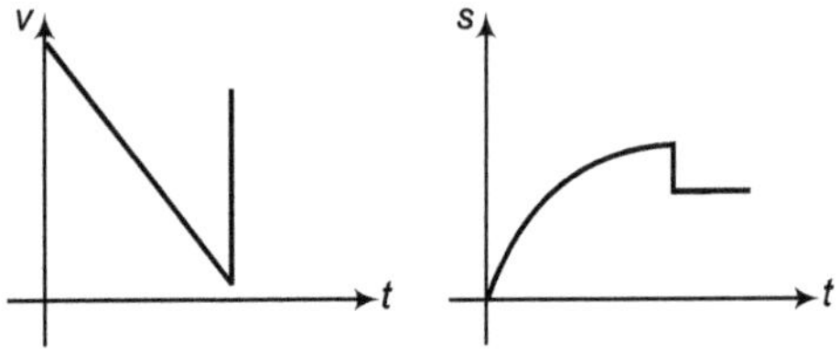

Fig. 3.14

- At one time, two values of velocity or displacement are not possible. Hence, the following graphs are not acceptable:

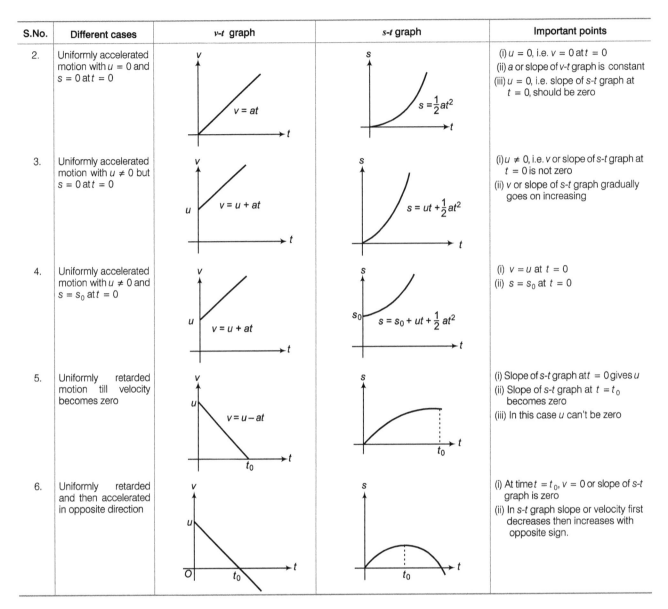

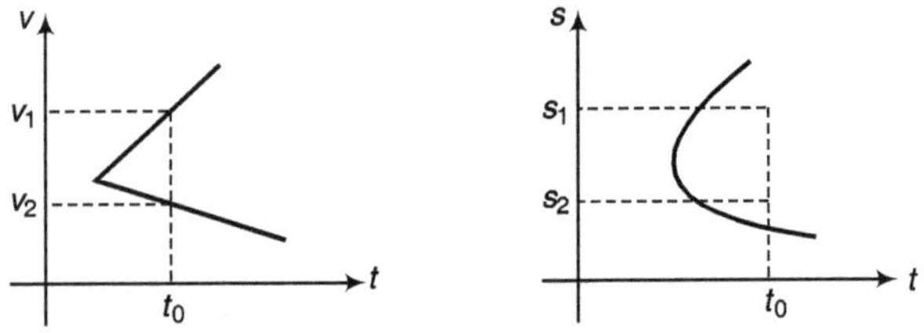

Fig. 3.15

- On time axis we always move forward.
- Different values of displacements in s-t graph corresponding to given v-t graph can be calculated just by calculating areas under v-t graph. There is no need of using equations like $v = u + at$, etc.

Example 3.13 *Displacement-time graph of a particle moving in a straight line is as shown in figure. State whether the motion is accelerated or not. Describe the motion in detail. Given* $s_0 = 20\,m$ *and* $t_0 = 4\,s$.

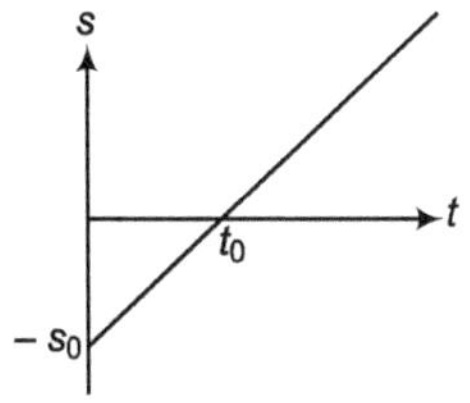

Fig. 3.16

Sol. Slope of s-t graph is constant. Hence, velocity of particle is constant. Further at time $t = 0$, displacement of the particle from the mean position is $-s_0$ or -20 m. Velocity of particle,

$$v = \text{slope} = \frac{s_0}{t_0} = \frac{20}{4} = 5\text{ m/s}$$

Motion of the particle is as shown in figure. At $t = 0$, particle is at -20 m and has a constant velocity of 5 m/s.
At $t_0 = 4$ s, particle will pass through $s = 0$ position.

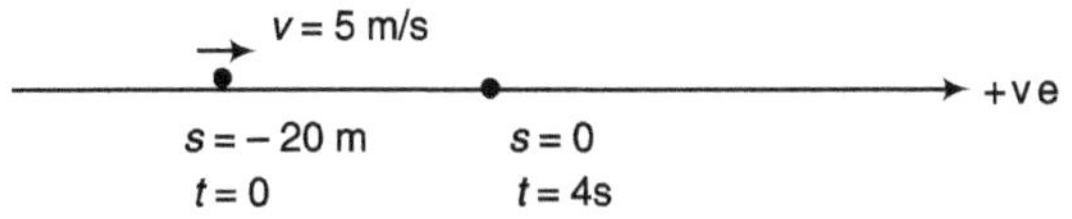

Fig. 3.17

Example 3.14 *Derive the following equations of motion for uniformly accelerated motion from velocity-time graph :*

(a) $v = u + at$ *(b)* $s = ut + \frac{1}{2}at^2$

(c) $v^2 - u^2 = 2as$

Sol. Equations of motion by graphical method. Consider an object moving along a straight line path with initial velocity u and uniform acceleration a. Its velocity-time graph is straight line.

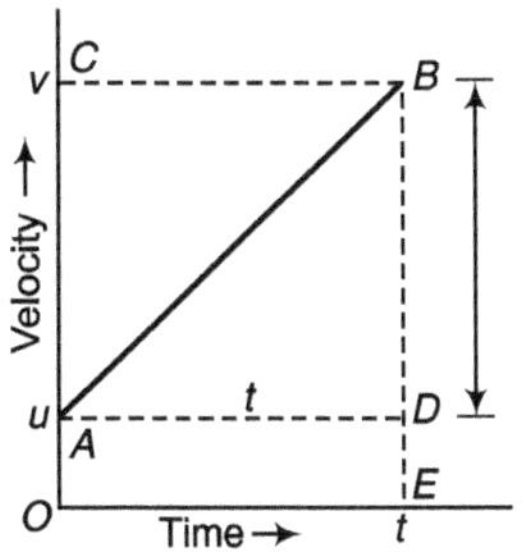

Fig. 3.18. Velocity-time graph for uniform acceleration

(a) We know that,
Acceleration = Slope of velocity-time graph AB

or $$a = \frac{DB}{AD} = \frac{v-u}{t}$$

or $v - u = at$ or $v = u + at$

This proves the first equation of motion.

(b) Acceleration

$$a = \frac{DB}{AD} = \frac{DB}{t} \quad \text{or} \quad DB = at$$

Distance travelled by the object in time t is

s = Area of the trapezium $OABE$

= Area of rectangle $OADE$ + Area of triangle ADB

$= OA \times OE + \frac{1}{2} DB \times AD$

$= ut + \frac{1}{2} at \times t$ or $s = ut + \frac{1}{2}at^2$

This proves the second equation of motion.

(c) Distance travelled by object in time t is

s = Area of trapezium $OABE$

$= \frac{1}{2}(EB + OA) \times OE = \frac{1}{2}(EB + ED) \times OE$

Acceleration,

a = Slope of velocity-time graph AB

or $a = \frac{DB}{AD} = \frac{EB - ED}{OE}$ or $OE = \frac{EB - ED}{a}$

$$s = \frac{1}{2}(EB + ED) \times \frac{(EB - ED)}{a}$$

$$= \frac{1}{2a}(EB^2 - ED^2) = \frac{1}{2a}(v^2 - u^2)$$

or $v^2 - u^2 = 2as$

This proves the third equation of motion.

Example 3.15 *Fig.* 3.19. *shows the distance-time graphs of two trains, which start moving simultaneously in the same direction. From the graphs, find*

(a) how much ahead of A and B when the motion starts ?

(b) what is the speed of B?

(c) when and where will A catch B?

(d) what is the difference between the speeds of A and B?

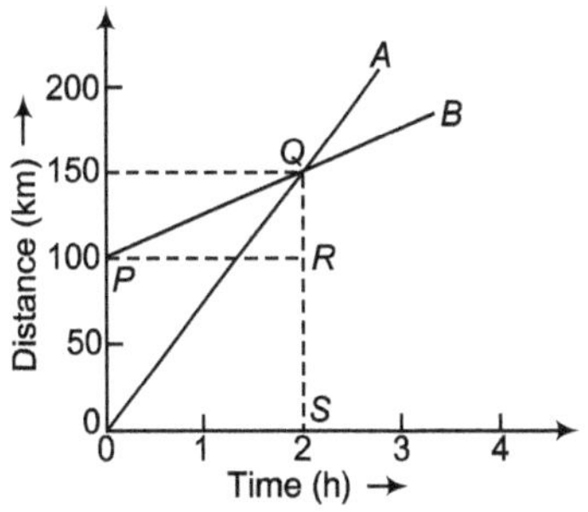

Fig. 3.19

Sol. (a) B is ahead of A by the distance $OP = 100$ km, when the motion starts.

(b) Speed of $B = \frac{QR}{PR} = \frac{150 - 100}{2 - 0}$

$= 25\text{ kmh}^{-1}$

(c) Since, the two graphs intersect at point Q, so A will catch B after 2 h and at a distance of 150 km from the origin.

(d) Speed of $A = \frac{QS}{OS} = \frac{150 - 0}{2 - 0} = 75\text{ kmh}^{-1}$

$\therefore$ Difference in speeds $= 75 - 25 = 50\text{ kmh}^{-1}$

Example 3.16 *A car accelerates from rest at a constant rate α for some time, after which it decelerates at a constant rate β, to come to rest. If the total time elapsed is t second evaluate (a) the maximum velocity reached and (b) the total distance travelled.*

Sol. (a) Let the car accelerates for time t_1 and decelerates for time t_2. Then,

$$t = t_1 + t_2 \qquad \text{...(i)}$$

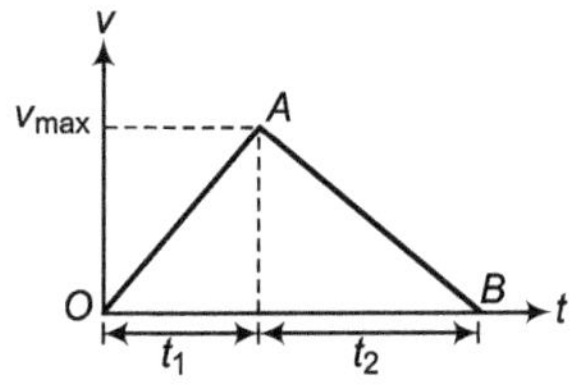

Fig. 3.20

and corresponding velocity-time graph will be as shown in Fig. 3.20.
From the graph,

$$\alpha = \text{slope of line } OA = \frac{v_{max}}{t_1}$$

or
$$t_1 = \frac{v_{max}}{\alpha} \qquad \text{...(ii)}$$

and
$$\beta = -\text{ slope of line } AB = \frac{v_{max}}{t_2}$$

or
$$t_2 = \frac{v_{max}}{\beta} \qquad \text{...(iii)}$$

From Eqs. (i), (ii) and (iii), we get

$$\frac{v_{max}}{\alpha} + \frac{v_{max}}{\beta} = t \quad \text{or} \quad v_{max}\left(\frac{\alpha + \beta}{\alpha\beta}\right) = t$$

$$\therefore \qquad v_{max} = \frac{\alpha\beta t}{\alpha + \beta}$$

(b) Total distance = displacement = area under v-t graph

$$= \frac{1}{2} \times t \times v_{max} = \frac{1}{2} \times t \times \frac{\alpha\beta t}{\alpha + \beta}$$

or
$$\text{Distance} = \frac{1}{2}\left(\frac{\alpha\beta t^2}{\alpha + \beta}\right)$$

- This problem can also be solved by using equations of motion ($v = u + at$ etc.). Try it yourself.

Example 3.17 *Acceleration-time graph of a particle moving in a straight line is shown in Fig. 3.21. Velocity of particle at time $t = 0$ is 2 m/s. Find velocity at the end of fourth second.*

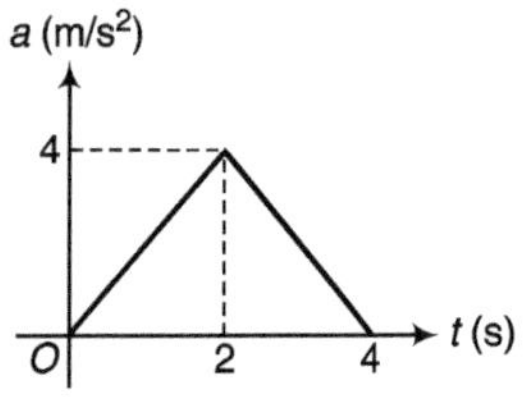

Fig. 3.21

Sol. Change in velocity = area under a-t graph

Hence,
$$v_f - v_i = \frac{1}{2}(4)(4) = 8 \text{ m/s}$$

$$\therefore \qquad v_f = v_i + 8 = (2 + 8) \text{ m/s} = 10 \text{ m/s}$$

Example 3.18 *A rocket is fired vertically upwards with a net acceleration of 4 m/s² and initial velocity zero. After 5 s its fuel is finished and it decelerates with g. At the highest point its velocity becomes zero. Then, it accelerates downwards with acceleration g and return back to ground. Plot velocity-time and displacement-time graphs for the complete journey. (Take $g = 10$ m/s²)*

Sol. In the graphs,

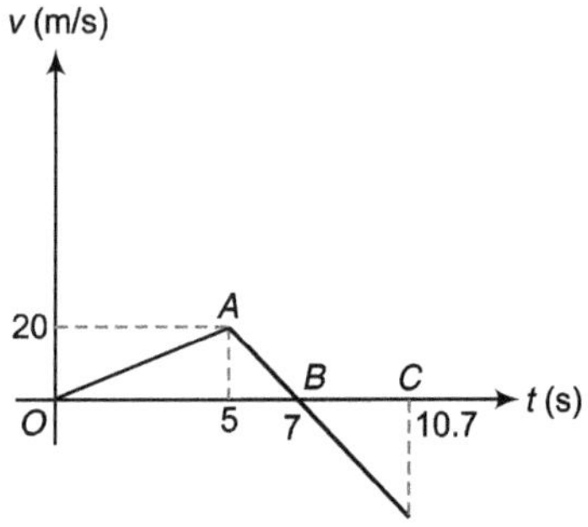

Fig. 3.22

$$v_A = at_{OA} = (4)(5) = 20 \text{ m/s}$$

$$v_B = 0 = v_A - gt_{AB}$$

$$\therefore \qquad t_{AB} = \frac{v_A}{g} = \frac{20}{10} = 2 \text{s}$$

$$\therefore \qquad t_{OAB} = (5 + 2) \text{s} = 7 \text{s}$$

Now, s_{OAB} = area under v-t graph between 0 to 7 s

$$= \frac{1}{2}(7)(20) = 70 \text{ m}$$

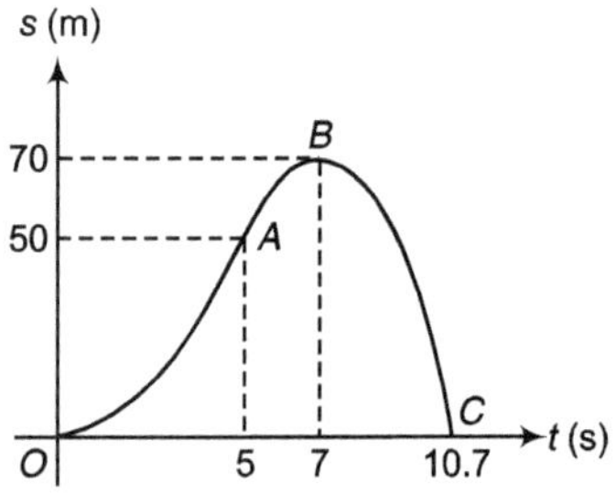

Fig. 3.23

Now,
$$|s_{OAB}| = |s_{BC}| = \frac{1}{2} g t_{BC}^2$$

$$\therefore \qquad 70 = \frac{1}{2}(10) t_{BC}^2$$

$$\therefore \qquad t_{BC} = \sqrt{14} = 3.7 \text{ s}$$

$$\therefore \qquad t_{OABC} = 7 + 3.7 = 10.7 \text{ s}$$

Also s_{OA} = area under v-t graph between OA

$$= \frac{1}{2}(5)(20) = 50 \text{ m}$$

Example 3.19 *Velocity-time graph of a particle moving in a straight line is shown in Fig. 3.24.*

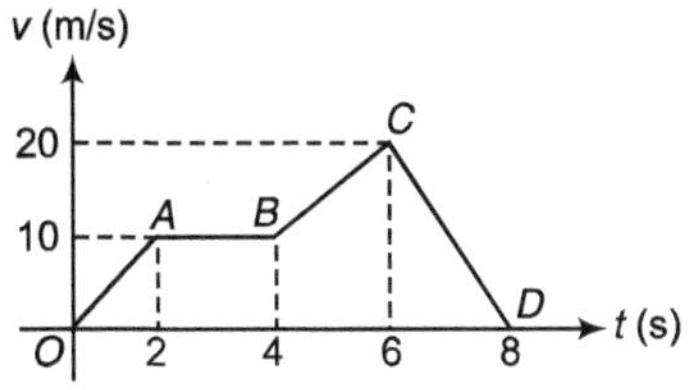

Fig. 3.24

Plot the corresponding displacement-time graph of the particle, if at time $t = 0$, displacement $s = 0$.

Sol. Displacement = Area under velocity-time graph

Hence, $s_{OA} = \frac{1}{2} \times 2 \times 10 = 10\,\text{m}$

$s_{AB} = 2 \times 10 = 20\,\text{m}$ or $s_{OAB} = 10 + 20 = 30\,\text{m}$

$s_{BC} = \frac{1}{2} \times 2(10 + 20) = 30\,\text{m}$

or $s_{OABC} = 30 + 30 = 60\,\text{m}$

and $s_{CD} = \frac{1}{2} \times 2 \times 20 = 20\,\text{m}$

or $s_{OABCD} = 60 + 20 = 80\,\text{m}$

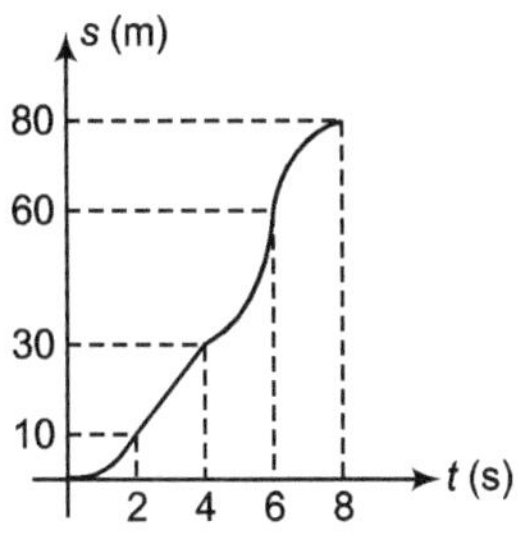

Fig. 3.25

Between 0 to 2 s and 4 to 6 s motion is accelerated, hence displacement-time graph is a parabola. Between 2 to 4 s, motion is uniform, so displacement-time graph will be a straight line. Between 6 to 8 s, motion is decelerated hence displacement-time graph is again a parabola but inverted in shape. At the end of 8 s velocity is zero, therefore, slope of displacement- time graph should be zero. The corresponding graph is shown in Fig. 3.25.

3.8 Relative Motion

The word 'relative' is a very general term, which can be applied to physical, non-physical, scalar or vector quantities. For example, my height is five feet and six inches while my wife's height is five feet and four inches. If I ask you how high I am relative to my wife, your answer will be two inches. What you did? You simply subtracted my wife's height from my height. The same concept is applied everywhere, whether it is a relative velocity, relative acceleration or anything else. So, from the above discussion we may now conclude that relative velocity of A with respect to B (written as $\mathbf{v}_{AB}$) is

$$\mathbf{v}_{AB} = \mathbf{v}_A - \mathbf{v}_B$$

Similarly, relative acceleration of A with respect to B is

$$\mathbf{a}_{AB} = \mathbf{a}_A - \mathbf{a}_B$$

If it is a one dimensional motion we can treat the vectors as scalars just by assigning the positive sign to one direction and negative to the other. So, in case of a one dimensional motion the above equations can be written as

$$\mathbf{v}_{AB} = \mathbf{v}_A - \mathbf{v}_B \quad \text{and} \quad \mathbf{a}_{AB} = \mathbf{a}_A - \mathbf{a}_B$$

Further, we can see that

$$\mathbf{v}_{AB} = -\,\mathbf{v}_{BA} \quad \text{or} \quad \mathbf{a}_{BA} = -\,\mathbf{a}_{AB}$$

Extra Edge

- By the concept of relative motion we convert a two body motion into one body motion. For example, if A and B, two bodies are in motion, then $\mathbf{v}_{AB}$ $(= \mathbf{v}_A - \mathbf{v}_B)$ means B is assumed to be at rest and A is moving with velocity $\mathbf{v}_{AB}$. Similarly in $\mathbf{v}_{BA}$, body A is assumed to be at rest and B moving with velocity $\mathbf{v}_{BA}$.

Example 3.20 *Seeta is moving due east with a velocity of 1 m/s and Geeta is moving due west with a velocity of 2 m/s. What is the velocity of Seeta with respect to Geeta?*

Sol. It is a one dimensional motion. So, let us choose the east direction as positive and the west as negative. Now, given that

v_S = velocity of Seeta = 1 m/s

and v_G = velocity of Geeta = −2 m/s

Thus, v_{SG} = velocity of Seeta with respect to Geeta

$= v_S - v_G = 1 - (-2) = 3\,\text{m/s}$

Hence, velocity of Seeta with respect to Geeta is 3 m/s due east.

Example 3.21 *Car A has an acceleration of $2\,m/s^2$ due east and car B, $4\,m/s^2$ due north. What is the acceleration of car B with respect to car A?*

Sol. It is a two dimensional motion. Therefore,

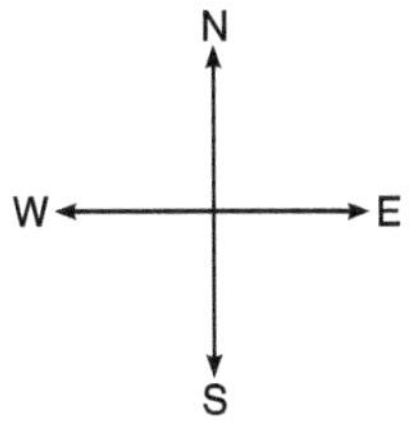

Fig. 3.26

$\mathbf{a}_{BA}$ = acceleration of car B with respect to car A

$= \mathbf{a}_B - \mathbf{a}_A$

Here $\mathbf{a}_B$ = acceleration of car B

$= 4\,\text{m/s}^2$ (due north)

and $\mathbf{a}_A$ = acceleration of car A

$= 2\,\text{m/s}^2$ (due east)

$$|\mathbf{a}_{BA}| = \sqrt{(4)^2 + (2)^2} = 2\sqrt{5}\,\text{m/s}^2$$

and $\alpha = \tan^{-1}\left(\frac{4}{2}\right) = \tan^{-1}(2)$

Thus, $\mathbf{a}_{BA}$ is $2\sqrt{5}$ m/s^2 at an angle of $\alpha = \tan^{-1}$ (2) from west towards north.

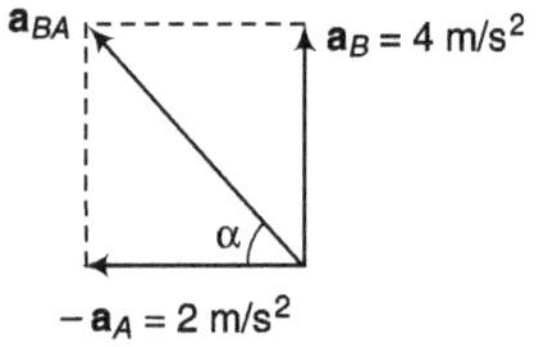

Fig. 3.27

Example 3.22 *Two particles A and B are moving along x-axis. Their x-t graphs are shown in figure below. Find velocity of A with respect to B.*

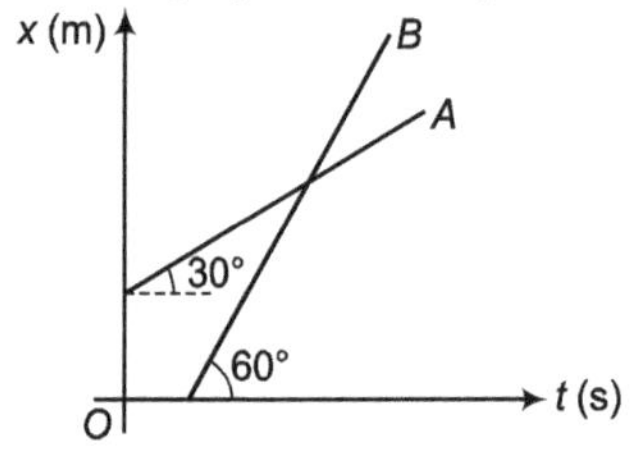

Fig. 3.28

Sol. Velocity of A,

$$v_A = \text{Slope of } x\text{-}t \text{ graph of } A$$
$$= \tan 30° = 0.58 \text{ m/s}$$

Velocity of B,

$$v_B = \text{Slope of } x\text{-}t \text{ graph of } B$$
$$= \tan 60° = 1.732 \text{ m/s}$$

Velocity of A with respect to B

$$v_{AB} = v_A - v_B = 0.58 - 1.732$$
$$= -1.152 \text{ m/s}$$

Negative sign indicates that A appears to be moving towards negative x-direction (to B) with velocity 1.152 m/s

Thus, $\mathbf{a}_{BA}$ is $2\sqrt{5}$ m/s^2 at an angle of $\alpha = \tan^{-1}$ (2) from west towards north.

The topic 'relative motion' is very useful in two and three dimensional motion. Questions based on relative motion are usually of following four types:

(a) Minimum distance between two bodies in motion
(b) River-boat problems
(c) Aircraft-wind problems
(d) Rain problems

Minimum Distance between Two Bodies in Motion

When two bodies are in motion, the questions like, the minimum distance between them or the time when one body overtakes the other can be solved easily by the principle of relative motion. In these type of problems, one body is assumed to be at rest and the relative motion of the other body is considered. By assuming so two body problem is converted into one body problem and the solution becomes easy. Following example will illustrate the statement.

Example 3.23 *Car A and car B start moving simultaneously in the same direction along the line joining them. Car A with a constant acceleration $a = 4$ m/s^2, while car B moves with a constant velocity $v = 1$ m/s. At time $t = 0$, car A is 10 m behind car B. Find the time when car A overtakes car B.*

Sol. Given, $u_A = 0$, $u_B = 1$ m/s,

$$a_A = 4 \text{ m/s}^2 \text{ and } a_B = 0$$

Assuming car B to be at rest, we have

$$u_{AB} = u_A - u_B = 0 - 1 = -1 \text{ m/s}$$
$$a_{AB} = a_A - a_B = 4 - 0 = 4 \text{ m/s}^2$$

Now, the problem can be assumed in simplified form as follows

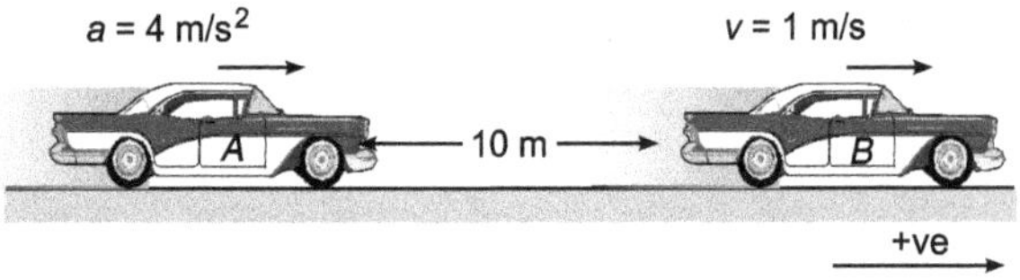

Fig. 3.29

Substituting the proper values in equation

$$s = ut + \frac{1}{2}at^2$$

We get $\quad 10 = -t + \frac{1}{2}(4)(t^2)$

or $\quad 2t^2 - t - 10 = 0$

or $\quad t = \dfrac{1 \pm \sqrt{1 + 80}}{4} = \dfrac{1 \pm \sqrt{81}}{4} = \dfrac{1 \pm 9}{4}$

or $\quad t = 2.5$ s and -2s

Ignoring the negative value, the desired time is 2.5 s.

- The above problem can also be solved without using the concept of relative motion as under.

At the time when A overtakes B,

$$s_A = s_B + 10$$

$\therefore \quad \frac{1}{2} \times 4 \times t^2 = 1 \times t + 10$

or $\quad 2t^2 - t - 10 = 0$

Which on solving gives $t = 2.5$ s and -2 s, the same as we found above.

As per my opinion, this approach (by taking absolute values) is more suitable in case of two body problem in one dimensional motion. Let us see one more example in support of it.

Example 3.24 *An open lift is moving upwards with velocity 10 m/s. It has an upward acceleration of 2 m/s^2. A ball is projected upwards with velocity 20 m/s relative to ground. Find*

(a) time when ball again meets the lift.
(b) displacement of lift and ball at that instant.
(c) distance travelled by the ball upto that instant.
(Take $g = 10$ m/s^2)

Sol. (a) At the time when ball again meets the lift,

$$s_L = s_B$$

$$\therefore \quad 10t + \frac{1}{2} \times 2 \times t^2 = 20t - \frac{1}{2} \times 10t^2$$

Solving this equation, we get

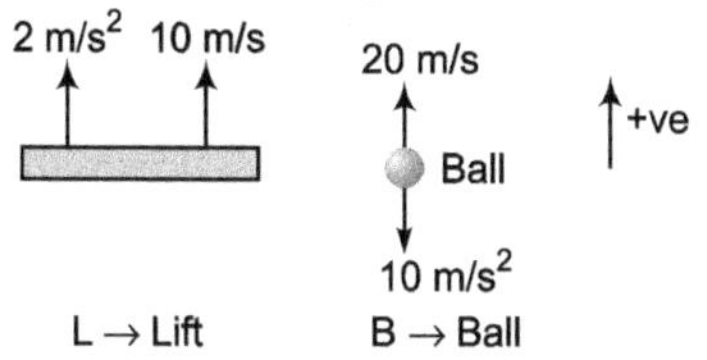

Fig. 3.30

$$t = 0 \quad \text{and} \quad t = \frac{5}{3}\,\text{s}$$

$\therefore$ Ball will again meet the lift after $\frac{5}{3}$ s.

(b) At this instant

$$s_L = s_B = 10 \times \frac{5}{3} + \frac{1}{2} \times 2 \times \left(\frac{5}{3}\right)^2$$

$$= \frac{175}{9}\,\text{m} = 19.4\,\text{m}$$

(c) For the ball $u \uparrow \downarrow a$. Therefore, we will first find t_0, the time when its velocity becomes zero.

$$t_0 = \left|\frac{u}{a}\right| = \frac{20}{10} = 2\,\text{s}$$

As $t \left(= \frac{5}{3}\,\text{s}\right) < t_0$, distance and displacement are equal.

or $\qquad d = 19.4$ m

Concept of relative motion is more useful in two body problem in two (or three) dimensional motion. This can be understood by the following example.

Example 3.25 *Two ships A and B are 10 km apart on a line running south to north. Ship A farther north is streaming west at 20 km/h and ship B is streaming north at 20 km/h. What is their distance of closest approach and how long do they take to reach it?*

Sol. Ships A and B are moving with same speed 20 km/h in the directions shown in figure. It is a two dimensional, two body problem with zero acceleration. Let us find $\mathbf{v}_{BA}$

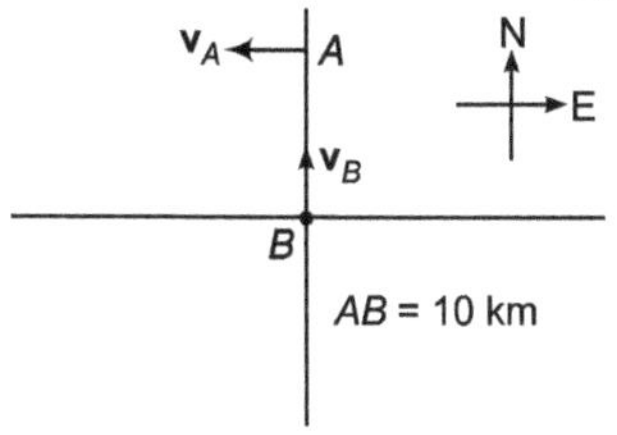

Fig. 3.31

$$\mathbf{v}_{BA} = \mathbf{v}_B - \mathbf{v}_A$$

Here, $\qquad |\mathbf{v}_{BA}| = \sqrt{(20)^2 + (20)^2}$

$$= 20\sqrt{2}\ \text{km/h}$$

i.e. $\mathbf{v}_{BA}$ is $20\sqrt{2}$ km/h at an angle of 45° from east towards north.

Thus, the given problem can be simplified as

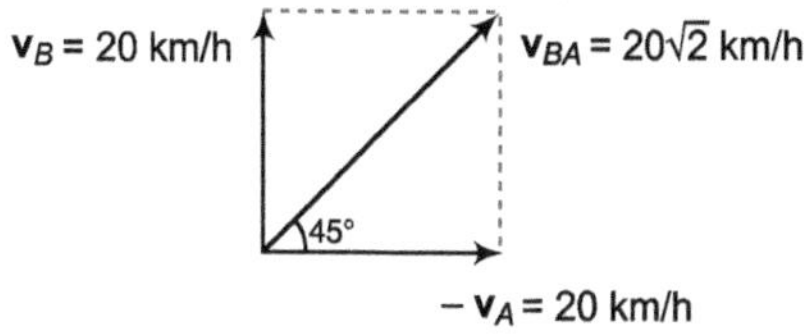

Fig. 3.32

A is at rest and B is moving with $\mathbf{v}_{BA}$ in the direction shown in Fig. 3.33.

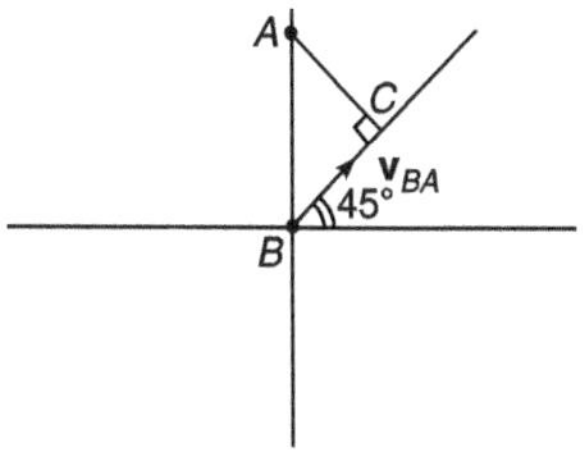

Fig. 3.33

Therefore, the minimum distance between the two is

$$s_{\min} = AC = AB \sin 45° = 10\left(\frac{1}{\sqrt{2}}\right)\text{km}$$

$= 5\sqrt{2}$ km and the desired time is

$$t = \frac{BC}{|\mathbf{v}_{BA}|} = \frac{5\sqrt{2}}{20\sqrt{2}} \qquad (BC = AC = 5\sqrt{2}\ \text{km})$$

$$= \frac{1}{4}\,\text{h} = 15\,\text{min}$$

River-Boat Problems

In river-boat problems, we come across the following three terms :

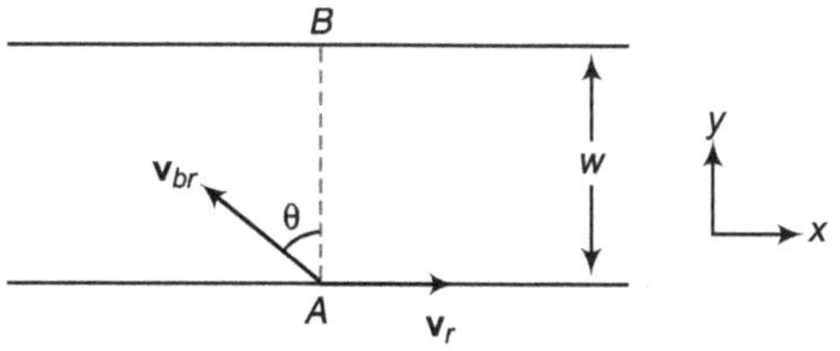

Fig. 3.34

$\mathbf{v}_r$ = absolute velocity of river,

$\mathbf{v}_{br}$ = velocity of boatman with respect to river or velocity of boatman in still water

and $\mathbf{v}_b$ = absolute velocity of boatman.

Here, it is important to note that $\mathbf{v}_{br}$ is the velocity of boatman with which he steers and $\mathbf{v}_b$ is the actual velocity of boatman relative to ground.

Further, $\qquad \mathbf{v}_b = \mathbf{v}_{br} + \mathbf{v}_r$

Now, let us derive some standard results and their special cases.

A boatman starts from point A on one bank of a river with velocity $\mathbf{v}_{br}$ in the direction shown in Fig. 3.34. River is flowing along positive x-direction with velocity $\mathbf{v}_r$. Width of the river is w. Then,

$$\mathbf{v}_b = \mathbf{v}_r + \mathbf{v}_{br}$$

Therefore, $v_{bx} = v_{rx} + v_{brx} = v_r - v_{br} \sin\theta$

and
$$v_{by} = v_{ry} + v_{bry} = 0 + v_{br} \cos\theta = v_{br}\cos\theta$$

Now, time taken by the boatman to cross the river is

$$t = \frac{w}{v_{by}} = \frac{w}{v_{br}\cos\theta}$$

or
$$t = \frac{w}{v_{br}\cos\theta} \qquad \text{...(i)}$$

Further, displacement along x-axis when he reaches on the other bank (also called drift) is

$$x = v_{bx}\, t = (v_r - v_{br}\sin\theta)\frac{w}{v_{br}\cos\theta}$$

or
$$x = (v_r - v_{br}\sin\theta)\frac{w}{v_{br}\cos\theta} \qquad \text{...(ii)}$$

Two special cases are

(i) **Condition when the boatman crosses the river in shortest interval of time**

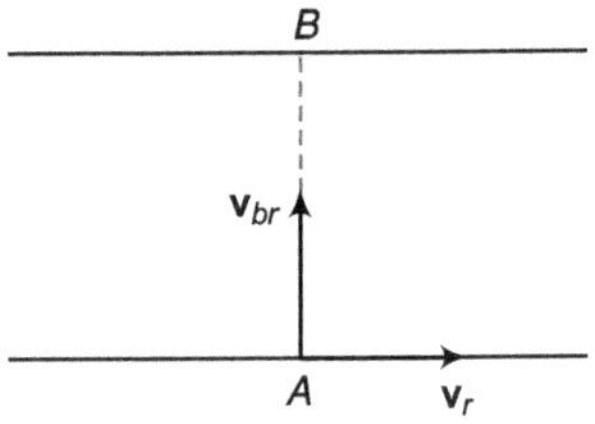

Fig. 3.35

From Eq. (i), we can see that time (t) will be minimum when $\theta = 0°$, i.e. the boatman should steer his boat perpendicular to the river current.

Also, $t_{min} = \dfrac{w}{v_{br}}$ as $\cos\theta = 1$

(ii) **Condition when the boatman wants to reach point *B*, i.e. at a point just opposite from where he started**

In this case, the drift (x) should be zero.

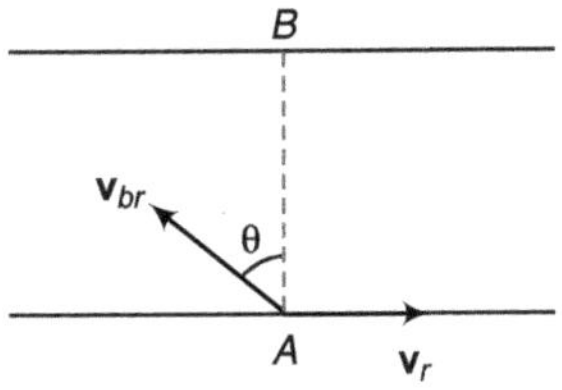

Fig. 3.36

$\therefore$ $x = 0$

or $(v_r - v_{br}\sin\theta)\dfrac{w}{v_{br}\cos\theta} = 0$

or $v_r = v_{br}\sin\theta$

or $\sin\theta = \dfrac{v_r}{v_{br}}$

or $\theta = \sin^{-1}\left(\dfrac{v_r}{v_{br}}\right)$

Hence, to reach point B the boatman should row at an angle $\theta = \sin^{-1}\left(\dfrac{v_r}{v_{br}}\right)$ upstream from AB.

Further, since $\sin\theta \not> 1$.

So, if $v_r \geq v_{br}$, the boatman can never reach at point B. Because if $v_r = v_{br}$, $\sin\theta = 1$ or $\theta = 90°$ and it is just impossible to reach at B if $\theta = 90°$. Moreover it can be seen that $v_b = 0$ if $v_r = v_{br}$ and $\theta = 90°$. Similarly, if $v_r > v_{br}$, $\sin\theta > 1$, i.e. no such angle exists. Practically it can be realized in this manner that it is not possible to reach at B, if river velocity (v_r) is too high.

Example 3.26 *A man can row a boat with 4 km/h in still water. If he is crossing a river where the current is 2 km/h.*

(a) In what direction will his boat be headed, if he wants to reach a point on the other bank, directly opposite to starting point?

(b) If width of the river is 4 km, how long will the man take to cross the river, with the condition in part (a)?

(c) In what direction should he head the boat, if he wants to cross the river in shortest time and what is this minimum time?

(d) How long will it take him to row 2 km up the stream and then back to his starting point?

Sol. (a) Given, that $v_{br} = 4$ km/h and $v_r = 2$ km/h

$$\therefore \quad \theta = \sin^{-1}\left(\frac{v_r}{v_{br}}\right) = \sin^{-1}\left(\frac{2}{4}\right) = \sin^{-1}\left(\frac{1}{2}\right) = 30°$$

Hence, to reach the point directly opposite to starting point he should head the boat at an angle of 30° with AB or 90° + 30° = 120° with the river flow.

(b) Time taken by the boatman to cross the river

w = width of river = 4 km

$v_{br} = 4$ km/h and $\theta = 30°$

$$\therefore \quad t = \frac{4}{4\cos 30°} = \frac{2}{\sqrt{3}}\text{ h}$$

(c) For shortest time $\theta = 0°$

and $t_{min} = \dfrac{w}{v_{br}\cos 0°} = \dfrac{4}{4} = 1$ h

Hence, he should head his boat perpendicular to the river current for crossing the river in shortest time and this shortest time is 1 h.

(d) $t = t_{CD} + t_{DC}$

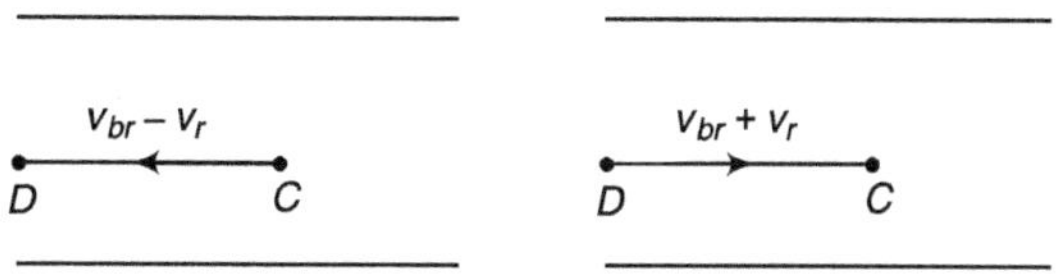

Fig. 3.37

or $$t = \frac{CD}{v_{br} - v_r} + \frac{DC}{v_{br} + v_r}$$

$$= \frac{2}{4-2} + \frac{2}{4+2} = 1 + \frac{1}{3} = \frac{4}{3}\ \text{h}$$

Aircraft Wind Problems

This is similar to river boat problem. The only difference is that $\mathbf{v}_{br}$ is replaced by $\mathbf{v}_{aw}$ (velocity of aircraft with respect to wind or velocity of aircraft in still air), $\mathbf{v}_r$ is replaced by $\mathbf{v}_w$ (velocity of wind) and $\mathbf{v}_r$ is replaced by $\mathbf{v}_a$ (absolute velocity of aircraft). Further, $\mathbf{v}_a = \mathbf{v}_{aw} + \mathbf{v}_w$. The following example will illustrate the theory.

Example 3.27 *An aircraft flies at 400 km/h in still air. A wind of $200\sqrt{2}$ km/h is blowing from the south. The pilot wishes to travel from A to a point B north-east of A. Find the direction he must steer and time of his journey if $AB = 1000$ km.*

Sol. Given that $v_w = 200\sqrt{2}$ km/h.

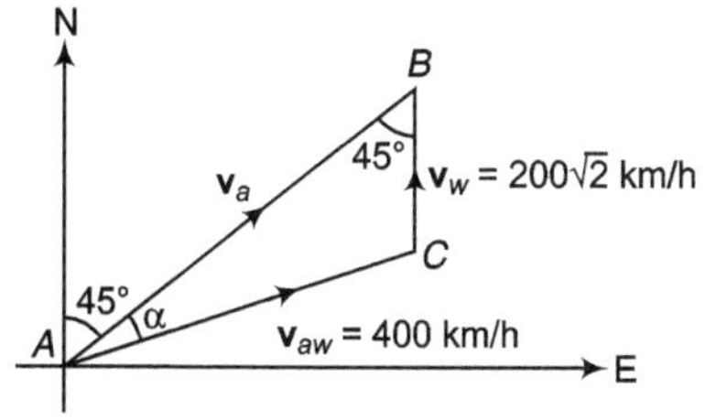

Fig. 3.38

$\mathbf{v}_{aw} = 400$ km/h and $\mathbf{v}_a$ should be along AB or in north-east direction. Thus, the direction of $\mathbf{v}_{aw}$ should be such as the resultant of $\mathbf{v}_w$ and $\mathbf{v}_{aw}$ is along AB or in north-east direction.

Let $\mathbf{v}_{aw}$ makes an angle α with AB as shown in Fig. 3.38. Applying sine law in triangle ABC, we get

$$\frac{AC}{\sin 45^\circ} = \frac{BC}{\sin\alpha}$$

or $$\sin\alpha = \left(\frac{BC}{AC}\right)\sin 45^\circ = \left(\frac{200\sqrt{2}}{400}\right)\frac{1}{\sqrt{2}} = \frac{1}{2}$$

$\therefore$ $$\alpha = 30^\circ$$

Therefore, the pilot should steer in a direction at an angle of $(45^\circ + \alpha)$ or 75° from north towards east.

Further, $$\frac{|\mathbf{v}_a|}{\sin(180^\circ - 45^\circ - 30^\circ)} = \frac{400}{\sin 45^\circ}$$

or $$|\mathbf{v}_a| = \frac{\sin 105^\circ}{\sin 45^\circ} \times (400)\ \text{km/h}$$

$$= \left(\frac{\cos 15^\circ}{\sin 45^\circ}\right)(400)\ \text{km/h}$$

$$= \left(\frac{0.9659}{0.707}\right)(400)\ \text{km/h}$$

$$= 546.47\ \text{km/h}$$

$\therefore$ The time of journey from A to B is

$$t = \frac{AB}{|\mathbf{v}_a|} = \frac{1000}{546.47}\ \text{h}$$

$$t = 1.83\ \text{h}$$

Rain Problems

In these type of problems, we again come across three terms $\mathbf{v}_r$, $\mathbf{v}_m$ and $\mathbf{v}_{rm}$, Here

$\mathbf{v}_r$ = velocity of rain

$\mathbf{v}_m$ = velocity of man (it may be velocity of cyclist or velocity of motorist also).

and $\mathbf{v}_{rm}$ = velocity of rain with respect to man.

Here, $\mathbf{v}_{rm}$ is the velocity of rain which appears to the man. Now, let us take one example of this.

Example 3.28 *To a man walking at the rate of 3 km/h the rain appears to fall vertically. When he increases his speed to 6 km/h it appears to meet him at an angle of 45° with vertical. Find the speed of rain.*

Sol. Let $\hat{\mathbf{i}}$ and $\hat{\mathbf{j}}$ be the unit vectors in horizontal and vertical directions respectively.

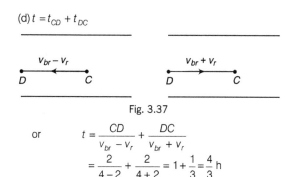

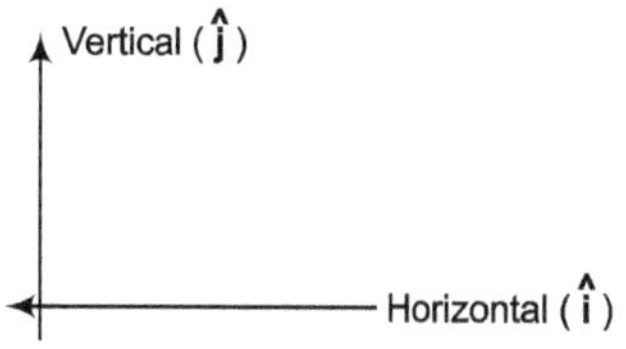

Fig. 3.39

Let velocity of rain

$$\mathbf{v}_r = a\hat{\mathbf{i}} + b\hat{\mathbf{j}} \quad \text{...(i)}$$

Then speed of rain will be

$$|\mathbf{v}_r| = \sqrt{a^2 + b^2} \quad \text{...(ii)}$$

In the first case, $\mathbf{v}_m$ = velocity of man = $3\hat{\mathbf{i}}$

$\therefore$ $$\mathbf{v}_{rm} = \mathbf{v}_r - \mathbf{v} = (a-3)\hat{\mathbf{i}} + b\hat{\mathbf{j}}$$

It seems to be in vertical direction. Hence,

$$a - 3 = 0 \quad \text{or} \quad a = 3$$

In the second case $\mathbf{v}_m = 6\hat{\mathbf{i}}$

$\therefore$ $$\mathbf{v}_{rm} = (a-6)\hat{\mathbf{i}} + b\hat{\mathbf{j}} = -3\hat{\mathbf{i}} + b\hat{\mathbf{j}}$$

This seems to be at 45° with vertical.

Hence, $|b| = 3$

Therefore, from Eq. (ii) speed of rain is

$$|\mathbf{v}_r| = \sqrt{(3)^2 + (3)^2} = 3\sqrt{2}\ \text{km/h}$$

Extra Knowledge Points

- If y (may be velocity, acceleration etc) is a function of time or $y = f(t)$ and we want to find the average value of y between a time interval of t_1 and t_2. Then,

$$< y >_{t_1 \text{ to } t_2} = \text{average value of } y \text{ between } t_1 \text{ and } t_2$$

$$= \frac{\int_{t_1}^{t_2} f(t)\, dt}{t_2 - t_1} \quad \text{or} \quad < y >_{t_1 \text{ to } t_2} = \frac{\int_{t_1}^{t_2} f(t)\, dt}{t_2 - t_1}$$

But if $f(t)$ is a linear function of t, then

$$y_{av} = \frac{y_f + y_i}{2}$$

Here, y_f = final value of y and

y_i = initial value of y

At the same time, we should not forget that

$$v_{av} = \frac{\text{total displacement}}{\text{total time}}$$

and $$a_{av} = \frac{\text{change in velocity}}{\text{total time}}$$

- **Example** In one dimensional uniformly accelerated motion, find average velocity from $t = 0$ to $t = t$.

Solution We can solve this problem by three methods.

Method 1. $v = u + at$

$$\therefore \quad < v >_{0-t} = \frac{\int_0^t (u + at)\, dt}{t - 0} = u + \frac{1}{2} at$$

Method 2. Since, v is a linear function of time, we can write,

$$v_{av} = \frac{v_f + v_i}{2} = \frac{(u + at) + u}{2} = u + \frac{1}{2} at$$

Method 3. $$v_{av} = \frac{\text{Total displacement}}{\text{Total time}} = \frac{ut + \frac{1}{2} at^2}{t} = u + \frac{1}{2} at$$

A particle is thrown upwards with velocity u. Suppose it takes time t to reach its highest point, then distance travelled in last second is independent of u.

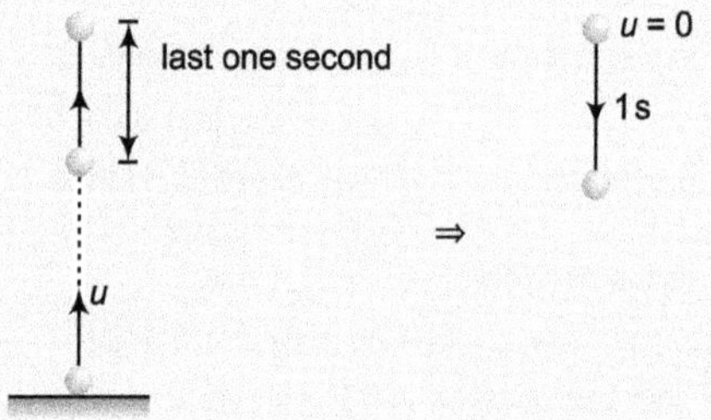

This is because this distance is equal to the distance travelled in first second of a freely falling object. Thus,

$$s = \frac{1}{2} g \times (1)^2 = \frac{1}{2} \times 10 \times 1 = 5 \text{ m}$$

Exercise A particle is thrown upwards with velocity u (> 20 m/s). Prove that distance travelled in last 2 s is 20 m.

- Suppose we have given velocity-time (v-t) graph. We want to plot corresponding displacement-time (s-t) graph then values of displacements at different times can be found just by adding the corresponding areas under v-t graph.
- The modulus of velocity is really the speed or $|\mathbf{v}| = v$
- Rate of change of velocity is acceleration, while rate of change of speed is the tangential acceleration (component of acceleration along velocity). Thus,

$$\frac{d\mathbf{v}}{dt} = \mathbf{a}, \text{ while } \frac{dv}{dt} = \frac{d|\mathbf{v}|}{dt} = a_t$$

Angle between velocity vector and acceleration vector **a** decides whether the speed of particle is increasing, decreasing or constant.

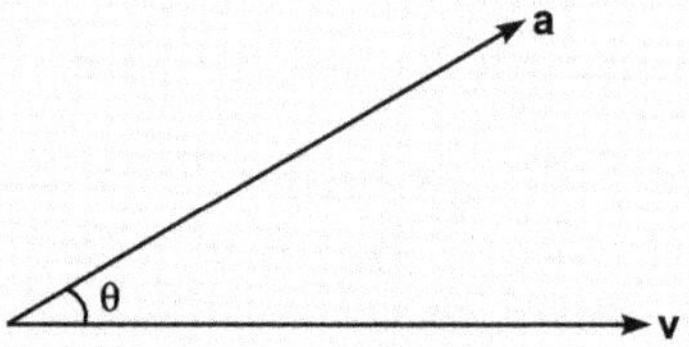

Speed increases if $0° \le \theta < 90°$

Speed decreases if $90° < \theta \le 180°$

Speed is constant if $\theta = 90°$

The angle θ between **v** and **a** can be obtained by the relation,

$$\theta = \cos^{-1}\left(\frac{\mathbf{v} \cdot \mathbf{a}}{va}\right)$$

Exercise Prove that speed of a particle increases if dot product of **v** and **a** is positive, speed decreases if the dot product is negative and speed remains constant, if dot product is zero.

- In the x-t graph of a particle moving along x-axis shown in figure average velocity of the particle between time interval t_1 and t_2 is :

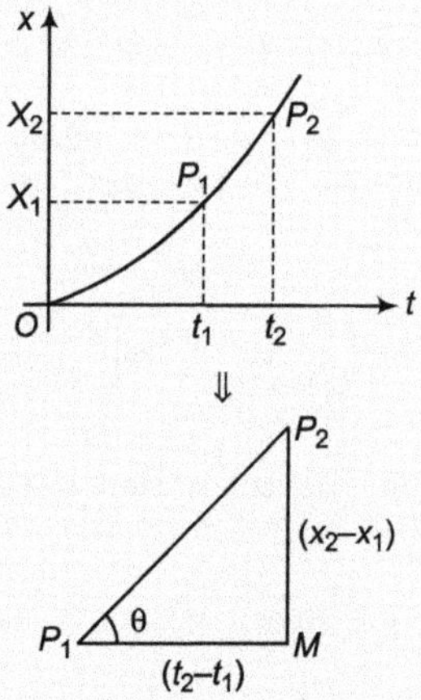

$$v_{av} = \frac{\Delta s}{\Delta t} = \frac{\Delta x}{\Delta t} = \frac{x_f - x_i}{t_2 - t_1} = \frac{x_2 - x_1}{t_2 - t_1} = \text{Slope of line}$$

$$P_1P_2 = \tan\theta$$

- From displacement time graph we can determine the sign of velocity its nature (whether speed of the particle is increasing, decreasing or constant) and sign of acceleration. For example

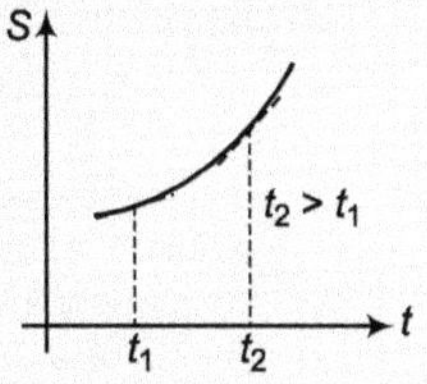

Slope of S-t graph = velocity.

Slopes at t_1 and t_2 both are positive. Hence, velocities at t_1 and t_2 are also positive. Further,

Slope at t_2 >Slope at t_1

$\therefore$ Velocity at t_2 >velocity at t_1 or,

speed of the particle is increasing, since speed is increasing, so acceleration should have the same sign as that of velocity. Or, acceleration is also positive.

Exercise What are the signs of accelerations in the three position-time graphs shown in figure.

Solution. (a) positive (b) negative (c) zero

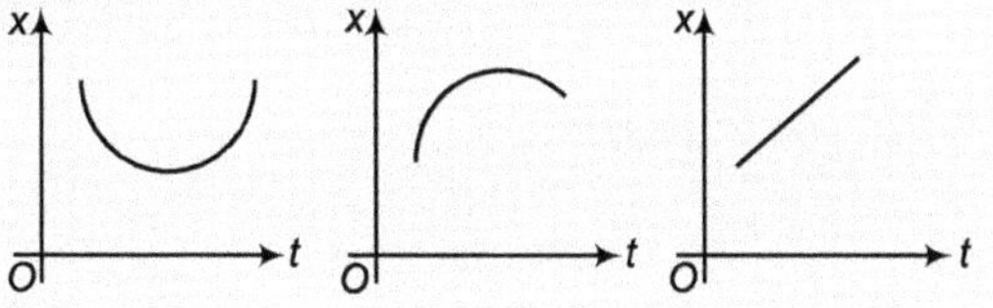

(a) (b) (c)

Reaction time

When a situation demands our immediate action, it takes some time before we really respond. Reaction time is the time a person takes to observe, think and act. For example, if a person is driving and suddenly a boy appears on the road, then the time elapsed before he slams the brakes of the car is the reaction time.

- If a particle is speeding up, acceleration is in the direction of velocity; if its speed is decreasing, acceleration is in the direction opposite to that of the velocity. This statement is independent of the choice of the origin and the axis.
- The sign of acceleration does not tell us whether the particle's speed is increasing or decreasing. Ths sign of acceleration depends on the choice of the positive direction of the axis. For example, if the vertically upward direction is chosen to be the positive direction, the acceleration due to gravity is negative. If a particle is falling under gravity, this acceleration, though negative, results in increase in speed. For a particle thrown upward, the same negative acceleration (of gravity) results in decrease in speed.
- The zero velocity of a particle at any instant does not necessarily imply zero acceleration at that instant. A particle may be momentarily at rest and yet have non-zero acceleration. For example, a particle thrown up has zero velocity at its highest point but the acceleration at that instant continues to be the acceleration due to gravity.

Chapter Summary with Formulae

- **Basic Definition**

(i) Displacement **a**

$$\mathbf{s} = \mathbf{r}_f - \mathbf{r}_i$$
$$= (x_f - x_i)\hat{\mathbf{i}} + (y_f - y_i)\hat{\mathbf{j}} + (z_f - z_i)\hat{\mathbf{k}}$$

(ii) Distance = actual path length

(iii) Average velocity

$$= \frac{\text{total displacement}}{\text{total time}} = \frac{s}{t}$$

(iv) Average speed

$$= \frac{\text{total distance}}{\text{total time}} = \frac{d}{t}$$

(v) Average acceleration

$$= \frac{\text{change in velocity}}{\text{time}} = \frac{\mathbf{v}_f - \mathbf{v}_i}{t}$$

(vi) Instantaneous velocity

$$= \frac{d\mathbf{s}}{dt} \text{ or } \frac{d\mathbf{r}}{dt}$$

(vii) Instantaneous acceleration

$$= \text{rate of change of velocity}$$
$$= \frac{d\mathbf{v}}{dt} = \frac{d^2\mathbf{s}}{dt^2} = \frac{d^2\mathbf{r}}{dt^2}$$

- **In One Dimensional Motion**

(viii) Instantaneous velocity

$$v = \frac{ds}{dt} \text{ or } \frac{dx}{dt}$$
$$= \text{slope of } v\text{-}t \text{ graph}$$

- **One Dimensional Motion with Uniform Acceleration**

(i) $v = u + at$

(ii) $s = ut + \frac{1}{2}at^2$

(iii) $s = s_0 + ut + \frac{1}{2}at^2$

(iv) $v^2 = u^2 + 2as$

(v) s_t = displacement (not distance) in t^{th} second

$$= (u + at) - \frac{a}{2}$$

- While using above equations, take a sign convention and substitute all vector quantities (v, u, a, s and s_t) with sign.
- In equation $s = ut + \frac{1}{2}at^2$, s is the displacement measured from the starting point.
- s_t is the displacement between $(t-1)$ and t seconds.

- **One Dimensional Motion with Non-uniform Acceleration**

(i) s-$t \rightarrow v$-$t \rightarrow a$-$t \rightarrow$ Differentiation

(ii) a-$t \rightarrow v$-$t \rightarrow s$-$t \rightarrow$ Integration

(iii) Equations of differentiation

$$v = \frac{ds}{dt} \quad \text{and} \quad = \frac{dv}{dt} = v.\frac{dv}{ds}$$

(iv) Equations of integration

$$\int ds = \int v dt, \int dv = \int a dt, \int v dv = \int a ds$$

In first integration equation v should be either a constant or function of t. In second equation 'a' should be either a constant or a function of t. Similarly, in third equation 'a' should be either a constant or function of s.

- **Two or Three Dimensional Motion with Uniform Acceleration**

(i) $\mathbf{v} = \mathbf{u} + \mathbf{a}t$

(ii) $\mathbf{s} = \mathbf{u}t + \frac{1}{2}\mathbf{a}t^2$

(iii) $\mathbf{v} \cdot \mathbf{v} = \mathbf{u} \cdot \mathbf{u} + 2\mathbf{a} \cdot \mathbf{s}$

- **Two or Three Dimensional Motion with Non-uniform Acceleration**

(i) $\mathbf{v} = \frac{d\mathbf{s}}{dt}$ or $\frac{d\mathbf{r}}{dt}$

(ii) $\mathbf{a} = \frac{d\mathbf{v}}{dt}$

(iii) $\int d\mathbf{v} = \int \mathbf{a} \cdot dt$

(iv) $\int d\mathbf{s} = \int \mathbf{v} \cdot dt$

- **Graphs**

(i) Uniform motion v = constant, $a = 0, s = vt$

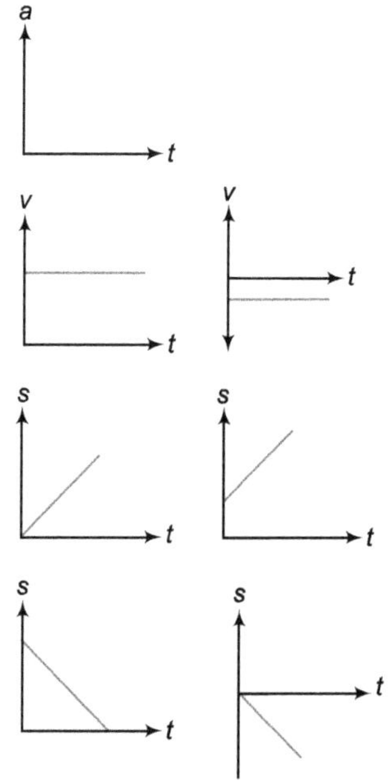

Since, $a = 0$, therefore slope of v-t graph $= 0$. Further, v = constant, therefore, slope of s-t graph = constant.

(ii) **Uniformly accelerated or retarded motion**

$a = \text{constant}$, $v = u \pm at$,

$$s = ut \pm \frac{1}{2}at^2$$

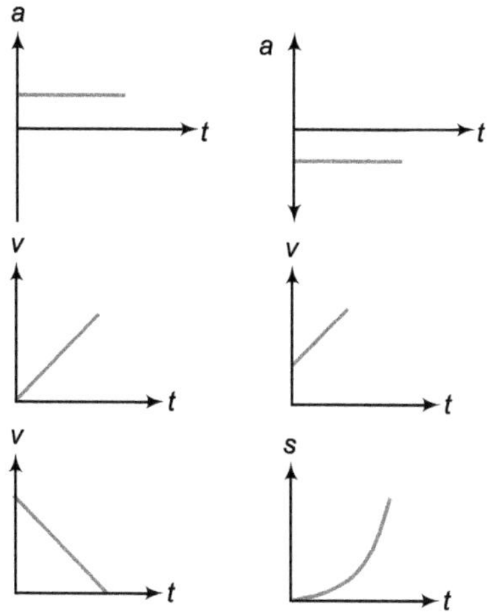

or

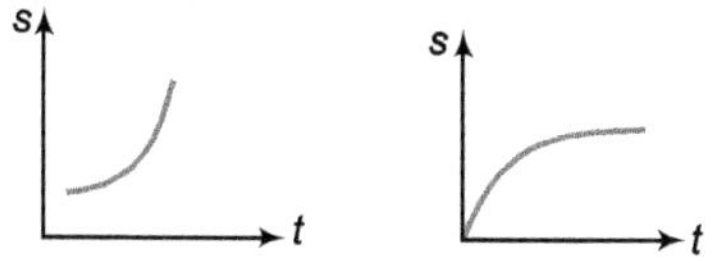

Since a = constant, therefore slope of v-t graph = constant.

Further, v is increasing or decreasing, therefore slope of s-t graph should either increase or decrease.

(iii) Slope of s-t graph = velocity

Slope of v-t graph = acceleration

Area under v-t graph = displacement and

Area under a-t graph = change in velocity

- **Relative Motion**

(i) $\mathbf{v}_{AB}$ = velocity of A with respect to $B = \mathbf{v}_A - \mathbf{v}_B$

(ii) $\mathbf{a}_{AB}$ = acceleration of A with respect to $B = \mathbf{a}_A - \mathbf{a}_B$.

In one dimensional motion take a sign convention.

In this case,

(iii) $v_{AB} = v_A - v_B$

(iv) $a_{AB} = a_A - a_B$

Additional Examples

Example 1. *A farmer has to go 500 m due north, 400 m due east and 200 m due south to reach his field. If he takes 20 min to reach the field,*

(a) what distance he has to walk to reach the field ?

(b) what is the displacement from his house to the field ?

(c) what is the average speed of farmer during the walk ?

(d) what is the average velocity of farmer during the walk ?

Sol. (a) Distance $= AB + BC + CD$

$$= (500 + 400 + 200)$$
$$= 1100 \text{ m}$$

(b) Displacement $= AD = \sqrt{(AB - CD)^2 + BC^2}$

$$= \sqrt{(500 - 200)^2 + (400)^2}$$
$$= 500 \text{ m}$$

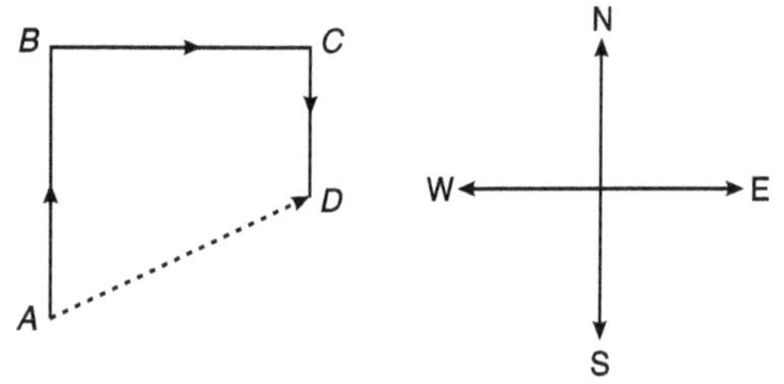

(c) Average speed $= \dfrac{\text{Total distance}}{\text{Total time}}$

$$= \frac{1100}{20} = 55 \text{ m/min}$$

(d) Average velocity $= \dfrac{AD}{t}$

$$= \frac{500}{20} = 25 \text{m/min (along } AD)$$

Example 2. *A particle starts with an initial velocity and passes successively over the two halves of a given distance with accelerations a_1 and a_2 respectively. Show that the final velocity is the same as if the whole distance is covered with a uniform acceleration $\dfrac{(a_1 + a_2)}{2}$.*

Sol. In the first case,

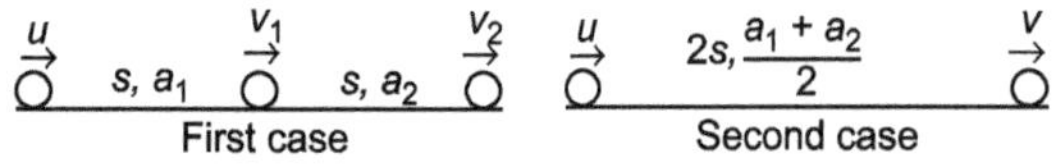

$$v_1^2 = u^2 + 2a_1 s \qquad \dots\text{(i)}$$
$$v_2^2 = v_1^2 + 2a_2 s \qquad \dots\text{(ii)}$$

Adding Eqs. (i) and (ii), we have

$$v_2^2 = u^2 + 2\left(\frac{a_1 + a_2}{2}\right)(2s) \qquad \dots\text{(iii)}$$

In the second case,

$$v^2 = u^2 + 2\left(\frac{a_1 + a_2}{2}\right)(2s) \qquad \dots\text{(iv)}$$

From Eqs. (iii) and (iv), we can see that

$$v_2 = v$$

Example 3. *In a car race, car A takes a time t less than car B at the finish and passes the finishing point with speed v more than that of the car B. Assuming that both the cars starts from rest and travel with constant acceleration a_1 and a_2 respectively. Show that $v = \sqrt{a_1 a_2}\, t$.*

Sol. Let A takes t_1 second, then according to the given problem B will take $(t_1 + t)$ seconds. Further, let v_1 be the velocity of B at finishing point, then velocity of A will be $(v_1 + v)$. Writing equations of motion for A and B.

$$v_1 + v = a_1 t_1 \qquad \dots\text{(i)}$$
$$v_1 = a_2(t_1 + t) \qquad \dots\text{(ii)}$$

From these two equations, we get

$$v = (a_1 - a_2)t_1 - a_2 t \qquad \dots\text{(iii)}$$

Total distance travelled by both the cars is equal.

or $\qquad s_A = s_B$

or $\qquad \dfrac{1}{2}a_1 t_1^2 = \dfrac{1}{2}a_2 (t_1 + t)^2$ or $t_1 = \dfrac{\sqrt{a_2}\, t}{\sqrt{a_1} - \sqrt{a_2}}$

Substituting this value of t_1 in Eq. (iii), we get the desired result.

or $\qquad v = (\sqrt{a_1 a_2})\, t$

Example 4. *A particle is moving with a velocity of $v = (3 + 6t + 9t^2)$ cm/s. Find out*

(a) the acceleration of the particle at $t = 3$ s.

(b) the displacement of the particle in the interval $t = 5$ s to $t = 8$ s.

Sol. (a) Acceleration of particle,

$$a = \frac{dv}{dt} = (6 + 18t) \text{ cm/s}^2$$

At $t = 3$ s, $\quad a = (6 + 18 \times 3) \text{ cm/s}^2 = 60 \text{cm/s}^2$

(b) Given, $\quad v = (3 + 6t + 9t^2)$

or $\quad \dfrac{ds}{dt} = (3 + 6t + 9t^2)$

or $\quad ds = (3 + 6t + 9t^2)\, dt$

$\therefore \quad \displaystyle\int_0^s ds = \int_5^8 (3 + 6t + 9t^2)\, dt$

$\therefore \quad s = [3t + 3t^2 + 3t^3]_5^8$

or $\quad s = 1287$ cm

Example 5. *The motion of a particle along a straight line is described by the function $x=(2t-3)^2$, where x is in metre and t is in second.*

(a) Find the position, velocity and acceleration at $t=2\,s$.

(b) Find velocity of the particle at origin.

Sol. (a) Position, $x=(2t-3)^2$

Velocity, $v=\dfrac{dx}{dt}=4\,(2t-3)\ \text{m/s}$

and acceleration,

$$a=\frac{dv}{dt}=8\ \text{m/s}^2$$

At $t=2$ s,

$$x=(2\times 2-3)^2=1.0\ \text{m}$$

$$v=4\,(2\times 2-3)=4\ \text{m/s}$$

and $a=8\ \text{m/s}^2$

(b) At origin, $x=0$

or $(2t-3)=0$

$\therefore$ $v=4\times 0=0$

Example 6. *An open elevator is ascending with zero acceleration. The speed $v=10$ m/s. A ball is thrown vertically up by a boy when he is at a height $h=10$ m from the ground. The velocity of projection is $v=30$ m/s with respect to elevator. Find*

(a) the maximum height attained by the ball.

(b) the time taken by the ball to meet the elevator again.

(c) time taken by the ball to reach the ground after crossing the elevator.

Sol. (a) Absolute velocity of ball = 40 m/s (upwards)

$\therefore$ $h_{max}=h_i+h_f$

Here, h_i = initial height = 10 m

and h_f = further height attained by ball

$$=\frac{u^2}{2g}=\frac{(40)^2}{2\times 10}=80\ \text{m}$$

$\therefore$ $h_{max}=(10+80)\ \text{m}=90\ \text{m}$

(b) The ball will meet the elevator again when displacement of lift = displacement of ball

or $$10\times t=40\times t-\frac{1}{2}\times 10\times t^2$$

or $t=6$ s

(c) Let t_0 be the total time taken by the ball to reach the ground. Then,

$$-10=40\times t_0-\frac{1}{2}\times 10\times t_0^2$$

Therefore, time taken by the ball to reach the ground after crossing the elevator $=(t_0-t)=2.24$ s

Example 7. *From an elevated point A, a stone is projected vertically upwards. When the stone reaches a distance h below A, its velocity is double of what it was at a height h above A. Show that the greatest height attained by the stone is $\frac{5}{3}h$.*

Sol. Let u be the velocity with which the stone is projected vertically upwards.

Given that, $v_{-h}=2v_h$

or $(v_{-h})^2=4v_h^2$

$\therefore$ $u^2-2g(-h)=4(u^2-2gh)$

$\therefore$ $u^2=\dfrac{10gh}{3}$

Now, $h_{max}=\dfrac{u^2}{2g}=\dfrac{5h}{3}$

Example 8. *A man crosses a river in a boat. If he cross the river in minimum time he takes 10 min with a drift 120 m. If he crosses the river taking shortest path, he takes 12.5 min, find*

(a) width of the river.

(b) velocity of the boat with respect to water.

(c) speed of the current.

Sol. Let v_r = velocity of river,

v_{br} = velocity of river in still water and

w = width of river

Given, $t_{min}=10$ min or $\dfrac{w}{v_{br}}=10$ ...(i)

Drift in this case will be,

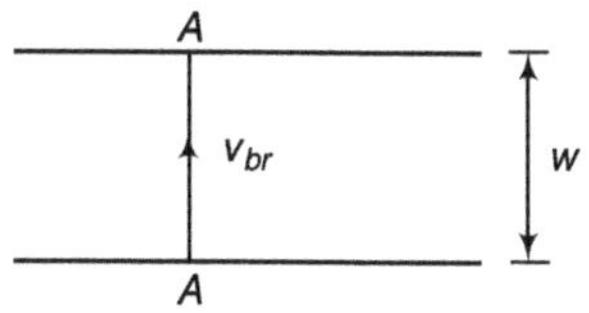

For minimum time

$x=v_r t\ \Rightarrow\ \therefore\ 120=10\,v_r$ ...(ii)

Shortest path is taken when v_b is along AB. In this case,

$$v_b=\sqrt{v_{br}^2-v_r^2}$$

Now, $12.5=\dfrac{w}{v_b}=\dfrac{w}{\sqrt{v_{br}^2-v_r^2}}$ ...(iii)

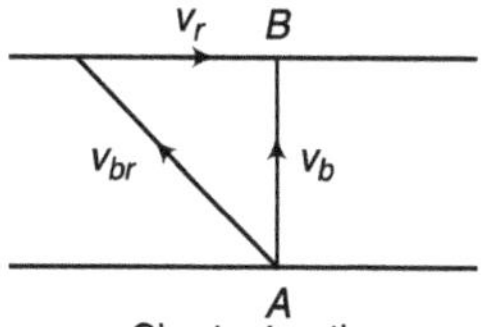

Shortest path

Solving these three equations, we get

$v_{br}=20$ m/min, $v_r=12$ m/min and $w=200$ m.

Example 9. *A man wants to reach point B on the opposite bank of a river flowing at a speed as shown in figure. What minimum speed relative to water should the man have so that he can reach point B ? In which direction should he swim?*

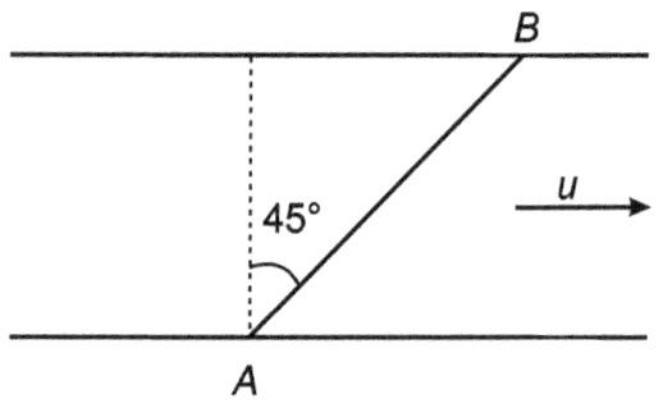

Sol. Let v be the speed of boatman in still water.

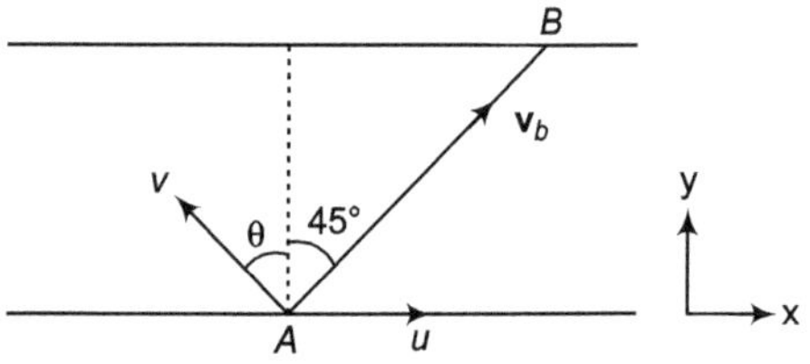

Resultant of v and u should be along AB. Components of $\mathbf{v}_b$ (absolute velocity of boatman) along x and y directions are,

$$v_x = u - v\sin\theta \quad \text{and} \quad v_y = v\cos\theta$$

Further, $\tan 45° = \dfrac{v_y}{v_x}$ or $1 = \dfrac{v\cos\theta}{u - v\sin\theta}$

$$\therefore \qquad v = \frac{u}{\sin\theta + \cos\theta} = \frac{u}{\sqrt{2}\sin(\theta + 45°)}$$

v is minimum at,

$$\theta + 45° = 90°$$

or $$\theta = 45°$$

and $$v_{\min} = \frac{u}{\sqrt{2}}$$

Example 10. *The acceleration versus time graph of a particle moving along a straight line is shown in the figure. Draw the respective velocity-time graph. (Assuming at $t = 0, v = 0$.)*

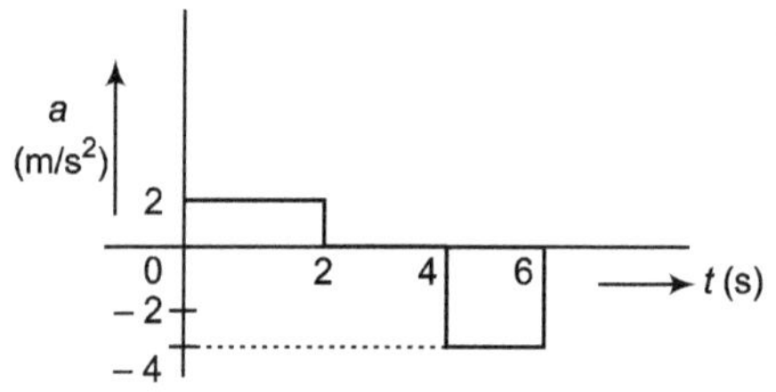

Sol. From $t = 0$ to $t = 2$ s, $a = +\,2$ m/s^2

$$\therefore \qquad v = at = 2t$$

or v-t graph is a straight line passing through origin with slope 2 m/s^2.

At the end of 2 s,

$$v = 2 \times 2 = 4 \text{ m/s}$$

From $t = 2$ to 4 s, $a = 0$.

Hence, $v = 4$ m/s will remain constant.

From $t = 4$ to 6 s, $a = -\,4$ m/s^2.

Hence, $\qquad v = u - at = 4 - 4t \qquad$ (with $t = 0$ at 4 s)

$v = 0$ at $t = 1$ s or at 5 s from origin.

At the end of 6 s (or $t = 2$ s) $v = -\,4$ m/s.

Corresponding v-t graph is as shown below.

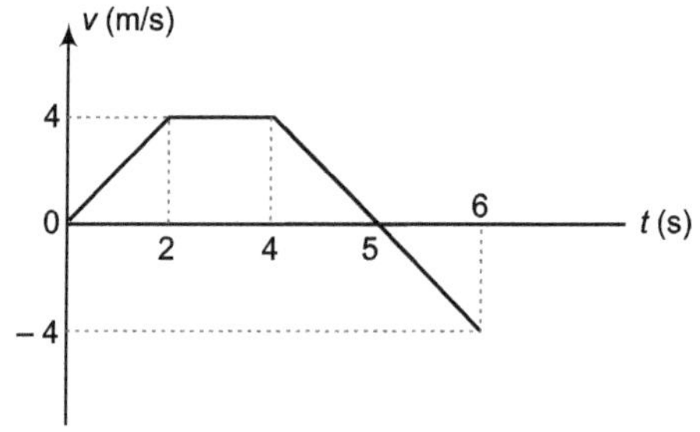

NCERT Selected Questions

Q 1. The position-time (x-t) graphs for two children A and B returning from their school O to their homes P and Q respectively are shown in figure. Choose the correct entries in the brackets below :

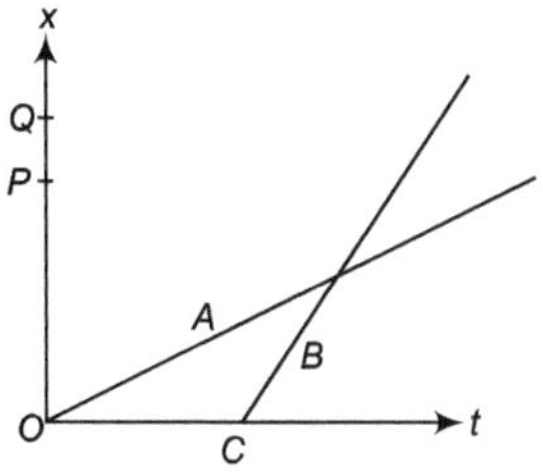

(a) (A/B) lives closer to the school than (B/A)
(b) (A/B) starts from school earlier than (B/A)
(c) (A/B) walks faster than (B/A)
(d) (A/B) overtakes (B/A) on the road (once/twice)

Sol. (a) It is clear from the graph that $OQ > OP$, so A lives closer to the school than B.

(b) The position-time (x-t) graph of A starts from the origin, so $x = 0$, $t = 0$ for A while the x-t graph of B starts from C which shows that B starts later than A after a time interval OC. So, A starts from school (O) earlier than B.

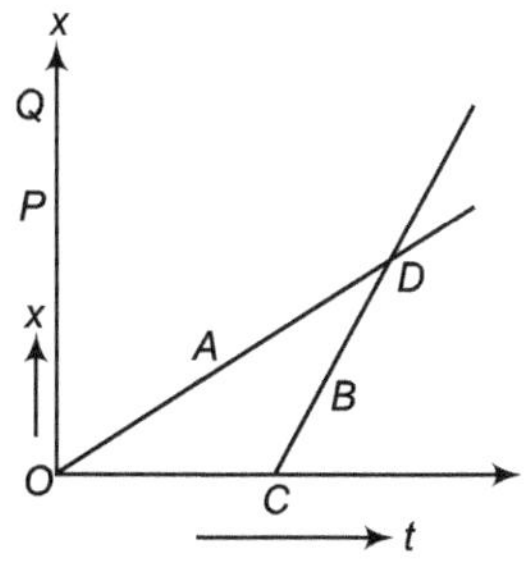

(c) The speed is represented by the magnitude of slope of the x-t graph. More steeper the graph, more will be the speed. As, the x-t graph of B is steeper than the x-t graph of A, so we conclude that B walks faster than A.

(d) As the x-t graphs for A and B intersect each other at point D only and B starts from the school later, so B over takes A on the road only once.

Q 2. A woman starts from her home at 9.00 am, walks with a speed of 5 kmh^{-1} on a straight road upto her office 2.5 km away, stays at the office upto 5.00 pm and returns home by an auto with a speed of 25 kmh^{-1}. Choose suitable scales and plot the x-t graph of her motion.

Sol. x-t graph of the motion of woman is shown in figure,

v_1 = speed of woman while walking at 5 kmh^{-1}

x = distance between home and office = 2.5 km

If t_1 is time taken to reach office then

$$t_1 = \frac{x}{v_1}$$

$$\therefore \quad t_1 = \frac{2.5}{5} = \frac{1}{2}\text{ h}$$

$$= 30 \text{ min}$$

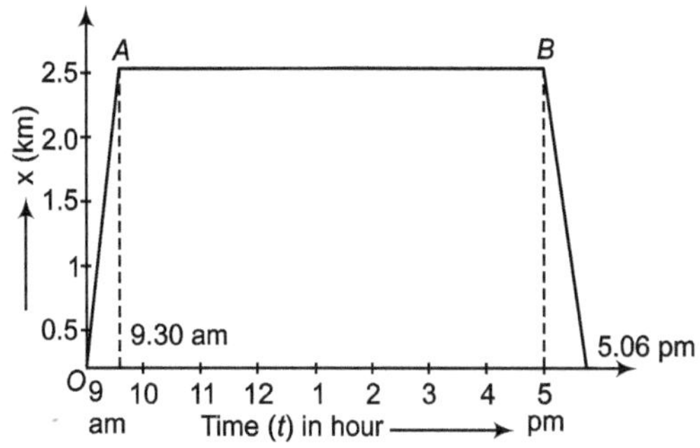

When she stays at her office from 9.30 am to 5.00 pm, then she is stationary.

On return journey, speed of auto, $v_2 = 25$ km/h.

$\therefore$ If t_2 is time taken by her in return journey from office to her home, then

$$t_2 = \frac{x}{v_2} = \frac{2.5}{25} = \frac{1}{10}\text{ h} = 6\text{ min}$$

Thus, she reaches back to her home at 5.06 pm.

Q 3. A drunkard walking in a narrow lane takes 5 steps forward and 3 steps backward, followed again by 5 steps forward and 3 steps backward and so on. Each step is 1 m long and requires 1 s. Plot the x-t graph of his motion. Determine graphically and otherwise how long the drunkard takes to fall in a pit 13 m away from the start.

Sol. The x-t graph of the drunkard is shown in figure.

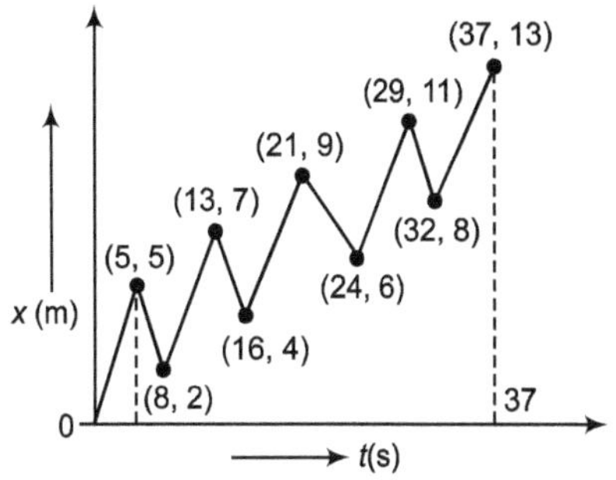

Length of each step = 1 m, time taken for each step = 1 s.

$\therefore$ Time taken to move by 5 steps = 5 s.

5 steps forward and 3 steps backward means that the net distance covered by him in first 8 steps is 2 m.

Distance covered by him in first 16 steps or $16\text{s} = 2 + 2 = 4\text{m}$.

Distance covered by the drunkard in first 24 s, i.e. 24 steps $= 2 + 2 + 2 = 6\text{m}$.

and distance covered in 32 steps i.e. 32 s = 8 m.

$\therefore$ Distance covered in first 37 steps = 8 + 5 = 13 m.

Distance of the pit from the start = 13 m.

$\therefore$ Total time taken by the drunkard to fall in the pit = 37 s.

***Q* 4.** A jet airplane travelling at a speed of $500\ \text{kmh}^{-1}$ ejects its products of combustion at the speed of $1500\ \text{kmh}^{-1}$ relative to the jet plane. What is the speed of the combustion with respect to an observer on the ground?

Sol. Let v_j, v_g and v_0 be the velocities of jet, ejected gases and observer on the ground respectively.

Let jet be moving towards right (+ ve direction).

$\therefore$ Ejected gases will move towards left (− ve direction).

$\therefore$ According to the statement

$$v_j - v_0 = v_j = 500\ \text{kmh}^{-1} \quad \text{...(i)}$$

As observer is at rest

$$v_g - v_j = -1500\ \text{kmh}^{-1} \text{ (given)} \quad \text{...(ii)}$$

$\therefore$ Adding Eqs. (i) and (ii), we get the speed of combustion products w.r.t. observer on the ground

$$(v_j - v_0)\ (v_g - v_j) = v_g - v_0 = 500 + (-1500)$$

or $$v_g - v_0 = -1000\ \text{kmh}^{-1}$$

−ve sign shows that relative velocity of the ejected gases w.r.t. observer is towards left or in a direction opposite to the motion of the jet plane.

***Q* 5.** A car moving along a straight highway with speed of $126\ \text{kmh}^{-1}$ is brought to a stop within a distance of 200 m. What is the retardation of the car (assumed uniform) and how long does it take for the car to stop?

Sol. Initial velocity of car,

$$u = 126 \times \frac{5}{18}\ \text{ms}^{-1} = 35\ \text{ms}^{-1} \quad \text{...(i)}$$

As the car finally comes to rest, $v = 0$

Distance covered, $s = 200$ m

Using the equation $v^2 - u^2 = 2as$

$$\therefore \quad a = \frac{v^2 - u^2}{2s} \quad \text{...(ii)}$$

Putting the values from Eq. (i) in Eq. (ii), we get

$$a = \frac{0 - (35)^2}{2 \times 200}$$

$$= \frac{35 \times 35}{400} = \frac{-49}{15}\ \text{ms}^{-2}$$

$$= -3.06\ \text{ms}^{-2}$$

−ve sign shows that car is retarded.

To find t, let us use the relation,

$$v = u + at$$

Here, $a = -3.06\ \text{ms}^{-2}$, $v = 0$, $u = 35$ m/s

$$\therefore \quad t = \frac{v - u}{a} = \frac{0 - 35}{-3.06} = 11.44\ \text{s}$$

$$\therefore \quad t = 11.44\ \text{s}$$

***Q* 6.** Two trains A and B of length 400 m each are moving on two parallel tracks with a uniform speed of $72\ \text{kmh}^{-1}$ in the same direction, with A ahead of B. The driver of B decides to overtake A and accelerates by $1\ \text{ms}^{-2}$. If after 50 s, the guard of B just brushes past the driver of A, what was the original distance between them?

Sol. Given that

$$u_A = u_B = 72\ \text{kmh}^{-1} = 72 \times \frac{5}{18} = 20\ \text{ms}^{-1}$$

Using the relations, $s = ut + \frac{1}{2}at^2$, we get

$$s_B = u_B t + \frac{1}{2}at^2$$

$$= 20 \times 50 + \frac{1}{2} \times 1 \times (50)^2$$

$$= 1000 + 1250 = 2250\ \text{m}$$

Also let s_A be the distance covered by the train A, then

$$s_A = u_A \times t \quad \text{or} \quad s_A = 20 \times 50 = 1000\ \text{m}$$

Original distance between the two trains $= s_B - s_A$

$$= 2250 - 1000$$

$$= 1250\ \text{m}$$

***Q* 7.** On a two lane road, car A is travelling with a speed of $36\ \text{kmh}^{-1}$. Two cars B and C approach car A in opposite directions with a speed of $54\ \text{kmh}^{-1}$ each. At a certain instant, when the distance AB is equal to AC, both being 1 km, B decides to overtake A before C does. What minimum acceleration of car B is required of avoid an accident?

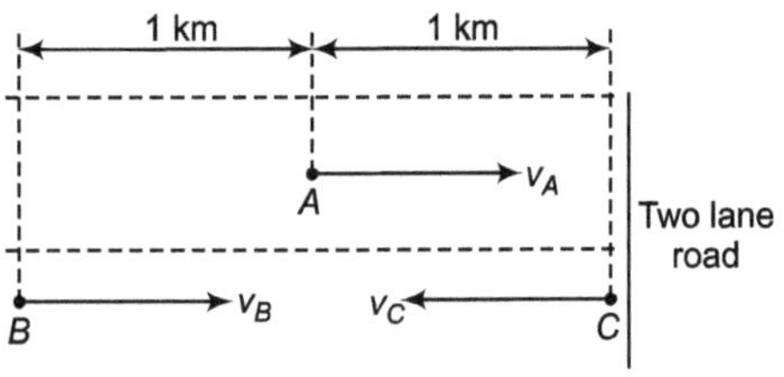

Sol. Speed of car A,

$$v_A = 36 \times \frac{5}{18} = 10\ \text{ms}^{-1}$$

Let v_B and v_C be the speeds of cars B and C

$$\therefore \quad v_B = v_C = 54\ \text{kmh}^{-1}$$

$$= 54 \times \frac{5}{18} = 15\ \text{ms}^{-1} \quad \text{(given)}$$

$\therefore$ Relative velocity of car B w.r.t. car A is given by

$$v_{BA} = v_B - v_A$$
$$= 15 - 10 = 5\ \text{ms}^{-1}$$

Also magnitude of relative velocity of car C w.r.t. car A is given by,

$$v_{CA} = v_C - (-v_A)$$
$$= 15 + 10 = 25\ \text{ms}^{-1}$$

$AB = AC = 1$ km (given) $= 1000$ m

Between cars A and C

$$t = \frac{s}{u} = \frac{AC}{v_{CA}} = \frac{1000}{25} = 40\ \text{s}$$

Let a = acceleration of car B for $t = 40$ s so it will cover 1000m in 40 s.

$\therefore$ Using the relation, $s = ut + \frac{1}{2}at^2$

We get,
$$AB = v_{BA}t + \frac{1}{2}at^2$$
$$1000 = 5 \times 40 + \frac{1}{2}a \times (40)^2$$
$$= 200 + a \times \frac{1600}{2}$$
$$800\,a = 800$$
$$a = 1\ \text{ms}^{-2}$$

Q 8. Two towns A and B are connected by a regular bus service with a bus leaving in either direction every T min. A man cycling with a speed of $20\ \text{kmh}^{-1}$ in the direction A to B notices that a bus goes past him every 18 min in the direction of his motion and every 6 min in opposite direction. What is the period T of the bus service and with what speed (assumed constant) do the buses play on the road?

Sol. Let the speed of each bus $= v_b\ \text{kmh}^{-1}$

and speed of cyclist $= v_c = 20\ \text{kmh}^{-1}$

Case I Relative speed of the buses playing in the direction of motion of cyclist i.e. from A to $B = v_b - v_c = (v_b - 20)\ \text{kmh}^{-1}$

Since, the bus goes past the cyclist every 18 min $\left(= \frac{18}{60}\text{ h}\right)$,

$\therefore$ Distance covered by the bus w.r.t. the cyclist

$$d_1 = (v_b - 20) \times \frac{18}{60}\text{ km} \quad \ldots\text{(i)}$$

Since, the bus leaves after every T min, so the distance covered by the bus in T min is given by

$$d_2 = v_b \times \frac{T}{60} \quad \ldots\text{(ii)}$$

$\therefore$ From Eqs. (i) and (ii), we get

$$(v_b - 20) \times \frac{18}{60} = v_b \times \frac{T}{60}$$

or
$$v_b - 20 = v_b \times \frac{T}{18} \quad \ldots\text{(iii)}$$

Case II Relative speed of the bus coming from town B to A w.r.t. cyclist $= (v_b + 20)\ \text{kmh}^{-1}$.

($\because$ Cyclist is moving from A to B).

Since, the bus goes past the cyclist after every 6 min.

$\therefore$ Distance covered by the bus w.r.t. the cyclist

$$d_3 = (v_b + 20) \times \frac{6}{60}\text{ km} \quad \ldots\text{(iv)}$$

Also distance covered by bus in T min is

$$d_4 = v_b \times \frac{T}{60}\text{ km} \quad \ldots\text{(v)}$$

$\therefore$ Equating Eqs. (iv) and (v), we get

$$(v_b + 20) \times \frac{6}{60} = v_b \times \frac{T}{60}$$

or
$$v_b + 20 = v_b \times \frac{T}{6} \quad \ldots\text{(vi)}$$

Dividing Eq. (vi) by Eq. (iii), we get

$$\frac{v_b + 20}{v_b - 20} = 3$$

or
$$v_b + 20 = 3\,v_b - 60$$

or
$$2v_b = 80 \quad \text{or} \quad v_b = 40\ \text{kmh}^{-1}$$

Putting the value of v_b in Eq. (iii), we get

$$40 - 20 = 40 \times \frac{T}{18}$$

or
$$20 = 40 \times \frac{T}{18}$$

or
$$T = 20 \times \frac{18}{40} = 9\ \text{min}$$

$\therefore$
$$v_b = 40\ \text{kmh}^{-1},$$
$$T = 9\ \text{min}$$

Q 9. A player throws a ball upwards with an initial speed of $29.4\ \text{ms}^{-1}$.

(a) What is the direction of acceleration during the upward motion of the ball?

(b) What are the velocity and acceleration of the ball at the highest point of its motion?

(c) Choose the $x = 0$ m and $t_0 = 0$ s to be the location and time of the ball at its highest point, vertically downward direction to the positive direction of x-axis, and give the signs, velocity and acceleration of the ball during its upward and downward motion.

Sol. (a) **Under gravity** The direction of acceleration due to gravity is always vertically downwards.

(b) At the highest point of its motion, its velocity becomes zero and the acceleration is equal to the acceleration due to gravity $= 9.8\ \text{ms}^{-2}$ in vertically downward direction.

(c) When the highest point is chosen as the location for $x = 0$ and $t = 0$ and vertically downward direction to be the positive direction of x-axis.

During upward motion, sign of velocity is negative and the sign of acceleration positive.

During downward motion, sign of velocity is positive and the sign of acceleration is also positive.

Q 10. Read each statement below carefully and state with reasons and examples, if it is true or false. A particle in one-dimensional motion

(a) with zero speed at an instant may have non-zero acceleration at that instant.

(b) with zero speed may have non-zero velocity.

(c) with constant speed must have zero acceleration.

(d) with positive value of acceleration must be speeding up.

Sol. (a) True, example : if the ball is thrown vertically upward, then it will have zero speed at the highest point and an acceleration of 9.8 ms^{-2} in downward direction.

(b) False, because speed is the magnitude of velocity.

(c) True, if a particle is moving with constant velocity, the speed is constant but acceleration is zero.

(d) True, if the positive direction of acceleration is along the direction of motion.

Q 11. A man walks on a straight road from his home to a market 2.5 km away with a speed of 5 kmh^{-1}. Finding the market closed, he instantly turns and walks back home with a speed of 7.5 kmh^{-1}. What is the

(a) magnitude of average velocity and

(b) average speed of the man over the interval of time

(i) 0 to 30 min

(ii) 0 to 50 min

(iii) 0 to 40 min

Sol. We know that average velocity $= \dfrac{\text{Total displacement}}{\text{Total time}}$

and average speed $= \dfrac{\text{Total distance}}{\text{Total time}}$

(i) **0 to 30 min interval**

Distance covered in going to market = 2.5 km

Speed = 5 kmh^{-1}

$\therefore$ Time taken to go to market = Distance /speed

$$= 2.5/5 = \frac{1}{2}\text{ h} = 30\text{ min}$$

$\therefore$ (a) Magnitude of average velocity $= \dfrac{\text{Displacement}}{\text{Time}}$

$$= \frac{2.5}{\frac{1}{2}\text{h}} = 5\text{ kmh}^{-1}$$

(b) Average speed $= \dfrac{2.5}{\frac{1}{2}\text{h}} = 5\text{ kmh}^{-1}$

(ii) **0 to 50 min interval**

Speed of return journey = 7.5 kmh^{-1}

Distance to be covered to reach home = 2.5 km

$\therefore$ Time taken in return journey $= \dfrac{\text{Distance}}{\text{Speed}}$

$$= \frac{2.5}{7.5} = \frac{1}{3}\text{ h}$$

$$= 20\text{ min.}$$

$\therefore$ Total time taken for the journey = 30 + 20 = 50 min

Now net displacement of the man is zero $= \dfrac{5}{6}$ h

$$\therefore \text{Average velocity} = \frac{0}{\frac{5}{6}} = 0$$

Total distance covered during to whole journey

$$= 2.5 + 2.5$$

$$= 5\text{ km}$$

$$\therefore \text{Average speed} = \frac{5}{(5/6)} = 6\text{ kmh}^{-1}$$

(iii) **0 to 40 min interval**

As the man takes 30 min; to go to market.

$\therefore$ Time for which he performed return journey

$$= 40 - 30 = 10\text{ min.}$$

Distance covered in 10 min

= Velocity of return journery × time

$$= 7.5 \times \frac{10}{60} = 1.25\text{ km}$$

$\therefore$ Net displacement of the man = 2.5 − 1.25 = 1.25 km

$$\therefore \text{Average velocity} = \frac{1.25}{\frac{2}{3}} = 1.25 \times \frac{3}{2}$$

$$= \frac{3.75}{2}$$

$$= 1.875\text{ kmh}^{-1}$$

and average speed

$$= \frac{\text{Total distance}}{\text{Total time}}$$

$$= \frac{2.5 + 1.25}{\frac{2}{3}}$$

$$= 3.75 \times \frac{3}{2}$$

$$= \frac{11.25}{2}$$

$$= 5.625\text{ kmh}^{-1}$$

Q 12. Look at the graphs (a) to (d) carefully and state with reasons, which of these cannot possibly represent one-dimensional motion of a particle ?

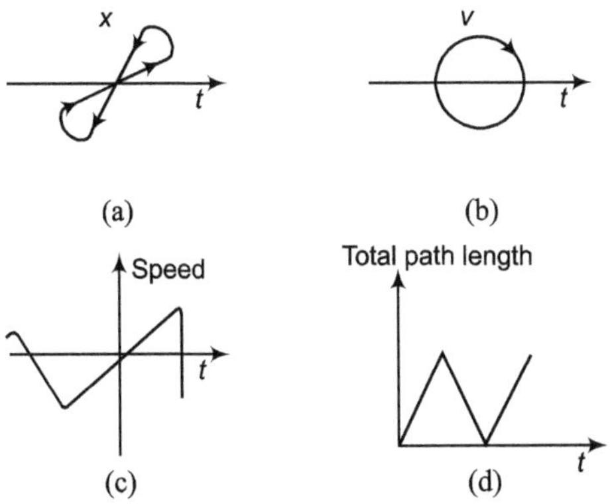

(a) (b) (c) (d)

Sol. (a) A line drawn for a given time parallel to position axis will cut the graph at two points which means that at a given instant of time, the particle will have to positions. Which is not possible. Hence, graph (a) is not possible.

(b) At a given instant of time, the particle will have two values of velocity in positive, as well as in negative direction which is not possible in one dimensional motion.

(c) Speed can never be negative. Hence, this graph is not possible.

(d) This does not represent one dimensional motion, as this graph tells that the total path length decreases after certain time but total path length of a moving particle can never decrease with time.

Q 13. Figure below shows the x-t plot of one dimensional motion of a particle. Is it correct to say from the graph that the particle moves in a straight line for $t<0$ and on a parabolic path for $t>0$? If not, suggest a suitable physical context for this graph.

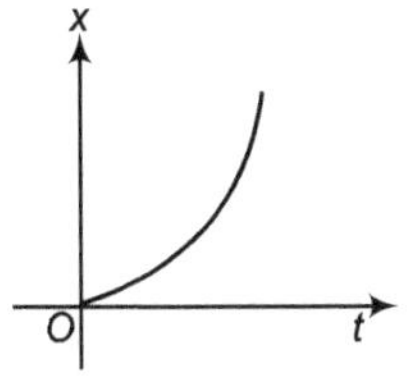

Sol. No, the x-t graph does not show the trajectory of the path of particle.

This may be x-t graph of one dimensional uniformly accelerated motion.

Q 14. A police van moving on a highway with a speed of 30 kmh^{-1} fires a bullet at a thief's car speeding away in the same direction with a speed of 192 kmh^{-1}. If the muzzle speed of the bullet is 150 ms^{-1}, with what speed does the bullet hit the thief's car?

Sol. Speed of thief's car, $v_t = 192 \text{ kmh}^{-1}$

$$= 192 \times \frac{5}{18} \text{ ms}^{-1} = \frac{160}{3} \text{ ms}^{-1}$$

Speed of police van, $v_p = 30 \text{ kmh}^{-1}$

$$= 30 \times \frac{5}{18} = \frac{25}{3} \text{ ms}^{-1}$$

Speed of the bullet with which it is fired = 150 ms^{-1}

$\therefore$ Actual speed of the bullet, v_b = speed of police van + 150

$$= \frac{25}{3} + 150$$

$$= \frac{475}{3} \text{ ms}^{-1}$$

$\therefore$ Speed of the bullet with which it hits the thief's car is

= Relative speed of the bullet w.r.t. thief 's car

$= v_{bt}$

i.e. $v_{bt} = v_b - v_t$

$$= \left(\frac{475}{3} - \frac{160}{3}\right) \text{ ms}^{-1} = \frac{315}{3} = 105 \text{ ms}^{-1}$$

Q 15. Suggest a suitable physical situation for each of the following graphs.

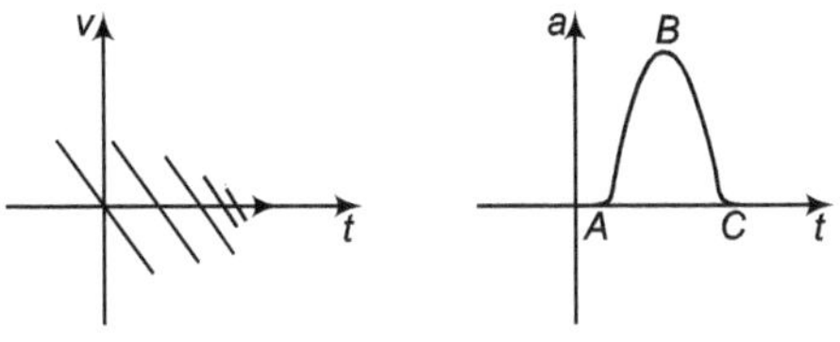

(a) (b)

Sol. (a) The graph represents the case of a ball thrown up with some initial velocity and rebounding from the floor with reduced speed.

(b) It is the acceleration-time graph. The graph represents a uniformly moving cricket ball turned back after hitting the bat for a very short time interval (represented by the portion ABC of the graph).

Q 16. A boy standing on a stationary lift (open from above) throws a ball upwards with the maximum initial speed he can, equal to 49 ms^{-1}. How much time does the ball take to return to his hands? If the lift starts moving up with a uniform speed of 5 ms^{-1} and the boy again throws the up with the maximum speed he can, how long does the ball take to return to his hands?

Sol. **Case I When the lift is stationary**

Let t be total time taken by ball in going vertically upward and coming down to the hands of the boy.

$$s = \text{total displacement} = 0$$

$\therefore$Using the relation, $s = ut + \frac{1}{2}at^2$, we get

$$0 = 49t + \frac{1}{2}(-9.8) \times t^2$$

or $\quad 49t = 4.9t^2$

or $\quad t = \frac{49}{4.9} = 10$ s.

Case II When the lift starts moving with uniform speed

As the lift starts moving upwards with uniform speed of 5 ms^{-1}, there is no change in the relative velocity of the ball w.r.t. the boy which remains 49 ms^{-1} due to the fact that there is no acceleration in the lift. i.e. initial velocity of the ball will remain 49 ms^{-1} only w.r.t. lift. Hence, the ball will naturally return back to the boy's hand after 10 s.

***Q* 17.** The speed-time graph of a particle moving along a fixed direction is shown below. Obtain the distance traversed by the particle between (a) $t = 0$ s to 10 s (b) $t = 2$ s to 6 s. What is the average speed of the particle over the intervals in (a) and (b)?

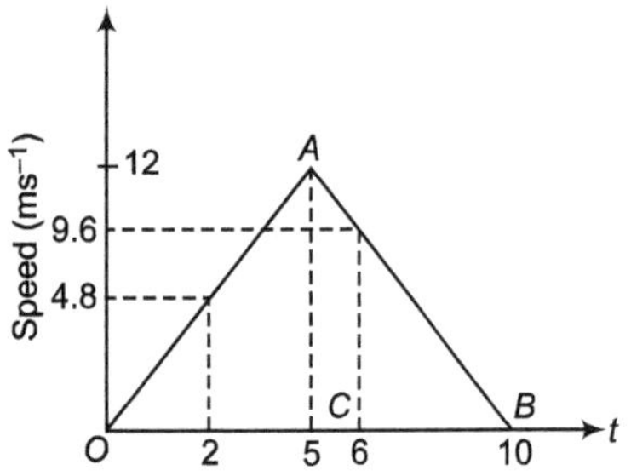

Sol. (a) (i) Distance covered by an object in a given time interval is equal to the area under speed-time graph.

$\therefore$ Distance traversed by the particle between the interval $t = 0$ to 10 s = area of the triangle OAB

$$= \frac{1}{2} \times 10 \times 12$$

$$= 60 \text{ m}$$

(ii) The average speed in a given time interval is given by

$$v_{av} = \frac{\text{Total distance covered in the given time interval}}{\text{Total time}}$$

$\therefore v_{av}$ for the interval 0 s to 10 s is given by

$$v_{av} = \frac{60}{10} = 6 \text{ ms}^{-1}$$

(b) (i) The distance covered from 2 s to 6 s is calculated in the following way.

Area of speed-time graph in given interval

= distance travelled

$$= \frac{1}{2}(3)(4.8 + 12) + \frac{1}{2}(1)(12 + 9.6)$$

$$= 36 \text{ m}$$

(ii) Average speed during this interval is given by

$$v_{av} = \frac{\text{Total distance traversed}}{\text{Total time}}$$

$$= \frac{36}{4} = 9 \text{ ms}^{-1}$$

Objective Problems

[Level 1]

Basic Definitions

1. A particle's velocity changes from $(2\hat{i} + 3\hat{j})$ m/s to $(2\hat{i} - 3\hat{j})$ m/s in 2 s. The acceleration in m/s^2 is
(a) $-(\hat{i} + 5\hat{j})$ (b) $(\hat{i} + 5\hat{j})/2$
(c) zero (d) $(-3\hat{j})$

2. A boy is running over a circular track with uniform speed of 10 m/s. Find the average velocity for movement of boy from A to B (in m/s)

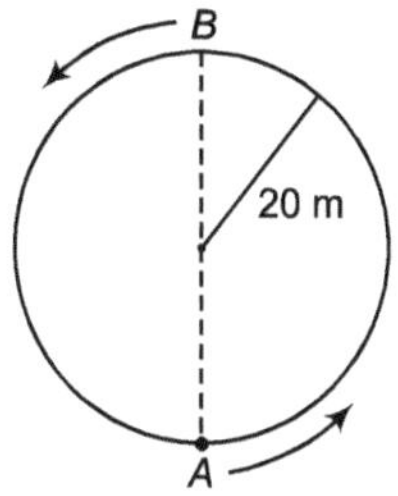

(a) $\frac{10}{\pi}$ (b) $\frac{40}{\pi}$ (c) 10 (d) None of these

3. A car has to cover the distance 60 km. If half of the total time it travels with speed 80 km/h and in rest half time its speed becomes 40 km/h, the average speed of car will be
(a) 60 km/h (b) 80 km/h (c) 120 km/h (d) 180 km/h

4. A person moves towards east for 3 m, then towards north for 4 m and then moves vertically up by 5 m. What is his distance now from the starting point?
(a) $5\sqrt{2}$ m (b) 5 m
(c) 10 m (d) 20 m

5. A point traversed 3/4 th of the circle of radius R in time t. The magnitude of the average velocity of the particle in this time interval is
(a) $\frac{\pi R}{t}$ (b) $\frac{3\pi R}{2t}$ (c) $\frac{R\sqrt{2}}{t}$ (d) $\frac{R}{\sqrt{2}t}$

6. A particle starts from the origin, goes along X-axis to the point (20 m, 0) and then returns along the same line to the point (−20 m, 0). The distance and displacement of the particle during the trip are
(a) 40 m, 0 (b) 40 m, 20 m
(c) 40 m, − 20 m (d) 60 m, − 20 m

7. An insect crawls a distance of 4 m along north in 10 s and then a distance of 3 m along east in 5 s. The average velocity of the insect is
(a) $\frac{7}{15}$ m/s (b) $\frac{1}{5}$ m/s (c) $\frac{1}{3}$ m/s (d) $\frac{4}{5}$ m/s

8. A wheel of radius 1 m rolls forward half a revolution on a horizontal ground. The magnitude of the displacement of the point of the wheel initially in contact with the ground is
(a) 2π
(b) $\sqrt{2}\pi$
(c) $\sqrt{\pi^2 + 4}$
(d) π

9. A particle is constrained to move on a straight line path. It returns to the starting point after 10 s. The total distance covered by the particle during this time is 30 m. Which of the following statements about the motion of the particle is false?
(a) Displacement of the particle is zero
(b) Average speed of the particle is 3 m/s
(c) Displacement of the particle is 30 m
(d) Both (a) and (b)

10. A particle, moving with uniform speed v, changes its direction by angle θ in time t. Magnitude of its average acceleration during this time is
(a) zero (b) $\frac{2v}{t}\sin\frac{\theta}{2}$
(c) $\frac{v\sqrt{2}}{t}$ (d) None of these

11. A point traversed half of the distance with a velocity v_0. The remaining part of the distance was covered with velocity v_1 for half the time and with velocity v_2 for the other half of the time. The mean velocity of the point averaged over the whole time of motion is
(a) $\frac{v_0 + v_1 + v_2}{3}$
(b) $\frac{2v_0 + v_1 + v_2}{3}$
(c) $\frac{v_0 + 2v_1 + 2v_2}{3}$
(d) $\frac{2v_0(v_1 + v_2)}{(2v_0 + v_1 + v_2)}$

12. During the first 18 min of a 60 min trip, a car has an average speed of 11 m/s. What should be the average speed for remaining 42 min so that car is having an average speed of 21 m/s for the entire trip?
(a) 25.3 m/s
(b) 29.2 m/s
(c) 31 m/s
(d) 35.6 m/s

Motion in 1-D with Constant Acceleration

13. The velocity v of a particle as a function of its position (x) is expressed as $v=\sqrt{c_1 - c_2 x}$, where c_1 and c_2 are positive constants. The acceleration of the particle is
(a) c_2 (b) $-\frac{c_2}{2}$ (c) $c_1 - c_2$ (d) $\frac{c_1 + c_2}{2}$

14. The displacement of a body in 8 s starting from rest with an acceleration of 20cm/s^2 is
(a) 64 m (b) 64 cm
(c) 640 cm (d) 0.064 m

15. A body falls from a height $h = 200$ m. The ratio of distance travelled in each 2 s, during $t = 0$ to $t = 6$ s of the journey is
(a) 1 : 4 : 9 (b) 1 : 2 : 4
(c) 1 : 3 : 5 (d) 1 : 2 : 3

16. A ball is released from height h and another from $2h$. The ratio of time taken by the two balls to reach the ground is
(a) $1:\sqrt{2}$ (b) $\sqrt{2}:1$
(c) 2 : 1 (d) 1 : 2

17. A body falling from the rest has a velocity v after it falls through a height h. The distance it has to fall down further for its velocity to become double, will be
(a) $8h$ (b) $6h$
(c) $3h$ (d) $5h$

18. The velocity of a particle moving in the positive direction of x-axis varies as $v = 5\sqrt{x}$. Assuming that at $t = 0$, particle was at $x = 0$. What is the acceleration of the particle?
(a) 12.5 m/s^2 (b) 7.5 m/s^2
(c) 5 m/s^2 (d) 2.5 m/s^2

19. A person throws balls into air after every second. The next ball is thrown when the velocity of the first ball is zero. How high do the ball rise above his hand?
(a) 2 m (b) 5 m (c) 8 m (d) 10 m

20. An object is moving with velocity 10 m/s. A constant force acts for 4 s on the object and gives it a speed of 2 m/s in opposite direction. The acceleration produced is
(a) 3 m/s^2 (b) -3 m/s^2
(c) 6 m/s^2 (d) -6 m/s^2

21. A ball is thrown vertically upward with a speed v from a height h metre above the ground. The time taken for the ball to hit ground is
(a) $\frac{v}{g}\sqrt{1-\frac{2hg}{v^2}}$ (b) $\frac{v}{g}\sqrt{1+\frac{2hg}{v^2}}$
(c) $\sqrt{1+\frac{2hg}{v^2}}$ (d) $\frac{v}{g}\left[1+\sqrt{1+\frac{2hg}{v^2}}\right]$

22. A helicopter, moving vertically upwards, releases a packet when it is at a certain height above the ground. The packet initially moves upwards for a time t_1 and then falls downwards for a t_2 until it reaches the ground. Then,
(a) $t_1 < t_2$ (b) $t_1 = t_2$
(c) $t_1 > t_2$ (d) Data insufficient

23. A particle starts with a velocity of 2 ms^{-1} and moves in a straight line with a retardation of 0.1 ms^{-2}. The first time at which the particle is 15 m from the starting point, is
(a) 10 s (b) 20 s
(c) 30 s (d) 40 s

24. If a stone is thrown up with a velocity of 9.8 m/s, then how much time will it take to come back?
(a) 1 s (b) 2 s
(c) 3 s (d) 4 s

25. The motion of a particle is described by the equation $v = at$. The distance travelled by the particle in the first 4 s
(a) $4a$ (b) $12a$
(c) $6a$ (d) $8a$

26. A particle starts from rest, accelerates at 2 m/s^2 for 10 s and then goes for constant speed for 30 s and then decelerates at 4 m/s^2 till it stops. What is the distance travelled by it?
(a) 750 m (b) 800 m
(c) 700 m (d) 850 m

27. A body dropped from the top of a tower covers a distance $7x$ in the last second of its journey, where x is the distance covered in first second. How much time does it take to reach the ground?
(a) 3 s (b) 4 s
(c) 5 s (d) 6 s

28. A particle returns to the starting point after 10 s. If the rate of change of velocity during the motion is constant, then its location after 7 s will be same as that after
(a) 1 s (b) 2 s
(c) 3 s (d) 4 s

29. A body is moving with uniform velocity of 8 ms^{-1}. When the body just crossed another body, the second one starts and moves with uniform acceleration of 4 m/s^2. The time after which two bodies meet, will be
(a) 2 s (b) 4 s
(c) 6 s (d) 8 s

30. Two bodies A and B start from rest from the same point with a uniform acceleration of 2 m/s^2. If B starts one second later, then the two bodies are separated, at the end of the next second, by
(a) 1 m (b) 2 m
(c) 3 m (d) 4 m

31. Velocity of a body moving along a straight line with uniform acceleration a reduces by $\frac{3}{4}$ of its initial velocity in time t_0. The total time of motion of the body till its velocity becomes zero is

(a) $\frac{4}{3}t_0$ (b) $\frac{3}{2}t_0$
(c) $\frac{5}{3}t_0$ (d) $\frac{8}{3}t_0$

32. A body travelling with uniform acceleration crosses two points A and B with velocities 20 m/s and 30 m/s respectively. The speed of the body at mid-point of A and B is

(a) 25 m/s (b) 25.5 m/s
(c) 24 m/s (d) $10\sqrt{6}$ m/s

33. A particle starting from rest with constant acceleration travels a distance x in first 2 s and a distance y in next 2 s, then

(a) $y = x$ (b) $y = 2x$
(c) $y = 3x$ (d) $y = 4x$

34. A man in a balloon rising vertically with an acceleration of 4.9 m/s^2 releases a ball 2 s after the balloon is let go from the ground. The greatest height above the ground reached by the ball is $(g = 9.8\text{ m/s}^2)$

(a) 14.7 m (b) 19.6 m
(c) 9.8 m (d) 24.5 m

35. A stone falls freely under gravity. The total distance covered by it in the last second of its journey equals the distance covered by it in first 3 s of its motion. The time for which the stone is in air is

(a) 5 s (b) 12 s
(c) 15 s (d) 8 s

36. The displacement of a particle moving in a straight line is described by the relation, $s = 6 + 12t - 2t^2$. Here, s is in metre and t is in second. The distance covered by particle in first 5 s is

(a) 20 m (b) 32 m
(c) 24 m (d) 26 m

37. The position of a particle along x-axis at time t is given by $x = 2 + t - 3t^2$. The displacement and the distance travelled in the interval $t = 0$ to $t = 1$ are respectively

(a) 2, 2 (b) −2, 2.5
(c) 0, 2 (d) −2, 2.1

38. A stone is allowed to fall freely from rest. The ratio of the times taken to fall through the first metre and the second metre distance is

(a) $\sqrt{2} - 1$
(b) $\sqrt{2} + 1$
(c) $\sqrt{2}$
(d) None of the above

39. A particle moves along x-axis as $x = 4(t-2) + a(t-2)^2$. Which of the following statements is true?

(a) The initial velocity of particle is 4
(b) The acceleration of particle is $2a$
(c) The particle is at origin at $t = 0$
(d) None of the above

40. A stone thrown upward with a speed u from the top of the tower reaches the ground with a speed $3u$. The height of the tower is

(a) $3u^2/g$ (b) $4u^2/g$
(c) $6u^2/g$ (d) $9u^2/g$

41. A particle is dropped under gravity from rest from a height h $(g = 9.8\text{ m/s}^2)$ and it travels a distance $9h/25$ in the last second, the height h is

(a) 100 m (b) 122.5 m
(c) 145 m (d) 167.5 m

42. An aeroplane is moving with a velocity u. It drops a packet from a height h. The time t taken by the packet in reaching the ground will be

(a) $\sqrt{\left(\frac{2g}{h}\right)}$ (b) $\sqrt{\left(\frac{2u}{g}\right)}$
(c) $\sqrt{\left(\frac{h}{2g}\right)}$ (d) $\sqrt{\left(\frac{2h}{g}\right)}$

43. When a ball is thrown up vertically with velocity v_0, it reaches a maximum height of h. If one wishes to triple the maximum height, then the ball should be thrown with velocity

(a) $\sqrt{3}\,v_0$ (b) $3\,v_0$
(c) $9\,v_0$ (d) $3/2\,v_0$

44. A body starts from rest with uniform acceleration a, its velocity after n second is v. The displacement of the body in last 3 s in (assume total time of journey from 0 to n second)

(a) $\frac{v(6n-9)}{2n}$ (b) $\frac{2v(6n-9)}{n}$
(c) $\frac{2v(2n+1)}{n}$ (d) $\frac{2v(n-1)}{n}$

45. A train accelerating uniformly from rest attains a maximum speed of 40 ms^{-1} in 20 s. It travels at this speed for 20 s and is brought to rest uniform retardation in further 40 s. What is the average velocity during this period?

(a) 80/3 ms^{-1} (b) 40 ms^{-1}
(c) 25 ms^{-1} (d) 30 ms^{-1}

46. A particle is thrown vertically upwards. Its velocity at half of the height is 10 m/s. Then, the maximum height attained by it is $(g = 10\text{ m/s}^2)$

(a) 16 m (b) 10 m
(c) 20 m (d) 40 m

47. Which of the following represents uniformly accelerated motion?

(a) $x = \sqrt{\frac{t+a}{b}}$ (b) $x = \frac{t+a}{b}$

(c) $t = \sqrt{\frac{x+a}{b}}$ (d) $x = \sqrt{t+a}$

48. Two particles A and B start from rest and move for equal time on a straight line. Particle A has an acceleration of $2\ \text{m/s}^2$ for the first half of the total time and $4\ \text{m/s}^2$ for the second half. The particle B has acceleration $4\ \text{m/s}^2$ for the first half and $2\ \text{m/s}^2$ for the second half. Which particle has covered larger distance?

(a) A
(b) B
(c) Both have covered the same distance
(d) Data insufficient

49. If a ball is thrown vertically upwards with speed u, the distance covered during the last t second of its ascent is

(a) $ut - (gt^2/2)$ (b) $(u + gt)\,t$
(c) ut (d) $gt^2/2$

50. A body thrown vertically up from the ground passes the height of 10.2 m twice in an interval of 10 s. What was its initial velocity?

(a) 52 m/s (b) 61 m/s
(c) 45 m/s (d) 26 m/s

51. A body is projected with a velocity u. It passes through a certain point above the ground after t_1 second. The time interval after which the body passes through the same point during the return journey is

(a) $\left(\frac{u}{g} - t_1^2\right)$ (b) $2\left(\frac{u}{g} - t_1\right)$

(c) $\left(\frac{u}{g} - t_1\right)$ (d) $\left(\frac{u^2}{g^2} - t_1\right)$

52. With what speed should a body be thrown upwards so that the distances traversed in 5th second and 6th second are equal?

(a) 5.84 m/s
(b) 49 m/s
(c) $\sqrt{98}$ m/s
(d) 98 m/s

53. A train accelerates from rest at a constant rate α for distance x_1 and time t_1. After that it retards to rest at constant rate β for distance x_2 and time t_2. Which of the following relations is correct?

(a) $\frac{x_1}{x_2} = \frac{\alpha}{\beta} = \frac{t_1}{t_2}$ (b) $\frac{x_1}{x_2} = \frac{\beta}{\alpha} = \frac{t_1}{t_2}$

(c) $\frac{x_1}{x_2} = \frac{\alpha}{\beta} = \frac{t_2}{t_1}$ (d) $\frac{x_1}{x_2} = \frac{\beta}{\alpha} = \frac{t_2}{t_1}$

54. A particle starts from rest and traverses a distance l with uniform acceleration, then moves uniformly over a further distance $2l$ and finally comes to rest after moving a further distance $3l$ under uniform retardation. Assuming entire motion to be rectilinear motion the ratio of average speed over the journey to the maximum speed on its ways is

(a) 1/5 (b) 2/5
(c) 3/5 (d) 4/5

55. A point moves with uniform acceleration and v_1, v_2 and v_3 denote the average velocities in the three successive intervals of time t_1, t_2 and t_3. Which of the following relation is correct?

(a) $(v_1 - v_2):(v_2 - v_3) = (t_1 - t_2):(t_2 + t_3)$
(b) $(v_1 - v_2):(v_2 - v_3) = (t_1 + t_2):(t_2 + t_3)$
(c) $(v_1 - v_2):(v_2 - v_3) = (t_1 - t_2):(t_1 - t_3)$
(d) $(v_1 - v_2):(v_2 - v_3) = (t_1 - t_2):(t_2 - t_3)$

56. A body is thrown vertically upwards from the top A of tower. It reaches the ground in t_1 sec. If it is thrown vertically downwards from A with the same speed it reaches the ground in t_2 sec. If it is allowed to fall freely from A, then the time it takes to reach the ground is given by

(a) $t = \frac{t_1 + t_2}{2}$ (b) $t = \frac{t_1 - t_2}{2}$

(c) $t = \sqrt{t_1 t_2}$ (d) $t = \sqrt{\frac{t_1}{t_2}}$

Motion in 1-D with Variable Acceleration

57. The position of a particle moving along the x-axis is expressed as $x = at^3 + bt^2 + ct + d$. The initial acceleration of the particle is

(a) $6a$ (b) $2b$
(c) $(a + b)$ (d) $(a + c)$

58. The acceleration a in m/s^2, of a particle is given by $a = 3t^2 + 2t + 2$, where t is the time. If the particle starts out with a velocity $v = 2$ m/s at $t = 0$, then the velocity at the end of 2 s is

(a) 12 m/s (b) 14 m/s (c) 16 m/s (d) 18 m/s

59. A particle initially at rest moves along the x-axis. Its acceleration varies with time as $a = 4t$. If it starts from the origin, the distance covered by it in 3 s is

(a) 12 m (b) 18 m
(c) 24 m (d) 36 m

60. The acceleration a (in ms^{-2}) of a body, starting from rest varies with time t (in second) according to the relation $a = 3t + 4$. The velocity of the body starting from rest at time $t = 2$ s will be

(a) $10\ \text{ms}^{-1}$ (b) $12\ \text{ms}^{-1}$
(c) $14\ \text{ms}^{-1}$ (d) $16\ \text{ms}^{-1}$

61. The displacement (x) of a particle depends on time t as $x = \alpha t^2 - \beta t^3$. Choose the incorrect statements from the following.

(a) The particle never returns to its starting point

(b) The particle comes to rest after time $\frac{2\alpha}{3\beta}$

(c) The initial velocity of the particle is zero

(d) The initial acceleration of the particle is zero

62. The displacement of a particle is given by $y = a + bt + ct^2 - dt^4$. The initial velocity and acceleration are respectively

(a) $b, -4d$ (b) $-b, -2c$

(c) $b, 2c$ (d) $2c, -4d$

63. The displacement of a particle starting from rest (at $t = 0$) is given by $s = 6t^2 - t^3$. The time in second at which the particle will attain zero velocity again, is

(a) 2

(b) 4

(c) 6

(d) 8

64. A particle moves along a straight line such that its displacement at any time t is given by $s = 3t^3 + 7t^2 + 14t + 5$. The acceleration of the particle at $t = 1$ s is

(a) 18 m/s^2 (b) 32 m/s^2

(c) 29 m/s^2 (d) 24 m/s^2

65. A particle moves along a straight line. Its position at any instant is given by $x = 32t - \frac{8t^3}{4}$, where x is in metre and t is in second. Find the acceleration of the particle at the instant when particle is at rest.

(a) -16 m/s^2 (b) -32 m/s^2

(c) 32 m/s^2 (d) 16 m/s^2

66. A body starts from rest, with uniform acceleration a. The acceleration of the body as function of time t is given by the equation $a = pt$, where p is a constant, then the displacement of the particle in the time interval $t = 0$ to $t = t_1$ will be

(a) $\frac{1}{2} pt_1^3$ (b) $\frac{1}{3} pt_1^2$

(c) $\frac{1}{2} pt_1^2$ (d) $\frac{1}{6} pt_1^3$

67. The acceleration of a particle is increasing linearly with time t as bt. The particle starts from the origin with an initial velocity v_0. The distance travelled by the particle in time t will be

(a) $v_0 t + \frac{1}{6} bt^3$ (b) $v_0 t + \frac{1}{3} bt^3$

(c) $v_0 t + \frac{1}{3} bt^2$ (d) $v_0 t + \frac{1}{2} bt^2$

68. The displacement of a particle moving in a straight line depends on time as $x = \alpha t^3 + \beta t^2 + \gamma t = \delta$.

The ratio of initial acceleration to its initial velocity depends

(a) only on α and γ (b) only on β and γ

(c) only on α and β (d) only on α

69. At time $t = 0$, a car moving along a straight line has a velocity of 16 m/s. It slows down with an acceleration of $-0.5t \text{ m/s}^2$, where t is in second. Mark the correct statement (s).

(a) The direction of velocity changes at $t = 8$ s

(b) The distance travelled in 4 s is approximately 59 m

(c) The distance travelled by the particle in 10 s is 94 m

(d) The velocity at $t = 10$ s is 8 m / s

Motion in 2 or 3-D with Constant or Variable Acceleration

70. A particle moves along the positive branch of the curve $y = \frac{x^2}{2}$ where $x = \frac{t^2}{2}$, x and y are measured in metres and t in second. At $t = 2$ s, the velocity of the particle is

(a) $2\hat{\mathbf{i}} - 4\hat{\mathbf{j}}$ m/s (b) $4\hat{\mathbf{i}} + 2\hat{\mathbf{j}}$ m/s

(c) $2\hat{\mathbf{i}} + 4\hat{\mathbf{j}}$ m/s (d) $4\hat{\mathbf{i}} - 2\hat{\mathbf{j}}$ m/s

71. The displacement of an object along the three axes are given by $x = 2t^2$, $y = t^2 - 4t$ and $z = 3t - 5$. The initial velocity of the particle is

(a) 10 unit (b) 12 unit

(c) 5 unit (d) 2 unit

72. A particle has an initial velocity of $3\hat{\mathbf{i}} + 4\hat{\mathbf{j}}$ and an acceleration of $0.4\hat{\mathbf{i}} + 0.3\hat{\mathbf{j}}$. Its speed after 10 s is

(a) 10 units

(b) 7 units

(c) $7\sqrt{2}$ units

(d) 8.5 units

73. A particle moves in the xy - plane according to the law $x = kt$, $y = kt\ (1 - \alpha t)$, where k and α are positive constants and t is time. The trajectory of the particle is

(a) $y = kx$ (b) $y = x - \frac{\alpha x^2}{k}$

(c) $y = -\frac{\alpha x^2}{k}$ (d) $y = \alpha x$

74. The position of a particle moving in the xy - plane at any time t is given by $x = (3t^2 - 6t)$ m, $y = (t^2 - 2t)$ m. Select the correct statement about the moving particle from the following.

(a) The acceleration of the particle is zero at $t = 0$ s

(b) The velocity of the particle is zero at $t = 0$ s

(c) The velocity of the particle is zero at $t = 1$ s

(d) The velocity and acceleration of the particle are never zero

75. Velocity and acceleration of a particle at some instant of time are $\mathbf{v} = (3\hat{\mathbf{i}} + 4\hat{\mathbf{j}})$ m/s and $\mathbf{a} = -(6\hat{\mathbf{i}} + 8\hat{\mathbf{j}})$ m/s^2 respectively. At the same instant particle is at origin. Maximum x-coordinate of particle will be

(a) 1.5 m (b) 0.75 m
(c) 2.25 m (d) 4.0 m

76. A particle's velocity changes from $(2\hat{\mathbf{i}} + 3\hat{\mathbf{j}})$ m/s into $(3\hat{\mathbf{i}} - 2\hat{\mathbf{j}})$ m/s in 2 s. If its mass is 1 kg, the acceleration (m/s^2) is

(a) $-(\hat{\mathbf{i}} + 5\hat{\mathbf{j}})$ (b) $(\hat{\mathbf{i}} + 5\hat{\mathbf{j}})/2$
(c) zero (d) $(\hat{\mathbf{i}} - 5\hat{\mathbf{j}})/2$

77. A particle has an initial velocity of $3\hat{\mathbf{i}} + 4\hat{\mathbf{j}}$ and an acceleration of $0.4\hat{\mathbf{i}} + 0.3\hat{\mathbf{j}}$. Its speed after 10 s is

(a) 10 units (b) 7 units
(c) $7\sqrt{2}$ units (d) 8.5 units

78. The position vector of a particle is

$$\mathbf{r} = a\cos\omega t\,\hat{\mathbf{i}} + \sin\omega t\,\hat{\mathbf{j}}$$

The velocity of the particle is

(a) parallel to position vector
(b) perpendicular to position vector
(c) directed towards origin
(d) directed away from the origin

79. A body is projected from origin such that its position vector varies with time as $\mathbf{r} = [3t\hat{\mathbf{i}} + (4t - 5t^2)\hat{\mathbf{j}}]$ m and t is time in second. Then,

(a) x-coordinate of particle is 2.4 m when y-coordinate is zero
(b) speed of projection is 5 m/s
(c) angle of projection with x-axis is $\tan^{-1}(4/3)$
(d) time when particle is again at x-axis is 0.8 s

Graphical Problems

80. A particle projected vertically upwards returns to the ground in time T. Which graph represents the correct variation of velocity (v) against time (t)?

(a)
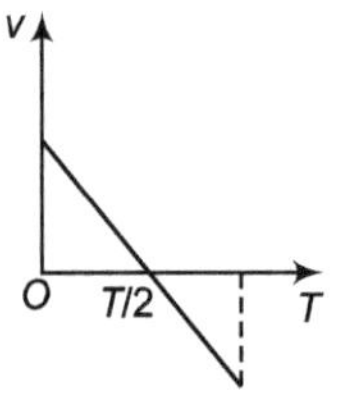

(b)
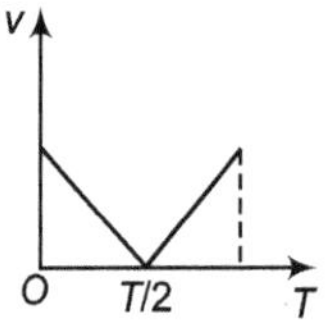

(c)
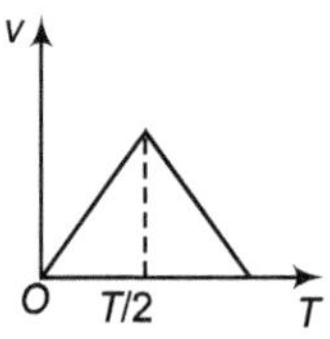

(d)
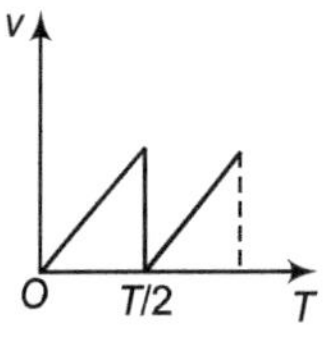

81. Which of the following graph correctly represents velocity-time relationship for a particle released from rest to fall freely under gravity?

(a)
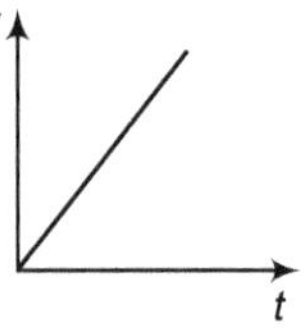

(b)
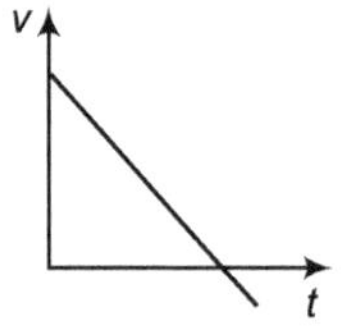

(c)
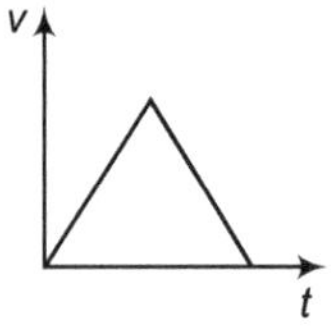

(d)
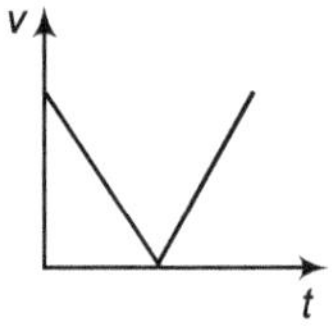

82. The velocity-time graph of a body is shown in figure. It implies that at point B

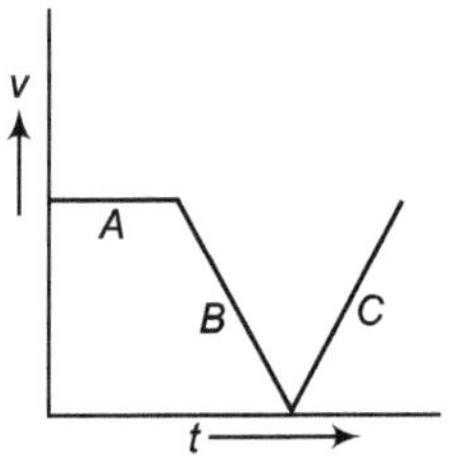

(a) the force is zero
(b) there is a force towards motion
(c) there is a force which opposes motion
(d) there is only gravitational force

83. The velocity-time graph for a particle moving along x-axis is shown in the figure. The corresponding displacement-time graph is correctly shown by

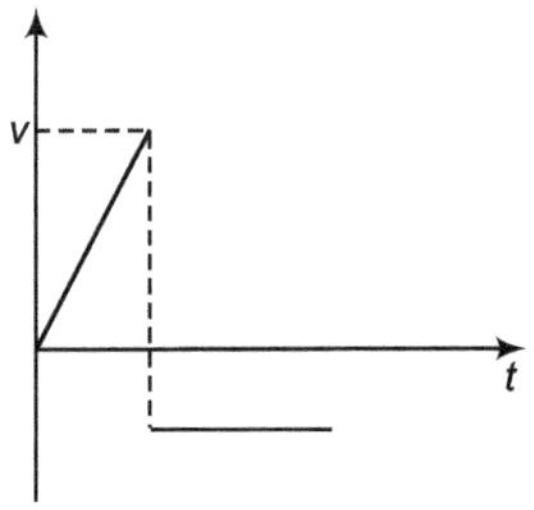

(a)
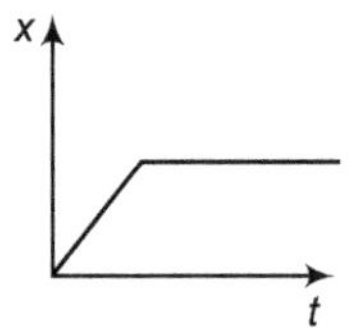

(b)
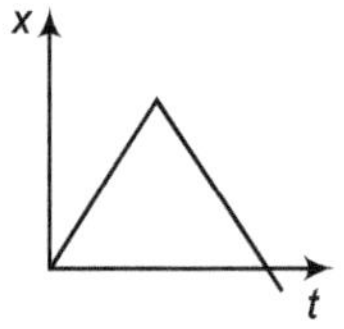

(c)
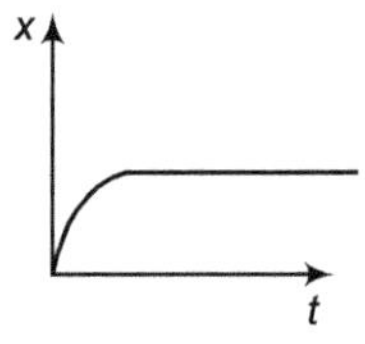

(d)
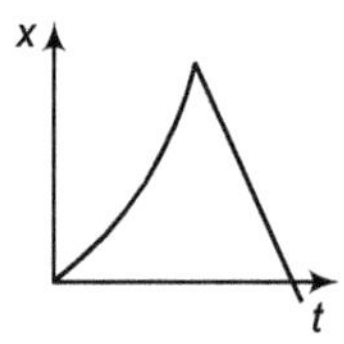

84. Acceleration *versus* time graph of a body starting from rest is shown in the figure. The velocity *versus* time graph of the body is given by

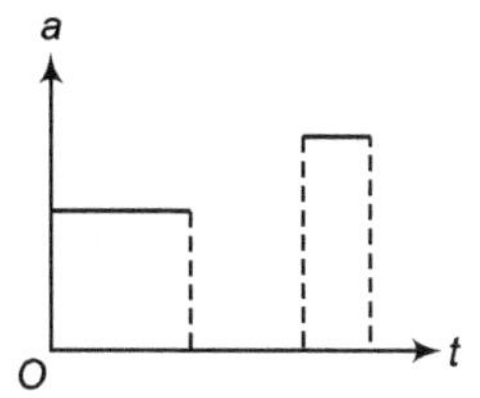

(a) 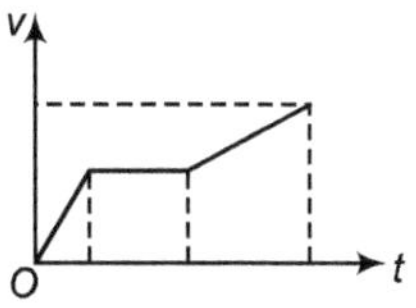

(b) 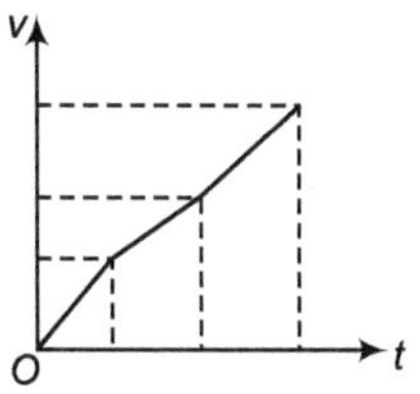

(c) 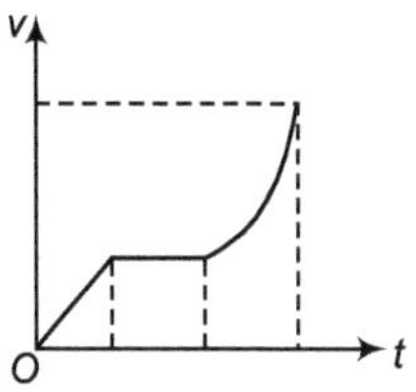

(d) 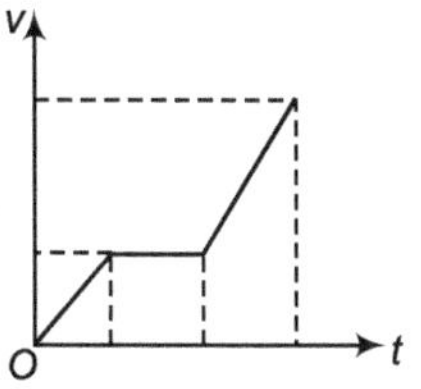

85. The acceleration of a train between two stations is shown in the figure. The maximum speed of the train is

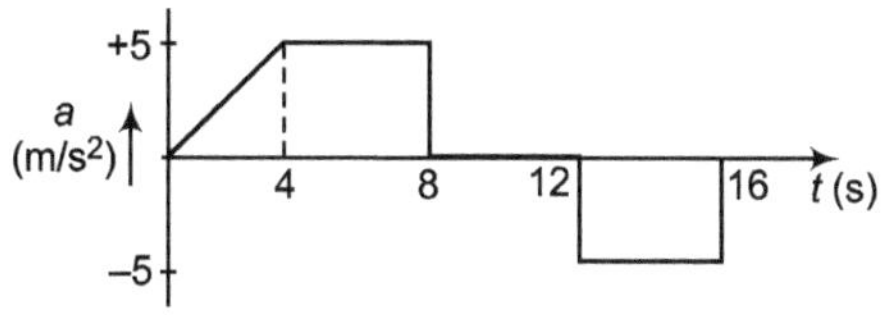

(a) 60 m/s (b) 30 m/s
(c) 120 m/s (d) 90 m/s

86. Which graph represents uniform motion?

(a) 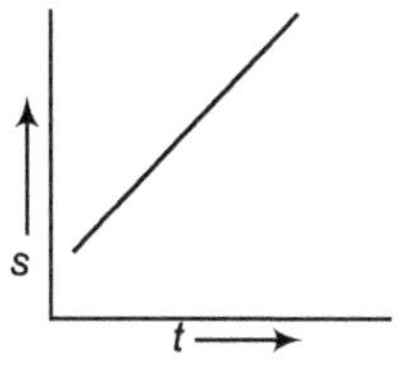

(b) 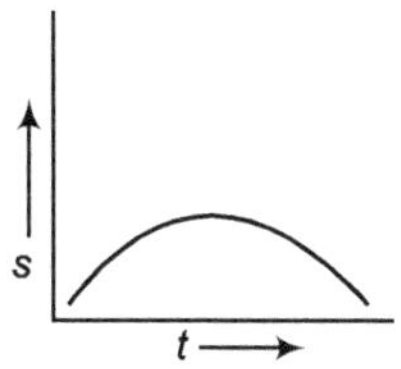

(c) 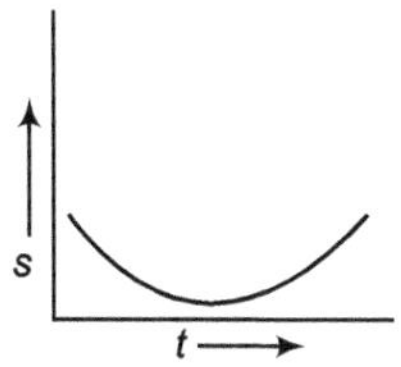

(d) None of these

87. The velocity-time graph of a body moving in a straight line is shown in the figure.

The displacement and distance travelled by the body in 6 s are respectively

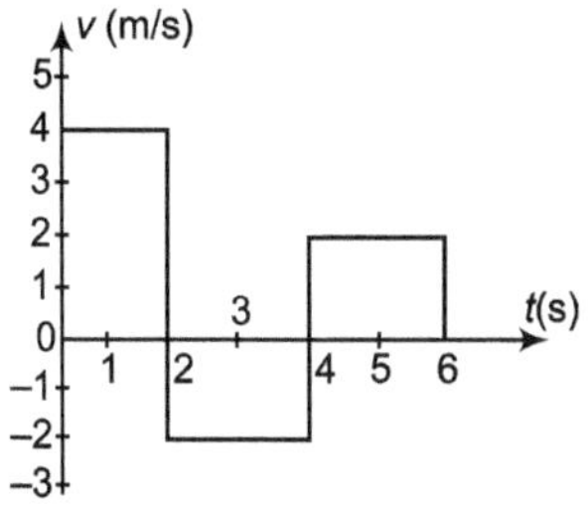

(a) 8 m, 16 m (b) 16 m, 32 m
(c) 16 m, 16 m (d) 8 m, 18 m

88. v^2 *versus* s-graph of a particle moving in a straight line is as shown in figure. From the graph some conclusions are drawn. State which statement is wrong?

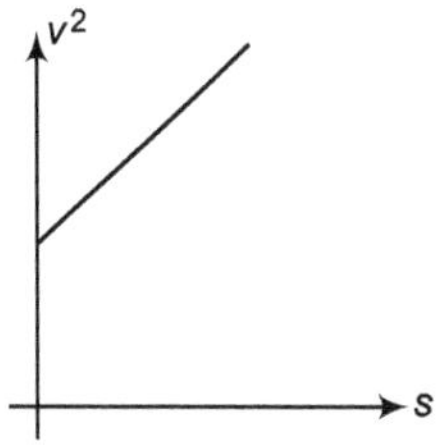

(a) The given graph shows a uniformly accelerated motion
(b) Initial velocity of particle is zero
(c) Corresponding s-t graph will be a parabola
(d) None of the above

89. A graph between the square of the velocity of a particle and the distance s moved by the particle is shown in the figure. The acceleration of the particle is

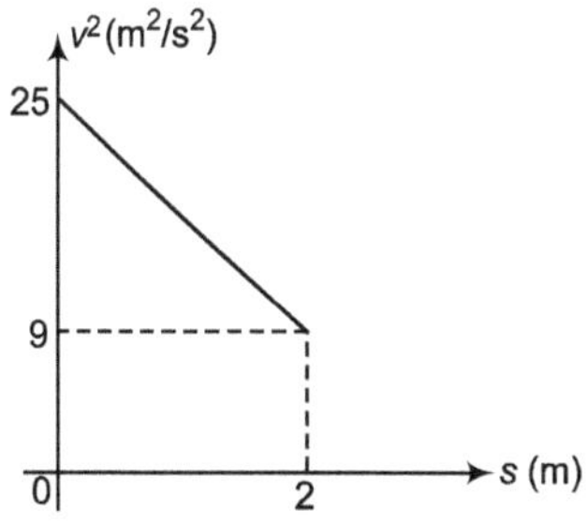

(a) $-8\ m/s^2$
(b) $4\ m/s^2$
(c) $-16\ m/s^2$
(d) None of the above

90. If the velocity v of a particle moving along a straight line decreases linearly with its displacement s from $20\ ms^{-1}$ to a value approaching zero at $s = 30$ m, then acceleration of the particle at $s = 15$ m is

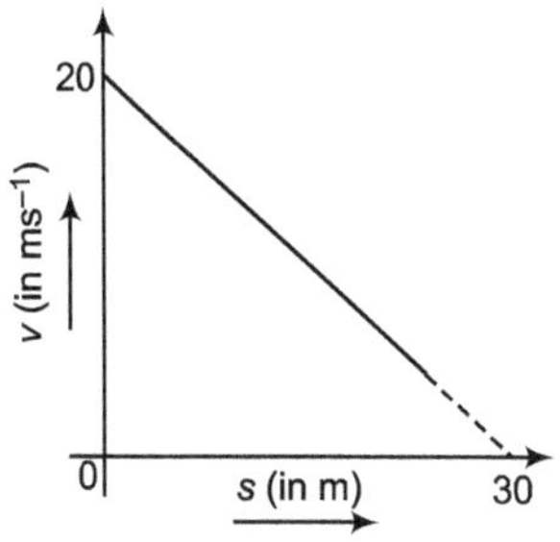

(a) $\frac{2}{3}\ ms^{-2}$ (b) $-\frac{2}{3}\ ms^{-2}$
(c) $\frac{20}{3}\ ms^{-2}$ (d) $-\frac{20}{3}\ ms^{-2}$

91. The distance-time graph of a particle at time t makes angle 45° with the time axis. After one second, it makes angle 60° with the time axis. What is the average acceleration of the particle?

(a) $\sqrt{3}-1$ (b) $\sqrt{3}+1$ (c) $\sqrt{3}$ (d) 1

92. Acceleration-time graph for a particle moving in a straight line is as shown in figure. Change in velocity of the particle from $t = 0$ to $t = 6$ s is

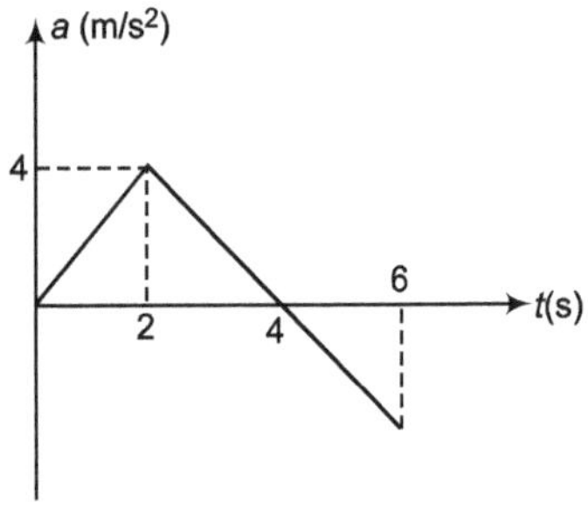

(a) 10 m/s (b) 4 m/s (c) 12 m/s (d) 8 m/s

93. The velocity-time graph is shown in the figure for a particle. The acceleration of the particle is

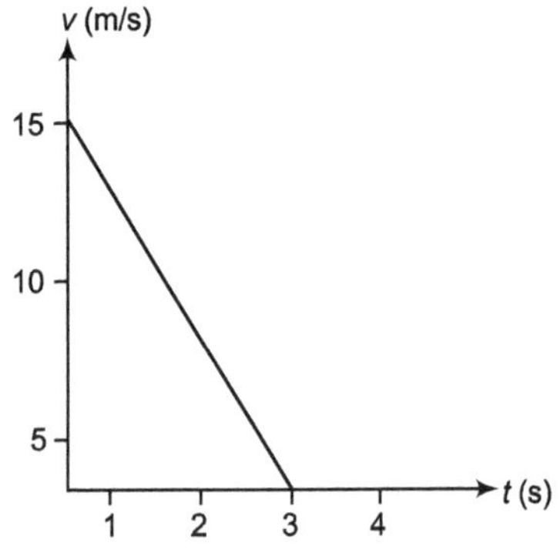

(a) 22.5 m/s^2 (b) 5 m/s^2 (c) -5 m/s^2 (d) -3 m/s^2

94. Figure shows the position-time (x-t) graph of the motion of two boys A and B returning from school O to their homes P and Q respectively. Which of the following statements is true?

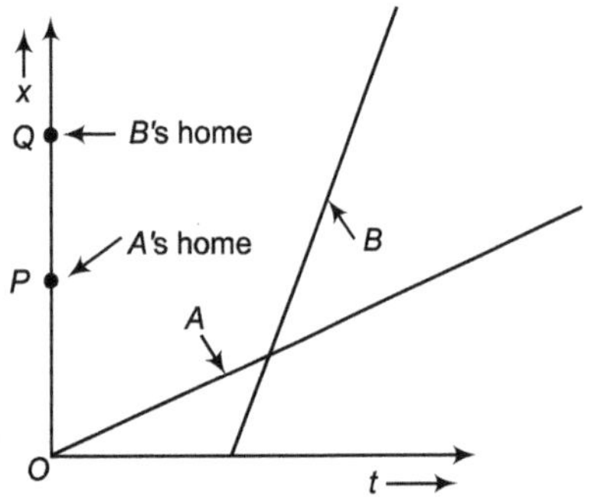

(a) A walks faster than B
(b) Both A and B reach home at the same time
(c) B starts for home earlier than A
(d) B overtakes A on his way to home

95. The x-t equation is given as $x = 2t + 1$. The corresponding v-t graph is

(a) a straight line passing through origin
(b) a straight line not passing through origin
(c) a parabola
(d) None of the above

96. The variation of velocity of a particle with time moving along a straight line is illustrated in the adjoining figure. The distance travelled by the particle in 4 s is

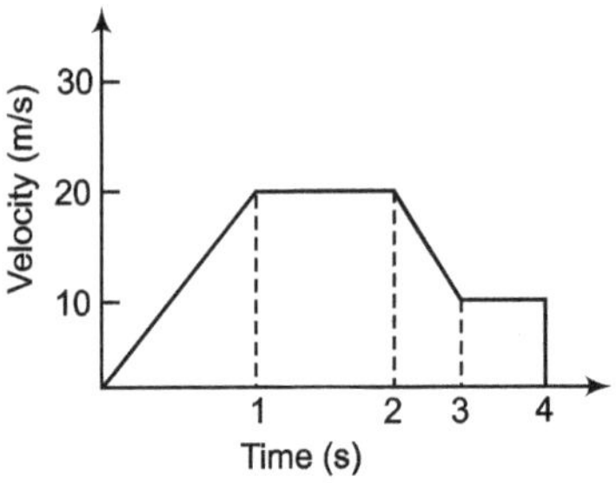

(a) 60 m (b) 55 m
(c) 25 m (d) 30 m

97. The v-t graph of a moving object is given in figure. The maximum acceleration is

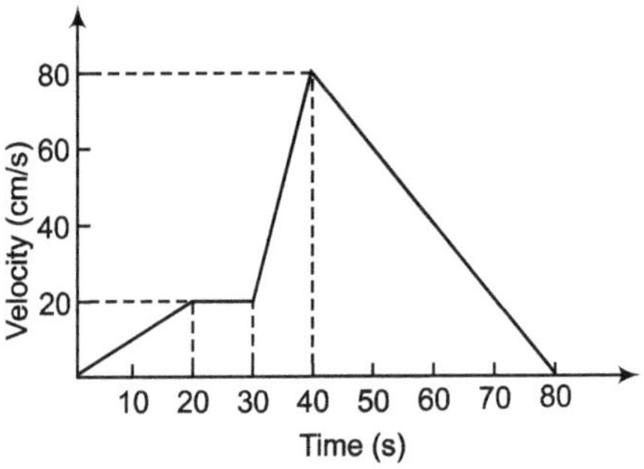

(a) 1 cm/s^2 (b) 2 cm/s^2
(c) 3 cm/s^2 (d) 6 cm/s^2

98. A lift is going up. The variation in the speed of the lift is as given in the graph. What is the height to which the lift takes the passengers?

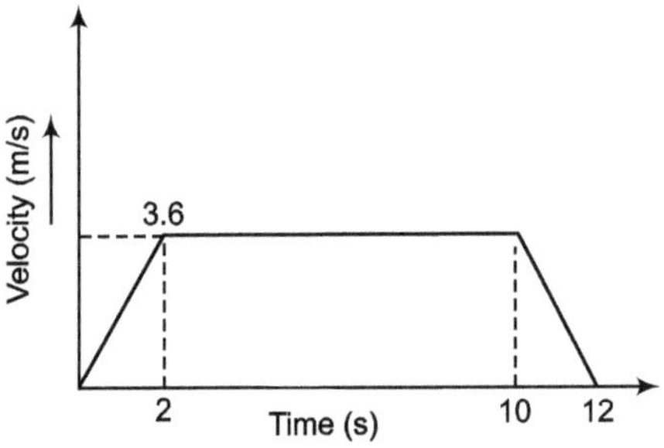

(a) 3.6 m
(b) 28.8 m
(c) 36.0 m
(d) Cannot be calculated from the above graph

99. A ball is dropped vertically from a height of above the ground. It hits the ground and bounces up vertically to a height $d/2$. Neglecting subsequent motion and air resistance, its velocity v varies with the height h above the ground is

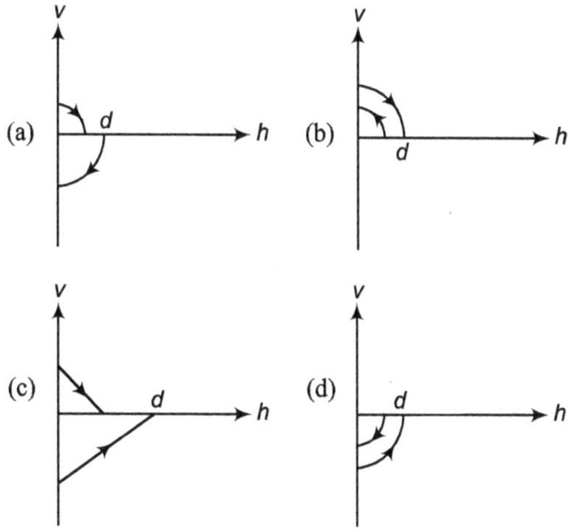

100. The given graph shows the variation of velocity with displacement. Which one of the graph given below correctly represents the variation of acceleration with displacement?

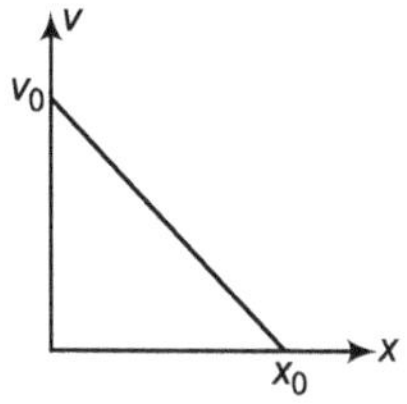

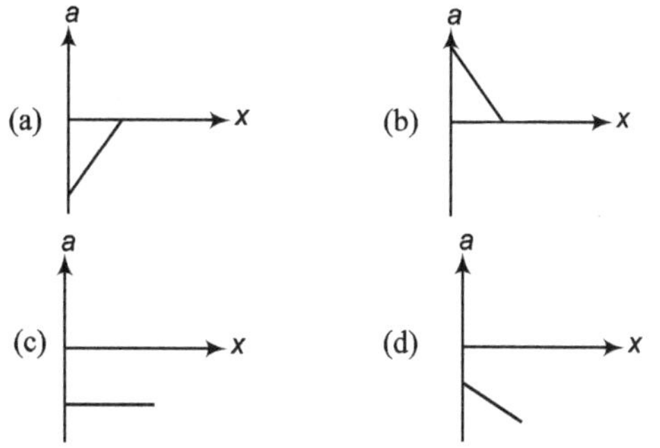

101. From the displacement-time graph find out the velocity of a moving body

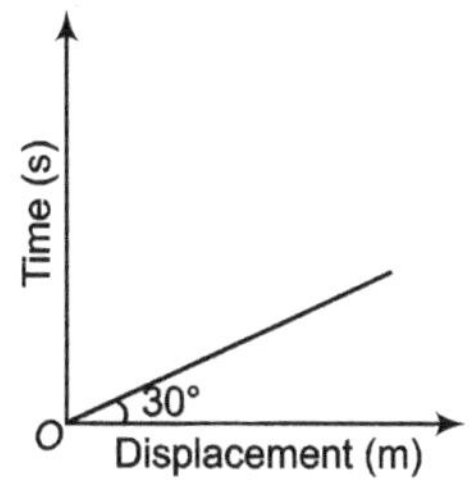

(a) $\frac{1}{\sqrt{3}}$ m/s (b) 3 m/s

(c) $\sqrt{3}$ m/s (d) $\frac{1}{3}$ m/s

102. The v-t plot of a moving object is shown in the figure. The average velocity of the object during the first 10 s, is

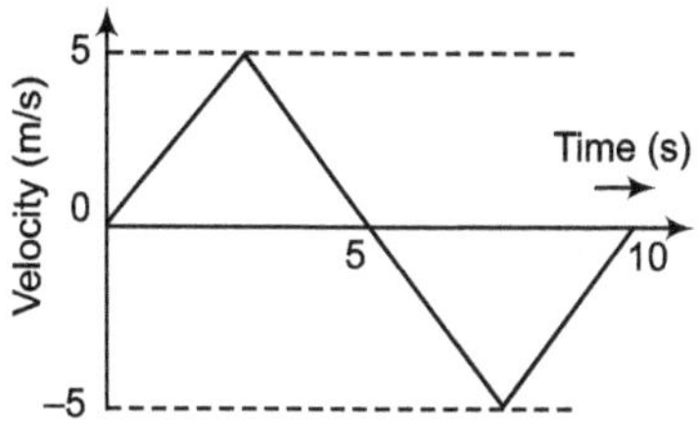

(a) zero (b) 2.5 ms^{-1}

(c) 5 ms^{-1} (d) 2 ms^{-1}

103. The acceleration-time $(a\text{-}t)$ graph of a particle moving in a straight line is as shown in the figure. The velocity-time graph of the particle would be

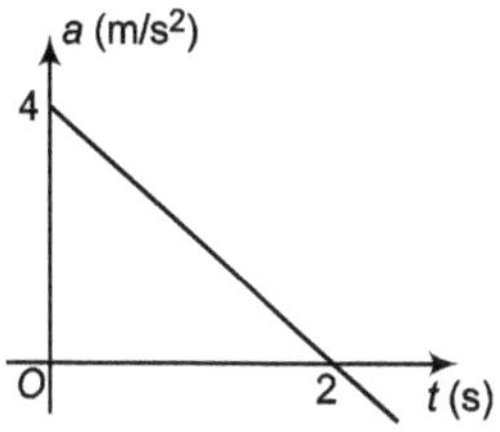

(a) a straight line (b) a parabola

(c) a circle (d) an ellipse

Relative Motion

104. A particle (A) moves due north at 3 km/h another particle (B) due west at 4 km/h. The relative velocity of A with respect to B is ($\tan 37° = 3/4$)

(a) 5 km/h, 37° north of east

(b) 5 km/h, 37° east of north

(c) $5\sqrt{2}$ km/h, 53° east of north

(d) $5\sqrt{2}$ km/h, 53° north of east

105. A river is flowing from west to east at a speed of 8 m per min. A man on the south bank of the river, capable of swimming at 20 m/min in still water, wants to swim across the river in the shortest time. He should swim in a direction

(a) due north (b) 30° east of north

(c) 30° west of north (d) 60° east of north

106. A man standing on a road has to hold his umbrella at 30° with the vertical to keep the rain away. He throws the umbrella and starts running at 10 km/h. He finds that raindrops are hitting his head vertically. What is the speed of rain with respect to ground?

(a) $10\sqrt{3}$ km/h

(b) 20 km/h

(c) $\frac{20}{\sqrt{3}}$ km/h

(d) $\frac{10}{\sqrt{3}}$ km/h

107. What are the speeds of two objects if, when they move uniformly towards each other, they get 4 m closer in each second and when they move uniformly in the same direction with the original speeds, they get 4.0 m closer each 10 s?
(a) 2.8 m/s and 1.2 m/s
(b) 5.2 m/s and 4.6 m/s
(c) 3.2 m/s and 2.1 m/s
(d) 2.2 m/s and 1.8 m/s

108. Two trains are each 50 m long moving parallel towards each other at speeds 10 m/s and 15 m/s respectively, at what time will they pass each other?
(a) 8 s (b) 4 s
(c) 2 s (d) 6 s

109. The rowing speed of a man relative to water is 5 km/h and the speed of water flow is 3 km/h. At what angle to the river flow should he head if he wants to reach a point on the other bank, directly opposite to starting point?
(a) 127° (b) 143°
(c) 120° (d) 150°

110. A stationary man observes that the rain is falling vertically downward. When he starts running with a velocity of 12 km/h, he observes that the rains is falling at an angle 60° with the vertical. The actual velocity of rain is
(a) $12\sqrt{3}$ km/h (b) $6\sqrt{3}$ km/h
(c) $4\sqrt{3}$ km/h (d) $2\sqrt{3}$ km/h

111. Two bodies are held separated by 9.8 m vertically one above the other. They are released simultaneously to fall freely under gravity. After 2 s, the relative distance between them is
(a) 4.9 m (b) 19.6 m
(c) 9.8 m (d) 39.2 m

112. A ball is dropped from the top of a building 100 m high. At the same instant, another ball is thrown upwards with a velocity of 40 ms^{-1} from the bottom of the building. The two balls will meet after
(a) 5 s (b) 2.5 s
(c) 2 s (d) 3 s

113. A 100 m long train crosses a man travelling at 5 km/h, in opposite direction , in 7.2 s, then the velocity of train is
(a) 40 km/h
(b) 25 km/h
(c) 20 km/h
(d) 45 km/h

114. A man is 25 m behind a bus, when bus starts accelerating at 2 m/s^2 and man starts moving with constant velocity of 10 m/ s. Time taken by him to board the bus is
(a) 2 s (b) 3 s
(c) 4 s (d) 5 s

115. Two stones are thrown up simultaneously with initial speeds of 'u_1 and u_2 ($u_2 > u_1$). They hit the ground after 6 s and 10 s respectively. Which graph in figure correctly represents the time variation of $\Delta x = (x_2 - x_1)$, the relative position of the second stone with respect to the first upto $t = 10$ s?
Assume that the stones do not rebound after hitting the ground.

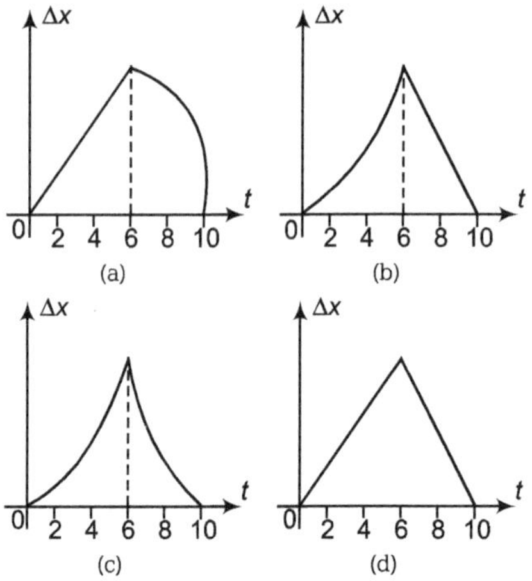

116. A boy is running on the plane road with velocity v with a long hollow tube in his hand. The water is failing vertically downwards with velocity u. At what angle to the vertical, he must incline the tube so that the water drops enter it without touching its sides?
(a) $\tan^{-1}\left(\frac{v}{u}\right)$ (b) $\sin^{-1}\left(\frac{v}{u}\right)$
(c) $\tan^{-1}\left(\frac{u}{v}\right)$ (d) $\cos^{-1}\left(\frac{v}{u}\right)$

117. Two points move in the same straight line starting at the same moment from the same point in it. The first moves with constant velocity u and the second with constant acceleration f. During the time elapses before the second catches, the first greatest distance between the particle is
(a) $\frac{u}{f}$ (b) $\frac{u^2}{2f}$ (c) $\frac{f}{2u^2}$ (d) $\frac{f}{u^2}$

118. A person walks up a stalled escalator in 90 s. When just standing on the same moving escalator, he is carried in 60 s. The time it would take him to walk up the moving escalator will be
(a) 27 s (b) 50 s (c) 18 s (d) 36 s

119. Particle A is moving along x-axis. At time $t = 0$, it has velocity of 10 m/s and acceleration -4 m/s^2. Particle B has velocity of 20 m/s and acceleration -2 m/s^2. Initially, both the particles are at origin. At time $t = 2$ s, distance between the two particles is
(a) 24 m (b) 36 m
(c) 20 m (d) 42 m

120. Driver of a train moving at a speed v_1 sights a freight train at a distance d ahead of him on the same track moving in the same direction with a slow speed v_2. He puts on the brakes and gives his train a constant deceleration α. Then, there will be no collision, if

(a) $d > \left(\frac{v_1 - v_2}{2\alpha}\right)$

(b) $d < \frac{(v_1 - v_2)^2}{2\alpha}$

(c) $d > \frac{(v_1 - v_2)^2}{2\alpha}$

(d) None of the above

121. The speed of a boat is 5 km/h in still water. It crosses a river of width 1 km along the shortest possible path in 15 min. Then, velocity of river will be

(a) 4.5 km/h (b) 4 km/h
(c) 1.5 km/h (d) 3 km/h

122. A ship X moving due north with speed v observes that another ship Y is moving due west with same speed v. The actual velocity of Y is

(a) $\sqrt{2v}$ towards south-west
(b) $\sqrt{2v}$ towards north-west
(c) $\sqrt{2v}$ towards south-east
(d) v towards north-east

[Level 2]

Only One Correct Option

1. A ball is dropped onto the floor from a height of 10 m. It rebounds to a height of 5 m. If the ball was in contact with the floor for 0.01 s, what was its average acceleration during contact? Take $g = 10\ \text{m/s}^2$.

(a) 2414 m/s^2
(b) 1735 m/s^2
(c) 3120 m/s^2
(d) 4105 m/s^2

2. Let **v** and **a** denote the velocity and acceleration respectively of a particle in the dimensional motion

(a) The speed of the particle decreases when $\mathbf{v} \cdot \mathbf{a} < 0$
(b) The speed of the particle increases when $\mathbf{v} \times \mathbf{a} > 0$
(c) The speed of the particle increases when $\mathbf{v} > \mathbf{a} = 0$
(d) The speed of the particle decreases when $|\mathbf{a}| < |\mathbf{a}|$

3. A stone is thrown vertically upward with an initial velocity v_0. The distance travelled in time $\frac{1.5 v_0}{g}$ is

(a) $\frac{v_0^2}{2g}$ (b) $\frac{3v_0^2}{8g}$
(c) $\frac{5v_0^2}{8g}$ (d) None of these

4. The vertical height of point P above the ground is twice that of Q. A particle is projected downward with a speed of 5 m/s from P and at the same time another particle is projected upward with the same speed from Q. Both particles reach the ground simultaneously, then

(a) $PQ = 30$ m
(b) time of flight of stones = 3 s
(c) Both (a) and (b) are correct
(d) Both (a) and (b) are wrong

5. The displacement x of a particle in a straight line motion is given by $x = 1 - t - t^2$. The correct representation of the motion is

(a)
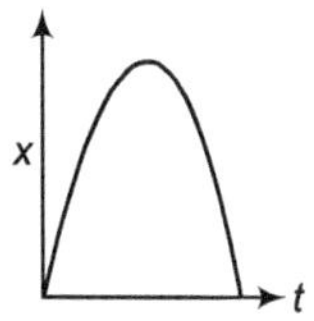

(b)
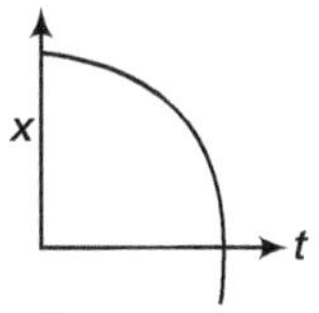

(c)
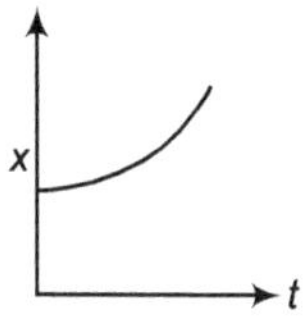

(d)
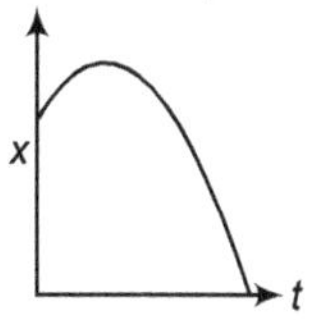

6. A swimmer crosses a river of width d flowing at velocity v. While swimming, he heads himself always at an angle of 120° with the river flow and on reaching the other end he finds a drift of $d/2$ in the direction of flow of river. The speed of the swimmer with respect to the river is

(a) $(2 - \sqrt{3})\, v$
(b) $2(2 - \sqrt{3})\, v$
(c) $4(2 - \sqrt{3})\, v$
(d) $(2 + \sqrt{3})\, v$

7. Three particles start simultaneously from a point on a horizontal smooth plane. First particle moves with speed v_1 towards east, second particle moves towards north with speed v_2 and third-one moves towards north-east. The velocity of the third particle, so that the three always lie on a line, is

(a) $\frac{v_1 + v_2}{\sqrt{2}}$ (b) $\sqrt{v_1 v_2}$
(c) $\frac{v_1 v_2}{v_1 + v_2}$ (d) $\sqrt{2}\frac{v_1 v_2}{v_1 + v_2}$

8. On a calm day, a boat can go across a lake and return in time T_0 at a speed V. On a rough day, there is uniform current at speed v to help the onward journey and impede the return journey. If the time taken to go across and return on the rough day be T, then T/T_0 is

(a) $1 - v^2/V^2$ (b) $\frac{1}{1 - v^2/V^2}$

(c) $1 + v^2/V^2$ (d) $\frac{1}{1 + v^2/V^2}$

9. A point moves in a straight line so that its displacement x at time t is given by $x^2 = t^2 + 1$. Its acceleration is

(a) $1/x$ (b) $1/x^3$

(c) $-1/x^2$ (d) $-1/x^3$

10. A car starts moving along a line, first with acceleration $a = 5\ \text{m/s}^2$ starting from rest then uniformly and finally decelerating at the same rate a comes to rest. The total time of motion is 25 s. The average speed during the time is $20\ \text{m/s}^2$. How long does the particle move uniformly?

(a) 10 s (b) 12 s (c) 20 s (d) 15 s

11. A ball is thrown vertically upwards with a speed u. It reaches a point B at a height h (lower than the maximum height) after time t_1. It returns to the ground after time t_2 from the instant it was at B during the upward journey. Then, $t_1\ t_2$ is equal to

(a) $2h/g$ (b) h/g

(c) $h/2g$ (d) $h/4g$

12. A target is made of two plates, one of wood and the other of iron. The thickness of the wooden plate is 4 cm and that of iron plate is 2 cm. A bullet fired goes through the wood first and then penetrates 1 cm into iron. A similar bullet fired with the same velocity from opposite direction goes through iron first and then penetrates 2 cm into wood. If a_1 and a_2 be the retardations offered to the bullet by wood and iron plates respectively, then

(a) $a_1 = 2a_2$ (b) $a_2 = 2a_1$

(c) $a_1 = a_2$ (d) Data insufficient

13. A body starts with an initial velocity of $10\ \text{ms}^{-1}$ and is moving along a straight line with constant acceleration. When the velocity of the particle is $50\ \text{ms}^{-1}$, the acceleration is reversed in direction. The velocity of the particle when it again reaches the starting point is

(a) $70\ \text{ms}^{-1}$ (b) $60\ \text{ms}^{-1}$ (c) $10\ \text{ms}^{-1}$ (d) $30\ \text{ms}^{-1}$

14. Two particles P and Q simultaneously start moving from point A with velocities 15 m/s and 20 m/s respectively. The two particles move with accelerations equal in magnitude but opposite in direction. When P overtakes Q at B, then its velocity is 30 m/s. The velocity of Q at point B will be

(a) 30 m/s (b) 5 m/s (c) 20 m/s (d) 15 m/s

15. If a particle takes t second less and acquires a velocity of $v\ \text{ms}^{-1}$ more in falling through the same distance on two planets, where the accelerations due to gravity are $2g$ and $8g$ respectively, then

(a) $v = 4gt$ (b) $v = 5gt$ (c) $v = 2gt$ (d) $v = 16gt$

16. Two particles start simultaneously from the same point and move along two straight lines, one with uniform velocity v and other with a uniform acceleration a. If α is the angle between the lines of motion of two particles, then the least value of relative velocity will be at time given by

(a) $\frac{v}{a}\sin\alpha$ (b) $\frac{v}{a}\cos\alpha$ (c) $\frac{v}{a}\tan\alpha$ (d) $\frac{v}{a}\cot\alpha$

17. Speed-time graph of two cars A and B approaching towards each other is shown in figure. Initial distance between them is 60 m. The two cars will cross each other after time

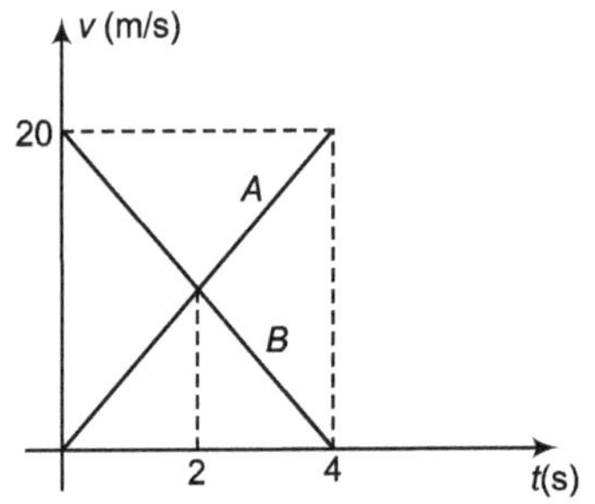

(a) 2 s (b) 3 s (c) 1.5 s (d) $\sqrt{2}$ s

18. The acceleration-time graph of a particle moving along a straight line is as shown in figure. At what time, the particle acquires its initial velocity?

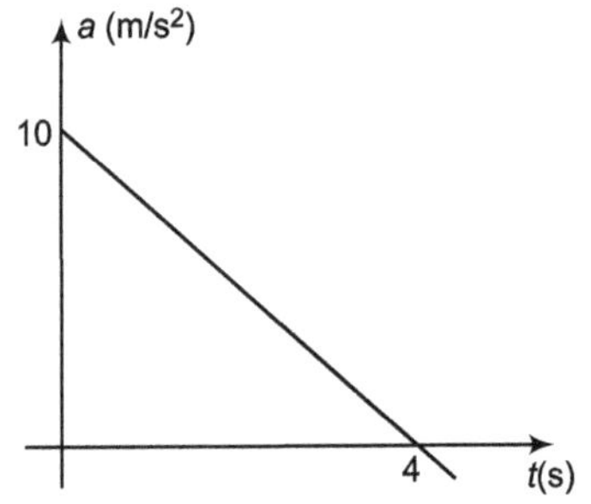

(a) 12 s (b) 5 s

(c) 8 s (d) 16 s

19. A street car moves rectilinearly from station A to the next station B with an acceleration varying according to the law $a = (b - cx)$, where b and c are constants and x is the distance from station A. The distance between the two stations and the maximum velocity are

(a) $x = 2b/c,\ v_{max} = \frac{b}{\sqrt{c}}$ (b) $x = \frac{c}{2b},\ v_{max} = b/c$

(c) $x = \frac{b}{2c},\ v_{max} = \frac{c}{\sqrt{a}}$ (d) $x = b/c,\ v_{max} = \frac{\sqrt{b}}{c}$

20. A lift performs the first part of its ascent with uniform acceleration a and the remaining with uniform retardation $2a$. If t is the time of ascent, find the depth of the shaft.

(a) $\frac{at^2}{4}$ (b) $\frac{at^2}{3}$

(c) $\frac{at^2}{2}$ (d) $\frac{at^2}{8}$

21. A particle starts moving from rest in a straight line with constant acceleration. After time t_0, acceleration changes its sign (just opposite to the initial direction), remaining the same in magnitude. Determine the time from the beginning of motion in which the particle returns to the initial position.

(a) $2t_0$ (b) $(2+\sqrt{2})\,t_0$

(c) $3t_0$ (d) $(2-\sqrt{2})\,t_0$

22. An elevator car whose floor to ceiling distance is 2.7 m starts ascending with a constant acceleration of $1.2\ \text{m/s}^2$. After 2 s of the start, a bolt falls from the ceiling of the car. The free fall time of the bolt is ($g = 9.8\ \text{m/s}^2$)

(a) $\sqrt{\frac{2.7}{9.8}}$ s (b) $\sqrt{\frac{5.4}{9.8}}$ s

(c) $\sqrt{\frac{5.4}{8.6}}$ s (d) $\sqrt{\frac{5.4}{11}}$ s

23. The displacement x of a particle varies with time t as $x = ae^{-\alpha t} + be^{\beta t}$, where a, b, α and β are positive constants. The velocity of the particle will

(a) go on decreasing with time
(b) be independent of α and β
(c) drop to zero when $\alpha = \beta$
(d) go on increasing with time

24. Two boys are standing at the ends A and B on ground, where $AB = a$. The boy at B starts running in a direction perpendicular to AB with velocity v_1. The boy at A starts running simultaneously with constant velocity v and catches the other boy in a time t, where t is

(a) $\frac{a}{\sqrt{v^2 + v_1^2}}$ (b) $\sqrt{\frac{a^2}{v^2 - v_1^2}}$

(c) $\frac{a}{(v - v_1)}$ (d) $\frac{a}{(v + v_1)}$

25. A man is, d distance behind a bus. The bus moves away from the man with an acceleration a. At the same time, man starts running towards bus with a constant velocity v.

(a) The man catches the bus, if $v \geq \sqrt{2\,ad}$
(b) If man just catches the bus the time of catching bus will be $t = \frac{v}{a}$
(c) If man just catches the bus, the time of catching bus will be $t = \frac{2v}{a}$
(d) The man will catch the bus, if $v \geq \sqrt{ad}$

26. A small electric car has a maximum constant acceleration of $1\ \text{m/s}^2$, a maximum constant deceleration of $2\ \text{m/s}^2$ and a maximum speed of 20 m/s. The amount of time it would take to drive this car 1 km starting from rest and finishing at rest is

(a) 15 s (b) 50 s

(c) 35 s (d) 65 s

27. The diagram shows the variation of $1/v$ (where, v is velocity of the particle) with respect to time. At time $t = 3$ s using the details given in the graph, the instantaneous acceleration will be equal to

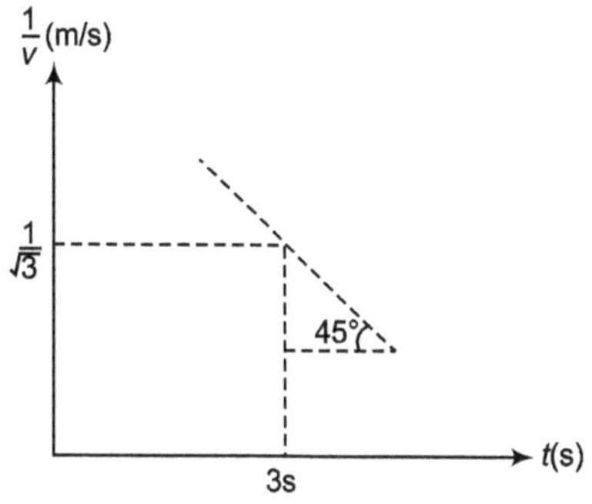

(a) $-2\ \text{m/s}^2$ (b) $+3\ \text{m/s}^2$

(c) $+5\ \text{m/s}^2$ (d) $-6\ \text{m/s}^2$

28. A particle starting from rest and moving with a uniform acceleration along a straight line covers distances a and b in successive intervals of p and q second. The acceleration of the particle is

(a) $\frac{a+b}{2\,(p+q)}$ (b) $\frac{2b}{(q+2p)\,q}$

(c) $\frac{2a}{p\,(p+2p)}$ (d) $\frac{a+b}{(p+q)^2}$

29. A rocket is fired vertically up from the ground with a resultant acceleration of $10\ \text{m/s}^2$. The fuel is finished in 1 min and it continues to move up ($g = 10\ \text{m/s}^2$)

(a) the maximum height reached by rocket from ground is 18 km
(b) the maximum height reached by rocket from ground is 36 km
(c) the time from initial in which rocket is again at ground is $(180 + 30\sqrt{2})$ s
(d) the time from initial in which rocket is again at ground is $(120 + 60\sqrt{2})$ s

30. Equation of motion of a body is $\frac{dv}{dt} = -4v + 8$, where v is the velocity in m/s and t is the time in second. Initial velocity of the particle was zero. Then,

(a) the initial rate of change of acceleration of the particle is $8\ \text{m/s}^3$
(b) the terminal speed is 2 m/s
(c) Both (a) and (b) are correct
(d) Both (a) and (b) are wrong

31. Among the four graph shown in the figure there is only one graph for which average velocity over the time interval $(0, T)$ can vanish for a suitably chosen T. Which one is it?

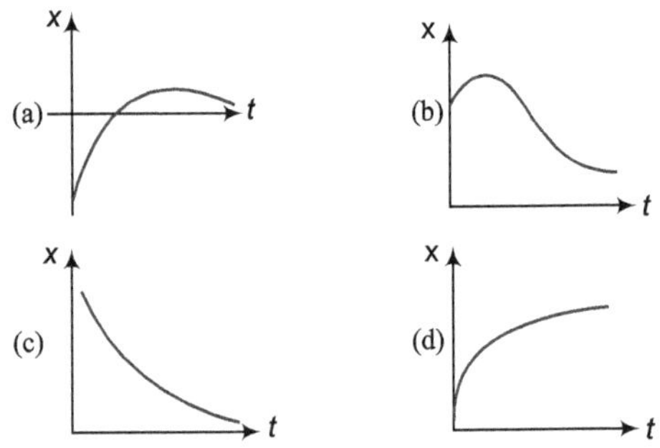

32. A lift is coming from 8th floor and is just about to reach 4th floor. Taking ground floor as origin and positive direction upwards for all quantities, which one of the following is correct?

(a) $x < 0, v < 0, a > 0$ (b) $x > 0, v < 0, a < 0$
(c) $x > 0, v < 0, a > 0$ (d) $x > 0, v > 0, a < 0$

33. A bird is tossing (flying to and fro) between two cars moving towards each other on a straight road. One car has a speed of 18 km/h while the other has the speed of 27 km/h. The bird starts moving from first car towards the other and is moving with the speed of 36 km/h and when the two cars were separated by 36 km. What is the total distance covered by the bird?

(a) 28.8 km (b) 36.4 km
(c) 58.2 km (d) None of these

34. A motor car moving at a speed of 72 km/h cannot come to a stop in less than 3.0 s while for a truck this time interval is 5.0 s. On a highway, the car is behind the truck both moving at 72 km/h. The truck gives a signal that it is going to stop at emergency. At what distance the car should be from the truck so that it does not bump onto (collide with) the truck. Human response time is 0.5 s.

(a) 6.75 m (b) 1.25 m
(c) 4.25 m (d) None of these

More than One Correct Options

1. A particle having a velocity $v = v_0$ at $t = 0$ is decelerated at the rate $|a| = \alpha\sqrt{v}$, where α is a positive constant.

(a) The particle comes to rest at $t = \dfrac{2\sqrt{v_0}}{\alpha}$
(b) The particle will come to rest at infinity
(c) The distance travelled by the particle before coming to rest is $\dfrac{2v_0^{3/2}}{\alpha}$
(d) The distance travelled by the particle before coming to rest is $\dfrac{2v_0^{3/2}}{3\alpha}$

2. At time $t = 0$, a car moving along a straight line has a velocity of 16 ms^{-1}. It slows down with an acceleration of $-0.5t$ ms^{-2}, where t is in second. Mark the correct statement (s).

(a) The direction of velocity changes at $t = 8$ s
(b) The distance travelled in 4 s is approximately 58.67 m
(c) The distance travelled by the particle in 10 s is 94 m
(d) The speed of particle at $t = 10$ s is 9 ms^{-1}

3. An object moves with constant acceleration **a**. Which of the following expressions are also constant ?

(a) $\dfrac{d|\mathbf{v}|}{dt}$ (b) $\left|\dfrac{d\mathbf{v}}{dt}\right|$
(c) $\dfrac{d(v^2)}{dt}$ (d) $\dfrac{d\left(\dfrac{\mathbf{v}}{|\mathbf{v}|}\right)}{dt}$

4. Ship A is located 4 km north and 3 km east of ship B. Ship A has a velocity of 20 kmh^{-1} towards the south and ship B is moving at 40 kmh^{-1} in a direction 37° north of east. X and Y-axes are along east and north directions, respectively.

(a) Velocity of A relative to B is $-32\hat{\mathbf{i}} - 44\hat{\mathbf{j}}$
(b) Position of A relative to B as a function of time is given by $\mathbf{r}_{AB} = (3 - 32t)\hat{\mathbf{i}} + (4 - 44t)\hat{\mathbf{j}}$
(c) Velocity of A relative to B is $32\hat{\mathbf{i}} - 44\hat{\mathbf{j}}$
(d) Position of A relative to B as a function of time is given by $(32t\hat{\mathbf{i}} - 44t\hat{\mathbf{j}})$

5. Starting from rest a particle is first accelerated for time t_1 with constant acceleration a_1 and then stops in time t_2 with constant retardation a_2. Let v_1 be the average velocity in this case and s_1 the total displacement. In the second case, it is accelerating for the same time t_1 with constant acceleration $2a_1$ and come to rest with constant retardation a_2 in time t_3. If v_2 is the average velocity in this case and s_2 the total displacement, then

(a) $v_2 = 2v_1$
(b) $2v_1 < v_2 < 4v_1$
(c) $s_2 = 2s_1$
(d) $2s_1 < s_2 < 4s_1$

6. A particle is moving along a straight line. The displacement of the particle becomes zero in a certain time ($t > 0$). The particle does not undergo any collision.

(a) The acceleration of the particle may be zero always
(b) The acceleration of the particle may be uniform
(c) The velocity of the particle must be zero at some instant
(d) The acceleration of the particle must change its direction

7. A particle is resting over a smooth horizontal floor. At $t = 0$, a horizontal force starts acting on it. Magnitude of the force increases with time according to law $F = \alpha t$, where α is a positive constant.

From figure, which of the following statements are correct ?

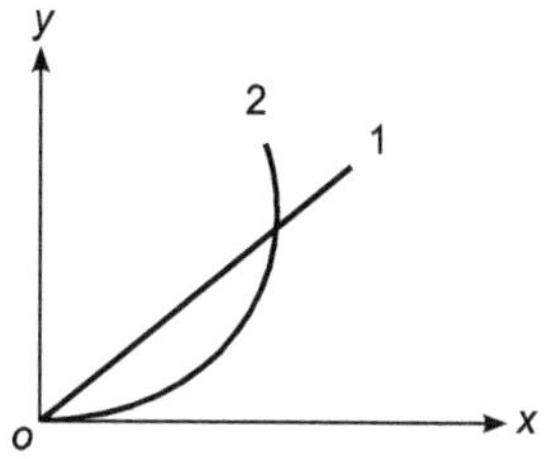

(a) Curve 1 can be the plot of acceleration against time
(b) Curve 2 can be the plot of velocity against time
(c) Curve 2 can be the plot of velocity against acceleration
(d) Curve 1 can be the plot of displacement against time

8. A train starts from rest at $S = 0$ and is subjected to an acceleration as shown in figure.

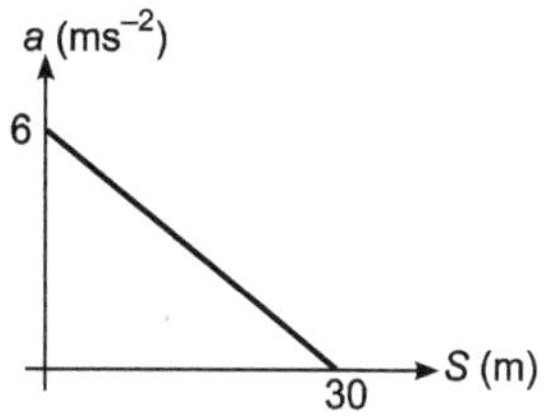

(a) Velocity at the end of 10 m displacement is 20 ms^{-1}
(b) Velocity of the train at $S = 10$ m is 10 ms^{-1}
(c) The maximum velocity attained by train is $\sqrt{180}$ ms^{-1}
(d) The maximum velocity attained by the train is 15 ms^{-1}

9. For a moving particle, which of the following options may be correct?

(a) $|\mathbf{v}_{av}| < v_{av}$
(b) $|\mathbf{v}_{av}| > v_{av}$
(c) $\mathbf{v}_{av} = 0$ but $v_{av} \neq 0$
(d) $\mathbf{v}_{av} \neq 0$ but $v_{av} = 0$

10. Identify the correct graph representing the motion of a particle along a straight line with constant acceleration with zero initial velocity

(a)

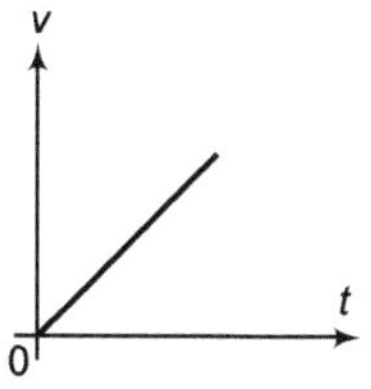

(b)

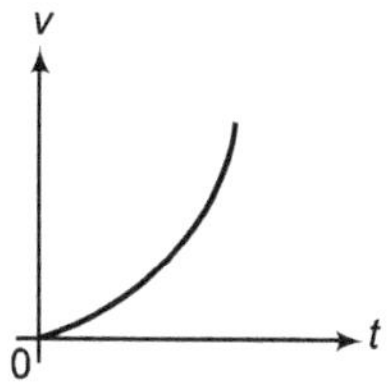

(c)

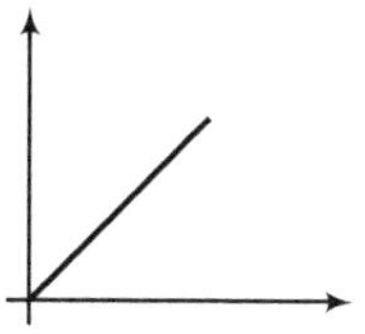

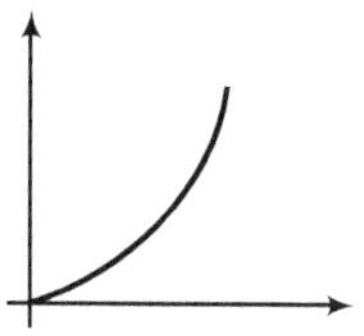

11. A man who can swim at a velocity v relative to water wants to cross a river of width b, flowing with a speed u.

(a) The minimum time in which he can cross the river is $\frac{b}{v}$
(b) He can reach a point exactly opposite on the bank in time $t = \frac{b}{\sqrt{v^2 - u^2}}$ if $v > u$
(c) He cannot reach the point exactly opposite on the bank if $u > v$
(d) He cannot reach the point exactly opposite on the bank if $v > u$

12. The figure shows the velocity (v) of a particle plotted against time (t).

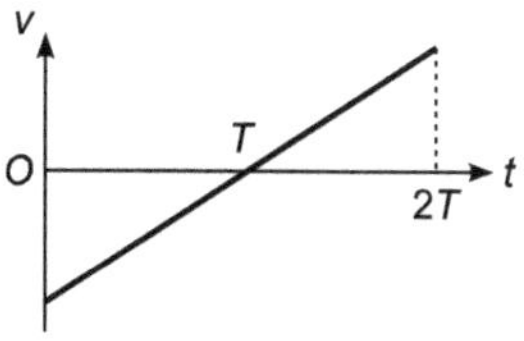

(a) The particle changes its direction of motion at some point
(b) The acceleration of the particle remains constant
(c) The displacement of the particle is zero
(d) The initial and final speeds of the particle are the same

13. The speed of a train increases at a constant rate α from zero to v and then remains constant for an interval and finally decreases to zero at a constant rate β. The total distance travelled by the train is l. The time taken to complete the journey is t. Then,

(a) $t = \frac{l(\alpha + \beta)}{\alpha\beta}$
(b) $t = \frac{l}{v} + \frac{v}{2}\left(\frac{1}{\alpha} + \frac{1}{\beta}\right)$
(c) t is minimum when $v = \sqrt{\frac{2l\alpha\beta}{(\alpha - \beta)}}$
(d) t is minimum when $v = \sqrt{\frac{2l\alpha\beta}{(\alpha + \beta)}}$

14. A particle moves in xy-plane and at time t is at the point $(t^2, t^3 - 2t)$, then which of the following is/are correct?

(a) At $t = 0$, particle is moving parallel to y-axis
(b) At $t = 0$, direction of velocity and acceleration are perpendicular
(c) At $t = \sqrt{\frac{2}{3}}$, particle is moving parallel to x-axis
(d) At $t = 0$, particle is at rest

15. A car is moving with uniform acceleration along a straight line between two stops X and Y. If its speeds at X and Y are 2 ms^{-1} and 14 ms^{-1}, then

(a) its speed at mid-point of XY is 10 ms^{-1}
(b) its speed at a point A such that $XA : AY = 1 : 3$ is 5 ms^{-1}
(c) the time to go from X to the mid-point of XY is double of that to go from mid-point to Y
(d) the distance travelled in first half of the total time is half of the distance travelled in the second half of the time

16. A graph of x *versus* t is shown in figure. Choose the correct alternatives given below

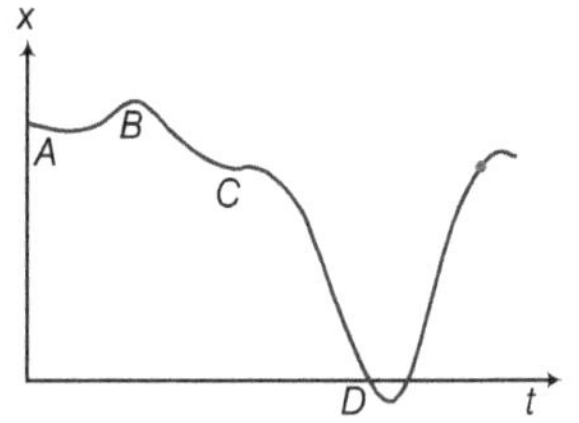

(a) The particle was released from rest at $t = 0$
(b) At B, the acceleration $a > 0$
(c) Average velocity for the motion between A and D is positive
(d) The speed at D exceeds that at E

17. A man is standing on top of a building 100m high. He throws two balls vertically, one at $t = 0$ and after a time interval (less than 2 s). The later ball is thrown at a velocity of half the first. The vertical gap between first and second ball is $+15$ m at $t = 2$ s. The gap is found to remain constant. Select the correct alternatives.
(a) The initial speeds of two balls (in m/s) are 10 and 20
(b) The initial speeds of two balls (in m/s) are 5 and 10
(c) The time interval between their throw is 0.5 s
(d) The time interval between their throw is 1 s

18. Following are four different relations about displacement, velocity and acceleration for the motion of a particle in general. Choose the incorrect one (s).
(a) $v_{av} = \frac{1}{2}[v(t_1) + v(t_2)]$
(b) $v_{av} = \frac{r(t_2) - r(t_1)}{t_2 - t_1}$
(c) $r = \frac{1}{2}(v(t_2) - v(t_1))(t_2 - t_1)$
(d) $a_{av} = \frac{v(t_2) - v(t_1)}{t_2 - t_1}$

Comprehension Based Questions

Passage I (Q.1 to 4)

An elevator without a ceiling is ascending up with an acceleration of 5 ms^{-2}. A boy on the elevator shoots a ball in vertical upward direction from a height of 2 m above the floor of elevator. At this instant the elevator is moving up with a velocity of 10 ms^{-1} and floor of the elevator is at a height of 50 m from the ground. The initial speed of the ball is 15 ms^{-1} with respect to the elevator. Consider the duration for which the ball strikes, the floor of elevator in answering following questions: ($g = 10\, ms^{-2}$)

1. The time in which the ball strikes the floor of elevator is given by
(a) 2.13 s (b) 2.0 s
(c) 1.0 s (d) 3.12 s

2. The maximum height reached by ball, as measured from the ground would be
(a) 73.65 m (b) 116.25 m
(c) 82.56 m (d) 63.25 m

3. Displacement of ball with respect to ground during its flight would be
(a) 16.25 m (b) 8.76 m
(c) 20.24 m (d) 30.56 m

4. The maximum separation between the floor of elevator and the ball during its flight would be
(a) 12 m (b) 15 m
(c) 9.5 m (d) 7.5 m

Passage II (Q.5 to 7)

A situation is shown in which two objects A and B start their motion from same point in same direction. The graph of their velocities against time is drawn. u_A and u_B are the initial velocities of A and B respectively. T is the time at which their velocities become equal after start of motion. You cannot use the data of one question while solving another question of the same set. So, all the questions are independent of each other.

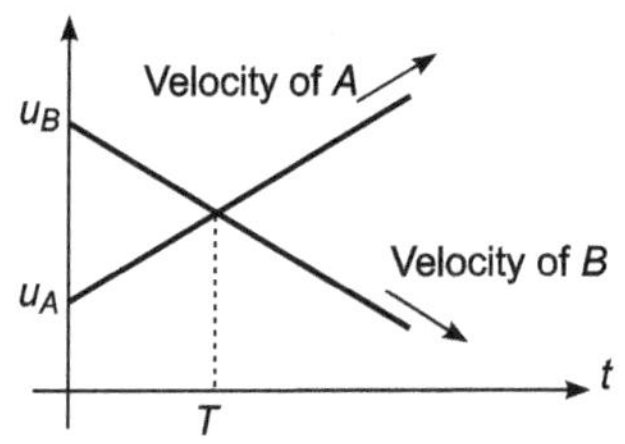

5. If the value of T is 4 s, then the time after which A will meet B is
(a) 12 s
(b) 6 s
(c) 8 s
(d) data insufficient

6. Let v_A and v_B be the velocities of the particles A and B respectively at the moment A and B meet after start of the motion. If $u_A = 5$ ms^{-1} and $u_B = 15$ ms^{-1}, then the magnitude of the difference of velocities v_A and v_B is
(a) 5 ms^{-1}
(b) 10 ms^{-1}
(c) 15 ms^{-1}
(d) data insufficient

7. After 10 s of the start of motion of both objects A and B, find the value of velocity of A, if $u_A = 6$ ms^{-1}, $u_B = 12$ ms^{-1} and at T velocity of A is 8 ms^{-1} and $T = 4$ s
(a) 12 ms^{-1}
(b) 10 ms^{-1}
(c) 15 ms^{-1}
(d) None of the above

Assertion and Reason

Directions (Q. Nos. 1-20) *These questions consists of two statements each printed as Assertion and Reason. While answering these question, you are required to choose any one of the following five responses*

(a) If both Assertion and Reason are correct and Reason is the correct explanation of Assertion
(b) If both Assertion and Reason are correct but Reason is not the correct explanation of Assertion
(c) If Assertion is true but Reason is false
(d) If Assertion is false but Reason is true
(e) If both Assertion and Reason are false.

1. **Assertion** A body is moving along a straight line such that its velocity varies with time as shown in figure. Magnitude of displacement of the body from $t = 0$ to $t = 12$ s is the same as the distance travelled by it in the given time duration.

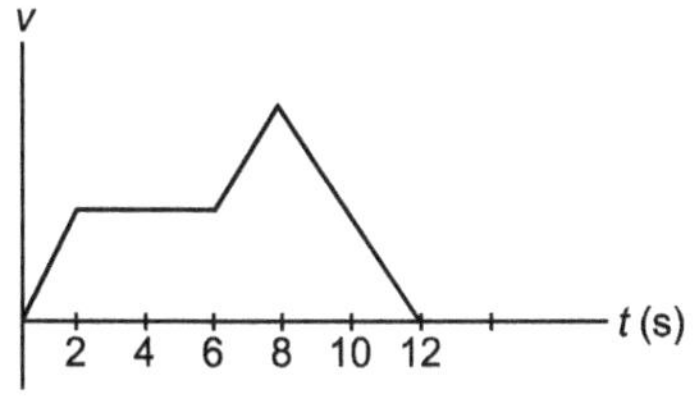

Reason For unidirectional motion of a body, $|\text{displacement}| = \text{distance}$

2. **Assertion** An object may have varying speed without having varying velocity.

Reason If the velocity is zero at an instant, the acceleration may not be zero at that instant.

3. **Assertion** A body of mass 4 kg has an initial velocity $5\hat{\mathbf{i}}$ m/s. It is subjected to a force of $4\hat{\mathbf{j}}$N. The displacement of the body from the origin after 4 s will be 21.5 m.

Reason The equation $\mathbf{v} = \mathbf{u} + \mathbf{a}\,t$ can be applied to obtain $\mathbf{v}$, if $\mathbf{a}$ is constant.

4. **Assertion** A body is momentarily at rest at the instant it reverses the direction.

Reason A body cannot have acceleration if its velocity is zero at a given instant of time.

5. **Assertion** On a curved path, average speed of a particle can never be equal to average velocity.

Reason Average speed is total distance travelled divided by total time. Whereas average velocity is final velocity, plus initial velocity divided by two.

6. **Assertion** Acceleration of a moving particle can change its direction without any change in direction of velocity.

Reason If the direction of change in velocity vector changes, the direction of acceleration vector also changes.

7. **Assertion** Plotting the acceleration-time graph from a given position-time graph of a particle moving along a straight line is possible.

Reason From position-time graph, sign of acceleration can be determined.

8. **Assertion** The v-t graph perpendicular to time axis is not possible in practice.

Reason Infinite acceleration cannot be realized in practice.

9. **Assertion** Magnitude of average velocity is equal to average speed, if velocity is constant.

Reason If velocity is constant, then there in no change in the direction of motion.

10. **Assertion** The average velocity of a particle having initial and final velocity $\mathbf{v}_1$ and $\mathbf{v}_2$ is $\mathbf{v}_1 + \mathbf{v}_2 / 2$.

Reason If $\mathbf{r}_1$ and $\mathbf{r}_2$ be the initial and final displacement in time t, then $\mathbf{v}_{av} = \dfrac{\mathbf{r}_1 - \mathbf{r}_2}{t}$.

11. **Assertion** If a particle is thrown upwards, then distance travelled in last second of upward journey is independent of the velocity of projection.

Reason In last second, distance travelled is 4.9 m. (Take $g = 9.8\ \text{m/s}^2$)

12. **Assertion** If acceleration of a particle moving in a straight line varies as $a \propto t^n$, then

$$s \propto t^{n+2}$$

Reason If a-t graph is a straight line,then s-t graph may be a parabola.

13. **Assertion** Distance between two particles moving with constant velocities always remains constant.

Reason In the above case relative motion between them is uniform.

14. **Assertion** Particle A is moving eastwards and particle B northwards with same speed. Then, velocity of A with respect to B is in south-east direction.

Reason Relative velocity between them is zero as their speeds are same.

15. **Assertion** In the v-t diagram shown in figure, average velocity between the interval $t = 0$ and $t = t_0$ is independent of t_0.

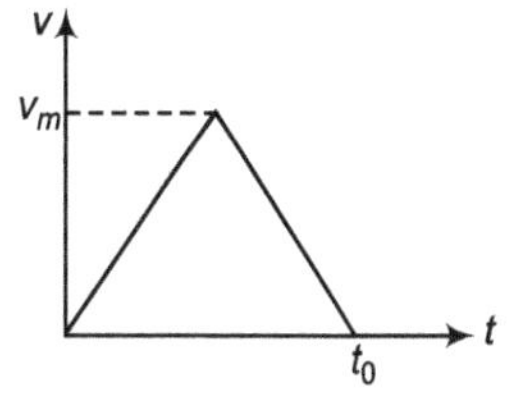

Reason Average velocity in the given interval is $\dfrac{1}{2} v_m$.

16. Assertion In the equation,

$$s = u + at - \frac{1}{2}a$$

s is the distance travelled in t th second.

Reason The above equation is dimensionally incorrect.

17. Assertion In the s-t graph shown in figure, velocity of particle is negative and acceleration is positive.

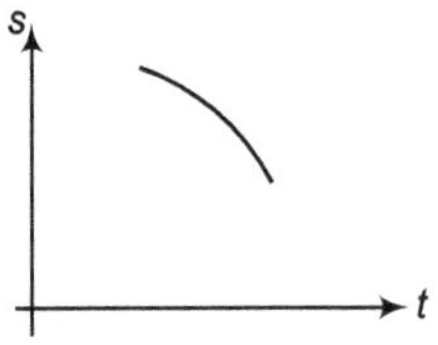

Reason Slope of s-t graph is negative and increasing in magnitude.

18. Assertion A lift is ascending with decreasing speed means acceleration of lift is downwards.

Reason A body always moves in the direction of its acceleration.

19. Assertion In the s-t diagram shown in figure, the body starts in positive direction but not form $s = 0$.

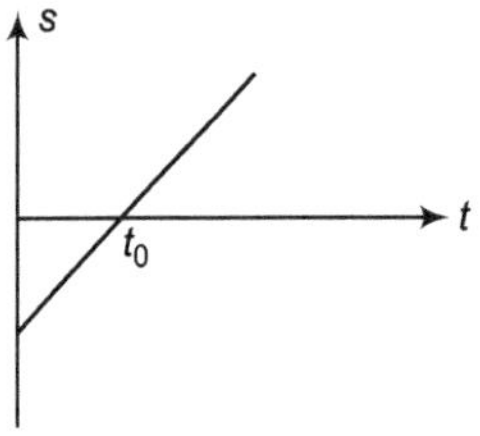

Reason At $t = t_0$, velocity of body changes its direction of motion.

20. Assertion If velocity-time equation of a particle moving in a straight line is quadratic, then displacement-time equation can't be linear.

Reason If displacement-time is quadratic, then velocity-time is linear.

Match the Columns

1. Match the following columns.

Column I	Column II
(A) $d\mathbf{v}/dt$	(p) Acceleration
(B) $d\lvert\mathbf{v}\rvert/dt$	(q) Magnitude of acceleration
(C) $d\mathbf{r}/dt$	(r) Velocity
(D) $\left\lvert\frac{d\mathbf{r}}{dt}\right\rvert$	(s) Magnitude of velocity

2. Velocity of a particle is in negative direction with constant acceleration in positive direction. Then, match the following columns.

Column I	Column II
(A) Velocity-time graph	(p) Slope $\to$ negative
(B) Acceleration-time graph	(q) Slope $\to$ positive
(C) Displacement-time graph	(r) Slope $\to$ zero
	(s) $\lvert$Slope$\rvert \to$ increasing

3. For the velocity-time graph shown in the figure, in a time interval from $t = 0$ to $t = 6$ s, match the following columns.

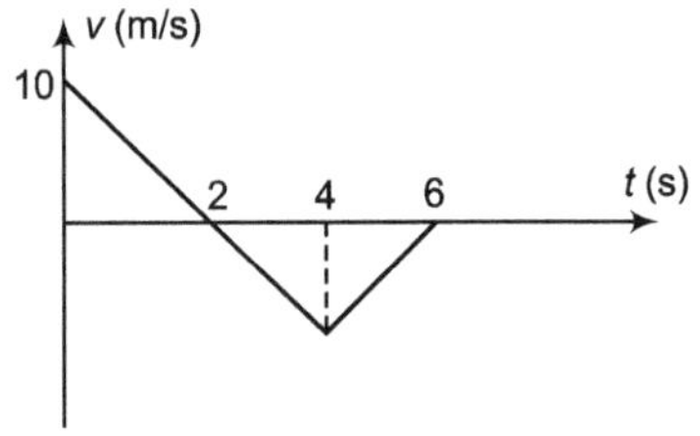

Column I	Column II
(A) Change in velocity	(p) – 5/3 SI unit
(B) Average acceleration	(q) – 20 SI unit
(C) Total displacement	(r) – 10 SI unit
(D) Acceleration at t = 3 s	(s) – 5 SI unit

4. Let us cell a motion, A when velocity is positive and increasing A^{-1} when velocity is negative and increasing. R when velocity is positive and decreasing and R^{-1} when velocity is negative and decreasing. Now, match the following two columns for the given s-t graph.

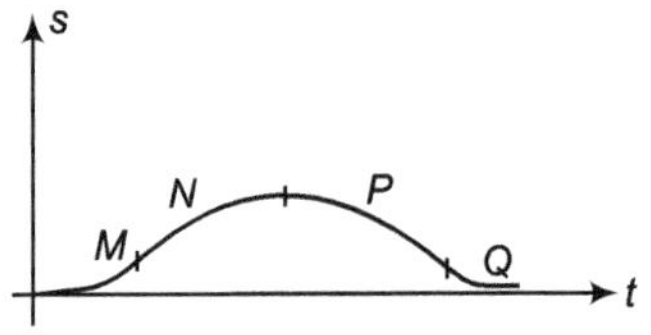

Column I		Column II	
(A)	M	(p)	A^{-1}
(B)	N	(q)	R^{-1}
(C)	P	(r)	A
(D)	Q	(s)	R

5. In the s-t equation ($s = 10 + 20t - 5t^2$), match the following columns.

Column I	Column II
(A) Distance travelled in 3 s	(p) –20 units
(B) Displacement in 1 s	(q) 15 units
(C) Initial acceleration	(r) 25 units
(D) Velocity at 4 s	(s) –10 units

6. A particle is rotating in a circle of radius 1m with constant speed 4 m/s. In time 1s, match the following (in SI units) columns.

Column I	Column II
(A) Displacement	(p) 8 sin 2
(B) Distance	(q) 4
(C) Average velocity	(r) 2 sin 2
(D) Average acceleration	(s) 4 sin 2

7. Match the following columns.

Column I	Column II
(A) Constant positive acceleration	(p) Speed may increase
(B) Constant negative acceleration	(q) Speed may decrease
(C) Constant displacement	(r) Speed is zero
(D) Constant slope of *a*-*t* graph	(s) Speed must increase

8. A balloon rises, up with constant net acceleration of $10\,m/s^2$. After 2 s a particle drops from the balloon. After further 2 s match the following columns. (Take $g = 10\,m/s^2$)

Column I	Column II
(A) Height of particle from ground	(p) Zero
(B) Speed of particle	(q) 10 SI unit
(C) Displacement of particle	(r) 40 SI unit
(D) Acceleration of particle	(s) 20 SI unit

9. Refer to the graph in figure. Match the following

Column I	Column II
(A) 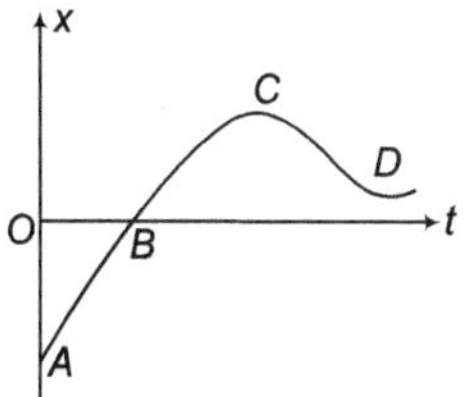	(p) has $v > 0$ and $a < 0$ throughout
(B) 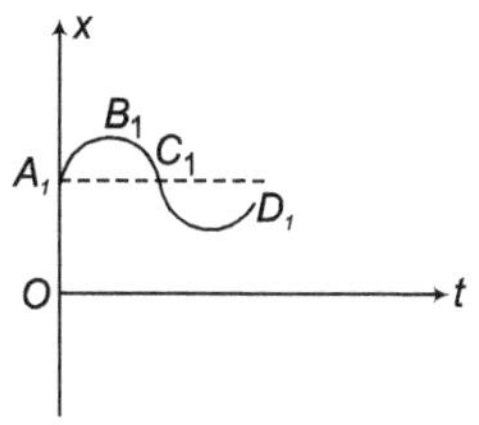	(q) has $x > 0$ throughout and has a point with $v = 0$ and a point with $a = 0$
(C) 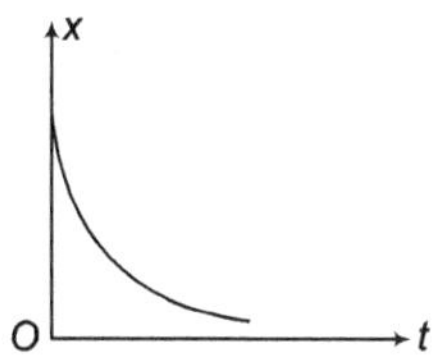	(r) has a point with zero displacement for $t > 0$
(D) 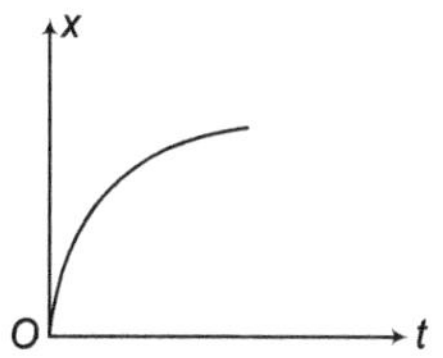	(s) has $v < 0$ and $a > 0$

Entrance Gallery

2014

1. A rocket is moving in a gravity free space with a constant acceleration of $2\ ms^{-2}$ along $+x$-direction (see figure). The length of a chamber inside the rocket is 4 m. A ball is thrown from the left end of the chamber in $+x$-direction with a speed of $0.3\ ms^{-1}$ relative to the rocket. At the same time, another ball is thrown in $-x$-direction with a speed of $0.2\ ms^{-1}$ from its right end relative to the rocket. The time (in seconds) when the two balls hit each other is **[JEE Advanced]**

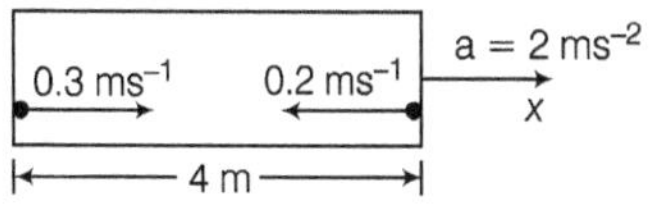

(a) 2 or 8
(b) 3
(c) 6
(d) 9

2. From a tower of height H, a particle is thrown vertically upwards with a speed u. The time taken by the particle, to hit the ground, is n times that taken by it to reach the highest point of its path. The relation between H, u and n is **[JEE Main]**

(a) $2gH = n^2u^2$
(b) $gH = (n-2)^2u^2$
(c) $2gH = nu^2(n-2)$
(d) $gH = (n-2)\,u^2$

3. The ratio of distance traversed in successive intervals of time when a body falls freely under gravity from certain height is **[Kerala CEE]**

(a) 1 : 2 : 3
(b) 1 : 5 : 9
(c) 1 : 3 : 5
(d) $\sqrt{1} : \sqrt{2} : \sqrt{3}$
(e) 1 : 4 : 9

4. A ball is dropped from the top of a tower of height 100 m and at the same time another ball is projected vertically upwards from ground with a velocity $25\,m^{-1}$. Then, the distance from the top of the tower at which the two balls meet, is **[Kerala CEE]**

(a) 68.4 m
(b) 48.4 m
(c) 18.4 m
(d) 28.4 m
(e) 78.4 m

5. A particle starting with certain initial velocity and uniform acceleration covers a distance of 12 m in first 3 s and a distance of 30 m in next 3 s. The initial velocity of the particle is **[Kerala CEE]**

(a) $3\ ms^{-1}$
(b) $2.5\ ms^{-1}$
(c) $2\ ms^{-1}$
(d) $1.5\ ms^{-1}$
(e) $1\ ms^{-1}$

6. A bullet when first into a target loses half of its velocity after penetrating 20 cm. Further distance of penetration before it comes to rest is [Kerala CEE]

(a) 6.66 cm (b) 3.33 cm
(c) 12.5 cm (d) 10 cm
(e) 5 cm

7. A car moves from A to B with a speed of 30 km/h and from B to A with a speed of 20 km/h. What is the average speed of the car? [Karnataka CET]

(a) 25 km/h (b) 24 km/h
(c) 50 km/h (d) 10 km/h

8. A body starts from rest and moves with constant acceleration for t. It travels a distance x_1 in first half of time and x_2 in next half of time, then [Karnataka CET]

(a) $x_2 = x_1$ (b) $x_2 = 2x_1$
(c) $x_2 = 3x_1$ (d) $x_2 = 4x_1$

9. A particle moves with constant acceleration along a straight line starting from rest. The percentage increase in its displacement during the 4th second compared to that in the 3rd is [WB JEE]

(a) 33% (b) 40%
(c) 66% (d) 77%

10. A cricket ball thrown across a field is at heights h_1 and h_2 from the point of projection at times t_1 and t_2 respectively after the throw. The ball is caught by a fielder at the same height as that of projection. The time of the ball in this journey is [WB JEE]

(a) $\dfrac{h_1t_2^2 - h_2t_1^2}{h_1t_2 - h_2t_1}$ (b) $\dfrac{h_1t_1^2 + h_2t_2^2}{h_2t_1 + h_1t_2}$

(c) $\dfrac{h_1t_2^2 + h_2t_1^2}{h_1t_2 + h_2t_1}$ (d) $\dfrac{h_1t_1^2 - h_2t_2^2}{h_1t_1 - h_2t_2}$

2013

11. A particle of mass 0.2 kg is moving in one dimension under a force that delivers a constant power 0.5 W to the particle. If the initial speed (in ms^{-1}) of the particle is zero, then the speed (in ms^{-1}) after 5 s is [JEE Advanced]

(a) 4 m/s (b) 2 m/s
(c) 1 m/s (d) 5 m/s

12. Two bullets are fired simultaneously, horizontally and with different speeds from the same place. Which bullet will hit the ground first? [Karnataka CET, OJEE]

(a) The faster one
(b) Depends on their mass
(c) The slower one
(d) Both will reach simultaneously

13. The bus moving with a speed of 42 km/h is brought to a stop by brakes after 6 m. If the same bus is moving at a speed of 90 km/h, then the minimum stopping distance is [J&K CET]

(a) 15.48 m (b) 18.64 m (c) 22.13 m (d) 27.55 m

2012

14. A person throws balls into air vertically upward in regular intervals of time of 1 s. The next ball is thrown when the velocity of the ball thrown earlier becomes zero. The height to which the balls rise is (assume, $g = 10\,m/s^2$) [Karnataka CET]

(a) 20 m (b) 5 m
(c) 10 m (d) 7.5 m

15. A ball is dropped from the roof of a tower of height h. The total distance covered by it in the last second of its motion is equal to the distance covered by it in first 3 s. What is the value of h? [OJEE]

(a) 125 m (b) 5 m
(c) 58 m (d) 250 m

16. The velocity of a particle is given by $v = \sqrt{180 - 16x}$ m/s. Its acceleration will be [OJEE]

(a) $-8\,m/s^2$
(b) $-6\,m/s^2$
(c) $0.4\,m/s^2$
(d) $5\,m/s^2$

17. A body moves in a plane, so that the displacement along X and Y-axes is $x = 3t^2$ and $y = 4t^2$, then velocity is [OJEE]

(a) $15\,t^2$
(b) $20\,t^2$
(c) $5\,t^2$
(d) None of the above

18. A particle is travelling along a straight line OX. The distance x (in metre) of the particle from O at a time t is given by $x = 37 + 27t - t^3$, where t is time in second. The distance of the particle from O when it comes to rest is [WB JEE]

(a) 81 m
(b) 91 m
(c) 101 m
(d) 111 m

19. A bullet on penetrating 30 cm into its target loses its velocity by 50%. What additional distance will it penetrate into the target before it comes to rest? [WB JEE]

(a) 30 cm
(b) 20 cm
(c) 10 cm
(d) 5 cm

20. From the top of a tower, 80 m high from the ground, a stone is thrown in the horizontal direction with a velocity of $8\,ms^{-1}$. The stone reaches the ground after a time t and falls at a distance of d from the foot of the tower are [WB JEE]

(a) 6 s, 64 m (b) 6 s, 48 m
(c) 4 s, 32 m (d) 4 s, 16 m

2011

21. An object moving with a speed of 6.25 m/s is decelerated at a rate given by $\frac{dv}{dt} = -2.5\sqrt{v}$, where v is the instantaneous speed. The time taken by the object, to come to rest, would be **[AIEEE]**

(a) 2 s (b) 4 s
(c) 8 s (d) 1 s

22. A bus begins to move with an acceleration of 1 ms^{-2}. A man who is 48 m behind the bus starts running at 10 ms^{-1} to catch the bus. The man will be able to catch the bus after **[Kerala CEE]**

(a) 6 s (b) 5 s (c) 3 s (d) 7 s
(e) 8 s

23. A particle is moving with constant acceleration from A to B in a straight line AB. If u and v are the velocities at A and B respectively. Then, its velocity at the mid-point C will be **[Kerala CEE]**

(a) $\left(\frac{u^2+v^2}{2u}\right)^2$

(b) $\frac{u+v}{2}$

(c) $\frac{v-u}{2}$

(d) $\sqrt{\frac{u^2+v^2}{2}}$

(e) $\sqrt{\frac{v^2-u^2}{2}}$

24. A particle crossing the origin of coordinates at time $t = 0$, moves in the xy-plane with a constant acceleration a in the y-direction. If its equation of motion is $y = bx^2$ (b is a constant), its velocity component in the x-direction is **[Kerala CEE]**

(a) $\sqrt{\frac{2b}{a}}$ (b) $\sqrt{\frac{a}{2b}}$

(c) $\sqrt{\frac{a}{b}}$ (d) $\sqrt{\frac{b}{a}}$

(e) $\sqrt{ab}$

25. A car moves a distance of 200 m. It covers first half of the distance at speed 60 kmh^{-1} and the second half at speed v. If the average speed is 40 kmh^{-1}, then value of v is **[Karnataka CET]**

(a) 30 kmh^{-1} (b) 13 kmh^{-1}
(c) 60 kmh^{-1} (d) 40 kmh^{-1}

26. The displacement-time graphs of two moving particles make angles of 30° and 45° with the X-axis. The ratio of their velocities is **[Karnataka CET]**

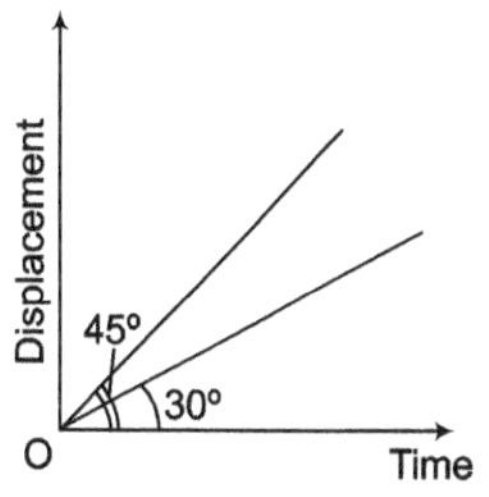

(a) $1 : \sqrt{3}$ (b) 1 : 2
(c) 1 : 1 (d) $\sqrt{3} : 2$

Answers

Level 1

Objective Problems

1. (d)	**2.** (d)	**3.** (a)	**4.** (a)	**5.** (c)	**6.** (d)	**7.** (c)	**8.** (c)	**9.** (c)	**10.** (b)
11. (a)	**12.** (a)	**13.** (b)	**14.** (c)	**15.** (c)	**16.** (a)	**17.** (c)	**18.** (a)	**19.** (b)	**20.** (b)
21. (d)	**22.** (a)	**23.** (a)	**24.** (b)	**25.** (d)	**26.** (a)	**27.** (b)	**28.** (c)	**29.** (b)	**30.** (c)
31. (a)	**32.** (b)	**33.** (c)	**34.** (a)	**35.** (a)	**36.** (d)	**37.** (d)	**38.** (b)	**39.** (b)	**40.** (b)
41. (b)	**42.** (d)	**43.** (a)	**44.** (a)	**45.** (c)	**46.** (b)	**47.** (c)	**48.** (b)	**49.** (d)	**50.** (a)
51. (b)	**52.** (b)	**53.** (b)	**54.** (c)	**55.** (b)	**56.** (c)	**57.** (b)	**58.** (d)	**59.** (b)	**60.** (c)
61. (a,d)	**62.** (c)	**63.** (b)	**64.** (b)	**65.** (b)	**66.** (d)	**67.** (a)	**68.** (b)	**69.** (a,b,c)	**70.** (c)
71. (c)	**72.** (c)	**73.** (b)	**74.** (c)	**75.** (b)	**76.** (d)	**77.** (c)	**78.** (b)	**79.** (a,b,c,d)	**80.** (a)
81. (a)	**82.** (c)	**83.** (d)	**84.** (d)	**85.** (b)	**86.** (a)	**87.** (a)	**88.** (b)	**89.** (d)	**90.** (d)
91. (a)	**92.** (b)	**93.** (c)	**94.** (d)	**95.** (b)	**96.** (b)	**97.** (d)	**98.** (c)	**99.** (a)	**100.** (a)
101. (c)	**102.** (a)	**103.** (b)	**104.** (a)	**105.** (a)	**106.** (b)	**107.** (d)	**108.** (b)	**109.** (a)	**110.** (c)
111. (c)	**112.** (b)	**113.** (d)	**114.** (d)	**115.** (a)	**116.** (a)	**117.** (b)	**118.** (d)	**119.** (a)	**120.** (c)
121. (d)	**122.** (b)								

Level 2

Only One Correct Option

1. (a)	**2.** (a,b)	**3.** (c)	**4.** (c)	**5.** (b)	**6.** (c)	**7.** (d)	**8.** (b)	**9.** (b)	**10.** (d)
11. (a)	**12.** (b)	**13.** (a)	**14.** (b)	**15.** (a)	**16.** (b)	**17.** (b)	**18.** (c)	**19.** (a)	**20.** (b)
21. (b)	**22.** (d)	**23.** (d)	**24.** (b)	**25.** (a,b)	**26.** (d)	**27.** (b)	**28.** (b)	**29.** (b,d)	**30.** (b)
31. (a)	**32.** (a)	**33.** (a)	**34.** (b)						

More than One Correct Options

1. (a,d)	**2.** (all)	**3.** (b)	**4.** (a,b)	**5.** (a,d)	**6.** (b,c)	**7.** (a,b)	**8.** (b,c)	**9.** (a,c)	**10.** (a,d)
11. (a,b,c)	**12.** (all)	**13.** (b,d)	**14.** (a,b,c)	**15.** (a,c)	**16.** (a,c,d)	**17.** (a,d)	**18.** (a,c)		

Comprehension Based Questions

1. (a)	**2.** (c)	**3.** (d)	**4.** (c)	**5.** (c)	**6.** (b)	**7.** (d)

Assertion and Reason

1. (a)	**2.** (d)	**3.** (b)	**4.** (c)	**5.** (c)	**6.** (a, b)	**7.** (d)	**8.** (a)	**9.** (a)	**10.** (d)
11. (a)	**12.** (b)	**13.** (d)	**14.** (c)	**15.** (a)	**16.** (e)	**17.** (d)	**18.** (c)	**19.** (c)	**20.** (b)

Match the Columns

1. (A → p, B → t, C → r, D → s)	**2.** (A → q, u, B → r, u, C → p, t)	**3.** (A → r, B → p, C → r, D → s)
4. (A → r, B → s, C → p, D → q)	**5.** (A → r, B → p, C → s, D → p)	**6.** (A → r, B → q, C → r, D → p)
7. (A → p, q, B → p, q, C → r, D → p, q)	**8.** (A → r, B → p, C → s, D → q)	**9.** (A → r, B → q, C → s, D → p)

Entrance Gallery

1. (a)	**2.** (c)	**3.** (c)	**4.** (e)	**5.** (e)	**6.** (a)	**7.** (b)	**8.** (c)	**9.** (b)	**10.** (a)
11. (a)	**12.** (d)	**13.** (d)	**14.** (b)	**15.** (a)	**16.** (a)	**17.** (d)	**18.** (b)	**19.** (c)	**20.** (a)
21. (a)	**22.** (e)	**23.** (d)	**24.** (b)	**25.** (a)	**26.** (a)				

Solutions

Level 1 : Objective Problems

1. $\mathbf{a} = \dfrac{\mathbf{v}_f - \mathbf{v}_i}{\Delta t} = \dfrac{(2\hat{i} - 3\hat{j}) - (2\hat{i} + 3\hat{j})}{2} = \dfrac{-6\hat{j}}{2} = -3\hat{j}\ \text{m/s}^2$

2. $v_{av} = \dfrac{\text{Displacement}}{\text{Time}} = \dfrac{2R}{(\pi R/v)} = \dfrac{2v}{\pi} = \dfrac{20}{\pi}\ \text{m/s}$

3. $v_{av} = \dfrac{\text{Total displacement}}{\text{Total time}}$

$$= \frac{80t + 40t}{2t} = 60\ \text{km/h}$$

4. Distance from starting point $= \sqrt{(3)^2 + (4)^2 + (5)^2}$

$$= 5\sqrt{2}\ \text{m}$$

5. $v_{av} = \dfrac{\text{Displacement}}{\text{Time}} = \dfrac{\sqrt{2}R}{t}$

7. $v_{av} = \dfrac{\text{Total displacement}}{\text{Total time}}$

$$= \frac{\sqrt{(4)^2 + (3)^2}}{10 + 5} = \frac{1}{3}\ \text{m/s}$$

8. Horizontal distance covered by the wheel in half revolution $= \pi R$

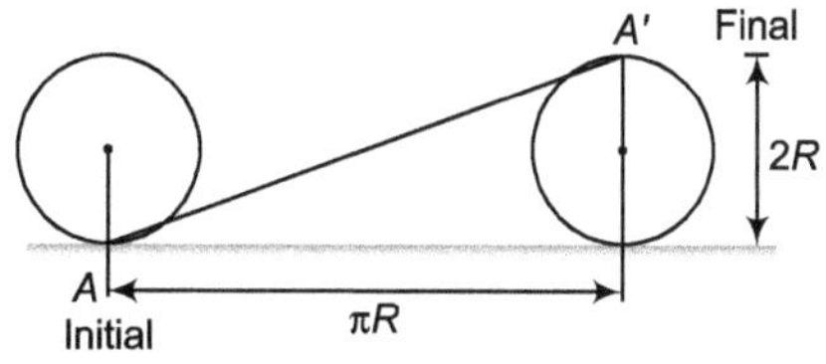

So, the displacement of the point which was initially in contact with ground $= AA'' = \sqrt{(\pi R)^2 + (2R)^2}$

$$= R\sqrt{\pi^2 + 4}$$

$$= \sqrt{\pi^2 + 4} \qquad \text{(as } R = 1\ \text{m)}$$

9. Displacement of the particle will be zero, because if comes back to its starting point.

$$\text{Average speed} = \frac{\text{Total distance}}{\text{Total time}}$$

$$= \frac{30\ \text{m}}{10\ \text{s}} = 3\ \text{m/s}$$

10. $a_{av} = \left|\dfrac{\Delta \mathbf{v}}{\Delta t}\right| = \dfrac{|\mathbf{v}_f - \mathbf{v}_i|}{t}$

$$= \frac{\sqrt{v^2 + v^2 - 2v.v.\cos\theta}}{t}$$

$$= \frac{2v}{t}\sin\theta/2$$

11. $v_1 t + v_2 t = s$

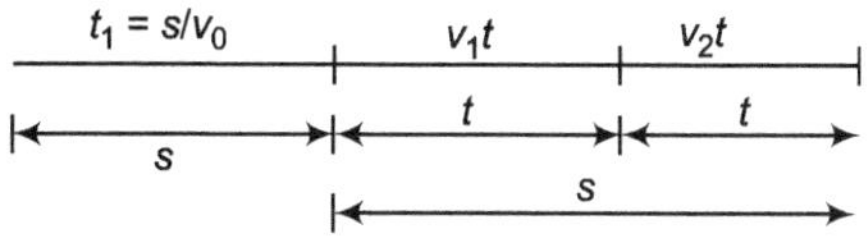

$$\therefore \qquad t = \frac{s}{v_1 + v_2}$$

Total time $= t_1 + 2t$ and total displacement $= 2s$

$$\text{Mean velocity} = \frac{\text{Total displacement}}{\text{Total time}}$$

$$= \frac{2s}{(s/v_0) + (2s/v_1 + v_2)}$$

$$= \frac{2}{(1/v_0) + (2/v_1 + v_2)}$$

$$= \frac{2v_0(v_1 + v_2)}{(2v_0 + v_1 + v_2)}$$

12. $21 = \dfrac{11 \times 18 + 42 \times v}{60}$

$\therefore\ v = 25.3\ \text{m/s}$

13. $v^2 = c_1 - c_2\, x$, comparing with $v^2 = u^2 + 2as$, we have $a = -\dfrac{c_2}{2}$

14. $s = \dfrac{1}{2}at^2 = \dfrac{1}{2} \times (0.2)(64) = 6.4\ \text{m} = 640\ \text{cm}$

15. $h_1 = \dfrac{1}{2} \times g \times (2)^2 = 2g$

$$h_2 = \frac{1}{2} \times g \times (4)^2 - 2g = 6g$$

$$h_3 = \frac{1}{2} \times g \times (6)^2 - 8g = 10g$$

$$\therefore \qquad h_1 : h_2 : h_3 = 1 : 3 : 5$$

16. $t = \sqrt{\dfrac{2h}{g}}$ or $t \propto \sqrt{h}$

17. $v = \sqrt{2gh}$ or $v \propto \sqrt{h}$

18. $v^2 = 25x$, comparing with $v^2 = 2as$, we have,

$a = 12.5\ \text{m/s}^2$

19. Time taken to reach maximum height is 1 s.

Height = free fall distance in 1 s $= \dfrac{1}{2}gt^2 = 5\ \text{m}$

20.
$$v = u + at$$
or
$$-2 = 10 + 4a$$
or
$$a = -3\ \text{m/s}^2$$

21. $-h = vt - \dfrac{1}{2}gt^2$ or $gt^2 - 2vt - 2h = 0$

$$\therefore \qquad t = \frac{2v + \sqrt{4v^2 + 8gh}}{2g}$$

$$= \frac{v}{g}\left[1 + \sqrt{1 + \frac{2hg}{v^2}}\right]$$

22.

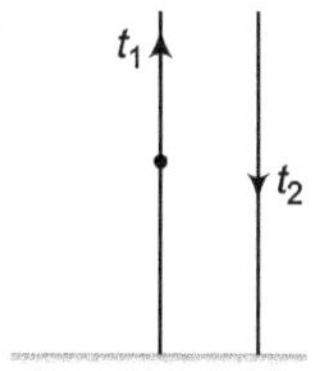

From figure, $t_1 < t_2$

23. $15 = 2t - \dfrac{1}{2} \times (0.1)t^2$ or $t = 10\ \text{s}$

24. $t = \dfrac{2u}{g}$

25. $s = \dfrac{1}{2}at^2$

26.
$$s = s_1 + s_2 + s_3$$
$$= \frac{1}{2}\times 2\times (10)^2 + (2)(10)(30) + \frac{1}{2}\times 4\times (5)^2$$
$$= 100 + 600 + 50$$
$$= 750\text{ m}$$

27. In first second distance travelled $x = \dfrac{1}{2}\times g\times t^2 = 5\text{ m}$

(for $g = 10\text{ m/s}^2$) in last second $7x = 35\text{ m}$

Now, $s_t\text{ th} = \left(u + at - \dfrac{1}{2}a\right)$

or $35 = 0 + 10\times t - \dfrac{1}{2}\times 10$

or $t = 4\text{ s}$

28.

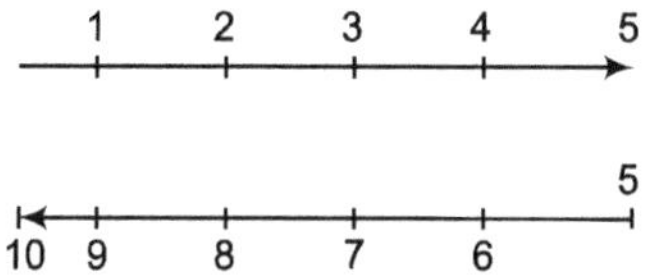

29. Displacements of both should be equal. Or,
$$8t = \frac{1}{2}\times 4\times t^2 \text{ or } t = 4\text{ s}$$

30. $s_B = \dfrac{1}{2}\times 2\times (1)^2 = 1\text{ m}$
$$s_A = \frac{1}{2}\times 2\times (2)^2 = 4\text{ m}$$
$\therefore$ $s_A - s_B = 3\text{ m}$

31. $\dfrac{u}{4} = u - at_0$

$\therefore$ $a = \dfrac{3u}{4t_0}$

or $\dfrac{u}{a} = \dfrac{4}{3}t_0$

Now, $0 = u - at$ or $t = \dfrac{u}{a} = \dfrac{4}{3}t_0$

32. $(30)^2 = (20)^2 + 2a(2s)$ or $2as = 250$

Now, $v^2 = (20)^2 + 2as = 650$

$\therefore$ $v = 25.5\text{ m/s}$

33. $\dfrac{1}{2}\times (a)(2)^2 = x$ or $2a = x$

Now, $\dfrac{1}{2}(a)(4)^2 - \dfrac{1}{2}(a)(2)^2 = 6a = 3x$

34. After 2 s $v = 4.9\times 2 = 9.8$

and $h = \dfrac{1}{2}\times 4.9\times (2)^2 = 9.8\text{ m}$

Greatest height, $h_{max} = h + \dfrac{v^2}{2g} = 14.7\text{ m}$

35. $s_t = u + at - \dfrac{1}{2}a$

or $0 + 10t - 5 = \dfrac{1}{2}\times 10\times (3)^2$

or $t = 5\text{ s}$

36. $v = \dfrac{ds}{dt} = 12 - 4t$

Comparing with $v = u + at$, $u = 12\text{ m/s}$ and $a = -4\text{ m/s}^2$

Velocity will become zero at $0 = 12 - 4t_0$ or $t_0 = 3\text{ s}$.

Since, the given time $t = 5\text{ s}$ is greater than $t_0 = 3\text{ s}$

distance > | displacement |

Distance $d = |s_{0-t_0}| + |s_{t_0-t}| = \dfrac{u^2}{2|a|} + \dfrac{1}{2}|a|(t - t_0)^2$
$$= \frac{(12)^2}{8} + \frac{1}{2}\times 4\times (2)^2 = 26\text{ m}$$

37. $x = 2 + t - 3t^2$, $v = \dfrac{dx}{dt} = 1 - 6t$, velocity will become zero at time, $0 = 1 - 6t_0$ or $t_0 = \dfrac{1}{6}\text{ s}$,

Since, the given time $t = 1\text{ s}$ is greater than $t_0 = \dfrac{1}{6}\text{ s}$,

distance > | displacement |

Displacement $s = x_f - x_i$
$$= (2 + 1 - 3) - (2 + 0 - 0) = -2\text{ m}$$
Distance $d = |s_{0-t_0}| + |s_{t-t_0}|$
$$= \frac{u^2}{2|a|} + \frac{1}{2}|a|(t - t_0)^2$$

Comparing $v = 1 - 6t$ with $v = u + ut$, we have
$u = 1\text{ m/s}$ and $a = -6\text{ m/s}^2$

$\therefore$ Distance $= \dfrac{(1)^2}{2\times 6} + \dfrac{1}{2}\times 6\times \left(1 - \dfrac{1}{6}\right)^2$
$$= 2.1\text{ m}$$

38. $1 = \dfrac{1}{2}gt_1^2$ or $t_1 = \sqrt{\dfrac{2}{g}}$
$$2 = \frac{1}{2}gt^2 \quad\text{or}\quad t = \sqrt{\frac{4}{g}}$$
but $t_2 = t - t_1 = \sqrt{\dfrac{4}{g}} - \sqrt{\dfrac{2}{g}}$

$\therefore$ $\dfrac{t_1}{t_2} = \dfrac{\sqrt{2/g}}{\sqrt{4/g} - \sqrt{2/g}}$
$$= \frac{\sqrt{2}}{2 - \sqrt{2}} = \frac{\sqrt{2}(2 + \sqrt{2})}{2}$$
$$= (\sqrt{2} + 1)$$

39. $x = 4(t - 2) + a(t - 2)^2$

At $t = 0$, $x = -8 + 4a = 4a - 8$
$$v = \frac{dx}{dt} = 4 + 2a(t - 2)$$
At $t = 0$, $v = 4 - 4a = 4(1 - a)$

But acceleration, $\dfrac{d^2x}{dt^2} = 2a$

40. $v^2 = u^2 + 2gh$

$\Rightarrow$ $(3u)^2 = (-u)^2 + 2gh$

$\Rightarrow$ $h = \dfrac{4u^2}{g} \downarrow +\text{ve}$

41. Let h distance is covered in t second

$\Rightarrow$ $h = \dfrac{1}{2}gt^2$

Distance covered in t^{th} second $= \frac{1}{2} g (2t - 1)$

$$\Rightarrow \quad \frac{9h}{25} = \frac{g}{2}(2t-1)$$

From above two equations, $h = 122.5$ m

42. The initial velocity of airplane is horizontal, Hence, the vertical component of velocity of packet will be zero.

$$\therefore \quad h = \frac{1}{2} g t^2$$

$$\text{So,} \quad t = \sqrt{\frac{2h}{g}}$$

43. $H_{max} \propto u^2$

$$\therefore \quad u \propto \sqrt{H_{max}}$$

i.e. to triple the maximum height, ball should be thrown with velocity $\sqrt{3}\, v_0$.

44. $v = an$

$$\therefore \quad a = \left(\frac{v}{n}\right)$$

$$\text{Now,} \quad S_n = \frac{1}{2} an^2 = \frac{1}{2}\left(\frac{v}{n}\right)(n^2) = \frac{vn}{2}$$

$$\text{and} \quad S_{n-3} = \frac{1}{2} a (n-3)^3 = \frac{1}{2}\left(\frac{v}{n}\right)(n-3)^2$$

$\therefore$ Displacement in last 3 s will be,

$$S = S_n - S_{n-3} = \frac{v}{2}\left[n - \frac{(n-3)^3}{n}\right] = \left(\frac{6n-9}{n}\right)\frac{v}{2}$$

45. $40 = (20)\, a_1$,

$$\therefore \quad a_1 = 2 \text{ m/s}^2$$

$$\text{Further} \quad 40 = (40)\, a_2$$

$$\therefore \quad a_2 = 1 \text{ m/s}^2$$

Therefore, acceleration is 2 m/s^2 and retardation 1 m/s^2.

$$\text{Now,} \quad s_1 = \frac{1}{2} a_1 t_1^2 = \frac{1}{2} \times 2 \times (20)^2 = 400 \text{ m}$$

$$s_2 = v_{max}\, t_2 = 40 \times 20 = 800 \text{ m}$$

$$s_3 = \frac{v_{max}^2}{2a_2} = \frac{(40)^2}{2 \times 1} = 800 \text{ m}$$

Now, average velocity

$$= \frac{\text{Total displacement}}{\text{Total time}} = \frac{400 + 800 + 800}{20 + 20 + 40} = 25 \text{ m/s}$$

46. $h = \frac{(10)^2}{2g} = 5$ m

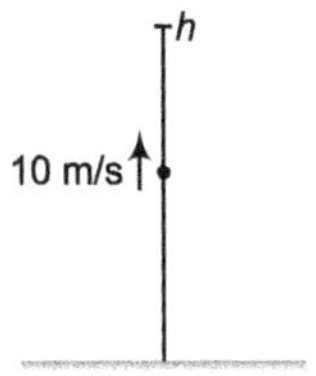

$\therefore$ Total height $= 2h = 10$ m

47. $t = \sqrt{\frac{x+a}{b}}$

$$\text{or} \quad (x + a) = bt^2$$

$$\text{or} \quad x = -a + bt^2$$

Comparing this equation with general equation of uniformly accelerated motion, $s = s_i + ut + \frac{1}{2} at^2$

we see that $s_i = -a, u = 0$ and acceleration $= 2b$.

48. Area under v-t graph gives displacement. We can see that area for B is greater than area of A. Hence, B has covered larger distance.

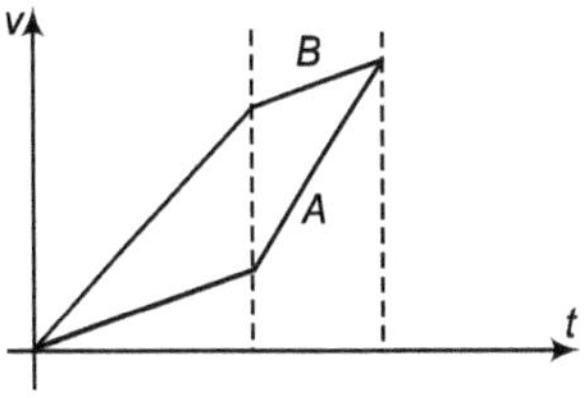

49. Let t second of upward journey = first t second of downward journey (with zero initial velocity)

$\therefore$ Desired distance $= \frac{1}{2} gt^2$

50. $t_{ABC} = 10$ s

$$\therefore \quad t_{AB} = 5 \text{ s}$$

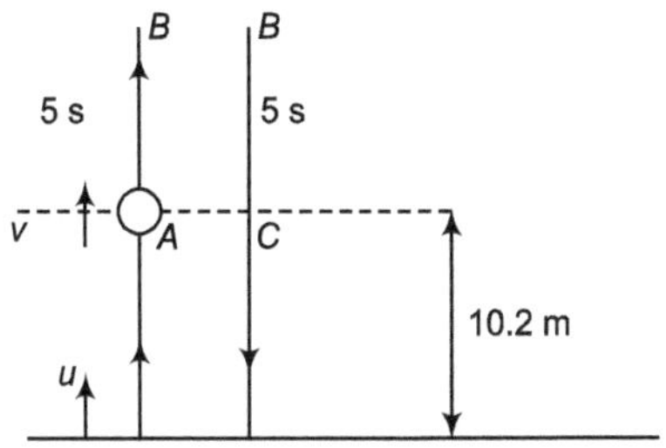

At B, velocity becomes zero. Hence, at A velocity should be 50 m/s.

$$\text{Now,} \quad (50)^2 = (u^2) - 2 \times 10 \times 10.2$$

$$\therefore \quad u = 52 \text{ m/s}$$

51. Velocity of particle of this instant will be $v = (u - gt_1)$

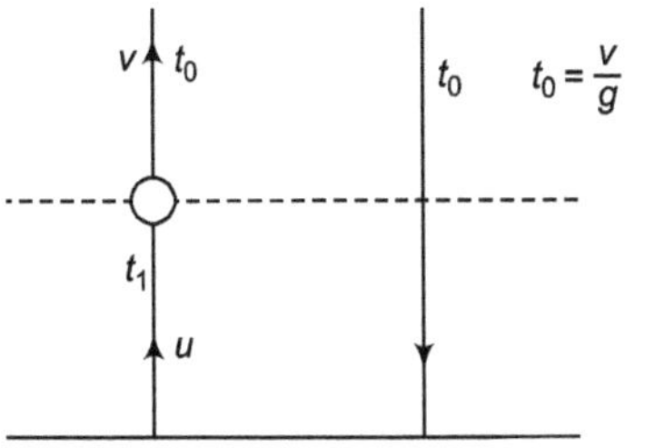

Now, the desired time interval will be $\frac{2v}{g}$ or

$$\frac{2(u - gt_1)}{g} = 2\left(\frac{u}{g} - t_1\right)$$

52. Total time taken in upward journey should be 5 s

$$\therefore \quad u = gt = 9.8 \times 5 = 49 \text{ m/s}$$

53. $x_1 = \frac{1}{2}\alpha t_1^2$ ...(i)

$$x_2 = \frac{1}{2}\beta t_2^2 \quad \text{...(ii)}$$

Further, $v_{max} = \alpha t_1$ ...(iii)

and $0 = v_{max} - \beta t_2$ ...(iv)

From Eqs. (iii) and (iv), we have

$$\frac{\alpha}{\beta} = \frac{t^2}{t_1}$$

Then, from Eqs. (i) and (ii),

$$\frac{x_1}{x_2} = \frac{t_1}{t_2} = \frac{\beta}{\alpha}$$

54. Let v_m be the maximum speed,

$$v_m = a_1 t_1 \text{ and } v_m = \sqrt{2a_1 l}$$

$$t_2 = \frac{2l}{v_m} \text{ and } v_m = a_2 t_3 = \sqrt{2a_2(3l)}$$

Now, average speed $v_{av} = \frac{l + 2l + 3l}{t_1 + t_2 + t_3}$

$$v_{av} = \frac{6l}{(v_m/a_1) + (2l/v_m) + (v_m/a_2)}$$

$$= \frac{6l}{\left(\frac{v_m}{v_m^2/2l}\right) + \left(\frac{2l}{v_m}\right) + \left(\frac{v_m}{v_m^2/6l}\right)}$$

$$= \frac{6l}{(10l/v_m)} = \frac{3v_m}{5}$$

$$\frac{v_{av}}{v_m} = \frac{3}{5}$$

55. Average velocity in uniformly accelerated motion is given by,

$$v_{av} = \frac{s}{t} = \frac{ut + \frac{1}{2}at^2}{t} = u + \frac{1}{2}at$$

Now, $v_1 = u + \frac{1}{2}at_1$

$$v_2 = (u + at_1) + \frac{1}{2}at_2$$

and $v_3 = u + a(t_1 + t_2) + \frac{1}{2}at_3$

$\therefore \quad (v_1 - v_2) : (v_2 - v_3) = (t_1 + t_2) : (t_2 + t_3)$

56. Taking downward direction as the positive direction

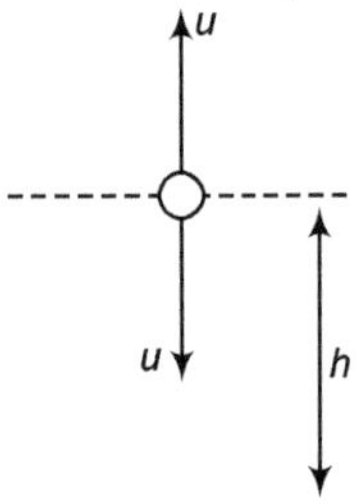

$$+h = -ut_1 + \frac{1}{2}gt_1^2 \quad \text{...(i)}$$

$$+h = ut_2 + \frac{1}{2}gt_2^2 \quad \text{...(ii)}$$

Multiplying t_2 by Eq. (i) and t_2 by Eq. (ii) and adding, we get

$$h(t_1 - t_2) = \frac{1}{2}gt_1t_2(t_1 + t_2)$$

or $h = \frac{1}{2}gt_1t_2$

For free fall from rest $h = \frac{1}{2}gt^2$

$\therefore \quad t^2 = t_1t_2$

$$t = \sqrt{t_1t_2}$$

57. $v = \frac{dx}{dt} = 3at^2 + 2bt + C$

$$a = \frac{dv}{dt} = 6at + 2b$$

At $t = 0, a = 2b$

58. $dv = a\,dt$

$\therefore \quad \int_2^v dv = \int_0^2 (3t^2 + 2t + 2)\,dt$

or $v = 18$ m/s

59. $dv = a\,dt$

$\therefore \quad \int_0^v dv = \int_0^t 4t\,dt$

or $v = 2t^2$

Now, $s = \int_0^3 v\,dt = \int_0^3 (2t^2)\,dt$

$$= \frac{2}{3}t^3 = 18 \text{ m}$$

60. $dv = a\,dt$

or $v = \int_0^2 (3t + 4)\,dt = 14$ m/s

61. $x = 0$ at $t = 0$ and $t = \alpha/\beta$. So, the particle returns to starting point at $t = \alpha/\beta$

$$v = \frac{dx}{dt} = 2\alpha t - 3\beta t^2$$

At $t = 0, v = 0$ i.e. initial velocity of particle is zero.

$$v = 0 \text{ at } t = 0 \text{ and } t = \frac{2\alpha}{2\beta}$$

Thus, the particle comes to rest after time

$$t = 2\alpha/3\beta$$

$$a = \frac{dv}{dt} = 2\alpha - 6\beta t$$

At $t = 0, a = 2a \neq 0$

62. $y = a + bt + ct^2 - dt^4$

$\therefore \quad v = \frac{dy}{dt} = b + 2ct - 4dt^3$

and $a = \frac{dv}{dt} = 2c - 12\,dt^2$

Hence, at $t = 0$, $v_{\text{initial}} = b$ and $a_{\text{initial}} = 2c$

63. $v = \frac{ds}{dt} = 12t - 3t^2$

Velocity is zero for $t = 0$

and $t = 4$ s

64. $v = \frac{ds}{dt} = 9t^2 + 14t + 14$

$$a = \frac{dv}{dt} = 18t + 14$$

At $t = 1$ s, $a = 32$ m/s^2

65. $v = \frac{dx}{dt} = 32 - 8t^2$

$v = 0$ at $t = 2\,s$ $\qquad a = \frac{dv}{dt} = -16t$

At 2 s, $\qquad a = -32\ m/s^2$

66. Integrate twice to convert a-t equation into s-t equation.

67. $a = bt$

$\therefore \quad \int_{v_0}^{v} dv = \int a dt = \int_a^t bt \cdot dt$

$\therefore \quad v = v_0 + \frac{bt^2}{2}$

Further integrating we get,

$$s = v_0 t + \frac{bt^3}{6}$$

68. $v = 3\alpha t^2 + 2\beta t + \gamma$

$v_{t=0} = v_i = \gamma$

$a = 6\alpha t + 2\beta : a_{t=0} = a_i = 2\beta$

$\therefore \quad \frac{v_i}{a_i} = \frac{\gamma}{2\beta}$

69. $a = -0.5t$ or $\frac{dv}{dt} = -0.5t$

$\therefore \quad \int_{16}^{v} dv = \int_0^t -(0.5t) \cdot dt$

$\therefore \quad v - 16 = -0.25t^2$

$\therefore \quad v = 16 - 0.25t^2$

Velocity will change its direction at the instant when $v = 0$

$\therefore \quad 0 = 16 - 0.25t^2$ or $t = 8\,s$

$$d_{4\,s} = S_{4\,s} = \int_0^4 v\, dt$$

$$\int_0^4 (16 - 0.25t^2)\, dt = 58.66\ m \approx 59\ m$$

Now, $\quad S_{8s} = \int_0^8 (16 - 0.25t^2)\, dt = 85.33\ m$

$S_{10s} = \int_0^{10} (16 - 0.25t^2)\, dt = 76.67\ m$

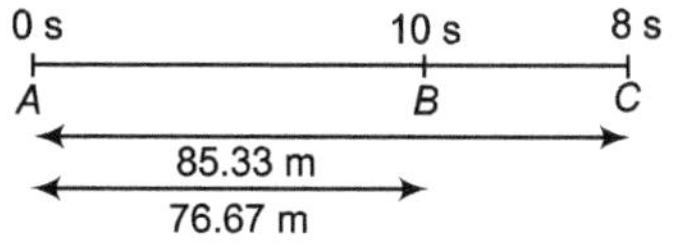

$d_{10s} = AC + BC = AC + (AC - BA)$

$= 2AC - BA = 94\ m$

$v_{10s} = 16 - (0.25)(10)^2 = -9\ m/s$

70. $v_x = \frac{dx}{dt} = \frac{2t}{2} = t$

At 2 s, $\qquad v_x = 2\ m/s$

Further, $\qquad y = \frac{x^2}{2} = \frac{(t^2/2)^2}{2} = \frac{t^4}{8}$

$v_y = \frac{dy}{dt} = \frac{t^3}{2}$

At $t = 2\,s$, $\qquad v_y = 4\ m/s$

71. $v_x = 4t, v_y = 2t - 4,\ v_z = 3$

At $t = 0, v_x = 0, v_y = -4$ and $v_z = 3$

$\therefore \quad v = \sqrt{v_x^2 + v_y^2} = 5$ units

72. $\mathbf{v} = \mathbf{u} + \mathbf{a}\,t = (3\hat{\mathbf{i}} + 4\hat{\mathbf{j}}) + (0.4\hat{\mathbf{i}} + 0.3\hat{\mathbf{j}})(10) = (7\hat{\mathbf{i}} + 7\hat{\mathbf{j}})$

$\therefore \quad v = \sqrt{(7)^2 + (7)^2}$

$= 7\sqrt{2}$ units

73. $x = kt,\ t = \frac{x}{k}$

Now, $\qquad y = k\left(\frac{x}{k}\right)\left(1 - \alpha \cdot \frac{x}{k}\right)$

or $\qquad y = x - \frac{\alpha x^2}{k}$

74. $v_x = \frac{dx}{dt} = 6t - 6,\ a_x = \frac{dv_x}{dt} = 6\ m/s^2$

$v_y = 2t - 2,\ a_y = \frac{dv_y}{dt} = 2\ m/s^2$

At $t = 1\,s$, v_x and v_y both are zero. Hence, net velocity is zero.

75. $u_x = 3\ m/s,\ a_x = -6\ m/s$

$\therefore \quad X_{max} = \frac{u_x^2}{2|a_x|} = \frac{9}{12} = 0.75\ m$

76. $\mathbf{a} = \frac{v_f - v_i}{t}$

77. $\mathbf{v} = \mathbf{u} + \mathbf{a}\,t = (3\hat{\mathbf{i}} + 4\hat{\mathbf{j}}) + (0.4\hat{\mathbf{i}} + 0.3\hat{\mathbf{j}})(10) = (7\hat{\mathbf{i}} + 7\hat{\mathbf{j}})$

$\therefore \quad v = \sqrt{(7)^2 + (7)^2}$

$= 7\sqrt{2}$ units

78. $\mathbf{r} = (a\cos\omega t)\,\hat{\mathbf{i}} + (a\sin\omega t)\,\hat{\mathbf{j}}$

$\therefore \mathbf{v} = \frac{d\mathbf{r}}{dt} = (-a\omega\sin\omega t)\,\hat{\mathbf{i}} + (a\omega\cos\omega t)\,\hat{\mathbf{j}}$

$\mathbf{r} \cdot \mathbf{v} = 0$

$\therefore \quad \mathbf{r} \perp \mathbf{v}$

79. Comparing with projectile motion,

$$\mathbf{r} = (u_x t)\,\hat{\mathbf{i}} + \left(u_y t - \frac{1}{2} g t^2\right)\hat{\mathbf{j}}$$

we have, $\qquad u_x = 3\ m/s, u_y = 4\ m/s$

and $\qquad g = 10\ m/s^2$

$T = \frac{2u_y}{g} = 0.8\,s$

(a) $R = u_x T = 2.4\ m$

(b) $u = \sqrt{u_x^2 + u_y^2} = 5\ m/s$

(c) $\theta = \tan^{-1}\left(\frac{u_y}{u_x}\right) = \tan^{-1}\left(\frac{4}{3}\right)$

(d) $T = 0.8\,s$

80. Velocity first decreases in upwards direction, then increases in downward direction.

81. Velocity will continuously increases, (starting from rest)

82. In region B velocity is decreasing.

83. Motion is first accelerated in positive direction, then uniform in negative direction.

84. Acceleration in second case is more. Hence slope is more.

85. v_{max} = Maximum positive area of a-t graph

= area between 0 and 8 s

= 30 m/s

86. Uniform motion means uniform velocity or constant slope of s-t graph.

87. Displacement = positive area − negative area

and distance = Σ(Areas)

88. For uniformly accelerate motion, $v^2 = u^2 + 2as$, i.e. v^2 *versus* s graph is a straight line with intercept u^2 and slope $2a$. Since, intercept in non-zero, initial velocity is non-zero.

89. $v^2 = u^2 + 2as$, slope $= 2a = -\frac{16}{2} = -8\ \text{m/s}^2$

or $a = -4\ \text{m/s}^2$

90. $a = v\left(\frac{dv}{ds}\right) = (10)\left(-\frac{20}{30}\right) = -\frac{20}{3}\ \text{m/s}^2$

91. v_i = slope of s-t graph $= \tan 45° = 1$ m/s

v_f = slope of s-t graph $= \tan 60° = \sqrt{3}$ m/s

Now $a_{av} = \frac{v_f - v_i}{\Delta t} = \frac{\sqrt{3}-1}{1} = (\sqrt{3}-1)$ units

92. Change in velocity = net area under a-t graph

93. a = slope of v-t graph $= -\frac{15}{3} = -5\ \text{m/s}^2$

94. Slope of B is more. Therefore, speed of B will be more.

95. $v = \frac{dx}{dt} = 2$ m/s = constant

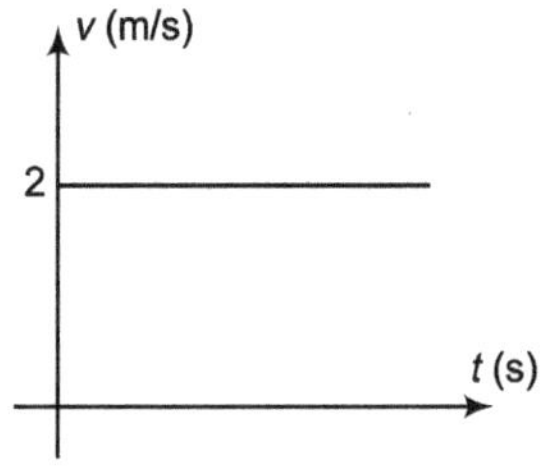

96. Distance = Area under v-t graph $= A_1 + A_2 + A_3 + A_4$

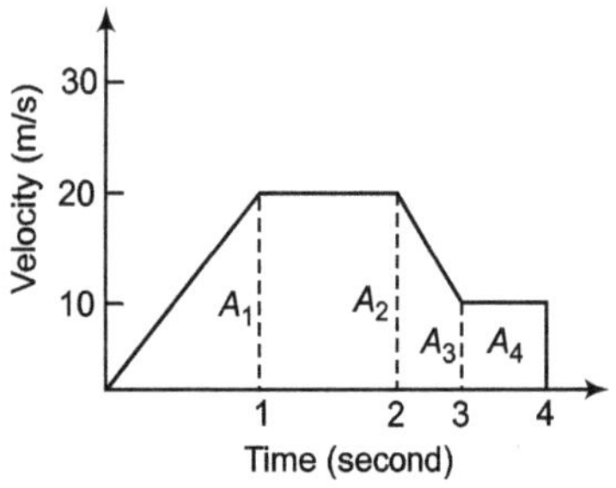

$= \frac{1}{2} \times 1 \times 20 + (20 \times 1) + \frac{1}{2}(20+10) \times 1 + (10 \times 1)$

$= 10 + 20 + 15 + 10 = 55$ m

97. Maximum acceleration means maximum change in velocity in minimum time interval.

In time interval $t = 30$ to $t = 40$ s,

$$a = \frac{\Delta v}{\Delta t} = \frac{80-20}{40-30} = \frac{60}{10} = 6\ \text{cm/s}^2$$

98. Area of trapezium $= \frac{1}{2} \times 3.6 \times (12+8) = 36.0$ m

99. For the given condition, initial height $h = d$ and velocity of the ball is zero. When the ball moves downward its velocity increases and it will be maximum when the ball hits the ground and just after the collision it becomes half and in opposite direction. As the ball moves upwards its velocity again decreases and becomes zero at height $d/2$. This explanation match with graph (a).

100. Given line have positive intercept but negative slope. So, its equation can be written as

$$v = -mx + v_0 \qquad \left[\text{where, } m = \tan\theta = \frac{v_0}{x_0}\right] \quad \text{...(i)}$$

By differentiating with respect to time, we get

$$\frac{dv}{dt} = -m\frac{dx}{dt} = -mv$$

Now, substituting the value of v from Eq. (i), we get

$$\frac{dv}{dt} = -m[-mx + v_0]$$

$$= m^2x - mv_0$$

$$\therefore \quad a = m^2x - mv_0$$

i.e. the graph between a and x should have positive slope but negative intercept on a-axis. So, graph (a) is correct.

101. In first instant, you will apply $v = \tan\theta$ and say

$$v = \tan 30° = \frac{1}{\sqrt{3}}\ \text{m/s}$$

But it is wrong because formula $v = \tan\theta$ is valid when angle is measured with time axis.

Here, angle is taken from displacement axis. So, angle from time axis $= 90° - 30° = 60°$

Now, $v = \tan 60° = \sqrt{3}$

102. Since, total displacement is zero, hence average velocity is also zero.

103. a-t equation from the given graph can be written as,

$$a = -2t + 4$$

or $dv = a \cdot dt = (-2t + 4)\,dt$

Integrating we get, $v = -t^2 + 4t$

v-t equation is quadratic. Hence, v-t graph is a parabola.

104. $\mathbf{v}_{AB} = \mathbf{v}_A - \mathbf{v}_B$

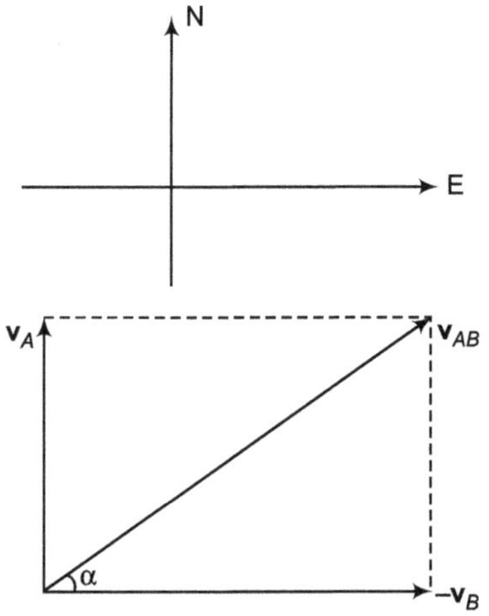

$$v_{AB} + \sqrt{v_A^2 + v_B^2} = 5\ \text{km/h}$$

$$\alpha = \tan^{-1}\left(\frac{v_A}{v_B}\right)$$

$$= \tan^{-1}\left(\frac{3}{4}\right) = 37°$$

105. For shortest time one should swim at right angles to river current.

106. Velocity of rain is at 30° with vertical. So, its horizontal component is $v_R \sin 30° = \frac{v_R}{2}$. When man starts walking with 10 km/h rain appears vertical. So, horizontal component $\frac{v_R}{2}$ is balanced by his speed of 10 km/h. Thus,

$$\frac{v_R}{2} = 10 \quad \text{or} \quad v_R = 20\,\text{km/h}$$

107. $$v_A + v_B = 4\,\text{m/s and } v_A - v_B = \frac{4.0}{10} = 0.4\,\text{m/s}$$

108. $$v_r = 25\,\text{m/s}$$

$$\therefore \quad t = \frac{50+50}{25} = 4\,\text{s}$$

109. $\sin\theta = \frac{v_r}{v_{br}} = \frac{3}{5}$

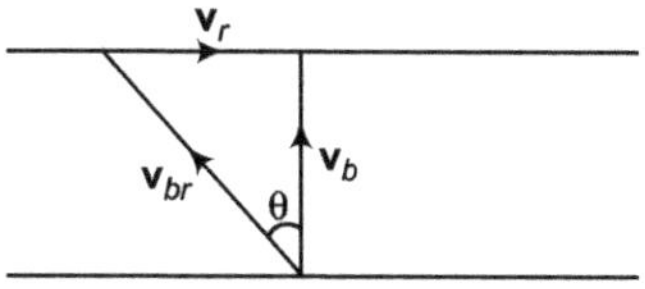

$$\therefore \quad \theta = 37°$$

The required angle is therefore $90° + \theta = 127°$

110. $\tan 60° = \frac{v_H}{v_V}$ or $\sqrt{3} = \frac{12}{v_V}$

$$\therefore \quad v_V = 4\sqrt{3}\,\text{km/h}$$

111. Both the particles, will fall same distance in same time interval.

So, the relative separation will remain unchanged.

112. Relative acceleration = 0, relative velocity is 40 m/s and relative separation is 100 m.

$$\therefore \quad t = \frac{100}{40} = 2.5\,\text{s}$$

113. $7.2 = \frac{100}{(v+5)(5/18)}$

or $$v = 45\,\text{km/h}$$

114. Displacement of man = displacement of bus + 25

or $$10t = \frac{1}{2} \times 2 \times t^2 + 25$$

or $$10t = t^2 + 25$$

Solving this equation, we get $t = 5\,\text{s}$.

115. Initially relative acceleration between them is zero, so distance between them will increase linearly. Later when one stone strikes the ground relative acceleration is g so distance will decrease parabolically.

116. $\tan\theta = \frac{v_H}{v_V} = \frac{v}{u}$

or $\theta = \tan^{-1}\left(\frac{v}{u}\right)$

117. The greatest distance, when velocities of both are equal.

or $$ft = u$$

$$\therefore \quad t = \frac{u}{f}$$

$$s_1 = ut = \frac{u^2}{f}$$

and $$s_2 = \frac{1}{2} ft^2 = \frac{u^2}{2f}$$

$$\therefore \quad s_{max} = s_1 - s_2 = \frac{u^2}{2f}$$

118. $v_1 = \frac{s}{90}, v_2 = \frac{s}{60}$

Now, $$t = \frac{s}{v_1 + v_2} = \frac{s}{\frac{s}{90} + \frac{s}{60}}$$

$$= \frac{90 \times 60}{90 + 60} = 36\,\text{s}$$

119. At 2 s

$$x_A = 10 \times 2 - \frac{1}{2} \times 4 \times (2)^2 = 12\,\text{m}$$

and $$x_B = 20 \times 2 - \frac{1}{2} \times 2 \times (2)^2$$

$$= 36\,\text{m}$$

$\therefore$ Distance between A and B at that instant is, 24 m.

120. Relative velocity of first train with respect to other

$$= (v_1 - v_2)$$

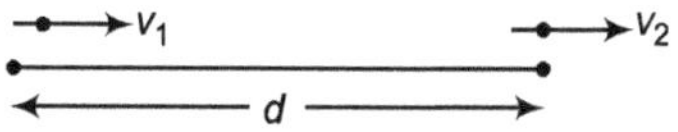

$$(0) = (v_1 - v_2)^2 - 2\alpha s$$

where, α = relative retardation

$$s = \frac{(v_1 - v_2)^2}{2\alpha}$$

For avoiding collision,

$$d > \frac{(v_1 - v_2)^2}{2\alpha}$$

121. $v_b = \frac{1}{1/4} = 4\,\text{km/h}$

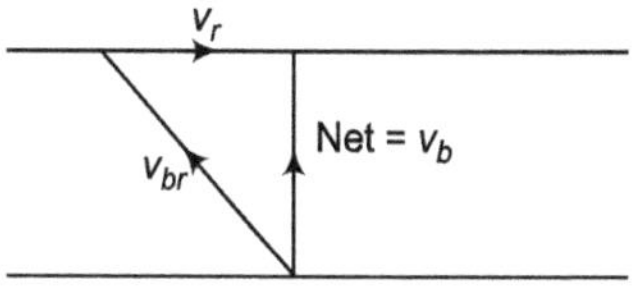

$$v_{br} = 5\,\text{km/h}$$

$$\therefore \quad v_r = 3\,\text{km/h}$$

122. $\mathbf{v}_X = -v\hat{\mathbf{j}}$

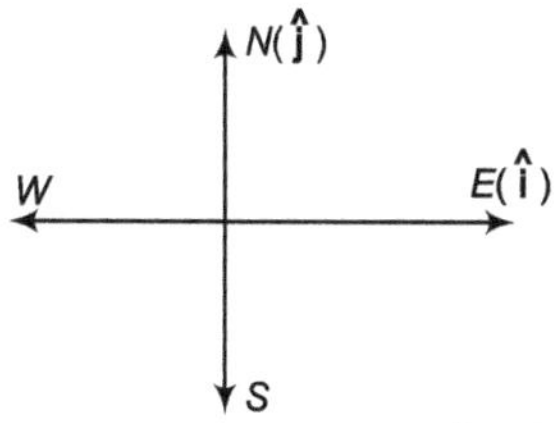

$$\therefore \quad \mathbf{v}_Y = \mathbf{v}_{YX} + \mathbf{v}_X = -v\hat{\mathbf{i}} + v\hat{\mathbf{j}}$$

$$\therefore \quad |\mathbf{v}_Y| = \sqrt{2}\,v$$

Direction of $\mathbf{v}_Y$ is north-west.

Level 2 : Only One Correct Option

1. $a_{av} = \frac{\mathbf{v}_f - \mathbf{v}_i}{\Delta t} = \frac{v_f - v_i}{\Delta t}$ (as they are in opposite direction)

$$= \frac{\sqrt{2gh_f} + \sqrt{2gh_i}}{\Delta t}$$

$$= \frac{\sqrt{2\times10\times5} + \sqrt{2\times10\times10}}{0.01}$$

$$\approx 2414 \text{ m/s}^2$$

2. $a_t = \frac{dv}{dt}$ = rate of change of speed $= a\cos\theta = \frac{\mathbf{a}\cdot\mathbf{v}}{v}$
= component of acceleration along velocity responsible for change in speed.

3. Velocity of particle will become zero in time $t_0 = \frac{v_0}{g}$

The given time $t = \frac{1.5v_0}{g}$ is greater than the time $t_0 = \frac{v_0}{g}$

Hence, distance > |displacement|

$$\text{Distance} = |S_{0-t_0}| + |S_{t-t_0}|$$

$$= \frac{v_0^2}{2g} + \frac{1}{2}g(t-t_0)^2$$

$$= \frac{5v_0^2}{8g}$$

4. $-h = 5t - 5t^2$ and $-2h = -5t - 5t^2$

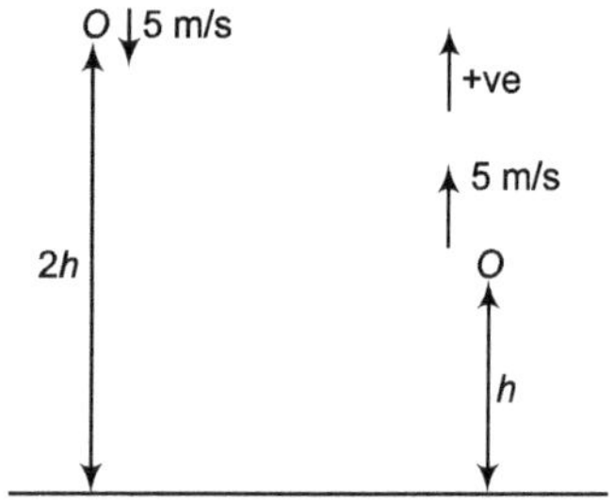

From these two equations we get, $t = 3$ s and h = distance between P and Q = 30 m.

5. $\frac{dx}{dt} = v = -1 - 2t$

Comparing with $v = u + at$, we have, $u = -1$ m/s and $a = -2$ m/s^2

At $t = 0, x = 1$ m. Then, u and a both are negative. Hence, x-coordinate of particle will go on decreasing.

6. $t = \frac{d}{v_r\cos 30°} = \frac{2d}{\sqrt{3}\,v_r}$

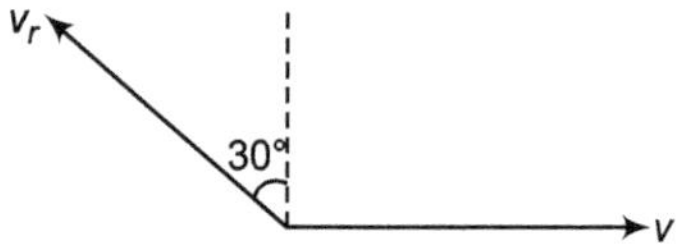

Now, drift $\frac{d}{2} = (v - v_r\sin 30°)\,t = (v - v_r/2)\left(\frac{2d}{\sqrt{3}\,v_r}\right)$

$$\therefore \quad \sqrt{3}\,v_r = 4v - 2v_r$$

$$\therefore \quad v_r = \left(\frac{4}{2+\sqrt{3}}\right)v = 4(2-\sqrt{3})\,v$$

7. Equation of line RS is $y = -mx + C$

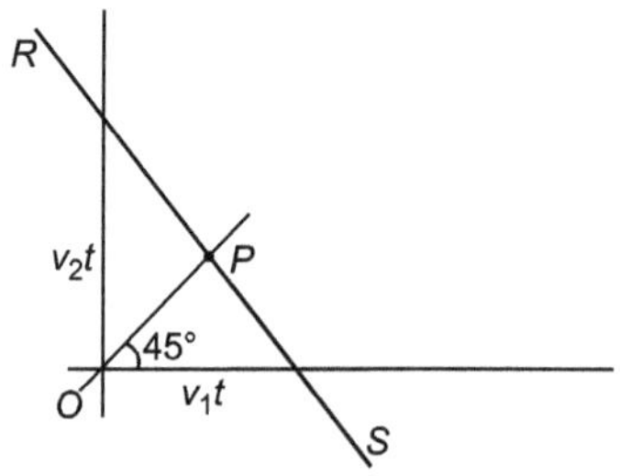

or $\quad y = -\left(\frac{v_2}{v_1}\right)x + v_2t$

or $\quad v_1y = -v_2x + v_1v_2t \quad \ldots\text{(i)}$

Equation of line OP is

$$y = x \quad \ldots\text{(ii)}$$

Point P is the point of intersection, we get

$$x_P = y_P = \frac{v_1v_2t}{v_1+v_2}$$

$$\therefore \quad OP = \sqrt{x_P^2 + y_P^2} = \frac{\sqrt{2}\,v_1v_2t}{v_1+v_2}$$

or $\quad v_3 = \frac{OP}{t} = \frac{\sqrt{2}\,v_1v_2}{v_1+v_2}$

8. $T_0 = \frac{2S}{V}$ and $T = \frac{S}{V+v} + \frac{S}{V-v} = \frac{2SV}{V^2-v^2}$

$$\therefore \quad T/T_0 = \frac{(2SV/V^2-v^2)}{2S/V} = \frac{V^2}{V^2-v^2} = \frac{1}{1-v^2/V^2}$$

9. $x^2 = t^2 + 1$ or $2x\cdot\frac{dx}{dt} = 2t$

or $\quad x\cdot\frac{dx}{dt} = t$

$$\therefore \quad \frac{dx}{dt} = \frac{t}{x} = \frac{t}{\sqrt{t^2+1}}$$

$$\frac{d^2x}{dt^2} = \frac{\sqrt{t^2+1} - \frac{t^2}{\sqrt{t^2+1}}}{(t^2+1)} = \frac{1}{(t^2+1)^{3/2}} = \frac{1}{x^3}$$

10. Average velocity = $\frac{\text{Total displacement}}{\text{Total time}}$

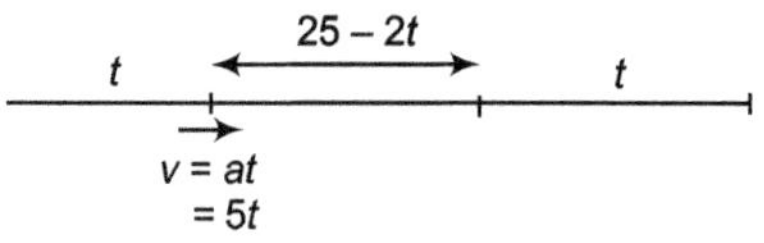

or $\quad 20 = \frac{\frac{1}{2}\times a\times t^2 + (at)(25-2t) + \frac{1}{2}\times a\times t^2}{25}$

Solving this equation with $a = 5$ m/s^2 we get, $t = 5$ s

Thus, the particle moved uniformly for $(25-2t)$ or 15 s.

11. Total time of journey is $(t_1 + t_2)$. Therefore, $\frac{2u}{g} = t_1 + t_2$ ...(i)

Further, $h = ut_1 - \frac{1}{2}gt_1^2$

or $h = g\frac{(t_1 + t_2)t_1}{2} - \frac{1}{2}g t_1^2$

or $gt_1t_2 = 2h$

$\therefore$ $t_1t_2 = \frac{2h}{g}$

12. $0 = u^2 - 2a_1(4) - 2a_2(1)$ ...(i)

$0 = u^2 - 2a_2(2) - 2a_1(2)$ ...(ii)

From these two equations, we get,

$$a_2 = 2a_1$$

13. $(50)^2 = (10)^2 + 2as$ ...(i)

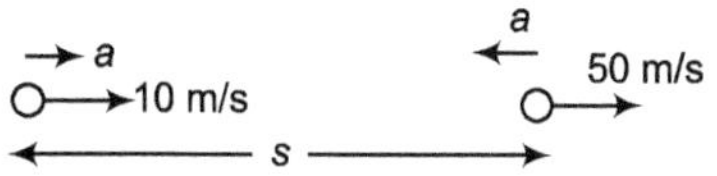

Now, $v^2 = (50)^2 + 2(-a)(-s) = (50)^2 + 2as$ ...(ii)

Solving these two equations, we get

$$v = 70 \text{ m/s}$$

14. For P, $30 = 15 + at$

or $at = 15$ m/s

For Q, $v = 20 - at$ or $v = 20 - 15 = 5$ m/s

15. Time $= \sqrt{\frac{2h}{g}}$ and velocity $= \sqrt{2gh}$

Now, $t = \sqrt{\frac{2h}{2g}} - \sqrt{\frac{2h}{8g}}$ and $v = \sqrt{16gh} - \sqrt{4gh}$

Dividing these two equations, we get

$$\frac{t}{v} = \frac{\sqrt{1/2g} - \sqrt{1/8g}}{\sqrt{8g} - \sqrt{2g}}$$

$$= \frac{2-1}{8g - 4g} = \frac{1}{4g} \quad \text{or} \quad v = 4gt$$

16. v_r is subtraction of vector. Hence,

$$v_r^2 = x \text{ (say)} = v^2 + (at)^2 - 2v(at)\cos\alpha$$

Now, v_r will be minimum when x is minimum. Hence,

$$\frac{dx}{dt} = 0 \text{ or } 2a^2t - 2va\cos\alpha = 0$$

$\therefore$ $t = \frac{v\cos\alpha}{a}$

17. Slope of v-t graph gives acceleration/retardation $|a| = 5\text{ m/s}^2$ for both of them. A has acceleration while B has retardation.

$$|s_A| + |s_B| = 60$$

or $\left\{\frac{1}{2} \times 5 \times t^2\right\} + \left\{20t - \frac{1}{2} \times 5 \times t^2\right\} = 60$

or $t = 3$ s

18. Area of a-t graph gives change in velocity. When net area will become zero, particle will acquire its initial velocity.

19. $a = (b - cx)$

or $v \cdot \frac{dv}{dx} = (b - cx)$ ...(i)

$\therefore$ $v \cdot dv = (b - cx)\,dx$

Integrating we get, $\frac{v^2}{2} = bx - \frac{cx^2}{2}$

$\therefore$ $v^2 = (2bx - cx^2)$ ...(ii)

At other station $v = 0$ or $x = \frac{2b}{c}$

Further, v is maximum, where $\frac{dv}{dx} = 0$

or $x = \frac{b}{c}$ [From Eq. (i)]

Substituting in Eq. (ii) we get,

$$v_m^2 = 2b \times \frac{b}{c} - c \times \frac{b^2}{c^2} = \frac{b^2}{c}$$

or $v_m = \frac{b}{\sqrt{c}}$

20. If t_0 is the time during acceleration, then $\frac{t_0}{2}$ will be the time during retardation.

Now, $t_0 + \frac{t_0}{2} = t$

$\therefore$ $t_0 = \frac{2t}{3}$ and $\frac{t_0}{2} = \frac{t}{3}$

$$s = \frac{1}{2}a\left(\frac{2t}{3}\right)^2 + \frac{1}{2} \times 2a \times \left(\frac{t}{3}\right)^2 = \frac{at^2}{3}$$

21. Taking, $s = ut + \frac{1}{2}at^2$, from point B onwards we have,

$$-\frac{1}{2}at_0^2 = (at_0)t - \frac{1}{2}at^2$$

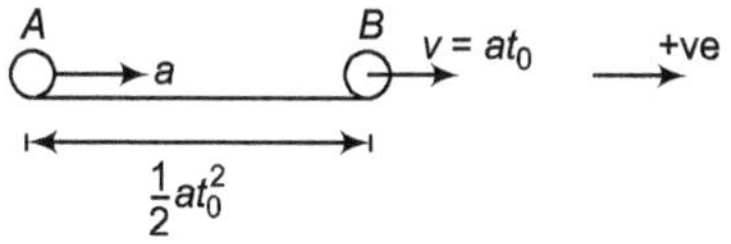

$\therefore$ $t^2 - 2t_0t - t_0^2 = 0$

$$t = \frac{2t_0 \pm \sqrt{4t_0^2 + 4t_0^2}}{2} = (t_0 + \sqrt{2}t_0)$$

$\therefore$ Total time $= t_{AB} + t = (2 + \sqrt{2})t_0$

22. Relative to lift

$$u_r = 0,$$

$$a_r = (9.8 + 1.2) = 11 \text{ m/s}^2$$

$$s_r = \frac{1}{2}a_rt^2$$

$\therefore$ $2.7 = \frac{1}{2} \times 11 \times t^2$

or $t = \sqrt{\frac{5.4}{11}}$ s

23. Given, $x = ae^{-\alpha t} + be^{\beta t}$

So, velocity $v = \frac{dx}{dt}$

$$= -a\alpha e^{-\alpha t} + b\beta e^{\beta t}$$

$$a = \frac{dv}{dt} = a\alpha^2e^{-\alpha t} + b\beta^2e^{\beta t}$$

$$= -\text{ve all the time.}$$

$\therefore$ v will go on increasing.

24. $v^2t^2 = v_1^2 t^2 + a^2$

$\therefore \qquad t = \sqrt{\dfrac{a^2}{v^2 - v_1^2}}$

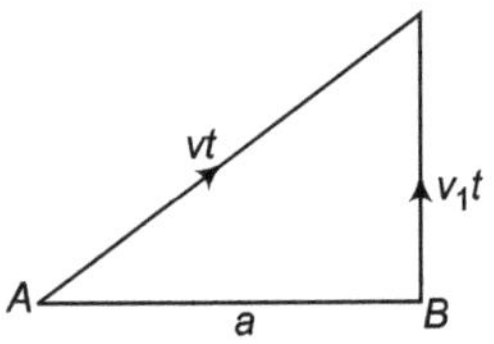

25. Let us see the relative motion, distance moved by the man before coming to rest

$$0 = v^2 - 2as \text{ or } s = \frac{v^2}{2a}$$

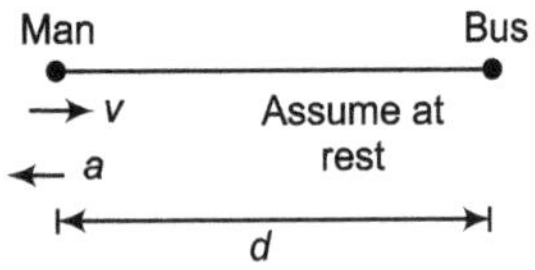

Now, man can catch the bus if,

$$s \geq d \text{ or } v \geq \sqrt{2ad}$$

If he just catches the bus, then

$$0 = v - at$$

or $\qquad t = \dfrac{v}{a}$

26. Car will acquire maximum speed i.e. 20 m/s in 20s with an acceleration of 1 m/s^2. In retardation, it will take 10s. Suppose it moves with constant speed of 20 m/s of t second. Then,

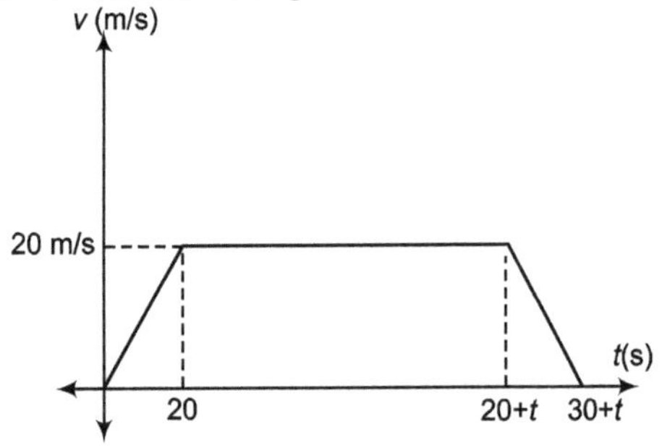

Area of v-t graph = Total displacement

or $\qquad \dfrac{1}{2}[t + 30 + t] \times 20 = 1000$

or $\qquad t = 35$ s

$\therefore$ Total time of journey $= (30 + t) = 65$ s

27. $\dfrac{1}{v}$ *versus* t equation from the given graph would be,

$$\frac{1}{v} = \left(3 + \frac{1}{\sqrt{3}}\right) - t \qquad (y = -mx + c)$$

or $\qquad v = \dfrac{1}{\left(3 + \dfrac{1}{\sqrt{3}}\right) - t}$

$\therefore \qquad a = \dfrac{dv}{dt} = \dfrac{1}{\left[\left(3 + \dfrac{1}{\sqrt{3}}\right) - t\right]^2}$

Putting $t = 3$ s we get, $a = 3$ m/s^2.

28. Let x be the acceleration. Then

$$a = \frac{1}{2} xp^2 \qquad \text{...(i)}$$

$$a + b = +\frac{1}{2} x (p + q)^2 \qquad \text{...(ii)}$$

Solving these two equations, we get

$$x = \frac{2b}{(q + 2p)\, q}$$

29. At 60 s

$$h_1 = \frac{1}{2} \times 10 \times (60)^2 = 18000 \text{ m}$$

$$v_1 = 10 \times 60 = 600 \text{ m/s}$$

After that

$$h_2 = \frac{v_1^2}{2g} = 18000 \text{ m}$$

$\therefore \qquad H_{max} = h_1 + h_2 = 36000 \text{ m} = 36 \text{ km}$

For time

$$-18000 = 600t - 5t^2$$

or $\qquad t^2 - 120t - 3600 = 0$

or $\qquad t = \dfrac{120 + \sqrt{14400 + 14400}}{2}$

$$= \frac{120 + 120\sqrt{2}}{2} = 60 + 60\sqrt{2}$$

$\therefore$ Total time $= 60 + (60 + 60\sqrt{2}) = (120 + 60\sqrt{2})$ s

30. $a = \dfrac{dv}{dt} = -4v + 8$

$\therefore \qquad \dfrac{da}{dt} = -4 \cdot \dfrac{dv}{dt} = -(-4v + 8)$

$$= 16v - 32$$

$$\left(\frac{da}{dt}\right)_{t=0} = -32 \text{ m/s}^3 \qquad (\text{as } v_i = 0)$$

At terminal velocity $a = 0$

$\therefore \qquad v = 2$ m/s

31. In graph (b) for one value of displacement there are two different points of time. Hence, for one time, the average velocity is positive and for other time is equivalent negative.

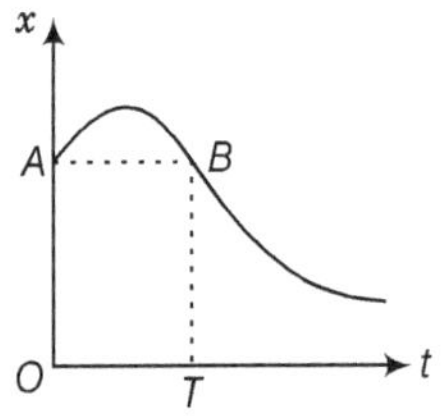

As there are opposite velocities in the interval 0 to T, hence average velocity can vanish in (b). This can be seen in the figure given

Here, $OA = BT$ (same displacement) for two different points of time.

32. As the lift is coming in downward directions displacement will be negative. We have to see whether the motion is accelerating or retarding.

We know that due to downward motion displacement will be negative. When the lift reaches 4th floor is about to stop hence, motion is retarding in nature hence, $x < 0$; $a > 0$.

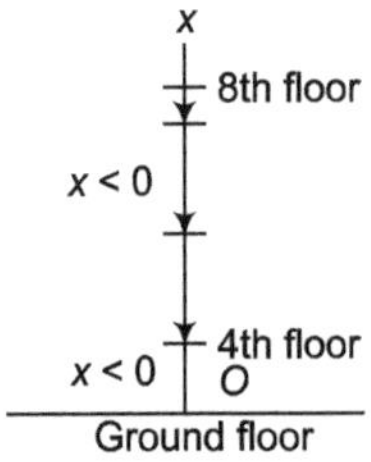

As displacement is in negative direction, velocity will also be negative i.e. $v < 0$.

This can be shown on the adjacent graph.

33. Given, speed of first car $= 18\,\text{km/h}$

Speed of second car $= 27\,\text{km/h}$

$\therefore$ Relative speed of each car w.r.t. each other

$$= 18 + 27 = 45\,\text{km/h}$$

Distance between the cars $= 36\,\text{km}$

$$\therefore \text{ Time of meeting the cars } (t) = \frac{\text{Distance between the cars}}{\text{Relative speed of cars}}$$

$$= \frac{36}{45} = \frac{4}{5}\,\text{h} = 0.8\,\text{h}$$

Speed of the bird $(v_b) = 36\,\text{km/h}$

$\therefore$ Distance covered by the bird $= v_b \times t = 36 \times 0.8 = 28.8\,\text{km}$

34. In this problem equations related to one dimensional motion will be applied for acceleration positive sign will be used and for retardation negative sign will be used.

Given, speed of car as well as truck $= 72\,\text{km/h}$

$$= 72 \times \frac{5}{18}\,\text{m/s} = 20\,\text{m/s}$$

Retarded motion for truck $v = u + a_t t$

$$0 = 20 + a_t \times 5 \quad \text{or} \quad a_t = -4\,\text{m/s}^2$$

Retarded motion for the car $v = u + a_c t$

$$0 = 20 + a_c \times 3 \quad \text{or} \quad a_c = -\frac{20}{3}\,\text{m/s}^2$$

Let car be at a distance x from truck, when truck gives the signal and t be the time taken to cover this distance.

As human response time is 0.5 s, therefore, time of retarded motion of car is $(t - 0.5)$ s.

Velocity of car after time t,

$$v_c = u - at = 20 - \left(\frac{20}{3}\right)(t - 0.5)$$

Velocity of truck after time t,

$$v_t = 20 - 4t$$

To avoid the car bump onto the truck,

$$v_c = v_t$$

$$20 - \frac{20}{3}(t - 0.5) = 20 - 4t$$

$$4t = \frac{20}{3}(t - 0.5) \Rightarrow t = \frac{5}{3}(t - 0.5)$$

$$3t = 5t - 2.5 \Rightarrow t = \frac{2.5}{2} = \frac{5}{4}\,\text{s}$$

Distance travelled by the truck in time t,

$$s_t = u_t t + \frac{1}{2} a_t t^2$$

$$= 20 \times \frac{5}{4} + \frac{1}{2} \times (-4) \times \left(\frac{5}{4}\right)^2$$

$$= 21.875\,\text{m}$$

Distance travelled by car in time $t =$ Distance travelled by car in 0.5 s (without retardation) + Distance travelled by car in $(t - 0.5)$ s (with retardation)

$$s_c = (20 \times 0.5) + 20\left(\frac{5}{4} - 0.5\right) - \frac{1}{2}\left(\frac{20}{3}\right)\left(\frac{5}{4} - 0.5\right)^2$$

$$= 23.125\,\text{m}$$

$$\therefore \quad s_c - s_t = 23.125 - 21.875 = 1.250\,\text{m}$$

Therefore, to avoid the bump onto the truck, the car must maintain a distance from the truck more than 1.250 m.

More than One Correct Options

1. (a) $a = -\alpha\sqrt{v}$

$$\therefore \quad \frac{dv}{dt} = -\alpha\sqrt{v}$$

$$\text{or} \quad \int_0^t dt = -\frac{1}{\alpha}\int_{v_0}^0 v^{-1/2} \cdot dv$$

$$\therefore \quad t = \frac{2\sqrt{v_0}}{\alpha}$$

(d) $a = -\alpha\sqrt{v}$

$$\therefore \quad v \cdot \frac{dv}{ds} = -\alpha\sqrt{v}$$

$$\therefore \quad \int_0^s ds = -\frac{1}{\alpha}\int_{v_0}^0 v^{1/2}\,dv$$

$$\therefore \quad s = \frac{2v_0^{3/2}}{3\alpha}$$

2. $a = -0.5t = \dfrac{dv}{dt}$

$$\therefore \quad \int_{16}^v dv = \int_0^t -0.5t\,dt$$

$$\therefore \quad v = 16 - 0.25t^2$$

$$s = \int v dt = 16t - \frac{0.25\,t^3}{3}$$

$$v = 0 \quad \text{when} \quad 16 - 0.25\,t^2 = 0$$

$$\text{or} \quad t = 8\,\text{s}$$

So, direction of velocity changes at 8 s. upto 8 s (i.e. at 4 s),

Distance = Displacement

$\therefore$ **At 4 s**

$$d = 16 \times 4 - \frac{0.25 \times (4)^3}{3} = 58.67\,\text{m}$$

$$S_{8\,s} = 16 \times 8 - \frac{(0.25)\,(8)^3}{3}$$

$$= 85.33\,\text{m}$$

$$S_{10\,s} = 16 \times 10 - \frac{(0.25)\,(10)^3}{3}$$

$$= 76.67\,\text{cm}$$

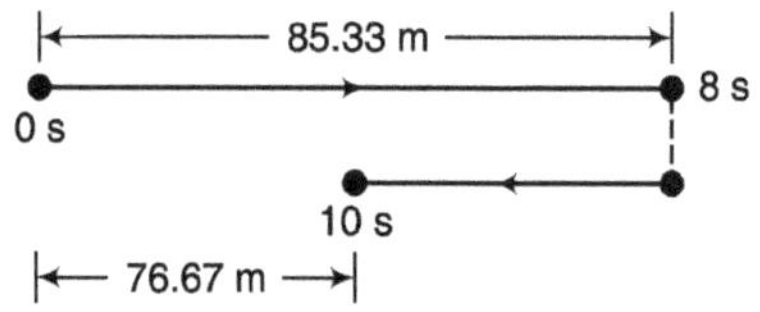

Distance travelled in 10s,

$$d = (85.33) + (85.33 - 76.67)$$

$$= 94\,\text{m}$$

At 10 s

$$v = 16 - 0.25\,(10)^2$$

$$= -9\,\text{m/s}$$

$$\therefore \quad \text{Speed} = 9\,\text{m/s}$$

3. If $\mathbf{a} =$ constant

Then $|\mathbf{a}|$ is also constant or $\left|\dfrac{d\mathbf{v}}{dt}\right| =$ constant

4. $\mathbf{r}_B = 0$ at $t = 0$

$$\mathbf{r}_A = 3\hat{\mathbf{i}} + 4\hat{\mathbf{j}}$$

at $t = 0$

$$\mathbf{v}_A = (-20\hat{\mathbf{j}})$$

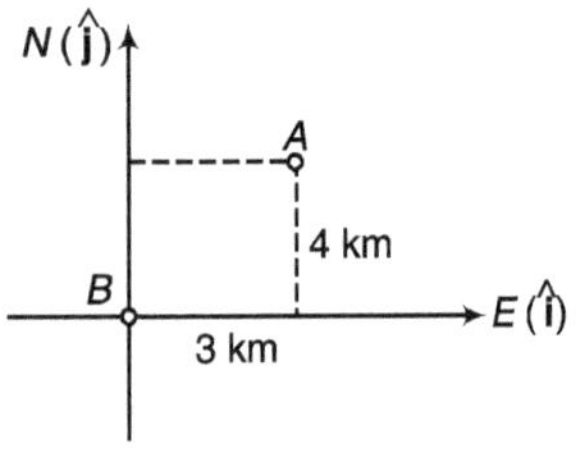

$$\mathbf{v}_B = (40\cos 37°)\hat{\mathbf{i}} + (40\sin 37°)\hat{\mathbf{j}}$$
$$= (32\hat{\mathbf{i}} + 24\hat{\mathbf{j}})$$
$$\mathbf{v}_{AB} = (-32\hat{\mathbf{i}} - 44\hat{\mathbf{j}})\ \text{km/h}$$

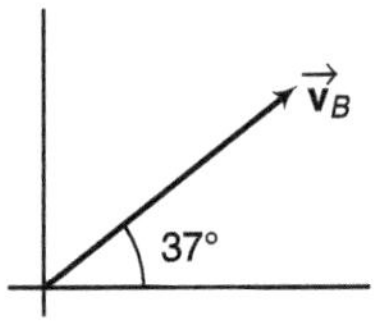

At time $t = 0$,

$$\mathbf{r}_{AB} = \mathbf{r}_A - \mathbf{r}_B = (3\hat{\mathbf{i}} + 4\hat{\mathbf{j}})$$

$\therefore$ At time $t = t$,

$$\mathbf{r}_{AB} = (\mathbf{r}_{AB}\ \text{at}\ t = 0) + \mathbf{v}_{AB}t$$
$$= (3 - 32t)\hat{\mathbf{i}} + (4 - 44t)\hat{\mathbf{j}}$$

5. $a_1 t_1 = a_2 t_2$

$$v_{\max} = a_1 t_1 = a_2 t_2$$
$$S_1 = \text{Area of } v\text{-}t \text{ graph}$$
$$= \frac{1}{2}(t_1 + t_2)(a_1 t_1)$$

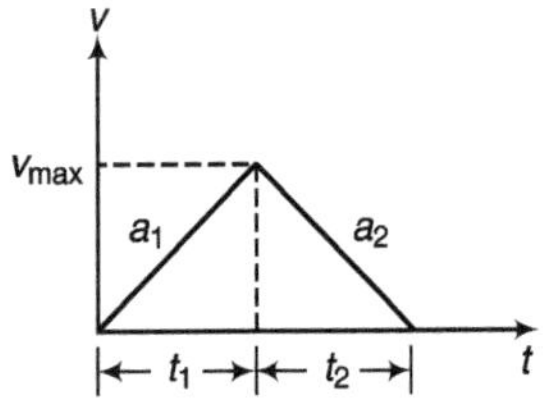

$$v_1 = \frac{S_1}{t_1 + t_2} = \frac{1}{2}a_1 t_1$$
$$S_2 = \frac{1}{2}(t_1 + t_3)v_{\max} = \frac{1}{2}(t_1 + 2t_2)(2a_1 t_1)$$
$$v_2 = \frac{S_2}{(t_1 + t_3)} = a_1 t_1$$

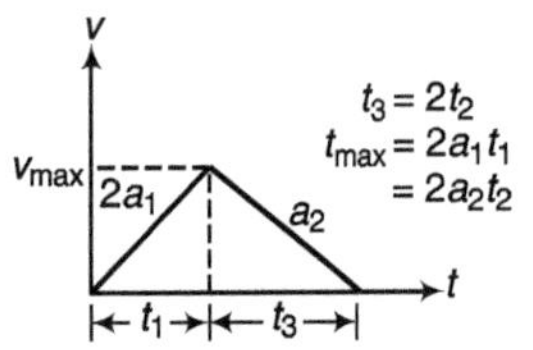

From the four relations, we can see that

$$v_2 = 2v_1$$

and $$2S_1 < S_2 < 4S_1$$

6. In the complete journey,

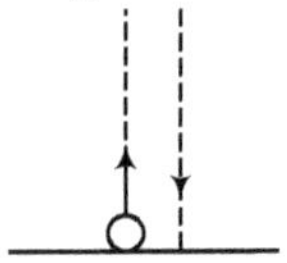

$S = 0$ and $a = \text{constant} = g$ (downwards)

7. $a = \frac{F}{m} = \frac{\alpha}{m}t$ ...(i)

or $a \propto t$

i.e. a-t graph is a straight line passing through origin.

If $u = 0$, then integration of Eq. (i) gives,

$$v = \frac{\alpha t^2}{2m} \quad \text{or} \quad v \propto t^2$$

Hence, in this situation (when $u = 0$) v-t graph is a parabola passing through origin.

8. a-s equation corresponding to given graph is,

$$a = 6 - \frac{s}{5}$$

$$\therefore \quad v \cdot \frac{dv}{ds} = \left(6 - \frac{s}{5}\right) \quad \text{or} \quad \int_0^v v\,dv = \int_0^s \left(6 - \frac{s}{5}\right) ds$$

or $$v = \sqrt{125 - \frac{s^2}{5}}$$

At $s = 10\,\text{m}, v = 10\ \text{m/s}$

Maximum values of v is obtained when

$$\frac{dv}{ds} = 0 \text{ which gives } s = 30\ \text{m}$$

$$\therefore \quad v_{\max} = \sqrt{12 \times 30 - \frac{(30)^2}{5}}$$
$$= \sqrt{180}\ \text{m/s}$$

9. $\mathbf{v}_{\text{av}} = \frac{\mathbf{s}}{t}$ and $v_{\text{av}} = \frac{d}{t}$

Now, $d \geq |\mathbf{s}|$

$\therefore$ $v_{\text{av}} \geq |\mathbf{v}_{\text{av}}|$

10. $v = at$ and $x = \frac{1}{2}at^2$ $(u = 0)$ i.e. v-t graph is a straight line passing through origin and x-t graph a parabola passing through origin.

11. For minimum time

$$\therefore \quad t_{\min} = \frac{b}{v}$$

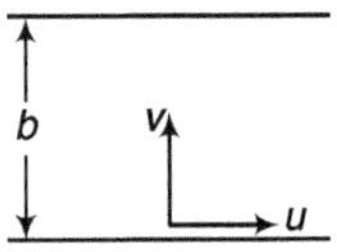

For reaching a point exactly opposite

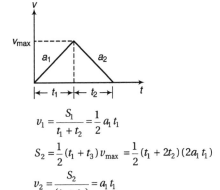

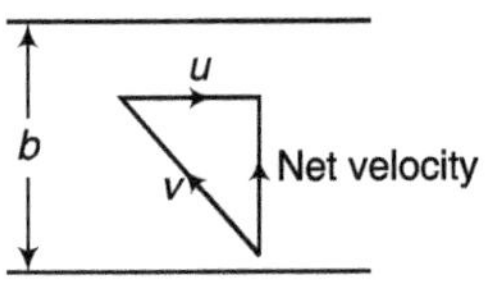

$$\text{Net velocity} = \sqrt{v^2 - u^2} \qquad (\text{but } v > u)$$

$$\therefore \quad t = \frac{b}{\text{net velocity}}$$

12. For $t<T, v=-\text{ve}$

For $t>T, v=+\text{ve}$

At $t=T, v=0$

$\therefore$ Particle changes direction of velocity at $t=T$

$S=$ Net area of v-t graph $=0$

$a=$ Slope of v-t graph $=$ constant

13. $v=\alpha t_1 \Rightarrow t_1=\frac{v}{\alpha}$

$v=\beta t_2 \Rightarrow t_2=\frac{v}{\beta}$

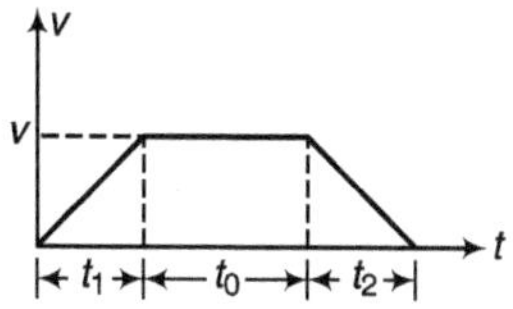

$\therefore \quad t_0=t-t_1-t_2=\left(t-\frac{v}{\alpha}-\frac{v}{\beta}\right)$

Now, $\quad l=\frac{1}{2}\alpha t_1^2+vt_0+\frac{1}{2}\beta t_2^2$

$$=\frac{1}{2}(\alpha)\left(\frac{v}{\alpha}\right)^2+v\left(t-\frac{v}{\alpha}-\frac{v}{\beta}\right)+\frac{1}{2}(\beta)\left(\frac{v}{\beta}\right)^2$$

$$=vt-\frac{v^2}{2\alpha}-\frac{v^2}{2\beta}$$

$$\therefore \quad t=\frac{l}{v}+\frac{v}{2}\left(\frac{1}{\alpha}+\frac{1}{\beta}\right)$$

For t to be minimum, its first derivation with respect to velocity be zero or

$$0=-\frac{l}{v^2}+\frac{\alpha+\beta}{2\alpha\beta}$$

$$\therefore \quad v=\sqrt{\frac{2l\alpha\beta}{\alpha+\beta}}$$

14. $x=t^2$

$\Rightarrow \quad v_x=\frac{dx}{dt}=2t \Rightarrow a_x=\frac{dv_x}{dt}=2$

$\Rightarrow \quad y=t^3-2t \Rightarrow v_y=\frac{dy}{dt}=3t^2-2$

$\Rightarrow \quad a_y=\frac{dv_y}{dt}=6t$

At $t=0$, $\quad v_x=0, v_y=-2, a_x=2$ and $a_y=0$

$\therefore \quad \mathbf{v}=-2\hat{\mathbf{j}}$ and $\mathbf{a}=2\hat{\mathbf{i}}$

or $\mathbf{v}\perp\mathbf{a}$

At $t=\sqrt{\frac{2}{3}}$, $\Rightarrow \quad v_y=0, v_x\neq 0$.

Hence, the particle is moving parallel to x-axis.

15. $(14)^2=(2)^2+2as$

$\therefore \quad 2as=192$ units

At mid point, $\quad v^2=(2)^2+2a\left(\frac{s}{2}\right)$

$$=4+\frac{192}{2}=100$$

$\therefore \quad v=10$ m/s

$XA:AY=1:3$

$\therefore \quad XA=\frac{1}{4}s$ and $AY=\frac{3}{4}s$

$$v_1^2=(2)^2+2a\left(\frac{s}{4}\right)=4+\frac{192}{4}=52$$

$\therefore \quad v_1=\sqrt{52}\neq 5$ m/s

$10=2+at_1 \qquad (v=u+at)$

$\therefore \quad t_1=\frac{8}{a} \quad 14=10+at_2$

$\therefore \quad t_2=\frac{4}{a}$ or $t_1=2t_2$

$S_1=(2t)+\frac{1}{2}a(t^2)=$ distance travelled in first half

$$S_2=2(2t)+\frac{1}{2}a(2t)^2$$

$S_3=S_2-S_1=$ distance travelled in second half

We can see that,

$$S_1\neq\frac{S_3}{2}$$

16. As per the diagram, at point A the graph is parallel to time axis hence, $v=\frac{dx}{dt}=0$. As the starting point is A hence, we can say that the particle is starting from rest.

At C, the graph changes slope, hence velocity also changes. As graph at C is almost parallel to time axis hence, we can say that velocity vanishes.

As direction of acceleration changes hence, we can say that it may be zero in between.

From the graph it is clear that

$$|\text{slope at } D|>|\text{slope at } E|$$

Hence, speed at D will be more than at E.

17. Let the speeds of the two balls (1and 2) be v_1 and v_2, where if $v_1=2v, v_2=v$

if y_1 and y_2 and the distance covered by the balls 1 and 2, respectively, before coming to rest, then

$$y_1=\frac{v_1^2}{2g}=\frac{4v^2}{2g} \text{ and } y_2=\frac{v_2^2}{2g}=\frac{v^2}{2g}$$

Since, $\quad y_1-y_2=15\text{m}, \frac{4v^2}{2g}-\frac{v^2}{2g}=15$ m

or $\quad \frac{3v^2}{2g}=15$ m

or $\quad v^2=\sqrt{5\text{m}\times(2\times10)}$ m/s^2

or $\quad v=10$ m/s

Clearly, $\quad v_1=20$ m/s and $v_2=10$ m/s

as $\quad y_1=\frac{v_1^2}{2g}=\frac{(20\text{m})^2}{2\times10\,\text{m}15}=20\text{m},$

$$y_2=y_1-15\text{ m}=5\text{ m}$$

If t_2 is the time taken by the ball 2 toner a distance of 5m, then from $y_2=v_2t-\frac{1}{2}gt_2^2$

$$5=10t_2-5t_2^2 \text{ or } t_2^2-2t_2+1=0,$$

where $\quad t_2=15$

Since, t_1 (time taken by ball 1 to cover distance of 20m) is 2s, time interval between the two throws

$$=t_1-t_2=2\text{s}-1\text{s}=1\text{s}$$

18. If an object undergoes a displacement Δr in time Δt, its average velocity is given by

$v = \dfrac{\Delta r}{\Delta t} = \dfrac{r_2 - r_1}{t_2 - t_1}$, where r_1 and r_2 are position vectors corresponding to time t_1 and t_2.

It the velocity of an object changes from v_1 to v_2 in time Δt. Average acceleration is given by

$$a_{av} = \frac{\Delta \mathbf{v}}{\Delta t} = \frac{v_2 - v_1}{t_2 - t_2}$$

But, when acceleration is non-uniform

$$v_{av} \neq \frac{v_1 + v_2}{2}$$

We can write $\Delta v = \dfrac{\Delta r}{\Delta t}$

Hence, $\Delta r = r_2 - r_1 = (v_2 - v_1)(t_2 - t_1)$

Comprehension Based Questions

1. Velocity of ball with respect to elevator is 15 m/s (up) and elevator has a velocity of 10 m/s (up). Therefore, absolute velocity of ball is 25 m/s (upwards). Ball strikes the floor of elevator if,

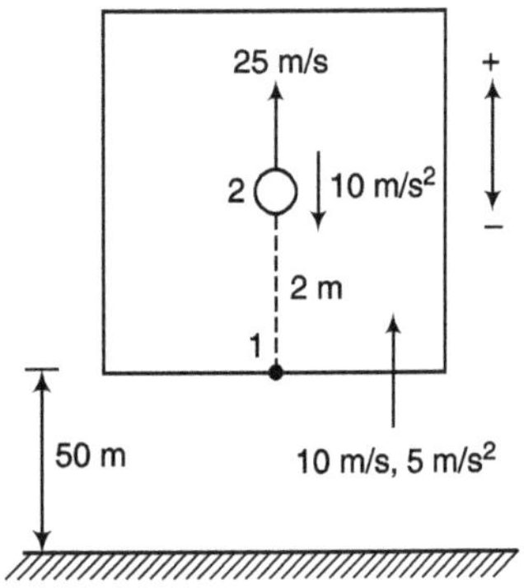

$$S_1 = S_2 + 2$$

$\therefore \quad 10t + 2.5t^2 = 25t - 5t^2 + 2$

Solving this equation we get,

$$t = 2.13 \text{ s}$$

2. If the ball does not collides, then it will reach its maximum height in time,

$$t_0 = \frac{u}{g} = \frac{25}{10} = 2.5 \text{ s}$$

Since, $t < t_0$, therefore as per the question ball is at its maximum height at 2.13 s.

$$h_{max} = 50 + 2 + 25 \times 2.13 - 5 \times (2.13)^2 = 82.56 \text{ m}$$

3. $S = 25 \times 2.13 - 5 \times (2.13)^2 = 30.56$ m

4. At maximum separation, their velocities are same

$\therefore \quad 25 - 10t = 10 + 5t$

or $\quad t = 1$ s

Maximum separation $= 2 + S_2 - S_1$

$= 2 + [25 \times 1 - 5 \times (1)^2] - [10 \times 1 + 2.5(1)^2]$

$= 9.5$ m

5. $u_A + a_A T = u_B - a_B T$

Putting $\quad T = 4$ s

we get $\quad 4(a_A + a_B) = u_B - u_A$

Now, $\quad S_A = S_B$

$\therefore \quad u_A t + \dfrac{1}{2} a_A t^2 = u_B t - \dfrac{1}{2} a_B t^2$

$\therefore \quad t = 2\dfrac{(u_B - u_A)}{(a_A + a_B)} = 2 \times 4 = 8$ s

6. $S_A = S_B$

$$5t + \frac{1}{2} a_A t^2 = 15t - \frac{1}{2} a_B t^2$$

or $\quad 10 + a_A t = 30 - a_B t$

$\therefore \quad (5 + a_A t) - (15 - a_B t) = 10$

or $\quad v_A - v_B = 10$ m/s

7. $8 = 6 + a_A T = 6 + 4a_A$

$\therefore \quad a_A = 0.5$ m/s

At 10 s,

$$v_A = u_A + a_A t = (6) + (0.5)(10) = 11 \text{ m/s}$$

Assertion and Reason

1. Velocity is always positive.

2. If speed varies, then velocity will definitely vary.

At highest point of a particle thrown upwards

$a \neq 0$ but $v = 0$.

3. $\mathbf{a} = \dfrac{\mathbf{F}}{m} = 1\hat{\mathbf{j}}$ N

$$\mathbf{s} = \mathbf{u}t + \frac{1}{2}\mathbf{a}t^2 = 20\hat{\mathbf{i}} + 8\hat{\mathbf{j}}$$

$\therefore \quad |\mathbf{s}| = 21.5$ m

4. When a particle is released from rest,

$v = 0$ but $a \neq 0$

5. On a curved path, distance > displacement

$\therefore$ Average speed > Average velocity

Further, average velocity $= \dfrac{\text{Total displacement}}{\text{Total time}}$

6. $\mathbf{a} = \dfrac{\mathbf{v}_f - \mathbf{v}_i}{\Delta t} = \dfrac{d\mathbf{v}}{dt}$ i.e. direction of acceleration is same as that of change in velocity vector, or in the direction of $\Delta \mathbf{v}$.

8. $a = \dfrac{dv}{dt}$ = slope of v-t graph.

Perpendicular to t-axis, slope $= \infty$

$\therefore \quad a = \infty$

11. Last second of upward journey is the first second of downward journey.

$\therefore$ Distance travelled $= \dfrac{1}{2} g t^2$

$= \dfrac{1}{2} \times 9.8 \times (1)^2 = 4.9$ m

12. By differentiating a-t equation two times we will get s-t equation. Further

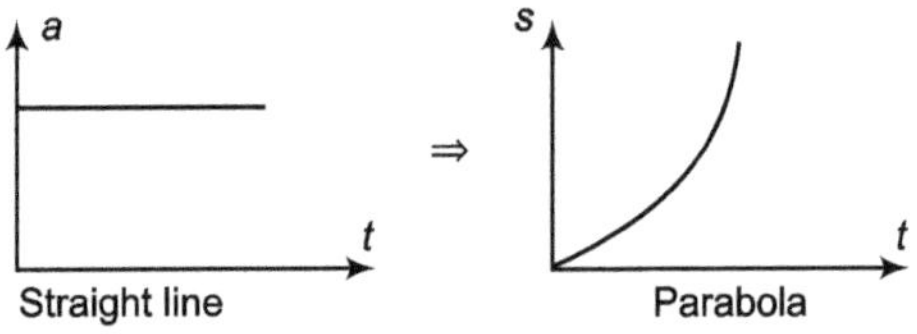

13. v_1 and v_2 = constant

But distance is increasing.

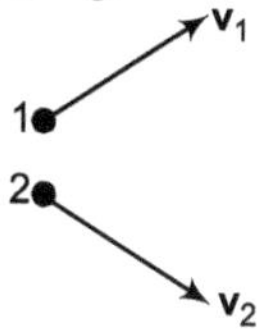

14.

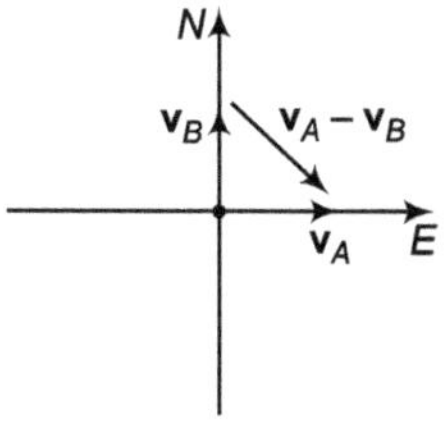

15. Average velocity $= \dfrac{\text{Displacement}}{\text{Time}} = \dfrac{\text{Area } v\text{ - }t \text{ graph}}{\text{Time}}$

$$= \frac{\frac{1}{2} v_m t_0}{t_0} = \frac{1}{2} v_m$$

16. S is displacement not distance.

17. Slope in negative. Therefore, velocity is negative.

Slope (therefore velocity) is increasing in magnitude. Therefore, acceleration is also negative.

18. Ascending means velocity is upwards. Speed is decreasing. It means acceleration is downwards.

Further, the body moves in the direction of velocity.

19. Slope of s-t graph = velocity = positive

At $t = 0,\ s \neq 0$

Further at $t = t_0 \Rightarrow s = 0, v \neq 0.$

20. $v\text{-}t \xrightarrow{\text{Intergration}} s\text{-}t$

$s\text{-}t \xrightarrow{\text{Differentiation}} v\text{-}t$

Match the Columns

1. $\dfrac{d|\mathbf{v}|}{dt}$ is rate of change of speed of the particle.

$\dfrac{d|\mathbf{r}|}{dt}$ is the rate by which distance of particle from the origin is changing.

2. Corresponding v-t, a-t and s-t graphs are as shown below

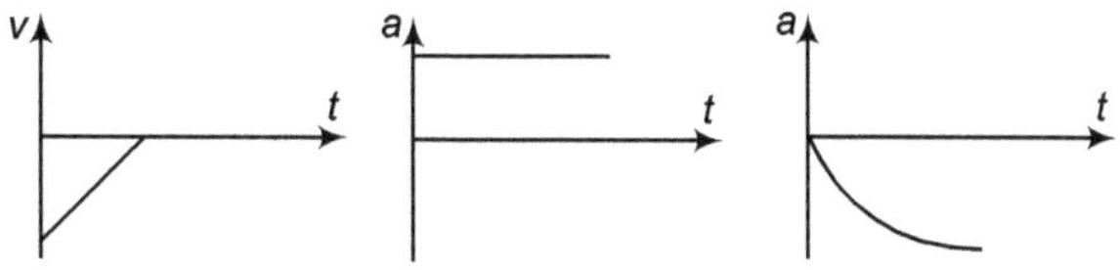

3. $v_i = +10$ m/s and $v_f = 0$

$\therefore$
$$\Delta v = v_f - v_i = -10 \text{ m/s}$$
$$a_{av} = \frac{\Delta u}{\Delta t} = \frac{-10}{6}$$
$$= \frac{-5}{3} \text{ m/s}^2$$

Total displacement = area under v-t graph (with sign)

And acceleration = slope of v-t graph

4. In motion M Slope of s-t graph is positive and increasing. Therefore, velocity of the particle is positive and increasing. Hence, it is A type motion. Similarly, N, P and Q can be observed from the slope.

5. Comparing the given equation with general equation of displacement, $s = s_0 + ut + \frac{1}{2}at^2$, we get

$$u = +\,20 \text{ unit} \quad \text{and} \quad a = -10 \text{ unit}$$

6. $\omega = \dfrac{v}{R} = \dfrac{4}{1} = 4$ rad/s

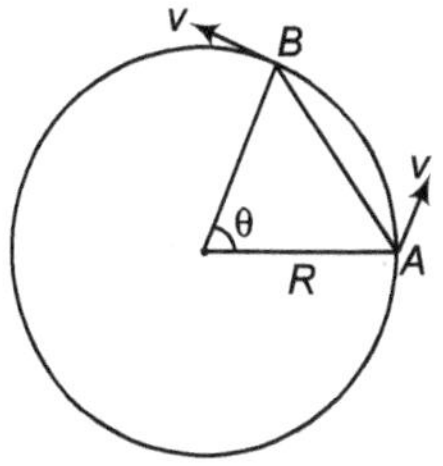

$$\theta = \omega t = 4 \text{ rad}$$

$$\text{Displacement} = AB = 2R\sin\frac{\theta}{2}$$
$$= 2\sin 2 \text{ unit}$$
$$\text{Distance} = vt = 4 \text{ unit}$$
$$\text{Average velocity} = \frac{\text{Displacement}}{\text{Time}} = 2\sin 2 \text{ unit}$$
$$\text{And average acceleration} = \frac{|\Delta \mathbf{v}|}{\Delta t}$$

From subtraction of vectors, we can find that change in velocity from A and B is $2v\sin\frac{\theta}{2} = 8\sin 2$ unit

$\therefore$ Average acceleration $= \dfrac{8\sin 2}{1}$

$$= 8\sin 2 \text{ unit}$$

7. With constant positive acceleration, speed will increase when velocity is positive, speed will decrease if velocity is negative.

Similarly, with constant negative acceleration speed will increase, if velocity is also negative and speed will decrease, if velocity is positive.

8. After 2s, velocity of balloon and hence the velocity of the particle will be 20 m/s (= at) and its height from the ground will be $20\text{ m}\left(=\frac{1}{2}at^2\right)$. Now, g will start acting on the particle.

9. We have to analyse slope of each curve i.e. $\dfrac{dx}{dt}$. For peak points $\dfrac{dx}{dt}$ will be zero as x is maximum at peak points.

For graph (A), there is a point (B) for which displacement is zero. So, a matches with (r).

In graph (B), x is positive (>0) throughout and at point $B_1, V = \dfrac{dx}{dt} = 0.$

since, at point of curvature changes $a = 0$, So b matches with (q)

In graph (C), slope $V = \dfrac{dx}{dt}$ is negative hence, velocity will be negative.
so matches wth (s).
In graph (D), as slope $V = \dfrac{dx}{dt}$ is positive hence, $V > 0$
Hence, d matches with (p).

Entrance Gallery

1. If balls are on floor or wall of rocket, then acceleration of rocket do not affect time, when balls hit each other. Balls hit after time t such that

$$0.3\,t + 0.2\,t = 4$$

$$\Rightarrow \quad t = \frac{4}{0.5} = 8\,\text{s} \quad \ldots\text{(i)}$$

If balls are not touching rocket wall or floor, then motion of left ball is affected by acceleration of rocket.
In small time 0.1s, left ball moves $0.3 \times 0.1 = 0.03$ m
Left wall moves $0\,(0.1) + \frac{1}{2}(2)\,(0.1)^2 = 0.01$ m
Initially, left ball is ahead of left wall at time t_1 given by

$$0\,(t_1) + \frac{1}{2}(2)\,(t_1)^2 = 0.3\,t \Rightarrow t_1 = 0.3\,\text{s}$$

Left wall and left ball touch and velocity of wall is $[0 + 2\,(0.3)] = 0.6\ \text{ms}^{-1}$, is more than velocity of ball.
Now, left wall takes the left ball along with it.
Left wall, left ball meet the right ball after time t_2 such that

$$4 - (0.3)\,(0.3) - (0.2)\,(0.3) = \left(0.6\,t_2 + \frac{1}{2}(2)\,t_2^2\right) + (0.2\,t_2)$$

$$t_2^2 + 0.8t_2 = 3.85$$

$$\Rightarrow \quad t_2^2 + 0.8t_2 + 0.16 = 3.85 + 0.16$$

$$\Rightarrow \quad (t_2 + 0.4)^2 = 4.01$$

$$\Rightarrow \quad t_2 = 2.0 - 0.4 = 1.6\,\text{s}$$

Total time $= t_1 + t_2 = 0.3\,\text{s} + 1.6\,\text{s} = 1.9\,\text{s} \approx 2\,\text{s}$ ...(ii)

2. 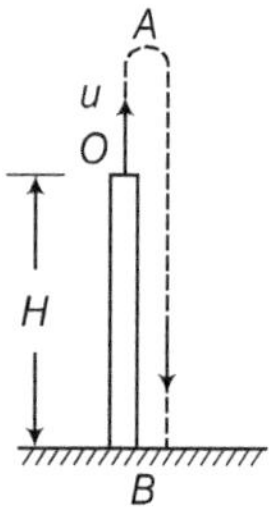

At A = highest point of path, velocity = 0

$$0 = u - gt$$

$$\Rightarrow \quad t = \frac{u}{g} = \text{time for } 0 \to A \text{ journey}$$

Given time for $0 \to A \to B$ journey

$$= nt \quad \ldots\text{(i)}$$

$0 \to A \to B$ journey

$$-H = u(nt) - \frac{1}{2}g\,(nt)^2$$

$$\Rightarrow \quad -H = un\left(\frac{u}{g}\right) - \frac{1}{2}gn^2\frac{u^2}{g^2}$$

$$\Rightarrow \quad -H = n\frac{u^2}{g} - \frac{n^2}{2}\frac{u^2}{g} \Rightarrow -2gH = 2nu^2 - n^2u^2$$

$$\Rightarrow \quad 2gH = n^2u^2 - 2nu^2$$

$$\Rightarrow \quad 2gH = nu^2\,(n-2)$$

3. $x = ut + \frac{1}{2}at^2$

For free fall starting from rest, $u = 0, a = g$

$$\Rightarrow \quad x = \frac{1}{2}gt^2$$

t	0	T	$2T$	$3T$
x	0	$\frac{g}{2}T^2$	$\frac{g}{2}(4T^2)$	$\frac{g}{2}(9T^2)$

Δt	0 to T	T to $2T$	$2T$ to $3T$
Δx	$\frac{g}{2}T^2$	$\frac{g}{2}(4-1)\,T^2$	$\frac{g}{2}(9-4)\,T^2 = \frac{g}{2}5T^2$

Required ratio

$$\frac{g}{2}T^2 : \frac{g}{2}3T^2 : \frac{g}{2}5T^2 = 1:3:5$$

4. 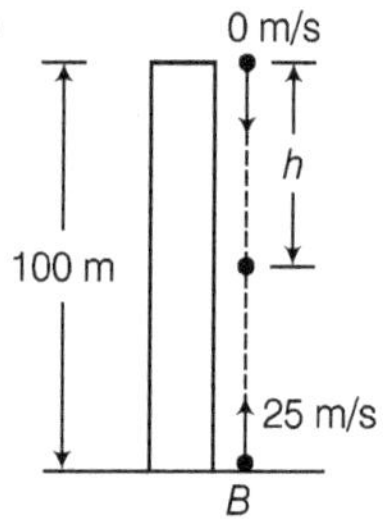

$$h = \frac{1}{2}gt^2 \quad \ldots\text{(i)}$$

$$\frac{1}{2}gt^2\left(25t - \frac{1}{2}gt^2\right) = 100 \Rightarrow t = \frac{100}{25} = 4\,\text{s} \quad \ldots\text{(ii)}$$

From Eqs. (i) and (ii), we get $h = \frac{1}{2}(9.8)\,(4)^2 = 78.4$ m

5. Given, $s_1 = 12\,\text{m}, t_1 = 3\,\text{s} \Rightarrow s_2 = 30 + 12 = 42\,\text{m}$

$$t_2 = 3 + 3 = 6\,\text{s} \Rightarrow s = ut + \frac{1}{2}at^2$$

$$\therefore \quad s_1 = ut_1 + \frac{1}{2}at_1^2 \Rightarrow 12 = 3 \times u + \frac{1}{2}a \times 3^2$$

$$12 = 3u + \frac{9}{2}a \quad \ldots\text{(i)}$$

Similarly, $\quad 42 = 6u + \frac{1}{2} \times a \times 36$ ...(ii)

On solving Eqs. (i) and (ii), we get

$$a = 2\,\text{m/s}^2 \quad \text{and} \quad u = 1\,\text{m/s}$$

6. Given, initial velocity, $u = v$
Final velocity, $v = v/2$
Distance, $s = 20$ cm
Let the further distance of penetration before it comes to rest be x.

$$v^2 = u^2 - 2as \Rightarrow \left(\frac{v}{2}\right)^2 = v^2 - 2a \times 20$$

$$40a = v^2 - \frac{v^2}{4} \Rightarrow 40a = \frac{3v^2}{4} \quad \ldots\text{(i)}$$

and $v^2 = u^2 + 2as$

$\Rightarrow$ $v^2 = 0 + 2a \times (20 + x)$

$$v^2 = 2 \times \frac{3}{160} v^2 (20 + x) \quad \text{[From Eq. (i)]}$$

$$1 = \frac{3}{80}(20 + x) \Rightarrow \frac{80}{3} = 20 + x$$

$$x = \frac{80}{3} - 20 \Rightarrow x = \frac{80 - 60}{3} = \frac{20}{3} = 6.66 \text{ m}$$

7. Let distance between A and $B = d$

t_1 = time for $A \to B$, journey $= \frac{d}{30}$

t_2 = time for $B \to A$, journey $= \frac{d}{20}$

Total distance travelled $= d + d = 2d$

Total time taken

$$= t_1 + t_2 = \frac{d}{30} + \frac{d}{20} = \frac{d}{10}\left(\frac{1}{3} + \frac{1}{2}\right) = \frac{d}{10}\left(\frac{5}{6}\right) = \frac{d}{12}$$

Average speed of car $= \frac{2d}{d/12} = 24$ km/h

8. $x = ut + \frac{1}{2}at^2$

For motion starting from rest, $u = 0$

$$x = \frac{1}{2}at^2$$

$$x_1 = \frac{1}{2}a\left(\frac{t}{2}\right)^2 = \frac{at^2}{8} \quad \text{...(i)}$$

$$x_1 + x_2 = \frac{1}{2}a(t)^2 = \frac{at^2}{2}$$

$$x_2 = \frac{at^2}{2} - \frac{at^2}{8} = \frac{3}{8}at^2 \quad \text{...(ii)}$$

From Eqs. (i) and (ii), we get

$$x_2 = 3x_1$$

9. $s_n = u(n) + \frac{1}{2}a(n)^2$ = distance travelled in n seconds

$$s_{n-1} = u(n-1) + \frac{1}{2}a(n-1)^2$$

$$\Rightarrow \quad s_{nth} = s_n - s_{n-1} = u + \frac{1}{2}a[n^2 - (n-1)^2]$$

$$\Rightarrow \quad s_{nth} = u + \frac{1}{2}a(2n - 1)$$

= distance travelled in nth second

$$s_{nth} = u + \frac{1}{2}a(2n-1)$$

$$s_{3rd} = 0 + \frac{1}{2}a(2 \times 3 - 1) = \frac{5}{2}a \quad (\text{for } n = 3 \text{ s})$$

$$s_{4th} = 0 + \frac{1}{2}a(2 \times 4 - 1) = \frac{7}{2}a \quad (\text{for } n = 4 \text{ s})$$

So, the percentage increase

$$\frac{s_{4th} - s_{3rd}}{s_{3rd}} \times 100 = \frac{\frac{7}{2}a - \frac{5}{2}a}{\frac{5}{2}a} \times 100$$

$$\frac{\frac{2a}{2}}{\frac{5}{2}a} \times 100 = 2 \times 20 = 40\%$$

10. For vertical motion, $h = ut - \frac{1}{2}gt^2$

$$h_1 = ut_1 - \frac{1}{2}gt_1^2 \quad \text{...(i)}$$

$$h_2 = ut_2 - \frac{1}{2}gt_2^2 \quad \text{...(ii)}$$

$$0 = uT - \frac{1}{2}gT^2 \Rightarrow T = \frac{2u}{g} \quad \text{...(iii)}$$

On multiplying Eq. (i) by t_2 and Eq. (ii) by t_1 and then subtracting Eq. (i) from Eq. (ii), we get

$$h_1t_2 - h_2t_1 = \frac{1}{2}g\,t_1t_2(t_2 - t_1) \Rightarrow g = \frac{2(h_1t_2 - h_2t_1)}{t_1t_2(t_2 - t_1)} \quad \text{...(iv)}$$

On multiplying Eq. (i) by t_2^2 and Eq. (ii) by t_1^2 and then subtracting Eq. (i) from Eq. (ii), we get

$$h_1t_2^2 - h_2t_1^2 = ut_1t_2(t_2 - t_1) \Rightarrow u = \frac{h_1t_2^2 - h_2t_1^2}{t_1t_2(t_2 - t_1)} \quad \text{...(v)}$$

From Eqs. (iii), (iv) and (v), we get

$$T = \frac{h_1t_2^2 - h_2t_1^2}{h_1t_2 - h_2t_1}$$

11. $v^2 - u^2 = 2ax$

$$\Rightarrow \quad \frac{mv^2}{2} - \frac{mu^2}{2} = max$$

where, x = displacement

v = final velocity

u = initial velocity

and m = mass of particle.

Final kinetic energy – Initial kinetic energy

= Force × displacement

$\Rightarrow$ Power × Time = Work

$\Rightarrow$ $(0.5)(5) = (0.2 \text{ kg})(ax)$

$\Rightarrow$ $ax = 12.5$ ...(i)

So, $v^2 - (0)^2 = 2(12.5)$

$\Rightarrow$ $v^2 = 25 \Rightarrow v = 5 \text{ ms}^{-1}$

12. Vertical motion for each bullet has same initial velocity (zero), same displacement and same acceleration.

From $x = ut + \frac{1}{2}at^2$ ...(i)

t will be same for both bullets.

Eq. (i) do not involve mass.

13. $u = 42 \times \frac{5}{18} = 11.66$ m/s and $v = 0$

So, $S_1 = 6$ m

when $u = 90 \times \frac{5}{18} = 25$ m/s and $v = 0$, then $S_2 = ?$

Case I $v^2 = u^2 + 2as$

$$0 = (11.66)^2 + 2a \times 5$$

$$a = \frac{-11.66 \times 11.66}{12}$$

$$a = -11.33 \text{ m/s}^2$$

Case II $v^2 = u^2 + 2as$

$$0 = (25)^2 + 2(-11.33) \times S_2$$

$$S_2 = \frac{625}{2 \times 11.23} = 27.5 \text{ m}$$

14.

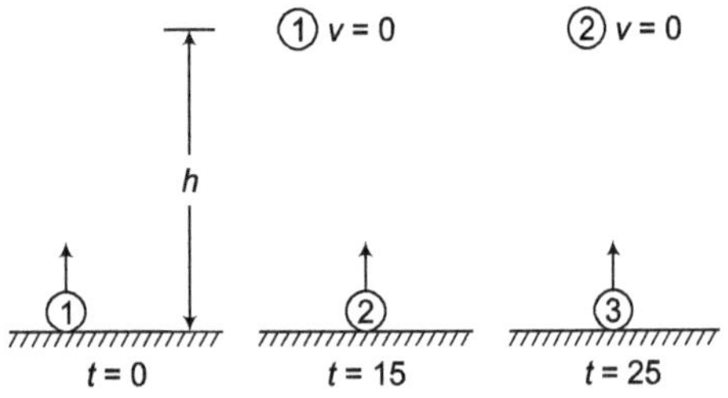

In upward motion, the velocity of ball 1 in 1s becomes zero

$$0 = u - g(1)$$

$\Rightarrow$ $u = g$

$\Rightarrow$ $h = ut - \frac{1}{2}gt^2$

$$h = g(1) - \frac{1}{2}g(1)^2$$

$$= \frac{g}{2} = \frac{10}{2} = 5\,\text{m}$$

15. If time for total journey $= t$

Distance covered in last 1s

$$= \frac{1}{2}gt^2 - \frac{1}{2}g(t-1)^2$$

$$= \frac{1}{2}g[t^2 - (t^2 - 2t + 1)]$$

$$= \frac{1}{2}g(2t - 1) \quad \ldots\text{(i)}$$

Distance covered in first 3 s

$$= \frac{1}{2}g(3)^2 = \frac{9}{2}g \quad \ldots\text{(ii)}$$

Given, $\frac{1}{2}g(2t-1) = \frac{9}{2}g$

$\Rightarrow$ $2t - 1 = 9$

$\Rightarrow$ $t = 5\,\text{s}$

Height $h = \frac{1}{2}gt^2 = \frac{1}{2}g(5)^2 = \frac{1}{2}(10)(25) = 125\,\text{m}$

16. The velocity of a particle is given by

$$\frac{dx}{dt} = v = \sqrt{180 - 16x}$$

$$a = \frac{dv}{dt} = \frac{1}{2}(180 - 16x)^{-1/2}(-16)\frac{dx}{dt}$$

$$a = -8 \times \frac{\sqrt{180 - 16x}}{\sqrt{180 - 16x}} = -8\,\text{m/s}^2$$

17. $x = 3t^2 \Rightarrow v_x = \frac{dx}{dt} = 3(2t) = 6t$

$y = 4t^2 \Rightarrow v_y = \frac{dy}{dt} = 4(2t) = 8t$

Velocity, $\mathbf{v} = v_x\hat{\mathbf{i}} + v_y\hat{\mathbf{j}} = 6t\,\hat{\mathbf{i}} + 8t\,\hat{\mathbf{j}}$

$$|\mathbf{v}| = \sqrt{(6t)^2 + (8t)^2} = 10t$$

18. Distance, $x = 37 + 27t - t^3$

Velocity, $v = \frac{dx}{dt} = 0 + 27(1) - 3t^2$

Time, when $v = 0 \Rightarrow 27 - 3t^2 = 0 \Rightarrow t^2 = 9 \Rightarrow t = 3\,\text{s}$

Distance at $t = 3\,\text{s}$,

$$x = 37 + 27(3) - (3)^3 = 37 + 81 - 27 = 37 + 54 = 91\,\text{m}$$

19. $v^2 = u^2 + 2ax$

Given, $v = \frac{u}{2}$ for $x = d_1 = 30\,\text{cm}$

Let $v = 0$ for $x = d_1 + d_2$

$\Rightarrow$ $\left(\frac{u}{2}\right)^2 = u^2 + 2ad_1$

$\Rightarrow$ $-\frac{3}{4}u^2 = 2ad_1 \quad \ldots\text{(i)}$

$$0^2 = u^2 + 2a(d_1 + d_2)$$

$\Rightarrow$ $-u^2 = 2a(d_1 + d_2) \quad \ldots\text{(ii)}$

On dividing Eq. (ii) by Eq. (i), we get

$$\frac{4}{3} = \frac{d_1 + d_2}{d_1} = 1 + \frac{d_2}{d_1}$$

$\Rightarrow$ $\frac{d_2}{d_1} = \frac{4}{3} - 1 = \frac{1}{3}$

$$d_2 = \frac{d_1}{3} = \frac{30\,\text{cm}}{3}$$

$$= 10\,\text{cm}$$

20. Horizontal motion, $h = 0 + \frac{1}{2}gt^2$

$$80 = \frac{1}{2} \times 10 \times t^2$$

$\Rightarrow$ $t^2 = 16$

$\Rightarrow$ $t = 4\,\text{s}$

Distance covered $= (8\,\text{ms}^{-1})(+\,5) = 32\,\text{m}$

21. $\frac{dv}{dt} = -2.5\sqrt{v}$

$\Rightarrow$ $\frac{dv}{\sqrt{v}} = -2.5\,dt$

$\Rightarrow$ $\int_{6.25}^{0} v^{-1/2}\,dv = -2.5\int_0^t dt$

$\Rightarrow$ $-2.5[t]_0^t = [2v^{1/2}]_{6.25}^0$

$\Rightarrow$ $t = 2\,\text{s}$

22. Given, $a = 1\,\text{m/s}^2$, $s = 48\,\text{m}$, $u = 10\,\text{m/s}$

By equation of motion, $s = ut + \frac{1}{2}at^2$

$\Rightarrow$ $48 = 10t + \frac{1}{2} \times 1 \times t^2$

$\Rightarrow$ $t = 8\,\text{s}$

23.

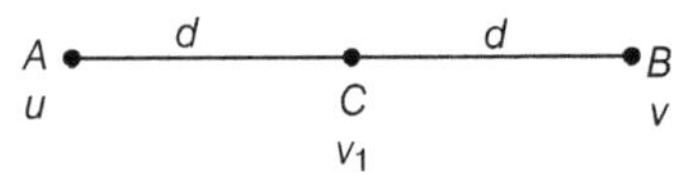

$$v_1^2 = u^2 + 2ad \quad \ldots\text{(i)}$$

$$v^2 = v_1^2 + 2ad \quad \ldots\text{(ii)}$$

Subtracting Eq. (i) and Eq. (ii), we get

$$v_1^2 - v^2 = u^2 - v_1^2$$

$\Rightarrow$ $2v_1^2 = u^2 + v^2$

$\Rightarrow$ $v_1 = \sqrt{\frac{u^2 + v^2}{2}}$

We have standard for this type of velocity

$$v_1 = \sqrt{\frac{u^2 + v^2}{2}}$$

24. Given, $y = bx^2$

$$\frac{dy}{dt} = 2bx\frac{dx}{dt} \quad \ldots\text{(i)}$$

$$\frac{dy}{dt} = u_y + at \quad (\because a_y = a)\ (\because v_y = u_y + a_y t)$$

$$u_y + at = 2bx\frac{dx}{dt}$$

$\Rightarrow$ $t = 0, x = 0$ imply $u_y = 0$

$$at\,dt = 2bx\,dx$$

Take integration on both sides, we get

$$\int at\,dt = \int 2bx\,dx$$

$$\frac{at^2}{2} = bx^2 + c \quad \ldots\text{(ii)}$$

At $t = 0, x = 0, c = 0$

Then, $\frac{at^2}{2} = bx^2$

$$\Rightarrow \quad x = \sqrt{\frac{at^2}{2b}} = t\sqrt{\frac{a}{2b}}$$

$$\therefore \quad v_x = \frac{dx}{dt} = \sqrt{\frac{a}{2b}}$$

25.

$$\text{Total time} = \frac{d}{v_1} + \frac{d}{v_2} = d\left(\frac{v_1 + v_2}{v_1 v_2}\right)$$

Total distance $= 2d$

$$\text{Average velocity} = \frac{2d}{d\left(\frac{v_1 + v_2}{v_1 v_2}\right)} = \frac{2v_2 v_1}{v_1 + v_2}$$

$$\text{Average velocity} = \frac{2v_1 v_2}{v_1 + v_2}$$

Given, $v_{av} = 40\,\text{km/h}, v_1 = 60\,\text{km/h}$ and $v_2 = ?$

$$\therefore \quad 40 = \frac{2 \times 60 \times v_2}{60 + v_2}$$

$$\Rightarrow \quad 60 + v_2 = 3v_2$$

$$v_2 = 30\,\text{km/h}$$

26. Slope of displacement-time graph is velocity

$$\frac{v_1}{v_2} = \frac{\tan\theta_1}{\tan\theta_2} = \frac{\tan 30^\circ}{\tan 45^\circ} = \frac{1}{\sqrt{3}}$$

$$\Rightarrow \quad v_1 : v_2 = 1 : \sqrt{3}$$

4

Projectile Motion

4.1 Projectile Motion

If a constant force (and hence constant acceleration) acts on a particle at an angle θ ($\neq 0°$ or $180°$) with the direction of its initial velocity ($\neq$ zero), the path followed by the particle is a parabola and the motion of the particle is called projectile motion. Projectile motion is a two-dimensional motion, i.e. motion of the particle is constrained in a plane.

Chapter Snapshot

- Projectile Motion
- Time of Flight (T)
- Horizontal Range (R)
- Maximum Height (H)

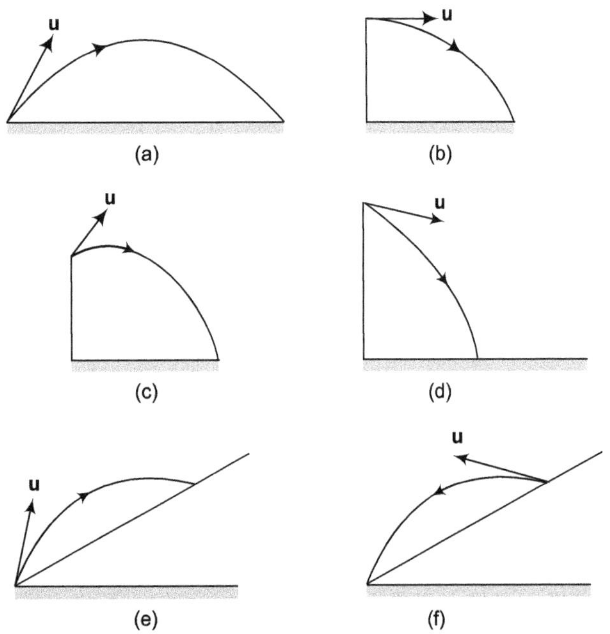

Fig. 4.1

When a particle is thrown obliquely near the earth's surface it moves in a parabolic path, provided the particle remains close to the surface of earth and the air resistance is negligible. This is an example of projectile motion. The different types of projectile motion we come across are shown in Fig. 4.1.

In all the above cases, acceleration g of the particle is downwards.

Let us first make ourselves familiar with certain terms used in projectile motion.

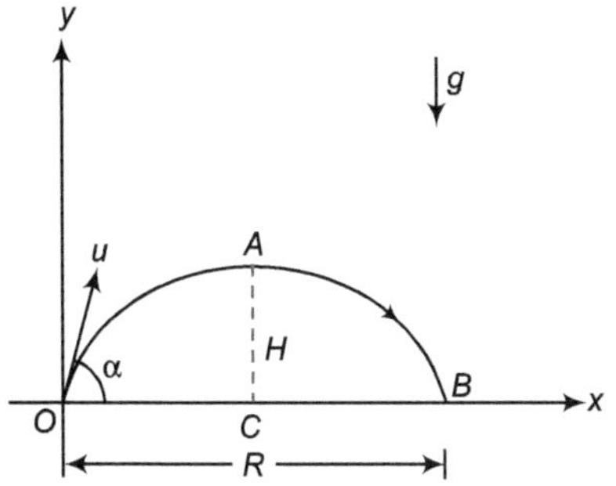

Fig. 4.2

Fig. 4.2 shows a particle projected from the point O with an initial velocity u at an angle α with the horizontal. It goes through the highest point A and falls at B on the horizontal surface through O. The point O is called the **point of projection**, the angle α is called the **angle of projection**, the distance OB is called the **horizontal range (*R*)** or simply range and the vertical height AC is called the maximum **height (*H*)**. The total time taken by the particle in describing the path OAB is called the **time of flight (*T*).**

As we have already discussed, projectile motion is a two-dimensional motion with constant acceleration (normally g). Problems related to projectile motion of any type can be solved by selecting two appropriate mutually perpendicular directions (x and y) and substituting the proper values in equations, i.e.

$$v_x = u_x + a_x t,$$
$$s_x = u_x t + \frac{1}{2} a_x t^2,$$
$$v_x^2 = u_x^2 + 2a_x s_x,$$
$$v_y = u_y + a_y t,$$
$$s_y = u_y t + \frac{1}{2} a_y t^2 \text{ and } v_y^2 = u_y^2 + 2a_y s_y$$

If any problem of projectile motion, we usually follow the three steps given below:

***Step* 1** Select two mutually perpendicular directions x and y.

***Step* 2** Write down the proper values of u_x, a_x, u_y and a_y with sign.

***Step* 3** Apply those equations from the six listed above which are required in the problem.

What should be the directions x and y or which equations are to be used, this you will learn after solving some problems of projectile motion. Using the above methodology, let us first prove the three standard results of time of flight (T), horizontal range (R) and the maximum height (H).

Time of Flight (*T*)

Refer Fig. 4.2. Here, x and y-axes are in the directions shown in figure. Axis x is along horizontal direction and axis y is vertically upwards. Thus,

$$u_x = u\cos\alpha,$$
$$u_y = u\sin\alpha,\ a_x = 0$$
and $$a_y = -g$$

At point B, $s_y = 0$. So, applying

$$s_y = u_y t + \frac{1}{2} a_y t^2, \text{ we have}$$
$$0 = (u\sin\alpha)\,t - \frac{1}{2} g t^2$$
$$\therefore \quad t = 0,\ \frac{2u\sin\alpha}{g}$$

Both $t = 0$ and $t = \dfrac{2u\sin\alpha}{g}$ correspond to the situation where, $s_y = 0$. The time $t = 0$ corresponds to point O and time $t = \dfrac{2u\sin\alpha}{g}$ corresponds to point B. Thus, time of flight of the projectile is given by

$$T = t_{OAB} \quad \text{or} \quad T = \frac{2u\sin\alpha}{g}$$

Horizontal Range (*R*)

Distance OB is the range R. This is also equal to the displacement of particle along x-axis in time $t = T$. Thus, applying $s_x = u_x t + \frac{1}{2} a_x t^2$,

we get $$R = (u\cos\alpha)\left(\frac{2u\sin\alpha}{g}\right) + 0$$

as $$a_x = 0 \text{ and } t = T$$

$$= \frac{2u\sin\alpha}{g}$$

$$\therefore \quad R = \frac{2u^2 \sin\alpha\cos\alpha}{g}$$
$$= \frac{u^2 \sin 2\alpha}{g}$$

or $$R = \frac{u^2 \sin 2\alpha}{g}$$

Here, two points are important regarding the range of a projectile.

(i) Range is maximum

where, $\sin 2\alpha = 1$ or $\alpha = 45°$ and this maximum range is

$$R_{\max} = \frac{u^2}{g} \qquad (\text{at } \alpha = 45°)$$

(ii) For given value of u range at α and range at $(90° - \alpha)$ are equal although times of flight and maximum heights may be different. Because,

$$R_{90° - \alpha} = \frac{u^2 \sin 2(90° - \alpha)}{g} = \frac{u^2 \sin (180° - 2\alpha)}{g} = \frac{u^2 \sin 2\alpha}{g} = R_\alpha$$

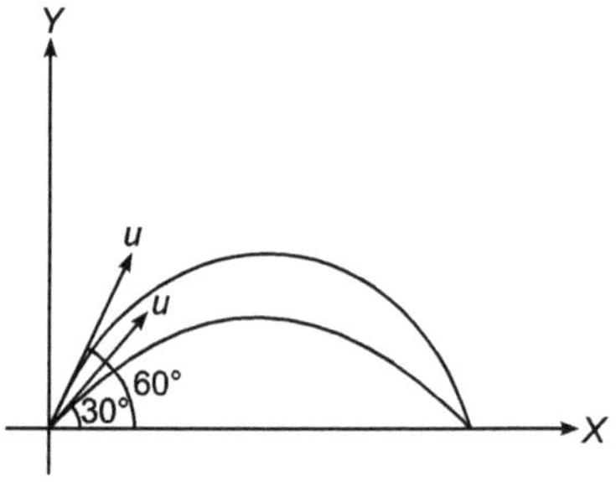

Fig. 4.3

So, $R_{30°} = R_{60°}$

or $R_{20°} = R_{70°}$

This is shown in Fig. 4.3.

Maximum Height (H)

At point A vertical component of velocity becomes zero, i.e., $v_y = 0$. Substituting the proper values in :

$$v_y^2 = u_y^2 + 2a_y s_y$$

we have, $0 = (u \sin \alpha)^2 + 2(-g)(H)$

$\therefore$ $$H = \frac{u^2 \sin^2 \alpha}{2g}$$

Example 4.1 *Find the angle of projection of a projectile for which the horizontal range and maximum height are equal.*

Sol. Given, $R = H$

$\therefore$ $\frac{u^2 \sin 2\alpha}{g} = \frac{u^2 \sin^2 \alpha}{2g}$

or $2 \sin \alpha \cos \alpha = \frac{\sin^2 \alpha}{2}$

or $\frac{\sin \alpha}{\cos \alpha} = 4$

or $\tan \alpha = 4$

$\therefore$ $\alpha = \tan^{-1}(4)$

Extra Knowledge Points

- As we have seen in the above derivations that $a_x = 0$, i.e., motion of the projectile in horizontal direction is uniform. Hence, horizontal component of velocity $u \cos \alpha$ does not change during its motion.
- Motion in vertical direction is first retarded, then accelerated in opposite direction. Because, u_y is upwards and a_y is downwards. Hence, vertical component of its velocity first decreases from O to A and then increases from A to B. This can be shown as in figure.

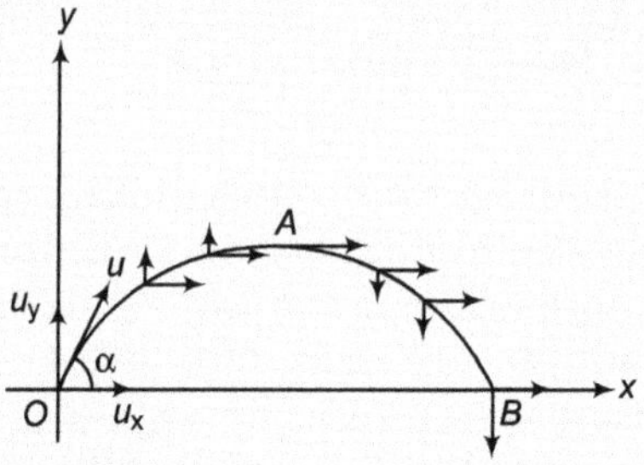

- The coordinates and velocity components of the projectile at time t are

$$x = s_x = u_x t = (u \cos \alpha) t$$
$$y = s_y = u_y t + \frac{1}{2} a_y t^2 = (u \sin \alpha) t - \frac{1}{2} g t^2$$
$$v_x = u_x = u \cos \alpha \text{ and } v_y = u_y + a_y t = u \sin \alpha - gt$$

Therefore, speed of projectile at time t is $v = \sqrt{v_x^2 + v_y^2}$ and the angle made by its velocity vector with positive x-axis is

$$\theta = \tan^{-1}\left(\frac{v_y}{v_x}\right)$$

- Equation of trajectory of projectile

$$x = (u \cos \alpha) t \Rightarrow \therefore t = \frac{x}{u \cos \alpha}$$

Substituting this value of t in, $y = (u \sin \alpha) t - \frac{1}{2} g t^2$, we get

$$y = x \tan \alpha - \frac{gx^2}{2u^2 \cos^2 \alpha} = x \tan \alpha - \frac{gx^2}{2u^2} \sec^2 \alpha = x \tan \alpha - \frac{gx^2}{2u^2}(1 + \tan^2 \alpha)$$

These are the standard equations of trajectory of a projectile. The equation is quadratic in x. This is why the path of a projectile is a parabola. The above equation can also be written in terms of range (R) of projectile as

$$y = x\left(1 - \frac{x}{R}\right) \tan \alpha$$

Now, let us take few examples based on the above theory.

Example 4.2 *Prove that the maximum horizontal range is four times the maximum height attained by the projectile; when fired at an inclination so as to have maximum horizontal range.*

Sol. For $\theta = 45°$, the horizontal range is maximum and is given by

$$R_{max} = \frac{u^2}{g}$$

Maximum height attained

$$H_{max} = \frac{u^2 \sin^2 45°}{2g}$$

$$= \frac{u^2}{4g} = \frac{R_{max}}{4}$$

or $$R_{max} = 4H_{max}$$

Example 4.3 *There are two angles of projection for which the horizontal range is the same. Show that the sum of the maximum heights for these two angles is independent of the angle of projection.*

Sol. There are two angles of projection α and $(90° - \alpha)$ for which the horizontal range R is same.

Now, $$H_1 = \frac{u^2 \sin^2 \alpha}{2g}$$

and $$H_2 = \frac{u^2 \sin^2 (90° - \alpha)}{2g}$$

$$= \frac{u^2 \cos^2 \alpha}{2g}$$

Therefore, $$H_1 + H_2 = \frac{u^2}{2g}(\sin^2 \alpha + \cos^2 \alpha) = \frac{u^2}{2g}$$

Clearly, the sum of the heights for the two angles of projection is independent of the angles of projection.

Example 4.4 *Show that there are two values of time for which a projectile is at the same height. Also show mathematically that the sum of these two times is equal to the time of flight.*

Sol. For vertically upward motion of a projectile,

$$y = (u \sin \alpha)t - \frac{1}{2}gt^2$$

or $$\frac{1}{2}gt^2 - (u \sin \alpha)t + y = 0$$

This is a quadratic equation in t. Its roots are

$$t_1 = \frac{u \sin \alpha - \sqrt{u^2 \sin^2 \alpha - 2gy}}{g}$$

and $$t_2 = \frac{u \sin \alpha + \sqrt{u^2 \sin^2 \alpha - 2gy}}{g}$$

$$\therefore \quad t_1 + t_2 = \frac{2u \sin \alpha}{g} = T \quad \text{(time of flight of the projectile)}$$

Example 4.5 *A projectile is fired horizontally with a velocity of 98 m/s from the top of a hill 490 m high. Find*

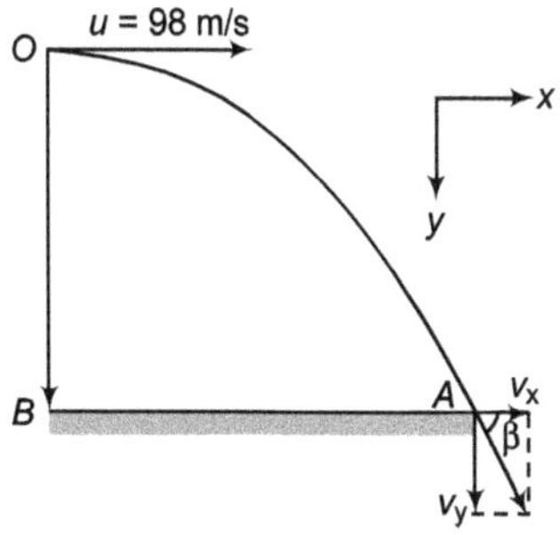

Fig. 4.4

(a) the time taken by the projectile to reach the ground,
(b) the distance of the point, where the particle hits the ground from foot of the hill and
(c) the velocity with which the projectile hits the ground. $(g = 9.8\ m/s^2)$

Sol. In this problem, we cannot apply the formulae of R, H and T directly. We will have to follow the three steps discussed in the theory. Here, it will be more convenient to choose x and y directions as shown in the figure.

Here, $u_x = 98$ m/s, $a_x = 0$, $u_y = 0$ and $a_y = g$.

(a) At A, $s_y = 490$ m. So, applying

$$s_y = u_y t + \frac{1}{2}a_y t^2$$

$$\therefore \quad 490 = 0 + \frac{1}{2}(9.8)t^2 \Rightarrow \therefore t = 10\,s$$

(b) $$BA = s_x = u_x t + \frac{1}{2}a_x t^2$$

or $BA = (98)(10) + 0$ or $BA = 980$ m

(c) $v_x = u_x = 98$ m/s

$$v_y = u_y + a_y t = 0 + (9.8)(10) = 98 \text{ m/s}$$

$$\therefore \quad v = \sqrt{v_x^2 + v_y^2} = \sqrt{(98)^2 + (98)^2} = 98\sqrt{2} \text{ m/s}$$

and $$\tan \beta = \frac{v_y}{v_x} = \frac{98}{98} = 1$$

$$\therefore \quad \beta = 45°$$

Thus, the projectile hits the ground with a velocity $98\sqrt{2}$ m/s at an angle of $\beta = 45°$ with horizontal.

Example 4.6 *A body is thrown horizontally from the top of a tower and strikes the ground after three seconds at an angle of 45° with the horizontal. Find the height of the tower and the speed with which the body was projected. (Take* $g = 9.8\ m/s^2$*)*

Sol. As shown in the Fig. 4.4,

$$u_y = 0$$

and $$a_y = g = 9.8 \text{ m/s}^2$$

$$s_y = u_y t + \frac{1}{2}a_y t^2$$

$$= 0 \times 3 + \frac{1}{2} \times 9.8 \times (3)^2$$

$$= 44.1 \text{ m}$$

Further, $v_y = u_y + a_y t = 0 + (9.8)(3)$

$$= 29.4 \text{ m/s}$$

As the resultant velocity v makes an angle of 45° with the horizontal so,

$$\tan 45° = \frac{v_y}{v_x}$$

or $1 = \frac{29.4}{v_x}$

$$v_x = 29.4 \text{ m/s}$$

Therefore, the speed with which the body was projected (horizontally) is 29.4 m/s.

Note Points

- Projectile motion is a two-dimensional motion with constant acceleration g. So, we can use $\mathbf{v} = \mathbf{u} + \mathbf{a}t, s = \mathbf{u}t + \frac{1}{2}\mathbf{a}t^2$, etc., in projectile motion as well.

Here, $\mathbf{u} = u\cos\alpha\hat{\mathbf{i}} + u\sin\alpha\hat{\mathbf{j}}$ and $\mathbf{a} = -g\hat{\mathbf{j}}$

Now, suppose we want to find velocity at time t.

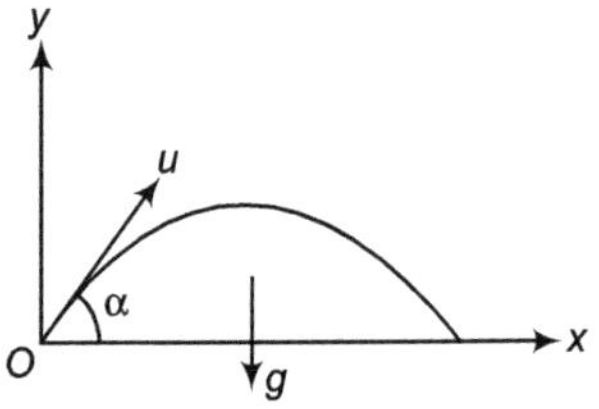

Fig. 4.5

$$\mathbf{v} = \mathbf{u} + \mathbf{a}t = (u\cos\alpha\hat{\mathbf{i}} + u\sin\alpha\hat{\mathbf{j}}) - gt\hat{\mathbf{j}}$$

or $\mathbf{v} = u\cos\alpha\hat{\mathbf{i}} + (u\sin\alpha - gt)\hat{\mathbf{j}}$

Similarly, displacement at time t will be

$$s = \mathbf{u}t + \frac{1}{2}\mathbf{a}t^2$$

$$= (u\cos\alpha\hat{\mathbf{i}} + u\sin\alpha\hat{\mathbf{j}})t - \frac{1}{2}gt^2\hat{\mathbf{j}}$$

$$= ut\cos\alpha\hat{\mathbf{i}} + \left(ut\sin\alpha - \frac{1}{2}gt^2\right)\hat{\mathbf{j}}$$

Extra Knowledge Points

- In projectile motion, the speed (and hence kinetic energy) is minimum at highest point.
  Speed = $(\cos\theta)$ times the speed of projection
  and kinetic energy $= (\cos^2\theta)$ times the initial kinetic energy
  Here, θ = angle of projection
- Path of a particle depends on the nature of acceleration and the angle between initial velocity **u** and acceleration **a**. Following are few paths which are observed frequently.
  (a) If **a** = constant and θ is either 0° or 180°, then path of the particle is a straight line.
  (b) If **a** = constant but θ is other then 0° or 180°, then path of the particle is parabola (as in projectile motion).
  (c) If $|\mathbf{a}|$ = constant and **a** is always perpendicular to velocity vector **v**, then path of the particle is a circle.
- In projectile motion, it is sometimes better to write the equations of H, R and T in terms of u_x and u_y as under.

$$T = \frac{2u_y}{g} \Rightarrow H = \frac{u_y^2}{2g} \text{ and } R = \frac{2u_x u_y}{g}$$

- In projectile motion $H = R$, when $u_y = 4u_x$ or $\tan\theta = 4$.

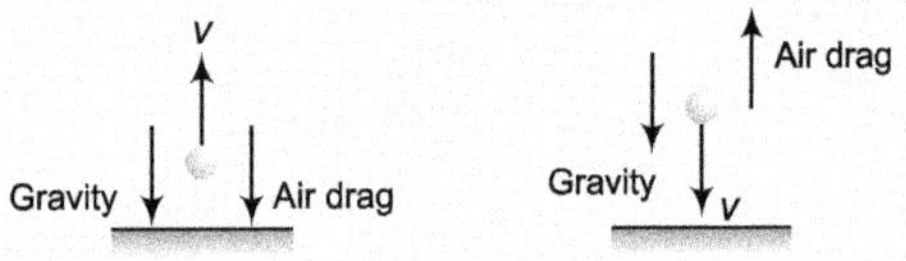

If a particle is projected vertically upwards, then during upward journey gravity forces (weight) and air drag both are acting downwards. Hence, |retardation| > $|g|$. During its downward journey, air drag is upwards while gravity is downwards. Hence, acceleration $< g$. Therefore, we may conclude that, time of ascent < time of descent.

- **Exercise** In projectile motion, if air drag is taken into consideration than state whether the H, R and T will increase, decrease or remain same

Chapter Summary with Formulae

(i) **Projectile Motion**

(a) $T = \frac{2u\sin\theta}{g} = \frac{2u_y}{g}$

(b) $H = \frac{u^2\sin^2\theta}{2g} = \frac{u_y^2}{2g}$

(c) $R = \frac{u^2\sin 2\theta}{g}$

$$= u_x T = \frac{2u_x u_y}{g}$$

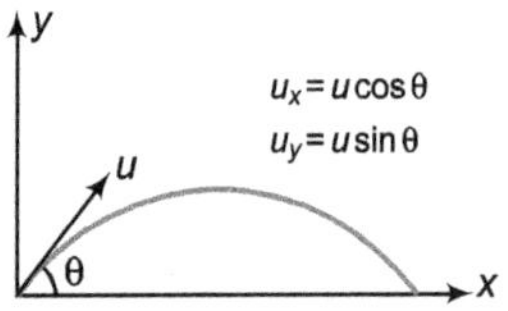

(d) $R_{max} = \frac{u^2}{g}$

$$\theta = 45°$$

(e) $R_\theta = R_{90°-\theta}$ for same value of u

(f) Equation of trajectory

$$y = x\tan\theta - \frac{gx^2}{2u^2\cos^2\theta}$$

$$= x\tan\theta - \frac{gx^2}{2u^2}(1 + \tan^2\theta).$$

$$= x\left(1 - \frac{x}{R}\right)\tan\theta$$

Additional Examples

Example 1. *Can there be motion in two dimensions with an acceleration only in one dimension ?*

Sol. Yes, in a projectile motion, the acceleration acts vertically downwards while the projectile follows a parabolic path.

Example 2. *A body projected horizontally from a tower moves with the same horizontal velocity, although it is under the action of force of gravity. Why?*

Sol. The force of gravity acts in the vertically downward direction and has no effect on the horizontal component of velocity and this makes the body to move with constant horizontal velocity.

Example 3. *What is the angle between the direction of velocity and acceleration at the highest point of a projectile path ?*

Sol. 90°. At the highest point the vertical component of velocity becomes zero, the projectile has only a horizontal velocity while the acceleration due to gravity acts vertically downwards.

Example 4. *A bullet is dropped from a certain height and at the same time, another bullet is fired horizontally from the same height. Which one will hit the ground earlier and why ?*

Sol. Since, the heights of both bullets from the ground is the same, so the time taken by both of them to reach the ground will be the same, because initial velocity component in vertical direction is zero for both of them.

Example 5. *Is the maximum height attained by projectile is largest when its horizontal range is maximum ?*

Sol. No, horizontal range is maximum when $\theta = 45^\circ$ and maximum height attained by projectile is largest when $\theta = 90^\circ$.

Example 6. *A person sitting in a moving train throws a ball vertically upwards. How does the ball appear to move to an observer (i) inside the train (ii) outside the train?*

Sol. (i) To the observer inside the train, the ball will appear to move straight vertically upwards and then downwards.

(ii) To the observer outside the train, the ball will appear to move along the parabolic path.

Example 7. *A boy is dropped freely from the window of a train. Will the time of the free fall be equal, if the train is stationary, the train moves with a constant velocity, the train moves with an acceleration ?*

Sol. The motion of the train does not affect the nature of the motion of the dropped body along the vertical. Therefore, in all the three cases, the time of free fall of the body will be equal.

Example 8. *A particle is projected from horizontal making an angle* 60° *with initial velocity* 40 *m/s. Find the time taken to the particle to make angle* 45° *from horizontal.*

Sol. At 45°, $v_x = v_y$ or $u_x = u_y - gt$

$$\therefore \quad t = \frac{u_y - u_x}{g} = \frac{40(\sin 60^\circ - \sin 30^\circ)}{9.8}$$

$$= 1.5 \text{ s}$$

Example 9. *A ball rolls off the edge of a horizontal table top* 4 *m high. If it strikes the floor at a point* 5 *m horizontally away from the edge of the table, what was its speed at the instant it left the table?*

Sol. Using $h = \frac{1}{2}gt^2$, we have

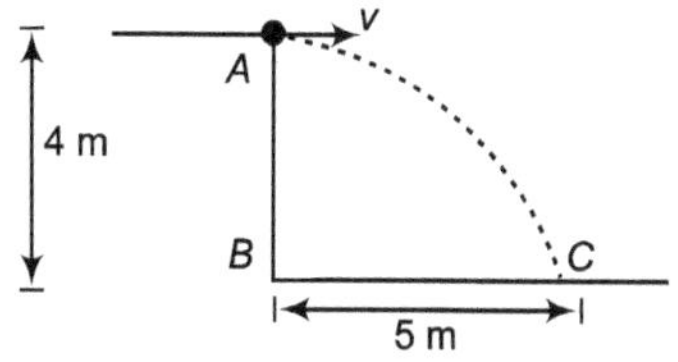

$$h_{AB} = \frac{1}{2}gt_{AC}^2 \quad \text{or} \quad t_{AC} = \sqrt{\frac{2h_{AB}}{g}}$$

$$= \sqrt{\frac{2 \times 4}{9.8}} = 0.9 \text{ s}$$

Further, $BC = vt_{AC}$

or $v = \frac{BC}{t_{AC}} = \frac{5.0}{0.9} = 5.55 \text{ m/s}$

Example 10. *An aeroplane is flying in a horizontal direction with a velocity* 600 *km/h at a height of* 1960 *m. When it is vertically above the point A on the ground, a body is dropped from it. The body strikes the ground at point B. Calculate the distance AB.*

Sol. From $h = \frac{1}{2} g t^2$,

We have $t_{OB} = \sqrt{\frac{2h_{OA}}{g}}$

$= \sqrt{\frac{2 \times 1960}{9.8}}$

= 20 s

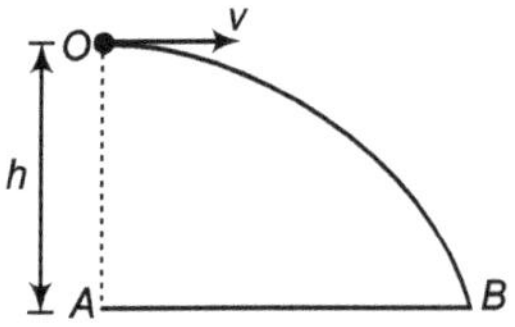

Horizontal distance, $AB = vt_{OB}$

$= \left(600 \times \frac{5}{18} \text{ m/s}\right)(20 \text{ s}) = 3333.33 \text{ m} = 3.33 \text{ km}$

Example 11. *A particle is projected with a velocity of 20 m/s at an angle of 30° to an inclined plane of inclination 30° to the horizontal. The particle hits the inclined plane at an angle of 30°, during its journey. Find the*

(a) time of impact,

(b) the height of the point of impact from the horizontal plane passing through the point of projection.

Sol. The particle hits the plane at 30° (the angle of inclination of plane). It means particle hits the plane horizontally.

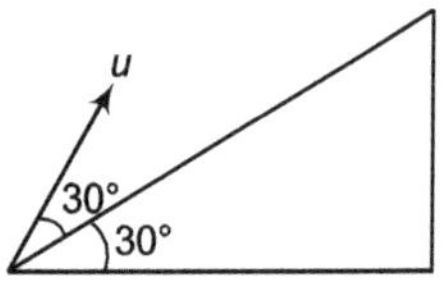

(a) $t = \frac{T}{2} = \frac{u \sin\theta}{g} = \frac{20 \sin(30° + 30°)}{9.8} = 1.76 \text{ s}$

(b) $H = \frac{u^2 \sin^2\theta}{2g} = \frac{(20)^2 \times \sin^2 60°}{2 \times 9.8} = 15.3 \text{ m}$

Example 12. *A particle is projected with velocity $2\sqrt{gh}$ so that it just clears two walls of equal height h, which are at a distance of 2h from each other. Show that the time interval of passing between the two walls is $2\sqrt{\frac{h}{g}}$.*

Sol. Let Δt be the time interval. Then,

$$2h = (u_x)(\Delta t) \quad \text{or} \quad u_x = \frac{2h}{\Delta t} \qquad \ldots\text{(i)}$$

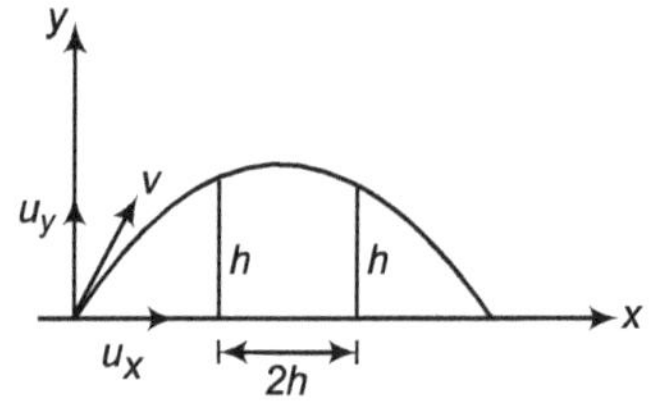

Further, $h = u_y t - \frac{1}{2} g t^2$ or $gt^2 - 2u_y t + 2h = 0$

$$\therefore \quad t_1 = \frac{2u_y + \sqrt{4u_y^2 - 8gh}}{2g}$$

$$\text{and} \quad t_2 = \frac{2u_y - \sqrt{4u_y^2 - 8gh}}{2g}$$

$$\Delta t = t_1 - t_2 = \frac{\sqrt{4u_y^2 - 8gh}}{g}$$

$$\text{or} \quad u_y^2 = \frac{g^2 (\Delta t)^2}{4} + 2gh \qquad \ldots\text{(ii)}$$

$$u_x^2 + u_y^2 = u^2 = (2\sqrt{gh})^2$$

$$\therefore \quad \frac{4h^2}{(\Delta t)^2} + \frac{g^2 (\Delta t)^2}{4} + 2gh = 4gh$$

$$\frac{g^2}{4} (\Delta t)^4 - 2gh (\Delta t)^2 + 4h^2 = 0$$

$$(\Delta t)^2 = \frac{2gh \pm \sqrt{4g^2 h^2 - 4g^2 h^2}}{g^2/2} = \frac{4h}{g}$$

$$\text{or} \quad \Delta t = 2\sqrt{\frac{h}{g}}$$

NCERT Selected Questions

Q 1. A bullet fired at an angle of 30° with the horizontal hits the ground 3 km away. By adjusting its angle of projection can one hope to hit a target 5 km away. Assume, the muzzle speed to be fixed and neglect air resistance.

Sol. Range, $R = \frac{u^2 \sin 2\theta}{g}$

$$\therefore \quad \frac{u^2}{g} = \frac{R}{\sin 2\theta} = \frac{3000}{\sin 60°} = \frac{3000 \times 2}{\sqrt{3}}$$

Maximum range is given by the formula

$$R_{max} = \frac{u^2}{g} = \frac{3000 \times 2}{\sqrt{3}}$$

= 3464 m

But required $R_{max} = 5000$ m

R_{max} with initial velocity of projection is short of 5000 m , so the bullet cannot hit the target 5 km away.

Objective Problems

[Level 1]

Direct Formula and Theory

1. A particle is projected at an angle of 45° with a velocity of 9.8 m/s. The horizontal range will be [take $g = 9.8\,m/s^2$]
(a) 9.8 m (b) 4.9 m (c) $\frac{9.8}{\sqrt{2}}$ (d) $9.8\sqrt{2}$

2. When a stone is projected which remains constant?
(a) Angular momentum
(b) Linear momentum
(c) Vertical component of velocity
(d) Horizontal component of velocity

3. A stone is projected with speed of 50 m/s at an angle of 60° with the horizontal. The speed of the stone at highest point of trajectory is
(a) 75 m/s (b) 25 m/s
(c) 50 m/s (d) cannot find

4. At the top of the trajectory of a projectile, the directions of its velocity and acceleration are
(a) parallel to each other
(b) anti-parallel to each other
(c) inclined to each other at an angle of 45°
(d) perpendicular to each other

5. A projectile, thrown with velocity v_0 at an angle α to the horizontal, has a range R. It will strike a vertical wall at a distance $R/2$ from the point of projection with a speed of
(a) v_0 (b) $v_0 \sin\alpha$ (c) $v_0 \cos\alpha$ (d) $\sqrt{\frac{gR}{2}}$

6. Two projectiles A and B are projected with same speed at angles 30° and 60° to be horizontal, then which one is wrong?
(a) $R_A = R_B$ (b) $H_B = 3H_A$
(c) $T_B = \sqrt{3}T_A$ (d) None of these

7. A projectile fired with initial velocity u at some angle θ has a range R. If the initial velocity be doubled at the same angle of projection, then the range will be
(a) $2R$ (b) $R/2$
(c) R (d) $4R$

8. If the initial velocity of a projection be doubled, keeping the angle of projection same, the maximum height reached by it will
(a) remain the same (b) be doubled
(c) become four times (d) be halved

9. An object is thrown along a direction inclined at an angle of 45° with the horizontal direction. The horizontal range of the particle is equal to
(a) vertical height
(b) twice the vertical height
(c) thrice the vertical height
(d) four times the vertical height

10. At the top of the trajectory of a particle, the acceleration is
(a) maximum (b) minimum
(c) zero (d) g

11. A football player throws a ball with a velocity of 50 m/s at an angle 30° from the horizontal. The ball remains in the air for ($g = 10\,m/s^2$).
(a) 2.5 s (b) 1.25 s
(c) 5 s (d) 0.625 s

12. For a projectile, the ratio of maximum height reached to the square of flight time is ($g = 10\,m/s^2$)
(a) 5 : 4 (b) 5 : 2
(c) 5 : 1 (d) 10 : 1

13. An object is projected at an angle of 45° with the horizontal. The horizontal range and the maximum height reached will be in the ratio
(a) 1 : 2 (b) 2 : 1
(c) 1 : 4 (d) 4 : 1

14. A bomb is dropped from an aeroplane moving horizontally at constant speed. If air resistance is taken into consideration, then the bomb
(a) falls on earth exactly below the aeroplane
(b) falls on the earth behind the aeroplane
(c) falls on the earth ahead of the aeroplane
(d) flies with the aeroplane

15. In the motion of a projectile falling freely under gravity, its (neglect air friction)
(a) total mechanical energy is conserved
(b) momentum is conserved
(c) mechanical energy and momentum both are conserved
(d) None is conserved

16. The horizontal range of a projectile is $4\sqrt{3}$ times of its maximum height. The angle of projection will be
(a) 60° (b) 37°
(c) 30° (d) 45°

Formulae Based Problems with Slight Modification

17. At what angle θ to the horizontal should an object is projected so that the maximum height reached is equal to the horizontal range?

(a) $\tan^{-1}(2)$ (b) $\tan^{-1}(4)$
(c) $\tan^{-1}(2/3)$ (d) $\tan^{-1}(3)$

18. The maximum range of a gun on horizontal terrain is 1.0 km. If $g = 10\,\text{m/s}^2$, what must be the muzzle velocity of the shell?

(a) 400 m/s (b) 200 m/s
(c) 100 m/s (d) 50 m/s

19. Two projectiles A and B are thrown from the same point with velocities v and $v/2$, respectively. If B is thrown at an angle 45° with horizontal, what is the inclination of A when their ranges are the same?

(a) $\sin^{-1}\left(\frac{1}{4}\right)$ (b) $\frac{1}{2}\sin^{-1}\left(\frac{1}{4}\right)$
(c) $2\sin^{-1}\left(\frac{1}{4}\right)$ (d) $\frac{1}{2}\sin^{-1}\left(\frac{1}{8}\right)$

20. Two stones having different masses m_1 and m_2 are projected at an angle α and $(90° - \alpha)$ with same speed from same point. The ratio of their maximum heights is

(a) $1:1$ (b) $1:\tan\alpha$
(c) $\tan\alpha:1$ (d) $\tan^2\alpha:1$

21. A body projected with velocity u at projection angle θ has horizontal range R. For the same velocity and projection angle, its range on the moon surface will be $(g_{\text{moon}} = g_{\text{earth}}/6)$

(a) $36R$ (b) $\frac{R}{36}$
(c) $\frac{R}{16}$ (d) $6R$

22. Three balls of same masses are projected with equal speeds at angle 15°, 45°, 75°, and their ranges are respectively R_1, R_2 and R_3, then

(a) $R_1 > R_2 > R_3$ (b) $R_1 < R_2 < R_3$
(c) $R_1 = R_2 = R_3$ (d) $R_1 = R_3 < R_2$

23. If 2 balls are projected at angles 45° and 60° and the maximum heights reached are same, what is the ratio of their initial velocities?

(a) $\sqrt{2}:\sqrt{3}$ (b) $\sqrt{3}:\sqrt{2}$
(c) $3:2$ (d) $2:3$

24. A projectile is thrown at an angle θ with the horizontal and its range is R_1. It is then thrown at an angle θ with vertical and the range is R_2, then

(a) $R_1 = 4R_2$ (b) $R_1 = 2R_2$
(c) $R_1 = R_2$ (d) data insufficient

25. A man can throw a stone such that it acquires maximum horizontal range 80 m. The maximum height to which it will rise for the same projectile in metre is

(a) 10 (b) 20
(c) 40 (d) 50

26. The ratio of the speed of a projectile at the point of projection to the speed at the top of its trajectory is x. The angle of projection with the horizontal is

(a) $\sin^{-1}(x)$ (b) $\cos^{-1}(x)$
(c) $\sin^{-1}(1/x)$ (d) $\cos^{-1}(1/x)$

27. The velocity at the maximum height of a projectile is half of its initial velocity of projection (u). Its range on horizontal plane is

(a) $\frac{3u^2}{g}$ (b) $\frac{3}{2}\cdot\frac{u^2}{g}$
(c) $\frac{u^2}{3g}$ (d) $\frac{\sqrt{3}}{2}\cdot\frac{u^2}{g}$

28. A projectile is thrown from a point in a horizontal plane such that the horizontal and vertical velocities are $9.8\,\text{ms}^{-1}$ and $19.6\,\text{ms}^{-1}$. It will strike the plane after covering distance of

(a) 39.2 m (b) 19.6 m
(c) 9.8 m (d) 4.9 m

29. A stone is projected in air. Its time of flight is 3 s and range is 150 m. Maximum height reached by the stone is $(g = 10\,\text{ms}^{-2})$

(a) 37.5 m (b) 22.5 m (c) 90 m (d) 11.25 m

30. The greatest height to which a man can throw a stone is h. The greatest distance to which he can throw it will be

(a) $\frac{h}{2}$ (b) h
(c) $2h$ (d) $3h$

31. The range of a projectile when launched at angle θ is same as when launched at angle 2θ. What is the value of θ?

(a) 15° (b) 30° (c) 45° (d) 60°

32. A boy throws a ball with a velocity u at an angle θ with the horizontal. At the same instant he starts running with uniform velocity to catch the ball before if hits the ground. To achieve this he should run with a velocity of

(a) $u\cos\theta$ (b) $u\sin\theta$
(c) $u\tan\theta$ (d) $u\sec\theta$

33. The range of a particle when launched at an angle of 15° with the horizontal is 1.5 km. What is the range of the projectile when launched at an angle of 45° to the horizontal?

(a) 1.5 km (b) 3.0 km
(c) 6.0 km (d) 0.75 km

34. Galileo writes that for angle of projection of a projectile at angle $(45° + \theta)$ and $(45° - \theta)$, the horizontal ranges described by the projectile are in the ratio of (if $\theta \le 45°$)
(a) 2 : 1 (b) 1 : 2
(c) 1 : 1 (d) 2 : 3

35. If time of flight of a projectile is 10 s. Range is 500 m. The maximum height attained by it will be
(a) 125 m (b) 50 m
(c) 100 m (d) 150 m

36. Four bodies P, Q, R and S are projected with equal velocities having angles of projection $15°, 30°, 45°$ and $60°$ with the horizontals respectively. The body having shortest range is
(a) P (b) Q (c) R (d) S

37. A stone is thrown at an angle θ to be the horizontal reaches a maximum height H. Then, the time of flight of stone will be
(a) $\sqrt{\frac{2H}{g}}$ (b) $2\sqrt{\frac{2H}{g}}$
(c) $\frac{2\sqrt{2H \sin\theta}}{g}$ (d) $\frac{\sqrt{2H \sin\theta}}{g}$

38. Two balls are thrown simultaneously from ground with same velocity of 10 m/s but different angles of projection with horizontal. Both balls fall at same distance $5\sqrt{3}$ m from point of projection. What is the time interval between balls striking the ground?
(a) $(\sqrt{3} - 1)$ s (b) $(\sqrt{3} + 1)$ s
(c) $\sqrt{3}$ s (d) 1 s

39. For a given velocity, a projectile has the same range R for two angles of projection. If t_1 and t_2 are the time of flight in the two cases, then $t_1\ t_2$ is equal to
(a) $\frac{2R}{g}$ (b) $\frac{R}{g}$
(c) $\frac{4R}{g}$ (d) $\frac{R}{2g}$

40. A projectile can have same range from two angles of projection with same initial speed. If h_1 and h_2 be the maximum heights, then
(a) $R = \sqrt{h_1 h_2}$ (b) $R = \sqrt{2h_1 h_2}$
(c) $R = 2\sqrt{h_1 h_2}$ (d) $R = 4\sqrt{h_1 h_2}$

Miscellaneous Problems

41. An arrow is shot into air. Its range is 200 m and its time of flight is 5 s. If $g = 10\ \text{m/s}^2$, then horizontal component of velocity and the maximum height will be respectively
(a) 20 m/s, 62.50 m (b) 40 m/s, 31.25 m
(c) 80 m/s, 62.5 m (d) None of these

42. A projectile is thrown upward with a velocity v_0 at an angle α to the horizontal. The change in velocity of the projectile when it strikes the same horizontal plane is
(a) $v_0 \sin\alpha$ vertically downwards
(b) $2v_0 \sin\alpha$ vertically downwards
(c) $2v_0 \sin\alpha$ vertically upwards
(d) zero

43. Two paper screens A and B are separated by a distance of 100 m. A bullet pierces A and then B. The hole in B is 10 cm below the hole in A. If the bullet is travelling horizontally at the time of hitting A, then the velocity of the bullet at A is
(a) 100 m/s (b) 200 m/s
(c) 600 m/s (d) 700 m/s

44. A piece of marble is projected from earth's surface with velocity of $19.6\sqrt{2}$ m/s at 45°. 2 s later its velocity makes an angle α with horizontal, where α is
(a) 45° (b) 30°
(c) 60° (d) 0°

45. A body is projected at an angle of 30° with the horizontal with momentum p. At its highest point, the magnitude of the momentum is
(a) $\frac{\sqrt{3}}{2}p$ (b) $\frac{2}{\sqrt{3}}p$
(c) p (d) $\frac{p}{2}$

46. A ball is projected upwards from the top of a tower with velocity $50\ \text{ms}^{-1}$ making an angle of 30° with the horizontal. If the height of the tower is 70 m, after what time from the instant of throwing, will the ball reach the ground? ($g = 10\ \text{ms}^{-2}$).
(a) 2 s (b) 5 s
(c) 7 s (d) 9 s

47. The equation of projectile is $y = \sqrt{3}x - \frac{g}{2}x^2$, the angle of its projection is
(a) 90° (b) zero
(c) 60° (d) 30°

48. A body is projected horizontally with a velocity of 4 m/s from the top of a high tower. The velocity of the body after 0.7 s is nearly (take $g = 10\ \text{m/s}^2$)
(a) 10 m/s (b) 8 m/s
(c) 19.2 m/s (d) 11 m/s

49. A cricket ball is hit for a six the bat at an angle of 45° to the horizontal with kinetic energy K. At the highest point, the kinetic energy of the ball is
(a) zero (b) K
(c) $K/2$ (d) $K/\sqrt{2}$

50. The height y and distance x along the horizontal for a body projected in the xy-plane are given by $y = 8t - 5t^2$ and $x = 6t$. The initial speed of projection is

(a) 8 m/s (b) 9 m/s
(c) 10 m/s (d) (10/3) m/s

51. An aeroplane is flying in horizontal direction with a velocity of 600 km/h and at a height of 1960 m. When it is vertically above a point A on the ground a body is dropped from it, the body strikes the ground at point B. Then, the distance AB will be

(a) 3.33 km (b) 4.33 km
(c) 5.33 km (d) 6.33 km

52. A stone is projected with a velocity $20\sqrt{2}$ m/s at an angle of 45° to the horizontal. The average velocity of stone during its motion from starting point to its maximum height is (take $g = 10\ \text{m/s}^2$)

(a) 20 m/s (b) $20\sqrt{5}$ m/s
(c) $5\sqrt{5}$ m/s (d) $10\sqrt{5}$ m/s

53. The maximum height attained by a projectile is increased by 10% by increasing its speed of projection, without changing the angle of projection. The percentage increases in the horizontal range will be

(a) 20% (b) 15%
(c) 10% (d) 5%

54. A ball is thrown up with a certain velocity at an angle θ to the horizontal. The kinetic energy KE of the ball varies with horizontal displacement x as

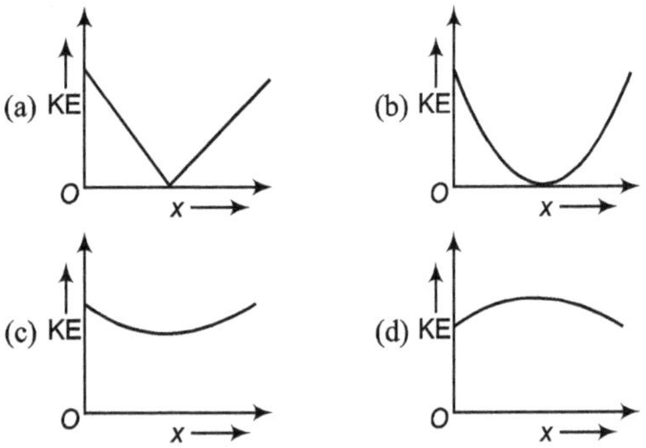

55. An object of mass m is projected with a momentum p at such an angle that its maximum height is $\frac{1}{4}$ th of its horizontal range. Its minimum kinetic energy in its path will be

(a) $\frac{p^2}{8m}$ (b) $\frac{p^2}{4m}$
(c) $\frac{3p^2}{4m}$ (d) $\frac{p^2}{m}$

56. From the top of a tower of height 40 m, a ball is projected upwards with a speed of 20 m/s at an angle of elevation of 30°. The ratio of the total time taken by the ball to hit the ground to its time of flight (time taken to come back to the same elevation) is (take $g = 10\ \text{m/s}^2$)

(a) 2 : 1 (b) 3 : 1
(c) 3 : 2 (d) 1.5 : 1

57. The equation of motion of a projectile is

$$y = 12x - \frac{3}{4}x^2$$

What is the range of the projectile?

(a) 12 m (b) 16 m
(c) 20 m (d) 24 m

58. A ball is thrown at different angles with the same speed u and from the same point. It has the same range in both cases. If y_1 and y_2 be the heights attained in the two cases, then $y_1 + y_2$ equals to

(a) $\frac{u^2}{g}$ (b) $\frac{2u^2}{g}$
(c) $\frac{u^2}{2g}$ (d) $\frac{u^2}{4g}$

59. A projectile is fired from level ground at an angle θ above the horizontal. The elevation angle ϕ of the highest point as seen from the launch point is related to θ by the relation.

(a) $\tan\phi = \frac{1}{4}\tan\theta$ (b) $\tan\phi = \tan\theta$
(c) $\tan\phi = \frac{1}{2}\tan\theta$ (d) $\tan\phi = 2\tan\theta$

60. Water is flowing from a horizontal pipe fixed at a height of 2 m from the ground. If it falls at a horizontal distance of 3 m as shown in figure, the speed of water when it leaves the pipe is (take $g = 9.8\ \text{ms}^{-2}$)

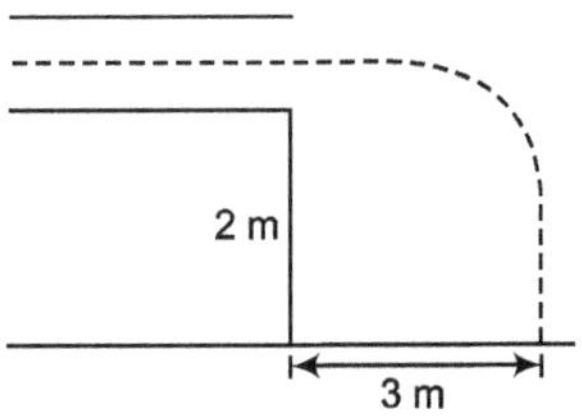

(a) $2.4\ \text{ms}^{-1}$ (b) $4.7\ \text{ms}^{-1}$
(c) $7.4\ \text{ms}^{-1}$ (d) $6.2\ \text{ms}^{-1}$

61. A particle is thrown with a speed u at an angle θ with the horizontal. When the particle makes an angle ϕ with the horizontal, its speed changes to v, where

(a) $v = u\cos\theta$ (b) $v = u\cos\theta\cos\phi$
(c) $v = u\cos\theta\sec\phi$ (d) $v = u\sec\theta\cos\phi$

62. A ball is thrown up with a certain velocity at an angle θ to the horizontal. The kinetic energy KE of the ball varies with height h as

(a) 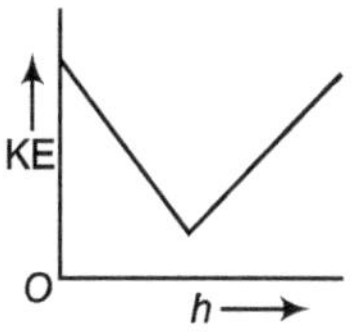

(b) 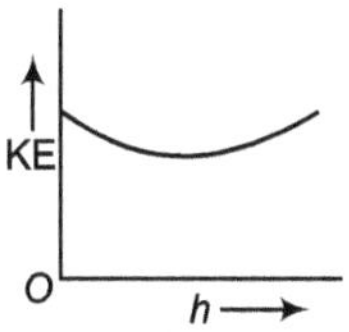

(c) 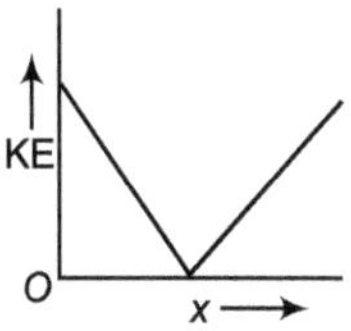

(d) 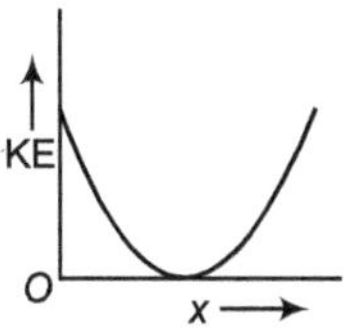

63. The equation of projectile is $Y = \sqrt{3X} - \frac{1}{2}gX^2$. The velocity of projection is

(a) 1 m/s (b) 2 m/s
(c) 3 m/s (d) 1.2 m/s

64. A ball of mass m is projected from the ground with an initial velocity u making an angle of θ with the horizontal. What is the change in velocity between the point of projection and the highest point?

(a) $u\cos\theta$ downward (b) $u\cos\theta$ upward
(c) $u\sin\theta$ upward (d) $u\sin\theta$ downward

65. A projectile is thrown at an angle θ such that it is just able to cross a vertical wall at its highest point as shown in the figure.
The angle θ at which the projectile is thrown is given by

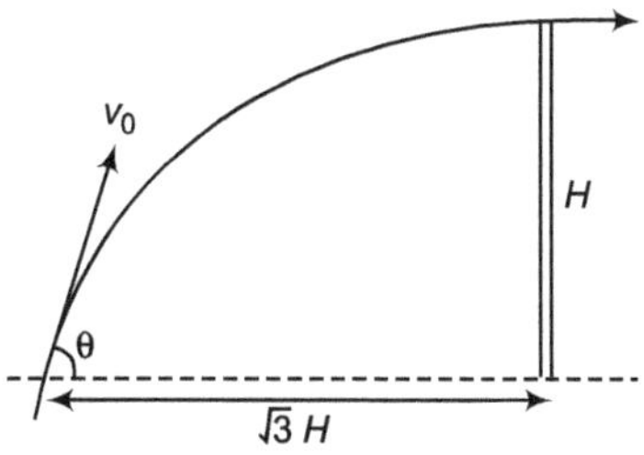

(a) $\tan^{-1}\left(\frac{1}{\sqrt{3}}\right)$ (b) $\tan^{-1}(\sqrt{3})$
(c) $\tan^{-1}\left(\frac{2}{\sqrt{3}}\right)$ (d) $\tan^{-1}\left(\frac{\sqrt{3}}{2}\right)$

66. A bomber plane moves horizontally with a speed of 500 m/s and a bomb released from it strikes the ground in 10 s. Angle at which it strikes the ground will be ($g = 10\ \text{m/s}^2$).

(a) $\tan^{-1}\left(\frac{1}{5}\right)$ (b) 60°
(c) 45° (d) $\tan^{-1}(5)$

67. A large number of bullets are fired in all directions with same speed v. What is the maximum area on the ground on which these bullets will spread?

(a) $\pi\frac{v^2}{g}$ (b) $\pi\frac{v^4}{g^2}$ (c) $\pi^2\frac{v^4}{g^2}$ (d) $\pi^2\frac{v^2}{g^2}$

68. A body of mass m is thrown upwards at an angle θ with the horizontal with velocity v. While rising up the velocity of the mass after t seconds will be

(a) $\sqrt{(v\cos\theta)^2 + (v\sin\theta)^2}$ (b) $\sqrt{(v\cos\theta - v\sin\theta)^2 - gt}$
(c) $\sqrt{v^2 + g^2t^2 - (2v\sin\theta)\,gt}$ (d) $\sqrt{v^2 + g^2t^2 - (2v\cos\theta)\,gt}$

69. Figure shows four paths for a kicked football. Ignoring the effects of air on the flight, rank the paths according to initial horizontal velocity component highest first.

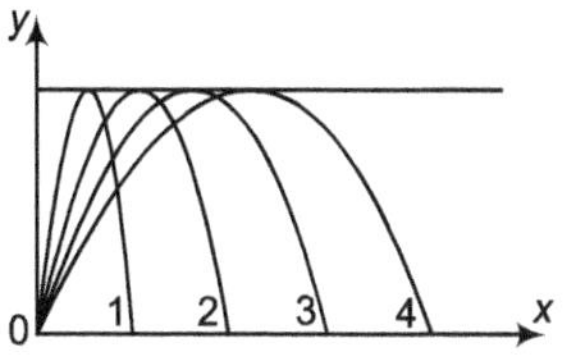

(a) 1, 2, 3, 4 (b) 2, 3, 4, 1
(c) 3, 4, 1, 2 (d) 4, 3, 2, 1

70. Which of the following is the graph between the height (h) of a projectile and time (t), when it is projected from the ground?

(a) 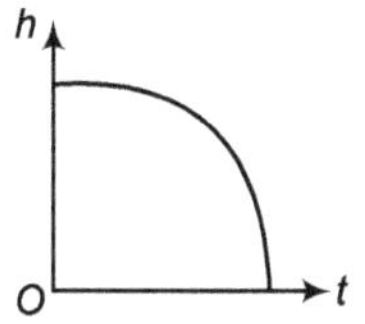

(b) 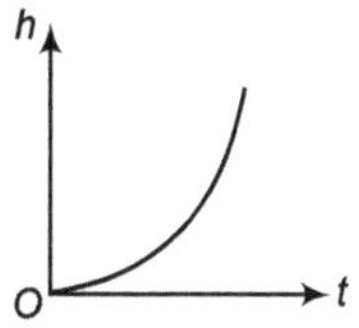

(c) 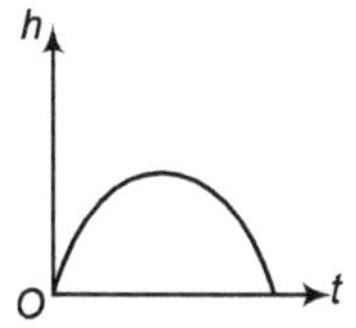

(d) 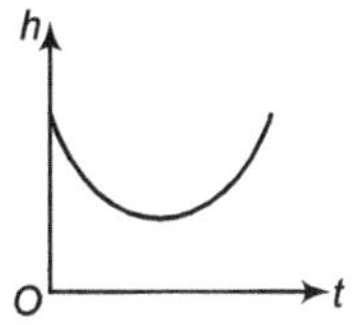

71. Which of the following is the altitude-time graph for a projectile thrown horizontally from the top of the tower?

(a) 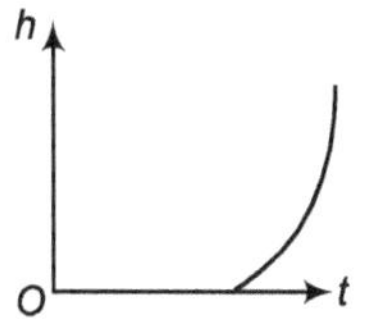

(b) 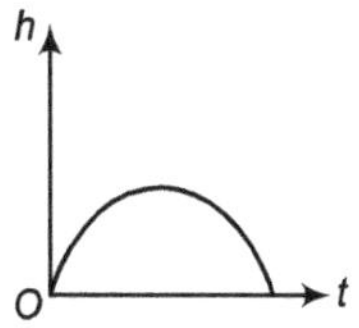

(c) 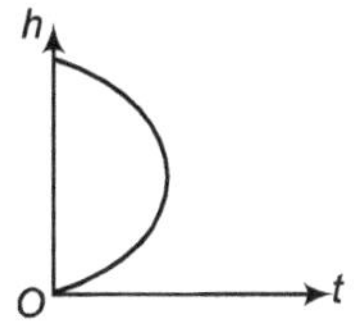

(d) 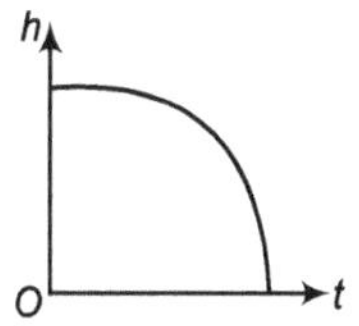

72. The initial velocity of a particle of mass 2 kg is $(4\hat{\mathbf{i}} + 4\hat{\mathbf{j}})$ m/s. A constant force of $-20\,\hat{\mathbf{j}}$ N is applied on the particle. Initially, the particle was at $(0, 0)$. Find the x-coordinate of the point where its y- coordinate is again zero.
(a) 3.2 m (b) 6 m (c) 4.8 m (d) 1.2 m

73. A particle is projected from the ground at an angle of θ with the horizontal with an initial speed of u. Time after which velocity vector of the projectile is perpendicular to the initial velocity.
(a) $u/g \sin\theta$ (b) $u/g \cos\theta$
(c) $2u/g \sin\theta$ (d) $2u \tan\theta$

74. A ball is projected with a velocity $20\sqrt{3}$ m/s at angle 60° to the horizontal. The time interval after which the velocity vector will make an angle 30° to the horizontal is (take $g = 10\,\text{m/s}^2$)
(a) 5 s (b) 2 s (c) 1 s (d) 3 s

75. A projectile is thrown with a velocity of 10 m/s at an angle of 60° with horizontal. The interval between the moments when speed is $\sqrt{5g}$ m/s is ($g = 10\,\text{m/s}^2$).
(a) 1 s (b) 3 s (c) 2 s (d) 4 s

76. A ball is projected from ground with a speed of 20 m/s at an angle of 45° with horizontal. There is a wall of 25 m height at a distance of 10 m from the projection point. The ball will hit the wall at a height of
(a) 5 m (b) 7.5 m
(c) 10 m (d) 12.5 m

77. Two particles are simultaneously projected in opposite directions horizontally from a given point in space, where gravity g is uniform. If u_1 and u_2 be their initial speeds, then the time t after which their velocities are mutually perpendicular is given by
(a) $\dfrac{\sqrt{u_1 u_2}}{g}$ (b) $\dfrac{\sqrt{u_1^2 + u_2^2}}{g}$
(c) $\dfrac{\sqrt{u_1 (u_1 + u_2)}}{g}$ (d) $\dfrac{\sqrt{u_2 (u_1 + u_2)}}{g}$

78. A ball is projected horizontally from the top of a tower with a velocity v_0. It will be moving at an angle of 60° with the horizontal after time.
(a) $\dfrac{v_0}{\sqrt{3}g}$ (b) $\dfrac{\sqrt{3}\,v_0}{g}$ (c) $\dfrac{v_0}{g}$ (d) $\dfrac{v_0}{2g}$

79. A ball rolls from the top of a stair way with a horizontal velocity u m/s. If the steps are h m high and b m wide, the ball will hit the edge of the nth step, if
(a) $n = \dfrac{2hu}{gb^2}$ (b) $n = \dfrac{2hu^2}{gb^2}$ (c) $n = \dfrac{2hu^2}{gb}$ (d) $n = \dfrac{hu^2}{gb^2}$

80. Two stones are projected so as to reach the same distance from the point of projection on a horizontal surface. The maximum height reached by one exceeds the other by an amount equal to half the sum of the height attained by them. Then, angle of projection of the stone which attains smaller height is
(a) 45° (b) 60°
(c) 30° (d) $\tan^{-1}(3/4)$

81. A projectile is thrown with an initial velocity of $(a\hat{\mathbf{i}} + b\hat{\mathbf{i}})\,\text{ms}^{-1}$. If the range of the projectile is twice the maximum height reached by it, then
(a) $a = 2b$ (b) $b = a$
(c) $b = 2a$ (d) $b = 4a$

82. A projectile A is thrown at an angle 30° to the horizontal from point P. At the same time another projectile B is thrown with velocity v_2 upwards from the point Q vertically below the highest point A would reach. For B to collide with A, the ratio $\dfrac{v_2}{v_1}$ should be

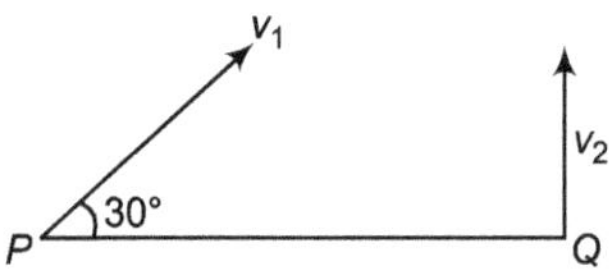

(a) $\dfrac{\sqrt{3}}{2}$ (b) 2
(c) $\dfrac{1}{2}$ (d) $\dfrac{2}{\sqrt{3}}$

83. A particle is projected with a velocity of 30 m/s, at an angle of $\theta_0 = \tan^{-1}\left(\dfrac{3}{4}\right)$. After 1 s, the particle is moving at an angle θ to the horizontal, where $\tan\theta$ will be equal to ($g = 10\,\text{m/s}^2$)
(a) 1 (b) 2
(c) $\dfrac{1}{2}$ (d) $\dfrac{1}{3}$

84. The equations of motion of a projectile are given by $x = 36t$ m and $2y = 96t - 9.8t^2$ m. The angle of projection is
(a) $\sin^{-1}\left(\dfrac{4}{5}\right)$ (b) $\sin^{-1}\left(\dfrac{3}{5}\right)$
(c) $\sin^{-1}\left(\dfrac{4}{3}\right)$ (d) $\sin^{-1}\left(\dfrac{3}{4}\right)$

85. A ball is thrown from a point with a speed v_0 at an angle of projection θ. From the same point and at the same instant a person starts running with a constant speed $v_0/2$ to catch the ball. Will the person be able to catch the ball? If yes, what should be the angle of projection?
(a) Yes, 60° (b) Yes, 30°
(c) No (d) Yes, 45°

[Level 2]

Only One Correct Option

1. A projectile has the maximum range 500 m. If the projectile is thrown up a smooth inclined plane of 30° with the same (magnitude) velocity, the distance covered by it along the inclined plane till it stops will be

(a) 250 m (b) 500 m
(c) 750 m (d) 1000 m

2. A projectile is fired at an angle of 30° to the horizontal such that the vertical component of its initial velocity is 80 m/s. Its time of flight is T. Its velocity at $t = \frac{T}{4}$ has a magnitude of nearly

(a) 200 m/s (b) 300 m/s
(c) 100 m/s (d) None of these

3. A very broad elevator is going up vertically with a constant acceleration of 2 m/s^2. At the instant when its velocity is 4 m/s a ball is projected from the floor of the lift with a speed of 4 m/s relative to the floor at an elevation of 30°. The time taken by the ball to return the floor is $(g = 10 \text{ m/s}^2)$

(a) $\frac{1}{2}$ s (b) $\frac{1}{3}$ s
(c) $\frac{1}{4}$ s (d) 1 s

4. A ball is dropped from a height of 49 m. The wind is blowing horizontally. Due to wind a constant horizontal acceleration is provided to the ball. Choose the correct statement (s). (Take $g = 9.8 \text{ m/s}^2$)

(a) Path of the ball is a straight line
(b) Path of the ball is a curved one
(c) The time taken by the ball to reach the ground is 3.16 s
(d) Actual distance travelled by the ball is more than 49 m

5. A particle moves along a parabolic path $y = -9x^2$ in such a way that the x component of velocity remains constant and has a value $\frac{1}{3}$ m/s. The acceleration of the particle is

(a) $\frac{1}{3} \text{ m/s}^2$ (b) 3 m/s^2 (c) $\frac{2}{3} \text{ m/s}^2$ (d) 2 m/s^2

6. Two particles A and B are projected simultaneously from a fixed point of the ground. Particle A is projected on a smooth horizontal surface with speed v, while particle B is projected in air with speed $\frac{2v}{\sqrt{3}}$. If particle B hits the particle A, the angle of projection of B with the vertical is

(a) 30° (b) 60°
(c) 45° (d) Both (a) and (b)

7. A ground to ground projectile is at point A at $t = \frac{T}{3}$, is at point B at $t = \frac{5T}{6}$ and reaches the ground at $t = T$. The difference in heights between points A and B is

(a) $\frac{gT^2}{6}$ (b) $\frac{gT^2}{12}$ (c) $\frac{gT^2}{18}$ (d) $\frac{gT^2}{24}$

8. A cart is moving horizontally along a straight line with a constant speed of 30 m/s. A projectile is to be fired from the moving cart in such a way that it will return to the cart (at the same point on cart) after the cart has moved 80 m. At what velocity (relative to the cart) must be projectile be fired? (Take $= 10 \text{ m/s}^2$)

(a) 10 m/s (b) $\frac{20}{3}$ m/s (c) $\frac{40}{3}$ m/s (d) $\frac{80}{3}$ m/s

9. A body of mass 1 kg is projected with velocity 50 m/s at an angle of 30° with the horizontal. At the highest point of its path a force 10 N starts acting on body for 5 s vertically upward besides gravitational force, what is horizontal range of the body? $(g = 10 \text{ m/s}^2)$

(a) $125\sqrt{3}$ m (b) $200\sqrt{3}$ m
(c) 500 m (d) $250\sqrt{3}$ m

10. A projectile is thrown with some initial velocity at an angle α to the horizontal. Its velocity when it is at the highest point is $(2/5)^{1/2}$ times the velocity when it is at height half of the maximum height. Find the angle of projection α with the horizontal.

(a) 30° (b) 45° (c) 60° (d) 37°

11. Balls A and B are thrown from two points lying on the same horizontal plane separated by a distance 120 m. Which of the following statement (s) is/are correct.

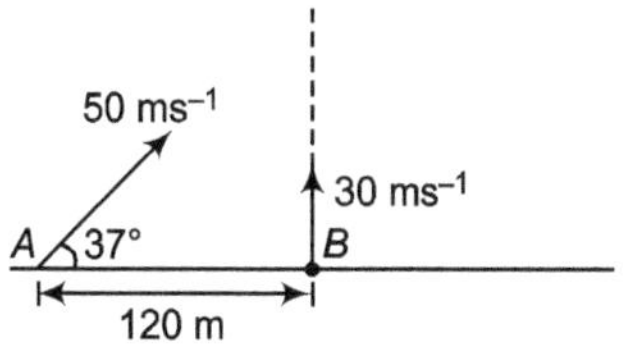

(a) The two balls can never meet
(b) The balls can meet, if the ball B is thrown 1 s later
(c) The two balls meet at a height of 45 m
(d) None of the above

12. Two second after projection, a projectile is travelling in a direction inclined at 30° to be horizontal. After 1 more second it is travelling horizontally. Then, $(g = 10 \text{ m/s}^2)$

(a) the velocity of projection is $20\sqrt{3}$ m/s
(b) the angle of projection is 30° with horizontal
(c) Both (a) and (b) are correct
(d) Both (a) and (b) are wrong

13. An object is projected with a velocity of 20 m/s making an angle of 45° with horizontal. The equation for trajectory is $h = Ax - Bx^2$, where h is height, x is horizontal distance. A and B are constants. The ratio $A : B$ is ($g = 10\,\text{m/s}^2$)

(a) 1 : 5 (b) 5 : 1
(c) 1 : 40 (d) 40 : 1

14. A projectile is launched with a speed of 10 m/s at an angle 60° with the horizontal from a sloping surface of inclination 30°. The range R is (Take $g = 10\ \text{m/s}^2$)

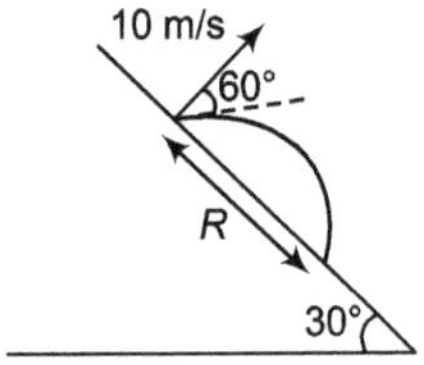

(a) 4.9 m (b) 13.3 m (c) 9.1 m (d) 12.6 m

15. If the instantaneous velocity of a particle projected as shown in figure is given by $\mathbf{v} = a\hat{\mathbf{i}} + (b - ct)\hat{\mathbf{j}}$, where $a, b,$ and c are positive constants, the range on the horizontal plane will be

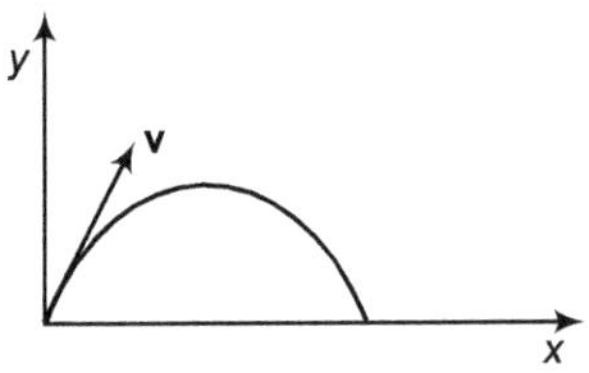

(a) $2ab/c$ (b) ab/c
(c) ac/b (d) $a/2bc$

16. A particle is projected from horizontal making an angle of 53° with initial velocity 100 m/s. The time taken by the particle to make angle 45° from horizontal is

(a) 14 s (b) 2.0 s (c) 12 s (d) 4 s

17. A ball is thrown from the ground to clear a wall 3 m high at a distance of 6 m and falls 18 m away from the wall, the angle of projection of ball is

(a) $\tan^{-1}\left(\frac{3}{2}\right)$ (b) $\tan^{-1}\left(\frac{2}{3}\right)$
(c) $\tan^{-1}\left(\frac{1}{2}\right)$ (d) $\tan^{-1}\left(\frac{3}{4}\right)$

18. Two particles are projected from the same point with same speed u at angles of projection α and β from horizontal. The maximum heights attained by them are h_1 and h_2 respectively, R is the range for both and t_1 and t_2 their times of flight respectively, then

(a) $\alpha + \beta = \frac{\pi}{2}$ (b) $R = 4\sqrt{h_1 h_2}$
(c) $\frac{t_1}{t_2} = \tan\alpha$ (d) $\tan\alpha = \sqrt{h_1 h_2}$

19. A man standing on a hill top projects a stone horizontally with speed v_0 as shown in figure. Taking the coordinates system as given in the figure. The coordinates of the point, where the stone will hit the hill surface are

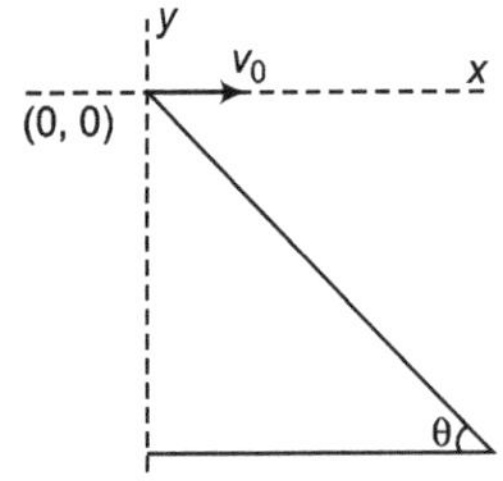

(a) $\left(\frac{2v_0^2\tan\theta}{g}, -\frac{2v_0^2\tan^2\theta}{g}\right)$ (b) $\left(\frac{2v_0^2}{g}, -\frac{2v_0^2\tan^2\theta}{g}\right)$
(c) $\left(\frac{2v_0^2\tan\theta}{g}, -\frac{2v_0^2}{g}\right)$ (d) $\left(\frac{2v_0^2\tan^2\theta}{g}, -\frac{2v_0^2\tan\theta}{g}\right)$

20. The horizontal range and maximum height attained by a projectile are R and H, respectively. If a constant horizontal acceleration $a = g/4$ is imparted to the projectile due to wind, then its horizontal range and maximum height will be

(a) $(R + H), \frac{H}{2}$ (b) $\left(R + \frac{H}{2}\right), 2H$
(c) $(R + 2H), H$ (d) $(R + H), H$

21. A stone is projected from a point on the ground so as to hit a bird on the top of a vertical pole of height h and then attain a maximum height $2h$ above the ground. If at the instant of projection the bird flies away horizontally with a uniform speed and if the stone hits the bird while descending, then the ratio of the speed of the bird to the horizontal speed of the stone is

(a) $\frac{\sqrt{2}}{\sqrt{2}+1}$ (b) $\frac{\sqrt{2}}{\sqrt{2}-1}$ (c) $\frac{1}{\sqrt{2}} + \frac{1}{2}$ (d) $\frac{2}{\sqrt{2}+1}$

22. A projectile is thrown with velocity u making angle θ with vertical. It just crosses the tops of two poles each of height h after 1 s and 3 s, respectively. The maximum height of projectile is

(a) 9.8 m (b) 19.6 m (c) 39.2 m (d) 4.9 m

23. A grasshopper can jump maximum distance 1.6 m. It negligible time of the ground. How far can it go in 10 s?

(a) $5\sqrt{2}$ m (b) $10\sqrt{2}$ m (c) $20\sqrt{2}$ m (d) $40\sqrt{2}$ m

More than One Correct Options

1. Two particles projected from the same point with same speed u at angles of projection α and β strike the horizontal ground at the same point. If h_1 and h_2 are the maximum heights attained by the projectile, R is the range for both and t_1 and t_2 are their times of flights, respectively , then

(a) $\alpha + \beta = \frac{\pi}{2}$ (b) $R = 4\sqrt{h_1 h_2}$
(c) $\frac{t_1}{t_2} = \tan\alpha$ (d) $\tan\alpha = \sqrt{\frac{h_1}{h_2}}$

2. A ball is dropped from a height of 49 m. The wind is blowing horizontally. Due to wind a constant horizontal acceleration is provided to the ball. Choose the correct statement (s). [Take $g = 9.8\text{ ms}^{-2}$]

(a) Path of the ball is a straight line
(b) Path of the ball is a curved one
(c) The time taken by the ball to reach the ground is 3.16 s
(d) Actual distance travelled by the ball is more then 49 m

3. A particle is projected from a point P with a velocity v at an angle θ with horizontal. At a certain point Q it moves at right angles to its initial direction. Then,

(a) velocity of particle at Q is $v\sin\theta$
(b) velocity of particle at Q is $v\cot\theta$
(c) time of flight from P to Q is $(v/g)\operatorname{cosec}\theta$
(d) time of flight from P to Q is $(v/g)\sec\theta$

4. At a height of 15 m from ground velocity of a projectile is $\mathbf{v} = (10\hat{\mathbf{i}} + 10\hat{\mathbf{j}})$. Here, $\hat{\mathbf{j}}$ is vertically upwards and $\hat{\mathbf{i}}$ is along horizontal direction then, $(g = 10\text{ ms}^{-2})$

(a) particle was projected at an angle of 45° with horizontal
(b) time of flight of projectile is 4 s
(c) horizontal range of projectile is 100 m
(d) maximum height of projectile from ground is 20 m

5. Which of the following quantities remain constant during projectile motion?

(a) Average velocity between two points
(b) Average speed between two points
(c) $\dfrac{d\mathbf{v}}{dt}$
(d) $\dfrac{d^2\mathbf{v}}{dt^2}$

6. In the projectile motion shown is figure, given $t_{AB} = 2$ s, then $(g = 10\text{ ms}^{-2})$

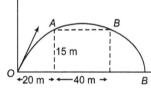

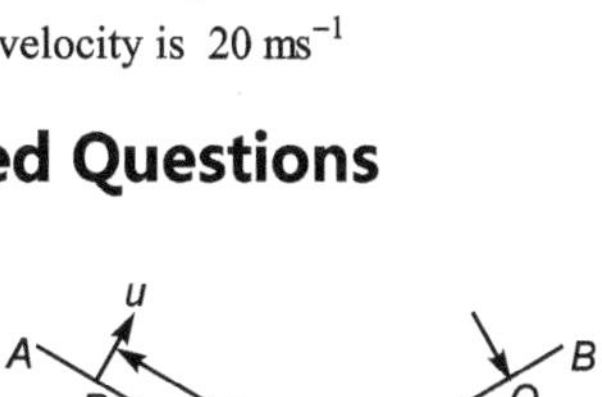

(a) particle is at point B at 3 s
(b) maximum height of projectile is 20 m
(c) initial vertical component of velocity is 20 ms^{-1}
(d) horizontal component of velocity is 20 ms^{-1}

Comprehension Based Questions

Passage (Q. 1 to 2)

Two inclined planes OA and OB intersect in a horizontal plane having their inclinations α and β with the horizontal as shown in figure. A particle is projected from point P with velocity u along a direction perpendicular to plane OA. The particle strikes plane OB perpendicularly at Q.

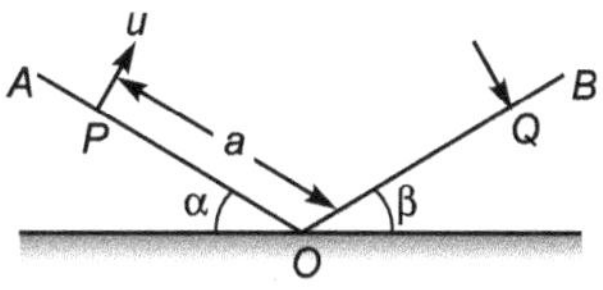

1. If $\alpha = 30°, \beta = 30°$, the time of flight from P to Q is

(a) $\dfrac{u}{g}$ (b) $\dfrac{\sqrt{3}u}{g}$ (c) $\dfrac{\sqrt{2}u}{g}$ (d) $\dfrac{2u}{g}$

2. If $\alpha = 30°, \beta = 30°$ and $a = 4.9$ m, the initial velocity of projection is

(a) 9.8 ms^{-1} (b) 4.9 ms^{-1} (c) $4.9\sqrt{2}$ ms^{-1} (d) 19.6 ms^{-1}

Assertion and Reason

Directions (Q. Nos. 1-13) *These questions consists of two statements each printed as Assertion and Reason. While answering these questions you are required to choose any one of the following five responses.*

(a) If both Assertion and Reason are correct and Reason is the correct explanation of Assertion
(b) If both Assertion and Reason are true but Reason is not the correct explanation of Assertion
(c) If Assertion is true but Reason is false
(d) If Assertion is false but Reason is true
(e) If both Assertion and Reason are false

1. Assertion In case of projectile motion, the magnitude of rate of change of velocity is variable.

Reason In projectile motion, magnitude of velocity first decreases and then increases during the motion.

2. Assertion A particle is projected with speed u at an angle θ with the horizontal. At any time during motion, speed of particle is v at angle α with the vertical, then $v\sin\alpha$ is always constant throughout the motion.

Reason In case of projectile motion, magnitude of radial acceleration at topmost point is maximum.

3. Assertion If in a projectile motion, we take air friction into consideration, then $t_{\text{ascent}} < t_{\text{descent}}$.

Reason During ascent magnitude of retardation is greater than magnitude of acceleration during descent.

4. Assertion In projectile motion, the angle between instantaneous velocity vector and acceleration vector can be anything between 0 to π (excluding the limiting case.)

Reason In projectile motion, acceleration vector is always pointing vertically downwards. (Neglect air friction).

5. Assertion In projectile motion if time of flight is made two times, then maximum height will become four times.

Reason $T \propto \sin\theta$ and $H \propto \sin^2\theta$

where, θ is angle of projection.

6. Assertion In projectile motion if time of flight is 4 s. Then, maximum height will be 20 m. $(g = 10\text{ m/s}^2)$

Reason Maximum height $= \dfrac{gT}{2}$.

7. Assertion For projection angle $\tan^{-1}(4)$, the horizontal and maximum height of a projectile are equal.

Reason The maximum range of projectile is directly proportional to square of velocity and inversely proportional to acceleration due to gravity.

8. Assertion A particle in xy-plane is governed by $x = a\sin\omega t$ and $y = a\cos\omega t$, where a as well as ω are constants, then the particle will have parabolic mation.

Reason A particle under the influence of mutually perpendicular velocities has parabolic motion.

9. Assertion Particle-1 is dropped from a tower and particle-2 is projected horizontal from the same tower. Then, both the particles reach the ground simultaneously.

Reason Both the particles strike the ground with different speeds.

10. Assertion If a particle is projected vertices upwards with velocity u, the maximum height attained by the particle is h_1. The same particle is projected at angle 30° from horizontal with the same speed u. Now, the maximum height is h_2. Thus, $h_1 = 4h_2$.

Reason In first case $v = 0$ at highest point and in second case $v \neq 0$ at highest point.

11. Assertion At highest point of a projectile dot product of velocity and acceleration is zero.

Reason At highest point velocity and acceleration are mutually perpendicular.

12. Assertion On the surface of moon, the value of g is $\frac{1}{6}$th the value on the surface of the earth. A particle is projected as projectile under similar condition on the surface of moon and on the surface of earth. Then, values of T, H and R on the surface of moon will become six times.

Reason T, H and $R \propto \frac{1}{g}$

13. Assertion At height 20 m from ground velocity of a projectile is $\mathbf{v} = (20\hat{\mathbf{i}} + 10\hat{\mathbf{j}})$ m/s. Here, $\hat{\mathbf{i}}$ is horizontal and $\hat{\mathbf{j}}$ is vertical. Then, the particle is at the same height after 4 s.

Reason Maximum height of particle from ground is 40 m. (Take $g = 10\,\text{m/s}^2$)

Match the Columns

1. Trajectory of particle in a projectile motion is given as $y = x - \frac{x^2}{80}$. Here, x and y are in metre. For this projectile motion match the following with $g = 10\,\text{m/s}^2$.

Column I	Column II
(A) Angle of projection	(p) 20 m
(B) Angle of velocity with horizontal after 4 s	(q) 80 m
(C) Maximum height	(r) 45°
(D) Horizontal range	(s) $\tan^{-1}\left(\frac{1}{2}\right)$

2. A particle is projected horizontally from a tower with velocity 10 m/s. Taking $g = 10\,\text{m/s}^2$. Match the following two columns at time $t = 1$ s.

Column I	Column II
(A) Horizontal component of velocity	(p) 5 SI unit
(B) Vertical component of velocity	(q) 10 SI unit
(C) Horizontal displacement	(r) 15 SI unit
(D) Vertical displacement	(s) 20 SI unit

3. A particle is projected from ground with velocity u at angle θ from horizontal. Match the following two columns.

Column I	Column II
(A) Average velocity between initial and final points	(p) $u\sin\theta$
(B) Change in velocity between initial and final points	(q) $u\cos\theta$
(C) Change in velocity between initial and final points	(r) zero
(D) Average velocity between initial and highest points	(s) None of the above

4. Given that u_x = horizontal component of initial velocity of a projectile, u_y = vertical component of initial velocity, R = horizontal range, T = time of flight and H = maximum height of projectile. Now match the following two columns.

Column I	Column II
(A) u_x is doubled, u_y is halved	(p) H will remain unchanged
(B) u_y is doubled u_x is halved	(q) R will remain unchanged
(C) u_x and u_y both are doubled	(r) R will become four times
(D) Only u_y is doubled	(s) H will become four times

5. Two particles are projected from a tower in opposite directions horizontally with speed 10 m/s each. At $t = 1$ s match the following two columns.

Column I	Column II
(A) Relative acceleration between two	(p) Zero
(B) Relative velocity between two	(q) 5 SI unit
(C) Horizontal distance between two	(r) 10 SI unit
(D) Vertical distance between two	(s) 20 SI unit

Entrance Gallery

2014

1. A particle of mass m is projected from the ground with an initial speed u_0 at an angle with the horizontal. At the highest point of its trajectory, it makes a completely inelastic collision with another identical particle, which was thrown vertically upward from the ground with the same initial speed u_0. The angle that the composite system makes with the horizontal immediately after the collision is **[JEE Advanced]**

(a) $\frac{\pi}{4}$ (b) $\frac{\pi}{4}+\alpha$
(c) $\frac{\pi}{4}-\alpha$ (d) $\frac{\pi}{2}$

2013

2. A projectile is given an initial velocity of $(\hat{\mathbf{i}}+2\hat{\mathbf{j}})$ m/s, where, $\hat{\mathbf{i}}$ is along the ground and $\hat{\mathbf{j}}$ is along the vertical. If $g = 10\,\text{m/s}^2$, then equation of its trajectory is **[JEE Main]**

(a) $y = x - 5x^2$ (b) $y = 2x - 5x^2$
(c) $4y = 2x - 5x^2$ (d) $4y = 2x - 25x^2$

3. A boy can throw a stone up to a maximum height of 10 m. The maximum horizontal distance that the boy can throw the same stone up to will be **[JEE Main]**

(a) $20\sqrt{2}$ m (b) 10 m
(c) $10\sqrt{2}$ m (d) 20 m

2012

4. A particle of mass m is projected with a velocity v making an angle of 30° with the horizontal. The magnitude of angular momentum of the projectile about the point of projection when the particle is at its maximum height h is **[AIEEE]**

(a) $\frac{\sqrt{3}}{2}\frac{mv^2}{g}$ (b) zero
(c) $\frac{mv^3}{\sqrt{2}g}$ (d) $\frac{\sqrt{3}}{16}\frac{mv^3}{g}$

2011

5. Two projectiles A and B thrown with speeds in the ratio $1:\sqrt{2}$ acquired the same heights. If A is thrown at an angle of 45° with the horizontal, the angle of projection of B will be **[Kerala CEE]**

(a) 0° (b) 60°
(c) 30° (d) 45°
(e) 15°

6. An aircraft is flying at a height of 3400 m above the ground. If the angle subtended at a ground observation point by the aircraft positions 10 s apart is 30°, then the speed of the aircraft is **[Kerala CEE]**

(a) 19.63 ms^{-1} (b) 1963 ms^{-1}
(c) 108 ms^{-1} (d) 196.3 ms^{-1}
(e) 10.8 ms^{-1}

7. The height y and the distance x along the horizontal plane of a projectile on a certain planet (with no surrounding atmosphere) are given by $y = 8t - 5t^2$ metre and $x = 6t$ metre, where t is in second. The velocity with which the projectile is projected, is **[Karnataka CET]**

(a) 14 ms^{-1} (b) 10 ms^{-1} (c) 8 ms^{-1} (d) 6 ms^{-1}

2010

8. A particle is moving with velocity $\mathbf{v} = k\,(y\hat{\mathbf{i}} + x\,\hat{\mathbf{j}})$, where k is a constant. The general equation for its path is **[AIEEE]**

(a) $y = x^2$ + constant (b) $y^2 = x$ + constant
(c) xy = constant (d) $y^2 = x^2$ + constant

9. A small particle of mass m is projected at an angle θ with the x-axis with an initial velocity v_0 in the xy-plane as shown in the figure. At a time $t < \frac{v_0 \sin\theta}{g}$, the angular momentum of the particle is **[AIEEE]**

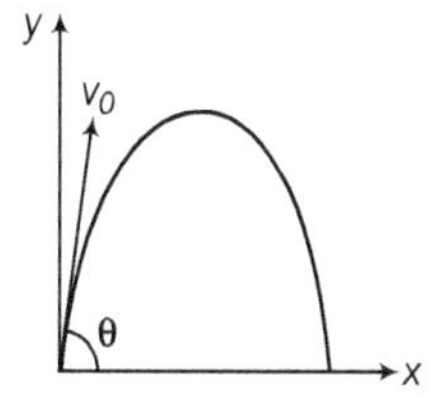

(a) $-mgv_0t^2\cos\theta\,\hat{\mathbf{j}}$ (b) $mgv_0t\cos\theta\,\hat{\mathbf{k}}$
(c) $-\frac{1}{2}mgv_0t^2\cos\theta\,\hat{\mathbf{k}}$ (d) $\frac{1}{2}mgv_0t^2\cos\theta\,\hat{\mathbf{i}}$

10. A boy throws a cricket ball from the boundary to the wicket-keeper. If the frictional force due to air cannot be ignored, the forces acting on the ball at the position X are respected by **[Karnataka CET]**

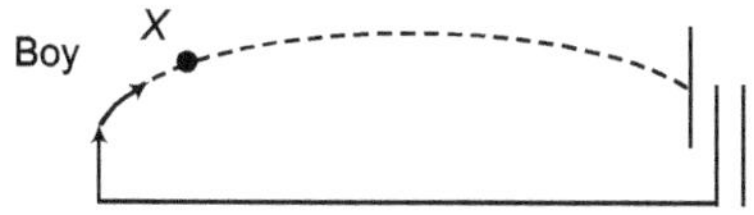

(a)  (b) 
(c)  (d) 

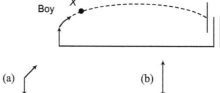

Answers

Level 1

Objective Problems

1. (a)	**2.** (d)	**3.** (b)	**4.** (d)	**5.** (c)	**6.** (d)	**7.** (d)	**8.** (c)	**9.** (d)	**10.** (d)
11. (c)	**12.** (a)	**13.** (d)	**14.** (b)	**15.** (a)	**16.** (c)	**17.** (b)	**18.** (c)	**19.** (b)	**20.** (d)
21. (b)	**22.** (d)	**23.** (b)	**24.** (c)	**25.** (b)	**26.** (d)	**27.** (d)	**28.** (a)	**29.** (d)	**30.** (c)
31. (b)	**32.** (a)	**33.** (b)	**34.** (c)	**35.** (a)	**36.** (a)	**37.** (b)	**38.** (a)	**39.** (a)	**40.** (d)
41. (b)	**42.** (b)	**43.** (d)	**44.** (d)	**45.** (a)	**46.** (c)	**47.** (c)	**48.** (b)	**49.** (c)	**50.** (c)
51. (a)	**52.** (d)	**53.** (c)	**54.** (c)	**55.** (b)	**56.** (a)	**57.** (b)	**58.** (c)	**59.** (c)	**60.** (b)
61. (c)	**62.** (a)	**63.** (b)	**64.** (d)	**65.** (c)	**66.** (a)	**67.** (b)	**68.** (c)	**69.** (d)	**70.** (c)
71. (d)	**72.** (a)	**73.** (a)	**74.** (b)	**75.** (a)	**76.** (b)	**77.** (a)	**78.** (b)	**79.** (b)	**80.** (c)
81. (c)	**82.** (c)	**83.** (d)	**84.** (a)	**85.** (a)					

Level 2

Only One Correct Option

1. (b)	**2.** (d)	**3.** (b)	**4.** (a,c,d)	**5.** (d)	**6.** (b)	**7.** (d)	**8.** (c)	**9.** (d)	**10.** (c)
11. (c)	**12.** (a)	**13.** (d)	**14.** (b)	**15.** (a)	**16.** (a.b)	**17.** (b)	**18.** (all)	**19.** (a)	**20.** (d)
21. (d)	**22.** (b)	**23.** (c)							

More than One Correct Options

1. (all)	**2.** (a,c,d)	**3.** (b,c)	**4.** (b,d)	**5.** (c,d)	**6.** (all)

Comprehension Based Questions

1. (b) **2.** (a)

Assertion and Reason

1. (d)	**2.** (b)	**3.** (a)	**4.** (a,b)	**5.** (a)	**6.** (c)	**7.** (b)	**8.** (e)	**9.** (b)	**10.** (b)
11. (a)	**12.** (a)	**13.** (b)							

Match the Columns

1. (A $\rightarrow$ r, B $\rightarrow$ r, C $\rightarrow$ p, D $\rightarrow$ q) **2.** (A $\rightarrow$ q, B $\rightarrow$ q, C $\rightarrow$ q, D $\rightarrow$ p) **3.** (A $\rightarrow$ q, B $\rightarrow$ s, C $\rightarrow$ p, D $\rightarrow$ p)

4. (A $\rightarrow$ q, B $\rightarrow$ q, r, C $\rightarrow$ r, s, D $\rightarrow$ s) **5.** (A $\rightarrow$ p, B $\rightarrow$ s, C $\rightarrow$ s, D $\rightarrow$ p)

Entrance Gallery

1. (a)	**2.** (b)	**3.** (d)	**4.** (d)	**5.** (c)	**6.** (d)	**7.** (b)	**8.** (d)	**9.** (c)	**10.** (c)

Solutions

Level 1 : Objective Problems

1. $R = \frac{u^2 \sin 2\theta}{g} = \frac{(9.8)^2 \sin 90°}{9.8} = 9.8$ m

3. $a_x = 0$

$\therefore u_x =$ constant

At highest point vertical component of velocity is zero. Only horizontal component of velocity is present.

4. Velocity is horizontal and acceleration is vertical.

5. Projectile will strike at highest point of its path with velocity $v_0 \cos\alpha$.

6. $T \propto \sin\theta,\ \frac{T_A}{T_B} = \frac{\sin 30°}{\sin 60°} = \frac{1}{\sqrt{3}}$ or $T_B = \sqrt{3}\, T_A$

$$H \propto \sin^2\theta,\ \frac{H_A}{H_B} = \frac{\sin^2 30°}{\sin^2 60°} = \frac{1}{3}$$

or $$H_B = 3H_A$$

$$R_\theta = R_{90° - \theta}$$

$\therefore$ $$R_A = R_B$$

7. $R = \frac{u^2 \sin 2\theta}{g}$

$\therefore$ $$R \propto u^2$$

If initial velocity be doubled, then range will become four times.

8. $H = \frac{u^2 \sin^2\theta}{2g}$

$\therefore$ $$H \propto u^2$$

If initial velocity be doubled, then maximum height reached by the projectile will become four times.

9. $R = 4H\cot\theta$ if $\theta = 45°$, then $R = 4H\cot 45° = 4H$

10. Acceleration throughout the projectile motion remains constant and equal to g.

11. Time of flight $= \frac{2u\sin\theta}{g} = \frac{2 \times 50 \times \sin 30°}{10} = 5$ s

12. $H = \frac{u^2 \sin^2\theta}{2g}$ and $T = \frac{2u\sin\theta}{g}$

So, $$\frac{H}{T} = \frac{u^2 \sin^2\theta/2g}{4u^2 \sin^2\theta/2g^2} = \frac{g}{8} = \frac{5}{4}$$

13. $R = 4H\cot\theta$

If $\theta = 45°$, then $4H \Rightarrow \frac{R}{H} = \frac{4}{1}$

14. Air resistance is in opposite direction of velocity.

15. An external force by gravity is present throughout the motion. So, momentum will not be conserved.

16. $R = 4\sqrt{3}H$

$$\frac{2u^2 \sin\theta\cos\theta}{g} = 4\sqrt{3}\left(\frac{u^2 \sin^2\theta}{2g}\right)$$

$$\tan\theta = \frac{1}{\sqrt{3}}$$

$\therefore$ $$\theta = 30°$$

17. $H = R$ or $\frac{u^2 \sin^2\theta}{2g} = \frac{u^2 \sin\theta\cos\theta}{g}$ or $\tan\theta = 4$

$\therefore$ $$\theta = \tan^{-1}(4)$$

18. $R_{max} = \frac{u^2}{g}$ at $\theta = 45°$

$\therefore$ $$u = \sqrt{g\,R_{max}} = 100 \text{ m/s}$$

19. $R_A = R_B$

$$\frac{v^2 \sin 2\theta}{g} = \frac{(v/2)^2 \sin 90°}{g}$$

$$\sin 2\theta = \frac{1}{2} \text{ or } \theta = \frac{1}{2}\sin^{-1}\left(\frac{1}{4}\right)$$

$\therefore$ $$\theta = \frac{1}{2}\sin^{-1}\left(\frac{1}{4}\right)$$

20. $H \propto \sin\alpha$

$\therefore$ $$\frac{H_1}{H_2} = \frac{\sin^2\alpha}{\sin^2(90° - \alpha)} = \tan^2\alpha$$

21. $R \propto \frac{1}{g}$

$\therefore$ $$R_{moon} = 6R_{earth}$$

22. At 45° range is maximum. At 15° and 75° ranges are equal as $R_\theta = R_{(90° - \theta)}$

23. $\frac{u_1^2 \sin^2 45°}{2g} = \frac{u_2^2 \sin^2 60°}{2g}$

$\therefore$ $$\frac{u_1}{u_2} = \frac{\sin 60°}{\sin 45°} = \frac{\sqrt{3}/2}{(1/\sqrt{2})} = \sqrt{3} : \sqrt{2}$$

24. $R_\theta = R_{(90° - \theta)}$

25. $\theta = 45°,\ R_{max} = \frac{u^2}{g} = 80$

$\therefore$ $$u^2 = 800 \text{ m/s}^2$$

Now, $$H = \frac{u^2 \sin^2 45°}{2g} = \frac{(800)(1/2)}{20}$$

$$= 20 \text{ m}$$

26. $\frac{u}{u\cos\theta} = x$ or $\cos\theta = \frac{1}{x} \Rightarrow \therefore\ \theta = \cos^{-1}\left(\frac{1}{x}\right)$

27. $u\cos\theta = \frac{u}{2}$

$\therefore\ \theta = 60°$

Now, $$R\frac{u^2 \sin 2\theta}{g} = \frac{u^2 \sin 120°}{g} = \frac{\sqrt{3}\,u^2}{2g}$$

28. $R = \frac{2u_x u_y}{g} = \frac{2 \times 9.8 \times 19.6}{9.8} = 39.2$ m

29. $T = \frac{2u_y}{g} = 3$

$\therefore$ $$u_y = 15 \text{ m/s}$$

Now, $$H = \frac{u_y^2}{2g} = \frac{(15)^2}{20} = 11.25 \text{ m}$$

30. $\frac{u^2}{2g}=h$ (given)

Now, $R_{max}=\frac{u^2 \sin 90^\circ}{g}=\frac{u^2}{g}=2h$

31. $2\theta=90^\circ-\theta$

or $3\theta=90^\circ$ or $\theta=30^\circ$

32. Velocity of boy should be equal to the horizontal component of velocity of ball.

33. $R_{15^\circ}=\frac{u^2 \sin(2\times15^\circ)}{g}=\frac{u^2}{2g}=1.5\text{ km}$

$R_{45^\circ}=\frac{u^2 \sin(2\times45^\circ)}{g}=\frac{u^2}{g}=1.5\times2=3\text{ km}$

34. For angle $(45^\circ-\theta)$, $R=\frac{u^2\sin(90^\circ-2\theta)}{g}=\frac{u^2\cos2\theta}{g}$

For angle $(45^\circ+\theta)$, $R=\frac{u^2\sin(90^\circ+2\theta)}{g}=\frac{u^2\cos2\theta}{g}$

35. $T=\frac{2u\sin\theta}{g}=10\text{ s}$

$\Rightarrow$ $u\sin\theta=50\text{ m/s}$

$\therefore$ $H=\frac{u^2\sin^2\theta}{2g}=\frac{(u\sin\theta)^2}{2g}$

$=\frac{50\times50}{2\times10}=125\text{ m}$

36. When the angle of projection is very far from 45°, then range will be minimum.

37. $H=\frac{u^2\sin^2\theta}{2g}$ and $T=\frac{2u\sin\theta}{g}\Rightarrow T^2=\frac{4u^2\sin^2\theta}{g^2}$

$\therefore$ $\frac{T^2}{H}=\frac{8}{g}\Rightarrow T=\sqrt{\frac{8H}{g}}=2\sqrt{\frac{2H}{g}}$

38. $5\sqrt{3}=\frac{(10)^2\sin2\theta}{g}$ or $\sin2\theta=\frac{\sqrt{3}}{2}$

$\therefore$ $2\theta=60^\circ$ or $\theta=30^\circ$

Two different angles of projection are therefore, θ and $(90^\circ-\theta)$ or 30° and 60°.

$T_1=\frac{2u\sin30^\circ}{g}=1\text{ s}$

$T_2=2\frac{u\sin60^\circ}{g}=\sqrt{3}\text{ s}$

$\therefore$ $\Delta t=T_2-T_1=(\sqrt{3}-1)\text{ s}$

39. $R=\frac{u^2\sin2\theta}{g}$ at angles θ and $(90^\circ-\theta)$

Now, $t_1=\frac{2u\sin\theta}{g}$

and $t_2=\frac{2u\sin(90^\circ-\theta)}{g}=\frac{2u\cos\theta}{g}$

$\therefore$ $t_1t_2=\frac{2}{g}\left[\frac{u^2\sin2\theta}{g}\right]=\frac{2R}{g}$

40. $R=\frac{u^2\sin\theta}{g}$ at angles θ and $(90^\circ-\theta)$

Now, $h_1=\frac{u^2\sin^2\theta}{2g}$

and $h_2=\frac{u^2\sin^2(90^\circ-\theta)}{2g}=\frac{u^2\cos^2\theta}{2g}$

$h_1h_2=\left(\frac{u^2\sin2\theta}{g}\right)^2\cdot\frac{1}{16}=\frac{R^2}{16}$

$\therefore$ $R=4\sqrt{h_1h_2}$

41. $T=\frac{2u_y}{g}$

$\therefore$ $u_y=\frac{gT}{2}=25\text{ m/s}$

Now, $H=\frac{u_y^2}{2g}=\frac{(25)^2}{20}$

$=31.25\text{ m}$

Further, $R=u_xT$

$\therefore$ $u_x=\frac{R}{T}$

$=40\text{ m/s}$

42. $\Delta\mathbf{a}=\mathbf{a}\,\Delta t$ (as $a=$ constant)

$=(-g\hat{\mathbf{j}})\left(\frac{2v_0\sin\alpha}{g}\right)=(-2v_0\,\alpha)\hat{\mathbf{j}}$

i.e. change in velocity is $2v_0\sin\alpha$, vertically downwards.

43. $h=\frac{1}{2}gt^2$ (in vertical direction)

$\therefore$ $t=\sqrt{\frac{2h}{g}}=\sqrt{\frac{2\times0.1}{10}}=0.141\text{ s}$

Now, in horizontal direction,

$v_x=\frac{S_x}{t}=\frac{100}{1.141}\approx700\text{ m/s}$

44. $u_x=19.6\text{ m/s}$ and $u_y=19.6\text{ m/s}$. After 2 s, vertical component of velocity v_y will become zero. So, particle is at its highest point.

45. At highest point velocity will remain $v\cos30^\circ$ or $\frac{\sqrt{3}v}{2}$.

Therefore, momentum will also remain $\frac{\sqrt{3}p}{2}$.

46. $s_y=u_yt+\frac{1}{2}a_yt^2$ or $-70=25t-5t^2$

Solving this equation, we get $t=7\text{ s}$.

47. Compare with $y=x\tan\theta=\frac{gx^2}{2u^2\cos^2\theta}$

$\tan\theta=\sqrt{3}$

$\therefore$ $\theta=60^\circ$

48. $v_y=gt=7\text{ m/s}, v_x=4\text{ m/s}$

$\therefore$ $v=\sqrt{v_x^2+v_y^2}\approx8\text{ m/s}$

49. At highest point speed will remain $\frac{1}{\sqrt{2}}$ times $(=u\cos45^\circ)$.

Therefore, kinetic energy will become $\frac{1}{2}$ times.

50. $v_y=\frac{dy}{dt}=8-10t, v_x=\frac{dx}{dt}=6$

At $t=0, v_y=8\text{ m/s}$ and $v_x=6\text{ m/s}$

$\therefore$ $v=\sqrt{v_x^2+v_y^2}=10\text{ m/s}$

51. $t=\sqrt{\frac{2h}{g}}=\sqrt{\frac{2\times1960}{9.8}}=20\,\text{s}$

$\therefore$ Horizontal distance $=600\times\frac{20}{60\times60}$ km $=3.33$ km

52. $T=(24\sin\theta/g)=4\,\text{s}, R=u^2\sin^2\theta/g=80\,\text{m}$

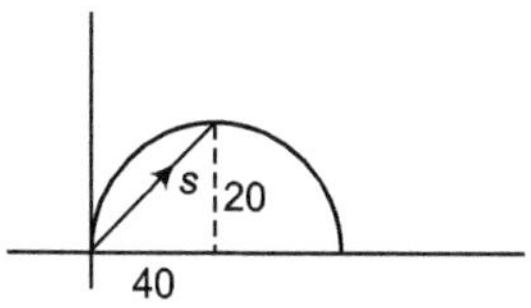

and $$H=\frac{u^2\sin^2\theta}{2g}=20\,\text{m}$$

Now, average velocity $=\frac{\text{Displacement}}{\text{Time}}$

$$=\frac{\sqrt{(20)^2+(40)^2}}{2}$$

$$=10\sqrt{5}\text{ m/s}$$

53. H and R both are proportional to u^2. Hence, percentage increases in horizontal range would also be 10%.

54. At highest point kinetic energy will be minimum but not zero.

55. $H=\frac{R}{4}$

$\therefore$ $$\frac{u^2\sin^2\theta}{2g}=\frac{2u^2\sin\theta\cos\theta}{4g}$$

or $$\tan\theta=1$$

$\therefore$ $$\theta=45°$$

At highest point momentum will remain $\frac{p}{\sqrt{2}}$.

$\therefore$ $$K=\frac{(p/\sqrt{2})^2}{2m}=\frac{p^2}{4m}$$

56. From $s_y=u_yt+\frac{1}{2}a_yt^2$ we have

$$-40=(20\sin30°)t-5t^2$$

or $$t^2-2t-8=0$$

Solving this, we get $t=4$ s

$$T=\frac{2u\sin\theta}{g}$$

$$=\frac{2\times20\times\sin30°}{10}=2\text{s}$$

$\therefore$ $$\frac{t}{T}=2$$

57. Put $y=0$,

$$0=12x-\frac{3}{4}x^2$$

$$x=16\,\text{m}$$

58. $y_1+y_2=\frac{u^2\sin^2\theta}{2g}+\frac{u^2\sin^2(90°-\theta)}{2g}=\frac{u^2}{2g}$

59. $\tan\phi=\frac{H}{R/2}=\frac{2H}{R}=\frac{(u^2\sin^2\theta/g)}{u^2\sin2\theta/g}$

$$=\frac{\sin^2\theta}{2\sin2\cos\theta}=\frac{\tan\theta}{2}$$

60. $t=\sqrt{\frac{2h}{g}}=\sqrt{\frac{4}{(9.8)}}=0.64$ s

Now, $$v=\frac{3}{t}=\frac{3}{0.64}=4.7\text{ m/s}$$

61. Horizontal component of velocity remains unchanged. Hence,

$$v\cos\phi=u\cos\theta$$

$\therefore$ $$v=u\cos\theta\sec\phi$$

62. $K=K_0-mgh$

Here, K = kinetic energy at height h,

K_0 = initial kinetic energy. Variation of K with h is linear. At highest point kinetic energy is not zero.

63. Comparing with $y=x\tan\theta-\frac{gx^2}{2u^2\cos^2\theta}$ we have,

$$\tan\theta=\sqrt{3}\quad\text{or}\quad\theta=60°$$

and $$u^2\cos^2\theta=1$$

or $$u=\sec\theta=\sec60°=2\text{ m/s}$$

64. $\Delta\mathbf{v}=\mathbf{a}\Delta t=\mathbf{a}\cdot\left(\frac{T}{2}\right)$

$$=(-g\hat{\mathbf{j}})\left(\frac{u\sin\theta}{g}\right)=(-u\sin\theta)\hat{\mathbf{j}}$$

Therefore, change in velocity is $u\sin\theta$ in downward direction.

65. $\frac{R/2}{H}=\frac{\sqrt{3}H}{H}=\sqrt{3}$

or $$\frac{(v_0^2\sin\theta\cos\theta)/g}{(v_0^2\sin^2\theta)/2g}=\sqrt{3}$$

$$2\cot\theta=\sqrt{3}$$

or $$\tan\theta=\frac{2}{\sqrt{3}}$$

or $$\theta=\tan^{-1}\left(\frac{2}{\sqrt{3}}\right)$$

66. Horizontal component of velocity $v_x=500$ m/s and vertical component of velocity while striking the ground.

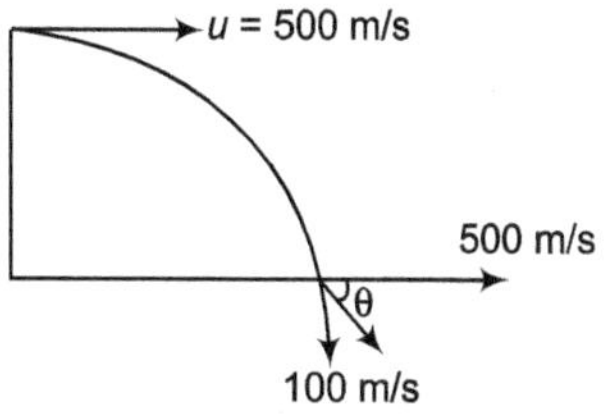

$$v_y=0+10\times10=100\text{ m/s}$$

$\therefore$ Angle with which it strikes the ground

$$\theta=\tan^{-1}\left(\frac{v_y}{y_x}\right)$$

$$=\tan^{-1}\left(\frac{100}{500}\right)=\tan^{-1}\left(\frac{1}{5}\right)$$

67. Area in which bullet will spread $=\pi r^2$

For maximum area, $r=R_{max}=\frac{v^2}{g}$ (when $\theta=45°$)

Maximum area $=\pi R_{max}^2=\pi\left(\frac{v^2}{g}\right)=\frac{\pi v^4}{g^2}$

68. Instantaneous velocity of rising mass after t s will be

$$v_t = \sqrt{v_x^2 + v_y^2}$$

where, $v_x = v\cos\theta$ = horizontal component of velocity

$v_y = v\sin\theta - gt$ = vertical component of velocity.

$$v_t = \sqrt{(v\cos\theta)^2 + (v\sin\theta - gt)^2}$$

$$v_t = \sqrt{v^2 + g^2t^2 - (2v\sin\theta)\,gt}$$

69. $R = \dfrac{u^2 \sin 2\theta}{g} = \dfrac{2u_x u_y}{g}$

$\therefore$ Range $\propto$ horizontal initial velocity component (u_x).

In path 4 range is maximum of football has maximum horizontal velocity component in this path.

72. Comparing with projectile motion we can see that it is like a projectile motion with $u_x = 4$ m/s, $u_y = 4$ m/s and $a_y = 10$ m/s^2.

x-coordinate = range $= \dfrac{2u_x u_y}{g}$

$$= \frac{2\times4\times4}{10} = 3.2 \text{ m}$$

73. $\mathbf{v} \perp \mathbf{u} = 0$ or $\mathbf{v}\cdot\mathbf{u} = 0$

or $(\mathbf{u} + \mathbf{a}\,t)\cdot\mathbf{u} = 0$

or $\mathbf{u}\cdot\mathbf{u} + (\mathbf{a}\cdot\mathbf{u})t = 0$

or $u^2 + gut\cos(90° + \theta) = 0$

(angle between **u** and **u** is $90° + \theta$)

or $u - gt\sin\theta = 0$

or $t = \dfrac{u}{g\sin\theta}$

74. $\tan 30° = \dfrac{v_y}{v_x} = \dfrac{u_y - gt}{u_x}$

$$= \frac{(20\sqrt{3}\sin 60°) - 10t}{(20\sqrt{3}\cos 60°)}$$

or $10 = 30 - 10t$

$\therefore$ $t = 2$ s

75. $v^2 = v_y^2 + v_x^2$ or $5g = (u_y - gt)^2 + u_x^2$

or $50 = (5\sqrt{3} - 10t)^2 + (5)^2$

$\therefore$ $(5\sqrt{3} - 10t) = \pm 5$

$$t_1 = \frac{5\sqrt{3}+15}{10} \text{ and } t_2 = \frac{5\sqrt{3}-5}{10}$$

$\therefore$ $t_1 - t_2 = 1$ s

76. $t = \dfrac{10}{20\cos 45°} = \dfrac{1}{\sqrt{2}}$ s

Now, $y = (20\sin 45°)\,t - \dfrac{1}{2}gt^2$

$$= 20\times\frac{1}{\sqrt{2}}\times\frac{1}{\sqrt{2}} - \frac{1}{2}\times10\frac{1}{2}$$

$= 7.5$ m

77. $\mathbf{v}_1 \perp \mathbf{v}_2$

$\therefore$ $\mathbf{v}_1\cdot\mathbf{v}_2 = 0$

or $(u_1\,\hat{\mathbf{i}} - gt\,\hat{\mathbf{j}})\cdot(-u_2\,\hat{\mathbf{i}} - gt\,\hat{\mathbf{j}}) = 0$

$\therefore$ $g^2t^2 = u_1u_2$

or $t = \dfrac{\sqrt{u_1u_2}}{g}$

78. $\tan 60° = \dfrac{v_V}{v_H} = \dfrac{gt}{v_0}$

$\therefore$ $t = \dfrac{\sqrt{3}\,v_0}{g}$

79. $nh = \dfrac{1}{2}gt^2$ and $nb = ut$

From these two equation we get, $n = \dfrac{2hu^2}{gb^2}$

80. $H_1 - H_2 = \dfrac{H_1 + H_2}{2}$ or $H_1 = 3H_2$

$\therefore$ $\dfrac{u^2\sin^2\theta}{2g} = 2\left\{\dfrac{u^2\sin^2(90° - \theta)}{2g}\right\}$

$\tan^2\theta = 3$

$\therefore$ $\tan\theta = \sqrt{3}$

or $\theta = 60°$

Therefore, the other angle is $(90° - \theta)$ or $30°$

81. $R = 2H$ or $\dfrac{2u_xu_y}{g} = \dfrac{2u_y^2}{2g}$

or $2u_x = y$ or $2a = b$

82. Vertical component of velocity of A should be equal to vertical velocity of B.

or $v_1\sin 30° = v_2$

or $\dfrac{v_1}{2} = v_2$

$\therefore$ $\dfrac{v_2}{v_1} = \dfrac{1}{2}$

83. $u_x = u\cos\theta_0 = 20\times\dfrac{4}{5} = 24$ m/s

and $u_y = u\sin\theta_0 = 30\times\dfrac{3}{5} = 18$ m/s

After 1 s, u_x will remain as it is u_y will decreases by 10 m/s or it will remain 8 m/s

$\therefore$ $\tan\theta = \dfrac{v_y}{v_x} = \dfrac{8}{24} = \dfrac{1}{3}$

84. $x = 36t$

$\therefore$ $v_x = \dfrac{dx}{dt} = 36$ m/s

$y = 48t - 4.9t^2$

$\therefore$ $v_y = 48 - 9.8t$

at $t = 0$ $v_x = 36$ m/s

and $v_y = 18$ m/s

So, angle of projection $\theta = \tan^{-1}\left(\dfrac{v_y}{v_x}\right)$

$$= \tan^{-1}\left(\frac{4}{3}\right)$$

or $\theta = \sin^{-1}\left(\dfrac{4}{5}\right)$

85. Person will catch the ball, if its velocity will be equal to horizontal component of velocity of the ball.

$$\frac{v_0}{2} = v_0\cos\theta$$

$\Rightarrow$ $\cos\theta = \dfrac{1}{2}$

$\Rightarrow$ $\theta = 60°$

Level 2 : Only One Correct Option

1. $\frac{u^2}{g} = R_{max} = 500$

Now, $$S = \frac{u^2}{2a} = \frac{u^2}{2g\sin 30°} = \frac{u^2}{g} = 500\text{ m}$$

2. $\tan 30° = \frac{u_y}{u_x}$

$\therefore$ $$\frac{1}{\sqrt{3}} = \frac{80}{u_x}$$

or $$u_x = \frac{80}{\sqrt{3}}\text{ m/s}$$

$$T = \frac{2u_y}{g} = \frac{2 \times 80}{10} = 16\text{ s}$$

At $$t = \frac{T}{4} = 4\text{ s}$$

$$v_x = u_x = \frac{80}{\sqrt{3}}\text{ m/s}$$

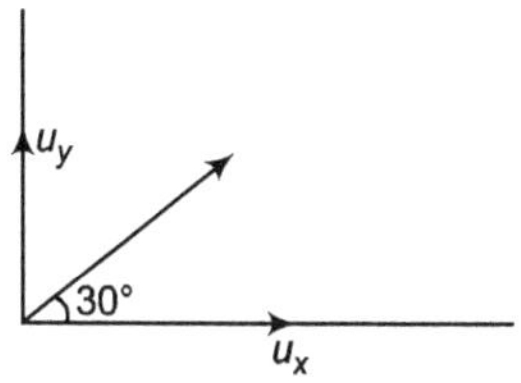

and $$v_y = u_y + a_y t$$

or $$v_y = 80 + (-10)(4) = 40\text{ m/s}$$

$\therefore$ $$\text{Speed} = \sqrt{(80\sqrt{3})^2 + (40)^2} = \frac{40}{\sqrt{13}}\text{ m/s}$$

3. Let us see the motion relative to elevator,

$$a_r = a_b - a_e = (-10) - (+2) = -12\text{ m/s}^2$$

Now, $$T = \frac{2u_y}{a_r} = \frac{2 \times u\sin\theta}{a_r}$$

$$= \frac{2 \times 4 \times \sin 30°}{12} = \frac{1}{3}\text{ s}$$

4. As initial velocity is zero. Particle will move in a straight line along a_{net}.

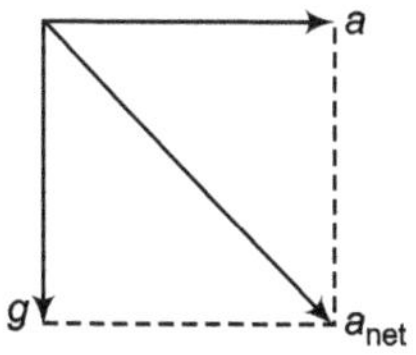

Further, $$t = \sqrt{\frac{2h}{g}} = \sqrt{\frac{2 \times 49}{9.8}} = \sqrt{10} = 3.16\text{ s}$$

5. Comparing with the trajectory of projectile in which particle is projected from certain height horizontally ($\theta = 0°$).

$$y = x\tan\theta - \frac{gx^2}{2u^2\cos^2\theta}$$

Putting $\theta = 0°$ and $g = a = 18u^2 = 2\text{ m/s}^2$

6. Their horizontal components should be same.

$\therefore$ $$\frac{2v}{\sqrt{3}} \cdot \cos\theta = v$$

or $$\theta = 60°$$

7. $T = \frac{2u_y}{g}$

$\therefore$ $$u_y = \frac{gT}{2}$$

$$h_A = u_y t_A - \frac{1}{2}g t_A^2$$

$$= u_y\left(\frac{2u_y}{3g}\right) - \frac{1}{2}g\left(\frac{2u_y}{3g}\right)^2$$

$$= \frac{4}{9}\frac{u^2{}_y}{g} = \left(\frac{4}{9g}\right)\left(\frac{gT}{2}\right)^2 = \frac{gT^2}{9}$$

$$h_B = u_y\left(\frac{5}{9} \times \frac{2u_y}{g}\right) - \frac{1}{2} \times g \times \left(\frac{5}{6} \times \frac{2u_y}{g}\right)^2$$

$$= \frac{5}{18}\frac{u_y^2}{g} = \frac{5}{18g}\left(\frac{gT}{2}\right)^2 = \frac{5}{12}gT^2$$

$\therefore$ $$h_A - h_B = \frac{gT^2}{24}$$

8. The time taken by cart to cover 80 m

$$\frac{s}{v} = \frac{80}{30} = \frac{8}{2}\text{ s}$$

The projectile must be fired (relative to cart) vertically upwards.

$$a = -g = -10\text{ ms}^{-2}, v' = 0 \text{ and } t = \frac{8/3}{2} = \frac{4}{3}\text{ s}$$

$\therefore$ $$v' = u + at \quad\text{or}\quad 0 = u - 10 \times \frac{4}{3}$$

or $$u = \frac{40}{3}\text{ ms}^{-1}$$

9. For 5 s weight of the body is balanced by the given force. Hence, it will move in a straight line as shown.

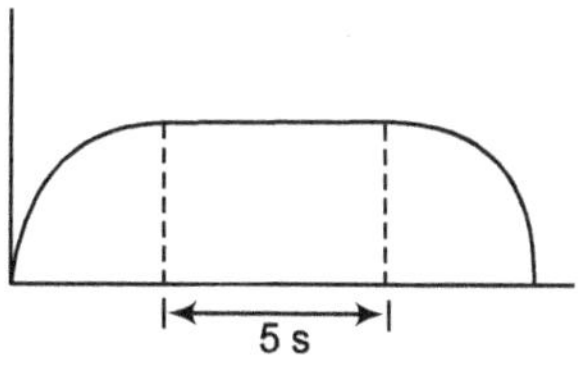

$$R = \frac{u^2\sin 2\theta}{g} + (u\cos\theta)(5)$$

$$= \frac{(50)^2 \cdot \sin 60°}{10} + (50 \times \cos 30°)(5)$$

$$= 250\sqrt{3}\text{ m}$$

10. $(u\cos\alpha) = \sqrt{\frac{2}{5}}\sqrt{(u\cos\alpha^2) + \{(u\sin\alpha)^2 - 2gh\}}$

Here, $h = \frac{H}{2} = \frac{u^2\sin^2\alpha}{4g}$

Solving this equation, we get $\alpha = 60°$

11. Two balls will meet if,

$$(50\cos 37°)t_A = 120 \text{ or } t_A = 3\text{ s}$$

Vertical component of A is also $50\sin 37°$ or 30 m/s, so they will meet if thrown simultaneously.

$$h_A = h_B$$

$$= 30 \times 3 - \frac{1}{2} \times 10 \times (3)^2$$

$$= 45\text{ m}$$

12. $T/2 = 2 + 1 = 3\text{s}$ or $T = 6\text{s}$

$\therefore \quad \frac{2u_y}{g} = 6$

$\therefore \quad u_y = 30 \text{ m/s}$

Further, $\tan 30° = \frac{v_y}{v_x} = \frac{u_y - gt}{u_x} = \frac{30 - 20}{u_x}$

or $\quad u_x = 10\sqrt{3}$ m/s

or $\quad u = \sqrt{u_x^2 + u_y^2}$

$= 20\sqrt{3}$ m/s

$\tan\theta = \frac{u_y}{u_x} = \frac{30}{10\sqrt{3}} = \sqrt{3}$

or $\quad \theta = 60°$

13. Standard equation of projectile motion

$$y = x\tan\theta - \frac{gx^2}{2u^2\cos^2\theta}$$

Comparing with given equation

$$A = \tan\theta \text{ and } B = \frac{g}{2u^2\cos^2\theta}$$

So $\quad \frac{A}{B} = \frac{\tan\theta \times 2u^2\cos^2\theta}{g} = 40$

(As $\theta = 45°, u = 20$ m/s, $g = 10$ m/s^2)

14. At B, $s_y = 0$

$\therefore \; u_y t + \frac{1}{2}a_y t^2 = 0 \quad$ or $\quad t = -\frac{2u_y}{a_y} = \frac{-2(10)}{-10\times\sqrt{3}/2} = \frac{4}{\sqrt{3}}$ s

Now, $\quad AB = R = \frac{1}{2}a_x t^2 = \frac{1}{2}\left(10 \times \frac{1}{2}\right)\left(\frac{16}{3}\right) = 13.33$ m

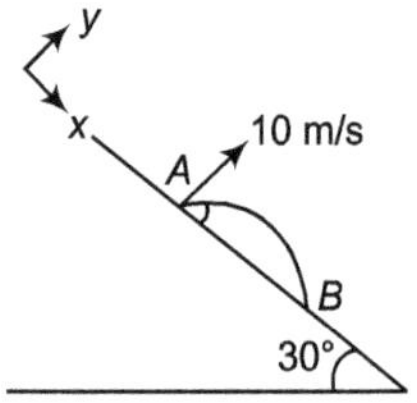

15. $u_x = a, u_y = b, g = c$

$$R = \frac{2u_x u_y}{g} = \frac{2ab}{c}$$

16. Component 60 m/s will remain unchanged. Velocity will make 45° with horizontal when vertical component also becomes $\pm$ 60 m/s.

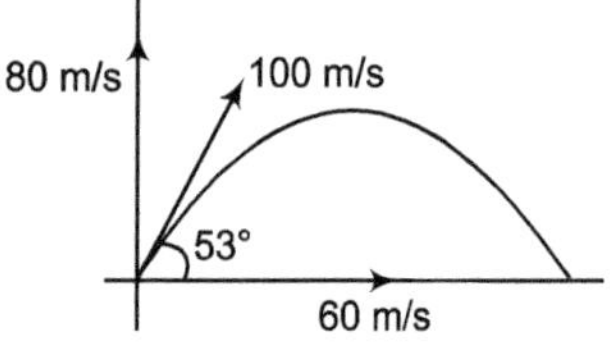

Using, $\quad v = u + at \quad$ (in vertical direction)

$+60 = 80 + (-10)t_1$

$\therefore \quad t_1 = 2\text{s}$

$-60 = 80 + (-10)t_2$

$\therefore \quad t_2 = 14$ s

17. $R = \frac{u^2\sin 2\theta}{g} = 24 \quad$ ...(i)

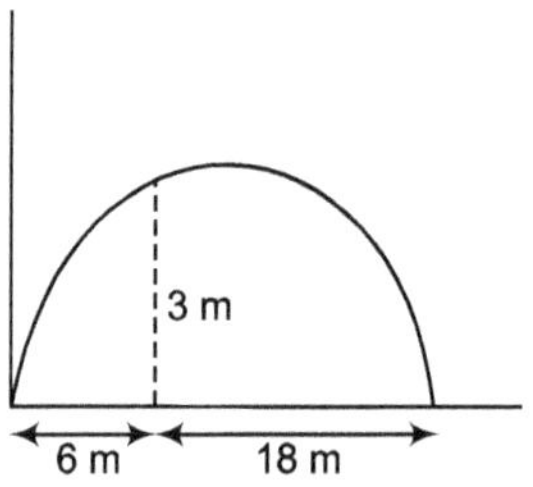

In $\quad y = x\tan\theta - \frac{gx^2}{2u^2\cos^2\theta}$

$3 = 6\tan\theta - \frac{36g}{2u^2\cos^2\theta} \quad$ ...(ii)

From Eq. (i), $\quad \frac{g}{u^2} = \frac{\sin 2\theta}{24} = \frac{\sin\theta\cos\theta}{12}$

Substituting in Eq. (ii), we have

$$3 = 6\tan\theta - \frac{3}{2}\tan\theta = \frac{9}{2}\tan\theta$$

$\therefore \quad \theta = \tan^{-1}\left(\frac{2}{3}\right)$

18. (a) Range becomes equal at complimentary angle.

Hence, $\quad \beta = 90° - \alpha$

(b) $h_1 = \frac{u^2\sin^2\alpha}{2g}$

$h_2 = \frac{u^2\cos^2\alpha}{2g} \quad$ (as $\beta = 90 - \alpha$)

$\therefore \quad 4\sqrt{h_1 h_2} = \frac{2u^2\sin\alpha\cos\alpha}{g}$

$= \frac{u^2\sin 2\alpha}{g} = R$

(c) $\frac{t_1}{t_2} = \frac{(2u\sin\alpha)/g}{(2u\cos\alpha/g)} = \tan\alpha$

(d) $\sqrt{\frac{h_1}{h_2}} = \tan\alpha$

19. $\frac{AB}{BC} = \frac{\frac{1}{2}gt^2}{v_0 t} = \tan\theta$

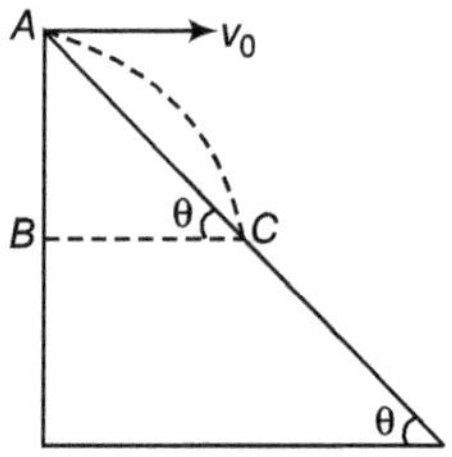

$\therefore \quad t = \frac{2v_0\tan\theta}{g}$

Now, x-coordinate $= v_0 t = \frac{2v_0^2\tan\theta}{g}$

and $\; y$-coordinate $= -\frac{1}{2}gt^2 = -\frac{2v_0^2\tan^2\theta}{g}$

20. $H' = H$ (as vertical component of acceleration has not changed)

$$R' = u_x T + \frac{1}{2} a_x T^2$$

$$= R + \frac{1}{2} \times \frac{g}{4} \times \frac{4u^2 \sin^2\theta}{g^2}$$

$$= R \frac{u^2 \sin^2\theta}{2g} = (R + H)$$

23. $R_{max} = \frac{u^2}{g} = 1.6\,\text{m}$ (at $\theta = 45°$)

or $u = \sqrt{16} = 4\,\text{m/s}$

Now, $T = \frac{2u_y}{g}$

$$= \frac{2(4/\sqrt{2})}{10} = \frac{4}{5\sqrt{2}}\,\text{s}$$

$\therefore$ Total distance travelled in 10 s $= 1.6 \times \frac{10}{4/5\sqrt{2}}$

$$= 20\sqrt{2}\,\text{m}$$

More than One Correct Options

1. $\alpha + \beta = 90°$ or $\beta = 90° - \alpha$

$$h_1 = \frac{u^2 \sin^2\alpha}{2g} \quad \text{and} \quad h_2 = \frac{u^2 \cos^2\alpha}{2g}$$

$$t_1 = \frac{2u\sin\alpha}{g} \quad \text{and} \quad t_2 = \frac{2u\cos\alpha}{g}$$

$$R_1 = R_2 = \frac{2u^2 \sin\alpha\cos\alpha}{g} = R$$

2. Since $u = 0$, motion of particle is a straight line in the direction of a_{net}.

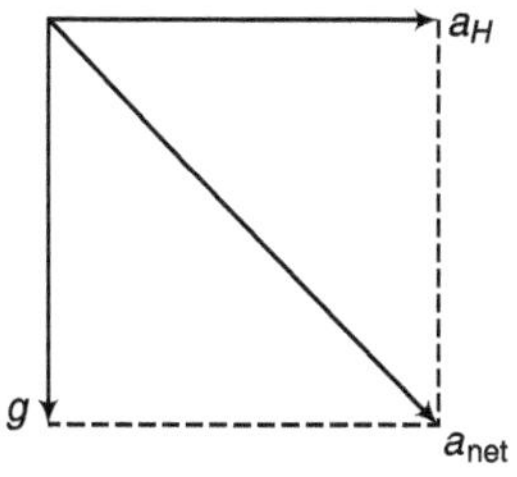

$$t = \sqrt{\frac{2h}{g}} = \sqrt{\frac{2 \times 49}{9.8}}$$

$$= 3.16\,\text{s}$$

3. Horizontal component of velocity remains unchanged

$\therefore$ $v\cos\theta = v'\cos(90° - \theta)$

or $v' = v\cot\theta$

In vertical (y) direction,

$$v_y = u_y + a_y t$$

$\therefore$ $t = \frac{v_y - u_y}{a_y}$

$$= \frac{-v^1 \sin(90 - \theta) - v\sin\theta}{-g}$$

$$= \frac{(v\cot\theta)\cdot\cos\theta + v\sin\theta}{g}$$

$$= \frac{v\,\text{cosec}\,\theta}{g}$$

4. $u_x = v_x = 10\,\text{m/s}$

$$u_y = \sqrt{v_y^2 + 2gh}$$

$$= \sqrt{(10)^2 + (2)(10)(15)}$$

$$= 20\,\text{m/s}$$

Angle of projection,

$$\theta = \tan^{-1}\left(\frac{u_y}{u_x}\right) = \tan^{-1}(2)$$

$$T = \frac{2u_y}{g} = \frac{(2)(20)}{10} = 4\,\text{s}$$

$$R = u_x T = (10)(4) = 40\,\text{m}$$

$$H = \frac{u_y^2}{2g} = \frac{(20)^2}{2 \times 10} = 20\,\text{m}$$

5. $\frac{d\mathbf{v}}{dt} = \mathbf{a} = \text{constant} = \mathbf{g}$

$$\frac{d^2\mathbf{v}}{dt^2} = \frac{d\mathbf{a}}{dt} = 0 = \text{constant}$$

6. Horizontal component of velocity remains unchanged

$$X_{OA} = 20\,\text{m} = \frac{X_{AB}}{2}$$

$\therefore$ $t_{OA} = \frac{t_{AB}}{2} = 1\,\text{s}$

For AB projectile

$$T = 2\,\text{s} = \frac{2u_y}{g}$$

$\therefore$ $u_y = 10\,\text{m/s}$

$$H = \frac{u_y^2}{2g} = \frac{(10)^2}{2 \times 10} = 5\,\text{m}$$

$\therefore$ Maximum height of total projectile,

$$= 15 + 5 = 20\,\text{m}$$

$$t_{OB} = t_{OA} + t_{AB} = 1 + 2 = 3\,\text{s}$$

For complete projectile

$$T = 2(t_{OA}) + t_{AB} = 4\,\text{s} = \frac{2u_y}{g}$$

$\therefore$ $u_y = 20\,\text{m/s}$

$$u_x = \frac{AB}{t_{AB}} = \frac{40}{2} = 20\,\text{m/s}$$

Comprehension Based Questions

1. At Q, component parallel to OB becomes zero

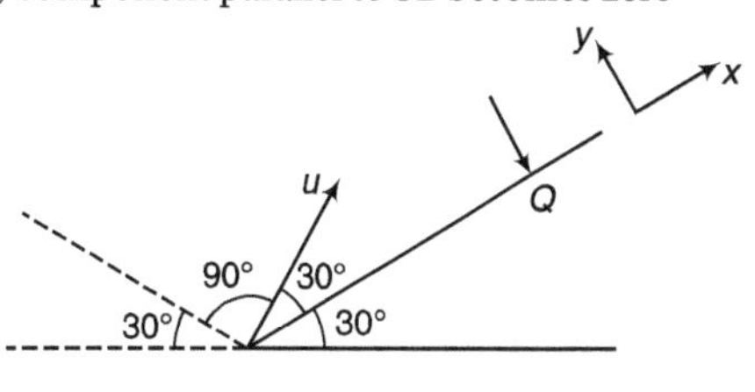

$$v_x = u_x + a_x t$$

$$0 = (u\cos 30°) + (-g\sin 30°)\,t$$

$$\therefore \quad t = \frac{u}{g}\cot 30° = \frac{\sqrt{3}u}{g}$$

2. $PQ = \text{range} = 2\,(PM) = 2a\cos 30°$

$$= (2)\,(4.9)\left(\frac{\sqrt{3}}{2}\right) = 4.9\sqrt{3}\text{ m}$$

$$= \frac{u^2 \sin 2\,(60°)}{9.8}$$

$$\therefore \quad 4.9\sqrt{3} = \frac{u^2\,(\sqrt{3}/2)}{9.8}$$

$$\therefore \quad u = 9.8\text{ m/s}$$

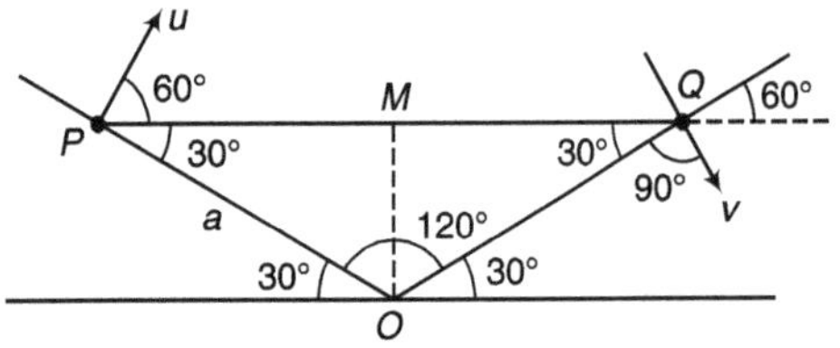

- Velocity at P is making an angle of 60° with horizontal and velocity at Q is making an angle of 60° with horizontal. That is the reason PQ = range. Because under this condition, points P and Q lie on same horizontal line.

Assertion and Reason

1. $\left|\frac{d\mathbf{v}}{dt}\right| = |\mathbf{a}| = 9.8\text{ m/s}^2 = \text{constant}$

2. **Assertion** $v\sin\alpha$ = horizontal component of velocity = constant

   Reason $a_r = \sqrt{g^2 - a_t^2}$

   At highest point $a_t = 0$. Therefore, a_r is maximum.

3. v = velocity

   w = weight and

   A = air resistance

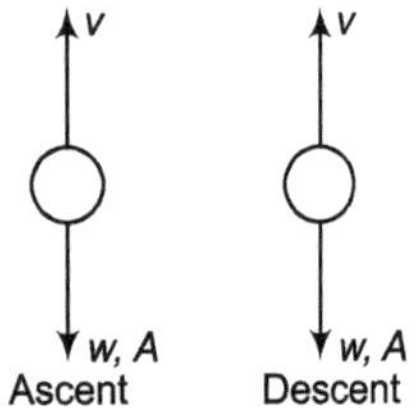

6. $H = \frac{u^2\sin^2\theta}{2g}$ and $T = \frac{2u\sin\theta}{g}$

$$\therefore \quad u\sin\theta = \frac{gT}{2}$$

$$\therefore \quad H = \frac{(g^2T^2/4)}{2g} = \frac{gT^2}{8}$$

9. $t_1 = t_2 = \sqrt{\frac{2h}{g}}$ (h = height of tower)

   $v_1 = \sqrt{2gh}$

   While $v_2 = \sqrt{v_1^2 + v_0^2}$ (v_0 = initial horizontal velocity)

10. $h_1 = \frac{u^2}{2g},\ h_2 = \frac{u^2\sin^2 30°}{2g} = \frac{u^2}{8g}$

11. Velocity is horizontal and acceleration is vertical.

13. $t = \frac{2u_y}{g} = \frac{2\times 20}{10} = 4\text{ s}$

$$H = 20 + \frac{u_y^2}{2g}$$

$$= 20 + \frac{(20)^2}{2\times 10} = 40\text{ m}$$

Match the Columns

1. Comparing with the standard equation of projectile,

$$y = x\tan\theta - \frac{gx^2}{2u^2\cos\theta}$$

   We get $\theta = 45°$

   and $u = 20\sqrt{2}$ m/s

   Time period of this projectile is 4 s. Hence, after 4 s velocity vector will again make 45° with horizontal.

Entrance Gallery

1. From momentum conservation equation, we have

$$\mathbf{p}_i = \mathbf{p}_f \quad\Rightarrow\quad m\,(u_0\cos\alpha)\hat{\mathbf{i}} + m\,(\sqrt{u_0^2 - 2gH})\hat{\mathbf{j}} = (2m)\,\mathbf{v} \qquad \text{...(i)}$$

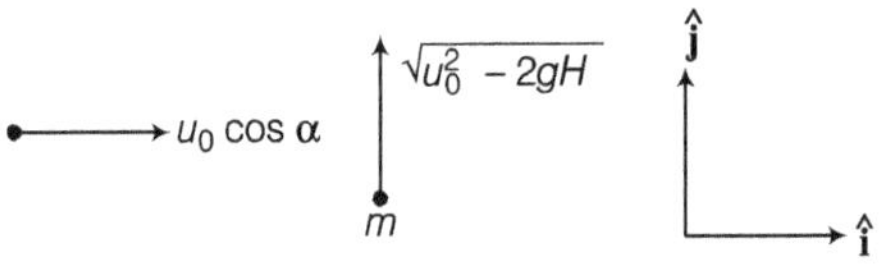

$$H = \frac{u_0^2\sin^2\alpha}{2g} \qquad \text{...(ii)}$$

   From Eqs. (i) and (ii), we get $\mathbf{v} = \frac{u_0\cos\alpha}{2}\hat{\mathbf{i}} + \frac{u_0\cos\alpha}{2}\hat{\mathbf{j}}$

   Since, both components of $\mathbf{v}$ are equal. Therefore, it is making 45° with horizontal.

2. Initial velocity $= (\hat{\mathbf{i}} + 2\hat{\mathbf{j}})$ m/s

   Magnitude of initial velocity, $u = \sqrt{(1)^2 + (2)^2} = \sqrt{5}$ m/s

   Equation of trajectory of projectile is

$$y = x\tan\theta - \frac{gx^2}{2u^2}(1 + \tan^2\theta) \qquad \left[\tan\theta = \frac{y}{x} = \frac{2}{1} = 2\right]$$

$$\therefore \quad y = x\times 2 - \frac{10\,(x)^2}{2(\sqrt{5})^2}[1 + (2)^2]$$

$$= 2x - \frac{10\,(x^2)}{2\times 5}(1 + 4) = 2x - 5x^2$$

3. Maximum speed with which the boy can throw stone is

$$u = \sqrt{2gh} = \sqrt{2\times 10\times 10} = 10\sqrt{2}\text{ m/s}$$

   Range is maximum when projectile is thrown at an angle of 45°. Thus,

$$R_{\max} = \frac{u^2}{g} = \frac{(10\sqrt{2})^2}{10} = 20\text{ m}$$

4. Angular momentum of the projectile

$$L = mv_h r_\perp = m(v\cos\theta)h$$

   (where, h is the maximum height)

$$\Rightarrow \quad = m(v\cos\theta)\left(\frac{v^2\sin^2\theta}{2g}\right)$$

$$L = \frac{mv^3\sin^2\theta\cos\theta}{2g} = \frac{\sqrt{3}\,mv^3}{16g}$$

5. Given, condition $h_1 = h_2$

Height is given as

$$H = \frac{u^2 \sin^2\theta}{2g}$$

$$u_1^2 \sin^2 45° = u_2^2 \sin^2\theta$$

$$\sin^2\theta = \frac{u_1^2}{u_2^2} \sin^2 45° = \frac{1}{2} \cdot \frac{1}{2} = \frac{1}{4}$$

$$\sin\theta = \frac{1}{2} \Rightarrow \theta = 30°$$

6. $\tan\theta = \frac{L}{H}$

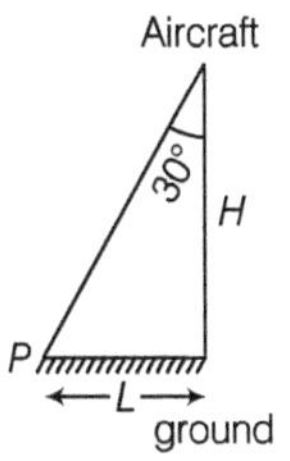

$$\tan 30° = \frac{10v}{3400}$$

$$\Rightarrow \quad v = \frac{340}{\sqrt{3}}$$

$$= 196.3 \text{ m/s}$$

7. Horizontal distance covered is

$$x = u\cos\theta(t) \quad \text{...(i)}$$

Given, $\quad x = 6t \quad$ ...(ii)

On comparing both Eqs. (i) and (ii), we get

$\Rightarrow \quad u\cos\theta = 6 \quad$ ...(iii)

Vertical distance covered is $y = u\sin\theta(t) - \frac{1}{2}gt^2 \quad$ ...(iv)

Given, $\quad y = 8t - 5t^2 \quad$ ...(v)

On comparing both Eqs. (iv) and (v), we get

$\Rightarrow \quad u\sin\theta = 8 \quad$ ...(vi)

From Eqs. (iii) and (vi), we get $u = 10 \text{ ms}^{-1}$

8. Velocity is given as $\mathbf{v} = ky\,\hat{\mathbf{i}} + kx\,\hat{\mathbf{j}}$

$$\frac{dx}{dt} = ky, \frac{dy}{dt} = kx$$

$$\frac{dy}{dx} = \frac{dy}{dt} \times \frac{dt}{dx} = \frac{kx}{ky}$$

$$y\,dy = x\,dx$$

$$y^2 = x^2 + \text{constant}$$

9. Angular momentum of a particle is given as $\mathbf{L} = m(\mathbf{r} \times \mathbf{v})$

$$\mathbf{L} = m\left[v_0 \cos\theta t\,\hat{\mathbf{i}} + (v_0 \sin\theta t - \frac{1}{2}gt^2)\,\hat{\mathbf{j}}\right]$$

$$\times [v_0 \cos\theta\,\hat{\mathbf{i}} + (v_0 \sin\theta - gt)\,\hat{\mathbf{j}}]$$

$$= mv_0 \cos\theta t\left[-\frac{1}{2}gt\right]\hat{\mathbf{k}} = -\frac{1}{2}mgv_0 t^2 \cos\theta\,\hat{\mathbf{k}}$$

10. The forces acting on the ball will be (i) in the direction opposite to its motion, i.e. frictional force and (ii) weight *mg*.

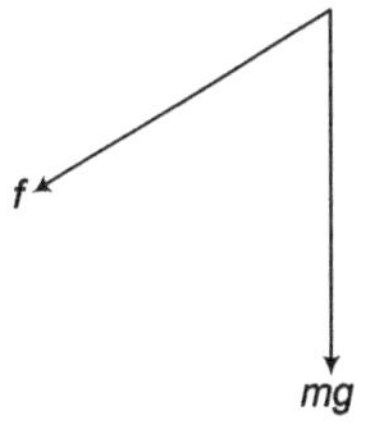

Hence, where f = frictional force

5

Laws of Motion

5.1 Types of Forces

There are basically three forces which are commonly encountered in mechanics.

Chapter Snapshot

- Types of Forces
- Free Body Diagram
- Equilibrium
- Newton's Laws of Motion
- Pulleys
- Constraint Equations
- Pseudo Force
- Friction

Field Forces

These are the forces in which contact between two objects is not necessary. Gravitational force between two bodies and electrostatic force between two charges are two examples of field forces. Weight ($w = mg$) of a body comes in this category.

Contact Forces

Two bodies in contact exert equal and opposite forces on each other. If the contact is frictionless, then contact force is perpendicular to the common surface and known as **normal reaction**.

If, however the objects are in rough contact and move (or have a tendency to move) relative to each other without losing contact, then **frictional force** arises which opposes such motion. Again each object exerts a frictional force on the other and the two forces are equal and opposite. This force is perpendicular to normal reaction. Thus, the contact force (F) between two objects is made up of two forces.

(i) Normal reaction (N) (ii) Force of friction (f)

and since these two forces are mutually perpendicular.

$$F = \sqrt{N^2 + f^2}$$

Consider two wooden blocks A and B being rubbed against each other.

In Fig. 5.1, A is being moved to the right while B is being moved leftward. In order to see more clearly which forces act on A and which on B, a second diagram is drawn showing a space between the blocks but they are still supposed to be in contact.

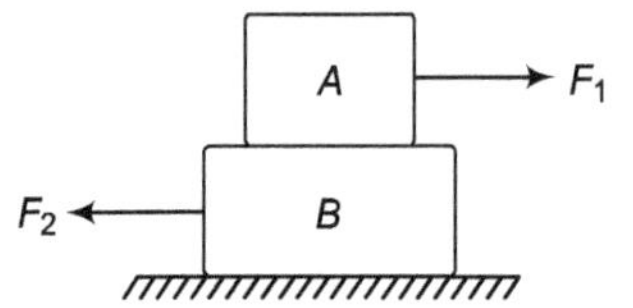

Fig. 5.1

In Fig. 5.2 the two normal reactions each of magnitude N are perpendicular to the surface of contact between the blocks and the two frictional forces each of magnitude f act along that surface, each in a direction opposing the motion of the block upon which it acts.

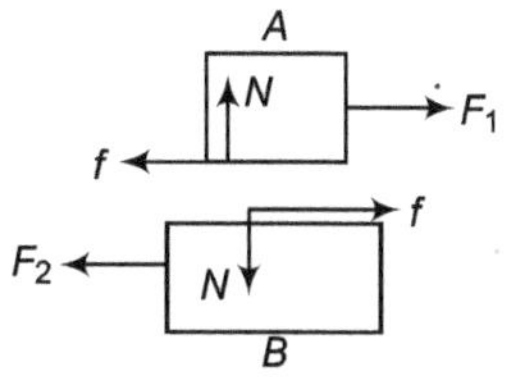

Fig. 5.2

Attachment to Another Body

Tension (T) in a string and **spring force** ($F = kx$) come in this group. Regarding the tension and string, the following three points are important to remember.

1. If a string is inextensible, the magnitude of acceleration of any number of masses connected through string is always same.
2. If a string is massless, the tension in it is same everywhere. However, if a string has a mass, tension at different points will be different.

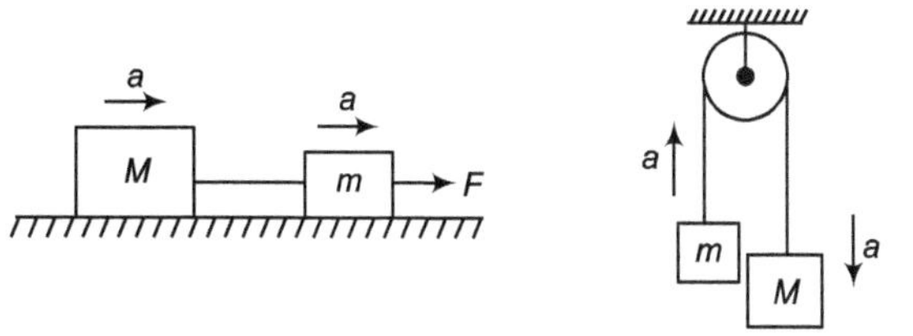

Fig. 5.3

3. If there is a friction between string and pulley, tension is different on two sides of the pulley, but if there is no friction between pulley and string, tension will be same on both sides of the pulley.

   Last two points can be understood in diagram as follows:

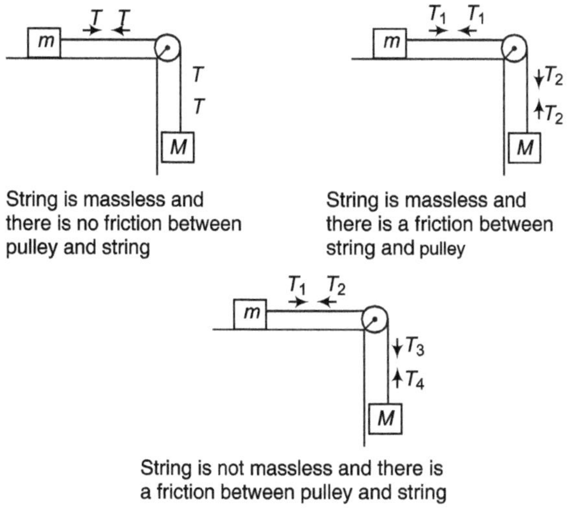

Fig. 5.4

Spring force ($F = kx$) has been discussed in detail in the chapter of work, energy and power.

5.2 Free Body Diagram

No system, natural or man made, consists of a single body alone or is complete in itself. A single body or a part of the system can, however, be isolated from the rest by appropriately accounting for its effect on the remaining system.

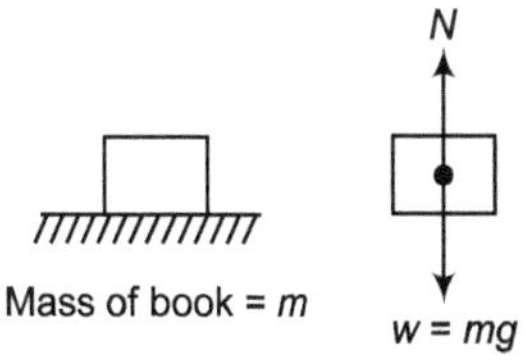

Fig. 5.5

A free body diagram (FBD) consists of a diagrammatic representation of a single body or a sub-system of bodies isolated from its surroundings showing all the forces acting on it.

Consider, for example, a book lying on a horizontal surface.

A free body diagram of the book alone would consist of its weight ($w = mg$), acting through the centre of gravity and the reaction (N) exerted on the book by the surface.

Example 5.1 *A cylinder of weight w is resting on a V-groove as shown in figure. Draw its free body diagram.*

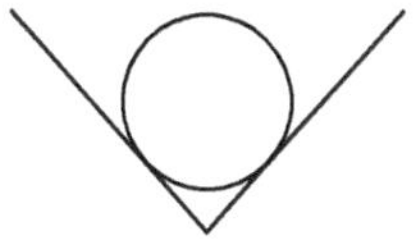
Fig. 5.6(a)

Sol. The free body diagram of the cylinder is as shown in Fig. 5.6(b)

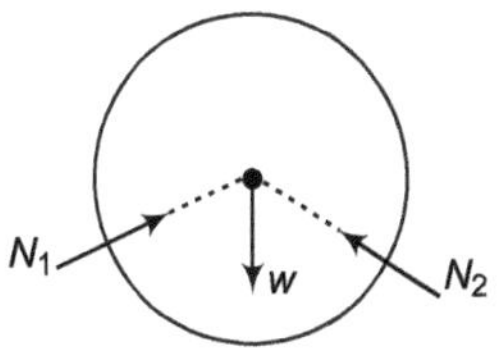

Fig. 5.6(b)

Here, w = weight of cylinder and N_1 and N_2 are the normal reactions between the cylinder and the two inclined walls.

Example 5.2 *Three blocks A, B and C are placed one over the other as shown in figure. Draw free body diagrams of all the three blocks.*

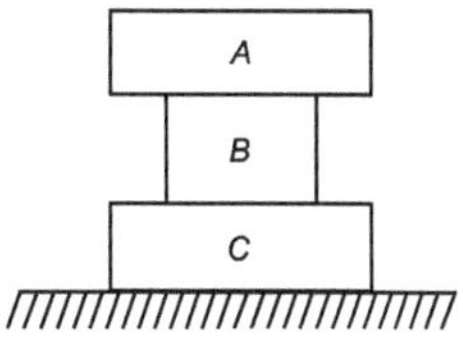

Fig. 5.7

Sol. Free body diagrams of *A*, *B* and *C* are shown below

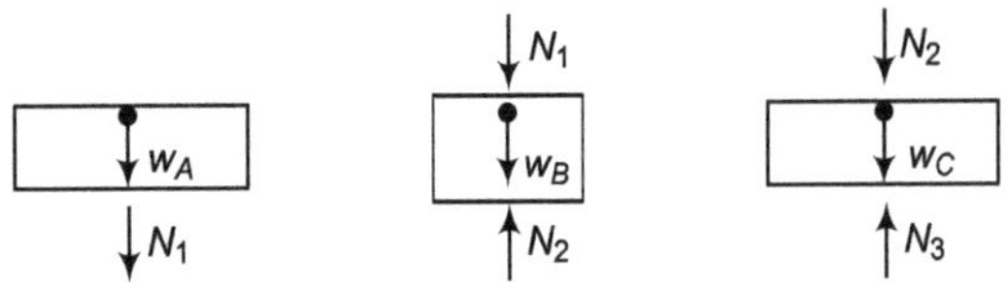

FBD of *A* FBD of *B* FBD of *C*

Fig. 5.8

Here, N_1 = normal reaction between *A* and *B*.
N_2 = normal reaction between *B* and *C*.
N_3 = normal reaction between *C* and ground.

Example 5.3 *A block of mass m is attached with two strings as shown in figure. Draw the free body diagram of the block.*

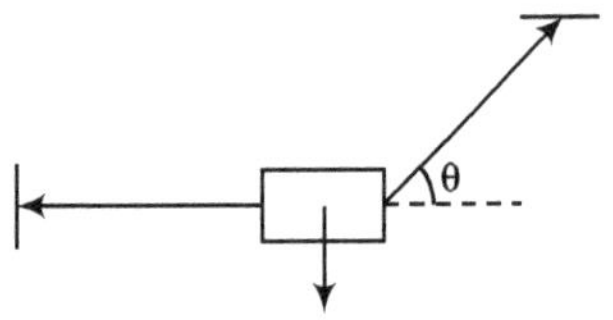

Fig. 5.9

Sol. The free body diagram of the block is as shown in Fig. 5.10.

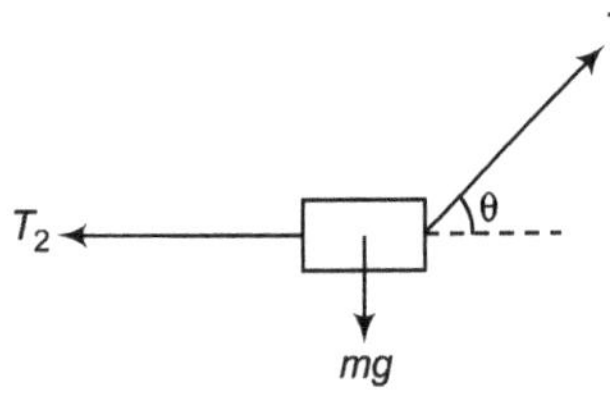

Fig. 5.10

5.3 Equilibrium

Forces which have zero linear resultant and zero turning effect will not cause any change in the motion of the object to which they are applied. Such forces (and the object) are said to be in equilibrium. For understanding the equilibrium of an object under two or more concurrent or coplanar forces let us first discuss the resolution of force and moment of a force about some point.

Resolution of a Force

When a force is replaced by an equivalent set of components, it is said to be resolved. One of the most useful ways in which to resolve a force is to choose only two components (although a force may be resolved in three or more components also) which are at right angles also. The magnitude of these components can be very easily found using trigonometry.

In Fig. 5.11,

$F_1 = F \cos\theta$ = component of **F** along *AC*

$F_2 = F \sin\theta$ = component of **F** perpendicular to *AC* or along *AB*

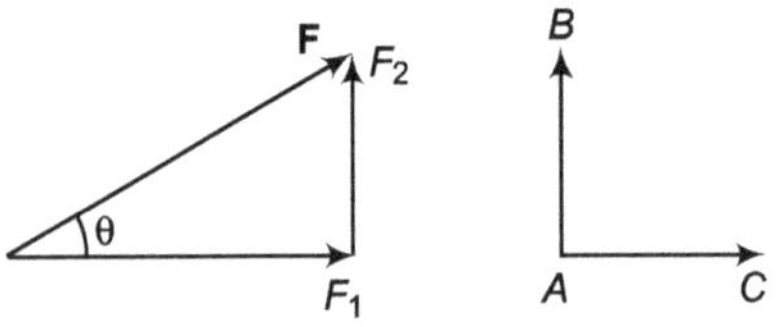

Fig. 5.11

Finding such components is referred to as resolving a force in a pair of perpendicular directions. Note that the component of a force in a direction perpendicular to itself is zero. For example, if a force of 10 N is applied on an object in horizontal direction, then its component along vertical is zero. Similarly, the component of a force in a direction parallel to the force is equal to the magnitude of the force. For example, component of the above force in the direction of force (horizontal) will be 10 N.

Example 5.4 *Resolve a weight of* 10 N *in two directions which are parallel and perpendicular to a slope inclined at* 30° *to the horizontal.*

Sol. Component perpendicular to the plane

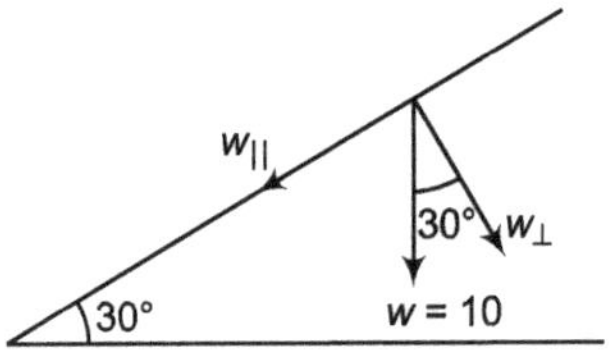

Fig. 5.12

$$w_\perp = w \cos 30^\circ$$
$$= (10)\frac{\sqrt{3}}{2} = 5\sqrt{3}\text{ N}$$

and component parallel to the plane

$$w_{||} = w \sin 30^\circ$$
$$= (10)\left(\frac{1}{2}\right) = 5\text{ N}$$

Example 5.5 *Resolve horizontally and vertically a force F* = 8 *N, which makes an angle of* 45° *with the horizontal.*

Sol. Horizontal component of **F** is

$$F_H = F \cos 45^\circ$$
$$= (8)\left(\frac{1}{\sqrt{2}}\right)$$
$$= 4\sqrt{2}\text{ N}$$

and vertical component of **F** is

$$F_V = F \sin 45^\circ$$
$$= (8)\left(\frac{1}{\sqrt{2}}\right)$$
$$= 4\sqrt{2}\text{ N}$$

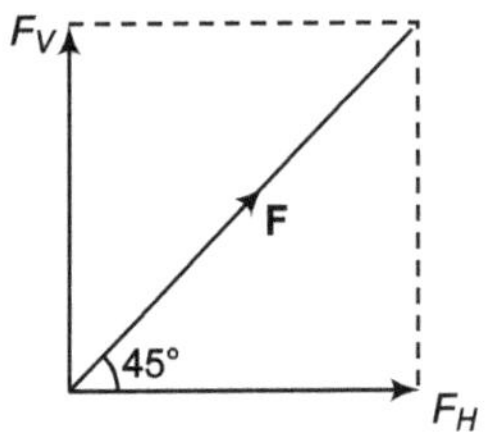

Fig. 5.13

Example 5.6 *A body is supported on a rough plane inclined at 30° to the horizontal by a string attached to the body and held at an angle of 30° to the plane. Draw a diagram showing the forces acting on the body and resolve each of these forces*
(a) horizontally and vertically,
(b) parallel and perpendicular to the plane.

Sol. The forces are :

The tension in the string T and the normal reaction with the plane N,
The weight of the body w and the friction f.

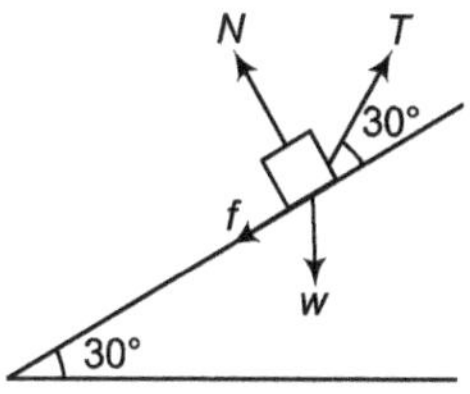

Fig. 5.14

(a) Resolving horizontally and vertically

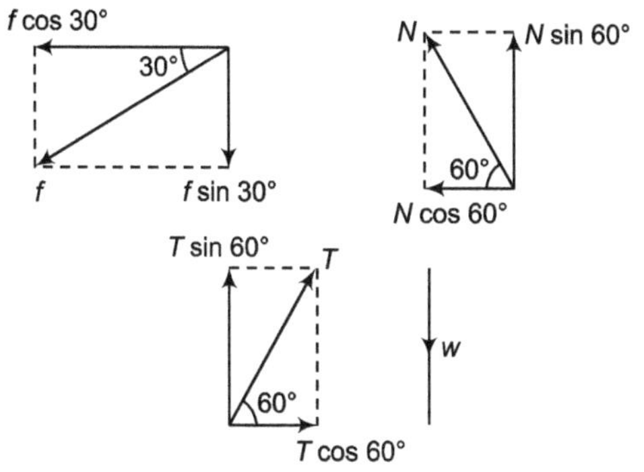

Fig. 5.15

(b) Resolving parallel and perpendicular to the plane

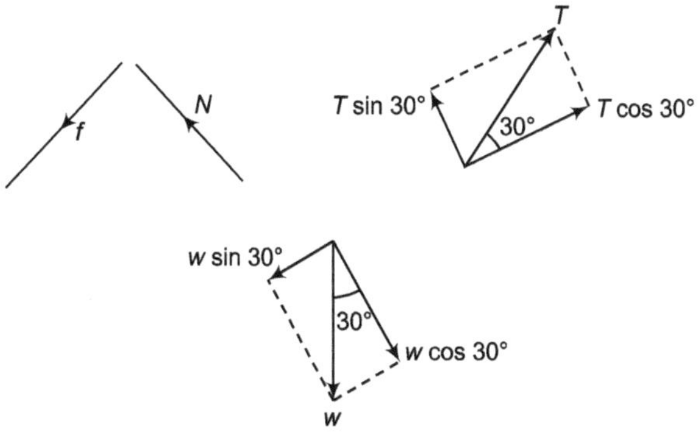

Fig. 5.16

Resolving horizontally and vertically in the senses OX and OY as shown, the components are

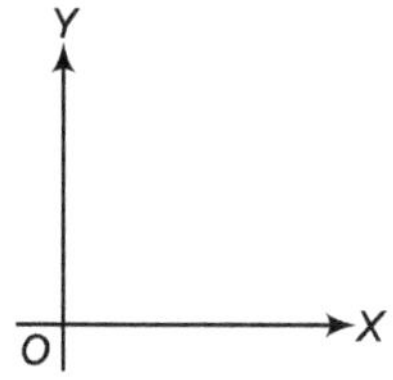

Fig. 5.17

Force	Components	
	Parallel to *OX* (horizontal)	Parallel to *OY* (vertical)
f	$-f \cos 30°$	$-f \sin 30°$
N	$-N \cos 60°$	$N \sin 60°$
T	$T \cos 60°$	$T \sin 60°$
w	0	$-w$

Resolving parallel and perpendicular to the plane in the senses OX' and OY' as shown, the components are

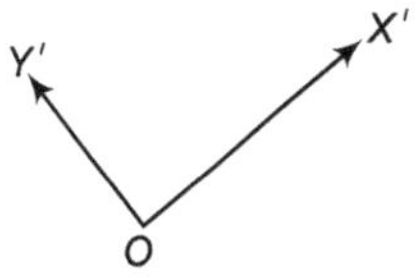

Fig. 5.18

Force	Components	
	Parallel to *OX′* (parallel to plane)	Parallel to *OY′* (perpendicular to plane)
f	$-f$	0
N	0	N
T	$T \cos 30°$	$T \sin 30°$
w	$-w \sin 30°$	$-w \cos 30°$

Moment of a Force

The general name given to any turning effect is **torque.** The magnitude of torque, also known as the moment of a force F is calculated by multiplying together the magnitude of the force and its perpendicular distance $r_\perp$ from the axis of rotation. This is denoted by C or τ (tau).

i.e. $$C = Fr_\perp$$

or $$\tau = Fr_\perp$$

Direction of Torque

The angular direction of a torque is the sense of the rotation it would cause.

Consider a lamina that is free to rotate in its own plane about an axis perpendicular to the lamina and passing through a point A on the lamina. In the diagram the moment about the axis of rotation of the force F_1 is $F_1 r_1$ anti-clockwise and the moment of the force F_2 is $F_2 r_2$ clockwise. A convenient way to differentiate between clockwise and anti-clockwise torques is to allocate a positive sign to one sense (usually, but not invariably, this is anti-clockwise) and negative sign to the other.

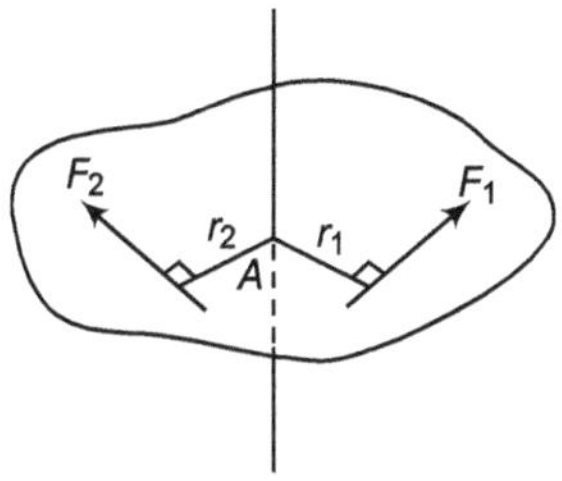

Fig. 5.19

With this convention, the moments of F_1 and F_2 are $+F_1 r_1$ and $-F_2 r_2$ (when using a sign convention in any problem it is advisable to specify the chosen positive sense).

Zero Moment

If the line of action of a force passes through the axis of rotation, its perpendicular distance from the axis is zero. Therefore, its moment about that axis is also zero.

- Later in the chapter of rotation, we will see that torque is a vector quantity.

Example 5.7 *ABCD is a square of side 2m and O is its centre. Forces act along the sides as shown in the diagram. Calculate the moment of each force about*

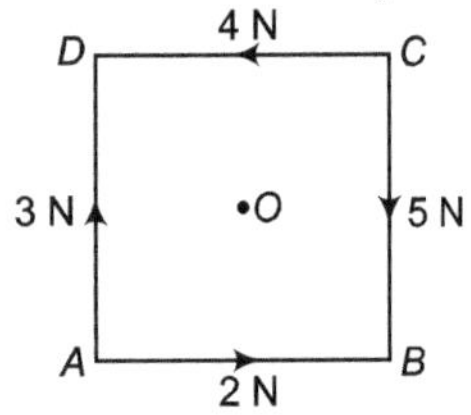

Fig. 5.20

(a) an axis through A and perpendicular to the plane of square.
(b) an axis through O and perpendicular to the plane of square.

Sol. Taking anti-clockwise moments as positive we have

(a)	Magnitude of force	2 N	5 N	4 N	3 N
	Perpendicular distance from *A*	0	2 m	2 m	0
	Moment about *A*	0	−10 N-m	+8 N-m	0

(b)	Magnitude of force	2 N	5 N	4 N	3 N
	Perpendicular distance from O	1 m	1 m	1 m	1 m
	Moment about O	+2 N-m	−5 N-m	+4 N-m	−3 N-m

Example 5.8 *Forces act as indicated on a rod AB which is pivoted at A. Find the anti-clockwise moment of each force about the pivot.*

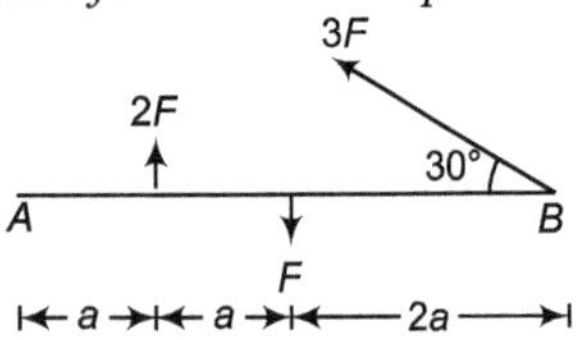

Fig. 5.21

Sol.

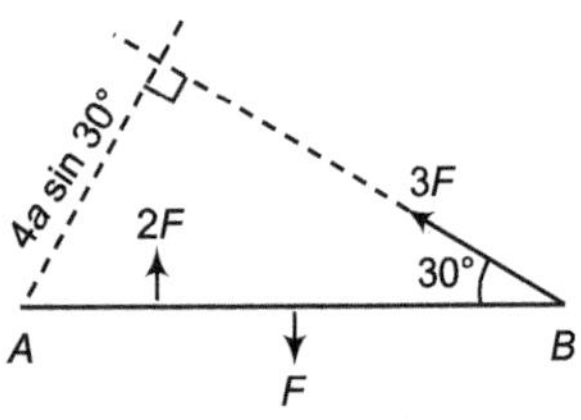

Fig. 5.22

Magnitude of force	2*F*	*F*	3*F*
Perpendicular distance from *A*	*a*	2*a*	4*a* sin 30° = 2*a*
Anti-clockwise moment about *A*	+ 2 *F a*	− 2 *F a*	+ 6 *F a*

Coplanar Forces in Equilibrium

When an object is in equilibrium under the action of a set of two or more coplanar forces, each of three factors which comprise the possible movement of the object must be zero, i.e. the object has

(i) no linear movement along any two mutually perpendicular directions *OX* and *OY*.

(ii) no rotation about any axis.

The set of forces must, therefore be, such that

(a) the algebraic sum of the components parallel to *OX* is zero or $\Sigma F_x = 0$

(b) the algebraic sum of the components parallel to *OY* is zero or $\Sigma F_y = 0$

(c) the resultant moment about any specified axis is zero or $\Sigma \tau_{\text{any axis}} = 0$

Thus, for the equilibrium of a set of two or more coplanar forces

$$\Sigma F_x = 0$$
$$\Sigma F_y = 0$$
and $$\Sigma \tau_{\text{any axis}} = 0$$

Using the above three conditions, we get only three sets of equations. So, in a problem number of unknown should not be more than three.

Example 5.9 *A rod AB rests with the end A on rough horizontal ground and the end B against a smooth vertical wall. The rod is uniform and of weight w. If the rod is in equilibrium in the position shown in figure. Find*
(a) frictional force at A,
(b) normal reaction at A,
(c) normal reaction at B.

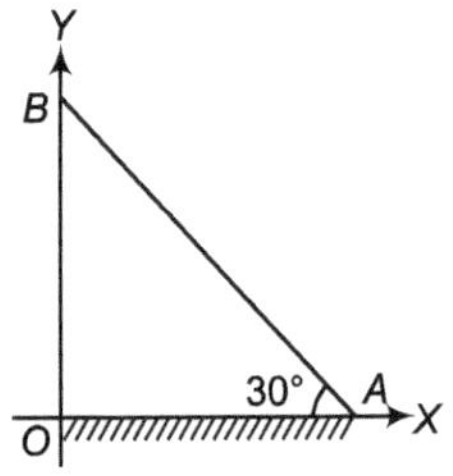

Fig. 5.23

Sol. Let length of the rod be 2*l*. Using the three conditions of equilibrium. Anti-clockwise moment is taken as positive.

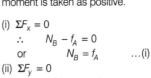

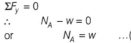

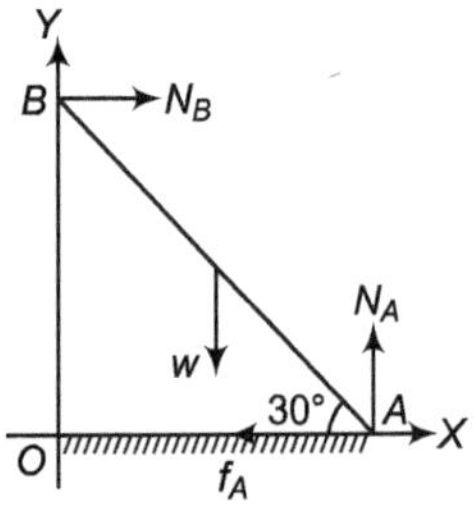

Fig. 5.24

(i) $\Sigma F_x = 0$

$\therefore \quad N_B - f_A = 0$

or $\quad N_B = f_A \quad$ ...(i)

(ii) $\Sigma F_y = 0$

$\therefore \quad N_A - w = 0$

or $\quad N_A = w \quad$ ...(ii)

(iii) $\Sigma\tau_0 = 0$

$\therefore \quad N_A (2l \cos 30°) - N_B(2l \sin 30°) - w(l \cos 30°) = 0$

or $\quad \sqrt{3}N_A - N_B - \frac{\sqrt{3}}{2} w = 0 \quad \ldots$(iii)

Solving these three equations, we get

(a) $f_A = \frac{\sqrt{3}}{2} w$

(b) $N_A = w$

(c) $N_B = \frac{\sqrt{3}}{2} w$

Equilibrium of Concurrent Coplanar Forces

If an object is in equilibrium under two or more concurrent coplanar forces, the algebraic sum of the components of forces in any two mutually perpendicular directions OX and OY should be zero, i.e. the set of forces must be such that

(a) the algebraic sum of the components parallel to OX is zero, i.e. $\Sigma F_x = 0$.

(b) the algebraic sum of the components parallel to OY is zero, i.e. $\Sigma F_y = 0$.

Thus, for the equilibrium of two or more concurrent coplanar forces

$$\Sigma F_x = 0$$
$$\Sigma F_y = 0$$

The third condition of zero moment about any specified axis is automatically satisfied, if the moment is taken about the point of intersection of the forces. So, here we get only two equations. Thus, number of unknowns in any problem should not be more than two.

Example 5.10 *An object is in equilibrium under four concurrent forces in the directions shown in figure. Find the magnitude of* $\mathbf{F}_1$ *and* $\mathbf{F}_2$.

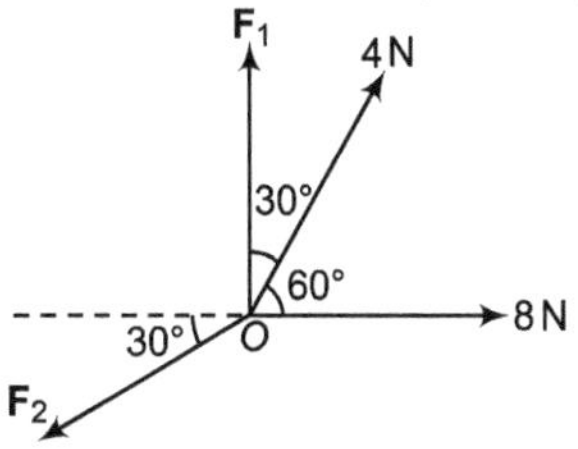

Fig. 5.25

Sol.

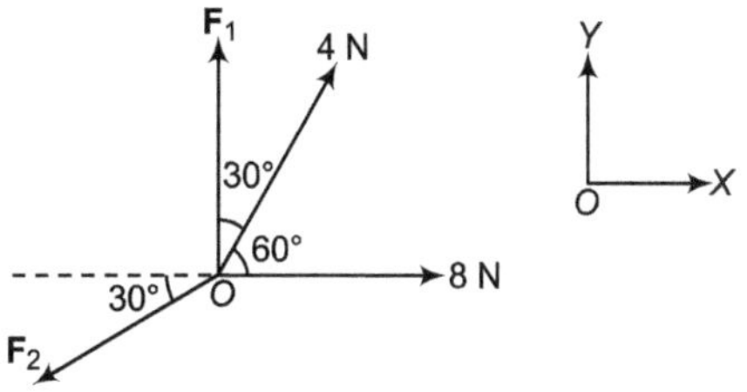

Fig. 5.26

The object is in equilibrium. Hence,

(i) $\Sigma F_x = 0$

$\therefore \quad 8 + 4\cos 60° - F_2 \cos 30° = 0$

or $\quad 8 + 2 - F_2 \frac{\sqrt{3}}{2} = 0$ or $F_2 = \frac{20}{\sqrt{3}}$ N

(ii) $\Sigma F_y = 0 \quad \therefore \quad F_1 + 4\sin 60° - F_2 \sin 30° = 0$

or $\quad F_1 + \frac{4\sqrt{3}}{2} - \frac{F_2}{2} = 0$

or $\quad F_1 = \frac{F_2}{2} - 2\sqrt{3}$

$\quad = \frac{10}{\sqrt{3}} - 2\sqrt{3}$

or $\quad F_1 = \frac{4}{\sqrt{3}}$ N

Lami's Theorem

If an object O is in equilibrium under three concurrent forces $\mathbf{F}_1$, $\mathbf{F}_2$ and $\mathbf{F}_3$ as shown in figure. Then,

$$\frac{F_1}{\sin\alpha} = \frac{F_2}{\sin\beta} = \frac{F_3}{\sin\gamma}$$

This property of three concurrent forces in equilibrium is known as Lami's theorem and is very useful method of solving problems related to three concurrent forces in equilibrium.

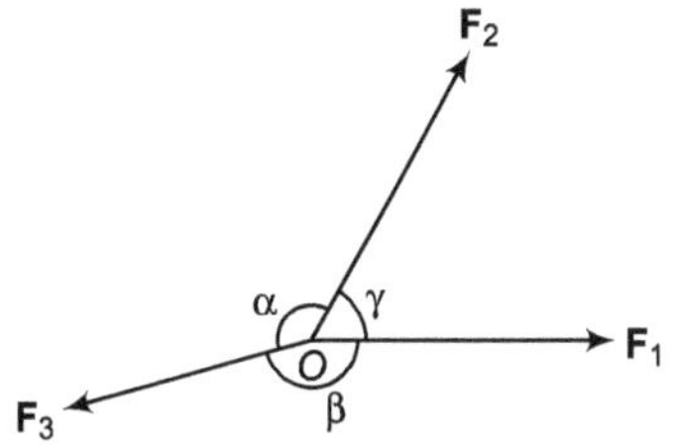

Fig. 5.27

Example 5.11 *One end of a string* 0.5 m *long is fixed to a point A and the other end is fastened to a small object of weight* 8 N. *The object is pulled aside by a horizontal force F, until it is* 0.3 m *from the vertical through A. Find the magnitudes of the tension T in the string and the force F.*

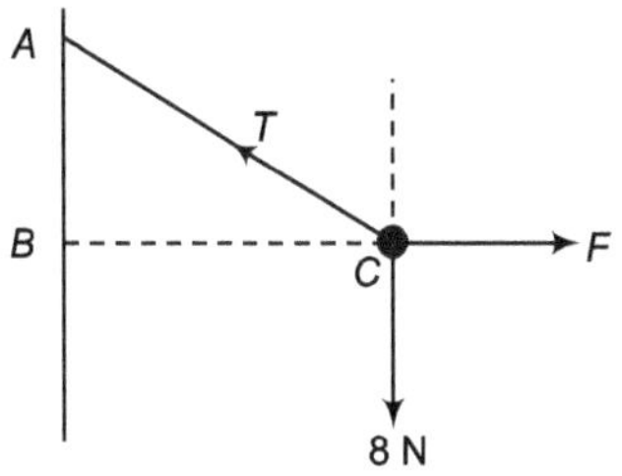

Fig. 5.28

Sol. $AC = 0.5$ m, $BC = 0.3$ m

$\therefore \quad AB = 0.4$ m

and if $\angle BAC = \theta$

Then, $\quad \cos\theta = \frac{AB}{AC} = \frac{0.4}{0.5} = \frac{4}{5}$

and $$\sin\theta = \frac{BC}{AC} = \frac{0.3}{0.5} = \frac{3}{5}$$

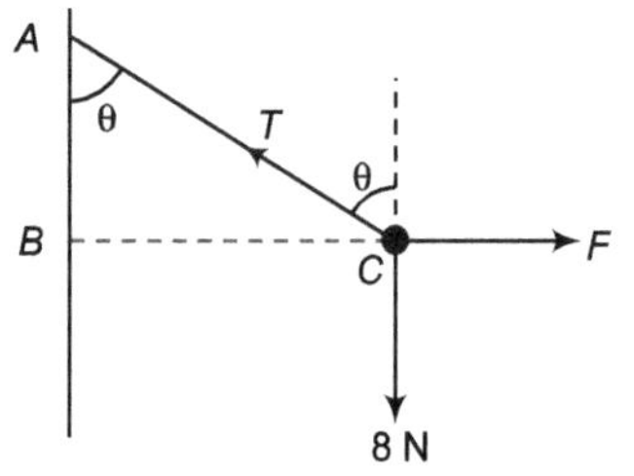

Fig. 5.29

Here, the object is in equilibrium under three concurrent forces. So, we can apply Lami's theorem.

or $$\frac{F}{\sin(180° - \theta)} = \frac{8}{\sin(90° + \theta)} = \frac{T}{\sin 90°}$$

or $$\frac{F}{\sin\theta} = \frac{8}{\cos\theta} = T$$

$\therefore$ $$T = \frac{8}{\cos\theta}$$

$$= \frac{8}{4/5} = 10\text{ N}$$

and $$F = \frac{8\sin\theta}{\cos\theta} = \frac{(8)(3/5)}{(4/5)} = 6\text{ N}$$

Example 5.12 *A block of mass m is at rest on a rough wedge as shown in figure. What is the force exerted by the wedge on the block?*

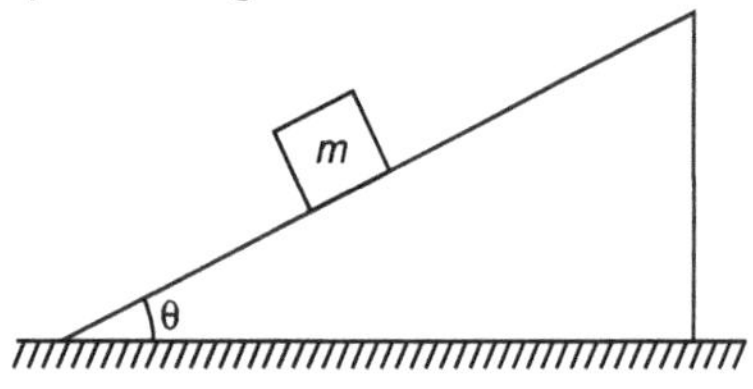

Fig. 5.30

Sol. Since, the block is permanently at rest, it is in equilibrium. Net force on it should be zero. In this case only two forces are acting on the block.

(1) Weight = mg (downwards)

(2) Contact force (resultant of normal reaction and friction force) applied by the wedge on the block.

For the block to be in equilibrium these two forces should be equal and opposite.

Therefore, force exerted by the wedge on the block is mg (upwards).

Note Points

- From Newton's third law of motion-force exerted by the block on the wedge is also mg but downwards.
- The result can also be obtained in a different manner. The normal force on the block is $N = mg\cos\theta$ and the friction force on the block is $f = mg\sin\theta$ (not $\mu mg\cos\theta$)

These two forces are mutually perpendicular.

$\therefore$ Net contact force would be $\sqrt{N^2 + f^2}$

or $\sqrt{(mg\cos\theta)^2 + (mg\sin\theta)^2}$ which is equal to mg.

5.4 Newton's Laws of Motion

It is interesting to read Newton's original version of the laws of motion.

Law I Every body continues in its state of rest or in uniform motion in a straight line unless it is compelled to change that state by forces impressed upon it.

Law II The change of motion is proportional to the magnitude of force impressed and is made in the direction of the straight line in which that force is impressed.

Law III To every action there is always an equal and opposite reaction or the mutual actions of two bodies upon each other are always directed to contrary parts.

The modern versions of these laws are :

1. A body continues in its initial state of rest or motion with uniform velocity unless acted on by an unbalanced external force.
2. The acceleration of a body is inversely proportional to its mass and directly proportional to the resultant external force acting on it, i.e.
$$\Sigma\mathbf{F} = \mathbf{F}_{net} = m\mathbf{a} \quad \text{or} \quad \mathbf{a} = \frac{\mathbf{F}_{net}}{m}$$
3. Forces always occur in pairs. If body A exerts a force on body B, an equal but opposite force is exerted by body B on body A.

Working with Newton's First and Second Laws

Normally, any problem relating to Newton's laws is solved in following four steps:

1. First of all we decide the system on which the laws of motion are to be applied. The system may be a single particle, a block or a combination of two or more blocks, two blocks connected by a string, etc. The only restriction is that all parts of the system should have the same acceleration.
2. Once the system is decided, we make the list of all the forces acting on the system. Any force applied by the system on other bodies is not included in the list of the forces.
3. Then, we make a free body diagram of the system and indicate the magnitude and directions of all the forces listed in step 2 in this diagram.
4. In the last step, we choose any two mutually perpendicular axes say x and y in the plane of the forces in case of coplanar forces. Choose the x-axis along the direction in which the system is known to have or is likely to have the acceleration. A direction perpendicular to it may be chosen as the y-axis. If the system is in equilibrium, any mutually perpendicular directions may be chosen. Write the components of all the forces along the x-axis and equate their sum to the product of the mass of the system and its acceleration,

i.e. $\Sigma F_x = ma$ ...(i)

This gives us one equation. Now, we write the components of the forces along the y-axis and equate the sum to zero. This gives us another equation, i.e.

$\Sigma F_y = 0$ ...(ii)

Note Points

- If the system is in equilibrium, we will write the two equations as

$$\Sigma F_x = 0 \text{ and } \Sigma F_y = 0$$

- If the forces are collinear, the second equation, i.e. $\Sigma F_y = 0$ is not needed.

Example 5.13 *Two blocks of mass 4 kg and 2 kg are placed side by side on a smooth horizontal surface as shown in the figure. A horizontal force of 20 N is applied on 4 kg block. Find*

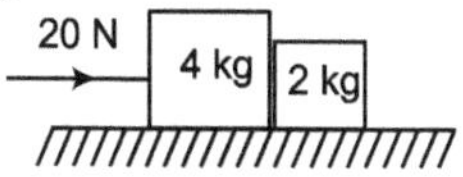

Fig. 5.31

(a) the acceleration of each block,

(b) the normal reaction between two blocks.

Sol. (a) Since, both the blocks will move with same acceleration (say a) in horizontal direction.

Let us take both the blocks as a system. Net external force on the system is 20 N in horizontal direction.

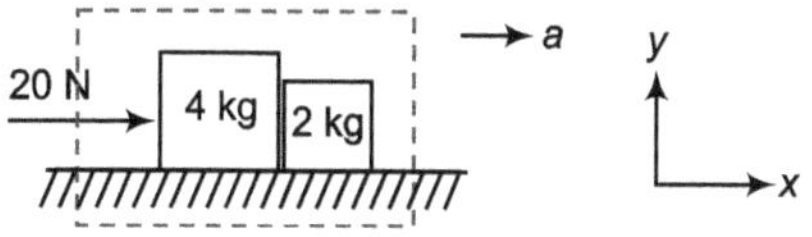

Fig. 5.32

Using $\Sigma F_x = ma_x$,

$$20 = (4+2)a = 6a \quad \text{or} \quad a = \frac{10}{3} \text{ m/s}^2$$

(b) The free body diagram of both the blocks are as shown in Fig. 5.33.

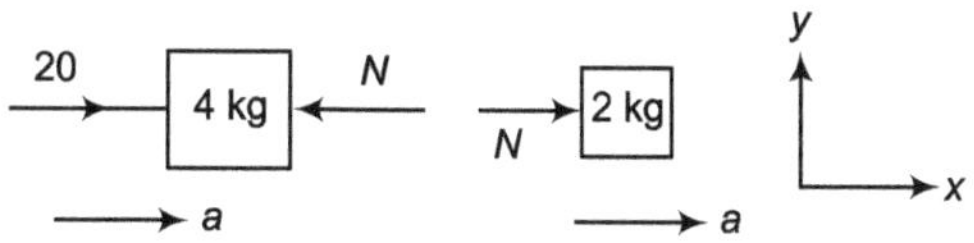

Fig. 5.33

Using $\Sigma F_x = ma_x$,

For 4 kg block, $20 - N = 4a = 4 \times \frac{10}{3}$

$$N = 20 - \frac{40}{3} = \frac{20}{3} \text{ N}$$

This can also be solved as under.

For 2 kg block, $N = 2a = 2 \times \frac{10}{3} = \frac{20}{3}$ N

Here, N is the normal reaction between the two blocks.

- In free body diagram of the blocks, we have not shown the forces acting on the blocks in vertical direction, because normal reaction between the blocks and acceleration of the system can be obtained without using $\Sigma F_y = 0$.

Extra Knowledge Points

- If a pulley is massless, net force on it is zero even if it is accelerated. For example in the following figure

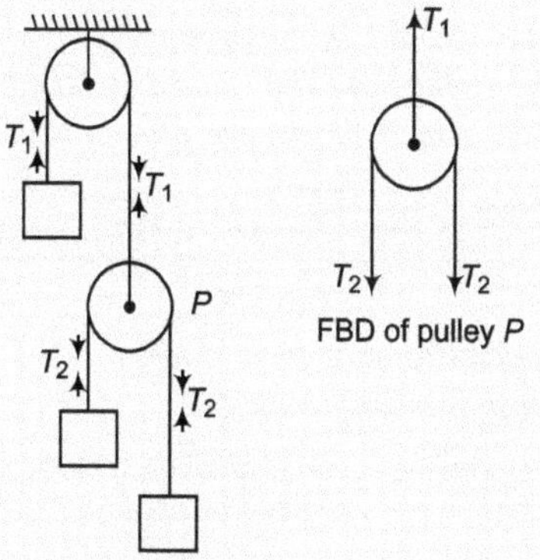

$T_1 = 2T_2$ whether the pulley is accelerated on not provided the pulley is massless. This is because wF mass x acceleration and wF will be zero, if pulley is massless.

I have found students often confused over the resolution of forces. If the body is in equilibrium, no matter in which direction you resolve the forces. Net force should be zero in any direction. If the body is accelerated, resolve in the direction of acceleration and its perpendicular. In perpendicular direction, net force is zero and in the direction of acceleration it is *mwF*.

Example 5.14 *Three blocks of mass 3 kg, 2 kg and 1 kg are placed side by side on a smooth surface as shown in figure. A horizontal force of 12 N is applied on 3 kg block. Find the net force on 2 kg block.*

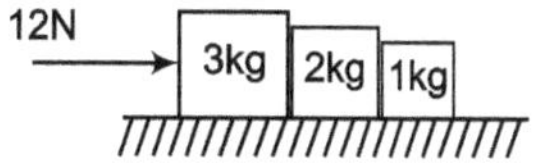

Fig. 5.34

Sol. Since, all the blocks will move with same acceleration (say a) in horizontal direction. Let us take all the blocks as a system.

Net external force on the system is 12 N in horizontal direction.

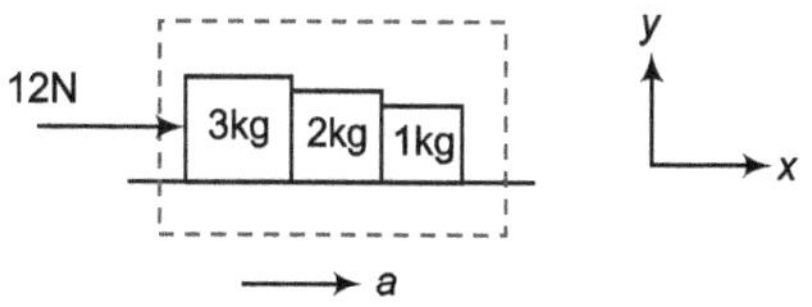

Fig. 5.35

Using $\Sigma F_x = ma_x$, we get

$$12 = (3+2+1)a = 6a \quad \text{or} \quad a = \frac{12}{6} = 2 \text{ m/s}^2$$

Now, let F be the net force on 2 kg block in x-direction, then using $\Sigma F_x = ma_x$ for 2 kg block, we get

$$F = (2)(2) = 4 \text{ N}$$

- Here, net force F on 2 kg block is the resultant of N_1 and N_2 $(N_1 > N_2)$
where, N_1 = normal reaction between 3 kg and 2 kg block
and N_2 = normal reaction between 2 kg and 1 kg block
Thus, $F = N_1 - N_2$

Example 5.15 *In the arrangement shown in figure. The strings are light and inextensible. The surface over which blocks are placed is smooth. Find*

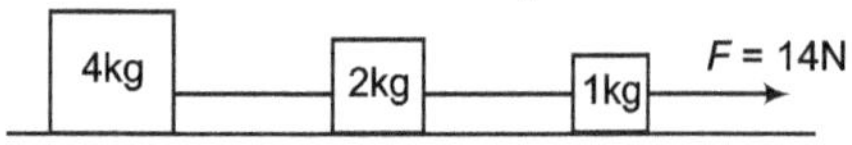

Fig. 5.36

(a) the acceleration of each block,
(b) the tension in each string.

Sol. (a) Let a be the acceleration of each block and T_1 and T_2 be the tensions, in the two strings as shown in figure.

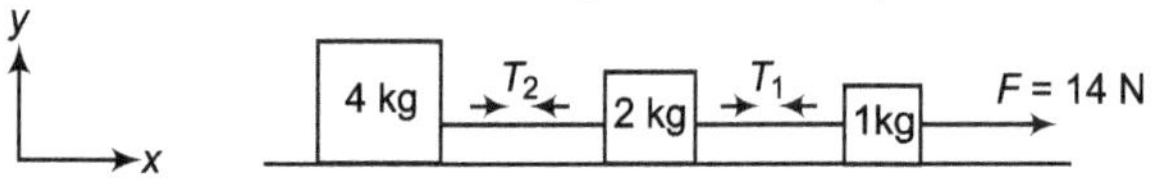

Fig. 5.37

Taking the three blocks and the two strings as the system.

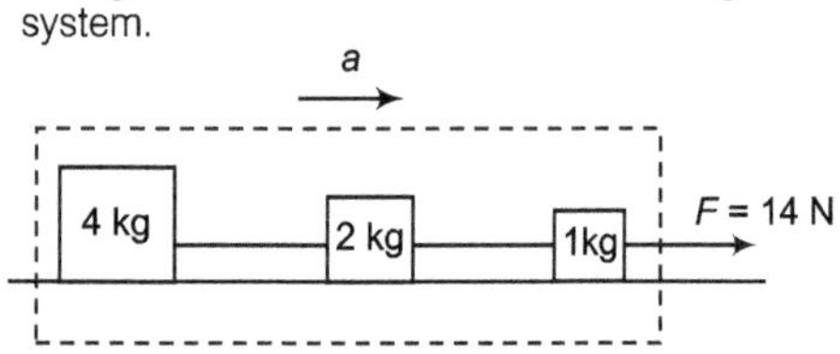

Fig. 5.38

Using $\Sigma F_x = ma_x$

or $14 = (4 + 2 + 1)a$

or $a = \frac{14}{7} = 2 \text{ m/s}^2$

(b) Free body diagram (showing the forces in x-direction only) of 4 kg block and 1 kg block are shown in figure.

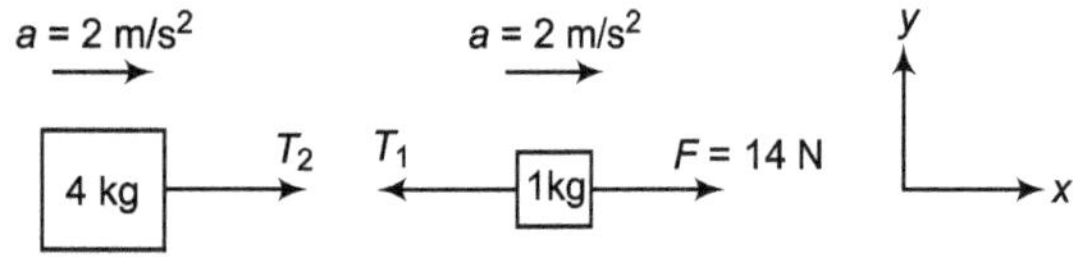

Fig. 5.39

Using $\Sigma F_x = ma_x$,

For 1 kg block, $F - T_1 = (1)(a)$

or $14 - T_1 = (1)(2) = (2)$

$\therefore$ $T_1 = 14 - 2 = 12 \text{ N}$

For 4 kg block, $T_2 = (4)(a)$

$\therefore$ $T_2 = (4)(2) = 8 \text{ N}$

Example 5.16 *Two blocks of mass 4 kg and 2 kg are attached by an inextensible light string as shown in figure. Both the blocks are pulled vertically upwards by a force $F = 120$ N. Find*
(a) the acceleration of the blocks,
(b) the tension in the string. (Take $g = 10 \text{ m/s}^2$)

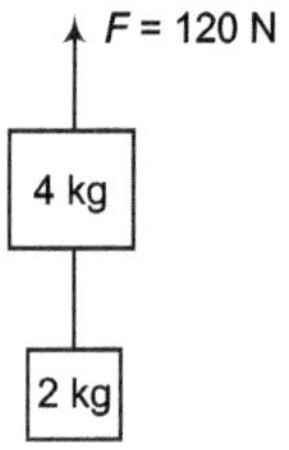

Fig. 5.40

Sol. (a) Let a be the acceleration of the blocks and T the tension in the string as shown in figure.

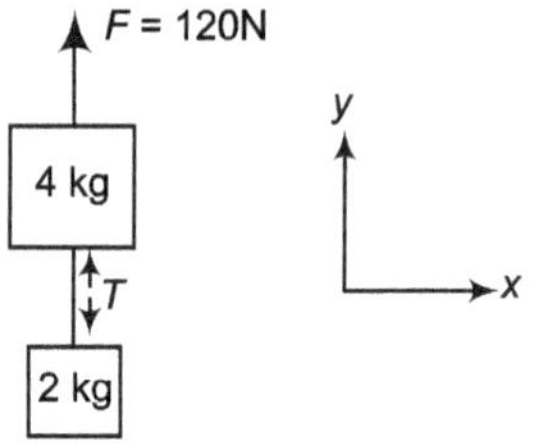

Fig. 5.41

Taking the two blocks and the string as the system.

Using $\Sigma F_y = ma_y$, we get

$F - 4g - 2g = (4 + 2)a$ or $120 - 40 - 20 = 6a$

or $60 = 6a \Rightarrow \therefore a = 10 \text{ m/s}^2$

(b) Free body diagram of 2 kg block is as shown in Fig. 5.42.

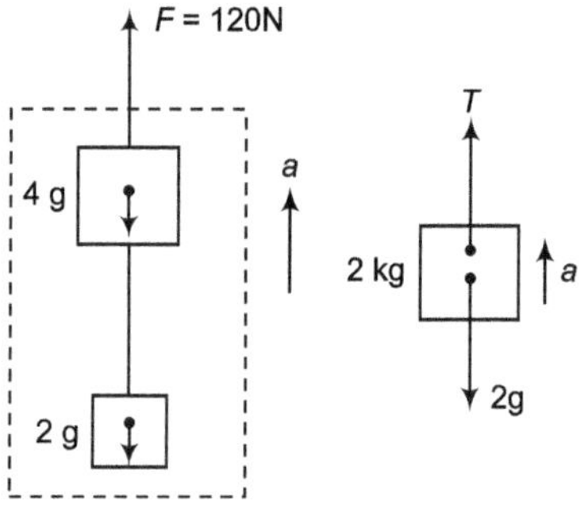

Fig. 5.42

Using $\Sigma F_y = ma_y$, we get

$T - 2g = 2a$ or $T - 20 = (2)(10)$

$\therefore$ $T = 40 \text{ N}$

- If the string is having some mass, tension in it is different at different points. Under such condition tension on the string at some point is calculated as under:

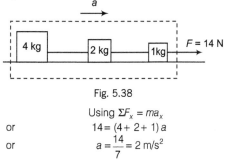

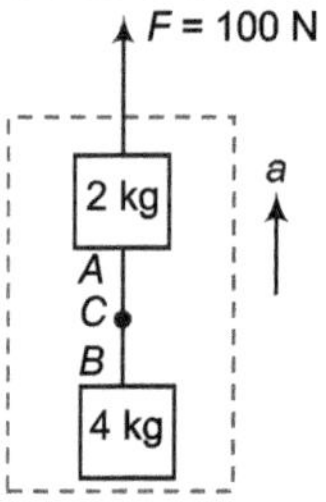

Fig. 5.43

In the above figure, the length of the string connecting the two blocks is 2 m and mass is 2 kg. Tension at A, B and C (centre point) can be calculated by considering the motion of system below A, B and C. For example

$$a = \frac{F - \text{weight of 2 kg} - \text{weight of 4 kg} - \text{weight of string}}{\text{mass of 2 kg} + \text{mass of 4 kg} + \text{mass of string}}$$

$$= \frac{100 - 20 - 40 - 20}{2 + 4 + 2} \qquad (g = 10 \text{ m/s}^2)$$

$$= \frac{20}{8} = 2.5 \text{ m/s}^2$$

Refer Fig. 5.44 (a)

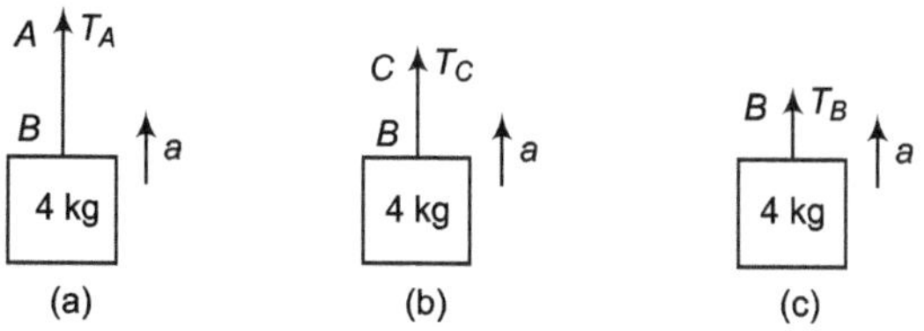

Fig. 5.44

$$T_A - m_{AB}\, g - 40 = (m_{AB} + 4)a$$

or $\quad T_A - 20 - 40 = (2 + 4)(2.5)$

or $\quad T_A = 75 \text{ N}$

Refer Fig. 5.44 (b)

$$T_C - m_{BC}\, g - 40 = (m_{BC} + 4)a$$

or $\quad T_C - 10 - 40 = (1 + 4)(2.5)$

or $\quad T_C = 62.5 \text{ N}$

Refer Fig. 5.44 (c) $\quad T_B - 40 = 4a$

or $\quad T_B = 40 + 4 \times 2.5$

or $\quad T_B = 50 \text{ N}$

5.5 Pulleys

As an author, I personally feel that problems based on pulleys become very simple using **pulling force** method. Now, let us see what is this pulling force method with the help of an example.

Suppose, two unequal masses m and $2m$ are attached to the ends of a light inextensible string which passes over a smooth massless pulley. We have to find the acceleration of the system. We can assume that the mass $2m$ is pulled downwards by a force equal to its weight, i.e. $2mg$. Similarly, the mass m is being pulled by a force of mg downwards. Therefore, net pulling force on the system is $2mg - mg = mg$ and total mass being pulled is $2m + m = 3m$.

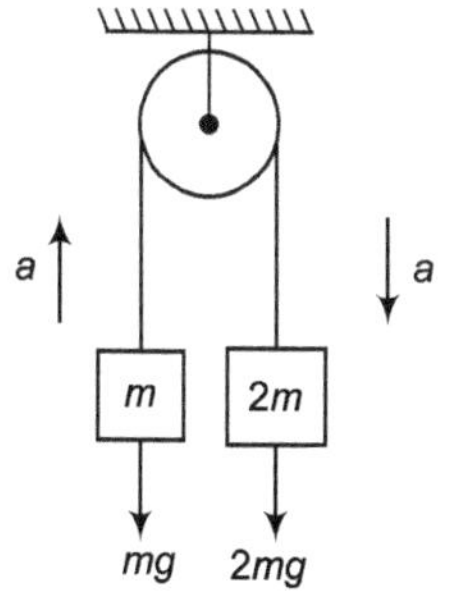

Fig. 5.45

$\therefore$ Acceleration of the system is

$$a = \frac{\text{Net pulling force}}{\text{Total mass to be pulled}}$$

$$= \frac{mg}{3m} = \frac{g}{3}$$

Let us now, take few examples based on pulling force method.

Example 5.17 *In the below figure, mass of A, B and C are 1 kg, 3 kg and 2 kg, respectively. Find*

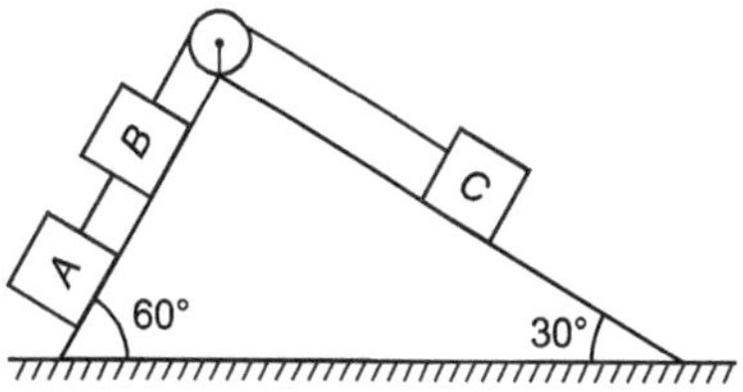

Fig. 5.46

(a) the acceleration of the system,
(b) tensions in the string.
Neglect friction. $(g = 10 \ m/s^2)$

Sol. (a) In this case, net pulling force

$$= m_A g \sin 60° + m_B g \sin 60° - m_C g \sin 30°$$

$$= (1)(10)\frac{\sqrt{3}}{2} + (3)(10)\left(\frac{\sqrt{3}}{2}\right) - (2)(10)\left(\frac{1}{2}\right)$$

$$= 21.17 \text{ N}$$

Total mass being pulled $= 1 + 3 + 2 = 6$ kg

$\therefore$ Acceleration of the system, $a = \dfrac{21.17}{6} = 3.53 \text{ m/s}^2$

(b) For the tension in the string between A and B,

FBD of A

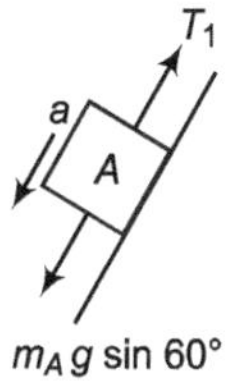

Fig. 5.47

$$m_A\, g \sin 60° - T_1 = (m_A)(a)$$

$$\therefore \quad T_1 = m_A\, g \sin 60° - m_A\, a$$

$$= m_A\, (g \sin 60° - a)$$

$$\therefore \quad T_1 = (1)\left(10 \times \frac{\sqrt{3}}{2} - 4.1\right)$$

$$= 5.13 \text{ N}$$

For the tension in the string between B and C,

FBD of C

$$T_2 - m_C\, g \sin 30° = m_C\, a$$

$$\therefore \quad T_2 = m_C (a + g \sin 30°)$$

$$\therefore \quad T_2 = 2\left[3.53 + 10\left(\frac{1}{2}\right)\right]$$

$$= 17.01 \text{ N}$$

5.6 Constraint Equations

These equations basically establish the relation between accelerations (or velocities) of different masses attached by string(s). Usually, it is observed that the number of constraint equations are as many as the number of strings in the system under consideration. From the following few examples, we can better understand the method.

Example 5.18 *Using constraint method find the relation between accelerations of 1 and 2.*

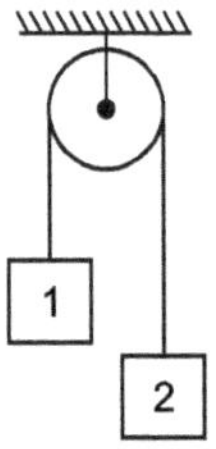

Fig. 5.49

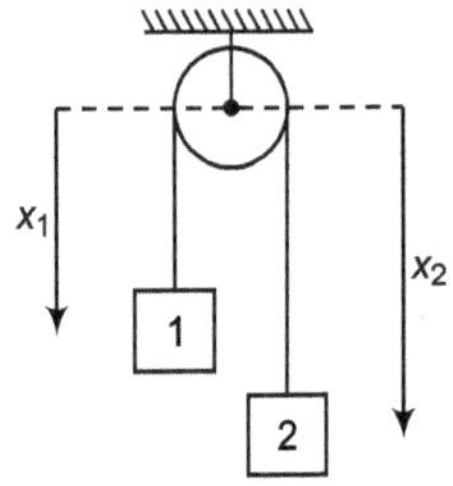

Fig.

Sol. At any instant of time, let x_1 and x_2 be the displacements of 1 and 2 from a fixed line (shown dotted).

Then, $x_1 + x_2 =$ constant.

or $x_1 + x_2 = l$ (length of string)

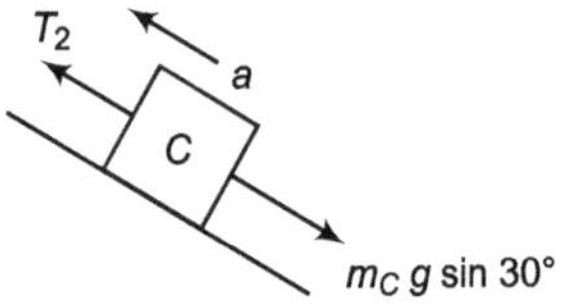

Fig. 5.48

Differentiating with respect to time, we have

$$v_1 + v_2 = 0$$

or $$v_1 = -v_2$$

Again differentiating with respect to time, we get

$$a_1 + a_2 = 0 \quad \text{or} \quad a_1 = -a_2$$

This is the required relation between a_1 and a_2, i.e. accelerations of 1 and 2 are equal but in opposite directions.

Example 5.19 *Find the constraint relation between a_1, a_2 and a_3.*

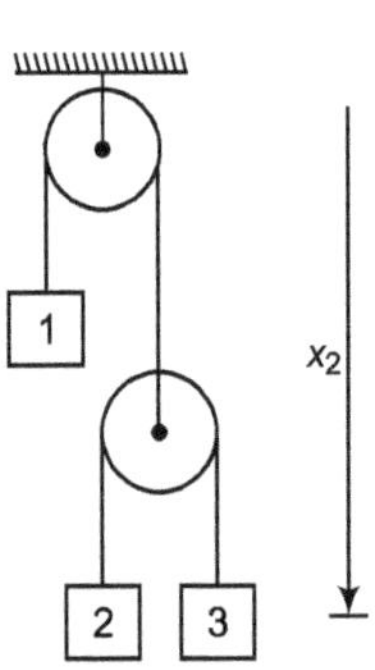

Fig. 5.51

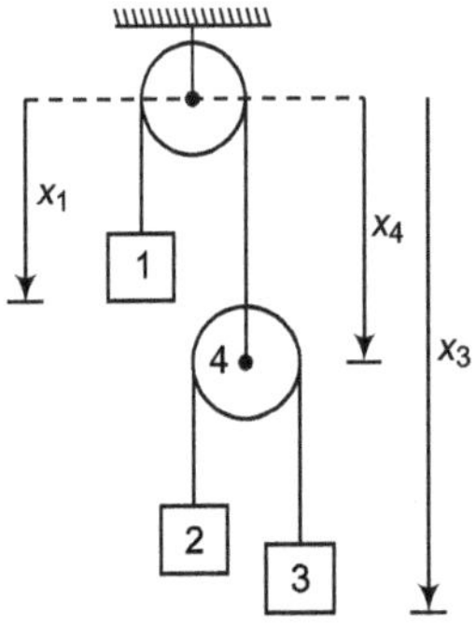

Fig. 5.52

Sol. Points 1, 2, 3 and 4 are movable. Let their displacements from a fixed line be x_1, x_2, x_3 and x_4.

We have

$x_1 + x_4 = l_1$ (length of first string)...(i)

and $(x_2 - x_4) + (x_3 - x_4) = l_2$ (length of second string)

or $x_2 + x_3 - 2x_4 = l_2$ ...(ii)

On double, differentiating with respect to time, we get

$a_1 + a_4 = 0$ ...(iii)

and $a_2 + a_3 - 2a_4 = 0$ ...(iv)

But since, $a_4 = -a_1$ [From Eq. (iii)]

We have $a_2 + a_3 + 2a_1 = 0$

This is the required constraint relation between a_1, a_2 and a_3.

Example 5.20 *Using constraint equations find the relation between a_1 and a_2.*

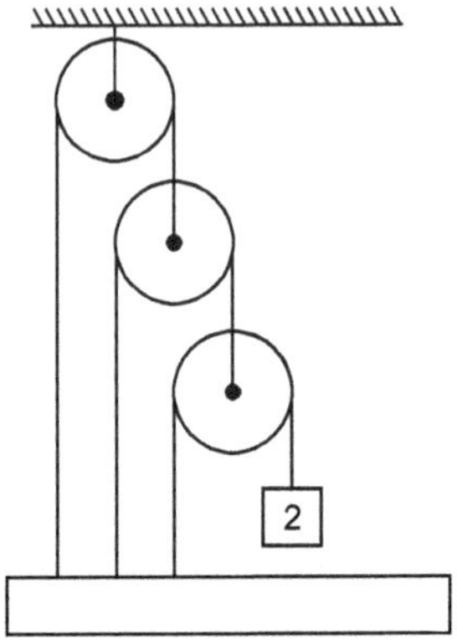

Fig. 5.53

Sol. Points 1, 2, 3 and 4 are movable. Let their displacements from a fixed line be x_1, x_2, x_3 and x_4.

$$x_1 + x_3 = l_1$$

$$(x_1 - x_3) + (x_4 - x_3) = l_2$$

$$(x_1 - x_4) + (x_2 - x_4) = l_3$$

On double differentiating with respect to time, we will get following three constraint relations,

$a_1 + a_3 = 0$ ...(i)

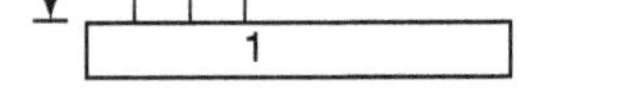

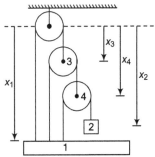

Fig. 5.54

$a_1 + a_4 - 2a_3 = 0$ ...(ii)

$a_1 + a_2 - 2a_4 = 0$ ...(iii)

Solving Eqs. (i), (ii) and (iii), we get

$$a_2 = -7a_1$$

which is the desired relation between a_1 and a_2.

Example 5.21 *At certain moment of time, velocities of 1 and 2 both are 1 m/s upwards. Find the velocity of 3 at that moment.*

Sol. In example 5.19, we have found

$$a_2 + a_3 + 2a_1 = 0$$

Similarly, we can find

$$v_2 + v_3 + 2v_1 = 0$$

Taking, upward direction as positive we are given

$$v_1 = v_2 = 1 \text{ m/s}$$

$$\therefore \quad v_3 = -3 \text{ m/s}$$

i.e. velocity of block 3 is 3 m/s (downwards).

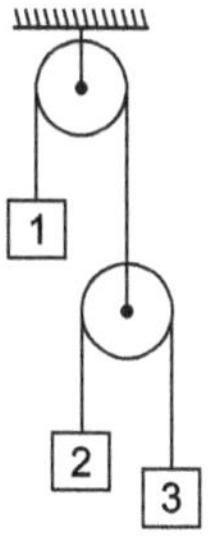

Fig. 5.55

5.7 Pseudo Force

Before studying pseudo force let us first discuss frame of reference. A system of coordinate axes which defines the position of a particle or an event in two or three dimensional space is called a frame of reference. The simplest frame of reference is, of course, the familiar cartesian system of coordinates, in which the position of the particle is specified by its three coordinates *x*, *y* and *z*. Frame of references are of two types

Inertial Frame of Reference

A non-accelerating frame of reference is called an inertial frame of reference. A frame of reference moving with a constant velocity is an inertial frame of reference.

Non-inertial Frame of Reference

An accelerating frame of reference is called a non-inertial frame of reference.

- A rotating frame of reference is a non-inertial frame of reference, because it is also an accelerating one.

Now, let us come to the pseudo force. Newton's first two laws hold good in an inertial frame only. However, we people spend most of our time on the earth which is an (approximate) inertial frame. We are so familiar with the Newton's laws that we will still like to use 'total force equals mass times acceleration' even when we use a non-inertial frame.

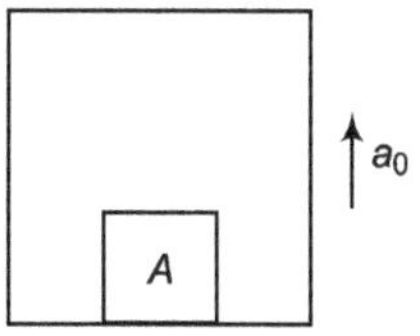

Fig. 5.56

This can be done, if we agree to call $(-m\mathbf{a}_0)$ a force acting on the particle. Then, while preparing the list of the forces acting on a particle *P*, we include all the (real) forces acting on *P* by all other objects and also include an imaginary force $(-m\mathbf{a}_0)$. Here, $\mathbf{a}_0$ is the acceleration of the non-inertial frame under consideration.

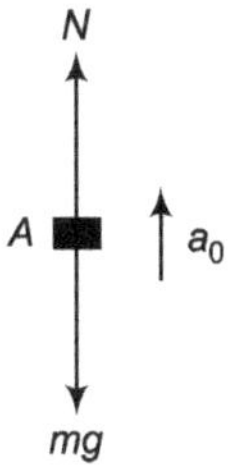

Fig. 5.57

After applying this additional imaginary force (called pseudo force) $(-m\mathbf{a}_0)$, we can now use 'total force equals mass time acceleration' even in non-inertial frames also. Now, with the help of a simple example let us see what problem arises, if we don't apply the pseudo force $(-m\mathbf{a}_0)$ while using $\mathbf{F} = m\mathbf{a}$ (second law) in non-inertial frame. Suppose a block *A* of mass *m* is placed on a lift ascending with an acceleration a_0. Let *N* be the normal reaction between the block and the floor of the lift. Free body diagram of *A* in ground frame of reference (inertial) is shown in Fig. 5.57.

$$\therefore \quad N - mg = ma_0$$

$$\text{or} \quad N = m(g + a_0) \quad \text{...(i)}$$

But if we draw the free body diagram of *A* with respect to the elevator (a non-inertial frame of reference) without applying the pseudo force, as shown in Fig. 5.58, we get

Fig. 5.58

$$N' - mg = 0$$

$$\text{or} \quad N' = mg \quad \text{...(ii)}$$

Since, $N' \neq N$, either of the equations is wrong. But if we apply a pseudo force in non-inertial frame of reference, N' becomes equal to *N* as shown in Fig. 5.59. Acceleration of block with respect to elevator is zero

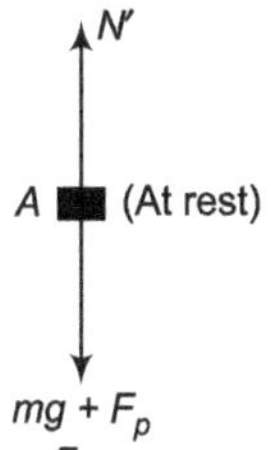

Here $F_P = ma_0$

Fig. 5.59

$$\therefore \quad N' - mg - ma_0 = 0$$

$$\text{or} \quad N' = m(g + a_0) \quad \text{...(iii)}$$

$$\therefore \quad N' = N$$

Pseudo force is given by $\mathbf{F}_P = -m\mathbf{a}_0$. Here, $\mathbf{a}_0$ is the acceleration of the non-inertial frame of reference and *m* the mass of the body under consideration. In the whole chapter, we will show the pseudo force by $\mathbf{F}_P$.

Thus, we may conclude that pseudo force is not a real force. When we draw the free body diagram of a mass with respect to an inertial frame of reference, we apply only the real forces (forces which are actually acting on the mass), but when the free body diagram is drawn from a non-inertial frame of reference a pseudo force (in addition to all real forces) has to be applied to make the equation $\mathbf{F} = m\mathbf{a}$, valid in this frame also.

- In case of rotating frame of reference, this pseudo force is called the centrifugal force when applied for centripetal acceleration. Let us take few examples of pseudo forces.

Example 5.22 *In the below figure, the coefficient of friction between wedge (of mass M) and block (of mass m) is μ.*

Find the minimum horizontal force F required to keep the block stationary with respect to wedge.

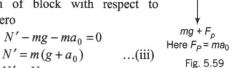

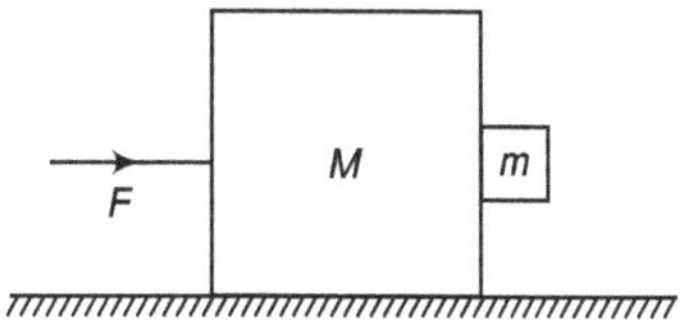

Fig. 5.60

Sol. Such problems can be solved with or without using the concept of pseudo force. Let us solve the problem by both the methods.

a = acceleration of (wedge + block) in horizontal direction

$$= \frac{F}{M+m}.$$

Inertial Frame of Reference (Ground)

FBD of block with respect to ground (only real forces have to be applied) with respect to ground block is moving with an acceleration a. Therefore,

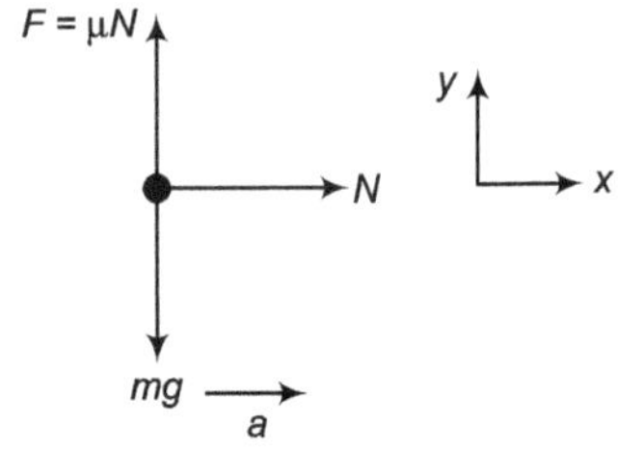

Fig. 5.61

$$\Sigma F_y = 0 \quad \text{and} \quad \Sigma F_x = ma$$
$$mg = \mu N \quad \text{and} \quad N = ma$$
$$\therefore \quad a = \frac{g}{\mu}$$
$$\therefore \quad F = (M+m)a = (M+m)\frac{g}{\mu}$$

Non-inertial Frame of Reference (Wedge)

FBD of m with respect to wedge (real + one pseudo force) with respect to wedge block is stationary.

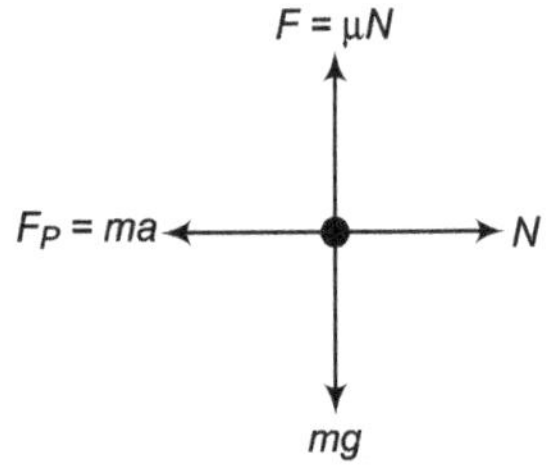

Fig. 5.62

$$\therefore \quad \Sigma F_x = 0 = \Sigma F_y$$
$$\therefore \quad mg = \mu N$$
$$\text{and} \quad N = ma$$
$$\therefore \quad a = \frac{g}{\mu}$$
$$\text{and} \quad F = (M+m)a = (M+m)\frac{g}{\mu}$$

From the above discussion, we can see that from both the methods results are same.

Example 5.23 *All surfaces are smooth in following figure. Find F such that block remains stationary with respect to wedge.*

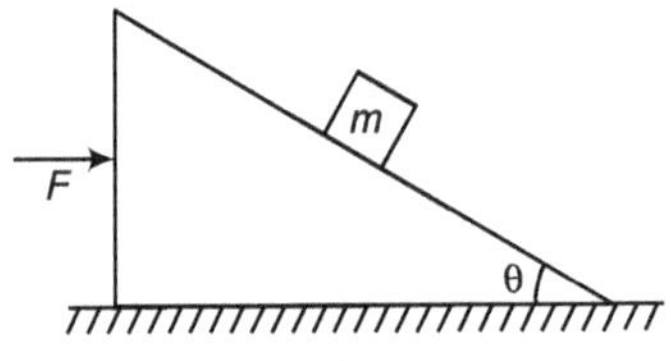

Fig. 5.63

Sol. Acceleration of (block + wedge)

$$a = \frac{F}{(M+m)}$$

Let us solve the problem by both the methods.

From Inertial Frame of Reference (Ground)

FBD of block w.r.t. ground (apply real forces) with respect to ground, block is moving with an acceleration a

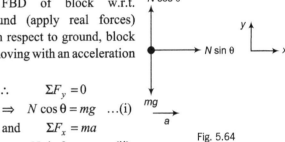

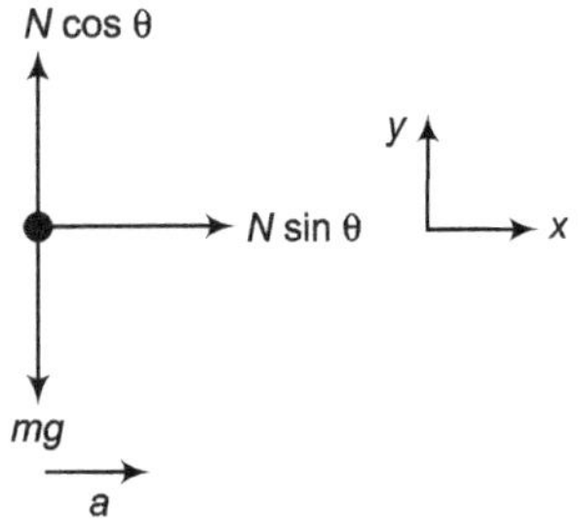

Fig. 5.64

$$\therefore \quad \Sigma F_y = 0$$
$$\Rightarrow \quad N\cos\theta = mg \quad \ldots\text{(i)}$$
$$\text{and} \quad \Sigma F_x = ma$$
$$\Rightarrow \quad N\sin\theta = ma \quad \ldots\text{(ii)}$$

From Eqs. (i) and (ii), we get

$$a = g\tan\theta$$
$$\therefore \quad F = (M+m)a = (M+m)\,g\tan\theta$$

From Non-inertial Frame of Reference (Wedge)

FBD of block w.r.t. wedge (real forces + pseudo force)

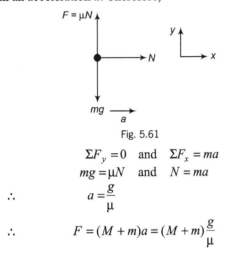

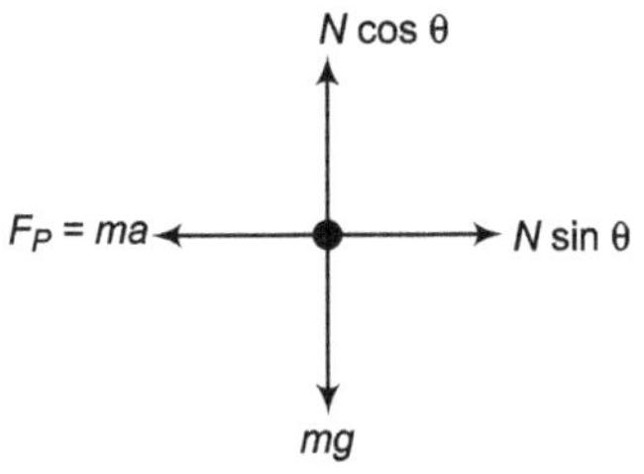

Fig. 5.65

w.r.t. wedge, block is stationary

$$\therefore \quad \Sigma F_y = 0 \Rightarrow N\cos\theta = mg \quad \ldots\text{(iii)}$$
$$\Sigma F_x = 0 \Rightarrow N\sin\theta = ma \quad \ldots\text{(iv)}$$

From Eqs. (iii) and (iv), we will get the same result i.e. $F = (M+m)\,g\tan\theta$.

Example 5.24 *A bob of mass m is suspended from the ceiling of a train moving with an acceleration a as shown in figure. Find the angle θ in equilibrium position.*

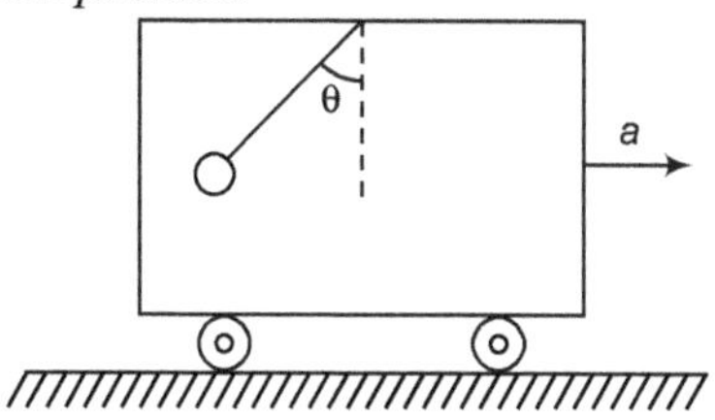

Fig. 5.66

Sol. This problem can also be solved by both the methods.

Inertial Frame of Reference (Ground)

FBD of bob w.r.t. ground (only real forces)

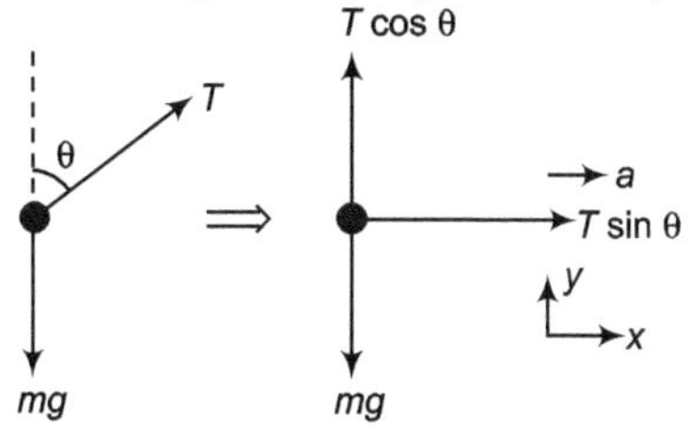

Fig. 5.67

with respect to ground, bob is also moving with an acceleration a

$$\therefore \quad \Sigma F_x = 0$$

$$\Rightarrow \quad T \sin\theta = ma \quad \ldots(\text{i})$$

and

$$\Sigma F_y = 0$$

$$\Rightarrow \quad T \cos\theta = mg \quad \ldots(\text{ii})$$

From Eqs. (i) and (ii), we get

$$\tan\theta = \frac{a}{g}$$

or

$$\theta = \tan^{-1}\left(\frac{a}{g}\right)$$

Non-inertial Frame of Reference (Train)

FBD of bob w.r.t. train (real forces + pseudo force)

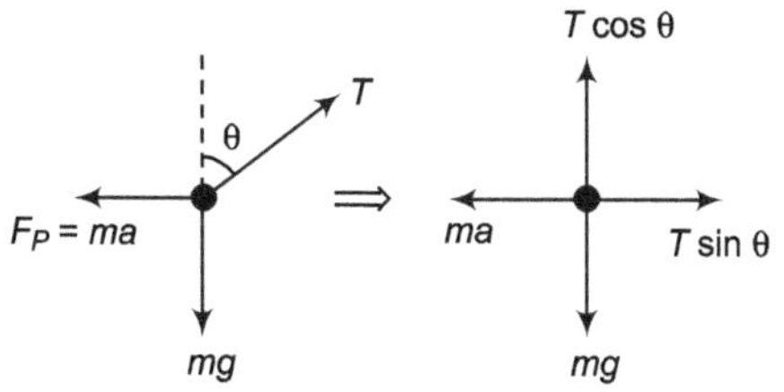

Fig. 5.68

with respect to train, bob is in equilibrium

$$\therefore \quad \Sigma F_x = 0$$

$$\Rightarrow \quad T \sin\theta = ma \quad \ldots(\text{iii})$$

$$\therefore \quad \Sigma F_y = 0$$

$$\Rightarrow \quad T \cos\theta = mg \quad \ldots(\text{iv})$$

From Eqs. (iii) and (iv), we get the same result, i.e.

$$\theta = \tan^{-1}\left(\frac{a}{g}\right)$$

Example 5.25 *In the below figure, a wedge is fixed to an elevator moving upwards with an acceleration a. A block of mass m is placed over the wedge. Find the acceleration of the block with respect to wedge. Neglect friction.*

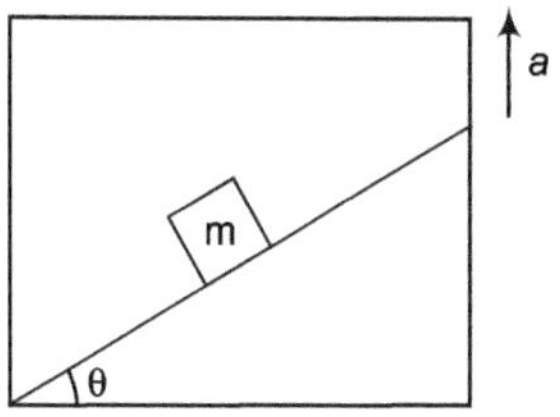

Fig. 5.69

Sol. Since, acceleration of block w.r.t. wedge (an accelerating or non-inertial frame of reference) is to be find out.

FBD of block w.r.t. wedge is shown in Fig. 5.70.

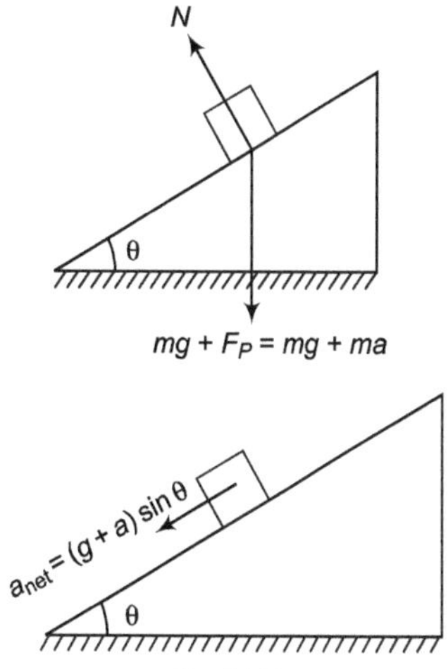

Fig. 5.70

The acceleration would had been $g \sin\theta$ (down the plane), if the lift were stationary or when only weight (i.e. mg) acts downwards.

Here, downward force is $m(g + a)$

So, acceleration of the block (of course w.r.t. wedge) will be $(g + a)\sin\theta$ down the plane.

5.8 Friction

As we have discussed in Article 5.1 friction is the parallel component of contact force between two bodies in contact. These forces are basically electromagnetic in nature. Friction can operate between a given pair of solids between a solid and a fluid or between a pair of fluids. Frictional force exerted by fluids is called **viscous force**. When two bodies slip over each other the force of friction is called kinetic friction, but when they do not slip but have a tendency to do so the force of friction is called static friction.

Regarding friction it is worth noting that

1. If a body is at rest and no pulling force is acting on it, force of friction on it is zero.
2. If a force is applied to pull the body and it does not move, the friction acts which is equal in magnitude and opposite in direction to the applied force, i.e. friction is self adjusting force. Further, as the body is at rest the friction is called static friction.
3. If the applied force is increased, the force of static friction also increases. If the applied force exceeds a certain (maximum) value, the body starts moving. This maximum force of static friction up to which body does not move is called limiting friction. Thus, static friction is a self adjusting force with an upper limit called limiting friction.
4. This limiting force of friction (f_L) is found experimentally to depend on normal reaction (N). Hence,
$$f_L \propto N$$
or $$f_L = \mu_s N$$
Here, μ_s is a dimensionless constant and called coefficient of static friction, which depends on nature of surfaces in contact.
5. If the applied force is further increased, the friction opposing the motion is called kinetic or sliding friction. Experimentally, it is well established that kinetic friction is lesser than limiting friction and is given by
$$f_k = \mu_k N$$
where, μ_k is coefficient of kinetic friction and less than μ_s.

Note Points

- In problems, if μ_s and μ_k are separately not given but only μ is given. Then, use
$$f_L = f_k = \mu N$$
- If more than two blocks are placed one over the other on a horizontal ground, then normal reaction between two blocks will be equal to the weight of the blocks over the common surface.

For example, N_1 = normal reaction between A and B
$= m_A g$
N_2 = normal reaction between B and C
$= (m_A + m_B)g$ and so on.

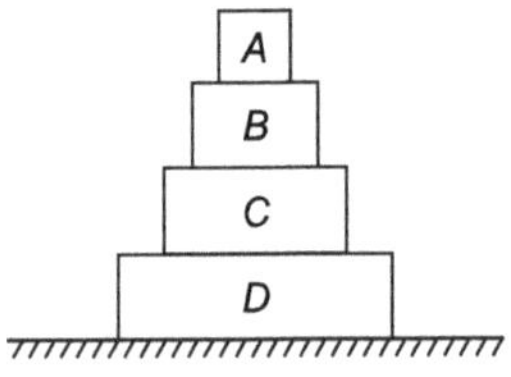

Fig. 5.71

The theory of static and kinetic friction can be better understood by the following simple example.

Example 5.26 *Suppose a block of mass 1 kg is placed over a rough surface and a horizontal force F is applied on the block as shown in figure. Now, let us see what are the values of force of friction f and acceleration of the block a, if the force F is gradually increased. Given, that* $\mu_s = 0.5$, $\mu_k = 0.4$ *and* $g = 10\,m/s^2$.

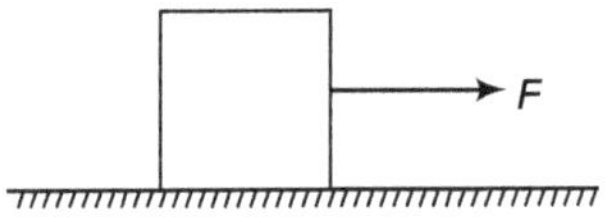

Fig. 5.72

Sol. Free body diagram of block is

$$\Sigma F_y = 0$$
$\therefore$ $$N - mg = 0$$
or $$N = mg = (1)(10) = 10\,\text{N}$$
$$f_L = \mu_s N = (0.5)(10) = 5\,\text{N}$$
and $$f_k = \mu_k N = (0.4)(10) = 4\,\text{N}$$

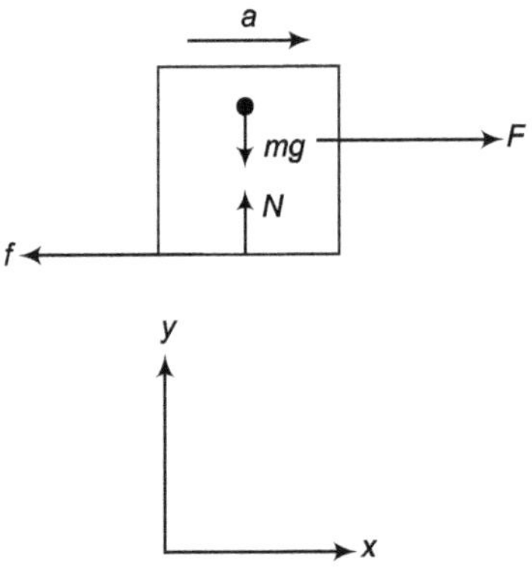

Fig. 5.73

Below is explained in tabular form, how the force of friction f depends on the applied force F.

F	f	$F_{net} = F - f$	Acceleration of block $a = \frac{F_{net}}{m}$	Diagram
0	0	0	0	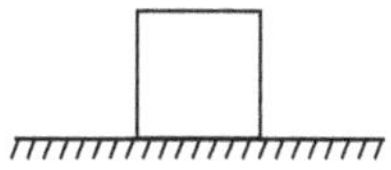
2 N	2 N	0	0	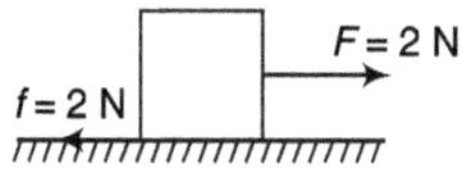
4 N	4 N	0	0	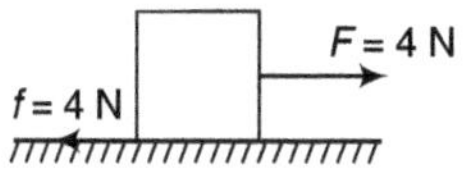
5 N	5 N	0	0	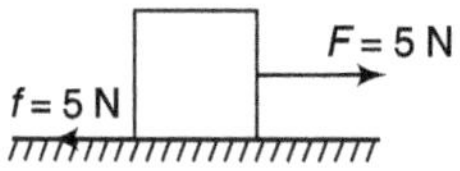
6 N	4 N	2 N	2 m/s^2	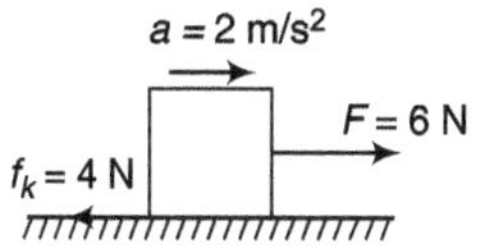
8 N	4 N	4 N	4 m/s^2	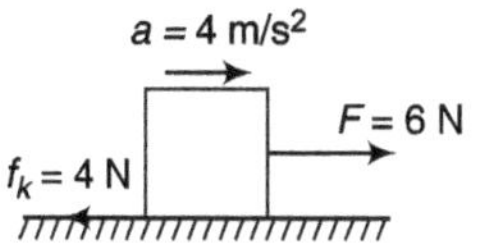

Graphically this can be understood as under

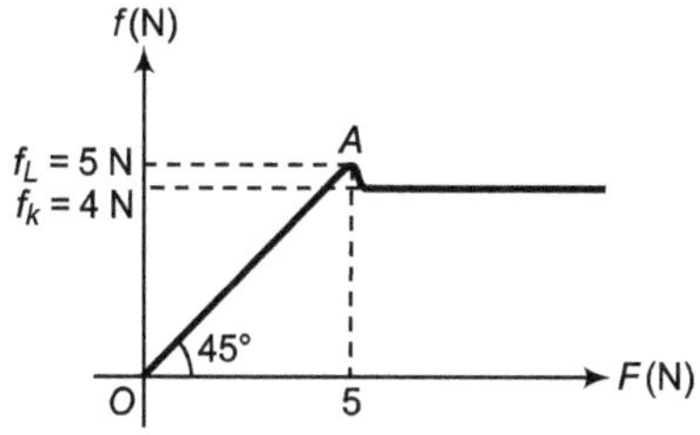

Fig. 5.74

Note that $f = F$ till $F \leq f_L$. Therefore, slope of line OA will be 1 ($y = mx$) or angle of line OA with F-axis is 45°.

- Hence, we will take coefficient of friction as μ unless and until specially mentioned in the question μ_s and μ_k, separately.

Angle of Friction (λ)

At a point of rough contact, where slipping is about to occur the two forces acting on each object are the normal reaction N and frictional force μN.

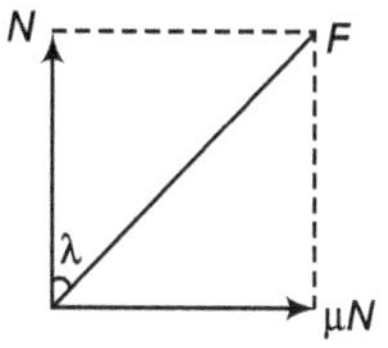

Fig. 5.75

The resultant of these two forces is F and it makes an angle λ with the normal, where

$$\tan \lambda = \frac{\mu N}{N} = \mu$$

or $$\lambda = \tan^{-1} (\mu) \quad \ldots(i)$$

This angle λ is called the **angle of friction**.

Angle of Repose (α)

Suppose a block of mass m is placed on an inclined plane whose inclination θ can be increased or decreased. Let, μ be the coefficient of friction between the block and the plane. At a general angle θ,

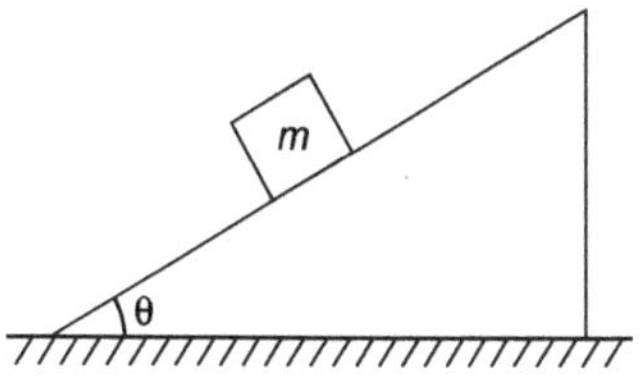

Fig. 5.76

Normal reaction, $N = mg \cos \theta$

Limiting friction

$$f_L = \mu N = \mu mg \cos \theta$$

and the driving force (or pulling force)

$$F = mg \sin \theta \quad \text{(Down the plane)}$$

From these three equations, we see that when θ is increased from 0° to 90°, normal reaction N and hence, the limiting friction f_L is decreased while the driving force F is increased. There is a critical angle called angle of repose (α) at which these two forces are equal. Now, if θ is further increased, then the driving force F becomes more than the limiting friction f_L and the block starts sliding.

Thus, $f_L = F$

at $\theta = \alpha$

or $\mu\, mg \cos\alpha = mg \sin\alpha$

or $\tan\alpha = \mu$

or $\alpha = \tan^{-1}(\mu)$ ...(ii)

From Eqs. (i) and (ii), we see that angle of friction (λ) is numerically equal to the angle of repose.

or $\lambda = \alpha$

From the above discussion, we can conclude that

if $\theta < \alpha$, $F < f_L$ the block is stationary

if $\theta = \alpha$, $F = f_L$ the block is on the verge of sliding

and if $\theta > \alpha$, $F > f_L$ the block slides down with acceleration.

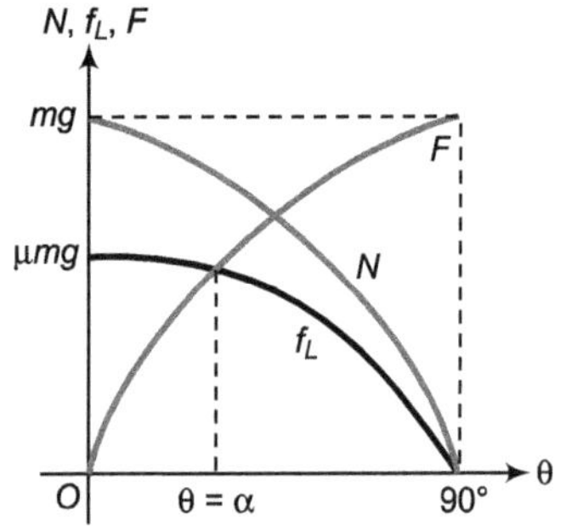

Fig. 5.77

$$a = \frac{F - f_L}{m}$$

$$= g(\sin\theta - \mu\cos\theta)$$

How, N, f_L and F varies with θ, this can be shown graphically as shown in Fig. 5.77

$N = mg\cos\theta$

or $N \propto \cos\theta$

$f_L = \mu mg\cos\theta$

or $f_L \propto \cos\theta$

$F = mg\sin\theta$

or $F \propto \sin\theta$

Normally, $\mu < 1$

So, $f_L < N$.

Extra Knowledge Points

- Friction always opposes the relative motion between two bodies in contact. They form an action and reaction pair, i.e. two equal and opposite forces act on two different bodies.

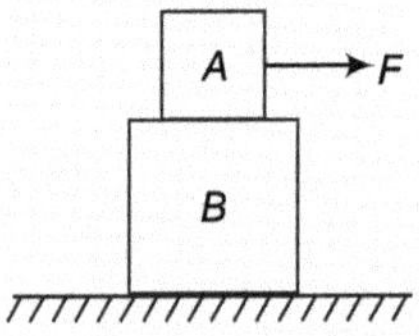

- The direction of friction force on each of them is such as it either stops the relative motion or attempts to do so. For example, if a force F is applied on block A of a two block system, the direction of frictional forces at different contacts on different bodies will be as shown.

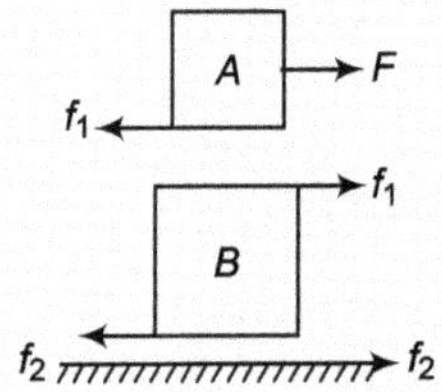

Here, f_1 = force of friction between A and B

and f_2 = force of friction between B and ground

- Force of friction $f = 0$, if no driving force is applied.

$f \le f_L \; (= \mu_s N)$

If driving force is applied but no relative motion is there and $f = \mu_k N$

If relative motion is there.

- A common mistake which the students are in hurry is that they always write $f_L = \mu mg$ (in case of horizontal ground) or $f_L = \mu mg\cos\theta$ (in inclined surface). The actual formula is $f_L = \mu N$. Here, N is equal to mg or $mg\cos\theta$ up to when no force is acting at some angle $(\neq 0°)$ with the plane.

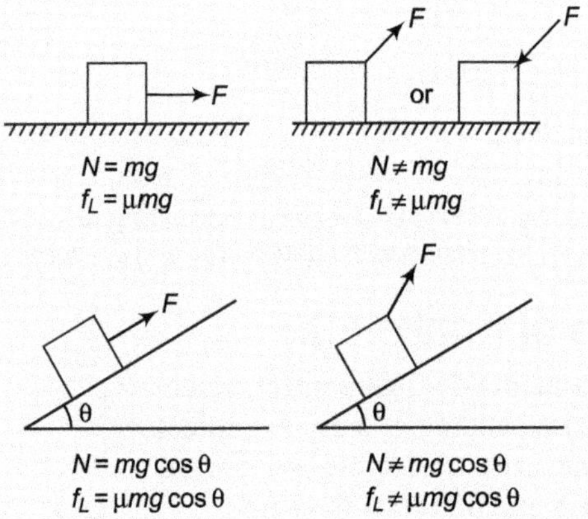

- A car (or any vehicle) accelerates and decelerates by friction. So, maximum acceleration or deceleration of a car on horizontal ground can be μg, unless some external force is applied.
- **Exercise** Think about maximum acceleration or retardation on an inclined road.
- The common problem which I feel, students often face is the resolution of forces. Following two rules can be made in this regard.

  Rule 1 If the body is in equilibrium, you can resolve the forces in any direction. Net force should be zero in all directions. A body moving with constant velocity is also in equilibrium.

  Rule 2 If the body is accelerated, resolve the forces along acceleration and perpendicular to it. Net force along acceleration $= ma$ and net force perpendicular to acceleration is zero.
- To find net force on a body find the acceleration of the body. Net force = mass × acceleration.

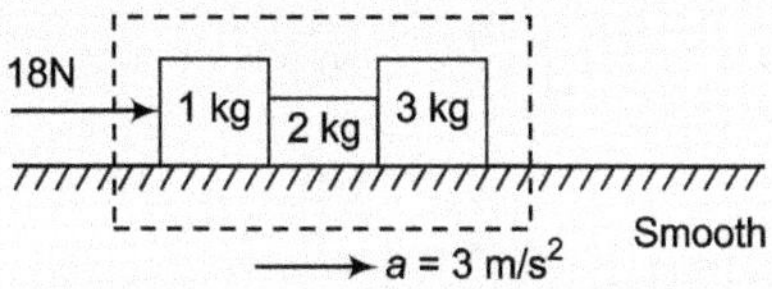

For example, in the figure shown,

$$a = \frac{18}{1+2+3} = 3\,\text{m/s}^2$$

Therefore, net force on 1 kg block $= 1 \times 3 = 3$ N.

This 3 N is the resultant of the applied force 18 N and normal reaction between 1 kg block and 2 kg block. Similarly, net force on 2 kg block $= 2 \times 3 = 6$ N.

This is the resultant of normal reactions (1 kg and 2 kg) and (2 kg and 3 kg) blocks.
- Force of friction does not oppose the motion of a body but it opposes the relative motion between two bodies in contact.

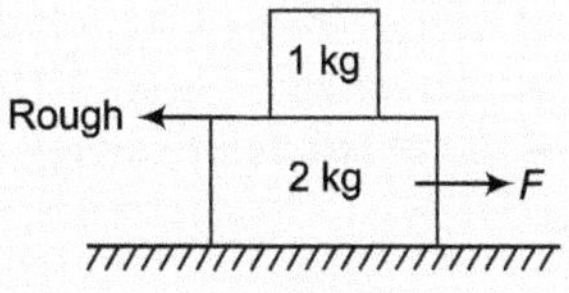

As far as motion of individual body is concerned it is sometimes friction which is responsible for its motion. For example, in the figure shown the 1 kg block moves with 2 kg block only due to friction.

Example 5.27 *A particle of mass 1 kg rests on rough contact with a plane inclined at 30° to the horizontal and is just about to slip. Find the coefficient of friction between the plane and the particle.*

Sol. The given angle 30° is really the angle of repose α. Hence,

$$\mu = \tan 30° = \frac{1}{\sqrt{3}}$$

Example 5.28 *A block of weight w rests on a horizontal plane with which the angle of friction is λ. A force P inclined at an angle θ to the plane is applied to the plane until it is on the point of moving. Find the value of θ for which the value of P will be least.*

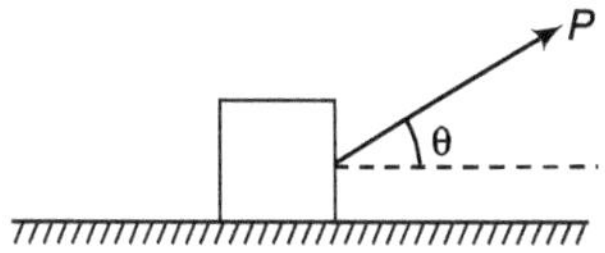

Fig. 5.78

Sol. In the limiting case, contact force F is inclined at λ to the normal. Only three forces act on the block. Applying Lami's theorem, we get

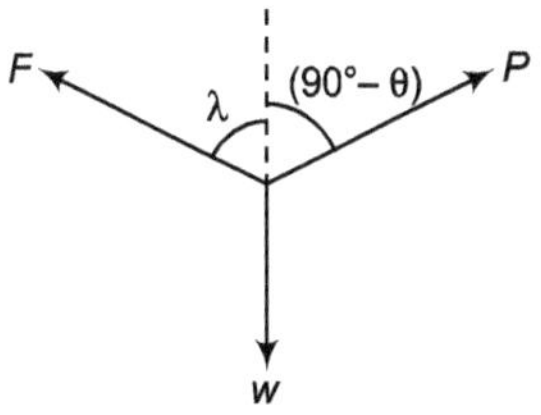

Fig. 5.79

$$\frac{P}{\sin(180° - \lambda)} = \frac{w}{\sin(90° - \theta + \lambda)} = \frac{F}{\sin(90° + \theta)}$$

or $$\frac{P}{\sin\lambda} = \frac{w}{\cos(\theta - \lambda)}$$

or $$P = \frac{w\sin\lambda}{\cos(\theta - \lambda)}$$

P will be least when $\cos(\theta - \lambda)$ is greatest, because w and λ are constant.

i.e. when $\cos(\theta - \lambda) = 1$

and $\theta - \lambda = 0°$ or $\theta = \lambda$

Additional Examples

Example 1. *A stone when thrown on a glass window smashes the windowpane to pieces, but a bullet from the gun passes through making a clean hole. Why?*

Sol. Due to its small speed, the stone remains in contact with the windowpane for a longer duration. It transfers its motion to the pane and breaks it into pieces. But the parcticles of windowpane near the hole are unable to share the fast motion of the bullet and so remain undisturbed.

Example 2. *If a ball is thrown up in a moving train, it comes back to the person's hands. Why?*

Sol. Both during its upward and downward motion, the ball continues to move (inertia of motion) with the same horizontal velocity as the train. In this period, the ball covers the same horizontal distance as the train and so it comes back to the thrower's hands.

Example 3. *Why we are hurt less when we jump on a muddy floor in comparision to a hard floor?*

Sol. When we jump on a muddy floor, the floor is carried in the direction of the jump and the time interval Δt for which force acts is increased. This decreases rate of change of momentum and hence the force of reaction. Therefore, we are hurt less.

Example 4. *Automobile tyres are generally provided with irregular projections over their surfaces. Why?*

Sol. Irregular projections increase the friction between the rubber tyres and the road. This provides a firm grip between the tyres and the road and prevents slipping.

Example 5. *Why frictional force gets increased when a surface is polished beyond a certain limit?*

Sol. When surfaces are highly polished, the area of contact between them increases. As a result of this, a large number of atoms and molecules lying on both the surfaces start exerting strong attractive forces on each other and therefore frictional force increases.

Example 6. *A rope is hanging from a tree as shown in figure. Equal mass of bananas are fastened to the higher end of the rope, and a monkey is hanging along the rope near its lower end. The monkey climbs along the rope, will it be able to eat the bananas?*

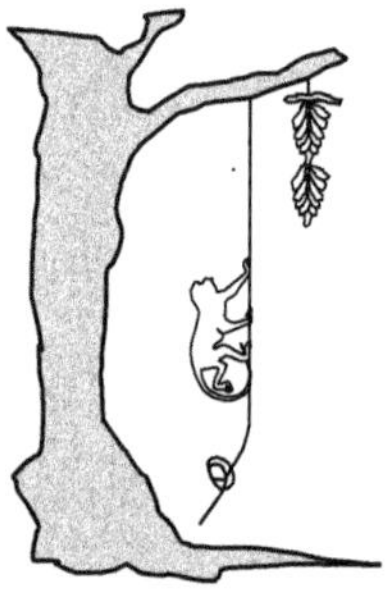

Sol. No. Force diagram on both sides is same. Tension (T) in upward direction and weight (mg) in downward direction. If $T > mg$, both will move up with same velocity and same acceleration.

Example 7. *Determine the tensions T_1 and T_2 in the strings as shown in figure.*

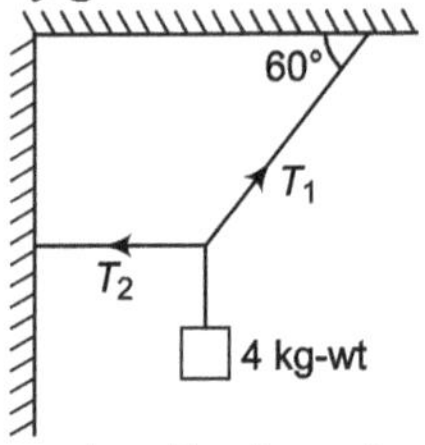

Sol. Resolving the tension T_1 along horizontal and vertical directions. As the body is in equilibrium,

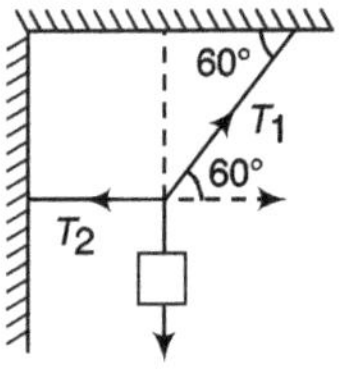

$$w = 4 \times 9.8 \text{ N}$$

$$T_1 \sin 60° = 4 \times 9.8 \text{ N} \quad \ldots\text{(i)}$$

$$T_1 \cos 60° = T_2 \quad \ldots\text{(ii)}$$

$$T_1 = \frac{4 \times 9.8}{\sin 60°}$$

$$= \frac{4 \times 9.8 \times 2}{\sqrt{3}} = 45.26 \text{ N}$$

$$T_2 = T_1 \cos 60°$$

$$= 45.26 \times 0.5 = 22.63 \text{ N}$$

Example 8. *A ball of mass 1 kg hangs in equilibrium from two strings OA and OB as shown in figure. What are the tensions in strings OA and OB? (Take $g = 10\ ms^{-2}$)*

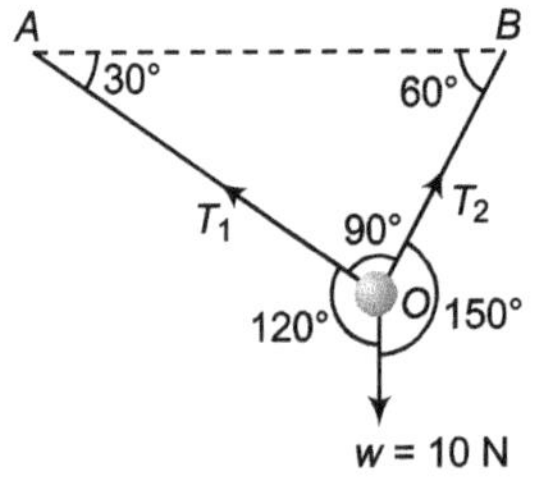

Sol. Various forces acting on the ball are as shown in figure. The three concurrent forces are in equilibrium. Using Lami's theorem,

$$\frac{T_1}{\sin 150°} = \frac{T_2}{\sin 120°} = \frac{10}{\sin 90°}$$

$$\text{or} \quad \frac{T_1}{\sin 30^\circ} = \frac{T_2}{\sin 60^\circ} = \frac{10}{1}$$

$$\therefore \quad T_1 = 10 \sin 30^\circ = 10 \times 0.5 = 5 \text{ N}$$

$$\text{and} \quad T_2 = 10 \sin 60^\circ = 10 \times \frac{\sqrt{3}}{2} = 5\sqrt{3} \text{ N}$$

Example 9. *A 4 m long ladder weighing 25 kg rests with its upper end against a smooth wall and lower end on rough ground. What should be the minimum coefficient of friction between the ground and the ladder for it to be inclined at 60° with the horizontal without slipping? (Take $g = 10\, ms^{-2}$)*

Sol. In figure, AB is a ladder of weight w which acts at its centre of gravity G.

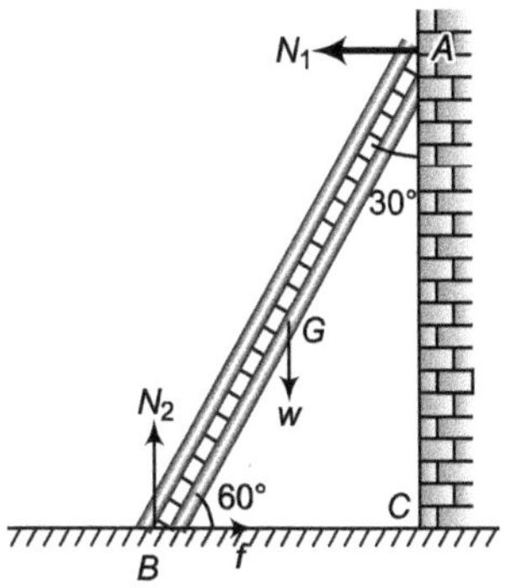

$$\angle ABC = 60^\circ$$
$$\therefore \quad \angle BAC = 30^\circ$$

Let N_1 be the reaction of the wall, and N_2 the reaction of the ground.

Force of friction f between the ladder and the ground acts along BC.

For horizontal equilibrium,

$$f = N_1 \quad \ldots\text{(i)}$$

For vertical equilibrium,

$$N_2 = w \quad \ldots\text{(ii)}$$

Taking moments about B, we get for equilibrium,

$$N_1(4 \cos 30^\circ) - w(2 \cos 60^\circ) = 0 \quad \ldots\text{(iii)}$$

Here, $w = 250$ N

Solving these three equations, we get

$$f = 72.17 \text{ N and } N_2 = 250 \text{ N}$$

$$\therefore \quad \mu = \frac{f}{N_2} = \frac{72.17}{250} = 0.288$$

Example 10. *A block of mass 25 kg is raised by a 50 kg man in two different ways as shown in figure. What is the action on the floor by the man in the two cases? If the floor yields to a normal force of 700 N, which mode should the man adopt to lift the block without the floor yielding? (Take $g = 9.8\, m/s^2$)*

Sol. In mode (a), the man applies a force equal to 25 kg weight in upward direction. According to Newton's third law of motion, there will be a downward force of reaction on the floor.

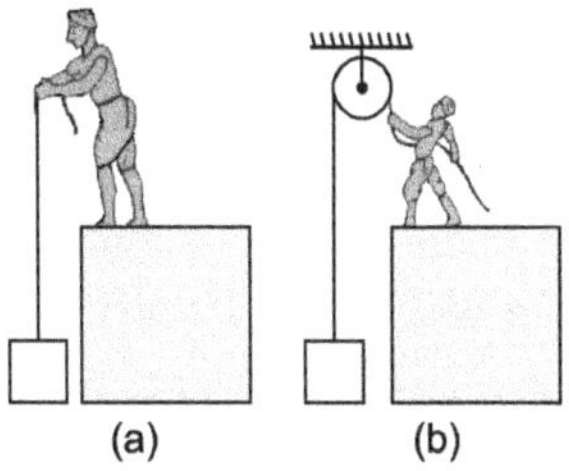

$\therefore$ Total action on the floor by the man

$$= 50 \text{ kg-wt} + 25 \text{ kg-wt} = 75 \text{ kg-wt} = 75 \times 9.8 \text{ N} = 735 \text{ N}$$

In mode (b), the man applies a downward force equal to 25 kg-wt. According to Newton's third law, the reaction will be in the upward direction.

$\therefore$ Total action on the floor by the man

$$= 50 \text{ kg-wt} - 25 \text{ kg-wt} = 25 \text{ kg-wt} = 25 \times 9.8 \text{ N} = 245 \text{ N}$$

As the floor yields to a downward force of 700 N, so the man should adopt mode (b).

Example 11. *A block of mass 200 kg is set into motion on a frictionless horizontal surface with the help of frictionless pulley and a rope system as shown in figure (a). What horizontal force F should be applied to produce in the block an acceleration of $1\, m/s^2$?*

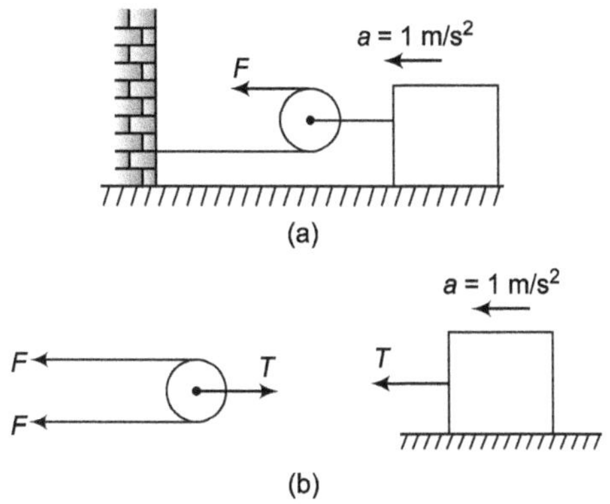

Sol. As shown in Fig. (b), when force F is applied at the end of the string, the tension in the lower part of the string is also F. If T is the tension in string connecting the pulley and the block, then

$$T = 2F$$

$$\text{But,} \quad T = ma = (200)(1) = 200 \text{ N}$$

$$\therefore \quad 2F = 200 \text{ N}$$

$$\text{or} \quad F = 100 \text{ N}$$

Example 12. *A block of mass 1 kg is pushed against a rough vertical wall with a force of 20 N, coefficient of static friction being $\frac{1}{4}$. Another horizontal force of 10 N is applied on the block in a direction parallel to the wall. Will the block move? If yes, in which direction? If no, find the frictional force exerted by the wall on the block.* ($g = 10\ m/s^2$)

Sol. Normal reaction on the block from the wall will be

$$N = F = 20\ \text{N}$$

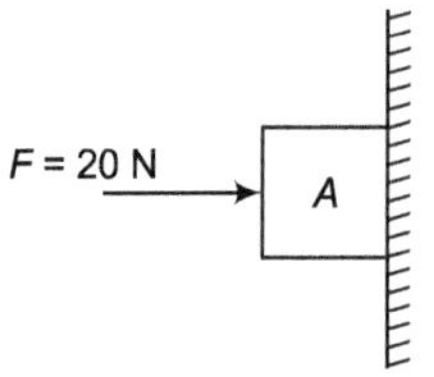

Therefore, limiting friction

$$f_L = \mu N = \left(\frac{1}{4}\right)(20) = 5\ \text{N}$$

Weight of the block is

$$w = mg = (1)(10) = 10\ \text{N}$$

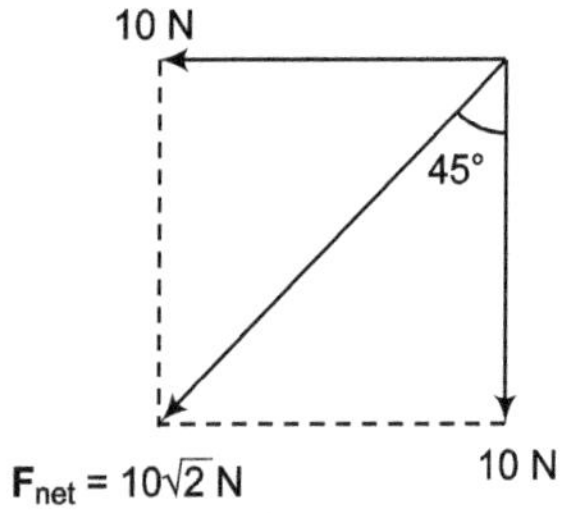

A horizontal force of 10 N is applied to the block. The resultant of these two forces will be $10\sqrt{2}$ N in the direction shown in figure. Since, this resultant is greater than the limiting friction. The block will move in the direction of $\mathbf{F}_{\text{net}}$ with acceleration.

$$a = \frac{F_{\text{net}} - f_L}{m}$$

$$= \frac{10\sqrt{2} - 5}{1} = 9.14\ \text{m/s}^2$$

Example 13. *Figure shows a man standing stationary with respect to a horizontal conveyor belt that is accelerating with 1 ms^{-2}. What is the net force on the man? If the coefficient of static friction between the man's shoes and the belt is 0.2, up to what acceleration of the belt can the man continue to be stationary relative to the belt? Mass of the man = 65 kg.* ($g = 9.8\ m/s^2$)

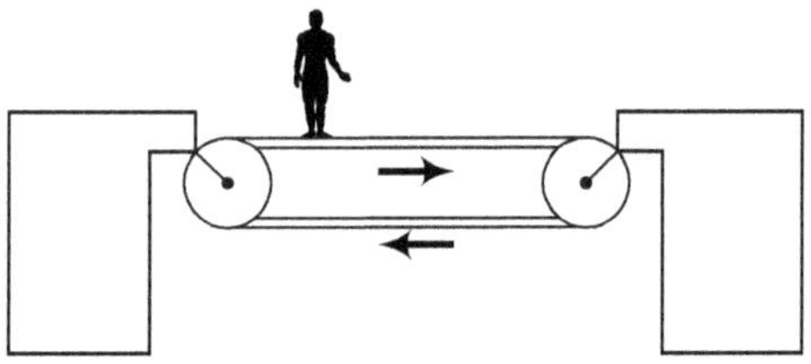

Sol. As the man is standing stationary w.r.t. the belt,

∴ Acceleration of the man = Acceleration of the belt

$$= a = 1\ \text{ms}^{-2}$$

Mass of the man,

$$m = 65\ \text{kg}$$

Net force on the man $= ma = 65 \times 1 = 65$ N

Given, coefficient of friction, $\mu = 0.2$

∴ Limiting friction, $f_L = \mu mg$

If the man remains stationary with respect to the maximum acceleration a_0 of the belt, then

$$ma_0 = f_L = \mu mg$$

∴ $$a_0 = \mu g$$

$$= 0.2 \times 9.8$$

$$= 1.96\ \text{ms}^{-2}$$

NCERT Selected Questions

***Q* 1.** Give the magnitude and direction of the net force acting on

(a) a drop of rain falling down with a constant speed.
(b) a cork of mass 10 g floating on water.
(c) a kite skillfully held stationary in the sky.
(d) a car moving with a constant velocity of 30 km/h on a rough road.

Sol. (a) Since, the drop of rain is falling with a constant speed, so the acceleration of the rain drop is zero. Hence, net force on the drop is zero.

(b) As the cork is floating in water, so the weight of the cork is balanced by upthrust. Therefore, net force on the cork is zero.

(c) Since, the kite is held stationary, so its acceleration is zero. Hence, the net force acting on the kite is zero.

(d) Since, the car is moving with a constant velocity, so its acceleration is zero. Hence, the net force acting on the car is zero.

***Q* 2.** A pebble of mass 0.05 kg is thrown vertically upwards. Give the direction and magnitude of the net force on the pebble

(a) during its upward motion.
(b) during its downward motion.
(c) at the highest point where it is momentarily at rest.

Do your answers change, if the pebble was thrown at an angle of 45° with the horizontal direction? Ignore air resistance.

Sol. (a) Net force on the pebble $= mg$ $(\because a = g)$

$$= 0.05 \times 9.8 = 0.49 \text{ N}$$

(Acts vertically downwards)

(b) Net force on the pebble $= mg$ $(\because a = g)$

$$= 0.05 \times 9.8 = 0.49 \text{ N}$$

(Acts vertically downwards)

(c) When the stone is at the highest point, then also the net force $(= mg)$ acts in vertically downward direction.

$\therefore$ Net force on the pebble $= mg$

$$= 0.05 \times 9.8 = 0.49 \text{ N}$$

If the pebble was thrown at an angle of 45° with the horizontal direction, then it will not affect the force on the pebble. Hence, our answers will not change in any case.

***Q* 3.** Give the magnitude and direction of the net force acting on a stone of mass 0.1 kg.

(a) Just after it is dropped from the window of a stationary train.
(b) Just after it is dropped from the window of a train running at a constant velocity of 36 km/h
(c) Just after it is dropped from the window of a train accelerating with 1 ms^{-2}

Sol. (a) $\therefore$ Net force on the stone $= mg$

$$= 0.1 \times 9.8 = 0.98 \text{ N}$$

and it acts vertically downwards.

(b) When the stone is dropped from the window of this train, then it is again falling freely, so

net force on the stone = its own weight

$$= mg$$
$$= 0.98 \text{ N}$$

and it acts vertically downwards

(c) Net force on stone is again mg or 0.98 N

***Q* 4.** A constant retarding force of 50 N is applied to a body of mass 20 kg moving initially with a speed of 15 ms^{-1}. How long does the body take to stop?

Sol. Using the relation $F = ma$,

We get $$a = \frac{F}{m}$$

or $$a = -\frac{50}{20}$$

$$= -2.5 \text{ ms}^{-2} \quad \text{(retardation)}$$

Using the relation $v = u + at$,

We get $$0 = 15 + (-2.5)\,t$$

or $$2.5\,t = 15$$

$\therefore$ $$t = \frac{15}{2.5} = 6 \text{ s}$$

***Q* 5.** A constant force acting on a body of mass 3.0 kg changes its speed from 2.0 ms^{-1} to 3.5 ms^{-1} in 25 s. The direction of motion of the body remains unchanged. What is the magnitude and direction of the force?

Sol. Using the relation $v = u + at$,

We get $$a = \frac{v-u}{t} = \frac{3.5-2}{25}$$

$$= \frac{1.5}{25} = \frac{15}{250} \text{ ms}^{-2}$$

$\therefore$ Using the relation $F = ma$,

We get the magnitude of the force as

$$F = 3 \times \frac{15}{250}$$

$$= 0.18 \text{ N}$$

The direction of the force is along the direction of motion.

***Q* 6.** A body of mass 5 kg is acted upon by two perpendicular forces 8 N and 6 N. Give the magnitude and direction of the acceleration of the body.

Sol. To find a let us first find net force on the body.

Let F be the net force acting on the body.

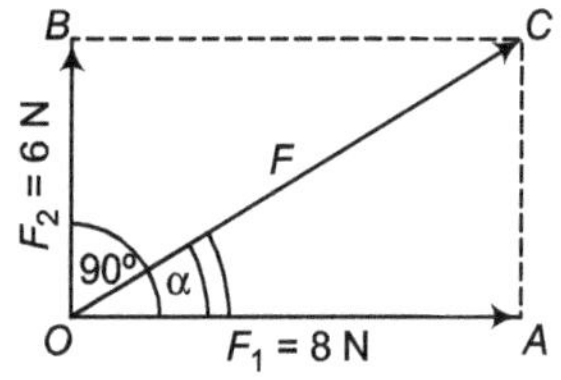

Magnitude of *F* The magnitude of F is given by

$$F = \sqrt{F_1^2 + F_2^2}$$
$$= \sqrt{8^2 + 6^2}$$
$$= 10 \text{ N}$$

Direction of *F* Let α be the angle made by F with F_1

Then, $\tan\alpha = \dfrac{F_2}{F_1} = \dfrac{6}{8} = 0.75$

or $\alpha = 37°$ with the direction of 8 N.

This is the direction of resultant force and hence the direction of acceleration of the body.

$$a = \frac{F}{m} = \frac{10}{5} = 2 \text{ ms}^{-2}$$

$\therefore$ Magnitude of acceleration $= 2 \text{ ms}^{-2}$ and acts in the direction of resultant force.

***Q* 7.** The driver of a three wheeler moving with a speed of 36 km/h sees a child standing in the middle of the road and brings his vehicle to rest in 4.0 s just in time to save the child. What is the average retarding force on the vehicle? The mass of the three wheeler is 400 kg and the mass of the driver is 65 kg.

Sol. Initial speed of the three wheeler,

$$u = 36 \text{ km}^{-1} = 36 \times \frac{5}{18} \text{ ms}^{-1}$$
$$= 10 \text{ ms}^{-1}$$

Final speed of the three wheeler,

$$v = 0$$

Time, $t = 4$ s

Total mass, m = mass of three wheeler + mass of driver

$$= 400 + 65 = 465 \text{ kg}$$

Using the relation $v = u + at$,

We get $a = \dfrac{v - u}{t} = \dfrac{0 - 10}{4}$

$$= -2.5 \text{ ms}^{-2}$$

– ve sign shows that it is retardation.

$\therefore$ $|\mathbf{F}| = m\,|\mathbf{a}|$

$$= 465 \times (2.5) = 1162.5 \text{ N}$$
$$= 1.16 \times 10^3 \text{ N} = 1.2 \times 10^3 \text{ N}$$

***Q* 8.** A rocket with a lift-off mass 20000 kg is blasted upwards with an initial acceleration of 5.0 ms^{-2}. Calculate the initial thrust (force) of the blast.

Sol. Let T be the initial thrust acting upward.

$\therefore$ Equation of motion becomes,

$$T - mg = ma \text{ or } T = mg + ma$$
$$= m\,(g + a) = 2 \times 10^4\,(9.8 + 5)$$
$$= 2.96 \times 10^5 \text{ N} \approx 3.0 \times 10^5 \text{ N}$$

***Q* 9.** A bob of mass 0.1 kg hung from the ceiling of a room by a string 2 m long is set into oscillation. The speed of the bob at its mean position is 1 ms^{-1}. What is the trajectory of the bob, if the string is cut when the bob is

(a) at one of its extreme positions.

(b) at its mean position.

Sol. (a) We know that at each extreme position, the instantaneous velocity of the bob is zero. If the string is cut at the extreme position, it is under the action of g only, hence the bob will fall extreme downwards.

(b) At mean position, the bob is having a velocity of 1 ms^{-1} along the tangent of the arc which is in the horizontal direction. If the string is cut at the mean position, the bob will behave as a horizontal projectile. Hence, it will follow a parabolic path.

***Q* 10.** A man of mass 70 kg stands on a weighing scale in a lift which is moving

(a) upwards with uniform speed of 10 ms^{-1}.

(b) downwards with a uniform acceleration of 5 ms^{-2}.

(c) upwards with uniform acceleration of 5 ms^{-2}.

What would be the readings on the scale in each case?

(d) What would be the reading if the lift mechanism failed and it falls down freely under gravity?

Sol. (a) When the speed of the lift is uniform, its acceleration $a = 0$. So, the weighing machine will show his weight, i.e.

$$N = w = mg = 70 \times 9.8 = 686 \text{ N}$$

$\therefore$ Reading of the scale = 686 N

= 70 kg

(b) When the lift moves downward with an acceleration

$$a = 5 \text{ ms}^{-2}$$

$\therefore$ Resultant force acts vertically downward

i.e. $F = mg - N$

or $ma = mg - N$

or $N = mg - ma = m(g - a)$

$$= 70 \times 4.8 = 336 \text{ N}$$

$\therefore$ Reading of the scale = 336 N

$$= \frac{336}{9.8} \text{ kg}$$
$$= 34.29 \text{ kg}$$

(c) In this case, the left is moving upwards with an acceleration,

$$a = 5\ \text{ms}^{-2}$$

$\therefore$ Resultant force acting is upward

$$F = N - mg$$

or
$$N = ma + mg$$
$$= m(g + a)$$
$$= 70\,(9.8 + 5) = 14.8 \times 70$$
$$= 1036\ \text{N}$$

$\therefore$ Reading of the scale = 1036 N

$$= \frac{1036}{9.8}\ \text{kg}$$
$$= 105.7\ \text{kg}$$

(d) In this case, the lift falls freely under gravity, i.e. with an acceleration

$$a = g = 9.8\ \text{ms}^{-2}$$

$\therefore$
$$N = m(g - a) = m(g - g) = 0$$

Thus, the reading of the scale is zero. This is the state of weightlessness.

Q 11. Two bodies of masses 10 kg and 20 kg respectively kept on a smooth, horizontal surface are tied to the ends of a light string. A horizontal force $F = 600$ N is applied to (i) A, (ii) B along the direction of string. What is the tension in the string in each case?

Sol.
$$F = 600\ \text{N}$$
$$m_1 = 10\ \text{kg}$$
$$m_2 = 20\ \text{kg}$$

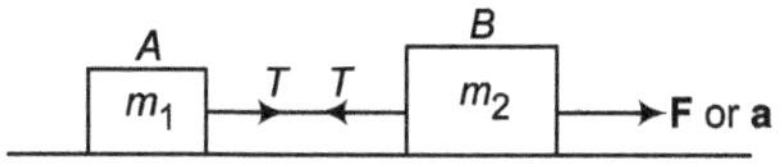

Let T be the tension in the string and a be the acceleration of the system, in the direction of force applied.

(a) When force is applied on the heavier mass.

Then, equations of motion of A and B are

$$m_1 a = T \quad \text{...(i)}$$
$$m_2 a = F - T \quad \text{...(ii)}$$

Solving the equations, we get

$$T = \frac{F}{3} = \frac{600}{3} = 200\ \text{N}$$

(b) When the force is applied on lighter mass, then

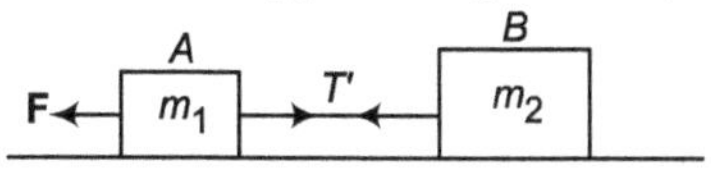

Let T be the tension in the string in this case.

Then, equations of motion of A and B are

$$F - T' = m_1 a \quad \text{...(iii)}$$
$$T' = m_2 a \quad \text{...(iv)}$$

Again solving the equations for given values, we have

or
$$T = \frac{2}{3}F = \frac{2}{3} \times 600 = 400\ \text{N}$$

Q 12. Two masses 8 kg and 12 kg are connected at the two ends of a light inextensible string that goes over a frictionless pulley. Find the acceleration of the masses and the tension in the string when the masses are released.

Sol. The equations of motion of m_1 and m_2 are given by

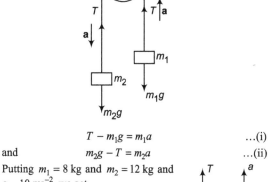

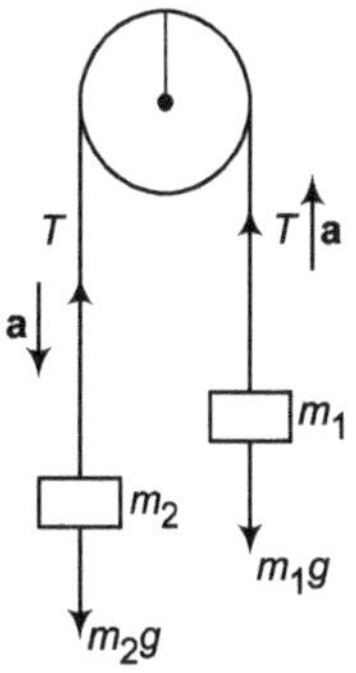

$$T - m_1 g = m_1 a \quad \text{...(i)}$$

and
$$m_2 g - T = m_2 a \quad \text{...(ii)}$$

Putting $m_1 = 8$ kg and $m_2 = 12$ kg and $g = 10\ \text{ms}^{-2}$, we get

$$a = 2\ \text{m/s}^2$$

and
$$T = 96\ \text{N}$$

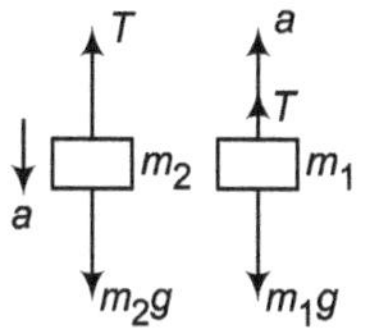

Q 13. Figure below shows a man standing stationary with respect to a horizontal conveyor belt that is accelerating with $1\ \text{ms}^{-2}$. What is the net force on the man? If the coefficient of static friction between the man's shoes and the belt is 0.2, up to what acceleration of the belt can the man continue to be stationary relative to the belt? (Mass of the man = 65 kg)

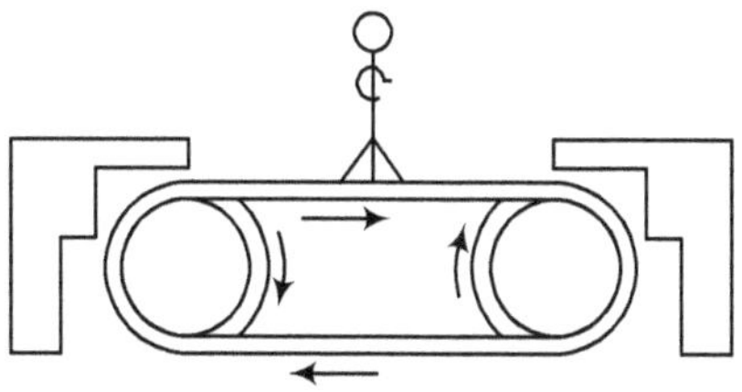

Sol. As the man is standing stationary w.r.t. the belt, so acceleration of man = acceleration of belt = $a = 1\ \text{ms}^{-2}$.

Mass of man, $m = 65$ kg

$\therefore$ Net force on the man, $F = ma = 65 \times 1 = 65$ N.

The direction of this force is in the direction of motion of the belt. Coefficient of static friction between man's shoes and belt.

$$\mu_s = 0.2$$

Let d be the acceleration of the belt up to which the man can continue to be stationary relative to the belt.

$\therefore$ md = maximum value of static friction

or
$$ma = \mu_s N = \mu_s mg$$

or
$$a = \mu_s g = 0.2 \times 9.8 = 1.96\ \text{ms}^{-2}$$

Q 14. A helicopter of mass 1000 kg rises with a vertical acceleration of 15 ms^{-2}. The crew and the passengers weigh 300 kg. Give the magnitude and direction of the

(a) force on floor by the crew and passengers.

(b) action of the rotor of the helicopter on the surrounding air.

(c) force on the helicopter due to the surrounding air, (Take $g = 10\ ms^{-2}$).

Sol. Mass of the helicopter, $m = 1000$ kg

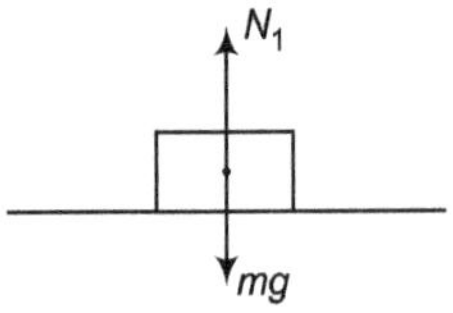

Mass of the crew and the passengers, $m = 300$ kg

Vertical acceleration of the helicopter, $a = 15\ ms^{-2}$

Acceleration due to gravity, $g = 10\ ms^{-2}$

(a) Let N_1 be the force by the floor on the crew and the passengers

$$N_1 - mg = ma$$

or $$N_1 = m(g + a) = 300(10 + 15)$$
$$= 300 \times 25 = 7500 \text{ N}$$

and it acts upwards.

By 3rd law, force on the floor by the crew and passengers = 7500N

(b) Let N_2 = force by air on the rotor (action) upwards

$$\therefore\ N_2 - (M + m)g = (M + m)a$$
$$N_2 = (M + m)(g + a)$$
$$= (1000 + 300)(10 + 15)$$
$$= 1300 \times 25 = 32500 \text{ N}$$

$\therefore$ By third law, force (action) of the rotor of helicopter on the surrounding air = 32500 N downwards.

(c) According to Newton's third law of motion, action and reaction are equal and opposite.

$\therefore$ Force on the helicopter due to surrounding air = 32500 Nand it acts upwards.

Q 15. A stream of water flowing horizontally with a speed of 15 ms^{-1} gushes out of a tube of cross-sectional area $10^{-2}\ m^2$, and hits a vertical wall nearby. What is the force exerted on the wall by the impact or water, assuming it does not rebound?

Sol. Area of cross-section of the tube, $a = 10^{-2}\ m^2$

Speed of stream of water = 15 ms^{-1}

V = volume of water coming out per second from the tube
$= a \times V = 15 \times 10^{-2}\ m^3s^{-1}$

Also we know that density of water, $\rho = 10^3\ kg\ m^{-3}$

$\therefore$ If m be the mass of water striking the wall per second, then

$$m = \rho V = 10^3 \times 15 \times 10^{-2}\ kg\ s^{-1} = 150\ kg\ s^{-1}$$

As on hitting the wall, water does not rebound, so

F = change in momentum per second

= mass of water flowing out per second × velocity

$= 150 \times 15 = 2250$ N

Q 16. Ten one-rupee coins are put on top of each other on a table. Each coin has mass *m*. Give the magnitude and direction of

(a) the force on the 7th coin (counted from the bottom) due to all the coins on its top.

(b) the force on the 7th coin by the eighth coin.

(c) the reaction of the 6th coin on the 7th coin.

Sol. (a) If F_1 be the force on 7th coin (counted from the bottom)
F_1 = weight of three coins above it = 3 mg (downward)

(b) F_8 = Force on 7th cion by 8th coin = weight of 8th coin + weight of two coins above the 8th coin $= mg + 2mg = 3mg$ (N) and it acts downwards.

(c) The sixth coin experiences force equal to weight of the four coins above it. Hence, reaction due to 6th coin on 7th coin = 4 mg (N) and it acts vertically upwards.

Q 17. A monkey of mass 40 kg climbs on a rope which can stand a maximum tension of 600 N. In which of the following cases will the rope break : the monkey

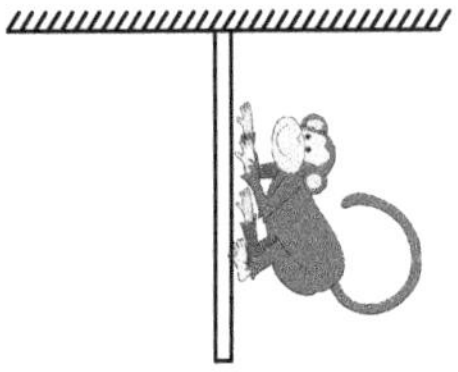

(a) climbs up with an acceleration of 6 ms^{-2}

(b) falls down with an acceleration of 4 ms^{-2}

(c) climbs up with a uniform speed of 5 ms^{-1}

(d) falls down the rope nearly freely under gravity

(Ignore the mass of the rope) and take $g = 10\ ms^{-2}$.

Sol. (a) When the monkey climbs up with $a = 6\ ms^{-2}$, the

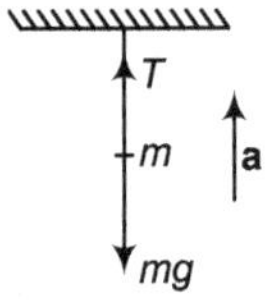

$\therefore$ Equation of motion is

$$T - mg = ma$$

$$\therefore\quad T = mg + ma = m(g + a)$$
$$= 40(10 + 6) = 40 \times 16$$
$$= 640 \text{ N}$$

Thus, $T > T_{max}$ hence the rope will break in this case.

(b) When the monkey falls down with $a = 4\ ms^{-2}$, then

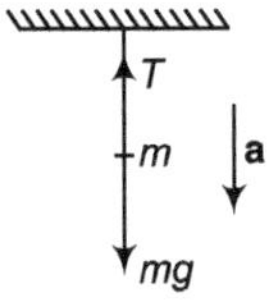

Equation of motion is

$$mg - T = ma \text{ or } T = mg - ma$$
$$= m(g - a) = 40(10 - 4)$$
$$= 40 \times 6 = 240 \text{ N}$$

As $T < T_{max}$, so the rope will not break in this case.

(c) When the monkey climbs up with a uniform speed of 5 ms^{-1}, then there is no acceleration, i.e. $a = 0$, hence

$$T = mg = 40 \times 10 = 400 \text{ N}$$

which is $< T_{max}$, so the string will not break.

(d) When the monkey falls freely, then $a = g$

$$\therefore\quad T = m(g - a) = m(g - g) = 0$$

i.e. the monkey is in the state of weightlessness and $T = 0$, hence the rope will not break.

Objective Problems

[Level 1]

Equilibrium of Forces

1. A rope of length L and mass M is hanging from a rigid support. The tension in the rope at a distance x from the rigid support is

(a) Mg (b) $\left(\frac{L-x}{L}\right) Mg$

(c) $\left(\frac{L}{L-x}\right) Mg$ (d) $\frac{x}{L} Mg$

2. The pulleys and strings shown in the figure are smooth and of negligible mass. For the system to remain in equilibrium, the angle θ should be

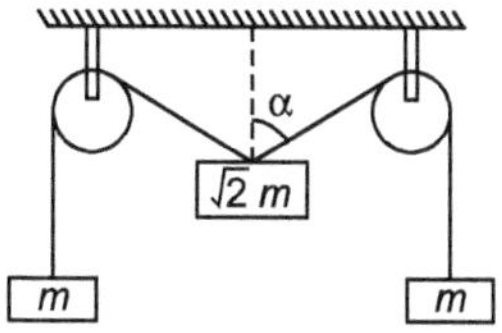

(a) $0°$ (b) $30°$
(c) $45°$ (d) $60°$

3. A weightless rod is acted upon by upward parallel forces of 4 N and 2 N magnitudes at ends A and B, respectively. The total length of the rod $AB = 3$ m. To keep the rod in equilibrium, a force of 6 N should act in the following manner

(a) downward at any point between A and B
(b) downward at the mid-point of AB
(c) downward at a point C such that $AC = 1$ m
(d) downward at a point C such that $BC = 1$ m

4. Ten coins are placed on top of each other on a horizontal table. If the mass of each coin is 10 g and acceleration due to gravity is $10\ \text{ms}^{-2}$, what is the magnitude and direction of the force on the 7th coin (counted from the bottom) due to all the coins above it?

(a) 0.3 N downwards
(b) 0.3 N upwards
(c) 0.7 N downwards
(d) 0.7 N upwards

5. An object is resting at the bottom of two strings which are inclined at an angle of 120° with each other. Each string can withstand a tension of 20 N. The maximum weight of the object that can be sustained without breaking the strings is

(a) 10 N (b) 20 N
(c) $20\sqrt{2}$ N (d) 40 N

6. A block of mass 10 kg is suspended by three strings as shown in the figure. The tension T_2 is

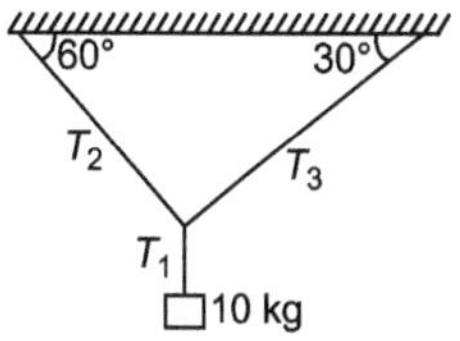

(a) 100 N (b) $\frac{100}{\sqrt{3}}$ N

(c) $\sqrt{3} \times 100$ N (d) $50\sqrt{3}$ N

7. A man of mass 50 kg stands on a frame of mass 30 kg. He pulls on a light rope which passes over a pulley. The other end of the rope is attached to the frame. For the system to be in equilibrium what force man must exert on the rope?

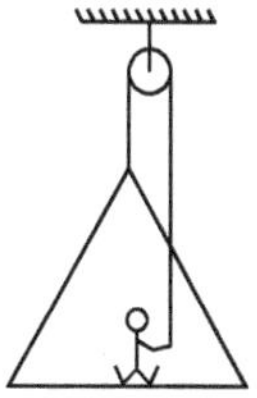

(a) 40 g (b) 80 g
(c) 30 g (d) 50 g

8. Three blocks are placed at rest on a smooth inclined plane with force acting on m_1 parallel to the inclined plane. Find the contact force between m_2 and m_3.

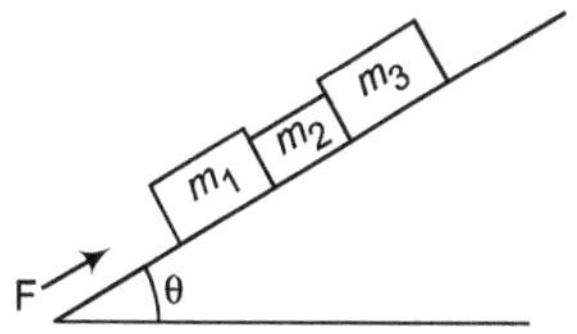

(a) $\frac{(m_1 + m_2 + m_3) F}{m_3}$ (b) $\frac{m_3 F}{m_1 + m_2 + m_3}$

(c) $F - (m_1 + m_2) g$ (d) None of these

9. Four forces act on a point object. The object will be in equilibrium, if

(a) all of them are in the same plane
(b) they are opposite to each other in pairs
(c) the sum of x, y and z-components of forces is zero separately
(d) they form a closed figure of 4 sides when added as per polygon law

10. A light string going over a clamped pulley of mass m supports a block of mass M as shown in the figure. The force on the pulley by the clamp is given by

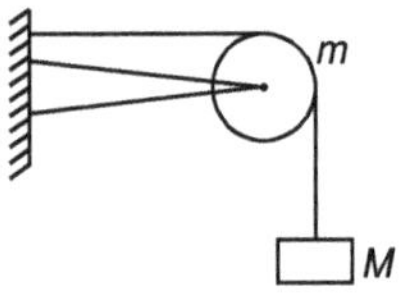

(a) $\sqrt{2}\,Mg$ (b) $\sqrt{2}\,mg$
(c) $g\sqrt{(M+m)^2+m^2}$ (d) $g\sqrt{(M+m)^2+M^2}$

11. A non-uniform rod AB of weight w is supported horizontally in a vertical plane by two light strings PA and QB as shown in the figure. G is the centre of gravity of the rod. If PA and QB make angles 30° and 60° respectively with the vertical, the ratio $\frac{AG}{GB}$ is

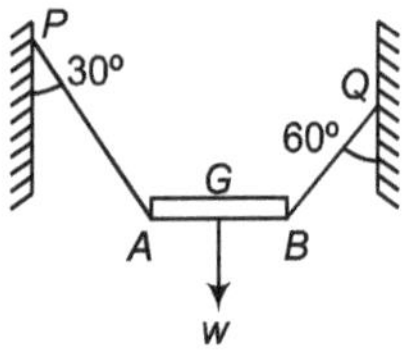

(a) $\frac{1}{2}$ (b) $\sqrt{3}$ (c) $\frac{1}{3}$ (d) $\frac{1}{\sqrt{3}}$

12. A 40 N block supported by two ropes. One rope is horizontal and the other makes an angle of 30° with the ceiling. The tension in the rope attached to the ceiling is approximately
(a) 80 N (b) 40 N
(c) $40\sqrt{3}$ N (d) $\frac{40}{\sqrt{3}}$ N

13. A ball of mass 1kg hangs in equilibrium from a two strings OA and OB as shown in figure. What are the tensions in strings OA and OB? (Take $g = 10\,\text{m/s}^2$)

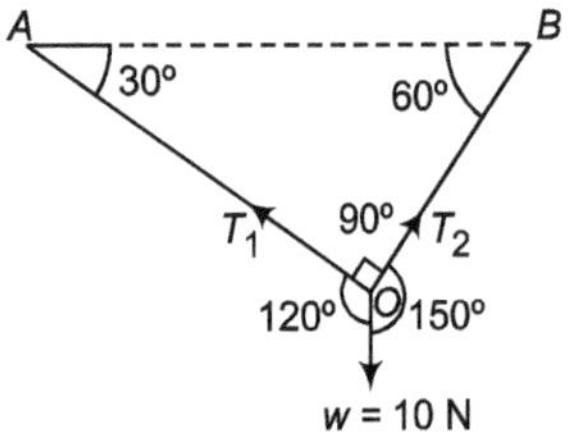

(a) 5 N, 5 N (b) $5\sqrt{3}$ N, $5\sqrt{3}$ N
(c) 5 N, $5\sqrt{3}$ N (d) $5\sqrt{3}$ N, 5N

Acceleration

14. A mass is hanging on a spring balance which is kept in a lift. The lift ascends with increasing speed. The spring balance will show in its reading
(a) increase
(b) decrease
(c) no change
(d) change will depend upon velocity

15. A body of mass 2 kg is hung on a spring balance mounted vertically in a lift. If the lift moves up with an acceleration equal to the acceleration due to gravity, the reading on the spring balance will be
(a) 2 kg (b) $2 \times g$ kg
(c) $4 \times g$ kg (d) 4 kg

16. A monkey is descending from the branch of a tree with a constant acceleration. If the breaking strength of the branch is 75% of the weight of the monkey, the minimum acceleration with which the monkey can slide down without breaking the branch is
(a) g (b) $3g/4$
(c) $g/2$ (d) $g/4$

17. The pendulum hanging from the ceiling of a railway carriage makes an angle 30° with the vertical when it is accelerating. The acceleration of the carriage is
(a) $\frac{\sqrt{3}}{2}g$ (b) $\frac{2}{\sqrt{3}}g$
(c) $g\sqrt{3}$ (d) $\frac{g}{\sqrt{3}}$

18. Two bodies of masses 5 kg and 3 kg respectively are connected to two ends of a light string passing over horizontal frictionless pulley. The tension in the string is (Take $g = 9.8\,\text{m/s}^2$)
(a) 60 N (b) 36.75 N
(c) 73.50 N (d) 18 N

19. Three blocks of masses m_1, m_2 and m_3 are connected by massless strings as shown on a frictionless table. They are pulled with a force $T_3 = 40$ N. If $m_1 = 10$ kg, $m_2 = 6$ kg and $m_3 = 4$ kg, the tension T_2 will be

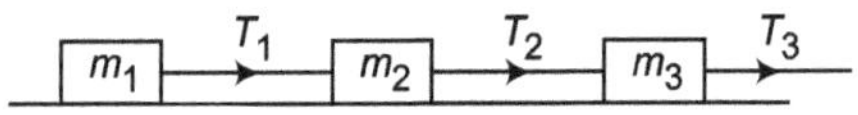

(a) 20 N (b) 40 N
(c) 10 N (d) 32 N

20. Three equal weights A, B and C of mass 2 kg each are hanging on a string passing over a fixed pulley which is frictionless as shown in figure. The tension in the string connecting weight B and C is

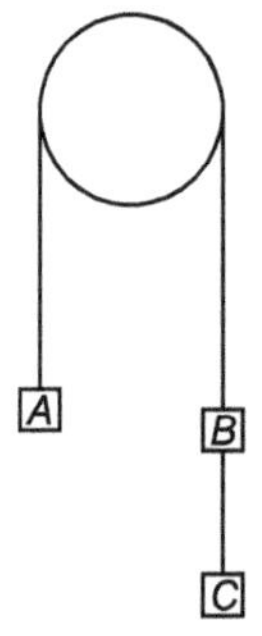

(a) zero (b) 13 N (c) 3.3 N (d) 19.6 N

21. Consider an elevator moving downwards with an acceleration a, the force exerted by a passenger of mass m on the floor of the elevator is

(a) ma (b) $ma - mg$
(c) $mg - ma$ (d) $mg + ma$

22. Two masses M_1 and M_2 are attached to the ends of a string which passes over a pulley attached to the top of an inclined plane. The angle of inclination of the plane is 30° and $M_1 = 10\,\text{kg}$, $M_2 = 5\,\text{kg}$. What is the acceleration of mass M_2?

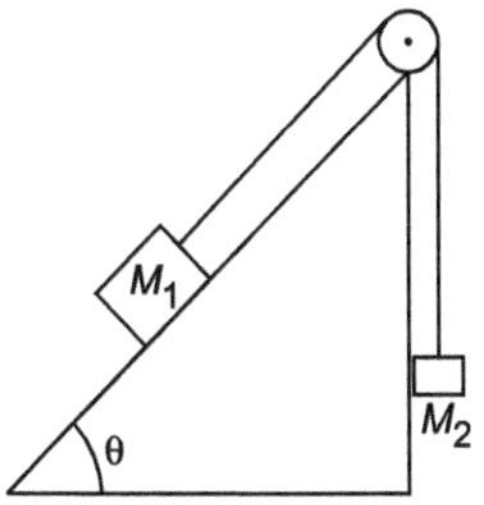

(a) $10\,\text{m/s}^2$ (b) $5\,\text{m/s}^2$
(c) Zero (d) Data insufficient

23. A force of 50 N is required to push a car on a level road with constant speed of $10\,\text{ms}^{-1}$. The mass of the car is 500 kg. What forces should be applied to make the car accelerate at $1\,\text{ms}^{-2}$?

(a) 450 N (b) 500 N
(c) 550 N (d) 2500 N

24. A 50 kg boy stands on a platform spring scale in a lift that is going down with a constant speed 3 m/s. If the lift is brought to rest by a constant deceleration in a distance of 9 m, what does the scale read during this period? (Take $g = 9.8\,\text{m/s}^2$)

(a) 500 N (b) 465 N
(c) 515 N (d) Zero

25. An elevator and its load have a total mass of 800 kg. If the elevator, originally moving downward at 10 m/s, is brought to rest with constant deceleration in a distance of 25 m, the tension in the supporting cable will be (Take $g = 10\,\text{m/s}^2$)

(a) 8000 N
(b) 6400 N
(c) 11200 N
(d) 9600 N

26. A block is placed on the top of a smooth inclined plane of inclination θ kept on the floor of a lift. When the lift is descending with a retardation a, the block is released. The acceleration of the block relative to the incline is

(a) $g\sin\theta$
(b) $a\sin\theta$
(c) $(g - a)\sin\theta$
(d) $(g + a)\sin\theta$

27. The acceleration of the 2 kg block if the free end of string is pulled with a force of 20 N as shown is

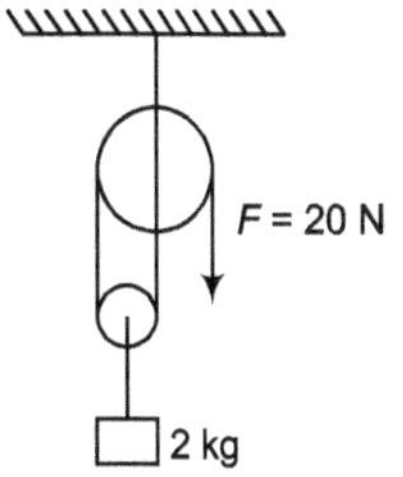

(a) zero (b) $10\,\text{m/s}^2$
(c) $5\,\text{m/s}^2$ upward (d) $5\,\text{m/s}^2$ downward

28. If the elevator in the shown figure is moving upwards with constant acceleration $1\,\text{m/s}^2$, the tension in the string connected to block A of mass 6 kg would be (Take $g = 10\,\text{m/s}^2$)

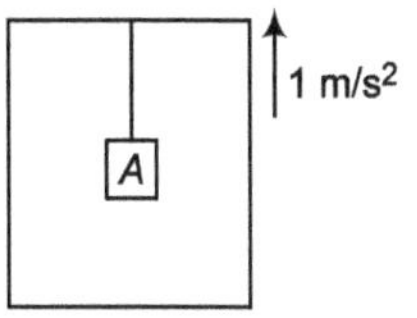

(a) 60 N (b) 66 N
(c) 54 N (d) 42 N

29. Two weights w_1 and w_2 are suspended from the ends of a light string passing over a smooth fixed pulley. If the pulley is pulled up at an acceleration g. The tension in the string will be

(a) $\dfrac{4w_1 w_2}{w_1 + w_2}$ (b) $\dfrac{2w_1 w_2}{w_1 + w_2}$
(c) $\dfrac{w_1 - w_2}{w_1 + w_2}$ (d) $\dfrac{w_1 w_2}{2(w_1 + w_2)}$

30. Two blocks, each having a mass M, rest on frictionless surfaces as shown in the figure. If the pulleys are light and frictionless and M on the incline is allowed to move down, then the tension in the string will be

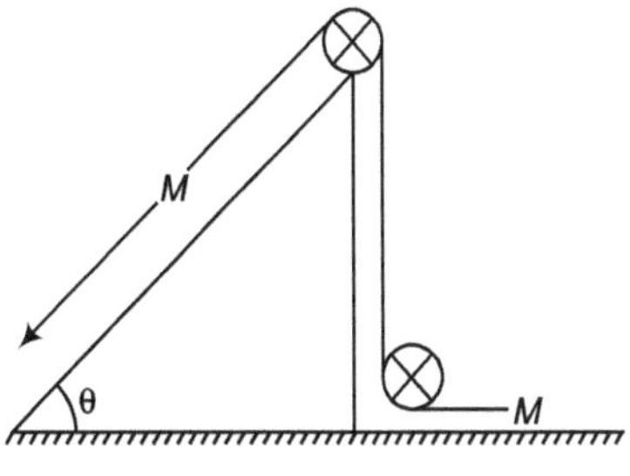

(a) $\dfrac{2}{3}Mg\sin\theta$
(b) $\dfrac{3}{2}Mg\sin\theta$
(c) $\dfrac{Mg\sin\theta}{2}$
(d) $2Mg\sin\theta$

31. When a force F acts on a body of mass m, the acceleration produced in the body is a. If three equal forces $F_1 = F_2 = F_3 = F$ act on the same body as shown in figure. The acceleration produced is

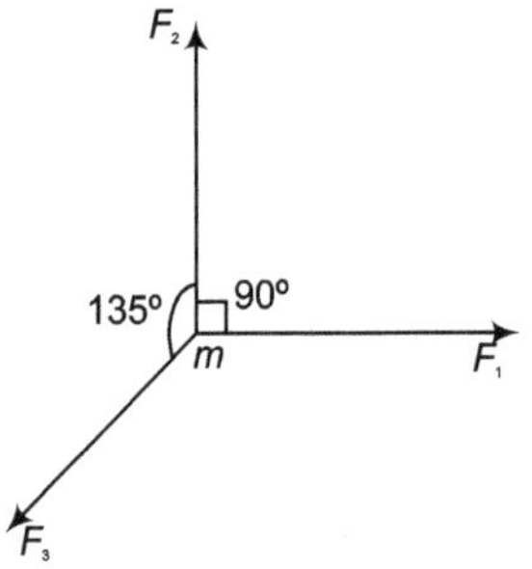

(a) $(\sqrt{2}-1)\,a$ (b) $(\sqrt{2}+1)\,a$
(c) $\sqrt{2}\,a$ (d) a

32. A balloon of weight w is falling vertically downward with a constant acceleration a ($< g$). The magnitude of the air resistance is

(a) w (b) $w\left(1+\frac{a}{g}\right)$
(c) $w\left(1-\frac{a}{g}\right)$ (d) $w\frac{a}{g}$

33. In the arrangement shown in the figure, the pulley has a mass $3m$. Neglecting friction on the contact surface, the force exerted by the supporting rope AB on the ceiling is

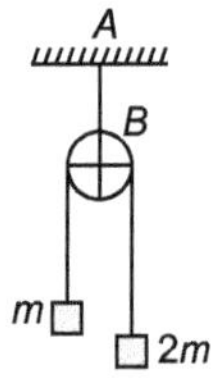

(a) 6 mg (b) 3 mg
(c) 4 mg (d) None of these

Friction

34. Two blocks are connected over a massless pulley as shown in figure. The mass of block A is 10 kg and the coefficient of kinetic friction is 0.2. Block A sliders down the incline at constant speed. The mass of block B in kg is

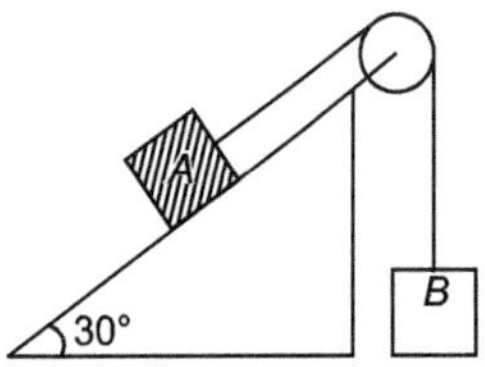

(a) 5.4 (b) 3.3
(c) 4.2 (d) 6.8

35. A block of mass 5 kg resting on a horizontal surface is connected by a cord, passing over a light frictionless pulley to a hanging block of mass 5 kg. The coefficient of kinetic friction between the block and the surface is 0.5. Tension in the cord is (Take $g = 9.8\ \text{m/s}^2$)

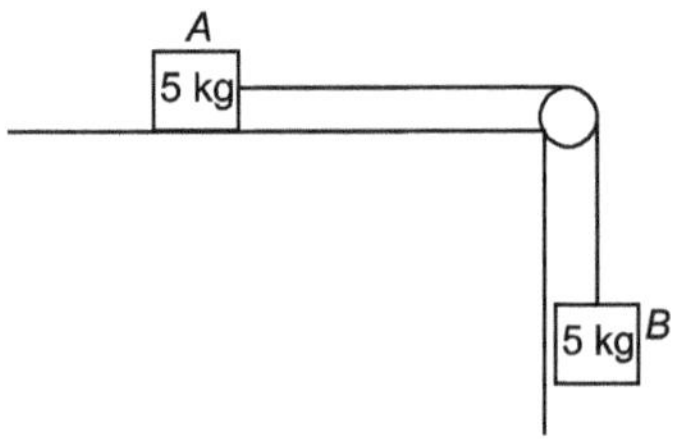

(a) 49 N (b) zero
(c) 36.75 N (d) 12.75

36. A wooden box of mass 8 kg slides down an inclined plane of inclination 30° to the horizontal with a constant acceleration of $0.4\ \text{m/s}^2$. What is the force of friction between the box and inclined plane? (Take $g = 10\ \text{m/s}^2$)

(a) 36.8 N (b) 76.8 N
(c) 65.6 N (d) None of these

37. A 30 kg block rests on a rough horizontal surface. A force of 200 N is applied on the block. The block acquires a speed of 4 m/s starting from rest in 2 s. What is the value of coefficient of friction?

(a) 10/3 (b) $\sqrt{3}/10$
(c) 0.47 (d) 0.184

38. Starting from rest, a body slides down a 45° inclined plane in twice the time it takes to slide down the same distance in the absence of friction. The coefficient of friction between the body and the inclined plane is

(a) 0.2 (b) 0.25
(c) 0.75 (d) 0.5

39. A car having a mass of 1000 kg is moving at a speed of 30 m/s. Brakes are applied to bring the car to rest. If the frictional force between the tyres and the road surface is 5000 N, the car will come to rest in

(a) 5 s (b) 10 s (c) 12 s (d) 6 s

40. A 100 N force acts horizontally on a block of mass 10 kg placed on a horizontal rough table of coefficient of friction $\mu = 0.5$. If g at the place is $10\ \text{ms}^{-2}$, the acceleration of the block is

(a) zero (b) $10\ \text{m/s}^2$
(c) $5\ \text{m/s}^2$ (d) $5.2\ \text{m/s}^2$

41. A block of mass 2 kg is placed on the floor. The coefficient of static friction is 0.4. If a force of 2.8 N is applied on the block parallel to the floor, the force of friction between the block and floor is (Take $g = 10\ \text{ms}^{-2}$)

(a) 2.8 N (b) 8 N (c) 2 N (d) zero

42. A mass placed on an inclined place is just in equilibrium. If μ is coefficient of friction of the surface, then maximum inclination of the plane with the horizontal is
(a) $\tan^{-1}\mu$ (b) $\tan^{-1}(\mu/2)$
(c) $\sin^{-1}\mu$ (d) $\cos^{-1}\mu$

43. A block of mass 0.1 kg is held against a wall applying a horizontal force of 5 N on the block. If the coefficient of friction between the block and the wall is 0.5, the magnitude of the frictional force acting on the block is
(a) 2.5 N (b) 0.98 N
(c) 4.9 N (d) 0.49 N

44. A block of mass 4 kg is placed on a rough horizontal plane. A time dependent force $F = kt^2$ acts on the block, where $k = 2\,\text{N/s}^2$. Coefficient of friction $\mu = 0.8$. Force of friction between block and the plane at $t = 2\,\text{s}$ is
(a) 8 N (b) 4 N
(c) 2 N (d) 32 N

45. A body is projected along a rough horizontal surface with a velocity 6 m/s. If the body comes to rest after travelling 9 m, then coefficient of sliding friction, is ($g = 10\,\text{m/s}^2$)
(a) 0.5 (b) 0.4
(c) 0.6 (d) 0.2

46. A block of mass m is given an initial downward velocity v_0 and left on an inclined place (coefficient of friction $= 0.6$). The block will

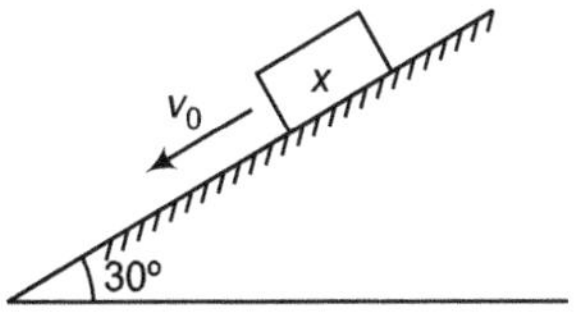

(a) continue of move down the plane with constant velocity v_0
(b) accelerate downward
(c) decelerate and come to rest
(d) first accelerate downward then decelerate

47. A block of weight 5 N is pushed against a vertical wall by a force 12 N. The coefficient of friction between the wall and block is 0.6. The magnitude of the force exerted by the wall on the block is

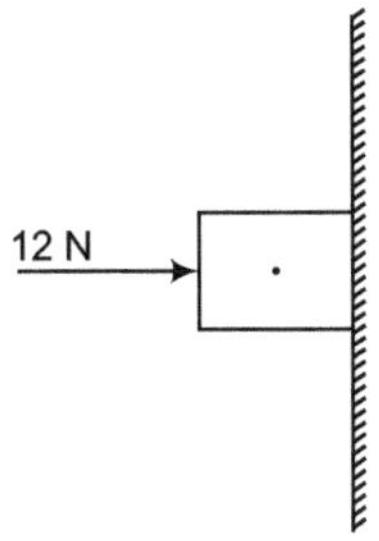

(a) 12 N (b) 5 N
(c) 7.2 N (d) 13 N

48. A body is projected up a 45° rough incline. If the coefficient of friction is 0.5, then the retardation of the block is
(a) $\frac{g}{2\sqrt{2}}$ (b) $\frac{g}{\sqrt{2}}$
(c) $\frac{3g}{2\sqrt{2}}$ (d) $\frac{g}{2}$

49. The coefficient of friction between the tyres and road is 0.4. The minimum distance covered before attaining a speed of $8\,\text{ms}^{-1}$ starting from rest is nearly ($g=10\,\text{m/s}^2$)
(a) 8.0 m (b) 4.0 m
(c) 10.0 m (d) 16.0 m

50. A block has been placed on an inclined plane. The slope angle θ of the place is such that the block slides down the plane at a constant speed. The coefficient of kinetic friction is equal to
(a) $\sin\theta$ (b) $\cos\theta$
(c) g (d) $\tan\theta$

51. A mass m rests on a horizontal surface. The coefficient of friction between the mass and the surface is μ. If the mass is pulled by a force F as shown in figure. The limiting friction between mass and the surface will be

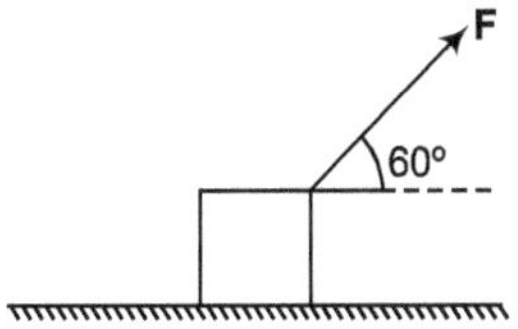

(a) μmg (b) $\mu\left[mg - \left(\frac{\sqrt{3}}{2}\right)F\right]$
(c) $\mu\left[mg - \left(\frac{F}{2}\right)\right]$ (d) $\mu\left[mg + \left(\frac{F}{2}\right)\right]$

52. A block of mass 5 kg is kept on a horizontal floor having coefficient of friction 0.09. Two mutually perpendicular horizontal forces of 3 N and 4 N act on this block. The acceleration of the block is ($g = 10\,\text{m/s}^2$)
(a) zero
(b) $0.1\,\text{m/s}^2$
(c) $0.2\,\text{m/s}^2$
(d) $0.3\,\text{m/s}^2$

53. A block of mass 3 kg is at rest on a rough inclined plane as shown in the figure. The magnitude of net force exerted by the surface on the block will be

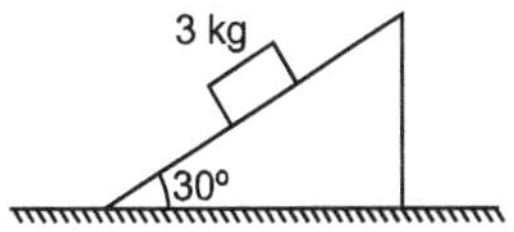

(a) $15\sqrt{3}$ N (b) 15 N
(c) 10 N (d) 30 N

54. A block of mass m is at rest on an inclined plane which is making angle θ with the horizontal. The coefficient of friction between the block and plane is μ. Then, frictional force acting between the surfaces is

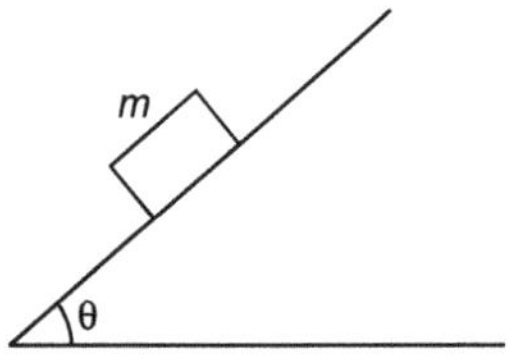

(a) $\mu\, mg$
(b) $\mu\, mg \sin\theta$
(c) $\mu\,(mg \sin\theta - mg\cos\theta)$
(d) $mg\sin\theta$

55. A block of mass 2 kg rests on a rough inclined plane making an angle of 30° with the horizontal. The coefficient of static friction between the block and the plane is 0.7. The frictional force on the block is
(a) 9.8 N
(b) $0.7\times 9.8\times\sqrt{3}$ N
(c) $9.8\times\sqrt{3}$ N
(d) 0.7×9.8 N

56. A minimum force F is applied to a block of mass 102 kg to prevent it from sliding on a plane with an inclination angle 30° with the horizontal. If the coefficients of static and kinetic friction between the block and the plane are 0.4 and 0.3 respectively, then the force F is
(a) 157 N
(b) 224 N
(c) 315 N
(d) zero

57. A uniform rope of length l lies on a table. If the coefficient of friction is μ, then the maximum length l_1 of the hanging part of the rope which can overhang from the edge of the table without sliding down is
(a) l/μ
(b) $l/(\mu + 1)$
(c) $\mu l/(\mu + 1)$
(d) $\mu l/(\mu - 1)$

58. An insect crawls up a hemispherical surface very slowly (see the figure). The coefficient of friction between the insect and the surface is $\frac{1}{3}$. If the line joining the centre of the hemispherical surface to the insect makes an angle α with the vertical, the maximum possible value of α is given by

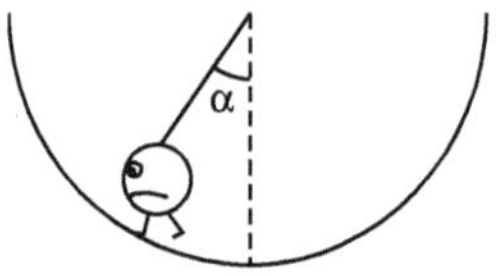

(a) $\cot\alpha = 3$ (b) $\tan\alpha = 3$
(c) $\sec\alpha = 3$ (d) $\operatorname{cosec}\alpha = 3$

59. The upper half of an inclined plane of inclination θ is perfectly smooth while the lower half rough. A block starting from rest at the top of the plane will again come to rest at the bottom if the coefficient of friction between the block and the lower half of the plane is given by
(a) $\mu = 2\tan\theta$ (b) $\mu = \tan\theta$
(c) $\mu = \frac{2}{\tan\theta}$ (d) $\mu = \frac{1}{\tan\theta}$

60. A block is gently placed on a conveyor belt moving horizontally with constant speed. After $t = 4$ s, the velocity of the block becomes equal to the velocity of the belt. If the coefficient of friction between the block and the belt is $\mu = 0.2$, then the velocity of the conveyor belt is
(a) $3\,\text{ms}^{-1}$ (b) $4\,\text{ms}^{-1}$
(c) 6ms^{-1} (d) 8ms^{-1}

61. The breaking strength of the cable used to pull a body is 40 N. A body of mass 8 kg is resting on a table of coefficient of friction $\mu = 0.20$. The maximum acceleration which can be produced by the cable is (Take $g = 10\,\text{ms}^{-2}$)
(a) 6ms^{-2} (b) $3\,\text{ms}^{-2}$
(c) 8ms^{-1} (d) 8ms^{-2}

62. A 13 m ladder is placed against a smooth vertical wall with its lower end 5m from the wall. What should be the minimum coefficient of friction between ladder and floor so that it remains in equilibrium?
(a) 0.36 (b) 0.72
(c) 0.21 (d) 0.52

63. A box of mass 8 kg is placed on a rough inclined plane of inclination θ. Its downward motion can be prevented by applying an upward pull F. And it can be made to slide upwards by applying a force $2F$. The coefficient of friction between the box and the inclined plane is
(a) $\frac{1}{3}\tan\theta$ (b) $3\tan\theta$
(c) $\frac{1}{2}\tan\theta$ (d) $2\tan\theta$

64. A block of mass m, lying on a rough horizontal plane, is acted upon by a horizontal force P and another force Q, inclined at an angle θ to the vertical upwards. The block will remain in equilibrium, if minimum coefficient of friction between it and the surface is
(a) $\frac{(P + Q\sin\theta)}{(mg - Q\cos\theta)}$
(b) $\frac{(P\cos\theta + Q)}{(mg - Q\sin\theta)}$
(c) $\frac{(P - Q\cos\theta)}{(mg + Q\sin\theta)}$
(d) $\frac{(P\sin\theta - Q)}{(mg - Q\cos\theta)}$

65. In the figure shown, if coefficient of friction is μ, then m_2 will start moving upwards if

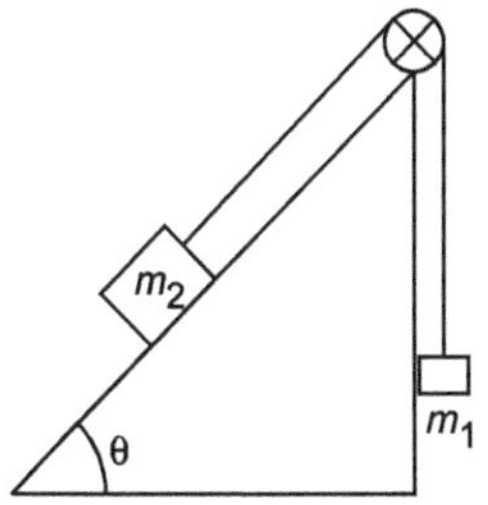

(a) $\frac{m_1}{m_2} > \sin\theta - \mu\cos\theta$ (b) $\frac{m_1}{m_2} > \sin\theta + \mu\cos\theta$

(c) $\frac{m_1}{m_2} > \mu\sin\theta - \cos\theta$ (d) $\frac{m_1}{m_2} > \mu\sin\theta + \cos\theta$

66. A block of mass M rests on a rough horizontal surface as shown. Coefficient of friction between the block and the surface is μ. A force $F = Mg$ acting at angle θ with the vertical side of the block pulls it. In which of the following cases, the block can be pulled along the surface?

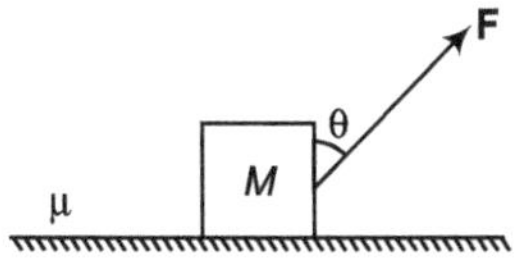

(a) $\tan\theta \geq \mu$ (b) $\tan\left(\frac{\theta}{2} \geq \mu\right)$

(c) $\cos\theta \geq \mu$ (d) $\cot\left(\frac{\theta}{2}\right) \geq \mu$

67. A man of mass 60 kg is standing on a horizontal conveyor belt. When the belt is given an acceleration of 1 ms^{-2}, the man remains stationary with respect to the moving belt. If $g = 10 ms^{-2}$, the net force acting on the man is

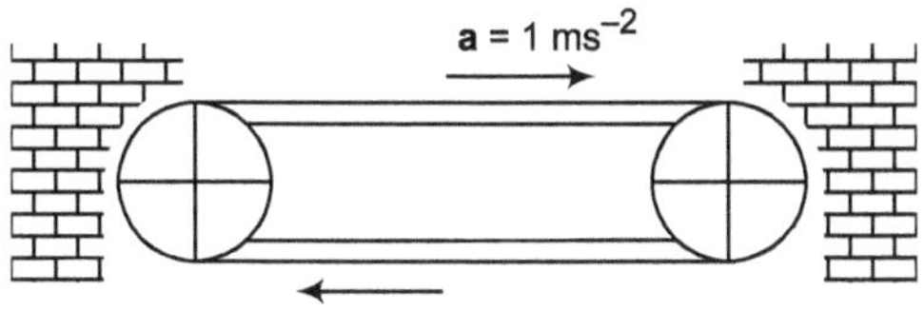

(a) zero (b) 120 N
(c) 60 N (d) 600 N

Miscellaneous Problems

68. Three masses of 6 kg, 4 kg and 2 kg are attached to a rigid support as shown in figure. If the string attached to the support breaks and the system falls freely, then the tension in the string connecting 4 kg and 2 kg mass is

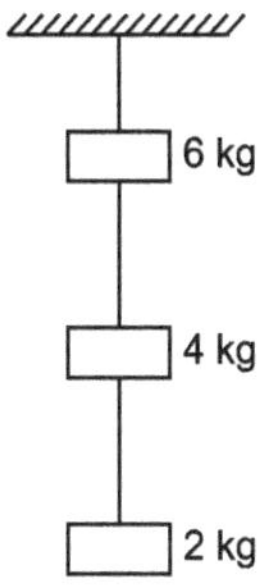

(a) zero (b) 8 kg-wt
(c) 12 kg-wt (d) 6 kg-wt

69. A ball of mass 0.5 kg moving with a velocity of 2 m/s strikes a wall normally and bounces back with the same speed. If the time of contact between the ball and the wall is one millisecond, the average force exerted by the wall on the ball is

(a) 2000 N (b) 1000 N
(c) 5000 N (d) 125 N

70. If a body loses half of its velocity on penetrating 3 cm in a wooden block, then how much will it penetrate more before coming to rest?

(a) 1cm (b) 2 cm
(c) 3 cm (d) 4 cm

71. A ball is dropped on to the floor from a height of 10 m. It rebounds to a height of 2.5 m. If the ball is in contact with floor for 0.01 s. What is the average acceleration during contact?

(a) 700 m/s^2 (b) 1400 m/s^2
(c) 2100 m/s^2 (d) 2800 m/s^2

72. Pushing force making an angle θ to the horizontal is applied on a block of weight w placed on a horizontal table. If the angle of friction is ϕ, the magnitude of force required to move the body is equal to

(a) $\frac{w\cos\theta}{\cos(\theta - \phi)}$ (b) $\frac{w\sin\theta}{\cos(\theta + \phi)}$

(c) $\frac{w\tan\phi}{\sin(\theta - \phi)}$ (d) $\frac{w\sin\phi}{\tan(\theta - \phi)}$

73. A block of mass m is placed on a smooth plane inclined at an angle θ with the horizontal. The force exerted by the plane on the block has a magnitude

(a) mg (b) $mg\sec\theta$
(c) $mg\cos\theta$ (d) $mg\sin\theta$

74. Two blocks are in contact on a frictionless table. One has mass m and the other $2m$. A force F is applied on $2m$ as shown in the figure. Now, the same force F is applied from the right on m. In the two cases, the ratio of force of contact between the two blocks will be

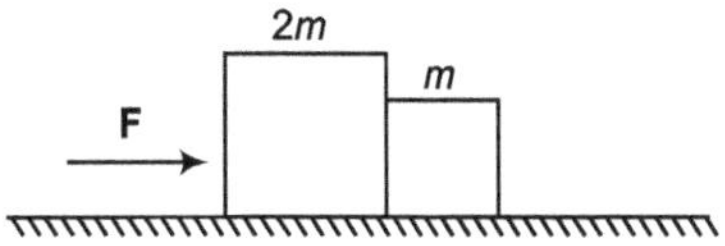

(a) same (b) 1: 2
(c) 2 : 1 (d) 1 : 3

75. A dynamometer D is attached to two bodies of masses $M = 6$ kg and $m = 4$ kg. Forces $F = 20$N and $f = 10$N are applied to the masses as shown. The dynamometer reads

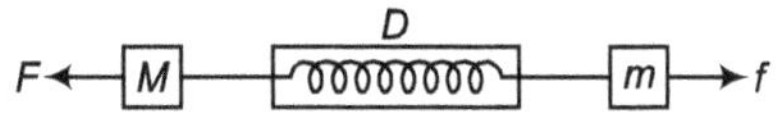

(a) 10N (b) 20 N
(c) 6 N (d) 14 N

76. A smooth inclined plane of length L, having an inclination θ with horizontal is inside a lift which is moving down with retardation a. The time taken by a block to slide down the inclined plane from rest will be

(a) $\sqrt{\frac{2L}{\sqrt{a\sin\theta}}}$ (b) $\sqrt{\frac{2L}{g\sin\theta}}$

(c) $\sqrt{\frac{2L}{(g-a)\sin\theta}}$ (d) $\sqrt{\frac{2L}{(g+a)\sin\theta}}$

77. Consider the shown arrangement. Assume all surfaces to be smooth. If N represents magnitudes of normal reaction between block and wedge, then acceleration of M along horizontal is equal to

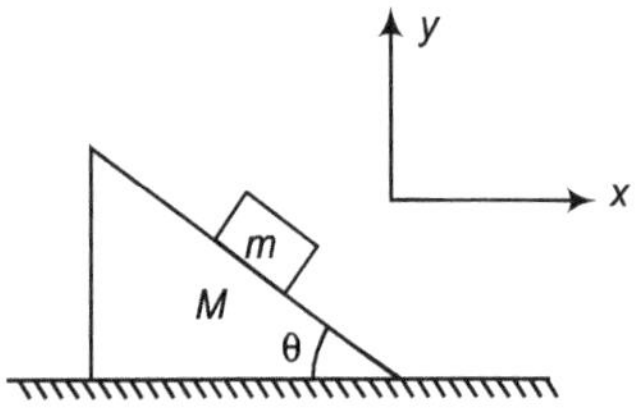

(a) $\frac{N\sin\theta}{M}$ (along + ve x-axis)

(b) $\frac{N\cos\theta}{M}$ (along − ve x-axis)

(c) $\frac{N\sin\theta}{M}$ (along − ve x-axis)

(d) $\frac{N\sin\theta}{m+M}$ (along − ve x-axis)

78. In the above problem, normal reaction between ground and wedge will have magnitude equal to

(a) $N\cos\theta + Mg$ (b) $N\cos\theta + Mg + mg$

(c) $N\cos\theta - Mg$ (d) $N\sin\theta + Mg + mg$

79. For the arrangement shown in the figure, the tension in the string is given by $\left(\sin 37° = \frac{3}{5}\right)$

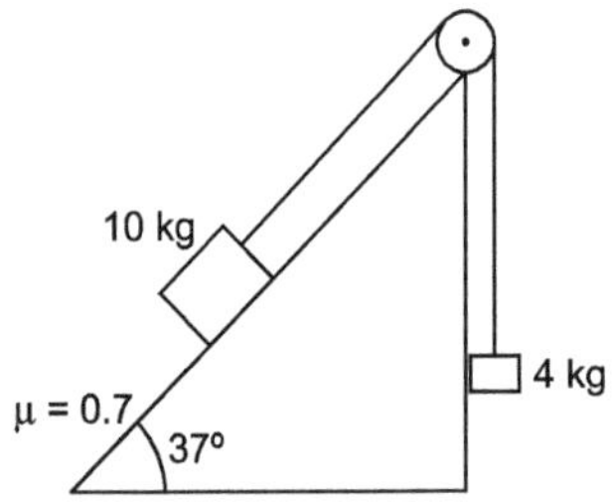

(a) 30 N (b) 40 N (c) 60 N (d) 30 N

80. A body of mass 10 kg is placed on rough surface, pulled by a force F making an angle of 30° above the horizontal. If the angle of friction is also 30°, then the minimum magnitude of force F required to move the body is equal to (Take $g = 10\,\text{m/s}^2$)

(a) 100 N (b) $50\sqrt{2}$ N (c) $100\sqrt{2}$ N (d) 50 N

81. There is no slipping between the two blocks. What is the force of friction between two blocks?

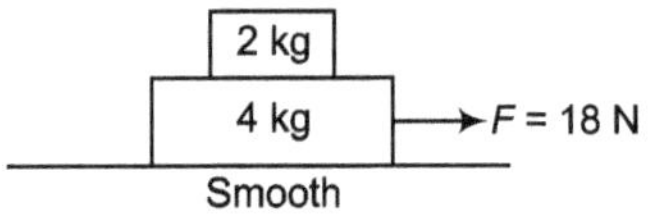

(a) Zero (b) 9 N (c) 12 N (d) 6 N

82. Two masses are connected by a string which passes over a pulley accelerating upwards at a rate A as shown. If a_1 and a_2 be the accelerations of bodies 1 and 2 respectively, then

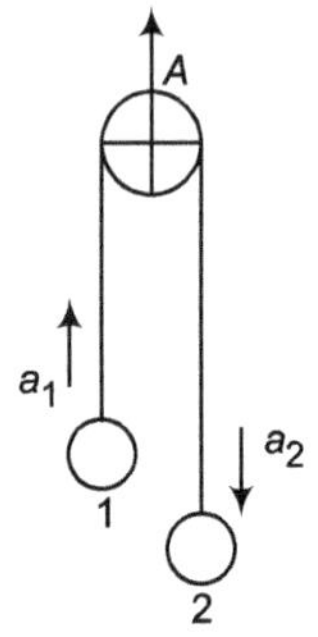

(a) $A = a_1 - a_2$ (b) $A = a_1 + a_2$

(c) $A = \frac{a_1 - a_2}{2}$ (d) $A = \frac{a_1 + a_2}{2}$

83. In the shown arrangement, mass of $A = 1\,\text{kg}$, mass of $B = 2\,\text{kg}$. Coefficient of friction between A and $B = 0.2$

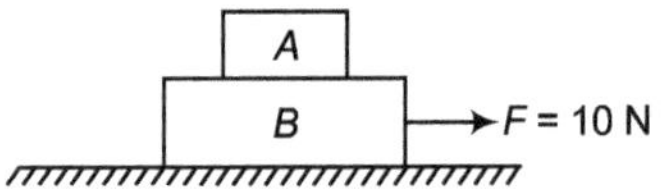

There is no friction between B and ground. The frictional force exerted by A on B equals

(a) 2 N (b) 3 N

(c) 4 N (d) 5 N

84. Find the force exerted by 5 kg block on floor of lift, as shown in figure. (Take $g = 10\,\text{m/s}^2$)

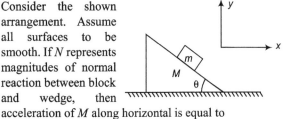

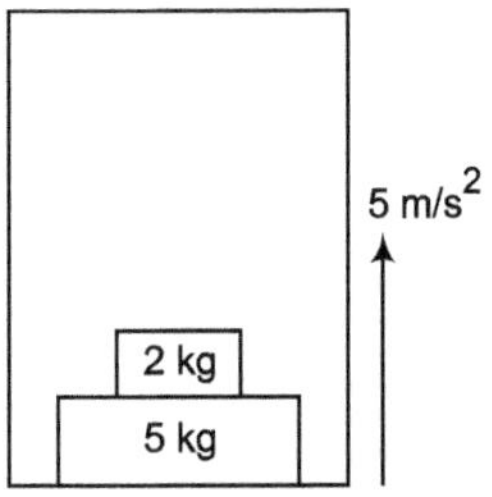

(a) 100 N (b) 115 N

(c) 105 N (d) 135 N

[Level 2]

Only One Correct Option

1. A 4 kg block A is placed on the top of 8 kg block B which rests on a smooth table.

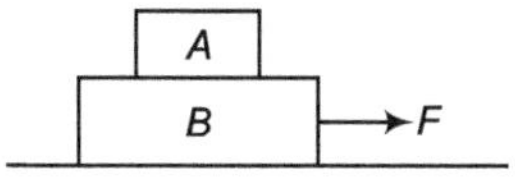

A just slips on B when a force of 12 N is applied on A. Then, the maximum horizontal force F applied on B to make both A and B move together, is

(a) 12 N (b) 24 N
(c) 36 N (d) 48 N

2. Find the value of friction forces between the blocks A and B and between B and ground. (Take $g = 10\ \text{m/s}^2$)

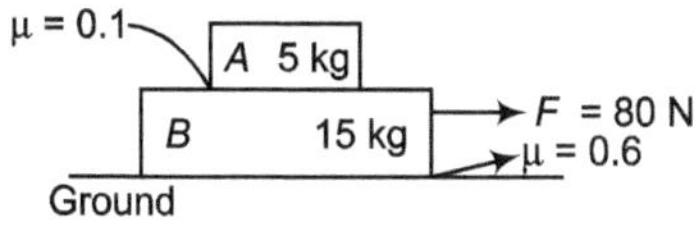

(a) 90 N, 5 N (b) 5 N, 90 N
(c) 5 N, 75 N (d) 0 N, 80 N

3. A block of mass m is placed on another block of mass M which itself is lying on a horizontal surface. The coefficient of friction between two blocks is μ_1 and that between the block of mass M and horizontal surface is μ_2. What maximum horizontal force can be applied to the lower block so that the two blocks move without separation?

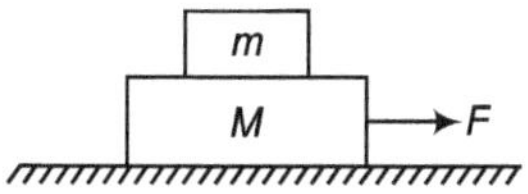

(a) $(M + m)(\mu_2 - \mu_1)g$ (b) $(M - m)(\mu_2 + \mu_1)g$
(c) $(M - m)(\mu_2 + \mu_1)g$ (d) $(M + m)(\mu_2 + \mu_1)g$

4. A block of mass m is placed on the top of another block of mass M as shown in the figure. The coefficient of friction between them is μ.

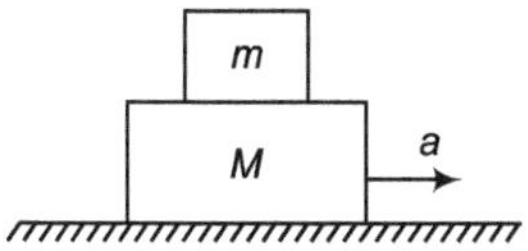

The maximum acceleration with which the block M may move so that m also moves along with it is

(a) μg (b) $\mu \frac{M}{m} g$
(c) $\mu \frac{m}{M} g$ (d) $\frac{g}{\mu}$

5. Block A of mass m rests on the plank B of mass $3m$ which is free to slide on a frictionless horizontal surface. The coefficient of friction between the block and plank is 0.2. If a horizontal force of magnitude $2mg$ is applied to the plank B, then acceleration of A relative to the plank and relative to the ground respectively, are

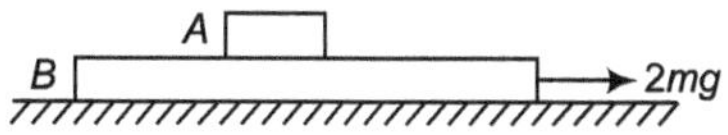

(a) $0, \frac{g}{2}$ (b) $0, \frac{2g}{3}$
(c) $\frac{3g}{5}, \frac{g}{5}$ (d) $\frac{2g}{5}, \frac{g}{5}$

6. A pivoted beam of negligible mass has a mass suspended from one end and an at wood's machine suspended from the other. The frictionless pulley has negligible mass and dimension.

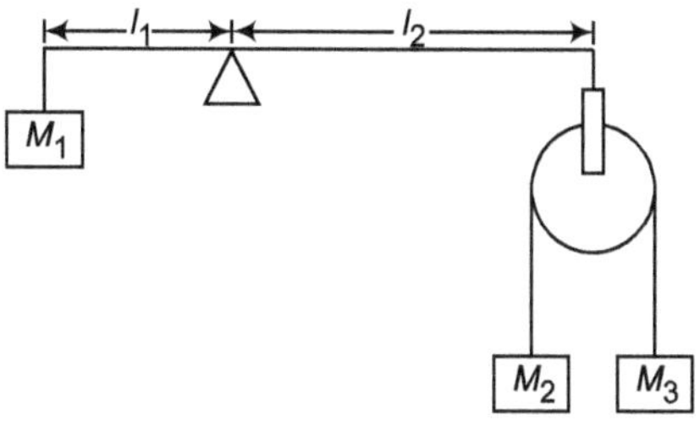

Gravity is directed downwards and $M_2 = 3M_3$, $l_2 = 3l_1$. Find the ratio $\frac{M_1}{M_2}$ which will ensure that the beam has no tendency to rotate just after the masses are released.

(a) $\frac{M_1}{M_2} = 2$ (b) $\frac{M_1}{M_2} = 3$
(c) $\frac{M_1}{M_2} = 4$ (d) None of these

7. If coefficient of friction between all surfaces is 0.4, then find the minimum force F to have equilibrium of the system. (Take $g = 10\text{m/s}^2$)

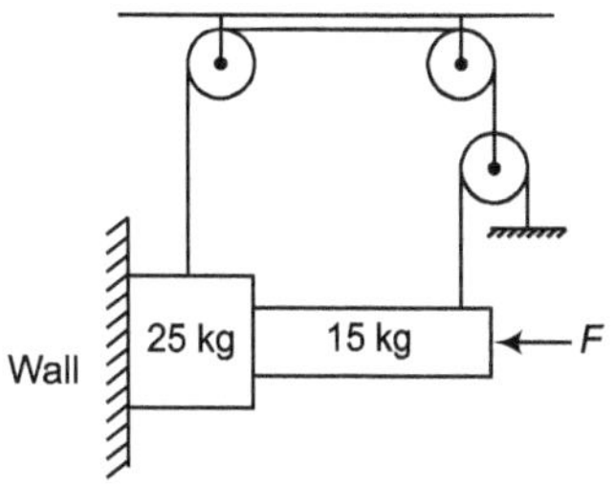

(a) 62.5 N (b) 150 N
(c) 135 N (d) 50 N

8. In the arrangement shown in figure, there is a friction force between the blocks of masses m and $2m$. The mass of the suspended block is m. The block of mass m is stationary with respect to block of mass $2m$. The minimum value of coefficient of friction between m and $2m$ is

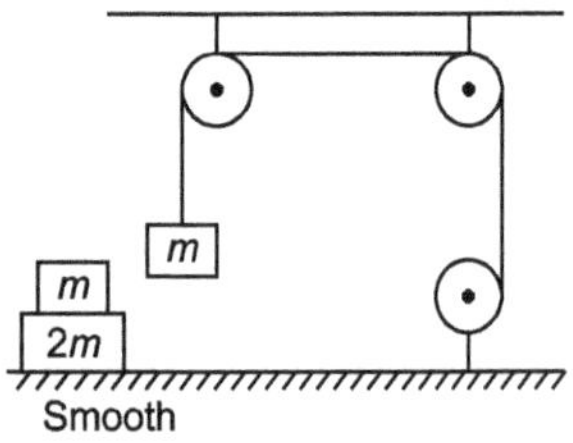

(a) $\frac{1}{2}$ (b) $\frac{1}{\sqrt{2}}$ (c) $\frac{1}{4}$ (d) $\frac{1}{3}$

9. A block of mass 5 kg is kept on a horizontal floor having coefficient of friction 0.09. Two mutually perpendicular horizontal forces of 3 N and 4 N act on this block. The acceleration of the block is ($g = 10 \text{m/s}^2$)

(a) zero (b) 0.1 m/s^2 (c) 0.2 m/s^2 (d) 0.3 m/s^2

10. Block B moves to the right with a constant velocity v_0. The velocity of body A relative to B is

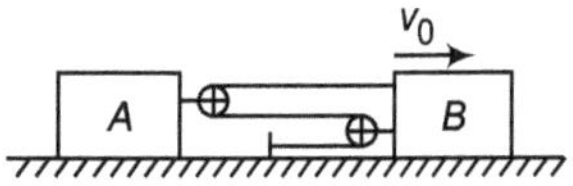

(a) $\frac{v_0}{2}$, towards left (b) $\frac{v_0}{2}$, towards right

(c) $\frac{3v_0}{2}$, towards left (d) $\frac{3v_0}{2}$, towards right

11. A mass of 3 kg descending vertically downwards supports a mass of 2 kg by means of a light string passing over a pulley. At the end of 5 s the string breaks . How much high from now the 2 kg mass will go? ($g = 10 \text{m/s}^2$)

(a) 4.9 m (b) 9.8 m
(c) 19.6 m (d) 2.45 m

12. The rear side of a truck is open and a box of mass 20 kg is placed on the truck 4 m away from the open end. $\mu = 0.15$ and $g = 10 \text{m/s}^2$. The truck starts from rest with an acceleration of 2m/s^2 on a straight road. The box will fall off the truck when it is at a distance from the starting point equal to

(a) 4 m (b) 8 m
(c) 16 m (d) 32 m

13. In the figure, pulleys are smooth and strings are massless, $m_1 = 1$ kg and $m_2 = \frac{1}{3}$ kg. To keep m_3 at rest, mass m_3 should be

(a) 1 kg
(b) $\frac{2}{3}$ kg
(c) $\frac{1}{4}$ kg
(d) 2 kg

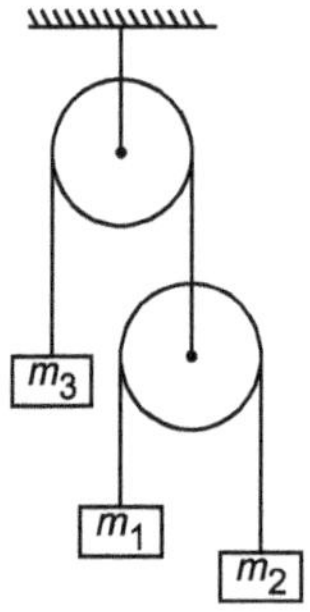

14. A pendulum of mass m hangs from a support fixed to a trolley. The direction of the string (i.e. angle θ) when the trolley rolls up a plane of inclination α with acceleration a is

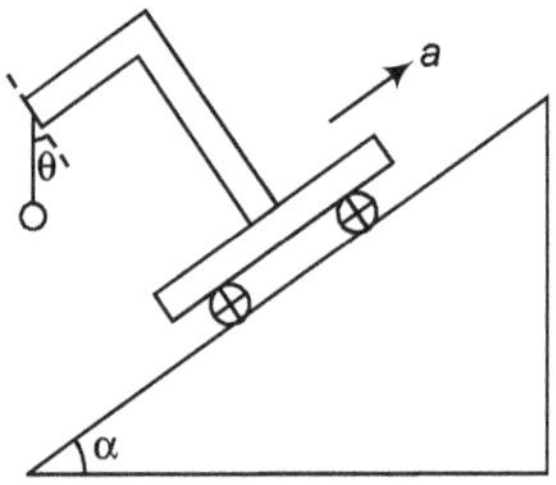

(a) 0 (b) $\tan^{-1}\alpha$

(c) $\tan^{-1}\left(\frac{a + g\sin\alpha}{g\cos\alpha}\right)$ (d) $\tan^{-1}\left(\frac{a}{g}\right)$

15. Two masses m and M are attached with strings as shown. For the system to be in equilibrium, we have

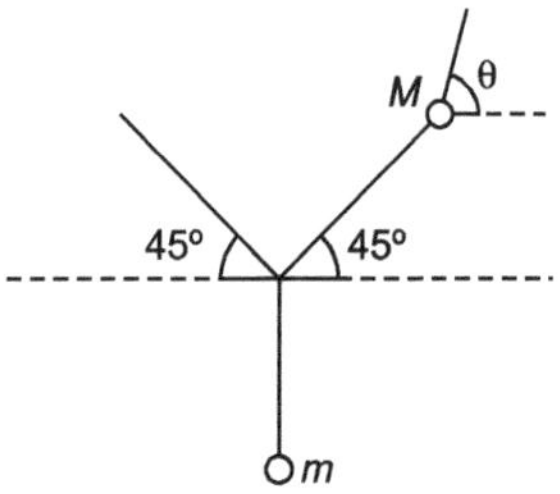

(a) $\tan\theta = 1 + \frac{2M}{m}$ (b) $\tan\theta = 1 + \frac{2m}{M}$

(c) $\tan\theta = 1 + \frac{M}{2m}$ (d) $\tan\theta = 1 + \frac{m}{2M}$

16. Two blocks of mass $m = 5$ kg and $M = 10$ kg are connected by a string passing over a pulley B as shown. Another string connects the centre of pulley B to the floor and passes over another pulley A as shown. An upward force F is applied at the centre of pulley A. Both the pulleys are massless. The accelerations of blocks m and M, if F is 300 N are

(Take $g = 10 \text{m/s}^2$)

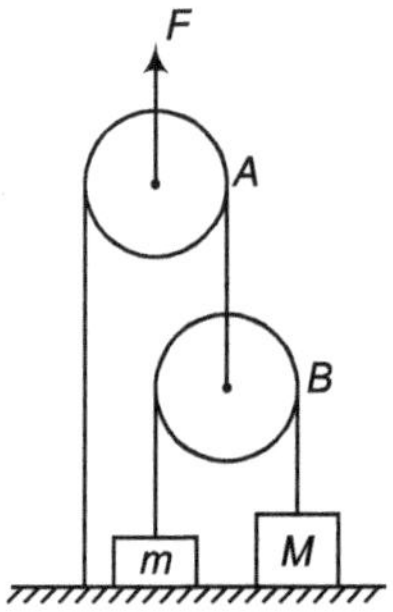

(a) 5 m/s^2, zero (b) zero, 5 m/s^2

(c) zero, zero (d) 5 m/s^2, 5 m/s^2

17. A body weighs 8 g when placed in one pan and 18 g when placed on the other pan of a false balance. If the beam is horizontal when both the pans are empty, the true weight of the body is

(a) 13 g (b) 12 g
(c) 15.5 g (d) 15 g

18. A man of mass m has fallen into a ditch of width d. Two of his friends are slowly pulling him out using a light rope and two fixed pulleys as shown in figure. Both the friends exert force of equal magnitudes F.

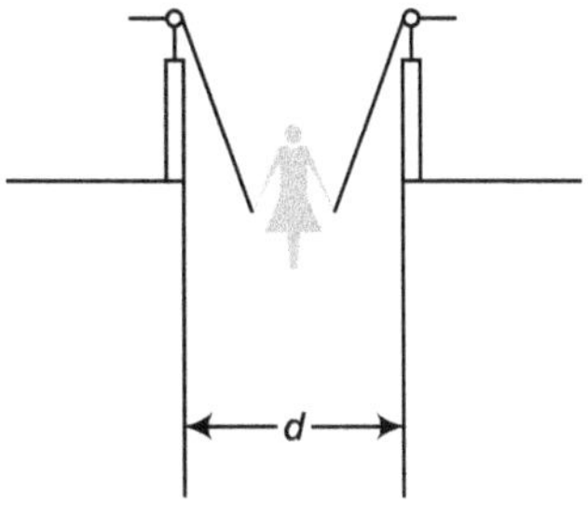

When the man is at a depth h, the value of F is

(a) $\frac{mg}{4h}\sqrt{d^2 + 4h^2}$ (b) hmg
(c) dmg (d) $\frac{mg}{2h}\sqrt{h^2 + d^2}$

19. A man of mass m stands on a platform of equal mass m and pulls himself by two ropes passing over pulleys as shown in figure. If he pulls each rope with a force equal to half his weight, his upward acceleration would be

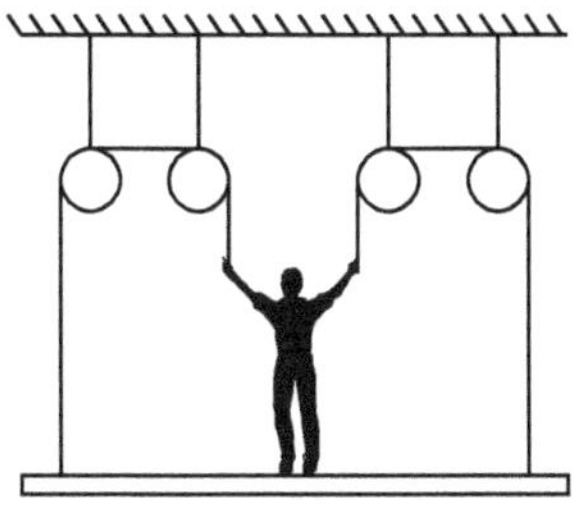

(a) $\frac{g}{2}$ (b) $\frac{g}{4}$
(c) g (d) zero

20. If the coefficient of friction between A and B is μ, the maximum acceleration of the wedge A for which B will remain at rest with respect to the wedge is

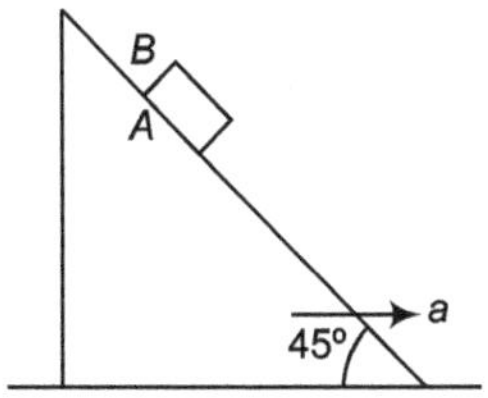

(a) μg (b) $g\left(\frac{1+\mu}{1-\mu}\right)$ (c) $g\left(\frac{1-\mu}{1+\mu}\right)$ (d) $\frac{g}{\mu}$

21. Two blocks A and B each of mass m are placed on a smooth horizontal surface. Two horizontal force F and $2F$ are applied on the blocks A and B respectively as shown in figure. The block A does not slide on block B.

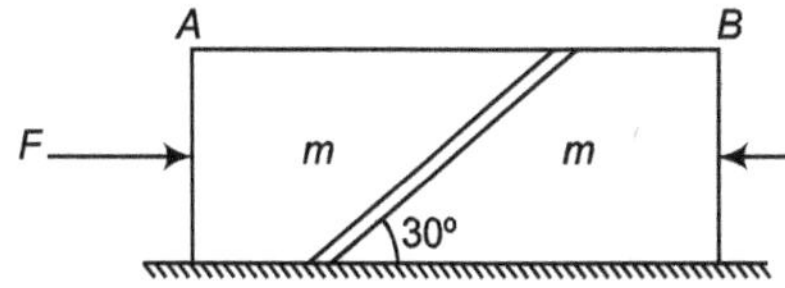

Then, the normal reaction acting between the two blocks is

(a) F (b) $\frac{F}{2}$ (c) $\frac{F}{\sqrt{3}}$ (d) $3F$

22. Two beads A and B move along a semicircular wire frame as shown in figure . The beads are connected by an inelastic string which always remains tight. At an instant the speed of A is u, $\angle BAC = 45°$ and $\angle BOC = 75°$, where O is the centre of the semicircular arc. The speed of bead B at that instant is

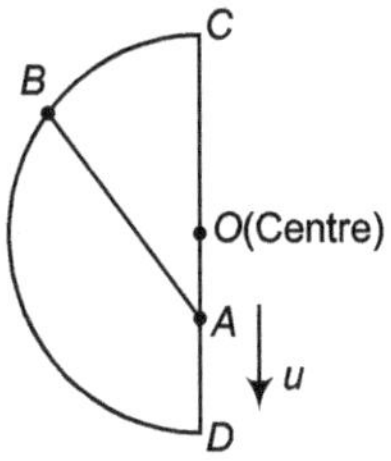

(a) $\sqrt{2}u$ (b) u (c) $\frac{u}{2\sqrt{2}}$ (d) $\sqrt{\frac{2}{3}}u$

23. In the figure shown, a person wants to raise a block lying on the ground to a height h. In both the cases, if time required is the same then in which case he has to exert more force. Assume pulleys and strings are light

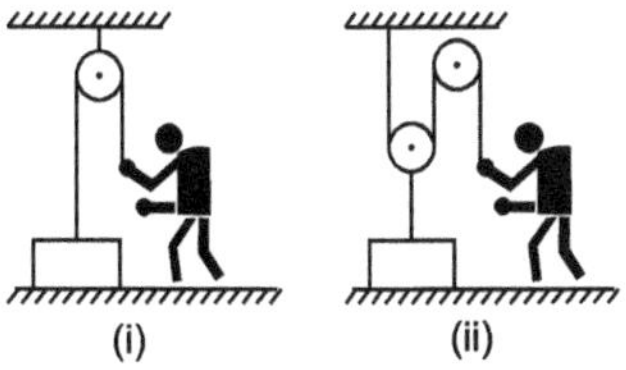

(a) (i) (b) (ii)
(c) same in both (d) Cannot be determined

24. Two blocks of masses m and $2m$ are placed one over the other as shown in figure.The coefficient of friction between m and $2m$ is μ and between $2m$ and ground is $\frac{\mu}{3}$.

If a horizontal force F is applied on upper block and T is tension developed in string, then choose the incorrect alternative.

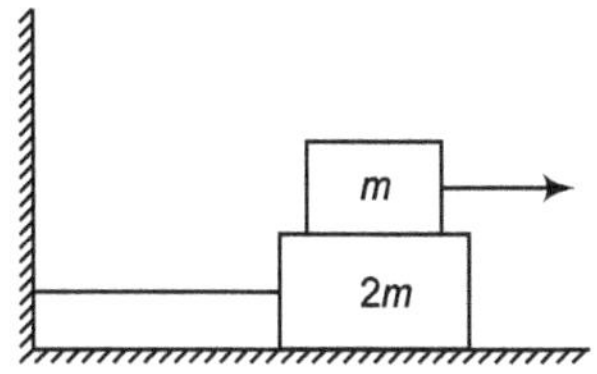

(a) If $F = \frac{\mu}{3}mg, T = 0$ (b) If $F = \mu mg, T = 0$
(c) If $F = 2\mu mg, T = \frac{\mu mg}{3}$ (d) If $F = 3\mu mg, T = 0$

25. Two blocks of masses 2 m and m are in equilibrium as shown in the figure . Now, the string between the blocks is suddenly broken. The accelerations of the blocks A and B respectively at that instant are

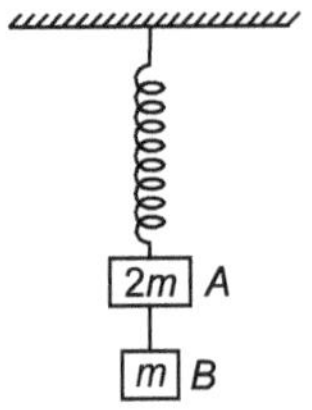

(a) g and g (b) g and $\frac{g}{2}$ (c) $\frac{g}{2}$ and g (d) $\frac{g}{2}$ and $\frac{g}{2}$

26. A block of mass m is placed on a wedge of mass $2m$ which rests on a rough horizontal surface. There is no friction between the block and the wedge. The minimum coefficient of friction between the wedge and the ground so that the wedge does not move is

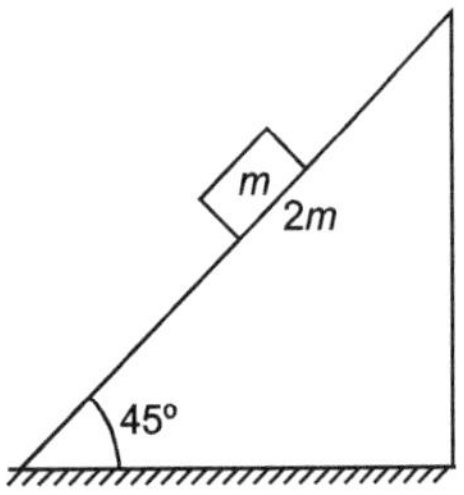

(a) 0.1 (b) 0.20 (c) 0.3 (d) 0.4

27. Two blocks of mass 5 kg and 3 kg are attached to the ends of a string passing over a smooth pulley fixed to the ceiling of an elevator. A man inside the elevator accelerated upwards, finds the acceleration of the blocks to be $\frac{9}{32}g$. The acceleration of the elevator is

(a) $\frac{g}{3}$ (b) $\frac{g}{4}$ (c) $\frac{g}{8}$ (d) $\frac{g}{6}$

28. The line of action of the resultant of two like parallel forces shifts by one-fourth of the distance between the forces when the two forces are interchanged. The ratio of the two forces is

(a) 1 : 2 (b) 2 : 3 (c) 3 : 4 (d) 3 : 5

29. Two masses A and B of 10 kg and 5 kg respectively are connected with a string passing over a frictionless pulley fixed at the corner of a table as shown in figure. The coefficient of friction of A with the table is 0.2. The minimum mass of C that may be placed on A to prevent it from moving is equal to

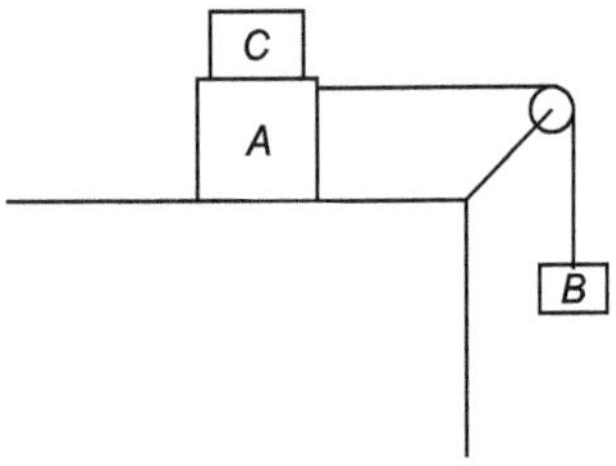

(a) 15 kg (b) 10 kg (c) 5 kg (d) 20 kg

30. A sphere of mass m is held between two smooth inclined walls. For $\sin 37° = \frac{3}{5}$, the normal reaction of the wall (2) is equal to

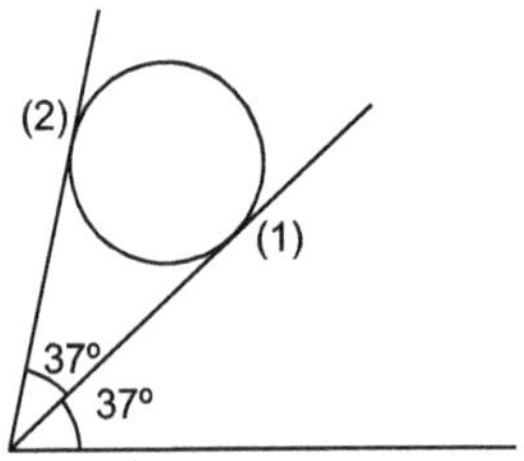

(a) $\frac{16mg}{25}$ (b) $\frac{25mg}{21}$ (c) $\frac{39mg}{21}$ (d) $\frac{Mg \sin\theta}{2}$

31. A block of mass m is kept on an inclined plane of a lift moving down with acceleration of 2 m/s^2. What should be the coefficient of friction to let the block move down with constant velocity relative to lift.

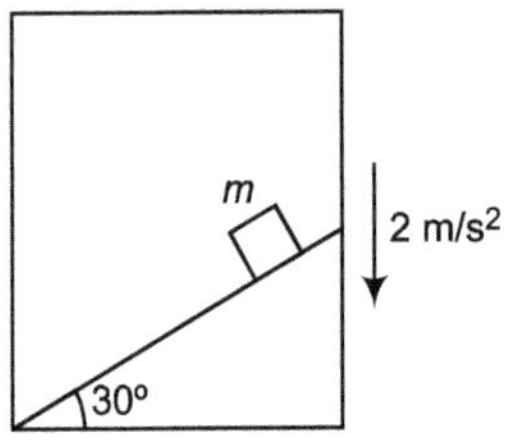

(a) $\mu = \frac{1}{\sqrt{3}}$ (b) $\mu = 0.4$ (c) $\mu = 0.8$ (d) $\mu = \frac{\sqrt{3}}{2}$

More than One Correct Options

1. Two blocks each of mass 1 kg are placed as shown. They are connected by a string which passes over a smooth (massless) pulley. There is no friction between m_1 and the ground. The coefficient of friction between m_1 and m_2 is 0.2. A force F is applied to m_2. Which of the following statement is/are correct?

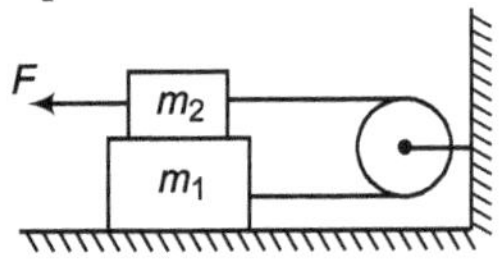

(a) The system will be in equilibrium if $F \leq 4$ N
(b) If $F > 4$ N, tension in the string will be 4 N
(c) If $F > 4$ N, the frictional force between the blocks will be 2 N
(d) If $F = 6$ N, tension in the string will be 3 N

2. Two particles A and B, each of mass m are kept stationary by applying a horizontal force $F = mg$ on particle B as shown in figure. Then,

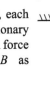

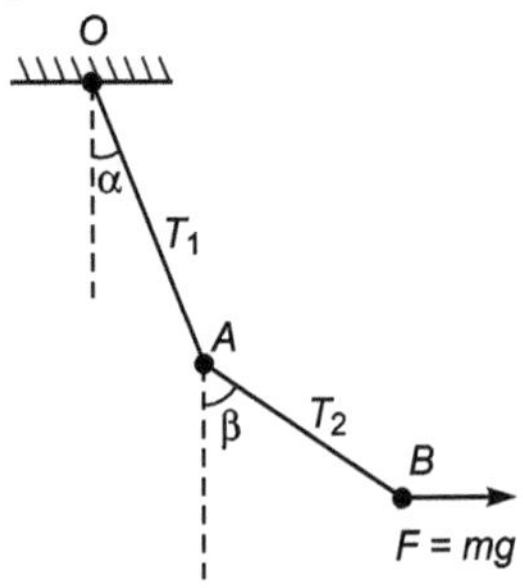

(a) $\tan\beta = 2\tan\alpha$
(b) $2T_1 = 5T_2$
(c) $\sqrt{2}\,T_1 = \sqrt{5}T_2$
(d) None of the above

3. The velocity-time graph of the figure shows the motion of a wooden block of mass 1 kg which is given an initial push at $t = 0$ along a horizontal table.

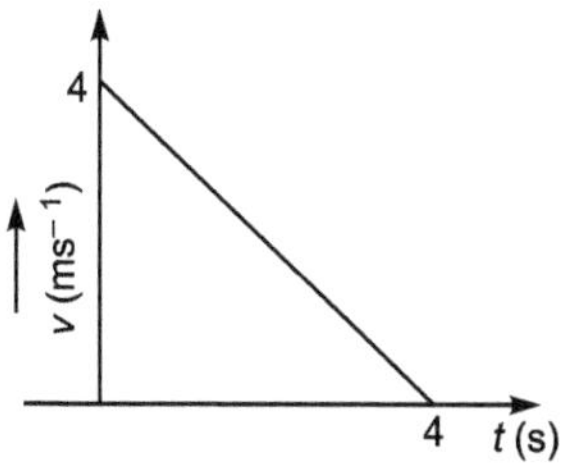

(a) The coefficient of friction between the block and the table is 0.1
(b) The coefficient of friction between the block and the table is 0.2
(c) If the table was half of its present roughness, the time taken by the block to complete the journey is 4 s
(d) If the table was half of its present roughness, the time taken by the block to complete the journey is 8 s

4. As shown in the figure, A is a man of mass 60 kg standing on a block of mass 40 kg kept on ground. The coefficient of friction between the feet of the man and the block is 0.3 and that between B and the ground is 0.2. If the person pulls the string with 125 N force, then

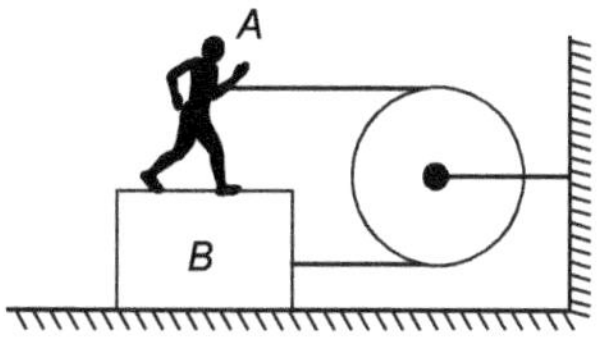

(a) B will slide on ground
(b) A and B will move with acceleration 0.5 ms^{-2}
(c) the force of friction acting between A and B will be 40 N
(d) the force of friction acting between A and B will be 180 N

5. In the figure shown, A and B are free to move. All the surfaces are smooth. Mass of A is m. Then,

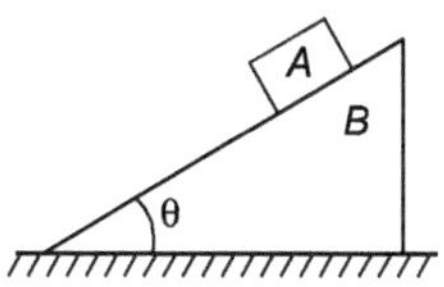

(a) the acceleration of A will be more than $g \sin\theta$
(b) the acceleration of A will be less than $g \sin\theta$
(c) normal reaction on A due to B will be more than $mg \cos\theta$
(d) normal reaction on A due to B will be less than $mg \cos\theta$

6. $M_A = 3$ kg, $M_B = 4$ kg, and $M_C = 8$ kg. μ between any two surfaces is 0.25. Pulley is frictionless and string is massless. A is connected to wall through a massless rigid rod.

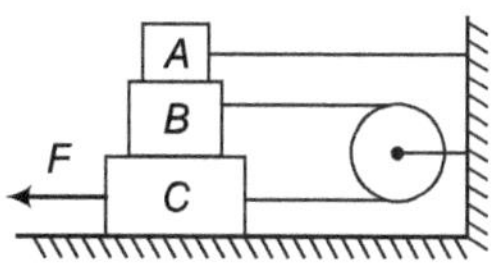

(a) the value of F to keep C moving with constant speed is 80 N
(b) the value of F to keep C moving with constant speed is 120 N
(c) if F is 200 N, then acceleration of B is 10 ms^{-2}
(d) to slide C towards left, F should be at least 50 N (Take $g = 10\ ms^{-2}$)

7. A man pulls a block of mass equal to himself with a light string. The coefficient of friction between the man and the floor is greater than that between the block and the floor

(a) if the block does not move, then the man also does not move
(b) the block can move even when the man is stationary
(c) if both move, then the acceleration of the block is greater than the acceleration of man
(d) if both move, then the acceleration of man is greater than the acceleration of block

8. A block of mass 1 kg is at rest relative to a smooth wedge moving leftwards with constant acceleration $a = 5\ ms^{-2}$. Let N be the normal reaction between the block and the wedge. Then, $(g = 10\,ms^{-2})$

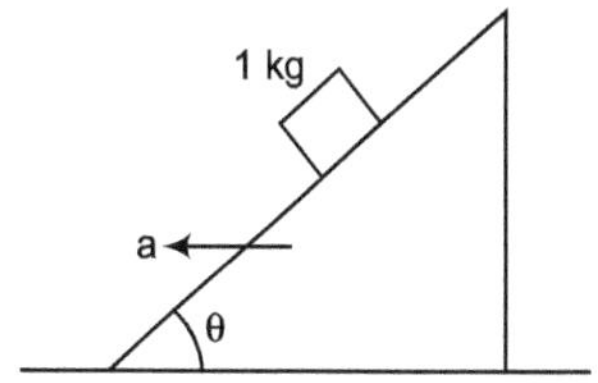

(a) $N = 5\sqrt{5}$ N
(b) $N = 15$ N
(c) $\tan\theta = \frac{1}{2}$
(d) $\tan\theta = 2$

9. For the given situation shown in figure, choose the correct options $(g = 10\,ms^{-2})$

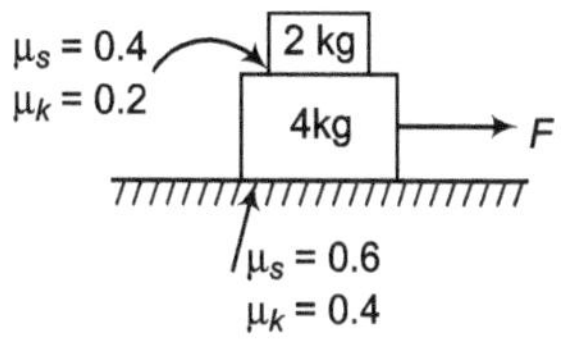

(a) At $t = 1$ s, force of friction between 2 kg and 4 kg is 2 N
(b) At $t = 1$ s, force of friction between 2 kg and 4 kg is zero
(c) At $t = 4$ s, force of friction between 4 kg and ground is 8 N
(d) At $t = 15$ s, acceleration of 2 kg is 1 ms^{-2}

10. In the figure shown, all the strings are massless and friction is absent everywhere. Choose the correct options.

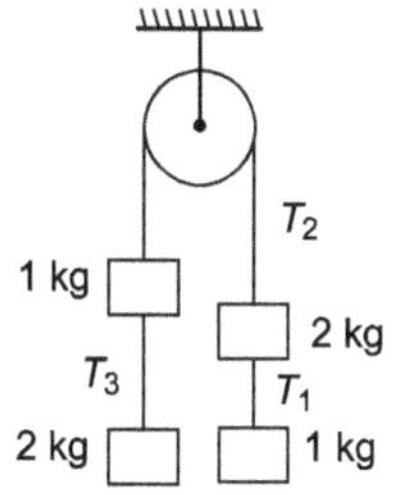

(a) $T_1 > T_3$
(b) $T_3 > T_1$
(c) $T_2 > T_1$
(d) $T_2 > T_3$

11. Force acting on a block *versus* time graph is as shown in figure. Choose the correct options. (Take $g = 10\,\text{ms}^{-2}$)

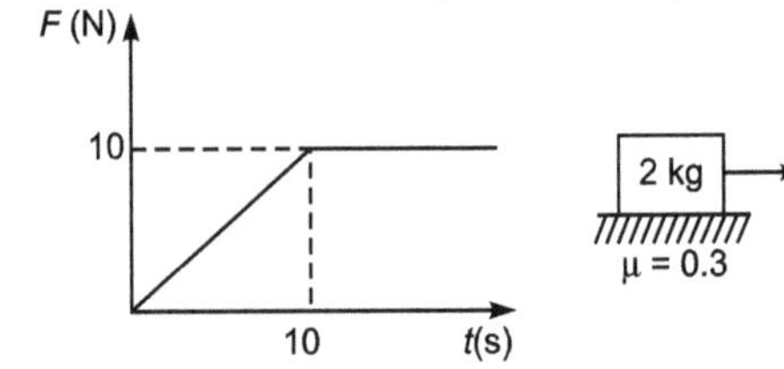

(a) At $t = 2$ s, force of friction is 2 N
(b) At $t = 8$ s, force of friction is 6 N
(c) At $t = 10$ s, acceleration of block is $2\ \text{ms}^{-2}$
(d) At $t = 12$ s, velocity of block is $8\ \text{ms}^{-1}$

12. For the situation shown in figure, mark the correct options.

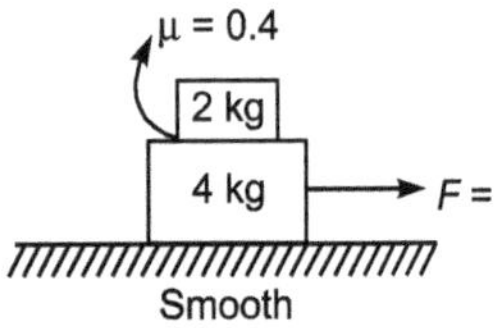

(a) At $t = 3$ s, pseudo force on 4 kg when applied from 2 kg is 4 N in forward direction
(b) At $t = 3$ s, pseudo force on 2 kg when applied from 4 kg is 2 N in backward direction
(c) Pseudo force does not make an equal and opposite pairs
(d) Pseudo force also makes a pair of equal and opposite forces

13. For the situation shown in figure, mark the correct options.

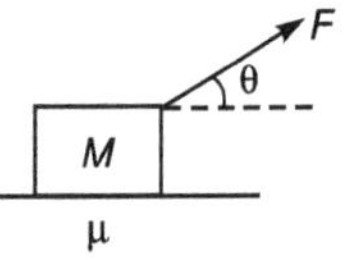

(a) Angle of friction is $\tan^{-1}(\mu)$
(b) Angle of repose is $\tan^{-1}(\mu)$
(c) At $\theta = \tan^{-1}(\mu)$, minimum force will be required to move the block
(d) Minimum force required to move the block is $\dfrac{\mu Mg}{\sqrt{1+\mu^2}}$

14. In figure, the coefficient of friction between the floor and the body B is 0.1. The coefficient of friction between the bodies B and A is 0.2. A force **F** is applied as shown on B. The mass of A is $m/2$ and of B is m. Which of the following statements are true?

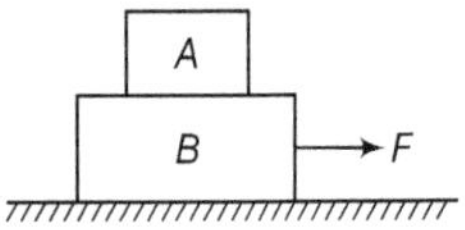

(a) The bodies will move together if $\mathbf{F} = 0.25$ mg
(b) The body A will slip with respect to B if $\mathbf{F} = 0.5$ mg
(c) The bodies will be at rest if $\mathbf{F} = 0.1$ mg
(d) The maximum value of **F** for which the two bodies will move together is 0.45 mg

15. Mass m_1 moves on a slope making an angle θ with the horizontal and is attached to mass m_2 by a string passing over a frictionless pulley as shown in figure. The coefficient of friction between m_1 and the sloping surface is μ. Which of the following statements are true?

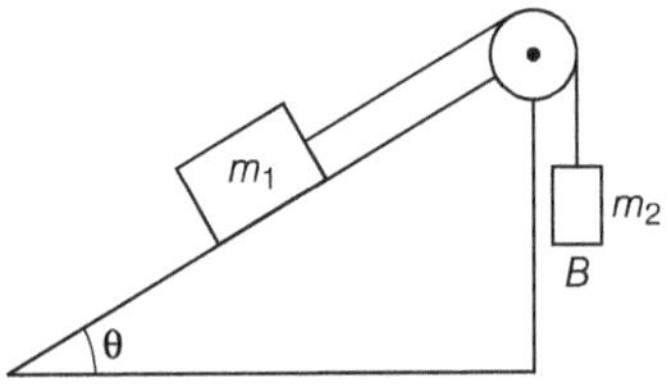

(a) If $m_2 > m_1 \sin\theta$, the body will move up the plane
(b) If $m_2 > m_1 (\sin\theta + \mu\cos\theta)$, the body will move up the plane
(c) If $m_2 < m_1 (\sin\theta + \mu\cos\theta)$, the body will move up the plane
(d) If $m_2 < m_1 (\sin\theta - \mu\cos\theta)$, the body will move down the plane

16. In figure, a body A of mass m slides on plane inclined at angle θ_1 to the horizontal and μ is the coefficient of friction between A and the plane. A is connected by a light string passing over a frictionless pulley to another body B, also of mass m, sliding on a frictionless plane inclined at an angle θ_2 to the horizontal. Which of the following statements are true?

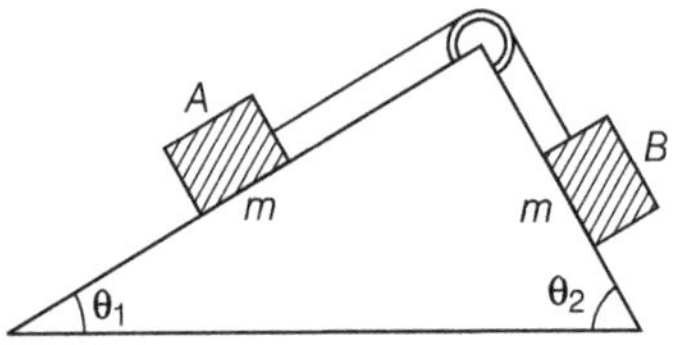

(a) A will never move up the plane
(b) A will just start moving up the plane when $\mu = \dfrac{\sin\theta_2 - \sin\theta_1}{\cos\theta_1}$
(c) For A to move up the plane, θ_2 must always be greater than θ_1
(d) B will always slide down with constant speed

Comprehension Based Questions

Passage I (Q.1 to 5)

A man wants to slide down a block of mass m which is kept on a fixed inclined plane of inclination 30° as shown in the figure. Initially, the block is not sliding.

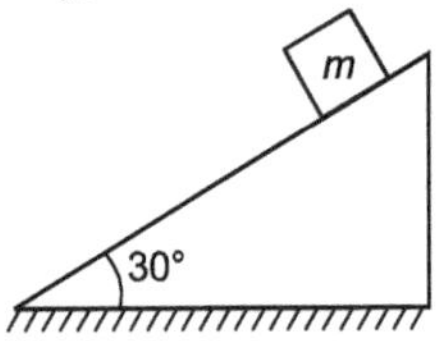

To just start sliding, the man pushes the block down the incline with a force F. Now, the block starts accelerating. To move it downwards with constant speed, the man starts pulling the block with same force. Surfaces are such that ratio of maximum static friction to kinetic friction is 2. Now, answer the following questions.

1. What is the value of F?
(a) $\frac{mg}{4}$ (b) $\frac{mg}{6}$ (c) $\frac{mg\sqrt{3}}{4}$ (d) $\frac{mg}{2\sqrt{3}}$

2. What is the value of μ_s, the coefficient of static friction?
(a) $\frac{4}{3\sqrt{3}}$ (b) $\frac{2}{3\sqrt{3}}$ (c) $\frac{3}{3\sqrt{3}}$ (d) $\frac{1}{2\sqrt{3}}$

3. If the man continues pushing the block by force F, its acceleration would be
(a) $\frac{g}{6}$ (b) $\frac{g}{4}$ (c) $\frac{g}{2}$ (d) $\frac{g}{3}$

4. If the man wants to move the block up the incline, what minimum force is required to start the motion?
(a) $\frac{2}{3}mg$ (b) $\frac{mg}{2}$ (c) $\frac{7mg}{6}$ (d) $\frac{5mg}{6}$

5. What minimum force is required to move it up the incline with constant speed?
(a) $\frac{2}{3}mg$ (b) $\frac{mg}{2}$ (c) $\frac{7mg}{6}$ (d) $\frac{5mg}{6}$

Passage II (Q.6 to 7)

A lift with a mass 1200 *kg is raised from rest by a cable with a tension* 1350 *g-N. After some time, the tension drops to* 1000 *g-N and the lift comes to rest at a height of* 25 *m above its initial point.* (1 g-N = 9.8 N)

6. What is the height at which the tension changes?
(a) 10.8 m (b) 12.5 m (c) 14.3 m (d) 16 m

7. What is greatest speed of lift?
(a) 9.8 ms^{-1} (b) 7.5 ms^{-1}
(c) 5.92 ms^{-1} (d) None of the above

Passage III (Q.8 to 9)

Blocks A and B shown in the figure are connected with a bar of negligible weight. A and B each has mass 170 *kg, the coefficient of friction between A and the plane is* 0.2 *and that between B and the plane is* 0.4 ($g = 10\ ms^{-2}$)

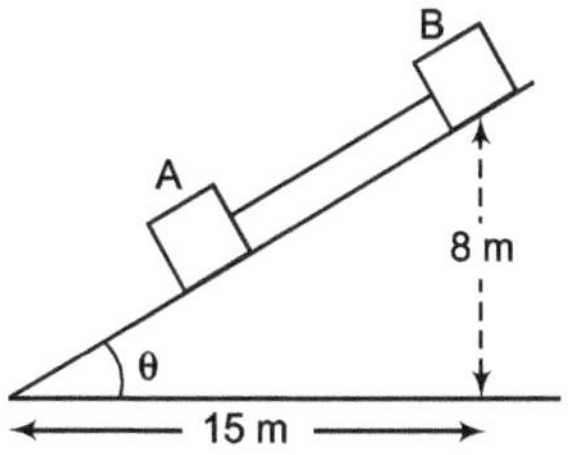

8. What is the total force of friction between the blocks and the plane?
(a) 900 N (b) 700 N (c) 600 N (d) 300 N

9. What is the force acting on the connecting bar?
(a) 140 N (b) 100 N
(c) 75 N (d) 125 N

Assertion and Reason

Directions (Q. Nos. 1-20) *These questions consists of two statements each printed as Assertion and Reason. While answering these questions you are required to choose any one of the following five responses.*

(a) If both Assertion and Reason are correct and Reason is the correct explanation of Assertion
(b) If both Assertion and Reason are true but Reason is not the correct explanation of Assertion
(c) If Assertion is true but Reason is false
(d) If Assertion is false but Reason is true
(e) If both Assertion and Reason are false

1. **Assertion** A block of weight 10 N is pushed against a vertical wall by a force of 15 N. The coefficient of friction between the wall and the block is 0.6. Then, the magnitude of maximum frictional force is 9 N.

   Reason For given system block will remain stationary.

2. **Assertion** The weighing machine measures the weight of a body.

   Reason Weightlessness means the absence of weight.

3. **Assertion** When a person walks on a rough surface, the net force exerted by surface on the person in the direction of his motion.

   Reason It is the force exerted by the road on the person that causes the motion.

4. **Assertion** A body of mass 10 kg is placed on a rough inclined surface ($\mu = 0.7$). The surface is inclined to horizontal at angle 30°. Acceleration of the body down the plane will be zero.

   Reason Work done by friction is always negative.

5. **Assertion** In the system of two blocks of equal masses as shown, the coefficient of friction between the blocks (μ_2) is less than coefficient of friction (μ_1) between lower block and ground.

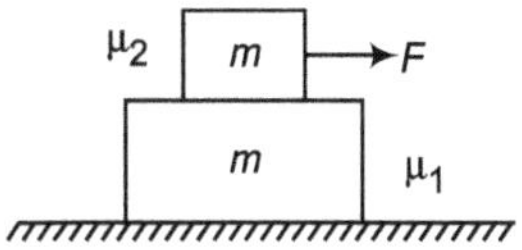

   For all values of force F applied on upper block lower block remains at rest.

   Reason Frictional force on lower block due to upper block is not sufficient to overcome the frictional force on lower block due to ground.

6. **Assertion** During free fall of a person, one feels weightlessness because his weight becomes zero.

   Reason He falls with an acceleration of g.

7. **Assertion** Earth is an inertial frame.

   Reason A frame in motion is sometimes an inertial frame and sometimes a non-inertial.

8. Assertion Friction opposes motion of a body.

Reason Static friction is self adjusting while kinetic friction is constant.

9. Assertion Friction always opposes the relative motion of two bodies.

Reason Without friction also, one can move on a smooth surface.

10. Assertion Static friction acting on a body is always greater than the kinetic friction acting on this body.

Reason Coefficient of static friction is more than the coefficient of kinetic friction.

11. Assertion A particle is thrown vertically upwards. If air resistance is taken into consideration, then retardation in upward journey is more than the acceleration in downward journey.

Reason Same mechanical energy is lost in the form of heat due to air friction.

12. Assertion A string has a mass m. If it is accelerated tension is non-uniform and if it is not accelerated tension is uniform.

Reason Tension force is an electromagnetic force.

13. Assertion If two equal and opposite forces act on a body, then it remains in equilibrium.

Reason Linear acceleration of body should be zero under the above condition.

14. Assertion In the figure shown, block of mass m is stationary with respect to lift . Force of friction acting on the block is greater than $mg \sin \theta$.

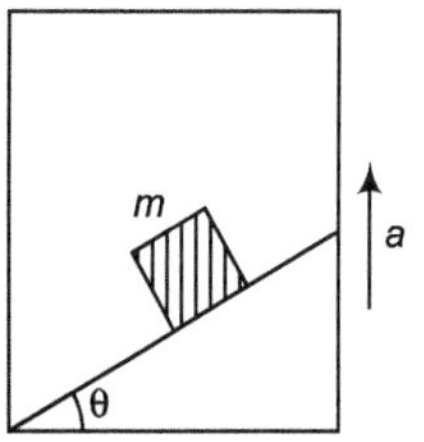

Reason If lift moves with constant velocity, then force of friction is equal to $mg \sin \theta$.

15. Assertion In the shown figure tension T connected to the ceiling is $20\text{ N} < T < 40\text{N}$

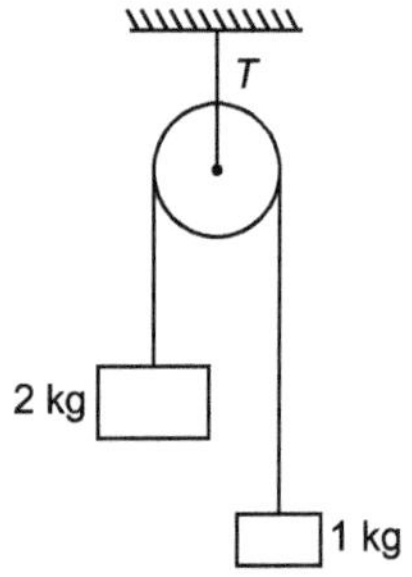

Reason System is not stationary.

16. Assertion Two forces are acting on a rope lying on a smooth table as shown in figure. In moving from A to B, tension on string decreases from $2F$ to F.

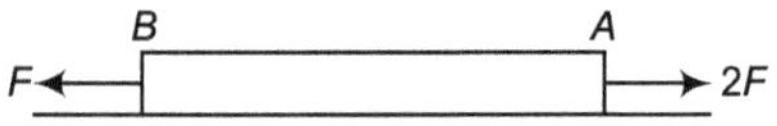

Reason Situation will become in determinant, if we take it a massless string.

17. Assertion A massless rod AB is suspended with the help of two strings as shown in figure . Tension on these two strings are T_1 and T_2. A force F is applied at distance x from end B . If x is decreased, then T_1 will decrease and T_2 will increase.

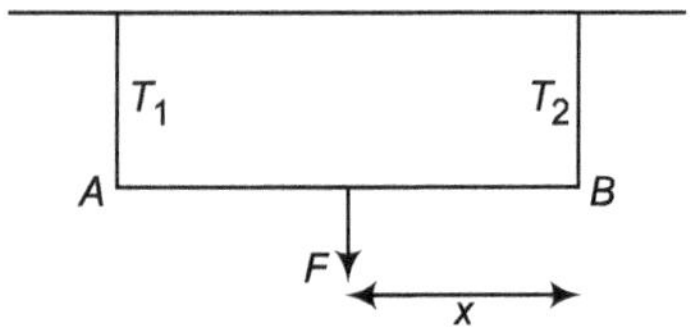

Reason When $x = 0, T_1 = 0$ and $T_2 = F$.

18. Assertion In the diagram shown in figure, string is massless and pulley is smooth, Net force on 1 kg block is $\frac{10}{3}$ N.

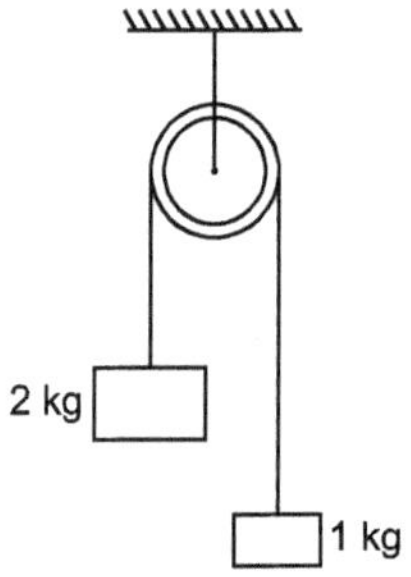

Reason Net force on both the blocks will be same.

19. Assertion Block A is resting on one corner of a box as shown in figure. Acceleration of box is $(2\hat{\mathbf{i}} + 2\hat{\mathbf{j}})\text{m/s}^2$. Let N_1 is the normal reaction on block from vertical wall and N_2 from ground of box. Then $\frac{N_1}{N_2} = \frac{1}{0}$.

(Neglect friction)

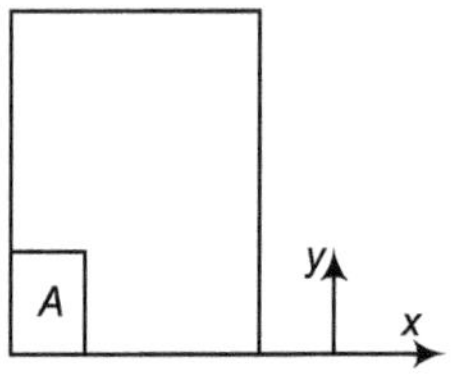

Reason $N_1 = 0$, if lift is stationary.

20. Assertion If vector sum of two or more than two forces is zero. Then, net moment of all the forces about any point is constant.

Reason Two equal and opposite forces make a couple. Moment of this couple = (force × distance between the forces).

Match the Columns

1. In the diagram shown in figure, match the following columns ($g = 10\ \text{m/s}^2$).

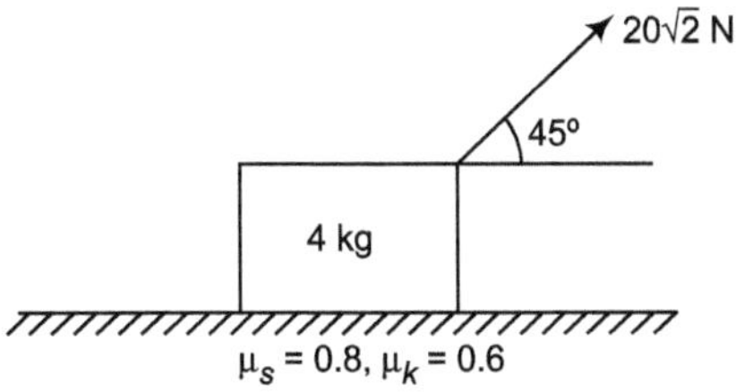

Column I	Column II
(A) Normal reaction	(p) 12 SI unit
(B) Force of friction	(q) 20 SI unit
(C) Acceleration of block	(r) zero
	(s) 2 SI unit

2. In the diagram shown in figure, all pulleys are smooth and massless and strings are light. Match the following columns.

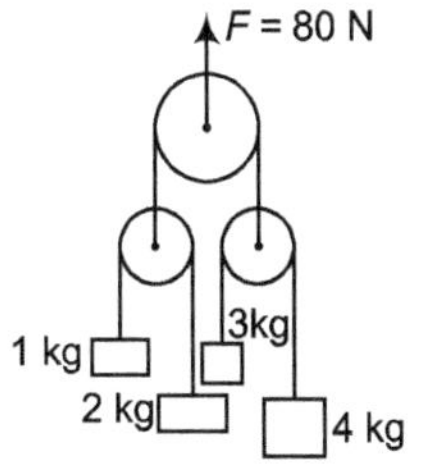

Column I	Column II
(A) 1 kg block	(p) will remain stationary
(B) 2 kg block	(q) will move down
(C) 3 kg block	(r) will move up
(D) 4 kg block	(s) 5 m/s^2

3. Match the following columns.

Column I	Column II
(A) Force of friction	(p) Opposes motion
(B) Normal reaction on a block kept on horizontal ground	(q) Opposes relative motion
	(r) Is always *mg*
	(s) May be equal to *mg*

4. A block of mass m is thrown upwards with some initial velocity as shown in figure. On the block

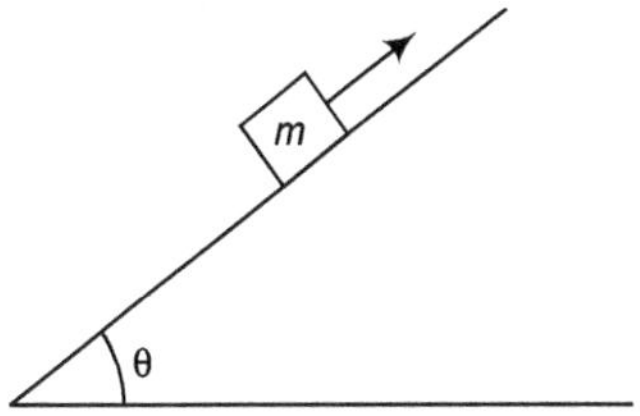

Column I	Column II
(A) Net force in horizontal direction	(p) Zero
(B) Net force in vertical direction	(q) $m(g\sin\theta + \mu g\cos\theta)$
(C) Net force along the plane	(r) $m(g\sin\theta\cos\theta + \mu g\cos^2\theta)$
(D) Net force perpendicular to plane	(s) $m(g\sin^2\theta + \mu g\sin\theta\cos\theta)$

5. In the diagram shown in figure, match the following columns. ($g = 10\,\text{m/s}^2$)

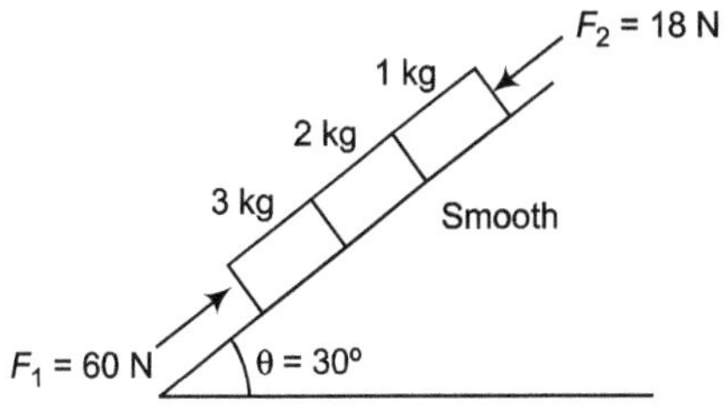

Column I	Column II
(A) Acceleration of 2 kg block	(p) 8 SI unit
(B) Net force on 3 kg block	(q) 25 SI unit
(C) Normal reaction between 2 kg and 1 kg	(r) 2 SI unit
(D) Normal reaction between 3 kg and 2 kg	(s) 45 N

6. Velocity of three particles A, B and C varies with time t as $\mathbf{v}_A = (2t\,\hat{\mathbf{i}} + 6\hat{\mathbf{j}})\,\text{m/s}$, $\mathbf{v}_B = (3\hat{\mathbf{i}} + 4\hat{\mathbf{j}})\,\text{m/s}$ and $\mathbf{v}_C = (6\hat{\mathbf{i}} - 4t\,\hat{\mathbf{j}})$. Regarding the pseudo force match the following columns.

Column I	Column II
(A) On *A* as observed by *B*	(p) Along positive *x*-direction
(B) On *B* as observed by *C*	(q) Along negative *x*-direction
(C) On *A* as observed by *C*	(r) Along positive *y*-direction
(D) On *C* as observed by *A*	(s) Along negative *y*-direction

7. In the diagram shown in figure, match the following columns.

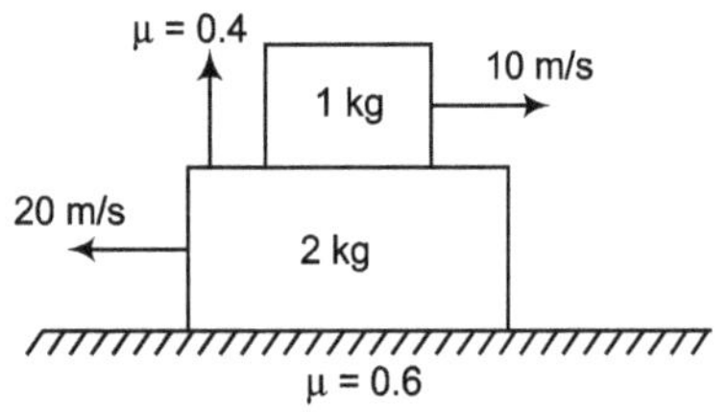

Column I	Column II
(A) Absolute acceleration of 1 kg block	(p) 11 m/s^2
(B) Absolute acceleration of 2 kg block	(q) 6 m/s^2
(C) Relative acceleration between the two	(r) 17 m/s^2
	(s) None of the above

Entrance Gallery

2014

1. A particle of mass m is at rest at the origin at time $t = 0$. It is subjected to a force, $F(t) = F_0 e^{-bt}$ in the x-direction. Its speed $v(t)$ is depicted by which of the following curves? **[JEE Main]**

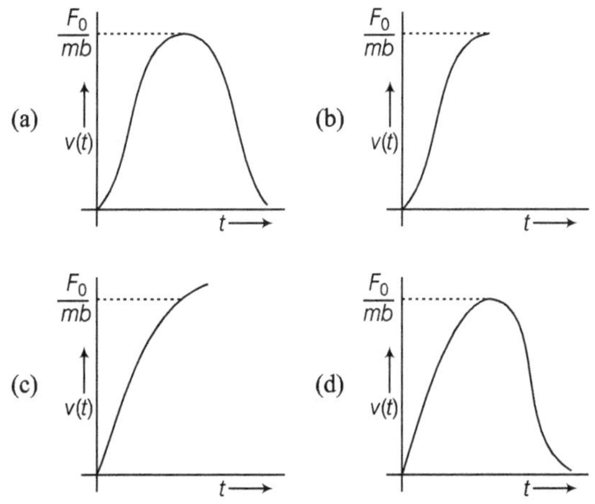

2. A block of mass m is placed on a surface with a vertical cross-section given by $y = x^3/6$. If the coefficient of friction is 0.5, the maximum height above the ground at which the block can be placed without slipping is **[JEE Main]**

(a) $\frac{1}{6}$ m (b) $\frac{2}{3}$ m (c) $\frac{1}{3}$ m (d) $\frac{1}{2}$ m

3. A box is lying on an inclined plane what is the coefficient of static friction if the box starts sliding when an angle of inclination is $60°$. **[Karnataka CET]**

(a) 1.173 (b) 1.732 (c) 2.732 (d) 1.677

4. To determine the coefficient of friction between a rough surface and a block, the surface is kept inclined at 45° and the block is released from rest. The block takes a time t in moving a distance d. The rough surface is then replaced by a smooth surface and the same experiment is repeated. The block now takes a time $t/2$ in moving down the same distance d. The coefficient of friction is **[WB JEE]**

(a) 3/4 (b) 5/4 (c) 1/2 (d) $1/\sqrt{3}$

5. A constant retarding force of 80 N is applied to a body of mass 50 kg which is moving initially with a speed of 20 m/s. What would be the time required by the body to come to rest? **[J&K CET]**

(a) 15 s (b) 14 s (c) 12.5 s (d) 18 s

6. An object is gently placed on a long converges belt moving with 11ms^{-1}. If the coefficient of friction is 0.4, then the block will slide in the belt up to a distance of **[J&K CET]**

(a) 10.21 m (b) 15.125 m (c) 20.3m (d) 25.6 m

2012

7. A small block of mass of 0.1 kg lies on a fixed inclined plane PQ which makes an angle θ with the horizontal. A horizontal force of 1 N acts on the block through its centre of mass as shown in the figure. The block remains stationary if (Take, $g = 10 \text{m/s}^2$) **[IIT JEE]**

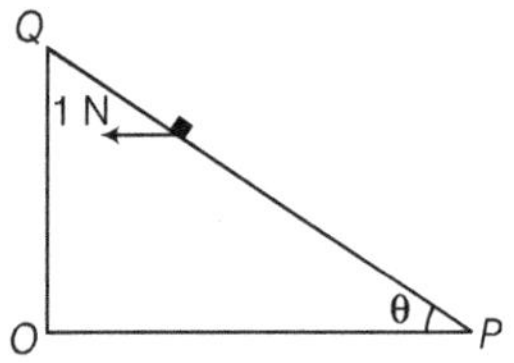

(a) $\theta = 45°$
(b) $\theta > 45°$ and a frictional force acts on the block towards P
(c) $\theta > 45°$ and a frictional force acts on the block towards Q
(d) $\theta < 45°$ and a frictional force acts on the block towards Q

8. A body of mass m is travelling with a velocity u. When a constant retarding force F is applied, it comes to rest after travelling a distance s_1. If the initial velocity is $2u$ with the same force F, the distance travelled before it comes to rest is s_2. Then, **[Karnataka CET]**

(a) $s_2 = 4s_1$ (b) $s_2 = 2s_1$ (c) $s_2 = \frac{s_1}{2}$ (d) $s_2 = s_1$

9. A block kept on a rough surface starts sliding when the inclination of the surface is θ with respect to the horizontal. The coefficient of static friction between the block and the surface is **[Karnataka CET]**

(a) $\sec\theta$ (b) $\sin\theta$ (c) $\tan\theta$ (d) $\cos\theta$

10. A box of mass 2 kg is placed on the roof of a car. The box would remain stationary until the car attains a maximum acceleration. Coefficient of static friction between the box and the roof of the car is 0.2 and $g = 10 \text{ms}^{-2}$. The maximum acceleration of the car, for the box to remain stationary, is **[WB JEE]**

(a) 8 ms^{-2} (b) 6 ms^{-2} (c) 4 ms^{-2} (d) 2 ms^{-2}

11. Three blocks of masses 4 kg, 2 kg and 1 kg, respectively are in contact on a frictionless table as shown in the figure. If a force of 14 N is applied on the 4 kg block, the contact force between the 4 kg and the 2 kg block will be **[WB JEE]**

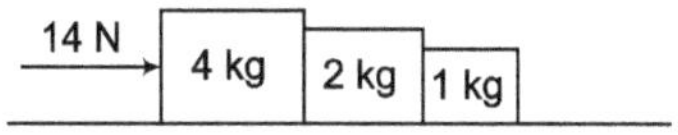

(a) 2 N (b) 6 N
(c) 8 N (d) 14 N

2011

12. Two masses $m_1 = 1\,\text{kg}$ and $m_2 = 2\,\text{kg}$ are connected by a light inextensible string and suspended by means of a weightless pulley as shown in the figure.

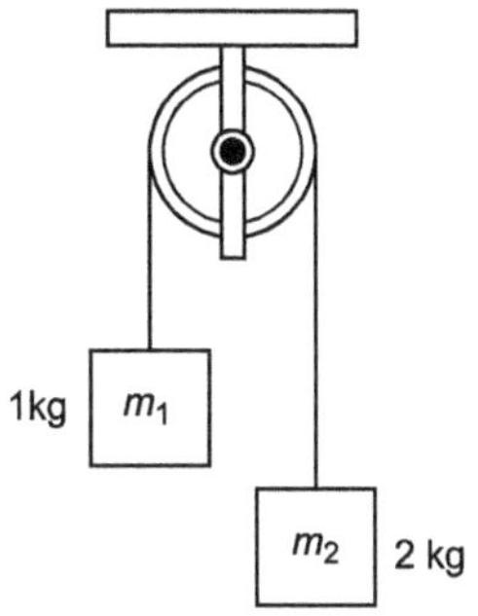

Assuming that both the masses start from rest, the distance travelled by the centre of mass in 2 s is (take, $g = 10\,\text{m/s}^2$) **[Kerala CEE]**

(a) $\frac{20}{9}$ m (b) $\frac{40}{9}$ m
(c) $\frac{2}{3}$ m (d) $\frac{1}{3}$ m
(e) 4 m

13. A block at rest slides down a smooth inclined plane which makes an angle $60°$ with the vertical and it reaches the ground in t_1 second. Another block is dropped vertically from the same point and reaches the ground in t_2 second. Then, the ratio of $t_1 : t_2$ is **[Kerala CEE]**

(a) 1 : 2 (b) 2 : 1
(c) 1 : 3 (d) $1 : \sqrt{2}$
(e) 3 : 1

14. The resultant of two forces acting at an angle of 120° is 10 kg-wt and is perpendicular to one of the forces. That force is **[Karnataka CET]**

(a) $\frac{10}{\sqrt{3}}$ kg-wt (b) 10 kg-wt
(c) $20\sqrt{3}$ kg-wt (d) $10\sqrt{3}$ kg-wt

15. Block A of mass 2 kg is placed over block B of mass 8 kg. The combination is placed over a rough horizontal surface. Coefficient of friction between B and the floor is 0.5. Coefficient of friction between A and B is 0.4. A horizontal force of 10 N is applied on block B.

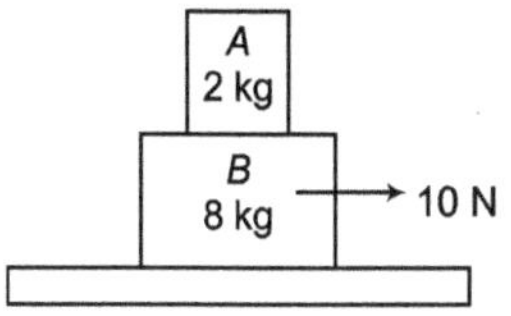

The force of friction between A and B is **[Karnataka CET]**

(a) zero (b) 50 N (c) 40 N (d) 100 N

2010

16. A block of mass m is on an inclined plane of angle θ. The coefficient of friction between the block and the plane is μ and $\tan\theta > \mu$. The block is held stationary by applying a force P parallel to the plane. The direction of force pointing up the plane is taken to be positive. As P is varied from $P_1 = mg\,(\sin\theta - \mu\cos\theta)$ to $P_2 = mg\,(\sin\theta + \mu\cos\theta)$, the frictional force f *versus* P graph will look like **[IIT JEE]**

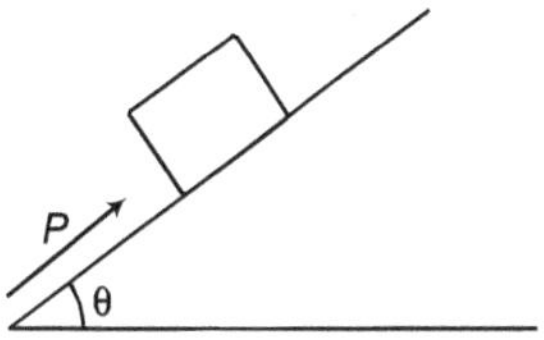

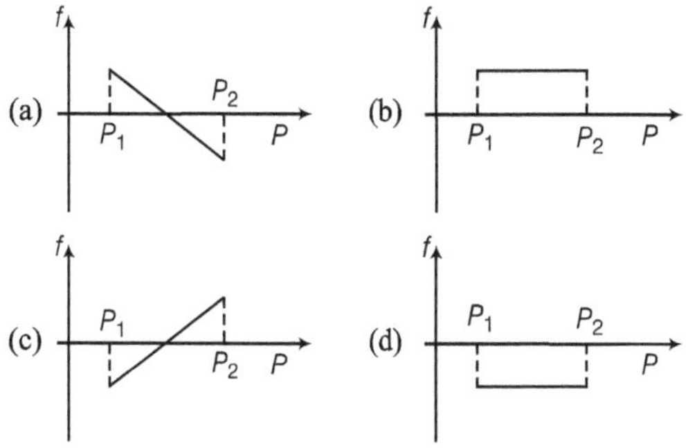

17. Two fixed frictionless inclined plane making an angle 30° and 60° with the vertical are shown in the figure. Two blocks A and B are placed on the two planes. What is the relative vertical acceleration of A with respect to B? **[AIEEE]**

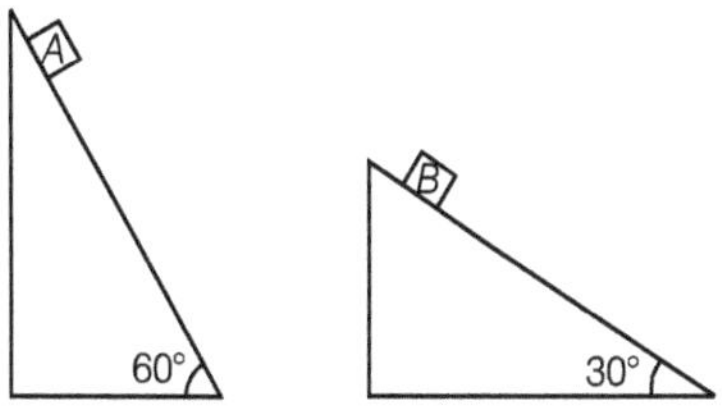

(a) 4.9 ms^{-2} in horizontal direction
(b) 9.8 ms^{-2} in vertical direction
(c) zero
(d) 4.9 ms^{-2} in vertical direction

Answers

Level 1

Objective Problems

1. (b)	**2.** (c)	**3.** (c)	**4.** (a)	**5.** (b)	**6.** (d)	**7.** (a)	**8.** (b)	**9.** (b,c,d)	**10.** (d)
11. (c)	**12.** (a)	**13.** (c)	**14.** (a)	**15.** (d)	**16.** (d)	**17.** (d)	**18.** (b)	**19.** (d)	**20.** (b)
21. (c)	**22.** (c)	**23.** (c)	**24.** (c)	**25.** (d)	**26.** (d)	**27.** (b)	**28.** (b)	**29.** (a)	**30.** (c)
31. (a)	**32.** (c)	**33.** (d)	**34.** (b)	**35.** (c)	**36.** (a)	**37.** (c)	**38.** (c)	**39.** (d)	**40.** (c)
41. (a)	**42.** (a)	**43.** (b)	**44.** (a)	**45.** (d)	**46.** (c)	**47.** (d)	**48.** (c)	**49.** (a)	**50.** (d)
51. (b)	**52.** (b)	**53.** (d)	**54.** (a)	**55.** (a)	**56.** (a)	**57.** (c)	**58.** (a)	**59.** (a)	**60.** (d)
61. (b)	**62.** (c)	**63.** (a)	**64.** (a)	**65.** (b)	**66.** (d)	**67.** (c)	**68.** (a)	**69.** (a)	**70.** (a)
71. (c)	**72.** (b)	**73.** (c)	**74.** (b)	**75.** (d)	**76.** (d)	**77.** (c)	**78.** (a)	**79.** (b)	**80.** (d)
81. (d)	**82.** (c)	**83.** (a)	**84.** (c)						

Level 2

Only One Correct Option

1. (b)	**2.** (d)	**3.** (d)	**4.** (a)	**5.** (d)	**6.** (b)	**7.** (a)	**8.** (c)	**9.** (b)	**10.** (b)
11. (a)	**12.** (c)	**13.** (a)	**14.** (c)	**15.** (a)	**16.** (a)	**17.** (b)	**18.** (a)	**19.** (d)	**20.** (b)
21. (d)	**22.** (a)	**23.** (a)	**24.** (c)	**25.** (b)	**26.** (b)	**27.** (c)	**28.** (d)	**29.** (a)	**30.** (d)
31. (a)									

More than One Correct Options

1. (a,c,d)	**2.** (a,c)	**3.** (a,d)	**4.** (a,b)	**5.** (a,d)	**6.** (a,c)	**7.** (a,b,c)	**8.** (a,c)	**9.** (b,c)	**10.** (b,c,d)
11. (all)	**12.** (b,c)	**13.** (all)	**14.** (all)	**15.** (b,d)	**16.** (b,c)				

Comprehension Based Questions

1. (b)	**2.** (a)	**3.** (d)	**4.** (c)	**5.** (d)	**6.** (c)	**7.** (c)	**8.** (a)	**9.** (a)

Assertion and Reason

1. (c)	**2.** (d)	**3.** (a,b)	**4.** (c)	**5.** (a)	**6.** (d)	**7.** (d)	**8.** (d)	**9.** (d)	**10.** (d)
11. (b)	**12.** (d)	**13.** (d)	**14.** (b)	**15.** (b)	**16.** (b)	**17.** (b)	**18.** (c)	**19.** (b)	**20.** (b)

Match the Columns

1. ($A \rightarrow q$, $B \rightarrow p$, $C \rightarrow s$)
2. ($A \rightarrow rt$, $B \rightarrow p$, $C \rightarrow q$, $D \rightarrow qs$)
3. ($A \rightarrow t$, $B \rightarrow s$)
4. ($A \rightarrow r$, $B \rightarrow s$, $C \rightarrow q$, $D \rightarrow p$)
5. ($A \rightarrow r$, $B \rightarrow t$, $C \rightarrow q$, $D \rightarrow t$)
6. ($A \rightarrow t$, $B \rightarrow r$, $C \rightarrow r$, $D \rightarrow q$)
7. ($A \rightarrow s$, $B \rightarrow p$, $C \rightarrow s$)

Entrance Gallery

1. (b)	**2.** (a)	**3.** (b)	**4.** (a)	**5.** (c)	**6.** (b)	**7.** (c)	**8.** (a)	**9.** (c)	**10.** (d)
11. (d)	**12.** (a)	**13.** (b)	**14.** (a)	**15.** (b)	**16.** (c)	**17.** (d)			

Solutions

Level 1 : Objective Problems

1. T_x = (mass of rope of length $L - x$) $\times g = \frac{M}{L}(L-x)g$

2. Equilibrium of m: $\quad T = mg \quad$ ... (i)

   Equilibrium of $\sqrt{2}m$: $2T\cos\theta = \sqrt{2}mg \quad$ ... (ii)

   Solving these two equations, we get $\theta = 45°$

3. Net moment about point C should be zero. Or

$$4x = 2(3 - x)$$

$$\therefore \quad x = 1\,\text{m}$$

4. Three coins are above the 7th coin. Therefore, force from 3 coins above it will be

   $3mg = 3(10\times10^{-3})(10) = 0.3\,\text{N}$ (in downward direction)

5. $2T\cos 60° = w$

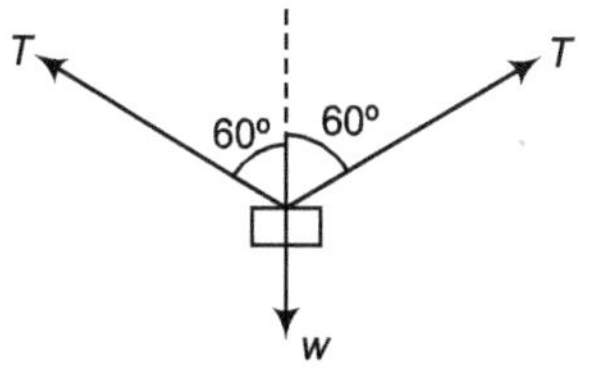

or $\quad w = T$

or $\quad w_{\text{max}} = T_{\text{max}} = 20\,\text{N}$

6. $\quad T_3 \sin 60° = T_2 \sin 30°$

$$\therefore \quad T_3 = \frac{T_2}{\sqrt{3}}$$

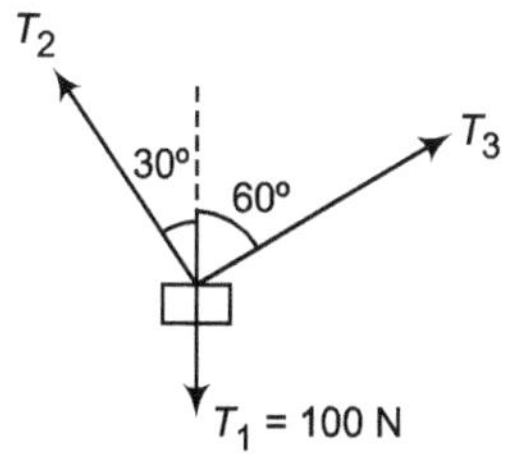

Now, $T_2\cos 30° + T_3\cos 60° = T_1 = 100$

or $\quad \frac{\sqrt{3}T_2}{12} + \frac{T_2}{\sqrt{3}} \times \frac{1}{2} = 100$

or $\quad \frac{8T_2}{4\sqrt{3}} = 100$ or $T_2 = 50\sqrt{3}\,\text{N}$

7. Force exerted by man on rope transfers to it in the form of tension.

   Net upwards force on the system is $2T$ or $2F$.

   Net downward force is $(50 + 30)\,g = 8g$.

   For equilibrium of system, $2F = 80g$ or $F = 40g$

8. Acceleration of system $a = \frac{\text{Net pushing force}}{\text{Total mass}}$

   or $\quad a = \frac{F - (m_1 + m_2 + m_3)g\sin\theta}{(m_1 + m_2 + m_3)}$

   Equation of motion for m_3

$$N - m_3 g\sin\theta = m_3 a$$

or $N = m_3 g\sin\theta + m_3\left\{\frac{F - (m_1 + m_2 + m_3)g\sin\theta}{(m_1 + m_2 + m_3)}\right\}$

$$= \frac{m_3 F}{m_1 + m_2 + m_3}$$

10. $T = mg$

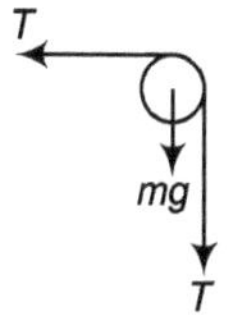

Force on the pulley (other than from clamp)

$$F_{\text{net}} = \sqrt{(T + mg)^2 + T^2}$$

$$= g\sqrt{(M + m)^2 + M^2}$$

Since, pulley is in equilibrium, clamp will exert the same amount of force in opposite direction. Or pulley will also exert this much force on clamp.

11. $\frac{T_1}{2} = \frac{\sqrt{3}T_2}{2}$

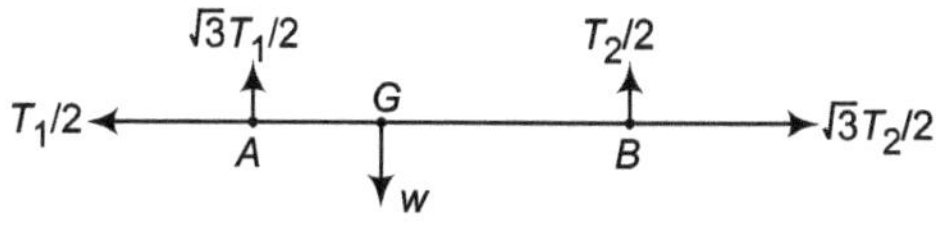

or $\quad T_1 = \sqrt{3}T_2 \quad$ ...(i)

$$\frac{\sqrt{3}T_1}{2} \times AG = \frac{T_2}{2} \times BG$$

$$\therefore \quad \frac{AG}{BG} = \frac{T_2}{\sqrt{3}T_1}$$

But $\frac{T_2}{T_1} = \frac{1}{\sqrt{3}}$ from Eq. (i)

Hence, $\quad \frac{AG}{BG} = \frac{1}{3}$

12. $T\sin 30° = w = 40$

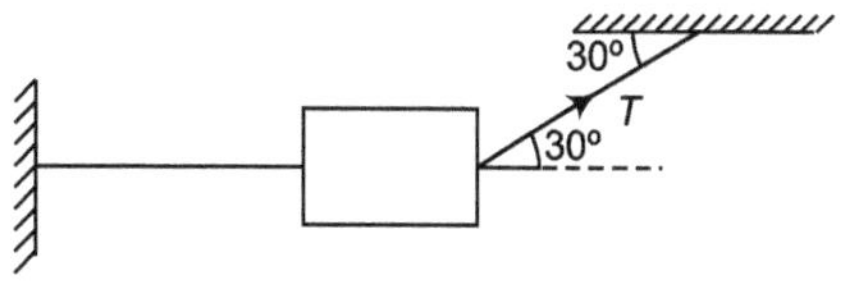

$$\therefore \quad \frac{T}{2} = 40$$

or $\quad T = 80\,\text{N}$

13. Apply Lami's theorem at O

$$\frac{T_1}{\sin 150°} = \frac{T_2}{\sin 120°} = \frac{10}{\sin 90°} = \frac{10}{1} = 10$$

$$\therefore \quad T_1 = 10\sin 150° = 10 \times \frac{1}{2} = 5\,\text{N}$$

$$T_2 = 10\sin 120°$$

$$= 10 \times \frac{\sqrt{3}}{2} = 5\sqrt{3}\,\text{N}$$

14. If lift is accelerating, reading will be more, if it is decelerating reading will be less and if it is moving with constant velocity reading will be same.

15. Suppose F is reading of spring balance. Then,

$$F - 2g = 2g \quad \text{or} \quad F = 4g$$

So, it is 4 kg or 4g-N.

Note That 1 kg = g-N

16.

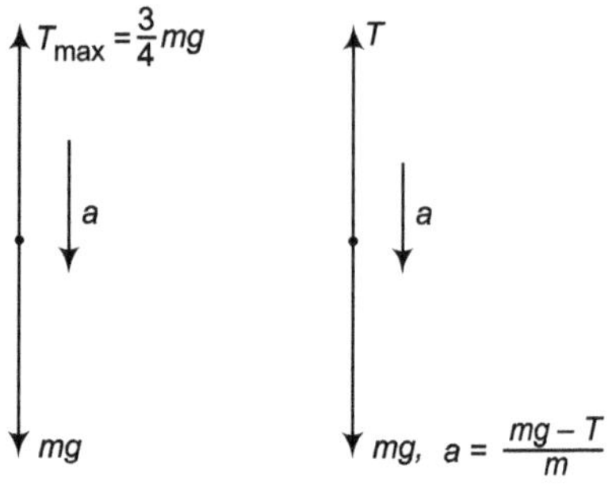

This is the minimum value of a, because if a is increased from this value T will be less than $\frac{3}{4}mg$. Which is less than T_{max}. Or we can say for minimum value of a, T should be maximum.

$$mg - \frac{3}{4}mg = ma$$

or
$$a = \frac{g}{4}$$

17. $T\cos\theta = mg$ and $T\sin\theta = ma$

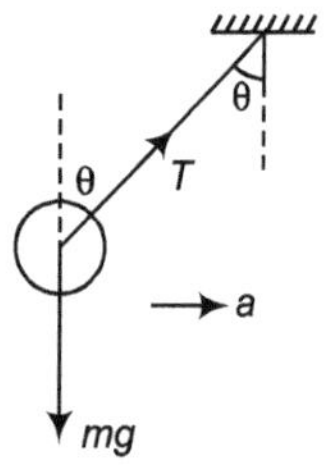

From these two equations, we get

$$a = g\tan\theta = g\tan 30° = \frac{g}{\sqrt{3}}$$

18. $a = \frac{\text{Net pulling force}}{\text{Total force}} = \frac{5g - 3g}{8} = \frac{g}{4}$

Equation of motion of 5 kg block gives

$$5g - T = 5a = \frac{5g}{4}$$

$\therefore$
$$T = \frac{15g}{4}$$
$$= 36.75 \text{ N}$$

19. Acceleration of system,

$$a = \frac{T_3}{m_1 + m_2 + m_3}$$
$$= \frac{40}{10 + 6 + 4}$$
$$= 2 \text{ m/s}^2$$

Equation of motion of m_3 is

$$T_3 - T_2 = m_3 a$$

$\therefore$
$$T_2 = T_3 - m_3 a = 40 - 4 \times 2 = 32 \text{N}$$

20. Acceleration of system,

$$a = \frac{\text{Net pulling force}}{\text{Total mass}}$$
$$= \frac{4g - 2g}{6} = \frac{g}{3}$$

Equation of motion of block C is

$$m_C g = T_{BC} = m_C a$$

$\therefore$
$$T_{BC} = m_C(g - a)$$
$$= 2\left(9.8 - \frac{9.8}{3}\right) = 13 \text{N}$$

21. $Mg - N = ma$ or $N = m(g - a)$

22. Since, $M_1 g\sin 30° = M_2 g$

$\therefore$ Net pulling force $= 0$

or acceleration of system $= 0$

Friction coefficient is not required in this question.

23. Maximum friction is 50 N.

Now,
$$F - f_{max} = ma$$

or
$$F = f_{max} + ma$$
$$= 50 + 500 \times 1$$
$$= 550 \text{N}$$

24. $0 = (3)^2 - 2(a)(9)$

$\therefore$
$$a = \frac{1}{2} = 0.5 \text{ m/s}^2 \qquad \text{(upwards)}$$
$$N = m(g + a)$$
$$= 50(9.8 + 0.5) = 515 \text{N}$$

25. $0 = (10)^2 - 2(a)(25)$

$\therefore$
$$a = 2 \text{m/s}^2 \qquad \text{(upwards)}$$

Now,
$$T - 800g = 800a$$

$\therefore$
$$T = 800(10 + 2) = 9600 \text{N}$$

26. Lift is descending with retardation. Therefore, acceleration a is upwards or pseudo force ma is upwards. Relative to lift force in downward direction is $m(g + a)$ in place of mg. Therefore in smooth plane acceleration will be $(g + a)\sin\theta$ not $g\sin\theta$.

27. Upward force on 2 kg block in upward direction will be 40 N ($= 2F$) in the form of tension.

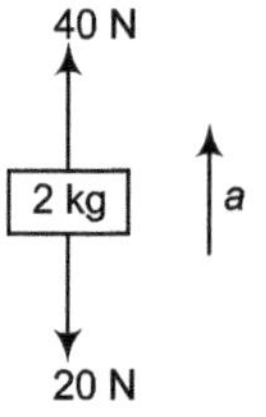

$\therefore$
$$a = \frac{40 - 20}{2} = 10 \text{ m/s}^2$$

28.
$$T - m_A g = m_A a$$

$\therefore$
$$T = m_A(g + a)$$
$$= 6(10 + 1) = 66 \text{N}$$

29. Writing equation of motion for two weights,

$$w_1 - T = \frac{w_1}{g}(a_r - a) \qquad \text{...(i)}$$

$$T - w_2 = \frac{w_2}{g}(a_r + a) \qquad \text{...(ii)}$$

Solving Eqs. (i) and (ii), we get

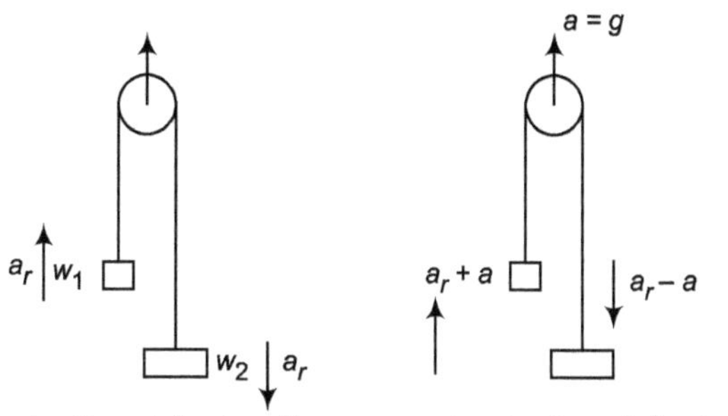

Acceleration relative to pulley   Acceleration relative to ground

$$T = \frac{4w_1w_2}{w_1 + w_2} \text{ with } a = g$$

30. Acceleration of system, $a = \frac{\text{Net pulling force}}{\text{Total mass}}$

$$= \frac{Mg\sin\theta}{2M}$$

$$a = \frac{1}{2}g\sin\theta$$

Now, the block on ground is moving due to tension

Hence, $T = Ma = \frac{Mg\sin\theta}{2}$

31. $F = ma$

Resultant of three forces $\mathbf{F}_1, \mathbf{F}_2$ and $\mathbf{F}_3$ will be $(\sqrt{2}-1)F$. Therefore, acceleration of body is also $(\sqrt{2}-1)a$.

32. Suppose air resistance is F (upwards), then from equation of motion of balloon, we have

$$w - F = \text{mass} \times \text{acceleration} = \frac{w}{g} \cdot a$$

$$\therefore \quad F = w\left(1 - \frac{a}{g}\right)$$

33. Acceleration of system,

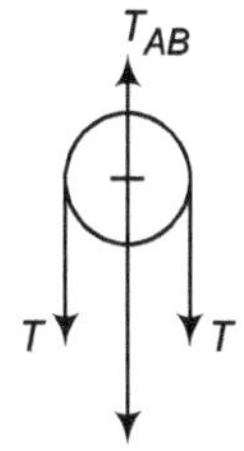

$$a = \frac{\text{Net pulling force}}{\text{Total mass}}$$

$$= \frac{2mg - mg}{3m} = \frac{g}{3}$$

Now, from equation of motion of m

$$T - mg = ma = \frac{mg}{3}$$

$$\therefore \quad T = \frac{4mg}{3}$$

For equilibrium of pulley

$$T_{AB} = 2T + \text{weight of pulley}$$

$$= \frac{8mg}{3} + 3mg = \frac{17mg}{3}$$

34. Net pulling force on the system should be zero as velocity is constant. Hence,

$$m_A g\sin 30° = \mu m_A g\cos 30° + m_B g$$

$$\therefore \quad m_B = \left(\frac{m_A}{2}\right) - \left(\frac{\mu m_A\sqrt{3}}{2}\right)$$

$$= 10\left[\frac{1}{2} - 0.2 \times \frac{\sqrt{3}}{2}\right] = 3.3 \text{ kg}$$

35. $a = \frac{\text{Net pulling force}}{\text{Total mass}} = \frac{5g - 0.5 \times 5 \times g}{10} = \frac{g}{4}$

Equation of motion for the hanging mass will be

$$5g - T = 5a = \frac{5g}{4}$$

$$\therefore \quad T = \frac{15}{4}g = \frac{15 \times 9.8}{4} = 36.75 \text{ N}$$

36. $mg\sin\theta - f = ma$ (f = force of friction)

or $f = m = (g\sin\theta - a)$

$$= 8\left(10 \times \frac{1}{2} - 0.4\right) = 36.8 \text{ N}$$

37. $a = \frac{v}{t} = \frac{4}{2} = 2 \text{ m/s}^2$

$$F - f = ma$$

or $200 - \mu \times 30 \times 10 = 30 \times 2$

$\therefore \quad \mu = 0.47$

38. $s = \frac{1}{2}at^2$ or $t = \sqrt{\frac{2s}{a}}$ or $t \propto \frac{1}{\sqrt{a}}$

$$\frac{t_1}{t_2} = \sqrt{\frac{a_2}{a_1}} \quad \text{or} \quad 2 = \sqrt{\frac{g\sin\theta}{g\sin\theta - \mu g\cos\theta}} = \sqrt{\frac{1}{1-\mu}}$$

as $\sin 45° = \cos 45°$

$\therefore \quad 1 - \mu = \frac{1}{4}$

$\therefore \quad \mu = 0.75$

39. Retardation, $a = \frac{F}{m} = 5 \text{ m/s}^2$

$$0 = 30 - 5t$$

$\therefore \quad t = 6 \text{ s}$

40. $a = \frac{F - \mu mg}{m} = \frac{100 - 0.5 \times 10 \times 10}{10} = 5 \text{ m/s}^2$

41. $f_{max} = \mu mg = 0.4 \times 2 \times 10 = 8 \text{ N}$

Since, the applied force is less than f_{max}, force of friction will be equal to the applied force or 2.8 N.

42. Maximum inclination of the plane with horizontal = angle of repose = $\tan^{-1}(\mu)$.

43. $N = F = 5 \text{N}$

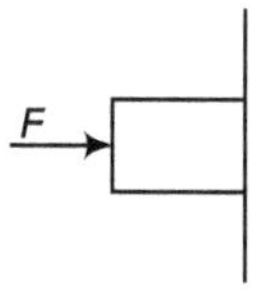

$\therefore \quad f_{max} = \mu N = 2.5 \text{ N}$

$$w = mg = 0.1 \times 9.8 = 0.98 \text{ N}$$

Since, $w < f_{max}$, force of friction will be 0.98 N.

44. $f_{max} = \mu mg = 0.8 \times 4 \times 10 = 32 \text{ N}$

At $t = 2 \text{ s}, F = kt^2 (2)(2)^2 = 8 \text{ N}$

Since, applied force $f < f_{max}$, force of friction will be 8 N.

45. Retardation, $a = \frac{\mu mg}{m} = \mu g = 10\mu$

Now, $0 = u^2 - 2as$

or $0 = (6^2) - 2(10\mu)(9)$

$\therefore \quad \mu = 0.2$

46. $f_{max} = \mu mg\cos\theta = 0.6 \times mg \times \frac{\sqrt{3}}{2} = 0.52mg$

$$mg\sin\theta = \frac{mg}{2} = 0.5mg$$

Since, $f_{max} > mg\sin\theta$

Block will decelerate and come to rest.

47. N = applied force = 12 N

$\therefore \quad f_{max} = \mu N = 7.2\,N$

Since, weight $w < f_{max}$

Force of friction $f = 5\,N$

$\therefore$ Net contact force $= \sqrt{N^2 + f^2} = \sqrt{(12)^2 + (5)^2} = 13\,N$

48. Acceleration, $a = (g\sin\theta + \mu g\cos\theta)$

$$= \frac{g}{\sqrt{2}} + 0.5 \times g \times \frac{1}{\sqrt{2}} = \frac{3g}{2\sqrt{2}}$$

49. Retardation, $a = \frac{\mu mg}{m} = \mu g = 4\,m/s^2$

$$0 = (8)^2 - 2(4)(s)$$

$\therefore \quad s = 8.0\,m$

50. Angle of plane is just equal to the angle of repose.

or $\quad \mu = \tan\theta$

51. $N = mg - F\sin 60° = mg - \frac{\sqrt{3}F}{2}$

$\therefore$ Limiting friction $= \mu N = \mu\left(mg - \frac{\sqrt{3}F}{2}\right)$

52. Net external force, $F = \sqrt{(4)^2 + (3)^2} = 5\,N$

Maximum friction $f_{max} = \mu mg = (0.09)(5)(10) = 4.5\,N$

Since, $F > f_{max}$ block will move with an acceleration,

$$a = \frac{F - f_{max}}{m} = \frac{5 - 4.5}{5} = 0.1\,m/s^2$$

53. Since, the block is at rest under two forces

(i) weight of block.

(ii) contact force from the plane (resultant of force of friction and normal reaction).

Contact force should be equal to weight (or 30 N) in upward direction. Because under two forces a body remains in equilibrium when both the forces are equal to magnitude but opposite in direction.

54. Block is at rest. Hence,

$$f = mg\sin\theta \neq \mu mg\cos\theta$$

55. Angle of repose, $\theta_r = \tan^{-1}(\mu_s) = \tan^{-1}(0.7)$

or $\quad \tan\theta_r = 0.7$

Angle of plane is $\theta = 30°$, $\tan\theta = \tan 30° = 0.577$

Since, $\quad \tan\theta < \tan\theta_r, \theta < \theta_r$

Block will not slide or $f = mg\sin\theta \neq \mu mg\cos\theta$

or $\quad f = (2)(9.8)\sin 30° = 9.8\,N$

56. $mg\sin\theta = (102)(10)\sin 30° = 510\,N$

$$\mu_s mg\cos\theta = (0.4)(102)(10)\cos 30° = 353\,N$$

$$F = 510 - 353 = 157\,N$$

57. In critical case, weight of hanging part = force of friction on the part of rope lying on table.

$\therefore \quad \frac{m}{l} \cdot l_1 g = \mu \frac{m}{l}(l - l_1)g$

Solving this we get, $\quad l_1 = \left(\frac{\mu}{1+\mu}\right)l$

58. α = angle of repose

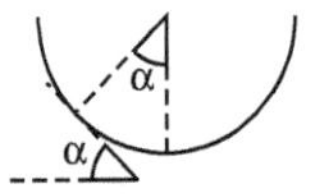

or $\quad \tan\alpha = \mu = \frac{1}{3}$

$\therefore \quad \cot\alpha = 3$

59. $v^2 = 2a_1 s$ $(a_1 = g\cos\theta)$

$0 = v^2 - 2a_2 s$ $(a_2 = \mu g\cos\theta - g\sin\theta)$

From these two equations, we see that $a_1 = a_2$

or $\quad g\sin\theta = \mu g\cos\theta - g\sin\theta$ or $2\tan\theta = \mu$

60. Due to friction $(a = \mu g)$, velocity of block will become equal to velocity of belt. Relative motion between two will stop.

$\therefore \quad v = at = \mu gt = 0.2 \times 10 \times 4 = 8\,m/s$

61. $a_{max} = \frac{T_{max} - \mu mg}{m} = \frac{40 - 0.2 \times 8 \times 10}{8} = 3\,m/s^2$

62. $N_2 = w$

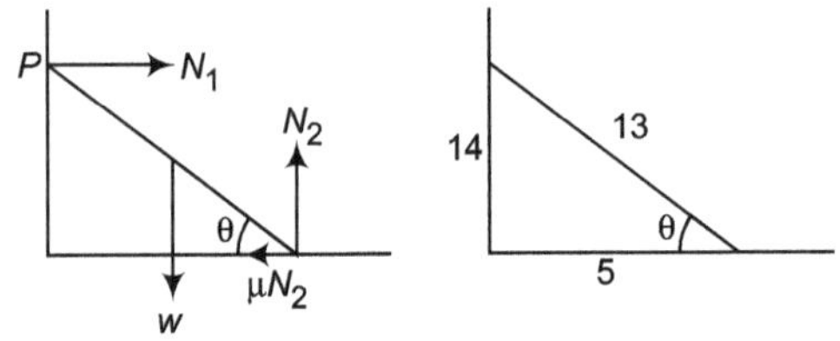

Net moments of all the forces about point P should be zero.

$\therefore \quad w\left(\frac{l}{2}\cos\theta\right) + \mu N_2(l\sin\theta) = N_2(l\cos\theta)$

or $\quad w \times \frac{l}{2} \times \frac{5}{13} + \mu wl \times \frac{12}{13} = w \times l \times \frac{5}{13}$

or $\quad 5 + 24\mu = 10$

$\therefore \quad \mu = 0.21$

63. $\quad F = mg\sin\theta - \mu mg\cos\theta \quad$ ... (i)

$2F = mg\sin\theta + \mu mg\cos\theta \quad$ ... (ii)

or $\quad 2mg\sin\theta - 2\mu mg\cos\theta = mg\sin\theta + \mu mg\cos\theta$

or $\quad \mu = \frac{1}{3}\tan\theta$

64. $\quad N = mg - Q\cos\theta$

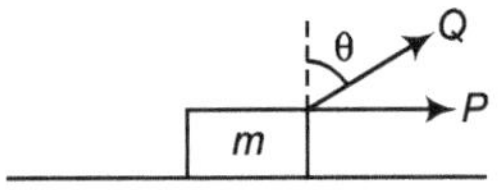

$P + Q\sin\theta = \mu N = \mu(mg - \theta\cos\theta)$

$\therefore \quad \mu = \frac{P + Q\sin\theta}{mg - Q\cos\theta}$

65. $m_1 g > m_2 g\sin\theta + \mu m_2 g\cos\theta$

or $\quad \frac{m_1}{m_2} > \sin\theta + \mu\cos\theta$

66. $N = Mg - F\cos\theta = Mg - Mg\cos\theta = 2g\sin^2\frac{\theta}{2}$

Further block can be pulled if

$$F\sin\theta \geq \mu\cos\theta$$

or $\quad 2Mg\sin\frac{\theta}{2} \cdot \cos\frac{\theta}{2} \geq 2\mu Mg\sin^2\frac{\theta}{2}$ or $\cot\frac{\theta}{2} \geq \mu$

67. F_{net} = mass × acceleration

68. In free fall, $T = 0$

69. $F_{av} = ma_{av} = m\frac{\Delta v}{\Delta t} = \frac{0.5[2-(-2)]}{10^{-3}} = 2000 + N$

70. Assuming the resistance force or retardation to be constant.

$$\left(\frac{v}{2}\right)^2 = v^2 - 2as_1 \quad \text{... (i)}$$

$$0 = \left(\frac{v}{2}\right)^2 - 2as_2 \quad \text{... (ii)}$$

Solving these two equations, we get $s_2 = \frac{s_1}{3} = 1\,cm$

71. $a_{av} = \frac{\Delta v}{\Delta t} = \frac{v_f + v_i}{\Delta t}$

$$= \frac{\sqrt{2gh_f} + \sqrt{2gh_i}}{\Delta t}$$

$$= \frac{\sqrt{2 \times 9.8 \times 2.5} + \sqrt{2 \times 9.8 \times 10}}{0.01}$$

$$= 2100 \text{ m/s}^2$$

72. $N = w + F\sin\theta$

$\therefore \quad f_{max} = \mu N = (\tan\phi)(w + F\sin\theta)$

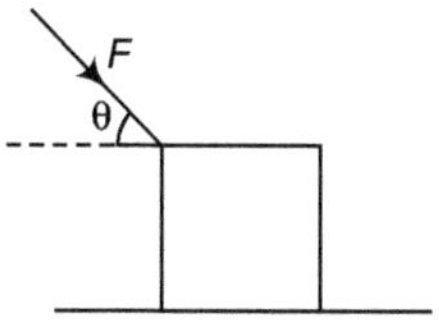

To move the body,

$$F\cos\theta = f_{max} = (\tan\phi)(w + F\sin\theta)$$

Solving this equation, we get

$$f = \frac{w\sin\theta}{\cos(\theta + \phi)}$$

73. Force of friction is zero. Only contact force is normal reaction which is $mg\cos\theta$.

74. Acceleration in both the cases will be same.

$$N_1 = ma \text{ but } N_2 = (2m)a$$

$$\therefore \quad \frac{N_1}{N_2} = \frac{1}{2}$$

75. Acceleration of system,

$$a = \frac{F - f}{M + m} = \frac{20 - 10}{6 + 4}$$

$= 1 \text{ m/s}^2$ (towards left)

Let F_0 be the reading of dynamometer, then the equation of motion of mass m would be

$$F_0 - f = ma \quad \text{or} \quad F_0 = f + ma = 10 + (4)(1)$$

$$= 14 \text{N}$$

76. Moving down with retardation a means, lift is accelerated upwards.With respect to lift pseudo force on the block will be ma in downward direction, where m is the mass of block. So, downward force mg on the block will be replaced by $m(g + a)$. Therefore, acceleration of block relative to plane will be

$a_r = (g + a)\sin\theta$ (down the plane)

From $L = \frac{1}{2}a_r t^2$ $\left(s = \frac{1}{2}at^2\right)$

$$t = \sqrt{\frac{2}{a_r}}$$

$$= \sqrt{\frac{2L}{(g + a)\sin\theta}}$$

77. Wedge moves due to horizontal component of normal reaction.

Thus, $a = \frac{N_H}{M} = \frac{N\sin\theta}{M}$ (along – ve x-axis)

78. Net force on M in vertical direction should be zero.

In vertically downwards, two forces $N\cos\theta$ and Mg are acting. Therefore, N' the normal reaction from ground should be equal to $N\cos\theta + Mg$.

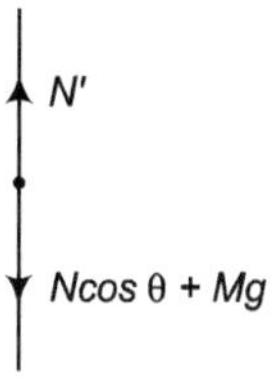

79. Net pulling force on the system $F = 10g\sin 37° - 4g = 20$ N

Maximum force of friction

$$f_{max} = \mu mg\cos 37° = 0.7 \times 10 \times 10 \times \frac{4}{5} = 56 \text{ N}$$

Since, $F < f_{max}$ system will not move. Equilibrium of 4 kg gives $T = 40$N.

80. $\mu = \tan 30° = \frac{1}{\sqrt{3}}$ (30° = angle of friction)

$$N = mg - F\sin 30°$$

$$= \left(100 - \frac{F}{2}\right)$$

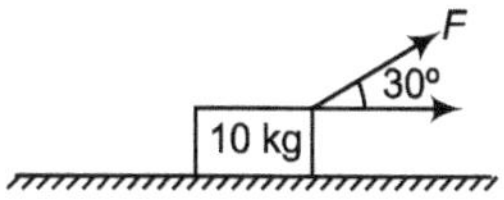

$$F\cos 30° = \mu N$$

or $\frac{\sqrt{3}F}{2} = \frac{1}{\sqrt{3}}\left(100 - \frac{F}{2}\right)$ or $\frac{3F}{2} = 100 - \frac{F}{2}$

or $F = 50$N

81. $a = \frac{18}{6} = 3 \text{ m/s}^2$

2 kg block moves by friction. Hence,

$$f = ma = 2 \times 3 = 6\text{N}$$

82. $x_P - x_1 + x_P - x_2 =$ length of string = constant

Differentiating twice with respect to time, we get

$$a_P = \frac{a_1 + a_2}{2}$$

Here, $a_P = A, a_1$ is positive and a_2 is negative. Hence,

$$A = \frac{a_1 - a_2}{2}$$

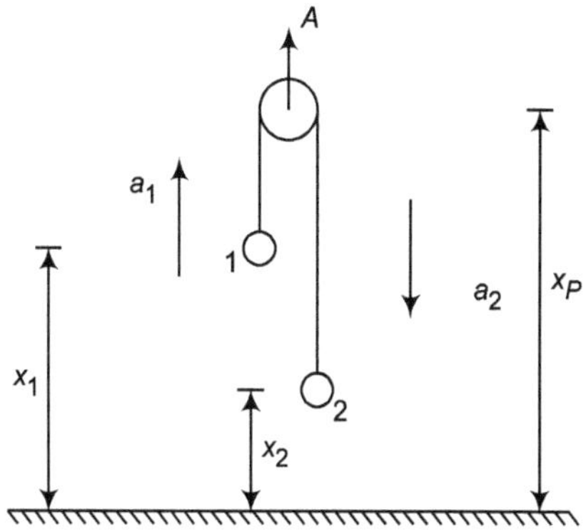

83. Block A moves due to friction . Maximum acceleration of A can be $\frac{f_{max}}{m}$ or $\frac{\mu mg}{m}$ or $\mu g = 0.2 \times 10 = 2 \text{ m/s}^2$. If both the blocks move together, then combined acceleration of A and B can be $\frac{10}{3}$ of 3.33 m/s^2. Since, this is more than the maximum acceleration of A. Slipping between them will take place and force of friction will be maximum or $\mu m_A g = 2$ N.

84.

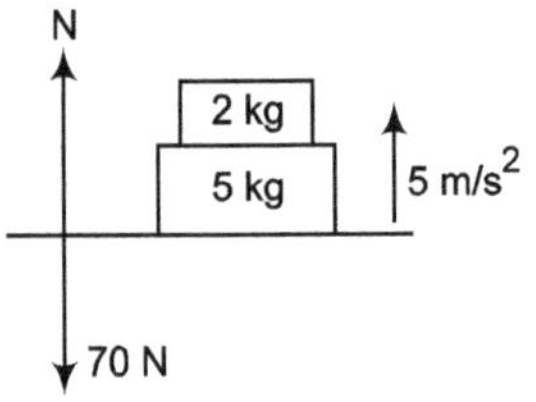

$$N - 70 = 7 \times 5 \qquad \dots \text{(i)}$$

$$\therefore \quad N = 105\,\text{N}$$

$$F = 300\,\text{N}$$

Level 2 : Only One Correct Option

1. When force is applied on A

$$\frac{F_A}{m_A + m_B} = \frac{\mu m_A g}{m_B} \qquad \dots \text{(i)}$$

When force is applied on B

$$\frac{F_B}{m_A + m_B} = \frac{\mu m_A g}{m_A} \qquad \dots \text{(ii)}$$

Dividing these two equations, we get

$$\frac{F_B}{F_A} = \frac{m_B}{m_A}$$

$$\therefore \quad F_B = \frac{m_B}{m_A} \cdot F_A = \frac{8}{4} \times 112 = 24\,\text{N}$$

2. Maximum force of friction between ground and B is

$$(15 + 5)(0.6)(10) = 120\,\text{N}$$

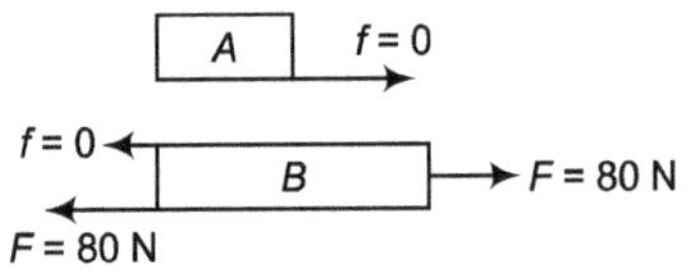

By applying 80 N force on B, blocks will not move. Hence, force diagram is as shown in above figure.

3. Maximum acceleration of m by friction = their maximum common acceleration

$$\therefore \quad \frac{\mu_1 m_1 g}{m_2} = \frac{F - \mu_2 (M + m) g}{(M + m)}$$

m will move by friction.

$$\therefore \quad F = (\mu_1 + \mu_2)(M + m) g$$

$$f_{\max} = \mu mg$$

$$\therefore \quad (a_m)_{\max} = \frac{\mu mg}{m} = \mu g$$

4. m will move by friction,

$$f_{\max} = \mu mg$$

$$\therefore \quad (a_m)_{\max} = \frac{\mu mg}{m} = \mu g$$

5. Block A moves due to friction. Maximum value of friction can be $\mu m_A g$. Therefore, maximum acceleration of A can be $\frac{\mu m_A g}{m_A}$ or $\mu g = 0.2g = \frac{g}{5}$. When force 2 mg is applied on lower block, common acceleration (if both move together) will be,

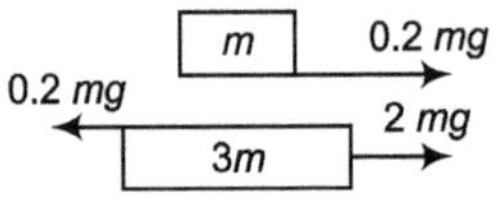

$$a = \frac{\text{Net force}}{\text{Total mass}} = \frac{2mg}{4m} = \frac{g}{2}$$

Since, $a = \frac{g}{2}$ is greater than maximum acceleration of A which can be given to it by friction.Therefore, slipping will take place.

$$a_A = 0.2g = \frac{g}{5}$$

$$a_B = \frac{2mg - 0.2mg}{3m}$$

$$= 0.6g = \frac{3g}{5}$$

$$a_{AB} = a_A - a_B = -\frac{2g}{5}$$

or $\frac{3g}{5}$ in backward direction.

6.

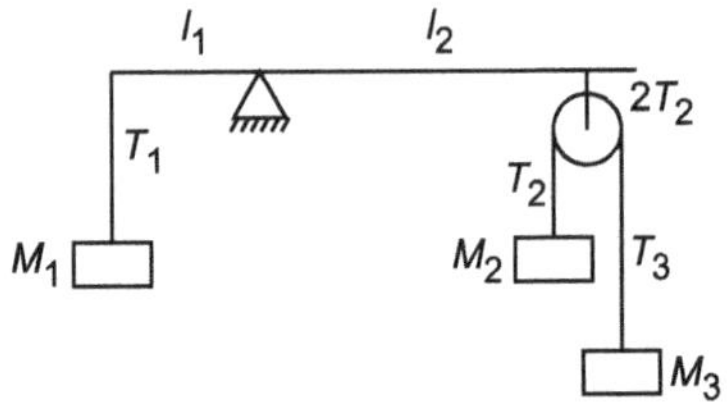

$$T_1 l_1 = 2T_2 l_2 \qquad \dots \text{(i)}$$

$$T_1 = M_1 g \qquad \dots \text{(ii)}$$

$$M_2 g - T_2 = M_2 \left(\frac{M_2 g - \frac{M_2}{3} \cdot g}{M_2 + \frac{M_2}{3}} \right)$$

$$\text{or} \quad T_2 = \frac{M_2 g}{2} \qquad \dots \text{(iii)}$$

Solving these three equations, we get

$$\frac{M_1}{M_2} = 3$$

7. $2T = 250$

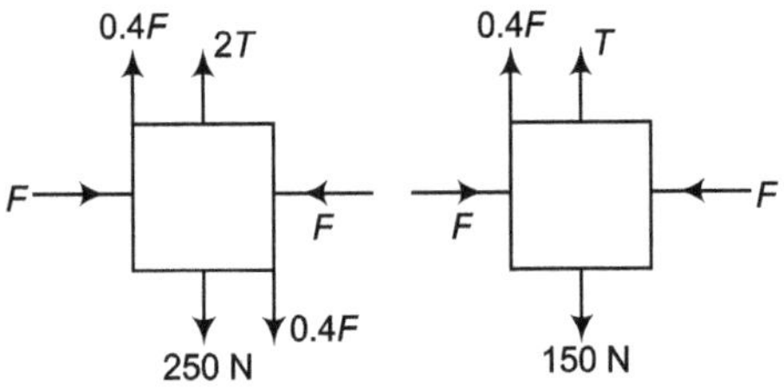

$$T = 125\,\text{N}$$

$$T + 0.4F = 150$$

$$\therefore \quad F = 62.5\,\text{N}$$

8. Maximum acceleration due to friction of mass m over mass $2m$ can be μg. Now, for the whole system

$$a = \frac{\text{Net pulling force}}{\text{Total mass}}$$

$$\therefore \quad \mu g = \frac{mg}{4m}$$

$$\text{or} \quad \mu = \frac{1}{4}$$

9. Net external force, $F=\sqrt{(4)^2+(3)^2}=5\,\text{N}$

Maximum friction, $f_{max}=\mu mg$

$$=(0.09)(5)(10)$$
$$=4.5\,\text{N}$$

Since, $F > f_{max}$, block will not move with an acceleration,

$$a=\frac{F-f_{max}}{m}$$
$$=\frac{5-4.5}{5}$$
$$=0.1\,\text{m/s}^2$$

10. $T_B=3T$ and $T_A=2A$

$\therefore$ $$v_A=\frac{T_B}{T_A}\cdot v_B=\frac{3}{2}v_0 \qquad \text{(towards right)}$$

$\therefore$ $$v_{AB}=\frac{3v_0}{2}-v_0=\frac{v_0}{2}, \qquad \text{(towards right)}$$

In such cases, velocity and acceleration are in increase ratio of tensions.

11. Acceleration of system before breaking the string was,

$$a=\frac{\text{Net pulling force}}{\text{Total mass}}$$
$$=\frac{3g-2g}{5}=\frac{g}{5}$$

After 5 s, velocity of system $v=at=\frac{g}{5}\times 5=g\,\text{m/s}$

Now, $$h=\frac{v^2}{2g}=\frac{g^2}{2g}=\frac{g}{2}=4.9\,\text{m}$$

12. Maximum acceleration of the box can be μg or $1.5\ \text{m/s}^2$, while acceleration of truck is $2\ \text{m/s}^2$. Therefore, relative acceleration of the box will be $a_r=0.5\,\text{m/s}^2$ (backward). It will fall off the truck in a time.

$$t=\sqrt{\frac{2l}{a_r}} \qquad \left(s=\frac{1}{2}at^2\right)$$
$$=\sqrt{\frac{2\times 4}{0.5}}=4\,\text{s}$$

Displacement of truck up to this instant is

$$s_r=\frac{1}{2}a_rt^2=\frac{1}{2}\times 2\times(4)^2=16\,\text{m}$$

13. m_3 is at rest. Therefore,

$$2T=m_3g \qquad \ldots\text{(i)}$$

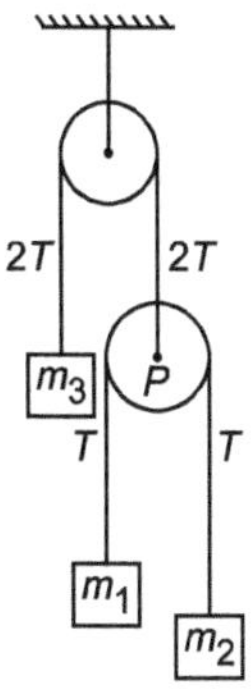

Further if m_3 is at rest, then pulley P is also at rest. Writing equations of motion.

$$m_1g-T=m_1a \qquad \ldots\text{(ii)}$$
$$T-m_2g=m_2a \qquad \ldots\text{(iii)}$$

Solving Eqs. (ii) and (iii), we get

$$m_3=1\,\text{kg}$$

14. $T\sin\theta-mg\sin\alpha=ma$

$$T\cos\theta=mg\cos\alpha$$

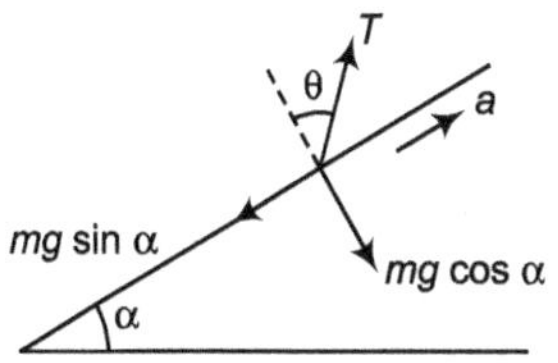

From these two equations, we get

$$\tan\theta=\frac{a+g\sin\alpha}{g\cos\alpha}$$
$$\theta=\tan^{-1}\left(\frac{a+g\sin\alpha}{g\cos\alpha}\right)$$

15. $2T_1\cos 45° mg$

$\therefore$ $$T_1=\frac{mg}{\sqrt{2}} \qquad \ldots\text{(i)}$$

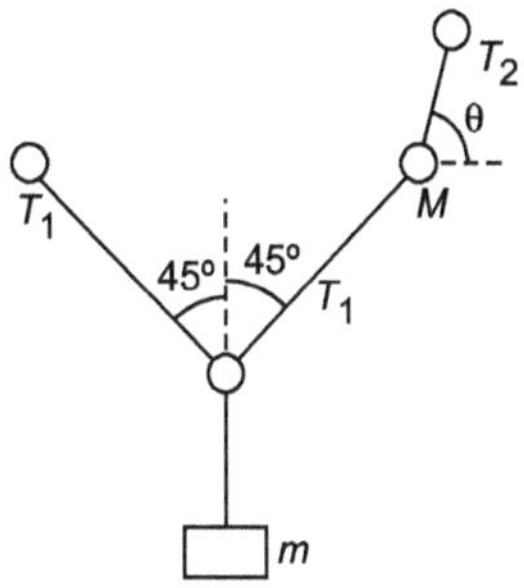

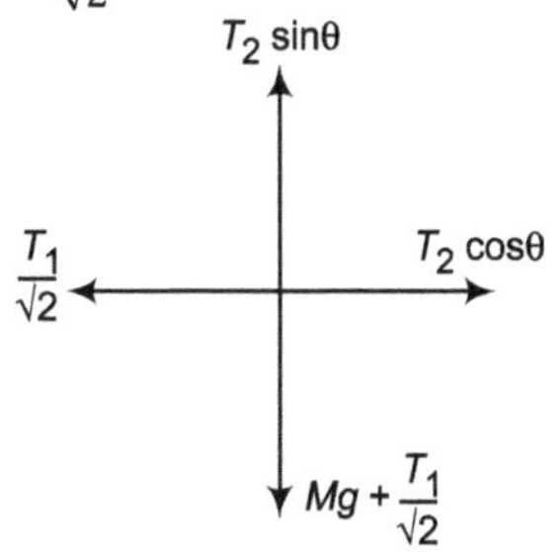

$$T_2\cos\theta=\frac{T_1}{\sqrt{2}}=\frac{mg}{2}$$
$$\tan\theta=\frac{mg+\frac{mg}{2}}{\frac{mg}{2}}=1+\frac{2M}{m}$$

16. $\frac{F}{4}=75\,\text{N}$ or $\frac{F}{4}<100\,\text{N}$

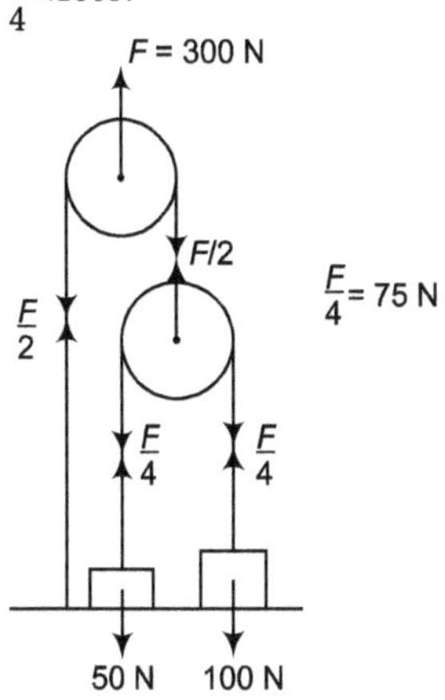

Therefore, $a_M=0$

$$a_m=\frac{75-50}{5}=5\,\text{m/s}^2$$

17. In false balance, $l_1 \neq l_2$

Moments about O should be zero.

$\therefore \qquad w_1 l_1 = w l_2 \quad \Rightarrow \quad w l_1 = w_2 l_2$

Dividing two equations, we get

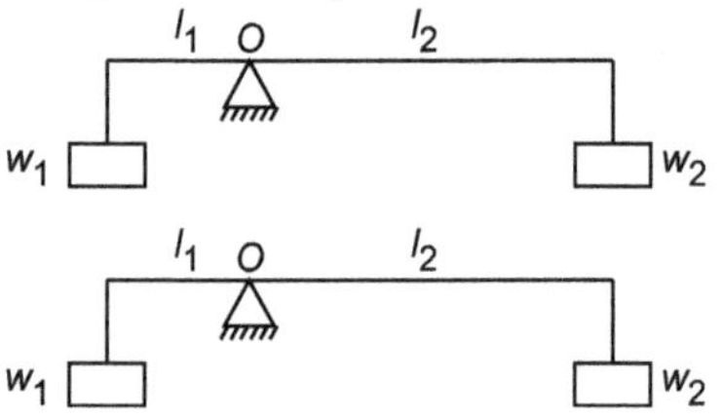

$$\frac{w_1}{w} = \frac{w}{w_2} \quad \text{or} \quad w = \sqrt{w_1 w_2}$$

$$= (\sqrt{8 \times 18})g = 12g$$

18. $2F\cos\theta = mg$

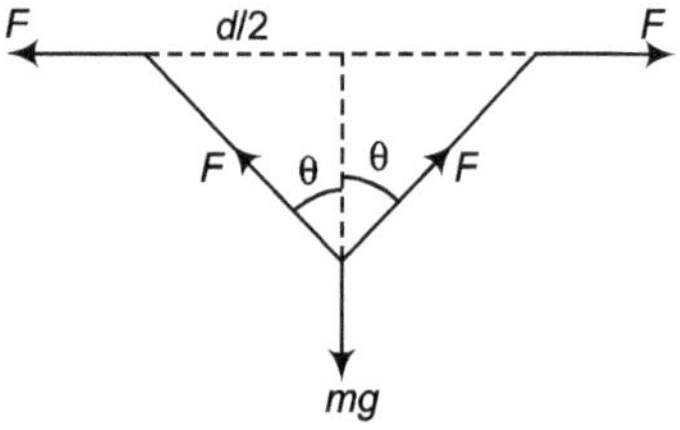

$$\therefore \qquad F = \frac{mg}{2\cos\theta} = \frac{mg\sqrt{h^2 + \frac{d^2}{4}}}{2h}$$

$$= \frac{mg}{4h}\sqrt{d^2 + 4h^2}$$

19. Total upward force $= 2\left(\frac{mg}{2}\right) = mg$

Total downward force is also mg

$\therefore \qquad F_{\text{net}} = 0 = a_{\text{net}}$

20. $N\sin\theta + \mu N\cos\theta = ma$

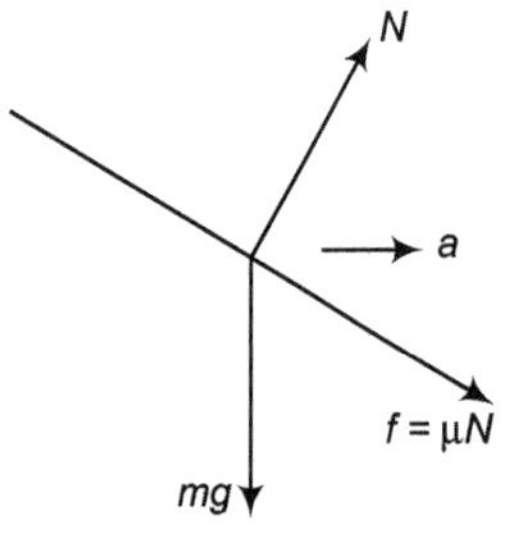

$$N\cos\theta = \mu N\sin\theta = mg$$

Putting $\theta = 45°$ and solving these two equations, we get

$$a = g\left(\frac{1+\mu}{1-\mu}\right)$$

21. $a = \dfrac{2F - F}{2m} = \dfrac{F}{2m}$ (towards left)

Writing equation of right hand side block

$$2F - N\sin 30° = ma = \frac{F}{2}$$

$$\frac{N}{2} = 2F - \frac{F}{2} = \frac{3F}{2}$$

$\therefore \qquad N = 3F$

22. To remain the string tight, component of velocities along the line joining A and B should be same.

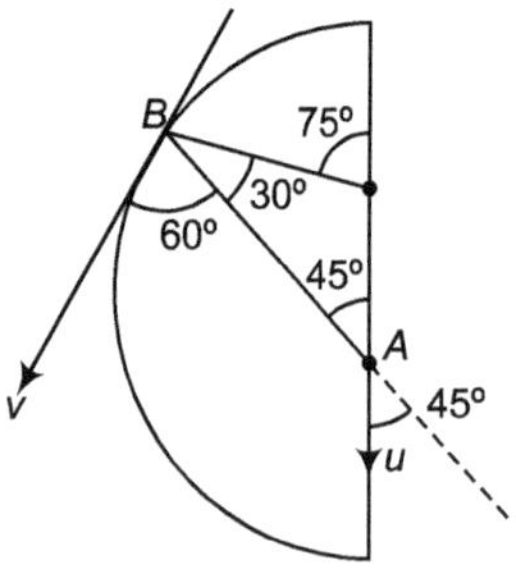

$\therefore \qquad u\cos 45° = u\cos 60°$

or $\qquad v = \sqrt{2}u$

23. In first case,

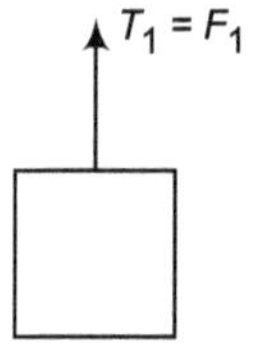

In second case,

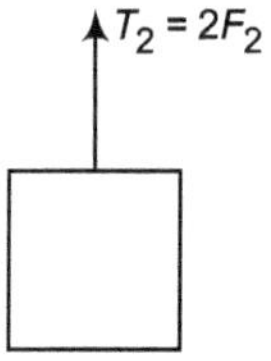

$\therefore$ In first case, person will have to apply more force.

24. f_1 = maximum value of friction between m and $2m = \mu mg$

f_2 = maximum value of friction between $2m$ and ground

$$= \frac{\mu}{3}(3m)g = \mu\, mg$$

Now, let us free body diagram of m and $2m$ in all four cases.

(a)

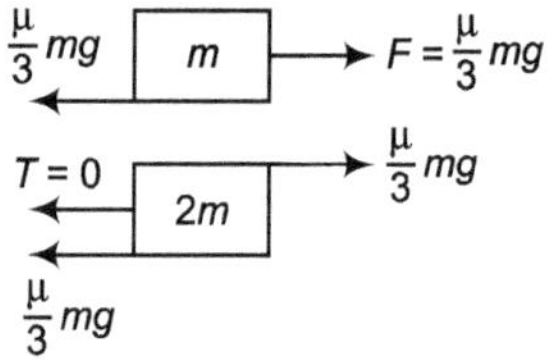

(b)

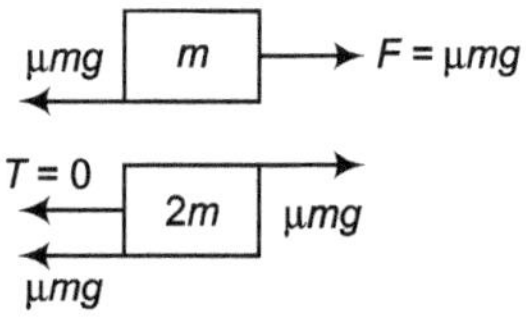

(c)

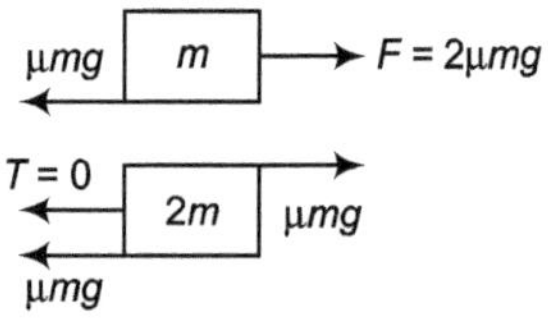

(d)

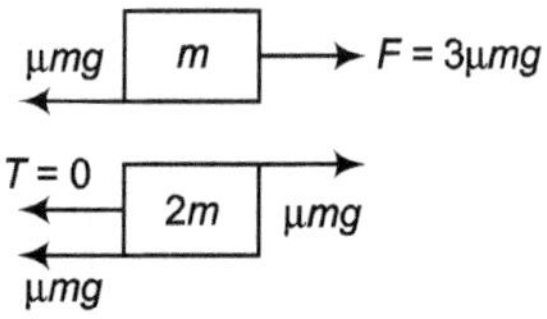

- Tension will be zero in all four cases.

25. $T = mg$

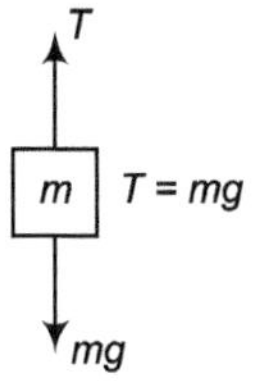

When the string breaks, T will suddenly disappear.

$\therefore \quad a_m = \dfrac{T}{m}$ (downwards)

$= \dfrac{mg}{m} = g$

$a_{2m} = \dfrac{T}{2m} = \dfrac{mg}{2m} = \dfrac{g}{2}$ (upwards)

26. $a = g\sin\theta = \dfrac{g}{\sqrt{2}}$

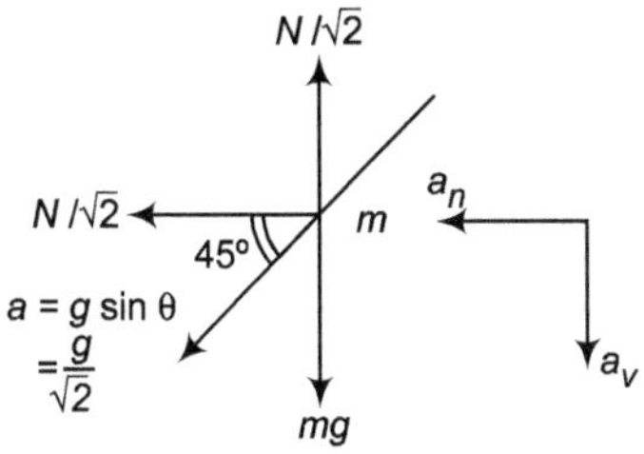

$$a_H = a_V = \frac{g}{\sqrt{2}} \times \frac{1}{\sqrt{2}} = \frac{g}{2}$$

$$\frac{N}{\sqrt{2}} = \frac{mg}{2} \quad \text{... (i)}$$

$$mg - \frac{N}{\sqrt{2}} = \frac{mg}{2} \quad \text{... (ii)}$$

$$N' = \frac{F}{\sqrt{2}} + 2mg \quad \text{... (iii)}$$

$$\mu N' = \frac{N}{\sqrt{2}} \quad \text{... (iv)}$$

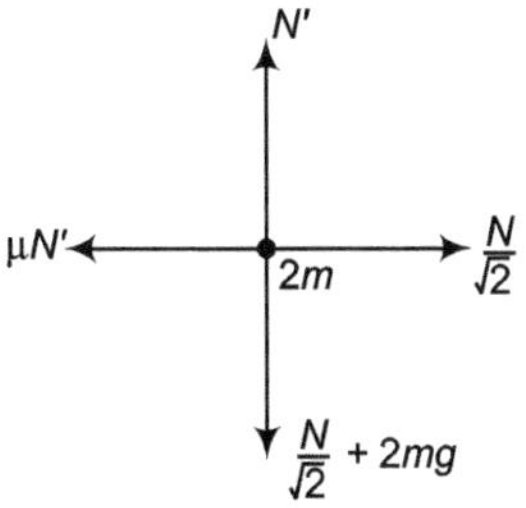

Solving above four equations, we get

$$\mu = 0.2$$

27. Let acceleration of lift is a upwards. Then, with respect to lift

$$a_r = \frac{\text{Net pulling force}}{\text{Total mass}}$$

$$= \frac{(5g + 5a) - (3g + 3a)}{8}$$

or $\quad \dfrac{9}{32}g = \dfrac{(g + a)}{4}$

$\therefore \quad a = \dfrac{g}{8}$

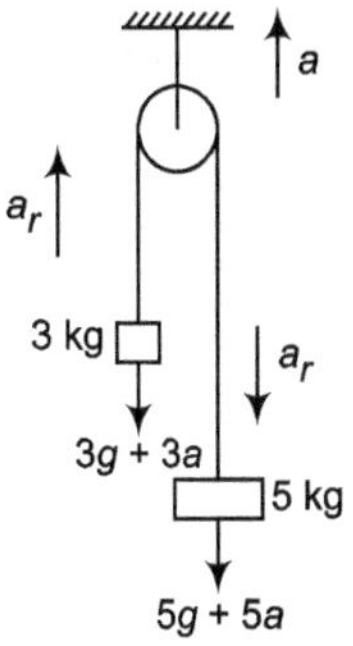

28. $F_1 r_1 = F_2 r_2$ (equating the moments about line of action or force)

$\therefore \quad \dfrac{r_1}{r_2} = \dfrac{F_2}{F_1}$

Let L be the distance between the forces. Then,

$$r_1 = \left(\frac{F_2}{F_1 + F_2}\right)L \quad \text{and} \quad r_2 = \left(\frac{F_1}{F_1 + F_2}\right)L$$

Now, given that $\quad r_1 = r_2 + \dfrac{L}{4}$

or $\quad \left(\dfrac{F_2}{F_1 + F_2}\right)L = \left(\dfrac{F_1}{F_1 + F_2}\right)L + \dfrac{L}{4}$

Solving this equations, we get

$$\frac{F_1}{F_2} = \frac{3}{5}$$

29. $\mu(m_A + m_C)g = m_B g$

$\therefore \quad m_C = \dfrac{m_B}{\mu} - m_A = \dfrac{5}{0.2} - 10 = 15\,\text{kg}$

30. $N_1 \sin 37° = N_2 \sin 74°$

or $\quad N_1 = 2N_2 \cos 37°$

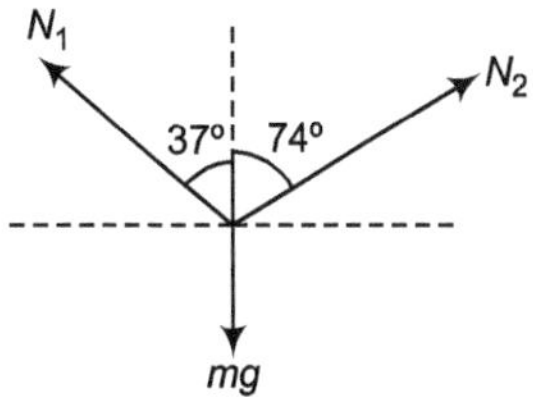

Now, $\quad N_1 \cos 37° + N_2 \cos 74° = mg$

or $\quad 2N_2 \cos^2 37° + N_2(2\cos^2 37° - 1) = mg$

or $\quad 2N_2\left(\dfrac{16}{25}\right) + N_2\left(\dfrac{32}{25} - 1\right) = mg$

or $\quad \dfrac{39}{25}N_2 = mg$

$\therefore \quad N_2 = \dfrac{25mg}{39}$

31. $m(g - a)\sin\theta = \mu m(g - a)\cos\theta$

or $\quad \mu = \tan\theta = \tan 30° = \dfrac{1}{\sqrt{3}}$

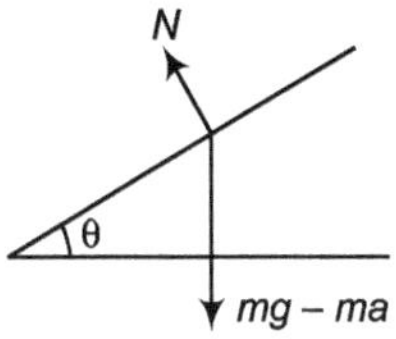

More than One Correct Options

1. Maximum value of friction between two blocks

$$f_{max} = 0.2 \times 1 \times 10 = 2\ \text{N}$$

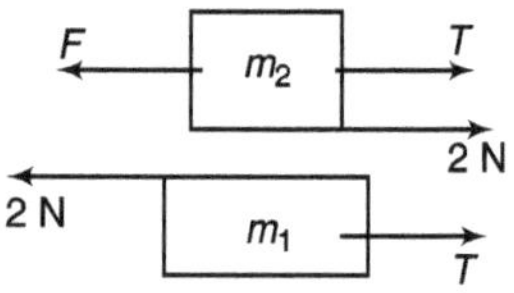

In critical case,

$$T = 2\ \text{N}$$
$$F = T + 2 = 4\ \text{N}$$

$\therefore$ System is in equilibrium if $f \le 4$ N

For $F > 4$ N

$$F - T + 2 = m_2 a = (1)(a) \quad ...(i)$$
$$T - 2 = m_1 a = (1)(a) \quad ...(ii)$$

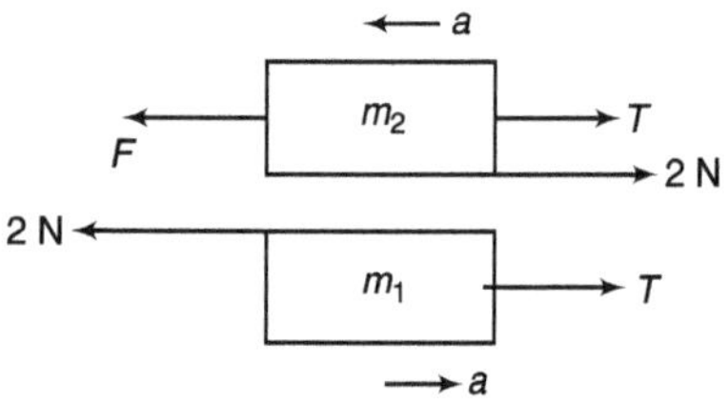

On solving these two equations, we get

$$T = \frac{F}{2}$$

When $F = 6\ \text{N},\ T = 3\ \text{N}$

2. Resultant of mg and mg is $\sqrt{2}\,mg$.

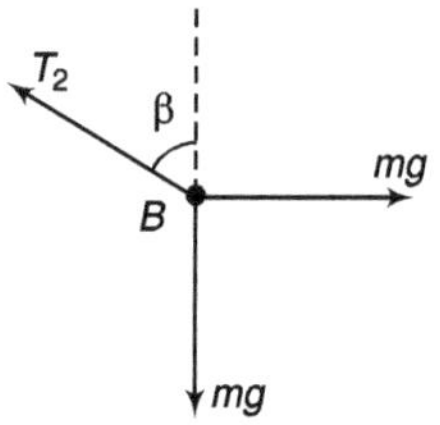

Therefore, T_2 should be equal and opposite of this. Or

$$T_2 = \sqrt{2}\,mg \quad ...(i)$$

Further,

$$T_2 \cos\beta = mg \quad ...(ii)$$

and $$T_2 \sin\beta = mg \quad ...(iii)$$

or $\sin\beta = \cos\beta \Rightarrow \beta = 45°$

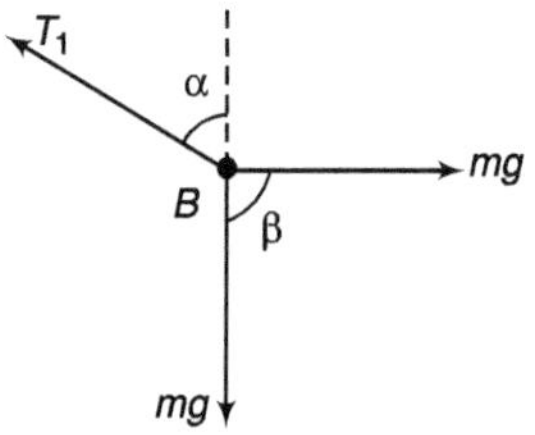

$$T_1 \cos\alpha = mg + T_2 \cos\beta$$
$$= mg + \sqrt{2}\,mg\left(\frac{1}{\sqrt{2}}\right)$$

or $$T_1 \cos\alpha = 2mg \quad ...(iv)$$

$$T_1 \sin\alpha = T_2 \sin\beta = \sqrt{2}\,mg\left(\frac{1}{\sqrt{2}}\right)$$

$\therefore$ $$T_1 \sin\alpha = mg \quad ...(v)$$

From Eqs. (iv) and (v), we get

$$\tan\alpha = \frac{1}{2} \quad \text{and} \quad T_1 = \sqrt{5}\,mg$$
$$\tan\beta = \tan 45° = 1 \quad \text{and} \quad T_2 = \sqrt{2}\,mg$$

$\therefore$ $$\tan\beta = 2\tan\alpha \quad \text{and} \quad \sqrt{2}\,T_1 = \sqrt{5}\,T_2$$

3. a = slope of v - t graph $= -1\ \text{m/s}^2$

$\therefore$ Retardation $= 1\ \text{m/s}^2 = \dfrac{\mu mg}{m} = \mu g$

or $$\mu = \frac{1}{g} = \frac{1}{10} = 0.1$$

If μ is half, then retardation a is also half. So, using

$$v = u - at \quad \text{or} \quad 0 = u - at$$

or $$t = \frac{u}{a} \quad \text{or} \quad t \propto \frac{1}{a}$$

we can see that t will be two times.

4. Maximum force of friction between A and B

$$(f_1)_{max} = 0.3 \times 60 \times 10 = 180\ \text{N}$$

Maximum force of friction between B and ground

$$(f_2)_{max} = 0.3 \times (60 + 40)\,g = 300\ \text{N}$$

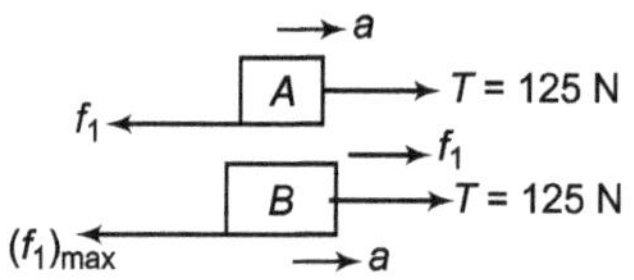

$$125 - f_1 = 60\,a \quad ...(i)$$
$$125 + f_1 - 300 = 40\,a \quad ...(ii)$$

On solving these two equations, we get

$$a = 0.5\ \text{m/s}^2$$

and $$f_1 = 95\ \text{N}$$

$f_1 = 95$ N is less than its maximum value of 180 N.

5. $a_x = \dfrac{mg\sin\theta}{m} = g\sin\theta$

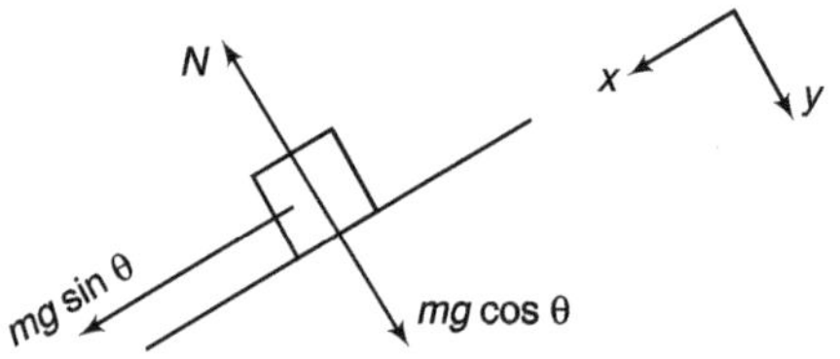

It is also moving in y -direction.

$\therefore$ $$mg\cos\theta > N$$
$$a_y = \frac{mg\cos\theta - N}{m}$$

Now, $$a = \sqrt{a_x^2 + a_y^2} > g\sin\theta$$

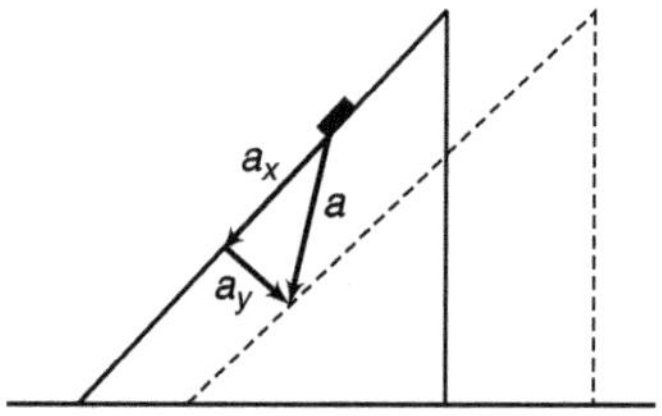

6. Maximum value of friction between A and B is

$$(f_1)_{max} = 0.25 \times 3 \times 10 = 7.5 \text{ N}$$

Maximum value of friction between B and C

$$(f_2)_{max} = 0.25 \times 7 \times 10 = 17.5 \text{ N}$$

and maximum value of friction between C and ground,

$$(f_3)_{max} = 0.25 \times 15 \times 10 = 37.5 \text{ N}$$

$$F_0 = \text{force on } A \text{ from rod}$$

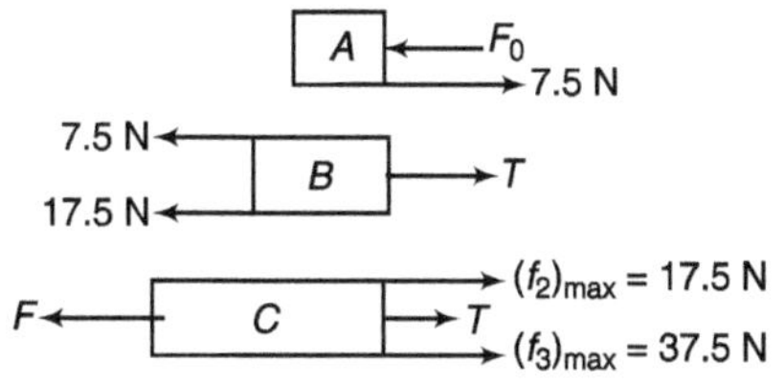

If C is moving with constant velocity, then B will also move with constant velocity.

For B, $\quad T = 17.5 + 7.5 = 25 \text{ N}$

For C, $\quad F = 17.5 + 25 + 37.5 = 80 \text{ N}$

For $F = 200$ N,

Acceleration of B towards right

= acceleration of C towards left

$= a$ (say)

Then $\quad T - 7.5 - 17.5 = 4a \quad$ ...(i)

$200 - 17.5 - 37.5 - T = 8a \quad$ ...(ii)

On solving these two equations, we get

$$a = 10 \text{ m/s}^2$$

7. Since, $\mu_1 > \mu_2$

$\therefore \quad (f_1)_{max} > (f_2)_{max}$

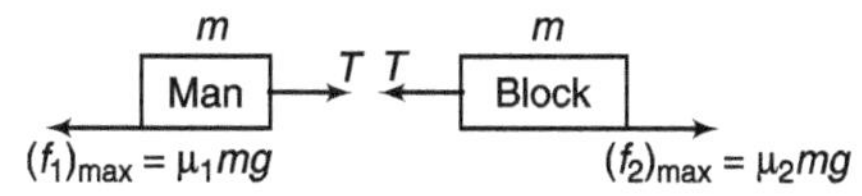

Further if both move,

$$a = \frac{T - \mu mg}{m}$$

μ of block is less. Therefore, its acceleration is more.

8. $N \cos\theta = mg = 10 \quad$ ...(i)

$N \sin\theta = ma = 5 \quad$ ...(ii)

On solving these two equations, we get

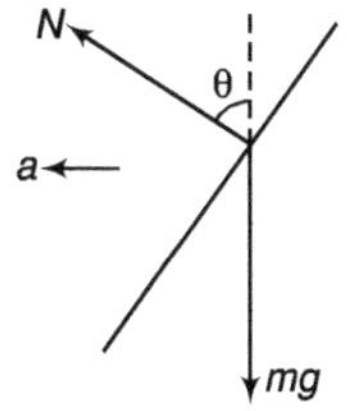

9. $f_1 \rightarrow$ force of friction between 2 kg and 4 kg

$f_2 \rightarrow$ force of friction between 4 kg and ground

$$(f_{S_1})_{max} = 0.4 \times 2 \times 10 = 8 \text{ N}$$

$$F_{K_1} = 0.2 \times 2 \times 10 = 4 \text{ N}$$

$$(f_{S_2})_{max} = 0.6 \times 6 \times 10 = 36 \text{ N}$$

$$F_{K_2} = 0.4 \times 6 \times 10 = 24 \text{ N}$$

At $t = 1$ s, $F = 2 \text{ N} < 36$ N, therefore system remains stationary and force of friction between 2 kg and 4 kg is zero.

At $t = 4$ s, $F = 8 \text{ N} < 36$ N. Therefore, system is again stationary and force of friction on 4 kg from ground is 8 N.

At $t = 15$ s, $F = 30 \text{ N} < 36$ N and system is stationary.

10. Net pulling force = 0

$$\Rightarrow \quad a = 0$$

$$T_1 = 1 \times g = 10 \text{ N}$$

$$T_3 = 2 \times g = 20 \text{ N}$$

$$T_2 = 20 + T_1 = 30 \text{ N}$$

11. $f_{max} = 0.3 \times 2 \times 10 = 6$ N

At $t = 2$ s, $F = 2 \text{ N} < f_{max}$

$\therefore \quad f = F = 2$ N

At $t = 8$ s, $F = 8 \text{ N} > f_{max}$

$\therefore \quad f = 6$ N

At $t = 10$ s, $F = 10 \text{ N} > f_{max}$

$\therefore \quad f = 6$ N

$$a = \frac{F - f}{m} = \frac{10 - 6}{2}$$

$$= 2 \text{ m/s}^2$$

$$F = f_{max} = 6 \text{ N at } 6 \text{ s}$$

For $6 \text{ s} \le t \le 10$ s,

$$a = \frac{F - f}{m} = \frac{t - 6}{2} = 0.5t - 3$$

$$\int_0^v dv = \int a\,dt = \int_6^{10} (0.5t - 3)\,dt$$

$$v = 4 \text{ m/s}$$

After 10 s,

$$a = \frac{F - f}{m} = \frac{10 - 6}{2} = 2 \text{ m/s}^2$$

$$= \text{constant}$$

$\therefore \quad v' = v + at$

$$= 4 + 2(10 - 10) = 8 \text{ m/s}$$

12. Maximum force of friction between 2 kg and 4 kg

$$= 0.4 \times 2 \times 10 = 8 \text{ N}$$

2 kg moves due to friction. Therefore, its maximum acceleration may be

$$a_{max} = \frac{8}{2} = 4 \text{ m/s}^2$$

Slip will start when their combined acceleration becomes 4 m/s^2.

$\therefore \quad a = \dfrac{F}{m} \quad$ or $\quad 4 = \dfrac{2t}{6} \quad$ or $\quad t = 12$ s

At $t = 3$ s

$$a_2 = a_4 = \frac{F}{m} = \frac{2t}{6}$$

$$= \frac{2 \times 3}{6} = 1 \text{ m/s}^2$$

Both a_2 and a_4 are towards right. Therefore, pseudo forces F_1 (on 2 kg from 4 kg) and F_2 (on 4 kg from 2 kg) are towards left.

$$F_1 = (2)(1) = 2 \text{ N}$$

$$F_2 = (4)(1) = 4 \text{ N}$$

From here, we can see that F_1 and F_2 do not make a pair of equal and opposite forces.

14. Consider the below diagram. Frictional force on $B(f_1)$ and frictional force on A (f_2) will be as shown.

Let A and B are moving together

$$a_{\text{common}} = \frac{F - f_1}{m_A + m_B} = \frac{F - f_1}{(m/2) + m} = \frac{2(F - f_1)}{3m}$$

Pseudo force on $A = (m_A) \times a_{\text{common}}$

$$= m_A \times \frac{2(F - f_1)}{3m} = \frac{m}{2} \times \frac{2(F - f_1)}{3m}$$

$$= \frac{(F - f_1)}{3}$$

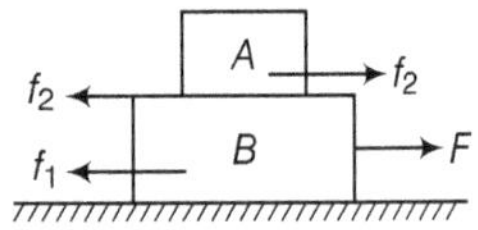

The force (F) will be maximum when

Pseudo force on $A =$ Frictional force on A

$$\Rightarrow \quad \frac{F_{\max} - f_1}{3} = \mu\, m_A g$$

$$= 0.2 \times \frac{m}{2} \times g = 0.1 \text{ mg}$$

$$\Rightarrow \quad F_{\max} = 0.3 \text{ mg} + f_1$$

$$= 0.3 \text{ mg} + (0.1)\frac{3}{2}\text{ mg} = 0.45 \text{ mg}$$

Hence, maximum force up to which bodies will move together is $F_{\max} = 0.45$ mg

(a) Hence, for $F = 0.25 \text{ mg} < F_{\max}$ bodies will move together.

(b) For $F = 0.5 \text{ mg} > F_{\max}$, body A will slip with respect to B.

(c) For $F = 0.5 \text{ mg} > F_{\max}$, bodies slip.

$$(f_1)_{\max} = \mu\, m_B g = (0.1) \times \frac{3}{2} m \times g = 0.15 \text{ mg}$$

$$(f_2)_{\max} = \mu\, m_A g = (0.2)\left(\frac{m}{2}\right)(g) = 0.1 \text{ mg}$$

Hence, minimum force required for movement of the system $(A + B)$

$$F_{\min} = (f_1)_{\max} + (f_2)_{\max}$$

$$= 0.15 \text{ mg} + 0.1 \text{ mg} = 0.25 \text{ mg}$$

(d) Maximum force for combined movement $F_{\max} = 0.45$ mg.

15. Let m_1, moves up the plane. Different forces involved are shown in the diagram.

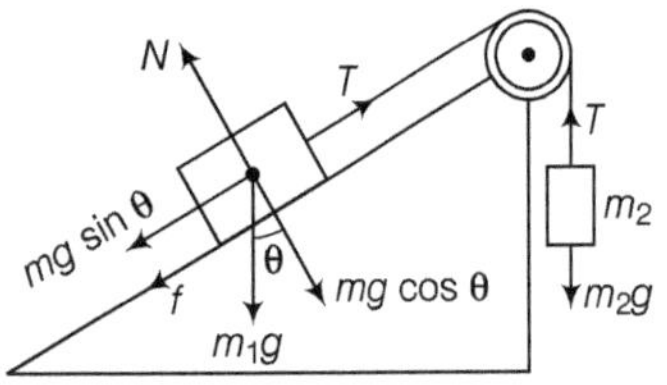

$N =$ Normal reaction

$f =$ Frictional force

$T =$ Tension in the string

$f = \mu N = \mu\, m_1 g \cos\theta$

For the system $(m_1 + m_2)$ to move up

$$m_2 g - (m_1 g \sin\theta + f) > 0$$

$$\Rightarrow \quad m_2 g - (m_1 g \sin\theta + \mu\, m_1 g \cos\theta) > 0$$

$$\Rightarrow \quad m_2 > m_1 (\sin\theta + \mu \cos\theta)$$

Hence, option (b) is correct.

Let the body moves down the plane, in this case f acts up the plane.

Hence, $\quad m_1 g \sin\theta - f > m_2 g$

$$\Rightarrow \quad m_1 g \sin\theta - \mu\, m_1 g \cos\theta > m_2 g$$

$$\Rightarrow \quad m_1 (\sin\theta - \mu \cos\theta) > m_2$$

$$\Rightarrow \quad m_2 < m_1 (\sin\theta - \mu \cos\theta)$$

Hence, option (d) is correct.

16. Let A moves up the plane frictional force on A will be downward as shown.

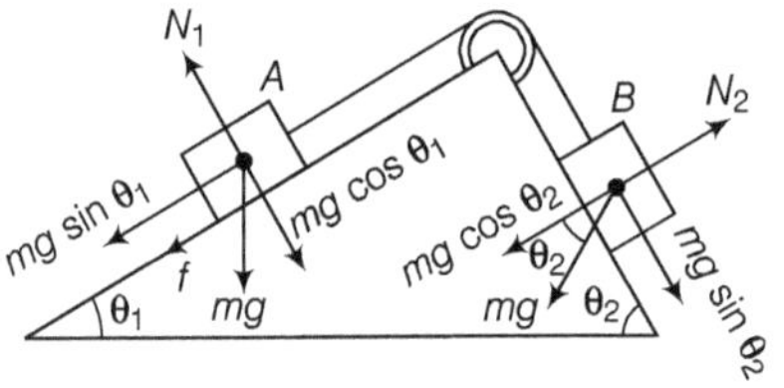

When A just starts moving up

$$mg \sin\theta_1 + f = mg \sin\theta_2$$

$$\Rightarrow \quad mg \sin\theta_1 + \mu mg \cos\theta_1 = mg \sin\theta_2$$

$$\Rightarrow \quad \mu = \frac{\sin\theta_2 - \sin\theta_1}{\cos\theta_1}$$

When A moves upwards

$$f = mg \sin\theta_2 - mg \sin\theta_1 > 0$$

$$\Rightarrow \quad \sin\theta_2 > \sin\theta_1$$

$$\Rightarrow \quad \theta_2 > \theta_1$$

Comprehension Based Questions

1. Let $\mu_K = \mu$, then $\mu_S = 2\mu$

According to first condition,

$$F + mg \sin\theta = \mu_S\, mg \cos\theta = 2\mu\, mg \cos\theta \quad \text{...(i)}$$

According to second condition,

$$mg \sin\theta = F + \mu_K\, mg \cos\theta$$

$$= F + \mu\, mg \cos\theta \quad \text{...(ii)}$$

Putting $\theta = 30°$, we get

$$F + mg/2 = 2\mu mg\left(\frac{\sqrt{3}}{2}\right)$$

or $$\sqrt{3}\mu\, mg = F + 0.5 mg \quad \text{...(iii)}$$

$$\frac{mg}{2} = F + \mu\, mg\left(\frac{\sqrt{3}}{2}\right)$$

or $$0.5\sqrt{3}\mu\, mg = 0.5 mg - F \quad \text{...(iv)}$$

Dividing Eqs. (iii) and (iv), we get

$$F = \frac{mg}{6}$$

2. Substituting value of F in Eq. (iii), we have

$$\mu = \frac{2}{3\sqrt{3}} = \mu_K$$

$$\therefore \quad \mu_S = 2\mu = \frac{4}{3\sqrt{3}}$$

3. $$a = \frac{F + mg \sin\theta - \mu_K\, mg \cos\theta}{m}$$

$$= \frac{(mg/6) + (mg/2) - \left(\frac{2}{3\sqrt{3}}\right) mg \left(\frac{\sqrt{3}}{2}\right)}{m} = \frac{g}{3}$$

4. $F' = mg\sin\theta + \mu_S mg\cos\theta$

$$= \frac{mg}{2} + \frac{4}{3\sqrt{3}} mg\left(\frac{\sqrt{3}}{2}\right) = \frac{7mg}{6}$$

5. $F'' = mg\sin\theta + \mu_K mg\cos\theta$

$$= (mg/2) + \left(\frac{2}{3\sqrt{3}}\right) mg\left(\frac{\sqrt{3}}{2}\right) = \frac{5mg}{6}$$

6. Acceleration, $a_1 = \dfrac{1350\times 9.8 - 1200\times 9.8}{1200} = 1.225\ \text{m/s}^2$

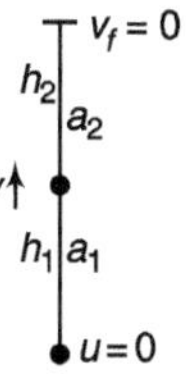

Retardation, $a_2 = \dfrac{1200g - 1000g}{1200} = 1.63\ \text{m/s}^2$

$$h_1 + h_2 = 25 \quad \text{...(i)}$$

$$v = \sqrt{2a_1 h_1}$$

or $2a_1 h_1 = 2a_2 h_2$

$\therefore \dfrac{h_1}{h_2} = \dfrac{a_2}{a_1} = \dfrac{1.63}{1.225} = 1.33$ ...(ii)

Solving these equations, we get

$$h_1 = 14.3\ \text{m}$$

7. $v = \sqrt{2a_1 h_1} = \sqrt{2\times 1.225\times 14.3}$

$= 5.92$ m/s **Ans.**

8. $\tan\theta = \dfrac{8}{15}$ $\therefore$ $\theta = \tan^{-1}\left(\dfrac{8}{15}\right) = 28°$

$$(f_A)_{max} = 0.2\times 170\times 10\times \cos 28°$$
$$= 300.2\ \text{N} \approx 300\ \text{N}$$
$$(f_B)_{max} = 0.4\times 170\times 10\times \cos 28°$$
$$= 600.4\ \text{N} \approx 600\ \text{N}$$

Now,

$$(m_A + m_B)\, g\sin\theta = (340)\,(10)\sin 28° = 1596\ \text{N}$$

Since, this is greater than $(f_A)_{max} + (f_B)_{max}$, therefore blocks slides downward and maximum force of friction will act on both surfaces

$\therefore$ $f_{total} = (f_A)_{max} + (f_B)_{max} = 900$ N

9. $a = \dfrac{(m_A + m_B)\, g\sin\theta - f_{total}}{m_A + m_B}$

$= \dfrac{1596 - 900}{340} = 2.1\ \text{m/s}^2$

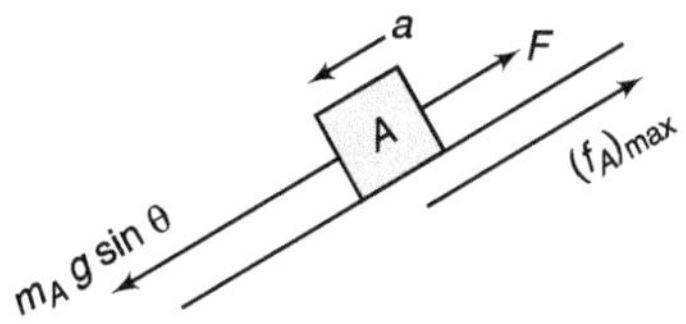

F = force on connecting bar

$$m_A\, g\sin\theta - F - (f_A)_{max} = m_A a$$

$\therefore$ $F = m_A g\sin\theta - (f_A)_{max} - m_A a$

$= 170\times 10\times \sin 28° - 300 - 170\times 2.1$

$= 141$ N

Assertion and Reason

1. Normal reaction = 15 N

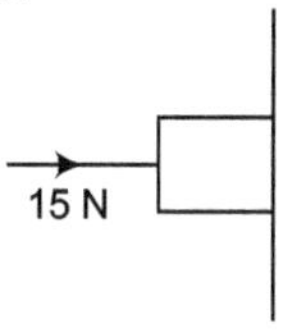

$$f_{max} = \mu N = 9\ \text{N}$$

Weight = 10 N

Since, weight $> f_{max}$

$\therefore$ 9 N friction will act and block will move downwards.

4. $mg\sin\theta = 10\times 10\times \dfrac{1}{2} = 50$ N

$\mu mg\cos\theta = 0.7\times 10\times 10\times \dfrac{\sqrt{3}}{2} = 60.62$ N

Since $\mu mg\cos\theta$ is more, block will remain stationary.

5. $(f_2)_{max} = \mu_2 mg$

$(f_1)_{max} = \mu_1 (2m) g$

Since, $(f_1)_{max} > (f_2)_{max}$

Lower block will not move at all.

7. Due to rotation of earth it is non-inertial. A frame moving with constant velocity is inertial.

8. Friction opposes the relative motion of the bodies in contact not the motion.

9. Friction opposes the relative motion of the bodies in contact. By throwing something backwards you can move forwards.

10. Static friction varies between zero to a maximum limiting value.

11. Retardation in upward journey, $a_1 = \dfrac{w + F}{m}$

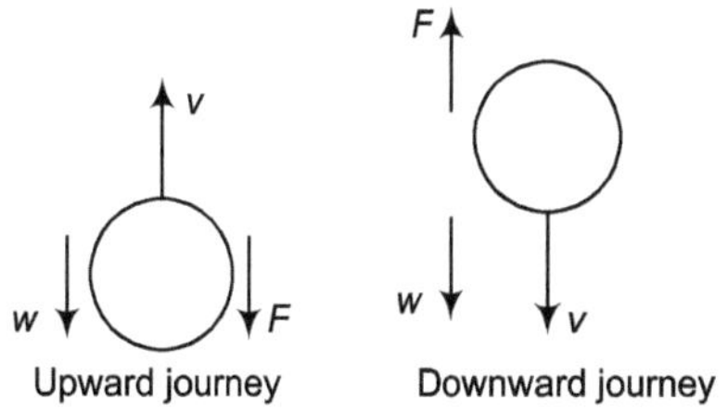

Acceleration in downward journey, $a_2 = \dfrac{w - F}{m}$

$\therefore$ $a_1 > a_2$

12. Tension is non-uniform even if string is not accelerated.

13. $\mathbf{F}_{net} = 0$

$\therefore$ $\mathbf{a}_{net} = 0$

But, torque may not be zero. Therefore, it can't be in rotational equilibrium.

14. $f - mg\sin\theta = ma\sin\theta$

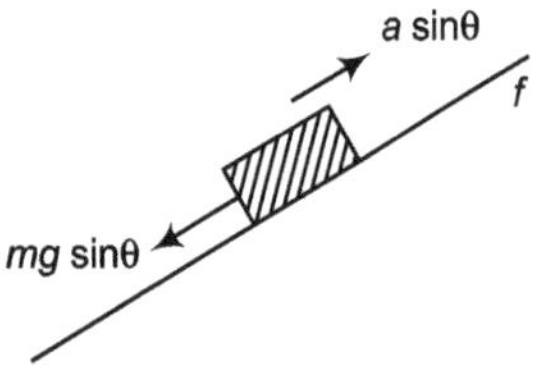

$\therefore \quad f = m(g + a)\sin\theta > mg\sin\theta$

15. Let T_0 is the tension in the string connecting the two blocks. Then,

$$T_0 - 10 = 1 \times a$$

$$\therefore \quad T_0 = 10\text{N}$$

$$20 - T_0 = 2 \times a$$

$$\therefore \quad T_0 < 20\text{N}$$

16. In case of massless string,

$$a = \frac{2F - F}{0} = \infty$$

17. Consider moment of forces about the point of application of force.

18. $F_{\text{net}} = ma$

19.
$$N_1 = ma_x - 2m$$

$$N_2 = m(g + a_y) = 12m$$

$$\therefore \quad \frac{N_1}{N_2} = \frac{1}{6}$$

Match the Columns

1. $N = mg - 20\sqrt{2}\sin 45° = 20\text{N}$

$$\mu_s N = 16\text{ N}$$

and $\quad \mu_k N = 12\text{ N}$

Since, $20\sqrt{2}\cos 45° > \mu_s N$ block will move and kinetic friction will act,

$$a = \frac{20\sqrt{2}\cos 45° - \mu_k N}{m}$$

$$= \frac{20 - 12}{4} = 2\text{ m/s}^2$$

2. Since, the pulleys are smooth, net force on each pulley should be zero. With this concept, tensions on all strings are shown below. Now, we can draw free body diagrams of all the four blocks.

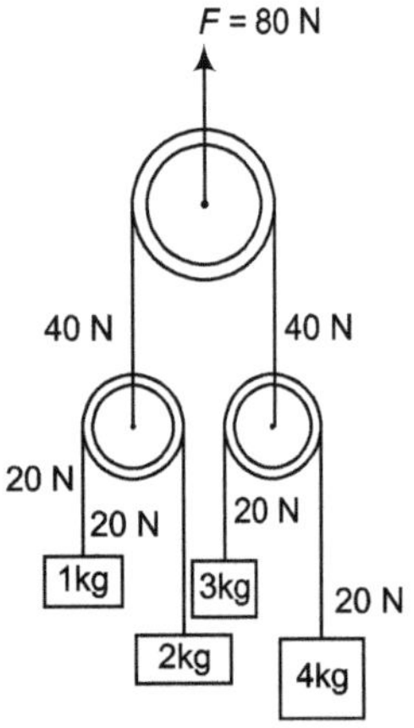

Now, we can draw free body diagrams of all the four blocks.

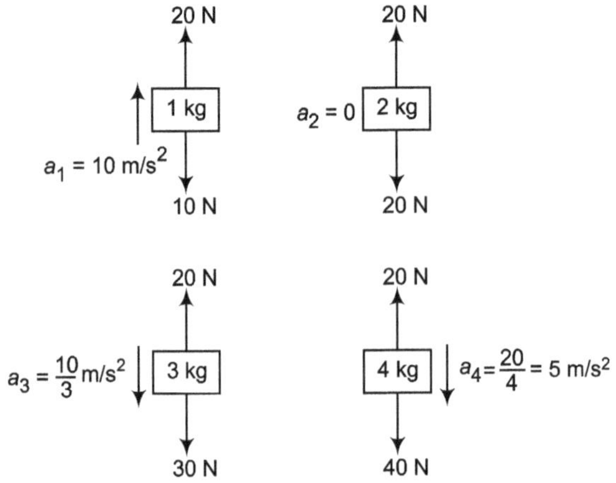

3. Force of friction opposes relative motion between two bodies in contact. It does not simply opposes the relative motion. Further, $N = mg$, only when no inclined force is acting on the block as shown in figure.

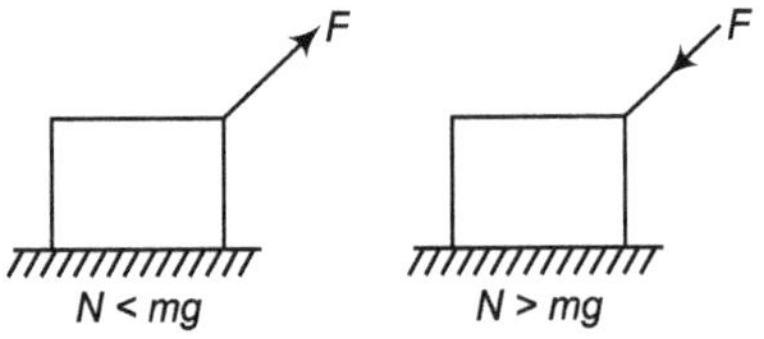

4. Force of friction and $mg\sin\theta$ both are downwards.

$\therefore$ Acceleration of the block is

$a = (g\sin\theta + \mu g\cos\theta)$ down the plane

Now, net force in any direction is equal to

$F = m$(component of acceleration that direction)

5. Acceleration of system,

$$a = \frac{60 - 18 - (m_1 + m_2 + m_3)g\sin 30°}{(m_1 + m_2 + m_3)} = 2\text{ m/s}^2$$

Net force on 3 kg block $= m_3 a = 6$ N

From free body diagram of 3 kg block, we have

$$N_{12} - m_1 g\sin 30° - 18 = m_1 a$$

$$\therefore \quad N_{12} = 25\text{N}$$

From free body diagram of 3 kg block, we have

$$60 - m_3 g\sin 30° - N_{32} = m_3 a$$

$$\therefore \quad N_{32} = 39\text{N}$$

6. $\mathbf{a}_A = \frac{d\mathbf{v}_A}{dt} = (2\hat{\mathbf{j}})\text{ m/s}^2$

$$\mathbf{a}_B = \frac{d\mathbf{v}_B}{dt} = 0$$

$$\mathbf{a}_C = \frac{d\mathbf{v}_C}{dt} = (-4\hat{\mathbf{j}})\text{ m/s}^2$$

Now, pseudo force in opposite direction of acceleration of frame from where object is observed.

7. Force diagram on both the blocks is as shown in figure.

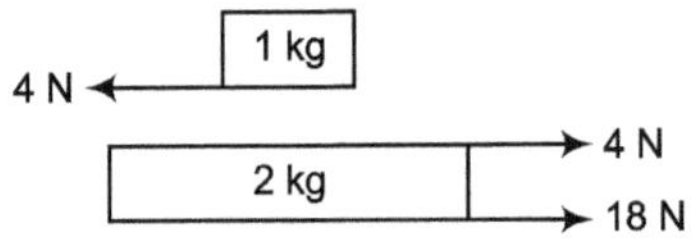

Entrance Gallery

1. As the force is exponentially decreasing, so its acceleration i.e. rate of increase of velocity will decrease with time. Thus, the graph of velocity will be an increasing curve with decreasing slope with time.

$$a = \frac{F}{m} = \frac{F_0}{m} e^{-bt} = \frac{dv}{dt}$$

$$\Rightarrow \quad \int_0^v dv = \int_0^t \frac{F_0}{m} e^{-bt}\, dt$$

$$\Rightarrow \quad v = \frac{F_0}{m}\left(\frac{1}{-b}\right) e^{-bt} \Big|_0^t$$

$$= \frac{F_0}{mb} e^{-bt} \Big|_t^0$$

$$v = \frac{F_0}{mb}(e^0 - e^{-bt}) = \frac{F_0}{mb}(1 - e^{-bt})$$

with $\quad v_{max} = \frac{F_0}{mb}$

2. A block of mass m is placed on a surface with a vertical cross-section, then

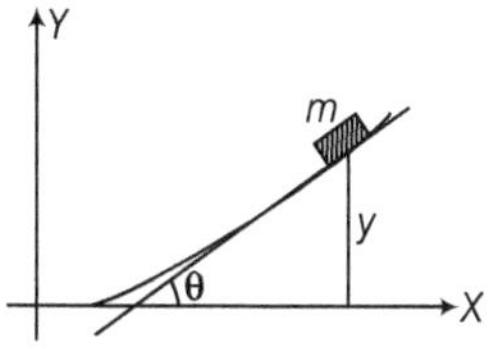

$$\tan\theta = \frac{dy}{dx} = \frac{d\left(\frac{x^3}{6}\right)}{dx} = \frac{x^2}{2}$$

At limiting equilibrium, we get

$$\mu = \tan\theta$$

$$0.5 = \frac{x^2}{2} \Rightarrow x^2 = 1 \Rightarrow x = \pm 1$$

Now, putting the value of x in $y = \frac{x^2}{6}$, we get

When, $x = 1$ $\quad y = \frac{(1)^3}{6} = \frac{1}{6}$ $\qquad$ When, $x = -1$ $\quad y = \frac{(-1)^3}{6} = \frac{-1}{6}$

So, the maximum height above the ground at which the block can be placed without slipping is $\frac{1}{6}$ m.

3. Angle of inclination, $\theta = 60°$

The formula of the coefficient of static friction μ is

$\mu = \tan\theta = \tan 60° = \sqrt{3}$

Thus, $\quad \mu = 1.732$

4. If the same wedge is made rough, then time taken by it to come down becomes n times more.

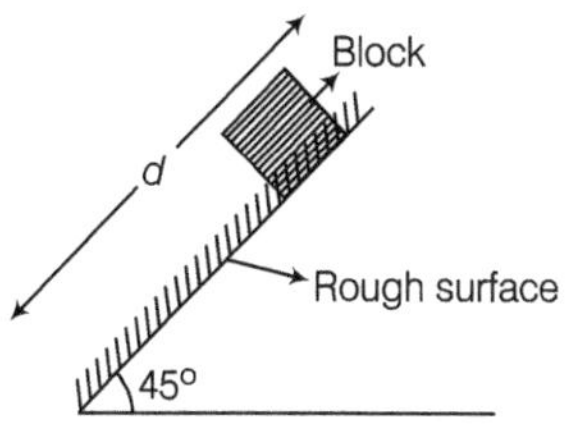

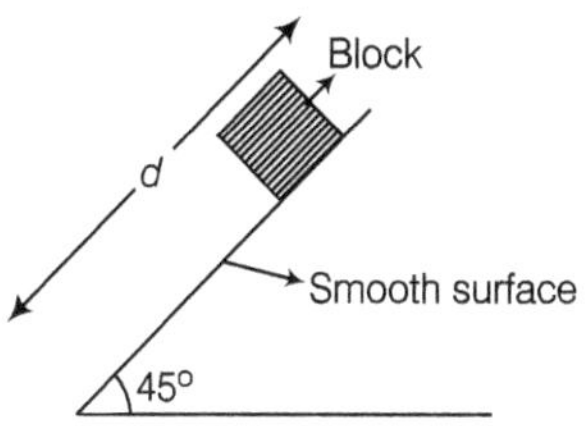

The coefficient of friction, $\mu = \left[1 - \frac{1}{n^2}\right]\tan\theta$

$$\Rightarrow \quad \mu = \left[1 - \frac{1}{2^2}\right]\tan 45°$$

$$\mu = \frac{4-1}{4}$$

$$\Rightarrow \quad \mu = \frac{3}{4}$$

5. Here, $\quad a = \frac{F}{m} = \frac{80}{50} = 1.6\ \text{m/s}^2$

$$u = 20\ \text{m/s and } v = 0$$

From equation of motion, $\quad v = u + at$

$$0 = 20 - 1.6 \times t$$

$$t = \frac{20}{1.6} = 12.5\ \text{s}$$

6. $u = 11\ \text{m/s}$

$\because \quad a = \mu g$

Given $\quad a = 0.4 \times 10 = 4\ \text{m/s}^2$

According to question, $v = 0$

So from, $\quad v^2 = u^2 + 2as$

$$0 = (11)^2 + 2(-4) \times s$$

$$s = \frac{11 \times 11}{8} = 15.125\ \text{m}$$

7. Weight, $w = mg = 0.1 \times 10 = 1$ N

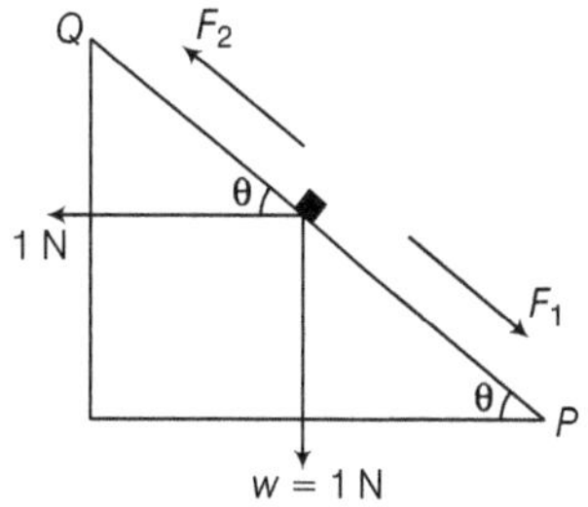

F_1 = component of weight $= 1 \cdot \sin\theta = \sin\theta$

F_2 = component of applied force $= 1 \cdot \cos\theta = \cos\theta$

Now, at $\theta = 45°, F_1 = F_2$ and block remains stationary without the help of friction.

For $\theta > 45°, F_1 > F_2$, so friction will act towards Q.

For $\theta < 45°, F_2 > F_1$ and friction will act towards P.

8. From equation of motion,

$$v^2 = u^2 + 2as \quad \text{...(i)}$$

where, v = final velocity

u = initial velocity

a = acceleration

and $\quad s$ = distance travelled.

Now,

Case I $\quad (0)^2 + (2u)^2 - 2a(s_1)$ $\qquad$ [using Eq. (i)]

$$u^2 = 2as_1 \quad \text{...(ii)}$$

Case II $\quad (0)^2 = (2u)^2 - 2as_2\ 4u^2 = 2as_2$ $\qquad$ ...(iii)

[using Eq. (i)]

Hence, from Eqs. (ii) and (iii), we get

$$s_2 = 4s_1$$

9. As, the block is at rest,

Balancing horizontal forces component on the block,

$$f = \mu mg \cos\theta = \mu N$$

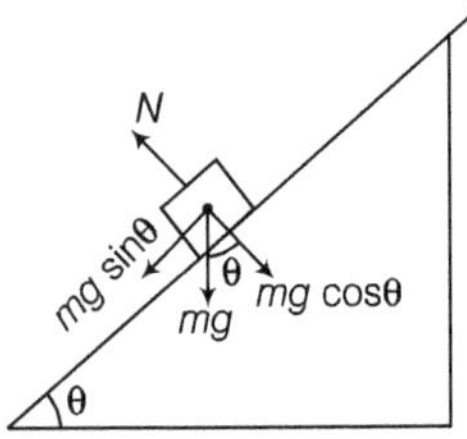

i.e. $mg\cos\theta(\mu_s) = mg\sin\theta$

$\Rightarrow$ $\tan\theta = \mu_s$

Coefficient of static friction, $\mu_s = \tan\theta$

10. Given, $m = 2\,\text{kg}, \mu = 0.2$ and $g = 10\,\text{ms}^{-2}$

Here, $ma = \mu mg$

$a = \mu g$

$a = 0.2 \times 10$

$\Rightarrow$ $a = 2\,\text{ms}^{-2}$

11. We know that, $F = ma$

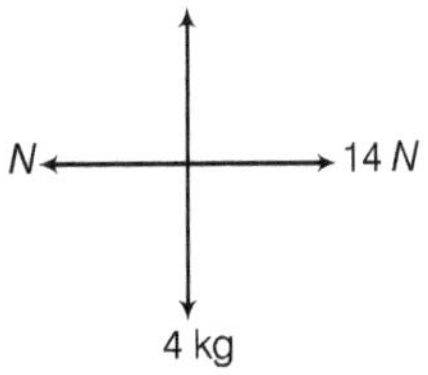

$$a = \frac{F}{m} = \frac{14}{17} = 2\,\text{ms}^{-2}$$

Hence, from the figure

$14 - N = 4a$

$14 - N = 8$

$\Rightarrow$ $N = 6\,\text{N}$

12. Given, $m_1 = 1\,\text{kg}, m_2 = 2\,\text{kg}$ and $g = 10\,\text{m/s}^2$

$$a = \left(\frac{m_2 - m_1}{m_1 + m_2}\right) g$$

$$= \left(\frac{2-1}{1+2}\right) 10 = \frac{10}{3}$$

$\because$ $s = ut + \frac{1}{2}at^2 \Rightarrow s = \frac{1}{2}at^2$ $[\because u = 0]$

$$= \frac{1}{2} \times \frac{10}{3} \times 4 = \frac{20}{3}$$

$$m = \frac{2 \times \frac{20}{3} - 1 \times \frac{20}{3}}{3} = \frac{20}{9}$$

13.

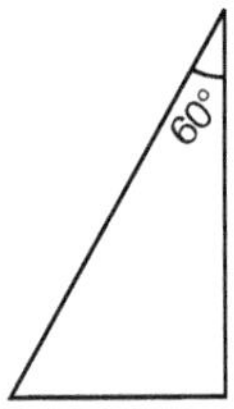

$$l = \frac{1}{2} g \cos 60° \, t_1^2 \quad \text{...(i)}$$

$$l\cos\theta = \frac{1}{2} g t_2^2 \quad \text{...(ii)}$$

$\therefore$ $\dfrac{t_1^2}{t_2^2} = \dfrac{1}{\cos^2 60°} = \dfrac{4}{1}$

$t_1 : t_2 = 2:1$

14. $\tan 30° = \dfrac{1}{\sqrt{3}} = \dfrac{x}{10}$

$\Rightarrow$ $x = \dfrac{10}{\sqrt{3}}$

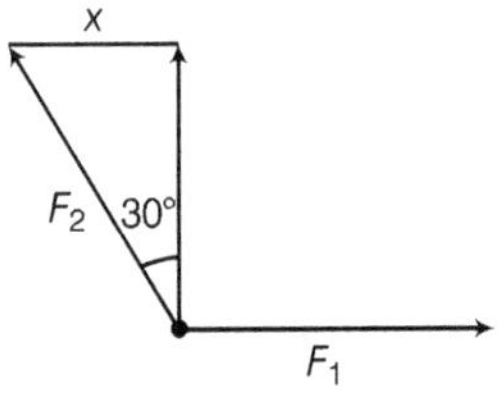

15. Net frictional force between block and surface is

$F = \mu R = 0.5 \times 10 \times 10 = 50\,\text{N}$

Applied force is 10 N and it is less than 50 N.

$\therefore$ System is at rest and no friction between A and B.

16. When, $P = mg(\sin\theta - \mu\cos\theta)$

Then, $f = \mu mg\cos\theta$ (upwards)

When $P = mg\sin\theta$,

Then, $f = 0$

and when $P = mg(\sin\theta + \mu\cos\theta)$

Then, $f = \mu\, mg\cos\theta$ (downwards)

17. $mg\sin\theta = ma$

$\therefore$ $a = g\sin\theta$

where, a is along the inclined plane.

$\therefore$ Vertical component of acceleration is

$$g\sin^2\theta.$$

$\therefore$ Relative vertical acceleration of A with respect to B is

$g(\sin^2 60° - \sin^2 30°) = \dfrac{g}{2} = 4.9\,\text{m s}^{-2}$ (in vertical direction)

6

Work, Energy and Power

Chapter Snapshot

- Introduction to Work
- Work Done By a Constant Force
- Work Done By a Variable Force
- Conservative and Non-conservative Force Field
- Kinetic Energy
- Work-Energy Theorem
- Potential Energy
- Law of Conservation of Mechanical Energy
- Three Types of Equilibrium
- Power

6.1 Introduction to Work

In our daily life 'work' has many different meanings. For example, Ram is working in a factory. The machine is in working order. Let us work out a plan for the next year, etc. In physics however, the term 'work' has a special meaning. In physics, work is always associated with a force and a displacement. We note that for work to be done, the force must act through a distance. Consider a person holding a weight a distance h off the floor as shown in figure. In everyday usage, we might say that the man is doing a work, but in our scientific definition, no work is done by a force acting on a stationary object. We could eliminate the effort of holding the weight by merely tying the string to some object and the weight could be supported with no help from us.

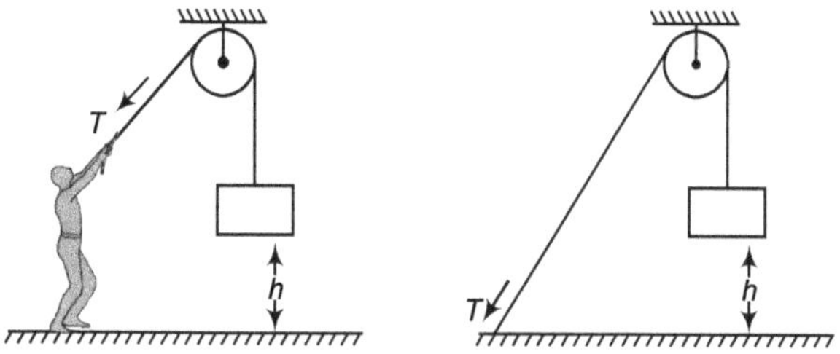

No work is done by the man holding the weight at a fixed position.
The same task could be accomplished by tying the rope to a fixed point.

Fig. 6.1

Let us now see what does 'work' mean in the language of physics.

6.2 Work Done by a Constant Force

Let us first consider the simple case of a constant force **F** acting on a body. Further, let us also assume that the body moves in a straight line in the direction of force. In this case, we define the work done by the force on the body as the product of the magnitude of the force **F** and the distance **S** through which the body moves.

e.g. the work W is given by

$$W = \mathbf{F} \cdot \mathbf{S}$$

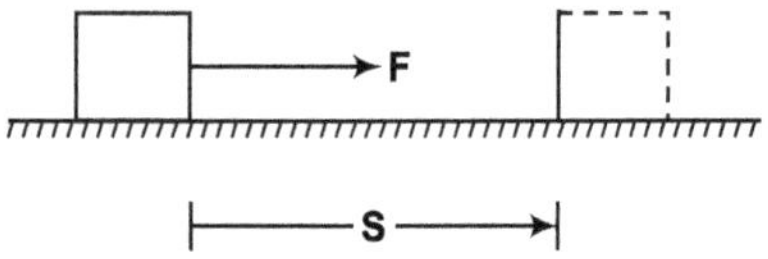

Fig. 6.2

On the other hand, in a situation when the constant force does not act along the same direction as the displacement of the body, the component of force **F** along the displacement **S** is effective in doing work.

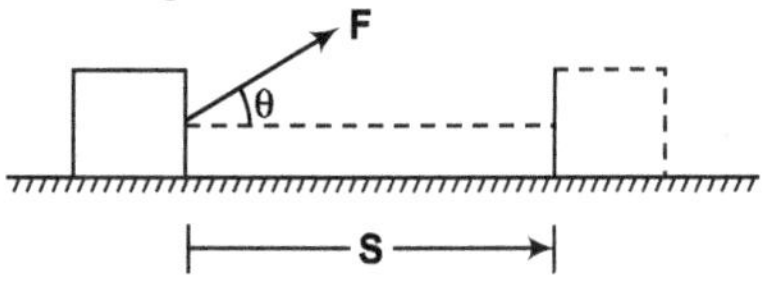

Fig. 6.3

Thus, in this case, work done by a constant force **F** is given by

$W =$ (component of force along the displacement) $\times$ (displacement)

or $\quad W = (F \cos\theta)(S)$

or $\quad W = \mathbf{F} \cdot \mathbf{S}$ (from the definition of dot product)

So, work done is a scalar or dot product of **F** and **S**.

Regarding work done it is worth noting that

1. Work can be positive, negative or even zero also, depending on the angle (θ) between the force vector **F** and displacement vector **S**. Work done by a force is zero when $\theta = 90°$, it is positive when $\theta < 90°$ and negative when $\theta > 90°$. For example, when a person lifts a body, the work done by the lifting force is positive (as $\theta = 0°$) but work done by the force of gravity is negative (as $\theta = 180°$). Similarly, work done by centripetal force is always zero (as $\theta = 90°$).
2. Work depends on frame of reference. With change of frame of reference inertial force does not change while displacement may change. So, the work done by a force will be different in different frames. For example, if a person is pushing a box inside a moving train, then work done as seen from the frame of reference of train is $\mathbf{F} \cdot \mathbf{S}$ while as seen from the ground it is $\mathbf{F} \cdot (\mathbf{S} + \mathbf{S}_0)$. Here, $\mathbf{S}_0$ is the displacement of train relative to ground.

Example 6.1 *A block of mass $m = 2$ kg is pulled by a force $F = 40$ N upwards through a height $h = 2$ m. Find the work done on the block by the applied force F and its weight mg. ($g = 10$ m/s^2)*

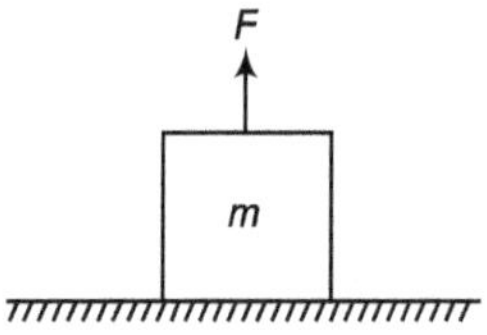

Fig. 6.4

Sol. Weight $= mg = (2)(10) = 20$ N

Work done by the applied force $W_F = Fh\cos 0°$

As the angle between force and displacement is 0°

or $\quad W_F = (40)(2)(1) = 80$ J

Similarly, work done by its weight

$W_{mg} = (mg)(h)\cos 180°$ or $W_{mg} = (20)(2)(-1) = -40$ J

Example 6.2 *Two unequal masses of 1 kg and 2 kg are attached at the two ends of a light inextensible string passing over a smooth pulley as shown in figure. If the system is released from rest, find the work done by string on both the blocks in 1 s. (Take $g = 10$ m/s^2)*

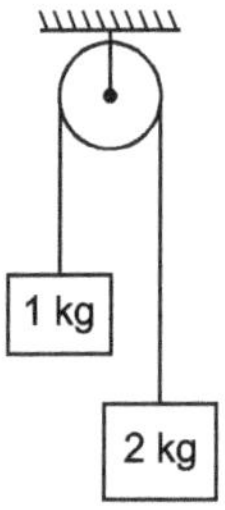

Fig. 6.5

Sol. Net pulling force on the system is

$$F_{net} = 2g - 1g = 20 - 10 = 10\text{ N}$$

Total mass being pulled

$$m = (1 + 2) = 3\text{ kg}$$

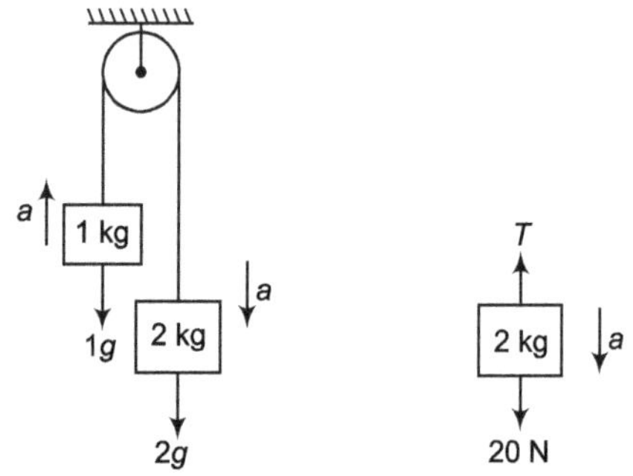

Fig. 6.6 (a) Fig. 6.6 (b)

Therefore, acceleration of the system will be

$$a = \frac{F_{net}}{m} = \frac{10}{3}\text{ m/s}^2$$

Displacement of both the blocks in 1 s is

$$S = \frac{1}{2}at^2 = \frac{1}{2}\left(\frac{10}{3}\right)(1)^2 = \frac{5}{3}\text{ m}$$

Free body diagram of 2 kg block is shown in Fig. 6.6 (b).

Using $\Sigma F = ma$, we get

$$20 - T = 2a = 2\left(\frac{10}{3}\right)$$

or

$$T = 20 - \frac{20}{3} = \frac{40}{3}\text{ N}$$

$\therefore$ Work done by string (tension) on 1 kg block in 1 s is

$$W_1 = (T)(S)\cos 0°$$

$$= \left(\frac{40}{3}\right)\left(\frac{5}{3}\right)(1)$$

$$= \frac{200}{9}\text{ J}$$

Similarly, work done by string on 2 kg block in 1 s will be

$$W_2 = (T)(S)(\cos 180°)$$

$$= \left(\frac{40}{3}\right)\left(\frac{5}{3}\right)(-1)$$

$$= -\frac{200}{9}\text{ J}$$

6.3 Work Done by a Variable Force

So far we have considered the work done by a force which is constant both in magnitude and direction. Let us now consider a force which acts always in one direction but whose magnitude may keep on varying. We can choose the direction of the force as x-axis. Further, let us assume that the magnitude of the force is also a function of x or say $F(x)$ is known to us. Now, we are interested in finding the work done by this force in moving a body from x_1 to x_2.

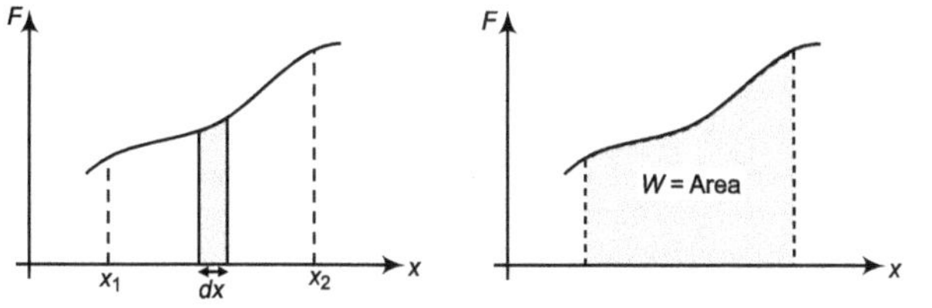

Fig. 6.7

Work done in a small displacement from x to $x + dx$ will be

$$dW = F \cdot dx$$

Now, the total work can be obtained by integration of the above elemental work from x_1 to x_2 or

$$W = \int_{x_1}^{x_2} dW = \int_{x_1}^{x_2} F \cdot dx$$

It is important to note that $\int_{x_1}^{x_2} F\, dx$ is also the area under F-x graph between $x = x_1$ to $x = x_2$.

Spring Force

An important example of the above idea is a spring that obeys Hooke's law. Consider the situation shown in figure.

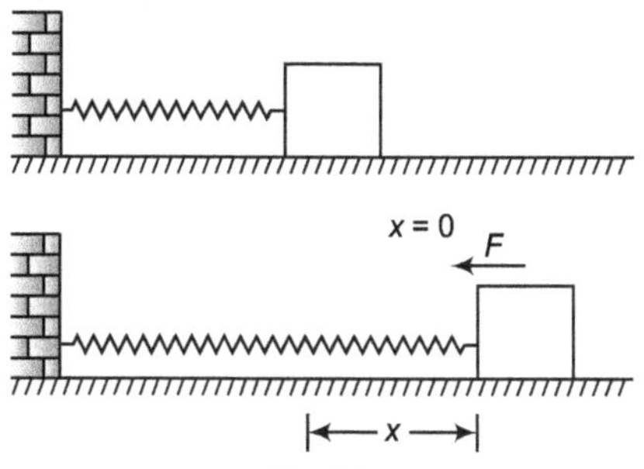

Fig. 6.8

One end of a spring is attached to a fixed vertical support and the other end to a block which can move on a horizontal table. Let $x = 0$ denote the position of the block when the spring is in its natural length. When the block is displaced by an amount x (either compressed or elongated) a restoring force (F) is applied by the spring on the block. The direction of this force F is always towards its mean position ($x = 0$) and the magnitude is directly proportional to x

$$F \propto x \qquad \text{(Hooke's law)}$$

$$\therefore \quad F = -kx \qquad \text{...(i)}$$

Here, k is a constant called force constant of spring and depends on the nature of spring. From Eq. (i), we see that F is a variable force and F-x graph is a straight line passing through origin with slope $= -k$. Negative sign in Eq. (i) implies that the spring force F is directed in a direction opposite to the displacement x of the block.

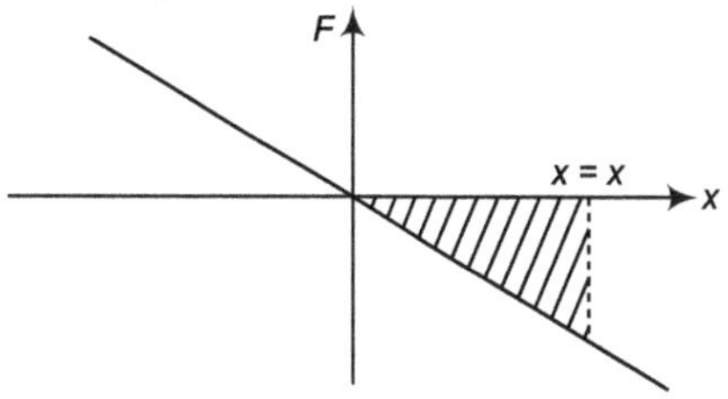

Fig. 6.9

Let us now find the work done by this force F when the block is displaced from $x = 0$ to $x = x$. This can be obtained either by integration or the area under F-x graph.

Thus, $$W = \int dW = \int_0^x F\, dx$$

$$= \int_0^x -kx\, dx = -\frac{1}{2} kx^2$$

Here, work done is negative, because force is in opposite direction of displacement.

Similarly, if the block moves from $x = x_1$ to $x = x_2$. The limits of integration are x_1 and x_2 and the work done is

$$W = \int_{x_1}^{x_2} -kx\, dx = \frac{1}{2} k (x_1^2 - x_2^2)$$

Example 6.3 *A force $F = (2 + x)$ acts on a particle in x-direction, where F is in newton and x in metre. Find the work done by this force during a displacement from $x = 1.0$ m to $x = 2.0$ m.*

Sol. As the force is variable, we shall find the work done in a small displacement from x to $x + dx$ and then integrate it to find the total work. The work done in this small displacement is

$$dW = F\, dx = (2 + x)\, dx$$

Thus, $$W = \int_{1.0}^{2.0} dW = \int_{1.0}^{2.0} (2 + x)\, dx$$

$$= \left[2x + \frac{x^2}{2}\right]_{1.0}^{2.0} = 3.5 \text{ J}$$

Example 6.4 *A force $F = -\frac{k}{x^2}$ $(x \neq 0)$ acts on a particle in x-direction. Find the work done by this force in displacing the particle from. $x = +a$ to $x = +2a$. Here, k is a positive constant.*

Sol. $W = \int F\,dx$

$$= \int_{+a}^{+2a} \left(\frac{-k}{x^2}\right) dx = \left[\frac{k}{x}\right]_{+a}^{+2a} = -\frac{k}{2a}$$

- It is important to note that work comes out to be negative which is quite obvious as the force acting on the particle is in negative x-direction $\left(F = -\frac{k}{x^2}\right)$ while displacement is along positive x-direction. (from $x = a$ to $x = 2a$)

Example 6.5 *A force F acting on a particle varies with the position x as shown in figure. Find the work done by this force in displacing the particle from*

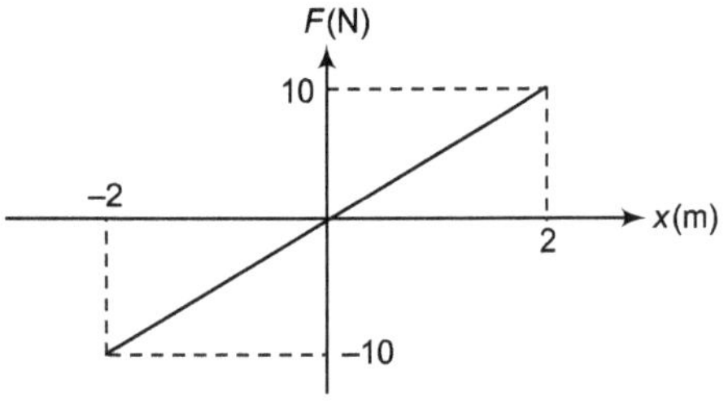

Fig. 6.10

(a) $x = -2\,m$ *to* $x = 0$ *(b)* $x = 0$ *to* $x = 2\,m$.

Sol. (a) From $x = -2$m to $x = 0$, displacement of the particle is along positive x-direction while force acting on the particle is along negative x-direction. Therefore, the work done is negative and given by the area under F-x graph.

$\therefore$ $$W = -\frac{1}{2}(2)(10) = -10\text{ J}$$

(b) From $x = 0$ to $x = 2$ m, displacement of particle and force acting on the particle both are along positive x-direction. Therefore, work done is positive and given by the area under F-x graph, or

$$W = \frac{1}{2}(2)(10) = 10\text{ J}$$

6.4 Conservative and Non-conservative Force Field

In Article 6.3, we considered the forces which were although variable but always directed in one direction. However, the most general expression for work done is

$$dW = \mathbf{F} \cdot d\mathbf{r}$$

and $$W = \int_{\mathbf{r}_i}^{\mathbf{r}_f} dW = \int_{\mathbf{r}_i}^{\mathbf{r}_f} \mathbf{F} \cdot d\mathbf{r}$$

Here, $d\mathbf{r} = dx\hat{\mathbf{i}} + dy\hat{\mathbf{j}} + dz\hat{\mathbf{k}}$

$\mathbf{r}_i$ = initial position vector

and $\mathbf{r}_f$ = final position vector

Conservative and non-conservative forces can be better understood after going through the following two examples.

Example 6.6 *An object is displaced from point* $A(2m, 3m, 4m)$ *to a point* $B(1m, 2m, 3m)$ *under a constant force* $\mathbf{F} = (2\hat{\mathbf{i}} + 3\hat{\mathbf{j}} + 4\hat{\mathbf{k}})$ *N. Find the work done by this force in this process.*

Sol. $W = \int_{\mathbf{r}_i}^{\mathbf{r}_f} \mathbf{F} \cdot d\mathbf{r}$

$$= \int_{(2m,\,3m,\,4m)}^{(1m,\,2m,\,3m)} (2\hat{i} + 3\hat{j} + 4\hat{k}) \cdot (dx\hat{i} + dy\hat{j} + dz\hat{k})$$

$$= [2x + 3y + 4z]_{(2m,\,3m,\,4m)}^{(1m,\,2m,\,3m)} = -9\text{ J}$$

Alternate Solution

Since, $\mathbf{F}$ = constant, we can also use

$W = \mathbf{F} \cdot \mathbf{S}$

Here, $\mathbf{S} = \mathbf{r}_f - \mathbf{r}_i$

$= (\hat{i} + 2\hat{j} + 3\hat{k}) - (2\hat{i} + 3\hat{j} + 4\hat{k})$

$= (-\hat{i} - \hat{j} - \hat{k})$

$\therefore$ $W = (2\hat{i} + 3\hat{j} + 4\hat{k}) \cdot (-\hat{i} - \hat{j} - \hat{k})$

$= -2 - 3 - 4 = -9$J

Example 6.7 *An object is displaced from position vector* $\mathbf{r}_1 = (2\hat{\mathbf{i}} + 3\hat{\mathbf{j}})$ *m to* $\mathbf{r}_2 = (4\hat{\mathbf{i}} + 6\hat{\mathbf{j}})$*m under a force* $\mathbf{F} = (3x^2\hat{\mathbf{i}} + 2y\hat{\mathbf{j}})$*N. Find the work done by this force.*

Sol. $W = \int_{\mathbf{r}_1}^{\mathbf{r}_2} \mathbf{F} \cdot d\mathbf{r}$

$$= \int_{\mathbf{r}_1}^{\mathbf{r}_2} (3x^2\hat{i} + 2y\hat{j}) \cdot (dx\hat{i} + dy\hat{j} + dz\hat{k})$$

$$= \int_{\mathbf{r}_1}^{\mathbf{r}_2} (3x^2 dx + 2y dy)$$

$$= [x^3 + y^2]_{(2,\,3)}^{(4,\,6)} = 83\text{ J}$$

In the above two examples, we saw that while calculating the work done we did not mention the path through which the object was displaced. Only initial and final coordinates were required. It shows that in both the examples, the work done is path independent or work done will be equal on whichever path we follow. Such forces in which work is path independent are known as **conservative forces.**

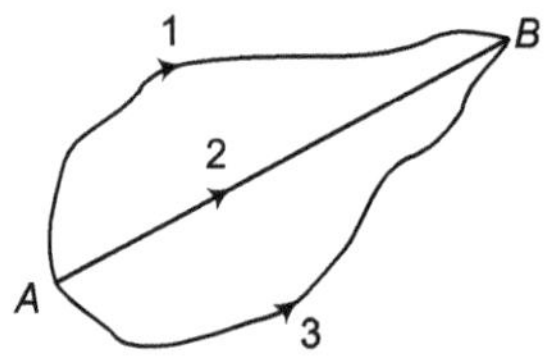

Fig. 6.11

Thus, if a particle or an object is displaced from position A to position B through three different paths under a conservative force field. Then

$$W_1 = W_2 = W_3$$

Further, it can be shown that work done in a closed path is zero under a conservative force field. ($W_{AB} = -W_{BA}$ or $W_{AB} + W_{BA} = 0$). Gravitational force, Coulomb's force are few examples of conservative forces. On the other hand, if the work is path dependent or $W_1 \neq W_2 \neq W_3$, the force is called a **non-conservative**. Frictional forces, viscous forces are non-conservative in nature. Work done in a closed path is not zero in a non-conservative force field.

6.5 Kinetic Energy

Kinetic energy (KE) is the capacity of a body to do work by virtue of its motion. If a body of mass m has a velocity v its kinetic energy is equivalent to the work which an external force would have to do to bring the body from rest upto its velocity v. The numerical value of the kinetic energy can be calculated from the formula.

$$\text{KE} = \frac{1}{2} mv^2$$

This can be derived as follows :

Consider a constant force F which acting on a mass m initially at rest, gives the mass a velocity v. If in reaching this velocity, the particle has been moving with an acceleration a and has been given a displacement s, then

$$F = ma \qquad \text{(Newton's law)}$$

$$v^2 = 2as$$

Work done by the constant force $= Fs$

or $$W = (ma)\left(\frac{v^2}{2a}\right) = \frac{1}{2} mv^2$$

But the kinetic energy of the body is equivalent to the work done in giving the body this velocity.

Hence, $$\text{KE} = \frac{1}{2} mv^2$$

Regarding the kinetic energy the following two points are important to note.

1. Since, both m and v^2 are always positive. KE is always positive and does not depend on the direction of motion of the body.
2. Kinetic energy depends on the frame of reference. For example, the kinetic energy of a person of mass m sitting in a train moving with speed v is zero in the frame of train but $\frac{1}{2} mv^2$ in the frame of the earth.

6.6 Work-Energy Theorem

This theorem is a very important tool that relates the works to kinetic energy. According to this theorem,

Work done by all the forces (conservative or non-conservative, external or internal) acting on a particle or an object is equal to the change in kinetic energy of it.

$$\therefore \qquad W_{\text{net}} = \Delta\text{KE} = K_f - K_i$$

Let, $\mathbf{F}_1, \mathbf{F}_2, \ldots$ be the individual forces acting on a particle. The resultant force is $\mathbf{F} = \mathbf{F}_1 + \mathbf{F}_2 + \ldots$ and the work done by the resultant force is

$$W = \int \mathbf{F} \cdot d\mathbf{r} = \int (\mathbf{F}_1 + \mathbf{F}_2 + \ldots) \cdot d\mathbf{r}$$
$$= \int \mathbf{F}_1 \cdot d\mathbf{r} + \int \mathbf{F}_2 \cdot d\mathbf{r} + \ldots$$

where, $\int \mathbf{F}_1 \cdot d\mathbf{r}$ is the work done on the particle by $\mathbf{F}$ and so on. Thus, work energy theorem can also be written as : work done by the resultant force is equal to the sum of the work done by the individual forces.

Regarding the work-energy theorem it is worth noting that

1. If W_{net} is positive, then $K_f - K_i =$ positive, i.e. $K_f > K_i$ or kinetic energy will increase and *vice-versa.*
2. This theorem can be applied to non-inertial frames also. In a non-inertial frame it can be written as : work done by all the forces (including the pseudo forces) = change in kinetic energy in non-inertial frame.

Let us take an example.

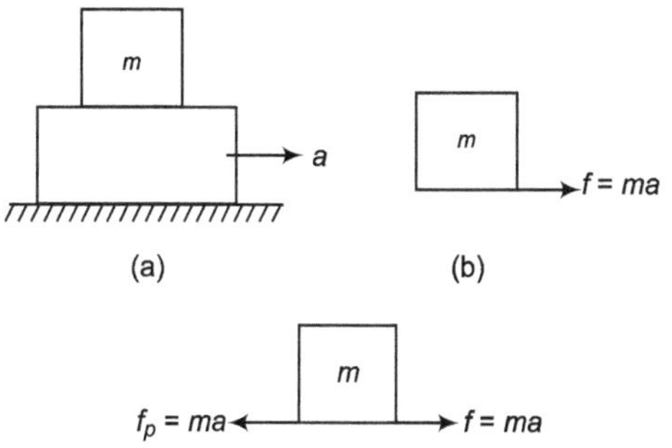

Fig. 6.12

Refer Fig. (a)

A block of mass m is kept on a rough plank moving with an acceleration a. There is no relative motion between block and plank. Hence, force of friction on block is $f = ma$ in forward direction.

Refer Fig. (b)

Horizontal forces on the block has been shown from ground (inertial) frame of reference.

If the plank moves a distance s on the ground the block will also move the same distance s as there is no slipping between the two.

Hence, work done by friction on the block (w.r.t. ground) is

$$W_f = fs = mas$$

From work-energy principle if v is the speed of block (w.r.t. ground).

$$\text{KE} = W_f$$

or $$\frac{1}{2}mv^2 = mas$$

or $$v = \sqrt{2as}$$

Thus, velocity of block relative to ground is $\sqrt{2as}$.

Refer Fig. (c)

Free body diagram of the block has been shown from accelerating frame (plank).

Here, f_p = pseudo force = ma

Work done by all the forces,

$$W = W_f + W_{fp} = mas - mas = 0$$

From work-energy theorem,

$$\frac{1}{2}mv_r^2 = W = 0$$

or $$v_r = 0$$

Thus, velocity of block relative to plank is zero.

Example 6.8 *An object of mass m is tied to a string of length l and a variable force F is applied on it which brings the string gradually at angle θ with the vertical. Find the work done by the force F.*

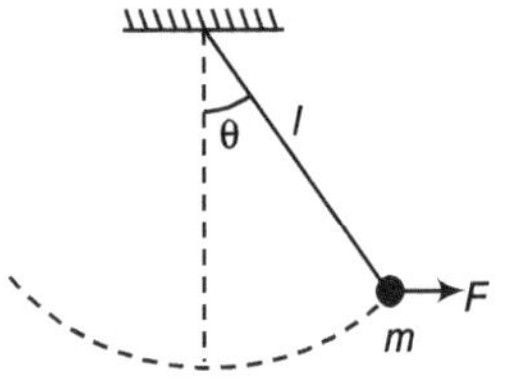

Fig. 6.13

Sol. In this case three forces are acting on the object

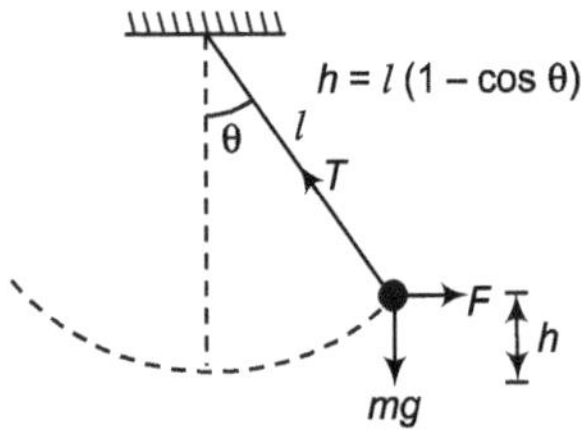

Fig. 6.14

1. Tension (T),
2. Weight (mg) and
3. Applied force (F)

Using work-energy theorem

$$W_{net} = \Delta\,\text{KE}$$

or $$W_T + W_{mg} + W_F = 0 \qquad \text{...(i)}$$

as $$\Delta\text{KE} = 0$$

because $$K_i = K_f = 0$$

Further, $W_T = 0$, as tension is always perpendicular to displacement.

$$W_{mg} = -mgh$$

or $$W_{mg} = -mgl\,(1 - \cos\theta)$$

Substituting these values in Eq. (i), we get

$$W_F = mgl\,(1 - \cos\theta)$$

- Here, the applied force F is variable. So if we do not apply the work energy theorem we will first find the magnitude of F at different locations and then integrate dW (= $\mathbf{F}\cdot d\mathbf{r}$) with proper limits.

6.7 Potential Energy

The energy possessed by a body or system by virtue of its position or configuration is known as the potential energy. For example, a block attached to a compressed or elongated spring possesses some energy called elastic potential energy. This block has a capacity to do work. Similarly, a stone when released from a certain height also has energy in the form of gravitational potential energy. Two charged particles kept at certain distance has electric potential energy.

Regarding the potential energy it is important to note that it is defined for a conservative force field only. For non-conservative forces it has no meaning. The change in potential energy (dU) of a system corresponding to a conservative internal force is given by

$$dU = -\mathbf{F}\cdot d\mathbf{r} = -dW \qquad \left(F = -\frac{dU}{dr}\right)$$

or $$\int_i^f dU = -\int_{\mathbf{r}_i}^{\mathbf{r}_f} \mathbf{F}\cdot d\mathbf{r}$$

or $$U_f - U_i = -\int_{\mathbf{r}_i}^{\mathbf{r}_f} \mathbf{F}\cdot d\mathbf{r}$$

We generally choose the reference point at infinity and assume potential energy to be zero there, i.e. if we take $r_i = \infty$ (infinite) and $U_i = 0$, then we can write

$$U = -\int_{\infty}^{\mathbf{r}} \mathbf{F}\cdot d\mathbf{r} = -W$$

or potential energy of a body or system is the negative of work done **by the conservative forces** in bringing it from infinity to the present position.

Regarding the potential energy it is worth noting that

1. Potential energy can be defined only for conservative forces and it should be considered to be a property of the entire system rather than assigning it to any specific particle.
2. Potential energy depends on frame of reference.

Now, let us discuss three types of potential energies which we usually come across.

Elastic Potential Energy

In Article 6.3, we have discussed the spring forces. We have seen there that the work done by the spring force (ofcourse conservative for an ideal spring) is $-\frac{1}{2}kx^2$, when the spring is stretched or compressed by an amount x from its unstretched position. Thus,

$$U = -W = -\left(-\frac{1}{2}kx^2\right)$$

or $$U = \frac{1}{2}kx^2 \quad (k = \text{spring constant})$$

Note that elastic potential energy is always positive.

Gravitational Potential Energy

The gravitational potential energy of two particles of masses m_1 and m_2 separated by a distance r is given by

$$U = -G\frac{m_1 m_2}{r}$$

Here, G = universal gravitation constant

$$= 6.67 \times 10^{-11} \frac{\text{N-m}^2}{\text{kg}^2}$$

If a body of mass m is raised to a height h from the surface of the earth, the change in potential energy of the system (earth + body) comes out to be

$$\Delta U = \frac{mgh}{\left(1 + \frac{h}{R}\right)} \quad (R = \text{ radius of the earth})$$

or $\Delta U \approx mgh$

if $h << R$

Thus, the potential energy of a body at height h, i.e. mgh is really the change in potential energy of the system for $h << R$. So be careful while using $U = mgh$, that h should not be too large. This we will discuss in detail in the chapter of gravitation.

Electric Potential Energy

The electric potential energy of two point charges q_1 and q_2 separated by a distance r in vacuum is given by

$$U = \frac{1}{4\pi\varepsilon_0} \cdot \frac{q_1 q_2}{r}$$

Here, $$\frac{1}{4\pi\varepsilon_0} = 9.0 \times 10^9 \frac{\text{N-m}^2}{\text{C}^2}$$

= constant

Extra Knowledge Points

- Change in potential energy is equal to the negative of work done by the conservative force ($\Delta U = -\Delta W$). If work done by the conservative force is negative change in potential energy will be positive or potential energy of the system will increase and *vice-versa*.

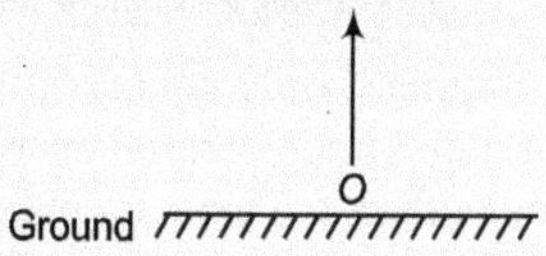

This can be understood by a simple example. Suppose a ball is taken from the ground to some height, work done by gravity is negative, i.e. change in potential energy should increase or potential energy of the ball will increase. Which happens so

$$\Delta W_{\text{gravity}} = -\text{ve}$$

$\therefore \Delta U = +\text{ve}$ $\quad (\Delta U = -\Delta W)$

or $\quad U_f - U_i = +\text{ve}$

- $F = -\frac{dU}{dr}$, i.e. conservative forces always act in a direction, where potential energy of the system is decreased. This can also be shown as in figure.

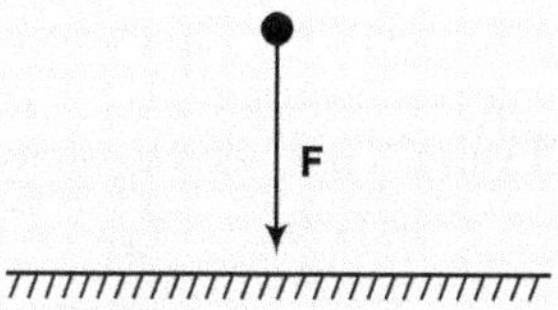

If a ball is dropped from a certain height. The force on it (its weight) acts in a direction in which its potential energy decreases.

- Suppose a particle is released from point A with $u = 0$.

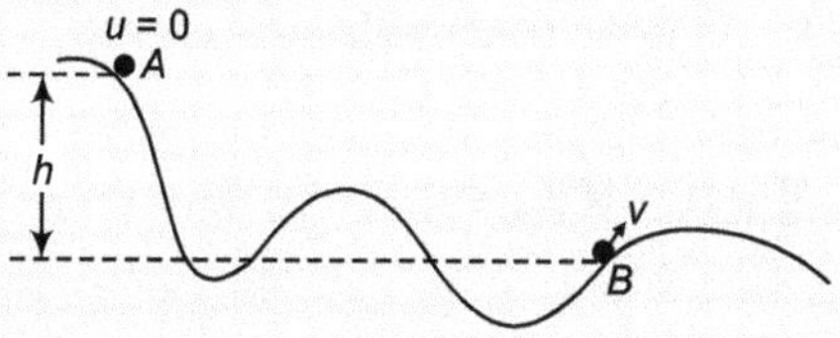

Friction is absent everywhere. Then velocity at B will be

$$v = \sqrt{2gh}$$

(irrespective of the track it follows from A to B)

Here, $\quad h = h_A - h_B$

- Suppose a car is moving with constant speed in the track as shown in figure. Then, the equations of motion are as under.

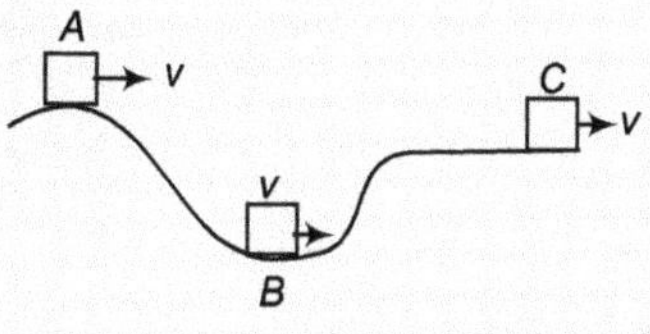

$$mg - N_A = \frac{mv^2}{R_A} \quad \text{or} \quad N_A = mg - \frac{mv^2}{R_A}$$

$$N_B - mg = \frac{mv^2}{R_B} \quad \text{or} \quad N_B = mg + \frac{mv^2}{R_B}$$

and $\quad N_C = mg$

Thus, we can say $\quad N_B > N_C > N_A$

6.8 Law of Conservation of Mechanical Energy

Suppose, only conservative forces operate on a system of particles and potential energy U is defined corresponding to these forces. There are either no other forces or the work done by them is zero. We have

$$U_f - U_i = -W$$

and $W = K_f - K_i$ (from work energy theorem)

then $U_f - U_i = -(K_f - K_i)$

or $U_f + K_f = U_i + K_i$ ...(i)

The sum of the potential energy and the kinetic energy is called the total mechanical energy. We see from Eq. (i), that the **total mechanical energy of a system remains constant if only conservative forces are acting on a system of particles and the work done by all other forces is zero.** This is called the conservation of mechanical energy.

The total mechanical energy is not constant, if non-conservative forces such as friction is acting between the parts of a system. However, the work energy theorem is still valid. Thus, we can apply

$$W_c + W_{nc} + W_{\text{ext}} = K_f - K_i$$

Here, $W_c = -(U_f - U_i)$

So, we get $W_{nc} + W_{\text{ext}} = (K_f + U_f) - (K_i + U_i)$

or $W_{nc} + W_{\text{ext}} = E_f - E_i$

Here, $E = K + U$ is the total mechanical energy.

Extra Knowledge Points

- If only conservative forces are acting on a system of particles and work done by any other external force is zero, then mechanical energy of the system will remain conserved. In this case some fraction of the mechanical energy will be decreasing while the other will be increasing. Problems can be solved by equating the magnitudes of the decrease and the increase. Let us see an example of this.

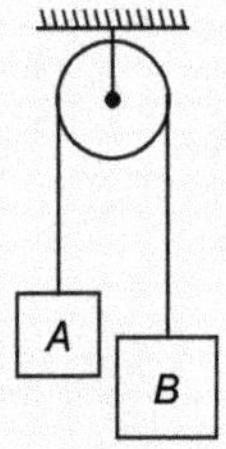

In the arrangement shown in figure string is light and inextensible and friction is absent everywhere. Find the speed of both the blocks after the block A has ascended a height of 1 m. Given that $m_A = 1$ kg and $m_B = 2$ kg ($g = 10$ m/s^2).

- ***Solution*** Friction is absent. Therefore, mechanical energy of the system will remain conserved. From constraint relations we see that speed of both the blocks will be same. Suppose it is v. Here, gravitational potential energy of 2 kg block is decreasing while gravitational potential energy of 1 kg block is increasing. Similarly, kinetic energy of both the blocks is also increasing. So we can write

Decrease in gravitational potential energy of 2 kg block = increase in gravitational potential energy of 1 kg block + increase in kinetic energy of 1 kg block + increase in kinetic energy of 2 kg block.

$\therefore$ $m_B gh = m_A gh + \frac{1}{2} m_A v^2 + \frac{1}{2} m_B v^2$

or $(2)(10)(1) = (1)(10)(1) + \frac{1}{2}(1)v^2 + \frac{1}{2}(2)v^2$

or $20 = 10 + 0.5\,v^2 + v^2$

or $1.5\,v^2 = 10$

$\therefore$ $v^2 = 6.67\ \text{m}^2/\text{s}^2$

or $v = 2.58$ m/s

- If some non-conservative forces such as friction are also acting on some parts of the system and work done by any other forces (excluding the conservative forces) is zero. Then, we can apply

$$W_{nc} = E_f - E_i$$

or $W_{nc} = (U_f - U_i) + (K_f - K_i)$

$= \Delta U + \Delta K$

i.e. work done by non-conservative forces is equal to the change in mechanical (potential + kinetic) energy. But note that here all quantities are to be substituted with sign. Let us see an example of this.

In the arrangement shown in figure, $m_A = 1$ kg, $m_B = 4$ kg. String is light and inextensible while pulley is smooth. Coefficient of friction between block A and the table is $\mu = 0.2$. Find the speed of both the blocks when block B has descended a height $h = 1$ m. (Take $g = 10$ m/s^2)

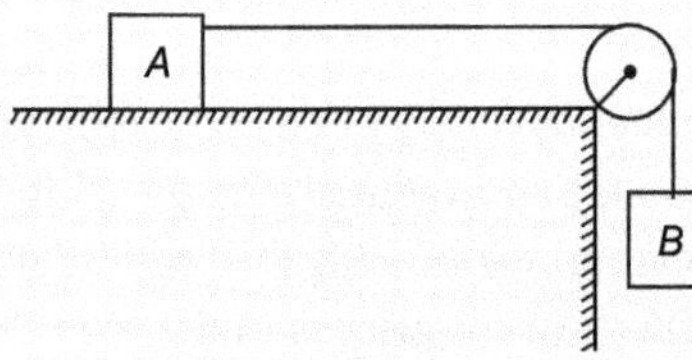

- ***Solution*** From constraint relation, we see that

$$v_A = v_B = v \quad \text{(say)}$$

Force of friction between block A and table will be

$$f = \mu m_A g = (0.2)(1)(10) = 2\ \text{N}$$

$\therefore$ $W_{nc} = \Delta U + \Delta K$

$\therefore$ $-fs = -m_B gh + \frac{1}{2}(m_A + m_B)v^2$

or $(-2)(1) = -(4)(10)(1) + \frac{1}{2}(4+1)v^2$

$-2 = -40 + 2.5\,v^2$ or $2.5\,v^2 = 38$

$\therefore$ $v^2 = 15.2\ \text{m}^2/\text{s}^2$

or $v = 3.9$ m/s

Example 6.9 *In the arrangement shown in figure $m_A = 4.0\,kg$ and $m_B = 1.0\,kg$. The system is released from rest and block B is found to have a speed 0.3 m/s after it has descended through a distance of 1 m. Find the coefficient of friction between the block and the table. Neglect friction elsewhere. (Take $g = 10\ m/s^2$)*

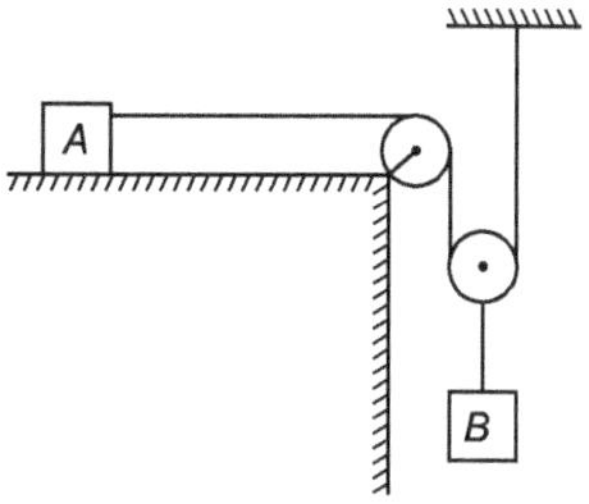

Fig. 6.16

Sol. From constraint relations, we can see that

$$v_A = 2\,v_B$$

Therefore, $v_A = 2(0.3) = 0.6$ m/s

as $v_B = 0.3$ m/s (given)

Applying $W_{nc} = \Delta U + \Delta K$

we get $-\mu\, m_A g S_A = -m_B g S_B + \frac{1}{2} m_A v_A^2 + \frac{1}{2} m_B v_B^2$

Here, $S_A = 2S_B = 2$ m as $S_B = 1$ m (given)

$\therefore\ -\mu(4.0)(10)\ (2) = -(1)(10)(1) + \frac{1}{2}(4)(0.6)^2 + \frac{1}{2}(1)\ (0.3)^2$

or $-80\,\mu = -10 + 0.72 + 0.045$

or $80\,\mu = 9.235$ or $\mu = 0.115$

6.9 Three Types of Equilibrium

A body is said to be in translatory equilibrium, if net force acting on the body is zero, i.e.

$$\mathbf{F}_{\text{net}} = 0$$

If the forces are conservative

$$F = -\frac{dU}{dr}$$

and for equilibrium $F = 0$.

So, $$-\frac{dU}{dr} = 0, \text{ or } \frac{dU}{dr} = 0$$

i.e. at equilibrium position slope of U-r graph is zero or the potential energy is optimum (maximum or minimum or constant). Equilibrium are of three types, i.e. the situation where, $F = 0$ and $\frac{dU}{dr} = 0$ can be obtained under three conditions. These are stable equilibrium, unstable equilibrium and neutral equilibrium. These three types of equilibrium can be better understood from the given three figures

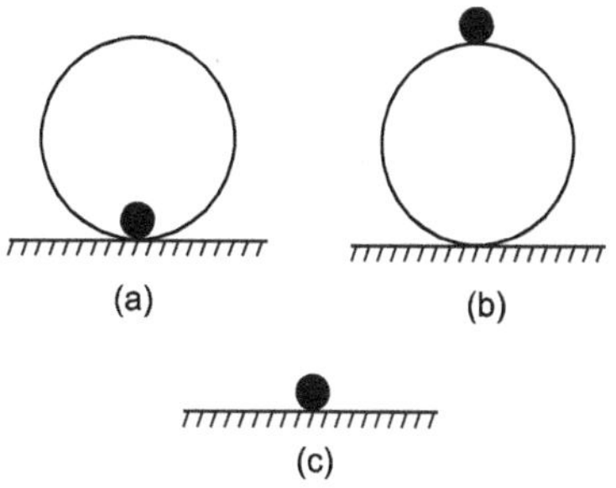

Fig. 6.16

Three identical balls are placed in equilibrium in positions as shown in Fig. (a), (b) and (c) respectively.

In Fig. (a), ball is placed inside a smooth spherical shell. This ball is in stable equilibrium position. In Fig. (b), the ball is placed over a smooth sphere. This is in unstable equilibrium position. In Fig. (c), the ball is placed on a smooth horizontal ground. This ball is in neutral equilibrium position.

The table given below explains what is the difference and what are the similarities between these three equilibrium positions in the language of physics.

Table 6.1

S. No.	Stable Equilibrium	Unstable Equilibrium	Neutral Equilibrium
1.	Net force is zero.	Net force is zero.	Net force is zero.
2.	$\frac{dU}{dr} = 0$ or slope of U-r graph is zero.	$\frac{dU}{dr} = 0$ or slope of U-r graph is zero.	$\frac{dU}{dr} = 0$ or slope of U-r graph is zero.
3.	When displaced from its equilibrium position a net restoring force starts acting on the body which has a tendency to bring the body back to its equilibrium position.	When displaced from its equilibrium position, a net force starts acting on the body which moves the body in the direction of displacement or away from the equilibrium position.	When displaced from its equilibrium position the body has neither the tendency to come back nor to move away from the original position.
4.	Potential energy in equilibrium position is minimum as compared to its neighbouring points or $\frac{d^2U}{dr^2}$ = positive	Potential energy in equilibrium position is maximum as compared to its neighbouring points or $\frac{d^2U}{dr^2}$ = negative	Potential energy remains contant even, if the body is displaced from its equilibrium position or $\frac{d^2U}{dr^2} = 0$
5.	When displaced from equilibrium position the centre of gravity of the body goes up.	When displaced from equilibrium position the centre of gravity of the body comes down.	When displaced from equilibrium position the centre of gravity of the body remains at the same level.

Extra Knowledge Points

- If we plot graphs between F and r or U and r, F will be zero at equilibrium while U will be maximum, minimum or constant depending on the type of equilibrium. This all is shown in figure.

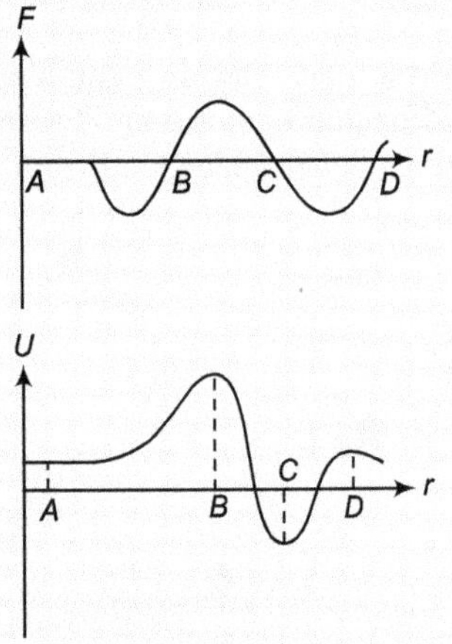

At point A, $F = 0$, $\frac{dU}{dr} = 0$, but U is constant. Hence, A is neutral equilibrium position. At points B and D, $F = 0$, $\frac{dU}{dr} = 0$ but U is maximum. Thus, these are the points of unstable equilibrium.

At point C, $F = 0$, $\frac{dU}{dr} = 0$, but U is minimum. Hence, point C is in stable equilibrium position.

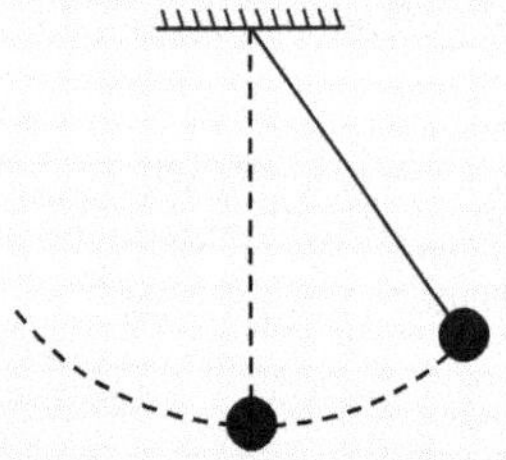

(a)

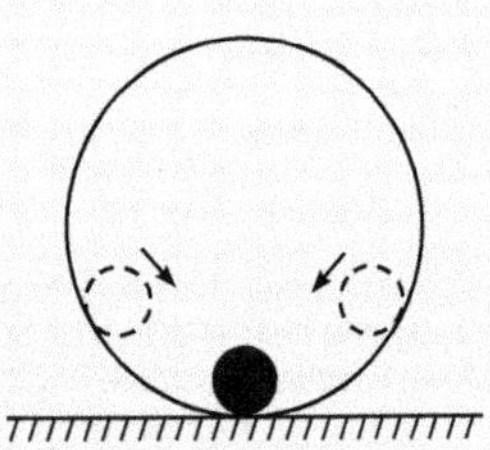

(b)

Oscillations of a body take place about stable equilibrium position. For example, bob of a pendulum oscillates about its lowest point which is also the stable equilibrium position of bob. Similarly, in Fig. (b), the ball will oscillate about its stable equilibrium position.

Example 6.10 *The potential energy of a conservative system is given by*

$$U = ax^2 - bx$$

(where, a and b are positive constants. Find the equilibrium position and discuss whether the equilibrium is stable, unstable or neutral.

Sol. In a conservative field

$$F = -\frac{dU}{dx}$$

$$\therefore \quad F = -\frac{d}{dx}(ax^2 - bx)$$

$$= b - 2ax$$

For equilibrium $F = 0$

or $\quad b - 2ax = 0$

$$\therefore \quad x = \frac{b}{2a}$$

From the given equation we can see that $\frac{d^2U}{dx^2} = 2a$ (positive), i.e. U is minimum.

Therefore, $x = \frac{b}{2a}$ is the stable equilibrium position.

6.10 Power

Power is the rate at which a force does work. If a force does 20 J of work in 10 s, the average rate at which it is working is 2 J/s or the average power is 2 W.

The work done by a force $\mathbf{F}$ in a small displacement $d\mathbf{r}$ is $dW = \mathbf{F} \cdot d\mathbf{r}$. Thus, the instantaneous power delivered by the force is

$$P = \frac{dW}{dt}$$

$$= \mathbf{F} \cdot \frac{d\mathbf{r}}{dt} = \mathbf{F} \cdot \mathbf{v}$$

$$= Fv \cos \theta$$

Thus, power is equal to the scalar product of force with velocity. It is zero, if force is perpendicular to velocity. For example, power of a centripetal force in a circular motion is zero.

Example 6.11 *A train has a constant speed of* 40 *m/s on a level road against resistive force of magnitude* 3×10^4 *N. Find the power of the engine.*

Sol. At constant speed, there is no acceleration, so the forces acting on the train are in equilibrium.

Therefore, $\quad F = R$

$\therefore \quad F = 3 \times 10^4$ N

or $\quad P = Fv$

We have, $\quad$ power $= 3 \times 10^4 \times 40$

$= 1.2 \times 10^6$ W

Example 6.12 *A train of mass 2.0×10^5 kg has a constant speed of 20 m/s up a hill inclined at $\theta = \sin^{-1}\left(\frac{1}{50}\right)$ to the horizontal when the engine is working at 8.0×10^5 W. Find the resistance to motion of the train. ($g = 9.8\ m/s^2$).*

Sol. Since, $P = Fv$

$$F = \frac{P}{v} = \frac{8.0 \times 10^5}{20} = 4.0 \times 10^4 \text{ N}$$

At constant speed, the forces acting on the train are in equilibrium. Resolving the forces parallel to the hill

$$F = R + (2.0 \times 10^5)\, g \times \frac{1}{50}$$

$$4.0 \times 10^4 = R + 39200 \quad \text{or} \quad R = 800 \text{ N}$$

Therefore, the resistance is 800 N.

Example 6.13 *A block of mass m is pulled by a constant power P placed on a rough horizontal plane. The friction coefficient between the block and surface is μ. Find the maximum velocity of the block.*

Sol. Power $P = F \cdot v$ = constant

$$F = \frac{P}{v} \quad \text{or} \quad F \propto \frac{1}{v}$$

as v increases, F decreases.

when $F = \mu mg$, net force on block becomes zero, i.e. it has maximum or terminal velocity

$$\therefore \quad P = (\mu mg)\, v_{max} \quad \text{or} \quad v_{max} = \frac{P}{\mu mg}$$

Extra Knowledge Points

- Work is a scalar quantity. It can be positive, negative or zero. The angle between **F** and **s** decides whether the work done is positive, negative or zero.

  If $0° \le \theta < 90°, W$ = positive

  If $\theta = 90°, \quad W = 0$ and

  If $90° < \theta \le 180°, W$ = negative
- The CGS unit of work is erg.

  $$1 \text{ J} = 10^7 \text{ erg}$$

  Power is also measured in horse power (HP)
- If only conservative forces are acting on a system, the mechanical energy of the system remains constant. Mechanical energy comprises of kinetic and potential. In potential usually gravitational and elastic comes in the question (as far as problems of work, power and energy are concerned). Now it may happen that some part of the energy might be decreasing while other part might be increasing. Energy conservation equation now can be written in two ways.

  First method Magnitude of decrease of energy = magnitude of increase of energy.

  Second method $E_i = E_f$ ($i \rightarrow$ initial and $f \rightarrow$ final)

  i.e. write down total initial mechanical energy on one side and total final mechanical energy on the other side. While writing gravitational potential energy we choose some reference point (where $h = 0$), but throughout the question this reference point should not change. Let us take a simple example.

  A ball of mass m is released from a height h as shown in figure. The velocity of particle at the instant when it strikes the ground can be found using energy conservation principle by following two methods.

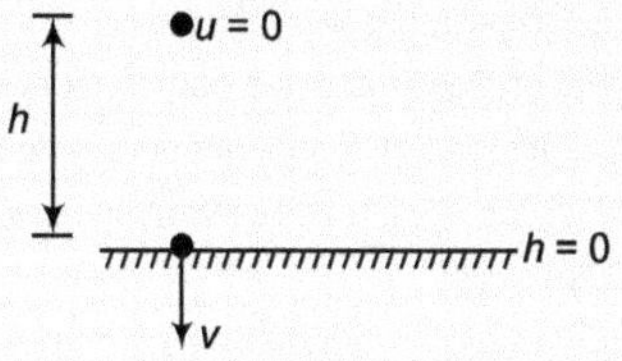

Method 1 Decrease in gravitational PE = increase in KE

or $$mgh = \frac{1}{2}mv^2$$

or $$v = \sqrt{2gh}$$

Method 2 $(PE + KE)_i = (PE + KE)_f$

For gravitational PE we take ground as the reference point.

$$\therefore \quad mgh + 0 = 0 + \frac{1}{2}mv^2$$

or $$v = \sqrt{2gh}$$

- If the system consists of frictional forces as well. Then, some mechanical energy will be lost in doing work against friction

  or $$E_f < E_i$$

  Now, suppose work done by friction is asked in the question, then find $E_f - E_i$ and if work done against friction is asked then write down $E_i - E_f$.

Chapter Summary with Formulae

- **Work Done**
  - (i) By a constant force

$$W = \mathbf{F}\cdot\mathbf{s} = \mathbf{F}\cdot(\mathbf{r}_f - \mathbf{r}_i) = Fs\cos\theta$$

= Force ×displacement in the direction of force.

  - (ii) By a variable force

$$W = \int_{x_i}^{x_f} F\cdot dx \quad \text{where } F = f(x)$$

  - (iii) **By area under F-x graph** If force is a function of x, we can find work done by area under F-x graph with projection along x-axis. In this method magnitude of work done can be obtained by area under F-x graph, but sign of work done should be decided by you. If force and displacement both are positive or negative, work done will be positive. If one is positive and other is negative, then work done will be negative.

- **Power of a Force**
  - (i) **Average power**

$$P_{av} = \frac{\text{Total work done}}{\text{Total time taken}} = \frac{W_{\text{Total}}}{t}$$

  - (ii) **Instantaneous power**

$$P_{ins} = \text{rate of doing work done}$$
$$= \frac{dW}{dt} = \mathbf{F}\cdot\mathbf{v} = Fv\cos\theta$$

- **Conservative and Non-conservative Forces**

  In case of conservative forces
  - (i) Work done is path independent.
  - (ii) In a closed path net work done is zero.
  - (iii) Potential energy is defined only for conservative forces.
  - (iv) If only conservative forces are acting on a system, its mechanical energy should remain constant.

  Forces which are not conservative are non-conservative.

- **Potential Energy**
  - (i) Potential energy is defined only for conservative forces.
  - (ii) In a conservative force field, difference in potential energy between two points is negative of work done by conservative forces in displacing the body (or system) from some initial position to final position. Hence,

$$\Delta U = -W \quad \text{or} \quad U_B - U_A = -W_{A\to B}$$

  - (iii) Absolute potential energy at a point can be defined with respect to a reference point where potential energy is assumed to be zero. Reference point corresponding to gravitational potential energy is assumed at infinity. Reference point corresponding to spring potential energy is taken at natural length points of spring.

  Now, negative of work done in displacement of body from reference point (say O) to the point under consideration (say P) is called absolute potential energy at P. Thus,

$$U_P = -W_{O\to P}$$

- **Spring Force**
  - (i) $F = -kx$ or $F \propto -x$
  - (ii) $U_x = \frac{1}{2}kx^2$
  - (iii)

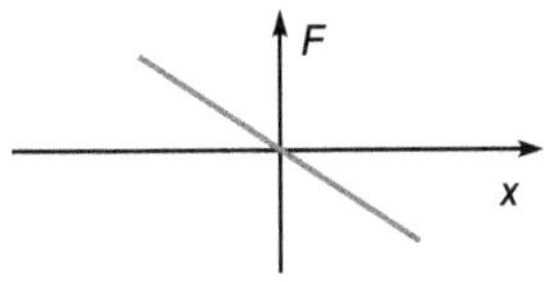

Additional Examples

Example 1. *How much work is done by a coolie walking on a horizontal platform with a load on his head ?*

Sol. Zero. In order to balance the load on his head, the coolie applies a force on it in the upward direction equal to its weight. His displacement is along the horizontal direction. Thus, the angle between force **F** and displacement **s** is 90°. Therefore work done, $W = Fs\cos\theta = Fs\cos 90° = 0$.

Example 2. *Can a body have energy without momentum ?*

Sol. Yes, there may be an internal energy in the body due to the thermal agitation of the particles of the body.

Example 3. *Can a body have momentum without energy ?*

Sol. No, if a body has momentum, it must be in motion and consequently possess kinetic energy.

Example 4. *A light body and a heavy body have the same momentum. Which one will have greater kinetic energy ?*

Sol. Kinetic energy, $K = \dfrac{P^2}{2m}$

For constant P, $K \propto \dfrac{1}{m}$

Thus, the lighter body has more kinetic energy than the heavier body.

Example 5. *A light body and a heavy body have the same kinetic energy. Which one will have the greater momentum?*

Sol. Momentum,

$$P = \sqrt{2mK}$$

i.e. $$P \propto \sqrt{m}$$

Thus, the heavier body has a greater momentum than the lighter one.

Example 6. *A lorry and a car with the same kinetic energy are brought to rest by the application of the brakes which provide equal retarding force. Which of them will come to rest in a shorter distance?*

Sol. From work-energy theorem,

Loss in KE of the vehicle

= Work done against retarding force

= Retarding force × distance travelled

$$\therefore \quad \text{Distance travelled} = \frac{\text{Loss in KE}}{\text{Retarding force}}$$

As both the kinetic energy and retarding force are same, so both, lorry and the car would come to rest in the same distance.

Example 7. *A small body of mass m is located on a horizontal plane at the point O. The body acquires a horizontal velocity v_0 due to friction. Find, the mean power developed by the friction force during the motion of the body, if the frictional coefficient $\mu = 0.27$, $m = 1.0$ kg and $v_0 = 1.5$ m/s.*

Sol. The body gains velocity due to friction. The acceleration due to friction.

$$a = \frac{\text{force of friction}}{\text{mass}} = \frac{\mu mg}{m} = \mu g$$

Further, $v_0 = at$

Therefore, $$t = \frac{v_0}{a} = \frac{v_0}{\mu g} \qquad \text{...(i)}$$

From work energy theorem,

work done by force of friction = change in kinetic energy

or $$W = \frac{1}{2}mv_0^2 \qquad \text{...(ii)}$$

$$\text{Mean power} = \frac{W}{t}$$

From Eqs. (i) and (ii), we get $P_{mean} = \dfrac{1}{2}\mu mg v_0$

Substituting the values, we have

$$P_{mean} = \frac{1}{2} \times 0.27 \times 1.0 \times 9.8 \times 1.5 \approx 2.0 \text{ W}$$

Example 8. *An object of mass 5 kg falls from rest through a vertical distance of 20 m and attaches a velocity of 10 m/s. How much work is done by the resistance of the air on the object? $(g = 10 \text{ m/s}^2)$*

Sol. Applying work-energy theorem,

work done by all the forces = change in kinetic energy

or $$W_{mg} + W_{air} = \frac{1}{2}mv^2$$

$$\therefore \quad W_{air} = \frac{1}{2}mv^2 - W_{mg}$$

$$= \frac{1}{2}mv^2 - mgh$$

$$= \frac{1}{2} \times 5 \times (10)^2 - (5) \times (10) \times (20)$$

$$= -750 \text{ J}$$

Example 9. *A rod of length 1.0 m and mass 0.5 kg fixed at one end is initially hanging, vertical. The other end is now raised until it makes an angle 60° with the vertical. How much work is required?*

Sol. For increase in gravitational potential energy of a rod we see the centre of the rod.

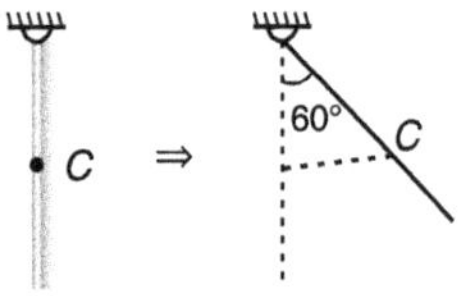

$$W = \text{change in potential energy}$$
$$= mg\frac{l}{2}(1 - \cos\theta)$$

Substituting the values, we have

$$W = (0.5)(9.8)\left(\frac{1.0}{2}\right)(1 - \cos 60°)$$
$$= 1.225 \text{ J}$$

Example 10. *A small mass m starts from rest and slides down the smooth spherical surface of R. Assume zero potential energy at the top. Find*

(a) the change in potential energy

(b) the kinetic energy

(c) the speed of the mass as a function of the angle θ made by the radius through the mass with the vertical.

Sol. In the figure, $h = R(1 - \cos\theta)$

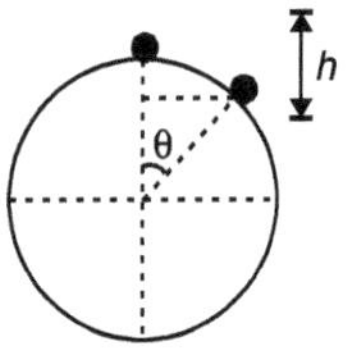

(a) As the mass comes down, potential energy will decrease. Hence,

$$\Delta U = -mgh = -mgR(1 - \cos\theta)$$

(b) Magnitude of decrease in potential energy = increase in kinetic energy

$\therefore$ Kinetic energy $= mgh = mgR(1 - \cos\theta)$

(c) $\frac{1}{2}mv^2 = mgR(1 - \cos\theta)$

$\therefore \quad v = \sqrt{2gR(1 - \cos\theta)}$

Example 11. *A smooth narrow tube in the form of an arc AB of a circle of centre O and radius r is fixed so that A is vertically above O and OB is horizontal. Particles P of mass m and Q of mass 2 m with a light inextensible string of length $\left(\frac{\pi r}{2}\right)$ connecting them are placed inside the tube with P at A and Q at B and released from rest. Assuming the string remains taut during motion, find the speed of particles when P reaches B.*

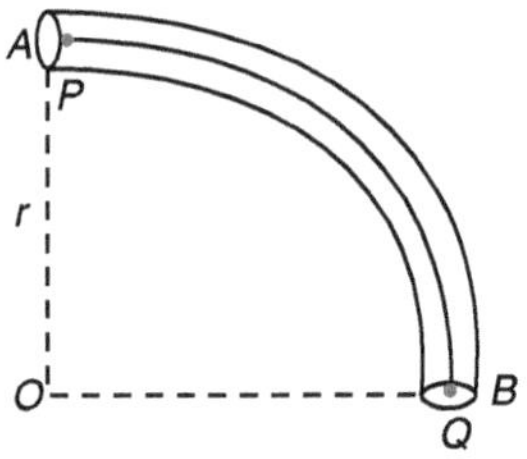

Sol. All surfaces are smooth. Therefore, mechanical energy of the system will remain conserved.

$\therefore$ Decrease in PE of both the blocks

= increase in KE of both the blocks

$$\therefore \quad (mgr) + (2mg)\left(\frac{\pi r}{2}\right) = \frac{1}{2}(m + 2m)v^2$$

or
$$v = \sqrt{\frac{2}{3}(1 + \pi)gr}$$

Example 12. *An automobile of mass m accelerates, starting from rest. The engine supplies constant power P, show that the velocity is given as a function by time by $v = \left(\frac{2Pt}{m}\right)^{1/2}$*

Sol. Power P = constant

Work done upto time t is $W = Pt$

From work energy theorem,

$$W = \Delta KE$$

or
$$Pt = \frac{1}{2}mv^2$$

$$\therefore \quad v = \left(\frac{2Pt}{m}\right)^{1/2}$$

NCERT Selected Questions

Q 1. The sign of work done by a force on a body is important to understand. State carefully if the following quantities are positive or negative

(a) work done by a man in lifting a bucket out of a well by means of a rope tied to the bucket.

(b) work done by gravitational force in the above case.

(c) work done by friction on a body sliding down an inclined plane.

(d) work done by an applied force on a body moving on a rough horizontal plane with uniform velocity.

(e) work done by the resistive force of air on a vibrating pendulum in bringing it to rest.

Sol. Work done by a force is given by $W = \mathbf{F} \cdot \mathbf{S} = FS\cos\theta$, where, θ = angle between F (force) and displacement S.

(a) To lift the bucket, force equal to or more than the weight of the bucket has to be applied vertically upwards and the bucket moves along the same direction, thus $\theta = 0$, so $W = FS$ is positive.

(b) Here, bucket moves in a direction opposite to the gravitational force which acts vertically downwards.

$\therefore \quad \theta = 180°$

So $\quad W = FS\cos 180°$

$= FS(-1) = -FS$

or $\quad W =$ negative.

(c) Work done by friction on a body sliding down an inclined plane is negative as friction is in opposite direction of motion, thus $\theta = 180°$

$\therefore \quad W = -FS.$

(d) As the body moves in the same direction in which the force is applied, so $\theta = 0$, thus $W = FS$ i.e. it is positive.

(e) Work done is negative as the direction of the resistive force of air on the vibrating pendulum is opposite to the direction of displacement (i.e. motion) of the bob.

Q 2. A body of mass 2 kg initially at rest moves under the action of an applied horizontal force of 7 N on a table with coefficient of kinetic friction $= 0.1$. Compute the

(a) work done by the applied force is 10 s.

(b) work done by friction in 10 s.

(c) work done by the net force on the body in 10 s.

(d) change in kinetic energy of the body in 10 s and interpret your results.

Sol. Force of friction, $f = u_k mg$

$= 0.1 \times 2 \times 9.8 = 1.96\text{ N}$

$\therefore \quad$ Net force $F' = F - f$

$= (7 - 1.96)\text{ N} = 5.04\text{ N}$

$\therefore$ Net acceleration with which the body moves,

$$a = \frac{F'}{m}$$

$$= \frac{5.04}{2} = 2.52\text{ ms}^{-2}$$

If S be the distance covered by the body in 10 s, then

$$S = \frac{1}{2}at^2 \text{ (as } u = 0)$$

$$= \frac{1}{2} \times 2.52 \times (10)^2$$

$$\therefore \quad S = 126\text{ m}$$

(a) Work done by the applied force,

$$W_1 = FS = 7 \times 126 = 882\text{ J}$$

(b) Work done by the force of friction

$$W_2 = -FS$$

$$= -1.96 \times 126$$

$$= -246.96\text{ J}$$

$$= -247\text{ J}$$

(c) Work done by the net force $W_3 = FS$

$$= 5.04 \times 126 = 635.04\text{ J}$$

(d) According to work energy theorem, change in kinetic energy = work done or change in KE = 635.04 J

Q 3. Underline the correct alternative.

(a) When a conservative force does positive work on a body, the potential energy of the body increases/decreases/ remain unaltered.

(b) Work done by a body against friction always results in a loss of its kinetic / potential energy.

Sol. (a) Potential energy of the body decreases.Because Work done by conservative forces

$$W = (-\text{ change in PE}) = -\Delta U.$$

If $W = +$ ve, then $\Delta U = -$ ve

(b) Work done by a body against friction results in a loss of its kinetic energy.

Q 4. State if each of the following statements are true or false. Give reasons for your answer.

(a) Work done in the motion of a body over a closed loop is zero for every force in nature.

(b) In an inelastic collision, the final kinetic energy is always less than the initial kinetic energy of the system.

Sol. (a) False. Work done in the motion of a body over a closed loop is zero only when the body moves under the action or conservative force.

(b) True. Usually but not always, because in an inelastic collision, some kinetic energy usually changes into some other forms of energy.

Q 5. A body is moving unidirectionally under the influence of a source of constant power. Its displacement in time t is proportional to

(a) $t^{1\backslash 2}$ (b) t

(c) $t^{3/2}$ (d) t^2

Sol. Let, the constant power P acts on the body of mass m for a time t to give it a velocity v.

$\therefore$ Its KE = work done = power × time

or $$\frac{1}{2}mv^2 = Pt$$

or $$v = \sqrt{\frac{2Pt}{m}} = \sqrt{\frac{2P}{m}}t^{1/2} \quad \ldots\text{(i)}$$

Also we know that

$$v = \frac{dx}{dt}$$

or $$dx = vdt \quad \ldots\text{(ii)}$$

If x be the displacement of the body, then

$$x = \int dx = \int v dt$$

$$= \sqrt{\frac{2P}{m}}\int t^{1/2}dt$$

$$= \sqrt{\frac{2P}{m}}\left(\frac{t^{\frac{1}{2}+1}}{\frac{1}{2}+1}\right)$$

$$= \sqrt{\frac{2P}{m}}\cdot\frac{t^{3/2}}{\left(\frac{3}{2}\right)}$$

$$= \frac{2}{3}\sqrt{\frac{2P}{m}}t^{3/2}$$

Here, P = constant and m is also constant for a body,

so $x = \text{constant} \times t^{3/2}$ or $x \propto t^{3/2}$.

Q 6. A body constrained to move along z-axis of a coordinate system is subjected to a constant force **F** given by

$$\mathbf{F} = -\hat{\mathbf{i}} + 2\hat{\mathbf{j}} + 3\hat{\mathbf{k}}\text{ N}$$

where $\hat{\mathbf{i}}$, $\hat{\mathbf{j}}$, and $\hat{\mathbf{k}}$ are the unit vector along the x, y and z-axis of the system respectively. What is the work done by this force in moving the body a distance of 4 m along the z-axis?

Sol. Since, the body is displaced 4 m along z-axis only

$\therefore$ $$\mathbf{S} = 4\hat{\mathbf{k}}$$

Also $$\mathbf{F} = -\hat{\mathbf{i}} + 2\hat{\mathbf{j}} + 3\hat{\mathbf{k}}$$

$\therefore$ Work done is given by

$$W = \mathbf{F}\cdot\mathbf{S}$$

$$= (-\hat{\mathbf{i}} + 2\hat{\mathbf{j}} + 3\hat{\mathbf{k}}).(0\hat{\mathbf{i}} + 0\hat{\mathbf{j}} + 4\hat{\mathbf{k}})$$

$$= (-1)\times 0 + 2\times 0 + 3\times 4 = 12\text{ J}$$

Q 7. A rain drop of radius 2 mm falls from a height of 500 m above the ground . It falls with decreasing acceleration (due to viscous resistance of the air) until at half its original height, it attains its maximum (terminal) speed, and moves with uniform speed thereafter. What is the work done by the gravitational force on the drop in the first and second half of its journey? What is the work done by the resistive force in the entire journey, if its speed on reaching the ground is 10 ms^{-1}?

Sol. Distance travelled in the half journey,

$$S = \frac{H}{2} = \frac{500}{2} = 250\text{m}$$

$\therefore$ Volume of the drop, $V = \frac{4}{3}\pi r^3$

or $$V = \frac{4}{3}\times\pi\times(2\times 10^{-3})^3$$

$$= 3.35\times 10^{-8}\text{ kg}$$

$\therefore$ If m be the mass of the rain drop, then

$$m = V\rho = 3.35\times 10^{-8}\times 10^3$$

$$= 3.35\times 10^{-5}\text{kg}$$

$\therefore$ Gravitational force on the drop,

$$F = mg = 3.35\times 10^{-5}\times 9.8\text{N}$$

$\therefore$ Work done by the gravitational force on the drop in the first half of its journey is given by

$$W = FS = 3.35\times 10^{-5}\times 9.8\times 250\text{ J}$$

$$= 0.082\text{ J}$$

Since, the distance travelled in the second half is also same, therefore, work done in the second half is also same i.e. 0.082 J.

Let us assume that the rain drop is initially at rest

i.e. $u = 0$

Final velocity on hitting the ground

$$v = 10\text{ ms}^{-1}$$

$\therefore$ Change in KE of the rain drop,

$$\Delta E_k = \frac{1}{2}mv^2 - \frac{1}{2}mu^2$$

$$= \frac{1}{2}mv^2 - 0$$

$$= \frac{1}{2}\times 3.35\times 10^{-5}\times(10)^2$$

$$= 1.675\times 10^{-3}\text{ J} = 0.00167\text{ J}$$

Work done by gravitational force

$$= mgh$$

$$= 3.35\times 10^{-5}\times 9.8\times 500$$

$$= 0.16415\text{J}$$

Work done by the resistive force in the entire journey

= change in KE – work done by gravitational force

= 0.001675 – 0.16415 = – 0.163 J

i.e. the work done by the resistive force is negative.

Q 8. A pump on the ground floor of a building can pump up water to fill a tank of value 30 m^3 in 15 min. If the tank is 40 m above the ground and the efficiency of the pump is 30%, how much electric power is consumed by the pump?

Sol. Mass of water to be pumped up,

$$m = \text{Volume} \times \text{density of water}$$
$$= 30 \times 10^3 = 3 \times 10^4 \text{kg}$$

Height of the tank, $h = 40$ m

∴ Work done by the pump to fill the tank,

$$W = mgh = 3 \times 10^4 \times 9.8 \times 40 \text{ J}$$
$$= 1.176 \times 10^7 \text{ J}$$

Time, $t = 15 \text{ min} = 15 \times 60 \text{s} = 900 \text{s}$

∴ Required average power,

$$P = \frac{W}{t} = \frac{1.176 \times 10^7}{900} = 13.07 \text{kW}$$

Percentage efficiency of pump = 30%

Now $$\eta\% = \frac{\text{Output power}}{\text{Input power}} \times 100$$

or Input power = power consumed by the pump

$$= \frac{\text{Output power}}{\eta\%} \times 100$$
$$= \frac{13.07}{30} \times 100 = 43.55 \text{kW}$$
$$= 43.6 \text{kW}$$

Q 9. The bob of a pendulum is released from a horizontal position A as shown in the figure. If the length of the pendulum is 1.5m, what is the speed with which the bob arrives at the lower most point B, given that it dissipated 5% of its initial energy against air resistance?

Sol. At point A, the energy of the pendulum is entirely PE. At point B, the energy of the pendulum is entirely KE. It means that as the bob of the pendulum lowers from A to B, PE is converted into KE. Thus, at B, KE = PE. But 5% of the PE is dissipated against air resistance.

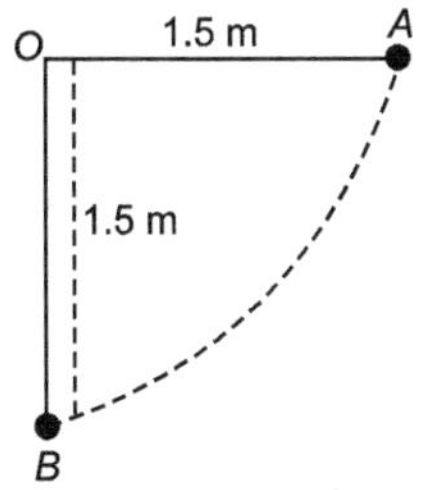

KE at B = 95% of PE at A ...(i)

∴ From Eq.(i), $$\frac{1}{2}mv^2 = \frac{95}{100}mgh$$

or $$v^2 = 2 \times \frac{95}{100}gh$$
$$= 2 \times \frac{95}{100} \times 9.8 \times 1.5$$
$$v = \sqrt{\frac{19 \times 9.8 \times 1.5}{10}}$$
$$= \sqrt{27.93}$$
$$= 5.285 \text{ ms}^{-1}$$

or $$v = 5.3 \text{ ms}^{-1}$$

Q 10. A person trying to loose weight (dieter) lifts a 10 kg mass 0.5 m 1000 times. Assume that the potential energy lost each time she lowers the mass is dissipated.

(a) How much work does she do against the gravitational force?

(b) Fat supplies 3.8×10^7 J of energy per kg, which is converted to mechanical energy with a 20% efficiency rate. How much fat will the dieter use up?

Sol. (a) Work done against gravitational force

$$= mngh$$
$$= 10 \times 1000 \times 9.8 \times 0.5$$
$$= 49000 \text{ J}$$

(b) Energy supplied by fat per kg = 3.8×10^7 J

i.e. 3.8×10^7 J energy is equivalent to 1 kg of fat

∴ 49×10^3 J energy will be equivalent to

$$= \frac{49 \times 10^3}{3.8 \times 10^7} \text{kg fat}$$

$$\text{Fat produced} = \frac{49}{3.8} \times 10^{-4} \text{kg}$$

$$\eta\% = 20\%$$

∴ $$\eta\% = \frac{\text{Fat produced}}{\text{Fats used up by dieter}} \times 100$$

or $$20 = 49 \times \frac{10^{-4}}{3.8} \times \frac{100}{\text{Fats used up by dieter}}$$

or fats used up by the dieter

$$= \frac{49 \times 10^{-4}}{3.8} \times \frac{100}{20}$$
$$= \frac{245}{3.8} \times 10^{-4}$$
$$= \frac{245}{38} \times 10^{-3} \text{kg}$$
$$= 6.447 \times 10^{-3} \text{kg}$$
$$= 6.45 \times 10^{-3} \text{kg}$$

Q 11. A large family uses 8 kW of power.

(a) Direct solar energy is incident on the horizontal surface at an average rate of 200 W/m^2. If 20% of this energy can be converted to useful electrical energy, how large an area is needed to supply 8 kW?

(b) Compare this area to that of the roof of a typical house.

Sol. Total power used by the family = 8k W = 8000 W

(a) Average rate of incidence of the direct solar energy on the horizontal surface = $200\ \text{W/m}^2$

Energy converted into useful electrical energy

$$= 20\% \text{ of } 200\,\text{W/m}^2 = \frac{20}{100} \times 200 = 40\,\text{W/m}^2$$

Total power required by family = 8000 W

$\therefore$ Area required to produce the same amount of electrical energy $= \dfrac{8000}{40} = 200\,\text{m}^2$

(b) In order to compare this area to that of the roof of a typical house, let a be the side of the roof

$\therefore$ Area of roof $= a \times a = a^2$

Thus, $a^2 = 200\ \text{m}^2$

or $a = \sqrt{200\,\text{m}^2} = 14.14\,\text{m}$

$\therefore$ Area of roof $= 14.14 \times 14.14\,\text{m}^2$

Thus, $200\ \text{m}^2$ is comparable to the roof of a large house of dimensions $14.14\,\text{m} \times 14.14\,\text{m} = 14\,\text{m} \times 14\,\text{m}$

***Q* 12.** A bolt of mass 0.3 kg falls from the ceiling of an elevator moving down with a uniform speed of $7\,\text{ms}^{-1}$. It hits the floor of the elevator (length of the elevator = 3 m) and does not rebound.

What is the heat produced by the impact? Would your answer be different, if the elevator were stationary?

Sol. Potential energy of the bolt at the ceiling

$$= mgh$$
$$= 0.3 \times 9.8 \times 3$$
$$= 8.82\ \text{J}$$

The bolt does not rebound, so the whole of the potential energy is converted into heat according to the law of conservation of energy.

Since, the value of acceleration due to gravity is the same in all inertial systems, therefore the answer will not change even, if the elevator is stationary or moving.

Objective Problems

[Level 1]

Work Done by Different Forces

1. A particle moves under a force $f = cx$ from $x = 0$ to $x = x_1$. The work done is
(a) cx_1^2 (b) $\frac{cx_1^2}{2}$
(c) zero (d) cx_1^3

2. A particle moves along the x-axis from $x = 0$ to $x = 5$ m under the influence of a force given by $F = 7 - 2x + 3x^2$. Work done in the process is
(a) 70 (b) 270
(c) 35 (d) 135

3. A body constrained to move in the y-direction, is subjected to a force $\mathbf{F} = (-2\hat{\mathbf{i}} + 15\hat{\mathbf{j}} + 6\hat{\mathbf{k}})$N. What is the work done by this force in moving the body through a distance of 10 m along the y-axis?
(a) 20 J (b) 150 J
(c) 160 J (d) 190 J

4. Work done by a force $\mathbf{F} = (\hat{\mathbf{i}} + 2\hat{\mathbf{j}} + 3\hat{\mathbf{k}})$N acting on a particle in displacing it from the point $\mathbf{r}_1 = \hat{\mathbf{i}} + \hat{\mathbf{j}} + \hat{\mathbf{k}}$ to the point $\mathbf{r}_2 = \hat{\mathbf{i}} - \hat{\mathbf{j}} + 2\hat{\mathbf{k}}$ is
(a) 3 J (b) −1 J
(c) zero (d) 2 J

5. A mass M is lowered with the help of a string by a distance x at a constant acceleration $\frac{g}{2}$. The magnitude of work done by the string will be
(a) Mgx (b) $\frac{1}{2}Mgx^2$
(c) $\frac{1}{2}Mgx$ (d) Mgx^2

6. The work done by pseudo forces is
(a) positive (b) negative
(c) zero (d) All of these

7. The work done by a spring force
(a) is always negative
(b) is always positive
(c) is always zero
(d) may be positive and negative

8. A force $(3\hat{\mathbf{i}} + 4\hat{\mathbf{j}})$ N acts on a body and displaces it by $(3\hat{\mathbf{i}} + 4\hat{\mathbf{j}})$m. The work done by the force is
(a) 10 J (b) 12 J (c) 16 J (d) 25 J

9. A ball is released from the top of a tower. The ratio of work done by force of gravity in first, second and third second of the motion of ball is
(a) 1:2:3
(b) 1:4:16
(c) 1:3:5
(d) 1:9:25

10. An object of mass 5 kg is acted upon by a force that varies with position of the object as shown. If the object starts out from rest at a point $x = 0$. What is its speed at $x = 50$ m?

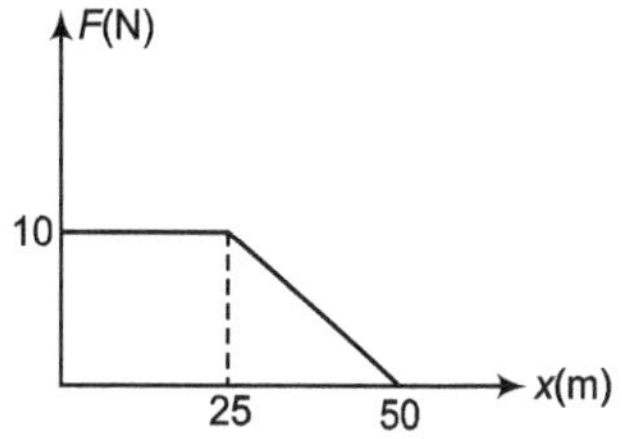

(a) 12.2 ms^{-1} (b) 18.2 ms^{-1}
(c) 16.4 ms^{-1} (d) 20.4 ms^{-1}

11. The force F acting on a particle is moving in a straight line as shown in figure. What is the work done by the force on the particle in the 1 m of the trajectory?

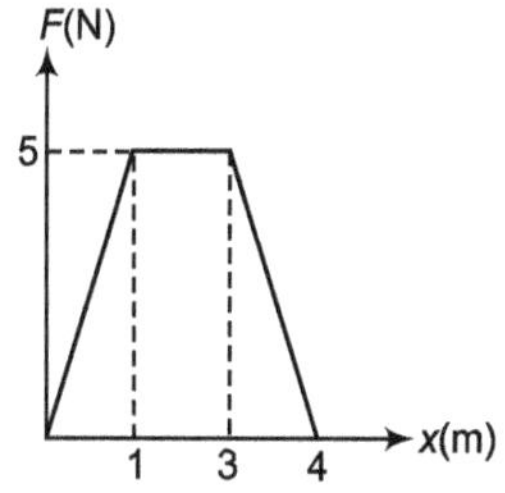

(a) 5 J (b) 10 J
(c) 15 J (d) 2.5 J

12. A position dependent force, $F = 7 - 2x + 3x^2$ N acts on a small body of mass 2 kg and displaces it from $x = 0$ to $x = 5$ m. The work done in joule is
(a) 35 (b) 70
(c) 135 (d) 270

13. Mark out the correct statement (s).
(a) Total work done by internal forces on a system is always zero
(b) Total work done by internal forces on a system may sometimes be zero
(c) Total work done by friction may sometimes be zero
(d) Total work done by friction is always zero

14. A body of mass 1 kg moves from points A (2m, 3m, 4m,) to B (3m, 2m, 5m,). During motion of body, a force $\mathbf{F} = (2\text{N})\hat{\mathbf{i}} - (4\text{N})\hat{\mathbf{j}}$ acts on it. The work done by the force on the particle during displacement is

(a) 6 J (b) 2 J (c) − 2 J (d) − 6 J

15. A force $F = Ay^2 + By + C$ acts on a body in the y-direction. The work done by this force during a displacement from $y = -a$ to $y = a$ is

(a) $\frac{2Aa^3}{3}$ (b) $\frac{2Aa^3}{3} + 2Ca$ (c) $\frac{2Aa^3}{3} + \frac{Ba^2}{2} + Ca$ (d) None of these

16. Force acting on a particle is $(2\hat{\mathbf{i}} + 3\hat{\mathbf{j}})\text{N}$. Work done by this force is zero, when a particle is moved on the line $3y + kx = 5$. Here, value of k is

(a) 2 (b) 4 (c) 6 (d) 8

17. A block of mass 10 kg is moving in x-direction with a constant speed of 10m/s. It is subjected to a retarding force $F = (-0.1x)\text{N}$ during its travel from $x = 20$ m to $x = 30$m. Its final kinetic energy will be

(a) 475 J (b) 450 J (c) 275 J (d) 250 J

Potential Energy, Kinetic Energy, Power, Work Energy Theorem and Types of Equilibrium

18. Under the action of a force, a 2 kg body moves such that its position x as a function of time t is given by $x = \frac{t^3}{3}$, where, x is in metre and t in second. The work done by the force in the first two seconds is

(a) 1600 J (b) 160 J (c) 16 J (d) 1.6 J

19. If the speed of a vehicle increase by 2m/s, its kinetic energy is doubled, then original speed of the vehicle is

(a) $(\sqrt{2}+1)$m/s (b) $2(\sqrt{2}-1)$m/s (c) $2(\sqrt{2}+1)$m/s (d) $\sqrt{2}(\sqrt{2}+1)$m/s

20. Unit of power is

(a) kilowatt hour (b) kilowatt per hour (c) kilowatt (d) erg

21. When work done by force of gravity is negative

(a) PE increase
(b) KE decreases
(c) PE remains constant
(d) PE decreases

22. A long spring is stretched by 2 cm. Its potential energy is U. If the spring is stretched by 10 cm, its potential energy would be

(a) $\frac{U}{25}$ (b) $\frac{U}{5}$ (c) $5U$ (d) $25U$

23. An engine develops 10 kW of power. How much time will it take to lift a mass of 200 kg to a height of 40 m? ($g = 10\text{ms}^{-2}$).

(a) 4 s (b) 5 s (c) 8 s (d) 10 s

24. Velocity-time graph of a particle of mass 2 kg moving in a straight line is as shown in figure. Work done by all the forces on the particle is

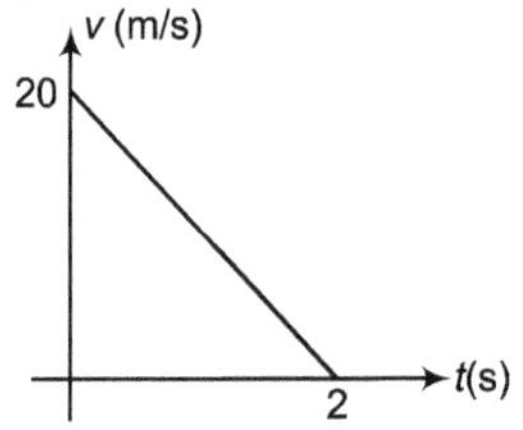

(a) 400 J (b) − 400 J (c) − 200 J (d) 200 J

25. A vehicle needs an engine of 7500 W to keep it moving with a constant velocity of 20 m/s on a horizontal surface. The force resisting the motion is

(a) 375 dyne (b) 375 N (c) 150000 dyne (d) 150000 N

26. A mass of 1 kg is acted upon by a single force $\mathbf{F} = (4\hat{\mathbf{i}} + 4\hat{\mathbf{j}})\text{N}$. Under this force it is displaced from (0, 0) to (1m, 1m). If initially the speed of the particle was 2m/s, its final speed should be

(a) 6 m/s (b) 4.5 m/s (c) 8 m/s (d) 4 m/s

27. A particle moves with a velocity $(5\hat{\mathbf{i}} - 3\hat{\mathbf{j}} + 6\hat{\mathbf{k}})$m/s under, the influence of a constant force $\mathbf{F} = (10\hat{\mathbf{i}} + 10\hat{\mathbf{j}} + 20\hat{\mathbf{k}})\text{N}$. The instantaneous power applied to the particle is

(a) 200 J/s (b) 40 J/s (c) 140 J/s (d) 170 J/s

28. Power applied to a particle varies with time as $P = (3t^2 - 2t + 1)$W, there t is in second. Find the change in its kinetic energy between time $t = 2$ s and $t = 4$ s.

(a) 32 J (b) 46 J (c) 61 J (d) 102 J

29. A particle is moving in a conservative force field from point A to point B. U_A and U_B are the potential energies of the particle at point A and B and W_C is the work done by conservative forces in the process of taking the particle from A and B.

(a) $W_C = U_B - U_A$ (b) $W_C = U_A - U_B$ (c) $U_A > U_B$ (d) $U_B > U_A$

30. A box is moved along a straight line by a machine delivering constant power. The distance moved by the body in time t is proportional to

(a) $t^{1/2}$ (b) $t^{3/4}$ (c) $t^{3/2}$ (d) t^2

31. In which of the following cases, the potential energy is defined?

(a) Both conservative and non-conservative forces
(b) Conservative force only
(c) Non-conservative force only
(d) Neither conservative nor-conservative forces

32. A particle of mass 0.01 kg travels with velocity given by $4\hat{\mathbf{i}} + 16\hat{\mathbf{k}}\ \text{ms}^{-1}$. After sometime, its velocity becomes $8\hat{\mathbf{i}} + 20\hat{\mathbf{j}}\ \text{ms}^{-1}$. The work done on particle during this interval of time is

(a) 0.32 J (b) 6.9 J (c) 9.6 J (d) 0.96 J

33. A running man has half the KE that a boy of half his mass. The man speeds up by 1 ms^{-1} and then has the same KE as that of boy. The original speeds of man and boy in ms^{-1} are

(a) $(\sqrt{2}+1), (\sqrt{2}-1)$ (b) $(\sqrt{2}+1), 2(\sqrt{2}+1)$
(c) $\sqrt{2}, \sqrt{2}$ (d) $(\sqrt{2}+1), 2(\sqrt{2}-1)$

34. A motor drives a body along a straight line with a constant force. The power P developed by the motor must vary with time t as

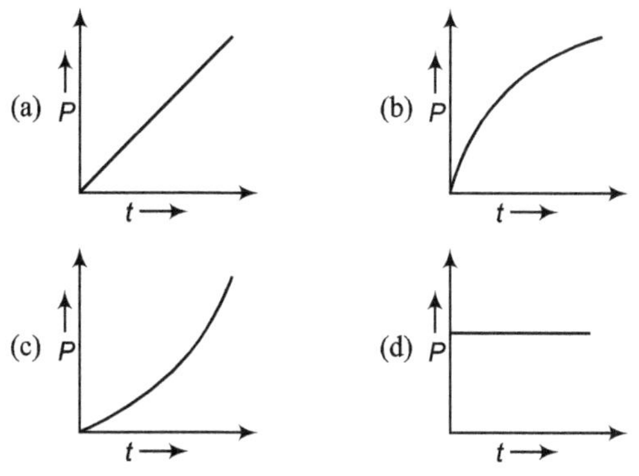

35. A 50 kg girl is swinging on a swing from rest. Then, the power delivered when moving with a velocity of 2 m/s upwards in a direction making an angle 60° with the verticle is

(a) 980 W (b) 490 W (c) $490\sqrt{3}$ W (d) 245 W

36. Power supplied to a particle of mass 2 kg varies with time as $P = \frac{3t^2}{2}$ W. Here, t is in second. If velocity of particle at $t = 0$ is $v = 0$. The velocity of particle at time $t = 2$ s will be

(a) 1 m/s (b) 4 m/s (c) 2 m/s (d) $2\sqrt{2}$ m/s

37. A particle moves in a straight line with retardation proportional to its displacement. Its loss of kinetic energy for any displacement x is proportional to

(a) x (b) x^2
(c) $\ln x$ (d) e^x

38. A body of mass 5 kg is raised vertically to a height of 10 m by a force 170 N. The velocity of the body at this height will be

(a) 9.8 m/s (b) 15 m/s
(c) 22 m/s (d) 37 m/s

39. A body of mass m thrown vertically upwards attains a maximum height h. At what height will its kinetic energy be 75% of its initial value?

(a) $\frac{h}{6}$ (b) $\frac{h}{5}$
(c) $\frac{h}{4}$ (d) $\frac{h}{3}$

40. A particle of mass m accelerating uniformly has velocity v at time t_1. What is work done in time t?

(a) $\frac{1}{2}\frac{mv^2}{t_1^2}t^2$ (b) $\frac{1}{2}\left(\frac{m}{t_1}\right)^2 t^2$
(c) $\frac{mv^2}{t_1^2}t^2$ (d) $\frac{2mv^2}{t_1^2}t^2$

41. A mass of 1 kg is acted upon by a single force $\mathbf{F} = (4\hat{\mathbf{i}} - 4\hat{\mathbf{j}})$N. Under this force it is displaced from (0, 0) to (1m, 1m). If initially, the speed of the particle was 2 m/s, its final speed is approximately

(a) 6.4 m/s (b) 4.5 m/s
(c) 8.6 m/s (d) 7.2 m/s

42. An engine exerts a force $\mathbf{F} = (20\hat{\mathbf{i}} - 3\hat{\mathbf{j}} + 5\hat{\mathbf{k}})$N and moves with velocity $\mathbf{v} = (6\hat{\mathbf{i}} + 20\hat{\mathbf{j}} - 3\hat{\mathbf{k}})$ m/s. The power of the engine (in watt) is

(a) 45 (b) 75
(c) 20 (d) 10

43. An object of mass m, initially at rest under the action of a constant force F attains a velocity v in time t. Then, the average power supplied to mass is

(a) $\frac{mv^2}{2t}$
(b) $\frac{Fv}{2}$
(c) Both are correct
(d) Both are wrong

44. A force F acting on a body depends on its displacement S as $F \propto S^{-1/3}$. The power delivered by F will depend on displacement as

(a) $S^{2/3}$ (b) $S^{-5/3}$
(c) $S^{1/2}$ (d) S^0

45. The force acting on a body moving along x-axis varies with the position of the particle as shown in the figure. The body is in stable equilibrium at

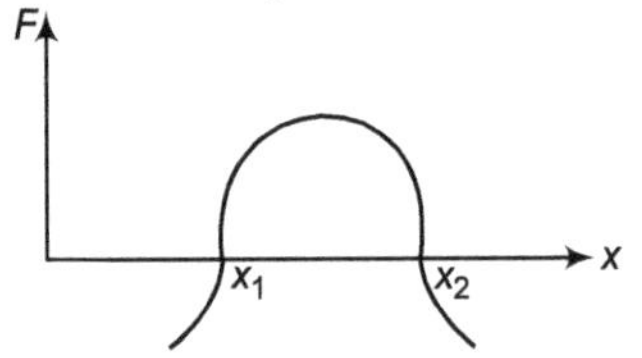

(a) $x = x_1$ (b) $x = x_2$
(c) Both x_1 and x_2 (d) Neither x_1 nor x_2

46. The given plot shows the variation of U, the potential energy of interaction between two particles with the distance separating them r.

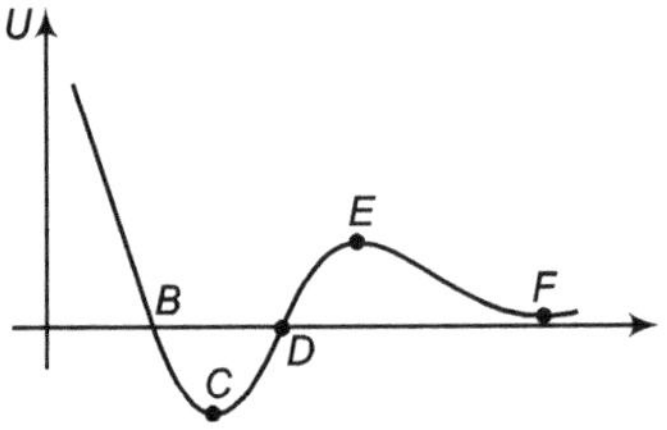

1. B and D are equilibrium points.
2. C is a point of stable equilibrium.
3. The force of interaction between the two particles is attractive between points C and D and repulsive between D and E.
4. The force of interaction between particles is repulsive between points E and F.

Which of the above statements are correct?

(a) 1 and 2 (b) 1 and 4
(c) 2 and 4 (d) 2 and 3

Smooth Surfaces where Mechanical Energy remains Conserved

47. Three particles A, B and C are projected from the top of a tower with the same speed. A is thrown straight upwards B straight down and C horizontally. They hit the ground with speeds v_A, v_B and v_C, then which of the following is correct?

(a) $v_A = v_B > v_C$ (b) $v_A = v_B = v_C$
(c) $v_A > v_B = v_C$ (d) $v_B > v_C > v_A$

48. A body of mass 0.1 kg moving with a velocity of 10 m/s hits a spring (fixed at the other end) of force constant 1000 N/m and comes to rest after compressing the spring. The compression of the spring is

(a) 0.01 m (b) 0.1 m
(c) 0.2 m (d) 0.5 m

49. A mass of 2 kg falls from a height of 40 cm on a spring with a force constant of 1960 N/m. The spring is compressed by (take $g = 9.8\,\text{m/s}^2$)

(a) 10 cm (b) 1.0 cm
(c) 20 cm (d) 5 cm

50. A toy gun uses a spring of very large value of force constant k. When charged before being triggered in the upward direction, the spring is compressed by a small distance x. If mass of shot is m, on being triggered it will go upto a height of

(a) $\dfrac{kx^2}{mg}$ (b) $\dfrac{x^2}{kmg}$
(c) $\dfrac{kx^2}{2mg}$ (d) $\dfrac{(kx)^2}{mg}$

51. A body is falling under gravity. When it loses a gravitational potential energy by U, its speed is v. The mass of the body shall be

(a) $\dfrac{2U}{v}$ (b) $\dfrac{U}{2v}$
(c) $\dfrac{2U}{v^2}$ (d) $\dfrac{U}{2v^2}$

52. A bead can slide on a smooth circular wire frame of radius r which is fixed in a vertical plane. The bead is displaced slightly from the highest point of the wire frame. The speed of the bead subsequently as a function of the angle θ made by the bead with the vertical line is

(a) $\sqrt{2gr}$ (b) $\sqrt{2gr(1-\sin\theta)}$
(c) $\sqrt{2gr(1-\cos\theta)}$ (d) $2\sqrt{gr}$

53. A body is moved along a straight line with constant power. The distance moved by the body in time t is proportional to

(a) $t^{3/2}$ (b) $t^{3/2}$
(c) $t^{1/4}$ (d) t^3

54. A particle is released from a height H. At certain height its kinetic energy is two times its potential energy. Height and speed of particle at that instant are

(a) $\dfrac{H}{3}, \sqrt{\dfrac{2gH}{3}}$ (b) $\dfrac{H}{3}, 2\sqrt{\dfrac{gH}{3}}$
(c) $\dfrac{2H}{3}, \sqrt{\dfrac{2gH}{3}}$ (d) $\dfrac{H}{3}, \sqrt{2gH}$

55. A uniform chain has a mass M and length L. It is placed on a frictionless table with length l_0 hanging over the edge. The chain begins to slide down . Then, the speed v with which the end slides down from the edge is given by

(a) $v = \sqrt{\dfrac{g}{L}(L + l_0)}$ (b) $v = \sqrt{\dfrac{g}{L}(L - l_0)}$
(c) $v = \sqrt{\dfrac{g}{L}(L^2 - l_0^2)}$ (d) $v = \sqrt{2g(L - l_0)}$

56. If a body of mass 200 g falls from a height 200 m and its total potential energy is converted into kinetic energy, at the point of contact of the body with the surface, then decrease in potential energy of the body at the contact is ($g = 10\,\text{m/s}^2$)

(a) 900 J (b) 600 J
(c) 400 J (d) 200 J

57. A stone of mass 2 kg is projected upwards with KE of 98 J. The height at which the KE of the body becomes half its original value, is given by (Take $g = 9.8\ \text{m/s}^2$)

(a) 5 m (b) 2.5 m
(c) 1.5 m (d) 0.5 m

58. The system shown in the figure is released from rest. At the instant when mass M has fallen through a distance h, the velocity of m will be

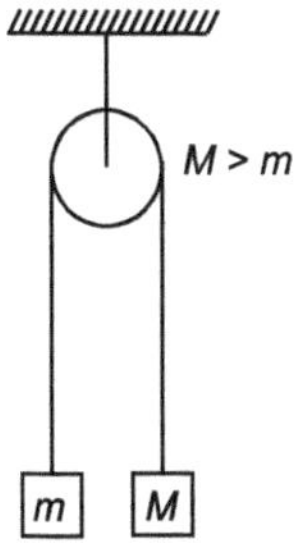

(a) $\sqrt{2gh}$ (b) $\frac{\sqrt{2ghM}}{m}$
(c) $\sqrt{\frac{2gh(M-m)}{m+M}}$ (d) $\sqrt{\frac{2gh(M+m)}{m-M}}$

59. If v be the instantaneous velocity of the body dropped from the top of a tower, when it is located at height h, then which of the following remains constant?

(a) $gh + v^2$ (b) $gh + \frac{v^2}{2}$ (c) $gh - \frac{v^2}{2}$ (d) $gh - v^2$

60. A 2.0 kg block is dropped from a height of 40 cm onto a spring of spring constant $k = 1960\ \text{N/m}$. Find the maximum distance the spring is compressed.

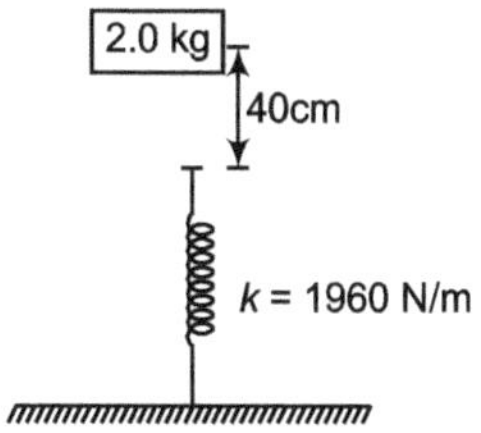

(a) 0.080 m (b) 0.20 m (c) 0.40 m (d) 0.10 m

Rough Surfaces where Mechanical Energy does not remain Constant

61. A pendulum of length 2 m left at A. When it reaches B, it loses 10% of its total energy due to air resistance. The velocity at B is

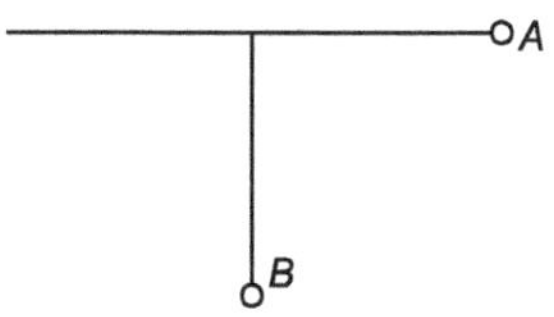

(a) 6 m/s (b) 1 m/s
(c) 2 m/s (d) 8 m/s

62. The net work done by kinetic friction

(a) is always negative
(b) is always zero
(c) may be negative and positive
(d) is always positive

63. A block of mass m is pulled along a horizontal surface by applying a force at an angle θ with the horizontal. If the block travels with a uniform velocity and has a displacement d and the coefficient of friction is μ, then the work done by the applied force is

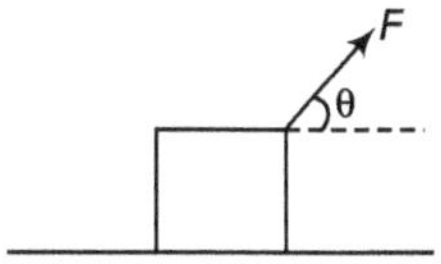

(a) $\frac{\mu mgd}{\cos\theta + \mu\sin\theta}$

(b) $\frac{\mu mgd\cos\theta}{\cos\theta + \mu\sin\theta}$

(c) $\frac{\mu mgd\sin\theta}{\cos\theta + \mu\sin\theta}$

(d) $\frac{\mu mgd\cos\theta}{\cos\theta - \mu\sin\theta}$

64. A horizontal force F pulls a 20 kg box at a constant speed along a rough horizontal floor. The coefficient of friction between the box and the floor is 0.25. The work done by force F on the block in displacing it by 2 m is

(a) 49 J (b) 98 J
(c) 147 J (d) 196 J

65. The coefficient of friction between the block and plank is μ and its value is such that block becomes stationary with respect to plank before it reaches the other end. Then,

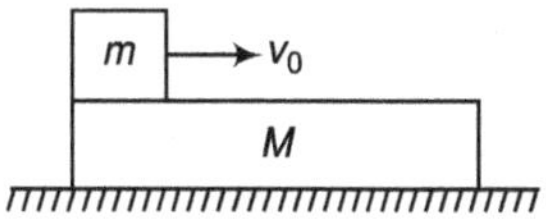

(a) the work done by friction on the block is negative
(b) the work done by friction on the plank is positive
(c) the net work done by friction is negative
(d) the net work done by the friction is zero

66. A plank of mass 10 kg and a block of mass 2 kg are placed on a horizontal plane as shown in the figure. There is no friction between plane and plank. The coefficient of friction between block and plank is 0.5. A force of 60 N is applied on plank horizontally. In first 2 s the work done by friction on the block is

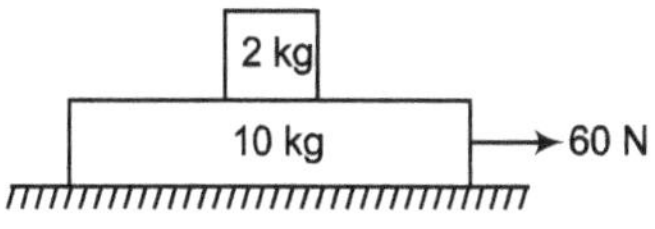

(a) – 100 J (b) 100 J
(c) zero (d) 200 J

67. A block of mass 5 kg slides down a rough inclined surface. The angle of inclination is 45°. The coefficient of sliding friction is 0.20 . When the block slides 10 cm, the work done on the block by force of friction is

(a) $-\frac{1}{\sqrt{2}}$ J (b) 1J
(c) $-\sqrt{2}$ J (d) −1J

68. A block of mass 1 kg slides down a rough inclined plane of inclination 60° starting from its top. If coefficient of kinetic friction is 0.5 and length of the plane $d = 1$ m, then work done against friction is

(a) 2.45 J (b) 4.9 J
(c) 9.8 J (d) 19.6 J

69. A particle moves on a rough horizontal ground with some initial velocity say v_0. If $\left(\frac{3}{4}\right)^{\text{th}}$ of its kinetic energy is lost due to friction in time t_0, then the coefficient of friction between the particle and the ground is

(a) $\frac{v_0}{2gt_0}$ (b) $\frac{v_0}{4gt_0}$
(c) $\frac{3v_0}{4gt_0}$ (d) $\frac{v_0}{gt_0}$

Miscellaneous Problems

70. If the linear momentum is increased by 50%, then kinetic energy will be increased by

(a) 50 % (b) 100 %
(c) 125 % (d) 25 %

71. A rod of mass m and length l is lying on a horizontal table. Work done in making it stand on one end will be

(a) mgl (b) $\frac{mgl}{2}$
(c) $\frac{mgl}{4}$ (d) $2mgl$

72. A body of mass m was slowly pulled up the hill by a force F which at each point was directed along the tangent of the trajectory. All surfaces are smooth. Find the work performed by this force.

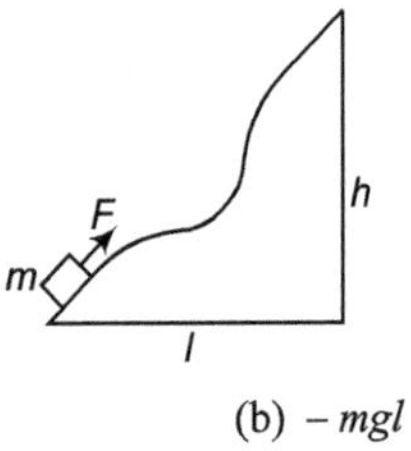

(a) mgl (b) $-mgl$
(c) mgh (d) zero

73. A particle is moved from (0, 0) to (a, a) under a force $\mathbf{F} = (3\hat{\mathbf{i}} + 4\hat{\mathbf{j}})$ from two paths. Path 1 is OP and path 2 is OQP. Let W_1 and W_2 be the work done by this force in these two paths. Then,

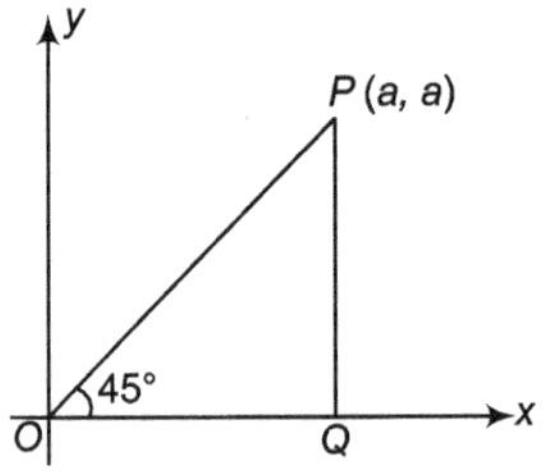

(a) $W_1 = W_2$ (b) $W_1 = 2W_2$
(c) $W_2 = 2W_1$ (d) $W_2 = 4W_1$

74. Two masses of 1 g and 4 g are moving with equal kinetic energies. The ratio of the magnitudes of their momenta is

(a) 4 : 1 (b) $\sqrt{2}$: 1
(c) 1 : 2 (d) 1 : 16

75. A uniform chain of length L and mass M is lying on a smooth table and one third of its length is hanging vertically down over the edge of the table . If g is acceleration due to gravity, the work required to pull the hanging part on the table is

(a) MgL (b) $\frac{MgL}{3}$
(c) $\frac{MgL}{9}$ (d) $\frac{MgL}{18}$

76. The net work done by the tension in the figure, when the bigger block of mass M touches the ground is

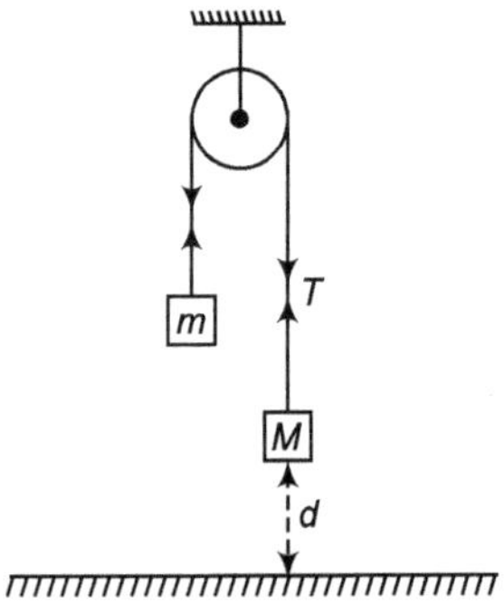

(a) $+Mgd$ (b) $-(M+m)gd$
(c) $-mgd$ (d) zero

77. An object of mass m is tied to a string of length L and a variable horizontal force is applied on it which starts at zero and gradually increases until the string makes an angle θ with the vertical. Work done by the force F is

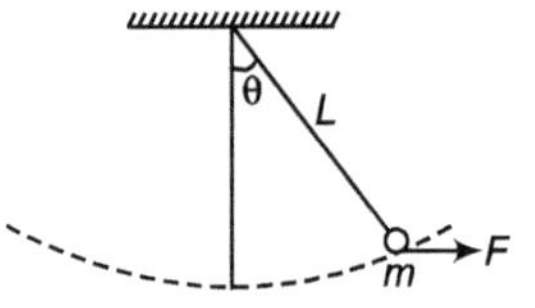

(a) $mgL(1-\sin\theta)$ (b) mgL
(c) $mgL(1-\cos\theta)$ (d) $mgL(1+\cos\theta)$

78. A particle at rest on a frictionless table is acted upon by a horizontal force which is constant in magnitude and direction. A graph is plotted of the work done on the particle W, against the speed of the particle v. If there are no frictional forces acting on the particle, the graph will look like

(a) 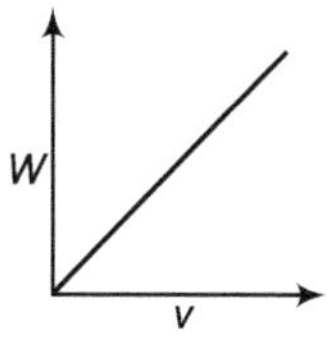

(b) 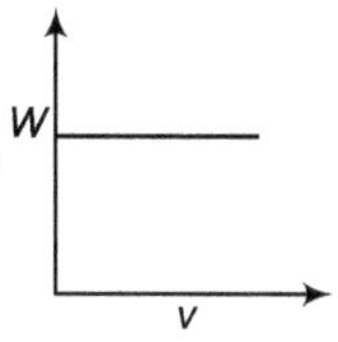

(c) 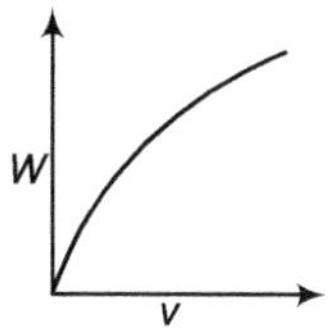

(d) 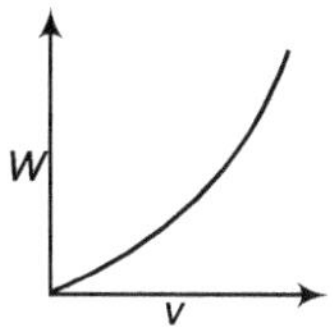

79. The energy required to accelerate a car from rest to 10 ms is W. The energy required to accelerate the car from $10\ \text{ms}^{-1}$ to $20\ \text{ms}^{-1}$ is

(a) W (b) $2W$
(c) $3W$ (d) $4W$

80. Given that the displacement of the body in metre is a function of time as follows

$$x = 2t^4 + 5$$

The mass of the body is 2 kg. What is the increase in its kinetic energy one second after the start of motion?

(a) 8 J (b) 16 J
(c) 32 J (d) 64 J

81. The momentum of a body is P and its kinetic energy is E. Its momentum becomes $2P$. Its kinetic energy will be

(a) $\frac{E}{2}$ (b) $3E$
(c) $2E$ (d) $4E$

82. A body moves from rest with a constant acceleration. Which one of the following graphs represents, the variation of its kinetic energy K with the distance travelled x?

(a) 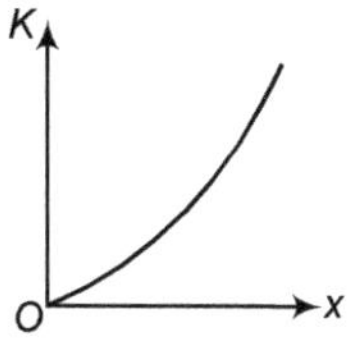

(b) 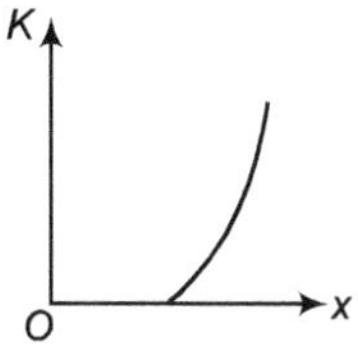

(c) 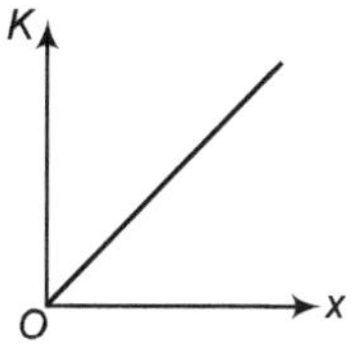

(d) 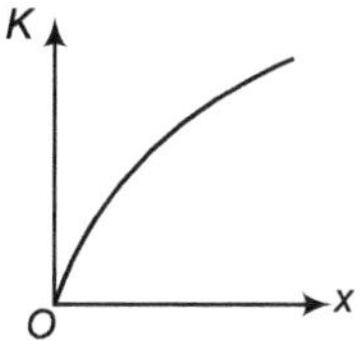

83. A body is attached to the lower end of a vertical spiral spring and it is gradually lowered to its equilibrium position. This stretches the spring by a length d. If the same body attached to the same spring is allowed to fall suddenly, what would be the maximum stretching in this case?

(a) d (b) $2d$
(c) $3d$ (d) $\frac{1}{2}d$

84. The pointer reading *versus* load graph for a spring balance is as shown

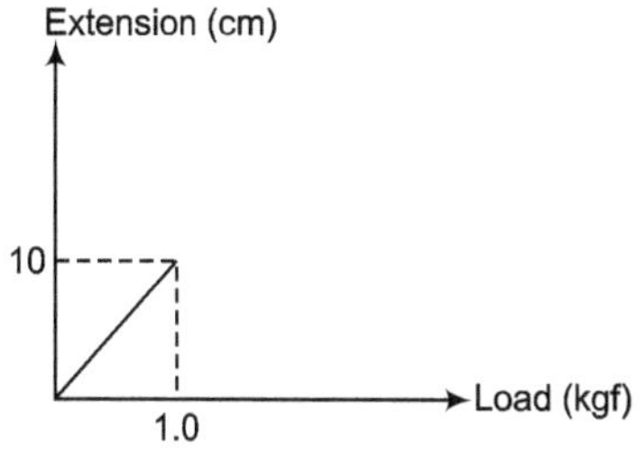

The spring constant is

(a) $\frac{15\ \text{kgf}}{\text{cm}}$ (b) $\frac{5\ \text{kgf}}{\text{cm}}$ (c) $\frac{0.1\ \text{kgf}}{\text{cm}}$ (d) $\frac{10\ \text{kgf}}{\text{cm}}$

85. v-t graph of an object of mass 1 kg is shown in figure. Select the wrong statement.

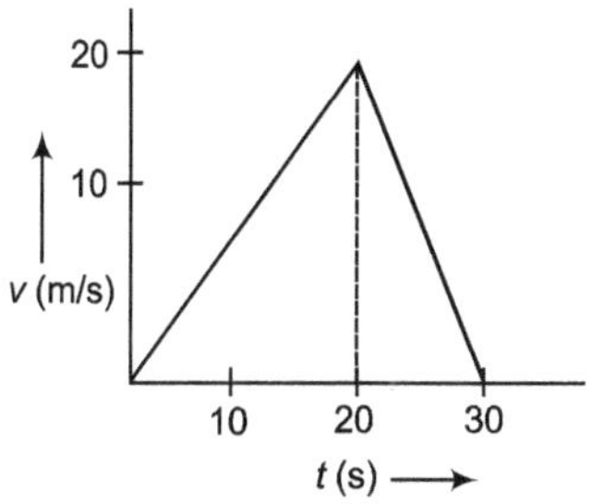

(a) Work done on the object in 30 s is zero
(b) The average acceleration of the object is zero
(c) The average velocity of the object is zero
(d) The average force on the object is zero

86. If v, p and E denote velocity, linear momentum and KE of the particle, then

(a) $p = \frac{dE}{dv}$ (b) $p = \frac{dE}{dt}$ (c) $p = \frac{dv}{dt}$ (d) $p = \frac{dE}{dv} \times \frac{dE}{dt}$

87. The force required to stretch a spring varies with the distance as shown in the figure . If the experiment is performed with the above spring of half the length, the line OA will

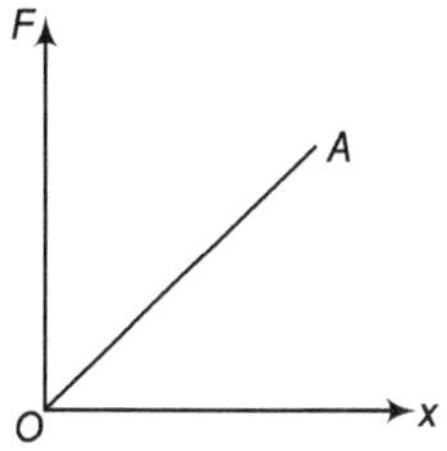

(a) shift towards F-axis (b) shift towards x-axis
(c) remain as it is (d) become double in length

88. Kinetic energy of a particle moving in a straight line varies with time t as $K = 4t^2$. The force acting on the particle
(a) is constant
(b) is increasing
(c) is decreasing
(d) first increases and then decreases

89. A car moving with a speed of 40 km/h can be stopped by applying brakes atleast 2m. If the car is moving with a speed of 80 km/h, the minimum stopping distance is
(a) 8 m
(b) 2 m
(c) 4 m
(d) 6 m

90. A ball is thrown vertically upwards with a velocity of 10 m/s. It returns to the ground with a velocity of 9 m/s. If $g = 9.8\text{ m/s}^2$, then the maximum height attained by the ball is nearly (assume air resistance to be uniform)
(a) 5.1 m (b) 4.1 m (c) 4.61 m (d) 5.0 m

91. An open knife edge of mass m is dropped from a height h on a wooden floor. If the blade penetrates upto the depth d into the wood, the average resistance offered by the wood to the knife edge is
(a) $mg\left(1+\frac{h}{d}\right)$ (b) $mg\left(1+\frac{h}{d}\right)^2$
(c) $mg\left(1-\frac{h}{d}\right)$ (d) $mg\left(1+\frac{d}{h}\right)$

[Level 2]

Only One Correct Option

1. A small body of mass m slides without friction from the top of a hemisphere of radius r. At what height will the body be detached from the centre of the hemisphere?

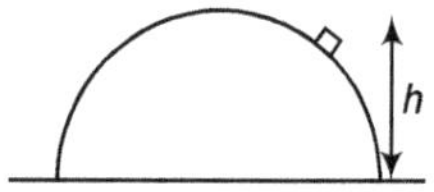

(a) $h = \frac{r}{2}$ (b) $h = \frac{r}{3}$
(c) $h = \frac{2r}{3}$ (d) $h = \frac{r}{4}$

2. A particle of mass 1 g executes an oscillatory motion on the concave surface of a spherical dish of radius 2 m placed on a horizontal plane. If the motion of the particle begins from a point on the dish at a height of 1 cm from the horizontal plane and the coefficient of friction is 0.01, the total distance covered by the particle before it comes to rest, is approximately
(a) 2.0 m (b) 10.0 m
(c) 1.0 m (d) 20.0 m

3. A mass-spring system oscillates such that the mass moves on a rough surface having coefficient of friction μ. It is compressed by a distance a from its normal length and on being released, it moves to a distance b from its equilibrium position. The decrease in amplitude for one half-cycle ($-a$ to b) is
(a) $\frac{\mu mg}{K}$ (b) $\frac{2\mu mg}{K}$ (c) $\frac{\mu g}{K}$ (d) $\frac{K}{\mu mg}$

4. A uniform flexible chain of mass m and length l hangs in equilibrium over a smooth horizontal pin of negligible diameter. One end of the chain is given a small vertical displacement so that the chain slips over the pin. The speed of chain when it leaves pin is
(a) $\sqrt{\frac{gl}{2}}$ (b) $\sqrt{gl}$
(c) $\sqrt{2gl}$ (d) $\sqrt{3gl}$

5. The potential energy of a particle of mass 1 kg is, $U = 10 + (x-2)^2$. Here, U is in joule and x in metre on the positive x-axis. Particle travels upto $x = +6$m. Choose the correct statement.
(a) On negative x-axis particle travels upto $x = -2$ m
(b) The maximum kinetic energy of the particle is 16 J
(c) Both (a) and (b) are correct
(d) Both (a) and (b) are wrong

6. A body is moving down an inclined plane of slope 37°. The coefficient of friction between the body and plane varies as $\mu = 0.3x$, where x is the distance travelled down the plane by the body. The body will have maximum speed. $\left(\sin 37° = \frac{3}{5}\right)$ $(g = 10\text{m/s}^2)$
(a) At $x = 1.16$m
(b) At $x = 2$m
(c) At bottom of plane
(d) At $x = 2.5$m

7. A force of $F = 0.5$ N is applied on lower block as shown in figure. The work done by lower block on upper block for a displacement of 3 m of the upper block with respect to ground is (Take $g = 10 \text{m/s}^2$)

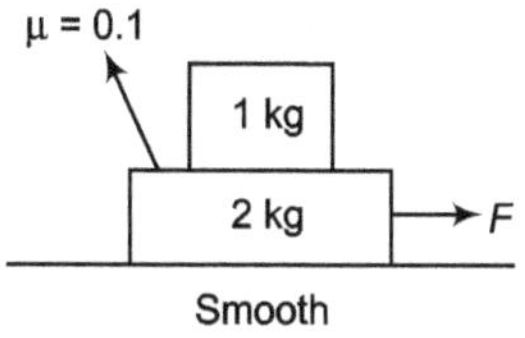

(a) -0.5 J (b) 0.5 J
(c) 2 J (d) -2 J

8. The potential energy between the atoms in a molecule is given by

$$U(x) = \frac{a}{x^{12}} - \frac{b}{x^6}$$

where, a and b positive constants and x is the distance between the atoms. The atom is in equilibrium when

(a) $x = 0$ (b) $x = \left(\frac{a}{2b}\right)^{\frac{1}{6}}$

(c) $x = \left(\frac{2a}{b}\right)^{\frac{1}{6}}$ (d) $x = \left(\frac{11a}{5b}\right)^{\frac{1}{6}}$

9. A bead of mass $\frac{1}{2}$ kg starts from rest from A to move in a vertical plane along a smooth fixed quarter ring of radius 5 m, under the action of a constant horizontal force $F = 5$ N as shown. The speed of bead as it reaches the point B is (Take $g = 10 \text{m/s}^2$)

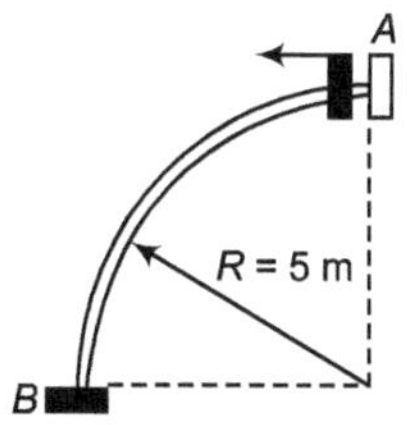

(a) 14.14 m/s (b) 7.07 m/s (c) 5 m/s (d) 25 m/s

10. A car of mass m is accelerating on a level smooth road under the action of a single force F. The power delivered to the car is constant and equal to P. If the velocity of the car at an instant is v, then after travelling how much distance it becomes double?

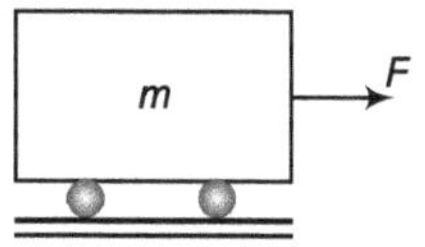

(a) $\frac{7mv^3}{3P}$ (b) $\frac{4mv^3}{3P}$

(c) $\frac{mv^3}{P}$ (d) $\frac{18mv^3}{7P}$

11. An ideal massless spring S can be compressed 1 m by a force of 100 N in equilibrium. The same spring is placed at the bottom of a frictionless inclined plane inclined at 30° to the horizontal. A 10 kg block M is released from rest at the top of the incline and is brought to rest momentarily after compressing the spring by 2 m. If $g = 10 \text{m/s}^2$, what is the speed of mass just before it touches the spring?

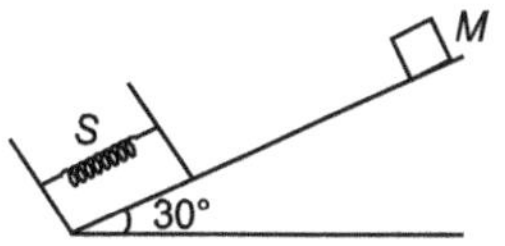

(a) $\sqrt{20}$ m/s (b) $\sqrt{30}$ m/s
(c) $\sqrt{10}$ m/s (d) $\sqrt{40}$ m/s

12. A pendulum of mass 1 kg and length $l = 1$ m is released from rest at angle $\theta = 60°$. The power delivered by all the forces acting on the bob at angle $\theta = 30°$ will be ($g = 10 \text{m/s}^2$).

(a) 13.4 W (b) 20.4 W
(c) 24.6 W (d) zero

13. A small block of mass m is kept on a rough inclined surface of inclination θ fixed in an elevator. The elevator goes up with a uniform velocity v and the block does not slide on the wedge. The work done by the force of friction on the block in a time t will be

(a) zero (b) $mgvt\cos^2\theta$

(c) $mgvt\sin^2\theta$ (d) $\frac{1}{2}mgvt\sin 2\theta$

14. In position A kinetic energy of a particle is 60 J and potential energy is – 20 J. In position B, kinetic energy is 100 J and potential energy is 40 J. Then, in moving the particle from A to B

(a) work done by conservative forces is – 60 J
(b) work done by external forces is 40 J
(c) net work done by all the forces is 40 J
(d) net work done by all the forces is 100 J

15. A block A of mass M rests on a wedge B of mass $2M$ and inclination θ. There is sufficient friction between A and B so that A does not slip on B. If there is no friction between B and ground, the compression in spring is

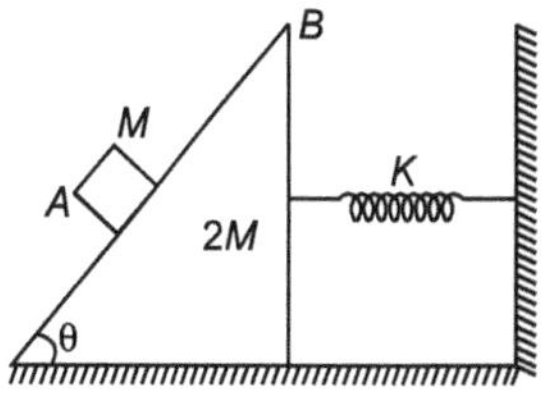

(a) $\frac{Mg\cos\theta}{K}$ (b) $\frac{Mg\cos\theta\sin\theta}{K}$

(c) $\frac{Mg\sin\theta}{K}$ (d) zero

16. A smooth chain AB of mass m rests against a surface in the form of a quarter of a circle of radius R. If it is released from rest, the velocity of the chain after it comes over the horizontal part of the surface is

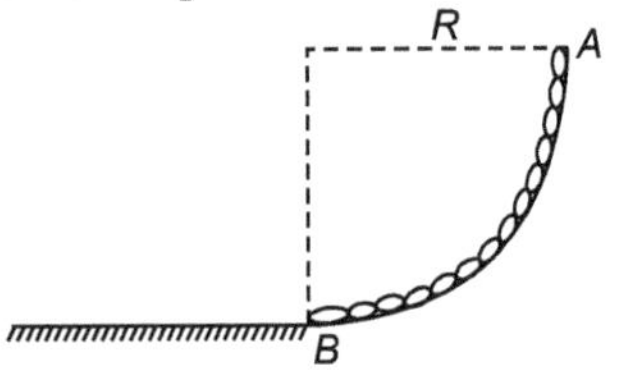

(a) $\sqrt{2gR}$ (b) $\sqrt{gR}$

(c) $\sqrt{2gR\left(1-\frac{2}{\pi}\right)}$ (d) $\sqrt{2gR(2-\pi)}$

17. Two inclined frictionless tracks, one gradual and the other steep meet at A from where two stones are allowed to slide down from rest, one on each track as shown in figure.

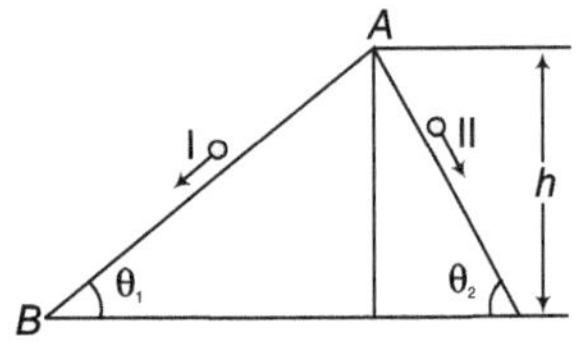

Which of the following statement is correct?

(a) Both the stones reach the bottom at the same time but not with the same speed
(b) Both the stones reach the bottom with the same speed and stone I reaches the bottom earlier than stone II
(c) Both the stones reach the bottom with the same speed and stone II reaches the bottom earlier than stone I
(d) Both the stones reach the bottom at different times and with different speeds

More than One Correct Options

1. The potential energy of a particle of mass 5 kg moving in x-y plane is given as $U=(7x+24y)$ J, x and y being in metre. Initially at $t=0$, the particle is at the origin (0, 0) moving with a velocity of $(8.6\hat{i}+23.2\hat{j})\ \text{ms}^{-1}$. Then,

(a) the velocity of the particle at $t=4$ s, is $5\ \text{ms}^{-1}$
(b) the acceleration of the particle is $5\ \text{ms}^{-2}$
(c) the direction of motion of the particle initially (at $t=0$) is at right angles to the direction of acceleration
(d) the path of the particle is circle

2. The potential energy of a particle is given by formula $U=100-5x+100x^2$, U and x are in SI units. If mass of the particle is 0.1 kg, then the magnitude of its acceleration

(a) at 0.05 m from the origin is $50\ \text{ms}^{-2}$
(b) at 0.05 m from the mean position is $100\ \text{ms}^{-2}$
(c) at 0.05 m from the origin is $150\ \text{ms}^{-2}$
(d) at 0.05 m from the mean position is $200\ \text{ms}^{-2}$

3. One end of a light spring of spring constant k is fixed to a wall and the other end is tied to a block placed on a smooth horizontal surface. In a displacement, the work done by the spring is $+\left(\frac{1}{2}\right)kx^2$. The possible cases are

(a) the spring was initially compressed by a distance x and was finally in its natural length
(b) it was initially stretched by a distance x and finally was in its natural length
(c) it was initially in its natural length and finally in a compressed position
(d) it was initially in its natural length and finally in a stretched position

4. Identify the correct statement about work energy theorem.

(a) Work done by all the conservative forces is equal to the decrease in potential energy
(b) Work done by all the forces except the conservative forces is equal to the change in mechanical energy
(c) Work done by all the forces is equal to the change in kinetic energy
(d) Work done by all the forces is equal to the change in potential energy

5. A disc of mass $3m$ and a disc of mass m are connected by a massless spring of stiffness k. The heavier disc is placed on the ground with the spring vertical and lighter disc on top. From its equilibrium position the upper disc is pushed down by a distance δ and released. Then,

(a) if $\delta > \frac{3\,mg}{k}$, the lower disc will bounce up
(b) if $\delta = \frac{2\,mg}{k}$, maximum normal reaction from ground on lower disc $= 6\,mg$
(c) if $\delta = \frac{2\,mg}{k}$, maximum normal reaction from ground on lower disc $= 4\,mg$
(d) if $\delta > \frac{4\,mg}{k}$, the lower disc will bounce up

6. In the adjoining figure block A is of mass m and block B is of mass 2 m. The spring has force constant k. All the surfaces are smooth and the system is released from rest with spring unstretched.

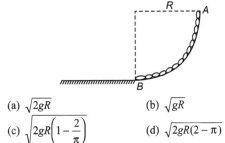

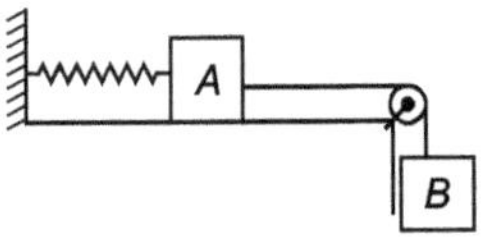

(a) The maximum extension of the spring is $\frac{4mg}{k}$
(b) The speed of block A, when the extension in spring is $\frac{2mg}{k}$, is $2g\sqrt{\frac{2m}{3k}}$
(c) Net acceleration of block B, when the extension in the spring is maximum, is $\frac{2}{3}g$
(d) Tension in the thread for extension of $\frac{2mg}{k}$ in spring is mg

7. If kinetic energy of a body is increasing, then

(a) work done by conservative forces must be positive
(b) work done by conservative forces may be positive
(c) work done by conservative forces may be zero
(d) work done by non-conservative forces may be zero

8. At two positions kinetic energy and potential energy of a particle are

$$K_1 = 10\,J:$$
$$U_1 = -20\,J,$$
$$K_2 = 20\ J,$$
$$U_2 = -10\,J$$

In moving from 1 to 2

(a) work done by conservative forces is positive
(b) work done by conservative forces is negative
(c) work done by all the forces is positive
(d) work done by all the forces is negative

9. Block A has no relative motion with respect to wedge fixed to the lift as shown in figure during motion-1 or motion-2

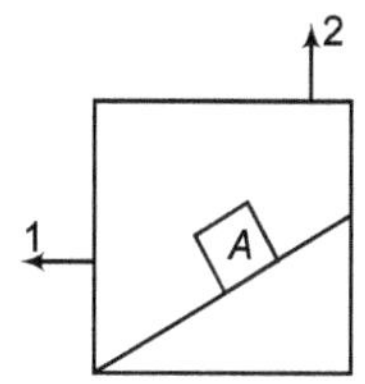

(a) work done by gravity on block A in motion-2 is less than in motion-1
(b) work done by normal reaction on block A in both the motions will be positive
(c) work done by force of friction in motion-1 may be positive
(d) work done by force of friction in motion-1 may be negative

10. A man of mass m, standing at the bottom of the stair case, of height L climbs it and stands at its top.

(a) Work done by all forces on man is equal to the rise in potential energy mgL
(b) Work done by all forces on man is zero
(c) Work done by the gravitational force on man is mgL
(d) The reaction force from a step does not do work, because the point of application of the force does not move while the force exists

Comprehension Based Questions

Passage (Q. 1 to 2)

The figure shows the variation of potential energy of a particle as a function of x, the x-coordinate of the region. It has been assumed that potential energy depends only on x. For all other values of x, U is zero, i.e. for $x < -10$ and $x > 15, U = 0$.

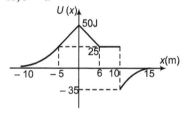

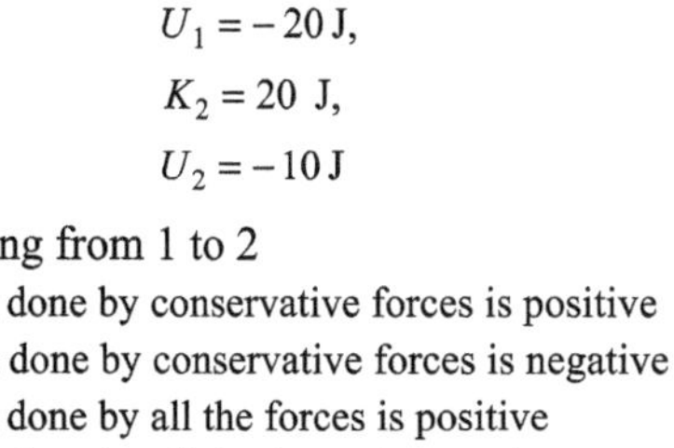

Based on above information answer the following questions.

1. If total mechanical energy of the particle is 25 J, then it can be found in the region

(a) $-10 < x < -5$ and $6 < x < 15$
(b) $-10 < x < 0$ and $6 < x < 10$
(c) $-5 < x < 6$
(d) $-10 < x < 10$

2. If total mechanical energy of the particle is -40 J, then it can be found in region

(a) $x < -10$ and $x > 15$
(b) $-10 < x < -5$ and $6 < x < 15$
(c) $10 < x < 15$
(d) It is not possible

Assertion and Reason

Directions (Q. Nos. 1-16) *These questions consists of two statements each printed as Assertion and Reason. While answering these questions you are required to choose any one of the following five responses.*

(a) If both Assertion and Reason are correct and Reason is the correct explanation of Assertion
(b) If both Assertion and Reason are true but Reason is not the correct explanation of Assertion
(c) If Assertion is true but Reason is false
(d) If Assertion is false but Reason is true
(e) If both Assertion and Reason are false

1. Assertion Consider a person of mass 80 kg who is climbing a ladder. In climbing up a vertical distance of 5 m, the contact force exerted by ladder on person's feet does 4000 J of work. (Consider $g = 10\,m/s^2$)

Reason Work done by a force F is defined as the dot product of force with the displacement of point of application of force.

2. Assertion The work done in bringing a body down from the top to the base along a frictionless inclined plane is the same as the work done in bringing it down from the vertical side.

Reason The gravitational force on the body along the inclined plane is the same as that along the vertical side.

3. **Assertion** An object A is dropped from the top of an incline at $t = 0$, as shown in figure. It will fall under gravity as indicated by the arrow. At the same time, i.e. $t = 0$, another object B begins to slide down the frictionless incline.

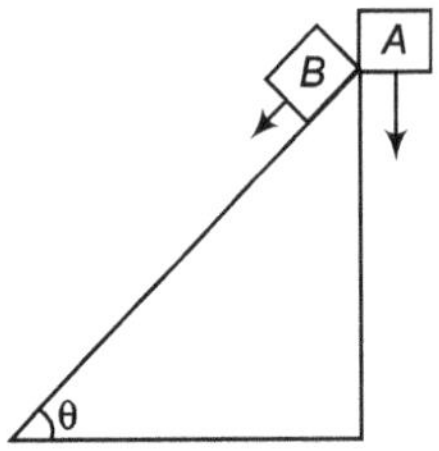

The two objects during their motion to the ground level will be travelling at equal speeds.

Reason Net force on both the objects during their motion is same.

4. **Assertion** In projectile motion, the rate of change in magnitude of potential energy of a particle first decreases and then increases during motion.

Reason In projectile motion, the rate of change in linear momentum of a particle remains constant during motion.

5. **Assertion** At any instant the magnitude of rate of change of potential energy of the projectile of mass 1 kg is numerically equal to magnitude of $\mathbf{a} \cdot \mathbf{v}$ (where, $\mathbf{a}$ is acceleration due to gravity and $\mathbf{v}$ is velocity at that instant).

Reason The graph representing power delivered by the gravitational force acting on the projectile with time will be straight line with negative slope.

6. **Assertion** A block of mass m starts moving on a rough horizontal surface with a velocity v. It stops due to friction between the block and the surface after moving through a certain distance. The surface is now tilted to an angle of 30° with the horizontal and the same block is made to go up on the surface with the same initial velocity v. The decrease in the mechanical energy in the second situation is smaller than that in the first situation.

Reason The coefficient of friction between the block and the surface decreases with the increase in the angle of inclination.

7. **Assertion** Surface between the blocks A and B is rough work done by friction on block B is always negative.

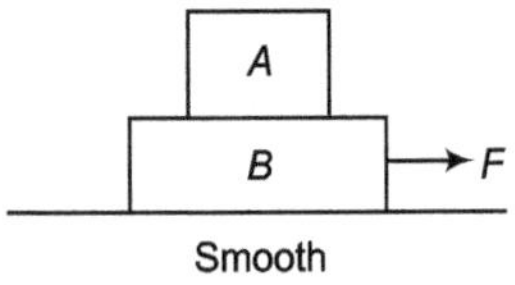

Reason Total work done by friction in both the blocks is always zero.

8. **Assertion** At stable equilibrium position of a body kinetic energy can't be zero. Because it is maximum.

Reason During oscillations of a body potential energy is maximum at stable equilibrium position.

9. **Assertion** A body is moving along x-axis. Force $F = -2x^2$ is acting on it work done by this force in moving the body from $x = -2$ to $x = +2$ is zero.

Reason From $x = -2$ to $x = 0$, work done is negative and from $x = 0$ to $x = +2$ work done is negative.

10. **Assertion** Spring force is a conservative force.

Reason Potential energy is defined only for conservative forces.

11. **Assertion** If work done by conservative force is negative, then potential energy associated with that force should increase.

Reason This is from the reaction

$$\Delta U = -W$$

Here, ΔU is change in potential energy and W is work done by conservative force.

12. **Assertion** Work done by constant force is equal to magnitude of force multiplied by displacement.

Reason Work done is scalar quantity. It may be positive, negative or zero.

13. **Assertion** Velocity of a block changes from $(2\hat{\mathbf{i}} + 3\hat{\mathbf{j}})$ m/s to $(-4\hat{\mathbf{i}} - 6\hat{\mathbf{j}})$ m/s. Then, work done by all the forces during this interval of time is positive.

Reason Speed of block is increasing

14. **Assertion** If a force is applied on a rigid body. Body is in motion but point of application of force is stationary. Then, work done by the force is zero.

Reason If body is moving, then point of application of force should also move.

15. **Assertion** Force applied on a block moving in one dimension is producing a constant power, then the motion should be uniformly accelerated.

Reason This constant power multiplied with time is equal to the change in kinetic energy.

16. **Assertion** Total work done by spring for may be positive, negative or zero.

Reason Direction of spring force is always towards mean position.

Match the Columns

1. Match the following columns.

Column I	Column II
(A) Work done by all forces	(p) Change in potential energy
(B) Work done by conservative forces	(q) Change in kinetic energy
(C) Work done by external forces	(r) Change in mechanical energy
	(s) None

2. A force $F = kx$ (where, k is a positive constant) is acting on a particle. Work done

Column I	Column II
(A) In displacement from $x = 2$ to $x = 4$	(p) Negative
(B) In displacing the body from $x = -4$ to $x = -2$	(q) Positive
(C) In displacing the body from $x = -2$ to $x = +2$	(r) zero

3. F-x and corresponding U-x graph are as shown in figures. Three points A, B and C in F-x graph may be corresponding to P, Q and R in the U-x graph. Match the following.

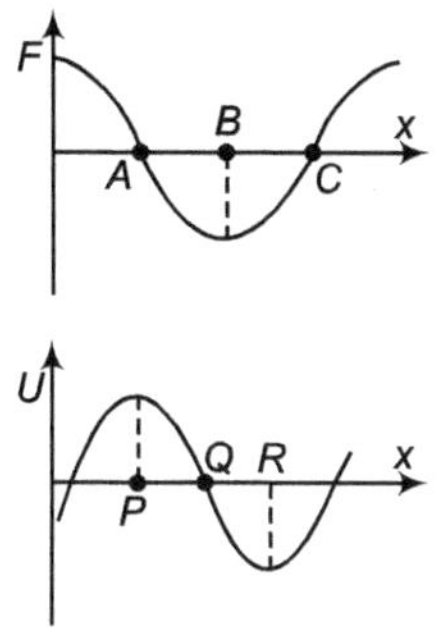

Column I		Column II	
(A)	A	(p)	P
(B)	B	(q)	Q
(C)	C	(r)	R
		(s)	None

4. A block of mass m is stationary with respect to a rough wedge as shown in figure. Starting from rest in time t, ($m = 1\text{kg}, \theta = 30°, a = 2\text{m/s}^2, t = 4\text{s}$) work done on block

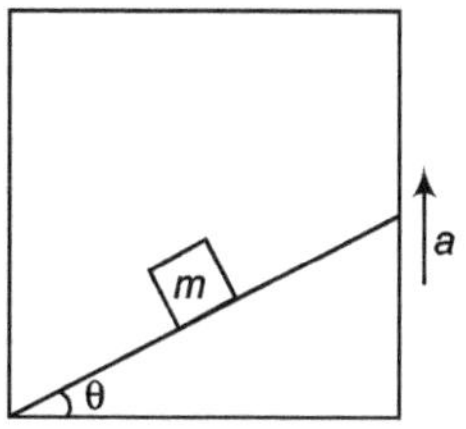

Column I	Column II
(A) By gravity	(p) 144 J
(B) By normal reaction	(q) 32 J
(C) By friction	(r) 56 J
(D) By all the forces	(s) 48 J
	(t) None

5. Acceleration *versus* x and potential energy *versus* x graph of a particle moving along x-axis is as shown in figure . Mass of the particle is 1 kg and velocity at $x = 0$ is 4 m/s. Match the following at $x = 8$ m

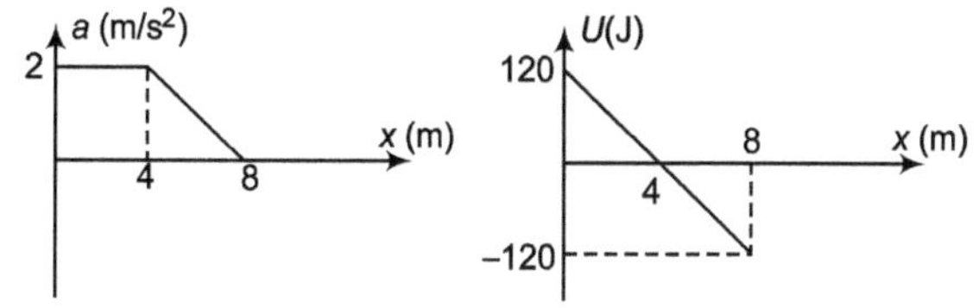

Column I	Column II
(A) Kinetic energy	(p) 120 J
(B) Work done by conservative forces	(q) 240 J
(C) Total work done	(r) 112 J
(D) Work done by external forces	(s) None

6. A body is moved along a straight line by a machine delivering a power proportional to time ($P \propto t$). Then, match the following.

Column I	Column II
(A) Velocity is proportional to	(p) t
(B) Displacement is proportional to	(q) t^2
(C) Work done is proportional to	(r) t^3

7. Match the following.

Column I	Column II
(A) Electrostatic potential energy	(p) Positive
(B) Gravitational potential energy	(q) Negative
(C) Elastic potential energy	(r) Zero
(D) Magnetic potential energy	(s) Not defined

Entrance Gallery

2014

1. Consider an elliptically shaped rail PQ in the vertical plane with $OP = 3$ m and $OQ = 4$ m. A block of mass 1 kg is pulled along the rail from P to Q with a force of 18 N, which is always parallel to line PQ (see figure). Assuming no frictional losses, the kinetic energy of the block when it reaches Q is $(n \times 10)$ J. The value of n is (take, acceleration due to gravity $= 10\text{ ms}^{-2}$)

[JEE Advanced]

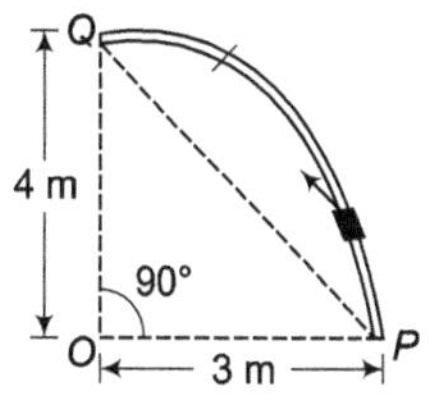

(a) 50 J
(b) 100 J
(c) 5×10^2
(d) 200 J

2. When a rubber band is stretched by a distance x, it exerts a restoring force of magnitude $F = ax + bx^2$, where a and b are constants. The work done in stretching the unstretched rubber band by L is **[JEE Main]**

(a) $aL^2 + bL^3$ (b) $\frac{1}{2}(aL^2 + bL^3)$

(c) $\frac{aL^2}{2} + \frac{bL^3}{3}$ (d) $\frac{1}{2}\left(\frac{aL^2}{2} + \frac{bL^3}{3}\right)$

3. A block of mass 0.18 kg is attached to a spring of force constant 2 N/m. The coefficient of friction between the block and the floor is 0.1. Initially the block is at rest and the spring is unstretched. An impulse is given to the block. The block sides a distance of 0.66 m and comes to rest for the first time the initial velocity of the blook in m/s is $v = N/10$, then N is **[J & K CET]**

(a) 2 (b) 3
(c) 4 (d) 6

2012

4. A spring stores 1 J of energy for a compression of 1 mm. The additional work to be done to compress it further by 1 mm is **[Kerala CEE]**

(a) 1 J (b) 2 J
(c) 3 J (d) 4 J
(e) 0.5 J

5. Two bodies of masses m_1 and m_2 are acted upon by constant force F for a time t. They start from rest and acquire kinetic energies E_1 and E_2, respectively. Then, $\frac{E_1}{E_2}$ is **[Karnataka CET]**

(a) $\frac{\sqrt{m_1 m_2}}{m_1 + m_2}$ (b) $\frac{m_1}{m_2}$

(c) $\frac{m_2}{m_1}$ (d) 1

6. Force of 50 N acting on a body at an angle θ with horizontal. If 150 J work is done by displacing it 3 m, then θ is **[OJEE]**

(a) 60° (b) 30°
(c) 0° (d) 45°

2011

7. At time $t = 0$ s particle starts moving along the x-axis. If its kinetic energy increases uniformly with time t, the net force acting on it must be proportional to **[AIEEE]**

(a) $\sqrt{t}$ (b) constant (c) t (d) $\frac{1}{\sqrt{t}}$

8. A ball dropped from a height of 2 m rebounds to a height of 1.5 m after hitting the ground. Then, the percentage of energy lost is **[Karnataka CET]**

(a) 25 (b) 30 (c) 50 (d) 100

9. A body of mass 5 kg is thrown vertically up with a kinetic energy of 490 J. The height at which the kinetic energy of the body becomes half of the original value is **[Karnataka CET]**

(a) 12.5 m (b) 10 m
(c) 2.5 m (d) 5 m

10. A particle is projected from the ground with a kinetic energy E at an angle of 60° with the horizontal. Its kinetic energy at the highest point of its motion will be **[WB JEE]**

(a) $E/\sqrt{2}$ (b) $E/2$
(c) $E/4$ (d) $E/8$

11. A cubical vessel of height 1 m is full of water. What is the amount of work done in pumping water out of the vessel? (take, $g = 10\,\text{ms}^{-2}$) **[WB JEE]**

(a) 1250 J (b) 5000 J
(c) 1000 J (d) 2500 J

12. A body of mass 6 kg is acted upon by a force which causes a displacement in it given by, $x = (t^2/4)$ m, where t is the time in second. The work done by the force in 2 s is **[WB JEE]**

(a) 12 J (b) 9 J
(c) 6 J (d) 3 J

13. A box is moved along a straight line by a machine delivering constant power. The distance moved by the body in time t is proportional to **[WB JEE]**

(a) $t^{1/2}$ (b) $t^{3/4}$ (c) $t^{3/2}$ (d) t^2

2010

14. A light inextensible string that goes over a smooth fixed pulley as shown in the figure connects two blocks of masses 0.36 kg and 0.72 kg. Taking, $g = 10\,\text{ms}^{-2}$, find the work done (in joule) by string on the block of mass 0.36 kg during the first second after the system is released from rest. **[IIT JEE]**

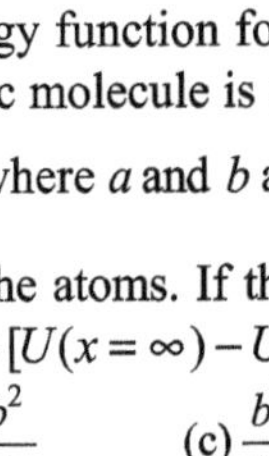

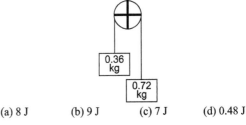

(a) 8 J (b) 9 J (c) 7 J (d) 0.48 J

15. The potential energy function for the force between two atoms in a diatomic molecule is approximately given by, $U(x) = \frac{a}{x^{12}} - \frac{b}{x^6}$, where a and b are constants and x is the distance between the atoms. If the dissociation energy of the molecule is $D = [U(x = \infty) - U_{\text{at equilibrium}}]$, D is **[AIEEE]**

(a) $\frac{b^2}{2a}$ (b) $\frac{b^2}{12a}$ (c) $\frac{b^2}{4a}$ (d) $\frac{b^2}{6a}$

16. A particle acted upon by constant forces $4\hat{i}+\hat{j}-3\hat{k}$ and $3\hat{i}+\hat{j}-\hat{k}$ is displaced from the point $\hat{i}+2\hat{j}+3\hat{k}$ to the point $5\hat{i}+4\hat{j}+\hat{k}$. The total work done by the forces in SI unit is **[Kerala CEE]**

(a) 20
(b) 40
(c) 50
(d) 30
(e) 35

17. The work done by a force acting on a body is as shown in the graph. The total work done in covering an initial distance of 20 m is **[Karnataka CET]**

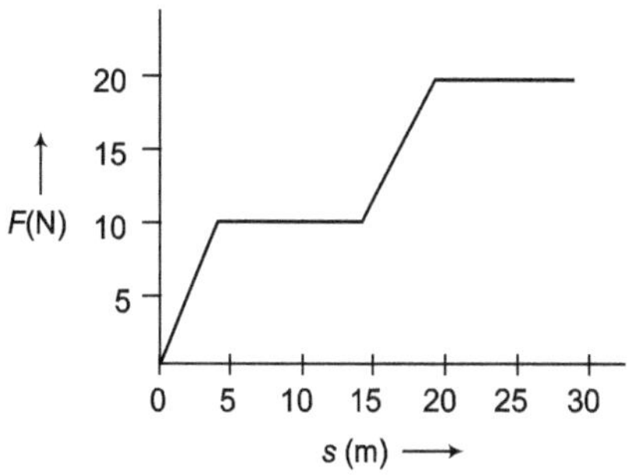

(a) 225 J (b) 200 J
(c) 400 J (d) 175 J

Answers

Level 1

Objective Problems

1. (b)	**2.** (d)	**3.** (b)	**4.** (b)	**5.** (c)	**6.** (d)	**7.** (d)	**8.** (d)	**9.** (c)	**10.** (a)
11. (d)	**12.** (c)	**13.** (b,c)	**14.** (a)	**15.** (b)	**16.** (a)	**17.** (a)	**18.** (c)	**19.** (c)	**20.** (c)
21. (a)	**22.** (d)	**23.** (c)	**24.** (b)	**25.** (b)	**26.** (b)	**27.** (c)	**28.** (b)	**29.** (b)	**30.** (c)
31. (b)	**32.** (d)	**33.** (b)	**34.** (a)	**35.** (c)	**36.** (c)	**37.** (b)	**38.** (c)	**39.** (b)	**40.** (a)
41. (b)	**42.** (a)	**43.** (c)	**44.** (d)	**45.** (b)	**46.** (c)	**47.** (b)	**48.** (b)	**49.** (a)	**50.** (c)
51. (c)	**52.** (c)	**53.** (c)	**54.** (b)	**55.** (c)	**56.** (c)	**57.** (b)	**58.** (c)	**59.** (b)	**60.** (d)
61. (a)	**62.** (c)	**63.** (b)	**64.** (b)	**65.** (a,b,c)	**66.** (b)	**67.** (a)	**68.** (a)	**69.** (a)	**70.** (c)
71. (b)	**72.** (c)	**73.** (a)	**74.** (c)	**75.** (d)	**76.** (d)	**77.** (c)	**78.** (d)	**79.** (c)	**80.** (d)
81. (d)	**82.** (c)	**83.** (b)	**84.** (c)	**85.** (c)	**86.** (a)	**87.** (a)	**88.** (a)	**89.** (a)	**90.** (c)
91. (a)									

Level 2

Only One Correct Option

1. (c)	**2.** (c)	**3.** (b)	**4.** (a)	**5.** (c)	**6.** (d)	**7.** (b)	**8.** (c)	**9.** (a)	**10.** (a)
11. (a)	**12.** (a)	**13.** (c)	**14.** (a,c)	**15.** (d)	**16.** (c)	**17.** (c)			

More than One Correct Options

1. (a,b)	**2.** (a,b,c)	**3.** (a,b)	**4.** (b,c)	**5.** (b,d)	**6.** (a)	**7.** (b,c,d)	**8.** (b,c)	**9.** (all)	**10.** (b,d)

Comprehension Based Questions

1. (a) **2.** (d)

Assertion and Reason

1. (d)	**2.** (b)	**3.** (c)	**4.** (b)	**5.** (c)	**6.** (c)	**7.** (c)	**8.** (d)	**9.** (e)	**10.** (a)
11. (a)	**12.** (d)	**13.** (a,b)	**14.** (c)	**15.** (d)	**16.** (b)				

Match the Columns

1. $A \to q, B \to s, C \to r$
2. $A \to q, B \to p, C \to r$
3. $A \to r, B \to s, C \to p$
4. $A \to t, B \to p, C \to s, D \to q$
5. $A \to r, B \to q, C \to p, D \to t$
6. $A \to p, B \to q, D \to q$
7. $A \to pqr, B \to qr, C \to pr, D \to pqr$

Entrance Gallery

1. (a)	**2.** (c)	**3.** (c)	**4.** (c)	**5.** (c)	**6.** (c)	**7.** (d)	**8.** (a)	**9.** (d)	**10.** (c)
11. (b)	**12.** (d)	**13.** (c)	**14.** (a)	**15.** (c)	**16.** (b)	**17.** (b)			

Solutions

Level 1 : Objective Problems

1. $W = \int_0^{x_1} cx \cdot dx = \frac{cx_1^2}{2}$

2. $W = \int_0^5 F dx = \int_0^5 (7 - 2x + 3x^2) dx$

$= [7x - x^2 + x^3]_0^5 = 135$ units

3. $W = (y\text{-component of force}) \times (\text{displacement along } y\text{-axis})$

$= 15 \times 10 = 150\,\text{J}$

4. $W = \mathbf{F} \cdot \mathbf{S} = \mathbf{F} \cdot (\mathbf{r}_2 - \mathbf{r}_1)$

$= (\hat{\mathbf{i}} + 2\hat{\mathbf{j}} + 3\hat{\mathbf{k}}) \cdot [(\hat{\mathbf{i}} - \hat{\mathbf{j}} + 2\hat{\mathbf{k}}) - (\hat{\mathbf{i}} + \hat{\mathbf{j}} + \hat{\mathbf{k}})]$

$= (\hat{\mathbf{i}} + 2\hat{\mathbf{j}} + 3\hat{\mathbf{k}}) \cdot (-2\hat{\mathbf{j}} + \hat{\mathbf{k}})$

$= -4 + 3 = -1\,\text{J}$

5. $Mg - T = \frac{Mg}{2}$ or $T = \frac{Mg}{2}$

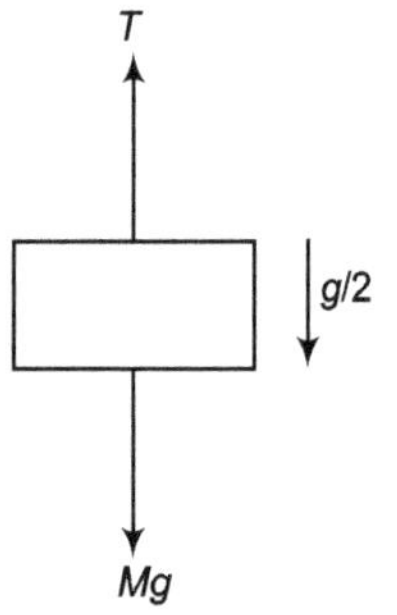

or magnitude of work done $= T \times x = \frac{Mgx}{2}$

7. Work done by spring force from A to B is negative. But work done by this force from B to A will be positive.

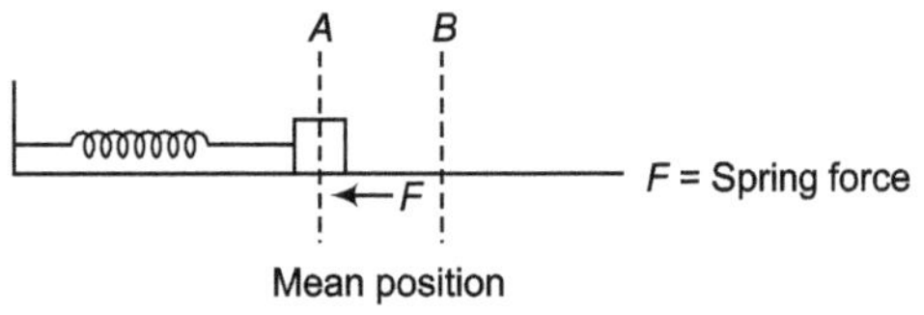

8. $W = \mathbf{F} \cdot \mathbf{S} = (3\hat{\mathbf{i}} + 4\hat{\mathbf{j}}) \cdot (3\hat{\mathbf{i}} + 4\hat{\mathbf{j}}) = 9 + 16 = 25\text{J}$

9. Ratio of displacement is

$\frac{1}{2}g(1)^2 \frac{1}{2}(g)(2)^2 - \frac{1}{2}g(1)^2 : \frac{1}{2}g(3)^2 - \frac{1}{2}g(2)^2$

or $\quad 1 : 3 : 5.$

Therefore, ratio of work done will also be $1 : 3 : 5$.

As $\quad W = mgs$

or $\quad W \propto s$

10. Change in kinetic energy = work done

= Area under F-x graph

$\therefore \quad \frac{1}{2} \times 5 \times v^2 = 10 \times 25 + \frac{1}{2} \times 25 \times 10$

$= 375$

$\therefore \quad v = 12.2\,\text{m/s}$

11. W = area under F-x graph.

12. $W = \int_0^5 F \cdot dx = \int_0^5 (7 - 2x + 3x^2) dx = 135\text{J}$

14. $\mathbf{r} = \mathbf{r}_B - \mathbf{r}_A = (\hat{\mathbf{i}} - \hat{\mathbf{j}} + \hat{\mathbf{k}})\text{m}$

$W = \mathbf{F} \cdot \mathbf{r} = 2 + 4 = 6\text{J}$

15. $W = \int_{-a}^{+a} F dy = \left[\frac{Ay^3}{3} + \frac{By^2}{2} + Cy\right]_{-a}^{+a} = \frac{2Aa^3}{3} + 2Ca$

16. Line and force should be perpendicular, as the work done is given to be zero.

$m_1 = \frac{F_y}{F_x} = \frac{3}{2} =$ slope of force

$m_2 = -\frac{k}{3} =$ slope of line

$m_1 m_2 = -1$

$\therefore \quad k = 2$

17. $K_f - K_i = W$

$\therefore \quad K_f = K_i + W$

$= \frac{1}{2} \times 10 \times (10)^2 + \int_{20}^{30} (-0.1\, x) dx$

$= 475\,\text{J}$

18. $v = \frac{dx}{dt} = t^2$

At $t = 0, v = 0,$

At $t = 2\text{s}, v = 4$ m/s

From work energy theorem,

W = change in kinetic energy

$= K_f - K_i$

$= \frac{1}{2} m(v_f^2 - v_i^2)$

$= \frac{1}{2} \times 2 \times (16 - 0)$

$= 16\text{J}$

19. $K_f = 2K_i$

or $\quad \frac{1}{2} m(v + 2)^2 = \left(\frac{1}{2} mv^2\right)(2)$

or $\quad v + 2 = \sqrt{2} v$

$v = \frac{2}{\sqrt{2} - 1} = 2(\sqrt{2} + 1)\,\text{m/s}$

21. When a body is moved upwards, work done by gravity is negative but potential energy increases.

22. $U = \frac{1}{2} Kx^2$ or $U \propto x^2$

Stretch is increased to 5 times. Therefore, stored potential energy will be increased by 25 times.

23. $P = \frac{W}{t}$

$\therefore \quad t = \frac{W}{P} = \frac{mgh}{P}$

$= \frac{200 \times 10 \times 40}{10 \times 10^3} = 8\,\text{s}$

24. Work done by all forces = change in kinetic energy

$$= \frac{1}{2}m(v_f^2 - v_i^2)$$

$$= \frac{1}{2} \times 2(0 - 400) = -400\text{J}$$

25. $P = F \cdot v$

$$\therefore \quad F = \frac{P}{v} = \frac{7500}{20} = 375\text{N}$$

26. Work done = change in kinetic energy.

$$\therefore \quad \int_{(0,0)}^{(1,1)} (4\hat{\mathbf{i}} + 4\hat{\mathbf{j}}) \cdot (dx\hat{\mathbf{i}} + dy\hat{\mathbf{j}}) = \frac{1}{2} \times 1 \times (v_f^2 - 2^2)$$

or $$8 = \frac{1}{2}(v_f^2 - 4)$$

or $$v_f = 4.5\text{m/s}$$

27. $P = \mathbf{F} \cdot \mathbf{v} = (10\hat{\mathbf{i}} + 10\hat{\mathbf{j}} + 20\hat{\mathbf{k}}) \cdot (5\hat{\mathbf{i}} - 3\hat{\mathbf{j}} + 6\hat{\mathbf{k}})$

$$= 50 - 30 + 120$$

$$= 140\text{J/s}$$

28. $\Delta\text{KE} = W = \int_2^4 P dt = \int_2^4 (3t^2 - 2t + 1)\, dt$

$$= [t^3 - t^2 + t]_2^4$$

$$= 46\text{J}$$

29. Work done by conservative force $= -\Delta U$

30. $$W = P \times t \qquad (P = \text{constant})$$

or $$\frac{1}{2}mv^2 = Pt \qquad \text{(From work-energy theorem)}$$

$$\therefore \quad v \propto t^{1/2}$$

Integrating, we get $s \propto t^{3/2}$.

32. $W = \Delta\text{KE} = \frac{1}{2}m(v_f^2 - v_i^2)$

$$= \frac{1}{2} \times 0.01[(64 + 400) - (16 + 256)]$$

$$= 0.96\text{J}$$

33. $\frac{1}{2}mv_1^2 = \frac{1}{2}\left[\frac{1}{2} \times \frac{m}{2} \times v_2^2\right]$

v_1 = speed of man and v_2 = speed of boy

Now, $$\frac{1}{2}m(v_1 + 1)^2 = \frac{1}{2} \times \frac{m}{2} \times v_2^2$$

Solving these two equations, we get

$$v_1 = (\sqrt{2} + 1)\,\text{m/s and } v_2 = 2(\sqrt{2} + 1)\,\text{m/s}$$

34. F = constant

$\therefore a$ = constant or $v = at$

Now, $P = F \cdot v = F \cdot at$ or $P \propto t$ (as F and a both are constant)

Hence, P-t graph is a straight line passing through origin.

35. Two forces are acting on bob, tension and weight. Power of tension will be zero and that of weight is,

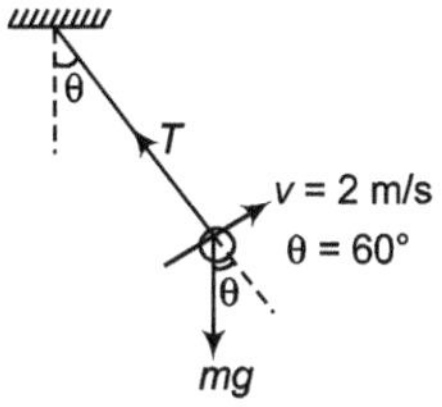

$$P = mgv\cos(90° + \theta)$$

$$= -mgv\sin 60°$$

$$= -50 \times 9.8 \times 2 \times \frac{\sqrt{3}}{2} = -490\sqrt{3}\,\text{W}$$

$\therefore$ Power delivered is $490\sqrt{3}$ W

36. $W = \int P dt$ or $\frac{1}{2}mv^2 = \int P dt = \int_0^2 \left(\frac{3t^2}{2}\right) dt = 4\text{J}$

$$\therefore \quad v = \sqrt{\frac{2 \times 4}{m}} = 2\text{m/s}$$

37. Compare with spring force,

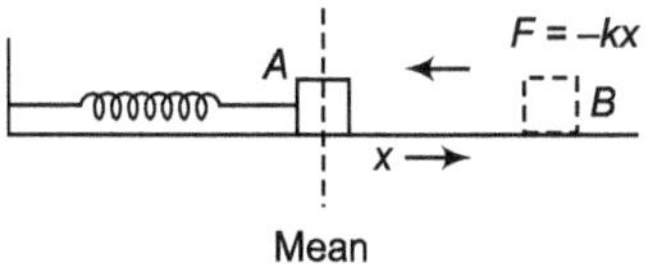

From A to B,

Loss in KE = gain in elastic PE

$$= \frac{1}{2}kx^2 \propto x^2$$

38. $a = \frac{F - W}{m} = \frac{170 - 50}{5} = 24\,\text{m/s}^2$

$$v = \sqrt{2as} = \sqrt{2 \times 24 \times 10}$$

$$= 21.9\text{ m/s}$$

$$= 22\text{ m/s}$$

39. $mgh = K$

Now, $$\frac{3}{4}K + mgh' = K \qquad \text{(at height } h')$$

$$\therefore \quad \frac{3}{4}mgh + mgh' = mgh$$

or $$h' = \frac{h}{4}$$

40. $v = at_1$

$$\therefore \quad a = \frac{v}{t_1}$$

$$W = \frac{1}{2}mv^2 = \frac{1}{2}m(at)^2$$

$$= \frac{1}{2}m\left(\frac{v}{t_1} \times t\right)^2$$

$$= \frac{1}{2}m\frac{v^2t^2}{t_1^2}$$

41. $\frac{1}{2}m(v_f^2 - v_i^2) = W = \mathbf{F} \cdot (\mathbf{r}_f - \mathbf{r}_i)$

42. $P = \mathbf{F} \cdot \mathbf{v} = (20\hat{\mathbf{i}} - 3\hat{\mathbf{j}} + 5\hat{\mathbf{k}}) \cdot (6\hat{\mathbf{i}} + 20\hat{\mathbf{j}} - 3\hat{\mathbf{k}})$

$$= 120 - 60 - 15 = 45\text{W}$$

43. P = Impulse $= Ft$

$$W = \text{KE} = \frac{P^2}{2m} = \frac{F^2t^2}{2m}$$

$$= \frac{\left(\frac{F}{m}\right)t \cdot Ft}{2}$$

$$= \frac{(at)Ft}{2} = \frac{Fvt}{2}$$

$$\therefore \quad \text{Average power} = \frac{W}{t} = \frac{F \cdot v}{2}$$

$$\text{Average power} = \frac{\text{KE}}{t} = \frac{mv^2}{2t}$$

44. $F \propto S^{-1/3}$ or $a \propto S^{-1/3}$

or $\quad vdv \propto S^{-1/3} ds \left(a = v \cdot \frac{dv}{ds} \right)$

Integrating we have,

$$v^2 \propto S^{2/3}$$

or $\quad v \propto S^{1/3}$

Now, power $P \propto F \cdot v \propto S^{-1/3} \cdot S^{1/3}$ or $P \propto S^0$

45. When displaced from x_2 in negative direction, force is positive. So this force is of restoring nature, or bringing the body back. Hence, at x_2, body is in stable equilibrium position.

46. In stable equilibrium, PE is minimum

$F = -\frac{dU}{dr} = -$ slope of v-r graph and negative force means attraction and positive force means repulsion.

47. Change in potential energy for all three particles is same. Hence, change in kinetic energy will also be same.

Or $\quad v_A = v_B = v_C$

48. Decrease in kinetic energy = increase in elastic potential energy.

$\therefore \quad \frac{1}{2}mv^2 = \frac{1}{2}Kx^2$

or $\quad x = \sqrt{\frac{m}{K}} \cdot v$

$$= \sqrt{\frac{0.1}{1000}} \times 10$$

$$= 0.1 \text{ m}$$

49. Decrease in gravitational potential energy

= increase in elastic potential energy

or $\quad mg(h + x) = \frac{1}{2}x^2$

or $\quad 2 \times 9.8(4.0 + x) = \frac{1}{2} \times 1960 \times x^2$

Solving this equation, we get

$$x = 0.1 \text{ m or } 10 \text{ cm.}$$

50. Increase in gravitational potential energy = decrease in elastic potential energy

$\therefore \quad mgh = \frac{1}{2}kx^2$

$\therefore \quad h = \frac{kx^2}{2mg}$

51. $\frac{1}{2}mv^2 = U$ or $m = \frac{2U}{v^2}$

52. $h = r(1 - \cos\theta)$

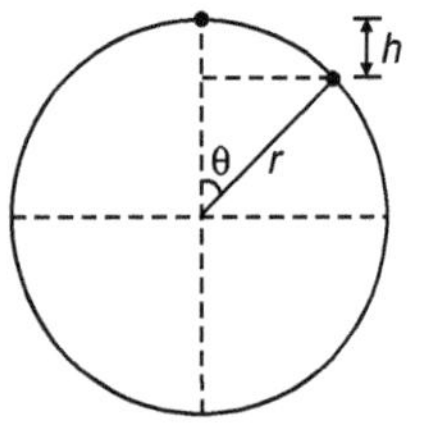

Now, decrease in PE = increase in KE

or $\quad mgr(1 - \cos\theta) = \frac{1}{2}mv^2$

or $\quad v = \sqrt{2gr(1 - \cos\theta)}$

53. $P =$ constant

$\therefore \quad W = \frac{1}{2}mv^2 = P \times t$

$\Rightarrow \quad v \propto t^{3/2}$ (integration)

54. $(mgH - mgh) = 2mgh$

$\therefore \quad h = \frac{H}{3}$

$$v = \sqrt{2g(H - h)}$$

$$= \sqrt{2g\left(\frac{2H}{3}\right)} = 2\sqrt{\frac{gH}{3}}$$

55. Decrease in gravitational PE = increase in kinetic energy

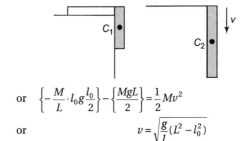

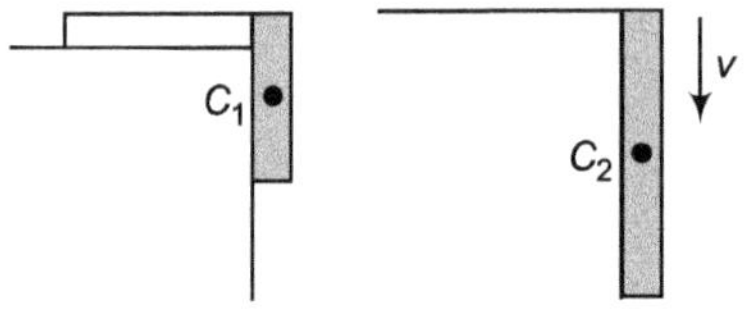

or $\quad \left\{-\frac{M}{L} \cdot l_0 g \frac{l_0}{2}\right\} - \left\{\frac{MgL}{2}\right\} = \frac{1}{2}Mv^2$

or $\quad v = \sqrt{\frac{g}{l}(L^2 - l_0^2)}$

56. Decrease in PE $= mgh$

57. At this height half is PE

$\therefore \quad mgh = \frac{98}{2}$

or $\quad 2 \times 9.8 \times h = 49$

or $\quad h = 2.5 \text{m}$

58. $\quad (M - m)gh = \frac{1}{2}(M + m)v^2$

$\therefore \quad v = \sqrt{\frac{2gh(M - m)}{M + m}}$

59. $mgh + \frac{1}{2}mv^2 =$ constant $= mgh$ ($H =$ initial height)

or $\quad gh + \frac{v^2}{2} =$ constant

60. Decrease in gravitational potential energy = increase in elastic potential energy

$\therefore \quad mg(h + x) = \frac{1}{2}Kx^2$

Solving we get $\quad x = 0.1$ m

61. $\frac{1}{2}mv^2 = (0.9)(mgh)$

$\therefore \quad v = \sqrt{1.8gh} = \sqrt{1.8 \times 10 \times 2}$

$$= 6 \text{ m/s}$$

63. $N = mg - F\sin\theta$

Block moves with uniform velocity. Hence, net force = 0

or, $\quad F\cos\theta = \mu N = \mu(mg - F\sin\theta)$

$\therefore \quad F = \frac{\mu mg}{\cos\theta + \mu\sin\theta}$

$$W = Fs\cos\theta = \frac{\mu\, mgd\cos\theta}{\cos\theta + \mu\sin\theta}$$

64. $F = \mu mg$ and $W = F \cdot S$

65.

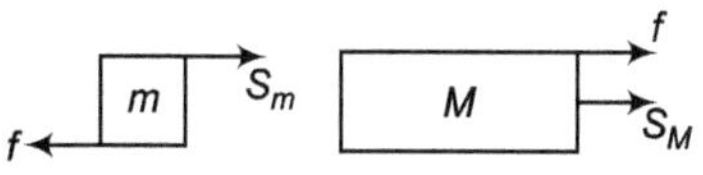

$$S_m \neq S_M$$

66. Maximum acceleration of 2 kg block due to friction can be μg or $5\,m/s^2$.

Combined acceleration, if both move together with same acceleration would be,

$$a = \frac{60}{12} = 5\,m/s^2$$

Since, both accelerations are equal, upper block will move with acceleration $5\,m/s^2$ due to friction.

In first two seconds, $S = \frac{1}{2}at^2 = \frac{1}{2} \times 5 \times 4 = 10m$

and force of friction $f = ma = 10N$

$\therefore \quad W_f = fs\cos 0^0 = 100\,J$

67. $W_f = fs\cos 180° = (\mu mg\cos\theta)(-1)(S)$

$$= -0.2 \times 5 \times 10 \times \frac{1}{\sqrt{2}} \times 0.1$$

$$= -\frac{1}{\sqrt{2}}\,J$$

68. Work done against friction $= (\mu mg\cos\theta)d$

69. $\frac{3}{4}$th is lost. Hence, left is $\frac{1}{4}$th, or,

$$v^2 = \frac{v_0^2}{u}$$

$\therefore \quad v = \frac{v_0}{2} = v_0 - at_0$

$$= v_0 - ugt_0$$

or $\quad u = \frac{v_0}{2gt_0}$

70. $K = \frac{P^2}{2m}, K' = \frac{(1.5P^2)}{2m}$

$$= 2.25\frac{P^2}{2m} = 2.25K$$

$\therefore$ Increase in kinetic energy is 125%.

71. Centre of mass of the rod moves a height $h = \frac{l}{2}$.

72. W = change in potential energy $= mgh$.

73. Given force is a constant force and work done by a constant force is always path independent.

74. $P = \sqrt{2Km}$ or $P \propto \sqrt{m}$

$\therefore \quad \frac{P_1}{P_2} = \sqrt{\frac{1}{4}} = \frac{1}{2}$

75. Mass $\frac{M}{3}$ has its centre of mass $\frac{L}{6}$ below the table surface.

$\therefore \quad W = mgh$

$$= \left(\frac{M}{3}\right)(g)\left(\frac{L}{6}\right)$$

$$= \frac{MgL}{18}$$

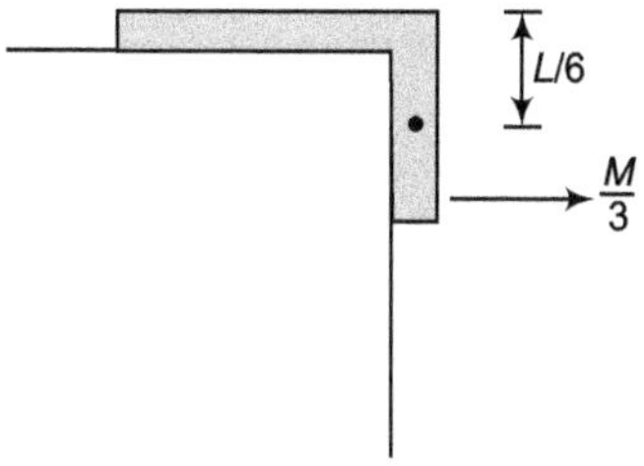

76. Work done by tension on M is negative (force and displacement are in opposite directions). But work done by tension on m is positive. Net work done will be zero.

77. W = change in potential energy $= mgh = mgL(1 - \cos\theta)$.

78. From work-energy theorem,

W = change in kinetic energy

or $\quad W = \frac{1}{2}mv^2$

$\therefore$ W-v graph is a parabola.

79. $W = \frac{1}{2} \times m \times (10)^2 = 50m$

$$W' = \frac{1}{2}m \times (400 - 100) = 150m = 3W$$

80. $v = \frac{dx}{dt} = 8t^3$

$$v_{0s} = 0, v_{1s} = 8\,m/s$$

$\therefore \quad \Delta KE = \frac{1}{2} \times 2 \times (64 - 0)\,J$

$$= 64\,J$$

81. $K = \frac{P^2}{2m}$ or $K \propto P^2$

When P is doubled, kinetic energy will become four times.

82. From work-energy theorem,

Work done = change in kinetic energy

$\therefore \quad Fx = K$ (as F = constant, because a = constant)

Therefore, K-x graph is a straight line passing through origin.

83. In equilibrium $kd = mg$ or $d = \frac{mg}{k}$

In allowed to fall suddenly, if does not stop in its equilibrium position. In that case,

decrease in gravitational PE = increase in elastic PE

or $\quad mgd' = \frac{1}{2}kd'^2$ or $d' = \frac{2mg}{k} = 2d$

84. $F = Kx$ or K = slope of F-x graph (F along y-axis)

Here, F is along x-axis.

So, $K = \frac{1.0}{10} = 0.1\frac{\text{kgf}}{\text{cm}}$

85. Average velocity $= \frac{\text{Total displacement}}{\text{Time}}$

$$= \frac{\text{Area under } v\text{ - }t \text{ graph}}{\text{Time}}$$

Since, area $\neq 0$

$\therefore$ Average velocity $\neq 0$

86. $E = \frac{1}{2} mv^2$

$\therefore \quad \frac{dE}{dv} = mv$

or $\quad P = \frac{dE}{dv} \quad$ (as $mv = P$)

87. $K = \frac{F}{x}$ = slope of F - x graph.

$$K \propto \frac{1}{l}$$

Length is reduced to half. Therefore, K will become two times. Slope will increase.

88. $K = 4t^2$ or $v^2 \propto t^2$

$\therefore \quad v \propto t$

v varies linearly with time when acceleration or force is constant.

89. Speed is doubled. Therefore, kinetic energy will become four times. Hence, minimum stopping distance will also become four times.

90. Total loss in friction $= K_i - K_f = \frac{1}{2} m(100 - 81)$

$= \frac{19}{2} m \quad$ (m = mass)

Half of the loss will be in upward journey.

Hence, $mgh + \frac{1}{2}\left(\frac{19}{2} m\right) = K_i = \frac{1}{2} m(100) = 50m$

$\therefore \quad h = 4.61 \text{ m}$

91. $F.d = mg(h + d)$

$\therefore F = mg\left(1 + \frac{h}{d}\right)$

Work done against resistance

= decrease in mechanical energy

Level 2 : Only One Correct Option

1. When released from top with zero velocity block leaves contact at $\cos\theta = \frac{2}{3}$

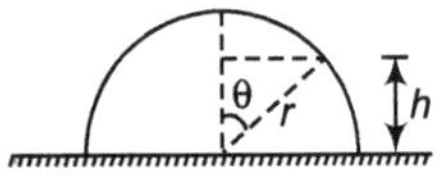

$\therefore \quad h = r\cos\theta = \frac{2r}{3}$

2. Since, the particle is released from a small height, θ (angle of radius with vertical) will be very small. Force of friction throughout the journey can be assumed to be μmg. Particle will finally come to rest when whole of its energy ($= mgh$) is lost in the work done against friction. Let particle stops after travelling a distance d. Then,

$$\mu mgd = mgh$$

or $\quad d = \frac{h}{\mu} = \frac{10^{-2}}{0.01} = 1.0 \text{ m}$

3. From $-a$ to b (ΔKE $= 0$ as mass will stop for a moment at extreme position) decrease in elastic potential energy = work done against friction.

$\therefore \quad \frac{1}{2} K a^2 - \frac{1}{2} K b^2 = \mu mg(a + b)$

or $\quad (a - b) = \frac{2\mu mg}{K}$

$\therefore$ Decrease in amplitude $= \frac{2\mu mg}{K}$

4. Decrease in gravitational potential energy = increase in kinetic energy.

Initially centre of mass of chain was at distance $\frac{l}{4}$ below the pin and in final position it is at distance $\frac{l}{2}$ below the pin.

Hence, centre of mass has descended $\frac{l}{4}$.

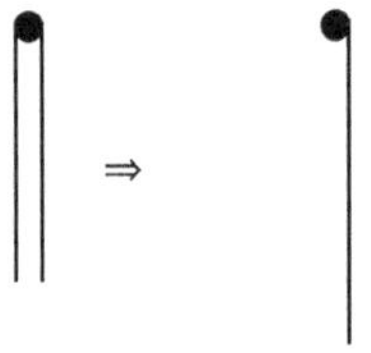

$\therefore \quad mg\frac{l}{4} = \frac{1}{2} mv^2$

or $\quad v = \sqrt{\frac{gl}{2}}$

5. At $x = 6$ m, $U = 26$ J (extreme position)

On the other side

$U = 26 = 10 + (x - 2)^2$ or $x - 2 = \pm 4$

$\therefore \quad x = 6 \text{ m}$ and $x = -2 \text{ m}$

Thus, $x = +6$ m and $x = -2$ m are the extreme positions. At $x = 2$ m, potential energy is minimum, which is 10 J. Hence, at this position kinetic energy will be maximum, which is equal to total mechanical energy ($= 26$ J) minus minimum potential energy ($= 10$ J). Thus maximum kinetic energy is 16 J.

6. Body will have maximum speed where,

$$Mg\sin\theta = \mu mg\cos\theta$$

or $\quad \sin 37° = (0.3x).\cos 37°$

or $\quad x = 2.5 \text{ m}$

7. Maximum acceleration of 1 kg block may be

$$a_{max} = \mu g = 1 \text{ m/s}^2$$

Common acceleration, without relative motion between two blocks may be,

$$a = \frac{0.5}{3} \text{ m/s}^2$$

Since, $\quad a < a_{max}$

There will be no relative motion and blocks will move with acceleration $\frac{0.5}{3}$ m/s^2.

Force of friction by lower block on upper block,

$$f = ma = (1)\left(\frac{0.5}{3}\right) = \frac{1}{6} \text{ N (towards right)}$$

$\therefore \quad W = f \times s = 0.5$ J

8. At equilibrium,

$$F = -\frac{dU}{dx} = 0$$

or $\quad (-12)ax^{-13} + (6bx^{-7}) = 0$

$\therefore \quad x = \left(\frac{2a}{b}\right)^{1/6}$

9. From work-energy theorem

$$W_F + W_{mg} + W_N = \frac{1}{2}mv^2$$

$$F \cdot R + mgR + 0 = \frac{1}{2}mv^2$$

$$5 \times 5 + \frac{1}{2} \times 10 \times 5 = \frac{1}{2} \times \frac{1}{2} \times v^2$$

$$\therefore \quad v = \sqrt{200} = 14.14 \text{ m/s}$$

10. $P = F.v = m\left(v \cdot \frac{dv}{ds}\right) \cdot v$

$$\therefore \quad \int_v^{2v} v^2 \cdot dv = \frac{P}{m} \cdot \int_0^s ds$$

$$\left[\frac{v^3}{3}\right]_v^{2v} = \frac{PS}{m} \quad \text{or} \quad S = \frac{7mv^3}{3P}$$

11. $F = kx$

$$\therefore \quad k = \frac{F}{x} = \frac{100}{1} \text{ N/m} = 100 \text{ N/m}$$

Now from energy conservation, between natural length of spring and its maximum compression state.

$$\frac{1}{2}mv^2 + mgh = \frac{1}{2}kx_{\text{max}}^2$$

$$\therefore \quad v = \sqrt{\frac{kx_{\text{max}}^2}{m} - 2gh}$$

$$= \sqrt{\frac{(100)(2)^2}{10} - (2)(10)\left(\frac{2}{2}\right)}$$

$$= \sqrt{20} \text{ m/s}$$

12. Power of tension = 0

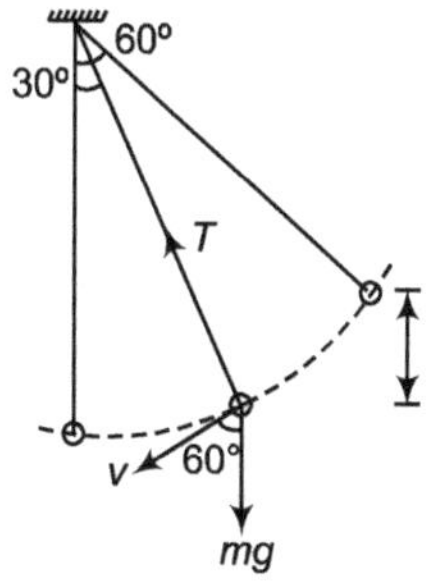

Power of $mg = (mg)(v)\cos 60°$

Here, $v = \sqrt{2gh}$ and $h = l(\cos 30° - \cos 60°)$

13. Block does not slide. Hence, the force of friction,

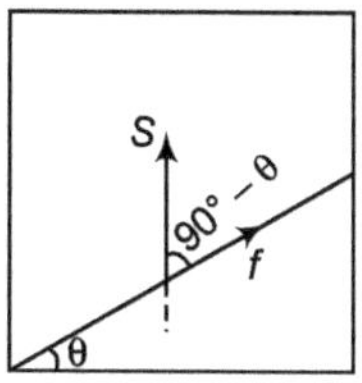

$$f = mg\sin\theta$$

In time t, displacement $S = vt$

$$\therefore \quad W_f = f \cdot s \cdot \cos(90° - \theta)$$

$$= (mg\sin\theta)(vt)(\sin\theta)$$

$$= mgvt\sin^2\theta$$

14. Work done by conservative forces $= U_i - U_f$

Work done by external forces $= E_f - E_i$

and net work done by all the forces $= K_f - K_i$

15. On M horizontal components of N and f are balanced (as Mg is vertical). Hence on $2M$ also they will be balanced.

$\therefore$ Horizontal Kx force on $2M$ should be zero.

16. $h = R(1 - \cos\theta)$

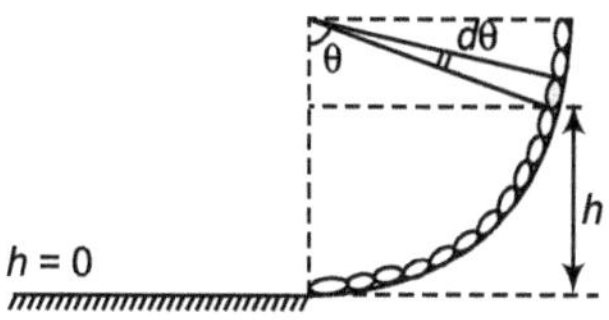

$$dm = \left(\frac{m}{\pi/2}\right) \cdot d\theta$$

$$= \frac{2md\theta}{\pi}$$

$$dv_i = (dm)gh$$

$$= \frac{2mgR(1 - \cos\theta)d\theta}{\pi}$$

$$\therefore \quad U_i = \int_0^{90°} dU_i$$

$$= mgR\left(1 - \frac{2}{\pi}\right)$$

$$U_f = 0$$

Decrease in PE = increase in KE.

17. As the given tracks are frictionless, hence, mechanical energy will be conserved. As both the tracks having common height, h.

From conservation of mechanical energy,

$$\frac{1}{2}mv^2 = mgh \quad \text{(for both tracks I and II)}$$

$$v = \sqrt{2gh}$$

Hence, speed is same for both stones.

For stone I, a_1 = acceleration along inclined plane $= g\sin\theta_1$

Similarly, for stone II $a_2 = g\sin\theta_2$ as $\theta_2 > \theta_1$ hence, $a_2 > a_1$.

And both length for track II is also less hence, stone II reaches earlier than stone I.

More than One Correct Options

1. $\mathbf{F} = -\left[\frac{\partial U}{\partial X}\hat{\mathbf{i}} + \frac{\partial U}{\partial y}\hat{\mathbf{j}}\right] = (-7\hat{\mathbf{i}} - 24\hat{\mathbf{j}})$ N

$$a = \frac{F}{m} = \left(-\frac{7}{5}\hat{\mathbf{i}} - \frac{24}{5}\hat{\mathbf{j}}\right) \text{m/s}$$

$$|\mathbf{a}| = \sqrt{\left(\frac{7}{5}\right)^2 + \left(\frac{24}{5}\right)^2} = 5 \text{ m/s}^2$$

Since, $\mathbf{a}$ = constant, we can apply,

$$\mathbf{v} = \mathbf{u} + \mathbf{a}t$$

$$= (8.6\hat{\mathbf{i}} + 23.2\hat{\mathbf{j}}) + \left(-\frac{7}{5}\hat{\mathbf{i}} - \frac{24}{5}\hat{\mathbf{j}}\right)(4)$$

$$= (3\hat{\mathbf{i}} + 4\hat{\mathbf{j}}) \text{ m/s}$$

$$|\mathbf{v}| = \sqrt{(3)^2 + (4)^2} = 5 \text{ m/s}$$

2. $F = -\dfrac{dU}{dX} = 5 - 200x$

Mean position is at $F = 0$

or at, $x = \dfrac{5}{200} = 0.025$ m

$$a = \frac{F}{m} = \frac{5 - 200x}{0.1} = (50 - 2000x) \quad ...(i)$$

At 0.05 m from the origin,

$x = +0.05$ m

or $x = -0.05$ m

Substituting in Eq. (i), We have

$|a| = 150 \text{ m/s}^2$

or $= 50 \text{ m/s}^2$

At 0.05 m from the mean position means,

$x = 0.075$

or $x = -0.025$ m

Substituting in Eq. (i) we have,

$|a| = 100 \text{ m/s}^2$

3. Spring force is always towards mean position. If displacement is also towards mean position, F and S will be of same sign and work done will be positive.
4. Work done by conservative force $= -\Delta U$

Work done by all the forces $= \Delta K$

Work done by forces other than conservative forces $= \Delta E$

5. (a) At equilibrium

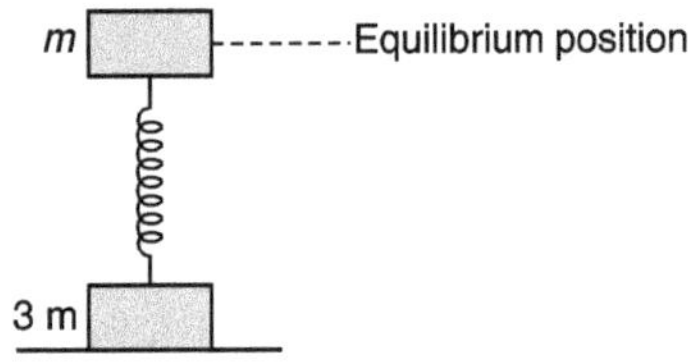

$k\delta_0 = mg$

or $\delta_0 = \dfrac{mg}{k}$

where, δ_0 = compression

(b) $\delta_{\text{Total}} = \delta + \delta_0 = \dfrac{3mg}{k}$

$$F_{\max} = k\delta_{\max} = k\left(\frac{3mg}{k}\right) = 3mg \quad \text{(downward)}$$

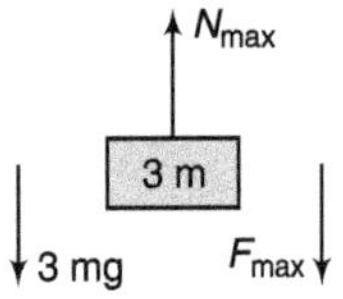

$\therefore \quad N_{\max} = 3mg + F_{\max} = 6mg$

(d) If $\delta > \dfrac{4mg}{k}$, then upper block will move a distance

$X > \dfrac{4mg}{k} - \delta_0$ or $X > \dfrac{3mg}{k}$ from natural length.

Hence in this case, extension

$X > \dfrac{3mg}{k}$

or $F = kx > 3mg$ (upwards on lower block)

So lower block will bounce up.

6. (a) Decrease in potential energy of B = increase in spring potential energy

$\therefore \quad 2mg\,X_m = \dfrac{1}{2}kX_m^2$

$\therefore \quad X_m = \dfrac{4mg}{k}$

(b) $E_i = E_f$

$$0 = \frac{1}{2}(m + 2m)v^2 + \frac{1}{2} \times k \times \left(\frac{2mg}{k}\right)^2 - (2mg)\left(\frac{2mg}{k}\right)$$

$\therefore \quad v = 2g\sqrt{\dfrac{m}{3k}}$

(c) $a = \dfrac{kX_m - 2mg}{2m}$ (upwards)

$$= \frac{k\left(\frac{4mg}{k}\right) - 2mg}{2m} = g$$

(d) $T - 2mg = ma$ ...(i)

$2mg - T = 2ma$ ...(ii)

Solving these two equations, we get

$a = 0$ and $T = 2mg$

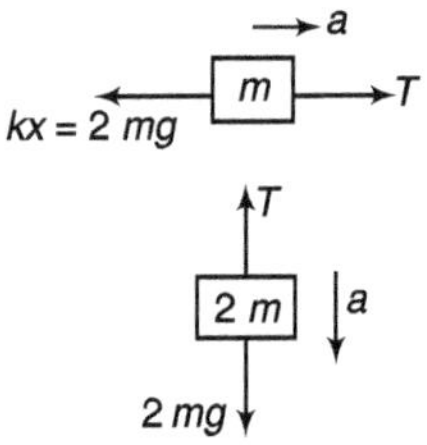

7. No solution is required.
8. Work done by conservative forces

$= U_i - U_f = -20 + 10 = -10$ J

Work done by all the forces,

$= K_f - K_i = 20 - 10 = 10$ J

9. (a) Work done by gravity in motion 1 is zero ($\theta = 90°$) and in motion 2 is negative ($\theta = 180°$).

(b) In both cases angle between N and S is acute.

(c) and (d) Depending on the value of acceleration in motion 1, friction may act up the plane or down the plane. Therefore, the angle between friction and displacement may be obtuse or acute. So, work done by friction may be negative or positive.

10. When a man of mass m climbs up the staircase of height L, work done by the gravitational force on the man is $-mgl$ work done by internal muscular forces will be mgL as the change in kinetic energy is almost zero.

Hence, total work done $= -mgL + mgL = 0$

As the point of application of the contact forces does not move, hence work done by reaction forces will be zero.

Comprehension Based Questions

1. $U = E - K = 25 - K$

Since, $K \geq 0$

$\therefore \quad U \leq 25$ J

2. $U = E - K = -40 - K$

Since, $K \geq 0$

$\therefore \quad U \leq -40$ J

Assertion and Reason

1. $W = 0$. That feet remains stationary, which is in contact with the ladder.

2. Change in potential energy is same.

3. Assertion Change in PE is same.

Reason Acceleration a is different. Hence, the net force is different.

4. Assertion $\frac{dU}{dt} = \frac{d}{dt}(mgh) = mg\frac{d}{dt}\left(u_y t - \frac{1}{2}gt^2\right)$

$$= mg(u_y - gt) = mgv_y$$

$\therefore$ $\left|\frac{dU}{dt}\right| = mg|v_y|$

$|v_y|$ first decreases then increase with time.

Reason $\frac{d\mathbf{P}}{dt} = \mathbf{F} = mg =$ constant

5. Assertion $\frac{dU}{dt} = \frac{d}{dt}(mgh) = mg\frac{d}{dt}\left(u_y t - \frac{1}{2}gt^2\right)$

$$= mg(u_y - gt) = mgv_y$$

$\therefore$ $\left|\frac{dv}{dt}\right| = mg|v_y|$

For $m = 1\,\text{kg}$

$$\left|\frac{dU}{dt}\right| = g(u_y - gt) \quad \text{...(i)}$$

$$\mathbf{a} \cdot \mathbf{v} = (-g\hat{\mathbf{j}}) \cdot [u_x\hat{\mathbf{i}} + (u_y - gt)\hat{\mathbf{j}}]$$

$$= g(u_y - gt) \quad \text{...(ii)}$$

From Eqs. (i) and (ii), we can see that two magnitudes are equal.

Reason $P = \mathbf{F} \cdot \mathbf{v} = (-mg\hat{\mathbf{i}}) \cdot [v_x\hat{\mathbf{i}} + (u_y - gt)\hat{\mathbf{j}}]$

$$= -mgu_y + mg^2t$$

So, P *versus* t graph has the positive slope.

6. Assertion Decrease in mechanical energy in case 1 will be

$$\Delta U_1 = \frac{1}{2}mv^2$$

But decrease in mechanical energy in case 2 will be

$$\Delta U_2 = \frac{1}{2}mv^2 - mgh$$

$\therefore$ $\Delta U_2 < \Delta U_1$

or assertion is correct.

Coefficient of friction is the mutual property of two surfaces. It does not depend on angle of inclination.

7. Work done by friction on A is positive and on B is negative. If there is no slipping between the two blocks, then $S_A = S_B$. Therefore, the net work done will be zero, otherwise not.

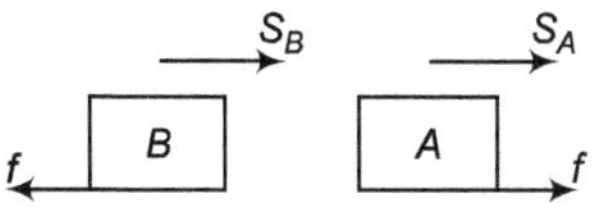

8. If oscillations are not taking place, then kinetic energy may be zero at stable equilibrium position.

9. Force is continuously negative. But displacement is positive (from $x = -2$ to $x = +2$)

$\therefore$ Work done is negative.

12. Work done by a constant force

= Force $\times$ displacement in the direction of force.

13. According to work-energy theorem, work done by all the forces = change in kinetic energy.

15. $W = Pt = \frac{1}{2}mv^2$

$$v \propto t^{1/2}$$

Differentiating we get,

$$a \propto t^{-1/2}$$

16. $W_{AC} = +$ ve

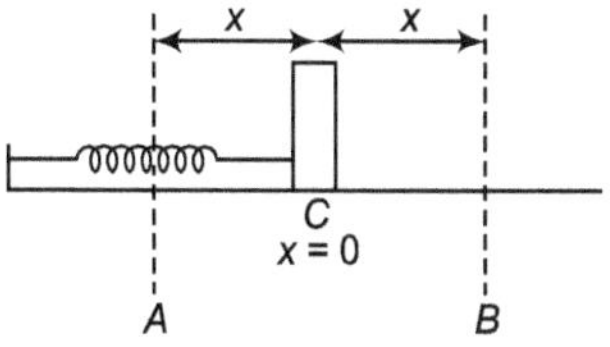

$W_{CB} = -$ ve and $W_{ACB} = 0$

Match the Columns

1. Work done by conservative forces in negative of change in potential energy.

2. From $x = 2$ to $x = 4$, force is positive and displacement is also positive. Hence, the work done is positive. Similar logic can be applied to other parts also.

3. A is the point of stable equilibrium, so potential energy is minimum. Similarly, point C is the unstable equilibrium position, where potential energy should be maximum.

4. In $t = 4\text{s}$,

$$v = at = 8\,\text{m/s} \quad \text{and}\; s = \frac{1}{2}at^2 = 16\,\text{m}$$

$$\text{KE} = \frac{1}{2}mv^2 = 32\,\text{J}$$

From work-energy theorem,

Work done by all the forces = ΔKE = 32J

Work done by gravity = $-mgh = -(1)(10)(16)$

$$= -160\,\text{J}$$

Writing equation of motion, we have, $\Sigma F_y = ma$

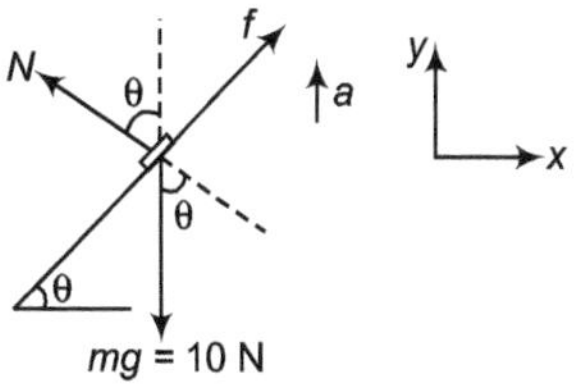

$$N\cos 30° + f\sin 30° - 10 = ma = 2$$

or $\sqrt{3}N + f = 24$ ...(i)

$$\Sigma F_x = 0$$

$\therefore$ $N\sin 30° = f\cos 30°$

or $N = \sqrt{3}f$ ...(ii)

Solving Eqs. (i) and (ii), we have

$$f = 6\,\text{N}$$

and $N = \frac{18}{\sqrt{3}} = 6\sqrt{3}\,N$

Now, $W_N = (N\cos\theta)(s)$

$= (6\sqrt{3})\left(\frac{\sqrt{3}}{2}\right)(16) = 144\,J$

$W_F = (f\sin\theta)(s)$

$= (6)\frac{1}{2}(16) = 48\,J$

5. $v_f - v_i$ = area of a-x graph $= 12\,m/s$

$\therefore \quad v_f = 12 + 4 = 16\,m/s$

$\Delta KE = \frac{1}{2}m(v_f^2 - v_i^2) = 120\,J$

Work done by all the forces $= \Delta KE = 120\,J$

$K_f = \frac{1}{2}mv_f^2 = 128\,J$

Work done by conservative forces $= U_i - U_f = 240\,J$

Work done by external force

= total work done – work done by conservative forces

$= -112\,J$

6. $P \propto t$

$$W = \int P dt = \int \alpha t\, dt \quad \text{or} \quad W \propto t^2$$

Since, work done is equal to change is KE

Hence, $v^2 \propto t^2$ or $v \propto t$

Further, $v = \frac{ds}{dt}$

$\therefore \quad \frac{ds}{dt} \propto t$ or $ds \propto t\,dt$

or $s \propto t^2$ (by integration)

Entrance Gallery

1. From work-energy theorem,

Work done by all forces = Change in kinetic energy

or $W_F + W_{mg} = K_f - K_i$

$18 \times 5 + (1 \times 10)(-4) = K_f$

$90 - 40 = K_f$

$\Rightarrow \quad K_f = 50\,J$

2. **Thinking Process** We know that change in potential energy of a system corresponding to a conservative internal force as,

$$U_f - U_i = -W = -\int_i^f F \cdot dr$$

Given, that $F = ax + bx^2$

We know that work done in stretching the rubber band by L is given by

$$|dW| = |Fdx| \Rightarrow |W| = \int_0^L (ax + bx^2)\,dx$$

$$= \left[\frac{ax^2}{2}\right]_0^L + \left[\frac{bx^3}{3}\right]_0^L$$

$$= \left[\frac{aL^2}{2} - \frac{a \times (0)^2}{2}\right] + \left[\frac{b \times L^3}{3} - \frac{b \times (0)^3}{3}\right]$$

$$= |W| = \frac{aL^2}{2} + \frac{bL^3}{3}$$

3. Here, $m = 0.18\,kg$,

$K = 2\,N/m \quad \mu = 0.1, x = 0.06\,m$

According to conservation of mechanical energy principle, we know

Decrease in mechanical energy = work done against friction

$$\frac{1}{2}mv^2 - \frac{1}{2}kx^2 = \mu mgh$$

$$v = \sqrt{\frac{2\mu mgx + kx^2}{m}}$$

Substituting the values of m, μ, g, x, we get

$$v = \sqrt{\frac{2 \times 0.1 \times 0.28 \times 9.8 \times 0.06 + 2 \times 0.06 \times 0.06}{0.18}}$$

$$v = \left(\frac{4}{10}\right) m/s$$

So, $N = 4$

4. Given, $W_1 = 1\,J, x = 1 \times 10^{-3}\,m$

Work done to compress it 1 mm,

$W_1 = \frac{1}{2}kx_1^2 \Rightarrow 1 = \frac{1}{2} \times k \times (1 \times 10^{-3})^2$

$\Rightarrow \quad k = 2 \times 10^6$

Work done to compress it further by 1 mm,

$W_2 = \frac{1}{2}kx_2^2 = \frac{1}{2} \times 2 \times 10^6 \times (1 \times 10^{-3} + 1 \times 10^{-3})^2$

$W_2 = 4 \times 10^{-6} \times 10^6 = 4\,J$

The additional work done $= W_2 - W_1 = 4 - 1 = 3\,J$

5. Momentum acquired by the bodies,

$p_1 = p_2 = Ft$

Now, their kinetic energies,

$E_1 = \frac{p^2}{2m_1}$ and $E_2 = \frac{p^2}{2m_2}$

$\therefore \quad \frac{E_1}{E_2} = \frac{m_2}{m_1}$

6. Given, $F = 50\,N$, $W = 150\,J, s = 3\,m$

Workdone, $W = Fs\cos\theta$

$150 = 50 \times 3 \times \cos\theta$

$\cos\theta = \frac{150}{150} = 1$

$\Rightarrow \quad \theta = 0°$

7. Given, $\frac{dK}{dt}$ = constant

$K \propto t \quad \left(\because K = \frac{1}{2}mv^2\right)$

$\frac{1}{2}mv^2 \propto t$

$\Rightarrow \quad v \propto \sqrt{t}$

$\therefore \quad P = Fv = \frac{dK}{dt}$ = constant

$F \propto \frac{1}{v}$

$\Rightarrow \quad F \propto \frac{1}{\sqrt{t}}$

8. Percentage of energy loss $= \frac{mg\,(2 - 1.5}{mgh} \times 100$

$= \frac{mg\,(0.5}{mg \times 2} \times 100 = 25\%$

9. According to law of conservation of energy,

$$\frac{1}{2}mu^2=\frac{1}{2}mv^2+mgh$$

$$490=245+5\times9.8\times h$$

$$h=\frac{245}{49}=5\,\text{m}$$

10. At the ground, $E=\frac{1}{3}mu^2$

At highest point, $E'=\frac{1}{2}m(u\cos60^\circ)^2$

$$E'=\frac{E}{4}$$

11. The total volume of water in vessel,

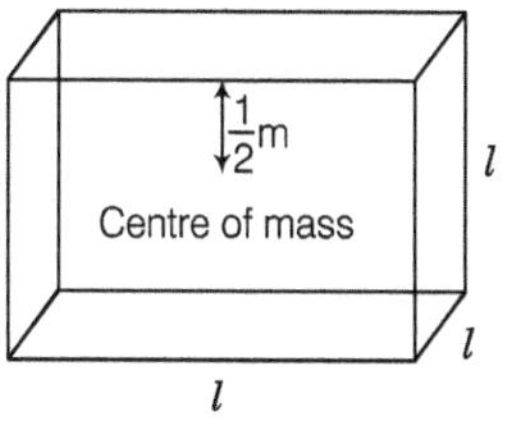

$V=l^3=1\,\text{m}^3$

$m=1\times1000=1000\,\text{kg}$

$W=mgh=1000\times10\times\frac{1}{2}=5000\,\text{J}$

12. Given, $m=6\,\text{kg},\ x=\frac{t^2}{4}$

$$\frac{dx}{dt}=v=\frac{t}{2}$$

Velocity at, $t=0\,\text{s}$

$\Rightarrow$ $v(0)=0$

$\therefore$ The initial kinetic energy, $K_i=\frac{1}{2}m(0)^2$

Velocity at $t=2\,\text{s},\ v(2)=\frac{2}{2}=1$

and the final kinetic energy,

$$K_f=\frac{1}{2}m(1)^2=\frac{1}{2}\times6\times1=3$$

According to work-energy theorem

Work, $W=K_f-K_i=3-0=3\,\text{J}$

13. We know that, $P=Fv=m\cdot\frac{dv}{dt}\cdot v$

$$\frac{vdv}{dt}=\frac{P}{m}$$

Integrating on both side, we get

$$\int v\,dv=\int\frac{Pdt}{m},\quad\Rightarrow\quad\frac{v^2}{2}=\frac{Pt}{m}$$

$$v=\sqrt{\frac{2P}{m}}t^{1/2},\quad\frac{dx}{dt}=\sqrt{\frac{2P}{m}}t^{1/2}$$

$$\int dx=\sqrt{\frac{2P}{m}}\int t^{1/2}dt,$$

$$x=\sqrt{\frac{2P}{m}}\frac{t^{3/2}}{3/2}=\frac{2}{3}\sqrt{\frac{2P}{m}}t^{3/2}$$

$\Rightarrow$ $x\propto t^{3/2}$

14. Let acceleration,

given, $m_1=0.36\,\text{kg},\ m_2=0.72\,\text{kg}$

and $g=10\,\text{ms}^{-1}$

$$a=\frac{\text{Net pulling force}}{\text{Total mass}}$$

$$=\frac{0.72g-0.36g}{0.72+0.36}=\frac{g}{3}$$

$$s=\frac{1}{2}at^2=\frac{1}{2}\left(\frac{g}{3}\right)(1)^2=\frac{g}{6}$$

$$T-0.36g=0.36a=0.36\left(\frac{g}{3}\right)$$

$\therefore$ $T=0.48\,\text{g}$

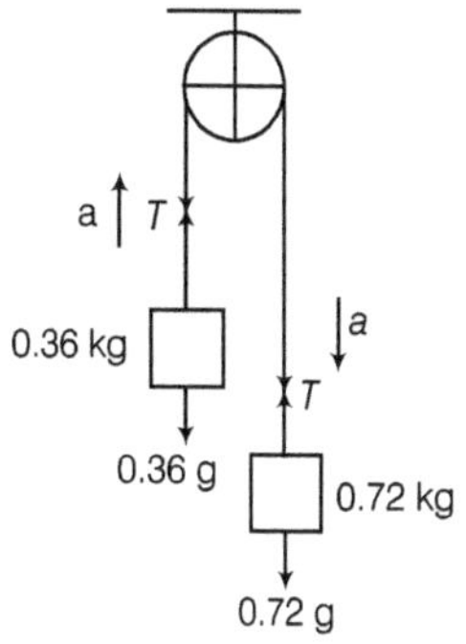

Now, $W_T=Ts\cos0^\circ$ (on 0.36 kg mass)

$$=(0.48g)\left(\frac{g}{6}\right)(1)=0.08(g^2)$$

$$=0.08(10)^2=8\,\text{J}$$

15. Given, potential energy function $U(x)=\frac{a}{x^{12}}-\frac{b}{x^6}$

$$U(x=\infty)=0$$

As, $F=-\frac{dU}{dx}=-\left[\frac{12a}{x^{13}}-\frac{6b}{x^7}\right]$

At equilibrium, $F=0$

$\therefore$ $x^6=\frac{2a}{b}$

$\therefore$ $U_{\text{equilibrium}}=\frac{a}{\left(\frac{2a}{b}\right)^{12}}-\frac{b}{\left(\frac{2a}{b}\right)^{8}}=\frac{-b^2}{4a}$

$\therefore$ $D=[U(x-\infty)-U_{\text{equilibrium}}]=\frac{b^2}{4a}$

16. Work done, $W=\mathbf{F}\cdot\mathbf{ds}=(\mathbf{F}_1+\mathbf{F}_2)\cdot(\mathbf{s}_2-\mathbf{s}_1)$

$$=\{(4\hat{\mathbf{i}}+\hat{\mathbf{j}}-3\hat{\mathbf{k}})+(3\hat{\mathbf{i}}+\hat{\mathbf{j}}-\hat{\mathbf{k}})\}$$

$$\{(5\hat{\mathbf{i}}+4\hat{\mathbf{j}}+\hat{\mathbf{k}})-(\hat{\mathbf{i}}+2\hat{\mathbf{j}}+3\hat{\mathbf{k}})\}=(7\hat{\mathbf{i}}+2\hat{\mathbf{j}}-4\hat{\mathbf{k}})\cdot(4\hat{\mathbf{i}}+2\hat{\mathbf{j}}-2\hat{\mathbf{k}})$$

$$=28+4+8=40\,\text{J}$$

17. Work done, $W=$ Area $ABCEFDA$

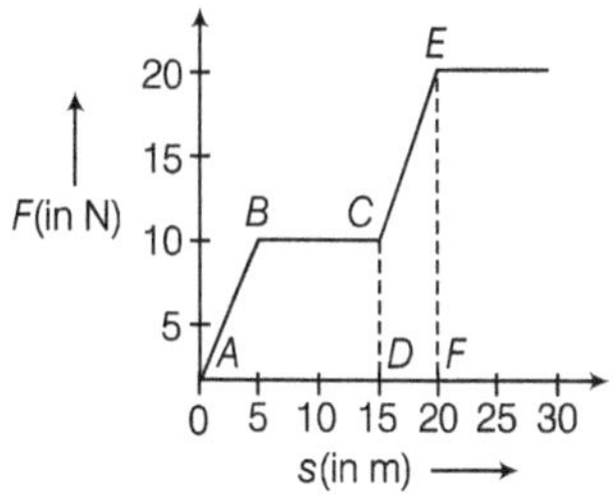

$=$ Area $ABCD+$ Area $CEFD$

$$=\frac{1}{2}\times(15+10)\times10+\frac{1}{2}\times(10+20)\times5$$

$$=125+75=200\,\text{J}$$

7

Circular Motion

7.1 Kinematics of Circular Motion

Angular Variables

Circular motion is a two dimensional motion or motion in a plane.

Suppose a particle P is moving in a circle of radius r and centre O.

The position of the particle P at a given instant may be described by the angle θ between OP and OX. This angle θ is called the **angular position** of the particle. As the particle moves on the circle its angular position θ changes. Suppose, the point rotates an angle $\Delta\theta$ in time Δt. The rate of change of angular position is known as the **angular velocity** (ω). Thus,

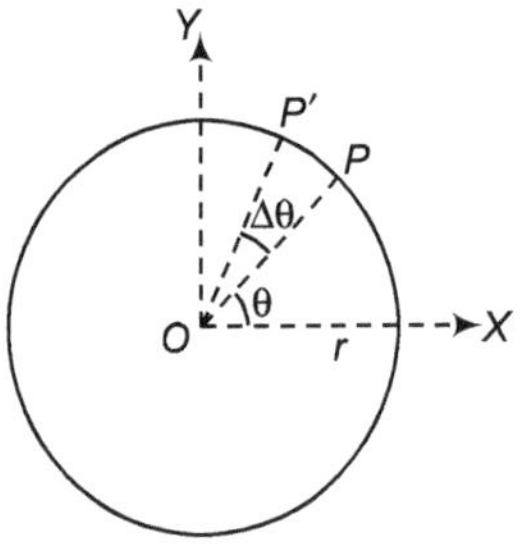

Fig. 7.1

$$\omega = \lim_{\Delta t \to 0} \frac{\Delta\theta}{\Delta t} = \frac{d\theta}{dt}$$

The rate of change of angular velocity is called the **angular acceleration** (α). Thus,

$$\alpha = \frac{d\omega}{dt} = \frac{d^2\theta}{dt^2}$$

If angular acceleration is constant, we have

$$\theta = \omega_0 t + \frac{1}{2}\alpha t^2$$

$$\omega = \omega_0 + \alpha t$$

and $$\omega^2 = \omega_0^2 + 2\alpha\theta$$

Here, ω_0 and ω are the angular velocities at time $t = 0$ and t and θ the angular position at time t. The linear distance PP' travelled by the particle in time Δt is

$$\Delta s = r\Delta\theta$$

or $$\lim_{\Delta t \to 0} \frac{\Delta s}{\Delta t} = r \lim_{\Delta t \to 0} \frac{\Delta\theta}{\Delta t}$$

Chapter Snapshot

- Kinematics of Circular Motion
- Dynamics of Circular Motion
- Motion in a Vertical Circle

or $$\frac{ds}{dt}=r\frac{d\theta}{dt}$$

or $$v=r\omega$$

Here, v is the linear velocity of the particle.

Following points are worthnoting regarding circular motion of any particle :

Velocity

A particle P in circular motion has two types of velocities and corresponding two speeds.

(i) **Linear velocity (v) and linear speed (v)** Linear velocity of particle is displacement of particle per unit time.

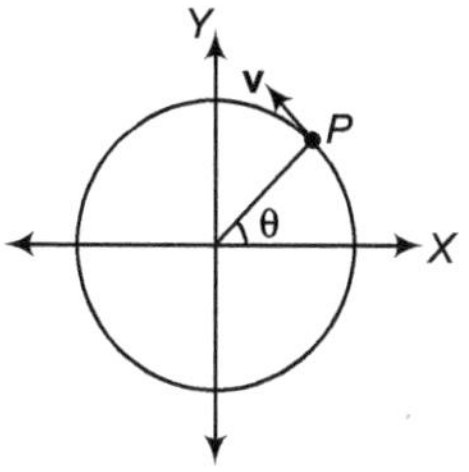

Fig. 7.2

or $$\mathbf{v}=\frac{d\mathbf{s}}{dt}$$

Magnitude of linear velocity is called linear speed, v which is distance travelled per unit time.

or $$v=|\mathbf{v}|=\left|\frac{d\mathbf{s}}{dt}\right|$$

(ii) **Angular velocity** (ω) **and angular speed** (ω_0) Small angular displacements $d\theta$ is a vector quantity. Rate of change of angular displacement is called angular velocity (ω). Thus,

$$\omega=\frac{d\theta}{dt} \qquad \text{(unit is rad/s)}$$

Magnitude of ω is called angular velocity. Thus,

$$\omega_0=|\omega|=\left|\frac{d\theta}{dt}\right|=\frac{d\theta}{dt}$$

Note that ω is an axial vector, direction of which is given by right hand screw law. For example, in the given figure ω is perpendicular to paper outwards. Relation between linear speed v and angular speed ω_0 is

$$v=r\omega_0$$

But relation between linear velocity **v** and angular velocity ω is given by

$$\mathbf{v}=\omega\times\mathbf{r}$$

Here, r is the radius but **r** is position vector of particle with respect to centre of circle.

Acceleration

Acceleration of the particle in circular motion has two components.

(i) **Tangential component** (a_t) This is the component of **a** in the direction of velocity, which is responsible for change in speed of particle. It is also equal to rate of change of speed. Hence,

$$a_t=\text{component of } \mathbf{a} \text{ along } \mathbf{v}$$
$$=\frac{dv}{dt}=\frac{d|\mathbf{v}|}{dt}$$

This component is tangential.

(ii) **Radial component** (a_r) This is component of **a** towards centre. This is responsible for change in direction of velocity. This is equal to $\frac{v^2}{r}$ or $r\omega^2$. Thus,

$$a_r=\frac{v^2}{r}=r\omega^2$$

Three Possible Types of Circular Motion

In circular motion direction of velocity definitely changes. Hence, a_r can never be zero. But speed may remain constant ($a_t=0$), may be increasing ($a_t=$ positive) or may be decreasing ($a_t=$ negative).

Accordingly we can classify circular motion in following three types.

(i) **Uniform circular motion** In this type of circular motion, speed of particle remains constant. Hence, $a_t=0$ and $a_r=\frac{v^2}{r}$ or $r\omega^2$. Thus, net acceleration is also equal to a_r. Angle between **v** and **a** is 90°.

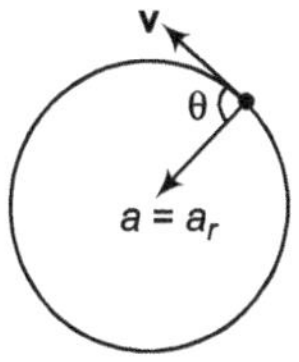

Fig. 7.3

$$\theta=90° \Rightarrow a=a_r=\frac{v^2}{r} \text{ or } r\omega^2$$

$$\Rightarrow \qquad v=\text{constant}$$

(ii) **Circular motion of increasing speed** In this type of circular motion speed of particle increases. Hence, a_t is positive or in the direction of velocity. Thus, net acceleration in this case will be $a=\sqrt{a_t^2+a_r^2}$. Angle between **a** and **v** will be acute.

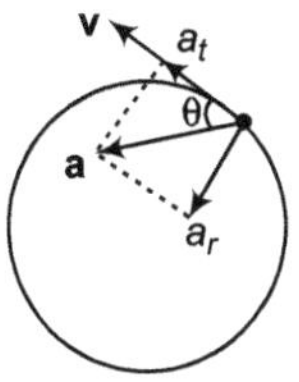

Fig. 7.4

$$\theta=\text{acute} \Rightarrow a=\sqrt{a_t^2+a_r^2}$$

where $$a_t=\frac{dv}{dt} \text{ or } \frac{d|\mathbf{v}|}{dt}$$

and $$a_r=\frac{v^2}{r} \text{ or } r\omega^2$$

(iii) **Circular motion of decreasing speed** In this type of circular motion speed of particle decreases. Hence, a_t is negative or in opposite direction of velocity. Thus, net acceleration in this case is also $a = \sqrt{a_t^2 + a_r^2}$ but angle between **v** and **a** is obtuse.

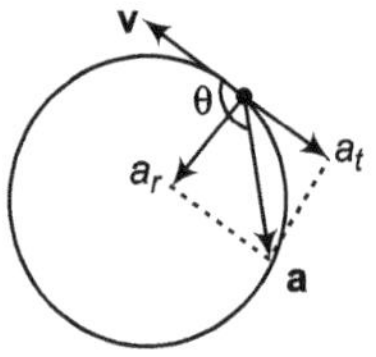

Fig. 7.5

$$\theta = \text{obtuse}$$
$$a = \sqrt{a_t^2 + a_r^2}$$

where, $Ta_t = \frac{dv}{dt}$ or $\frac{d|\mathbf{v}|}{dt}$

and $a_r = \frac{v^2}{r}$ or $r\omega^2$

- That radial acceleration (a_r) is also sometimes called normal acceleration (a_n).

Example 7.1 *A particle moves in a circle of radius 0.5 m with a linear speed of 2 m/s. Find its angular speed.*

Sol. The angular speed is

$$\omega = \frac{v}{r} = \frac{2}{0.5} = 4 \text{ rad/s}$$

Example 7.2 *A particle moves in a circle of radius 0.5 m at a speed that uniformly increases. Find the angular acceleration of particle, if its speed changes from 2.0 m/s to 4.0 m/s in 4.0 s.*

Sol. The tangential acceleration of the particle is

$$a_t = \frac{dv}{dt} = \frac{4.0 - 2.0}{4.0} = 0.5 \text{ m/s}^2$$

The angular acceleration is

$$\alpha = \frac{a_t}{r} = \frac{0.5}{0.5} = 1 \text{ rad/s}^2$$

Example 7.3 *The speed of a particle moving in a circle of radius $r = 2$ m varies with time t as $v = t^2$ where, t is in second and v in m/s. Find the radial, tangential and net acceleration at $t = 2$ s.*

Sol. Linear speed of particle at $t = 2$ s is

$$v = (2)^2 = 4 \text{ m/s}$$

∴ Radial acceleration

$$a_r = \frac{v^2}{r} = \frac{(4)^2}{2} = 8 \text{ m/s}^2$$

The tangential acceleration is

$$a_t = \frac{dv}{dt} = 2t$$

∴ Tangential acceleration at $t = 2$ s is

$$a_t = (2)(2) = 4 \text{ m/s}^2$$

∴ Net acceleration of particle at $t = 2$s is

$$a = \sqrt{(a_r)^2 + (a_t)^2} = \sqrt{(8)^2 + (4)^2} \text{ or } a = \sqrt{80} \text{ m/s}^2$$

- On any curved path (not necessarily a circular one) the acceleration of the particle has two components a_t and a_r in two mutually perpendicular directions. Component of **a** along **v** is a_t and perpendicular to **v** is a_r. Thus,

$$|\mathbf{a}| = \sqrt{a_t^2 + a_r^2}$$

7.2 Dynamics of Circular Motion

If a particle moves with constant speed in a circle, motion is called uniform circular motion. In uniform circular motion a resultant non-zero force acts on the particle. This is because a particle moving in a circle is accelerated even, if speed of the particle is constant. This acceleration is due to the change in direction of the velocity vector. As we have seen in article 7.1 that in uniform circular motion tangential acceleration (a_t) is zero. The acceleration of the particle is towards the centre and its magnitude is $\frac{v^2}{r}$. Here, v is the speed of the particle and r the radius of the circle. The direction of the resultant force F is, towards centre and its magnitude is

$$F = ma \quad \text{or} \quad F = \frac{mv^2}{r} \quad \text{or} \quad F = mr\omega^2 \quad (\text{as } v = r\omega)$$

Here, ω is the angular speed of the particle. This force F is called the centripetal force. Thus, a centripetal force of magnitude $\frac{mv^2}{r}$ is needed to keep the particle moving in a circle with constant speed. This force is provided by some external source such as friction, magnetic force, Coulomb force, gravitation, tension, etc.

- I have found students often confused over the centripetal force. They think that this force acts on a particle moving in a circle. This force does not act but required for moving in a circle which is being provided by the other forces acting on the particle. Let, us take an example. Suppose a particle of mass m is moving in a vertical circle with the help of a string of length l fixed at point O. Let v be the speed of the particle at lowest position. When I ask the students what forces are acting on the particle in this position ? They immediately say, three forces are acting on the particle (1) tension T (2) weight mg and (3) centripetal force $\frac{mv^2}{l}$ ($r = l$).

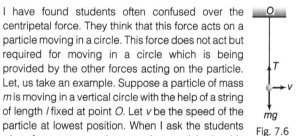

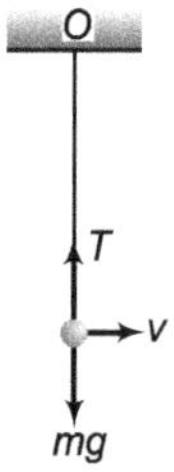

Fig. 7.6

However, they are wrong. Only first two forces T and mg are acting

on the particle. Third force $\frac{mv^2}{l}$ is required for circular motion which is being provided by T and mg. Thus, the resultant of these two forces is $\frac{mv^2}{l}$ towards O. Or we can write

$$T - mg = \frac{mv^2}{l}$$

Circular Turning of Roads

When vehicles go through turnings, they travel along a nearly circular arc. There must be some force which will produce the required centripetal acceleration. If the vehicles travel in a horizontal circular path, this resultant force is also horizontal. The necessary centripetal force is being provided to the vehicles by following three ways.

1. By friction only.
2. By banking of roads only.
3. By friction and banking of roads both.

In real life the necessary centripetal force is provided by friction and banking of roads both. Now let us write equations of motion in each of the three cases separately and see what are the constraints in each case.

By Friction Only

Suppose a car of mass m is moving at a speed v in a horizontal circular arc of radius r. In this case, the necessary centripetal force to the car will be provided by force of friction f acting towards centre.

Thus, $$f = \frac{mv^2}{r}$$

Further, limiting value of f is μN

or $$f_L = \mu N = \mu mg \qquad (N = mg)$$

Therefore, for a safe turn without sliding

$$\frac{mv^2}{r} \le f_L$$

or $$\frac{mv^2}{r} \le \mu mg$$

or $$\mu \ge \frac{v^2}{rg}$$

or $$v \le \sqrt{\mu rg}$$

Here, two situations may arise. If μ and r are known to us, the speed of the vehicle should not exceed $\sqrt{\mu rg}$ and if v and r are known to us, the coefficient of friction should be greater than $\frac{v^2}{rg}$.

- You might have seen that if the speed of the car is too high, car starts skidding outwards. With this radius of the circle increases or the necessary centripetal force is reduced (centripetal force $\propto \frac{1}{r}$).

By Banking of Roads Only

Friction is not always reliable at circular turns, if high speeds and sharp turns are involved. To avoid dependence on friction, the roads are banked at the turn so that the outer part of the road is some what lifted compared to the inner part.

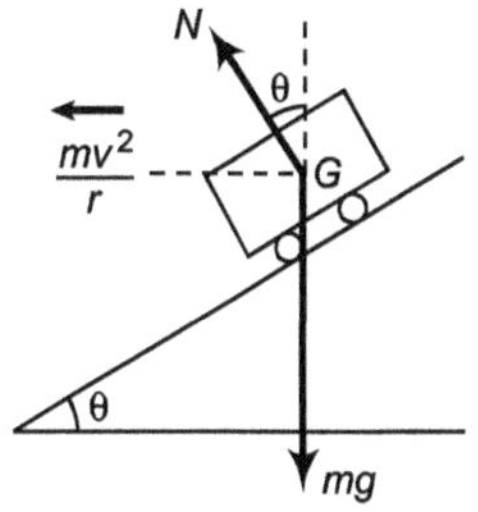

Fig. 7.7

Applying Newton's second law along the radius and the first law in the vertical direction, then

$$N \sin\theta = \frac{mv^2}{r}$$

and $$N \cos\theta = mg$$

From these two equations, we get

$$\tan\theta = \frac{v^2}{rg} \qquad \text{...(i)}$$

or $$v = \sqrt{rg \tan\theta} \qquad \text{...(ii)}$$

- This is the speed at which car does not slide down even if track is smooth. If track is smooth and speed is less than $\sqrt{rg\tan\theta}$, vehicle will move down so that r gets decreased and if speed is more than this vehicle will move up.

Conical Pendulum

If a small particle of mass m tied to a string is whirled in a horizontal circle, as shown in Fig. 7.8. The arrangement is called the **conical pendulum**. In case of conical pendulum the vertical component of tension balances the weight while its horizontal component provides the necessary centripetal force. Thus,

$$T \sin\theta = \frac{mv^2}{r} \qquad \text{...(i)}$$

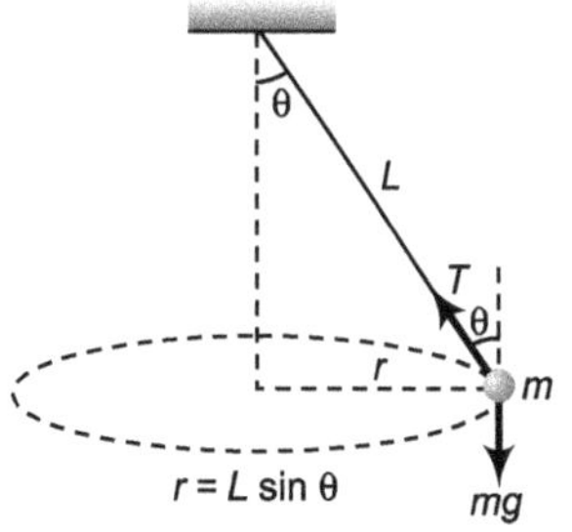

Fig. 7.8

and $$T\cos\theta = mg \qquad \text{...(ii)}$$

From these two equations, we can find

$$v = \sqrt{rg\tan\theta}$$

$\therefore$ Angular speed $\omega = \frac{v}{r} = \sqrt{\frac{g \tan \theta}{r}}$

So the time period of pendulum is

$$T = \frac{2\pi}{\omega} = 2\pi \sqrt{\frac{r}{g \tan \theta}}$$

$$= 2\pi \sqrt{\frac{L \cos \theta}{g}}$$

or $$T = 2\pi \sqrt{\frac{L \cos \theta}{g}}$$

- This is similar to the case, when necessary centripetal force to vehicles is provided by banking. The only difference is that the normal reaction is being replaced by the tension.

'Death Well' or Rotor

In case of 'death well' a person drives a bicycle on a vertical surface of a large wooden well while in case of a rotor at a certain angular speed of rotor a person hangs resting against the wall without any support from the bottom. In death well walls are at rest and person revolves while in case of rotor person is at rest and the walls rotate.

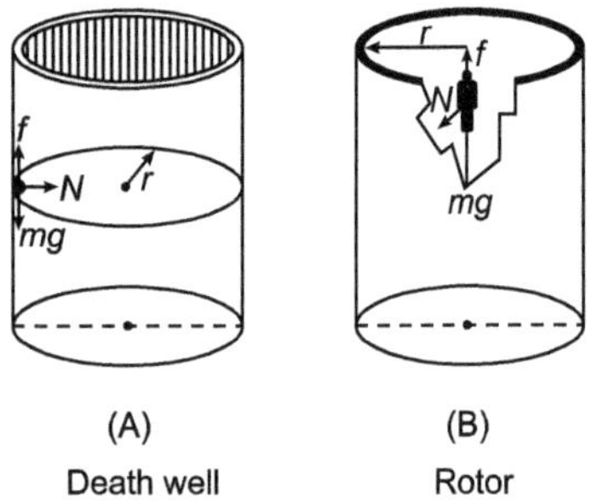

Fig. 7.9

In both cases friction balances the weight of person while reaction provides the centripetal force for circular motion, i.e.

$$f = mg$$

and $$N = \frac{mv^2}{r} = mr\omega^2 \qquad (v = r\omega)$$

A Cyclist on the Bend of a Road

In figure,

$$F = \sqrt{N^2 + f^2}$$

When the cyclist is inclined to the centre of the rounding of its path, the resultant of N, f and mg is directed horizontally to the centre of the circular path of the cycle.

This resultant force imparts a centripetal acceleration to the cyclist.

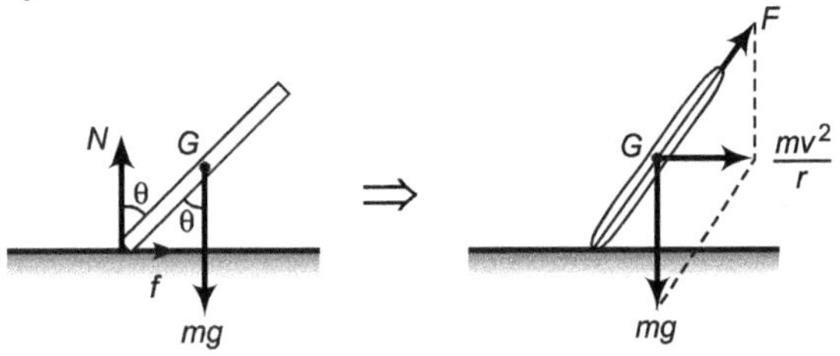

Fig. 7.10

Resultant of N and f, i.e. F should pass through G, the centre of gravity of cyclist (for complete equilibrium, rotational as well as translational). Hence,

$$\tan \theta = \frac{f}{N}, \text{ where } f = \frac{mv^2}{r}$$

and $$N = mg \Rightarrow \therefore \tan \theta = \frac{v^2}{rg}$$

Centrifugal Force

Newton's laws are valid only in inertial frames. In non-inertial frames a pseudo force $-m\mathbf{a}$ has to be applied on a particle of mass m ($\mathbf{a}$ = acceleration of frame of reference). After applying the pseudo force one can apply Newton's laws in their usual form. Now suppose a frame of reference is rotating with constant angular velocity ω in a circle of radius r. Then it will become a non-inertial frame of acceleration $r\omega^2$ towards the centre.

Now, if we see an object of mass m from this frame then obviously a pseudo force of magnitude $mr\omega^2$ will have to be applied to this object in a direction away from the centre. This pseudo force is called the **centrifugal force.** After applying this force we can now apply Newton's laws in their usual form. Following example will illustrate the concept more clearly.

Example 7.4 *A particle of mass m is placed over a horizontal circular table rotating with an angular velocity ω about a vertical axis passing through its centre. The distance of the object from the axis is r. Find the force of friction f between the particle and the table.*

Sol. Let us solve this problem from both frames. The one is a frame fixed on ground and the other is a frame fixed on table itself.

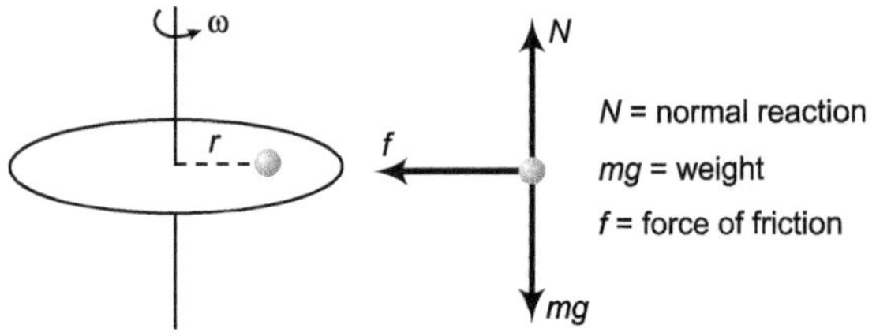

Fig. 7.11

From Frame of Reference Fixed on Ground (Inertial)

Here, N will balance its weight and the force of friction f will provide the necessary centripetal force.

Thus, $$f = mr\omega^2$$

From Frame of Reference Fixed on Table Itself (Non-inertial)

In the free body diagram of particle with respect to table, in addition to above three forces (N, mg and f) a pseudo force of magnitude $mr\omega^2$ will have to be applied in a direction away from the centre. But one thing should be clear that in this frame the particle is in equilibrium, i.e. N will balance its weight in vertical direction while f will balance the pseudo force in horizontal direction.

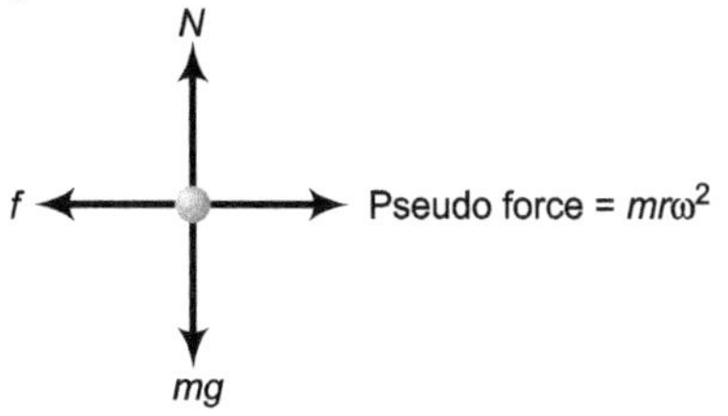

Fig. 7.12

or $$f = mr\omega^2$$

Thus, we see that f comes out to be $mr\omega^2$ from both the frames.

Now, let us take few more examples of circular motion.

Example 7.5 *A simple pendulum is constructed by attaching a bob of mass m to a string of length L fixed at its upper end. The bob oscillates in a vertical circle. It is found that the speed of the bob is v when the string makes an angle α with the vertical. Find the tension in the string and the magnitude of net force on the bob at that instant.*

Sol. (i) The forces acting on the bob are

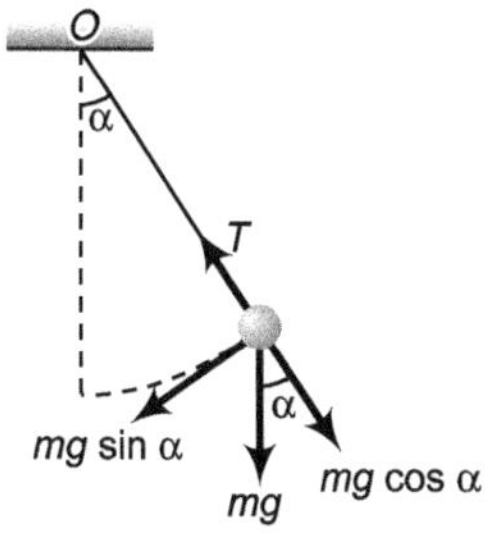

Fig. 7.13

(a) the tension T

(b) the weight mg

As the bob moves in a circle of radius L with centre at O. A centripetal force of magnitude $\frac{mv^2}{L}$ is required towards O. This force will be provided by the resultant of T and $mg\cos\alpha$. Thus,

$$T - mg\cos\alpha = \frac{mv^2}{L}$$

or $$T = m\left(g\cos\alpha + \frac{v^2}{L}\right)$$

(ii) $$|\mathbf{F}_{net}| = \sqrt{(mg\sin\alpha)^2 + \left(\frac{mv^2}{L}\right)^2}$$

$$= m\sqrt{g^2\sin^2\alpha + \frac{v^4}{L^2}}$$

Example 7.6 *A hemispherical bowl of radius R is rotating about its axis of symmetry which is kept vertical. A small ball kept in the bowl rotates with the bowl without slipping on its surface. If the surface of the bowl is smooth and the angle made by the radius through the ball with the vertical is α. Find the angular speed at which the bowl is rotating.*

Sol. Let ω be the angular speed of rotation of the bowl. Two forces are acting on the ball.

1. Normal reaction N
2. Weight mg

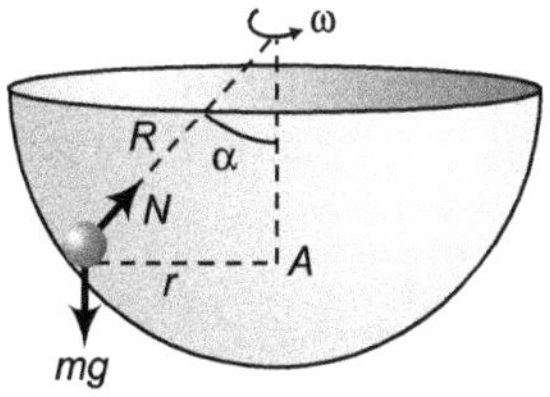

Fig.7.14

The ball is rotating in a circle of radius $r (= R\sin\alpha)$ with centre at A at an angular speed ω. Thus,

$$N\sin\alpha = mr\omega^2$$

$$= mR\omega^2\sin\alpha \quad \text{...(i)}$$

and $$N\cos\alpha = mg \quad \text{...(ii)}$$

Dividing Eqs. (i) by (ii), we get

$$\frac{1}{\cos\alpha} = \frac{\omega^2 R}{g}$$

$$\therefore \quad \omega = \sqrt{\frac{g}{R\cos\alpha}}$$

7.3 Motion in a Vertical Circle

Suppose a particle of mass m is attached to an inextensible light string of length R. The particle is moving in a vertical circle of radius R about a fixed point O. It is imparted a velocity u in horizontal direction at lowest point A. Let v be its velocity at point B of the circle as shown in figure. Here

$$h = R(1 - \cos\theta) \quad \text{...(i)}$$

From conservation of mechanical energy

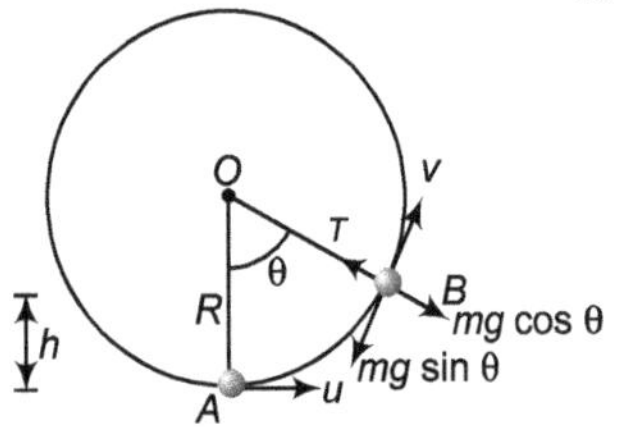

Fig. 7.15

$$\frac{1}{2}m(u^2 - v^2) = mgh \quad \text{or} \quad v^2 = u^2 - 2gh \quad \text{...(ii)}$$

The necessary centripetal force is provided by the resultant of tension T and $mg\cos\theta$

$$\therefore \quad T - mg\cos\theta = \frac{mv^2}{R} \quad \text{...(iii)}$$

Now, following three conditions arise depending on the value of u.

Condition of Looping the Loop ($u \geq \sqrt{5\,gR}$)

The particle will complete the circle, if the string does not slack even at the highest point ($\theta = \pi$). Thus, tension in the string should be greater than or equal to zero ($T \geq 0$) at $\theta = \pi$. In critical case substituting $T = 0$ and $\theta = \pi$ in Eq. (iii), we get

$$mg = \frac{mv^2_{\min}}{R} \text{ or } v^2_{\min} = gR \text{ or } v_{\min} = \sqrt{gR}$$

(at highest point)

Substituting $\theta = \pi$ in Eq. (i), $h = 2R$

Therefore, from Eq. (ii)

$$u^2_{\min} = v^2_{\min} + 2gh$$

or $$u^2_{\min} = gR + 2g(2R) = 5\,gR \text{ or } u_{\min} = \sqrt{5\,gR}$$

Thus, if $u \geq \sqrt{5gR}$, the particle will complete the circle. At $u = \sqrt{5gR}$, velocity at highest point is $v = \sqrt{gR}$ and tension in the string is zero.

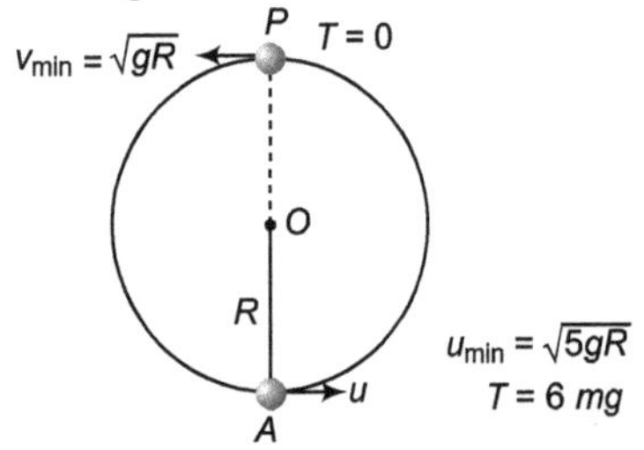

Fig. 7.16

Substituting $\theta = 0°$ and $v = \sqrt{5gR}$ in Eq. (iii), we get $T = 6\,mg$ or in the critical condition tension in the string at lowest position is $6\,mg$. This is shown in figure.

If $u < \sqrt{5gR}$, following two cases are possible.

Condition of Leaving the Circle ($\sqrt{2gR} < u < \sqrt{5gR}$)

If $u < \sqrt{5gR}$, the tension in the string will become zero before reaching the highest point. From Eq. (iii), tension in the string becomes zero ($T = 0$)

where, $\cos\theta = \frac{-v^2}{Rg}$ or $\cos\theta = \frac{2gh - u^2}{Rg}$

Substituting, this value of $\cos\theta$ in Eq. (i), we get

$$\frac{2gh - u^2}{Rg} = 1 - \frac{h}{R}$$

or $$h = \frac{u^2 + Rg}{3g} = h_1 \text{ (say)} \quad \text{...(iv)}$$

or we can say that at height h_1 tension in the string becomes zero. Further, if $u < \sqrt{5gR}$, velocity of the particle becomes zero, when

$$0 = u^2 - 2gh$$

or $$h = \frac{u^2}{2g} = h_2 \text{ (say)} \quad \text{...(v)}$$

i.e. at height h_2 velocity of particle becomes zero.

Now, the particle will leave the circle, if tension in the string becomes zero but velocity is not zero. or $T = 0$ but $v \neq 0$. This is possible only when

$$h_1 < h_2 \quad \text{or} \quad \frac{u^2 + Rg}{3g} < \frac{u^2}{2g}$$

or $$2u^2 + 2Rg < 3u^2 \text{ or } u^2 > 2Rg$$

or $$u > \sqrt{2Rg}$$

Therefore, if $\sqrt{2gR} < u < \sqrt{5gR}$, the particle leaves the circle.

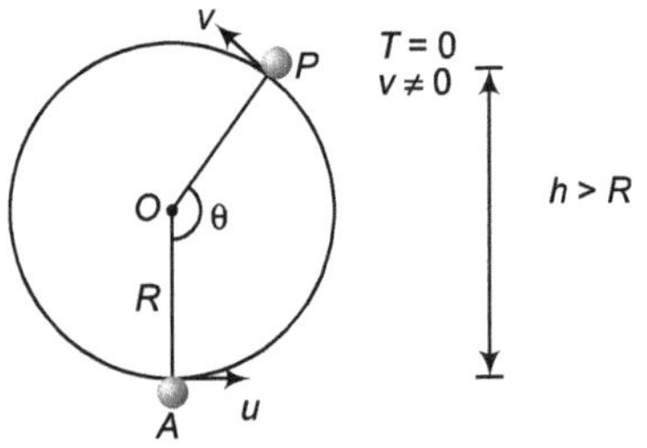

Fig. 7.17

From Eq. (iv), we can see that $h > R$, if $u^2 > 2gR$. Thus, the particle, will leave the circle when $h > R$ or $90° < \theta < 180°$. This situation is shown in the figure.

$$\sqrt{2gR} < u < \sqrt{5gR}$$

or $$90° < \theta < 180°$$

- That after leaving the circle, the particle will follow a parabolic path.

Condition of Oscillation ($0 < u \leq \sqrt{2gR}$)

The particle will oscillate, if velocity of the particle becomes zero but tension in the string is not zero. or $v = 0$, but $T \neq 0$.

This is possible when $h_2 < h_1$

$$\text{or} \qquad \frac{u^2}{2g} < \frac{u^2 + Rg}{3g}$$

$$\text{or} \qquad 3u^2 < 2u^2 + 2Rg$$

$$\text{or} \qquad u^2 < 2Rg$$

$$\text{or} \qquad u < \sqrt{2Rg}$$

Moreover, if $h_1 = h_2, u = \sqrt{2Rg}$ and tension and velocity both becomes zero simultaneously.

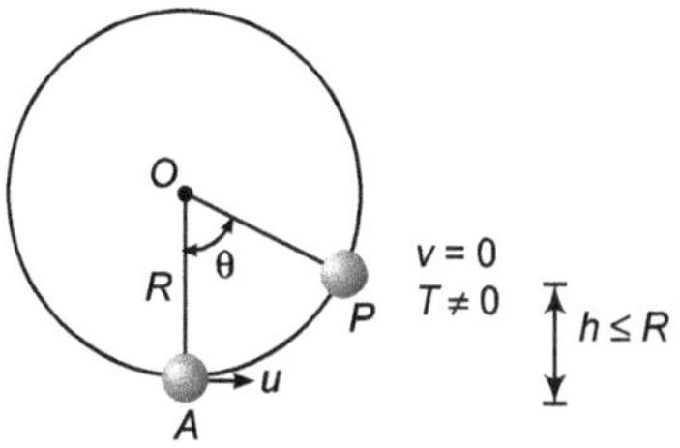

Fig. 7.18

Further, from Eq. (iv), we can see that $h \leq R$, if $u \leq \sqrt{2Rg}$. Thus, for $0 < u \leq \sqrt{2gR}$, particle oscillates in lower half of the circle ($0° < \theta \leq 90°$). This situation is shown in the figure.

$$0 < u \leq \sqrt{2gR}$$

$$\text{or} \qquad 0° < \theta \leq 90°$$

- The above three conditions have been derived for a particle moving in a vertical circle attached to a string. The same conditions apply, if a particle moves inside a smooth spherical shell of radius R. The only difference is that the tension is replaced by the normal reaction N.

Condition of Looping the Loop is $u \geq \sqrt{5gR}$

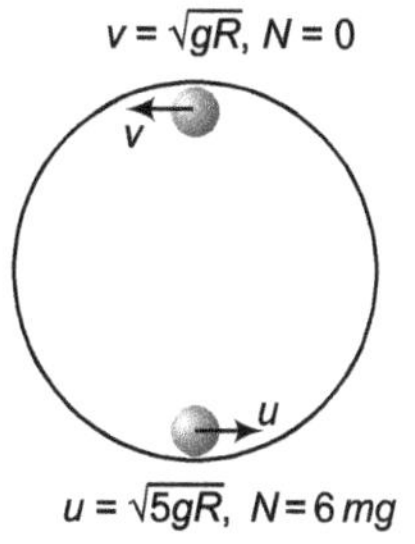

Fig. 7.19

Condition of Leaving the Circle $\sqrt{2gR} < u < \sqrt{5gR}$

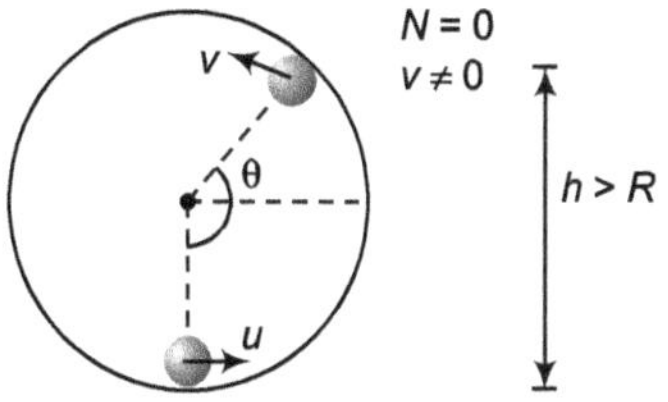

Fig. 7.20

Condition of Oscillation is $0 < u \leq \sqrt{2gR}$

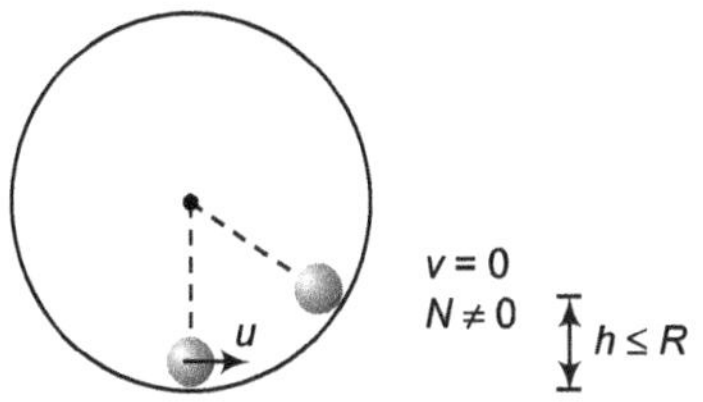

Fig. 7.21

Motion of a Ball over a Smooth Solid Sphere

Suppose a small ball of mass m is given a velocity v over the top of a smooth sphere of radius R. The equation of motion for the ball at the topmost point will be

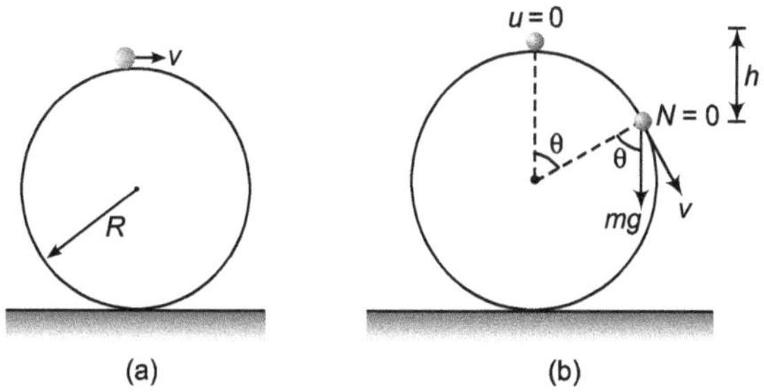

Fig. 7.22

$$mg - N = \frac{mv^2}{R}$$

$$\text{or} \qquad N = mg - \frac{mv^2}{R}$$

From this equation we see that value of N decreases as v increases. Minimum value of N can be zero. Hence,

$$0 = mg - \frac{mv_{\max}^2}{R}$$

$$\text{or} \qquad v_{\max} = \sqrt{Rg}$$

From here we can conclude that ball will lose contact with the sphere right from the beginning, if velocity of ball at topmost point $v > \sqrt{Rg}$. If $v < \sqrt{Rg}$ it will definitely lose contact but after moving certain distance over the sphere. Now let us find the angle θ where the ball loses contact with the sphere if velocity at topmost point is just zero. Fig. 7.22 (b)

$$h = R(1 - \cos\theta) \quad \ldots(\text{i})$$

$$v^2 = 2gh \quad \ldots(\text{ii})$$

$$mg\cos\theta = \frac{mv^2}{R} \quad (\text{as } N = 0) \quad \ldots(\text{iii})$$

Solving Eqs. (i), (ii) and (iii), we get

$$\theta = \cos^{-1}\left(\frac{2}{3}\right) = 48.2^\circ$$

Thus, the ball can move on the sphere maximum upto $\theta = \cos^{-1}\left(\frac{2}{3}\right)$.

Exercise : Find angle θ where the ball will lose contact with the sphere if velocity at topmost point is $u = \frac{v_{\text{max}}}{2} = \frac{\sqrt{gR}}{2}$.

$$\theta = \cos^{-1}\left(\frac{3}{4}\right) = 41.4^\circ$$

Hint : Only Eq. (ii) will change as,

$$v^2 = u^2 + 2gh \quad (u \neq 0)$$

Example 7.7 *A heavy particle hanging from a fixed point by a light inextensible string of length l is projected horizontally with speed $\sqrt{gl}$. Find the speed of the particle and the inclination of the string to the vertical at the instant of the motion when the tension in the string is equal to the weight of the particle.*

Sol. Let $T = mg$ at angle θ as shown in figure.

$$h = l(1 - \cos\theta) \quad \ldots(\text{i})$$

Applying conservation of mechanical energy between points A and B, we get

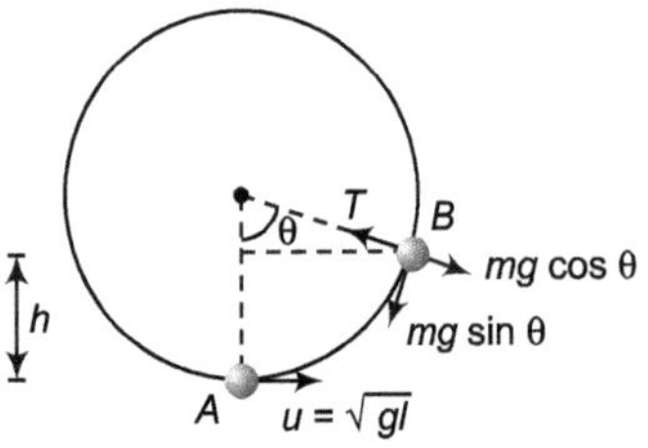

Fig. 7.23

$$\frac{1}{2}m(u^2 - v^2) = mgh$$

Here, $u^2 = gl$ ...(ii)

and v = speed of particle in position B

$\therefore$ $v^2 = u^2 - 2gh$ ...(iii)

Further, $T - mg\cos\theta = \frac{mv^2}{l}$

or $mg - mg\cos\theta = \frac{mv^2}{l}$ $(T = mg)$

or $v^2 = gl(1 - \cos\theta)$ ...(iv)

Substituting values of v^2, u^2 and h from Eqs. (iv), (ii) and (i) in Eq. (iii), we get

$$gl(1 - \cos\theta) = gl - 2gl(1 - \cos\theta)$$

or $\cos\theta = \frac{2}{3}$

or $\theta = \cos^{-1}\left(\frac{2}{3}\right)$

Substituting $\cos\theta = \frac{2}{3}$ in Eq. (iv), we get

$$v = \sqrt{\frac{gl}{3}}$$

Extra Knowledge Points

- In circular motion linear velocity is always tangential to the circle (either clockwise or anti-clockwise).

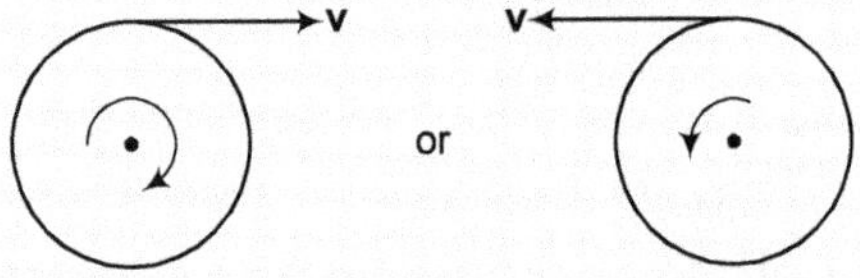

(a) (b)

- In uniform circular motion acceleration (also called the centripetal acceleration) is always towards the centre $\left(= \frac{v^2}{r} \text{ or } r\omega^2\right)$.
- If a particle of mass m is connected to a light rod and whirled in a vertical circle of radius R, then to complete the circle, the minimum velocity of the particle at the bottommost point is not $\sqrt{5gR}$. Because in this case, velocity of the particle at the topmost point can be zero also. Using conservation of mechanical energy between points A and B as shown in figure, we get

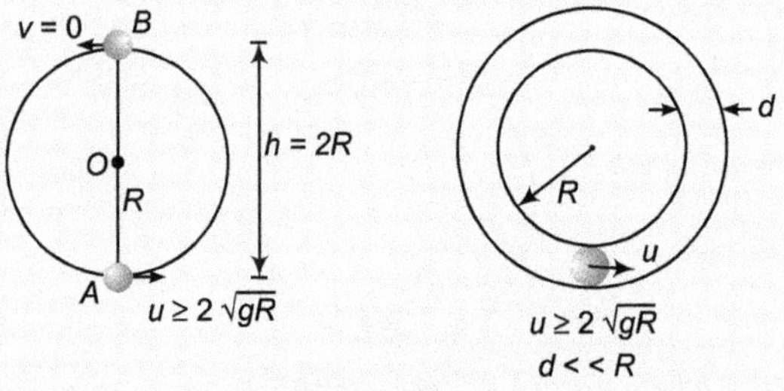

$$\frac{1}{2}m(u^2 - v^2) = mgh$$

or $$\frac{1}{2}mu^2 = mg(2R) \qquad \text{(as } v = 0\text{)}$$

$\therefore$ $$u = 2\sqrt{gR}$$

Therefore, the minimum value of u in this case is $2\sqrt{gR}$.Same is the case when a particle is compelled to move inside a smooth vertical tube as shown in figure.

- In uniform circular motion although the speed of the particle remains constant yet the particle is accelerated due to change in direction of velocity. Therefore the forces acting on the particle in uniform circular motion can be resolved in two directions one along the radius (parallel to acceleration) and another perpendicular to radius (perpendicular to acceleration). Along the radius net force should be equal to $\frac{mv^2}{R}$ and perpendicular to it net force should be zero.
- Oscillation of a pendulum is part of a circular motion. At point A and C since velocity is zero, net centripetal force will be zero. Only tangential force is present. From A and B or C to B speed of the bob increases. Therefore, tangential force is parallel to velocity. From B to A or B to C speed of the bob decreases. Hence, tangential force is antiparallel to velocity.

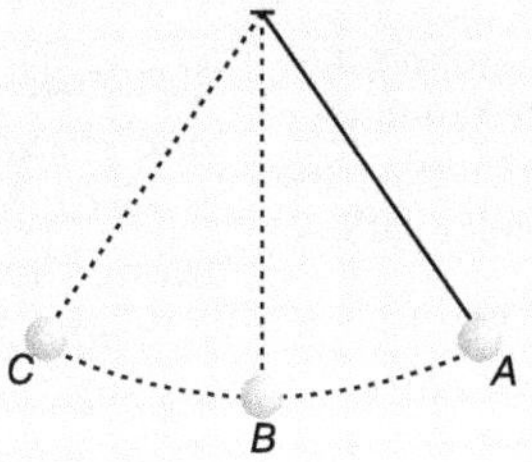

Chapter Summary with Formulae

(i) **Circular Motion** A particle in circular motion may have two types of speeds.

(a) Linear speed v and

(b) Angular speed ω

These two are related by the equation

$$v = R\omega \qquad (R = \text{radius of circular path})$$

(ii) Centripetal acceleration in uniform circular motion :

$$a = \frac{v^2}{R} = R\omega^2$$

Direction of this acceleration is always towards centreand lying in the plane of circle.

(iii) Linear velocity in circular motion is always tangential to the circle and lying in the plane of circle.

(iv) Angular velocity is perpendicular to plane of circle given by screw law.

(v) $\mathbf{v} = \boldsymbol{\omega} \times \mathbf{r}$

(vi) In uniform circular motion body is accelerated.

(vii) In uniform circular motion linear speed and angular velocity are constants. Linear velocity and linear acceleration keep on changing, due to change in their directions.

Additional Examples

Example 1. *For uniform circular motion, does the direction of centripetal acceleration depends upon the sense of rotation ?*

Sol. The direction of centripetal acceleration does not depend on the clockwise or anti-clockwise sense of rotation of the body. It always acts along the radius towards the centre of the circle.

Example 2. *A stone tied to the end of a string is whirled in a circle. If the string breaks, the stone files away tangentially. Why ?*

Sol. The instantaneous velocity of the stone going around the circular path is always along tangent to the circle. When the string breaks, the centripetal force ceases. Due to inertia, the stone continues its motion along tangent to the circular path.

Example 3. *A particle moves in a circle of radius 2.0 cm at a speed given by $v = 4t$, where v is in cm/s and t in seconds.*

(a) Find the tangential acceleration at $t = 1$ s.

(b) Find total acceleration at $t = 1$ s.

Sol. (a) Tangential acceleration

$$a_t = \frac{dv}{dt} \text{ or } a_t = \frac{d}{dt}(4t) = 4 \text{ cm/s}^2$$

i.e. a_t is constant or tangential acceleration at $t = 1$ s is 4 cm/s^2.

(b) Normal acceleration

$$a_n = \frac{v^2}{R} = \frac{(4t)^2}{R} \text{ or } a_n = \frac{16t^2}{2.0} = 8.0\,t^2$$

At $t = 1$ s, $\quad a_n = 8.0$ cm/s^2

$\therefore$ Total acceleration $a = \sqrt{a_t^2 + a_n^2}$

or $\quad a = \sqrt{(4)^2 + (8)^2}$

$$= \sqrt{80} = 4\sqrt{5} \text{ cm/s}^2$$

Example 4. *A particle is projected with a speed u at an angle θ with the horizontal. What is the radius of curvature of the parabola traced out by the projectile at a point, where the particle velocity makes an angle $\frac{\theta}{2}$ with the horizontal?*

Sol. Let v be the velocity at the desired point. Horizontal component of velocity remains unchanged. Hence,

$$v \cos \theta/2 = u \cos \theta$$

$$\therefore \quad v = \frac{u \cos \theta}{\cos \theta/2} \quad \text{... (i)}$$

Radial acceleration is the component of acceleration perpendicular to velocity or

$$a_n = g \cos\left(\frac{\theta}{2}\right) \Rightarrow \therefore \frac{v^2}{R} = g \cos\left(\frac{\theta}{2}\right) \quad \text{... (ii)}$$

Substituting value of v from Eq. (i) in Eq. (ii), we have radius of curvature,

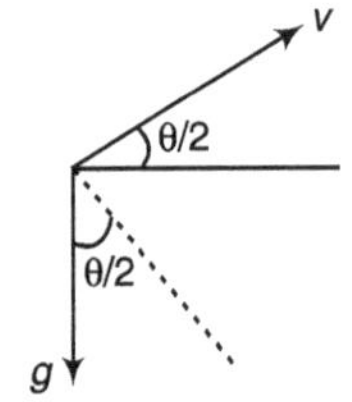

$$R = \frac{\left[\dfrac{u \cos \theta}{\cos\left(\dfrac{\theta}{2}\right)}\right]^2}{g \cos\left(\dfrac{\theta}{2}\right)} = \frac{u^2 \cos^2 \theta}{g \cos^3\left(\dfrac{\theta}{2}\right)}$$

Example 5. *Two particles A and B start at the origin O and travel in opposite directions along the circular path at constant speeds $v_A = 0.7 m/s$ and $v_B = 1.5 m/s$, respectively. Determine the time when they collide and the magnitude of the acceleration of B just before this happens.*

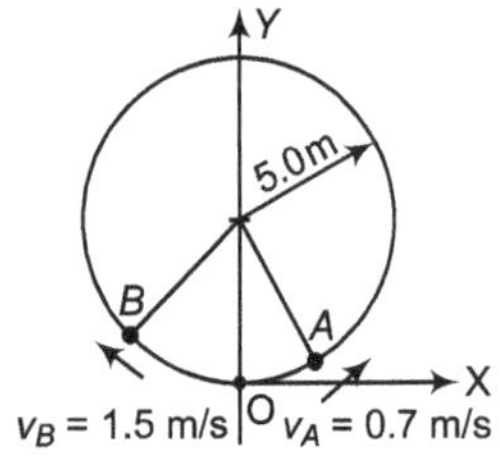

Sol. $1.5t + 0.7t = 2\pi R = 10\pi$

$$\therefore \quad t = \frac{10\pi}{2.2} = 14.3 \text{ s} \Rightarrow a = \frac{v_B^2}{R} = 0.45 \text{ m/s}^2$$

Example 6. *A boy whirls a stone in a horizontal circle of radius 1.5 m and at height 2.0 m above level ground. The string breaks, and the stone flies off horizontally and strikes the ground after travelling a horizontal distance of 10 m. What is the magnitude of the centripetal acceleration of the stone while in circular motion?*

Sol. $t = \sqrt{\frac{2h}{g}} = \sqrt{\frac{2 \times 2}{9.8}} = 0.64$ s

$$v = \frac{10}{t} = 15.63 \text{ m/s}$$

$$\therefore \quad a = \frac{v^2}{R} = 163 \text{ m/s}^2$$

Example 7. *A car is travelling along a circular curve that has a radius of 50 m. If its speed is 16 m/s and is increasing uniformly at 8 m/s^2, determine the magnitude of its acceleration at this instant.*

Sol. $a = \sqrt{a_t^2 + a_n^2} = \sqrt{\left(\frac{dv}{dt}\right)^2 + \left(\frac{v^2}{R}\right)^2}$

$$= \sqrt{(8)^2 + \left(\frac{256}{50}\right)^2} = 9.5 \text{ m/s}^2$$

NCERT Selected Questions

Q 1. A stone tied to the end of a string 80 cm long is whirled in a horizontal circle with a constant speed. If the stone makes 14 revolutions in 25 s, what is the magnitude and direction of acceleration of the stone?

Sol. Radius of the horizontal circle, $r = 80\,\text{cm} = 0.80\text{ m}$

$$n = 14$$
$$t = 25\,\text{s}$$

Angular speed of revolution of the stone,

$$\omega = \frac{\theta}{t} = \frac{2\pi n}{t} = 2\pi\left(\frac{n}{t}\right)$$
$$= 2\times\frac{22}{7}\times\left(\frac{14}{25}\right)$$
$$= \frac{88}{25}\text{ rads}^{-1}$$

∴ Magnitude of acceleration produced in the stone will be equal to the magnitude of centripetal acceleration

$$= r\omega^2$$
$$= 0.80\times\left(\frac{88}{25}\right)^2$$
$$= 9.91\text{ ms}^{-2}$$

The direction of the acceleration is towards the centre.

Q 2. An aircraft executes a horizontal loop of radius 1 km with a speed of 900 kmh^{-1}. Compare its centripetal acceleration with the acceleration due to gravity.

Sol. Here, $r = 1\,\text{km} = 1000\text{ m}$

$$v = 900\text{ kmh}^{-1} = 900\times\frac{1000}{3600}\text{ ms}^{-1}$$
$$= 250\text{ ms}^{-1}$$

The centripetal acceleration of the aircraft is given by

$$a = \frac{v^2}{r} = \frac{(250)^2}{1000} = \frac{62500}{1000}$$
$$= 62.5\text{ ms}^{-2}$$

Acceleration due to gravity,

$$g = 9.8\text{ ms}^{-2}$$

$$\therefore \quad \frac{\text{Centripetal acceleration}}{\text{Acceleration due to gravity}} = \frac{a}{g} = \frac{62.5}{9.8}$$

or $\quad \frac{a}{g} = 6.4 \quad$ or $\quad a = 6.4\,g$

Q 3. A cyclist is ridding with a speed of 27 kmh^{-1}. As he approaches a circular turn on the road of radius 80 m, he applies brakes and reduces his speed at the constant rate of 0.5 ms^{-1} every second. What is the magnitude and direction of the net acceleration of the cyclist on the circular turn?

Sol. Speed of cyclist $v = 27\text{ kmh}^{-1} = 27\times\frac{5}{18}\text{ ms}^{-1} = \frac{15}{2}\text{ ms}^{-1}$

Radius of circular turn, $r = 80$ m.

Here, the cyclist will be acted upon by two accelerations a_c and a_t.

Centripetal acceleration (a_c) is given by

$$a_c = \frac{v^2}{r} = \frac{\left(\frac{15}{2}\right)^2}{80} = 0.703\text{ ms}^{-2}$$

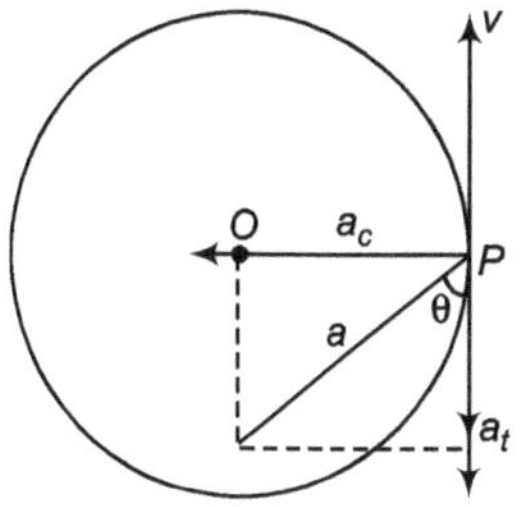

Let P be the point at which the cyclist applies brakes, then the tangential acceleration a_t (which will be negative) will act opposite to velocity of cyclist and is given by

$$a_t = \frac{\text{Change in velocity}}{\text{Time}}$$
$$= 0.5\text{ ms}^{-2}$$

As both a_c and a_t act along the radius of circle and tangent to the circle respectively, so a_c and a_t are mutually perpendicular, thus the magnitude of the net acceleration a, is given by

$$a = \sqrt{a_c^2 + a_t^2}$$
$$= \sqrt{(0.703)^2 + (0.5)^2}$$
$$= 0.863\text{ ms}^{-2}$$

Direction of a Let θ be the angle made by the net acceleration with the velocity of cyclist, then

$$\tan\theta = \frac{a_c}{a_t} = \frac{0.703}{0.5}$$
$$= 1.406$$
$$\therefore \quad \theta = 54.44°$$

Q 4. Read each statement below carefully and state with reasons, if it is true or false.

(a) The net acceleration of a particle in the circular motion is always along the radius of the circle towards the centre.

(b) The velocity vector of a particle at a point is always along the tangent to the path of the particle at that point.

(c) The acceleration vector of a particle in uniform circular motion averaged over one cycle is a null vector.

Sol. (a) The statement is false since, the centripetal acceleration is towards the centre only in the case of uniform circular motion (constant speed) for which it is true.

(b) True, the velocity of a particle is always along the tangent to the path of the particle at that point either in rectilinear or circular or curvilinear motion.

(c) True, because in one cycle change in velocity = 0

***Q* 5.** One end of a string of length l is connected to a particle of mass m and the other to a small peg on smooth horizontal table. If the particle moves in a circle with speed v, the net force on the particle (directed towards the centre) is

(i) T (ii) $T - \frac{mv^2}{l}$

(iii) $T + \frac{mv^2}{l}$ (iv) 0

Sol. (i) The net force on the particle is T, because tension T provides the necessary centripetal force, i.e. $T = \frac{mv^2}{l}$

***Q* 6.** A stone of mass 0.25 kg tied to the end of a string is whirled round in a circle of radius 1.5 m with speed 40 rev/min in a horizontal plane. What is the tension in the string? What is the maximum speed with which the stone can be whirled around, if the string can withstand a maximum tension of 200 N?

Sol. Frequency of revolution of stone,

$$f = 40 \text{rev/min} = \frac{40}{60} \text{rev/s}$$

Angular speed of the stone, $\omega = 2\pi f$

$$= 2\pi \times \frac{40}{60} = \frac{4\pi}{3} \text{rad s}^{-1}$$

The centripetal force is provided by the tension (T) in the string

i.e. $$T = \frac{mv^2}{r} = mr\omega^2$$

$$= 0.25 \times 1.5 \times \left(\frac{4\pi}{3}\right)^2 \text{N}$$

$$= 6.58 \text{ N} \approx 6.6 \text{ N}$$

As the string can withstand a maximum tension of 200 N

$\therefore$ $$T_{max} = \frac{mv_{max}^2}{r}$$

or $$v_{max} = \sqrt{\frac{rT_{max}}{m}}$$

$$= \sqrt{\frac{1.5 \times 200}{0.25}}$$

$$= 34.64 \text{ ms}^{-1}$$

$$\approx 35.0 \text{ ms}^{-1}$$

$\therefore$ $$T = 6.6 \text{ N},$$

$$v_{max} = 35.0 \text{ ms}^{-1}$$

***Q* 7.** A stone of mass m tied to the end of a string revolves in a vertical circle of radius R. The net forces at the lowest and highest points of the circle directed vertically downwards are (choose the correct alternative).

Lowest point	Highest point
(a) $mg - T_1$	$mg + T_2$
(b) $mg + T_1$	$mg - T_2$
(c) $mg + T_1 - (mv_1^2)/R$	$mg - T_2 + (mv_1^2)/R$
(d) $mg - T_1 - (mv_1^2)/R$	$mg + T_2 + (mv_1^2)/R$

Here, T_1, T_2 and (v_1, v_2) denote the tension in the string (and the speed of the stone) at the lowest and the highest point respectively.

Sol. In the figure shown, here L and H show the lowest and highest points respectively.

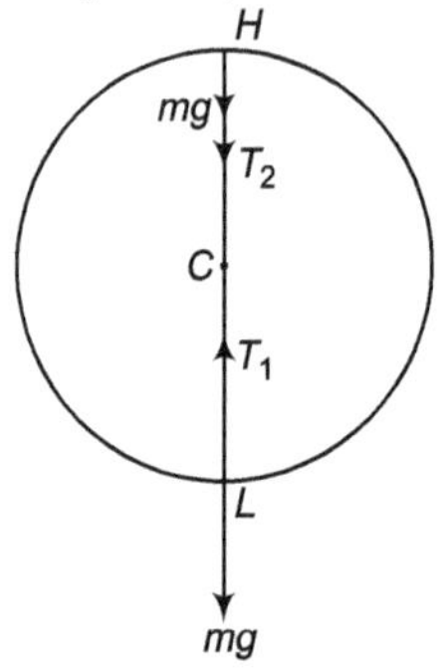

At point *L* T_1 acts towards the centre of the circle and mg acts vertically downward.

$\therefore$ Net force on the stone at the lowest point in the downward direction = $mg - T_1$

At point *H* Both T_2 and mg act vertically downward towards the centre of the circle.

$\therefore$ Net force on the stone at the highest point in the downward direction = $T_2 + mg$

So, option (a) is the correct alternative.

***Q* 8.** An aircraft executes a horizontal loop at a speed of 720 km/h with its wings banked at 15°. What is the radius of the loop?

Sol. Speed of aircraft, $v = 720 \text{ kmh}^{-1}$

$$= 720 \times \frac{5}{18} = 200 \text{ ms}^{-1}$$

For a banked curve, $\tan\theta = \frac{v^2}{rg}$

or $$r = \frac{v^2}{g \tan\theta}$$

$$= \frac{(200)^2}{9.8 \times \tan 15°}$$

$$= 15.24 \times 10^3 \text{ m}$$

$$= 15.24 \text{ km}$$

Q 9. A train rounds a circular bend of radius 30 m at a speed of 54 kmh^{-1}. The mass of the train is 10^6 kg. What is the angle of banking required to prevent wearing out of the rail?

Sol. Radius of circular bend, $r = 30$ m

Speed of train $= v = 54\ kmh^{-1}$

$$= 54 \times \frac{5}{18}$$

$$= 15\ ms^{-1}$$

The angle of banking is given by

$$\tan\theta = \frac{v^2}{rg}$$

$$= (15)^2 / 30 \times 9.8$$

$$= 0.7653$$

$$\therefore \quad \theta = 37°25' = 37.42°$$

Q 10. You may have seen in a circus, a motorcyclist driving in vertical loop inside a 'death well' (a hollow spherical chamber with holes, so that spectators can watch from outside). What is the minimum speed required at the uppermost position to perform a vertical loop, if the radius of the chamber is 25 m?

Sol. When the motorcyclist is at the highest point of the 'death well', the normal reaction N on him by the ceiling of the chamber acts downwards. His weight mg also acts downwards.

$$\therefore \quad N + mg = \frac{mv^2}{r} \quad \ldots\text{(i)}$$

where, v = speed of the motorcyclist,

m = mass of the motorcyclist (mass of motorcycle + driver)

The minimum speed required to perform a vertical loop is given by Eq. (i), when $N = 0$

$$mg = \frac{mv_{min}^2}{r}$$

$$\therefore \quad v_{min} = \sqrt{gr} \quad \ldots\text{(ii)}$$

Here, r = radius of the loop

$$g = 9.8\ ms^{-2}$$

$\therefore$ From Eq. (ii), $v_{min} = \sqrt{9.8 \times 25}$

$$= 15.65\ ms^{-1}$$

$$\approx 16\ ms^{-1}$$

Q 11. A 70 kg man stands in contact against the inner wall of a hollow cylindrical drum of radius 3 m rotating about its vertical axis with 200 rev/min. The coefficient of friction between the wall and his clothing is 0.15. What is the minimum rotational speed of the cylinder to enable the man to remain stuck to the wall (without falling) when the floor is suddenly removed?

Sol. The cylinder being vertical, the normal reaction of the wall on the man acts horizontally and provides the necessary centripetal force

$$\therefore \quad N = mr\omega^2 \quad \ldots\text{(i)}$$

The frictional force f, acting upwards balances his weight

i.e. $f = mg$ ...(ii)

The man will remain stuck to the wall after the floor is removed, i.e. he will continue to rotate with the cylinder without slipping

if $f \le \mu N$

or $mg \le \mu m r\omega^2$

or $\omega^2 \ge \frac{g}{\mu r}$

or $\omega \ge \sqrt{\frac{g}{\mu r}}$ ...(iii)

$\therefore$ The minimum angular speed of rotation of the cylindrical drum is given by

$$\omega_{min} = \sqrt{\frac{g}{\mu r}} \quad \ldots\text{(iv)}$$

Given $\mu = 0.15$, $r = 3$ m, $g = 9.8\ ms^{-2}$ ...(v)

$\therefore$ From Eqs. (iv) and (v), we get

$$\omega_{min} = \sqrt{\frac{9.8}{0.15 \times 3}}$$

$$= 4.67\ \text{rad/s} \approx 5\ \text{rad/s}$$

Objective Problems

[Level 1]

Kinematics of Circular Motion

1. A particle is moving on a circular path of 10 m radius. At any instant of time its speed is 5 m/s and the speed is increasing at a rate of 2 m/s^2. At this instant the magnitude of the net acceleration will be
(a) 3.2 m/s^2 (b) 2 m/s^2
(c) 2.5 m/s^2 (d) 4.3 m/s^2

2. In a typical projectile motion tangential acceleration at the topmost point P of the trajectory is

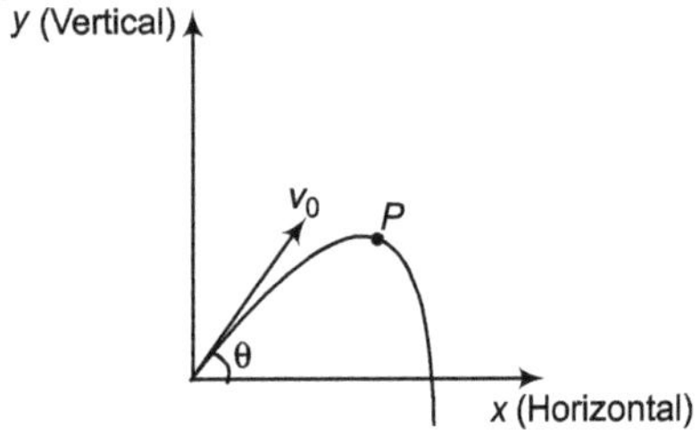

(a) g (b) $g \cos\theta$
(c) 0 (d) None of these

3. A point on the rim of a flywheel has a peripheral speed of 10 m/s at an instant when it is decreasing at the rate of 60 m/s^2. If the magnitude of the total acceleration of the point at this instant is 100 m/s^2, the radius of the flywheel is
(a) 1.25 m (b) 12.5 m
(c) 25 m (d) 2.5 m

4. A particle is moving on a circular track of radius 30 cm with a constant speed of 6 m/s. Its acceleration is
(a) zero (b) 120 m/s^2
(c) 1.2 m/s^2 (d) 36 m/s^2

5. In case of a uniform circular motion, velocity and acceleration are
(a) perpendicular (b) in same direction
(c) in opposite direction (d) not related to each other

6. Let a_r and a_t represent radial and tangential accelerations. The motion of a particle may be circular, if
(a) $a_r = 0, a_t = 0$ (b) $a_r = 0, a_t \neq 0$
(c) $a_r \neq 0, a_t = 0$ (d) None of these

7. The speed of a particle moving in a circle is increasing. The dot product of its acceleration and velocity is
(a) negative
(b) zero
(c) positive
(d) may be positive or negative

8. The ratio of angular speeds of minute hand and hour hand of a watch is
(a) 1 : 12 (b) 6 : 1
(c) 12 : 1 (d) 1 : 6

9. The angular speed of a flywheel making 120 rev/min is
(a) 2π rad/s (b) $4\pi^2$ rad/s
(c) π rad/s (d) 4π rad/s

10. A particle moves with constant speed v along a circular path of radius r and completes the circle in time T. The acceleration of the particle is
(a) $\frac{2\pi v}{T}$ (b) $\frac{2\pi r}{T}$
(c) $\frac{2\pi r^2}{T}$ (d) $\frac{2\pi v^2}{T}$

11. A body is moving in a circular path with acceleration a. If its speed gets doubled, find the ratio of acceleration after and before the speed is changed
(a) 1 : 4 (b) 1 : 2
(c) 2 : 1 (d) 4 : 1

12. The circular orbit of two satellites have radii r_1 and r_2 respectively ($r_1 < r_2$). If angular velocities of satellites are same, then their centripetal accelerations are related as
(a) $a_1 > a_2$ (b) $a_1 = a_2$
(c) $a_1 < a_2$ (d) Data insufficient

13. The wheel of a toy car rotates about a fixed axis. It slows down from 400 rps to 200 rps in 2 s. Then, its angular retardation in rad/s^2 is (rps = revolutions per second)
(a) 200π (b) 100
(c) 400π (d) None of these

14. A car is circulating on a circular path of radius r. At some instant its velocity is v and rate of increase of speed is a. The resultant acceleration of the car will be
(a) $\sqrt{\frac{v^2}{a^2} + r^2}$ (b) $\sqrt{\frac{v^2}{r} + a}$
(c) $\sqrt{\frac{v^4}{r^2} + a^2}$ (d) $\left(\frac{v^2}{r} + a\right)$

15. A point starts from rest and moves along a circular path with a constant tangential acceleration. After one rotation, the ratio of its radial acceleration to its tangential acceleration will be equal to
(a) 1 (b) 2π
(c) $\frac{1}{2}\pi$ (d) 4π

16. A particle moves in a circular path of radius R with an angular velocity $\omega = a - bt$, where a and b are positive constants and t is time. The magnitude of the acceleration of the particle after time $\frac{2a}{b}$ is

(a) $\frac{a}{R}$ (b) a^2R

(c) $R(a^2 + b)$ (d) $R\sqrt{a^4 + b^2}$

17. A car wheel is rotated to uniform angular acceleration about its axis. Initially its angular velocity is zero. It rotates through an angle θ_1 in the first 2 s. In the next 2 s, it rotates through an additional angle θ_2, the ratio of $\frac{\theta_2}{\theta_1}$ is

(a) 1 (b) 2

(c) 3 (d) 4

Dynamics of Circular Motion

18. A coin, placed on a rotating turn-table slips, when it is placed at a distance of 9 cm from the centre. If the angular velocity of the turn-table is trippled, it will just slip, if its distance from the centre is

(a) 27 cm (b) 9 cm

(c) 3 cm (d) 1 cm

19. If mass, speed and radius of the circle of a particle moving uniformly in a circular path are all increased by 50%, the necessary force required to maintain the body moving in the circular path will have to be increased by

(a) 225% (b) 125%

(c) 150% (d) 100%

20. A ball is placed on a smooth inclined plane of inclination $\theta = 30°$ to the horizontal, which is rotating at frequency 0.5 Hz about a vertical axis passing through its lower end. At what distance from the lower end does the ball remain at rest?

(a) 0.87 m (b) 0.33 m (c) 0.5 m (d) 0.67 m

21. A person wants to drive on the vertical surface of a large cylindrical wooden 'well' commonly known as 'death well' in a circus. The radius of the well is R and the coefficient of friction between the tyres of the motorcycle and the wall of the well is μ_s. The minimum speed, the motorcycle must have in order to prevent slipping, should be

(a) $\sqrt{\frac{Rg}{\mu_s}}$ (b) $\sqrt{\frac{\mu_s}{Rg}}$ (c) $\sqrt{\frac{\mu_s g}{R}}$ (d) $\sqrt{\frac{R}{\mu_s g}}$

22. The maximum tension that an inextensible ring of radius 1m and mass density 0.1 kg m^{-1} can bear is 40 N. The maximum angular velocity with which it can be rotated in a circular path is

(a) 20 rad/s (b) 18 rad/s

(c) 16 rad/s (d) 15 rad/s

23. A motorcyclist wants to drive on the vertical surface of wooden 'well' of radius 5 m, with a minimum speed of $5\sqrt{5}$ m/s. The minimum value of coefficient of friction between the tyres and the wall of the well must be (Take $g = 10$ m/s^2)

(a) 0.10 (b) 0.20

(c) 0.30 (d) 0.40

24. A particle of mass 2 kg is moving along a circular path of radius 1 m. If its angular speed is 2π rad s^{-1}, the centripetal force on it is

(a) 4π N (b) 8π N

(c) $4\pi^4$ N (d) $8\pi^2$ N

25. Two particles of equal masses are revolving in circular paths of radii r_1 and r_2 respectively with the same speed. The ratio of their centripetal forces is

(a) $\frac{r_2}{r_1}$ (b) $\sqrt{\frac{r_2}{r_1}}$

(c) $\left(\frac{r_1}{r_2}\right)^2$ (d) $\left(\frac{r_2}{r_1}\right)^2$

26. A mass of 100 g is tied to one end of a string 2 m long. The body is revolving in a horizontal circle making a maximum of 200 rev/min. The other end of the string is fixed at the centre of the circle of revolution. The maximum tension that the string can bear is (approximately)

(a) 8.76 N (b) 8.94 N

(c) 89.42 N (d) 87.64 N

27. A particle of mass m is executing uniform circular motion on a path of radius r. If p is the magnitude of its linear momentum. The radial force acting on the particle is

(a) pmr (b) $\frac{rm}{p}$

(c) $\frac{mp^2}{r}$ (d) $\frac{p^2}{rm}$

28. A mass of 2 kg is whirled in a horizontal circle by means of a string at an initial speed of 5 rev/min. Keeping the radius constant the tension in the string is doubled. The new speed is nearly

(a) $\frac{5}{\sqrt{2}}$ rpm (b) 10 rpm

(c) $10\sqrt{2}$ rpm (d) $5\sqrt{2}$ rpm

29. A stone of mass of 16 kg is attached to a string 144 m long and is whirled in a horizontal circle. The maximum tension the string can withstand is 16 N. The maximum velocity of revolution that can be given to the stone without breaking it, will be

(a) 20 ms^{-1} (b) 16 ms^{-1}

(c) 14 ms^{-1} (d) 12 ms^{-1}

30. Three identical particles are joined together by a thread as shown in figure. All the three particles are moving in a horizontal plane. If the velocity of the outermost particle is v_0, then the ratio of tensions in the three sections of the string is

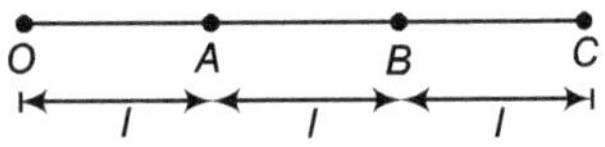

(a) 3 : 5 : 7 (b) 3 : 4 : 5
(c) 7 : 11 : 6 (d) 3 : 5 : 6

31. Toy cart tied to the end of an unstretched string of length a, when revolved moves in a horizontal circle of radius $2a$ with a time period T. Now, the toy cart is speeded up until it moves in a horizontal circle of radius $3a$ with a period T. If Hooke's law $(F = kx)$ holds, then

(a) $T' = \sqrt{\frac{3}{2}}T$ (b) $T' = \left(\frac{\sqrt{3}}{2}\right)T$
(c) $T' = \left(\frac{3}{2}\right)T$ (d) $T' = T$

32. A national roadway bridge over a canal is in the form of an arc of a circle of radius 49 m. What is the maximum speed with which a car can move without leaving the ground at the highest point? (Take $g = 9.8\,\text{m/s}^2$)

(a) 19.6 m/s (b) 40 m/s
(c) 22 m/s (d) None of these

Vertical Circular Motion

33. A car when passes through a convex bridge exerts a force on it which is equal to

(a) $Mg + \frac{Mv^2}{r}$ (b) $\frac{Mv^2}{r}$
(c) $Mg - \frac{Mv^2}{r}$ (d) None of these

34. A block of mass m at the end of a string is whirled round in a vertical circle of radius R. The critical speed of the block at top of its swing below which the string would slacken before the block reaches the bottom is?

(a) $\sqrt{5\,Rg}$ (b) $\sqrt{3\,Rg}$
(c) $\sqrt{2\,Rg}$ (d) $\sqrt{Rg}$

35. A sphere is suspended by a thread of length l. What minimum horizontal velocity has to be imparted to the ball for it to reach the height of the suspension?

(a) $\sqrt{5\,gl}$ (b) $2\,gl$
(c) $\sqrt{gl}$ (d) $\sqrt{2gl}$

36. A pendulum bob on a 2m string is displaced 60° from the vertical and then released. What is the speed of the bob as it passes through the lowest point in its path?

(a) $\sqrt{2}$ m/s (b) $\sqrt{9.8}$ m/s
(c) 4.43 m/s (d) $\frac{1}{\sqrt{2}}$ m/s

37. A particle is moving in a vertical circle. The tensions in the string when passing through two positions at angles 30° and 60° from vertical (lowest position) are T_1 and T_2 respectively, then

(a) $T_1 = T_2$ (b) $T_2 > T_1$
(c) $T_1 > T_2$ (d) Data insufficient

38. A body of mass 1 kg is moving in a vertical circular path of radius 1 m. The difference between the kinetic energies at its highest and lowest positions is

(a) 20 J (b) 10 J
(c) $4\sqrt{5}$ J (d) $10(\sqrt{5}-1)$ J

39. A particle of mass m is being circulated on a vertical circle of radius r. If the speed of particle at the highest point be v, then

(a) $mg = \frac{mv^2}{r}$ (b) $mg > \frac{mv^2}{r}$
(c) $mg \le \frac{mv^2}{r}$ (d) $mg \ge \frac{mv^2}{r}$

40. The string of a pendulum is horizontal. The mass of bob attached to it is m. Now, the string is released. The tension in the string in the lowest positions, is

(a) mg (b) $2\,mg$
(c) $3\,mg$ (d) $4\,mg$

41. A bucket full of water is rotated in a vertical circle of radius R. If the water does not split out, the speed of the bucket at topmost point will be

(a) $\sqrt{Rg}$ (b) $\sqrt{5gR}$
(c) $\sqrt{2Rg}$ (d) $\sqrt{\left(\frac{R}{g}\right)}$

42. A pendulum bob has a speed of 3 m/s at its lowest position. The pendulum is 0.5 m long. The speed of the bob, when string makes an angle of 60° to the vertical is $(g = 10\,\text{m/s}^2)$

(a) 2 m/s (b) $\frac{1}{2}$ m/s
(c) 1 m/s (d) 2.5 m/s

43. A simple pendulum of length l has a maximum angular displacement θ. The maximum kinetic energy of the bob of mass m will be

(a) $mgl\,(1-\cos\theta)$ (b) $mgl\cos\theta$
(c) $mgl\sin\theta$ (d) None of these

44. A small ball is pushed from a height h along a smooth hemispherical bowl of radius R. With what speed should the ball be pushed so that it just reaches the top of the opposite end of the bowl?

(a) $\sqrt{2gh}$
(b) $\sqrt{2g\,(R+h)}$
(c) $\sqrt{2g\,(R-h)}$
(d) None of the above

45. A child is swinging a swing. Minimum and the maximum heights of swing from the earth's surface are 0.75 m and 2 m respectively. The maximum velocity of this swing is $(g = 10\,\text{m/s}^2)$

(a) 5 m/s (b) 10 m/s
(c) 15 m/s (d) 20 m/s

46. A small body of mass m slides without friction from the top of a hemisphere of radius r. At what height will the body be detached from the centre of the hemisphere?

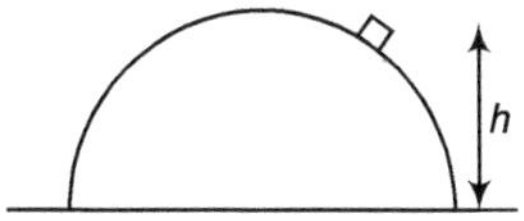

(a) $h = \frac{r}{2}$ (b) $h = \frac{r}{3}$
(c) $h = \frac{2r}{3}$ (d) $h = \frac{r}{4}$

47. A frictionless track $ABCDE$ ends in a circular loop of radius R. A body slides down the track from point A which is at height $h = 5\,\text{cm}$. Maximum value of R for a body to complete the loop successfully is

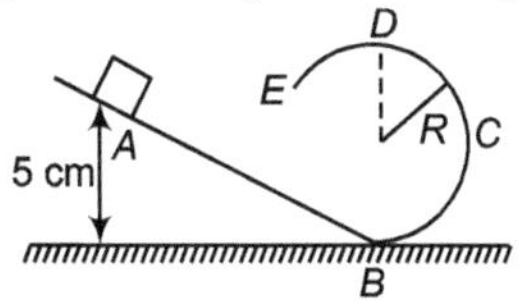

(a) 2 cm (b) $\frac{10}{3}$ cm
(c) $\frac{15}{4}$ cm (d) $\frac{18}{3}$ cm

48. A small sphere of mass m is suspended by a thread of length l. It is raised upto the height of suspension with thread fully stretched and released. Then, the maximum tension in thread will be

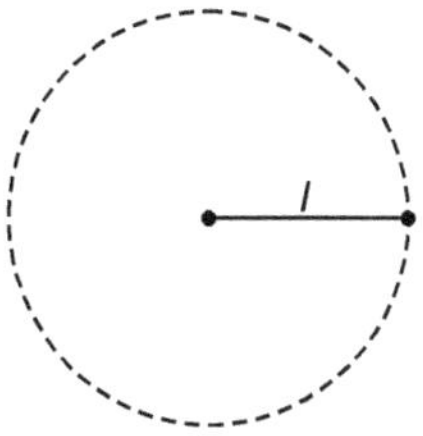

(a) mg (b) $2\,mg$
(c) $3\,mg$ (d) $6\,mg$

49. A stone of mass 1 kg is tied to the end of a string 1 m long. It is whirled in a vertical circle. The velocity of the stone at the bottom of the circle is just sufficient to take it to the top of circle without slackening of the string. What is the tension in the string at the top of the circle? (Take $g = 10\,\text{ms}^{-2}$)

(a) Zero (b) 1 N (c) $\sqrt{10}$ N (d) 10 N

50. A block of mass m slides from the rim of a hemispherical bowl of radius R. The velocity of the block at the bottom will be

(a) $\sqrt{3Rg}$ (b) $\sqrt{2gR}$
(c) $\sqrt{Rg}$ (d) $\sqrt{4gR}$

51. A body is moving in a vertical circle of radius r such that the string is just taut at its highest point. The speed of the particle when the string is horizontal is

(a) $\sqrt{gr}$ (b) $\sqrt{2gR}$ (c) $\sqrt{3gr}$ (d) $\sqrt{4gR}$

52. A stone is attached to one end of a string and rotated in a vertical circle. If string breaks at the position of maximum tension, it will break at

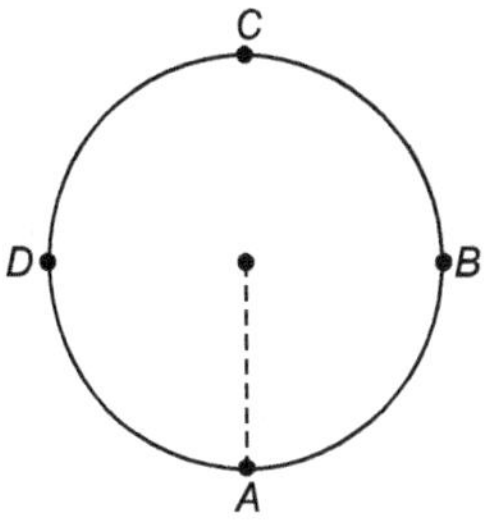

(a) A (b) B (c) C (d) D

Miscellaneous Problems

53. A particle moves in a uniform circular motion. Choose the wrong statement.

(a) The particle moves with constant speed
(b) The acceleration is always normal to the velocity
(c) The particle moves with uniform acceleration
(d) The particle moves with variable velocity

54. A circular disc of radius R is rotating about its axis O with a uniform angular velocity ω rad/s as shown in the figure. The magnitude of the relative velocity of point A relative to point B on the disc is

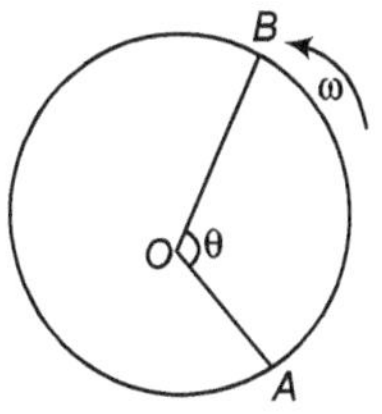

(a) zero (b) $R\omega \sin\left(\frac{\theta}{2}\right)$
(c) $2R\omega \sin\left(\frac{\theta}{2}\right)$ (d) $\sqrt{3}\,R\omega \sin\left(\frac{\theta}{2}\right)$

55. A circular curve of a highway is designed for traffic moving at 72 km/h. If the radius of the curved path is 100 m, the correct angle of banking of the road should be

(a) $\tan^{-1}\left(\frac{2}{5}\right)$ (b) $\tan^{-1}\left(\frac{3}{5}\right)$ (c) $\tan^{-1}\left(\frac{1}{5}\right)$ (d) $\tan^{-1}\left(\frac{1}{4}\right)$

56. A stone is tied with a string and is rotated in a circle horizontally. When the string suddenly breaks, the stone will move
(a) tangential to the motion
(b) away from the centre
(c) towards the centre
(d) None of the above

57. When the angular velocity of a uniformly rotating body has increased thrice, the resultant of forces applied to it increases by 60 N. Find the accelerations of the body in the two cases. The mass of the body, $m = 3$ kg
(a) $2.5 \text{ ms}^{-2}, 7.5 \text{ ms}^{-2}$
(b) $7.5 \text{ ms}^{-2}, 22.5 \text{ ms}^{-2}$
(c) $5 \text{ ms}^{-2}, 45 \text{ ms}^{-2}$
(d) $2.5 \text{ ms}^{-2}, 22.5 \text{ ms}^{-2}$

58. A particle is moving along a circular path of radius 5 m with a uniform speed 5 ms^{-1}. What will be the average acceleration when the particle completes half revolution?
(a) Zero (b) 10 ms^{-2}
(c) $10\pi \text{ ms}^{-2}$ (d) $\frac{10}{\pi} \text{ ms}^{-2}$

59. If the banking angle of curved road is given by $\tan^{-1}\left(\frac{3}{5}\right)$ and the radius of curvature of the road is 6m, then the safe driving speed should not exceed ($g = 10 \text{ m/s}^{-2}$)
(a) 86.4 km/h (b) 43.2 km/h
(c) 21.6 km/h (d) 30.4 km/h

60. A particle moving along a circular path due to a centripetal force having constant magnitude is an example of motion with
(a) constant speed and velocity
(b) variable speed and variable velocity
(c) variable speed and constant velocity
(d) constant speed and variable velocity

61. A particle moves with constant angular velocity in a circle. During the motion its
(a) energy is conserved
(b) momentum is conserved
(c) energy and momentum both are conserved
(d) None of the above

62. A car is moving on a circular path and takes a turn. If R_1 and R_2 be the reactions on the inner and outer wheels respectively, then
(a) $R_1 = R_2$ (b) $R_1 < R_2$
(c) $R_1 > R_2$ (d) $R_1 \geq R_2$

63. An unbanked curve has a radius of 60 m. The maximum speed at which a car can make a turn, if the coefficient of static friction is 0.75 is
(a) 2.1 m/s (b) 14 m/s
(c) 21 m/s (d) 7 m/s

64. A particle is moving in a circle with uniform speed v. In moving from a point to another diametrically opposite point
(a) the momentum changes by mv
(b) the momentum changes by $2\,mv$
(c) the kinetic energy changes by $\left(\frac{1}{2}\right) mv^2$
(d) the kinetic energy changes by mv^2

65. An object is moving in a circle of radius 100 m with a constant speed of 31.4 m/s. What is its average speed for one complete revolution?
(a) Zero
(b) 31.4 m/s
(c) 3.14 m/s
(d) $\sqrt{2} \times 31.4$ m/s

66. In 1.0 s, a particle goes from point A to point B, moving in a semicircle of radius 1.0 m (see figure). The magnitude of the average velocity is

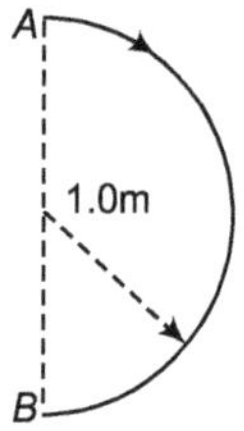

(a) 3.14 m/s (b) 2.0 m/s
(c) 1.0 m/s (d) zero

67. A cyclist starts from the centre O of a circular park of radius 1 km, reaches the edge P of the park, then cycles along the circumference and returns to the centre along QO as shown in the figure. If the round trip takes 10 min., the net displacement and average speed of the cyclist (in m and km per hour) are

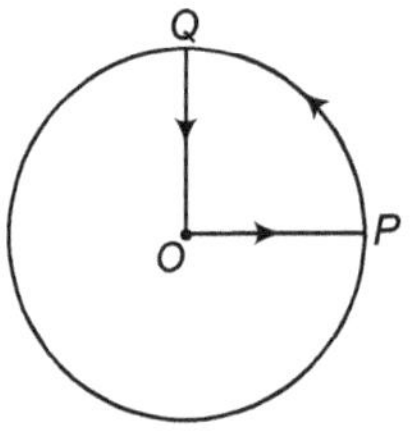

(a) 0, 1 (b) $\frac{\pi + 4}{2}, 4$
(c) $21.4, \frac{\pi + 4}{2}$ (d) 0, 21.4

68. A particle of mass m is circulating on a circle of radius r having angular momentum L about centre. Then, the centripetal force will be
(a) $\frac{L^2}{mr}$ (b) $\frac{L^2}{mr^2}$
(c) $\frac{L^2}{mr^3}$ (d) $\frac{L}{mr^2}$

69. Three particles A, B and C move in a circle of radius $r = \frac{1}{\pi}$ m, in anti-clockwise direction with speed 1 m/s, 2.5 m/s and 2 m/s respectively. The initial positions of A, B and C are as shown in figure.

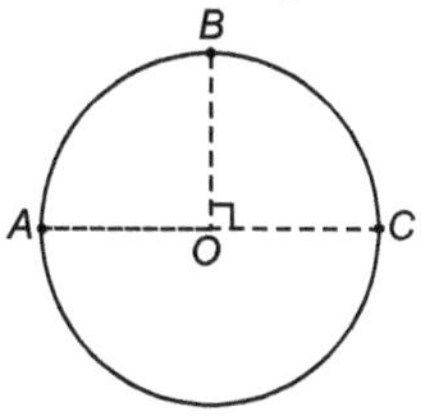

The ratio of distance travelled by B and C by the instant A, B and C meet for the first time is
(a) 3 : 2
(b) 5 : 4
(c) 3 : 5
(d) 3 : 7

70. A motorcyclist moving with a velocity of 72 km/h on a flat road takes a turn on the road at a point, where the radius of curvature of the road is 20 m. The acceleration due to gravity is 10 m/s^2. In order to avoid sliding, he must bend with respect to the vertical plane by an angle
(a) $\theta = \tan^{-1}(4)$
(b) $\theta = 45°$
(c) $\theta = \tan^{-1}(2)$
(d) $\theta = \tan^{-1}(6)$

71. A train has to negotiate a curve of radius 400 m. By how much height should the outer rail be raised with respect to inner rail for a speed of 48 km/h? The distance between the rails is 1 m
(a) 4.4 cm
(b) 9 cm
(c) 2.2 cm
(d) 3.3 cm

72. A string of length l fixed at one end carries a mass m at the other end. The strings makes $\frac{2}{\pi}$ rev/s around the axis through the fixed end as shown in the figure, the tension in the string is

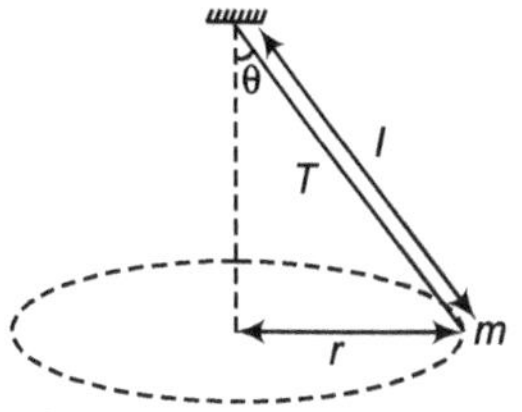

(a) 16 ml
(b) 4 ml
(c) 8 ml
(d) 2 ml

[Level 2]

Only One Correct Option

1. A heavy particle is tied to the end A of a string of length 1.6 m. Its other end O is fixed. It revolves as a conical pendulum with the string making 60° with the horizontal. Then,
(a) its period of revolution is $4\pi/7$ s
(b) the tension in the string is $\frac{2}{\sqrt{3}}$ times the weight of the particle
(c) the speed of the particle is $2.8\sqrt{3}$ m/s
(d) the centripetal acceleration of the particle is $9.8/\sqrt{3}$ m/s^2

2. A bullet of mass m moving with a horizontal velocity u strikes a stationary wooden block of mass M suspended by a string of length $L = 50$ cm. The bullet emerges out of the block with speed $\frac{u}{4}$. If $M = 6$m, the minimum value of u so that the block can complete the vertical circle (Take $g = 10\ m/s^2$)
(a) 10 m/s
(b) 20 m/s
(c) 30 m/s
(d) 40 m/s

3. Two bodies of mass m and $4m$ are attached to a spring as shown in the figure. The body of mass m hanging from a string of length l is executing periodic motion with amplitude $\theta = 60°$ while other body is at rest on the surface. The minimum coefficient of friction between the mass $4m$ and the horizontal surface must be

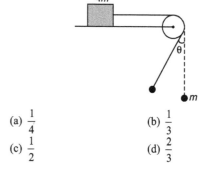

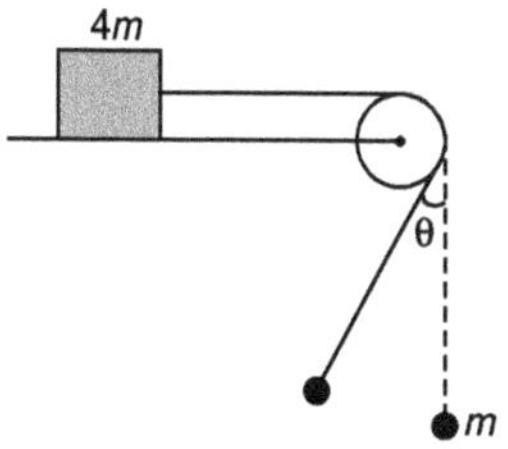

(a) $\frac{1}{4}$
(b) $\frac{1}{3}$
(c) $\frac{1}{2}$
(d) $\frac{2}{3}$

4. A mass is attached to the end of a string of length l which is tied to a fixed point O. The mass is released from the initial horizontal position of the string. Below the point O at what minimum distance a peg P should be fixed so that the mass turns about P and can describe a complete circle in the vertical plane?
(a) $\left(\frac{3}{5}\right)l$
(b) $\left(\frac{2}{5}\right)l$
(c) $\frac{l}{3}$
(d) $\frac{2l}{3}$

5. The bob of a 0.2 m pendulum describes an arc of circle in a vertical plane. If the tension in the cord is $\sqrt{3}$ times, the weight of the bob when the cord makes an angle 30° with the vertical, the acceleration of the bob in that position is

(a) g (b) $\frac{g}{2}$ (c) $\frac{\sqrt{3}\,g}{2}$ (d) $\frac{g}{4}$

6. An automobile enters a turn of radius R. If the road is banked at an angle of 45° and the coefficient of friction is 1, the minimum speed with which the automobile can negotiate the turn without skidding is

(a) $\sqrt{\frac{rg}{2}}$ (b) $\frac{\sqrt{rg}}{2}$
(c) $\sqrt{rg}$ (d) zero

7. A jeep runs around a curve of radius 0.3 km at a constant speed of 60 ms^{-1}. The jeep covers a curve of 60° arc

(a) resultant change in velocity of jeep is 60 ms^{-1}
(b) instantaneous acceleration of jeep is 12 ms^{-2}
(c) average acceleration of jeep is approximately 11.5 ms^{-2}
(d) instantaneous and average acceleration are same in this case

8. A stone is tied to a string of length l and is whirled in a vertical circle with the other end of the string as the centre. At a certain instant of time, the stone is at its lowest position and has a speed u. The magnitude of the change in velocity as it reaches a position, where the string is horizontal (g being acceleration due to gravity) is

(a) $\sqrt{2(u^2 - gl)}$ (b) $\sqrt{u^2 - gl}$
(c) $u - \sqrt{u^2 - 2gl}$ (d) $\sqrt{2gl}$

9. A stone of mass 1 kg tied to a light inextensible string of length $L = \frac{10}{3}$ m, whirling in a circular path in a vertical plane. The ratio of maximum tension to the minimum tension in the string is 4. If g is taken to be 10 m/s^2, the speed of the stone at the highest point of the circle is

(a) 10 m/s (b) $5\sqrt{2}$ m/s
(c) $10\sqrt{3}$ m/s (d) 20 m/s

10. A wet open umbrella is held vertical and whirled about the handle at a uniform rate of 21 rev in 44 s. If the rim of the umbrella is a circle of 1 m in diameter and the height of the rim above the floor is 4.9 m. The locus of the drop on floor is a circle of radius

(a) $\sqrt{2.5}$ m (b) 1 m (c) 3 m (d) 1.5 m

11. A 50 kg girl is swinging on a swing from rest. Then, the power delivered when moving with a velocity of 2 m/s upwards in a direction making an angle 60° with the vertical is

(a) 980 W (b) 490 W (c) $490\sqrt{3}$ W(d) 245 W

12. A particle of mass 1 g executes an oscillatory motion on the concave surface of a spherical dish of radius 2 m placed on a horizontal plane. If the motion of the particle begins from a point on the dish at a height of 1 cm from the horizontal plane and the coefficient of friction is 0.01, the total distance covered by the particle before it comes to rest, is approximately

(a) 2.0 m (b) 10.0 m
(c) 1.0 m (d) 20.0 m

13. A particle suspended by a light inextensible thread of length l is projected horizontally from its lowest position with velocity $\sqrt{7gl/2}$. The string will slack after swinging through an angle equal to

(a) 30° (b) 90°
(c) 120° (d) 150°

14. A particle moves from rest at A on the surface of a smooth circular cylinder of radius r as shown in figure. At B it leaves the cylinder. The equation relating α and β is

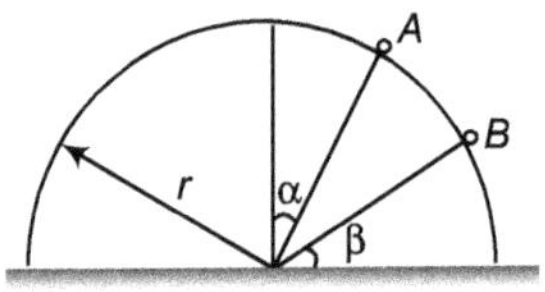

(a) $3\sin\alpha = 2\cos\beta$ (b) $2\sin\alpha = 3\cos\beta$
(c) $3\sin\beta = 2\cos\alpha$ (d) $2\sin\beta = 3\cos\alpha$

15. A ball suspended by a thread swings in a vertical plane so that its acceleration in the extreme position and lowest position are equal. The angle θ of thread deflection in the extreme position will be

(a) $\tan^{-1}(2)$
(b) $\tan^{-1}(\sqrt{2})$
(c) $\tan^{-1}\left(\frac{1}{2}\right)$
(d) $2\tan^{-1}\left(\frac{1}{2}\right)$

16. A body of mass m hangs at one end of a string of length l, the other end of which is fixed. It is given a horizontal velocity so that the string would just reach, where it makes an angle of 60° with the vertical. The tension in the string at bottommost point position is

(a) 2 mg (b) mg
(c) 3 mg (d) $\sqrt{3}$ mg

17. A simple pendulum oscillates in a vertical plane. When it passes through the bottommost point, the tension in the string is 3 times the weight of the pendulum bob. What is the maximum displacement of the pendulum of the string with respect to the vertical?

(a) 30° (b) 45°
(c) 60° (d) 90°

18. A boy whirls a stone in a horizontal circle of radius 1.5 m and at height 2.0 m above level ground. The string breaks and the stone flies off tangentially and strikes the ground after travelling a horizontal distance of 10 m. What is the magnitude of the centripetal acceleration of the stone while in circular motion?

(a) 163 m/s^2 (b) 64 m/s^2
(c) 15.63 m/s^2 (d) 125 m/s^2

19. A stone is rotated in a vertical circle. Speed at bottommost point is $\sqrt{8gR}$, where R is the radius of circle. The ratio of tension at the top and the bottom is

(a) 1 : 2 (b) 1 : 3
(c) 2 : 3 (d) 1 : 4

20. The kinetic energy K of a particle moving along a circle of radius R depends on the distance covered s as $K = as^2$. The force acting on the particle is

(a) $\frac{2as^2}{R}$ (b) $2as\left(1+\frac{s^2}{R^2}\right)^{1/2}$

(c) $as\left(1+\frac{s^2}{R^2}\right)^{1/2}$ (d) None of these

21. A conical pendulum of length L makes an angle θ with the vertical. The time period will be

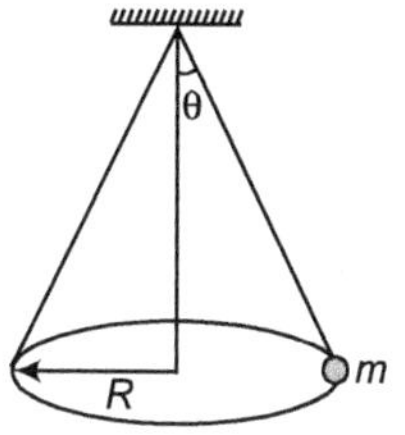

(a) $2\pi\sqrt{\frac{L\cos\theta}{g}}$ (b) $2\pi\sqrt{\frac{L}{g\cos\theta}}$

(c) $2\pi\sqrt{\frac{L\tan\theta}{\gamma}}$ (d) $2\pi\sqrt{\frac{L}{g\tan\theta}}$

22. A particle starts travelling on a circle with constant tangential acceleration. The angle between velocity vector and acceleration vector, at the moment when particle complete half the circular track, is

(a) $\tan^{-1}(2\pi)$ (b) $\tan^{-1}(\pi)$
(c) $\tan^{-1}(3\pi)$ (d) zero

23. A simple pendulum is vibrating with an angular amplitude of 90° as shown in the figure. For what value of α, is the acceleration directed?

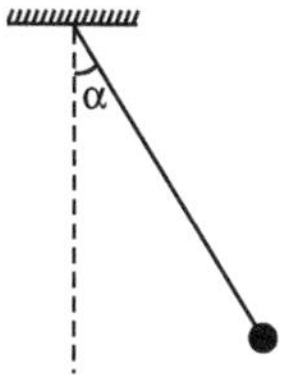

(i) Vertically upwards
(ii) Horizontally
(iii) Vertically downwards

(a) 0°, $\cos^{-1}\left(\frac{1}{\sqrt{3}}\right)$, 90° (b) 90°, $\cos^{-1}\left(\frac{1}{\sqrt{3}}\right)$, 0°

(c) 0°, $\cos^{-1}\sqrt{3}$, 90° (d) $\cos^{-1}\left(\frac{1}{\sqrt{3}}\right)$, 90°, 0°

More than One Correct Options

1. A ball tied to the end of the string swings in a vertical circle under the influence of gravity.

(a) When the string makes an angle 90° with the vertical, the tangential acceleration is zero and radial acceleration is somewhere between minimum and maximum
(b) When the string makes an angle 90° with the vertical, the tangential acceleration is maximum and radial acceleration is somewhere between maximum and minimum
(c) At no place in circular motion, tangential acceleration is equal to radial acceleration
(d) When radial acceleration has its maximum value, the tangential acceleration is zero

2. A small spherical ball is suspended through a string of length l. The whole arrangement is placed in a vehicle which is moving with velocity v. Now, suddenly the vehicle stops and ball starts moving along a circular path. If tension in the string at the highest point is twice the weight of the ball, then (Assume that the ball completes the vertical circle)

(a) $v=\sqrt{5gl}$
(b) $v=\sqrt{7gl}$
(c) velocity of the ball at highest point is $\sqrt{gl}$
(d) the velocity of the ball at the highest point is $\sqrt{3gl}$

3. A particle is describing circular motion in a horizontal plane in contact with the smooth surface of a fixed right circular cone with its axis vertical and vertex down. The height of the plane of motion above the vertex is h and the semi-vertical angle of the cone is α. The period of revolution of the particle

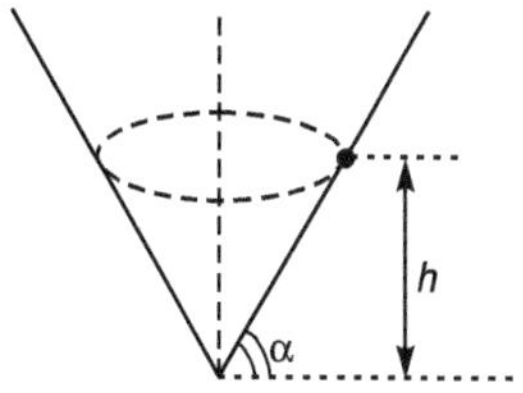

(a) increases as h increases
(b) decreases as h decreases
(c) increases as α increases
(d) decreases as α increases

4. In circular motion of a particle,

(a) particle cannot have uniform motion
(b) particle cannot have uniformly accelerated motion
(c) particle cannot have net force equal to zero
(d) particle cannot have any force in tangential direction

5. A smooth cone is rotated with an angular velocity ω as shown. A block A is placed at height h. A block has no motion relative to cone. Choose the correct options, when ω is increased.

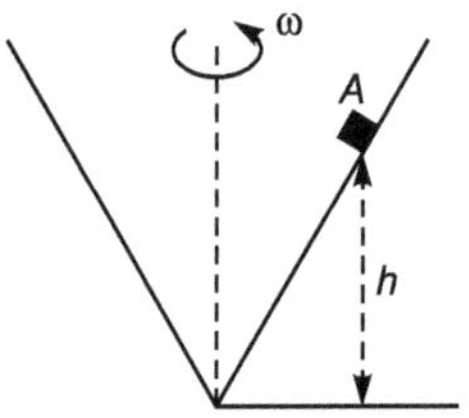

(a) Net force acting on block will increase
(b) Normal reaction acting on block will increase
(c) h will increase
(d) Normal reaction will remain unchanged

Comprehension Based Questions

Passage I (Q. 1 to 2)

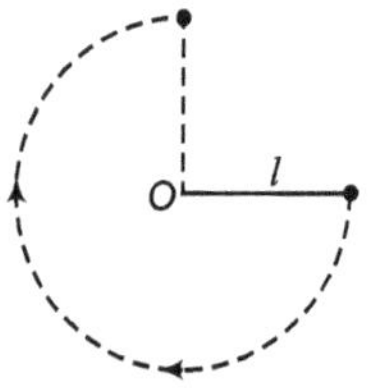

A ball with mass m is attached to the end of a rod of mass M and length l. The other end of the rod is pivoted so that the ball can move in a vertical circle. The rod is held in the horizontal position as shown in the figure and then given just enough a downward push so that the ball swings down and just reaches the vertical upward position having zero speed there. Now answer the following questions

1. The change in potential energy of the system (ball + rod) is

(a) mgl (b) $(M+m)gl$
(c) $\left(\frac{M}{2}+m\right)gl$ (d) $\frac{(M+m)}{2}gl$

2. The initial speed given to the ball is

(a) $\sqrt{\frac{Mgl+2mgl}{m}}$ (b) $\sqrt{2gl}$
(c) $\sqrt{\frac{2Mgl+mgl}{m}}$ (d) None of these

● Attempt the above question after studying chapter of rotational motion.

Passage II (Q. 3 to 5)

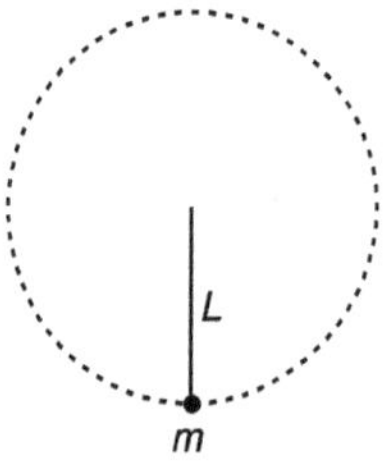

A small particle of mass m attached with a light inextensible thread of length L is moving in a vertical circle. In the given case particle is moving in complete vertical circle and ratio of its maximum to minimum velocity is 2 : 1.

3. Minimum velocity of the particle is

(a) $4\sqrt{\frac{gL}{3}}$ (b) $2\sqrt{\frac{gL}{3}}$ (c) $\sqrt{\frac{gL}{3}}$ (d) $3\sqrt{\frac{gL}{3}}$

4. The kinetic energy of particle at the lower most position is

(a) $\frac{4mgL}{3}$ (b) $2mgL$ (c) $\frac{8mgL}{3}$ (d) $\frac{2mgL}{3}$

5. Velocity of particle when it is moving vertically downward is

(a) $\sqrt{\frac{10gL}{3}}$ (b) $2\sqrt{\frac{gL}{3}}$
(c) $\sqrt{\frac{8gL}{3}}$ (d) $\sqrt{\frac{13gL}{3}}$

Assertion and Reason

Directions (Q. Nos. 1-16) *These questions consists of two statements each printed as Assertion and Reason. While answering these questions you are required to choose any one of the following five responses.*

(a) In both Assertion and Reason are correct and Reason is the correct explanation of Assertion.
(b) It both Assertion and Reason are correct but Reason is not the correct explanation of Assertion.
(c) If Assertion is true but Reason is false.
(d) If Assertion is false but Reason is true.
(e) If both Assertion and Reason are false.

1. Assertion One end of a massless rod of length l is hinged so that it is free to rotate in a vertical plane about a horizontal axis. If a particle is attached to the other end of the rod, then the minimum speed at lower most position of the particle is $\sqrt{5gl}$ to complete the circular motion.

Reason Work done by centripetal force on the particle is always zero.

2. Assertion When water in a bucket is whirled fast overhead, the water does not fall out at the top of the circular path.

Reason The centripetal force in this position on water is more than the weight of water.

3. Assertion A ball tied by thread is undergoing circular motion (of radius R) in a vertical plane. (Thread always remains in vertical plane). The difference of maximum and minimum tension in thread is independent of speed (u) of ball at the lowest position $(u > \sqrt{5gR})$.

Reason For a ball of mass m tied by thread undergoing vertical circular motion (of radius R), difference in maximum and minimum magnitude of centripetal acceleration of the ball is independent of speed (u) of ball at the lowest position $(u > \sqrt{5gR})$.

4. **Assertion** A car moves along a road with uniform speed. The path of car lies in vertical plane and is shown in figure. The radius of curvature (R) of the path is same everywhere. If the car does not loose contact with road at the highest point, it can travel the shown path without loosing contact with road anywhere else.

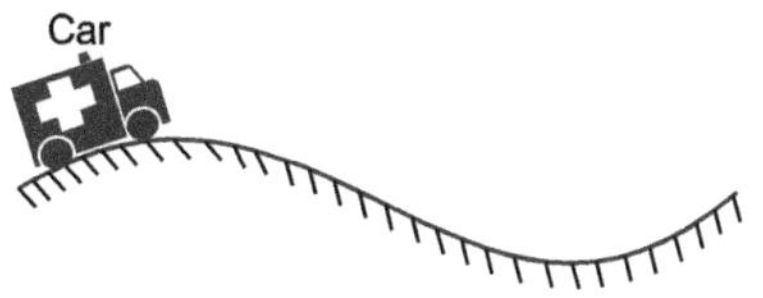

Reason For car to loose contact with road, the normal reaction between car and road should be zero.

5. **Assertion** Uniform circular motion is uniformly accelerated motion.

Reason Acceleration in uniform circular motion is always towards centre.

6. **Assertion** In vertical circular motion speed of a body cannot remain constant.

Reason In moving upwards work done by gravity is negative.

7. **Assertion** In circular motion acceleration of particle is not always towards centre.

Reason If speed of particle is not constant acceleration is not towards centre.

8. **Assertion** In circular motion average speed and average velocity are never equal.

Reason In any curvilinear path these two are never equal.

9. **Assertion** In circular motion dot product of v and ω is always zero.

Reason ω is always perpendicular to the plane of the circular motion.

10. **Assertion** Centripetal force $\frac{mv^2}{R}$ acts on a particle rotating in a circle.

Reason Summation of net forces acting on the particle is equal to $\frac{mv^2}{R}$ in the above case.

11. **Assertion** A particle is rotating in a circle of radius 1 m. At some given instant its speed is 2 m/s. Then acceleration of particle at the given instant is 4 m/s^2.

Reason Centripetal acceleration at this instant is 4 m/s^2 towards centre of circle.

12. **Assertion** If a particle is rotating in a circle, then centrifugal force is acting on the particle in radially outward direction.

Reason Centrifugal force is equal and opposite to the centripetal force.

13. **Assertion** Angle (θ) between **a** and **v** in circular motion is

$$0^\circ < \theta < 180^\circ$$

Reason Angle between any two vectors lies in the above range.

14. **Assertion** A small block of mass m is rotating in a circle inside a smooth cone as shown in figure. In this case the normal reaction,

$$N \neq mg \cos \theta$$

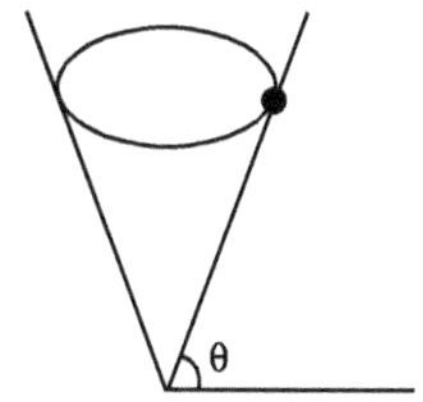

Reason In this case acceleration of the block is not along the surface of cone. It is horizontal.

15. **Assertion** When a car takes a circular turn on a horizontal road, then normal reaction on inner wheels is always less than the normal reaction on outer wheels.

Reason This is for rotational equilibrium of car.

16. **Assertion** On every satellite of the earth we feel weightlessness.

Reason The gravitational force acting on it by the earth is completely utilized in providing the necessary centripetal force.

Match the Columns

1. Three balls each of mass 1 kg are attached with three strings each of length 1 m as shown in figure. They are rotated in a horizontal circle with angular velocity $\omega = 4$ rad/s about point O. Match the following columns.

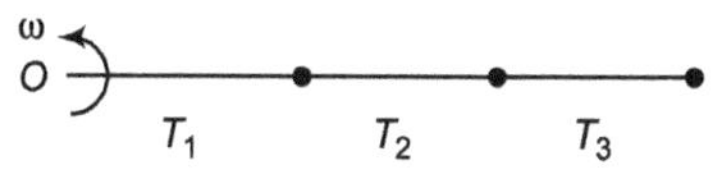

Column I	Column II
(A) T_1	(p) Maximum
(B) T_2	(q) Minimum
(C) T_3	(r) 80 N
	(s) 48 N
	(t) 90 N

2. A particle is suspended from a string of length R. It is given a velocity $u = 3\sqrt{gR}$ at the bottom.

Match the following columns.

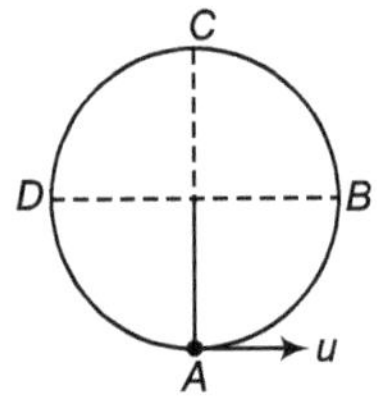

Column I	Column II
(A) Velocity at B	(p) 7 mg
(B) Velocity at C	(q) $\sqrt{5gR}$
(C) Tension in string at B	(r) $\sqrt{7gR}$
(D) Tension in string at C	(s) 5 mg
	(t) None

3. In the system shown in figure, mass m is released from rest from position A. Suppose potential energy of m at point A with respect to point B is E. Dimensions of m are negligible and all surfaces are smooth. When mass m reaches at point B.

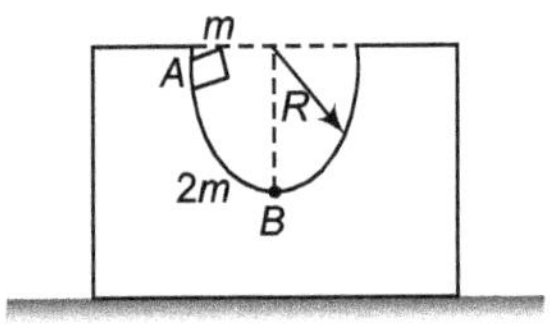

Column I	Column II
(A) Kinetic energy of m	(p) $E/3$
(B) Kinetic energy of $2m$	(q) $\frac{2E}{3}$
(C) Momentum of m	(r) $\sqrt{\frac{4}{3}mE}$
(D) Momentum of $2m$	(s) $\sqrt{\frac{2}{3}mE}$
	(t) None

4. A pendulum is released from point A as shown in figure. At some instant net force on the bob is making an angle θ with the string. Then, match the following columns.

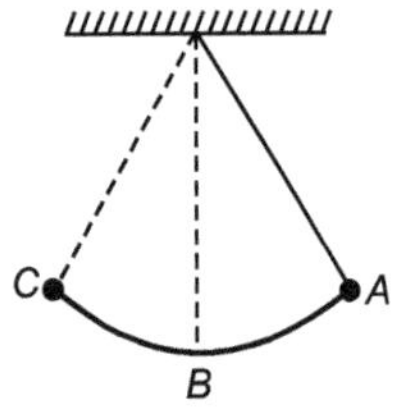

Column I	Column II
(A) For $\theta = 30°$	(p) Particle may be moving along BA
(B) For $\theta = 120°$	(q) Particle may be moving along CB
(C) For $\theta = 90°$	(r) Particle is at A
(D) For $\theta = 0°$	(s) Particle is at B
	(t) None

Entrance Gallery

2014

1. A bob of mass m attached to an inextensible string of length l is suspended from a vertical support. The bob rotates in a horizontal circle with an angular speed ω rad/s about the vertical support. About the point of suspension, **[JEE Main]**
   (a) angular momentum is conserved
   (b) angular momentum changes in magnitude but not in direction
   (c) angular momentum changes in direction but not in magnitude
   (d) angular momentum changes both in direction and magnitudes

2. A particle is moving uniformly in a circular path of radius r. When it moves through an angular displacement θ, then the magnitude of the corresponding linear displacement will be **[WB JEE]**
   (a) $2r\cos\left(\frac{\theta}{2}\right)$ (b) $2r\cot\left(\frac{\theta}{2}\right)$
   (c) $2r\tan\left(\frac{\theta}{2}\right)$ (d) $2r\sin\left(\frac{\theta}{2}\right)$

2013

3. The work done on a particle of mass m by a force,

$$K\left[\frac{x}{(x^2+y^2)^{3/2}}\hat{\mathbf{i}}+\frac{y}{(x^2+y^2)^{3/2}}\hat{\mathbf{j}}\right]$$

(K being a constant of appropriate dimension), when the particle is taken from the point $(a, 0)$ to the point $(0, a)$ along a circular path of radius a about the origin in the x-y plane is **[JEE Main]**
   (a) $\frac{2K\pi}{a}$ (b) $\frac{K\pi}{a}$
   (c) $\frac{K\pi}{2a}$ (d) zero

4. A hoop of radius r and mass m rotating with an angular velocity ω_0 is placed on a rough horizontal surface. The initial velocity of the centre of the hoop is zero. What will be the velocity of the centre of the hoop when it ceases to slip? **[JEE Main]**
   (a) $\frac{r\omega_0}{4}$ (b) $\frac{r\omega_0}{3}$ (c) $\frac{r\omega_0}{2}$ (d) $r\omega_0$

2012

5. Two cars of masses m_1 and m_2 are moving in circles of radii r_1 and r_2 respectively. Their speeds are such that they make complete circles in the same time t. The ratio of their centripetal accelerations is **[AIEEE]**
   (a) $m_1r_1 : m_2r_2$ (b) $m_1 : m_2$
   (c) $r_1 : r_2$ (d) $1:1$

2011

6. A ball of mass 0.5 kg is attached to the end of a string having length 0.5 m as shown in figure. The ball is rotated on a horizontal circular path about vertical axis. The maximum tension that the string can bear is 324 N. The maximum possible value of angular velocity of ball (in rad/s) is **[IIT JEE]**

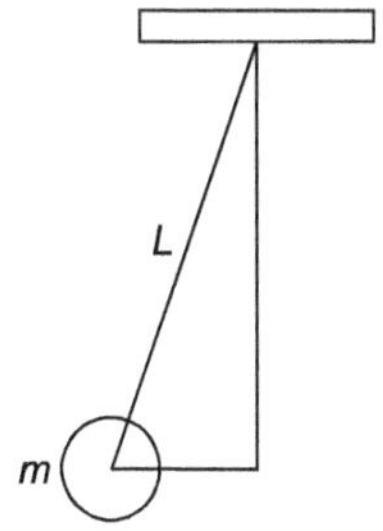

(a) 9 (b) 18 (c) 27 (d) 36

7. A bridge is in the form of a semicircle of radius 40 m. The greatest speed with which a motor cycle can cross the bridge without leaving the ground at the highest point is (take, $g = 10\,ms^{-2}$) and (frictional force is negligibly small) **[Kerala CEE]**

(a) 40 ms^{-1} (b) 20 ms^{-1} (c) 30 ms^{-1} (d) 15 ms^{-1}
(e) 25 ms^{-1}

8. A particle of mass m is moving in a horizontal circle of radius r, under a centripetal force $F = \frac{k}{r^2}$, where k is a constant. **[Kerala CEE]**

(a) The potential energy of the particle is zero
(b) The potential energy of the particle is $\frac{k}{r}$
(c) The total energy of the particle is $-\frac{k}{2r}$
(d) The kinetic energy of the particle is $-\frac{k}{r}$
(e) The potential energy of the particle is $-\frac{k}{2r}$

9. A particle is moving with a constant speed v in a circle. What is the magnitude of average after half rotation? **[WB JEE]**

(a) $2v$ (b) $2\frac{v}{\pi}$
(c) $\frac{v}{2}$ (d) $\frac{v}{2\pi}$

2010

10. A point P moves in counter-clockwise direction on a circular path as shown in the figure. The movement of P is such that it sweeps out a length $s = t^3 + 5$, where s is in metre and t is in second. The radius of the path is 20 m. The acceleration of P, when $t = 2\,s$ is nearly **[AIEEE]**

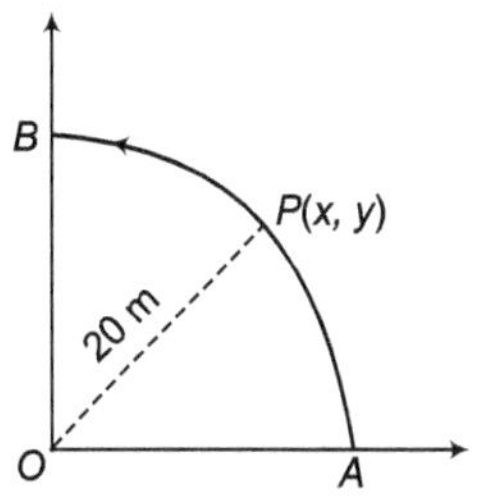

(a) 13 ms^{-2} (b) 12 ms^{-2}
(c) 7.2 ms^{-2} (d) 14 ms^{-2}

11. For a particle in uniform circular motion, the acceleration **a** at a point $P(R,\theta)$ on the circle of radius R is (here θ, is measured from the x-axis) **[AIEEE]**

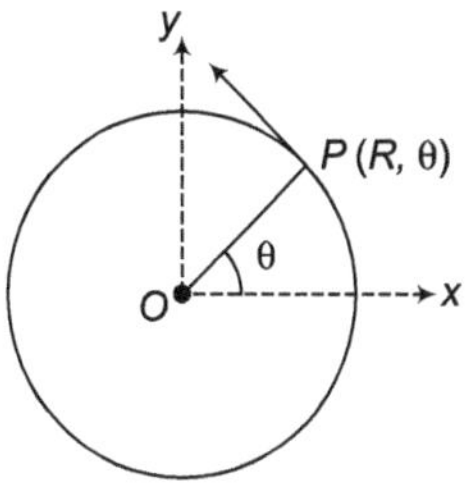

(a) $-\frac{v^2}{R}\cos\theta\,\hat{i} + \frac{v^2}{R}\sin\theta\,\hat{j}$ (b) $-\frac{v^2}{R}\sin\theta\,\hat{i} + \frac{v^2}{R}\cos\theta\,\hat{j}$
(c) $-\frac{v^2}{R}\cos\theta\,\hat{i} - \frac{v^2}{R}\sin\theta\,\hat{j}$ (d) $\frac{v^2}{R}\hat{i} + \frac{v^2}{R}\hat{j}$

12. If α is angular acceleration, ω is angular velocity and a is the centripetal acceleration, then which of the following is true? **[MHT CET]**

(a) $\alpha = \frac{\omega a}{v}$ (b) $\alpha = \frac{v}{\omega a}$
(c) $\alpha = \frac{va}{\omega}$ (d) $\alpha = \frac{a}{\omega v}$

13. If KE of the particle of mass m performing UCM in a circle of radius r is E. Find the acceleration of the particle **[MHT CET]**

(a) $\frac{2E}{mr}$ (b) $\left(\frac{2E}{mr}\right)^2$
(c) $2Emr$ (d) $\frac{4E}{mr}$

14. A particle of mass m is circulating on a circle of radius r having angular momentum L, then the centripetal force will be **[MHT CET]**

(a) L^2/mr (b) L^2m/r
(c) L^2/mr^3 (d) L^2/mr^2

Answers

Level 1

Objective Problems

1. (a)	**2.** (c)	**3.** (a)	**4.** (b)	**5.** (a)	**6.** (c)	**7.** (c)	**8.** (c)	**9.** (d)	**10.** (a)
11. (d)	**12.** (c)	**13.** (a)	**14.** (c)	**15.** (d)	**16.** (d)	**17.** (c)	**18.** (d)	**19.** (b)	**20.** (d)
21. (a)	**22.** (a)	**23.** (d)	**24.** (d)	**25.** (a)	**26.** (d)	**27.** (d)	**28.** (d)	**29.** (d)	**30.** (d)
31. (b)	**32.** (c)	**33.** (c)	**34.** (d)	**35.** (d)	**36.** (c)	**37.** (c)	**38.** (a)	**39.** (d)	**40.** (c)
41. (a)	**42.** (a)	**43.** (a)	**44.** (c)	**45.** (a)	**46.** (c)	**47.** (a)	**48.** (c)	**49.** (a)	**50.** (b)
51. (c)	**52.** (a)	**53.** (c)	**54.** (c)	**55.** (a)	**56.** (a)	**57.** (d)	**58.** (d)	**59.** (c)	**60.** (b)
61. (a)	**62.** (b)	**63.** (c)	**64.** (b)	**65.** (b)	**66.** (b)	**67.** (d)	**68.** (c)	**69.** (b)	**70.** (c)
71. (a)	**72.** (a)								

Level 2

Only One Correct Option

1. (b,d)	**2.** (d)	**3.** (c)	**4.** (a)	**5.** (a)	**6.** (d)	**7.** (a,b,c)	**8.** (a)	**9.** (a)	**10.** (a)
11. (c)	**12.** (c)	**13.** (c)	**14.** (c)	**15.** (d)	**16.** (a)	**17.** (d)	**18.** (a)	**19.** (b)	**20.** (b)
21. (b)	**22.** (a)	**23.** (a)							

More than One Correct Options

1. (b,d) **2.** (b,d) **3.** (a,d) **4.** (a,b,d) **5.** (a,b)

Comprehension Based Questions

1. (c) **2.** (d) **3.** (b) **4.** (c) **5.** (a)

Assertion and Reason

1. (d)	**2.** (a)	**3.** (a)	**4.** (d)	**5.** (d)	**6.** (d)	**7.** (a)	**8.** (a)	**9.** (a)	**10.** (e)
11. (d)	**12.** (d)	**13.** (c)	**14.** (a)	**15.** (a)	**16.** (d)				

Match the Columns

1. (A → p, B → r, C → qs) **2.** (A → r, B → q, C → p, D → t) **3.** (A → q, B → p, C → r, D → r)
4. (A → qp, B → t, C → r, D → s)

Entrance Gallery

1. (c)	**2.** (d)	**3.** (d)	**4.** (c)	**5.** (c)	**6.** (d)	**7.** (b)	**8.** (a)	**9.** (b)	**10.** (d)
11. (c)	**12.** (a)	**13.** (a)	**14.** (c)						

Solutions

Level 1 : Objective Problems

1. $a=\sqrt{a_t^2+a_n^2}$

a_t = rate of change of speed = 2 m/s²

$$a_n=\frac{v^2}{R}=\frac{(5)^2}{10}=2.5\ \text{m/s}^2$$

$\therefore$ $a=\sqrt{a_t^2+a_n^2}=\sqrt{(2)^2+(2.5)^2}=3.2\ \text{m/s}^2$

2. At highest point of a projectile, tangential component of acceleration is zero as acceleration is vertically downwards (g) and velocity is horizontal or the angle between acceleration and velocity is 90°.

3. $a_n=\sqrt{a^2-a_t^2}=\sqrt{(100)^2-(60)^2}=80\ \text{m/s}^2$

Now $a_n=\frac{v^2}{R}$ or $R=\frac{v^2}{a_n}=\frac{(10)^2}{80}=1.25$ m

4. $a_t=0, a=a_n=\frac{v^2}{R}=\frac{(6)^2}{0.3}=120\ \text{m/s}^2$

5. Velocity is tangential and acceleration is radial.

6. In circular motion, a_r can never be zero.

7. Speed of particle is increasing, means tangential component of acceleration is positive.

8. $\omega_{\text{min}}=\frac{2\pi}{60}$ rad/min and $\omega_{\text{hr}}=\frac{2\pi}{12\times 60}$ rad/min

$\therefore$ $\frac{\omega_{\text{min}}}{\omega_{\text{hr}}}=\frac{2\pi/60}{2\pi/12\times 60}$

9. 120 rev/min $=120\times\frac{2\pi}{60}$ rad/s $=4\pi$ rad/s

10. Acceleration $=\omega^2 r=\frac{v^2}{r}=\omega v=\frac{2\pi}{T}v$

11. $a=\frac{v^2}{R}$ or $a\propto v^2$

12. $a=r\omega^2$ or $a\propto r$

13. $\alpha=\frac{\Delta\omega}{\Delta t}$ but $\omega=2\pi f$

$\therefore$ $\alpha=\frac{2\pi\Delta f}{\Delta t}=\frac{2\pi\times 200}{2}$

$=(200\pi)$ rad/s²

14. $a=\sqrt{a_n^2+a_t^2}=\sqrt{\left(\frac{v^2}{r}\right)^2+a^2}=\sqrt{\frac{v^2}{r^2}+a^2}$

15. $\frac{a_n}{a_t}=\frac{v^2/R}{a}$ (let $a_r=a$)

Here, $vb^2=2al=2a(2\pi R)=4\pi R$.

Therefore, the ratio is $\frac{4\pi}{l}$.

16. $\alpha=\frac{d\omega}{dt}=-b$

$a_t=R\alpha=-Rb$ ...(i)

At $t=\frac{2a}{b}$, $\omega=-a$

$a_n=R\omega^2=Ra^2$ ...(ii)

Now, $a=\sqrt{a_t^2+a_n^2}$

$=R\sqrt{a^4+b^2}$

17. α = constant

$\therefore$ $\theta=\frac{1}{2}\propto t^2$

$\therefore$ $\theta\propto t^2$

or $\frac{\theta_2+\theta_1}{\theta_1}=\left(\frac{2+2}{2}\right)^2=4$

or $\frac{\theta_2}{\theta_1}=3$

18. Necessary centripetal force to the coin is provided by friction. Thus,

$$mr\omega_{\text{max}}^2=\mu mg \quad\text{or}\quad r=\frac{\mu g}{\omega_{\text{max}}^2}$$

ω_{max} is made three times. Therefore, distance from centre r will remain $\frac{1}{9}$ times.

19. $F=\frac{mv^2}{r}, F'=\frac{(1.5m)(1.5v)^2}{(1.5r)}=2.25\frac{mv^2}{r}=2.25F$

Therefore, F has to be increased by 125%.

20. $\omega=2\pi f=2\pi(0.5)=\pi$ rad/s

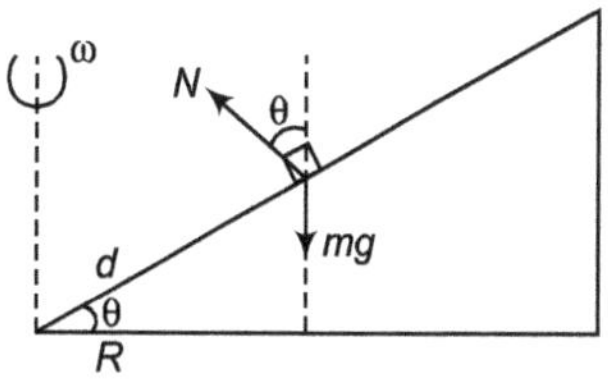

$N\sin\theta=mR\omega^2$

$N\cos\theta=mg$

or $\tan\theta=\frac{R\omega^2}{g}$

$\therefore$ $R=\frac{g\tan\theta}{\omega^2}$

$=\frac{10\cdot\tan 30^\circ}{(\pi^2)}\approx\frac{1}{\sqrt{3}}$ m

Now, $d=\frac{R}{\cos\theta}=\frac{1/\sqrt{3}}{\cos\theta}$

$=\frac{1/\sqrt{3}}{\sqrt{3}/2}=\frac{2}{3}=0.67$ m

21. $N=\frac{mv^2}{R}, mg=\mu_s N=\frac{\mu_s mv^2}{R}$

$\therefore$ $v=\sqrt{\frac{Rg}{\mu_s}}$

22. To find tension in the ring, let us take an arc which subtends angle $2(d\theta)$ at centre. Tangential components of T cancel out each other, while inward components provide the necessary centripetal force. Thus,

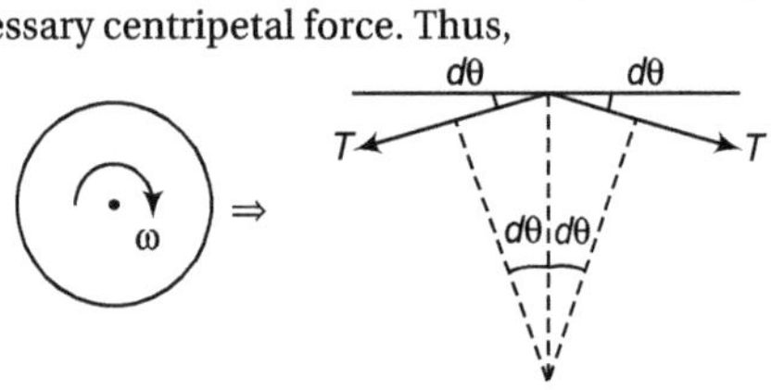

$$2T\sin(d\theta) = (dm)\,R\omega^2$$
$$= (\lambda 2R\cdot d\theta)(R\omega^2)$$

Here, $\lambda \to$ linear mass of density.

For small angles, $\sin d\theta \approx d\theta$

$\therefore$ $$2Td\theta = 2\lambda R^2\omega^2 d\theta$$

or $$T = \lambda R^2\omega^2$$

$$\omega_{max} = \sqrt{\frac{T_{max}}{\lambda}\cdot\frac{1}{R}}$$
$$= \sqrt{\frac{40}{0.1}\times\frac{1}{1}}$$
$$= 20\,\text{rad/s}$$

23. $N = \dfrac{mv^2}{R}, mg = \mu N = \dfrac{\mu mv^2}{R}$

$\therefore$ $$\mu = \frac{Rg}{v^2} = \frac{5\times 10}{(5\sqrt{5})^2} = 0.4$$

24. Centripetal force $= mR\omega^2 = (2)(1)(2\pi)^2 = 8\pi^2$ N

25. $F = \dfrac{mv^2}{r}$. If m and v are constants, then $F \propto \dfrac{1}{r}$

$\therefore$ $$\frac{F_1}{F_2} = \left(\frac{r_2}{r_1}\right)$$

26. Maximum tension $= m\omega^2 r = m\times 4\pi^2\times n^2\times r$

By substituting the values, we get $T_{max} = 87.64$ N

27. Radial force $= \dfrac{mv^2}{r} = \dfrac{m}{r}\left(\dfrac{p}{m}\right)^2 = \dfrac{p^2}{mr}$

28. Tension in the string $T = m\omega^2 r = 4\pi^2 n^2 mr$

$\therefore$ $$T \propto n^2$$

$\Rightarrow$ $$\frac{n_2}{n_1} = \sqrt{\frac{T_2}{T_1}}$$

$\Rightarrow$ $$n_2 = 5\sqrt{\frac{2T}{T}} = 5\sqrt{2}\,\text{rpm}$$

29. Maximum tension $= \dfrac{mv^2}{r} = 16$ N

$\Rightarrow$ $$\frac{16\times v^2}{144} = 16$$

$\Rightarrow$ $$v = 12\,\text{m/s}$$

30. Let ω is the angular speed of revolution.

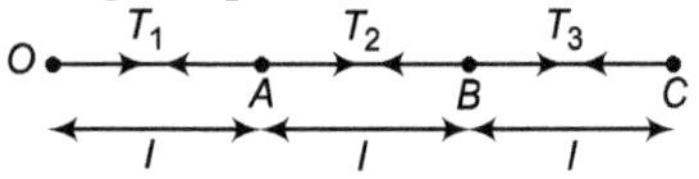

$$T_3 = m\omega^3(3l)$$
$$T_2 - T_3 = m\omega^2\,2l$$

$\Rightarrow$ $$T_2 = m\omega^2(5l)$$
$$T_1 - T_2 = m\omega^2 l$$

$\Rightarrow$ $$T_1 = m\omega^2(6l)$$
$$T_3 : T_2 : T_1 = 3:5:6$$

31. $Kx = \dfrac{mv^2}{r}$

$$v = \sqrt{\frac{Kxr}{m}}$$
$$T = \frac{2\pi r}{v} = 2\pi\sqrt{\frac{mr}{Kx}}$$

$\therefore$ $$T \propto \sqrt{\frac{r}{x}}$$
$$T' = T\sqrt{\frac{r'x}{rx'}} = T\sqrt{\frac{3a\times a}{2a\times 2a}} = \frac{\sqrt{3}}{2}T$$

32. $v_{max} = \sqrt{gR} = \sqrt{9.8\times 49} = 22$ m/s

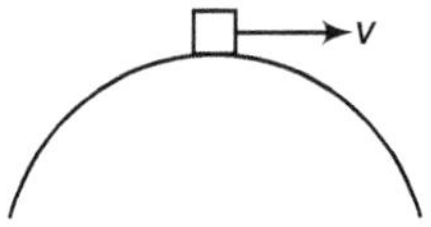

33. $Mg - N = \dfrac{Mv^2}{r}$

or $$N = Mg - \frac{Mv^2}{r}$$

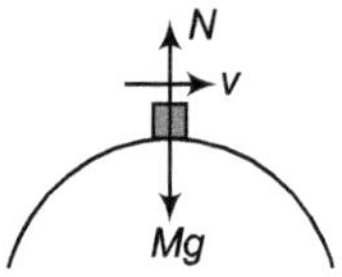

34. $mg = \dfrac{mv_{min}^2}{R}$ or $v_{min} = \sqrt{Rg}$ (at topmost point)

35. Kinetic energy given to a sphere at lowest point= potential energy at the height of suspension

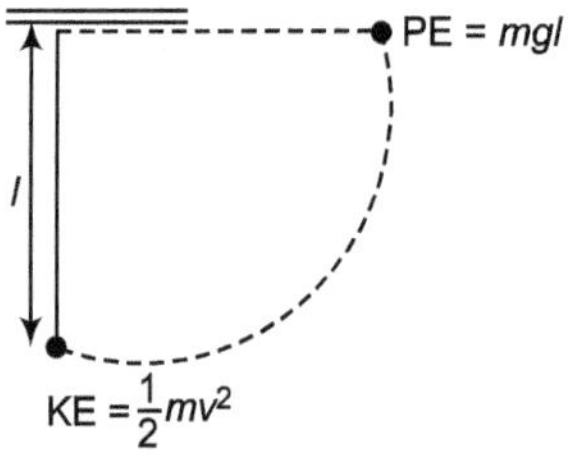

$\Rightarrow$ $$\frac{1}{2}mv^2 = mgl$$

$\therefore$ $$v = \sqrt{2gl}$$

36. $h = l(1-\cos 60°) = \dfrac{l}{2} = 1$ m

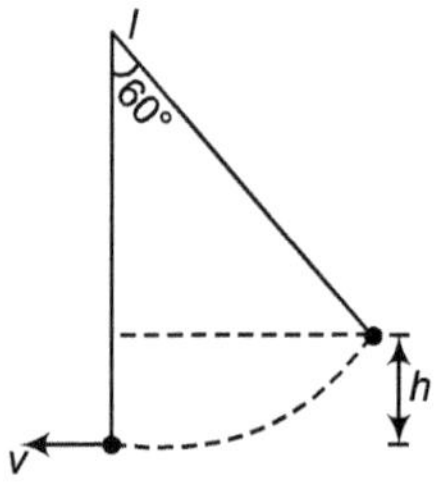

$$v = \sqrt{2gh} = \sqrt{2\times 9.8\times 1} = 4.43\,\text{m/s}$$

37. Tension, $T = \frac{mv^2}{r} + mg\cos\theta$

For, $\theta = 30°$,

$$T_1 = \frac{mv_1^2}{r} + mg\cos 30°$$

$$\theta = 60°,\ T_2 = \frac{mv_2^2}{r} + mg\cos 60°$$

$\therefore$ $T_1 > T_2$ as $v_1 > v_2$

and $\cos 30° > \cos 60°$

38. Difference in kinetic energy

$$= 2mgr$$
$$= 2 \times 1 \times 10 \times 1$$
$$= 20\,\text{J}$$

39. Otherwise the particle will fall down.

40. $v^2 = 2gh = 2gR$

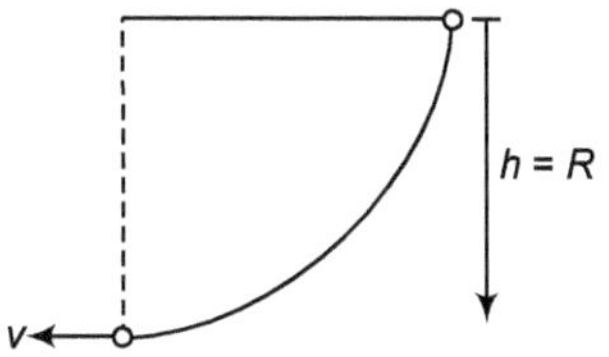

$$T - mg = \frac{mv^2}{R}$$

Substituting value of v^2, we get

$$T = 3mg$$

41. Minimum velocity at topmost point is $\sqrt{Rg}$.

42. $h = R(1 - \cos 60°) = \frac{R}{2}$

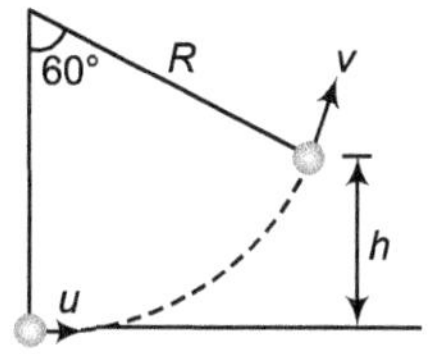

$$v = \sqrt{u^2 - 2gh}$$

43. $h = l(1 - \cos\theta)$

$$v_m^2 = 2gh$$
$$= 2gl(1 - \cos\theta)$$

$\therefore$ $K_m = \frac{1}{2}mv_m^2$

$$= mgl(1 - \cos\theta)$$

44. $h' = R - h$

$\therefore$ $v_{\min} = \sqrt{2gh'}$

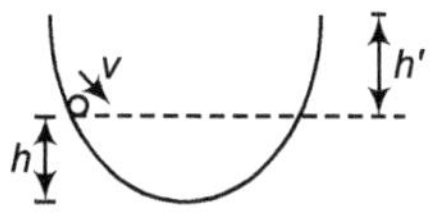

45. $v = \sqrt{2gh} = \sqrt{2gh(h_{\max} - h_{\min})}$

$$= \sqrt{2 \times 10 \times 1.25}$$
$$= 5\,\text{m/s}$$

46. When released from top with zero velocity block leaves contact at

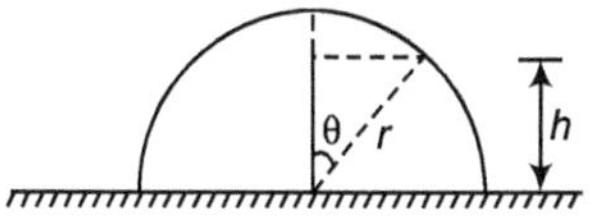

$$\cos\theta = \frac{2}{3}$$

$\therefore$ $h = r\cos\theta = \frac{2r}{3}$

47. $\sqrt{2gh} = \sqrt{5gR}$

$\therefore$ $R = \frac{2h}{5} = \frac{2 \times 5}{5} = 2\,\text{cm}$

48. $T - mg = \frac{mv^2}{l}$ or $T = 3mg$

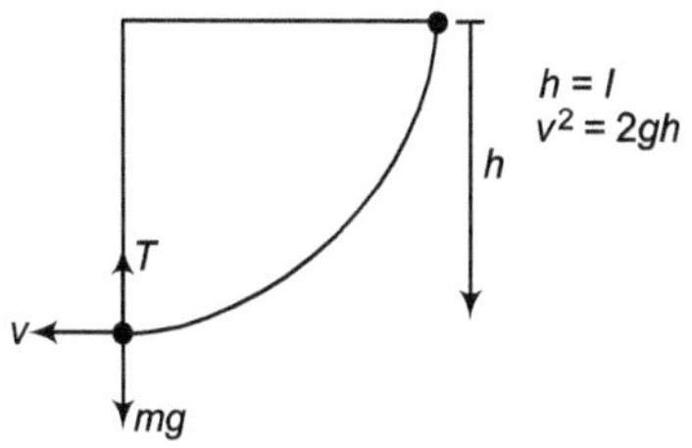

Tension will be maximum at bottommost point.

49. In critical case tension at topmost point is zero.

50. $v = \sqrt{2gh} = \sqrt{2gR}$

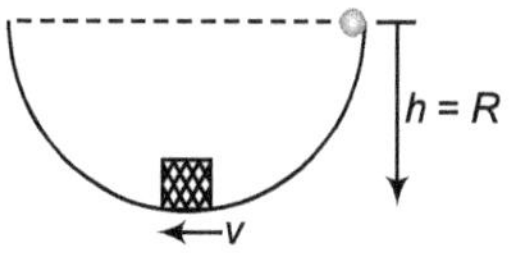

51. $v = \sqrt{u^2 - 2gh} = \sqrt{5gr - 2gr} = \sqrt{3gr}$

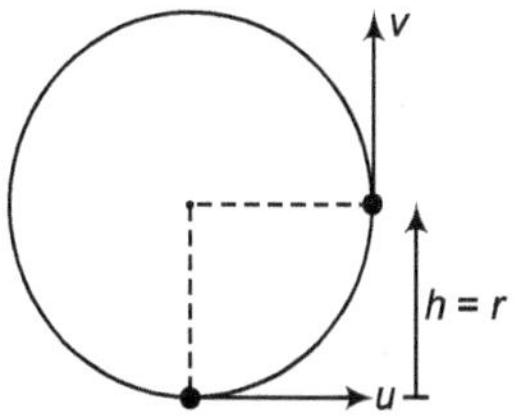

52. Maximum tension will be at bottommost point.

54. $|\mathbf{v}_{AB}| = |\mathbf{v}_A - \mathbf{v}_B| = \sqrt{(R\omega)^2 + (R\omega)^2 - 2(R\omega)(R\omega)\cos\theta}$

$$= 2R\omega\sin\frac{\theta}{2}$$

55. $v = 72\,\text{km/h} = 72 \times \frac{5}{18}\,\text{m/s} = 20\,\text{m/s}$

$$\theta = \tan^{-1}\left(\frac{v^2}{Rg}\right)$$
$$= \tan^{-1}\left(\frac{400}{100 \times 10}\right) = \tan^{-1}\left(\frac{2}{5}\right)$$

65. As the speed is constant throughout the circular motion therefore, its average speed is equal to instantaneous speed.

66. Average velocity $= \frac{\text{Total displacement}}{\text{Total time}} = \frac{2\,\text{m}}{1\,\text{s}} = 2\,\text{ms}^{-1}$

67. Average speed $= \dfrac{(2R + \pi R/2)\text{ km}}{(1/6)\text{ h}}$

$$= (12R + 3\pi R)\text{ km/h}$$

$$= (12 + 3\pi)\frac{\text{km}}{\text{h}} \qquad (\text{as } R = 1\text{ km})$$

$$= 21.4\text{ km/h}$$

68. $L = mvr$

$\therefore$ $$v = \frac{L}{mr}$$

$$F = \frac{mv^2}{r} = \frac{L^2}{mr^3}$$

69. $\dfrac{d_B}{d_C} = \dfrac{v_B t}{v_C t} = \dfrac{v_B}{v_C} = \dfrac{2.5}{2} = \dfrac{5}{4}$

70. $v = 72\text{ km/h} = 72 \times \dfrac{5}{18}\text{ m/s} = 20\text{ m/s}$

$$\tan\theta = \frac{v^2}{Rg}$$

71. $v = 48 \times \dfrac{5}{18}\text{ m/s} = 13.33$

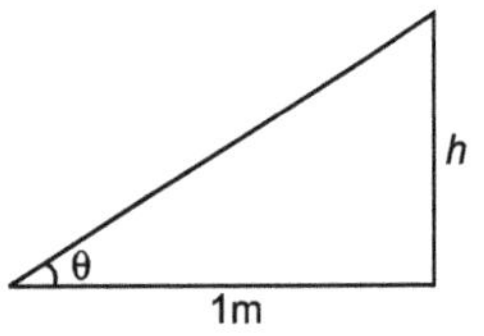

$$\tan\theta = \frac{v^2}{Rg} = 0.044$$

$$0.044 = \frac{h}{1}$$

$\therefore$ $$h = 0.044\text{ m} = 4.4\text{ cm}$$

72. $$T\sin\theta = mr\omega^2$$

or $$T\sin\theta = m(l\sin\theta)\,\omega^2$$

$\therefore$ $$T = ml\omega^2 = ml\,(2\pi f)^2$$

$$= ml\left(2\pi \times \frac{2}{\pi}\right)^2 = 16ml$$

Level 2 : Only One Correct Option

1. $R = 1.6\cos 60° = 0.8\text{ m}$

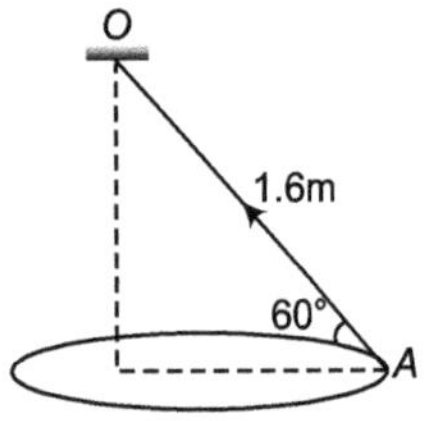

$$T\sin 60° = mg$$

or $$T = \frac{2mg}{\sqrt{3}}$$

$$T\cos 60° = \frac{mv^2}{R}$$

$\therefore$ $$\frac{v^2}{R} = \frac{T}{2m} = \frac{g}{\sqrt{3}} = \frac{9.8}{\sqrt{3}}\text{ m/s}^2$$

$$v = \sqrt{\frac{9.8}{\sqrt{3}} \times 0.8}\text{ m/s}$$

$$\text{Time period} = 2\pi\frac{R}{v}$$

2. From momentum conservation,

$$mu = m\frac{u}{4} + (6m)\sqrt{5 \times 10 \times 0.5}$$

Solving we get, $u = 40\text{ m/s}$

3. $h = l(1 - \cos 60°) = \dfrac{l}{2}$, $v^2 = 2gh = gl$

Now, $$T_{\max} - mg = \frac{mv^2}{l} \qquad (\text{at bottommost point})$$

$\therefore$ $$T_{\max} = 2mg = \mu_s\,(4mg)$$

$\therefore$ $$\mu_s = 0.5$$

4. $v^2 = 2gl = 5gR$

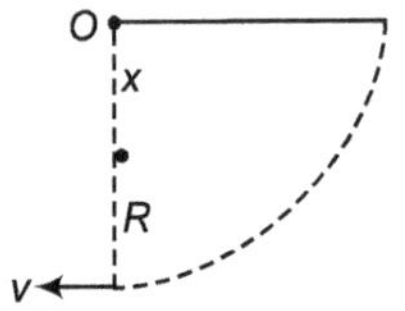

$\therefore$ $$R = 0.4l \quad \text{or} \quad x = l - R = 0.6l$$

5. $a = \sqrt{a_n^2 + a_t^2} = \sqrt{\left(\dfrac{T - mg\cos 30°}{m}\right)^2 + (g\sin 30°)^2}$

$$= g\sqrt{\left(3 - \frac{\sqrt{3}}{2}\right)^2 + \left(\frac{1}{4}\right)} = g$$

6. $\dfrac{N}{\sqrt{2}} - \dfrac{\mu N}{\sqrt{2}} = \dfrac{mv_{\min}^2}{R}$

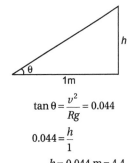

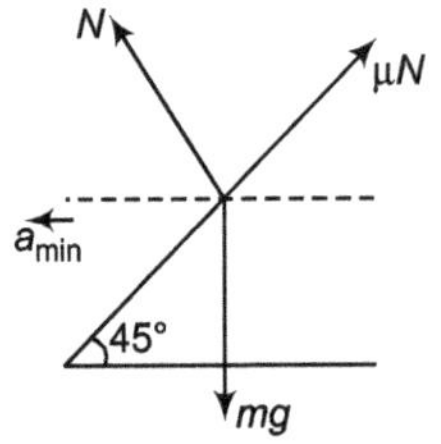

$\Rightarrow$ $$v_{\min} = 0, \text{ as } \mu = 1$$

7. $\Delta t = \dfrac{\text{Distance travelled}}{\text{Speed}}$

$$= \frac{(2\pi R/6)}{v} = \frac{3.14 \times 300}{60 \times 3} = 5.23\text{ s}$$

(a) $|\Delta \mathbf{v}| = |\mathbf{v}_f| - |\mathbf{v}_i|$

$$= \sqrt{v^2 + v^2 - 2v \cdot v\cos 60°}$$

$$= 2v\sin 30° = 60\text{ m/s}$$

(b) $a_i = \dfrac{v^2}{R} = 12\text{ m/s}^2$

(c) $|\mathbf{a}_{\text{av}}| = \dfrac{|\Delta \mathbf{v}|}{\Delta t} = \dfrac{60}{5.23} = 11.5\text{ m/s}^2$

8. $h = l$

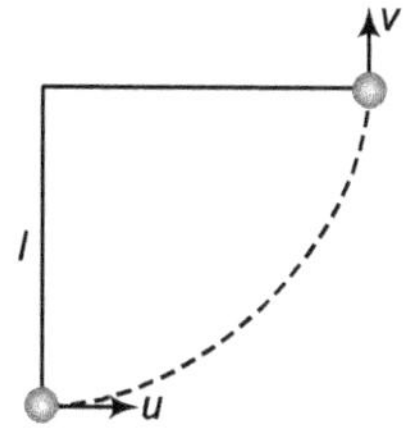

$$v = \sqrt{u^2 - 2gh} = \sqrt{u^2 - 2gl}$$

$$|\Delta \mathbf{v}| = |\mathbf{v}_f - \mathbf{v}_i|$$

= subtraction of vectors at 90°

$$= \sqrt{v^2 + u^2 - 2vu \cdot \cos 90°}$$

$$= \sqrt{v^2 + u^2} = \sqrt{(u^2 - 2gl) + u^2}$$

$$= \sqrt{2(u^2 - gl)}$$

9. Minimum tension is at topmost point (speed $= v$) and maximum tension at bottommost point (speed $= u$).

$$\frac{T_{\max}}{T_{\min}} = \frac{mg + \dfrac{mu^2}{L}}{-mg + \dfrac{mv^2}{L}} = 4$$

$$u^2 = v^2 + 2g(2L)$$

Solving we get, $v = 10 \text{ m/s}$

10. $v = \dfrac{d}{t_1} = \dfrac{21 \times 2\pi \times 0.5}{44} = 1.5 \text{ m/s}$

$$t_2 = \sqrt{\frac{2h}{g}} = \sqrt{\frac{2 \times 4.9}{9.8}} = 1 \text{ s}$$

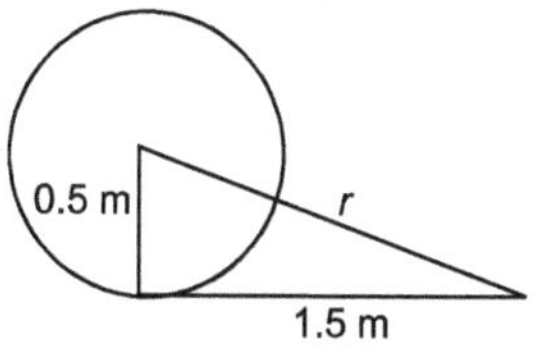

Horizontal distance travelled by drop $= vt_2 = 1.5$ m

$\therefore \quad r = \sqrt{(1.5)^2 + (0.5)^2} = \sqrt{2.5}$ m

11. Two forces are acting on bob, tension and weight. Power of tension will be zero and that of weight is,

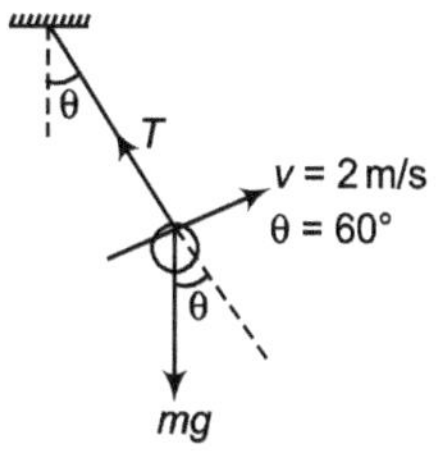

$$P = mgv \cos(90° + \theta)$$

$$= -mgv \sin 60°$$

$$= -50 \times 9.8 \times 2 \times \frac{\sqrt{3}}{2}$$

$$= -490\sqrt{3} \text{ W}$$

$\therefore$ Power delivered is $490\sqrt{3}$ W.

12. Since, the particle is released from a small height, θ (angle of radius with vertical) will be very small.

Force of friction throughout the journey can be assumed to be μmg. Particle will finally come to rest when whole of its energy ($= mgh$) is lost in the work done against friction. Let particle stops after travelling a distance d. Then,

$$\mu mgd = mgh$$

or $\quad d = \dfrac{h}{\mu} = \dfrac{10^{-2}}{0.01} = 1.0$ m

13. $h = l + l\sin\theta = l(1 + \sin\theta)$

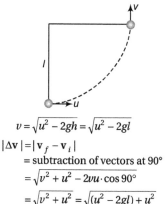

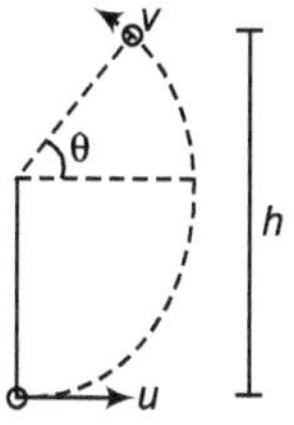

$$v^2 = u^2 - 2gh$$

$$= u^2 - 2gl(1 + \sin\theta)$$

String will slack where, component of weight towards centre is just equal to centripetal force or

$$mg\sin\theta = \frac{mv^2}{l} = \frac{m}{l}[u^2 - 2gl(1 + \sin\theta)]$$

Substituting $u^2 = \dfrac{7gl}{2}$

we get, $\quad \sin\theta = \dfrac{1}{2}$ or $\theta = 30°$

$\therefore$ The desired angle is 90° + 30° or 120°.

14. $h_{AB} = (r\cos\alpha - r\sin\beta)$

Velocity of particle at B

$$v = \sqrt{2gh_{AB}} = \sqrt{2g(r\cos\alpha - r\sin\beta)}$$

Particle will leave contact at B, if component of weight is just equal to centripetal force (towards centre).

or $\quad mg\sin\beta = \dfrac{mv^2}{r}$

or $\quad \sin\beta = 2\cos\alpha - 2\sin\beta$

$\therefore \quad 3\sin\beta = 2\cos\alpha$

15. $g\sin\theta = \dfrac{v^2}{R} = \dfrac{2gh}{R} = \dfrac{2gR(1 - \cos\theta)}{R}$

or $\quad \sin\theta = 2(1 - \cos\theta)$

$$2\sin\frac{\theta}{2}\cos\frac{\theta}{2} = 2\left(2\sin^2\frac{\theta}{2}\right)$$

$\therefore \quad \tan\dfrac{\theta}{2} = \dfrac{1}{2}$

$$\frac{\theta}{2} = \tan^{-1}\left(\frac{1}{2}\right)$$

or $\quad \theta = 2\tan^{-1}\left(\dfrac{1}{2}\right)$

- In extreme position of pendulum only tangential component of acceleration ($a_t = g\sin\theta$) is present. In lowest position, only normal acceleration ($a_n = v^2/R$) is present.

16. When body is released from the position p (inclined at angle θ from vertical) then velocity at mean position

$$v = \sqrt{2gl(1 - \cos\theta)}$$

$\therefore$ Tension at the lowest point $= mg + \dfrac{mv^2}{l}$

$$= mg + \frac{m}{l}[2gl(1 - \cos 60°)]$$

$$= mg + mg = 2mg$$

17. Tension at mean position,

$$mg + \frac{mv^2}{r} = 3mg$$

$$v = \sqrt{2gl} \qquad \ldots\text{(i)}$$

and if the body displace by angle θ with the vertical, then

$$v = \sqrt{2gl\,(1-\cos\theta)} \qquad \ldots\text{(ii)}$$

Comparing Eqs. (i) and (ii),

$$\cos\theta = 0$$

$$\Rightarrow \qquad \theta = 90^\circ$$

18. $t = \sqrt{\dfrac{2l}{g}} = \sqrt{\dfrac{2\times 2.0}{9.8}} = 0.64$ s

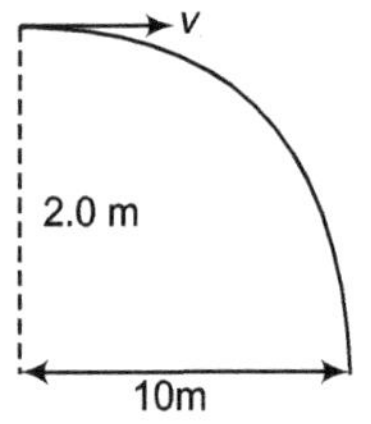

$$v = \frac{10}{t} = 15.625\,\frac{\text{m}}{\text{s}}$$

$$a = \frac{v^2}{R} = 163\ \text{m/s}^2$$

19. $T_{\text{top}} = 3mg$ and $T_{\text{bottom}} = 9mg$

20. $\frac{1}{2}mv^2 = as^2$

or $\qquad F_n = \dfrac{mv^2}{R} = \dfrac{2as^2}{R} \qquad \ldots\text{(i)}$

Further, $\qquad v = \sqrt{\dfrac{2a}{m}}\cdot s$

or $\qquad a_t = \dfrac{dv}{dt} = \sqrt{\dfrac{2a}{m}}\cdot\dfrac{ds}{dt} = \sqrt{\dfrac{2a}{m}}\cdot v$

$$= \sqrt{\frac{2a}{m}}\cdot\sqrt{\frac{2a}{m}}\cdot s = \frac{2as}{m}$$

$\therefore \qquad F_t = ma_t = 2as \qquad \ldots\text{(ii)}$

$\therefore \qquad F_{\text{net}} = \sqrt{F_n^2 + F_t^2}$

$$= 2as\sqrt{1+\frac{s^2}{R^2}}$$

21. $T\cos\theta = mg,\ T\sin\theta = mR\omega^2$

or $\qquad \tan\theta = \dfrac{R\omega^2}{g} = \dfrac{R(2\pi/T)^2}{g}$

$\therefore \qquad T = 2\pi\sqrt{\dfrac{R}{g\tan\theta}}$

But $R = L\sin\theta$

$\therefore \qquad T = 2\pi\sqrt{\dfrac{L}{g\cos\theta}}$

22. $v = \sqrt{2a_t s} = \sqrt{2a_t\,(\pi R)}$

$\therefore \qquad a_n = \dfrac{v^2}{R} = 2\pi a_t$

or $\qquad \dfrac{a_n}{a_t} = 2\pi$

$$\tan\theta = \frac{a_n}{a_t} = 2\pi$$

$\therefore \qquad \theta = \tan^{-1}(2\pi)$

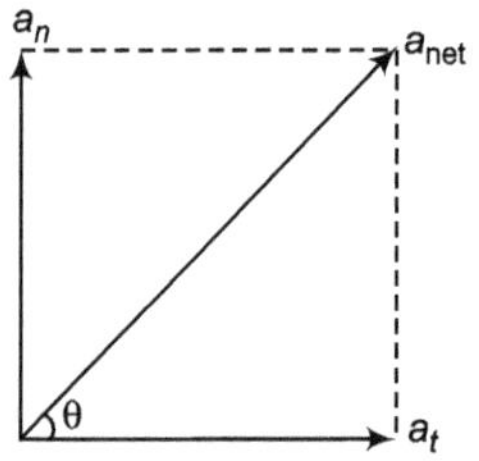

23. When a is horizontal

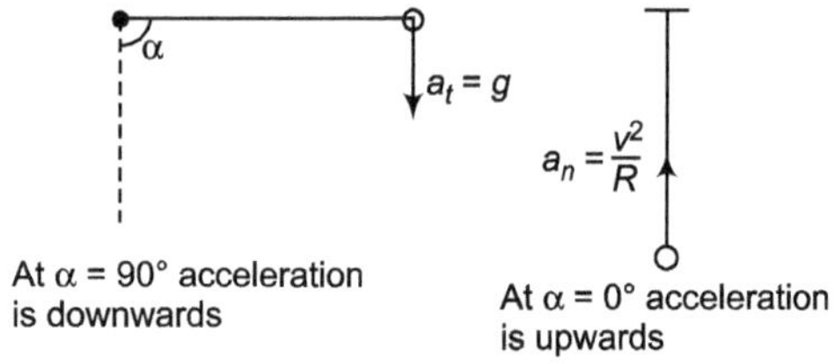

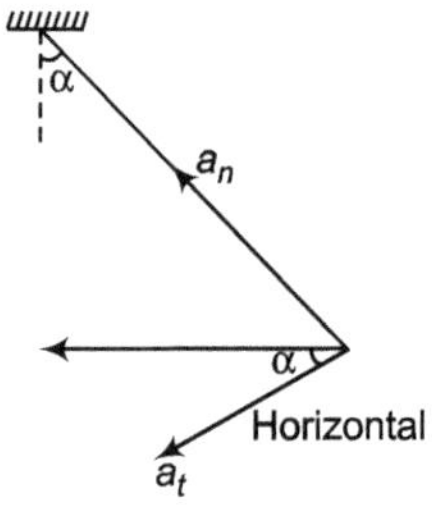

$$\tan\alpha = \frac{a_n}{a_t} = \frac{v^2/l}{g\sin\alpha} = \frac{2gh/l}{g\sin\alpha}$$

$$= \frac{2gl\cos\alpha/l}{g\sin\alpha}$$

$$= 2\cot\alpha$$

$$\tan\alpha = \sqrt{2} \quad\text{or}\quad \cos\alpha = \frac{1}{\sqrt{3}}$$

$\therefore \qquad \alpha = \cos^{-1}\left(\dfrac{1}{\sqrt{3}}\right)$

More than One Correct Options

1. Radial acceleration is given by,

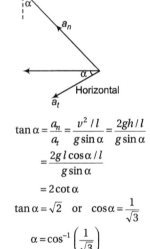

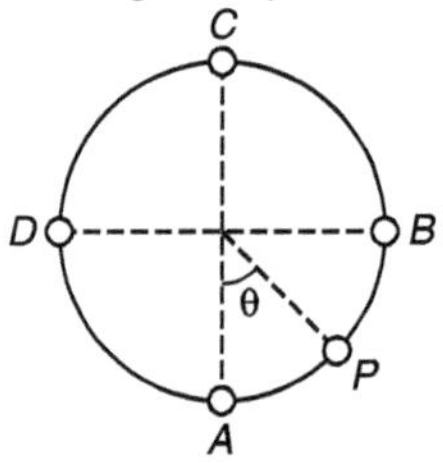

$$a_r = \frac{v^2}{R}$$

At A, speed is maximum.

Therefore, a_r is maximum. At C, speed is minimum. Therefore, a_r is minimum. Tangential acceleration is $g\sin\theta$.

At point $B, \theta = 90^\circ$.

Therefore, tangential acceleration is maximum $(= g)$.

2. $T + mg = \frac{mu^2}{l}$

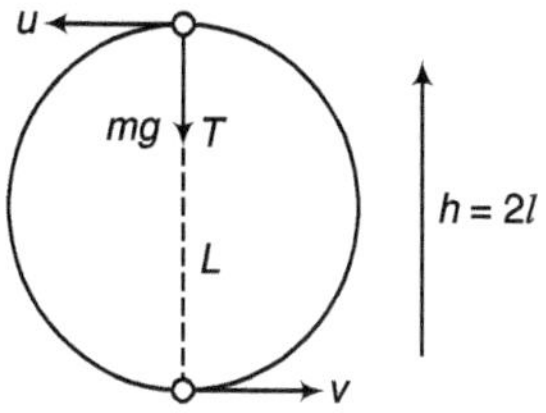

$\therefore \quad 2mg + mg = \frac{mu^2}{l}$ or $u = \sqrt{3gl}$

$$v^2 = u^2 + 2gh = 3gl + 2g(2l)$$
$$= 7gl$$
$$\therefore \quad v = \sqrt{7gl}$$

3. $R = h\cot\alpha$

$$N\cos\alpha = mg \quad \text{...(i)}$$
$$N\sin\alpha = \frac{mv^2}{R} \quad \text{...(ii)}$$

Solving these two equations, we get

$$v = \sqrt{Rg\tan\alpha}$$
$$= \sqrt{(h\cot\alpha)(g\tan\alpha)}$$
$$= \sqrt{gh}$$

Now, $T = \frac{2\pi R}{v} = \frac{2\pi h\cot\alpha}{\sqrt{gh}}$

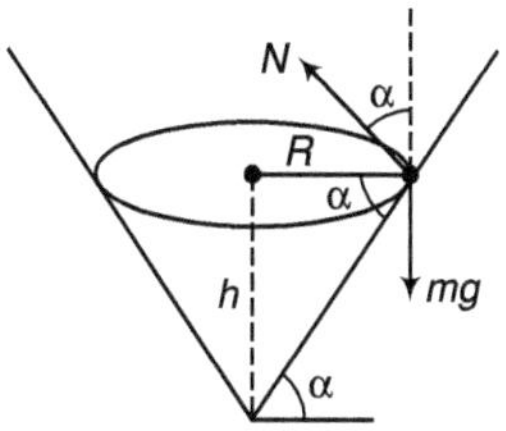

$$= 2\pi\sqrt{\frac{h}{g}}\cot\alpha$$

$\therefore \quad T \propto \sqrt{h}$ and $T \propto \cot\alpha$

4. No solution is required.

5. $N\cos\theta = mg$ ...(i)

$$N\sin\theta = mR\omega^2 \quad \text{...(ii)}$$

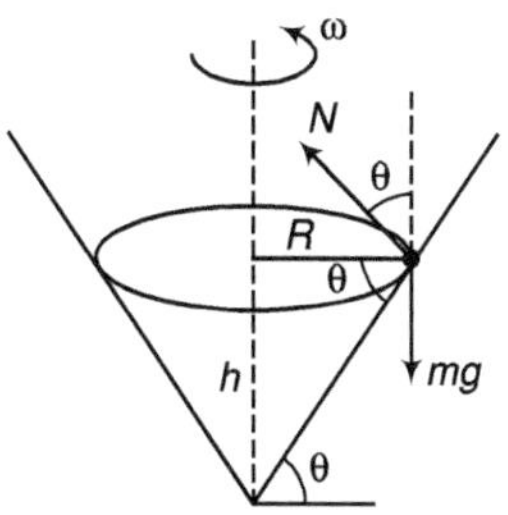

Net force is the resultant of N and mg.

From Eq. (ii), we can see that N will increase with increase in the value of ω.

If N will increase then net force will also increase.

On solving these two equations, we get

$$\omega = \sqrt{\frac{g}{R}\tan\theta}$$

But $R = \frac{h}{\tan\theta}$

$\therefore \quad \omega = \sqrt{\frac{g}{h}}\tan\theta$

or $\omega \propto \frac{1}{\sqrt{h}}$ (θ = constant)

Comprehension Based Questions

1. $\Delta U = \Delta U_{rod} + \Delta U_{ball}$

$$= Mg\left(\frac{l}{2}\right) + mgl$$
$$= \left(\frac{M}{2} + m\right)gl$$

2. $\omega = \frac{v}{l}$

Now decrease in rotational kinetic energy = increase in potential energy

$$\therefore \quad \frac{1}{2}I\omega^2 = \left(\frac{M}{2} + m\right)gl$$

or $\frac{1}{2}\left[\frac{Ml^2}{3} + ml^2\right]\left(\frac{v}{l}\right)^2 = \left(\frac{M}{2} + m\right)gl$

$$\therefore \quad v = \sqrt{\frac{\left(\frac{M}{2} + m\right)gl}{\left(\frac{M}{6} + \frac{m}{2}\right)}}$$

3. Maximum velocity is at bottommost point and minimum velocity is at topmost point.

$$\frac{\sqrt{u_{min}^2 + 2g(2L)}}{u_{min}} = \frac{2}{1}$$

On solving, we get

$$u_{min} = 2\sqrt{\frac{gL}{3}}$$

4. $u_{max} = 2u_{min} = 4\sqrt{\frac{gL}{3}}$

$$\therefore \quad K_{max} = \frac{1}{2}$$
$$mu_{max}^2 = \frac{8mgL}{3}$$

5. $v = \sqrt{u_{max}^2 - 2g(L)}$

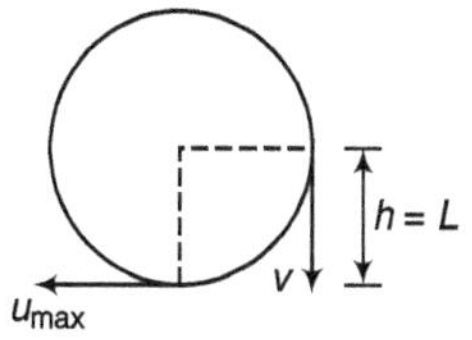

$$= \sqrt{\frac{16gL}{3} - 2gL} = \sqrt{\frac{10gL}{3}}$$

Assertion and Reason

1. In case of massless rod, $v_{min} = \sqrt{4gl}$.

3. Let the minimum and maximum tensions be T_{max} and T_{min} and the minimum and maximum speed be u and v.

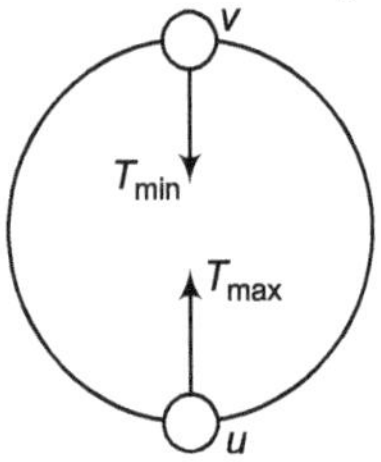

$$\therefore \quad T_{max} = \frac{mu^2}{R} + mg$$

$$T_{min} = \frac{mv^2}{R} - mg$$

$$\therefore \quad \Delta T = m\left(\frac{u^2}{R} - \frac{v^2}{R}\right) + 2mg$$

From conservation of energy, $\frac{u^2}{R} - \frac{v^2}{R} = 4g$

If is independent of u and $\Delta T = 6\,mg$.

$\therefore$ Reason is correct explanation of assertion.

4. The normal reaction is not least at topmost point, hence assertion is false.

5. Direction of acceleration continuously changes.

9. v is always perpendicular to ω.

10. $\frac{mv^2}{R}$ does not act. But this much force is required.

11. $a_c = \frac{v^2}{R} = \frac{(2)^2}{1} = 4\,\text{m/s}^2$

But no information is given for tangential acceleration a_t.

12. Centripetal force acts when the particle is observed from a rotating non-inertial frame.

13. Angle between any two vectors lies in the following range,

$$0° \le \theta \le 180°.$$

14. In vertical direction,

$$N\cos\theta = mg$$

In horizontal direction

$$N\sin\theta = \frac{mv^2}{R}$$

Match the Columns

1. $T_3 = (1)(3)(4)^2 = 48\,\text{N}\ (F = mR\omega^2)$

$$T_2 - T_3 = (1)(2)(4)^2$$

$$\therefore \quad T_2 = 80\,\text{N}$$

$$T_1 - T_2 = (1)(1)(4)^2$$

$$\therefore \quad T_1 = 96\,\text{N}$$

2. $v_B^2 = u_A^2 - 2gh_{AB} = (9gR) - (2gR) = 7gR$

$$\therefore \quad v_B = \sqrt{7gR}$$

Further, $T_B + \frac{mv_B^2}{R} = 7mg$

Again, $v_C^2 = v_A^2 - 2gh_{AC}$

$$= (9gR) - 2(2R) = 5gR$$

$$\therefore \quad v_C = \sqrt{5gR}$$

Further, $T_C + mg = \frac{mv_C^2}{R}$

$$\therefore \quad T_C = 4mg$$

3. Apply conservation of linear momentum in horizontal direction and conservation of mechanical energy.

4. Angle between net force and the string can never be obtuse. It is 90° at A, 0° at B and acute in between.

Entrance Gallery

1. Angular momentum of the pendulum about the suspension point O is

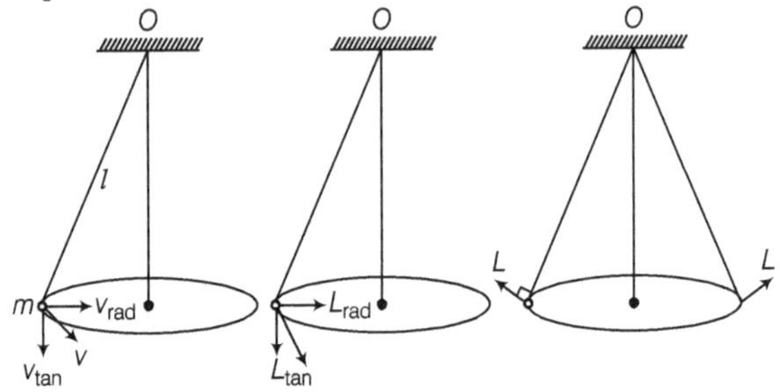

Then, **v** can be resolved into two components, radial component v_{rad} and tangential component v_{tan}. Due to v_{rad}, L will be tangential and due to v_{tan}, L will be radially outwards as shown above.

So, net angular momentum will be as shown in figure, whose magnitude will be constant ($|L| = mvl$). But its direction will change as shown in the figure.

$$\mathbf{L} = m(\mathbf{r} \times \mathbf{v})$$

where, **r** = position of particle from point O.

2. Suppose a particle is moving in a circular path of radius r as shown in the figure.

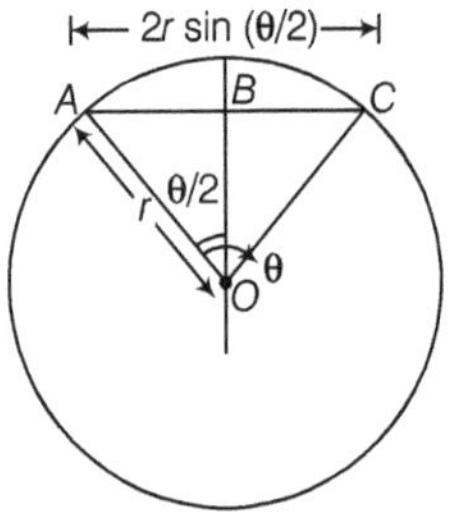

In ΔAOB, $\sin\left(\frac{\theta}{2}\right) = \frac{AB}{AO}$ $\quad (\because AO = r)$

$$AB = AO\sin\left(\frac{\theta}{2}\right) \Rightarrow AB = r\sin\left(\frac{\theta}{2}\right)$$

$$AC = AB + BC = r\sin\left(\frac{\theta}{2}\right) + r\sin\left(\frac{\theta}{2}\right) \quad (\because AB = BC)$$

$$AC = 2r\sin\left(\frac{\theta}{2}\right)$$

So, the magnitude of the corresponding linear displacement will be $2r\sin\left(\frac{\theta}{2}\right)$.

3. A circle with radius r as shown below, such that the position of point P from centre O, i.e. $\mathbf{OP} = \mathbf{r}$ is given as,

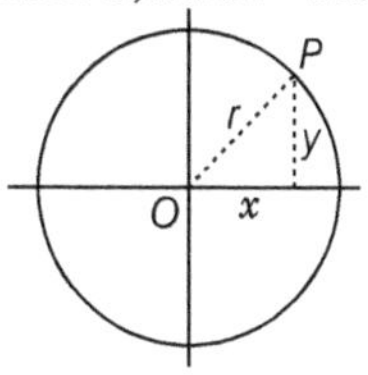

$$\mathbf{r} = \mathbf{OP} = x\hat{\mathbf{i}} + y\hat{\mathbf{j}}$$

$$\mathbf{F} = \frac{k}{(x^2+y^2)^{3/2}}(x\hat{\mathbf{i}} + y\hat{\mathbf{j}}) = \frac{k}{r^3}(\mathbf{r})$$

Since, **F** is along **r** or in radial direction. Therefore, work done is zero.

4.

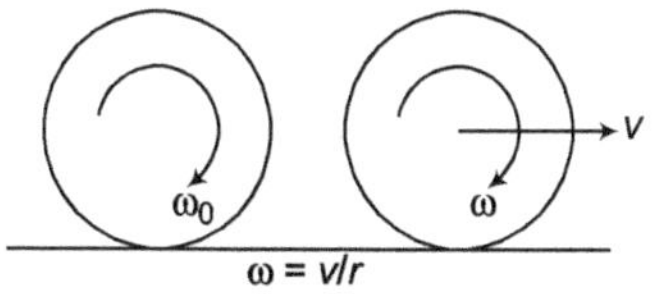

From conservation of angular momentum about bottommost point.

$$mr^2\omega_0 = mvr + mr^2 \times \frac{v}{r} \Rightarrow v = \frac{\omega_0 r}{2}$$

5. As, centripetal acceleration is given as, $a_c = \dfrac{v^2}{r}$

For first body of mass m_1, $a_{c_1} = \dfrac{v_1^2}{r_1}$

For second body of mass m_2, $a_{c_2} = \dfrac{v_2^2}{r_2}$

Also time to complete one revolution by both body is same. Hence,

$$\frac{2\pi r_1}{v_1} = \frac{2\pi r_2}{v_2} \Rightarrow \frac{v_1}{v_2} = \frac{r_1}{r_2} \quad \text{...(i)}$$

i.e. $$a_{c_1} : a_{c_2} = \frac{v_1^2}{r_1} \times \frac{r_2}{v_2^2} \quad \text{[From Eq. (i)]}$$

$$= \frac{r_1^2}{r_2^2} \times \frac{r_2}{r_1} = \frac{r_1}{r_2} = r_1 : r_2$$

6. $T\cos\theta$ component will cancel mg.

$T\sin\theta$ component will provide necessary centripetal force to the ball towards centre C.

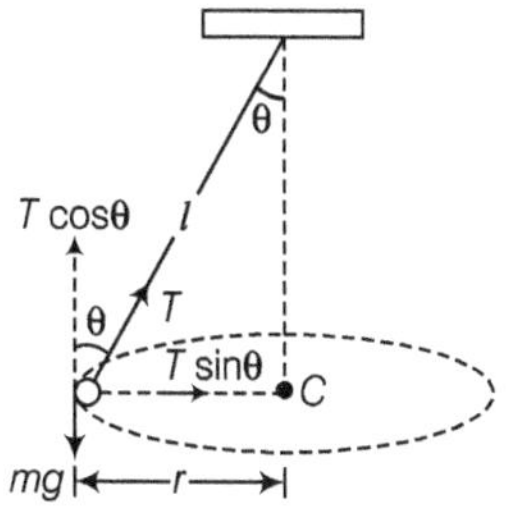

$\therefore$ $$T\sin\theta = mr\omega^2 = m(l\sin\theta)\omega^2$$

or $$T = ml\omega^2 \Rightarrow \omega = \sqrt{\frac{T}{ml}} \text{ rad/s}$$

or $$\omega_{max} = \sqrt{\frac{T_{max}}{ml}} = \sqrt{\frac{324}{0.5 \times 0.5}} = 36 \text{ rad/s}$$

7. Given, $r = 40\,\text{m}$ and $g = 10\,\text{m/s}^2$

We have, $$v = \sqrt{gr}$$
$$= 10 \times 40 = \sqrt{400}$$
$$= 20\,\text{m/s}$$

8. For horizontal planes potential energy remains constant equal to zero, if we assumes surface to be the zero level.

9. Time, $T = \dfrac{2\pi r}{v}$

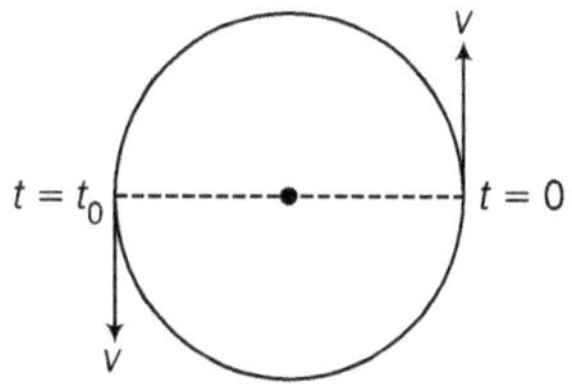

and $$t_0 = \frac{T}{2} = \frac{\pi r}{v}$$

$\therefore$ $$v_{av} = \frac{2r}{\pi r / v} = \frac{2v}{\pi}$$

10. Given, $s = t^3 + 5$

$\therefore$ Speed, $v = \dfrac{ds}{dt} = 3t^2$

and rate of change of speed, $a_t = \dfrac{dv}{dt} = 6t$

$\therefore$ Tangential acceleration at $t = 2\,\text{s}$

$$a_t = 6 \times 2 = 12\,\text{ms}^{-2}$$

and at $t = 2\,\text{s}$, $v = 3(2)^2 = 12\,\text{ms}^{-1}$

$\therefore$ Centripetal acceleration, $a_c = \dfrac{v^2}{R} = \dfrac{144}{20}\,\text{ms}^{-2}$

$\therefore$ Net acceleration $= a_t^2 + a_c \approx 14\,\text{ms}^{-2}$

11. For a particle in uniform circular motion,

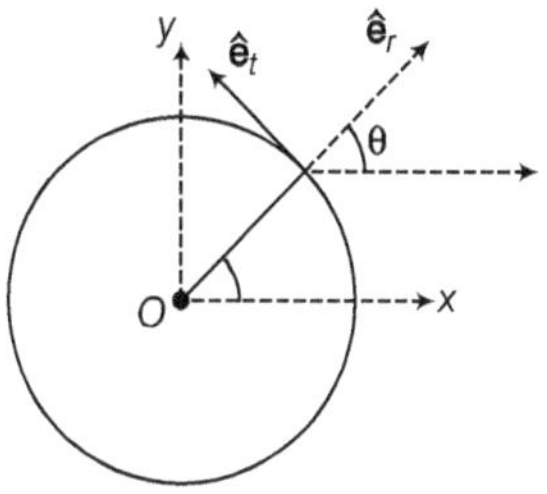

where, $\hat{\mathbf{e}}_r$ is radial unit vector and $\hat{\mathbf{e}}_t$ is tangential unit vector. As, $\hat{\mathbf{e}}_r = \hat{\mathbf{i}}\cos\theta + \hat{\mathbf{j}}\sin\theta$

$\mathbf{a} = \dfrac{v^2}{R}$ towards centre of circle, i.e. $(-\hat{\mathbf{e}}_r)$

$$\mathbf{a} = \frac{v^2}{R}(-\cos\theta\,\hat{\mathbf{i}} - \sin\theta\,\hat{\mathbf{j}})$$

or $$\mathbf{a} = -\frac{v^2}{R}\cos\theta\,\hat{\mathbf{i}} - \frac{v^2}{R}\sin\theta\,\hat{\mathbf{j}}$$

12. Centripetal acceleration, $a = \dfrac{v}{t}$

Angular acceleration, $\alpha = \dfrac{\omega}{t} = \dfrac{\omega v}{vt} = \dfrac{\omega a}{v}$

13. Kinetic energy, $E = \dfrac{1}{2}mv^2$ or $\dfrac{1}{2}mr\dfrac{v^2}{r} = E$

or $$\frac{1}{2}mra = E$$

or $$a = \frac{2E}{mr}$$

14. Angular momentum $L = r \times p = r \times m \times v$

$$v = \frac{L}{mr} \quad \text{...(i)}$$

Now, as centripetal force, $F_c = \dfrac{mv^2}{r}$ ...(ii)

Substituting the value of v from Eq. (i) in Eq. (ii), we get

$$F_c = \frac{m}{r}\left[\frac{L}{mr}\right]^2 = \frac{L^2}{mr^3}$$

8

COM, Conservation of Linear Momentum, Impulse and Collision

Chapter Snapshot

- Centre of Mass
- Law of Conservation of Linear Momentum
- Variable Mass
- Impulse
- Collision

8.1 Centre of Mass

When we consider the motion of a system of particles, there is one point in it which behaves as though the entire mass of the system (i.e. the sum of the masses of all the individual particles) is concentrated there and its motion is the same as would ensue if the resultant of all the forces acting on all the particles were applied directly to it. This point is called the centre of mass (CM) of the system. The concept of CM is very useful in solving many problems, in particular, those concerned with collision of particles.

Position of Centre of Mass

First of all, we find the position of CM of a system of particles. Just to make the subject easy we classify a system of particles in three groups:

1. System of two particles,
2. System of a large number of particles,
3. Continuous bodies.

Now, let us take them separately.

Position of CM of Two Particles

Centre of mass of two particles of mass m_1 and m_2 separated by a distance of d lies in between the two particles. The distance of centre of mass from any of the particle (r) is inversely proportional to the mass of the particle (m).

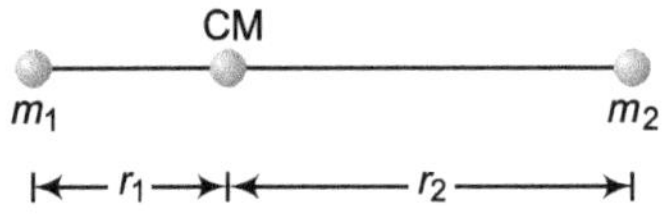

Fig. 8.1

i.e. $$r \propto \frac{1}{m} \quad \text{or} \quad \frac{r_1}{r_2} = \frac{m_2}{m_1}$$

or $$m_1 r_1 = m_2 r_2$$

$$r_1 = \left(\frac{m_2}{m_2 + m_1}\right) d$$

and $$r_2 = \left(\frac{m_1}{m_1 + m_2}\right) d$$

Here, r_1 = distance of CM from m_1

and r_2 = distance of CM from m_2

From the above discussion, we see that

$r_1 = r_2 = \frac{d}{2}$ if $m_1 = m_2$, i.e. CM lies midway between the two particles of equal masses.

Similarly, $r_1 > r_2$ if $m_1 < m_2$ and $r_1 < r_2$ if $m_1 > m_2$, i.e. CM is nearer to the particle having larger mass.

Example 8.1 *Two particles of mass 1 kg and 2 kg are located at $x = 0$ and $x = 3$ m. Find the position of their centre of mass.*

Sol. Since, both the particles lie on x-axis, the CM will also lie on x-axis. Let the CM is located at $x = x$, then

r_1 = distance of CM from the particle of mass 1 kg = x

and r_2 = distance of CM from the particle of mass 2 kg $= (3 - x)$

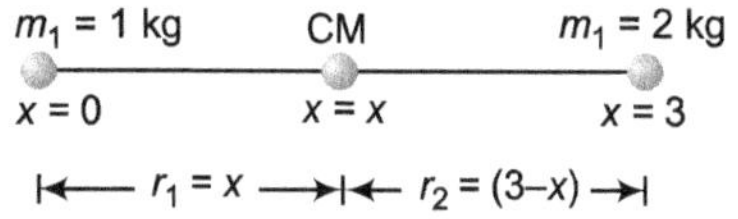

Fig. 8.2

Using, $\frac{r_1}{r_2} = \frac{m_2}{m_1}$ or $\frac{x}{3 - x} = \frac{2}{1}$ or $x = 2$ m

Thus, the CM of the two particles is located at $x = 2$ m.

Position of CM of a Large Number of Particles

If we have a system consisting of n particles, of mass $m_1, m_2, \ldots, m_n$ with $\mathbf{r}_1\ \mathbf{r}_2, \ldots, \mathbf{r}_n$ as their position vectors at a given instant of time. The position vector $\mathbf{r}_{CM}$ of the CM of the system at that instant is given by

$$\mathbf{r}_{CM} = \frac{m_1 \mathbf{r}_1 + m_2 \mathbf{r}_2 + \ldots + m_n \mathbf{r}_n}{m_1 + m_2 + \ldots + m_n} = \frac{\sum_{i=1}^{n} m_i \mathbf{r}_i}{\sum_{i=1}^{n} m_i}$$

or $$\mathbf{r}_{CM} = \frac{\sum_{i=1}^{n} m_i \mathbf{r}_i}{M}$$

Here, $M = m_1 + m_2 + \ldots + m_n$ and $\Sigma m_i \mathbf{r}_i$ is called the first moment of the mass.

Further, $$\mathbf{r}_i = x_i \hat{\mathbf{i}} + y_i \hat{\mathbf{j}} + z_i \hat{\mathbf{k}}$$

and $$\mathbf{r}_{CM} = x_{CM} \hat{\mathbf{i}} + y_{CM} \hat{\mathbf{j}} + z_{CM} \hat{\mathbf{k}}$$

So, the cartesian coordinates of the CM will be

$$x_{CM} = \frac{m_1 x_1 + m_2 x_2 + \ldots + m_n x_n}{m_1 + m_2 + \ldots + m_n} = \frac{\sum_{i=1}^{n} m_i x_i}{\Sigma m_i}$$

or $$x_{CM} = \frac{\sum_{i=1}^{n} m_i x_i}{M}$$

Similarly, $$y_{CM} = \frac{\sum_{i=1}^{n} m_i y_i}{M} \text{ and } z_{CM} = \frac{\sum_{i=1}^{n} m_i z_i}{M}$$

Example 8.2 *The position vector of three particles of mass $m_1 = 1$ kg, $m_2 = 2$ kg and $m_3 = 3$ kg are $\mathbf{r}_1 = (\hat{\mathbf{i}} + 4\hat{\mathbf{j}} + \hat{\mathbf{k}})$ m, $\mathbf{r}_2 = (\hat{\mathbf{i}} + \hat{\mathbf{j}} + \hat{\mathbf{k}})$ m and $\mathbf{r}_3 = (2\hat{\mathbf{i}} - \hat{\mathbf{j}} - 2\hat{\mathbf{k}})$ m respectively. Find the position vector of their centre of mass.*

Sol. The position vector of CM of the three particles will be given by

$$\mathbf{r}_{CM} = \frac{m_1\mathbf{r}_1 + m_2\mathbf{r}_2 + m_3\mathbf{r}_3}{m_1 + m_2 + m_3}$$

Substituting the values, we get

$$\mathbf{r}_{CM} = \frac{1(\hat{\mathbf{i}} + 4\hat{\mathbf{j}} + \hat{\mathbf{k}}) + 2(\hat{\mathbf{i}} + \hat{\mathbf{j}} + \hat{\mathbf{k}}) + 3(2\hat{\mathbf{i}} - \hat{\mathbf{j}} - 2\hat{\mathbf{k}})}{1 + 2 + 3}$$

$$= \frac{9\hat{\mathbf{i}} + 3\hat{\mathbf{j}} - 3\hat{\mathbf{k}}}{6}$$

$$\mathbf{r}_{CM} = \frac{1}{2}(3\hat{\mathbf{i}} + \hat{\mathbf{j}} - \hat{\mathbf{k}})\text{ m}$$

Example 8.3 *Four particles of mass 1 kg, 2 kg, 3 kg and 4 kg are placed at the four vertices A, B, C and D of a square of side 1 m. Find the position of centre of mass of the particles.*

Sol. Assuming D as the origin, DC as x-axis and DA as y-axis, we have

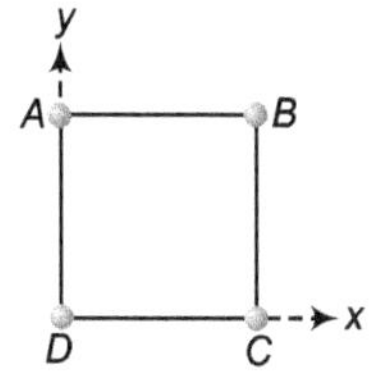

Fig. 8.3

$m_1 = 1$ kg, $(x_1, y_1) = (0, 1\text{ m})$

$m_2 = 2$ kg, $(x_2, y_2) = (1\text{ m}, 1\text{ m})$

$m_3 = 3$ kg, $(x_3, y_3) = (1\text{ m}, 0)$

and $m_4 = 4$ kg, $(x_4, y_4) = (0, 0)$

Coordinates of their CM are

$$x_{CM} = \frac{m_1 x_1 + m_2 x_2 + m_3 x_3 + m_4 x_4}{m_1 + m_2 + m_3 + m_4}$$

$$= \frac{1(0) + 2(1) + 3(1) + 4(0)}{1 + 2 + 3 + 4}$$

$$= \frac{5}{10} = \frac{1}{2}\text{ m} = 0.5\text{ m}$$

Similarly,

$$y_{CM} = \frac{m_1y_1 + m_2y_2 + m_3y_3 + m_4y_4}{m_1 + m_2 + m_3 + m_4}$$

$$= \frac{1(1) + 2(1) + 3(0) + 4(0)}{1 + 2 + 3 + 4}$$

$$= \frac{3}{10}\text{ m} = 0.3\text{ m}$$

$\therefore \quad (x_{CM}, y_{CM}) = (0.5\text{m}, 0.3\text{m})$

Thus, position of CM of the four particles is as shown in figure.

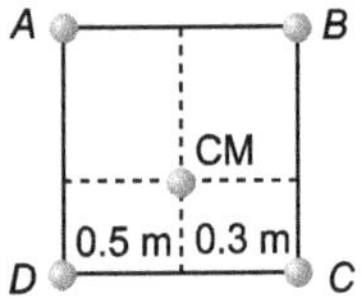

Fig. 8.4

Position of CM of Continuous Bodies

If we consider the body to have continuous distribution of matter, then summation in the formula of CM is replaced by integration. Suppose x, y and z are the coordinates of a small element of mass dm, we write the coordinates of CM as

$$x_{CM} = \frac{\int x\,dm}{\int dm} = \frac{\int x\,dm}{M}$$

$$y_{CM} = \frac{\int y\,dm}{\int dm} = \frac{\int y\,dm}{M}$$

and $$z_{CM} = \frac{\int z\,dm}{\int dm} = \frac{\int z\,dm}{M}$$

Let us take an example.

Centre of Mass of a Uniform Rod

Suppose a rod of mass M and length L is lying along the x-axis with its one end at $x = 0$ and the other at $x = L$.

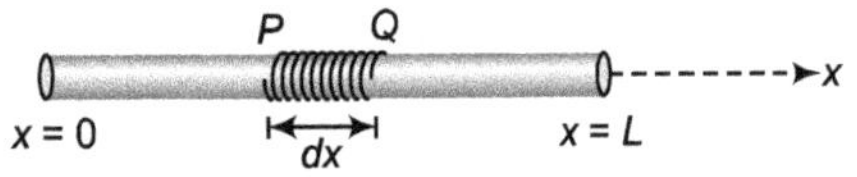

Fig. 8.5

Mass per unit length of the rod $= \frac{M}{L}$

Hence, the mass of the element PQ of length dx situated at $x = x$ is $dm = \frac{M}{L}dx$

The coordinates of the element PQ are $(x, 0, 0)$. Therefore, x-coordinate of CM of the rod will be

$$x_{CM} = \frac{\int_0^L x\,dm}{\int dm}$$

$$= \frac{\int_0^L (x)\left(\frac{M}{L}dx\right)}{M}$$

$$= \frac{1}{L}\int_0^L x\,dx = \frac{L}{2}$$

The y-coordinate of CM is

$$y_{CM} = \frac{\int y\,dm}{\int dm} = 0 \qquad (\text{as, } y = 0)$$

Similarly, $\quad z_{CM} = 0$

i.e. the coordinates of CM of the rod are $\left(\frac{L}{2}, 0, 0\right)$ or it lies at the centre of the rod.

Proceeding in the similar manner, we can find the CM of certain rigid bodies. Centre of mass of some well known rigid bodies are given below:

1. Centre of mass of a uniform rectangular, square or circular plate lies at its centre.

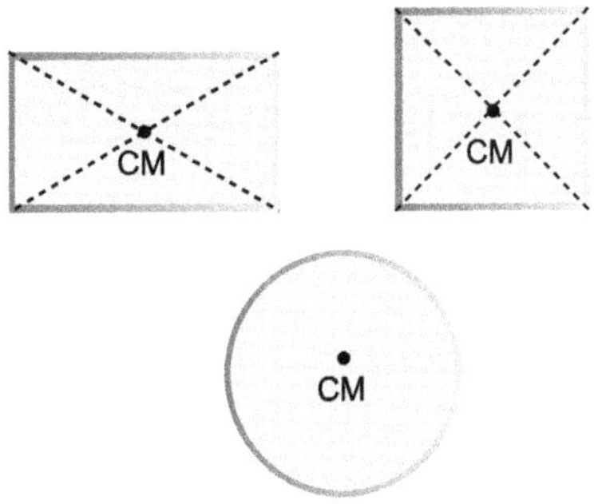

Fig. 8.6

2. Centre of mass of a uniform **semicircular ring** lies at a distance of $h = \frac{2R}{\pi}$ from its centre, on the axis of symmetry where R is the radius of the ring.

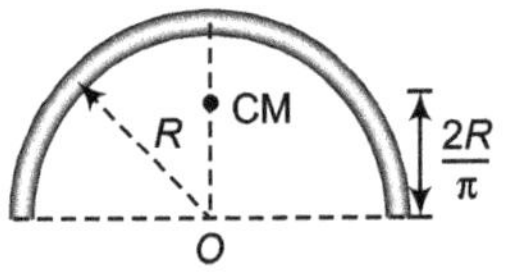

Fig. 8.7

3. Centre of mass of a uniform **semicircular disc** of radius R lies at a distance of $h = \frac{4R}{3\pi}$ from the centre on the axis of symmetry as shown in Fig. 8.8.

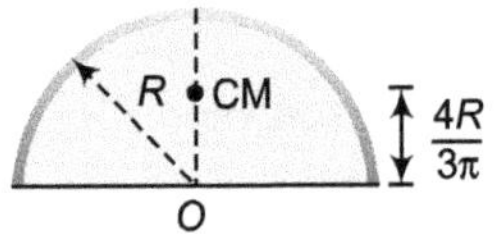

Fig. 8.8

4. Centre of mass of a **hemispherical shell** of radius R lies at a distance of $h = \frac{R}{2}$ from its centre on the axis of symmetry as shown in figure.

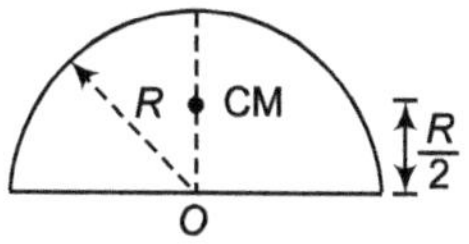

Fig. 8.9

5. Centre of mass of a **solid hemisphere** of radius R lies at a distance of $h = \frac{3R}{8}$ from its centre on the axis of symmetry.

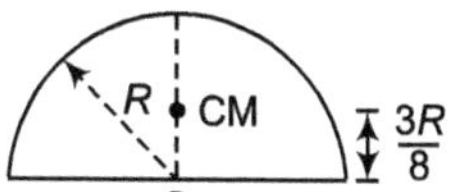

Fig. 8.10

Example 8.4 *Find the position of centre of mass of the uniform lamina as shown in figure.*

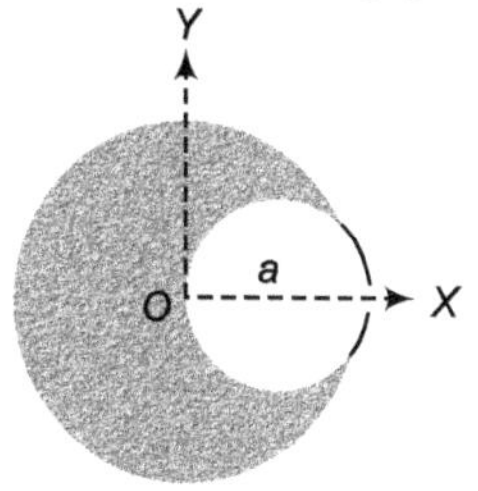

Fig. 8.11

Sol. Here, A_1 = area of complete circle $= \pi a^2$

A_2 = area of small circle

$$= \pi\left(\frac{a}{2}\right)^2 = \frac{\pi a^2}{4}$$

(x_1, y_1) = coordinates of centre of mass of small circle = (0, 0)

(x_2, y_2) = coordinates of centre of mass of small circle

$$= \left(\frac{a}{2}, 0\right)$$

Using, $$x_{CM} = \frac{A_1x_1 - A_2x_2}{A_1 - A_2}$$

We get, $$x_{CM} = \frac{-\frac{\pi a^2}{4}\left(\frac{a}{2}\right)}{\pi a^2 - \frac{\pi a^2}{4}} = \frac{-\left(\frac{1}{8}\right)}{\left(\frac{3}{4}\right)} a = -\frac{a}{6}$$

and $y_{CM} = 0$ as y_1 and y_2 both are zero. Therefore, coordinates of CM of the lamina as shown in above figure are $\left(-\frac{a}{6}, 0\right)$.

Extra Knowledge Points

- For a laminar type (2-dimensional) body the formulae for finding the position of centre of mass are as follows:

(i) $$\mathbf{r}_{CM} = \frac{A_1\mathbf{r}_1 + A_2\mathbf{r}_2 + \ldots + A_n\mathbf{r}_n}{A_1 + A_2 + \ldots + A_n}$$

(ii) $$x_{CM} = \frac{A_1x_1 + A_2x_2 + \ldots + A_nx_n}{A_1 + A_2 + \ldots + A_n}$$

$$y_{CM} = \frac{A_1y_1 + A_2y_2 + \ldots + A_ny_n}{A_1 + A_2 + \ldots + A_n}$$

and $$z_{CM} = \frac{A_1z_1 + A_2z_2 + \ldots + A_nz_n}{A_1 + A_2 + \ldots + A_n}$$

Here, A stands for the area.

- If some mass or area is removed from a rigid body, then the position of centre of mass of the remaining portion is obtained from the following formulae:

(i) $$\mathbf{r}_{CM} = \frac{m_1\mathbf{r}_1 - m_2\mathbf{r}_2}{m_1 - m_2} \quad \text{or} \quad \mathbf{r}_{CM} = \frac{A_1\mathbf{r}_1 - A_2\mathbf{r}_2}{A_1 - A_2}$$

(ii) $$x_{CM} = \frac{m_1x_1 - m_2x_2}{m_1 - m_2} \quad \text{or} \quad x_{CM} = \frac{A_1x_1 - A_2x_2}{A_1 - A_2}$$

$$y_{CM} = \frac{m_1y_1 - m_2y_2}{m_1 - m_2} \quad \text{or} \quad y_{CM} = \frac{A_1y_1 - A_2y_2}{A_1 - A_2}$$

and $$z_{CM} = \frac{m_1z_1 - m_2z_2}{m_1 - m_2} \quad \text{or} \quad z_{CM} = \frac{A_1z_1 - A_2z_2}{A_1 - A_2}$$

Here, m_1, A_1, $\mathbf{r}_1$, x_1, y_1 and z_1 are the values for the whole mass while m_2, A_2, $\mathbf{r}_2$, x_2, y_2 and z_2 are the values for the mass which has been removed. Let us see two examples in support of the above theory.

Students are often confused over the problems of centre of mass. They cannot answer even the basic problems of CM. For example, let us take a simple problem : two particles one of mass 1 kg and the other of 2 kg are projected simultaneously with the same speed from the roof of a tower, the one of mass 1 kg vertically upwards and the other vertically downwards. What is the acceleration of centre of mass of these two particles? When I ask this question in my first class of centre of mass, three answers normally come among the students g, $\frac{g}{3}$ and zero. The correct answer is g. Because

$$\mathbf{a}_{CM} = \frac{m_1\mathbf{a}_1 + m_2\mathbf{a}_2}{m_1 + m_2}$$

Here, $\mathbf{a}_1 = \mathbf{a}_2 = g$ (downwards)

$$\therefore \quad \mathbf{a}_{CM} = \frac{(1)(g) + (2)(g)}{1 + 2} = g \text{ (downwards)}$$

The idea behind this is that apply the basic equations when asked anything about centre of mass. Just as a revision I am writing below all the basic equations of CM at one place.

$$\mathbf{r}_{CM} = \frac{m_1\mathbf{r}_1 + m_2\mathbf{r}_2 + \dots + m_n\mathbf{r}_n}{m_1 + m_2 + \dots + m_n}$$

$$x_{CM} = \frac{m_1x_1 + m_2x_2 + \dots + m_n x_n}{m_1 + m_2 + \dots + m_n}$$

$$y_{CM} = \frac{m_1y_1 + m_2y_2 + \dots + m_n y_n}{m_1 + m_2 + \dots + m_n}$$

$$z_{CM} = \frac{m_1z_1 + m_2z_2 + \dots + m_n z_n}{m_1 + m_2 + \dots + m_n}$$

$$\mathbf{v}_{CM} = \frac{m_1\mathbf{v}_1 + m_2\mathbf{v}_2 + \dots + m_n\mathbf{v}_n}{m_1 + m_2 + \dots + m_n}$$

$$\mathbf{p}_{CM} = \mathbf{p}_1 + \mathbf{p}_2 + \dots + \mathbf{p}_n$$

$$\mathbf{a}_{CM} = \frac{m_1\mathbf{a}_1 + m_2\mathbf{a}_2 + \dots + m_n\mathbf{a}_n}{m_1 + m_2 + \dots + m_n}$$

and $\mathbf{F}_{CM} = \mathbf{F}_1 + \mathbf{F}_2 + \dots + \mathbf{F}_n$

- If a projectile explodes in air in different parts, the path of the centre of mass remains unchanged. This is because during explosion no external force (except gravity) acts on the centre of mass. The situation is as shown in figure.

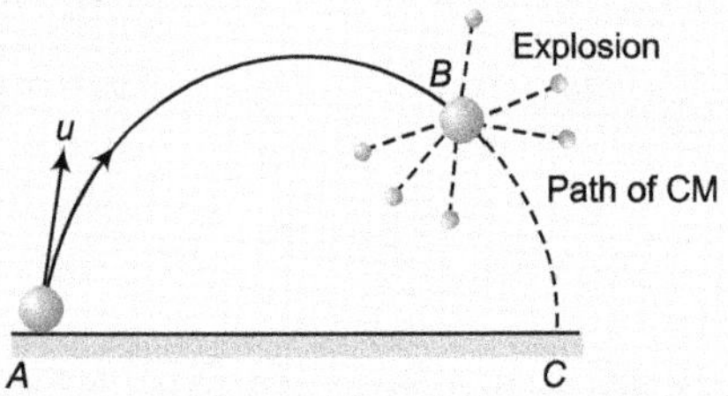

Path of CM is *ABC*, even though the different parts travel in different directions after explosion.Suppose a system consists of more than one particle (or bodies). Net external force on the system in a particular direction is zero. Initially the centre of mass of the system is at rest, then obviously the centre of mass will not move along that particular direction even though some particles (or bodies) of the system may move along that direction. The following example will illustrate the above theory.

Motion of the Centre of Mass

Let us consider the motion of a system of n particles of individual masses $m_1, m_2, \dots, m_n$ and total mass M. It is assumed that no mass enters or leaves the system during its motion, so that M remains constant. Then, as we have seen, we have the relation

$$\mathbf{r}_{CM} = \frac{m_1\mathbf{r}_1 + m_2\mathbf{r}_2 + \dots m_n\mathbf{r}_n}{m_1 + m_2 + \dots + m_n} = \frac{m_1\mathbf{r}_1 + m_2\mathbf{r}_2 + \dots + m_n\mathbf{r}_n}{M}$$

or $M\mathbf{r}_{CM} = m_1\mathbf{r}_1 + m_2\mathbf{r}_2 + \dots + m_n\mathbf{r}_n$

Differentiating this expression with respect to time t, we have

$$M\frac{d\mathbf{r}_{CM}}{dt} = m_1\frac{d\mathbf{r}_1}{dt} + m_2\frac{d\mathbf{r}_2}{dt} + \dots + m_n\frac{d\mathbf{r}_n}{dt}$$

Since, $\frac{d\mathbf{r}}{dt}$ = velocity

Therefore, $M\mathbf{v}_{CM} = m_1\mathbf{v}_1 + m_2\mathbf{v}_2 + \dots + m_n\mathbf{v}_n$ ...(i)

or velocity of the CM is

$$\mathbf{v}_{CM} = \frac{m_1\mathbf{v}_1 + m_2\mathbf{v}_2 + .. + m_n\mathbf{v}_n}{M}$$

or $$\mathbf{v}_{CM} = \frac{\sum_{i=1}^{n} m_i\mathbf{v}_i}{M}$$

Further, $m\mathbf{v}$ = momentum of a particle $\mathbf{p}$. Therefore, Eq. (i) can be written as

$$\mathbf{p}_{CM} = \mathbf{p}_1 + \mathbf{p}_2 + \dots + \mathbf{p}_n \quad \text{or} \quad P_{CM} = \sum_{i=1}^{n} P_i$$

Differentiating Eq. (i) with respect to time t, we get

$$M\frac{d\mathbf{v}_{CM}}{dt} = m_1\frac{d\mathbf{v}_1}{dt} + m_2\frac{d\mathbf{v}_2}{dt} + \dots + m_n\frac{d\mathbf{v}_n}{dt}$$

or $M\mathbf{a}_{CM} = m_1\mathbf{a}_1 + m_2\mathbf{a}_2 + \dots + m_n\mathbf{a}_n$ ...(ii)

or $$\mathbf{a}_{CM} = \frac{m_1\mathbf{a}_1 + m_2\mathbf{a}_2 + \dots + m_n\mathbf{a}_n}{M}$$

or $$\mathbf{a}_{CM} = \frac{\sum_{i=1}^{n} m_i\mathbf{a}_i}{M}$$

Further, in accordance with Newton's second law of motion $\mathbf{F} = m\mathbf{a}$. Hence, Eq. (ii) can be written as

$$\mathbf{F}_{CM} = \mathbf{F}_1 + \mathbf{F}_2 + \dots + \mathbf{F}_n \quad \text{or} \quad \mathbf{F}_{CM} = \sum_{i=1}^{n} \mathbf{F}_i$$

Thus, as pointed out earlier also, the centre of mass of a system of particles moves as though it were a particle of mass equal to that of the whole system with all the external forces acting directly on it.

Example 8.5 *Two particles A and B of mass 1 kg and 2 kg respectively are projected in the directions shown in figure with speeds $u_A = 200$ m/s and $u_B = 50$ m/s. Initially they were 90 m apart. Find the maximum height attained by the centre of mass of the particles. Assume acceleration due to gravity to be constant.* ($g = 10\, m/s^2$).

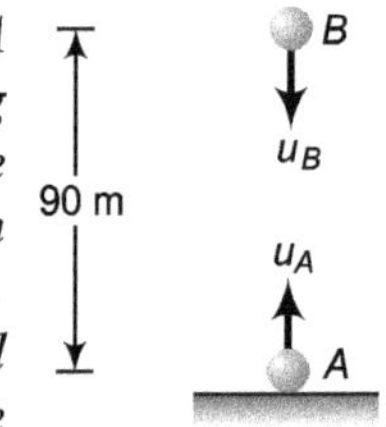

Fig. 8.12

Sol. Using, $m_A r_A = m_B r_B$

or $1(r_A) = 2(r_B)$

or $r_A = 2r_B$ ...(i)

and $r_A + r_B = 90$ m ...(ii)

Solving these two equations, we get

$r_A = 60$ m and $r_B = 30$ m

i.e. CM is at height 60 m from the ground at time $t = 0$.

Further, $\mathbf{a}_{CM} = \dfrac{m_A\mathbf{a}_A + m_B\mathbf{a}_B}{m_A + m_B}$

$= g = 10\ \text{m/s}^2$ (downwards)

as $\mathbf{a}_A = \mathbf{a}_B = g$ (downwards)

$$\mathbf{u}_{CM} = \frac{m_A\mathbf{u}_A + m_B\mathbf{u}_B}{m_A + m_B} = \frac{1(200) - 2(50)}{1+2}$$

$= \dfrac{100}{3}$ m/s (upwards)

Let h be the height attained by CM beyond 60 m. Using

$$v_{CM}^2 = u_{CM}^2 + 2a_{CM}h$$

or $0 = \left(\dfrac{100}{3}\right)^2 - (2)(10)h$

or $h = \dfrac{(100)^2}{180}$

$= 55.55$ m

Therefore, maximum height attained by the centre of mass is

$$H = 60 + 55.55 = 115.55\ \text{m}$$

Example 8.6 *In the arrangement shown in figure, $m_A = 2\,kg$ and $m_B = 1\,kg$. String is light and inextensible. Find the acceleration of centre of mass of both the blocks. Neglect friction everywhere.*

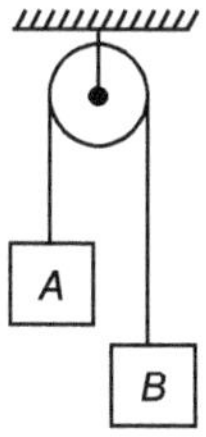

Fig. 8.13

Sol. Net pulling force on the system is $(m_A - m_B)g$

or $(2-1)g = g$

Total mass being pulled is $m_A + m_B$ or 3 kg

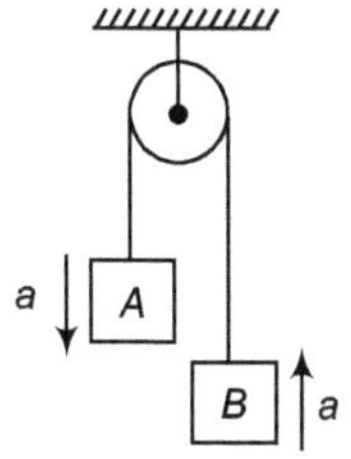

Fig. 8.14

$\therefore$ $a = \dfrac{\text{Net pulling force}}{\text{Total mass}} = \dfrac{g}{3}$

Now, $\mathbf{a}_{CM} = \dfrac{m_A\mathbf{a}_A + m_B\,\mathbf{a}_B}{m_A + m_B}$

$= \dfrac{2(a) - 1(a)}{1+2} = \dfrac{a}{3} = \dfrac{g}{9}$ (downwards)

Alternate Method

Free body diagram of block A is shown in figure.

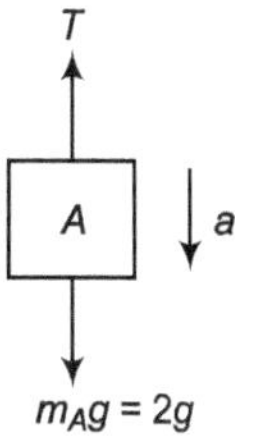

Fig. 8.15

$2g - T = m_A(a)$ or $T = 2g - m_Aa$

$= 2g - (2)\left(\dfrac{g}{3}\right) = \dfrac{4g}{3}$

Free body diagrams of A and B both are as shown in Fig. 8.16.

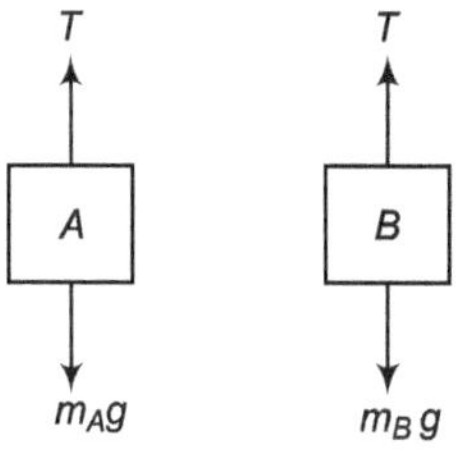

Fig. 8.16

$$\mathbf{a}_{CM} = \frac{\text{Net force on both the blocks}}{m_A + m_B} = \frac{(m_A + m_B)g - 2T}{2+1}$$

$= \dfrac{3g - \dfrac{8g}{3}}{3} = \dfrac{g}{9}$ (downwards)

8.2 Law of Conservation of Linear Momentum

The product of mass and the velocity of a particle is defined as its linear momentum (**p**). So,

$$\mathbf{p} = m\mathbf{v}$$

The magnitude of linear momentum may be written as

$$p = mv$$

or $p^2 = m^2v^2$

$= 2m\left(\dfrac{1}{2}mv^2\right)$

$= 2mK$

Thus, $p = \sqrt{2Km}$

or $K = \dfrac{p^2}{2m}$

Here, K is the kinetic energy of the particle. In accordance with Newton's second law,

$$\mathbf{F} = m\mathbf{a} = m\frac{d\mathbf{v}}{dt} = \frac{d(m\mathbf{v})}{dt} = \frac{d\mathbf{p}}{dt}$$

Thus, $$\mathbf{F} = \frac{d\mathbf{p}}{dt}$$

In this case, the external force applied to a particle (or a body) be zero, we have

$$\mathbf{F} = \frac{d\mathbf{p}}{dt} = 0 \quad \text{or} \quad \mathbf{p} = \text{constant}$$

Showing that in the absence of an external force, the linear momentum of a particle (or the body) remains constant. This is called the law of conservation of linear momentum. This law may be extended to a system of particles or to the centre of mass of a system of particles. For example, for a system of particles it takes the form.

If net force (or the vector sum of all the forces) on a system of particles is zero, the vector sum of linear momentum of all the particles remain conserved, or

If $$\mathbf{F} = \mathbf{F}_1 + \mathbf{F}_2 + \mathbf{F}_3 + \ldots + \mathbf{F}_n = 0$$

Then, $$\mathbf{p}_1 + \mathbf{p}_2 + \mathbf{p}_3 + \ldots + \mathbf{p}_n = \text{constant}$$

The same is the case for the centre of mass of a system of particles, i.e. if

$$\mathbf{F}_{CM} = 0, \quad \mathbf{p}_{CM} = \text{constant}.$$

Thus, the law of conservation of linear momentum can be applied to a single particle, to a system of particles or even to the centre of mass of the particles.

The law of conservation of linear momentum enables us to solve a number of problems which cannot be solved by a straight application of the relation $\mathbf{F} = m\mathbf{a}$.

For example, suppose a particle of mass m initially at rest, suddenly explodes into two fragments of masses m_1 and m_2 which fly apart with velocities $\mathbf{v}_1$ and $\mathbf{v}_2$, respectively. Obviously, the forces resulting in the explosion of the particle must be internal forces, since no external force has been applied. In the absence of the external forces, therefore, the momentum must remain conserved and we should have

$$m\mathbf{v} = m_1\mathbf{v}_1 + m_2\mathbf{v}_2$$

Since, the particle was initially at rest, $\mathbf{v} = 0$ and therefore,

$$m_1\mathbf{v}_1 + m_2\mathbf{v}_2 = 0$$

$$\mathbf{v}_1 = -\frac{m_2}{m_1}\mathbf{v}_2 \quad \text{or} \quad \frac{|\mathbf{v}_1|}{|\mathbf{v}_2|} = \frac{m_2}{m_1}$$

Showing at once that the velocities of the two fragments must be inversely proportional to their masses and in opposite directions along the same line. This result could not possibly be arrived at from the relation $\mathbf{F} = m\mathbf{a}$, since we know nothing about the forces that were acting during the explosion. Nor, could we derive it from the law of conservation of energy.

Example 8.7 *A wooden plank of mass 20 kg is resting on a smooth horizontal floor. A man of mass 60 kg starts moving from one end of the plank to the other end. The length of the plank is 10 m. Find the displacement of the plank over the floor when the man reaches the other end of the plank.*

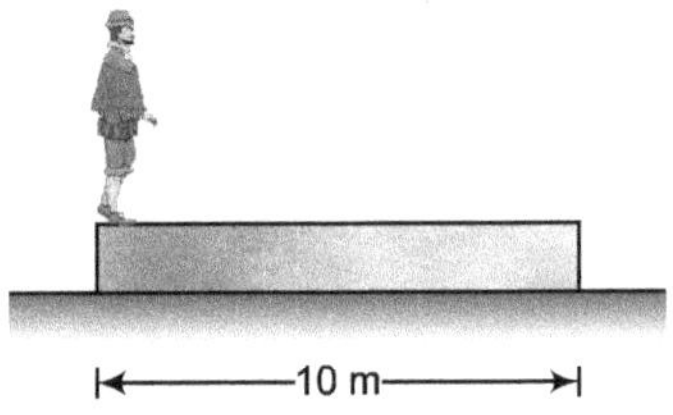

Fig. 8.17

Sol. Here the system is man + plank. Net force on this system in horizontal direction is zero and initially the centre of mass of the system is at rest. Therefore, the centre of mass does not move in horizontal direction.

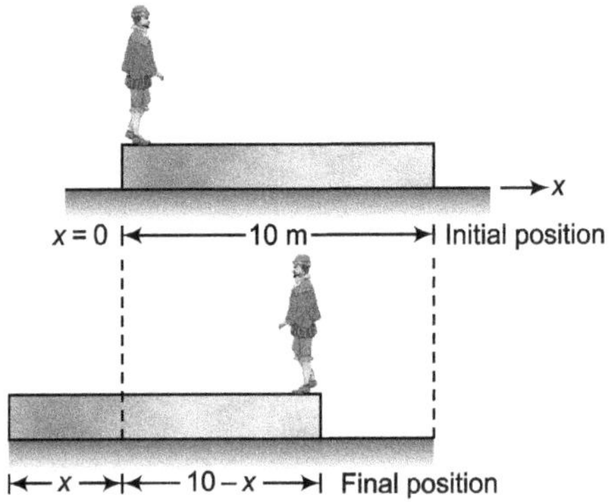

Fig. 8.18

Let x be the displacement of the plank. Assuming the origin, i.e. $x = 0$ at the position shown in figure.

As we said earlier also, the centre of mass will not move in horizontal direction (x-axis). Therefore, for centre of mass to remain stationary,

$$x_i = x_f$$

$$\frac{60(0) + 20\left(\frac{10}{2}\right)}{60 + 20} = \frac{60(10 - x) + 20\left(\frac{10}{2} - x\right)}{60 + 20}$$

or $$\frac{5}{4} = \frac{6(10 - x) + 2\left(\frac{10}{2} - x\right)}{8}$$

$$= \frac{60 - 6x + 10 - 2x}{8}$$

or $$5 = 30 - 3x + 5 - x$$

or $$4x = 30$$

or $$x = \frac{30}{4}\ \text{m}$$

or $$x = 7.5\ \text{m}$$

- The centre of mass of the plank lies at its centre.

Example 8.8 *A man of mass m_1 is standing on a platform of mass m_2 kept on a smooth horizontal surface. The man starts moving on the platform with a velocity v_r relative to the platform. Find the recoil velocity of platform.*

Sol. Absolute velocity of man $= v_r - v$, where v = recoil velocity of platform. Taking the platform and the man as a system, net external force on the system in horizontal direction is zero. The linear momentum of the system remains constant. Initially both the man and the platform were at rest.

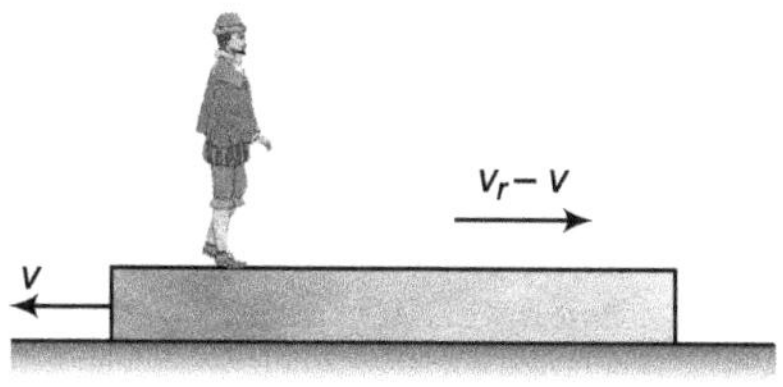

Fig. 8.19

Hence, $0 = m_1(v_r - v) - m_2 v \Rightarrow \quad \therefore \quad v = \frac{m_1 v_r}{m_1 + m_2}$

Example 8.9 *A gun (mass = M) fires a bullet (mass = m) with speed v_r relative to barrel of the gun which is inclined at an angle of 60° with horizontal. The gun is placed over a smooth horizontal surface. Find the recoil speed of gun.*

Sol. Let the recoil speed of gun is v. Taking gun + bullet as the system. Net external force on the system in horizontal direction is zero. Initially the system was at rest. Therefore, applying the principle of conservation of linear momentum in horizontal direction, we get

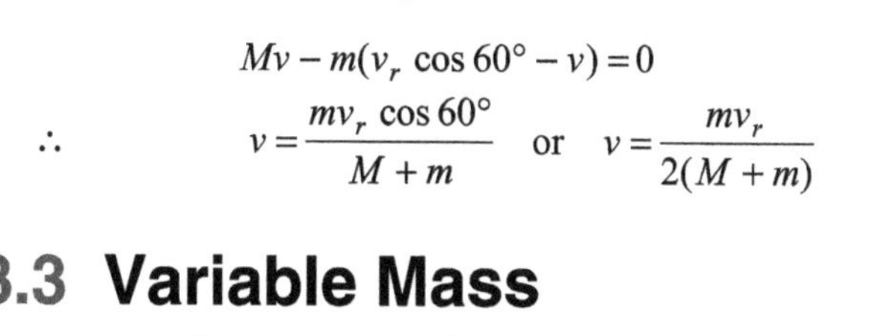

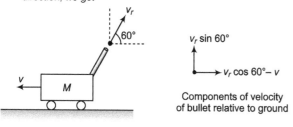

Fig. 8.20

$$Mv - m(v_r \cos 60° - v) = 0$$

$$\therefore \quad v = \frac{mv_r \cos 60°}{M + m} \quad \text{or} \quad v = \frac{mv_r}{2(M + m)}$$

8.3 Variable Mass

In our discussion of the conservation of linear momentum, we have so far dealt with systems whose mass remains constant. We now consider those systems whose mass is variable, i.e. those in which mass enters or leaves the system. A typical case is that of the rocket from which hot gases keep on escaping, thereby continuously decreasing its mass.

In such problems, you have nothing to do but apply a thrust force ($\mathbf{F}_t$) to the main mass in addition to the all other forces acting on it. This thrust force is given by,

$$\mathbf{F}_t = \mathbf{v}_{\text{rel}} \left(\frac{dm}{dt}\right)$$

Here, $\mathbf{v}_{\text{rel}}$ is the velocity of the mass gained or mass ejected relative to the main mass. In case of rocket, this is sometimes called the exhaust velocity of the gases. $\frac{dm}{dt}$ is the rate at which mass is increasing or decreasing.

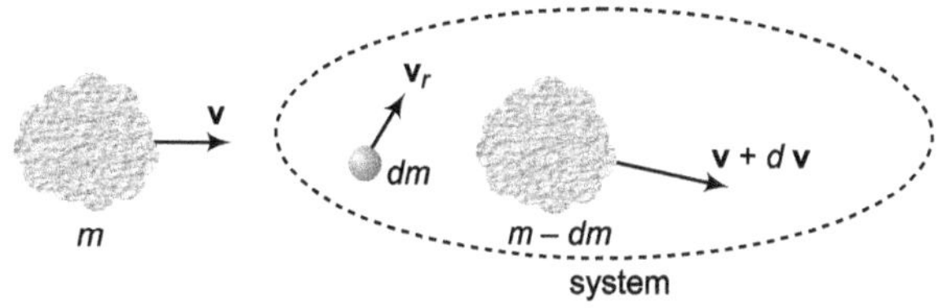

Fig. 8.21

The expression for the thrust force can be derived from the conservation of linear momentum in the absence of any external forces on a system as follows :

Suppose at some moment $t = t$ mass of a body is m and its velocity is $\mathbf{v}$. After some time at $t = t + dt$ its mass becomes $(m - dm)$ and velocity becomes $\mathbf{v} + d\mathbf{v}$. The mass dm is ejected with relative velocity $\mathbf{v}_r$. Absolute velocity of mass dm is therefore $(\mathbf{v}_r + \mathbf{v} + d\mathbf{v})$. If no external forces are acting on the system, then linear momentum of the system will remain conserved, or

Hence, $$\mathbf{p}_i = \mathbf{p}_f$$

or $$m\mathbf{v} = (m - dm)(\mathbf{v} + d\mathbf{v}) + dm(\mathbf{v}_r + \mathbf{v} + d\mathbf{v})$$

or $$m\mathbf{v} = m\mathbf{v} + md\mathbf{v} - dm\mathbf{v} - (dm)(d\mathbf{v}) + dm\mathbf{v} + \mathbf{v}_r\, dm + (dm)(d\mathbf{v})$$

$$\therefore \quad md\mathbf{v} = -\mathbf{v}_r\, dm \quad \text{or} \quad m\left(\frac{d\mathbf{v}}{dt}\right) = \mathbf{v}_r\left(\frac{dm}{dt}\right)$$

Here, $m\left(\frac{d\mathbf{v}}{dt}\right)$ = thrust force ($\mathbf{F}_t$)

and $-\frac{dm}{dt}$ = rate at which mass is ejecting

Problems Related to Variable Mass can be Solved in Following Three Steps

1. Make a list of all the forces acting on the main mass and apply them on it.
2. Apply an additional thrust force $\mathbf{F}_t$ on the mass, the magnitude of which is $\left|\mathbf{v}_r\left(\pm\frac{dm}{dt}\right)\right|$ and direction is given by the direction of $\mathbf{v}_r$ in case the mass is

increasing and otherwise the direction of $-\mathbf{v}_r$ if it is decreasing.

3. Find net force on the mass and apply

$$\mathbf{F}_{\text{net}} = m\frac{d\mathbf{v}}{dt} \quad (m = \text{mass at that particular instant})$$

Rocket Propulsion

Let m_0 be the mass of the rocket at time $t = 0$. m its mass at any time t and v its velocity at that moment. Initially let us suppose that the velocity of the rocket is u.

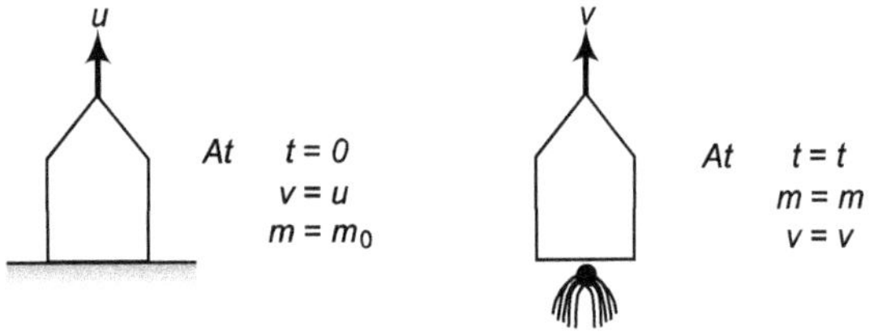

Fig. 8.22

Further, let $\left(\frac{-dm}{dt}\right)$ be the mass of the gas ejected per unit time and v_r the exhaust velocity of the gases. Usually $\left(\frac{-dm}{dt}\right)$ and v_r are kept constant throughout the journey of the rocket. Now, let us write few equations which can be used in the problems of rocket propulsion. At time $t = t$,

1. Thrust force on the rocket

$$F_t = v_r\left(-\frac{dm}{dt}\right) \quad \text{(upwards)}$$

2. Weight of the rocket

$$W = mg \quad \text{(downwards)}$$

3. Net force on the rocket

$$F_{\text{net}} = F_t - W \quad \text{(upwards)}$$

or $$F_{\text{net}} = v_r\left(\frac{-dm}{dt}\right) - mg$$

4. Net acceleration of the rocket $a = \frac{F}{m}$

or $$\frac{dv}{dt} = \frac{v_r}{m}\left(\frac{-dm}{dt}\right) - g$$

or $$dv = v_r\left(\frac{-dm}{m}\right) - g\,dt$$

or $$\int_u^v dv = v_r\int_{m_0}^{m}\frac{-dm}{m} - g\int_0^t dt$$

or $$v - u = v_r\ln\left(\frac{m_0}{m}\right) - gt$$

Thus, $$v = u - gt + v_r\ln\left(\frac{m_0}{m}\right) \quad \ldots\text{(i)}$$

Note Points

- $F_t = v_r\left(-\frac{dm}{dt}\right)$ is upwards, as v_r is downwards and $\frac{dm}{dt}$ is negative.
- If gravity is ignored and initial velocity of the rocket $u = 0$, Eq. (i) reduces to $v = v_r\ln\left(\frac{m_0}{m}\right)$.

Example 8.10 *(a) A rocket set for vertical firing weighs 50 kg and contains 450 kg of fuel. It can have a maximum exhaust velocity of 2 km/s. What should be its minimum rate of fuel consumption*

(i) to just lift it off the launching pad.

(ii) to give it an acceleration of 20 m/s^2.

(b) What will be the speed of the rocket when the rate of consumption of fuel is 10 kg/s after whole of the fuel is consumed? (Take $g = 9.8\ m/s^2$)

Sol. (a) (i) To just lift it off the launching pad

weight = thrust force

or $$mg = v_r\left(\frac{-dm}{dt}\right)$$

or $$\left(\frac{-dm}{dt}\right) = \frac{mg}{v_r}$$

Substituting the values, we get

$$\left(\frac{-dm}{dt}\right) = \frac{(450 + 50)(9.8)}{2\times 10^3} = 2.45 \text{ kg/s}$$

(ii) Net acceleration $a = 20\ m/s^2$

$\therefore$ $$ma = F_t - mg$$

or $$a = \frac{F_t}{m} - g$$

or $$a = \frac{v_r}{m}\left(\frac{-dm}{dt}\right) - g$$

This gives $$\left(-\frac{dm}{dt}\right) = \frac{m(g + a)}{v_r}$$

Substituting the values, we get

$$\left(-\frac{dm}{dt}\right) = \frac{(450 + 50)(9.8 + 20)}{2\times 10^3}$$

$$= 7.45 \text{ kg/s}$$

(b) The rate of fuel consumption is 10 kg/s. So, the time for the consumption of entire fuel is

$$t = \frac{450}{10} = 45 \text{ s}$$

Using Eq. (i), i.e. $v = u - gt + v_r\ln\left(\frac{m_0}{m}\right)$

Here, $u = 0$, $v_r = 2\times 10^3$ m/s

$m_0 = 500$ kg

and $m = 50$ kg

Substituting the values, we get,

$$v = 0 - (9.8)(45) + (2\times 10^3)\ln\left(\frac{500}{50}\right)$$

or $$v = -441 + 4605.17$$

or $$v = 4164.17 \text{ m/s} \quad \text{or} \quad v = 4.164 \text{ km/s}$$

8.4 Impulse

Consider a constant force **F** which acts for a time t on a body of mass m thus. changing its velocity from **u** to **v**. Because the force is constant, the body will travel with constant acceleration **a**. Force is where

$$\mathbf{F} = m\mathbf{a} \quad \text{and} \quad \mathbf{a}t = \mathbf{v} - \mathbf{u}$$

Hence, $\dfrac{\mathbf{F}}{m}t = \mathbf{v} - \mathbf{u}$ or $\mathbf{F}t = m\mathbf{v} - m\mathbf{u}$

The product of constant force **F** and the time t for which it acts is called the impulse (**J**) of the force and this is equal to the change in linear momentum which it produces.

Thus, impulse $(\mathbf{J}) = \mathbf{F}t = \Delta\mathbf{p} = \mathbf{p}_f - \mathbf{p}_i$

Instantaneous Impulse There are many occasions when a force acts for such a short time that the effect is instantaneous, e.g. a bat striking a ball. In such cases, although the magnitude of the force and the time for which it acts may each be unknown but the value of their product (i.e. impulse) can be known by measuring the initial and final momenta. Thus, we can write

$$\mathbf{J} = \int \mathbf{F}dt = \Delta\mathbf{p} = \mathbf{p}_f - \mathbf{p}_i$$

Regarding the impulse, it is important to note that impulse applied to an object in a given time interval can also be calculated from the area under force-time (F-t) graph in the same time interval.

Example 8.11 *A truck of mass 2×10^3 kg travelling at 4 m/s is brought to rest in 2 s when it strikes a wall. What force (assume constant) is exerted by the wall?*

Sol. Using impulse = change in linear momentum.

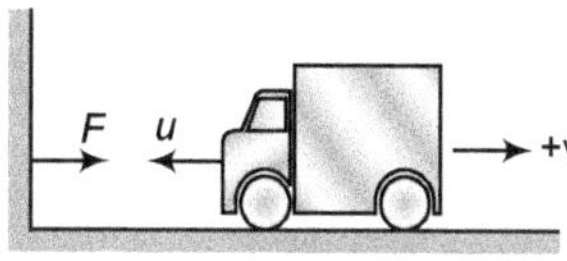

Fig. 8.23

We have $F \cdot t = mv_f - mv_i = m(v_f - v_i)$

or $F(2) = 2 \times 10^3 [0 - (-4)]$

or $2F = 8 \times 10^3$

or $F = 4 \times 10^3$ N

Example 8.12 *A ball of mass m, travelling with velocity $2\mathbf{i} + 3\mathbf{j}$ receives an impulse $-3m\hat{\mathbf{i}}$. What is the velocity of the ball immediately afterwards?*

Sol. Using, $\mathbf{J} = m(\mathbf{v}_f - \mathbf{v}_i)$

$-3m\hat{\mathbf{i}} = m[\mathbf{v}_f - (2\hat{\mathbf{i}} + 3\hat{\mathbf{j}})]$

or $\mathbf{v}_f = -3\hat{\mathbf{i}} + (2\hat{\mathbf{i}} + 3\hat{\mathbf{j}})$

or $\mathbf{v}_f = -\hat{\mathbf{i}} + 3\hat{\mathbf{j}}$

- The velocity component in the direction of $\hat{\mathbf{j}}$ is unchanged. This is because there is no impulse component in this direction.

Example 8.13 *A particle of mass 2 kg is initially at rest. A force starts acting on it in one direction whose magnitude changes with time. The force-time graph is as shown in figure.*

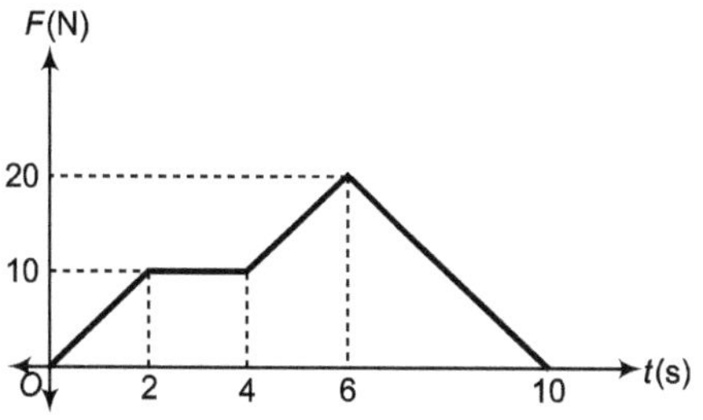

Fig. 8.24

Find the velocity of the particle at the end of 10 s.

Sol. Using impulse = change in linear momentum (or area under F-t graph)

We have $m(v_f - v_i) = \text{Area}$

or $2(v_f - 0) = \frac{1}{2} \times 2 \times 10 + 2 \times 10 + \frac{1}{2} \times 2 \times (10 + 20) + \frac{1}{2} \times 4 \times 20$

$= 10 + 20 + 30 + 40$

or $2v_f = 100$

$\therefore$ $v_f = 50$ m/s

8.5 Collision

Contrary to the meaning of the term collision in our everyday life, in physics it does not necessarily mean one particle 'striking' against other. Indeed two particles may not even touch each other and may still be said to collide. All that is implied is that as the particles approach each other,

(i) an impulse (a large force for a relatively short time) acts on each colliding particles.

(ii) the total momentum of the particles remain conserved.

The collision is in fact a redistribution of total momentum of the particles. Thus, law of conservation of linear momentum is indispensible in dealing with the phenomenon of collision between particles. Consider a situation shown in figure.

Two blocks of masses m_1 and m_2 are moving with velocities v_1 and $v_2 (< v_1)$ along the same straight line in a smooth horizontal surface. A spring is attached to the block of mass m_2. Now, let us see what happens during the collision between two particles.

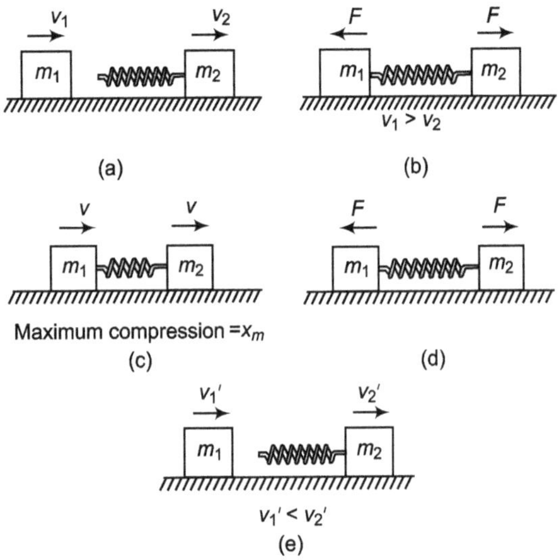

Fig. 8.25

Fig. (a) Block of mass m_1 is behind m_2. Since, $v_1 > v_2$, the blocks will collide after some time.

Fig. (b) The spring is compressed. The spring force $F\ (= kx)$ acts on the two blocks in the directions shown in figure. This force decreases the velocity of m_1 and increases the velocity of m_2.

Fig. (c) The spring will compress till velocity of both the blocks become equal. So, at maximum compression (say x_m) velocities of both the blocks are equal (say v).

Fig. (d) Spring force is still in the directions shown in figure, i.e. velocity of block m_1 is further decreased and that of m_2 is increased. The spring now starts relaxing.

Fig. (e) The two blocks are separated from one another. Velocity of block m_2 becomes more than the velocity of block m_1, i.e. $v_2' > v_1'$.

Equations which can be Used in the Above Situation

Assuming spring to be perfectly elastic following, two equations can be applied in the above situation.

(i) In the absence of any external force on the system the linear momentum of the system will remain conserved before, during and after collision, i.e.

$$m_1 v_1 + m_2 v_2 = (m_1 + m_2) v = m_1 v_1' + m_2 v_2' \qquad \ldots\text{(i)}$$

(ii) In the absence of any dissipative forces, the mechanical energy of the system will also remain conserved, i.e.

$$\frac{1}{2} m_1 v_1^2 + \frac{1}{2} m_2 v_2^2 = \frac{1}{2}(m_1 + m_2) v^2 + \frac{1}{2} k x_m^2 = \frac{1}{2} m_1 v_1'^2 + \frac{1}{2} m_2 v_2'^2 \qquad \ldots\text{(ii)}$$

- In the above situation, we have assumed spring to be perfectly elastic, i.e. it regains its original shape and size after the two blocks are separated. In actual practice, there is no such spring between the two blocks. During collision both the blocks (or bodies) are a little bit deformed. This situation is similar to the compression of the spring. Due to deformation two equal and opposite forces act on both the blocks. These two forces redistribute their linear momentum in such a manner that both the blocks are separated from one another. The collision is said to be elastic if both the blocks regain their original shape and size completely after they are separated. On the other hand if the blocks do not return to their original form the collision is said to be inelastic. If the deformation is permanent and the blocks move together with same velocity after the collision, the collision is said to be perfectly inelastic.

Example 8.14 *Two blocks A and B of equal mass $m = 1.0$ kg are lying on a smooth horizontal surface as shown in figure. A spring of force constant $k = 200$ N/m is fixed at one end of block A. Block B collides with block A with velocity $v_0 = 2.0$ m/s. Find the maximum compression of the spring.*

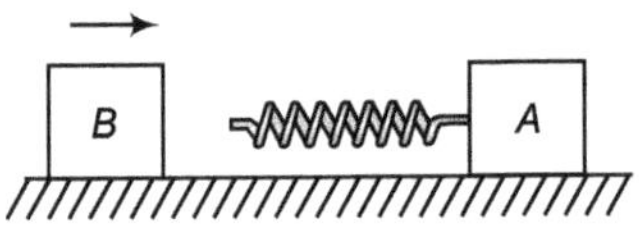

Fig. 8.26

Sol. At maximum compression (x_m) velocity of both the blocks is same, say it is v. Applying conservation of linear momentum, we have,

$$(m_A + m_B) v = m_B v_0$$

or $$(1.0 + 1.0) v = (1.0) v_0$$

or $$v = \frac{v_0}{2} = \frac{2.0}{2} = 1.0 \text{ m/s}$$

Using conservation of mechanical energy, we have

$$\frac{1}{2} m_B v_0^2 = \frac{1}{2}(m_A + m_B) v^2 + \frac{1}{2} k x_m^2$$

Substituting the values, we get

$$\frac{1}{2} \times (1) \times (2.0)^2 = \frac{1}{2} \times (1.0 + 1.0) \times (1.0)^2 + \frac{1}{2} \times (200) \times x_m^2$$

or $$2 = 1.0 + 100 x_m^2$$

or $$x_m = 0.1 \text{ m} = 10.0 \text{ cm}$$

Types of Collision

Collision between two bodies may be classified in two ways :

1. Elastic collision and inelastic collision.
2. Head on collision or oblique collision.

As discussed earlier also collision between two bodies is said to be **elastic** if both the bodies come to their original shape and size after the collision, i.e. no fraction of mechanical energy remains stored as deformation potential energy in the bodies. Thus, in addition to the linear momentum, kinetic energy also remains conserved before and after collision. On the other hand, in an **inelastic** collision, the colliding bodies do not return to their original shape and size completely after collision and some part of the mechanical energy of the system goes to the deformation potential energy. Thus, only linear momentum remains conserved in case of an inelastic collision.

Further, a collision is said to be head on (or direct) if the directions of the velocity of colliding objects are along the line of action of the impulses, acting at the instant of collision. If just before collision, at least one of the colliding objects was moving in a direction different from the line of action of the impulses, the collision is called oblique or indirect.

Problems related to oblique collision are usually not asked in any medical entrance test. Hence, only head on collision is discussed below.

Head on Elastic Collision

Let the two balls of mass m_1 and m_2 collide each other elastically with velocities v_1 and v_2 in the directions shown in Fig. 8.27. Their velocities become v_1' and v_2' after the collision along the same line. Applying conservation of linear momentum, we get

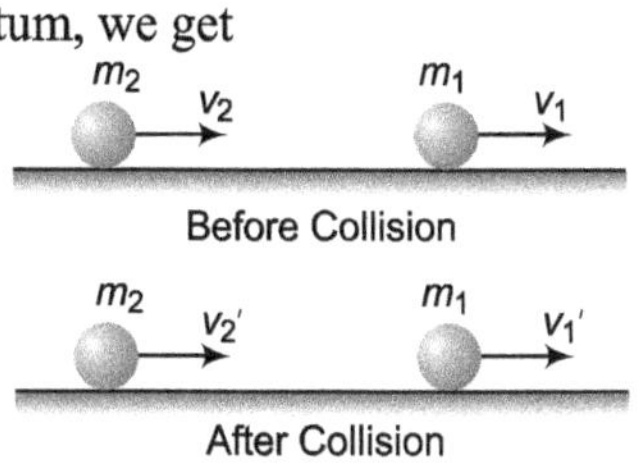

Fig. 8.27

$$m_1 v_1 + m_2 v_2 = m_1 v_1' + m_2 v_2' \qquad \ldots(i)$$

In an elastic collision kinetic energy before and after collision is also conserved. Hence,

$$\frac{1}{2} m_1 v_1^2 + \frac{1}{2} m_2 v_2^2 = \frac{1}{2} m_1 v_1'^2 + \frac{1}{2} m_2 v_2'^2 \qquad \ldots(ii)$$

Solving Eqs. (i) and (ii) for v_1' and v_2', we get

$$v_1' = \left(\frac{m_1 - m_2}{m_1 + m_2}\right) v_1 + \left(\frac{2m_2}{m_1 + m_2}\right) v_2 \qquad \ldots(iii)$$

and
$$v_2' = \left(\frac{m_2 - m_1}{m_1 + m_2}\right) v_2 + \left(\frac{2m_1}{m_1 + m_2}\right) v_1 \qquad \ldots(iv)$$

Special Cases

1. If $m_1 = m_2$, then from Eqs. (iii) and (iv), we can see that

   $$v_1' = v_2 \text{ and } v_2' = v_1$$

   i.e. when two particles of equal mass collide elastically and the collision is head on, they exchange their velocities., e.g.

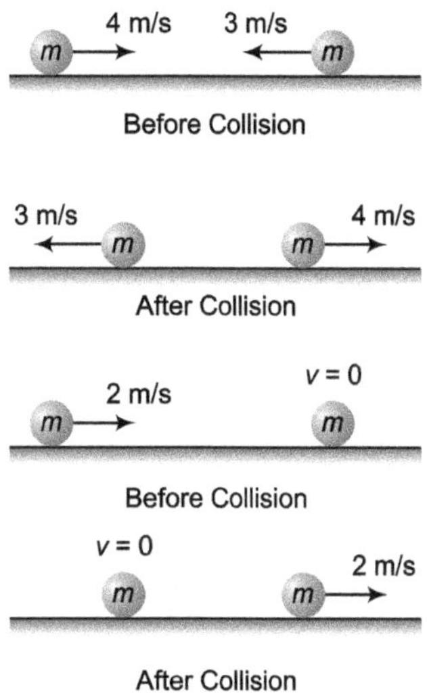

Fig. 8.28

2. If $m_1 >> m_2$ and $v_1 = 0$.

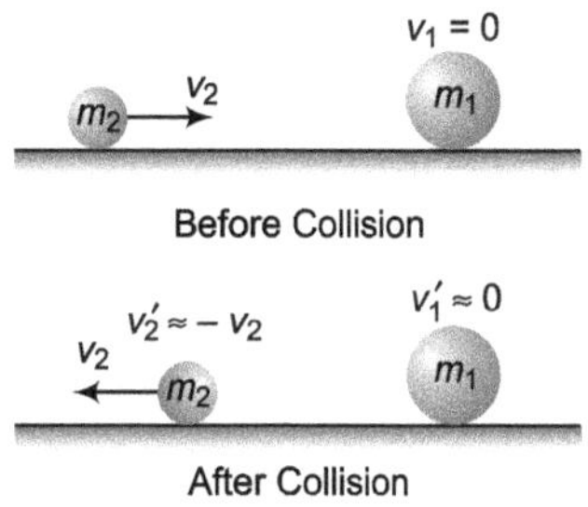

Fig. 8.29

Then, $\dfrac{m_2}{m_1} \approx 0$

With these two substitutions $\left(v_1 = 0 \text{ and } \dfrac{m_2}{m_1} = 0\right)$

We get the following two results

$$v_1' \approx 0 \quad \text{and} \quad v_2' \approx -v_2$$

i.e. the particle of mass m_1 remains at rest while the particle of mass m_2 bounces back with same speed v_2.

3. If $m_2 >> m_1$
   and $v_1 = 0$

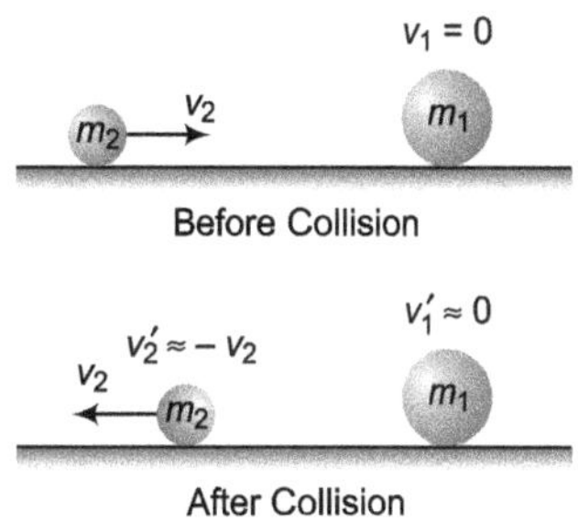

Fig. 8.30

With the substitution $\frac{m_1}{m_2} \approx 0$ and $v_1 = 0$, we get the results

$$v_1' \approx 2v_2 \quad \text{and} \quad v_2' \approx v_2$$

i.e. the mass m_1 moves with velocity $2v_2$ while the velocity of mass m_2 remains unchanged.

- It is important to note that Eqs. (iii) and (iv) and their three special cases can be used only in case of a head on elastic collision between two particles. I have found that many students apply these two equations even if the collision is inelastic and do not apply these relations where clearly a head on elastic collision is given in the problem.

Example 8.15 *Two particles of mass m and 2m moving in opposite directions collide elastically with velocities v and 2v. Find their velocities after collision.*

Sol. Here, $v_1 = -v, v_2 = 2v, m_1 = m$ and $m_2 = 2m$.

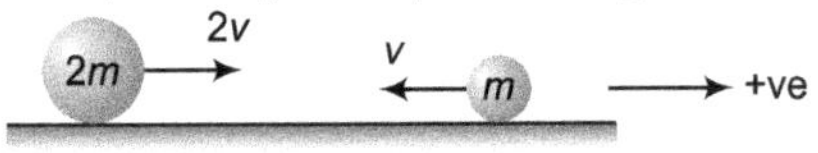

Fig. 8.31

Substituting these values in Eqs. (iii) and (iv), we get

$$v_1' = \left(\frac{m-2m}{m+2m}\right)(-v) + \left(\frac{4m}{m+2m}\right)(2v)$$

or $$v_1' = \frac{v}{3} + \frac{8v}{3} = 3v$$

and $$v_1' = \left(\frac{2m-m}{m+2m}\right)(2v) + \left(\frac{2m}{m+2m}\right)(-v)$$

or $$v_1' = \frac{2}{3}v - \frac{2}{3}v = 0$$

Fig. 8.32

i.e. the second particle (of mass $2m$) comes to a rest while the first (of mass m) moves with velocity $3v$ in the direction shown in Fig. 8.32.

Example 8.16 *Two pendulum bobs of mass m and 2m collide elastically at the lowest point in their motion. If both the balls are released from a height H above the lowest point, to what heights do they rise for the first time after collision?*

Sol. Given, $m_1 = m, m_2 = 2m,$

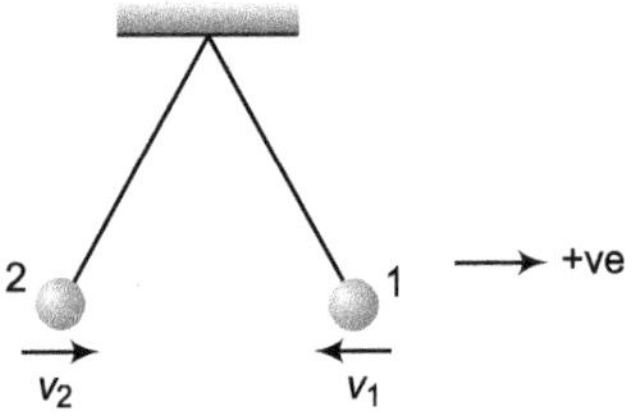

Fig. 8.33

$$v_1 = -\sqrt{2gH} \quad \text{and} \quad v_2 = \sqrt{2gH}$$

Since, the collision is elastic. Using Eqs. (iii) and (iv) discussed in the theory the velocities after collision are

$$v_1' = \left(\frac{m-2m}{m+2m}\right)(-\sqrt{2gH}) + \left(\frac{4m}{m+2m}\right)\sqrt{2gH}$$

$$= \frac{\sqrt{2gH}}{3} + \frac{4\sqrt{2gH}}{3}$$

$$= \frac{5}{3}\sqrt{2gH}$$

and $$v_2' = \left(\frac{2m-m}{m+2m}\right)(\sqrt{2gH}) + \left(\frac{2m}{m+2m}\right)(-\sqrt{2gH})$$

$$= \frac{\sqrt{2gH}}{3} - \frac{2\sqrt{2gH}}{3}$$

$$= -\frac{\sqrt{2gH}}{3}$$

i.e. the velocities of the balls after the collision are as shown in figure.

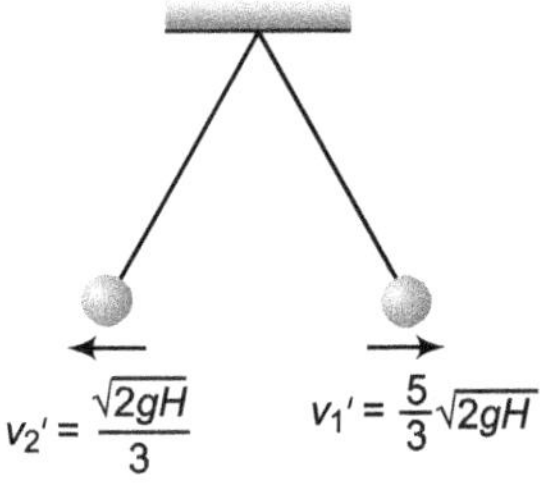

Fig. 8.34

Therefore, the heights to which the balls rise after the collision are

$$h_1 = \frac{(v_1')^2}{2g} \qquad \text{(using, } v^2 = u^2 - 2gh)$$

or $$h_1 = \frac{\left(\frac{5}{3}\sqrt{2gH}\right)^2}{2g} \quad \text{or} \quad h_1 = \frac{25}{9}H$$

and $$h_2 = \frac{(v_2')^2}{2g} \quad \text{or} \quad h_2 = \frac{\left(\frac{\sqrt{2gH}}{3}\right)^2}{2g}$$

or $$h_2 = \frac{H}{9}$$

- Since, the collision is elastic, mechanical energy of both the balls will remain conserved, or

$$E_i = E_f$$

$$\Rightarrow \quad (m + 2m)\,gH = mgh_1 + 2mgh_2$$

$$\Rightarrow \quad 3mgH = (mg)\left(\frac{25}{9}H\right) + (2mg)\left(\frac{H}{9}\right)$$

$$\Rightarrow \quad 3mgH = 3mgH$$

Head on Inelastic Collision

As we have discussed earlier also, in an inelastic collision, the particles do not regain their shape and size completely after collision. Some fraction of mechanical energy is retained by the colliding particles in the form of deformation potential energy. Thus, the kinetic energy of the particles no longer remains conserved. However, in the absence of external forces, law of conservation of linear momentum still holds good.

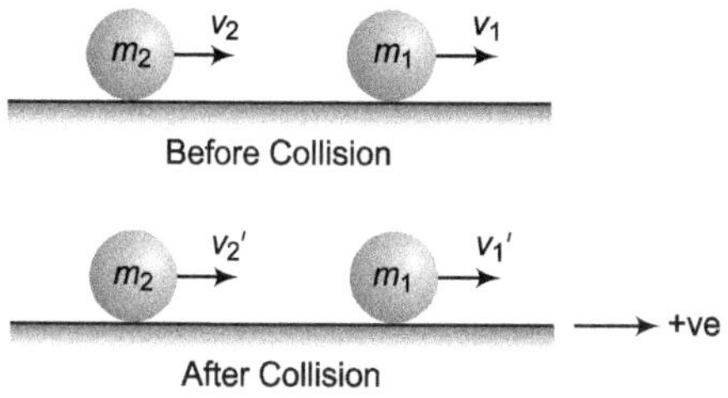

Fig. 8.35

Suppose the velocities of two particles of mass m_1 and m_2 before collision be v_1 and v_2 in the directions shown in figure. Let v_1' and v_2' be their velocities after collision. The law of conservation of linear momentum gives

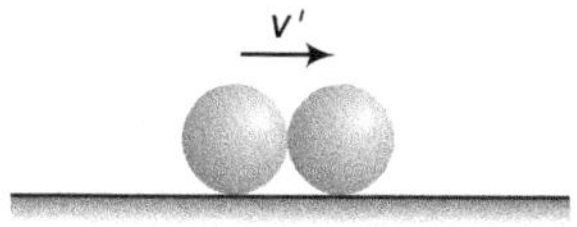

Fig. 8.36

$$m_1 v_1 + m_2 v_2 = m_1 v_1' + m_2 v_2' \qquad \ldots(v)$$

Collision is said to be perfectly inelastic if both the particles stick together after collision and move with same velocity, say v' as shown in figure. In this case, Eq. (v) can be written as

$$m_1 v_1 + m_2 v_2 = (m_1 + m_2) v'$$

or $$v' = \frac{m_1 v_1 + m_2 v_2}{m_1 + m_2} \qquad \ldots(vi)$$

Newton's Law of Restitution

When two objects are in direct (head on) impact, the speed with which they separate after impact is usually less than or equal to their speed of approach before impact.

Experimental evidence suggests that the ratio of these relative speeds is constant for two given set of objects. This property formulated by Newton, is known as the law of restitution and can be written in the form

$$\frac{\text{Separation speed}}{\text{Approach speed}} = e \qquad \ldots(vii)$$

The ratio e is called the coefficient of restitution and is constant for two particular objects.

In general, $$0 \le e \le 1$$

$e = 0$, for completely inelastic collision, as both the objects stick together. So, their separation speed is zero or $e = 0$ from Eq. (vii).

$e = 1$, for an elastic collision, as we can show from Eqs. (iii) and (iv), that

$$v_1' - v_2' = v_2 - v_1$$

or separation speed = approach speed

or $$e = 1$$

Let us now find the velocities of two particles after collision if they collide directly and the coefficient of restitution between them is given as e.

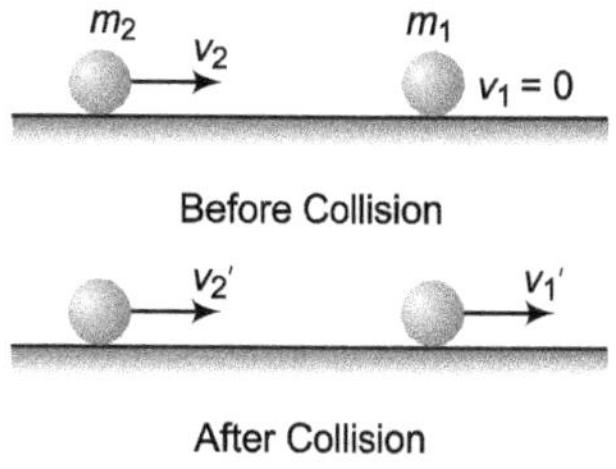

Fig. 8.37

Applying conservation of linear momentum

$$m_1 v_1 + m_2 v_2 = m_1 v_1' + m_2 v_2' \qquad \ldots(viii)$$

Further, separation speed = e (approach speed)

or $$v_1' - v_2' = e(v_2 - v_1) \qquad \ldots(ix)$$

Solving Eqs. (viii) and (ix), we get

$$v_1' = \left(\frac{m_1 - e m_2}{m_1 + m_2}\right) v_1 + \left(\frac{m_2 + e m_2}{m_1 + m_2}\right) v_2 \qquad \ldots(x)$$

and $$v_2' = \left(\frac{m_2 - e m_1}{m_1 + m_2}\right) v_2 + \left(\frac{m_1 + e m_1}{m_1 + m_2}\right) v_1 \qquad \ldots(xi)$$

Special Cases

1. If collision is elastic, i.e. $e = 1$, then

$$v_1' = \left(\frac{m_1 - m_2}{m_1 + m_2}\right) v_1 + \left(\frac{2m_2}{m_1 + m_2}\right) v_2$$

and $$v_2' = \left(\frac{m_2 - m_1}{m_1 + m_2}\right) v_2 + \left(\frac{2m_1}{m_1 + m_2}\right) v_1$$

which are same as Eqs. (iii) and (iv).

2. If collision is perfectly inelastic, i.e. $e = 0$, then

$$v_1' = v_2' = \frac{m_1 v_1 + m_2 v_2}{m_1 + m_2} = v' \text{ (say)}$$

which is same as Eq. (vi).

3. If $m_1 = m_2$ and $v_1 = 0$, then

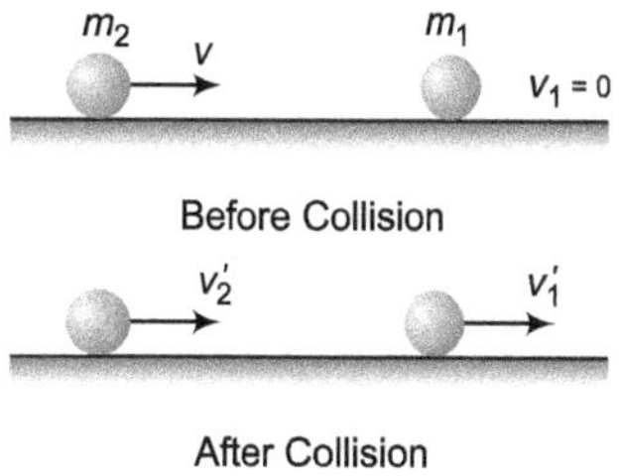

Fig. 8.38

$$v_1' = \left(\frac{1+e}{2}\right) v_2 \text{ and } v_2' = \left(\frac{1-e}{2}\right) v_2$$

Note Points

- If mass of one body is very-very greater than that of the other, then after collision velocity of heavy body does not change appreciably. (Whether the collision is elastic or inelastic).

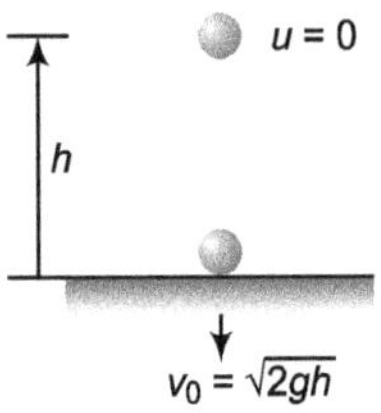

Fig. 8.39

- In the situation shown in figure, if e is the coefficient of restitution between the ball and the ground, than after nth collision with the floor the speed of ball will remain $e^n v_0$ and it will go upto a height $e^{2n} h$ or,

$$v_n = e^n v_0 = e^n \sqrt{2gh} \quad \text{and} \quad h_n = e^{2n} h$$

Example 8.17 *A ball of mass m moving at a speed v makes a head on collision with an identical ball at rest. The kinetic energy of the balls after the collision is 3/4 th of the original. Find the coefficient of restitution.*

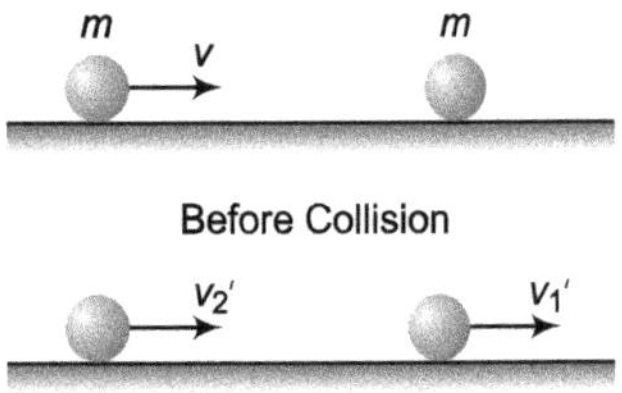

Fig. 8.40

Sol. As, we have seen in the above discussion, that under the given conditions

$$v_1' = \left(\frac{1+e}{2}\right) v \quad \text{and} \quad v_2' = \left(\frac{1-e}{2}\right) v$$

Given that $K_f = \frac{3}{4} K_i$ or $\frac{1}{2} m v_1'^2 + \frac{1}{2} m v_2'^2 = \frac{3}{4}\left(\frac{1}{2} m v^2\right)$

Substituting the value, we get

$$\left(\frac{1+e}{2}\right)^2 + \left(\frac{1-e}{2}\right)^2 = \frac{3}{4}$$

or $(1+e)^2 + (1-e)^2 = 3$ or $2 + 2e^2 = 3$

or $e^2 = \frac{1}{2}$ or $e = \frac{1}{\sqrt{2}}$

Example 8.18 *A ball is moving with velocity 2 m/s towards a heavy wall moving towards the ball with speed 1m/s as shown in figure. Assuming collision to be elastic, find the velocity of ball immediately after the collision.*

Fig. 8.41

Sol. The speed of wall will not change after the collision. So, let v be the velocity of the ball after collision in the direction shown in figure. Since, collision is elastic ($e = 1$),

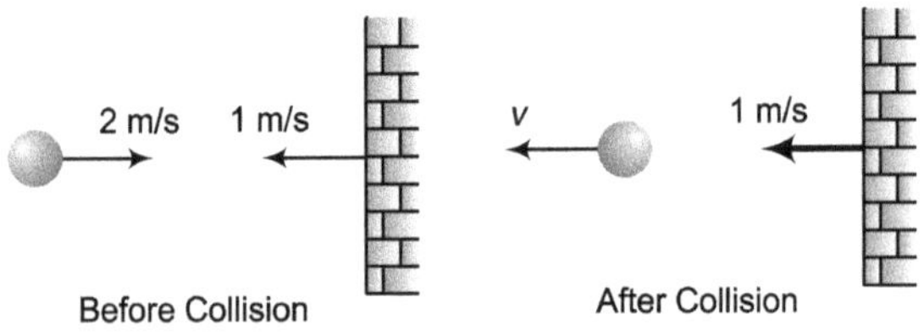

Fig. 8.42

separation speed = approach speed

or $v - 1 = 2 + 1$

or $v = 4$ m/s

Example 8.19 *After perfectly inelastic collision between two identical particles moving with same speed in different directions, the speed of the particles becomes half the initial speed. Find the angle between the two before collision.*

Sol. Let θ be the desired angle. Linear momentum of the system will remain conserved. Hence

$$p^2 = p_1^2 + p_2^2 + 2p_1 p_2 \cos\theta$$

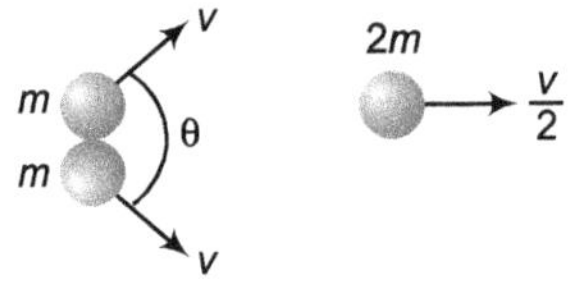

Fig 8.43

or $\left\{2m\left(\frac{v}{2}\right)\right\}^2 = (mv)^2 + (mv)^2 + 2(mv)(mv)\cos\theta$

or $1 = 1 + 1 + 2\cos\theta$ or $\cos\theta = -\frac{1}{2}$

$\therefore$ $\theta = 120°$

Extra Knowledge Points

- During collision if mass of one body is very much greater than the mass of the other body then the velocity of heavy body remains almost unchanged after collision, whether the collision is elastic or inelastic.
- Net force on a system is zero, it does not mean that centre of mass is at rest. It might be moving with constant velocity.
- Centre of mass of a rigid body is not necessarily the geometric centre of the rigid body.
- If two many particles are in air, then relative acceleration between any two is zero but acceleration of their centre of mass is g downwards.
- Coefficient of restitution is the mutual intrinsic properties of two bodies. Its value varies from 0 to 1.

Chapter Summary with Formulae

- **Position of Centre of Mass**

  (i) Two point masses

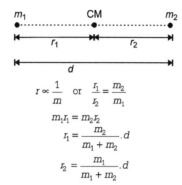

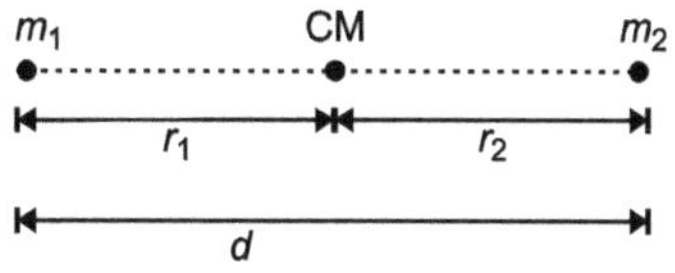

$$r \propto \frac{1}{m} \quad \text{or} \quad \frac{r_1}{r_2} = \frac{m_2}{m_1}$$

$\therefore$
$$m_1 r_1 = m_2 r_2$$
$$r_1 = \frac{m_2}{m_1 + m_2} . d$$
$$r_2 = \frac{m_1}{m_1 + m_2} . d$$

  (ii) More than two point masses

  (a) $\mathbf{r}_{CM} = \dfrac{m_1 \mathbf{r}_1 + m_2 \mathbf{r}_2}{m_1 + m_2}$

  (b) $X_{CM} = \dfrac{m_1 x_1 + m_2 x_2}{m_1 + m_2} \Rightarrow Y_{CM} = \dfrac{m_1 y_1 + m_2 y_2}{m_1 + m_2} \Rightarrow Z_{CM} = \dfrac{m_1 z_1 + m_2 z_2}{m_1 + m_2}$

  (iii) More than two rigid bodies

  (a) Centre of mass of symmetrical rigid body (like sphere, disc, cube etc.) lies at its geometric centre.

  (b) For two or more than two rigid bodies we can use,

$$\mathbf{r}_{CM} = \frac{m_1 \mathbf{r}_1 + m_2 \mathbf{r}_2}{m_1 + m_2} \Rightarrow X_{CM} = \frac{m_1 x_1 + m_2 x_2}{m_1 + m_2}$$

$$Y_{CM} = \frac{m_1 y_1 + m_2 y_2}{m_1 + m_2} \quad \text{and} \quad Z_{CM} = \frac{m_1 z_1 + m_2 z_2}{m_1 + m_2}$$

  (c) If three dimensional rigid body has uniform density, then mass in above formulae can be replaced by volume (V)

$$\mathbf{r}_{CM} = \frac{V_1 \mathbf{r}_1 + V_2 \mathbf{r}_2}{V_1 + V_2}$$

  (d) In case of two dimensional body, mass can be replaced by area (A).

$$\mathbf{r}_{CM} = \frac{A_1 \mathbf{r}_1 + A_2 \mathbf{r}_2}{A_1 + A_2}$$

  (e) If some portion is removed from the body. Then,

$$\mathbf{r}_{CM} = \frac{A_1 \mathbf{r}_1 - A_2 \mathbf{r}_2}{A_1 - A_2}$$ (in case of two dimensional body)

  Here, A_1 = Area of whole body (without removing)

  $\mathbf{r}_1$ = Position vector of its centre of mass

  A_1 = Area of removed portion

  $\mathbf{r}_2$ = Position vector of centre of mass of removed portion

- **Other Formulae of Centre of Mass**

  (i) $\mathbf{F}_{CM} = \mathbf{F}_1 + \mathbf{F}_2$

  (ii) $\mathbf{p}_{CM} = \mathbf{p}_1 + \mathbf{p}_2$

  (iii) $\mathbf{v}_{CM} = \dfrac{m_1 \mathbf{v}_1 + m_2 \mathbf{v}_2}{m_1 + m_2}$

  (iv) $\mathbf{a}_{CM} = \dfrac{m_1 \mathbf{a}_1 + m_2 \mathbf{a}_2}{m_1 + m_2}$

  (v) $\mathbf{s}_{CM} = \dfrac{m_1 \mathbf{s}_1 + m_2 \mathbf{s}_2}{m_1 + m_2}$

Additional Examples

Example 1. *Define centre of mass.*

Sol. The point in a system, where the whole mass of the system can be supposed to be concentrated is called centre of mass of the system.

Example 2 *Should the centre of mass of a body necessarily lie inside the body?*

Sol. Not necessarily. For example, the centre of mass of a ring lies at the centre of the ring, i.e. at a point, where actually there is no mass.

Example 3. *What is the difference between centre of gravity and centre of mass?*

Sol. The centre of gravity of a body is a point, where the whole weight of the body may be supposed to act. Further, the total gravitational torque on the body about its centre of gravity is always zero. The centre of mass of a body is a point, where, the whole mass of the body can be supposed to be concentrated. The motion of the body under the action of external forces can be studied by studying the motion of centre of mass, when all the external forces are applied directly on it. The centre of mass and centre of gravity are two different concepts. However, the centre of gravity of the body coincides with the centre of mass in uniform gravity or gravity-free space.

Example 4. *If an external force can only change the state of motion of CM of a body, how does the internal force of the brakes bring a car to rest?*

Sol. Actually, it is not the external force which brings the car to rest. The internal force of the brakes on the wheel locks the wheel. Now a large frictional force comes into play between the wheels and the ground. This force is external to the system and brings the car to rest.

Example 5. *Two men stand facing each other on two boats floating on still water at a distance apart. A rope is held at its ends by both. The two boats are found to meet always at the same point, whether each man pulls separately or both pull together, why? Will the time taken be different in the two cases? Neglect friction.*

Sol. Two men on the boats floating on water constitute a single system. So, the forces applied by the two men are internal forces. Whether each man pulls separately or both pull together, the centre of mass of the system of boats remains fixed due to the absence of any external force. Consequently, the two boats meet at a fixed point, which is the centre of mass of the system.

Example 6. *The friction coefficient between the horizontal surface and each of the block shown in the figure is 0.2. The collision between the blocks is perfectly elastic. Find the separation between them when they come to rest.*
(Take $g = 10\ m/s^2$)

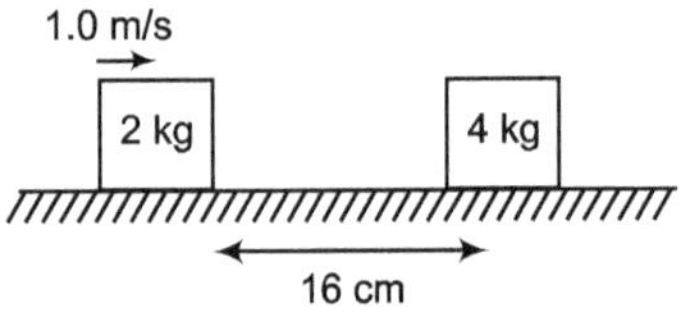

Sol. Velocity of first block before collision,

$$v_1^2 = 1^2 - 2(2) \times 0.16$$
$$= 1 - 0.64$$
$$v_1 = 0.6\ \text{m/s}$$

By conservation of momentum, $2 \times 0.6 = 2v_1' + 4v_2'$

also $v_2' - v_1' = v_1$ for elastic collision

It gives $\quad v_2' = 0.4\ \text{m/s}$

$\quad v_1' = -0.2\ \text{m/s}$

Now, distance moved after collision

$$s_1 = \frac{(0.4)^2}{2 \times 2} \text{ and } s_2 = \frac{(0.2)^2}{2 \times 2}$$

$\therefore \quad s = s_1 + s_2 = 0.05\ \text{m}$

$= 5\ \text{cm}$

Example 7. *A pendulum bob of mass 10^{-2} kg is raised to a height 5×10^{-2} m and then released. At the bottom of its swing, it picks up a mass 10^{-3} kg. To what height will the combined mass rise?*

Sol. Velocity of pendulum bob in mean position

$$v_1 = \sqrt{2gh} = \sqrt{2 \times 10 \times 5 \times 10^{-2}}$$
$$= 1\ \text{m/s}$$

When the bob picks up a mass 10^{-3} kg at the bottom, then by conservation of linear momentum the velocity of coalesced mass is given by

$$m_1 v_1 + m_2 v_2 = (m_1 + m_2)\, v$$
$$10^{-2} + 10^{-3} \times 0 = (10^{-2} + 10^{-3})\, v$$

or $$v = \frac{10^{-2}}{1.1 \times 10^{-2}} = \frac{10}{11}\ \text{m/s}$$

Now $$h = \frac{v^2}{2g} = \frac{(10/11)^2}{2 \times 10}$$
$$= 4.1 \times 10^{-2}\ \text{m}$$

Example 8. *Three identical balls, ball I, ball II and ball III are placed on a smooth floor on a straight line at the separation of* 10 *m between balls as shown in figure. Initially balls are stationary.*

Ball I is given velocity of 10 *m/s towards ball II, collision between ball I and II is inelastic with coefficient of restitution* 0.5 *but collision between ball II and III is perfectly elastic. What is the time interval between two consecutive collisions between balls I and II ?*

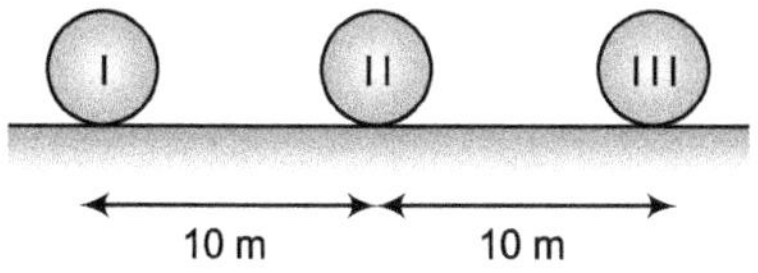

Sol. Let velocity of I ball and II ball after collision be v_1 and v_2

$$v_2 - v_1 = 0.5 \times 10 \quad \ldots\text{(i)}$$

$$mv_2 + mv_1 = m \times 10 \quad \ldots\text{(ii)}$$

$$\Rightarrow \quad v_2 + v_1 = 10$$

Solving Eqs. (i) and (ii)

$$v_1 = 2.5 \text{ m/s},$$

$$v_2 = 7.5 \text{ m/s}$$

Ball II after moving 10 m collides with ball III elastically and stops. But ball I moves towards ball II. Time taken between two consecutive collisions

$$\frac{10}{7.5} = \frac{10 - 10 \times \frac{2.5}{7.5}}{2.5}$$

$$= 4 \text{ s}$$

Example 9. *A plank of mass* 5 *kg is placed on a frictionless horizontal plane. Further a block of mass* 1 *kg is placed over the plank. A massless spring of natural length* 2 *m is fixed to the plank by its one end. The other end of spring is compressed by the block by half of spring's natural length. They system is now released from the rest. What is the velocity of the plank when block leaves the plank ?*

(The stiffness constant of spring is 100 *N/m)*

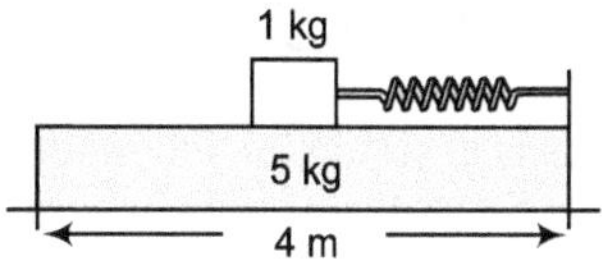

Sol. Let the velocity of the block and the plank, when the block leaves the spring be u and v respectively.

By conservation of energy $\frac{1}{2}kx^2 = \frac{1}{2}mu^2 + \frac{1}{2}Mv^2$

(M = mass of the plank, m = mass of the block)

$$\Rightarrow \quad 100 = u^2 + 5v^2 \quad \ldots\text{(i)}$$

By conservation of momentum

$$mu + Mv = 0$$

$$u = -5v \quad \ldots\text{(ii)}$$

Solving Eqs. (i) and (ii), we get

$$30\,v^2 = 100 \Rightarrow v = \sqrt{\frac{10}{3}} \text{ m/s}$$

From this moment until block falls, both plank and block keep their velocity constant.

Thus, when block falls, velocity of plank $= \sqrt{\frac{10}{3}}$ m/s.

Example 10. *A particle of mass* 2 *kg moving with a velocity* $5\mathbf{i}$ *m/s collides head-on with another particle of mass* 3 *kg moving with a velocity* $-2\mathbf{i}$ *m/s. After the collision the first particle has speed of* 1.6 *m/s in negative x direction. Find*

(a) velocity of the centre of mass after the collision.
(b) velocity of the second particle after the collision.
(c) coefficient of restitution.

Sol. (a) $\mathbf{v}_c = \frac{m_1\mathbf{u}_1 + m_2\mathbf{u}_2}{m_1 + m_2} = 0.8\,\hat{\mathbf{i}}$ m/s

(b) $\mathbf{v}_1 = -1.6\,\hat{\mathbf{i}}$ m/s

From CM, $m_1\mathbf{u}_1 + m_2\mathbf{u}_2 = m_1\mathbf{v}_1 + m_2\mathbf{v}_2$

$$\Rightarrow \quad v_2 = 2.4\,\hat{\mathbf{i}} \text{ m/s}$$

(c) $e = \frac{v_2 - v_1}{u_1 - u_2} = \frac{4}{7}$

NCERT Selected Questions

***Q* 1.** A nucleus is at rest in the laboratory frame of reference. Show that if it disintegrates into two smaller nuclei the products must be emitted in opposite directions.

Sol. Let m = mass of the nucleus at rest.

$$\mathbf{u} = \text{its initial velocity} = 0$$

Also let m_1, m_2 be the masses of the two smaller product nuclei and $\mathbf{v}_1, \mathbf{v}_2$ be their respective velocities.

If $\mathbf{p}_i$ and $\mathbf{p}_f$ be the initial and final momentum of the nucleus and the two nuclei respectively, then

$$\mathbf{p}_i = m\mathbf{u} = 0 \qquad \text{...(i)}$$

and $$\mathbf{p}_f = m_1\mathbf{v}_1 + m_2\mathbf{v}_2 \qquad \text{...(ii)}$$

Now, according to the law of conservation of linear momentum, we know that

$$\mathbf{p}_i = \mathbf{p}_f$$

or $$0 = m_1\mathbf{v}_1 + m_2\mathbf{v}_2$$

or $$m_2\mathbf{v}_2 = -m_1\mathbf{v}_1$$

or $$\mathbf{v}_2 = -\frac{m_1\mathbf{v}_1}{m_2} \qquad \text{...(iii)}$$

The negative sign in Eq. (iii) shows that $\mathbf{v}_1$ and $\mathbf{v}_2$ are in opposite directions, i.e. the two smaller nuclei are emitted in opposite directions.

***Q* 2.** Two billiard balls each of mass 0.05 kg moving in opposite directions with speed 6 ms^{-1} collide and rebound with the same speed. What is the impulse imparted to each ball due to the other?

Sol. ∴ Initial momentum of each ball

$$\mathbf{p}_i = m\mathbf{v} = (0.05)\,(6)$$
$$= 0.30 \text{ kg ms}^{-1}$$

As after collision, the direction of velocity of each ball is reversed on rebounding.

∴ Final momentum of each ball = $m(-\mathbf{v})$

$$\mathbf{p}_f = 0.05\,(-6)$$
$$= -0.30 \text{ kg ms}^{-1}$$

∴ Impulse imparted of each ball

$$= \mathbf{p}_f - \mathbf{p}_i$$
= change in momentum of each ball
$$= -0.30 - (0.30)$$
$$= -0.60 \text{ kg ms}^{-1}$$

or magnitude of impulse imparted by one ball due to collision with the other = 0.6 kg ms^{-1}. The two impulses are opposite in direction.

***Q* 3.** A shell of mass 0.020 kg is fired by a gun of mass 100 kg. If the muzzle speed of the shell is 8.0 ms^{-1}, then what is the recoil speed of the gun?

Sol. Here, mass of gun, $M = 100$ kg

Mass of shell, $m = 0.02$ kg

Speed of shell, $v = 80 \text{ ms}^{-1}$

and let recoil speed of gun = V

As, initially both the gun and the shell are at rest before firing, so the initial momentum $(\mathbf{p}_i)$ of the system = 0

Final momentum $(\mathbf{p}_f)$ of the system after firing

$$= MV + mv = 100\,V + 0.02 \times 80$$

∴ According to the law of conservation of linear momentum,

$$\mathbf{p}_i = \mathbf{p}_f \quad \text{or} \quad 0 = 100\,V + 1.6$$

or $$V = -\frac{1.6}{100} = 0.016 \text{ ms}^{-1}$$

∴ Recoil speed of gun = V = 0.016 ms^{-1} = 1.6 cms^{-1}

***Q* 4.** (a) The rate of change of total momentum of a many particle system is proportional to the external force/sum of the internal forces on the system.

(b) In an inelastic collision of two bodies, the quantities which do not change after the collision are the total kinetic energy/total linear momentum/total energy of the system of two bodies.

Sol. (a) Internal forces cannot change the total or net momentum of a system. Hence, the rate of change of total momentum of many particle system is proportional to the external force on the system.

(b) In an inelastic collision of two bodies, the quantities which do not change after the collision are total linear momentum and total energy of the system of two bodies (if the system is isolated). The total KE of the system, is not conserved as it may change to an equivalent amount of energy in some other form.

***Q* 5.** State if each of the following statements are true or false. Give reasons for your answer.

(a) In an elastic collision of two bodies, the momentum and energy of each body is conserved.

(b) In an inelastic collision, the final kinetic energy is always less than the initial kinetic energy of the system.

Sol. (a) False. The total momentum and energy of the system are conserved and not of each body.

(b) True. Usually but not always because in an inelastic collision, some kinetic energy usually changes into some other forms of energy.

***Q* 6.** Answer carefully, with reasons.

(a) In an elastic collision of two billiard balls, is the total kinetic energy conserved during the short time of collision of the balls (i.e. when they are in contact)?

(b) Is the total linear momentum conserved during the short time of an elastic collision of two balls?

(c) What are the answers to (a) and (b) for an inelastic collision?

Sol. (a) No, the total kinetic energy is not conserved during the given elastic collision because a part of the kinetic energy is used in deforming the balls in that short interval for which they are in contact during collision and gets converted into potential energy. In an elastic collision, the KE before and after collision is same.

(b) Yes, the total linear momentum is conserved during the short time of an elastic collision of two balls.

(c) In an inelastic collision, total KE is not conserved during collision and even after collision. The total linear momentum is however conserved during as well as after collision.

***Q* 7.** A molecule in a gas container hits a horizontal wall with speed 200 ms^{-1} and angle 30° with the normal and rebounds with the same speed. Is momentum conserved in the collision? Is the collision elastic or inelastic?

Sol. Yes, linear momentum is always conserved whether the collision is elastic or inelastic. Thus, in the given situation, momentum of system is conserved.

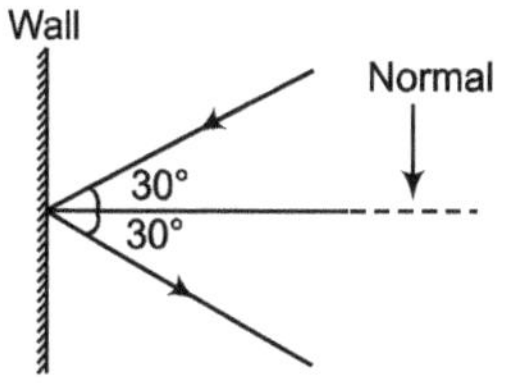

As, the speed of the molecule remains the same before and after the collision, so kinetic energy of the molecule is also conserved.

Hence, the collision is elastic.

***Q* 8.** Two identical ball bearings in contact with each other and resting on a frictionless table are hit head-on by another ball bearing of the same mass moving initially with a speed v. If the collision is elastic, which of the following is a possible result after collision?

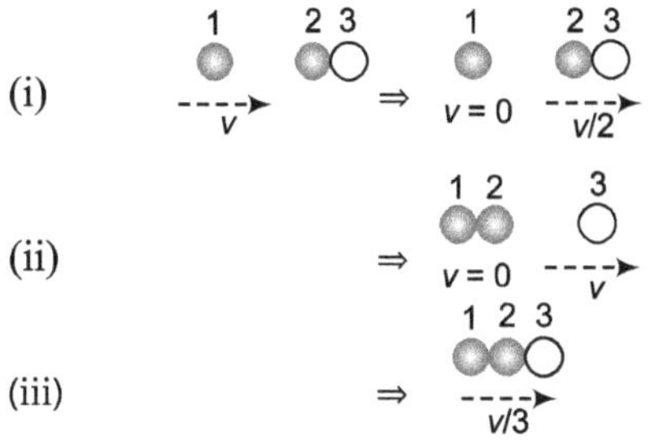

Sol. Before collision, total KE of the system $= \frac{1}{2}mv^2$ ...(i)

After collision, KE of the system

In case (i) $\quad \text{KE} = \frac{1}{2}2m\left(\frac{v}{2}\right)^2 = \frac{1}{4}mv^2$ ...(ii)

In case (ii) $\quad \text{KE} = \frac{1}{2}mv^2$ ...(iii)

In case (iii) $\quad \text{KE} = \frac{1}{2}(3m)\left(\frac{v}{3}\right)^2 = \frac{1}{6}mv^2$ ...(iv)

Thus, from the above equations we observe that the KE is conserved only in case (ii), hence case (ii) is the only possible result after collision.

***Q* 9.** The bob A of a pendulum released from 30° to the vertical hits another bob B of the same mass at rest on a table as shown in figure given below. How high does the bob A rise after the collision? Neglect the size of the bobs and assume the collision to be elastic.

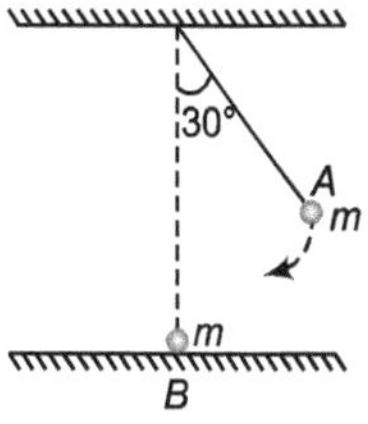

Sol. In elastic head on collision, when two equal masses collide each other, then they exchange their velocities.

In the present case, bob A is moving with certain speed and bob B is at rest. Hence, after collision, bob A comes to rest and the bob B starts moving with the speed of the bob A. The bob A transfers whole of its momentum to ball B and hence bob A, will not rise at all after the collision.

***Q* 10.** A trolley of mass 300 kg carrying a sand bag of 25 kg is moving uniformly with a speed of 27 km/h on a frictionless track. After a while, sand starts leaking out of a hole on the floor of trolley at the rate of 0.05 $kg\,s^{-1}$. What is the speed of the trolley after the entire sand bag is empty?

Sol. The trolley carrying a sand bag is moving with a speed of 27 kmh^{-1}. It means that the system of trolley and sand bag is not acted upon by any external force. If the sand leaks out of a hole on the floor of trolley, it does not give rise to any external force on the trolley.

Therefore, speed of the trolley will not change when the sand is leaking out or even after the sand bag becomes empty. This is because the weight of sand is leaks by taking away the same momentum.

$\therefore$ Speed = 27 km/h

Q 11. A bullet of mass 0.012 kg and horizontal speed $70\,\text{ms}^{-1}$ strikes a block of wood of mass 0.4 kg and instantly comes to rest with respect to the block. The block is suspended from the ceiling by means of a thin wire. Calculate the height to which the block rises. Also, estimate the amount of heat produced in the block.

Sol. Let final speed of bullet and block after collision which move together = v

According to the law of conservation of momentum,

initial momentum of (bullet + block) before collision = final momentum of bullet and block after collision

or $\quad 0.012\times 70 + 0.4\times 0 = (0.012+0.4)\,v$

or $\quad 0.012\times 70 = 0.412\,v \quad$ or $\quad v = \dfrac{0.012\times 70}{0.412}$

or $\quad v = 2.04\ \text{ms}^{-1}$

Let h = height through which the block rises after collision.

Then, $\quad h = \dfrac{v^2}{2g} = \dfrac{(2.04)^2}{2\times 9.8}$

$= 0.212\ \text{m} = 21.2\ \text{cm}$

Now, heat produced in the block = loss of KE

= KE of bullet − KE of block and bullet

$= \dfrac{1}{2}\times 0.012\times (70)^2 - \dfrac{1}{2}(0.412)\times(2.04^2)$

$= 28.543\ \text{J}$

Q 12. A trolley of mass 200 kg moves with a uniform speed of 36 km/h on frictionless track. A child of mass 20 kg runs on the trolley from one end to the other (10 m away) with a speed of $4\,\text{ms}^{-1}$ relative to the trolley in a direction opposite to the trolley's motion, and jumps out of the trolley. What is the final speed of the trolley? How much has the trolley moved from the time the child begins to run?

Sol. Mass of the trolley, $m_1 = 200\,\text{kg}$

Speed of the trolley, $v = 36\,\text{kmh}^{-1} = 36\times\dfrac{5}{18} = 10\ \text{ms}^{-1}$

Mass of the child, $m_2 = 20\,\text{kg}$

If p_i be the initial momentum of the system before the child starts running.

Then, $\quad p_i = (m_1 + m_2)\,v$

$= (200+20)\times 10$

$= 220\times 10$

$= 2200\ \text{kgms}^{-1} \quad$ ...(i)

When the child starts running, with a velocity of $4\ \text{ms}^{-1}$ in a direction opposite to trolley, then let v' be the final speed of the trolley (w.r.t. earth)

∴ Speed of the child relative to earth $= v' - 4$

∴ Momentum of the system when the child is running,

$p_f = m_1v' + 20(v'-4)$

$= 200v' + 20(v'-4) \quad$ ...(ii)

As, no external force is applied on the system

∴ $\quad p_f = p_i$

or $\quad 200v' + 20v' - 80 = 2200$

or $\quad 220v' = 2200 + 80 = 2280$

or $\quad v' = \dfrac{2280}{220} = 10.36\ \text{ms}^{-1}$

Time taken by the child to run a distance of 10 m over the trolley,

$t = \dfrac{10\ \text{m}}{4\ \text{ms}^{-1}} = 2.5\ \text{s}$

If x be the distance moved by the trolley in this time, then

$x = v'\times t = 10.36\times 2.5$

$= 25.9\ \text{m}$

Q 13. Give the location of the centre of mass of a (i) sphere (ii) cylinder (iii) ring and (iv) cube each of uniform mass density. Does the centre of mass of a body necessarily lie inside the body?

Sol. (i) Centre of sphere

(ii) Mid- point of axis of symmetry of the cylinder, i.e. its geometrical centre.

(iii) Centre of ring.

(iv) Point of intersection of diagonals, i.e. at its geometrical centre.

No, in some cases, centre of mass may lie outside.

Q 14. In the HCl molecule, the separation between the nuclei of the two atoms is about $1.27\ \text{Å}\ (1\ \text{Å} = 10^{-10}\ \text{m})$. Find the approximate location of the CM of the molecule, given that a chlorine atom is about 35.5 times as massive as a hydrogen atom and nearly all the mass of an atom is concentrated in its nucleus.

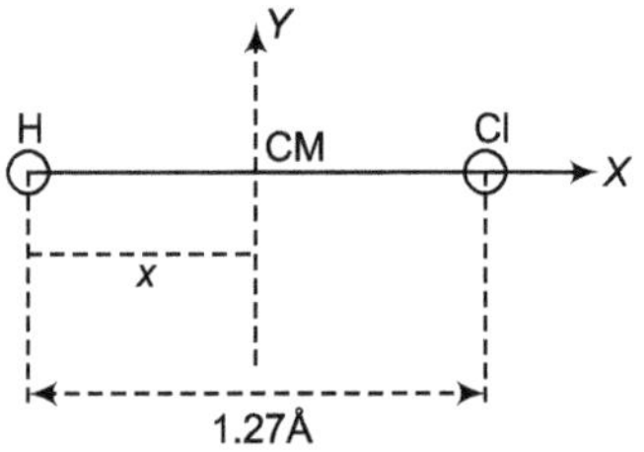

Sol. Let CM be at a distance of x from H-atom.

If x_1 and x_2 be the position vectors of the H and Cl atoms w.r.t. the CM as origin, then

$x_{CM} = 0 = \dfrac{m_1x_1 + m_2x_2}{m_1 + m_2} \quad$ ...(i)

or $\quad m_1x_1 + m_2x_2 = 0$

or $\quad m(-x) + 35.5\,m\,(1.27 - x) = 0$

or $\quad mx = 35.5\,m\,(1.27 - x)$

or $\quad x + 35.5x = 35.5\times 1.27$

∴ $\quad x = \dfrac{35.5\times 1.27}{36.5} = 1.235\ \text{Å} \approx 1.24\ \text{Å}$

From H-atom and on the line joining H and Cl atoms.

Q 15. A child sits stationary at one end of a long trolley moving uniformly with a speed v on a smooth horizontal floor. If the child gets up and runs about on the trolley in any manner, what is the speed of the CM of the (trolley + child) system?

Sol. The speed of the centre of mass of the system and child remains unchanged. It is because the state of system can change only under the effect of an external force and in this case no external force is acting. The forces involved in running on the trolley are from within the system, i.e. internal forces.

Q 16. From a uniform disc of radius R, a circular section of radius $R/2$ is cut out. The centre of the hole is at $R/2$ from the centre of the original disc. Locate the centre of gravity of the resulting flat body.

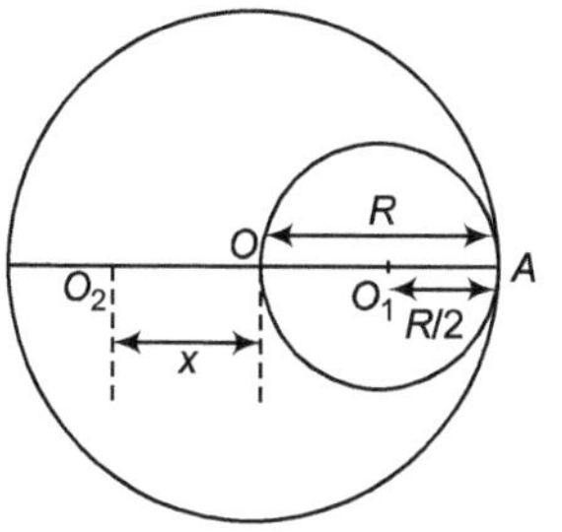

Sol. Here, radius of original disc = R

Radius of removed circular section = $\frac{R}{2}$

If A and a be their respective areas, then

$$A = \pi R^2 \quad \text{and} \quad a = \pi\left(\frac{R}{2}\right)^2 = \frac{\pi R^2}{4}$$

Here O is the centre of the original disc.
and O_1 is the centre of the removed circular hole.
Also let O_2 = centre of the remaining portion.
ρ = mass per unit area of the disc.

If m and m_2 be the masses of the original disc and the removed portion respectively, then

$$m = \rho A = \pi R^2 \rho$$

and
$$m_2 = \rho a = \frac{\pi R^2}{4}\rho$$

If m_1 be the mass of remaining portion, then

$$m_1 = m - m_2 = \pi R^2 \rho - \frac{\pi R^2 \rho}{4}$$

$$= \frac{3}{4}\pi R^2 \rho$$

Let O be the origin.

Then,
$$x_{CM} = \frac{m_1 x_1 + m_2 x_2}{m_1 + m_2} \quad \text{...(i)}$$

$$0 = \frac{\left(\frac{3}{4}\pi R^2 \rho\right)(-x) + \left(\frac{\pi R^2}{4}\rho\right)\left(\frac{R}{2}\right)}{\left(\frac{3}{4}\pi R^2 \rho + \frac{\pi R^2}{4}\rho\right)}$$

$\therefore$
$$x = \frac{R}{6}$$

Objective Problems

[Level 1]

Position of Centre of Mass

1. A uniform square plate $ABCD$ has a mass of 10 kg. If two points masses of 5 kg each are placed at the corners C and D as shown in the adjoining figure, then the centre of mass shifts to the mid-point of

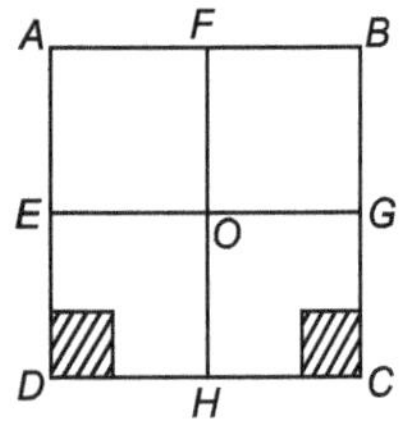

(a) OH (b) DH (c) OG (d) OF

2. In carbon monoxide molecules, the carbon and the oxygen atoms are separated by distance 1.2×10^{-10} m. The distance of the centre of mass, from the carbon atom is

(a) 0.48×10^{-10} m (b) 0.51×10^{-10} m
(c) 0.56×10^{-10} m (d) 0.69×10^{-10} m

3. A square plate of side 20 cm has uniform thickness and density. A circular part of diameter 8 cm is cut out symmetrically and shown in figure. The position of centre of mass of the remaining portion is

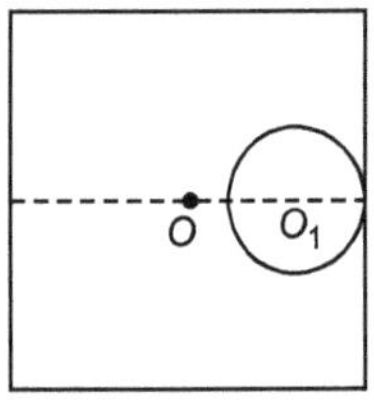

(a) at O_1
(b) at O
(c) 0.54 cm from O on the left hand side
(d) None of the above

4. A circular plate of diameter d is kept in contact with a square plate of edge d as shown in figure. The density of the material and the thickness are same everywhere. The centre of mass of the composite system will be

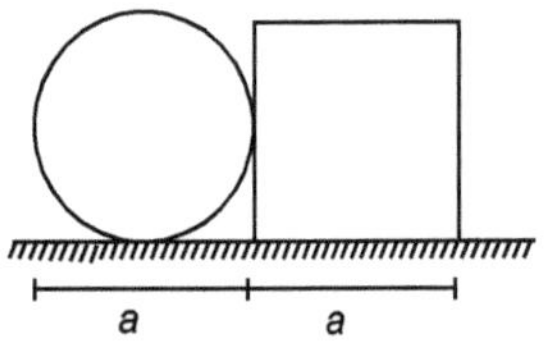

(a) inside the circular plate (b) inside the square plate
(c) at the point of contact (d) outside the system

5. Four particles of mass $m_1 = 2m$, $m_2 = 4m$, $m_3 = m$ and m_4 are placed at four corners of a square. What should be the value of m_4 so that the centre of mass of all the four particles are exactly at the centre of the square?

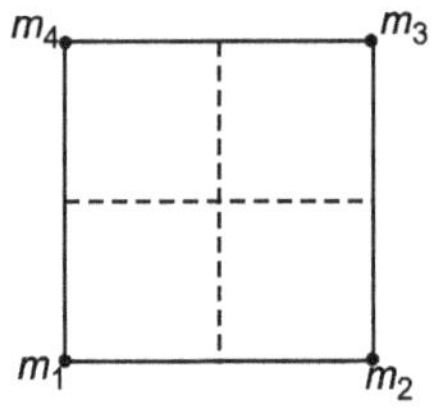

(a) $2m$ (b) $8m$
(c) $6m$ (d) None of these

6. Three rods of the same mass are placed as shown in the figure. What will be the coordinate of centre of mass of the system?

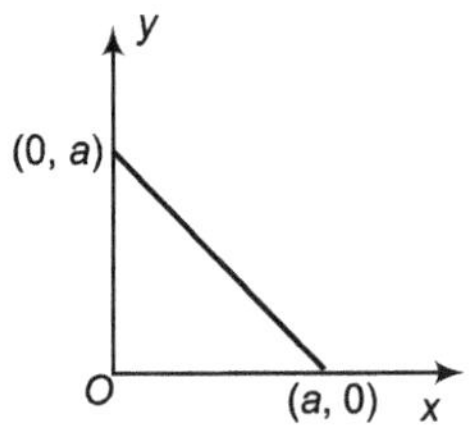

(a) $\left(\frac{a}{2}, \frac{a}{2}\right)$ (b) $\left(\frac{a}{\sqrt{2}}, \frac{a}{\sqrt{2}}\right)$
(c) $\frac{2a}{3}, \frac{2a}{3}$ (d) $\left(\frac{a}{3}, \frac{a}{3}\right)$

7. Centre of mass of three particles of masses 1 kg, 2 kg and 3 kg lies at the point (1, 2, 3) and centre of mass of another system of particles of total mass 3 kg lies at the point (−1, 3, −2). Where should we put a particle of mass 5 kg so that the centre of mass of the entire system lies at the centre of mass of Ist system?

(a) (0, 0, 0)
(b) (1, 3, 2)
(c) (−1, 2, 3)
(d) None of the above

8. A uniform metal disc of radius R is taken and out of it a disc of diameter R is cut off from the end. The centre of mass of the remaining part will be

(a) $\frac{R}{4}$ from the centre (b) $\frac{R}{3}$ from the centre
(c) $\frac{R}{5}$ from the centre (d) $\frac{R}{6}$ from the centre

9. A uniform circular disc of radius a is taken. A circular portion of radius b has been removed from it as shown in the figure. If the centre of hole is at a distance c from the centre of the disc, the distance x_2 of the centre of mass of the remaining part from the initial centre of mass O is given by

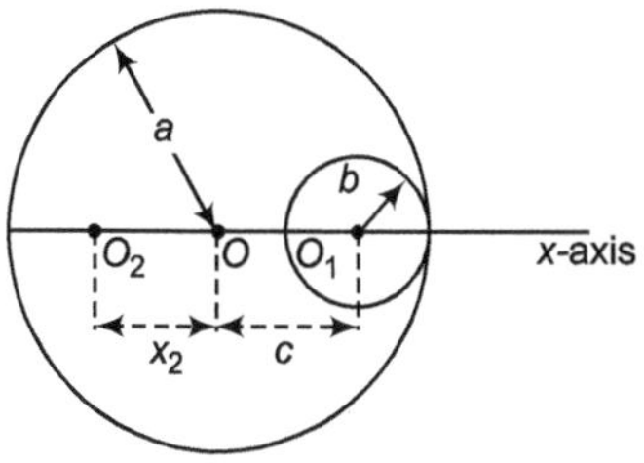

(a) $\frac{\pi b^2}{(a^2 - c^2)}$ (b) $\frac{cb^2}{(a^2 - b^2)}$

(c) $\frac{\pi c^2}{(a^2 - b^2)}$ (d) $\frac{ca^2}{(c^2 - b^2)}$

10. Four rods AB, BC, CD and DA have mass m, $2m$, $3m$ and $4m$ respectively. The centre of mass of all the four rods

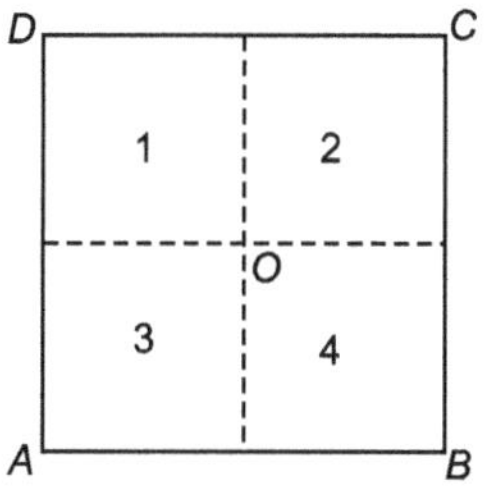

(a) lie in region
(b) lie in region 2
(c) lie in region 3
(d) lie at O

11. All the particles of a body are situated at a distance R from the origin. The distance of centre of mass of the body from the origin is

(a) $= R$ (b) $\leq R$
(c) $> R$ (d) $\geq R$

12. A uniform metal rod of length 1 m is bent at 90° so as to form two arms of equal length. The centre of mass of this bent rod is

(a) on the bisector of the angle, $\left(\frac{1}{\sqrt{2}}\right)$ m from vertex

(b) on the bisector of angle, $\left(\frac{1}{2\sqrt{2}}\right)$ m from vertex

(c) on the bisector of the angle, $\left(\frac{1}{2}\right)$ m from vertex

(d) on the bisector of the angle, $\left(\frac{1}{4\sqrt{2}}\right)$ m from vertex

13. Three identical spheres, each of mass 1 kg are placed touching each other with their centres on a straight line. Their centre are marked P, Q and R, respectively. The distance of centre of mass of the system from P is

(a) $\frac{PQ + PR + QR}{3}$ (b) $\frac{PQ + PR}{3}$

(c) $\frac{PQ + QR}{3}$ (d) None of these

14. A circular ring of mass 6 kg and radius a is palced such that its centre lies at the origin. Two particles of masses 2 kg each are placed at the intersecting points of the circle with positive x-axis and positive y-axis. Then, the angle made by the position vector of centre of mass of entire system with x-axis is

(a) 45° (b) 60°
(c) $\tan^{-1}\left(\frac{4}{5}\right)$ (d) 30°

15. From a circular disc of radius R, a square is cut out with a radius as its diagonal. The centre of mass of remainder is at a distance (from the centre)

(a) $\frac{R}{(4\pi - 2)}$ (b) $\frac{R}{2\pi}$

(c) $\frac{R}{(\pi - 2)}$ (d) $\frac{R}{(2\pi - 2)}$

Motion of Centre of Mass

16. Two particles A and B initially at rest, move towards each other under a mutual force of attraction. At the instant when the speed of A is v and the speed of B is $2v$, the speed of centre of mass of the system is

(a) zero (b) v
(c) 1.5 v (d) $3v$

17. Consider a system of two identical particles. One of the particles is at rest and the other has an acceleration a. The centre of mass has an acceleration

(a) zero (b) $\frac{1}{2}$ a
(c) a (d) 2 a

18. Two blocks of masses 10 kg and 4 kg are connected by a spring of negligible mass and placed on a frictionless horizontal surface. An impulse gives a velocity of 14 m/s to the heavier block in the direction of the lighter block. The velocity of the centre of mass is

(a) 30 m/s (b) 20 m/s
(c) 10 m/s (d) 5 m/s

19. A body falling vertically downwards under gravity breaks in two parts of unequal masses. The centre of mass of the two parts taken together shifts horizontally towards

(a) heavier piece
(b) lighter piece
(c) does not shift horizontally
(d) depends on the vertical velocity at the time of breaking

20. A ball kept in a closed box moves in the box making collisions with the walls. The box is kept on a smooth surface. The velocity of the centre of mass

(a) of the box remains constant
(b) of the box plus the ball system remains constant
(c) of the ball remains constant
(d) of the ball relative to the box remains constant

21. Two balls are thrown simultaneously in air. The acceleration of the centre of mass of the two balls while in air
(a) depends on the direction of the motion of the balls
(b) depends on the masses of the two balls
(c) depends on the speeds of two balls
(d) is equal to g

22. A man of mass M stands at one end of a plank of length L which lies at rest on a frictionless surface. The man walks to the other end of the plank. If the mass of the plank is $\frac{M}{3}$, the distance that the man moves relative to the ground is
(a) $\frac{3L}{4}$ (b) $\frac{L}{4}$ (c) $\frac{4L}{5}$ (d) $\frac{L}{3}$

23. In a gravity free space, a man of mass M standing at a height h above the floor, throws a ball of mass m straight down with a speed u. When the ball reaches the floor, the distance of the man above the floor will be
(a) $h\left(1+\frac{m}{M}\right)$ (b) $\left(1+\frac{M}{m}\right)h$
(c) h (d) $\frac{m}{M}h$

24. Two particles of equal mass have coordinates (2m, 4m, 6m) and (6m, 2m, 8m). Of these one particle has a velocity $\mathbf{v}_1 = (2\hat{\mathbf{i}})$ m/s and another particle has velocity $\mathbf{v}_2 = (2\hat{\mathbf{j}})$ m/s at time $t = 0$. The coordinate of their centre of mass at time $t = 1$ s will be
(a) (4 m, 4 m, 7 m) (b) (5 m, 4 m, 7 m)
(c) (2 m, 4 m, 6 m) (d) (4 m, 5 m, 4 m)

25. A ball falls freely from a height of 45 m. When the ball is at a height of 25 m it explodes into two equal pieces. One of them moves horizontally with a speed of $10\ \text{ms}^{-1}$. The distance between the two pieces on the ground is
(a) 20 m (b) 30 m
(c) 40 m (d) 60 m

26. A metre stick is placed vertically at the origin on a frictionless surface. A gentle push in positive x direction is given to the top most point of the rod, when it has fallen completely x-coordinate of centre of rod is at
(a) origin (b) -0.5 m
(c) -1 m (d) $+0.5$ m

27. Two blocks of masses 10 kg and 30 kg are placed along a vertical line. The first block is raised through a height of 7 cm. By what distance should the second mass be moved to raise the centre of mass by 1 cm?
(a) 2 cm upward
(b) 1 cm upward
(c) 2 cm downward
(d) 1 cm downward

28. In a free space, a rifle of mass M shoots a bullet of mass m at a stationary block of mass M distance D away from it. When the bullet has moved through a distance d towards the block, the centre of mass of the bullet-block system is at a distance of
(a) $\frac{(D-d)\,m}{M+m}$ from the bullet
(b) $\frac{md+MD}{M+m}$ from the block
(c) $\frac{2md+MD}{M+m}$ from the block
(d) $\frac{(D-d)\,M}{M+m}$ from the bullet

29. Two particles P and Q initially at rest, move towards each other under a mutual force of attraction. At the instant when the speed of P is v and the speed of Q is $2v$, the speed of the centre of mass of the system is
(a) $3v$ (b) $1.5\ v$
(c) zero (d) data insufficient

30. Blocks A and B are resting on a smooth horizontal surface given equal speeds of 2 m/s in opposite sense as shown in the figure.

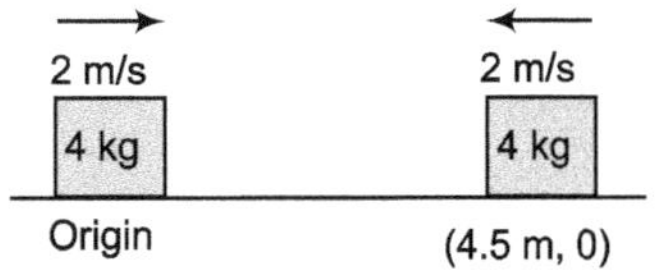

At $t = 0$, the position of blocks are shown, then the coordinates of centre of mass $t = 3$ s will be
(a) (1, 0) (b) (3, 0)
(c) (5, 0) (d) (2.25, 0)

31. Two balls of equal mass are projected from a tower simultaneously with equal speeds. One at angle θ above the horizontal and the other at the same angle θ below the horizontal. The path of the centre of mass of the two balls is
(a) a vertical straight line
(b) a horizontal straight line
(c) a straight line at an angle α ($<\theta$) with horizontal
(d) a parabola

32. A cracker is thrown into air with a velocity of 10 m/s at an angle of 45° with the vertical. When it is at a height of (1/2) m from the ground, it explodes into a number of pieces which follow different parabolic paths. What is the velocity of centre of mass, when it is at a height of 1 m from the ground? ($g = 10\ \text{m/s}^2$)
(a) $4\sqrt{5}\ \text{ms}^{-1}$
(b) $2\sqrt{5}\ \text{ms}^{-1}$
(c) $5\sqrt{4}\ \text{ms}^{-1}$
(d) $5\ \text{ms}^{-1}$

33. Two bodies having masses m_1 and m_2 and velocities $\mathbf{u}_1$ and $\mathbf{u}_2$ collide and form a composite system. If $m_1\mathbf{u}_1 + m_2\mathbf{u}_2 = 0\ (m_1 \neq m_2)$, then velocity of composite system will be

(a) $\mathbf{u}_1 - \mathbf{u}_2$ (b) $\mathbf{u}_1 + \mathbf{u}_2$
(c) $\frac{\mathbf{u}_1 + \mathbf{u}_2}{2}$ (d) zero

34. An isolated particle of mass m is moving in horizontal plane (x-y), along the x-axis, at a certain height above the ground, it suddenly explodes into two fragments of masses $\frac{m}{4}$ and $\frac{3m}{4}$. An instant later, the smaller fragment is at $y = +15$ cm. The larger fragment at this instant is at

(a) $y = -5$ cm (b) $y = +20$ cm
(c) $y = +5$ cm (d) $y = -20$ cm

35. A man of mass m is standing on a plank of equal mass m resting on a smooth horizontal surface. The man starts moving on the plank with speed u relative to the plank.

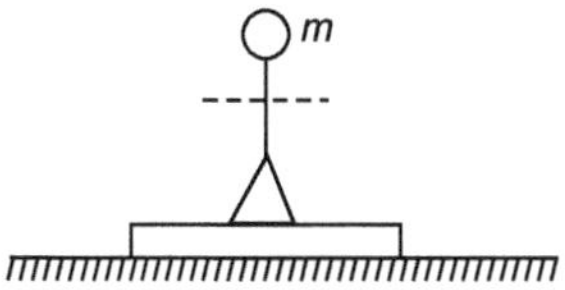

The speed of the man relative to the ground is

(a) $2u$ (b) $\frac{u}{2}$
(c) zero (d) $\frac{u}{4}$

36. A body of mass a moving with a velocity b strikes a body of mass c and gets embedded into it. The velocity of the system after collision is

(a) $\frac{a+c}{ab}$ (b) $\frac{ab}{a+c}$ (c) $\frac{a}{b+c}$ (d) $\frac{a}{a+b}$

37. A man of mass m moves with a constant speed on a plank of mass M and length l kept initially at rest on a frictionless horizontal surface, from one end to the other in time t. The speed of the plank relative to ground while man is moving, is

(a) $\frac{l}{t}\left(\frac{M}{m}\right)$ (b) $\frac{l}{t}\left(\frac{m}{m+M}\right)$
(c) $\frac{l}{t}\left(\frac{M}{M+m}\right)$ (d) None of these

Conservation Laws, Impulse and Variable Mass System

38. A bullet of mass 5 g is fired at a velocity of 900 ms^{-1} from a rifle of mass 2.5 kg. What is the recoil velocity of the rifle?

(a) 0.9 ms^{-1} (b) 180 ms^{-1}
(c) 900 ms^{-1} (d) 1.8 ms^{-1}

39. A stationary bomb explodes into two parts of masses 3 kg and 1 kg. The total KE of the two parts after explosion is 2400 J. The KE of the smaller part is

(a) 600 J (b) 1800 J
(c) 1200 J (d) 2160 J

40. When a ball of mass 5 kg hits a bat with a velocity 3 m/s, in positive direction and it moves back with a velocity 4 m/s, find the impulse in SI units.

(a) 5 (b) 15
(c) 25 (d) 35

41. One projectile moving with velocity v in space, gets burst into 2 parts of masses in the ratio 1 : 3. The smaller part becomes stationary. What is the velocity of the other part?

(a) $4v$ b) v (c) $\frac{4v}{3}$ (d) $\frac{3v}{4}$

42. A bomb of mass 9 kg explodes into two pieces of mass 3 kg and 6 kg. The velocity of 3 kg mass is 16 m/s. The velocity of 6 kg mass is

(a) 4 m/s (b) 8 m/s (c) 16 m/s (d) 32 m/s

43. A body at rest breaks into two pieces of equal masses. The parts will move

(a) in same direction
(b) along different lines
(c) in opposite directions with equal speeds
(d) in opposite directions with unequal speeds

44. The magnitude of the impulse developed by a mass of 0.2 kg which changes its velocity from $5\hat{\mathbf{i}} - 3\hat{\mathbf{j}} + 7\hat{\mathbf{k}}$ m/s to $2\hat{\mathbf{i}} + 3\hat{\mathbf{j}} + \hat{\mathbf{k}}$ m/s is

(a) 2.7 N-s (b) 1.8 N-s (c) 0.9 N-s (d) 3.6 N-s

45. Gravels are dropped on a conveyor belt at the rate of 0.5 kg/s. The extra force required in newtons to keep the belt moving at 2 m/s is

(a) 1 (b) 2 (c) 4 (d) 0.5

46. A particle of mass 15 kg has an initial velocity $\mathbf{v}_i = \hat{\mathbf{i}} - 2\hat{\mathbf{j}}$ m/s. It collides with another body and the impact time is 0.1 s, resulting in a velocity $\mathbf{v}_f = 6\hat{\mathbf{i}} + 4\hat{\mathbf{j}} + 5\hat{\mathbf{k}}$ m/s after impact. The average force of impact on the particle is

(a) $15\,|5\hat{\mathbf{i}} + 6\hat{\mathbf{j}} + 5\hat{\mathbf{k}}|$ (b) $15\,|5\hat{\mathbf{i}} + 6\hat{\mathbf{j}} - 5\hat{\mathbf{k}}|$
(c) $150\,|5\hat{\mathbf{i}} - 6\hat{\mathbf{j}} + 5\hat{\mathbf{k}}|$ (d) $150\,|5\hat{\mathbf{i}} + 6\hat{\mathbf{j}} + 5\hat{\mathbf{k}}|$

47. Two trains A and B are running in the same direction on parallel rails such that A is faster than B. Packets of equal weight are transferred between them. What will happen due to this

(a) A will be accelerated but B will be retarded
(b) B will be accelerated but A will be retarded
(c) there will be no change in A but B will be accelerated
(d) there will be no change in B but A will be accelerated

48. A shell of mass m is moving horizontally with velocity v_0 and collides with the wedge of mass M just above point A, as shown in the figure. As a consequence, wedge starts to move towards left and the shell returns with a velocity in x-y plane. The principle of conservation of momentum can be applied for

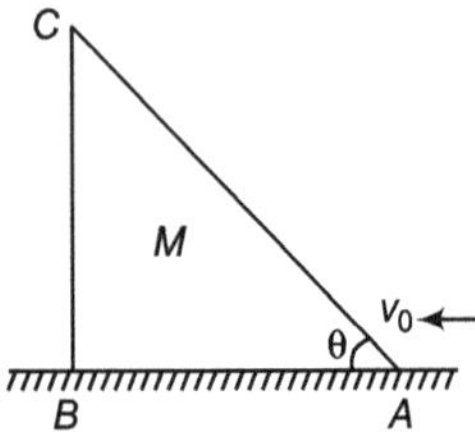

(a) system $(m + M)$ along any direction
(b) system $(m + M)$ along vertical
(c) system $(m + M)$ horizontally
(d) None of the above

49. A shell is fired from cannon with a velocity v at an angle θ with the horizontal direction. At the highest point in its path it explodes into two pieces of equal mass. One of the pieces retraces its path to the cannon and the speed (in ms^{-1}) of the other piece immediately after the explosion is

(a) $3v\cos\theta$ (b) $2v\cos\theta$ (c) $\frac{3}{2}v\cos\theta$ (d) $\sqrt{\frac{3}{2}}v\cos\theta$

50. In a two block system an initial velocity v_0 with respect to ground is given to block A

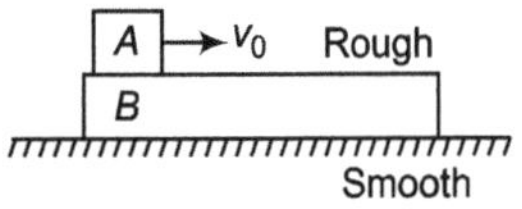

(a) the momentum of block A is not conserved
(b) the momentum of system of blocks A and B is conserved
(c) the increase in momentum of B is equal to the decrease in momentum of block A
(d) All of the above

51. If the net external forces acting on the system of particles is zero, then which of the following may vary?
(a) Momentum of the system (b) Velocity of centre of mass
(c) Position of centre of mass (d) None of these

52. For a rocket moving in free space, the fraction of mass to be disposed off to attain a speed equal to two times the exhaust speed is given by (given $e^2 = 7.4$)
(a) 0.63 (b) 0.37
(c) 0.50 (d) 0.86

53. An object of mass $3m$ splits into three equal fragments. Two fragments have velocities $v\hat{\mathbf{j}}$ and $v\hat{\mathbf{i}}$. The velocity of the third fragment is

(a) $v(\hat{\mathbf{j}} - \hat{\mathbf{i}})$ (b) $v(\hat{\mathbf{i}} - \hat{\mathbf{j}})$ (c) $-v(\hat{\mathbf{i}} + \hat{\mathbf{j}})$ (d) $\frac{v(\hat{\mathbf{i}} + \hat{\mathbf{j}})}{\sqrt{2}}$

54. A mass of 100 g strikes the wall with speed 5 m/s at an angle as shown in figure and it rebounds with the same speed. If the contact time is 2×10^{-3} s, what is the force applied on the mass by the wall

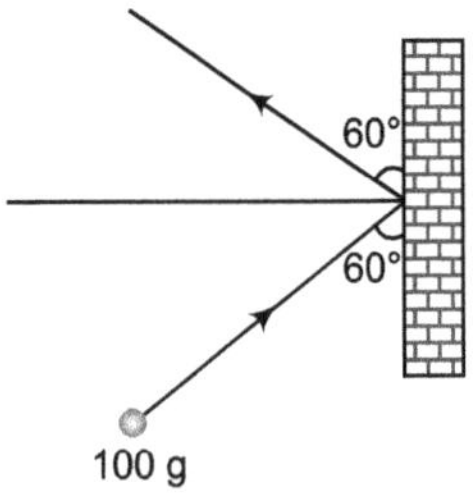

(a) $250\sqrt{3}$ N to right
(b) 250 N to right
(c) $250\sqrt{3}$ N to left
(d) 250 N to left

55. A cannon ball is fired with a velocity 200 m/s at an angle of 60° with the horizontal. At the highest point of its flight, it explodes into 3 equal fragments, one going vertically upwards with a velocity 100 m/s, the second one falling vertically downwards with a velocity 100 m/s. The third fragment will be moving with a velocity
(a) 100 m/s in the horizontal direction
(b) 300 m/s in the horizontal direction
(c) 300 m/s in a direction making an angle of 60° with the horizontal
(d) 200 m/s in a direction making an angle of 60° with the horizontal

56. A particle of mass m moving with velocity u makes an elastic one dimensional collision with a stationary particle of mass m. They are in contact for a brief time T. Their force of interaction increases from zero to F_0 linearly in time $\frac{T}{2}$ and decreases linearly to zero in further time $\frac{T}{2}$. The magnitude of F_0 is

(a) $\frac{mu}{T}$ (b) $\frac{2mu}{T}$
(c) $\frac{mu}{2T}$ (d) None of these

57. A ball of mass m moving with velocity v_0 collides a wall as shown in figure. After impact it rebounds with a velocity $\frac{3}{4}v_0$. The impulse acting on ball during impact is

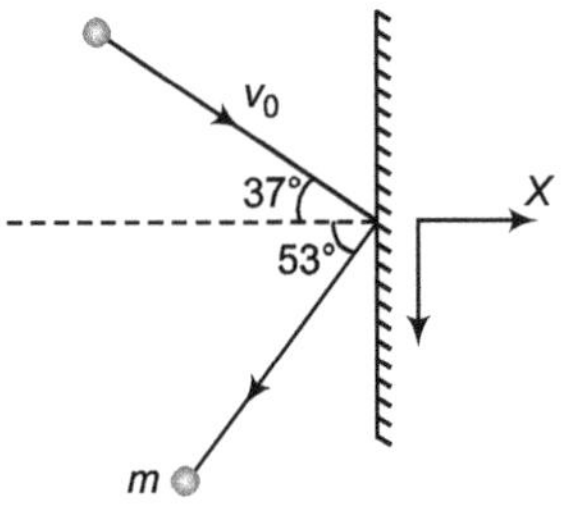

(a) $-\frac{m}{2}v_0\hat{\mathbf{j}}$ (b) $-\frac{3}{4}mv_0\hat{\mathbf{i}}$
(c) $-\frac{5}{4}mv_0\hat{\mathbf{i}}$ (d) None of these

58. A unidirectional force F varying with time t as shown in the figure acts on a body initially at rest for a short duration $2T$. Then, the velocity acquired by the body is

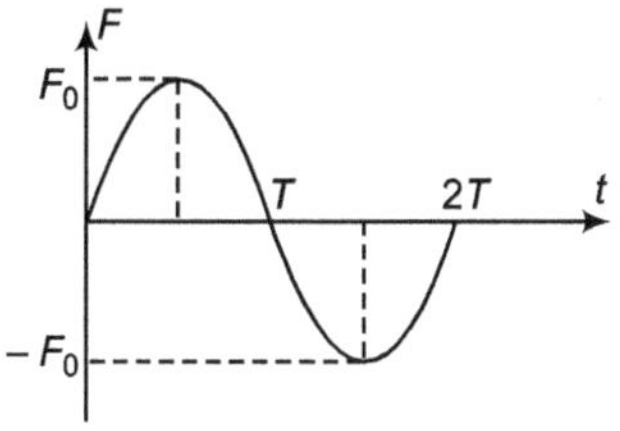

(a) $\frac{\pi F_0 T}{4m}$ (b) $\frac{\pi F_0 T}{2m}$
(c) $\frac{F_0 T}{4m}$ (d) zero

59. In the figure given the position-time graph of a particle of mass 0.1 kg is shown in the given figure. The impulse at $t = 2\,\text{s}$ is

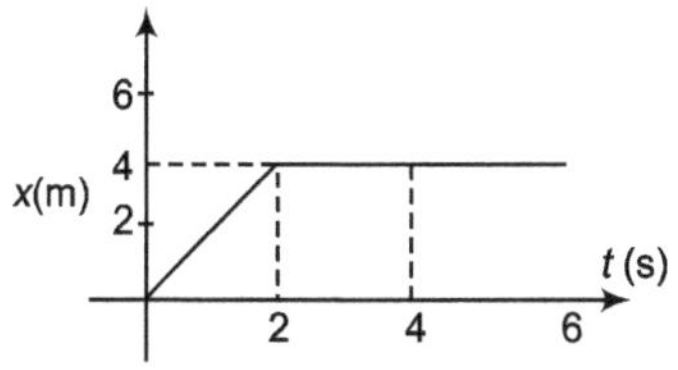

(a) 0.2 kg-ms^{-1} (b) −0.2 kg-ms^{-1}
(c) 0.1 kg-ms^{-1} (d) −0.4 ms^{-1}

60. A player takes 0.1 s in catching a ball of mass 150 g moving with velocity of 20 m/s. The force imparted by the ball on the hands of the player is
(a) 0.3 N (b) 3 N (c) 30 N (d) 300 N

61. If two balls, each of mass 0.06 kg, moving in opposite directions with speed 4 m/s collide and rebound with the same speed, then the impulse imparted to each ball due to the other is
(a) 0.92 kg-m/s (b) 0.24 kg-m/s
(c) 0.48 kg-m/s (d) 0.52 kg-m/s

62. A batsman hits a 150 g ball moving horizontally at 20 m/s back to bowler at 12 m/s. If the contact of bat and ball lasted for 0.04 s, then the average force exerted by the bat on the ball is
(a) 30 N (b) 120 N
(c) 150 N (d) 120×10^3 N

63. An open railroad car of mass M is moving with initial velocity v_0 on a straight horizontal frictionless track. It suddenly starts raining at time $t = 0$. The raindrops fall vertically with velocity v and add a mass of μ kg/s of water. Determine the velocity v of car after t seconds
(a) $\frac{Mv_0}{M - \mu t}$ (b) $\frac{\mu t v_0}{M + \mu t}$
(c) $\frac{Mv_0}{M + \mu t}$ (d) None of these

64. A rocket of mass 6000 kg is fired from earth with exhaust speed of gases 1 km/s. What should be the rate of ejection of gas, so that it moves with initial acceleration of 20 m/s^2? (take $g = 10\,\text{m/s}^2$)
(a) 60 kg/s (b) 180 kg/s
(c) 200 kg/s (d) 120 kg/s

65. A rocket with a lift-off mass 3.5×10^4 kg is blasted upwards with an initial net acceleration of $10\,\text{m/s}^2$. Then, the initial thrust of the blast is
(a) 3.5×10^5 N (b) 7.0×10^5 N
(c) 14.0×10^5 N (d) 17.5×10^5 N

Collision

66. In an elastic collision
(a) both momentum and KE are conserved
(b) only momentum is conserved
(c) only KE is conserved
(d) Neither KE nor momentum is conserved

67. A bullet of mass m and velocity v is fired into a block of mass M and sticks to it. The final velocity of the system equals
(a) $\frac{M}{m + M} \cdot v$ (b) $\frac{m}{m + M} \cdot v$
(c) $\frac{m + M}{m} \cdot v$ (d) None of these

68. Two perfectly elastic particles A and B of equal mass travelling along the line joining them with velocities 15 m/s and 10 m/s. After collision, their velocities will be
(a) 10 m/s, 10 m/s
(b) 15 m/s, 15 m/s
(c) 10 m/s, 15 m/s
(d) 15 m/s, 10 m/s

69. A body of mass m moving with velocity v collides head on with another body of mass $2m$ which is initially at rest. The ratio of KE of colliding body before and after collision will be

A	B
(i) 0	25
(ii) 5	20
(iii) 10	15
(iv) 20	5

(a) 1 : 1 (b) 2 : 1
(c) 4 : 1 (d) 9 : 1

70. A block of mass m moving at a velocity v collides with another block of mass $2m$ at rest. The lighter block comes to rest after collision. Find the coefficient of restitution.
(a) $\frac{1}{2}$ (b) 1
(c) $\frac{1}{3}$ (d) $\frac{1}{4}$

71. A steel ball strikes a steel plate at an angle θ with the vertical. If the coefficient of restitution is e, the angle at which the rebound will take place is

(a) θ
(b) $\tan^{-1}\left[\frac{\tan\theta}{e}\right]$
(c) $e\tan\theta$
(d) $\tan^{-1}\left[\frac{e}{\tan\theta}\right]$

72. Two ice skaters A and B approach each other at right angles. Skater A has a mass 30 kg and velocity 1 m/s skater B has a mass 20 kg and velocity 2 m/s. They meet and cling together. Their final velocity of the couple is

(a) 2 m/s (b) 1.5 m/s (c) 1 m/s (d) 2.5 m/s

73. The collision of two balls of equal mass takes place at the origin of coordinates. Before collision, the components of velocities are ($v_x = 50$ cm/s, $v_y = 0$) and ($v_x = -40$ cm/s and $v_y = 30$ cm/s). The first ball comes to rest after collision. The velocity components (v_x and v_y respectively) of the second ball are

(a) 10 and 30 cm/s
(b) 30 and 10 cm/s
(c) 5 and 15 cm/s
(d) 15 and 5 cm/s

74. If a body of mass m collides head on, elastically with velocity u with another identical body at rest. After collision velocity of the second body will be

(a) zero
(b) u
(c) $2u$
(d) data insufficient

75. Two balls of equal mass have a head on collision with speed 4 m/s each travelling in opposite directions. If the coefficient of restitution is 1/2, the speed of each ball after impact will be

(a) 1 m/s
(b) 2 m/s
(c) 3 m/s
(d) data insufficient

76. A ball hits the floor and rebounds after an inelastic collisions. In this case

(a) the momentum of the ball just after the collision is the same as that just before the collision
(b) the mechanical energy of the ball remains the same in the collision
(c) the total momentum of the ball and the earth is conserved
(d) the total energy of the ball and the earth is conserved

77. Two vehicles of equal masses are moving with same speed v on two roads inclined at an angle θ. They collide inelastically at the junction and, then move together. The speed of the combination is

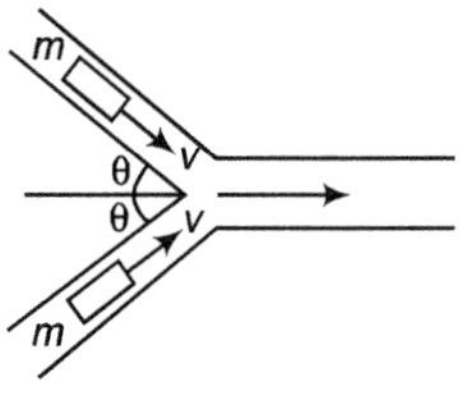

(a) $v\cos\theta$ (b) $2v\cos\theta$ (c) $\frac{3}{2}v\cos\theta$ (d) $\frac{v}{2}\cos\frac{\theta}{2}$

78. A sphere of mass m moving with a constant velocity u hits another stationary sphere of same mass. If e is the coefficient of restitution, then ratio of velocities of the two spheres $\frac{v_1}{v_2}$ after collision will be

(a) $\frac{1-e}{1+e}$ (b) $\frac{1+e}{1-e}$ (c) $\frac{e+1}{e-1}$ (d) $\frac{e-1}{e+1}$

79. A smooth sphere of mass M moving with velocity u directly collides elastically with another sphere of mass m at rest. After collision their final velocities are V and v respectively. The value of v is

(a) $\frac{2uM}{m}$ (b) $\frac{2um}{M}$ (c) $\frac{2u}{1+\frac{m}{M}}$ (d) $\frac{2u}{1+\frac{M}{m}}$

80. A metal ball falls from a height of 32 m on a steel plate. If the coefficient of restitution is 0.5 to what height will the ball rise after second bounce?

(a) 2 m
(b) 4 m
(c) 8 m
(d) 16 m

81. A body of mass m_1 moving with a velocity 3 ms^{-1} collides with another body at rest of mass m_2. After collision the velocities of the two bodies are 2 ms^{-1} and 5 ms^{-1} respectively along the direction of motion of m_2. The ratio $\frac{m_1}{m_2}$ is

(a) $\frac{5}{12}$ (b) 5 (c) $\frac{1}{5}$ (d) $\frac{12}{5}$

82. The bob A of a simple pendulum is released when the string makes an angle of 45° with the vertical. It hits another bob B of the same material and same mass kept at rest on a table. If the collision is elastic, then

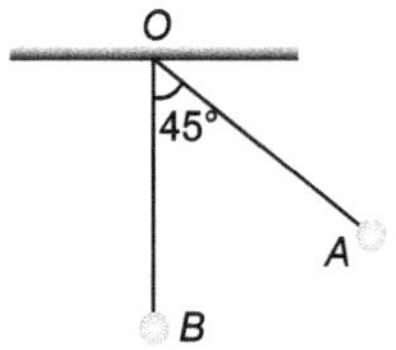

(a) both A and B rise to the same height
(b) both A and B come to rest at B
(c) both A and B move with the same velocity of A
(d) A comes to rest and B moves with the velocity of A

83. A body of mass M_1 collides elastically with another body of mass M_2 at rest. There is maximum transfer of energy when

(a) $M_1 > M_2$
(b) $M_1 < M_2$
(c) $M_1 = M_2$
(d) same for all values of M_1 and M_2

84. A mass m moves with a velocity v and collides inelastically with another identical mass. After collision, the 1st mass moves with velocity $\frac{v}{\sqrt{3}}$ in a direction perpendicular to the initial direction of motion. Find the speed of the 2nd mass after collision

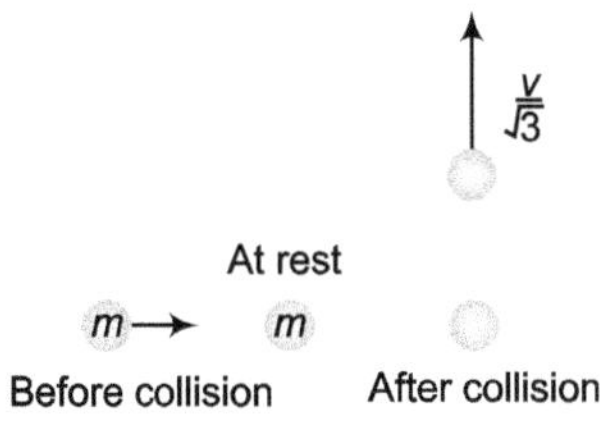

(a) $\frac{2}{\sqrt{3}}v$ (b) $\frac{v}{\sqrt{3}}$

(c) v (d) $\sqrt{3}v$

85. Two identical balls marked 2 and 3, in contact with each other and at rest on a horizontal frictionless table are hit head on by another identical ball marked 1 moving initially with a speed v as shown in figure. Assume the collision is elastic, then mark out the correct statement.

(a) Ball 2 comes to rest and ball 2 and 3 move with speed $\frac{v}{2}$ each

(b) Ball 1 and 2 come to rest and ball 3 moves with speed v

(c) Each ball moves with speed $\frac{v}{3}$

(d) None of the above

86. Two balls of equal masses have a head on collision with speed 6 m/s each. If the coefficient of restitution is 1/3, the speed of each ball after impact will be

(a) 18 m/s
(b) 2 m/s
(c) 6 m/s
(d) data insufficient

87. The two diagrams show the situations before and after a collision between two spheres A and B of equal radii moving along the same straight line on a smooth horizontal surface. The coefficient of restitution e is

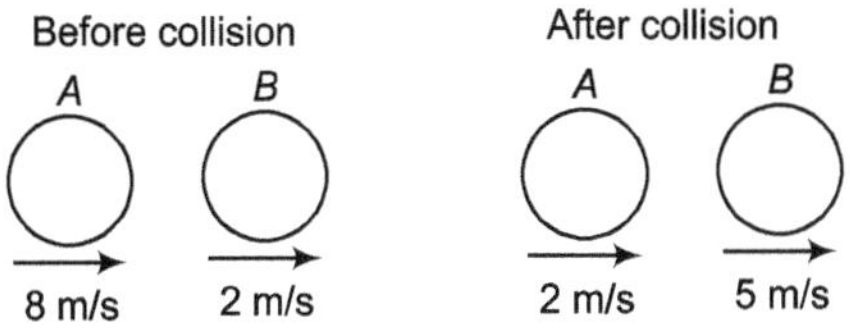

(a) $\frac{1}{3}$ (b) $\frac{1}{2}$

(c) $\frac{2}{3}$ (d) $\frac{3}{4}$

Miscellaneous Problems

88. A mass m with velocity u strikes a wall normally and returns with the same speed. What is magnitude of the change in momentum of the body when it returns

(a) 4 mu (b) mu (c) 2 mu (d) zero

89. Consider the following two statements.

A. Linear momentum of the system remains constant.
B. Centre of mass of the system remains at rest.

(a) A implies B and B implies A
(b) A does not imply B and B does not imply A
(c) A implies B but B does not imply A
(d) B implies A but A does not imply B

90. Consider the following two statements.

A. Linear momentum of a system of particles is zero.
B. Kinetic energy of a system of particles is zero.

(a) A implies B and B implies A
(b) A does not imply B and B does not imply A
(c) A implies B but B does not imply A
(d) B implies A but A does not imply B

91. A particle of mass m has momentum p. Its kinetic energy will be

(a) mp (b) p^2m (c) $\frac{p^2}{m}$ (d) $\frac{p^2}{2m}$

92. A machine gun fires a steady stream of bullets at the rate of n per minute into a stationary target in which the bullets get embedded. If each bullet has a mass m and arrives at the target with a velocity v, the average force on the target is

(a) 60 mnv (b) $\frac{60\,v}{mn}$ (c) $\frac{mn\,v}{60}$ (d) $\frac{m\,v}{60n}$

93. 10000 small balls, each weighing 1 g, strikes one square cm of area per second with a velocity 100 m/s in a normal direction and rebound with the same velocity. The value of pressure on the surface will be

(a) 2×10^3 N/m^2 (b) 2×10^5 N/m^2

(c) 10^7 N/m^2 (d) 2×10^7 N/m^2

94. A particle is projected from a point at an angle with the horizontal. At any instant t, if p is the linear momentum and E the kinetic energy, then which of the following graph is/are correct?

(a)
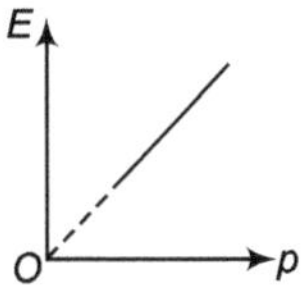

(b)
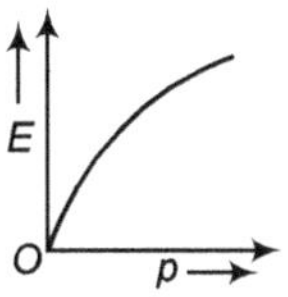

(c)
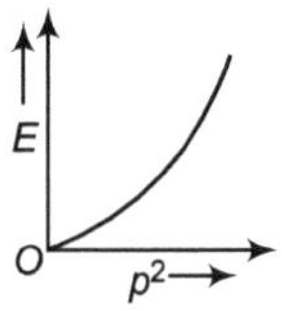

(d)
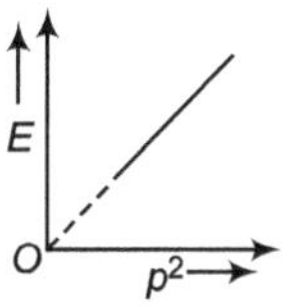

95. A 2 kg block of wood rests on a long table top. A 5 g bullet moving horizontally with a speed of 150 m/s is shot into the block and sticks to it. The block then slides 2.7 m along the table top and comes to a stop. The force of friction between the block and the table is

(a) 0.052 N (b) 3.63 N (c) 2.50 N (d) 1.04 N

96. A projectile of mass m is fired with velocity v from a point P, as shown in the figure. Neglecting air resistance, the magnitude of the change in momentum between the points P and arriving at Q is

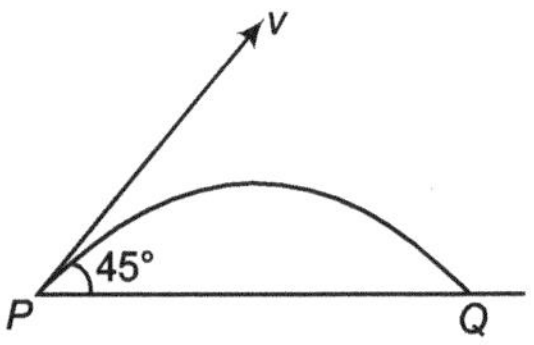

(a) zero (b) $\frac{mv}{\sqrt{2}}$ (c) $mv\sqrt{2}$ (d) $2mv$

97. If the KE of a body is increased by 300%, its momentum will increase by

(a) 100% (b) 150% (c) $\sqrt{300}$% (d) 175%

98. If the linear momentum is increased by 50%, the kinetic energy will increase by

(a) 50% (b) 100% (c) 125% (d) 25%

99. If the kinetic energy of a body increases by 0.1%, the percent increase of its momentum will be

(a) 0.05% (b) 0.1% (c) 1.0% (d) 10%

100. A particle of mass m at rest is acted upon by a constant force F for a time t. Its kinetic energy after an interval t is

(a) $\frac{F^2t^2}{m}$ (b) $\frac{F^2t^2}{2m}$ (c) $\frac{F^2t^2}{3m}$ (d) $\frac{Ft}{2m}$

101. Two identical blocks A and B, each of mass m resting on smooth floor are connected by a light spring of natural length L and spring constant K with the spring at its natural length. A third identical block C (mass m) moving with a speed v along the line joining A and B collides with A, the maximum compression in the spring is

(a) $v\sqrt{\frac{m}{2k}}$ (b) $m\sqrt{\frac{v}{2k}}$

(c) $\sqrt{\frac{mv}{k}}$ (d) $\frac{mv}{2k}$

102. A bullet moving with a speed of $100\,\text{ms}^{-1}$ can just penetrate two planks of equal thickness. Then, the number of such planks penetrated by the same bullet when the speed is doubled will be

(a) 4 (b) 8

(c) 6 (d) 10

103. A ball is projected vertically down with an initial velocity from a height of 20 m onto a horizontal floor. During the impact it loses 50% of its energy and rebounds to the same height. The initial velocity of its projection is

(a) 20 ms^{-1} (b) 15 ms^{-1} (c) 10 ms^{-1} (d) 5 ms^{-1}

104. A block C of mass m is moving with velocity v_0 and collides elastically with block A of mass m and connected to another block B of mass 2 m through spring of spring constant k. What is k if x_0 is compression of spring when velocity of A and B is same?

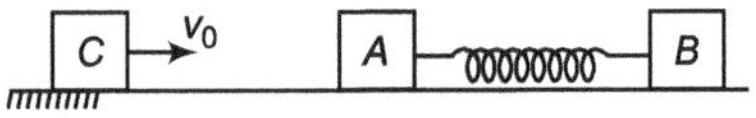

(a) $\frac{mv_0^2}{x_0^2}$ (b) $\frac{mv_0^2}{2x_0^2}$ (c) $\frac{3mv_0^2}{2x_0^2}$ (d) $\frac{2mv_0^2}{3x_0^2}$

105. A bullet of mass m is fired into a block of wood of mass M which hangs on the end of pendulum and gets embedded into it. When the bullet strikes the wooden block, the pendulum starts to swing with maximum rise R. Then, the velocity of the bullet is given by

(a) $\frac{M}{m+M}\sqrt{2gR}$ (b) $\frac{M+m}{m}\sqrt{2gR}$

(c) $\frac{M}{m}\sqrt{2gR}$ (d) None of these

[Level 2]

Only One Correct Options

1. A block A of mass M moving with speed u collides elastically with block B of mass m which is connected to block C of mass m with a spring.

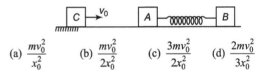

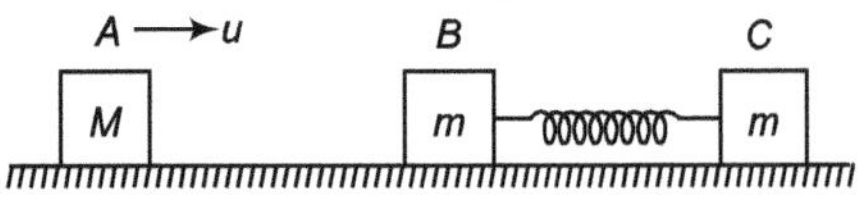

When the compression in spring is maximum the velocity of block C with respect to block A is (Neglect the friction everywhere)

(a) zero (b) $\left(\frac{M}{M+m}\right)u$

(c) $\left(\frac{m}{M+m}\right)u$ (d) $\left(\frac{m}{M}\right)u$

2. A particle A of mass m initially at rest slides down a height of 1.25 m on a frictionless ramp, collides with and sticks to an identical particle B of mass m at rest as shown in the figure.

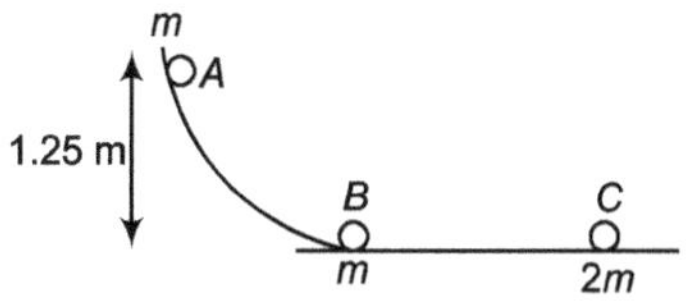

Then, particles A and B together collide elastically with particle C of mass 2 m at rest. The speed of particle C after the collision with combined body $(A + B)$ would be $(g = 10\ \text{m/s}^2)$

(a) 2.0 m/s (b) 1.25 m/s
(c) 2.5 m/s (d) 5 m/s

3. A bullet is fired from a gun. The force on the bullet is given by $F = 600 - 2 \times 10^5 t$

where, F is in newton and t in second. The force on the bullet becomes zero as soon as it leaves the barrel. What is the average impulse imparted to the bullet?

(a) 8 N-s (b) Zero (c) 0.9 N-s (d) 1.8 N-s

4. At high altitude, a body explodes at rest into two equal fragments with one fragment receiving horizontal velocity of 10 m/s. Time taken by the velocity vectors of the fragments to make 90° is $(g = 10\ \text{m/s}^2)$

(a) 0.5 s (b) 4 s (c) 2 s (d) 1 s

5. A train of mass M is moving on a circular track of radius R with constant speed v. The length of the train is half of the perimeter of the track. The linear momentum of the train will be

(a) $\pi M v$ (b) $\dfrac{2Mv}{\pi}$
(c) $\dfrac{\pi M v}{2}$ (d) Mv

6. A pendulum consists of a wooden bob of mass m and length l. A bullet of mass m_1 is fired towards the pendulum with a speed v_1 and it emerges from the bob with speed $\frac{v_1}{3}$. The bob just completes motion along a vertical circle. Then, v_1 is

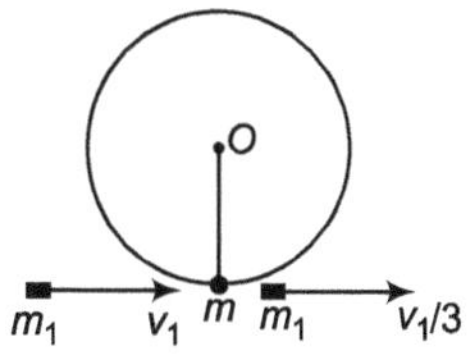

(a) $\dfrac{m}{m_1}\sqrt{5gl}$ (b) $\dfrac{3m}{2m_1}\sqrt{5gl}$
(c) $\dfrac{2}{3}\left(\dfrac{m}{m_1}\right)\sqrt{5gl}$ (d) $\left(\dfrac{m_1}{m}\right)\sqrt{gl}$

7. n elastic balls are placed at rest on a smooth horizontal plane which is circular at the ends with radius r as shown in the figure. The masses of the balls are m, $\frac{m}{2}$, $\frac{m}{2^2}, \ldots, \frac{m}{2^{n-1}}$ respectively. What is the minimum velocity which should be imparted to the first ball of mass m such that this nth ball will complete the vertical circle?

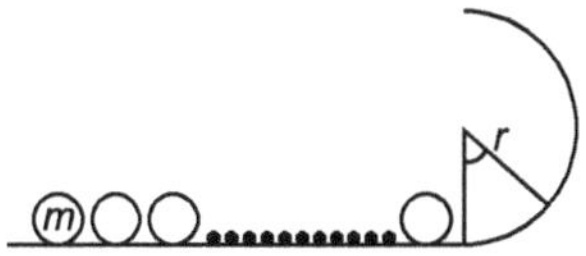

(a) $\left(\dfrac{3}{4}\right)^{n-1}\sqrt{5gr}$ (b) $\left(\dfrac{4}{3}\right)^{n-1}\sqrt{5gr}$
(c) $\left(\dfrac{3}{2}\right)^{n-1}\sqrt{5gr}$ (d) $\left(\dfrac{2}{3}\right)^{n-1}\sqrt{5gr}$

8. A particle of mass m moving with speed v hits elastically another stationary particle of mass 2 m inside a smooth horizontal circular tube of radius r. The time after which the second collision will take place is

(a) $\dfrac{2\pi r}{v}$ (b) $\dfrac{4\pi r}{v}$
(c) $\dfrac{3\pi r}{2v}$ (d) $\dfrac{\pi r}{v}$

9. Two blocks of mass m and $2m$ are kept on a smooth horizontal surface. They are connected by an ideal spring of force constant k. Initially the spring is unstretched. A constant force is applied to the heavier block in the direction shown in figure. Suppose at time t displacement of smaller block is x, then displacement of the heavier block at this moment would be

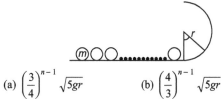

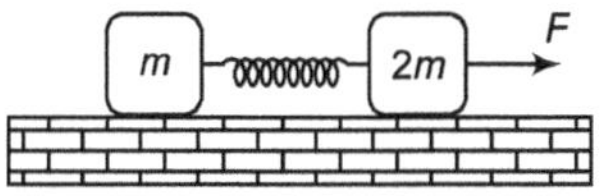

(a) $\dfrac{x}{2}$ (b) $\dfrac{Ft^2}{6m} + \dfrac{x}{3}$
(c) $\dfrac{x}{3}$ (d) $\dfrac{Ft^2}{4m} - \dfrac{x}{2}$

10. A bullet of mass 20 g and moving with 600 m/s collides with a block of mass 4 kg hanging with the string of length 0.4 m. What is velocity of bullet when it comes out of block, if block rises to height 0.2 m after collision?

(a) 200 m/s (b) 150 m/s
(c) 400 m/s (d) 300 m/s

11. A particle of mass 1 kg is thrown vertically upward with speed 100 m/s. After 5 s, it explodes into two parts. One part of mass 400 g comes back with speed 25 m/s, what is the velocity of other part just after explosion?

(a) 100 m/s upward (b) 600 m/s upward
(c) 100 m/s downward (d) 300 m/s upward

12. A ball falling freely from a height of 4.9 m/s, hits a horizontal surface. If $e = \frac{3}{4}$, then the ball will hit the surface, second time after

(a) 1.0 s (b) 1.5 s
(c) 2.0 s (d) 3.0 s

13. A disc of mass 10 *g* is kept floating horizontally by throwing 10 marbles per second against it from below. If the mass of each marble is 5 *g*. What will be velocity with which the marble are striking the disc? Assume that the marble strikes the disc normally and rebound downwards with the same speed.

(a) 2.98 m/s (b) 0.98 m/s
(c) 0.49 m/s (d) 1.96 m/s

14. If a man of mass *M* jumps to the ground from a height *h* and it moves a small distance *x* inside the ground, the average force acting on him from ground is

(a) $\frac{Mgh}{x}$ (b) $\frac{Mgx}{h}$
(c) $Mg\left(\frac{h}{x}\right)^2$ (d) None of these

15. A machine gun fires a bullet of mass 40 g with a velocity $1200\,\text{ms}^{-1}$. The man holding it, can exert a maximum force of 144 N on the gun. How many bullets can be fired per second at the most?

(a) One (b) Four
(c) Two (d) Three

16. A straight rod of length *L* has one of its ends at the origin and the other at $x = L$. If the mass per unit length of the rod is given by Ax, where *A* is constant, where is its mass centre?

(a) $\frac{L}{3}$ (b) $\frac{L}{2}$ (c) $\frac{2L}{3}$ (d) $\frac{3L}{4}$

17. In a one dimensional collision between two identical particle *A* and *B*. *B* is stationary and *A* has momentum *p* before impact. During impact *B* gives an impulse *J* to *A*. Then, coefficient of restitution between the two is

(a) $\frac{2J}{p} - 1$ (b) $\frac{2J}{p} + 1$
(c) $\frac{J}{p} + 1$ (d) $\frac{J}{p} - 1$

18. Three identical blocks *A, B* and *C* are placed on horizontal frictionless surface. The blocks *B* and *C* are at rest. But *A* is approaching towards *B* with a speed 10 m/s.

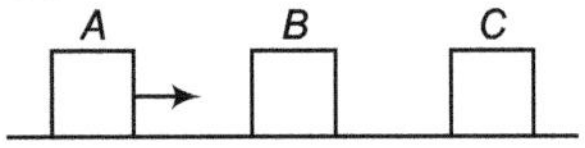

The coefficient of restitution for all collision is 0.5. The speed of the block *C* just after collision is approximately

(a) 5.6 m/s (b) 6.4 m/s (c) 3.2 m/s (d) 4.6 m/s

19. A particle of mass *m* moving with a velocity $(3\hat{\mathbf{i}} + 2\hat{\mathbf{j}})$ m/s collides with a stationary body of mass *M* and finally moves with a velocity $(-2\hat{\mathbf{i}} + \hat{\mathbf{j}})$ m/s. If $\frac{m}{M} = \frac{1}{13}$, then

(a) the impulse is $\pm m(5\hat{\mathbf{i}} + \hat{\mathbf{j}})$
(b) the velocity of the *M* is $\frac{1}{13}(5\hat{\mathbf{i}} + \hat{\mathbf{j}})$
(c) Both (a) and (b) are wrong
(d) Both (a) and (b) are correct

20. A small ball rolls off the top landing of the staircase. It strikes the mid-point of the first step and then the mid-point of the second step. The steps are smooth and identical in height and width. The coefficient of restitution between the ball and the first step is

(a) 1 (b) $\frac{3}{4}$ (c) $\frac{1}{2}$ (d) $\frac{1}{4}$

21. You are supplied with three identical rods of same length and mass. If the length of each rod is 2π. Two of them are converted into rings and then placed over the third rod as shown in figure. If point *A* is considered as origin of the coordinate system, the coordinate of the centre of mass will be (you may assume *AB* as *x*-axis of the coordinate system)

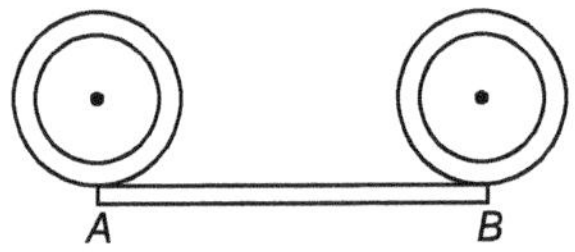

(a) $\left(\frac{\pi}{2}, \frac{1}{3}\right)$ (b) $\left(\frac{\pi}{2}, \frac{2}{3}\right)$ (c) $\left(\pi, \frac{1}{3}\right)$ (d) $\left(\pi, \frac{2}{3}\right)$

22. A mass of 10 g moving horizontally with a velocity of 100 cm/s strikes a pendulum bob of mass 10 g. Length of string is 50 cm. The two masses stick together. The maximum height reached by the system now is $(g = 10\,\text{m/s}^2)$

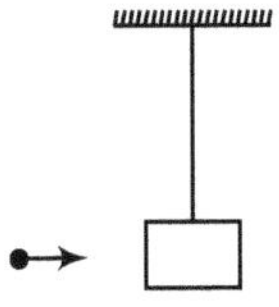

(a) 7.5 cm (b) 5 cm
(c) 2.5 cm (d) 1.25 cm

23. Both the blocks as shown in the given arrangement are given together a horizontal velocity towards right. If a_{CM} be the subsequent acceleration of the centre of mass of the system of blocks, then a_{CM} equals

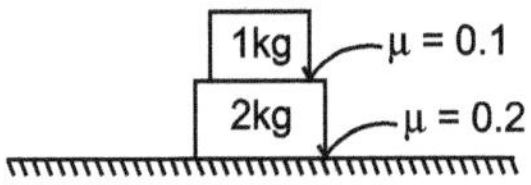

(a) zero (b) $\frac{5}{3}\,\text{m/s}^2$ (c) $\frac{7}{3}\,\text{m/s}^2$ (d) $2\,\text{m/s}^2$

24. An object comprises of a uniform ring of radius R and its uniform chord AB (not necessarily made of the same material) as shown in figure. Which of the following can not be the centre of mass of the object?

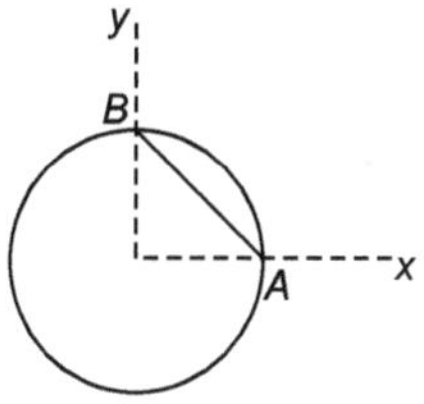

(a) $\left(\frac{R}{3}, \frac{R}{3}\right)$ (b) $\left(\frac{R}{\sqrt{2}}, \frac{R}{\sqrt{2}}\right)$

(c) $\left(\frac{R}{4}, \frac{R}{4}\right)$ (d) None of these

25. A girl throws a ball with initial velocity v at an inclination of 45°. The ball strikes the smooth vertical wall at a horizontal distance d from the girl and after rebounding returns to her hand. What is the coefficient of restitution between wall and the ball?

(a) $v^2 - gd$ (b) $\frac{gd}{v^2 - gd}$

(c) $\frac{gd}{v^2}$ (d) $\frac{v^2}{gd}$

More than One Correct Option

1. A particle of mass m, moving with velocity v collides a stationary particle of mass $2m$. As a result of collision, the particle of mass m deviates by 45° and has final speed of $\frac{v}{2}$. For this situation, mark out the correct statement (s).

(a) The angle of divergence between particles after collision is $\frac{\pi}{2}$

(b) The angle of divergence between particles after collision is less than $\frac{\pi}{2}$

(c) Collision is elastic

(d) Collision is inelastic

2. A pendulum bob of mass m connected to the end of an ideal string of length l is released from rest from horizontal position as shown in the figure. At the lowest point, the bob makes an elastic collision with a stationary block of mass $5m$, which is kept on a frictionless surface. Mark out the correct statement(s) for the instant just after the impact.

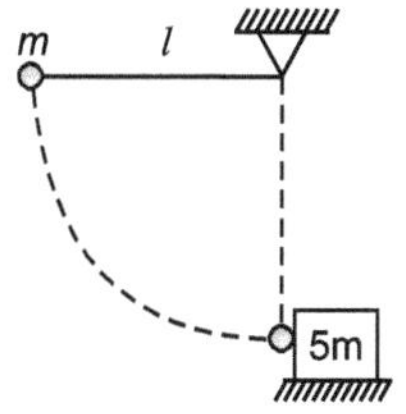

(a) Tension in the string is $\frac{17}{9}mg$

(b) Tension in the string is $3\,mg$

(c) The velocity of the block is $\frac{\sqrt{2gl}}{3}$

(d) The maximum height attained by the pendulum bob after impact is (measured from the lowest position) $\frac{4l}{9}$

3. A particle of mass m strikes a horizontal smooth floor with a velocity u making an angle θ with the floor and rebound with velocity v making an angle ϕ with the floor. The coefficient of restitution between the particle and the floor is e. Then,

(a) the impulse delivered by the floor to the body is $mu(1+e)\sin\theta$

(b) $\tan\phi = e\tan\theta$

(c) $v = u\sqrt{1-(1-e^2)\sin^2\theta}$

(d) the ratio of the final kinetic energy to the initial kinetic energy is $\cos^2\theta + e^2\sin^2\theta$

4. A particle of mass m moving with a velocity $(3\hat{\mathbf{i}} + 2\hat{\mathbf{j}})\,\text{ms}^{-1}$ collides with another body of mass M and finally moves with velocity $(-2\hat{\mathbf{i}} + \hat{\mathbf{j}})\,\text{ms}^{-1}$. Then, during the collision

(a) impulse received by m is $m(5\hat{\mathbf{i}} + \hat{\mathbf{j}})$

(b) impulse received by m is $m(-5\hat{\mathbf{i}} - \hat{\mathbf{j}})$

(c) impulse received M is $m(-5\hat{\mathbf{i}} - \hat{\mathbf{j}})$

(d) impulse received by M is $m(5\hat{\mathbf{i}} + \hat{\mathbf{j}})$

5. All surfaces shown in figure are smooth. System is released from rest. x and y components of acceleration of COM are

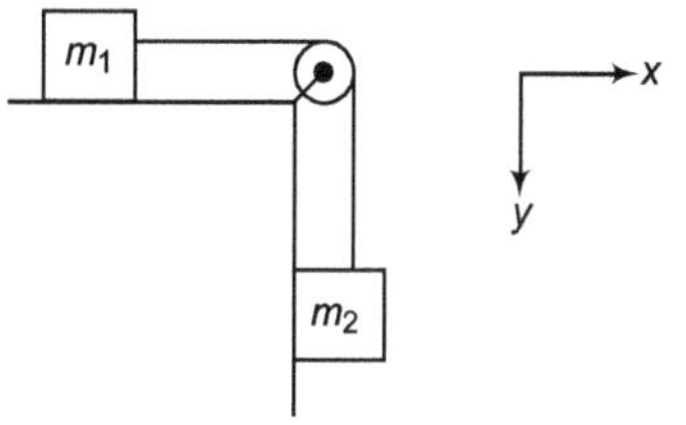

(a) $(a_{cm})_x = \frac{m_1 m_2 g}{m_1 + m_2}$ (b) $(a_{cm})_x = \frac{m_1 m_2 g}{(m_1 + m_2)^2}$

(c) $(a_{cm})_y = \left(\frac{m_2}{m_1 + m_2}\right)^2 g$ (d) $(a_{cm})_y = \left(\frac{m_2}{m_1 + m_2}\right) g$

6. A block of mass m is placed at rest on a smooth wedge of mass M placed at rest on a smooth horizontal surface.As the system is released

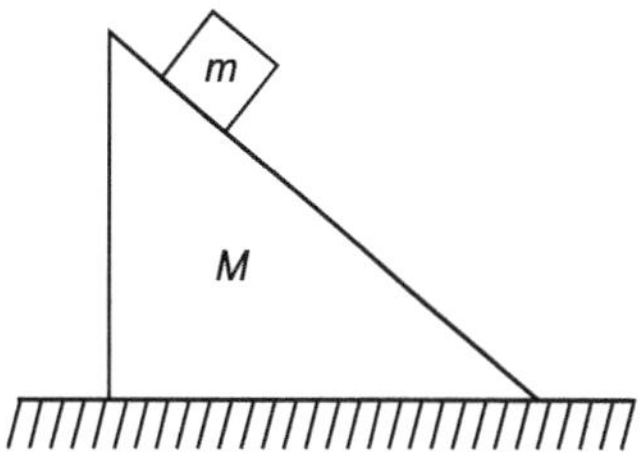

(a) the COM of the system remains stationary

(b) the COM of the system has an acceleration g vertically downward

(c) momentum of the system is conserved along the horizontal direction

(d) acceleration of COM is vertically downward $(a < g)$

7. In the figure shown, coefficient of restitution between A and B is $e = \frac{1}{2}$, then

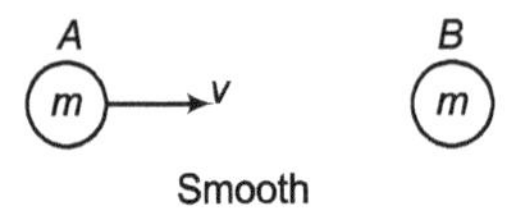

(a) velocity of B after collision is $\frac{v}{2}$

(b) impulse between two during collision is $\frac{3}{4} mv$

(c) loss of kinetic energy during the collision is $\frac{3}{16} mv^2$

(d) loss of kinetic energy during the collision is $\frac{1}{4} mv^2$

8. In case of rocket propulsion, choose the correct options.

(a) Momentum of system always remains constant
(b) Newton's third law is applied
(c) If exhaust velocity and rate of burning of mass is kept constant, then acceleration of rocket will go on increasing
(d) Newton's second law can be applied

Comprehension Based Questions

Passage I (Q. 1 to 2)

A block of mass 2 *kg is attached with a spring of spring constant* 4000 Nm^{-1} *and the system is kept on smooth horizontal table. The other end of the spring is attached with a wall. Initially spring is stretched by* 5 *cm from its natural position and the block is at rest. Now, suddenly an impulse of* 4 *kg-ms*$^{-1}$ *is given to the block towards the wall.*

1. Find the velocity of the block when spring acquires its natural length.

(a) 5 ms^{-1} (b) 3 ms^{-1}
(c) 6 ms^{-1} (d) None of these

2. Approximate distance travelled by the block when it comes to rest for a second time (not including the initial one) will be (Take $\sqrt{45} = 6.70$)

(a) 30 cm (b) 25 cm
(c) 40 cm (d) 20 cm

Passage II (Q. 3 to 7)

A uniform bar of length $12L$ *and mass* $48m$ *is supported horizontally on two smooth tables as shown in figure. A small moth (an insect) of mass* $8m$ *is sitting on end A of the rod and a spider (an insect) of mass* $16m$ *is sitting on the other end B. Both the insects moving towards each other along the rod with moth moving at speed* $2v$ *and the spider at half this speed (absolute).*

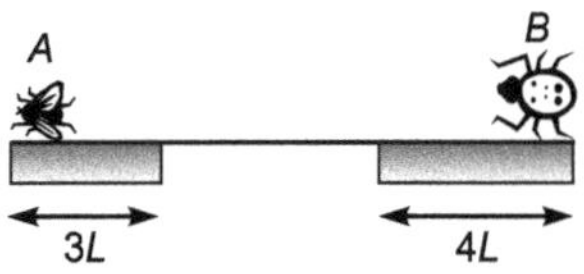

They meet at a point P on the rod and the spider eats the moth. After this the spider moves with a velocity $\frac{v}{2}$ *relative to the rod towards the end A. The spider takes negligible time in eating on the other insect. Also, let* $v = \frac{L}{T}$, *where T is a constant having value 4 s.*

3. Displacement of the rod by the time the insect meet the moth is

(a) $\frac{L}{2}$ (b) L (c) $\frac{3L}{4}$ (d) zero

4. The point P is at

(a) the centre of the rod
(b) the edge of the table supporting the end B
(c) the edge of the table supporting end A
(d) None of the above

5. The speed of the rod after the spider eats up the moth and moves towards A is

(a) $\frac{v}{2}$ (b) v
(c) $\frac{v}{6}$ (d) $2v$

6. After starting from end B of the rod the spider reaches the end A at a time

(a) 40 s (b) 30 s (c) 80 s (d) 10 s

7. By what distance the centre of mass of the rod shifts during this time?

(a) $\frac{8L}{3}$ (b) $\frac{4L}{3}$
(c) L (d) $\frac{L}{3}$

Assertion and Reason

Directions (Q. Nos. 1-23) *These questions consist of two statements each printed as Assertion and Reason. While answering these questions you are required to choose any one of the following five responses.*

(a) If both Assertion and Reason are correct and Reason is the correct explanation of Assertion
(b) If both Assertion and Reason are true but Reason is not the correct explanation of Assertion
(c) If Assertion is true but Reason is false
(d) If Assertion is false but Reason is true
(e) If both Assertion and Reason are false

1. Assertion In head inelastic collision, the final momentum is less than the initial momentum.

Reason For inelastic collision, $0 \le e < 1$. Hence, the magnitude of relative velocity of separation after collision is less than relative velocity of approach before collision.

2. **Assertion** A projectile gets exploded at its highest point. All the pieces get only horizontal velocities. The centre of mass will always fall at a point which is farther than the point where the projectile would have fallen in unexploded condition.

   Reason The weight of the projectile is the external force for projectile.

3. **Assertion** If a projectile explodes in mid air, linear momentum of centre of mass of different fragments remains constant.

   Reason When net force on a system of particles is zero, linear momentum of the system remains constant.

4. **Assertion** A rocket moves forward by pushing the surrounding air backwards.

   Reason It derives the necessary thrust to move forward according to Newton's third law of motion.

5. **Assertion** A body is thrown with a velocity u inclined to the horizontal at some angle. It moves along a parabolic path and falls to the ground. Linear momentum of the body, during its motion, will remain conserved.

   Reason Throughout the motion of the body, a constant force acts on it.

6. **Assertion** In an elastic collision between two bodies, the magnitude of relative velocity of the bodies after collision is equal to the magnitude of relative velocity before collision.

   Reason In an elastic collision, the linear momentum of the system is conserved.

7. **Assertion** When a body dropped from a height explodes in mid air, its centre of mass keeps moving in vertically downward direction.

   Reason Explosion occurs under internal forces only. External force is zero.

8. **Assertion** The centre of mass of an electron and proton, when released from rest moves faster towards proton.

   Reason Proton is heavier than electron.

9. **Assertion** The relative velocity of the two particles in head-on elastic collision is unchanged both in magnitude and direction.

   Reason The relative velocity is unchanged in magnitude but gets reversed in direction.

10. **Assertion** A given force applied in turn to a number of different masses by cause the same rate of change in momentum in each but not the same acceleration to all.

    Reason $\mathbf{F} = \dfrac{d\mathbf{p}}{dt}$ and $\mathbf{a} = \dfrac{\mathbf{F}}{m}$

11. **Assertion** In an elastic collision between two bodies, the relative speed of the bodies after collision is equal to the relative speed before the collision.

    Reason In an elastic collision, the linear momentum of the system is conserved.

12. **Assertion** A moving ball having an inelastic collision with a moving wall can have larger kinetic energy after collision.

    Reason During a collision between two bodies, transfer of energy may takes place from one body to another.

13. **Assertion** A rocket launched vertically upward explodes at the highest point it reaches. The explosion produces three fragments with non-zero initial velocity. Then, the initial velocity vectors of all the three fragments are in one plane.

    Reason For sum of momentum of three particles, all the three momentum vectors must be coplanar.

14. **Assertion** If net force on a system is zero, then momentum of every individual body remains constant.

    Reason If momentum of a system is constant, then kinetic energy of the system may change.

15. **Assertion** Two blocks of masses m_A and m_B $(> m_A)$ are thrown towards each other with same speed over a rough ground. The coefficient of friction of both the blocks with ground is same. Initial velocity of CM is towards left.

    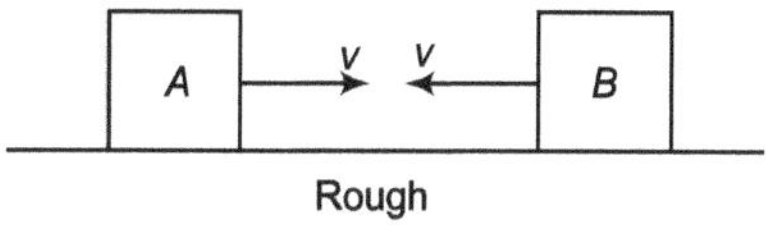

    Reason Initial acceleration of centre of mass is towards right.

16. **Assertion** In inelastic collision linear momentum of system does not remain constant during collision. But before collision and after collision it is constant.

    Reason In elastic collision, momentum remains constant during collision also.

17. **Assertion** If a projectile explodes in mid air, then no external force acts on the projectile during explosion.

    Reason Centre of mass in this case follows the same path.

18. **Assertion** Two identical spherical spheres are half filled with two liquids of densities ρ_1 and ρ_2 $(> \rho_1)$. The centre of mass of both the spheres lie at same level.

    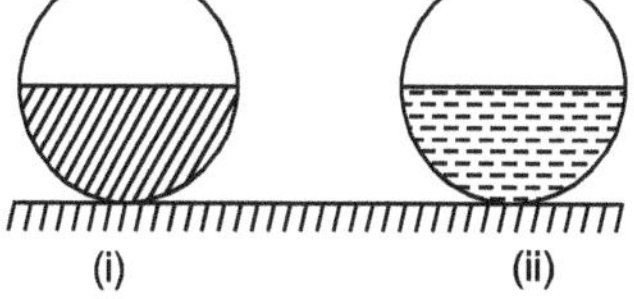

    Reason The centre of mass will lie at centre of the sphere.

19. Assertion Two spherical bodies of mass ratio 1 : 2 travel towards each other (starting from rest) under the action of their mutual gravitational attraction. Then, the ratio of their kinetic energies at any instant is 2 : 1.

Reason At any instant their momenta are same.

20. Assertion Two bodies moving in opposite directions with same magnitude of linear momentum collide each other. Then, after collision both the bodies will come to rest.

Reason Linear momentum of the system of bodies is zero.

21. Assertion Two blocks *A* and *B* are connected at the two ends of an ideal spring as shown in figure. Initially spring was relaxed. Now, block *B* is pressed. Linear momentum of the system will not remain constant till the spring reaches its initial natural length.

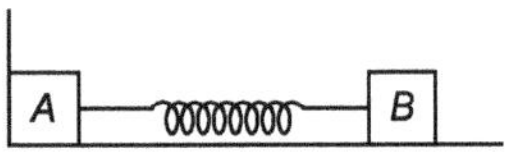

Reason An external force will act from the wall on block *A*.

22. Assertion In the figure shown, linear momentum of system (of blocks *A* and *B*) moves towards right during motion of block *A* over the block *B*.

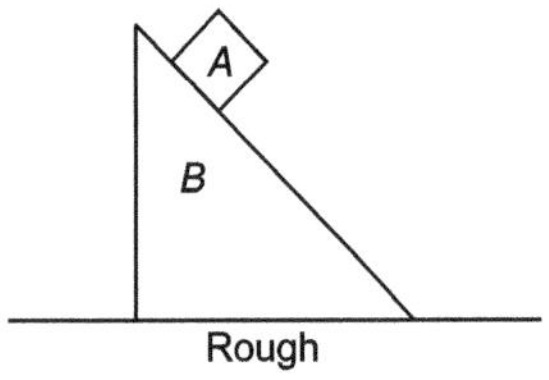

Reason Force of friction will act on *B* towards right (from ground).

23. Assertion Centre of mass and centre of gravity of a body will coincide on moon.

Reason There is no gravity on moon.

Match the Columns

1. If net force on a system of particles is zero, then match the following columns.

Column I	Column II
(A) Acceleration of centre of mass	(p) Constant
(B) Momentum of centre of mass	(q) Zero
(C) Velocity of centre of mass	(r) May be zero
(D) Velocity of an individual particle of the system	(s) May be constant

2. In the arrangement shown in figure match the following

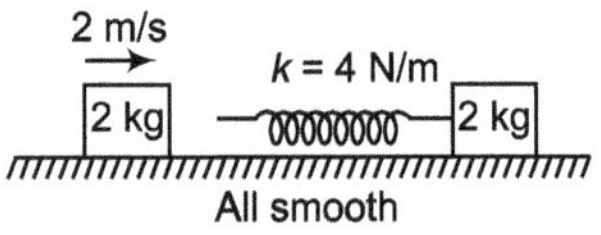

Column I	Column II
(A) Velocities of centre of mass	(p) 2 SI unit
(B) Velocity of combined mass when compression in the spring is maximum	(q) 1 SI unit
(C) Maximum compression in the spring	(r) 4 SI unit
(D) Maximum potential energy stored in the spring	(s) 0.5 SI unit

3. A particle of mass *m*, kinetic energy *K* and momentum *p* collides head on elastically with another particle of mass 2 *m* at rest. Match the following columns (after collision).

Column I	Column II
(A) Momentum of first particle	(p) $3/4\ p$
(B) Momentum of second particle	(q) $-K/9$
(C) Kinetic energy of first particle	(r) $-p/3$
(D) Kinetic energy of second particle	(s) $8K/9$

4. Match the following columns.

(p = momentum of particle, K = kinetic energy of particle)

Column I	Column II
(A) p is increased by 200%, corresponding change in K	(p) 800%
(B) K is increased by 300%, corresponding change in p	(q) 200%
(C) p is increased by 1%, corresponding change in K	(r) 0.5%
(D) K is increased by 1%, corresponding change in p	(s) 2%

5. Four point masses are placed at four corners of a square of side 4 m as shown in the figure. Match the following columns.

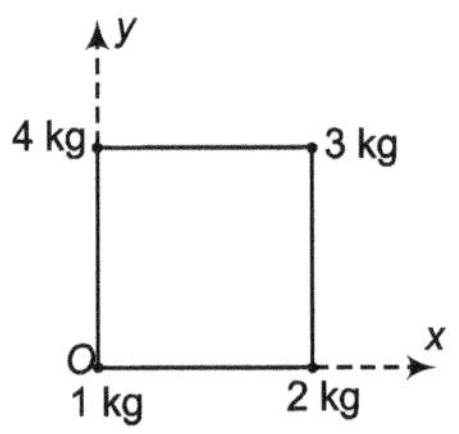

Column I	Column II
(A) *x*-coordinate of centre of mass of 4 kg and 2 kg	(p) (7/2) m
(B) *x*-coordinate of centre of mass of 4 kg, 2 kg and 3 kg	(q) 4/3 m
(C) *y*-coordinate of centre of mass of 1 kg, 4 kg and 3 kg	(r) 3 m
(D) *y*-coordinate of centre of mass of 1 kg and 3 kg	(s) (20/9) m

6. In a two block system shown in figure match the following.

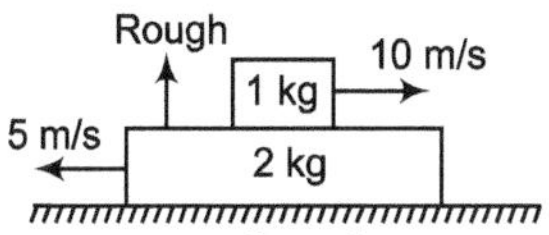

Column I	Column II
(A) Velocity of centre of mass	(p) Keep on changing all the time
(B) Momentum of centre of mass	(q) First decreases then become zero
(C) Momentum of 1 kg block	(r) Zero
(D) Kinetic energy of 2 kg block	(s) Constant

7. A particle of mass 1 kg is projected upwards with velocity 60 m/s. Another particle of mass 2 kg is just dropped from a certain height. After 2 s, match the following

[Take $g = 10 \text{m/s}^2$]

Column I	Column II
(A) Acceleration of CM	(p) Zero
(B) Velocity of CM	(q) 10 SI unit
(C) Displacement of CM	(r) 20 SI unit
	(s) None

8. In the diagram shown in figure mass of both the balls is same. Match the following columns.

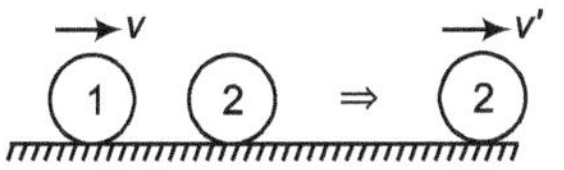

Column I	Column II
(A) For $v' = v$	(p) $e = 0$
(B) For $v' = v/2$	(q) $e = 1$
(C) For $v' = 3/4\, v$	(r) $e = 1/2$
	(s) Data is insufficient

9. A particle of mass 1 kg has velocity $\mathbf{v}_1 = (2t)\hat{\mathbf{i}}$ and another particle of mass 2 kg has velocity $\mathbf{v}_2 = (t^2)\hat{\mathbf{j}}$. Match the following columns.

Column I	Column II
(A) Net force on centre of mass at 2 s	(p) $\frac{20}{9}$ unit
(B) Velocity of centre of mass at 2 s	(q) $\sqrt{68}$ unit
(C) Displacement of centre of mass in 2 s	(r) $\sqrt{80}/3$ unit
	(s) None

Entrance Gallery

2014

1. In elastic collision, **[Kerala CEE]**

(a) both momentum and kinetic energy are conserved
(b) neither momentum nor kinetic energy is conserved
(c) only momentum is conserved
(d) only kinetic energy is conserved
(e) forces involved in the interaction and non-conservative

2. A smooth massless string passes over a smooth fixed pulley. Two masses m_1 and m_2, $(m_1 > m_2)$ are tied at the two ends of the string. The masses are allowed to move under gravity starting from rest. The total external force acting on the two masses is **[WB JEE]**

(a) $(m_1 + m_2)g$ (b) $\frac{(m_1 - m_2)^2}{m_1 + m_2} g$

(c) $(m_1 - m_2)g$ (d) $\frac{(m_1 + m_2)^2}{m_1 - m_2} g$

2013

3. A particle of mass m is attached to one end of a massless spring of force constant k, lying on a frictionless horizontal plane. The other end of the spring is fixed. The particle starts moving horizontally from its equilibrium position at times, $t = 0$ with an initial velocity u_0. When the speed of the particle is $0.5\,u_0$, it collides elastically with a rigid wall. After this collision, **[JEE Advanced]**

(a) the speed of the particle when it returns to its equilibrium position is u_0
(b) the time at which the particle passes through the equilibrium position for the first time is $t = \pi\sqrt{\frac{m}{k}}$
(c) the time at which the maximum compression of the spring occurs is $t = \frac{4\pi}{3}\sqrt{\frac{m}{k}}$
(d) the time at which the particle passes through the equilibrium position for the second time is $t = \frac{5\pi}{3}\sqrt{\frac{m}{k}}$

4. A bob of mass m, suspended by a string of length l_1, is given a minimum velocity required to complete a full circle in the vertical plane. At the highest point, it collides elastically with another bob of mass m suspended by a string of length l_2, which is initially at rest. Both the strings are massless and inextensible. If the second bob, after collision acquires the minimum speed required to complete a full circle in the vertical plane, the ratio $\frac{l_1}{l_2}$ is **[JEE Advanced]**

(a) 1 (b) 3 (c) 4 (d) 5

5. This question has Statement I and Statement II, of the four choices given after the statements, choose the one that best describes the two statements. [JEE Main]

Statement I A point particle of mass m moving with speed v collides with stationary point particle of mass M. If the maximum energy loss possible is given as,

$$f\left(\frac{1}{2}mv^2\right), \text{ then } f=\left(\frac{m}{M+m}\right).$$

Statement II Maximum energy loss occurs when the particles get stuck together as a result of the collision.

(a) Statement I is correct, Statement II is correct, and Statement II is the correct explanation of Statement I
(b) Statement I is correct, Statement II is correct, but Statement II is incorrect explanation of Statement I
(c) Statement I is correct, Statement II is incorrect
(d) Statement I is incorrect, Statement II is correct

2012

Paragraph

The general motion of a rigid body can be considered to be a combination of (i) a motion of its centre of mass about an axis and (ii) its motion about an instantaneous axis passing through the centre of mass.

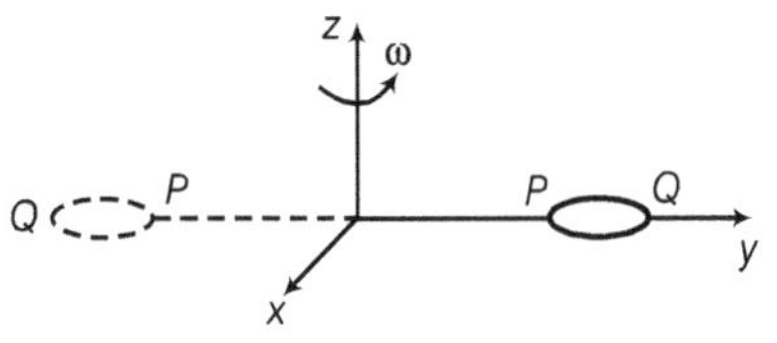

These axes need not be stationary. Consider, for example, a thin uniform disc welded (rigidly fixed) horizontally at its rim to a massless stick, as shown in the figure.

When the disc-stick system is rotated about the origin on a horizontal frictionless plane with angular speed ω, the motion at any instant can be taken as a combination of (i) a rotation of the centre of mass of the disc about the z-axis, and (ii) a rotation of the disc through an instantaneous vertical axis passing through its centre of mass (as is seen from the changed orientation of points P and Q).

Both these motions have the same angular speed ω this case.

Now, consider two similar systems as shown in the figure.

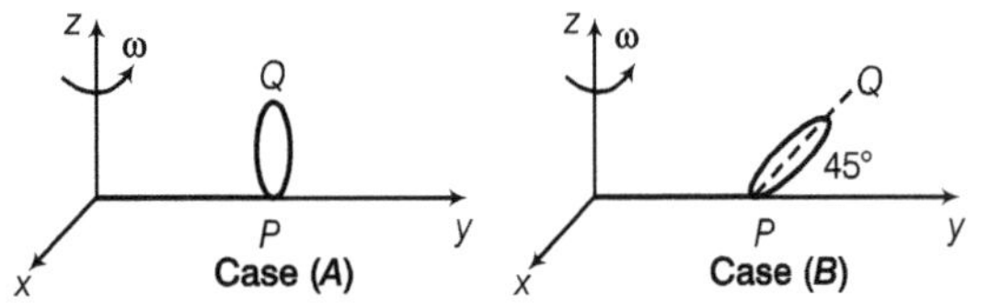

Case (*A*) the disc with its face vertical and parallel to x-z plane,

Case (*B*) the disc with its face making an angle of 45° with x-y plane and its horizontal diameter parallel to x-axis. In both the cases, the disc is welded at point P and the systems are rotated with constant angular speed ω about the z-axis.

6. Which of the following statements regarding the angular speed about the instantaneous axis (passing through the centre of mass) is correct? [IIT JEE]

(a) It is $\sqrt{2}\omega$ for both the cases
(b) It is ω for case (A) and $\omega/\sqrt{2}$ for case (B)
(c) It is ω for case (A) and $\sqrt{2}\omega$ for case (B)
(d) It is ω for both the cases

7. Which of the following statements about the instantaneous axis (passing through the centre of mass) is correct?

(a) It is vertical for both the cases A and B [IIT JEE]
(b) It is vertical for case A and is at 45° to the x-z plane and lies in the plane of the disc for case B
(c) It is horizontal for case A and is at 45° to the x-z plane and is normal to the plane of the disc for case B
(d) It is vertical for case A and is at 45° to the x-z plane and is normal to the plane of the disc for case B

8. Two solid cylinders P and Q of same mass and same radius start rolling down a fixed inclined plane from the same height at the same time. Cylinder P has most of its mass concentrated near its surface, while Q has most of its mass concentrated near the axis. Which statements (s) is (are) correct? [IIT JEE]

(a) Both cylinders P and Q reach the ground at the same time
(b) Cylinder P has larger linear acceleration than cylinder Q
(c) Both cylinders reach the ground with same translational kinetic energy
(d) Cylinder Q reaches the ground with larger angular speed

2011

9. A thin ring of mass 2 kg and radius 0.5 m is rolling without slipping on a horizontal plane with velocity 1 m/s. A small ball of mass 0.1 kg, moving with velocity 20 m/s in the opposite direction, hits the ring at a height of 0.75 m and goes vertically up with velocity 10 m/s. Immediately after the collision, [IIT JEE]

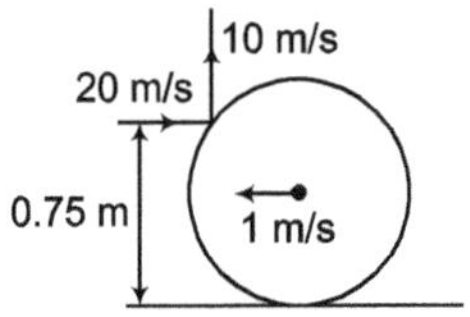

(a) the ring has pure rotation about its stationary CM
(b) the ring comes to a complete stop
(c) friction between the ring the ground is to the left
(d) there is no friction between the ring and the ground

10. A ball of mass 0.2 kg rests on a vertical post of height 5 m. A bullet of mass 0.01 kg, travelling with a velocity v m/s in a horizontal direction, hits the centre of the ball. After the collision, the ball and bullet travel independently. The ball hits the ground at a distance of 20 m and the bullet at a distance of 100 m from the foot of the post. The initial velocity v of the bullet is **[IIT JEE]**

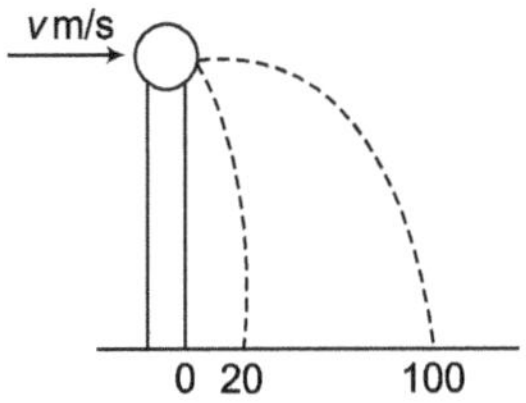

(a) 250 m/s (b) $250\sqrt{2}$ m/s
(c) 400 m/s (d) 500 m/s

11. A stationary bomb explodes into three pieces. One piece of 2 kg mass moves with a velocity of $8\,ms^{-1}$ at right angles to the other piece of mass 1 kg moving with a velocity of $12\,ms^{-1}$. If the mass of the third piece is 0.5 kg, then its velocity is **[Kerala CEE]**

(a) $10\,ms^{-1}$ (b) $20\,ms^{-1}$
(c) $30\,ms^{-1}$ (d) $40\,ms^{-1}$
(e) $50\,ms^{-1}$

12. A cricket ball of mass 0.25 kg with speed 10 m/s collides with a bat and returns with same speed with in 0.01s. The force acted on bat is **[WB JEE]**

(a) 25N (b) 50N (c) 250N (d) 500N

2010

13. A point mass of 1 kg collides elastically with a stationary point mass of 5 kg. After their collision, the 1 kg mass reverses its direction and moves with a speed of $2\,ms^{-1}$. Which of the following statement(s) is/are correct for the system of these two masses? **[IIT JEE]**

(a) Total momentum of the system is 3 kg-ms^{-1}
(b) Momentum of 5 kg mass after collision is 4 kg-ms^{-1}
(c) Kinetic energy of the centre of mass is 0.75 J
(d) Total kinetic energy of the system is 4 J

14. Statement I Two particles moving in the same direction do not lose all their energy in a completely inelastic collision.

Statement II Principle of conservation of momentum holds true for all kinds of collisions. **[AIEEE]**

(a) Statement I is correct, Statement II is correct, Statement II is the correct explanation of Statement I
(b) Statement I is correct Statement II is correct, Statement II incorrect explanation of Statement I
(c) Statement I is incorrect, Statement II is correct
(d) Statement I is correct, Statement II is incorrect

15. A T shaped object with dimensions shown in the figure, is lying on a smooth floor. A force **F** is applied at the point P parallel to AB, such that the object has only the translational motion without rotation. Find the location of P with respect to C. **[AIEEE]**

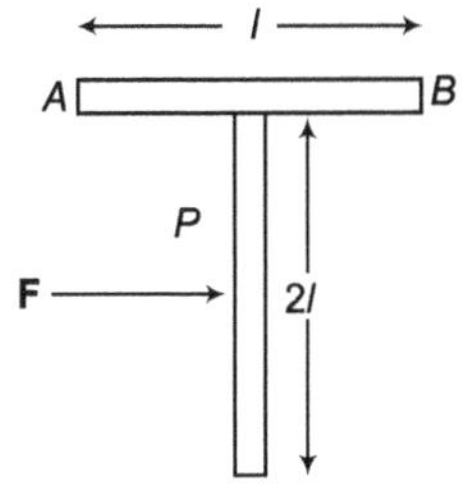

(a) $\frac{2}{3}l$ (b) $\frac{3}{2}l$ (c) $\frac{4}{3}l$ (d) l

16. The figure shows the position-time (x-t) graph of one-dimensional motion of a body of mass 0.4 kg. The magnitude of each impulse is **[AIEEE]**

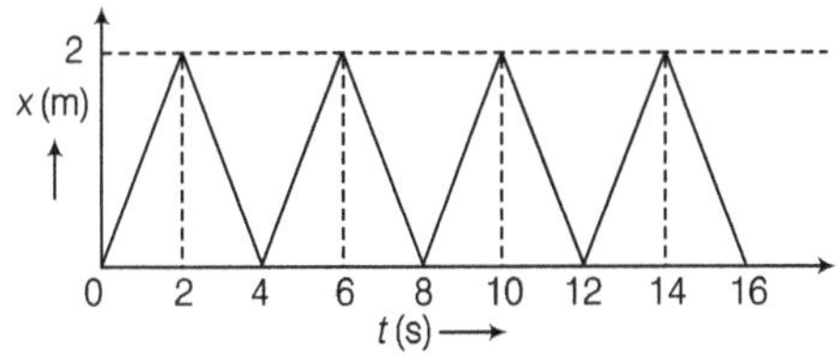

(a) 0.4 N-s (b) 0.8 N-s
(c) 1.6 N-s (d) 0.2 N-s

Answers

Level 1

Objective Problems

1. (a)	**2.** (d)	**3.** (d)	**4.** (b)	**5.** (d)	**6.** (d)	**7.** (d)	**8.** (d)	**9.** (b)	**10.** (a)
11. (b)	**12.** (d)	**13.** (b)	**14.** (a)	**15.** (a)	**16.** (a)	**17.** (b)	**18.** (c)	**19.** (c)	**20.** (b)
21. (d)	**22.** (b)	**23.** (a)	**24.** (b)	**25.** (a)	**26.** (a)	**27.** (d)	**28.** (d)	**29.** (c)	**30.** (d)
31. (d)	**32.** (a)	**33.** (d)	**34.** (a)	**35.** (b)	**36.** (b)	**37.** (b)	**38.** (d)	**39.** (b)	**40.** (d)
41. (c)	**42.** (b)	**43.** (c)	**44.** (b)	**45.** (a)	**46.** (d)	**47.** (b)	**48.** (c)	**49.** (a)	**50.** (d)
51. (d)	**52.** (d)	**53.** (c)	**54.** (d)	**55.** (b)	**56.** (b)	**57.** (c)	**58.** (d)	**59.** (b)	**60.** (c)
61. (c)	**62.** (b)	**63.** (c)	**64.** (b)	**65.** (b)	**66.** (a)	**67.** (b)	**68.** (c)	**69.** (d)	**70.** (a)
71. (b)	**72.** (c)	**73.** (c)	**74.** (b)	**75.** (b)	**76.** (c)	**77.** (a)	**78.** (b)	**79.** (c)	**80.** (a)
81. (b)	**82.** (d)	**83.** (c)	**84.** (a)	**85.** (b)	**86.** (b)	**87.** (b)	**88.** (c)	**89.** (d)	**90.** (d)
91. (d)	**92.** (c)	**93.** (d)	**94.** (d)	**95.** (a)	**96.** (c)	**97.** (a)	**98.** (c)	**99.** (a)	**100.** (b)
101. (a)	**102.** (b)	**103.** (a)	**104.** (d)	**105.** (b)					

Level 2

Only One Correct Option

1. (c)	**2.** (c)	**3.** (c)	**4.** (d)	**5.** (b)	**6.** (b)	**7.** (a)	**8.** (a)	**9.** (d)	**10.** (a)
11. (a)	**12.** (b)	**13.** (b)	**14.** (a)	**15.** (d)	**16.** (c)	**17.** (a)	**18.** (a)	**19.** (d)	**20.** (b)
21. (d)	**22.** (d)	**23.** (d)	**24.** (b)	**25.** (b)					

More than One Correct Options

1. (b,d)	**2.** (a,c,d)	**3.** (all	**4.** (b,d)	**5.** (b,c)	**6.** (c,d)	**7.** (b,c)	**8.** (b,c,d)

Comprehension Based Questions

1. (b)	**2.** (b)	**3.** (d)	**4.** (b)	**5.** (c)	**6.** (c)	**7.** (a)

Assertion and Reason

1. (d)	**2.** (d)	**3.** (d)	**4.** (d)	**5.** (d)	**6.** (b)	**7.** (c)	**8.** (d)	**9.** (d)	**10.** (a)
11. (d)	**12.** (a,b)	**13.** (a)	**14.** (d)	**15.** (b)	**16.** (d)	**17.** (d)	**18.** (e)	**19.** (a,b)	**20.** (d)
21. (a)	**22.** (a)	**23.** (e)							

Match the Columns

1. $(A \rightarrow q, B \rightarrow r, C \rightarrow r, D \rightarrow s)$	**2.** $(A \rightarrow q, B \rightarrow q, C \rightarrow q, D \rightarrow p)$	**3.** $(A \rightarrow r, B \rightarrow t, C \rightarrow t, D \rightarrow s)$
4. $(A \rightarrow p, B \rightarrow t, C \rightarrow s, D \rightarrow r)$	**5.** $(A \rightarrow q, B \rightarrow s, C \rightarrow p, D \rightarrow r)$	**6.** $(A \rightarrow r,s, B \rightarrow rs, C \rightarrow q, D \rightarrow q)$
7. $(A \rightarrow q, B \rightarrow p, C \rightarrow r)$	**8.** $(A \rightarrow q, B \rightarrow p, C \rightarrow r)$	**9.** $(A \rightarrow q, B \rightarrow r, C \rightarrow p)$

Entrance Gallery

1. (a)	**2.** (b)	**3.** (d)	**4.** (d)	**5.** (a)	**6.** (d)	**7.** (b)	**8.** (a)	**9.** (a,c)	**10.** (d)
11. (d)	**12.** (d)	**13.** (b)	**14.** (b)	**15.** (c)	**16.** (b)				

Solutions

Level 1 : Objective Problems

1. Centre of mass of square plate is at O. Centre of mass of two masses of 5 kg each is at H. Hence, centre of mass of the whole system is at mid-point of OH.

2. Distance distributes in inverse ratio of masses. Hence,

$$r_c = d\left(\frac{m_0}{m_0 + m_c}\right) = 1.2\times10^{-10}\left(\frac{16}{16+12}\right)$$

$$= 0.69\times 10^{-10}\ \text{m}$$

3. $A_1x_1 = A_2x_2$ or $x_1 = \frac{A_2}{A_1}\cdot x_2$

$$= \frac{\left(\frac{\pi}{4}\right)(8)^2}{(20)^2}\times 6 = 0.12\ \text{cm from } O.$$

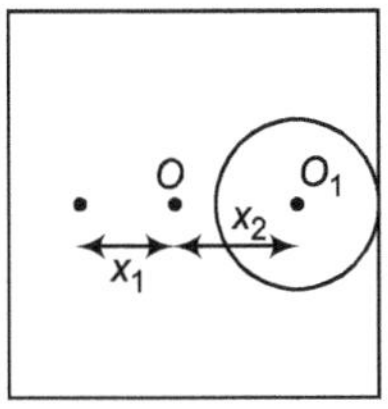

4. Area of circle, $\frac{\pi}{4}a^2 = A_1$, area of square $= a^2 = A_2$.

Since, $A_2 > A_1$, centre of mass will lie inside the square plate.

5. $x_{CM} = 0$

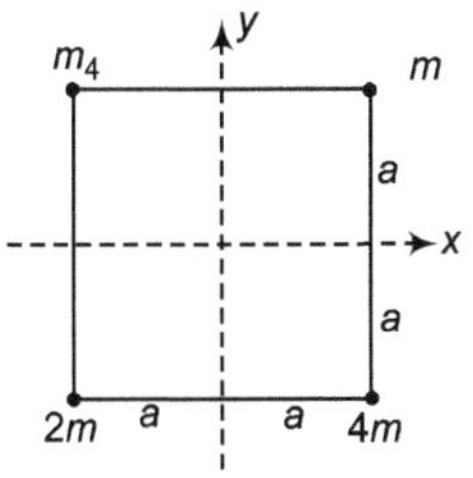

or $$\frac{m_1x_1 + m_2x_2 + m_3x_3 + m_4x_4}{m_1 + m_2 + m_3 + m_4} = 0$$

or $(2m)(-a) + 4m(a) + m(a) + m_4(-a) = 0$

or $m_4 = 3m$

Similarly, $y_{CM} = 0$

or $(2m)(-a) + 4m(-a) + m(a) + m_4(a) = 0$

or $m_4 = 5m$

Since, value of m_4 are different to satisfy both $x_{CM} = 0$ and $y_{CM} = 0$.

Hence, it is not possible.

6. $$X_{CM} = \frac{m_1x_1 + m_2x_2 + m_3x_3}{m_1 + m_2 + m_3}$$

$$= \frac{m(0) + (m)\left(\frac{a}{2}\right) + m\left(\frac{a}{2}\right)}{m + m + m} = \frac{a}{3}$$

Similarly, $y_{CM} = \frac{a}{3}$

7. Centre of mass of Ist system already lies at (1, 2, 3). Therefore, centre of mass of 3 kg and 5 kg should lie at (1, 2, 3).

$\therefore$ $$\frac{3(-\hat{\mathbf{i}} + 3\hat{\mathbf{j}} - 2\hat{\mathbf{k}}) + 5\mathbf{r}_5}{(3+5)} = (\hat{\mathbf{i}} + 2\hat{\mathbf{j}} + 3\hat{\mathbf{k}})$$

Solving this we get, $\mathbf{r}_5 = \frac{11}{5}\hat{\mathbf{i}} + \frac{7}{5}\hat{\mathbf{j}} + 6\hat{\mathbf{k}}$

i.e. 5 kg mass should be kept at (11/5, 7/5, 6)

8. Centre of mass of complete disc should lie at point O. C_1 is the position of centre of mass of remaining portion and C_2 the position of centre of mass of the removed disc.

$\therefore$ x (Area of remaining portion) $= \frac{R}{2}$ (Area of removed disc)

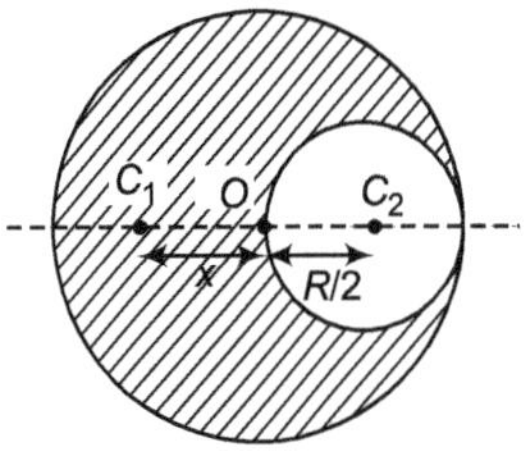

$\therefore$ $$x\left[\pi R^2 - \frac{\pi R^2}{4}\right] = \frac{R}{2}\left[\frac{\pi R^2}{4}\right]$$

$\therefore$ $$x = \frac{R}{6}$$

9. Centre of mass of whole system was at point O. Hence,

x_2 (area of remaining portion) $= c$ (area of removed disc)

$\therefore$ $$x_2(\pi a^2 - \pi b^2) = c(\pi b^2)$$

$\therefore$ $$x_2 = \frac{cb^2}{a^2 - b^2}$$

10. $C_1 \rightarrow$ Position of centre of mass of rods AB and CD (nearer to CD, as it is heavy)

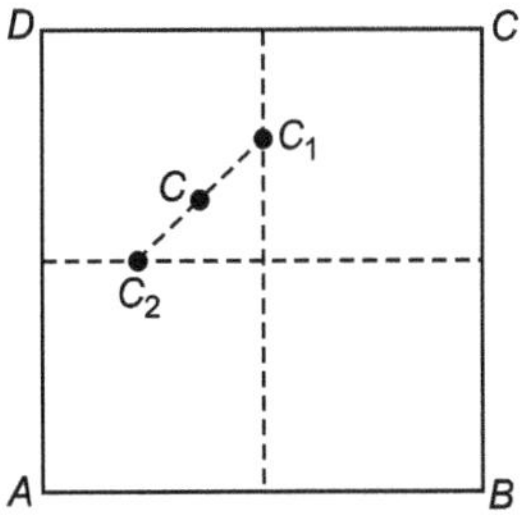

$C_2 \rightarrow$ Position of centre of mass of rods BC and DA.

$C \rightarrow$ Overall centre of mass of all four rods.

11. For a single particle distance of centre of mass from origin is R. For more than one particles distance $\leq R$.

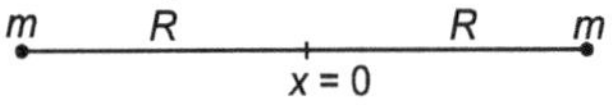

For example for two particles of equal mass, kept as shown in figure, distance = 0.

12. $OC_1 = \frac{1}{4}$ m

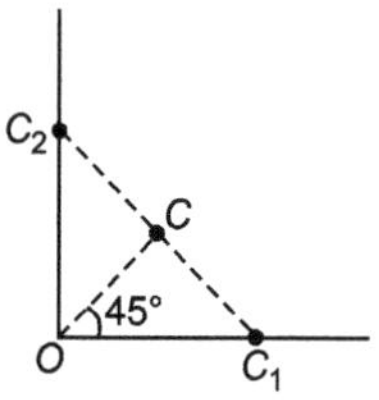

$\therefore$ $$OC = OC_1 \cos 45°$$
$$= \frac{1}{4\sqrt{2}} \text{ m}$$

13. X_{CM} (from P) $= \dfrac{m \times O + m \times PQ + m \times PR}{m + m + m}$ $(m = 1 \text{ kg})$

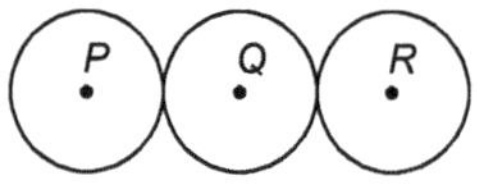

or $$X_{CM} = \frac{PQ + PR}{3}$$

14. P is the position of centre of mass of particle at 2 and 3. Q is position of centre of mass of all three particles.

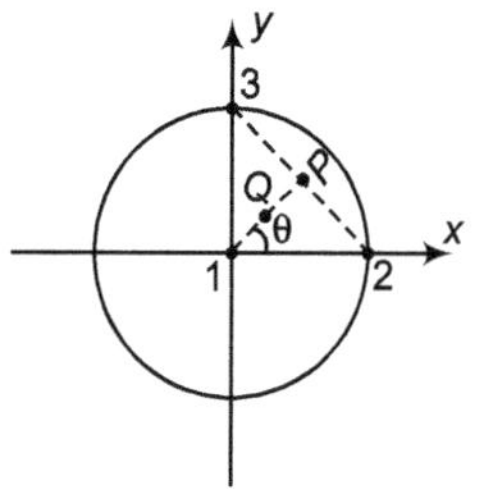

$$\tan\theta = \frac{y_{CM}}{x_{CM}} = \frac{m_1y_1 + m_2y_2 + m_3y_3}{m_1x_1 + m_2x_2 + m_3x_3}$$
$$= \frac{6 \times 0 + 2 \times 0 + 2 \times a}{6 \times 0 + 2 \times a + 2 \times 0}$$
$$= 1$$

or $$\theta = 45°$$

15. $A_1(CC_1) = A_2(CC_2)$

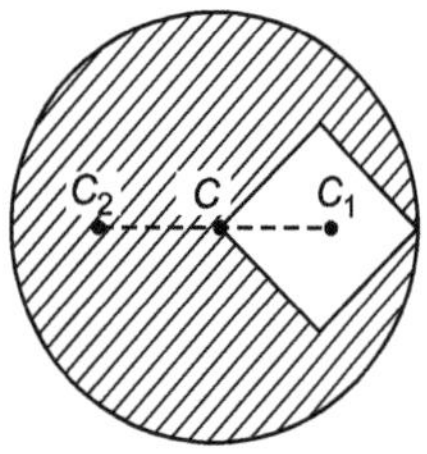

Side of square will be $\frac{R}{\sqrt{2}}$

$\therefore$ $$CC_2 = \frac{A_1}{A_2}(CC_1)$$
$$= \frac{(R/\sqrt{2})^2}{\pi R^2 - (R/\sqrt{2})^2}\left(\frac{R}{2}\right)$$
$$= \left(\frac{R}{4\pi - 2}\right)$$

16. $P_i = 0$

$\therefore$ $P_f = 0$ or centre of mass should be at rest at all instants.

17. $\mathbf{a}_{CM} = \dfrac{(m)(0) + (m)(\mathbf{a})}{m + m} = \dfrac{1}{2}\mathbf{a}$

18. $v_{CM} = \dfrac{10(14) + 4(0)}{10 + 14} = 10$ m/s

19. Centre of mass does not change its path during explosion. Therefore, it will keep on falling vertically.

20. Net external force is zero. Hence, velocity of CM of the box and ball system will remain constant.

21. Both the balls in air have acceleration g downwards. Hence, the acceleration of their centre of mass will also be g downwards.

22. Let plank moves x distance in opposite direction. Then, displacement of man relative to ground will be, $(L - x)$. Applying

$$m_R x_R = m_L x_L \quad \text{or} \quad M(L - x) = \frac{M}{3}x$$

Solving this equation we get, $x = \frac{3L}{4}$

$\therefore$ Displacement of man relative to ground $= L - \frac{3L}{4} = \frac{L}{4}$

23. Centre of mass will remain at height h.

$\therefore$ $$h_{CM} = \frac{m \times 0 + MH}{m + M} = h$$

$\therefore$ $$H = h\left(1 + \frac{m}{M}\right)$$

24. After 1 s, coordinates of first particle will become (4m, 4m, 6m) and co-ordinates of second particle will become (6m, 4m, 8m).

$\therefore$ $$X_{CM} = \frac{4 + 6}{2} = 5\text{m}$$
$$Y_{CM} = \frac{4 + 4}{2} = 4\text{m}$$

and $$Z_{CM} = \frac{6 + 8}{2} = 7\text{m}$$

- In this problem, answer is independent of the fact whether the first particle has velocity $\mathbf{v}_1$ or $\mathbf{v}_2$. Think why?

25. Remaining time for the pieces to reach the ground will be,

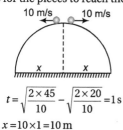

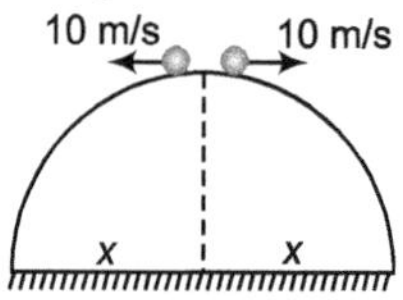

$$t = \sqrt{\frac{2 \times 45}{10}} - \sqrt{\frac{2 \times 20}{10}} = 1\text{ s}$$
$$x = 10 \times 1 = 10\text{ m}$$

$\therefore$ Distance between two pieces will be 20 m.

26. In the absence of friction, centre of mass will not move in x-direction.

27. $y_{CM} = \dfrac{m_1y_1 + m_2y_2}{m_1 + m_2}$

or $$+1 = \frac{(10)(7) + (30)y_2}{10 + 30}$$

$\therefore$ $$y_2 = -1\text{ cm}$$

28. Distance between bullet and block at this instant is $(D-d)$

Distance from bullet $= \dfrac{M}{M+m}(D-d)$

And distance from block $= \dfrac{m}{M+m}(D-d)$

29. Initially centre of mass was at rest. Hence, at any instant centre of mass will be at rest as net external force is zero.

30. At $t=0$, centre of mass is at mid- point or at (2.25 m, 0)

Velocity of centre of mass is zero. Hence, centre of mass will remain at this position all the time.

31. Vertical component of velocity of CM is zero. Horizontal component of velocity of CM is non-zero. Acceleration of CM is g downwards. Hence, path is a parabola as show in figure.

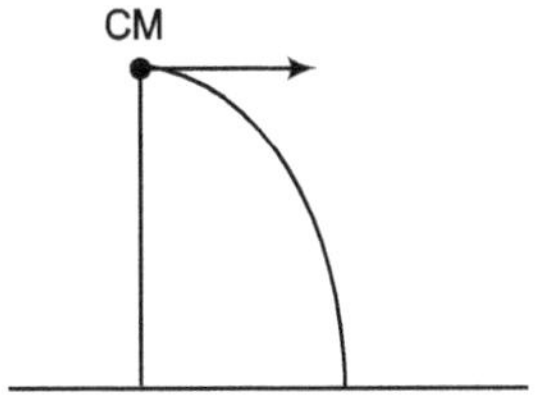

32. During explosion of a projectile path of centre of mass does not change. At height 1 m,

$v=\sqrt{u^2-2gh}$ (from conservation of mechanical energy)

$=4\sqrt{5}\ \text{m/s}$

33. $\mathbf{u}_{CM}=\dfrac{m_1\mathbf{u}_1+m_2\mathbf{u}_2}{m_1+m_2}=0$

as $m_1\mathrm{u}_1+m_2\mathrm{u}_2=0$ is given. Hence, velocity of composite system will be zero.

34. Centre of mass will not move along y-axis.

or $\quad Y_{CM}=0$ (always)

35. $m(u-v)=mv$

$\therefore \quad v=\dfrac{u}{2}$

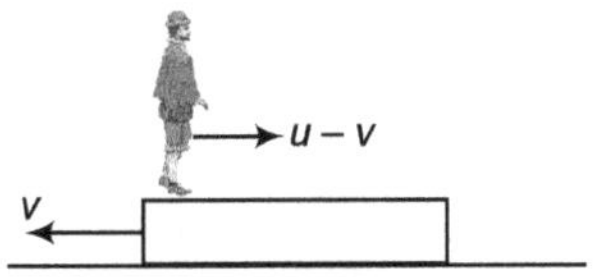

m (u – v) = mv

$\therefore$ Velocity of man relative to ground

$=u-v=\dfrac{u}{2}$

36. $\quad m_1v_1+m_2v_2=(m_1+m_2)v$

$\therefore \quad a\cdot b+c\cdot 0=(a+c)v$

$\Rightarrow \quad v=\dfrac{ab}{a+c}$

37. $v_r=\dfrac{l}{t}$. Now, $m(v_r-v)=Mv$ (v = speed of plank)

$\therefore \quad v=\dfrac{mv_r}{M+m}$

$=\dfrac{l}{t}\left(\dfrac{m}{M+m}\right)$

38. $m_1v_1=m_2v_2$ or $v_2=\dfrac{m_1v_1}{m_2}$

$=\dfrac{5\times10^{-3}\times 900}{2.5}=1.8\ \text{m/s}$

39. $K_1+K_2=2400 \quad$ ...(i)

$p_1=p_2$

$\therefore \quad \sqrt{2K_1m_1}=\sqrt{2K_2m_2}$

or $\quad \dfrac{K_1}{K_2}=\dfrac{m_2}{m_1}=\dfrac{3}{1} \quad$ ...(ii)

Solving these two equations, we get

K_1 = Kinetic energy of smaller part = 1800 J

40. Impulse = change in linear momentum

$=p_f-p_i$

$=-5\times4-5\times3$

$=-35$ N-s

$\therefore$ Magnitude of impulse in SI Units = 35

41. $mv=\left(\dfrac{3}{4}m\right)v'$

$\therefore v'=\dfrac{4v}{3}$

42. $p_3=p_6$ or $3\times16=6\times v$ or $v=8$ m/s. As $p_i=0$ or both the pieces should have equal and opposite momentum.

43. $p_i=0$

$\therefore p_f=0$, i.e. both the pieces should have equal and opposite momentum.

44. Impulse $\mathbf{J}=\Delta\mathbf{p}=m(\mathbf{v}_f-\mathbf{v}_i)$

$0.2[(2\hat{\mathbf{i}}+3\hat{\mathbf{j}}+\hat{\mathbf{k}})-(5\hat{\mathbf{i}}-3\hat{\mathbf{j}}+7\hat{\mathbf{k}})]$

$=(-0.6\hat{\mathbf{i}}+1.2\hat{\mathbf{j}}-1.2\hat{\mathbf{k}})$

$\therefore \quad J=\sqrt{(0.6)^2+(1.2)^2+(1.2)^2}$

$=1.8$ N-s

45. F = thrust force $=v_r\left(\dfrac{dm}{dt}\right)$

$=2\times0.5$

$=1$ N

46. $\mathbf{F}=\dfrac{\Delta\mathbf{p}}{\Delta t}=\dfrac{m(\mathbf{v}_f-\mathbf{v}_i)}{4t}$

$=\dfrac{15[(6\hat{\mathbf{i}}+4\hat{\mathbf{j}}+5\hat{\mathbf{k}})-(\hat{\mathbf{i}}-2\hat{\mathbf{j}})]}{0.1}$

$=150(5\hat{\mathbf{i}}+6\hat{\mathbf{j}}+5\hat{\mathbf{k}})$

47. Packet from A falls with greater momentum on B. Therefore, B is slightly accelerated.

48. In horizontal direction, net force on the system is zero.

49. From conservation of linear momentum,

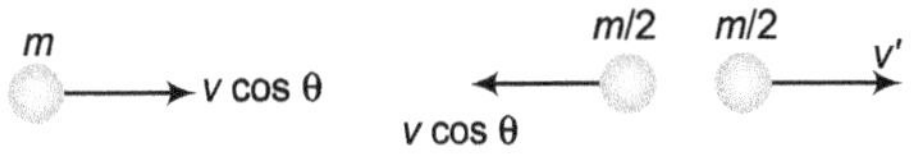

$m(v\cos\theta)=\dfrac{m}{2}v'-\dfrac{m}{2}v\cos\theta$

$\therefore \quad v=3v\cos\theta$

50. Force of friction on A is backwards and force of friction on B is forwards. Net external force on the system is zero. Hence, momentum of system will remain conserved. As the momentum of system is conserved, increase in momentum of B is equal to decrease in momentum of A.

51. The centre of mass under the given condition may be at rest or may be moving with constant velocity.

52. $v = v_r \ln\left(\frac{m_0}{m}\right)$

or $2v_r = v_r \ln\left(\frac{m_0}{m}\right)$

$\therefore$ $2 = \ln\left(\frac{m_0}{m}\right)$

$\therefore$ $\frac{m_0}{m} = e^2 = 7.4$

or $\frac{m}{m_0} = 0.14$

This is fraction of mass remaining. Hence, fraction of mass disposed off $= 0.86$

53. Initial momentum of $3m$ mass $= 0$ ...(i)

Due to explosion, this mass splits into three fragments of equal masses.

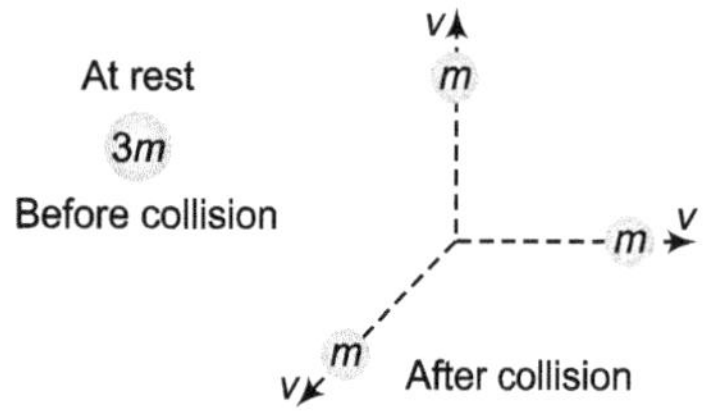

Final momentum of system

$= m\mathbf{v} + mv\hat{\mathbf{i}} + mv\hat{\mathbf{j}}$ ...(ii)

From law of conservation of linear momentum

$m\mathbf{v} = mv\hat{\mathbf{i}} + mv\hat{\mathbf{j}} = 0$

$\Rightarrow$ $\mathbf{v} = -v(\hat{\mathbf{i}} + \hat{\mathbf{j}})$

54. Force = Rate of change of momentum

Initial momentum $\mathbf{p}_1 = mv\sin\theta\hat{\mathbf{i}} + mv\cos\theta\hat{\mathbf{j}}$

Final momentum $\mathbf{p}_2 = -mv\sin\theta\hat{\mathbf{i}} + mv\cos\theta\hat{\mathbf{j}}$

$\therefore$ $\mathbf{F} = \frac{\Delta\mathbf{p}}{\Delta t}$

$= \frac{-2mv\sin\theta}{2\times10^{-3}}$

Substituting $m = 0.1$ kg, $v = 5$ m/s, $\theta = 60°$

Force on the ball $\mathbf{F} = -250\sqrt{3}$ N

Negative sign indicates direction of the force.

55. $\mathbf{p}_i = \mathbf{p}_f$ (at highest point)

$(3m)(100\hat{\mathbf{i}}) = m(100\hat{\mathbf{j}}) - m(100\hat{\mathbf{j}}) + m(\mathbf{v})$

$\therefore$ $\mathbf{v} = 300\hat{\mathbf{i}}$ (of third part)

56. In one dimensional elastic collision between two equal masses velocities are interchanged. Therefore, change in linear momentum of any of the particle will be mu. Now, impulse or area under F-t graph gives the change in linear momentum.

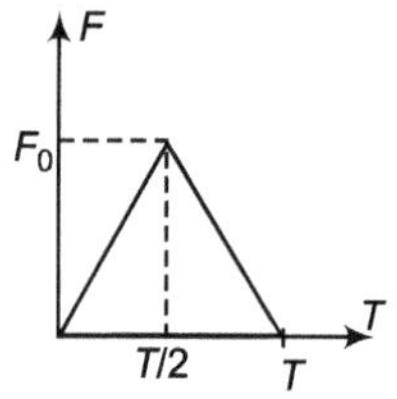

$\therefore$ $\frac{1}{2}F_0T = mu$ or $F_0 = \frac{2mu}{T}$

57. $\mathbf{v}_i = v_0\cos 37°\,\hat{\mathbf{i}} + v_0\sin 37°\,\hat{\mathbf{j}} = \frac{4}{5}v_0\hat{\mathbf{i}} + \frac{3}{5}v_0\hat{\mathbf{j}}$

$\mathbf{v}_f = -\frac{3}{4}v_0\cos 53°\,\hat{\mathbf{i}} + \frac{3}{4}v_0\sin 53°\,\hat{\mathbf{j}}$

$= -\frac{9}{20}v_0\hat{\mathbf{i}} + \frac{3}{5}v_0\hat{\mathbf{j}}$

$\mathbf{J} = m(\mathbf{v}_f - \mathbf{v}_i) = -\frac{5}{4}mv_0\hat{\mathbf{i}}$

58. Net impulse = area under F-t graph $= 0$

$\therefore$ p or $v = 0$

59. Just before 2 s, v_i = slope of x-t graph = 2 m/s

Just after 2 s, v_f = slope of x-t graph $= 0$

$\therefore$ Impulse at 2 s

$= \Delta p = p_f - p_i = m(v_f - v_i) = -0.2$ kg-m/s

60. $F = \frac{\Delta p}{\Delta t}$

$\therefore$ $F = \frac{0.150\times(20-0)}{0.1} = 30\text{N}$

61. Impulse $= \Delta p = p_1 - p_2 = mv_1 - mv_2$

$= m(v_1 - v_2)$

$= 0.06[4-(-4)]$

$= 0.06\times 8 = 0.48$ kg-m/s

62. $F = \frac{m\Delta v}{\Delta t}$

$= \frac{0.150[20-(-12)]}{0.04} = \frac{0.150\times 32}{0.04} = 120\text{N}$

63. $V_r = v$ (Opposite to the motion of railroad car)

$\therefore$ $F_t = -\mu v = (M+\mu t)\cdot\frac{dv}{dt}$

$\therefore$ $\frac{dt}{M+\mu t} = -\mu\left(\frac{dv}{v}\right)$

$\therefore$ $\int_0^t \frac{dty}{M+\mu t} = -\frac{1}{\mu}\int_{v_0}^{v}\frac{dv}{v}$

$\therefore$ $\frac{1}{\mu}\ln\left(\frac{M+\mu t}{M}\right) = \frac{1}{\mu}\ln\left(\frac{v_0}{v}\right)$

Solving we get, $v = \frac{Mv_0}{M+\mu t}$

64. $F - mg = ma$ (F = thrust force)

or $F = m(g+a)$

$V_r\left(-\frac{dm}{dt}\right) = m(g+a)$

$\therefore$ $\left(-\frac{dm}{dt}\right) = \frac{m(g+a)}{V_r}$

$= \frac{6000\times 30}{1000} = 180$ kg/s

65. $F - mg = ma \, (F = \text{thrust})$

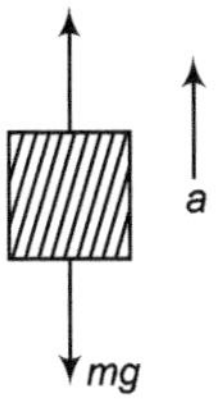

$$F = m(g + a)$$
$$= (3.5 \times 10^4)\,(10 + 10)$$
$$= 7.0 \times 10^5 \text{ N}$$

67. $mv = (m + M)\,v'$

$\therefore \quad v' = \left(\dfrac{m}{m + M}\right) v$

68. In perfectly elastic collision between two bodies of equal masses, velocities are exchanged.

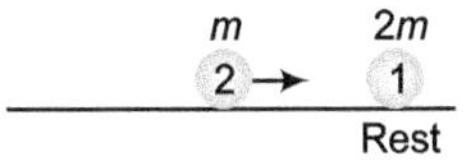

69. $v_2' = \left(\dfrac{m_2 - m_1}{m_2 + m_1}\right) v_2 + \left(\dfrac{2m_1}{m_1 + m_2}\right) v_1$

$$v_1 = 0$$

$\therefore \quad \dfrac{v_2}{v_2'} = \left(\dfrac{m_2 + m_1}{m_2 - m_1}\right)$

$$= \left(\dfrac{m + 2m}{m - 2m}\right) = -3$$

$\therefore \quad \dfrac{K_2}{K_2'} = \left(\dfrac{v_2}{v_2'}\right)^2 = 9$

70. From conservation of linear momentum, we can see that velocity of $2m$ will become $\dfrac{v}{2}$ after collision (as mass is doubled)

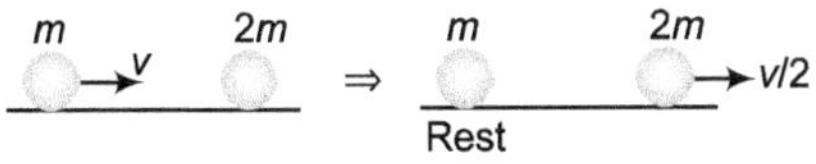

Now, $\quad e = \dfrac{\text{Relative velocity of separation}}{\text{Relative velocity of approach}}$

$$= \frac{v/2}{v} = \frac{1}{2}$$

71. $\tan\theta' = \dfrac{v\sin\theta}{ev\cos\theta} = \dfrac{\tan\theta}{e}$

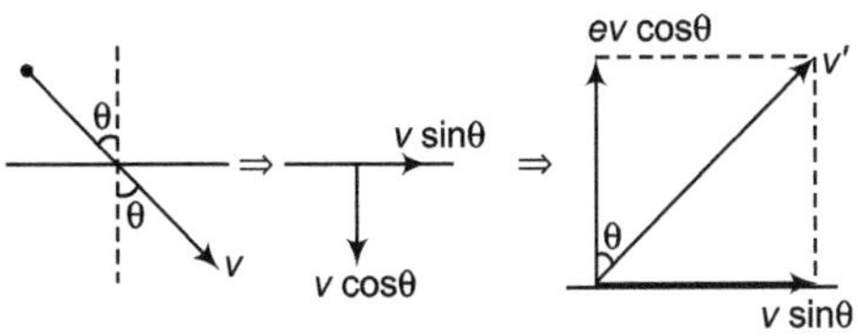

or $\quad \theta = \tan^{-1}\left(\dfrac{\tan\theta}{e}\right)$

72. Mass of combined body = 50 kg

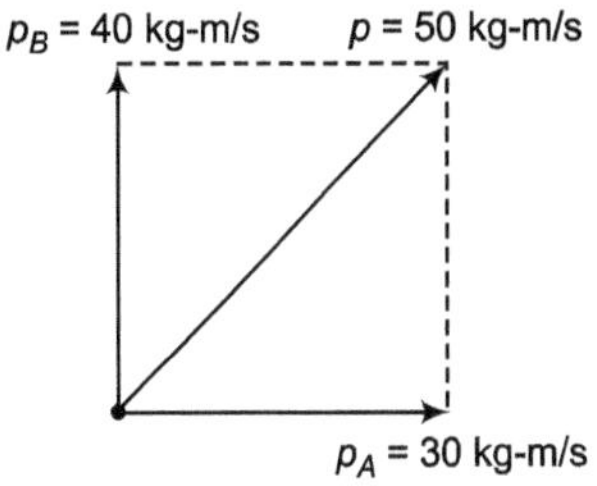

$\therefore$ Velocity of combined body $= \dfrac{50 \text{ kg-m/s}}{5 \text{ kg}} = 1 \text{ m/s}$

73. $v_x = \dfrac{m \times 50 + m(-40)}{m + m} = 5 \text{ cm/s}$

$v_y = \dfrac{m(0) + m(30)}{m + m} = 15 \text{ cm/s}$

74. Between elastic collision of two identical masses, velocities are interchanged.

75. Relative speed of approach is 8 m/s, $e = \dfrac{1}{2}$. Therefore, relative speed of separation will be 4 m/s.

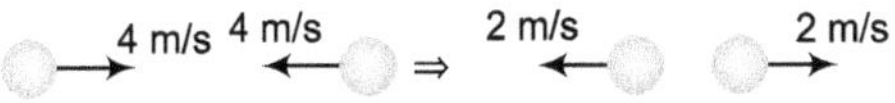

76. Net force on ball and earth system is zero.

77. $2mv\cos\theta = (2m)\,v'$

$\therefore \quad v' = v\cos\theta$

78. From conservation of linear momentum,

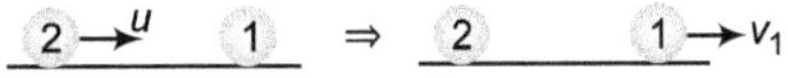

$$mu = mv_1 + mv_2$$

or $\quad u = v_1 + v_2 \qquad \text{...(i)}$

From definition of e,

$$v_1 - v_2 = eu \qquad \text{...(ii)}$$

Solving these two equations, we get

$$v_1 = \left(\frac{1+e}{2}\right) u$$

and $\quad v_2 = \left(\dfrac{1-e}{2}\right) u$

$\therefore \quad \dfrac{v_1}{v_2} = \left(\dfrac{1+e}{1-e}\right)$

79. $v_2 = \left(\dfrac{m_2 - m_1}{m_1 + m_2}\right) u_2 + \dfrac{2m_1 u_1}{m_1 + m_2}$

$$= \frac{2Mu}{M + m}$$

$$= \frac{2u}{1 + \dfrac{m}{M}}$$

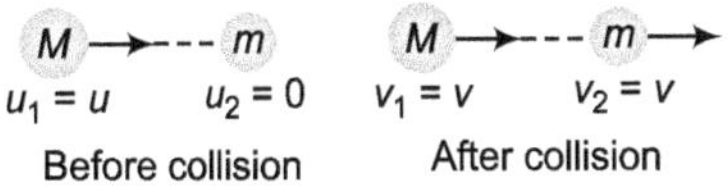

80. $h_n = he^{2n} = 32\left(\dfrac{1}{2}\right)^4 = \dfrac{32}{16} = 2m$ (here, $n = 2, e = 1/2$)

81. If target is at rest, then final velocity of bodies are

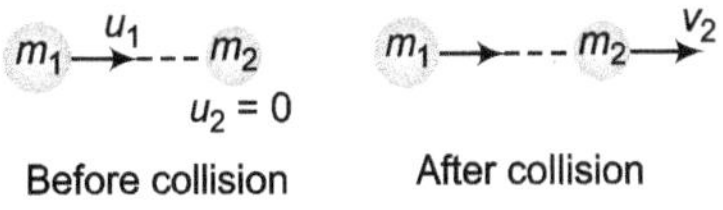

$$v_1 = \left(\frac{m_1 - m_2}{m_1 + m_2}\right)u_1 \quad \text{...(i)}$$

and
$$v_2 = \frac{2m_1u_1}{m_1 + m_2} \quad \text{...(ii)}$$

From Eqs. (i) and (ii), $\frac{v_1}{v_2} = \frac{m_1 - m_2}{2m_1} = \frac{2}{5} \Rightarrow \frac{m_1}{m_2} = 5$

82. Due to the same mass of A and B as well as due to elastic collision velocities of spheres get interchanged after the collision.

84. Let mass A moves with velocity v and collides inelastically with mass B, which is at rest.

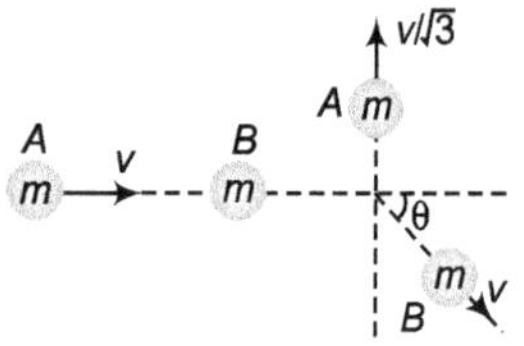

According to problem, mass A moves in a perpendicular direction and let the mass B moves at angle θ with the horizontal with velocity v.

Initially in horizontal momentum of system

$$\text{(before collision)} = mv \quad \text{...(i)}$$

Final horizontal momentum of system

$$\text{(after collision)} = mv\cos\theta \quad \text{...(ii)}$$

From the conservation of linear momentum

$$mv = mV\cos\theta \Rightarrow v = v\cos\theta \quad \text{...(iii)}$$

Initial perpendicular direction momentum of system (before collision) is zero.

Final momentum of system in this direction is

$$\frac{mv}{\sqrt{3}} - mV\sin\theta$$

From conservation of momentum

$$\frac{mv}{\sqrt{3}} - mV\sin\theta = 0$$

$$\Rightarrow \quad \frac{v}{\sqrt{3}} = V\sin\theta \quad \text{...(iv)}$$

By solving Eqs. (iii) and (iv), we have

$$v^2 + \frac{v^2}{3} = V^2(\sin^2\theta + \cos^2\theta)$$

$$\Rightarrow \quad \frac{4v^2}{3} = V^2 \Rightarrow V = \frac{2}{\sqrt{3}}v$$

85. In elastic collision, both linear momentum and kinetic energy should remain constant. This is possible in option (b).

86. Relative speed of approach is 12 m/s.

$$e = \frac{1}{3}$$

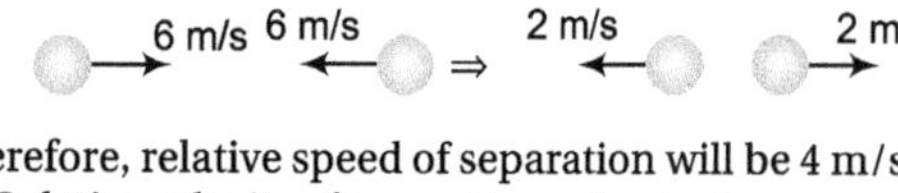

Therefore, relative speed of separation will be 4 m/s.

87. $e = \frac{\text{Relative velocity of separation}}{\text{Relative velocity of approach}} = \frac{5-2}{8-2} = \frac{1}{2}$

88. $\Delta p = p_f - p_i = -mu - mu = -2mu$ but $|\Delta p| = 2mu$

89. If CM is at rest, it definitely means momentum of the system is constant. But momentum of the system is constant, it does not mean centre of mass is at rest.

90. Kinetic energy of the system is zero, it definitely means momentum is zero. But momentum of the system is zero, it does not mean kinetic energy is zero.

Example $\quad O \to v$

$v \leftarrow O$

$p = O$

but $\quad K = 2\left(\frac{1}{2}mv^2\right) = mv^2.$

91. $K = \frac{p^2}{2m}$ and $p = \sqrt{2Km}$ are two standard results.

92. F = rate of change of linear momentum

In 1 s, $\frac{n}{60}$ bullets are embedded. Momentum of each bullet is mv.

$$\therefore \quad F = \left(\frac{n}{60}\right)mv$$

93. In 1 cm^2 area, 10^4 balls are striking per second. Therefore, in 1 m^2 area, 10^8 balls will strike per unit time.

Change in momentum of each ball per second will be $2mu$.

$$\therefore \quad \text{Pressure} = \frac{F}{A} = \frac{(\Delta p/\Delta t)}{A}$$

$$= 10^8 \times 2mu$$

$$= 10^8 \times 2 \times 10^{-3} \times 100$$

$$= 2 \times 10^7 \text{ N/m}^2$$

94. $p^2 = 2Em$ or $p^2 \propto E$

i.e. p^2 *versus* E graph is a straight line passing through origin.

95. Velocity of block just after collision will be

$$v = \frac{5 \times 10^{-3} \times 150}{(2 + 5 \times 10^{-3})}$$

(From conservation of linear momentum)

$$= 0.374 \text{ m/s}$$

Let, F be the force of friction. Then, work done against friction = initial kinetic energy

or $\quad F \times 2.7 = \frac{1}{2} \times 2.005 \times (0.374)^2$

or $\quad F = 0.052$ N

96. $\Delta \mathbf{p} = \mathbf{p}_f - \mathbf{p}_i = m(\mathbf{v}_f - \mathbf{v}_i)$

$$= m\left[\left(\frac{v}{\sqrt{2}}\hat{\mathbf{i}} - \frac{v}{\sqrt{2}}\hat{\mathbf{j}}\right) - \left(\frac{v}{\sqrt{2}}\hat{\mathbf{i}} + \frac{v}{\sqrt{2}}\hat{\mathbf{j}}\right)\right]$$

$$= -\sqrt{2}\,mv\hat{\mathbf{j}}$$

$$\therefore \quad |\Delta \mathbf{p}| = \sqrt{2}\,mv$$

97. Let initial kinetic energy, $E_1 = E$

Final kinetic energy, $\quad E_2 = E + 300\%$ of $E = 4E$

As $p \propto \sqrt{E}$

$$\Rightarrow \quad \frac{p_2}{p_1} = \sqrt{\frac{E_2}{E_1}} = \sqrt{\frac{4E}{E}} = 2$$

$$\Rightarrow \quad p_2 = 2p_1 \Rightarrow \quad p_2 = p_1 + 100\% \text{ of } p_1$$

i.e. Momentum will increase by 100%.

98. Let $p_1 = p, p_2 = p_1 + 50\%$ of $p_1 = p_1 + \frac{p_1}{2} = \frac{3p_1}{2}$

$$E \propto p^2$$

$$\Rightarrow \quad \frac{E_2}{E_1} = \left(\frac{p_2}{p_1}\right)^2$$

$$= \left(\frac{3p_1/2}{p_1}\right)^2 = \frac{9}{4}$$

$$\Rightarrow \quad E_2 = 2.25\,E_2 = E_1 + 1.25E_1$$

$$\therefore \quad E_2 = E_1 + 125\% \text{ of } E_1$$

i.e. Kinetic energy will increase by 125%.

99. $p = \sqrt{2mE}$

$$\therefore \quad p \propto \sqrt{E}$$

Percentage increase in $p \approx \frac{1}{2}$ (percentage increase in E)

$$= \frac{1}{2}(0.1\%) = 0.05\%$$

100. Kinetic energy $E = \frac{p^2}{2m}$

$$= \frac{(Ft)^2}{2m} = \frac{F^2t^2}{2m} \qquad (\text{As } p = Ft)$$

101. After striking with A, the block C comes to rest and block A moves with velocity v, when compression in spring is maximum both A and B will be moving with common velocity v.

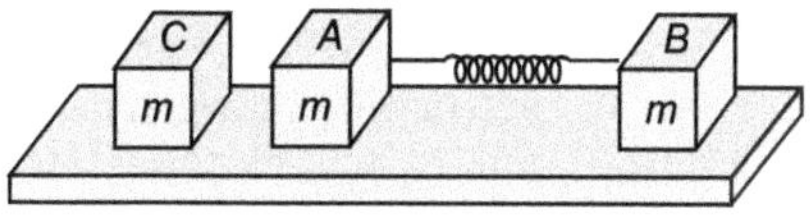

From law of conservation of linear momentum,

$$mv = (m + m)\,V \Rightarrow V = \frac{v}{2}$$

From law of conservation of energy,

KE of block C = KE of system + PE of system

$$\frac{1}{2}mv^2 = \frac{1}{2}(2m)\,v^2 + \frac{1}{2}kx^2$$

$$\Rightarrow \quad \frac{1}{2}mv^2 = \frac{1}{2}(2m)\left(\frac{v}{2}\right)^2 + \frac{1}{2}kx^2$$

$$\Rightarrow \quad kx^2 = \frac{1}{2}mv^2$$

$$\Rightarrow \quad x = v\sqrt{\frac{m}{2k}}$$

102. Let the thickness of each of plank is s. If the initial speed of bullet is 100 m/s, then it stops by covering a distance $2s$.

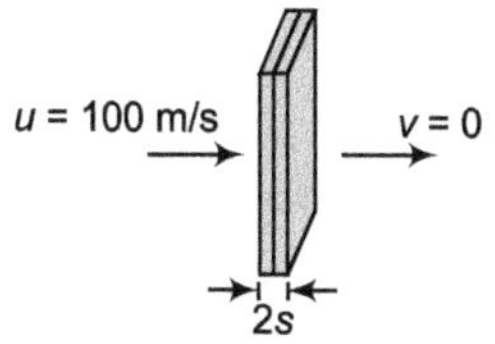

By applying $v^2 = u^2 - 2as \Rightarrow 0 = u^2 - 2as$

$$s = \frac{u^2}{2a}$$

or $\quad s \propto u^2 \quad$ (If retardation is constant)

If the speed of the bullet is double, then bullet will cover four times distance before coming to rest

i.e. $\quad s_2 = 4(s_1) = 4(2s) \Rightarrow s_2 = 8s$

So, number of planks required = 8

103. Let ball is projected vertically downward with velocity v from height h.

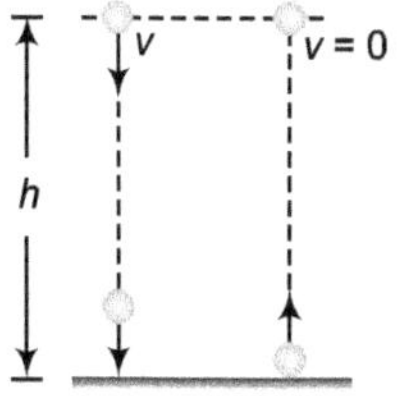

Total energy at point $A = \frac{1}{2}mv^2 + mgh$

During collision loss of energy is 50% and the ball rise up to same height. It means it posses only potential energy at same level.

$$50\%\left(\frac{1}{2}mv^2 + mgh\right) = mgh$$

$$\frac{1}{2}\left(\frac{1}{2}mv^2 + mgh\right) = mgh$$

$$v = \sqrt{2gh} = \sqrt{2 \times 10 \times 20}$$

$$\therefore \quad v = 20\,\text{m/s}$$

104. Their common velocity would be

$$v = \frac{mv_0}{m + 2m} = \frac{v_0}{3}$$

Now, applying conservation of mechanical energy

$$\frac{1}{2}mv_0^2 = \frac{1}{2}kx_0^2 + \frac{1}{2}(3m)\left(\frac{v_0}{3}\right)^2$$

$$k = \frac{2}{3}\frac{mv_0^2}{x_0^2}$$

105. $v' = \sqrt{2gR}$

From conservation of linear momentum,

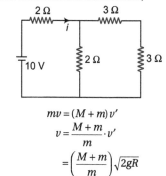

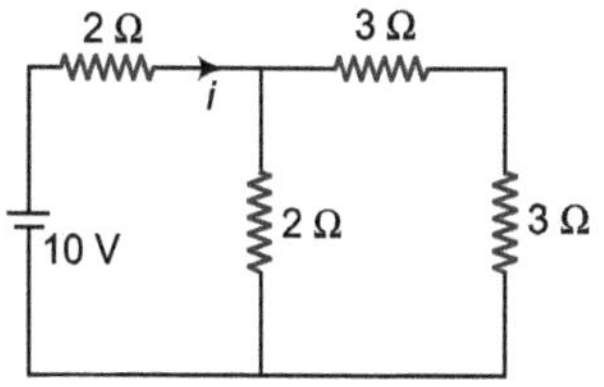

$$mv = (M + m)\,v'$$

$$\therefore \quad v = \frac{M + m}{m} \cdot v'$$

$$= \left(\frac{M + m}{m}\right)\sqrt{2gR}$$

Level 2 : Only One Correct Option

1. $v_A = \left(\frac{M - m}{M + m}\right)u$

Velocity of C at maximum compression

$$v_C = \frac{1}{2}\left[\frac{2M}{M + m}\right]u = \left(\frac{M}{M + m}\right)u$$

$$\therefore \quad v_{CA} = v_C - v_A = \left(\frac{m}{M + m}\right)u$$

2. Velocity of A just before collision

$$= \sqrt{2gh} = \sqrt{2 \times 10 \times 1.25}$$
$$= 5 \text{ m/s}$$

Velocity of $(A + B)$ just after collision $= \frac{5}{2} = 2.5 \text{ m/s}$

(From conservation of linear momentum)

In elastic collision between two bodies of equal masses velocities are interchanged.

Hence, velocity of C will become 2.5 m/s.

3. $F = 0$ at $t = \frac{600}{2 \times 10^5}$ s $= 0.003$ s

$$\text{Impulse} = \int_0^{0.003} F\,dt = 0.9 \text{ N-s}$$

4. At 1 s

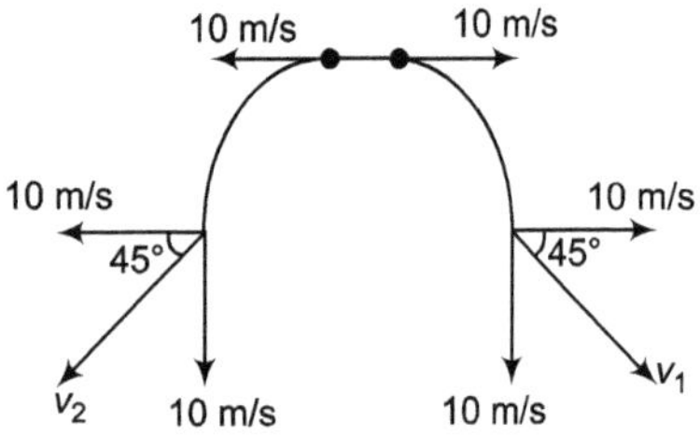

5. $p_{\text{net}} = \int dp \sin\theta = \int_0^{\pi} \left(\frac{M}{\pi} \cdot d\theta\right) v \cdot \sin\theta = \frac{2Mv}{\pi}$

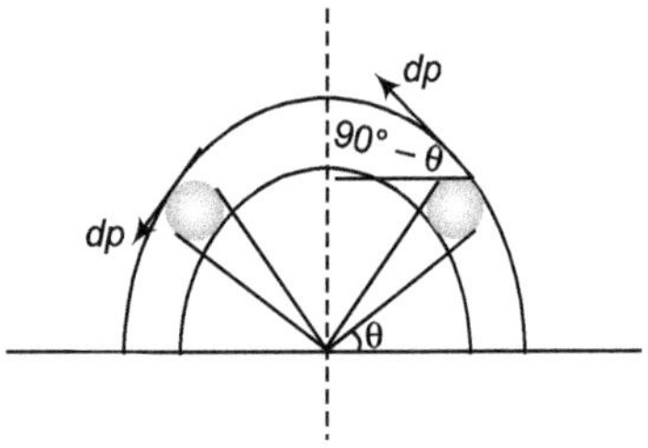

6. From conservation of linear momentum,

$$m_1 v_1 = m\sqrt{5gl} + m_1 \frac{v_1}{3}$$

or
$$v_1 = \frac{3}{2} \frac{m}{m_1} \sqrt{5gl}$$

7. In head-on elastic collision

$$v'_1 = \left(\frac{m/2 - m}{m/2 + m}\right)(0) + \left[\frac{2(m)}{m/2 + m}\right](v) = \frac{4}{3} v$$

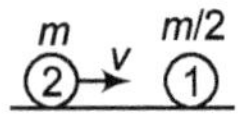

Then
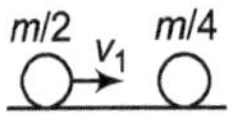

Finally, $v'_n = \left(\frac{4}{3}\right)^{n-1} \cdot v = \sqrt{5gr}$

$$\therefore \quad v = \left(\frac{3}{4}\right)^{n-1} \sqrt{5gr}$$

8. Relative speed of separation = relative speed of approach = v

$\therefore$ Time of next collision $= \frac{2\pi r}{v}$

9. $X_{\text{CM}} = \frac{mX_m + 2mX_{2m}}{2m}$

$$\therefore \quad \frac{1}{2} a_{\text{CM}} t^2 = \frac{x + 2x_{2m}}{3}$$

$$\therefore \quad \frac{3}{2}\left(\frac{F}{3m}\right) t^2 = x + 2x_m$$

$$\therefore \quad X_{2m} = \frac{Ft^2}{4m} - \frac{x}{2}$$

10. Velocity of block just after collision $= \sqrt{2gh}$

$$= \sqrt{2 \times 10 \times 0.2}$$
$$= 2 \text{ m/s}$$

Now, applying conservation of linear momentum just before and just after collision.

$$0.02 \times 600 = 4 \times 2 = 0.02 \times v$$
$$\therefore \quad v = 200 \text{ m/s}$$

11. Velocity of particle after 5 s

$$v = u - gt$$
$$= 100 - 10 \times 5$$
$$= 100 - 50$$
$$= 50 \text{ m/s (upwards)}$$

Conservation of linear momentum gives

$$Mv = m_1 v_1 + m_2 v_2 \quad \text{...(i)}$$

Taking upward direction positive,

$$v_1 = -25 \text{ m/s}, v = 50 \text{ m/s}$$
$$M = 1 \text{ kg}, m_1 = 400 \text{ g} = 0.4 \text{ kg}$$
$$m_2 = M - m_1 = 1 - 0.4$$
$$= 0.6 \text{ kg}$$

From Eq. (i),

$$1 \times 50 = 0.4 \times (-25) + 0.6\, v_2$$

or
$$v_2 = 100 \text{ m/s (upwards)}$$

12. Velocity on hitting the surface

$$= \sqrt{2 \times 9.8 \times 49} = 9.8 \text{ m/s}$$

Velocity after first bounce

$$v = \frac{3}{4} \times 9.8$$

Time taken from first bounce to the second bounce $= \frac{2v}{g}$

$$= 2 \times \frac{3}{4} \times 9.8 \times \frac{1}{9.8} = 1.5 \text{ s}$$

13. $\mathbf{F} = \frac{\Delta \mathbf{p}}{\Delta t} = \frac{\Delta(mn\mathbf{v})}{\Delta t}$

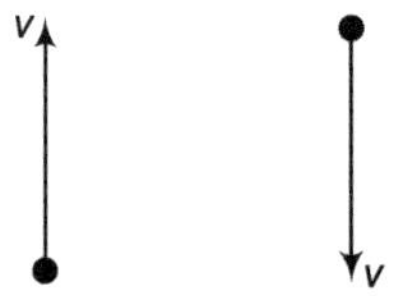

$$|\Delta v| = 2v$$

Here, m = mass of one marble = 5 g

$\frac{n}{\Delta t}$ = number of molecules striking per second = 10

$$\therefore \quad Mg = m\left(\frac{n}{\Delta t}\right) |\Delta \mathrm{v}|$$

Here, M = mass of disc

$$(10 \times 10^{-3})\,(9.80) = (5 \times 10^{-3})\,(10)\,2v$$

or
$$v = 0.98 \text{ m/s}$$

14. $$Mgh = Fx$$

$\therefore$ $$F = \frac{Mgh}{x}$$

15. $F = \frac{\Delta p}{\Delta t} = n(mv)$

Here, n = number of bullets fired per second.

$\therefore$ $$n = \frac{F}{mv} = \frac{144}{0.04 \times 1200} = 3$$

16. $X_{CM} = \frac{\int X\, dm}{\int dm}$

$$= \frac{\int_0^L (Ax) \cdot x \cdot dx}{\int_0^L (Ax)\, dx} = \frac{2L}{3}$$

17. $e = \frac{RSOS}{RSOA}$

$$= \frac{J + (J - p)}{p} = \frac{2J}{p} - 1$$

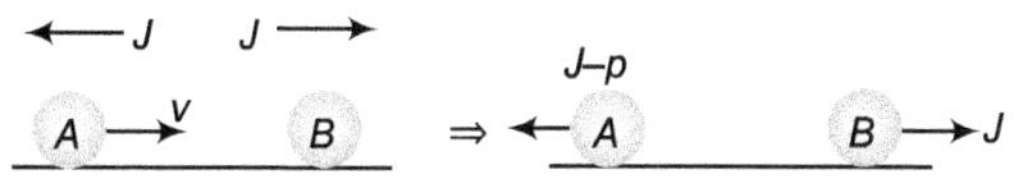

18. Between A and B

$$m \times 10 = mv_A + mv_B \qquad \text{...(i)}$$

$$e = \frac{1}{2} = \frac{v_B - v_A}{10}$$

or $$v_B - v_A = 5 \qquad \text{...(ii)}$$

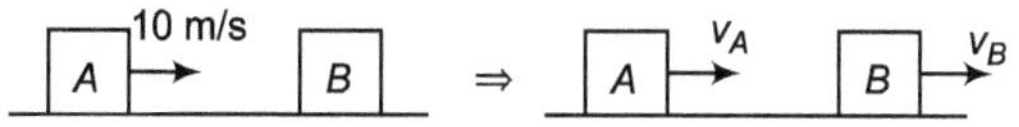

Solving Eqs. (i) and (ii), we get

$$v_B = 7.5 \text{ m/s}$$

Hence, A has given 75% of its speed to B will also transfer its 75% speed to C.

$\therefore$ $$v_C = \frac{75}{100} \times 7.5 = 5.625 \text{ m/s}$$

19. $m(3\hat{\mathbf{i}} + 2\hat{\mathbf{j}}) = m(-2\hat{\mathbf{i}} + \hat{\mathbf{j}}) + M\mathbf{v}$

(Put $M = 13m$)

$\therefore$ $$\mathbf{v} = \frac{(5\hat{\mathbf{i}} + \hat{\mathbf{j}})}{13}$$

$$\mathbf{J}_m = \mathbf{P}_f - \mathbf{P}_i \text{ of } m$$

$$= [(-2\hat{\mathbf{i}} + \hat{\mathbf{j}}) - (3\hat{\mathbf{i}} + 2\hat{\mathbf{j}})] - m(5\hat{\mathbf{i}} + \hat{\mathbf{j}})$$

$\therefore$ $$\mathbf{J}_M = +m(5\hat{\mathbf{i}} + \hat{\mathbf{j}})$$

20.

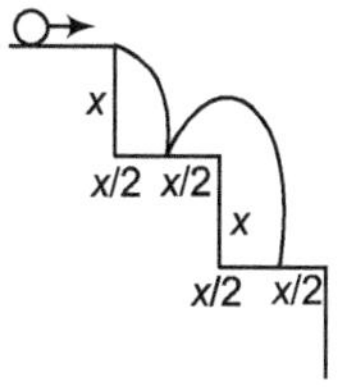

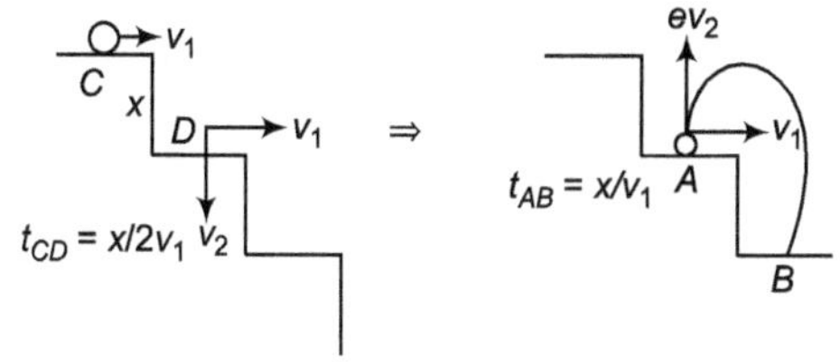

C to D $v_2 = \sqrt{2gx} = gt \Rightarrow \therefore\ t = \frac{v_2}{g}$

From A to B, time will become two times.

Applying $s = ut + \frac{1}{2}at^2$ in vertical direction, we have

$$-x = (ev_2)(2t) - \frac{1}{2} \times g \times (2t)^2$$

$$-x = \frac{2ev_2^2}{g} - \frac{2v_2^2}{g}$$

$$-x = 2e(2x) - 2(2x)$$

$\therefore$ $$e = \frac{3}{4}$$

21. $2\pi R = 2\pi \Rightarrow \therefore\ R = 1$

$$y_{CM} = \frac{m \times 0 + m \times 1 + m \times 1}{m + m + m} = \frac{2}{3},$$

$$x_{CM} = \frac{m(\pi) + m(0) + m(2\pi)}{m + m + m} = \pi$$

22. From conservation of linear momentum, velocity of combined mass just after collision will be 50 cm/s as mass has doubled.

Now, $$H = \frac{u^2}{2g} = \frac{(0.5)^2}{20} \text{ m} = 1.25 \text{ cm}$$

23. $a_{CM} = \frac{\text{External force}}{\text{Total mass}}$

$$= \frac{\text{Force of friction from ground}}{\text{Total mass}}$$

$$= \frac{0.2 \times (2 + 1)\,(10)}{1 + 2} = 2 \text{ m/s}^2$$

24. The centre of mass of the object must lie on the line segment joining (0, 0) and $(R/2, R/2)$. Here (0, 0) is the centre of mass of the ring and $(R/2, R/2)$ is the centre of mass of the chord.

25. $T = \frac{d}{v/\sqrt{2}} + \frac{d}{ev/\sqrt{2}} = \left(1 + \frac{1}{e}\right)\frac{\sqrt{2}d}{v}$

or $$\frac{2v/\sqrt{2}}{g} = \left(1 + \frac{1}{e}\right)\frac{\sqrt{2}d}{v}$$

or $$e = \frac{gd}{v^2 - gd}$$

More than One Correct Options

1. Momentum remains conserved in any type of equation

$$\mathbf{p}_i = \mathbf{p}_f$$

$$\therefore \quad (mv\hat{\mathbf{i}}) = \left(\frac{mv}{2\sqrt{2}}\hat{\mathbf{i}} + \frac{mv}{2\sqrt{2}}\hat{\mathbf{j}}\right) + (2m)\,\mathbf{v}$$

$\mathbf{v}$ = velocity of mass 2m

$$= 0.32\hat{\mathbf{i}} - 0.35\hat{\mathbf{j}}$$

$$= (0.32v)\,\hat{\mathbf{i}} - (0.18v)\,\hat{\mathbf{j}}$$

$$\mathbf{v} = 0.37v$$

$$\tan\theta = \frac{0.18v}{0.32v} = 0.5625$$

$$\therefore \quad \theta = 29.35°$$

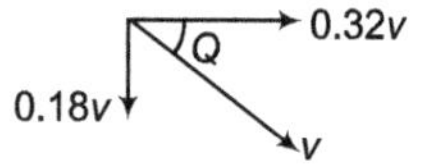

Since, $(\theta + 45°) < 90°$. Therefore, the angle of divergence between particles after collision is less than 90°.

Further, $K_i = \frac{1}{2}mv^2$ and

$$K_f = \frac{1}{2}m\left(\frac{v}{2}\right)^2 + \frac{1}{2}(2m)(0.37v)^2$$

$$K_f < K_i$$

Therefore, collision is inelastic.

2. Just after collision,

$$v_m = \left(\frac{m-5m}{m+5m}\right)\sqrt{2gl} = -\frac{2}{3}\sqrt{2gl}$$

$$v_{5m} = \left(\frac{2\times m}{m+5m}\right)\sqrt{2gl} = \frac{\sqrt{2gl}}{3}$$

$$T - mg = \frac{mv_m^2}{l} = \frac{m}{l}\left(\frac{8gl}{9}\right)$$

$$\therefore \quad T = \frac{17}{9}mg$$

$$h_m = \frac{v_m^2}{2g} = \frac{4l}{9}$$

3. $u\cos\theta = v\cos\phi$ ...(i)

$$v\sin\phi = eu\sin\theta$$

or $\quad eu\sin\theta = v\sin\phi$ ...(ii)

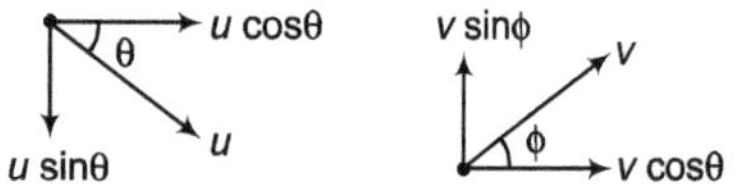

From Eqs. (i) and (ii), we can see that,

$$\tan\phi = e\tan\theta$$

Momentum or velocity changes only in vertical direction.

$$\therefore \quad |\text{Impulse}| = |\Delta P|$$

$$= m(u\sin\theta + eu\sin\theta)$$

$$= m(1+e)\,u\sin\theta$$

$$v = \sqrt{(v\cos\phi)^2 + (v\sin\phi)^2}$$

$$= \sqrt{(u\cos\theta)^2 + (eu\sin\theta)^2}$$

$$= \sqrt{u^2(\cos^2\theta + e^2\sin^2\theta)}$$

$$= u\sqrt{1-(1-e^2)\sin^2\theta}$$

$$\frac{K_f}{K_i} = \frac{\frac{1}{2}mv^2}{\frac{1}{2}mu^2} = \frac{v^2}{u^2} = \cos^2\theta + e^2\sin^2\theta$$

4. Impulse $= \Delta\mathbf{p} = m(\mathbf{v}_f - \mathbf{v}_i)$

$\therefore$ Impulse recieved by m

$$= m[(-2\hat{\mathbf{i}} + \hat{\mathbf{j}}) - (3\hat{\mathbf{i}} + 2\hat{\mathbf{j}})] = m(-5\hat{\mathbf{i}} - \hat{\mathbf{j}})$$

Impulse recieved by M = – (impulse received by m)

$$= m(5\hat{\mathbf{i}} + \hat{\mathbf{j}})$$

5. $a = \dfrac{\text{Net pulling force}}{\text{Total mass}} = \dfrac{m_2 g}{m_1 + m_2}$

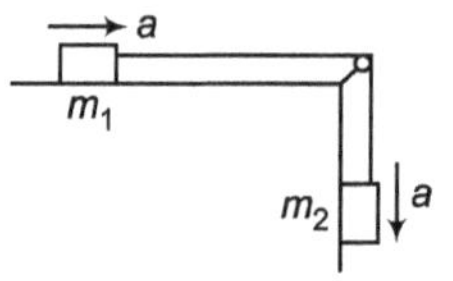

Now, $\quad (a_{cm})_x = \dfrac{m_1 a}{m_1 + m_2} = \dfrac{m_1 m_2 g}{(m_1 + m_2)^2}$

$$(a_{cm})_y = \frac{m_2 a}{m_1 + m_2} = \left(\frac{m_2}{m_1 + m_2}\right)^2 g$$

6. Out of two blocks, one block of mass m is moving in vertical direction also (downwards). Therefore, COM is moving vertically system is not conserved in vertical direction.

7. $|\text{Impulse}| = |\Delta p_1 \text{ or } \Delta p_2|$

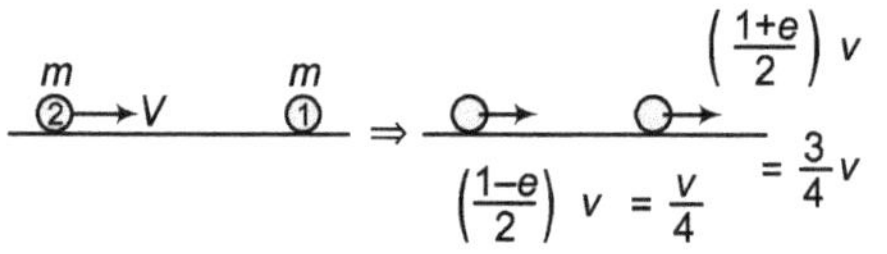

$$= m\left(\frac{3}{4}v\right) = \frac{3mv}{4}$$

Loss of kinetic energy $= K_i - K_f$

$$= \frac{1}{2}mv^2 - \left[\frac{1}{2}m\left(\frac{3v}{4}\right)^2 + \frac{1}{2}m\left(\frac{v}{4}\right)^2\right] = \frac{3}{16}mv^2$$

8. External force gravity acts on system. Therefore, momentum of system is not conserved.

Mass keep on decreasing. Therefore, acceleration will keep on increasing.

Comprehension Based Questions

1. Impulse $= mu$

$$\therefore \quad u = \frac{\text{Impulse}}{m} = \frac{4}{2} = 2\text{ m/s}$$

Now, $\quad E_i = E_f$

$$\therefore \quad \frac{1}{2}\times 2\times(2)^2 + \frac{1}{2}\times 4000\times(0.05)^2 = \frac{1}{2}\times 2\times v^2$$

Solving we get,

$$v = 3\text{ m/s}$$

2. Again using,

$$E_i = E_f$$

$$\therefore \quad \frac{1}{2}\times 2\times(2)^2 + \frac{1}{2}\times 4000\times(0.05)^2 = \frac{1}{2}\times 4000\times x^2$$

$\therefore \quad x = 0.067 \text{ m} = 6.7 \text{ m} = \text{compression}$

$\therefore \quad d = 5 \text{ cm} + 6.7 \text{ cm} + 6.7 \text{ cm} + 6.7 \text{ cm} \approx 25 \text{ cm}$

3. $p_i = p_f$

$\therefore \quad 0 = (8m)(2v) - (16m)(v) + (48m)v'$

Here, v' = absolute speed of rod $= 0$

$\therefore$ Displacement of rod = 0

4. $X_A + X_B = 12L$

$\therefore \quad 2vt + vt = 12L$

$\therefore \quad vt = 4L \Rightarrow X_B = vt = 4L$

5. $(24m)\left(\frac{v}{2} - u\right) = (48m)u$

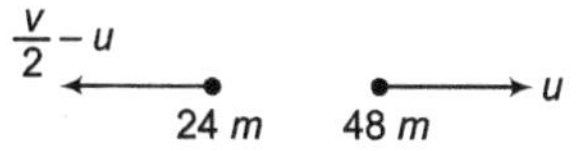

$\therefore \quad u = \frac{v}{6}$

6. $t = t_1 + t_2$

$$= \left(\frac{4L}{v}\right) + \left(\frac{12L - 4L}{v/2}\right) = \frac{20L}{v}$$

But $\quad \frac{L}{v} = T = 4 \text{ s}$

$\therefore \quad t = 80 \text{ s}$

7. Till t_1, rod is stationary. For time t_2, rod is moving with absoulte speed $u (= v/6)$

$\therefore$ Displacement of rod $= \left(\frac{v}{6}\right)t_2$

$$= \left(\frac{v}{6}\right)\left(\frac{16L}{v}\right) = \frac{8L}{3}$$

Assertion and Reason

2. It will fall at the same point.

3. Path of centre of mass remains unchanged.

4. Rocket does not push the surrounding air, but Newton's third law acts between rocket and the exhaust gases.

7. Weight is the external force.

8. Centre of mass remains stationary.

9. $e = \frac{|RVOS|}{|RVOA|}$

In elastic collision, $e = 1$

$\therefore \quad |RVOS| = |RVOA|$

11. In elastic collision magnitude of relative velocity of separation = magnitude of relative velocity of approach.

Relative speed after collision = Relative speed before collision

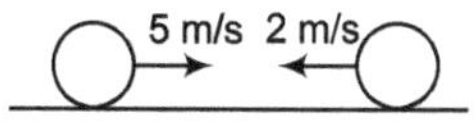

For example in the shown figure,

$|RVOA| = 7 \text{ m/s}$

But, relative speed $= 3 \text{ m/s}$

14. Two bodies are released from rest in space. Net force on the system is zero. Momentum of system is constant. But momentum of individual body is not constant. Further, kinetic energy of system is also increasing.

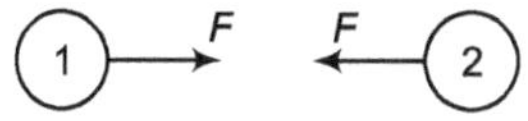

15. $v_{CM} = \frac{m_A(+v) + m_B(-v)}{m_A + m_B} = -\text{ve}$

Force of friction, $f_B > f_A$

f_B is acting towards right and f_A towards left.

16. Momentum is always constant during any type of collision.

17. Gravity forces act during explosion.

18. Mass of second liquid is more. Therefore, centre of mass of second liquid is at lower heights.

19. $K = \frac{p^2}{2m}$

As momentum is equal and opposite.

$\therefore \quad K \propto \frac{1}{m}$

20. Only in case of perfectly inelastic collision they will come to rest.

21. Till then spring will remain compressed. Therefore, a force will act on block A from the wall.

22. Force of friction will act on B towards right (from ground).

23. Atmosphere is absent on the surface of moon. Gravity is present.

Match the Columns

2. $v_{CM} = \frac{m_1v_1 + m_2v_2}{m_1 + m_2} = 1 \text{ m/s}$

During maximum compression also, velocity of combined mass is 1 m/s

Now, $\quad U_{max} = K_i - K_f$

$$= \frac{1}{2} \times 2(2)^2 - \frac{1}{2} \times 4 \times (1)^2 = 2\text{J}$$

From $\quad \frac{1}{2}KX_{max}^2 = 2\text{J}$

We have, $\quad X_{max} = 1 \text{ m}$

3. $p_1 + p_2 = p \quad$ ...(i)

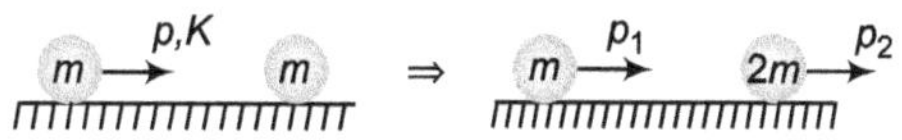

Further, $\quad K_1 + K_2 = K$

or $\quad \frac{p_1^2}{2m} + \frac{p_2^2}{4m} = \frac{p^2}{2m}$

or $\quad 2p_1^2 + p_2^2 = 2p^2 \quad$ ...(ii)

Solving these two equations, we get

$$p_2 = \frac{4}{3}p \text{ and } p_1 = -\frac{p}{3}, K_1 = \frac{K}{9}$$

and $\quad K_2 = \frac{8K}{9}$

4. (a) $K = \frac{p^2}{2m}$

$$K' = \frac{(3p)^2}{2m} = 9\left(\frac{p^2}{2m}\right) = 9K$$

$\therefore$ % increase in K = 800%

(b) $p=\sqrt{2km}$

$$p'=\sqrt{2(4K)m}=2\sqrt{2km}=2p$$

$\therefore$ % increase in $p=100\%$

(c) $p=\sqrt{2km}=(2km)^{1/2}$

% increase in $p=\frac{1}{2}$ (% increase in K) $=0.5\%$

5. Apply $X_{CM}=\dfrac{m_1x_1+\ldots+m_nx_n}{m_1+\ldots+m_n}$

and $\quad Y_{CM}=\dfrac{m_1y_1+\ldots+m_ny_n}{m_1+\ldots+m_n}$

6. $v_{CM}=\dfrac{m_1v_1+m_2v_2}{m_1+m_2}=\dfrac{(1)(10)+(2)(-5)}{3}=0$

Similarly, $\quad p_{CM}=0$

Net force on the system is zero, hence, Vcm and Pcm will remain constant.

Velocity of 1 kg and 2 kg blocks keep on changing initially and finally both of them stop as Vcm was zero.

7. From $a_{CM}=\dfrac{m_1a_1+m_2a_2}{m_1+m_2}$

$$=\frac{(1)(-10)+2(-10)}{3}=-10\text{ m/s}^2$$

$$u_{CM}=\frac{m_1u_1+m_2u_2}{m_1+m_2}=\frac{(1)(60)+(2)(0)}{3}$$

$$=+20\text{ m/s}$$

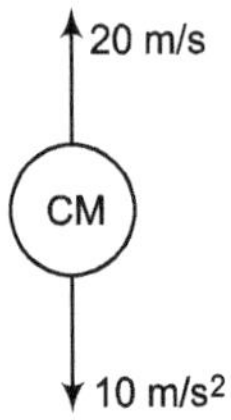

8. When $e=1$, collision is elastic and equal masses exchange their velocities.

For $e=0$, collision is perfectly inelastic. Hence, velocity of each will remain half.

In the last case, when $v_2=\frac{3}{4}v$

Then, $v_1=\frac{v}{4}$ (from conservation of momentum)

$$\therefore \quad e=\frac{\frac{3v}{4}-\frac{v}{4}}{v}=\frac{1}{2}$$

9. $\mathbf{F}_{CM}=\mathbf{F}_1+\mathbf{F}_2=m_1a_1+m_2a_2=(2\hat{\mathbf{i}}+8\hat{\mathbf{j}})$

$$\therefore \quad |\mathbf{F}_{CM}|=\sqrt{4+64}$$

$$\text{CM}=\sqrt{68}\text{ unit}$$

$$\mathbf{v}_{CM}=\frac{m_1\mathbf{v}_1+m_2\mathbf{v}_2}{m_1+m_2}$$

$$=\frac{(1)(4\hat{\mathbf{i}})+(2)(4\hat{\mathbf{j}})}{3}=\frac{4\hat{\mathbf{i}}+8\hat{\mathbf{j}}}{3}$$

$$\therefore \quad |\mathbf{v}_{CM}|=\frac{1}{3}\sqrt{16+64}=\frac{\sqrt{80}}{3}\text{ unit}$$

$$\mathbf{s}_1=\int_0^2 \mathbf{v}_1\,dt=(4\hat{\mathbf{i}})$$

$$\mathbf{s}_2=\int_0^2 \mathbf{v}_2\,dt=\left(\frac{8}{3}\hat{\mathbf{j}}\right)$$

Now, $\quad \mathbf{s}_{CM}=\dfrac{m_1\mathbf{s}_1+m_2\mathbf{s}_2}{m_1+m_2}$

$$=\frac{(1)(4\hat{\mathbf{i}})+2\left(\frac{8}{3}\hat{\mathbf{j}}\right)}{3}=\left(\frac{4}{3}\hat{\mathbf{i}}+\frac{16}{9}\hat{\mathbf{j}}\right)$$

$$\therefore \quad |\mathbf{s}_{CM}|=\sqrt{\frac{16}{9}+\frac{256}{81}}=\frac{20}{9}$$

Entrance Gallery

1. In elastic collision, both momentum and kinetic energy are conserved.

2. We know that,

$$\text{Acceleration, } a_{CM}=\left(\frac{m_1-m_2}{m_1+m_2}\right)^2\times g \qquad (\because m_1>m_2)$$

So, resultant external force,

$$F=(m_1+m_2)a_{CM}=(m_1+m_2)\times\left(\frac{m_1-m_2}{m_1+m_2}\right)^2\times g$$

$$F=\frac{(m_1-m_2)^2}{(m_1+m_2)}\times g$$

3. (a) At equilibrium $(t=0)$, particle has maximum velocity u_{max}. Therefore, velocity at time t can be written as,

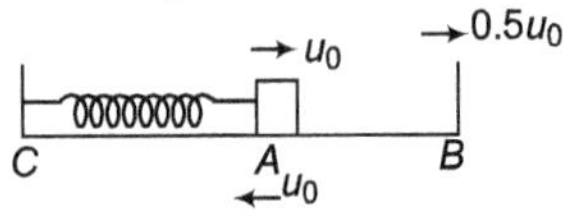

$$u=u_{max}\cos\omega t+u_0\cos\omega t \Rightarrow u=0.5u_0=u_0\cos\omega t$$

$$\therefore \quad \omega t=\frac{\pi}{3} \Rightarrow \frac{2\pi}{T}t=\frac{\pi}{3} \Rightarrow t=\frac{T}{6}$$

(b) $t=t_{AB}+t_{BA}=\dfrac{T}{6}+\dfrac{T}{6}=\dfrac{T}{3}=\dfrac{2\pi}{3}\sqrt{\dfrac{m}{k}}$

(c) $t=t_{AB}+t_{BA}+t_{AC}=\dfrac{T}{6}+\dfrac{T}{6}+\dfrac{T}{4}=\dfrac{7}{12}T=\dfrac{7\pi}{6}\sqrt{\dfrac{m}{k}}$

(d) $t=t_{AB}+t_{BA}+t_{AC}+t_{CA}$

$$=\frac{T}{6}+\frac{T}{6}+\frac{T}{4}+\frac{T}{4}=\frac{5}{6}T=\frac{5\pi}{3}\sqrt{\frac{m}{k}}$$

4. Velocity of first bob at highest point,

$$v_1=\sqrt{gR}=\sqrt{gl_1} \quad \text{(to just complete the vertical circle)}$$

Velocity of second bob just after elastic collision

= Velocity of second bob at the bottom most point

$$=\sqrt{5gl_2} \Rightarrow \frac{l_1}{l_2}=5$$

5. Maximum energy loss

$$=\frac{p^2}{2m}-\frac{p^2}{2(m+M)} \qquad \left(\because \text{KE}=\frac{p^2}{2m}\right)$$

Before collision, the mass is m and after collision mass is $m+M$.

$$=\frac{p^2}{2m}\left[\frac{M}{(m+M)}\right]=\frac{1}{2}mv^2\left\{\frac{M}{m+M}\right\}=\frac{1}{2}mv^2f$$

$$\therefore \quad \left(f=\frac{M}{m+M}\right)$$

Sol. 6 and 7

(*i*) Every particle of the disc is rotating in a horizontal circle.

(*ii*) Actual velocity of any particle is horizontal.

(*iii*) Magnitude of velocity of any particle is given by

$$v=r\omega$$

where, r is the perpendicular distance of that from actual axis of rotation (z-axis).

(*iv*) When it is broken into two parts, then actual velocity of any particle is resultant of two velocities,

$$v_1=r_1\omega_1 \text{ and } v_2=r_2\omega_2$$

Here,

r_1 = perpendicular distance of centre of mass from z-axis

ω_1 = angular speed of rotation of centre of mass from z-axis

r_2 = distance of particle from centre of mass

ω_2 = angular speed of rotation of the disc about the axis passing through centre of mass

(v) Net v will be horizontal, if v_1 and v_2 both are horizontal. Further, v_1 is already horizontal, because centre of mass is rotating about a vertical z-axis. To make v_2 also be vertical.

8. Along y -axis, momentum remains zero. Here,

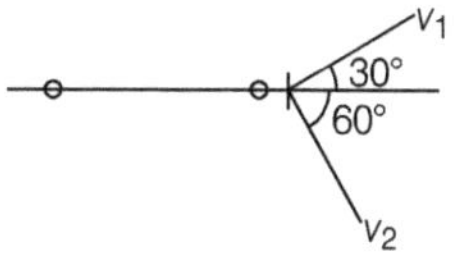

$$4v_1 \sin 30° = v_2 \sin 60°$$

$$\frac{v_1}{v_2} = \frac{\sqrt{3}}{4}$$

9. The data is incomplete. Let us assume that friction from ground on ring is not impulsive during impact.

From linear momentum, conservation is in the horizontal direction, we have

–ve
+ve

$$(-2\times 1) + (0.1\times 20) = (0.1\times 0) + (2\times v)$$

Here, v is the velocity of centre of mass of ring after impact.

Solving the above equation, we have

$$v = 0$$

Thus, centre of mass becomes stationary.

Linear impulse during impact

(*i*) In horizontal direction, $J_1 = \Delta p = 0.1\times 20 = 2$ N-s

(*ii*) In vertical direction, $J_2 = \Delta p = 0.1\times 10 = 1$ N-s

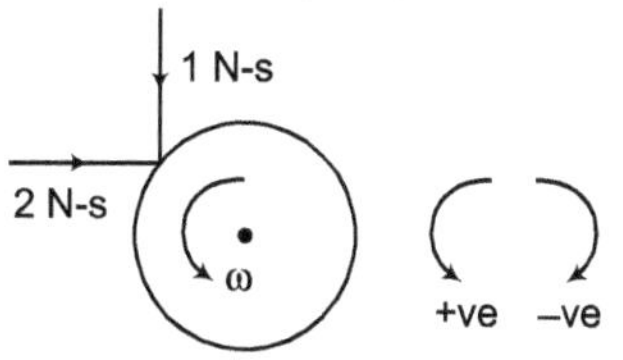

Writing the equation (about CM),

angular impulse = change in angular momentum

We have,

$$1\times\left(\frac{\sqrt{3}}{2}\times\frac{1}{2}\right) - 2\times 0.5\times\frac{1}{2} = 2\times(0.5)^2\left[\omega - \frac{1}{0.5}\right]$$

Solving this equation, ω comes out to be positive or ω anti-clockwise. So just after collision rightwards slipping is taking place. Hence, friction is leftwards. Therefore, option (c) is also correct.

10. Time taken by the bullet and ball to strike the ground is

$$t = \sqrt{\frac{2h}{g}} = \sqrt{\frac{2\times 5}{10}} = 1 \text{ s}$$

Let v_1 and v_2 are the velocities of ball and bullet after collision. Then, applying

$$x = vt$$

We have, $20 = v_1 \times 1$ or $v_1 = 20$ m/s

$100 = v_2 \times 1$ or $v_2 = 100$ m/s

Now, from conservation of linear momentum before and after collision, we have

$$0.01v = (0.2\times 20) + (0.01\times 100)$$

On solving, we get

$$v = 500 \text{ m/s}$$

11. Momentum of third piece,

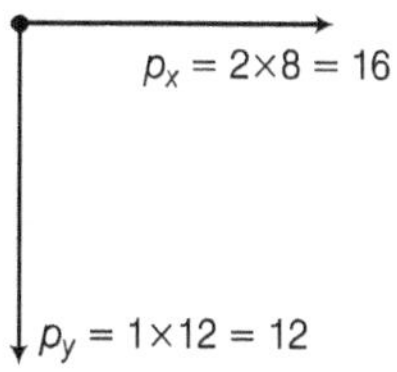

$p_x = 2\times 8 = 16$
$p_y = 1\times 12 = 12$

$$p = \sqrt{p_x^2 + p_y^2} = \sqrt{(16)^2 + (12)^2}$$

$$p = 20 \text{ kg-m/s}$$

$$v = \frac{p}{m} = \frac{20}{0.5} = 40 \text{ m/s}$$

12. Change in momentum

Force acted on the bat

$$\Delta p = 2mv = 2\times 0.25\times 10 = 5 \text{ kg-m/s}$$

$$F = \frac{\Delta p}{\Delta t} = \frac{5}{0.01} = 500 \text{ N}$$

13. Momentum of 5kg mass after the collision is 4 kg m/s.

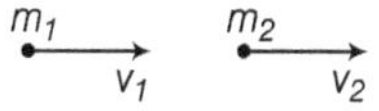

14. If it is a completely inelastic collision then,

$$m_1v_1 + m_2v_2 = m_1v + m_2v$$

$$v = \frac{m_1v_1 + m_2v_2}{m_1 + m_2}$$

$$\text{KE} = \frac{p_1^2}{2m_1} + \frac{p_2^2}{2m_2}$$

As, p_1 and p_2 both simultaneously cannot be zero therefore, total kinetic energy cannot be lost.

15. For translatory motion the force should be applied on the centre of mass of the body so, we have to calculate the location of centre of mass of T shaped object.

Let mass of rod AB is m so the mass of rod CD will be $2m$.

Let y_1 is the centre of mass of rod AB and y_2 is the centre of mass of rod CD. We can consider that whole mass of the rod is placed at their respective centre of mass i.e. mass m is placed at y_1 and mass $2m$ is placed at y_2.

Taking point C at the origin position vector of points y_1 and y_2 an be written as

$$\mathbf{r}_1 = 2l\,\hat{\mathbf{j}}, \mathbf{r}_2 = l\,\hat{\mathbf{j}}$$

and $m_1 = m$ and $m_2 = 2m$

Position vector of centre of mass of the system,

$$\mathbf{r}_{CM} = \frac{m_1\mathbf{r}_1 + m_2\mathbf{r}_2}{m_1 + m_2} = \frac{m2l\,\hat{\mathbf{j}} + 2ml\,\hat{\mathbf{j}}}{m + 2m} = \frac{4ml\,\hat{\mathbf{j}}}{3m} = \frac{4l\,\hat{\mathbf{j}}}{3}$$

16. From the graph, it is a straight line so, it represents uniform motion. Because of impulse direction of velocity changes as can be seen from the slope of the graph.

Initial velocity, $\mathbf{v}_1 = \frac{2}{2} = 1 \text{ ms}^{-1}$

Final velocity, $\mathbf{v}_2 = \frac{-2}{2} = -1 \text{ ms}^{-1}$

$$\mathbf{p}_i = m\mathbf{v}_1 = 0.4 \text{ N-s}$$

$$\mathbf{p}_f = m\mathbf{v}_2 = -0.4 \text{ N-s}$$

$$\mathbf{J} = \mathbf{p}_f - \mathbf{p}_i = -0.4 - 0.4 = -0.8 \text{ N-s}$$

$\therefore$ $|\mathbf{J}| = 0.8$ N-s

9

Rotation

9.1 Moment of Inertia

Like the centre of mass, the moment of inertia is a property of an object that is related to its mass distribution. The moment of inertia (denoted by I) is an important quantity in the study of system of particles which are rotating. The role of the moment of inertia in the study of rotational motion is analogous to that of mass in the study of linear motion. Moment of inertia gives a measurement of the resistance of a body to a change in its rotational motion. If a body is at rest, the larger the moment of inertia of a body, the more difficult it is to put that body into rotational motion. Similarly, the larger the moment of inertia of a body, the more difficult it is to stop its rotational motion. The moment of inertia is calculated about some axis (usually the rotational axis) and it depends on the mass as well as its distribution about that axis.

Chapter Snapshot

- Moment of Inertia
- Torque
- Rotation of a Rigid Body about a Fixed Axis
- Angular Momentum
- Conservation of Angular Momentum
- Combined Translational and Rotational Motion of a Rigid Body
- Uniform Pure Rolling
- Accelerated Pure Rolling
- Angular Impulse

Moment of Inertia of a Single Particle

For a very simple case, the moment of inertia of a single particle about an axis is given by

$$I = mr^2 \qquad \ldots\text{(i)}$$

Here, m is the mass of the particle and r its distance from the axis under consideration.

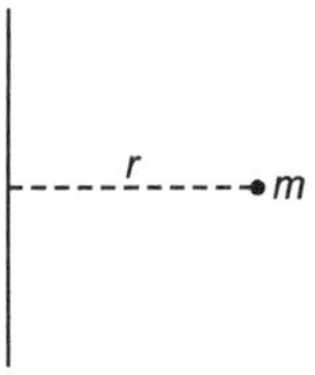

Fig. 9.1

Moment of Inertia of a System of Particles

The moment of inertia of a system of particles about an axis is given by

$$I = \sum_i m_i r_i^2 \qquad \ldots\text{(ii)}$$

where, r_i is the perpendicular distance from the axis to the ith particle, which has a mass m_i.

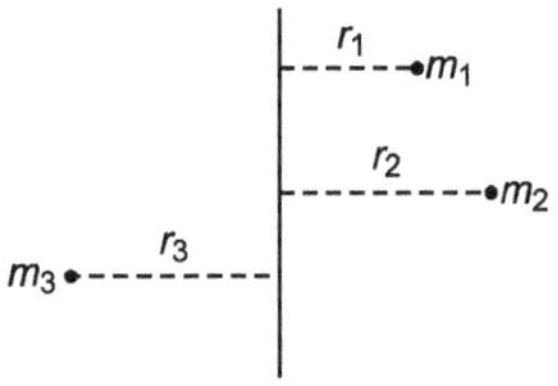

Fig. 9.2

Moment of Inertia of Rigid Bodies

For a continuous mass distribution such as found in a rigid body, we replace the summation of Eq. (ii) by an integral. If the system is divided into infinitesimal elements of mass dm and if r is the distance from a mass element to the axis of rotation, the moment of inertia is

$$I = \int r^2 \, dm$$

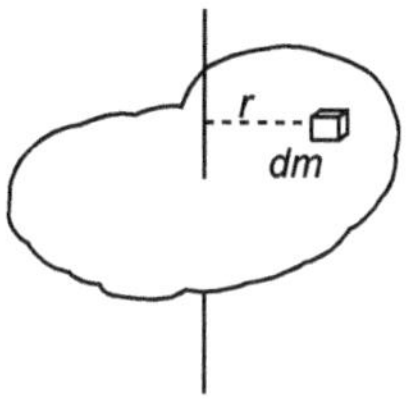

Fig. 9.3

where, integral is taken over the system.

Moment of Inertia of a Uniform Cylinder

Let us find the moment of inertia of a uniform cylinder about an axis through its centre of mass and perpendicular to its base. Mass of the cylinder is M and radius is R.

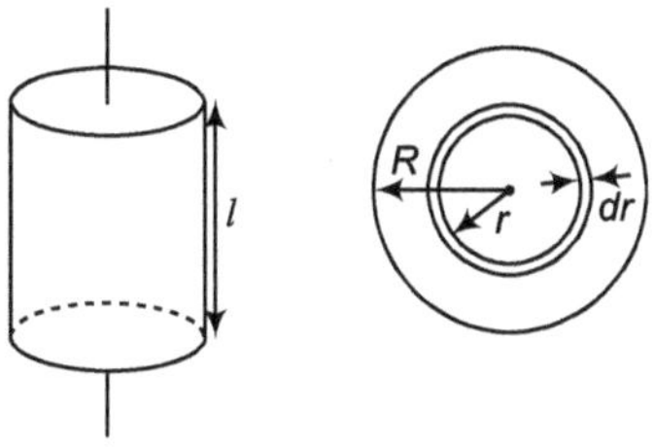

(a) (b)

Fig. 9.4

We first divide the cylinder into annular shells of width dr and length l as shown in figure. The moment of inertia of one of these shells is

$$dI = r^2 dm = r^2 \, (\rho dV)$$

Here, ρ = density of cylinder

and dV = volume of shell $= 2\pi rl \, dr$

$\therefore \quad dI = 2\pi\rho l \, r^3 dr$

The cylinder's moment of inertia is found by integrating this expression between 0 and R

So, $$I = 2\pi\rho l \int_0^R r^3 dr = \frac{\pi\rho l}{2} R^4 \qquad \text{...(iii)}$$

The density ρ of the cylinder is the mass divided by the volume.

$$\therefore \quad \rho = \frac{M}{\pi R^2 l} \qquad \text{...(iv)}$$

From Eqs. (iii) and (iv), we have

$$I = \frac{1}{2} MR^2$$

Proceeding in the similar manner we can find the moment of inertia of certain rigid bodies about some given axis. Moments of inertia of several rigid bodies with symmetry are listed in Table. 9.1.

In all cases except (f), the rotational axis AA' passes through the centre of mass.

Table 9.1

(a) Very thin circular hoop

$I = MR^2$

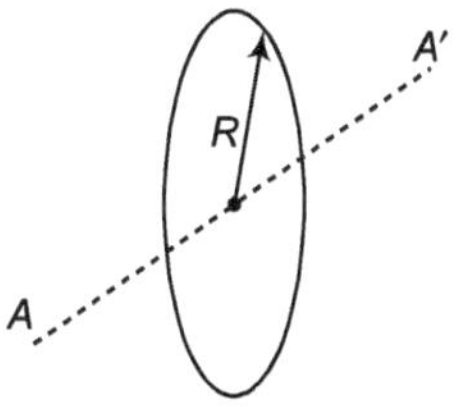

(b) Uniform circular hoop

$I = M \dfrac{R_1^2 + R_2^2}{2}$

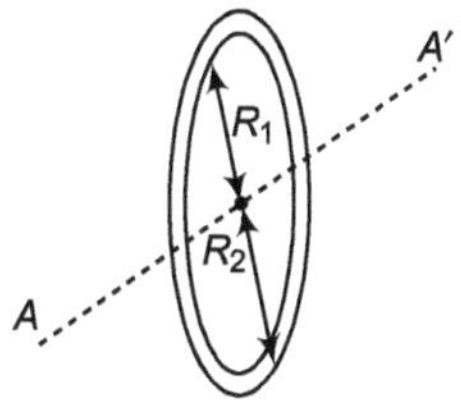

(c) Uniform solid cylinder

$I = \dfrac{1}{2} MR^2$

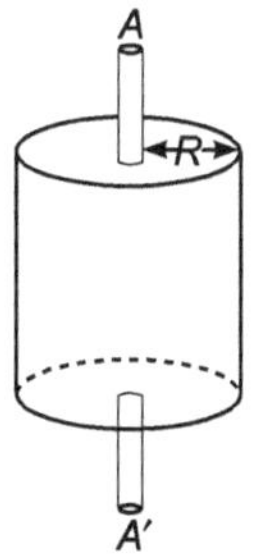

(d) Uniform solid sphere

$I = \dfrac{2}{5} MR^2$

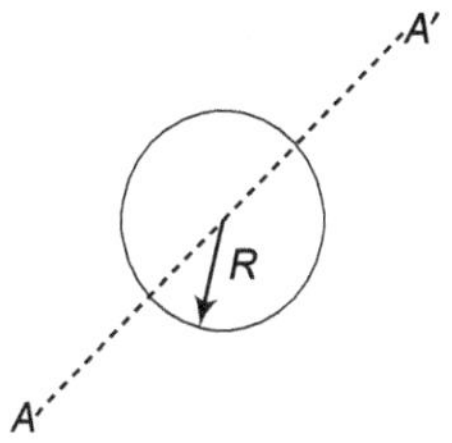

(e) Uniform thin rod

$I = \dfrac{1}{12} Ml^2$

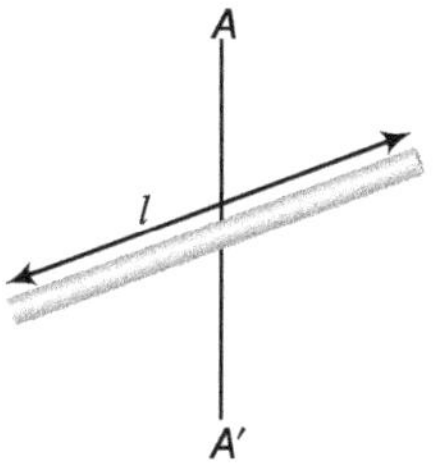

(f) Uniform thin rod

$I = \dfrac{1}{3} Ml^2$

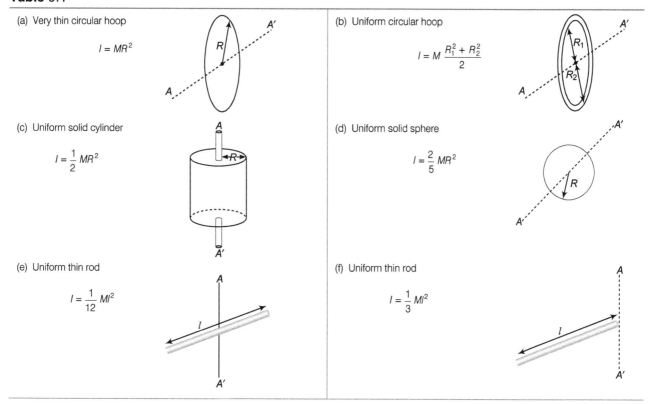

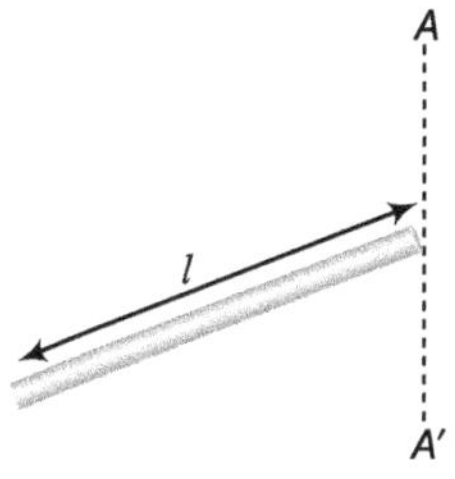

(g) Very thin spherical shell

$$I = \frac{2}{3} MR^2$$

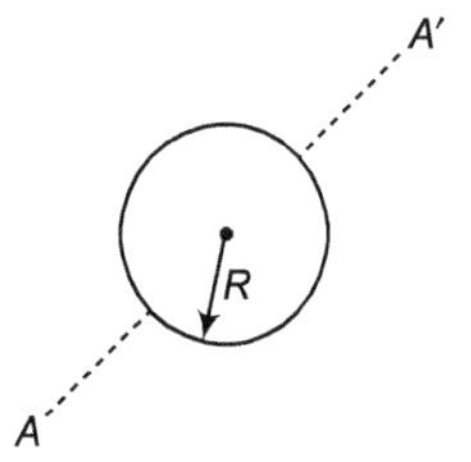

(h) Thin circular sheet

$$I = \frac{1}{4} MR^2$$

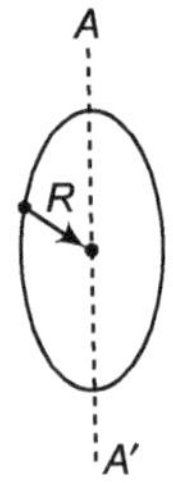

(i) Thin rectangular sheet

$$I = M\left(\frac{a^2 + b^2}{12}\right)$$

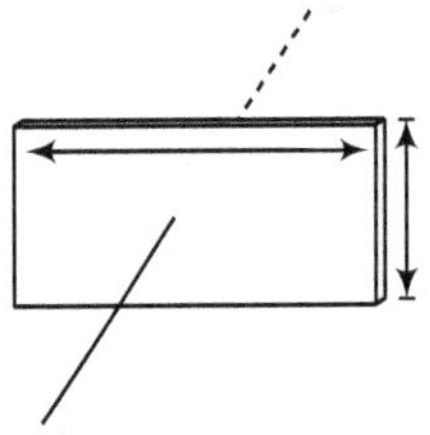

(j) Uniform right cone

$$I = \frac{3}{10} MR^2$$

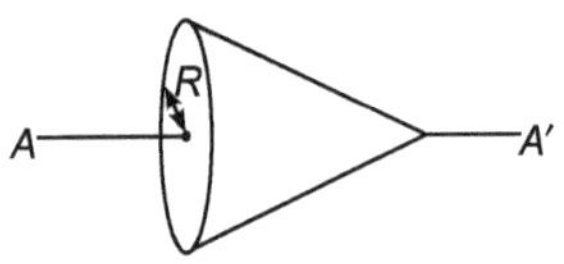

Theorem on Moment of Inertia

There are two important theorems on moment of inertia, which in some cases, enable the moment of inertia of a body to be determined about an axis, if its moment of inertia about some other axis is known. Let us now discuss both of them.

Theorem of Parallel Axis

A very useful theorem, called the parallel axis theorem relates the moment of inertia of a rigid body about two parallel axes, one of which passes through the centre of mass.

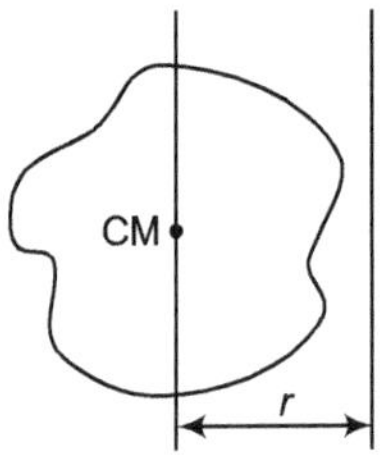

Fig. 9.5

Two such axes are shown in figure for a body of mass M. If r is the distance between the axes and I_{CM} and I are the respective moments of inertia about them, these moments are related by

$$I = I_{CM} + Mr^2$$

- From the above theorem we can see that among too many parallel axes moment of inertia is least about an axis which passes through centre of mass. *e.g.* I_2 is least among I_1, I_2 and I_3. Similarly, I_5 is least among I_4, I_5 and I_6.

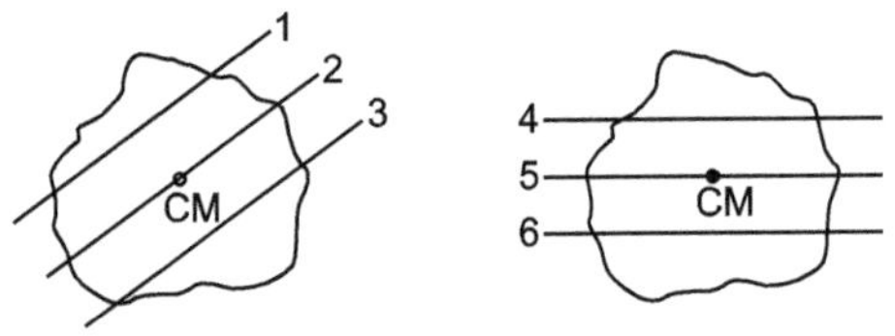

Fig. 9.6

Theorem of Perpendicular Axis

This theorem is applicable only to the plane bodies (two dimensional). The theorem states that the moment of inertia of a plane lamina about an axis perpendicular to the plane of the lamina is equal to the sum of the moments of inertia of the lamina about two axes perpendicular to each other, in its own plane and intersecting each other at the point, where the perpendicular axis passes through it. Let x and y axes be chosen in the plane of the body and z-axis perpendicular to this plane, three axes being mutually perpendicular, then the theorem states that

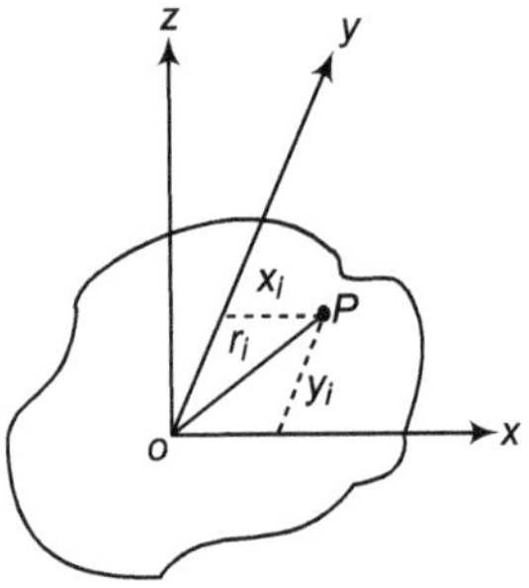

Fig. 9.7

$$I_z = I_x + I_y$$

Radius of Gyration

Radius of gyration (K) of a body about an axis is the effective distance from this axis, where the whole mass can be assumed to be concentrated so that the moment of inertia remains the same. Thus,

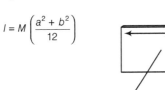

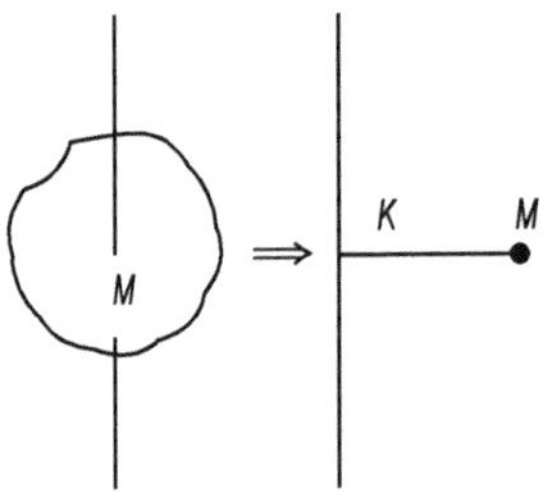

Fig. 9.8

$$I = MK^2 \quad \text{or} \quad K = \sqrt{\frac{I}{M}}$$

e.g. Radius of gyration of a disc about an axis perpendicular to its plane and passing through its centre of mass is

$$K = \sqrt{\frac{\frac{1}{2}MR^2}{M}} = \frac{R}{\sqrt{2}}$$

Example 9.1 *Three rods each of mass m and length l are joined together to form an equilateral triangle as shown in figure. Find the moment of inertia of the system about an axis passing through its centre of mass and perpendicular to the plane of the triangle.*

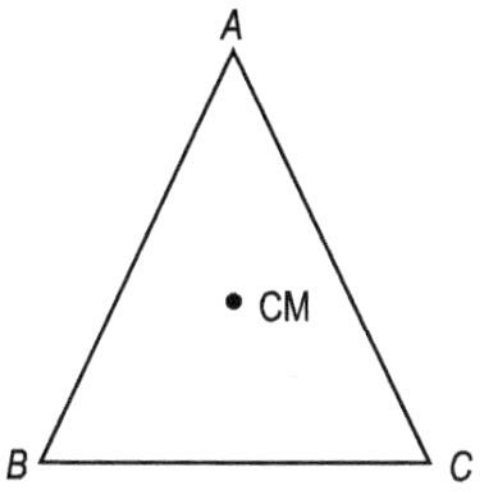

Fig. 9.9

Sol. Moment of inertia of rod BC about an axis perpendicular to plane of triangle ABC and passing through the mid-point of rod BC (i.e. D) is

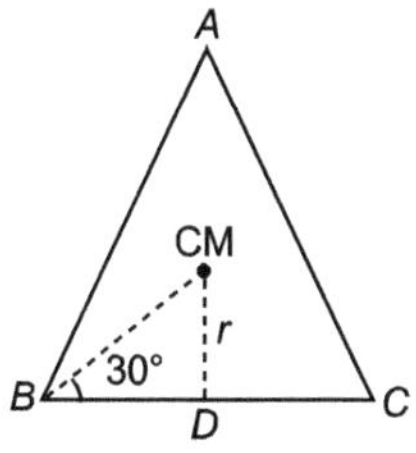

Fig. 9.10

$$I_1 = \frac{ml^2}{12}$$

From theorem of parallel axis, moment of inertia of this rod about the asked axis is

$$I_2 = I_1 + mr^2$$

$$= \frac{ml^2}{12} + m\left(\frac{l}{2\sqrt{3}}\right)^2$$

$$= \frac{ml^2}{6}$$

∴ Moment of inertia of all the three rods is

$$I = 3I_2 = 3\left(\frac{ml^2}{6}\right)$$

$$= \frac{ml^2}{2}$$

Extra Knowledge Points

- Theorem of parallel axis is applicable for any type of rigid body whether it is a two dimensional or three dimensional, while the theorem of perpendicular axis is applicable for laminar type or two dimensional bodies only.
- In theorem of perpendicular axis, the point of intersection of the three axes (x, y and z) may be any point on the plane of body (it may even lie outside the body). This point may or may not be the centre of mass of the body.
- Moment of inertia of a part of a rigid body (symmetrically cut from the whole mass) is the same as that of the whole body. e.g. in Fig. (a), moment of inertia of the section shown (a part of a circular disc) about an axis perpendicular to its plane and passing through point O is $\frac{1}{2}MR^2$ as the moment of inertia of the complete disc is also $\frac{1}{2}MR^2$. This can be shown as in figure.

  Suppose, the given section is $\frac{1}{n}$ th part of the disc, then mass of the disc will be nM.

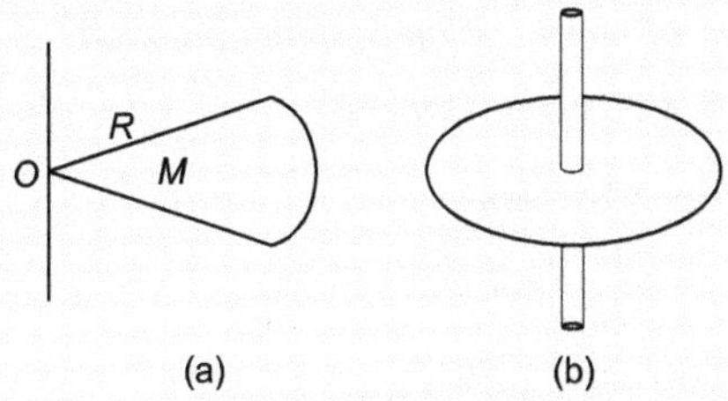

$$I_{\text{disc}} = \frac{1}{2}(nM)R^2$$

$$\therefore \quad I_{\text{section}} = \frac{1}{n}I_{\text{disc}} = \frac{1}{2}MR^2$$

- If a rigid body is rotating about a fixed axis with angular speed ω, all the particles in rigid body rotate same angle in same interval of time, i.e. their angular speed is same (ω). They rotate in different circles of different radii. The planes of these circles are perpendicular to the rotational axis. Linear speeds of different particles are different. Linear speed of a particle situated at a distance r from the rotational axis is

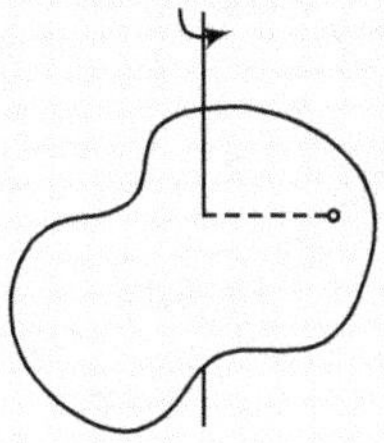

Example 9.2 *Find the moment of inertia of a solid sphere of mass M and radius R about an axis xx' shown in figure.*

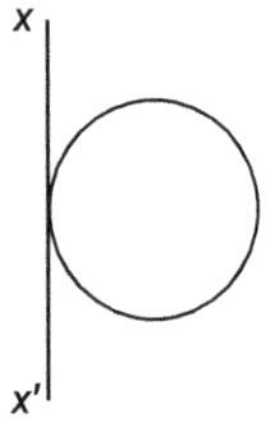

Fig. 9.11

Sol. From theorem of parallel axis,

$$I_{xx'} = I_{CM} + Mr^2$$
$$= \frac{2}{5}MR^2 + MR^2$$
$$= \frac{7}{5}MR^2$$

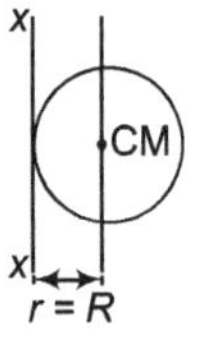

Fig. 9.12

Example 9.3 *Consider a uniform rod of mass m and length 2l with two particles of mass m each at its ends. Let AB be a line perpendicular to the length of the rod and passing through its centre. Find the moment of inertia of the system about AB.*

Sol. MI of the system about AB

$$I_{AB} = I_{rod} + I_{both\ particles}$$
$$= \frac{m(2l)^2}{12} + 2\,(ml^2)$$
$$= \frac{7}{3}ml^2$$
$$v = r\omega$$
or $\quad v \propto r$

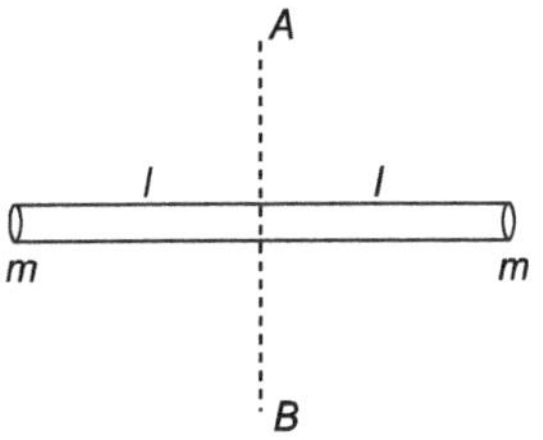

Fig. 9.13

9.2 Torque

Suppose a force **F** is acting on a particle P and let **r** be the position vector of this particle about some reference point O. The torque of this force **F**, about O is defined as

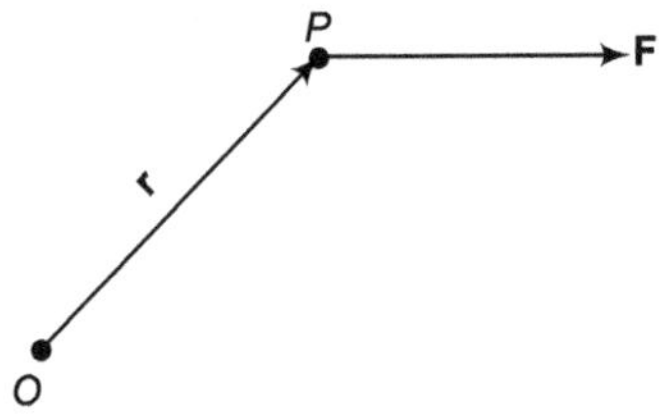

Fig. 9.14

$$\boldsymbol{\tau} = \mathbf{r} \times \mathbf{F}$$

This is a vector quantity having its direction perpendicular to both **r** and **F**, according to the rule of cross product.

Torque of a Force about a Line

Consider a rigid body rotating about a fixed axis AB. Let **F** be a force acting on the body at point P. Take the origin O somewhere on the axis of rotation. The torque of **F** about O is

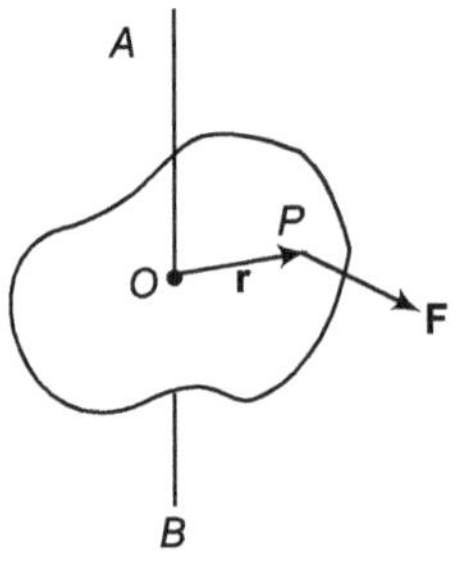

Fig. 9.15

$$\boldsymbol{\tau} = \mathbf{r} \times \mathbf{F}$$

Its component along AB is called the torque of **F** about AB.

Example 9.4 *A small ball of mass 1.0 kg is attached to one end of a 1.0 m long massless string and the other end of the string is hung from a point. When the resulting pendulum is 30° from the vertical, what is the magnitude of torque about the point of suspension? (Take $g = 10\ m/s^2$)*

Sol. Two forces are acting on the ball

(i) Tension (T) (ii) Weight (mg)

Torque of tension about point O is zero, as it passes through O.

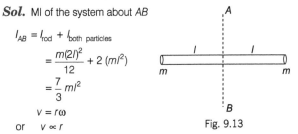

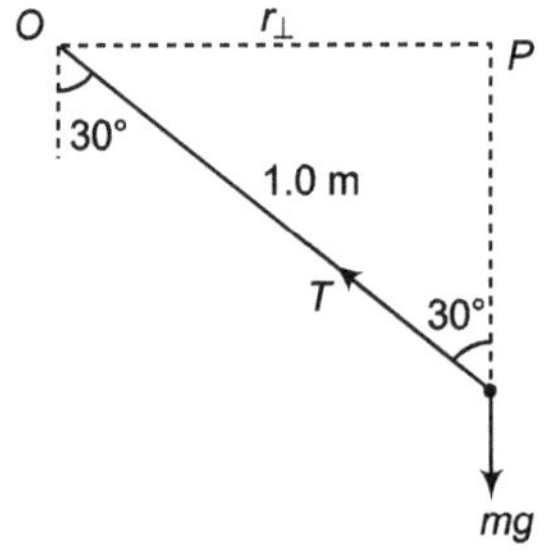

Fig. 9.16

$$\tau_{mg} = F \times r_{\perp}$$

Here, $\quad r_{\perp} = OP = 1.0 \sin 30°$
$$= 0.5\ \text{m}$$

$\therefore \quad \tau_{mg} = (mg)(0.5)$
$$= (1)(10)(0.5)$$
$$= 5\ \text{N-m}$$

Example 9.5 *Find the torque of a force* $\mathbf{F} = (\hat{\mathbf{i}} + 2\hat{\mathbf{j}} - 3\hat{\mathbf{k}})\,\text{N}$ *about a point O. The position vector of point of application of force about O is* $\mathbf{r} = (2\hat{\mathbf{i}} + 3\hat{\mathbf{j}} - \hat{\mathbf{k}})\,\text{m}$.

Sol. Torque, $\tau = \mathbf{r} \times \mathbf{F}$

$$= \begin{vmatrix} \hat{i} & \hat{j} & \hat{k} \\ 2 & 3 & -1 \\ 1 & 2 & -3 \end{vmatrix} = \hat{i}(-9+2) + \hat{j}(-1+6) + \hat{k}(4-3)$$

or $$\tau = (-7\hat{i} + 5\hat{j} + \hat{k})\text{N-m}$$

9.3 Rotation of a Rigid Body about a Fixed Axis

When a body is rotating about a fixed axis, any point P located in the body travels along a circular path. Before, analysing the circular motion of point P, we will first study the angular motion properties of a rigid body.

Angular Motion

Since, a point is without dimension, it has no angular motion. Only lines or bodies undergo angular motion. Let us consider the angular motion of a radial line r located with the shaded plane.

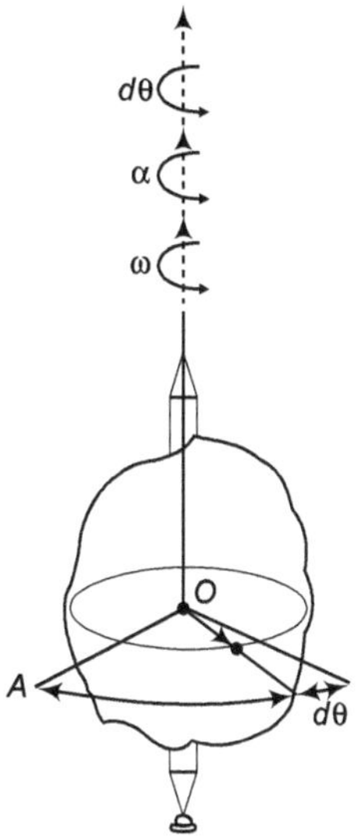

Fig. 9.17

Angular Position

The angular position of r is defined by the angle θ, measured between a fixed reference line OA and r.

Angular Displacement

The change in the angular position, often measured as a differential $d\theta$ is called the angular displacement. (Finite angular displacements are not vector quantities, although differential rotations $d\theta$ are vectors). This vector has a magnitude $d\theta$ and the direction of $d\theta$ is along the axis.

Specifically, the direction of $d\theta$ is determined by right hand rule; that is, the fingers of the right hand are curled with the sense of rotation, so that in this case the thumb or $d\theta$ points upward.

Angular Velocity

The time rate of change in the angular position is called the angular velocity ω. Thus,

$$\omega = \frac{d\theta}{dt} \qquad \ldots(\text{i})$$

It is expressed here in scalar form, since its direction is always along the axis of rotation, i.e. in the same direction as $d\theta$.

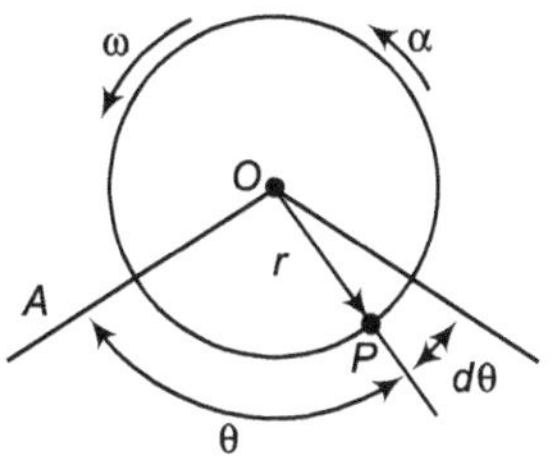

Fig. 9.18

Angular Acceleration

The angular acceleration α measures the time rate of change of the angular velocity. Hence, the magnitude of this vector may be written as,

$$\alpha = \frac{d\omega}{dt} \qquad \ldots(\text{ii})$$

It is also possible to express α as,

$$\alpha = \frac{d^2\theta}{dt^2}$$

The line of action of α is the same as that for ω, however its sense of direction depends on whether ω is increasing or decreasing with time. In particular if ω is decreasing, α is called an angular deceleration and therefore, has a sense of direction which is opposite to ω.

Torque and Angular Acceleration for a Rigid Body

The angular acceleration of a rigid body is directly proportional to the sum of the torque components along the axis of rotation. The proportionality constant is the inverse of the moment of inertia about that axis or

$$\alpha = \frac{\Sigma\tau}{I}$$

Thus, for a rigid body we have the rotational analog of Newton's second law

$$\Sigma\tau = I\alpha \qquad \ldots(\text{iii})$$

Following two points are important regarding the above equation

(i) The above equation is valid only for rigid bodies. If the body is not rigid like a rotating tank of water, the angular acceleration α is different for different particles.

(ii) The sum $\Sigma\tau$ in the above equation includes only the torques of the external forces, because all the internal torques add to zero.

Rotation with Constant Angular Acceleration

If the angular acceleration of the body is constant, then Eqs. (i) and (ii) when integrated yield a set of formulae which relate the body's angular velocity, angular position and time. These equations are similar to equations used for rectilinear motion. Table given below compares the linear and angular motion with constant acceleration.

Table 9.2

Straight line motion with constant linear acceleration	Fixed axis rotation with constant angular acceleration
a = constant	α = constant
$v = u + at$	$\omega = \omega_0 + \alpha t$
$s = s_0 + ut + \frac{1}{2}at^2$	$\theta = \theta_0 + \omega_0 t + \frac{1}{2}\alpha t^2$
$v^2 = u^2 + 2a(s - s_0)$	$\omega^2 = \omega_0^2 + 2\alpha(\theta - \theta_0)$

Here, θ_0 and ω_0 are the initial values of the body's angular position and angular velocity respectively.

Motion of Point *P*

As the rigid body as shown in figure rotates, point P travels along a circular path of radius r and centre at point O.

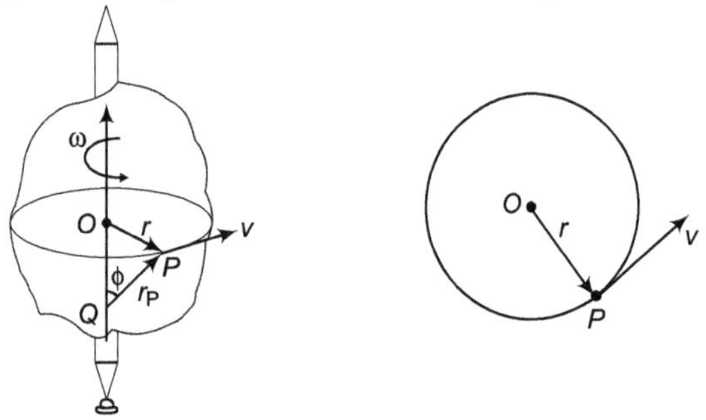

Fig. 9.19

Position

The position of P is defined by the position vector **r**, which extends from O to P.

Velocity

The velocity of P has a magnitude given by,

$$v = \omega r \qquad \text{...(iv)}$$

As shown in figures the direction of velocity **v** is tangential to the circular path.

Kinetic Energy of a Rigid Body Rotating About a Fixed Axis

Suppose a rigid body is rotating about a fixed axis with angular speed ω. Then, kinetic energy of the rigid body will be

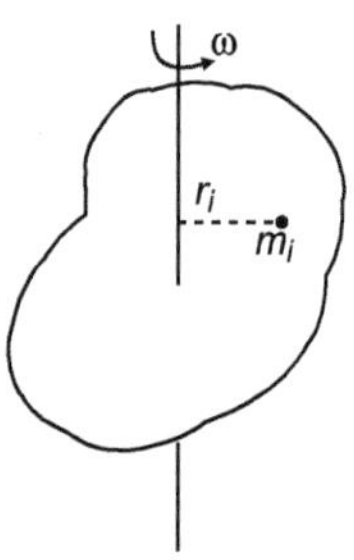

Fig. 9.20

$$K = \sum_i \frac{1}{2} m_i v_i^2 = \sum_i \frac{1}{2} m_i (\omega r_i)^2$$

$$= \frac{1}{2}\omega^2 \sum_i m_i r_i^2 = \frac{1}{2} I\omega^2 \qquad (\text{as } \sum_i m_i r_i^2 = I)$$

Thus, $\quad \text{KE} = \frac{1}{2} I\omega^2$

Sometimes it is called the rotational kinetic energy.

Example 9.6 *The angular position of a point on the rim of a rotating wheel is given by* $\theta = 4t - 3t^2 + t^3$, *where* θ *is in radians and t is in seconds. What are the angular velocities at*

(a) $t = 2.0$ s *and*

(b) $t = 4.0$ s

(c) What is the average angular acceleration for the time interval that begins at $t = 2.0$ s *and ends at* $t = 4.0$ s*?*

(d) What are the instantaneous angular acceleration at the beginning and the end of this time interval?

Sol. Angular velocity

$$\omega = \frac{d\theta}{dt} = \frac{d}{dt}(4t - 3t^2 + t^3)$$

or $\quad \omega = 4 - 6t + 3t^2$

(a) At $t = 2.0$ s,

$$\omega = 4 - 6 \times 2 + 3(2)^2$$

or $\quad \omega = 4$ rad/s

(b) At $t = 4.0$ s,

$$\omega = 4 - 6 \times 4 + 3(4)^2$$

or $\quad \omega = 28$ rad/s

(c) Average angular acceleration

$$\alpha_{av} = \frac{\omega_f - \omega_i}{t_f - t_i} = \frac{28 - 4}{4 - 2}$$

or $\quad \alpha_{av} = 12 \text{ rad/s}^2$

(d) Instantaneous angular acceleration is

$$\alpha = \frac{d\omega}{dt} = \frac{d}{dt}(4 - 6t + 3t^2)$$

or $\quad \alpha = -6 + 6t$

At $t = 2.0$ s,

$$\alpha = -6 + 6 \times 2 = 6 \text{ rad/s}^2$$

At $t = 4.0$ s

$$\alpha = -6 + 6 \times 4 = 18 \text{ rad/s}^2$$

Example 9.7 *A solid fly wheel of* 20 *kg mass and* 120 *mm radius revolves at* 600 *rev/min. With what force must a brake lining be pressed against it for the flywheel to stop in* 3 *s, if the coefficient of friction is* 0.1*?*

Sol. $n_0 = 600$ rev/min (revolutions per minute)

$$= \frac{600}{60} \text{ rev/min}$$

$= 10$ rev/s (revolutions per second)

So, initial angular velocity,

$$\omega_0 = 2\pi n_0 = (2\pi)(10) = 20\pi \text{ rad/s}$$

Let α be the constant angular retardation, then applying

$$\omega = \omega_0 - \alpha t$$

or $$0 = (20\,\pi) - 3(\alpha)$$

or $$\alpha = \frac{20}{3}\pi \text{ rad/s}^2$$

Further, $$\alpha = \frac{\tau}{I}$$

Here, $$\tau = \mu NR \qquad (R = \text{radius})$$

or $$\tau = \mu FR \qquad (F = \text{applied force})$$

as $$N = F \text{ and } I = \frac{1}{2}mR^2$$

From the above equations,

$$\frac{20}{3}\pi = \frac{\mu FR}{\frac{1}{2}mR^2} = \frac{2\mu F}{mR}$$

or $$F = \frac{10\pi mR}{3\mu}$$

Substituting the values, we have

$$F = \frac{10 \times 22 \times 20 \times 0.12}{3 \times 7 \times 0.1}$$

or $$F = 251.43 \text{ N}$$

Example 9.8 *A wheel rotates around a stationary axis so that the rotation angle θ varies with time as $\theta = at^2$, where $\alpha = 0.2$ rad/s^2. Find the magnitude of net acceleration of the point A at the rim at the moment $t = 2.5$ s, if the linear velocity of the point A at this moment is $v = 0.65$ m/s.*

Sol. Instantaneous angular velocity at time t is

$$\omega = \frac{d\theta}{dt} = \frac{d}{dt}(at^2)$$

or $$\omega = 2at = 0.4t \qquad (\text{as } a = 0.2 \text{ rad/s}^2)$$

Further, instantaneous angular acceleration is

$$\alpha = \frac{d\omega}{dt} = \frac{d}{dt}(0.4t)$$

or $$\alpha = 0.4 \text{ rad/s}^2$$

Angular velocity at $t = 2.5$ s is

$$\omega = 0.4 \times 2.5 = 1.0 \text{ rad/s}$$

Further, radius of the wheel

$$R = \frac{v}{\omega}$$

or $$R = \frac{0.65}{1.0} = 0.65 \text{ m}$$

Now, magnitude of total acceleration is

$$a = \sqrt{a_n^2 + a_t^2}$$

Here, $$a_n = R\omega^2 = (0.65)(1.0)^2 = 0.65 \text{ m/s}^2$$

and $$a_t = R\alpha = (0.65)(0.4) = 0.26 \text{ m/s}^2$$

$\therefore$ $$a = \sqrt{(0.65)^2 + (0.26)^2}$$

or $$a = 0.7 \text{ m/s}^2$$

9.4 Angular Momentum

A mass moving in a straight line has linear momentum (**p**). When a mass rotates about some point/axis, there is momentum associated with rotational motion called the angular momentum (**L**). Just as net external force is required to change the linear momentum of an object a net external torque is required to change the angular momentum of an object. The angular momentum is classified in following two types

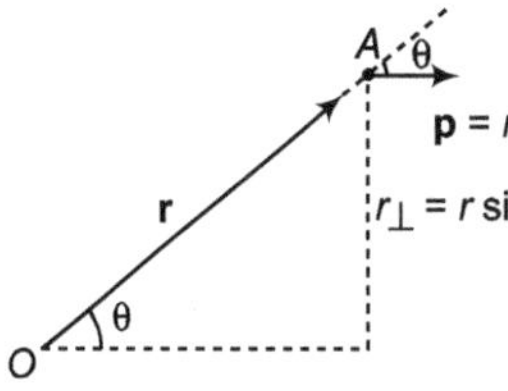

Fig. 9.21

Angular Momentum of a Particle about Some Point

Suppose a particle A of mass m is moving with linear momentum $\mathbf{p} = m\,\mathbf{v}$. Its angular momentum **L** about point O is defined as

$$\mathbf{L} = \mathbf{r} \times \mathbf{p} = \mathbf{r} \times (m\,\mathbf{v}) = m(\mathbf{r} \times \mathbf{v})$$

Here, **r** is the radius vector of particle A about O at that instant of time. The magnitude of **L** is

$$L = mvr \sin\theta = mvr_\perp$$

Here, $r_\perp = r\sin\theta$ is the perpendicular distance of line of action of velocity **v** from point O. The direction of **L** is same as that of $\mathbf{r} \times \mathbf{v}$.

- The angular momentum of a particle about a line (say AB) is the component along AB of the angular momentum of the particle about any point (say O) on the line AB. This component is independent of the choice of point O, so far as it is chosen on the line AB.

Example 9.9 *A particle of mass m is moving along the line $y = b$, $z = 0$ with constant speed v. State whether the angular momentum of particle about origin is increasing, decreasing or constant.*

Sol. $|\mathbf{L}| = mvr\sin\theta = mvr_\perp = mvb$

$\therefore$ $|\mathbf{L}|$ = constant as m, v and b all are constants.

Direction of $\mathbf{r} \times \mathbf{v}$ also remains the same. Therefore, angular momentum of particle about origin remains constant with due course of time.

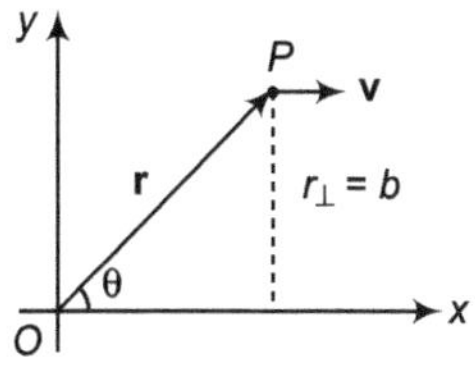

Fig. 9.22

- In this problem, $|\mathbf{r}|$ is increasing, θ is decreasing but $r\sin\theta$, i.e. b remains constant. Hence, the angular momentum remains constant.

Example 9.10 *A particle of mass m is projected from origin O with speed u at an angle θ with positive x-axis. Positive y-axis is in vertically upward direction. Find the angular momentum of particle at any time t about O before the particle strikes the ground again.*

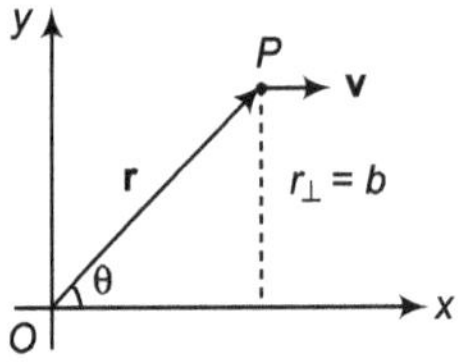

Fig. 9.23

Sol. $\mathbf{L} = m(\mathbf{r} \times \mathbf{v})$

Here, $\mathbf{r}(t) = x\hat{\mathbf{i}} + y\hat{\mathbf{j}}$

$$= (u\cos\theta)t\,\hat{\mathbf{i}} + \left(ut\sin\theta - \frac{1}{2}gt^2\right)\hat{\mathbf{j}}$$

and $\mathbf{v}(t) = v_x\hat{\mathbf{i}} + v_y\hat{\mathbf{j}}$

$$= (u\cos\theta)\hat{\mathbf{i}} + (u\sin\theta - gt)\hat{\mathbf{j}}$$

$\therefore$ $\mathbf{L} = m(\mathbf{r} \times \mathbf{v})$

$$= m\begin{vmatrix} \hat{\mathbf{i}} & \hat{\mathbf{j}} & \hat{\mathbf{k}} \\ (u\cos\theta)t & (u\sin\theta)t - \frac{1}{2}gt^2 & 0 \\ u\cos\theta & u\sin\theta - gt & 0 \end{vmatrix}$$

$$= m\left[(u^2\sin\theta\cos\theta)t - (u\cos\theta)gt^2 - (u^2\sin\theta\cos\theta)t + \frac{1}{2}(u\cos\theta)gt^2\right]\hat{\mathbf{k}}$$

$$= -\frac{1}{2}m(u\cos\theta)\,gt^2\,\hat{\mathbf{k}}$$

Angular Momentum of a Rigid Body Rotating about a Fixed Axis

Suppose a particle P of mass m is going in a circle of radius r and at some instant the speed of the particle is v. For finding the angular momentum of the particle about the axis of rotation, the origin may be chosen anywhere on the axis. We choose it at the centre of the circle. In this case, $\mathbf{r}$ and $\mathbf{p}$ are perpendicular to each other and $\mathbf{r} \times \mathbf{p}$ is along the axis. Thus, component of $\mathbf{r} \times \mathbf{p}$ along the axis is mvr itself. The angular momentum of the whole rigid body about AB is the sum of components of all particles, i.e.

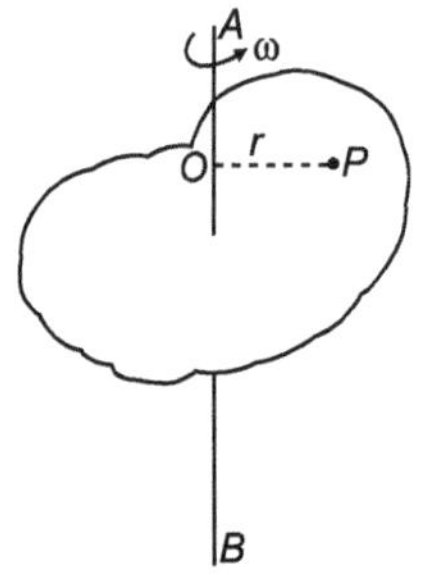

Fig. 9.24

$$L = \sum_i m_i r_i v_i$$

Here, $v_i = r_i\omega$

$\therefore$ $L = \sum_i m_i r_i^2 \omega$

or $L = \omega \sum_i m_i r_i^2$

or $L = I\omega$ (as $\sum_i m_i r_i^2 = I$)

Here, I is the moment of inertia of the rigid body about AB.

9.5 Conservation of Angular Momentum

As we have seen in Article 9.4, the angular momentum of a particle about some reference point O is defined as,

$$\mathbf{L} = \mathbf{r} \times \mathbf{p} \quad \text{...(i)}$$

Here, $\mathbf{p}$ is the linear momentum of the particle and $\mathbf{r}$ its position vector with respect to the reference point O. Differentiating Eq. (i) with respect to time, we get

$$\frac{d\mathbf{L}}{dt} = \mathbf{r} \times \frac{d\mathbf{p}}{dt} + \frac{d\mathbf{r}}{dt} \times \mathbf{p} \quad \text{...(ii)}$$

Here, $\frac{d\mathbf{p}}{dt} = \mathbf{F}$ and $\frac{d\mathbf{r}}{dt} = \mathbf{v}$ (velocity of particle)

Hence, Eq. (ii) can be rewritten as,

$$\frac{d\mathbf{L}}{dt} = \mathbf{r} \times \mathbf{F} + \mathbf{v} \times \mathbf{p}$$

Now, $\mathbf{v} \times \mathbf{p} = 0$, because $\mathbf{v}$ and $\mathbf{p}$ are parallel to each other and the cross product of two parallel vectors is zero. Thus,

$$\frac{d\mathbf{L}}{dt} = \mathbf{r} \times \mathbf{F} = \tau \quad \text{or} \quad \tau = \frac{d\mathbf{L}}{dt} \quad \text{...(iii)}$$

Which states that the time rate of change of angular momentum of a particle about some reference point in an inertial frame of reference is equal to the net torques acting on it. This result is rotational analog of the equation $\mathbf{F} = \frac{d\mathbf{p}}{dt}$, which states that the time rate of change of the linear momentum of a particle is equal to the force acting on it. Eq. (iii) like all vector equations, is equivalent to three scalar equations, namely

$$\tau_x = \left(\frac{dL}{dt}\right)_x, \quad \tau_y = \left(\frac{dL}{dt}\right)_y \quad \text{and} \quad \tau_z = \left(\frac{dL}{dt}\right)_z$$

The same equation can be generalized for a system of particles as, $\tau_{\text{ext}} = \frac{d\mathbf{L}}{dt}$. According to which the time rate of change of the total angular momentum of a system of particles about some reference point of an inertial frame of reference is equal to the sum of all external torques (of course the vector sum) acting on the system about the same reference point.

Now, suppose that $\tau_{\text{ext}} = 0$, then $\frac{d\mathbf{L}}{dt} = 0$, so that $\mathbf{L} = \text{constant}$.

"When the resultant external torque acting on a system is zero, the total vector angular momentum of the system remains constant. This is the principle of the conservation of angular momentum.

For a rigid body rotating about an axis (the z-axis, say) that is fixed in an inertial reference frame, we have

$$L_z = I\omega$$

It is possible for the moment of inertia I of a rotating body to change by rearrangement of its parts. If no net external torque acts, then L_z must remain constant and if I does change, there must be a compensating change in ω. The principle of conservation of angular momentum in this case is expressed as

$$I\omega = \text{constant} \qquad \ldots\text{(iv)}$$

Example 9.11 *A wheel of moment of inertia I and radius R is rotating about its axis at an angular speed ω_0. It picks up a stationary particle of mass m at its edge. Find the new angular speed of the wheel.*

Sol. Net external torque on the system is zero. Therefore, angular momentum will remain conserved. Thus,

$$I_1\omega_1 = I_2\omega_2 \quad \text{or} \quad \omega_2 = \frac{I_1\omega_1}{I_2}$$

Here, $I_1 = I,\ \omega_1 = \omega_0,\ I_2 = I + mR^2$

$$\therefore \quad \omega_2 = \frac{I\omega_0}{I + mR^2}$$

9.6 Combined Translational and Rotational Motion of a Rigid Body

Up until now we have considered only bodies rotating about some fixed axis. Let us consider combined translational and rotational motion of a rigid body.

In such problems if two things are known,

(i) velocity of centre of mass (v_{CM})

(ii) angular velocity of the rigid body (ω).

The motion of whole rigid body can be described.

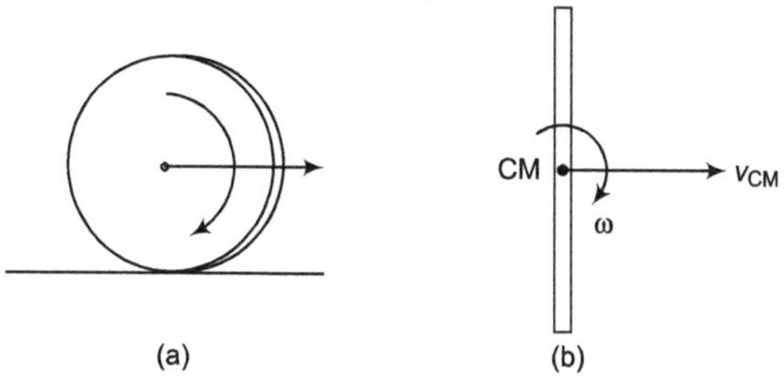

Fig. 9.25

For example, let the velocity of centre of mass of a rigid body shown in figure is v and angular velocity of the rigid body is ω. Then, velocity of any point P on the rigid body can be obtained as,

$$\mathbf{v}_p = \mathbf{v}_{CM} + \mathbf{v}_{p,CM}$$

Here, $|\mathbf{v}_{CM}| = v$

and $\mathbf{v}_{p,CM} = r\omega$ in a direction perpendicular to line CP.

Thus, the velocity of point P is the vector sum of $\mathbf{v}_{CM}$ and $\mathbf{v}_{p,CM}$ as shown in figure.

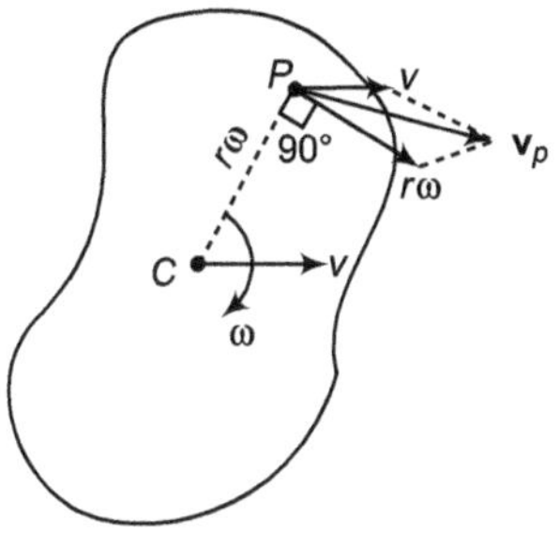

Fig. 9.26

Kinetic energy of rigid body in combined translational and rotational motion.

Here, two energies are associated with the rigid body. One is translational $\left(= \frac{1}{2} m v_{CM}^2\right)$ and another is rotational $\left(= \frac{1}{2} I_{CM}\,\omega^2\right)$. Thus, total kinetic energy of the rigid body is

$$K = \frac{1}{2} m v_{CM}^2 + \frac{1}{2} I_{CM}\,\omega^2$$

9.7 Uniform Pure Rolling

Pure rolling means no relative motion (or no slipping) at point of contact between two bodies.

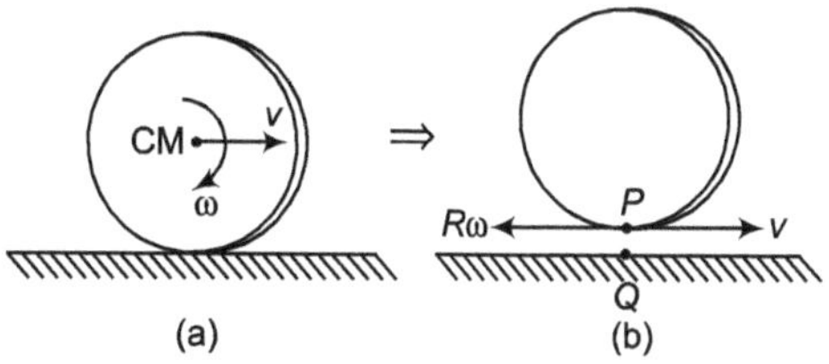

Fig. 9.27

For example, consider a disc of radius R moving with linear velocity v and angular velocity ω on a horizontal ground. The disc is said to be moving without slipping, if velocities of points P and Q (shown in Fig. b) are equal,

$$v_P = v_Q$$

or $\quad v - R\omega = 0 \quad$ or $\quad v = R\omega$

If $v_P > v_Q$ or $v > R\omega$, the motion is said to be forward slipping and if $v_P < v_Q$ or $v < R\omega$, the motion is said the backward slipping (or sometimes called forward slipping).

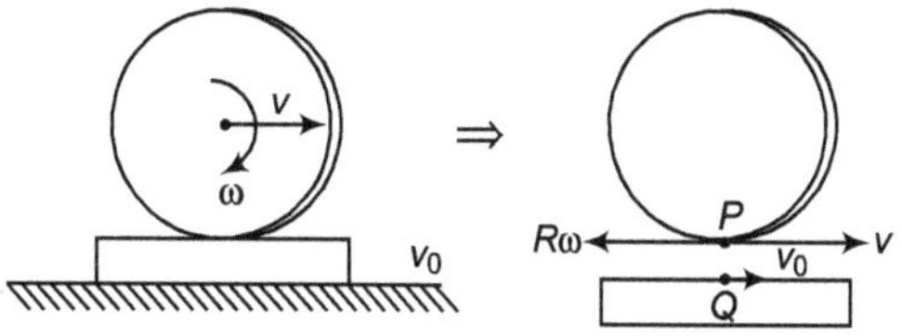

Fig. 9.28

Thus, $v = R\omega$ is the condition of pure rolling on a stationary ground. Sometimes it is simply said rolling. Suppose, the base over which the disc in rolling, is also moving with some velocity (say v_0), then in that case condition of pure rolling is different.

For example, in the above figure,

$$v_P = v_Q \quad \text{or} \quad v - R\omega = v_0$$

Thus, in this case $v - R\omega \neq 0$, but $v - R\omega = v_0$. By uniform pure rolling we mean that v and ω are constant. They are neither increasing nor decreasing.

Example 9.12 *A disc of radius R start at time $t = 0$ moving along the positive x-axis with linear speed v and angular speed ω. Find the x and y coordinates of the bottommost point at any time t.*

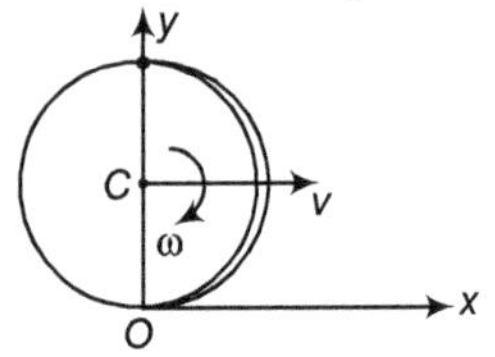

Fig. 9.29

Sol. At time t the bottommost point will rotate an angle $\theta = \omega t$ with respect to the centre of the disc C. The centre C will travel a distance $s = vt$.

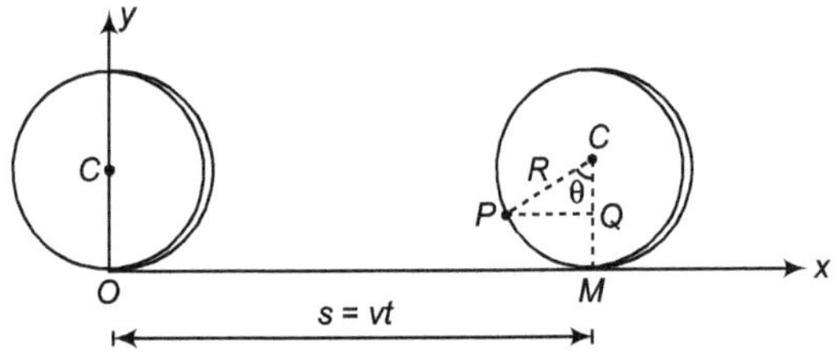

Fig. 9.30

In the figure, $PQ = R \sin\theta = R \sin\omega t$

and $CQ = R\cos\theta = R\cos\omega t$

Coordinates of point P at time t are,

$$x = OM - PQ = vt - R\sin\omega t$$

and $y = CM - CQ$

$$= R - R\cos\omega t$$

$$\therefore \quad (x, y) \equiv (vt - R\sin\omega t, R - R\cos\omega t)$$

9.8 Accelerated Pure Rolling

So far we were discussing the uniform pure rolling in which v and ω were constants. Now, suppose an external force is applied to the rigid body, the motion will no longer remain uniform. The condition of pure rolling on a stationary ground is

$$v = R\omega$$

Differentiating this equation with respect to time we have

$$\frac{dv}{dt} = R\frac{d\omega}{dt} \quad \text{or} \quad a = R\alpha$$

Thus, in addition to $v = R\omega$ at every instant of time, linear acceleration = $R \times$ angular acceleration or $a = R\alpha$ for pure rolling to take place. Here, friction plays an important role in maintaining the pure rolling. The friction may sometimes act in forward direction, sometimes in backward direction or under certain conditions it may be zero. Here, we should not forget the basic nature of friction, which is a self adjusting force (upto a certain maximum limit) and which has a tendency to stop the relative motion between two bodies in contact. Let us take an example illustrating the above theory.

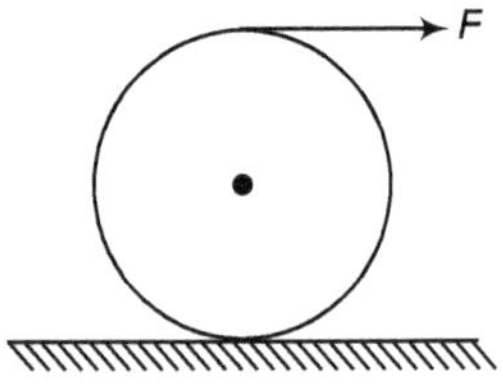

Fig. 9.31

Suppose a force F is applied at the topmost point of a rigid body of radius R, mass M and moment of inertia I about an axis passing through the centre of mass. Now, the applied force F can produce by itself

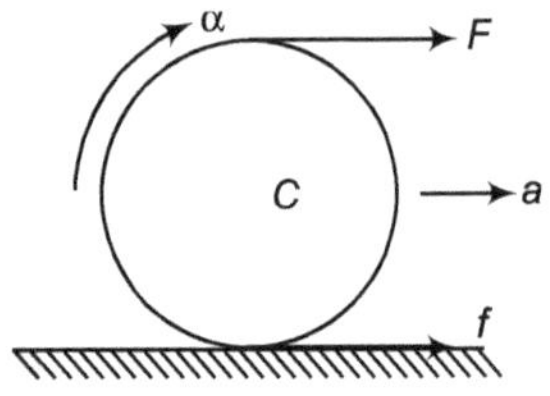

Fig. 9.32

(i) a linear acceleration a and

(ii) an angular acceleration α.

If $a = R\alpha$, then there is no need of friction and force of friction $f = 0$. If $a < R\alpha$, then to support the linear motion the force of friction f will act in forward direction. Similarly, if $a > R\alpha$, then to support the angular motion the force of friction will act in backward direction. So, in this case force of friction will be either backward, forward or even zero also. It all depends on M, I and R. For calculation you can choose any direction of friction. Let we assume it in forward direction,

Let, a = linear acceleration, α = angular acceleration then,

$$a = \frac{F_{\text{net}}}{M} = \frac{F + f}{M} \quad \ldots\text{(i)}$$

$$\alpha = \frac{\tau_c}{I} = \frac{(F - f)R}{I} \quad \ldots\text{(ii)}$$

For pure rolling to take place,

$$a = R\alpha \quad \ldots\text{(iii)}$$

Solving Eqs. (i), (ii) and (iii), we get

$$f = \frac{(MR^2 - I)}{(MR^2 + I)} . F \quad \ldots\text{(iv)}$$

From Eq. (iv) following conclusions can be drawn

(i) If $I = MR^2$ (e.g. in case of a ring)

$$f = 0$$

if a force F is applied on the top of a ring, the force of friction will be zero and the ring will roll without slipping.

(ii) If $I < MR^2$, (e.g. in case of a solid sphere or a hollow sphere), f is positive force of friction will be forward.

(iii) If $I > MR^2$, f is negative force of friction will be backwards. Although under no condition $I > MR^2$. (Think why?). So, force of friction is either in forward direction or zero.

Here, it should be noted that the force of friction f obtained in Eq. (iv) should be less than the limiting friction (μMg), for pure rolling to take place. Further, we saw that, if $I < MR^2$, force of friction acts in forward direction. This is because α is more, if I is small $\left(\alpha = \frac{\tau}{I}\right)$ i.e. to support the linear motion force of friction is in forward direction.

- It is often said that rolling friction is less than the sliding friction. This is because the force of friction calculated by equation number (iv) normally comes less than the sliding friction ($\mu_k N$) and even sometimes it is in forward direction, i.e. it supports the motion.

There are certain situations in which the direction of friction is fixed. For example, in the following situations the force of friction is backward.

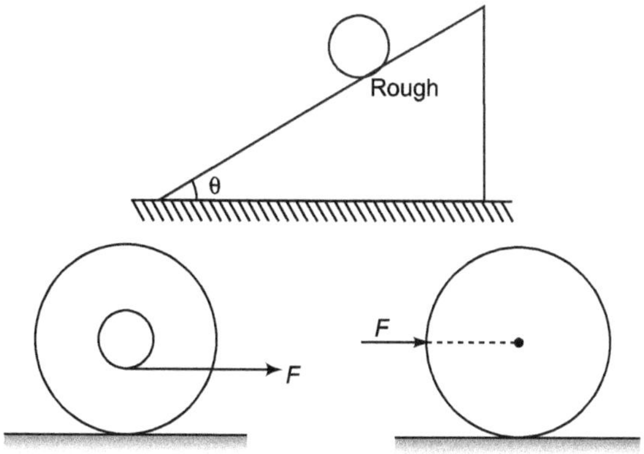

Fig. 9.33

Rolling on Rough Inclined Plane

As we said earlier also, force of friction in this case will be backward. Equations of motion are

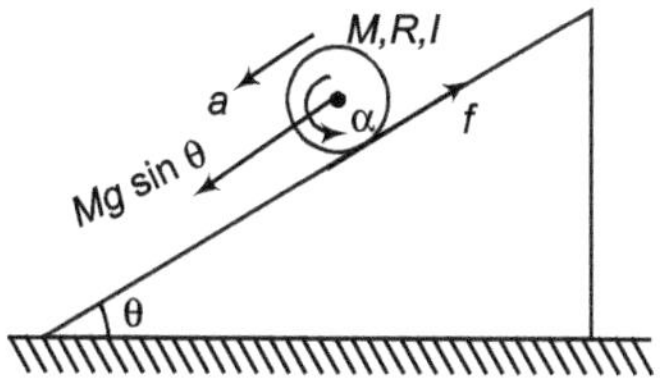

Fig. 9.34

$$a = \frac{Mg\sin\theta - f}{M} \quad \ldots(\text{i})$$

$$\alpha = \frac{fR}{I} \quad \ldots(\text{ii})$$

For pure rolling to take place,

$$a = R\alpha \quad \ldots(\text{iii})$$

Solving Eqs. (i), (ii) and (iii), we get

$$f = \frac{Mg\sin\theta}{1 + \frac{MR^2}{I}} \quad \ldots(\text{iv})$$

and

$$a = \frac{g\sin\theta}{1 + \frac{I}{MR^2}} \quad \ldots(\text{v})$$

From Eq. (v), we can see that if a solid sphere and a hollow sphere of same mass and radius are released from a rough inclined plane, the solid sphere reaches the bottom first because

$$I_{\text{solid}} < I_{\text{hollow}}$$

or $$a_{\text{solid}} > a_{\text{hollow}}$$

$\therefore$ $$t_{\text{solid}} < t_{\text{hollow}}$$

Further, the force of friction calculated in Eq. (iv) for pure rolling to take place should be less than or equal to the maximum friction $\mu\, Mg\cos\theta$.

or $$\frac{Mg\sin\theta}{1 + \frac{MR^2}{I}} \le \mu\, Mg\cos\theta$$

or $$\mu \ge \frac{\tan\theta}{1 + \frac{MR^2}{I}}$$

Example 9.13 *In the arrangement shown in figure the mass of the uniform solid cylinder of radius R is equal to m and the masses of two bodies are equal to m_1 and m_2. The thread slipping and the friction in the axle of the cylinder are supposed to be absent. Find the angular acceleration of the cylinder and the ratio of tensions $\frac{T_1}{T_2}$ of the vertical sections of the thread in the process of motion.*

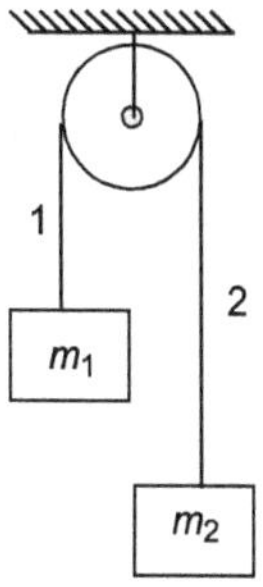

Fig. 9.35

Sol. Let α = angular acceleration of the cylinder

and a = linear acceleration of two bodies

Equations of motion are,

For mass m_1, $T_1 - m_1 g = m_1 a$ ...(i)

For mass m_2, $m_2g - T_2 = m_2a$ ...(ii)

For cylinder, $\alpha = \dfrac{(T_2 - T_1)R}{\frac{1}{2}mR^2}$ ...(iii)

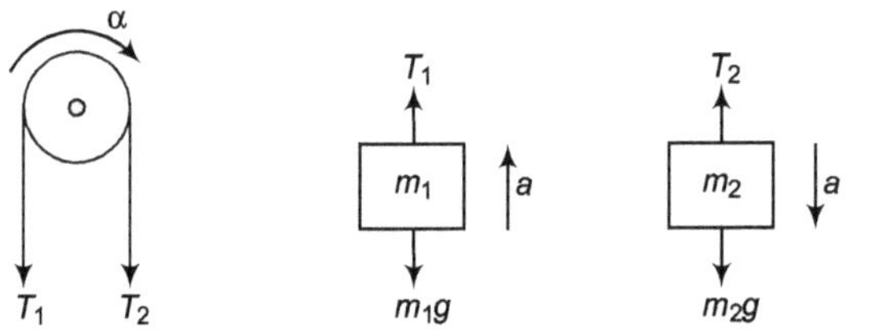

Fig. 9.36

For no slipping condition

$$a = R\alpha \quad ...(iv)$$

Solving these equations, we get

$$\alpha = \frac{2(m_2 - m_1)g}{(2m_1 + 2m_2 + m)R}$$

and $\dfrac{T_1}{T_2} = \dfrac{m_1(m + 4m_2)}{m_2(m + 4m_1)}$

Example 9.14 *Consider the arrangement shown in figure. The string is wrapped around a uniform cylinder which rolls without slipping. The other end of the string is passed over a massless, frictionless pulley to a falling weight. Determine the acceleration of the falling mass m in terms of only the mass of the cylinder M, the mass m and g.*

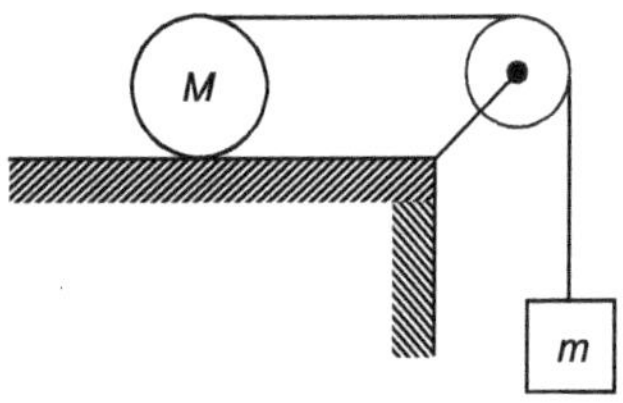

Fig. 9.37

Sol. Let T be the tension in the string and f the force of (static) friction, between the cylinder and the surface

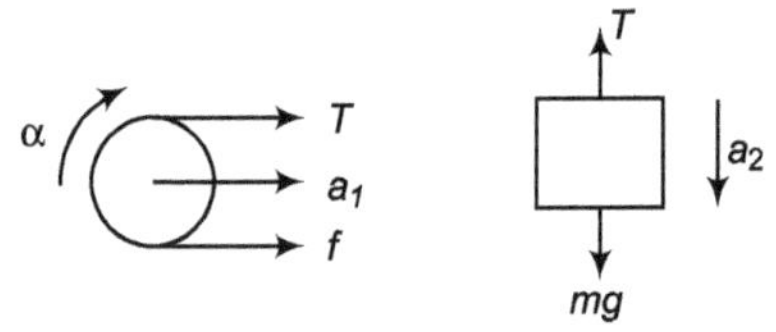

Fig. 9.38

a_1 = acceleration of centre of mass of cylinder towards right,
a_2 = downward acceleration of block m,
α = angular acceleration of cylinder (clockwise)
Equations of motion are

For block, $mg - T = ma_2$ ...(i)

For cylinder, $T + f = Ma_1$ ...(ii)

$$\alpha = \frac{(T - f)R}{\frac{1}{2}MR^2} \quad ...(iii)$$

The string attaches the mass m to the highest point of the cylinder, hence

$$v_m = v_{CM} + R\omega$$

Differentiating the above equation, we get

$$a_2 = a_1 + R\alpha \quad ...(iv)$$

We also have (for rolling without slipping)

$$a_1 = R\alpha \quad ...(v)$$

Solving these equations, we get

$$a_2 = \frac{8mg}{3M + 8m}$$

Alternate Solution (Energy Method)

Since, there is no slipping at all contacts mechanical energy of the system will remain conserved.

$\therefore$ Decrease in gravitational potential energy of block m in time t = increase in translational kinetic energy of block + increase in rotational as well as translational kinetic energy of cylinder.

$$\therefore \quad mgh = \frac{1}{2}mv_2^2 + \frac{1}{2}I\omega^2 + \frac{1}{2}Mv_1^2$$

or
$$mg\left(\frac{1}{2}a_2t^2\right) = \frac{1}{2}m(a_2t)^2 + \frac{1}{2}\left(\frac{1}{2}MR^2\right)(\alpha t)^2 + \frac{1}{2}M(a_1t)^2 \quad ...(vi)$$

Solving Eqs. (iv), (v) and (vi), we get the same result.

9.9 Angular Impulse

The angular impulse of a torque in a given time interval is defined as $\int_{t_1}^{t_2} \tau \, dt$

Here, τ is the resultant torque acting on the body. Further, since

$$\tau = \frac{d\mathbf{L}}{dt}$$

$$\therefore \quad \tau \, dt = d\mathbf{L}$$

or
$$\int_{t_1}^{t_2} \tau \, dt = \text{angular impulse} = \mathbf{L}_2 - \mathbf{L}_1$$

Thus, the angular impulse of the resultant torque is equal to the change in angular momentum. Let us take an example based on the angular impulse.

Chapter Summary with Formulae

- **Moment of Inertia**

(i) **Thin rod**

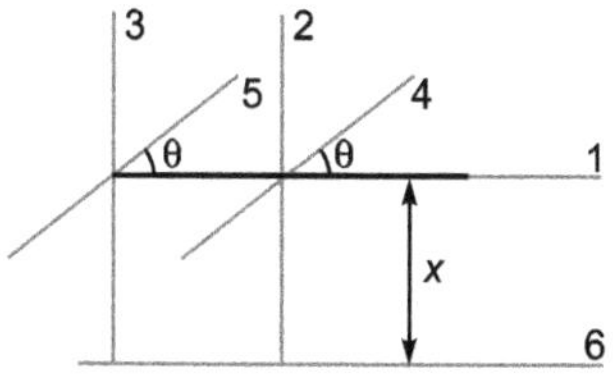

$$I_1 = 0,\ I_2 = \frac{ml^2}{12},\ I_3 = \frac{ml^2}{3},$$

$$I_4 = \frac{ml^2}{12}\sin^2\theta,\ I_5 = \frac{ml^2}{3}\sin^2\theta,$$

$$I_6 = mx^2$$

(ii) **Circular disc**

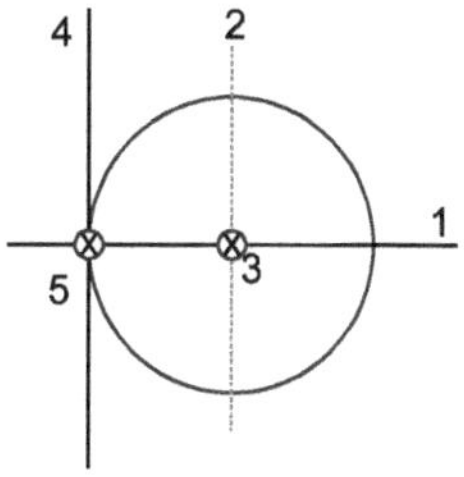

$$I_1 = I_2 = \frac{mR^2}{4}$$

$$I_3 = I_1 + I_2 = \frac{mR^2}{2}$$

$$I_4 = I_2 + mR^2 = \frac{5}{4}mR^2$$

$$I_5 = I_3 + mR^2 = \frac{3}{2}mR^2$$

(iii) **Circular ring**

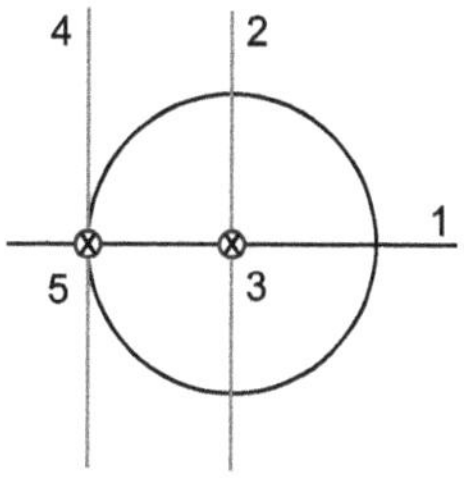

$$I_1 = I_2 = \frac{mR^2}{2}$$

$$I_3 = I_1 + I_2 = mR^2$$

$$I_4 = I_2 + mR^2 = \frac{3}{2}mR^2$$

$$I_5 = I_3 + mR^2 = 2mR^2$$

(iv) **Rectangular slab**

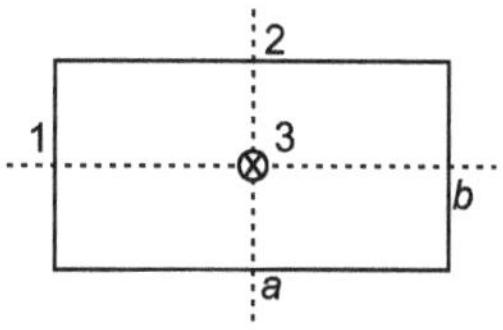

$$I_1 = \frac{mb^2}{12}$$

$$I_2 = \frac{ma^2}{12}$$

$$I_3 = I_1 + I_2 = \frac{m}{12}(a^2 + b^2)$$

(v) **Square slab**

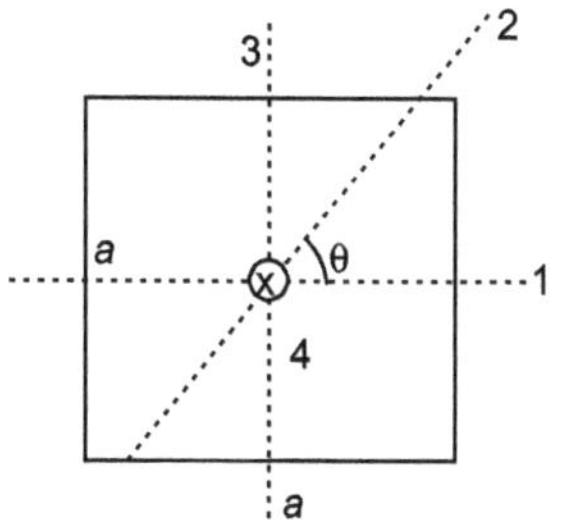

$$I_1 = I_2 = I_3 = \frac{ma^2}{12}$$

$$I_4 = I_1 + I_3 = \frac{ma^2}{6}$$

(vi) **Solid sphere**

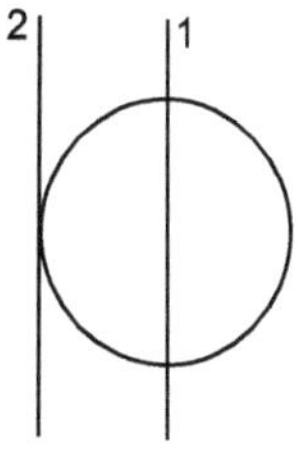

$$I_1 = \frac{2}{5}mR^2$$

$$I_2 = I_1 + mR^2 = \frac{7}{5}mR^2$$

(vii) **Hollow sphere**

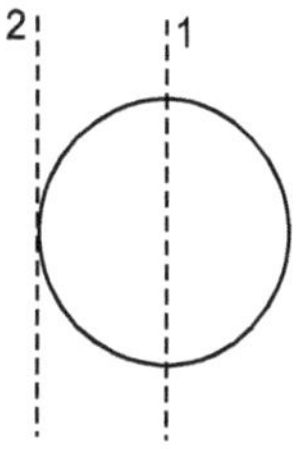

$$I_1 = \frac{2}{3}mR^2$$

$$I_2 = I_1 + mR^2 = \frac{5}{3}mR^2$$

- **Two Theorems**

(i) **Theorem of parallel axes**

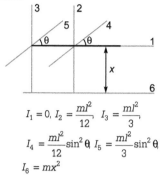

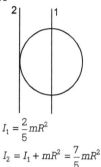

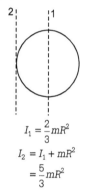

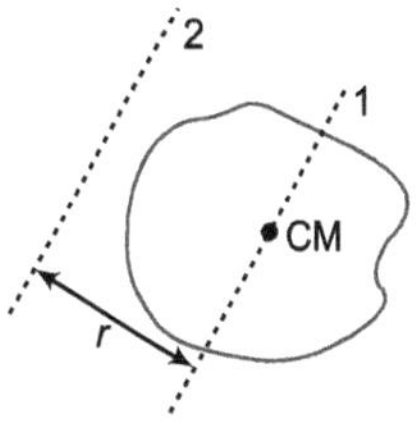

$$I = I_{CM} + mr^2 \quad \text{or} \quad I_2 = I_1 + mr^2$$

(ii) **Theorem of perpendicular axes** This theorem is applicable only for a two dimensional body of negligible thickness. If x and y are two perpendicular axes lying in the plane of body. z is the axis perpendicular to plane of body and passing through point of intersection of x and y, then

$$I_z = I_x + I_y$$

- **Radius of gyration** (K) It is an imaginary distance from the axis at which whole mass of the rigid body, if kept as a point mass, moment of inertia remains unchanged.

Thus, $$I = mK^2 \quad \text{or} \quad K = \sqrt{\frac{I}{m}}$$

- **Three type of motions of a rigid body** A rigid body is made up of many point masses (or particles). If distance between any two particles remains constant, body is said to be rigid body. A rigid body may have either of the following three type of motions

(i) only translational motion
(ii) only rotational motion
(iii) both rotational and translational motion.

Only in case of translational motion velocity and acceleration of all particles are same. In rotational or rotational plus translational motion different particles have different displacements, different velocities and different accelerations.

Angular Momentum (L)

Angular momentum can be defined in following three ways :

(i) **Angular momentum of a particle (A) in motion, about a fixed point (O)**

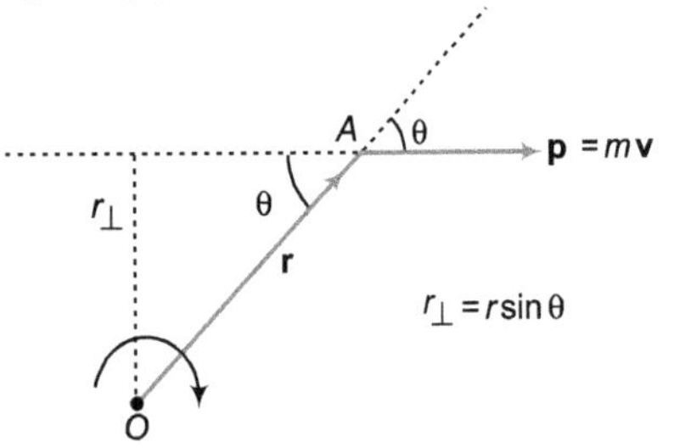

Suppose a particle A has a linear momentum $\mathbf{p} = m\mathbf{v}$ as shown.

Its position vector about a fixed point O at this instant is $\mathbf{r}$. Then angular momentum of particle A about point O will be,

$$\mathbf{L} = \mathbf{r} \times \mathbf{p} = \mathbf{r} \times (m\mathbf{v}) = m\,(\mathbf{r} \times \mathbf{v})$$

Magnitude of L is $L = mvr\sin\theta = mvr_\perp$, where θ is the angle between r and p. Further, $r_\perp = r\sin\theta$ is the perpendicular distance on line of action of p (or v) from point O. Direction of L will be given by right hand screw law. In the shown figure direction of L is perpendicular to paper inwards.

If the particle passes through point O, $r_\perp = 0$. Therefore, angular momentum is zero.

(ii) **Angular momentum of a rigid body in pure rotation about axis of rotation**

If a rigid body is in pure rotation about a fixed axis, then angular momentum of rigid body about this axis will be given by $L = I\omega$

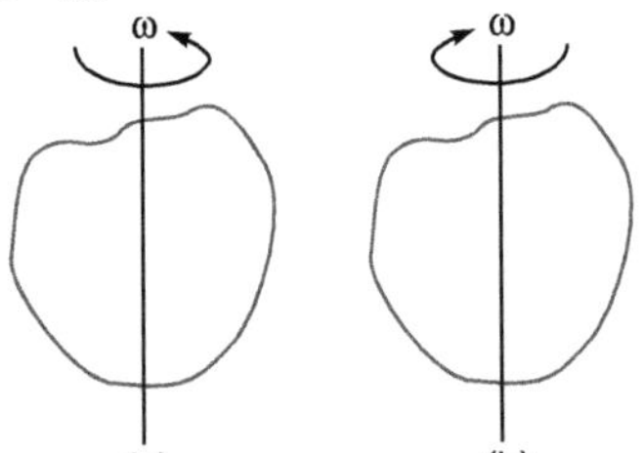

This is actually component of total angular momentum about axis of rotation. Direction of this component is again given by right hand screw law. In figure (a), this is along the axis in upward direction. In figure (b) this is along the axis in downward direction.

(iii) **Angular momentum of a rigid body in rotation plus translation motion about a general axis**

Suppose a rigid body is in rotation and translational motion. Velocity of its centre of mass is v and angular velocity of rigid body is ω. We want to find angular momentum of rigid body about an axis passing from a general point O and perpendicular to plane of paper.

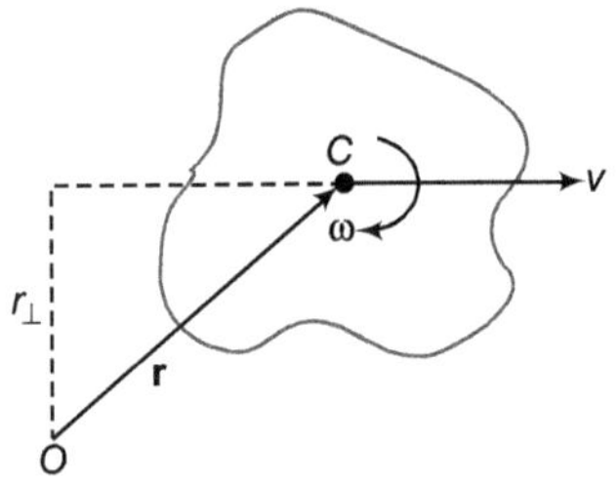

This will contain two terms

(a) $I_c\omega$

(b) $mv_c r_\perp = mvr_\perp$

From right hand screw law, we can see that $I_c\omega$ and $mvr_\perp$ both terms are perpendicular to the paper in inward direction. Hence, they are added.

or $$L_{\text{Total}} = I_c\omega + mvr_\perp$$

In a problem, if these two terms are in opposite directions, then they will be subtracted.

Equations of Motion of Pure Rotational Motion of a Rigid Body about a Fixed Axis

(i) In pure rotational motion all particles (except those lying on the axis) of the rigid body rotate in circles. The planes of these circles are mutually parallel. The centres of circles lie on the axis. Radii of different circles are different. Angular velocity corresponding to all particles in circular motion is same. This is also called angular velocity of rigid body. Plane of every circle is perpendicular to the axis of rotation.

(ii) Velocity of any particle P is, $v = r\omega$, tangential to its own circle. Since, ω is same for all particles we can say that, $v \propto r$.

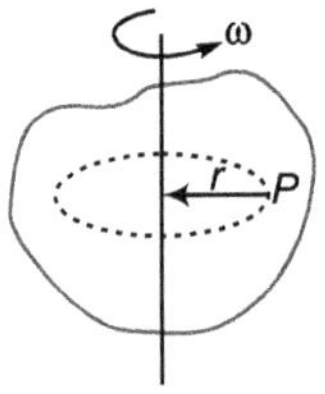

(iii) Acceleration of particle P will have two components a_n and a_t (as it is rotating in a circle). In the figure,

$$a_n = r\omega^2 \text{ or } \frac{v^2}{r} \quad \text{(towards centre)}$$

$$a_t = r\alpha = r \cdot \frac{d\omega}{dt} \quad \text{(tangential to the circle)}$$

$$\therefore \quad a = \sqrt{a_n^2 + a_t^2}$$

(iv) If $\alpha \left(= \frac{d\omega}{dt} = \frac{\tau}{I} \right)$ is constant, then

$$\omega = \omega_0 + \alpha t,$$

$$\theta = \omega_0 t + \frac{1}{2}\alpha t^2$$

and $$\omega^2 = \omega_0^2 + 2\alpha\theta$$

Here, ω is angular velocity at time t and ω_0, initial angular velocity.

If α is not constant, then we will have to go for differentiation or integration.

The basic equations of differentiation or integration are,

$$\omega = \frac{d\theta}{dt}, \alpha = \frac{d\omega}{dt} = \omega \cdot \frac{d\omega}{d\theta}$$

(equations of differentiation)

$$\int d\theta = \int \omega dt, \int d\omega = \int \alpha dt, \int \omega \, d\omega = \int \alpha \cdot d\theta$$

(equations of integration)

(v) Number of rotations made by rigid body,

$$N = \frac{\text{Angle rotated}}{2\pi} = \frac{\theta}{2\pi}$$

- **Rotational Plus Translational Motion of a Rigid Body**

A complex motion of rotation plus translation can be simplified by considering,

(i) the translational motion of centre of mass of the rigid body and

(ii) rotation about centre of mass.

As discussed earlier also, in this type of motion velocities of different points of the rigid body are different.

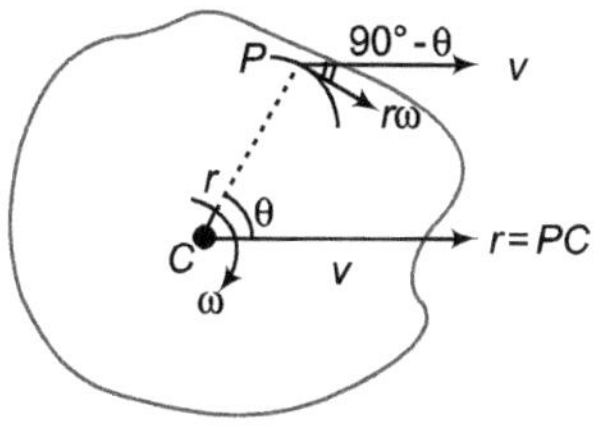

- To find velocity of a general point (say P) we will be required following two things

(i) velocity of centre of mass of the rigid body v,

(ii) angular velocity of rigid body ω.

Now, velocity of point P is the vector sum of two terms v and $r\omega$. Here, v is common for all points, while $r\omega$ is different for different points, as r is different.

In the figure shown,

$$v_P = \sqrt{v^2 + (r\omega)^2 + 2(v)(r\omega)\cos(90° - \theta)}$$

$$= \sqrt{v^2 + r^2\omega^2 + 2vr\omega\sin\theta}$$

- To find acceleration of point P we will be required following three things

(i) acceleration of centre of mass of the rigid body a

(ii) angular velocity of rigid body ω and

(iii) angular acceleration $\left(\alpha = \frac{d\omega}{dt}\right)$ of the rigid body.

- Now, acceleration of point P is the vector sum of three terms

(i) a

(ii) $a_n = r\omega^2$ (acting towards centre O)

(iii) $a_t = r\alpha$ (acting tangentially)

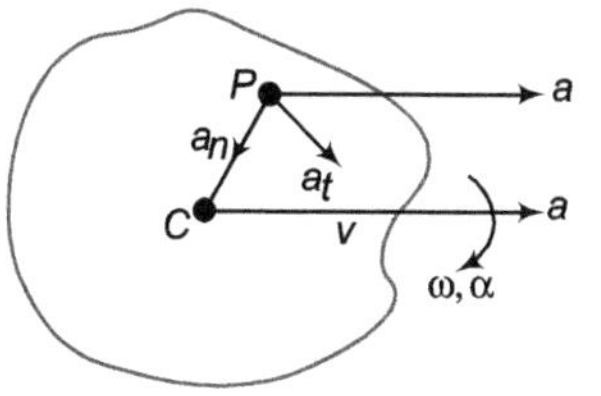

Here, a is common for all points, while a_n and a_t are different.

- **Pure Rolling**

Pure rolling (also called rolling without slipping) may be of two types

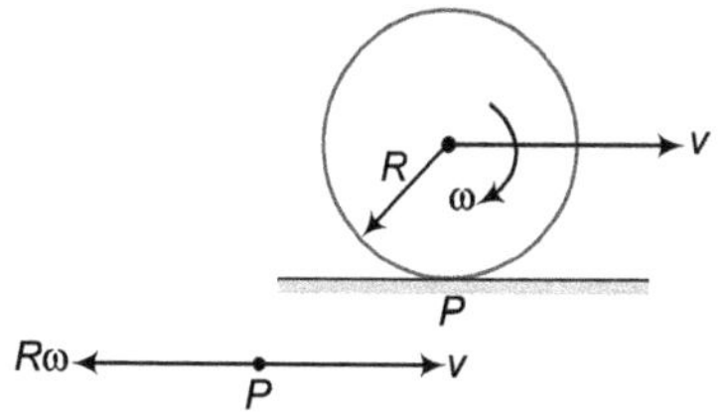

(i) **Uniform pure rolling** In which v and ω remain constant.

Condition of pure rolling is $v = R\omega$. In this case, bottommost point of the spherical body is at rest. It has no slipping with its contact point on ground. Because ground point is also at rest.

If $v > R\omega$, then net velocity of point P is in the direction of v (in the direction of motion of body). This is called forward slipping.

If $v < R\omega$, then net velocity of point P is in opposite direction of v. This is called backward slipping.

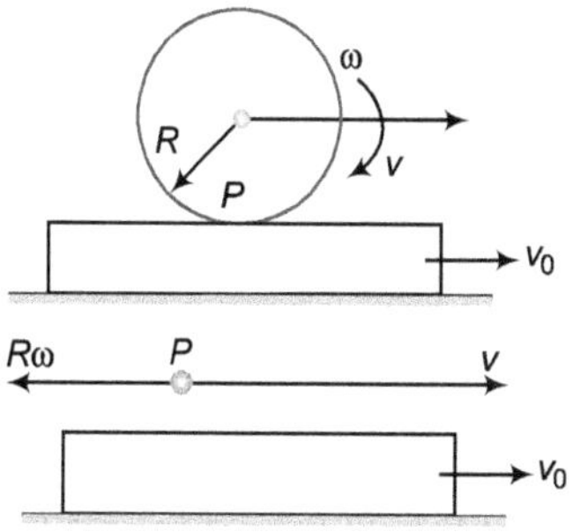

If a spherical body is rolling over a plank, condition for no slipping between spherical body and plank is,

$$v - R\omega = v_0$$

(ii) **Accelerated pure rolling** If v and ω are not constant then, $a = R\alpha$ is an additional condition for pure rolling on horizontal ground, which takes place in the presence of some external forces.

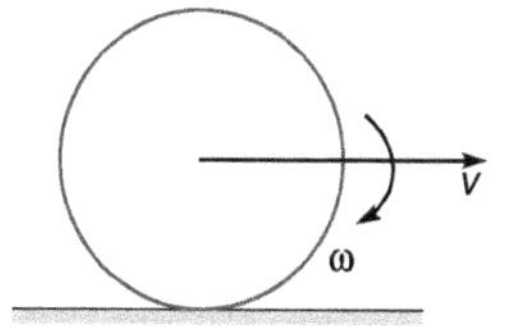

Here, friction plays very important role. Magnitude and direction of friction is so adjusted that equation $a = R\alpha$ is satisfied. If friction is insufficient for satisfying the equation $a = R\alpha$, slipping (either forward or backward) will occur and kinetic friction will act.

- **Motion of a Spherical Body on Rough Inclined Surface**

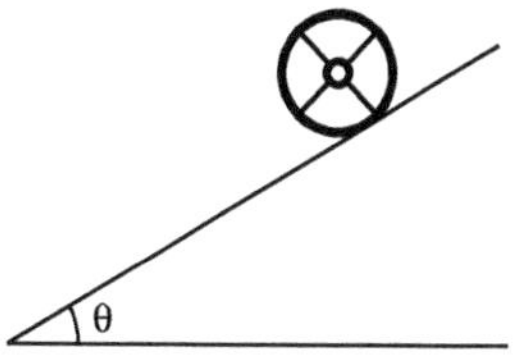

(i) Minimum value of coefficient of friction required for pure rolling,

$$\mu_{min} = \frac{\tan\theta}{1 + \frac{mR^2}{I}}$$

(ii) If $\mu = 0$, body will slip downwards (only translational motion) with an acceleration,

$$a_1 = g\sin\theta$$

(iii) If $\mu > \mu_{min}$, body will roll down without slipping with an acceleration,

(rotation + translation both)

$$a_2 = \frac{g\sin\theta}{1 + \frac{I}{mR^2}}$$

(iv) In the above case (when $\mu > \mu_{min}$) force of friction will act upwards. Magnitude of this force is,

$$f = \frac{mg\sin\theta}{1 + \frac{mR^2}{I}}$$

(v) If $\mu < \mu_{min}$, body will roll downwards with forward slipping. Maximum friction will act in this case. The acceleration of body is

$$a_3 = g\sin\theta - \mu g\cos\theta$$

Angular Impulse

Linear impulse when multiplied by perpendicular distance gives angular impulse. Angular impulse is equal to change in angular momentum.

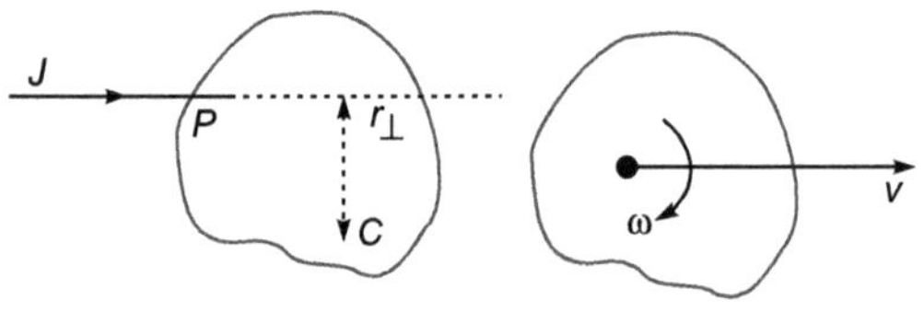

A rigid body is kept over a smooth table. It is hit at point P by a linear impulse J at a perpendicular distance $r_\perp$ from C as shown. Since, it is hit at some perpendicular distance from C its motion is rotational plus translational. Velocity of centre of mass will be given by,

$$v = \frac{J}{m} \qquad \text{(as } J = mv\text{)}$$

Angular velocity of rigid body is,

$$\omega = \frac{J \times r_\perp}{I}$$

(as $J \times r_\perp$ = angular impulse = change in angular momentum = $I\omega$).

Additional Examples

Example 1. *Define radius of gyration.*

Sol. It is defined as the distance from the axis of rotation at which, if whole mass of the body were supposed to be concentrated, the moment of inertia would be same as with the actual distribution of the mass of body in the form of the constituting particles.

Example 2. *There is a stick half of which is wooden and half is of steel. It is pivoted at the wooden end and a force is applied at the steel end at right angles to its length. Next, it is pivoted at the steel end and the same force is applied at the wooden end. In which case is the angular acceleration more and why ?*

Sol. We know, $\tau = I\alpha$

or $\alpha = \tau / I$

In the second case, half of the stick made of steel (heavier material) lies at the lesser distance from the axis of rotation. Therefore, MI of the stick will be lesser than that in the first case. Hence, the angular acceleration in the second case will be more than that in the former case.

Example 3. *How will you distinguish between a hard boiled egg and a raw egg by spinning each on a table top ?*

Sol. The egg which spins faster will be the hard boiled egg. It is because, a hard boiled egg will spin (rotate) more or less as a rigid body, whereas a raw egg will not do so. In case of a raw egg, its matter in the liquid state moves away from the axis of rotation, thereby increasing its moment of inertia. As MI of the raw egg is more, the angular acceleration produced will be lesser, when the same torque is applied in both the cases to set them spinning.

Example 4. *The angular velocity of revolution of the earth around the sun increases, when it comes closer to the sun. Why?*

Sol. The earth revolves around the sun in elliptical orbit with the sun at one of the two focii of the elliptical orbit. Therefore, moment of inertia of the earth about an axis through the sun keeps on changing due to change in its distance from the sun. Since, no external torque acts on the earth, its angular momentum ($I\omega$) must remain conserved. Thus, angular velocity of the earth increases, when moment of inertia of the earth decreases and *vice-versa.*

Example 5. *Why in hand driven grinding machine, handle is put near the circumference of the stone or wheel ?*

Sol. For a given force, torque can be increased, if the perpendicular distance of the point of application of the force from the axis of rotation is increased. Hence, the handle put near the circumference produces maximum torque.

Example 6. *About which axis would a uniform cube have a minimum rotational inertia ?*

Sol. About a diagonal, because the mass is more concentrated about a diagonal.

Example 7. *A disc is recast into a thin walled cylinder of same radius. Which will have larger moment of inertia?*

Sol. Hollow cylinder will have larger moment of inertia, because most of its mass is located at comparatively larger distance from the axis of rotation.

Example 8. *Why there are two propellers in a helicopter ?*

Sol. If there were only one propeller in the helicopter, then due to conservation of angular momentum, the helicopter itself would have turned in the opposite direction.

Example 9. *A thin wheel can stay up right on its rim for a considerable length of time when rolled with a considerable velocity, while it falls from its upright position at the slightest disturbance when stationary. Give reason.*

Sol. When the wheel is rolling upright, it has angular momentum in the horizontal direction, i.e. along the axis of the wheel. If slight disturbing force is applied on the wheel then its torque just changes the direction of angular momentum.

The wheel, topples (or falls) due to this torque when wheel is at rest.

Example 10. *If angular momentum is conserved in a system whose moment of inertia is decreased, will its rotational kinetic energy be conserved ?*

Sol. Here, $L = I\omega =$ constant

Rotational KE is given by

$$K = \frac{1}{2} I\omega^2 = \frac{1}{2}\frac{I^2\omega^2}{I} = \frac{1}{2}\cdot\frac{L^2}{I}$$

For constant L, $K \propto \frac{1}{I}$

So, when the moment of inertia decreases, the rotational KE increases. Hence, rotational KE is not conserved.

Example 11. *If no external torque acts on a body, will its angular velocity remain constant? Give reason.*

Sol. When no external torque acts on a body, its angular momentum remains constant. But

$$L = I\omega$$

Clearly, the angular velocity ω will remain constant only if the moment of inertia I of the body also remains constant.

Example 12. *A disc starts rotating with constant angular acceleration of π rad/s^2 about a fixed axis perpendicular to its plane and through its centre. Find*

(a) the angular velocity of the disc after 4 s,

(b) the angular displacement of the disc after 4 s.

Sol. Here, $\alpha = \pi$ rad/s^2

$$\omega_0 = 0, \; t = 4 \text{ s}$$

(a) $\omega_{(4\text{ s})} = 0 + (\pi \text{ rad/s}^2) \times 4 \text{ s} = 4\pi$ rad/s

(b) $\theta_{(4\text{ s})} = 0 + \frac{1}{2}(\pi \text{ rad/s}^2) \times (16 \text{ s}^2) = 8\pi$ rad

Example 13. *A wheel rotates with an angular acceleration given by $\alpha = 4at^3 - 3bt^2$, where t is the time and a and b are constants. If the wheel has initial angular speed ω_0, write the equations for the*

(i) angular speed, (ii) angle displacement.

Sol. (i) $\alpha = \dfrac{d\omega}{dt}$

$\Rightarrow \quad d\omega = \alpha \, dt$

$\Rightarrow \quad \int_{\omega_0}^{\omega} d\omega = \int_0^t \alpha \, dt = \int_0^t (4at^3 - 3bt^2) \, dt$

$\Rightarrow \quad \omega = \omega_0 + at^4 - bt^3$

(ii) Further, $\omega = \dfrac{d\theta}{dt}$

$\Rightarrow \quad d\theta = \omega \, dt$

$\Rightarrow \quad \int_0^{\theta} d\theta = \int_0^t \omega \, dt = \int_0^t (\omega_0 + at^4 - bt^3) \, dt$

$\Rightarrow \quad \theta = \omega_0 t + \dfrac{at^5}{5} - \dfrac{bt^4}{4}$

Example 14. *A particle of mass m is projected with velocity v at an angle θ with the horizontal. Find its angular momentum about the point of projection when it is at the highest point of its trajectory.*

Sol.

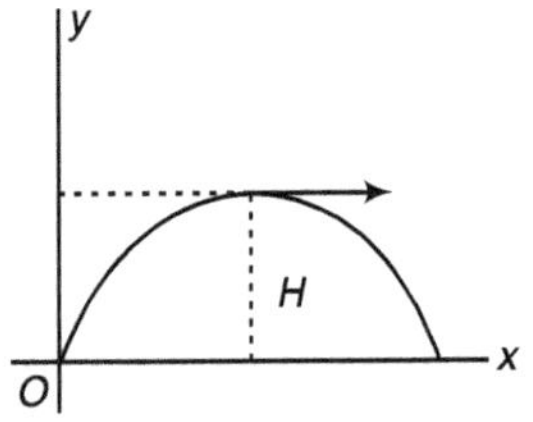

At the highest point it has only horizontal velocity $v_x = v\cos\theta$.

Length of the perpendicular to the horizontal velocity from O is the maximum height, where

$$H_{\max} = \frac{v^2 \sin^2 \theta}{2g}$$

$\Rightarrow$ Angular momentum $L = \dfrac{mv^3 \sin^2 \theta \cos\theta}{2g}$

Example 15. *A solid cylinder of mass m and radius r starts rolling down an inclined plane of inclination θ. Friction is enough to prevent slipping. Find the speed of its centre of mass when its centre of mass has fallen a height h.*

Sol. Considering the two shown positions of the cylinder. As it does not slip hence, total mechanical energy will be conserved.

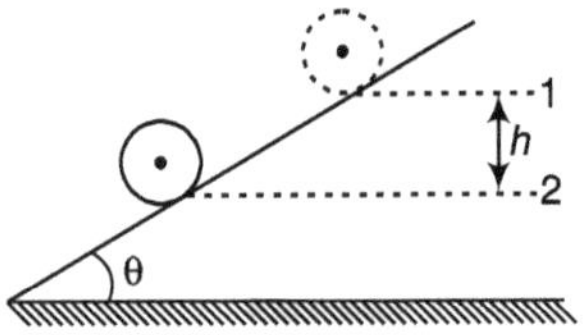

Energy at position 1 is

$$E_1 = mgh$$

Energy at position 2 is

$$E_2 = \frac{1}{2} mv_{CM}^2 + \frac{1}{2} I_{CM} \, \omega^2$$

$\because \quad \dfrac{v_{CM}}{r} = \omega$

and $\quad I_{CM} = \dfrac{mr^2}{2}$

$\Rightarrow \quad E_2 = \dfrac{3}{4} mv_{CM}^2$

From condemnation of energy, $E_1 = E_2$

$\Rightarrow \quad v_{CM} = \sqrt{\dfrac{4}{3} gh}$

Example 16. *If the radius of the earth contracts to half of its present value without change in its mass, what will be the new duration of the day?*

Sol. Present angular momentum of the earth

$$L_1 = I\omega = \frac{2}{5} MR^2 \omega$$

New angular momentum, because of change in radius

$$L_2 = \frac{2}{5} M\left(\frac{R}{2}\right)^2 \omega'$$

If external torque is zero, then angular momentum must be conserved

$$L_1 = L_2$$

$$\frac{2}{5} MR^2 \omega = \frac{1}{4} \times \frac{2}{5} MR^2 \omega'$$

i.e. $\quad \omega' = 4\,\omega$

$$T' = \frac{1}{4} T = \frac{1}{4} \times 24$$

$$= 6 \text{ h}$$

Example 17. *A solid ball of radius* 0.2m *and mass* 1 *kg is given an instantaneous impulse of* 50 *N-s at point P as shown in figure. Find the number of rotations made by the ball about its diameter before* *ground. The ball is kept on smooth surface initially.*

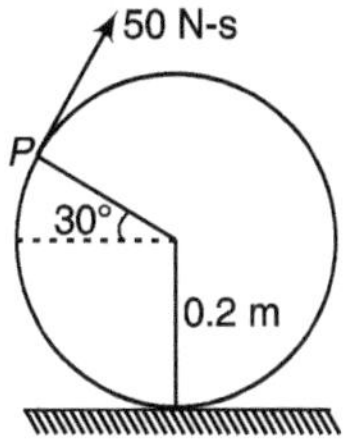

Sol. Impulse gives translational velocity

$$u = \frac{\text{Impulse}}{\text{Mass}}$$
$$= 50 \text{ m/s}$$
$$T = \text{time of flight of projectile}$$
$$= \frac{2u\sin\theta}{g} = \frac{2 \times 50 \times \sin 60°}{10}$$
$$= 5\sqrt{3}\text{ s}$$

Impulse give angular impulse also

$$\omega = \frac{\text{Impulse} \times R}{I}$$

or

$$\omega = \frac{\text{Impulse} \times R}{\frac{2}{5}mR^2}$$

Number of rotations,

$$n = \frac{\omega T}{2\pi}$$
$$= \frac{3125\sqrt{3}}{2\pi}$$

Example 18. *A sphere of mass m attached to a spring on incline as shown in figure is held in unstretched position of spring. Suddenly sphere is left free what is the maximum extension of spring, if friction allows only rolling of sphere about horizontal diametre?*

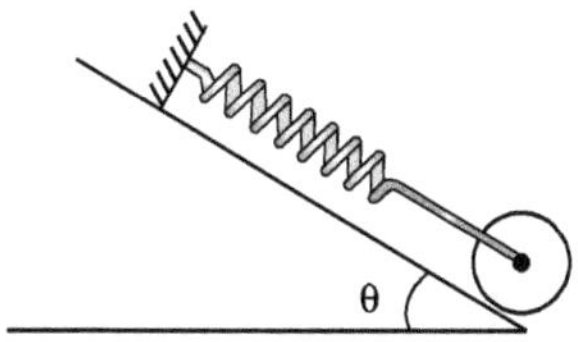

Sol. At the point of maximum extension sphere is at rest. No work is done by frictional force in rolling.

$\Rightarrow$ Loss in gravitational PE = Gain in PE of spring

$$\Rightarrow \quad \frac{1}{2}kx^2 = mgx\sin\theta$$
$$\Rightarrow \quad x = \frac{2mg\sin\theta}{k}$$

Example 19. *A horizontal force F acts on the sphere at its centre as shown in figure. Coefficient of friction between ground and sphere is μ. What is maximum value of F, for which there is no slipping?*

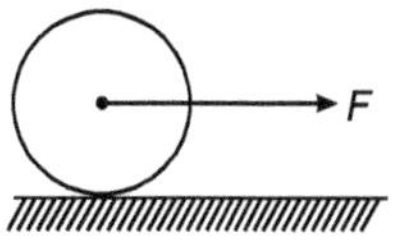

Sol.

$$F - f = Ma$$
$$f \cdot R = \frac{2}{5}MR^2\frac{a}{R} \Rightarrow f = \frac{2}{5}Ma$$
$$\Rightarrow \quad f = \frac{2}{7}F \Rightarrow \frac{2}{7}F \le \mu mg$$
$$\Rightarrow \quad F \le \frac{7}{2}\mu mg$$

Example 20. *A tangential force F acts at the top of a thin spherical shell of mass m and radius R. Find the acceleration of the shell, if it rolls without slipping.*

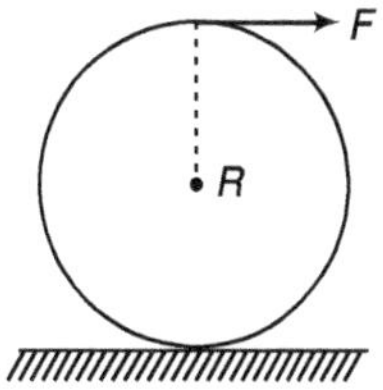

Sol. Let f be the force of friction between the shell and the horizontal surface.

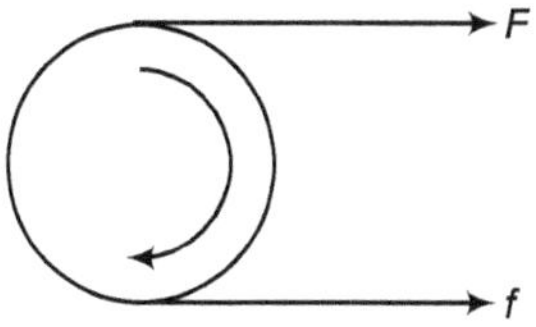

For translational motion

$$F + f = ma \qquad \ldots \text{(i)}$$

For rotational motion,

$$FR - fR = I\alpha = I\frac{a}{R}$$

$(\because a = R\alpha$ for pure rolling$)$

$$\Rightarrow \quad F - f = I\frac{a}{R^2} \qquad \ldots \text{(ii)}$$

Adding Eqs. (i) and (ii)

$$2F = \left(m + \frac{I}{R^2}\right)a = \left(m + \frac{2}{3}m\right)a = \frac{5}{3}ma$$

or

$$F = \frac{5}{6}ma \qquad \left[\because I_{\text{shell}} = \frac{2}{3}mR^2\right]$$
$$\Rightarrow \quad a = \frac{6F}{5m}$$

NCERT Selected Questions

Q 1. (a) Find the moment of inertia of a sphere about a tangent to the sphere, given that the MI of the sphere about any of its diameter to be $2MR^2/5$, where, M is the mass of the sphere and R is the radius of the sphere.

(b) Given that the MI of a disc of mass M and radius R about any of its diameters to the $MR^2/4$, find its MI about an axis normal to the disc and passing through a point on its edge.

Sol. (a)

$$I_{AB} = \frac{2}{5}MR^2 \quad \text{...(i)}$$

Let, CD be a tangent of the sphere parallel to the diameter AB of the sphere.

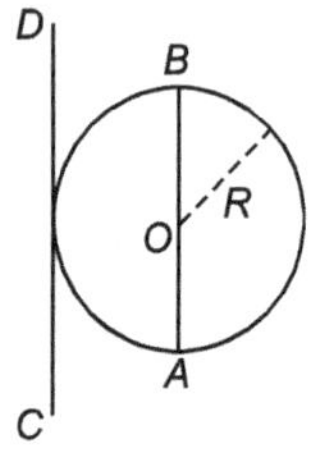

∴ Distance between the two parallel axis is R. If I_{CD} be its MI about CD axis, then according to the theorem of parallel axis,

$$I_{CD} = I_{AB} + MR^2$$

$$= \frac{2}{5}MR^2 + MR^2$$

$$= \frac{7}{5}MR^2$$

(b) Let EF be an axis perpendicular to the plane of disc and passing through its CM. Clearly, DG is the axis normal to the disc and passing through a point D on its edge.

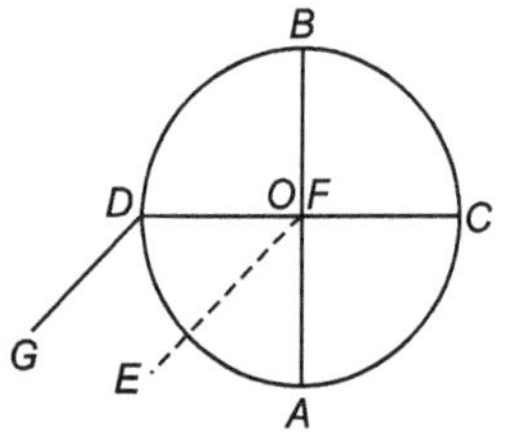

Clearly, the axis DG is parallel to the axis EF.

∴ If I_{EF} be the MI of the disc about EF axis.

Then, according to theorem of perpendicular axes,

$$I_{EF} = I_{AB} + I_{CD}$$

$$= \frac{MR^2}{4} + \frac{MR^2}{4}$$

Here, perpendicular distance between EF and DG axis $= R$

∴ If I_{DG} be the MI of the disc about the required axis, then according to theorem of parallel axes

$$I_{DG} = I_{EF} + MR^2$$

$$= \frac{1}{2}MR^2 + MR^2$$

$$= \frac{3}{2}MR^2$$

Q 2. Torques of equal magnitude are applied to a hollow cylinder and a solid sphere both having the same mass and radius. The cylinder is free to rotate about its standard axis of symmetry and the sphere is free to rotate about an axis passing through its centre. Which of the two will acquire a greater angular speed after a given time?

Sol. Let I_1 and I_2 be the moments of inertia of the hollow cylinder about its axis of symmetry and solid sphere about its axis through its centre respectively.

Then, $$I_1 = MR^2 \quad \text{...(i)}$$

and $$I_2 = \frac{2}{5}MR^2 \quad \text{...(ii)}$$

Let $\tau =$ magnitude of the torque applied on each of them.

If α_1 and α_2 be the angular accelerations produced in the cylinder and sphere respectively, then

$$\tau = I_1\alpha_1$$

and $$\tau = I_2\alpha_2$$

∴ $$I_1\alpha_1 = I_2\alpha_2$$

or $$\frac{\alpha_1}{\alpha_2} = \frac{I_2}{I_1}$$

$$= \frac{\frac{2}{5}MR^2}{MR^2} = \frac{2}{5}$$

or $$\alpha_2 = \frac{5}{2}\alpha_1$$

$$= 2.5\,\alpha_1 \quad \text{...(iii)}$$

If ω_1 and ω_2 be the angular speed of the cylinder and sphere after a time t, then

$$\omega_1 = \omega_0 + \alpha_1 t \quad \text{...(iv)}$$

and $$\omega_2 = \omega_0 + \alpha_2 t$$

$$= \omega_0 + 2.5\,\alpha_1 t \quad \text{...(v)}$$

where, $\omega_0 =$ initial angular speed

∴ From Eqs. (iv) and (v), it is clear that

$$\omega_2 > \omega_1$$

∴ The sphere will acquire more angular speed as compared to that of the cylinder after a given time.

Q 3. A solid cylinder of mass 20 kg rotates about its axis with angular speed 100 rad s^{-1}. The radius of the cylinder is 0.25 m. What is the kinetic energy associated with the rotation of the cylinder? What is the magnitude of angular momentum of the cylinder about its axis?

Sol. If I be the MI of the cylinder about its axis,

then $$I = \frac{1}{2}MR^2 = \frac{1}{2} \times 20 \times (0.25)^2$$

$$= 0.625 \text{ kg-m}^2$$

$\therefore$ KE associated with the rotating cylinder is given by,

$$KE = \frac{1}{2} I\omega^2 = \frac{1}{2} \times 0.625 \times (100)^2$$

$$= 3125 \text{ J}$$

Also using the relation,

or $L = \sqrt{2I \text{ KE}}$

We have,

$L = \sqrt{2 \times 0.625 \times 3125}$ or $L = 62.5$ J-s

***Q* 4.** (a) A child stands at the centre of a turn table with his two arms outstretched. The turn table is set rotating with an angular speed of 40 rpm. How much is the angular speed of the child, if he folds his hands back and thereby reduces his moments of inertia to 2/5 times the initial value? Assume that the turn table rotates without friction.

(b) Show that the child's new kinetic energy of rotation is more than the initial kinetic energy of rotation. How do you account for this increase in kinetic energy?

Sol. (a) $I_2 = \frac{2}{5} I_1$ (Given)

$$f_1 = 40 \text{rpm} = \frac{40}{60} \text{rps}$$

$$\therefore \quad \omega_1 = 2\pi f_1 = \frac{2\pi \times 40}{60} \text{ rad s}^{-1}$$

$$= \frac{4}{3} \pi \text{ rad s}^{-1}$$

Let, ω_1 and ω_2 be the angular speed of the child with outstretched and folding arms respectively.

$\therefore$ According to the law of conservation of angular momentum, we get

$$I_1\omega_1 = I_2\omega_2$$

or $\omega_2 = \frac{I_1}{I_2} \omega_1$ or $\omega_2 = \frac{I_1}{\frac{2}{5} I_2} \times \frac{4\pi}{3}$

$$= \frac{5}{2} \times \frac{4\pi}{3} = \frac{10\pi}{3} \text{ rad s}^{-1}$$

$\therefore$ Frequency of revolution f_2 is given by

$$f_2 = \frac{\omega_2}{2\pi} = \frac{10\pi}{3 \times 2\pi} \text{ rps}$$

$$= \frac{5}{3} \times 60 \text{ rpm} = 100 \text{rpm}$$

$$\therefore \quad f_2 = 100 \text{ rpm}$$

(b) Initial KE of rotation $= \frac{1}{2} I_1 \omega_1^2$

Final KE of rotation $= \frac{1}{2} I_2 \omega_2^2$

$$= \frac{1}{2} \left(\frac{2}{5} I_1\right) \left(\frac{5}{2}\right)^2 \omega_1^2 \qquad \left(\because \omega_2 = \frac{5}{2} \omega_1\right)$$

$$= \frac{2}{5} \times \left(\frac{5}{2}\right)^2 \times \left(\frac{1}{2} I_1\omega_1^2\right) = \frac{5}{2} \left(\frac{1}{2} I_1\omega_1^2\right)$$

$$\therefore \quad \frac{\text{Final KE of rotation}}{\text{Initial KE of rotation}} = \frac{\frac{5}{2}\left(\frac{1}{2} I_1\omega_1^2\right)}{\left(\frac{1}{2} I_1\omega_1^2\right)} = \frac{5}{2}$$

Clearly, new KE of rotation (when child folds his hands back) is $\frac{5}{2}$ times greater than the initial KE of rotation.

The increase in KE of rotation on folding back his hands appears due to the use of child's own internal energy i.e. muscular energy (used in folding the arms) to increase the rotational kinetic energy.

***Q* 5.** A rope of negligible mass is wound round a hollow cylinder of mass 3 kg and radius 40cm. What is the angular acceleration of the cylinder, if the rope is pulled with a force of 30 N? What is the linear acceleration of the rope? Assuming that there is no slipping.

Sol. If I be the MI of the hollow cylinder about its axis, then

$$I = MR^2 = 3(0.4)^2 = 0.48 \text{ kg-m}^2$$

If τ = torque acting on the cylinder, then

$$\tau = FR = 30 \times 0.4 = 12 \text{N-m}$$

$$\therefore \quad \alpha = \frac{\tau}{I} = \frac{12}{0.48} = 25 \text{ rad s}^{-2}$$

$$\therefore \quad a = R\alpha = 0.4 \times 25 = 10 \text{m/s}^2$$

***Q* 6.** To maintain a rotor at a uniform angular speed of 200 rad s^{-1}, an engine needs to transmit a torque of 180 N-m. What is the power required by the engine?

Sol. Here, $\omega = 200$ rad s^{-1}

$$\tau = 180 \text{N-m}$$

Using the relation, $P = \tau\omega$, we get

$$P = 180 \times 200$$

$$= 36000 \text{ W} = 36 \text{ kW}$$

***Q* 7.** A solid sphere rolls down two different inclined planes of the same heights but different angles of inclination.

(a) Will it reach the bottom with the same speed in each case?

(b) Will it take longer to roll down one plane than the other?

Sol. (a) As work done by friction in case of pure rolling = 0

Therefore, in both cases mechanical energy will remain constant.

Kinetic energy in reaching bottom in both cases is *mgh.*

(b) Yes, it will take longer time down one plane than the other. It will be longer for the plane having smaller angle of inclination.

***Q* 8.** A hoop of radius 2 m weight 100 kg. It rolls along a horizontal floor so that its centre of mass has a speed of 20 cm^{-1}. How much work has to be done to stop it?

Sol. If ω be the angular velocity of the CM of the hoop, then

$$\omega = \frac{v}{r} = \frac{0.20}{2} = 0.10 \text{ rad s}^{-1}$$

$$I = mr^2 = 100 \times (2)^2 = 400 \text{ kg-m}^2$$

∴ Total KE of ring = Rotational KE + Translational KE

or $$E = \frac{1}{2} I\omega^2 + \frac{1}{2} mv^2$$

or $$E = \frac{1}{2} \times 400 \times (0.10)^2 + \frac{1}{2} \times 100 \times (0.20)^2$$

$$= 4 \text{ J}$$

Work done to stop it = Total KE of the hoop = 4 J

***Q* 9.** The oxygen molecule has a mass of 5.30×10^{-26} kg and a moment of inertia of 1.94×10^{-46} kg-m^2 about an axis through its centre perpendicular to the lines joining the two atoms. Suppose, the mean speed of such a molecule in a gas is 500 ms^{-1} and KE of rotation is $\frac{2}{3}$ of its KE of translation. Find the average angular velocity of the molecule.

Sol. According to the given condition,

KE of rotation = $\frac{2}{3}$ × KE of translation

or $$\frac{1}{2} I\omega^2 = \frac{2}{3}\frac{1}{2} mv^2 \text{ or } \omega = \sqrt{\frac{2}{3}\frac{mv^2}{I}}$$

$$= \sqrt{\frac{2}{3} \times \frac{5.30 \times 10^{-26} \times (500)^2}{1.94 \times 10^{-46}}}$$

$$= 6.75 \times 10^{12} \text{ rad s}^{-1}$$

***Q* 10.** A cylinder rolls up an inclined plane of angle of inclination 30°. At the bottom of the inclined plane, the centre of mass of the cylinder has a speed of 5 ms^{-1}.

(a) How far will the cylinder go up the plane?

(b) How long will it take to return to the bottom?

Sol. (a) Retardation, $a = -\dfrac{g \sin\theta}{1 + \dfrac{I}{mR^2}}$

Let the cylinder be solid, then

$$I = \frac{1}{2} mR^2$$

∴ $$a = -\frac{g \sin\theta}{1 + \frac{1}{2}} = -\frac{2}{3} \times 9.8 \times \frac{1}{2}$$

$$= -\frac{9.8}{3} \text{ ms}^{-2}$$

Using the relation, $v^2 - u^2 = 2as$, we get

$$s = \frac{v^2 - u^2}{2a} = \frac{0 - 5^2}{2\left(-\frac{9.8}{3}\right)}$$

$$= \frac{25}{2 \times 9.8} \times 3 = 3.83 \text{ m}$$

(b) Let T = time taken by the cylinder to return to the bottom $T = 2t$, where t = time of ascending or descending

∴ $$a = \frac{g \sin\theta}{1 + \dfrac{I}{mR^2}} = \frac{9.8}{3} \text{ ms}^{-2}$$

$$s = 3.83 \text{ m}$$

Here, initial velocity = 0

∴ Using the relation, $s = ut + \frac{1}{2} at^2$, we get

$$t = \sqrt{\frac{2s}{a}} = \sqrt{\frac{2 \times 3.83}{\left(\frac{9.8}{3}\right)}} = 1.53 \text{ s}$$

∴ $$T = 2 \times 1.53 = 3.06 \text{ s} \approx 3 \text{ s}$$

***Q* 11.** A disc rotating about its axis with angular speed ω_0 is placed lightly [without any translational push] on a perfectly frictionless table. Will the disc roll on the surface?

Sol. The disc cannot roll on smooth surface, if it is not placed on rolling condition.

***Q* 12.** Read each statement below carefully and state with reasons, if it is true or false.

(a) The instantaneous speed of the point of contact during rolling is zero.

(b) For perfect rolling motion, work done against friction is zero, if it is taking place on stationary ground.

(c) A wheel moving down a perfectly frictionless inclined plane will undergo slipping (not rolling) motion.

Sol. (a) It is true. A rolling body can be imagined to be rotating about an axis passing through the point of contact of the body with the ground and hence its instantaneous speed is zero, if it is taking place on stationary ground.

(b) It is true. If rolling is taking place on stationary ground. In a perfect rolling motion, the work done against friction is zero,because for perfect rolling, the point of contact should be momentarily at rest. So, no work is done against friction at this point.

(c) True. On an inclined plane a body rolls due to the force of friction acting on it. If the wheel is moving down a perfectly frictionless inclined plane, it will be under the effect of its weight only. Since, the weight of the wheel acts along the vertical through its CM, the wheel will not rotate.

It will keep on slipping, i.e. in the absence of friction no rolling will take place and the body slips due to translational force.

Objective Problems

[Level 1]

Moment of Inertia

1. Moment of inertia of a body depends upon
(a) axis of rotation
(b) torque
(c) angular momentum
(d) angular velocity

2. Analogue of mass in rotational motion is
(a) moment of inertia
(b) angular momentum
(c) gyration
(d) None of the above

3. One circular ring and one circular disc both having the same mass and radius. The ratio of their moments of inertia about the axes passing through their centres and perpendicular to planes will be
(a) 1 : 1 (b) 2 : 1
(c) 1 : 2 (d) 4 : 1

4. The radius of gyration of a solid sphere of radius R about a is tangential
(a) $\sqrt{\frac{7}{5}}R$ (b) $\sqrt{\frac{2}{5}}R$
(c) $\sqrt{\frac{5}{7}}R$ (d) R

5. The ratio of the radii of gyration of a hollow sphere and a solid sphere of the same radii about a tangential axis is
(a) $\sqrt{\frac{7}{3}}$ (b) $\frac{5}{\sqrt{21}}$
(c) $\sqrt{\frac{21}{5}}$ (d) $\frac{25}{9}$

6. What is the moment of inertia of a solid sphere of density ρ and radius R about its diameter?
(a) $\frac{105}{176}R^5\rho$ (b) $\frac{176}{105}R^5\rho$
(c) $\frac{105}{176}R^2\rho$ (d) $\frac{176}{105}R^2\rho$

7. Let I_A and I_B be moments of inertia of a body about two axes A and B respectively. The axis A passes through the centre of mass of the body but B does not
(a) $I_A < I_B$
(b) $I_A < I_B$, whether the axes are parallel or not parallel
(c) if the axes are parallel $I_A < I_B$
(d) if the axes are not parallel $I_A \geq I_B$

8. Three thin rods each of length L and mass M are placed along x, y and z-axis such that one end of each rod is at origin. The moment of inertia of this system about z-axis is
(a) $\frac{2}{3}ML^2$ (b) $\frac{4ML^2}{3}$
(c) $\frac{5ML^2}{3}$ (d) $\frac{ML^2}{3}$

9. The ratio of the radii of gyration of a circular disc and a circular ring of the same radii about a tangential axis perpendicular to plane of disc or ring is
(a) 1 : 2 (b) $\sqrt{5}:\sqrt{6}$
(c) 2 : 3 (d) $\frac{\sqrt{3}}{2}$

10. Figure represents the MI of the solid sphere about an axis parallel to the diameter of the solid sphere and at a distance x from it. Which one of the following represents the variations of I with x?

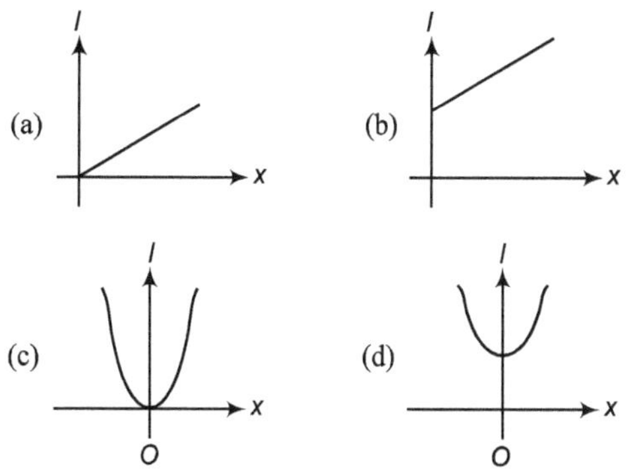

11. If I_1 is the moment of inertia of a thin rod about an axis perpendicular to its length and passing through its centre of mass and I_2 is the moment of inertia of the ring about an axis perpendicular to plane of ring and passing through its centre formed by bending the rod, then
(a) $\frac{I_1}{I_2}=\frac{3}{\pi^2}$ (b) $\frac{I_1}{I_2}=\frac{2}{\pi^2}$ (c) $\frac{I_1}{I_2}=\frac{\pi^2}{2}$ (d) $\frac{I_1}{I_2}=\frac{\pi^2}{3}$

12. A rod is placed along the line $y = 2x$ with its centre at origin. The moment of inertia of the rod is maximum about
(a) x-axis (b) y-axis
(c) z-axis (d) Data insufficient

13. The moment of inertia of a thin rectangular plate *ABCD* of uniform thickness about an axis passing through the centre *O* and perpendicular to the plane of the plate is

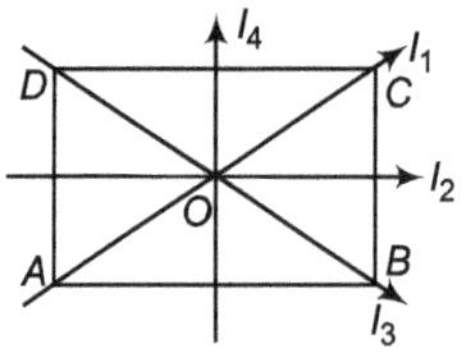

(a) $I_1 + I_2$ (b) $I_2 + I_4$
(c) $I_1 + I_3$ (d) $I_1 + I_2 + I_3 + I_4$

14. A wheel comprises of a ring of radius *R* and mass *M* and three spokes of mass *m* each. The moment of inertia of the wheel about its axis is

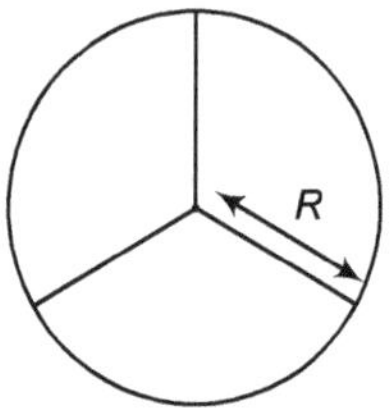

(a) $\left(M + \frac{m}{4}\right) R^2$ (b) $(M + m) R^2$
(c) $(M + 3m) R^2$ (d) $\left(\frac{M + m}{2}\right) R^2$

15. Moment of inertia of a rod of mass *m* and length *l* about its one end is *I*. If one-fourth of its length is cut away, then moment of inertia of the remaining rod about its one end will be

(a) $\frac{3}{4} I$ (b) $\frac{9}{16} I$ (c) $\frac{27}{64} I$ (d) $\frac{I}{16}$

16. For the uniform *T* shaped structure, with mass 3*M*, moment of inertia about an axis normal to the plane and passing through *O* would be

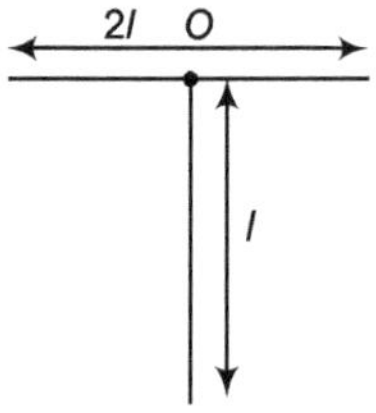

(a) $\frac{2}{3} Ml^2$ (b) Ml^2
(c) $\frac{Ml^2}{3}$ (d) None of these

17. The moment of inertia of a solid cylinder of mass *M*, length 2*R* and radius *R* about an axis passing through the centre of mass and perpendicular to the axis of the cylinder is I_1 and about an axis passing through one end of the cylinder and perpendicular to the axis of cylinder is I_2

(a) $I_2 - I_1 = MR^2$ (b) $I_2 = I_1$
(c) $\frac{I_2}{I_1} = \frac{19}{12}$ (d) $I_1 - I_2 = MR^2$

18. A particle of mass 1 kg is kept at (1 m, 1 m, 1 m). The moment of inertia of this particle about *z*-axis would be

(a) 1 kg-m^2 (b) 2 kg-m^2
(c) 3 kg-m^2 (d) None of these

19. Two uniform, thin identical rods each of mass *M* and length *l* are joined together to form a cross. What will be the moment of inertia of the cross about an axis passing through the point at which the two rods are joined and perpendicular to the plane of the cross

(a) $\frac{Ml^2}{12}$ (b) $\frac{Ml^2}{6}$
(c) $\frac{Ml^2}{4}$ (d) $\frac{Ml^2}{3}$

20. A square lamina is as shown in figure. The moment of inertia of the frame about the three axes shown in figure are I_1, I_2 and I_3 respectively. Select the correct alternative.

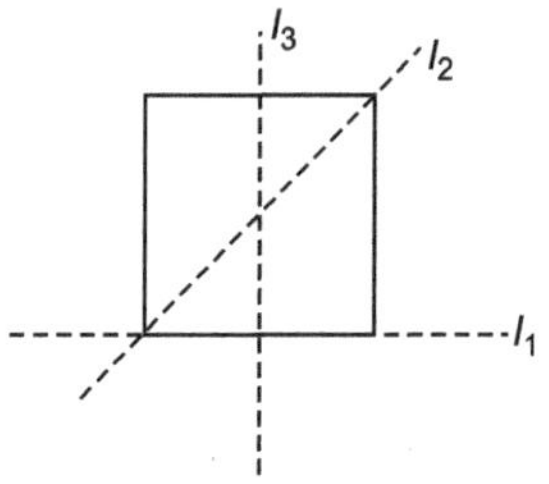

(a) $I_2 = I_3 > I_1$ (b) $I_1 > I_2 > I_3$
(c) $I_2 = I_3 < I_1$ (d) $I_1 < I_2 < I_3$

21. Moment of a force of magnitude 10 N acting along positive *y*-direction at point (2m, 0, 0) about the point (0, 1m, 0) in N-m is

(a) 10 (b) 20 (c) $10\sqrt{2}$ (d) 30

22. One circular ring and one circular disc both having the same mass and radius. The ratio of their moments of inertia about the axes passing through their centres and perpendicular to planes will be

(a) 1 : 1 (b) 2 : 1
(c) 1 : 2 (d) 4 : 1

23. The radius of gyration of a uniform rod of length *L* about an axis passing through its centre of mass is

(a) $\frac{L}{2\sqrt{3}}$ (b) $\frac{L^2}{12}$
(c) $\frac{L}{\sqrt{3}}$ (d) $\frac{L}{\sqrt{2}}$

24. Five particles of masses 2 kg each are attached to the rim of a circular disc of radius 0.1 m and negligible mass. Moment of inertia of the system about the axis passing through the centre of the disc and perpendicular to its plane is

(a) 1 kg-m^2 (b) 0.1 kg-m^2
(c) 2 kg-m^2 (d) 0.2 kg-m^2

25. ABC is a right angled triangular plate of uniform thickness. The sides are such that $AB > BC$ as shown in figure. I_1, I_2 and I_3 are moments of inertia about AB, BC and AC respectively. Then, which of the following relation is correct?

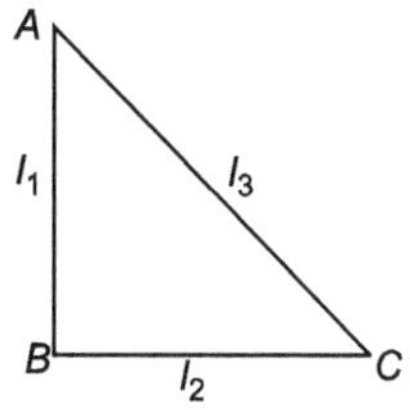

(a) $I_1 = I_2 = I_3$ (b) $I_2 > I_1 > I_3$
(c) $I_3 < I_2 < I_1$ (d) $I_3 > I_1 > I_2$

26. The moment of inertia of a cube of mass m and side a about one of its edges is equal to

(a) $\frac{2}{3}ma^2$ (b) $\frac{4}{3}ma^2$ (c) $3ma^2$ (d) $\frac{8}{3}ma^2$

27. The moment of inertia of a semicircular ring of mass M and radius R about an axis which is passing through its centre and at an angle θ with the line joining its ends as shown in figure is

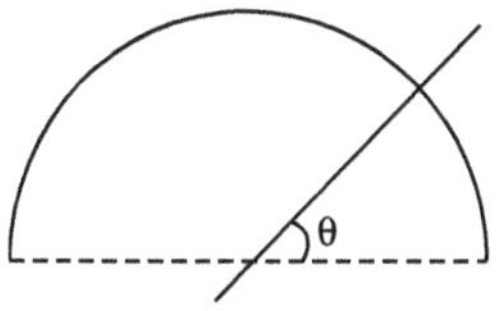

(a) $\frac{MR^2}{4}$ at $\theta = 0°$ (b) $\frac{MR^2}{2}$ if $\theta = 0°$
(c) $\frac{MR^2}{2}$ if $\theta = 45°$ (d) $\frac{MR^2}{2}$ if $\theta = 90°$

28. A square is made by joining four rods each of mass M and length L. Its moment of inertia about an axis PQ, in its plane and passing through one of its corner is

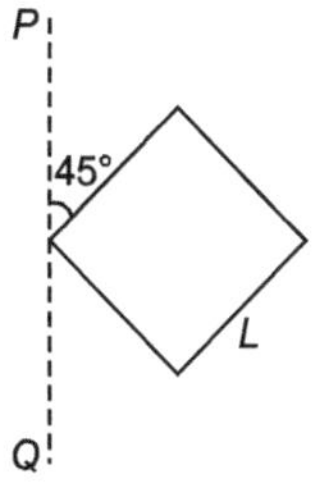

(a) $6ML^2$ (b) $\frac{4}{3}ML^2$ (c) $\frac{8}{3}ML^2$ (d) $\frac{10}{3}ML^2$

29. From a uniform square plate of side a and mass m, a square portion $DEFG$ of side $\frac{a}{2}$ is removed. Then, the moment of inertia of remaining portion about the axis AB is

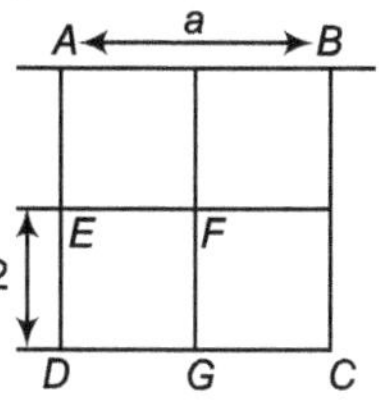

(a) $\frac{7ma^2}{16}$ (b) $\frac{3ma^2}{16}$ (c) $\frac{3ma^2}{4}$ (d) $\frac{9ma^2}{16}$

30. A semicircular 2-dimensional plate of mass m has radius r and centre C. Its centre of mass is at a distance x from C. Its moment of inertia about an axis through its centre of mass and perpendicular to its plane is

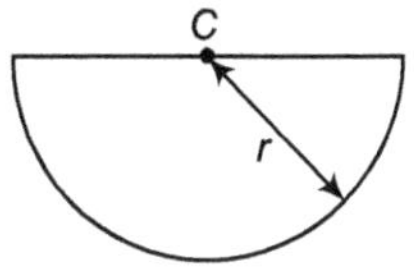

(a) $\frac{1}{2}mr^2$ (b) $\frac{1}{4}mr^2$
(c) $\frac{1}{2}mr^2 + mx^2$ (d) $\frac{1}{2}mr^2 - mx^2$

31. Two point masses m_1 and m_2 and attached from two ends of a rod of negligible mass. The system is rotated about an axis perpendicular to the length of the rod. The minimum moment of inertia of the system is

(a) about centre of mass
(b) $\frac{m_1 m_2 l^2}{m_1 + m_2}$
(c) Both (a) and (b)
(d) None of the above

32. The ratio of the radii of gyration of a circular disc and a circular ring of the same radius about a tangential axis in the plane is

(a) $\sqrt{3} : \sqrt{4}$ (b) $\sqrt{5} : \sqrt{6}$
(c) $\sqrt{6} : \sqrt{5}$ (d) $\sqrt{4} : \sqrt{3}$

Torque, Angular Momentum, Conservation of Angular Momentum and Angular Velocity

33. If a person standing on a rotating disc stretches out his hands, the angular speed will

(a) increase (b) decrease
(c) remain same (d) None of these

34. The angular momentum of a system of particles is conserved

(a) when no external force acts upon the system
(b) when no external torque acts upon the system
(c) when no external impulse acts upon the system
(d) when axis of rotation remains same

35. Angular momentum is

(a) moment of momentum
(b) product of mass and angular velocity
(c) product of moment of inertia and velocity
(d) moment in angular motion

36. If the radius of the earth contracts $\frac{1}{n}$ of its present day value, the length of the day will be approximately

(a) $\frac{24}{n}$ h (b) $\frac{24}{n^2}$ h
(c) $24n$ h (d) $24n^2$ h

37. A thin circular ring of mass M and radius R is rotating about its axis with a constant angular velocity ω. Two objects each of mass m are attached gently to the ring. The wheel now rotates with an angular velocity

(a) $\frac{\omega M}{(m+M)}$ (b) $\frac{\omega (M-2m)}{(M+2m)}$

(c) $\frac{\omega M}{(M+2m)}$ (d) $\frac{\omega (M+2m)}{M}$

38. A particle of mass 5 g is moving with a uniform speed of $3\sqrt{2}$ cm/s in the x-y plane along the line $y = 2\sqrt{5}$ cm. The magnitude of its angular momentum about the origin in g-cm^2/s is

(a) zero (b) 30

(c) $30\sqrt{2}$ (d) $30\sqrt{10}$

39. A particle of mass $m = 5$ units is moving with a uniform speed $v = 3\sqrt{2}$ units in the x-y plane along the line $y = x + 4$. The magnitude of the angular momentum about origin is

(a) zero (b) 60 unit

(c) 7.5 unit (d) $40\sqrt{2}$ unit

40. A uniform disc of radius a and mass m, is rotating freely with angular speed ω in a horizontal plane about a smooth fixed vertical axis through its centre. A particle, also of mass m, is suddenly attached to the rim of the disc and rotates with it. The new angular speed is

(a) $\frac{\omega}{6}$ (b) $\frac{\omega}{3}$

(c) $\frac{\omega}{2}$ (d) $\frac{\omega}{5}$

41. When a body is projected at an angle with the horizontal in the uniform gravitational field of the earth, the angular momentum of the body about the point of projection, as it proceeds along its path

(a) remains constant

(b) increases

(c) decreases

(d) initially decreases and after its highest point increases.

42. A force $\mathbf{F} = a\hat{\mathbf{i}} + 3\hat{\mathbf{j}} + 6\hat{\mathbf{k}}$ is acting at a point $\mathbf{r} = 2\hat{\mathbf{i}} - 6\hat{\mathbf{j}} - 12\hat{\mathbf{k}}$. The value of a for which angular momentum about origin is conserved is

(a) zero (b) 1 (c) −1 (d) 2

43. A thin uniform circular disc of mass M and radius R is rotating in a horizontal plane about an axis passing through its centre and perpendicular to its plane with an angular velocity ω. Another disc of same dimensions but of mass $\frac{1}{4}M$ is placed gently on the first disc coaxially. The angular velocity of the system is

(a) $\frac{2}{3}\omega$ (b) $\frac{4}{5}\omega$ (c) $\frac{3}{4}\omega$ (d) $\frac{1}{3}\omega$

44. The torque of a force $\mathbf{F} = -6\hat{\mathbf{i}}$ acting at a point $\mathbf{r} = 4\hat{\mathbf{j}}$ about origin will be

(a) $-24\hat{\mathbf{k}}$ (b) $24\hat{\mathbf{k}}$

(c) $24\hat{\mathbf{j}}$ (d) $24\hat{\mathbf{i}}$

45. A particle of mass 2 kg located at the position $(\hat{\mathbf{i}} + \hat{\mathbf{j}})$ m has a velocity $2(+\hat{\mathbf{i}} - \hat{\mathbf{j}} + \hat{\mathbf{k}})$ m/s. Its angular momentum about z-axis in kg-m^2/s is

(a) +4 (b) +8

(c) −4 (d) −8

46. A wheel of radius 20 cm forces applied to it as shown in the figure. The torque produced by the forces 4 N at A, 8 N at B, 6 N at C and 9 N at D at angles indicated is

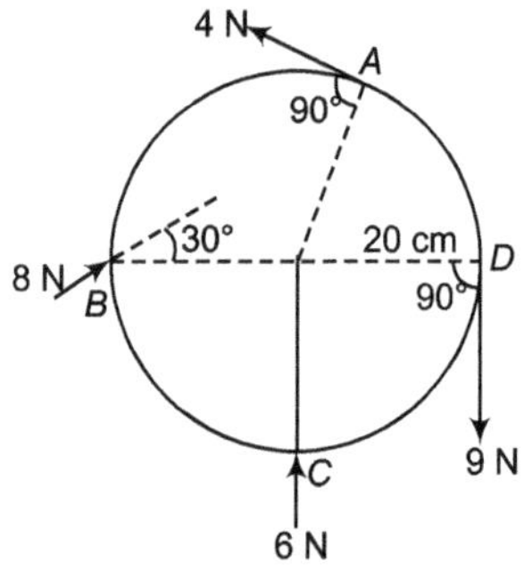

(a) 5.4 N-m anti-clockwise (b) 1.80 N-m clockwise

(c) 2.0 N-m clockwise (d) 3.6 N-m clockwise

47. The torque of force $\mathbf{F} = -3\hat{\mathbf{i}} + \hat{\mathbf{j}} + 5\hat{\mathbf{k}}$ acting on a point $\mathbf{r} = 7\hat{\mathbf{i}} + 3\hat{\mathbf{j}} + \hat{\mathbf{k}}$ about origin will be

(a) $14\hat{\mathbf{i}} - 38\hat{\mathbf{j}} + 16\hat{\mathbf{k}}$ (b) $4\hat{\mathbf{i}} - 4\hat{\mathbf{j}} + 6\hat{\mathbf{k}}$

(c) $-14\hat{\mathbf{i}} - 38\hat{\mathbf{j}} + 16\hat{\mathbf{k}}$ (d) $-21\hat{\mathbf{i}} - 3\hat{\mathbf{j}} + 5\hat{\mathbf{k}}$

48. O is the centre of an equilateral triangle ABC. F_1, F_2 and F_3 are three forces acting along the sides AB, BC and AC as shown in figure.

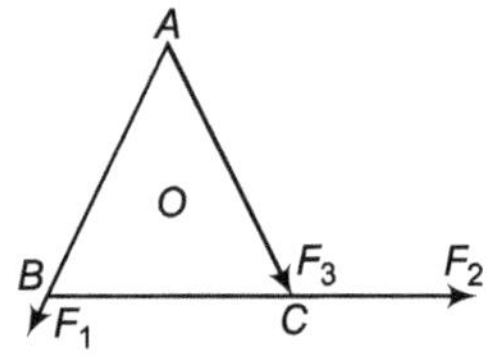

What should be the magnitude of F_3 so that the total torque about O is zero?

(a) $\frac{(F_1 + F_2)}{2}$ (b) $(F_1 - F_2)$

(c) $(F_1 + F_2)$ (d) $2(F_1 + F_2)$

49. A diver in a swimming pool bends his head before diving, because it

(a) decreases his moment of inertia

(b) decreases his angular velocity

(c) increases his moment of inertia

(d) decreases his linear velocity

50. A particle of mass m is projected with a velocity v making an angle of 45° with the horizontal. The magnitude of angular momentum of projectile about the point of projection when the particle is at its maximum height h is

(a) zero (b) $\frac{mvh}{\sqrt{2}}$

(c) mvh (d) $\sqrt{2}\,mvh$

51. A sphere rolls without slipping on a rough horizontal surface with centre of mass speed v_0. If mass of the sphere is M and its radius is R, then what is the angular momentum of the sphere about the point of contact?

(a) $\frac{5}{2}Mv_0R$ (b) $\frac{7}{5}Mv_0R$

(c) $\frac{3}{5}Mv_0R$ (d) $\frac{1}{2}Mv_0R$

Pure Rotational Motion

52. The figure shows the angular velocity *versus* time graph of a flywheel. The angle, in radians through which the flywheel turns during 25 s is

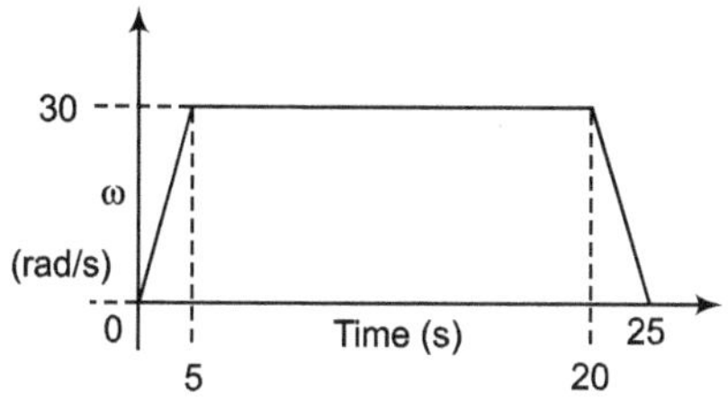

(a) 120 (b) 480
(c) 600 (d) 750

53. A body is in pure rotation. The linear speed v of a particle, the distance r of the particle from the axis and the angular velocity ω of the body are related as $\omega = \frac{v}{r}$. Thus,

(a) $\omega \propto \frac{1}{r}$ (b) $\omega \propto r$

(c) $\omega = 0$ (d) ω is independent of r

54. The motor of an engine is rotating about its axis with an angular velocity of 100 rpm. It comes to rest is 15 s, after being switched off. Assuming constant angular deceleration. What are the numbers of revolutions made by it before coming to rest?

(a) 12.5 (b) 40 (c) 32.6 (d) 15.6

55. A wheel is subjected to uniform angular acceleration about its axis. Initially its angular velocity is zero. In the first 2 s, it rotates through an angle θ_1, in the next 2 s, it rotates through an angle θ_2. The ratio of $\frac{\theta_2}{\theta_1}$, is

(a) 1 (b) 2
(c) 3 (d) 5

56. A wheel is rotating at 900 rpm about its axis. When power is cut-off it comes to rest in 1 min. The angular retardation in rad/s^2 is

(a) $\frac{\pi}{2}$ (b) $\frac{\pi}{4}$

(c) $\frac{\pi}{6}$ (d) $\frac{\pi}{8}$

57. If the equation for the displacement of a particle moving on a circular path is given by $\theta = 2t^3 + 0.5$, where θ is in radian and t is in second, then the angular velocity of the particle after 2 s is

(a) 8 rad/s (b) 12 rad/s
(c) 24 rad/s (d) 36 rad/s

58. A constant torque of 1000 N-m turns a wheel of moment of inertia 200 kg-m^2 about an axis through its centre. Its angular velocity after 3 s is

(a) 1 rad/s (b) 5 rad/s
(c) 10 rad/s (d) 15 rad/s

59. A table fan, rotating at a speed of 2400 rpm is switched off and the resulting variation of the rpm with time is shown in the figure. The total number of revolutions of the fan before it comes to rest is

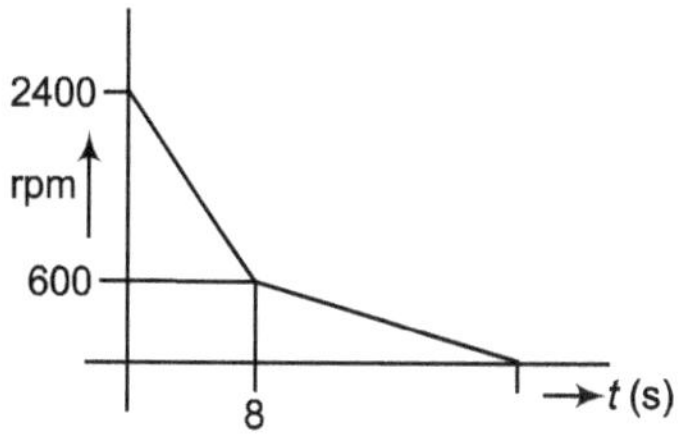

(a) 420 (b) 190
(c) 280 (d) 380

60. A wheel is at rest. Its angular velocity increases uniformly and becomes 80 rad/s after 5 s. The total angular displacement is

(a) 800 rad (b) 400 rad
(c) 200 rad (d) 100 rad

61. A rigid body rotates about a fixed axis with variable angular velocity equal to $\alpha - \beta t$, at the time t, where α, β are constants. The angle through which it rotates before its stops

(a) $\frac{\alpha^2}{2\beta}$ (b) $\frac{\alpha^2 - \beta^2}{2\alpha}$

(c) $\frac{\alpha^2 - \beta^2}{2\beta}$ (d) $\frac{(\alpha - \beta)\alpha}{2}$

62. A flywheel having a radius of gyration of 2 m and mass 10 kg rotates at an angular speed of 5 rad/s about an axis perpendicular to it through its centre. The kinetic energy of rotation is

(a) 500 J (b) 2000 J
(c) 1000 J (d) 250 J

63. A wheel is rotating at the rate of 33 rpm. If it comes to stop in 20 s. Then, the angular retardation will be

(a) π rad/s (b) 11π rad/s^2

(c) $\frac{\pi}{200}$ rad/s^2 (d) $\frac{11\pi}{200}$ rad/s^2

Rotational and Translational Motion

64. A sphere of moment of inertia I rolls down an inclined plane without slipping. The ratio of the rotational kinetic energy to the translational kinetic energy is nearly

(a) $\frac{2}{5}$ (b) $\frac{7}{5}$

(c) $\frac{7}{2}$ (d) $\frac{5}{2}$

65. Total KE of sphere of mass M rolling with velocity v is

(a) $\frac{7}{10}Mv^2$ (b) $\frac{5}{6}Mv^2$

(c) $\frac{7}{5}Mv^2$ (d) $\frac{10}{7}Mv^2$

66. A sphere cannot roll without applying external force on

(a) a smooth inclined surface
(b) a smooth horizontal surface
(c) a rough inclined surface
(d) a rough horizontal surface

67. The centre of a wheel rolling on a plane surface moves with a speed v_0. A particle on the rim of the wheel at the same level as the centre will be moving at speed

(a) zero (b) v_0

(c) $\sqrt{2}v_0$ (d) $2v_0$

68. A solid sphere, a hollow sphere and a disc, all having same mass and radius, are placed at the top of a smooth incline and released. Least time will be taken in reaching the bottom by

(a) the solid sphere (b) the hollow sphere

(c) the disc (d) all will take same time

69. A solid sphere, a hollow sphere and a disc, all having same mass and radius, are placed at the top of an inclined plane and released. The friction coefficients between the objects and the incline are same and not sufficient to allow pure rolling. Least time will be taken in reaching the bottom by

(a) the solid sphere
(b) the hollow sphere
(c) the disc
(d) all will take same time

70. In the previous question, the smallest kinetic energy at the bottom of the incline will be achieved by

(a) the solid sphere
(b) the hollow sphere
(c) the disc
(d) all will achieve same kinetic energy

71. A wheel of radius R rolls on the ground with a uniform velocity v. The velocity of topmost point relative to the bottommost point is

(a) v (b) $2v$

(c) $\frac{v}{2}$ (d) zero

72. A sphere can roll on a surface inclined at an angle θ, if the friction coefficient $\mu > (2/7) g \sin\theta$. Now, suppose the friction coefficient is $(1/7) g \sin\theta$ and the sphere is released from rest on the incline

(a) it will stay at rest
(b) it will make pure translational motion
(c) it will translate and rotate about the centre
(d) the angular momentum of the sphere about its centre will remain constant.

73. A solid homogeneous sphere is moving on a rough horizontal surface, partly rolling and partly sliding. During this kind of motion of the sphere

(a) total KE is conserved
(b) angular momentum of the sphere about the point of contact with the plane is conserved
(c) only the rotational KE about centre of mass is conserved
(d) angular momentum about centre of mass is conserved.

74. A spherical ball rolls on a table without slipping. The fraction of its total energy associated with rotation is

(a) $\frac{3}{5}$ (b) $\frac{2}{7}$ (c) $\frac{2}{5}$ (d) $\frac{3}{7}$

75. The least coefficient of friction for an inclined plane inclined at an angle α with horizontal, in order that a solid cylinder will roll down without slipping is

(a) $\frac{2}{3}\tan\alpha$ (b) $\frac{2}{7}\tan\alpha$

(c) $\frac{1}{3}\tan\alpha$ (d) $\frac{4}{3}\tan\alpha$

76. A string of negligible thickness is wrapped several times around a cylinder kept on a rough horizontal surface. A man standing at a distance l from the cylinder holds one end of the string and pulls the cylinder towards him. There is no slipping anywhere. The length of the string passed through the hand of the man while the cylinder reaches his hands is

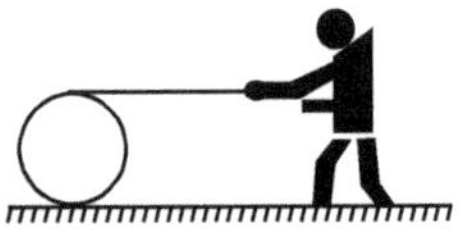

(a) l (b) $2l$

(c) $3l$ (d) $4l$

77. Two uniform solid spheres having unequal masses and unequal radii are released from rest from the same height on a rough incline. If the spheres roll without slipping

(a) the heavier sphere reaches the bottom first
(b) the bigger sphere reaches the bottom first
(c) the two spheres reach the bottom together
(d) the information given is not sufficient to tell which sphere will reach the bottom first

78. The ratio of the time taken by a solid sphere and that taken by a disc of the same mass and radius to roll down a rough inclined plane from rest from the same height is

(a) 15 : 14 (b) $\sqrt{15}:\sqrt{14}$
(c) 14 : 15 (d) $\sqrt{14}:\sqrt{15}$

79. An inclined plane makes an angle of 30° with the horizontal. A solid sphere rolling down this inclined plane from rest without slipping has a linear acceleration equal to

(a) $\frac{g}{3}$ (b) $\frac{2g}{3}$
(c) $\frac{5g}{7}$ (d) $\frac{5g}{14}$

80. A solid cylinder and a solid sphere both having the same mass and radius are released from a rough inclined plane. Both roll without slipping. Then,

(a) the force of friction that acts on the two is the same
(b) the force of friction is greater in case of a sphere than for a cylinder
(c) the force of friction is greater in case of a cylinder than for a sphere
(d) the force of friction will depend on the nature of the surface of the body that is moving and that of the inclined surface and is independent of the shape and size of the moving body

81. A solid sphere of mass M rolls without slipping on an inclined plane of inclination θ. What should be the minimum coefficient of friction, so that the sphere rolls down without slipping?

(a) $\frac{2}{5}\tan\theta$ (b) $\frac{2}{7}\tan\theta$
(c) $\frac{5}{7}\tan\theta$ (d) $\tan\theta$

82. A solid sphere of mass m rolls without slipping on an inclined plane of inclination θ. The linear acceleration of the sphere is

(a) $\frac{7}{5}g\sin\theta$ (b) $\frac{2}{7}g\sin\theta$
(c) $\frac{3}{7}g\sin\theta$ (d) $\frac{5}{7}g\sin\theta$

83. A ring is kept on a rough inclined surface. But the coefficient of friction is less than the minimum value required for pure rolling. At any instant of time let K_T and K_R be the translational and rotational kinetic energies of the ring, then

(a) $K_R = K_T$ (b) $K_R > K_T$
(c) $K_T > K_R$ (d) $K_R = 0$

84. A sphere is rolling down a plane of inclination θ to the horizontal. The acceleration of its centre down the plane is

(a) $g\sin\theta$
(b) less than $g\sin\theta$
(c) greater than $g\sin\theta$
(d) zero

85. The speed of a uniform spherical shell after rolling down an inclined plane of vertical height h from rest, is

(a) $\sqrt{\frac{10gh}{7}}$ (b) $\sqrt{\frac{6gh}{5}}$
(c) $\sqrt{\frac{4gh}{5}}$ (d) $\sqrt{2gh}$

86. A disc of radius R rolls on a rough horizontal surface. The distance covered by the point A in one revolution is

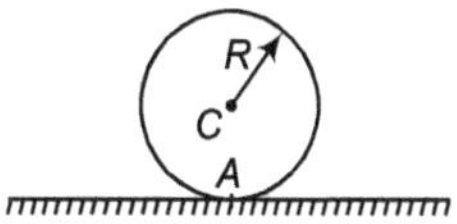

(a) $2\pi R$ (b) $2R$
(c) $8R$ (d) πR

87. A wheel of bicycle is rolling without slipping on a level road. The velocity of the centre of mass is v_{CM} then true statement is

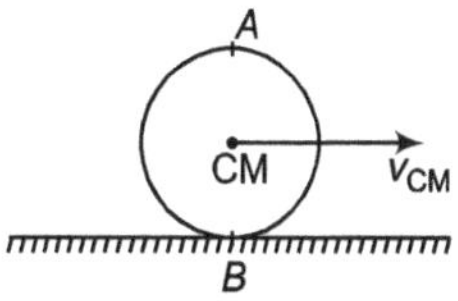

(a) the velocity of point A is $2v_{CM}$ and velocity of point B is zero
(b) the velocity of point A is zero and velocity of point B is $2v_{CM}$
(c) the velocity of point A is $2v_{CM}$ and velocity of point B is $-v_{CM}$
(d) the velocities of both A and B are v_{CM}

88. A disc is rolling without slipping on a horizontal surface with C, as its centre and Q and P the two points equidistant from C. Let v_P, v_Q and v_C be the magnitudes of velocities of points P, Q and C respectively, then

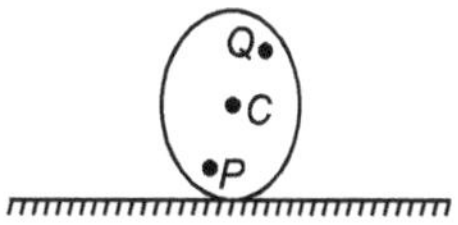

(a) $v_Q > v_C > v_P$ (b) $v_Q < v_C < v_P$
(c) $v_Q = v_P, v_C = \frac{1}{2}v_P$ (d) $v_Q < v_C > v_P$

Miscellaneous Problems

89. A rigid body rotates with an angular momentum L. If its rotational kinetic energy is made 4 times, its angular momentum will become

(a) $4L$ (b) $16L$
(c) $\sqrt{2}\,L$ (d) $2L$

90. A constant torque acting on a uniform circular wheel changes its angular momentum from A_0 to $4A_0$ in 4 s. The magnitude of this torque is

(a) $\frac{3A_0}{4}$ (b) $4A_0$
(c) A_0 (d) $12A_0$

91. A uniform rod of mass 2 kg and length 1 m lies on a smooth horizontal plane. A particle of mass 1 kg moving at a speed of 2 m /s perpendicular to the length of the rod strikes it at a distance $\frac{1}{4}$ m from the centre and stops. What is the angular velocity of the rod about its centre just after the collision?

(a) 3 rad /s (b) 4 rad /s
(c) 1 rad /s (d) 2 rad /s

92. A ring and a disc of different masses are rotating with the same kinetic energy. If we apply a retarding torque τ on the ring, it stops after making n revolutions. After how many revolutions will the disc stop, if the retarding torque on it is also τ?

(a) $\frac{n}{2}$ (b) n
(c) $2n$ (d) Data insufficient

93. A thin bar of mass m and length l is free to rotate about a fixed horizontal axis through a point at its end. The bar is brought to a horizontal position ($\theta = 90°$) and then released. The angular velocity when it reaches the lowest point is

(a) directly proportional to its length and inversely proportional to its mass
(b) independent of mass and inversely proportional to the square root of its length
(c) dependent only upon the acceleration due to gravity and the mass of the bar
(d) directly proportional to its length and inversely proportional to the acceleration due to gravity

94. The rotational KE of a body is E and its moment of inertia is I. The angular momentum is

(a) EI (b) $2\sqrt{(EI)}$
(c) $\sqrt{(2EI)}$ (d) $\frac{E}{I}$

95. A disc is free to rotate about a smooth horizontal axis passing through its centre of mass. A particle is fixed at the top of the disc. A slight push is given to the disc and it starts rotating. During the process

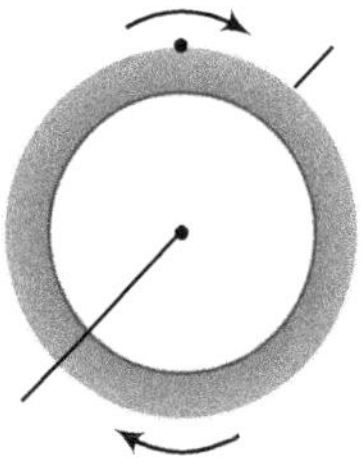

(a) only mechanical energy is conserved
(b) only angular momentum (about the axis of rotation) is conserved
(c) both mechanical energy and angular momentum are conserved
(d) neither the mechanical energy nor the angular momentum are conserved

96. Work done by friction in case of pure rolling

(a) is always zero
(b) is always positive
(c) is always negative
(d) may be positive, negative or zero

97. A particle is moving in a circular orbit with constant speed. Select wrong alternate.

(a) Its linear momentum is conserved
(b) Its angular momentum is conserved
(c) It is moving with variable velocity
(d) It is moving with variable acceleration

98. A fly wheel is in the form of a uniform circular disc of radius 1 m and mass 2 kg. The work which must be done on it to increase its frequency of rotation from 5 rev s^{-1} to 10 rev s^{-1} is approximately

(a) 1.5×10^2 J (b) 3.5×10^2 J
(c) 1.5×10^3 J (d) 3.0×10^3 J

99. If **F** be a force acting on a particle having the position vector **r** and τ be the torque of this force about the origin, then

(a) $\mathbf{r} \times \tau = 0$ and $\mathbf{F} \times \tau = 0$ (b) $\mathbf{r} \times \tau = 0$ and $\mathbf{F} \times \tau \neq 0$
(c) $\mathbf{r} \times \tau \neq 0$ and $\mathbf{F} \times \tau \neq 0$ (d) $\mathbf{r} \times \tau \neq 0$ and $\mathbf{F} \times \tau = 0$

100. A body is under the action of two equal and oppositely directed forces and the body is rotating with constant non-zero angular acceleration. Which of the following cannot be the separation between the lines of action of the forces?

(a) 1 m (b) 0.4 m (c) 0.25 m (d) Zero

101. A particle P is moving in a circle of radius a with a uniform speed u. C is the centre of the circle and AB is a diameter. The angular velocities of P about A and C are in the ratio

(a) 1 : 1 (b) 1 : 2
(c) 2 : 1 (d) 4 : 1

102. A particle performs uniform circular motion with an angular momentum L. If the frequency of the particle motion is doubled, the angular momentum becomes

(a) $2L$ (b) $4L$ (c) $\frac{L}{2}$ (d) $\frac{L}{4}$

103. A rod of uniform mass and of length L can freely rotate in a vertical plane about an axis passing through O. The angular velocity of the rod when it falls from position P to P' through an angle α is

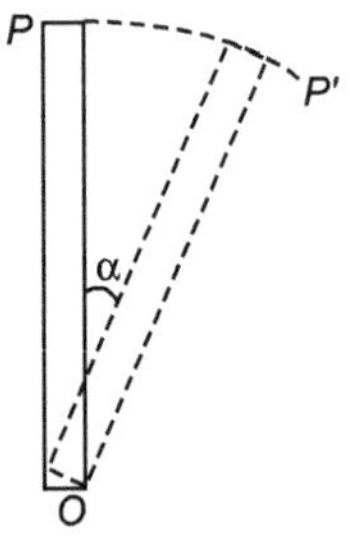

(a) $\sqrt{\frac{6g}{5L}} \sin \alpha$
(b) $\sqrt{\frac{6g}{L}} \sin \frac{\alpha}{2}$
(c) $\sqrt{\frac{6g}{L}} \cos \frac{\alpha}{2}$
(d) $\sqrt{\frac{6g}{L}} \sin \alpha$

[Level 2]

Only One Correct Option

1. A ball rolls without slipping. The radius of gyration of the ball about an axis passing through its centre of mass is K. If radius of the ball be R, then the fraction of total energy associated with its rotational energy will be

(a) $\frac{K^2}{K^2+R^2}$ (b) $\frac{R^2}{K^2+R^2}$

(c) $\frac{K^2+R^2}{R^2}$ (d) $\frac{K^2}{R^2}$

2. A particle of mass m is projected with velocity u at an angle of θ with the horizontal. The initial angular momentum of the particle about the highest point of its trajectory is equal to

(a) $\frac{mu^3 \sin^2\theta \cos\theta}{3g}$

(b) $\frac{3\,mu^3 \sin^2\theta \cos\theta}{2g}$

(c) $\frac{mu^3 \sin^2\theta \cos\theta}{2g}$

(d) None of the above

3. A uniform rod of mass 2 kg and length 1 m lies on a smooth horizontal plane. A particle of mass 1 kg moving at a speed of 2 m/s perpendicular to the length of the rod strikes it at a distance $\frac{1}{4}$ m from the centre and stops. What is the angular velocity of the rod about its centre just after the collision?

(a) 3 rad/s (b) 4 rad/s

(c) 1 rad/s (d) 2 rad/s

4. A cord is wound around the circumference of wheel of radius r. The axis of the wheel is horizontal and MI is I. A weight mg is attached to the end of the cord and falls from rest. After falling through a distance h, the angular velocity of the wheel will be

(a) $\sqrt{\frac{2gh}{I+mr^2}}$ (b) $\left(\frac{2mgh}{I+mr^2}\right)^{1/2}$

(c) $\left(\frac{2mgh}{I+2mr^2}\right)^{1/2}$ (d) $\sqrt{2gh}$

5. The speed of a homogeneous solid sphere after rolling down an inclined plane of vertical height h, from rest without sliding is

(a) $\sqrt{gh}$ (b) $\sqrt{\left(\frac{g}{5}\right)gh}$

(c) $\sqrt{\left(\frac{4}{3}\right)gh}$ (d) $\sqrt{\left(\frac{10}{7}\right)gh}$

6. A solid sphere rolls without slipping and presses a spring of spring constant k as shown in figure. Then, the compression in the spring will be

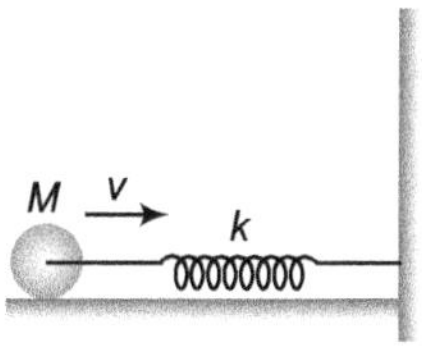

(a) $v\sqrt{\frac{2M}{3k}}$ (b) $v\sqrt{\frac{2M}{5k}}$

(c) $v\sqrt{\frac{5k}{7M}}$ (d) $v\sqrt{\frac{7M}{5k}}$

7. A portion of a ring of radius R has been removed as shown in figure. Mass of the remaining portion is m. Centre of the ring is at origin O. Let I_A and I_O be the moment of inertias passing through points

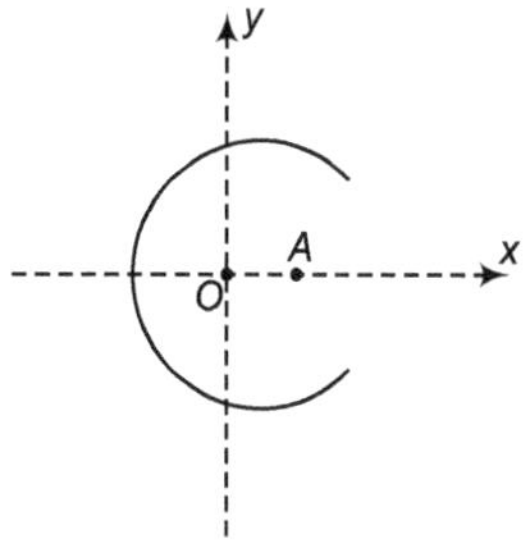

A and O are perpendicular to the plane of the ring. Then,

(a) $I_O = mR^2$ (b) $I_O = I_A$

(c) $I_O > I_A$ (d) $I_A > I_O$

8. Consider three solid spheres, sphere (i) has radius r and mass m, sphere (ii) has radius r and mass $3m$, sphere (iii) has radius $3r$ and mass m. All can be placed at the same point on the same inclined plane, where they will roll without slipping to the bottom. If allowed to roll down the incline, then at the bottom of the incline

(a) sphere (i) will have the largest speed

(b) sphere (ii) will have the largest speed

(c) sphere (ii) will have the largest kinetic energy

(d) all the spheres will have equal speeds

9. A circular disc of mass m and radius R rests flat on a horizontal frictionless surface. A bullet, also of mass m and moving with a velocity v, strikes the disc and gets embedded in it.

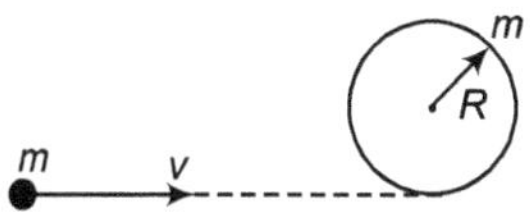

The angular velocity with which the system rotates after the bullet strikes the hoop is

(a) $\frac{v}{2R}$ (b) $\frac{v}{3R}$

(c) $\frac{2v}{3R}$ (d) $\frac{3v}{4R}$

10. A small pulley of radius 20 cm and moment of inertia 0.32 kg-m^2 is used to hang a 2 kg mass with the help of massless string. If the block is released, for no slipping condition acceleration of the block will be

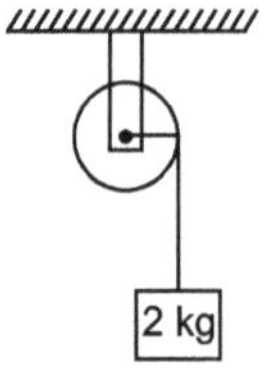

(a) 2 m/s^2 (b) 4 m/s^2
(c) 1 m/s^2 (d) 3 m/s^2

11. A disc of mass m_0 rotates freely about a fixed horizontal axis through its centre. A thin cotton pad is fixed to its rim, which can absorb water?

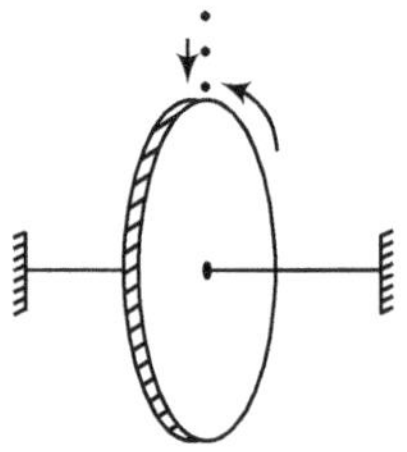

The mass of water dripping onto the pad is μ per second. After what time will the angular velocity of the disc get reduced to half of its initial value?

(a) $\frac{2m_0}{\mu}$ (b) $\frac{3m_0}{\mu}$ (c) $\frac{m_0}{\mu}$ (d) $\frac{m_0}{2\mu}$

12. A solid sphere and a solid cylinder of same mass are rolled down on two inclined planes of heights h_1 and h_2 respectively. If at the bottom of the plane the two objects have same linear velocities, then the ratio of $h_1 : h_2$ is

(a) 2 : 3 (b) 7 : 5
(c) 14 : 15 (d) 15 : 14

13. If I_1 is the moment of inertia of a thin rod about an axis perpendicular to its length and passing through its centre of mass and I_2 is the moment of inertia of the ring about an axis perpendicular to plane of ring and passing through its centre formed by bending the rod, then

(a) $\frac{I_1}{I_2} = \frac{3}{\pi^2}$ (b) $\frac{I_1}{I_2} = \frac{2}{\pi^2}$
(c) $\frac{I_1}{I_2} = \frac{\pi^2}{2}$ (d) $\frac{I_1}{I_2} = \frac{\pi^2}{3}$

14. A horizontal disc rotates freely about a vertical axis through its centre. A ring, having the same mass and radius as the disc is now gently placed on the disc. After some time, the two rotate with a common angular velocity

(a) some friction exists between the disc and the ring
(b) the angular momentum of the disc plus ring is conserved
(c) the final common angular velocity is $\frac{2}{3}$rd of the initial angular velocity of the disc
(d) $\frac{2}{3}$rd of the initial kinetic energy is converted into heat

15. A uniform rod AB of length 7m is undergoing combined rotational and translational motion such that at some instant of time, velocities of its end points A and centre C are both perpendicular to the rod and opposite in direction, having magnitudes 11 m/s and 3 m/s respectively as shown in the figure.

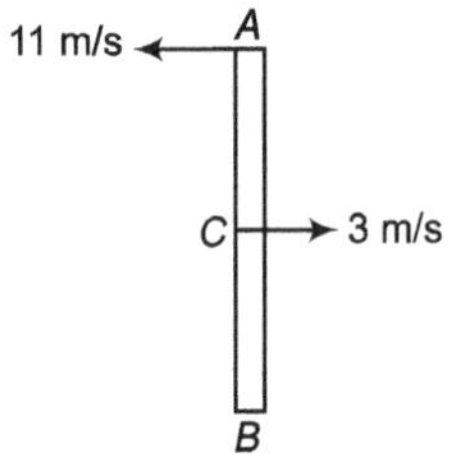

Velocity of centre C and angular velocity of the rod remain constant

(a) acceleration of point A is 56 m/s^2
(b) acceleration of point B is 56 m/s^2
(c) at the instant shown in the figure acceleration of point B is more than that of point A
(d) angular velocity of the rod is 4 rad/s

16. A rod of length L whose lower end is fixed along the horizontal plane starts to topple from the vertical position. The velocity of the upper end of the rod when it hits the ground is

(a) $\sqrt{3gL}$ (b) $\sqrt{2gL}$ (c) $\sqrt{gL}$ (d) $\sqrt{5gL}$

17. A disc of mass m and radius R is rolling on horizontal ground with linear velocity v. What is the angular momentum of the disc about an axis passing through bottommost point and perpendicular to the plane of motion?

(a) $\frac{3}{2}mvR$ (b) mvR (c) $\frac{1}{2}mvR$ (d) $\frac{4}{3}mvR$

18. A force F is applied on the top of a cube as shown in figure. The coefficient of friction between the cube and the ground is μ. If F is gradually increased, the cube will topple before sliding, if

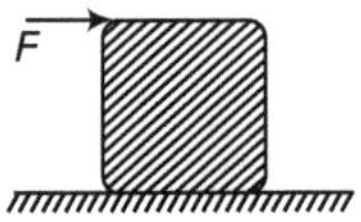

(a) $\mu > 1$ (b) $\mu < \frac{1}{2}$ (c) $\mu > \frac{1}{2}$ (d) $\mu < 1$

19. Two uniform rods of equal length but different masses are rigidly joined to form an L-shaped body, which is then pivoted as shown in figure. If in equilibrium the body is in the shown configuration, ratio M/m will be

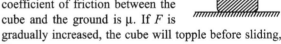

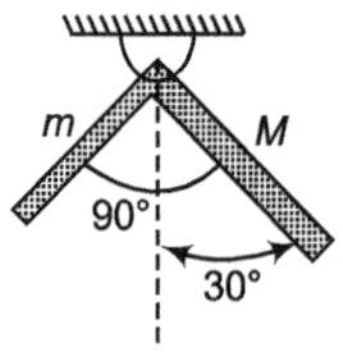

(a) 2 (b) 3 (c) $\sqrt{2}$ (d) $\sqrt{3}$

20. The figure shows a uniform rod lying along the x-axis. The locus of all the points lying on the x-y plane, about which the moment of inertia of the rod is same as that about O is

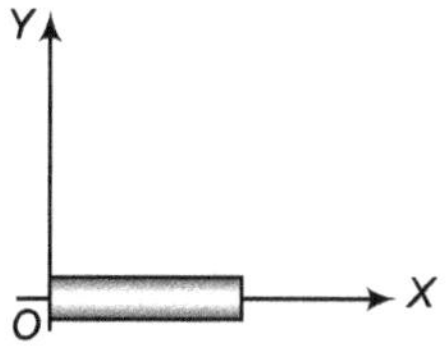

(a) an ellipse (b) a circle
(c) a parabola (d) a straight line

21. Two discs have same mass and thickness. Their materials are of densities d_1 and d_2. The ratio of their moments of inertia about an axis passing through the centre and perpendicular to the plane is

(a) $d_1 : d_2$ (b) $d_2 : d_1$ (c) $\left(\frac{d_1}{d_2}\right)^2$ (d) $\left(\frac{d_2}{d_1}\right)^2$

22. A solid sphere of mass 2 kg rolls up a 30° incline with an initial speed of 10 m/s. The maximum height reached by the sphere is ($g = 10\,\text{m/s}^2$)

(a) 3.5 m (b) 7.0 m
(c) 10.5 m (d) 14.0 m

23. A hole of radius $R/2$ is cut from a thin circular plate of radius R and mass M. The moment of inertia of the plate about an axis through O perpendicular to the x-y plane (i.e. about the z-axis) is

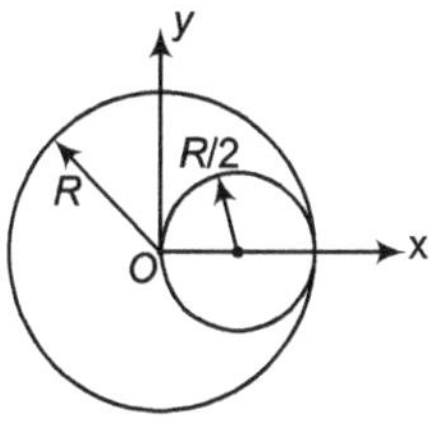

(a) $\frac{5}{7}MR^2$ (b) $\frac{7}{12}MR^2$ (c) $\frac{13}{32}MR^2$ (d) $\frac{13}{24}MR^2$

24. If a disc of mass m and radius r is reshaped into a ring or radius $2r$, the mass remaining the same, the radius of gyration about centroidal axis perpendicular to plane goes up by a factor of

(a) $\sqrt{2}$ (b) 2
(c) $2\sqrt{2}$ (d) 4

25. A spool is pulled horizontally by two equal and opposite forces as shown in the figure. Which of the following statements is correct?

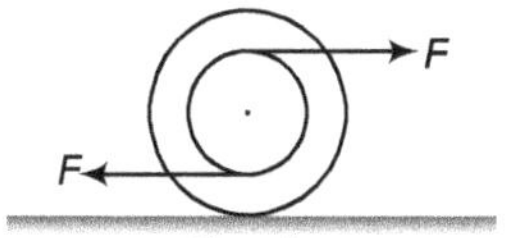

Rough

(a) The centre of mass moves towards left
(b) The centre of mass moves towards right
(c) The centre of mass remains stationary
(d) The net torque about the centre of mass of the spool is zero.

26. Two men each of mass m stand on the rim of a horizontal circular disc, diametrically opposite to each other. The disc has a mass M and is free to rotate about a vertical axis passing through its centre of mass. Each mass start simultaneously along the rim clockwise and reaches their original starting points on the disc. The angle turned through by the disc with respect to the ground (in radian) is

(a) $\frac{8m\pi}{4m+M}$ (b) $\frac{2m\pi}{4m+M}$
(c) $\frac{m\pi}{M+m}$ (d) $\frac{4m\pi}{2M+m}$

27. A ring of mass m and radius R has three particles attached to the ring as shown in the figure. The centre of the ring has a speed v_0. The kinetic energy of the system in case of no slipping is

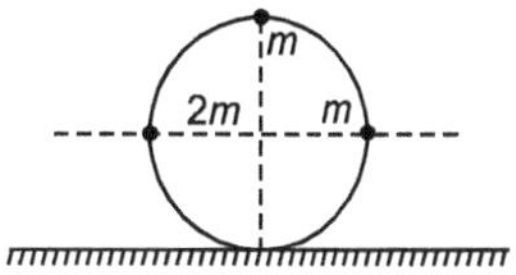

(a) $6mv_0^2$
(b) $12mv_0^2$
(c) $4mv_0^2$
(d) $8mv_0^2$

More than One Correct Options

1. A mass m of radius r is rolling horizontally without any slip with a linear speed v. It then rolls upto a height given by $\frac{3}{4}\frac{v^2}{g}$

(a) the body is identified to be a disc or a solid cylinder
(b) the body is a solid sphere
(c) moment of inertia of the body about instantaneous axis of rotation is $\frac{3}{2}mr^2$
(d) moment of inertia of the body about instantaneous axis of rotation is $\frac{7}{5}mr^2$

2. Four identical rods each of mass m and length l are joined to form a rigid square frame. The frame lies in the x-y plane, with its centre at the origin and the sides parallel to the x and y-axes. Its moment of inertia about

(a) the x-axis is $\frac{2}{3}ml^2$
(b) the z-axis is $\frac{4}{3}ml^2$
(c) an axis parallel to the z-axis and passing through a corner is $\frac{10}{3}ml^2$
(d) one side is $\frac{5}{3}ml^2$

3. A uniform circular ring rolls without slipping on a horizontal surface. At any instant, its position is as shown in the figure. Then

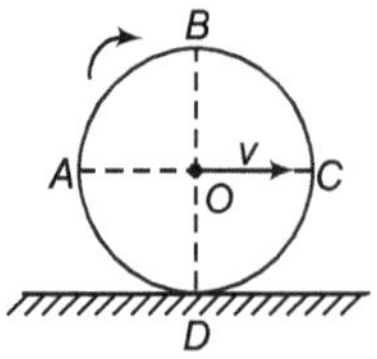

(a) section ABC has greater kinetic energy than section ADC
(b) section BC has greater kinetic energy than section CD
(c) the section BC has the same kinetic energy as section DA
(d) the sections CD and DA have the same kinetic energy

4. A cylinder of radius R is to roll without slipping between two planks as shown in the figure. Then

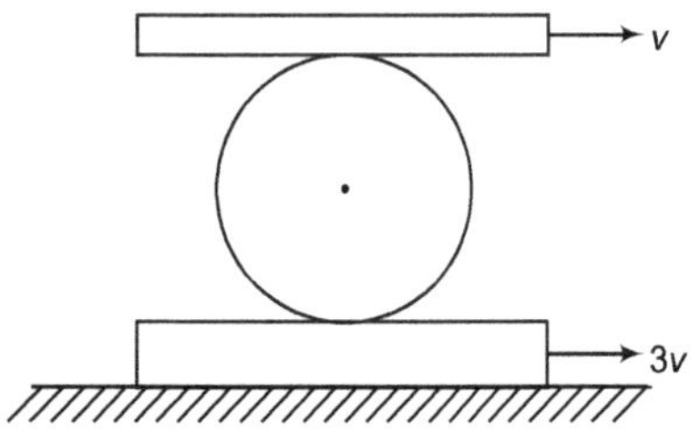

(a) angular velocity of the cylinder is $\frac{v}{R}$ counter clockwise
(b) angular velocity of the cylinder is $\frac{2v}{R}$ clockwise
(c) velocity of centre of mass of the cylinder is v towards left
(d) velocity of centre of mass of the cylinder is $2v$ towards right

5. A uniform rod of mass $m = 2$ kg and length $l = 0.5$ m is sliding along two mutually perpendicular smooth walls with the two ends P and Q having velocities $v_P = 4$ m/s and $v_Q = 3$ m/s as shown in figure. Then,

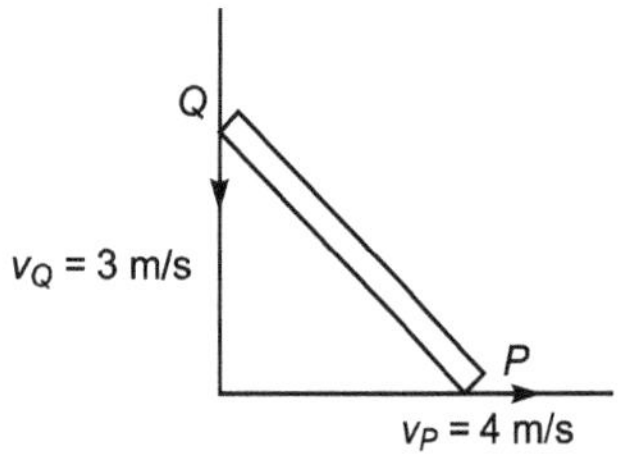

(a) The angular velocity of rod, $\omega = 10$ rad/s, counter clockwise
(b) The angular velocity of rod, $\omega = 5.0$ rad/s, counter clockwise
(c) The velocity of centre of mass of rod, $v_{cm} = 2.5$ m/s
(d) The total kinetic energy of rod, $K = \frac{25}{3}$ J

6. A wheel is rolling without slipping on a horizontal plane with velocity v and acceleration a of centre of mass as shown in figure. Acceleration at

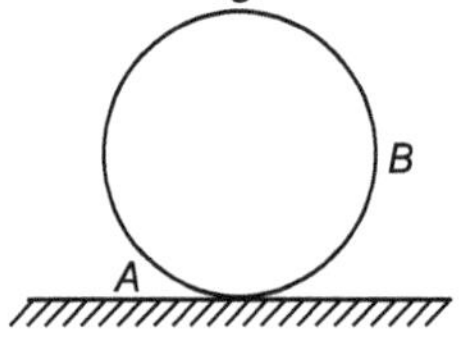

(a) A is vertically upwards
(b) B may be vertically downwards
(c) C cannot be horizontal
(d) a point on the rim may be horizontal leftwards

7. A uniform rod of length l and mass $2\,m$ rests on a smooth horizontal table. A point mass m moving horizontally at right angles to the rod with velocity v collides with one end of the rod and sticks it. Then

(a) angular velocity of the system after collision is $\frac{2}{5}\frac{v}{l}$
(b) angular velocity of the system after collision is $\frac{v}{2l}$
(c) the loss in kinetic energy of the system as a whole as a result of the collision is $\frac{3}{10}mv^2$
(d) the loss in kinetic energy of the system as a whole as a result of the collision is $\frac{7mv^2}{24}$

8. A non-uniform ball of radius R and radius of gyration about geometric centre $= R/2$, is kept on a frictionless surface. The geometric centre coincides with the centre of mass. The ball is struck horizontally with a sharp impulse $= J$. The point of application of the impulse is at a height h above the surface. Then

(a) the ball with slip on surface for all cases
(b) the ball will roll purely, if $h = 5R/4$
(c) the ball will roll purely, if $h = 3R/2$
(d) there will be no rotation, if $h = R$

9. A hollow spherical ball is given an initial push up an incline of inclination angle α. The ball rolls purely. Coefficient of static friction between ball and incline $= \mu$. During its upwards journey

(a) friction acts up along the incline
(b) $\mu_{min} = (2\tan\alpha)/5$
(c) friction acts down along the incline
(d) $\mu_{min} = (2\tan\alpha)/7$

10. A uniform disc of mass m and radius R rotates about a fixed vertical axis passing through its centre with angular velocity ω. A particle of same mass m and having velocity of $2\omega R$ towards centre of the disc collides with the disc moving horizontally and sticks to its rim.

(a) The angular velocity of the disc will become $\omega/3$
(b) The angular velocity of the disc will become $5\,\omega/3$
(c) The impulse on the particle due to disc is $\frac{\sqrt{37}}{3}m\omega R$
(d) The impulse on the particle due to disc is $2m\omega R$

11. The end B of the rod AB which makes angle θ with the floor is being pulled with a constant velocity v_0 as shown in figure. The length of the rod is l.

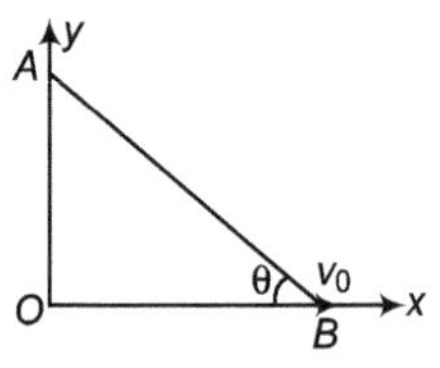

(a) At $\theta = 37°$ velocity of end A is $\frac{4}{3}v_0$ downwards
(b) At $\theta = 37°$ angular velocity of rod is $\frac{5v_0}{3l}$
(c) Angular velocity of rod is constant
(d) Velocity of end A is constant

Comprehension Based Questions

Passage I (Q. 1 to 3)

Consider a uniform disc of mass m, radius r, rolling without slipping on a rough surface with linear acceleration a and angular acceleration α due to an external force **F** *as shown in the figure. Coefficient of friction is μ*

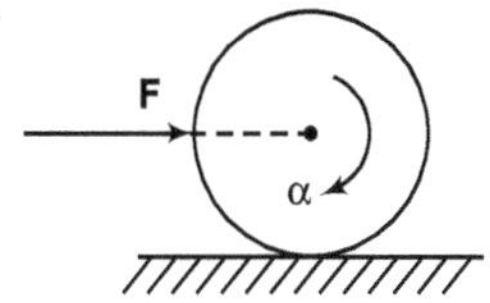

1. The work done by the frictional force at the instant of pure rolling is

(a) $\frac{\mu mgat^2}{2}$ (b) $\mu mgat^2$

(c) $\mu mg\frac{at^2}{\alpha}$ (d) zero

2. The magnitude of frictional force acting on the disc is

(a) ma (b) μmg (c) $\frac{ma}{2}$ (d) zero

3. Angular momentum of the disc will be conserved about

(a) centre of mass
(b) point of contact
(c) a point at a distance $3R/2$ vertically above the point of contact
(d) a point at a distance $4R/3$ vertically above the point of contact

Passage II (Q. 4 to 6)

A tennis ball, starting from rest, rolls down the hill in the drawing. At the end of the hill the ball becomes airborne, leaving at an angle of 37° with respect to the ground. Treat the ball as a thin-walled spherical shell.

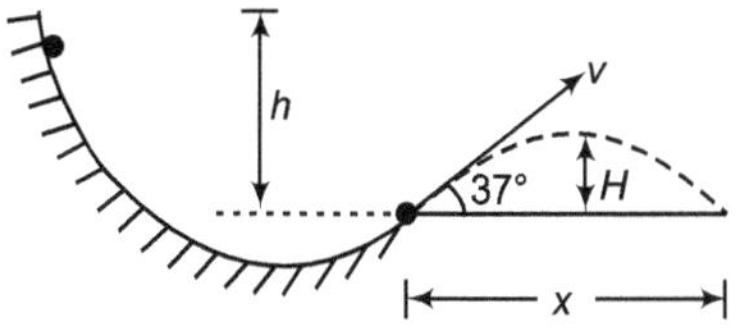

4. The velocity of projection v is

(a) $\sqrt{2gh}$ (b) $\sqrt{\frac{10}{7}gh}$

(c) $\sqrt{\frac{5}{7}gh}$ (d) $\sqrt{\frac{6}{5}gh}$

5. Maximum height reached by ball H above ground is

(a) $\frac{9h}{35}$ (b) $\frac{18h}{35}$ (c) $\frac{18h}{25}$ (d) $\frac{27h}{125}$

6. Range x of the ball is

(a) $\frac{144}{125}h$ (b) $\frac{48}{25}h$

(c) $\frac{48}{35}h$ (d) $\frac{24}{7}h$

Assertion and Reason

Directions (Q. Nos. 1-20) *These questions consists of two statements each printed as Assertion and Reason. While answering these questions you are required to choose any one of the following five responses.*

(a) If both Assertion and Reason are correct and Reason is the correct explanation of Assertion
(b) If both Assertion and Reason are true but Reason is not the correct explanation of Assertion
(c) If Assertion is true but Reason is false
(d) If Assertion is false but Reason is true
(e) If both Assertion and Reason are false

1. **Assertion** Moment of inertia about an axis passing through the centre of mass is always minimum.

Reason Theorem of parallel axis can be applied for 2-D as well as 3-D bodies.

2. **Assertion** A sphere is placed in pure rolling condition over a rough inclined surface. Then, force of friction will act in downward direction.

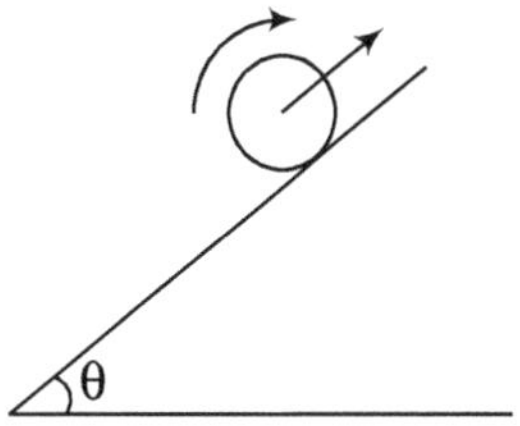

Reason Angular acceleration (actually retardation) due to friction is anti-clockwise.

3. **Assertion** A ring and a disc of same mass and radius begin to roll without slipping from the top of an inclined surface at $t = 0$. The ring reaches the bottom of incline in time t_1 while the disc reaches the bottom in time t_2, then $t_1 < t_2$.

Reason Disc will roll down the plane with more acceleration, because of its lesser value of moment of inertia.

4. **Assertion** A body is moving along a circle with a constant speed. Its angular momentum about the centre of the circle remains constant.

Reason In this situation, a constant non-zero torque acts on the body.

5. **Assertion** A solid and a hollow sphere both of equal masses and radii are put on a rough surface after rotating with some angular velocity say ω_0. Coefficient of friction for both the spheres and ground is same. Solid sphere will start pure rolling first.

Reason Radius of gyration of hollow sphere about an axis passing through its centre of mass is more.

6. Assertion Speed of any point or rigid body executing rolling motion can be calculated by expression $v = r\omega$, where r is distance of point from instantaneous centre of rotation.

Reason Rolling motion of rigid body can be considered as a pure rotation about instantaneous centre of rotation.

7. Assertion The condition of equilibrium for a rigid body is

Translational equilibrium $\Sigma F = 0$ and

Rotational equilibrium $\Sigma\tau = 0$

Reason A rigid body must be in equilibrium under the action of two equal and opposite forces.

8. Assertion The angular velocity of a rigid body in motion is defined for the whole body.

Reason All points on a rigid body performing pure rotational motion are having same angular velocity.

9. Assertion A uniform disc of radius R is performing impure rolling motion on a rough horizontal plane as shown in figure. After some time the disc comes to rest. It is possible only when $v_0 = \dfrac{\omega_0 R}{2}$.

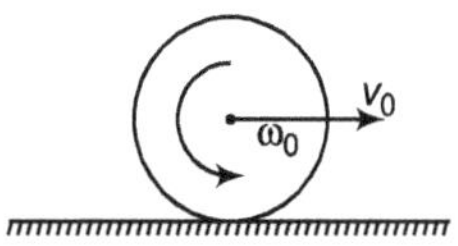

Reason For a body performing pure rolling motion, the angular momentum is conserved about any point in space.

10. Assertion A solid sphere and a ring of same mass and radius are released simultaneously from the top of an inclined surface. The two objects roll down the plane without slipping. They reach the bottom of the incline with equal linear speeds.

Reason Decrease in potential energy for both is the same.

11. Assertion If a particle is rotating in a circle, then angular momentum about any point is mvR.

Reason In circular motion, angular momentum about centre is always constant.

12. Assertion A solid sphere cannot roll without slipping on smooth horizontal surface.

Reason If the sphere is left free on smooth inclined surface, it cannot roll without slipping.

13. Assertion Two axes AB and CD are as shown in figure. Given figure is of a semi circular ring. As the axis moves from AB towards CD, moment of inertia first decreases then increases.

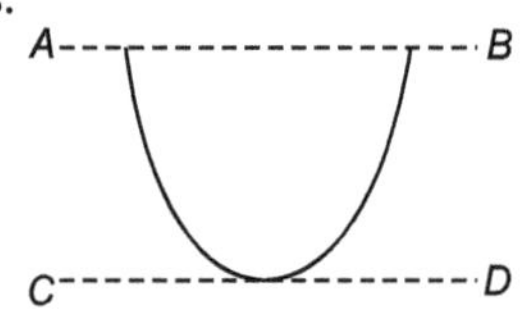

Reason Centre of mass lies somewhere between AB and CD.

14. Assertion If a particle moves with a constant velocity, then angular momentum of this particle about any point remains constant.

Reason Angular momentum has the units of Planck's constant.

15. Assertion Moment of inertia about an axis passing through centre of mass is maximum.

Reason Theorem of parallel axis can be applied only for two dimensional body of negligible thickness.

16. Assertion If we draw a circle around the centre of mass of a rigid body, then moment of inertia about all parallel axes passing through this circle has a constant value.

Reason Dimensions of radius of gyration are $[M^0LT^0]$.

17. Assertion In rotational plus translational motion of a rigid body different particles of the rigid body may have different velocities but they will have same accelerations.

Reason Translational motion of a particle is equivalent to the translational motion of a rigid body.

18. Assertion Two identical solid spheres are rotated from rest to same angular velocity ω about two different axes as shown in figure. More work will have to be done to rotate the sphere in case-2.

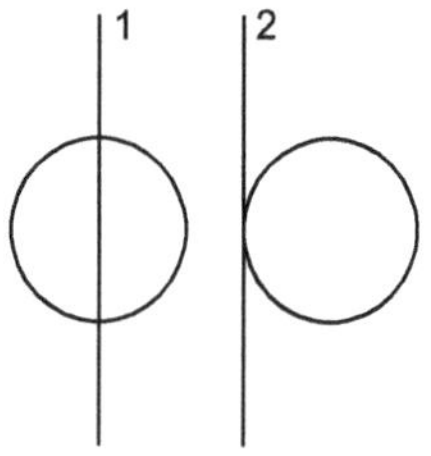

Reason Moment of inertia in case-2 is more.

19. Assertion Angular momentum of sun and planet system about any point remains constant.

Reason Two equal and opposite forces will act on them. Net torque of those two set of forces about any point is zero.

20. Assertion Two identical spherical balls are released from two inclined plane. First is sufficiently rough and second is smooth. Both the balls will have same kinetic energy on reaching the bottom.

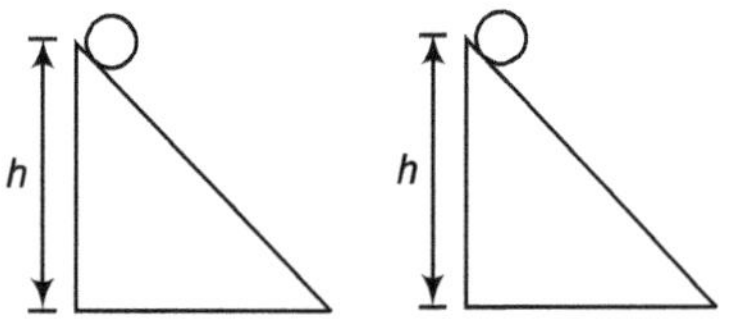

Reason Linear velocity of second ball will be more.

Match the Columns

1. Four rods of equal length l and mass m each from a square as shown in figure. Moment of inertia about three axes 1, 2 and 3 are say I_1, I_2 and I_3. Then, match the following columns

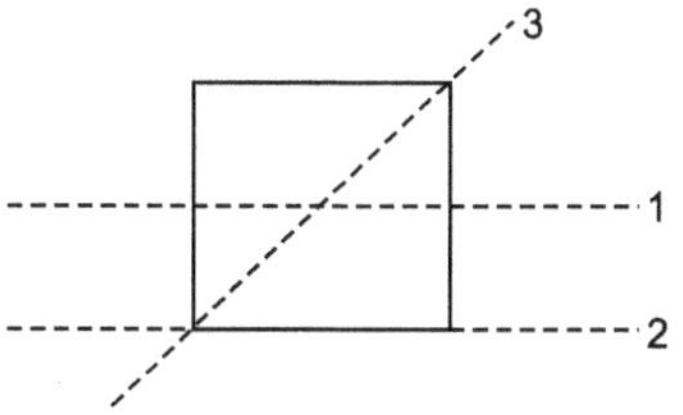

Column I		Column II	
(A)	I_1	(p)	$\frac{4}{3}ml^2$
(B)	I_2	(q)	$\frac{2}{3}ml^2$
(C)	I_3	(r)	$\frac{1}{2}ml^2$
		(s)	None of the above

2. A disc rolls on ground without slipping. Velocity of centre of mass is v. There is a point P on circumference of disc at angle θ. Suppose, v_P is the speed of this point. Then, match the following columns

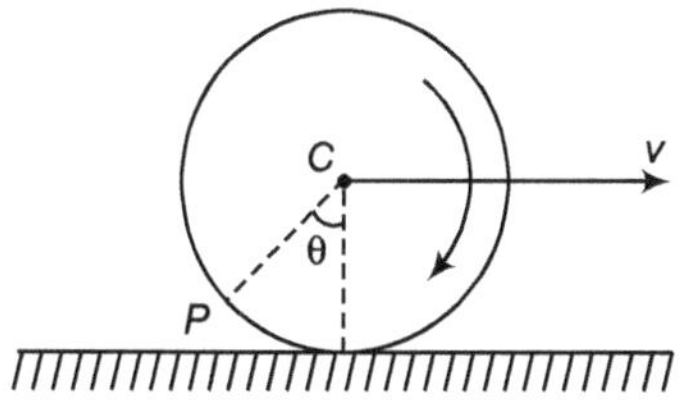

Column I		Column II	
(A)	If $\theta = 60°$	(p)	$v_P = \sqrt{2}\,v$
(B)	If $\theta = 90°$	(q)	$v_P = v$
(C)	If $\theta = 120°$	(r)	$v_P = 2v$
(D)	If $\theta = 180°$	(s)	$v_P = \sqrt{3}\,v$

3. If radius of the earth is reduced to half without changing its mass, then match the following columns

Column I	Column II
(A) Angular momentum of the earth	(p) Will become two times
(B) Time period of rotation of the earth	(q) Will become four times
(C) Rotational kinetic energy of the earth	(r) Will remain constant
	(s) None of the above

4. A semicircular ring has mass m and radius R as shown in figure. Let I_1, I_2, I_3 and I_4 be the moments of inertias of the four axes as shown in figure. Axis 1 passes through centre and is perpendicular to plane of ring. Then, match the following columns

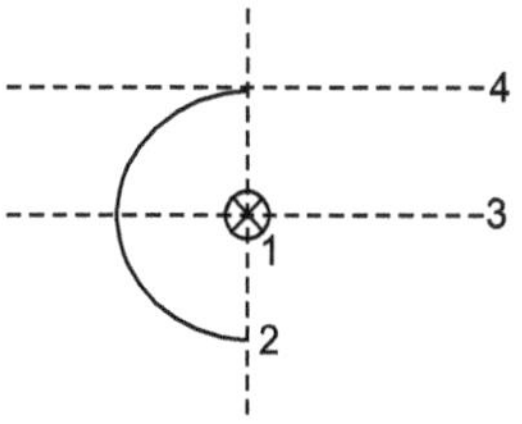

Column I		Column II
(A)	I_1	(p) $\frac{mR^2}{2}$
(B)	I_2	(q) $\frac{3}{2}mR^2$
(C)	I_3	(r) mR^2
(D)	I_4	(s) Data is insufficient

5. A solid sphere is rotating about an axis as shown in figure. An insect follows the dotted path on the circumference of sphere as shown. Then, match the following columns

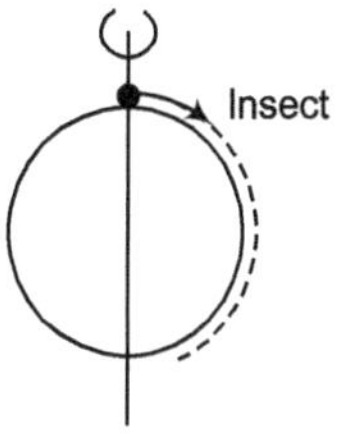

Column I	Column II
(A) Moment of inertia	(p) Will remain constant
(B) Angular velocity	(q) Will first increase, then decrease
(C) Angular momentum	(r) Will first decrease, then increase
(D) Rotational kinetic energy	(s) Will continuously decrease

6. A uniform cube of mass m and side a is placed on a frictionless horizontal surface. A vertical force F is applied to the edge as shown in figure. Match the following (most appropriate choice)

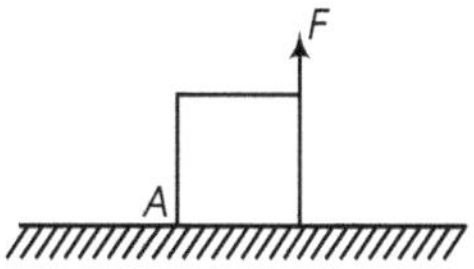

Column I	Column II
(a) $mg/4 < F < mg/2$	(p) Cube will move up.
(b) $F > mg/2$	(q) Cube will not exhibit motion.
(c) $F > mg$	(r) Cube will begin to rotate and slip at A.
(d) $F = mg/4$	(s) Normal reaction effectively at $a/3$ from A, no motion.

7. A uniform sphere of mass m and radius R is placed on a rough horizontal surface (figure). The sphere is struck horizontally at a height h from the floor. Match the following

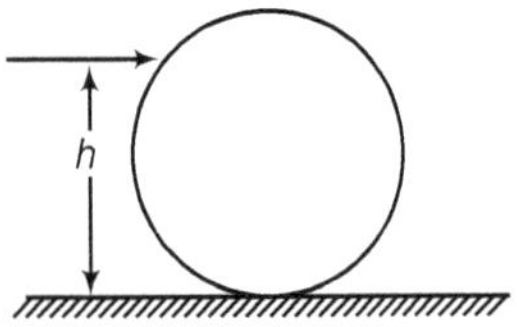

Column I	Column II
(a) $h = R/2$	(p) Sphere rolls without slipping with a constant velocity and no loss of energy.
(b) $h = R$	(q) Sphere spins clockwise, loses energy by friction.
(c) $h = 3R/2$	(r) Sphere spins anti-clockwise, loses energy by friction.
(d) $h = 7R/5$	(s) Sphere has only a translational motion, loses energy by friction.

Entrance Gallery

2014

1. A force $\mathbf{F} = 5\hat{\mathbf{i}} + 2\hat{\mathbf{j}} - 5\hat{\mathbf{k}}$ acts on a particle whose position vector is $\mathbf{r} = \hat{\mathbf{i}} - 2\hat{\mathbf{j}} + \hat{\mathbf{k}}$. What is the torque about the origin? **[Karnataka CET]**

(a) $8\hat{\mathbf{i}} + 10\hat{\mathbf{j}} + 12\hat{\mathbf{k}}$ (b) $8\hat{\mathbf{i}} + 10\hat{\mathbf{j}} - 12\hat{\mathbf{k}}$
(c) $8\hat{\mathbf{i}} - 10\hat{\mathbf{j}} - 8\hat{\mathbf{k}}$ (d) $10\hat{\mathbf{i}} - 10\hat{\mathbf{j}} - \hat{\mathbf{k}}$

2. A body having a moment of inertia about its axis of rotation equal to 3 kg-m^2 is rotating with angular velocity of 3 rad s^{-1}. Kinetic energy of this rotating body is same as that of a body of mass 27 kg moving with a velocity v. The value of v is **[Karnataka CET]**

(a) 1 ms^{-1} (b) 0.5 ms^{-1} (c) 2 ms^{-1} (d) 1.5 ms^{-1}

3. A solid uniform sphere resting on a rough horizontal plane is given a horizontal impulse directed through its centre so that it starts sliding with an initial velocity v_0. When it finally starts rolling without sliping the speed of its centre is **[WB JEE]**

(a) $\frac{2}{7}v_0$ (b) $\frac{3}{7}v_0$ (c) $\frac{5}{7}v_0$ (d) $\frac{6}{7}v_0$

2013

4. A uniform circular disc of mass 50 kg and radius 0.4 m is rotating with an angular velocity of 10 rad/s about its own axis, which is vertical. Two uniform circular rings, each of mass 6.25 kg and radius 0.2 m, are gently placed symmetrically on the disc in such a manner that they are touching each other along the axis of the disc and are horizontal. Assume that the friction is large enough such that the rings are at rest relative to the disc and the system rotates about the original axis. The new angular velocity (in rad/s^{-1}) of the system is **[JEE Advanced]**

(a) 8 (b) 7 (c) 9 (d) 11

5. A sphere rolls down an inclined plane of inclination θ. What is the acceleration as the sphere reaches bottom?. **[O JEE]**

(a) $\frac{5}{7}g\sin\theta$ (b) $\frac{3}{5}g\sin\theta$
(c) $\frac{2}{7}g\sin\theta$ (d) $\frac{2}{5}g\sin\theta$

6. A particle is moving in a circle with uniform speed v. In moving from a point to another diametrically opposite point **[O JEE]**

(a) the momentum changes by mv
(b) the momentum changes by $2mv$
(c) the kinetic energy changes by $(1/2)\,mv^2$
(d) the kinetic energy changed by mv^2

2012

7. A lamina is made by removing a small disc of diameter $2R$ from a bigger disc of uniform mass density and radius $2R$, as shown in the figure. The moment of inertia of this lamina about axes passing through O and P is I_O and I_P, respectively. Both these axes are perpendicular to the plane of the lamina. The ratio $\frac{I_P}{I_O}$ to the nearest integer is **[IIT JEE]**

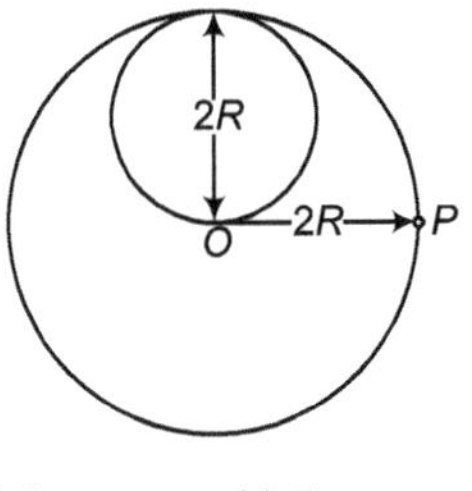

(a) 3 (b) 4 (c) 5 (d) 6

8. A small mass m is attached to a massless string whose other end is fixed at P as shown in the figure. The mass is undergoing circular motion in the x-y plane with centre at O and constant angular speed ω. If the angular momentum of the system, calculated about O and P are denoted by L_O and L_P respectively, then **[IIT JEE]**

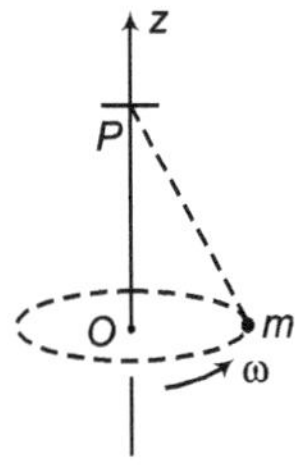

(a) L_O and L_P do not vary with time
(b) L_O varies with time while L_P remains constant
(c) L_O remains constant while L_P varies with time
(d) L_O anad L_P both vary with time.

9. A thin uniform rod, pivoted at O, is rotating in the horizontal plane with constant angular speed ω, as shown in the figure. At time $t = 0$, a small insect starts from O and moves with constant speed v w.r.t. the rod towards the other end. It reaches the end of the rod at $t = T$ and stops. The angular speed of the system remains

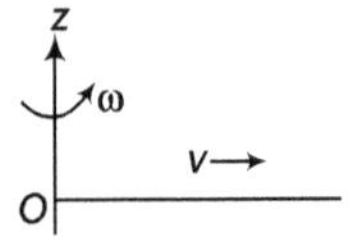

ω throughout. The magnitude of the torque $|\tau|$ on the system about O, as a function of time is best represented by which plot ? **[IIT JEE]**

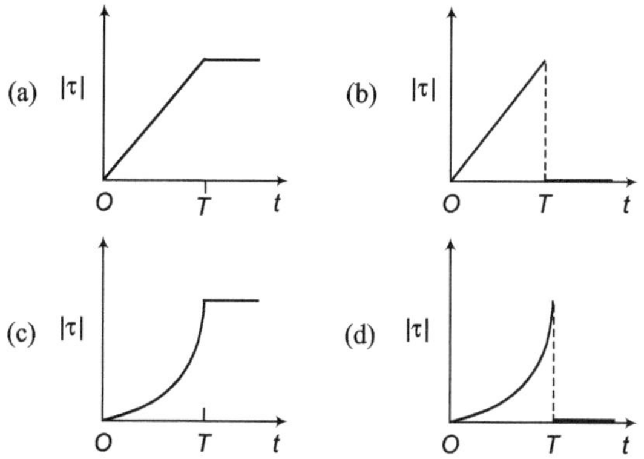

10. Two identical discs of same radius R are rotating about their axes in opposite directions with the same constant angular speed ω. The discs are in the same horizontal plane. At time $t = 0$, the points P and Q are facing each other as shown in the figure.The relative speed between the two points P and Q is v_r. In one time period (T) of rotation of the discs, v_r as a function of time is best represented by **[IIT JEE]**

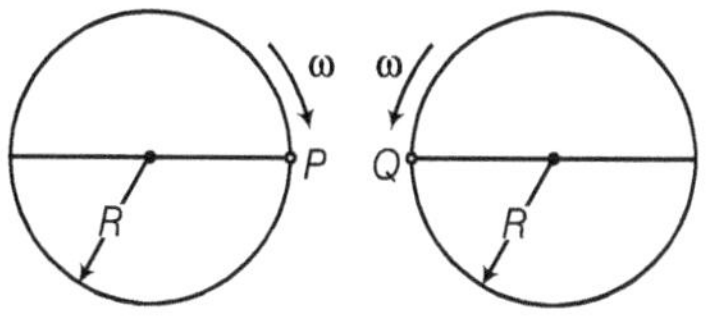

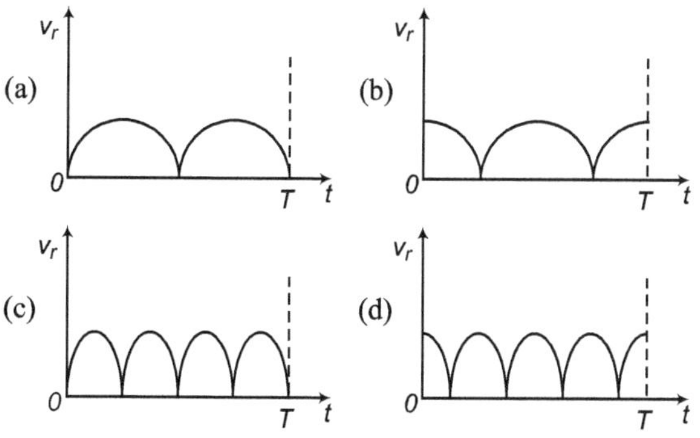

11. A solid cylinder rolls from an inclined plane of height h. The velocity of body at bottom is **[O JEE]**

(a) $\sqrt{\frac{2gh}{3}}$ (b) $\sqrt{\frac{4gh}{3}}$

(c) $\sqrt{\frac{3gh}{2}}$ (d) $\sqrt{gh}$

12. A wire is wound on a hollow cylinder of radius 40 cm. Mass of the cylinder is 3 kg. Force of 30 N is applied on wire. The angular acceleration is **[O JEE]**

(a) 25 rad/s^2 (b) 30 rad/s^2

(c) 35 rad/s^2 (d) 40 rad/s^2

2011

13. A thin horizontal circular rotating about a vertical axis passing through its centre. An insect is at rest at a point near the rim of the disc. The insect now moves along a diameter of the disc to reach its other end. During the journey of the insect, the angular speed of the disc

(a) continuously decreases **[IIT JEE]**

(b) continuously increases

(c) first increases and then decreases

(d) remains unchanged

14. A body is rotating with angular velocity 30 rad s^{-1}. If its kinetic energy is 360 J, then its moment of inertia is

(a) 0.8 kg-m^2 **[IIT JEE]**

(b) 0.4 kg-m^2

(c) 1 kg-m^2

(d) 1.2 kg-m^2

15. A pulley of radius 2 m is rotating about its axis by a force $F = (20t - 5t^2)$ N (where, t is measured in sec) applied tangentially. If the moment of inertia of the pulley about its axis of rotation is 10 kg - m^2, the number of rotations made by the pulley before its direction of motion, if reserved, is **[AIEEE]**

(a) more than 3 but less than 6

(b) more than 6 but less than 9

(c) more than 9

(d) less than 6

16. A mass m hangs with the help of a string wrapped around a pulley on a frictionless bearing. The pulley has mass m and radius R. Assuming pulley to be a perfect uniform circular disc, the acceleration of the mass m, if the string does not slip on the pulley, is **[AIEEE]**

(a) g (b) $\frac{2}{3}g$

(c) $\frac{g}{3}$ (d) $\frac{3}{2}g$

17. A ring starts to roll down the inclined plane of height h without slipping. The velocity with which it reaches the ground is **[Kerala CEE]**

(a) $\sqrt{\frac{10gh}{7}}$ (b) $\sqrt{\frac{4gh}{7}}$

(c) $\sqrt{\frac{4gh}{3}}$ (d) $\sqrt{2gh}$

(e) $\sqrt{gh}$

18. The angular momentum of a particle describing uniform circular motion is L. If its kinetic energy is halved and angular velocity doubled, its new angular momentum is **[Kerala CEE]**

(a) $4L$ (b) $\frac{L}{4}$ (c) $\frac{L}{2}$ (d) $2L$

(e) $\frac{L}{8}$

19. A solid sphere of mass m rolls down an inclined plane without slipping, starting from rest at the top of an inclined plane. The linear speed of the sphere at the bottom of the inclined plane is v. The kinetic energy of the sphere at the bottom is **[Karnataka CET]**

(a) $\frac{7}{10}mv^2$ (b) $\frac{2}{5}mv^2$
(c) $\frac{5}{3}mv^2$ (d) $\frac{1}{2}mv^2$

2010

20. A binary star consists of two stars A (mass $2.2M_S$) and B (mass $11M_S$), where M_S is the mass of the sun. They are separated by distance d and are rotating about their centre of mass, which is stationary. The ratio of the total angular momentum of the binary star to the angular momentum of star B about the centre of mass is **[IIT JEE]**

(a) 7 (b) 6 (c) 9 (d) 10

21. From a circular ring of mass M and radius R, an arc corresponding to a 90° sector is removed. The moment of inertia of the remaining part of the ring about an axis passing through the centre of the ring and perpendicular to the plane of the ring is k times MR^2. Then, the value of k is **[Kerala CEE]**

(a) $\frac{3}{4}$ (b) $\frac{7}{8}$ (c) $\frac{1}{4}$ (d) 1
(e) $\frac{1}{8}$

22. A wheel of moment of inertia 2.5 kg-m^2 has an initial angular velocity of 40 rad s^{-1}. A constant torque of 10 N-m acts on the wheel. The time during which the wheel is accelerated to 60 rad s^{-1} is **[Kerala CEE]**

(a) 4 s (b) 6 s (c) 5 s (d) 2.5 s
(e) 4.5 s

23. When a disc is rotating with angular velocity ω, a particle situated at a distance of 4 cm just begins to slip. If the angular velocity is doubled, at what distance will the particle start to slip? **[MHT CET]**

(a) 1 cm (b) 2 cm
(c) 3 cm (d) 4 cm

24. The moment of inertia of a thin uniform rod of length L and mass M about an axis passing through a point at a distance of 1/3 from one of its ends and perpendicular to the rod is **[MHT CET]**

(a) $\frac{ML^2}{12}$ (b) $\frac{ML^2}{9}$
(c) $\frac{7ML^2}{48}$ (d) $\frac{ML^2}{48}$

25. Moment of inertia of a rod of mass M and length L about an axis passing through a point midway between centre and end is **[MHT CET]**

(a) $\frac{ML^2}{6}$ (b) $\frac{ML^2}{12}$
(c) $\frac{7ML^2}{24}$ (d) $\frac{7ML^2}{48}$

26. From a disc of radius R, a concentric circular portion of radius r is cut out so as to leave an annular disc of mass M. The moment of inertia of this annular disc about the axis perpendicular to its plane and passing through its centre of gravity is **[MHT CET]**

(a) $\frac{1}{2}M(R^2+r^2)$

(b) $\frac{1}{2}M(R^2-r^2)$

(c) $\frac{1}{2}M(R^4+r^4)$

(d) $\frac{1}{2}M'(R^4-r^4)$

27. Which relation is not correct of the following? **[MHT CET]**

(a) Torque = moment of inertia × angular acceleration
(b) Torque = dipole moment × magnetic induction
(c) Moment of inertia = torque/angular acceleration
(d) Linear momentum = moment of inertia × angular velocity

28. Moment of inertia of a disc about a diameter is I. Find the moment of inertia of disc about an axis perpendicular to its plane and passing through its rim? **[MHT CET, J&K CET]**

(a) $6I$ (b) $4I$
(c) $2I$ (d) $8I$

Answers

Level 1

Objective Problems

1. (d)	**2.** (b)	**3.** (b)	**4.** (a)	**5.** (b)	**6.** (b)	**7.** (c)	**8.** (a)	**9.** (d)	**10.** (d)
11. (d)	**12.** (c)	**13.** (b)	**14.** (b)	**15.** (c)	**16.** (b)	**17.** (a)	**18.** (b)	**19.** (b)	**20.** (c)
21. (b)	**22.** (b)	**23.** (a)	**24.** (b)	**25.** (b)	**26.** (a)	**27.** (b,c,d)	**28.** (c)	**29.** (b)	**30.** (d)
31. (c)	**32.** (b)	**33.** (b)	**34.** (b)	**35.** (a)	**36.** (b)	**37.** (c)	**38.** (d)	**39.** (b)	**40.** (b)
41. (b)	**42.** (c)	**43.** (b)	**44.** (a)	**45.** (d)	**46.** (b)	**47.** (a)	**48.** (c)	**49.** (a)	**50.** (b)
51. (b)	**52.** (c)	**53.** (d)	**54.** (a)	**55.** (c)	**56.** (a)	**57.** (c)	**58.** (d)	**59.** (c)	**60.** (c)
61. (a)	**62.** (a)	**63.** (d)	**64.** (a)	**65.** (a)	**66.** (a)	**67.** (c)	**68.** (d)	**69.** (d)	**70.** (b)
71. (b)	**72.** (c)	**73.** (b)	**74.** (b)	**75.** (c)	**76.** (b)	**77.** (c)	**78.** (d)	**79.** (d)	**80.** (c)
81. (b)	**82.** (d)	**83.** (c)	**84.** (b)	**85.** (b)	**86.** (c)	**87.** (a)	**88.** (a)	**89.** (d)	**90.** (a)
91. (a)	**92.** (b)	**93.** (b)	**94.** (c)	**95.** (a)	**96.** (d)	**97.** (a)	**98.** (c)	**99.** (a)	**100.** (d)
101. (b)	**102.** (a)	**103.** (b)							

Level 2

Only One Correct Option

1. (a)	**2.** (c)	**3.** (a)	**4.** (b)	**5.** (d)	**6.** (d)	**7.** (a,d)	**8.** (c,d)	**9.** (a)	**10.** (a)
11. (d)	**12.** (c)	**13.** (c)	**14.** (c)	**15.** (d)	**16.** (a,b,d)	**17.** (a,b,d)	**18.** (a)	**19.** (a)	**20.** (c)
21. (d)	**22.** (b)	**23.** (b)	**24.** (b)	**25.** (c)	**26.** (c)	**27.** (b)			

More than One Correct Options

1. (a,c)	**2.** (all)	**3.** (a,b,d)	**4.** (a,d)	**5.** (a,c,d)	**6.** (all)	**7.** (a,c)	**8.** (b,d)	**9.** (a,b)	**10.** (a,c)
11. (a,b)									

Comprehension Based Questions

1. (d)	**2.** (c)	**3.** (c)	**4.** (d)	**5.** (d)	**6.** (a)

Assertion and Reason

1. (d)	**2.** (d)	**3.** (d)	**4.** (c)	**5.** (a)	**6.** (a)	**7.** (c)	**8.** (b)	**9.** (c)	**10.** (d)
11. (e)	**12.** (d)	**13.** (a)	**14.** (b)	**15.** (b)	**16.** (b)	**17.** (d)	**18.** (a)	**19.** (a)	**20.** (b)

Match the Columns

1. ($A \rightarrow q$, $B \rightarrow s$, $C \rightarrow p$) **2.** ($A \rightarrow q$, $B \rightarrow p$, $C \rightarrow s$, $D \rightarrow r$) **3.** ($A \rightarrow r$, $B \rightarrow s$, $C \rightarrow q$) **4.** ($A \rightarrow r$, $B \rightarrow p$, $C \rightarrow p$, $D \rightarrow q$)
5. ($A \rightarrow q$, $B \rightarrow r$, $C \rightarrow p$, $D \rightarrow r$) **6.** ($A \rightarrow q$, $B \rightarrow r$, $C \rightarrow p$, $D \rightarrow s$) **7.** ($A \rightarrow r$, $B \rightarrow s$, $C \rightarrow p$, $D \rightarrow p$)

Entrance Gallery

1. (a)	**2.** (a)	**3.** (c)	**4.** (a)	**5.** (a)	**6.** (b)	**7.** (a)	**8.** (c)	**9.** (a)	**10.** (a)
11. (b)	**12.** (a)	**13.** (c)	**14.** (a)	**15.** (a)	**16.** (b)	**17.** (e)	**18.** (b)	**19.** (a)	**20.** (b)
21. (a)	**22.** (c)	**23.** (a)	**24.** (b)	**25.** (d)	**26.** (b)	**27.** (d)	**28.** (b)		

Solutions

Level 1 : Objective Problems

3. $I_{\text{Ring}} = mR^2, I_{\text{disc}} = \frac{1}{2} mR^2$

4. K = radius of gyration

$$= \sqrt{\frac{I}{m}} = \sqrt{\frac{\frac{7}{5} mR^2}{m}} = \sqrt{\frac{7}{5}} R$$

5. $I_H = \sqrt{\frac{I}{m}} = \sqrt{\frac{\frac{2}{3} mR^2 + mR^2}{m}} = \sqrt{\frac{5}{3}} R$

$$I_S = \sqrt{\frac{I}{m}} = \sqrt{\frac{\frac{2}{5} mR^2 + mR^2}{m}} = \sqrt{\frac{7}{5}} R$$

$$\therefore \qquad \frac{I_H}{I_S} = \sqrt{\frac{25}{21}} = \frac{5}{\sqrt{21}}$$

6. $I = \frac{2}{5} mR^2 = \frac{2}{5}\left[\frac{4}{3}\pi R^3 \rho\right] R^2 = \frac{176}{105} \rho R^5$

7. I_{CM} is less than I about any other axis not passing through centre of mass but only when two axes are parallel.

8. Moment of inertia of the rod lying along z-axis will be zero. Of the rods along x and y-axis will be $\frac{ML^2}{3}$ each. Hence, total moment of inertia is $\frac{2}{3} ML^2$.

9. Radius of gyration, $K = \sqrt{\frac{I}{m}}$

$$K_{\text{disc}} = \sqrt{\frac{\frac{1}{2} mR^2 + mR^2}{m}} = \sqrt{\frac{3}{2}} R$$

$$K_{\text{ring}} = \sqrt{\frac{mR^2 + mR^2}{m}} = \sqrt{2} R$$

$$\therefore \qquad \frac{K_{\text{disc}}}{K_{\text{ring}}} = \frac{\sqrt{3/2}}{\sqrt{2}} = \frac{\sqrt{3}}{2}$$

10. $I = I_{\text{CM}} + mx^2$

i.e. I-x graph is a parabola not passing through origin.

11. $l = 2\pi R$

$$\therefore \qquad R = \frac{l}{2\pi}$$

$$I_1 = \frac{ml^2}{12}$$

$$I_2 = mR^2 = \frac{ml^2}{4\pi^2}$$

$$\therefore \qquad \frac{I_1}{I_2} = \frac{\pi^2}{3}$$

12. Moment of inertia depends on the distribution of mass about the axis.

13. Apply theorem of perpendicular axis.

14. $I = I_{\text{ring}} + 3I_{\text{spoke}}$

$$= MR^2 + 3\left(\frac{mR^2}{3}\right) = (M + m)R^2$$

15. $I = \frac{ml^2}{3}$, $I' = \frac{(3m/4)(3l/4)^2}{3}$

$$= \frac{27}{67}\left(\frac{ml^2}{3}\right) = \frac{27}{64} I$$

16. $I = 3\left[\frac{Ml^2}{3}\right] = Ml^2$

17. From theorem of parallel axes,

$$I_2 = I_1 + MR^2$$

18. Perpendicular distance from z-axis would be

$$\sqrt{(1)^2 + (1)^2} = \sqrt{2}\, m$$

$$\therefore \qquad I = mr^2 = (1)(\sqrt{2})^2$$

$$= 2 \text{ kg-m}^2$$

19. $I = 4\left[\frac{M}{2}\frac{(l/2)^2}{3}\right] = \frac{Ml^2}{6}$

20. For square lamina $I_2 = I_3$. This value will be less than I_1, because mass is nearer to axis in this case.

21. Moment = Force $\times r_\perp = 10 \times 2 = 20$ N-m

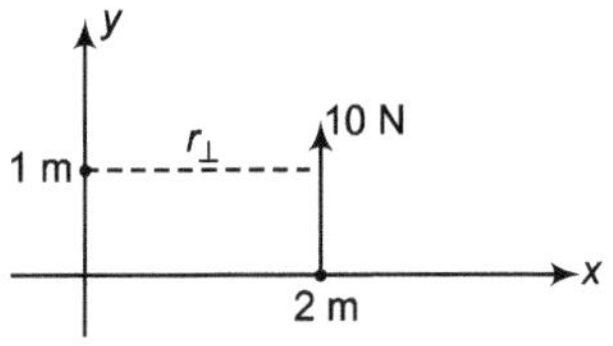

22. The desired ratio is $\frac{mR^2}{\frac{1}{2} mR^2} = 2:1$

23. $K = \sqrt{\frac{I}{m}} = \sqrt{\frac{(mL^2/12)}{m}} = \frac{L}{2\sqrt{3}}$

24. $I = 5mR^2 = 5 \times 2 \times (0.1)^2 = 0.1 \text{ kg-m}^2$

25. For a given body mass is same, so it will depend only on the distribution of mass about the axis.

The mass is farthest from axis BC, so I_2 is maximum. Mass is nearest to axis AC, so I_3 is minimum.

Hence, the correct sequence will be

$$I_2 > I_1 > I_3.$$

26. From theorem of perpendicular axes, we have

$$I = I_C + m\left(\frac{a}{\sqrt{2}}\right)^2$$

$$= \left[\frac{ma^2}{12} + \frac{ma^2}{12}\right] + \frac{ma^2}{2}$$

$$= \frac{2}{3} ma^2$$

27. If we complete the ring, its mass becomes 2 m.

$\therefore \quad I_{\text{Whole ring}} = \frac{1}{2}(2M)(R^2) = MR^2$

$\therefore \quad I_{\text{Half ring}} = \frac{1}{2}(MR^2)$

This value is independent of angle θ.

28.

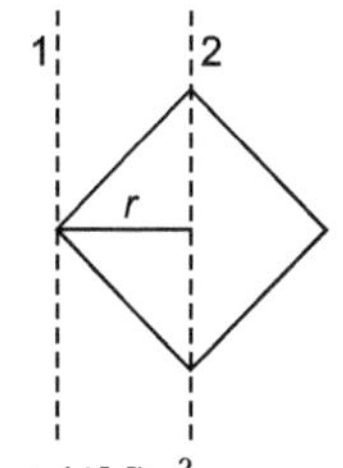

$$I_1 = I_2 + (4M)r^2$$

$$= 4\left[\frac{ML^2}{3}\sin^2 45°\right] + 4M\left[\frac{\sqrt{2}\,L}{2}\right]^2$$

$$= \frac{8}{3}ML^2$$

29. $I = \left[\frac{ma^2/4}{12\times 4} + \frac{m}{4}\frac{a^2}{\times 16}\right] + \left[\frac{ma^2/4}{12\times 4} + \frac{ma^2}{16\times 4}\right] + \left[\frac{ma^2/4}{12\times 4} + \frac{m}{4}\left(\frac{3a}{4}\right)^2\right]$

$$= \frac{3ma^2}{16}$$

30. $\{I_C\}_{\text{circular}} = \frac{1}{2}(2m)r^2 = mr^2$

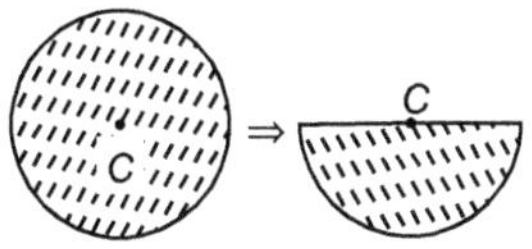

$\therefore \quad \{I_C\}_{\text{semicircle}} = \frac{1}{2}mr^2$

$$= I_{\text{CM}} + mx^2$$

$\therefore \quad I_{\text{CM}} = \frac{1}{2}mr^2 - mx^2$

31. Minimum moment of inertia is about centre of mass.

$$I_{\min} = m_1r_1^2 + m_2r_2^2$$

$$= m_1\left(\frac{m_2}{m_1+m_2}l\right)^2 + m_2\left(\frac{m_1}{m_1+m_2}l\right)^2$$

$$= \left(\frac{m_1m_2}{m_1+m_2}\right)l^2$$

32. $K_1 = \sqrt{\frac{I_1}{m_1}} = \sqrt{\frac{(5/4)m_1R^2}{m_1}} = \sqrt{\frac{5}{4}}R$

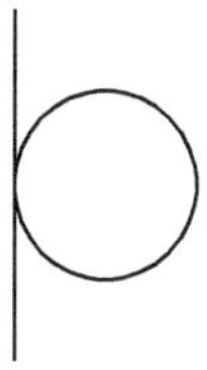

$$K_2 = \sqrt{\frac{I_2}{m_2}}$$

$$= \sqrt{\frac{(3/2)m_2R^2}{m_2}} = \sqrt{\frac{3}{2}}R$$

$\therefore \quad \frac{K_1}{K_2} = \sqrt{\frac{5}{6}}$

33. $I\omega = \text{constant}\ \omega \propto \frac{1}{I}$

On stretching hands, I will increase, therefore ω will decrease.

34. $\tau = \frac{d\mathbf{L}}{dt}$. If $\tau = 0, \mathbf{L} = \text{constant}$

35. $\mathbf{L} = \mathbf{r}\times\mathbf{p}$

36. $I\omega = \text{constant}$

or $\quad \frac{R^2}{T} = \text{constant} \quad \left(\text{as } I \propto R^2 \text{ and } \omega \propto \frac{1}{T}\right)$

$\therefore \quad T \propto R^2$

As $R' = \frac{1}{n}R$

$\therefore \quad T' = \frac{T}{n^2} = \frac{24}{n^2}\text{ h}$

37. $\omega_1 I_1 = \omega_2 I_2$

$\therefore \quad \omega_2 = \frac{\omega_1 I_1}{I_2}$

or $\quad \omega_2 = \omega\left[\frac{MR^2}{MR^2 + 2mR^2}\right]$

$$= \omega\left(\frac{M}{M+2m}\right)$$

38. $L = mvr_1 = (5)(3\sqrt{2})(2\sqrt{5}) = 30\sqrt{10}\text{ g-cm}^2/\text{s}$

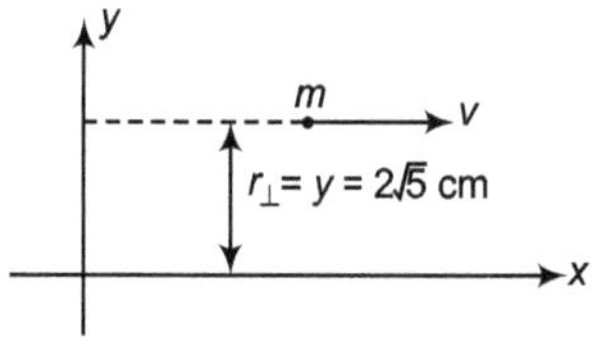

39. $r_\perp = 4\sin 45° = 2\sqrt{2}$ unit

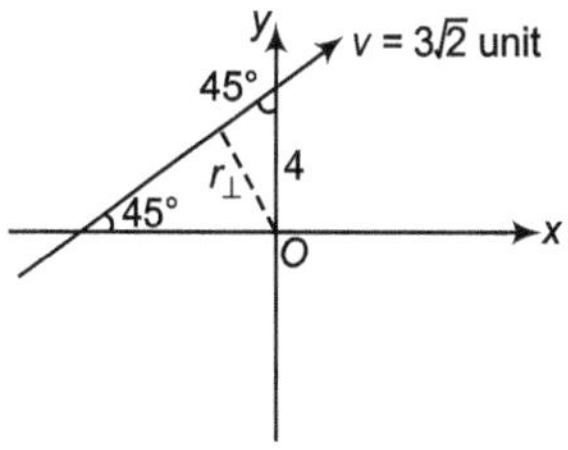

$$L = mvr_\perp = 5(3\sqrt{2})(2\sqrt{2}) = 60\text{ unit}$$

40. $I_1\omega_1 = I_2\omega_2$

$\therefore \quad \omega_2 = \frac{I_1}{I_2}\cdot\omega_1$

$$= \frac{(1/2ma^2)}{(1/2ma^2) + ma^2}\cdot\omega = \frac{\omega}{3}$$

41. $\tau_0 = mg \times r_\perp$

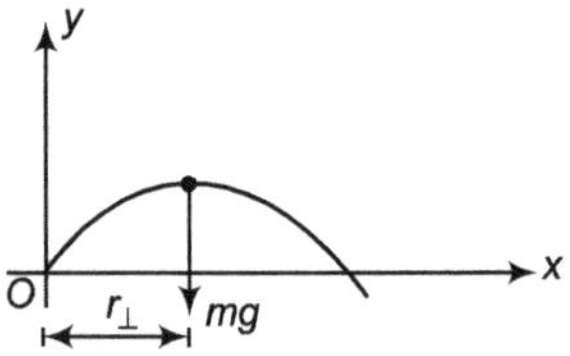

As $r_\perp$ is continuously increasing or torque is continuously increasing on the particle. Hence, angular momentum is continuously increasing.

42. $\tau = \mathbf{r} \times \mathbf{F}$ should be zero. Or $\mathbf{r}$ should be parallel to $\mathbf{F}$.

43. $I_1\omega_1 = I_2\omega_2$ or $\omega_2 = \frac{I_1}{I_2}\omega_1$

$$= \frac{(1/2MR^2)}{(1/2MR^2) + \frac{1}{2}(1/4MR^2)}\omega$$

$$= \frac{4}{5}\omega$$

44. $\tau = \mathbf{r} \times \mathbf{F}$

45. About origin, angular momentum will be

$$m(\mathbf{r} \times \mathbf{v}) = 2\begin{vmatrix} \hat{\mathbf{i}} & \hat{\mathbf{j}} & \hat{\mathbf{k}} \\ 1 & 1 & 0 \\ 2 & -2 & 2 \end{vmatrix}$$

$$= 4\hat{\mathbf{i}} - 4\hat{\mathbf{j}} - 8\hat{\mathbf{k}}$$

Therefore, component of angular moment about z-axis would be $-8 \text{ kg-m}^2/\text{s}$.

46. Force 8 N can also be resolved in tangential and radial directions.

$$\tau_{\text{net}} = (9 + 8\sin 30° - 4)(0.2) \text{ N-m clockwise}$$
$$= 1.8 \text{ N-m clockwise}$$

47. $\tau = \mathbf{r} \times \mathbf{F}$

48. For total torque about O to be zero,

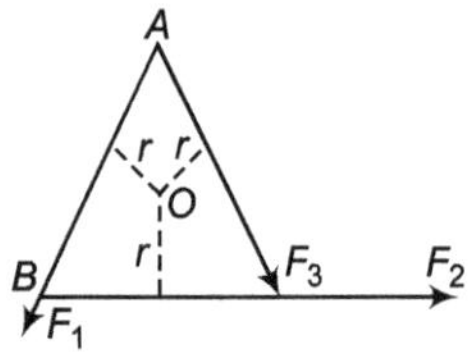

$$F_1r + F_2r - F_3r = 0$$

$\Rightarrow$ $F_3 = F_1 + F_2$

50. $L = mvr_\perp$

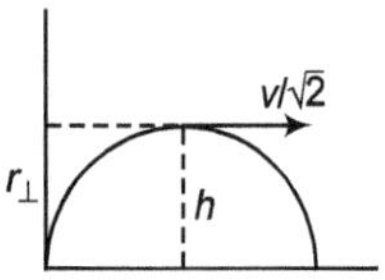

51. About bottommost point,

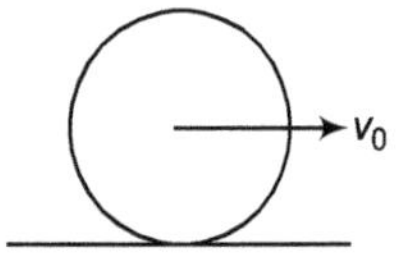

$$L = MRv_0 + \frac{2}{5}MR^2\left(\frac{v_0}{R}\right)$$
$$= \frac{7}{5}MRv_0$$

52. Angle rotated θ = area under ω-t graph.

53. If r, the distance of point from axis will increase, then corresponding v of the point will also increase accordingly. Hence, ω is constant.

54. $0 = \omega_0 - \alpha t$

$\therefore$ $$\alpha = \frac{\omega_0}{t} = \frac{(100 \times 2\pi)/60}{15}$$
$$= 0.7 \text{ rad/s}^2$$

Now, angle rotated before coming to rest $\theta = \frac{\omega_0^2}{2\alpha}$

or $$\theta = \frac{(100 \times 2\pi/60)^2}{2 \times 0.7}$$
$$= 78.33 \text{ rad}$$

or number of rotations, $n = \frac{\theta}{2\pi} = 12.5$

55. $\theta_1 = \frac{1}{2}(\alpha)(2)^2 = 2\alpha$,

$$\theta_2 = \frac{1}{2}\alpha(4)^2 - \frac{1}{2}\alpha(2)^2 = 6\alpha$$

$\therefore$ $$\frac{\theta_2}{\theta_1} = \frac{3}{1}$$

56. $\omega_0 = \frac{900 \times 2\pi}{60} \text{ rad/s} = 30\pi \text{ rad/s}$

Now, $$0 = \omega_0 - \alpha t$$

or $$\alpha = \frac{\omega_0}{t} = \frac{30\pi}{60} \text{ rad/s}^2$$
$$= \frac{\pi}{2} \text{ rad/s}^2$$

57. $\omega = \frac{d\theta}{dt} = 6t^2$, At $t = 2$ s

$$\omega = 6(2)^2 = 24 \text{ rad/s}$$

58. Angular acceleration, $\alpha = \frac{\tau}{I} = \frac{1000}{200} = 5 \text{ rad/s}^2$

$$\omega = \alpha t = (5)(3) = 15 \text{ rad/s}$$

59. Area under n-t graph will give total number of revolutions before coming to stop.

Area = Total number of revolutions

$$= \frac{1}{2} \times 8 \times \left(\frac{1800}{60}\right) + 8 \times \frac{600}{60} + \frac{1}{2} \times 16 \times \left(\frac{600}{60}\right)$$
$$= 120 + 80 + 80$$
$$= 280$$

60. $\omega = \alpha t$

$\therefore$ $$\alpha = \frac{\omega}{t}$$
$$= \frac{80}{5} = 16 \text{ rad/s}^2$$
$$\theta = \frac{1}{2}\alpha t^2$$
$$= \frac{1}{2}(16)(5)^2$$
$$= 200 \text{ rad}$$

61. $\omega = \alpha - \beta t$. Comparing with $\omega = \omega_0 - \alpha t$

Initial angular velocity $= \alpha$

Angular retardation $= \beta$

$\therefore$ Angle rotated before it stops is $\frac{\alpha^2}{2\beta}$

from $$0 = \alpha^2 - 2\beta\theta$$

62. $I = mK^2 = 10(2)^2 = 40\ \text{kg-m}^2$

$$K_R = \frac{1}{2} I\omega^2 = \frac{1}{2}(40)(5)^2 = 500\ \text{J}$$

63. $\alpha = \frac{\omega_0}{t}$ $(\because 0 = \omega_0 - \alpha t)$

$$= \frac{2\pi \times 33/60}{20} = \frac{11\pi}{200}\ \text{rad/s}^2$$

64. $\frac{K_R}{K_T} = \frac{I}{mR^2} = \frac{2}{5}$

65. $\frac{K_R}{K_T} = \frac{2}{5}$

$$\therefore \quad K = \frac{7}{5} K_T = \frac{7}{5}\left(\frac{1}{2} Mv^2\right) = \frac{7}{10} Mv^2$$

66. On smooth horizontal surface, if $v = R\omega$, but external force $= 0$. Then, it can roll without slipping.

67. $v_P = r\omega = (\sqrt{2}R)\omega = \sqrt{2}\, v_0$

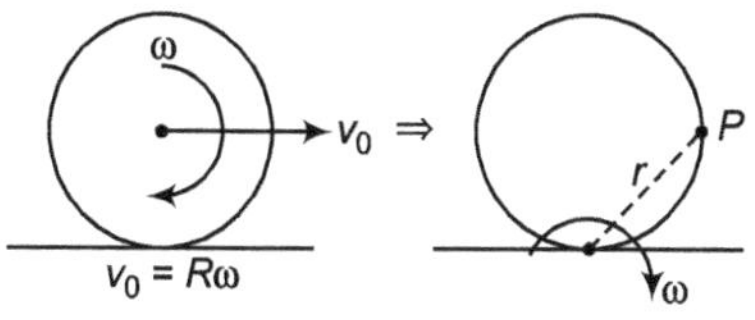

68. Acceleration of all bodies on smooth surface $= g\sin\theta$ (equal).

$\therefore$ All bodies will take equal time.

69. For all bodies $a = g\sin\theta - \mu g\cos\theta$

Hence, all will take equal time.

70. Translational kinetic energy of all bodies is same. As a is same, but moment of inertia of hollow sphere is maximum. Hence, rotational kinetic energy for hollow sphere will be least, as its angular acceleration will be least. Because force of friction for all bodies is same (maximum). Hence, total kinetic energy of hollow sphere will be minimum.

71. $v_P = r\omega = (2R)\omega = 2v$

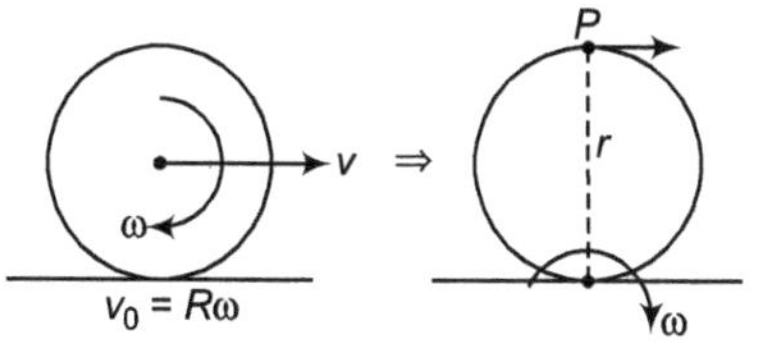

72. $\mu < \mu_{min}$ but $\mu \neq 0$. Therefore, sphere will roll with forward slipping.

73. Since, the sphere is sliding, maximum force of friction will act passing through the bottommost point. Torque of this force about point of contact is zero. Hence, angular momentum will remain conserved.

74. For a sphere $\frac{K_R}{K_T} = \frac{2}{5}$ in case of pure rolling.

Hence, $$\frac{K_R}{K} = \frac{2}{7}$$

75. $\mu_{min} = \frac{\tan\alpha}{1 + \frac{mR^2}{I}} = \frac{\tan\alpha}{1+2} = \frac{1}{3}\tan\alpha$ $\quad \left(\text{as } I = \frac{1}{2} mR^2\right)$

76. In case of pure rolling, velocity of topmost point is 2 times the velocity of centre of mass.

77. In case of pure rolling,

$$a = \frac{g\sin\theta}{1 + \frac{I}{mR^2}}; \quad \frac{I}{mR^2} = \frac{2}{5}$$

$$\therefore \quad a = \frac{5}{7} g\sin\theta$$

Since, a is independent of m and R, both spheres reach the bottom together.

78. In case of pure rolling, $a = \frac{g\sin\theta}{1 + \frac{I}{mR^2}}$

$\frac{I}{mR^2} = \frac{2}{5}$ for solid sphere

and $\frac{I}{mR^2} = \frac{1}{2}$ for a disc.

$\therefore$ For a solid sphere, $a_1 = \frac{5}{7} g\sin\theta$

and for a hollow sphere, $a_2 = \frac{2}{3} g\sin\theta$

From $$s = \frac{1}{2}at^2,\ t = \sqrt{\frac{2s}{a}}$$

or $$\frac{t_1}{t_2} = \sqrt{\frac{a_2}{a_1}}$$

$$= \sqrt{\frac{2/3}{5/7}} = \sqrt{\frac{14}{15}}$$

79. $a = \frac{g\sin\theta}{1 + \frac{I}{mR^2}}$, $\frac{I}{mR^2} = \frac{2}{5}$ for solid sphere

$$\therefore \quad a = \frac{g\sin 30^\circ}{1 + \frac{2}{5}} = \frac{5g}{14}$$

80. In case of pure rolling,

Force of friction, $f = \frac{mg\sin\theta}{1 + \left(\frac{mR^2}{I}\right)}$

Since, moment of inertia of disc is more, force of friction on cylinder will be more.

81. $\mu_{min} = \frac{\tan\theta}{1 + \frac{mR^2}{I}} = \frac{\tan\theta}{1 + \frac{5}{2}} = \frac{2}{7}\tan\theta$

82. $a = \frac{g\sin\theta}{1 + \frac{I}{mR^2}}$

$$= \frac{g\sin\theta}{1 + \frac{2}{5}} = \frac{5}{7} g\sin\theta$$

83. Friction in this case is backwards. If it is insufficient, it is a case of forward slipping or $K_T > K_R$.

84. $a = \dfrac{g\sin\theta}{1+\dfrac{I}{mR^2}}$ or $a < g\sin\theta$

85. In pure rolling mechanical energy remains conserved. Therefore, at bottommost point, total kinetic energy will be *mgh*.

Ratio of rotational to translational kinetic energy will be $\dfrac{2}{3}$.

$$\therefore \quad K_T = \frac{3}{5}(mgh)$$

$$= \frac{1}{2}mv^2$$

$$\therefore \quad v = \sqrt{\frac{6gh}{5}}$$

86. Path of *A* is cycloid and distance travelled in one rotation is 8*R*.

88. $v = r\omega$ or $v \propto r$

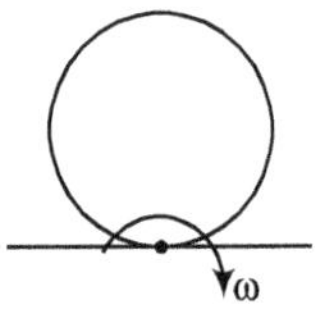

$$r_Q > r_C > r_P$$

$$\therefore \quad v_Q > v_C > v_P$$

89. $L = \sqrt{2KI}$ (Just like $p = \sqrt{2km}$)

K is made 4 times. Hence, *L* will become 2 times as $L \propto \sqrt{K}$.

90. Angular impulse = Change in angular momentum

$$\therefore \quad \tau \times t = L_f - L_i$$

$$\text{or} \quad 4\tau = 4A_0 - A_0$$

$$\therefore \quad \tau = \frac{3A_0}{4}$$

91. $L_i - l_f$ (about centre of mass of the rod)

$$mvr = \frac{ML^2}{12}\cdot\omega$$

$$\therefore \quad \omega = \frac{12mvr}{ML^2}$$

$$= \frac{(12)(1)(2)\left(\frac{1}{4}\right)}{(2)(1)^2} = 3\,\text{rad/s}$$

92. Work done by retarding torque = change in kinetic energy

Since, kinetic energy of both are equal, both will rotate same revolutions.

93. Decrease in gravitational potential energy = Increase in rotational kinetic energy about an axis passing through end point.

$$\therefore \quad mg\frac{l}{2} = \frac{1}{2}\left(\frac{ml^2}{3}\right)\omega^2$$

$$\Rightarrow \quad \omega = \sqrt{\frac{3g}{l}} \quad \text{or} \quad \omega \mu \sqrt{\frac{g}{l}}$$

94. $E = \dfrac{1}{2}I\omega^2$

$$\omega = \sqrt{\frac{2E}{I}}$$

Now, $\quad L = I\omega = \sqrt{2EI}$

95. Angular momentum will not remain conserved due to a torque of weight of particle about axis of rotation.

96. It depends whether pure rolling is taking place on ground or over a platform.

97. Direction of linear velocity always keeps on changing. Hence, linear momentum is varying.

98. W = change in rotational kinetic energy $= \dfrac{1}{2}I(\omega_f^2 - \omega_i^2)$

$$= \frac{1}{2}\left(\frac{1}{2}mR^2\right)[(2\pi f_f)^2 - (2\pi f_i)^2]$$

$$= \frac{1}{4}\times 2\times(1)^2\times 4\times\pi^2\,(100-25)$$

$$= 1.48\times10^3\ \text{J} \approx 1.5\times10^3\ \text{J}$$

99. $\boldsymbol{\tau} = \mathbf{r}\times\mathbf{F}$, i.e. $\boldsymbol{\tau}$ is perpendicular to both $\mathbf{r}$ and $\mathbf{F}$

$$\therefore \quad \boldsymbol{\tau}\cdot\mathbf{r} = 0 \quad \text{and} \quad \boldsymbol{\tau}\cdot\mathbf{F} = 0$$

100. Otherwise net torque or net angular acceleration will become zero.

101. From the property of a circle, if an arc subtends an angle θ at any point *A* on circumference, then it will subtend an angle 2θ at centre *C*. So in the same time interval a particle rotating in a circle turn double the angle with respect to centre point *C* compared to point *A*.

$$\therefore \quad \omega_C = 2\omega_A$$

102. $L = I\omega = I(2\pi f)$

Frequency *f* is doubled. Hence, angular momentum will become 2*L*.

103. Decrease in gravitational PE = Increase in rotational KE

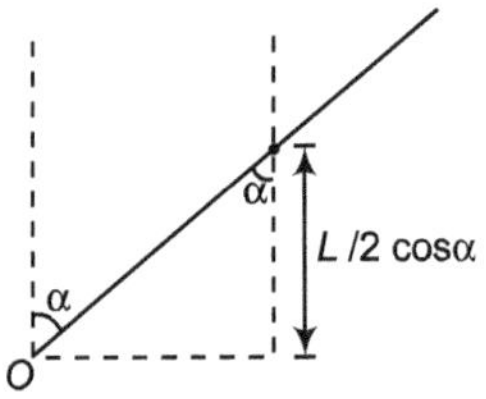

$$\therefore \quad mgh = \frac{1}{2}I\omega^2$$

$$mg\left[\frac{L}{2} - \frac{L}{2}\cos\alpha\right] = \frac{1}{2}\left(\frac{mL^2}{3}\right)\omega^2$$

$$\therefore \quad \omega = \sqrt{\frac{3g}{L}(1-\cos\alpha)}$$

$$= \sqrt{\frac{3g}{L}\left(2\sin^2\frac{\alpha}{2}\right)} = \sqrt{\frac{6g}{L}}\sin\frac{\alpha}{2}$$

Level 2 : Only One Correct Option

1. $\dfrac{K_R}{K_T} = \dfrac{I}{mR^2}$, if $v = R\omega$ in case of pure rolling $= \dfrac{K^2}{R^2}$

$$\therefore K_R = \left(\frac{K^2}{K^2+R^2}\right)K_{\text{Total}} \quad \text{or} \quad \frac{K_R}{K_{\text{Total}}} = \frac{K^2}{K^2+R^2}$$

2. $L_P = (mu\sin\theta)\left(\frac{R}{2}\right)(mu\cos\theta)(H)$

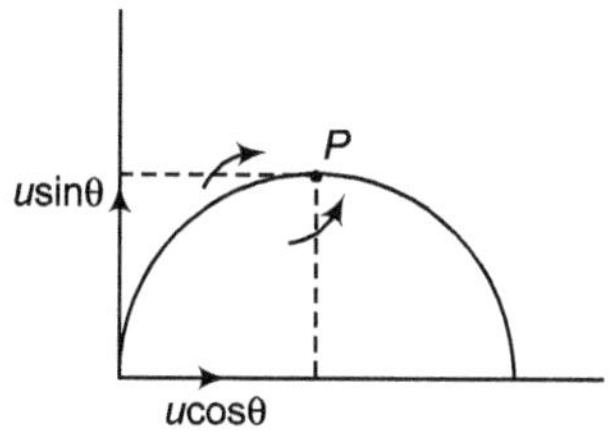

$$= (mu\sin\theta)\left(\frac{u\sin\theta}{g}\right)(u\cos\theta) = (mu\cos\theta)\left(\frac{u^2\sin^2\theta}{2g}\right)$$

$$= \frac{mu^3\sin^2\theta\cos\theta}{2g}$$

3. From $L_i = L_f$ about centre of rod.

$$m \cdot vr_\perp = I_{CM} \cdot \omega$$

$$\Rightarrow \quad (1)(2)\left(\frac{1}{4}\right) = \frac{1}{12} \times (2)(1)^2 \cdot \omega$$

$$\therefore \quad \omega = 3 \text{ rad/s}$$

4. $v = r\omega$

Decrease in gravitational potential energy = Increase in kinetic energy

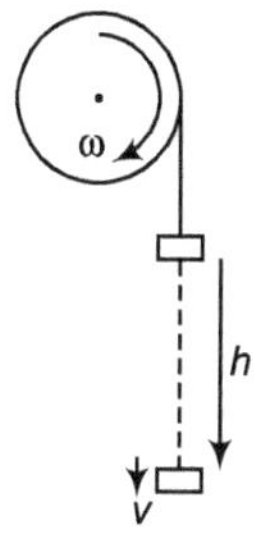

$$\therefore \quad mgh = \frac{1}{2}I\omega^2 + \frac{1}{2}mv^2$$

$$= \frac{1}{2}I\omega^2 + \frac{1}{2}m(r\omega)^2$$

$$\therefore \quad \omega = \sqrt{\frac{2mgh}{I + mr^2}}$$

5. At bottommost point, total kinetic energy will be *mgh*. Ratio of rotational to translational kinetic energy will be $\frac{2}{5}$.

$$\therefore \quad K_T = \frac{5}{7}mgh = \frac{1}{2}mv^2$$

$$\therefore \quad v = \sqrt{\frac{10}{7}gh}$$

6. In case of pure rolling, ratio of rotational to translational kinetic energy is $\frac{2}{5}$. Therefore, total kinetic energy is $\frac{7}{5}$ times the translational kinetic energy. At maximum compression whole of energy is elastic potential. Hence,

$$\frac{7}{5}\left(\frac{1}{2}Mv^2\right) = \frac{1}{2}kx_{max}^2$$

$$\therefore \quad x_{max} = v\sqrt{\frac{7M}{5k}}$$

7. Whole mass has equal distance from the centre O. Hence, $I_0 = mR^2$. Further centre of mass of the remaining portion will be to the left of point O. More the distance of axis from centre of mass, more is the moment of inertia.

Hence, $I_A > I_0$.

8. $a = \dfrac{g\sin\theta}{1 + \dfrac{I}{mr^2}}$

For a sphere, $\dfrac{I}{mr^2} = \dfrac{2}{5}$

Hence, $a = \frac{5}{7}g\sin\theta = \text{constant}$.

Hence, speed of all spheres is same at the bottom. Sphere (ii) has the largest mass. Hence, it will have the maximum kinetic energy.

9. Conserved angular momentum about common centre of mass.

$$mv\frac{R}{2} = I_{Total}\,\omega = \left[m\left(\frac{R}{2}\right)^2 + \frac{1}{2}mR^2 + m\left(\frac{R}{2}\right)^2\right]\omega$$

$$\therefore \quad \omega = \frac{v}{2R}$$

10. $TR = I\alpha, TR = I\dfrac{a}{R}$ or $T = \dfrac{Ia}{R^2}$

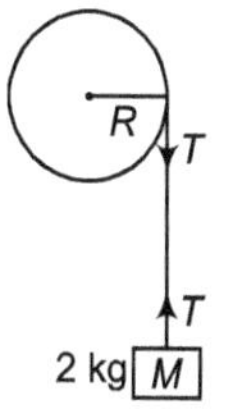

$$Mg - T = Ma$$

or $$Mg = Ma + T = \left(M + \frac{I}{R^2}\right)a$$

or $$a = \frac{Mg}{\left(M + \frac{I}{R^2}\right)} = \frac{2 \times 10}{\left(2 + \frac{0.32}{0.04}\right)} = 2 \text{ m/s}^2$$

11. $I_1\omega_1 = I_2\omega_2$, $\omega_2 = \frac{1}{2}\omega_1$

$$\therefore \quad I_2 = 2I_1$$

$$\frac{1}{2}m_0R^2 + (\mu t)R^2 = 2\left[\frac{1}{2}m_0R^2\right]$$

$$\therefore \quad t = \frac{m_0}{2\mu}$$

12. $(K_T)_{sphere} = (K_T)_{cylinder}$ (m and v are same)

$$\therefore \quad \frac{5}{7}(mgh_1) = \frac{2}{3}(mgh_2)$$

$$\therefore \quad \frac{h_1}{h_2} = \frac{14}{15}$$

13. $l = 2\pi R \Rightarrow \therefore R = \dfrac{l}{2\pi}$

$$I_1 = \frac{ml^2}{12}$$

$$I_2 = mR^2 = \frac{ml^2}{4\pi^2} \Rightarrow \therefore \frac{I_1}{I_2} = \frac{\pi^2}{3}$$

14. $I_1\omega_1 = I_2\omega_2$

$\therefore \quad \frac{1}{2}(mR^2)\omega_0 = \left(\frac{1}{2}mR^2 + mR^2\right)\omega_2$

$$\omega_2 = \frac{\omega_0}{3}$$

$$K_i = \frac{1}{2}I_1\omega_1^2 = \frac{1}{2}\left(\frac{1}{2}mR^2\right)\omega_0^2 = \frac{1}{4}mR^2\omega_0^2$$

$$K_f = \frac{1}{2}\left(\frac{1}{2}mR^2 + mR^2\right)\left(\frac{\omega_0}{3}\right)^2 = \frac{1}{12}mR^2\omega_0^2 = \frac{K_i}{3}$$

$\therefore \quad \frac{2K_i}{3}$ is converted into heat.

15. Relative velocity of A with respect to C perpendicular. to AC is 14 m/s.

$\therefore \quad \omega = \frac{14}{AC} = \frac{14}{7/2} = 4$ rad/s

A and α are zero. Hence, acceleration of any point is

$$a_n = r\omega^2$$

$\therefore \quad |\mathbf{a}_A| = |\mathbf{a}_B|$

$$= r\omega^2 = \frac{l}{2}\cdot\omega^2 = 56 \text{ m/s}^2$$

16. Decrease in gravitational potential energy = Increase in rotational kinetic energy about point O.

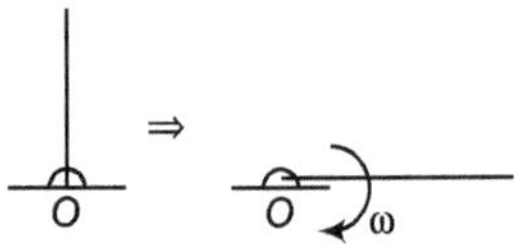

or $mg\frac{L}{2} = \frac{1}{2}\left(\frac{mL^2}{3}\right)\omega^2$ or $\omega = \sqrt{\frac{3g}{L}}$

Now, $\quad v = r\omega \quad$ (in pure rotation)

$\therefore \quad v = L\sqrt{\frac{3g}{L}} = \sqrt{3gL}$

17. $L = mvR + I_C\omega = mvR + \frac{1}{2}mvR = \frac{3}{2}mvR$

18. Cube will slide, if $F > \mu mg$ ...(i)

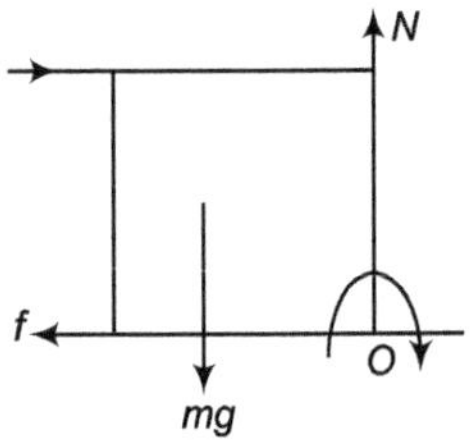

Cube will topple about O, if

$$\tau_F > \tau_{mg}$$

$\therefore \quad F\cdot a > mg\left(\frac{a}{2}\right)$

or $\quad F > \frac{1}{2}mg$ ...(ii)

If $\mu > \frac{1}{2}$, Eq. (ii) will be satisfied earlier.

19. Net torque about O should be zero

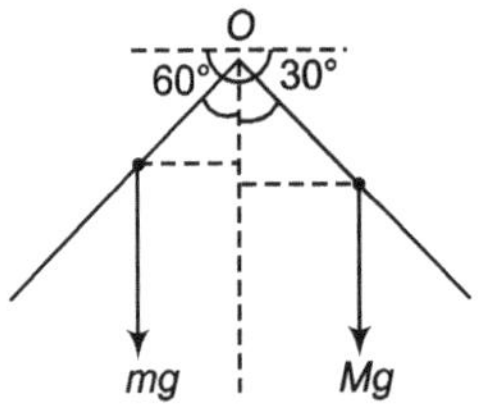

$\therefore \quad Mg\frac{l}{2}\sin 30° = mg\frac{l}{2}\sin 60°$

$$\frac{M}{m} = \frac{\sin 60°}{\sin 30°} = \sqrt{3}$$

20. From theorem of parallel axes, moment of inertia about an axis passing through a distance $\frac{l}{2}$ from centre in all directions will have moment of inertia $\frac{ml^2}{3}$. All such points at same distance from centre will lie in a circle.

21. $(\pi R^2 td) = m$

$\therefore \quad R^2 = \frac{m}{\pi td}$

Now, $I = \frac{1}{2}mR^2 = \frac{m^2}{2\pi td}$ or $I \propto \frac{1}{d}$

$\therefore \quad \frac{I_1}{I_2} = \frac{d_2}{d_1}$

22. In case of pure rolling, $\frac{K_R}{K_T} = \frac{2}{5}$

$\therefore$ Total kinetic energy $= \frac{7}{5}$ times the translational kinetic energy. At highest point whole of the kinetic energy will be converted into potential energy.

$\therefore \quad mgh = \frac{7}{5}\left(\frac{1}{2}mv^2\right)$

or $\quad h = \frac{7v^2}{10g} = \frac{7(10)^2}{10\times 10} = 7$ m

23. Moment of inertia of remaining portion

$$I = \frac{1}{2}MR^2 - \left[\frac{1}{2}\left(\frac{M}{4}\right)\left(\frac{R}{2}\right)^2 + \left(\frac{M}{4}\right)\left(\frac{R}{2}\right)^2\right] = \frac{13}{32}MR^2$$

24. $K_1 = \sqrt{\frac{I_1}{m}} = \sqrt{\frac{\frac{1}{2}mr^2}{m}} = \frac{r}{\sqrt{2}}$

$$K_2 = \sqrt{\frac{I_2}{m}} = \sqrt{\frac{m(2r)^2}{m}} = 2r$$

$\therefore \quad \frac{K_2}{K_1} = 2\sqrt{2}$

25. About bottommost point net torque is clockwise. Hence, centre of mass moves with acceleration $a = R\alpha$ towards right.

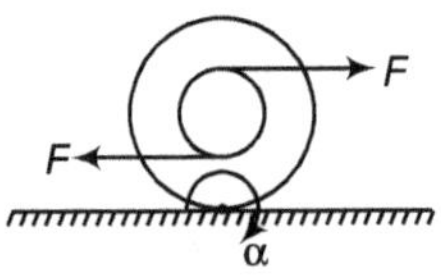

26. Let disc rotates anti-clockwise by an angle θ. Then, angular displacement of each men is $(2\pi - \theta)$ in clockwise direction. In the absence of external torque.

Applying : $\Sigma I_A \theta_A = \Sigma I_C \theta_C$ (A = anti-clockwise, C = clockwise)

$$\frac{1}{2}MR^2 \cdot \theta = 2(mR^2)(2\pi - \theta)$$

or
$$\theta = \frac{8m\pi}{4m + M}$$

27. In case of a ring, its translational and rotational kinetic energies are equal. Hence, total kinetic energy of ring will be mv_0^2.

Kinetic energy of system

$$= mv_0^2 + \frac{1}{2}(2m)(\sqrt{2}v_0)^2 + \frac{1}{2} \times m \times (2v_0)^2 + \frac{1}{2} \times m \times (\sqrt{2}v_0)^2$$

$$= 6mv_0^2$$

More than One Correct Options

1. $\frac{K_R}{K_T} = \frac{I}{mr^2}$

$$\therefore \quad K_T = \left(\frac{mr^2}{I + mr^2}\right) K_{\text{Total}}$$

or
$$K_{\text{Total}} = \left(\frac{I + mr^2}{mr^2}\right)\left(\frac{1}{2}mv^2\right)$$

$$= mgh = mg\left(\frac{3}{4}\frac{v^2}{g}\right)$$

Solving, we get $I = \frac{1}{2}mr^2$

So, it is either solid cylinder or disc.

2. (a) $I_x = I_y = 2\left[\frac{ml^2}{12} + m\left(\frac{l}{2}\right)^2\right]$

$$= \frac{2}{3}ml^2$$

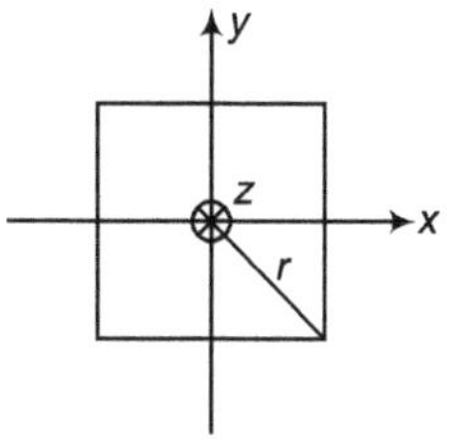

(b) $I_z = I_x + I_y = \frac{4}{3}ml^2$

(c) $I = I_z + (4m)r^2$

$$= \frac{4}{3}ml^2 + (4m)\left(\frac{l}{\sqrt{2}}\right)^2$$

$$= \frac{10}{3}ml^2$$

(d) $I = 2\frac{ml^2}{3} + ml^2$

$$= \frac{5}{3}ml^2$$

3. $O \rightarrow$ Instantaneous axis of rotation

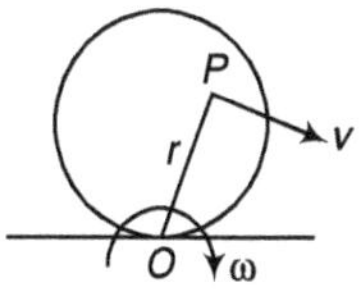

$$v = r\omega$$

or
$$v \propto r$$

More the value of r from O more is the speed of point P.

4. $v_0 + R\omega_0 = v$ ...(i)

$$v_0 - R\omega_0 = 3v \quad \text{...(ii)}$$

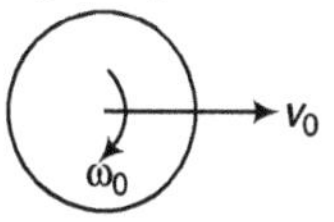

Solving these two equations, we get

$$v_0 = 2v \quad \text{and} \quad \omega_0 = -\frac{v}{R}$$

5. $x^2 + y^2 = l^2 = (0.25)\,m^2$

$$y\omega = 4 \quad \text{...(i)}$$

$$x\omega = 3 \quad \text{...(ii)}$$

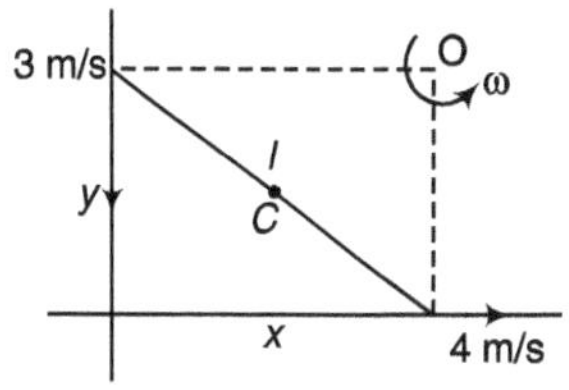

Squaring and adding, we get

$$(x^2 + y^2)\omega^2 = 25 \quad \text{or} \quad (0.25)\omega^2 = 25$$

$$\therefore \quad \omega = 10 \text{ rad/s}$$

$$OC = \frac{l}{2} = 0.25 \text{ m}$$

$$v_{\text{CM}} = (OC)\,\omega = 2.5 \text{ m/s}$$

$$K_{\text{Total}} = \frac{1}{2}mv_{\text{CM}}^2 + \frac{1}{2}I_{\text{CM}}\,\omega^2$$

$$= \frac{1}{2} \times 2 \times (2.5)^2 + \frac{1}{2}\left(\frac{2 \times 0.25}{12}\right)(10)^2$$

$$= \frac{25}{3}\text{ J}$$

6. Net acceleration of any point on the rim is vector sum of a, $R\omega^2$ and $R\alpha$ with $a = R\alpha$

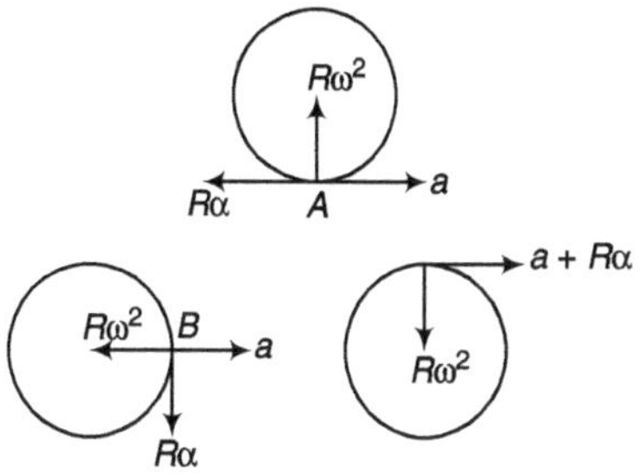

$a_A = R\omega^2 \rightarrow$ vertically upwards

If $a = R\omega^2$, a_B is vertically downwards and so on.

7. C_1 is centre of mass of rod,

C_2 is centre of mass of both

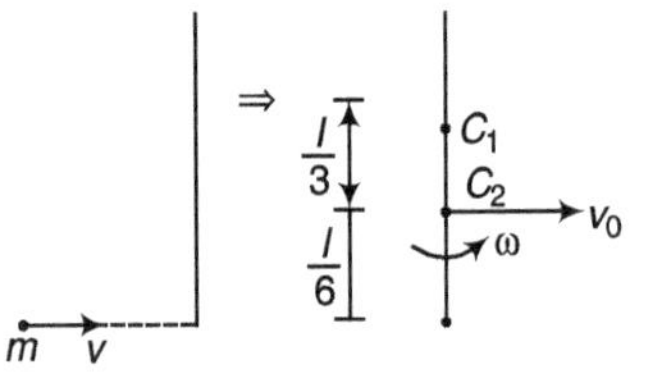

$$P_i = P_f$$

$\therefore$ $mv = 3mv_0$

$\Rightarrow$ $v_0 = \frac{v}{3}$ ...(i)

$L_i = L_f$ about C_2 we have,

$$mv\frac{l}{6} = I_{C_2}\omega$$

$$= \left[\left(\frac{2ml^2}{12}\right) + 2m\left(\frac{l}{3}\right)^2 + m\left(\frac{l}{6}\right)^2\right]$$

$\therefore$ $\omega = \frac{2}{5}\frac{v}{l}$

$$K_i = \frac{1}{2}mv^2$$

$$K_f = \frac{1}{2}(3m)\left(\frac{v}{3}\right)^2 + \frac{1}{2}I_{C_2}\omega^2$$

$$= \frac{1}{6}mv^2 + \frac{1}{2}\left(\frac{5}{12}ml^2\right)\left(\frac{2}{5}\frac{v}{l}\right)^2$$

$$= \frac{1}{5}mv^2$$

$\therefore$ Loss of kinetic energy $= K_i - K_f$

$$= \frac{3}{10}mv^2$$

8. $I = mK^2$ $\left(\text{where, } K = \frac{R}{2} = \text{radius of gyration}\right)$

$$= m\left(\frac{R}{2}\right)^2$$

$$= \frac{1}{4}mR^2$$

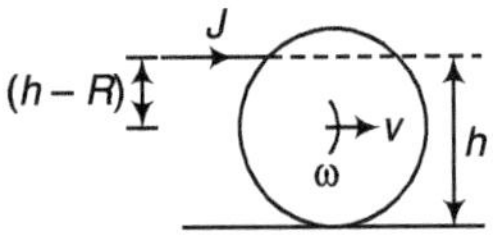

Ball will roll purely, if

$$v = R\omega$$

$\therefore$ $\left(\frac{J}{m}\right) = R\left[\frac{J(h-R)}{(1/4)mR^2}\right]$

Solving this equation, we get

$$h = \frac{5R}{4}$$

Further, if ball is struck at centre of mass, there will be no rotation only translation.

9. Friction always acts upwards. If the ball moves upwards, it becomes a case of retardation with pure rolling.

$$\mu_{min} = \frac{\tan\alpha}{1 + \frac{mR^2}{I}}$$

$$= \frac{\tan\alpha}{1 + \frac{3}{2}} = \frac{2}{5}\tan\alpha$$

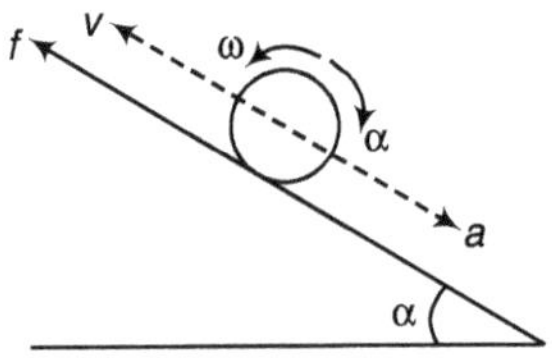

10. $I_1\omega_1 = I_2\omega_2$

$\therefore$ $\omega_2 = \frac{I_1}{I_2}\cdot\omega_1$

$$= \frac{(1/2mR^2)}{(1/2mR^2 + mR^2)}\cdot\omega$$

$$= \frac{\omega}{3}$$

$$p_i = mv_i = m(2\omega R) = 2m\omega R$$

$$p_f = mv_f = m\left(\frac{\omega}{3}R\right) = \frac{1}{3}m\omega R$$

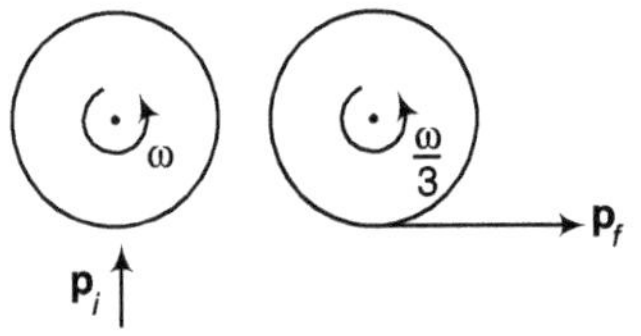

Impulse, $J = |\mathbf{p}_f - \mathbf{p}_i|$

$$= \sqrt{p_i^2 + p_f^2} \quad (\text{as } \theta = 90°)$$

$$= \frac{\sqrt{37}}{3}m\omega R$$

11. Velocity component along $AB = 0$

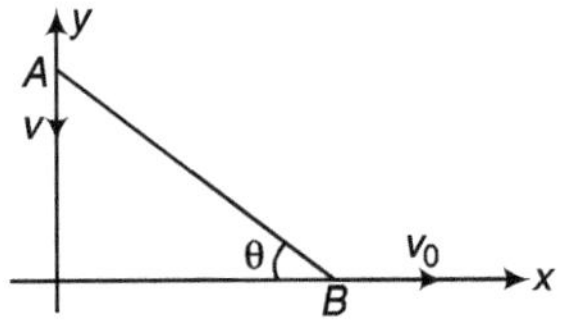

$\therefore$ $v_0\cos\theta = v\sin\theta$

or $v = v_0\cot\theta = f_1(\theta)$

At $\theta = 37°, v = \frac{4}{3}v_0$

$$\omega = \frac{\text{Relative velocity} \perp \text{to } AB}{l}$$

$$= \frac{v_0\sin\theta + v\cos\theta}{l} = f_2(\theta)$$

$$= \frac{(v_0)(3/5) + (4/3v_0)(4/5)}{l}$$

$$= \frac{5v_0}{3l}$$

Comprehension Based Questions

1. Bottommost point where friction actually acts is at rest.

2. $\alpha = \dfrac{F \cdot R}{\frac{3}{2}mR^2} = \dfrac{2F}{3mR}$

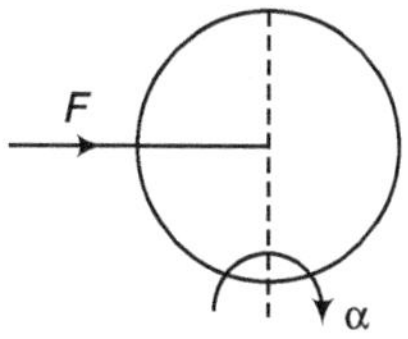

$$a = R\alpha = \frac{2F}{3m}$$

$$F - f = ma = \frac{2F}{3}$$

$$\therefore \quad f = \frac{F}{3} = \frac{1}{2}ma$$

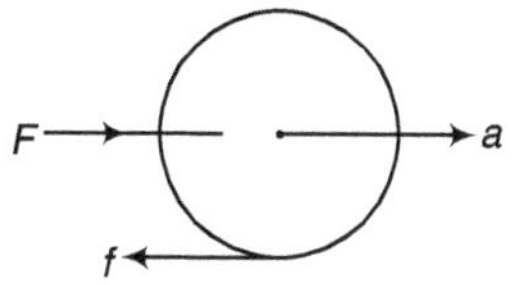

3. About point P net torque of F and f is zero. So, angular momentum is conserved.

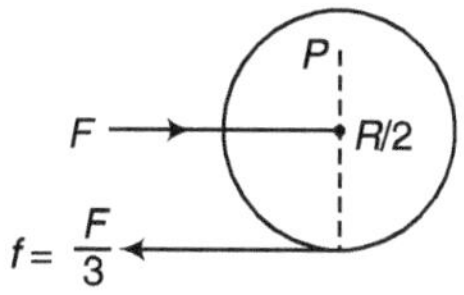

4. K_{Total} after falling a height h is mgh

$$\frac{K_R}{K_T} = \frac{2}{3}$$

$$\therefore \quad K_T = \frac{1}{2}mv^2 = \left(\frac{3}{5}\right)(mgh)$$

$$\therefore \quad v = \sqrt{\frac{6gh}{5}}$$

5. $H = \dfrac{v^2 \sin^2 \sin 37°}{2g} = \dfrac{(6gh/5)(3/5)^2}{2g} = \dfrac{27h}{125}$

6. $R = x = \dfrac{2u^2 \sin 37° \cos 37°}{g}$

$$= \frac{(2)(6gh/5)(3/5)(4/5)}{g} = \frac{144}{125}h$$

Assertion and Reason

1. When comparing between parallel axes moment of inertia about an axis passing through centre of mass is minimum.

2. Linear acceleration due to $mg\sin\theta$ is downwards. Hence, for pure rolling to continue force of friction should act upwards so that angular acceleration is anti-clockwise.

3. $a = \dfrac{g\sin\theta}{1 + \dfrac{I}{mR^2}}$

4. L about centre $= mvR =$ constant

$$\tau_{centre} = 0$$

5. Pure rolling will start when

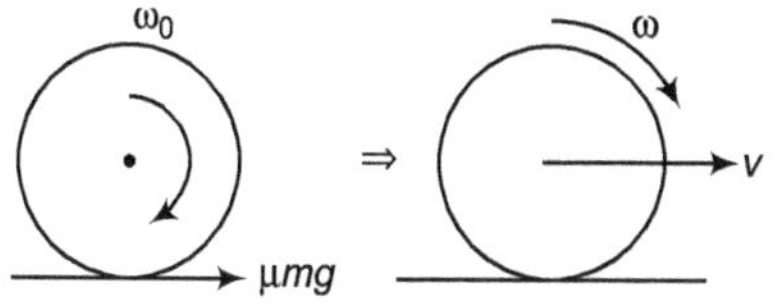

$$v = R\omega, \mu g t = R\left[\omega_0 - \frac{\mu mgR}{I}t\right]$$

$$\therefore \quad t = \frac{R\omega_0}{\mu g + \dfrac{\mu mgR}{I}}; I_{solid} < I_{hollow}$$

$$\therefore \quad t_{solid} < t_{hollow}$$

7. Under the action of two equal and opposite forces body is in translational equilibrium, not in rotational equilibrium.

9. Angular momentum about bottommost point will be

$$L = mv_0R - \frac{1}{2}mR^2\omega_0$$

If $L = 0$ or $v_0 = \dfrac{\omega_0 R}{2}$, disc will come to rest after some time.

10. $\dfrac{1}{2}mv^2 + \dfrac{1}{2}I\omega^2 = mgh$ (for both)

I is different for two bodies.

Hence, v and ω will be different, as $v = R\omega$.

11. Angular momentum about centre is mvR.

12. Sphere can roll without slipping on surface, if $v = r\omega$

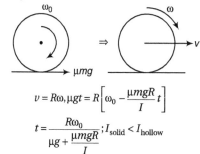

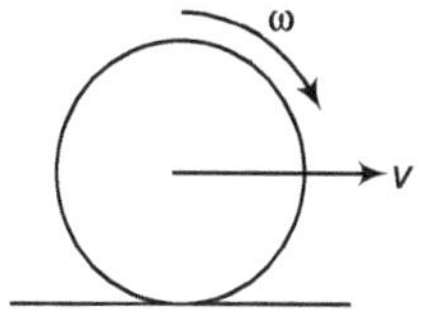

14. $L = mvr\sin\theta$ or $mvr_\perp$

In case of constant velocity $m : v$ and $r_\perp$ all are constant. Therefore, angular momentum is constant.

Further, $L = n\dfrac{h}{2\pi}$ (in Bohr's theory).

Therefore L and h have same units.

15. When compared with many parallel axes, I_{CM} is least. But, it is wrong to say that I_{CM} is least.

16. $I = I_C + mr^2$

Radius of gyration is a distance.

20. In first case, work done by friction will be zero. That's why mechanical energy will remain conserved. Further, in second case, linear velocity will be more because this ball has only translational kinetic energy.

Match the Columns

1. $I_1 = 2\left(\dfrac{ml^2}{2}\right) + 2(m)\left(\dfrac{l}{2}\right)^2 = \dfrac{2}{3}ml^2$

$$I_2 = 0 + 2\left(\frac{ml^2}{3}\right) + ml^2 = \left(\frac{5}{3}\right)ml^2$$

$$I_3 = 4\left[\frac{ml^2}{3}\sin^2 45°\right] = \frac{2}{3}ml^2 = I_1$$

Note $I_1 = I_3$ (think why?)

2. In general as $v_P = 2v \sin\left(\frac{\theta}{2}\right)$.
3. In the absence of external torque, angular momentum remains constant.

$$L = I\omega = \text{constant}$$
$$I \propto R^2$$

$\therefore$ I will remain $\frac{1}{4}$th or ω will become 4 times.

Therefore, time period (T) will remain $\frac{1}{4}$th

Further, $K = \frac{L^2}{2I}$

Since, angular momentum is constant and I has become $\frac{1}{4}$th.

Therefore, kinetic energy will become 4 times.

4. First find I for complete ring, by doubling the given mass. Then, the desired value is half the calculated value.
5. $\tau = 0$

$\therefore$ $L = \text{constant}, K = \frac{L^2}{2I}$ and $I\omega = \text{constant}$,

Insect first moves away from the axis, the towards it. Hence, I will first increase and then decrease.

6. Consider the below diagram

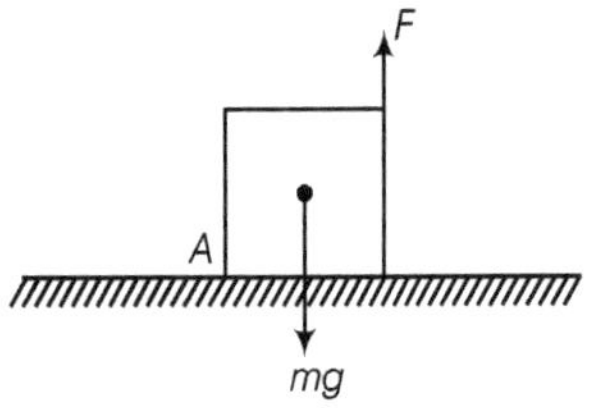

Moment of the force F about point A, $\tau_1 = F \times a$ (anti-clockwise)

Moment of weight mg of the cube about point A,

$$\tau_2 = mg \times \frac{a}{2} \text{ (clockwise)}$$

Cube will not exhibit motion, if $\tau_1 = \tau_2$

($\because$ In this case, both the torque will cancel the effect of each other)

$\therefore$ $F \times a = mg \times \frac{a}{2}$

$\Rightarrow$ $F = \frac{mg}{2}$

Cube will rotate only when, $\tau_1 > \tau_2$

$\Rightarrow$ $F \times a > mg \times \frac{a}{2} \Rightarrow F > \frac{mg}{2}$

Let normal reaction is acting at $\frac{a}{3}$ from point A, then

$mg \times \frac{a}{3} = F \times a$ or $F = \frac{mg}{3}$ (For no motion)

When $F = \frac{mg}{4}$ which is less than $\frac{mg}{3}$,. there will be no motion. $\left(F < \frac{mg}{3}\right)$

$\therefore$ A→q B→r C→p D→s

7. Consider the diagram where a sphere of m and radius R, struck horizontally at height h above the floor

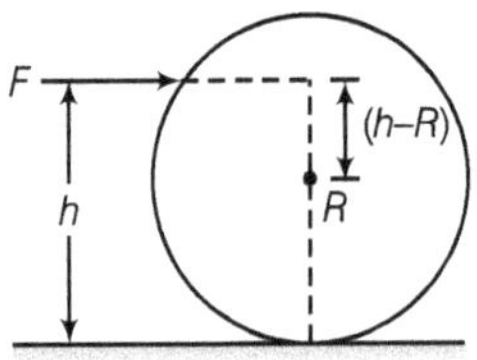

The sphere will roll without slipping when $\omega = \frac{v}{r}$, where v is linear velocity and ω is angular velocity of the sphere.

Now, angular momentum of sphere, about centre of mass [We are applying conservation of angular momentum just before and after struck]

$$mv(h - R) = I\omega = \left(\frac{2}{5}mR^2\right)\left(\frac{v}{R}\right)$$

$$\Rightarrow \quad mv(h - R) = \frac{2}{5}mvR$$

$$h - R = \frac{2}{5}R \Rightarrow h = \frac{7}{5}R$$

Therefore, the sphere will roll without slipping with a constant velocity and hence, no loss of energy, so D→p

Torque due to applied force, F about centre of mass

$$\tau = F(h - R) \quad \text{(clockwise)}$$

For $\tau = 0$, $h = R$, sphere will have only translational motion. It would lose energy by friction.

Hence, B→s

The sphere will spin clockwise when $\tau > 0 \Rightarrow h > R$

Therefore, C→q

The sphere will spin anti-clockwise when

$\tau < 0 \Rightarrow h < R$, A→r

Entrance Gallery

1. Given, $\mathbf{F} = 5\hat{\mathbf{i}} + 2\hat{\mathbf{j}} - 5\hat{\mathbf{k}}$ and $\mathbf{r} = \hat{\mathbf{i}} - 2\hat{\mathbf{j}} + \hat{\mathbf{k}}$

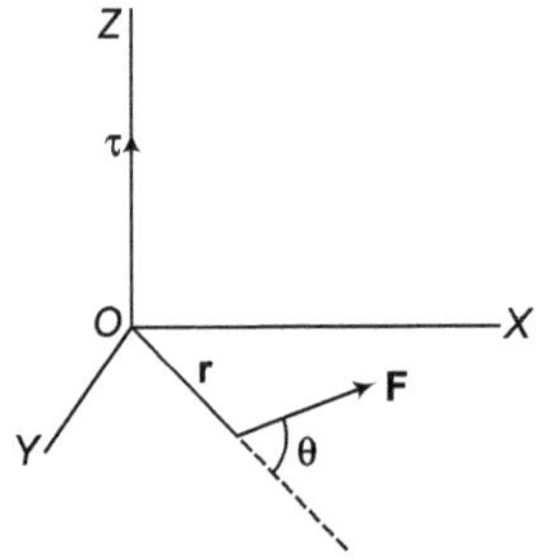

We know that, $\tau = \mathbf{r} \times \mathbf{F}$

So, torque about the origin will be given by

$$= \begin{vmatrix} \hat{\mathbf{i}} & \hat{\mathbf{j}} & \hat{\mathbf{k}} \\ 1 & -2 & 1 \\ 5 & +2 & -5 \end{vmatrix}$$

$$= \hat{\mathbf{i}}(10 - 2) - \hat{\mathbf{j}}(-5 - 5) + \hat{\mathbf{k}}(2 + 10) = 8\hat{\mathbf{i}} + 10\hat{\mathbf{j}} + 12\hat{\mathbf{k}}$$

2. We know that kinetic energy, $K = \frac{1}{2}mv^2 = \frac{1}{2}I\omega^2$

where, $m = 27$ kg (mass of the body)

$\omega = 3$ rad/s (angular velocity)

$$I = 3\ \text{kg-m}^2 \qquad \text{(moment of inertia)}$$

We have, $mv^2 = I\omega^2$

$$v^2 = \frac{I\omega^2}{m} \Rightarrow v^2 = \frac{3 \times 3^2}{27}$$

$$v^2 = \frac{27}{27} = 1 \Rightarrow v = \sqrt{1} = 1\ \text{ms}^{-1}$$

3. Let the final velocity be v.

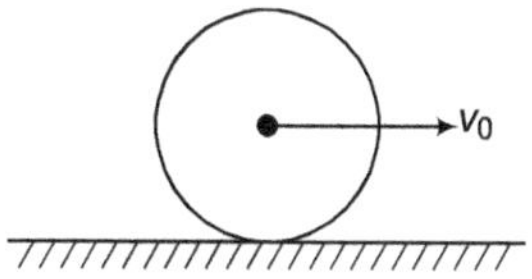

So, angular momentum will remain conserved along point of contact.

By conservation of angular momentum,

Angular momentum will remain conserved along point of contact,

$$I\omega = \text{constant}$$

$$mv_0 r = mvr + \frac{2}{5}mr^2 \times \omega \qquad \left(\because \omega = \frac{v}{r}\right)$$

$$mv_0 r = mvr + \frac{2}{5}mr^2\left(\frac{v}{r}\right)$$

$$\Rightarrow \quad v_0 = v + \frac{2}{5}v$$

$$v_0 = \frac{7}{5}v$$

$$\Rightarrow \quad v = \frac{5}{7}v_0$$

4. We know that angular momentum is given by $L = I\omega$.

$I_1\omega_1 = I_2\omega_2$

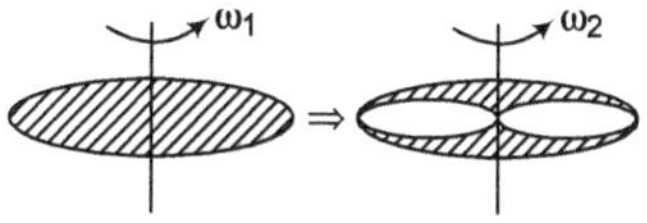

[Disc → M, R Ring → m, r]

Then,

$$\therefore \quad \omega_2 = \left(\frac{I_1}{I_2}\right)\omega_1 = \left[\frac{\frac{1}{2}MR^2}{\frac{1}{2}MR^2 + 2(mr^2)}\right]\omega_1$$

$$= \left[\frac{50(0.4)^2}{50(0.4)^2 + 8 \times (6.25) \times (0.2)^2}\right](10)$$

$$= \left[\frac{8}{10}\right] \times 10 = 8\ \text{rad/s}$$

5. Acceleration of the sphere,

$$a = \frac{g\sin\theta}{1 + \frac{v^2}{r^2}}$$

$$= \frac{g\sin\theta}{1 + \frac{2}{5}}$$

$$a = \frac{5}{7}g\sin\theta$$

6. Initial velocity, $v_1 = v$

Final velocity, $v_2 = -v$

Initial momentum, $p_1 = mv$

Final momentum, $p_2 = m(-v) = -mv$,

Change in momentum, $\Delta p = p_1 - p_2$

$$= mv - (-mv) = 2mv$$

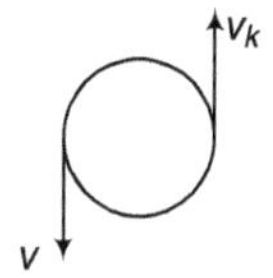

7. Here, T = total portion, R = remaining portion and C = cavity and let σ = mass per unit area.

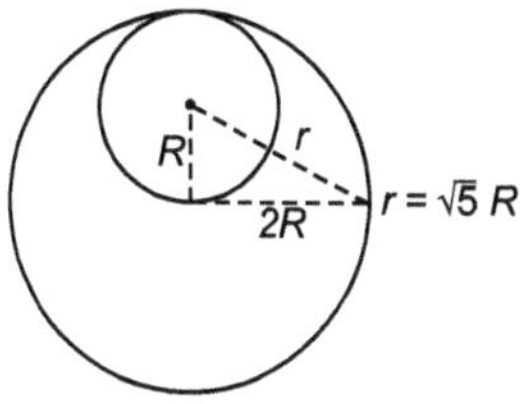

Then, $m_T = \pi(2R)^2\,\sigma = 4\pi R^2\sigma$

$$m_c = \pi(R)^2\,\sigma = \pi R^2\sigma$$

For I_P, $I_R = I_T - I_C = \frac{3}{2}m_T\,(2R)^2 - \left[\frac{1}{2}m_C R^2 + m_C r^2\right]$

$$= \frac{3}{2}(4\pi R^2\sigma)(4R^2) - \left[\frac{1}{2}(\pi R^2\sigma)R^2 + (\pi R^2\sigma)(5R^2)\right]$$

$$= (18.5\,\pi R^4\sigma)$$

For I_O, $I_R = I_T - I_C = \frac{1}{2}m_T\,(2R)^2 - \frac{3}{2}m_C R^2$

$$= \frac{1}{2}(4\pi R^2\sigma)(4R^2) - \frac{3}{2}(\pi R^2\sigma)(R^2)$$

$$= 6.5\,\pi R^4\sigma$$

$$\therefore \quad \frac{I_P}{I_O} = \frac{18.5\pi R^4\sigma}{6.5\pi R^4\sigma}$$

$$= 2.846$$

Therefore, the nearest integer is 3.

8. Angular momentum of a particle about a point is given by

$$\mathbf{L} = \mathbf{r} \times \mathbf{p} = m\,(r \times v)$$

For $\mathbf{L}_O$,

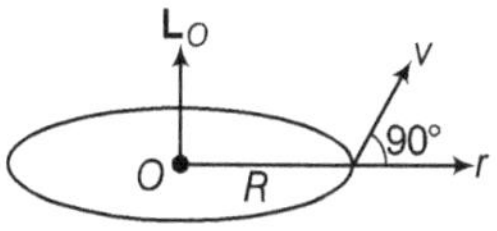

$|\mathbf{L}_O| = (mvr\sin\theta) = m(R\omega)(R)\sin 90° = mR^2\omega$ = constant.

Direction of $\mathbf{L}_O$ is always upwards. Therefore, complete $\mathbf{L}_O$ is constant, both in magnitude as well as direction.

For $\mathbf{L}_P$,

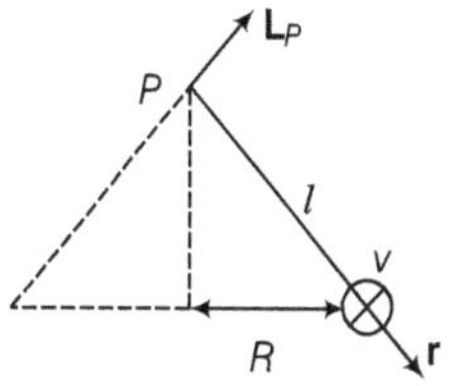

$$|\mathbf{L}_P| = (mvr\sin\theta) = (m)(R\omega)(l)\sin 90°$$

$$= (mRl\omega)$$

Magnitude of $\mathbf{L}_P$ will remain constant but direction of $\mathbf{L}_O$ keeps on changing.

9. Angular moment, $|\mathbf{L}|$ or $L = I\omega$ (about axis of rod)

Here, $m =$ mass of insect

$\therefore \quad L = (I_{rod} + mv^2t^2)\omega$

Now, $|\tau| = \dfrac{dL}{dt} = (2mv^2t\omega)$ or $|\tau| \propto t$

i.e. the graph is straight line passing through the origin.

After time T,

$$L = \text{constant}\,|\tau| \quad \text{or} \quad \frac{dL}{dt} = 0$$

10. Language of question is wrong, because relative speed is not the correct word. Relative speed between two is always zero. The correct word is magnitude of relative velocity.

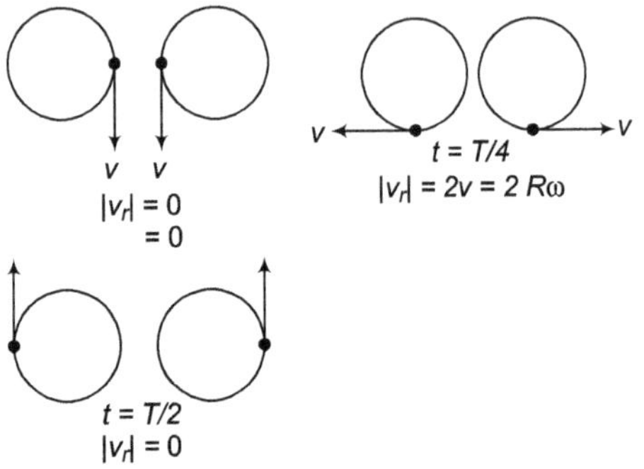

$$v = R\omega$$

Corresponding to above values the correct graph is (a).

11. The velocity of solid cylinder,

$$v = \sqrt{\frac{2gh}{1 + \frac{k^2}{R}}}$$

For solid cylinder, $\dfrac{k^2}{R} = 1/2$

$$\Rightarrow \quad v = \sqrt{\frac{4gh}{3}}$$

12. Centripetal force, $F = \dfrac{mv^2}{r}$

$F = ma \Rightarrow 30 = 3 \times a \Rightarrow a = 10\,\text{rad/s}^2$

The angular acceleration given that

$$\alpha = \frac{a}{r} = \frac{10}{40 \times 10^{-2}} = 25\,\text{rad/s}^2$$

13. Moment of inertia $= \dfrac{1}{2}MR^2 + mx^2$

where, $m =$ mass of insect

$x =$ distance of insect from centre

Clearly, as the insect moves along the diameter of the disc, mass of inertia first decreases, then increases.

By conservation of angular momentum, angular speed first increases, then decreases.

14. Rotational kinetic energy, $K_R = \dfrac{1}{2}I\omega^2$

$\therefore$ Its moment of inertia $= \dfrac{2K_R}{\omega^2} = \dfrac{2 \times 360}{(30)^2} = 0.8\,\text{kg-m}^2$

15. For, $F = 20t - 5t^2$

$$\Rightarrow \quad \omega = \frac{FR}{I} = 4t - t^2 \Rightarrow \frac{d\theta}{dt} = 4t - t^2$$

$$\Rightarrow \quad \int_0^{\omega} d\theta = \int_0^t (4t - t^2)\,dt$$

$$\Rightarrow \quad \omega = 2t^2 - \frac{t^3}{3}$$

When direction is reversed,

$$\omega = 0, \text{ i.e. } t = 0, 6\,\text{s}$$

Now, $d\theta = \omega\,dt$

$$\int_0^{\theta} d\theta = \int_0^6 \left(2t^2 - \frac{t^3}{3}\right) dt$$

$$\Rightarrow \quad \theta = \left[\frac{2t^3}{3} - \frac{t^4}{12}\right]_0^6$$

$$\Rightarrow \quad \theta = 144 - 108 = 36\,\text{rad}$$

$\therefore$ Number of rotations,

$$n = \frac{\theta}{2\pi} = \frac{36}{2\pi} < 6$$

16. For the motion of the block,

$$mg - T = ma \qquad \ldots\text{(i)}$$

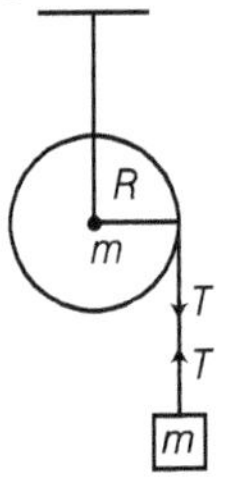

For the rotation of the pulley,

$$\tau = TR = I\alpha$$

$$\Rightarrow \quad T = \frac{1}{2}mR\alpha \qquad \ldots\text{(ii)}$$

As string does not slip on the pulley

then, $a = R\alpha$ ...(iii)

On solving Eqs. (i), (ii) and (iii), we get

$$a = \frac{2g}{3}$$

17. For a ring, $K^2 = r^2$, then

$$v^2 = \frac{2gh}{1 + \frac{K^2}{r^2}} \Rightarrow v^2 = \frac{2gh}{2} = gh \Rightarrow v = \sqrt{gh}$$

18. We know that,

$$L = I\omega \qquad \ldots\text{(i)}$$

$$L^2 = 2KI$$

From Eq. (i), we get

$$L^2 = 2K\frac{L}{\omega} \Rightarrow L = \frac{2K}{\omega} \Rightarrow L' = \frac{2\left(\frac{K}{2}\right)}{2\omega} = \frac{L}{4}$$

19. Total KE at bottom $= \dfrac{1}{2}mv^2\left[1 + \dfrac{K^2}{R^2}\right]$

$$= \frac{1}{2}mv^2\left[1 + \frac{2}{5}\right] = \frac{7}{10}mv^2$$

20. $\frac{L_{total}}{L_B} = \frac{(I_A + I_B)\omega}{I_B \cdot \omega}$ (as ω will be same in both cases)

$= \frac{I_A}{I_B} + 1 = \frac{m_A r_A^2}{m_B r_B^2} + 1 = \frac{r_A}{r_B} + 1$ (as, $m_A r_A = m_B r_B$)

$= \frac{11}{2.2} + 1 = 6$ $\left(\text{as } r \propto \frac{1}{m}\right)$

21. The moment of inertia of ring $= MR^2$

The moment of inertia of removed sector $= \frac{1}{4} MR^2$

The moment of inertia of remaining part

$$= MR^2 - \frac{1}{4} MR^2 = \frac{3}{4} MR^2 \quad \ldots\text{(i)}$$

According to question, the moment of inertia of the remaining part $= kMR^2$ ...(ii)

On comparing Eq. (i) and Eq. (ii), we get

$$k = \frac{3}{4}$$

22. Given, moment of inertia $= 2.5 \text{ kgm}^{-2}$

$\omega = 40 \text{ rad s}^{-1}$, $\omega = 60 \text{ rad s}^{-1}$, $\tau = 10 \text{ N-m}$

As, $\tau = I\alpha$

$\Rightarrow$ $10 = 2.5 \times \alpha$

$\Rightarrow$ $\alpha = 4 \text{ rad s}^{-2}$

Now, $\omega = \omega_0 + \alpha t$

$\Rightarrow$ $60 = 40 + 4 \times t$

$\Rightarrow$ $20 = 4t$

$\Rightarrow$ $t = 5 \text{ s}$

23. Angular velocity $= \omega$

Centripetal force, $F = mr\omega^2$ or $r \propto \frac{1}{\omega^2}$

$\therefore$ $\frac{r_1}{r_2} = \frac{\omega_2^2}{\omega_1^2}$

or $\frac{4}{r_2} = \frac{4\omega^2}{\omega^2}$

or $r_2 = 1 \text{ cm}$

24. $I_{CM} = \frac{ML^2}{12}$ (about middle-point)

$\therefore$ $I = I \quad + Mx^2$

$$= \frac{ML^2}{12} + M\left(\frac{L}{6}\right)^2$$

$$I = \frac{ML^2}{9}$$

25. From theorem of parallel axes,

$$I = I_{CM} + Mx^2 = \frac{ML^2}{12} + M\left[\frac{L}{4}\right]^2$$

$$= \frac{ML^2}{12} + \frac{ML^2}{16} = \frac{7ML^2}{48}$$

26. Mass for unit area $= \frac{M}{\pi(R^2 - r^2)}$

$\therefore$ Mass of whole disc $= \frac{M \cdot \pi r R^2}{\pi(R^2 - r^2)} = \frac{MR^2}{(R^2 - r^2)}$

MI of disc of radius, $r = \frac{1}{2} \frac{M\pi r^2}{\pi(R^2 - r^2)} \cdot r^2 = \frac{1}{2} \frac{Mr^4}{(R^2 - r^2)}$

MI of whole disc $= \frac{1}{2} \frac{Mr^2 \cdot r^2}{(R^2 - r^2)} = \frac{Mr^4}{2(R^2 - r^2)}$

$\therefore$ MI of annular disc $= \frac{M}{2}\left[\frac{R^4 - r^4}{R^2 - r^2}\right]$

$$= \frac{M(R^2 - r^2)}{2}$$

27. The correct relations all as below:

Torque = Moment of inertia × angular acceleration

Torque = dipole moment × magnetic field

Moment of inertia = torque ÷ angular acceleration

Given, momentum = mass × velocity

28. Moment of inertia of a disc about a diameter is given by

$$\frac{1}{4} MR^2 = I$$

$\Rightarrow$ $MR^2 = 4I$ (given)

10

Gravitation

10.1 Introduction

Why are planets, moon and the sun all nearly spherical? Why do some Earth satellites circle the Earth in 90 min, while the moon takes 27 days for the trip? And why don't satellites fall back to the Earth? The study of gravitation provides the answers for these and many related questions.

Gravitation is one of the four classes of interactions found in nature. These are given below:

(i) The gravitational force

(ii) The electromagnetic force

(iii) The strong nuclear force (also called the hadronic forces)

(iv) The weak nuclear forces

Although, of negligible importance in the interactions of elementary particles, gravity is of primary importance in the interactions of large objects. It is gravity that holds the universe together.

In this chapter, we will learn the basic laws that govern gravitational interactions.

Chapter Snapshot

- Introduction
- Newton's Law of Gravitation
- Acceleration due to Gravity
- Gravitational Field
- Gravitational Potential
- Gravitational Potential Energy
- Binding Energy
- Motion of Satellites
- Kepler's Laws

10.2 Newton's Law of Gravitation

Along with his three laws of motion, Newton published the law of gravitation in 1687. According to him, every particle of matter in the universe attracts every other particle with a force that is directly proportional to the product of the masses of the particles and inversely proportional to the square of the distance between them.

Thus, the magnitude of the gravitational force F between two particles m_1 and m_2 placed at a distance r is

$$F \propto \frac{m_1 m_2}{r^2} \quad \text{or} \quad F = G\frac{m_1 m_2}{r^2}$$

Here, G is a universal constant called gravitational constant whose magnitude is

$$G = 6.67\times 10^{-11} \text{ N-m}^2/\text{kg}^2$$
$$= 6.67\times 10^{-8} \text{ dyne/cm}^2\text{-g}^2$$
$$[G] = [\text{M}^{-1}\text{L}^3\text{T}^{-2}]$$

The direction of the force F is along the line joining the two particles.

Following three points are important regarding the gravitational force:

(i) Unlike the electrostatic force, it is independent of the medium between the particles.

(ii) It is conservative in nature.

(iii) It expresses the force between two point masses (of negligible volume). However, for external points of spherical bodies the whole mass can be assumed to be concentrated at its centre of mass.

Gravity

In Newton's law of gravitation, gravitation is the force of attraction between any two bodies. If one of the bodies is Earth, then the gravitation is called **gravity**. Hence, gravity is the force by which Earth attracts a body towards its centre. It is a special case of gravitation.

Example 10.1 *Spheres of the same material and same radius r are touching each other. Show that gravitational force between them is directly proportional to* r^4.

Sol.

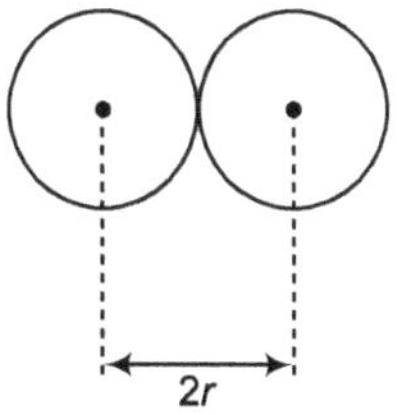

Fig. 10.1

$$m_1 = m_2 = (\text{volume}) \times (\text{density})$$

$$= \left(\frac{4}{3}\pi r^3\right)\rho$$

$$\therefore \quad F = \frac{Gm_1m_2}{r^2}$$

$$= \frac{G\left(\frac{4}{3}\pi r^3\right)\left(\frac{4}{3}\pi r^3\right)\rho^2}{r^2}$$

or $$F \propto r^4$$

10.3 Acceleration due to Gravity

When a body is dropped from a certain height above the ground, it begins to fall towards the earth under gravity. The acceleration produced in the body due to gravity is called the acceleration due to gravity. It is denoted by g. Its value close to the earth's surface is 9.8 m/s^2.

Suppose that the mass of the earth is M, its radius is R, then the force of attraction acting on a body of mass m close to the surface of the earth is

$$F = \frac{GMm}{R^2}$$

According to Newton's second law, the acceleration due to gravity

$$g = \frac{F}{m} = \frac{GM}{R^2}$$

This expression is free from m. If two bodies of different masses are allowed to fall freely, they will have the same acceleration, i.e. if they are allowed to fall from the same height, they will reach the earth simultaneously.

Variation in the Value of g

The value of g varies from place to place on the surface of earth. It also varies as we go above or below the surface of Earth. Thus, value of g depends on the following factors:

Shape of the Earth

The earth is not a perfect sphere. It is somewhat flat at the two poles. The equatorial radius is approximately 21 km more than the polar radius. And since,

$$g = \frac{GM}{R^2} \quad \text{or} \quad g \propto \frac{1}{R^2}$$

The value of g is minimum at the equator and maximum at the poles.

Height above the Surface of Earth

The force of gravity on an object of mass m at a height h above the surface of the earth is

$$F = \frac{GMm}{(R+h)^2}$$

$\therefore$ Acceleration due to gravity at this height will be

$$g' = \frac{F}{m} = \frac{GM}{(R+h)^2}$$

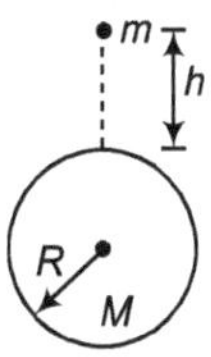

Fig. 10.2

This can also be written as

$$g' = \frac{GM}{R^2\left(1+\frac{h}{R}\right)^2}$$

or $$g' = \frac{g}{\left(1+\frac{h}{R}\right)^2} \quad \text{as} \quad \frac{GM}{R^2} = g$$

Thus, $$g' < g$$

i.e. the value of acceleration due to gravity g goes on decreasing as we go above the surface of the earth. Further,

$$g' = g\left(1+\frac{h}{R}\right)^{-2}$$

or $$g' \approx g\left(1-\frac{2h}{R}\right),$$

if $$h << R$$

Depth below the Surface of Earth

Let an object of mass m is situated at a depth h below the earth's surface. Its distance from the centre of the earth is $(R-h)$. This mass is situated at the surface of the inner solid sphere and lies inside the outer spherical shell. The gravitational force of attraction on a mass inside a spherical shell is always zero. Therefore, the object experiences gravitational attraction only due to inner solid sphere.

The mass of this sphere is

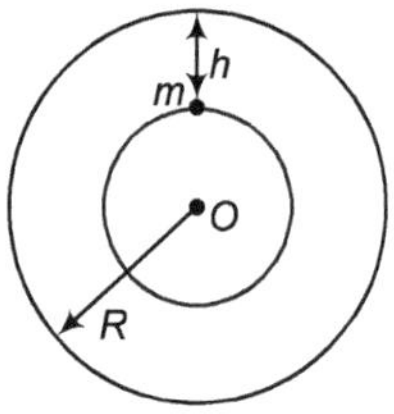

Fig. 10.3

$$M' = \left(\frac{M}{4/3\,\pi R^3}\right)\frac{4}{3}\pi\,(R-h)^3$$

or $$M' = \frac{(R-h)^3}{R^3}M$$

$$F = \frac{GM'm}{(R-h)^2} = \frac{GMm(R-h)}{R^3}$$

and $$g' = \frac{F}{m}$$

Substituting the values, we get

$$g' = g\left(1-\frac{h}{R}\right) \text{ i.e. } g' < g$$

- We can see from this equation that $g' = 0$ at $h = R$, i.e. acceleration due to gravity is zero at the centre of the earth.

Thus, the variation in the value of g with r (the distance from the centre of Earth) is as follows:

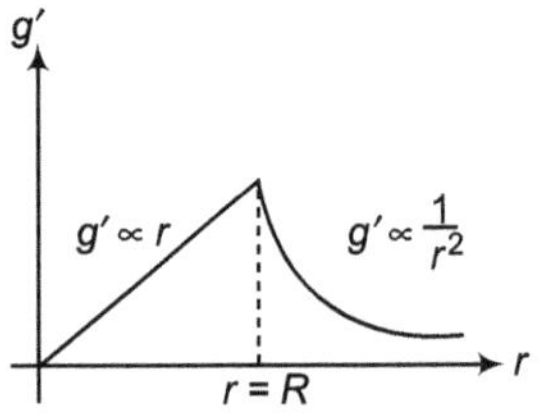

Fig. 10.4

For $r \le R$,

$$g' = g\left(1-\frac{h}{R}\right) = \frac{gr}{R}$$

as $$R - h = r \quad \text{or} \quad g' \propto r$$

For $r > R$, $$g' = \frac{g}{\left(1+\frac{h}{R}\right)^2} = \frac{gR^2}{r^2}$$

or $$g' \propto \frac{1}{r^2}$$

Extra Knowledge Points

- Acceleration due to Moon's gravity on Moon's surface is $\frac{g_e}{6}$, because

$$\frac{M_m}{R_m^2} \approx \frac{1}{6}\frac{M_e}{R_e^2} \qquad \left(g = \frac{GM}{R^2}\right)$$

While acceleration due to Earth's gravity on Moon's surface is approximately $\frac{g_e}{(60)^2}$ or $\frac{g_e}{3600}$. This is because distance of Moon from the Earth's centre is approximately equal to 60 times the radius of the Earth and $g \propto \frac{1}{r^2}$. This can be understood from the figure.

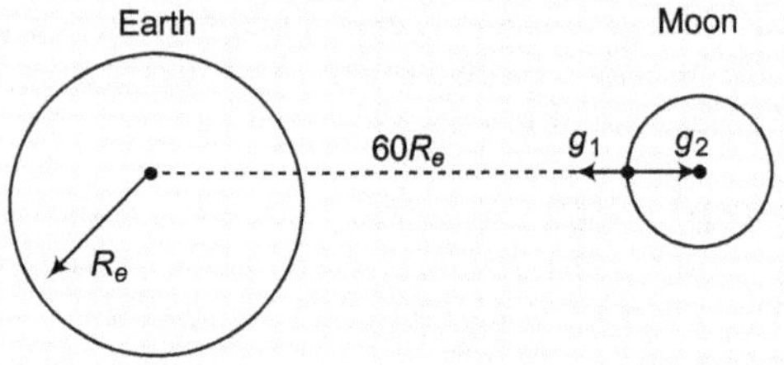

$$g_1 = \frac{g_e}{(60)^2} \quad \text{while} \quad g_2 = \frac{g_e}{6}$$

Axial Rotation of Earth

Let us consider a particle P at rest on the surface of the Earth, in latitude ϕ. Then, the pseudo force acting on the particle is $mr\omega^2$ in outward direction. The true acceleration g is acting towards the centre O of the Earth. Thus, the effective acceleration g' is the resultant of g and $r\omega^2$. After derivation we can find the following relation

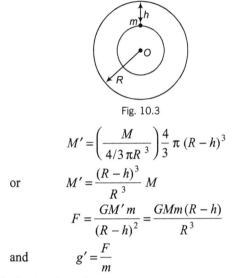

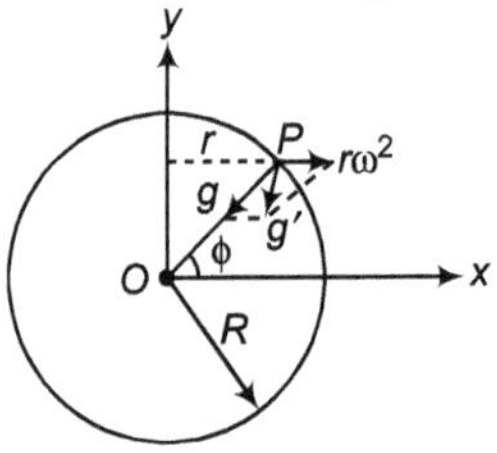

Fig. 10.5

$$g' = g - R\omega^2\cos^2\phi$$

Following conclusions can be drawn from the above discussion

(i) The effective value of g is not truely vertical.

(ii) The effect of centrifugal force due to rotation of Earth is to reduce the effective value of g.

(iii) At equators $\phi = 0°$.

Therefore, $g' = g - R\omega^2$ and at poles $\phi = 90°$,

Therefore, $g' = g$

Thus, at equator g' is minimum while at poles g' is maximum.

Example 10.2 *Assuming Earth to be a sphere of uniform mass density, how much would a body weigh half way down the centre of the Earth, if it weighed 100 N on the surface?*

Sol. Given, $mg = 100\text{ N}$

$$g' = g\left(1 - \frac{h}{R}\right)$$

$$\frac{h}{R} = \frac{1}{2}$$

$$\therefore \quad g' = g\left(1 - \frac{1}{2}\right) = \frac{g}{2}$$

$$\therefore \quad mg' = \frac{mg}{2} = \frac{100}{2} = 50\text{ N}$$

Example 10.3 *Suppose the Earth increases its speed of rotation. At what new time period will the weight of a body on the equator becomes zero? (Take $g = 10\ m/s^2$ and radius of Earth $R = 6400\ km$)*

Sol. The weight will become zero when

$$g' = 0$$

or $$g - R\omega^2 = 0 \quad (\text{on the equator } g' = g - R\omega^2)$$

or $$\omega = \sqrt{\frac{g}{R}}$$

$$\therefore \quad \frac{2\pi}{T} = \sqrt{\frac{g}{R}} \quad \text{or} \quad T = 2\pi\sqrt{\frac{R}{g}}$$

Substituting the values,

$$T = \frac{2\pi\sqrt{\frac{6400 \times 10^3}{10}}}{3600}\text{ h}$$

or $$T = 1.4\text{ h}$$

Thus, the new time period should be 1.4 h instead of 24 h for the weight of a body to be zero on the equator.

10.4 Gravitational Field

The space around a body in which any other body experiences a force of attraction is called the gravitational field of the first body.

The force experienced (both in magnitude and direction) by a unit mass placed at a point in a gravitational field is called the gravitational field strength or intensity of gravitational field at that point. Usually, it is denoted by **E**. Thus,

$$\mathbf{E} = \frac{\mathbf{F}}{m}$$

In Article 10.3, we have seen that acceleration due to gravity g is also $\frac{\mathbf{F}}{m}$. Hence, for the Earth's gravitational field g and **E** are same. Unless the effect of rotation of Earth about its own axis is to be considered. The E *versus* r (the distance from the centre of Earth) graph are same as that of g' *versus* r graph.

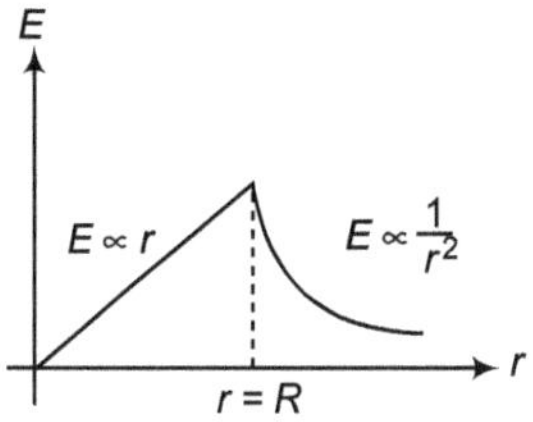

Fig. 10.6

Field due to a Point Mass

Suppose, a point mass M is placed at point O. We want to find the intensity of gravitational field **E** at a point P, a distance r from O. Magnitude of force F acting on a particle of mass m placed at P is

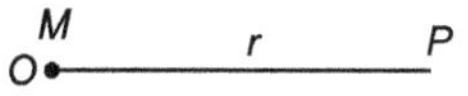

Fig. 10.7

$$F = \frac{GMm}{r^2}$$

$$\therefore \quad \mathbf{E} = \frac{F}{m} = \frac{GM}{r^2} \quad \text{or} \quad \mathbf{E} = \frac{GM}{r^2}$$

The direction of the force F and hence of E is from P to O as shown in Fig. 10.7

Gravitational Field due to a Uniform Solid Sphere

Field at an External Point

A uniform sphere may be treated as a single particle of same mass placed at its centre for calculating the gravitational field at an external point. Thus,

$$E(r) = \frac{GM}{r^2}$$

For $$r \geq R \quad \text{or} \quad E(r) \propto \frac{1}{r^2}$$

Here, r is the distance of the point from the centre of the sphere and R the radius of sphere.

Field at an Internal Point

The gravitational field due to a uniform sphere at an internal point is proportional to the distance of the point from the centre of the sphere. At the centre itself, it is zero and at surface it is $\frac{GM}{R^2}$, where R is the radius of the sphere. Thus,

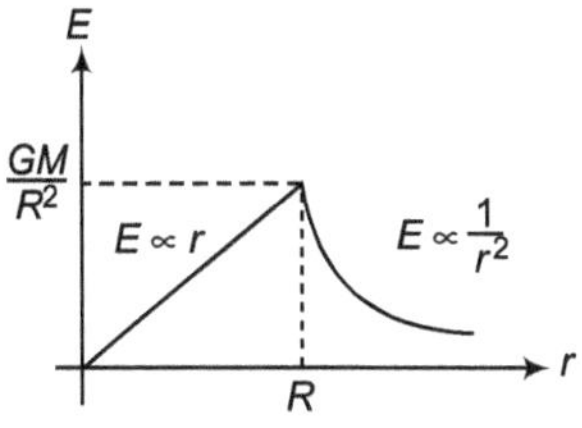

Fig. 10.8

$$E(r) = \frac{GM}{R^3}r,$$

for $$r \leq R \quad \text{or} \quad E(r) \propto r$$

Hence, E *versus* r graph is as shown in Fig. 10.8.

Field due to a Uniform Spherical Shell

At an External Point

For an external point, the shell may be treated as a single particle of same mass placed at its centre. Thus, at an external point the gravitational field is given by

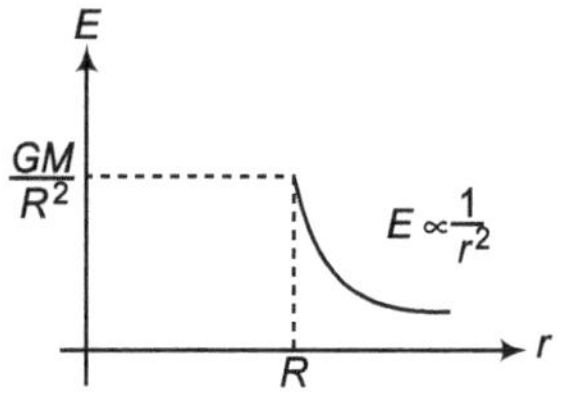

Fig. 10.9

$$E(r) = \frac{GM}{r^2}, \text{ for } r \geq R$$

At $r = R$ (the surface of shell)

$$E = \frac{GM}{R^2}$$

and otherwise $E \propto \frac{1}{r^2}$

At an Internal Point

The field inside a uniform spherical shell is zero.

Thus, *E versus r* graph is as shown in Fig. 10.9.

Field due to a Uniform Circular Ring at a Point on its Axis

Field strength at a point *P* on the axis of a circular ring of radius *R* and mass *M* is given by

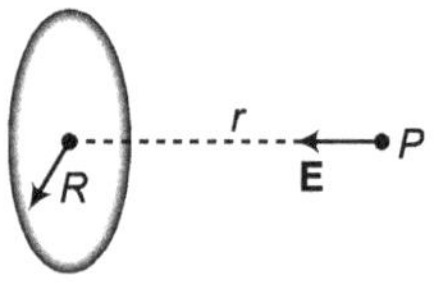

Fig. 10.10

$$E(r) = \frac{GMr}{(R^2 + r^2)^{3/2}}$$

This is directed towards the centre of the ring. It is zero at the centre of the ring and maximum at $r = \frac{R}{\sqrt{2}}$ (can be obtained by putting $\frac{dE}{dr} = 0$). Thus, *E-r* graph is as shown in Fig. 10.11.

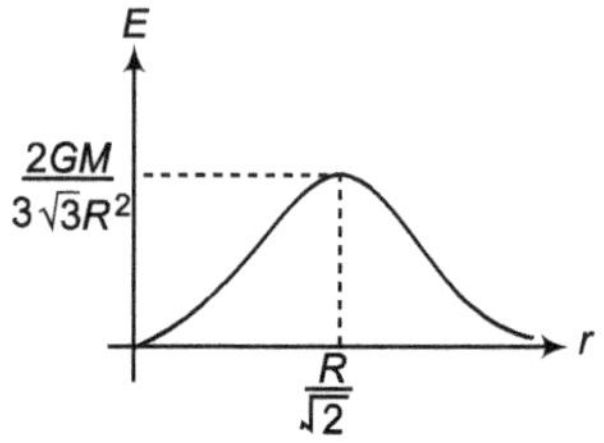

Fig. 10.11

The maximum value is $E_{max} = \frac{2GM}{3\sqrt{3}R^2}$.

10.5 Gravitational Potential

If a body is moved in a gravitational field from one place to the other either work is done against the gravitational attraction or it is obtained.

The work done in bringing a unit mass from infinity to a point in the gravitational field is called the 'gravitational potential' at that point.

This work is obtained (not done) by the agent in bringing the mass. The gravitational potential is denoted by *V*. So, let *W* joule of work is obtained in bringing a test mass *m* from infinity to some point, then gravitational potential at that point will be

$$V = \frac{W}{m}$$

Since, work is obtained, it is negative. Hence, gravitational potential is always negative.

Potential due to a Point Mass

Suppose a point mass *M* is situated at a point *O*. We want to find the gravitational potential due to this mass at a point *P* a distance *r* from *O*. For this let us find work done in taking the unit mass from *P* to infinity. This will be

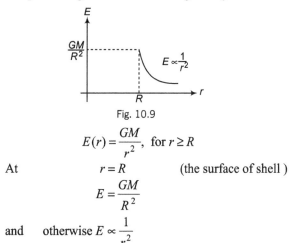

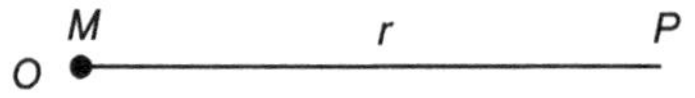

Fig. 10.12

$$W = \int_r^\infty F\,dr$$

$$= \int_r^\infty \frac{GM}{r^2}\,dr = \frac{GM}{r}$$

Hence, the work done in bringing unit mass from infinity to *P* will be $-\frac{GM}{r}$. Thus, the gravitational potential at *P* will be

$$V = -\frac{GM}{r}$$

Potential due to a Uniform Solid Sphere

Potential at an External Point

The gravitational potential due to a uniform sphere at an external point is same as that due to a single particle of same mass placed at its centre. Thus,

$$V(r) = -\frac{GM}{r}, \text{ where } r \geq R$$

At the surface, $r = R$

and $V = -\frac{GM}{R}$

Potential at an Internal Point

At some internal point, potential at a distance r from the centre is given by

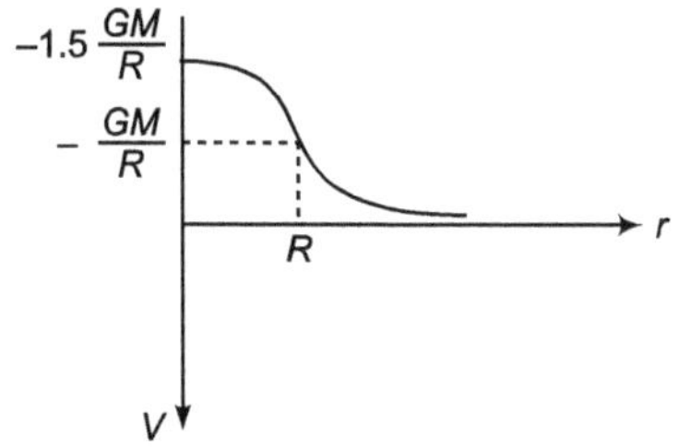

Fig. 10.13

$$V(r) = -\frac{GM}{R^3}(1.5R^2 - 0.5r^2) \quad r \le R$$

At $r = R$, $V = -\dfrac{GM}{R}$

while at $r = 0$, $V = -\dfrac{1.5\,GM}{R}$

i.e. at the centre of the sphere, the potential is 1.5 times the potential at surface. The variation of V *versus* r graph is as shown in Fig. 10.13.

Potential due to a Uniform Thin Spherical Shell

Potential at an External Point

To calculate the potential at an external point, a uniform spherical shell may be treated as a point mass of same magnitude at its centre. Thus, potential at a distance r is given by

$$V(r) = -\frac{GM}{r}, \text{ where } r \ge R$$

At $r = R$, $V = -\dfrac{GM}{R}$

Potential at an Internal Point

The potential due to a uniform spherical shell is constant throughout at any point inside the shell and this is equal to $-\dfrac{GM}{R}$. Thus, V-r graph for a spherical shell is as shown in Fig. 10.14.

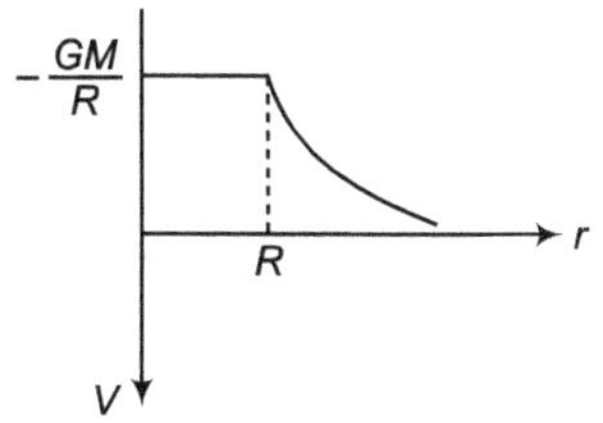

Fig. 10.14

Potential due to a Uniform Ring at a Point on its Axis

The gravitational potential at a distance r from the centre on the axis of a ring of mass M and radius R is given by

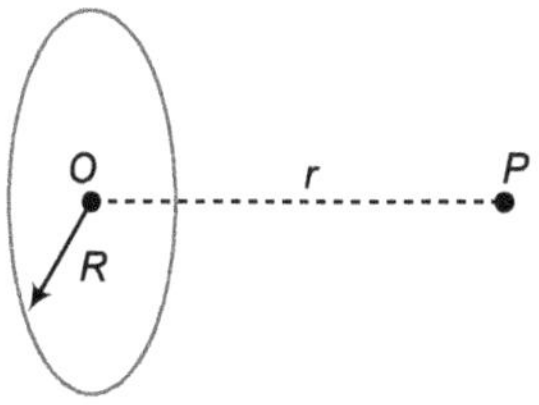

Fig. 10.15

$$V(r) = -\frac{GM}{\sqrt{R^2 + r^2}}, \quad 0 \le r \le \infty$$

At $r = 0$, $V = -\dfrac{GM}{R}$, i.e. at the centre of the ring gravitational potential is $-\dfrac{GM}{R}$.

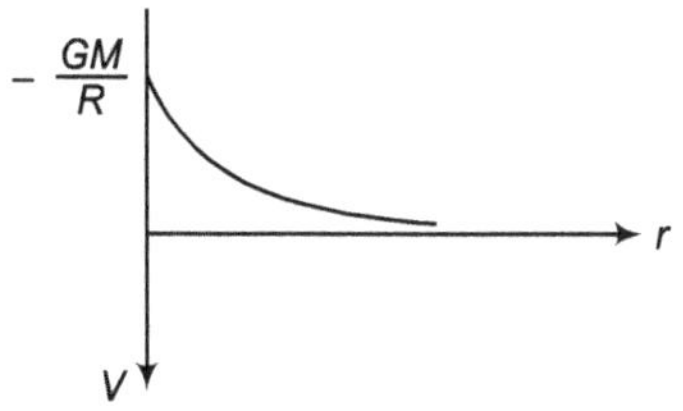

Fig. 10.16

The V-r graph is as shown in Fig. 10.16.

Example 10.4 *Two concentric spherical shells have masses m_1 and m_2 and radii r_1 and r_2 ($r_2 > r_1$). What is the force exerted by this system on a particle of mass m_3, if it is placed at a distance r ($r_1 < r < r_2$) from the centre?*

Sol. The outer shell will have no contribution in the gravitational field at point P.

$$\therefore \quad E_P = \frac{Gm_1}{r^2}$$

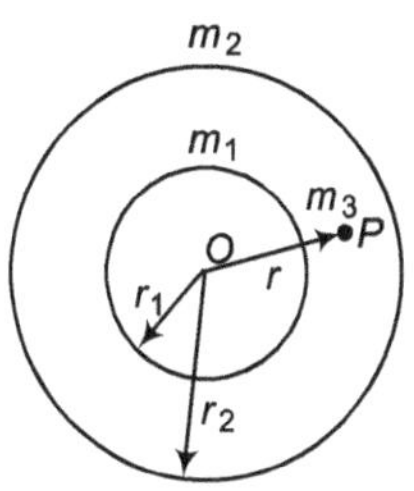

Fig. 10.17

Thus, force on mass m_3 placed at P is

$$F = (m_3 E_P) \quad \text{or} \quad F = \frac{Gm_1 m_3}{r^2}$$

The field $\mathbf{E}_P$ and the force $\mathbf{F}$ both are towards centre O.

Example 10.5 *A particle of mass 1 kg is kept on the surface of a uniform sphere of mass 20 kg and radius 1.0 m. Find the work to be done against the gravitational force between them to take the particle away from the sphere.*

Sol. Potential at the surface of sphere,

$$V = -\frac{GM}{R} = -\frac{(6.67\times 10^{-11})(20)}{1}\ \text{J/kg}$$
$$= -1.334\times 10^{-9}\ \text{J/kg}$$

i.e. 1.334×10^{-9} J work is obtained to bring a mass of 1 kg from infinity to the surface of sphere. Hence, the same amount of work will have to be done to take the particle away from the surface of sphere. Thus,

$$W = 1.334\times 10^{-9}\ \text{J}$$

10.6 Gravitational Potential Energy

The concept of potential energy has already been discussed in the chapter of work, energy and power. The word potential energy is defined only for a conservative force field. There we have discussed that the change in potential energy (dU) of a system corresponding to a conservative internal force is given by

$$dU = -\mathbf{F}\cdot d\mathbf{r}$$

or
$$\int_i^f dU = -\int_{\mathbf{r}_i}^{\mathbf{r}_f} \mathbf{F}\cdot d\mathbf{r}$$

or
$$U_f - U_i = -\int_{\mathbf{r}_i}^{\mathbf{r}_f} \mathbf{F}\cdot d\mathbf{r}$$

We generally choose the reference point at infinity and assume potential energy to be zero there, i.e. if we take $r_i = \infty$ (infinite) and $U_i = 0$, then we can write

$$U = -\int_{\infty}^{\mathbf{r}} \mathbf{F}\cdot d\mathbf{r} = -W$$

or potential energy of a body or system is negative of work done by the conservative forces in bringing it from infinity to the present position.

Gravitational Potential Energy of a two Particle System

The gravitational potential energy of two particles of masses m_1 and m_2 separated by a distance r is given by

$$U = -\frac{Gm_1m_2}{r}$$

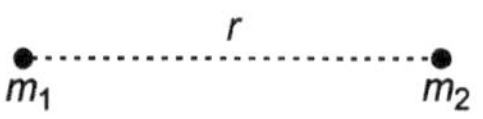

Fig. 10.18

This is actually the negative of work done in bringing those masses from infinity to a distance r by the gravitational forces between them.

Gravitational Potential Energy for a System of Particles

The gravitational potential energy for a system of particles (say m_1, m_2, m_3 and m_4) is given by

$$U = -G\left[\frac{m_4m_3}{r_{43}} + \frac{m_4m_2}{r_{42}} + \frac{m_4m_1}{r_{41}} + \frac{m_3m_2}{r_{32}} + \frac{m_3m_1}{r_{31}} + \frac{m_2m_1}{r_{21}}\right]$$

Thus, for a n particle system there are $\frac{n(n-1)}{2}$ pairs and the potential energy is calculated for each pair and added to get the total potential energy of the system.

Gravitational Potential Energy of a Body on Earth's Surface

The gravitational potential energy of mass m in the gravitational field of mass M at a distance r from it is

$$U = -\frac{GMm}{r}$$

The Earth behaves for all external points as, if its mass M were concentrated at its centre. Therefore, a mass m near Earth's surface may be considered at a distance R (the radius of Earth) from M. Thus, the potential energy of m due to Earth will be

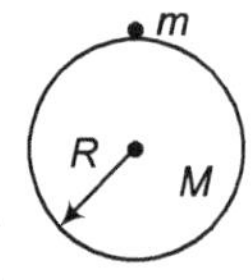

Fig. 10.19

$$U = -\frac{GMm}{R}$$

10.7 Binding Energy

Total mechanical energy (potential + kinetic) of a closed system is negative. The modulus of this total mechanical energy is known as the binding energy of the system. This is the energy due to which system is closed or different parts of the system are bound to each other.

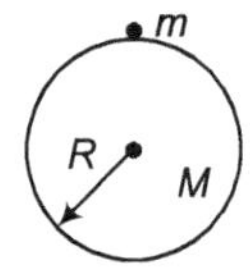

Fig. 10.20

Suppose, the mass m is placed on the surface of Earth. The radius of the Earth is R and its mass is M. Then, the kinetic energy of the particle $K = 0$ and potential energy of the particle is $U = -\frac{GMm}{R}$.

Therefore, the total mechanical energy of the particle is

$$E = K + U = 0 - \frac{GMm}{R}$$

or
$$E = -\frac{GMm}{R}$$

$\therefore$ Binding energy $= |E| = \frac{GMm}{R}$

It is due to this energy, the particle is attached with the Earth. If minimum this much energy is supplied to the particle in any form (normally kinetic) the particle no longer remains bound to the Earth. It goes out of the gravitational field of Earth.

Escape Velocity

As we discussed the binding energy of a particle on the surface of Earth kept at rest is $\frac{GMm}{R}$. If this much energy in the form of kinetic energy is supplied to the particle, it leaves the gravitational field of the Earth. So, if v_e is the escape velocity of the particle, then

$$\frac{1}{2}mv_e^2 = \frac{GMm}{R}$$

or $$v_e = \sqrt{\frac{2GM}{R}} \quad \text{or} \quad v_e = \sqrt{2gR}$$

as $$g = \frac{GM}{R^2}$$

Substituting the value of $g\ (9.8\ \text{m/s}^2)$ and $R\ (6.4 \times 10^6\ \text{m})$, we get

$$v_e \approx 11.2\ \text{km/s}$$

Thus, the minimum velocity needed to take a particle infinitely away from the earth is called the escape velocity. On the surface of the earth, its value is 11.2 km/s.

Extra Knowledge Points

- Let us find the difference in potential energy of a mass m in two positions shown in figure. The potential energy of the mass on the surface of Earth (at B) is

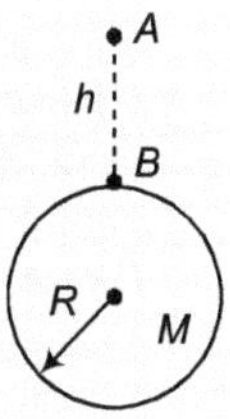

$$U_B = -\frac{GMm}{R}$$

and potential energy of mass m at height h above the surface of Earth (at A) is

$$U_A = -\frac{GMm}{R+h} \qquad (U_A > U_B)$$

$$\therefore \quad \Delta U = U_A - U_B = -\frac{GMm}{R+h} - \left(-\frac{GMm}{R}\right)$$

$$= GMm\left(\frac{1}{R} - \frac{1}{R+h}\right) = \frac{GMmh}{R(R+h)}$$

$$= \frac{GMmh}{R^2\left(1+\frac{h}{R}\right)} \qquad \left(\frac{GM}{R^2} = g\right)$$

$$\therefore \quad \Delta U = \frac{mgh}{1+\frac{h}{R}}$$

For $h << R$, $\Delta U \approx mgh$

Thus, what we read the mgh is actually, the difference in potential energy (not the absolute potential energy), that too for $h << R$.

- Maximum height attained by a particle. Suppose a particle of mass m is projected vertically upwards with a speed v and we want to find the maximum height h attained by the particle. Then, we can use conservation of mechanical energy, i.e.

Decrease in kinetic energy = increase in gravitational potential energy of particle.

$$\therefore \quad \frac{1}{2}mv^2 = \Delta U$$

or $$\frac{1}{2}mv^2 = \frac{mgh}{1+\frac{h}{R}}$$

Solving this, we get

$$h = \frac{v^2}{2g - \frac{v^2}{R}}$$

From this we can see that

(i) if $v = v_e$ or $v^2 = v_e^2 = 2gR$, $h = \infty$ and if

(ii) v is small, $h = \frac{v^2}{2g}$

Both the results are quite obvious.

Example 10.6 *Three masses of* 1 kg, 2 kg *and* 3 kg *are placed at the vertices of an equilateral triangle of side* 1 m. *Find the gravitational potential energy of this system.*
(Take, $G = 6.67 \times 10^{-11}$ *N-m²/kg²)*

Sol. $U = -G\left(\frac{m_3m_2}{r_{32}} + \frac{m_3m_1}{r_{31}} + \frac{m_2m_1}{r_{21}}\right)$

Here, $r_{32} = r_{31} = r_{21} = 1.0$ m

$m_1 = 1$ kg

$m_2 = 2$ kg

and $m_3 = 3$ kg

Substituting in above, we get

$$U = -(6.67 \times 10^{-11})\left(\frac{3\times 2}{1} + \frac{3\times 1}{1} + \frac{2\times 1}{1}\right)$$

or $$U = -7.337 \times 10^{-10}\ \text{J}$$

Example 10.7 *Calculate the escape velocity from the surface of Moon. The mass of the Moon is* 7.4×10^{22} *kg and radius* $= 1.74 \times 10^6$ *m.*

Sol. Escape velocity from the surface of Moon is

$$v_e = \sqrt{\frac{2GM_m}{R_m}}$$

Substituting the values, we have

$$v_e = \sqrt{\frac{2\times 6.67\times 10^{-11} \times 7.4\times 10^{22}}{1.74\times 10^6}}$$

$$= 2.4 \times 10^3\ \text{m/s} \quad \text{or} \quad 2.4\ \text{km/s}$$

Example 10.8 *A particle is projected from the surface of Earth with an initial speed of 4.0 km/s. Find the maximum height attained by the particle. Radius of Earth = 6400 km, and g = 9.8 m/s².*

Sol. The maximum height attained by the particle is

$$h = \frac{v^2}{2g - \frac{v^2}{R}}$$

Substituting the values, we have

$$h = \frac{(4.0\times 10^3)^2}{2\times 9.8 - \frac{(4.0\times 10^3)^2}{6.4\times 10^6}}$$

$$= 9.35\times 10^5 \text{ m}$$

or $h \approx 935$ km

10.8 Motion of Satellites

Just as the planets revolve around the sun, in the same way few celestial bodies revolve around these planets. These bodies are called **satellites**. For example, the Moon, is a satellite of the Earth. Artificial satellites are launched from the Earth. Such satellites are used for telecommunication, weather forecast and other applications. The path of these satellites are elliptical with the centre of the Earth at a focus. However, the difference in major and minor axes is so small that they can be treated as nearly circular for not too sophisticated calculations. Let us derive certain characteristics of the motion of satellites by assuming the orbit to be perfectly circular.

Orbital Speed

The necessary centripetal force to the satellite is being provided by the gravitational force exerted by the Earth on the satellite. Thus,

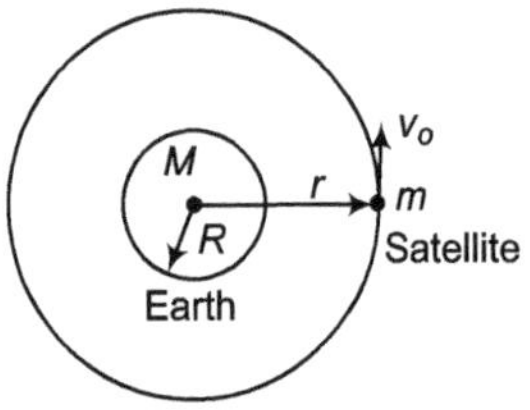

Fig. 10.22

$$\frac{mv_o^2}{r} = \frac{GMm}{r^2}$$

$$\therefore \quad v_o = \sqrt{\frac{GM}{r}} \quad \text{or} \quad v_o \propto \frac{1}{\sqrt{r}}$$

Hence, the orbital speed (v_o) of the satellite decreases as the orbital radius (r) of the satellite increases. Further, the orbital speed of a satellite close to the Earth's surface ($r \approx R$) is

$$v_o = \sqrt{\frac{GM}{R}} = \sqrt{gR} = \frac{v_e}{\sqrt{2}}$$

Substituting the values of G, M and R. Then,

$$v_e = 11.2 \text{ km/s}$$

$$v_o = 7.9 \text{ km/s}$$

Period of Revolution

The period of revolution (T) is given by

$$T = \frac{2\pi r}{v_o} \quad \text{or} \quad T = \frac{2\pi r}{\sqrt{\frac{GM}{r}}}$$

$$\text{or} \quad T = 2\pi\sqrt{\frac{r^3}{GM}} \quad \text{or } T = 2\pi\sqrt{\frac{r^3}{gR^2}} \quad (\text{as } GM = gR^2)$$

From this expression of T, we can make the following conclusions.

(i) $T \propto r^{3/2}$

or $T^2 \propto r^3$

(which is also the Kepler's third law)

(ii) Time period of a satellite very close to the Earth's surface ($r \approx R$) is

$$T = 2\pi\sqrt{\frac{R}{g}}$$

Substituting the values, we get

$$T \approx 84.6 \text{ min}$$

(iii) Suppose, the height of a satellite is such that the time period of the satellite is 24 h and it moves in the same sense as the earth. The satellite will always be overhead a particular place on the equator. As seen from the earth, this satellite will appear to be stationary. Such a satellite is called a **geostationary satellite**. Putting $T = 24$ h in the expression of T, the radius of geostationary satellite comes out to be $r = 4.2\times 10^4$ km. The height above the surface of the earth is about 3.6×10^4 km.

Energy of Satellite

The potential energy of the system is

$$U = -\frac{GMm}{r}$$

The kinetic energy of the satellite is

$$K = \frac{1}{2}mv_o^2$$

$$= \frac{1}{2}m\left(\frac{GM}{r}\right)$$

or

$$K = \frac{1}{2}\frac{GMm}{r}$$

The total energy is

$$E = K + U = -\frac{GMm}{2r}$$

or $$E = -\frac{GMm}{2r}$$

This energy is constant and negative, i.e. the system is closed. The farther the satellite from the Earth the greater its total energy.

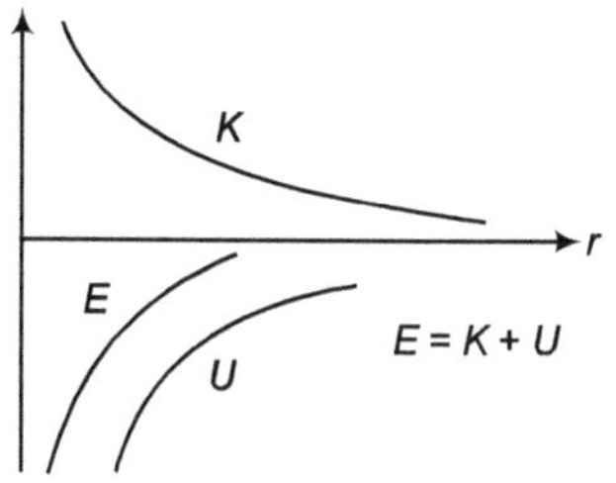

Fig. 10.23

Extra Knowledge Points

- Total energy of a closed system is always negative. For example, energy of planet-sun, satellite-Earth or electron-nucleus system are always negative.
- If the law of force obeys the inverse square law .

$$\left(F \propto \frac{1}{r^2},\ F = \frac{-dU}{dr}\right)$$

$$K = \frac{|U|}{2} = |E|$$

It is also same for electron-nucleus system, because there also, the electrostatic force $F_e \propto \frac{1}{r^2}$.

- The areal velocity of a planet is constant (Kepler's second law) and is given by

$$\frac{dA}{dt} = \frac{L}{2m}$$

Here, L is the angular momentum of the planet about the sun.

Example 10.9 *A spaceship is launched into a circular orbit close to the Earth's surface. What additional velocity has now to be imparted to the spaceship in the orbit to overcome the gravitational pull? Radius of Earth = 6400 km, $g = 9.8\ m/s^2$.*

Sol. The speed of the spaceship in a circular orbit close to the Earth's surface is given by

$$v_o = \sqrt{gR}$$

and escape velocity is given by

$$v_e = \sqrt{2gR}$$

∴ Additional velocity required to escape

$$v_e - v_o = \sqrt{2gR} - \sqrt{gR}$$

$$= (\sqrt{2} - 1)\sqrt{gR}$$

Substituting the values of $\sqrt{2}$ g and R, we get

$$v_e - v_o = 3.278 \times 10^3 \text{ m/s}$$

10.9 Kepler's Laws

Kepler discovered three empirical laws that accurately described the motions of the planets. The three laws may be stated as

(i) Each planet moves in an elliptical orbit, with the sun at one focus of the ellipse. This law is also known as the **law of elliptical orbits** and obviously gives the shape of the orbits of the planets round the sun.

(ii) The radius vector, drawn from the sun to a planet, sweeps out equal areas in equal time, i.e. its areal velocity (or the area swept out by it per unit time) is constant.

This is referred to as the **law of areas** and gives the relationship between the orbital speed of the planet and its distance from the sun.

(iii) The square of the planet's time period is proportional to the cube of the semi-major axis of its orbit. This is known as the **harmonic law** and gives the relationship between the size of the orbit of a planet and its time of revolution.

Chapter Summary with Formulae

1. **Gravitational Force between Two Point Masses is**

$$F = G\frac{m_1 m_2}{r^2}$$

2. **Acceleration Due to Gravity**

(i) On the surface of Earth $g = \frac{GM}{R^2} = 9.81 \text{ ms}^{-2}$

(ii) At height h from the surface of the Earth,

$$g' = \frac{g}{\left(1 + \frac{h}{R}\right)^2} \approx g\left(1 - \frac{2h}{R}\right)$$

if $h << R$

(iii) At depth d from the surface of the Earth,

$$g' = g\left(1 - \frac{d}{R}\right)$$

$g' = 0$ if $d = R$, i.e. at centre of the Earth

3. **Field Strength**

(i) Gravitational field strength at a point in gravitational field is defined as

$\mathbf{E} = \frac{\mathbf{F}}{m}$ = gravitational force per unit mass.

(ii) **Due to a point mass**

$$E = \frac{GM}{r^2} \quad \text{(towards the mass)}$$

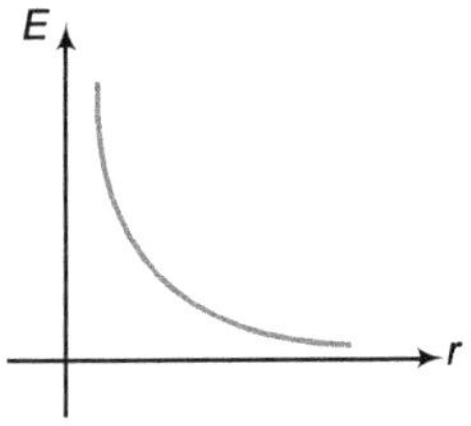

or $$E = \frac{1}{r^2}$$

(iii) Due to a solid sphere

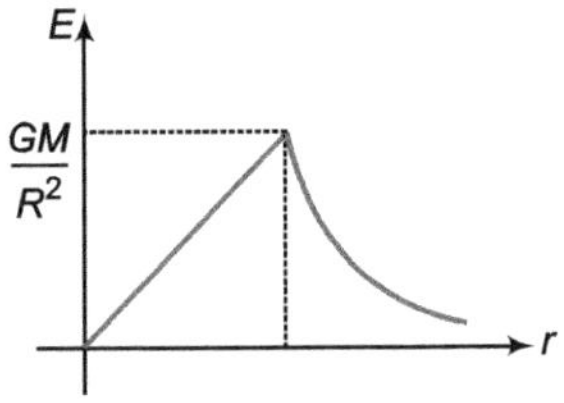

Inside points, $E_i = \frac{GM}{R^3} r$

At $r = 0, E = 0$ (at centre),

At $r = R, E = \frac{GM}{R^2}$, i.e. on surface

Outside points, $E_o = \frac{GM}{r^2}$ or $E_o \propto \frac{1}{r^2}$

At $r = R, E = \frac{GM}{R^2}$, i.e. on surface

As, $R \to \infty, E \to 0$

On the surface, E-r graph is continuous.

(iv) **Due to a spherical shell**

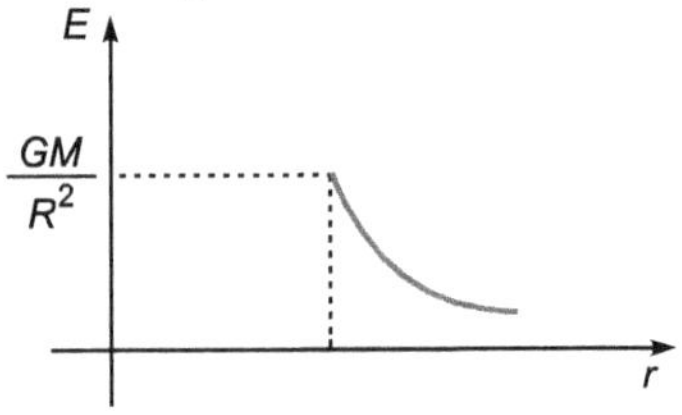

Inside points, $E_i = 0$

Outside points, $E_o = \frac{GM}{r^2}$

Just outside the surface, $E = \frac{GM}{R^2}$

On the surface, E - r graph is discontinuous.

4. **Gravitational Potential**

(i) Gravitational potential at a point in a gravitational field is defined as the negative of work done by gravitational force in moving a unit mass from infinity to that point. Thus,

$$V_P = -\frac{W_{\infty \to P}}{m}$$

(ii) **Due to a point mass**

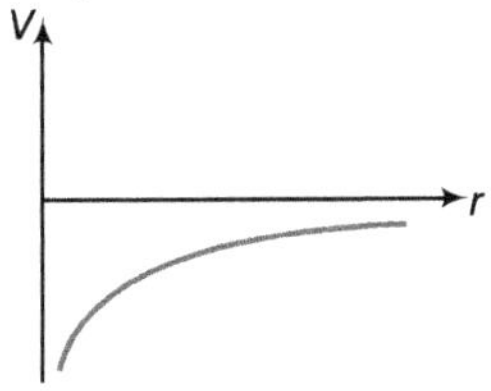

$$V = -\frac{Gm}{r}$$

$V \to -\infty$

as $r \to 0$ and $V \to 0$

as $r \to \infty$

(iii) **Due to a solid sphere**

Inside points, $V_i = -\frac{GM}{R^3}(1.5R^2 - 0.5r^2)$

At $r = R$, $V = -\frac{GM}{R}$, i.e. on surface

At $r = 0$, $V = -1.5\frac{GM}{R}$, i.e. at centre

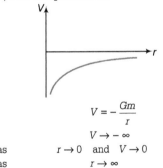

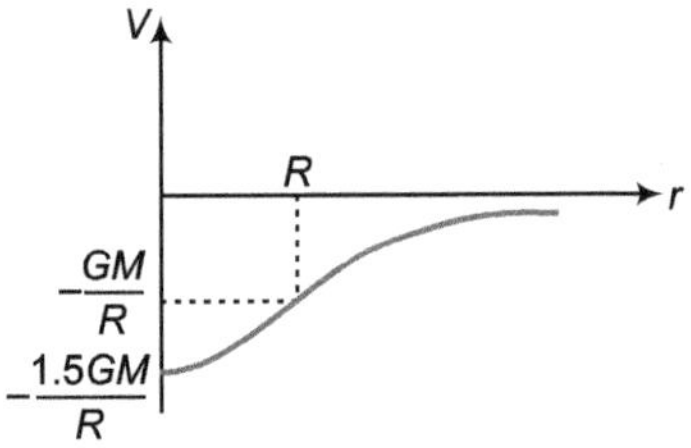

V-r graph is parabolic for inside points and potential at centre is 1.5 times the potential at surface.

Outside points, $V_o = -\frac{GM}{r}$

At $r = R$, $V = -\frac{GM}{R}$, i.e. on surface

As $r \to \infty, V \to 0$

(iv) **Due to a spherical shell**

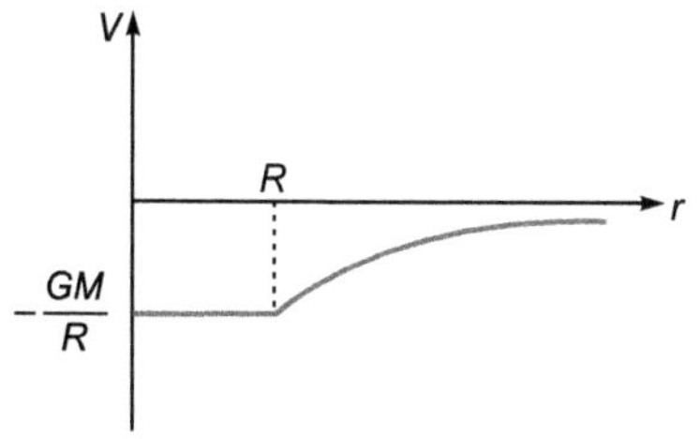

Inside points, $V_i = -\frac{GM}{R}$ = constant

Outside points, $V_o = -\frac{GM}{r}$

5. **Gravitational Potential Energy**

(i) This is negative of work done by gravitational forces in making, the system from infinite separation to the present position.

(ii) Gravitational potential energy of two point masses is

$$U = -\frac{Gm_1m_2}{r}$$

(iii) For finding gravitational potential energy is more than two point masses we have to make pairs of masses. Neither of the pair should be repeated. For example, in case of four point masses,

$$U = -G\left[\frac{m_4m_3}{r_{43}} + \frac{m_4m_2}{r_{42}} + \frac{m_4m_1}{r_{41}} + \frac{m_3m_2}{r_{32}} + \frac{m_3m_1}{r_{31}} + \frac{m_2m_1}{r_{21}}\right]$$

For n point masses, total number of pairs will be $\frac{n(n-1)}{2}$.

(iv) If a point mass m is placed on the surface of Earth, the potential energy here is

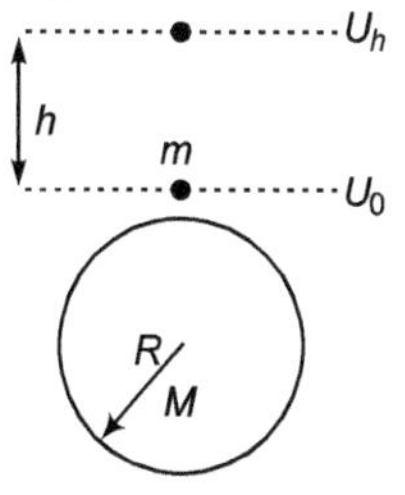

$$U_0 = -\frac{GMm}{R}$$

and potential energy at height h is

$$U_h = -\frac{GMm}{(R+h)}$$

The difference in potential energy would be

$$\Delta U = U_h - U_0 \quad \text{or} \quad \Delta U = \frac{mgh}{1 + \frac{h}{R}}$$

If $h << R$, $\Delta U \approx mgh$

6. **Escape Velocity**

(i) From the surface of Earth,

$$v_e = \sqrt{2gR} = \sqrt{\frac{2GM}{R}} \quad \left(\text{as } g = \frac{GM}{R^2}\right)$$

$$\approx 11.2 \text{ kms}^{-1}$$

(ii) Escape velocity does not depend upon the angle at which particle is projected from the surface.

7. **Motion of Satellites**

(i) Orbital speed, $v_o = \sqrt{\frac{GM}{r}}$

(ii) Time period, $T = \frac{2\pi}{\sqrt{GM}} r^{3/2}$

(iii) Kinetic energy, $K = \frac{GMm}{2r}$

(iv) Potential energy, $U = -\frac{GMm}{r}$

(v) Total mechanical energy,

$$E = -\frac{GMm}{2r}$$

Near the surface of the Earth, $r \approx R$ and

$$v_o = \sqrt{\frac{GM}{R}} = \sqrt{gR} = 7.9 \text{ kms}^{-1}.$$

This is the maximum speed of Earth's satellite.

Time period of such a satellite would be

$$T = \frac{2\pi}{\sqrt{GM}} R^{3/2} = 2\pi\sqrt{\frac{R}{g}}$$

$$= 84.6 \text{ min}$$

This is the minimum time period of any Earth's satellite.

8. **Kepler's Laws**

Kepler's three empirical laws describe the motion of the planets.

First law Each planet moves in an elliptical orbit with the sun at one focus of the ellipse.

Second law The radius vector drawn from the sun to a planet, sweeps out equal areas in equal time interval, i.e. areal velocity is constant. This law is derived from law of conservation of angular momentum.

$$\frac{dA}{dt} = \frac{L}{2m} = \text{constant}.$$

Here, L is angular momentum and m is mass of planet.

Third law $T^2 \propto r^3$, where r is semi-major axis of elliptical path.

Circle is a special case of an ellipse. Therefore, second and third laws can also be applied for circular path. In third law, r is radius of circular path.

Additional Examples

Example 1. *Moon has got no atmosphere on it. Why?*

Sol. The escape velocity in case of Moon is 2.38 km/s and is comparable to the rms velocity of all the constituents of atmosphere like O_2, N_2, CO_2 and water vapours.

Example 2. *Moon is continuously revolving around the Earth without falling towards it. Why?*

Sol. It is on account of the fact that the gravitational force of attraction between Moon and Earth is used up in providing, the necessary centripetal force for keeping the Moon in an orbit around the Earth.

Example 3. *A person sitting in a satellite of Earth feels weightlessness but a person standing on Moon does not feel weightlessness though Moon is also a satellite of Earth.*

Sol. When a person is sitting in an artificial satellite of Earth, the gravitational attraction on him due to Earth provides the necessary centripetal force. But when he is standing on Moon, the gravitational attraction acting on him due to Moon, accounts for his weight on Moon, because mass of Moon is large and it applies considerable force on person.

Example 4. *Is it possible to shield a body from gravitational effect?*

Sol. No, it is not possible to shield a body from gravitational effects.

Example 5. *What is the difference between inertial mass and gravitational mass of a body?*

Sol. The inertial mass of a body is a measure of its inertia and is given by the ratio of the external force applied on it to the acceleration produced in it on the other hand. The gravitational mass of a body is a measure of the gravitational pull acting on it due to the Earth. This is the ratio of gravitational force acting on it and the acceleration produced by it.

Example 6. *If suddenly the gravitational force of attraction between the Earth and a satellite revolving around it becomes zero, what will happen to the satellite?*

Sol. If the gravitational force of the Earth suddenly becomes zero, the satellite will stop revolving around the Earth and it will move in a direction tangential to its original orbit with a speed with which it was revolving around the Earth.

Example 7. *What is the fractional decrease in the value of free-fall acceleration g for a particle when it is lifted from the surface to an elevation h ? (h < < R)*

Sol. $\because g = \dfrac{GM}{R^2}$

$$\frac{dg}{dR} = \frac{-2GM}{R^3}$$

$$\Rightarrow \quad \frac{dg}{h} = \frac{-2GM}{R^2}\cdot\frac{1}{R} \Rightarrow \frac{dg}{g} = -2\left(\frac{h}{R}\right)$$

Example 8. *Two concentric shells of mass M_1 and M_2 are concentric as shown. Calculate the gravitational force on m due to M_1 and M_2 at points P,Q and R.*

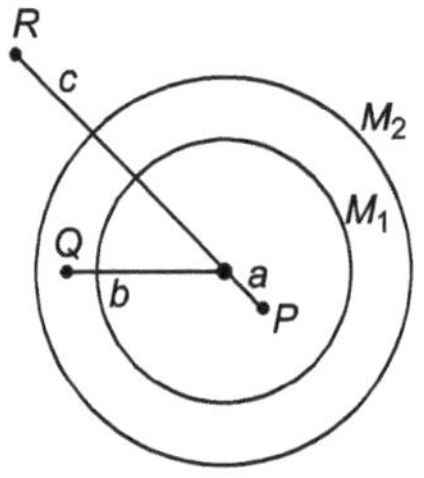

Sol. At P, $\quad F = 0$

At Q, $\quad F = \dfrac{GM_1m}{b^2}$

At R, $\quad F = \dfrac{G(M_1 + M_2)m}{c^2}$

Example 9. *Assuming the radius of the Earth to be 6.4×10^6 m.*

(a) Calculate the time period T of a satellite for equatorial orbit at 1.4×10^3 km above the surface of the Earth.

(b) What is the speed of the satellite in this orbit?

Sol. (a) $r = R_{\text{earth}} + h$

$$= (6.4\times 10^6 + 1.4\times 10^6)\ \text{m} \Rightarrow r = 7.8\times 10^6\ \text{m}$$

$$T = \left[\frac{4\pi^2 r^3}{GM_{\text{earth}}}\right]$$

$$= 6831\ \text{s}$$

(b) Speed of satellite, $v = \sqrt{\dfrac{GM_{\text{earth}}}{r}}$

$$= 7174\ \text{m/s}$$

Example 10. *Three particles each of mass m are located at the vertices of an equilateral triangle of side a. At what speed must they move, if they all revolve under the influence of their gravitational force of attraction in a circular orbit, circumscribing the triangle while still preserving the equilateral triangle?*

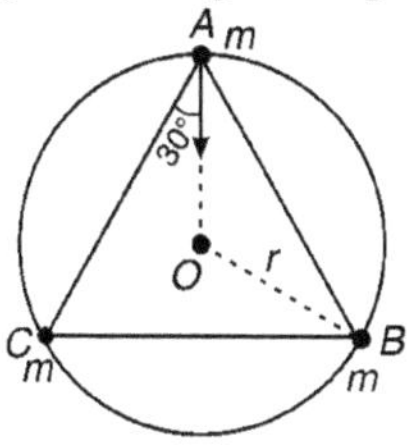

Sol. $F_A = F_{AB} + F_{AC}$

$$= 2\left[\frac{GM^2}{a^2}\right]\cos 30°$$

$$= \left[\frac{GM^2}{a^2}\cdot\sqrt{3}\right]$$

$$r = \frac{a}{\sqrt{3}}$$

Now, $$\frac{mv^2}{r} = F$$

or $$\frac{mv^2\sqrt{3}}{a} = \frac{GM^2}{a^2}\sqrt{3}$$

$\therefore$ $$v = \sqrt{\frac{GM}{a}}$$

Example 11. *A planet of mass m revolves in an elliptical orbit around the sun so that its maximum and minimum distances from the sun are equal to r_a and r_p respectively. Find the angular momentum of this planet relative to the sun.*

Sol. Using conservation of angular momentum,

$$mv_pr_p = mv_ar_a$$

As, velocities are perpendicular to the radius vectors at apogee and perigee.

$\Rightarrow$ $$v_pr_p = v_ar_a$$

Using conservation of energy,

$$-\frac{GMm}{r_p} + \frac{1}{2}mv_p^2 = \frac{-GMm}{r_a} + \frac{1}{2}mv_a^2$$

By solving, the above equations,

$$v_p = \sqrt{\frac{2GMr_a}{r_p(r_p + r_a)}}$$

$$L = mv_pr_p$$

$$= m\sqrt{\frac{2GMr_pr_a}{(r_p + r_a)}}$$

Example 12. *Three concentric shells of masses M_1, M_2 and M_3 having radii a,b and c respectively are situated as shown in figure. Find the force on a particle of mass m.*

(a) When the particle is located at Q.

(b) When the particle is located at P.

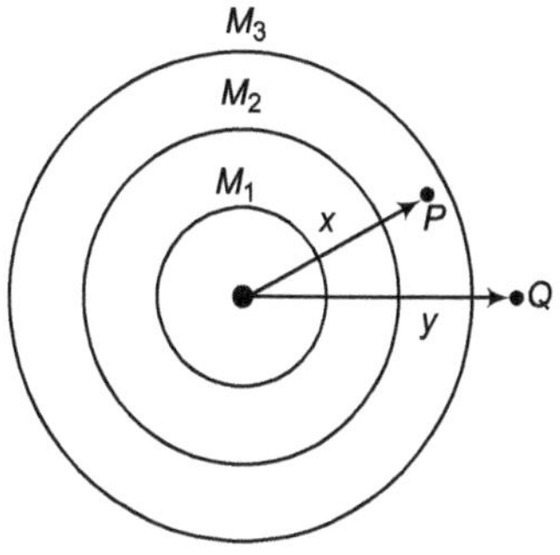

Sol. Attraction at an external point due to spherical shell of mass M is $\left(\frac{GM}{r^2}\right)$ while at an internal point is zero.

(a) Point is external to shell M_1, M_2 and M_3,

So, force at Q will be

$$F_Q = \frac{GM_1m}{y^2} + \frac{GM_2m}{y^2} + \frac{GM_3m}{y^2}$$

$$= \frac{Gm}{y^2}(M_1 + M_2 + M_3)$$

(b) Force at P will be

$$F_P = \frac{GM_1m}{x^2} + \frac{GM_2m}{x^2} + 0$$

$$= \frac{Gm}{x^2}(M_1 + M_2)$$

Example 13. *If the radius of the Earth contracts to half of its present value without change in its mass, what will be the new duration of the day?*

Sol. Present angular momentum of the Earth

$$L_1 = I\omega = \frac{2}{5}MR^2\omega$$

New angular momentum, because of change in radius

$$L_2 = \frac{2}{5}M\left(\frac{R}{2}\right)^2\omega'$$

If external torque is zero, then angular momentum must be conserved

$$L_1 = L_2$$

$$\frac{2}{5}MR^2\omega = \frac{1}{4}\times\frac{2}{5}MR^2\omega'$$

i.e. $$\omega' = 4\omega$$

$$T' = \frac{1}{4}T = \frac{1}{4}\times 24 = 6\text{ h}$$

NCERT Selected Questions

Q 1. Answer the following:

(a) You can shield a charge from electrical forces by putting it inside a hollow conductor. Can you shield a body from gravitational influence of nearby matter by putting it inside a hollow sphere or by some other means?

(b) An astronaut inside a small spaceship orbiting around the earth cannot detect gravity. If the space station orbiting around the earth has a large size, can he hope to detect gravity?

Sol. (a) No, induction effects take place in electrostatics not in gravitation.

(b) Yes, he can hope to detect gravity. If the size of the spaceship is extremely large, then the magnitude of the gravity will become appreciable and hence the gravitational effect of the spaceship may become measurable.

Q 2. Choose the correct alternatives.

(a) Acceleration due to gravity increases/decreases with increasing altitude.

(b) Acceleration due to gravity increases/decreases with increasing depth (assume the earth to be a sphere of uniform density).

(c) Acceleration due to gravity is independent of the mass of the earth/mass of the body.

(d) The formula $-GMm\left(\frac{1}{r_2}-\frac{1}{r_1}\right)$ is more or less accurate than the formula $mg(r_2-r_1)$ for the difference of potential energy between two points r_2 and r_1 distance away from the centre of the earth.

Sol. (a) Acceleration due to gravity decreases with increasing altitude.

(b) Acceleration due to gravity decreases with increasing depth.

(c) Acceleration due to gravity is independent of the mass of the body.

(d) The formula $-GMm\left(\frac{1}{r_2}-\frac{1}{r_1}\right)$ is more accurate than the formula $mg(r_2-r_1)$ for the difference of potential energy between two points r_1 and r_2 distance away from the centre of the earth.

Q 3. Suppose there existed a planet that went around the sun twice as fast as the earth. What would be its orbital size as compared to that of the earth?

Sol. From the relation, $T^2 \propto R^3$ we have

$$\left(\frac{T'}{T}\right)^{2/3}=\left(\frac{r'}{r}\right)$$

or $$r'=\left(\frac{1}{2}\right)^{2/3} r$$

or $$r'=\left(\frac{1}{4}\right)^{1/3} r$$

$$=\frac{1}{(4)^{1/3}}r=\frac{r}{(4)^{1/3}}=\frac{r}{1.59}=0.63r$$

Therefore, orbital size of planet = 0.63 times the orbital size of the earth.

Q 4. If one of the satellites of the jupiter has an orbital period of 1.769 days and the radius of the orbit is 4.22×10^8 m. Show that mass of the jupiter is about one thousandth that of the sun.

Sol. Using the relation,

$$\frac{GM}{r^3}=\omega^2,$$

$$\frac{GM_J}{r^3}=\left(\frac{2\pi}{T}\right)^2$$

or $$M_J=\frac{4\pi^2r^3}{T^2G}$$

$$=\frac{4\times9.87\times(4.22\times10^8)^3}{(15.2841\times10^4)^2\times6.67\times10^{-11}}$$

$$=1.9\times10^{27}\text{ kg}$$

$$\simeq 2\times10^{27}\text{ kg}$$

$\therefore$ $$\frac{M_J}{M_S}=\frac{2\times10^{27}}{2\times10^{30}}\approx\frac{1}{1000}$$

or $$M_J\approx\frac{1}{1000}M_S$$

i.e. mass of the jupiter is about one thousandth of the mass of the sun.

Q 5. Let us assume that our galaxy consists of 2.5×10^{11} stars each of one solar mass. How long will a star at a distance of 50000 light years from the galactic centre take to complete one revolution? Take the diameter of the Milky way to be 10^5 light years.

Sol. One solar mass = 2×10^{30} kg

1 light year = 9.46×10^{15} m

Let M = Mass of stars in the galaxy

$= 2.5\times10^{11}\times2\times10^{30}$ kg

$= 5\times10^{41}$ kg

r = Radius of orbit of a star

= Distance of a star from galactic centre

$= 50000$ light years

$= 50000 \times 9.46 \times 10^{15}$m

Diameter of Milky way = 10^5 light years

Using the relation,

$$\frac{GM}{r^3} = \omega^2 = \left(\frac{2\pi}{T}\right)^2, \text{ we get}$$

$$T^2 = \frac{4\pi^2 r^3}{GM} = \frac{4 \times 9.87 \times (5 \times 9.46 \times 10^{19})^3}{6.67 \times 10^{-11} \times 5 \times 10^{41}}$$

$$= 12527.5 \times 10^{28} \text{ s}^2$$

$$\therefore \quad T = 111.93 \times 10^{14} \text{ s}$$

$$= \frac{111.93 \times 10^{14}}{365 \times 24 \times 3600} \text{ yr} = 3.55 \times 10^8 \text{ yr}$$

Q 6. Choose the correct alternative.

(a) If the zero of potential energy is at infinity, the total energy of an orbiting satellite is negative of the kinetic/potential energy.

(b) The energy required to rocket an orbiting satellite out of the earth's gravitational influence is more/less than the energy required to project a stationary object at the same height (as the satellite) out of the Earth's influence.

Sol. (a) Kinetic energy (b) Less.

Q 7. Does the escape velocity of a body from the earth depend on (a) the mass of the body, (b) the height from where it is projected, (c) the direction of projection?

Sol. (a) No, we know that the escape velocity of the body is given by

$v_e = \sqrt{\frac{2GM}{R}}$, where M and R are mass and radius of the earth. Thus clearly, it does not depend on the mass of the body.

(b) Yes.

(c) No, it does not depend on the direction of projection.

Q 8. A comet orbits the sun in highly elliptical orbit. Does the comet has a constant (a) linear speed, (b) angular speed, (c) angular momentum, (d) kinetic energy, (e) potential energy and (f) total energy throughout its orbit? Neglect any mass loss of the comet when it comes very close to the sun.

Sol. (a) The comet moves faster when it is close to the sun and moves slower when it is farther away from the sun. Hence, the linear speed of the comet does not remain constant.

(b) Angular speed of the comet is not constant.

(c) The angular momentum of the comet remains constant.

(d) Since, the linear speed of the comet around the sun changes continuously, so its kinetic energy also changes continuously.

(e) Potential energy depends upon the distance between the sun and the comet, hence PE changes in the elliptical orbit as the distance between sun and the comet changes continuously.

(f) As total energy is the sum of KE and PE, the total energy always remains constant.

Q 9. A rocket is fired from the earth towards the sun. At what distance from the earth's centre is the gravitational force on the rocket zero? Mass of the sun = 2×10^{30} kg, mass of the earth = 6.0×10^{24} kg. Neglect the effect of the other planets, etc. (Orbital radius = 1.5×10^{11} m)

Sol. Let P be a point at a distance r from the earth's centre where gravitational force due to sun and the earth are equal and opposite and hence gravitational force on the rocket is zero.

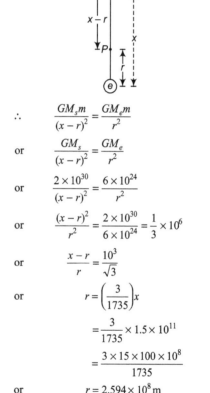

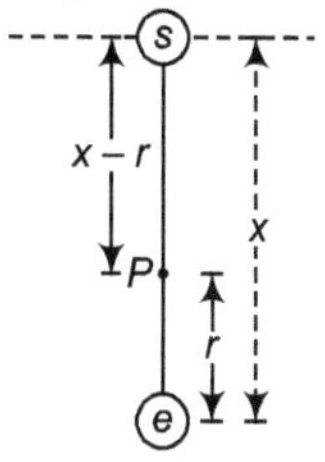

$$\therefore \quad \frac{GM_s m}{(x-r)^2} = \frac{GM_e m}{r^2}$$

$$\text{or} \quad \frac{GM_s}{(x-r)^2} = \frac{GM_e}{r^2}$$

$$\text{or} \quad \frac{2 \times 10^{30}}{(x-r)^2} = \frac{6 \times 10^{24}}{r^2}$$

$$\text{or} \quad \frac{(x-r)^2}{r^2} = \frac{2 \times 10^{30}}{6 \times 10^{24}} = \frac{1}{3} \times 10^6$$

$$\text{or} \quad \frac{x-r}{r} = \frac{10^3}{\sqrt{3}}$$

$$\text{or} \quad r = \left(\frac{3}{1735}\right)x$$

$$= \frac{3}{1735} \times 1.5 \times 10^{11}$$

$$= \frac{3 \times 15 \times 100 \times 10^8}{1735}$$

$$\text{or} \quad r = 2.594 \times 10^8 \text{ m}$$

$$= 2.6 \times 10^8 \text{ m from the earth.}$$

Q 10. How will you weigh the sun, i.e. estimate its mass? You will need to know the period of one of its planets and the radius of the planetary orbit. The mean orbital radius of the Earth around the Sun is 1.5×10^8 km. Estimate the mass of the Sun.

Sol. We know that the earth revolves around the sun in an orbit of radius 1.5×10^{11}m and completes one revolution around the sun in 365 days.

$\therefore$ R = Radius of orbit of the earth $= 1.5 \times 10^{11}$ m

T = Time period of earth around the sun

$= 365$ days $= 365 \times 24 \times 60 \times 60$ s

$G = 6.67 \times 10^{-11}\ \text{Nm}^2\text{kg}^{-2}$

We know that

$$\frac{R}{T^2} = \frac{GM_s}{4\pi^2R^2}$$

or $$M_s = \frac{4\pi^2R^3}{T^2G}$$

$$= \frac{4 \times 9.87 \times (1.5 \times 10^{11})^3}{(365 \times 24 \times 60 \times 60)^2 \times 6.67 \times 10^{-11}}\ \text{kg}$$

or $$M_s = 2.0 \times 10^{30}\ \text{kg}$$

Q 11. A saturn year is 29.5 times the earth year. How far is the saturn from the sun, if the earth is 1.50×10^8 km away from the sun?

Sol. According to the Kepler's third law,

$$T^2 \propto R^3$$

$\therefore$ $$\frac{T_s^2}{T_e^2} = \frac{R_s^3}{R_e^3} \text{ or } \left(\frac{T_s}{T_e}\right)^2 = \left(\frac{R_s}{R_e}\right)^3$$

Now, $T_s = 29.5\,T_e$ (given) or $\frac{T_s}{T_e} = 29.5$

and R_e = Distance of earth from sun $= 1.5 \times 10^8$ km

$$(29.5)^2 = \left(\frac{R_s}{1.5 \times 10^8}\right)^3$$

or $$R_s^3 = (29.5)^2 \times (1.5 \times 10^8)^3$$

or $$R_s^3 = 2.937 \times 10^{27}\ \text{km}^3$$

$\therefore$ $$R_s = (2.937 \times 10^{27}\ \text{km})^{1/3}$$

$$= 1.43 \times 10^9\ \text{km} = 1.43 \times 10^{12}\ \text{m}$$

Q 12. A body weighs 63 N on the surface of the earth. What is the gravitational force on it due to the earth at a height equal to half of the radius of the earth?

Sol. $g_h = g\left(1 + \frac{h}{R}\right)^{-2}$

Here, $$h = \frac{R}{2}$$

$\therefore$ $$g_h = g\left(1 + \frac{R/2}{R}\right)^{-2}$$

or $$g_h = \frac{4}{9}g$$

$$w = mg = 63\text{N} \quad \text{...(given)}$$

and $$w_h = mg_h = m \times \frac{4}{9}g = \frac{4}{9}mg$$

$$= \frac{4}{9} \times 63 = 28\ \text{N}$$

$$w_h = 28\ \text{N}$$

Q 13. Assuming the earth to be a sphere of uniform mass density, how much would a body weigh half way down to the centre of the earth, if it weighed 250 N on the surface?

Sol. $g_d = g\left(1 - \frac{d}{R}\right)$

Here, $$d = \frac{R}{2}$$

$$g_d = g\left(1 - \frac{R/2}{R}\right) = g\left(1 - \frac{1}{2}\right)$$

$$= \frac{g}{2}$$

$\therefore$ $$w_d = mg_d = m \times \frac{g}{2}$$

$$= \frac{1}{2}mg = \frac{1}{2}w$$

$$= \frac{1}{2} \times 250$$

$$= 125\ \text{N}$$

$\therefore$ Weight of the body half way down to the centre of earth $= 125$ N.

Q 14. A rocket is fired vertically with a speed of 5 km s^{-1} from the earth's surface. How far from the earth does the rocket go before returning to the earth? Mass of the earth $= 6.0 \times 10^{24}$ kg, mean radius of earth $= 6.4 \times 10^6$ m, $G = 6.67 \times 10^{-11}\ \text{Nm}^2\text{kg}^{-2}$

Sol. $g = \frac{GM}{R^2} = 9.8\ \text{m/s}^2$

Maximum height is given by

$\therefore$ $$h = \frac{Rv^2}{2gR - v^2}$$

Here, $v = 5\,\text{km s}^{-1} = 5000\ \text{ms}^{-1}$ (given)

$$R = 6.4 \times 10^6\ \text{m}$$

Putting these values, we get

$$h = \frac{6.4 \times 10^6 \times (5 \times 10^3)^2}{2 \times 9.8 \times 6.4 \times 10^6 - (5 \times 10^3)^2}$$

$$= 1.6 \times 10^6\text{m} = 1600\ \text{km}$$

$\therefore$ Distance from centre of the earth

$$= R + h = 6.4 \times 10^6 + 1.6 \times 10^6$$

$$= 8.0 \times 10^6\text{m}$$

Q 15. The escape speed of a projectile on the earth's surface is 11.2 km s^{-1}. A body is projected out with thrice this speed. What is the speed of the body far away from the earth? Ignore the presence of the sun and other planets.

Sol. According to the principle of conservation of the energy

Initial KE + Initial PE = Final KE + Final PE

or $$\frac{1}{2}mv^2 - \frac{GMm}{R} = \frac{1}{2}mv'^2 + 0$$

or $$\frac{1}{2}mv'^2 = \frac{1}{2}mv^2 - \frac{GMm}{R} \quad \text{...(i)}$$

Also let, v_e = escape velocity

$$\frac{1}{2}mv_e^2 = \frac{GMm}{R} \quad \text{...(ii)}$$

∴ From Eqs. (i) and (ii), we get

$$\frac{1}{2}mv'^2 = \frac{1}{2}mv^2 - \frac{1}{2}mv_e^2 \quad \text{...(iii)}$$

or $$v'^2 = v^2 - v_e^2$$

Now, $$\left.\begin{array}{l} v_e = 11.2 \text{ kms}^{-1} \\ v = 3v_e \end{array}\right\} \quad \text{...(iv) (given)}$$

∴ From Eqs. (iii) and (iv), we get

$$v'^2 = (3v_e)^2 - v_e^2$$
$$= 9v_e^2 - v_e^2 = 8v_e^2$$
$$= 8 \times (11.2)^2$$

or $$v' = \sqrt{8 \times (11.2)^2} = \sqrt{8} \times 11.2$$
$$= 2 \times 1.414 \times 11.2$$
$$= 31.68 \text{ kms}^{-1}$$

∴ Speed of the body far away from the earth,

$$v' = 31.68 \text{ kms}^{-1} = 31.7 \text{ kms}^{-1}$$

Q 16. Two stars each of 1 solar mass (2×10^{30} kg) are approaching towards each other for a head on collision. When they are at a distance 10^9 km, their speeds are negligible. What is the speed with which they collide? The radius of each star is 10^4 km. Assume the stars to remain undistorted untill they collide. (Use the known value of G)

Sol. Here, mass of each star, $M = 2 \times 10^{30}$ kg = one solar mass.

Initial distance between two stars, $r = 10^9$ km $= 10^{12}$m

Size of each star = radius of each star, $R = 10^4$ km $= 10^7$ m

Let v = speed with which they just collide with each other.

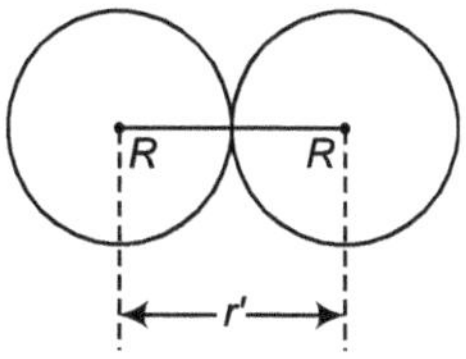

∴ Total final KE of stars at the time of collision

$$= \frac{1}{2}Mv^2 + \frac{1}{2}Mv^2 = Mv^2$$

∴ According to the law of conservation of energy.

Total initial energy = Final KE + Final PE

or $$-\frac{GM^2}{r} = Mv^2 - \frac{GM^2}{2R}$$

or $$v^2 = GM\left(\frac{1}{2R} - \frac{1}{r}\right)$$

$$= 6.67 \times 10^{-11} \times 2 \times 10^{30}\left(\frac{1}{2 \times 10^7} - \frac{1}{10^{12}}\right)$$
$$= 13.34 \times 10^9 \,(5 \times 10^{-8} - 10^{-12})$$
$$= 13.34 \times 10^{19} \times 4.9999 \times 10^{-8}$$

or $$v = \sqrt{6.6699 \times 10^{12}} \text{ or } v = 2.583 \times 10^6 \text{ms}^{-1}$$
$$= 2.6 \times 10^6 \text{ms}^{-1}$$

Q 17. Two heavy spheres each of mass 100 kg and radius 0.10 m are placed 1.0 m apart on a horizontal table. What is the gravitational force and potential at the mid-point of the line joining the centres of the spheres? Is an object placed at that point in equilibrium? If so, is the equilibrium stable or unstable?

Sol.

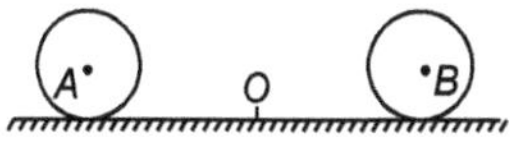

Net force on any mass placed at the centre of two spheres is zero.

If V be the total gravitational potential at point O, i.e. mid-point of the line joining the centres of two spheres, then

$$V = V_A + V_B$$
$$= -\frac{G \times 100}{(0.5)} + \left(-\frac{G \times 100}{(0.5)}\right)$$
$$= -\frac{2 \times G \times 100}{0.5}$$
$$= \frac{-2 \times 6.67 \times 10^{-11} \times 100}{0.5}$$
$$= -2.7 \times 10^{-8} \text{ J kg}^{-1}$$
$$= -27 \times 10^{-9} \text{ Jkg}^{-1}$$

An object placed at the mid-point is in an unstable equilibrium.

Q 18. As you have learnt in the text, a geostationary satellite orbits the earth at a height of nearly 36000 km from the surface of the earth. What is the potential due to the earth's gravity at the site of this satellite? (Take the potential energy at infinite to be zero.) Mass of earth $= 6.0 \times 10^{24}$kg, radius = 6400 km.

Sol. Mass of the earth, $M = 6.0 \times 10^{24}$ kg

Radius of the earth, $R = 6400$ km

Distance of the satellite from earth's surface, $h = 36000$ km

∴The gravitational potential due to earth's gravity at the site of the satellite $$= \frac{-GM}{R+h} = -\frac{6.67 \times 10^{-11} \times 6 \times 10^{24}}{(6400 + 3600) \times 10^3}$$
$$= -\frac{40.02 \times 10^{13}}{425 \times 10^5} \text{Jkg}^{-1}$$
$$= -9.43 \times 10^6 \text{ Jkg}^{-1}$$

Objective Problems

[Level 1]

Gravitational Force, Acceleration due to Gravity

1. The mass of a planet is twice the mass of the earth and diameter of the planet is thrice the diameter of the earth, then the acceleration due to gravity on the planet's surface

(a) $g/2$ (b) $2g$ (c) $2g/9$ (d) $3g/\sqrt{2}$

2. If radius of the earth is R, then the height h at which the value of g becomes one-fourth, is

(a) $2R$ (b) $3R$ (c) R (d) $4R$

3. Weight of a body is maximum at

(a) poles (b) equator
(c) centre of the earth (d) at latitude 45°

4. The depth d, at which the value of acceleration due to gravity becomes $\frac{1}{n}$ times the value at the surface, is

(R = radius of the earth)

(a) $\frac{R}{n}$ (b) $R\left(\frac{n-1}{n}\right)$
(c) $\frac{R}{n^2}$ (d) $R\left[\frac{n}{n+1}\right]$

5. If a planet consists of a satellite whose mass and radius were both half that of the earth, the acceleration due to gravity at its surface would be

(a) 4.9 m/s^2 (b) 9.8 m/s^2
(c) 19.6 m/s^2 (d) 29.4 m/s^2

6. Two point masses each equal to 1 kg attract one another with a force of 10^{-9} kg-wt. The distance between the two point masses is approximately

($G = 6.6 \times 10^{-11}$ MKS units)

(a) 8 cm (b) 0.8 cm
(c) 80 cm (d) 0.08 cm

7. A body has a weight 72 N. When it is taken to a height $h = R$ = radius of the earth, it would weight

(a) 72 N (b) 36 N
(c) 18 N (d) zero

8. The angular speed of the earth in rad/s so that the object on equator may appear weightlessness, is (radius of earth = 6400 km)

(a) 1.23×10^{-3}
(b) 6.20×10^{-3}
(c) 1.56
(d) 1.23×10^{-5}

9. The weight of a body at the centre of the earth is

(a) zero
(b) infinite
(c) same as on the surface of the earth
(d) None of the above

10. If the change in the value of g at a height h above the surface of earth is the same as at a depth d below it (both h and d are much smaller than the radius of the earth), then

(a) $d = h$ (b) $d = 2h$
(c) $d = h/2$ (d) $d = h^2$

11. The rotation of the earth having radius R about its axis speeds up to a value such that a man at latitude angle 60° feels weightlessness. The duration of the day in such a case is

(a) $\pi\sqrt{\frac{R}{g}}$ (b) $\frac{\pi}{2}\sqrt{\frac{R}{g}}$
(c) $\frac{\pi}{3}\sqrt{\frac{R}{g}}$ (d) $\pi\sqrt{\frac{g}{R}}$

12. Gravitational force between a point mass m and M separated by a distance is F. Now, if a point mass $2m$ is placed next to m in contact with it, the force on M due to m and the total force on M are

(a) $2F, F$ (b) F, 2F
(c) $F, 3F$ (d) F, F

13. The weights of an object in a coal mine, at sea level and at top of mountains are w_1, w_2 and w_3 respectively, then

(a) $w_1 < w_2 > w_3$ (b) $w_1 = w_2 = w_3$
(c) $w_1 < w_2 < w_3$ (d) $w_1 > w_2 > w_3$

14. When a body is taken from the equator to the poles, its weight

(a) remains constant
(b) increases
(c) decreases
(d) increases at N-pole and decreases at S-pole

15. If the earth suddenly shrinks (without changing mass) to half of its present radius, the acceleration due to gravity will be

(a) $g/2$ (b) $4g$
(c) $g/4$ (d) $2g$

16. The diameters of two plantes are in the ratio 4:1 and their mean densities in the ratio 1:2. The acceleration due to gravity on the planets will be in ratio

(a) 1 : 2 (b) 2 : 3
(c) 2 : 1 (d) 4 : 1

17. If density of the earth increased 4 times and its radius become half of what it is, our weight will?
(a) Be four times its present value
(b) Be doubled
(c) Remain same
(d) Be halved

18. The acceleration due to gravity near the surface of a planet of radius R and density d is proportional to
(a) $\frac{d}{R^2}$ (b) dR^2 (c) dR (d) $\frac{d}{R}$

19. If earth is supposed to be sphere of radius R, if $g_{30°}$ is value of acceleration due to gravity at latitude of 30° and g at the equator, the value of $g - g_{30°}$ is
(a) $\frac{1}{4}\omega^2 R$ (b) $\frac{3}{4}\omega^2 R$
(c) $\omega^2 R$ (d) $\frac{1}{2}\omega^2 R$

20. If M is the mass of the earth and R its radius, the ratio of the gravitational acceleration and the gravitational constant is
(a) $\frac{R^2}{M}$ (b) $\frac{M}{R^2}$
(c) MR^2 (d) $\frac{M}{R}$

21. A mass M is split into two parts, m and $(M - m)$, which are then separated by a certain distance. What ratio of m/M maximizes the gravitational force between the two parts?
(a) 1/3 (b) 1/2
(c) 1/4 (d) 1/5

22. Assuming, the earth to have a constant density, point out which of the following curves show the variation of acceleration due to gravity from the centre of earth to the points far away from the surface of earth

(a) 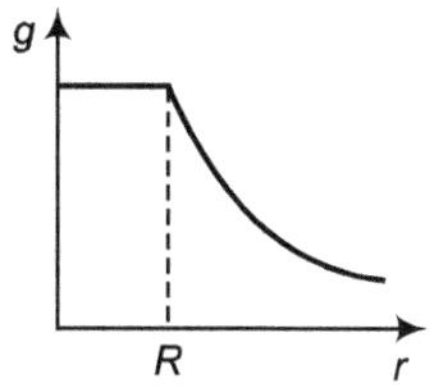

(b) 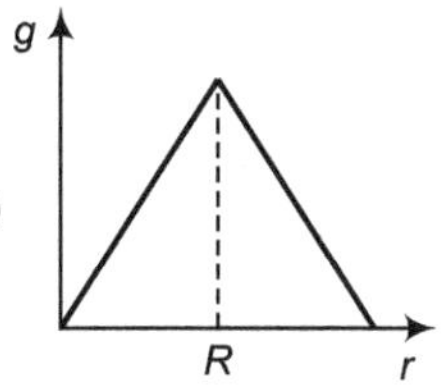

(c) 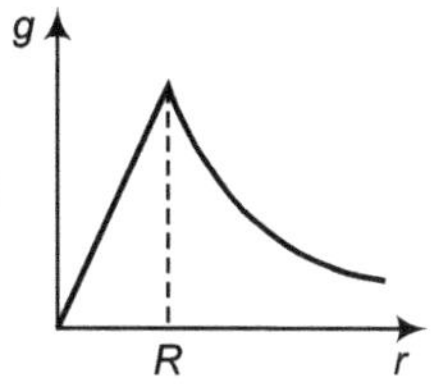

(d) None of these

23. The height above the surface of the earth, where acceleration due to gravity is $\frac{1}{64}$ of its value at surface of the earth is approximately
(a) 45×10^6 m (b) 54×10^6 m
(c) 102×10^6 m (d) 72×10^6 m

24. The angular velocity of rotation of a star (of mass M and radius R) at which the matter will start escaping from its equator is
(a) $\sqrt{\frac{2GR}{M}}$ (b) $\sqrt{\frac{GM}{R^3}}$
(c) $\sqrt{\frac{GR}{M}}$ (d) $\sqrt{\frac{2GM^2}{R}}$

25. At what depth below the surface of the earth acceleration due to gravity will be half its value 1600 km above the surface of the earth?
(a) 4.3×10^6 m (b) 2.4×10^6 m
(c) 3.2×10^6 m (d) 1.6×10^6 m

Gravitational Field and Potential

26. Unit of gravitational potential is
(a) Joule (b) Joule/kilogram
(c) Joule kilogram (d) None of these

27. A mass m is placed inside a hollow sphere of mass M as shown in figure. The gravitation force on mass m is

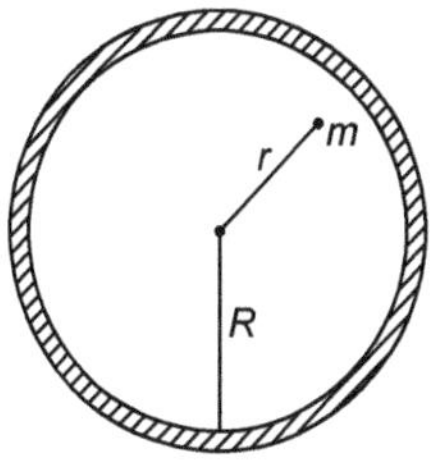

(a) $\frac{GMm}{R^2}$ (b) $\frac{GMm}{r^2}$
(c) $\frac{GMm}{(R-r)^2}$ (d) zero

28. For a uniform ring of mass M and radius R at its centre,
(a) field and potential both are zero
(b) field is zero but potential is $\frac{GM}{R}$
(c) field is zero but potential is $-GM/R$
(d) magnitude of field is $\frac{GM}{R^2}$ and potential $-\frac{GM}{R}$

29. Two bodies of masses m and M are placed a distance d apart. The gravitational potential at the position, where the gravitational field due to them is zero, is V. Then,
(a) $V = -\frac{G}{d}(m + M)$ (b) $V = -\frac{Gm}{d}$
(c) $V = -\frac{GM}{d}$ (d) $V = -\frac{G}{d}(\sqrt{m} + \sqrt{M})^2$

30. A thin rod of length L is bent to form a semicircle. The mass of rod is M. What will be the gravitational potential at the centre of the circle?
(a) $-\frac{GM}{L}$ (b) $-\frac{GM}{2\pi L}$
(c) $-\frac{\pi GM}{2L}$ (d) $-\frac{\pi GM}{L}$

31. A uniform solid sphere of mass m and radius r is surrounded symmetrically by a uniform thin spherical shell of radius $2r$ and mass m.

(a) The gravitational field at a distance of $1.5r$ from the centre is $\frac{2}{9}\frac{Gm}{r^2}$

(b) The gravitational field at a distance of $2.5r$ from centre is $\frac{8}{25}\frac{Gm}{r^2}$

(c) The gravitational field at a distance of $1.5r$ from centre is zero

(d) The gravitational field between the sphere and spherical shell is uniform

32. Inside a uniform shell,

(a) potential energy is zero (b) potential is zero
(c) potential is constant (d) All of these

33. At what height the gravitational field reduces by 75%, the gravitational field at the surface of earth?

(a) R (b) $2R$ (c) $3R$ (d) $4R$
(where, R is the radius of the earth)

34. If V is the gravitational potential on the surface of the earth, then what is its value at the centre of the earth?

(a) $2V$ (b) $3V$
(c) $\frac{3}{2}V$ (d) $\frac{2}{3}V$

35. 3 particles each of mass m are kept at vertices of an equilateral triangle of side L. The gravitational field at centre due to these particles is

(a) zero
(b) $\frac{3GM}{L^2}$
(c) $\frac{9GM}{L^2}$
(d) $\frac{12GM}{\sqrt{3}L^2}$

36. The diagram showing the variation of gravitational potential of the earth with distance from the centre of the earth is

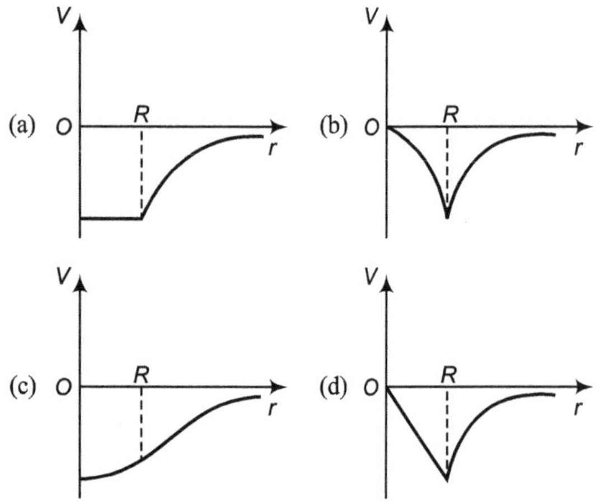

37. By which curve will be variation of gravitational potential of a hollow sphere of radius R with distance be depicted?

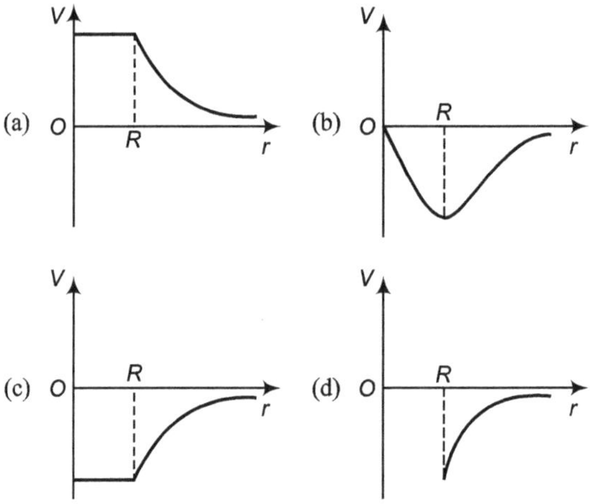

38. Which one of the following graphs represents correctly represent the variation of the gravitational field (E) with the distance (r) from the centre of a spherical shell of mass M radius R?

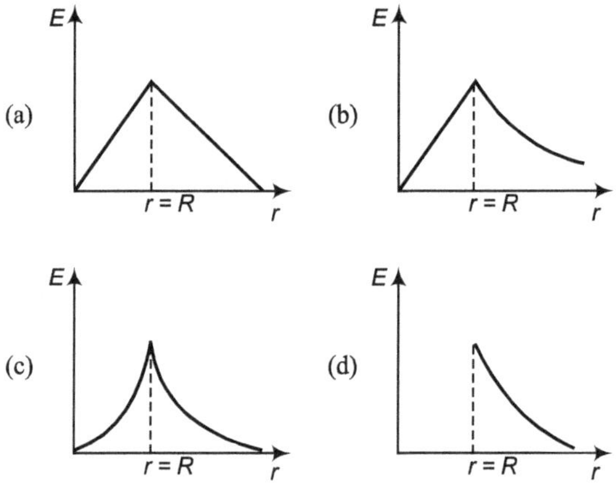

Gravitational Potential Energy, Energy Conservation and Escape Velocity

39. In a gravitational field, if a body is bound with the earth, then total mechanical energy has

(a) positive
(b) zero
(c) negative
(d) may be positive, negative or zero

40. The velocity with which a projectile must be fired to escape from the earth does not depend upon

(a) mass of the earth
(b) mass of the projectile
(c) radius of the earth
(d) None of the above

41. When a body is lifted from surface of the earth height equal to radius of the earth, then the change in its PE is

(a) mgR (b) $2mgR$ (c) $\frac{1}{2}mgR$ (d) $4mgR$

42. Escape velocity is given by

(a) $\sqrt{2gR}$ (b) $\sqrt{gR}$ (c) $\sqrt{2}gR$ (d) $2\sqrt{gR}$

43. Escape velocity on the earth is 11.2 km/s. What would be the escape velocity on a planet whose mass is 1000 times and radius is 10 times that of the earth?

(a) 112 km/s (b) 11.2 km/s
(c) 1.12 km/s (d) 3.7 km/s

44. The escape velocity from the earth is v_e. A body is projected with velocity $2v_e$. With what constant velocity will it move in the inter planetary space?

(a) v_e (b) $\sqrt{2}v_e$
(c) $\sqrt{3}v_e$ (d) $\sqrt{5}v_e$

45. The escape velocity for a body of mass 1 kg from the earth's surface is 11.2 kms^{-1}. The escape velocity for a body of mass 100 kg would be

(a) $11.2 \times 10^2 \text{ kms}^{-1}$ (b) 112 kms^{-1}
(c) 11.2 kms^{-1} (d) $11.2 \times 10^{-2} \text{ kms}^{-1}$

46. When escape velocity is given to a particle on surface of the earth, its total energy is

(a) zero (b) greater than zero
(c) less than zero (d) $-\frac{GMm}{2R}$

where, M = mass of the earth, m = mass of particle and R = radius of the earth.

47. The escape velocity of a body projected vertically upward from the earth's surface is 11.2 km/s. If the body is projected in a direction making 30° angle to the vertical, its escape velocity in this case will be

(a) 11.2 km/s (b) $\frac{11.2}{2}$ km/s
(c) $11.2 \times \frac{\sqrt{3}}{2}$ km/s (d) $\frac{11.2}{3}$ km/s

48. The escape velocity of a particle of mass m varies as

(a) m^2 (b) m (c) m^0 (d) m^{-1}

49. The ratio of the radius of the earth to that of the moon is 10. The ratio of the acceleration due to gravity on the earth to that on the moon is 6. The ratio of the escape velocity from the earth's surface to that from the moon is

(a) 4 (b) 6
(c) 12 (d) None of these

50. With what velocity should a particle be projected so that its height becomes equal to radius of the earth?

(a) $\left(\frac{GM}{R}\right)^{1/2}$ (b) $\left(\frac{8GM}{R}\right)^{1/2}$
(c) $\left(\frac{2GM}{R}\right)^{1/2}$ (d) $\left(\frac{4GM}{R}\right)^{1/2}$

51. At what angle with the horizontal should a projectile be fired with the escape velocity to enable it escape from gravitational pull of the earth?

(a) Less than 45° (b) 45°
(c) More than 45° (d) Any angle

52. Energy required in moving a body of mass m from a distance $2R$ to $3R$ from centre of the earth of mass M is

(a) $\frac{GMm}{12R^2}$ (b) $\frac{GMm}{3R^2}$
(c) $\frac{GMm}{8R}$ (d) $\frac{GMm}{6R}$

53. The kinetic energy needed to project a body of mass m from the earth surface (radius R) to infinity is

(a) $\frac{mgR}{2}$ (b) $2mgR$
(c) mgR (d) $\frac{mgR}{2}$

54. Gas escapes from the surface of a planet, because it acquires an escape velocity. The escape velocity will depend on which of the following factors:

I. Mass of the planet
II. Mass of the particle escaping
III. Temperature of the planet
IV. Radius of the planet

Select the correct answer from the codes given below

(a) I and II
(b) II and IV
(c) I and IV
(d) I, III and IV

55. What is the energy required to launch a m kg satellite from the earth's surface in a circular orbit at an altitude of $2R$ (R = radius of the earth)

(a) $\frac{2}{3}mgR$ (b) mgR
(c) $\frac{5}{6}mgR$ (d) $\frac{1}{3}mgR$

Motion of Satellites in Circular Orbits and Planets in Elliptical Orbits

56. A satellite of the earth is revolving in a circular orbit with a uniform speed v. If gravitational force suddenly disappears, the satellite will

(a) continue to move with speed v along the original orbit
(b) move with the velocity v tangentially to the original orbit
(c) fall downward with increasing velocity
(d) ultimately come to rest somewhere on the original orbit

57. Kepler's law states that square of the time period of any planet about the sun is directly proportional to

(a) R (b) $\frac{1}{R}$
(c) R^3 (d) $\frac{1}{R^3}$

58. A body is projected from earth's surface to become its satellite, its time period of revolution will not depend upon
(a) mass of the earth
(b) its own mass
(c) gravitational constant
(d) radius of orbit

59. Satellite is revolving around the earth. If its radius of orbit is increased to 4 times of the radius of geostationary satellite, what will become its time period?
(a) 8 days (b) 4 days
(c) 2 days (d) 16 days

60. The ratio of mean distances of three planets from the sun are 0.5 : 1 : 1 : 5, then the square of time periods are in the ratio of
(a) 1 : 4 : 9 (b) 1 : 9 : 4
(c) 1 : 8 : 27 (d) 2 : 1 : 3

61. A planet moves around the sun. It is closest to sun at a distance d_1 and have velocity v_1.At farthest distance d_2 its speed will be
(a) $\frac{d_1^2 v_1}{d_2^2}$ (b) $\frac{d_2 v_1}{d_1}$
(c) $\frac{d_1 v_1}{d_2}$ (d) $\frac{d_2^2 v_1}{d_1^2}$

62. Two satellites A and B, ratio of masses 3 : 1 are in circular orbits of radii r and $4r$. Then, the ratio of total mechanical energy of A and B is
(a) 1:3 (b) 3:1 (c) 3:4 (d) 12:1

63. The ratio of distance of two satellites from the centre of earth is 1:4 . The ratio of their time periods of rotation will be
(a) 1:4 (b) 4:1 (c) 1:8 (d) 8:1

64. For a satellite in elliptical orbit which of the following quantities does not remain constant?
(a) Angular momentum
(b) Momentum
(c) Areal velocity
(d) Total energy

65. The orbital velocity of an artificial satellite in a circular orbit just above the earth's surface is v. For a satellite orbiting at an altitude of half the earth's radius, the orbital velocity is
(a) $\frac{3}{2}v$ (b) $\sqrt{\frac{3}{2}}v$
(c) $\sqrt{\frac{2}{3}}v$ (d) $\frac{2}{3}v$

66. In case of an orbiting satellite, if the radius of orbit is decreased
(a) its kinetic energy decreases
(b) its potential energy increases
(c) Both (a) and (b) are correct
(d) Both (a) and (b) are wrong

67. The period of revolution of planet A round the sun is 8 times that of B. The distance of A from the sun is how many times greater than that of B from the sun?
(a) 5 (b) 4 (c) 3 (d) 2

68. Kepler's second law is based on
(a) newton's first law
(b) newton's second law
(c) special theory of relativity
(d) conservation of angular momentum

69. An artificial satellite moving in a circular orbit around the earth has total mechanical energy E_0.
Its potential energy is
(a) $-2E_0$ (b) $1.5E_0$ (c) $2E_0$ (d) E_0

70. An artificial satellite moving in a circular orbit around the earth has a total (kinetic + potential) energy E_0. Its potential energy and kinetic energy respectively are
(a) $2E_0$ and $-2E_0$ (b) $-2E_0$ and E_0
(c) $2E_0$ and $-E_0$ (d) $-2E_0$ and $-E_0$

71. If mean radius of earth is R, its angular velocity is ω, and the acceleration due to gravity at the surface of the earth is g, then the cube of the radius of the orbit of geostationary satellite will be
(a) $\frac{R^2 g}{\omega^2}$ (b) $\frac{R^2\omega^2}{g}$
(c) $\frac{R^2 g}{\omega}$ (d) $\frac{Rg}{\omega^2}$

72. An earth's satellite is moved from one stable circular orbit to another higher stable circular orbit. Which one of the following quantities increase for the satellite as a result of this change.
(a) Angular momentum (b) Kinetic energy
(c) Angular velocity (d) Linear orbital speed

73. When a planet moves around sun, its
(a) areal velocity is constant
(b) linear velocity is constant
(c) angular velocity is constant
(d) All of the above

74. The orbital velocity of a body close to the earth's surface is
(a) 8 km/s (b) 11.2 km/s
(c) 3×10^8 m/s (d) 2.2×10^3 km/s

75. For a satellite orbiting very close to earth's surface, total energy is
(a) zero (b) $\frac{GMm}{R}$
(c) $-\frac{GMm}{R}$ (d) $-\frac{GMm}{2R}$

76. Two identical satellites are orbiting at distances R and $7R$ from the surface of the earth, R being the radius of the earth. The ratio of their
(a) kinetic energies is 4 (b) potential energies is 4
(c) total energies is 4 (d) All of these

77. The distance of two planets from the sun are 10^{13} and 10^{12} m, respectively. The ratio of the periods of the planet is

(a) 100 (b) $\frac{1}{\sqrt{10}}$

(c) $\sqrt{10}$ (d) $10\sqrt{10}$

78. The period of a satellite in a circular orbit around a planet is independent of

(a) the mass of the planet
(b) the radius of the planet
(c) the mass of the satellite
(d) All the three parameters (a), (b) and (c)

79. Two satellites A and B go round a planet P in circular orbits having radii $4R$ and R, respectively. If the speed of the satellite A is $3v$, the speed of the satellite B will be

(a) $12v$ (b) $6v$ (c) $\frac{4}{3}v$ (d) $\frac{3}{2}v$

80. A satellite is revolving in circular orbit of radius r around the earth of mass M. Time of revolution of satellite is

(a) $T \propto \frac{r^5}{GM}$

(b) $T \propto \sqrt{\frac{r^3}{GM}}$

(c) $T \propto \sqrt{\frac{\frac{r}{GM^2}}{3}}$

(d) $T \propto \sqrt{\frac{\frac{r^3}{GM}}{4}}$

81. Two satellites A and B, ratio of masses 3 : 1 are in circular orbits of radii r and $4r$. Then, the ratio of total mechanical energy of A to B is

(a) 1:3 (b) 3:1 (c) 3:4 (d) 12:1

82. Which of the following quantities does not depend upon the orbital radius of a satellite?

(a) $\frac{T}{R}$ (b) $\frac{T^2}{R}$

(c) $\frac{T^2}{R^2}$ (d) $\frac{T^2}{R^3}$

(R = radius of orbit)

83. A satellite moves round the earth in a circular orbit of radius R making one revolution per day. A second satellite moving in a circular orbit, moves round the earth once in 8 days. The radius of the orbit of the second satellite is

(a) $8R$ (b) $4R$
(c) $2R$ (d) R

84. Kepler discovered

(a) laws of motion
(b) laws of rotational motion
(c) laws of planetary motion
(d) laws of curvilinear motion

Miscellaneous Problems

85. The acceleration due to gravity g and mean density of earth ρ are related by which of the following relations? (G = gravitational constant and R = radius of the earth)

(a) $\rho = \frac{4\pi g R^2}{3G}$ (b) $\rho = \frac{4\pi g R^3}{3G}$

(c) $\rho = \frac{3g}{4\pi GR}$ (d) $\rho = \frac{3g}{4\pi GR^3}$

86. What is not conserved in case of celestial bodies revolving around sun all the time?

(a) Kinetic energy (b) Mass
(c) Angular momentum (d) None of these

87. The distance of the centres of moon and the earth is D. The mass of the earth is 81 times the mass of the moon. At what distance from the centre of the earth, the gravitational force on a particle will be zero?

(a) $\frac{D}{2}$ (b) $\frac{2D}{3}$ (c) $\frac{4D}{3}$ (d) $\frac{9D}{10}$

88. The rotation of the earth about its axis speeds up such that a man on the equator becomes weightless. In such a situation, what would be the duration of one day?

(a) $2\pi\sqrt{\frac{R}{g}}$ (b) $\frac{1}{2\pi}\sqrt{\frac{R}{g}}$

(c) $2\pi\sqrt{Rg}$ (d) $\frac{1}{2\pi}\sqrt{Rg}$

89. A body of mass m is kept at a small height h above the ground. If the radius of the earth is R and its mass is M, the potential energy of the body and the earth system (with $h = \infty$ being the reference position) is

(a) $\frac{GMm}{R} + mgh$ (b) $\frac{-GMm}{R} + mgh$

(c) $\frac{GMm}{R} - mgh$ (d) $-\frac{GMm}{R} - mgh$

90. A simple pendulum has a time period T_1 when on the earth's surface and T_2 when taken to a height R above the earth's surface, where R is the radius of the earth. The value of $\frac{T_2}{T_1}$ is

(a) 1 (b) $\sqrt{2}$ (c) 4 (d) 2

91. A particle is projected vertically upwards from the surface of earth (radius R) with a kinetic energy equal to half of the minimum value needed for it to escape. The height to which it rises above the surface of earth is

(a) R (b) $2R$
(c) $3R$ (d) $4R$

92. Suppose, the gravitational attraction varies inversely as the distance from the earth. The orbital velocity of a satellite in such a case varies as nth power of distance, where n is equal to

(a) -1
(b) zero
(c) $+1$
(d) $+2$

93. Consider the two identical particles shown in the given figure. They are released from rest and may move towards each other under influence of mutual gravitation force

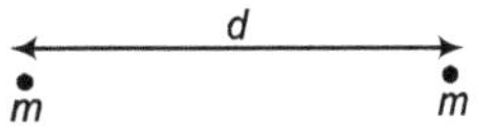

(i) The velocity of the centre of mass of the two particle system
(a) is zero
(b) is constant ($\neq 0$)
(c) increases as the separation decreases
(d) None of the above

(ii) Speed of each particle, when the separation reduces to half its initial value
(a) $\sqrt{\frac{Gm}{d}}$ (b) $\sqrt{\frac{2Gm}{d}}$
(c) $\sqrt{\frac{Gm}{2d}}$ (d) None of these

94. If G is universal gravitational constant and g is acceleration due to gravity, then the unit of the quantity $\frac{G}{g}$ is
(a) kg-m^2 (b) kg/m
(c) kg/m^2 (d) m^2/kg

95. If gravitational attraction between two points masses be given by $F = G\frac{m_1 m_2}{r^n}$. Then, the period of a satellite in a circular orbit will be proportional to
(a) $r^{\frac{n-1}{2}}$ (b) $r^{\frac{n+1}{2}}$
(c) $r^{\frac{n}{2}}$ (d) independent of n

96. The centripetal force on a satellite orbiting round the earth and the gravitational force of the earth acting on the satellite both equal F. The net force on the satellite is
(a) zero (b) F
(c) $F\sqrt{2}$ (d) $2F$

97. Reason of weightlessness in a satellite is
(a) zero gravity
(b) no atmosphere
(c) zero reaction force by satellite surface
(d) None of the above

98. Assume that the acceleration due to gravity on the surface of the moon is 0.2 times the acceleration due to gravity on the surface of the earth. If R_e is the maximum range of a projectile on the earth's surface, what is the maximum range on the surface of the moon for the same velocity of projection?
(a) $0.2\ R_e$
(b) $2\ R_e$
(c) $0.5\ R_e$
(d) $5\ R_e$

● Speed of projectile is not very high.

99. If orbital velocity of planet is given by $v = G^a M^b R^c$, then
(a) $a = \frac{1}{3}, b = \frac{1}{3}, c = -\frac{1}{3}$
(b) $a = \frac{1}{2}, b = \frac{1}{2}, c = -\frac{1}{2}$
(c) $a = \frac{1}{2}, b = -\frac{1}{2}, c = \frac{1}{2}$
(d) $a = \frac{1}{2}, b = -\frac{1}{2}, c = -\frac{1}{2}$

100. Suppose a smooth tunnel is dug along a straight line joining two points on the surface of the earth and a particle is dropped from rest at its one end. Assume that mass of the earth is uniformly distributed over its volume. Then,
(a) the particle will emerge from the other end with velocity $\sqrt{\frac{GM_e}{2R_e}}$, where M_e and R_e are earth's mass and radius respectively
(b) the overlapping will come to rest at centre of the tunnel
(c) potential energy of the particle will be equal to zero at centre of tunnel, if it is along a diameter
(d) acceleration of the particle will be proportional to its distance from mid-point of the tunnel

101. A planet is revolving round the sun in an elliptical orbit, If v is the velocity of the planet when its position vector from the sun is r, then areal velocity of the planet is
(a) $|\mathbf{v} \times \mathbf{r}|$
(b) $2|\mathbf{r} \times \mathbf{v}|$
(c) $\left|\frac{1}{2}(\mathbf{r} \times \mathbf{v})\right|$
(d) None of the above

102. The magnitude of gravitational potential energy of a body at a distance r from the centre of earth is u. Its weight at a distance $2r$ from the centre of earth is
(a) $\frac{u}{r}$ (b) $\frac{u}{4r}$ (c) $\frac{u}{2r}$ (d) $\frac{4r}{u}$

103. Two satellites of same mass are launched in the same orbit of radius r around the earth so as to rotate opposite to each other. If they collide inelastically and stick together as wreckage, the total energy of the system just after collision is
(a) $-\frac{2GMm}{r}$
(b) $-\frac{GMm}{r}$
(c) $\frac{GMm}{2r}$
(d) zero

104. Earth orbiting satellite will escape, if
(a) its speed is increased by 41%
(b) its KE is doubled
(c) Both (a) and (b) are correct
(d) Both (a) and (b) are wrong

105. A planet of mass m is in an elliptical orbit about the sun with an orbital period T. If A be the area of orbit, then its angular momentum would be

(a) $\frac{2mA}{T}$ (b) mAT (c) $\frac{mA}{2T}$ (d) $2mAT$

106. What additional velocity must be given to a satellite orbiting around earth with radius R to become free from the earth's gravitational field? Mass of earth is M.

(a) $\sqrt{\frac{GM}{R}}(\sqrt{2}-1)$ (b) $\sqrt{\frac{GM}{2R}}(\sqrt{2}-1)$

(c) $\sqrt{\frac{GM}{R}}(\sqrt{3}-1)$ (d) $\sqrt{\frac{GM}{R}}(\sqrt{2}+1)$

107. What impulse need to be given to a body of mass m, released from the surface of earth along a straight tunnel passsing through centre of earth, at the centre of earth, to bring it to rest? (Mass of earth M, radius of earth R).

(a) $m\sqrt{\frac{GM}{R}}$

(b) $\sqrt{\frac{GMm}{R}}$

(c) $m\sqrt{\frac{GM}{2R}}$

(d) Zero

[Level 2]

Only One Correct Option

1. A rocket is launched vertical from the surface of the earth of radius R with an initial speed v. If atmospheric resistance is neglected, the maximum height attained by the rocket is

(a) $h=\frac{R}{\left(\frac{2gR}{v^2}-1\right)}$ (b) $h=\frac{R}{\left(\frac{2gR}{v^2}+1\right)}$

(c) $h=\frac{R^2}{\left(\frac{2gR}{v^2}-1\right)}$ (d) $h=\frac{R^2}{\left(\frac{2gR}{v^2}+1\right)}$

2. Suppose, the gravitational force varies inversely as the nth power of distance. Then, the time period of a planet in circular orbit of radius r around the sun will be proportional to

(a) $r^{\frac{1}{2}(n+1)}$ (b) $r^{\frac{1}{2}(n-1)}$

(c) r^n (d) $r^{\frac{1}{2}(n-2)}$

3. A solid sphere of mass M and radius R has a spherical cavity of radius $\frac{R}{2}$ such that the centre of cavity is at a distance $\frac{R}{2}$ from the centre of the sphere. A point mass m is placed inside the cavity at a distance $\frac{R}{4}$ from the centre of sphere. The gravitational force on mass m is

(a) $\frac{11GMm}{R^2}$ (b) $\frac{14GMm}{R^2}$

(c) $\frac{GMm}{2R^2}$ (d) $\frac{GMm}{R^2}$

4. The required kinetic energy of an object of mass m so that it may escape, will be

(a) $\frac{1}{4}mgR$ (b) $\frac{1}{2}mgR$

(c) mgR (d) $2mgR$

5. A body is projected vertically upwards from the surface of earth with a velocity equal to half the escape velocity. If R be the radius of earth, maximum height attained by the body from the surface of the earth is

(a) $\frac{R}{6}$

(b) $\frac{R}{3}$

(c) $\frac{2R}{3}$

(d) R

6. Pertaining to two planets, the ratio of escape velocities from respective surfaces is 1:2, the ratio of the time period of the same simple pendulum at their respective surfaces is 2:1 (in same order). Then, the ratio of their average densities is

(a) 1:1 (b) 1:2

(c) 1:4 (d) 8:1

7. Four equal masses (each of mass M) are placed at the corners of a square of side a. The escape velocity of a body from the centre O of the square is

(a) $4\sqrt{\frac{2GM}{a}}$ (b) $\sqrt{\frac{8\sqrt{2}GM}{a}}$

(c) $\frac{4GM}{a}$ (d) $\sqrt{\frac{4\sqrt{2}GM}{a}}$

8. A point $P(R\sqrt{3},0,0)$ lies on the axis of a ring of mass M and radius R. The ring is located in y-z plane with its centre at origin O. A small particle of mass m starts from P and reaches O under gravitational attraction only. Its speed at O will be

(a) $\sqrt{\frac{GM}{R}}$ (b) $\sqrt{\frac{Gm}{R}}$

(c) $\sqrt{\frac{GM}{2R}}$ (d) $\sqrt{\frac{Gm}{\sqrt{2}R}}$

9. Two spherical bodies of masses m and $5m$ and radii R and $2R$ respectively are released in free space with initial separation between their centres equal to $12\ R$. If they attract each other due to gravitational force only, then the distance covered by smaller sphere just before collision will be

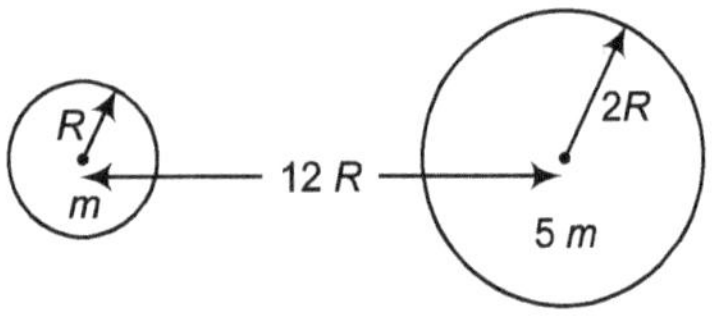

(a) $5\ R$ (b) $7.5\ R$
(c) $2.5\ R$ (d) $6\ R$

10. Energy of a satellite in circular orbit is E_0. The energy required to move the satellite to a circular orbit of 3 times the radius of the initial orbit is

(a) $\frac{2}{3}E_0$ (b) $2E_0$ (c) $\frac{E_0}{3}$ (d) $\frac{3}{2}E_0$

11. Two identical thin rings each of radius R are coaxially placed at a distance R. If the rings have a uniform mass distribution and each has mass m_1 and m_2 respectively, then the work done in moving a mass m from centre of one ring to that of the other is

(a) zero (b) $\frac{Gm(m_1 \pm m_2)(\sqrt{2}-1)}{\sqrt{2}R}$

(c) $\frac{Gm\sqrt{2}(m_1 \pm m_2)}{R}$ (d) $\frac{Gmm_1(\sqrt{2} \pm 1)}{m_2 R}$

12. A person brings a mass of 1 kg infinity to a point A. Initially, the mass was at rest but it moves at a speed of 2 m/s as it reached A. The work done by the person on the mass is –3 J . The potential at A is

(a) – 3 J/kg (b) – 12 J/kg
(c) – 5 J/kg (d) None of these

13. A satellite is moving in a circular orbit round the earth with a diameter of orbit $2R$. At a certain point a rocket fixed to the satellite is fired such that it increases the velocity of the satellite tangentially. The resulting orbit of the satellite would be

(a) same as before
(b) circular orbit with diameter greater than $2\ R$
(c) elliptical orbit with minimum distance from the earth equal to R
(d) elliptical orbit with maximum distance from the earth equal to R

14. A particle would take a time t to move down a straight tube from the surface of the earth (supposed to be a homogeneous sphere) to its centre. If gravity were to remain constant, then the time would be t. The ratio of $\frac{t}{t'}$ will be

(a) $\frac{\pi}{2\sqrt{2}}$ (b) $\frac{\pi}{\sqrt{3}}$ (c) $\sqrt{2}\pi$ (d) $\frac{\pi}{\sqrt{2}}$

15. Two particles of mass m and M are initially at rest at infinite distance. Find their relative velocity of approach due to gravitational attraction, when d is their separation at any instant

(a) $\sqrt{\frac{2G(M+m)}{d}}$ (b) $\sqrt{\frac{G(M+m)}{d}}$

(c) $\sqrt{\frac{G(M+m)}{2d}}$ (d) $\sqrt{\frac{G(M+m)}{4d}}$

16. An earth satellite of mass m revolves in a circular orbit at a height h from the surface of the earth. R is the radius of the earth and g is acceleration due to gravity at the surface of the earth. The velocity of the satellite in the orbit is given by

(a) $\frac{gR^2}{R+h}$ (b) gR

(c) $\frac{gR}{R+h}$ (d) $\sqrt{\frac{gR^2}{R+h}}$

17. The orbital angular momentum of a satellite revolving at a distance r form the centre is L. If the distance is increased to $16r$, then the new angular momentum will be

(a) $16\ L$ (b) $64\ L$
(c) $\frac{L}{4}$ (d) $4\ L$

18. The ratio of energy required to raise a satellite to a height h above the earth surface to that required to put it into the orbit is

(a) $h:2R$ (b) $2h:R$
(c) $R:h$ (d) $h:R$

19. A body which is initially at rest at a height R above the surface of the earth of radius R, falls freely towards the earth, then its velocity on reaching the surface of the earth is

(a) $\sqrt{(2gR)}$ (b) $\sqrt{(gR)}$

(c) $\sqrt{\frac{3}{2}gR}$ (d) $\sqrt{(4gR)}$

20. What is the energy required to launch a m kg satellite from the earth's surface in a circular orbit at an altitude of $2R$? (R = radius of the earth)

(a) $\frac{2}{3}mgR$ (b) mgR

(c) $\frac{5}{6}mgR$ (d) $\frac{1}{3}mgR$

21. A body of supercondense material with mass twice the mass of earth but size very small compared to the size of earth starts from rest from $h << R$ above the earth's surface. It reaches earth in time

(a) $t=\sqrt{\frac{h}{g}}$ (b) $t=\sqrt{\frac{2h}{g}}$

(c) $t=\sqrt{\frac{2h}{3g}}$ (d) $t=\sqrt{\frac{4h}{3g}}$

22. A solid sphere of uniform density and radius R applies a gravitational force of attraction equal to F_1 on a particle placed at a distance $2R$ from the centre of the sphere. A spherical cavity of radius $\frac{R}{2}$ is now made in the sphere as shown in the figure. The sphere with cavity now applies a gravitational force F_2 on the same particle. The ratio $\frac{F_2}{F_1}$ is

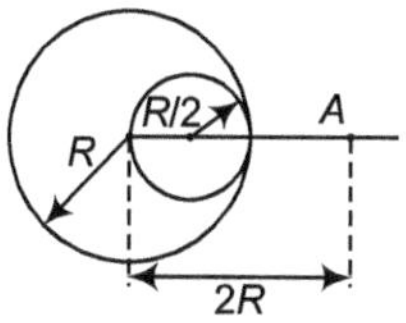

(a) $\frac{5}{9}$ (b) $\frac{7}{8}$ (c) $\frac{3}{4}$ (d) $\frac{7}{9}$

23. The magnitudes of the gravitational force at distances r_1 and r_2 from the centre of a uniform sphere of radius R and mass M are F_1 and F_2, respectively. Then, (more than one are correct)

(a) $\frac{F_1}{F_2} = \frac{r_1}{r_2}$, if $r_1 < R$ and $r_2 < R$

(b) $\frac{F_1}{F_2} = \frac{r_1^2}{r_2^2}$, if $r_1 > R$ and $r_2 > R$

(c) $\frac{F_1}{F_2} = \frac{r_2}{r_1}$, if $r_1 < R$ and $r_2 < R$

(d) $\frac{F_1}{F_2} = \frac{r_2^2}{r_1^2}$, if $r_1 > R$ and $r_2 > R$

24. The escape velocity from earth is v_e. A body is projected with velocity $2v_e$. With what constant velocity will it move in the inter planetary space?

(a) v_e (b) $\sqrt{2}v_e$ (c) $\sqrt{3}v_e$ (d) $\sqrt{5}v_e$

25. Suppose, the gravitational attraction varies inversely as the distance from the earth. The orbital velocity of a satellite in such a case varies as nth power of distance, where n is equal to

(a) -1 (b) zero (c) $+1$ (d) $+2$

26. Let E be the energy required to raise a satellite to height h above the earth's surface and E' be the energy required to put the same satellite into orbit at that height. Then, $\frac{E}{E'}$ is equal to

(a) $\frac{2h}{(R+2h)}$ (b) $\frac{2h}{(2R+3h)}$ (c) $\frac{R}{R+h}$ (d) $\frac{2R}{2h+R}$

27. Two spheres of masses m and $2m$ are separated by distance d. A particle of mass $\frac{m}{5}$ is projected straight from $2m$ towards m with a velocity v_0. Which of the following statements is correct?

(a) Velocity of the particle decreases constantly

(b) Velocity of the particle increases constantly

(c) Acceleration of the particle may become momentarily zero

(d) The particle may retrace its path depending on value of v_0

28. A satellite is revolving round the earth with orbital speed v_o. If it stops suddenly, the speed with which it will strike the surface of earth would be (v_e = escape velocity of a particle on earth's surface.)

(a) $\frac{v_e^2}{v_o}$ (b) $2v_o$ (c) $\sqrt{v_e^2 - v_o^2}$ (d) $\sqrt{v_e^2 - 2v_o^2}$

29. Two particles of equal mass m go round a circle of radius R under the action of their mutual gravitational attraction. The speed of each particle is

(a) $v = \sqrt{\frac{Gm}{R}}$ (b) $v = \sqrt{\frac{Gm}{2R}}$

(c) $v = \frac{1}{2}\sqrt{\frac{Gm}{R}}$ (d) $v = \sqrt{\frac{4Gm}{R}}$

30. Four particles, each of mass M, move along a circle of radius R under the action of their mutual gravitational attraction. The speed of each particle is

(a) $\frac{GM}{R}$ (b) $\sqrt{2\sqrt{2}\frac{GM}{R}}$

(c) $\sqrt{\frac{GM}{R}(2\sqrt{2}+1)}$ (d) $\sqrt{\frac{GM}{R}\left(\frac{2\sqrt{2}+1}{4}\right)}$

31. Suppose a vertical tunnel is along the diameter of earth, assumed to be a sphere of uniform mass density ρ. If a body of mass m is thrown in this tunnel, its acceleration at a distance y from the centre is given by

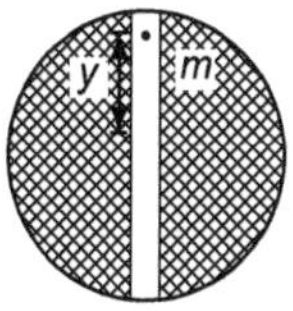

(a) $\frac{4\pi}{3}G\rho y m$ (b) $\frac{3}{4}\pi\rho y$

(c) $\frac{4}{3}\pi\rho y$ (d) $\frac{4}{3}\pi G\rho y$

32. Gravitational field at the centre of a semicircle formed by a thin wire AB of mass M and length l is

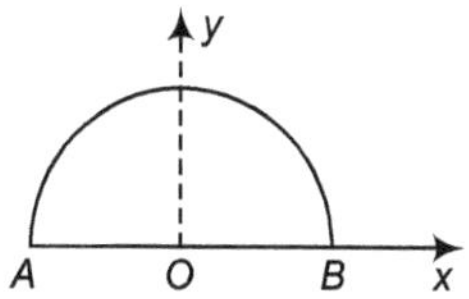

(a) $\frac{GM}{l^2}$ along x-axis (b) $\frac{GM}{\pi l^2}$ along y-axis

(c) $2\pi\frac{GM}{l^2}$ along x-axis (d) $\frac{2\pi GM}{l^2}$ along y-axis

33. A mass m is at a distance a from one end of a uniform rod of length l and M. The gravitational force on the mass due to the rod is

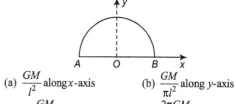

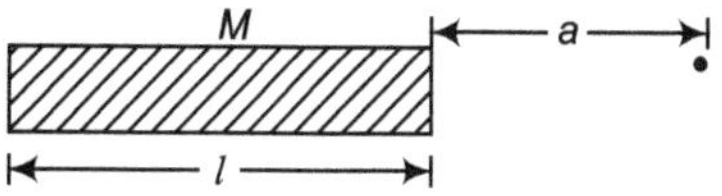

(a) $\frac{GMm}{(a+l)}$ (b) $\frac{GmM}{a(l+a)}$

(c) $\frac{GMm}{a}$ (d) $\frac{GmM}{2(l+a)}$

34. A uniform ring of mass m is lying at a distance $\sqrt{3}\,a$ from the centre of a sphere of mass M just over the sphere (where, a is the radius of the ring as well as that of the sphere). Then, magnitude of gravitational force between them is

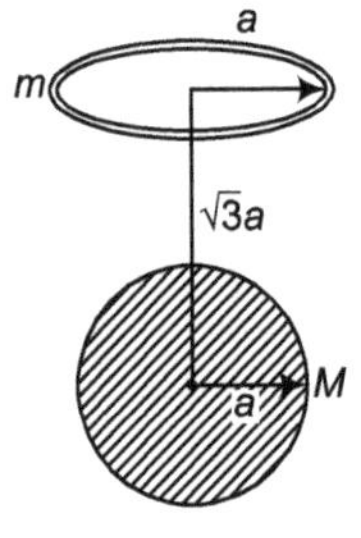

(a) $\dfrac{GMm}{8a^2}$ (b) $\dfrac{GMm}{\sqrt{3}a^2}$

(c) $\sqrt{3}\dfrac{GMm}{a^2}$ (d) $\sqrt{3}\dfrac{GMm}{8a^2}$

35. A ring of mass m_1 and radius R is fixed in space at some location. An external agent brings a point mass m_2 from infinity to centre of the ring .Work done by the external agent will be

(a) $-\dfrac{GM_1m_2}{R}$ (b) $\dfrac{GM_1m_2}{R}$

(c) $\dfrac{G\sqrt{m_1^2+m_2^2}}{R}$ (d) $\dfrac{GM_1m_2}{R(m_1^2+m_2^2)}$

36. Energy required in moving a body of mass m from a distance $2R$ to $3R$ from centre of earth of mass M is

(a) $\dfrac{GMm}{12R^2}$ (b) $\dfrac{GMm}{3R^2}$

(c) $\dfrac{GMm}{8R}$ (d) $\dfrac{GMm}{6R}$

37. A planet of mass m moves around the sun of mass M in an elliptical orbit. The maximum and minimum distance of the planet from the sun are r_1 and r_2, respectively. The time period of the planet is proportional to

(a) $r_1^{3/2}$ (b) $(r_1+r_2)^{3/2}$

(c) $(r_1-r_2)^{3/2}$ (d) $r_1^{3/2}$

38. A body attains a height equal to the radius of the earth. The velocity of the body with which it was projected is

(a) $\sqrt{\dfrac{GM}{R}}$ (b) $\sqrt{\dfrac{2GM}{R}}$

(c) $\sqrt{\dfrac{1}{4}\dfrac{GM}{R}}$ (d) $\sqrt{\dfrac{GM}{2R}}$

39. If the mass of moon is $\dfrac{M}{81}$, where M is the mass of earth, find the distance of the point, where gravitational field due to earth and moon cancel each other, from the centre of moon. Given that distance between centres of earth and moon is $60\,R$, where R is the radius of earth

(a) $4\,R$ (b) $8\,R$
(c) $12\,R$ (d) $6\,R$

40. The minimum energy required to launch a m kg satellite from the earth's surface in a circular orbit at an altitude of $2R$ where R is the radius of earth, will be

(a) $\dfrac{1}{6}mgR$ (b) $\dfrac{5}{6}mgR$ (c) $\dfrac{2}{3}mgR$ (d) $\dfrac{1}{5}mgR$

41. Three point masses each of mass m rotate in a circle of radius r with constant angular velocity ω due to their mutual gravitational attraction. If at any instant, the masses are on the vertex of an equilateral triangle of side a, then the value of ω is

(a) $\sqrt{\dfrac{Gm}{a^3}}$ (b) $\sqrt{\dfrac{3Gm}{a^3}}$

(c) $\sqrt{\dfrac{Gm}{3a^3}}$ (d) None of these

42. A sphere of mass M and radius R_2 has a concentric cavity of radius R_1 as shown in figure. The force F exerted by the sphere on a particle of mass m located a distance r from the centre of sphere varies as $(0 \le r \le \infty)$

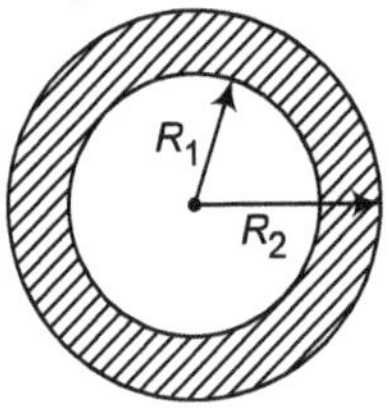

(a)
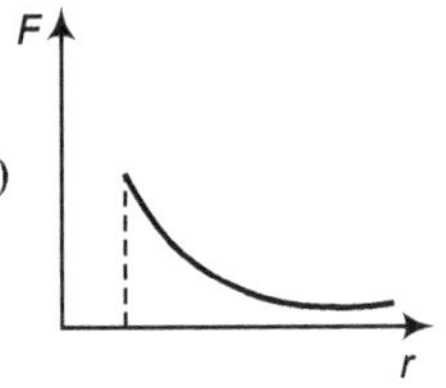

(b)
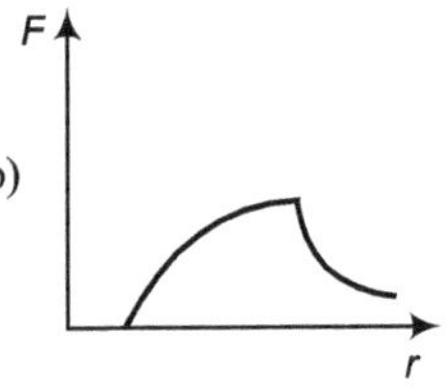

(c)
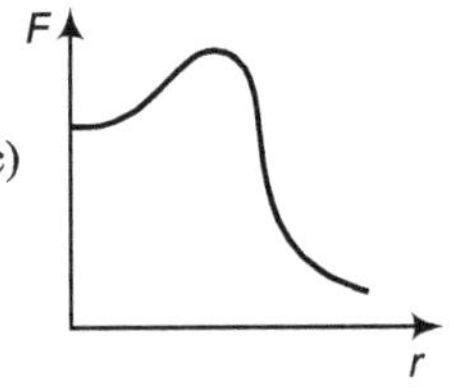

(d)
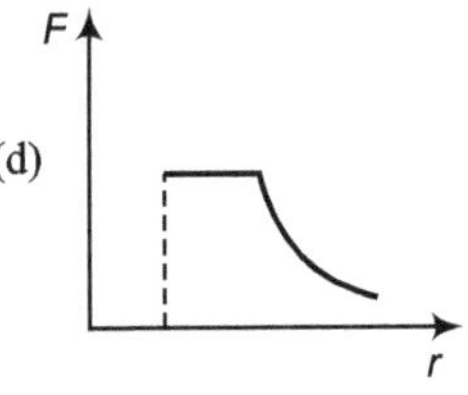

43. If an artificial satellite is moving in a circular orbit around the earth with a speed equal to half the magnitude of the escape velocity from the earth, the height of the satellite above the surface of the earth is

(a) $2\,R$ (b) $\dfrac{R}{2}$ (c) R (d) $\dfrac{R}{4}$

More than One Correct Options

1. Three planets of same density have radii R_1, R_2 and R_3 such that $R_1 = 2R_2 = 3R_3$. The gravitational field at their respective surfaces are g_1, g_2 and g_3 and escape velocities from their surfaces are v_1, v_2 and v_3, then

(a) $g_1/g_2 = 2$ (b) $g_1/g_3 = 3$
(c) $v_1/v_2 = 1/4$ (d) $v_1/v_3 = 3$

2. For a geostationary satellite orbiting around the earth identify the necessary condition.
(a) It must lie in the equatorial plane of the earth
(b) Its height from the surface of the earth must be 36000 km
(c) Its period of revolution must be $2\pi\sqrt{\frac{R}{g}}$, where R is the radius of the earth
(d) Its period of revolution must be 24 h

3. A ball of mass m is dropped from a height h equal to the radius of the earth above the tunnel dug through the earth as shown in the figure. Choose the correct options.

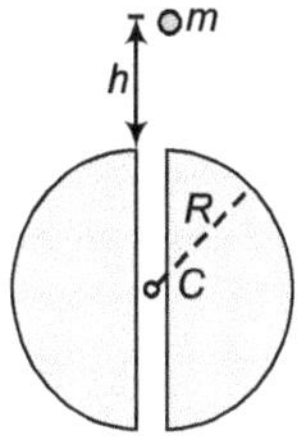

(a) Particle will oscillate through the earth to a height h on both sides
(b) Particle will execute simple harmonic motion
(c) Motion of the particle is periodic
(d) None of the above

4. Two point masses m and $2m$ are kept at points A and B as shown. E represents magnitude of gravitational field strength and V the gravitational potential. As, we move from A to B

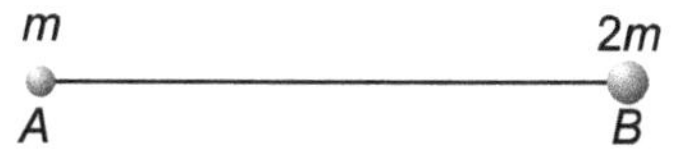

(a) E will first decrease then increases
(b) E will first increase then decreases
(c) V will first decrease then increases
(d) V will first increase then decreases

5. Two spherical shells have mass m and $2m$ as shown. Choose the correct options.

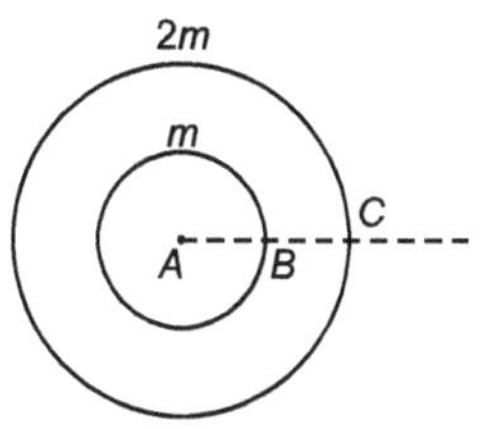

(a) Between A and B gravitational field strength is zero
(b) Between A and B gravitational potential is constant
(c) There will be two points one lying between B and C and other lying between C and infinity, where gravitational field strength are same
(d) There will be a point between B and C, where gravitational potential will be zero

6. Four point masses are placed at four corners of a square as shown. When positions of m and $2m$ are interchanged,

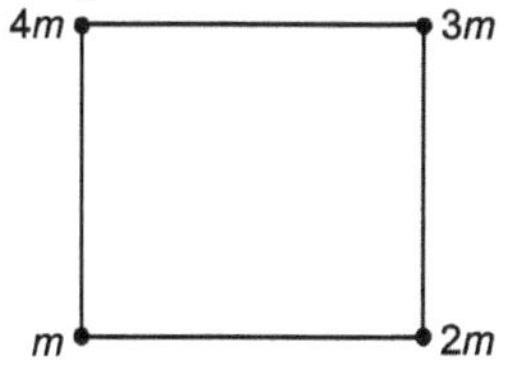

(a) gravitational field strength at centre will increase
(b) gravitational field strength at centre will decrease
(c) gravitational potential at centre will remain unchanged
(d) gravitational potential at centre will decrease

7. Two identical particles 1 and 2 are projected from surface of the earth with same velocities in the directions shown in figure.

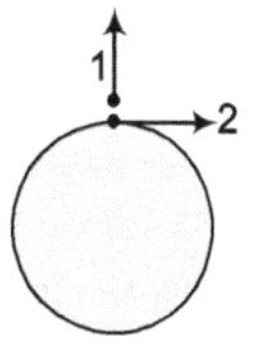

(a) Both the particles will stop momentarily (before striking with ground) at different times
(b) Particle 2 will rise up to lesser height compared to particle 2
(c) Minimum speed of particle 2 is more than that of particle 1
(d) Particle-1 will strike the ground earlier

8. A planet is moving round the sun in an elliptical orbit as shown. As, the planet moves from A to B

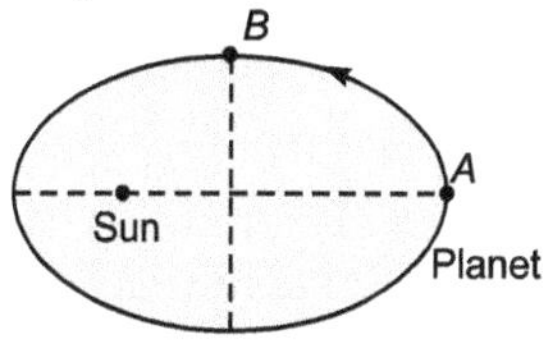

(a) its kinetic energy will decrease
(b) its potential energy will remain unchanged
(c) its angular momentum about centre of sun will remain unchanged
(d) its speed is minimum at A

9. A satellite of mass m is just placed over the surface of the earth. In this position, mechanical energy of satellite is E_1. Now, it starts orbiting round the earth in a circular path at height h = radius of the earth. In this position, kinetic energy, potential energy and total mechanical energy of satellite are K_2, U_2 and E_2, respectively. Then,
(a) $U_2 = \frac{E_1}{2}$ (b) $E_2 = \frac{E_1}{4}$ (c) $K_2 = -E_2$ (d) $K_2 = -\frac{U_2}{2}$

10. A satellite is revolving round the earth in circular orbit
(a) if mass of the earth is made four times, keeping other factors constant, orbital speed of satellite will become two times
(b) corresponding to change in part (a), times period of satellite will remain half
(c) when value of G is made two times orbital speed increases and time period decreases
(d) G has no effect on orbital speed and time period

11. If the mass of the sun were ten times smaller and gravitational constant G were ten times larger in magnitude. Then,

(a) walking on ground would become more difficult
(b) the acceleration due to gravity on the earth will not change
(c) raindrops will fall much faster
(d) airplanes will have to travel much faster

12. If the sun and the planets carried huge amounts of opposite charges,

(a) all three of Kepler's laws would still be valid
(b) only the third law will be valid
(c) the second law will not change
(d) the first law will still be valid

13. There have been suggestions that the value of the gravitational constant G becomes smaller when considered over very large time period (in billions of years) in the future. If that happens, for our earth,

(a) nothing will change
(b) we will become hotter after billions of years
(c) we will be going around but not strictly in closed orbits
(d) after sufficiently long time we will leave the solar system

Assertion and Reason

Directions (Q. Nos. 1-20) *These questions consist of two statements each printed as assertion and reason. While answering these questions, you are required to choose anyone of the following five responses.*

(a) If both Assertion and Reason are true and Reason is the correct explanation of Assertion.
(b) If both Assertion and Reason are true but Reason is not the correct explanation of Assertion.
(c) If Assertion is true but Reason is false.
(d) If Assertion is false but Reason is true.
(e) If both Assertion and Reason are false.

1. Assertion If gravitational potential at some point is zero, then gravitational field strength at that point will also be zero.

Reason Except at infinity gravitational potential due to a system of point masses at some finite distance can't be zero.

2. Assertion Gravitational force between two masses in air is F. If they are immersed in water, force will remain F.

Reason Gravitational force does not depend on the medium between the masses.

3. Assertion The centre of semicircular ring of mass m and radius R is the origin O. The potential at origin is $-\frac{Gm}{R}$.

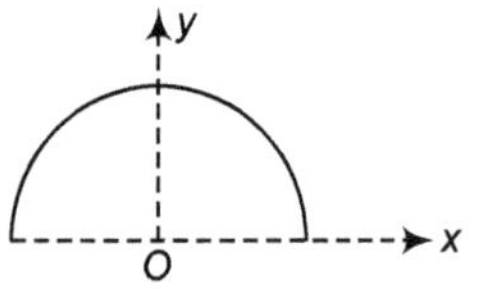

Reason The gravitational field strength is towards y-axis.

4. Assertion Four point masses each of mass m are placed at points 1, 2, 3 and 6 of a regular hexagon of side a. Then, the gravitational field at the centre of hexagon is $\frac{Gm}{a^2}$.

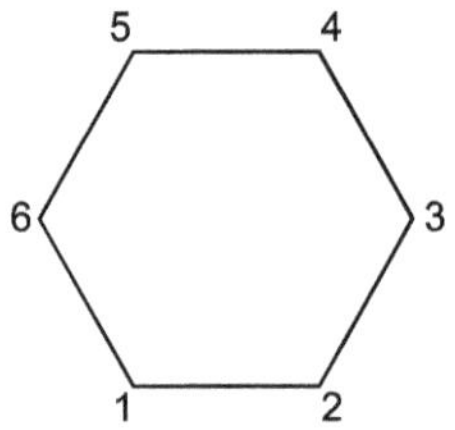

Reason The field strength due to masses at 3 and 6 are cancelled out.

5. Assertion The field strength at the centre of a ring is zero.

Reason At the centre of the ring, slope of v-r graph is zero.

6. Assertion Angular momentum of a planet is constant about any point.

Reason Force acting on the planet is a central force.

7. Assertion The binding energy of a satellite does not depend upon the mass of the satellite.

Reason Binding energy is the negative value of total energy of satellite.

8. Assertion If the product of surface area and density is same for two planets, escape velocities will be same for both.

Reason Product of surface area and density is proportional to the mass of the planet per unit radius of the planet.

9. Assertion Kepler's laws for planetary motion are consequence of Newton's laws.

Reason Kepler's laws can be derived by using Newton's laws.

10. Assertion The centres of two cubes of masses m_1 and m_2 are separated by a distance r. The gravitational force between these two cubes will be $\frac{Gm_1 m_2}{r^2}$.

Reason According to Newton's law of gravitation, gravitational force between two point masses m_1 and m_2 separated by a distance r is $\frac{Gm_1 m_2}{r^2}$.

11. Assertion Mass of the rod AB is m_1 and of particle P is m_2. Distance between centre of rod and particle is r. Then, the gravitational force between the rod and the particle is

$$F = \frac{Gm_1 m_2}{r^2}$$

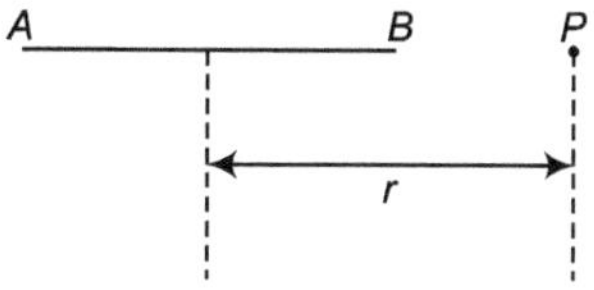

Reason The relation $F = \frac{Gm_1 m_2}{r^2}$ can be applied directly only two find force between two particles.

12. **Assertion** Two spherical shells have masses m_1 and m_2. Their radii are r_1 and r_2. Let r be the distance of a point from centre. Then, gravitational field strength and gravitational potential both are equal to zero for $0 < r < r_1$

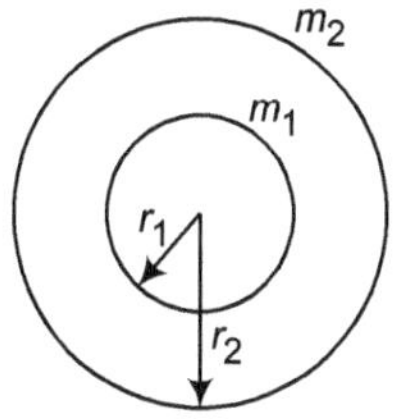

Reason In the region $r_1 < r < r_2$, gravitational field strength due to m_2 is zero. But gravitational potential due to m_2 is constant (but non-zero).

13. **Assertion** If radius of the earth suddenly shrinks to half its present without changing its mass value, then the period of an earth's satellite will not change.

Reason Time period of a satellite does not depond upon the mass of the earth.

14. **Assertion** On two sides of a point mass, gravitational field strength is same at same distance.

Reason As we move away from a point mass value of gravitational potential decreases.

15. **Assertion** Areal velocity of a planet around sun will become two times if mass of planet is halved.

Reason Areal velocity $= \frac{L}{2m}$, where L is angular momentum of planet about centre of sun.

16. **Assertion** Let W_1 is the work done in taking away a satellite from the surface of the earth to its orbit and then W_2 the work done in rotating the satellite in circular orbit there. Then,

$$W_1 = W_2$$

Reason $W_1 = W_2 = \frac{GMm}{2r}$

17. **Assertion** Plane of a satellite always passes through the centre of the earth.

Reason Gravitational force on satellite is always towards centre of the earth.

18. **Assertion** On earth's satellite, we feel weightlessness. Moon is also satellite of the earth. But on the surface of moon we do not feel weightlessness.

Reason Gravitational force by the earth on us on the surface of moon is zero. But, gravitational force by moon on us on its surface is non-zero.

19. **Assertion** Gravitational potential and gravitational potential energy both are related to the work done by gravitational force in the gravitational field.

Reason Gravitational field strength is related to the gravitational force in gravitational field.

20. **Assertion** If a particle is projected from the surface of the earth with velocity equal to escape velocity, then total mechanical energy is zero.

Reason Total mechanical energy of any closed system is always negative.

Match the Columns

1. On the surface of the earth, acceleration due gravity is g and gravitational potential is V. Match the following columns.

Column I	Column II
(A) At height $h = R$, value of g	(p) Decreases by a factor $\frac{1}{4}$
(B) At depth $h = \frac{R}{2}$, value of g	(q) Decreases by a factor $\frac{1}{2}$
(C) At height $h = R$, value of V	(r) Increases by a factor $\frac{11}{8}$
(D) At depth $h = \frac{R}{2}$, value of V	(s) Increases by a factor 2
	(t) None of the above

2. A particle is projected from the surface of the earth with speed v. Suppose it travels a distance x when its speed becomes v to $\frac{v}{2}$ and y when speed changes from $\frac{v}{2}$ to 0. Similarly, the corresponding times are suppose t_1 and t_2. Then,

Column I		Column II	
(A)	$\frac{x}{y}$	(p)	$= 1$
(B)	$\frac{t_1}{t_2}$	(q)	> 1
		(r)	< 1

3. Density of a planet is two times the density of the earth. Radius of this planet is half. Match the following (as compared to the earth) columns.

Column I	Column II
(A) Acceleration due to gravity on this planet's surface	(p) Half
(B) Gravitational potential on the surface	(q) Same
(C) Gravitational potential at centre	(r) Two times
(D) Gravitational field strength at centre	(s) Four times

4. In elliptical orbit of a planet, as the planet moves from apogee position to perigee position, match the following columns.

Column I	Column II
(A) Speed of planet	(p) Remains same
(B) Distance of planet from centre of sun	(q) Decreases
(C) Potential energy	(r) Increases
(D) Angular momentum about centre of sun	(s) Cannot say

5. Match the following columns.

Column I	Column II
(A) Kepler's first law	(p) $T^2 \propto r^3$
(B) Kepler's second law	(q) Areal velocity is constant
(C) Kepler's third law	(r) Orbit of planet is elliptical

6. Let V and E denote the gravitational potential and gravitational field at a point. Then, match the following columns.

Column I	Column II
(A) $E = 0, V = 0$	(p) At centre of spherical shell
(B) $E \neq 0, V = 0$	(q) At centre of solid sphere
(C) $V \neq 0, E = 0$	(r) At centre of circular ring
(D) $V \neq 0, E \neq 0$	(s) At centre of two point masses of equal magnitude

7. Two concentric spherical shells are as shown in figure. Match the following columns.

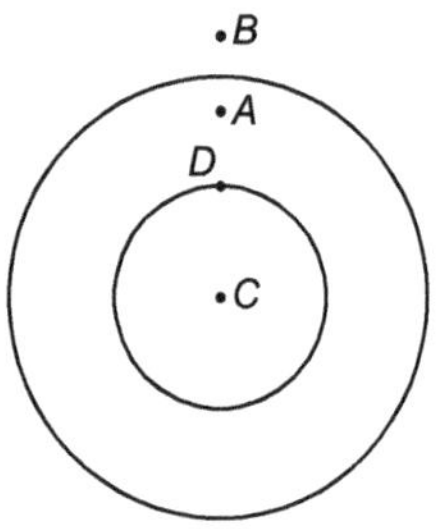

Column I	Column II
(A) Potential at A	(p) greater than B
(B) Gravitational field at A	(q) less than B
(C) As one moves from C to D	(r) Potential remains constant
(D) As one moves from D to A	(s) Gravitational field decreases
	(t) None of the above

8. Match the following columns.

Column I	Column II
(A) Kinetic energy of a particle in gravitational field is increasing	(p) Work done by gravitational force should be positive
(B) Potential energy of a particle in gravitational field is increasing	(q) Work done by external force should be non-zero
(C) Mechanical energy of a particle in gravitational field is increasing	(r) Work done by gravitational force should be negative
	(s) Cannot say anything

9. Match the following columns.

Column I	Column II
(A) Time period of an earth satellite in circular orbit	(p) Independent of mass of satellite
(B) Orbital velocity of satellite	(q) Independent of radius of orbit
(C) Mechanical energy of satellite	(r) Independent of mass of the earth
	(s) None of the above

10. If the earth decreases its rotational speed. Match the following columns.

Column I	Column II
(A) Value of g at pole	(p) Will remain same
(B) Value of g at equator	(q) Will increase
(C) Distance of geostationary satellite	(r) Will decrease
(D) Energy of geostationary satellite	(s) Cannot say

11. Match the following columns. (for a satellite in circular orbit)

Column I	Column II
(A) Kinetic energy	(p) $-\frac{GMm}{2r}$
(B) Potential energy	(q) $\sqrt{\frac{GM}{r}}$
(C) Total energy	(r) $-\frac{GMm}{r}$
(D) Orbital energy	(s) $\frac{GMm}{2r}$

Entrance Gallery

2014

1. A planet of radius $R = \frac{1}{10} \times$ (radius of the earth) has the same mass density as the earth. Scientist dig a well of depth $\frac{R}{5}$ on it and lower and wire of the same length and of linear mass density 10^{-3} kgm^{-1} into it. If the wire is not touching anywhere, the force applied at the top of the wire by a person holding it in place is (take the radius of earth $= 6 \times 10^6$ m and the acceleration due to gravity on the earth is 10 ms^{-2}). **[JEE Advanced]**

(a) 96 N (b) 108 N (c) 120 N (d) 150 N

2. Four particles, each of mass M and equidistant from each other, move along a circle of radius R under the action of their mutual gravitational attraction. The speed of each particle is **[JEE Main]**

(a) $\sqrt{\frac{GM}{R}}$ (b) $\sqrt{2\sqrt{2}\frac{GM}{R}}$

(c) $\sqrt{\frac{GM}{R}(1+2\sqrt{2})}$ (d) $\frac{1}{2}\sqrt{\frac{GM}{R}(1+2\sqrt{2})}$

3. If a body of mass m has to be taken from the surface to the earth to a height $h = R$, then the amount of energy required is (R = radius of the earth) **[Kerala CEE]**

(a) mgR (b) $\frac{mgR}{3}$ (c) $\frac{mgR}{2}$ (d) $\frac{mgR}{12}$

(e) $\frac{mgR}{9}$

4. The total energy of an artificial satellite of mass m revolving in a circular orbit around the earth with a speed v is **[Kerala CEE]**

(a) $\frac{1}{2}mv^2$ (b) $\frac{1}{4}mv^2$ (c) $-\frac{1}{4}mv^2$ (d) $-mv^2$

(e) $-\frac{1}{2}mv^2$

5. What is a period of revolution of the earth satellite? Ignore the height of satellite above the surface of the earth.

Given, the value of gravitational acceleration $g = 10\ \text{ms}^{-2}$, radius of the earth $R_e = 6400$ km.

(Take, $\pi = 3.14$) **[Karnataka CET]**

(a) 85 min (b) 156 min (c) 83.73 min (d) 90 min

6. What happens to the acceleration due to gravity with the increase in altitude from the surface of the earth?

(a) Increases **[J&K CET]**

(b) Decreases

(c) First decreases and then increases

(d) Remains same

7. A body weighs 45 N on the surface of the earth. What is the gravitational force on it due to the earth at a height equal to half of the radius of the earth? **[J&K CET]**

(a) 20 N (b) 45 N

(c) 40 N (d) 90 N

2013

8. Two bodies each of mass M are kept fixed with a separation $2L$. A particle of mass m is projected from the mid-point of the line joining their centres perpendicular to the line. The gravitational constant is G. The correct statement(s) is (are) **[JEE Advanced]**

(a) The minimum initial velocity of the mass m to escape the gravitational field of the two bodies is $4\sqrt{\frac{GM}{L}}$

(b) The minimum initial velocity of the mass m to escape the gravitational field of the two bodies is $2\sqrt{\frac{GM}{L}}$

(c) The minimum initial velocity of the mass m to escape the gravitational field of the two bodies is $\sqrt{\frac{2GM}{L}}$

(d) The energy of the mass m remains constant

9. What is the minimum energy required to launch a satellite of mass m from the surface of a planet of mass M and R in a circular orbit an altitude of $2R$? **[JEE Main]**

(a) $\frac{5\,GmM}{6R}$ (b) $\frac{2\,GmM}{3R}$ (c) $\frac{GmM}{2R}$ (d) $\frac{GmM}{3R}$

2012

10. Two spherical planets P and Q have the same uniform density ρ, masses M_P and M_Q and surface areas A and $4A$, respectively. A spherical planet R also has uniform density ρ and its mass is $(M_P + M_Q)$. The escape velocities from the planets P, Q and R, are v_P, v_Q and v_R, respectively. Then, **[JEE Main]**

(a) $v_Q > v_R > v_P$ (b) $v_R > v_Q > v_P$

(c) $v_R/v_P = 3$ (d) $v_P/v_Q = \frac{1}{2}$

2011

11. A satellite is moving with a constant speed v in a circular orbit about the earth. An object of mass m is ejected from the satellite such that it just escapes from the gravitational pull of the earth. At the time of its ejection, the kinetic energy of the object is **[IIT JEE]**

(a) $\frac{1}{2}mv^2$ (b) mv^2 (c) $\frac{3}{2}mv^2$ (d) $2mv^2$

12. Two particles of equal mass m go around a circle of radius R under the action of their mutual gravitational attraction. The speed of each particle with respect to their centre of mass is **[AIEEE]**

(a) $\sqrt{\frac{Gm}{R}}$ (b) $\sqrt{\frac{Gm}{4R}}$

(c) $\sqrt{\frac{GM}{3R}}$ (d) $\sqrt{\frac{Gm}{2R}}$

13. Two bodies of masses m and $4m$ are placed at a distance r. The gravitational potential at a point on the line joining them where the gravitational field is zero, is **[AIEEE]**

(a) $-\frac{4Gm}{r}$ (b) $-\frac{6Gm}{r}$

(c) $-\frac{9Gm}{r}$ (d) zero

14. A body is projected with a velocity of 2×11.2 km/s from the surface of earth. The velocity of the body when it escapes the gravitational pull of the earth is **[Kerala CEE]**

(a) $\sqrt{3} \times 11.2$ km/s (b) 11.2 km/s

(c) $\sqrt{2} \times 11.2$ km/s (d) 0.5×11.2 km/s

(e) 2×11.2 km/s

15. A satellite is launched into a circular orbit of radius R around of the earth. A second satellite is launched into an orbit of radius $4R$. The ratio of their respective periods is **[Kerala CEE]**

(a) 4 : 1 (b) 1 : 8 (c) 8 : 1 (d) 1 : 4

(e) 1 : 2

16. If the earth were to suddenly contract to $\frac{1}{n}$ th of its present radius without any change in its mass, the duration of the new day will be nearly **[WB JEE]**

(a) $\frac{24}{n}$ h (b) $24n$ h

(c) $\frac{24}{n^2}$ h (d) $24n^2$ h

17. If g is the acceleration due to gravity on the surface of the earth, the gain in potential energy of an object of mass m raised from the earth's surface to a height equal to the radius R of the earth is **[WB JEE]**

(a) $\frac{mgR}{4}$ (b) $\frac{mgR}{2}$
(c) mgR (d) $2mgR$

2010

18. A thin uniform angular disc (see figure) of mass M has outer radius $4R$ and inner radius $3R$. The work required to take a unit mass from point P on its axis to infinity is

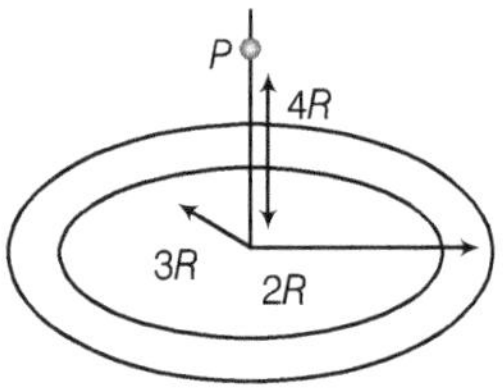

[IIT JEE]

(a) $\frac{2GM}{7R}(4\sqrt{2}-5)$
(b) $-\frac{2GM}{7R}(4\sqrt{2}-5)$
(c) $\frac{GM}{4R}$
(d) $\frac{2GM}{5R}(\sqrt{2}-1)$

19. Gravitational acceleration on the surface of a planet is $\frac{\sqrt{6}}{11}g$, where g is the gravitational acceleration on the surface of the earth. The average mass density of the planet is $\frac{2}{3}$ times that of the earth. If the escape speed on the surface of the earth is taken on be 11 kms^{-1}, the escape speed on the surface of the planet in kms^{-1} will be **[IIT JEE]**

(a) 5 (b) 7
(c) 3 (d) 11

20. The height at which the acceleration due to gravity becomes $\frac{g}{9}$ (where, g = the acceleration due to gravity on the surface of the earth) in terms of R, the radius of the earth, is **[AIEEE]**

(a) $2R$ (b) $\frac{R}{\sqrt{3}}$
(c) $\frac{R}{2}$ (d) $\sqrt{2}R$

21. The ratio of radii of the earth to another planet is $\frac{2}{3}$ and the ratio of their mean densities is $\frac{4}{5}$. If an astronaut can jump to a maximum height of 1.5 m on the earth, with the same effort, the maximum height he can jump on the planet is **[Kerala CEE]**

(a) 1 m (b) 0.8 m (c) 0.5 m (d) 1.25 m
(e) 2 m

22. At what depth below the surface of the earth, the value of g is the same as that at a height of 5 km? **[Kerala CEE]**

(a) 1.25 km (b) 2.5 km (c) 5 km (d) 7.5 km
(e) 10 km

23. A body is at rest on the surface of the earth. Which of the following statement is correct? **[OJEE]**

(a) No force is acting on the body
(b) Only weight of the body acts on it
(c) Net downward force is equal to the net upward force
(d) None of the above statement is correct

24. The density of the earth in terms of acceleration due to gravity (g), radius of the earth (R) and universal gravitational constant (G) is **[OJEE]**

(a) $\frac{4\pi RG}{3g}$ (b) $\frac{3\pi RG}{4g}$
(c) $\frac{4g}{3\pi RG}$ (d) $\frac{3g}{4\pi RG}$

25. A body is taken to a height of nR from the surface of the earth. The ratio of the acceleration due to gravity on the surface to that at the altitude is **[MHT CET]**

(a) $(n+1)^2$ (b) $(n+1)^{-2}$ (c) $(n+1)^{-1}$ (d) $(n+1)$

26. If the density of the earth is doubled keeping radius constant, find the new acceleration due to gravity? (Take, $g = 9.8\ m/s^2$) **[MHT CET]**

(a) 9.8 m/s^2 (b) 19.6 m/s^2
(c) 4.9 m/s^2 (d) 39.2 m/s^2

27. If ρ is the density of the planet, the time period of near by satellite is given by **[MHT CET]**

(a) $\sqrt{\frac{4\pi}{3G\rho}}$ (b) $\sqrt{\frac{4\pi}{G\rho}}$
(c) $\sqrt{\frac{3\pi}{G\rho}}$ (d) $\sqrt{\frac{\pi}{G\rho}}$

28. In a satellite, if the time of revolution is T, then KE is proportional to **[MHT CET]**

(a) $\frac{1}{T}$ (b) $\frac{1}{T^2}$
(c) $\frac{1}{T^3}$ (d) $T^{-2/3}$

Answers

Level 1

Objective Problems

1.	(c)	**2.**	(c)	**3.**	(a)	**4.**	(b)	**5.**	(c)	**6.**	(a)	**7.**	(c)	**8.**	(a)	**9.**	(a)	**10.**	(b)
11.	(a)	**12.**	(c)	**13.**	(a)	**14.**	(b)	**15.**	(b)	**16.**	(c)	**17.**	(b)	**18.**	(c)	**19.**	(b)	**20.**	(b)
21.	(b)	**22.**	(c)	**23.**	(a)	**24.**	(b)	**25.**	(a)	**26.**	(b)	**27.**	(d)	**28.**	(c)	**29.**	(d)	**30.**	(d)
31.	(b)	**32.**	(c)	**33.**	(a)	**34.**	(c)	**35.**	(a)	**36.**	(c)	**37.**	(c)	**38.**	(d)	**39.**	(c)	**40.**	(b)
41.	(c)	**42.**	(a)	**43.**	(a)	**44.**	(c)	**45.**	(c)	**46.**	(a)	**47.**	(a)	**48.**	(c)	**49.**	(d)	**50.**	(a)
51.	(d)	**52.**	(d)	**53.**	(c)	**54.**	(c)	**55.**	(c)	**56.**	(b)	**57.**	(c)	**58.**	(b)	**59.**	(a)	**60.**	(c)
61.	(c)	**62.**	(d)	**63.**	(c)	**64.**	(b)	**65.**	(c)	**66.**	(d)	**67.**	(b)	**68.**	(d)	**69.**	(c)	**70.**	(c)
71.	(a)	**72.**	(a)	**73.**	(a)	**74.**	(a)	**75.**	(d)	**76.**	(d)	**77.**	(d)	**78.**	(c)	**79.**	(b)	**80.**	(b)
81.	(d)	**82.**	(d)	**83.**	(b)	**84.**	(c)	**85.**	(c)	**86.**	(a)	**87.**	(d)	**88.**	(a)	**89.**	(b)	**90.**	(d)
91.	(a)	**92.**	(b)	**93.**	(a,a)	**94.**	(d)	**95.**	(b)	**96.**	(b)	**97.**	(c)	**98.**	(d)	**99.**	(b)	**100.**	(d)
101.	(c)	**102.**	(b)	**103.**	(a)	**104.**	(c)	**105.**	(a)	**106.**	(a)	**107.**	(a)						

Level 2

Only One Correct Option

1.	(a)	**2.**	(a)	**3.**	(c)	**4.**	(c)	**5.**	(b)	**6.**	(c)	**7.**	(b)	**8.**	(a)	**9.**	(b)	**10.**	(a)
11.	(b)	**12.**	(c)	**13.**	(c)	**14.**	(a)	**15.**	(a)	**16.**	(d)	**17.**	(d)	**18.**	(b)	**19.**	(b)	**20.**	(c)
21.	(c)	**22.**	(d)	**23.**	(a,d)	**24.**	(c)	**25.**	(b)	**26.**	(a)	**27.**	(c,d)	**28.**	(d)	**29.**	(c)	**30.**	(d)
31.	(d)	**32.**	(d)	**33.**	(b)	**34.**	(d)	**35.**	(a)	**36.**	(d)	**37.**	(b)	**38.**	(a)	**39.**	(d)	**40.**	(b)
41.	(b)	**42.**	(b)	**43.**	(c)														

More than One Correct Options

1.	(a,b,d)	**2.**	(a,b,d)	**3.**	(a,c,d)	**4.**	(a,d)	**5.**	(a,b,c)	**6.**	(a,c)	**7.**	(b,c,d)	**8.**	(c,d)	**9.**	(all)	**10.**	(a,b)
11.	(a,c,d)	**12.**	(a,c,d)	**13.**	(c,d)														

Assertion and Reason

1.	(d)	**2.**	(a)	**3.**	(b)	**4.**	(d)	**5.**	(a,b)	**6.**	(d)	**7.**	(d)	**8.**	(a)	**9.**	(d)	**10.**	(d)
11.	(e)	**12.**	(d)	**13.**	(b)	**14.**	(e)	**15.**	(a)	**16.**	(e)	**17.**	(a)	**18.**	(c)	**19.**	(b)	**20.**	(b)

Match the Columns

1. (A $\rightarrow$ q, B $\rightarrow$ q, C $\rightarrow$ s) **2.** (A $\rightarrow$ r, B $\rightarrow$ r) **3.** (A $\rightarrow$ q, B $\rightarrow$ p, C $\rightarrow$ p, D $\rightarrow$ q) **4.** (A $\rightarrow$ r, B $\rightarrow$ q, C $\rightarrow$ q, D $\rightarrow$ p)
5. (A $\rightarrow$ r, B $\rightarrow$ q, C $\rightarrow$ p) **6.** (A $\rightarrow$ t, B $\rightarrow$ t, C $\rightarrow$ p,q,r,s, D $\rightarrow$ t) **7.** (A $\rightarrow$ q, B $\rightarrow$ t, C $\rightarrow$ r, D $\rightarrow$ s) **8.** (A $\rightarrow$ s, B $\rightarrow$ r, C $\rightarrow$ q)
9. (A $\rightarrow$ p, B $\rightarrow$ p, C $\rightarrow$ s) **10.** (A $\rightarrow$ p, B $\rightarrow$ q, C $\rightarrow$ q, D $\rightarrow$ q) **11.** (A $\rightarrow$ s, B $\rightarrow$ r, C $\rightarrow$ p, D $\rightarrow$ q)

Entrance Gallery

1.	(b)	**2.**	(d)	**3.**	(c)	**4.**	(e)	**5.**	(c)	**6.**	(b)	**7.**	(a)	**8.**	(b)	**9.**	(a)	**10.**	(b)
11.	(b)	**12.**	(b)	**13.**	(c)	**14.**	(a)	**15.**	(b)	**16.**	(c)	**17.**	(b)	**18.**	(a)	**19.**	(c)	**20.**	(a)
21.	(c)	**22.**	(e)	**23.**	(e)	**24.**	(d)	**25.**	(a)	**26.**	(b)	**27.**	(c)	**28.**	(d)				

Solutions

Level 1 : Objective Problems

1. $g = \frac{GM}{R^2}$ or $g \propto \frac{M}{R^2}$

Mass is 2 times and diameter is 3 times. Hence, $g' = \frac{2g}{9}$

2. With height, $g' = \frac{g}{\left(1 + \frac{h}{R}\right)^2}$ ... (i)

Given $g' = \frac{g}{4}$, substituting in Eq. (i) we get, $h = R$.

3. At poles value of g is maximum. There is no effect of rotation of the earth.

4. At depth, $g' = g\left(1 - \frac{h}{R}\right)$ or $g\left(1 - \frac{d}{R}\right)$

$$\frac{g'}{g} = \frac{1}{n} = \left(1 - \frac{d}{R}\right) \text{ or } d = R\left(\frac{n-1}{n}\right)$$

5. $g = \frac{GM}{R^2}$ or $g \propto \frac{M}{R^2}$

6. $F = \frac{Gm_1m_2}{r^2}$

$$\therefore \quad r = \sqrt{\frac{Gm_1m_2}{F}} = \sqrt{\frac{6.67 \times 10^{-11} \times 1 \times 1}{9.8 \times 10^{-9}}}$$

$$= 0.082 \text{ m} = 8.2 \text{ cm}$$

7. $g' = \frac{g}{\left(1 + \frac{h}{R}\right)^2}$ (at $h = R$)

$$= \frac{g}{4}$$

$$\therefore \quad W' = \frac{72}{4} = 18 \text{N}$$

8. At equator, $g' = g - R\omega^2$

$$0 = g - R\omega^2$$

$$\therefore \quad \omega = \sqrt{\frac{g}{R}} = \sqrt{\frac{9.8}{6400 \times 10^3}}$$

$$= 1.23 \times 10^{-3} \text{ rad/s}$$

9. At centre of earth value of g is zero. Hence, weight is zero.

10. $g\left(1 - \frac{2h}{R}\right) = \left(1 - \frac{d}{R}\right)$ (at $h << R$)

$$\text{or} \quad 1 - \frac{2h}{R} = 1 - \frac{d}{R}$$

$$\therefore \quad d = 2h$$

11. $0 = g - R\omega^2 \cos^2 60°$

$$\text{or} \quad \omega^2 = \frac{4g}{R}$$

$$\text{or} \quad \omega^2 = 2\sqrt{\frac{g}{R}}$$

$$\frac{2\pi}{T} = 2\sqrt{\frac{g}{R}}$$

$$\therefore \quad T = \pi\sqrt{\frac{R}{g}}$$

12. $F = \frac{Gm_1m_2}{r^2}$ or $F \propto m_1m_2$.

On M due to masses m force is F and due to mass $2m$ force is $2F$. Therefore, the net force is $3F$.

13. In a coal mine and at the top of a mountain value of g is less, hence apparent weight is less.

14. Because acceleration due to gravity increases.

15. $g = \frac{GM}{R^2}$. If radius shrinks to half of its present value, then g will become four times.

16. $$g = \frac{4}{3}G\pi R\rho$$

$$\Rightarrow \quad \frac{g_1}{g_2} = \frac{\rho_1 R_1}{\rho_2 R_2} = \frac{1}{2} \times \frac{4}{1} = \frac{2}{1}$$

17. $g \propto \rho R$

18. $g = \frac{4}{3}\pi\rho G R$

$\Rightarrow g \propto dR$ ($\rho = d$ given in the problem)

19. Acceleration due to gravity at latitude ϕ is given by

$$g' = g - R\omega^2 \cos^2 \phi$$

$$\text{At} \quad 30°, g_{30°} = g - R\omega^2 \cos^2 30° = g - \frac{3}{4}R\omega^2$$

$$\therefore \quad g - g_{30°} = \frac{3}{4}\omega^2 R$$

20. Acceleration due to gravity, $g = \frac{GM}{R^2}$

$$\therefore \quad \frac{g}{G} = \frac{M}{R^2}$$

21. $F = \frac{Gm(M - m)}{r^2}$

For maximum force, $\frac{dF}{dm} = 0$

$$\Rightarrow \quad \frac{d}{dm}\left(\frac{GmM}{r^2} - \frac{Gm^2}{r^2}\right) = 0$$

$$\Rightarrow \quad M - 2m = 0 \Rightarrow \frac{m}{M} = \frac{1}{2}$$

22. $g \propto r$ (if $r < R$) and $g \propto \frac{1}{r^2}$ (if $r < R$)

23. $g' = \frac{g}{\left(1 + \frac{h}{R}\right)^2}$

$$\text{Given,} \quad g' = \frac{g}{64}$$

$$\therefore \quad 1 + \frac{h}{r} = 8 \quad \text{or} \quad \frac{h}{R} = 7$$

$$\therefore \quad h = 7R = 44.8 \times 10^6 \text{ m}$$

24. At equator $g' = g - R\omega^2$. Matter will start escaping from equator when $g' = 0$

$$\therefore \quad \omega = \sqrt{\frac{g}{R}} = \sqrt{\frac{GM}{R^3}}$$

25. $g\left(1-\frac{d}{R}\right)=\frac{1}{2}\frac{g}{\left(1+\frac{h}{R}\right)^2}$

or $$1-\frac{d}{R}=\frac{1}{2\left(1+\frac{1600}{6400}\right)^2}=\frac{16}{50}$$

$$\frac{d}{R}=\frac{34}{50}$$

$\therefore$ $$d=(6.4\times10^6\text{ m})\left(\frac{34}{50}\right)$$

$$=4.352\times10^6\text{ m}$$

26. Potential is work done per unit mass. Hence, unit is J/kg.

27. Inside a shell, field strength is zero. Therefore, force on a particle is zero.

29. $E_{\text{net}}=0$

$\therefore$ $\frac{Gm}{x^2}=\frac{GM}{(d-x)^2}$, x is distance from m

$\therefore$ $$\frac{x}{d-x}=\frac{\sqrt{m}}{\sqrt{M}}$$

$$x=\frac{\sqrt{m}}{\sqrt{m}+\sqrt{M}}\cdot d$$

and $$d-x=\frac{\sqrt{M}}{\sqrt{m}+\sqrt{M}}\cdot d$$

$$V=-\frac{Gm}{x}-\frac{GM}{d-x}$$

$$=-\frac{Gm(\sqrt{m}+\sqrt{m})}{\sqrt{m}\cdot d}-\frac{GM\sqrt{m}+\sqrt{M}}{\sqrt{M}\cdot d}$$

$$=-\frac{G}{d}(\sqrt{m}+\sqrt{M})^2$$

30. $$\pi R=L$$

$\therefore$ $$R=\frac{L}{\pi}$$

$$V=-\frac{GM}{R}=-\frac{\pi GM}{L}$$

31. At distance $2.5\,r$ or $\frac{5}{2}r$

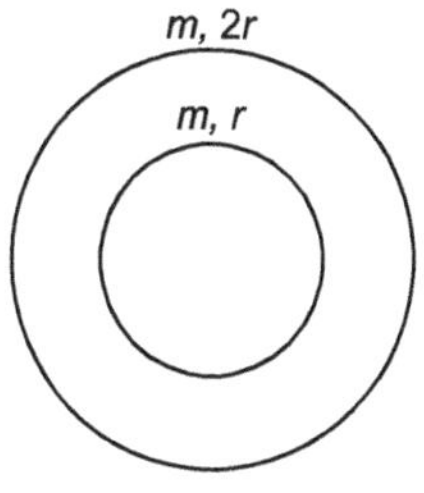

$$E=\frac{G(m+m)}{\left(\frac{5}{2}r\right)^2}=\frac{8Gm}{25r^2}$$

33. Reduces by 75%, means 25% is left.

$$\frac{GM}{r^2}=\frac{1}{4}\left(\frac{GM}{R^2}\right)$$

or $$r=2R$$

$\therefore$ $$h=r-R$$

$$=2R-R=R$$

34. At centre, $V_c=-\frac{3}{2}\frac{GM}{R}$, on surface $V_s=-\frac{Gm}{R}$

$\therefore$ $$V_c=\frac{3}{2}V_s$$

35. Due to three particles, net intensity at the centre $I=I_A+I_B+I_C=0$. Because, these three intensities are equal in magnitude and the angle between each other is 120°.

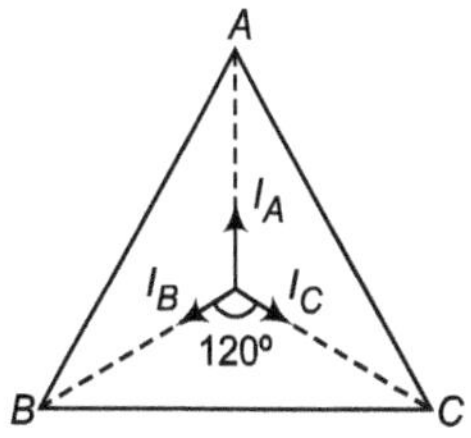

36. $V_{\text{in}}=\frac{-Gm}{2R}\left[3-\left(\frac{r}{R}\right)^2\right], V_{\text{surface}}=\frac{-GM}{R}, V_{\text{out}}=\frac{-GM}{r}$

37. For hollow sphere,

$$V_{\text{in}}=\frac{-GM}{R}, V_{\text{surface}}=\frac{-GM}{R}, V_{\text{out}}=\frac{-GM}{r}$$

i.e. potential remains constant inside the sphere and it is equal to potential at the surface and increase when the point moves away from the surface of sphere.

38. Intensity will be zero inside the spherical shell.

$$E=0 \text{ up to } r=R$$

and $$E\propto\frac{1}{r^2}, \text{ when } r>R$$

39. Total mechanical energy of any closed system is always negative.

40. $v_e=\sqrt{2gR}$, v_e is independent of mass of projectile.

41. $\Delta U=\frac{mgh}{\left(1+\frac{h}{R}\right)}$, $h=R$ (given)

$\therefore$ $$\Delta U=\frac{mgR}{2}$$

43. $v_e=\sqrt{2gR}=\sqrt{2\frac{GM}{R^2}\cdot R}$ or $v_e\propto\sqrt{\frac{M}{R}}$

Mass is 1000 times and radius is 10 times. Therefore, escape velocity will become 10 times.

44. $U_i+K_i=U_f+K_f$

$$-\frac{GMm}{R}+\frac{1}{2}m(2v_e)^2=0+\frac{1}{2}mv^2$$

or $$-\frac{GM}{R}+2v_e^{\,2}=\frac{1}{2}v^2$$

or $$-\frac{2GM}{R}+\frac{8GM}{R}=v^2$$

or $$v=\sqrt{\frac{6GM}{R}}=\sqrt{3\left(\frac{2GM}{R}\right)}$$

$$=\sqrt{3(2gR)}=\sqrt{3}v_e$$

45. Escape velocity is independent of mass of projectile which is projected.

46. At $v<v_e$, total energy is negative.

$v=v_e$, total energy is zero and

$v=v_e$, total energy is positive

47. Escape velocity is independent of angle at which particle is projected.

48. Escape velocity is independent of mass of particle which is projected.

49. $v = \sqrt{2gR}$

$$\therefore \quad \frac{v_e}{v_m} = \sqrt{\frac{g_e}{g_m} \times \frac{R_e}{R_m}} = \sqrt{\frac{6}{1} \times \frac{10}{1}} = \sqrt{60}$$

50. Decrease in kinetic energy = increase in potential energy

$$\therefore \quad \frac{1}{2}mv^2 = \frac{mgh}{1+\frac{h}{R}}, \quad \text{(given, } h = R)$$

$$\therefore \quad v^2 = \frac{2gR}{2} = gR = \frac{GM}{R}$$

$$\text{or} \quad v = \sqrt{\frac{GM}{R}}$$

51. Escape velocity is independent of the angle at which it is projected.

52. Change in potential energy in displacing a body from r_1 to r_2 is given by

$$\Delta U = GMm\left[\frac{1}{r_1} - \frac{1}{r_2}\right] = GMm\left(\frac{1}{2R} - \frac{1}{3R}\right) = \frac{GMm}{6R}$$

53. $\frac{1}{2}mv_e^2 = \frac{1}{2}m \times 2gR = mgR$

54. $v_e = \sqrt{\frac{2GM}{R}}$

i.e. escape velocity depends upon the mass and radius of the planet.

55. E = Energy of satellite – energy of satellite on surface of earth

$$= -\frac{GMm}{2(3R)} - \left[-\frac{GMm}{R}\right] = \frac{5}{6}\frac{GMm}{R} = \frac{5}{6}mgR \quad \left(\text{as } \frac{GM}{R} = gR\right)$$

56. In circular path, if necessary centripetal force disappears, body moves in tangential path.

58. $T = \frac{2\pi (r)^{3/2}}{\sqrt{GM}}$, T is independent of m, the mass of satellite.

59. $T \propto r^{3/2}$

$$\therefore \quad \frac{T_2}{T_1} = \left(\frac{r_2}{r_1}\right)^{3/2}$$

$$\text{or} \quad T_2 = T_1 (4)^{3/2} = 8T_1 = 8 \text{ days}$$

60. $T^2 \propto r^3, r_1 : r_2 : r_3 = \frac{1}{2} : 1 : \frac{3}{2}$

$$\therefore \quad T_1^2 : T_2^2 : T_3^2 = \frac{1}{8} : 1 : \frac{27}{8} = 1 : 8 : 27$$

61. $d_1 v_1 = d_2 v_2$ (from conservation of angular momentum at these two points)

$$\therefore \quad v_2 = \frac{d_1 v_1}{d_2}$$

62. Energy of satellite is given by

$$E = -\frac{GMm}{2r}$$

$$\text{or} \quad E \propto \frac{m}{r}$$

63. $T \propto r^{3/2}$

64. In elliptical orbit velocity, both in magnitude and direction keep on changing.

65. $v = \sqrt{\frac{GM}{r}}$

$$\text{or} \quad v \propto \frac{1}{\sqrt{r}}$$

$$\frac{v_2}{v_1} = \sqrt{\frac{r_1}{r_2}}$$

$$\therefore \quad v_2 = \sqrt{\frac{r_1}{r_2}} \cdot v = \sqrt{\frac{R}{R+\frac{R}{2}}} \cdot v = \sqrt{\frac{2}{3}}\, v$$

66. $K = \frac{GMm}{2r}, U = -\frac{GMm}{r}, E = -\frac{GMm}{2r}$

If r is decreased, K will increase but U and E will decrease.

67. $T \propto r^{3/2}$

68. $\frac{dA}{dt} = \frac{L}{2m}$

69. Potential energy is two times the mechanical energy.

70. $U = 2E$ and $K = -E$

$$\therefore \quad U = 2E_0 \text{ and } K = -E_0$$

71. $T = \frac{2\pi r^{3/2}}{\sqrt{GM}}$ or $\sqrt{GM} = \frac{2\pi}{T} \cdot r^3$

$$\frac{2\pi}{T} = \omega$$

$$\therefore \quad \omega r^{3/2} = \sqrt{GM}$$

$$\text{or} \quad r^3 = \frac{GM}{\omega^2} = \frac{gR^2}{\omega^2}$$

72. $L = mvr$ or $L \propto vr$

$$v \propto \frac{1}{\sqrt{r}}$$

$$\therefore \quad L \propto r^{1/2}$$

i.e. with increase in r, L will increase.

75. Energy of a satellite is given by $E = -\frac{GMm}{2r}$

For a satellite very close to the earth, $r = R$

$$\therefore \quad E = -\frac{GMm}{2R}$$

76. $r_1 = R + R = 2R$ and $r_2 = 7R + R = 8R$

$$\text{Now,} \quad K = \frac{GMm}{2r}$$

$$\frac{K_1}{K_2} = \frac{r_2}{r_1} = 4$$

$$U = -\frac{GMm}{r}, \frac{U_1}{U_2} = \frac{r_2}{r_1} = 4$$

$$\text{and} \quad E = -\frac{GMm}{2r} \text{ or } \frac{U_1}{U_2} = \frac{r_2}{r_1} = 4$$

77. $T \propto r^{3/2}$

$\therefore \quad \dfrac{T_1}{T_2} = \left(\dfrac{10^{13}}{10^{12}}\right)^{3/2} = 10\sqrt{10}$

79. $v = \sqrt{\dfrac{GM}{R}}$

$\Rightarrow \quad \dfrac{v_A}{v_B} = \sqrt{\dfrac{R_B}{R_A}} = \sqrt{\dfrac{R}{4R}} = \dfrac{1}{2}$

$\therefore \quad \dfrac{v_A}{v_B} = \dfrac{3v}{v_B} = \dfrac{1}{2}$

$\therefore \quad v_B = 6v$

81. Total mechanical energy of satellite

$$E = \frac{-GMm}{2r}$$

$\Rightarrow \quad \dfrac{E_A}{E_B} = \dfrac{m_A}{m_B} \times \dfrac{r_B}{r_A}$

$\Rightarrow \quad \dfrac{3}{1} \times \dfrac{4r}{r} = \dfrac{12}{1}$

82. $T^2 \propto R^3$

$\therefore \quad \dfrac{T^2}{R^3} = \text{constant}$

83. Given that, $T_1 = 1$ day and $T_2 = 8$ days

$\therefore \quad \dfrac{T_2}{T_1} = \left(\dfrac{r_2}{r_1}\right)^{3/2}$

$\Rightarrow \quad \dfrac{r_2}{r_1} = \left(\dfrac{T_2}{T_1}\right)^{2/3} = \left(\dfrac{8}{1}\right)^{2/3} = 4$

$\Rightarrow \quad r_2 = 4r_1 = 4R$

85. $g = \dfrac{GM}{R^2} = \dfrac{G\left(\dfrac{4}{3}\pi R^3\right)\rho}{R^2}$

$\therefore \quad \rho = \dfrac{g}{G \cdot 4\pi \dfrac{R}{3}} = \dfrac{3g}{4\pi GR}$

86. Total mechanical energy is conserved, not the kinetic energy.

87. Force will be zero at the point of zero intensity

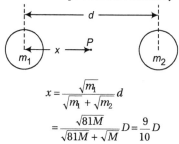

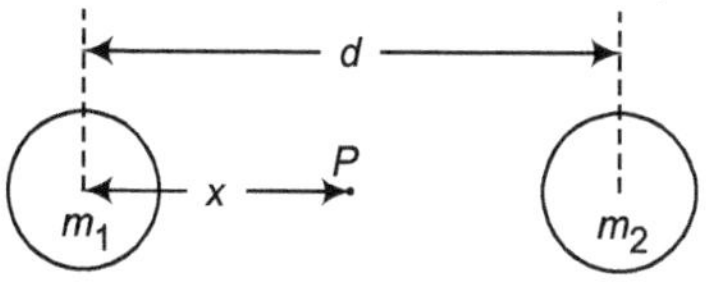

$$x = \frac{\sqrt{m_1}}{\sqrt{m_1} + \sqrt{m_2}} d$$

$$= \frac{\sqrt{81M}}{\sqrt{81M} + \sqrt{M}} D = \frac{9}{10} D$$

88. At equator, $g' = g - R\omega^2 = 0$

$\therefore \quad \omega = \sqrt{\dfrac{g}{R}}$

$$\frac{2\pi}{T} = \sqrt{\frac{g}{R}}$$

$\therefore \quad T = 2\pi\sqrt{\dfrac{R}{g}}$

89. On surface of earth, $U = -\dfrac{GMm}{R}$

At height, $h (<< R)$, increase in potential energy is mgh

$\therefore \quad U_h = -\dfrac{GMm}{R} + mgh$

90. $T = 2\pi\sqrt{\dfrac{l}{g}} \propto \dfrac{1}{\sqrt{g}}$

$\therefore \quad \dfrac{T_2}{T_1} = \sqrt{\dfrac{g_1}{g_2}} = \sqrt{\dfrac{g}{\dfrac{g}{\left(1+\dfrac{h}{R}\right)^2}}} = 2 \qquad (\text{at } h = R)$

91. Decrease in kinetic energy = increase in PE

$\therefore \quad \dfrac{1}{2} m \left(\dfrac{v_e}{\sqrt{2}}\right)^2 = \dfrac{mgh}{1+\dfrac{h}{R}}$

or $\quad \dfrac{v_e^2}{4} = \dfrac{gh}{1+\dfrac{h}{R}}$

or $\quad \dfrac{2gR}{4} = \dfrac{gh}{1+\dfrac{h}{R}}$ or $\dfrac{R}{2} = \dfrac{h}{1+\dfrac{h}{R}}$

Solving this equation, we get $h = R$

Note Kinetic energy is half the value required to escape.

Therefore, speed is $\dfrac{1}{\sqrt{2}}$ times the value required to escape.

92. $F = \dfrac{k}{r} \qquad (k = \text{constant})$

$\therefore \quad \dfrac{mv^2}{r} = \dfrac{k}{r}$ or $v \propto r^0$

93. (i) Net force on the system is zero. Hence, velocity of centre of mass at any instant will remain zero. As, initially they were at rest.

(ii) Increase in kinetic energy of both particles = decrease in gravitational potential energy

$\therefore \quad 2\left(\dfrac{1}{2}mv^2\right) = U_i - U_f = -\dfrac{Gmm}{d} + \dfrac{Gmm}{\dfrac{d}{2}} = \dfrac{Gm^2}{d}$

$\therefore \quad v = \sqrt{\dfrac{Gm}{d}}$

94. $g = \dfrac{GM}{R^2}$ or $\dfrac{G}{g} = \dfrac{R^2}{m}$

$\therefore \quad \dfrac{G}{g}$ will have the units $\dfrac{\text{m}^2}{\text{kg}}$.

95. $\dfrac{mv^2}{r} = \dfrac{Gm_1m_2}{r^n}$ or $v \propto r^{\left(\frac{1-n}{2}\right)}$

Now, $\quad T = \dfrac{2\pi r}{v}$

or $\quad T \propto \dfrac{r}{v}$

$\therefore \quad T \propto \dfrac{r}{r^{\left(\frac{1-n}{2}\right)}}$

$$T \propto r^{\frac{1+n}{2}}$$

96. Actually, gravitational force provides the centripetal force.

98. Range of projectile, $R = \frac{u^2 \sin 2\theta}{g}$

If u and θ are constant, then $R \propto \frac{1}{g}$

$$\frac{R_m}{R_e} = \frac{g_e}{g_m}$$

$$\Rightarrow \quad \frac{R_m}{R_e} = \frac{1}{0.2}$$

$$\Rightarrow \quad R_m = \frac{R_e}{0.2}$$

$$\Rightarrow \quad R_m = 5R_e$$

99. $v = \sqrt{\frac{GM}{R}} = G^{1/2} M^{1/2} R^{-1/2}$

101. $\frac{dA}{dt} = \frac{vr\sin\theta}{2} = \frac{1}{2}|\mathbf{r} \times \mathbf{v}|$

102. $u = \frac{GMm}{r}$

At distance $2r$, $E = \frac{GM}{(2r)^2} = \frac{GM}{4r^2} = \frac{u}{4mr}$

Now, $\quad W = mE$

or $\quad mg = \frac{u}{4r}$

103. KE = 0, only PE is present.

$$\therefore \quad E = U = 2\left(-\frac{GMm}{r}\right) = -\frac{2GMm}{r}$$

104. $v_e = \sqrt{2} v_0 = 1.414 v_0$

or orbital speed is to be increased by 41%.

Further, speed is to increase $\sqrt{2}$ times. Or kinetic energy is to increase two times.

105. $\frac{A}{T} = \frac{L}{2m}$ or $L = \frac{2mA}{T}$

106. $\Delta v = v_e - v_0 = \sqrt{2gR} - \sqrt{gR}$

$$= (\sqrt{2} - 1)\sqrt{gR} = (\sqrt{2} - 1)\sqrt{\frac{GM}{R}}$$

107. $\frac{1}{2}mv^2 = m\left[-\frac{GM}{R} + \frac{3GM}{2R}\right] = \frac{GMm}{2R}$

(increase in KE = decrease in PE)

or $\quad v = \sqrt{\frac{GM}{R}}$

Momentum, $mv = m\sqrt{\frac{GM}{R}}$

$\therefore$ Impulse required $= m\sqrt{\frac{GM}{R}}$

Level 2 : Only One Correct Option

1. $\frac{1}{2}mv^2 = \frac{mgh}{1 + \frac{h}{R}}$

$$\therefore \quad h = \frac{v^2}{2g - \frac{v^2}{R}} = \frac{R}{\left(\frac{2gR}{v^2}\right) - 1}$$

2. $\frac{mv^2}{r} \propto r^{-n}$

$$\therefore \quad v \propto r^{(1-n)/2}$$

$$T = \frac{2\pi r}{v} \quad \text{or} \quad T \propto rv^{-1}$$

or $\quad T \propto r \cdot r^{(n-1)/2}$

or $\quad T \propto r^{(n+1)/2}$

3. Field strength is uniform inside the cavity. Let us find at its centre.

$E_T = E_P + E_C$ (T = Total, R = Remaining, C = Cavity)

$$\therefore \quad E_R = E_T - E_C = \frac{GM}{R^3}\frac{R}{2} - 0 = \frac{GM}{2R^2}$$

$$\therefore \quad F = mF_R = \frac{GMm}{2R^2}$$

4. $\text{KE} = \frac{1}{2}mv_e^2 = \frac{1}{2}m(2gR) = mgR$

5. $\frac{1}{2}m\left(\frac{v_e}{2}\right)^2 = \frac{mgh}{1 + \frac{h}{R}}$

or $\quad \frac{v_e^2}{8} = \frac{gh}{1 + \frac{h}{R}}$

or $\quad \frac{2gR}{8} = \frac{gh}{1 + \frac{h}{R}}$

Solving, we get $h = \frac{R}{3}$

6. $T = 2\pi\sqrt{\frac{l}{g}}$

or $\quad T \propto \frac{1}{\sqrt{g}}$

$$\frac{T_1}{T_2} = \sqrt{\frac{g_2}{g_1}} = \frac{2}{1}$$

$$\therefore \quad \frac{g_2}{g_1} = 4$$

$$v_e = \sqrt{2gR} \quad \text{or} \quad v_e \propto \sqrt{gR}$$

$$\frac{v_{e1}}{v_{e2}} = \sqrt{\frac{g_1}{g_2} \cdot \frac{R_1}{R_2}} = 1 : 2$$

$$\therefore \quad \sqrt{\frac{1}{4} \times \frac{R_1}{R_2}} = \frac{1}{2}$$

$$\therefore \quad \frac{R_1}{R_2} = 1$$

Further, $\quad g = \frac{Gm}{R^2} = \frac{G\left(\frac{4}{3}\pi R^3 \rho\right)}{R^2}$

or $\quad g \propto R\rho$

$$\therefore \quad \frac{g_1}{g_2} = \frac{R_1}{R_2} \cdot \frac{\rho_1}{\rho_2}$$

or $\quad \left(\frac{1}{4}\right) = (1)\frac{\rho_1}{\rho_2}$

or $\quad \frac{\rho_1}{\rho_2} = \frac{1}{4}$

7. Potential at centre $= -\frac{4GM}{r}$

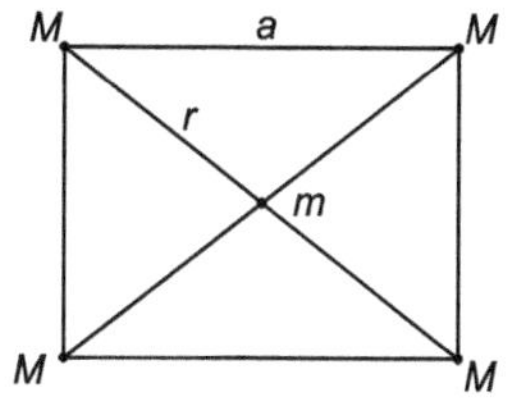

$$= \frac{-4GM}{a/\sqrt{2}}$$

$$= -\frac{4\sqrt{2}GM}{a}$$

Potential energy $= \frac{-4\sqrt{2}GMm}{a}$

Binding energy $= \frac{4\sqrt{2}GMm}{a}$

$\therefore \quad \frac{1}{2}mv_e^2 = \frac{4\sqrt{2}GMm}{a}$

or $\quad v_e = \sqrt{\frac{8\sqrt{2}GM}{a}}$

8. Increase in kinetic energy = decrease in potential energy

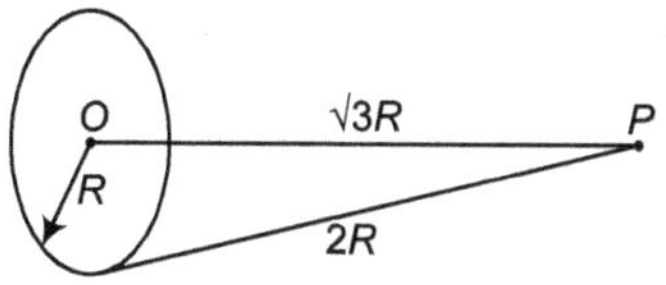

or $\quad \frac{1}{2}mv^2 = U_i - U_f = m(V_i - V_f)$

or $\quad v = \sqrt{2(V_i - V_f)}$

$$= \sqrt{2\left[-\frac{GM}{2R} + \frac{GM}{R}\right]}$$

$$= \sqrt{\frac{GM}{R}}$$

9. Just before collision total distance travelled by both $= 12R - R - 2R = 9R$. Now, let smaller sphere travels a distance x. Then,

$$m(x) = 5m(9R - x)$$

$\therefore \quad x = 7.5R$

10. $E_i = E_0 = -\frac{GMm}{2r}$

or $\quad E_f = -\frac{GMm}{2(3r)} = -\frac{GMm}{6r}$

$\therefore \quad W = E_f - E_i$

$$= -\frac{GMm}{3r} = \frac{2}{3}E_0$$

11. $W = U_2 - U_1$

$$= mv_2 - mv_1$$

$$= m\left[\left\{-\frac{Gm_2}{R} - \frac{Gm_1}{\sqrt{2}R}\right\} - \left\{-\frac{Gm_1}{R} - \frac{Gm_2}{\sqrt{2}R}\right\}\right]$$

$$= \frac{Gm(m_1 - m_2)}{\sqrt{2}R}(\sqrt{2} - 1)$$

12. $W = E_A - E_\infty$

$$= (U_A + K_A) - (U_\infty + K_\infty)$$

$\therefore \quad = \left\{(1)V_A + \frac{1}{2} \times (1)(2)^2\right\} - (0 + 0)$

$\therefore \quad V_A = -5 \text{ J/kg}$

13. The initial position will become the perigee (nearest) position.

14. $t = \frac{T}{4} = \frac{2\pi\sqrt{\frac{R}{g}}}{4} = \frac{\pi}{2}\sqrt{\frac{R}{g}}$

$$t' = \sqrt{\frac{2R}{g}}$$

$\therefore \quad \frac{t}{t'} = \frac{\pi}{2\sqrt{2}}$

15. $\frac{1}{2}\mu v_r^2 = U_i - U_f$

Here, $\quad \mu =$ reduced mass $= \frac{mM}{M + m}$

$\therefore \quad \frac{1}{2}\left(\frac{mMv_r^2}{m + M}\right) = 0 - \left(-\frac{GMm}{d}\right)$

$\therefore \quad v_r = \sqrt{\frac{2G(M + m)}{d}}$

16. $v_0 = \sqrt{\frac{GM}{r}} = \sqrt{\frac{gR^2}{R + h}}$

17. $L = mvr = m\sqrt{\frac{GM}{r}}r = m\sqrt{GMr}$

$\therefore \quad L \propto \sqrt{r}$

18. $E_1 = \Delta U = \left(\frac{mgh}{1 + \frac{h}{R}}\right)$

E_2 = Energy of satellite – energy of satellite on surface of earth.

$$= -\frac{GMm}{2(R + h)} + \frac{GMm}{R}$$

$$= mgR\left[1 - \frac{1}{2\left(1 + \frac{h}{R}\right)}\right]$$

$$= \frac{mgR\left(\frac{2h}{R} + 1\right)}{2\left(1 + \frac{h}{R}\right)}$$

$\therefore \quad \frac{E_1}{E_2} = \frac{mgh}{1 + \frac{h}{R}} \times \frac{2\left(1 + \frac{h}{R}\right)}{mgR} = \frac{2h}{R}$

19. Increase in kinetic energy = Decrease in potential energy

$\therefore \quad \frac{1}{2}mv^2 = \frac{mgR}{1 + \frac{R}{R}} = \frac{mgR}{2} \qquad \left(\Delta U = \frac{mgh}{1 + \frac{h}{R}}\right)$

20. E = Energy of satellite – energy of satellite on surface of the earth

$$= -\frac{GMm}{2(3R)} - \left[-\frac{GMm}{R}\right]$$

$$= \frac{5}{6}\frac{GMm}{R}$$

$$= \frac{5}{6}mgR \qquad \left(\text{as} \frac{GM}{R} = gR\right)$$

21. Acceleration of super dense material $= g$

but acceleration of earth $= 2g$ (towards the mass)

$\therefore$ Relative acceleration $= 3g$

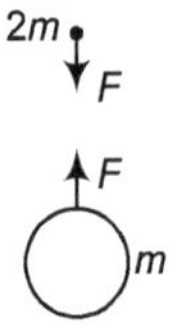

Now, $h = \frac{1}{2} a_r t^2$

or $h = \frac{1}{2}(3g)t^2$

or $t = \sqrt{\frac{2h}{3g}}$

22. $F_1 = \frac{GMm}{(2R)^2} = \frac{GMm}{4R^2}$

$$F_2 = F_1 - F_{\text{cavity}}$$

$$= \frac{GMm}{4R^2} - \frac{G\left(\frac{M}{8}\right)(m)}{\left(\frac{3}{2}R\right)^2}$$

$$= \frac{7GMm}{36R^2}$$

$\therefore$ $\frac{F_2}{F_1} = \frac{7}{9}$

23. $g = \frac{4}{3}\pi\rho G r$

$\therefore$ $g \propto r$, if $r < R$

$$g = \frac{GM}{r^2}$$

$\therefore$ $g \propto \frac{1}{r^2}$, if $r > R$

If $r_1 < R$ and $r_2 < R$, then $\frac{F_1}{F_2} = \frac{g_1}{g_2} = \frac{r_1}{r_2}$

If $r_1 > R$ and $r_2 > R$, then $\frac{F_1}{F_2} = \frac{g_1}{g_2} = \left(\frac{r_2}{r_1}\right)^2$

24. $U_i + K_i = U_f + K_f$

$$-\frac{GMm}{R} + \frac{1}{2}m(2v_e)^2 = 0 + \frac{1}{2}mv^2$$

or $-\frac{GM}{r} + 2v_e^2 = \frac{1}{2}v^2$

or $-\frac{2GM}{R} + \frac{8GM}{R} = v^2$

or $v = \sqrt{\frac{6GM}{R}}$

$$= \sqrt{3\left(\frac{2GM}{R}\right)} = \sqrt{3(2gR)}$$

$$= \sqrt{3}v_e$$

25. $F = \frac{k}{r}$ (k = constant)

$\therefore$ $\frac{mv^2}{r} = \frac{k}{r}$ or $v \propto r^0$

26. $E = \Delta U = \frac{mgh}{1 + \frac{h}{R}} = \frac{mghR}{(R+h)}$

E' = energy of satellite – energy of satellite on surface of earth

$$= -\frac{GMm}{2(R+h)} - \left(-\frac{GMm}{R}\right)$$

$$= mgR^2\left[\frac{1}{R} - \frac{1}{2(R+h)}\right] = \frac{mg(R+2h)R}{2(R+h)}$$

Now, $\frac{E}{E'} = \frac{2h}{(R+2h)}$

27. Acceleration between them is zero, where force between $2m$ and $\frac{m}{5}$ is equal to the force between m and $\frac{m}{5}$ or the net force on $\frac{m}{5}$ is zero. Suppose, this point is P. Now if velocity v_0 is less than value necessary to cross point P, it will retrace its path otherwise not.

28. $U_2 = 2\left(\frac{1}{2}mv^2\right)$

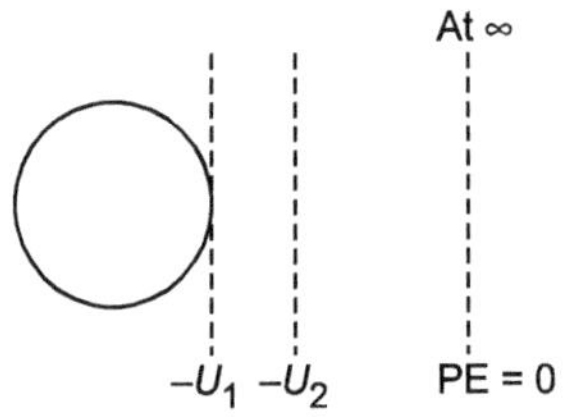

$\Rightarrow$ $v_0^2 = \frac{t}{2}$

Now, $\frac{1}{2}mv^2 = -U_2 + U_1$

$$= U_1 - U_2 = \frac{mv_e^2}{2} - mv_o^2$$

$\therefore$ $v = \sqrt{v_e^2 - 2v_o^2}$

29. $\frac{G \cdot m \cdot m}{4R^2} = \frac{mv^2}{(R)}$

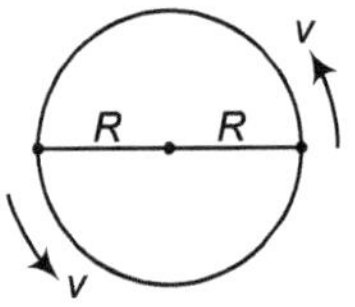

$\therefore$ $v = \frac{1}{2}\sqrt{\frac{Gm}{R}}$

30. Side of square : $r = 2R\cos 45° = \sqrt{2}R$

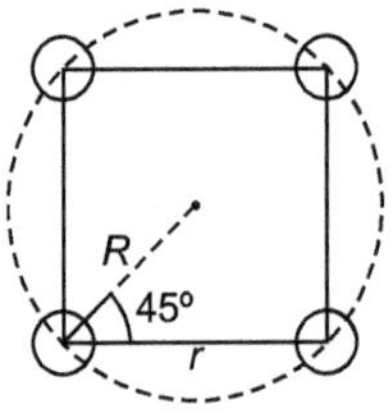

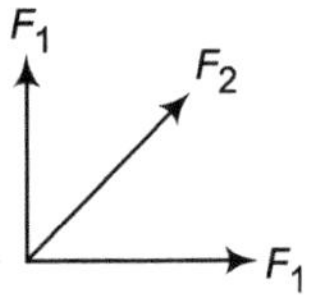

Now, $2F_1\cos 45° + F_2 = \frac{Mv^2}{R}$

$$v=\sqrt{\frac{R}{M}(\sqrt{2}F_1+F_2)}$$

$$v=\sqrt{\frac{R}{M}\left[\sqrt{2}\frac{GM.M}{(\sqrt{2}R)^2}+\frac{GMM}{(2R)^2}\right]}$$

$$=\sqrt{\frac{GM}{R}\left(\frac{2\sqrt{2}+1}{4}\right)}$$

31. $a=g'=g\left(1-\frac{h}{R}\right)=g\left(\frac{R-h}{R}\right)=\frac{g}{R}\cdot y$

$$=\frac{\left(\frac{GM}{R^2}\right)}{R}\cdot y$$

$$=\frac{G}{R^3}\left(\frac{4}{3}\pi R^3\rho\right)\cdot y=\frac{4}{3}\pi\rho G y$$

32. $E=2\int_0^{\pi/2} dE\sin\theta$ (along Y- axis)

$$=2\int_0^{\pi/2}\frac{G\left(\frac{M}{l}\cdot R\cdot d\theta\right)}{R^2}.\sin\theta$$

$$=\frac{2GM}{lR}$$

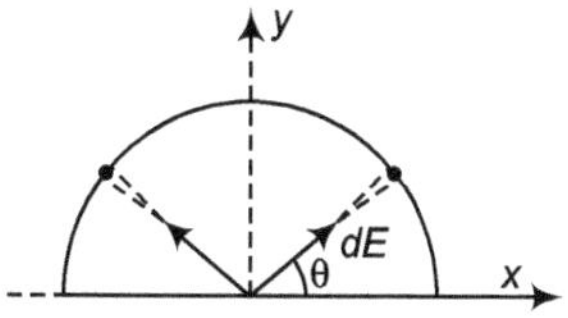

Now, $l=\pi R$

$\therefore$ $R=\frac{l}{\pi}$

Substituting, we get

$$E=\frac{2\pi GM}{l^2}$$

33. $dF=\frac{GmdM}{x^2}$

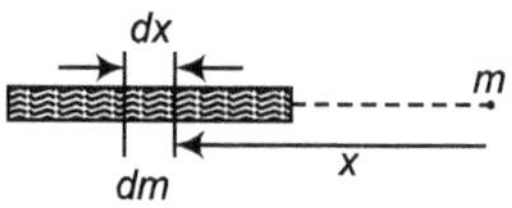

$\therefore$ $F=\int_{x=a}^{x=a+l}dF$

$$=\int_a^{a+l}\frac{G\cdot m\left(\frac{M}{l}\cdot dx\right)}{x^2}$$

$$=\frac{GMm}{a(a+l)}$$

34. $F=\int\limits_{\text{(whole ring)}} dF\sin 60°$

$$=\int\limits_{\text{(whole ring)}}\frac{G.M(dm)}{(2a^2)}\cdot\frac{\sqrt{3}}{2}$$

$$=\frac{\sqrt{3}GMm}{8a^2}\quad (\text{as}\int\limits_{\text{(whole ring)}} dm=m)$$

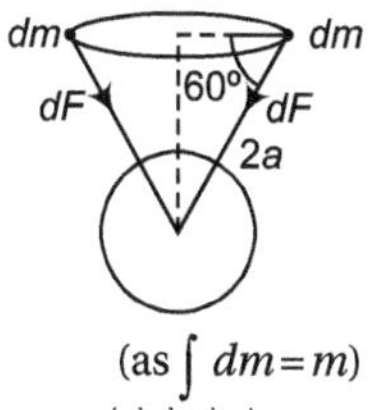

35. $W=\Delta U=U_f-U_i=U_f$ (as $U_i=0$)

36. Change in potential energy in displacing a body from r_1 to r_2 is given by

$$\Delta U=GMm\left[\frac{1}{r_1}-\frac{1}{r_2}\right]$$

$$=GMM\left(\frac{1}{2R}-\frac{1}{3R}\right)$$

$$=\frac{GMm}{6R}$$

37. Semimajor axis

$$a=\frac{r_1+r_2}{2}$$

Now, $T^2\propto a^3$

$\Rightarrow$ $T\propto a^{3/2}\propto\left(\frac{r_1+r_2}{2}\right)^{3/2}$

$\Rightarrow$ $T\propto(r_1+r_2)^{3/2}$

38. $\frac{1}{2}mv^2=\frac{mgh}{1+\frac{h}{R}}$

Putting $h=R$, we get

$$v=\sqrt{gR}=\sqrt{\frac{GM}{R}}$$

39. $\frac{\frac{GM}{81}}{r^2}=\frac{GM}{(60R-r)^2}$

Solving, we get

$$60R-r=9r$$

$\therefore$ $r=6R$

40. W = Energy of satellite – potential energy of satellite on surface of earth.

$$=-\frac{GMm}{2(3R)}+\frac{GMm}{R}$$

$$=\frac{5}{6}\frac{GMm}{R}=\frac{5}{6}mgR$$

41. $r=\frac{a}{\sqrt{3}}$

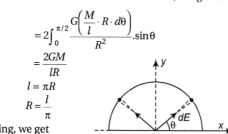

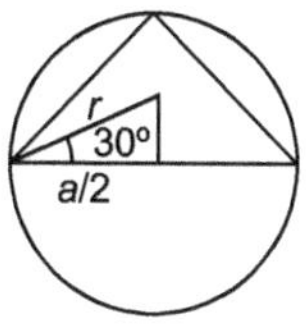

$$\sqrt{3}F=mr\omega^2$$

$$\sqrt{3}\left(\frac{Gmm}{a^2}\right)=m\frac{a}{\sqrt{3}}.\omega^2$$

$\therefore$ $\omega=\sqrt{\frac{3Gm}{a^3}}$

42. Inside the shell force, will be zero (according to Gauss theorem). Then, force will increase and then decrease.

43. $\sqrt{\frac{GM}{r}}=\frac{1}{2}\sqrt{2gR}=\sqrt{\frac{gR}{2}}=\sqrt{\frac{GM}{2R}}$

$\therefore$ $r=2R$

or $h=r-R=R$

More than One Correct Options

1. $g=\frac{GM}{R^2}=\frac{G\left(\frac{4}{3}\pi R^3\rho\right)}{R^2}$

or $g\propto R$ (as ρ is same)

$$v_e=\sqrt{2gR}=\sqrt{\frac{2GM}{R}}$$

$$=\sqrt{\frac{2G\left(\frac{4}{3}\pi R^3\rho\right)}{R}}$$

So, v_e or $v\propto R$ (as ρ is same).

2. No solution is required.

3. $K_i+U_i=K_f+U_f$

$\therefore$ $0-\frac{GMm}{2R}=\frac{1}{2}mv^2-\frac{3}{2}\frac{GMm}{R}$

or $v=\sqrt{\frac{2Gm}{R}}$

4. E due to point mass is $E = \dfrac{Gm}{r^2}$.

As $r \to 0, E \to \infty$

So, just over the point masses, $E = \infty$. Hence, in moving from one point mass to other point mass, E first decreases and then increases.

V due to a point mass is

$$V = -\frac{Gm}{r}$$

As $r \to 0, V \to -\infty$

So, just over the point mass, V is $-\infty$. Hence, in moving from one point mass to other point mass, V first increases and then decreases.

5. Inside a shell, $V =$ constant and $E = 0$.

Between A and B, $E_{net} = 0$, $V_{net} =$ constant, because these points inside both shells.

Between B and C

E of $m \neq 0$, E of $2m = 0$

V of $m \neq$ constant, V of $2m =$ constant.

Beyond C

E and V due to both shells are neither zero nor constant.

6. $E = \dfrac{Gm}{r^2}$

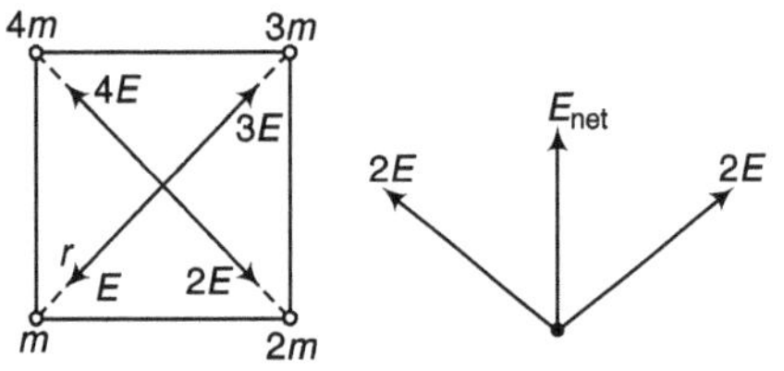

$$E'_{net} = 2\sqrt{2}\,E = \frac{2\sqrt{2}\,Gm}{r^2}$$

$$E'_{net} = \sqrt{(9E)^2 + E^2}$$

$$= \sqrt{10}\,E$$

$$= \frac{\sqrt{10}\,Gm}{r^2}$$

$$E'_{net} > E_{net}$$

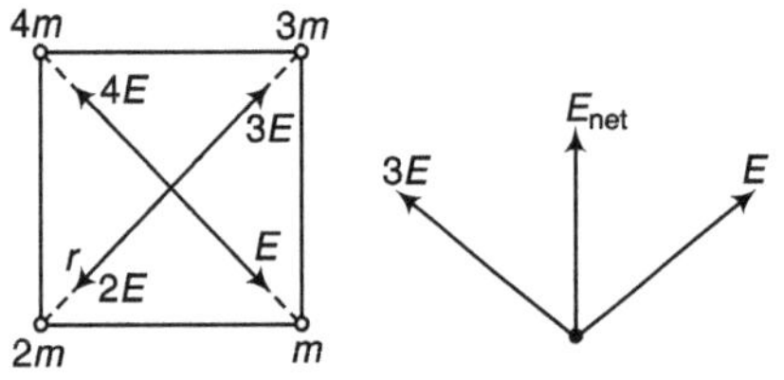

Potential in both cases is

$$V_{net} = -\frac{G}{r}(m + 2m + 3m + 4m)$$

$$= -\frac{10\,Gm}{r}$$

7. At highest point velocity of particle 1will become zero. But velocity of particle 2 is non-zero.

8. At maximum distance (at A) kinetic energy is minimum. But, angular momentum about centre of sun always remains constant.

9. $E_1 = -\dfrac{GMm}{R}$

$$U_2 = -\frac{GMm}{2R} \qquad (r = R + R = 2R)$$

$$K_2 = \frac{GMm}{4R} \text{ and } E_2 = -\frac{GMm}{4R}$$

10. $v = \sqrt{\dfrac{GM}{r}}$ or $v \propto \sqrt{GM}$

$$T = \frac{2\pi}{\sqrt{GM}} r^{3/2}$$

or $$T \propto \frac{1}{\sqrt{GM}}$$

11. Given, $G' = 10G$

Consider the below diagram.

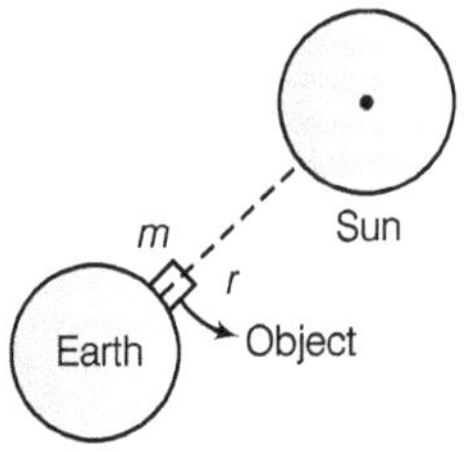

Force on the object due to the earth $= \dfrac{G'M_e m}{R^2} = \dfrac{10GM_e m}{R^2}$

$[\because G' = 10G \text{ given}]$

$$= 10\left(\frac{GM_e m}{R^2}\right) \qquad \left[\because g = \frac{GM_e}{R^2}\right]$$

$$= (10g)\,m = 10\,mg \qquad \text{...(i)}$$

Force on the object due to the sun, $F = \dfrac{GM_s' m}{r^2}$

$$= \frac{G(M_s)m}{10r^2} \qquad \left[\because M_s' = \frac{M_s}{10} \text{ (given)}\right]$$

As $r >> R$ (radius of the earth) $\Rightarrow F$ will be very small.

So, the effect of the sun will be neglected.

Now, as $g' = 10g$

Hence, weight of person

$$= mg' = 10\,mg \qquad \text{[from Eq. (i)]}$$

i.e. gravity pull on the person will increase. Due to it, walking on ground would become more difficult.

Critical velocity, v_c is proportional to g *i.e.*,

$$v_c \propto g$$

As, $g' > g$

$\Rightarrow$ $v_c' > v_c$

Hence, rain drops will fall much faster.

To overcome the increased gravitational force of the earth, the aeroplanes will have to travel much faster.

12. Due to huge amount of opposite charges on the sun and the earth's electrostatic force of attraction will be large. Gravitational force is also attractive in nature have both forces will be added.

Both the forces obey inverse square law and are central forces. As both the forces are of same nature, hence all the three **Kepler's laws** will be valid.

13. We know that gravitational force between the earth and the sun.

$F_G = \dfrac{GMm}{r^2}$, where M is mass of the sun and m is mass of the earth.

When G decreases with time, the gravitational force F_G will become weaker with time. As F_G is changing with time. Due to it, the earth will be going around the sun not strictly in closed orbit and radius also increases, since the attraction force is getting weaker.

Hence, after long time the earth will leave the solar system.

Assertion and Reason

1. Gravitational potential due to a point mass at some finite distance is always negative.

4. Field strengths due to masses at 1 and 2 are acting at 60°. Therefore, resultant is $\dfrac{\sqrt{3}Gm}{a^2}$.

6. Angular momentum is constant about the centre of sun.

8. $v_e = \sqrt{2gR} = \sqrt{\dfrac{2GM}{R}}$

$\therefore \quad v_e \propto \sqrt{\dfrac{M}{R}}$

$$(4\pi R^2)\cdot\delta = (4\pi R^2)\frac{M}{\frac{4}{3}\pi R^3} \propto \frac{M}{R}$$

10. In case of cube, the formula cannot be applied directly.

11. In the three cases, $F = \dfrac{Gm_1m_2}{r^2}$ can be applied directly.

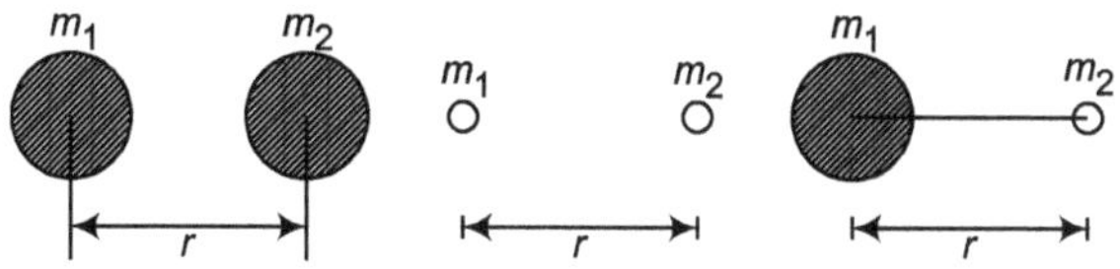

12. In the region, $0 < r < r_1$.

$F = 0$ but $v =$ constant $+ 0$

13. $T = \dfrac{2\pi}{\sqrt{GM}} r^{3/2}$

Time period does not depend upon the radius of earth r but it depends upon the mass of earth M.

14. On two sides of a point mass, direction of gravitational field strength will be different. Further, gravitational potential is negative. Hence, it increases in moving away from the point mass.

15. $\dfrac{dA}{dt} = \dfrac{L}{2m} = \text{constant} = \dfrac{mvr\sin\theta}{2m} = \dfrac{vr\sin\theta}{2}$

i.e. $\dfrac{dA}{dt}$ is independent of m.

16. $W_2 = \dfrac{1}{2}mv^2 = \dfrac{1}{2}m\left(\sqrt{\dfrac{GM}{r}}\right)^2 = \dfrac{GMm}{2r}$

$$W_1 = \Delta U = \frac{mgh}{1+\frac{h}{R}} = \frac{mgRh}{R+h}$$

$$= \frac{m\left(\frac{GM}{R^2}\right)Rh}{r} = \left(\frac{GMm}{rR}\right)h$$

18. Gravitational force by earth is utilized in providing the necessary centripetal force for rotating round the earth. That's why this force in not felt to us. But gravitational force by moon is unutilized. That's why it is felt.

Match the Columns

1. $g = \dfrac{GM}{R^2}, V = -\dfrac{GM}{R}$

At height $\quad h = R,$

$$g' = \frac{g}{1+\frac{h}{R}} = \frac{g}{2} \Rightarrow V' = -\frac{GM}{2R}$$

i.e. decrease by a factor $\dfrac{1}{2}$ and V' increase by a factor.

At depth $\quad h = \dfrac{R}{2},$

$$g' = g\left(1 - \frac{h}{R}\right) = g\left(1 - \frac{1}{2}\right) = \frac{g}{2}$$

$$V' = -\frac{GM}{R^3}[1.5R^2 - 0.5r^2]$$

$$= -\frac{GM}{R^3}\left[1.5R^2 - \frac{1}{8}R^2\right]$$

$$= -\frac{11}{8}\frac{GM}{R}$$

i.e. g' decrease by a factor $\dfrac{1}{2}$ and V' also decreases by a factor $\dfrac{11}{8}$.

2. Near the surface of earth retardation of the particle will be more.

3. $g = \dfrac{Gm}{R^2} = \dfrac{G\left(\frac{4}{3}\pi R^3\right)\rho}{R^2}$ or $g \propto \rho R$

At centre, $\quad V = -\dfrac{Gm}{R} = -\dfrac{G\left(\frac{4}{3}\pi R^3\rho\right)}{R}$

or $\quad V \propto \rho R^2$

At centre potential is 1.5 times potential at centre. At centre of a solid sphere field strength is zero.

4. At perigee position, planet is nearest to sun.

7. Inside a shell, $V = -\dfrac{GM}{R} =$ constant

and $E = 0$

Outside the shell, $V = -\dfrac{GM}{r}$ and $E = \dfrac{GM}{r^2}$

As r increases, V increases and E decreases.

9. $T = \dfrac{2}{\sqrt{GM}} r^{3/2}, v_0 = \sqrt{\dfrac{GM}{r}}$ and $E = -\dfrac{GMm}{2r}$

10. Due to rotation of earth,

$$g' = g - R\omega^2\cos^2\phi$$

At pole there is no effect of rotation of earth.

As $\phi = 90°$

At equator g' will increase, if ω decreases.

Further, T will increase with decrease in ω.

From Kepler's third law, r should also increase.

$E = -\dfrac{GMm}{2r}$, so with increase in r, E also increases.

Entrance Gallery

1. Given, $R_p = \frac{R_e}{10}$, $\rho_p = \rho_e = \rho$

Length of the wire, $l = \frac{R_p}{5}$ m

Density of the wire, $\rho_w = 10^{-3}$ kg / m^3

Mass of the wire, $m_p = \frac{R_p}{5} \times 10^{-5}$ kg

The force applied at the top of the wire by a person holding it in place = weight of the wire.

The weight of the wire will act at the centre of the gravity of the wire which is located at a distance of $\frac{R_p}{5 \times 2} = \frac{R_p}{10}$ from the surface of the planet.

Now, calculating the effective value of acceleration due to gravity g_p' at the depth $d = \frac{R}{10}$ for the planet.

$$g_p' = g_p\left(1 - \frac{d}{R_p}\right) \Rightarrow g_p = \frac{g_e}{10} \quad (\because g \times \rho R)$$

$$\therefore g_p = \frac{g_e}{10}\left(1 - \frac{R_p}{10 \times R_p}\right) \Rightarrow g_p' = \frac{9g_e}{100}$$

Weight of the wire $= mg_p'$

$$= \frac{R_p}{5} \times 10^{-3} \times \frac{9 \times g_e}{100}$$

$$= \frac{6 \times 10^5 \times 10^{-3}}{5} \times \frac{9 \times 10}{100} = \frac{60 \times 9}{5} = 108 \text{ N}$$

2. Resultant force acting on mass M due to other three masses

$$\text{F} = \frac{F}{\sqrt{2}} + \frac{F}{\sqrt{2}} + F \Rightarrow \frac{F}{\sqrt{2}} + \frac{F}{\sqrt{2}} + F = \frac{Mv^2}{R}$$

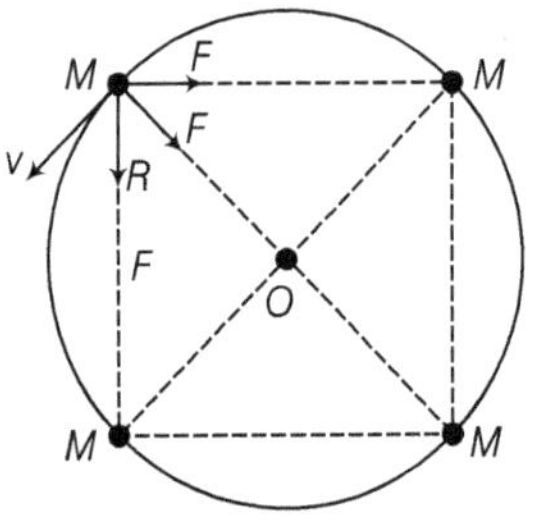

$$\frac{2 \times GM^2}{\sqrt{2}(R\sqrt{2})^2} + \frac{GM^2}{4R^2} = \frac{Mv^2}{R}$$

$$\frac{GM^2}{R}\left[\frac{1}{4} + \frac{1}{\sqrt{2}}\right] = Mv^2$$

$$v = \sqrt{\frac{GM}{R}\left(\frac{\sqrt{2}+4}{4\sqrt{2}}\right)} = \frac{1}{2}\sqrt{\frac{GM}{R}(1 + 2\sqrt{2})}$$

3. We know that gravitational potential energy,

$$U = -\frac{GM_e m}{R} \quad \text{...(i)}$$

and gravitational kinetic energy,

$$K = \frac{1}{2} \cdot \frac{GM_e m}{R} \quad \text{...(ii)}$$

$\therefore$ Total energy of a body is

$$E = U + K = -\frac{GM_e m}{R} + \frac{GM_e m}{2R}$$

$$= -\frac{2GM_e m + GM_e m}{2R} = -\frac{GM_e m}{2R}$$

But, acceleration due to gravity (g) in terms of gravitational constant (G) is

$$g = \frac{GM_e}{R^2} \quad \text{...(iii)}$$

$$\therefore \quad -\frac{GM_e}{R^2} \times R^2 \times \frac{m}{2R} = g \times R^2 \times \frac{m}{2R} \quad \text{[From Eq. (iii)]}$$

(Cancellation of negative sign, because energy can never be negative) $= \frac{gmR}{2} = \frac{mgR}{2}$

4. Total energy of the satellite is

$$E = -\frac{1}{2}\frac{GM_e m}{R_e}$$

where, $K = \frac{1}{2}\frac{GM_e m}{R_e}$

$\therefore$ Total energy = – Kinetic energy

$$E = -\frac{1}{2}mv^2$$

Putting the value of KE in the form of mass of a satellite m and speed v.

5. Given, $R_e = 6400 \text{ km} = 6.4 \times 10^6$ m

$\pi = 3.14, g = 10 \text{ m/s}^2$

We know that the period of revolution of the earth satellite

$$T = 2\pi\sqrt{\frac{(R_e + h)^3}{gR_e^2}} \quad [\text{if } h << R_e, \text{then } (R_e + h = R_e)]$$

So, $T = 2\pi\sqrt{\frac{R_e^3}{gR_e^2}} = 2\pi\sqrt{\frac{R_e}{g}} = 2 \times 3.14\sqrt{\frac{6.4 \times 10^6}{10}}$

$$= 2 \times 3.14 \times 0.8 \times 10^3 = 5.024 \times 10^3 = 5024 \text{ s}$$

and $T = \frac{5024}{60} = 83.73$ min

6. The acceleration due to gravity decreases with the increase in altitude from the surface of the earth. According to the given relation,

$$g' = \frac{g}{\left(1 + \frac{h}{R}\right)^2}$$

7. Weight of the body at the surface of the earth $= mg = 45$ N

Using $g' = \frac{g}{\left(1 + \frac{h}{R}\right)^2} \Rightarrow mg' = \frac{mg}{\left(1 + \frac{R}{2R}\right)^2} \quad \left[h = \frac{R}{2}\right]$

$$w' = \frac{45}{\left(1 + \frac{1}{2}\right)^2} = \frac{4 \times 45}{9} \quad (\therefore w = mg)$$

$w' = 20$ N

8.

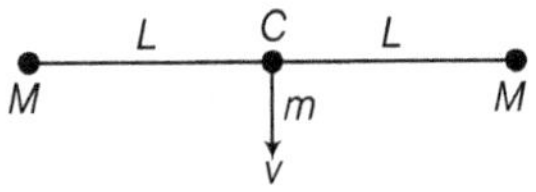

Let v is minimum velocity. From energy conservation,

$$U_c + K_c = U_\infty + K_\infty$$

$$\therefore \quad mV_c + \frac{1}{2}mv^2 = 0 + 0$$

$$\therefore \quad v = \sqrt{-2V_c} = \sqrt{(-2)\left(\frac{-2GM}{L}\right)} = 2\sqrt{\frac{GM}{L}}$$

9. E = Energy of satellite – energy of mass on the surface of planet

$$= -\frac{GMm}{2r} - \left(-\frac{GMm}{R}\right)$$

Here, $r = R + 2R = 3R$

Substituting the value of r in above equation,

$$E = \frac{5GMm}{6R}$$

10. Surface area of Q is four times. Therefore, radius of Q is two times. Volume is eight times. Therefore, mass of Q is also eight times.

So, let $M_P = M$ and $R_P = r$

Then, $M_Q = 8M$ and $R_Q = 2r$

Now, mass of R is $(M_P + M_Q)$ or $9M$. Therefore, radius of R is $(9)^{1/3}r$. Now, escape velocity from the surface of a planet is given by

$$v = \sqrt{2\frac{GM}{r}} \quad (r = \text{radius of that planet})$$

$$\therefore \quad v_P = \sqrt{\frac{2GM}{r}} \Rightarrow v_Q = \sqrt{\frac{2G(8M)}{(2r)}}$$

$$v_R = \sqrt{\frac{2G(9M)}{(9)^{1/3}r}}$$

From here, we can see that, $\frac{v_P}{v_Q} = \frac{1}{2}$ and $v_R > v_Q > v_P$.

11. In circular orbit of a satellite of potential energy

$$= -2 \times (\text{kinetic energy}) = -2 \times \frac{1}{2}mv^2 = -mv^2$$

Just to escape from the gravitational pull, its total mechanical energy should be zero. Therefore, its kinetic energy should be $+mv^2$.

12. Gravitational force provides necessary centripetal force

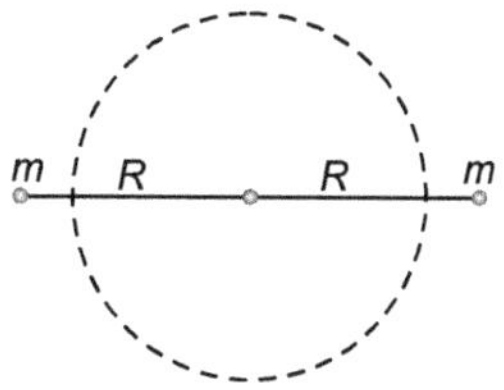

$$\frac{Gm^2}{(2R)^2} = \frac{mv^2}{R} \Rightarrow v = \sqrt{\frac{Gm}{4R}}$$

13. Let gravitational field is zero at P as shown in figure.

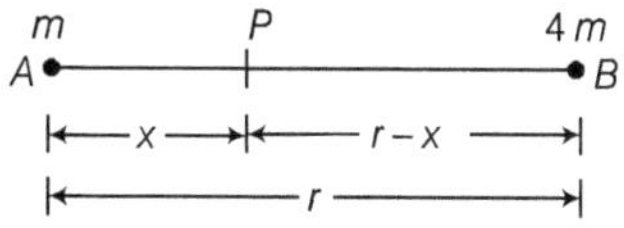

$$\therefore \quad \frac{Gm}{x^2} = \frac{G(4m)}{(r-x)^2} \Rightarrow 4x^2 = (r-x)^2$$

$$\Rightarrow \quad 2x = r - x$$

$$\Rightarrow \quad x = \frac{r}{3}$$

$\therefore$ The gravitational potential

$$\therefore \quad V_p = -\frac{Gm}{x} - \frac{G(4m)}{r-x} = -\frac{3Gm}{r} - \frac{6Gm}{r} = -\frac{9Gm}{r}$$

14. $\text{KE} = \frac{1}{2}mv_0^2 - \frac{1}{2}m(11.2)^2 \quad [v_0 = 2 \times 11.2 \text{ km/s}]$

$$= \frac{1}{2}m(2 \times 11.2)^2 - \frac{1}{2}m(11.2)^2$$

$$\frac{1}{2}mv^2 = 3 \times \frac{1}{2}m \times (11.2)^2$$

$$v = \sqrt{3} \times 11.2 \text{ km/s}$$

15. Given, $R_1 = r$ and $R_2 = 4r$

We have $T^2 \propto R^3$

$$\therefore \quad \frac{T_1^2}{T_2^2} = \frac{(r)^3}{(4r)^3} \quad \text{or} \quad \frac{T_1}{T_2} = \frac{1}{8}$$

16. From the conservation of angular momentum,

$$I_1\omega_1 = I_2\omega_2$$

$$\frac{2}{5}MR^2\left(\frac{2\pi}{T_1}\right) = \frac{2}{5}M \cdot \frac{R^2}{n^2}\left(\frac{2\pi}{T_2}\right) \Rightarrow T_2 = \frac{T_1}{n^2} = \frac{24}{n^2}$$

17. Potential energy, $\Delta U = \frac{mgh}{1 + \frac{h}{R}} = \frac{mgh}{1 + \frac{R}{R}} = \frac{mgR}{2}$

18. $W = \Delta U = U_f - U_i = U_\infty - U_P = -U_P = -mV_P = -V_P$ (as $m = 1$)

Potential at point P will be obtained by integration as given below.

Let dM be the mass of small ring as shown in figure.

$$dM = \frac{M}{\pi(4R)^2 - \pi(3R)^2}(2\pi r)\,dr = \frac{2Mr\,dr}{7R^2}$$

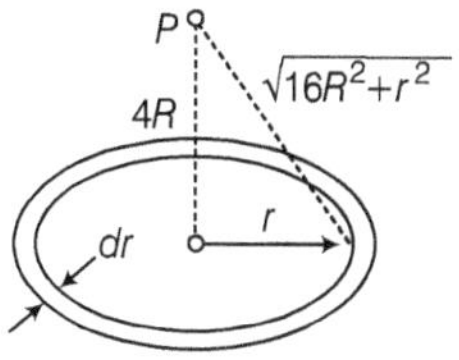

$$dV_p = -\frac{G \cdot dM}{\sqrt{16R^2 + r^2}}$$

$$= -\frac{2GM}{7R^2}\int_{3R}^{4R}\frac{r}{\sqrt{16R^2 + r^2}} \cdot dr$$

$$= -\frac{2GM}{7R}(4\sqrt{2} - 5)$$

$$\therefore \quad W = +\frac{2GM}{7R}(4\sqrt{2} - 5)$$

19. Acceleration due to gravity,

$$g = \frac{GM}{R^2} = \frac{G\left(\frac{4}{3}\pi R^3\right)\rho}{R^2}$$

or $\quad g \propto \rho R$

or $\quad R \propto \frac{g}{\rho}$

Now escape velocity, $v_e = \sqrt{2gR}$

or $\quad v_e \propto \sqrt{gR}$

or $\quad v_e \propto \sqrt{g \times \frac{g}{\rho}} \propto \sqrt{\frac{g^2}{\rho}}$

$$\therefore \quad (v_e)_{\text{planet}} = (9.8)\sqrt{\frac{6}{121} \times \frac{3}{2}}$$

$$= 3 \text{ kms}^{-1}$$

20. $g' = \frac{GM}{(R+h)^2}$

Acceleration due to gravity at height h,

$$\Rightarrow \quad \frac{g}{9} = \frac{GM}{R^2} \cdot \frac{R^2}{(R+h)^2} = g\left(\frac{R}{R+h}\right)^2$$

$$\Rightarrow \quad \frac{1}{9} = \left(\frac{R}{R+h}\right)^2 \quad \Rightarrow \quad \frac{R}{R+h} = \frac{1}{3}$$

$$\Rightarrow \quad 3R = R + h \quad \Rightarrow \quad 2R = h$$

21. Given, $\frac{R_e}{R_p} = \frac{2}{3}$ and $\frac{d_e}{d_p} = \frac{4}{5}$

As, $MG = gR_e^2$ and $M = d_e \times \frac{4}{3}\pi R_e^3$

$$\therefore \quad d_e \times \frac{4}{3}\pi R_e \times G = g_e \quad \text{...(i)}$$

Similarly, for planet, $d_p \times \frac{4}{3}\pi R_p G = g_p$ ...(ii)

Dividing Eq. (i) by Eq. (ii), we get

$$\frac{g_e}{g_p} = \frac{R_e}{R_p} \times \frac{d_e}{d_p} \quad \Rightarrow \quad \frac{g_e}{g_p} = \frac{2}{3} \times \frac{4}{5} = \frac{8}{15} = 0.5 \text{ m}$$

22. Acceleration due to gravity at depth d below the surface of the earth

$$g_d = g\left(1 - \frac{d}{R}\right)$$

Acceleration due to gravity at height h from the surface of the earth

$$g_h = g\left(g - \frac{2h}{R}\right)$$

Given, $g_h = g_d$

$$\therefore \quad \frac{2h}{R} = \frac{d}{R}$$

$$d = 2h = 10 \text{ km} \qquad (\text{given, } h = 5 \text{ km})$$

23. The net force on the body is zero. Weight of the body is balanced by the reaction of the ground.

24. Acceleration due to gravity,

$$g = \frac{GM}{R^2} = \frac{G}{R^2} \times \frac{4}{3}\pi R^3 \rho$$

$$\therefore \quad \rho = \frac{3g}{4\pi GR}$$

25. Acceleration due to gravity at a height above the earth surface

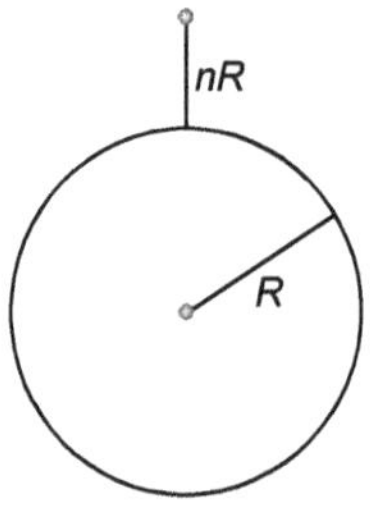

$$g' = g\left(\frac{R}{R+h}\right)^2$$

$$\frac{g}{g'} = \left(\frac{R+h}{R}\right)^2$$

$$\frac{g}{g'} = \left(\frac{R+nR}{R}\right)^2$$

$$\frac{g}{g'} = (1+n)^2$$

26. Acceleration due to gravity, $g = \frac{4}{3}\pi\rho GR$

or $\quad g \propto \rho, \quad \frac{g_1}{g_2} = \frac{\rho_1}{\rho_2}$

$$\frac{g_1}{g_2} = \frac{\rho}{2\rho} \qquad [\because \rho_2 = 2\rho]$$

$$g_2 = g_1 \times 2 = 9.8 \times 2 \quad \Rightarrow \quad g_2 = 19.6 \text{ m/s}^2$$

27. Time period of near by satellite,

$$T = 2n\sqrt{\frac{r^3}{GM}} = 2\pi\sqrt{\frac{R^3}{GM}}$$

$$= \frac{2\pi (R^3)^{1/2}}{\left[G\frac{4}{3}\pi R^3 \rho\right]^{1/2}} = \sqrt{\frac{3\pi}{G\rho}}$$

28. Velocity of satellite, $v = \sqrt{\frac{GM}{r}}$, $\quad KE \propto v^2 \propto \frac{1}{r}$

and $\quad T^2 \propto r^3$,

So, $KE \propto T^{-2/3}$.

11

Simple Harmonic Motion

11.1 Introduction

A particle has oscillatory (vibrational) motion when it moves periodically about stable equilibrium position. The motion of a pendulum is oscillatory. A weight attached to a stretched spring, once it is released, starts oscillating.

When the particle is moved away from the equilibrium position and released, a force comes into play to pull it back toward equilibrium. But by the time it gets there, it has picked up some kinetic energy, so it overshoots, stopping somewhere on the other side and is again pulled back toward equilibrium.

Of all the oscillatory motions, the most important is called **simple harmonic motion (SHM)**. In this type of oscillatory motion, displacement, velocity, acceleration and force all vary (w.r.t. time) in a way that can be described by either the sine or the cosine function collectively called **sinusoids**. Any oscillatory motion that cannot be described, so simply is called **anharmonic oscillation.**

Besides being the simplest motion to describe and analyze, it constitutes a rather accurate description of many oscillations found in nature.

Understanding periodic motion will be essential for our later study of waves, sound, alternating electric currents and light.

Chapter Snapshot

- Introduction
- The Causes of Oscillation
- Kinematics of Simple Harmonic Motion
- Force and Energy in Simple Harmonic Motion
- Relation between Simple Harmonic Motion and Uniform Circular Motion
- Method of Finding Time Period of a Simple Harmonic Motion
- Vector Method of Combining Two or More Simple Harmonic Motions in Same Direction

11.2 The Causes of Oscillation

Consider a particle free to move on x-axis, is being acted upon by a force given by

$$F = -kx^n$$

Here, k is a positive constant.

Now, following cases are possible depending on the value of n.

(i) If n is an even integer ($0, 2, 4, \ldots$ etc), force is always along negative x-axis, whether x is positive or negative. Hence, the motion of the particle is not oscillatory. If the particle is released from any position on the x-axis (except at $x = 0$) a force in negative direction of x-axis acts on it and it moves rectilinearly along negative x-axis.

(ii) If n is an odd integer (1, 3, 5, … etc), force is along negative x-axis for $x > 0$, along positive x-axis for $x < 0$ and zero for $x = 0$. Thus, the particle will oscillate about stable equilibrium position, $x = 0$. The force in this case is called the restoring force. Of these, if $n = 1$, i.e. $F = -kx$ the motion is said to be SHM.

11.3 Kinematics of Simple Harmonic Motion

A particle has simple harmonic motion along an axis OX when its displacement x relative to the origin of the coordinate system is given as a function of time by the relation,

$$x = A\cos(\omega t + \phi)$$

The quantity $(\omega t + \phi)$ is called the **phase angle** or simply the phase of the SHM and ϕ is the initial phase, i.e. the phase at $t = 0$. Although, we have defined SHM in terms of a cosine function, it may just as well be expressed in terms of a sine function. The only difference between the two forms is an initial phase difference of $\frac{\pi}{2}$. Since, the cosine (or sine) function varies between a value of -1 and $+1$, the displacement of the particle varies between $x = -A$ and $x = +A$. The maximum displacement from the origin A, is the amplitude of the SHM. The cosine or sine function repeats itself every time the angle ωt increases by 2π. Thus, the displacement of the particle repeats itself after a time interval of $\frac{2\pi}{\omega}$. Therefore, SHM is periodic, and its **period** is

$$T = \frac{2\pi}{\omega}$$

The **frequency** ν of a SHM is equal to the number of complete oscillations per unit time. Thus,

$$\nu = \frac{1}{T} = \frac{\omega}{2\pi}$$

and is measured in hertz. The quantity ω, called the angular frequency of the oscillating particle is related to the frequency by the relation similar to the equation for a circular motion, namely

$$\omega = \frac{2\pi}{T} = 2\pi\nu$$

The velocity of the particle is

$$v = \frac{dx}{dt} = -\omega A\sin(\omega t + \phi)$$

which varies periodically between the values $+\omega A$ and $-\omega A$. Similarly, the acceleration is given by $a = \frac{dv}{dt} = -\omega^2 A\cos(\omega t + \phi) = -\omega^2 x$ and therefore, varies periodically between the values $+\omega^2 A$ and $-\omega^2 A$. This expression also indicates that,

In SHM the acceleration is proportional and opposite to the displacement.

In figures x, v and a as functions of time are illustrated.

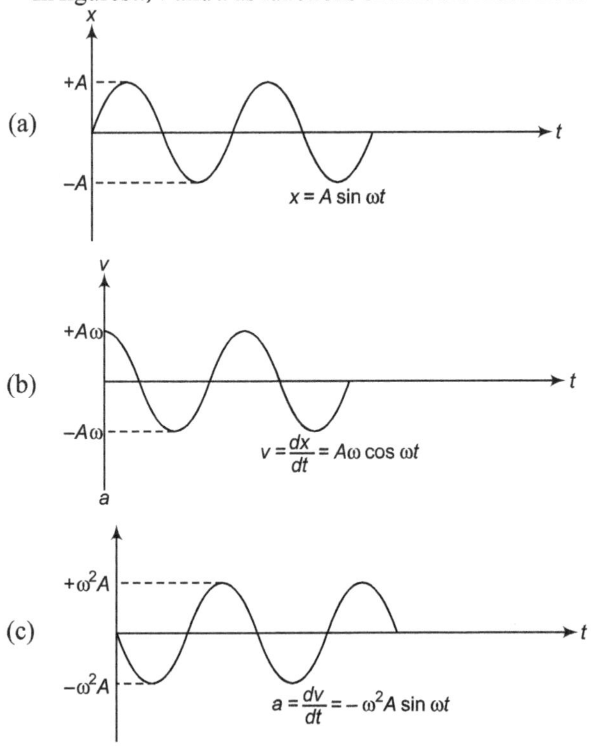

Fig. 11.1 Graphs of (a) displacement, (b) velocity and (c) acceleration *vs* time in SHM

11.4 Force and Energy in Simple Harmonic Motion

In the above article, we found that the acceleration of a body in SHM is $a = -\omega^2 x$. Applying the equation of motion $\mathbf{F} = m\mathbf{a}$, we have

$$F = -m\omega^2 x = -kx$$

where, $$\omega = \sqrt{\frac{k}{m}}$$

Thus, in SHM the force is proportional and opposite to the displacement.

That is when the displacement is to the right (positive) the force points to the left and when the displacement is to the left (negative) the force points to the right. Thus, the force is always pointing toward the origin O. Such type of force appears when an elastic body such as a spring is deformed. That is why the constant $k = m\omega^2$ is sometimes called the **elastic constant**. Further, since

$$\omega = \frac{2\pi}{T} = \sqrt{\frac{k}{m}} \Rightarrow T = 2\pi\sqrt{\frac{m}{k}} \text{ and } f = \frac{1}{2\pi}\sqrt{\frac{k}{m}}$$

Kinetic Energy

The kinetic energy of the particle is

$$K = \frac{1}{2} mv^2 = \frac{1}{2} m A^2 \sin^2 (\omega t + \phi)$$

Since, $\sin^2 \theta = 1 - \cos^2 \theta$

and using $x = A \cos (\omega t + \phi)$ for the displacement, we can also express the kinetic energy as

$$K = \frac{1}{2} m\omega^2 A^2 [1 - \cos^2 (\omega t + \phi)]$$

which can be written as

$$K = \frac{1}{2} m\omega^2 (A^2 - x^2) = \frac{1}{2} k(A^2 - x^2)$$

From this expression we can see that, the kinetic energy is maximum at the centre $(x = 0)$ and zero at the extremes of oscillation $(x = \pm A)$.

Potential Energy

To obtain the potential energy we use the relation,

$$F = -\frac{dU}{dx}$$

or $\frac{dU}{dx} = kx$ (as $F = -kx$)

$\therefore$ $\int_0^U dU = \int_0^x kx \, dx$

$\therefore$ $U = \frac{1}{2} kx^2 = \frac{1}{2} m\omega^2 x^2$

Thus, the potential energy has a minimum value at the centre $(x = 0)$ and increases as the particle approaches either extreme of the oscillation $(x = \pm A)$.

Total Energy

Total energy can be obtained by adding potential and kinetic energies. Therefore,

$$E = K + U$$

$$= \frac{1}{2} m\omega^2 (A^2 - x^2) + \frac{1}{2} m\omega^2 x^2$$

$$= \frac{1}{2} m\omega^2 A^2$$

or $E = \frac{1}{2} kA^2$ (as $m\omega^2 = k$)

Which is a **constant** quantity. This was to be expected since, the force is conservative.

Therefore, we may conclude that, during an oscillation, there is a continuous exchange of kinetic and potential energies. While moving away from the equilibrium position, the potential energy increases at the expense of the kinetic energy. When the particle moves towards the equilibrium position, the reverse happens.

Figure shows the variation of total energy (E), potential energy (U) and kinetic energy (K) with displacement (x).

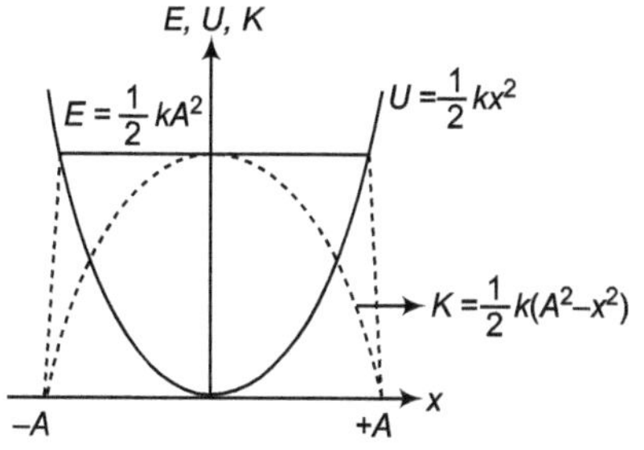

Fig. 11.2

Extra Knowledge Points

- In SHM, $F = -kx$ or $a = -\omega^2 x$, i.e. F-x graph or a-x graph is a straight line passing through origin with negative slope. The corresponding graphs are shown below:

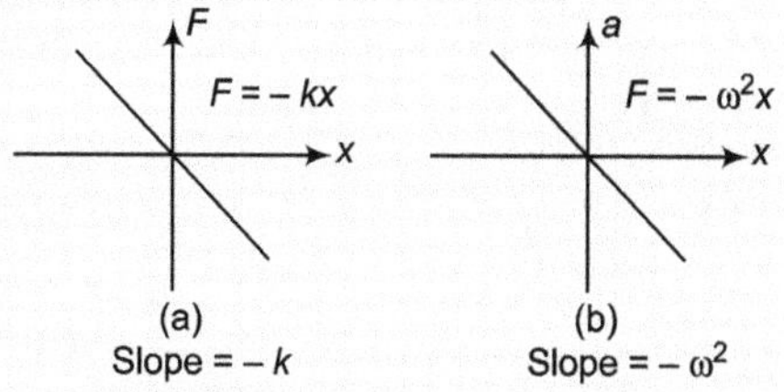

- Any function of t, say $y = y(t)$ oscillates simple harmonically if, $\frac{d^2y}{dt^2} \propto -y$ or we can say, if above condition is satisfied, y will oscillate simple harmonically.
- All sine and cosine functions of t are simple harmonic in nature, i.e. for the function

  $$y = A \sin (\omega t \pm \phi) \quad \text{or} \quad y = A \cos (\omega t \pm \phi)$$

  $\frac{d^2y}{dt^2}$ is directly proportional to $-y$. Hence, they are simple harmonic in nature.
- Kinetic energy *versus* time equation can also be written as

  $$K = \frac{1}{4} mA^2\omega^2 [1 - \cos 2(\omega t + \phi)]$$

  This function is also periodic with angular frequency 2ω. Thus, kinetic energy in SHM is also periodic with double the frequency, then that of x, v and a. But these oscillations are not simple harmonic in nature, because $\frac{d^2K}{dt^2}$ is not proportional to $-K$. But,

  $$K - \frac{1}{4} mA^2\omega^2 = -\frac{1}{4} mA^2\omega^2 \cos 2(\omega t + \phi) = K_0 \text{ (say)}$$

  Here, K_0 is simply a cosine function of time. So, K_0 will oscillate simple harmonically with angular frequency 2ω. Same is the case with potential energy function. U also oscillate with angular frequency 2ω but the oscillations are not simple harmonic in nature.

Total energy does not oscillate.

It is constant. Thus,

$x \rightarrow$ oscillates simple harmonically with angular frequency ω.

$v \rightarrow$ oscillates simple harmonically with angular frequency ω.

$a \rightarrow$ oscillates simple harmonically with angular frequency ω.

$K \rightarrow$ oscillates with angular frequency 2ω but not simple harmonically.

$U \rightarrow$ oscillates with angular frequency 2ω but not simple harmonically.

$E \rightarrow$ does not oscillate.

- In the above discussion, we have read that potential energy is zero at mean position and maximum at extreme positions and kinetic energy is maximum at mean position and zero at extreme positions. But the correct statement is like this,

  At mean position $\rightarrow$ K is maximum and U is minimum (it may be zero also, but it is not necessarily zero).

  At extreme positions $\rightarrow$ K is zero and U is maximum.

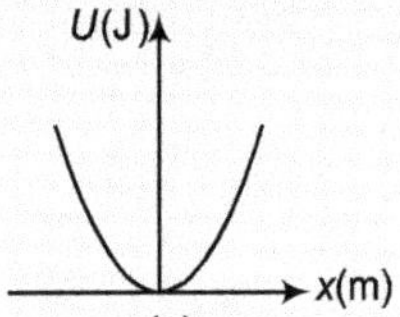

(a)

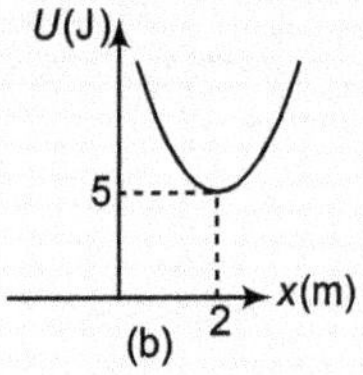

(b)

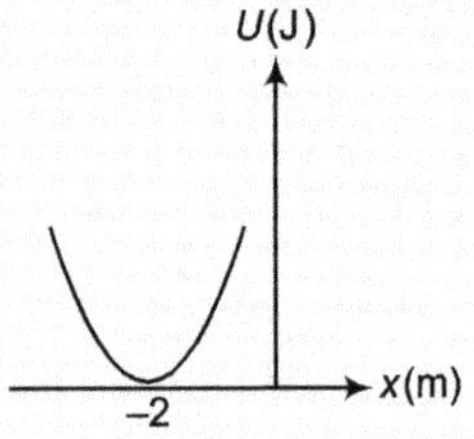

(c)

Thus, in Fig. (a), oscillations will take place about the mean position $x = 0$ and minimum potential energy at mean position is zero.

In Fig. (b), mean position is at $x = 2\,\text{m}$ and the minimum potential energy in this position is 5 J.

In Fig. (c), mean position is at $x = -2\,\text{m}$ and the minimum potential energy in this position is again zero.

- A function $f(t)$ is said to be periodic of time period T, if $f(t+T) = f(t)$

  All sine or cosine functions of time are periodic. Thus,

$$Y = A \sin \omega t$$

or $A \cos \omega t$ is periodic, of time period $T = \dfrac{2\pi}{\omega}$.

How the different physical quantities (e.g. displacement, velocity, acceleration, kinetic energy, etc) vary with time or displacement are listed below in tabular form.

Table 11.1

S. No.	Name of the equation	Expression of the equation	Remarks
1.	Displacement- time	$x = A \cos (\omega t + \phi)$	x varies between $+A$ and $-A$
2.	Velocity- time $\left(v = \dfrac{dx}{dt}\right)$	$v = -A\omega \sin (\omega t + \phi)$	v varies between $+A\omega$ and $-A\omega$
3.	Acceleration- time $\left(a = \dfrac{dv}{dt}\right)$	$a = -A\omega^2 \cos (\omega t + \phi)$	a varies between $+A\omega^2$ and $-A\omega^2$
4.	Kinetic energy- time $\left(K = \dfrac{1}{2} mv^2\right)$	$K = \dfrac{1}{2} mA^2\omega^2 \sin^2 (\omega t + \phi)$	K varies between 0 and $\dfrac{1}{2} mA^2\omega^2$
5.	Potential energy-time $\left(U = \dfrac{1}{2} m\,\omega^2 x^2\right)$	$K = \dfrac{1}{2} m\omega^2 A^2 \cos^2 (\omega t + \phi)$	U varies between $\dfrac{1}{2} mA^2\omega^2$ and 0
6.	Total energy- time $(E = K + U)$	$E = \dfrac{1}{2} m\,\omega^2 A^2$	E is constant
7.	Velocity- displacement	$v = \omega\sqrt{A^2 - x^2}$	$v = 0$ at $x = \pm A$ and at $x = 0, v = \pm A\omega$
8.	Acceleration-displacement	$a = -\omega^2 x$	$a = 0$ at $x = 0$ and $a = \pm\omega^2 A$ at $x = \mp A$
9.	Kinetic energy-displacement	$K = \dfrac{1}{2} m\,\omega^2 (A^2 - x^2)$	$K = 0$ at $x = \pm A$ and $K = \dfrac{1}{2} m\,\omega^2 A^2$ at $x = 0$
10.	Potential energy-displacement	$U = \dfrac{1}{2} m\,\omega^2 x^2$	$U = 0$ at $x = 0$ and $U = \dfrac{1}{2} m\,\omega^2 A^2$ at $x = \pm A$
11.	Total energy-displacement	$E = \dfrac{1}{2} m\,\omega^2 A^2$	E is constant

From the above table we see that x, v and a are sine or cosine functions of time. So, they all oscillate simple harmonically with same angular frequency ω.

Phase difference between x and a is π and between any other two is $\dfrac{\pi}{2}$.

Example 11.1 *Find the period of the function,*

$$y = \sin \omega t + \sin 2\omega t + \sin 3\omega t$$

Sol. The given function can be written as

$$y = y_1 + y_2 + y_3$$

Here, $y_1 = \sin \omega t, \quad T_1 = \frac{2\pi}{\omega}$

$y_2 = \sin 2\omega t, \quad T_2 = \frac{2\pi}{2\omega} = \frac{\pi}{\omega}$

and $y_3 = \sin 3\,\omega t, \quad T_3 = \frac{2\pi}{3\,\omega}$

$\therefore \quad T_1 = 2T_2$ and $T_1 = 3T_3$

So, the time period of the given function is T_1 or $\frac{2\pi}{\omega}$.

- Because in time $T = \frac{2\pi}{\omega}$, first function completes one oscillation, the second function two oscillations and the third, three.

Example 11.2 *A linear harmonic oscillator has a total mechanical energy of 200 J. Potential energy of it at mean position is 50 J. Find*

(i) the maximum kinetic energy,
(ii) the minimum potential energy,
(iii) the potential energy at extreme positions.

Sol. At mean position, potential energy is minimum and kinetic energy is maximum.

Hence, $U_{min} = 50$ J (at mean position)

and $K_{max} = E - U_{min}$

$= 200 - 50 = 150$ J (at mean position)

At extreme positions, kinetic energy is zero and potential energy is maximum.

$\therefore \quad U_{max} = E = 200$ J (at extreme position)

Example 11.3 *The potential energy of a particle oscillating on x-axis is given as*

$$U = 20 + (x - 2)^2$$

Here, U is in joules and x in metres. Total mechanical energy of the particle is 36 J.

(a) State whether the motion of the particle is simple harmonic or not.
(b) Find the mean position.
(c) Find the maximum kinetic energy of the particle.

Sol. (a) $F = -\frac{dU}{dx} = -2(x - 2)$

By assuming $x - 2 = X$, we have $F = -2X$

Since, $F \propto -X$

The motion of the particle is simple harmonic.

(b) The mean position of the particle is $X = 0$ or $x - 2 = 0$, which gives $x = 2$ m

(c) Maximum kinetic energy of the particle is

$K_{max} = E - U_{min} = 36 - 20$

$= 16$ J

- U_{min} is 20 J at mean position or at $x = 2$ m.

11.5 Relation between Simple Harmonic Motion and Uniform Circular Motion

Consider a particle Q, moving on a circle of radius A with constant angular velocity ω. The projection of Q on a diameter BC is P. It is clear from the figure that as Q moves around the circle of the projection P oscillates between B and C. The angle that the radius OQ makes with the x-axis is $\theta = \omega t + \phi$. Here, ϕ is the angle made by the radius OQ with the x-axis at time $t = 0$. Further,

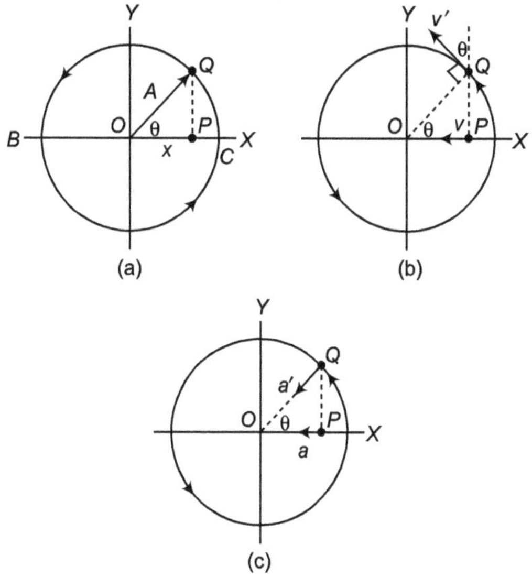

Fig. 11.3 Relation between SHM and uniform circular motion. (a) position, (b) velocity and (c) acceleration

$$OP = OQ \cos \theta$$

or $$x = A \cos (\omega t + \phi)$$

In other words, P moves with SHM. That is

When a particle moves with uniform circular motion, its projection on a diameter moves with SHM.

The velocity of Q is perpendicular to OQ and has a magnitude of velocity $v' = \omega A$. The component of v' along the x-axis is

$$v = -v' \sin \theta$$

or $$v = -\omega A \sin (\omega t + \phi)$$

which is also the velocity of P. The acceleration of Q is centripetal and has a magnitude, $a' = \omega^2 A$.

The component of a' along the x-axis is

$$a = -a' \cos \theta \quad \text{or} \quad a = -\omega^2 A \cos (\omega t + \phi)$$

Which again coincides with the acceleration of P.

11.6 Method of Finding Time Period of a Simple Harmonic Motion

There are basically following two methods of finding time period of a SHM. These are the restoring force or torque method and the energy method.

Restoring Force or Torque Method

The following steps are usually followed in this method:

Step 1 Find the stable equilibrium position which usually is also known as the mean position. Net force or torque on the particle in this position is zero. Potential energy is minimum.

Step 2 Displace the particle from its mean position by a small displacement x (in case of a linear SHM) or θ (in case of an angular SHM).

Step 3 Find net force or torque in this displaced position.

Step 4 Show that this force or torque has a tendency to bring the particle back to its mean position and magnitude of force or torque is proportional to displacement, i.e.

$$F \propto -x \quad \text{or} \quad F = -kx \qquad \ldots\text{(i)}$$

or $$\tau \propto -\theta \quad \text{or} \quad \tau = -k\theta \qquad \ldots\text{(ii)}$$

This force or torque is also known as restoring force or restoring torque.

Step 5 Find linear acceleration by dividing Eq. (i) by mass m or angular acceleration by dividing Eq. (ii) by moment of inertia I. Hence,

$$a = -\frac{k}{m}x = -\omega^2 x \quad \text{or} \quad \alpha = -\frac{k}{I}\theta = -\omega^2\theta$$

Step 6 Finally,

$$\omega = \sqrt{\left|\frac{a}{x}\right|} \quad \text{or} \quad \sqrt{\left|\frac{\alpha}{\theta}\right|}$$

or $$\frac{2\pi}{T} = \sqrt{\left|\frac{a}{x}\right|} \quad \text{or} \quad \sqrt{\left|\frac{\alpha}{\theta}\right|}$$

$\therefore$ $$T = 2\pi\sqrt{\left|\frac{x}{a}\right|} \quad \text{or} \quad 2\pi\sqrt{\left|\frac{\theta}{\alpha}\right|}$$

Energy Method

Repeat step 1 and step 2 as in method 1. Find the total mechanical energy (E) in the displaced position. Since, mechanical energy in SHM remains constant.

$$\frac{dE}{dt} = 0$$

By differentiating the energy equation with respect to time and substituting $\frac{dx}{dt} = v, \frac{d\theta}{dt} = \omega, \frac{dv}{dt} = a$ and $\frac{d\omega}{dt} = \alpha$ we come to step 5. The remaining procedure is same.

Note Points

- E usually consists of following terms
  (a) Gravitational PE,
  (b) Elastic PE,
  (c) Electrostatic PE,
  (d) Rotational KE and
  (e) Translational KE.
- For gravitational PE, choose the reference point ($h = 0$) at mean position.

Now, let us take few examples of finding time period (T) of certain simple harmonic motions.

The Simple Pendulum

An example of SHM is the motion of a pendulum. A simple pendulum is defined as a particle of mass m suspended from a point O by a string of length l and of negligible mass.

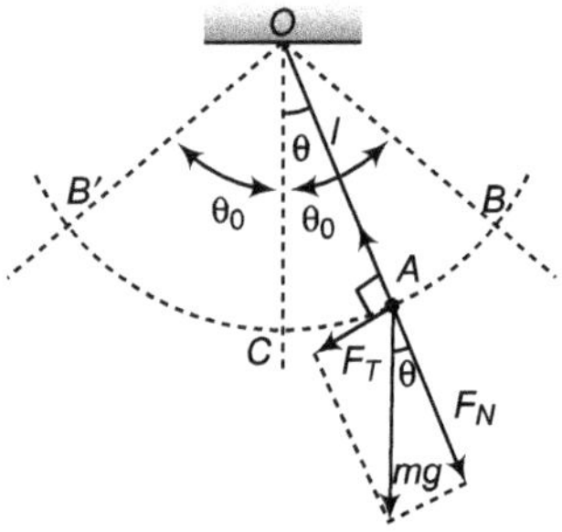

Fig. 11.4

When the particle is pulled aside to position B, so that the string makes an angle θ_0 with the vertical OC and then released, the pendulum will oscillate between B and the symmetric position B'. The oscillatory motion is due to the tangential component F_T of the weight mg of the particle. This force F_T is maximum at B and B', and zero at C. Thus, we can write

$$F_T = -mg\sin\theta$$

Here, minus sign appears because it is opposite to the displacement.

$$x = CA$$

$\therefore$ $$ma_T = -mg\sin\theta \qquad \ldots\text{(i)}$$

Here, $$a_T = l\alpha \qquad \left(\text{where, } \alpha = \frac{d^2\theta}{dt^2}\right)$$

and $$\sin\theta \approx \theta \qquad \text{for small oscillations}$$

$\therefore$ $$ml\alpha = -mg\theta$$

or $$\alpha = -\left(\frac{g}{l}\right)\theta \quad \text{or} \quad \left|\frac{\theta}{\alpha}\right| = \frac{l}{g}$$

$\therefore$ $$T = 2\pi\sqrt{\left|\frac{\theta}{\alpha}\right|} \quad \text{or} \quad T = 2\pi\sqrt{\frac{l}{g}}$$

- Note that the period is independent of the mass of the pendulum.

Energy Method

Let us derive the same expression by energy method. Suppose ω be the angular velocity of particle at angular displacement θ about point O. Then, total mechanical energy of particle in position A is

$$E = \frac{1}{2} I\omega^2 + mg(h_A - h_C)$$

or $$E = \frac{1}{2}(ml^2)\,\omega^2 + mgl(1 - \cos\theta)$$

E is constant therefore, $\dfrac{dE}{dt} = 0$

or $$0 = ml^2\omega\left(\frac{d\omega}{dt}\right) + mgl\sin\theta\left(\frac{d\theta}{dt}\right)$$

Putting $\dfrac{d\theta}{dt} = \omega$, $\dfrac{d\omega}{dt} = \alpha$ and $\sin\theta \approx \theta$, we get the same expression, *viz.*

$$\alpha = -\left(\frac{g}{l}\right)\theta$$

$$\therefore \quad T = 2\pi\sqrt{\left|\frac{\theta}{\alpha}\right|} \quad \text{or} \quad T = 2\pi\sqrt{\frac{l}{g}}$$

Following points should be remembered in case of a simple pendulum.

1. **For** large amplitudes the approximation $\sin\theta \approx \theta$ is not valid and the calculation of the period is more complex.
2. If the time period of a simple pendulum is 2 s, it is called **seconds pendulum.**
3. If length of the pendulum is large, g no longer remain vertical but will be directed towards the centre of the earth and expression for time period is given by

$$T = 2\pi\sqrt{\frac{1}{g\left(\frac{1}{l} + \frac{1}{R}\right)}}$$

Here, R is the radius of earth. From this expression, we can see that,

(a) if $l << R$, $\frac{1}{l} >> \frac{1}{R}$ and $T = 2\pi\sqrt{\frac{l}{g}}$

(b) as $l \to \infty$, $\frac{1}{l} \to 0$ and $T = 2\pi\sqrt{\frac{R}{g}}$

and substituting the value of R and g, we get $T = 84.6$ min.

Note Points

- In physics, this 84.6 min come in following four places:
  (i) Time period of a satellite close to earth's surface is 84.6 min.
  (ii) Time period of a pendulum of infinite length is 84.6 min.
  (iii) If a tunnel is dug along any chord of the earth and a particle is released from the surface of earth along this tunnel, then motion of this particle is simple harmonic and time period of this is also 84.6 min.
  (iv) If length of day on earth becomes 84.6 min, one will feel weightless on equator

4. Time period of a simple pendulum depends on acceleration due to gravity, g $\left(\text{as } T \propto \frac{1}{\sqrt{g}}\right)$, so take $|\mathbf{g}_{\text{eff}}|$ in $T = 2\pi\sqrt{\frac{l}{g}}$. Following two cases are possible:

(i) If a simple pendulum is in a carriage which is accelerating with acceleration $\mathbf{a}$, then

$$\mathbf{g}_{\text{eff}} = \mathbf{g} - \mathbf{a}$$

e.g. if the acceleration $\mathbf{a}$ is upwards, then

$$|\mathbf{g}_{\text{eff}}| = g + a$$

and $$T = 2\pi\sqrt{\frac{l}{g + a}}$$

If the acceleration $\mathbf{a}$ is downwards, then $(g > a)$

$$|\mathbf{g}_{\text{eff}}| = g - a \quad \text{and} \quad T = 2\pi\sqrt{\frac{l}{g - a}}$$

If the acceleration $\mathbf{a}$ is in horizontal direction, then

$$|\mathbf{g}_{\text{eff}}| = \sqrt{a^2 + g^2}$$

In a freely falling lift, $\mathbf{g}_{\text{eff}} = 0$ and $T = \infty$, i.e. the pendulum will not oscillate.

(ii) If in addition to gravity one additional force $\mathbf{F}$, (e.g. electrostatic force $\mathbf{F}_e$) is also acting on the bob, then in that case,

$$\mathbf{g}_{\text{eff}} = \mathbf{g} + \frac{\mathbf{F}}{m}$$

Here, m is the mass of the bob.

Example 11.4 *A simple pendulum of length l is suspended from the ceiling of a cart which is sliding without friction on an inclined plane of inclination θ. What will be the time period of the pendulum?*

Sol. Here, point of suspension has an acceleration. $\mathbf{a} = g\sin\theta$ (down the plane). Further, $\mathbf{g}$ can be resolved into two components $g\sin\theta$ (along the plane) and $g\cos\theta$ (perpendicular to plane).

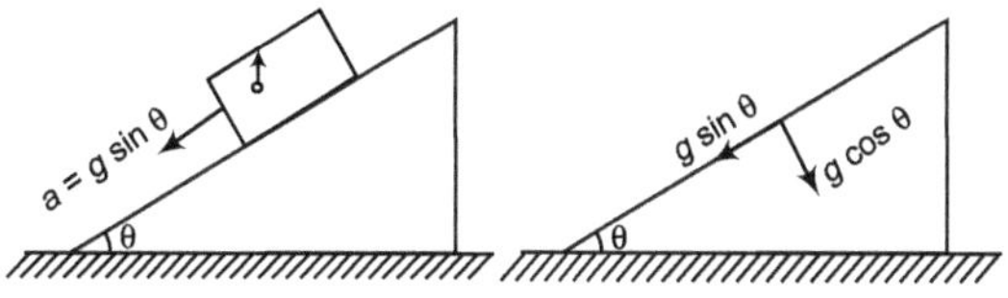

Fig. 11.5

$$\therefore \quad \mathbf{g}_{\text{eff}} = \mathbf{g} - \mathbf{a}$$

$$= g \cos\theta \quad \text{(perpendicular to plane)}$$

$$\therefore \quad T = 2\pi\sqrt{\frac{l}{|g_{\text{eff}}|}}$$

$$= 2\pi\sqrt{\frac{l}{g\cos\theta}}$$

- If $\theta = 0°$, $T = 2\pi\sqrt{\frac{l}{g}}$ which is quiet obvious.

Example 11.5 *A simple pendulum consists of a small sphere of mass m suspended by a thread of length l. The sphere carries a positive charge q. The pendulum is placed in a uniform electric field of strength E directed vertically upwards. With what period will pendulum oscillate, if the electrostatic force acting on here is less than the gravitational force?*

Sol. The two forces acting on the bob are shown in figure

g_{eff} in this case will be $\frac{w - F_e}{m}$

or $$g_{\text{eff}} = \frac{mg - qE}{m}$$

$$= g - \frac{qE}{m}$$

$$\therefore \quad T = 2\pi\sqrt{\frac{l}{g_{\text{eff}}}}$$

$$= 2\pi\sqrt{\frac{l}{g - \frac{qE}{m}}}$$

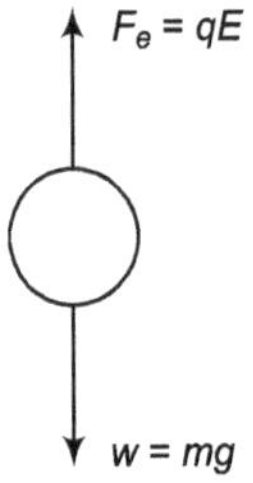

Fig. 11.6

Spring-block System

Suppose a mass m is attached to the free end of a massless spring of spring constant k, with its other end fixed to a rigid support.

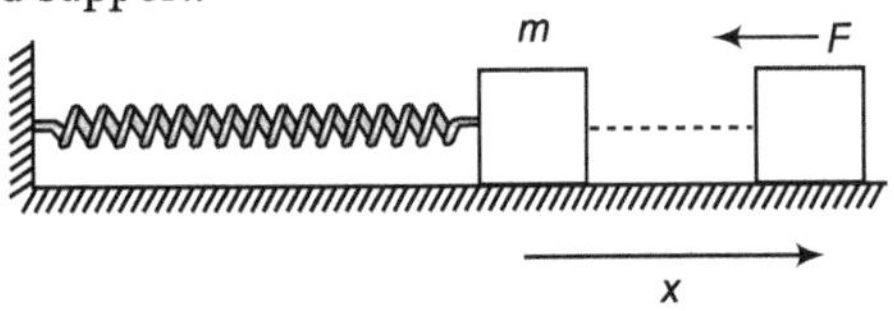

Fig. 11.7

If the mass be displaced through a distance x, as shown a linear restoring force,

$$F = -kx \quad \text{...(i)}$$

starts acting on the mass, tending to bring it back into its original position. The negative sign simply indicates that it is directed oppositely to the displacement of the mass.

Eq. (i) can be written as

$$ma = -kx \quad \text{...(ii)}$$

or $$\left|\frac{x}{a}\right| = \frac{m}{k}$$

$$\therefore \quad T = 2\pi\sqrt{\left|\frac{x}{a}\right|} \quad \text{or} \quad T = 2\pi\sqrt{\frac{m}{k}}$$

Energy Method

The time period of the spring-block system can also be obtained by the energy method. Let v be the speed of the mass in displaced position. Then, total mechanical energy of the spring-block system is

E = kinetic energy of the block + elastic potential energy

or $$E = \frac{1}{2}mv^2 + \frac{1}{2}kx^2$$

Since, E is constant

$$\frac{dE}{dt} = 0 \quad \text{or} \quad 0 = mv\left(\frac{dv}{dt}\right) + kx\left(\frac{dx}{dt}\right)$$

Substituting, $\frac{dx}{dt} = v$ and $\frac{dv}{dt} = a$

We have $ma = -kx$

$$\therefore \quad T = 2\pi\sqrt{\left|\frac{x}{a}\right|} = 2\pi\sqrt{\frac{m}{k}}$$

Following points are important for a spring block system.

(i) Although, we have considered a horizontal system. But the same is true for vertical system also.

$$T = 2\pi\sqrt{\frac{m}{k}}$$

Fig. 11.8

(ii) In case of a vertical spring-block system, time period can also be written as

$$T = 2\pi\sqrt{\frac{l}{g}}$$

Here, l = extension in the spring when the mass m is suspended from the spring.

This can be seen as under

$$kl = mg \quad \text{(in equilibrium position)}$$

$$\therefore \quad \frac{m}{k} = \frac{l}{g} \quad \text{or} \quad T = 2\pi\sqrt{\frac{m}{k}} = 2\pi\sqrt{\frac{l}{g}}$$

(iii) **Equivalent force constant (k)** If a spring pendulum is constructed by using two springs and a mass, the following three situations are possible.

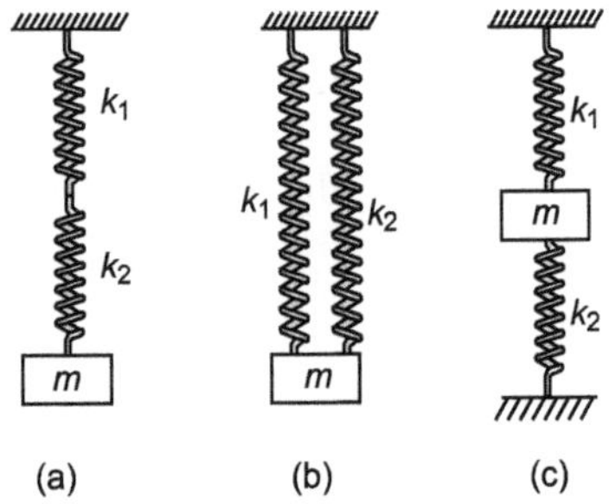

Fig. 11.9

Refer Fig. (a)

In this case,

$$\frac{1}{k}=\frac{1}{k_1}+\frac{1}{k_2} \quad \text{or} \quad k=\frac{k_1k_2}{k_1+k_2}$$

$$\therefore \quad T=2\pi\sqrt{\frac{m}{k}}=2\pi\sqrt{\frac{m(k_1+k_2)}{k_1k_2}}$$

Refer Fig. (b) and (c)

In both the cases,

$$k=k_1+k_2 \quad \text{or} \quad T=2\pi\sqrt{\frac{m}{k}}=2\pi\sqrt{\frac{m}{k_1+k_2}}$$

(iv) If spring has a mass m_s and a mass m is suspended from it, then time period is given by

$$T=2\pi\sqrt{\frac{m+\frac{m_s}{3}}{k}}$$

(v) If two masses m_1 and m_2 are connected by a spring and made to oscillate on horizontal surface, then time period is given by

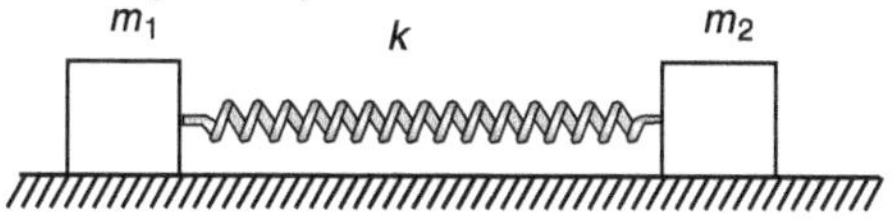

Fig. 11.10

$$T=2\pi\sqrt{\frac{\mu}{k}}$$

Here, $\mu=$ reduced mass $=\dfrac{m_1m_2}{m_1+m_2}$

(vi) The force constant (k) of a spring is inversely proportional to the length of the spring, i.e.

$$k\propto\frac{1}{\text{Length of spring}}$$

This can be visualized as under

A spring of length l and spring constant k can be supposed to be made up by two springs in series of length $\frac{l}{2}$ and force constant $2k$. In series,

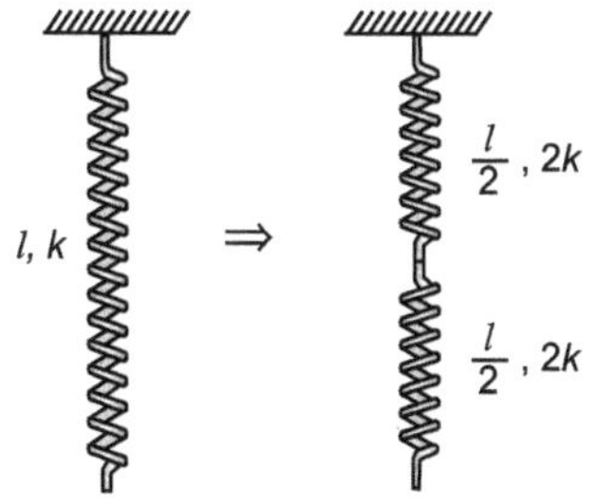

Fig. 11.11

$$k_{\text{eff}}=\frac{(2k)(2k)}{2k+2k}=k$$

Example 11.6 *A block with a mass of 3.00 kg is suspended from an ideal spring having negligible mass and stretches the spring 0.2 m.*

(a) What is the force constant of the spring?

(b) What is the period of oscillation of the block, if it is pulled down and released?

Sol. (a) In equilibrium,

$$kl=mg$$

$$\therefore \quad k=\frac{mg}{l}$$

Substituting the proper values, we have

$$k=\frac{(3.00)(9.8)}{0.2}=147\text{ N/m}$$

(b) $$T=2\pi\sqrt{\frac{l}{g}}=2\pi\sqrt{\frac{0.2}{9.8}}=0.897\text{ s}$$

Example 11.7 *A block with mass M attached to a horizontal spring with force constant k is moving with simple harmonic motion having amplitude A_1. At the instant when the block passes through its equilibrium position a lump of putty with mass m is dropped vertically on the block from a very small height and sticks to it.*

(a) Find the new amplitude and period.

(b) Repeat part (a) for the case in which the putty is dropped on the block when it is at one end of its path.

Sol. (a) Before the lump of putty is dropped the total mechanical energy of the block and spring is

$$E_1=\frac{1}{2}kA_1^2$$

Since, the block is at the equilibrium position, $U=0$, and the energy is purely kinetic. Let v_1 be the speed of the block at the equilibrium position, we have

$$E_1=\frac{1}{2}Mv_1^2=\frac{1}{2}kA_1^2$$

$$\therefore \quad v_1=\sqrt{\frac{k}{M}}A_1$$

During the process momentum of the system in horizontal direction is conserved. Let v_2 be the speed of the combined mass, then

$$(M+m)v_2=Mv_1$$

$$\therefore \quad v_2=\frac{M}{M+m}v_1$$

Now, let A_2 be the amplitude afterwards. Then,

$$E_2=\frac{1}{2}kA_2^2$$

$$=\frac{1}{2}(M+m)v_2^2$$

Substituting the proper values, we have

$$A_2=A_1\sqrt{\frac{M}{M+m}}$$

- $E_2<E_1$, as some energy is lost into heating up the block and putty.

Further, $$T_2=2\pi\sqrt{\frac{M+m}{k}}$$

(b) When the putty drops on the block, the block is instantaneously at rest. All the mechanical energy is stored in the spring as potential energy. Again the momentum in horizontal direction is conserved during the process. But, now it is zero just before and after putty is dropped. So, in this case, adding the extra mass of the putty has no effect on the mechanical energy, i.e.

$$E_2 = E_1 = \frac{1}{2} kA_1^2$$

and the amplitude is still A_1. Thus,

$$A_2 = A_1$$

and $$T_2 = 2\pi \sqrt{\frac{M+m}{k}}$$

The Physical Pendulum

The physical pendulum is just a rigid body, of whatever shape, capable of oscillating about a horizontal axis passing through it. For small oscillations, the motion of a physical pendulum is almost as easy as for a simple pendulum. Figure shows a body of irregular shape pivoted at O so, that it can oscillate without friction about an axis passing through O.

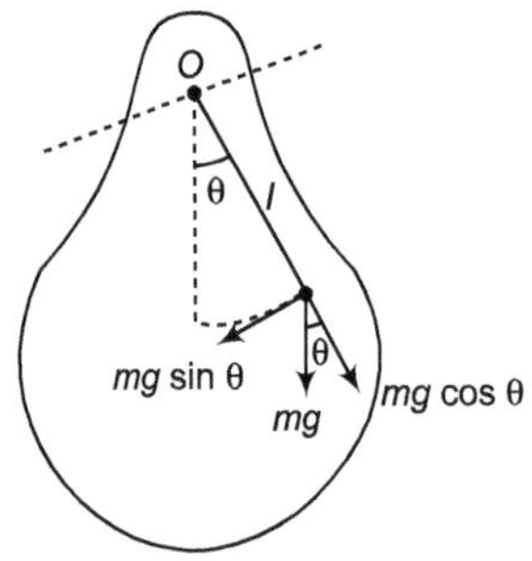

Fig. 11.12

In equilibrium the centre of gravity (G) in directly below O. In the position shown in figure, the body is displaced from equilibrium by an angle θ. The distance from O to the centre of gravity is l. The moment of inertia of the body about the axis of rotation through O is I and the total mass is m. In the displaced position, the weight mg causes a restoring torque,

$$\tau = -(mg)(l \sin\theta)$$

The negative sign shows that the restoring torque is clockwise when the displacement is counterclockwise and *vice-versa*.

For small oscillations,

$$\sin\theta \approx \theta \quad \text{and} \quad \Sigma\tau = I\alpha$$

$$\therefore \quad -(mgl)\theta = I\alpha$$

As α is proportional to $-\theta$, the motion is simple harmonic, the time period of which is

$$T = 2\pi\sqrt{\left|\frac{\theta}{\alpha}\right|} \quad \text{or} \quad T = 2\pi\sqrt{\frac{I}{mgl}}$$

Example 11.8 *A uniform circular disc of radius R oscillates in a vertical plane about a horizontal axis. Find the distance of the axis of rotation from the centre for which the period is minimum. What is the value of this period?*

Sol. The time period of a compound pendulum is the minimum when its length is equal to the radius of gyration about its centre of gravity, i.e. $l = K$.

Since, the moment of inertia of a disc about an axis perpendicular to its plane and passing through its centre is equal to,

$$I = MK^2 = \frac{1}{2}MR^2 \quad \therefore \quad K = \frac{R}{\sqrt{2}}$$

Thus, the disc will oscillate with the minimum time period when the distance of the axis of rotation from the centre is $\frac{R}{\sqrt{2}}$.

And the value of this minimum time period will be

$$T_{min} = 2\pi\sqrt{\frac{\frac{2R}{\sqrt{2}}}{g}} = 2\pi\sqrt{\frac{\sqrt{2}R}{g}}$$

or $$T_{min} \approx 2\pi\sqrt{\frac{1.414R}{g}}$$

Example 11.9 *Find the period of small oscillations of a uniform rod with length l, pivoted at one end.*

Sol. $T = 2\pi\sqrt{\frac{I_0}{mg(OG)}}$

Here, $$I_0 = \frac{1}{3}ml^2$$

and $$OG = \frac{l}{2}$$

$$\therefore \quad T = 2\pi\sqrt{\frac{\left(\frac{1}{3}ml^2\right)}{(m)(g)\left(\frac{l}{2}\right)}}$$

or $$T = 2\pi\sqrt{\frac{2l}{3g}}$$

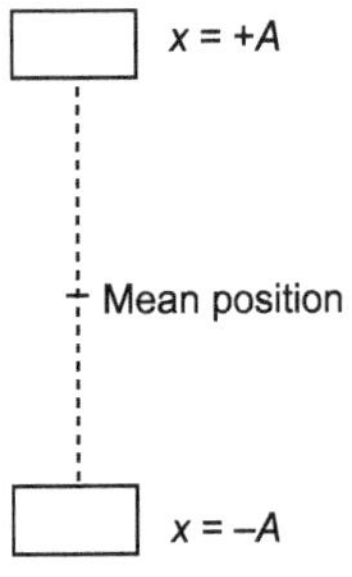

Fig. 11.13

Oscillations of a Fluid Column

Initially, the level of liquid in both the columns is same. The area of cross-section of the tube is uniform. If the liquid is depressed by x in one limb, it will rise by x along the length of the tube is the other limb. Here, the restoring force is provided by the hydrostatic pressure difference.

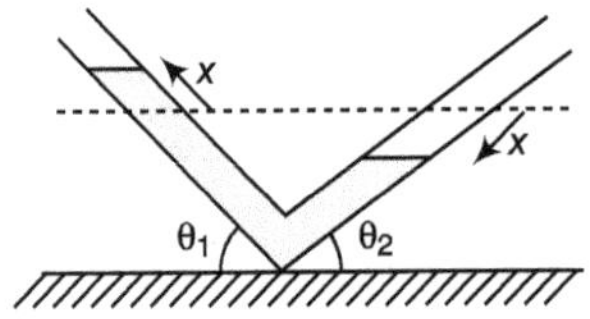

Fig. 11.14

$$\therefore \quad F = -(\Delta p)A = -(h_1 + h_2)\rho g A$$
$$= -\rho g A(\sin\theta_1 + \sin\theta_2)x$$

Let, m be the mass of the liquid in the tube. Then,

$$ma = -\rho g A(\sin\theta_1 + \sin\theta_2)x$$

Since, F or a is proportional to $-x$, the motion of the liquid column is simple harmonic in nature, time period of which is given by

$$T = 2\pi\sqrt{\left|\frac{x}{a}\right|}$$

or
$$T = 2\pi\sqrt{\frac{m}{\rho g A(\sin\theta_1 + \sin\theta_2)}}$$

- For a U-tube, if the liquid is filled to a height l, $\theta_1 = 90° = \theta_2$ and $m = 2\,(lA\rho)$

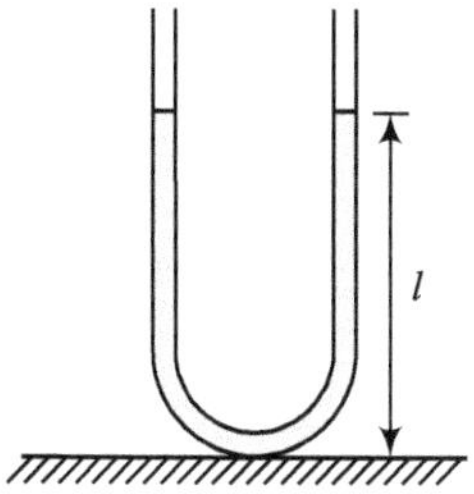

Fig. 11.15

So,
$$T = 2\pi\sqrt{\frac{l}{g}}$$

Thus, we see that the expression $T = 2\pi\sqrt{l/g}$ comes in picture at three places.

(i) Time period of a simple pendulum for small oscillations.
(ii) Time period of a spring-block system in vertical position.
(iii) Time period of a liquid column in a U-tube filled to a height (l). But, (l) has different meanings at different places.

11.7 Vector Method of Combining Two or More Simple Harmonic Motions in Same Direction

A simple harmonic motion is produced when a force (called restoring force) proportional to the displacement acts on a particle. If a particle is acted upon by two such forces, the resultant motion of the particle is a combination of two simple harmonic motions. Suppose the two individual motions are represented by

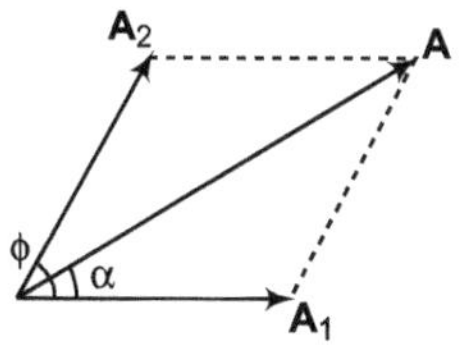

Fig. 11.16

$$x_1 = A_1 \sin \omega t$$

and
$$x_2 = A_2 \sin(\omega t + \phi)$$

Both the simple harmonic motions have same angular frequency ω.

The resultant displacement of the particle is given by

$$x = x_1 + x_2$$
$$= A_1 \sin \omega t + A_2 \sin(\omega t + \phi)$$
$$= A \sin(\omega t + \alpha)$$

Here,
$$A = \sqrt{A_1^2 + A_2^2 + 2A_1A_2\cos\phi}$$

and
$$\tan\alpha = \frac{A_2 \sin\phi}{A_1 + A_2\cos\phi}$$

Thus, we can see that this is similar to the vector addition. The same method of vector addition can be applied to the combination of more than two simple harmonic motions.

Example 11.10 *Find the displacement equation of the simple harmonic motion obtained by combining the motions.*

$$x_1 = 2\sin\omega t, \quad x_2 = 4\sin\left(\omega t + \frac{\pi}{6}\right)$$

and
$$x_3 = 6\sin\left(\omega t + \frac{\pi}{3}\right)$$

Sol. The resultant equation is

$$x = A\sin(\omega t + \phi)$$
$$\Sigma A_x = 2 + 4\cos 30° + 6\cos 60°$$
$$= 8.46$$

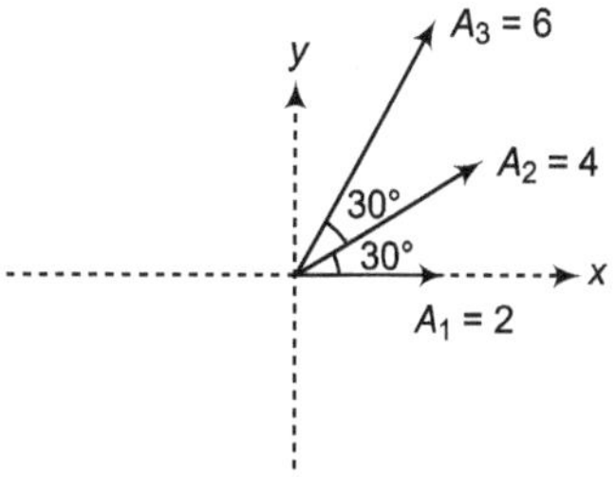

Fig. 11.17

and
$$\Sigma A_y = 4\sin 30° + 6\cos 30°$$
$$= 7.2$$

$\therefore$
$$A = \sqrt{(\Sigma A_x)^2 + (\Sigma A_y)^2}$$
$$= \sqrt{(8.46)^2 + (7.2)^2}$$
$$= 11.25$$

and
$$\tan\phi = \frac{\Sigma A_y}{\Sigma A_x}$$
$$= \frac{7.2}{8.46} = 0.85$$

or
$$\phi = \tan^{-1}(0.85) = 40.4°$$

Thus, the displacement equation of the combined motion is

$$x = 11.25\sin(\omega t + \phi)$$

where, $\phi = 40.4°$

Extra Knowledge Points

- **Lissajous figure**

Suppose two forces act on a particle, the first alone would produce a simple harmonic motion in x-direction given by

$$x = a \sin \omega t$$

and the second would produce a simple harmonic motion in y-direction given by

$$y = b \sin (\omega t + \phi)$$

The amplitudes a and b may be different and their phases differ by ϕ. The frequencies of the two simple harmonic motions are assumed to be equal. The resultant motion of the particle is a combination of the two simple harmonic motions.

Depending on the value of ϕ and relation between a and b, the particle follows different paths. Given below are few special cases

- **Case 1 (When $\phi = 0°$)** When the phase difference between two simple harmonic motions is 0°, i.e.

$$x = a \sin \omega t$$

$$\Rightarrow \quad \sin \omega t = \frac{x}{a} \quad \text{...(i)}$$

$$y = b \sin \omega t$$

$$\Rightarrow \quad \sin \omega t = \frac{y}{b} \quad \text{...(ii)}$$

From Eqs. (i) and (ii), we get

$$\frac{x}{a} = \frac{y}{b}$$

or

$$y = \left(\frac{b}{a}\right) x$$

which is equation of a straight line with slope $\frac{b}{a}$. Thus, the path of the particle is a straight line. As a special case $y = x$, if $a = b$ or slope is 1.

- **Case 2 $\left(\text{When } \phi = \frac{\pi}{2}\right)$** When the phase difference is $\frac{\pi}{2}$, i.e.

$$x = a \sin \omega t$$

$$\Rightarrow \quad \sin \omega t = \frac{x}{a} \quad \text{...(iii)}$$

$$y = b \sin\left(\omega t + \frac{\pi}{2}\right) = b \cos \omega t$$

$$\Rightarrow \quad \cos \omega t = \frac{y}{b} \quad \text{...(iv)}$$

Squaring and adding Eqs. (iii) and (iv), we get

$$\frac{x^2}{a^2} + \frac{y^2}{b^2} = 1$$

Table 11.2

S. No.	ϕ	$\frac{b}{a}$	Corresponding Lissajous figure
1.	0°	2	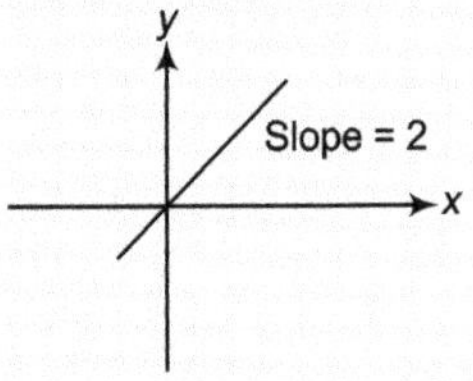 Straight line with slope 2
2.	0°	1	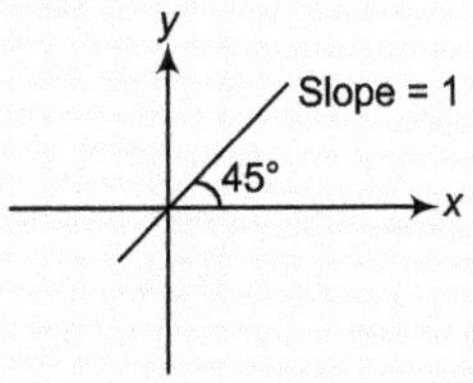 Straight line with slope 1
3.	90°	2	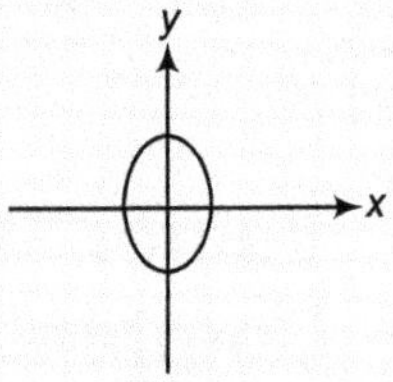 Ellipse
4.	90°	1	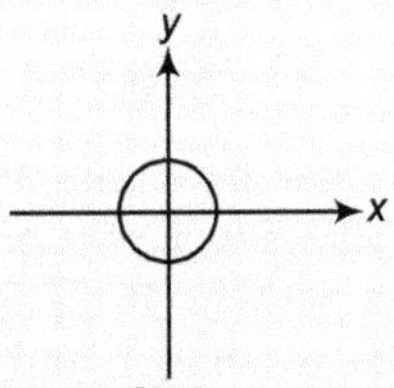 Circle

Chapter Summary with Formulae

- **Different Equations in SHM**

(i) $F = -kx$

(ii) $a = \frac{F}{m} = -\left(\frac{k}{m}\right)x = -\omega^2 x$

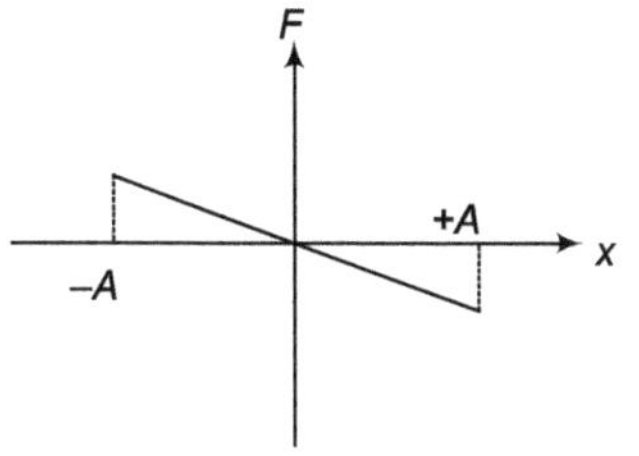

Slope $= -k$

(iii) $\omega = \sqrt{\frac{k}{m}}$ = angular frequency of SHM

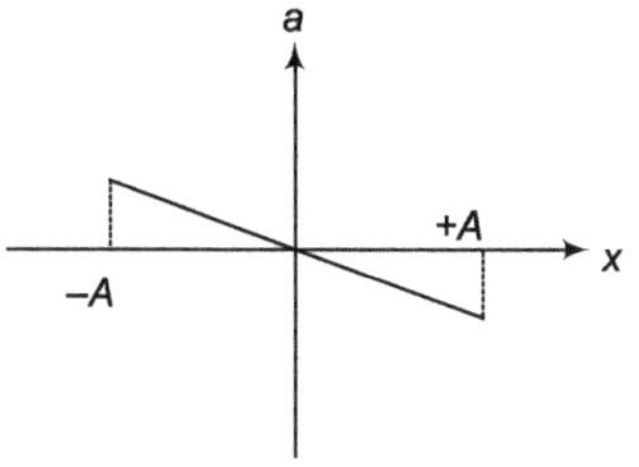

Slope $= -\omega^2$

(iv) $\frac{d^2x}{dt^2} = -\omega^2 x$

(v) $F \propto -x$ or $a \propto -x$ is the sufficient and necessary condition for a periodic motion to be simple harmonic.

(vi) General solution of differential equation

$$\frac{d^2x}{dt^2} = -\omega^2 x \text{ is}$$

$$x = A\sin(\omega t \pm \phi)$$

or $$x = A\cos(\omega t \pm \phi)$$

Here, x is displacement from mean position (not x-coordinate), A is amplitude of oscillation, ϕ is phase angle at $t = 0$ (also called initial phase) and $(\omega t \pm \phi)$ is phase angle at a general time t, also called instantaneous phase.

(vii) If $x = A\sin\omega t$, then

$$v = \frac{dx}{dt} = \omega A\cos\omega t$$

and $$a = \frac{dv}{dt} = -\omega^2 A\sin\omega t$$

From these three equations, we can see that x-t, v-t and a-t all three functions oscillate simple harmonically with same angular frequency ω. Here, x oscillates between $+A$ and $-A$, v between $+\omega A$ and $-\omega A$ and a between $+\omega^2 A$ and $-\omega^2 A$.

Phase difference between x-t and v-t functions or between v-t and a-t functions is $\frac{\pi}{2}$. But, phase difference between x-t and a-t function is π.

(viii) $v = \omega\sqrt{A^2 - x^2}$

$v = 0$ at $x = \pm A$ or at extreme positions

$v = \pm\omega A$, at $x = 0$ or at mean position.

$|v|_{max} = \omega A$ at mean position and $|v|_{min} = 0$, at extreme positions.

(ix) $a = -\omega^2 x$

$|a|_{min} = 0$ at $x = 0$, at mean position.

$|a|_{max} = \omega^2 A$ at $x = \pm A$, at extreme positions.

(x) $PE = U_0 + \frac{1}{2}kx^2$, $KE = \frac{1}{2}k(A^2 - x^2)$

and $TE = PE + KE = U_0 + \frac{1}{2}kA^2$.

Here, U_0 is minimum potential energy at mean position and $\frac{1}{2}kA^2$ or $\frac{1}{2}m\omega^2 A^2$ is called energy of oscillation.

This much work is done on the system when displaced from mean position to extreme position. This much energy keeps on oscillating between potential and kinetic during oscillation.

- **At Mean Position**

$F = 0$, $a = 0$ potential energy is minimum (this minimum value may be zero also) and speed or kinetic energy is maximum.

- **At Extreme Positions**

Speed or kinetic energy is zero. Force, acceleration and potential energy is maximum.

Table 11.3

S. No.	Physical quantity	At mean position	At extreme position	At general point
1.	Speed	ωA	zero	$\omega\sqrt{A^2 - x^2}$
2.	Acceleration	zero	$\pm\omega^2 A$	$-\omega^2 x$
3.	Force	zero	$\pm kA$	$-kx$
4.	Kinetic energy	$\frac{1}{2}kA^2 = \frac{1}{2}m\omega^2 A^2$	zero	$\frac{1}{2}k(A^2 - x^2)$
5.	Potential energy	U_0	$U_0 + \frac{1}{2}kA^2$	$U_0 + \frac{1}{2}kx^2$
6.	Total mechanical energy	$U_0 + \frac{1}{2}kA^2$	$U_0 + \frac{1}{2}kA^2$	$U_0 + \frac{1}{2}kA^2$

U_0 or minimum potential energy at mean position may be zero also.

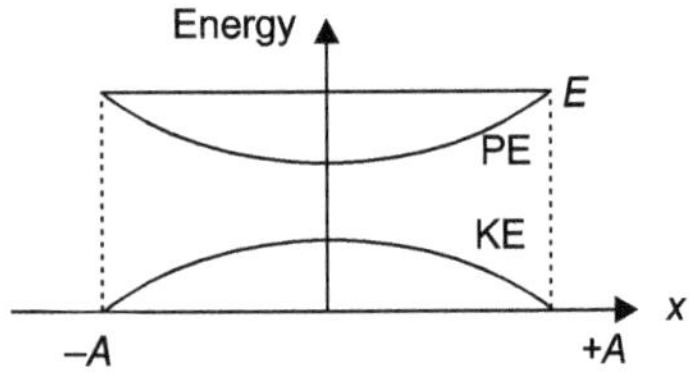

- Potential energy *versus* x or kinetic energy *versus* x graph is parabola. While total energy *versus* x graph is a straight line, as it remains constant.
- 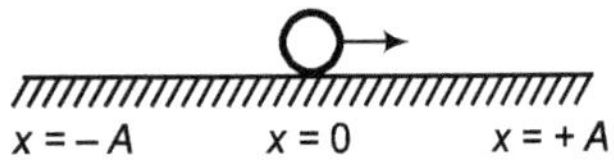

$t = 0$

$x = A \sin\omega t$

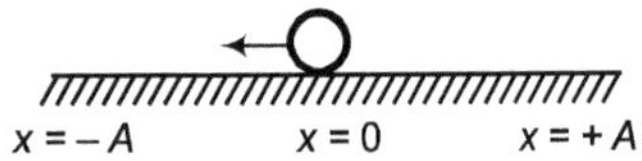

$t = 0$

$x = -A \sin\omega t$

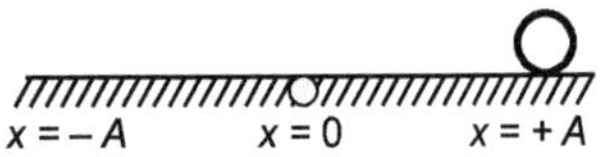

$t = 0$

$x = A \cos\omega t$

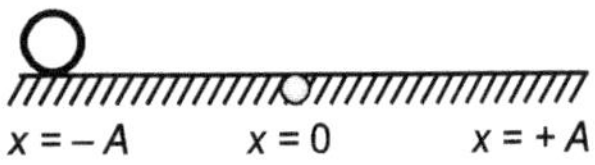

$t = 0$

$x = -A \cos\omega t$

- If $x = A \sin\omega t$. Then,

  $v = \omega A \cos\omega t$ and $a = -\omega^2 A \sin\omega t$

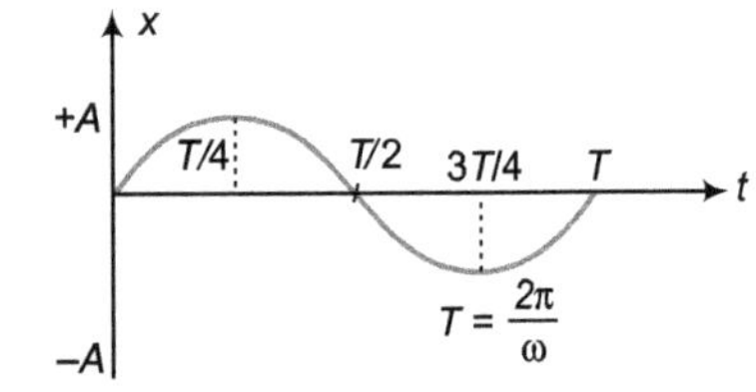

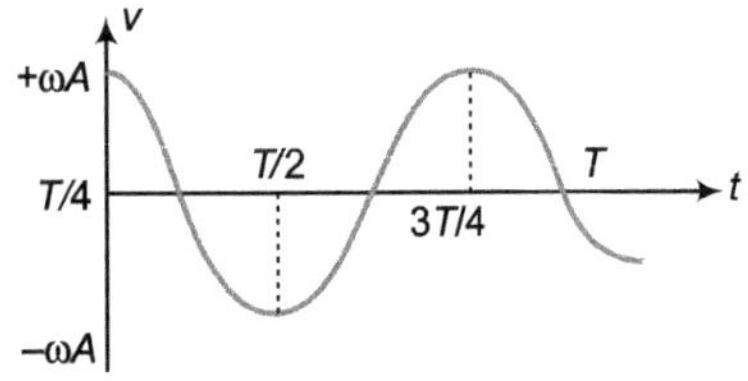

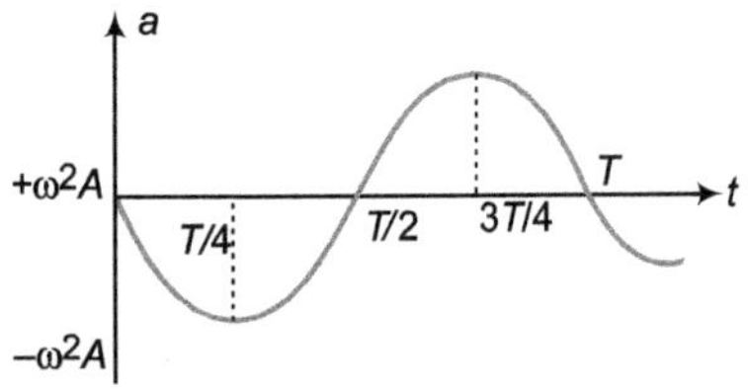

- **Spring Block System**

  (i) $\omega = \sqrt{\frac{k}{m}}, T = \frac{2\pi}{\omega} = 2\pi\sqrt{\frac{m}{k}},$

  $f = \frac{1}{T} = \frac{1}{2\pi}\sqrt{\frac{k}{m}}$

  (ii) 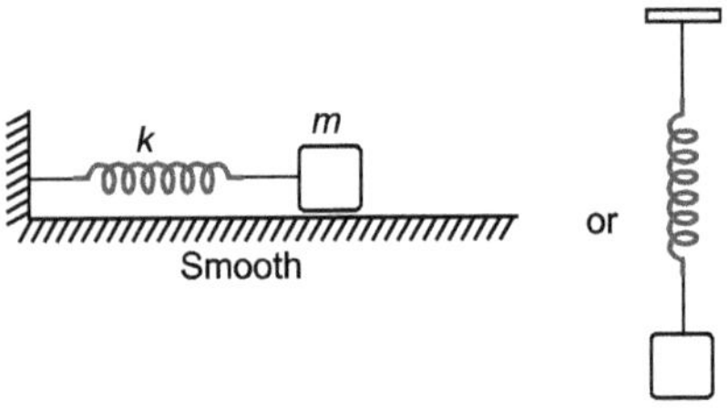

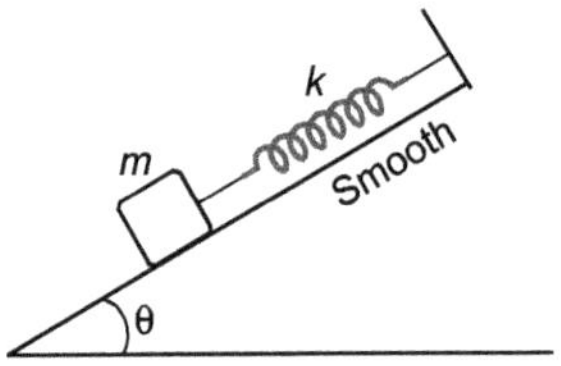

or

In all three cases, $T = 2\pi\sqrt{\frac{m}{k}}$

(iii) 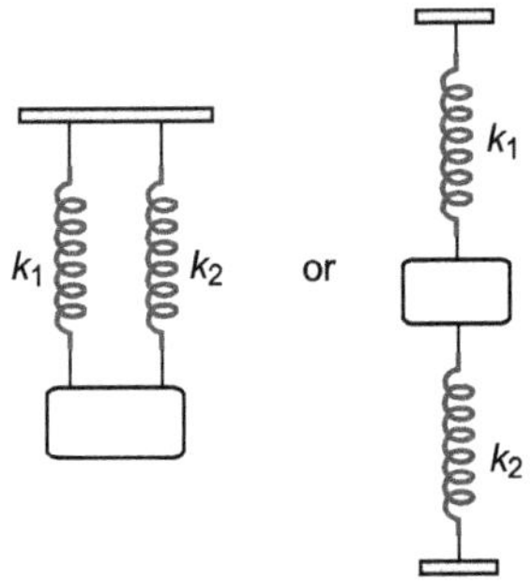

In both cases, $k_e = k_1 + k_2$

(iv) $k_e = \frac{k_1 k_2}{k_1 + k_2}$

$\frac{1}{k_e} = \frac{1}{k_1} + \frac{1}{k_2}$

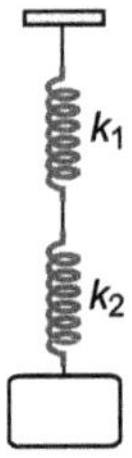

- **Pendulum**

  (i) Only small oscillations of a pendulum are simple harmonic in nature. Time period of which is given by $T = 2\pi\sqrt{\frac{l}{g}}.$

  (ii) Second's pendulum is one whose time period is 2 s and length is 1 m.

Additional Examples

Example 1. *In an SHM, can velocity and displacement be in the same direction?*

Sol. Yes. When a particle moves from mean position to extreme position, velocity and displacement are in the same direction.

Example 2. *Can velocity and acceleration be in the same direction in an SHM?*

Sol. Yes. When a particle moves from extreme to mean position, velocity and acceleration are in the same direction.

Example 3. *Can displacement and acceleration be in the same direction in an SHM?*

Sol. No. In an SHM, acceleration is always opposite to displacement.

Example 4. *What is the most important characteristic of a simple harmonic motion?*

Sol. The most important characteristic of SHM is that acceleration is directly proportional to the displacement and is directed towards the mean position. Mathematically, $a = -\omega^2 x$.

Example 5. *At what points, the energy of the simple harmonic oscillator is entirely potential?*

Sol. At the extreme positions, energy of simple harmonic oscillator is entirely potential. At these positions, the velocity of the oscillator is zero and as such it has no kinetic energy.

Example 6. *Can we use a pendulum watch in an artificial satellite?*

Sol. Inside an artificial satellite, a body is in a state of weightlessness and the effective value of g is zero. The time period of the simple pendulum is given by $T = 2\pi\sqrt{l/g}$. As $g = 0, T = \infty$. Thus, inside a satellite, the pendulum does not oscillate at all. Hence, we cannot use a pendulum watch in an artificial satellite.

Example 7. *Will a pendulum clock gain or lose time when taken to the top of a mountain?*

Sol. On the top of a mountain, the value of g is less than that on the surface of Earth. With a decrease in the value of g, time period of the simple pendulum increases at the top of the mountain and accordingly the pendulum loses time.

Example 8. *What will be the change in time period of a loaded spring, when taken to moon?*

Sol. There will be no change in the time period of the loaded spring. The time period of a loaded spring is independent of acceleration due to gravity.

Example 9. *A girl is swinging in the sitting position. How will the period of the swing change if she stands up?*

Sol. The girl and the swing together constitute a pendulum of time period,

$$T = 2\pi\sqrt{\frac{l}{g}}$$

As the girl stands up, her centre of gravity is raised. The distance between the point of suspension and the centre of gravity, decreases, i.e. length l decreases. Hence, the time period T decreases.

Example 10. *The length of a second's pendulum on the surface of earth is 1 m. What will be the length of a second's pendulum on the surface of moon?*

Sol. $T = 2\pi\sqrt{\frac{l}{g}}$. In both the cases, T is same and $T \propto \sqrt{l/g}$

On the moon, the value of acceleration due to gravity is one-sixth of that on the surface of earth. So, the length of second's pendulum is $\frac{1}{6}$ m.

Example 11. *The bob of simple pendulum is made of wood. What will be the effect on the time period if the wooden bob is replaced by an identical bob of iron?*

Sol. There will be no effect because the time period does not depend upon the nature of material of the bob.

Example 12. *A spring having a force constant k is divided into three equal parts. What would be the force constant for each individual part?*

Sol. Force constant of the spring $k = \frac{F}{x}$, where F is the restoring force. When the spring is divided into three parts, the displacement for the same force reduces to $x/3$, therefore, the force constant for each individual part is

$k' = \frac{F}{x/3} = 3\left(\frac{F}{x}\right) = 3k$. Therefore, we can say that

$$\text{Force constant of a spring} \propto \frac{1}{\text{Length of spring}}$$

Example 13. *There are two springs, one delicate and another stiffer one. Which spring will have a greater frequency of oscillation for a given load?*

Sol. Frequency, $f = \frac{1}{2\pi}\sqrt{\frac{k}{m}}$

Force constant k is larger for the stiffer spring, so its frequency of oscillation will be greater than that of delicate spring.

Example 14. *A particle executes simple harmonic motion about the point $x = 0$. At time $t = 0$, it has displacement $x = 2$ cm and zero velocity. If the frequency of motion is $0.25\ s^{-1}$, find (a) the period, (b) angular frequency, (c) the amplitude, (d) maximum speed, (e) the displacement at $t = 3$ s and (f) the velocity at $t = 3$ s.*

Sol. (a) Period $T = \frac{1}{f} = \frac{1}{0.25\ s^{-1}} = 4$ s

(b) Angular frequency,

$$\omega = \frac{2\pi}{T} = \frac{2\pi}{4} = \frac{\pi}{2}\ \text{rad/s} = 1.57\ \text{rad/s}$$

(c) Amplitude is the maximum displacement from mean position. Hence, $A = 2 - 0 = 2$ cm.

(d) Maximum speed

$$v_{max} = A\omega = 2 \cdot \frac{\pi}{2} = \pi\ \text{cm/s} = 3.14\ \text{cm/s}$$

(e) The displacement is given by

$$x = A \sin(\omega t + \phi)$$

Initially, at $t = 0$, $x = 2$ cm, then

$$2 = 2 \sin \phi$$

or $\sin \phi = 1 = \sin 90°$

or $\phi = 90°$

Now, at $t = 3$s

$$x = 2 \sin\left(\frac{\pi}{2} \times 3 + \frac{\pi}{2}\right) = 0$$

(f) Velocity at $x = 0$ is v_{max}, i.e. 3.14 cm/s.

Example 15. *If an SHM is represented by the equation $x = 10 \sin\left(\pi t + \frac{\pi}{6}\right)$, in SI units determine its amplitude, time period and maximum velocity v_{max}?*

Sol. Comparing the above equation with

$$x = A \sin(\omega t + \phi)$$

we get,

$$A = 10\ \text{m}$$

$$\omega = \pi\ s^{-1} \quad \text{and} \quad \phi = \frac{\pi}{6}$$

$\because$ $$T = \frac{2\pi}{\omega} \Rightarrow T = 2\text{s}$$

$$v_{max} = \omega A = 10\pi\ \text{m/s}$$

Example 16. *A particle executes SHM with a time period of 4 s. Find the time taken by the particle to go directly from its mean position to half of its amplitude.*

Sol. $x = A \sin(\omega t + \phi)$

At $t = 0$, $x = 0$

$\Rightarrow$ $A \sin \phi = 0$

or $\phi = 0$

Hence, $x = A \sin \omega t$

or $\frac{A}{2} = A \sin \omega t$

or $\frac{1}{2} = \sin \omega t$

$$\omega t = \sin^{-1}\left(\frac{1}{2}\right) = \frac{\pi}{6}$$

$$t = \frac{\pi}{6\omega} = \frac{\pi T}{6(2\pi)}$$

As $$\omega = \frac{2\pi}{T} \Rightarrow t = \frac{T}{12} = \frac{1}{3}\ \text{s}$$

Example 17. *Two particles move parallel to x-axis about the origin with the same amplitude and frequency. At a certain instant they are found at distance $\frac{A}{3}$ from the origin on opposite sides but their velocities are found to be in the same direction. What is the phase difference between the two?*

Sol. Let equations of two SHM be

$$x_1 = A \sin \omega t \quad \text{...(i)}$$

$$x_2 = A \sin(\omega t + \phi) \quad \text{...(ii)}$$

Give that $\frac{A}{3} = A \sin \omega t$

and $-\frac{A}{3} = A \sin(\omega t + \phi)$

Which gives $\sin \omega t = \frac{1}{3}$ ...(iii)

$$\sin(\omega t + \phi) = -\frac{1}{3} \quad \text{...(iv)}$$

From Eq. (iv), $\sin \omega t \cos \phi + \cos \omega t \sin \phi = -\frac{1}{3}$

$$\Rightarrow \frac{1}{3} \cos \phi + \sqrt{1 - \frac{1}{9}} \sin \phi = -\frac{1}{3}$$

Solving this equation, we get

or $$\cos \phi = -1, \frac{7}{9}$$

$$\Rightarrow \phi = \pi \text{ or } \cos^{-1}\left(\frac{7}{9}\right)$$

Differentiating Eqs. (i) and (ii), we obtain

$$v_1 = A\omega \cos \omega t$$

and $$v_2 = A\omega \cos(\omega t + \phi)$$

If we put $\phi = \pi$, we find v_1 and v_2 are of opposite signs. Hence, $\phi = \pi$ is not acceptable.

$\therefore$ $$\phi = \cos^{-1}\left(\frac{7}{9}\right)$$

Example 18. *A particle executes SHM*

(a) What fraction of total energy is kinetic and what fraction is potential when displacement is one half of the amplitude.

(b) At what value of displacement are the kinetic and potential energies equal.

Sol. We know that $E_{total} = \frac{1}{2} m\omega^2 A^2$

$$KE = \frac{1}{2} m\omega^2 (A^2 - x^2)$$

and $$U = \frac{1}{2} m\omega^2 x^2$$

(a) When $$x = \frac{A}{2}$$

$$KE = \frac{1}{2} m\omega^2 \frac{3A^2}{4}$$

$$\Rightarrow \quad \frac{KE}{E_{total}} = \frac{3}{4}$$

At $$x = \frac{A}{2},$$

$$U = \frac{1}{2} m\omega^2 \frac{A^2}{4}$$

$$\Rightarrow \quad \frac{PE}{E_{total}} = \frac{1}{4}$$

(b) Since, $K = U$

$$\frac{1}{2} m\omega^2 (A^2 - x^2) = \frac{1}{2} m\omega^2 x^2$$

or $$2x^2 = A^2 \text{ or } x = \frac{A}{\sqrt{2}} = 0.707A$$

Example 19. *A spring mass system is hanging from the ceiling of an elevator in equilibrium. The elevator suddenly starts accelerating upwards with acceleration a, find*

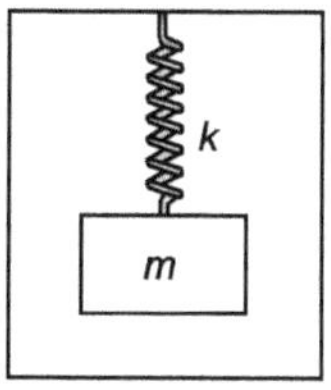

(a) the frequency and
(b) the amplitude of the resulting SHM.

Sol. (a) Frequency $= 2\pi\sqrt{\frac{m}{k}}$

(Frequency is independent of g in spring)

(b) Extension in spring in equilibrium

$$\text{initial} = \frac{mg}{k}$$

Extension in spring in equilibrium in accelerating lift

$$= \frac{m(g+a)}{k}$$

$$\therefore \quad \text{Amplitude} = \frac{m(g+a)}{k} - \frac{mg}{k} = \frac{ma}{k}$$

Example 20. *A ring of radius r is suspended from a point on its circumference. Determine its angular frequency of small oscillations.*

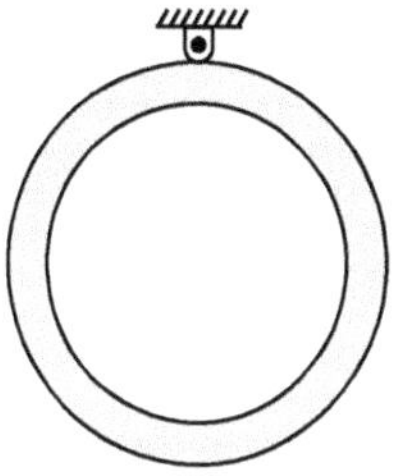

Sol. It is a physical pendulum, the time period of which is

$$T = 2\pi\sqrt{\frac{I}{mgl}}$$

Here, I = moment of inertia of the ring about point of suspension
$= mr^2 + mr^2 = 2mr^2$

and l = distance of point of suspension from centre of gravity $= r$

$$\therefore \quad T = 2\pi\sqrt{\frac{2mr^2}{mgr}} = 2\pi\sqrt{\frac{2r}{g}}$$

$\therefore$ Angular frequency, $\omega = \frac{2\pi}{T}$

or $$\omega = \sqrt{\frac{g}{2r}}$$

NCERT Selected Questions

Q 1. Which of the following examples represent periodic motion?

(a) A swimmer completing one (return) trip from one bank of a river to the other and back.

(b) A freely suspended bar magnet displaced from its N-S direction and released.

(c) A hydrogen molecule rotating about its centre of mass.

(d) An arrow released from a bow.

Sol. (a) It is not a periodic motion. Though, the motion of a swimmer is to and fro but will not have a fixed period.

(b) It is a periodic motion because a freely suspended magnet if once displaced from N-S direction and let it go, oscillates about this position after a fixed time interval.

(c) It is a periodic motion.

(d) It is not a periodic motion.

Q 2. Which of the following examples represent (nearly) simple harmonic motion and which represent periodic, but not simple harmonic motion?

(a) The rotation of earth about its axis.

(b) Motion of an oscillating mercury column in a U-tube.

(c) Motion of a ball bearing inside a smooth curved bowl, when released from a point slightly above the lower most point.

Sol. (a) It is periodic but not simple harmonic motion, as it is not to and fro motion about a fixed point.

(b) It is SHM.

(c) It is SHM.

Q 3. Which of the following functions of time represent (i) Simple harmonic motion (ii) periodic but not simple harmonic and (iii) non-periodic motion? Give period for each case of periodic motion (ω is any positive constant)

(a) $\sin \omega t - \cos \omega t$

(b) $\sin^3 \omega t$

(c) $3 \cos\left(\frac{\pi}{4} - 2\omega t\right)$

(d) $\cos \omega t + \cos 3\omega t + \cos 5\omega t$

(e) $\exp(-\omega^2 t^2)$

(f) $1 + \omega t + \omega^2 t^2$

Sol. (a) $\sin \omega t - \cos \omega t = \sqrt{2}\left(\frac{1}{\sqrt{2}} \sin \omega t - \frac{1}{\sqrt{2}} \cos \omega t\right)$

$$= \sqrt{2}\left(\sin \omega t \cos \frac{\pi}{4} - \cos \omega t \sin \frac{\pi}{4}\right)$$

$$= \sqrt{2} \sin\left(\omega t - \frac{\pi}{4}\right)$$

$\therefore$ It represents simple harmonic with a period, $T = \frac{2\pi}{\omega}$

(b) $\sin^3 \omega t = \frac{1}{4}(3 \sin \omega t - \sin 3\omega t)$

$(\because \sin 3A = 3 \sin A - 4 \sin^3 A)$

Here, each term $\sin \omega t$ and $\sin 3\omega t$ individually represents SHM but it is not outcome of superposition of two SHMs, so it will represent only periodic but not simple harmonic motion. Its time period $= \frac{2\pi}{\omega}$.

(c) $3 \cos\left(\frac{\pi}{4} - 2\omega t\right) = 3 \cos\left(2\omega t - \frac{\pi}{4}\right)$

$[\because \cos(-\theta) = \cos\theta]$

It represents simple harmonic motion and its time period is $\frac{2\pi}{2\omega} = \frac{\pi}{\omega}$.

(d) $\cos \omega t + \cos 3\omega t + \cos 5\omega t$

It represents periodic but not simple harmonic motion. Its time period is $\frac{2\pi}{\omega}$. It can be noted that each term represents a periodic function with a different angular frequency. Since, period is the least interval of time after which a function repeats its value, $\cos \omega t$ has a period $T = \frac{2\pi}{\omega}$, $\cos 3\omega t$ has a period $\frac{2\pi}{3\omega} = \frac{T}{3}$, and $\cos 5\omega t$ has a period $\frac{2\pi}{5\omega} = \frac{T}{5}$. The last two terms repeat after any integral multiple of their period. Thus, each term in the sum repeats itself after T, and hence the sum is a periodic function with a period $\frac{2\pi}{\omega}$.

(e) $\exp(-\omega^2 t^2)$. It is an exponential function which decreases with time and tends to zero as $t \to \infty$ and thus never repeats itself. Therefore, it represents non-periodic motion.

(f) $1 + \omega t + \omega^2 t^2$. It also represents non-periodic motion.

Q 4. A particle is in linear simple harmonic motion between two points A and B, 10 cm apart. Take the direction from A to B as the positive direction and give the signs of velocity, acceleration and force on the particle when it is

(a) at the end A.

(b) at the end B.

(c) at the mid-point of AB going towards A.

(d) at 2 cm away from B going towards A.

(e) at 3 cm away from A going towards B.

Sol.

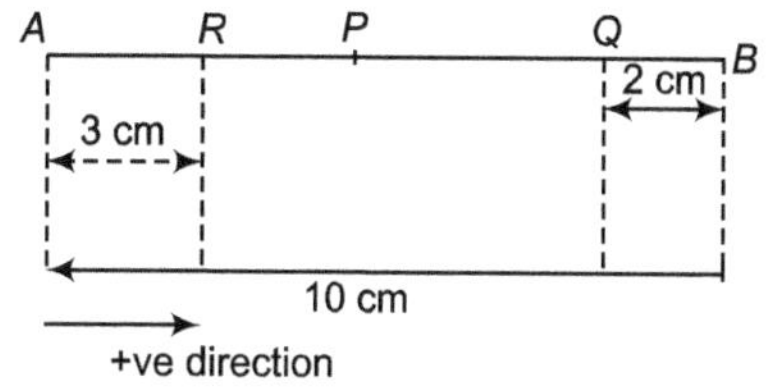

Points A and B are the extreme points and point P is the mean position.

(a) At the end A, the particle executing SHM is momentarily at rest being at its extreme position of motion, hence its velocity is zero, acceleration is +ve as it is directed towards P. Force is also +ve as the force is also directed towards P.

Ans. 0, +, +

(b) At end B, velocity is zero. Acceleration and force are negative as they are directed towards P, i.e. along negative direction.

Ans. 0, –, –

(c) At the mid-point of AB going towards A, the particle is at the mean position P, with a tendency to move towards A, i.e. along negative direction. Hence, velocity is –ve and both acceleration and force are zero.

Ans. –, 0, 0

(d) At 2 cm away from B going towards A, the particle is at point Q with a tendency to move along QP which is negative direction. Hence, velocity, acceleration and force all are –ve, i.e.

Ans. –, –, –

(e) At 3 cm away from A going towards B. The particle is now at R moving towards B, i.e. in +ve direction. So, velocity, acceleration and force all are +ve.

Ans. +, +, +

***Q* 5.** Which of the following relationships between the acceleration a and the displacement x of a particle involve simple harmonic motion?

(a) $a = 0.7x$

(b) $a = -200x^2$

(c) $a = -10x$

(d) $a = 100x^3$

Sol. A particle is said to be executing SHM if

$$a = -\omega^2 x \text{ or } a \propto -x \qquad \text{...(i)}$$

(a) $a = 0.7x$, does not satisfy Eq.(i), so, it does not represent SHM.

(b) $a = -200x^2$, does not satisfy Eq.(i), hence it does not represent SHM.

(c) $a = -10x$, satisfies Eq. (i), so, it represents SHM.

(d) $a = 100x^3$, again does not represent SHM.

***Q* 6.** A spring having a spring constant $1200\ \text{Nm}^{-1}$ is mounted on a horizontal table as shown. A mass of 3.0 kg is attached to the free end of the spring. The mass is then pulled sideways to a distance of 0.2 cm and released. Determine

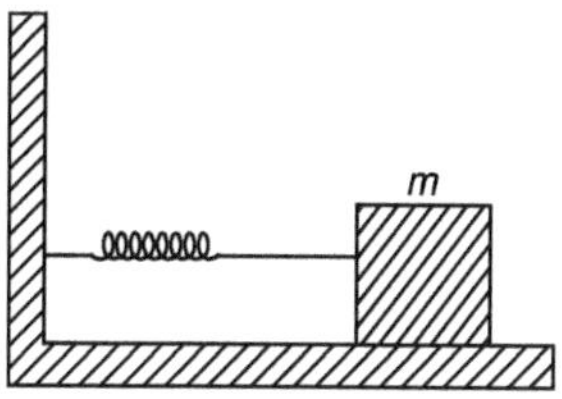

(i) the frequency of oscillations.

(ii) the maximum acceleration of the mass and

(iii) the maximum speed of the mass?

Sol. (i) Frequency,

$$f = \frac{1}{2\pi}\sqrt{\frac{k}{m}} = \frac{1}{2 \times 3.142} \times \sqrt{\frac{1200}{3}} = 3.18 \text{ Hz}$$

(ii) The acceleration is given by

$$a = -\omega^2 x = -\frac{k}{m}x \text{ or } |a_{max}| = \frac{k}{m}|x_{max}|$$

i.e. acceleration will be maximum, when x is maximum.

i.e. $x_{max} = A = 0.02\,\text{m}$

$$\therefore \quad a = \frac{1200}{3} \times 0.02 = 8.0 \text{ ms}^{-2}$$

(iii) The maximum speed of the mass is given by

$$v = A\omega = A\sqrt{\frac{k}{m}} = 0.02 \times \sqrt{\frac{1200}{3}} = 0.40 \text{ ms}^{-1}$$

***Q* 7.** In previous question, let us take the position of mass when the spring is unstretched as $x = 0$, and the direction from left to right as the positive direction of x-axis. Give x as a function of time t for the oscillating mass if at the moment we start the stop watch ($t = 0$), the mass is

(a) at the mean position.

(b) at the maximum stretched position.

(c) at the maximum compressed position.

In what way do these different functions for SHM differ from each other in frequency, in amplitude or the initial phase?

Sol. Here, $A = 2 \text{ cm} = 0.02 \text{ m}$

$$k = 1200\,\text{N/m}$$

$$\omega = \sqrt{\frac{k}{m}} = \sqrt{\frac{1200}{3}} = 20 \text{ s}^{-1}$$

(a) When the mass starts from mean position towards positive direction,

$$x = A \sin \omega t$$

$$\therefore \quad x = 2 \sin 20t$$

(b) In the maximum stretched position,

$$x = A \cos \omega t$$

$$\therefore \quad x = 2 \cos 20t$$

(c) In the maximum compressed position,

$$x = -A\cos\omega t = -2\cos 20t$$

The functions differ only in initial phase because their amplitudes ($A = 2$ cm) and periods are same, i.e.

$$T = \frac{2\pi}{\omega} = \frac{2\pi}{20} = \frac{\pi}{10}\text{ s}$$

Q 8. Figure below corresponds to two circular motions. The radius of the circle, the period of revolution, the initial position, and the sense of revolution (i.e. clockwise or anti-clockwise) are indicated on each figure. Obtain the corresponding simple harmonic motion of the x-projection of the radius vector of the revolving particle P, in each case.

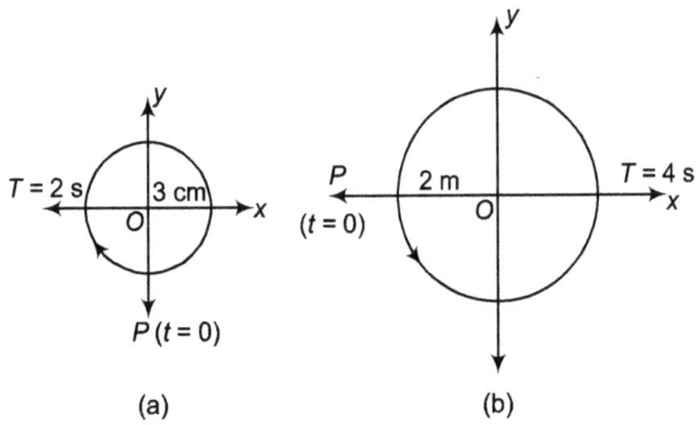

Sol. (a) Here, at $t = 0$, OP makes an angle $\frac{\pi}{2}$ with x-axis. Since, motion is clockwise, so $\phi = \frac{+\pi}{2}$ rad. Thus, the x-projection of OP at time t will give us the equation of SHM given by

$$x = A\cos\left(\frac{2\pi t}{T} + \phi\right)$$

$$= 3\cos\left[\frac{2\pi t}{2} + \frac{\pi}{2}\right] \quad (\because A = 3\text{ cm}, T = 2\text{ s})$$

or $$x = 3\cos\left(\pi t + \frac{\pi}{2}\right)$$

$$= -3\sin \pi t \quad (x \text{ is in cm})$$

$\therefore$ $$x = -3\sin \pi t \quad (\text{cm})$$

(b) $T = 4$ s, $A = 2$ m

At $t = 0$, OP makes an angle π with the positive direction of x-axis, i.e. $\phi = +\pi$.

Thus, the x-projection of OP at time t will give us the equation of SHM given by

$$x = A\cos\left(\frac{2\pi}{T}t + \phi\right)$$

$$= 2\cos\left(\frac{2\pi}{4}t + \pi\right)$$

$$= -2\cos\left(\frac{\pi}{2}t\right)$$

or $$x = -2\cos\left(\frac{\pi}{2}t\right)\text{ (m)}$$

Q 9. The piston in the cylinder of a locomotive has a stroke (twice the amplitude) of 1.0 m. If the piston moves with simple harmonic motion with an angular frequency of 200 rad/min, what is its maximum speed?

Sol. Stroke of piston = 2 times the amplitude

$\therefore$ $$A = \text{amplitude} = \frac{1}{2}\text{ m} \quad \text{(given)}$$

Angular frequency, $\omega = 200$ rad/min

$$v_{max} = \omega A = 200 \times \frac{1}{2} = 100\text{ m/min}$$

Q 10. The acceleration due to gravity on the surface of moon is 1.7 ms^{-2}. What is the time period of a simple pendulum on the surface of moon if its time period on the surface on earth is 3.5 s? (g on the surface of earth is 9.8 ms^{-2}.)

Sol. Using the formula

$$T = 2\pi\sqrt{\frac{l}{g}}, \text{ we get}$$

For earth, $$T_e = 2\pi\sqrt{\frac{l}{g_e}} \quad \text{...(i)}$$

and for moon, $$T_m = 2\pi\sqrt{\frac{l}{g_m}} \quad \text{...(ii)}$$

Dividing Eq. (ii) by Eq. (i), we get

$$\frac{T_m}{T_e} = \sqrt{\frac{g_e}{g_m}} = \sqrt{\frac{9.8}{1.7}}$$

or $$T_m = 3.5 \times \sqrt{\frac{98}{17}}$$

$\therefore$ $$T_m = 8.4\text{ s}$$

Q 11. A simple pendulum of length l and having a bob of mass M is suspended in a car. The car is moving on a circular track of radius R with a uniform speed v. If the pendulum makes small oscillations in a radial direction about its equilibrium position, what will be its time period?

Sol. Point of suspension has an acceleration $\mathbf{a}_c$ in horizontal direction.

Hence, $$\mathbf{a} = \mathbf{g} - \mathbf{a}_c$$

$$-\mathbf{a}_c = \frac{v^2}{R}$$

If T be the time period of oscillation of the pendulum, then

$$T = 2\pi\sqrt{\frac{l}{a}} = 2\pi\sqrt{\frac{l}{\sqrt{g^2 + \frac{v^4}{R^2}}}}$$

or $$T = 2\pi\sqrt{\frac{l}{\left(g^2 + \frac{v^4}{R^2}\right)^{\frac{1}{2}}}}$$

Objective Problems

[Level 1]

Basic Equations in SHM

1. Two bodies M and N of equal masses are suspended from two separate massless springs of spring constants k_1 and k_2, respectively. If the two bodies oscillate vertically such that their maximum velocities are equal, the ratio of the amplitude of vibration of M to that on N is

(a) $\frac{k_1}{k_2}$ (b) $\sqrt{\frac{k_1}{k_2}}$

(c) $\frac{k_2}{k_1}$ (d) $\sqrt{\frac{k_2}{k_1}}$

2. A body is executing SHM with an amplitude of 0.1 m. Its velocity while passing through the mean position is 3 m/s. Its frequency in Hz is

(a) 15π (b) $\frac{15}{\pi}$

(c) 30π (d) 25π

3. Two particles are executing SHMs. The equations of their motions are

$$y_1 = 10\sin\left(\omega t + \frac{\pi}{4}\right)$$

$$y_2 = 5\sin\left(\omega t + \frac{\sqrt{3}\,\pi}{4}\right)$$

What is the ratio of their amplitudes?

(a) 1 : 1 (b) 2 : 1

(c) 1 : 2 (d) None of these

4. What is the maximum acceleration of the particle doing the SHM?

$$y = 2\sin\left[\frac{\pi t}{2} + \phi\right] \text{ where } y \text{ is in cm.}$$

(a) $\frac{\pi}{2}$ cm/s^2 (b) $\frac{\pi^2}{2}$ cm/s^2

(c) $\frac{\pi}{4}$ cm/s^2 (d) $\frac{\pi^2}{4}$ cm/s^2

5. The amplitude and the time period in a SHM is 0.5 cm and 0.4 s, respectively. If the initial phase is $\frac{\pi}{2}$ rad, then the equation of SHM will be

(a) $y = 0.5 \sin 5\pi t$

(b) $y = 0.5 \sin 4\pi t$

(c) $y = 0.5 \sin 2.5\pi t$

(d) $y = 0.5 \cos 5\pi t$

6. A particle is executing simple harmonic motion with a period of T seconds and amplitude a metre. The shortest time it takes to reach a point $\frac{a}{\sqrt{2}}$ m from its mean position in seconds is

(a) T (b) $\frac{T}{4}$

(c) $\frac{T}{8}$ (d) $\frac{T}{16}$

7. A particle executes a simple harmonic motion of time period T. Find the time taken by the particle to go directly from its mean position to half the amplitude.

(a) $\frac{T}{2}$ (b) $\frac{T}{4}$

(c) $\frac{T}{8}$ (d) $\frac{T}{12}$

8. A particle executing SHM of amplitude 4 cm and $T = 4$ s. The time taken by it to move from positive extreme position to half the amplitude is

(a) 1 s (b) $\frac{1}{3}$ s

(c) $\frac{2}{3}$ s (d) $\sqrt{\frac{3}{2}}$ s

9. Two simple harmonic motions are represented by the equations $y_1 = 0.1\sin\left(100\pi t + \frac{\pi}{3}\right)$ and $y_2 = 0.1\cos \pi t$.

The phase difference of velocity of particle 1 with respect to the velocity of particle 2 at time $t = 0$ is

(a) $-\frac{\pi}{3}$ (b) $\frac{\pi}{6}$ (c) $-\frac{\pi}{6}$ (d) $\frac{\pi}{3}$

10. A particle executing simple harmonic motion has an amplitude of 6 cm. Its acceleration at a distance of 2 cm from the mean position is 8 cm/s^2. The maximum speed of the particle is

(a) 8 cm/s (b) 12 cm/s

(c) 16 cm/s (d) 24 cm/s

11. A particle executes simple harmonic motion with an amplitude of 4 cm. At the mean position the velocity of the particle is 10 cm/s. The distance of the particle from the mean position when its speed becomes 5 cm/s is

(a) $\sqrt{3}$ cm (b) $\sqrt{5}$ cm

(c) $2\sqrt{3}$ cm (d) $2\sqrt{5}$ cm

12. The maximum velocity of a simple harmonic motion represented by $y = 3\sin\left(100t + \frac{\pi}{6}\right)$ is given by

(a) 300 units (b) $\frac{3\pi}{6}$ units

(c) 100 units (d) $\frac{\pi}{6}$ units

13. Velocity at mean position of a particle executing SHM is v. Velocity of the particle at a distance equal to half of the amplitude will be

(a) $\frac{v}{2}$ (b) $\frac{v}{\sqrt{2}}$

(c) $\frac{\sqrt{3}}{2}v$ (d) $\frac{\sqrt{3}}{4}v$

14. Two massless springs of force constants k_1 and k_2 are joined end to end. The resultant force constant k of the system is

(a) $\frac{k_1 + k_2}{k_1k_2}$ (b) $k_1 + k_2$

(c) $\frac{k_1k_2}{k_1 + k_2}$ (d) $\frac{k_1k_2}{k_1 - k_2}$

15. A light spring of constant k is cut into two equal parts. The spring constant of each part is

(a) k (b) $2k$

(c) $\frac{k}{2}$ (d) $4k$

16. A body is vibrating in simple harmonic motion. If its acceleration is 12 cm/s^2 at a displacement 3 cm from the mean position, then time period is

(a) 6.28 s (b) 3.14 s (c) 1.57 s (d) 2.57 s

17. The period of a particle executing SHM is 8 s. At $t = 0$, it is at the mean position. The ratio of the distances covered by the particle in the 1st second to the 2nd second is

(a) $\frac{1}{\sqrt{2}+1}$ (b) $\sqrt{2}$

(c) $\frac{1}{\sqrt{2}}$ (d) $\sqrt{2}+1$

18. The maximum acceleration of a body moving in SHM is a_0 and maximum velocity is v_0. The amplitude is given by

(a) $\frac{v_0^2}{a_0}$ (b) a_0v_0

(c) $\frac{a_0^2}{v_0}$ (d) $\frac{1}{a_0v_0}$

19. The displacement of a particle executing SHM is given by $y = 0.25\sin 200t$ cm. The maximum speed of the particle is

(a) 200 cm s^{-1} (b) 100 cm s^{-1}

(c) 50 cm s^{-1} (d) 5.25 cm s^{-1}

20. The displacement of a particle executing SHM is given by $x = 0.01\sin 100\pi(t + 0.05)$. The time period is

(a) 0.01 s (b) 0.02 s

(c) 0.1 s (d) 0.2 s

21. A body is executing simple harmonic motion with an angular frequency of 2 rad/s. The velocity of the body at 20 mm displacement, when the amplitude of the motion is 60 mm is

(a) 131 mm/s (b) 118 mm/s

(c) 113 mm/s (d) 90 mm/s

Energy in SHM

22. Force constant of a weightless spring is 16 N/m. A body of mass 1.0 kg suspended from it is pulled down through 5 cm from its mean position and then released. The maximum kinetic energy of the body will be

(a) 2×10^{-2} J (b) 4×10^{-2} J

(c) 8×10^{-2} J (d) 16×10^{-2} J

23. Energy of particle executing SHM depends upon

(a) amplitude only

(b) amplitude and frequency

(c) velocity only

(d) frequency only

24. A particle undergoing SHM has the equation $x = A\sin(\omega t + \phi)$, where x represents the displacement of the particle. The kinetic energy oscillates with time period

(a) $\frac{2\pi}{\omega}$ (b) $\frac{\pi}{\omega}$

(c) $\frac{4\pi}{\omega}$ (d) None of these

25. A particle of mass 0.10 kg executes SHM with an amplitude 0.05 m and frequency 20 vib/s. Its energy of oscillation is

(a) 2 J (b) 4 J

(c) 1 J (d) zero

26. In SHM, for how many times potential energy is equal to kinetic energy during one complete period?

(a) 1 (b) 2

(c) 4 (d) 8

27. Amplitude of a particle in SHM is 6 cm. If instantaneous potential energy is half the total energy, then distance of particle from its mean position is

(a) 3 cm (b) 4.2 cm

(c) 5.8 cm (d) 6 cm

28. A body of mass 1 kg is executing simple harmonic motion. Its displacement y (cm) at t seconds is given by $y = 6\sin\left(100t + \frac{\pi}{4}\right)$. Its maximum kinetic energy is

(a) 6 J (b) 18 J (c) 24 J (d) 36 J

29. In a simple harmonic oscillator, at the mean position
(a) kinetic energy is minimum, potential energy is maximum
(b) both kinetic and potential energies are maximum
(c) kinetic energy is maximum, potential energy is minimum
(d) both kinetic and potential energies are minimum

30. A body executes simple harmonic motion. The potential energy (PE), the kinetic energy (KE) and total energy (TE) are measured as a function of displacement x. Which of the following statements is true?
(a) PE is maximum when $x = 0$
(b) KE is maximum when $x = 0$
(c) TE is zero when $x = 0$
(d) KE is maximum when x is maximum

31. The total energy of a particle, executing simple harmonic motion is
(a) $\propto x$ (b) $\propto x^2$
(c) independent of x (d) $\propto x^{1/2}$

32. A body of mass 1 kg is executing simple harmonic motion. Its displacement y (cm) at t seconds is given by $y = 6\sin\left(100t + \frac{\pi}{4}\right)$. Its maximum kinetic energy is
(a) 6 J
(b) 18 J
(c) 24 J
(d) 36 J

33. Two simple harmonic motions $y_1 = A\sin\omega t$ and $y_2 = A\cos\omega t$ are superimposed on a particle of mass m. The total mechanical energy of the particle is
(a) $\frac{1}{2}m\omega^2A^2$ (b) $m\omega^2A^2$
(c) $\frac{1}{4}m\omega^2A^2$ (d) zero

34. A particle starts SHM from the mean position. Its amplitude is a and total energy E. At one instant its kinetic energy is $\frac{3E}{4}$. Its displacement at that instant is
(a) $\frac{a}{\sqrt{2}}$ (b) $\frac{a}{2}$
(c) $\sqrt{3}\frac{a}{2}$ (d) zero

Time Period and Frequency in SHM

35. The time period of a simple pendulum inside a stationary lift is $\sqrt{5}$ s. What will be the time period when the lift moves upwards with an acceleration $\frac{g}{4}$?
(a) $\sqrt{5}$ s
(b) $2\sqrt{5}$ s
(c) $(2+\sqrt{5})$ s
(d) 2 s

36. A rectangular block of mass m and area of cross-section A floats in a liquid of density ρ. If it is given small vertical displacement from equilibrium, it undergoes oscillation with a time period T, then
(a) $T \propto m$ (b) $T \propto \rho$
(c) $T \propto \frac{1}{A}$ (d) $T \propto \sqrt{\frac{1}{\rho}}$

37. A cabin is falling freely under gravity, what is the time period of a pendulum attached to its ceiling
(a) zero (b) ∞
(c) 1 s (d) 2 s

38. If both spring constants k_1 and k_2 are increased to $4k_1$ and $4k_2$ respectively, what will be the new frequency, if f was the original frequency?

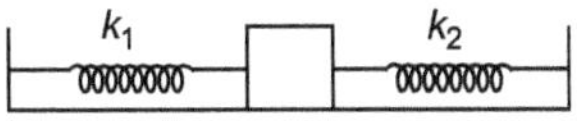

(a) f (b) $2f$ (c) $\frac{1}{2}f$ (d) $4f$

39. Frequency of oscillation is proportional to

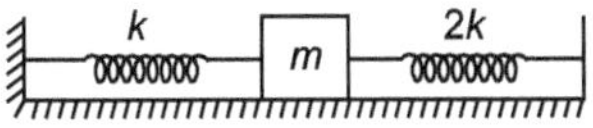

(a) $\sqrt{\frac{3k}{m}}$ (b) $\sqrt{\frac{k}{m}}$
(c) $\sqrt{\frac{2k}{m}}$ (d) $\sqrt{\frac{m}{3k}}$

40. The mass and the radius of a planet are twice that of earth. Then, period of oscillation of a second's pendulum on that planet will be
(a) $\frac{1}{\sqrt{2}}$ s (b) $2\sqrt{2}$ s
(c) 1 s (d) $\frac{1}{2}$ s

41. A mass m is suspended from a spring. Its frequency of oscillation is f. The spring is cut into two halves and the same mass is suspended from one of the two pieces of the spring. The frequency of oscillation of the mass will be
(a) $\sqrt{2}f$ (b) $\frac{f}{2}$ (c) f (d) $2f$

42. A block of mass 0.2 kg, which slides without friction on a 30° incline, is connected to the top of the incline by a massless spring of force constant 80 N/m as shown in figure. If the block is pulled slightly from its mean position, what is the period of oscillations?

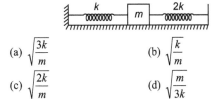

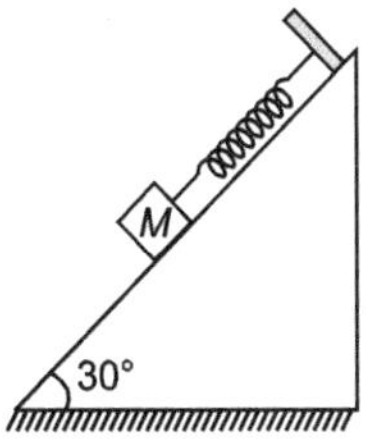

(a) π s (b) $\frac{\pi}{10}$ s (c) $\frac{2\pi}{5}$ s (d) $\frac{\pi}{2}$ s

43. A horizontally placed spring mass system has time period T. The same system is now placed on a car moving with acceleration a in horizontal direction. Then,
(a) time period will increase
(b) time period will decrease
(c) time period will remain constant
(d) no conclusion can be drawn

44. A particle executes linear simple harmonic motion with an amplitude of 2 cm. When the particle is at 1 cm from the mean position, the magnitude of its velocity is equal to that of its acceleration. Then, its time period in seconds is
(a) $\frac{1}{2\pi\sqrt{3}}$ (b) $2\pi\sqrt{3}$
(c) $\frac{2\pi}{\sqrt{3}}$ (d) $\frac{\sqrt{3}}{2\pi}$

45. The equation of motion of a particle is $\frac{d^2y}{dt^2} + Ky = 0$, where K is a positive constant. The time period of the motion is given by
(a) $\frac{2\pi}{K}$ (b) $2\pi K$
(c) $\frac{2\pi}{\sqrt{K}}$ (d) $2\pi\sqrt{K}$

46. The displacement x (in metre) of a particle in, simple harmonic motion is related to time t (in seconds) as

$$x = 0.01\cos\left(\pi t + \frac{\pi}{4}\right)$$

The frequency of the motion will be
(a) 0.5 Hz (b) 1.0 Hz
(c) $\frac{\pi}{2}$ Hz (d) π Hz

47. A simple pendulum is made of a body which is a hollow sphere containing mercury suspended by means of a wire. If a little mercury is drained off, the period of pendulum will
(a) remain unchanged (b) increase
(c) decrease (d) become erratic

48. A man measures the period of a simple pendulum inside a stationary lift and finds it to be T second. If the lift accelerates upwards with an acceleration $\frac{g}{4}$, then the period of the pendulum will be
(a) T (b) $\frac{T}{4}$
(c) $\frac{2T}{\sqrt{5}}$ (d) $2T\sqrt{5}$

49. If the metal bob of a simple pendulum is replaced by a wooden bob, then its time period will
(a) increase
(b) decrease
(c) remain the same
(d) may increase or decrease

50. A simple pendulum is attached to the roof of a lift. If time period of oscillation, when the lift is stationary is T, then frequency of oscillation, when the lift falls freely, will be
(a) zero (b) infinite
(c) $\frac{1}{T}$ (d) None of these

51. A particle is attached to a vertical spring and is pulled down a distance 0.04 m below its equilibrium position and is released from rest. The initial upward acceleration of the particle is 0.30 m/s^2. The period of the oscillation is
(a) 4.08 s (b) 1.92 s
(c) 3.90 s (d) 2.29 s

52. Three masses 0.1 kg, 0.3 kg and 0.4 kg are suspended at end of a spring. When the 0.4 kg mass is removed, the system oscillates with a period 2 s. When the 0.3 kg mass is also removed, the system will oscillate with a period
(a) 1 s (b) 2 s (c) 3 s (d) 4 s

53. The time period of a simple pendulum of infinite length is
(a) infinite (b) $2\pi\sqrt{\frac{R}{g}}$
(c) $2\pi\sqrt{\frac{g}{R}}$ (d) $\frac{1}{2\pi}\sqrt{\frac{R}{g}}$

54. A disc of radius R and mass M is pivoted at the rim and is set for small oscillations about an axis perpendicular to plane of disc. If a simple pendulum has to have the same time period as that of the disc, the length of the pendulum should be
(a) $\frac{5}{4}R$ (b) $\frac{2}{3}R$ (c) $\frac{3}{4}R$ (d) $\frac{3}{2}R$

55. A simple pendulum is suspended from the roof of a trolley which moves, in a horizontal direction with an acceleration a, then the time period T is given by $T = 2\pi\sqrt{\frac{l}{g'}}$, where g' is equal to
(a) g (b) $g - a$
(c) $g + a$ (d) $\sqrt{g^2 + a^2}$

56. Five identical springs are used in the three configurations as shown in figure. The time periods of vertical oscillations in configurations (a), (b) and (c) are in the ratio

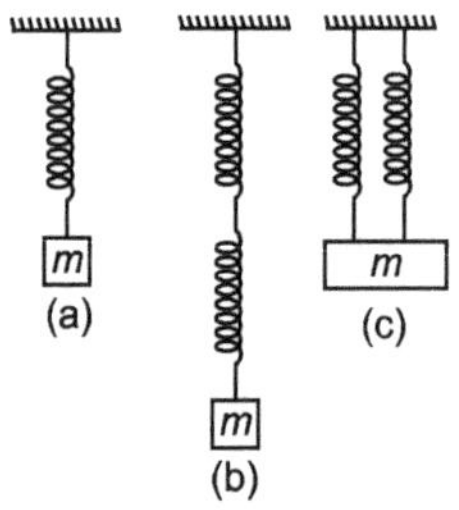

(a) $1 : \sqrt{2} : \frac{1}{\sqrt{2}}$
(b) $2 : \sqrt{2} : \frac{1}{\sqrt{2}}$
(c) $\frac{1}{\sqrt{2}} : 2 : 1$
(d) $2 : \frac{1}{\sqrt{2}} : 1$

57. A uniform spring whose unstressed length is l has a force constant k. The spring is cut into two pieces of unstressed lengths l_1 and l_2, where $l_1 = nl_2$, where n being an integer. Now, a mass m is made to oscillate with first spring. The time period of its oscillation would be

(a) $T = 2\pi\sqrt{\dfrac{mn}{k(n+1)}}$ (b) $T = 2\pi\sqrt{\dfrac{m}{k(n+1)}}$

(c) $T = 2\pi\sqrt{\dfrac{m}{nk}}$ (d) $T = 2\pi\sqrt{\dfrac{m(n+1)}{nk}}$

58. Two masses m_1 and m_2 are suspended together by a massless spring of spring constant k. When the masses are in equilibrium m_1 is removed without distributing the system, then the angular frequency of oscillation will be

(a) $\sqrt{\dfrac{k}{m_1 + m_2}}$ (b) $\sqrt{\dfrac{k}{m_2} + m_1}$

(c) $\sqrt{\dfrac{k}{m_1}}$ (d) $\sqrt{\dfrac{k}{m_2}}$

59. Two identical springs of spring constant k each are connected in series and parallel as shown in figure. A mass M is suspended from them. The ratio of their frequencies of vertical oscillation will be

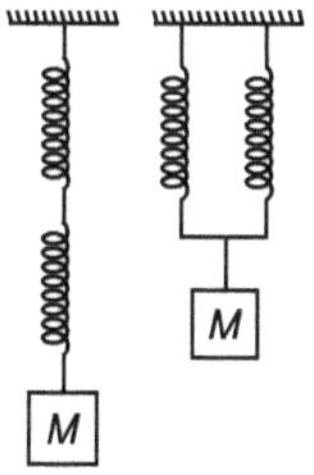

(a) 1 : 2 (b) 2 : 1
(c) 4 : 1 (d) 1 : 4

60. A point mass m is suspended at the end of a massless wire of length L and cross-sectional area A. If Y is the Young's modulus of the wire. Then, the frequency of the oscillation for the simple harmonic oscillation along the vertical direction is

(a) $\dfrac{1}{2\pi}\sqrt{\dfrac{LA}{mY}}$ (b) $\dfrac{1}{2\pi}\sqrt{\dfrac{LAm}{Y}}$

(c) $\dfrac{1}{2\pi}\sqrt{\dfrac{YA}{mL}}$ (d) $\dfrac{1}{2\pi}\sqrt{\dfrac{mY}{AL}}$

61. A thin uniform rod of length l is pivoted at its upper end. It is free to swing in a vertical plane. Its time period for oscillations of small amplitude is

(a) $2\pi\sqrt{\dfrac{l}{g}}$ (b) $2\pi\sqrt{\dfrac{2l}{3g}}$

(c) $2\pi\sqrt{\dfrac{3l}{2g}}$ (d) $2\pi\sqrt{\dfrac{l}{3g}}$

Superposition of Two or More than Two SHMs and Graphical Problems

62. The motion of a particle is given $x = A\sin\omega t + B\cos\omega t$. The motion of the particle is

(a) not simple harmonic
(b) simple harmonic with amplitude $A + B$
(c) simple harmonic with amplitude $\dfrac{(A+B)}{2}$
(d) simple harmonic with amplitude $\sqrt{A^2 + B^2}$

63. In case of a simple pendulum, time period *versus* length is depicted by

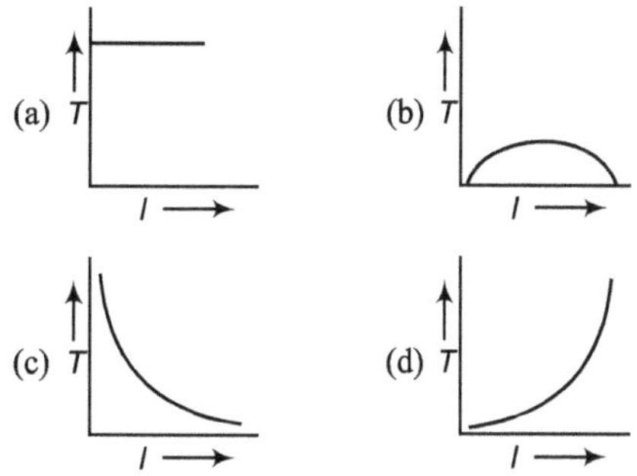

64. The displacement-time graph of a particle executing SHM is shown in figure. Which of the following statements is false?

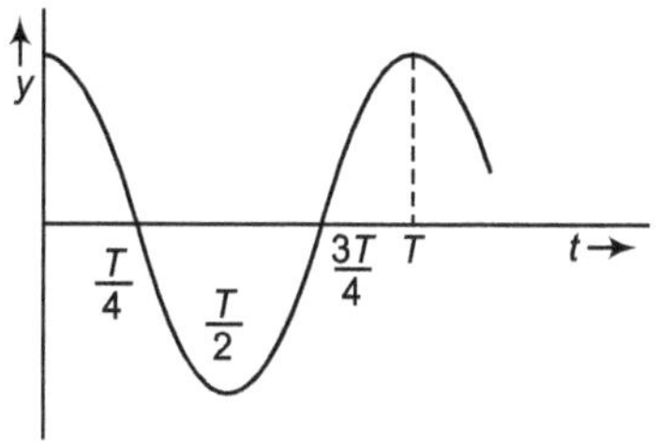

(a) The acceleration is maximum at $t = T$
(b) The force is zero at $t = \dfrac{3T}{4}$
(c) The potential energy equals the total oscillation energy at $t = \dfrac{T}{2}$
(d) None of the above

65. The displacement equation of a particle is $x = 3\sin 2t + 4\cos 2t$. The amplitude and maximum velocity will be respectively

(a) 5, 10 (b) 3, 2
(c) 4, 2 (d) 3, 4

66. The acceleration a of a particle undergoing SHM is shown in the figure. Which of the labelled points corresponds to the particle being at $-x_{max}$?

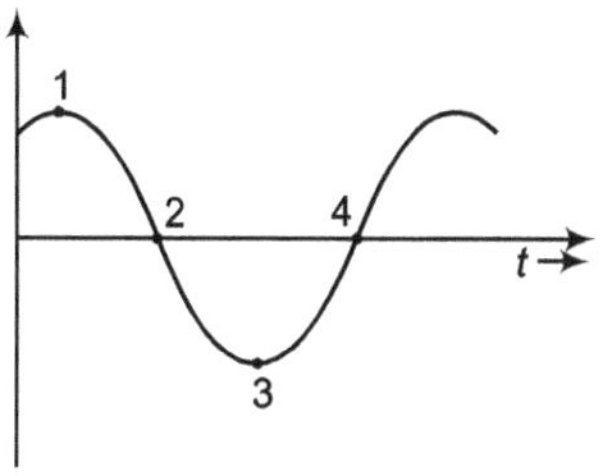

(a) 4
(b) 3
(c) 2
(d) 1

67. The displacement-time graph of a particle executing SHM is as shown in the figure.

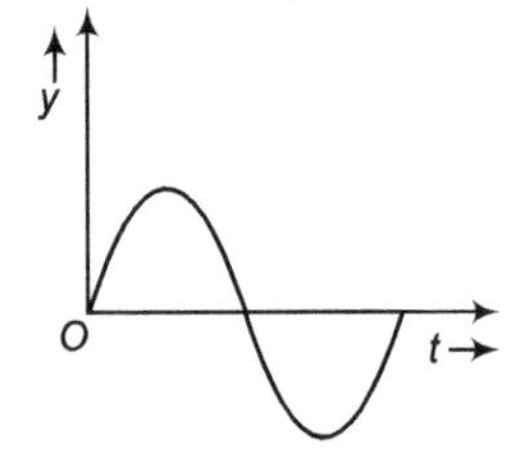

The corresponding force-time graph of the particle is

(a)
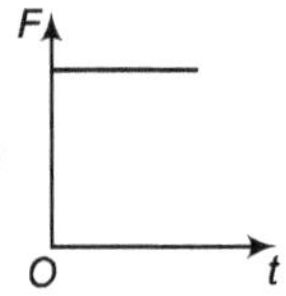

(b)
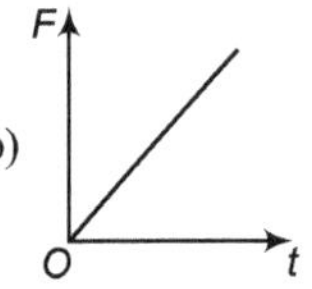

(c)
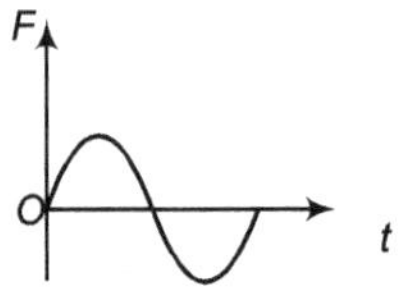

(d)
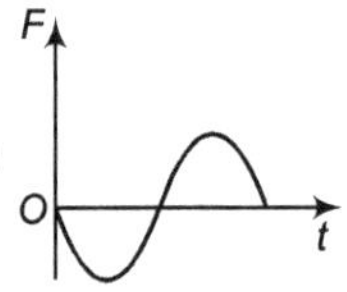

68. The graph shows the variation of displacement of a particle executing SHM with time. We infer from this graph that

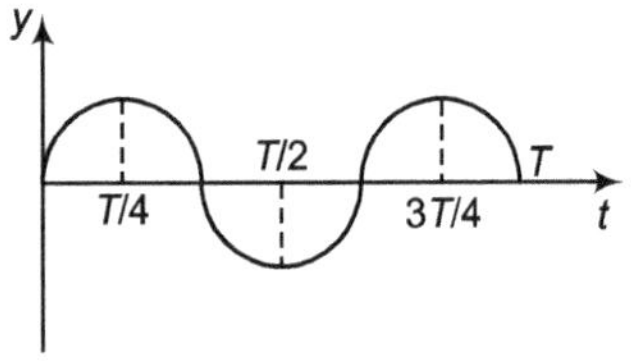

(a) the force is zero at time $\frac{3T}{4}$

(b) the velocity is maximum at time $\frac{T}{2}$

(c) the acceleration is maximum at time T

(d) the PE is equal to total energy at time $\frac{T}{2}$

69. The velocity-time diagram of a harmonic oscillator is shown in the below figure. The frequency of oscillation is

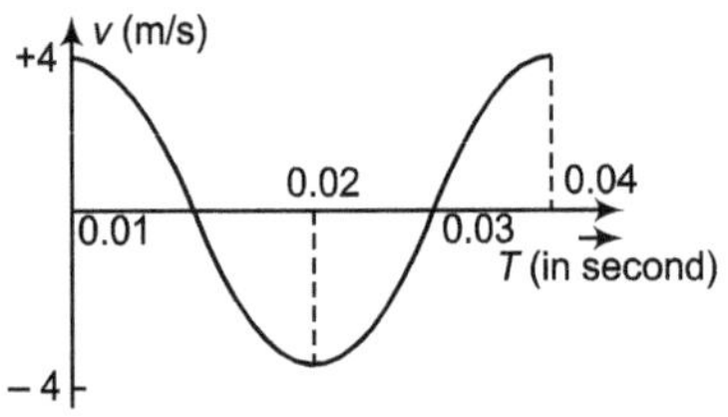

(a) 25 Hz
(b) 50 Hz
(c) 12.25 Hz
(d) 33.3 Hz

70. A body of mass 0.1 kg executes simple harmonic motion (SHM) about $x = 0$ under the influence of a force shown in figure.

The period of the SHM is

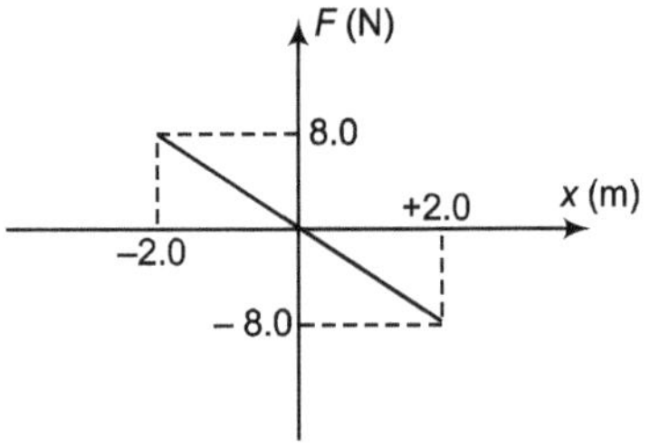

(a) 1.05 s
(b) 0.52 s
(c) 0.25 s
(d) 0.31 s

71. In order that the resultant path on superimposing two mutually perpendicular SHM be a circle, the conditions are that

(a) the amplitudes on both SHM should be equal and they should have a phase difference of $\frac{\pi}{2}$

(b) the amplitudes should be in the ratio 1 : 2 and the phase difference should be zero

(c) the amplitudes should be in the ratio 1 : 2 and the phase difference should be $\frac{\pi}{2}$

(d) the amplitudes should be equal and the phase difference should be zero

72. Lissajous figure obtained by combining

$$x = a \sin \omega t \text{ and } y = b \sin \left(\omega t + \frac{\pi}{2}\right)$$

will be

(a) an ellipse
(b) a straight line
(c) a circle
(d) a parabola

73. Two particles are executing simple harmonic motion. At an instant of time t their displacements are

$$y_1 = a \cos (\omega t) \text{ and } y_2 = a \sin (\omega t)$$

Then, the phase difference between y_1 and y_2 is

(a) 120°
(b) 90°
(c) 180°
(d) zero

74. A particle of mass m oscillates with simple harmonic motion between points x_1 and x_2, the equilibrium position being at O. Its potential energy is plotted. It will be as given below in the graph

(a)
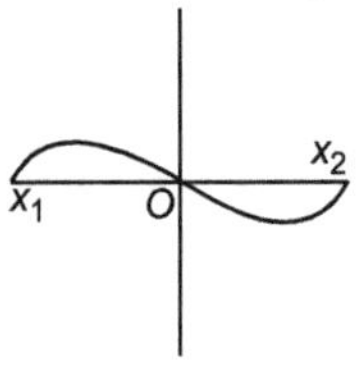

(b)
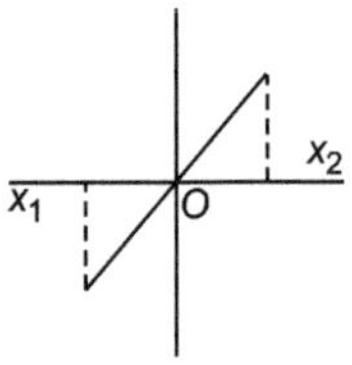

(c)
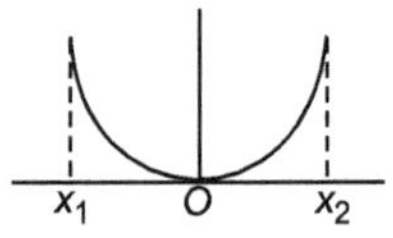

(d)
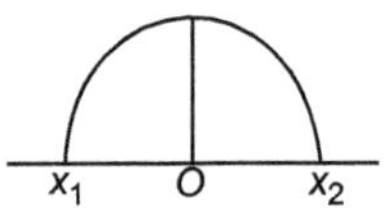

Miscellaneous Problems

75. In a spring-mass system, the length of the spring is L, and it has a mass M attached to it and oscillates with an angular frequency ω. The spring is then cut into two parts, one (a) with relaxed length αL and the other (b) with relaxed length $(1-\alpha)L$. The force constants of the two springs A and B are

(a) $\frac{k}{1-\alpha}$ and $\frac{k}{\alpha}$ (b) $\frac{k}{\alpha}$ and $\frac{k}{1-\alpha}$
(c) $\alpha k, (1-\alpha)k$ (d) k and k

76. A light spring of constant k is cut into two equal parts. The spring constant of each part is

(a) k (b) $2k$
(c) $\frac{k}{2}$ (d) $4k$

77. A light spring of force constant $8\,\text{Nm}^{-1}$ is cut into two equal halves and the two are connected in parallel; the equivalent force constant of the system is

(a) $16\,\text{Nm}^{-1}$ (b) $32\,\text{Nm}^{-1}$ (c) $8\,\text{Nm}^{-1}$ (d) $24\,\text{Nm}^{-1}$

78. Two springs of spring constants k each are connected in series, then what is the net spring constant?

(a) k (b) $\frac{k}{2}$
(c) $2k$ (d) $3\frac{k}{2}$

79. Two point masses of 3.0 kg and 6.0 kg are attached to opposite ends of horizontal spring whose spring constant is $300\,\text{Nm}^{-1}$ as shown in the figure. The natural vibration frequency of the system is approximately

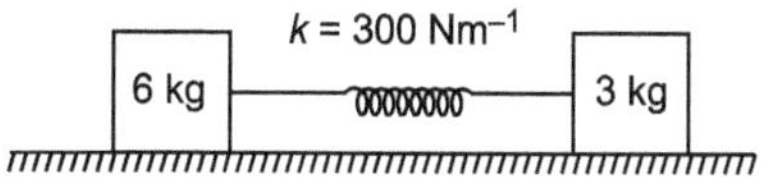

(a) 4 Hz (b) 3 Hz (c) 2 Hz (d) 1 Hz

80. In simple harmonic motion, the particle is

(a) always accelerated
(b) always retarded
(c) alternately accelerated and retarded
(d) neither accelerated nor retarded

81. The dimensions of mass ÷ force constant of SHM are same as that of

(a) time (b) time^2
(c) acceleration (d) $\frac{1}{\text{acceleration}}$

82. The length of second's pendulum on the earth is about 1 m. What should be the length of second's pendulum on the moon?

(a) 1 m (b) $\frac{1}{6}$ m
(c) 6 m (d) 36 m

83. A mass of 0.2 kg is attached to the lower end of a massless spring of force constant 200 N/m, the upper end of which is fixed to a rigid support. Which of the following statement is true?

(a) In equilibrium, the spring will be stretched by 1 cm
(b) If the mass is raised till the spring becomes unstretched and then released, it will go down by 2 cm before moving upwards
(c) The frequency of oscillation will be nearly 5 Hz
(d) All of the above

84. In SHM, phase difference between displacement and velocity is ϕ_1 and that between displacement and acceleration is ϕ_2, then

(a) $\phi_2 = 2\phi_1$ (b) $\phi_2 = \phi_1$
(c) $\phi_1 = 2\phi_2$ (d) None of these

85. Which of the following quantities is always negative in SHM?

(a) $\mathbf{F}\cdot\mathbf{a}$ (b) $\mathbf{v}\cdot\mathbf{s}$
(c) $\mathbf{a}\cdot\mathbf{s}$ (d) $\mathbf{F}\cdot\mathbf{v}$

Here, **s** is displacement from mean position.

86. A small ball is dropped from a certain height on the surface of a non-viscous liquid of density less than the density of ball. The motion of the ball is

(a) SHM
(b) periodic but not SHM
(c) not periodic
(d) SHM for half a period and non-periodic for rest half of the period

87. When a particle executes SHM there is always a constant ratio between its displacement and

(a) velocity
(b) acceleration
(c) mass
(d) time period

88. A mass $M = 5$ kg is attached to a spring as shown in the figure and held in position so that the spring remains unstretched. The spring constant is 200 N/m. The mass M is then released and begins to undergo small oscillations. The amplitude of oscillation is

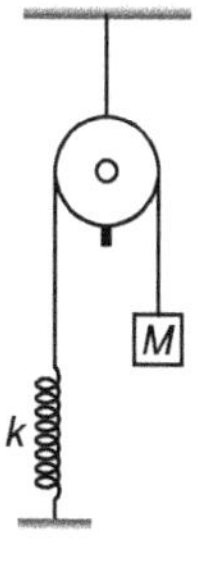

(a) 0.5 m
(b) 0.25 m
(c) 0.2 m
(d) 0.1 m

89. Under the action of a force $F = -kx^3$, the motion of a particle is (k = a positive constant)

(a) simple harmonic motion
(b) uniformly accelerated motion
(c) not periodic
(d) periodic but not simple harmonic

90. The phase (at a time t) of a particle in simple harmonic motion tells

(a) only the position of the particle at time t
(b) only the direction of motion of the particle at time t
(c) both the position and direction of motion of the particle at time t
(d) neither the position of the particle nor its direction of motion at time t

91. A particle is moving with constant angular velocity along the circumference of a circle. Which of the following statements is true?

(a) The particle executes SHM
(b) The projection of the particle on any one of the diameters executes SHM
(c) The projection of the particle on any of the diameters executes SHM
(d) None of the above

92. A particle executing simple harmonic motion along y-axis has its motion described by the equation $y = A\sin(\omega t) + B$. The amplitude of the simple harmonic motion is

(a) A
(b) B
(c) $A + B$
(d) $\sqrt{A+B}$

93. Which of the following is a necessary and sufficient condition for SHM?

(a) Constant period
(b) Constant acceleration
(c) Proportionality between acceleration and velocity
(d) Proportionality between restoring force and displacement from equilibrium position

94. In simple harmonic motion, the ratio of acceleration of the particle to its displacement at any time is a measure of

(a) spring constant
(b) angular frequency
(c) (angular frequency)^2
(d) restoring force

95. A tunnel has been dug through the centre of the earth and a ball is released in it. It will reach the other end of the tunnel after

(a) 84.6 min
(b) 42.3 min
(c) 1 day
(d) will not reach the other end

96. What is constant in SHM?

(a) Restoring force
(b) Kinetic energy
(c) Potential energy
(d) Periodic time

97. The displacement x (in metre) of a particle in, simple harmonic motion is related to time t (in second) as

$$x = 0.01\cos\left(\pi t + \frac{\pi}{4}\right)$$

The frequency of the motion will be

(a) 0.5 Hz (b) 1.0 Hz
(c) $\frac{\pi}{2}$ Hz (d) π Hz

98. In a simple pendulum, the period of oscillation T is related to length of the pendulum l as

(a) $\frac{l}{T}$ = constant (b) $\frac{l^2}{T}$ = constant
(c) $\frac{l}{T^2}$ = constant (b) $\frac{l^2}{T^2}$ = constant

99. The ratio of frequencies of two pendulums are 2 : 3, then their lengths are in ratio

(a) $\sqrt{\frac{2}{3}}$ (b) $\sqrt{\frac{3}{2}}$
(c) $\frac{4}{9}$ (d) $\frac{9}{4}$

100. Molten-wax of mass m drops on a block of mass M, which is oscillating on a frictionless table as shown. Select the incorrect option.

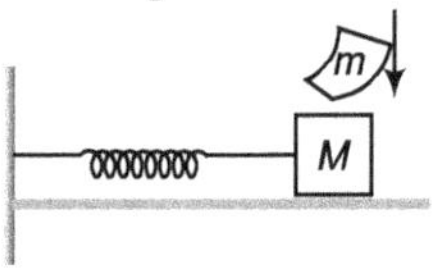

(a) If the collision takes place at extreme position, amplitude does not change
(b) If the collision takes place at mean position, amplitude decreases
(c) If the collision takes place at mean position, time period decreases
(d) If the collision takes place at extreme position, time period increases

101. A particle executes SHM with amplitude of 20 cm and time period of 12 s. What is the minimum time required for it to move between two points 10 cm on either side of the mean position?

(a) 1 s (b) 2 s (c) 3 s (d) 4 s

102. The maximum acceleration of a particle in SHM is made two times keeping the maximum speed to be constant. It is possible when

(a) amplitude of oscillation is doubled while frequency remains constant
(b) amplitude is doubled while frequency is halved
(c) frequency is doubled while amplitude is halved
(d) frequency of oscillation is doubled while amplitude remains constant

[Level 2]

Only One Correct Option

1. A body of mass 0.01 kg executes simple harmonic motion (SHM) about $x = 0$ under the influence of a force shown in figure. The period of the SHM is

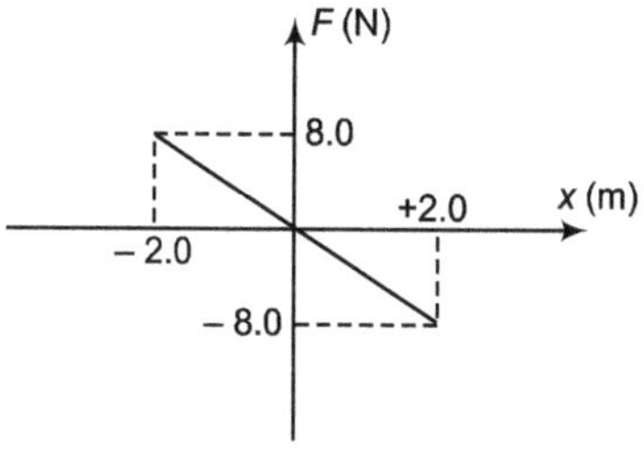

(a) 1.05 s (b) 0.52 s
(c) 0.25 s (d) 0.31 s

2. The vertical motion of a ship at sea is described by the equation $\frac{d^2x}{dt^2} = -4x$, where x is the vertical height of the ship (in metre) above its mean position. If it oscillates through a height of 1 m,

(a) its maximum vertical speed will be 1 m/s
(b) its maximum vertical speed will be 2 m/s
(c) its greatest vertical acceleration is 2 m/s^2
(d) its greatest vertical acceleration is 1 m/s^2

3. Period of small oscillations in the two cases shown in figure is T_1 and T_2 respectively, then

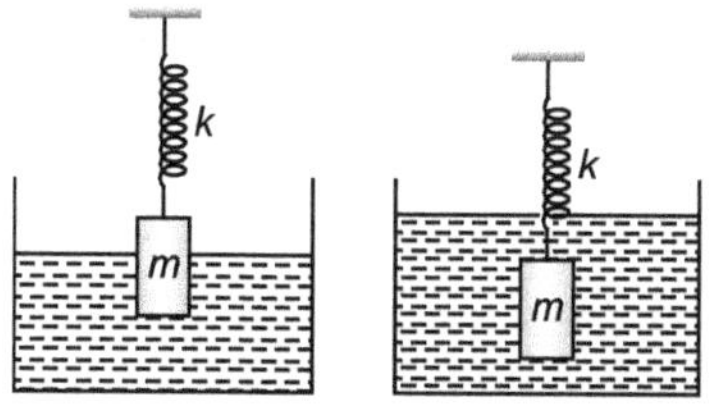

(a) $T_1 = T_2$
(b) $T_1 < T_2$
(c) $T_1 > T_2$
(d) Cannot say anything

4. The potential energy of a particle of mass 2 kg in SHM is $(9x^2)$ J. Here, x is the displacement from mean position. If total mechanial energy of the particle is 36 J, then maximum speed of the particle is

(a) 4 m/s
(b) 2 m/s
(c) 6 m/s
(d) 10 m/s

5. The variation of PE of harmonic oscillator is as shown in figure. The spring constant is

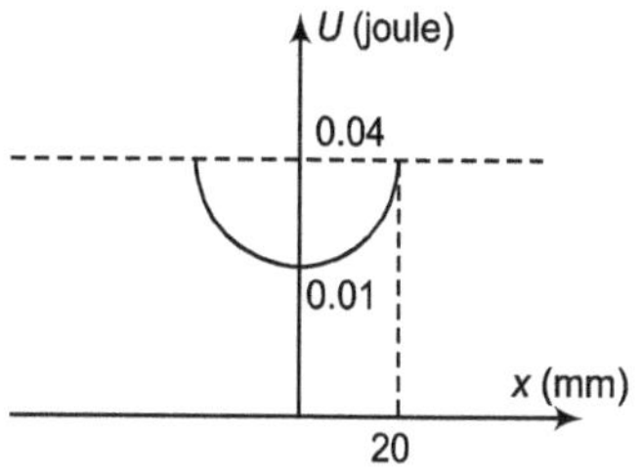

(a) 1×10^2 N/m (b) 1.5×10^2 N/m
(c) 2×10^2 N/m (d) 3×10^2 N/m

6. A particle of mass 0.1 kg is executing SHM of amplitude 0.1 m. When the particle passes through the mean position, its KE is 8×10^{-3} J. Find the equation of motion of the particle if the initial phase of oscillation is 45°.

(a) $y = 0.1 \cos\left(3t + \frac{\pi}{4}\right)$ (b) $y = 0.1 \sin\left(6t + \frac{\pi}{4}\right)$
(c) $y = 0.1 \sin\left(4t + \frac{\pi}{4}\right)$ (d) $y = 0.1 \cos\left(4t + \frac{\pi}{4}\right)$

7. The system shown in figure is in equilibrium. The mass of the container with liquid is M, density of liquid in the container is ρ and the volume of the block is V. If the container is now displaced downwards through a distance x_0 and released such that the block remains well inside the liquid then during subsequent motion,

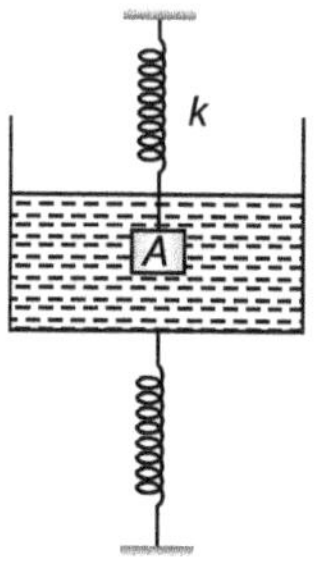

(a) time period of SHM of the container will be $2\pi\sqrt{\frac{M}{k}}$
(b) time period of SHM of the container will be $2\pi\sqrt{\frac{M + \rho V}{k}}$
(c) amplitude of SHM of container is x_0
(d) amplitude of SHM of the container is $2x_0$

8. A particle is in linear SHM of amplitude A and time period T. If v refers to its average speed during any interval of $\frac{T}{3}$, then the maximum possible value of v is

(a) $\frac{3\sqrt{3}}{T}A$ (b) $\frac{\sqrt{3}A}{T}$
(c) $\frac{2\sqrt{3}}{T}A$ (d) $\frac{3A}{T}$

9. A simple pendulum is suspended from the ceiling of a car and its period of oscillation is T when the car is at rest. The car starts moving on a horizontal road with a constant acceleration g (equal to the acceleration due to gravity, in magnitude) in the forward direction. To keep the time period same, the length of the pendulum

(a) will have to be increased by $\sqrt{2}\, l$
(b) will have to be increased by $(\sqrt{2}-1)\, l$
(c) will have to be decreased by $\sqrt{2}\, l$
(d) will have to be decreased by $(\sqrt{2}-1)\, l$

10. A particle of mass m is dropped from a great height h above the hole in the earth dug along its diameter.

(a) The motion of the particle is simple harmonic
(b) The motion of the particle is periodic
(c) The speed of the particle at the centre of earth equals $\sqrt{\frac{2GM}{(R+h)}}$, where R and M are the radius and mass of the earth respectively
(d) The speed of the particle at the centre of earth equals $\sqrt{\frac{GM(R+3h)}{R(R+h)}}$, where R and M are the radius and mass of the earth respectively

11. A solid cube floats in water half immersed and has small vertical oscillations of time period $\frac{\pi}{5}$ s. Its mass (in kg) is (Take $g = 10\,\text{m/s}^2$)

(a) 4 (b) 2
(c) 1 (d) 0.5

12. Maximum kinetic energy of a particle of mass 1 kg in SHM is 8 J. Time period of SHM is 4 s. Maximum potential energy during the motion is 10 J. Then,

(a) amplitude of oscillations is approximately 2.53 m
(b) minimum potential energy of the particle is 2 J
(c) maximum acceleration of the particle is approximately 6.3 m/s^2
(d) minimum kinetic energy of the particle is 2 J

13. The potential energy of a particle of mass 0.1 kg moving along the x-axis, is given by $U = 5x(x-4)$ J, where x is in metres. Choose the wrong option.

(a) The speed of the particle is maximum at $x = 2$ m
(b) The particle executes simple harmonic motion
(c) The period of oscillation of the particle is $\frac{\pi}{5}$ s
(d) None of the above

14. The speed (v) of a particle moving along a straight line, when it is at a distance (x) from a fixed point on the line, is given by $v^2 = 144 - 9x^2$. Select wrong alternate.

(a) Displacement of the particle $\le$ distance moved by it
(b) The magnitude of acceleration at a distance 3 units from the fixed point is 27 units
(c) The motion is simple harmonic with $T = \frac{\pi}{3}$ units
(d) The maximum displacement from the fixed point is 4 units

15. Two pendulums of time periods 3 s and 7 s, respectively start oscillating simultaneously from two opposite extreme positions. After how much time they will be in same phase?

(a) $\frac{21}{8}$ s (b) $\frac{21}{4}$ s (c) $\frac{21}{2}$ s (d) $\frac{21}{10}$ s

16. A particle under the action of a SHM has a period of 3 s and under the effect of another it has a period 4 s. What will be its period under the combined action of both the SHM's in the same direction?

(a) 7 s (b) 5 s
(c) 2.4 s (d) 0.4 s

17. A particle of mass m is attached to three identical springs A, B and C each of force constant k as shown in figure. If the particle of mass m is pushed slightly against the spring A and released, then the time period of oscillation is

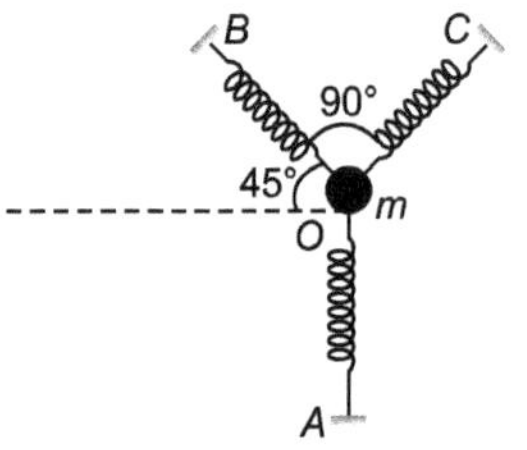

(a) $2\pi\sqrt{\frac{2m}{k}}$ (b) $2\pi\sqrt{\frac{m}{2k}}$
(c) $2\pi\sqrt{\frac{m}{k}}$ (d) $2\pi\sqrt{\frac{m}{3k}}$

18. In SHM, potential energy of a particle at mean position is E_1 and kinetic energy is E_2, then

(a) $E_1 = E_2$
(b) total potential energy at $x = \frac{\sqrt{3}A}{2}$ is $E_1 + \frac{3E_2}{4}$
(c) total kinetic energy at $x = \frac{\sqrt{3}A}{2}$ is $\frac{3E_2}{4}$
(d) total kinetic energy at $x = \frac{A}{\sqrt{2}}$ is $\frac{E_2}{4}$

19. A particle performs SHM in a straight line. In the first second, starting from rest, it travels a distance a and in the next second it travels a distance b in the same side of mean position. The amplitude of the SHM is

(a) $a - b$ (b) $\frac{2a-b}{3}$
(c) $\frac{2a^2}{3a-b}$ (d) None of these

20. A block of mass 100 g attached to a spring of spring constant 100 N/m is lying on a frictionless floor as shown. The block is moved to compress the spring by 10 cm and then released. If the collisions with the wall in front are elastic, then the time period of the motion is

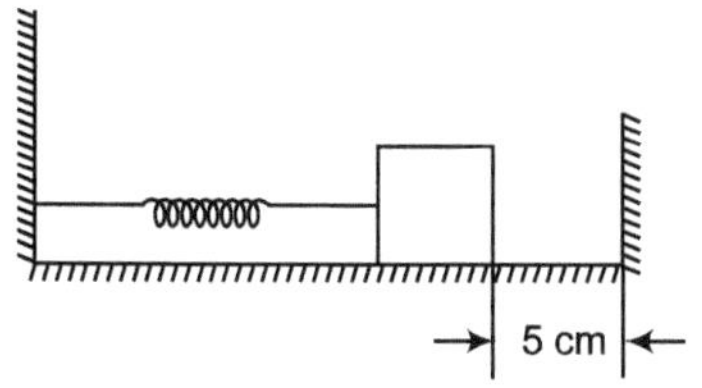

(a) 0.2 s (b) 0.1 s
(c) 0.15 s (d) 0.132 s

21. In the figure, the block of mass *m*, attached to the spring of stiffness k is in contact with the completely elastic wall, and the compression in the spring is *e*. The spring is compressed further by *e* by displacing the block towards left and is then released. If the collision between the block and the wall is completely elastic, then the time period of oscillations of the block will be

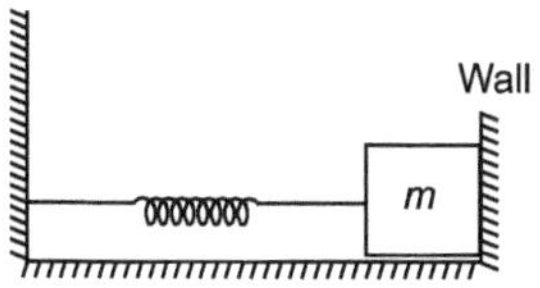

(a) $\frac{2\pi}{3}\sqrt{\frac{m}{k}}$ (b) $2\pi\sqrt{\frac{m}{k}}$ (c) $\frac{\pi}{3}\sqrt{\frac{m}{k}}$ (d) $\frac{\pi}{6}\sqrt{\frac{m}{k}}$

22. A cubical block of mass M vibrates horizontally with amplitude of 4.0 cm and a frequency of 2.0 Hz. A small block of mass m is palced on the bigger block. In order that the smaller block does not slide on the bigger block, the minimum value of the coefficient of static friction between the two blocks is

(a) 0.36 (b) 0.40
(c) 0.64 (d) 0.72

23. A spring has a natural length of 50 cm and a force constant of $2.0 \times 10^3\ \text{Nm}^{-1}$. A body of mass 10 kg is suspended from it and the spring is stretched. If the body is pulled down to a length of 58 cm and released, it executes simple harmonic motion. What is the net force on the body when it is at its lowermost position of its oscillation? (Take $g = 10\ \text{ms}^{-2}$).

(a) 20 N (b) 40 N
(c) 60 N (d) 80 N

24. Two masses 8 kg and 4 kg are suspended together by a massless spring of spring constant 1000 N/m. When the masses are in equilibrium 8 kg is removed without disturbing the system. The amplitude of oscillation is

(a) 0.5 m (b) 0.08 m
(c) 0.4 m (d) 0.04 m

25. The period of oscillation of a simple pendulum of length L suspended from the roof of a vehicle which moves without friction down an inclined plane of inclination α is, given by

(a) $2\pi\sqrt{\frac{L}{(g\cos\alpha)}}$ (b) $2\pi\sqrt{\frac{L}{(g\sin\alpha)}}$

(c) $2\pi\sqrt{\frac{L}{g}}$ (d) $2\pi\sqrt{\frac{L}{(g\tan\alpha)}}$

26. A horizontal platform with an object placed on it is executing SHM in the vertical direction. The amplitude of oscillation is 4×10^{-3} m. What must be least period of these oscillations, so that the object is not detached from the platform? (Take $g = 10\,\text{m/s}^2$)

(a) $\frac{\pi}{25}$ s (b) $\frac{\pi}{5}$ s

(c) $\frac{\pi}{10}$ s (d) $\frac{\pi}{50}$ s

27. A block of mass *m*, attached to a spring of spring constant *k*, oscillates on a smooth horizontal table. The other end of the spring is a fixed to a wall. The block has a speed v when the spring is at its natural length. Before coming to an instantaneous rest, if the block moves a distance x from the mean position, then

(a) $x = \sqrt{\frac{m}{k}}$ (b) $x = \frac{1}{v}\sqrt{\frac{m}{k}}$

(c) $x = v\sqrt{\frac{m}{k}}$ (d) $x = \sqrt{\frac{mv}{k}}$

28. One end of a long metallic wire of length L is tied to the ceiling. The other end is tied to massless spring of spring constant *k*. A mass *m* hangs freely from the free end of the spring. The area of cross-section and Young's modulus of the wire are A and *Y*, respectively. If the mass is slightly pulled down and released, it will oscillate with a time period T equal to

(a) $2\pi\sqrt{\frac{m}{k}}$ (b) $2\pi\sqrt{\frac{(YA + kL)}{YAk}}$

(c) $2\pi\sqrt{\frac{(YA + kL)}{Ak}}$ (d) $2\pi\sqrt{\frac{(Y + kL)}{YAk}}$

29. A mass M is attached to a horizontal spring of force constant k fixed one side to a rigid support as shown in figure. The mass oscillates on a frictionless surface with time period T and amplitude A. When the mass is in equilibrium position, another mass m is gently placed on it. What will be the new amplitude of oscillations?

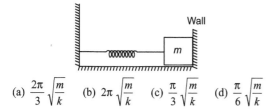

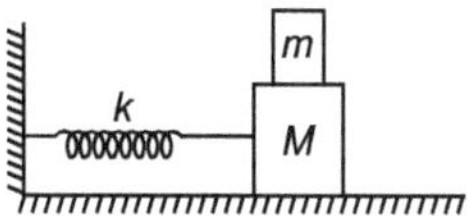

(a) $\sqrt{\frac{(M + m)}{M}}\,A$ (b) $\sqrt{\frac{(M - m)}{M}}\,A$

(c) $\sqrt{\frac{M}{(M + m)}}\,A$ (d) $\sqrt{\frac{M}{(M - m)}}\,A$

30. A linear harmonic oscillator of force constant 2×10^6 N/m and amplitude 0.01 m has a total mechanical energy of 160 J. Its

(a) maximum potential energy is 160 J
(b) maximum potential energy is 100 J
(c) minimum potential energy is zero
(d) minimum potential energy is 100 J

31. A mass is suspended separately by two springs of spring constants k_1 and k_2 in successive order. The time periods of oscillations in the two cases are T_1 and T_2, respectively. If the same mass be suspended by connecting the two springs in parallel, (as shown in figure) then the time period of oscillations is T. The correct relation is

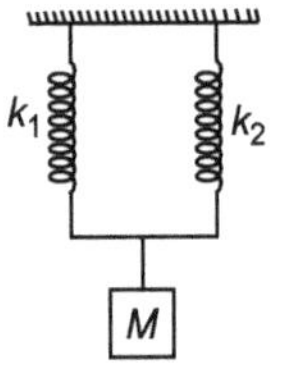

(a) $T^2=T_1^2+T_2^2$ (b) $T^{-2}=T_1^{-2}+T_2^{-2}$
(c) $T^{-1}=T_1^{-1}+T_2^{-1}$ (d) $T=T_1+T_2$

32. Two particles execute SHM of the same amplitude and frequency along the same straight line. They pass one another when going in opposite directions each time their displacement is half their amplitude. What is the phase difference between them?

(a) 60° (b) 30°
(c) 90° (d) 120°

33. A particle of mass 2 kg moves in simple harmonic motion and its potential energy U varies with position x as shown. The period of oscillation of the particle is

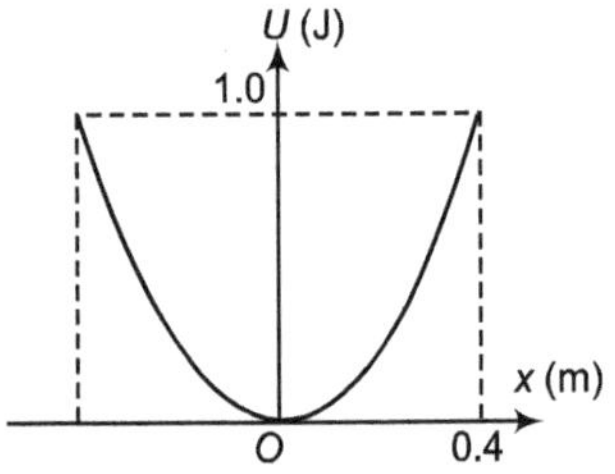

(a) $\dfrac{2\pi}{5}$ s (b) $\dfrac{2\sqrt{2}\,\pi}{5}$ s
(c) $\dfrac{\sqrt{2}\,\pi}{5}$ s (d) $\dfrac{4\pi}{5}$ s

34. Two simple harmonic motions are represented by the following equations

$$y_1=40\sin\omega t$$

and $$y_2=10(\sin\omega t+c\cos\omega t)$$

If their displacement amplitudes are equal, then the value of c (in appropriate units) is

(a) $\sqrt{13}$ (b) $\sqrt{15}$
(c) $\sqrt{17}$ (d) 4

35. A mass M is suspended from a massless spring. An additional mass m stretches the spring further by a distance x. The combined mass will oscillate with a period

(a) $2\pi\sqrt{\left\{\dfrac{(M+m)x}{mg}\right\}}$ (b) $2\pi\sqrt{\left\{\dfrac{mg}{(M+m)x}\right\}}$

(c) $2\pi\sqrt{\left\{\dfrac{(M+m)}{mgx}\right\}}$ (d) $\dfrac{\pi}{2}\sqrt{\left\{\dfrac{mg}{(M+m)x}\right\}}$

More than One Correct Options

1. A simple pendulum with a bob of mass m is suspended from the roof of a car moving with horizontal acceleration a. Then,

(a) the string makes an angle of $\tan^{-1}(a/g)$ with the vertical
(b) the string makes an angle of $\sin^{-1}\left(\dfrac{a}{g}\right)$ with the vertical
(c) the tension in the string is $m\sqrt{a^2+g^2}$
(d) the tension in the string is $m\sqrt{g^2-a^2}$

2. A particle starts from a point P at a distance of $A/2$ from the mean position O and travels towards left as shown in the figure. If the time period of SHM, executed about O is T and amplitude is A, then the equation of the motion of particle is

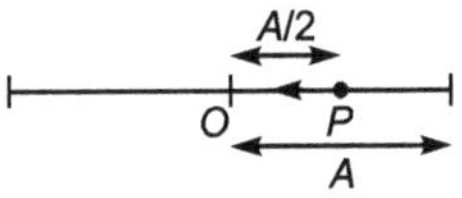

(a) $x=A\sin\left(\dfrac{2\pi}{T}t+\dfrac{\pi}{6}\right)$ (b) $x=A\sin\left(\dfrac{2\pi}{T}t+\dfrac{5\pi}{6}\right)$

(c) $x=A\cos\left(\dfrac{2\pi}{T}t+\dfrac{\pi}{6}\right)$ (d) $x=A\cos\left(\dfrac{2\pi}{T}t+\dfrac{\pi}{3}\right)$

3. A spring has natural length 40 cm and spring constant 500 N/m. A block of mass 1 kg is attached at one end of the spring and other end of the spring is attached to a ceiling. The block is released from the position, where the spring has length 45 cm. Then,

(a) the block will perform SHM of amplitude 5 cm
(b) the block will have maximum velocity $30\sqrt{5}$ cm/s
(c) the block will have maximum acceleration 15 m/s^2
(d) the minimum elastic potential energy of the spring will be zero

4. The system shown in the figure can move on a smooth surface. They are initially compressed by 6 cm and then released.

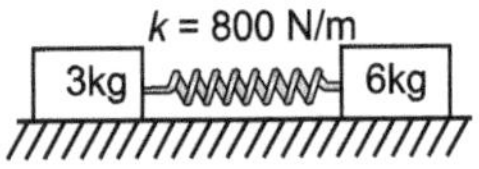

(a) The system performs, SHM with time period $\dfrac{\pi}{10}$ s
(b) The block of mass 3 kg performs SHM with amplitude 4 cm
(c) The block of mass 6 kg will have maximum momentum of 2.40 kg m/s
(d) The time periods of two blocks are in the ratio of $1:\sqrt{2}$

5. The displacement-time graph of a particle executing SHM is shown in figure. Which of the following statement(s) is/are true?

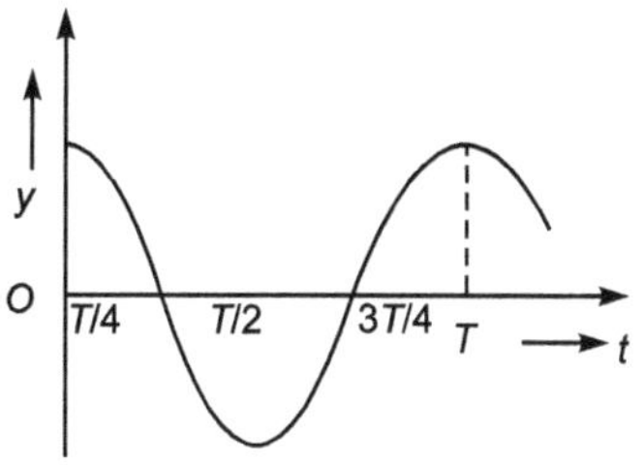

(a) The velocity is maximum at $t = T/2$
(b) The acceleration is maximum at $t = T$
(c) The force is zero at $t = 3T/4$
(d) The kinetic energy equals total oscillation energy at $t = T/2$

6. For a particle executing SHM, $x =$ displacement from mean position, $v =$ velocity and $a =$ acceleration at any instant, then
(a) v-x graph is a circle (b) v-x graph is an ellipse
(c) a-x graph is a straight line (d) a-x graph is a circle

7. The acceleration of a particle is $a = -100x + 50$. It is released from $x = 2$. Here, a and x are in SI units. Then,
(a) the particle will perform SHM of amplitude 2 m
(b) the particle will perform SHM of amplitude 1.5 m
(c) the particle will perform SHM of time period 0.63 s
(d) the particle will have a maximum velocity of 15 m/s

8. Two particles are performing SHM in same phase. It means that
(a) the two particles must have same distance from the mean position simultaneously
(b) two particles may have same distance from the mean position simultaneously
(c) the two particles must have maximum speed simultaneously
(d) the two particles may have maximum speed simultaneously

9. A particle moves along y-axis according to the equation

$$y\,(\text{in cm}) = 3\sin 100\pi t + 8\sin^2 50\pi t - 6.$$ Then,

(a) the particle performs SHM
(b) the amplitude of the particle's oscillation is 5 cm
(c) the mean position of the particle is at $y = -2$ cm
(d) the particle does not perform SHM

10. The displacement-time graph of a particle executing SHM is shown in figure. Which of the following statement(s) is/are true?

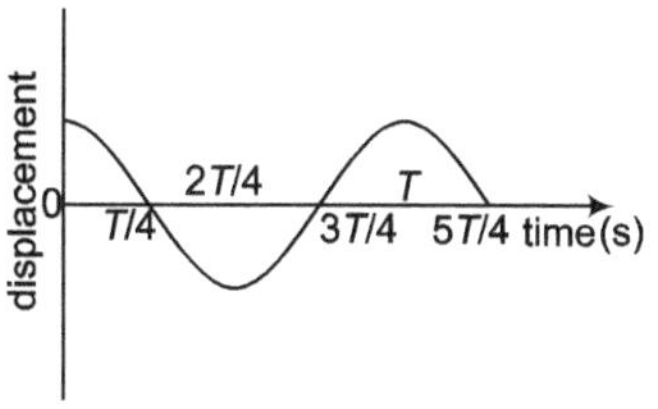

(a) The force is zero at $t = \frac{3T}{4}$
(b) The acceleration is maximum at $t = \frac{4T}{4}$
(c) The velocity is maximum at $t = \frac{T}{4}$
(d) The PE is equal to KE of oscillation at $t = \frac{T}{2}$

11. A body is performing SHM, then its
(a) average total energy per cycle is equal to its maximum kinetic energy
(b) average kinetic energy per cycle is equal to half of its maximum kinetic energy
(c) mean velocity over a complete cycle is equal to $\frac{2}{\pi}$ times of its maximum velocity
(d) root mean square velocity is $\frac{1}{\sqrt{2}}$ times of its maximum velocity

12. A particle is in linear simple harmonic motion between two points. A and B, 10 cm apart (figure) take the direction from A to B as the positive direction and choose the correct statements.

$$AO = OB = 5\,\text{cm} \;\Rightarrow\; BC = 8\,\text{cm}$$

(a) The sign of velocity, acceleration and force on the particle when it is 3 cm away from A going towards B are positive
(b) The sign of velocity of the particle at C going towards B is negative
(c) The sign of velocity, acceleration and force on the particle when it is 4 cm away from B going towards A are negative
(d) The sign of acceleration and force on the particle when it is at points B is negative

Comprehension Based Questions

Passage (Q. 1 to 2)

A 2 kg block hangs without vibrating at the bottom end of a spring with a force constant of 400 N/m. The top end of the spring is attached to the ceiling of an elevator car. The car is rising with an upward acceleration of $5\,\text{m/s}^2$ when the acceleration suddenly ceases at time $t = 0$ and the car moves upward with constant speed ($g = 10\,\text{m/s}^2$)

1. What is the angular frequency of oscillation of the block after the acceleration ceases?
(a) $10\sqrt{2}$ rad/s (b) 20 rad/s
(c) $20\sqrt{2}$ rad/s (d) 32 rad/s

2. The amplitude of the oscillation is
(a) 7.5 cm (b) 5 cm (c) 2.5 cm (d) 1 cm

Assertion and Reason

Direction (Q. Nos. 1-20) *These questions consist of two statements each printed as Assertion and Reason. While answering these questions you are required to choose anyone of the following five responses.*

(a) If both Assertion and Reason are true and Reason is the correct explanation of Assertion.
(b) If both Assertion and Reason are true but Reason is not the correct explanation of Assertion.
(c) If Assertion is true but Reason is false.
(d) If Assertion is false but Reason is true.
(e) If both Assertion and Reason are false.

1. **Assertion** In SHM, v-x graph is an ellipse, where v is velocity and x is displacement from mean position.

Reason Equation between v and x is $\frac{v^2}{\omega^2}+\frac{x^2}{1}=A^2$, which is the equation of an ellipse.

2. **Assertion** In $x = 3 + 4\cos\omega t$, amplitude of oscillation is 4 units.

Reason Mean position is at $x = 3$.

3. **Assertion** Bob is released from rest from position A. Given, θ_0 very small. Angular velocity of bob about point O is maximum and $\left(\sqrt{\frac{g}{l}}\right)\theta_0$ at point O.

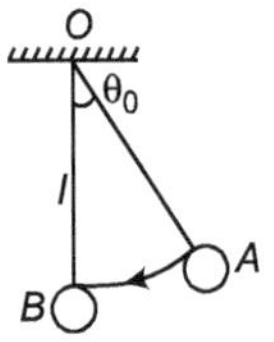

Reason For small angular amplitudes motion of bob is simple harmonic.

4. **Assertion** If a pendulum is suspended in a lift and lift accelerates upwards, then its time period will decrease.

Reason Value of effective value of g will be

$$g_e = g + a$$

5. **Assertion** Mean position of SHM is the stable equilibrium position.

Reason In stable equilibrium, position potential energy is minimum.

6. **Assertion** In $x = A\cos\omega t$ equation of SHM, x is measured from the extreme position, $x = +A$.

Reason At time $t = 0$, particle is at $x = +A$.

7. **Assertion** In a spring block system of length of spring and mass both are halved, time period of oscillation will remain unchanged.

Reason Angular frequency of SHM is $\omega = \sqrt{\frac{k}{m}}$ where, k is spring constant and m is mass of block.

8. **Assertion** Displacement-time equation of a particle moving along x-axis is $x = 4 + 6\sin\omega t$. Under this situation, motion of particle is not simple harmonic.

Reason $\frac{d^2x}{dt^2}$ for the given equation is not proportional to $-x$.

9. **Assertion** Simple harmonic motion is an example of one dimensional motion with non-uniform acceleration.

Reason In simple harmonic motion, acceleration varies with displacement linearly.

10. **Assertion** A particle performing SHM at certain instant is having velocity v. It again acquires a velocity v for the first time after a time interval of T second. Then, the time period of oscillation is T second.

Reason A particle performing SHM can have same velocity at two instants in one cycle.

11. **Assertion** Velocity *versus* displacement (from the mean position) graph in SHM is a parabola.

Reason $v = \omega\sqrt{A^2 - x^2}$ is quadratic equation in v and x.

12. **Assertion** In SHM, to find time taken in moving from one point to another point we cannot apply the relation,

$$\text{Time} = \frac{\text{Distance}}{\text{Speed}}$$

Reason In SHM, speed is not constant.

13. **Assertion** In the x-t graph of a particle in SHM, acceleration of particle at time t_0 is positive but velocity is negative.

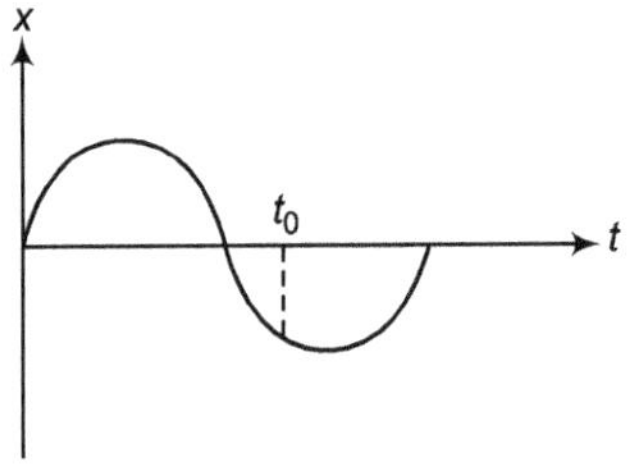

Reason $a \propto -x$ and velocity is slope of x-t graph.

14. **Assertion** x-t equation of a particle in SHM is given as

$$x = A\cos\omega t$$

At time $t = \frac{\pi}{2\omega}$, potential energy is minimum.

Reason In the given equation, the minimum potential energy is zero.

15. **Assertion** In $x = 5 - 4\sin\omega t$ motion of body is SHM about the mean position $x = 5$.

Reason Amplitude of oscillations is 9.

16. **Assertion** Time period of a spring-block system is T. If length of spring is decreased, time period will decrease.

Reason If length is decreased, then the block will have to travel less distance and it will take less time.

17. Assertion If two simple harmonic motions are imposed on a body in mutually perpendicular directions, then the resultant path is a straight line if they have same angular frequency.

Reason Resultant path can be obtained by eliminating t and finding x-y relation.

18. Assertion If amplitude of SHM is increased, time period of SHM will increase.

Reason If amplitude is increased, body have to travel more distance in one complete oscillation.

19. Assertion In the equation.

$$x = 3A \sin \omega t + 4A \cos \omega t,$$

maximum speed is $5\omega A$.

Reason The given equation can be written as

$$x = 5A \sin (\omega t + 53°)$$

20. Assertion Average kinetic energy in one oscillation during SHM of a body is $\frac{1}{4} m\omega^2 A^2$.

Reason Maximum kinetic energy is $\frac{1}{2} m\omega^2 A^2$.

Match the Columns

1. In SHM, match the following columns.

Column I	Column II
(A) Displacement and velocity	(p) Phase difference is zero
(B) Displacement and acceleration	(q) Phase difference is $\frac{\pi}{2}$
(C) Velocity and acceleration	(r) Phase difference is π

2. In the equation $y = A \sin \left(\omega t + \frac{\pi}{4}\right)$, match the following columns. For $x = \frac{A}{2}$

Column I	Column II
(A) Kinetic energy	(p) Half the maximum value
(B) Potential energy	(q) 3/4 times the maximum value
(C) Acceleration	(r) 1/4 times the maximum value
	(s) Cannot say anything

3. In spring-block system, match the following columns.

Column I	Column II
(A) If k (the spring constant) is made 4 times	(p) Speed will become 16 times
(B) If m (the mass of block) is made 4 times	(q) Potential energy will become 4 times
(C) If k and m both are made 4 times	(r) Kinetic energy will remain unchanged
	(s) None

4. In $y = A \sin \omega t + A \sin \left(\omega t + \frac{2\pi}{3}\right)$, match the following.

Column I	Column II
(A) Motion	(p) is periodic but not SHM
(B) Amplitude	(q) is SHM
(C) Initial phase	(r) A
(D) Maximum velocity	(s) $\frac{\pi}{3}$
	(t) $\omega A/2$
	(u) None

5. In SHM, match the following columns.

Column I	Column II
(A) Acceleration-displacement graph	(p) Parabola
(B) Velocity-acceleration graph	(q) Straight line
(C) Velocity	(r) Circle
	(s) None

6. Velocity-time graph of a particle in SHM is as shown in figure. Match the following columns.

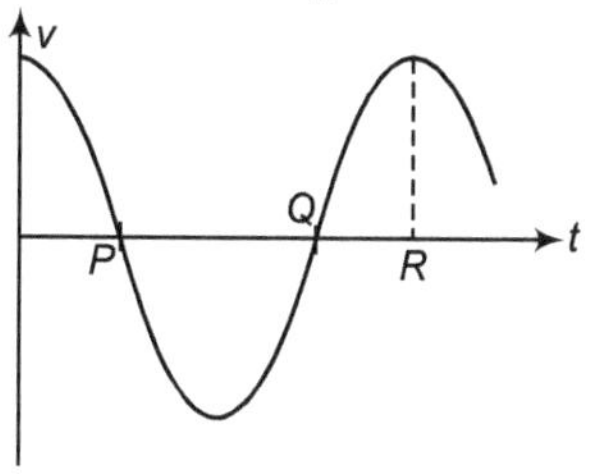

Column I	Column II
(A) At P	(p) Particle is at $x = -A$
(B) At Q	(q) Acceleration of particle is maximum
(C) At R	(r) Displacement of particle is zero
	(s) Acceleration of particle is zero
	(t) None

7. A uniform rod of length l is suspended from a point P and the rod and is made to undergo small oscillations. Match the following columns for the time period.

Column I	Column II
(A) If P is the centre of mass	(p) Zero
(B) If P is the end point	(q) $2\pi \sqrt{\frac{l}{3g}}$
(C) Length of simple pendulum having the time period equal to that of the rod when P is end point	(r) $\frac{l}{3}$
	(s) $\frac{2l}{3}$
	(t) None

8. In the two block-spring system, force constant of spring is $k = 6\,\text{N/m}$. Spring is stretched by 12 cm and then left. Match the following columns.

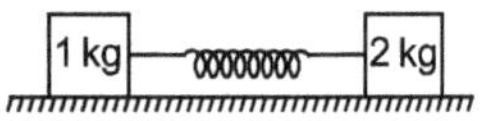

Column I	Column II
(A) Angular frequency of oscillation	(p) 4.8×10^{-3} SI unit
(B) Maximum kinetic energy of 1 kg	(q) 3 SI unit
(C) Maximum kinetic energy of 2 kg	(r) 2.4×10^{-3} SI unit
	(s) None

9. In case of second's pendulum, match the following columns. (Consider shape of earth also).

Column I	Column II
(A) At pole	(p) $T > 2$ s
(B) On a satellite	(q) $T < 2$ s
(C) At mountain	(r) $T = 2$ s
(D) At centre of earth	(s) $T = 0$
	(t) $T = \infty$

10. *F-x* and *x-t* graph of a particle in SHM are as shown in figure. Match the following columns.

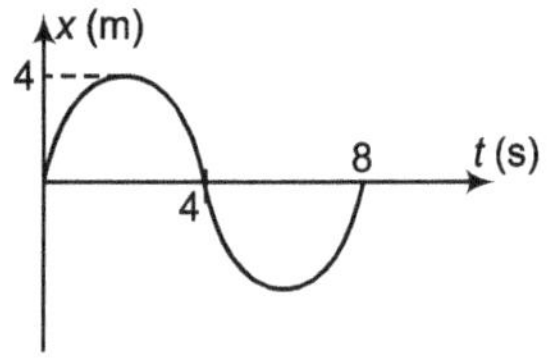

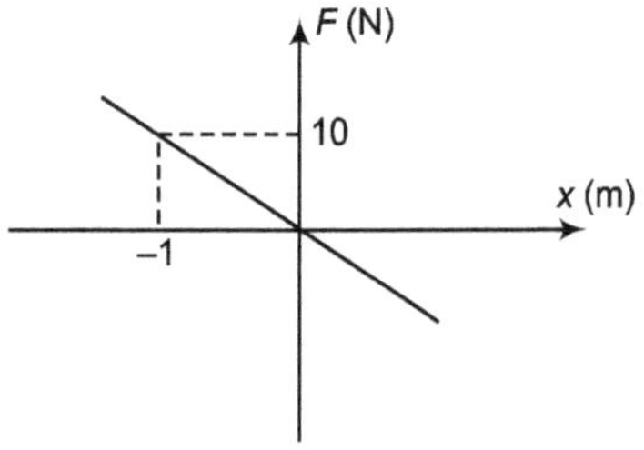

Column I	Column II
(A) Mass of the particle	(p) $\pi/2$ SI unit
(B) Maximum kinetic energy of particle	(q) $(160/\pi^2)$ SI unit
(C) Angular frequency of particle	(r) (8.0×10^{-3}) SI unit
	(s) None

11. *x-t* equation of a particle in SHM is given as $x = 1.0 \sin(12\pi t)$ in SI units. Potential energy at mean position is zero. Mass of particle is $\frac{1}{4}$ kg. Match the following columns.

Column I	Column II
(A) Frequency with which kinetic energy oscillates	(p) 1/2 SI unit
(B) Speed of particle is maximum at time *t*	(q) $18\pi^2$ SI unit
(C) Maximum potential energy	(r) 12 SI unit
(D) Force constant *k*	(s) $36\pi^2$ SI unit

Entrance Gallery

2014

1. A particle moves with simple harmonic motion in a straight line. In first τ sec, after starting from rest it travels a distance a and in next τ sec, it travels $2a$, in same direction, then **[JEE Main]**

(a) amplitude of motion is $3a$
(b) time period of oscillations is 8π
(c) amplitude of motion is $4a$
(d) time period of oscillations is 6π

2. A 10 kg metal block is attached to a spring of spring constant $1000\,\text{Nm}^{-1}$. A block is displaced from equilibrium position by 10 cm and released. The maximum acceleration of the block is **[Karnataka CET]**

(a) $10\ \text{ms}^{-2}$ (b) $100\ \text{ms}^{-2}$ (c) $200\ \text{ms}^{-2}$ (d) $0.1\ \text{ms}^{-2}$

3. The displacement of a particle in a periodic motion is given by $y = 4\cos^2\left(\frac{t}{2}\right)\sin(1000t)$. This displacement may be considered as the result of superposition of n independent harmonic oscillations. Here, n is **[WB JEE]**

(a) 1 (b) 2 (c) 3 (d) 4

4. When a particle executing SHM oscillates with a frequency ν, then the kinetic energy of particle **[WB JEE]**

(a) changes periodically with a frequency of ν
(b) changes periodically with a frequency of 2ν
(c) changes periodically with a frequency of $\nu/2$
(d) remains constant

5. A spring balance has a scale that reads from 0 to 60 kg. The length of the scale is 30 cm. A body suspended from this balance and when displaced are released, oscillates with a period of 0.8 s, what is the weight of the body when oscillating? **[J&K CET]**

(a) 350.67 N (b) 540.11 N
(c) 311.24 N (d) 300.5 N

2013

6. The amplitude of a damped oscillator decreases to 0.9 times its original magnitude is 5 s. In another 10 s, it will decrease to α times its original magnitude, where α equals to **[JEE Main]**

(a) 0.7 (b) 0.81 (c) 0.729 (d) 0.6

2012

7. A small block is connected to one end of a massless spring of unstretched length 4.9 m. The other end of the spring (see the figure) is fixed. The system lies on a horizontal frictionless surface. The block is stretched by 0.2 m and released from rest at $t = 0$. Then, it executes simple harmonic motion with angular frequency $\omega = \pi/3$ rad/s. Simultaneously at $t = 0$, a small pebble is projected with speed v from point P at an angle of 45° as shown in the figure. Point P is at a horizontal distance of 10 m from point O. If the pebble hits the block at $t = 1$ s, the value of v is (Take, $g = 10\text{ m/s}^2$) **[JEE Main]**

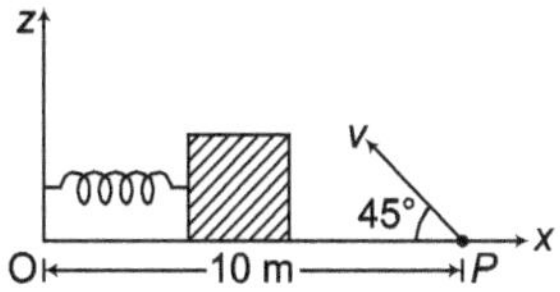

(a) $\sqrt{50}$ m/s (b) $\sqrt{51}$ m/s (c) $\sqrt{52}$ m/s (d) $\sqrt{53}$ m/s

8. This question has Statement I and Statement II. Of the four choices given after the statements, choose the one that best describes the two statements.

If two springs S_1 and S_2 of force constants k_1 and k_2, respectively are stretched by the same force, it is found that more work is done on the spring S_1 than on spring S_2.

Statement I If stretched by the same amount, work done on S_1 will be more than that on S_2.

Statement II $k_1 < k_2$ **[AIEEE]**

(a) Statement I is incorrect, Statement II is correct
(b) Statement I correct, Statement II is incorrect
(c) Statement I is correct, Statement II is correct, Statement II is the correct explanation for Statement I
(d) Statement I is correct, Statement II is correct, Statement II is incorrect explanation of Statement I.

9. If a simple pendulum has significant amplitude (up to a factor of $1/e$ of original) only in the period between $t = 0$ s to $t = \tau$ s, then τ may be called the average life of the pendulum. When the spherical bob of the pendulum suffers a retardation (due to viscous drag) proportional to its velocity with b as the constant of proportionality, the average life time of the pendulum is (assuming damping is small) in sec **[AIEEE]**

(a) $\frac{0.693}{b}$ (b) b (c) $\frac{1}{b}$ (d) $\frac{2}{b}$

10. For a body performing SHM, at a distance $A/\sqrt{2}$, then correct relation between KE and PE will be **[O JEE]**

(a) KE is equal to PE (b) KE is 2 times of PE
(c) KE is 3 times of PE (d) KE is half of PE

11. When a spring is stretched by 10 cm, the potential energy stored is E. When the spring is stretched by 10 cm more, the potential energy stored in the spring becomes **[WB JEE]**

(a) $2E$ (b) $4E$
(c) $6E$ (d) $10E$

2011

12. A metal rod of length L and mass m is pivoted at one end. A thin disc of mass M and radius R ($< L$) is attached at its centre to the free end of the rod. Consider two ways the disc is attached. **Case *A*-**the disc is not free to rotate about its centre. **Case *B*-**the disc is free to rotate about its centre. The rod disc system performs SHM in vertical plane after being released from the same displaced position. Which of the following statement(s) is/are correct? **[IIT JEE]**

(a) Restoring torque in case A = Restoring torque in case B
(b) Restoring torque in case A < Restoring torque in case B
(c) Angular frequency for case A < Angular frequency for case B
(d) Angular frequency for case A < Angular frequency for case B

13. The x-t graph of a particle undergoing simple harmonic motion is shown below. The acceleration of the particle at $t = \frac{4}{3}$ s is **[JEE Main]**

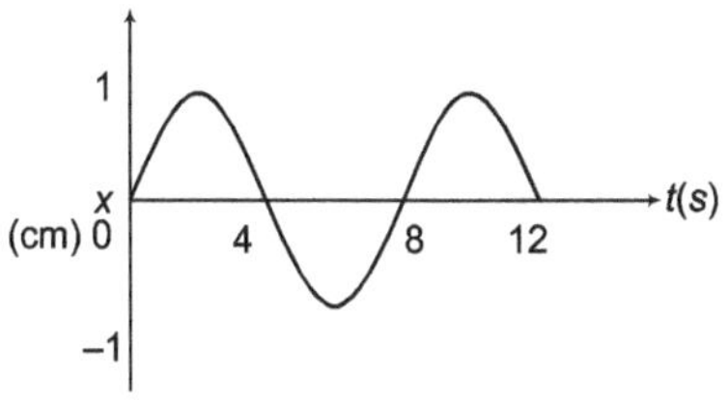

(a) $\frac{\sqrt{3}}{32}\pi^2\text{ cms}^{-2}$ (b) $-\frac{\pi^2}{32}\text{ cms}^{-2}$
(c) $\frac{\pi^2}{32}\text{ cms}^{-2}$ (d) $-\frac{\sqrt{3}}{32}\pi^2\text{ cms}^{-2}$

14. A mass M, attached to a horizontal spring, executes SHM with amplitude A_1. When the mass M passes through its mean position, then a smaller mass m is placed over it and both of them move together with amplitude A_2. The ratio of $\left(\frac{A_1}{A_2}\right)$ is **[AIEEE]**

(a) $\frac{M+m}{M}$ (b) $\left(\frac{M}{M+m}\right)^{1/2}$
(c) $\left(\frac{M+m}{M}\right)^{1/2}$ (d) $\frac{M}{M+m}$

15. Two particles are executing simple harmonic motion of the same amplitude A and frequency ω along the X-axis. Their mean position is separated by distance x_0 ($x_0 > A$). If the maximum separation between them is $(x_0 + A)$, the phase difference between their motions is **[AIEEE]**

(a) $\frac{\pi}{3}$ (b) $\frac{\pi}{4}$
(c) $\frac{\pi}{6}$ (d) $\frac{\pi}{2}$

16. A wooden cube (density of wood d) of side l floats in a liquid of density ρ with its upper and lower surfaces horizontal. If the cube is pushed slightly down and released, it performs simple harmonic motion of period T, then T is equal **[AIEEE]**

(a) $2\pi\sqrt{\frac{l\rho}{(\rho-d)g}}$ (b) $2\pi\sqrt{\frac{ld}{\rho g}}$

(c) $2\pi\sqrt{\frac{l\rho}{dg}}$ (d) $2\pi\sqrt{\frac{ld}{(\rho-d)g}}$

17. A body of mass 4 kg hangs from a spring and oscillates with a period of 0.5 s on the removal of the body, the spring is shortened by **[Kerala CEE]**

(a) 6.2 cm (b) 0.63 cm (c) 6.25 cm
(d) 6.3 cm (e) 0.625 cm

18. The amplitude of a damped oscillator becomes 1/3rd in 2 s. If its amplitude after 6 s is $1/n$ times the original amplitude, the value of n is **[Kerala CEE]**

(a) 3^2 (b) $3\sqrt{2}$ (c) $3\sqrt{3}$
(d) 2^3 (e) 3^3

19. A particle executing a simple harmonic motion has a period of 6 s. The time taken by the particle to move from the mean position to half the amplitude starting from the mean position is **[Karnataka CET]**

(a) $\frac{1}{4}$ s (b) $\frac{3}{4}$ s

(c) $\frac{1}{2}$ s (d) $\frac{3}{2}$ s

20. The displacement of a particle in SHM various according to the relation $x = 4(\cos\pi t + \sin\pi t)$. The amplitude of the particle is **[Karnataka CET]**

(a) -4 (b) 4
(c) $4\sqrt{2}$ (d) 8

21. If two springs A and B with spring constants $2k$ and k, are stretched separately by same suspended weight, then the ratio between the work done in stretching A and B is **[Karnataka CET]**

(a) 1 : 2 (c) 1 : 4 (b) 1 : 3 (d) 4 : 1

22. A body is vibrating in simple harmonic motion. If its acceleration is 12 cms^{-2} at a displacement 3 cm, then time period is **[O JEE]**

(a) 6.28 s (b) 3.14 s (c) 1.57 s (d) 2.57 s

23. A particle of mass m is located in a one-dimensional potential field where potential energy is given by $V(x) = A(1-\cos px)$, where A and p are constants. The period of small oscillations of the particle is **[WB JEE]**

(a) $2\pi\sqrt{\frac{m}{Ap}}$ (b) $2\pi\sqrt{\frac{m}{Ap^2}}$

(c) $2\pi\sqrt{\frac{m}{A}}$ (d) $\frac{1}{2\pi}\sqrt{\frac{AR}{m}}$

24. Two identical springs are connected to mass m as shown, (k = spring constant).If the period of the configuration in Fig. (i) is 2 s, the period of the configuration in Fig. (ii) is **[WB JEE]**

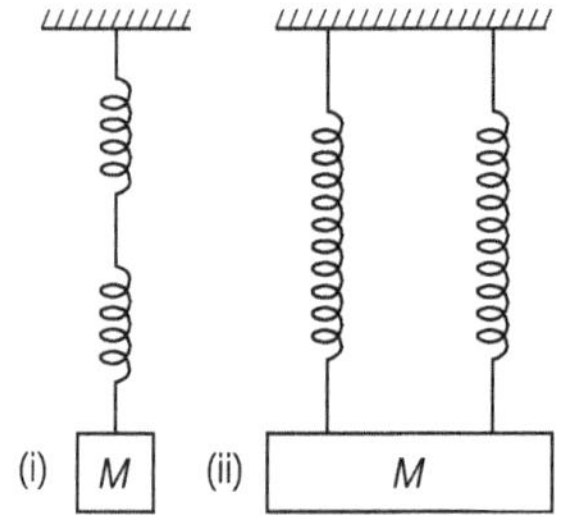

(a) $2\sqrt{2}$ s (b) 1s (c) $\frac{1}{\sqrt{2}}$ s (d) $\sqrt{2}$ s

25. The period of oscillation of a simple pendulum of length l suspended from the roof of a vehicle, which moves without friction down an inclined plane of inclination α is given by **[WB JEE]**

(a) $2\pi\sqrt{\frac{l}{g\cos\alpha}}$ (b) $2\pi\sqrt{\frac{l}{g\sin\alpha}}$

(c) $2\pi\sqrt{\frac{l}{g}}$ (d) $2\pi\sqrt{\frac{l}{g\tan\alpha}}$

26. For a particle in SHM, if the amplitude of the displacement is a and the amplitude of velocity is v, the amplitude of acceleration is **[MHT CET]**

(a) va (b) $\frac{v^2}{a}$

(c) $\frac{v^2}{2a}$ (d) $\frac{v}{a}$

27. The average acceleration of a particle performing SHM over one complete oscillation is **[MHT CET]**

(a) $\frac{\omega^2 A}{2}$ (b) $\frac{\omega^2 A}{\sqrt{2}}$

(c) zero (d) $A\omega^2$

28. U is of an oscillating particle and F is the force acting on it at a given instant. Which of the following is correct? **[MHT CET]**

(a) $\frac{U}{F} + x = 0$ (b) $\frac{2U}{F} + x = 0$

(c) $\frac{F}{U} + x = 0$ (d) $\frac{F}{2U} + x = 0$

29. A particle performing SHM has time period $\frac{2\pi}{\sqrt{3}}$ and path length 4 cm. The displacement from mean position at which acceleration is equal to velocity is **[MHT CET]**

(a) zero
(b) 0.5 cm
(c) 1 cm
(d) 1.5 cm

30. If a simple pendulum oscillates with an amplitude of 50 mm and time period of 2 s, then its maximum velocity is [MHT CET]

(a) 0.10 ms^{-1} (b) 0.15 ms^{-1} (c) 0.8 ms^{-1} (d) 0.26 ms^{-1}

31. Ratio of kinetic energy at mean position to potential energy at $A/2$ of a particle performing SHM is [MHT CET]

(a) 2 : 1 (b) 4 : 1 (c) 8 : 1 (d) 1 : 1

2010

32. Two simple harmonic motions are represented by, $y_1 = 5[\sin 2\pi t + \sqrt{3}\cos 2\pi t]$ and $y_2 = 5\sin\left(2\pi t + \frac{\pi}{4}\right)$

The ratio of their amplitudes is [Karnataka CET]

(a) 1 : 1 (b) 2 : 1
(c) 1 : 3 (d) $\sqrt{3}$: 1

Answers

Level 1

Objective Problems

1. (d) **2.** (b) **3.** (b) **4.** (b) **5.** (d) **6.** (c) **7.** (d) **8.** (c) **9.** (c) **10.** (b)
11. (c) **12.** (a) **13.** (c) **14.** (c) **15.** (b) **16.** (b) **17.** (d) **18.** (a) **19.** (c) **20.** (b)
21. (c) **22.** (a) **23.** (b) **24.** (b) **25.** (a) **26.** (c) **27.** (b) **28.** (b) **29.** (c) **30.** (b)
31. (c) **32.** (b) **33.** (b) **34.** (c) **35.** (d) **36.** (d) **37.** (b) **38.** (c) **39.** (a) **40.** (b)
41. (a) **42.** (b) **43.** (c) **44.** (c) **45.** (c) **46.** (a) **47.** (b) **48.** (c) **49.** (c) **50.** (a)
51. (d) **52.** (a) **53.** (b) **54.** (d) **55.** (d) **56.** (a) **57.** (a) **58.** (d) **59.** (a) **60.** (c)
61. (b) **62.** (b) **63.** (d) **64.** (d) **65.** (a) **66.** (d) **67.** (d) **68.** (d) **69.** (a) **70.** (d)
71. (a) **72.** (a) **73.** (b) **74.** (c) **75.** (b) **76.** (b) **77.** (b) **78.** (b) **79.** (c) **80.** (c)
81. (b) **82.** (b) **83.** (d) **84.** (a) **85.** (c) **86.** (c) **87.** (b) **88.** (b) **89.** (d) **90.** (c)
91. (c) **92.** (a) **93.** (d) **94.** (c) **95.** (b) **96.** (d) **97.** (c) **98.** (d) **99.** (d) **100.** (c)
101. (b) **102.** (c)

Level 2

Only One Correct Option

1. (d) **2.** (b) **3.** (b) **4.** (c) **5.** (b) **6.** (c) **7.** (b,c) **8.** (a) **9.** (b) **10.** (b,d)
11. (a) **12.** (a,b,c) **13.** (d) **14.** (c) **15.** (a) **16.** (c) **17.** (b) **18.** (b) **19.** (c) **20.** (d)
21. (a) **22.** (c) **23.** (c) **24.** (b) **25.** (a) **26.** (a) **27.** (c) **28.** (c) **29.** (c) **30.** (c)
31. (b) **32.** (b) **33.** (d) **34.** (b) **35.** (b)

More than One Correct Options

1. (a,c) **2.** (b,d) **3.** (b,c,d) **4.** (a,b,c) **5.** (b,c) **6.** (b,c) **7.** (b,c,d) **8.** (b,c) **9.** (a,b,c) **10.** (a,b,c)
11. (a,b,d) **12.** (a,c,d)

Comprehension Based Questions

1. (a) **2.** (c)

Assertion and Reason

1. (a) **2.** (b) **3.** (b) **4.** (a) **5.** (a) **6.** (d) **7.** (d) **8.** (d) **9.** (d) **10.** (d)
11. (e) **12.** (a) **13.** (a) **14.** (c) **15.** (c) **16.** (c) **17.** (d) **18.** (d) **19.** (a,b) **20.** (b)

Match the Columns

1. (A→q, B→r, C→q) **2.** (A→q, B→s, C→p) **3.** (A→q, B→r, C→s) **4.** (A→q, B→r, C→s, D→u)
5. (A→q, B→s, C→s, D→s) **6.** (A→q, B→p,q, C→r,s) **7.** (A→t, B→t, C→s) **8.** (A→q, B→s, C→p)
9. (A→q, B→t, C→p, D→q) **10.** (A→q, B→s, C→s) **11.** (A→r, B→p, C→q, D→s)

Entrance Gallery

1. (d) **2.** (a) **3.** (c) **4.** (b) **5.** (c) **6.** (c) **7.** (a) **8.** (a) **9.** (d) **10.** (b)
11. (b) **12.** (d) **13.** (d) **14.** (c) **15.** (a) **16.** (b) **17.** (a) **18.** (e) **19.** (c) **20.** (c)
21. (a) **22.** (b) **23.** (b) **24.** (b) **25.** (a) **26.** (b) **27.** (c) **28.** (b) **29.** (c) **30.** (b)
31. (b) **32.** (b)

Solutions

Level 1 : Objective Problems

1. $$\omega_1 A_1 = \omega_2 A_2$$

or $$\frac{A_1}{A_2} = \frac{\omega_2}{\omega_1} = \sqrt{\frac{k_2}{k_1}}$$

as $$\omega = \sqrt{\frac{k}{m}}$$

2. $\omega A = 3\,\text{m/s}$ or $\omega = \frac{3}{A} = \frac{3}{0.1} = 30\,\text{rad/s}$

$$f = \frac{\omega}{2\pi} = \frac{30}{2\pi} = \frac{15}{\pi}\,\text{Hz}$$

3. There is no effect of phase angle on amplitude. Hence,

$$\frac{A_1}{A_2} = \frac{10}{5} = \frac{2}{1}$$

4. $\omega = \frac{\pi}{2}\,\text{rad/s}$ and $A = 2\,\text{cm}$

$\therefore$ Maximum acceleration $= \omega^2 A = \frac{\pi^2}{2}\,\text{cm/s}^2$

5. $y = a\sin(\omega t + \phi) = a\sin\left(\frac{2\pi}{T}t + \phi\right)$

$\Rightarrow$ $$\psi = 0.5\,\sigma\iota\nu\left(\frac{2\pi}{0.4}\tau + \frac{\pi}{2}\right)$$

$$y = 0.5\sin\left(5\pi t + \frac{\pi}{2}\right)$$

$$= 0.5\cos 5\pi t$$

6. $y = a\sin\frac{2\pi}{T}t$

$\Rightarrow$ $$\frac{a}{\sqrt{2}} = a\sin\frac{2\pi}{T}\cdot t$$

$\Rightarrow$ $$\sin\frac{2\pi}{T}t = \frac{1}{\sqrt{2}} = \sin\frac{\pi}{4}$$

$\Rightarrow$ $$\frac{2\pi}{T}t = \frac{\pi}{4}$$

$\Rightarrow$ $$t = \frac{T}{8}$$

7. $y = A\sin\omega t \Rightarrow \frac{A}{2} = A\sin\frac{2\pi t}{T}$

$\Rightarrow$ $$\tau = \frac{T}{12}$$

8. Equation of motion, $y = a\cos\omega t$

$\Rightarrow$ $$\frac{a}{2} = a\cos\omega t$$

$\Rightarrow$ $$\omega t = \frac{\pi}{3}$$

$\Rightarrow$ $$\frac{2\pi t}{T} = \frac{\pi}{3}$$

$\Rightarrow$ $$t = \frac{\frac{\pi}{3}\times T}{2\pi}$$

$$= \frac{4}{3\times 2} = \frac{2}{3}\,\text{s}$$

9. $v_1 = \frac{dy_1}{dt} = 0.1\times 100\pi\cos\left(100\pi t + \frac{\pi}{3}\right)$

$$v_2 = \frac{dy_2}{dt} = -0.1\,\pi\sin\pi t$$

$$= 0.1\,\pi\cos\left(\pi t + \frac{\pi}{2}\right)$$

Phase difference of velocity of first particle with respect to the velocity of 2nd particle at $t = 0$ is

$$\Delta\phi = \phi_1 - \phi_2 = \frac{\pi}{3} - \frac{\pi}{2} = -\frac{\pi}{6}$$

10. $a = \omega^2 y \Rightarrow \omega = \sqrt{\frac{a}{y}} = \sqrt{\frac{8}{2}} = 2\,\text{rad/s}$

Now, $$v_{max} = A\omega = 6\times 2 = 12\,\text{cm/s}$$

11. $v_{max} = a\omega \Rightarrow \omega = \frac{v_{max}}{a} = \frac{10}{4}$

Now, $$v = \omega\sqrt{a^2 - y^2}$$

$\Rightarrow$ $$v^2 = \omega^2(a^2 - y^2)$$

$\Rightarrow$ $$y^2 = a^2 - \frac{v^2}{\omega^2}$$

$\Rightarrow$ $$y = \sqrt{a^2 - \frac{v^2}{\omega^2}}$$

$$= \sqrt{4^2 - \frac{5^2}{(10/4)^2}} = 2\sqrt{3}\,\text{cm}$$

12. $v_{max} = a\omega = 3\times 100 = 300$ units

13. Velocity in mean position $v = a\omega$, velocity at a distance of half amplitude

$$v' = \omega\sqrt{a^2 - y^2} = \omega\sqrt{a^2 - \frac{a^2}{4}}$$

$$= \frac{\sqrt{3}}{2}a\omega = \frac{\sqrt{3}}{2}v$$

14. $\frac{1}{k} = \frac{1}{k_1} + \frac{1}{k_2}$

15. $k \propto \frac{1}{l}$

16. $$|a| = \omega^2 x$$

$$12 = \omega^2(3)$$

$\therefore$ $$\omega = 2\,\text{rad/s} = \frac{2\pi}{T}$$

$\therefore$ $$T = \pi\ \text{second}$$

or 3.14 s

17. $\omega = \frac{2\pi}{T} = \frac{\pi}{4}$ s. Therefore, $y = A\sin\frac{\pi}{4}t$. Now, put $x = 1$ s and then $x = 2$ s.

18. $$\omega^2 A = a_0 \quad \text{...(i)}$$

$$\omega A = v_0 \quad \text{...(ii)}$$

Solving we get, $$A = \frac{v_0^2}{a_0}$$

19. $v_{max} = \omega A = (200)\,(0.25) = 50\,cm/s$

20. $\omega = 100\pi$

$\therefore \quad T = \frac{2\pi}{\omega} = 0.02\,s$

21. $v = \omega\sqrt{A^2 - x^2}$

$= 2\sqrt{(60)^2 - (20)^2} = 113\,mm/s$

22. $\omega = \sqrt{\frac{k}{m}} = \sqrt{\frac{16}{1}}$

$= 4\,rad/s$

Now, $K_{max} = \frac{1}{2}m\omega^2A^2$

$= \frac{1}{2}\times 1\times(4)^2\times(5\times10^{-2})^2$

$= 2\times10^{-2}\,J$

23. Energy in SHM, $E = U_{mean} + \frac{1}{2}m\omega^2A^2$

i.e. E depends on ω and A both.

24. Frequency of oscillation of kinetic energy is doubled, i.e. 2ω

$\therefore \quad T = \frac{2\pi}{2\omega} = \frac{\pi}{\omega}$

25. $E = \frac{1}{2}m\omega^2A^2 = \frac{1}{2}m(2\pi f)^2A^2$

$= 2\pi^2f^2mA^2$

$= 2\pi^2(20)^2\,(0.1)\,(0.05)^2$

$= 2\,J$ (approximately)

26. $U = K$ or $U = \frac{E}{2}$

$\therefore \quad \frac{1}{2}Kx^2 = \frac{1}{2}\left(\frac{1}{2}KA^2\right)$

i.e. at $x = \pm\frac{A}{\sqrt{2}}, U = K$

and this situation will occur for four times in one complete period.

27. If at any instant displacement is y, then it is given that

$U = \frac{1}{2}\times E$

$\Rightarrow \quad \frac{1}{2}m\omega^2y^2 = \frac{1}{2}\times\left(\frac{1}{2}m\omega^2a^2\right)$

$\Rightarrow \quad y = \frac{a}{\sqrt{2}} = \frac{6}{\sqrt{2}}$

$= 4.2\,cm$

28. $a = 6\,cm, \omega = 100\,rad/s$

$K_{max} = \frac{1}{2}m\omega^2a^2$

$= \frac{1}{2}\times 1\times(100)^2\times(6\times10^{-2})^2$

$= 18\,J$

30. In SHM at mean position, kinetic energy will be maximum and potential energy will be minimum. Total energy is always constant.

31. Total energy $= \frac{1}{2}m\omega^2a^2 =$ constant

32. $a = 6\,cm, \omega = 100\,rad/s$

$K_{max} = \frac{1}{2}m\omega^2a^2$

$= \frac{1}{2}\times 1\times(100)^2\times(6\times10^{-2})^2 = 18\,J$

33. $y = y_1 + y_2 = \sqrt{2}A\sin\left(\omega t + \frac{\pi}{4}\right)$

$E = \frac{1}{2}m\omega^2(\sqrt{2}A)^2 = m\omega^2A^2$

34. $KE = \frac{1}{2}K(A^2 - x^2) \quad (a = A)$

Put $\frac{1}{2}KA^2 = E$ and $KE = \frac{3E}{4}$

35. $T = 2\pi\sqrt{\frac{l}{g}}$ and $T' = 2\pi\sqrt{\frac{l}{\left(g + \frac{g}{4}\right)}}$

or $T' = \frac{2}{\sqrt{5}}T = 2\,s$

36. $T = 2\pi\sqrt{\frac{m}{\rho Ag}}$

37. $T = 2\pi\sqrt{\frac{l}{g_e}}$. Here, g_e = effective value of g, which is zero in a freely falling lift. Hence, $T = \infty$

38. $f = \frac{1}{2\pi}\sqrt{\frac{k}{m}}$ or $f \propto \sqrt{k}$

$k_i = k_1 + k_2$

and $k_f = 4(k_1 + k_2) = 4k_i$

39. $f = \frac{1}{2\pi}\sqrt{\frac{k}{m}}$ or $f \propto \sqrt{\frac{k}{m}}$

Here, value of k_e will become $3k$.

$\therefore \quad f \propto \sqrt{\frac{3k}{m}}$

40. $g \propto \frac{M}{R^2}$, mass and radius both are doubled. Hence, g will become 1/2 times.

Now, $T \propto \frac{1}{\sqrt{g}}$ and time period of second's pendulum is 2 s on earth. Hence, time period of second's pendulum on this planet will become $2\sqrt{2}$ s.

41. $k \propto \frac{1}{l}$ and $f \propto \sqrt{k}$

$\therefore \quad f \propto \frac{1}{\sqrt{l}}$

42. $T = 2\pi\sqrt{\frac{m}{k}} = 2\pi\sqrt{\frac{0.2}{80}} = \frac{\pi}{10}\,s$

43. In case of spring-block system, time period does not change because restoring force is still kx.

44. Velocity, $v = \omega\sqrt{A^2 - x^2}$ and acceleration $= \omega^2x$

Now given, $\omega^2x = \omega\sqrt{A^2 - x^2}$

$\Rightarrow \quad \omega^2\cdot 1 = \omega\sqrt{2^2 - 1^2}$

$\Rightarrow \quad \omega = \sqrt{3}$

$\therefore \quad T = \frac{2\pi}{\omega} = \frac{2\pi}{\sqrt{3}}$

45. On comparing with standard equation $\frac{d^2y}{dt^2}+\omega^2 y=0$,

We get $\omega^2=K \Rightarrow \omega=\frac{2\pi}{T}=\sqrt{K}$

$\Rightarrow \quad T=\frac{2\pi}{\sqrt{K}}$

46. Comparing given equation with standard equation, $x=a\cos(\omega t+\phi)$ we get, $a=0.01$ and $\omega=\pi$

$\Rightarrow \quad 2\pi n=\pi \Rightarrow n=0.5\,\text{Hz}$

47. When a little mercury is drained off, the position of centre of mass of ball falls (with respect to fixed end) so that effective length of pendulum increases hence T increases.

48. In stationary lift, $T=2\pi\sqrt{\frac{l}{g}}$

In upward moving lift,

$$T'=2\pi\sqrt{\frac{l}{(g+a)}}$$

(a = acceleration of lift)

$$\Rightarrow \frac{T'}{T}=\sqrt{\frac{g}{g+a}}=\sqrt{\frac{g}{\left(g+\frac{g}{4}\right)}}=\sqrt{\frac{4}{5}}$$

$$\Rightarrow T'=\frac{2T}{\sqrt{5}}$$

49. Remains the same because time period of simple pendulum T is independent of mass of the bob.

50. When lift falls freely, effective acceleration and frequency of oscillations will be zero.

$$g_{\text{eff}}=0 \Rightarrow T'=\infty,$$

Hence, frequency = 0

51. $A=0.04\,\text{m}$, $\omega^2 A=0.3\,\text{m/s}^2$

$\therefore \quad \omega=2.74\,\text{rad/s}$

Now, $T=\frac{2\pi}{\omega}=2.29.\,\text{s}$

52. $T=2\pi\sqrt{\frac{m}{k}}$ or $T\propto\sqrt{m}$

Mass has reduced to $\frac{1}{4}$th the initial mass. Hence, time period will remain half.

54. $2\pi\sqrt{\frac{I}{MgL}}=2\pi\sqrt{\frac{l}{g}}$

or $l=\frac{I}{ML}=\frac{(3/2\,MR^2)}{MR}=\frac{3}{2}R$

55. $g'=\sqrt{g^2+a^2}$

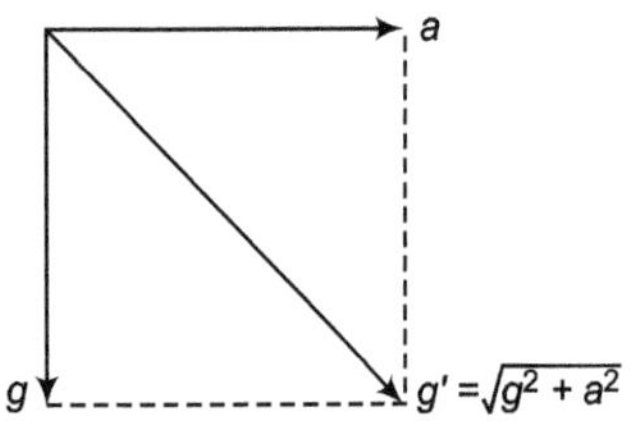

56. $T_a=2\pi\sqrt{\frac{m}{k}}$

$$T_b=2\pi\sqrt{\frac{m}{(k/2)}}$$

$$T_c=2\pi\sqrt{\frac{m}{2k}}$$

57. $k\propto\frac{1}{l} \Rightarrow \frac{l_1}{l_2}=n$

$\therefore \quad l_1=\left(\frac{n}{n+1}\right)l$

or $\quad k'=\left(\frac{n+1}{n}\right)k$

$\therefore \quad T=2\pi\sqrt{\frac{mn}{k(n+1)}}$

58. $\omega=\sqrt{\frac{k}{m_2}}$, here $m_1=m_2$

59. $k_s=\frac{k\times k}{k+k}=\frac{k}{2}$

$$k_p=k+k=2k$$

$$\therefore \frac{n_s}{n_p}=\frac{\frac{1}{2\pi}\sqrt{\frac{k_s}{M}}}{\frac{1}{2\pi}\sqrt{\frac{k_p}{M}}}=\sqrt{\frac{k_s}{k_p}}=\sqrt{\frac{k/2}{2k}}=\frac{1}{2}$$

60. $k=\frac{YA}{L}$

$\therefore \quad f=\frac{1}{2\pi}\sqrt{\frac{k}{m}}=\frac{1}{2\pi}\sqrt{\frac{YA}{mL}}$

61. $T=2\pi\sqrt{\frac{I}{mgl/2}}=2\pi\sqrt{\frac{(ml^2/3)}{(mgl/2)}}$

$$=2\pi\sqrt{\frac{2l}{3g}}$$

62. $x=A\sin\omega t+B\cos\omega t$

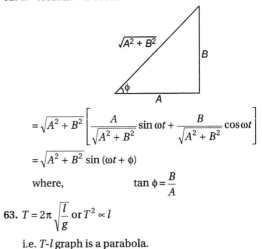

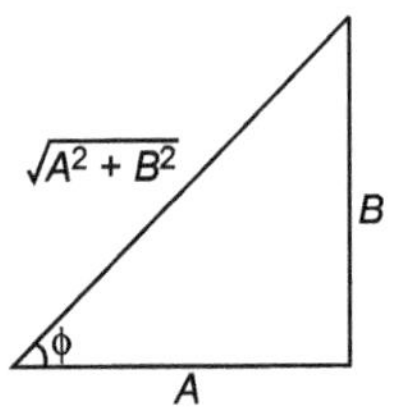

$$=\sqrt{A^2+B^2}\left[\frac{A}{\sqrt{A^2+B^2}}\sin\omega t+\frac{B}{\sqrt{A^2+B^2}}\cos\omega t\right]$$

$$=\sqrt{A^2+B^2}\sin(\omega t+\phi)$$

where, $\tan\phi=\frac{B}{A}$

63. $T=2\pi\sqrt{\frac{l}{g}}$ or $T^2\propto l$

i.e. T-l graph is a parabola.

arc = sin θ

64. Acceleration is maximum when displacement is maximum. Force is zero when displacement is zero and potential energy is maximum when displacement is maximum.

65. $x = 3\sin 2t + 4\cos 2t$. From given equation,

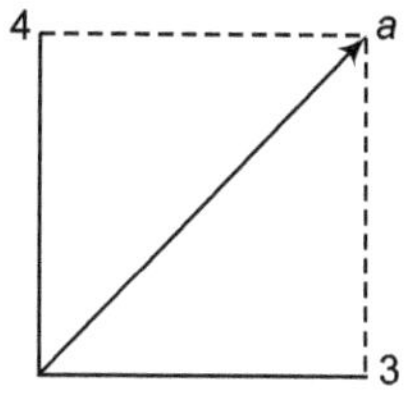

$$a_1 = 3, a_2 = 4$$

and $\phi = \frac{\pi}{2}$

$\therefore$ $a = \sqrt{a_1^2 + a_2^2}$

$= \sqrt{3^2 + 4^2} = 5$

$\Rightarrow$ $v_{max} = a\omega = 5 \times 2 = 10$

66. Using acceleration, $A = -\omega^2 x$

At $-x_{max}$, A will be maximum and positive.

67. Acceleration $= -\omega^2 y$. So, $F = -m\omega^2 y$

y is sinusoidal function.

So, F will also be sinusoidal function with phase difference π.

68. At time $\frac{T}{2}; v = 0$

$\therefore$ Total energy = Potential energy

69. $f = \frac{1}{T} = \frac{1}{0.04} = 25\,\text{Hz}$

70. From graph, slope $K = \frac{F}{x} = \frac{8}{2} = 4$

$$T = 2\pi\sqrt{\frac{m}{k}}$$

$\Rightarrow$ $T = 2\pi\sqrt{\frac{0.01}{4}} = 0.31\,\text{s}$

71. Suppose $x = A\sin\omega t$ and $y = A\cos\omega t$

Then, by squaring and adding these two equations we get

$$x^2 + y^2 = A^2$$

72. $x = a\sin\omega t$ or $\sin\omega t = \frac{x}{a}$ ...(i)

$$y = b\sin\left(\omega t + \frac{\pi}{2}\right) \quad \text{or} \quad y = b\cos\omega t$$

or $\cos\omega t = \frac{y}{b}$ ...(ii)

Squaring and adding Eqs. (i) and (ii), we get

$\frac{x^2}{a^2} + \frac{y^2}{b^2} = 1$ which is the equation of ellipse.

73. $y_1 = a\cos\omega t = a\sin\left(\omega t + \frac{\pi}{2}\right)$

$$y_2 = a\sin\omega t$$

$\therefore$ $\Delta\phi = \omega t + \frac{\pi}{2} - \omega t$

$= \frac{\pi}{2} = 90°$

74. Potential energy graph is parabolic with its minimum value at mean position.

75. Force constant, $k \propto \frac{1}{\text{Length of spring}}$

$\therefore$ $\frac{k_A}{k} = \frac{L}{\alpha L}$

$\therefore$ $k_A = \frac{k}{\alpha}$

Similarly, $k_B = \frac{k}{1-\alpha}$

76. $k \propto \frac{1}{l}$

77. $k \propto \frac{1}{l}$ hence, k of individual halves will be 16 N/m. When they are connected in parallel, effective value of k will become 32 N/m.

78. $\frac{1}{k_e} = \frac{1}{k} + \frac{1}{k}$ or $k_e = \frac{k}{2}$

79. Reduced mass of two blocks,

$$\mu = \frac{m_1 m_2}{m_1 + m_2} = 2\,\text{kg}$$

Now, $f = \frac{1}{2\pi}\sqrt{\frac{k}{\mu}}$

$= \frac{1}{2\pi}\sqrt{\frac{300}{2}}$

$\approx 2\,\text{Hz}$

80. From mean position to extreme position body is decelerated and from extreme position to mean position body is accelerated.

81. $T = 2\pi\sqrt{\frac{m}{k}}$

$\therefore$ Dimensionally $\frac{m}{k}$ is equal to T^2.

82. $T = 2\pi\sqrt{\frac{l}{g}}$. For T to be constant value of $\frac{l}{g}$ should remain constant. So, if $g_{moon} = \frac{g_{earth}}{6}$, then $l_{moon} = \frac{l_{earth}}{6}$.

83. At equilibrium, $x = \frac{mg}{k} = \frac{0.2 \times 10}{200} = 0.01\text{ m} = 1\text{ cm}$

$$f = \frac{1}{2\pi}\sqrt{\frac{k}{m}}$$

$= \frac{1}{2\pi}\sqrt{\frac{200}{0.2}} \approx 5.0\,\text{Hz}$

84. $\phi_1 = \frac{\pi}{2}$

and $\phi_2 = \pi$

85. **a** and **s** are always in opposite directions.

86. After attaining terminal velocity, ball will move with constant velocity.

87. $a = -\omega^2 x$

$\therefore$ $\frac{x}{a} = -\frac{1}{\omega^2} = \text{constant}$

88. $A = \frac{mg}{k} = \frac{5 \times 10}{200} = 0.25\,\text{m}$

89. At $x=0, F=0$. For $x>0$, force is in negative direction and for $x<0$, force is in positive x-direction. Therefore, motion of the particle is periodic about mean position $x=0$.

92. The amplitude is a maximum displacement from the mean position.

93. $F=-kx$

94. $a=-\omega^2 x \Rightarrow \left|\frac{a}{x}\right|=\omega^2$

95. Ball will execute SHM inside the tunnel with time period

$$T=2\pi\sqrt{\frac{R}{g}}=84.63\,\text{min}$$

Hence, time to reach the ball from one end to the other end of the tunnel, $t=\frac{84.63}{2}=42.3\,\text{min}$.

98. $T=2\pi\sqrt{\frac{l}{g}} \Rightarrow \frac{\lambda}{T^2}=\frac{\gamma}{4\pi^2}=\text{constant}$

99. Frequency, $n\propto\frac{1}{\sqrt{l}} \Rightarrow \frac{n_1}{n_2}=\sqrt{\frac{l_2}{l_1}}$

$$\Rightarrow \qquad \frac{l_2}{l_1}=\frac{n_2^2}{n_1^2}=\frac{3^2}{2^2}=\frac{9}{4}$$

100. $T=2\pi\sqrt{\frac{m}{K}}$ or $T\propto\sqrt{m}$

Mass is increasing therefore time period will increase.

101. $X=A\sin\omega t$ or $10=20\sin\omega t$

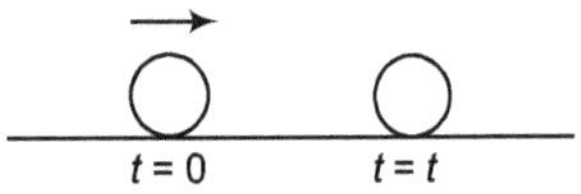

$$\therefore \qquad \omega t=\left(\frac{\pi}{6}\right)$$

$$t=\frac{\pi}{6\omega}$$

The desired time will be $2t$.

or $$\frac{\pi}{3\omega}=\frac{\pi}{3(2\pi/T)}=\frac{T}{6}=\frac{12}{6}=2\,\text{s}$$

102. $\omega A=\text{constant}$ and $\omega^2 A$ is made two times.

This is possible when ω is doubled and A is halved.

Level 2 : Only One Correct Option

1. From graph, slope $k=\frac{F}{x}=\frac{8}{2}=4$

$$T=2\pi\sqrt{\frac{m}{k}} \Rightarrow T=2\pi\sqrt{\frac{0.1}{4}}=0.31\,\text{s}$$

2. Comparing with $\frac{d^2x}{dt^2}=-\omega^2 x$

We have $\omega=2\,\text{rad/s}$

$$v_{\max}=\omega A=2\,\text{m/s}$$

$$a_{\max}=\omega^2 A=4\,\text{m/s}^2$$

3. $T_1=2\pi\sqrt{\frac{m}{k+\rho Ag}}$ but $T_2=2\pi\sqrt{\frac{m}{k}}$.

Hence, $\qquad T_1<T_2$

4. At $x=0, U=0$. Therefore, total mechanical energy is equal to the maximum kinetic energy.

$$\therefore \qquad \frac{1}{2}mv_{\max}^2=36$$

or $$v_{\max}=\sqrt{\frac{72}{m}}=\sqrt{\frac{72}{2}}=6\,\text{m/s}$$

5. $\frac{1}{2}kA^2=(0.04-0.01)\,\text{J}=0.03\,\text{J}$

$$\therefore \qquad k=\frac{0.06}{A^2}=\frac{0.06}{(0.02)^2}=150\,\text{N/m}$$

6. $\frac{1}{2}m\omega^2A^2=8\times10^{-3}\,\text{J}$

$$\therefore \qquad \frac{1}{2}\times0.1\times\omega^2\times(0.1)^2=8\times10^{-3}$$

or $$\omega=4\,\text{rad/s}$$

7. $M_e=\text{Mass of container}+\frac{\text{Upthrust}}{g}=M+\frac{\rho Vg}{g}=M+\rho V$

8. $\frac{T}{3}$ means 120° in reference circle. Maximum possible value of v is near mean position between P and Q.

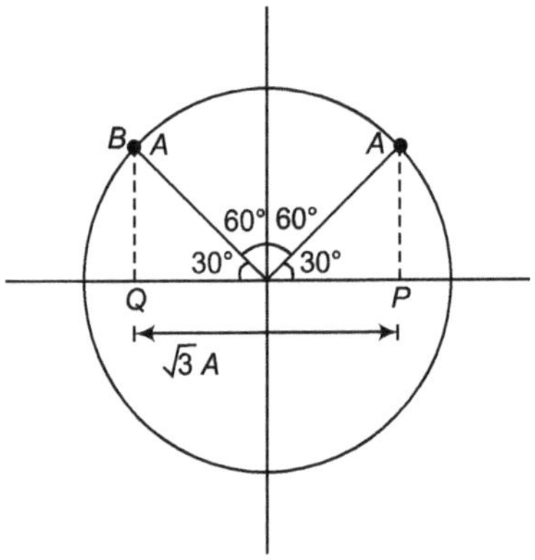

$$\therefore \qquad v=\frac{\text{Distance}}{\text{Time}}$$

$$=\frac{\sqrt{3}A}{(T/3)}=\frac{3\sqrt{3}A}{T}$$

9. $\therefore \quad l'=\sqrt{2}\,l$

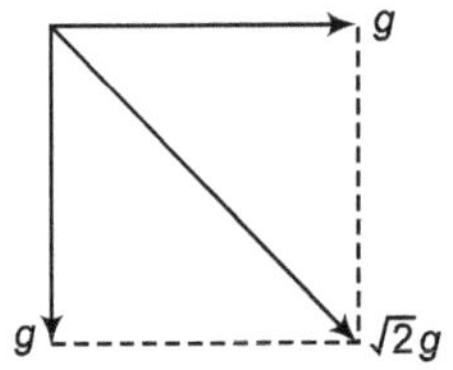

10. The motion is simple harmonic only inside earth. Further

$$\frac{1}{2}mv^2=-\frac{GMm}{(R+h)}+\frac{3}{2}\frac{GMm}{R}$$

$$v=\sqrt{\frac{GM(R+3h)}{R(R+h)}}$$

11. Since, the cube is half immersed. The density of cube should be half the density of water, i.e. $500\,\text{kg/m}^3$.

$$T = 2\pi\sqrt{\frac{m}{\rho_w A g}} \qquad (k = \rho_w A g)$$

$$\frac{\pi}{5} = 2\pi\sqrt{\frac{a^3 \times \rho}{\rho_w \times a^2 \times g}}$$

$\therefore$ $$\frac{a\rho}{g\rho_w} = \frac{1}{100}$$

$\therefore$ $$\frac{a}{20} = \frac{1}{100}$$

or $$a = (0.2)\text{ m}$$

Now, $$m = a^3\rho = 4\text{ kg}$$

12. Maximum kinetic energy = energy of oscillation in SHM

$\therefore$ $$8 = \frac{1}{2}kA^2$$

$\therefore$ $$kA^2 = 16 \qquad \text{...(i)}$$

Further, $$2\pi\sqrt{\frac{m}{k}} = 4$$

$\therefore$ $$\frac{1}{k} = \frac{4}{\pi^2}$$

or $$k = \frac{\pi^2}{4} \qquad \text{...(ii)}$$

From Eqs. (i) and (ii), we get

$$k \approx 2.4\text{ N/m} \quad \text{and} \quad A \approx 2.53\text{ m}$$

Maximum acceleration of the particle will be

$$a_{max} = \omega^2 A = \frac{k}{m}A$$

$$= \left(\frac{2.5}{1}\right)(2.53) \approx 6.3\text{ m/s}^2$$

$$= \frac{k}{m}\cdot A = \frac{2.5}{1} \times 2.53$$

$$= 6.3\text{ m/s}^2$$

13. $U = 5x^2 - 20x, \dfrac{dU}{dx} = 10x - 20,$

$$F = -\frac{dU}{dx} = (-10x + 20)$$

Let us suppose, $x = (X + 2)$. Then, $F = -10X$, $F = 0$ at $X = 0$ or $x = 2$, i.e. $x = 2$ m is the mean position about which particle is in SHM. $k = 10$

$\therefore$ $$T = 2\pi\sqrt{\frac{m}{k}} = 2\pi\sqrt{\frac{0.1}{10}} = \frac{\pi}{5}\text{ s.}$$

At mean position kinetic energy is maximum and force acting on particle in SHM is variable.

14. $v = \omega\sqrt{A^2 - x^2}$ (in SHM)

or $v^2 = \omega^2 A^2 - \omega^2 x^2$

Comparing the given equation with this equation, we get

$$\omega^2 = 9$$

$\therefore$ $$\omega = 3 = \frac{2\pi}{T}$$

$\therefore$ $$T = \frac{2\pi}{3}\text{ units}, \omega^2 A^2 = 144$$

In SHM,

$\therefore$ $$A = 4\text{ units}, |a| = \omega^2 x = (9)(3) = 27\text{ units}$$

Displacement $\le$ distance

15. $$y_1 = A\sin\left(\omega_1 t + \frac{\pi}{2}\right)$$

$$y_2 = A\sin\left(\omega_2 t + \frac{\pi}{2}\right)$$

Now, $\omega_1 t + \dfrac{\pi}{2} = \omega_2 t - \dfrac{\pi}{2}$

or $$t = \frac{\pi}{\omega_2 - \omega_1}$$

$$= \frac{\pi}{(2\pi/T_2) - (2\pi/T_1)}$$

$$= \frac{T_1 T_2}{2(T_1 - T_2)}$$

$$= \frac{3 \times 7}{2(7-3)} = \frac{21}{8}\text{ s}$$

16. $T = 2\pi\sqrt{\dfrac{m}{k}}$

or $$T \propto \frac{1}{\sqrt{k}}$$

T has increased $\dfrac{4}{3}$ times. Hence, $k' = \dfrac{9}{16}k$

When both are combined,

$$k_{net} = k + k' = \frac{25}{16}k$$

or $$\frac{25}{16}\text{ times}$$

Hence, new time period will become 4/5 times of 3 s or 2.4 s.

17.

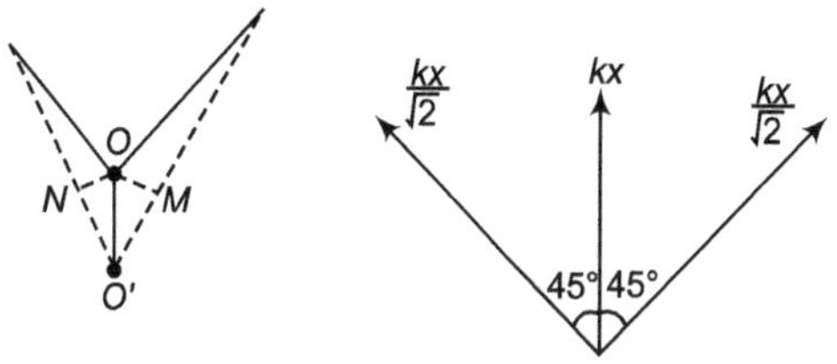

$$OO' = x\text{ (say)}$$

Then, $$O'M = O'N \approx \frac{x}{\sqrt{2}}$$

i.e. elongation in spring B and C is $\dfrac{x}{\sqrt{2}}$, while compression in spring is x.

Net restoring force,

$$F = -\left[kx + \frac{2kx}{\sqrt{2}}\cos 45°\right] = -2kx$$

$\therefore$ $$a = \frac{F}{m} = -\frac{2k}{m}x$$

$$T = 2\pi\sqrt{\left|\frac{x}{a}\right|} = 2\pi\sqrt{\frac{m}{2k}}$$

18. $U = U_{mean} + \dfrac{1}{2}kX^2$

Given, $U_{mean} = E_1$ and $\dfrac{1}{2}kA^2 = E_2$ = maximum kinetic energy at mean position.

$\therefore$ $$X = \frac{\sqrt{3}A}{2}, U = E_1 + \frac{1}{2}k\left(\frac{\sqrt{3}A}{2}\right)^2$$

$$= E_1 + \frac{3}{4}E_2$$

19. Particle starts from rest. Hence, $x = A\cos\omega t$

$$a = A - A\cos(\omega \times 1)$$

or $$\cos\omega = \frac{A-a}{A} = \left(1 - \frac{a}{A}\right)$$

$$a + b = A - A\cos(\omega \times 2)$$
$$= A - A(2\cos^2\omega - 1)$$
$$= 2A - 2A\left(1 - \frac{a}{A}\right)^2$$

Solving we get, $$A = \frac{2a^2}{3a-b}$$

20. $T = 2\pi\sqrt{\frac{m}{k}} = 2\pi\sqrt{\frac{0.1}{100}} = 0.198\text{ s}$

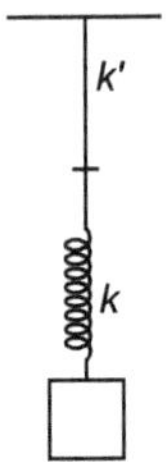

From $x = A\sin\omega t$ or $5 = 10\sin\omega t$

$$\omega t = \frac{\pi}{6}$$

or $$\frac{2\pi}{T}t = \frac{\pi}{6}$$

or $$2t = \frac{T}{6}$$

The desired time in the question will be $\frac{T}{2} + 2t$

or $$\frac{T}{2} + \frac{T}{6} = \frac{2T}{3}$$
$$= \left(\frac{2}{3}\right)(0.198) = 0.132\text{ s}$$

21. From e to $2e$ or $\frac{A}{2}$ to A, $t = \frac{T}{6}$

$\therefore$ Time period of oscillation $= 2t = \frac{T}{3} = \frac{2\pi}{3}\sqrt{\frac{m}{k}}$

22. $\mu g >$ maximum acceleration in SHM $> \omega^2 A$

$\therefore$ $$\mu > \frac{\omega^2 A}{g}$$

23. At its lowermost point spring is stretched by 8 cm or 8×10^{-2} m.

$\therefore$ $$F_{\text{net}} = kx - mg$$
$$= (2 \times 10^3 \times 8 \times 10^{-2}) - (10 \times 10) = 60\text{ N}$$

24. $x_{12} = \frac{12 \times 10}{1000} = 0.12\text{ m}$ $\qquad \left(x = \frac{mg}{k}\right)$

$$x_4 = \frac{4 \times 10}{1000} = 0.04\text{ m}$$

$\therefore$ $$A = x_{12} - x_4 = 0.08\text{ m}$$

25. $\mathbf{g}_e = \mathbf{g} - \mathbf{a}$

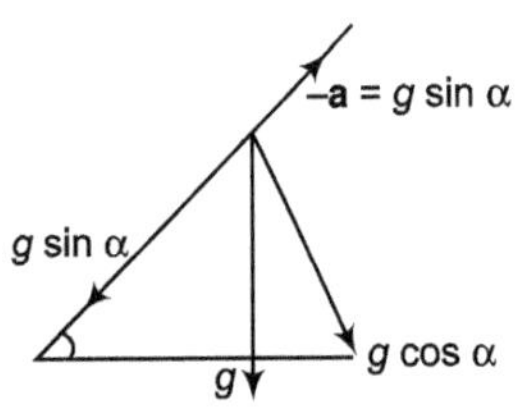

where, $\mathbf{a}$ = acceleration of roof

$$|\mathbf{g}_e| = g\cos\alpha$$

$\therefore$ $$T = 2\pi\sqrt{\frac{L}{g\cos\alpha}}$$

26. $$\omega^2 A = g$$
$$\omega = \sqrt{\frac{g}{A}} \quad \text{or} \quad T = 2\pi\sqrt{\frac{A}{g}}$$

27. $\frac{1}{2}mv^2 = \frac{1}{2}kx^2$

$\therefore$ $x = v\sqrt{\frac{m}{k}}$

28. $k' = \frac{YA}{L}$

$$k_e = \frac{kk'}{k+k'}$$
$$T = 2\pi\sqrt{\frac{m}{k_e}}$$

29. $(M)(\omega A) = (M+m)v$ $\qquad (P_i = P_f)$

$\therefore$ $$v = \frac{M\omega A}{(M+m)} = \omega' A'$$

$\therefore$ $$A' = \left(\frac{M}{M+m}\right)\cdot\frac{\omega}{\omega'}$$

Now, $$\omega = \sqrt{\frac{k}{m}}$$

or $$\propto \frac{1}{\sqrt{m}}$$

$\therefore$ $$\frac{\omega}{\omega'} = \sqrt{\frac{M+m}{M}}$$

$\therefore$ $$A' = \left(\sqrt{\frac{M}{m+M}}\right)A$$

30. $\frac{1}{2}kA^2 = \frac{1}{2} \times 2 \times 10^6 \times (0.01)^2 = 100\text{ J}$

At mean position

$U = 60\text{ J}$

and $K = 100\text{ J}$

At extreme positions

$U = 160\text{ J}$

and $K = 0$

31. $T = 2\pi\sqrt{\frac{m}{k}}$ or $T \propto \frac{1}{\sqrt{k}}$ or $k = \frac{\alpha}{T^2}$

Now, $$k_e = k_1 + k_2$$

$\therefore$ $$\frac{\alpha}{T^2} = \frac{\alpha}{T_1^2} + \frac{\alpha}{T_2^2}$$

or $$T^{-2} = T_1^{-2} + T_2^{-2}$$

32.

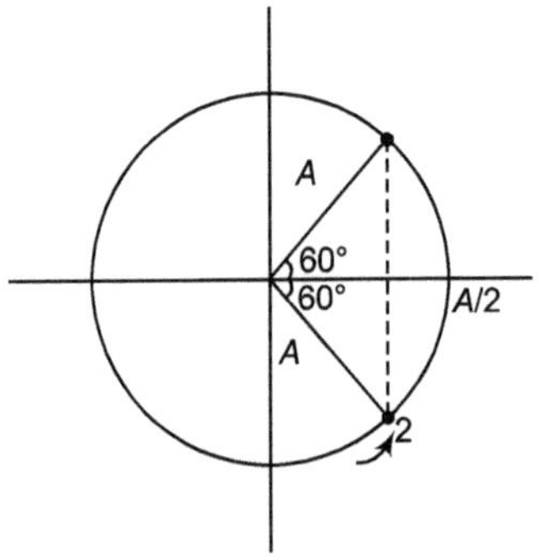

33. $\frac{1}{2}kA^2 = 1.0\,J$

$\therefore \quad k = \frac{2}{A^2} = \frac{2}{(0.4)^2} = \frac{25}{2}\,N/m$

$$T = 2\pi\sqrt{\frac{m}{k}} = 2\pi\sqrt{\frac{2}{(25/2)}}$$

$$= \frac{4\pi}{5}\,s$$

34. For y_2,

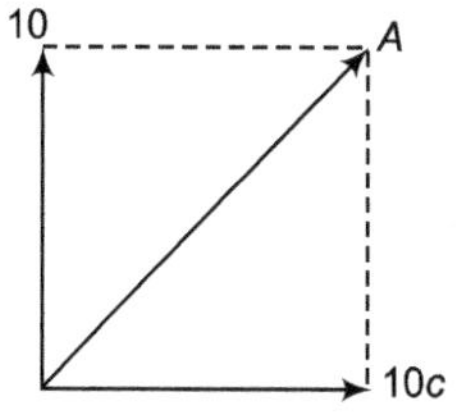

$$A = \sqrt{(10)^2 + (10c)^2} \quad \text{or} \quad 10\sqrt{1+c^2} = 40$$

$\therefore \quad c = \sqrt{15}$

35. $mg = kx$

$\therefore \quad k = \frac{mg}{x}$

$\therefore \quad T = 2\pi\sqrt{\frac{M+m}{k}} = 2\pi\sqrt{\frac{(M+m)x}{mg}}$

More than One Correct Options

1. $\tan\theta = \frac{ma}{mg} = \frac{a}{g}$

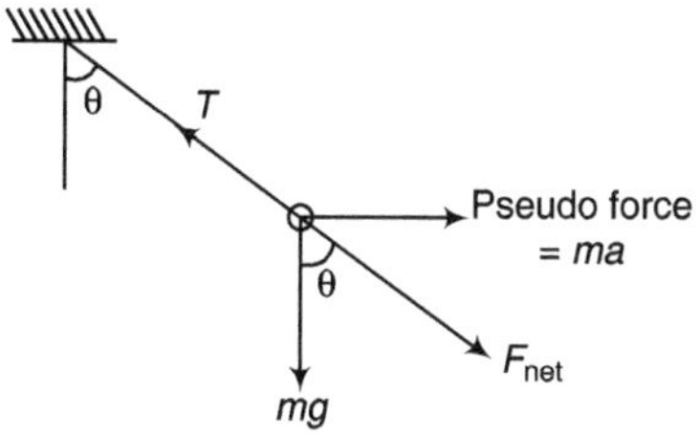

$$\theta = \tan^{-1}\left(\frac{a}{g}\right)$$

$$T = F_{net}$$

$$= m\sqrt{a^2 + g^2}$$

2. $X = A\sin(\omega t + 150°)$

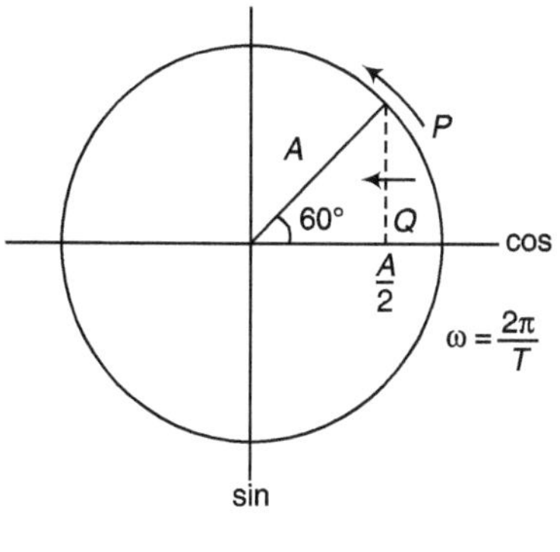

$$= A\sin\left(\frac{2\pi}{T}t + \frac{5\pi}{6}\right)$$

or $\quad X = A\cos(\omega t + 60°)$

$$= A\cos\left(\frac{2\pi}{T}t + \frac{\pi}{3}\right)$$

3. $kx_0 = mg$

$\therefore \quad x_0 = \frac{mg}{k} = \frac{1\times 10}{500}$

$$= 0.02\text{ m} = 2\text{ cm}$$

So, equilibrium is obtained after an extension of 2 cm of at a length of 42 cm. But, it is released from a length of 45 cm.

$\therefore \quad A = 3\text{ cm} = 0.03\text{ m}$

(b) $v_{max} = \omega A = \sqrt{\frac{k}{m}}A$

$$= \left(\sqrt{\frac{500}{1}}\right)(0.03) = 0.3\sqrt{5}\text{ m/s}$$

$$= 30\sqrt{5}\text{ cm/s}$$

(c) $a_{max} = \omega^2 A$

$$= \left(\frac{k}{m}\right)(A) = \left(\frac{500}{1}\right)(0.03) = 15\text{ m/s}^2$$

(d) Mean position is at 42 cm length and amplitude is 3 cm. Hence, block oscillates between 45 cm length and 39 cm. Natural length 40 cm lies in between these two, where elastic potential energy = 0.

4. μ = Reduced mass

$$= \frac{m_1 m_2}{m_1 + m_2} = 2\text{ kg}$$

$$T = 2\pi\sqrt{\frac{\mu}{k}} = 2\pi\sqrt{\frac{2}{800}}$$

$$= \frac{\pi}{10}\,s = T_3 = T_6$$

$$A \propto \frac{1}{m}$$

$\therefore \quad \frac{A_3}{A_6} = \frac{6}{3} = \frac{2}{1}$

$\therefore \quad A_3 = 4\text{ cm} \quad \text{and} \quad A_6 = 2\text{ cm}$

$$(P_6)_{max} = m_6 (v_6)_{max}$$

$$= m_6(\omega A_6)$$

$$= (6)\left(\frac{2\pi}{\pi/10}\right)(2\times 10^{-2})$$

$$= 2.4\text{ kg/m}$$

5. v or KE = 0 at $y = \pm A$

v or KE = maximum at $y = 0$

F or a is maximum at $y = \pm A$

F or a is zero at $y = 0$

6. $v = \omega\sqrt{A^2 - x^2}$

$\therefore \quad \frac{v^2}{\omega^2} + \frac{X^2}{(1)^2} = A^2$

i.e. v - x graph is an ellipse.

$$a = -\omega^2 x$$

i.e. a - x graph is a straight line passing through origin with negative slope.

7. $a = 0$ at $x = 0.5$ m and particle is released from $x = 2$ m

Hence,

$$A = 2 - 0.5 = 1.5 \text{ m}$$
$$\omega^2 = 100$$
$$\therefore \quad \omega = 10 \text{ rad/s}$$
$$T = \frac{2\pi}{\omega} = \frac{2\pi}{10} = 0.63 \text{ s}$$
$$v_{max} = \omega A = (10)(1.5) = 15 \text{ m/s}$$

8. Two particles shown in figure are in same phase, although they have distances from the mean position at $t = 0$

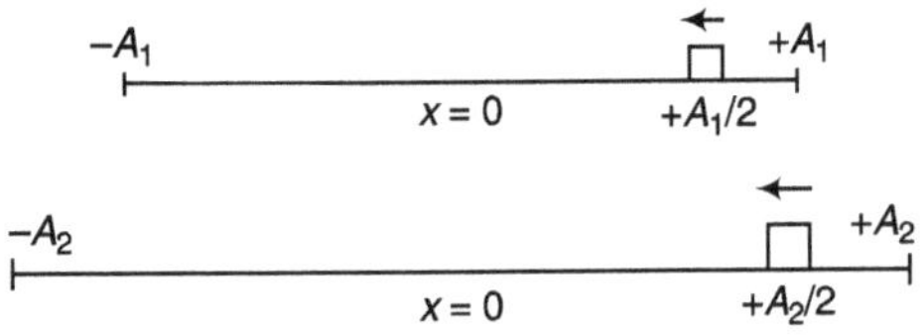

9. The given equation can be written as

$$y = 3\sin 100\,\pi t + (4 - 4\cos 100\,\pi t) - 6$$
$$= 3\sin 100\,\pi t + 4\sin(100\,\pi t + \pi/2) - 2$$
$$\text{or} \quad y = 5\sin(100\,\pi - 53°) - 2$$

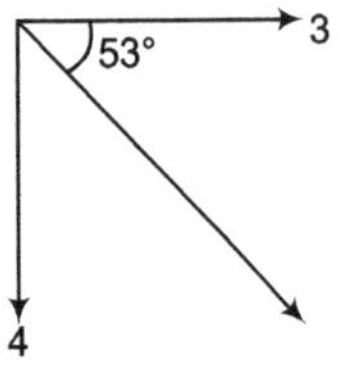

$$y_{max} = 5 - 2 = 3$$
$$y_{min} = -5 - 2 = -7$$
$$\text{Mean position} = \frac{y_{max} + y_{min}}{2} = -2 \text{ cm}$$

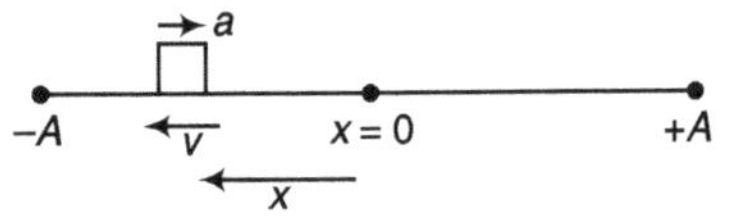

$\therefore$ Velocity is negative.

(d) At mean position, velocity is maximum. Just after few seconds acceleration becomes positive. So, displacement will become negative. Hence, the velocity should be negative.

10. Consider the diagram

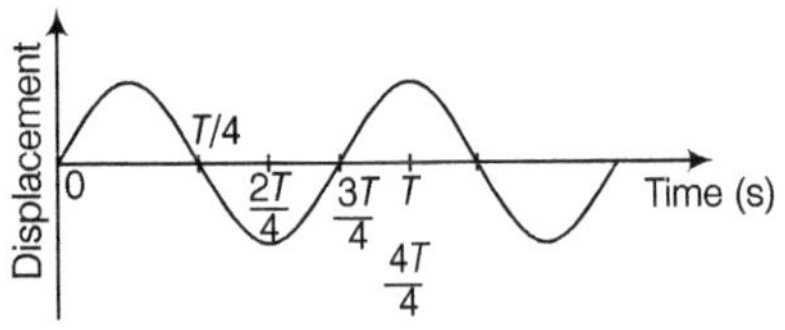

From the given diagram, it is clear that

(a) At $t = \frac{3T}{4}$, the displacement of the particle is zero.

Hence, the particle executing SHM will be at mean position, i.e. $x = 0$. So, acceleration is zero and force is also zero.

(b) At $t = \frac{4T}{4}$, displacement is maximum, i.e. extreme position, so acceleration is maximum.

(c) At $t = \frac{T}{4}$, corresponds to mean position, so velocity will be maximum at this position.

(d) At $t = \frac{2T}{4} = \frac{T}{2}$, corresponds to extreme position, so KE = 0 and PE = maximum.

11. Let the equation of a SHM is represented as $x = a\sin\omega t$

Assume mass of the body is m.

(a) Total mechanical energy of the body at any time t is

$$E = \frac{1}{2}m\omega^2 a^2 \quad \text{...(i)}$$

Kinetic energy at any instant t is

$$K = \frac{1}{2}mv^2 = \frac{1}{2}m\left[\frac{dx}{dt}\right]^2 \quad \left[\because v = \frac{dx}{dt}\right]$$
$$= \frac{1}{2}m\omega^2 a^2 \cos^2\omega t \quad [\because x = a\sin\omega t]$$
$$\Rightarrow \quad K_{max} = \frac{1}{2}m\omega^2 a^2 = E$$
$$[\because \text{ for } k_{max}, \cos\omega t = 1] \quad \text{...(ii)}$$

(b) KE at any instant t is

$$K = \frac{1}{2}m\omega^2 a^2 \cos^2\omega t$$
$$(K_{av}) \text{ for a cycle} = \frac{1}{2}m\omega^2 a^2 [(\cos^2\omega t)_{av}] \text{ for a cycle}$$
$$= \frac{1}{2}m\omega^2 a^2\left[\frac{0+1}{2}\right]$$
$$= \frac{1}{4}m\omega^2 a^2 = \frac{K_{max}}{2} \quad \text{[from Eq. (ii)]}$$

(c) Velocity $= v = \frac{dx}{dt} = a\omega\cos\omega t$

$$v_{mean} = \frac{v_{max} + v_{min}}{2}$$
$$= \frac{a\omega + (-a\omega)}{2} = 0 \quad \text{[For a complete cycle]}$$
$$v_{max} \neq v_{mean}$$

(d) $v_{rms} = \sqrt{\frac{v_1^2 + v_2^2}{2}} = \sqrt{\frac{0 + a^2\omega^2}{2}} = \frac{a\omega}{\sqrt{2}}$

$$\Rightarrow \quad v_{rms} = \frac{v_{max}}{\sqrt{2}}$$

12. Consider the diagram.

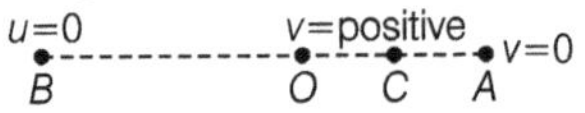

(a) When the particle is 3cm away from A going towards B, velocity is towards AB i.e. positive.

In SHM, acceleration is always towards mean position O in this case.

Hence, it is positive.

(b) When the particle is at C, velocity is towards B hence positive.

(c) When the particle is 4 cm away from B going towards A velocity is negative and acceleration is towards mean position O hence negative.

(d) Acceleration is always towards mean position O. When the particle is at B, acceleration and force are towards BA that is negative.

Comprehension Based Questions

1. $\omega = \sqrt{\frac{k}{m}} = \sqrt{\frac{400}{2}} = 10\sqrt{2}$ rad/s

2. $Kx_0 = ma$

$$x_0 = \frac{ma}{K} = \frac{2 \times 5}{400}\text{ m}$$
$$= 0.025\text{ m} = 2.5\text{ cm}$$
$$\therefore \quad A = x_0 = 2.5\text{ cm}$$

Assertion and Reason

1. $v = \omega\sqrt{A^2 - x^2}$

$$\therefore \quad \frac{v^2}{\omega^2} + \frac{x^2}{1} = A^2$$

2. $X = x - 3 = 4\cos\omega t$

$X = 0$ at $x = 3$

3. Maximum angular velocity $= \omega\theta_0$

$$= (2\pi f)\theta_0 = (2\pi)\left(\frac{1}{2\pi}\sqrt{\frac{g}{l}}\right)\theta_0 = \left(\sqrt{\frac{g}{l}}\right)\theta_0$$

4. $T = 2\pi\sqrt{\frac{l}{g_e}}$

Here, $g_e = g + a$, if the lift accelerates upwards.

6. In any equation of SHM, x is always measured from the mean position.

7. $k \propto \frac{1}{\text{length of spring}}$

If length is halved, k will become two times.

From $\quad T = 2\pi\sqrt{\frac{m}{k}}$

Time period T will remain half for the given conditions.

8. Motion of particle is simple harmonic, but mean position is at $x = 4$. Amplitude is 6. The particle will oscillate between $x = 10$ and $x = -2$.

9. Simple harmonic motion is not always one-dimensional. In case of angular simple harmonic motion it is 2-D also.

10. $t_{PQ} = T$ (given)

But t_{PQ} is not the time period of oscillation.

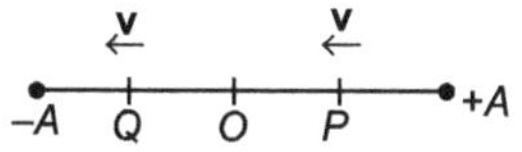

11. v *versus* x graph is an ellipse.

13. At given time x is negative. Therefore, acceleration is positive ($a \propto -x$). Further slope is negative. Therefore, velocity is negative.

14. At given time $x = 0$, i.e. body is at mean position. At mean position potential energy is always minimum. But this minimum potential energy may or may not be zero.

15. Amplitude is 4.

16. Force constant of a spring, $k \propto \frac{1}{l}$ and $T \propto \frac{1}{\sqrt{k}}$

$$\therefore \quad T \propto \sqrt{l}$$

18. Time period does not depend on the amplitude of oscillation.

19. 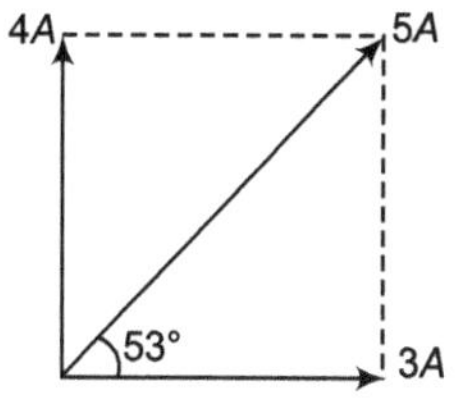

20. If $v = \omega A\sin\omega t$

Then, $\quad v_{max} = \omega A$

$$\therefore \quad K_{max} = \frac{1}{2}mv_{max}^2 = \frac{1}{2}m\omega^2A^2$$

Further, $\quad \langle K\rangle = \langle\frac{1}{2}m\omega^2A^2\sin^2\omega t\rangle$

$$= \frac{1}{2}m\omega^2A^2\langle\sin^2\omega t\rangle$$

But, $\quad \langle\sin^2\omega t\rangle = \frac{1}{2}$ in one oscillation

$$\therefore \quad \langle K\rangle = \frac{1}{4}m\omega^2A^2$$

Match the Columns

1. Suppose $\quad x = A\sin\omega t$

Then, $\quad v = \frac{dx}{dt} = \omega A\cos\omega t$

and $\quad a = \frac{dv}{dt} = -\omega^2A\sin\omega t$

2. $v = \omega\sqrt{A^2 - \frac{A^2}{4}} = \frac{\sqrt{3}}{2}\omega A = \frac{\sqrt{3}}{2}v_{max}$

$$\therefore \quad K = \left(\frac{\sqrt{3}}{2}\right)^2 \text{ times, } K_{max} = \frac{3}{4}K_{max}$$

About PE, it will be $\frac{1}{4}$ times its maximum value only when potential energy at mean position is zero.

3. $\omega = \sqrt{\frac{k}{m}}, v = \omega\sqrt{A^2 - X^2}, \text{PE} = \frac{1}{2}kx^2$

4. The given equation,

$y = A\sin\omega t + A\sin\left(\omega t + \frac{2\pi}{3}\right)$ can also be written as

$$y = 2A\sin\left(\omega t + \frac{\pi}{3}\right)\cdot\cos\left(\frac{\pi}{3}\right)$$
$$= A\sin\left(\omega t + \frac{\pi}{3}\right)$$

Now, we can see that this is SHM with amplitude A and initial phase $\frac{\pi}{3}$.

5. $a = -\omega^2x$, i.e. a-x graph is straight line passing through origin.

If $\quad v = v_0\sin\omega t$

then, $\quad a = \frac{dv}{dt} = v_0\omega\cos\omega t$

$$= v_0\omega\sqrt{1 - \sin^2\omega t}$$
$$= v_0\omega\sqrt{1 - \frac{v^2}{v_0^2}}$$
$$a = \omega\sqrt{v_0^2 - v^2}$$

So, a-v graph is neither a straight line nor a parabola. Further, acceleration and velocity time graphs are sine or cosine functions.

6. At P, velocity is zero, but just after few seconds velocity is negative. Therefore, at P particle is at $x = +A$. At this point acceleration of the particle will be maximum. On similar ground we can say that

At Q, $x = -A$

At R, v is maximum, i.e. particle is at mean position, where acceleration is also zero.

7. $T = 2\pi\sqrt{\frac{I}{mgl}}$. Here l is distance of point P from centre of mass.

When P is the centre, $l = 0$. Therefore, $T = \infty$, i.e. rod will not oscillate.

When P is end point, $I = \frac{ml^2}{3}$

$$\therefore \quad T = 2\pi\sqrt{\frac{2l}{3g}}$$

$$\text{or} \quad 2\pi\sqrt{\frac{2l}{3g}} = 2\pi\sqrt{\frac{L}{g}} \quad \text{or} \quad L = \frac{2l}{3}$$

8. $\omega = \sqrt{\frac{k}{\mu}}$

Here, μ = reduced mass

$$= \frac{m_1 m_2}{m_1 + m_2} = \frac{2}{3}\text{kg}$$

$$\therefore \quad \omega = \sqrt{\frac{6}{2/3}} = 3\,\text{rad/s}$$

Amplitude 12 cm distributes in inverse ratio of mass.

$$\therefore \quad A_1 = 8\,\text{cm and } A_2 = 4\,\text{cm}$$

Now, maximum kinetic energy

$$K_1 = \frac{1}{2} m_1 \omega A_1^2$$

$$= \frac{1}{2} \times 1 \times 3 \times (8 \times 10^{-2})^2 = 9.6\,\text{mJ}$$

$$K_2 = \frac{1}{2} \times 2 \times 3 \times (4 \times 10^{-2})^2 = 4.8\,\text{mJ}$$

9. On a satellite and at centre of earth $g' = 0$

$$\therefore \quad T = \infty$$

At pole, value of g is more than the normal value. Hence, $T < 2\,\text{s}$

10. $T = 8\,\text{s}$

$$\therefore \quad \omega = \frac{2\pi}{T} = \frac{\pi}{4}\,\text{rad/s}$$

$K = -$ slope of F-x graph $= 10\,\text{N/m}$

From, $T = 2\pi\sqrt{\frac{m}{k}}$,

We have $8 = 2\pi\sqrt{\frac{m}{10}}$

$$\therefore \quad m = \frac{160}{\pi^2}\,\text{kg}$$

$$K_{max} = \frac{1}{2}\pi\omega^2 A^2 = \frac{1}{2} \times \frac{160}{\pi^2} \times \frac{\pi^2}{16} \times 4$$

$$= 8.0 \times 10^{-3}\,\text{J}$$

Entrance Gallery

1. In SHM, a particle starts from rest, we have

i.e. $x = A\cos\omega t$, at $t = 0, x = A$

When $t = \tau$, then

$$x = A - a \quad \text{...(i)}$$

When $t = 2\tau$, then

$$x = A - 3a \quad \text{...(ii)}$$

On comparing Eqs. (i) and (ii), we get

$$A - a = A\cos\omega\tau$$

$$A - 3a = A\cos 2\omega\tau$$

As, $\cos 2\omega\tau = 2\cos^2\omega\tau - 1$

$$\Rightarrow \quad \frac{A - 3a}{A} = 2\left(\frac{A - a}{A}\right)^2 - 1$$

$$= \frac{2A^2 + 2a^2 - 4Aa - A^2}{A^2}$$

$$A^2 - 3aA = A^2 + 2a^2 - 4Aa \Rightarrow 2a^2 = aA$$

Now, $A - a = A\cos\omega\tau$ $[\because \tau = t]$

$$\Rightarrow \quad \cos\omega t = 1/2 \Rightarrow \frac{2\pi}{T}t = \frac{\pi}{3} \Rightarrow T = 6\pi$$

2. We know that spring represents SHM.

So, the restoring force is proportional to displacement

i.e. $$F = -m\omega^2 y \quad \text{...(i)}$$

$$F = -ky \quad \text{...(ii)}$$

where, k = force constant of the spring

$$m = 10\,\text{kg}, \quad k = 1000\,\text{N/m}$$

$$A = 10\,\text{cm} = 0.1\,\text{m}$$

On comparing both equations, we get

$$\omega^2 = \frac{k}{m} \Rightarrow \omega = \sqrt{\frac{k}{m}} = \sqrt{\frac{1000}{10}} = 10\,\text{rad/s}$$

and acceleration in SHM

$$a_{max} = -\omega^2 \cdot y \text{ where, } \omega^2 \text{ is constant}$$

$$= -10^2 \times (0.1) = -10\,\text{m/s}^2 = 10\,\text{ms}^{-2}$$

3. Given, $y = 4\cos^2\left(\frac{t}{2}\right) \cdot \sin(1000t)$

$$= 2 \times 2\cos^2\left(\frac{t}{2}\right) \cdot \sin(1000t) \quad \left[\because 2\cos^2\frac{t}{2} = (1 + \cos t)\right]$$

$$= 2(1 + \cos t)\sin(1000t)$$

$$= 2\sin(1000\,t) + 2\cos t \cdot \sin(1000\,t)$$

$$= 2\sin(1000\,t) + 2\sin(1000\,t) \cdot \cos t$$

$$= 2\sin(1000\,t) + \sin(1000\,t + t) + \sin(1000\,t - t)$$

$$[\because 2\sin A \cdot \cos B = \sin(A + B) + \sin(A - B)]$$

$$= 2\sin(1000\,t) + \sin(1001\,t) + \sin(999\,t).$$

So, $n = 3$

4. We know that,

Kinetic energy, $K = \frac{1}{2}mu^2$

where, $u = \frac{dy}{dt} = \omega a\cos\omega t$

So, $K = \frac{1}{2}m\omega^2 a^2\cos^2\omega t$

Hence, kinetic energy varies periodically with double the frequency of SHM. So, when a particle executing SHM oscillates with a frequency ν, then the kinetic energy of particle changes periodically with a frequency of 2ν.

5. Time period of the spring mass system,

$$T = 2\pi\sqrt{\frac{m}{k}} \quad \ldots\text{(i)}$$

From spring, $mg = kx$

$$\Rightarrow \quad k = \frac{mg}{x} = \frac{60 \times 9.8}{30 \times 10^{-2}} = 1960\ \text{N/m} \quad \ldots\text{(ii)}$$

From Eqs. (i) and (ii), we get

$$0.8 = 2\pi\sqrt{\frac{m}{1960}} \Rightarrow \frac{0.8 \times 0.8 \times 1960}{4p^2} = m$$

$$m = 31.81\ \text{kg} \Rightarrow w = mg$$

$$\therefore \quad w = 31.81 \times 9.8 = 311.24\ \text{N}$$

6. Amplitude decreases exponentially. In 5 s, it remains 0.9 times. Therefore, in total 15 s it will remains (0.9) (0.9) (0.9) = 0.729 times its original value.

7. Time of flight of projectile,

$$t = \frac{2v\sin\theta}{g} \quad (\theta = 45°)$$

$$\therefore \quad v = \frac{gt}{2\sin\theta} = \frac{10 \times 1}{2 \times 1/\sqrt{2}} = \sqrt{50}\ \text{m/s}$$

8. As, no relation between k_1 and k_2 is given in the question, that is why, nothing can be predicted about Statement I. But as in Statement II, $k_1 < k_2$.

Then, for same force

$$W = F \cdot x = F \cdot \frac{F}{k} = \frac{F^2}{k} \Rightarrow W \propto \frac{1}{k}, \text{ i.e. } W_1 > W_2$$

But for same displacement,

$$W = F \cdot x = \frac{1}{2}kx \cdot x = \frac{1}{2}kx^2 \Rightarrow W \propto k, \text{i.e. } W_1 < W_2$$

Thus, in the correct of Statement II, Statement I is incorrect.

9. For damped harmonic motion,

$$ma = -kx - mbv \quad \text{or} \quad ma + mbv + kx = 0$$

Solution to above equation is

$$x = A_0 e^{-\frac{bt}{2}} \sin(\omega t) \quad \text{with} \quad \omega^2 = \frac{k}{m} - \frac{b^2}{4}$$

where, amplitude drops exponentially with time.

i.e. $A_\tau = A_0 e^{-\frac{b\tau}{2}}$

Average time τ is that duration when amplitude drops by 63%, i.e. becomes A_0/e.

Thus, $A_\tau = \frac{A_0}{e} = A_0 e^{-\frac{b\tau}{2}}$ or $\frac{b\tau}{2} = 1$ or $\tau = \frac{2}{b}$

10. Potential energy PE of body $= \frac{1}{2}m\omega^2 y^2$

$$\text{PE at } A/\sqrt{2} = \frac{1}{2}m\omega^2\left(\frac{A}{\sqrt{2}}\right)^2 = \frac{1}{4}m\omega^2 A^2$$

Kinetic energy of a body $= \frac{1}{2}m\omega^2(A^2 - y^2)$

$$\text{KE at } A/\sqrt{2} = \frac{1}{2}m\omega^2\left[A^2 - \left(\frac{A}{\sqrt{2}}\right)^2\right] = \frac{1}{4}m\omega^2 A^2$$

11. For 1st situation, $E = \frac{1}{2}k(10 \times 10^{-2})^2 \quad \ldots\text{(i)}$

For 2nd situation, $E' = \frac{1}{2}k(20 \times 10^{-2})^2 \quad \ldots\text{(ii)}$

On solving Eqs. (i) and (ii), we get $E' = 4E$

12. $\tau_A = \tau_B = (mg\frac{L}{2}\sin\theta + MgL\sin\theta)$

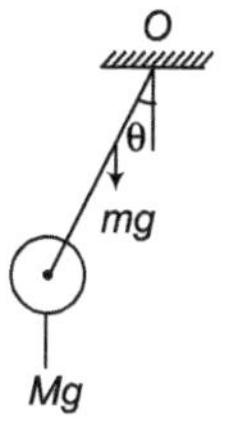

= Restoring torque about point O

In case A, moment of inertia will be more.

Hence, angular acceleration ($\alpha = \tau/I$) will be less.

Therefore, angular frequency will be less.

13. Given, $T = 8\text{s}, \omega = \frac{2\pi}{T} = \left(\frac{\pi}{4}\right)\text{rads}^{-1}$

We have, $\quad x = A\sin\omega t$

$$\therefore \quad a = -\omega^2 x = -\left(\frac{\pi^2}{16}\right)\sin\left(\frac{\pi}{4}t\right)$$

On substituting $t = \frac{4}{3}$s, we get

$$a = -\left(\frac{\sqrt{3}}{32}\pi^2\right)\text{cm s}^{-2}$$

14. At mean position, $F_{\text{net}} = 0$

$\therefore$ By conservation of linear momentum,

$$Mv_1 = (M + m)v_L$$

$$\Rightarrow \quad M\omega_1 A_1 = (M + m)\omega_2 A_2$$

$$\omega_1 = \sqrt{\frac{k}{m}} \Rightarrow \omega_2 = \sqrt{\frac{k}{M + m}}$$

On solving, $\quad \frac{A_1}{A_2} = \sqrt{\frac{M + m}{M}}$

15. $x_1 = A\sin(\omega t + \phi)$

$$x_2 = A\sin(\omega t + \phi)$$

$$\Rightarrow \quad x_2 - x_1 = A[\sin(\omega t + \phi_2) - \sin(\omega t + \phi_1)]$$

$$= 2A\cos\left(\frac{2\omega t + \phi_1 + \phi_2}{2}\right)\sin\left(\frac{\phi_2 - \phi_1}{2}\right)$$

The resultant motion can be treated as a simple harmonic motion with amplitude $2A\sin\left(\frac{\phi_2 - \phi_1}{2}\right)$.

Given, maximum distance between the particles $= x_0 + A$

$\therefore$ Amplitude of resultant SHM $= x_0 + A - x_0 = A$

$$\Rightarrow \quad 2A\sin\left(\frac{\phi_2 - \phi_1}{2}\right) = A$$

$$\phi_2 - \phi_1 = \frac{\pi}{3}$$

16. Let at any instant, cube is at a depth x from the equilibrium position, then net force acting on the cube = upthrust on the portion of length x

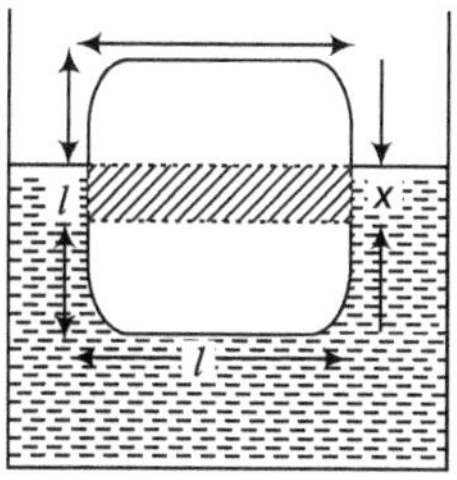

$$F = -\rho l^2 xg = -\rho l^2 gx \quad \ldots\text{(i)}$$

Negative sign shows that, force is opposite to x. Hence, equation of SHM

$$F = -kx \quad \ldots\text{(ii)}$$

On comparing Eqs. (i) and (ii), we get

$$k = \rho l^2 g$$

$$\Rightarrow \quad T = 2\pi\sqrt{\frac{m}{k}}$$

$$= 2\pi\sqrt{\frac{l^3 d}{\rho l^2 g}} = 2\pi\sqrt{\frac{ld}{\rho g}}$$

17. Time period, $T = 2\pi\sqrt{\frac{m}{k}} \Rightarrow mg = kx$

$\therefore \quad T = 2\pi\sqrt{\frac{x}{g}} \Rightarrow (0.5)^2 = 4\pi^2 \times \frac{x}{10}$

$\frac{(0.5)^2 \times 9.8}{4 \times 3.14 \times 3.14} = x \Rightarrow x = 0.0621 \text{ m} = 6.2 \text{ cm}$

18. We have $A, A_0 = e^{bt/2m}$

In this case after 6 s, amplitude becomes $\frac{1}{27}$ times.

19. Displacement, $y = A\sin\left(\frac{2\pi}{T}\right)t \quad \left(\because \omega = \frac{2\pi}{T}\right)$

$\frac{A}{2} = A\sin\left(\frac{2\pi}{T}\right)t \Rightarrow \sin^{-1}\left(\frac{1}{2}\right) = \frac{2\pi}{T}t$

$\Rightarrow \quad \frac{\pi}{6} = \frac{2\pi}{T}t \Rightarrow t = \frac{T}{12} = \frac{6}{12} = \frac{1}{2} \text{ s}$

20. The given relation is given that

$$x = 4(\cos \pi t + \sin \pi t)$$

On comparing, we get

$$R\sin\delta = 4 \text{ and } R\cos\delta = 4$$

On squaring both the equations given above and adding them, we get

$$R^2(\sin^2\delta + \cos^2\delta) = 32 \Rightarrow R = 4\sqrt{2}$$

21. Weight of spring A, $mg = 2kx_A$

Weight of spring B, $mg = kx_B$

$\therefore \quad \frac{x_A}{x_B} = \frac{1}{2}$

$\therefore \quad W = Fx \Rightarrow \frac{W_A}{W_B} = \frac{Fx_A}{Fx_B} = \frac{1}{2}$

22. Here, $a = 12 \text{ cms}^{-2}$, $x = 3 \text{ cm}$

In SHM, acceleration, $a = -\omega^2 x$

$\therefore$ Magnitude of acceleration, $a = \omega^2 x$

(discarding off negative sign)

$\therefore \quad \omega^2 = \frac{a}{x}$ or $\omega = \sqrt{\frac{a}{x}}$ or $\frac{2\pi}{T} = \omega = \sqrt{\frac{a}{x}}$

or $\quad T = 2\pi\sqrt{\frac{x}{a}} = 2\pi\sqrt{\frac{3}{12}} = \pi \text{ s} = 3.14 \text{ s}$

23. Given, $V_x = A(1 - \cos px)$

$$F = -\frac{dV}{dx} = -Ap\sin px$$

For small (x), $F = -Ap^2 x \Rightarrow a = -\frac{Ap^2}{m}x$

$$a = \omega^2 x \Rightarrow \omega = \sqrt{\frac{Ap^2}{m}}$$

$\therefore \quad T = 2\pi\sqrt{\frac{m}{Ap^2}}$

24. Time period, $T = 2\pi\sqrt{\frac{m}{k}}$

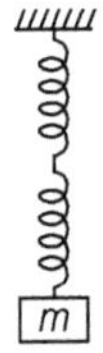

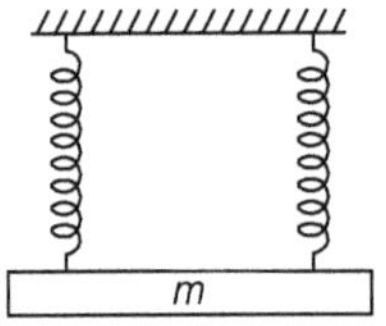

$\therefore \quad \frac{T_1}{T_2} = \sqrt{\frac{k_2}{k_1}}$

$\Rightarrow \quad \frac{2}{T} = \sqrt{\frac{2k}{\frac{k}{2}}} = 2, T = 1 \text{ s}$

25. Time period, $T = 2\pi\sqrt{\frac{l}{g_{\text{eff}}}}$

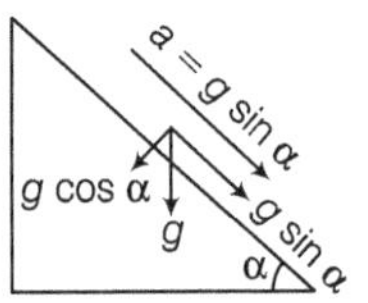

$$T = 2\pi\sqrt{\frac{l}{g\cos\alpha}}$$

26. Maximum velocity, $v_{\max} = a\omega$ and

Maximum acceleration, $= \omega^2 a = \left(\frac{v}{a}\right)^2 a = \frac{v^2}{a}$

27. The average acceleration of a particle performing SHM over one complete oscillation is zero.

28. The potential energy, $U = \frac{1}{2}kx^2$

$2U = kx^2 \Rightarrow 2U = -Fx \quad (\because F = -kx)$

or $\quad \frac{2U}{F} = -x$ or $\frac{2U}{F} + x = 0$

29. Velocity, $v = \omega\sqrt{A^2 - x^2}$

and acceleration $= \omega^2 x$

Given, $\omega\sqrt{A^2 - x^2} = \omega^2 x$ or $\sqrt{A^2 - x^2} = \omega x$

Given, $\quad T = \frac{2\pi}{\sqrt{3}}$ and $\omega = \frac{2\pi}{T} = \sqrt{3}$

Substituting the value of ω in Eq.(i), we get

$$\sqrt{A^2 - x^2} = \sqrt{3}x \Rightarrow A = 2x$$

As, amplitude $= \frac{\text{Path length}}{2} \times 2 \text{ cm} \Rightarrow x = 1 \text{ cm}$

30. Maximum velocity, $v_{\max} = a\omega = a \times \frac{2\pi}{T}$

$$= (50 \times 10^{-3}) \times \frac{2\pi}{2} = 0.15 \text{ ms}^{-1}$$

31. Kinetic energy, $\text{KE} = \frac{1}{2}m\omega^2 A^2 - y^2$

At mean position, $y = 0$

$$\text{KE} = \frac{1}{2}m\omega^2(A^2)$$

Potential energy, $U = \frac{1}{2}m\omega^2 y^2 \quad \ldots\text{(i)}$

Potential energy, U at $y = \frac{A}{2}$

$\Rightarrow \quad U = \frac{1}{2}\frac{mA^2}{4}\omega^2 \quad \ldots\text{(ii)}$

On dividing Eq. (i) by Eq. (ii), we get

$$\frac{\text{KE}}{U} = \frac{4}{1}$$

$\Rightarrow \quad \text{KE } U = 4:1$

32. Given that, $y_1 = 5[\sin 2\pi t + \sqrt{3}\cos 2\pi t]$

$$= 10\left[\frac{1}{2}\sin 2\pi t + \frac{\sqrt{3}}{2}\cos 2\pi t\right]$$

$$= 10\left[\cos\frac{\pi}{3}\sin 2\pi t + \sin\frac{\pi}{3}\cos 2\pi t\right] = 10\sin\left[\left(2\pi t + \frac{\pi}{3}\right)\right]$$

$\Rightarrow A_1 = 10$

Similarly, $\quad y_2 = 5\sin\left(2\pi t + \frac{\pi}{4}\right) \Rightarrow A_2 = 5$

Hence, $\quad \frac{A_1}{A_2} = \frac{10}{5} = \frac{2}{1}$

12

Elasticity

12.1 Introduction

Whenever a load is attached to a thin hanging wire, it elongates and the load moves downwards (sometimes through a negligible distance). The amount by which the wire elongates depends upon the amount of load and the nature of wire material. Cohesive force, between the molecules of the hanging wire offer resistance against the deformation, and the force of resistance increases with the deformation. The process of deformation stops, when the force of resistance is equal to the external force (i.e. the load attached). Sometimes, the force of resistance offered by the molecules is less than the external force. In such a case, the deformation continues until failure takes place.

Thus, we may conclude that, if some external force is applied to a body it has two effects on it, namely

(i) deformation of the body

(ii) internal resistance (restoring) forces are developed.

Chapter Snapshot

- Introduction
- Elasticity
- Stress and Strain
- Hooke's Law and the Modulus of Elasticity
- The Stress-Strain Curve
- Potential Energy Stored in a Stretched Wire
- Thermal Stresses and Strains

12.2 Elasticity

As we have already discussed that whenever a single force (or a system of forces) acts on a body it undergoes some deformation and the molecules offer some resistance to the deformation. When the external force is removed, the force of resistance also vanishes and the body returns back to its original shape. But it is only possible, if the deformation is within a certain limit. Such a limit is called elastic limit. This property of materials of returning back to their original position is called the elasticity.

A body is said to be perfectly elastic, if it returns back completely to its original shape and size after removing the external force(s). If a body remains in the deformed state and does not even partially regain its original shape after the removal of the deforming forces, it is called a perfectly inelastic or plastic body. Quite often, when the external forces are removed, the body partially regains the original shape. Such bodies are partially elastic. If the force acting on the body is increased and the deformation exceeds the elastic limit, the body loses to some extent, its property of elasticity. In this case, the body will not return to its original shape and size even after removal of the external force. Some deformation is left permanently.

12.3 Stress and Strain

Stress

When an external force is applied to a body, then at each cross-section of the body an internal restoring force is developed which tends to restore the body to its original state. The internal restoring force per unit area of cross-section of the deformed body is called stress. It is usually denoted by σ (sigma).

Thus, $$\text{Stress } \sigma = \frac{\text{Restoring force}}{\text{Area}}$$

Depending upon the way, the deforming forces are applied to a body, there are three types of stress: longitudinal stress, shearing stress and volume stress.

Longitudinal and Shearing Stress

The body of figure is in static equilibrium under an arbitrary set of external forces. In Fig. 12.1(b), we see the same body with an imaginary sectional cut at CC'. Since, each of the two individual parts of the body is also in static equilibrium, both internal forces and internal torques are developed at the cross-section. Those on the right portion are due to the left portion and *vice-versa*. On the left portion, the normal and tangential components of the internal forces are $\mathbf{F}_n$ and $\mathbf{F}_t$ respectively, and the net internal torque is τ. From Newton's third law, the right portion is subjected at this same cross-section to force components $-\mathbf{F}_n$ and ... $-\mathbf{F}_t$ and the torque $-\tau$. We define the normal stress or longitudinal stress over the area as

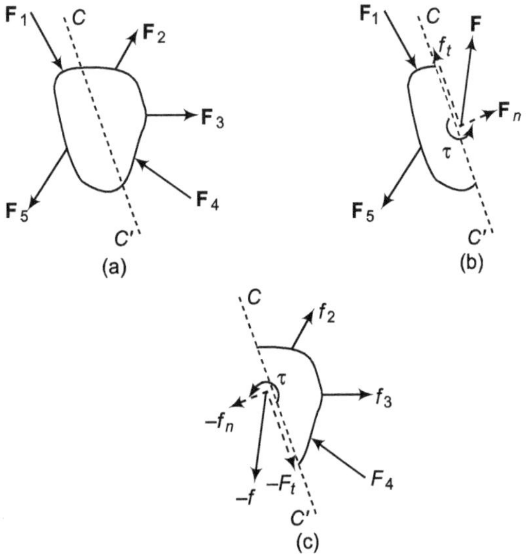

Fig. 12.1

$$\sigma_n = \frac{f_n}{A}$$

and the tangential stress or shearing stress over the area as

$$\sigma_t = \frac{f_t}{A}$$

Here, A is the cross-section area of the body at CC'. The longitudinal stress can be of two types. The two parts of the body on two sides of a cross-section may pull each other. The longitudinal stress is, then called the tensile stress. This is the case when a rod or a wire is stretched by equal and opposite forces. In case of tensile stress in a wire or a rod, the force $\mathbf{F}_n$ is just the tension.

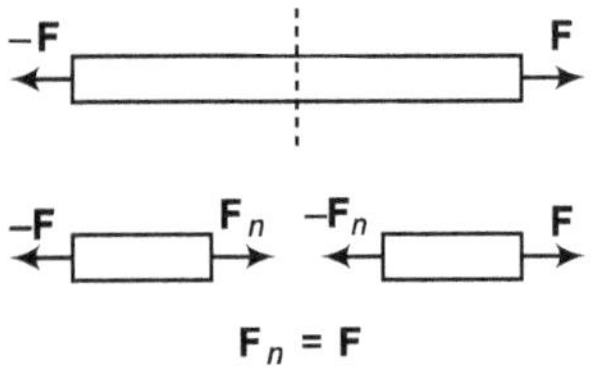

Fig. 12.2

If the rod is pushed at the two ends with equal and opposite forces, it will be under compression. Taking any cross-section of the rod the two parts on the two sides push each other. The longitudinal stress in this case is called the compressive stress.

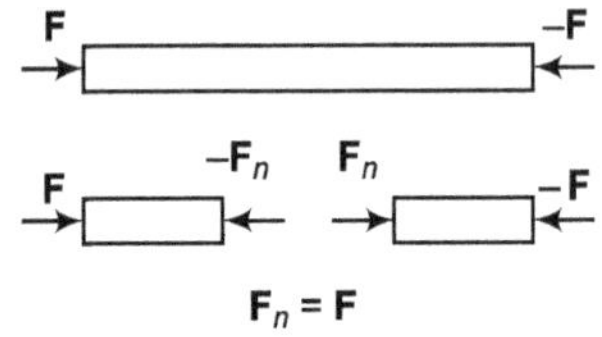

Fig. 12.3

Volume Stress

When a body is acted upon by forces in such a manner that,

(i) the force at any point is normal to the surface.

(ii) the magnitude of the force on any small area is proportional to the area.

The force per unit area is then called the volume stress, i.e.

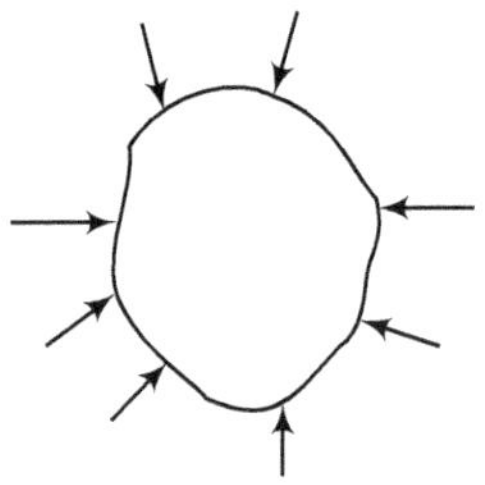

Fig. 12.4

$$\sigma_V = \frac{F}{A}$$

which is same as the pressure. This is the case when a body is immersed in a liquid.

Strain

When the size or shape of a body is changed under an external force, the body is said to be strained. The change occurred in the unit size of the body is called strain. Usually, it is denoted by ε. Thus,

$$\varepsilon = \frac{\Delta x}{x}$$

Here, Δx is the change (may be in length, volume, etc.) and x the original value of the quantity in which change has occurred.

For example, when the length of a suspended wire increases under an applied load, the value of strain is,

$$\varepsilon = \frac{\Delta l}{l}$$

and it is called **longitudinal strain.**

Similarly, if the change has occurred in the volume of a body, it is called **volumetric strain** and is given by

$$\varepsilon = \frac{\Delta V}{V}$$

Shearing Strain

This type of strain is produced when a shearing stress is present.

Consider a body of square cross-section *ABCD*. Four forces of equal magnitude *F* are applied as shown in figure. Net resultant force and net torque is zero. Hence, the body is in translational as well as rotational equilibrium. Because of the forces, the shape of the cross-section changes from a square to a parallelogram.

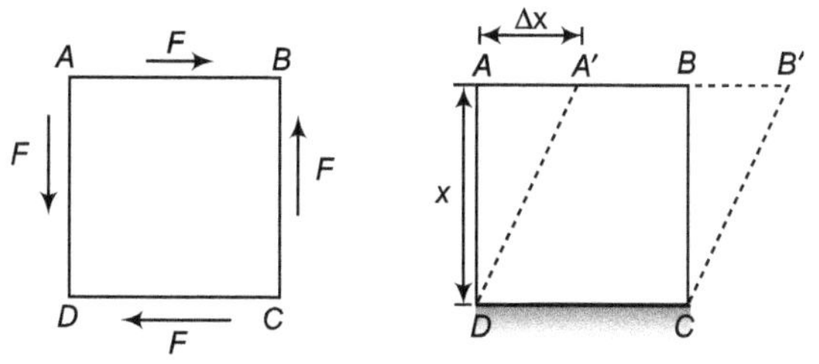

Fig. 12.5

We define the shearing strain as the displacement of a layer divided by its distance from the fixed layer. Thus, shearing strain

$$\varepsilon = \frac{\Delta x}{x}$$

12.4 Hooke's Law and the Modulus of Elasticity

According to Hooke's law,

"For small deformation, the stress in a body is proportional to the corresponding strain."

i.e. stress $\propto$ strain or strees = (E) (strain)

Here, $E = \frac{\text{stress}}{\text{strain}}$ is a constant called the modulus of elasticity. Depending upon the nature of force applied on the body, the modulus of elasticity is classified in following three types:

Young's Modulus of Elasticity (Y)

When a wire is acted upon by two equal and opposite forces in the direction of its length, the length of the body is changed. The change in length per unit length $\left(\frac{\Delta l}{l}\right)$ is icalled the longitudinal strain and the restorng force (which is equal to the applied force in equilibrium) per unit area of cross-section of the wire is called the longitudinal stress.

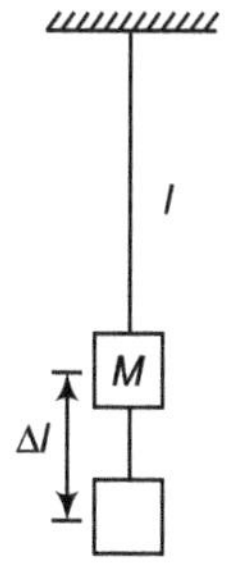

Fig. 12.6

For small change in the length of the wire, the ratio of the longitudinal stress to the corresponding strain is called the Young's modulus of elasticity (Y) of the wire. Thus,

$$Y = \frac{\frac{F}{A}}{\frac{\Delta l}{l}} \quad \text{or} \quad Y = \frac{Fl}{A\Delta l}$$

Let there be a wire of length l and radius r. Its one end is clamped to a rigid support and a mass M is attached at the other end. Then,

$$F = Mg$$

and $$A = \pi r^2$$

Substituting in above equation, we have

$$Y = \frac{Mgl}{(\pi r^2)\Delta l}$$

Bulk Modulus of Elasticity (B)

When a uniform pressure (normal force) is applied all over the surface of a body, the volume of the body changes. The change in volume per unit volume of the body is called the 'volume strain' and the normal force acting per unit area of the surface (pressure) is called the normal stress or volume stress. For small strains, the ratio of the volume stress to the volume strain is called the 'Bulk modulus' of the material of the body. It is denoted by B. Then,

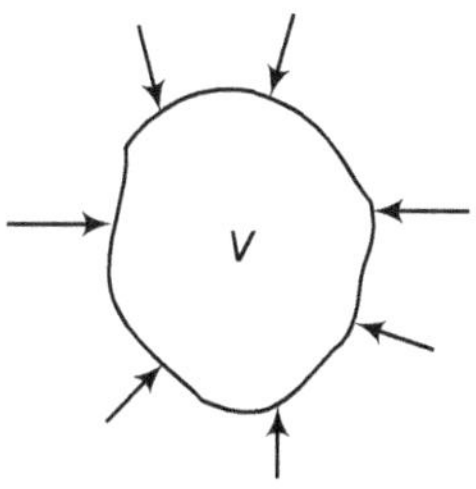

Fig. 12.7

$$B = \frac{-p}{\Delta V / V}$$

Here, negative sign implies that when the pressure increases volume decreases and *vice-versa.*

Compressibility

The reciprocal of the bulk modulus of the material of a body is called the 'compressibility' of that material. Thus,

$$\text{Compressibility} = \frac{1}{B}$$

Modulus of Rigidity (η)

When a body is acted upon by an external force tangential to a surface of the body, the opposite surface being kept fixed, it suffers a change in shape, its volume remaining unchanged. Then, the body is said to be sheared.

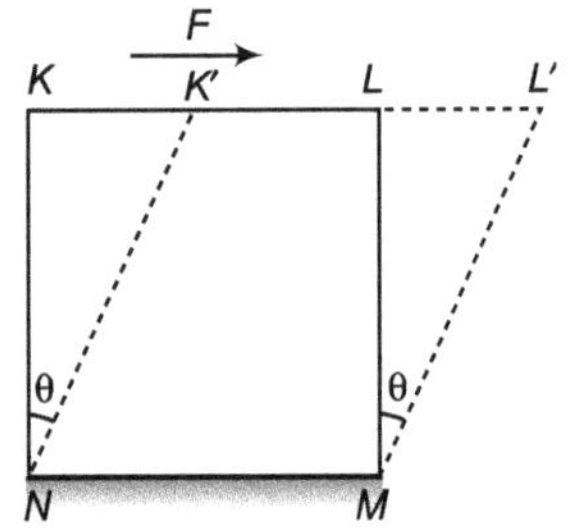

Fig. 12.8

The ratio the displacement of a layer in the direction of the tangential force and the distance of the layer from the fixed surface is called the shearing strain and the tangential force acting per unit area of the surface is called the "shearing stress".

For small strain the ratio of the shearing stress to the shearing strain is called the "modulus of rigidity" of the material of the body. It is denoted by η.

Thus, $$\eta = \frac{F/A}{KK'/KN}$$

Here, $$\frac{KK'}{KN} = \tan\theta \approx \theta$$

∴ $$\eta = \frac{F/A}{\theta}$$

or $$\eta = \frac{F}{A\theta}$$

12.5 The Stress-Strain Curve

A plot of normal stress (either tensile or compressive) *versus* normal strain for a typical solid is shown in figure.

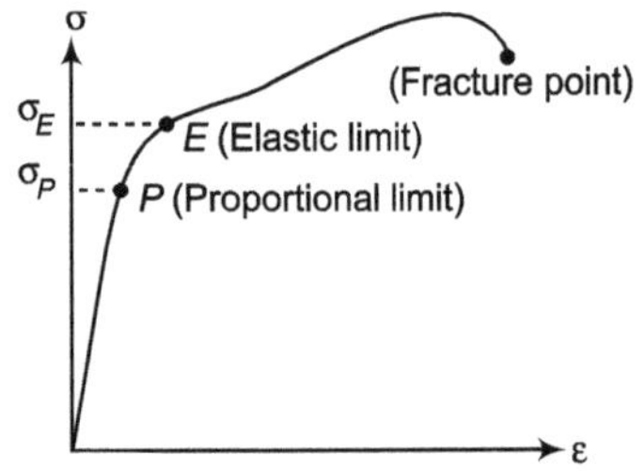

The stress-strain curve for a typical solid

Fig. 12.9

The strain is directly proportional to the applied stress for values of stress up to σ_P. In this linear region, the material returns to its original size, when the stress is removed. Point P is known as the proportional limit of the solid. For stresses between σ_P and σ_E, where point E is called the elastic limit, the material also returns to its original size.

However, notice that stress and strain are not proportional in this region. For deformations beyond the elastic limit, the material does not return to its original size, when the stress is removed, it is permanently distorted. Finally, further stretching beyond the elastic limit leads to the eventual fracture of the solid. The proportionality constant for linear region or the slope of stress-strain curve in this curve is called the Young's modulus of elasticity Y.

Extra Knowledge Points

- Modulus of elasticity E (whether it is Y, B or η) is given by
$$E = \frac{\text{stress}}{\text{strain}}$$
Following conclusions can be made from the above expression :
  (i) $E \propto$ stress (for same strain), i.e. if we want the equal amount of strain in two different materials, the one which needs more stress is having more E.
  (ii) $E \propto \frac{1}{\text{strain}}$ (for same stress), i.e. if the same amount of stress is applied on two different materials, the one having the less strain is having more E. Rather we can say that, the one which offers more resistance to the external forces is having greater value of E. So, we can see that modulus of elasticity of steel is more than that of rubber or
$$E_{\text{steel}} > E_{\text{rubber}}$$
  (iii) $E =$ stress for unit strain $\left(\frac{\Delta x}{x} = 1 \text{ or } \Delta x = x\right)$, i.e. suppose, the length of a wire is 2 m, then the Young's modulus of elasticity (Y) is the stress applied on the wire to stretch the wire by the same amount of 2 m.
- The material which has smaller value of Y is more ductile, i.e. it offers less resistance in framing it into a wire. Similarly, the material having the smaller value of B is more malleable. Thus, for making wire we choose a material having less value of Y.
- A solid will have all the three modulii of elasticity Y, B and η. But in case of a liquid or a gas only B can be defined as a liquid or a gas cannot be framed into a wire or no shear force can be applied on them.
- For a liquid or a gas,
$$B = \left(\frac{-dp}{dV/V}\right)$$
So, instead of p we are more interested in change in pressure dp.
- In case of a gas,
$B = Xp$ in the process $pV^X =$ constant

For example, for $X = 1$,

or pV = constant (isothermal process) $B = p$.

i.e. isothermal bulk modulus of a gas (denoted by B_T) is equal to the pressure of the gas at that instant of time or

$$B_T = p$$

Similarly, for $x = \gamma = \frac{C_p}{C_V}$ or pV^γ = constant (adiabatic process)

$$B = \gamma p$$

i.e. adiabatic bulk modulus of a gas (denoted by B_s) is equal to γ times the pressure of the gas at that instant of time or $B_s = \gamma p$

- For a gas, $B \propto p$ whether it is an isothermal process or an adiabatic process. Physically, this can be understood as under :

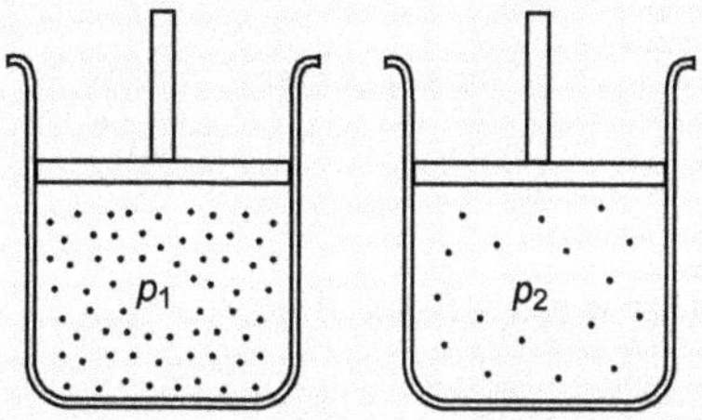

Suppose, we have two containers A and B. Some gas is filled in both the containers. But the pressure in A is more than the pressure in B, i.e.

$$p_1 > p_2$$

So, bulk modulus of A should be more than the bulk modulus of B,

or $$B_1 > B_2$$

and this is quiet obvious, because it is more difficult to compress the gas in chamber A, i.e. it provides more resistance to the external forces. And as we have said in point number 1(ii) the modulus of elasticity is greater for a substance which offers more resistance to external forces.

- If a spring is stretched or compressed by an amount Δl, then restoring force produced in it is

$$F_s = k\,\Delta l \qquad \text{...(i)}$$

Here, k = force constant of spring.

Similarly, if a wire is stretched by an amount Δl, the restoring force produced in it is

$$F = \left(\frac{YA}{l}\right)\Delta l \qquad \text{...(ii)}$$

Comparing Eqs. (i) and (ii), we can see that force constant of a wire is

$$k = \frac{YA}{l} \qquad \text{...(iii)}$$

i.e. a wire is just like a spring of force constant $\frac{YA}{l}$. So, all formulae which we use in case of a spring can be applied to a wire also.

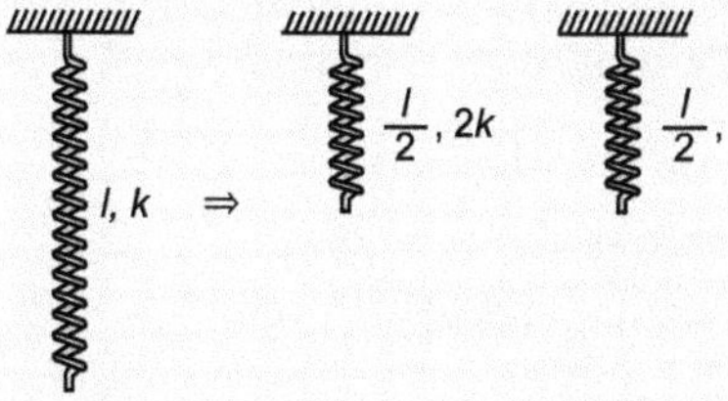

From Eq. (iii), we may also conclude that force constant of a spring is inversely proportional to the length of the spring l or

$$k \propto \frac{1}{l}$$

i.e. if a spring is cut into two equal pieces its force constant is doubled.

- When a pressure (dp) is applied on a substance, its density is changed. The change in density can be calculated as under

$$\rho = \frac{\text{mass}}{\text{volume}} \qquad (\rho = \text{density})$$

or $$\rho \propto \frac{1}{V} \qquad (\text{mass = constant})$$

$$\frac{\rho'}{\rho} = \frac{V}{V'} = \frac{V}{V + dV}$$

or $$\rho' = \rho\left(\frac{V}{V + dV}\right)$$

$$= \rho\left(\frac{V}{V - (dp/B)V}\right)$$

as $$B = -\frac{dp}{dV/V}$$

$$\rho' = \frac{\rho}{1 - \frac{dp}{B}}$$

- From this expression, we can see that ρ' increases as pressure is increased (dp is positive) and *vice-versa*.

Example 12.1 *Determine the elongation of the steel bar* 1 m *long and* 1.5 cm^2 *cross-sectional area when subjected to a pull of* 1.5×10^4 N. *(Take* $Y = 2.0 \times 10^{11}$ N/m^2*)*

Sol. $Y = \frac{F/A}{\Delta l / l}$

$$\therefore \quad \Delta l = \frac{Fl}{AY}$$

Substituting the values,

$$\Delta l = \frac{(1.5 \times 10^4)(1.0)}{(1.5 \times 10^{-4})(2.0 \times 10^{11})}$$

$$= 0.5 \times 10^{-3}\ \text{m}$$

or $$\Delta l = 0.5\ \text{mm}$$

Example 12.2 *A bar of mass m and length l is hanging from point A as shown in figure. Find the increase in its length due to its own weight. The Young's modulus of elasticity of the wire is Y and area of cross-section of the wire is A.*

Fig. 12.10

Sol. Consider a small section dx of the bar at a distance x from B. The weight of the bar for a length x is

$$w = \left(\frac{mg}{l}\right)x$$

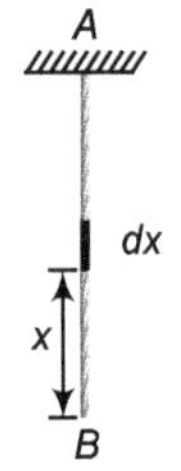

Fig. 12.11

Elongation in section dx will be

$$dl = \left(\frac{w}{AY}\right)dx = \left(\frac{mg}{lAY}\right)x\,dx$$

Total elongation in the bar can be obtained by integrating this expression for $x = 0$ to $x = l$.

$$\therefore \quad \Delta l = \int_{x=0}^{x=l} dl = \left(\frac{mg}{lAY}\right)\int_0^l x\,dx$$

or $$\Delta l = \frac{mgl}{2AY}$$

12.6 Potential Energy Stored in a Stretched Wire

When a wire is stretched, work is done against the inter-atomic forces. This work is stored in the wire in the form of elastic potential energy. Suppose on applying a force F on a wire of length l, the increase in length is Δl. The area of cross-section of the wire is A. The potential energy stored in the wire should be

$$U = \frac{1}{2}K(\Delta l)^2$$

Here, $$K = \frac{YA}{l}$$

$$\therefore \quad U = \frac{1}{2}\frac{YA}{l}(\Delta l)^2$$

Elastic potential energy per unit volume of the wire is

$$u = \frac{U}{\text{volume}}$$

or $$u = \frac{\frac{1}{2}\frac{YA}{l}(\Delta l)^2}{Al}$$

or $$u = \frac{1}{2}\left(\frac{\Delta l}{l}\right)\left(Y\cdot\frac{\Delta l}{l}\right)$$

or $$u = \frac{1}{2}(\text{strain})\,(Y \times \text{strain})$$

or $$u = \frac{1}{2}(\text{strain} \times \text{stress})$$

12.7 Thermal Stresses and Strains

Whenever there is some increase or decrease in the temperature of the body, it causes the body to expand or contract. If the body is allowed to expand or contract freely, with the rise or fall of the temperature, no stresses are induced in the body. But, if the deformation of the body is prevented, some stresses are induced in the body. Such stresses are called thermal stresses or temperature stresses. The corresponding strains are called thermal strains or temperature strains.

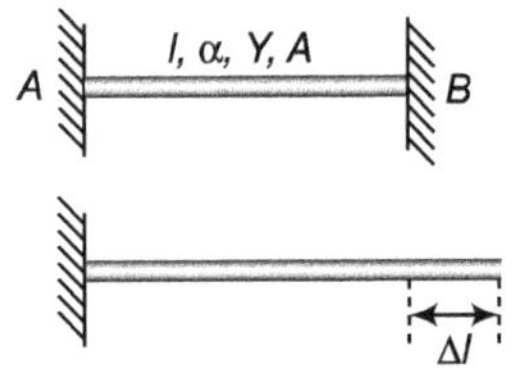

Fig. 12.12

Consider a rod AB fixed at two supports as shown in figure.

Let l = length of rod,

A = area of cross-section of the rod,

Y = Young's modulus of elasticity of the rod

and α = thermal coefficient of linear expansion of the rod.

Let the temperature of the rod is increased by an amount t. The length of the rod would had increased by an amount Δl, if it were not fixed at two supports. Here,

$$\Delta l = l\alpha t$$

But since the rod is fixed at the supports, a compressive strain will be produced in the rod. Because at the increased temperature, the natural length of the rod is $l + \Delta l$, while being fixed at two supports its actual length is l.

Hence, thermal strain $\varepsilon = \frac{\Delta l}{l} = \frac{l\alpha t}{l} = \alpha t$ or

$\varepsilon = \alpha t$

Therefore, thermal stress

$$\sigma = Y\varepsilon \qquad (\text{stress} = Y \times \text{strain})$$

or $$\sigma = Y\alpha t$$

or force on the supports,

$$F = \sigma A = YA\alpha t$$

This force F is in the direction shown below:

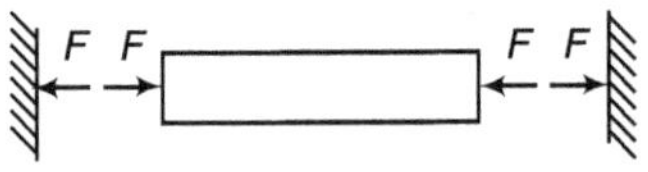

Fig. 12.13

Extra Knowledge Points

- **Poisson's ratio**

When a longitudinal force is applied on a wire, its length increases but its radius decreases. Thus, two strains are produced by a single force :

(i) Longitudinal strain $= \frac{\Delta l}{l}$ and

(ii) Lateral strain $= \frac{\Delta R}{R}$

The ratio of these two strains is called the Poisson's ratio.

Thus, the Poisson's ratio

$$\sigma = \frac{\text{Lateral strain}}{\text{Longitudinal strain}} = -\frac{\Delta R / R}{\Delta l / l}$$

Following points are worthnothing in case of Poisson's ratio :

(i) Negative sign in σ indicates that radius of the wire decreases as the length increases.

(ii) Theoretical value of σ lies between -1 and $+\frac{1}{2}$.

(iii) Practical value of σ lies between 0 and $+\frac{1}{2}$.

- **Relation between Y, B, η and σ**

Following are some relations between the four

(a) $B = \frac{Y}{3(1-2\sigma)}$

(b) $\eta = \frac{Y}{2(1+\sigma)}$

(c) $\sigma = \frac{3B - 2\eta}{2\eta + 6B}$

(d) $\frac{9}{Y} = \frac{1}{B} + \frac{3}{\eta}$

- **Streamline and Turbulent flow**

The flow of a fluid is said to be steady, if at any given point the velocity of each particle passing through that point remains constant. Although velocity at different points may be different. That is, at some other point the particle may have a different velocity, but every other particle which passes the second point behaves exactly as the previous particle that has just passed that point.

Each particle follows a smooth path and the paths of the particles do not cross each other. The path followed by a fluid particle under a steady flow is called a streamline.

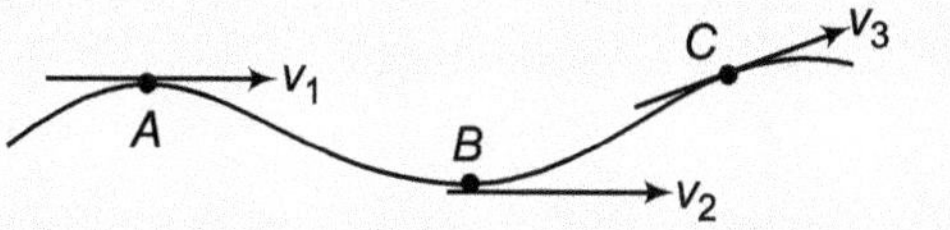

In the streamline ABC shown in above figure, velocity of fluid particles at A is v_1 at B is v_2 and at C is v_3.

Steady flow is achieved when the speed of flow is not very high. Beyond a limiting value, called critical speed, flow is no longer streamlined it becomes turbulent.

- **Reynold's number**

When speed of fluid flow is high, flow of fluid becomes turbulent. Turbulent flow is less likely for viscous fluids flowing at low speeds. Reynold's number (R_e) is a dimensionless number, whose value gives us an idea whether the flow would be laminar (streamline) or turbulent.

- $R_e = \frac{\rho v d}{\eta}$

Here, ρ = density of fluid,
v = speed of fluid,
d = dimension of fluid and
η = viscosity of fluid.

For $R_e < 1000$, flow is streamline or laminar.

For $R_e \to 1000$ to 2000, flow is unsteady and for $R_e > 2000$, flow is turbulent. Turbulence dissipates kinetic energy usually in the form of heat. Racing cars and aeroplanes are designed to minimize turbulence. But, sometimes turbulence is desirable. Turbulence promotes mixing.

For example, blades of a kitchen mixer induce turbulent flow and provide thick milk shakes.

Chapter Summary with Formulae

1. Stress $= \frac{F}{A}$ = restoring force per unit area.
2. Strain $= \frac{\Delta x}{x}$ = change per unit original. It may be $\frac{\Delta l}{l}$ or $\frac{\Delta V}{V}$, etc.
3. Modulus of Elasticity $$E = \frac{\text{Stress}}{\text{Strain}}$$
4. Materials which offer more resistance to external deforming forces have higher value of modulus of elasticity.
5. Young's Modulus of Elasticity
$$Y = \frac{F/A}{\Delta l/l} = \frac{Fl}{A\Delta l}$$
$$F = Mg \quad \text{and} \quad A = \pi r^2$$
6. Bulk Modulus of Elasticity $$B = \frac{F/A}{\Delta V/V} = -\frac{\Delta p}{\Delta V/V} \quad \text{or} \quad -\frac{dp}{\Delta V/V}$$
7. Shear Modulus of Elasticity of Modulus of Rigidity
$$G \text{ or } \eta = \frac{F/A}{\theta} = \frac{F}{A\theta}$$

   where, $\theta \approx \tan\theta = \frac{DE}{DF}$ or $\frac{BC}{BA}$

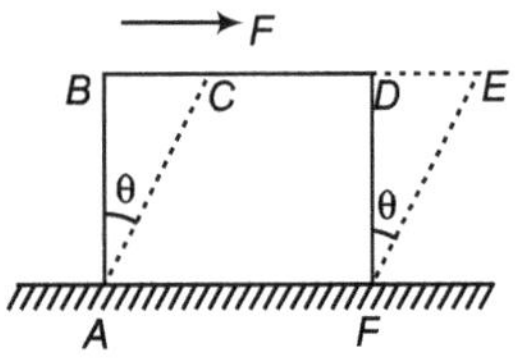

8. Solids have all three modulii of elasticities, Young's modulus, bulk modulus and shear modulus. Whereas liquids and gases have only bulk modulus.
9. Every wire is like a spring whose force constant is equal to,
$$k = \frac{YA}{l} \quad \text{or} \quad k \propto \frac{1}{l}$$
10. Potential energy stored in a stretched wire
$$U = \frac{1}{2} k (\Delta l)^2 = \frac{1}{2}\left(\frac{YA}{l}\right)(\Delta l)^2$$
11. Potential energy stored per unit volume (also called energy density) in a stretched wire,
$$U = \frac{1}{2} \times \text{Stress} \times \text{Strain}$$
12. Change in Length of a Wire
$$\Delta l = \frac{Fl}{AY}$$

   Here, F is tension in the wire. If wire is having negligible mass, tension is a uniform throughout the wire and change in length is obtained directly. Otherwise by integration.
13. In case of solids and liquids bulk modulus is almost constant. In case of a gas, it is process dependent.
   In isothermal process, $B = B_T = p$
   In adiabatic process, $B = B_S = \gamma p$
   Here, p is pressure of the gas and γ is called adiabatic exponent of the gas.
14. Compressibility $= \frac{1}{B}$
15. When pressure is applied on a substance, its volume decreases, while mass remains constant. Hence, its density will increase,
$$\rho' = \frac{\rho}{1 - \Delta p/B} \quad \text{or} \quad \rho' \approx \rho\left(1 + \frac{\Delta p}{B}\right)$$

   If $\frac{\Delta p}{B} << 1$

Additional Examples

Example 1. *State Hooke's law.*

Sol. Within elastic limit, stress is directly proportional to strain.

Example 2. *What are the factors on which modulus of elasticity of a material depends?*

Sol. Nature of the material and the manner in which it is deformed.

Example 3. *The stress versus strain graphs for wires of two materials A and B are as shown in figure.*

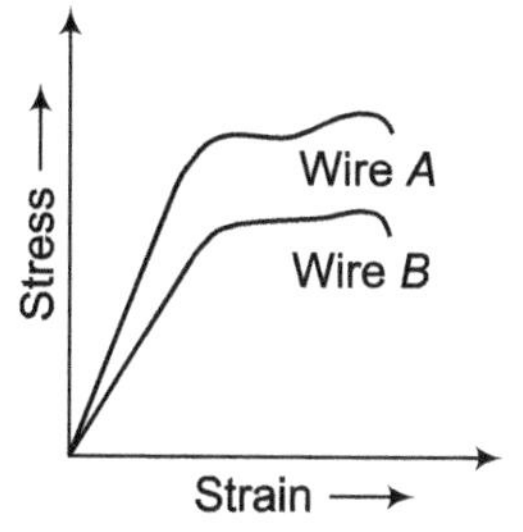

(*a*) *Which material is more ductile?*

(*b*) *Which material is more brittle?*

Sol. (a) Material *A* is more ductile. It is because, the material *A* has greater plastic range (portion of graph between the elastic limit and breaking point).

(b) Material *B* is more brittle. It is because, the material z*A* has lesser plastic range.

Example 4. *Why do we prefer steel to copper in the manufacturing of spring?*

Sol. A better spring will be the one, in which a large restoring force is developed on being deformed. This, in turn, depends upon the elasticity of the material of the spring. As Young's modulus of steel is greater than that of copper, steel is preferred to manufacture a spring.

Example 5. *A cable is cut to half of its original length. Why this change has no effect on the maximum load, the cable can support?*

Sol. The breaking stress is a constant for a given material. We know that :

Breaking load = breaking stress × area of cross-section.

When the cable is cut to half of its length, the area of cross-section does not change. Hence, there is no effect on the maximum load (breaking load), the cable can support.

Example 6. *The breaking force for a wire is F. What will be the breaking force for (a) two parallel wires of the same size (b) for a single wire of double the thickness?*

Sol. (a) When two wires of the same size are suspended in parallel, a force equal to $2F$ has to be applied on the parallel combination, so that a force F equal to the breaking force for the wire acts on each of the two wires. Hence, the breaking force in this case is $2F$.

(b) Breaking force = (Breaking stress) × (area of C.S.)

$\therefore$ Breaking force $\propto$ area of C.S.

If the wire is of double the thickness, i.e. of double the diameter, the area of C.S. will be four times. Hence, breaking force will be $4F$.

Example 7. *A cable is replaced by another of the same length and material of twice diameter?*

(*a*) *How does this affect elongation under a given load?*

(*b*) *How many times will be the maximum load supported by the later as compared to the former?*

Sol. (a) We know, $Y = \dfrac{FL}{(A)(\Delta l)} = \dfrac{4FL}{(\pi d^2)(\Delta l)}$,

where d is the diameter of the cable.

Therefore, $$\Delta l = \frac{4\,FL}{\pi d^2 Y}$$

As $\Delta l \propto 1/d^2$,

Hence, the extension will become one-fourth on replacing the cable by another cable of double the diameter.

(b) Breaking load = breaking stress × area of cross-section.

When the cable of double the diameter is used, area of cross-section will become four times. Since, the breaking stress is a constant for the given material, such a cable will be able to support a load four times as large as the former cable can support.

Example 8. *Why are the bridges declared unsafe after long use?*

Sol. A bridge undergoes continuous alternating strains a large number of times everyday. Therefore, such a long use of the bridge results in the loss of its elastic strength. In other words, after a long use, the strain produced is quite large for a given stress and may lead to the collapse of the bridge. This is the reason why, the bridges are declared unsafe after a long use.

Example 9. *What is a perfectly elastic body? Give an example.*

Sol. If, on removal of deforming force, a body completely regains its original configuration, then it is said to be perfectly elastic. Example is quartz.

Example 10. *What is a perfectly plastic body? Give an example.*

Sol. If, on removal of deforming force, a body does not regain its original configuration even a little, then it is said to be perfectly plastic. Example is putty.

Example 11. *Is it possible to double the length of a metallic wire by applying a force over it?*

Sol. No, it is not possible, because within elastic limit strain is only of the order of 10^{-3}. Wires actually break much before it if they are stretched to double the length.

Example 12. *Stress and pressure both are force per unit area. Then, in what respect does stress differ from pressure?*

Sol. Pressure is the external force per unit area, while stress is the internal restoring force (which comes into play in a deformed body) per unit area of the body.

Example 13. *Which is more elastic-water or air?*

Sol. Water is more elastic than air. Air can be easily compressed while water is incompressible and bulk modulus is reciprocal of compressibility.

Example 14. *The ratio stress/strain remains constant for a small deformation. What happens to this ratio if deformation is made very large?*

Sol. When the deforming force exceeds the elastic limit, the strain increases more rapidly than stress. Hence, the ratio of stress/strain decreases.

Example 15. *A steel wire of length 4 m and diameter 5 mm is stretched by 5 kg-wt. Find the increase in its length, if the Young's modulus of steel is* 2.4×10^{12} *dyne/cm*^{2}.

Sol. Hence, $l = 4\text{ m} = 400\text{ cm}, \; 2r = 5\text{ mm}$

or $r = 2.5\text{ mm} = 0.25\text{ cm}$

$f = 5\text{ kg-wt} = 5000\text{ g-wt}$

$= 5000\times 980\text{ dyne}$

$\Delta l = ?, \; Y = 2.4\times 10^{12}\text{ dyne/cm}^2$

As $$Y = \frac{F}{\pi r^2}\times\frac{l}{\Delta l}$$

or $$\Delta l = \frac{Fl}{\pi r^2 Y}$$

$$= \frac{(5000\times 980)\times 400}{(22/7)\times(0.25)^2\times 2.4\times 10^{12}}$$

$= 0.0041\text{ cm}$

Example 16. *A cable is replaced by another one of same length and material but twice the diameter. How will this effect the elongation under a given load ? How does this effect the maximum load it can support without exceeding the elastic limit?*

Sol. Young's modulus

$$Y = \frac{Mgl}{\pi r^2\cdot\Delta l} = \frac{Mgl}{\pi\left(\frac{D}{2}\right)^2\Delta l} = \frac{4Mgl}{\pi D^2\cdot\Delta l}$$

where, D is the diameter of the wire.

$\therefore$ Elongation, $\Delta l = \dfrac{4Mgl}{\pi D^2\cdot Y}$

i.e. $\Delta l \propto \dfrac{1}{D^2}$

Clearly, if the diameter is doubled, the elongation will become one-fourth.

Also, maximum stress, $\sigma_m = \dfrac{M_m g}{(\pi D^2/4)}$ or $M_m \propto D^2$

Clearly, if the diameter is doubled, the wire can support four times the original load.

Example 17. *The bulk modulus of water is* 2.3×10^9 *N/m*^{2}.

(a) Find its compressibility.

(b) How much pressure in atmospheres is needed to compress a sample of water by 0.1%.

Sol. Here, $B = 2.3\times 10^9\text{ N/m}^2 = \dfrac{2.3\times 10^9}{1.01\times 10^5}$

$= 2.27\times 10^4\text{ atm}$

(a) Compressibility $= \dfrac{1}{B} = \dfrac{1}{2.27\times 10^4}$

$= 4.4\times 10^{-5}\text{ atm}^{-1}$

(b) Here, $\dfrac{\Delta V}{V} = -0.1\% = -0.001$

Required increase in pressure,

$$\Delta p = B\times\left(-\frac{\Delta V}{V}\right)$$

$= 2.27\times 10^4\times 0.001$

$= 22.7\text{ atm}$

Example 18. *A steel wire* 4.0 *m in length is stretched through* 2.0 *mm. The cross-sectional area of the wire is* 2.0 *mm*^{2}. *If Young's modulus of steel is* 2.0×10^{11} *N/m*^{2}. *Find*

(a) the energy density of wire,

(b) the elastic potential energy stored in the wire.

Sol. Here, $l = 4.0\text{ m}, \; \Delta l = 2\times 10^{-3}\text{ m}$

$A = 2.0\times 10^6\text{ m}^2$

$Y = 2.0\times 10^{11}\text{ N/m}^2$

(a) The energy density of stretched wire,

$$u = \frac{1}{2}\times\text{stress}\times\text{strain}$$

$$= \frac{1}{2}\times Y\times(\text{strain})^2$$

$$= \frac{1}{2} \times 2.0 \times 10^{11} \times \left(\frac{2 \times 10^{-3}}{4} \right)^2$$

$$= 2.5 \times 10^4 \text{ J/m}^3$$

(b) Elastic potential energy = energy density × volume

$$= 2.5 \times 10^4 \times (2.0 \times 10^{-6}) \times 4.0 \text{ J} = 0.20 \text{ J}$$

Example 19. *Find the greatest length of steel wire that can hang vertically without breaking. Breaking stress of steel* $= 8.0 \times 10^8$ N/m^2. *Density of steel* $= 8.0 \times 10^3$ kg/m^3. *(Take* $g = 10$ m/s^2*)*

Sol. Let l be the length of the wire that can hang vertically without breaking. Then, the maximum tension on it is at the topmost point and it is equal to its own weight. If, A is the area of cross-section and ρ the density, then

Maximum stress, $\sigma_m = \dfrac{\text{weight}}{A}$ $\quad \left(\text{Stress} = \dfrac{\text{force}}{\text{area}} \right)$

or $\quad \sigma_m = \dfrac{(Al\rho)g}{A}$

$\therefore \quad l = \dfrac{\sigma_m}{\rho g}$

Substituting the values, we get

$$l = \frac{8.0 \times 10^8}{(8.0 \times 10^3)(10)}$$

$$= 10^4 \text{ m}$$

Example 20. *What is the density of lead under a pressure of* 2.0×10^8 N/m^2, *if the bulk modulus of lead is* 8.0×10^9 N/m^2 *and initially the density of lead is* 11.4 g/cm^3*?*

Sol. The changed density,

$$\rho' = \frac{\rho}{1 - \dfrac{dp}{B}}$$

Substituting the values, we have

$$\rho' = \frac{11.4}{1 - \dfrac{2.0 \times 10^8}{8.0 \times 10^9}} \quad \text{or} \quad \rho' = 11.69 \text{ g/cm}^3$$

Example 21. *(a) A wire* 4 m *long and* 0.3 mm *in diameter is stretched by a force of* 100 N. *If extension in the wire is* 0.3 mm, *calculate the potential energy stored in the wire. (b) Find the potential energy stored in a wire of cross-section* 1 mm² *and length* 2 m *through* 0.1 mm. *Young's modulus for the material of wire is* 2.0×10^{11} N/m^2.

Sol. (a) Energy stored

$$U = \frac{1}{2} (\text{stress})(\text{strain})(\text{volume})$$

or $\quad U = \dfrac{1}{2} \left(\dfrac{F}{A} \right) \left(\dfrac{\Delta l}{l} \right) (Al)$

$$= \frac{1}{2} F \cdot \Delta l$$

$$= \frac{1}{2} (100)(0.3 \times 10^{-3})$$

$$= 0.015 \text{ J}$$

(b) Potential energy stored

$$= \frac{1}{2} k (\Delta l)^2$$

$$= \frac{1}{2} \left(\frac{YA}{l} \right) (\Delta l)^2 \qquad \left(\text{as } k = \frac{YA}{l} \right)$$

Substituting the values, we have

$$U = \frac{1}{2} \frac{(2.0 \times 10^{11})(10^{-6})}{(2)} (0.1 \times 10^{-3})^2$$

$$= 5.0 \times 10^{-4} \text{ J}$$

Example 22. *A rubber cord has a cross-sectional area* 1 mm² *and total unstretched length* 10.0 cm. *It is stretched to* 12.0 cm *and then released to project a missile of mass* 5.0 g. *Taking Young's modulus Y for rubber as* 5.0×10^8 N/m^2. *Calculate the velocity of projection.*

Sol. Equivalent force constant of rubber cord.

$$k = \frac{YA}{l}$$

$$= \frac{(5.0 \times 10^8)(1.0 \times 10^{-6})}{(0.1)}$$

$$= 5.0 \times 10^3 \text{ N/m}$$

Now, from conservation of mechanical energy, elastic potential energy of cord

= kinetic energy of missile

$\therefore \quad \dfrac{1}{2} k (\Delta l)^2 = \dfrac{1}{2} mv^2$

$\therefore \quad v = \left(\sqrt{\dfrac{k}{m}} \right) \Delta l$

$$= \left(\sqrt{\frac{5.0 \times 10^3}{5.0 \times 10^{-3}}} \right) (12.0 - 10.0) \times 10^{-2}$$

$$= 20 \text{ m/s}$$

Example 23. *A light rod of length 2.00 m is suspended from the ceiling horizontally by means of two vertical wires of equal length tied to its ends. One of the wires is made of steel and is of cross-section 10^{-3} m² and the other is of brass of cross-section 2×10^{-3} m². Find out the position along the rod at which a weight may be hung to produce,*

(a) equal stresses in both wires,

(b) equal strains on both wires.

Young's modulus for steel is 2×10^{11} N/m² and for brass is 10^{11} N/m².

Sol. (a) Given, s

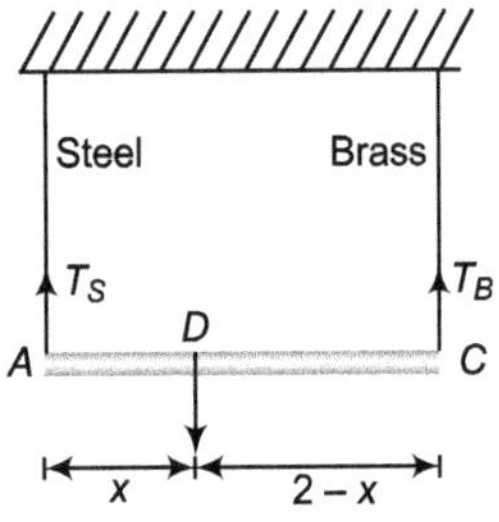

$$\therefore \quad \frac{T_S}{A_S} = \frac{T_B}{A_B}$$

$$\therefore \quad \frac{T_S}{T_B} = \frac{A_S}{A_B}$$

$$= \frac{10^{-3}}{2\times10^{-3}} = \frac{1}{2} \quad ...(i)$$

As the system is in equilibrium, taking moments about D, we have

$$T_S\cdot x = T_B\,(2-x)$$

$$\therefore \quad \frac{T_S}{T_B} = \frac{2-x}{x} \quad ...(ii)$$

From Eqs. (i) and (ii), we get

$$x = 1.33 \text{ m}$$

(b) $\text{Strain} = \dfrac{\text{Stress}}{Y}$

Given, strain in steel = strain in brass

$$\therefore \quad \frac{T_S/A_S}{Y_S} = \frac{T_B/A_B}{Y_B}$$

$$\therefore \quad \frac{T_S}{T_B} = \frac{A_S Y_S}{A_B Y_B}$$

$$= \frac{(1\times10^{-3})(2\times10^{11})}{(2\times10^{-3})(10^{11})} = 1 \quad ...(iii)$$

From Eqs. (ii) and (iii), we have

$$x = 1.0 \text{ m}$$

Example 24. *A steel rod of cross-sectional area 16 cm² and two brass rods each of cross-sectional area 10 cm² together support a load of 5000 kg as shown in figure. Find the stress in the rods. Take Y for steel $= 2.0\times10^6$ kg/cm² and for brass $= 1.0\times10^6$ kg/cm².*

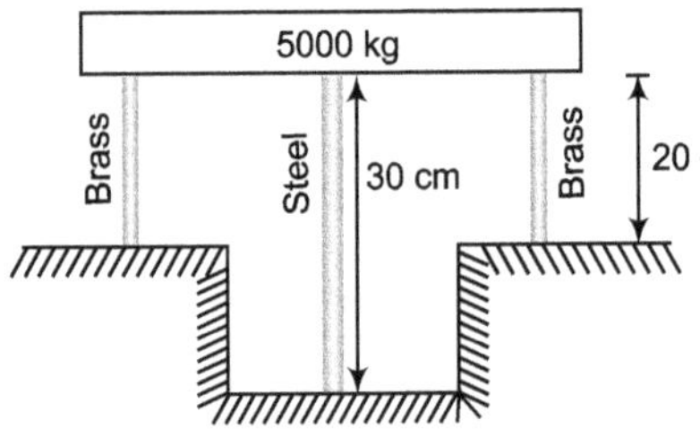

Sol. Given, area of steel rod

$$A_S = 16 \text{ cm}^2$$

Area of two brass rods

$$A_B = 2\times10 = 20 \text{ cm}^2$$

Load, $F = 5000$ kg

Y for steel $Y_S = 2.0\times10^6$ kg/cm²

Y for brass $Y_B = 1.0\times10^6$ kg/cm²

Length of steel rod, $l_S = 30$ cm

Length of brass rod, $l_B = 20$ cm

Let σ_S = stress in steel and σ_B = stress in brass

Decrease in length is given by

$$\Delta l = (l)\,(\text{strain}) = (l)\,\frac{(\text{stress})}{r}$$

Now, given that

decrease in length of steel rod = decrease in length of brass rod

or
$$\frac{\sigma_S}{Y_S}\times l_S = \frac{\sigma_B}{Y_B}\times l_B$$

or
$$\sigma_S = \frac{Y_S}{Y_B}\times\frac{l_B}{l_S}\times\sigma_B$$

$$= \frac{2.0\times10^6}{1.0\times10^6}\times\frac{20}{30}\times\sigma_B$$

$$\therefore \quad \sigma_S = \frac{4}{3}\sigma_B \quad ...(i)$$

Now, using the relation,

$$F = \sigma_S A_S + \sigma_B A_B$$

or
$$5000 = \sigma_S\times16 + \sigma_B\times20 \quad ...(ii)$$

Solving Eqs. (i) and (ii), we get

$$\sigma_B = 120.9 \text{ kg/cm}^2 \text{ and } \sigma_S = 161.2 \text{ kg/cm}^2$$

NCERT Selected Questions

Q 1. A steel wire of length 4.7 m and cross-sectional area 3.0×10^{-5} m^2 stretches by the same amount as a copper wire of length 3.5 m and cross-sectional area 4.0×10^{-5} m^2 under a given load. What is the ratio of the Young's modulus of steel to that of copper?

Sol. Let Y_1 and Y_2 be the Young's modulus of steel and copper wires, respectively.

$$\therefore \quad Y_1 = \frac{F_1 \times l_1}{A_1 \times \Delta l_1}$$

$$= \frac{F \times 4.7}{3 \times 10^{-5} \times \Delta l} \quad \text{...(i)}$$

and

$$Y_2 = \frac{F_2 \times l_2}{A_2 \times \Delta l_2}$$

$$= \frac{F \times 3.5}{4 \times 10^{-5} \times \Delta l} \quad \text{...(ii)}$$

Dividing Eq. (i) by Eq. (ii), we get

$$\frac{Y_1}{Y_2} = \frac{F \times 4.7}{3 \times 10^{-5} \times \Delta l} \times \frac{4 \times 10^{-5} \times \Delta l}{F \times 3.5}$$

$$= 1.79 \approx 1.8$$

Q 2. Figure shows the strain-stress curve for a given material. What are (a) Young's modulus and (b) approximate yield strength for his material?

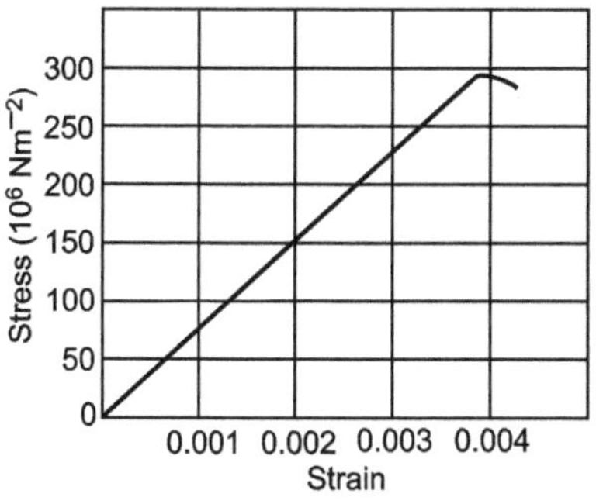

Sol. From the given graph for a stress of 150×10^6 Nm^{-2}, the strain is 0.002.

(a) $\therefore$ Young's modulus of the material Y is given by

$$Y = \frac{\text{Stress}}{\text{Strain}} = \frac{150\times10^6}{0.002}$$

$$= \frac{150 \times 10^6}{2 \times 10^{-3}}$$

$$= 75 \times 10^9 \text{ N/m}^2$$

$$= 7.5 \times 10^{10} \text{ N/m}^2$$

(b) Yield strength of a material is defined as the maximum stress it can sustain.

$\therefore$ From the given graph, the approximate yield strength of the given material = 300×10^6 N/m^2 = 3×10^8 N/m^2.

Q 3. The stress-strain graph for materials A and B are shown in Fig. (a) and Fig. (b).

The graph are drawn to the same scale.

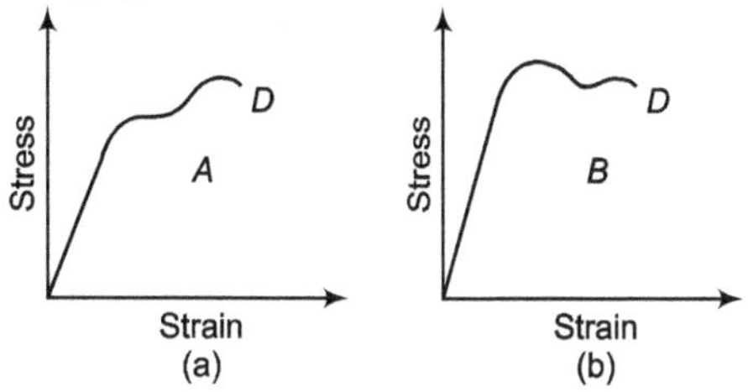

(a) Which of the materials has greater Young's modulus?

(b) Which of the two is the stronger material?

Sol. (a) From the two graphs, we note that for a given strain, stress for B is more than that of A. Hence, Young's modulus $\left(= \frac{\text{stress}}{\text{strain}}\right)$ is greater for B than that of A.

(b) Strength of a material is determined by the amount of stress (load) required to cause breaking or fracture of the material corresponding the breaking point.

$\therefore$ Material A is stronger than B as it can withstand more load without breaking than the material B corresponding to point D.

Q 4. Read the following two statements below carefully and state, if it is true or false.

(a) The Young's modulus of rubber is greater than that of steel.

(b) The stretching of a coil is determined by its shear modulus.

Sol. (a) False. This is because if steel and rubber wires of same length and area of cross-section are subjected to same deforming force, then the extension produced in steel is less than the extension produced in rubber. For producing same strain in steel and rubber more stress is required in case of steel.

(b) True.

Q 5. Two wires of diameter 0.25 cm, one made of steel and other made of brass are loaded as shown in figure. The unloaded length of steel wire is 1.5 m and that of brass wire is 1.0 m. Young's modulus of steel is 2.0×10^{11} Pa and that of brass is 0.91×10^{11} Pa. Compute the elongations of steel and brass wires. (1 Pa = 1 Nm^{-2}).

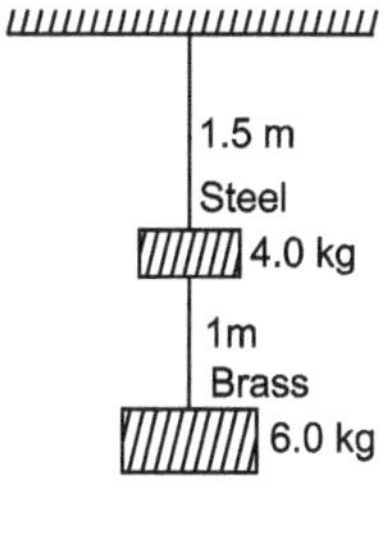

Sol. For steel wire,

Total force, $F_1 = 10 \times 9.8$ N

$l_1 = 1.5$ m

$$\Delta l_1 = \frac{F_1 \times l_1}{\pi r_1^2 \times Y_1}$$

$$= \frac{(10 \times 9.8) \times (1.5) \times 7}{22 \times (0.125 \times 10^{-2})^2 \times 2 \times 10^{11}}$$

$= 1.5 \times 10^{-4}$ m

For brass wire, $F_2 = 6 \times 9.8$ N

$$\therefore \quad \Delta l_2 = \frac{F_2 \times l_2}{\pi r_2^2 \times Y_2}$$

$$= \frac{(6 \times 9.8) \times 1.0 \times 7}{22 \times (0.125 \times 10^{-2})^2 \times 0.91 \times 10^{11}}$$

$= 1.3 \times 10^{-4}$ m

Q 6. Four identical hollow cylindrical columns of mild steel support a big structure of mass 50000 kg. The inner and outer radii of each column are 30 cm and 60 cm, respectively. Assuming, the load distribution to be uniform, calculate the compressional strain of each column. Young's modulus, $Y = 2.0 \times 10^{11}$ Pa.

Sol. Total weight of the structure to be supported by four columns

$= Mg$

$= 50000 \times 9.8$ N

Since, this weight is to be supported by 4 columns.

$\therefore$ Compressional force on each column (F) is given by

$$F = \frac{Mg}{4} = \frac{50000 \times 9.8}{4}$$

Inner radius of a column, $r_1 = 30$ cm $= 0.3$ m

Outer radius of a column, $r_2 = 60$ cm $= 0.6$ m

$\therefore$ Area of cross-section of each column is given by

$$A = \pi (r_2^2 - r_1^2)$$

$$= \pi [(0.6)^2 - (0.3)^2]$$

$$= 0.27 \pi \text{m}^2$$

Young's modulus, $Y = 2 \times 10^{11}$ Pa

$$\therefore \quad Y = \frac{\text{Compressional force/Area}}{\text{Compressional strain}}$$

$$= \frac{F/A}{\text{Compressional strain}}$$

or compressional strain of each column

$$= \frac{F}{AY} = \frac{50000 \times 9.8 \times 7}{4 \times 0.27 \times 22 \times 2 \times 10^{11}}$$

$= 0.722 \times 10^{-6}$

Q 7. A piece of copper having a rectangular cross-section of 15.2 mm $\times$ 19.1 mm is pulled with 44500 N force, producing only elastic deformation. Calculate the resulting strain? (Y for copper $= 1.1 \times 10^{11}$ Nm^{-2}).

Sol. Here, $Y = 1.1 \times 10^{11}$ Nm^{-2}

A = Area of cross-section

$= 15.2$ mm $\times$ 19.1 mm

$= 15.2 \times 10^{-3}$ m $\times$ 19.1×10^{-3} m

Force, $F = 44500$ N

$$\therefore \quad Y = \frac{\text{Stress}}{\text{Strain}}$$

$$\text{Strain} = \frac{\text{Stress}}{Y} = \frac{F}{AY}$$

$$\text{Longitudinal strain} = \frac{44500}{15.2 \times 19.1 \times 10^{-6} \times 1.1 \times 10^{11}}$$

$= 0.139$

Q 8. A steel cable with a radius of 1.5 cm supports a chair lift. If the maximum stress is not to exceed 10^8 Nm^{-2}, what is the maximum load the cable can support?

Sol. Here, radius of steel cable,

$r = 1.5$ cm $= 1.5 \times 10^{-2}$ m

Maximum stress $= 10^8$ N/m^2

$\therefore$ Area of cross-section of cable

$$A = \pi r^2$$

$$= \pi (1.5 \times 10^{-2})^2$$

$$\text{Maximum stress} = \frac{\text{Maximum force}}{\text{Area of cross - section}}$$

or Maximum force

= Maximum stress $\times$ area of cross-section

$= 10^8 \times \pi \times (1.5 \times 10^{-2})^2 = 7.1 \times 10^4$ N

or Maximum load the cable can withstand $= 7.1 \times 10^4$ N.

Q9. A rigid bar of mass 15 kg is supported symmetrically by three wires each 2 m long. Those at each end are of copper and the middle one is of iron. Determine the ratios of their diameters, if each is to have the same tension.

Sol. Let Y_1 and Y_2 be the Young's modulus of copper and iron wires, respectively.

$\therefore$ $Y_1 = 110 \times 10^9$ Nm^{-2}

$Y_2 = 190 \times 10^9$ Nm^{-2}

Also let A_1 and A_2 be the areas of cross-section of copper and iron wires, respectively. If d_1 and d_2 be their respective diameters.

Then, $$A_1 = \frac{\pi d_1^2}{4}$$

and $$A_2 = \frac{\pi d_2^2}{4}$$

$$\therefore \quad \frac{A_1}{A_2} = \frac{d_1^2}{d_2^2} = \left(\frac{d_1}{d_2}\right)^2$$

$L = 2$ m

Let Δl be the extension produced in each wire.

Let F = Tension produced in each wire.

$\therefore$ From the relation, $Y = \frac{\text{stress}}{\text{strain}}$, we get

Strain for copper wire $= \frac{F/A_1}{Y_1}$

and strain for iron wire $= \frac{F/A_2}{Y_2}$

As the bar is supported symmetrically.

$\therefore$ The two strains are equal

$$\therefore \quad \frac{F}{A_1Y_1} = \frac{F}{A_2Y_2}$$

$$\text{or} \quad A_1Y_1 = A_2Y_2$$

$$\text{or} \quad \frac{A_1}{A_2} = \frac{Y_2}{Y_1}$$

$$\text{or} \quad \frac{\pi d_1^2/4}{\pi d_2^2/4} = \frac{Y_2}{Y_1}$$

$$\text{or} \quad \frac{d_1}{d_2} = \sqrt{\frac{Y_2}{Y_1}} = \sqrt{\frac{190 \times 10^9}{110 \times 10^9}}$$

$$\text{i.e.} \quad \frac{d_1}{d_2} = 1.31$$

$$\text{or} \quad d_1 : d_2 = 1.31 : 1$$

***Q* 10.** A 14.5 kg mass, fastened to one end of a steel wire of unstretched length 1 m is whirled in a vertical circle with an angular velocity of 2 rev/s at the bottom of the circle. The cross-sectional area of the wire is 0.065 cm^2. Calculate the elongation of the wire, when the mass is at the lowest point of its path.

Sol. Mass attached to one end of steel wire,

$$m = 14.5 \text{ kg}$$

Length of steel wire, $l = 1$ m

Frequency, $f = 2$ rev/s $= 2$ rps

$\therefore$ Angular frequency, $\omega = 2\pi f = 2\pi \times 2 = 4\pi$ rad/s

Area of cross-section of the wire,

$$A = 0.065 \text{ cm}^2$$

$$= 0.065 \times 10^{-4} \text{ m}^2$$

Y for steel $= 2 \times 10^{11}$ Pa

The stretching force developed in the wire due to the rotation of the mass,

$$mg + ml\omega^2 = 2432 \text{ N}$$

$\therefore$ Using the relation, $Y = \frac{\text{Stress}}{\text{Strain}} = \frac{F/A}{\Delta l / l}$, we get

$$\Delta l = \frac{F}{A} \times \frac{l}{Y}$$

We get
$$\Delta l = \frac{(2432)(1)}{(0.065 \times 10^{-4})(2 \times 10^{11})}$$

$$= 1.87 \times 10^{-3} \text{ m}$$

***Q* 11.** Compute the bulk modulus of water from the following data : initial volume = 100.0 L, pressure increase = 100.0 atm (1 atm = 1.013×10^5 Pa). Final volume = 100.5 L. Compare the bulk modulus of water with that of air (at constant temperature). Explain in simple terms why the ratio is so large?

Sol. Here, $p = 100$ atm

$$= 100 \times 1.013 \times 10^5 \text{ Pa} \quad (\because 1 \text{ atm} = 1.013 \times 10^5 \text{ Pa})$$

Initial volume, $V_1 = 100 \text{ L} = 100 \times 10^{-3} \text{ m}^3$

Final volume, $V_2 = 100.5 \text{L} = 100.5 \times 10^{-3} \text{ m}^3$

$$\Delta V = \text{change in volume} = V_2 - V_1$$

$$= (100.5 - 100) \times 10^{-3} \text{ m}^3$$

$$= 0.5 \times 10^{-3} \text{ m}^3$$

$\therefore$ From the formula, $B = \frac{p}{\Delta V / V}$, we get

$$B_w = \frac{pV}{\Delta V}$$

$$\text{e.g.} \quad = \frac{100 \times 1.013 \times 10^5 \times 100 \times 10^{-3}}{0.5 \times 10^{-3}}$$

$$\text{or} \quad B_w = 2.026 \times 10^9 \text{ Pa}$$

Bulk modulus of air at STP is given by

$$B_{\text{air}} = 1.0 \times 10^{-4} \text{ GPa}$$

$$= 1 \times 10^{-4} \times 10^9 \text{ Pa}$$

$$= 10^5 \text{ Pa}$$

$$\frac{B_w}{B_{\text{air}}} = \frac{2.026 \times 10^9}{10^5}$$

$$= 20260$$

The ratio is too large. This is due to the fact that the strain for air is much larger than for water at the same pressure. In other words, the intermolecular distances in case of liquid are very small as compared to the corresponding distances in the case of gases.

***Q* 12.** What is the density of water at a depth, where pressure is 80.0 atm, given that its density at the surface is 1.03×10^3 kg - m^{-3}? Compressibility of water is 45.8×10^{-11} Pa^{-1}.

Sol. Here, $p = 80.0 \text{ atm} = 80 \times 1.013 \times 10^5$ Pa

Compressibility $= \frac{1}{B} = 45.8 \times 10^{-11} \text{ Pa}^{-1}$

Density of water at the surface.

$$\rho = 1.03 \times 10^3 \text{ kg - m}^{-3}$$

Let ρ be the density of water at a given depth.

If V and V' be the volumes of certain mass M of the water at the surface and at a given depth, then

$$V = \frac{M}{\rho}$$

$$\text{and} \quad V' = \frac{M}{\rho'}$$

$\therefore$ Change in volume, $\Delta V = V - V' = M\left(\frac{1}{\rho} - \frac{1}{\rho'}\right)$

$\therefore$ Volumetric strain, $\frac{\Delta V}{V} = M\left(\frac{1}{\rho} - \frac{1}{\rho'}\right) \times \frac{\rho}{M}$

$$= \left(1 - \frac{\rho}{\rho'}\right)$$

or $$\frac{\Delta V}{V} = 1 - \frac{1.03 \times 10^3}{\rho'}$$

Also, we know that bulk modulus of water is given by the formula,

$$B = \frac{p}{\left(\frac{\Delta V}{V}\right)} = \frac{pV}{\Delta V}$$

$\therefore$ Compressibility $= \frac{1}{B} = \frac{\Delta V}{pV} = \frac{1}{p}\left(\frac{\Delta V}{V}\right)$

or $45.8 \times 10^{-11} = \frac{1}{80 \times 1.013 \times 10^5} \times \left(1 - \frac{1.03 \times 10^3}{\rho'}\right)$

Solving, we get,

$$\rho' = 1.034 \times 10^3 \text{ kg-m}^{-3}$$

Q 13. Compute the fractional change in volume of a glass slab, when subjected to a hydraulic pressure of 10 atm.

Sol. Here, $p = 10$ atm $= 10 \times 1.013 \times 10^5$ Pa

Bulk modulus for glass slab $= 37 \times 10^9$ Nm^{-2}

Using the relation $B = \frac{p}{\Delta V / V}$, we get

$$\frac{\Delta V}{V} = \frac{p}{B}$$

$$= \frac{10 \times 1.013 \times 10^5}{37 \times 10^9}$$

$$= \frac{1.013}{37 \times 10^3}$$

$$= 0.0274 \times 10^{-3}$$

Q 14. Determine the volume contraction of a solid copper cube, 10 cm on an edge, when subjected to a hydraulic pressure of 7×10^6 Pa. B for copper $= 140 \times 10^9$ Pa.

Sol. Here, $L = 10$ cm 0.1 m

B = bulk modulus of Cu

$= 140 \times 10^9$ Pa

$p = 7 \times 10^6$ Pa

$\therefore$ $V = L^3 = (0.1)^3 = 0.001$ m^3

From the relation $B = -\frac{p}{\Delta V / V}$, we get

$$\Delta V = -\frac{pV}{B}$$

$$= -\frac{7 \times 10^6 \times 0.001}{140 \times 10^9} \text{ m}^3$$

$$= -0.05 \text{ cm}^3$$

Here, –ve sign shows volume contraction.

Q 15. How much should the pressure on a litre of water be changed to compress it by 0.10%?

Sol. Here, $V = 1$ L

$$\Delta V = -0.10\% \text{ of } V$$

$$= -\frac{0.10}{100} \times 1$$

$$= -\frac{1}{1000} \text{ L}$$

Let Δp = change in pressure required for compression of 1 L of water.

B = bulk modulus of water $= 2.2 \times 10^9$ Nm^{-2}

Using the relation, $B = -\frac{\Delta p}{\Delta V / V}$, we get

$$\Delta p = -B \cdot \frac{\Delta V}{V}$$

$$= 2.2 \times 10^9 \times \frac{\left(\frac{1}{1000}\right)}{1}$$

$$= 2.2 \times 10^6 \text{ Nm}^{-2}$$

Q 16. The Marina trench is located in the Pacific ocean, and at one place, it is nearly 11km beneath the surface of water. The water pressure at the bottom of the trench is about 1.1×10^8 Pa. A steel ball of initial volume 0.32 m^3 is dropped into the ocean and falls to the bottom of trench. What is the change in the volume of the ball when it reaches to the bottom?

Sol. Pressure exerted by a 11 km column of water on the bottom of the trench,

$$p = h\rho g$$

$$= 11 \times 10^3 \times 10^3 \times 10 \text{ Pa}$$

$$= 1.1 \times 10^8 \text{ Pa}$$

$$V = 0.32 \text{ m}^3$$

B for steel $= 1.6 \times 10^{11}$ Nm^{-2}

Here, $p = 1.1 \times 10^8$ Pa

$$|\Delta V| = \frac{pV}{B}$$

$$= \frac{1.1 \times 10^8 \times (0.32)}{1.6 \times 10^{11}}$$

$$= 2.2 \times 10^{-4} \text{ m}^3$$

Objective Problems

[Level 1]

1. Young's modulus of the material of a wire of length L and radius r is Y N/m^2. If the length is reduced to $\frac{L}{2}$ and radius to $\frac{r}{2}$, the Young's modulus will be

(a) Y (b) $2Y$
(c) $\frac{Y}{4}$ (d) $\frac{Y}{2}$

2. Strain energy per unit volume in a stretched string is

(a) $\frac{1}{2}$ stress × strain (b) stress × strain
(c) (stress × strain)^2 (d) stress/strain

3. The Young's modulus of a wire is numerically equal to the stress which will

(a) not change the length of the wire
(b) double the length of the wire
(c) increase the length by 50%
(d) change the radius of the wire to half

4. The longitudinal extension of any elastic material is very small. In order to have an appreciable change, the material must be in the form of

(a) long thick wire (b) short thick wire
(c) long thin wire (d) short thin wire

5. The Young's modulus of a perfectly rigid body

(a) is zero
(b) is unity
(c) is infinity
(d) may have any finite non-zero value

6. In practice, Poisson's ratio σ lies between

(a) $-\infty$ to $+\infty$ (b) 0 and $+\infty$
(c) 0 and 0.5 (d) 0.5 and 1.0

7. A cable that can support a load W is cut into two equal parts. The maximum load that can be supported by either part is

(a) $\frac{W}{4}$ (b) $\frac{W}{2}$ (c) W (d) $2W$

8. The only elastic modulus that applies to fluids is

(a) Young's modulus (b) Shear modulus
(c) Modulus of rigidity (d) Bulk modulus

9. For steel $Y = 2\times10^{11}$ N/m^2. The force required to double the length of a steel wire of area 1 cm^2 is

(a) 2×10^7 N (b) 2×10^6 N
(c) 2×10^8 N (d) 2×10^5 N

10. When a metal wire elongates by hanging a load Mg on it, the gravitational potential energy of mass M decreases by Mgl. This energy appears

(a) as elastic potential energy completely
(b) as thermal energy completely
(c) half as elastic potential energy and half as thermal energy
(d) as kinetic energy of the load completely

11. Longitudinal stress of 1 kg/mm^2 is applied on a wire. The percentage increase in length is ($Y = 10^{11}$ N/m^2)

(a) 0.002 (b) 0.001 (c) 0.003 (d) 0.01

12. When a pressure of 100 atm is applied on a spherical ball, then its volume reduces to 0.01%. The bulk modulus of the material of the rubber in dyne/cm^2 is

(a) 10×10^{12} (b) 100×10^{12} (c) 1×10^{12} (d) 20×10^{12}

13. A uniform cube is subjected to volume compression. If each side is decreased by 1%, then bulk strain is

(a) 0.01 (b) 0.06
(c) 0.02 (d) 0.03

14. A ball falling in a lake of depth 200 m shows 0.01% decrease in its volume at the bottom. What is the bulk modulus of the material of the ball?

(a) 19.6×10^8 N/m^2 (b) 19.6×10^9 N/m^2
(c) 19.6×10^{10} N/m^2 (d) 19.6×10^{12} N/m^2

15. The Young's modulus of a wire is Y. If the energy per unit volume is E, then the strain will be

(a) $\sqrt{\frac{2E}{Y}}$ (b) $E\sqrt{2Y}$ (c) EY (d) $\frac{E}{Y}$

16. The bulk modulus for an incompressible liquid is

(a) zero (b) unity
(c) infinity (d) between 0 and 1

17. A steel wire of cross-sectional area 3×10^{-6} m^2 can withstand a maximum strain of 10^{-3}. Young's modulus of steel is 2×10^{11} N/m^2. The maximum mass this wire can hold is

(a) 40 kg (b) 60 kg (c) 80 kg (d) 100 kg

18. When a force is applied at one end of an elastic wire, it produces a strain E in the wire. If Y is the Young's modulus of the material of the wire, the amount of energy stored per unit volume of the wire is given by

(a) $Y\times E$ (b) $\frac{1}{2}(Y\times E)$ (c) $Y\times E^2$ (d) $\frac{1}{2}(Y\times E^2)$

[Level 2]

Only One Correct Option

1. In the given figure, if the dimensions of the wires are the same and materials are different, Young's modulus is more for

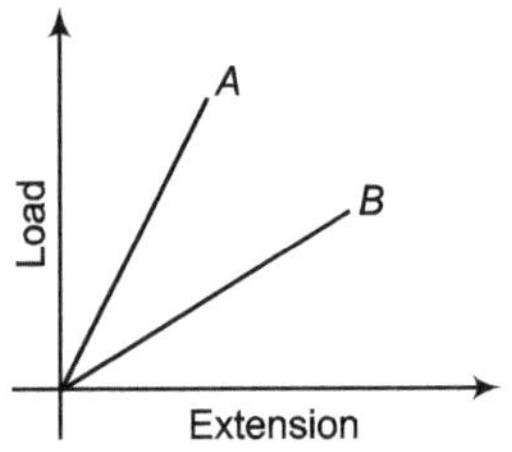

(a) A (b) B
(c) Both (d) None of these

2. Young's modulus of rubber is $10^4\ \text{N/m}^2$ and area of cross-section is $2\ \text{cm}^2$. If force of 2×10^5 dynes is applied along its length, then its initial length l becomes

(a) $3l$ (b) $4l$
(c) $2l$ (d) None of these

3. An elastic material with Young's modulus Y is subjected to a tensile stress S, elastic energy stored per unit volume of the material is

(a) $\frac{YS}{2}$ (b) $\frac{S^2}{Y}$
(c) $\frac{S^2}{2Y}$ (d) $\frac{S}{2Y}$

4. The bulk modulus of water is $2.1\times10^9\ \text{N/m}^2$. The pressure required to increase the density of water by 0.1% is

(a) $2.1\times10^3\ \text{N/m}^2$
(b) $2.1\times10^6\ \text{N/m}^2$
(c) $2.1\times10^5\ \text{N/m}^2$
(d) $2.1\times10^7\ \text{N/m}^2$

5. A thick rope of rubber of density $1.5\times10^3\ \text{kg/m}^3$ and Young's modulus $5\times10^6\ \text{N/m}^2$, 8 m in length, when hung from ceiling of a room, the increase in length due to its own weight is

(a) 96×10^{-3} m (b) 19.2×10^{-5} m
(c) 9.4 cm (d) 9.6 mm

6. Four wires of the same material are stretched by the same load. The dimensions of the wires are as given below. The one which has the maximum elongation is of

(a) diameter 1 mm and length 1 m
(b) diameter 2 mm and length 2 m
(c) diameter 0.5 mm and length 0.5 m
(d) diameter 3 mm and length 3 m

7. A metal block is experiencing an atmospheric pressure of $10^5\ \text{N/m}^2$. When the same block is placed in a vacuum chamber, the fractional change in its volume (the bulk modulus of metal is $1.25\times10^{11}\ \text{N/m}^2$)

(a) 4×10^{-7}
(b) 2×10^{-7}
(c) 8×10^{-7}
(d) 1×10^{-7}

8. A brass rod of length 2 m and cross-sectional area $2.0\ \text{cm}^2$ is attached end to end to a steel rod of length L and cross-sectional area $1.0\ \text{cm}^2$. The compound rod is subjected to equal and opposite pulls of magnitude 5×10^4 N at its ends. If the elongations of the two rods are equal, the length of the steel rod (L) is ($Y_{\text{brass}} = 1.0\times10^{11}\ \text{N/m}^2$ and $Y_{\text{steel}} = 2.0\times10^{11}\ \text{N/m}^2$)

(a) 1.5 m
(b) 1.8 m
(c) 1 m
(d) 2 m

9. An elevator cable is to have a maximum stress of $7\times10^7\ \text{N/m}^2$ to allow for appropriate safety factors. Its maximum upward acceleration is $1.5\ \text{m/s}^2$. If the cable has to support the total weight of 2000 kg of a loaded elevator, the area of cross-section of the cable should be

(a) $3.22\ \text{cm}^2$ (b) $2.38\ \text{cm}^2$
(c) $0.32\ \text{cm}^2$ (d) $8.23\ \text{cm}^2$

10. A uniform steel bar of cross-sectional area A and length L is suspended so that it hangs vertically. The stress at the middle point of the bar is (ρ is the density of steel)

(a) $\frac{L}{2A}\rho g$ (b) $\frac{L\rho g}{2}$
(c) $\frac{LA}{\rho g}$ (d) $L\rho g$

11. One end of a uniform wire of length L and weight w is attached rigidly to a point in roof and a weight w_1 is suspended from its lower end. If S is the area of cross-section of the wire, the stress in the wire at a height of $\frac{3L}{4}$ from its lower end is

(a) $\frac{w_1}{S}$ (b) $\frac{w_1 + \frac{w}{4}}{S}$
(c) $\left(\frac{w_1 + \frac{3w}{4}}{S}\right)$ (d) $\frac{w_1 + w}{S}$

12. Two wires of same material and length but diameter in the ratio 1 : 2 are stretched by the same force. The potential energy per unit volume for the two wires when stretched will be in the ratio

(a) 16 : 1 (b) 4 : 1
(c) 2 : 1 (d) 1 : 1

13. The length of an elastic string is a metre when the tension is 4 N, and b metre when the tension is 5 N. The length, in metre, when the tension is 9 N, is

(a) $(a + b)$ (b) $(4b - 5a)$
(c) $(5b - 4a)$ (d) $(9b - 9a)$

14. The length of a metal wire is l_1, when the tension in it is T_1 and is l_2 when the tension is T_2. The unstretched length of the wire is

(a) $\frac{l_1T_1 + l_2T_2}{T_1 + T_2}$ (b) $\frac{l_1 + l_2}{2}$
(c) $\frac{l_1T_2 - l_2T_1}{T_2 - T_1}$ (d) $\frac{l_1T_2 - l_2T_1}{T_2 + T_1}$

15. The Young's modulus of brass and steel are respectively 1.0×10^{10} N/m^2 and 2×10^{10} N/m^2. A brass wire and a steel wire of the same length are extended by 1 mm under the same force, the radii of brass and steel wires are R_B and R_S, respectively. Then,

(a) $R_S = \sqrt{2}\, R_B$ (b) $R_S = \frac{R_B}{\sqrt{2}}$
(c) $R_S = 4R_B$ (d) $R_S = \frac{R_B}{4}$

16. A block of weight w produces an extension of 9 cm, when it is hung by an elastic spring of length 60 cm and is cut into two parts one of length 40 cm and the other of length 20 cm. The same load w hangs in equilibrium supported by both parts as shown in figure. The extension in cm now is

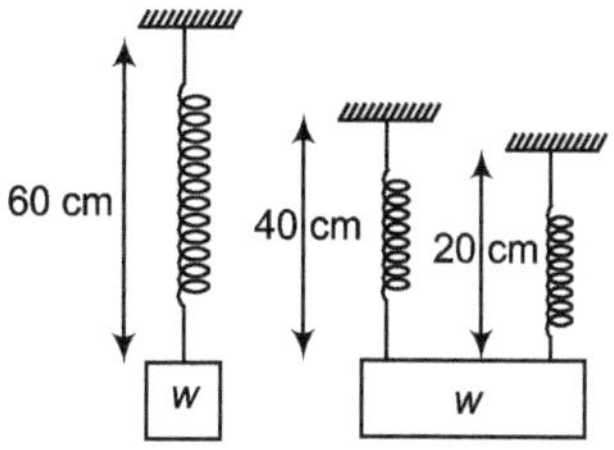

(a) 9 (b) 6 (c) 3 (d) 2

17. Two wires of the same material (Young's modulus = Y) and same length L but radii R and $2R$ respectively, are joined end to end and a weight w is suspended from the combination as shown in the figure. The elastic potential energy in the system is

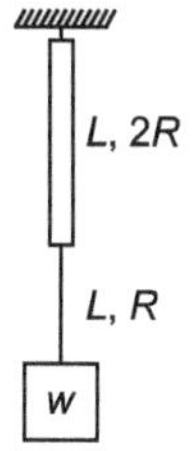

(a) $\frac{3w^2L}{4\pi R^2Y}$ (b) $\frac{3w^2L}{8\pi R^2Y}$
(c) $\frac{5w^2L}{8\pi R^2Y}$ (d) $\frac{w^2L}{\pi R^2Y}$

18. A uniform elastic plank moves due to a constant force F placed over a smooth surface. The area of end face is S and Young's modulus of the material is E. What is average strain produced in the direction of the force?

(a) $\frac{F}{SE}$ (b) $\frac{F}{2SE}$ (c) $\frac{F}{4SE}$ (d) Zero

19. What is the approximate change in density of water in a lake at a depth of 400 m below the surface? The density of water at the surface is 1030 kg/m^3 and bulk modulus of water is 2×10^9 N / m^2

(a) 4 kg/m^3 (b) 2 kg/m^3
(c) 6 kg/m^3 (d) 8 kg/m^3

20. A wire is elongated by 2 mm when a brick is suspended from it. When the brick is immersed in water, the wire contracts by 0.6 mm. What is the density of brick?

(a) 3333 kg/m^3 (b) 4210 kg/m^3
(c) 5000 kg/m^3 (d) 2000 kg/m^3

21. A load suspended by a massless spring produces an extension of x cm in equilibrium. When it is cut into two unequal parts, the same load produces an extension of 7.5 cm when suspended by the larger part of length 60 cm. When it is suspended by the smaller part, the extension is 5.0 cm. Then,

(a) $x = 12.5$
(b) $x = 3.0$
(c) the length of the original spring is 90 cm
(d) the length of the original spring is 80 cm

22. A copper wire ($Y = 10^{11}$ N / m^2) of length 8 m and a steel wire ($Y = 2\times 10^{11}$ N / m^2) of length 4 m, each of 0.5 cm^2 cross- section are fastened end to end and stretched with a tension of 500 N.

(a) Elongation in copper wire is 0.8 mm
(b) Elongation in steel is $\frac{1}{4}$th the elongation in copper wire
(c) Total elongation is 1.0 mm
(d) All of the above

23. A uniform cylinder rod of length L, cross-sectional area A and Young's modulus Y is acted upon by the forces shown in the figure. The elongation of the rod is

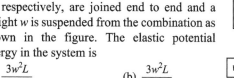

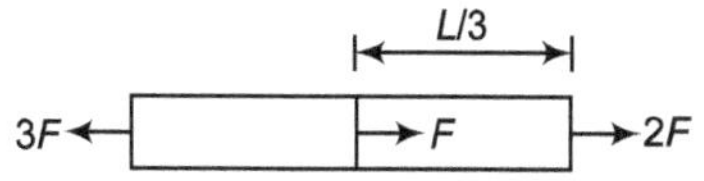

(a) $\frac{3FL}{5AY}$ (b) $\frac{2FL}{5AY}$ (c) $\frac{3FL}{8AY}$ (d) $\frac{8FL}{3AY}$

24. A stress of 10^6 N / m^2 is required for breaking a material. If the density of the material is 3×10^3 kg/m^3, then what should be the length of the wire made of this material so that it breaks under its own weight?

(a) 10 m (b) 33.3 m
(c) 5 m (d) 66.6 m

25. A uniform pressure p is exerted on all sides of a solid cube at temperature $t°$C. By what amount should the temperature be raised in order to bring the volume back to that it had been before the pressure was applied.

(Linear expansivity of the material of the cube is α and the bulk modulus of elasticity is β)

(a) $\frac{p\alpha}{\beta}$ (b) $\frac{3p\alpha}{\beta}$
(c) $\frac{p}{\alpha\beta}$ (d) $\frac{p}{3\alpha\beta}$

26. The length of an iron wire is L and area of cross-section is A. The increase in length is l on applying the force F on its two ends. Which of the statements is correct?
(a) Increase in length is inversely proportional to its length L
(b) Increase in length is proportional to area of cross-section A
(c) Increase in length is inversely proportional to A
(d) Increase in length is proportional to Young's modulus

27. The temperature of a wire of length 1 m and area of cross-section $1\,\text{cm}^2$ is increased from 0°C to 100°C. If the rod is not allowed to increase in length, the force required will be ($\alpha = 10^{-5}/°\text{C}$ and $Y = 10^{11}\ \text{N/m}^2$)
(a) 10^3 N (b) 10^4 N
(c) 10^5 N (d) 10^9 N

28. Two wires of copper having the length in the ratio 4 : 1 and their radii ratio as 1 : 4 are stretched by the same force. The ratio of longitudinal strain in the two will be
(a) 1 : 16 (b) 16 : 1
(c) 1 : 64 (d) 64 : 1

29. The interatomic distance for a metal is 3×10^{-10} m. If the interatomic force constant is 3.6×10^{-9} N/Å, then the Young's modulus in N/m^2 will be
(a) 1.2×10^{11} (b) 4.2×10^{11}
(c) 10.8×10^{8} (d) 2.4×10^{10}

30. A force of 20 N is applied at one end of a wire of length 2 m and having area of cross-section $10^{-2}\ \text{m}^2$. The other end of the wire is rigidly fixed. Its coefficient of linear expansion of the wire is $\alpha = 8 \times 10^{-6}/°\text{C}$ and Young's modulus $Y = 2.2 \times 10^{11}\ \text{N/m}^2$ and its temperature is increased by 5°C, then the increase in the tension of the wire will be
(a) 4.2 N (b) 4.4 N
(c) 2.4 N (d) 8.8 N

31. Two wires of same diameter of the same material having the length l and $2l$. If the force F is applied on each, the ratio of the work done in the two wires will be
(a) 1 : 2
(b) 1 : 4
(c) 2 : 1
(d) 1 : 1

32. A 5 m long wire is fixed to the ceiling. A weight of 10 kg is hung at the lower end is 1 m above the floor. The wire was elongated by 1 mm. The energy stored in the wire due to stretching is
(a) 0.01 J (b) 0.05 J
(c) 0.02 J (d) 0.04 J

33. The stress *versus* strain graphs for wires of two materials A and B are as shown in the figure. If Y_A and Y_B are the Young's modulii of the materials, then

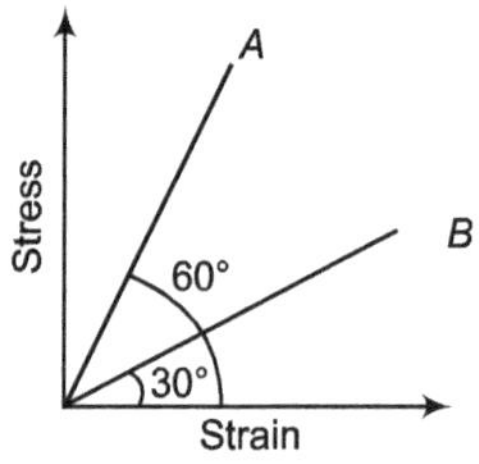

(a) $Y_B = 2Y_A$ (b) $Y_A = Y_B$
(c) $Y_B = 3Y_A$ (d) $Y_A = 3Y_B$

34. The load *versus* elongation graph for four wires of the same material is shown in the figure. The thickest wire is represented by the line

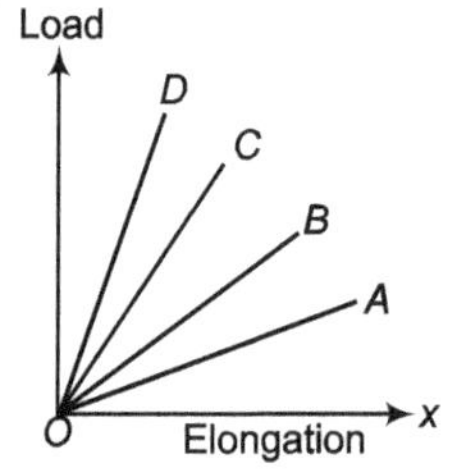

(a) OD (b) OC
(c) OB (d) OA

35. The strain-stress curves of three wires of different materials are shown in the figure. P, Q and R are elastic limits of the wires. The figure shown that

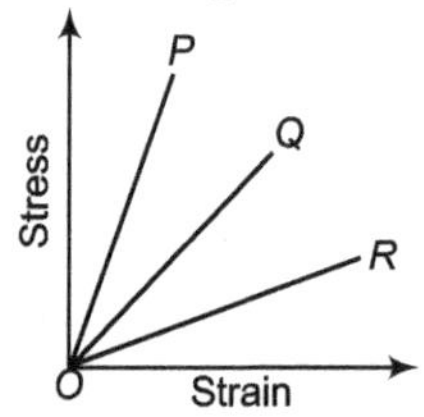

(a) elasticity of wire P is maximum
(b) elasticity of wire Q is maximum
(c) tensile strength of R is maximum
(d) None of the above

36. The density of a metal at normal pressure is ρ. Its density when it is subjected to an excess pressure p is ρ'. If B is the bulk modulus of the metal, the ratio $\frac{\rho'}{\rho}$ is

(a) $\frac{1}{1-\frac{p}{B}}$ (b) $1+\frac{B}{p}$ (c) $\frac{1}{1-\frac{B}{p}}$ (d) $1+\frac{p}{B}$

37. When a steel wire fixed at one end is pulled by a constant force F at its other end, its length increases by l. Which of the following statements is not correct?

(a) Work done by the external force is Fl
(b) Some heat is produced in the wire in the process
(c) The elastic potential energy of the wire is $\frac{Fl}{2}$
(d) The heat produced is equal to half of the elastic potential energy stored in the wire

38. A uniform rod of mass m, length L, area of cross- section A and Young's modulus Y hangs from the ceiling. Its elongation under its own weight will be

(a) zero (b) $\frac{mgL}{2AY}$
(c) $\frac{mgL}{AY}$ (d) $\frac{2mgL}{AY}$

More than One Correct Options

1. A metal wire of length L, area of cross-section A and modulus Y is stretched by a variable force F such that F is always slightly greater than the elastic forces of resistance in the wire. When the elongation of the wire is l

(a) the work done by F is $\frac{YAl^2}{2L}$
(b) the work done by F is $\frac{YAl^2}{L}$
(c) the elastic potential energy stored in the wire is $\frac{YAl^2}{2L}$
(d) the elastic potential energy stored in the wire is $\frac{YAl^2}{4L}$

2. Two wires A and B of same length are made of same material. The figure represents the load F *versus* extension Δx graph for the two wires. Then,

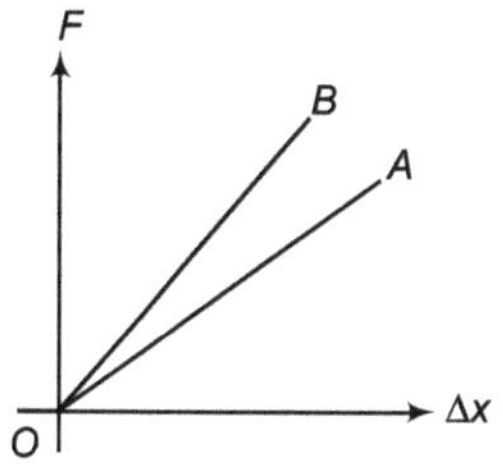

(a) the cross-sectional area of A is greater than that of B
(b) the elasticity of B is greater than that of A
(c) the cross-sectional area of B is greater than that of A
(d) the elasticity of A is greater than that of B

3. A body of mass M is attached to the lower end of a metal wire, whose upper end is fixed. The elongation of the wire is l.

(a) Loss in gravitational potential energy of M is Mgl
(b) The elastic potential energy stored in the wire is Mgl
(c) The elastic potential energy stored in the wire is $\frac{1}{2}Mgl$
(d) Heat produced is $\frac{1}{2}Mgl$

4. The stress-strain graphs for two materials are shown in figure. (assume same scale)

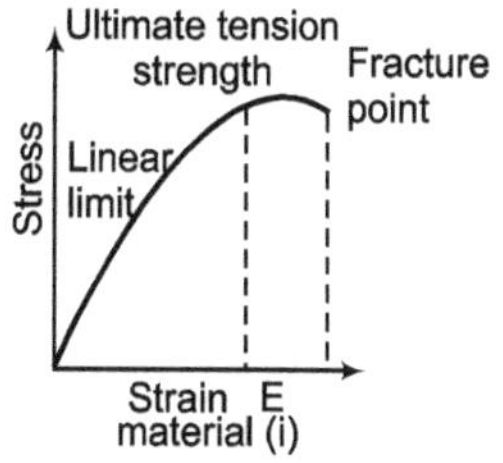

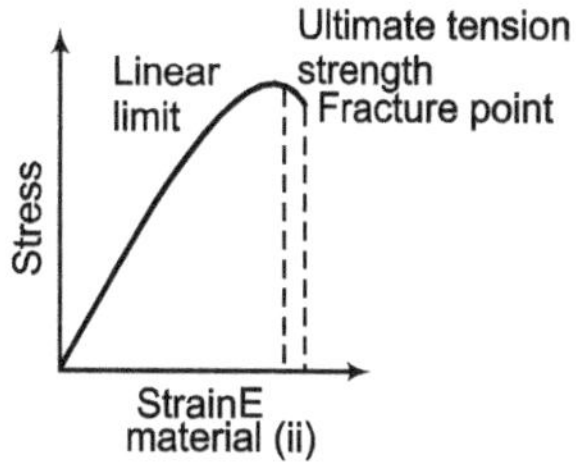

(a) Material (ii) is more elastic than material (i) and hence material (ii) is more brittle
(b) Material (i) and (ii) have the same elasticity and the same brittleness
(c) Material (ii) is elastic over a larger region of strain as compared to (i)
(d) Material (ii) is more brittle than material (i)

5. A rod of length l and negligible mass is suspended at its two ends by two wires of steel (wire A) and aluminium (wire B) of equal lengths (figure). The cross-sectional areas of wires A and B are 1.0 mm^2 and 2.0 mm^2, respectively.
($Y_{Al} = 70\times10^9 Nm^{-2}$ and $Y_{steel} = 200\times 10^9 Nm^{-2}$)

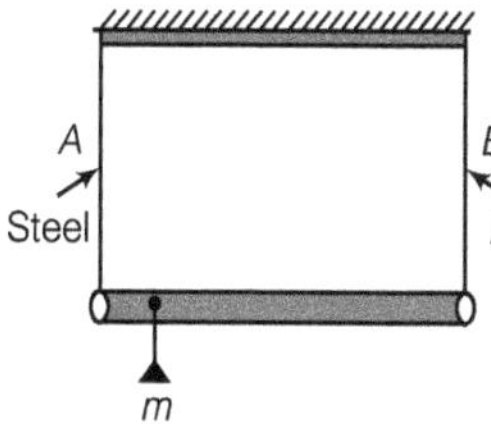

(a) Mass m should be suspended close to wire A to have equal stresses in both the wires
(b) Mass m should be suspended close to B to have equal stresses in both the wires
(c) Mass m should be suspended at the middle of the wires to have equal stresses in both the wires
(d) Mass m should be suspended close to wire A to have equal strain in both wires

6. A copper and a steel wire of the same diameter are connected end to end. A deforming force F is applied to this composite wire which causes a total elongation of 1 cm. The two wires will have

(a) the same stress
(b) different streses
(c) the same strain
(d) different strains

Assertion and Reason

Directions (Q. Nos. 1-14) *These question consists of two statements each printed as assertion and reason. While answering these questions you are required to choose anyone of the following five responses.*

(a) If both Assertion and Reason are true and Reason is the correct explanation of Assertion.
(b) If both Assertion and Reason are true but Reason is not the correct explanation of Assertion.
(c) If Assertion is true but Reason is false.
(d) If Assertion is false but Reason is true.
(e) If both Assertion and Reason are false.

1. Assertion Up to elastic limit of a stress-strain curve, the steel wire tends to regain its original shape when stress is removed.

Reason Within elastic limit, the wire follows Hooke's law.

2. Assertion Young's modulus of elasticity is not defined for liquids.

Reason Liquids cannot be stretched as wires.

3. Assertion Incompressible liquids have finite value of bulk modulus of elasticity.

Reason Compressibility is inverse of bulk modulus of elasticity.

4. Assertion If length of a wire is halved, its Young's modulus of elasticity will become two times.

Reason The ratio of longitudinal stress and longitudinal strain is called Young's modulus of elasticity.

5. Assertion The materials having low value of Young's modulus of elasticity are more ductile.

Reason If Young's modulus is less, they can be easily stretched as wires.

6. Assertion Bulk modulus of elasticity of gases is process dependent.

Reason More the pressure of gas more is the bulk modulus of elasticity of gas.

7. Assertion Stress and modulus of elasticity have the same dimensions.

Reason Strain is dimensionless.

8. Assertion If radius of cylindrical wire is doubled, then this wire can bear four times stress.

Reason By doubling the radius, the area of cross-section will become four times.

9. Assertion If a wire is stretched, only half of the work done in stretching the wire remains stored as elastic potential energy.

Reason Potential energy stored in the wire is $\frac{1}{2}(\text{stress}) \times (\text{strain})$

10. Assertion Bulk modulus of an incompressible fluid is infinite.

Reason Density of incompressible fluid remains constant.

11. Assertion If value of Young's modulus of elasticity of a material is less, then it means material is ductile.

Reason If a material is easily stretchable, then its value of Young's modulus of elasticity should be less.

12. Assertion Modulus of elasticity does depend upon the dimensions of material.

Reason Modulus of elasticity is a material property.

13. Assertion Up to the elastic limit strain $\propto$ stress.

Reason Up to elastic limit material returns to its original shape and size. When external force is removed.

14. Assertion Adiabatic bulk modulus of an ideal gas is more than its isothermal bulk modulus.

Reason Both the modulii are proportional to the pressure of gas at that moment.

Match the Columns

1. Match the following columns.

Column I	Column II
(A) Steel	(p) Young's modulus of elasticity
(B) Water	(q) Bulk modulus of elasticity
(C) Hydrogen gas filled in a chamber	(r) Shear modulus of elasticity

2. Match the following columns.

Column I	Column II
(A) Coefficient of viscosity	(p) $[M^2L^{-1}T^{-2}]$
(B) Surface tension	(q) $[ML^0T^{-2}]$
(C) Modulus of elasticity	(r) $[ML^{-1}T^2]$
(D) Energy per unit volume of a fluid	(s) None

3. Match the following columns.

Column I	Column II
(A) Stress × strain	(p) J
(B) $\frac{YA}{l}$	(q) N/m
(C) Yl^3	(r) J/m^3
(D) $\frac{Fl}{AY}$	(s) m

Entrance Gallery

2014

1. One end of a horizontal thick copper wire of length $2L$ and radius $2R$ is welded to an end of another horizontal thin copper wire of length L and radius R. When the arrangement is stretched by applying forces at two ends, the ratio of the elongation in the thin wire to that in the thick wire is **[JEE Advanced]**

(a) 0.25 (b) 0.50
(c) 2.00 (d) 4.00

2. The pressure that has to be applied to the ends of a steel wire of length 10 cm to keep its length constant when its temperature is raised by 100°C is (For steel, Young's modulus is 2×10^{11} Nm^{-2} and coefficient of thermal expansion is 1.1×10^{-5} K^{-1}) **[JEE Main]**

(a) 2.2×10^{8} Pa (b) 2.2×10^{9} Pa
(c) 2.2×10^{7} Pa (d) 2.2×10^{6} Pa

3. The length of the wire is increased by 2% by applying a load of 2.5 kg-wt. What is the linear strain produced in the wire? **[J&K CET]**

(a) 0.1 (b) 0.01 (c) 0.2 (d) 0.02

2013

4. If in a wire of Young's modulus Y, longitudinal strain X is produced, then the value of potential energy stored in its unit volume will be **[Karnataka CET]**

(a) $0.5\,YX^2$ (b) $0.5\,Y^2X$
(c) $2YX^2$ (d) YX^2

2011

5. The average depth of Indian ocean is about 3000 m. The fractional compression, $\frac{\Delta V}{V}$ of water at the bottom of the ocean (given that the Bulk modulus of the water $= 2.2\times 10^{9}\,Nm^{-2}$ and $g = 10\ ms^{-2}$) is **[Kerala CEE]**

(a) 0.82% (b) 0.91% (c) 1.36% (d) 1.24%
(e) 1.52%

6. Identify the incorrect statement. **[Kerala CEE]**

(a) Young's modulus and shear modulus are relevant only for solids
(b) Bulk modulus is relevant for solids, liquids and gases
(c) Alloys have larger values of Young's modulus than metals
(d) Metals have larger values of Young's modulus than elastomers
(e) Stress is not a vector quantity

7. The following four wires of length L and radius r are made of the same material. Which of these will have the largest extension, when the same tension is applied? **[Karnataka CET]**

(a) $L = 400$ cm, $r = 0.8$ mm (b) $L = 300$ cm, $r = 0.6$ mm
(c) $L = 200$ cm, $r = 0.4$ mm (d) $L = 100$ cm, $r = 0.2$ mm

8. A material has Poisson's ratio 0.50. If a uniform rod of it suffers a longitudinal strain of 2×10^{-3}, then the percentage change in volume is **[WB JEE]**

(a) 0.6 (b) 0.4
(c) 0.2 (d) zero

2010

9. A 0.1 kg mass is suspended from a wire of negligible mass. The length of the wire is 1 m and its cross-sectional area is 4.9×10^{-7} m^2. If the mass is pulled a little in the vertically downward direction and released, it performs simple harmonic motion of angular frequency

140 rad s^{-1}. If the Young's modulus of the material of the wire is $n\times 10^{9}$ Nm^{-2}, the value of n is **[IIT JEE]**

(a) 4 (b) 2
(c) 4.5 (d) 5

10. Two wires are made of the same material and have the same volume. However, wire 1 has cross-sectional area A and wire 2 has cross-sectional area $3A$. If the length of wire 1 increases by Δx on applying force F, how much force is needed to stretch wire 2 by the same amount?

(a) F (b) $4F$ **[AIEEE]**
(c) $6F$ (d) $9F$

11. Which of the following substances has the highest elasticity? **[Karnataka CET]**

(a) Sponge (b) Steel
(c) Rubber (d) Copper

12. There is no change in the volume of a wire due to the change in its length on stretching. The Poisson's ratio of the material of the wire is **[OJEE]**

(a) $+\frac{1}{2}$ (b) $-\frac{1}{2}$
(c) $+\frac{1}{4}$ (d) $-\frac{1}{4}$

13. The increase in pressure required to decrease the 200 L volume of a liquid by 0.008% in kPa is (Bulk modulus of the liquid = 2100 MPa) **[MHT CET]**

(a) 8.4 (b) 84
(c) 92.4 (d) 168

14. Four wires of the same material are stretched by the same load. Which one of them will elongate most if their dimensions are as follows? **[MHT CET]**

(a) $L = 100$ cm, $r = 1$ mm (b) $L = 200$ cm, $r = 3$ mm
(c) $L = 300$ cm, $r = 3$ mm (d) $L = 400$ cm, $r = 4$ mm

15. Which of the following relations is true? **[MHT CET]**

(a) $Y = 2\eta\,(1 - 2\sigma)$ (b) $Y = 2\eta\,(1 + 2\sigma)$
(c) $Y = 2\eta\,(1 - \sigma)$ (d) $(1 + \sigma)2\eta = Y$

16. For a given material, the Young's modulus is 2.4 times that of rigidity modulus. Its Poisson's ratio is **[MHT CET]**

(a) 2.4 (b) 1.2 (c) 0.4 (d) 0.2

Answers

Level 1

Objective Problems

1. (a)	**2.** (a)	**3.** (b)	**4.** (c)	**5.** (c)	**6.** (c)	**7.** (c)	**8.** (d)	**9.** (a)	**10.** (c)
11. (b)	**12.** (c)	**13.** (d)	**14.** (a)	**15.** (a)	**16.** (c)	**17.** (b)	**18.** (d)		

Level 2

Only One Correct Option

1. (a)	**2.** (c)	**3.** (c)	**4.** (b)	**5.** (c)	**6.** (c)	**7.** (c)	**8.** (d)	**9.** (a)	**10.** (b)
11. (c)	**12.** (a)	**13.** (c)	**14.** (c)	**15.** (b)	**16.** (d)	**17.** (c)	**18.** (b)	**19.** (b)	**20.** (b)
21. (a)	**22.** (d)	**23.** (d)	**24.** (b)	**25.** (d)	**26.** (c)	**27.** (b)	**28.** (b)	**29.** (a)	**30.** (b)
31. (a)	**32.** (b)	**33.** (d)	**34.** (a)	**35.** (d)	**36.** (a)	**37.** (d)	**38.** (b)		

More than One Correct Options

1. (a,c)	**2.** (c)	**3.** (a,c,d)	**4.** (c,d)	**5.** (b,d)	**6.** (a,d)

Assertion and Reason

1. (c)	**2.** (a)	**3.** (b)	**4.** (d)	**5.** (a)	**6.** (b)	**7.** (a,b)	**8.** (d)	**9.** (c)	**10.** (b)
11. (a)	**12.** (a)	**13.** (d)	**14.** (b)						

Match the Columns

1. (A→p,q,r; B→q; C→q) **2.** (A→s; B→q; C→r; D→r) **3.** (A→r; B→q; C→p; D→s)

Entrance Gallery

1. (c)	**2.** (a)	**3.** (d)	**4.** (a)	**5.** (c)	**6.** (e)	**7.** (d)	**8.** (d)	**9.** (a)	**10.** (d)
11. (b)	**12.** (b)	**13.** (b)	**14.** (a)	**15.** (d)	**16.** (d)				

Solutions

Level 1 : Objective Problems

1. Young's modulus is property of material.

3. $Y = \dfrac{\text{Stress}}{\text{Strain}}$, Y = Stress for unit strain or $\Delta l = l$

4. $\Delta l = \dfrac{Fl}{AY}$ or $\Delta l \propto \dfrac{l}{A}$

5. For perfectly rigid body, $\Delta l = 0$

$\therefore$ Strain $= 0$ or $Y = \infty$

7. $(\text{Stress})_1 = (\text{Stress})_2$

Maximum load does not depend on the length.

9. Stress required to double the length is called Young's modulus.

$\therefore \quad Y = \dfrac{F}{A}$

or $\quad F = Y \cdot A = 2 \times 10^{11} \times 10^{-4} = 2 \times 10^7$ N

11. Longitudinal strain $= \dfrac{\text{stress}}{Y} = \dfrac{10^6}{10^{11}} = 10^{-5}$

Percentage increase in length $= 10^{-5} \times 100 = 0.001\%$

12. $B = \dfrac{100}{0.01/100} = 10^6$ atm $= 10^{11}$ N/m^2 $= 10^{12}$ dyne/cm^2

13. If side of the cube is L, then $V = L^3$

$\Rightarrow \quad \dfrac{dV}{V} = 3\dfrac{dL}{L}$

$\therefore$ % change in volume $= 3 \times$ (% change in length) $= 3 \times 1\% = 3\%$

$\therefore$ Bulk strain, $\dfrac{\Delta V}{V} = 0.03$

14. $B = \dfrac{\Delta p}{\Delta V/V} = \dfrac{h\rho g}{0.1/100} = \dfrac{200 \times 10^3 \times 9.8}{1/1000}$
$= 19.6 \times 10^8$ N/m^2

15. Energy per unit volume $= \dfrac{1}{2} \times Y \times (\text{strain})^2$

$\therefore \quad$ Strain $= \sqrt{\dfrac{2E}{Y}}$

16. $B = \dfrac{-\Delta p}{(\Delta V / V)}$

For incompressible liquid, $\Delta V = 0$

$\therefore \quad B = \infty$

17. Maximum stress $= Y \times$ (maximum strain)

$$\frac{Mg}{A} = 2 \times 10^{11} \times 10^{-3} = 2 \times 10^{8}$$

$$\therefore \quad M = \frac{2 \times 10^{8} \times 3 \times 10^{-6}}{10} = 60 \text{ kg}$$

18. Energy per unit volume $= \dfrac{1}{2} \times$ stress $\times$ strain

$$= \frac{1}{2} \times (Y \times \text{strain}) \times \text{strain}$$

$$= \frac{1}{2} \times Y \times E^2$$

Level 2 : Only One Correct Option

1. $F = \left(\dfrac{YA}{l}\right) \cdot \Delta l$

i.e. F-Δl graph is a straight line with slope $\dfrac{YA}{l}$ or slope proportional to Y.

$$(\text{Slope})_A > (\text{slope})_B$$

$$\therefore \quad Y_A > Y_B$$

2. $\Delta l = \dfrac{Fl}{AY} = \dfrac{(2)l}{(10^4)(2 \times 10^{-4})} = l$

$\therefore$ New length will become $2l$.

3. Strain $= \dfrac{\text{Stress}}{Y} = \dfrac{S}{Y}$

Now, energy stored per unit volume $= \dfrac{1}{2} \times$ stress $\times$ strain

$$= \frac{1}{2} \times S \times \frac{S}{Y} = \frac{S^2}{2Y}$$

4. $d\rho = \dfrac{\rho}{B} \cdot dp$

or $\quad \dfrac{d\rho}{\rho} = \dfrac{dp}{B}$

$$\frac{0.1}{100} = \frac{dp}{2.1 \times 10^{9}}$$

$$\therefore \quad dp = 2.1 \times 10^{6} \text{ N/m}^2$$

5. Due to own weight, $\Delta l = \dfrac{mgl}{2AY}$

$$= \frac{(lA\rho)\, gl}{2AY} = \frac{l^2 \rho g}{2Y}$$

$$= \frac{(8)^2 (1.5 \times 10^{3})(9.8)}{2 \times 5 \times 10^{6}}$$

$$= 9.4 \times 10^{-2} \text{ m} = 9.4 \text{ cm}$$

6. $\Delta l = \dfrac{Fl}{AY}$

or $\quad \Delta l \propto \dfrac{l}{d^2} \left(A = \dfrac{\pi}{4} d^2\right)$

Therefore, Δl will be maximum for that wire for which $\dfrac{l}{d^2}$ is maximum.

8. $(\Delta l)_b = (\Delta l)_s$

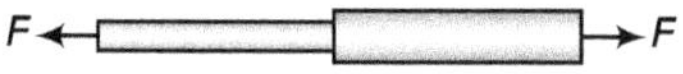

or $\quad \left(\dfrac{Fl}{AY}\right)_b = \left(\dfrac{Fl}{AY}\right)_s \qquad (F_b = F_s)$

or $\quad \left(\dfrac{l}{AY}\right)_b = \left(\dfrac{l}{AY}\right)_s$

$$\therefore \quad l_s = \left(\frac{A_s Y_s}{A_b Y_b}\right) l_b$$

$$= \left(\frac{1.0 \times 2.0 \times 10^{11}}{2.0 \times 1.0 \times 10^{11}}\right) (2\text{m}) = 2 \text{ m}$$

9. $T_{max} = m(g + a) = (2000)(9.8 + 1.5) = 22600$ N

Maximum stress $= \dfrac{T_{max}}{\text{Area}}$

$$\therefore \quad \text{Area} = \frac{T_{max}}{\text{Maximum stress}}$$

$$= \frac{22600}{7 \times 10^{7}}$$

$$= 3.22 \times 10^{-4} \text{ m}^2 = 3.22 \text{ cm}^2$$

10. At middle, T = weight of half the length of steel bar

$$= \left(\frac{L}{2}\right) A\rho g$$

$$\therefore \quad \text{Stress} = \frac{T}{A} = \frac{L\rho g}{2}$$

11. At length $\dfrac{3L}{4}$ from lower end, tension in the wire.

$$T = \text{suspended load} + \frac{3}{4} \times \text{weight of wire}$$

$$= w_1 + \frac{3w}{4}$$

$$\therefore \quad \text{Stress} = \frac{T}{S} = \frac{w_1 + \dfrac{3w}{4}}{S}$$

12. $\Delta l = \dfrac{Fl}{AY} \Rightarrow \dfrac{\Delta l}{l} = \left(\dfrac{F}{AY}\right)$

Potential energy per unit volume $= \dfrac{1}{2} \times$ stress $\times$ strain

$$u = \frac{1}{2} \times \frac{F}{A} \times \frac{F}{AY}$$

or $\quad u \propto \dfrac{1}{A^2}$ or $u \propto \dfrac{1}{d^4}$

$$\therefore \quad \frac{u_1}{u_2} = \left(\frac{d_2}{d_1}\right)^4 = \frac{16}{1}$$

13. Let l be the natural length and $K\left(= \dfrac{YA}{l}\right)$ be the force constant of wire. Then,

$$a = l + \frac{4}{K} \qquad \left(F = K\,\Delta l \text{ or } \Delta l = \frac{F}{K}\right)$$

and $\quad b = l + \dfrac{5}{K}$

or $\quad \dfrac{1}{K} = (b - a)$

and $\quad l = (5a - 4b)$

Now when $\quad T = 9\,\text{N}$

$$l' = l + \frac{9}{K}$$

$$= (5a - 4b) + 9(b - a)$$

$$= (5b - 4a)$$

14. Let natural length be l and force constant is $K\left(= \frac{YA}{l}\right)$. Then,

$$l_1 = l + \frac{T_1}{K} \qquad \left(F = K\Delta l \text{ or } \Delta l = \frac{F}{K}\right)$$

and $$l_2 = l + \frac{T_2}{K}$$

Solving these two equations, we get

$$l = \frac{l_1T_2 - l_2T_1}{T_2 - T_1}$$

15. $\Delta l = \dfrac{Fl}{AY} = \dfrac{Fl}{\pi R^2 Y}$

Δl, F and l are same

Hence, $R^2Y =$ constant

or $$\frac{R_S}{R_B} = \sqrt{\frac{Y_B}{Y_S}} = \sqrt{\frac{1}{2}}$$

or $$R_S = \frac{R_B}{\sqrt{2}}$$

16. Force constant, $k \propto \dfrac{1}{l}$

Let k be the force constant of original spring.

Then, $$k_{40} = \frac{60}{40}k = \frac{3}{2}k$$

and $$k_{20} = \frac{60}{20}k = 3k$$

$$k_{\text{net}} = k_{40} + k_{20} = \frac{9k}{2}$$

Now, $$\Delta l = \frac{F}{k} \propto \frac{1}{k}$$

$\therefore$ $$\Delta l' = \left(\frac{k}{9k/2}\right)(9\text{ cm})$$

$$= 2\text{ cm}$$

17. $\Delta l_1 = \dfrac{wL}{(4\pi R^2)y}$, $\Delta l_2 = \dfrac{wL}{\pi R^2 Y}$

$\therefore$ $$U = \frac{1}{2}K_1(\Delta l_1)^2 + \frac{1}{2}K_2(\Delta l_2)^2 \qquad \left(K = \frac{YA}{L}\right)$$

$$= \frac{1}{2} \times \frac{Y(4\pi R^2)}{L} \times \left[\frac{wL}{4\pi R^2 Y}\right]^2 + \frac{1}{2} \times \frac{Y(\pi R^2)}{L} \times \left[\frac{wL}{\pi R^2 Y}\right]^2$$

$$= \frac{5w^2L}{8\pi R^2 Y}$$

18. Tension at distance x from end A, $T_x = F - \dfrac{F}{L}x$

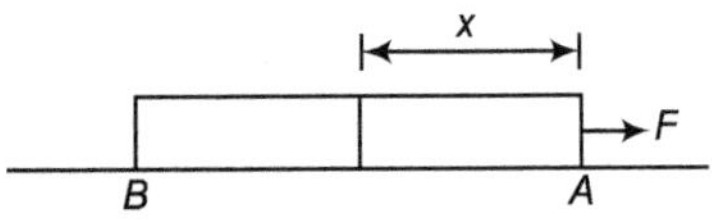

Total change in length,

$$\Delta l = \int_0^L \frac{T_x dx}{ES} = \frac{FL}{2ES}$$

$\therefore$ Average strain $= \dfrac{\Delta l}{L} = \dfrac{F}{2ES}$

19. $d\rho = \dfrac{\rho}{B} \cdot dp$

$$dp = \rho g h = (1030)(10)(400)$$

$$= 4.12 \times 10^6\ \text{N/m}^2$$

$\therefore$ $$d\rho = \frac{1030}{2 \times 10^9} \times 4.12 \times 10^6 = 2.12\ \text{kg/m}^3$$

20. $\Delta l \propto F$

$$\frac{\Delta l_1}{\Delta l_2} = \frac{F_1}{F_2} = \frac{\text{Weight}}{\text{Weight} - \text{upthrust}}$$

$$\frac{2}{1.4} = \frac{V\rho g}{V\rho g - V\rho_w g}$$

or $$\rho = \frac{2}{0.6}\rho_w$$

$$= \frac{2}{0.6} \times 1000$$

$$= 3333\ \text{kg/m}^3$$

21. $\Delta l = \dfrac{Fl}{AY}$ (can also be applied for a spring)

$\therefore$ $$\Delta l \propto l$$

$$\frac{7.5}{5.0} = \frac{60}{l_2}$$

$\therefore$ $$l_2 = 40\text{ cm}$$

$\therefore$ Length of original spring is $(60 + 40)$ cm $= 100$ cm

Now, $$\frac{x}{7.5} = \frac{100}{60}$$

$\therefore$ $$x = 12.5\text{ cm}$$

22. $(\Delta l)_C = \left(\dfrac{Fl}{AY}\right)_C = \dfrac{500 \times 8}{0.5 \times 10^{-4} \times 10^{11}}$

$$= 0.8 \times 10^{-3}\ \text{m} = 0.8\ \text{mm}$$

$$(\Delta l)_S = \left(\frac{Fl}{AY}\right)_S$$

$$= \frac{500 \times 4}{0.5 \times 10^{-4} \times 2 \times 10^{11}}$$

$$= 0.2 \times 10^{-3}\ \text{m}$$

$$= 0.2\ \text{mm}$$

$$(\Delta l)_S = \frac{1}{4}(\Delta l)_C$$

and $$\Delta l = 0.8 + 0.2 = 1.0\ \text{mm}$$

23. The free body diagrams of two parts are shown in figure.

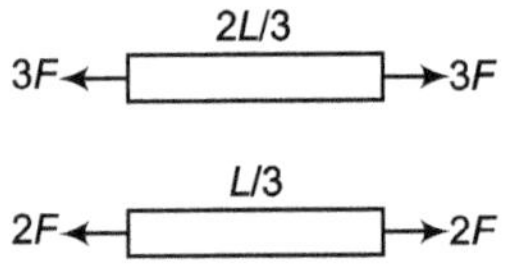

Both parts are stretched. Therefore, total elongation

$$\Delta l = \Delta l_1 + \Delta l_2$$
$$= \frac{3F\left(\frac{2L}{3}\right)}{AY} + \frac{2F\left(\frac{L}{3}\right)}{AY}$$
$$= \frac{8FL}{3AY}$$

24. $\frac{mg}{A}$ = maximum breaking stress

or $\frac{(Al\rho)g}{A} = \sigma_{max}$

$\therefore$ $l_{max} = \frac{\sigma_{max}}{\rho g}$

$$= \frac{10^6}{(3\times10^3)(10)}$$
$$= 33.33 \text{ m}$$

25. $\left(\frac{\Delta V}{V}\right)_1 = \gamma\Delta t = 3\alpha(\Delta t)$ (by change in temperature)

and $\left(\frac{\Delta V}{V}\right)_2 = \frac{p}{\beta}$ (by pressure)

$\left(\frac{\Delta V}{V}\right)_1 = \left(\frac{\Delta V}{V}\right)_2$ or $3\alpha(\Delta t) = \frac{p}{\beta}$

$\therefore$ $\Delta t = \frac{p}{3\alpha\beta}$

26. $\Delta l = \frac{FL}{YA} \Rightarrow l \propto \frac{1}{A}$

27. F = force developed = $YA\alpha(\Delta\theta)$

$= 10^{11} \times 10^{-4} \times 10^{-5} \times 100 = 10^4$ N

28. Strain $\propto$ stress $\propto \frac{F}{A}$

Ratio of strain $= \frac{A_2}{A_1} = \left(\frac{r_2}{r_1}\right)^2 = \left(\frac{4}{1}\right)^2 = \frac{16}{1}$

29. $Y = \frac{3.6\times10^{-9} \text{ N/Å}}{3\times10^{-10} \text{ m}} = 1.2\times10^{11} \text{N/m}^2$

30. Increase in tension of wire $= YA\alpha\Delta\theta$

$= 8\times10^{-6}\times2.2\times10^{11}\times10^{-2}\times10^{-4}\times5$

$= 8.8$ N

31. $W = \frac{1}{2}Fl$

$\therefore$ $W \propto l$ (F is constant)

$\therefore$ $\frac{W_1}{W_2} = \frac{l_1}{l_2} = \frac{l}{2l} = \frac{1}{2}$

32. $W = \frac{1}{2}\times F\times l = \frac{1}{2}mgl$

$$= \frac{1}{2}\times10\times10\times1\times10^{-1}$$
$$= 0.05 \text{ J}$$

33. $\frac{Y_A}{Y_B} = \frac{\tan\theta_A}{\tan\theta_B} = \frac{\tan 60°}{\tan 30°} = \frac{\sqrt{3}}{1/\sqrt{3}} = 3$

$\Rightarrow$ $Y_A = 3Y_B$

34. $l = \frac{FL}{AY}$

$\therefore$ $l \propto \frac{1}{r^2}$ (Y, L and F are constants)

i.e for the same load, thickest wire will show minimum elongation. So, graph D represents the thickest wire.

35. As, stress is shown on x-axis and strain on y-axis.

So, we can say that $Y = \cot\theta = \frac{1}{\tan\theta} = \frac{1}{\text{slope}}$

So, elasticity of wire P is minimum and of wire R is maximum.

36. $\rho = \frac{m}{V}$ or $\rho \propto \frac{1}{V}$ (as m = constant)

$\therefore$ $\frac{\rho'}{\rho} = \frac{V}{V_1}$

$$= \frac{V}{V+dV}\left(B = -\frac{p}{dV/V}\right)$$
$$= \frac{V}{V - \frac{Vp}{B}} = \frac{1}{1-\frac{p}{B}}$$

37. The heat produced is equal to the elastic potential energy stored in the wire.

38. Tension at distance x from the bottom,

$$T = \left(\frac{mg}{L}\right)x\text{dt}$$

Now, $dl = \frac{T\cdot dx}{AY}$

$\therefore$ $\Delta l = \int_0^L dl = \frac{mgL}{2AY}$

More than One Correct Options

1. W_F = elastic potential energy stored in the wire

$$= \frac{1}{2}k(\Delta l)^2 = \frac{1}{2}\left(\frac{YA}{L}\right)(l)^2 = \frac{YAl^2}{2L}$$

2. $\Delta x = \frac{Fl}{AY}$

$\therefore$ $F = \left(\frac{AY}{L}\right)\Delta x$

i.e. F *versus* Δx graph is a straight line of slope $\frac{YA}{L}$.

$$(\text{Slope})_B > (\text{Slope})_A$$

$\therefore$ $\left(\frac{YA}{L}\right)_B > \left(\frac{YA}{L}\right)_A$ or $(A)_B > (A)_A$

They are of same material.

Hence, $Y_B = Y_A$

3. Half of energy is lost in heat and rest half is stored as elastic potential energy.

$$l = \frac{Mgl}{AY} \quad \text{...(i)}$$
$$U = \frac{1}{2}Kl^2 = \frac{1}{2}\left(\frac{YA}{L}\right)l^2 \quad \text{...(ii)}$$

From Eqs. (i) and (ii), we can prove that

$$U = \frac{1}{2}Mgl$$

4. It is clear from the two graphs, the ultimate tensile strength for material (ii) is greater, hence material (ii) is elastic over larger region as compared to material (i).

For material (ii) fracture point is nearer, hence it is more brittle.

5. Let the mass is placed at x from the end B.

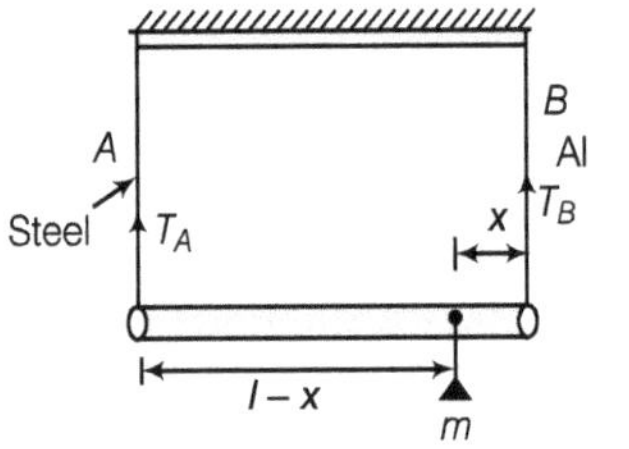

Let T_A and T_B be the tensions in wire A and wire B, respectively.

For the rotational equilibrium of the system,

$$\Sigma\tau = 0 \quad \text{(Total torque = 0)}$$

$$\Rightarrow \quad T_B x - T_A(l - x) = 0$$

$$\Rightarrow \quad \frac{T_B}{T_A} = \frac{l - x}{x} \quad \text{...(i)}$$

$$\text{Stress in wire } A = S_A = \frac{T_A}{a_A}$$

$$\text{Stress in wire } B = S_B = \frac{T_B}{a_B}$$

where, a_A and a_B are cross-sectional areas of wire A and B, respectively.

By question, $a_B = 2a_A$

Now, **for equal stress** $\quad S_A = S_B$

$$\Rightarrow \quad \frac{T_A}{a_A} = \frac{T_B}{a_B} \Rightarrow \frac{T_B}{T_A} = \frac{a_B}{a_A} = 2$$

$$\Rightarrow \quad \frac{l - x}{x} = 2 \Rightarrow \frac{l}{x} - 1 = 2$$

$$\Rightarrow \quad x = \frac{l}{3}$$

$$\Rightarrow \quad l - x = l - l/3 = \frac{2l}{3}$$

Hence, mass m should be placed closer to B.

For equal strain, $\quad (\text{strain})_A = (\text{strain})_B$

$$\Rightarrow \quad \frac{(Y_A)}{S_A} = \frac{Y_B}{S_B} \quad \text{(where, } Y_A \text{ and } Y_B \text{ are Young, modulii)}$$

$$\Rightarrow \quad \frac{Y_{\text{steel}}}{T_A/a_A} = \frac{Y_{\text{Al}}}{T_B/a_B}$$

$$\Rightarrow \quad \frac{Y_{\text{steel}}}{Y_{\text{Al}}} = \frac{T_A}{T_B} \times \frac{a_B}{a_A} = \left(\frac{x}{l - x}\right)\left(\frac{2a_A}{a_A}\right)$$

$$\Rightarrow \quad \frac{200 \times 10^9}{70 \times 10^9} = \frac{2x}{l - x}$$

$$\Rightarrow \quad \frac{20}{7} = \frac{2x}{l - x}$$

$$\Rightarrow \quad \frac{10}{7} = \frac{x}{l - x}$$

$$\Rightarrow \quad 10l - 10x = 7x$$

$$\Rightarrow \quad 17x = 10l$$

$$\Rightarrow \quad x = \frac{10l}{17}$$

$$l - x = l - \frac{10l}{17} = \frac{7l}{17}$$

Hence, mass m should be placed closer to wire A.

6. Consider the diagram where a deforming force F is applied to the combination.

For steel wire, $Y_{\text{steel}} = \dfrac{\text{Stress}}{\text{Strain}} = \dfrac{F/A}{\text{Strain}}$

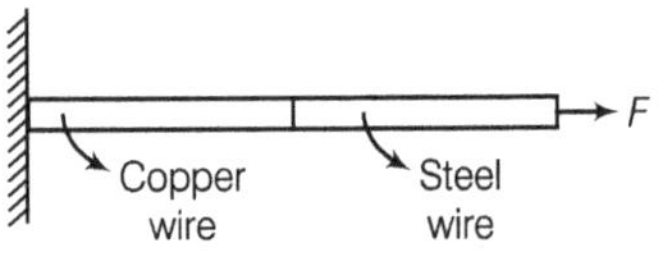

where, F is tension in each wire and A is cross-section area of each wires.

As F and A are same for both the wires, hence, stress will be same for both the wires.

$$(\text{Strain})_{\text{steel}} = \frac{\text{Stress}}{Y_{\text{steel}}}, \; (\text{Strain})_{\text{copper}} = \frac{\text{Stress}}{Y_{\text{copper}}}$$

As, $\quad Y_{\text{steel}} \neq Y_{\text{copper}}$

Hence, the two wires will have different strains.

Assertion and Reason

1. Hooke's law, stress = Strain

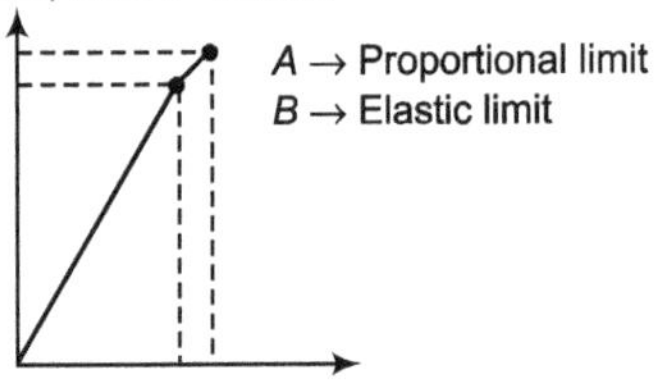

Is obey up to proportional limit.

3. Bulk modulus, $B = \dfrac{-\Delta p}{\Delta V / V}$

For incompressible fluids, $\Delta V = 0$

$\therefore \quad B = \infty$

4. Young's modulus is the property of material. It does not depend on length of the wire.

6. If pressure of gas is more, it is more difficult to compress a gas. Therefore, bulk modulus is high.

8. Load bearing capacity will become four times not the stress bearing capacity.

9. Potential energy per unit volume is $\frac{1}{2}$ (stress) × strain.

10. Incompressible fluid is that which cannot be compressed ($\Delta V = 0$) by applying pressure on it.

14. $B_S = \gamma p$ = Adiabatic bulk modulus of elasticity.

$B_T = p$ = Isothermal bulk modulus of elasticity

$B_S > B_T$ and both are proportional to p.

Entrance Gallery

1. The elongation in the thin wire

$$\because \quad \Delta l = \frac{FL}{AY} = \frac{FL}{(\pi r^2)Y} \Rightarrow \Delta l \propto \frac{L}{A}$$

$$\therefore \quad \frac{\Delta l_1}{\Delta l_2} = \frac{L/R^2}{2L/(2R)^2} = 2$$

2. According to Hooke's law,

Young's modulus, $Y = \dfrac{\text{Tensile stress}}{\text{Tensile strain}} = \dfrac{F/A}{\Delta L/L}$

If the rod is compressed, then compressive stress and strain appear. Their ratio Y is same as that for tensile case.

Given, length of a steel wire, $L = 10$ cm,

Temperature, $\Delta\theta = 100°C$,

$$Y = 2 \times 10^{11} \text{ N/m}^2 \quad \text{and} \quad \alpha = 1.1 \times 10^{-5} \text{ K}^{-1}$$

As, length is constant.

$$\therefore \quad \text{Strain} = \frac{\Delta L}{L} = \alpha\Delta\theta$$

Now, pressure = stress = $Y \times$ strain

$$= 2 \times 10^{11} \times 1.1 \times 10^{-5} \times 100$$
$$= 2.2 \times 10^8 \text{ Pa}$$

3. So, percentage change in length of wire $= \frac{\Delta l}{l} \times 100\%$

$$\frac{\Delta l}{l} \times 100 = 2\% \Rightarrow \frac{\Delta l}{l} = \frac{2}{100} = 0.02$$

4. Given, Young's modulus = Y, longitudinal strain = X

The potential energy stored is given by

$$= \frac{1}{2} \times \text{Stress} \times \text{Strain}$$
$$= \frac{1}{2} \times Y \times X \times X = \frac{1}{2} YX^2 = 0.5\, YX^2 \quad (\because \text{Stress} = Y \times \text{Strain})$$

5. We know that $B = \frac{\Delta p}{\Delta V/V}$

$$\Rightarrow \quad \frac{\Delta V}{V} = \frac{\Delta p}{B} = \frac{\rho g h}{B} = \frac{10^3 \times 10 \times 3000}{2.2 \times 10^9} = 1.36\ \%$$

6. Metals have larger values of Young's modulus, than elastomers, and the alloys having high densities, i.e. alloys have larger values of Young's modulus than metals.

7. Young's modulus, $Y = \frac{F}{A} \times \frac{L}{l}$

$$Y = \frac{F}{\pi r^2} \times \frac{L}{l} \Rightarrow \text{Extension, } l \propto \frac{L}{r^2}$$

Option (d) has the largest extension, when the same tension is applied.

8. Poisson's ratio, $\sigma = \frac{\text{Lateral strain}}{\text{Longitudinal strain}}$

$$= \frac{\frac{\Delta D}{D}}{\frac{\Delta L}{L}} = \frac{\frac{\Delta r}{r}}{\frac{\Delta L}{L}} \quad \text{or} \quad \frac{\Delta r}{r} = \sigma \times \frac{\Delta L}{L}$$
$$= 0.5 \times 2 \times 10^{-3} = 10^{-3}$$

Volume of rod, $V = \pi r^2 L$

$$V + \Delta V = \pi (r - \Delta r)^2 (L + \Delta L)$$

Neglecting Δr^2 and $\Delta r, \Delta L$, we get

$$\text{or} \quad \Delta V/V = \pi r^2 \Delta L / \pi r^2 L - \frac{2\pi r L\, \Delta r}{\pi r^2 L}$$
$$\Rightarrow \quad \frac{\Delta V}{V} = \frac{\Delta L}{L} - 2\frac{\Delta r}{r}$$
$$= 2 \times 10^{-3} - 2 \times 10^{-3} = 0$$

9. For SHM, we know that $\omega = \sqrt{\frac{k}{m}} = \sqrt{\frac{YA}{lm}}$

$$= \sqrt{\frac{(n \times 10^9)(4.9 \times 10^{-7})}{1 \times 0.1}}$$

Given, $\omega = 140 \text{ rads}^{-1}$ in above equation, we get

$$n = 4$$

10. As wires are of same material, we can write $A_1 l_1 = A_2 l_2$

$$\Rightarrow \quad l_2 = \frac{A_1 l_1}{A_2} = \frac{A \times l_1}{3A} = \frac{l_1}{3} \Rightarrow \frac{l_1}{l_2} = 3$$

Now,
$$\Delta x_1 = \frac{F_1}{AY} l_1 \quad \ldots \text{(i)}$$
$$\Delta x_2 = \frac{F_2}{3AY} l_2 \quad \ldots \text{(ii)}$$

Here,
$$\Delta x_1 = \Delta x_2 \Rightarrow \frac{F_2}{3AY} l_2 = \frac{F_1}{AY} l_1$$
$$F_2 = 3F_1 \times \frac{l_1}{l_2}$$
$$= 3F_1 \times 3 = 9F$$

11. Out of the given substances, steel has greater value of Young's modulus. Therefore, steel has highest elasticity.

12. Volume of cylindrical wire, $V = \frac{\pi x^2 L}{4} \quad \ldots \text{(i)}$

where, x is the diameter of wire.

Differentiating both sides of Eq. (i), we get

$$\frac{dV}{dx} = \frac{\pi}{4}\left[2xL + x^2 \cdot \frac{dL}{dx}\right]$$

Also, volume remains constant.

$$\therefore \quad \frac{dV}{dx} = 0 \Rightarrow 2xL + x^2 \frac{dL}{dx} = 0$$
$$\Rightarrow \quad 2xL = -x^2 \frac{dL}{dx} \Rightarrow \frac{\frac{dx}{x}}{\frac{dL}{L}} = -\frac{1}{2}$$

Poisson's ratio $= -\frac{1}{2}$

13. Bulk modulus, $K = \frac{\Delta p}{\Delta V} V$

$$\Delta p = \frac{K \Delta V}{V} \Rightarrow \Delta p = \frac{2100 \times 10^6 \times 0.008}{200} = 84 \text{ kPa}$$

14. As, $\Delta L = \frac{FL}{AY}$

Because, wires of the same material are stretched by the same load. So, F and Y will be constant.

$$\therefore \quad \Delta L \propto \frac{L}{\pi r^2}$$
$$\Delta L_1 = \frac{100}{\pi \times (1 \times 10^{-3})^2} = \frac{100}{\pi \times 10^{-6}} = \frac{100}{\pi} \times 10^6$$
$$\therefore \quad \Delta L_2 = \frac{200}{\pi \times (3 \times 10^{-3})^2} = \frac{200}{\pi \times 9 \times 10^{-6}} = \frac{22.2}{\pi} \times 10^6$$
$$\therefore \quad \Delta L_3 = \frac{300}{\pi \times (3 \times 10^{-3})^2} = \frac{300}{\pi \times 9 \times 10^{-6}} = \frac{33.3}{\pi} \times 10^6$$
$$\therefore \quad \Delta L_4 = \frac{400}{\pi \times (4 \times 10^{-3})^2} = \frac{400}{\pi \times 16 \times 10^{-6}} = \frac{25}{\pi} \times 10^6$$

We can see that, $L = 100$ cm and $r = 1$ mm will elongate most.

15. We know that

$$Y = (1 + \sigma) 2\eta$$

16. As, $\quad Y = 2\eta(1 + \sigma)$

$$\Rightarrow \quad 2.4\eta = 2\eta(1 + \sigma)$$
$$\Rightarrow \quad 1.2 = 1 + \sigma$$
$$\Rightarrow \quad \sigma = 0.2$$

13

Fluid Mechanics

Chapter Snapshot

- Definition of a Fluid
- Density of a Liquid
- Pressure in a Fluid
- Archimedes' Principle
- Flow of Fluids
- Bernoulli's Equation
- Applications Based on Bernoulli's Equation
- Viscosity
- Stoke's Law and Terminal Velocity
- Surface Tension

13.1 Definition of a Fluid

Fluid mechanics deals with the behaviour of fluids at rest and in motion. Fluids comprise the liquid and gas (or vapour) phases of the physical forms in which matter exists. The distinction between a fluid and the solid state of matter is clear, if you compare fluid and the solid behaviour. A solid deforms when a shear stress is applied but it does not continue to increase with time. However, if a shear stress is applied to a fluid, then deformation continues to increase as long as the stress is applied. We may alternatively define a fluid as a substance that **cannot sustain a shear stress when at rest.**

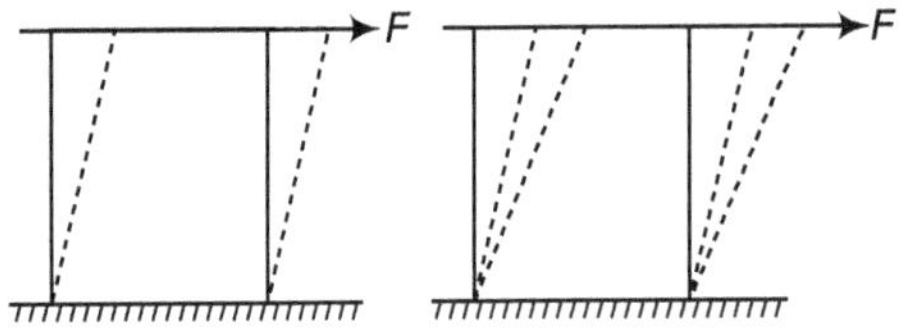

Fig. 13.1 Behaviour of a solid and a fluid, under the action of a constant shear force

In the present chapter, we shall deal with liquids. An ideal liquid is incompressible and non-viscous in nature. An incompressible liquid means the density of the liquid is constant, it is independent of the variations in pressure. A non-viscous liquid means that, parts of the liquid in contact do not exert any tangential force on each other. Thus, there is no friction between the adjacent layers of a liquid. The force by one part of the liquid on the other part is **perpendicular to the surface of contact.**

13.2 Density of a Liquid

Density (ρ) of any substance is defined as the mass per unit volume and it can be expressed as

$$\rho = \frac{\text{Mass}}{\text{Volume}} \quad \text{or} \quad \rho = \frac{m}{V}$$

Relative Density (RD)

In case of a liquid, sometimes an another term **relative density** (RD) is defined. It is the ratio of density of the substance to the density of water at 4°C. Hence,

$$\text{RD} = \frac{\text{Density of substance}}{\text{Density of water at } 4°\text{C}}$$

Relative density is a pure ratio. So, it has no units. It is also sometimes referred as specific gravity.

Density of water at 4°C in CGS is 1 g/cm^3. Therefore, numerically the RD and density of substance (in CGS) are equal. In SI units the density of water at 4°C is 1000 kg/m^3.

Example 13.1 *Relative density of an oil is 0.8. Find the absolute density of oil in CGS and SI units.*

Sol. Density of oil (in CGS) = (RD) g/cm^3

$$= 0.8 \text{ g/cm}^3 = 800 \text{ kg/m}^3$$

Density of a Mixture of Two or More Liquids

Here, we have two cases :

Case 1 Suppose two liquids of densities ρ_1 and ρ_2 having masses m_1 and m_2 are mixed together. Then, the density of the mixture will be

$$\rho = \frac{\text{Total mass}}{\text{Total volume}} = \frac{(m_1 + m_2)}{(V_1 + V_2)} = \frac{(m_1 + m_2)}{\left(\dfrac{m_1}{\rho_1} + \dfrac{m_2}{\rho_2}\right)}$$

If $m_1 = m_2$, then $\rho = \dfrac{2\rho_1\rho_2}{\rho_1 + \rho_2}$

Case 2 If two liquids of densities ρ_1 and ρ_2 having volumes V_1 and V_2 are mixed together, then the density of mixture is

$$\rho = \frac{\text{Total mass}}{\text{Total volume}} = \frac{m_1 + m_2}{V_1 + V_2} = \frac{\rho_1 V_1 + \rho_2 V_2}{V_1 + V_2}$$

If $V_1 = V_2$, then $\rho = \dfrac{\rho_1 + \rho_2}{2}$

Effect of Temperature on Density

As the temperature of a liquid is increased, the mass remains the same while the volume is increased and hence, the density of the liquid decreases (as $\rho \propto 1/V$). Thus,

$$\frac{\rho'}{\rho} = \frac{V}{V'} = \frac{V}{V + dV} = \frac{V}{V + V\gamma\Delta\theta}$$

or $$\frac{\rho'}{\rho} = \frac{1}{1 + \gamma\Delta\theta}$$

Here, γ = thermal coefficient of volume expansion

and $\Delta\theta$ = rise in temperature

$$\therefore \quad \rho' = \frac{\rho}{1 + \gamma\Delta\theta}$$

Effect of Pressure on Density

As pressure is increased, volume decreases and hence density will increase. Thus,

$$\rho \propto \frac{1}{V}$$

$$\therefore \quad \frac{\rho'}{\rho} = \frac{V}{V'} = \frac{V}{V + dV} = \frac{V}{V - \left(\dfrac{dp}{B}\right)V} \quad \text{or} \quad \frac{\rho'}{\rho} = \frac{1}{1 - \dfrac{dp}{B}}$$

Here, dp = change in pressure,

and B = bulk modulus of elasticity of the liquid

Therefore, $$\rho' = \frac{\rho}{1 - dp/B}$$

13.3 Pressure in a Fluid

When a fluid (either liquid or gas) is at rest, it exerts a force perpendicular to any surface in contact with it, such as a container wall or a body immersed in the fluid.

While the fluid as a whole is at rest, the molecules that makes up the fluid are in motion, the force exerted by the fluid is due to the molecules colloiding with their surroundings.

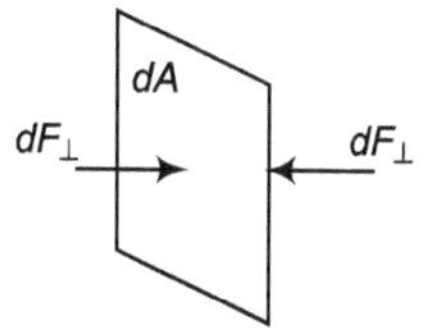

Fig.13.2

If we think of an imaginary surface within the fluid, then fluid on the two sides of the surface exerts equal and opposite forces on the surface, otherwise the surface would accelerate and the fluid would not remain at rest.

Consider a small surface of area dA centered on a point on the fluid, the normal force exerted by the fluid on each side is $dF_\perp$. The pressure p is defined at that point as the normal force per unit area, i.e.

$$p = \frac{dF_\perp}{dA}$$

If the pressure is the same at all points of a finite plane surface with an area A, then

$$p = \frac{F_\perp}{A}$$

where, $F_\perp$ is the normal force on one side of the surface. The SI unit of pressure is pascal, where

$$1 \text{ pascal} = 1\text{Pa} = 1.0 \text{ N/m}^2$$

One unit used principally in meterology is the bar which is equal to 10^5 Pa.

$$1 \text{ bar} = 10^5 \text{ Pa}$$

Atmospheric Pressure (p_0)

It is pressure of the earth's atmosphere. This changes with weather and elevation. Normal atmospheric pressure at sea level (an average value) is 1.013×10^5 Pa. Thus,

$$1 \text{ atm} = 1.013 \times 10^5 \text{ Pa}$$

- Fluid pressure acts perpendicular to any surface in the fluid no matter how that surface is oriented. Hence, pressure has no intrinsic direction of its own and it is a **scalar quantity.** By contrast, force is a vector with a definite direction.

Absolute Pressure and Gauge Pressure

The excess pressure above atmospheric pressure is usually called gauge pressure and the total pressure is called absolute pressure. Thus,

Gauge pressure = absolute pressure − atmospheric pressure

Absolute pressure is always greater than or equal to zero. While gauge pressure can be negative also.

Variation in Pressure with Depth

If the weight of the fluid can be neglected, then pressure in a fluid is the same throughout its volume. But often the fluid's weight is not negligible and under such condition pressure increases with increasing depth below the surface.

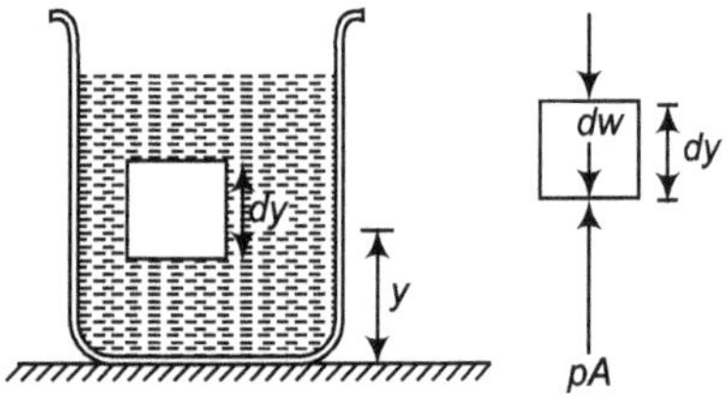

Fig. 13.3

Let us now derive a general relation between the pressure p at any point in a fluid at rest and the elevation y of that point. We will assume that the density ρ and the acceleration due to gravity g are the same throughout the fluid. If the fluid is in equilibrium, then every volume element is in equilibrium.

Consider a thin element of fluid with height dy. The bottom and top surfaces each have area A and they are at elevations y and $y + dy$ above some reference level, where $y = 0$. The weight of the fluid element is

$$dw = (\text{volume})\,(\text{density})\,(g) = (A\,dy)(\rho)(g)$$

or $\quad dw = \rho g A\,dy$

What are the other forces in Y-direction on this fluid element? Call the pressure at the bottom surface p, the total Y-component of upward force is pA. The pressure at the top surface is $p + dp$ and the total y-component of downward force on the top surface is $(p + dp)A$. The fluid element is in equilibrium, so the total y-component of force including the weight and the forces at the bottom and top surfaces must be zero.

$$\Sigma F_y = 0$$

$\therefore \quad pA - (p + dp)A - \rho g A\,dy = 0$

or $$\frac{dp}{dy} = -\rho g \qquad \ldots\text{(i)}$$

This equation shows that when y increases, p decreases, i.e. as we move upward in the fluid, pressure decreases.

If p_1 and p_2 be the pressures at elevations y_1 and y_2 and if ρ and g are constant, then integrating Eq. (i), we get

$$\int_{p_1}^{p_2} dp = -\rho g \int_{y_1}^{y_2} dy$$

or $$p_2 - p_1 = -\rho g (y_2 - y_1) \qquad \ldots\text{(ii)}$$

It's often convenient to express Eq. (ii) in terms of the depth below the surface of a fluid. Take point 1 at depth h below the surface of fluid and let p represents pressure at this point. Take point 2 at the surface of the fluid, where the pressure is p_0 (subscript zero for zero depth). The depth of point 1 below the surface is

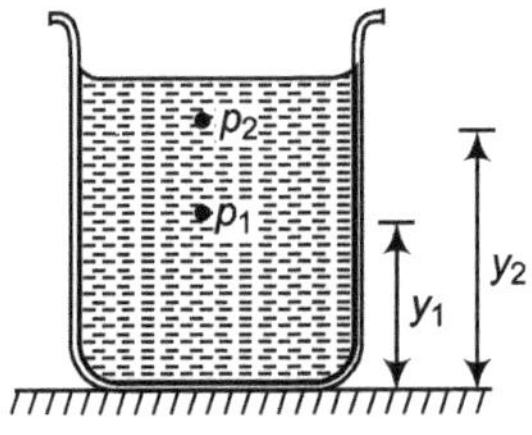

Fig. 13.4

$$h = y_2 - y_1$$

and Eq. (ii) becomes

$$p_0 - p = -\rho g (y_2 - y_1) = -\rho g h$$

$\therefore$ $$p = p_0 + \rho g h \qquad \ldots\text{(iii)}$$

Thus, pressure increases linearly with depth, if ρ and g are uniform. A graph between p and h is shown below.

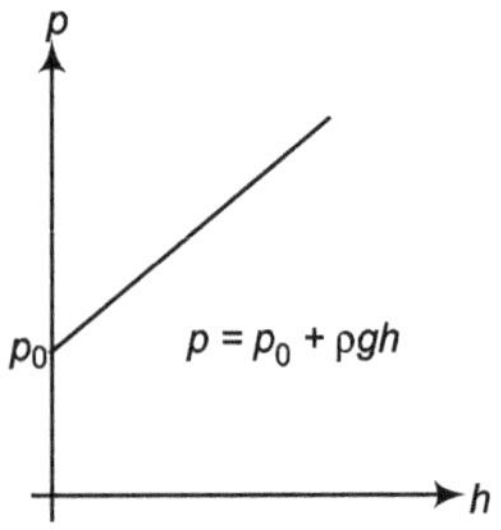

Fig. 13.5

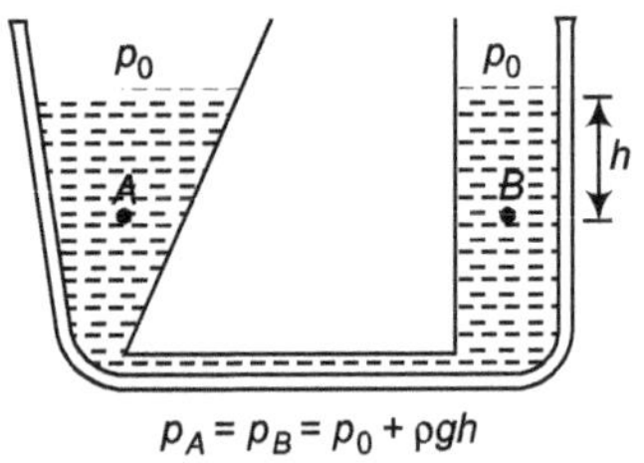

Fig. 13.6

Further, the pressure is the same at any two points at the same level in the fluid. The shape of the container does not matter.

Pascal's Law

It states that "pressure applied to an enclosed fluid is transmitted undiminished to every portion of the fluid and the walls of the containing vessel".

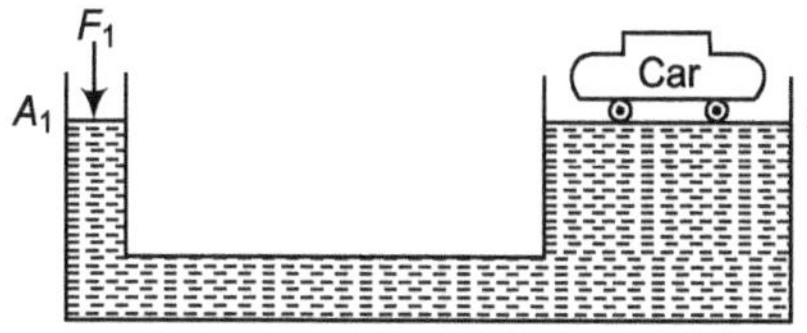

Fig. 13.7

A well known application of Pascal's law is the hydraulic lift used to support or lift heavy objects. It is schematically illustrated in figure.

A piston with small cross-section area A_1 exerts a force F_1 on the surface of a liquid such as oil. The applied pressure $p=\frac{F_1}{A_1}$ is transmitted through the connecting pipe to a larger piston of area A_2. The applied pressure is the same in both cylinders, so

$$p_2=\frac{A_2}{A_1}=\frac{F_2}{A_2} \text{ or } F_2=\frac{A_2}{A_1}.F_1$$

Now, since $A_2 > A_1$, therefore, $F_2 > F_1$. Thus, hydraulic lift is a force multiplying device with a multiplication factor equal to the ratio of the areas of the two pistons. Dentist's chairs, car lifts and jacks, many elevators and hydraulic brakes all use this principle.

Example 13.2 *Figure shows a hydraulic press with the larger piston of diameter 35 cm at a height of 1.5 m relative to the smaller piston of diameter 10 cm. The mass on the smaller piston is 20 kg. What is the force exerted on the load by the larger piston? The density of oil in the press is 750 kg/m^3. (Take $g=9.8$ m/s^2)*

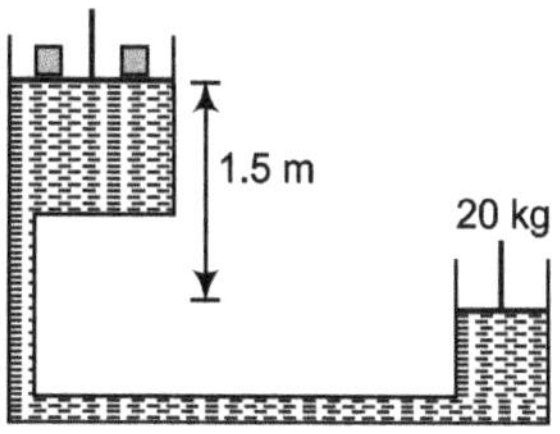

Fig. 13.8

Sol. Pressure on the smaller piston

$$=\frac{20\times 9.8}{\pi\times(5\times 10^{-2})^2}\text{ N/m}^2$$

Pressure on the larger piston

$$=\frac{F}{\pi\times(17.5\times 10^{-2})^2}\text{ N/m}^2$$

The difference between the two pressures = $h\rho g$

where, $h=1.5$ m

and $\rho=750$ kg/m^3

Thus, $$\frac{20\times 9.8}{\pi\times(5\times 10^{-2})^2}-\frac{F}{\pi\times(17.5\times 10^{-2})^2}=1.5\times 750\times 9.8$$

which gives, $F=1.3\times 10^3$ N

- Atmospheric pressure is common to both pistons and has been ignored.

Extra Knowledge Points

- At same point on a fluid pressure is same in all directions. In the figure,

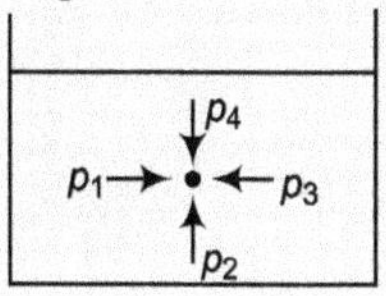

$$p_1=p_2=p_3=p_4$$

- Forces acting on a fluid in equilibrium have to be perpendicular to its surface. Because, it cannot sustain the shear stress.
- In the same liquid pressure will be same at all points at the same level.

For example, in the figure

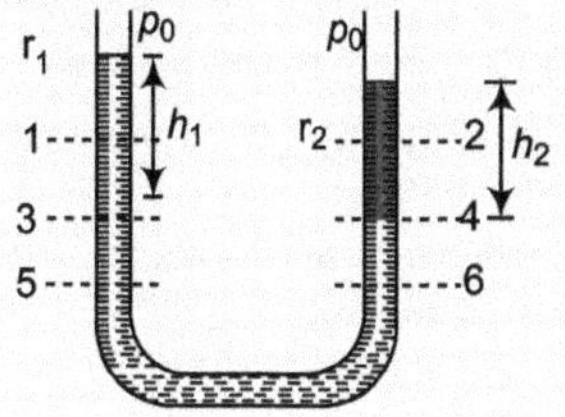

$$p_1\neq p_2, p_3=p_4 \text{ and } p_5=p_6$$

Further, $p_3=p_4$

$\therefore$ $p_0+\rho_1 gh_1=p_0+\rho_2 gh_2$

or $\rho_1 h_1=\rho_2 h_2$ or $h\propto\frac{1}{\rho}$

- **Barometer** It is a device used to measure atmospheric pressure.

In principle, any liquid can be used to fill the barometer, but mercury is the substance of choice, because its great density makes possible an instrument of reasonable size.

$$p_1=p_2$$

Here, p_1 = atmospheric pressure (p_0)

and $p_2=0+\rho gh=\rho gh$

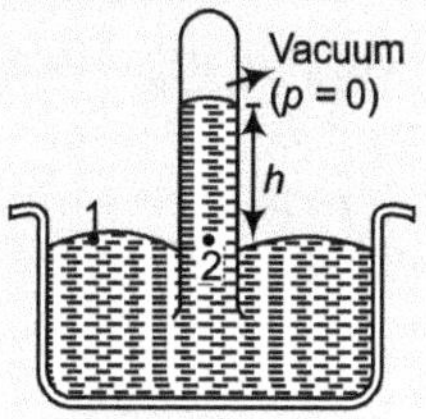

Here, ρ = density of mercury

$\therefore$ $p_0=\rho gh$

Thus, the mercury barometer reads the atmospheric pressure (p_0) directly from the height of the mercury column.

For example, if the height of mercury in a barometer is 760 mm, then atmospheric pressure will be,

$$p_0 = \rho g h = (13.6 \times 10^3)(9.8)(0.760)$$
$$= 1.01 \times 10^5 \text{ N/m}^2$$

- **Manometer** It is a device used to measure the pressure of a gas inside a container.

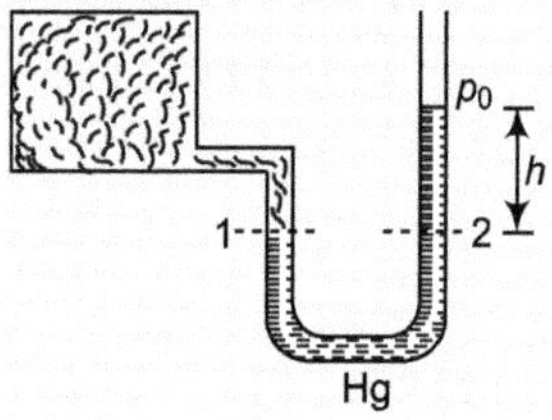

The U-shaped tube often contains mercury.

$$p_1 = p_2$$

Here, p_1 = pressure of the gas in the container (p),

and p_2 = atmospheric pressure (p_0) + $\rho g h$

$$\therefore \quad p = p_0 + \rho g h$$

This can also be written as

$$p - p_0 = \text{gauge pressure} = \rho g h$$

Here, ρ is the density of the liquid used in U-tube.

Thus, by measuring h we can find absolute (or gauge) pressure in the vessel.

- **Free body diagram of a liquid** The free body diagram of the liquid (showing the vertical forces only) is shown in Fig.(b). For the equilibrium of liquid.

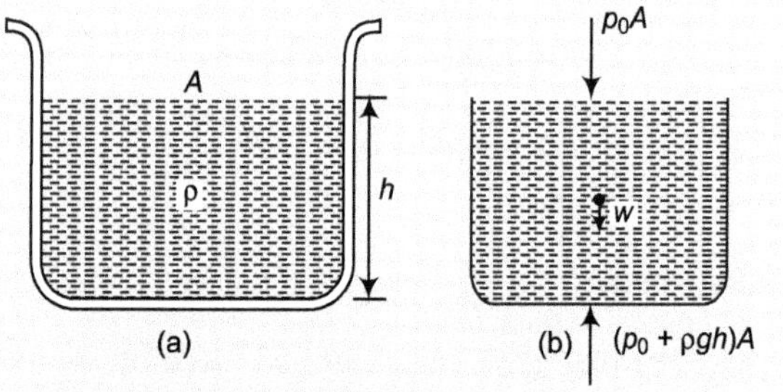

Net downward force = net upward force

$$\therefore \quad p_0 A + w = (p_0 + \rho g h)A$$

or $w = \rho g h A$

Example 13.3 *A glass full of water up to a height of* 10 cm *has a bottom of area* 10 cm^2, *top of area* 30 cm^2 *and volume* 1 *L.*

(a) Find the force exerted by the water on the bottom.

(b) Find the resultant force exerted by the sides of the glass on the water.

(c) If the glass is covered by a jar and the air inside the jar is completely pumped out, what will be the answers to parts (a) and (b)?

(d) If a glass of different shape is used provided the height, the bottom area and the volume are unchanged, will the answers to parts (a) and (b) change.

(Take $g = 10\, m/s^2$ *density of water* $= 10^3\, kg/m^3$ *and atmospheric pressure* $= 1.01 \times 10^5\, N/m^2$*)*

Sol. (a) Force exerted by the water on the bottom

$$F_1 = (p_0 + \rho g h)A_1 \qquad \ldots(i)$$

Here, p_0 = atmospheric pressure
$= 1.01 \times 10^5$ N/m²

ρ = density of water = 10^3 kg/m³

$g = 10$ m/s²,

$h = 10$ cm = 0.1 m

and A_1 = area of base = 10 cm² = 10^{-3} m²

Substituting in Eq. (i), we get

$$F_1 = (1.01 \times 10^5 + 10^3 \times 10 \times 0.1) \times 10^{-3}$$

or $F_1 = 102$ N (downwards)

(b) Force exerted by atmosphere on water

$$F_2 = (p_0)A_2$$

Here, A_2 = area of top = 30 cm² = 3×10^{-3} m²

$$\therefore \quad F_2 = (1.01 \times 10^5)(3 \times 10^{-3})$$
$$= 303 \text{ N (downwards)}$$

Force exerted by bottom on the water

$$F_3 = -F_1 \quad \text{or} \quad F_3 = 102 \text{ N (upwards)}$$

Weight of water w = (volume)(density)(g)

$$= (10^{-3})(10^3)(10)$$
$$= 10 \text{ N (downwards)}$$

Let F be the force exerted by side walls on the water (upwards). Then, from equilibrium of water.

Net upward force = net downward force

or $F + F_3 = F_2 + W$

$\therefore$ $F = F_2 + W - F_3 = 303 + 10 - 102$

or $F = 211$ N (upwards)

(c) If the air inside the jar is completely pumped out,

$$F_1 = (\rho g h)A_1 \qquad (\text{as } p_0 = 0)$$
$$= (10^3)(10)(0.1)(10^{-3})$$
$$= 1 \text{ N (downwards)}$$

In this case, $F_2 = 0$

and $F_3 = 1$ N (upwards)

$\therefore$ $F = F_2 + W - F_3 = 0 + 10 - 1$
$= 9$ N (upwards)

(d) No, the answer will remain the same. Because the answers depend upon p_0, ρ, g, h, A_1 and A_2.

Example 13.4 *Two vessels have the same base area but different shapes. The first vessel takes twice the volume of water that the second vessel requires to fill up to a particular common height. Is the force exerted by water on the base of the vessel the same in the two cases? If so, why do the vessels filled with water to that same height give different readings on a weighing scale?*

Sol. Pressure (and therefore force) on the two equal base areas are identical. But force is exerted by water on the sides of the vessels also, which has a non-zero vertical component, when the sides of the vessel are not perfectly normal to the base. This net vertical component of force by water on the sides of the vessel is greater for the first vessel than the second. Hence, the vessels weigh different when on the base is the same in the two cases.

13.4 Archimedes' Principle

If a heavy object is immersed in water, it seems to weigh less than when it is in air. This is because the water exerts an upward force called **buoyant force.** It is equal to the weight of the fluid displaced by the body.

A body wholly or partially submerged in a fluid is buoyed up by a force equal to the weight of the displaced fluid.

This result is known as **Archimedes' principle.**

Thus, the magnitude of buoyant force (F) is given by,

$$F = V_i \rho_L g$$

Here, V_i = immersed volume of solid,

ρ_L = density of liqid,

and g = acceleration due of gravity

Proof

Consider an arbitrarily shaped body of volume V placed in a container filled with a fluid of density ρ_L. The body is shown completely immersed, but complete immersion is not essential to the proof. To begin with, imagine the situation before the body was immersed. The region now occupied by the body was filled with fluid, whose weight was $V\rho_L g$. Because the fluid as a whole was in hydrostatic equilibrium, the net upwards force (due to difference in pressure at different depths) on the fluid in that region was equal to the weight of the fluid occuping that region.

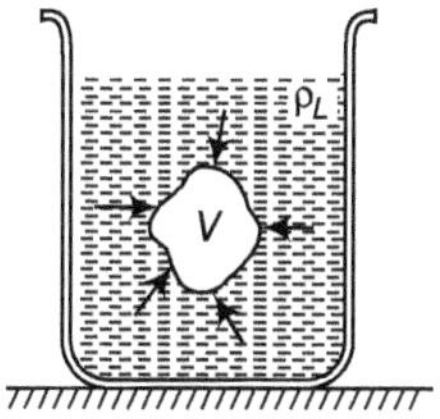

Fig. 13.9

Now, consider what happens when the body has displaced the fluid. The pressure at every point on the surface of the body is unchanged from the value at the same location when the body was not present. This is because the pressure at any point depends only on the depth of that point below the fluid surface.

Hence, the net force exerted by the surrounding fluid on the body is exactly the same as that exerted on the region before the body was present. But we know the latter to be $V\rho_L g$, the weight of the displaced fluid. Hence, this must also be the buoyant force exerted on the body. Archimedes' principle is thus, proved.

Law of Floatation

Consider an object of volume V and density ρ_S floating in a liquid of density ρ_L. Let V_i be the volume of object immersed in the liquid.

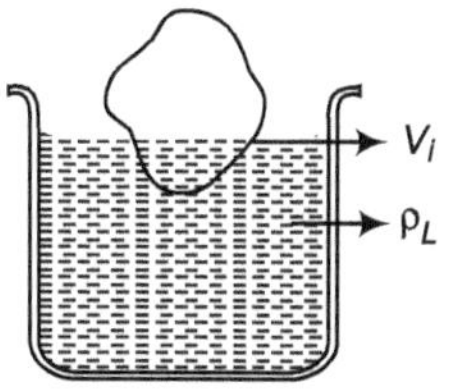

Fig. 13.10

For equilibrium of object,

Weight = Upthrust

$$\therefore \quad V\rho_S g = V_i \rho_L g$$

$$\therefore \quad \frac{V_i}{V} = \frac{\rho_S}{\rho_L} \qquad \text{...(i)}$$

This is the fraction of volume immersed in liquid.

Percentage of volume immersed in liquid

$$= \frac{V_i}{V} \times 100 = \frac{\rho_S}{\rho_L} \times 100$$

Three possibilities may now arise:

(i) If $\rho_S < \rho_L$, then only fraction of body will be immersed in the liquid. This fraction will be given by the above equation.

(ii) If $\rho_S = \rho_L$, then whole of the rigid body will be immersed in the liquid. Hence, the body remains floating in the liquid wherever it is left.

(iii) If $\rho_S > \rho_L$, then body will sink.

Apparent Weight of a Body Inside a Liquid

If a body is completely immersed in a liquid its effective weight gets decreased. The decrease in its weight is equal to the upthrust on the body. Hence,

$$W_{app} = W_{actual} - \text{upthrust} \quad \text{or} \quad W_{app} = V\rho_S g - V\rho_L g$$

Here, V = total volume of the body

ρ_S = density of body

and ρ_L = density of liquid

Thus, $W_{app} = Vg(\rho_S - \rho_L)$

If the liquid in which body is immersed is water, then

$$\frac{\text{Weight in air}}{\text{Decrease in weight}} = \text{Relative density of body,}$$

This can be shown as under

$$\frac{\text{Weight in air}}{\text{Decrease in weight}} = \frac{\text{Weight in air}}{\text{Upthrust}} = \frac{V\rho_S g}{V\rho_w g}$$

$$\frac{\rho_S}{\rho_w} = \text{RD}$$

Buoyant Force in Accelerating Fluids

Suppose a body is dipped inside a liquid of density ρ_L placed in an elevator moving with an acceleration **a**. The buoyant force F in this case becomes,

$$F = V\rho_L g_{\text{eff}}$$

Here, $g_{\text{eff}} = |\mathbf{g} - \mathbf{a}|$

For example, if the lift is moving upwards with an acceleration a, then value of g_{eff} is $g + a$ and if it is moving downwards with acceleration a, then g_{eff} is $g - a$. In a freely falling lift g_{eff} is zero (as $a = g$) and hence, net buoyant force is zero. This is why, in a freely falling vessel filled with some liquid, the air bubbles do not rise up (which otherwise move up due to buoyant force). The above result can be derived as follows.

Suppose a body is dipped inside a liquid of density ρ_L in an elevator moving up with an acceleration a. As, we done earlier also, replace the body into the liquid by the same liquid of equal volume. The replaced liquid is at rest with respect to the elevator. Thus, this replaced liquid is also moving up with an acceleration a together with the rest of the liquid.

The forces acting on the replaced liquid are:

(i) the buoyant force F, and

(ii) the weight mg of the substituted liquid.

From Newton's second law,

$$F - mg = ma \quad \text{or} \quad F = m(g + a)$$

Here, $m = V\rho_L$

$\therefore$ $F = V\rho_L (g + a) = V\rho_L g_{\text{eff}}$

where, $g_{\text{eff}} = g + a$

Example 13.5 *Density of ice is 900 kg/m^3. A piece of ice is floating in water (of density 1000 kg/m^3). Find the fraction of volume of the piece of ice outside the water.*

Sol. Let V be the total volume and V_i be the volume of ice piece immersed in water. For equilibrium of ice piece,

weight = upthrust

$\therefore$ $V\rho_i g = V_i \rho_w g$

Here, ρ_i = density of ice = 900 kg/m^3

and ρ_w = density of water = 1000 kg/m^3

Substituting in above equation, we get

$$\frac{V_i}{V} = \frac{900}{1000} = 0.9$$

i.e. the fraction of volume outside the water,

$$f = 1 - 0.9 = 0.1$$

Example 13.6 *A piece of ice is floating in a glass vessel filled with water. How will the level of water in the vessel change when the ice melts?*

Sol. Let m be the mass of ice piece floating in water.

In equilibrium, weight of ice piece = upthrust

or $mg = V_i \rho_w g$ or $V_i = \dfrac{m}{\rho_w}$ ...(i)

Here, V_i is the volume of ice piece immersed in water.

When the ice melts, let V be the volume of water formed by m mass of ice. Then

$$V = \frac{m}{\rho_w} \qquad \text{...(ii)}$$

From Eqs. (i) and (ii), we see that $V_i = V$

Hence, the level will not change.

Example 13.7 *A piece of ice having a stone frozen in it floats in a glass vessel filled with water. How will the level of water in the vessel change when the ice melts?*

Sol. Let, m_1 = mass of ice, m_2 = mass of stone,

ρ_S = density of stone and ρ_w = density of water.

In equilibrium, when the piece of ice floats in water,

weight of (ice + stone) = upthrust

$$(m_1 + m_2)g = V_i \rho_w g$$

$\therefore$ $V_i = \dfrac{m_1}{\rho_w} + \dfrac{m_2}{\rho_w}$ ...(i)

Here, V_i = volume of ice immersed.

When the ice melts, m_1 mass of ice converts into water and stone of mass m_2 is completely submerged.

Volume of water formed by m_1 mass of ice,

$$V_1 = \frac{m_1}{\rho_w}$$

Volume of stone (which is also equal to the volume of water displaced)

$$V_2 = m_2/\rho_s$$

Since, $\rho_S > \rho_w 0$

Therefore, $V_1 + V_2 < V_i$

or, the level of water will decrease.

Example 13.8 *An ornament weighing 50 g in air weights only 46 g is water. Assuming that some copper is mixed with gold to prepare the ornament. Find the amount of copper in it. Specific gravity of gold is 20 that of copper is 10.*

Sol. Let m be the mass of the copper in ornament. Then, mass of gold in it is $(50 - m)$.

Volume of copper $V_1 = \dfrac{m}{10}$ $\left(\text{volume} = \dfrac{\text{mass}}{\text{density}}\right)$

and volume of gold $V_2 = \dfrac{50 - m}{20}$

When immersed in water ($\rho_w = 1$ g/cm^3)

Decrease in weight = upthrust

$\therefore$ $(50 - 46)g = (V_1 + V_2)\rho_w g$

or $4 = \dfrac{m}{10} + \dfrac{50 - m}{20}$ or $80 = 2m + 50 - m$

$\therefore$ $m = 30$ g

Example 13.9 *The tension in a string holding a solid block below the surface of a liquid (of density greater than that of solid) as shown in figure is T_0, when the system is at rest. What will be the tension in the string, if the system has an upward acceleration a ?*

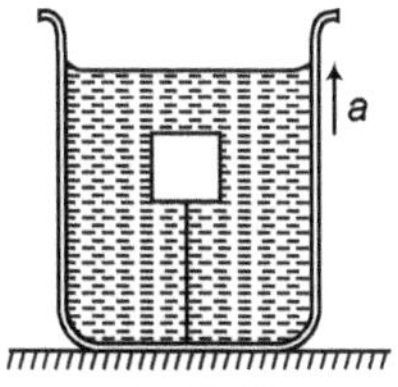

Fig. 13.11

Sol. Let m be the mass of block.

Initially for the equilibrium of block,

$$F = T_0 + mg \qquad \ldots(\text{i})$$

Here, F is the upthrust on the block.

When the lift is accelerated upwards, g_{eff} becomes $g + a$ instead of g. Hence,

$$F' = F\left(\frac{g+a}{g}\right) \qquad \ldots(\text{ii})$$

From Newton's second law,

$$F' - T - mg = ma \qquad \ldots(\text{iii})$$

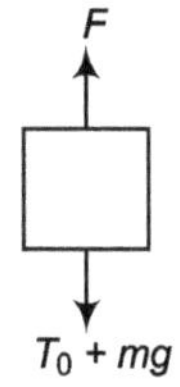

Fig. 13.12

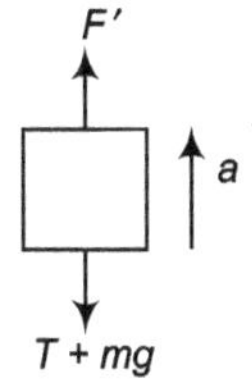

Fig. 13.13

Solving Eqs. (i), (ii) and (iii), we get

$$T = T_0\left(1 + \frac{a}{g}\right)$$

Extra Knowledge Points

- **Torque due to hydrostatic forces** Consider a dam in which water of density ρ is filled upto a height H. We are interested in finding the torque of hydrostatic forces on wall AB about point B. For this, consider a small length RS equal to dh of the wall AB at a depth h below the free surface of the dam. Further, let us assume a unit width perpendicular to paper inwards. Pressure on RS from left side is $p_0 + \rho gh$ and from right side is p_0.

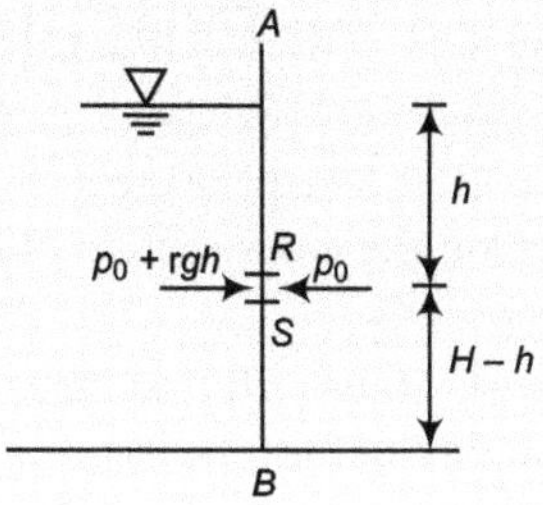

Area of RS, $\quad dA = (1)\,dh = dh$

Excess pressure $p = \rho gh$

$\therefore \qquad$ Net force $F = p\,dA = \rho gh\,dh$

Perpendicular distance of this force from point B is,

$$r_\perp = H - h$$

$\therefore$ Torque of this force about B, $d\tau = Fr_\perp$

or $$d\tau = \rho gh(H-h)dh$$

Therefore, net torque $\tau = \int_0^H d\tau = \int_0^H \rho gh\,(H-h)\,dh$

$$\therefore \qquad \tau = \frac{\rho g H^3}{6}$$

This is the torque of hydrostatic forces per unit width of the wall.

Note

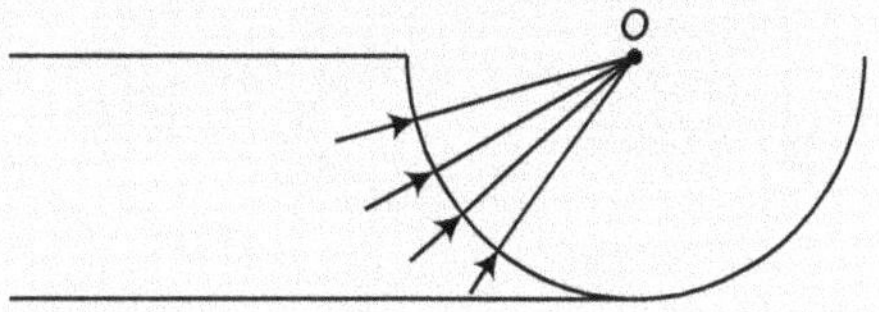

In the figure shown, torque of hydrostatic force about point O, the centre of a semicylindrical (or hemispherical) gate is zero as the hydrostatic force at all points passes through point O.

13.5 Flow of Fluids

Steady Flow

If the velocity of fluid particles at any point does not vary with time, then flow is said to be steady. Steady flow is also called streamlined or laminar flow. The velocity at different points may be different. Hence, in the figure,

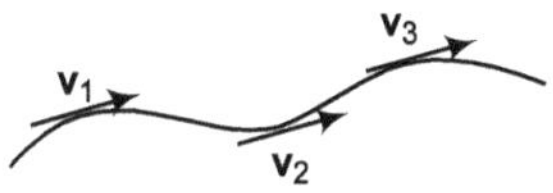

Fig. 13.14

$$\mathbf{v}_1 = \text{constant}, \quad \mathbf{v}_2 = \text{constant}, \quad \mathbf{v}_3 = \text{constant},$$

but $$\mathbf{v}_1 \neq \mathbf{v}_2 \neq \mathbf{v}_3$$

Principle of Continuity

It states that, when an incompressible and non-viscous liquid flows in a stream lined motion through a tube of non-uniform cross-section, then the product of the area of cross-section and the velocity of flow is same at every point in the tube.

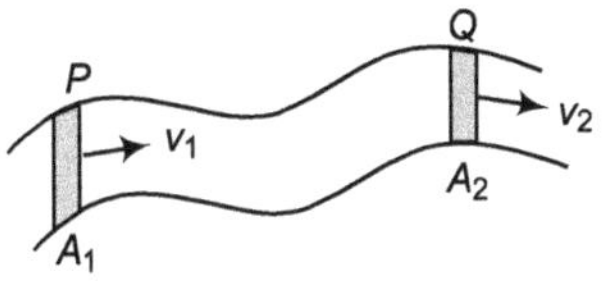

Fig. 13.15

Thus, $$A_1 v_1 = A_2 v_2 \text{ or } Av = \text{constant},$$

or $$v \propto \frac{1}{A}$$

This is basically the law of conservation of mass in fluid dynamics.

Proof

Let us consider two cross-sections P and Q of area A_1 and A_2 of a tube through which a fluid is flowing. Let v_1 and v_2 be the speeds at these two cross-sections. Then, being an incompressible fluid, mass of fluid going through P in a time interval Δt = mass of fluid passing through Q in the same interval of time Δt.

$\therefore \quad A_1 v_1 \rho \Delta t = A_2 v_2 \rho \Delta t$

or $\quad A_1 v_1 = A_2 v_2$

Therefore, the velocity of the liquid is smaller in the wider parts of the tube and larger in the narrower parts.

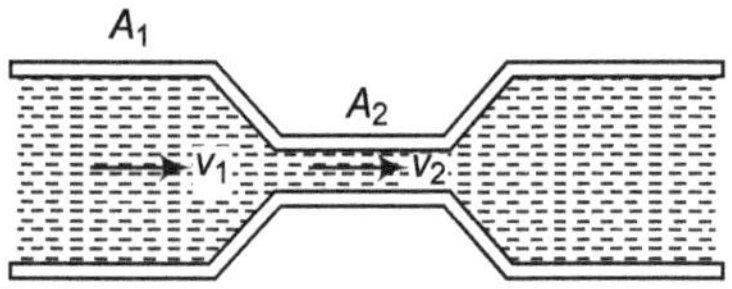

Fig. 13.16

or $\quad v_2 > v_1$

as $\quad A_2 < A_1$

- The product Av is the volume flow rate $\frac{dV}{dt}$, the rate at which volume crosses a section of the tube. Hence,

$$\frac{dV}{dt} = \text{volume flow rate} = Av$$

The mass flow rate is the mass flow per unit time through a cross-section. This is equal to density (ρ) times the volume flow rate $\frac{dV}{dt}$.

We can generalize the continuity equation for the case in which the fluid is not incompressible. If ρ_1 and ρ_2 are the densities at sections 1 and 2, then

$$\rho_1 A_1 v_1 = \rho_2 A_2 v_2$$

So, this is the continuity equation for a compressible fluid.

Example 13.10 *Water is flowing through a horizontal tube of non-uniform cross-section. At a place the radius of the tube is 1.0 cm and the velocity of water is 2 m/s. What will be the velocity of water, where the radius of the pipe is 2.0 cm ?*

Sol. Using equation of continuity,

$$A_1 v_1 = A_2 v_2$$

$$v_2 = \left(\frac{A_1}{A_2}\right) v_1$$

or $$v_2 = \left(\frac{\pi r_1^2}{\pi r_2^2}\right) v_1 = \left(\frac{r_1}{r_2}\right)^2 v_1$$

Substituting the values, we get

$$v_2 = \left(\frac{1.0 \times 10^{-2}}{2.0 \times 10^{-2}}\right)^2 (2)$$

or $$v_2 = 0.5 \text{ m/s}$$

13.6 Bernoulli's Equation

Bernoulli's equation relates the pressure, flow speed and height for flow of an ideal (incompressible and non-viscous) fluid. The pressure of a fluid depends on height as in the static situation, and it also depends on the speed of flow.

The dependence of pressure on speed can be understood from the continuity equation. When an incompressible fluid flows along a tube with varying cross-section, its speed must change, and so, an element of fluid must have an acceleration. If the tube is horizontal, then force that causes this acceleration has to be applied by the surrounding fluid. This means that the pressure must be different in regions of different cross-section.

When a horizontal flow tube narrows and a fluid element speeds up, it must be moving towards a region of lower pressure in order to have a net forward force to accelerate it. If the elevation also changes, this causes additional pressure difference.

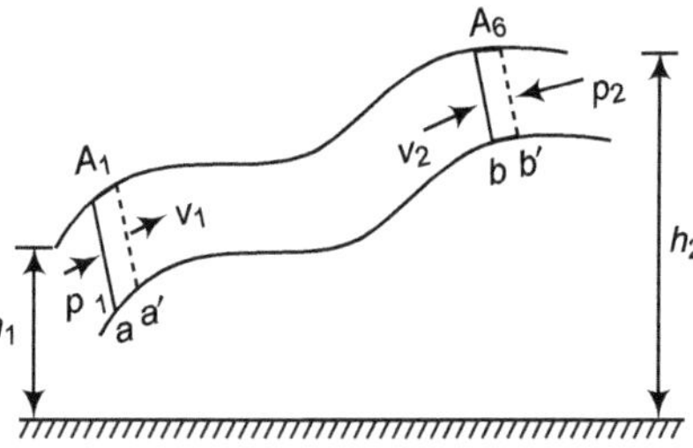

Fig. 13.17

To derive Bernoulli's equation, we apply the work energy theorem to the fluid in a section of the fluid element. Consider the element of fluid that at some initial time lies between two cross-sections a and b. The speeds at the lower and upper ends are v_1 and v_2.

In a small time interval the fluid that is initially at a moves to a' a distance $aa' = ds_1 = v_1\, dt$ and the fluid that is initially at b moves to b' a distance $bb' = ds_2 = v_2\, dt$. The cross-section areas at the two ends are A_1 and A_2 as shown in the figure. The fluid is incompressible, hence by the continuity equation, the volume of fluid dV passing through any cross-section during time dt is the same.

That is, $\quad dV = A_1 ds_1 = A_2 ds_2$

Work Done on the Fluid Element

Let us calculate the work done on this fluid element during time interval dt. The pressure at the two ends are p_1 and p_2, the force on the cross-section at a is $p_1 A_1$ and the force at b is $p_2 A_2$.

The net work done dW on the element by the surrounding fluid during this displacement is,

$$dW = p_1 A_1 ds_1 - p_2 A_2 ds_2 = (p_1 - p_2) dV \qquad \text{...(i)}$$

The second term is negative, because the force at b opposes the displacement of the fluid.

This work dW is due to forces other, than the conservative force of gravity, so it equals the change in total mechanical energy (kinetic plus potential). The mechanical energy for the fluid between sections a and b does not change.

Change in Potential Energy

At the beginning of dt the potential energy for the mass between a and a' is $dmgh_1 = \rho dVgh_1$. At the end of dt the potential energy for the mass between b and b' is $dmgh_2 = \rho dVgh_2$. The net change in potential energy dU during dt is,

$$dU = \rho (dV) g (h_2 - h_1) \quad \text{...(ii)}$$

Change in Kinetic Energy

At the beginning of dt the fluid between a and a' has volume $A_1 ds_1$, $\rho A_1 ds_1$ and kinetic energy $\frac{1}{2}\rho (A_1 ds_1) v_1^2$. At the end of dt the fluid between b and b' has kinetic energy $\frac{1}{2}\rho (A_2 ds_2) v_2^2$. The net change in kinetic energy dK during time dt is,

$$dK = \frac{1}{2}\rho (dV)(v_2^2 - v_1^2) \quad \text{...(iii)}$$

Combining Eqs. (i), (ii) and (iii) in the energy equation,

$$dW = dK + dU$$

We obtain,

$$(p_1 - p_2) dV = \frac{1}{2}\rho dV (v_2^2 - v_1^2) + \rho (dV) g (h_2 - h_1)$$

or $$p_1 - p_2 = \frac{1}{2}\rho (v_2^2 - v_1^2) + \rho g (h_2 - h_1) \quad \text{...(iv)}$$

This is Bernoulli's equation. It states that the work done on a unit volume of fluid by the surrounding fluid is equal to the sum of the changes in kinetic and potential energies per unit volume that occur during the flow. We can also express Eq. (iv) in a more convenient form as,

Bernoulli's equation

$$p_1 + \rho g h_1 + \frac{1}{2}\rho v_1^2 = p_2 + \rho g h_2 + \frac{1}{2}\rho v_2^2$$

The subscripts 1 and 2 refer to any two points along the flow tube, so we can also write

$$p + \rho g h + \frac{1}{2}\rho v^2 = \text{constant}$$

- When the fluid is not moving ($v_1 = 0 = v_2$). Bernoulli's equation reduces to

$$p_1 + \rho g h_1 = p_2 + \rho g h_2$$

$$\therefore \quad p_1 - p_2 = \rho g (h_2 - h_1)$$

This is the pressure relation we derived for a fluid at rest.

Extra Knowledge Points

- **Energy of a flowing fluid** There are following three types of energies in a flowing fluid.
  (i) **Pressure energy** If p is the pressure on the area A of a fluid, and the liquid moves through a distance l due to this pressure, then

  Pressure energy of liquid = work done
  = force × displacement
  = pAl

  The volume of the liquid is Al.

  ∴ Pressure energy per unit volume of liquid $= \frac{pAl}{Al} = p$

  (ii) **Kinetic energy** If a liquid of mass m and volume V is flowing with velocity v, then the kinetic energy is $\frac{1}{2}mv^2$.

  ∴ Kinetic energy per unit volume of liquid $= \frac{1}{2}\left(\frac{m}{V}\right)v^2 = \frac{1}{2}\rho v^2$

  Here, ρ is the density of liquid.

  (iii) **Potential energy** If a liquid of mass m is at a height h from the reference line ($h = 0$), then its potential energy is mgh.

  ∴ Potential energy per unit volume of the liquid $= \left(\frac{m}{V}\right)gh = \rho g h$

  Thus, the Bernoulli's equation

  $$p + \frac{1}{2}\rho v^2 + \rho g h = \text{constant (J/m}^3)$$

  can also be written as3

  Sum of total energy per unit volume (pressure + kinetic + potential) is constant for an ideal fluid.

- **Pressure head, velocity head and gravitational head of a flowing liquid**
  Dividing the Bernoulli's equation by ρg, we have

  $$\frac{p}{\rho g} + \frac{v^2}{2g} + h = \text{constant (J/m}^3)$$

  In this expression, $\frac{p}{\rho g}$ is called the pressure head, $\frac{v^2}{2g}$ the velocity head and h the gravitational head. The SI unit of each of these three is metre. Therefore, Bernoulli's equation may also be stated as,
  Sum of pressure head, velocity head and gravitational head is constant for an ideal fluid.

Example 13.11 *Calculate the rate of flow of glycerine of density $1.25 \times 10^3 \ kg/m^3$ through the conical section of a pipe, if the radii of its ends are $0.1 \ m$ and $0.04 \ m$ and the pressure drop across its length is $10 N/m^2$.*

Sol. From continuity equation,

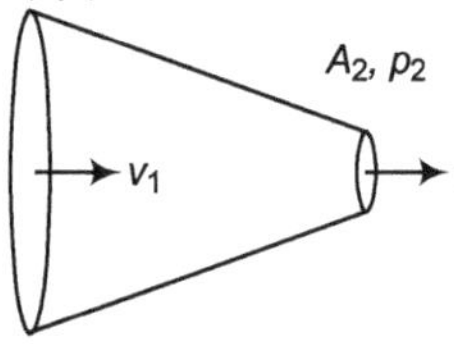

Fig. 13.18

$$A_1v_1 = A_2v_2$$

or $$\frac{v_1}{v_2} = \frac{A_2}{A_1} = \frac{\pi r_2^2}{\pi r_1^2} = \left(\frac{r_2}{r_1}\right)^2 = \left(\frac{0.04}{0.1}\right)^2 = \frac{4}{25} \qquad \ldots(i)$$

From Bernoulli's equation,

$$p_1 + \frac{1}{2}\rho v_1^2 = p_2 + \frac{1}{2}\rho v_2^2 \quad \text{or} \quad v_2^2 - v_1^2 = \frac{2(p_1 - p_2)}{\rho}$$

or $$v_2^2 - v_1^2 = \frac{2 \times 10}{1.25 \times 10^3} = 1.6 \times 10^{-2} \text{ m}^2/\text{s}^2 \qquad \ldots(ii)$$

Solving Eqs. (i) and (ii), we get

$$v_2 \approx 0.128 \text{ m/s}$$

$\therefore$ Rate of volume flow through the tube

$$Q = A_2v_2 = (\pi r_2^2)v_2 = \pi(0.04)^2(0.128)$$
$$= 6.43 \times 10^{-4} \text{ m}^2/\text{s}^2$$

13.7 Applications Based on Bernoulli's Equation

Venturimeter

Figure shows a venturimeter used to measure flow speed in a pipe. We apply Bernoulli's equation to the wide (point 1) and narrow (point 2) parts of the pipe, with $h_1 = h_2$

$$p_1 + \frac{1}{2}\rho v_1^2 = p_2 + \frac{1}{2}\rho v_2^2$$

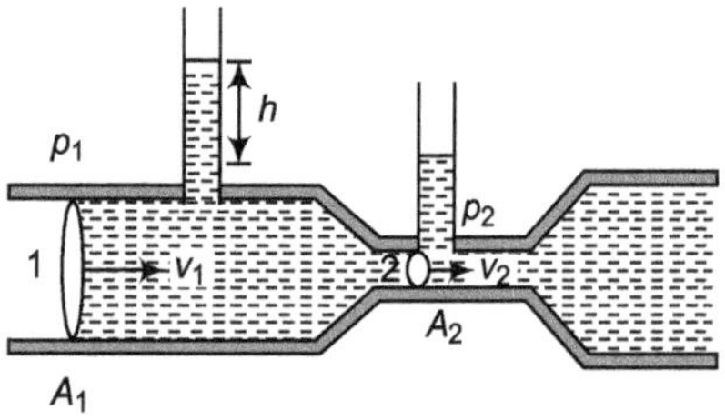

Fig. 13.19

From the continuity equation $v_2 = \dfrac{A_1 v_1}{A_2}$

Substituting and rearranging, we get

$$p_1 - p_2 = \frac{1}{2}\rho v_1^2\left(\frac{A_1^2}{A_2^2} - 1\right) \qquad \ldots(i)$$

Because, A_1 is greater than A_2, v_2 is greater than v_1 and hence the pressure p_2 is less than p_1. A net force to the right accelerates the fluid as it enters the narrow part of the tube (called throat) and a net force to the left slows as it leaves. The pressure difference is also equal to ρgh, where h is the difference in liquid level in the two tubes. Substituting in Eq. (i), we get

$$v_1 = \sqrt{\frac{2gh}{\left(\frac{A_1}{A_2}\right)^2 - 1}}$$

Note Points

- The discharge or volume flow rate can be obtained as,

$$\frac{dV}{dt} = A_1v_1 = A_1\sqrt{\frac{2gh}{\left(\frac{A_1}{A_2}\right)^2 - 1}}$$

- The venturi effect can be used to give a qualitative understanding of the lift of an air plane wing and the path of a pitcher's curve ball. An airplane wing is designed, so that air moves faster over the top of the wing that it does under the wing, thus, making the air pressure less on top, than underneath. This difference in pressure results in a net force upward on the wing.

Speed of Efflux

Suppose, the surface of a liquid in a tank is at a height h from the orifice O on its sides, through which the liquid issues out with velocity v. The speed of the liquid coming out is called the speed of efflux. If the dimensions of the tank be sufficiently large, then velocity of the liquid at its surface may be taken to be zero and since the pressure there as well as at the orifice O is the same *viz* atmospheric it plays no part in the flow of the liquid, which thus, occurs purely in consequence of the hydrostatic pressure of the liquid itself. So that, considering a tube of flow, starting at the liquid surface and ending at the orifice, as shown in figure. Applying Bernoulli's equation, we have

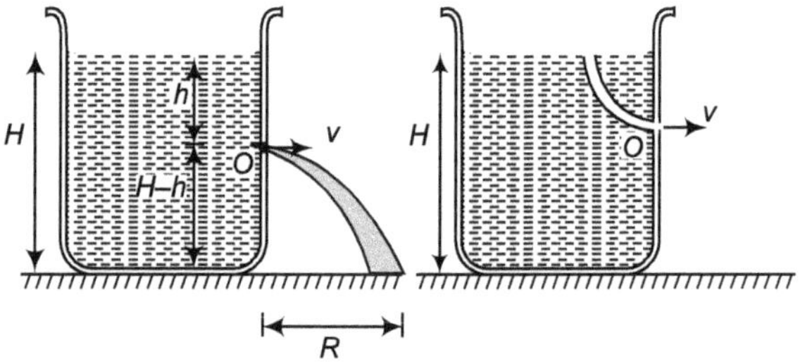

Fig. 13.20

Total energy per unit volume of the liquid at the surface

$$= \text{KE} + \text{PE} + \text{pressure energy}$$
$$= 0 + \rho gh + p_0 \qquad \ldots(i)$$

and total energy per unit volume at the orifice

$$= \text{KE} + \text{PE} + \text{pressure energy}$$
$$= \frac{1}{2}\rho v^2 + 0 + p_0 \qquad \ldots(ii)$$

Since, the total energy of the liquid must remain constant in steady flow, in accordance with Bernoulli's equation, we have

$$\rho gh + p_0 = \frac{1}{2}\rho v^2 + p_0$$

or $$v = \sqrt{2gh}$$

Evangelista Torricelli showed that this velocity is the same as the liquid will attain in falling freely through the vertical height (h) from the surface to the orifice. This is known as **Torricelli's theorem** and may be stated as, "The velocity of efflux of a liquid issuing out of an orifice is the same as it would attain, if allowed to fall freely through the vertical height between the liquid surface and orifice."

Range (R)

Let us find the range R on the ground.

Considering the vertical motion of the liquid,

$$(H-h) = \frac{1}{2}gt^2 \text{ or } t = \sqrt{\frac{2(H-h)}{g}}$$

Now, considering the horizontal motion,

$$R = vt$$

or $$R = (\sqrt{2gh})\left(\sqrt{\frac{2(H-h)}{g}}\right)$$

or $$R = 2\sqrt{h(H-h)}$$

From the expression of R, following conclusions can be drawn,

(i) $R_h = R_{H-h}$

as $$R_h = 2\sqrt{h(H-h)}$$

and $$R_{H-h} = 2\sqrt{(H-h)h}$$

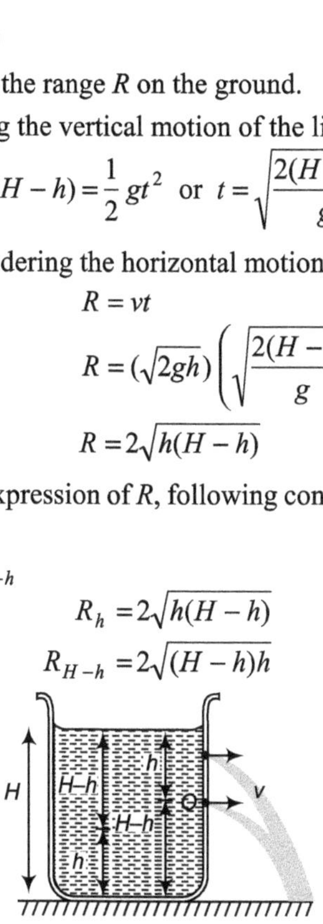

Fig. 13.21

This can be shown as in Fig. 13.21.

(ii) R is maximum at $h = \frac{H}{2}$ and $R_{max} = H$.

$$R^2 = 4(Hh - h^2)$$

For R to be maximum, $\frac{dR^2}{dh} = 0$

or $$H - 2h = 0$$

or $$h = \frac{H}{2}$$

That is, R is maximum at $h = \frac{H}{2}$

and $$R_{max} = 2\sqrt{\frac{H}{2}\left(H - \frac{H}{2}\right)} = H$$

Time taken to Empty a Tank

We are here interested in finding the time required to empty a tank, if a hole is made at the bottom of the tank.

Consider a tank filled with a liquid of density ρ up to a height H. A small hole of area of cross-section a is made at the bottom of the tank. The area of cross-section of the tank is A.

Let at some instant of time, the level of liquid in the tank is y. Velocity of efflux at this instant of time would be,

$$v = \sqrt{2gy}$$

Now, at this instant volume of liquid coming out of the hole per second is $\left(\frac{dV_1}{dt}\right)$.

Volume of liquid coming down in the tank per second is $\left(\frac{dV_2}{dt}\right)$.

$$\frac{dV_1}{dt} = \frac{dV_2}{dt}$$

$\therefore$ $$av = A\left(-\frac{dy}{dt}\right)$$

$\therefore$ $$a\sqrt{2gy} = A\left(-\frac{dy}{dt}\right)$$

or $$\int_0^t dt = -\frac{A}{a\sqrt{2g}}\int_H^0 y^{-1/2}\,dy \qquad \text{...(i)}$$

$\therefore$ $$t = \frac{2A}{a\sqrt{2g}}[\sqrt{y}]_0^H$$

$\therefore$ $$t = \frac{A}{a}\sqrt{\frac{2H}{g}}$$

Example 13.12 *A tank is filled with a liquid up to a height H. A small hole is made at the bottom of this tank. Let t_1 be the time taken to empty first half of the tank and t_2 the time taken to empty rest half of the tank. Then, find $\frac{t_1}{t_2}$.*

Sol. Substituting the proper limits in Eq. (i), derived in the theory we have,

$$\int_0^{t_1} dt = -\frac{A}{a\sqrt{2g}}\int_H^{H/2} y^{-1/2}\,dy$$

or $$t_1 = \frac{2A}{a\sqrt{2g}}[\sqrt{y}]_{H/2}^{H}$$

or $$t_1 = \frac{2A}{a\sqrt{2g}}\left[\sqrt{H} - \sqrt{\frac{H}{2}}\right]$$

or $$t_1 = \frac{A}{a}\sqrt{\frac{H}{g}}(\sqrt{2}-1) \quad \text{...(ii)}$$

Similarly, $$\int_0^{t_2} dt = -\frac{A}{a\sqrt{2g}}\int_{H/2}^{0} y^{-1/2}\,dy$$

or $$t_2 = \frac{A}{a}\sqrt{\frac{H}{g}} \quad \text{...(iii)}$$

From Eqs. (ii) and (iii), we get

$$\frac{t_1}{t_2} = \sqrt{2} - 1$$

or $$\frac{t_1}{t_2} = 0.414$$

- From here we see that $t_1 < t_2$. This is because initially, the pressure is high and the liquid comes out with greater speed.

13.8 Viscosity

Viscosity is internal friction in a fluid. Viscous forces opposes the motion of one portion of a fluid relative to the other.

The simplest example of viscous flow is motion of a fluid between two parallel plates.

The bottom plate is stationary and the top plate moves with constant velocity **v**. The fluid in contact with each surface has same velocity at that surface. The flow speeds of intermediate layers of fluid increase uniformly from bottom to top, as shown by arrows. So, the fluid layers slide smoothly over one another.

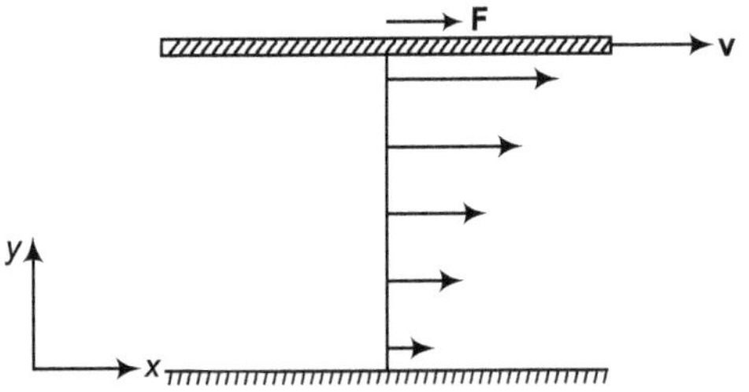

Fig. 13.22

According to Newton, the frictional force F (or viscous force) between two layers depends upon the following factors:

(i) Force F is directly proportional to the area (A) of the layers in contact, i.e.

$$F \propto A$$

(ii) Force F is directly proportional to the velocity gradient $\left(\frac{dv}{dy}\right)$ between the layers. Combining these two, we have

$$F \propto A\frac{dv}{dy} \quad \text{or} \quad F = -\eta A\frac{dv}{dy}$$

Here, η is constant of proportionality and is called **coefficient of viscosity.** Its value depends on the nature of the fluid. The negative sign in the above equation shows that the direction of viscous force F is opposite to the direction of relative velocity of the layer.

The SI unit of η is N-s/m^2. It is also called decapoise or pascal second. Thus,

1 decapoise = 1 N-s/m^2 = 1 Pa-s = 10 poise

Dimensions of η are $[ML^{-1}T^{-1}]$.

Coefficient of viscosity of water at 10°C is $\eta = 1.3 \times 10^{-3}$ N-s / m^2. Experiments show that coefficient of viscosity of a liquid decreases as its temperature rises.

Example 13.13 *A plate of area 2 m^2 is made to move horizontally with a speed of 2 m/s by applying a horizontal tangential force over the free surface of a liquid. If the depth of the liquid is 1 m and the liquid in contact with the bed is stationary. Coefficient of viscosity of liquid is 0.01 poise. Find the tangential force needed to move the plate.*

Sol. Velocity gradient $= \frac{\Delta v}{\Delta y} = \frac{2-0}{1-0} = 2\frac{\text{m/s}}{\text{m}}$

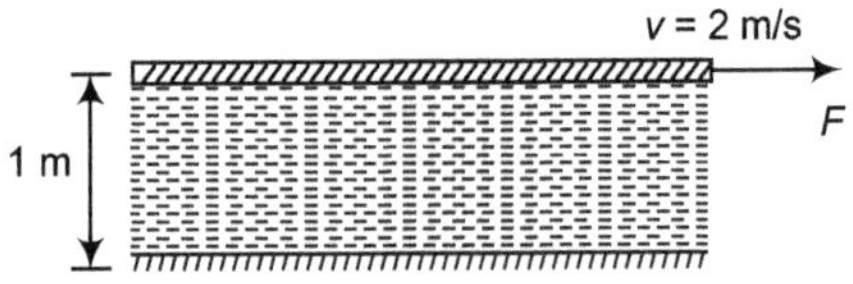

Fig. 13.23

From, Newton's law of viscous force,

$$|F| = \eta A\frac{\Delta v}{\Delta y} = (0.01 \times 10^{-1})(2)(2) = 4 \times 10^{-3}\text{ N}$$

So, to keep the plate moving, a force of 4×10^{-3} N must be applied.

Flow of Liquid through a Cylindrical Pipe

Figure shows the flow speed profile for laminar flow of a viscous fluid in a long cylindrical pipe. The speed is greatest along the axis and zero at the pipe walls. The flow speed v at a distance r from the axis of a pipe of radius R is,

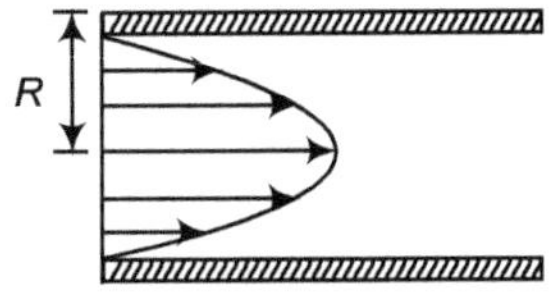

Fig. 13.24

$$v = \frac{p_1 - p_2}{4\eta L}(R^2 - r^2)$$

where, p_1 and p_2 are the pressure at the two ends of a pipe with length L. The flow is always in the direction of decreasing pressure.

From the above equation, we can see that v-r graph is a parabola.

$$v = 0 \quad \text{at} \quad r = R \quad \text{(along the walls)}$$

and $$v = \frac{(p_1 - p_2)R^2}{4\eta L} = v_{max} \quad \text{at } r = 0 \text{ (along the axis)}$$

Volume Flow Rate $\left(Q \text{ or } \frac{dV}{dt}\right)$

To find the total volume flow rate through the pipe, we consider a ring with inner radius r, outer radius-$r + dr$ and cross-sectional area $dA = 2\pi r\, dr$. The volume flow rate through this element is $v\, dA$. The total volume flow rate is found by integrating from $r = 0$ to $r = R$. The result is,

$$Q = \frac{dV}{dt}$$

$$= \frac{\pi}{8}\left(\frac{R^4}{\eta}\right)\left(\frac{p_1 - p_2}{L}\right)$$

The relation was first derived by Poiseulle and is called Poiseulle's equation.

Extra Knowledge Points

- Poiseulle's equation can also be written as,

$$Q = \frac{p_1 - p_2}{\left(\frac{8\eta L}{\pi R^4}\right)} = \frac{\Delta p}{X}$$

Here, $$X = \frac{8\eta L}{\pi R^4}$$

This equation can be compared with the current equation through a resistance, i.e.

$$i = \frac{\Delta V}{R}$$

Here, ΔV = potential difference

and R = electrical resistance

For current flow through a resistance, potential difference is a
requirement similarly for flow of liquid through a pipe pressure difference is must.

Problems of series and parallel combination of pipes can be solved in the similar manner as is done in case of an electrical circuit. The only difference is,

(i) Potential difference (ΔV) is replaced by the pressure difference (Δp).

(ii) The electrical resistance $R\left(= \rho\frac{L}{A}\right)$ is replaced by $X\left(= \frac{8\eta L}{\pi R^4}\right)$.

(iii) The electrical current i is replaced by volume flow rate Q or $\frac{dV}{dt}$. The following example will illustrate the above theory.

Example 13.14 *A liquid is flowing through horizontal pipes as shown in fig4ure.*

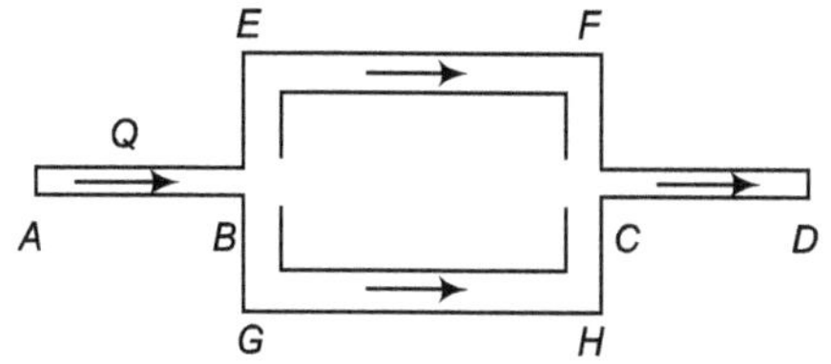

Fig. 13.25

Length of different pipes has the following ratio

$$L_{AB} = L_{CD} = \frac{L_{EF}}{2} = \frac{L_{GH}}{2}$$

Similarly, radii of different pipes has the ratio

$$R_{AB} = R_{EF} = R_{CD} = \frac{R_{GH}}{2}$$

Pressure at A is $2p_0$ and pressure at D is p_0. The volume flow rate through the pipe AB is Q. Find,

(a) volume flow rates through EF and GH

(b) pressure at E and F.

Sol. The equivalent electrical circuit can be drawn as under,

$$X \propto \frac{L}{R^4} \qquad \left(\text{as } X = \frac{8\eta L}{\pi R^4}\right)$$

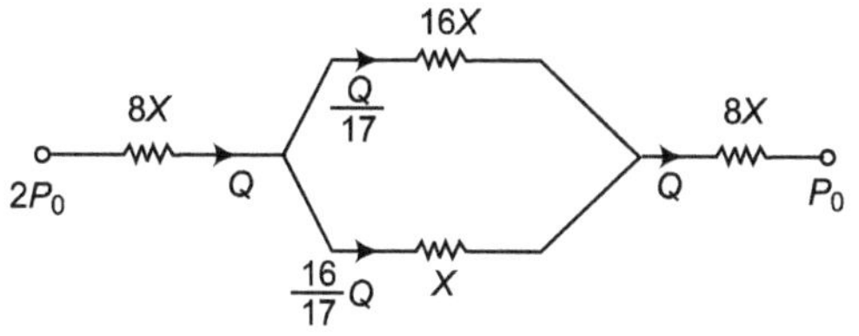

Fig. 13.26

$$\therefore \quad X_{AB} : X_{CD} : X_{EF} : X_{GH}$$

$$= \frac{\frac{1}{2}}{\left(\frac{1}{2}\right)^4} : \frac{\left(\frac{1}{2}\right)}{\left(\frac{1}{2}\right)^4} : \frac{(1)}{\left(\frac{1}{2}\right)^4} : \frac{(1)}{(1)^4}$$

$$= 8 : 8 : 16 : 1$$

(a) As, the current is distributed in the inverse ratio of the resistance (in parallel). The Q will be distributed in the inverse ratio of X.

Thus, volume flow rate through EF will be $\frac{Q}{17}$ and that from GH will be $\frac{16}{17}Q$.

(b) $$X_{net} = 8X + \left[\frac{(16X)(X)}{(16X) + (X)}\right] + 8X = \frac{288}{17}X$$

$$\therefore \quad Q = \frac{\Delta p}{X_{net}} \qquad \left(\text{as } i = \frac{\Delta V}{R}\right)$$

$$= \frac{(2p_0 - p_0)}{\frac{288}{17}X} = \frac{17p_0}{288X}$$

Now, let p_1 be the pressure at E, then

$$2p_0 - p_1 = 8QX = \frac{8 \times 17p_0}{288}$$

$$\therefore \quad p_1 = \left(2 - \frac{17 \times 8}{288}\right) p_0 = 1.53 P_0$$

Similarly, if p_2 be the pressure at F, then

$$p_2 - p_0 = 8QX$$

$$\therefore \quad p_2 = p_0 + \frac{8 \times 17}{288} p_0$$

$$\text{or} \quad p_2 = 1.47 p_0$$

13.9 Stoke's Law and Terminal Velocity

When an object moves through a fluid, it experiences a viscous force which acts in opposite direction of its velocity. The mathematics of the viscous force for an irregular object is difficult, we will consider here only the case of a small sphere moving through a fluid.

The formula for the viscous force on a sphere was first derived by the English physicist G. Stokes in 1843. According to him, a spherical object of radius r moving at velocity v experiences a viscous force given by

$$F = 6\pi\eta r v \quad (\eta = \text{coefficient of viscosity})$$

This law is called **Stoke's law.**

Terminal Velocity (v_T)

Consider a small sphere falling from rest through a large column of viscous fluid. The forces acting on the sphere are:

(i) Weight w of the sphere acting vertically downwards.

(ii) Upthrust F_t acting vertically upwards.

(iii) Viscous force F_v acting vertically upwards, i.e. in a direction opposite to velocity of the sphere.

Initially, $F_v = 0$ (as $v = 0$)

and $w > F_t$

and the sphere accelerates downwards. As the velocity of the sphere increases, F_v increases. Eventually a stage in reached when

$$w = F_t + F_v \quad \text{...(i)}$$

After this net force on the sphere is zero and it moves downwards with a constant velocity called **terminal velocity (v_T).**

Substituting proper values in Eq. (i), we have

$$\frac{4}{3}\pi r^3 \rho g = \frac{4}{3}\pi r^3 \sigma g + 6\pi\eta r v_T \quad \text{...(ii)}$$

Here, ρ = density of sphere, σ = density of fluid

and η = coefficient of viscosity of fluid

From Eq. (ii), we get

$$v_T = \frac{2}{9}\frac{r^2(\rho - \sigma)g}{\eta}$$

Figure shows the variation of the velocity v of the sphere with time

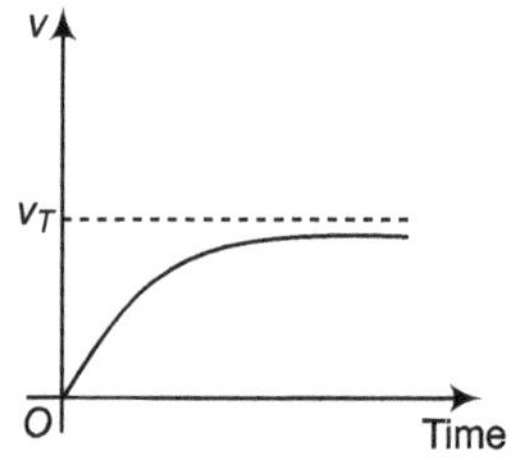

Fig. 13.28

- From the above expression we can see that terminal velocity of a spherical body is directly proportional to the difference in the densities of the body and the fluid ($\rho - \sigma$). If the density of fluid is greater than that of body (i.e. $\sigma > \rho$), then terminal velocity is negative. This means that the body instead of falling, moves upward. This is why air bubbles rise up in water.

Example 13.15 *Two spherical raindrops of equal size are falling vertically through air with a terminal velocity of* 1 m/s. *What would be the terminal speed, if these two drops were to coalesce to form a large spherical drop.*

Sol. $v_T \propto r^2$ ...(i)

Let r be the radius of small raindrops and R the radius of large drop. Equating the volumes, we have

$$\frac{4}{3}\pi R^3 = 2\left(\frac{4}{3}\pi r^3\right)$$

$$\therefore \quad R = (2)^{1/3}.r \quad \text{or} \quad \frac{R}{r} = (2)^{1/3}$$

$$\therefore \quad \frac{v_T'}{v_T} = \left(\frac{R}{r}\right)^2 = (2)^{2/3}$$

$$\therefore \quad v_T' = (2)^{2/3} v_T = (2)^{2/3}(1.0) \text{ m/s} = 1.587 \text{ m/s}$$

13.10 Surface Tension

A needle can made to float on a water surface, if it is placed there carefully. The forces that support the needle are not buoyant forces but are due to surface tension. The surface of a liquid behaves like a membrane under tension. The molecules of the liquid exert attractive forces on each other. There is zero net force on a molecule inside the volume of the liquid.

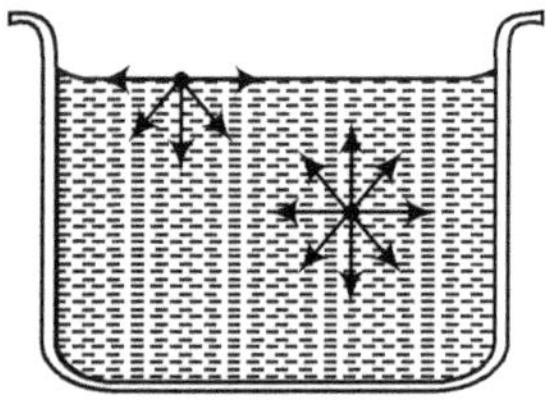

Fig. 13.29

But a surface molecule is drawn into the volume. Thus, the liquid tends to minimize its surface area, just as a stretched membrane does.

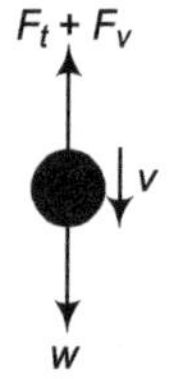

Fig. 13.27

Freely falling raindrops are spherical, because a sphere has a smaller surface area for a given volume, than any other shape. Hence, the surface tension can be defined as the property of a liquid at rest by virtue of which its free surface behaves like a stretched membrane under tension and tries to occupy as small area as possible.

Let an imaginary line AB be drawn in any direction in a liquid surface. The surface on either side of this line exerts a pulling force on the surface on the other side. This force is at right angles to the line AB. The magnitude of this force per unit length of AB is taken as a measure of the surface tension of the liquid. Thus, if F be the total force acting on either side of the line AB of length L, then the surface tension is given by,

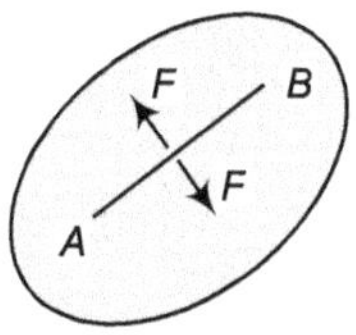

Fig. 13.30

$$T = \frac{F}{L}$$

Hence, the surface tension of a liquid is defined as the force per unit length in the plane of the liquid surface, acting at right angles on either side of an imaginary line drawn on that surface.

Few Examples of Surface Tension

Example *Take a ring of wire and dip it in a soap solution. When the ring is taken out, a soap film is formed. Place a loop of thread gently on the soap film. Now, prick a hole inside the loop.*

The thread is radially pulled by the film surface outside and it takes a circular shape.

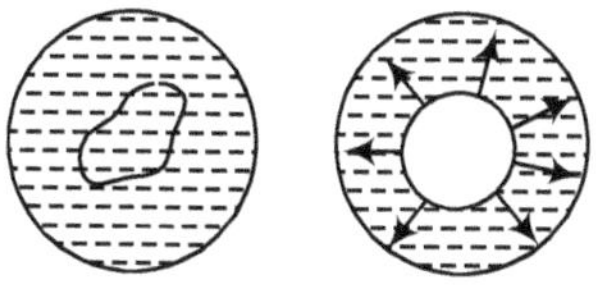
Fig. 13.31

Reason Before the pricking, there were surfaces both inside and outside the thread loop. Surfaces on both sides pull it equally and the net force is zero. Once the surface inside was punctured, the outside surface pulled the thread to take the circular shape, so that area outside the loop becomes minimum (because for given perimeter area of circle is maximum).

Example *A piece of wire is bent into a U-shape and a second piece of wire slides on the arms of the U. When the apparatus is dipped into a soap solution and removed, a liquid film is formed. The film exerts a surface tension force on the slider and if the frame is kept in a horizontal position, the slider quickly slides towards the closing arm of the frame. If the frame is kept vertical, one can have some weight to keep it in equilibrium. This shows that the soap surface in contact the slider pulls it parallel to the surface.*

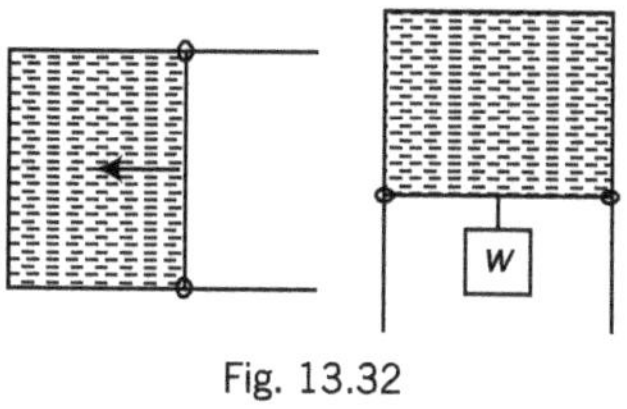

Fig. 13.32

- The surface tension of a particular liquid usually decreases as temperature increases. To wash clothing thoroughly, water must be forced through the tiny spaces between the fibres. This requires increasing the surface area of the water, which is difficult to do because of surface tension. Hence, hot water and soapy water is better for washing.

Surface Energy

When the surface area of a liquid is increased, the molecules from the interior rise to the surface. This requires work against force of attraction of the molecules just below the surface. This work is stored in the form of potential energy. Thus, the molecules in the surface have some additional energy due to their position. This additional energy per unit area of the surface is called 'surface energy'. The surface energy is related to the surface tension as discussed below :

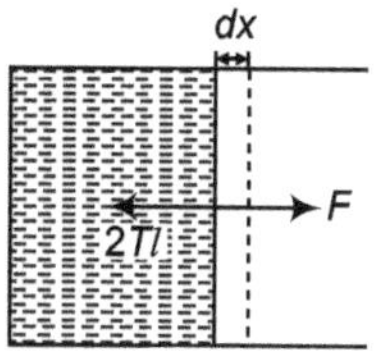

Fig. 13.33

Let a liquid film be formed on a wire frame and a straight wire of length l can slide on this wire frame as shown in figure. The film has two surfaces and both the surfaces are in contact with the sliding wire and hence, exert forces of surface tension on it. If T be the surface tension of the solution, then each surface will pull the wire parallel to itself with a force Tl. Thus, net force on the wire due to both the surfaces is $2Tl$. One has to apply an external force F equal and opposite to it to keep the wire in equilibrium. Thus,

$$F = 2Tl$$

Now, suppose the wire is moved through a small distance dx, the work done by the force is

$$dW = F\ dx = (2Tl)\ dx$$

But $(2l)(dx)$ is the total increase in area of both the surfaces of the film. Let it be dA. Then

$$dW = T\ dA \quad \text{or} \quad T = \frac{dW}{dA}$$

Thus, the surface tension T can also be defined as the work done in increasing the surface area by unity.

Further, since there is no change in kinetic energy, the work done by the external force is stored as the potential energy of the new surface.

$$\therefore \qquad T = \frac{dU}{dA} \qquad (\text{as } dW = dU)$$

Thus, the surface tension of a liquid is equal to the surface energy per unit surface area.

Example 13.16 *How much work will be done in increasing the soap bubble from 2 cm to 5 cm. Surface tension of soap solution is* 3.0×10^{-2} *N/m.*

Sol. Soap bubble has two surfaces.

Hence, $W = T\,\Delta A$

Here, $\Delta A = 2[4\pi\{(2.5 \times 10^{-2})^2 - (1.0 \times 10^{-2})^2\}]$

$= 1.32 \times 10^{-2}\ \text{m}^2$

$\therefore\quad W = (3.0 \times 10^{-2})(1.32 \times 10^{-2})$ J

$= 3.96 \times 10^{-4}$ J

Example 13.17 *Calculate the energy released when* 1000 *small water drops each of same radius* 10^{-7} *m coalesce to form one large drop. The surface tension of water is* 7.0×10^{-2} *N/m.*

Sol. Let r be the radius of smaller drops and R of bigger one. Equating the initial and final volumes, we have,

$$\frac{4}{3}\pi R^3 = (1000)\left(\frac{4}{3}\pi r^3\right)$$

or $R = 10r = (10)(10^{-7})$ m or $R = 10^{-6}$ m

Further, the water drops have only one free surface.

Therefore, $\Delta A = 4\pi R^2 - (1000)(4\pi r^2)$

$= 4\pi\ [(10^{-6})^2 - (10^3)(10^{-7})^2]$

$= -\ 36\pi(10^{-12})\ \text{m}^2$

Here, negative sign implies that surface area is decreasing. Hence, energy released in the process.

$U = T|\Delta A| = (7 \times 10^{-2})(36\pi \times 10^{-12})$ J

$= 7.9 \times 10^{-12}$ J

Excess Pressure inside a Bubble or Liquid Drop

Surface tension causes a pressure difference between the inside and outside of a soap bubble or a liquid drop.

Excess Pressure inside Soap Bubble

A soap bubble consists of two spherical surface films with a thin layer of liquid between them. Because of surface tension, the film tend to contract in an attempt to minimize their surface area. But as the bubble contracts, it compresses the inside air, eventually increasing the interior pressure to a level that prevents further contraction.

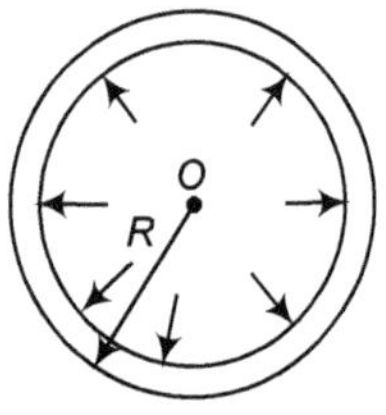

Fig. 13.34

We can derive an expression for the excess pressure inside a bubble in terms of its radius R and the surface tension T of the liquid.

Each half of the soap bubble is in equilibrium. The lower half is shown in figure. The forces at the flat circular surface where this half joins the upper half are

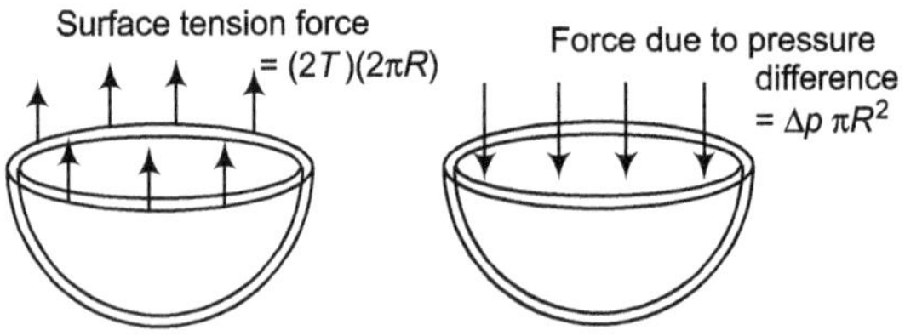

Fig. 13.35

(i) The upward force of surface tension. The total surface tension force for each surface (inner and outer) is $T(2\pi R)$, for a total of $(2T)(2\pi R)$

(ii) Downward force due to pressure difference.

The magnitude of this force is $(\Delta p)(\pi R^2)$. In equilibrium, these two forces have equal magnitude.

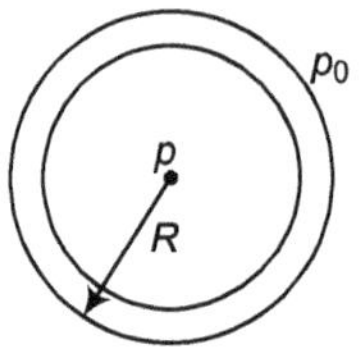

Fig. 13.36

$$\therefore\quad (2T)(2\pi R) = (\Delta p)(\pi R^2)$$

$$\text{or}\quad \Delta p = \frac{4T}{R}$$

- Suppose, the pressure inside the air bubble is p, then

$$p - p_0 = \frac{4T}{R}$$

Excess Pressure inside a Liquid Drop

A liquid drop has only one surface film. Hence, the surface tension force is $T(2\pi R)$, half that for a soap bubble. Thus, in equilibrium,

$$T(2\pi R) = \Delta p(\pi R^2) \ \text{or}\ \Delta p = \frac{2T}{R}$$

Note Points

- If we have an air bubble inside a liquid, a single surface is formed. There is air on the concave side and liquid on the convex side. The pressure in the concave side (that is in the air) is greater than the pressure in the convex side (that is in the liquid) by an amount $\frac{2T}{R}$.

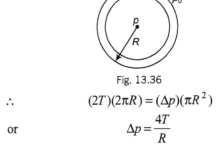

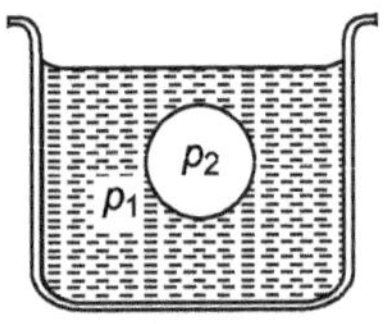

Fig. 13.37

$\therefore\quad p_2 - p_1 = 2T/R$

The above expression has been written by assuming p_1 to be constant from all sides of the bubble. For small size bubbles this can be assumed.

- From the above discussion, we can make a general statement. The pressure on the concave side of a spherical liquid surface is greater than the convex side by $2T/R$.

Example 13.18 *What should be the pressure inside a small air bubble of 0.1 mm radius situated just below the water surface? Surface tension of water* $= 7.2 \times 10^{-2}$ *N/m and atmospheric pressure* $= 1.013 \times 10^5$ *N/m²*.

Sol. Surface tension of water $T = 7.2 \times 10^{-2}$ N/m

Radius of air bubble $R = 0.1$ mm $= 10^{-4}$ m

The excess pressure inside the air bubble is given by,

$$p_2 - p_1 = \frac{2T}{R}$$

∴ Pressure inside the air bubble, $p_2 = p_1 + 2T/R$

Substituting the values, we have

$$p_2 = (1.013 \times 10^5) + \frac{(2 \times 7.2 \times 10^{-2})}{10^{-4}}$$

$$= 1.027 \times 10^5 \text{ N/m}^2$$

Shape of Liquid Surface

The surface of a liquid when meets a solid, such as the wall of a container, it usually curves up or down near the solid surface. The angle θ at which it meets the surface is called the **contact angle.** The curved surface of the liquid is called meniscus. The shape of the meniscus (convex or concave) is determined by the relative strengths of what are called the cohesive and adhesive forces. The force between the molecules of the same material is known as **cohesive force** and the force between the molecules of different kinds of material is called **adhesive force.**

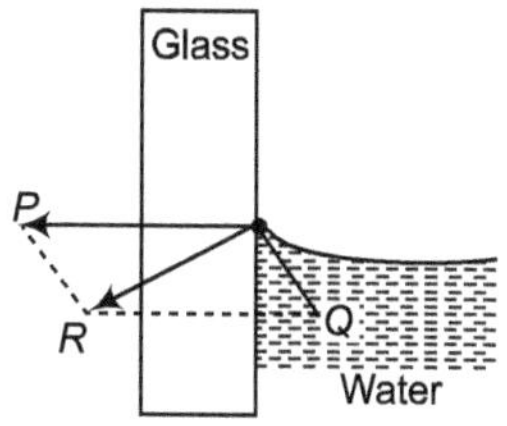

Fig. 13.38

When the adhesive force (p) between solid and liquid molecules is more than the cohesive force (Q) between liquid-liquid molecules (as with water and glass), shape of the meniscus is concave and the angle of contact θ is less than 90°. In this case the liquid wets or adheres to the solid surface. The resultant (R) of p and Q passes through the solid.

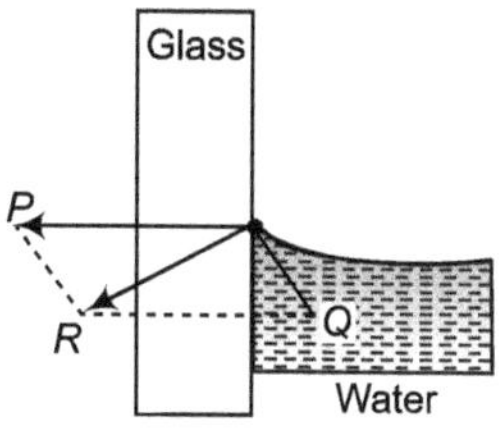

Fig. 13.39

On the other hand, when $p < Q$ (as with glass and mercury), shape of the meniscus is convex and the angle of contact $\theta > 90°$. The resultant (R) of p and Q in this case passes through the liquid.

Let us now see why the liquid surface bends near the contact with a solid. A liquid in equilibrium cannot sustain tangential stress. The resultant force on any small part of the surface layer must be perpendicular to the surface at that point. Basically three forces are acting on a small part of the liquid surface near its contact with solid. These forces are:

(i) p, attraction due to the molecule of the solid surface near it.

(ii) Q, attraction due to liquid molecules near this part, and

(iii) the weight w of the part considered.

We have considered very small part, so weight of that part can be ignored for better understanding. As we have seen in the last figures, to make the resultant (R) of p and Q perpendicular to the liquid surface, the surface becomes curved (convex or concave).

- The angle of contact between water and clean glass is zero and that between mercury and clean glass is 137°.

Capillarity

Surface tension causes elevation or depression of the liquid in a narrow tube. This effect is called capillarity.

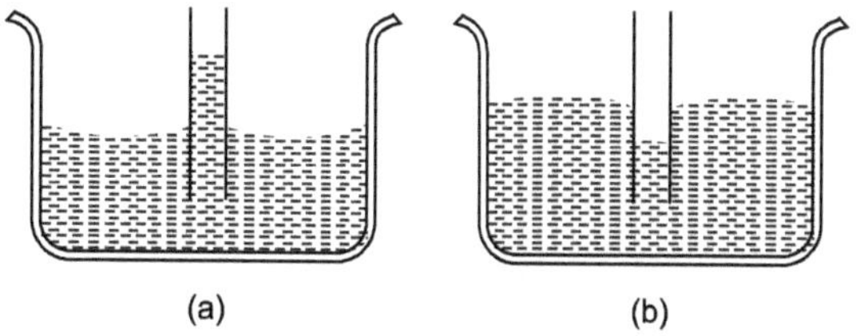

Fig. 13.40

When a glass capillary tube (A tube of very small diameter is called a capillary tube) open at both ends is dipped vertically in water, the water in the tube will rise above the level of water in the vessel as shown in Fig. (a). In case of mercury, the liquid is depressed in the tube below the level of mercury in the vessel as shown in Fig. (b).

When the contact angle is less than 90°, the liquid rises in the tube. For a nonwetting liquid angle of contact is greater than 90° and the surface is depressed, pulled down by the surface tension forces.

Explanation

When a capillary tube is dipped in water, the water meniscus inside the tube is concave. The pressure just below the meniscus is less than the pressure just above it by $\frac{2T}{R}$, where, T is the surface tension of water and R is the radius of curvature of the meniscus. The pressure on the surface of water is p_0, the atmospheric pressure. The pressure just below the plane surface of water outside the tube is also p_0, but that just below the meniscus inside the tube is $p_0 - \frac{2T}{R}$. We know that pressure at all points in the same level of water must be the same. Therefore, to make up the deficiency of pressure $\frac{2T}{R}$ below the meniscus water begins to flow from outside into the tube. The rising of water in the capillary stops at a certain height h. In this position, the pressure of water column of height h becomes equal to $\frac{2T}{R}$, i.e.

$$h\rho g = \frac{2T}{R} \quad \text{or} \quad h = \frac{2T}{Rrg}$$

If r is the radius of the capillary tube and θ the angle of contact, then

$$R = \frac{r}{\cos\theta}$$

$$\therefore \quad h = \frac{2T\cos\theta}{r\rho g}$$

Alternative Proof for the Formula of Capillary Rise

As we have already seen, when the contact angle is less than 90°, the total surface tension force just balances the extra weight of the liquid in the tube.

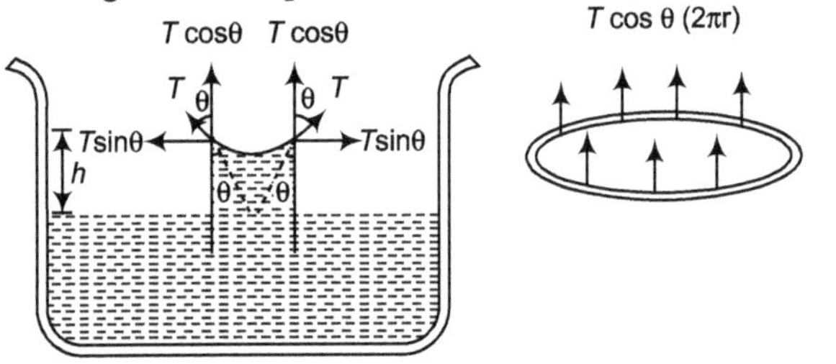

Fig. 13.41

The water, meniscus in the tube is along a circle of circumference $2\pi r$ which is in contact with the glass. Due to the surface tension of water, a force equal to T per unit length acts at all points of the circle. If the angle of contact is θ, then this force is directed inward at an angle θ from the wall of the tube. In accordance with Newton's third law, the tube exerts an equal and opposite force T per unit length on the circumference of the water meniscus.

This force which is directed outward, can be resolved into two components. $T\cos\theta$ per unit length acting vertically upward and $T\sin\theta$ per unit length acting horizontally outward. Considering the entire circumference $2\pi r$, for each horizontal component $T\sin\theta$ there is an equal and opposite component and the two neutralise each other. The vertical components being in the same direction are added up to give a total upward force $(2\pi r)(T\cos\theta)$. It is this force which supports the weight of the water column so raised. Thus,

$$(T\cos\theta)(2\pi r) = \text{Weight of the liquid column}$$

$$= (\pi r^2 \rho g h)$$

$$\therefore \quad h = \frac{2T\cos\theta}{r\rho g}$$

The result has following notable features:

(i) If the contact angle θ is greater than 90°, then term $\cos\theta$ is negative and hence, h is negative. The expression, then gives the depression of the liquid in the tube.

(ii) Suppose a capillary tube is held vertically in a liquid which has a concave meniscus, then capillary rise is given by,

$$h = \frac{2T\cos\theta}{r\rho g} = \frac{2T}{R\rho g} \quad \left(\text{as } R = \frac{r}{\cos\theta}\right)$$

$$\text{or} \quad hR = 2T/\rho g$$

When the length of the tube is greater than h, then liquid rises in the tube, so as to satisfy the above relation. But, if the length of the tube is insufficient (i.e. less than h) say h', the liquid does not emerge in the form of a fountain from the upper end (because it will violate the law of conservation of energy) but the angle made by the liquid surface and hence, the R changes in such a way that the force $2\pi r T\cos\theta$ equals the weight of the liquid raised. Thus

$$2\pi r T\cos\theta' = \pi r^2 \rho g h', \quad h' = \frac{2T\cos\theta'}{r\rho g}$$

$$\text{or} \quad h' = \frac{2T}{R'\rho g} \quad \text{or} \quad h'R' = \frac{2T}{\rho g}$$

From Eqs. (ii) and (iii)

$$hR = h'R' = \frac{2T}{\rho g}$$

Example 13.19 *A capillary tube whose inside radius is 0.5 mm is dipped in water having surface tension 7.0×10^{-2} N/m. To what height is the water raised above the normal water level? Angle of contact of water with glass is 0°. Density of water is 10^3 kg/m³ and $g = 9.8$ m/s²*

Sol. $h = \frac{2T\cos\theta}{r\rho g}$

Substituting the proper values, we have

$$h = \frac{(2)(7.0\times10^{-2})\cos 0^\circ}{(0.5\times10^{-3})(10^3)(9.8)} = 2.86\times10^{-2}\ \text{m} = 2.86\ \text{cm}$$

Example 13.20 *A glass tube of radius 0.4 mm is dipped vertically in water. Find upto what height the water will rise in the capillary? If the tube in inclined at an angle of 60° with the vertical, how much length of the capillary is occupied by water. Surface tension of water = 7.0×10^{-2}N/m, density of water = 10^3 kg/m³.*

Sol. For glass-water, angle of contact θ = 0°.

Now, $$h = \frac{2T\cos\theta}{r\rho g} = \frac{(2)(7.0\times10^{-2})\cos 0^\circ}{(0.4\times10^{-3})(10^3)(9.8)}$$

$$= 3.57\times10^{-2}\ \text{m} = 3.57\ \text{cm}$$

$$l = \frac{h}{\cos 60^\circ} = \frac{3.57}{1/2} = 7.14\ \text{cm}$$

Extra Knowledge Points

- **Stream line and turbulent flow** The flow of a fluid is said to be steady, if at any given point, the velocity of each particle passing through that point remains constant. Although velocity at different points may be different. That is, at some other point the particle may have a different velocity, but every other particle which passes the second point behaves exactly as the previous particle that has just passed that point.
  Each particle follows a smooth path and the paths of the particles do not cross each other. The path followed by a fluid particle under a steady flow is called a streamline.

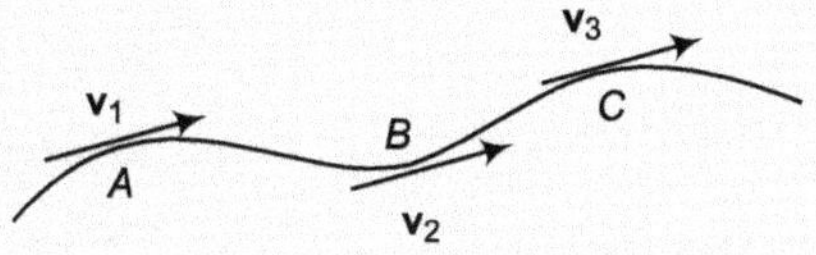

  In the streamline *ABC* shown is figure, velocity of fluid particles at *A* is v_1, at *B* is v_2 and at *C* is v_3.
  Steady flow is achieved when the speed of flow is not very high. Beyond a limiting value, called critical speed flow is no longer streamlined it becomes turbulent.

- **Reynolds number** When speed of fluid flow is high, flow of fluid becomes turbulent. Turbulent flow is less likely for viscous fluids flowing at low speeds.
  Reynolds number (R_e) is a dimensionless number, whose value gives us an idea whether the flow would be laminar (streamlined)or turbulent.

$$R_e = \frac{\rho v d}{\eta}$$

  Here, ρ = density of fluid
  v = speed of fluid
  d = dimensions of fluid and
  η = viscosity of fluid

  For $R_e < 1000$, flow is streamline or laminar.

  For $R_e \to 1000$ to 2000, flow is unsteady and for $R_e > 2000$, flow is turbulent. Turbulence dissipates kinetic energy usually in the form of heat. Racing cars and aeroplanes are designed to minimize turbulence. But sometimes turbulence is desirable. Turbulence promotes mixing. For example, blades of a kitchen mixer induce turbulent flow and provide thick milk shakes.

Chapter Summary with Formulae

(i) **Upthrust** $F = V_i \rho_l g_e$

(ii) When a solid whose density is less than the density of liquid floats in it then, some fraction of solid remains immersed in the liquid. In this case,

(a) Weight = Upthrust

(b) Fraction of volume immersed

$$f = \frac{\rho_s}{\rho_l}$$

(iii) When a solid whose density is more than the density of liquid is completely immersed in it, then upthrust acts on its 100% volume and apparent weight is less than its actual weight.

$$w_{app} = w - F$$

Here, F = Upthrust on 100% volume of solid.

(iv) **Relative Density (Specific Gravity) of any Substance**

$$\text{RD} = \frac{\text{Density of that substance}}{\text{Density of water}} = \frac{\text{Weight in air}}{\text{Change in weight in water}}$$

(v) **Effect of Temperature on Density**

$$\rho' = \frac{\rho}{1 + \gamma\Delta\theta}$$

(vi) **Effect of Pressure on Density**

$$\rho' = \frac{\rho}{1 - \dfrac{\Delta p}{B}}$$

(vii) 1 Pa = 1 Nm^{-2}, 1 Bar = 10^5 Pa, 1 atm = 1.013×10^5 Pa

Gauge pressure
= absolute pressure − atmospheric pressure.

(viii) Pressure at depth h below the surface of water,

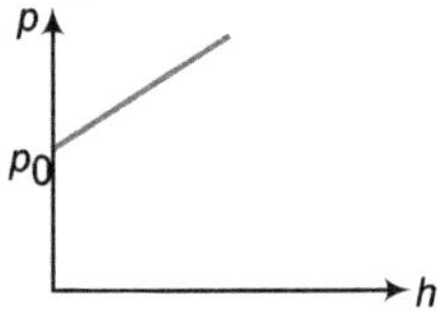

$$p = p_0 + \rho g h$$

Change in pressure per unit depth,

$$\frac{dp}{dh} = \rho g$$

(ix) **Volume Flow Rate**

$$Q = Av \text{ or } \frac{dV}{dt} = Av$$

(x) **Continuity Equation**

$$Q_1 = Q_2 \quad \text{or} \quad \frac{dV_1}{dt} = \frac{dV_2}{dt}$$

or $\quad A_1 v_1 = A_2 v_2$ or Av = constant or $v \propto \dfrac{1}{A}$

(xi) **Bernoulli's Equation**

$$p + \rho g h + 1/2\, \rho v^2 = \text{constant}$$

or $p_1 + \rho g h_1 + \dfrac{1}{2}\rho v_1^2 = p_2 + \rho g h_2 + \dfrac{1}{2}\rho v_2^2$

(xii) (a) $v = \sqrt{2gh}$

(b) $t = \sqrt{\dfrac{2(H-h)}{g}}$

(c) $R = vt = 2\sqrt{h(H-h)}$

(d) $R_{max} = H$ at $h = H/2$

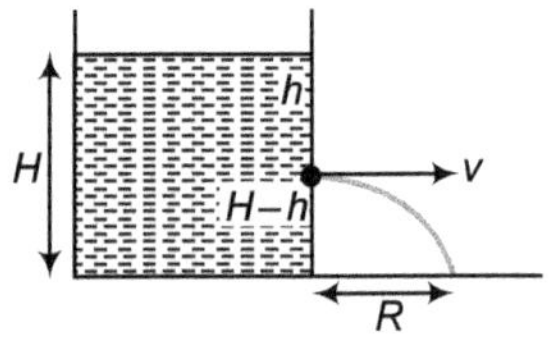

(xiii) **Barometer**

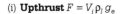

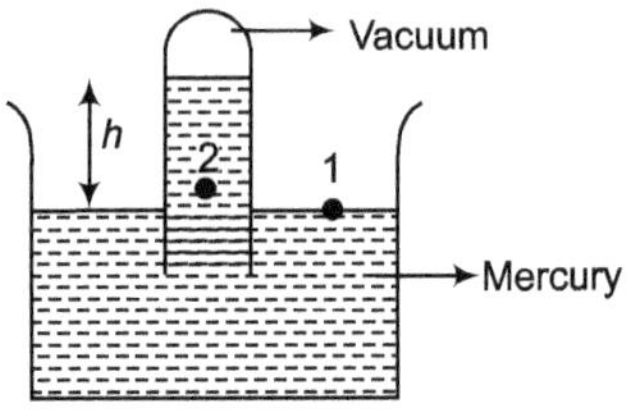

$$p_1 = p_2$$

$\therefore \quad p_0 = h\rho g$ or $h = \dfrac{p_0}{\rho g}$

h is approximately 76 cm of mercury.

(xiv) **Viscosity**

(a) $F = -\eta A \dfrac{dv}{dy}$ $\quad \left(\dfrac{dv}{dy} = \text{velocity gradient}\right)$

(b) $F = 6\pi\eta r v$ = For spherical ball

(c) $v_T = \dfrac{2}{9}\dfrac{r^2(\rho - \sigma)g}{\eta}$ or $v_T \propto r^2$

Here, ρ = density of ball and

σ = density of viscous medium in which ball is moving.

(xv) **Surface Tension**

(a) $T = \dfrac{F}{l} = \dfrac{\Delta W}{\Delta A}$ or $\Delta W = T \times \Delta A$

(b) $\Delta p = \dfrac{2T}{R}$ for single surface and $\dfrac{4T}{R}$ for double surface

(c) Capillary rise or fall,

$$h = \frac{2T}{R\rho g} = \frac{2T\cos\theta}{r\rho g} \quad \left(\text{as } R = \frac{r}{\cos\theta}\right)$$

(xvi) The unit of the pressure is the pascal (Pa). It is the same as Nm^{-2}. Other common units of pressure are

1 atm = 1.01×10^5 Pa

1 bar = 10^5 Pa

1 torr = 133 Pa = 0.133 kPa

1 mm of Hg = 1 torr = 133 Pa

(xvii) The onset of turbulence in a fluid is determined by a dimensionless parameter is called the Reynold number given by

$$R_e = \rho v d / \eta$$

where, d is a typical geometrical length associated with the fluid flow and the other symbols have their usual meaning.

Additional Examples

Example 1. *Why are the sleepers used below the rails?*

Sol. Sleepers are placed below the rails to increase area (A) and to consequently decrease the pressure (p) due to the weight (F) of train on the rails as $p = F/A$.

Example 2. *What will be the weight of a weightless bag when it is filled with water and weighed in water ?*

Sol. Let V be the volume of the bag.

Weight of the bag filled with water of density ρ is $W = V\rho g$

Force of buoyancy acting on the (weightless) when in water i.e.

$$F = \text{weight of water displaced by the bag} = V\rho g$$

Apparent weight of the bag in water

$$= w - F = V\rho g - V\rho g = 0$$

Example 3. *Explain why does a boat rise as it enters the sea from a river?*

Sol. For the boat to float, $W = F$

If V_r and V_s are the volumes of the portions of the boat inside river water and sea water and ρ_r, ρ_s are the densities of river and sea water respectively,

$$F = V_r \rho_r g = V_s \rho_s g$$

or

$$\frac{V_s}{V_r} = \frac{\rho_r}{\rho_s}$$

As, $\rho_r < \rho_s$ therefore, $V_s < V_r$.

Obviously, the boat rises as it enters sea.

Example 4. *A bucket of water rests on a scale. Does the scale reading change when a lead block is suspended from a thread and lowered into water where it is held submerged without touching the bottom or the sides of the bucket?*

Sol. Water exerts an upthrust force on the block from Newton's third law, there must be an equal but opposite force pushing down. The scale reading will increase by an amount equal to the buoyant force.

Example 5. *Two streamlines cannot cross each other. Explain, why?*

Sol. The tangent at any point on a streamline gives the direction of flow of liquid molecules at that point. In case, the two streamlines cross each other, it would mean that the liquid molecule can have two velocities along the two different directions, which is against the definition of streamline motion. Hence, two streamlines cannot cross each other.

Example 6. *The three vessels shown in figure are filled to the same height with water. The three vessels have the same base area. In which vessel, will the force on the base be minimum?*

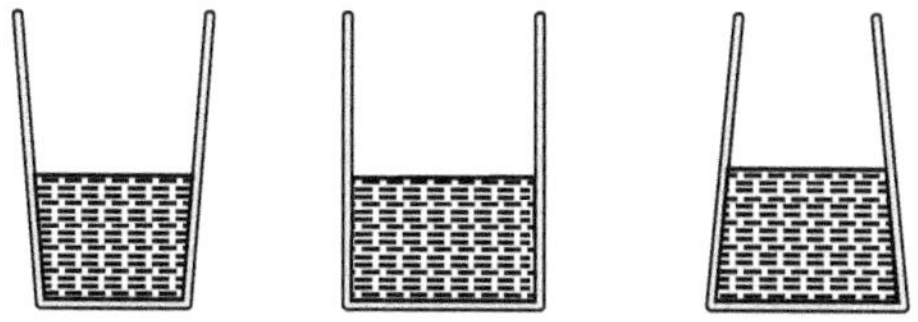

Sol. Since, the height of water column is same in the three vessels, the pressure at the base will be same. As, the three vessels have the same base area, the force will also be the same in the three cases.

Example 7. *To empty an oil tin, two holes are made. Why ?*

Sol. If one hole is made in the tin, then oil will not come out. It is because, as the oil comes out, the pressure inside the tin becomes less than the atmospheric pressure. Due to this, the oil cannot come out. On the other hand, when two holes are made in the tin, air keeps on entering the tin through the other.

Example 8. *Why is it easier to swim in sea water than in the river water?*

Sol. The density of sea water (salt dissolved in water) is greater than that of the river water. Therefore, upthrust on the swimmer is greater in sea water than in river water. Hence, it is easier to swim in sea water than in river water.

Example 9. *A balloon filled with helium does not rise in air indefinitely but halts after a certain height (neglect winds). Explain, why?*

Sol. Initially, the balloon filled with helium rises in air as the weight of the air displaced by the balloon is greater than the weight of the helium gas and the balloon. We know that the density of air decreases with height. Therefore, the balloon halts after attaining a height at which the density of air decreases to a value, such that the weight of the air displaced just equal to the weight of gas and the balloon.

Example 10. *As soon as parachute of a falling soldier opens, his acceleration decreases and soon becomes zero. Why?*

Sol. Viscous force keeps on increasing with increase in velocity. As a result, its acceleration decreases. However, it soon attains terminal velocity. On attaining the terminal velocity, the soldier falls with a constant velocity i.e. the acceleration of the soldier becomes zero.

Example 11. *When air is blown in between the two balls suspended from a string such that they do not touch each other, then balls come nearer to each other, instead of moving away. Why?*

Sol. When air is blown between the two suspended balls, then kinetic energy of the air between the balls increases. From Bernoulli's principle, pressure decreases. Due to pressure difference on the two sides of the two balls, the balls come nearer to each other.

Example 12. *If a small ping pong ball is placed in a vertical jet of air or water, it will rise to a certain height above the nozzle and stay at that level. Explain.*

Sol. Due to high velocity of the jet of water, pressure on the side of the ping pong ball, that faces the water jet, decreases. On the other side of the ball, the pressure is still equal to the atmospheric pressure. Due to the difference in pressure on the two sides, the ball get pushed towards the jet of water.

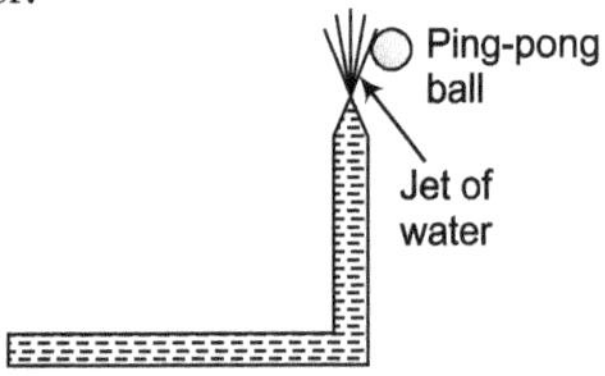

The ball does not fall down as it is constantly pressed against the water jet due to difference of pressure on the two sides of the ball.

Example 13. *End of a glass tube becomes round on heating. Explain.*

Sol. When an end of a glass tube is heated, the glass at that end melts. In molten (liquid) state, in an attempt to acquire minimum surface area due to the property of surface tension, the end of the glass tube becomes spherical i.e. round.

Example 14. *Why does a small piece of camphor dance on the water surface?*

Sol. When camphor dissolves in water, then surface tension of water decreases. Due to irregular shape of the camphor piece, it may dissolve more at one end than at the other end. Thus, surface tension of water will decrease by unequal amounts at the different ends of the camphor piece. It produces a resultant force on the camphor piece and it starts moving erom one place to another place.

Example 15. *A large bubble is formed at one end of a capillary tube and a small one at the other end. Which one will grow if they are connected?*

Sol. Since, the excess of pressure inside a bubble is inversely proportional to radius of the bubble $\left(\Delta p = \frac{4T}{R}\right)$, the pressure inside the bigger bubble will be less than that inside the smaller bubble. As air will flow from a region of greater pressure to the region of lower pressure, the bigger bubble will grow at the expense of the smaller bubble.

Example 16. *For the arrangement shown in the figure, what is the density of oil?*

Sol. $p_0 + \rho_w gl = p_0 + \rho_{oil} (l + d) g$

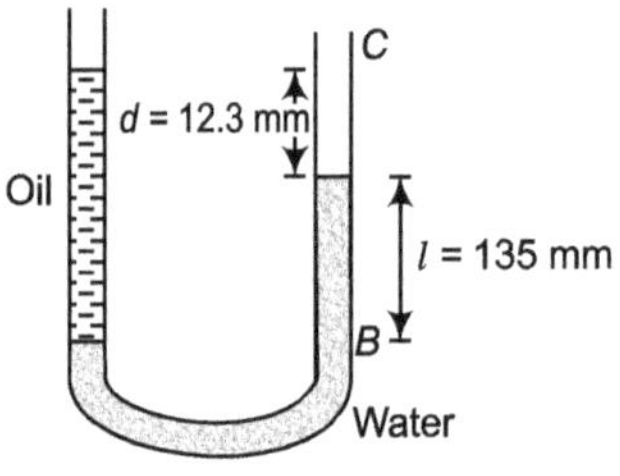

$$\Rightarrow \quad \rho_{oil} = \frac{\rho_w l}{l + d}$$

$$= \frac{1000 \times (135)}{(135 + 12.3)} = 916 \text{ kg/m}^3$$

Example 17. *A boat floating in a water tank is carrying a number of stones. If the stones were unloaded into water, what will happen to the water level?*

Sol. Let weight of boat $= W$ and weight of stone $= w$.

Assuming density of water = 1 g/cc

Volume of water displaced initially $= (w + W)$

Later, volume displaced $= \left(W + \frac{w}{\rho}\right)$

($\because \rho =$ density of stones)

$\Rightarrow$ Water level comes down.

Example 18. *Figure shows how the stream of water emerging from a faucet necks down as it falls. The area changes from A_0 to A through a fall of h. At what rate does the water flow from the tap?*

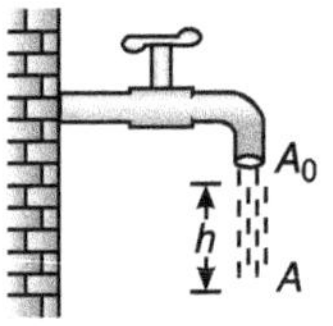

Sol. $A_0 v_0 = Av$

$$v^2 = v_0^2 + 2gh$$

$$v_0 = \sqrt{\frac{2ghA^2}{A_0^2 - A^2}}$$

or $$R = A_0 v_0 = A_0 \sqrt{\frac{2gh\, A^2}{A_0^2 - A^2}}$$

Example 19. *A bent tube is lowered into the stream as shown in the figure. The velocity of the stream relative to the tube is equal to v. The closed upper end of the tube located at the height h_0. To what height h will the water jet spurt ?*

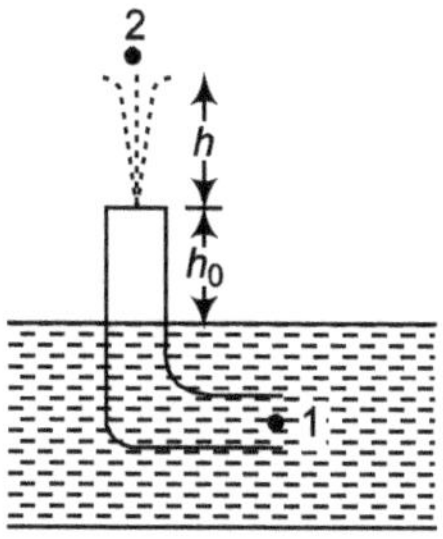

Sol. Let tube's entrance be a depth y below the surface. Take point 1 at entry, 2 at the maximum height of the fountain. Applying Bernoulli's theorem,

$$p_1 + \rho g h_1 + \frac{1}{2}\rho v_1^2 = p_2 + \rho g h_2 + \frac{1}{2}\rho v_2^2$$

Taking, $h_1 = 0,\ \ h_2 = (y + h_0 + h),\ \ v_1 = v,\ \ v_2 = 0,$

Substituting these values, we get

$$p_0 + \rho g y + \frac{1}{2}\rho v^2 = p_0 + \rho g (y + h_0 + h)$$

$$\Rightarrow \quad \frac{1}{2}\rho v^2 = \rho g (h_0 + h)$$

$$\Rightarrow \quad h = \left(\frac{v^2}{2g} - h_0\right)$$

Example 20. *A solid floats in a liquid of different materials. Carry out an analysis to see whether the level of liquid in the container will rise or fall when the solid melts.*

Sol. Let M = Mass of the floating solid

ρ_1 = density of liquid formed by melting of the solid

ρ_2 = density of the liquid in which the solid is floating

The mass of liquid displaced by the solid is M. Hence, the volume of liquid displaced is $\frac{M}{\rho_2}$.

When the solid melts, then volume occupied by it is M/ρ_1. Hence, the level of liquid in container will rise or fall according as

$$\frac{M}{\rho_1} > \text{ or } < \frac{M}{\rho_2}$$

i.e. $$\rho_1 < \text{ or } > \rho_2$$

There will be no change in the level, if $\rho_1 = \rho_2$. In case of ice floating in water $\rho_1 = \rho_2$ and hence, the level of water remains unchanged when ice melts.

Example 21. *An iron casting containing a number of cavities weigths 6000 N in air and 4000 N in water. What is the volume of the cavities in the casting? Density of iron is 7.87 g/cm^3.*
(Take $g = 9.8$ m/s^2 and density of water $= 10^3$ kg/m^3)

Sol. Let v be the volume of cavities and V be the volume of solid iron. Then,

$$V = \frac{\text{mass}}{\text{density}} = \left(\frac{6000/9.8}{7.87 \times 10^3}\right) = 0.078\ \text{m}^3$$

Further, decrease in weight = upthrust

$$\therefore \quad (6000 - 4000) = (V + v)\rho_w g$$

or $$2000 = (0.078 + v) \times 10^3 \times 9.8$$

or $$0.078 + v \approx 0.2$$

$$\therefore \quad v = 0.12\ \text{m}^3$$

Example 22. *Water rises in a capillary tube to a height of 2.0 cm. In another capillary tube whose radius is one third of it, how much the water will rise?*

Sol. $h = \dfrac{2T\cos\theta}{r\rho g}$

$$\therefore \quad hr = \frac{2T\cos\theta}{rg} = \text{constant}$$

$$h_1 r_1 = h_2 r_2 \quad \text{or} \quad h_2 = \frac{h_1 r_1}{r_2}$$

Substituting the values, we get

$$h_2 = (2.0)(3) \qquad \left(\frac{r_2}{r_1} = \frac{1}{3}\right)$$

$$= 6.0\ \text{cm}$$

Example 23. *Mercury has an angle of contact of 120° with glass. A narrow tube of radius 1.0 mm made of this glass is dipped in a trough containing mercury. By what amount does the mercury dip down in the tube relative to the liquid surface outside. Surface tension of mercury at the temperature of the experiment is 0.5 N/m and density of mercury is 13.6×10^3 kg/m^3. (Take $g = 9.8$ m/s^2)*

Sol. $h = \dfrac{2T\cos\theta}{r\rho g}$

Substituting the values, we get

$$h = \frac{2 \times 0.5 \times \cos 120^\circ}{10^{-3} \times 13.6 \times 10^3 \times 9.8} = -3.75 \times 10^{-3}\ \text{m}$$

or $$h = -3.75\ \text{mm}.$$

- Here, negative sign implies that mercury suffers capillary depression.

Example 24. *Two narrow bores of radius 3.0 mm and 6.0 mm are joined together to form a U-shaped tube open at both ends. If the U-tube contains water, what is the difference in its levels in the two limbs of the tube. Surface tension of water is 7.3×10^{-2} N/m. (Take the angle of contact to be zero and density of water to be 10^3 kg/m³, $g = 9.8\ m/s^2$)*

Sol. $h\rho g = \Delta p = \dfrac{2T\cos\theta}{r_1} - \dfrac{2T\cos\theta}{r_2}$

or $$h = \frac{2T\cos\theta}{\rho g}\left(\frac{r_2 - r_1}{r_1 r_2}\right)$$

Substituting the values, we have

$$h = \frac{2 \times 7.3 \times 10^{-2} \times \cos 0°}{10^3 \times 9.8}\left(\frac{6.0 - 3.0}{6.0 \times 3.0}\right) \times \frac{1}{10^{-3}}$$

$$= 2.48 \times 10^{-3}\ \text{m} = 2.48\ \text{mm}$$

Example 25. *With what terminal velocity will an air bubble 0.8 mm in diameter rise in a liquid of viscosity 0.15 N-s/m² and specific gravity 0.9. Density of air is 1.293 kg/m³.*

Sol. The terminal velocity of the bubble is given by,

$$v_T = \frac{2}{9}\frac{r^2(\rho - \sigma)g}{\eta}$$

Here, $r = 0.4 \times 10^{-3}$ m, $\sigma = 0.9 \times 10^3$ kg/m³

$\rho = 1.293$ kg/m³, $\eta = 0.15$, Ns/m²

and $g = 9.8$ m/s²

Substituting the values we have,

$$v_T = \frac{2}{9} \times \frac{(0.4 \times 10^{-3})^2(1.293 - 0.9 \times 10^3) \times 9.8}{0.15}$$

$$= -0.0021\ \text{m/s} \quad \text{or} \quad v_T = -0.21\ \text{cm/s}$$

- Here, negative sign implies that the bubble will rise up.

Example 26. *A spherical ball of radius 3.0×10^{-4} m and density 10^4 kg/m³ falls freely under gravity through a distance h before entering a tank of water. If after entering the water the velocity of the ball does not change, then find the value of h. Viscosity of water is 9.8×10^{-6} N-s/m².*

Sol. Before entering the water the velocity of ball is $\sqrt{2gh}$. If after entering the water this velocity does not change, then this value should be equal to the terminal velocity. Therefore,

$$\sqrt{2gh} = \frac{2}{9}\frac{r^2(\rho - \sigma)g}{\eta}$$

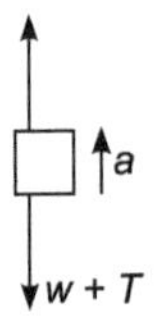

$$\therefore \quad h = \frac{\left\{\dfrac{2}{9}\dfrac{r^2(\rho - \sigma)g}{\eta}\right\}^2}{2g}$$

$$= \frac{2}{81} \times \frac{r^4(\rho - \sigma)^2 g}{\eta^2}$$

$$= \frac{2}{81} \times \frac{(3 \times 10^{-4})^4(10^4 - 10^3)^2 \times 9.8}{(9.8 \times 10^{-6})^2}$$

$$= 1.65 \times 10^3\ \text{m}$$

Example 27. *A solid ball of density half that of water falls freely under gravity from a height of 19.6 m and then enters water. Up to what depth will the ball go. How much time will it take to come again to the water surface. Neglect air resistance and viscosity effects in water. ($g = 9.8\ m/s^2$)*

Sol. $v = \sqrt{2gh} = \sqrt{2 \times 9.8 \times 19.6} = 19.6$ m/s

Let ρ be the density of ball and 2ρ the density of water. Net retardation inside the water,

$$a = \frac{\text{upthrust - weight}}{\text{mass}}$$

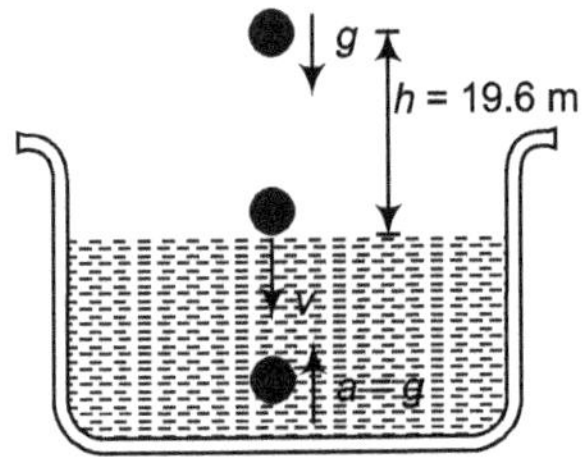

$$= \frac{V(2\rho)g - V(\rho)(g)}{V(\rho)g} \quad (V = \text{volume of ball})$$

$$= g = 9.8\ \text{m/s}^2$$

Hence, the ball will go up to the same depth 19.6 m below the water surface.

Further, time taken by the ball to come back to water surface is

$$t = 2\left(\frac{v}{a}\right) = 2\left(\frac{19.6}{9.8}\right) = 4\ \text{s}$$

Example 28. *A block of mass 1 kg and density 0.8 g/cm³ is held stationary with the help of a string as shown in figure. The tank is accelerating vertically upwards with an acceleration $a = 1.0$ m/s². Find,*

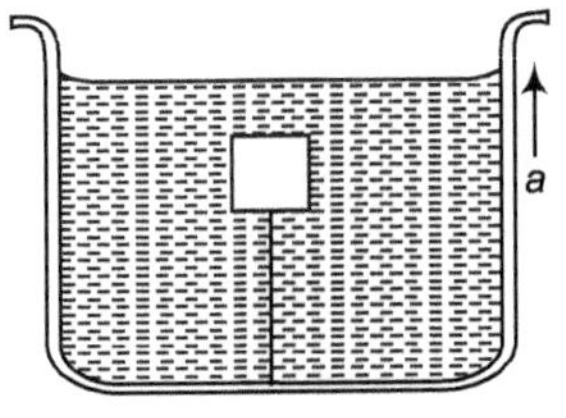

(a) the tension in the string,
(b) if the string is now cut find the acceleration of block. (Take $g = 10\ m/s^2$ and density of water $= 10^3$ kg/m³)

Sol. (a) Free body diagram of the block is shown in above figure.

In the figure,

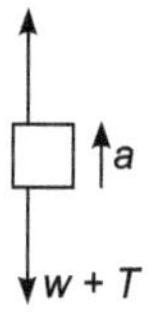

$$F = \text{Upthrust force}$$
$$= V\rho_\omega(g + a)$$
$$= \left(\frac{\text{Mass of block}}{\text{Density of block}}\right)\rho_w\,(g + a)$$
$$= \left(\frac{1}{800}\right)(1000)(10 + 1) = 13.75 \text{ N}$$
$$w = mg = 10 \text{ N}$$

Equation of motion of the block is

$$F - T - w = ma$$
$$\therefore \quad 13.75 - T - 10 = 1 \times 1$$
$$\therefore \quad T = 2.75 \text{ N}$$

(b) When the string is cut $T = 0$

$$\therefore \quad a = \frac{F - w}{m}$$
$$= \frac{13.75 - 10}{1} = 3.75 \text{ m/s}^2$$

Example 29. *Two separate air bubbles (radii* 0.004 *m and 0 m) formed of the same liquid (surface tension* 0.07 *N/m) come together to form a double bubble. Find the radius and the sense of curvature of the internal film surface common to both the bubbles.*

Sol. $p_1 = p_0 + \dfrac{4T}{r_1}$

$$p_2 = p_0 + \frac{4T}{r_2}$$

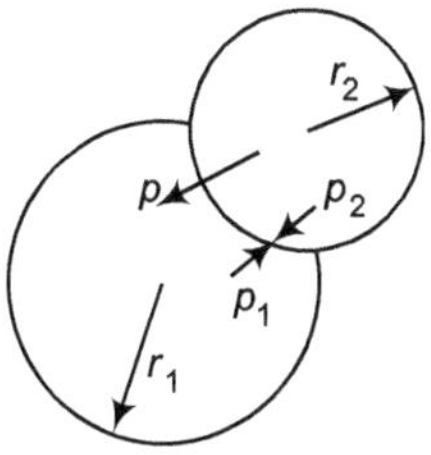

$$r_2 < r_1$$
$$\therefore \quad p_2 > p_1$$

i.e. pressure inside the smaller bubble will be more. The excess pressure

$$p = p_2 - p_1 = 4T\left(\frac{r_1 - r_2}{r_1 r_2}\right) \quad \text{...(i)}$$

This excess pressure acts from concave to convex side, the interface will be concave towards smaller bubble and convex towards larger bubble. Let R be the radius of interface, then

$$p = \frac{4T}{R} \quad \text{...(ii)}$$

From Eqs. (i) and (ii), we get

$$R = \frac{r_1 r_2}{r_1 - r_2} = \frac{(0.004)(0.002)}{(0.004 - 0.002)} = 0.004 \text{ m}$$

Example 30. *Under isothermal condition two soap bubbles of radii r_1 and r_2 coalesce to form a single bubble of radius r. The external pressure is p_0. Find the surface tension of the soap in terms of the given parameters.*

Sol. As mass of the air is conserved,

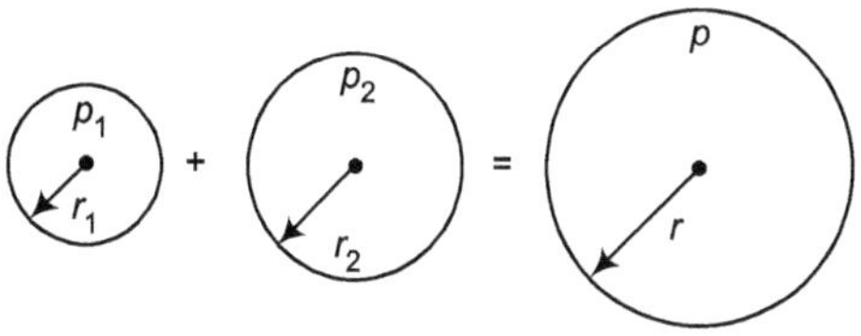

$$\therefore \quad n_1 + n_2 = n \qquad (\text{as } pV = nRT)$$
$$\therefore \quad \frac{p_1V_1}{RT_1} + \frac{p_2V_2}{RT_2} = \frac{pV}{RT}$$

As temperature is constant,

$$T_1 = T_2 = T$$
$$\therefore \quad p_1V_1 + p_2V_2 = pV$$
$$\therefore \quad \left(p_0 + \frac{4S}{r_1}\right)\left(\frac{4}{3}\pi r_1^3\right) + \left(p_0 + \frac{4S}{r_2}\right)\left(\frac{4}{3}\pi r_2^3\right)$$
$$= \left(p_0 + \frac{4S}{r}\right)\left(\frac{4}{3}\pi r^3\right)$$

Solving this, we get

$$S = \frac{p_0(r^3 - r_1^3 - r_2^3)}{4(r_1^2 + r_2^2 - r^2)}$$

- To avoid confusion with the temperature surface tension here is represented by S.

NCERT Selected Questions

Q 1. Explain why?

(a) The blood pressure in humans is greater at the feet than at the brain.

(b) Atmospheric pressure at a height of about 6 km decreases to nearly half its value at the sea level though the height of the atmosphere is more than 100 km.

(c) Hydrostatic pressure is a scalar quantity even though pressure is force divided by area.

Sol. (a) The height of the blood column in the human body is more at the feet as compared to that at the brain. Consequently, the blood pressure in humans is greater at the feet than at the brain.

(b) Density of air is maximum near the surface of the earth and decreases rapidly with height.

(c) Pressure is transmitted equally in all directions inside the liquid. Thus, there is no fixed direction for the pressure due to liquid. Hence, hydrostatic pressure is a scalar quantity.

Q 2. Fill in the blanks using the word(s) from the list appended with each statement.

(a) Surface tension of liquids generally .... with temperature. (increases/decreases)

(b) Viscosity of gases .... with temperature, whereas viscosity of liquids .... with temperature. (increases/decreases)

(c) For solids with elastic modulus of rigidity, the shearing force is proportional to ...., while for fluids it is proportional to .... ,(shear strain/rate of shear strain)

(d) For a fluid in a steady flow, the increase in flow speed at a constriction follows from ...., while the decrease of pressure there follows from .... ,(conservation of mass /Bernoulli's principle)

(e) For the model of a plane in a wind tunnel, turbulence occurs at a.............. speed than the critical speed for turbulence for an actual plane. (greater/smaller)

Sol. (a) decreases

(b) increases,decreases

(c) shear strain, rate of shear strain

(d) conservation of mass, Bernoulli's principle

(e) greater

Q 3. Explain why?

(a) To keep a piece of paper horizontal, you should blow over, not under it.

(b) When we try to close a water tap with our fingers, fast jets of water gush through the openings between our fingers.

(c) A fluid flowing out of a small hole in a vessel results in a backward thrust on the vessel.

Sol. (a) If we blow over a piece of paper, velocity of air above the paper becomes more than that below it. In accordance with Bernoulli's theorem $(p + \frac{1}{2}\rho v^2 = \text{constant})$, its pressure above the paper decreases. Due to greater value of pressure below the piece of paper it remains horizontal and does not fall.

(b) This can be explained from the equation of continuity $A_1 v_1 = A_2 v_2$. As, we try to close a water tap with our fingers, the area of cross-section of the outlet of water jet is reduced considerably as the openings between our fingers provide constriction (i.e. regions of smaller area). Thus velocity of water increases greatly and fast jets of water come through the openings between our fingers.

(c) When a fluid is flowing out of a small hole in a vessel, it acquires a large velocity and hence possesses large momentum. Since, no external force is acting on the system, a backward velocity must be attained by the vessel (according to the of conservation of momentum).

Q 4. A 50 kg girl wearing high heel shoes balances on a single heel. The heel is circular with a diameter 1.0 cm. What is the pressure exerted on the horizontal floor?

Sol. $p = \frac{F}{A}$ or $p = \frac{50 \times 9.8}{3.142 \times 25 \times 10^{-6}} = 6.24 \times 10^6$ Pa

$p = 6.24 \times 10^6$ Pa.

Q 5. Torricelli's barometer used mercury. Pascal duplicated it using French wine of density 984 kg m^{-3}. Determine the height of the wine column for normal atmospheric pressure.

Sol. p = normal atmospheric pressure $= 1.013 \times 10^5$ Pa

Let h be the height of the French wine column.

Then $h \rho_w g = p$

or $$h = \frac{p}{\rho_w g} = \frac{1.013 \times 10^5}{984 \times 9.8} = 10.5 \text{ m}$$

Q 6. A vertical off-shore structure is built to withstand a maximum stress of 10^9 Pa. Is the structure suitable for putting up on top of an oil well in the ocean? Take the depth of the ocean to be roughly 3 km and ignore ocean currents.

Sol. If p be the pressure exerted by this water column at the given depth.

Then, $p = h\rho g = 3 \times 10^3 \times 10^3 \times 9.8$

$= 29.4 \times 10^6 \text{ Pa} \approx 3 \times 10^7 \text{ Pa}$

Since, $3 \times 10^7 \text{ Pa} < 10^9 \text{ Pa}$

Thus, we conclude that the structure is suitable as the stress applied by it is much lesser than the maximum stress it can withstand.

Q 7. A hydraulic automobile lift is designed to lift cars with a maximum mass of 3000 kg. The area of cross-section of the piston carrying the load is 425 cm^2. What maximum pressure would the smaller piston have to bear?

Sol. The maximum force which the bigger piston can bear,

$$F = 3000 \text{ kgf} = 3000 \times 9.8 \text{ N}$$

Area of piston, $A = 425 \text{ cm}^2 = 425 \times 10^{-4} \text{m}^2$

If p=maximum pressure on the bigger piston.

Then, $p = \dfrac{F}{A} = \dfrac{3000 \times 9.8}{425 \times 10^{-4}}$

$= 6.92 \times 10^5 \text{ Pa}$

Since, the liquid transmits pressure equally in all directions, hence, the maximum pressure the smaller piston can bear is 6.92×10^5 Pa.

Q 8. A U-tube contains water and methylated spirit separated by mercury. The mercury columns in the two arms are in level with 10.0 cm of water in one arm and 12.5 cm of spirit in the other. What is the specific gravity of spirit?

Sol. $h_1 \rho_1 g = h_2 \rho_2 g$

or $\rho_2 = \dfrac{h_1 \rho_1}{h_2} = \dfrac{10 \times 1}{12.5}$

$= \dfrac{4}{5} = 0.8 \text{ g cm}^{-3}$

Now, specific gravity of spirit

$= \dfrac{\text{density of spirit}}{\text{density of water}}$

$= \dfrac{0.8 \text{ gcm}^{-3}}{1 \text{ gcm}^{-3}} = 0.800$

Q 9. In the previous problem, if 15.0 cm of water and spirit each are further poured into the respective arms of the tube, what is the difference in the levels of mercury in the two arms? (Specific gravity of mercury = 13.6).

Sol. Pressure at A = Pressure at B

or $p_0 + h_w \rho_w g = p_0 + h_s \rho_s g + h_m \rho_m g$

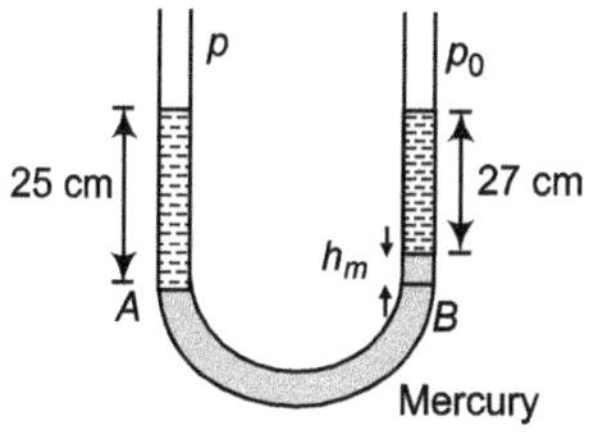

or $h_w \rho_w = h_s \rho_s + h_m \rho_m$

$\therefore$ $h_m = \dfrac{h_w \rho_w - h_s \rho_s}{\rho_m}$

Substituting the values, we get

or $h_m = 0.221 \text{ cm}$

Q 10. Can Bernoulli's equation be used to describe the flow of water through a rapid river? Explain.

Sol. No, Bernoulli's equation cannot be used to describe the flow of water through a rapid river. This is due to the reason that Bernoulli's equation applies to streamlined flow and in a rapid river, the flow of water is not streamlined.

Q 11. Does it matter, if one uses gauge instead of absolute pressures in applying Bernoulli's equation? Explain.

Sol. No, it does not matter, if one uses gauge pressure instead of absolute pressure in applying Bernoulli's equation unless the atmospheric pressure at the two points, where Bernoulli's equation is applied are significantly different.

Q 12. In a test experiment on a model aeroplane in a wind tunnel, the flow speeds on the upper and lower surfaces of the wing are 70 ms^{-1} and 63 ms^{-1} respectively. What is the lift on the wing if its area is 2.5 m^2? Take the density of air to be 1.3 kgm^{-3}.

Sol. The level of the upper and lower surfaces of the wings from the ground may be taken same, i.e.

$$h_1 = h_2$$

From Bernoulli's theorem,

$$p_1 + \rho g h_1 + \frac{1}{2}\rho v_1^2 = p_2 + \rho g h_2 \ \frac{1}{2}\rho v_2^2$$

or $p_2 - p_1 = \dfrac{1}{2}\rho(v_1^2 - v_2^2)$ ...(i)

This pressure difference provides the lift to the aeroplane i.e. lift on the wing.

Thus, if F be the lift on the wing, then

$F = (p_2 - p_1) \times A$

$= \dfrac{1}{2}\rho(v_1^2 - v_2^2) \times A$

$= \dfrac{1}{2} \times 1.3 \times (70^2 - 63^2) \times 2.5$

$= 1.5 \times 10^3 \text{ N}$

Q 13. Fig. (a) and Fig. (b) refer to the steady flow of a (non-viscous) liquid. Which of the two figures is incorrect? Why?

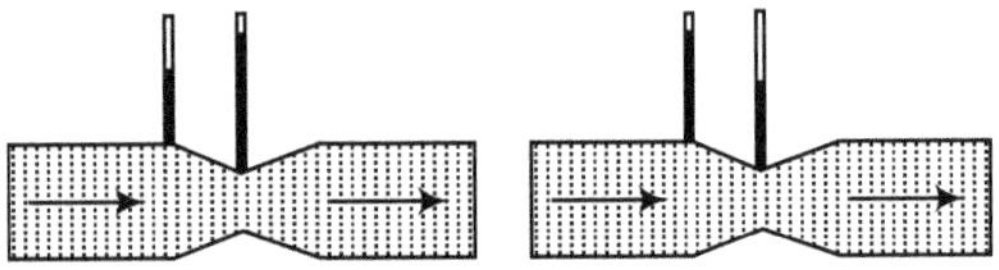

(a) (b)

Sol. Fig. (a) is incorrect. This is because at a section of smaller cross-section, the flow speed is larger due to equation of continuity. Hence, the according to Bernoulli's theorem, pressure in the narrow part must be low. But in this figure the pressure at narrow part is shown to be larger.

Q 14. The cylindrical tube of a spray pump has a cross-section of $8.0\,cm^2$ one end of which has 40 fine holes each of diameter 1.0 mm. If the liquid flow inside the tube is 1.5 m/min. What is the speed of ejection of the liquid through the holes?

Sol. Area of corss-section of tube,

$$a_1 = 8.0\text{ cm}^2 = 8\times10^{-4}\text{ m}^2$$

Number of fine holes = 40

$\therefore$ Radius of each hole $= \frac{d}{2} = 0.5\times10^{-3}$ m

$\therefore$ Area of cross-section of each hole,

$$a = \pi r^2 = \pi(0.5\times10^{-3})^2\text{ m}^2$$

$\therefore$ Area of cross-section of 40 holes,

$$a_2 = 40\,a = 40\,\pi\,(0.5\times10^{-3})^2\text{ m}^2$$

Velocity of flow of liquid in the tube,

$$v_1 = 1.5\text{ m/min} = \frac{1.5}{60}\text{ ms}^{-1}$$

Let v_2 be the velocity of ejection of the fluid. Then, according to equation of continuity,

$$a_1 v_1 = a_2 v_2$$

or $$v_2 = \frac{a_1 v_2}{a_2}$$

$$= \frac{8\times10^{-4}}{40\pi(0.5\times10^{-3})^2}\times\frac{1.5}{60}$$

$$= 0.64\text{ ms}^{-1}$$

Q 15. A U-shaped wire is dipped in a soap solution and removed. The thin soap film formed between the wire and a light slider supports a weight of 1.5×10^2 N (which includes the small weight of the slider). The length of the slider is 30 cm. What is the surface tension of the film?

Sol. A soap film has two free surfaces, so total length of the film to be supported, $l = 2\times30\text{ cm} = 0.60$ m

Let T = surface tension of the film

If F = total force on the slider due to surface tension.

Then, $$F = T\times2l = T\;\times0.6\text{ N}$$

$$w = 1.5\times10^{-2}\text{ N}$$

In equilibrium position, the force F on the slider due to surface tension must be balanced by the weight (w) supported by the slider.

i.e. $$F = w = mg$$

or $$T\times0.6 = 1.5\times10^{-2}$$

$\therefore$ $$T = \frac{1.5\times10^{-2}}{0.6} = 2.5\times10^{-2}\text{ Nm}^{-1}$$

Q 16. What is the pressure inside the drop of mercury of radius 3.00 mm at room temperature? Surface tension of mercury at that temperature (20°C) is $4.65\times10^{-1}\text{ Nm}^{-1}$. The atmospheric pressure is 1.01×10^5 Pa. Also give the excess pressure inside the drop.

Sol. $$p_{in} - p_{out} = \frac{2T}{r} = \frac{2\times4.65\times10^{-1}}{3\times10^{-3}} = 310\text{ Pa}$$

$\therefore$ $$p_{in} = p_{out} + 310 = 310 + 1.01\times10^5\text{ Pa}$$

$$= 1.01\times10^5 + 0.00310\times10^5$$

$$= 1.01310\times10^5\text{ Pa}$$

Q 17. What is the excess pressure inside a bubble of soap solution of radius 5.00 mm, given that the surface tension of soap solution at the temperature (20°C) is $2.50\times10^{-2}\text{ Nm}^{-1}$? If an air bubble of the same dimension were formed at a depth of 40.0 cm inside a container containing the soap solution (of relative density 1.20), what would be the pressure inside the bubble? (1 atmospheric pressure is 1.01×10^5 Pa.)

Sol. Excess of pressure inside the soap bubble is given by,

$$p_i - p = \frac{4T}{r}$$

$$= \frac{4\times2.5\times10^{-2}}{5.0\times10^{-3}}$$

$$= 20\text{ Pa}$$

Excess pressure inside the air bubble in the soap solution is

$$p_i - p = \frac{2T}{r}$$

$$= \frac{2\times2.5\times10^{-2}}{5.0\times10^{-3}}$$

$$= 10\text{ Pa}$$

Now, pressure outside the air bubble at a depth of 40 cm is

p = Atmospheric pressure + pressure due to 40 cm

$= 1.01\times10^5 + 0.4\times1.2\times10^3\times9.8$

$= 1.05704\times10^5$ Pa $\qquad (\because p = h\rho g)$

$\approx 1.06\times10^5$ Pa

$\therefore$ Pressure inside the air bubble,

$$p_1 = p + \frac{2T}{r} = (1.06\times10^5 + 10)\text{ Pa}$$

Q 18. A tank with a square base of area $1.0\,m^2$ is divided by a vertical partition in the middle. The bottom of the partition has a small hinged door of area $20\,cm^2$. The tank is filled with water in one compartment, and an acid (of relative density 1.7) in the other, both to a height of 4.0 m. Compute the force necessary to keep the door close.

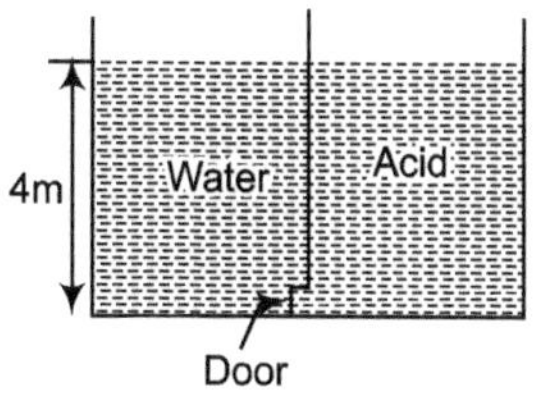

Sol. (a) If p_1 be the pressure exerted by water at the door provided at the bottom, then

$$p_1 = h_1 \rho_1 g = 4.0 \times 10^3 \times 9.8 = 39.2 \times 10^3 \text{ Pa}$$

(b) If P_2 be the pressure exerted by acid at the door provided at the bottom, then

$$p_2 = h_2 \rho_2 g = 4.0 \times 1.7 \times 9.8 \times 10^3 = 66.64 \times 10^3 \text{ Pa}$$

$\therefore$ Difference of pressure

$$\Delta p = p_2 - p_1 = 66.64 \times 10^3 - 39.2 \times 10^3$$
$$= 27.44 \times 10^3 \text{ Pa}$$

Area of door, $A = 20\text{ cm}^2 = 20 \times 10^{-4}\text{ m}^2$

$\therefore$ Force on the door = difference in pressure $\times$ area

$$= \Delta p \times A = 27.44 \times 10^3 \times 20 \times 10^{-4} = 54.88 \text{ N} \approx 55 \text{ N}$$

To keep the door closed, a force equal to 55 N should be applied horizontally on the door from compartment containing water.

Q 19. Two vessels have the same base area but different shapes. The first vessel takes twice the volume of water that the second vessel requires to fill upto a particular common height. Is the force exerted by the water on the base of the vessel the same in the two cases? If so, why do the vessels filled with water to that same height give different readings on a weighing scale?

Sol. Since, the pressure depends upon the height of water column and the height of the water column in the two vessels of different shapes is the same, hence there will be same pressure due to water on the base of each vessel. As the base area of each vessel is same, hence there will be equal force acting on the two base areas due to water pressure. But the two vessels have different weights of liquids. That is why, the two vessels filled with water to same vertical height show different readings on a weighing machine.

Q 20. In deriving Bernoulli's equation, we equated the work done on the fluid in the tube to its change in the potential and kinetic energy.

(a) How does the pressure change as the fluid moves along the tube, if dissipative forces are present?

(b) Do the dissipative forces become more important as the fluid velocity increases? Discuss qualitatively.

Sol. (a) If dissipating forces are present, then a part of work done will be used in overcoming these forces during the flow of fluid. So, there shall be greater drop of pressure as the fluid move along the tube.

(b) Yes, the dissipative forces become more important as the fluid velocity increases.

According to the equation

$$F = -\eta A\, dv/dy$$

Clearly as v increases, velocity gradient increases and hence, viscous drag i.e. dissipative force also increases.

Q 21. In Millikan's oil drop experiment, what is the terminal speed of an uncharged drop of radius of 2.0×10^{-5} m and density $1.2 \times 10^3\text{ kg m}^{-3}$? Take the viscosity of air at the temperature of the experiment to be $1.8 \times 10^{-5}\,\frac{\text{N-s}}{\text{m}^2}$. How much is the viscous force on the drop at that speed? Neglect buoyancy of the drop due to air?

Sol. Terminal velocity, $v_T = \frac{2}{9} r^2 \frac{(\rho - \rho_0)g}{\eta}$

Since, the buoyancy of the drop due to air is to be neglected,

$\therefore \quad \rho_0 = 0$ for air

$$\therefore \quad v_r = \frac{2 r^2 \rho g}{9 \eta} = \frac{2}{9} \times \frac{(2 \times 10^{-5})^2 \times (1.2 \times 10^3)}{1.8 \times 10^{-5}} \times 9.8$$
$$= 5.8 \times 10^{-2}\text{ ms}^{-1} = 5.8\text{ cms}^{-1}$$

The viscous force on the drop according to Stoke's law is given by

$$F = 6\pi\eta r v_T$$
$$= 6 \times 3.142 \times (1.8 \times 10^{-5}) \times (2 \times 10^{-5}) \times (5.8 \times 10^{-2})$$
$$= 3.93 \times 10^{-10}\text{ N}$$

Q 22. Mercury has an angle of contact equal to 140° with soda lime glass. A narrow tube of radius 1.00 mm made of this glass is dipped in a trough containing mercury. By what amount does the mercury dip down in the tube relative to the liquid surface outside? Surface tension of mercury at the temperature of the experiment is 0.465 Nm^{-1}. Density of mercury $= 13.6 \times 10^3\text{ kgm}^{-3}$.

Sol. Using the formula, $h = \frac{2T\cos\theta}{r\rho g}$, we get

$$h = \frac{2 \times 0.465 \times \cos 140°}{10^{-3} \times 13.6 \times 10^3 \times 9.8} = \frac{2 \times 0.465 \times (-0.7660)}{10^{-3} \times 13.6 \times 10^3 \times 9.8}$$
$$= -5.34 \times 10^{-3}\text{ m} = -5.34\text{ mm}$$

Here, negative sign shown that the mercury level is depressed in the tube relative to the mercury surface outside.

i.e. depression = 5.34 mm

Objective Problems

[Level 1]

Upthrust and Concept of Floating

1. A body floats in a liquid contained in a beaker. The whole system shown in figure is falling under gravity, the upthrust on the body due to the liquid is

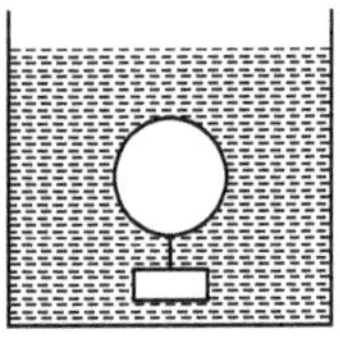

(a) zero
(b) equal to weight of the body in air
(c) equal to weight of liquid displaced
(d) equal to the weight of the immersed part of the body

2. A raft of wood of mass 120 kg floats in water. The weight that can be put on the raft to make it just sink should be ($d_{raft} = 600\,kg/m^3$)
(a) 80 kg (b) 50 kg
(c) 60 kg (d) 30 kg

3. A balloon has volume of $1000\,m^3$. It is filled with hydrogen ($\rho = 0.09g/L$). If the density of air is 1.29 g/L, then it can lift a total weight of
(a) 600 kg (b) 1200 kg
(c) 300 kg (d) 1800 kg

4. A boat having a length 3 m and breadth 2 m is floating on a lake. The boat sinks by 1 cm, when a man gets on it. The mass of the man is
(a) 60 kg (b) 72 kg
(c) 52 kg (d) 65 kg

5. A small block of wood of relative density 0.5 is submerged in water. When the block is released, it states moving upwards, then acceleration of the block is ($g = 10\,ms^{-2}$)
(a) $5\,ms^{-2}$
(b) $10\,ms^{-2}$
(c) $7.5\,ms^{-2}$
(d) $15\,ms^{-2}$

6. The reading of spring balance when a block is suspended from it in air, is 6 N. This reading is changed to 40 N, when the block is immersed in water. The specific gravity of the block is
(a) 3 (b) 2
(c) 6 (d) $\frac{3}{2}$

7. A tank contains water on top of mercury as shown in figure. A cubical block of side 10 cm is in equilibrium inside the tank. The depth of the block inside mercury is (RD of the material of block = 8.56, RD of mercury = 13.6)

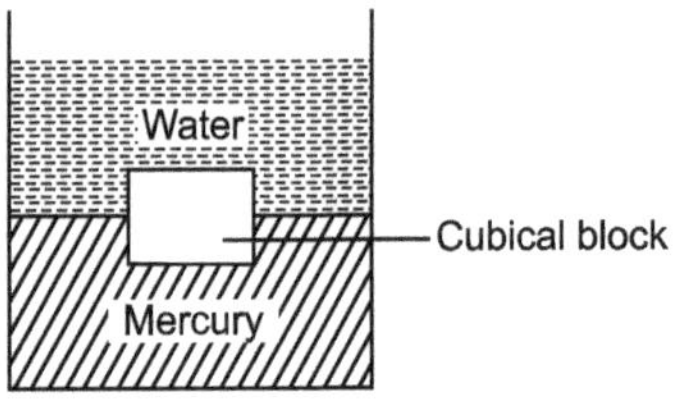

(a) 6 cm (b) 5 cm (c) 7 cm (d) 8 cm

8. An object of weight w and density ρ is submerged in liquid of density σ, its apparent weight will be
(a) $(\rho - \sigma)$ (b) $(\rho - \sigma)/w$
(c) $w\left(1 - \frac{\sigma}{\rho}\right)$ (d) $w\left(1 - \frac{\rho}{\sigma}\right)$

9. A block of wood is floating on the surface of water in a beaker. The beaker is covered with a bell jar and the air is evacuated. What will happen to the block?
(a) Sink a little (b) Rise a little
(c) Remain unchanged (d) Sink completely

10. A solid of density D is floating in a liquid of density d. If v is the volume of solid submerged in the liquid and V is the total volume of the solid, then $\frac{v}{V}$ is equal to
(a) $\frac{d}{D}$ (b) $\frac{D}{d}$
(c) $\frac{D}{(D+d)}$ (d) $\frac{D+d}{D}$

11. A metallic sphere floats in immiscible mixture of water (density $10^3\,kg/m^3$) and a liquid (density $8\times 10^3\,kg/m^3$) such that its $\left(\frac{2}{3}\right)$ part is in water and $\left(\frac{1}{3}\right)$ part in the liquid. The density of the metal is
(a) $\frac{5000}{3}\,kg/m^3$ (b) $\frac{10000}{3}\,kg/m^3$
(c) $5000\,kg/m^3$ (d) $2000\,kg/m^3$

12. A body measures 5 N in air and 2 N when put in water. The buoyant force is
(a) 7 N (b) 9 N
(c) 3 N (d) None of these

13. The relative density of ice is 0.9 and that of sea water is 1.125. What fraction of the whole volume of an iceberg appears above the surface of the sea?
(a) 1/5 (b) 2/5
(c) 3/5 (d) 4/5

14. The reading of spring balance when a block is suspended from it in air is 60 N. This reading is changed to 40 N, when the block is immersed in water. The specific gravity of the block is
(a) 3 (b) 2 (c) 6 (d) 3/2

15. A cubical block is floating in a liquid with half of its volume immersed in the liquid. When the whole system accelerates upwards with acceleration of $\frac{g}{3}$, then fraction of volume immersed in the liquid will be

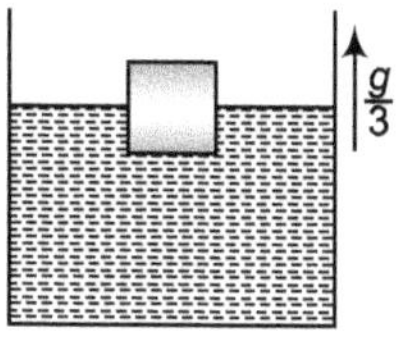

(a) 1/2 (b) 3/8 (c) 2/3 (d) 3/4

16. A raft of mass $M = 600$ kg floats in calm water with 7 cm submerged. When a man stands on the raft, 8.4 cm are submerged, the man's mass is
(a) 30 kg (b) 60 kg
(c) 90 kg (d) 120 kg

17. An ice-cube of density 900 kg/m^3 is floating in water of density 1000 kg/m^3. The percentage of volume of ice cube outside the water is
(a) 20% (b) 80% (c) 10% (d) 90%

18. A beaker containing water is kept on a spring scale. The mass of water and beaker is 5 kg. A block of mass 2 kg and specific gravity 10 is suspended by means of thread from a spring balance as shown in the figure. The readings of scales S_1 and S_2 are respectively (Take $g = 10\ \text{ms}^{-2}$).

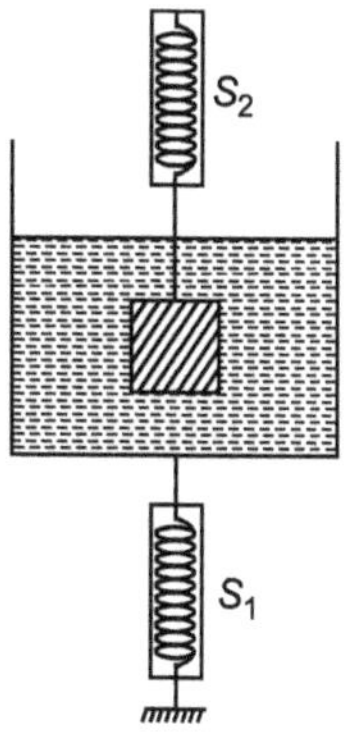

(a) 52 N and 20 N (b) 50 N and 18 N
(c) 52 N and 18 N (d) 52 N and 22 N

19. A cubical block of steel of each side equal to l is floating on mercury in vessel. The densities of steel and mercury are ρ_s and ρ_m. The height of the block above the mercury level is given by
(a) $l\left(1+\frac{\rho_s}{\rho_m}\right)$ (b) $l\left(1-\frac{\rho_s}{\rho_m}\right)$
(c) $l\left(1+\frac{\rho_m}{\rho_s}\right)$ (d) $l\left(1-\frac{\rho_m}{\rho_s}\right)$

20. A balloon has volume of 1000 m^3. It is filled with hydrogen ($\rho = 0.009$ g / L). If the density of air is 1.29 g / L, then it can lift a total weight of
(a) 600 kg (b) 1200 kg
(c) 300 kg (d) 1800 kg

Fluid Pressure and Pressure Force

21. Assuming that the atmosphere has the same density anywhere as at sea level ($\rho = 1.3$ kg / m^3) and g to be constant ($g = 10$ m/s^2). What should be the approximate height of atmosphere ($p_0 = 1.01 \times 10^5$ N/ m^2)
(a) 6 km (b) 8 km
(c) 12 km (d) 18 km

22. In a U-tube experiment, a column *AB* of water is balanced by a column *CD* of paraffin. The relative density of paraffin is

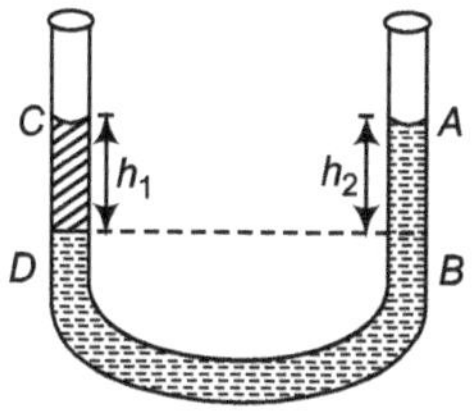

(a) $\frac{h_2}{h_1}$ (b) $\frac{h_1}{h_2}$
(c) $\frac{h_2 - h_1}{h_1}$ (d) $\frac{h_2}{h_1 + h_2}$

23. A U-tube of uniform cross-section shown in figure is partially filled with liquid I. Another liquid II which does not mix with I is poured into one side. The liquid levels of the two sides is found the same, while the level of liquid I has risen by 2 cm. If the specific gravity of liquid I is 1.1, then specific gravity of liquid II must be

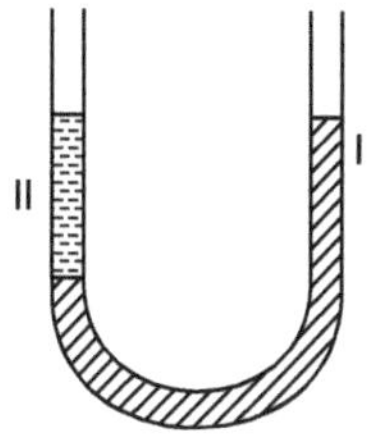

(a) 1.2 (b) 1.1
(c) 1.3 (d) 1.0

24. A closed rectangular tank is completely filled with water and is accelerated horizontally with an acceleration towards right. Pressure is

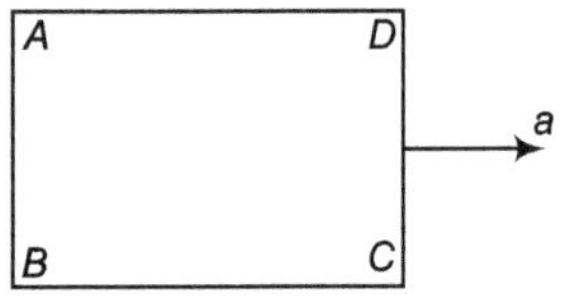

(i) maximum at, and (ii) minimum at

(a) (i) B (ii) D
(b) (i) C (ii) D
(c) (i) B (ii) C
(d) (i) B (ii) A

25. A uniformly tapering vessel is filled with a liquid of density 900 kgm^{-3}. The force that acts on the base of the vessel due to the liquid (excluding atmospheric force) is ($g = 10\ \text{m/s}^2$)

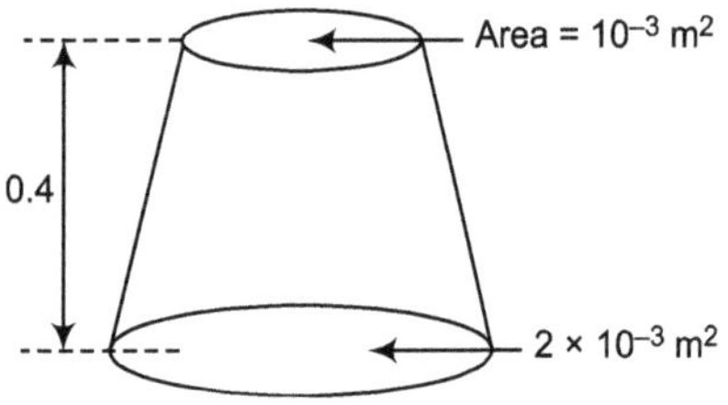

(a) 3.6 N
(b) 7.2 N
(c) 9.0 N
(d) 12.6 N

26. For the arrangement shown in the figure, the force at the bottom of the vessel is

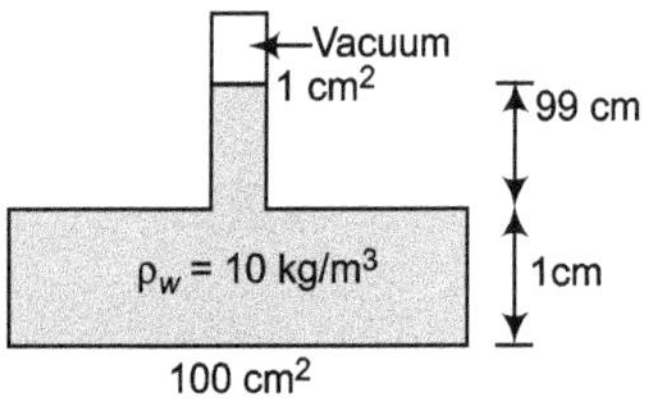

(a) 200 N (b) 100 N (c) 20 N (d) 2 N

27. An object of uniform density is allowed to float in water kept in a beaker. The object has triangular cross-section as shown in the figure. If the water pressure measured at the three point A, B and C below the object are p_A, p_B and p_C respectively. Then

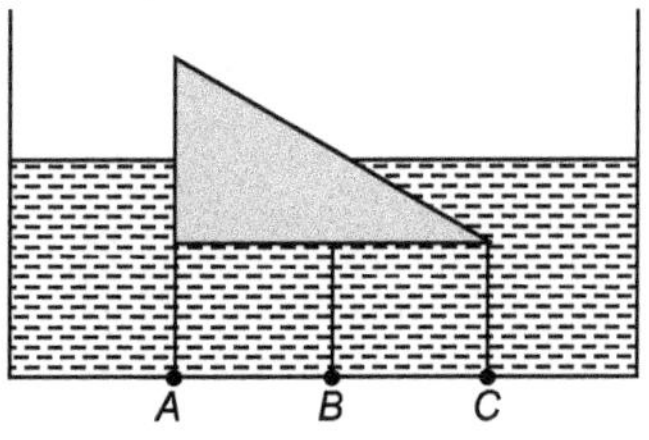

(a) $p_A > p_B > p_C$
(b) $p_A > p_B < p_C$
(c) $p_A = p_B = p_C$
(d) $p_A = p_B < p_C$

28. A cylindrical tank contains water up to a height H. If the tank is accelerated upwards with acceleration a, then pressure at the point A is p_1. If the tank is accelerated downwards with acceleration a the pressure at A is p_2. Then

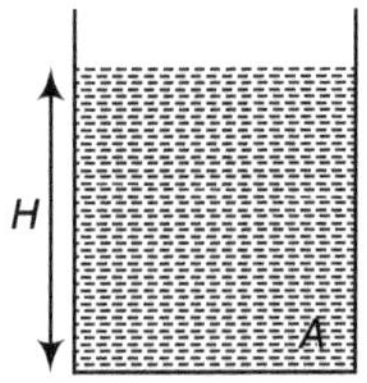

(a) $p_1 < p_2$
(b) $p_1 = p_1$
(c) $p_1 > p_2$
(d) data insufficient

29. In the given figure shown.

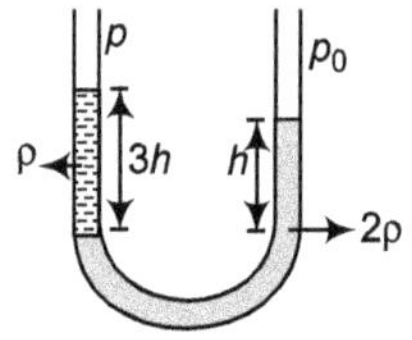

(a) $p_1 > p_0$
(b) $p > p_0$
(c) $p = p_0$
(d) $p = 0$

30. A piston of cross-sectional area $100\,\text{cm}^2$ is used in a hydraulic pressure to exert a force of 10^7 dyne on the water. The cross-sectional area of the other piston which support a truck of mass 2000 kg is

(a) $9.8 \times 10^2\ \text{cm}^2$
(b) $9.8 \times 10^3\ \text{cm}^2$
(c) $1.96 \times 10^3\ \text{cm}^2$
(d) $1.96 \times 10^4\ \text{cm}^2$

31. The liquid inside the container has density ρ. Choose the correct options.

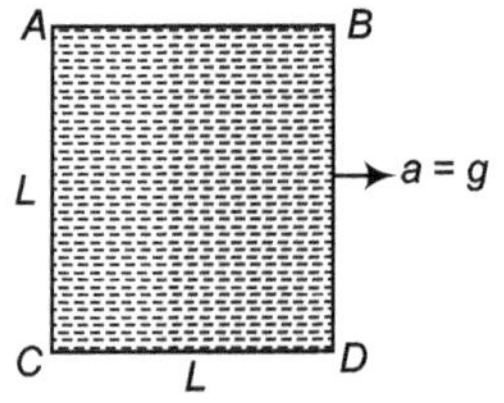

(a) $p_A - p_C = \rho g L$
(b) $p_C - p_B = 2\rho g L$
(c) $p_C - p_D = \rho g L$
(d) $p_A - p_D = 0$

Continuity Equation, Bernoulli's Equation and Torricelli's Equation

32. Bernoulli's theorem is a consequence of

(a) conservation of mass
(b) conservation of energy
(c) conservation of linear momentum
(d) conservation of angular momentum

33. The velocity of efflux of a liquid through an orifice in the bottom of a tank does not depend upon
(a) density of liquid
(b) height of the liquid column above orifice
(c) acceleration due to gravity
(d) None of the above

34. A hole is made at the bottom of the tank filled with water (density $= 1000 \text{ kg/m}^3$). If the total pressure at the bottom of the tank is three atmospheres (1 atmosphere $= 10^5 \text{ N/m}^2$), then the velocity of efflux is
(a) $\sqrt{400}$ m/s (b) $\sqrt{200}$ m/s
(c) $\sqrt{600}$ m/s (d) $\sqrt{500}$ m/s

35. Water from a tap emerges vertically down with an initial speed of 1.0 ms^{-1}. The cross-sectional area of tap is 10^{-4} m^2. Assume that the pressure is constant throughout the stream of water and that the flow is steady. The cross-sectional area of the stream 0.15 m below the tap is
(a) $5.0 \times 10^{-4} \text{ m}^2$
(b) $1.0 \times 10^{-5} \text{ m}^2$
(c) $5.0 \times 10^{-5} \text{ m}^2$
(d) $2.0 \times 10^{-5} \text{ m}^2$

36. The lift of an aeroplane is based on
(a) Torricelli's theorem (b) Bernoulli's principle
(c) law of gravitation (d) continuity equation

37. If the velocity head of a stream of water is equal to 10 cm, then its speed of flow is approximately
(a) 1.0 m/s (b) 1.4 m/s (c) 140 m/s (d) 10 m/s

38. A tank is filled to a height H. The range of water coming out of a hole which is a depth $H/4$ from the surface of water level is
(a) $\frac{2H}{\sqrt{3}}$ (b) $\frac{\sqrt{3}H}{2}$
(c) $\sqrt{3}H$ (d) $\frac{3H}{4}$

39. The level of water in a tank is 5 m high. A hole of area 1 cm^2 is made at the bottom of the tank. The rate of leakage of water from the hole is ($g = 10 \text{ m/s}^2$)
(a) $10^{-3} \text{ m}^3/\text{s}$
(b) $10^{-4} \text{ m}^3/\text{s}^3$
(c) $10 \text{ m}^3/\text{s}$
(d) $10^{-2} \text{ m}^3/\text{s}$

40. A water tank standing on the floor has two small holes punched in the vertical wall one above the other. The holes are 2.4 cm and 7.6 cm above the floor. If the jest of water from the holes hit the floor at the same point, then the height of water in the tank is
(a) 10 cm (b) 5 cm
(c) 20 cm (d) 4.8 cm

41. Water is flowing through two horizontal pipes of different diameters which are connected together. The diameters of the two pipes are 3 cm and 6 cm respectively. If the speed of water in the narrower tube is 4 m/s. Then, the speed of water in the wider tube is
(a) 16 m/s (b) 1 m/s (c) 4 m/s (d) 2 m/s

42. Water flows along a horizontal pipe of non-uniform cross-section. The pressure is 1 cm of Hg, where the velocity is 35 cm/s. At a point where the velocity is 65 cm/s, the pressure will be
(a) 0.89 of Hg (b) 0.62 cm of Hg
(c) 0.5 cm of Hg (d) 1 cm of Hg

43. In the given figure, the velocity v_3 will be

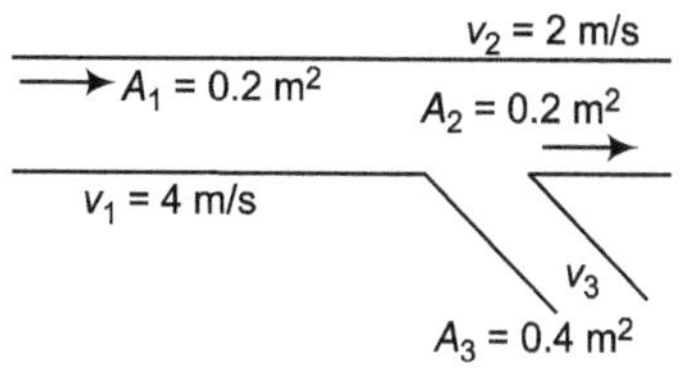

(a) 2 m/s (b) 4 m/s
(c) 1 m/s (d) 3 m/s

44. The pressure of water in a pipe when tap is closed is $5.5 \times 10^5 \text{ N/m}^2$. When tap gets open, pressure reduces to $5 \times 10^5 \text{ N/m}^2$. The velocity with which water comes out on opening the tap is
(a) 10 m/s (b) 5 m/s
(c) 20 m/s (d) 15 m/s

45. Air stream flows horizontally past an aeroplane wing of surface area 4 m^2. The speed of air over the top surface is 60 m/s and under the bottom surface is 40 m/s. The force of lift on the wing is (density of air $= 1 \text{ kg/m}^3$)
(a) 800 N (b) 1000 N
(c) 4000 N (d) 3200 N

46. If cross- sectional area of limb I is A_1 and that of limb II is A_2, then velocity of the liquid in the tube will be, (cross-sectional area of tube is very small)

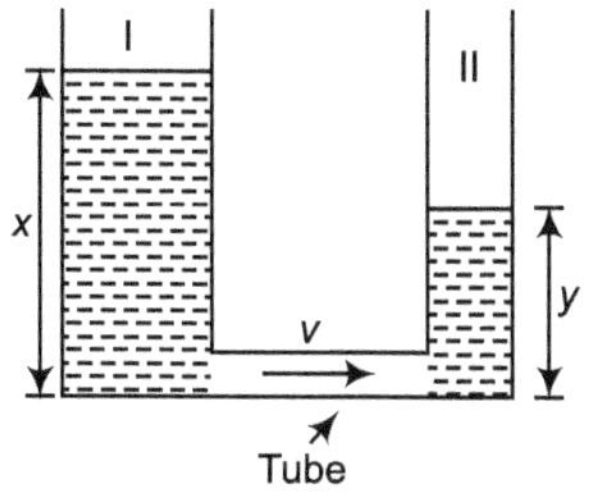

(a) $\sqrt{2g(x-y)}$
(b) $\frac{A_1}{A_2}\sqrt{2g(x-y)}$
(c) $\frac{A_2}{A_1}\sqrt{2g(x-y)}$
(d) None of the above

47. There are two hole O_1 and O_2 in a tank of height H. The water emerging from O_1 and O_2 strikes the ground at the same points, as shown in figure. Then

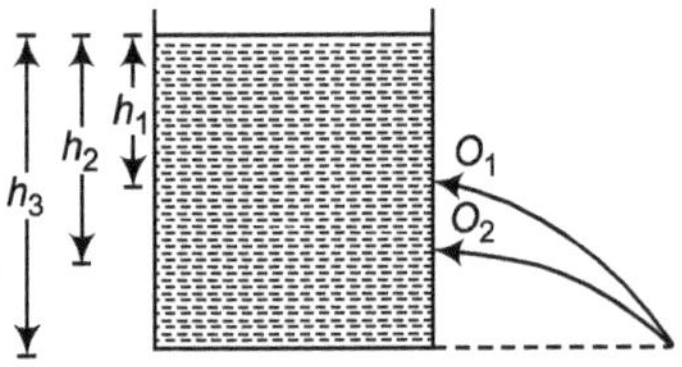

(a) $H = h_1 + h_2$ (b) $H = h_2 - h_1$
(c) $H = \sqrt{h_1 h_2}$ (d) None of these

48. A cylindrical vessel is filled with a liquid up to a height H. A small hole is made in the vessel at a distance y below, the liquid surface as shown in figure. The liquid emerging from the hole strike the ground at distance x

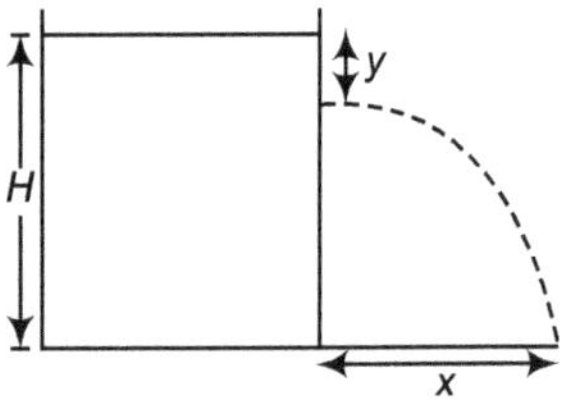

(a) x is equal if hole is at depth y or $H - y$
(b) x is maximum for $y = \frac{H}{2}$
(c) Both (a) and (b) are correct
(d) Both (c) and (d) are wrong

49. Air is blown through a pipe AB at a rate of 15 L/min. The cross-sectional area of the broad portion of the pipe AB is $2\,\text{cm}^2$ and that of the narrow portion is $0.5\,\text{cm}^2$. The difference in water level h is (density of air $= 1.32\,\text{kg/m}^3$)

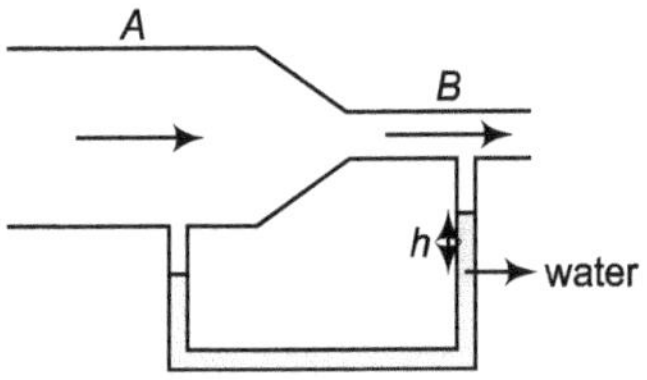

(a) 16 mm (b) 1.5 mm
(c) 10 mm (d) 3.2 mm

50. A tank if filled with water up to height H. When a hole is made at a distance h below the level of water. What will be the horizontal range of water jet?

(a) $2\sqrt{h(H-h)}$ (b) $4\sqrt{h(H+h)}$
(c) $4\sqrt{h(H-h)}$ (d) $2\sqrt{h(H+h)}$

51. A tank is filled to a height H. The range of water coming out of a hole which is a depth $H/4$ from the surface of water level is

(a) $\frac{2H}{\sqrt{3}}$ (b) $\frac{\sqrt{3}H}{2}$ (c) $\sqrt{3}H$ (d) $\frac{3H}{4}$

52. There is hole of area a at the bottom of a cylindrical of area A. Water is filled upto a height h and water flows out in t second. If water is filled to a height $4h$, then it will flow out in time

(a) $\frac{t}{4}$ (b) $2t$ (c) $4t$ (d) $\frac{t}{2}$

53. A container has a small hole at its bottom. Area of cross-section of the hole is A_1 and that of the container is A_2. Liquid is poured in the container at a constant rate $Q\,\text{m}^3/\text{s}$. The maximum level of liquid in the container will be

(a) $\frac{Q^2}{2g\,A_1A_2}$ (b) $\frac{Q^2}{2g\,A_1^2}$ (c) $\frac{Q}{2g\,A_1A_2}$ (d) $\frac{Q^2}{2g\,A_2^2}$

Viscosity

54. The ratio of the terminal velocities of two drops of radii R and $R/2$ is

(a) 2 (b) 1
(c) 1/2 (d) 4

55. An air bubble rises from the bottom of a lake of large depth. The rising speed of air bubble will

(a) go on increasing till it reaches surface
(b) go on decreasing till it reaches surface
(c) increase in the beginning, then will become constant
(d) be constant all throughout

56. Units of coefficient of viscosity are

(a) Nms^{-1} (b) Nm^2s^{-1}
(c) Nm^{-2}s (d) None of these

57. The terminal velocity v of a small steel ball of radius r falling under gravity through a column of viscous liquid of coefficient of viscosity η depends on mass of the ball m, acceleration due to gravity g, coefficient of viscosity η and radius r. Which of the following relations is dimensionally correct?

(a) $v \propto \frac{mgr}{\eta}$ (b) $v \propto mg\eta r$
(c) $v \propto \frac{mg}{r\eta}$ (d) $v \propto \frac{\eta mg}{\eta}$

58. Two equal drops of water are falling through air with a steady velocity v. If the drops coalesce, then new velocity will be

(a) $2v$ (b) $\sqrt{2}v$
(c) $2^{2/3}v$ (d) $\frac{v}{\sqrt{2}}$

59. As the temperature of water increases, its viscosity

(a) remains unchanged
(b) decreases
(c) increases
(d) increases or decreases depending on the external pressure

60. The rate of flow of liquid in a tube of radius r, length l, whose ends are maintained at a pressure difference p is $V = \frac{\pi Q\, pr^4}{\eta l}$, where η is coefficient of the viscosity and Q is

(a) 8 (b) $\frac{1}{8}$

(c) 16 (d) $\frac{1}{16}$

61. Two capillary tubes of the same length but different radii r_1 and r_2 are fitted in parallel to the bottom of a vessel. The pressure head is p. What should be the radius of a single tube that can replace the two tubes so that the rate of flow is same as before?

(a) $r_1 + r_2$ (b) $\frac{r_1 r_1}{r_1 + r_2}$

(c) $\frac{r_1 + r_2}{2}$ (d) None of these

62. Two capillaries of same length and radii in the ratio 1:2 are connected in series. A liquid flows through them in streamlined condition. If the pressure across the two extreme ends of the combination is 1 m of water, then pressure difference across first capillary is

(a) 9.4 m (b) 4.9 m (c) 0.49 m (d) 0.94 m

63. Water flows in a streamlined manner through a capillary tube of radius a, the pressure difference being p and the rate of flow Q. If the radius is reduced to $a/2$ and the pressure increased to $2p$, then rate of flow becomes

(a) $4Q$ (b) Q (c) $\frac{Q}{4}$ (d) $\frac{Q}{8}$

64. A viscous fluid is flowing through a cylindrical tube. The velocity distribution of the fluid is best represented by the diagram

(a) 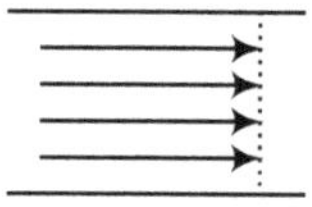

(b) 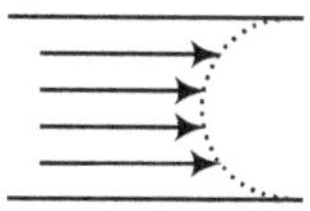

(c) 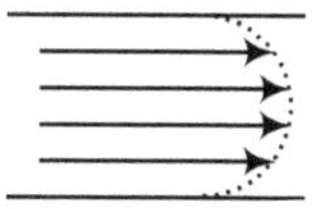

(d) None of these

65. Two capillary of lengths L and $2L$ and of radius R and $2R$ are connected in series. The net rate of flow of fluid through them will be (given rate of the flow through single capillary, $X = \pi p R^4 / 8\eta L$)

(a) $\frac{8}{9}X$ (b) $\frac{9}{8}X$

(c) $\frac{5}{7}X$ (d) $\frac{7}{5}X$

66. A lead shot of 1 mm diameter falls through a long column of glycerine. The variation of its velocity v, with distance covered is represented by

(a) 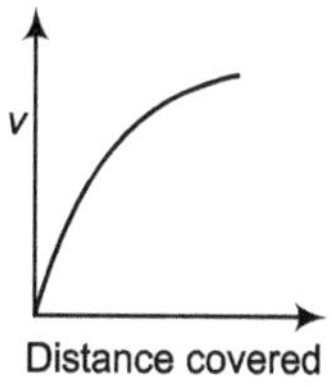

(b) 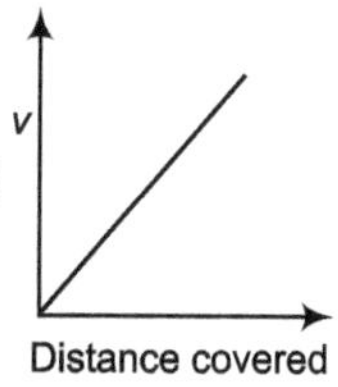

(c) 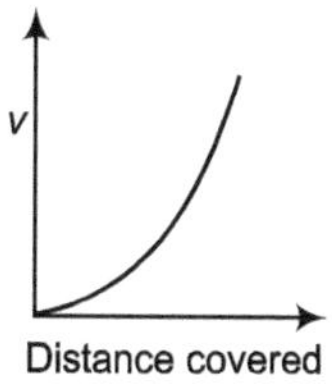

(d) 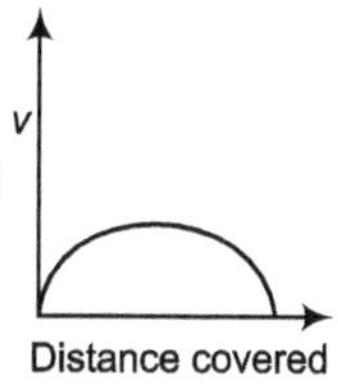

67. Water flows through a frictionless duct with a cross-section varying as shown in figure. Pressure p at points along the axis is represented by

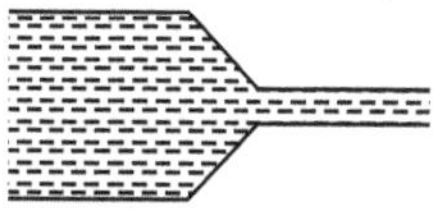

(a) 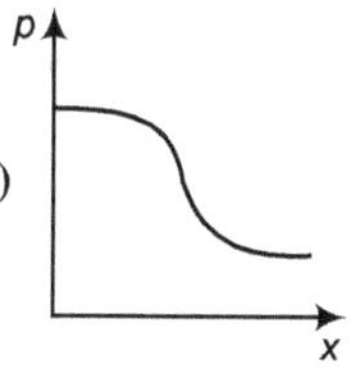

(b) 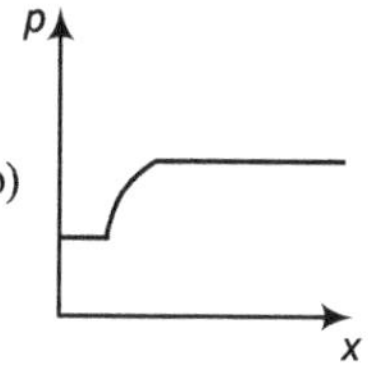

(c) 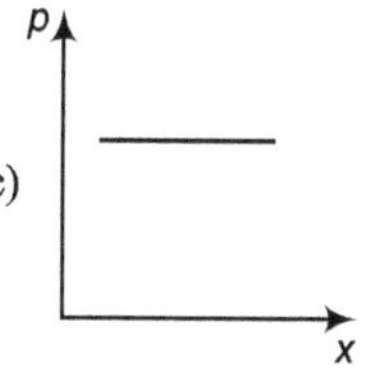

(d) 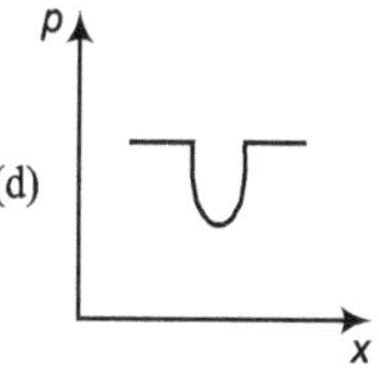

68. From amongst the following curves, which one shows the variation of the velocity v with time t for a small sized spherical body falling vertically in a long column of a viscous liquid

(a) 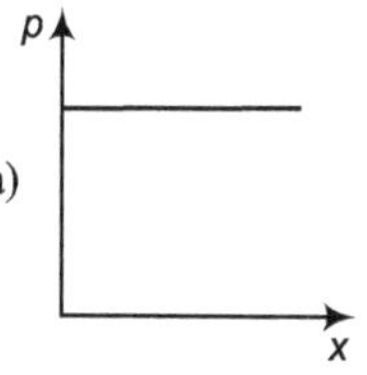

(b) 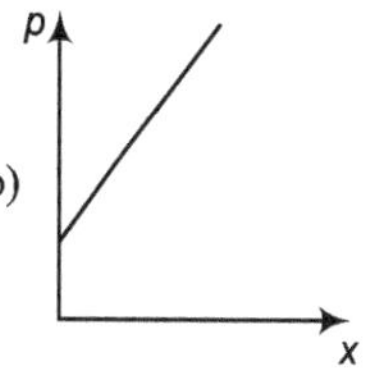

(c) 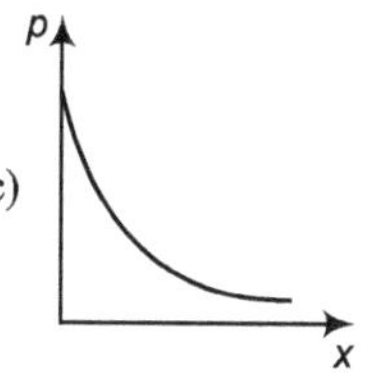

(d) 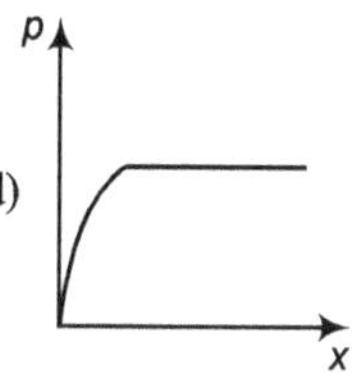

Surface Tension

69. The water droplets in free fall are spherical due to
(a) gravity
(b) viscosity
(c) surface tension
(d) intermolecular attraction

70. Two small drops of mercury each of radius r form a single large drop. The ratio of surface energy before and after this change is
(a) $2 : 2^{2/3}$ (b) $2^{2/3} : 1$
(c) 2:1 (d) 1:2

71. A water proofing agent changes, the angle of contact
(a) from acute to 90°
(b) from obtuse to 90°
(c) from an acute to obtuse value
(d) from an obtuse to acute value

72. With the rise in temperature the angle of contact
(a) increases
(b) decreases
(c) remains constant
(d) sometimes increases and sometimes decreases

73. If the angle of contact is less than 90°, then pressure just inside the surface of a meniscus
(a) is less than atmospheric pressure
(b) is greater than atmospheric pressure
(c) is same as the atmospheric pressure
(d) None of the above

74. Pressure inside two soap bubbles are 1.01 and 1.02 atmospheres. Ratio of their volumes is
(a) 2:1 (b) 4:1
(c) 8:1 (d) 12:1

75. Which of the following is not the unit of surface tension?
(a) Newton/metre
(b) Joule / $(\text{metre})^2$
(c) kg/ $(\text{second})^2$
(d) Watt/metre

76. Surface tension is due to
(a) friction forces between molecules
(b) cohesive forces between molecules
(c) adhesive forces between molecules
(d) gravitational forces

77. Coatings used on raincoat are waterproof, because
(a) water is absorbed by the coating
(b) cohesive force becomes greater
(c) water is not scattered away by the coating
(d) angle of contact decreases

78. If temperature increases, then surface tension of a liquid
(a) increases
(b) decreases
(c) remains the same
(d) first increases, then decreases

79. The spiders and insects move and run on the surface of water without sinking, because
(a) elastic membrane is formed on water due to property of surface tension
(b) spiders and insects are lighter
(c) spiders and insects swim on water
(d) spiders and insects experience upthrust

80. Small droplets of liquid are usually more spherical in shape than larger drops of the same liquid, because
(a) force of surface tension is equal and opposite to the force of gravity
(b) force of surface tension predominates the force of gravity
(c) force of gravity predominates the force of surface tension
(d) force of gravity and force of surface tension act in the same direction and are equal

81. Hair of shaving brush cling together when it is removed from water due to
(a) force of attraction between hair
(b) surface tension
(c) viscosity of water
(d) characteristic property of hair

82. A square frame of side L is dipped in a liquid. On taking out, a membrane is formed. If the surface tension of the liquid is T, then force acting on one side of the frame will be
(a) $2\,TL$ (b) $4\,TL$ (c) TL (d) $\frac{TL}{2}$

83. Water does not wet an oily glass, because
(a) cohesive force of oil > adhesive force between oil and glass
(b) cohesive force of oil > cohesive force of water
(c) oil repels water
(d) cohesive force of water > adhesive force between water and oil molecules

84. Which of the fact is not due to surface tension?
(a) Dancing of a camphor piece over the surface of water
(b) Small mercury drop itself becomes spherical
(c) A liquid surface comes at rest after stirring
(d) Mercury does not wet the glass vessel

85. The property of surface tension is obtained in
(a) solids, liquids and gases (b) liquids
(c) gases (d) matter

86. If two glass plates are quite nearer to each other in water, then there will be force of
(a) attraction
(b) repulsion
(c) attraction or repulsion
(d) Neither attraction nor repulsion

87. On mixing the salt in water, the surface tension of water will
(a) increase (b) decrease
(c) remain unchanged (d) None of these

88. A 10 cm long wire is placed horizontally on the surface of water and is gently pulled up with a force of 2×10^2 N to keep the wire in equilibrium. The surface tension in Nm^{-1} of water is
(a) 0.1 (b) 0.2 (c) 0.001 (d) 0.002

89. The dimensions of surface tension are
(a) $[MLT^{-1}]$ (b) $[ML^2T^{-2}]$ (c) $[ML^0T^{-2}]$ (d) $[ML^{-1}T^{-2}]$

90. If work W is done in blowing a bubble of radius R from soap solution, then the work done in blowing a bubble of radius $2R$ from the same solution is
(a) $W/2$ (b) $2W$ (c) $4W$ (d) $2\frac{1}{3}W$

91. If two identical mercury drops are combined to form a single drop, then its temperature will
(a) decrease (b) increase
(c) remains the same (d) None of these

92. The surface tension of a soap solution is 2×10^{-2} N / m. To blow a bubble of radius 1 cm, the work done is
(a) $4\pi\times 10^{-6}$J (b) $8\pi\times 10^{-6}$J
(c) $12\pi\times 10^{-6}$J (d) $16\pi\times 10^{-6}$J

93. The surface tension of liquid at its boiling point
(a) becomes zero
(b) becomes infinity
(c) is equal to the value at room temperature
(d) is half to the value at the room temperature

94. Which of the following statements are true in case, when two water drops coalesce and make a bigger drop?
(a) energy is released
(b) energy is absorbed
(c) the surface area of the bigger drop is greater than the sum of the surface areas of both the drops
(d) the surface area of the bigger drop is smaller than the sum of the surface areas of both the drops

95. If two soaps bubbles of equal radii r coalesce, then the radius of curvature of interface between two bubble will be
(a) r (b) 0 (c) infinity (d) $\frac{r}{2}$

96. A liquid does not wet the sides of a solid, if the angle of contact is
(a) zero (b) obtuse (c) acute (d) 90°

97. When the temperature is increased the angle of contact of a liquid?
(a) Increases
(b) Decreases
(c) Remains the same
(d) First increases and then decreases

98. The angle of contact between glass and mercury is
(a) 0° (b) 30°
(c) 90° (d) 135°

99. A liquid rises in a vertical tube. The relation between the weight of the liquid in the tube, surface tension of the liquid T and radius of the tube r is given by, if the angle of contact is zero
(a) $w=\pi r^2 T$ (b) $w=2\pi rT$
(c) $w=2r^2\pi T$ (d) $w=\frac{3}{4}\pi r^3 T$

100. A glass plate is partly dipped vertically in the mercury and angle of contact is measured. If the plate is inclined, then the angle of contact will
(a) increase (b) remain unchanged
(c) increase or decrease (d) decrease

101. If a water drop is kept between two glass plates, then its shape is

(a) 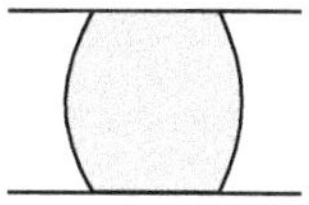 (b) 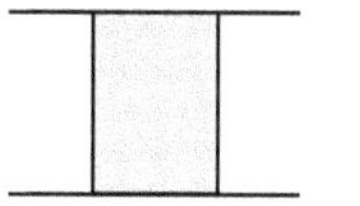

(c) 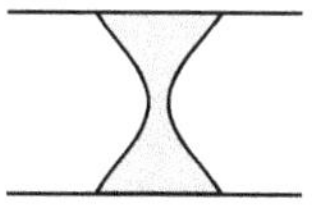 (d) None of these

102. A liquid wets a solid completely. The meniscus of the liquid in a sufficiently long tube is
(a) flat
(b) concave
(c) convex
(d) cylindrical

103. If two soap bubble of different radii are in communication with each other
(a) air flows from larger bubble into the smaller one
(b) the size of the bubbles remains the same
(c) air flows from the smaller bubble into the large one and the larger bubble grows at the expense of the smaller one
(d) None of the above

104. The surface tension of soap solution is 25×10^{-3} Nm^{-1}. The excess pressure inside a soap bubble of diameter 1 cm is
(a) 10 Pa (b) 20 Pa
(c) 5 Pa (d) None of these

105. When two soap bubbles of radius r_1 and r_2 $(r_2 > r_1)$ coalesce, then radius of curvature of common surface is
(a) $r_2 - r_1$ (b) $\frac{r_2 - r_1}{r_1 r_2}$
(c) $\frac{r_1 r_2}{r_2 - r_1}$ (d) $r_2 + r_1$

106. The excess pressure due to surface tension in a spherical liquid drop of radius r is directly proportional to
(a) r (b) r^2
(c) r^{-1} (d) r^{-2}

107. A long cylindrical glass vessel has a small hole of radius r at its bottom. The depth to which the vessel can be lowered vertically in the deep water bath (surface tension T) without any water entering inside is

(a) $4T/\rho rg$
(b) $3T/\rho rg$
(c) $2T/\rho rg$
(d) $T/\rho rg$

108. Two soap bubbles of radii r_1 and r_2 equal to 4 cm and 5 cm are touching each other over a common surface S_1S_2 (shown in figure). Its radius will be

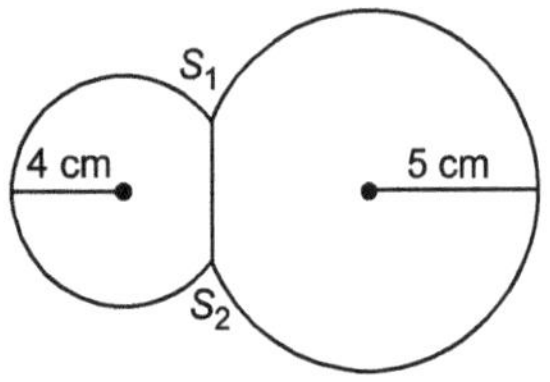

(a) 4 cm
(b) 20 cm
(c) 5 cm
(d) 4.5 cm

109. A vessel, whose bottom has round holes with diameter of 1.0 mm, is filled with water. The maximum height to which the water can be filled without leakage is (surface tension of water = 75 dyne/cm, $g = 1000 \text{ cm/s}^2$)

(a) 100 cm (b) 75 cm
(c) 50 cm (d) 30 cm

110. A soap bubble is blown with the help of a mechanical pump at the mouth of a tube. The pump produces a constant increase per minute in the radius of the bubble, irrespective of its internal pressure. The graph between the excess pressure inside, the soap bubble and time t will be

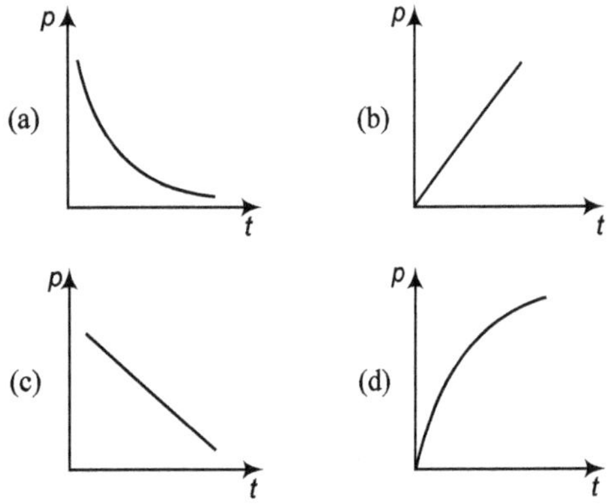

111. Which graph represents the variation of surface tension with temperature over small temperature ranges for water?

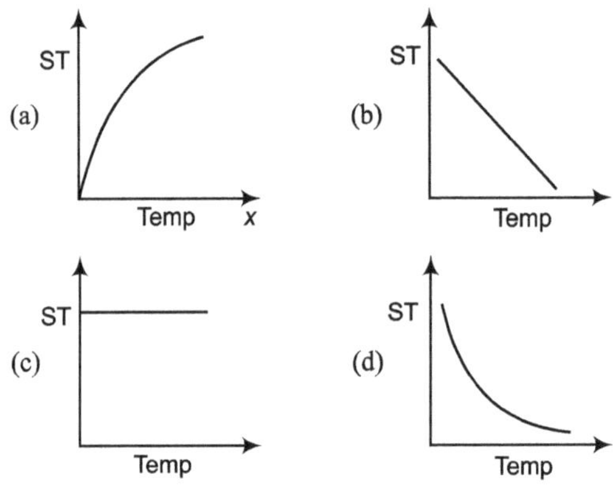

112. A thread is tied slightly loose to a wire frame as in figure and the frame is dipped into a soap solution and taken out. The frame is completely covered with the film. When A is pricked

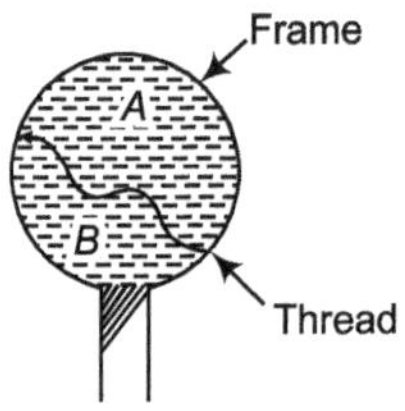

(a) thread will become concave on seeing from side A
(b) thread will become concave on seeing from side B
(c) thread will become straight
(d) thread will remain as it is

Capillary Rise or Fall

113. The liquid in the capillary tube will rise, if the angle of contact is

(a) 0°
(b) 90°
(c) obtuse
(d) acute

114. If a capillary tube is dipped into liquid and the levels of the liquid inside and outside are same, then the angle of contact is

(a) 0°
(b) 90°
(c) 45°
(d) 30°

115. Water rises to a height of 30 mm in a capillary tube. If the radius of the capillary tube is made $\frac{3}{4}$th of its previous value. The height to which the water will rise in the tube is

(a) 30 mm (b) 20 mm
(c) 40 mm (d) 10 mm

116. When two capillary tubes of different diameters are dipped vertically, then rise of the liquid is

(a) same in both the tubes
(b) more in the tube of larger diameter
(c) less in the tube of smaller diameter
(d) more in the tube of smaller diameter

117. Due to capillary action, a liquid will rise in a tube, if the angle of contact is

(a) acute (b) obtuse
(c) 90° (d) zero

118. Two parallel glass plates are dipped partly in the liquid of density d keeping them vertical. If the distance between the plates is x surface tension for the liquid is T and angle of contact θ, then rise of liquid between the plates due to capillary will be

(a) $\frac{T\cos\theta}{xd}$ (b) $\frac{2T\cos\theta}{xdg}$
(c) $\frac{2T}{xdg\cos\theta}$ (d) $\frac{T\cos\theta}{xdg}$

119. Water rises in a capillary tube to a certain height such that the upward force due to surface tension is balanced by 75×10^{-4} N force due to the weight of the liquid. If the surface tension of water is 6×10^{-2} N/m, then inner circumference of the capillary must be

(a) 1.25×10^{-2} m (b) 0.50×10^{-2} m
(c) 6.5×10^{-2} m (d) 12.5×10^{-2} m

120. Two capillary tubes P and Q are dipped in water. The height of water level in capillary P is $2/3$ to the height in Q capillary. The ratio of their diameters is

(a) 2:3 (b) 3:2
(c) 3:4 (d) 4:3

121. Water rises up to 10 cm height in a long capillary tube. If this tube is immersed in water so that the height above the water surface is only 8 cm, then

(a) water flows out continuously from the upper end
(b) water rises upto upper end and forms a spherical surface
(c) water only rises upto 6 cm height
(d) water does not rise at all

122. The correct relation is

(a) $r=\frac{2T\cos\theta}{hdg}$ (b) $r=\frac{hdg}{2T\cos\theta}$
(c) $r=\frac{2T\,dgh}{\cos\theta}$ (d) $r=\frac{T\cos\theta}{2hdg}$

123. In a surface tension experiment with capillary tube, water rises upto 0.1 m. If the same experiment is repeated on an artificial satellite, while is revolving around the earth, then water will rise in the capillary tube up to a height of

(a) 0.1 m
(b) 0.2 m
(c) 0.98 m
(d) full length of the capillary tube

124. The correct curve between the height or depression h of liquid in a capillary tube and its radius is

(a)
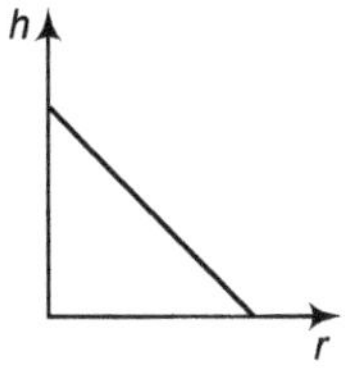

(b)
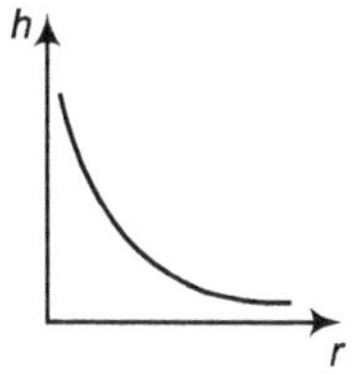

(c)
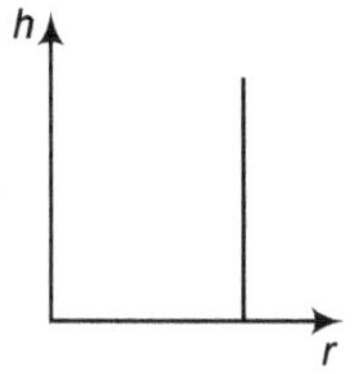

(d)
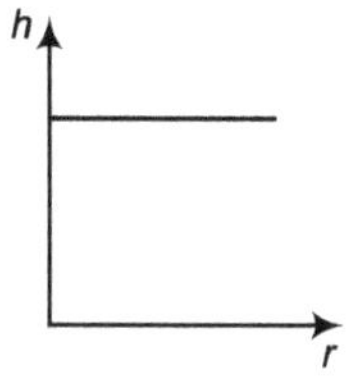

Miscellaneous Problems

125. A tank full of water has a small hole at its bottom. Let t_1 be the time taken to empty, the first half of the tank and t_2 be the time needed to empty the rest half of the tank, then

(a) $t_1 = t_2$ (b) $t_1 > t_2$
(c) $t_1 < t_2$ (d) $t_1 = 0.523t_2$

126. The density of ice is $x\,\text{g/cm}^3$ and that of water is $y\,\text{g/cm}^3$, when m gram of ice melts, then the change in volume is

(a) $m(y-x)$ (b) $\frac{y-x}{m}$
(c) $my(y-x)$ (d) $\frac{m}{y}-\frac{m}{x}$

127. A ball of relative density 0.8 falls into water from a height of 2 m. The depth to which the ball will sink is (neglect viscous forces)

(a) 8 m (b) 2 m
(c) 6 m (d) 16 m

128. A cylindrical vessel of 92 cm height is kept filled up to the brim. It has four holes 1, 2, 3, 4 which are respectively at heights of 20 cm, 30 cm, 46 cm and 80 cm from the horizontal floor. The water falling at the maximum horizontal distance from the vessel comes from

(a) hole number 4
(b) hole number 3
(c) hole number 2
(d) hole number 1

129. Two substances of relative densities ρ_1 and ρ_2 are mixed in equal volume and the relative density of mixture is 4. When they are mixed in equal masses, then relative density of the mixture is 3. The values of ρ_1 and ρ_2 are

(a) $\rho = 6$ and $\rho_2 = 2$ (b) $\rho_1 = 3$ and $\rho_2 = 5$
(c) $\rho_1 = 12$ and $\rho_2 = 4$ (d) None of these

130. A solid shell loses half its weight in water. Relative density of shell is 0.5, what fraction of its volume is hollow?

(a) $\frac{3}{5}$ (b) $\frac{2}{5}$
(c) $\frac{1}{5}$ (d) $\frac{4}{5}$

131. An iron block is on a boat which floats in a pond. The block is thrown into the water. The level of water in the pond will be

(a) equal to the earlier level
(b) less than the earlier level
(c) more than the earlier level
(d) depends on how large the block is

132. The pressure of the gas in a cylindrical chamber is p_0. The vertical force exerted by the gas on its hemispherical end is

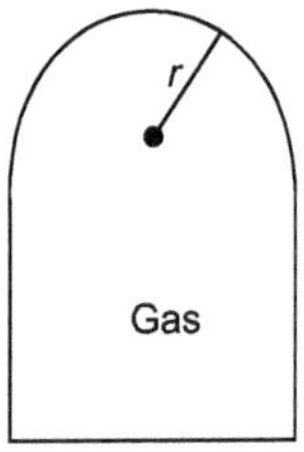

(a) $p_0 r^2$ (b) $4p_0\pi r^2$
(c) $2p_0\pi r^2$ (d) $p_0\pi r^2$

133. A metal sphere connected by a string is dipped in a liquid of density ρ as shown in figure. The pressure at the bottom of the vessel will be, (p_0 = atmospheric pressure)

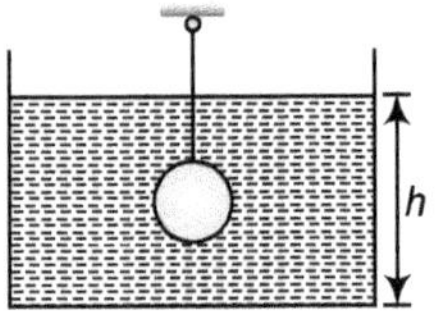

(a) $p = p_0 + \rho gh$ (b) $p > p_0 + \rho gh$
(c) $p < p_0 + \rho gh$ (d) p_0

134. A tank filled with water has two taps to exhaust and pour. A hollow spherical ball is half submerged in water. Through one tap, water is taken out and through another tap, a liquid of density double the density of water is poured in tank such that volume of liquid in tank remains constant. Sphere will

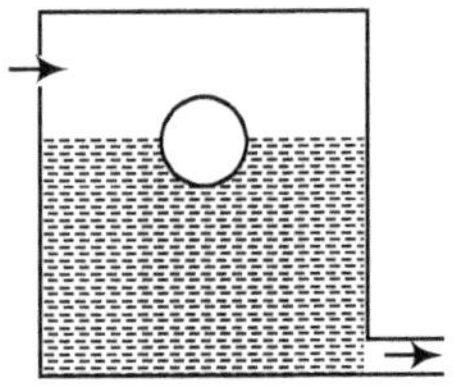

(a) go down
(b) go up
(c) maintain same height
(d) sink to bottom

135. An air bubble of radius 1 mm is formed inside water at a depth 10 m below free surface (where, air pressure is 10^5 N/m^2). The pressure inside the bubble is (surface tension of water = 2×10^{-7} N / m).

(a) 2.28×10^5 N/m^2 (b) 2.0028×10^5 N/m^2
(c) 2.14×10^5 N/m^2 (d) 2.0014×10^5 N/m^2

[Level 2]

Only One Correct Option

1. A block of mass 4 kg and volume 5×10^{-4} m^3 is suspended by a spring balance in a lift which is accelerating. The apparent weight shown by the spring balance is 3 kg. Now, the block is immersed in water in a container inside the lift. The apparent weight in kg shown by the spring balance is

(a) 2.375 (b) 2.625 (c) 2.5 (d) 3.125

2. A ball of mass 1 kg falls from a height of 5m above the free surface of water. The relative density of the solid ball is $s = \frac{2}{3}$. The ball travels a distance of 2 m under water and becomes stationary. The work done by the resistive forces of water is

(a) −50 J (b) −20 J
(c) −40 J (d) −30 J

3. A cubical block of wood of specific gravity 0.5 and chunk of concrete of specific gravity 2.5 are fastened together. The ratio of the mass of wood to the mass of concrete, which makes the combination to float with its entire volume submerged under water is

(a) 1/5 (b) 1/3
(c) 3/5 (d) 2/3

4. A cylinder of mass M and density d_1 hanging from a string, is lowered into a vessel of cross-sectional area A, containing a liquid of density d_2 ($d_2 < d_1$) until it is fully immersed. The increase in pressure at the bottom of the vessel is

(a) $\frac{Md_2g}{d_1A}$ (b) $\frac{Mg}{A}$
(c) $\frac{Md_1g}{d_2A}$ (d) zero

5. Two identical cylindrical vessels, each of base area A, have their bases at the same horizontal level. They contain a liquid of density ρ. In one vessel the height of the liquid is h_1 and in the other h_2 ($> h_1$). When the two vessels are connected, the word done by gravity in equalizing the levels is

(a) $2\rho Ag(h_2 - h_1)^2$ (b) $\rho Ag(h_2 - h_1)^2$
(c) $\frac{1}{2}\rho Ag(h_2 - h_1)^2$ (d) $\frac{1}{4}\rho Ag(h_2 - h_1)^2$

6. A small ball of density ρ is immersed in a liquid of density $\sigma(\sigma > \rho)$ to a depth h and then released. The height above the surface of water up to which the ball will jump is

(a) $\frac{\sigma h}{\rho}$ (b) $\left(\frac{\sigma h}{\rho} - 1\right)h$
(c) $\left(1 - \frac{\sigma h}{\rho}\right)h$ (d) $\frac{\rho h}{\sigma}$

7. A U-tube of base length l filled with same volume of two liquids of densities ρ and 2ρ is moving with an acceleration a on the horizontal plane. If the height difference between the two surfaces (open to atmosphere) becomes zero, then height h is given by

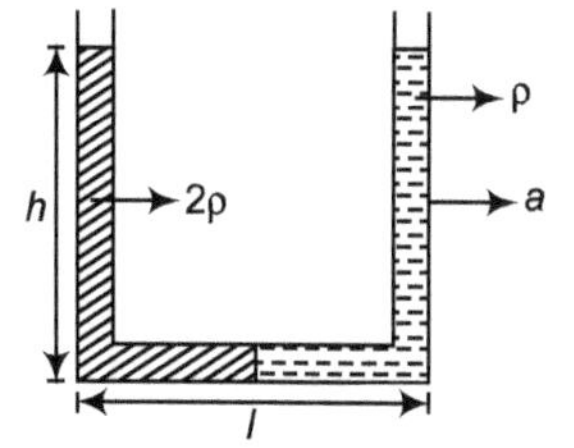

(a) $\frac{al}{g}$ (b) $\frac{3al}{2g}$
(c) $\frac{2al}{3g}$ (d) $\frac{al}{2g}$

8. A body of volume V and density ρ is initially submerged in a non-viscous liquid of density σ (> ρ). If it is rises by itself through a height h in the liquid. Its kinetic energy will

(a) increase by $hV(\sigma - \rho)g$
(b) increase by $hV(\rho + \sigma)g$
(c) increase by $\frac{hV\rho g}{\sigma}$
(d) decrease by $\frac{hV\rho g}{\sigma}$

9. A cubical block of side 10 cm floats at the interface of an oil and water. The pressure above that of atmosphere at the lower face of the block is

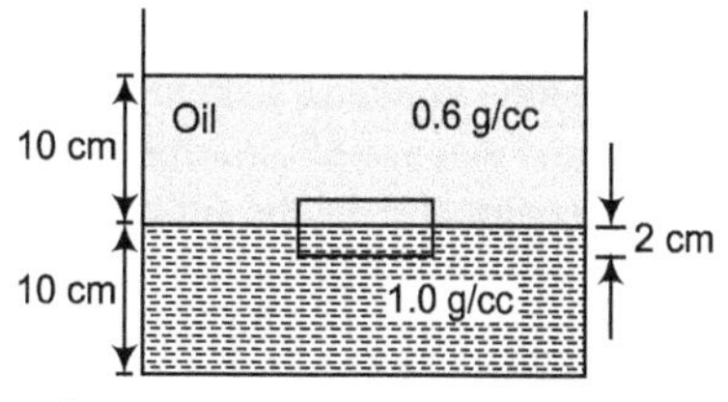

(a) 200 N/m^2 (b) 680 N/m^2
(c) 400 N/m^2 (d) 800 N/m^2

10. A sphere of solid material of specific gravity 8 has a concentric spherical cavity and just sinks in water. Then, the ratio of the radius of the cavity to the outer radius of the sphere must be

(a) $\frac{\sqrt[3]{3}}{2}$ (b) $\frac{\sqrt[3]{5}}{2}$
(c) $\frac{\sqrt[3]{7}}{2}$ (d) $\frac{2}{\sqrt[3]{7}}$

11. A large block of ice 10 cm thick with a vertical hole drilled through it is floating in a lake. The minimum length of the rope required to scoop out a bucket full of water through the hole is (density of ice $= 0.9\ g/cm^3$)

(a) 0.5 m (b) 1.0 m
(c) 1.2 m (d) 1.8 m

12. A large tank is filled with water (density $= 10^3\ kg/m^3$). A small hole is made at a depth 10 m below water surface. The range of water issuing out of the hole is R on ground. What extra pressure must be applied on the water surface so that the range become $2R$? (take 1 atm $= 10^5$ Pa and $g = 10\ m/s^2$)

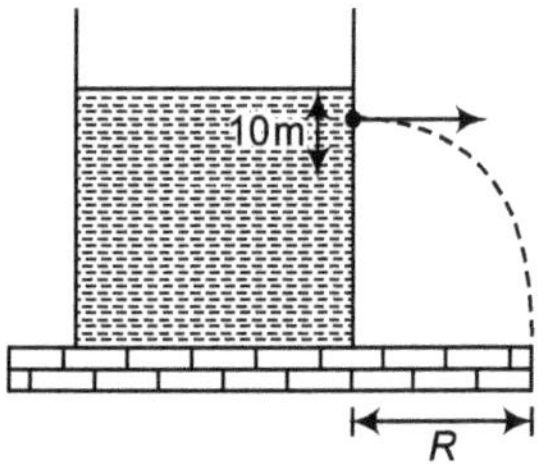

(a) 1 atm (b) 2 atm
(c) 4 atm (d) 3 atm

13. A wooden plank of length 1m and uniform cross-section is hinged at one end to the bottom of a tank as shown in the figure. The tank is filled with water up to a height of 0.5m. The specific gravity of the plank is 0.5. The angle θ made by the plank in equilibrium position is

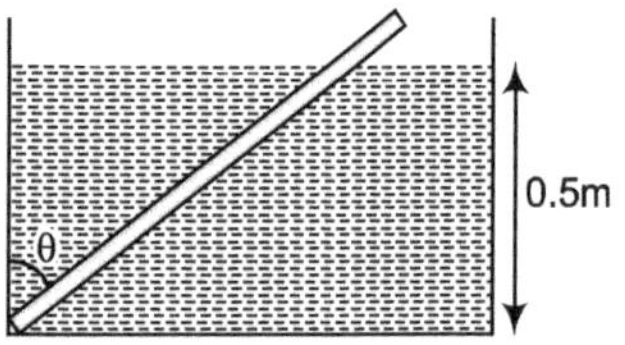

(a) 30° (b) 45°
(c) 60° (d) 90°

14. An open U-tube contains mercury. When 11.2 cm of water is poured into one of the arms of the tube, how high does the mercury rise in the other arm from its initial level?

(a) 0.82 cm (b) 1.35 cm
(c) 0.41 cm (d) 2.32 cm

15. A body of density ρ is dropped from rest from a height h into a lake of density $\sigma (\sigma > \rho)$. The maximum depth the body sinks inside the liquid is (neglect viscous effect of liquid)

(a) $\frac{h\rho}{\sigma - \rho}$
(b) $\frac{h\sigma}{\sigma - \rho}$
(c) $\frac{h\rho}{\sigma}$
(d) $\frac{h\sigma}{\rho}$

16. A liquid stands at the plane level in U-tube when at rest. If areas of cross-section of both the limbs are equal, what will be the difference in heights h of the liquid in the two limbs of U-tube, when the system is given an acceleration a in horizontal direction towards right as shown in the figure?

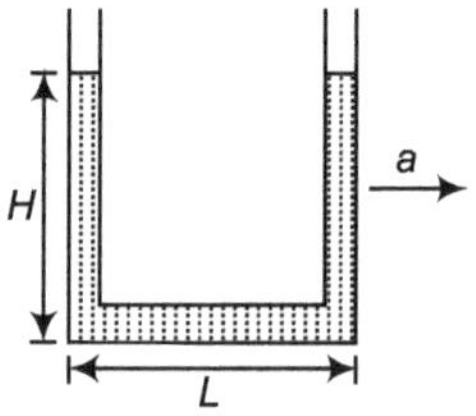

(a) $\frac{g}{a}\frac{L^2}{H}$ (b) $\frac{La}{g}$ (c) $\frac{L^2}{H}\frac{a}{g}$ (d) $\frac{Lg}{a}$

17. A capillary tube is dipped in a liquid. Let pressure at point A, B and C be p_A, p_B, p_C respectively, then

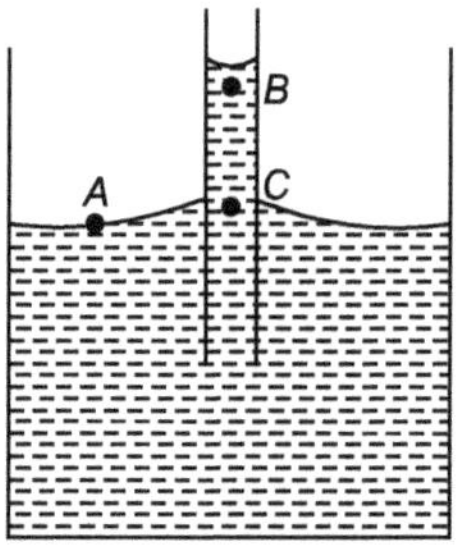

(a) $p_A = p_B = p_C$ (b) $p_A = p_B < p_C$
(c) $p_A = p_C < p_B$ (d) $p_A = p_C > p_B$

18. A small ball (mass m) falling under gravity in a viscous medium experiences a drag force proportional to the instantaneous speed u such that $F_{drag} = ku$. Then the terminal speed of ball within viscous medium is

(a) $\frac{k}{mg}$ (b) $\frac{mg}{k}$
(c) $\sqrt{\frac{mg}{k}}$ (d) None of these

19. A candle of diameter d is floating on a liquid in a cylindrical container of diameter $D (D >> d)$ as shown in figure. If it is burning at the rate of $2\ cm/h$, then the top of the candle will

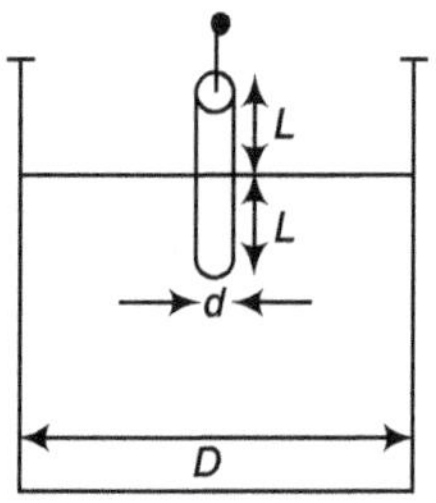

(a) remain at the same height (b) fall at the rate 1 cm/h
(c) fall at the rate of 2 cm/h (d) go up at the rate of 1 cm/h

20. A container has two immiscible liquids of densities ρ_1 and ρ_2 ($> \rho_1$). A capillary tube of radius r is inserted in the liquid so that its bottom reaches up to the denser liquid. The denser liquid rises in the capillary and attains a height h from the interface of the liquids, which is equal to the column length of the lighter liquid. Assuming angle of contact to be zero, the surface tension of heavier liquid is

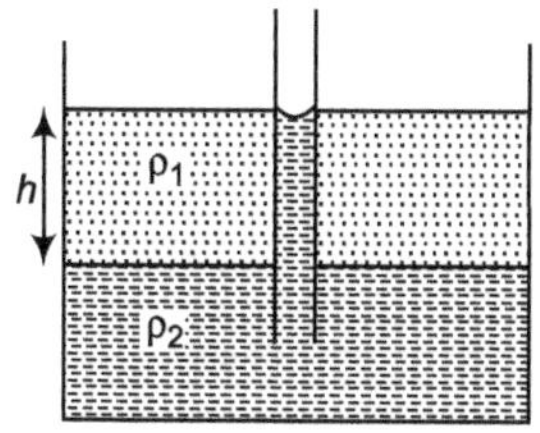

(a) $2\pi r \rho_2 gh$ (b) $\frac{\rho_2 rgh}{2}$
(c) $\frac{r}{2}(\rho_2 - \rho_1)gh$ (d) $2\pi r(\rho_2 - \rho_1)gh$

21. A spherical object of mass 1 kg and radius 1m is falling vertically downward inside a viscous liquid in a gravity free space. At a certain instant the velocity of the sphere is 2 m/s. If the coefficient of viscosity of the liquid is $\frac{1}{6\pi}$ SI units, then velocity of ball will become 0.5 m / s after a time.
(a) ln 4 s (b) 2 ln 4 s
(c) 3 ln 4 s (d) 3 ln 2s

22. If a capillary tube of radius r is immersed in water, then mass of water risen in capillary is M. If the radius of capillary be doubles, then mass of water risen in the capillary will be
(a) $M/2$ (b) M
(c) $2M$ (d) $4M$

23. A wooden block of mass 8 kg is tied to a string attached to the bottom of the tank. In the equilibrium the block is completely immersed in water. If relative density of wood is 0.8 and $g = 10\ ms^{-2}$, then tension T in the string is
(a) 120 N
(b) 100 N
(c) 80 N
(d) 20 N

24. A metal ball immersed in alcohol weighs w_1 at 0° C and w_2 at 59° C. The coefficient of cubical expansion of the metal is less than that of alcohol. Assuming that the density of the metal is large compared to that of alcohol, it can be shown that
(a) $w_1 > w_2$
(b) $w_1 = w_2$
(c) $w_1 < w_2$
(d) $w_1 = (w_1/2)$

25. The volume of an air bubble becomes three times as it rises from the bottom of a lake to its surface. Assuming temperature to be constant and atmospheric pressure to be 75 cm of Hg and the density of water to be 1/10 of the density of the mercury, the depth of the lake is
(a) 5 m (b) 10 m
(c) 15 m (d) 20 m

26. A barometer kept in an elevator reads 76 cm when it is at rest. If the elevator goes up with increasing speed, then reading will be
(a) zero (b) 76 cm
(c) > 76 cm (d) <76 cm

27. The surface energy of a liquid drop is E. It is sprayed into 1000 equal droplets. Then its surface energy becomes
(a) E (b) 10 E (c) 100 E (d) 1000 E

28. An open tank containing non-viscous liquid to a height of 5 m is placed over the ground. A heavy spherical ball falls from height 40 m over the ground in the tank. Ignoring air resistance find the height to which ball will go back. Collision between ball and bottom of tank is perfectly elastic
(a) 45 m (b) 35 m
(c) 40 m (d) 20 m

29. A large open tank has two holes in the wall. One is a square hole of side L at a depth y from the top and the other is a circular hole of radius R at a depth $4y$ from the top. When the tank is completely filled with water, then quantities of water flowing out per second from holes are the same. Then R is equal to
(a) $L/\sqrt{2\pi}$
(b) $2\pi L$
(c) L
(d) $L/2\pi$

30. A pump is designed as a horizontal cylinder with a piston area A and an outlet orifice arranged near the axis of the cylinder. Find the velocity of outflow of liquid from pump, if the piston moves with a constant velocity under the action of a constant force F. The density of liquid is ρ

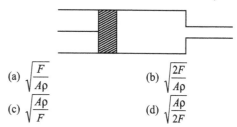

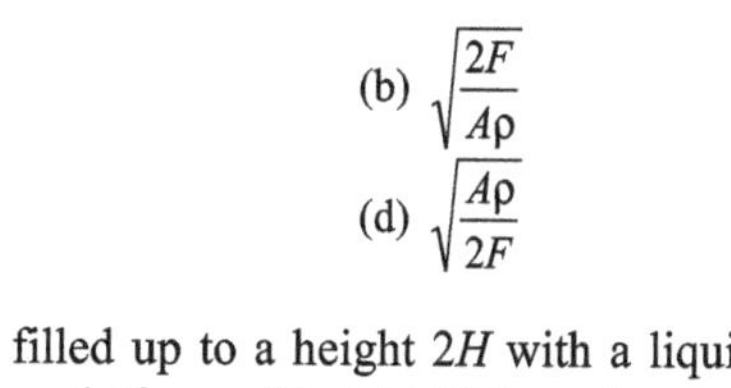

(a) $\sqrt{\frac{F}{A\rho}}$ (b) $\sqrt{\frac{2F}{A\rho}}$
(c) $\sqrt{\frac{A\rho}{F}}$ (d) $\sqrt{\frac{A\rho}{2F}}$

31. A tank is filled up to a height $2H$ with a liquid and is placed on a platform of height H from the ground. The distance x from the ground, where a small hole is punched to get the maximum range R is
(a) H (b) 1.25 H
(c) 1.5 H (d) 2 H

32. A piece of steel has a weight w in air, w_1 when completely immersed in water and w_2 when completely immersed in an unknown liquid. The relative density (specific gravity) of liquid is

(a) $\frac{w - w_1}{w - w_2}$ (b) $\frac{w - w_2}{w - w_1}$

(c) $\frac{w_1 - w_2}{w - w_1}$ (d) $\frac{w_1 - w_2}{w - w_2}$

33. Two cylinders of same cross-section and length L but made of two materials of densities d_1 and d_2 are connected together to a form a cylinder of length $2L$. The combination floats in a liquid of density d with a length $L/2$ above the surface of the liquid. If $d_1 < d_2$, then

(a) $d_1 < \frac{3}{4} d$ (b) $\frac{d}{2} > d_1$

(c) $\frac{d}{4} > d_1$ (d) $d_1 > \frac{d}{4} d$

More than One Correct Options

1. A large wooden plate of area $10\,m^2$ floating on the surface of a river is made to move horizontally with a speed of 2 m/s by applying a tangential force. River is 1 m deep and the water in contact with the bed is stationary. Then choose correct statement.
(coefficient of viscosity of water $= 10^{-3}$ N-s/m^2)

(a) Velocity gradient is $2\,s^{-1}$
(b) Velocity gradient is $1\,s^{-1}$
(c) Force required to keep the plate moving with constant speed is 0.02 N
(d) Force required to keep the plate moving with constant speed is 0.01 N

2. Choose the correct options.

(a) Viscosity of liquids increases with temperature
(b) Viscosity of gases increases with temperature
(c) Surface tension of liquids decreases with temperature
(d) For angle of contact $\theta = 0°$, liquid neither rises nor falls on capillary

3. A plank is floating in a non-viscous liquid as shown in the figure. Choose the correct options.

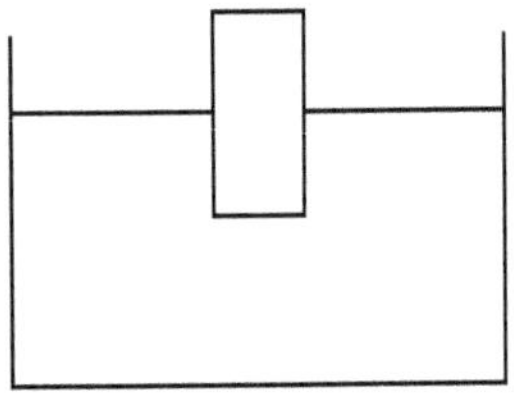

(a) Equilibrium of plank is stable in vertical direction
(b) For small oscillations of plank in vertical direction motion is simple harmonic
(c) Even, if oscillations are large, motion is simple harmonic till it is not fully immersed
(d) On vertical displacement motion is periodic but not simple harmonic

4. A non-viscous incompressible liquid is flowing from a horizontal pipe of non-uniform cross-section as shown in figure. Choose the correct options.

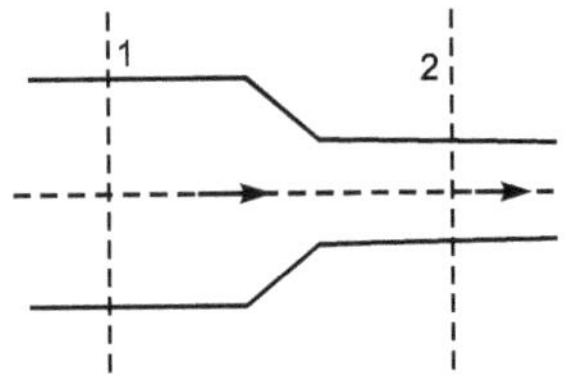

(a) Speed of liquid at section 2 is more
(b) Volume of liquid flowing per second from section 2 is more
(c) Mass of liquid flowing per second at both the sections is same
(d) Pressure at section 2 is less

5. A plank is floating in a liquid as shown in the figure. Fraction f of its volume is immersed. Choose the correct options.

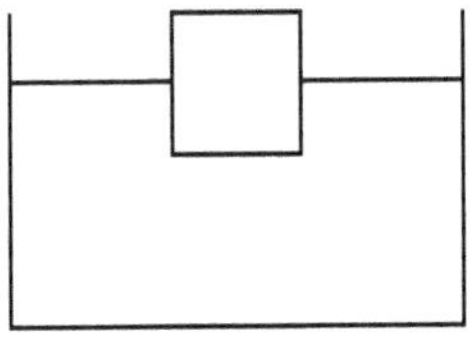

(a) If the system is taken to a place, where atmospheric pressure is more, f will increase
(b) In above condition f will remain unchanged
(c) If temperature is increased and expansion of only liquid is considered f will increase
(d) If temperature is increased and expansion of only plank is considered f will decrease

6. In two figures

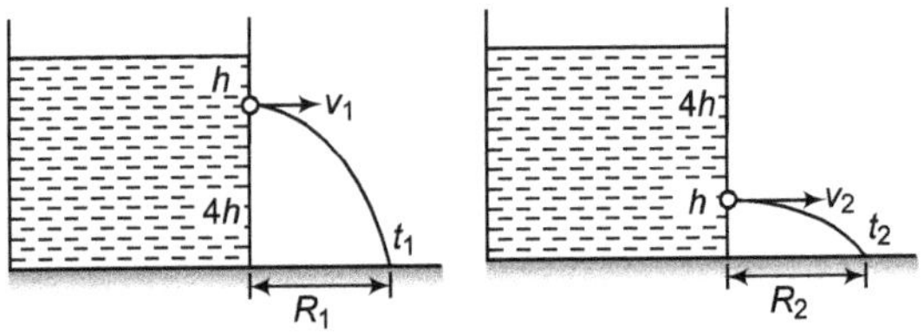

(a) $v_1/v_2 = 1/2$
(b) $t_1/t_2 = 2/1$
(c) $R_1/R_2 = 1$
(d) $v_1/v_2 = 1/4$

7. A liquid is filled in a container as shown in figure. Container is accelerated towards right. There are four points A, B, C and D in the liquid. Choose the correct options.

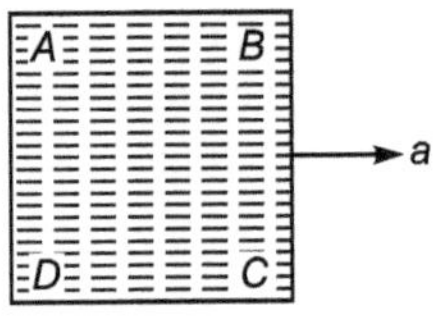

(a) $P_A > P_B$ (b) $P_C > P_A$
(c) $P_D > P_B$ (d) $P_A > P_C$

8. A ball of density ρ is dropped from a height on the surface of a non-viscous liquid of density 2ρ. Choose the correct options.

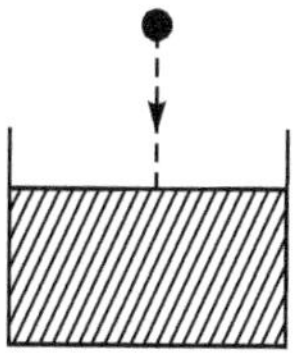

(a) Motion of ball is periodic but not simple harmonic
(b) Acceleration of ball in air and in liquid are equal
(c) Magnitude of upthrust in the liquid is two times the weight of ball
(d) Net force on ball in air and in liquid are equal and opposite

9. Two holes 1 and 2 are made at depths h and $16h$, respectively. Both the holes are circular but radius of hole 1 is two times.

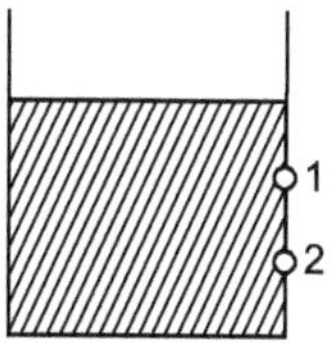

(a) Initially equal volumes of liquid will flow from both the holes in unit time
(b) Initially more volume of liquid will flow from hole 2 per unit time
(c) After some time more volume of liquid will flow from hole 1 per unit time
(d) After some time more volume of liquid will flow from hole 2 per unit time

10. A solid sphere, a cone and a cylinder are floating in water. All have same mass, density and radius. Let f_1, f_2 and f_3 are the fraction of their volumes inside the water and h_1, h_2 and h_3 depths inside water. Then,

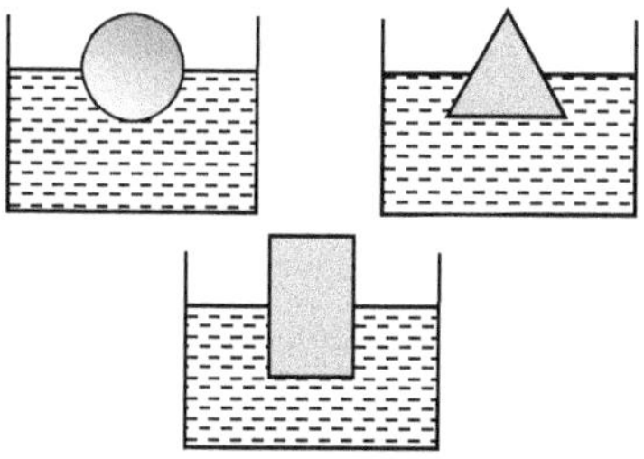

(a) $f_1 = f_2 = f_3$ (b) $f_3 > f_2 > f_1$
(c) $h_3 < h_1$ (d) $h_3 < h_2$

11. Streamline flow is more likely for liquids with
(a) high density (b) high viscosity
(c) low density (d) low viscosity

Comprehension Based Questions

Passage I (Q.1 to 3)

The spouting can is something used to demonstrate the variation of pressure with depth. When the corks are removed from the tubes in the side of the can, water flows out with a speed that depends on the depth. In a certain can, three tubes T_1, T_2 and T_3 are set at equal distances a above the base of the can. When water contained in this can is allowed to come out of the tubes the distances on the horizontal surface are measured as x_1, x_2 and x_3

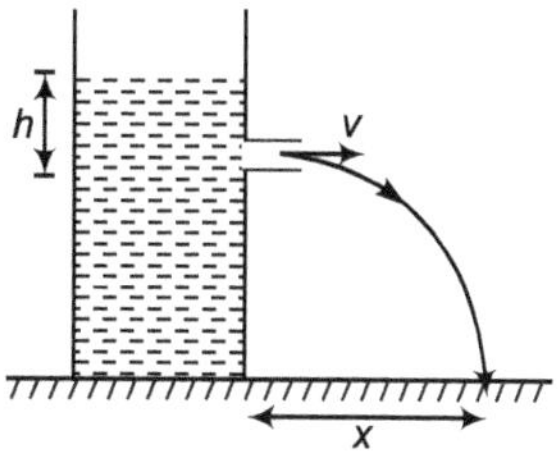

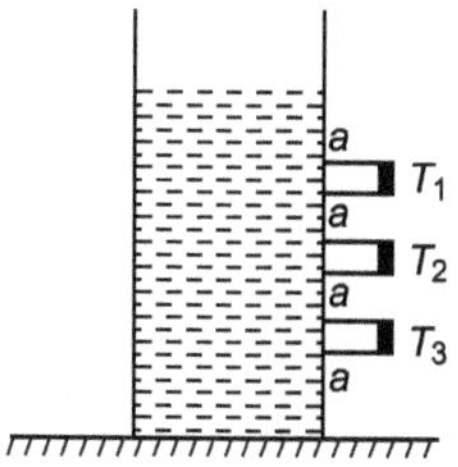

1. Speed of efflux is
(a) $\sqrt{3gh}$ (b) $\sqrt{2gh}$ (c) $\sqrt{gh}$ (d) $\frac{1}{2}\sqrt{2gh}$

2. Distance x_3 is given by
(a) $\sqrt{3}\,a$ (b) $\sqrt{2}\,a$ (c) $\frac{1}{2}\sqrt{3}\,a$ (d) $2\sqrt{3}\,a$

3. The correct sketch is

(a)
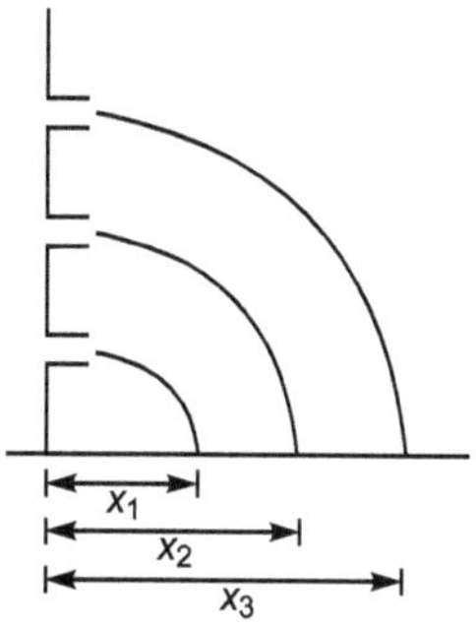

(b)
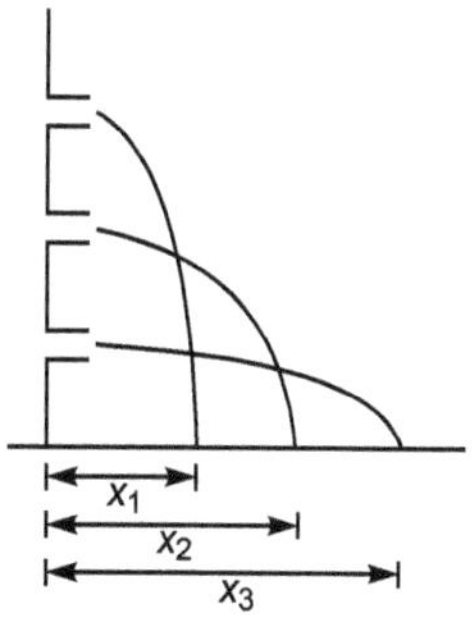

(c)
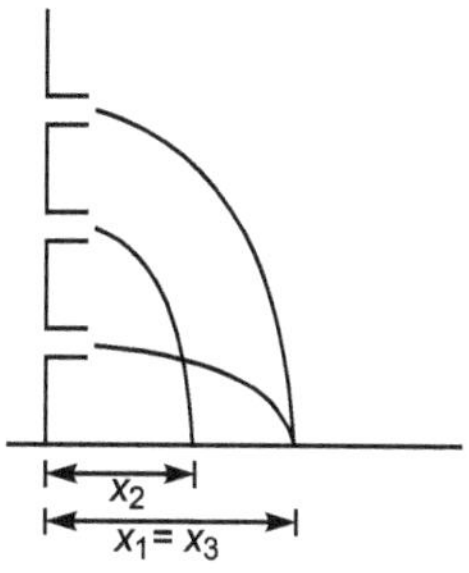

(d) None of the above

Passage II (Q. 4 to 7)

A container of large uniform cross-sectional area A resting on a horizontal surface, holds, two immiscible, non-viscous and incompressible liquids of densities d and 2d each of height $H/2$ as shown in the figure. The lower density liquid is open to the atmosphere having pressure P_0. A homogeneous solid cylinder of length $L(L< H/2)$, cross-sectional area $A/5$ is immersed such that it floats with its axis vertical at the liquid-liquid interface with length $L/4$ in the denser liquid.

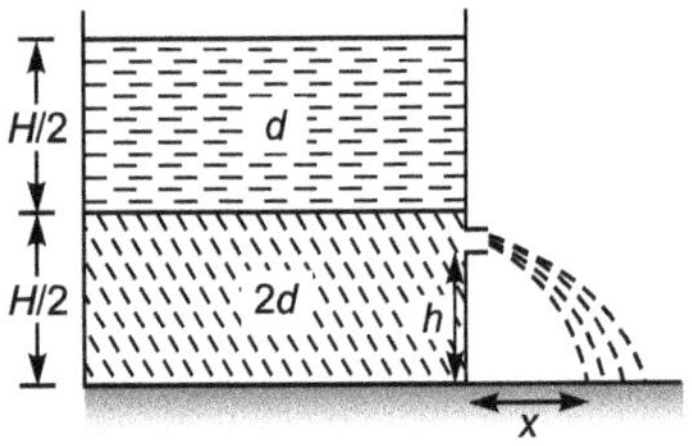

The cylinder is then removed and the original arrangement is restored. A tiny hole of area $s(s<< A)$ is punched on the vertical side of the container at a height $h(h< H/2)$. As a result of this, liquid starts flowing out of the hole with a range x on the horizontal surface.

4. The density D of the material of the floating cylinder is
(a) $5d/4$ (b) $3d/4$
(c) $4d/5$ (d) $4d/3$

5. The total pressure with cylinder, at the bottom of the container is
(a) $p_0 + \frac{(6L+H)}{4} dg$
(b) $p_0 + \frac{(L+6H)}{4} dg$
(c) $p_0 + \frac{(L+3H)}{4} dg$
(d) $p_0 + \frac{(L+2H)}{4} dg$

6. The initial speed of efflux without cylinder is
(a) $v = \sqrt{\frac{g}{3}[3H+4h]}$
(b) $v = \sqrt{\frac{g}{2}[4H-3h]}$
(c) $v = \sqrt{\frac{g}{2}[3H-4h]}$
(d) None of the above

7. The horizontal distance travelled by the liquid, initially, is
(a) $\sqrt{(3H+4h)h}$
(b) $\sqrt{(3h+4H)h}$
(c) $\sqrt{(3H-4h)\,h}$
(d) $\sqrt{(3H-3h)\,h}$

Assertion and Reason

Direction (Q. Nos. 1-20) *These questions consists of two statements each printed as assertion and reason. While answering these questions you are required to choose any one of the following five responses.*
(a) If both Assertion and Reason are correct and Reason is the correct explanation of Assertion
(b) If both Assertion and Reason are true but Reason is not the correct explanation of Assertion
(c) If Assertion is true but Reason is false
(d) If Assertion is false but Reason is true
(e) If both Assertion and Reason are false

1. Assertion When an ideal fluid flows through a pipe of non-uniform cross-section, then pressure is more at that section, where area is more if the pipe is horizontal.

Reason According to Bernoulli's theorem speed at broader cross-section will be less.

2. Assertion A ball is released from the bottom of a tank filled with a liquid. It moves upwards. In moving upwards upthrust will decrease.

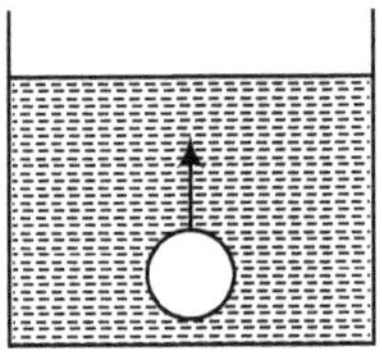

Reason Density of ball is less than the density of liquid.

3. Assertion A wooden plank is floating in two liquids as shown. Net force applied by liquid-1 on plank is zero.

Reason Contribution in upthrust due to liquid 1 is $V_1\rho_1 g$. Where, V_1 = volume immersed in liquid 1 and ρ_1 is the density of liquid 1.

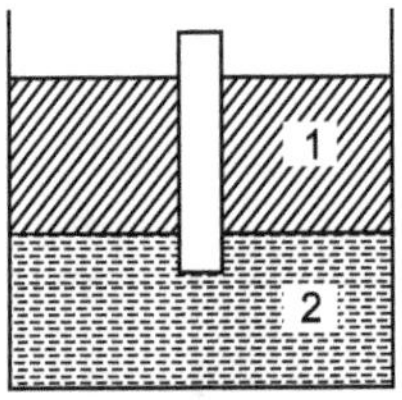

4. Assertion In the siphon figure shown, $p_1 = p_2$.

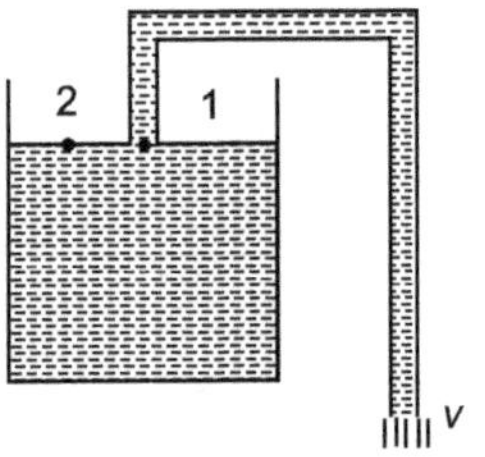

Reason Pressure at 1 is less than the atmospheric pressure.

5. **Assertion** If ice is floating in water and it melts, then level of water remains unchanged.

   Reason When the ice is floating, weight of liquid displaced is equal to the weight of ice.

6. **Assertion** An ice ball is floating in water. Some stone pieces are embedded inside the ice. When ice will melt, level of water will fall.

   Reason In floating condition, stone pieces will displace more liquid compared to the condition when they sink.

7. **Assertion** In a freely falling liquid container, upthrust force is zero.

   Reason In freely falling case value of effective value of g is zero.

8. **Assertion** Bulk modulus of an incompressible liquid is infinite.

   Reason Compressibility is inverse of bulk modulus.

9. **Assertion** Density of an incompressible liquid is constant.

   Reason An ideal fluid is incompressible.

10. **Assertion** A solid object of iron is dipped in water, both are at same temperature of 2°C. If the temperature of water is increased by 2°C, then the buoyancy force action of the object will increase.

    Reason If we increase the temperature of water from 2°C to 4°C, then density of water will increase. Ignore expansion of solid sphere.

11. **Assertion** A solid sphere and a hollow sphere both of same material are immersed in a liquid, then change in weight in both the spheres will be same.

    Reason Upthrust depends upon the volume of the solid immersed not the mass.

12. **Assertion** If water is filled in a balloon and this is immersed in water itself. Then volume of water displaced is equal to the volume of water filled in the balloon.

    Reason Volume of a liquid displaced is equal to the volume of solid immersed in that liquid.

13. **Assertion** At same level of same liquid pressure is always same.

    Reason When any fluid travels from a region of higher pressure to lower pressure (at same levels) it gains same speed.

14. **Assertion** A solid is floating in a liquid. If temperature is increased and expansion of solid is ignored, then fraction of volume immersed will increase.

    Reason By increasing the temperature density of liquid will decrease.

15. **Assertion** If angle of contact is 0°, then liquid will neither rise nor fall in a capillary.

    Reason When angle of contact is 0°, surface is neither convex nor concave inside the capillary. It is flat.

16. **Assertion** Small water drops are spherical while bigger water drops are not.

    Reason In small water drops surface tension forces dominate while in bigger water drops gravity forces dominate.

17. **Assertion** Bulk modulus of an ideal fluid is infinite.

    Reason An ideal fluid is incompressible.

18. **Assertion** Deep inside a liquid density is more than the density on surface.

    Reason Density of liquid increases with increase in depth.

19. **Assertion** A solid is floating in a liquid of density ρ_1. When the solid melts its density becomes ρ_2 in liquid state. If $\rho_1 > \rho_2$ level of liquids will increase after melting.

    Reason In liquid state volume always increases after a solid melts.

20. **Assertion** Mass of solid floating in liquid is m_1 and mass of liquid is m_2. Base area is A. Then pressure at bottom is $p_0 + \dfrac{(m_1 + m_2)g}{A}$.

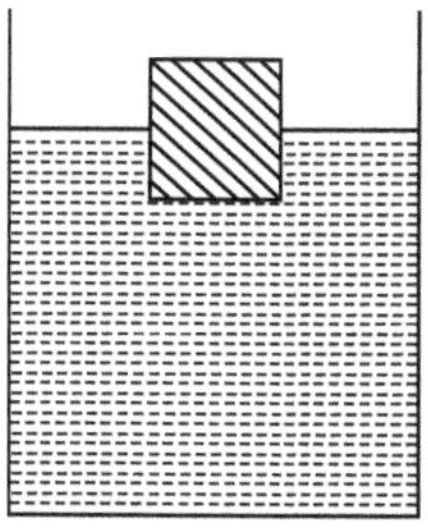

    Reason Upward force on liquid from base of vessel is pA, where p is pressure at bottom.

Match the Columns

1. A tube is inverted in a mercury vessel as shown in figure. If pressure p is increased, then

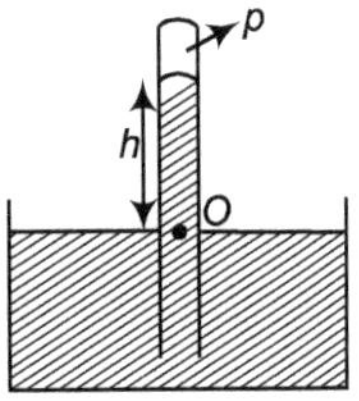

Column I	Column II
(A) Height h	(p) Will increase
(B) Pressure at O	(q) Will decrease
(C) Pressure at 1 cm above	(r) Will remain same

2. In the figure shown, velocity of liquid which comes out is v, time of liquid to fall to ground is t and range on ground is R. If the vessel is taken to a mountain, match the following, (consider all cases which might possible)

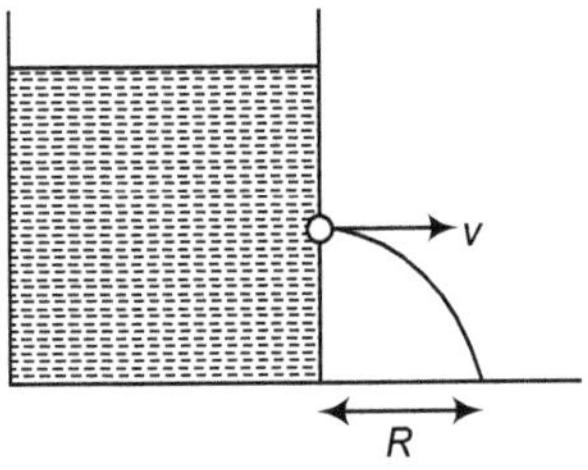

Column I	Column II
(A) v	(p) Will increase
(B) t	(q) Will decrease
(C) R	(r) Will remain same

3. A liquid is flowing through a pipe of non-uniform cross-section. At a point where area of cross-section of the pipe is less, match the following columns.

Column I	Column II
(A) Volume of liquid flowing per second	(p) Is less
(B) Speed of liquid	(q) Is more
(C) Pressure of liquid	(r) Is same

4. There are two points A and B inside a liquid as shown in figure. Now, the vessel starts moving upwards with an acceleration a. Match the following columns.

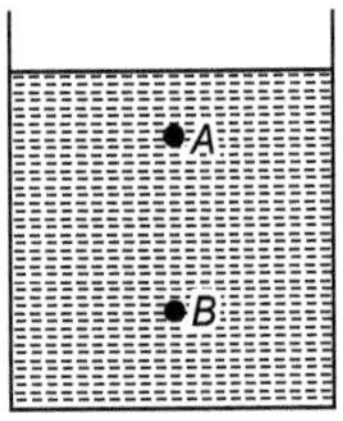

Column I	Column II
(A) Pressure B will	(p) Increase
(B) Pressure difference between A and B will	(q) Decrease
(C) Upthrust on an object inside the vessel will	(r) Remain same

5. Two soap bubbles coalesce to form a single large drop. Match the following columns.

Column I	Column II
(A) Surface energy in the process will	(p) Increase
(B) Temperature of the drop will	(q) Decrease
(C) Pressure inside the soap bubble will	(r) Remain same

6. A cube is floating in liquid as shown in figure. Match the following columns.

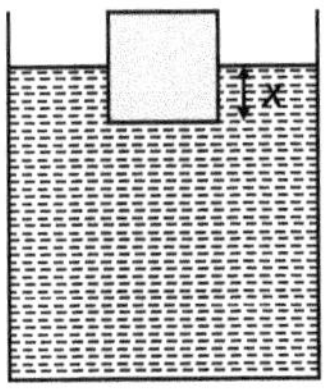

Column I	Column II
(A) If density of liquid decreases x will	(p) Increase
(B) If size of cube is increased x will	(q) Decrease
(C) If the whole system is accelerated upwards x will	(r) Remain same

Entrance Gallery

2014

1. There is a circular tube in a vertical plane. Two liquids which do not mix and of densities d_1 and d_2 are filled in the tube. Each liquid subtends 90° angle at centre. Radius joining their interface makes an angle α with vertical. Ratio d_1/d_2 is **[JEE Main]**

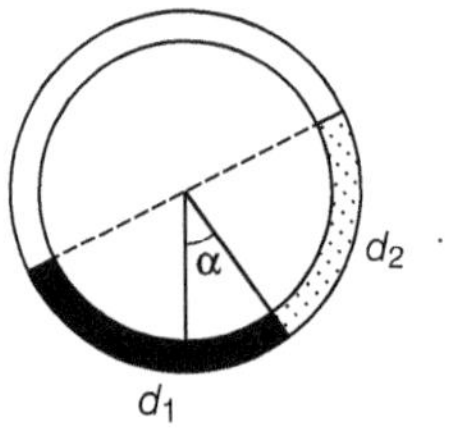

(a) $\dfrac{1+\sin\alpha}{1-\sin\alpha}$ (b) $\dfrac{1+\cos\alpha}{1-\cos\alpha}$ (c) $\dfrac{1+\tan\alpha}{1-\tan\alpha}$ (d) $\dfrac{1+\sin\alpha}{1-\cos\alpha}$

2. On heating water, bubbles beings formed at the bottom of the vessel detach and rise. Take the bubbles to be spheres of radius R and making a circular contact of radius r with the bottom of the vessel. If $r << R$ and the surface tension of water is T, value of r just before bubbles detach is (density of water is ρ) **[JEE main]**

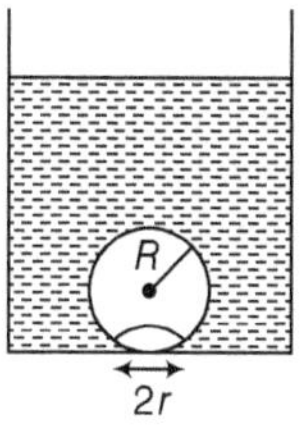

(a) $R^2\sqrt{\frac{2\rho_w g}{3T}}$ (b) $R^2\sqrt{\frac{\rho_w g}{6T}}$
(c) $R^2\sqrt{\frac{\rho_w g}{T}}$ (d) $R^2\sqrt{\frac{3\rho_w g}{T}}$

3. A flow of liquid is streamline, if the Reynold's number is **[Karnataka CET]**
(a) less than 1000
(b) greater than 1000
(c) between 2000 to 3000
(d) between 4000 to 5000

4. A drop of some liquid of volume 0.04 cm^3 is placed on the surface of a glass slide. Then, another glass slide is placed on it in such a way that the liquid forms a thin layer of area 20 cm^2 between the surfaces of the two slides. To separate the slides a force of 16×10^5 dyne has to be applied normal to the surfaces. The surface tension of the liquid is (in dyne - cm^{-1}). **[WB JEE]**
(a) 60 (b) 70
(c) 80 (d) 90

5. Which one of the following equation is Torricelli's law? **[J&K CET]**
(a) $p = \rho gh$ (b) $v = \sqrt{2hg}$
(c) $\eta Re = \rho vd$ (d) $S\ (2dl) = Fd$

2013

6. Assume that a drop of liquid evaporates by decrease in its surface energy, so that its temperature remains unchanged. What should be the minimum radius of the drop for this to be possible?

The surface tension is T, density of liquid is ρ and L is its latent heat of vaporisation. **[JEE Main]**
(a) $\rho L/T$ (b) $\sqrt{T/\rho L}$
(c) $T/\rho L$ (d) $2T/\rho L$

7. A rain drop of radius 0.3 mm has a terminal velocity of 1 ms^{-1} in air. The viscosity of air is 18×10^{-5} poise. The viscous force on the drop will be **[Karnataka CET]**
(a) 16.95×10^{-9}N (b) 1.695×10^{-9}N
(c) 10.17×10^{-9}N (d) 101.73×10^{-9}N

2012

8. Energy of bubble is E. If one thousand such bubbles coalesce to form a large bubble then its energy is E_1, then ratio E and E_1 is equal to **[O JEE]**
(a) $\frac{1}{100}$ (b) $\frac{1000}{1}$
(c) $\frac{1}{10}$ (d) $\frac{10}{1}$

2011

9. Two solid spheres A and B of equal volumes but of different densities d_A and d_B are connected by a string. They are fully immersed in a fluid of density d_F. They get arranged into an equilibrium state as shown in the figure with a tension in the string. The arrangement is possible only if **[IIT JEE]**

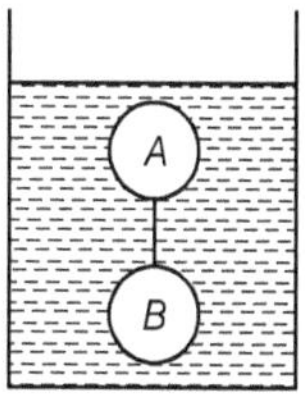

(a) $d_A < d_F$ (b) $d_B > d_F$
(c) $d_A > d_F$ (d) $d_A + d_B = 2d_F$

10. If a ball of steel (density $\rho = 7.8\ g\ cm^{-3}$) attains a terminal velocity of 10 cms^{-1} when falling in a tank of water (coefficient of viscosity $\eta_{water} = 8.5 \times 10^{-4}$ Pa-s) then its terminal velocity in glycerine ($\rho = 1.2\ g\ cm^{-3}$, $\eta = 13.2$ Pa-s) would be nearly **[AIEEE]**
(a) $1.6 \times 10^{-5}\ cm s^{-1}$ (b) $6.25 \times 10^{-4}\ cm s^{-1}$
(c) $6.45 \times 10^{-4}\ cm s^{-1}$ (d) $1.5 \times 10^{-5}\ cm s^{-1}$

11. Water is flowing continuously from a tap having an internal diameter 8×10^{-3} m. The water velocity as it leaves the tap is 0.4 m/s. The diameter of the water stream at a distance 2×10^{-1} m below the tap is close to **[AIEEE]**
(a) 7.5×10^{-3}m (b) 9.6×10^{-3}m
(c) 3.6×10^{-3}m (d) 5.0×10^{-3}m

12. Two mercury drops (each of radius r) merge to form a bigger drop. The surface energy of the bigger drop, if T is the surface tension is **[AIEEE]**
(a) $2^{5/3}\pi r^2 T$ (b) $4\pi r^2 T$
(c) $2\pi r^2 T$ (d) $2^{8/3}\pi r^2 T$

13. The terminal speed of a sphere of gold (density = 19.5 kg m^{-3}) is 0.2 ms^{-1} in a viscous liquid (density = 1.5 kg m^{-3}). Then, the terminal speed of a sphere of silver (density = 10.5 kg m^{-3}) of the same size in the same liquid is **[Kerala CEE]**
(a) 0.1 ms^{-1} (b) 1.133 ms^{-1}
(c) 0.4 ms^{-1} (d) 0.2 ms^{-1}
(e) 0.3 ms^{-1}

14. A large open tank has two holes in its wall. One is a square hole of side a at a depth of x from the top and the other is a circular hole of radius r at a depth $4x$ from the top. When the tank is completely filled with water, the quantities of water flowing out per second from both holes are the same. Then, r is equal to **[Kerala CEE]**

(a) $2\pi a$ (b) a (c) $\frac{a}{2\pi}$ (d) $\frac{a}{\pi}$

15. Ice pieces are floating in a beaker A containing water and also in a beaker B containing miscible liquid of specific gravity 1.2. When ice melts, then level of **[Kerala CEE]**

(a) water increases in A
(b) water decreases in A
(c) liquid in B decreases
(d) liquid in B increases
(e) water in A and liquid in B remains unaltered

16. Eight equal drops of water are falling through air with a steady velocity of 10 cms^{-1}. If the drops combine to form a single drop big in size, then the terminal velocity of this big drop is **[Karnataka CET]**

(a) 80 cms^{-1} (b) 30 cms^{-1} (c) 10 cms^{-1} (d) 40 cms^{-1}

17. An object weighs m_1 in a liquid of density d_1 and that in liquid of density d_2 is m_2. The density d of the object is **[WB JEE]**

(a) $d = \frac{m_2 d_2 - m_1 d_1}{m_2 - m_1}$ (b) $d = \frac{m_1 d_1 - m_2 d_2}{m_2 - m_1}$

(c) $d = \frac{m_2 d_1 - m_1 d_2}{m_1 - m_2}$ (d) $d = \frac{m_1 d_2 - m_2 d_1}{m_1 - m_2}$

18. A body floats in water with 40% of its volume outside water. When the same body floats in an oil, 60% of its volume remains outside oil. The relative density of oil is **[WB JEE]**

(a) 0.9 (b) 1.0 (c) 1.2 (d) 1.5

19. With an increase in temperature, surface tension of liquid (except molten copper and cadmium) **[MHT CET]**

(a) increases
(b) remain same
(c) decreases
(d) first decreases then increases

20. On the surface of the liquid in equilibrium, molecules of the liquid possess **[MHT CET]**

(a) maximum potential energy
(b) minimum potential energy
(c) maximum kinetic energy
(d) minimum kinetic energy

2010

21. A ball is made of a material of density ρ, where $\rho_{oil} < \rho < \rho_{water}$ with ρ_{oil} and ρ_{water} representing the densities of oil and water respectively. The oil and water are immiscible. If the above ball is in equilibrium in a mixture of this oil and water, which of the following pictures represents its equilibrium position? **[AIEEE]**

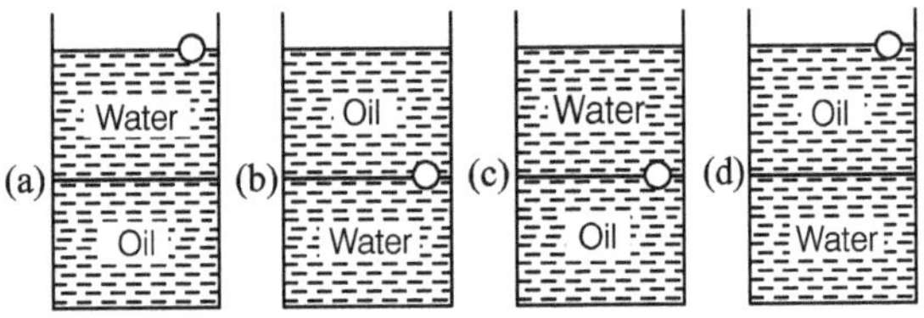

22. Three liquids of equal masses are taken in three identical cubical vessels A, B and C. Their densities are ρ_A, ρ_B and ρ_C respectively but $\rho_A < \rho_B < \rho_C$. The force exerted by the liquid on the base of the cubical vessel is **[Karnataka CET]**

(a) maximum in vessel C (b) minimum in vessel C
(c) the same in all the vessels (d) maximum in vessel A

Answers

Level 1

Objective Problems

1. (a)	**2.** (a)	**3.** (b)	**4.** (a)	**5.** (b)	**6.** (a)	**7.** (a)	**8.** (c)	**9.** (c)	**10.** (b)
11. (b)	**12.** (c)	**13.** (a)	**14.** (a)	**15.** (a)	**16.** (d)	**17.** (c)	**18.** (c)	**19.** (b)	**20.** (b)
21. (b)	**22.** (a)	**23.** (b)	**24.** (a)	**25.** (b)	**26.** (b)	**27.** (c)	**28.** (c)	**29.** (a)	**30.** (d)
31. (b,c,d)	**32.** (b)	**33.** (a)	**34.** (a)	**35.** (c)	**36.** (b)	**37.** (b)	**38.** (b)	**39.** (a)	**40.** (a)
41. (b)	**42.** (a)	**43.** (c)	**44.** (a)	**45.** (c)	**46.** (a)	**47.** (a)	**48.** (c)	**49.** (b)	**50.** (a)
51. (b)	**52.** (b)	**53.** (b)	**54.** (d)	**55.** (c)	**56.** (c)	**57.** (c)	**58.** (c)	**59.** (b)	**60.** (b)
61. (d)	**62.** (d)	**63.** (d)	**64.** (c)	**65.** (a)	**66.** (a)	**67.** (a)	**68.** (d)	**69.** (c)	**70.** (a)
71. (c)	**72.** (c)	**73.** (a)	**74.** (c)	**75.** (d)	**76.** (b)	**77.** (b)	**78.** (b)	**79.** (a)	**80.** (b)
81. (b)	**82.** (a)	**83.** (d)	**84.** (c)	**85.** (b)	**86.** (a)	**87.** (a)	**88.** (a)	**89.** (c)	**90.** (c)
91. (b)	**92.** (d)	**93.** (d)	**94.** (a,d)	**95.** (c)	**96.** (b)	**97.** (b)	**98.** (d)	**99.** (b)	**100.** (b)
101. (c)	**102.** (b)	**103.** (c)	**104.** (b)	**105.** (c)	**106.** (c)	**107.** (c)	**108.** (b)	**109.** (d)	**110.** (a)
111. (b)	**112.** (a)	**113.** (a,d)	**114.** (b)	**115.** (c)	**116.** (d)	**117.** (a)	**118.** (b)	**119.** (d)	**120.** (b)
121. (b)	**122.** (a)	**123.** (d)	**124.** (b)	**125.** (c)	**126.** (d)	**127.** (a)	**128.** (b)	**129.** (a)	**130.** (a)
131. (b)	**132.** (d)	**133.** (a)	**134.** (b)	**135.** (d)					

Level 2

Objective Problems

1. (b)	**2.** (c)	**3.** (c)	**4.** (a)	**5.** (d)	**6.** (b)	**7.** (b)	**8.** (a)	**9.** (d)	**10.** (c)
11. (b)	**12.** (d)	**13.** (b)	**14.** (c)	**15.** (a)	**16.** (b)	**17.** (d)	**18.** (d)	**19.** (b)	**20.** (c)
21. (a)	**22.** (c)	**23.** (d)	**24.** (c)	**25.** (c)	**26.** (d)	**27.** (b)	**28.** (c)	**29.** (a)	**30.** (b)
31. (c)	**32.** (b)	**33.** (a)							

More than One Correct Options

1. (a,c)	**2.** (b,c)	**3.** (a,b,c)	**4.** (a,c,d)	**5.** (b,c,d)	**6.** (a,b,c)	**7.** (a,c)	**8.** (a,c,d)	**9.** (a,d)	**10.** (a,c,d)
11. (b,c)									

Comprehension Based Questions

1. (b)	**2.** (d)	**3.** (d)	**4.** (a)	**5.** (b)	**6.** (c)	**7.** (c)

Assertion and Reason

1. (c)	**2.** (d)	**3.** (b)	**4.** (d)	**5.** (b)	**6.** (a)	**7.** (a)	**8.** (b)	**9.** (b)	**10.** (a)
11. (a)	**12.** (b)	**13.** (d)	**14.** (a)	**15.** (e)	**16.** (a)	**17.** (a)	**18.** (a)	**19.** (c)	**20.** (b)

Match the Columns

1. $A \to q, B \to r, C \to r$ **2.** $A \to q, B \to p, C \to r$ **3.** $A \to r, B \to q, C \to p$ **4.** $A \to p, B \to p, C \to p$ **5.** $A \to q, B \to p, C \to q$
6. $A \to p, B \to p, C \to r$

Entrance Gallery

1. (c)	**2.** (a)	**3.** (c)	**4.** (c)	**5.** (b)	**6.** (d)	**7.** (d)	**8.** (a)	**9.** (a,b,d)	**10.** (b)
11. (c)	**12.** (d)	**13.** (a)	**14.** (e)	**15.** (e)	**16.** (d)	**17.** (d)	**18.** (d)	**19.** (c)	**20.** (a)
21. (b)	**22.** (c)								

Solutions

Level 1 : Objective Problems

1. In a freely falling vessel upthrust is zero, because effective value of g is zero.

2. $(m_{raft} + m)g = V_{raft}\,\rho_w\, g$

$\therefore \quad m = V_{raft}\,\rho_w - m_{raft}$

$= \frac{120}{600} \times 1000 - 120 = 80$ kg

3. $1\ \text{g/L} = 1\ \text{kg/m}^3$

Upthrust = total downward force

$\therefore \quad 1000 \times 1.29 \times g = (1000 \times 0.09 \times g + mg)$

$\therefore \quad m = 1200$ kg

4. Weight of man = extra upthrust

$\therefore \quad mg = (3 \times 2 \times 10^{-2})\,(10)^3\,(10)$

or $\quad m = 60$ kg

5. $a = \frac{\text{upthrust} - \text{weight}}{\text{mass}} = \frac{V\rho_w g - V\rho g}{V\rho}$

$= \left(\frac{\rho_w - \rho}{\rho}\right) g = \left(\frac{1 - 0.5}{0.5}\right)(10) = 10\ \text{m/s}^2$

6. Specific gravity or relative density

$= \frac{\text{weight in air}}{\text{change in weight in water}} = \frac{60}{20} = 3$

7. Weight = upthrust from mercury + upthrust from water

$\therefore \quad 10 \times 8.56 = (x \times 13.6) + (10 - x) \times 1.0$

Here, x = depth inside mercury

Solving we get $\quad x = 6$ cm

8. $w_{app} = w -$ upthrust

$= w - \left(\frac{w/g}{\rho}\right)\sigma g = w\left(1 - \frac{\sigma}{\rho}\right)$

9.

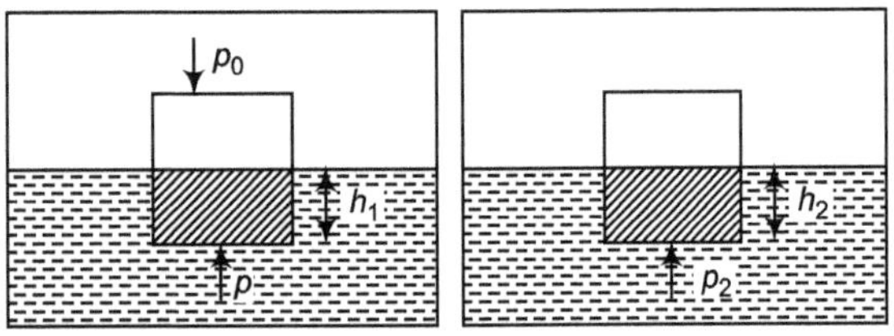

In the first case,

Initially $(p - p)A$ = weight of block

or $\quad [(p_0 + \rho g h_1) - p_0]\,A = w$

or $\quad h_1 = \frac{w}{A\rho g}$ ...(i)

In the second case

$p_2 A = w$

or $\quad (\rho g h_2)A = w$

or $\quad h_2 = \frac{w}{\rho A g}$ ...(ii)

From Eqs. (i) and (ii), we see that $h_1 = h_2$

10. w = Upthrust or $VDg = vdg$ or $\frac{v}{V} = \frac{D}{d}$.

11. Weight of sphere = upthrust from water + upthrust from liquid

$\therefore \quad V\rho g = \left(\frac{2}{3}V \times 10^3 \times g\right) + \left(\frac{V}{3} \times 8 \times 10^3 \times g\right)$

or $\quad \rho = \frac{10000}{3}\ \text{kg/m}^3$

12. $F = w_{air} - w_{water}$ or Δw

$= 5 - 2 = 3$ N

13. Fraction immersed $= \frac{\rho_{solid}}{\rho_{liquid}} = \frac{0.9}{1.125} = 0.8.$

Therefore, fraction outside water = 0.2.

14. Specific gravity or relative density

$= \frac{\text{weight in air}}{\text{change in weight in water}} = \frac{60}{20} = 3$

15. Let x be the fraction of volume immersed in first case and y the fraction in second case.

Then in first case, w = Upthrust

or $\quad V\rho_b g = (xV)\rho_l g \quad$ or $\quad x = \frac{\rho_b}{\rho_l}$ ...(i)

In second case,

Upthrust − weight = mass × acceleration

or $\quad (yV)\rho_l(g + g/3) - (V\rho_b)g = (V\rho_b)\left(\frac{g}{3}\right)$

or $\quad y = \frac{\rho_b}{\rho_l}$ ...(ii)

From Eqs. (i) and (ii), we see that $x = y$ or fraction of volume immersed does not change by the acceleration of vessel.

16. Upthrust on 7 cm = 600 kg

$\therefore$ Upthrust on further 1.4 cm $= \frac{600}{7} \times 1.4 = 120$ kg

17. Percentage volume outside the water

$= \frac{\rho_{water} - \rho_{ice}}{\rho_{water}} \times 100 = \frac{1000 - 900}{1000} \times 100 = 10\%$

18. Upthrust $= \frac{2}{10} \times 1 \times 10 = 2$ N

Reading of S_1 will increase by 2N while that of S_2 will decrease by 2 N.

19. $f_i = \frac{\rho_{solid}}{\rho_{liquid}} = \frac{\rho_s}{\rho_m}$

$\therefore \quad h_i = \frac{\rho_s}{\rho_m} . l$

$\therefore$ Height above the mercury level,

$= \frac{\rho_s}{\rho_m} l = \left(1 - \frac{\rho_s}{\rho_m}\right) l$

20. $1\ \text{g/c} = 1\ \text{kg/m}^3$

Upthrust = total downward force

$\therefore \quad 1000 \times 1.29 \times g = (1000 \times 0.09 \times g + mg)$

$\therefore \quad m = 1200$ kg

21. 1 atmosphere $= 1.01 \times 10^5 \text{ N/m}^2 = \rho g h$

$\therefore \quad h = \dfrac{1.01 \times 10^5}{1.3 \times 10} = 7769 \text{ m} \approx 8 \text{ km}$

22. $p_D = p_B$

$\therefore \quad p_0 + \rho_1 g h_1 = p_0 + \rho_w g h_2$

or $\dfrac{\rho_1}{\rho_w}$ = relative density of paraffin $= \dfrac{h_2}{h_1}$

23. Equating pressure at 1 and 2, we see that $\rho_1 = \rho_2$.

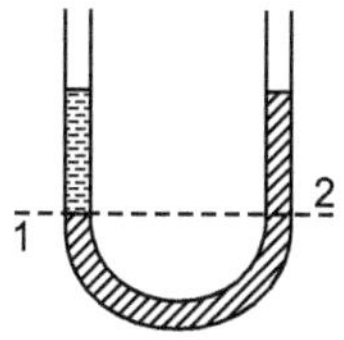

24. $\dfrac{dp}{dx} = -\rho a$ (along horizontal towards right) i.e. in the direction of acceleration along horizontal pressure will decrease of $p_{AB} > p_{CD}$.

$\dfrac{dp}{dy} = -\rho g$ (along vertically upwards direction pressure will decrease .)

Hence, pressure at B is maximum and pressure at D is minimum.

25. $F = p \times (\rho g h)$ (Area of base)

$= 900 \times 10 \times 0.4 \times 2 \times 10^{-3} = 7.2 \text{ N}$

26. $F = p \times A = (\rho g h_{\text{total}}) A = (10^3 \times 10 \times 1.0)(100 \times 10^{-4}) = 100 \text{ N}$

27. In horizontal direction, pressure remains constant unless it is accelerated.

28. When moving upwards $\dfrac{dp}{dh} = -\rho(g + a)$ and when moving downwards $\dfrac{dp}{dh} = -\rho(g - a)$

In first case pressure decreases with h more rapidly.

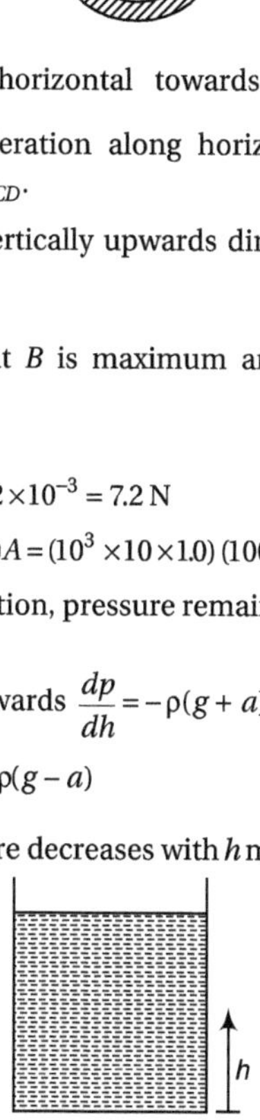

29. Pressure at boundary of two liquids will be same.

$\therefore \quad p + \rho g (3h) = p + 2\rho g h$

$\therefore \quad p = p_0 - \rho g h$ or $p < p_0$

30. $\dfrac{F_1}{A_1} = \dfrac{F_2}{A_2}$ (Pascal's law)

$\therefore \quad A_2 = \dfrac{F_2}{F_1} . A_1 = \left(\dfrac{2000 \times 9.8}{10^2} \right) (100) = 1.96 \times 10^4 \text{ cm}^2$

31. $\dfrac{\Delta p}{\Delta x} = \rho a = \rho g$ (in horizontal direction)

Pressure decreases in the direction of acceleration in horizontal.

$\dfrac{\Delta p}{\Delta x} = \rho g$ (in vertical direction)

Pressure increases with depth in vertical direction.

33. Velocity of efflux $= \sqrt{2gh}$, which is independent of density of liquid.

34. Difference in pressure energy = difference in kinetic energy

$\therefore \quad (p_0 - p_0) = \dfrac{1}{2} \rho v^2$

or $\quad v = 2\sqrt{\dfrac{p_0}{\rho}} = 2\sqrt{\dfrac{10^5}{10^3}} = 20 \text{ m/s}$ or $\sqrt{400}$ m/s

35. Decrease in potential energy = increase in kinetic energy

$\therefore \quad \rho g h = \dfrac{1}{2} \rho (v_f^2 - v_i^2)$

or $\quad 2(10)(0.15) = v_f^2 - (1.0)^2$

or $\quad v_f = 2 \text{ m/s}$

Now, from continuity equation,

$A_1 v_1 = A_2 v_2$ or $A \propto \dfrac{1}{v}$

Velocity has become two times. Hence, area of cross-section will remain half.

37. $\dfrac{1}{2} \rho v^2 = \rho g h \quad \therefore v = \sqrt{2gh} = \sqrt{2 \times 10 \times 0.1} = \sqrt{2}$ m/s

38. $v = 2\sqrt{H(H-h)} = 2\sqrt{\dfrac{H}{4} \times \dfrac{3H}{4}} = \dfrac{\sqrt{3}H}{2}$

39. Rate of leakage of water from the hole

$= Av = A\sqrt{2gh} = 10^{-4} \sqrt{2 \times 10 \times 5} = 10^{-3} \text{ m}^3/\text{s}$

40. $R_h = R_{H-h}$.

Since, range from both the holes is same

$h = 2.4 \text{ cm}$ and $H - h = 7.6 \text{ cm}$

Adding these two, we get $H = 10$ cm.

41. From continuity equation, $v \propto \dfrac{1}{A}$

or $\quad v \propto \dfrac{1}{d^2}$ (d = diameter of pipe)

42. $p_1 + \dfrac{1}{2} \rho v_1^2 = p_2 + \dfrac{1}{2} \rho v_2^2$

or $\quad (10^{-2} \times 13.6 \times 10^3 \times 10) + \dfrac{1}{2} \times 10^3 \times (0.35)^2$

$= p_2 + \dfrac{1}{2} \times 10^3 \times (0.65)^2$

or $\quad p_2 = 1210 \text{ N/m}^2 = \dfrac{1210 \times 10^2}{13.6 \times 10^3 \times 10}$ cm of Hg

$= 0.89$ cm of Hg

43. $A_1 v_1 = A_2 v_2 + A_3 v_3$

$\therefore \quad v_3 = \dfrac{A_1 v_1 - A_2 v_2}{A_3} = \dfrac{(0.2 \times 4) - (0.2 \times 2)}{0.4} = 1 \text{ m/s}$

44. Decrease in pressure energy = increase in kinetic energy

or $\quad \Delta p = \dfrac{1}{2} \rho v^2$

$\therefore \quad v = \sqrt{\dfrac{2(\Delta p)}{\rho}} = \sqrt{\dfrac{2 \times 0.5 \times 10^5}{10^3}} = 10 \text{ m/s}$

45. Pressure difference $\Delta p = \Delta \text{KE} = \dfrac{1}{2} \times \rho (v_1^2 - v_2^2)$

or $\quad \Delta p = \dfrac{1}{2} \times 1.0 \times (3600 - 1600) = 1000 \text{ N/m}^2$

$F = (\Delta p) \times A = 4000 \text{ N}$

46. $p_2 = p_3$ and $p_1 > p_2$

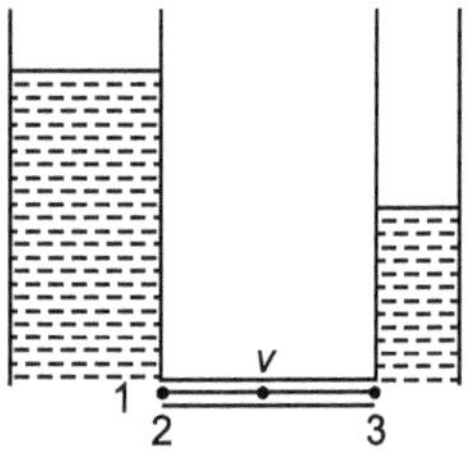

$$p_1 = p_0 + \rho g x$$
$$p_2 = p_0 + \rho g y$$
$$\therefore \quad \Delta p = p_1 - p_2 = \rho g\,(x - y)$$

Between 1 and 2,

Difference in pressure energy = difference in kinetic energy

or $$\rho g\,(x - y) = \frac{1}{2}\rho v^2$$

or $$v = \sqrt{2g(x - y)}$$

47. $R_h = R_{H-h}$ or $h_1 = H - h_2$ or $h_1 + h_2 = H$

48. At $h = \frac{H}{2}$, maximum range is obtained, which is equal to H.

Further $$R_h = R_{H-h}.$$

49. $Q = A_1 v_1 = A_2 v_2 = 15\,\text{L/min} = \frac{15 \times 10^{-3}}{60} = 2.5 \times 10^{-4}\ \text{m}^3/\text{s}$

$$v_1 = \frac{2.5 \times 10^{-4}}{2 \times 10^{-4}} = 1.25\ \text{m/s}$$

$$v_2 = \frac{2.5 \times 10^{-4}}{0.5 \times 10^{-4}} = 5.0\ \text{m/s}$$

Now, difference in pressure = difference in kinetic energy

$$\rho_w g h = \frac{1}{2}\rho_a\,(v_2^2 - v_1^2)$$

$$\therefore \quad h = \frac{\rho_a (v_2^2 - v_1^2)}{2\rho_w g} = 1.5 \times 10^{-3}\ \text{m} \approx 1.5\,\text{mm}$$

or $$h\rho_w g = \frac{1}{2}\rho_{\text{air}}\,(v_2^2 - v_1^2)$$

or $$h = \frac{\rho_{\text{air}}}{2\rho_w g}(v_2^2 - v_1^2) = \frac{1.32}{2 \times 10^3 \times 10}[(5)^2 - (1.25)^2]$$

$$= 1.5 \times 10^{-3}\ \text{m} \approx 1.5\,\text{mm}$$

50. $R = 2\sqrt{h_{\text{top}} \times h_{\text{bottom}}} = 2\sqrt{h(H - h)}$

51. $v = 2\sqrt{H(H - h)} = 2\sqrt{\frac{H}{3} \times \frac{3H}{4}} = \frac{\sqrt{3}H}{2}$

52. Time to empty a tank, $t = \frac{A}{a}\sqrt{\frac{2H}{g}}$

or $$t \propto \sqrt{H}$$

53. Level in the container will become maximum when rate of inflow = rate of outflow. Or,

$$Q = A_1 v = A_1\sqrt{2gh_{\max}}$$

$$\therefore \quad h_{\max} = \frac{Q^2}{2gA_1^2}$$

54. $v_T \propto r^2$

$$\therefore \quad \frac{(v_T)_1}{(v_T)_2} = \left(\frac{R}{R/2}\right)^2 = 4$$

55. Velocity becomes constant when terminal velocity is attained.

56. $F = 6\pi\eta r v$ or units of η are similar to that of $\frac{F}{rv}$ N-s/m^2

57. $F = 6\pi\eta r v$, i.e., v has the dimensions of $\frac{F}{\eta r}$ or $\frac{mg}{\eta r}$.

58. Let r be the radius of smaller drop and R the radius of bigger drop. By equating the volumes, we have

$$2\left(\frac{4}{3}\pi r^3\right) = \frac{4}{3}\pi R^3$$

or $$R = (2)^{1/3}.r$$

Now, terminal velocity $\propto$ (radius)^2

$$\therefore \quad v' = (2)^{2/3} v$$

61. $V = V_1 + V_2$

$$\Rightarrow \quad \frac{\pi p r^4}{8\eta l} = \frac{p r_1^4}{8\eta l} + \frac{\pi p r_2^4}{8}$$

$$\Rightarrow \quad r^4 = r_1^4 + r_2^4$$

$$\therefore \quad r = (r_1^4 + r_2^4)^{1/4}$$

62. Given, $l_1 = l_2 = 1\text{m}$ and $\frac{r_1}{r_2} = \frac{1}{2}$

$$V = \frac{\pi p_1 r_1^4}{8\eta l} = \frac{\pi p_2 r_2^4}{8\eta l}$$

$$\Rightarrow \quad \frac{p_1}{p_2} = \left(\frac{r_2}{r_1}\right) = 16$$

$$\Rightarrow \quad p_1 = 16 p_2$$

Since, both tubes are connected in series, hence pressure difference across combination,

$$p = p_1 + p_2 \Rightarrow p_1 = \frac{p_1}{16} \Rightarrow p_1 = \frac{16}{17} = 0.94\,\text{m}$$

63. $V = \frac{\pi p r^4}{8\eta l}$

$$\therefore \quad V \propto p r^4 \qquad (\eta \text{ and } l \text{ are constants})$$

$$\therefore \quad \frac{V_2}{V_1} = \left(\frac{p_2}{p_1}\right)\left(\frac{r_2}{r_1}\right)^4 = 2 \times \left(\frac{1}{2}\right)^4 = \frac{1}{8}$$

$$\therefore \quad V_2 = \frac{Q}{8}$$

65. Fluid resistance is given by $R = \frac{8\eta l}{\pi r^4}$.

When two capillary tubes of same size are joined in parallel, then equivalent fluid resistance is

$$R_e = R_1 + R_2 = \frac{8\eta l}{\pi r^4} + \frac{8\eta \times 2l}{\pi (2R)^4} = \left(\frac{8\eta l}{\pi r^4}\right) \times \frac{9}{8}$$

Equivalent resistance becomes 9/8 times so rate of flow will be $8/9\,X$.

67. When cross-section of duct is decreased, the velocity of water increased in accordance with Bernoulli's theorem, the pressure p decreased at that place.

70. Suppose, R be the radius of bigger drop. Then, by equating volumes, we have,

$$2(4/3\pi r^3) = (4/3)\pi R^3 \quad \text{or} \quad R = (2)^{1/3} r$$

Now, surface energy $\propto$ surface area

$$\therefore \quad \frac{U_1}{U_2} = \frac{A_1}{A_2} = \frac{2r^2}{R^2} = \frac{2}{2^{2/3}}$$

73. $p_1 < p_2$ for $\theta < 90°$

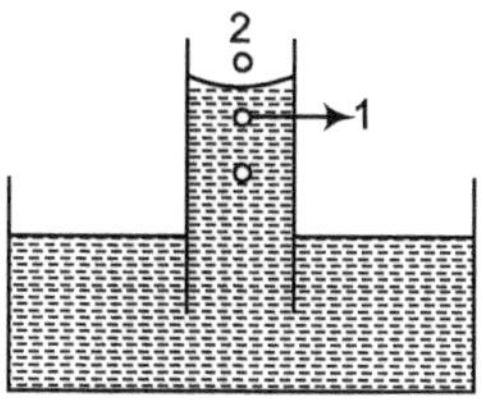

74. $\Delta p_1 = 0.01$ atm

$$\Delta p_1 = 0.01 \text{ atm}$$
$$\Delta p = \frac{4T}{R} \text{ or } \Delta p \propto 1/R$$
$$\therefore \quad \frac{R_1}{R_2} = \frac{\Delta p_2}{\Delta p_1} = \frac{2}{1}$$
$$\frac{V_1}{V_2} = \left(\frac{R_1}{R_2}\right)^3 = \frac{8}{1}$$

75. Surface tension $T = F/l$ and $\frac{\Delta W}{\Delta A}$, i.e. unit of surface tension is $\frac{\text{N}}{\text{m}}$ or $\frac{\text{J}}{\text{m}^2}$.

$$\frac{\text{N}}{\text{m}} = \frac{\text{kg-m/s}^2}{\text{m}} = \frac{\text{kg}}{\text{s}^2}$$

79. Weight of spiders or insects can be balanced by vertical component of force due to surface tension.

82. Force on each side $= 2TL$ (due to two surface).

84. This happens due to viscosity.

86. Due to force of attraction it is not easier to separate the two glass plates.

87. Soluble impurities increases the surface tension.

88. $T = \frac{F}{2l} = \frac{2 \times 10^{-2}}{2 \times 10 \times 10^{-2}} = 0.1$ N/m

89. $T = \frac{F}{l} = \frac{[\text{MLT}^{-2}]}{[\text{L}]} = [\text{ML}^0\text{T}^{-2}]$

90. $W = 8\pi R^2 T$

$$\therefore \quad W \propto R^2$$

If the radius becomes double, then work done will become four times.

91. Surface energy of combined drop will be lowered, so excess surface energy will raise the temperature of the drop.

92. $W = 8\pi R^2 T = 8 \times \pi \times (10^{-2})^2 \times 2 \times 10^{-2} = 16\pi \times 10^{-6}$ J

95. $r = \frac{r_1 r_2}{r_2 - r_1} = \infty$

97. Cohesive force decreases. So, angle of contact decreases.

101. Angle of contact is acute.

103. Since, $\Delta p \propto \frac{1}{R}$

104. Express pressure $\Delta p = \frac{4T}{r} = \frac{4 \times 2 \times 25 \times 10^{-3}}{1 \times 10^{-2}}$

$$= 20 \text{ N/m}^2 = 20 \text{ Pa} \qquad (\text{as } r = d/2)$$

107. $h\rho g = \frac{2T}{r} \Rightarrow h = \frac{2T}{r\rho g}$

108. $r = \frac{r_1 r_2}{r_1 - r_2} = \frac{5 \times 4}{5 - 4} = 20$ cm

109. $h = \frac{2T}{rdg} \Rightarrow h = \frac{2 \times 75}{0.005 \times 1 \times 10^3} = 30$ cm

110. $\Delta p = \frac{4T}{r}$

$$\therefore \quad \Delta p \propto \frac{1}{r}$$

As radius of soap bubble increases with time. $\therefore \Delta p \propto \frac{1}{t}$

111. $T_c = T_0(1 - at)$ i.e. surface tension decreases linearly with increase in temperature.

112. B will keep its area as minimum as possible.

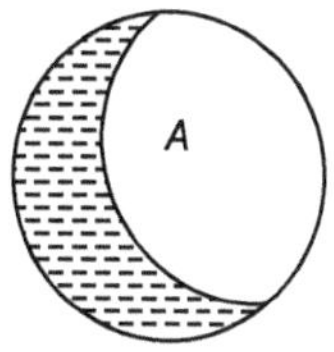

115. $h_1 r_1 = h_2 r_2$

116. $h = \frac{2T\cos\theta}{rdg}$ $\therefore h \propto \frac{1}{r}$. If θ is less, then h will be more.

117. $h = \frac{2T\cos\theta}{rdg}$. If θ is less than 90°, then h will be positive.

118. Let the width of each plate is b and due to surface tension liquid will rise up to height h, then upward force due to surface tension $= 2Tb\cos\theta$ ...(i)

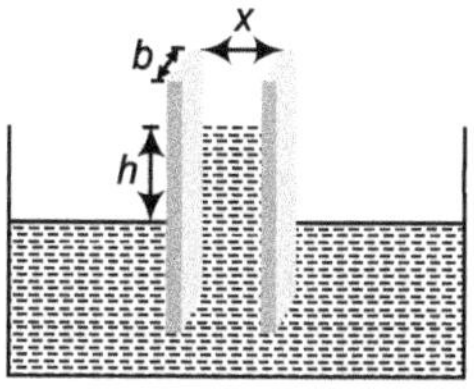

Weight of the liquid rises in between the plates

$$= Vdg = (bxh)dg \qquad \text{...(ii)}$$

Equating Eqs. (i) and (ii), we get

$$2Tb\cos\theta = bxhdg$$
$$\therefore \quad h = \frac{2T\cos\theta}{xdg}$$

119. $6 \times 10^{-2} \times$ circumference $=$ Force

$$\therefore \quad \text{Circumference} = \frac{75 \times 10^4}{6 \times 10^2} = 12.5 \times 10^2 \text{ m}$$

120. $r \propto \frac{1}{h} \Rightarrow \frac{r_P}{r_Q} = \frac{h_Q}{h_P} = \frac{h}{\frac{2}{3}h} = \frac{3}{2}$

124. $h = \frac{2T\cos\theta}{rdg}$

$$\therefore \quad h \propto \frac{1}{r}$$

So, the graph between h and r will be rectangular hyperbola.

125. Initially due to more pressure at the hole velocity of efflux will be more. So, the liquid drains out more quickly.

126. Volume of ice > Volume of water

$$\Delta V = V_w - V_i = \frac{m}{y} - \frac{m}{x}$$

127. Velocity of ball just before striking water surface $v = \sqrt{2gh}$

or $v = \sqrt{4g}$ m/s ($h = 2$m)

Retardation in water $a = \frac{\text{Upthrust} - \text{Weight}}{\text{Mass}}$

or $a = \frac{V\rho_w g - V\rho g}{V\rho}$

$= \left(\frac{1-0.8}{0.8}\right) g = g/4$

Now, depth to which ball will sink is

$$d = \frac{v^2}{2a} = \frac{4g}{2g/4} = 8\text{m}$$

128. When hole is made at centre of vessel, maximum range is obtained.

129. $4 = \frac{\rho_1 V + \rho_2 V}{2V}$ or $\rho_1 + \rho_2 = 8$ ...(i)

$$3 = \frac{m+m}{(m/\rho_1)+(m/\rho_2)}$$

or $\frac{\rho_1\rho_2}{\rho_1+\rho_2} = \frac{3}{2}$ ...(ii)

Solving these two equations, we get $\rho_1 = 6$ and $\rho_2 = 2$

130. Loss of weight = Upthrust

$\frac{W}{2}$ = total volume $\times \rho_w \times g$

or $\frac{(V - V_C) 5 \times \rho_w \times g}{2} = V \times \rho_w \times g$ ($\because V_C$ = volume of cavity)

or $V - V_C = \frac{2}{5}V$

$V_C = \frac{3}{5}V$

131. When iron block was floating, it displaces V_1 volume of water such that weight of this V_1 volume of water is equal to weight of block. But since density of iron is more than density of water, this volume V_1 will be greater than volume V_2 of iron block.

When iron block sinks in the pond, it displaces water of volume equal to its own volume V_2.

Since, $V_1 > V_2$, displaced volume in first case was more than displaced volume in second case. Hence, level will fall.

132. Force exerted by gas on its hemispherical end = Pressure of gas × projected area = $p_0(\pi r^2)$.

133. Upthrust on sphere from the liquid makes equal and opposite pair of forces. Hence, there will be no effect on the pressure at bottom of vessel. Or $p = p_0 + \rho gh$

134. W = Upthrust = $V_i \rho l g$ or $V_i = \frac{W}{\rho l g}$

If density of liquid ρl is increased, then immersed volume V_i will decrease or the ball will go up.

135. Pressure inside the bubble

$= \text{Pressure outside it} + \frac{2T}{r}$

$= (p_0 + \rho gh) + \frac{2T}{r}$

$= 10^5 + (10^3 \times 10 \times 10) + \frac{2 \times 7 \times 10^{-2}}{10^{-3}}$

$= 2.0014 \times 10^5$ N/m^2

Level 2 : Only One Correct Option

1. $3g = 4(g - a)$

$\therefore \quad g - a = \frac{3g}{4}$

Upthrust = $(5 \times 10^{-4})(10^3)\left(\frac{3g}{4}\right)$ N = 0.375 kg

Apparent weight = (3 – 0.375) = 2.625 kg

2. From work-energy theorem work done by all the forces = change in kinetic energy.

$\therefore$ (Work done by gravity for 7 m) + (work done by upthrust for 2 m) + (work done by resistive forces) = 0 as Δ KE = 0

$\therefore \quad (1)(10)(7)\cos 0° + \left(\frac{1}{2/3 \times \rho_w}\right)(\rho_w g)$

$(2)\cos 80°$ + work done by resistive forces = 0

$\therefore$ Work done by resistive forces = – 40 J

3. w = Upthrust

or $(m_1 + m_2) g = [(m_1/0.5 \times \rho_w) + (m_2/2.5 \times \rho_w)]\rho_w g$

or $m_1 + m_2 = 2m_1 + 0.4m_2$

or $m_1 = 0.6m_2$

$\therefore \quad \frac{m_1}{m_2} = 0.6 = \frac{3}{5}$

4. Increase in pressure

$= \frac{\text{Force exerted by the cylinder on the liquid}}{A} = \frac{\text{Upthrust}}{A}$

5. Common height after they are connected can be determined by equating the volumes. Hence,

$(A_1 + A_2)h = A_1h_1 + A_2h_2$ $\quad (A_1 = A_2 = A)$

$\therefore \quad h = \frac{h_1 + h_2}{2}$

Work done by gravity = $-\Delta U = U_i - U_f$

$= \left\{Ah_1\rho.g.\frac{h_1}{2} + Ah_2\rho.g.\frac{h_2}{2}\right\} - A(h_1 + h_2)\rho.g.\frac{(h_1+h_2)}{4}$

$= \frac{\rho Ag}{4}(h_2 - h_1)^2$

6. $2a_1S_1 = 2a_2S_2$

Here, a_1 = acceleration inside the liquid

$= \frac{\text{upthrust} - \text{weight}}{\text{mass}} = \frac{V\sigma g - V\rho g}{\rho} = \left(\frac{\sigma}{\rho} - 1\right)g$

a_2 = retardation in air = g

$\therefore \quad S_2 = \frac{a_1}{a_2}.S_1 = \left(\frac{\sigma}{\rho} - 1\right)g$

7. Writing pressure equation from one end to other end of tube

$p_0 + 2\rho gh - 2\rho.\frac{l}{2}.a - \rho\frac{l}{2}.a - \rho gh = p_0$

$\therefore \quad h = \frac{3al}{2g}$

8. Work done by all the forces = change in kinetic energy.

Two forces are acting, weight and upthrust.

9. $p = p_0 + \rho_1 gh_1 + \rho_2 gh_2$

or $p - p_0 = \rho_1 gh_1 + \rho_2 gh_2$

$= (600)(10)(0.1) + (1000)(10)(0.02)$

$= 800$ N/m^2

10. Weight = upthrust

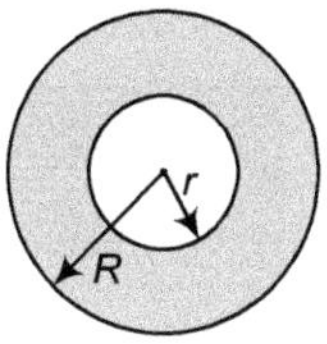

$$\left(\frac{4}{3}\pi R^3-\frac{4}{3}\pi r^3\right)(8\times\rho_w\times g)=\left(\frac{4}{3}\pi R^3\right)\rho_w g$$

or $$8R^3-8r^3=R^3$$

$\therefore$ $$8r^3=7R^3 \text{ or } \frac{r}{R}=\left(\frac{7}{8}\right)^{1/3}=\frac{\sqrt[3]{7}}{2}$$

11. In equilibrium, weight = upthrust i.e. $V\rho_i g=V_i\rho_w g$

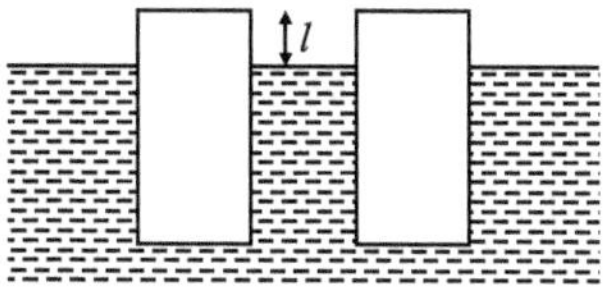

or $$\frac{V_i}{V}=\frac{\rho_i}{\rho_w}=\frac{0.9}{1.0}=0.9$$

i.e. 90% or 9 m of ice block is inside the water. Therefore, minimum length l required to scoop out a bucket of water is $10-9=1.0$ m

12. $\Delta p=\frac{1}{2}\rho v^2$ or $v=\sqrt{\frac{2(\Delta p)}{\rho}}$

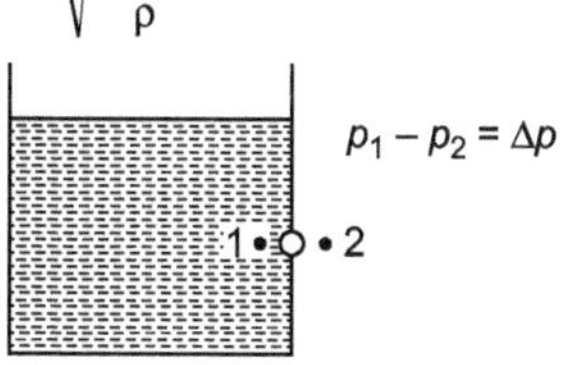

$$R=vt \text{ or } R\propto\sqrt{\Delta p}$$

To make R two times Δp should be 4 times. Initially

$$\Delta p=\rho_w gh=10^3\times10\times10=10^5\text{ N/m}^2=1\text{ atm}$$

In second case, $\rho_w gh$ + extra pressure = $4(\Delta p)$ = 4 atm

or 1 atm + extra pressure = 4 atm

$\therefore$ Extra pressure = 3 atm.

13. About hinge clockwise torque of hinge should be equal to anti-clockwise torque of upthrust. Immersed length will be $0.5\sec\theta$.

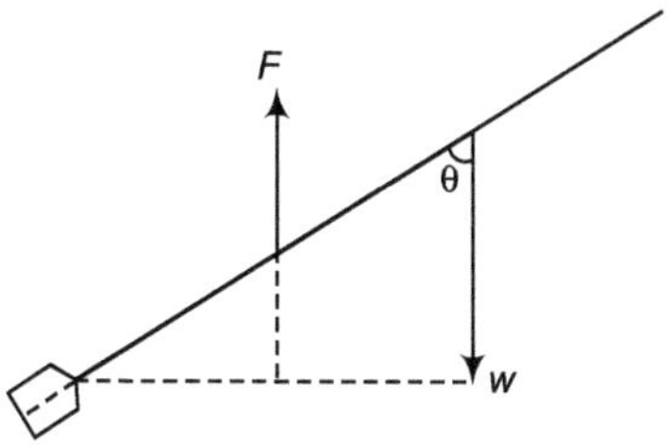

$\therefore$ $$w=\left(\frac{1}{2}\sin\theta\right)=F\left(\frac{0.5\sec\theta}{2}\right)\sin\theta$$

or $$w=0.5\,F\sec\theta$$

or $(1\times A\times0.5\times\rho_w\times g)=0.5\,(0.5\sec\theta)\times A\times1.0\times\rho_w\times g\times\sec\theta$

$$\sec^2\theta=2 \text{ or } \theta=45°$$

14. $(11.2)(1)=(13.6)(2x)$

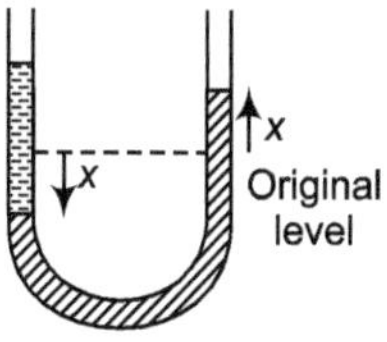

$\therefore$ $$nx=0.41\text{ cm}$$

15. $v=\sqrt{2gh}$

Retardation in liquid $a=\frac{V\sigma g-V\rho g}{V}=\left(\frac{\sigma-\rho}{\rho}\right)g$

$\therefore$ $$d_{max}=\frac{v^2}{2a}=\frac{2gh}{2=\frac{\rho h}{\pi-\rho}}=\frac{\rho h}{\sigma-\rho}$$

16. $\frac{dp}{dX}=\rho a$

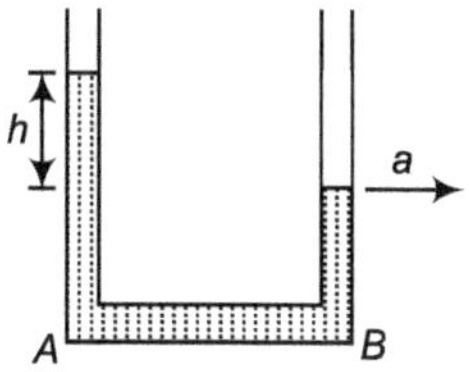

$\therefore$ $$\Delta p_{AB}=\rho aL=\rho gh$$

$\therefore$ $$h=\frac{La}{g}$$

17. $p_A=p_C=p_0$ and $p_B=p_0-\frac{2T}{R}$

18. Weight − upthrust = drag force.

But no information is given regarding the upthrust.

19. Fraction of length will remain unchanged.

$\frac{y_i}{y}$ = contact = $\frac{1}{2}$ in the given question.

$\therefore$ If total melting rate is 2 cm/h, then top of candle will fall at the rate of 1 cm/h.

20. Let us write pressure equation in path $ABCDE$.

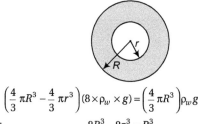

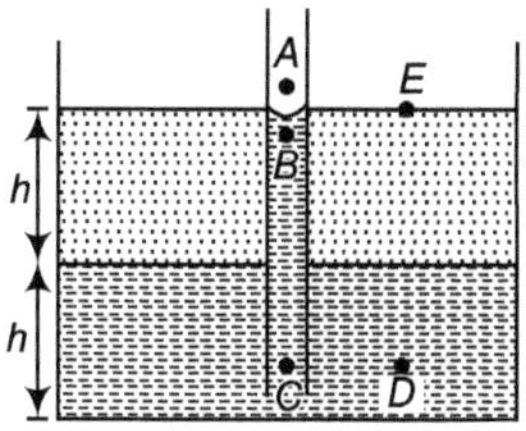

$$p_0-\frac{2I}{r}+\rho_2 g(h+h')-\rho_2 gh'-\rho_1 gh=p_0$$

$\therefore$ $$T=\frac{r}{2}(\rho_2-\rho_1)\,gh$$

21. $F=(6\pi\eta rv)=6\pi\left(\frac{1}{6\pi}\right)(1)(v)=v$

$\therefore$ Retardation $a=-\frac{F}{m}=-v$ or $\frac{dv}{dt}=-v$

$$\int_0^t dt=-\int_2^{0.5}\frac{dv}{v}$$

$$t=\ln(4)\text{ s}$$

22. $h=\frac{2T\cos\theta}{r\rho g}$ or $h\propto\frac{1}{r}$

$$M=(\pi r^2h)\rho \text{ or } m\propto r^2h$$

or $$M\propto r$$

23. Upthrust $F = w + T$

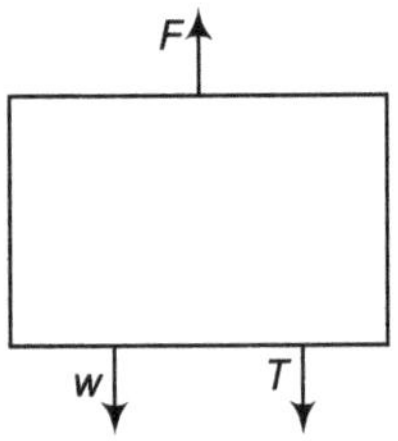

$$T = F - w = V\rho_w g - V\rho_b g = (\rho_w - \rho_b)Vg$$
$$= (1000 - 800)\left(\frac{8}{800}\right)(10) = 20\text{ N}$$

24. Upthrust = (volume of metal ball) × (density of liquid) × g with increase in temperature volume of ball will increase and density of liquid will decrease.

But coefficient of cubical expansion of liquid is more.

Hence, second effect is more dominating. Therefore, upthrust at higher temperatures will be less or apparent weight will be more.

25. $p_1V_1 = p_2V_2$

$$(75 + h)V = (75)(3V)$$

$\therefore$ $h = 150$ cm of Hg $= 1.5$ m of Hg or 15 m of water.

26. Initially p_0A supports the weight of mercury column. While in second case it provides acceleration too.

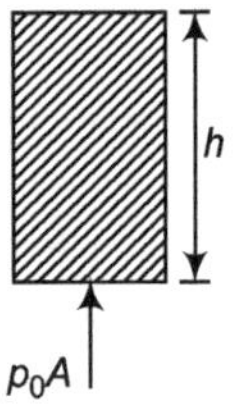

27. Let r be the radius of smaller drops and R the radius of bigger drop. Then, by equating volume we have

$$1000\left(\frac{4}{3}\pi r^3\right) = \frac{4}{3}\pi R^3 \text{ or } r = \frac{R}{10}$$

Surface energy $\propto$ surface area $S_i = 4\pi R^2$

or $\quad S_f = 1000(4\pi r^2) = 10(4\pi R^2) = 10\,S_i$

As surface area has increased 10 times, surface energy will also become ten times.

28. From conservation of mechanical energy.

29. $A_1v_1 = A_2v_2$

or $\quad A_1 = \sqrt{2gh_2} = A_2\sqrt{2gh}$

or $\quad A_1^2h_1 = A_2^2h_2$

$$(L^2)^2(y) = (\pi R^2)^2(4y)$$

or $\quad R = \frac{L}{\sqrt{2\pi}}$

30. $\Delta p = \frac{F}{A}$. Difference in pressure energy is equal to difference in kinetic energy.

$\therefore$ $\quad \frac{F}{A} = \frac{1}{2}\rho v^2$ or $v = \sqrt{\frac{2F}{\rho A}}$

31. We can assume total height of liquid to be equal to $3H$. Maximum range is obtained when hole is drilled at centre.

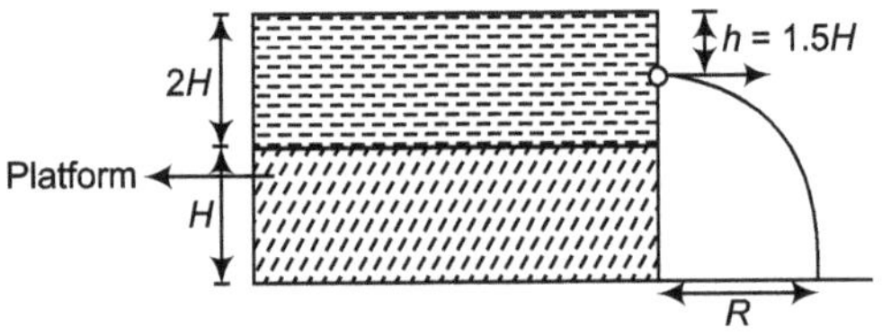

32. Relative density of steel

$$= \frac{\text{weight in air}}{\text{change in weight in water}} = \frac{w}{w - w_1}$$

Now, change in weight in the given liquid
= upthrust in this liquid

or $\quad w - w_2 = \left\{\frac{w/g}{(w/w - w_1)\rho_w}\right\}d_2 \times \rho_w \times g$

$\therefore$ $\quad d_2 = \frac{w - w_2}{w - w_1}$ = relative density of given liquid.

33. Weight of combined cylinder = upthrust

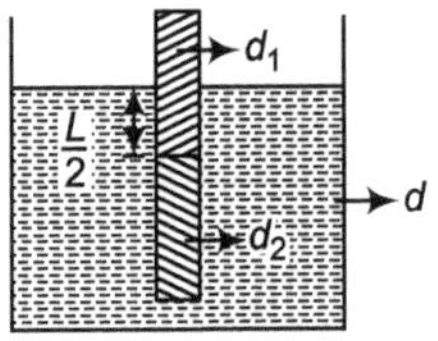

$\therefore$ $\quad d_1LAg + d_2LAg = d\left(\frac{3L}{2}\right)Ag$

or $\quad d_1 + d_2 = \frac{3}{2}d$

But $\quad d_1 < d_2$

$$d_1 < \frac{3}{4}d_2$$

More than One Correct Options

1. Velocity gradient $= \frac{\Delta v}{\Delta h} = \frac{2\text{m/s}}{1\text{ m}} = 2\text{s}^{-1}$

$$F = \eta A\frac{\Delta v}{\Delta h} = (10^{-3})(10)(2) = 0.02\text{ N}$$

2. For contact angle $\theta = 90°$, liquid neither rises nor falls.

3. Restoring force $= -(\rho Ag)x$ or $F \propto -x$

This is just like a spring-block system of force constant

$$K = \rho Ag$$

4. From continuity equation, Av = constant

$$v_2 > v_2 \quad \text{as} \quad A_2 < A_1$$

From Bernoulli's equation,

$$p + \frac{1}{2}\rho v^2 = \text{constant} \qquad (\text{as } h = \text{constant})$$

$$p_2 < p_1 \quad \text{as} \quad v_2 > v_2$$

5. Fraction of volume immersed,

$$f = \frac{\rho_s}{\rho_l}$$

This fraction is independent of atmospheric pressure. With increase in temperature ρ_s and ρ_l both will decrease.

6. $v = \sqrt{2gh_T} : t = \sqrt{\frac{2h_B}{g}}, R = 2\sqrt{h_T h_B}$

Here, h_T = distance of hole from top surface of liquid and h_B = distance of hole from bottom surface.

7. Pressure increase with depth in vertical direction and in horizontal direction it increases in opposite direction of acceleration based on this concept pressure is maximum at point D and minimum at B.

8. In air, $a_1 = g$ (downwards)

In liquid, $a_2 = \frac{\text{Upthrust} - \text{Weight}}{\text{Mass}} = \frac{(V)(2\rho)(g) - (V)(\rho)(g)}{(V\rho g)}$

$= g$ (upwards)

$\therefore \quad a_1 \neq a_2$

9. Initially

$$\frac{dV_1}{dt} = v_1 a_1 = (\sqrt{2gh})(\pi)(2R)^2 \quad \text{...(i)}$$

$$\frac{dV_2}{dt} = v_2 a_2 = (\sqrt{2g(16h)})(\pi)(2R)^2 \quad \text{...(ii)}$$

From Eqs. (i) and (ii), we can see that $\frac{dV_1}{dt} = \frac{dV_2}{dt}$

After some time v_1 and v_2 both will decrease, but decrease in the value of v_1 is more dominating. So,

$$a_1 v_1 \text{ or } \frac{dV_1}{dt} < a_2 v_2 \text{ or } \frac{dV_2}{dt}$$

10. Fraction of volume immersed is given by

$$f = \frac{\rho_s}{\rho_l}$$

ρ_s and ρ_e are same. Hence,

$$f_1 = f_2 = f_3$$

11. Base area in third case is uniform. Hence, h_3 is minimum.

Streamline flow is more likely for liquids having low density. We know that greater the coefficient of viscosity of a liquid more will be velocity gradient hence each line of flow can be easily differentiated. Also higher the coefficient of viscosity lower will be Reynolds number, hence flow more like to be streamline.

Comprehension Based Questions

1. Not required.

2. $x_3 = 2\sqrt{h_{\text{top}} \times h_{\text{bottom}}} = 2\sqrt{3a \times a} = 2\sqrt{3}a$

3. $x_1 = 2\sqrt{a \times 3a} = 2\sqrt{3}a$

$$x_2 = 2\sqrt{2a \times 2a} = 2\sqrt{4}a$$

$$x_3 = 2\sqrt{3}a$$

$\therefore \quad x_1 = x_3 < x_2$

4. Weight = upthrust

$$\therefore \quad \left(L\frac{A}{5}Dg\right) = \frac{L}{4} \times \frac{A}{5} \times 2d \times g + \frac{3L}{4} \times \frac{A}{5} \times d \times g$$

$$\therefore \quad D = \frac{5d}{4}$$

5. $pA = p_0 A$ + weight of two liquids + weight of cylinder

$$= p_0 A + \left(\frac{H}{2}\right)(A)(d)(g) + \left(\frac{H}{2}\right)(A)(2d)(g) + L\left(\frac{A}{5}\right)\left(\frac{5d}{4}\right)g$$

$$\therefore \quad p = p_0 + \frac{(L + 6H)}{4} dg$$

6. Applying Bernoulli's equation just inside and just outside the hole,

$$p_0 + \left(\frac{H}{2}\right)(d)g + \left(\frac{H}{2} - H\right)(2d)(g) = \frac{1}{2}(2d)v^2 + p_0$$

$$\therefore \quad v = \sqrt{\frac{g}{2}(3H - 4h)}$$

7. $t = \sqrt{\frac{2h}{g}}$

$$\therefore \quad x = vt = \sqrt{(3H - 4H)h}$$

Assertion and Reason

1. Reason is correct due to continuity equation.
2. Upthrust will remain same.
4. Fluid at 1 is moving while fluid at 2 is stationary. Therefore, according to Bernoulli's theorem, $p_1 < p_2$. Here $p_2 = p_0$ Hence, $p_1 < p_0$.
10. Density of water at 4°C is maximum.
13. At same level of same liquid pressures may be different, if at one point speed is zero and at other point speed is non-zero.
14. Fraction of volume immersed,

$$f = \frac{\rho_s}{\rho_l}$$

If ρ_l is decreased, then f will increase.

15. Capillary rises when $\theta = 0°$

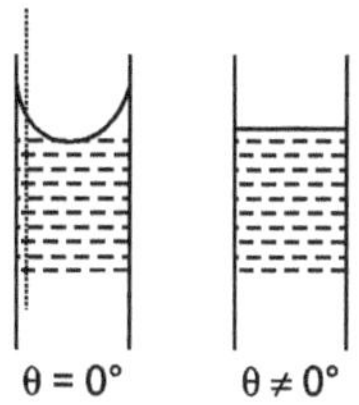

18. With increase in depth pressure increases. Therefore, volume of given mass of liquid decreases. Hence, density increases.
19. In floating condition, $mg = V_1 \rho_1 g$

(V_1 = volume immersed or volume displaced)

$$\therefore \quad V_1 = \frac{m}{\rho_1}$$

When the solid melts, volume of liquid formed,

$$V_2 = \frac{m}{\rho_2}$$

Since $\quad \rho_1 < \rho_2$

$\therefore \quad V_2 > V_1$

20. Equate net downward force and net upward force on block + liquid system.

Match the Columns

1. Pressure at O is the atmospheric pressure which will remain same.

2. $v = \sqrt{2gh}, t = \sqrt{\frac{2h}{g}}$ and $R = 2\sqrt{h(H - h)}$

At mountain, value of g will be less. Hence, v will decrease, t t will increase and R will remain unchanged.

3. From, $A_1v_1 = A_2v_2$ or $v \propto \frac{1}{A}$

At a point where area of cross-section is less, volume of liquid flowing per second is same but speed is more. Therefore, here kinetic energy is more or the pressure will be less.

4. At depth h $\quad p = p_0 + \rho(g + a)h$

$$\Delta p = \rho(g + a)\,\Delta h$$

Upthrust will also increase, because upthrust

$$F = V_i\,(g + a)\rho$$

Here, V_1 is the immersed volume.

5. When two soap bubbles coalesce to from a larger drop, radius increases. Hence, the excess pressure $\frac{4T}{R}$ will decrease. Further, surface area will decrease. Hence, the surface energy will decrease or the temperature will increase.

6. In equilibrium upthrust = weight

If density of liquid is decreasing, more volume should be immersed in the liquid, so that upthrust remains unchanged and it balances the weight.

Entrance Gallery

1. Equating pressure at A, we get

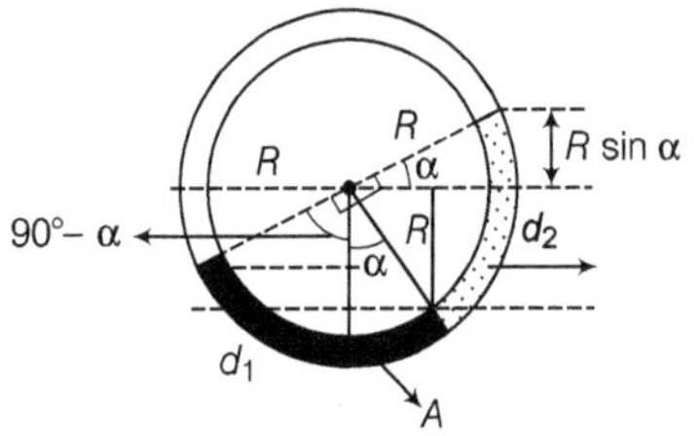

$$R\sin\alpha\, d_2 + R\cos\alpha\, d_2 + R(1 - \cos\alpha)\, d_1 = R\,(1 - \sin\alpha)\, d_1$$

$$(\sin\alpha + \cos\alpha)\, d_2 = d_1(\cos\alpha - \sin\alpha) \;\Rightarrow\; \frac{d_1}{d_2} = \frac{1 + \tan\alpha}{1 - \tan\alpha}$$

2. The bubble will detach, if

buoyant force ≥ surface tension force

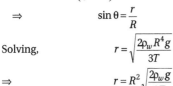

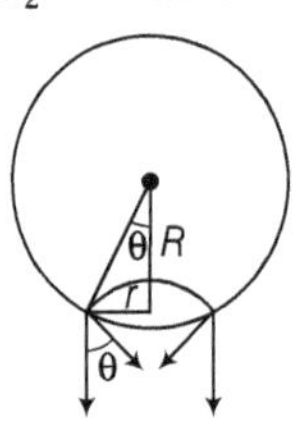

$$\frac{4}{3}\pi R^3 \rho_w g \ge \int T \times dl \sin\theta$$

$$(\rho_w)\left(\frac{4}{3}\pi R^3\right)g \ge (T)(2\pi r)\sin\theta$$

$\Rightarrow$ $$\sin\theta = \frac{r}{R}$$

Solving, $$r = \sqrt{\frac{2\rho_w R^4 g}{3T}}$$

$\Rightarrow$ $$r = R^2\sqrt{\frac{2\rho_w g}{3T}}$$

3. Reynold's number is a pure number and it is equal to the ratio of the inertial force per unit area to the viscous force per unit area for a flowing fluid.

Reynold's number, $R_e = \frac{v_c \cdot \rho \cdot r}{\eta}$

where, ρ = density of the liquid

v_c = critical velocity

η = coefficient of viscosity of liquid

r = radius of capillary tube.

Fact

I. For pure water flowing in a cylindrical pipe, Re is about 1000. When $0 < R_e < 2000$, the flow of liquid is streamlined.

II. When $2000 < R_e < 3000$, the flow of liquid is variable between streamlined and turbulent.

III. When $K > 3000$, the flow of liquid is turbulent.

4. Let, thickness of layer be x.

So, $\quad$ volume V = area $\times\, x$

$$V = A \times x \;\Rightarrow\; x = V/A \qquad (\because x = 2r)$$

$\therefore$ $$2r = \frac{V}{A} \;\Rightarrow\; r = \frac{V}{2A} \qquad \ldots\text{(i)}$$

and $$\Delta p = \frac{T}{r}$$

We know that, $F = \Delta p \times A = \frac{T}{r} \times A$

$$F = \frac{T}{\left(\frac{V}{2A}\right)} \times A \qquad \text{[from Eq. (i)]}$$

$$T = \frac{F \times V}{2A^2}$$

where, $F = 16 \times 10^5$ dyne, $V = 0.04\ \text{cm}^3$

$$A = 20\ \text{cm}^2, T = \frac{16 \times 10^5 \times 0.04}{2 \times 20^2} = \frac{8 \times 10^5 \times 4}{20^2 \times 100} = \frac{8 \times 10^5 \times 4}{400 \times 100}$$

$$= 8 \times 10^5 \times 10^{-4}$$

$$= 80\ \text{dyne/cm}$$

$$= 80\ \text{dyne cm}^{-1}$$

5. According to Torricelli's law,

For an open container, the velocity of efflux is given by,

$$v = \sqrt{2gh}$$

6. Decrease in surface energy = Heat required in vaporisation

$\therefore$ $$T\,(dS) = L(dm)$$

$\therefore$ $$T(2)\,(4\pi r)dr = L(4\pi r^2 dr)\rho$$

$$x = V/A$$

$$r = \frac{2T}{\rho L}$$

7. The formula for viscous force is

$$F = 6\pi\eta r v$$

Given, viscosity

$$\eta = 18 \times 10^{-5}\ \text{poise}$$

$$= 18 \times 10^{-6}\ \text{kg m}^{-1}\text{s}^{-1}$$

Radius, $\quad r = 0.3\ \text{mm} = 0.3 \times 10^{-3}\ \text{m}$

Velocity, $\quad v = 1\ \text{ms}^{-1}$

So, $F = 6 \times 3.14 \times 18 \times 10^{-6} \times 0.3 \times 10^{-3} \times 1 = 101.73 \times 10^{-9}\ \text{N}$

8. As volume remain constant, therefore $R = n^{1/3}r$.

$$\frac{\text{Energy of one small drop}}{\text{Energy of one big drop}} = \frac{E}{E_1} = \frac{(4\pi r^2)T}{(4\pi R^2)T}$$

$\Rightarrow$ $$\frac{E}{E_1} = \frac{r^2}{R^2} = \frac{r^2}{(n^{1/3r})^2} = \frac{1}{n^{2/3}}$$

Given, $\quad n = 1000$

So, $$\frac{E}{E_1} = \frac{1}{(1000)^{2/3}} = \frac{1}{100}$$

9. $F =$ upthrust $= Vd_F g$

Equilibrium of A, $Vd_F g = T + w_A = T + Vd_A g$ ...(i)

Equilibrium of B, $Vd_B g = T + Vd_F g$ ...(ii)

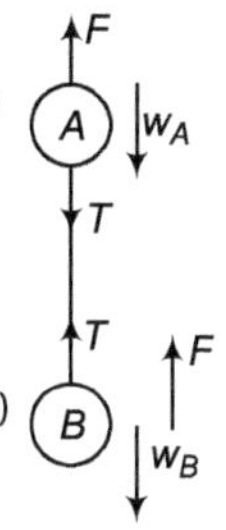

Adding Eqs. (i) and (ii), we get

$$2d_F = d_A + d_B$$

$\therefore$ Option (d) is correct

From Eq. (i), we can see that,

$$d_F > d_A \qquad (\text{as}, T > 0)$$

$\therefore$ Option (a) is correct.

From Eq. (ii), we can see that $d_B > d_F$

$\therefore$ Option (b) is correct.

$\therefore$ Correct options are (a), (b) and (d).

10. We know that, $v \propto \dfrac{\rho - \rho_0}{\eta} \Rightarrow \dfrac{v_2}{v_1} = \dfrac{\rho - \rho_{02}}{\rho - \rho_{01}} \times \dfrac{\eta_1}{\eta_2}$

or $v_2 = \dfrac{7.8 - 1.2}{7.8 - 1} \times \dfrac{8.5 \times 10^{-4} \times 10}{13.2} = 6.25 \times 10^{-4} \text{cm/s}$

11. From Bernoulli's theorem,

$$\rho g h = \frac{1}{2}\rho(v_2^2 - v_1^2) \Rightarrow gh = \frac{1}{2}v_1^2\left[\left(\frac{v_2}{v_1}\right)^2 - 1\right]$$

$$\Rightarrow \quad gh = \frac{1}{2}v_1^2\left[\left(\frac{A_1}{A_2}\right)^2 - 1\right] \qquad (\because A_1 v_1 = A_2 v_2)$$

$$\Rightarrow \left(\frac{A_1}{A_2}\right)^2 = 1 + \frac{2hg}{v_1^2} \Rightarrow \left(\frac{D_1}{D_2}\right)^4 = 1 + \frac{2gh}{v_1^2}$$

$$\Rightarrow \quad D_2 = \frac{D_1}{\left(1 + \dfrac{2gh}{v_1^2}\right)^{1/4}} = \frac{8 \times 10^{-3}}{\left[1 + \dfrac{2 \times 10 \times 0.2}{(0.4)^2}\right]^{1/4}} = 3.6 \times 10^{-3} \text{ m}$$

12. Let R be the radius of the bigger drop, then volume of bigger drop $= 2 \times$ volume of small drop

$$\frac{4}{3}\pi R^3 = 2 \times \frac{4}{3}\pi r^3 \Rightarrow R = 2^{1/3} r$$

Surface energy of bigger drop,

$$E = 4\pi R^2 T \Rightarrow E = 4 \times 2^{2/3}\pi r^2 T = 2^{8/3}\pi r^2 T$$

13. Velocity, $v = \dfrac{2}{9}\dfrac{(\rho - \sigma) r^2 g}{\eta}$

$$\therefore \quad v \propto (\rho - \sigma) \Rightarrow \frac{v_1}{v_2} = \frac{(\rho_1 - \sigma)}{(\rho_2 - \sigma)}$$

$$\Rightarrow \quad \frac{0.2}{v_2} = \frac{19.5 - 1.5}{10.5 - 1.5} \text{ or } v_2 = 0.1 \text{ m/s}$$

14. We have, $v^2 = \rho g h \Rightarrow a^2\sqrt{\rho g h} = \pi r^2 \sqrt{\rho g h} \times 2 \Rightarrow r = \dfrac{a}{\sqrt{2\pi}}$

15. If we have m gram of ice, which is floating in a liquid of density 1.2 and 9 L will displace volume $\dfrac{m_{cc}}{1.2} < m_{cc}$. After melting it occupies m_{cc}.

16. We know that,

$$v \propto r^2 \text{ and } 8\left(\frac{4}{3}\pi r_1^3\right) = \frac{4}{3}\pi r_2^3$$

$$\therefore \quad \frac{v_1}{v_2} = \frac{r_1^2}{r_2^2} \Rightarrow \frac{10}{v_2} = \frac{r^2}{8^{2/3} r^2} = \frac{1}{4} \Rightarrow v_2 = 40 \text{ cm/s}$$

17. $V(d - d_1)g = m_1 g$

and $\quad V(d - d_2)g = m_2 g \Rightarrow \dfrac{d - d_1}{d - d_2} = \dfrac{m_1}{m_2}$

$$\Rightarrow \quad d = \frac{m_1 d_2 - m_2 d_1}{m_1 - m_2}$$

18. $V\sigma g = 0.6 V\sigma_1 g$ and $V\sigma g = 0.4 V\sigma_2 g$

$$\therefore \quad 1 = \frac{6}{4}\frac{\sigma_1}{\sigma_2} \Rightarrow \frac{\sigma_2}{\sigma_1} = \frac{3}{2} = 1.5$$

19. The surface tension of liquid decreases with rise of temperature. The surface tension of liquid is zero at its boiling point and it vanishes at critical temperature. At critical temperature intermolecular forces for liquid and gases becomes equal and liquid can expand without any restriction. For small temperature differences, the variation in surface tension with temperature is linear and is given by relation

$$T_t = T_0(1 - \alpha t)$$

where, T_t and T_0 are the surface tensions at t°C and 0°C, respectively and α is the temperature coefficient of surface tension.

20. On the surface of the liquid in equilibrium molecules of the liquid possess maximum potential energy.

21. $\rho \; < \rho < \rho$. Oil is the least denser of them, so it should settle at the top with water at the base. Now, the ball is denser than oil but less denser than water. So, it will sink through oil but will not sink in water.

So, it will stay at the oil water interface.

22. Force exerted by the liquid on the base of the vessel is $F = mg$.

Here, $\quad m_A = m_B = m_C$

$\therefore \quad F_A = F_B = F_C$

14

Thermometry, Thermal Expansion and **Kinetic Theory** of **Gases**

14.1 Thermometers and the Celsius Temperature Scale

Chapter Snapshot

- Thermometers and the Celsius Temperature Scale
- The Constant Volume Gas Thermometer and the Absolute Temperature Scale
- Quantity of Heat
- Thermal Expansion
- Concept of an Ideal Gas
- Gas Laws
- Ideal Gas Equation
- Degree of Freedom (*f*)
- Internal Energy of an Ideal Gas
- Law of Equipartition of Energy
- Molar Heat Capacity
- Kinetic Theory of Gases

Thermometers are those devices which are used to measure temperatures. All thermometers are based on the principle that some physical properties of a system change as the system's temperature changes. Some physical properties that change with temperature are:

1. the volume of a liquid,
2. the length of a solid,
3. the pressure of a gas at constant volume,
4. the volume of a gas at constant pressure and
5. the electric resistance of a conductor.

A common thermometer in everyday use consists of a mass of liquid, usually mercury or alcohol that expands in a glass capillary tube when heated. In this case, the physical property is the **change in volume of the liquid**. Any temperature change is proportional to the change in length of the liquid column.

The thermometer can be calibrated accordingly. On the celsius temperature scale, a thermometer is usually calibrated between 0°C (called the ice point of water) and 100°C (called the steam point of water).

Once the liquid levels in the thermometer have been established at these two points, the distance between the two points is divided into 100 equal segments to create the celsius scale. Thus, each segment denotes a change in temperature of one celsius degree (1°C). A practical problem in this type of thermometer is that readings may vary for two different liquids. When one thermometer reads a temperature, for example 40°C the other may indicate a slightly different value.

These discrepancies between thermometers are especially large at temperatures far from the calibration points. To surmount this problem, we need a universal thermometer whose readings are independent of the substance used in it. The gas thermometer used in the next article meets this requirement.

14.2 The Constant Volume Gas Thermometer and The Absolute Temperature Scale

The physical property used by the constant volume gas thermometer is the **change in pressure of a gas at constant volume.**

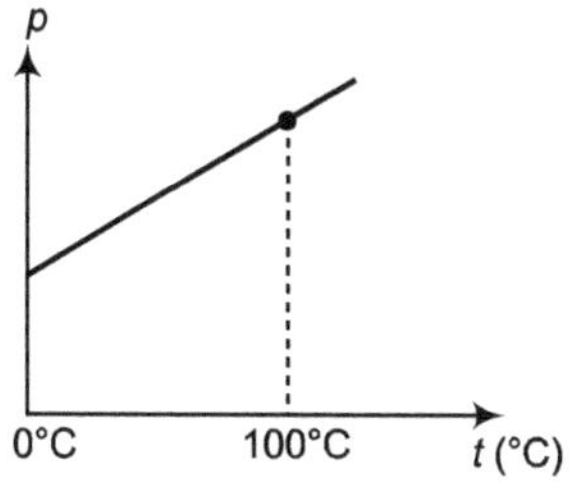

Fig. 14.1

The pressure *versus* temperature graph for a typical gas taken with a constant volume is shown in above figure. The two dots represent the two reference temperatures namely, the ice and steam points of water. The line connecting them serves as a calibration curve for unknown temperatures. Experiment shows that the thermometer readings are nearly independent of the type of gas used as long as the gas pressure is low and the temperature is well above the point at which the gas liquefied.

If you extend the curves shown in figure toward negative temperatures, you find, in every case, that the pressure is zero when the temperature is –273.15°C. This significant temperature is used as the basis for the **absolute temperature scale,** which sets –273.15°C as its zero point.

This temperature is often referred to as **absolute zero.** The size of a degree on the absolute temperature scale is identical to the size of a degree on the celsius scale. Thus, the conversion between these temperatures is

$$T_C = T - 273.15 \quad \ldots\text{(i)}$$

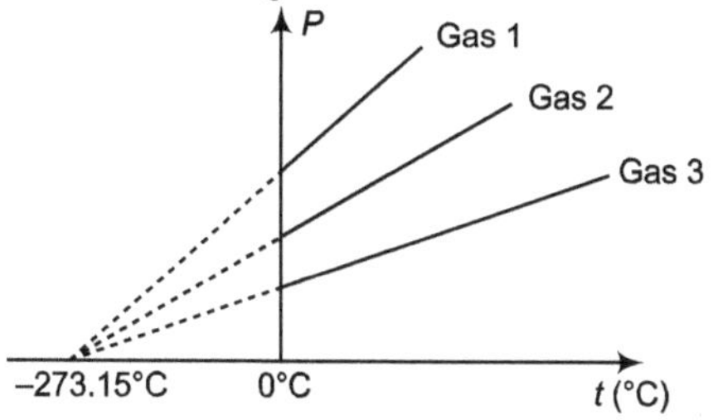

Fig. 14.2

In 1954, by the International committee on weights and measures, the **triple point of water** was chosen as the reference temperature for this new scale. The triple point of water is the single combination of temperature and pressure at which liquid water, gaseous water and ice (solid water) coexist in equilibrium.

This **triple point** occurs at a temperature of approximately 0.01°C and a pressure of 4.58 mm of mercury. On the new scale, which uses the unit **kelvin,** the temperature of water at the triple point was set at 273.16 K, abbreviated as 273.16 K. (No degree sign is used with the unit kelvin).

This new absolute temperature scale (also called the kelvin scale) employs the SI unit of absolute temperature, the kelvin which is defined to be, $\frac{1}{273.16}$ of the difference between absolute zero and the temperature of the triple point of water.

The Celsius, Fahrenheit and Kelvin Temperature Scales

Eq. (i) shows the relation between the temperatures in celsius scale and kelvin scale. Because the size of a degree is the same on the two scales, a temperature difference of 10°C is equal to a temperature difference of 10 K. The two scales differ only in the choice of the zero point. The ice point temperature on the kelvin scale, 273.15 K, corresponds to 0.00°C and the kelvin steam point 373.15 K, is equivalent to 100.00°C.

A common temperature scale in everyday use in US is the **fahrenheit scale**. The ice point in this scale is 32°F and the steam point is 212°F. The distance between these two points are divided into 180 equal parts. The relation between celsius scale and fahrenheit scale is as derived below:

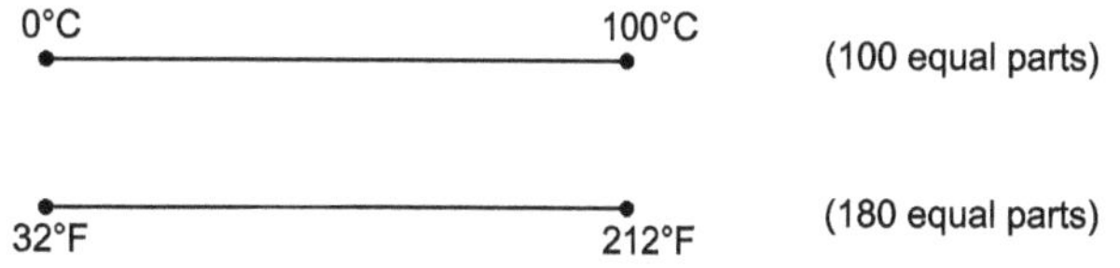

Fig. 14.3

$$100 \text{ parts of celsius scale} = 180 \text{ parts of fahrenheit scale}$$

$$\therefore \; 1 \text{ part of celsius scale} = \frac{9}{5} \text{ parts of fahrenheit scale}$$

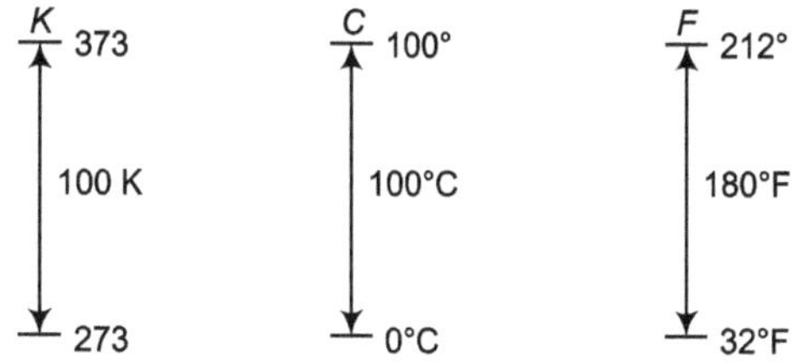

Relation among Kelvin, Celsius and Fahrenheit temperature scales

Fig. 14.4

Hence, $$T_F = 32 + \frac{9}{5} T_C \quad \ldots\text{(ii)}$$

Further, $$\Delta T_C = \Delta T = \frac{5}{9} \Delta T_F \quad \ldots\text{(iii)}$$

Example 14.1 *Express a temperature of 60°F in degree celsius and in kelvin.*

Sol. Substituting, T_F = 60°F in Eq. (ii)

$$T_C = \frac{5}{9}(T_F - 32°)$$
$$= \frac{5}{9}(60° - 32°)$$
$$= 15.55°C$$

From Eq. (i), we get

$$T = T_C + 273.15$$
$$= 15.55°C + 273.15$$
$$= 288.7 \text{ K}$$

Example 14.2 *The temperature of an iron piece is heated from 30°C to 90°C. What is the change in its temperature on the fahrenheit scale and on the kelvin scale?*

Sol. ΔT_C = 90°C − 30°C = 60°C

Using Eq. (iii),

$$\Delta T_F = \frac{9}{5}\Delta T_C = \frac{9}{5}(60°C) = 108°F$$

and $$\Delta T = \Delta T_C = 60 \text{ K}$$

Extra Knowledge Points

Different Thermometers

- **Thermometric property** It is the property that can be used to measure the temperature. It is represented by any physical quantity such as length, volume, pressure and resistance, etc. Which varies linearly with a certain range of temperature. Let X denotes the thermometric physical quantity and X_0, X_{100} and X_t be its values at 0°C, 100°C and t°C respectively. Then,

$$t = \left(\frac{X_t - X_0}{X_{100} - X_0}\right) \times 100°C$$

(i) **Constant volume gas thermometer** The pressure of a gas at constant volume is the thermometric property. Therefore,

$$t = \left(\frac{p_t - p_0}{p_{100} - p_0}\right) \times 100°C$$

(ii) **Platinum resistance thermometer** The resistance of a platinum wire is the thermometric property. Hence,

$$t = \left(\frac{R_t - R_0}{R_{100} - R_0}\right) \times 100°C$$

(iii) **Mercury thermometer** In this thermometer, the length of a mercury column from some fixed point is taken as thermometric property. Thus,

$$t = \left(\frac{l_t - l_0}{l_{100} - l_0}\right) \times 100°C$$

- Two other thermometers, commonly used are thermocouple thermometer and **total radiation pyrometer**.
- **Total radiation pyrometer** is used to measure very high temperatures. When a body is at a high temperature, it glows brightly and the radiation emitted per second from unit area of the surface of the body is proportional to the fourth power of the absolute temperature of the body. If this radiation is measured by some device, the temperature of the body is calculated. This is the principle of a total radiation pyrometer. The main advantage of this thermometer is that the experimental body is not kept in contact with it. Hence, there is no definite higher limit of its temperature range. It can measure temperature from 800°C to 3000°C – 4000°C. However, it cannot be used to measure temperatures below 800°C because at low temperatures the emission of radiation is so poor that it cannot be measured directly.
- Ranges of different thermometers.

Thermometer	Lower limit	Upper limit
Mercury thermometer	− 30°C	300°C
Gas thermometer	− 268°C	1500°C
Platinum resistance thermometer	− 200°C	1200°C
Thermocouple thermometer	− 200°C	1600°C
Radiation thermometer	800°C	No limit

- **Reaumer scale** Other than celsius, fahrenheit and kelvin temperature scales reaumer scale was designed by Reaumer in 1730. The lower fixed point is 0°R representing melting point of ice. The upper fixed point is 80°R, which represents boiling point of water. The distance between the two fixed points is divided into 80 equal parts. Each part represents 1°R. If T_C, T_F and T_R are temperature values of a body on celsius scale, fahrenheit scale and reaumer scale respectively, then,

$$\frac{T_C - 0}{100} = \frac{T_F - 32}{180} = \frac{T_R - 0}{80}$$

- A substance is found to exist in three states solid, liquid and gas. For each substance there is a set of temperature and pressure at which all the three states may coexist. This is called **triple point** of that substance. For water, the values of pressure and temperature corresponding to triple point are 4.58 mm of Hg and 273.16 K.

14.3 Quantity of Heat

When a cold body is brought in contact with a hot body, the cold body warms up and the hot body cools down as they approach thermal equilibrium. Fundamentally, a transfer of energy takes place from one substance to the other. This type of energy transfer that takes place slowly because of a temperature difference is called **heat flow** or **heat transfer** and energy transfer in this way is called **heat**.

Water can be warmed up by vigorous stirring with a pedal wheel. The pedal wheel adds energy to the water by doing work on it. The same temperature change can also be caused by putting the water in contact with some hotter body.

Hence, this interaction must also involve an energy exchange. Before exploring the relation between heat and mechanical energy let us define a unit of **quantity of heat.**

One **calorie** (1 cal) is defined as the amount of heat required to raise the temperature of 1 g of water from 14.5°C to 15.5°C.

Experiments have shown that.

$$1 \text{ cal} = 4.186 \text{ J}$$

Similarly, $1 \text{ kcal} = 1000 \text{ cal} = 4186 \text{ J}$

The calorie is not a fundamental SI unit.

14.4 Thermal Expansion

Most substances expand when they are heated. Thermal expansion is a consequence of the change in average separation between the constituent atoms of an object. Atoms of an object can be imagined to be connected to one another by stiff springs as shown in figure.

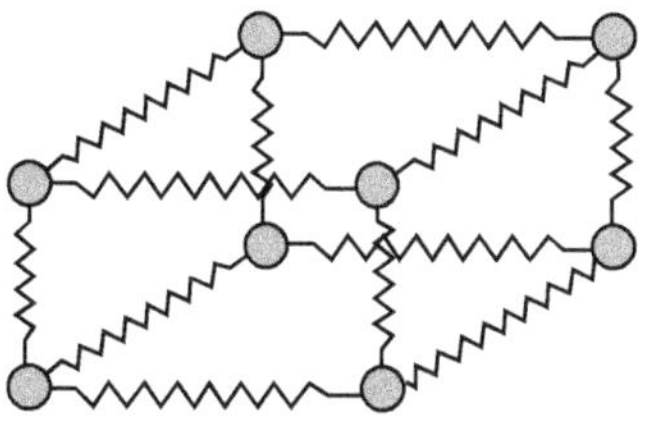

Fig. 14.5

At ordinary temperatures, the atoms in a solid oscillate about their equilibrium positions with an amplitude of approximately 10^{-11} m. The average spacing between the atoms is about 10^{-10} m. As the temperature of solid increases, the atoms oscillate with greater amplitudes as a result the average separation between them increases, consequently the object expands.

Linear Expansion

Suppose that the temperature of a thin rod of length l is changed from T to $T+\Delta T$. It is found experimentally that, if ΔT is not too large, the corresponding change in length Δl of the rod is directly proportional to ΔT and l. Thus,

$$\Delta l \propto \Delta T \quad \text{and} \quad \Delta l \propto l$$

Introducing a proportionality constant α (which is different for different materials) we may write Δl as

$$\Delta l = l\alpha\Delta T \qquad \text{...(i)}$$

Here, the constant α is called the **coefficient of linear expansion** of the material of the rod and its units are K^{-1} or $[(°C)^{-1}]$. Remember that $\Delta T = \Delta T_C$.

Actually, α does depend slightly on the temperature, but its variation is usually small enough to be negligible, even over a temperature range of 100°C. We will always assume that α is a constant.

Volume Expansion

Because the linear dimensions of an object change with temperature, it follows that surface area and volume change as well. Just as with linear expansion, experiments show that, if the temperature change ΔT is not too great (less than 100°C or so), the increase in volume ΔV is proportional to both the temperature change ΔT and the initial volume V.

Thus, $\Delta V \propto \Delta T$ and $\Delta V \propto V$

Introducing a proportionality constant γ, write ΔV as

$$\Delta V = V \times \gamma \times \Delta T \qquad \text{... (ii)}$$

Here, γ is called the **coefficient of volume expansion.** The units of γ are K^{-1} or $(°C)^{-1}$.

Relation between γ and α

For an isotropic solid (which has the same value of α in all directions) $\gamma = 3\alpha$. To see that $\gamma = 3\alpha$ for a solid, consider a cube of length l and volume $V = l^3$.

When the temperature of the cube is increased by dT, the side length increases by dl and the volume increases by an amount dV is given by

$$dV = \left(\frac{dV}{dl}\right) \cdot dl = 3l^2 \cdot dl$$

Now, $dl = l\alpha dT$

$\therefore$ $dV = 3l^3\alpha . dT$

$= (3\alpha) VdT$

This is consistent with Eq. (ii), $dV = \gamma VdT$, only if

$$\gamma = 3\alpha \qquad \text{...(iii)}$$

Average values of α and γ for some materials are listed in Table 14.1. You can check the relation $\gamma = 3\alpha$ for the materials given in the table.

Table 14.1

Material	α [K^{-1} or $(°C)^{-1}$]	γ [K^{-1} or $(°C)^{-1}$]
Steel	1.2×10^{-5}	3.6×10^{-5}
Copper	1.7×10^{-5}	5.1×10^{-5}
Brass	2.0×10^{-5}	6.0×10^{-5}
Aluminium	2.4×10^{-5}	7.2×10^{-5}

The Anomalous Expansion of Water

Most liquids also expand when their temperatures increase. Their expansion can also be described by Eq. (ii). The volume expansion coefficients for liquids are about 100 times larger than those for solids.

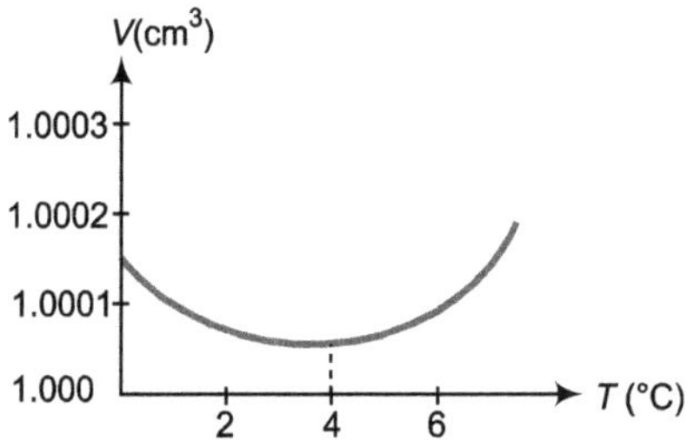

Fig. 14.6

Some substances contract when heated over a certain temperature range. The most common example is water.

Figure shows how the volume of 1 g of water varies with temperature at atmospheric pressure. The volume decreases as the temperature is raised from 0°C to about 4°C, at which point the volume is a minimum and the density is a maximum $(1000\ \text{kg/m}^3)$. Above 4°C, water expands with increasing temperature like most substances.

This anomalous behaviour of water causes ice to form first at the surface of a lake in cold weather. As winter approaches, the water temperature increases initially at the surface. The water there sinks because of its increased density. Consequently, the surface reaches 0°C first and the lake becomes covered with ice. Aquatic life is able to survive the cold winter as the lake bottom remains unfrozen at a temperature of about 4°C.

Extra Knowledge Points

- If a solid object has a hole in it, what happens to the size of the hole, when the temperature of the object increases ? A common misconception is that, if the object expands, the hole will shrink because material expands into the hole. But the truth is that, if the object expands, the hole will expand too, because every linear dimension of an object changes in the same way when the temperature changes.

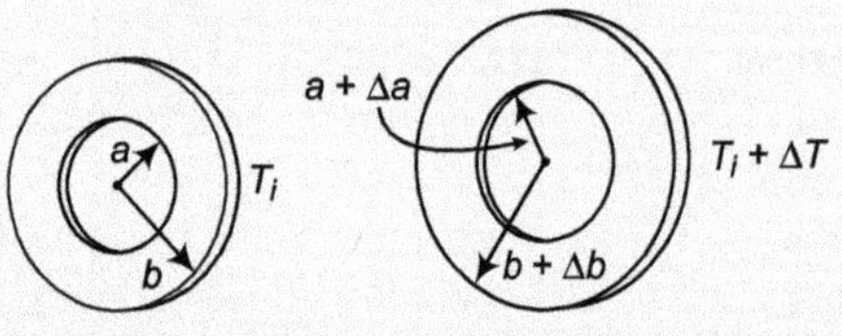

- **Expansion of a bimetallic strip**

As, Table 14.1 indicates, each substance has its own characteristic average coefficient of expansion. For example, when the temperatures of a brass rod and a steel rod of

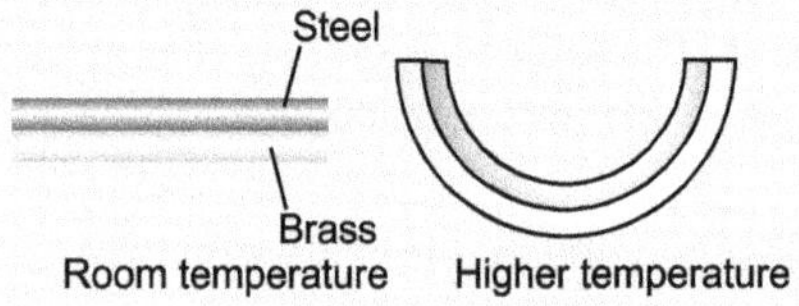

equal length are raised by the same amount from some common initial value, the brass rod expands more than the steel rod because brass has a greater average coefficient of expansion than steel. Such type of bimetallic strip is found in practical devices such as thermostats to break or make electrical contact.

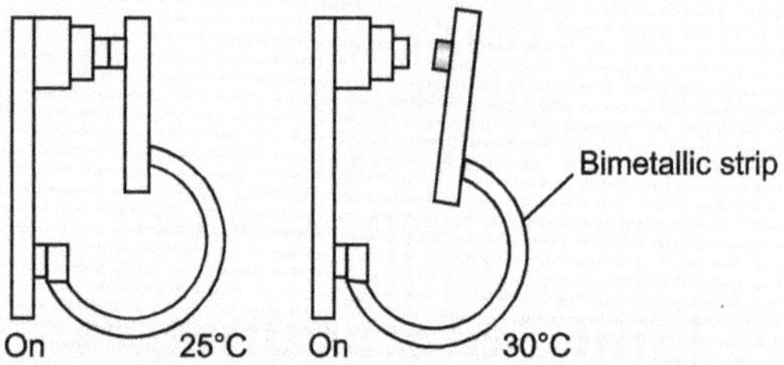

- **Variation of density with temperature**

Most substances expand when they are heated, i.e. volume of a given mass of a substance increases on heating, so the density should decrease $\left(\text{as } \rho \propto \frac{1}{V}\right)$.

Let us see how the density ρ varies with increase in temperature.

$$\rho = \frac{m}{V} \quad \text{or} \quad \rho \propto \frac{1}{V} \qquad \text{(for a given mass)}$$

$$\therefore \quad \frac{\rho'}{\rho} = \frac{V}{V'} = \frac{V}{V + \Delta V} = \frac{V}{V + \gamma V \Delta T} = \frac{1}{1 + \gamma \Delta T}$$

$$\therefore \quad \rho' = \frac{\rho}{1 + \gamma \Delta T}$$

This expression can also be written as

$$\rho' = \rho (1 + \gamma \Delta T)^{-1}$$

As γ is small, $(1 + \gamma \Delta T)^{-1} \approx 1 - \gamma \Delta T$ $\rho' \approx \rho (1 - \gamma \Delta T)$

- **Effect of temperature on upthrust**

When a solid body is completely immersed in a liquid, its apparent weight gets decreased due to an upthrust acting on it by the liquid. The apparent weight is given by

$$w_{app} = w - F$$

Here, $F = \text{upthrust} = V_S \rho_L g$

where, V_S = volume of solid and ρ_L = density of liquid.

Now, as the temperature is increased V_S increases while ρ_L decreases. So, F may increase or decrease (or may remain constant also) depending upon the condition that which factor dominates on the other. We can write

$$F \propto V_S \rho_L \quad \text{or} \quad \frac{F'}{F} = \frac{V_S'}{V_S} \cdot \frac{\rho_L'}{\rho_L} = \frac{(V_S + \Delta V_S)}{V_S} \cdot \left(\frac{1}{1 + \gamma_L \Delta T}\right)$$

$$= \left(\frac{V_S + \gamma_S V_S \Delta T}{V_S}\right)\left(\frac{1}{1+\gamma_L \Delta T}\right)$$

or $$F' = F\left(\frac{1+\gamma_S \Delta T}{1+\gamma_L \Delta T}\right)$$

Now, if $\gamma_S > \gamma_L, F' > F$ or $w'_{app} < w_{app}$ and *vice-versa.*

And if $\gamma_S = \gamma_L, F' = F$ or $w'_{app} = w_{app}$

- **Effect of temperature on the time period of a pendulum**

The time period of a simple pendulum is given by

$$T = 2\pi\sqrt{\frac{l}{g}} \quad \text{or} \quad T \propto \sqrt{l}$$

As the temperature is increased length of the pendulum and hence, time period gets increased or a pendulum clock becomes slow and it loses the time.

$$\frac{T'}{T} = \sqrt{\frac{l'}{l}} = \sqrt{\frac{l+\Delta l}{l}}$$

Here, we put $\Delta l = l\alpha\Delta\theta$ in place of $l\alpha\Delta T$, so as to avoid the confusion with change in time period. Thus,

$$\frac{T'}{T} = \sqrt{\frac{l + l\alpha\Delta\theta}{l}} = (1+\alpha\Delta\theta)^{1/2}$$

or $T' \approx T\left(1 + \frac{1}{2}\alpha\Delta\theta\right)$ or $\Delta T = T' - T = \frac{1}{2}T\alpha\Delta\theta$

Time lost in time t (by a pendulum clock whose actual time period is T and the changed time period at some higher temperature is T') is

$$\Delta t = \left(\frac{\Delta T}{T'}\right)t$$

Similarly, if the temperature is decreased the length and hence, the time period gets decreased. A pendulum clock in this case runs fast and it gains the time.

$$\frac{T'}{T} = \sqrt{\frac{l'}{l}} = \sqrt{\frac{l - l\alpha\Delta\theta}{l}} \approx 1 - \frac{1}{2}\alpha\Delta\theta$$

or $$T' = T\left(1 - \frac{1}{2}\alpha\Delta\theta\right)$$

$$\Delta T = T - T' = \frac{1}{2}T\alpha\,\Delta\theta$$

and time gained in time t is the same, i.e. $\Delta t = \left(\frac{\Delta T}{T'}\right)t$

At some higher temperature a scale will expand and scale reading will be lesser than true values, so that

true value = scale reading $(1 + \alpha\Delta T)$

Here, ΔT is the temperature difference.

However, at lower temperature scale reading will be more or true value will be less.

- When a rod whose ends are rigidly fixed such as to prevent from expansion or contraction undergoes a change in temperature, thermal stresses are developed in the rod.

This is because, if the temperature is increased, the rod has a tendency to expand but since, it is fixed at two ends, the rod exerts a force on supports.

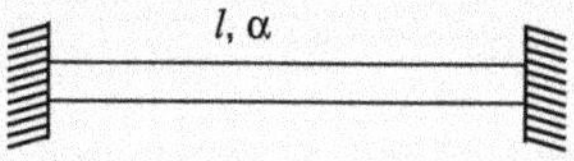

Thermal strain $= \frac{\Delta l}{l} = \alpha \cdot \Delta T$

So, thermal stress $= (\gamma)$(thermal strain)

$$= Y\alpha\Delta T$$

or force on supports; $F = A\,(\text{stress}) = YA\,\alpha\Delta T$

Here, Y = Young's modulus of elasticity of the rod.

$$F = YA\alpha\Delta T$$

- **Expansion of liquid**

For heating a liquid, it has to be put in some container. When the liquid is heated, the container will also expand. We define coefficient of apparent expansion of a liquid as the apparent increase in volume per unit original volume per °C rise in temperature. It is represented by γ_a. Thus,

$$\gamma_a = \gamma_r - \gamma_g$$

Here, γ_r = coefficient of real expansion of a liquid and

γ_g = coefficient of cubical expansion of the container.

Example 14.3 *A steel ruler exactly 20 cm long is graduated to give correct measurements at 20° C.*

(a) Will it give readings that are too long or too short at lower temperatures?

(b) What will be the actual length of the ruler be when it is used in the desert at a temperature of 40° C? $(\alpha_{steel} = 1.2 \times 10^{-5}\,°C^{-1})$

Sol. (a) If the temperature decreases, the length of the ruler also decreases through thermal contraction. Below 20°C, each centimetre division is actually somewhat shorter than 1.0 cm, so the steel ruler gives readings that are too long.

(b) At 40°C, the increase in length of the ruler is

$$\Delta l = l\alpha\Delta T = (20)(1.2 \times 10^{-5})(40° - 20°)$$

$$= 0.48 \times 10^{-2} \text{ cm}$$

$\therefore$ The actual length of the ruler is

$$l' = l + \Delta l = 20.0048 \text{ cm}$$

Example 14.4 *Find the coefficient of volume expansion for an ideal gas at constant pressure.*

Sol. For an ideal gas, $pV = nRT$

As p is constant, we have

$$p\,dV = nR\,dT$$

$$\therefore \quad \frac{dV}{dT} = \frac{nR}{p} \quad \text{or} \quad \gamma = \frac{1}{V}\cdot\frac{dV}{dT} = \frac{nR}{pV} = \frac{nR}{nRT} = \frac{1}{T}$$

$$\therefore \quad \gamma = \frac{1}{T}$$

Example 14.5 *The scale on a steel metre stick is calibrated at 15°C. What is the error in the reading of 60 cm at 27°C?*
$(\alpha_{steel} = 1.2 \times 10^{-5} {}^\circ C^{-1})$

Sol. At higher temperatures actual reading is more than the scale reading. The error in the reading will be

$$\Delta l = (\text{scale reading})(\alpha)(\Delta T)$$
$$= (60)(1.2 \times 10^{-5})(27^\circ - 15^\circ)$$
$$= 0.00864 \text{ cm}$$

Example 14.6 *A second's pendulum clock has a steel wire. The clock is calibrated at 20°C. How much time does the clock lose or gain in one weak when the temperature is increased to 30° C?*
$(\alpha_{steel} = 1.2 \times 10^{-5} {}^\circ C^{-1})$

Sol. The time period of second's pendulum is 2 s. As the temperature increases length and hence, time period increases. Clock becomes slow and it loses the time. The change in time period is

$$\Delta T = \frac{1}{2} T\alpha\Delta\theta$$
$$= \left(\frac{1}{2}\right)(2)(1.2 \times 10^{-5})(30^\circ - 20^\circ)$$
$$= 1.2 \times 10^{-4} \text{ s}$$

$\therefore$ New time period is

$$T' = T + \Delta T$$
$$= (2 + 1.2 \times 10^{-4})$$
$$= 2.00012 \text{ s}$$

$\therefore$ Time lost in one week

$$\Delta t = \left(\frac{\Delta T}{T'}\right) t$$
$$= \frac{(1.2 \times 10^{-4})}{(2.00012)}(7 \times 24 \times 3600)$$
$$= 36.28 \text{ s}$$

14.5 Concept of an Ideal Gas

A gas has no shape and size and can be contained in a vessel of any size or shape. It expands indefinitely and uniformly to fill the available space. It exerts pressure on its surroundings.

The gases whose molecules are point masses (mass without volume) and do not attract each other are called **ideal** or **perfect** gases. It is a hypothetical concept which can't exist in reality.

The gases such as hydrogen, oxygen or helium which cannot be liquefied easily are called **permanent** gases. An actual gas behaves as ideal gas most closely at low pressure and high temperature.

14.6 Gas Laws

Assuming permanent gases to be ideal, through experiments, it was established that gases irrespective of their nature obey the following laws:

Boyle's Law

According to this law, for a given mass of a gas the volume of a gas at constant temperature (called **isothermal** process) is inversely proportional to its pressure, i.e.

$$V \propto \frac{1}{p} \; (T = \text{constant}) \text{ or } pV = \text{constant} \quad \text{or } p_i V_i = p_f V_f$$

Thus, *p-V* graph in an isothermal process is a rectangular hyperbola. Or *pV versus p* or *V* graph is a straight line parallel to *p* or *V* axis.

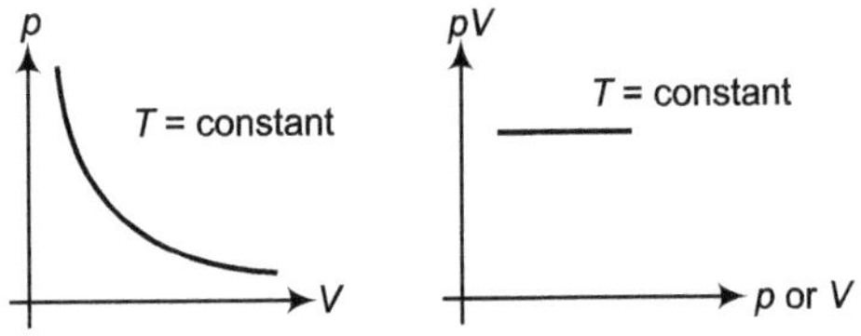

Fig. 14.7

Charles' Law

According to this law, for a given mass of a gas the volume of a gas at constant pressure (called **isobaric** process) is directly proportional to its absolute temperature, i.e.

$$V \propto T \qquad (p = \text{constant})$$

$$\text{or} \quad \frac{V}{T} = \text{constant} \quad \text{or} \quad \frac{V_i}{T_i} = \frac{V_f}{T_f}$$

Thus, *V-T* graph in an isobaric process is a straight line passing through origin. Or *V/T versus V* or *T* graph is a straight line parallel to *V* or *T*-axes.

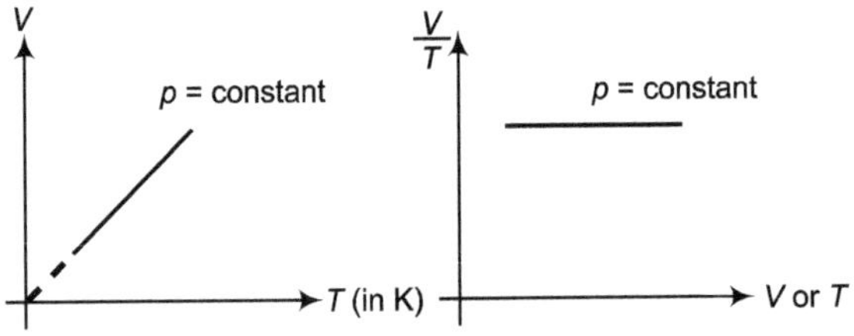

Fig. 14.8

Gay Lussac's Law or Pressure Law

According to this law, for a given mass of a gas the pressure of a gas at constant volume (called **isochoric process**) is directly proportional to its absolute temperature, i.e.

$$p \propto T \; (V = \text{constant})$$

or $$\frac{p}{T} = \text{constant}$$

or $$\frac{p_i}{T_i} = \frac{p_f}{T_f}$$

Thus, p-T graph in an isochoric process is a straight line passing through origin or p/T *versus* p or T graph is a straight line parallel to p or T axis.

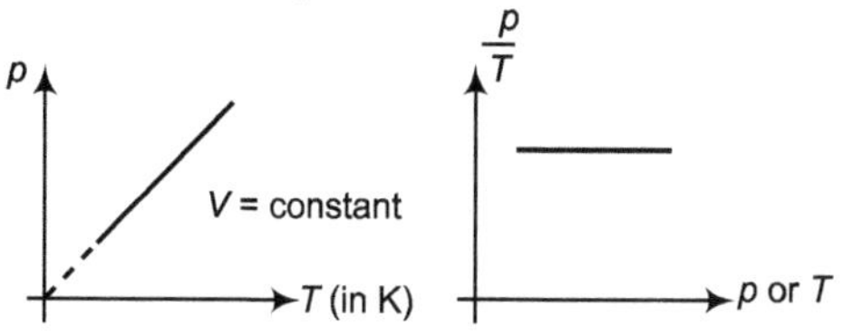

Fig. 14.9

Avogadro's Law

According to this law, at same temperature and pressure equal volumes of all gases contain equal number of molecules.

14.7 Ideal Gas Equation

All the above four laws can be written in one single equation known as ideal gas equation. According to this equation.

$$pV = nRT = \frac{m}{M} RT$$

In this equation, n = number of moles of the gas

$$= \frac{m}{M}$$

where, m = total mass of the gas,

M = molecular mass of the gas,

and R = universal gas constant.

$= 8.31$ J/mol-K

$= 2.0$ cal/mol-K

The above four laws can be derived from this single equation. For example, for a given mass of a gas (m = constant)

pV = constant at constant temperature (Boyle's law)

$\frac{p}{T}$ = constant at constant volume (Pressure law)

$\frac{V}{T}$ = constant at constant pressure (Charles' law)

and if p, V and T are constants, then n = constant for all gases.

And since, equal number of moles contain equal number of molecules. So, at constant pressure, volume and temperature all gases will contain equal number of molecules. Which is nothing but Avogadro's law.

Extra Knowledge Points

- In our previous discussion, we have read Charles' law and pressure law in absolute temperature scale. In centigrade scale, these laws are as under

Charles' law

When a given mass of a gas is heated at constant pressure, then for each 1°C rise in temperature the volume of the gas increases by a fraction α of its volume at 0°C. Thus, if the volume of a given mass of a gas at 0°C is V_0, then on heating at constant pressure to t°C its volume will increase by $V_0 \alpha t$. Therefore, if its volume at t°C be V_t, then

$$V_t = V_0 + V_0 \alpha t \quad \text{or} \quad V_t = V_0 (1 + \alpha t)$$

Here, α is called the 'volume coefficient' of the gas. For all gases the experimental value of α is nearly $\frac{1}{273}$ / °C.

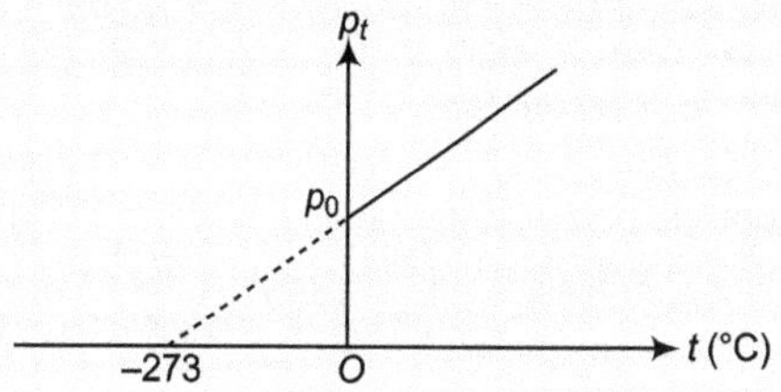

$$\therefore \quad V_t = V_0 \left(1 + \frac{t}{273}\right)$$

Thus, V_t *versus* t graph is a straight line with slope $\frac{V_0}{273}$ and positive intercept V_0.

Further, $V_t = 0$ at $t = -273$°C.

Pressure law

According to this law, when a given mass of a gas is heated at constant volume then for each 1°C rise in temperature, the pressure of the gas increases by a fraction β of its pressure at 0°C. Thus, if the pressure of a given mass of a gas at 0°C be p_0, then on heating at constant volume to t°C, its pressure will increase by $p_0 \beta t$. Therefore, if its pressure at t°C be p_t, then

$$p_t = p_0 + p_0 \beta t \quad \text{or} \quad p_t = p_0 (1 + \beta t)$$

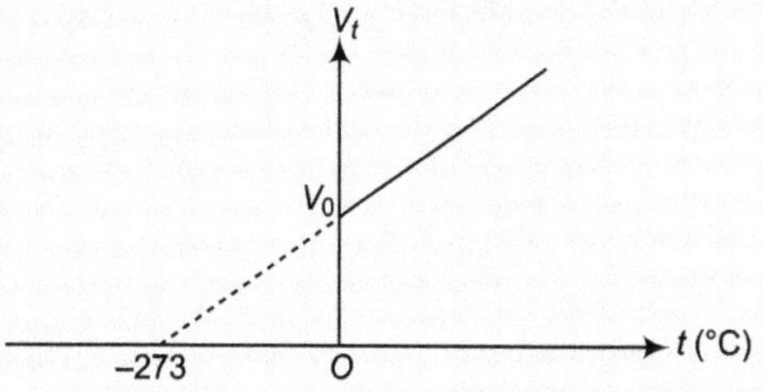

Here, β is called the pressure coefficient of the gas. For all gases the experimental value of β is also $\frac{1}{273}$ / °C.

$$\therefore \quad p_t = p_0 \left(1 + \frac{t}{273}\right)$$

The p_t *versus* t graph is as shown in figure.

- The above forms of Charles' law and pressure law can be simply expressed in terms of absolute temperature.
- Let at constant pressure, the volume of a given mass of a gas at 0°C, t_1°C and t_2°C be V_0, V_1 and V_2 respectively. Then,

$$V_1 = V_0\left(1+\frac{t_1}{273}\right) = V_0\left(\frac{273+t_1}{273}\right)$$

$$V_2 = V_0\left(1+\frac{t_2}{273}\right) = V_0\left(\frac{273+t_2}{273}\right)$$

$$\therefore \quad \frac{V_1}{V_2} = \frac{273+t_1}{273+t_2} = \frac{T_1}{T_2}$$

where, T_1 and T_2 are the absolute temperatures corresponding to t_1°C and t_2°C. Hence,

$$\frac{V_1}{T_1} = \frac{V_2}{T_2} \quad \text{or} \quad \frac{V}{T} = \text{constant} \quad \text{or} \quad V \propto T$$

This is the form of Charles' law which we have already studied in article 14.6. In the similar manner, we can prove the pressure law.

- Under isobaric conditions (p = constant), V-T graph is a straight line passing through origin (where T is in kelvin). The slope of this line is $\left(\frac{nR}{p}\right)$ as $V = \left(\frac{nR}{p}\right)T$ or slope of the line is directly proportional to $\frac{n}{p}$.

$$\text{Slope} = \frac{nR}{p} \quad \text{or} \quad \text{slope} \propto \frac{n}{p}$$

Similarly, under isochoric conditions (V = constant), p-T graph is a straight line passing through origin whose slope is $\frac{nR}{V}$ or slope is directly proportional to $\frac{n}{V}$.

- **Density of a gas** The ideal gas equation is

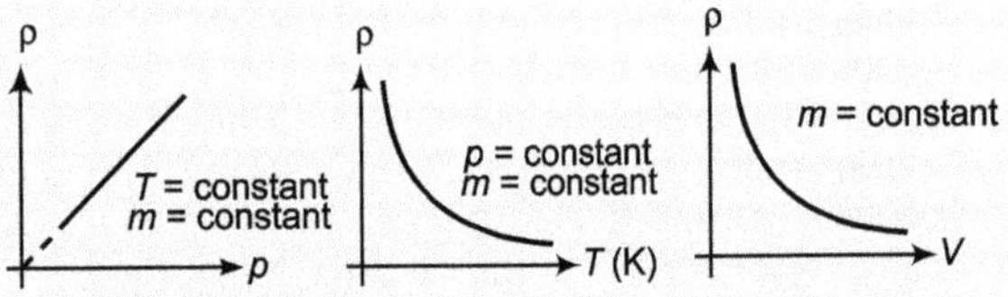

$$pV = nRT = \frac{m}{M}RT$$

$$\therefore \quad \frac{m}{V} = \rho = \frac{pM}{RT} \qquad (\text{As}, \rho = \text{density})$$

$$\therefore \quad \rho = \frac{pM}{RT}$$

- From this equation, we can see that ρ-p graph is straight line passing through origin at constant temperature ($\rho \propto p$) for a given gas and ρ-T graph is rectangular hyperbola at constant pressure $\left(\rho \propto \frac{1}{T}\right)$. Similarly, for a given mass of a gas ρ-V graph is a rectangular hyperbola $\left(\rho \propto \frac{1}{V}\right)$.

Example 14.7 *p-V diagrams of same mass of a gas are drawn at two different temperatures T_1 and T_2. Explain whether $T_1 > T_2$ or $T_2 > T_1$.*

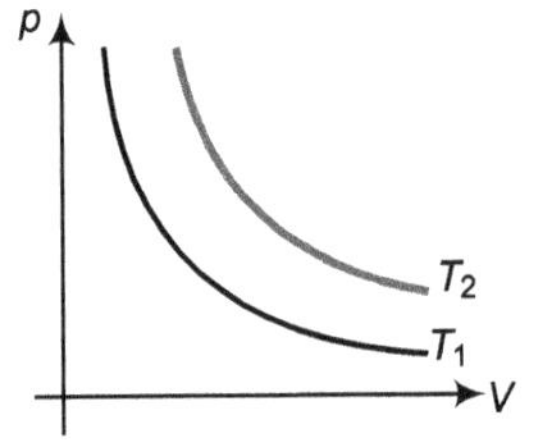

Fig. 14.10

Sol. The ideal gas equation is $pV = nRT$ or $T = \frac{pV}{nR}$

$T \propto pV$, if number of moles of the gas are kept constant. Here, mass of the gas is constant, which implies that number of moles are constant, i.e. $T \propto pV$. In the given diagram product of p and V for T_2 is more than T_1 at all points (keeping either p or V same for both graphs). Hence,

$$T_2 > T_1$$

Example 14.8 *The p-V diagrams of two different masses m_1 and m_2 are drawn (as shown) at constant temperature T. State whether $m_1 > m_2$ or $m_2 > m_1$?*

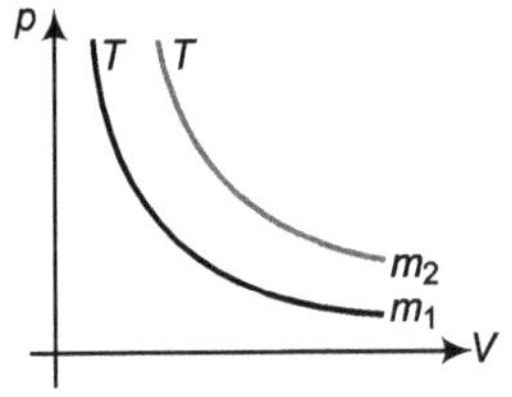

Fig. 14.11

Sol. $pV = nRT = \frac{m}{M}RT \Rightarrow \therefore \quad m = (pV)\left(\frac{M}{RT}\right)$

or $\qquad m \propto pV \qquad$ (if T = constant)

From the graph we can see that $p_2V_2 > p_1V_1$ (for same p or V). Therefore,

$$m_2 > m_1$$

Example 14.9 *The p-T graph for the given mass of an ideal gas is shown in figure. What inference can be drawn regarding the change in volume (whether it is constant, increasing or decreasing)?*

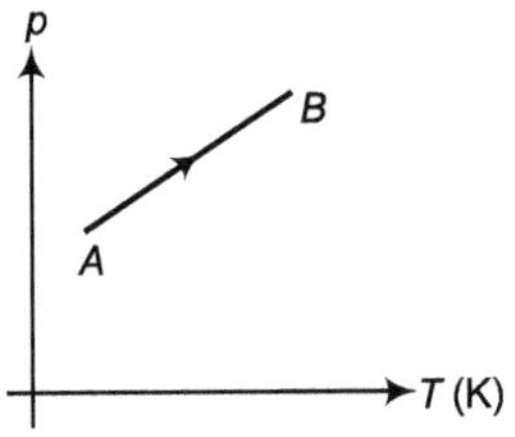

Fig. 14.12

HOW TO PROCEED *Definitely, it is not constant. Because when volume is constant, p-T graph is a straight line passing through origin. The given line does not pass through origin, hence volume is not constant.*

$$V = (nR)(T/P)$$

Now, to see the volume of the gas we will have to see whether $\frac{T}{p}$ is increasing or decreasing.

Sol. From the given graph, we can write the *p-T* equation as

$$p = aT + b \qquad (y = mx + c)$$

Here, *a* and *b* are positive constants. Further,

$$\frac{p}{T} = a + \frac{b}{T}$$

Now, $T_B > T_A \quad \therefore \quad \frac{b}{T_B} < \frac{b}{T_A}$

or $\left(\frac{p}{T}\right)_B < \left(\frac{p}{T}\right)_A$ or $\left(\frac{T}{p}\right)_B > \left(\frac{p}{T}\right)_A$

or $V_B > V_A$

Thus, as we move from *A* to *B*, volume of the gas is increasing.

Example 14.10 *p-V diagram of n moles of an ideal gas is as shown in figure. Find the maximum temperature between A and B.*

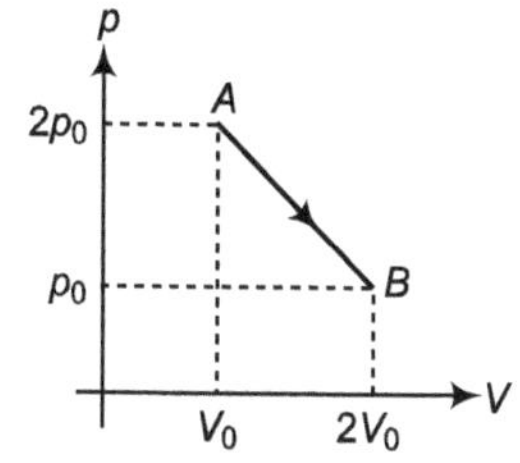

Fig. 14.13

HOW TO PROCEED *For given number of moles of a gas,*

$$T \propto pV \qquad (pV = nRT)$$

Although $(pV)_A = (pV)_B$ or $T_A = T_B$, still it is not an isothermal process. Because in isothermal process p-V graph is a rectangular hyperbola while it is a straight line. So, to see the behaviour of temperature first we will find either T-V equation or T-p equation and from that equation we can judge how the temperature varies. From the graph first, we will write p-V equation, then we will convert it either in T-V equation or in T-p equation.

Sol. From the graph, the *p-V* equation can be written as

$$p = -\left(\frac{p_0}{V_0}\right)V + 3p_0 \qquad (y = -mx + c)$$

or $$pV = -\left(\frac{p_0}{V_0}\right)V^2 + 3p_0V$$

or $$nRT = 3p_0V - \left(\frac{p_0}{V_0}\right)V^2 \qquad (\text{as } pV = nRT)$$

or $$T = \frac{1}{nR}\left[3p_0V - \left(\frac{p_0}{V_0}\right)V^2\right]$$

This is the required *T-V* equation. This is quadratic in *V*. Hence, *T-V* graph is a parabola. Now, to find maximum or minimum value of *T* we can substitute.

$$\frac{dT}{dV} = 0$$

or $$3p_0 - \left(\frac{2p_0}{V_0}\right)V = 0 \quad \text{or} \quad V = \frac{3}{2}V_0$$

Further $\frac{d^2T}{dV^2}$ is negative at $V = \frac{3}{2}V_0$

Hence, *T* is maximum at $V = \frac{3}{2}V_0$ and this maximum value is

$$T_{max} = \frac{1}{nR}\left[(3p_0)\left(\frac{3V_0}{2}\right) - \left(\frac{p_0}{V_0}\right)\left(\frac{3V_0}{2}\right)^2\right] \text{ or } T_{max} = \frac{9p_0V_0}{4nR}$$

Thus, *T-V* graph is as shown in the following figure.

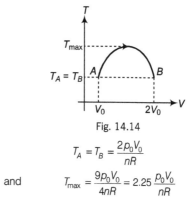

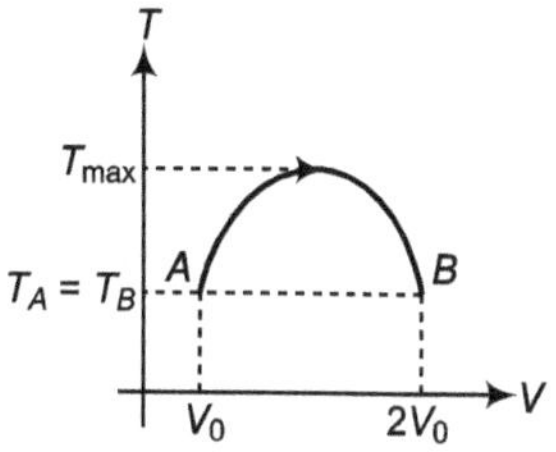

Fig. 14.14

$$T_A = T_B = \frac{2p_0V_0}{nR}$$

and $$T_{max} = \frac{9p_0V_0}{4nR} = 2.25\frac{p_0V_0}{nR}$$

14.8 Degree of Freedom (*f*)

The term degree of freedom refers to the number of possible independent ways in which a system can have energy.

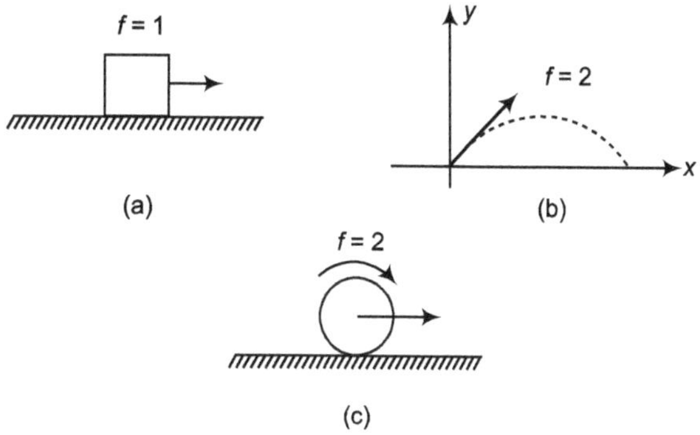

Fig. 14.15

For example In Fig. (a), block has one degree of freedom, because it is confined to move in a straight line and has only one translational degree of freedom.

In Fig. (b), the projectile has two degrees of freedom because it is confined to move in a plane and so it has two translational degrees of freedom.

In Fig. (c), the sphere has two degrees of freedom one rotational and another translational.

Similarly, a particle free to move in space will have three translational degrees of freedom.

Degree of Freedom of Gas Molecules

A gas molecule can have following types of energies :

(i) translational kinetic energy
(ii) rotational kinetic energy
(iii) vibrational energy (potential + kinetic)

Vibrational Energy

The forces between different atoms of a gas molecule may be visualised by imagining every atom as being connected to its neighbours by springs. Each atom can vibrate along the line joining the atoms. Energy associated with this is called **vibrational energy**.

Degree of Freedom of Monoatomic Gas

A monoatomic gas molecule (like He) consists of a single atom. It can have translational motion in any direction in space. Thus, it has 3 translational degrees of freedom.

$$f = 3 \qquad \text{(all translational)}$$

It can also rotate but due to its small moment of inertia, rotational kinetic energy is neglected.

Degree of Freedom of a Diatomic and Linear Polyatomic Gas

The molecules of a diatomic and linear polyatomic gas (like O_2, CO_2 and H_2) cannot only move bodily but also rotate about any one of the three coordinate axes as shown in figure. However, its moment of inertia about the axis joining the two atoms (x-axis) is negligible. Hence, it can have only two rotational degrees of freedom. Thus, a diatomic molecule has 5 degrees of freedom 3 translational and 2 rotational.

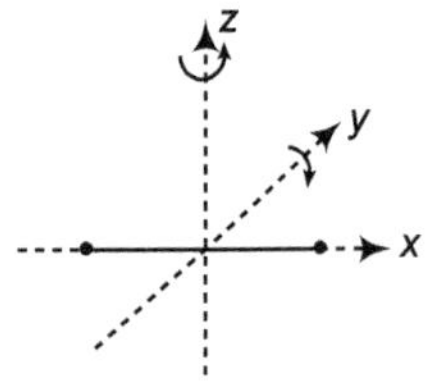

Fig. 14.16

At sufficiently high temperatures it has vibrational energy as well providing it two more degrees of freedom (one vibrational kinetic energy and another vibrational potential energy). Thus, at high temperatures a diatomic molecule has 7 degrees of freedom, 3 translational, 2 rotational and 2 vibrational. Thus,

$$f = 5$$

(3 translational + 2 rotational) at room temperatures

and $f = 7$

(3 translational + 2 rotational + 2 vibrational) at high temperatures

Degree of Freedom of Non-linear Polyatomic Gas

A non-linear polyatomic molecule (such as NH_3) can rotate about any coordinate axes. Hence, it has 6 degrees of freedom 3 translational and 3 rotational. At room temperatures a polyatomic gas molecule has vibrational energy greater than that of a diatomic gas. But at high enough temperatures it is also significant. So, it has 8 degrees of freedom 3 rotational 3 translational and 2 vibrational. Thus,

(3 translational + 3 rotational) at room temperatures

$$f = 6 \text{ and } f = 8$$

(3 translational + 3 rotational + 2 vibrational) at high temperatures

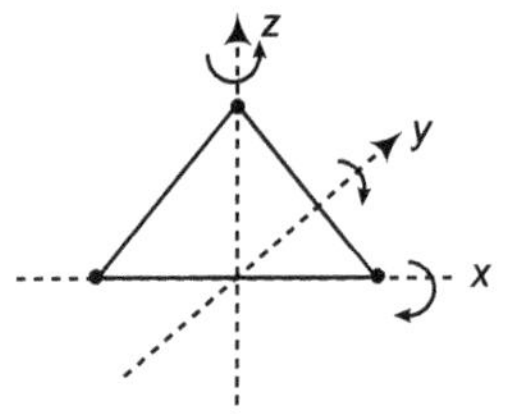

Fig. 14.17

Degree of Freedom of a Solid

An atom in a solid has no degrees of freedom for translational and rotational motion. At high temperatures due to vibration along 3 axes it has $3 \times 2 = 6$ degrees of freedom. $f = 6$ (all vibrational) at high temperatures

Note Points

- Degrees of freedom of a diatomic and polyatomic gas depends on temperature and since there is no clear cut demarcation line above which vibrational energy become significant. Moreover, this temperature varies from gas to gas. On the other hand, for a monoatomic gas there is no such confusion. Degree of freedom here is 3 at all temperatures. Unless and until stated in the question you can take $f = 3$ for monoatomic gas, $f = 5$ for a diatomic gas and $f = 6$ for a non-linear polyatomic gas.
- When a diatomic or polyatomic gas dissociates into atoms, it behaves as a monoatomic gas. Whose degrees of freedom are changed accordingly.

14.9 Internal Energy of an Ideal Gas

Suppose a gas is contained in a closed vessel as shown in figure. If the container as a whole is moving with some speed, then this motion is called the **ordered motion** of the gas. Source of this motion is some external force.

The *zig-zag* motion of gas molecules within the vessel is known as the **disordered motion**. This motion is directly related to the temperature of the gas. As the temperature is increased, the disordered motion of the gas molecules gets fast.

The internal energy (U) of the gas is concerned only with its disordered motion. It is in no way concerned with its ordered motion. When the temperature of the gas is increased, its disordered motion and hence its internal energy is increased.

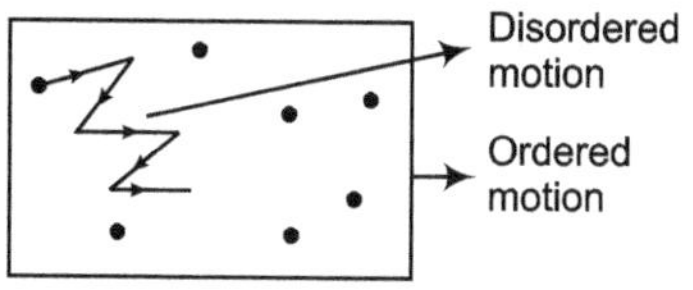

Fig. 14.18

Intermolecular forces in an ideal gas is zero. Thus, PE due to intermolecular forces of an ideal gas is zero. A monoatomic gas is having a single atom. Hence, its vibrational energy is zero. For dia and polyatomic gases vibrational energy is significant only at high temperatures. So, they also have only translational and rotational KE. We may thus, conclude that at room temperature the internal energy of an ideal gas (whether it is mono, dia or poly) consists of only translational and rotational KE. Thus, U (of an ideal gas) $= K_T + K_R$ at room temperatures.

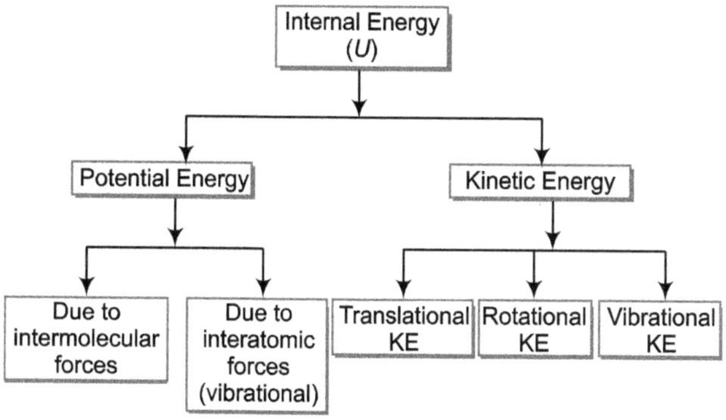

Later in the next article, we will see that K_T (translational KE) and K_R (rotational KE) depends on T only. They are directly proportional to the absolute temperature of the gas. Thus, internal energy of an ideal gas depends only on its absolute temperature (T) and is directly proportional to T.

or $$U \propto T$$

14.10 Law of Equipartition of Energy

An ideal gas is just like an ideal father. As an ideal father distributes whole of its assets equally among his children. Same is the case with an ideal gas. It distributes its internal energy equally in all degrees of freedom. In each degree of freedom energy of one mole of an ideal gas is $\frac{1}{2}RT$, where T is the absolute temperature of the gas. Thus, if f be the number of degrees of freedom, the internal energy of 1 mole of the gas will be $\frac{f}{2}RT$ or internal energy of n moles of the gas will be $n/2fRT$. Thus,

$$U = \frac{n}{2}fRT \qquad \text{...(i)}$$

For a monoatomic gas, $f = 3$

Therefore, $$U = \frac{3}{2}RT$$

(for 1 mole of a monoatomic gas)

For a dia and linear polyatomic gas at low temperatures, $f = 5$, so

$$U = \frac{5}{2}RT \qquad \text{(for 1 mole)}$$

and for non-linear polyatomic gas at low temperatures, $f = 6$, so

$$U = \frac{6}{2}RT = 3RT \qquad \text{(for 1 mole)}$$

- From Eq. (i), we can see that internal energy of an ideal gas depends only on its temperature and which is directly proportional to its absolute temperature T. In an isothermal process, T = constant. Therefore, the internal energy of the gas does not change or $dU = 0$.

14.11 Molar Heat Capacity

"Molar heat capacity C is the heat required to raise the temperature of 1 mole of a gas by 1°C (or 1 K)." Thus,

$$C = \frac{\Delta Q}{n\Delta T} \quad \text{or} \quad \Delta Q = nC\Delta T$$

For a gas, the value of C depends on the process through which its temperature is raised.

For example, in an isothermal process $\Delta T = 0$ or $C_{iso} = \infty$. In an adiabatic process (we will discuss it later), $\Delta Q = 0$. Hence, $C_{adi} = 0$. Thus, molar heat capacity of a gas varies from 0 to ∞ depending on the process. In general, experiments are made either at constant volume or at constant pressure. In case of solids and liquids, due to small thermal expansion, the difference in measured values of molar heat capacities is very small and is usually neglected. However, in case of gases molar heat capacity at constant volume C_V is quite different from that at constant pressure C_p. Later in the next chapter, we will derive the following relations, for an ideal gas

$$C_V = \frac{dU}{dT}$$
$$= \frac{f}{2}R = \frac{R}{\gamma - 1}$$
$$C_p = C_V + R$$
$$\gamma = \frac{C_p}{C_V} = 1 + \frac{2}{f}$$

Here, U is the internal energy of one mole of the gas. The most general expression for C in the process pV^x = constant is

$$C = \frac{R}{\gamma - 1} + \frac{R}{1 - x} \text{ (we will derive it later)}$$

For example For isobaric process

$$p = \text{constant} \quad \text{or} \quad x = 0$$

and $$C = C_p = \frac{R}{\gamma - 1} + R = C_V + R$$

For isothermal process, pV = constant or $x = 1$

$\therefore$ $$C = \infty$$

For adiabatic process, pV^γ = constant or $x = \gamma$

$\therefore$ $$C = 0$$

Values of f, U, C_V, C_p and γ for different gases are shown in Table 14.2.

Table 14.2

Nature of gas	f	$U = \frac{f}{2}RT$	$C_V = dU/dT = \frac{f}{2}R$	$C_p = C_V + R$	$\gamma = C_p/C_V = 1 + \frac{2}{f}$
Monoatomic	3	$\frac{3}{2}RT$	$\frac{3}{2}R$	$\frac{5}{2}R$	1.67
Dia and linear polyatomic	5	$\frac{5}{2}RT$	$\frac{5}{2}R$	$\frac{7}{2}R$	1.4
Non-linear polyatomic	6	$3RT$	$3R$	$4R$	1.33

14.12 Kinetic Theory of Gases

We have studied the mechanics of single particle. When we approach the mechanics associated with the many particles in systems such as gases, liquids and solids, we are faced with analyzing the dynamics of a huge number of particles. The dynamics of such many particle systems is called **statistical mechanics**.

The game involved in studying a system with a large number of particles is similar to what happens after every physics test. Of course we are interested in our individual marks, but we also want to know the class average.

The kinetic theory that we study in this article is a special aspect of the statistical mechanics of large number of particles. We begin with the simplest model for a monoatomic ideal gas, a dilute gas whose particles are single atoms rather than molecules.

Macroscopic variables of a gas are pressure, volume and temperature and microscopic properties are speed of gas molecules and momentum of molecules, etc. Kinetic theory of gases relates the microscopic properties to macroscopic properties. Further more, the kinetic theory provides us with a physical basis for our understanding of the concept of pressure and temperature.

The Ideal Gas Approximation

We make the following assumptions while describing an ideal gas

1. The number of particles in the gas is very large.
2. The volume V containing the gas is much larger than the total volume actually occupied by the gas particles themselves.
3. The dynamics of the particles is governed by Newton's laws of motion.
4. The particles are equally likely to be moving in any direction.
5. The gas particles interact with each other and with the walls of the container only via elastic collisions.
6. The particles of the gas are identical and indistinguishable.

The Pressure of an Ideal Gas

Consider an ideal gas consisting of N molecules in a container of volume V. The container is a cube with edges of length d. Consider the collision of one molecule moving with a velocity v towards the right hand face of the cube. The molecule has velocity components v_x, v_y and v_z. Previously, we used m to represent the mass of a sample, but in this article we shall use m to represent the mass of one molecule.

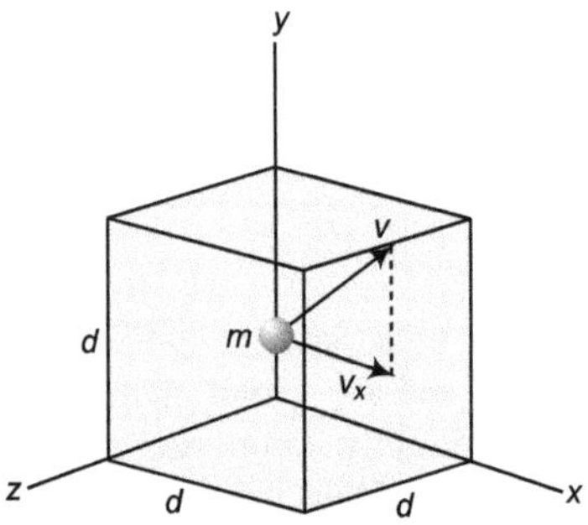

A cubical box with sides of length d containing an ideal gas. The molecule shown moves with velocity v

Fig. 14.19

As the molecule collides with the wall elastically its x-component of velocity is reversed, while its y and z-component of velocity remain unaltered. Because the x-component of the momentum of the molecule is mv_x before the collision and $-mv_x$ after the collision, the change in momentum of the molecule is

$$\Delta p_x = -mv_x - (mv_x) = -2mv_x$$

Applying impulse = change in momentum to the molecule

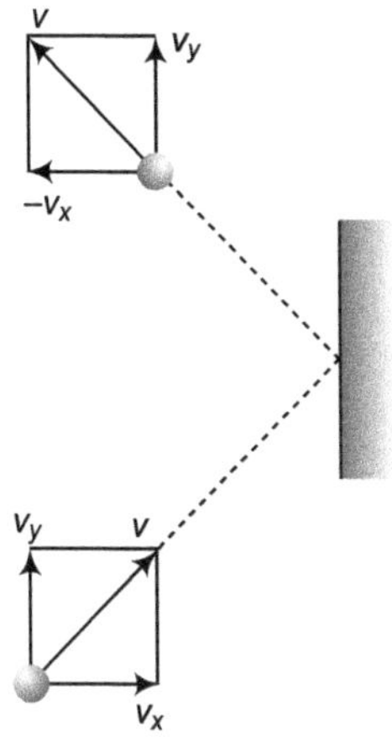

A molecule makes an elastic collision with the wall of the container. Its x-component of momentum is reversed, while its y-component remains unchanged. In this construction, we assume that the molecule moves in the x-y plane.

Fig. 14.20

$$F\Delta t = \Delta p_x = -2mv_x$$

where, F is the magnitude of the average force exerted by the wall on the molecule in time Δt. For the molecules to collide twice with the same wall, it must travel a distance $2d$ in the x-direction. Therefore, the time interval between two collisions with the same wall is $\Delta t = \frac{2d}{v_x}$. Over a time interval that is long compared with Δt, the average force exerted on the molecules for each collision is

$$F = \frac{-2mv_x}{\Delta t} = \frac{-2mv_x}{2d/v_x} = \frac{-mv_x^2}{d}$$

According to Newton's third law, the average force exerted by the molecule on the wall is $\frac{mv_x^2}{d}$.

Each molecule of the gas exerts a force on the wall. We find the total force exerted by all the molecules on the wall by adding the forces exerted by the individual molecules.

$$\therefore \quad F_{\text{wall}} = \frac{m}{d}(v_{x_1}^2 + v_{x_2}^2 + \ldots + v_{x_N}^2)$$

This can also be written as

$$F_{\text{wall}} = \frac{Nm}{d}\bar{v}_x^2$$

where, $$\bar{v}_x^2 = \frac{v_{x_1}^2 + v_{x_2}^2 + \ldots + v_{x_N}^2}{N}$$

Since, the velocity has three components v_x, v_y and v_z, we can have

$$\bar{v}^2 = \bar{v}_x^2 + \bar{v}_y^2 + \bar{v}_z^2 \quad (\text{as } v^2 = v_x^2 + v_y^2 + v_z^2)$$

Because the motion is completely random, the average values $\bar{v}_x^2$, $\bar{v}_y^2$ and $\bar{v}_z^2$ are equal to each other. So,

$$\bar{v}^2 = 3\bar{v}_x^2 \quad \text{or} \quad \bar{v}_x^2 = \frac{1}{3}\bar{v}^2$$

Therefore, $$F_{\text{wall}} = \frac{N}{3}\left(\frac{m\bar{v}^2}{d}\right)$$

$\therefore$ Pressure on the wall

$$p = \frac{F_{\text{wall}}}{A} = \frac{F_{\text{wall}}}{d^2} = \frac{1}{3}\left(\frac{N}{d^3}m\bar{v}^2\right)$$

$$= \frac{1}{3}\left(\frac{N}{V}\right)m\bar{v}^2$$

$$= \frac{2}{3}\left(\frac{N}{V}\right)\left(\frac{1}{2}m\bar{v}^2\right)$$

$$\therefore \quad p = \frac{1}{3}\frac{mN}{V}\bar{v}^2 = \frac{2}{3}\left(\frac{N}{V}\right)\left(\frac{1}{2}m\bar{v}^2\right) \quad \ldots\text{(i)}$$

This result indicates that the pressure is proportional to the number of molecules per unit volume (N/V) and to the average translational kinetic energy of the molecules $1/2\, m\bar{v}^2$. This result relates the large scale quantity (macroscopic) of pressure to an atomic quantity (microscopic) the average value of the square of the molecular speed. The above equation varifies some features of pressure with which you are probably familiar. One way to increase the pressure inside a container is to increase the number of molecules per unit volume in the container.

The meaning of the absolute temperature

Rewriting Eq. (i) in the more familiar form

$$pV = \frac{2}{3}N\left(\frac{1}{2}m\bar{v}^2\right)$$

Let us now compare it with the ideal gas equation

$$pV = nRT$$

$$nRT = \frac{2}{3}N\left(\frac{1}{2}m\bar{v}^2\right)$$

Here, $$n = \frac{N}{N_A} \quad (N_A = \text{Avogadro's number})$$

$\therefore \qquad T = \frac{2}{3}\left(\frac{N_A}{R}\right)\left(\frac{1}{2}m\bar{v}^2\right)$

or $\qquad T = \frac{2}{3k}\left(\frac{1}{2}m\bar{v}^2\right) \qquad \ldots(\text{ii})$

where, k is **Boltzmann constant** which has the value

$$k = \frac{R}{N_A} = 1.38 \times 10^{-23} \text{ J/K}$$

By rearranging Eq. (ii) we can relate the translational molecular kinetic energy to the temperature

$$\frac{1}{2}mv^2 = \frac{3}{2}kT$$

That is, the average translational kinetic energy per molecule is $\frac{3}{2}kT$. Because $\bar{v}_x^2 = \frac{1}{3}\bar{v}^2$, it follows that

$$\frac{1}{2}m\bar{v}_x^2 = \frac{1}{2}kT$$

In the similar manner it follows that

$$\frac{1}{2}m\bar{v}_y^2 = \frac{1}{2}kT \text{ and } \frac{1}{2}m\bar{v}_z^2 = \frac{1}{2}kT$$

Thus, in each translational degree of freedom one gas molecule has an energy $\frac{1}{2}kT$. One mole of a gas has N_A number of molecules. Thus, one mole of the gas has an energy $\frac{1}{2}(kN_A)T = \frac{1}{2}RT$ in each degree of freedom. Which is nothing but the law of equipartition of energy. The total translational kinetic energy of one mole of an ideal gas is therefore, $\frac{3}{2}RT$.

$$(\text{KE})_{\text{trans}} = \frac{3}{2}RT \qquad \text{(of one mole)}$$

Root Mean Square Speed

The square root of $\bar{v}^2$ is called the root mean square (rms) speed of the molecules. From Eq. (ii) we obtain, for the rms speed

$$v_{\text{rms}} = \sqrt{\bar{v}^2} = \sqrt{\frac{3kT}{m}}$$

Using $\qquad k = \frac{R}{N_A}, \; mN_A = M$

and $\qquad \frac{RT}{M} = \frac{p}{\rho}$

We can write,

$$v_{\text{rms}} = \sqrt{\frac{3kT}{m}} = \sqrt{\frac{3RT}{M}} = \sqrt{\frac{3p}{\rho}}$$

Mean Speed of Average Speed

The particles of a gas have a range of speeds. The average speed is found by taking the average of the speeds of all the particles at a given instant. Remember that the speed is a positive scalar since, it is the magnitude of the velocity.

$$v_{\text{av}} = \frac{v_1 + v_2 + \ldots + v_N}{N}$$

From Maxwellian speed distribution law, can show that

$$v_{\text{av}} = \sqrt{\frac{8kT}{\pi m}} = \sqrt{\frac{8RT}{\pi M}} = \sqrt{\frac{8p}{\pi\rho}}$$

Most Probable Speed

This is defined as the speed which is possessed by maximum fraction of total number of molecules of the gas. For example, if speeds of 10 molecules of a gas are, 1, 2, 2, 3, 3, 3, 4, 5, 6, 6 km/s, then the most probable speed is 3 km/s, as maximum fraction of total molecules possess this speed. Again from Maxwellian speed distribution law (out of JEE syllabus)

$$v_{\text{mp}} = \sqrt{\frac{2kT}{m}} = \sqrt{\frac{2RT}{M}} = \sqrt{\frac{2p}{\rho}}$$

Note Points

- In the above expressions of v_{rms}, v_{av} and v_{mp}, M is the molar mass in kg/mol. For example, molar mass of hydrogen is 2×10^{-3} kg/mol.
- $v_{\text{rms}} > v_{\text{av}} > v_{\text{mp}}$ (RAM)
- $v_{\text{rms}} : v_{\text{av}} : v_{\text{mp}} = \sqrt{3} : \sqrt{\frac{8}{\pi}} : \sqrt{2}$ and since, $\frac{8}{\pi} \approx 2.5$, we have $v_{\text{rms}} : v_{\text{av}} : v_{\text{mp}} = \sqrt{3} : \sqrt{2.5} : \sqrt{2}$

Example 14.11 *A tank used for filling helium balloons has a volume of 3.0 m^3 and contains 2.0 mol of helium gas at 20.0°C. Assuming that the helium behaves like an ideal gas.*

(a) What is the total translational kinetic energy of the molecules of the gas?

(b) What is the average kinetic energy per molecule?

Sol. (a) Using $(\text{KE})_{\text{trans}} = \frac{3}{2}nRT$

with $n = 2.0$ mol and $T = 293$ K, we find that

$$(\text{KE})_{\text{trans}} = \frac{3}{2}(2.0)(8.31)(293) = 7.3 \times 10^3 \text{ J}$$

(b) The average kinetic energy per molecule is $\frac{3}{2}kT$.

or $\qquad \frac{1}{2}m\bar{v}^2 = \frac{1}{2}m\bar{v}_{\text{rms}}^2 = \frac{3}{2}kT$

$$= \frac{3}{2}(1.38 \times 10^{-23})(293) = 6.07 \times 10^{-21} \text{ J}$$

Example 14.12 *Consider an 1100 particles gas system with speeds distribution as follows :*

1000 particles each with speed 100 m/s,
2000 particles each with speed 200 m/s,
4000 particles each with speed 300 m/s,
3000 particles each with speed 400 m/s, and
1000 particles each with speed 500 m/s.
Find the average speed, and rms speed.

Sol. The average speed is

$$v_{av} = \frac{(1000)(100) + (2000)(200) + (4000)(300) + (3000)(400) + (1000)(500)}{1100}$$

$= 309$ m/s

The rms speed is

$$v_{rms} = \sqrt{\frac{(1000)(100)^2 + (2000)(200)^2 + (4000)(300)^2 + (3000)(400)^2 + (1000)(500)^2}{1100}}$$

$= 328$ m/s

- Here, $\frac{v_{rms}}{v_{av}} \neq \sqrt{\frac{3}{8/\pi}}$ as values and gas molecules are arbitrarily taken.

Example 14.13 *Calculate the change in internal energy of 3.0 mol of helium gas when its temperature is increased by 2.0 K.*

Sol. Helium is a monoatomic gas. Internal energy of n moles of the gas is,

$$U = \frac{3}{2}nRT \Rightarrow \therefore \Delta U = \frac{3}{2}nR(\Delta T)$$

Substituting the values,

$$\Delta U = \left(\frac{3}{2}\right)(3)(8.31)(2.0) = 74.8 \text{ J}$$

Example 14.14 *In a crude model of a rotating diatomic molecule of chlorine (Cl_2), the two Cl atoms are 2.0×10^{-10} m apart and rotate about their centre of mass with angular speed $\omega = 2.0 \times 10^{12}$ rad/s. What is the rotational kinetic energy of one molecule of Cl_2, which has a molar mass of 70.0 g/mol ?*

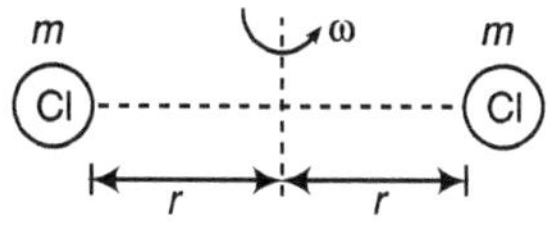

Fig. 14.21

Sol. Moment of inertia, $I = 2(mr^2) = 2mr^2$

Here, $m = \frac{70 \times 10^{-3}}{2 \times 6.02 \times 10^{23}} = 5.81 \times 10^{-26}$ kg

and $r = \frac{2.0 \times 10^{-10}}{2} = 1.0 \times 10^{-10}$ m

$\therefore$ $I = 2(5.81 \times 10^{-26})(1.0 \times 10^{-10})^2$

$= 1.16 \times 10^{-45}$ kg-m^2

$\therefore$ $K_R = \frac{1}{2}I\omega^2$

$= \frac{1}{2} \times (1.16 \times 10^{-45}) \times (2.0 \times 10^{12})^2$

$= 2.32 \times 10^{-21}$ J

- At $T = 300$ K, rotational KE should be equal to $\frac{1}{2}kT$

$= \frac{1}{2} \times (1.38 \times 10^{-23}) \times (300) = 2.07 \times 10^{-21}$ J

Example 14.15 *Prove that the pressure of an ideal gas is numerically equal to two-third of the mean translational kinetic energy per unit volume of the gas.*

Sol. Translational KE per unit volume

$$E = \frac{1}{2}(\text{mass per unit volume})(\overline{v^2})$$

$$= \frac{1}{2}(\rho)\left(\frac{3p}{\rho}\right) = \frac{3}{2}p \text{ or } p = \frac{2}{3}E$$

- Students are advised to remember this result. In this expression, E is the translational KE per unit volume.

Extra Knowledge Point

- Pressure exerted by an ideal gas is numerically equal to two-third of the mean kinetic energy of translation per unit volume of the gas. Thus,

$$p = \frac{2}{3}E$$

- **Mean Free Path**

Every gas consists of a very large number of molecules. These molecules are in a state of continuous rapid and random motion. They undergo perfectly elastic collisions against one another. Therefore, path of a single gas molecule consists of a series of short *zig-zag* paths of different lengths.

The mean free path of a gas molecule is the average distance between two successive collisions.

It is represented by λ.

$$\lambda = \frac{kT}{\sqrt{2}\pi\sigma^2 p}$$

Here, σ = diameter of the molecule

k = Boltzmann constant

- **Avogadro's Hypothesis**

At constant temperature and pressure equal volumes of different gases contain equal number of molecules. In 1 g-mol of any gas, there are 6.02×10^{23} molecules of that gas. This is called Avogadro's number. Thus,

$$N = 6.02 \times 10^{23}/\text{g-mol}$$

Therefore, the number of molecules in mass m of the substance

Number of molecules $= nN = \frac{m}{M} \times N$

- **Dalton's Law of Partial Pressure**

According to this law, if the gases filled in a vessel do not react chemically, then the combined pressure of all the gases is due to the partial pressure of the molecules of the individual gases. If $p_1, p_2, \ldots$ represent the partial pressures of the different gases, then the total pressure is

$$p = p_1 + p_2 \ldots$$

- **van der Waals' Equation**

Experiments have proved that real gases deviate largely from ideal behaviour. The reason of this deviation is two wrong assumptions in the kinetic theory of gases.

(i) The size of the molecules is much smaller in comparison to the volume of the gas, hence, it may be neglected.

(ii) Molecules do not exert intermolecular force on each other.

- van der Waals' made corrections for these assumptions and gave a new equation. This equation is known as van der Waals' equation for real gases.

(i) **Correction for the finite size of molecules**
Molecules occupy some volume. Therefore, the volume in which they perform thermal motion is less than the observed volume of the gas. It is represented by $(V - b)$. Here, b is a constant which depends on the effective size and number of molecules of the gas. Therefore, we should use $(V - b)$ in place of V in gas equation.

(ii) **Correction for intermolecular attraction** Due to the intermolecular force between gas molecules the molecules which are very near to the wall experiences a net inward force. Due to this inward force there is a decrease in momentum of the particles of a gas. Thus, the pressure exerted by real gas molecules is less than the pressure exerted by the molecules of an ideal gas. So, we use $\left(p + \frac{a}{V^2}\right)$ in place of p in gas equation.

Here, again a is a constant.
van der Waals' equation of state for real gases thus becomes,

$$\left(p + \frac{a}{V^2}\right)(V - b) = RT$$

- **Critical Temperature, Pressure and Volume**

Gases cannot be liquefied above a temperature called critical temperature (T_C), however large the pressure may be. The pressure required to liquefy the gas at critical temperature is called critical pressure (p_C) and the volume of the gas at critical temperature and pressure is called critical volume (V_C). Value of critical constants in terms of van der Waals' constants a and b are as under

$$V_C = 3b$$

$$p_C = \frac{a}{27b^2} \quad \text{and} \quad T_C = \frac{8a}{27Rb}$$

- Further, $\frac{RT_C}{p_C V_C} = \frac{8}{3}$ is called critical coefficient and is same for all gases.

Chapter Summary with Formulae

- **Thermal Expansion**
  (i) $\Delta l = l\alpha\Delta T$, $\Delta s = s\beta\Delta T$ and $\Delta V = V\gamma\Delta T$
  (ii) $\beta = 2\alpha$ and $\gamma = 3\alpha$ for isotropic medium.
  (iii) **Thermal Stress** If temperature of a rod fixed at both ends is increased, then thermal stresses are developed in the rod.

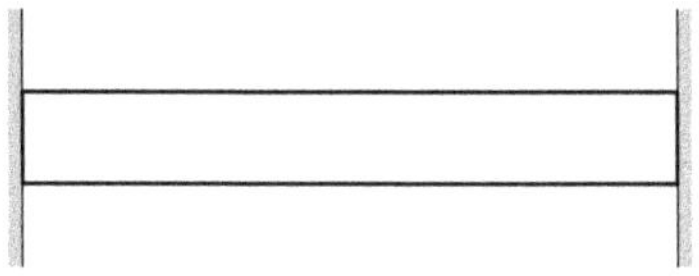

Fig. 14.22

Rod applies this much force on wall to expand. In turn, wall also exerts equal and opposite pair of encircled forces on rod. Due to this pair of forces only, we can say that rod is compressed.

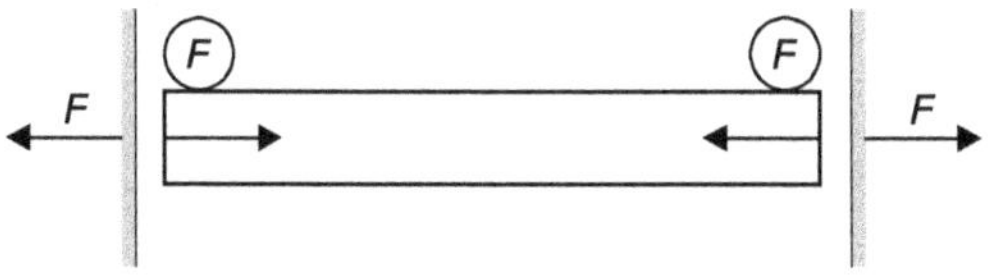

Fig. 14.23

- **Kinetic theory of gases**
  (i) f = degree of freedom
  = 3 for monatomic gas
  = 5 for diatomic and linear polyatomic gas
  = 6 for non-linear polyatomic gas

Note Points

- Vibrational degree of freedom is not taken into consideration.
- Translational degree of freedom for any type of gas is three.

  (ii) Total internal energy of a gas is

$$U = \frac{nf}{2} RT$$

Here, n = total number of gram moles

  (iii) $C_V = \frac{dU}{dT}$

[where, U = internal energy of one mole of a gas $= \frac{f}{2} RT$]

$$\therefore \quad C_V = \frac{f}{2} R = \frac{R}{\gamma - 1}$$

  (iv) $C_p = C_V + R = \left(1 + \frac{f}{2}\right) R = \left(\frac{\gamma}{\gamma - 1}\right) R$

  (v) $\gamma = \frac{C_p}{C_V} = 1 + \frac{2}{f}$

  (vi) Internal energy of 1 mole in one degree of freedom of any gas is $\frac{1}{2} RT$.

  (vii) Translational kinetic energy of one mole of any type of gas is $\frac{3}{2} RT$.

  (viii) Rotational kinetic energy of 1 mole of monatomic gas is zero of dia or linear polyatomic gas is $\frac{2}{2} RT$ or RT, of non-linear polyatomic gas is $\frac{3}{2} RT$.

  (ix) Mixture of non-reactive gases
  (a) $n = n_1 + n_2$
  (b) $p = p_1 + p_2$
  (c) $U = U_1 + U_2$
  (d) $\Delta U = \Delta U_1 + \Delta U_2$
  (e) $C_V = \frac{n_1 C_{V_1} + n_2 C_{V_2}}{n_1 + n_2}$
  (f) $C_p = \frac{n_1 C_{p_1} + n_2 C_{p_2}}{n_1 + n_2} = C_V + R$
  (g) $\gamma = \frac{C_p}{C_V}$ or $\frac{n}{\gamma - 1} = \frac{n_1}{\gamma_1 - 1} + \frac{n_2}{\gamma_2 - 1}$
  (h) $M = \frac{n_1 M_1 + n_2 M_2}{n_1 + n_2}$

- **Kinetic theory of gases**
  (i) $pV = nRT = \frac{m}{M} RT$
  (m = mass of gas in gms)
  (ii) Density, $\rho = \frac{m}{V}$ (general)

$$= \frac{pM}{RT} \qquad \text{(for an ideal gas)}$$

  (iii) **Gas laws**
  (a) **Boyle's law** is applied when T = constant or process is isothermal. In this condition,

$$pV = \text{constant}$$

or $$p_1V_1 = p_2V_2$$

or $$p \propto \frac{1}{V}$$

  (b) **Charles' law** is applied when p = constant or process is isobaric. In this condition $\frac{V}{T}$ = constant or

$$\frac{V_1}{T_1} = \frac{V_2}{T_2} \quad \text{or} \quad V \propto T$$

  (c) **Pressure law or Gay Lussac's law** is applied when V = constant or process is isochoric.
  In this condition,

$$\frac{p}{T} = \text{constant or } \frac{p_1}{T_1} = \frac{p_2}{T_2}$$

or $$p \propto T$$

(iv) Four speeds, $v = \sqrt{\frac{ART}{M}} = \sqrt{\frac{AkT}{m}} = \sqrt{\frac{Ap}{\rho}}$

Here, m = mass of one gas molecule.

$A = 3$ for rms speed of gas molecules

$= \frac{8}{\pi} \approx 2.5$ for average speed of gas molecules

$= 2$ for most probable speed of gas molecules

$\gamma = \frac{C_p}{C_V}$ for speed of sound in a gas

(v) $p = \frac{1}{3}\frac{mn}{V} v_{rms}^2$

Here, m = mass of one gas molecule and n = total number of molecules.

(vi) $p = \frac{2}{3}E$.

Here, E = total translational kinetic energy per unit volume

(vii) f = degree of freedom

= 3 for monatomic gas

= 5 for diatomic and linear polyatomic gas

= 6 for non-linear polyatomic gas

Note Points

- (a) Vibrational degree of freedom is not taken into consideration.
- (b) Translational degree of freedom for any type of gas is three.

(viii) Total internal energy of a gas is

$$U = \frac{nf}{2}RT$$

Here, n = total number of gram moles

(ix) $C_V = \frac{dU}{dT}$ (where U = internal energy of one mole of a gas $= \frac{f}{2}RT$)

$$\therefore \quad C_V = \frac{f}{2}R = \frac{R}{\gamma - 1}$$

(x) $C_p = C_V + R = \left(1 + \frac{f}{2}\right)R = \left(\frac{\gamma}{\gamma - 1}\right)R$

(xi) $\gamma = \frac{C_p}{C_V} = 1 + \frac{2}{f}$

(xii) Internal energy of 1 mole in one degree of freedom of any gas is $\frac{1}{2}RT$.

(xiii) Translational kinetic energy of one mole of any type of gas is $\frac{3}{2}RT$.

(xiv) Rotational kinetic energy of 1 mole of monatomic gas is zero of dia or linear polyatomic gas is $\frac{2}{2}RT$.

or RT, of non-linear polyatomic gas is $\frac{3}{2}RT$.

(xv) **Mixture of non-reactive gases**

(a) $n = n_1 + n_2$

(b) $p = p_1 + p_2$

(c) $U = U_1 + U_2$

(d) $\Delta U = \Delta U_1 + \Delta U_2$

(e) $C_V = \frac{n_1 C_{V_1} + n_2 C_{V_2}}{n_1 + n_2}$

(f) $C_p = \frac{n_1 C_{p_1} + n_2 C_{p_2}}{n_1 + n_2} = C_V + R$

(g) $\gamma = \frac{C_p}{C_V}$ or $\frac{n}{\gamma - 1} = \frac{n_1}{\gamma_1 - 1} + \frac{n_2}{\gamma_2 - 1}$

(h) $M = \frac{n_1 M_1 + n_2 M_2}{n_1 + n_2}$

(xvi) An ideal gas equation connecting pressure (p), volume (V) and absolute temperature (T) is

$$pV = nRT = kNT$$

where, n is the number of moles and N is the number of molecules, R and k are universal constants.

$R = 8.314 \text{ J mol}^{-1}\text{ K}^{-1}$, $k = 1.38 \times 10^{-23} \text{ J K}^{-1}$

Real gases satisfy the ideal gas equation only approximately, more so at low pressures and high temperatures.

(xvii) Kinetic theory of an ideal gas gives the relation

$$p = \frac{1}{3} nm v_{rms}^2$$

where, n is number density of molecules or number of molecules per unit volume m is the mass of one molecule and v_{rms}^2 is the root mean of square speed.

(xviii) The translational kinetic energy

$$E = \frac{3}{2}kNT$$

Here, N = total number of molecules

This leads to a relation $pV = \frac{2}{3}E$

Additional Examples

Example 1. *Why are gas thermometers more sensitive than mercury thermometers?*

Sol. The coefficient of expansion of a gas is very large as compared to the coefficient of expansion of mercury. For the same temperature range, a gas would undergo a much larger change in volume as compared to mercury.

Example 2. *Why is a constant volume gas thermometer preferred as a standard thermometer than a constant pressure gas thermometer?*

Sol. This is because the changes in pressure can be measured with greater accuracy than changes in volume.

Example 3. *Do all solids expand on heating? If not, give an example.*

Sol. No. Camphor contracts on heating.

Example 4. *Is the temperature coefficient of resistance always positive?*

Sol. No. Temperature coefficient α is positive for metals and alloys and negative for semiconductors and insulators.

Example 5. *Why a small gap is left between the iron rails of railway tracks?*

Sol. If no gap is left between the iron rails, the rails may bend due to expansion in summer and the train may get derailed.

Example 6 *Pendulum clocks generally run fast in winter and slow in summer. Why?*

Sol. The time period of a simple pendulum is given by

$$T = 2\pi\sqrt{\frac{l}{g}} \quad \text{i.e. } T \propto \sqrt{l}$$

In winter, l decreases with the fall in temperature, so T decreases and clocks run fast. In summer, l increases with the increase in temperature, so T increases and clock runs slow.

Example 7 *Why iron rims are heated red hot before being put on the cart wheels?*

Sol. The iron ring to be put on the rim of a cart wheel is always of slightly smaller diameter than that of the wheel. When the iron ring is heated to become red hot, it expands and slips on to the wheel easily. When it is cooled, it contracts and grips the wheel firmly.

Example 8 *Explain why a beaker filled with water at 4° C overflows if the temperature is decreased or increased?*

Sol. It is because of the anomalous expansion of water. Water has a maximum density at 4°C . Therefore, water expands whether it is heated above 4°C or cooled below 4°C.

Example 9. *If an electric fan is switched on in a closed room, will the air of the room be cooled? If not, why do we feel cold?*

Sol. The air will not be cooled. In fact, it will get heated up due to the increase in the speed of its molecules. We feel cold due to faster evaporation of sweat.

Example 10. *Two large holes are cut in a metal sheet. If the sheet is heated, how will the diameters of the holes change?*

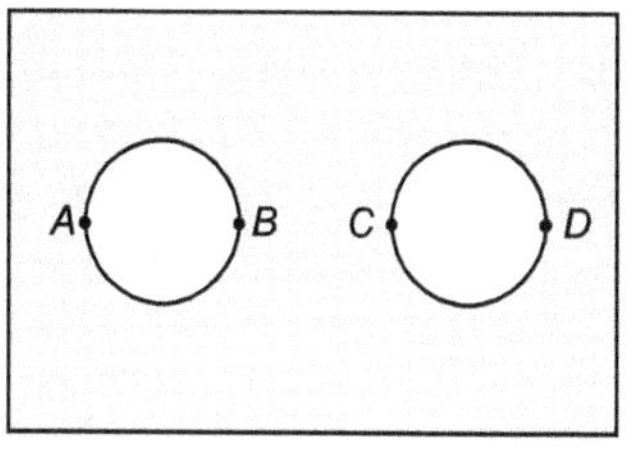

Sol. When a body is heated, the distance between any of its two points increases. Hence, the diameters AB and CD of the two holes will increase.

Example 11. *In example* 11, *will the distance between the two holes increase or decrease on heating?*

Sol. When the metal sheet is heated, it expands as a whole. Therefore, the holes will increase in diameter as well as move outwards. The distance BC between the two holes thus increases.

Example 12. *Out of the parameters-temperature, pressure, work and volume, which parameter does not characterize the thermodynamic state of matter?*

Sol. Work

Example 13. *What is the nature of total internal energy possessed by molecules of an ideal gas?*

Sol. Since, in an ideal gas there is no molecular attraction, the potential energy of the molecules is zero. Therefore, internal energy of an ideal gas is only kinetic in nature.

Example 14. *A piece of metal is hammered. Does its internal energy increase?*

Sol. The work done during hammering gets converted into heat energy. Due to this, its internal energy increases.

Example 15. *What are the conditions for thermodynamic equilibrium?*

Sol.
1. Temperature of every part of the system be the same.
2. There should be no net unbalanced force on a part or whole of the system.
3. There should be no changes due to chemical reactions.

Example 16. *A gas has two specific heats whereas a liquid and a solid have only one. Why?*

Sol. When solids and liquids are heated, there is only a slight change in their volume and as such these possess only one specific heat, i.e. specific heat at constant volume. But in case of gases, pressure and volume both change and as such these possess two principal specific heats, one at constant pressure and one at constant volume.

Example 17. *How does the internal energy of an ideal gas differ from that of real gas?*

Sol. The internal energy of an ideal gas consists of only the kinetic energy of the particles. But for real gases it consists of both the kinetic as well as potential energies.

Example 18. *Out of a solid, liquid and gas of the same mass and at the same temperature, which one has the greatest internal energy? Which one least? Justify .*

Sol. The gas has greatest internal energy because the potential energy (which is negative) of the molecules is very small. On the other hand, the (negative) potential energy of the molecules of a solid is very large, hence the internal energy of a solid is least.

Example 19. *Two vessels separately contain* $500\ cm^3$ *of hydrogen and* $500\ cm^3$ *of oxygen at N.T.P. Which will have larger number of molecules? Give reasons.*

Sol. Both the gases will have equal number of molecules. It is in accordance with Avogadro's hypothesis.

Example 20. *Does the value of degree of freedom of a gas molecule change with rise in temperature?*

Sol. Yes.

Example 21. *Cooking gas containers are kept in a lorry moving with uniform speed. What will be the effect on temperature of the gas molecules inside?*

Sol. As the lorry is moving at a uniform speed, the translational motion of the gas molecules will not be affected. Hence, the temperature of the gas molecules will remain the same.

Example 22. *Molecular motion ceases at zero kelvin. Explain.*

Sol. All molecular motion ceases at 0 K, i.e. at absolute zero. According to the kinetic interpretation of temperature, absolute temperature $\propto$ mean kinetic energy of molecules

Therefore, at temperature = 0 K, kinetic energy = 0

Hence, at 0 K, the velocity of molecules also become zero.

Example 23. *When a gas expands at constant temperature, why does the pressure decrease? Explain it on the basis of kinetic theory of gases.*

Sol. When the gas expands at constant temperature, the average kinetic energy of the gas molecules remains the same. However, due to increase in volume of the gas, separation between the molecules increases.

As a result, the number of molecules colliding per second against the walls decreases. Consequently, less momentum is transferred to the walls of the container per second. Hence, the pressure exerted by the gas decreases.

Example 24. *Why temperature less than absolute zero is not possible?*

Sol. According to the kinetic interpretation of temperature, absolute temperature $\propto$ mean kinetic energy of molecules As the temperature of a gas is decreased, the average speed of the gas molecules also decreases.

At absolute zero the average speed of the gas molecules becomes zero, i.e. energy is possible. Hence, the temperature of the gas cannot be decreased to a temperature below absolute zero.

Example 25. *Briefly explain, why there is practically no atmosphere on the surface of the moon?*

Sol. The molecules of a gas are always in random motion and they possess velocities of the order of a few $km\ s^{-1}$. The velocity of the molecules is quite below the escape velocity on the surface of the earth and hence they are retained in the earth's atmosphere.

But at the surface of the moon, the escape velocity is very low ($\approx 2.5\ ms^{-1}$) and the gas molecules having velocity greater than this value are bound to leave the moon's atmosphere.

Example 26. *Why do the gases at low temperature and high pressure show large deviations from ideal behaviour?*

Sol. At low temperature and high pressure, the intermolecular attractions become appreciable. Moreover, the volume occupied by the gas molecules cannot be neglected in comparison to the volume of the gas. Hence, the real gases show large deviations from ideal gas behaviour.

Example 29. *When an automobile travels for a long distance, the air pressure in the tyres increases slightly. Why?*

Sol. Due to the friction between the tyres and the road, the tyres get heated. The temperature of air inside the tyres increases. Consequently, the air pressure in the tyres increases slightly.

Example 28. *In the kinetic theory of gases, why do we not take into account the changes in gravitational potential energy of the molecule?*

Sol. The changes in gravitational potential energy are negligibly small as compared to the mean kinetic energy of molecules.

Example 29. *Find the rms speed of hydrogen molecules at room temperature $(= 300\,K)$.*

Sol. Mass of 1 mole of hydrogen gas $= 2\,g = 2 \times 10^{-3}$ kg

$$\Rightarrow \quad v_{rms} = \sqrt{\frac{3RT}{M}} = \sqrt{\frac{3 \times 8.31 \times 300}{2 \times 10^{-3}}}$$

$$= 1.93 \times 10^3 \text{ m/s}$$

Example 30. *4 g hydrogen is mixed with 11.2 L of He at STP in a container of volume 20 L. If the final temperature is 300 K, find the pressure.*

Sol. 4 g hydrogen = 2 moles of hydrogen

11.2 L He at STP $= \frac{1}{2}$ mole of He

$$p = p_H + p_{He} = (n_H + n_{He})\frac{RT}{V}$$

$$= \left(2 + \frac{1}{2}\right)\frac{8.31 \times (300\text{ K})}{(20 \times 10^{-3})\text{m}^3}$$

$$= 3.12 \times 10^5 \text{ N/m}^2$$

Example 31. *Find the average kinetic energy per molecule at temperature T for an equimolar mixture of two ideal gases A and B, where A is monoatomic and B is diatomic.*

Sol. Number of degrees of freedom per molecule for $A = 3$

Number of degrees of freedom per molecule for $B = 5$

Since, the mixture is equimolar, the average kinetic energy per molecule will be the simple average of the two values, i.e. $\left(\frac{3+5}{2}\right)kT = 4kT$,

where k is Boltzmann constant.

Example 32. *The steam point and the ice point of a mercury thermometer are marked as 80° and 10°. At what temperature on centigrade scale the reading of this thermometer will be 59° ?*

Sol. Let the relation between the thermometer reading and centigrade be $y = ax + b$

Given, at $x = 100$, $y = 80$ and at $x = 0$, $y = 10$

$\therefore \quad 80 = 100\,a + b, 10 = b$

$\Rightarrow \quad a = 0.7$

Now, we have to find x when $y = 59$

$\therefore \quad 59 = 0.7x + b \Rightarrow x = 70$

$\therefore$ The answer is 70° C

Example 33. *A glass vessel of volume V_0 is completely filled with a liquid and its temperature is raised by ΔT. What volume of the liquid will overflow? Coefficient of linear expansion of glass $= \alpha_g$ and coefficient of volume expansion of the liquid $= \gamma_l$.*

Sol. Volume of the liquid overflown

= Increase in the volume of the liquid

− Increase in the volume of the container

$$= [V_0(1 + \gamma_l \Delta T) - V_0] - [V_0(1 + \gamma_g \Delta T) - V_0]$$

$$= V_0 \Delta T(\gamma_l - \gamma_g) = V_0 \Delta T(\gamma_l - 3\alpha_g) \qquad (\because \gamma_g \approx 3\alpha_g)$$

Example 34. *Find the temperature at which oxygen molecules would have the same rms speed as of hydrogen molecules at 300 K.*

Sol. If T be the corresponding temperature,

$$\sqrt{\frac{3RT}{M_O}} = \sqrt{\frac{3R(300)}{M_H}} \Rightarrow T = (300)\left(\frac{M_O}{M_H}\right)$$

$$= 4800 \text{ K}$$

Example 35. *A platinum resistance thermometer reads 0°C when its resistance is 80 Ω and 100°C when its resistance is 90 Ω. Find the temperature at which the resistance is 86 Ω.*

Sol. The temperature on the platinum scale is

$$t = \frac{R_t - R_0}{R_{100} - R_0} \times 100° = \frac{86 - 80}{90 - 80} \times 100° \text{ C}$$

$$= 60° \text{ C}$$

Example 36. *A sphere of diameter 7 cm and mass 266.5 g floats in a bath of liquid. As the temperature is raised, the sphere just sinks at a temperature of 35°C. If the density of the liquid at 0° C is 1.527 gm/cm³, find the coefficient of cubical expansion of the liquid.*

Sol. The sphere will sink in the liquid at 35°C, when its density becomes equal to the density of liquid at 35° C.

The density of sphere,

$$\rho_{35} = \frac{266.5}{\frac{4}{3} \times \left(\frac{22}{7}\right) \times \left(\frac{7}{2}\right)^3} = 1.483 \text{ g/cm}^3$$

Now,
$$\rho_0 = \rho_{35}[1 + \gamma \Delta T]$$

$$1.527 = 1.483[1 + \gamma \times 35]$$

$$1.029 = 1 + \gamma \times 35$$

$$\gamma = \frac{1.029 - 1}{35} = 0.00083 \text{ /°C}$$

Example 37. *A light steel wire of length l and area of cross-section A is hanging vertically downward with a ceiling. It cools to the room temperature 30°C from the initial temperature 100°C. Calculate the weight which should be attached at its lower end such that its length remains same. Young's modulus of steel is Y and coefficient of linear expansion is α.*

Sol.
$$l\alpha \Delta T = \frac{wl}{AY}$$

$$\therefore \quad w = AY\alpha\Delta T = 70A\alpha Y$$

Example 38. *An air bubble starts rising from the bottom of a lake. Its diameter is 3.6 mm at the bottom and 4 mm at the surface. The depth of the lake is 250 cm and the temperature at the surface is 40°C. What is the temperature at the bottom of the lake? Given atmospheric pressure = 76 cm of Hg and $g = 980\ cm/s^2$.*

Sol. At the bottom of the lake, volume of the bubble

$$V_1 = \frac{4}{3}\pi r_1^3 = \frac{4}{3}\pi (0.18)^3 \text{ cm}^3$$

Pressure on the bubble, p_1 = atmospheric pressure + pressure due to a column of 250 cm of water

$$= 76 \times 13.6 \times 980 + 250 \times 1 \times 980$$
$$= (76 \times 13.6 + 250)\, 980 \text{ dyne/cm}^2$$

At the surface of the lake, volume of the bubble

$$V_2 = \frac{4}{3}\pi r_2^3 = \frac{4}{3}\pi (0.2)^3 \text{ cm}^3$$

Pressure on the bubble,

$$p_2 = \text{atmospheric pressure}$$
$$= (76 \times 13.6 \times 980) \text{ dyne/cm}^2$$
$$T_2 = 273 + 40° \text{C} = 313° \text{ K}$$

Now, $$\frac{p_1V_1}{T_1} = \frac{p_2V_2}{T_2}$$

or $$\frac{(76 \times 13.6 + 250)980 \times \left(\frac{4}{3}\right)\pi (0.18)^3}{T_1} = \frac{(76 \times 13.6) \times 980 \left(\frac{4}{3}\right)\pi (0.2)^3}{313}$$

or $$T_1 = 283.37 \text{ K}$$

$\therefore$ $$T_1 = 283.37 - 273 = 10.37° \text{ C}$$

Example 39. *Given, Avogadro's number $N = 6.02 \times 10^{23}$ and Boltzmann constant $k = 1.38 \times 10^{-23}$ J/K. Calculate*

(i) the average kinetic energy of translation of the molecules of an ideal gas at 0°C and at 100°C.

(ii) also calculate the corresponding energies per mole of the gas.

Sol. (i) According to the kinetic theory, the average kinetic energy of translation per molecule of an ideal gas at kelvin temperature T is $\left(\frac{3}{2}\right)kT$, where k is Boltzman's constant.

At 0° C (T = 273 K), the kinetic energy of translation

$$= \frac{3}{2}kT = \frac{3}{2} \times (1.38 \times 10^{-23}) \times 273$$
$$= 5.65 \times 10^{-21} \text{ J/molecule}$$

At 100° C (T = 373 K), the energy is

$$\frac{3}{2} \times (1.38 \times 10^{-23}) \times 373 = 7.72 \times 10^{-21} \text{ J/mol}$$

(i) 1 mole of gas contains N $(= 6.02 \times 10^{23})$ mol

Therefore, at 0° C, the kinetic energy of translation of 1 mole of the gas is

$$= (5.65 \times 10^{-21})\ (6.02 \times 10^{23})$$
$$\approx 3401 \text{ J/mol and at } 100° \text{ C}$$

The kinetic energy of translation of 1 mol of gas is

$$= (7.72 \times 10^{-21})(6.02 \times 10^{23}) \approx 4647 \text{ J/mol}$$

Example 40. *One mole of an ideal monoatomic gas is taken at a temperature of 300 K. Its volume is doubled keeping its pressure constant. Find the change in internal energy.*

Sol. Since, pressure is constant

$\therefore$ $$V \propto T$$

$\therefore$ $$\frac{V_i}{T_i} = \frac{V_f}{T_f}$$

$\therefore$ $$T_f = \frac{V_f}{V_i} T_i$$

$\Rightarrow$ $$T_f = 2T_i = 600 \text{ K}$$

$\therefore$ $$\Delta U = \frac{f}{2} n \cdot R\Delta T$$
$$= \frac{3}{2} R(600 - 300) = 450\, R$$

NCERT Selected Questions

Q 1. The triple point of neon and carbon dioxide are 24.57 K and 216.55 K, respectively. Express these temperatures on the celsius and fahrenheit scales.

Sol. **On Celsius Scale**

$$\frac{C-0}{100-0} = \frac{T-273.15}{100}$$

or $\quad C = T - 273.15$

For neon, $\quad t_1\,°C = 24.57 - 273.15 = -248.48°C$

For CO_2, $\quad t_2\,°C = 216.55 - 273.15 = -56.6°C$

On Fahrenheit Scale

$$\frac{F-32}{180} = \frac{T-273.15}{100}$$

For neon, $\quad F_1 = (T_1 - 273.15)\times\frac{9}{5} + 32$

$$= (24.57 - 273.15)\times\frac{9}{5} + 32$$

$$= -248.58 \times \frac{9}{5} + 32$$

$$= -415.44\ °F$$

For CO_2, $\quad F_2 = (T_2 - 273.15)\times\frac{9}{5} + 32$

$$= (216.55 - 273.15)\frac{9}{5} + 32$$

$$= -56.6 \times \frac{9}{5} + 32$$

$$= -69.88\ °F$$

Q 2. Two absolute scales A and B have triple points of water defined to be 200 A and 350 B. What is the relation between T_A and?T_B

Sol. Here, triple point of water on absolute scale A = 200 A

and triple point of water on absolute scale B = 350 B

Also, we know that triple point of water on absolute scale = 273.16K

Thus, it follows that temperature 200 A and 350 B on absolute scale A are equivalent to temperature 273.16 on absolute scale, i.e. the absolute scales measure the triple point as 200 A and 350 B.

∴ Size of one degree of kelvin scale on absolute scale

$$A = \frac{273.16}{200} \quad \text{or} \quad 1A = \frac{273.16}{200}\text{ K}$$

and size of one degree of Kelvin scale onabsolute scale

$$B = \frac{273.16}{350} \quad \text{or} \quad 1B = \frac{273.16}{350}\text{ K}$$

Value of temperature T_A on absolute scale

$$= \frac{273.16}{200}\times T_A$$

and value of temperature T_B on absolute scale

$$= \frac{273.16}{350}\times T_B$$

As T_A and T_B are the same temperature.

∴ $$\frac{273.16}{200}\times T_A = \frac{273.16}{350}\times T_B$$

or $$T_A = \frac{200}{350}\times T_B = \frac{4}{7}T_B$$

Q 3. A steel tape 1 m long is correctly calibrated for a temperature of 27.0 °C. The length of steel rod measured by this tape is found to be 63.0 cm on a hot day when the temperature is 45.0 °C. What is the actual length of the steel rod on that day? What is the length of the same steel rod on a day when temperature is 27.0°C? Coefficient of linear expansion of steel = $1.20\times 10^{-5}\ K^{-1}$.

Sol. Change in length, $\Delta l = \alpha\, l_0\, \Delta t = 1.2\times 10^{-5}\times 63\times 18$ cm

$(\because \Delta T = 45 - 27 = 18\ °C)$

$= 0.0136$ cm

∴ The actual length of the steel rod at 45° C.

$= l_0 + \Delta l\ \ 63 + 0.0136 = 63.0136$ cm

The data is correct only up to three significant figures. The actual length of the rod is 63.0 cm, but the change in length of the rod is 0.0136 cm.

On a day when temperature is 27 °C, the size of 1 cm mark on the steel tape will be exactly 1 cm (due to calibration of steel tape at 27 °C)

∴ Length of rod at 27°C = 63.0× 1 cm = 63.0 cm

Q 4. A large steel wheel is to be fitted on a shaft of the same material. At 27°C, the outer diameter of the shaft is 8.70 cm and the diameter of the central hole of the wheel is 8.69 cm. The shaft is cooled using 'dry ice'. At what temperature of the shaft does the wheel slip on the shaft? Assume the coefficient on linear expansion of the steel to be constant over the required temperature range

$$\alpha_{\text{steel}} = 1.20\times 10^{-5}\ K^{-1}$$

Sol. When the shaft is cooled, its linear dimension, i.e. its diameter decreases according to the formula

$$d_2 = d_1\,[1 + \alpha\,(T_1 - T_1)]$$

$$T_1 = 27°C = 300\text{ K}$$

$$8.69 = 8.70\,[1 + 1.20\times 10^{-5}\,(T_2 - 300)]$$

or $$T_2 - 300 = \frac{8.69 - 8.70}{8.70\times 1.20\times 10^5} = -95.78\text{ K}$$

or $$T_2 = 300 - 95.78 = 204.22\text{ K}$$

$$= 204.22 - 273.15 = -68.93°C$$

or $$T_2 \approx -69°C$$

Q 5. A hole is drilled in a copper sheet. The diameter of the hole is 4.24 cm at 27.0°C. What is the change in the diameter of the hole when the sheet is heated to 227°C? Coefficient of linear expansion of copper $= 1.70 \times 10^{-5}\,°C^{-1}$.

Sol. Let β be the coefficient of superficial expansion of copper, then $\beta = 2\alpha = 2 \times 1.7 \times 10^{-5} = 3.4 \times 10^{-5}\,°C^{-1}$

If S_1, S_2 be the surface areas of the hole at 27°C and 227 °C, then

$$S_1 = \frac{\pi d_1^2}{4} = \frac{\pi}{4} \times (4.24)^2$$

$$= 4.494\,\pi \text{ cm}^2$$

$\therefore$ $$S_2 = S_1 (1 + \beta\,\Delta t)$$

$$= 4.494\,\pi\,(1 + 3.40 \times 10^{-5} \times 200)$$

or $$S_2 = 4.494\,\pi \times 1.0068 = 4.525\,\pi \text{ cm}^2$$

or $$\frac{\pi d_2^2}{4} = 4.525\,\pi$$

or $$d_2 = \sqrt{4.525 \times 4} = 4.2544 \text{ cm}$$

$\therefore$ $$\Delta d = d_2 - d_1$$

$$= 4.2544 - 4.24 = 0.0144 \text{ cm}$$

or $$\Delta d = 1.44 \times 10^{-2} \text{ cm}$$

Q 6. A brass wire 1.8 m long at 27 °C is held taut with little tension between two rigid supports. If the wire is cooled to a temperature of −39°C, what is the tension developed in the wire, if its diameter is 2.0 mm? Coefficient of linear expansion of brass $= 2.0 \times 10^{-5}\,K^{-1}$. Young's modulus of brass $= 0.91 \times 10^{11}$ Pa.

Sol. If A be the area of cross-section of the wire, then

$$A = \frac{\pi d^2}{4}$$

$$= \frac{\pi}{4} \times (2.0 \times 10^{-3})^2$$

$$= 3.142 \times 10^{-6} \text{ m}^2$$

If F be the tension developed in the wire, then using the relation

$$Y = \frac{F/A}{\Delta l / l}, \text{ we get}$$

$$F = \frac{YA\Delta l}{l} \quad \text{...(i)}$$

Now, $$\Delta l = l\alpha\,\Delta t = 1.8 \times 2 \times 10^{-5} \times (-66)$$

$$= -0.00237 \text{ m} = -0.0024 \text{ m}$$

– ve sign shows that the length decreases.

Putting values of Y, A, Δl and l in Eq. (i), we get

$$F = \frac{0.91 \times 10^{11} \times 3.142 \times 10^{16} \times 24 \times 10^{-4}}{1.8} = 381 \text{ N}$$

Q 7. A brass rod of length 50 cm and diameter 3.0 mm is joined to a steel rod of the same length and diameter. What is the change in length of the combined rod at 250°C, if the original lengths are at 40°C? Is there a 'thermal stress' developed at the junction? The ends of the rod are free to expand. (coefficient of linear expansion of brass $= 2.0 \times 10^{-5}\,K^{-1}$, steel $= 1.2 \times 10^{-5}\,K^{-1}$.)

Sol. **For brass rod,**

$$\Delta t = t_2 - t_1 = 250 - 40 = 210°C$$

If l_2 be its length at t_2°C, then

$$l_2 = l_1 (1 + \alpha\Delta t)$$

$$= 50 (1 + 2 \times 10^{-5} \times 210)$$

$$= 50.21 \text{ cm}$$

$\therefore$ $$\Delta l_{brass} = l_2 - l_1 = 50.21 - 50$$

$$= 0.21 \text{ cm}$$

For steel rod,

$$\Delta t' = t_2' - t_1' = 250 - 40 = 210°C$$

If l_2' be the length of the steel rod at 250°C, then

$$l_2' = l_1' (1 + \alpha\,\Delta t')$$

$$= 50 (1 + 1.2 \times 10^{-5} \times 210)$$

$$= 50.126 \text{ cm}$$

$\therefore$ $$\Delta l_{steel} = l_2' - l_1' = 50.126 - 50$$

$$= 0.126 \text{ cm} \approx 0.13 \text{ cm}$$

$\therefore$ The length of the combined rod at 250°C

$$= l_2 + l'_2$$

$$= 50.21 + 50.126$$

$$= 100.336 \text{ cm}$$

and length of the combined rod at

$$40°C = l_1 + l'_1$$

$$= 50 + 50 = 100 \text{ cm}$$

$\therefore$ Total change in length of the combined rod

$$= 100.336 - 100$$

$$= 0.336 \text{ cm}$$

$$\approx 0.34 \text{ cm}$$

No thermal stress is developed at the junction since the rods can freely expand.

Q 8. The coefficient of volume expansion of glycerine is $49 \times 10^{-5}\,K^{-1}$. What is the fractional change in its density for a 30°C rise in temperatures?

Sol. Let V_0 be the initial volume of glycerine, i.e. at 0°C (dry)

If V_t be its volume at 30°C

Then, $$V_t = V_0 (1 + \gamma\Delta t)$$

$$= V_0 (1 + 49 \times 10^{-5} \times 30)$$

$$= V_0 (1 + 0.01470) = 1.01470\,V_0$$

or $$\frac{V_0}{V_t} = \frac{1}{1.01470} \quad \text{...(i)}$$

Let ρ_0 and ρ_t be the initial and final densities of the glycerine.

Then initial density, $\rho_0 = \frac{m}{V_0}$ and final density, $\rho_t = \frac{m}{V_t}$

where, $m =$ mass of glycerine.

$$\frac{\Delta\rho}{\rho_0} = \text{Fractional change in density}$$

$$= \frac{\rho_t - \rho_0}{\rho_0} = \frac{m\left(\frac{1}{V_t} - \frac{1}{V_0}\right)}{\frac{m}{V_0}} = \left(\frac{V_0}{V_t} - 1\right)$$

or $$\frac{\Delta\rho}{\rho_0} = \left(\frac{1}{1.01470} - 1\right) = -0.0145$$

Here, negative sign shows that the density decreases with the rise in temperature.

$\therefore$ $$\frac{\Delta\rho}{\rho_0} = 0.0145 = 1.45 \times 10^{-2} \approx 1.5 \times 10^{-2}$$

Q 9. Estimate the fraction of molecular volume to the actual volume occupied by oxygen gas at STP. Take the radius of an oxygen molecule to be roughly 3 Å.

Sol. A gram mole of oxygen gas occupies a volume at STP,

$$V = 22400 \times 10^{-6}\ \text{m}^3$$

$r =$ Radius of an oxygen molecule $= 3\text{Å} = 3 \times 10^{-10}$ m

Molecular volume of one mole of oxygen, V'

$= $ Volume of one molecule of $O_2 \times$ Avogadro's number (N)

$$= \frac{4}{3}\pi r^3 \times N$$

or $$V' = \frac{4}{3}\pi \times (3 \times 10^{-10})^3 \times 6.023 \times 10^{23}$$

$\therefore$ Fraction of molecular volume to the actual volume occupied by oxygen

$$= \frac{V'}{V} = \frac{\text{Molecular volume of one mole } O_2}{\text{Gram molar volume of } O_2 \text{ at STP}}$$

$$= \frac{\frac{4}{3}\pi \times (3 \times 10^{-10})^3 \times 6.023 \times 10^{23}\ \text{m}^3}{22400 \times 10^{-6}\ \text{m}^3}$$

$$= 4 \times 10^{-4}$$

Q 10. Molar volume is the volume occupied by 1 mole of any (ideal) gas at standard temperature and pressure (STP-1 atm pressure, 0°C). Show that it is 22.4 L.

Sol. At STP,

$$p = 1\ \text{atm} = 1.013 \times 10^5\ \text{Nm}^{-2}$$

$$T = 273\ \text{K}$$

$\therefore$ Using the gas equation,

$pV = nRT$, we get

$$V = \frac{nRT}{p} = \frac{1 \times 8.31 \times 273}{1.013 \times 10^5}$$

$$= 22.4 \times 10^{-3}\ \text{m}^3 = 22.4\ \text{L}$$

Q 11. An oxygen cylinder of volume 30 L has an initial gauge pressure of 15 atm and a temperature of 27°C. After some oxygen is withdrawn from the cylinder, the gauge pressure drops to 11 atm and its temperature drops to 17°C. Estimate the mass of oxygen taken out of the cylinder. ($R = 8.31\ \text{JK}^{-1}\ \text{mol}^{-1}$, molecular mass of $O_2 = 32$ u.)

Sol. If n_1 be the moles of oxygen gas contained in the cylinder, then using the gas equation $pV = nRT$, we get

$$n_1 = \frac{p_1 V_1}{RT_1}$$

$$= \frac{15 \times 1.013 \times 10^5 \times (30 \times 10^{-3})}{8.31 \times 300}$$

$$= 18.253$$

For oxygen, molecular weight

$$m = 32\text{g} = 32 \times 10^{-3}\ \text{kg}$$

$\therefore$ $$m_1 = n_1 M$$

$$m = 584.1 \times 10^{-3}\ \text{kg}$$

Finally, in the oxygen cylinder let n_2 moles of oxygen are left

Then, $$n_2 = \frac{p_2 V_2}{RT_2}$$

$$= \frac{(11 \times 1.013 \times 10^5) \times (30 \times 10^{-3})}{8.31 \times 290}$$

$$= 13.847$$

$\therefore$ Final mass of oxygen gas in the cylinder,

$$m_2 = 13.847 \times 32 \times 10^{-3}$$

$$= 453.1 \times 10^{-3}\ \text{kg}$$

$\therefore$ Mass of oxygen taken out of the cylinder

$$= m_1 - m_2$$

$$= (584.1 - 453.1) \times 10^{-3}\ \text{kg}$$

$$= 141 \times 10^{-3}\ \text{kg}$$

$$= 0.141\ \text{kg}$$

Q 12. An air bubble of volume 1.0 cm^3 rises from the bottom of a lake 40 m deep at a temperature of 12°C. To what volume does it grow when it reaches the surface, which is at a temperature of 35°C?

Sol. When the air bubble is at a depth of 40 m,

Then, $$p_1 = 1\ \text{atm} + 40\ \text{m depth of water}$$

$$= 1\ \text{atm} + h_1 \rho g$$

$$= 1.013 \times 10^5 + 40 \times 10^3 \times 9.8$$

$$= 493000\ \text{Pa} = 4.93 \times 10^5\ \text{Pa}$$

When the air bubble reaches at the surface of lake, then

$$T_2 = 35°\text{C}$$

$$= 35 + 273 = 308\ \text{K}$$

$$p_2 = 1\ \text{atm} = 1.013 \times 10^5\ \text{Pa}$$

Using the equation $\frac{p_1V_1}{T_1} = \frac{p_2V_2}{T_2}$, we get

$$V_2 = \frac{p_1V_1}{T_1} \times \frac{T_2}{p_2}$$

$$= \frac{4.93 \times 10^5 \times 1 \times 10^{-6} \times 308}{285 \times 1.013 \times 10^5}$$

$$= 5.275 \times 10^{-6} \text{ m}^3$$

$$\approx 5.3 \times 10^{-6} \text{ m}^3$$

Q 13. Estimate the total number of air molecules (inclusive of oxygen, nitrogen, water vapour and other constituents) in a room of capacity 25.0 m^3 at a temperature of 27°C and 1 atm pressure.

Sol. $k = 1.38 \times 10^{-23} \text{ JK}^{-1}$

$p = 1 \text{ atm} = 1.013 \times 10^5 \text{ Pa}$

Now, the gas equation $pV = nRT$ may be written as

$$p = \frac{nRT}{V} = \frac{n}{V}(Nk)T \qquad \left(\because \frac{R}{N} = k\right)$$

$$= (nN)\frac{kT}{V} = N'\frac{kT}{V}$$

where, $N' = nN =$ total number of air molecules in the given gas.

$$\therefore \quad N' = \frac{pV}{kT} = \frac{(1.013 \times 10^5) \times 25}{1.38 \times 10^{-23} \times 300} = 6.10 \times 10^{26}$$

Q 14. Estimate the average thermal energy of helium atom at (i) room temperature (27°C), (ii) the temperature on the surface of the sun (6000 K), (iii) the temperature of 10 million kelvin (the typical core temperature in the case of a star).

Sol. According to kinetic theory of gases, the average kinetic energy of the gas at a temperature T is given by

$$E = \frac{3}{2}kT$$

Here, $k = 1.38 \times 10^{-23} \text{ JK}^{-1}$

(i) $T = 27°\text{C} = 273 + 27 = 300 \text{ K}$

$$E = \frac{3}{2} \times 1.38 \times 10^{-23} \times 300$$

$$= 621 \times 10^{-23} \text{ J} = 6.21 \times 10^{-21} \text{ J}$$

(ii) $T = 6000 \text{ K}$

$$\therefore \quad E = \frac{3}{2} \times 1.38 \times 10^{-23} \times 6000 = 1.24 \times 10^{-19} \text{ J}$$

(iii) $T = 10 \times 10^6 \text{ K} = 10^7 \text{ K}$

$$\therefore \quad E = \frac{3}{2} \times 1.38 \times 10^{-23} \times 10^7$$

$$= 2.07 \times 10^{-16} \text{ J} \approx 2.1 \times 10^{-16} \text{ J}$$

Q 15. Three vessels of equal capacity have gases at the same temperature and pressure. The first vessel contains neon (monoatomic gas) the second contains chlorine (diatomic) and the third contains uranium hexafluoride (polyatomic). Do the vessels contain equal number of respective molecules? Is the root mean square speed of molecules the same in the three cases? If not, in which case is v_{rms} the largest?

Sol. Yes. According to Avogadro's hypothesis, if the condition of temperature and pressure remain same, the equal volumes of all gases have equal number of molecules. So, the number of molecules in the three vessels containing different gases must be same i.e. the three vessels contain equal number of respective molecules.

No. rms speed of molecules is not same in three cases as explained below.

We know that the rms velocity of a gas is given by

$$v_{rms} = \sqrt{\frac{3RT}{M}} \text{ or } v_{rms} \propto \frac{1}{\sqrt{M}},$$

where, $M =$ molecular mass and is different for different gases, therefore v_{rms} of three gases will be different.

As v_{rms} is inversely proportional to mass of molecules of the gas and since neon is the lightest of the three gases, therefore v_{rms} for neon gas is largest.

Q 16. At what temperature is the root mean square speed of an atom in an argon gas cylinder equal to the rms speed of a helium gas atom at −20° C? (Atomic mass of Ar = 39.9 u of He = 4.0 u)

Sol. Let v_1 and v_2 be the rms speed of argon and helium gas atoms at temperature T_1 and T_2, respectively.

Here, $M_1 = 39.9 \times 10^{-3}$ kg

$M_2 = 4.0 \times 10^{-3}$ kg

$T_2 = -20 + 273 = 253$ K

Now, we know that rms speed is given by $v = \sqrt{\frac{3RT}{M}}$

$$\therefore \quad v_1 = \sqrt{\frac{3RT_1}{M_1}} \text{ and } v_2 = \sqrt{\frac{3RT_2}{M_2}}$$

As $v_1 = v_2$ (given)

$$\therefore \quad \sqrt{\frac{3RT_1}{M_1}} = \sqrt{\frac{3RT_2}{M_2}} \text{ or } \frac{T_1}{M_1} = \frac{T_2}{M_2}$$

or $$T_1 = \frac{M_1}{M_2} \times T_2 = \frac{39.9 \times 10^{-3}}{4.0 \times 10^{-3}} \times 253$$

or $$T_1 = 2523.7 \text{ K} \approx 2524 \text{ K}$$

Objective Problems

[Level 1]

Temperature Scales, Thermometers and Thermal Expansion

1. The absolute zero temperature in fahrenheit scale is

(a) – 273°F (b) – 32°F
(c) – 460 °F (d) – 132 °F

2. A faulty thermometer has its fixed points marked 5 and 95. If the temperature of a body as shown on the celsius scale is 40, then its temperature shown on this faulty thermometer is

(a) 39° (b) 40°
(c) 41° (d) 44.4°

3. A steel rod of diameter 1 cm is clamped firmly at each end when its temperature is 25°C so that it cannot contract on cooling.The tension in the rod at 0°C is approximately ($\alpha = 10^{-5}/°C, Y = 2\times10^{11}$ N/m^2)

(a) 4000 N
(b) 7000 N
(c) 7400 N
(d) 4700 N

4. The temperature of a physical pendulum, whose time period is T, is raised by $\Delta\theta$. The change in its time period is

(a) $\frac{1}{2}\alpha T\Delta\theta$ (b) $2\alpha T\Delta\theta$
(c) $\frac{1}{2}\alpha\Delta\theta$ (d) $2\alpha\Delta\theta$

5. On heating a liquid having coefficient of volume expansion α in a container having coefficient of linear expansion $\alpha/2$, the level of the liquid in the container would

(a) rise
(b) fall
(c) remains almost stationary
(d) cannot be predicted

6. Two rods of length l_1 and l_2 are made of materials whose coefficient of linear expansions are α_1 and α_2, respectively. If the difference between two lengths is independent of temperature, then

(a) $\frac{l_1}{l_2} = \frac{\alpha_1}{\alpha_2}$
(b) $\frac{l_1}{l_2} = \frac{\alpha_2}{\alpha_1}$
(c) $l_2^2\alpha_1 = l_1^2\alpha_2$
(d) $\frac{\alpha_1^2}{l_1} = \frac{\alpha_2^2}{l_2}$

7. If two rods of lengths L and $2L$ having coefficient of linear expansion α and 2α respectively, are connected end-on-end, the average coefficient of linear expansion of the composite rod, equals

(a) $\frac{3}{2}\alpha$ (b) $\frac{5}{2}\alpha$
(c) $\frac{5}{3}\alpha$ (d) None of these

8. A solid ball of metal has a spherical cavity inside it. If the ball is heated, the volume of the cavity will

(a) increase
(b) decrease
(c) remain unchanged
(d) data insufficient

9. The radius of a ring is R and its coefficient of linear expansion is α. If the temperature of ring increases by θ, then its circumference will increase by

(a) $\pi R\alpha\theta$ (b) $2\pi R\alpha\theta$
(c) $\pi R\alpha\frac{\theta}{2}$ (d) $\pi R\alpha\frac{\theta}{4}$

10. Two rods of different materials having coefficients of thermal expansions α_1, α_2 and Young's moduli Y_1, Y_2 respectively are fixed between two rigid massive walls. The rods are heated such that they undergo the same increase in temperature. There is no bending of the rods. If $\alpha_1 : \alpha_2 = 2:3$, the thermal stresses developed in the two rods are equal provided $Y_1 : Y_2$ is equal to

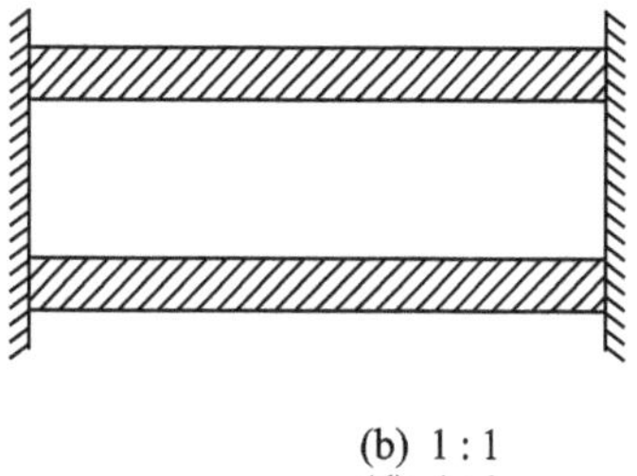

(a) 2 : 3 (b) 1 : 1
(c) 3 : 2 (d) 4 : 9

11. The freezing point on a thermometer is marked as – 20° and the boiling point as 130°. A temperature of human body (34° C) on this thermometer will be read as

(a) 31° (b) 51°
(c) 20° (d) None of these

12. On which of the following scales of temperature, the temperature is never negative?

(a) Celsius (b) Fahrenheit
(c) Reaumur (d) Kelvin

13. The temperature of a body on kelvin scale is found to be x K. When it is measured by fahrenheit thermometer, it is found to be $x°$F, then the value of x is

(a) 40 (b) 313
(c) 574.25 (d) 301.25

14. A device used to measure very high temperature is

(a) pyrometer (b) thermometer
(c) bolometer (d) calorimeter

15. A uniform metal rod is used as a bar pendulum. If the room temperature rises by 10° C, and the coefficient of linear expansion of the metal of the rod is 2×10^{-6} per° C, the period of the pendulum will have percentage increase of

(a) -2×10^{-3} (b) -1×10^{-3}
(c) 2×10^{-3} (d) 1×10^{-3}

16. In a vertical U-tube containing a liquid, the two arms are maintained at different temperatures t_1 and t_2. The liquid columns in the two arms have heights l_1 and l_2, respectively. The coefficient of volume expansion of the liquid is equal to

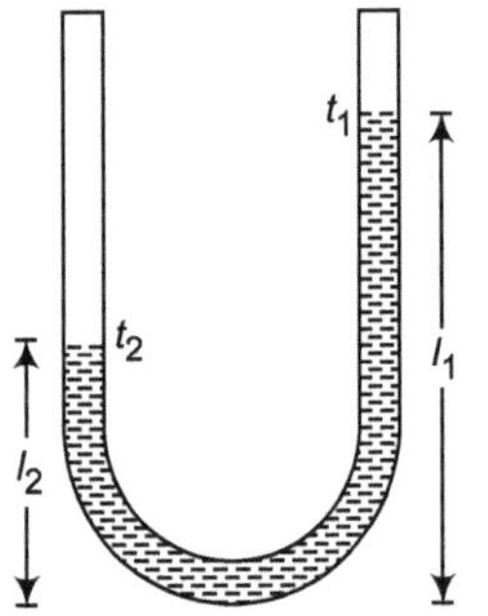

(a) $\frac{l_1 - l_2}{l_2 t_1 - l_1 t_2}$ (b) $\frac{l_1 - l_2}{l_1 t_1 - l_2 t_2}$
(c) $\frac{l_1 + l_2}{l_2 t_1 + l_1 t_2}$ (d) $\frac{l_1 + l_2}{l_1 t_1 + l_2 t_2}$

17. The coefficient of linear expansion of crystal in one direction is α_1 and that in other two directions perpendicular to it is α_2. The coefficient of cubical expansion is

(a) $\alpha_1 + \alpha_2$ (b) $2\alpha_1 + \alpha_2$
(c) $\alpha_1 + 2\alpha_2$ (d) None of these

18. The absolute zero temperature in fahrenheit scale is

(a) $-237°$F (b) $-32°$F
(c) $-460°$F (d) $-132°$F

19. The reading of air thermometer at 0° C and 100° C are 50 cm and 75 cm of mercury column, respectively. The temperature at which its reading is 80 cm of mercury column is

(a) 105°C (b) 110°C
(c) 115°C (d) 120°C

20. Bimetal strips are used for

(a) metal thermometers
(b) opening of closing electrical circuits
(c) thermostats
(d) All of the above

Gas Laws and Kinetic Theory of Gases

21. What will be the temperature when the rms velocity is double of that at 300 K?

(a) 300 K (b) 600 K (c) 900 K (d) 1200 K

22. By what factor the rms velocity will change, if the temperature is raised from 27°C to 327°C?

(a) $\sqrt{2}$ (b) 2 (c) $2\sqrt{2}$ (d) 1

23. If Maxwell distribution is valid and v_p denotes the most probable speed, v the average speed and v_{rms} the root-mean-square speed, then

(a) $v < v_p < v_{rms}$ (b) $v < v_{rms} < v_p$
(c) $v_p < v < v_{rms}$ (d) $v_p < v_{rms} < v$

24. The root mean square (rms) speed of oxygen molecules (O_2) at a certain absolute temperature is v. If the temperature is doubled and the oxygen gas dissociates into atomic oxygen, the rms speed would be

(a) v (b) $\sqrt{2}\,v$ (c) $2\,v$ (d) $2\sqrt{2}\,v$

25. The number of molecules in 1 cc of water is closed to

(a) 6×10^{23} (b) 22.4×10^{24} (c) $\frac{10^{23}}{3}$ (d) 10^{23}

26. 16 g of oxygen, 14 g of nitrogen and 11 g of carbon dioxide are mixed in an enclosure of volume 5 L and temperature 27°C. The pressure exerted by the mixture is

(a) 4×10^5 Nm^{-2} (b) 5×10^5 Nm^{-2}
(c) 6×10^5 Nm^{-2} (d) 9×10^5 Nm^{-2}

27. 22 g of CO_2 at 27°C is mixed in a closed container with 16 g of O_2 at 37°C. If both gases are considered as ideal kinetic theory gases, then the temperature of the mixture is

(a) 28.5°C (b) 30.5°C (c) 31.5°C (d) 33.0°C

28. The ratio of the speed of sound in nitrogen gas to that in helium gas, at 300 K is

(a) $\sqrt{\left(\frac{2}{7}\right)}$ (b) $\sqrt{\left(\frac{1}{7}\right)}$ (c) $\frac{\sqrt{3}}{5}$ (d) $\frac{\sqrt{6}}{5}$

29. The velocities of three molecules are $3v, 4v$ and $5v$ respectively. Their rms speed will be

(a) $\sqrt{\frac{50}{3}}\,v$ (b) $\sqrt{\frac{5}{2}}\,v$ (c) $\frac{7}{2}\,v$ (d) $\frac{5}{2}\,v$

30. The number of molecules per unit volume of a gas is given by

(a) $\frac{p}{kT}$ (b) $\frac{kT}{p}$ (c) $\frac{p}{RT}$ (d) $\frac{RT}{p}$

31. The temperature at which the root mean square speed of a gas will be half its value at 0°C is (assume the pressure remains constant)

(a) −86.4°C (b) −204.75°C
(c) −104.75°C (d) −68.25°C

32. Four molecules of a gas have speeds 1, 2, 3 and 4 km s^{-1}. The value of the root mean square speed of the gas molecules is

(a) $\frac{1}{2}\sqrt{15}\ \text{kms}^{-1}$ (b) $\frac{1}{2}\sqrt{10}\ \text{kms}^{-1}$
(c) $2.5\ \text{kms}^{-1}$ (d) $\sqrt{\frac{15}{2}}\ \text{kms}^{-1}$

33. In the gas below $pV = RT, V$ refers to the volume of

(a) any amount of gas
(b) one gram of gas
(c) one gram mole of gas
(d) one litre of the gas

34. In the adjacent V-T diagram, what is the relation between p_1 and p_2?

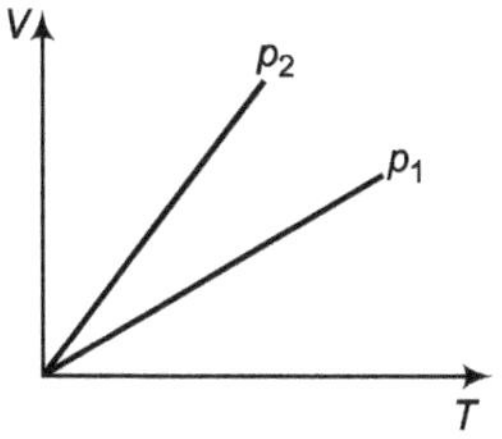

(a) $p_2 = p_1$ (b) $p_2 > p_1$
(c) $p_2 < p_1$ (d) Cannot be predicted

35. The molecules of a given mass of gas have rms speed $200\ \text{ms}^{-1}$ at 27°C and $10^5\ \text{Nm}^{-2}$ pressure. When the absolute temperature is doubled and the pressure is halved, the rms speed of the molecules of the same gas is

(a) $200\ \text{ms}^{-1}$ (b) $400\ \text{ms}^{-1}$
(c) $200\sqrt{2}\ \text{ms}^{-1}$ (d) $400\sqrt{2}\ \text{ms}^{-1}$

36. The average speed of molecules in a gas is given by

(a) $\sqrt{\frac{3p}{\rho}}$ (b) $\sqrt{\frac{\gamma p}{\rho}}$
(c) $\sqrt{\frac{2p}{\rho}}$ (d) $\sqrt{\frac{8p}{\pi\rho}}$

37. The most probable speed of molecules in a gas is given by

(a) $\sqrt{\frac{3p}{\rho}}$ (b) $\sqrt{\frac{8p}{\pi\rho}}$
(c) $\sqrt{\frac{2p}{\rho}}$ (d) $\sqrt{\frac{\gamma p}{\rho}}$

38. A perfect gas at 27°C is heated at constant pressure so as to triple its volume. The temperature of the gas will be

(a) 81°C (b) 900°C
(c) 627°C (d) 450°C

39. The below figure shows graph of pressure and volume of a gas at two temperatures T_1 and T_2. Which of the following is correct?

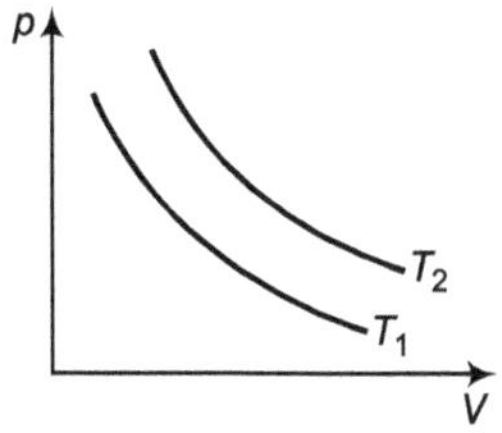

(a) $T_1 > T_2$
(b) $T_1 = T_2$
(c) $T_1 < T_2$
(d) Nothing can be said about temperatures

40. The figure shows pressure *versus* density graph for an ideal gas at two temperatures T_1 and T_2.

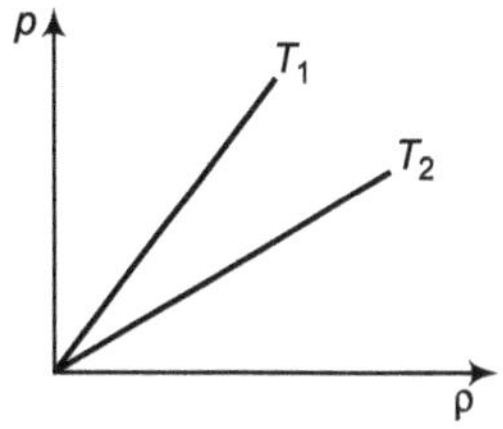

The relation between T_1 and T_2 is

(a) $T_1 > T_2$
(b) $T_1 = T_2$
(c) $T_1 < T_2$
(d) Nothing can be predicted

41. From the p-T graph what conclusion can be drawn?

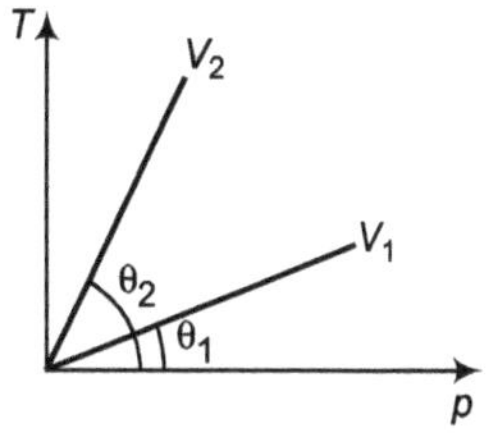

(a) $V_2 = V_1$ (b) $V_2 < V_1$
(c) $V_2 > V_1$ (d) Nothing can be predicted

42. pV *versus* T graph of equal masses of H_2, He and O_2 is shown in figure. Choose the correct alternative.

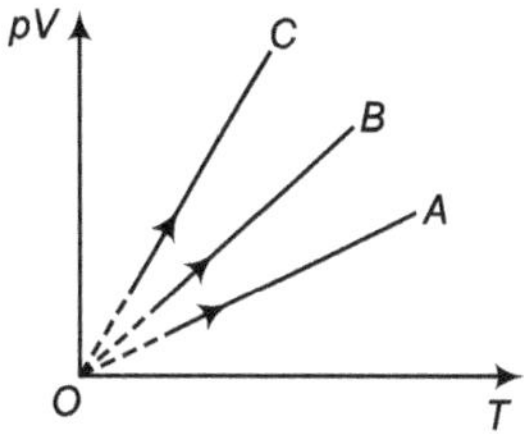

(a) A corresponds to H_2, B to He and C to O_2
(b) A corresponds to He, B to H_2 and C to O_2
(c) A corresponds to He, B to O_2 and C to H_2
(d) A corresponds to O_2, B to H_2 and C to He

43. Pressure *versus* temperature graph of an ideal gas at constant volume V is shown by the straight line A. Now, mass of the gas is doubled and volume is halved, then the corresponding pressure *versus* temperature graph will be shown by the line

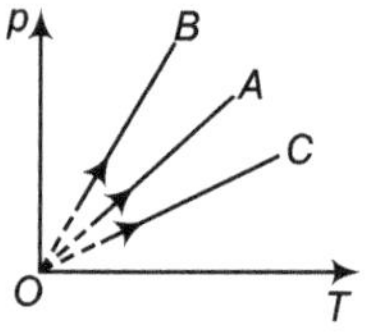

(a) A (b) B
(c) C (d) None of these

44. Which one of the following graphs represent the behaviour of an ideal gas at constant temperature?

(a)
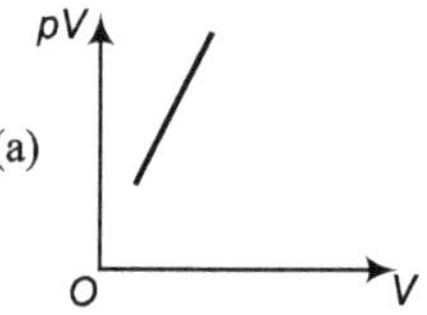

(b)
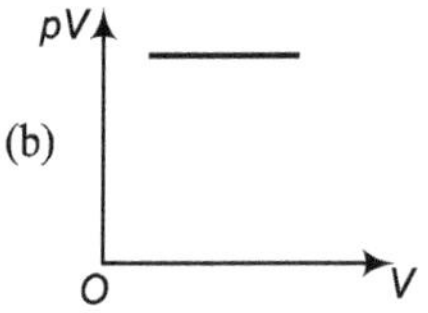

(c)
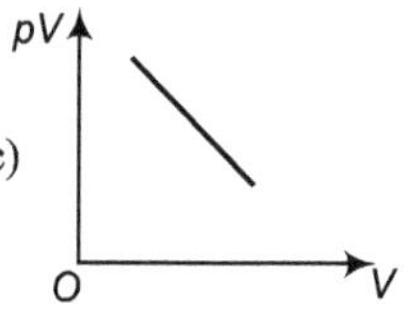

(d)
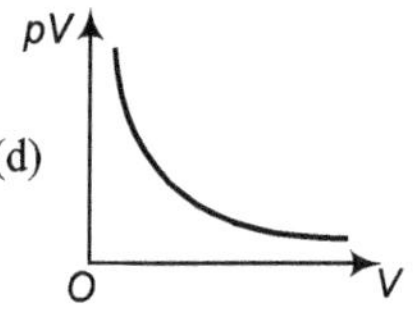

45. The curve between absolute temperature and v_{rms}^2 is

(a)
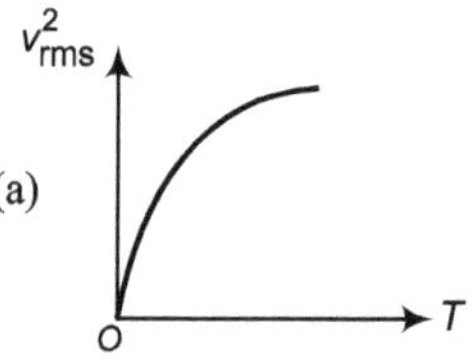

(b)
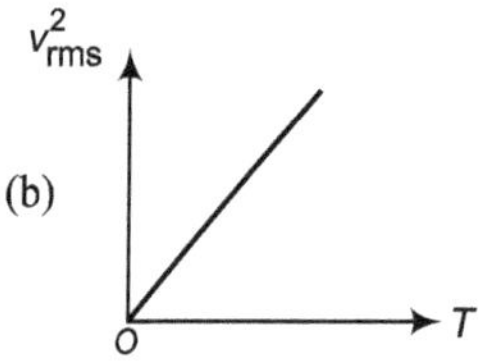

(c)
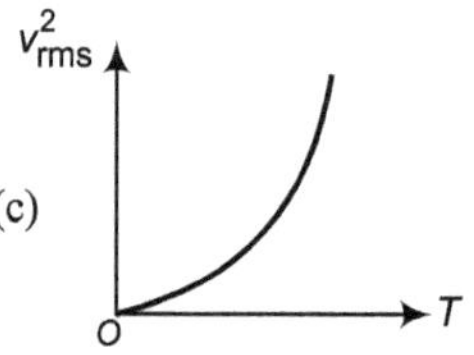

(d)
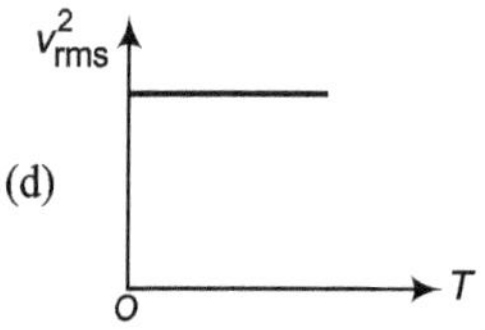

46. Volume-temperature graph at constant pressure for a monoatomic gas (V in m^3, T in °C) is

(a)
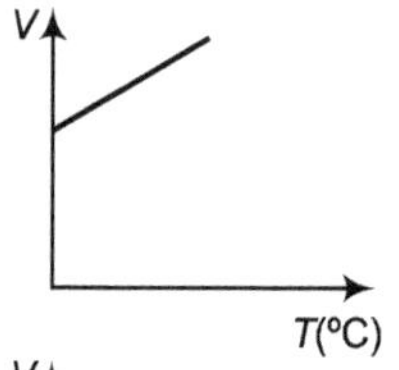

(b)
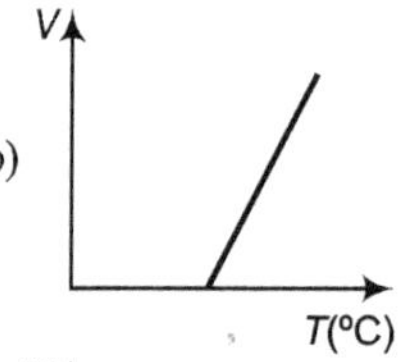

(c)
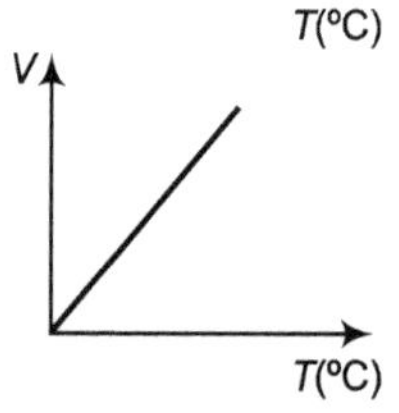

(d)
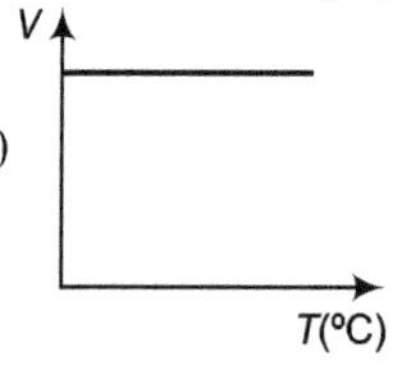

47. By what factor, the rms velocity will change, if the temperature is raised from 27°C to 327°C?

(a) $\sqrt{2}$ (b) 2
(c) $2\sqrt{2}$ (d) 1

48. Two different gases of molecular masses M_1 and M_2 are at the same temperature. What is the ratio of their mean square speeds?

(a) M_1/M_2 (b) M_2/M_1
(c) $\sqrt{M_1/M_2}$ (d) $\sqrt{M_2/M_1}$

49. In a mixture of gases, the average number of degree of freedoms per molecule is 6. The rms speed of the molecule of the gas is c. The velocity of sound in the gas is

(a) $\frac{c}{\sqrt{3}}$ (b) $\frac{c}{\sqrt{2}}$
(c) $\frac{2c}{3}$ (d) $\frac{3c}{3}$

Degree of Freedom, Internal Energy, Law of Equipartition of Energy and Molar Heat Capacity

50. What is the degree of freedom in case of a monoatomic gas?

(a) 1 (b) 3
(c) 5 (d) None of these

51. Two gases are at absolute temperatures 300 K and 350 K respectively. Ratio of average kinetic energy of their molecules is

(a) 7 : 6 (b) 6 : 7 (c) 36 : 49 (d) 49 : 36

52. The specific heat at constant pressure is greater than that of the same gas at constant volume because

(a) at constant volume work is done in expanding the gas
(b) at constant pressure work is done in expanding the gas
(c) the molecular attraction increases more at constant pressure
(d) the molecular vibration increases more at constant pressure

53. The average translational kinetic energy of O_2 (molar mass 32) at a particular temperature is 0.048 eV. The average translational kinetic energy of N_2 (molar mass 28) molecules in eV at the same temperature is

(a) 0.0015 (b) 0.003
(c) 0.048 (d) 0.768

54. Each molecule of a gas has f degrees of freedom. The ratio $\frac{C_p}{C_V} = \gamma$ for the gas is

(a) $1 + \frac{f}{2}$ (b) $1 + \frac{1}{f}$ (c) $1 + \frac{2}{f}$ (d) $\frac{f}{2}$

55. Two moles of argon are mixed with one mole of hydrogen, then $\frac{C_p}{C_V}$ for the mixture is nearly

(a) 1.2 (b) 1.3 (c) 1.4 (d) 1.5

56. The ratio of specific heats γ of an ideal gas is given by

(a) $\dfrac{1}{1-\dfrac{R}{C_p}}$ (b) $1+\dfrac{R}{C_V}$

(c) $\dfrac{C_p}{C_p - R}$ (d) All of these

57. A gas has volume V and pressure p. The total translational kinetic energy of all the molecules of the gas is

(a) $\frac{3}{2}pV$ only if the gas is monoatomic

(b) $\frac{3}{2}pV$ only if the gas is diatomic

(c) $\frac{3}{2}pV$ in all cases

(d) None of the above

58. The number of translational degree of freedom for a diatomic gas is

(a) 2 (b) 3 (c) 5 (d) 6

59. A vessel contains a mixture of one mole of oxygen and two moles of nitrogen at 300 K. The ratio of the average rotational kinetic energy per O_2 molecule to that per N_2 molecules is

(a) 1 : 1 (b) 1 : 2

(c) 2 : 1 (d) 8 : 7

60. Internal energy of two moles of an ideal gas at a temperature of 127°C is 1200 R. Then, the specific heat of the gas at constant pressure is

(a) 0.5 R (b) 0.1 R

(c) 1.5 R (d) 2.5 R

61. Choose the correct option related to the specific heat at constant volume C_V

(a) $C_V = \frac{1}{2} f R$ (b) $C_V = \dfrac{R}{\gamma - 1}$

(c) $C_V = \dfrac{C_p}{\gamma}$ (d) All of these

62. The graph which represents the variation of mean kinetic energy of molecules with temperature $t°$ C is

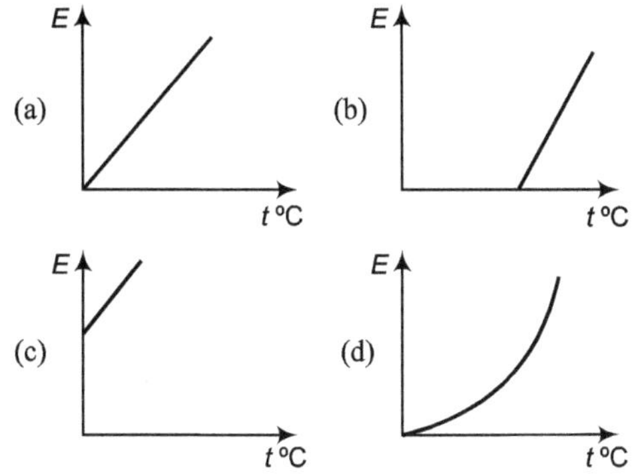

63. The mean kinetic energy of one mole of gas per degree of freedom is

(a) $\frac{1}{2}kT$ (b) $\frac{3}{2}kT$

(c) $\frac{3}{2}RT$ (d) $\frac{1}{2}RT$

64. The degrees of freedom of a molecule of a non-linear triatomic gas is (ignore vibrational motion)

(a) 2 (b) 4

(c) 6 (d) 8

65. A gas mixture consists of 2 moles of O_2 and 4 moles of Ar at temperature T. Neglecting all vibrational modes, the total internal energy of the system is

(a) $4RT$ (b) $15RT$

(c) $9RT$ (d) $11RT$

66. The ratio $\dfrac{C_p}{C_V} = \gamma$ for a gas. Its molecular weight is M. Its specific heat capacity at constant pressure is

(a) $\dfrac{R}{\gamma - 1}$ (b) $\dfrac{\gamma R}{\gamma - 1}$

(c) $\dfrac{\gamma R}{M(\gamma - 1)}$ (d) $\dfrac{\gamma RM}{(\gamma - 1)}$

67. At 27°C temperature the kinetic energy of an ideal gas is E_1. If the temperature is increased to 327°C, then the kinetic energy will be

(a) $\dfrac{E_1}{\sqrt{2}}$ (b) $\sqrt{2}\,E_1$

(c) $2E_1$ (d) $\dfrac{E_1}{2}$

68. The translational kinetic energy of 1 g molecule of a gas, at temperature 300 K is ($R = 8.31$ J/mol K)

(a) 3.4×10^3 J

(b) 2.97×10^3 J

(c) 1.2×10^2 J

(d) 0.66×10^3 J

69. For a gas, if the ratio of specific heats at constant pressure p and constant volume V is γ, then the value of degree of freedom is

(a) $\dfrac{\gamma + 1}{\gamma - 1}$ (b) $\dfrac{\gamma - 1}{\gamma + 1}$

(c) $\frac{1}{2}(\gamma - 1)$ (d) $\dfrac{2}{\gamma - 1}$

70. If the ratio of specific heats of a gas at constant pressure to that at constant volume is γ, the change in internal energy of a gas, when the volume changes from V to $2V$ at constant pressure p is

(a) $\dfrac{R}{(\gamma - 1)}$ (b) pV

(c) $\dfrac{pV}{(\gamma - 1)}$ (d) $\dfrac{\gamma pV}{(\gamma - 1)}$

Miscellaneous Problems

71. A gas at absolute temperature 300 K has pressure $= 4 \times 10^{-10}\ \text{N/m}^2$.

Boltzmann constant, $k = 1.38 \times 10^{-23}$ J/K. The number of molecules per cm^3 is of the order of

(a) 100 (b) 10^5 (c) 10^8 (d) 10^{11}

72. The energy density $\frac{u}{V}$ of an ideal diatomic gas is related to its pressure p as

(a) $\frac{u}{V} = 3p$
(b) $\frac{u}{V} = \frac{3}{2}p$
(c) $\frac{u}{V} = \frac{p}{3}$
(d) $\frac{u}{V} = \frac{5}{2}p$

73. If a 5 kg body falls to the ground from a height of 30 m and if all of its mechanical energy is converted into heat. The heat produced will be $(g = 10\ \text{m/s}^2)$

(a) 359 cal
(b) 150 cal
(c) 60 cal
(d) 254 cal

74. A cylinder contains 20 kg of N_2 gas $(M = 28\ \text{kg/kmol})$ at a pressure of 5 atm. The mass of hydrogen $(M = 2\ \text{kg/kmol})$ at a pressure of 3 atm contained in the same cylinder at same temperature is

(a) 1.08 kg
(b) 0.86 kg
(c) 0.68 kg
(d) 1.68 kg

75. If the density of a gas at NTP is $1.3\ \text{kg/m}^3$ and velocity of sound in it is 330 m/s. The number of degrees of freedom of gas molecule is

(a) 2 (b) 3 (c) 6 (d) 5

76. A gas is found to obey the law p^2V = constant. The initial temperature and volume are T_0 and V_0. If the gas expands to a volume $3V_0$, its final temperature becomes

(a) $\frac{T_0}{3}$
(b) $\frac{T_0}{\sqrt{3}}$
(c) $3T_0$
(d) None of these

77. Identify which pair of state parameters can completely describe the system.

(a) p and V
(b) p and ρ
(c) p and U
(d) All of these

78. 5 L of benzene weight

(a) more in summer than in winter
(b) more in winter than in summer
(c) equal in winter than in summer
(d) None of the above

79. A beaker is completely filled with water at 4° C. It will overflow if

(a) heated above 4°C
(b) cooled below 4°C
(c) both heated and cooled above and below 4°C, respectively
(d) None of the above

80. Calorie is defined as the amount of heat required to raise temperature of 1g of water by 1°C and it is defined under which of the following conditions?

(a) From 14.5°C to 15.5°C at 760 mm of Hg
(b) From 98.5°C to 99.5°C at 760 mm of Hg
(c) From 13.5°C to 14.5°C at 76 mm of Hg
(d) From 3.5°C to 4.5°C at 76 mm of Hg

81. A block of mass 100 g slides on a rough horizontal surface. If the speed of the block decreases from 10 m/s and 5 m/s, the thermal energy developed in the process is

(a) 3.75 J
(b) 37.5 J
(c) 0.375 J
(d) 0.75 J

82. SI unit of universal gas constant is

(a) cal/°C
(b) J/mol
(c) $\text{J/mol}^{-1}\text{K}^{-1}$
(d) J/kg

83. The density of a gas at normal pressure and 27°C temperature is 24 units. Keeping the pressure constant, the density at 127°C will be

(a) 6 units
(b) 12 units
(c) 18 units
(d) 24 units

84. The gas equation $\frac{pV}{T}$ = constant is true for a constant mass of an ideal gas undergoing

(a) isothermal change
(b) adiabatic change
(c) isobaric change
(d) any type of change

85. A vessel contains 1 mole of O_2 gas (molar mass 32) at a temperature T. The pressure of the gas is p. An identical vessel containing one mole of He gas (molar mass 4) at temperature $2T$ has a pressure of

(a) $p/8$
(b) p
(c) $2p$
(d) $8p$

86. When volume of an ideal gas is increased two times and temperature is decreased half of its initial temperature, then pressure becomes

(a) 2 times
(b) 4 times
(c) $\frac{1}{4}$ times
(d) $\frac{1}{2}$ times

87. Relationship between p, V and E for a gas is (E = total translational kinetic energy)

(a) $p = \frac{3}{2}EV$
(b) $V = \frac{2}{3}Ep$
(c) $pV = \frac{3}{2}E$
(d) $pV = \frac{2}{3}E$

88. Pressure *versus* temperature graph of an ideal gas is as shown in figure. Density of the gas at point A is ρ_0. Density at B will be

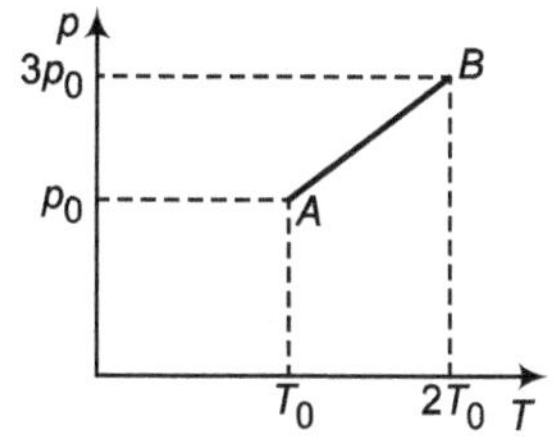

(a) $\frac{3}{4}\rho_0$ (b) $\frac{3}{2}\rho_0$ (c) $\frac{4}{3}\rho_0$ (d) $2\rho_0$

89. Two different isotherms representing the relationship between pressure p and volume V at a given temperature of the same ideal gas are shown for masses m_1 and m_2, then

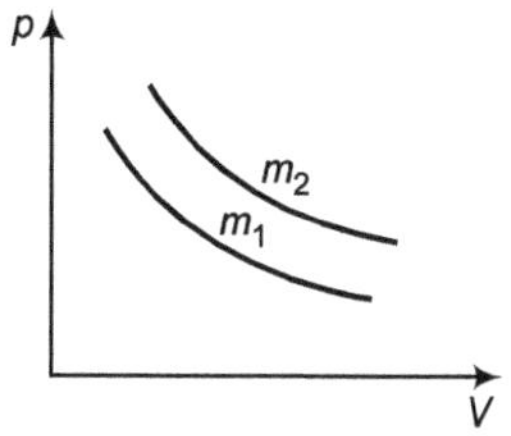

(a) $m_1 > m_2$ (b) $m_1 = m_2$
(c) $m_1 < m_2$ (d) Nothing can be predicted

90. Two different masses m and $3m$ of an ideal gas are heated separately in a vessel of constant volume. The pressure p and absolute temperature T graphs for these two cases are shown in the figure as A and B. The ratio of slopes of curves B to A is

(a) 3 : 1
(b) 1 : 3
(c) 9 : 1
(d) 1 : 9

91. Each atom of mass m of a monoatomic gas has got three degrees of freedom. The rms velocity of these atoms is v at temperature T. For a diatomic molecule of mass m and temperature T which has got five degrees of freedom, rms velocity of molecule is

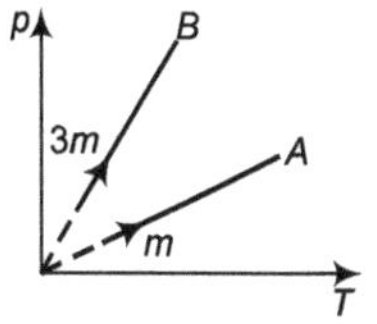

(a) $\sqrt{\frac{5}{3}}\,v$
(b) $\sqrt{\frac{3}{5}}\,v$
(c) v
(d) None of the above

[Level 2]

Only One Correct Option

1. The mass of hydrogen molecule is 3.32×10^{-27} kg . If 10^{23} hydrogen molecules strike per second at $2\,\text{cm}^2$ area of a rigid wall at an angle of 45° from the normal and rebound back with a speed of 1000 m/s, then the pressure exerted on the wall is

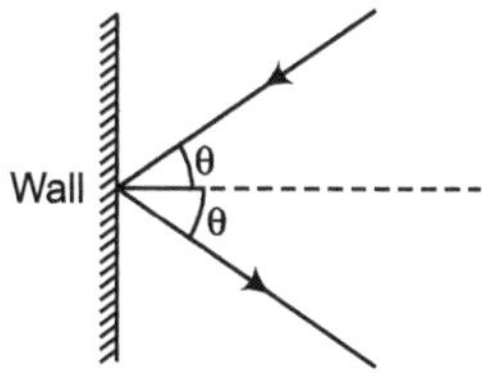

(a) 2.34×10^3 Pa
(b) 0.23×10^6 Pa
(c) 0.23×10^3 Pa
(d) 23.4×10^3 Pa

2. A ring shaped tube contains two ideal gases with equal masses and relative molar masses $M_1 = 32$ and $M_2 = 28$. The gases are separated by one fixed partition p and another movable stopper S which can move freely without friction inside the ring.

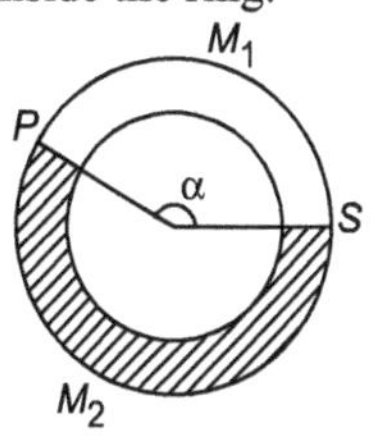

The angle α in equilibrium as shown in the figure (in degrees) is

(a) 291 (b) 219
(c) 129 (d) 192

3. A cylindrical tube of uniform cross-sectional area A is fitted with two air tight frictionless pistons.

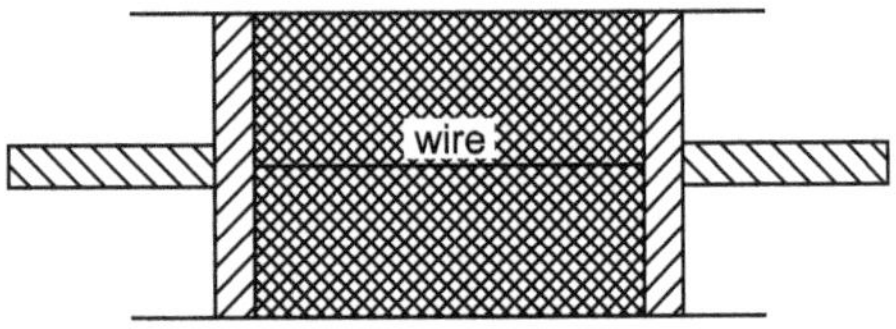

The pistons are connected to each other by a metallic wire. Initially, the pressure of the gas is p_0 and temperature is T_0, atmospheric pressure is also p_0. Now, the temperature of the gas is increased to $2T_0$, the tension in the wire will be

(a) $2p_0A$ (b) p_0A (c) $\frac{p_0A}{2}$ (d) $4\,p_0A$

4. If the density of a gas at NTP is $1.3\ \text{kg/m}^3$ and velocity of sound in it is 330 m/s. The number of degrees of freedom of gas molecule is

(a) 2 (b) 3 (c) 6 (d) 5

5. A vertical cylinder closed at both ends is fitted with a smooth piston dividing the volume into two parts each containing one mole of air. At the equilibrium temperature of 320 K, the upper and lower parts are in the ratio 4 : 1. The ratio will become 3 : 1 at a temperature of

(a) 450 K (b) 228 K (c) 420 K (d) 570 K

6. The expansion of an ideal gas of mass m at a constant pressure p is given by the straight line B. Then, the expansion of the same ideal gas of mass $2m$ at a pressure $2p$ is given by the straight line.

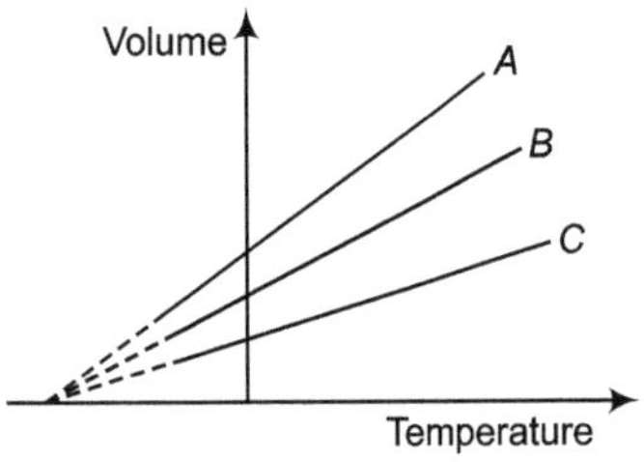

(a) C (b) A
(c) B (d) None of these

7. Pressure *versus* temperature graph of an ideal gas of equal number of moles of different volumes are plotted as shown in figure. Choose the correct alternative.

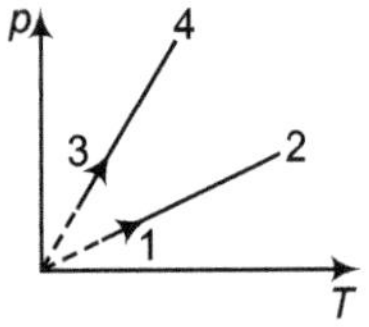

(a) $V_1 = V_2,\ V_3 = V_4$ and $V_2 > V_3$
(b) $V_1 = V_2,\ V_3 = V_4$ and $V_2 < V_3$
(c) $V_1 = V_2 = V_3 = V_4$
(d) $V_4 > V_3 > V_2 > V_1$

8. A cyclic process 1-2-3-4-1 is depicted on V-T diagram. The p-T diagram and p-V diagrams for this cyclic process are given below. Select the incorrect choices.

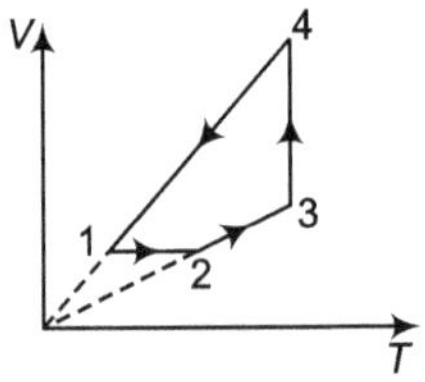

(a)
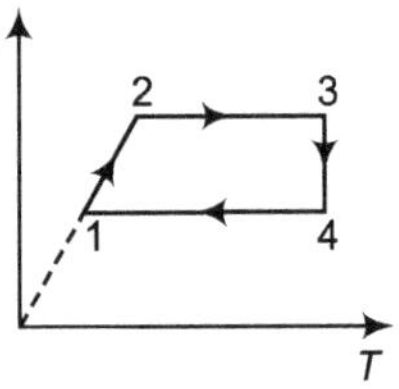

(b)
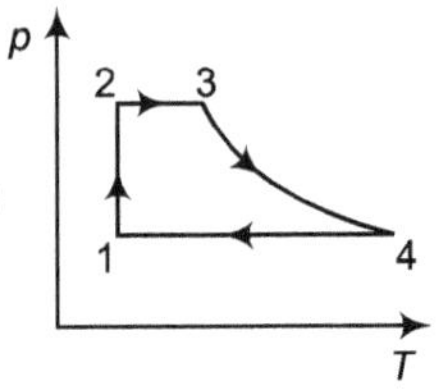

(c)
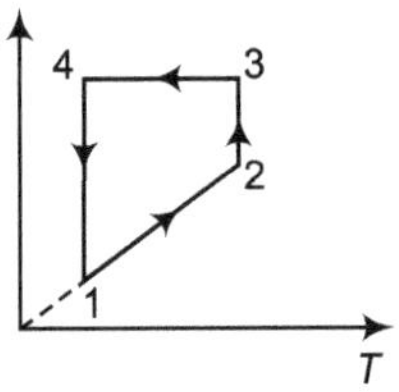

(d) None of these

9. The given curve represents the variations of temperature as a function of volume for one mole of an ideal gas. Which of the following curves best represent the variation of pressure as a function of volume?

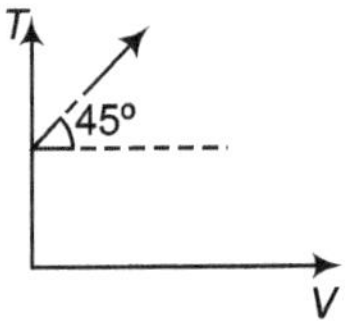

(a)
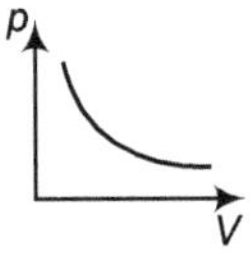

(b)
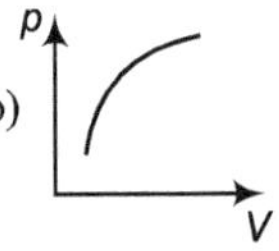

(c)
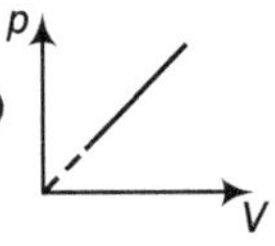

(d)
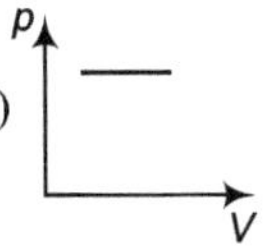

10. A cyclic process $ABCD$ is shown in the p-V diagram. Which of the following curves represent the same process?

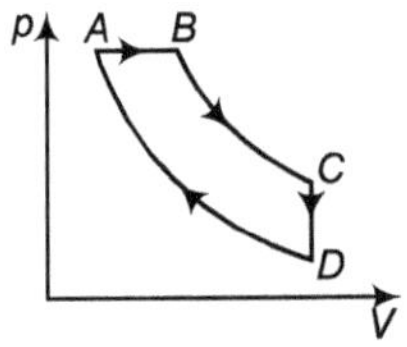

(a) 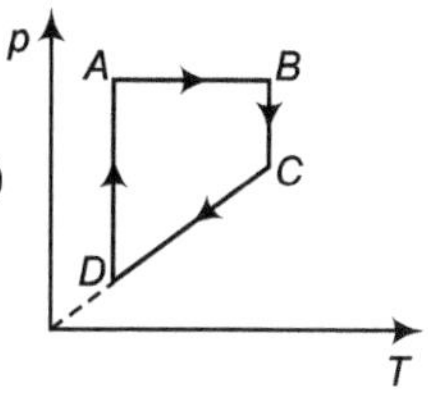

(b) 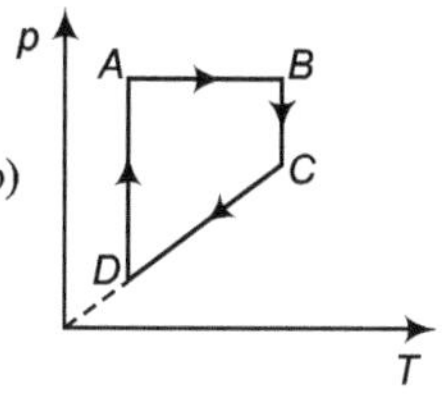

(c) 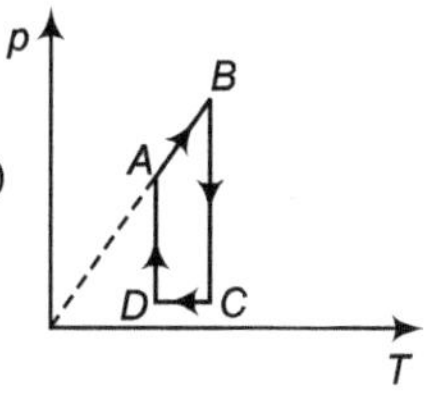

(d) 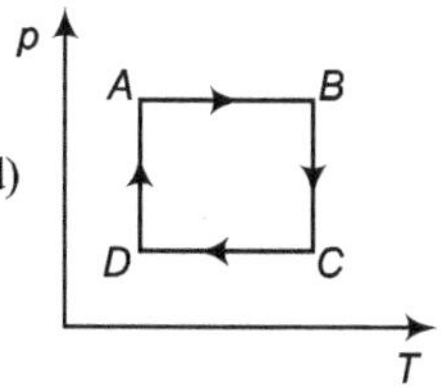

Note: *BC* and *DA* are parts of rectangular hyperbola.

11. Variation of internal energy with density of one mole of monoatomic gas is depicted in the below figure, corresponding variation of pressure with volume can be depicted as (Assuming the curve is rectangular hyperbola)

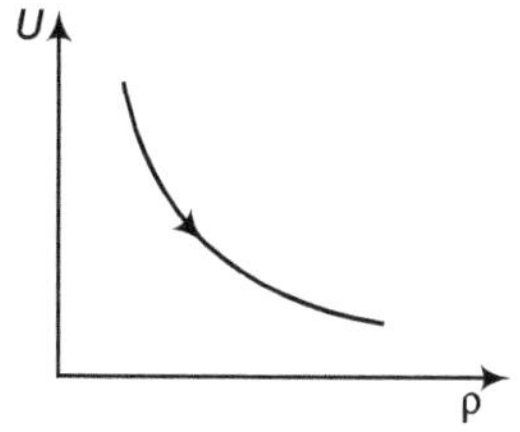

(a) 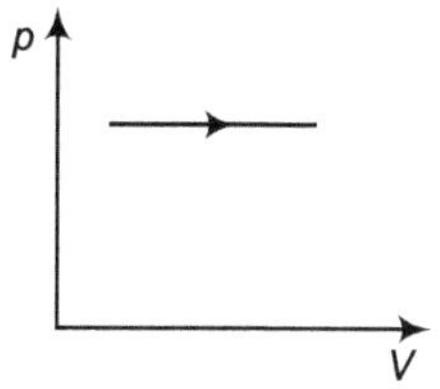

(b) 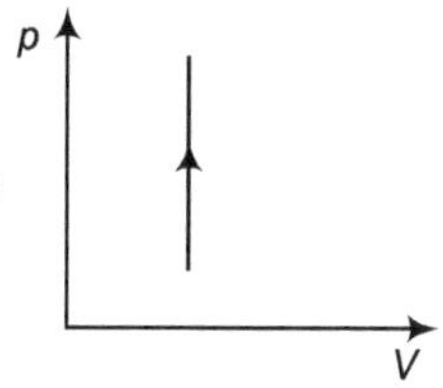

(c) 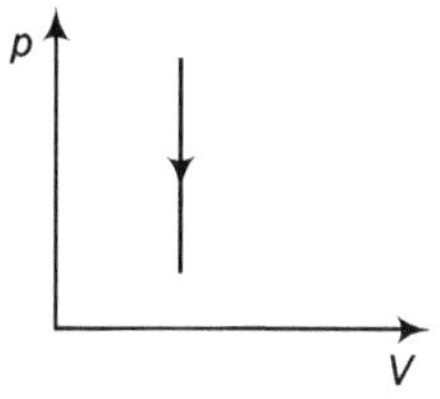

(d) 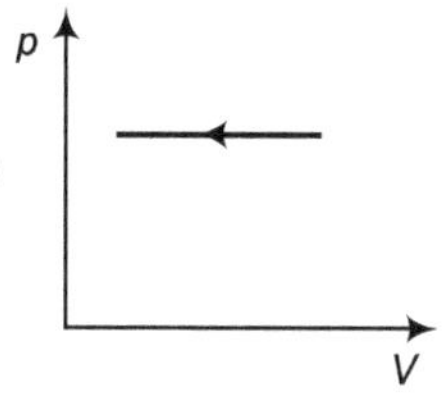

12. A gas at absolute temperature 300 K has pressure $p = 4 \times 10^{-10}$ N/m^2. Boltzmann constant $k = 1.38 \times 10^{-23}$ J/K. The number of molecules per cm^3 is of the order of

(a) 100 (b) 10^5
(c) 10^8 (d) 10^{11}

13. The number of molecules in 1 cc of water is closed to

(a) 6×10^{23}
(b) 22.4×10^{24}
(c) $\frac{10^{23}}{3}$
(d) 10^{23}

14. The number of molecules per unit volume of a gas is given by

(a) $\frac{p}{kT}$ (b) $\frac{kT}{p}$
(c) $\frac{p}{RT}$ (d) $\frac{RT}{p}$

15. A gas has volume V and pressure p. The total translational kinetic energy of all the molecules of the gas is

(a) $\frac{3}{2}pV$ only if the gas is monoatomic
(b) $\frac{3}{2}pV$ only if the gas is diatomic
(c) $> \frac{3}{2}pV$ if the gas is diatomic
(d) $\frac{3}{2}pV$ in all cases

16. Two identical containers joined by a small pipe initially contain the same gas at pressure p_0 and absolute temperature T_0. One container is now maintained at the same temperature while the other is heated to $2T_0$. The common pressure of the gases will be

(a) $\frac{3}{2}p_0$ (b) $\frac{4}{3}p_0$
(c) $\frac{5}{3}p_0$ (d) $2p_0$

17. A cylindrical steel plug is inserted into a circular hole of diameter 2.60 cm in a brass plate. When the plug and the plates are at a temperature of 20°C, the diameter of the plug is 0.010 cm smaller than that of the hole. The temperature at which the plug will just fit in it is

$$\left(\text{Given, } \alpha_{\text{steel}} = \frac{11 \times 10^{-6}}{°\text{C}} \text{ and } \alpha_{\text{brass}} = \frac{19 \times 10^{-6}}{°\text{C}}\right)$$

(a) – 48°C (b) – 20°C
(c) – 10°C (d) – 458°C

18. A piece of metal weigh 46 g in air. When it is immersed in a liquid of specific gravity 1.24 at 27°C, it weigh 30 g. When the temperature is raised to 42°C, the metal piece weigh 30.5 g. If the specific gravity of the liquid at 42°C is 1.20, the coefficient of linear expansion of the metal is

(a) $\frac{1.4 \times 10^{-5}}{°\text{C}}$ (b) $\frac{2.3 \times 10^{-5}}{°\text{C}}$
(c) $\frac{4.3 \times 10^{-5}}{°\text{C}}$ (d) $\frac{3.4 \times 10^{-5}}{°\text{C}}$

19. A chamber containing a gas was evacuated till the vacuum attained was 10^{-14} m of Hg. If the temperature of the chamber was 30°C, the number of molecules that remains in it per cubic metre is

(a) 3.2×10^{11} (b) 3.2×10^{12}
(c) 2.3×10^{12} (d) 2.3×10^{10}

20. The apparent coefficient of expansion of a liquid when heated in a copper vessel is C and when heated in a silver vessel is S. If A is the linear coefficient of expansion of copper, then the linear coefficient of expansion of silver is

(a) $\frac{C+S-3A}{3}$ (b) $\frac{C+3A-S}{3}$
(c) $\frac{S+3A-C}{3}$ (d) $\frac{C+S+3A}{3}$

21. Three rods of equal length l are joined to form an equilateral ΔPQR. O is the mid-point of PQ. Distance OR remains same for small change in temperature. Coefficient of linear expansion for PR and RQ is same, i.e. α_2 but that for PQ is α_1. Then,

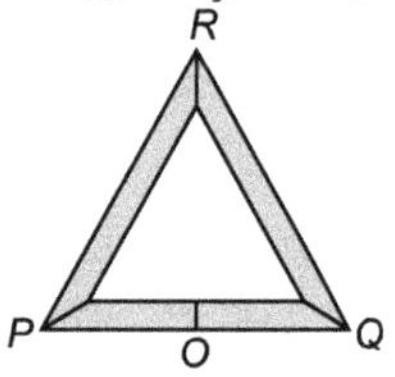

(a) $\alpha_2 = 3\alpha_1$ (b) $\alpha_2 = 4\alpha_1$ (c) $\alpha_1 = 3\alpha_2$ (d) $\alpha_1 = 4\alpha_2$

22. An ideal gas is initially at temperature T and volume V. Its volume is increased by ΔV due to an increase in temperature ΔT, pressure remaining constant. The quantity $\delta = \Delta V/(V\,\Delta T)$ varies with temperature as

(a)
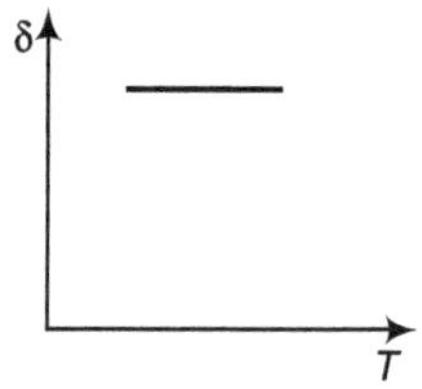

(b)
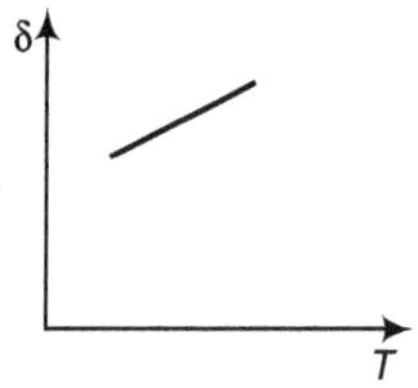

(c)
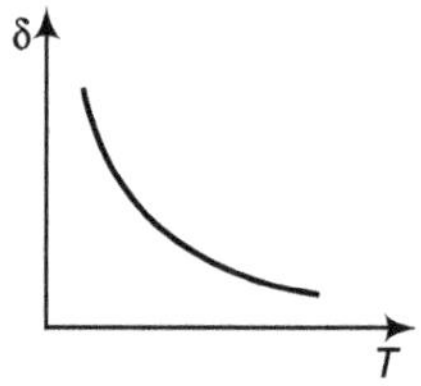

(d)
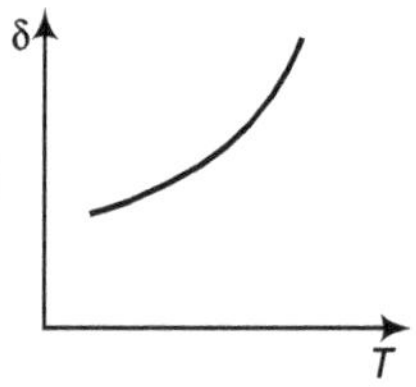

23. p-T diagram was obtained when a given mass of gas was heated. During the heating process from the state 1 to state 2, the volume

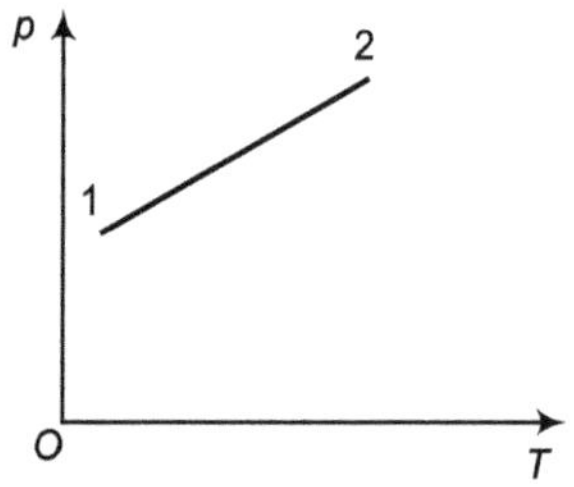

(a) remain constant (b) decreases
(c) increases (d) changes erratically

24. p-V diagram was obtained from state 1 to state 2 when a given mass of a gas is subjected to temperature changes. During this process, the gas is

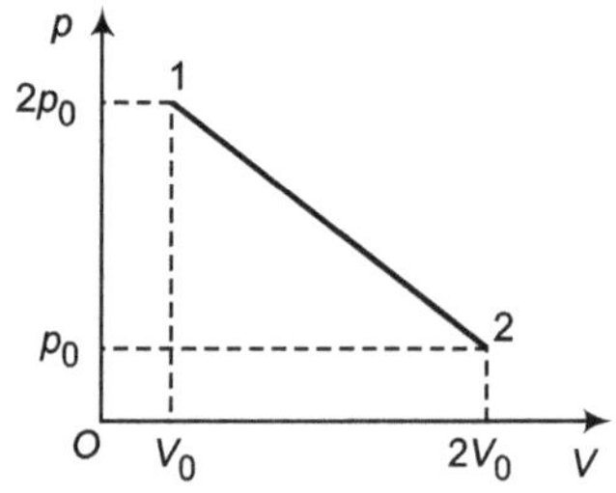

(a) heated continuously
(b) cooled continuously
(c) heated in the beginning and cooled towards the end
(d) cooled in the beginning and heated towards the end

25. 1 mole of H_2 gas is contained in a box of volume $V = 1.00\text{ m}^3$ at $T = 300$ K. The gas is heated to a temperature of $T = 3000$ K and the gas gets converted to a gas of hydrogen atoms. The final pressure would be (considering all gases to be ideal)

(a) same as the pressure initially
(b) 2 times the pressure initially
(c) 10 times the pressure initially
(d) 20 times the pressure initially

More than One Correct Options

1. During an experiment, an ideal gas is found to obey a condition $\frac{p^2}{\rho}$ = constant. (ρ = density of the gas). The gas is initially at temperature T, pressure p and density ρ. The gas expands such that density changes to $\rho/2$.

(a) The pressure of the gas changes to $\sqrt{2}\,p$
(b) The temperature of the gas changes to $\sqrt{2}\,T$
(c) The graph of the above process on p-T diagram is parabola
(d) The graph of the above process on p-T diagram is hyperbola

2. During an experiment, an ideal gas is found to obey a condition Vp^2 = constant. The gas is initially at a temperature T, pressure p and volume V. The gas expands to volume $4V$.

(a) The pressure of gas changes to $\frac{p}{2}$
(b) The temperature of gas changes to $4T$
(c) The graph of the above process on p-T diagram is parabola
(d) The graph of the above process on p-T diagram is hyperbola

3. Find the correct options.

(a) Ice point in Fahrenheit scale is 32°F
(b) Ice point in Fahrenheit scale is 98.8°F
(c) Steam point in Fahrenheit scale is 212°F
(d) Steam point in Fahrenheit scale is 252°F

4. In the T-V diagram shown in figure, choose the correct options for the process 1-2.

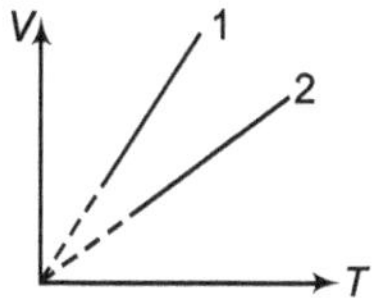

(a) Density of gas has reduced to half
(b) Temperature of gas has increased to two times
(c) Internal energy of gas has increased to four times
(d) T-V graph is a parabola passing through origin

5. Choose the wrong options.

(a) Translational kinetic energy of all ideal gases at same temperature is same
(b) In one degree of freedom, all ideal gases have internal energy = $1/2RT$
(c) Translational degree of freedom of all ideal gases is three
(d) Translational kinetic energy of one mole of all ideal gases is $\frac{3}{2}RT$

6. Along the line-1, mass of gas is m_1 and pressure is p_1. Along the line-2 mass of same gas is m_2 and pressure is p_2. Choose the correct options.

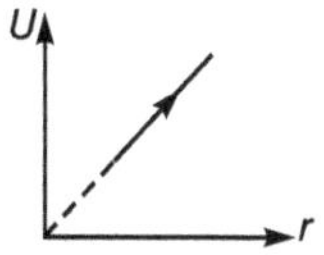

(a) m_1 may be less than m_2 (b) m_2 may be less than m_1
(c) p_1 may be less than p_2 (d) p_2 may be less than p_1

7. Choose the correct options.

(a) In $p = \frac{m}{M}RT$, m is mass of gas per unit volume
(b) In $pV = \frac{m}{M}RT$, m is mass of one molecule of gas
(c) In $p = \frac{1}{3}\frac{mN}{V}v_{rms}^2$, m is total mass of gas.
(d) In $v_{rms} = \sqrt{\frac{3kT}{m}}$, m is mass of one molecule of gas

Assertion and Reason

Direction (Q. Nos. 1-14) *These questions consist of two statements each printed as assertion and reason. While answering these questions, you are required to choose anyone of the following five responses.*

(a) If both Assertion and Reason are true and Reason is the correct explanation of Assertion.
(b) If both Assertion and Reason are true but Reason is not the correct explanation of Assertion.
(c) If Assertion is true but Reason is false.
(d) If Assertion is false but Reason is true.
(e) If both Assertion and Reason are false.

1. Assertion Pressure of a gas is 2/3 times translational kinetic energy of gas molecules.

Reason Translational degree of freedom of any type of gas is three, whether the gas is monoatomic, diatomic or polyatomic.

2. Assertion Straight line on V-T graph represents isobaric process.

Reason If $V \propto T$, then p = constant, i.e. process is isobaric.

3. Assertion At the same temperature and pressure equal volumes of all gases contain equal number of molecules.

Reason In 1 L at NTP, total number of molecules are 6.02×10^{23}.

4. Assertion In isochoric process, ρ-V graph is straight line parallel to ρ-axis.

Reason In isochoric process density ρ remains constant.

5. Assertion A metallic rod is fixed from two ends as shown in figure. When the temperature is increased compressive stresses are developed in the rod.

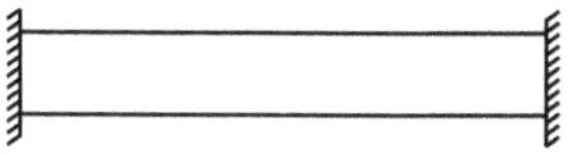

Reason At higher temperature, natural length of the rod will be more.

6. Assertion Pressure of a gas is given as $p = \frac{2}{3}E$.

Reason In the above expression, E represents kinetic energy of the gas per unit volume.

7. Assertion Total internal energy of oxygen gas at a given temperature is E of this energy $\frac{3}{5}E$ is translational kinetic energy and $\frac{2}{5}E$ is rotational kinetic energy.

Reason Potential energy of an ideal gas is zero.

8. Assertion Total kinetic energy of any gas at temperature T would be $\frac{1}{2}mv_{rms}^2$

Reason Translational kinetic energy of any type of gas at temperature T would be $\frac{3}{2}RT$ of one mole.

9. Assertion Degree of freedom of a monoatomic gas is always three, whether we consider vibrational effects or not.

Reason At all temperatures (low or high), vibrational kinetic energy of an ideal gas is zero.

10. Assertion An actual gas behaves as an ideal gas most closely at low pressure and high temperature.

Reason At low pressure and high temperature real gases obey the gas laws.

11. Assertion At triple point, three states (solid, liquid and gas) may co-exist simultaneously.

Reason For water, the values of pressure and temperature corresponding to triple point are 10 mm of Hg and 273.16 K.

12. Assertion If a gas chamber containing a gas is moved translationally, then temperature of gas will increase.

Reason Total kinetic energy of the gas molecules will increase by the translational motion of gas chamber.

13. Assertion In summers, a metallic scale will read more than the actual.

Reason In summer, length of metallic scale will increase.

14. Assertion Any straight line on V-T diagram represents isobaric process.

Reason In isobaric process, if V is doubled, then T will also become two times.

Match the Columns

1. In the process $T \propto \frac{1}{V}$, pressure of the gas increases from p_0 to $4p_0$. Match the following columns.

Column I	Column II
(A) Temperature of the gas	(p) Positive
(B) Volume of the gas	(q) Negative
(C) Work done by the gas	(r) Two times
(D) Heat supplies to the gas	(s) Cannot say anything
	(t) None of the above

2. For a monoatomic gas at temperature T, match the following columns.

Column I	Column II
(A) Speed of sound	(p) $\sqrt{2RT/M}$
(B) rms speed of gas molecules	(q) $\sqrt{8RT/\pi M}$
(C) Average speed of gas molecules	(r) $\sqrt{3RT/M}$
(D) Most probable speed of gas molecules	(s) $\sqrt{5RT/3M}$

3. Match the following columns.

Column I	Column II
(A) In $p = \frac{2}{3}E$, E is	(p) Change in internal energy in only isochoric process
(B) In $U = 3RT$ for an momoatomic gas U is	(q) Translational kinetic energy of unit volume
(C) In $W = p(V_f - V_i)$, W (R) is	(r) Internal energy of one mole
(D) In $\Delta U = nC_V \Delta T$, ΔU is	(s) Work done in isobaric process
	(t) None of the above

4. In the V-T graph shown in the figure match, the following columns.

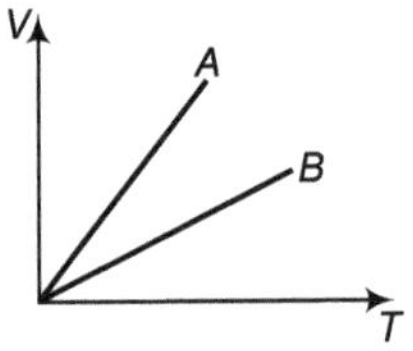

Column I	Column II
(A) Gas A is ... and gas B is ...	(p) Monoatomic, diatomic
(B) p_A/p_B is	(q) Diatomic, monoatomic
(C) n_A/n_B is	(r) >1
	(s) <1
	(t) Cannot say anything

5. Match the following columns.

Column I	Column II
(A) Adiabatic bulk modulus	(p) $-\frac{p}{V}$
(B) Slope of p-V graph in isothermal process	(q) $\frac{2}{\gamma - 1}$
(C) Degree of freedom	(r) γp
(D) Molar heat capacity at constant pressure divided by R	(s) $\frac{\gamma}{\gamma - 1}$

Entrance Gallery

2014

1. If m represents the mass of each molecule of a gas and T its absolute temperature, then the root mean square velocity of the gaseous molecule is proportional to **[Kerala CEE]**

(a) mT (b) $m^{1/2}T^{1/2}$ (c) $m^{-1/2}T$
(d) $m^{-1/2}T^{1/2}$ (e) $m^{-1}T$

2. A molecule of a gas has six degrees of freedom. Then, the molar specific heat of the gas at constant volume is **[Kerala CEE]**

(a) $\frac{R}{2}$ (b) R (c) $\frac{3R}{2}$
(d) $3R$ (e) $5R$

3. Total number of degrees of freedom of a rigid diatomic molecule is **[Kerala CEE]**

(a) 3 (b) 6 (c) 5
(d) 2 (e) 7

4. Which of the given substances A, B and C have more specific heat? **[Kerala CEE]**

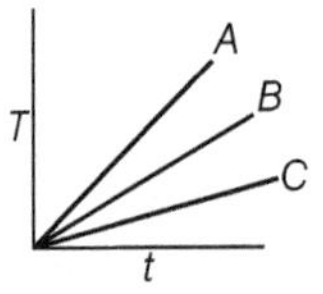

(a) A (b) B
(c) C (d) Both (a) and (b)
(e) All are equal

2013

5. Two non-reactive monoatomic ideal gases have their atomic masses in the ratio 2 : 3. The ratio of their partial pressure, when enclosed in a vessel kept at a constant temperature is 4 : 3.
The ratio of their densities is [JEE Advanced]
(a) 1 : 4 (b) 1 : 2 (c) 6 : 9 (d) 8 : 9

6. The velocity of 4 gas molecules are given by 1 km/s, 3 km/s, 5 km/s and 7 km/s. Calculate the difference between average and rms velocity. [MHT CET]
(a) 0.338 (b) 0.438 (c) 0.583 (d) 0.683

2012

7. In a mercury thermometer, the ice point (lower fixed point) is marked as 10°C and the steam point (upper fixed point) is marked as 130° C. At 40°C temperature, what will this thermometer read? [WB JEE]
(a) 78°C (b) 66°C
(c) 62°C (d) 58°C

2011

8. An aluminium sphere of 20 cm diameter is heated from 0°C to 100° C. Its volume changes by (given that coefficient of linear expansion for aluminium $\alpha_{A_1} = 23 \times 10^{-6}/°\text{C}$) [AIEEE]
(a) 28.9 cc (b) 2.89 cc
(c) 9.28 cc (d) 4.98 cc

9. A metal rod of Young's modulus Y and coefficient of thermal expansion α is held at its two ends such that its length remains invariant. If its temperature is raised by $t°\text{C}$, the linear stress developed in it, is [AIEEE]
(a) $\frac{\alpha\Delta t}{Y}$ (b) $\frac{Y}{\alpha E}$ (c) $Y\alpha\Delta t$ (d) $\frac{I}{Y\alpha\Delta t}$

10. A container with insulating walls is divided into two equal parts by a partition fitted with a valve. One part is filled with an ideal gas at a pressure p and temperature T, whereas the other part is completely evacuated. If the valve is suddenly opened, then the pressure and temperature of the gas will be [AIEEE]
(a) $\frac{p}{2}, T$ (b) $\frac{p}{2}, \frac{T}{2}$ (c) p, T (d) $p, \frac{T}{2}$

11. Three perfect gases at absolute temperatures T_1, T_2 and T_3 are mixed. The masses of molecules are m_1, m_2 and m_3 and the number of molecules are n_1, n_2 and n_3 respectively. Assuming no loss of energy, the final temperature of the mixture is [AIEEE]
(a) $\frac{n_1T_1 + n_2T_2 + n_3T_3}{n_1 + n_2 + n_3}$ (b) $\frac{n_1T_1^2 + n_2T_2^2 + n_3T_3^3}{n_1T_1 + n_2T_2 + n_3T_3}$
(c) $\frac{n_1^2T_1^2 + n_2^2T_2^2 + n_3^2T_3^2}{n_1T_1 + n_2T_2 + n_3T_3}$ (d) $\frac{T_1 + T_2 + T_3}{3}$

12. The ratio of the molar heat capacities of a diatomic gas at constant pressure to that at constant volume is [Kerala CEE]
(a) 7/2 (b) 3/2 (c) 3/5
(d) 7/5 (e) 5/2

13. A perfect gas at 27°C is heated at constant pressure so as to double its volume. The increase in temperature of the gas will be [Karnataka CET]
(a) 300°C (b) 54°C (c) 327°C (d) 600°C

14. Two temperature scales A and B are related by $\frac{A-42}{110} = \frac{B-72}{220}$. At which temperature two scales have the same reading? [WB JEE]
(a) – 42°C (b) – 72°C (c) 12°C (d) 40°C

15. Air inside a closed container is saturated with water vapour. The air pressure is p and the saturated vapour pressure of water is $\bar{p}$. If the mixture is compressed to one half of its volume by maintaining temperature constant, the pressure becomes [WB JEE]
(a) $2(p+\bar{p})$ (b) $2p+\bar{p}$ (c) $(p+\bar{p})/2$ (d) $p+2\bar{p}$

2010

16. A real gas behaves like an ideal gas, if its [IIT JEE]
(a) pressure and temperature are both high
(b) pressure and temperature are both low
(c) pressure is high and temperature is low
(d) pressure is low and temperature is high

17. The temperature of a gas contained in a closed vessel of constant volume increases by 1°C when the pressure of the gas is increased by 1%. The initial temperature of the gas is [Karnataka CET]
(a) 100 K (b) 273°C (c) 100°C (d) 200 K

18. One mole of a monoatomic ideal gas is mixed with one mole of a diatomic ideal gas. The molar specific heat of the mixture at constant volume is [O JEE]
(a) $(3/2)R$ (b) $(5/2)R$ (c) $2R$ (d) $4R$

19. To what temperature should the hydrogen at 327°C be cooled at constant pressure, so that the root mean square velocity of its molecules becomes half of its previous value? [MHT CET]
(a) – 123°C (b) 123°C
(c) – 100°C (d) 0°C

20. At what temperature rms speed of air molecules is doubled of that at NTP? [MHT CET]
(a) 819°C (b) 719°C
(c) 909°C (d) None of these

21. Kinetic energy per unit volume is E. The pressure exerted by the gas is given by [MHT CET]
(a) $\frac{E}{3}$ (b) $\frac{2E}{3}$ (c) $\frac{3E}{2}$ (d) $\frac{E}{2}$

Answers

Level 1

Objective Problems

1. (c)	**2.** (c)	**3.** (a)	**4.** (a)	**5.** (b)	**6.** (b)	**7.** (c)	**8.** (a)	**9.** (b)	**10.** (c)
11. (a)	**12.** (d)	**13.** (c)	**14.** (a)	**15.** (d)	**16.** (a)	**17.** (c)	**18.** (c)	**19.** (d)	**20.** (d)
21. (d)	**22.** (a)	**23.** (c)	**24.** (c)	**25.** (c)	**26.** (c)	**27.** (c)	**28.** (c)	**29.** (a)	**30.** (a)
31. (b)	**32.** (d)	**33.** (c)	**34.** (c)	**35.** (c)	**36.** (d)	**37.** (c)	**38.** (c)	**39.** (c)	**40.** (a)
41. (c)	**42.** (a)	**43.** (b)	**44.** (b)	**45.** (b)	**46.** (a)	**47.** (a)	**48.** (d)	**49.** (c)	**50.** (b)
51. (b)	**52.** (b)	**53.** (c)	**54.** (c)	**55.** (d)	**56.** (d)	**57.** (c)	**58.** (b)	**59.** (a)	**60.** (d)
61. (d)	**62.** (c)	**63.** (d)	**64.** (c)	**65.** (d)	**66.** (b)	**67.** (c)	**68.** (a)	**69.** (d)	**70.** (c)
71. (b)	**72.** (d)	**73.** (a)	**74.** (b)	**75.** (d)	**76.** (d)	**77.** (d)	**78.** (b)	**79.** (c)	**80.** (a)
81. (a)	**82.** (c)	**83.** (c)	**84.** (d)	**85.** (c)	**86.** (c)	**87.** (d)	**88.** (b)	**89.** (c)	**90.** (a)
91. (c)									

Level 2

Only One Correct Option

1. (a)	**2.** (d)	**3.** (b)	**4.** (d)	**5.** (a)	**6.** (c)	**7.** (a)	**8.** (a,b)	**9.** (a)	**10.** (a)
11. (d)	**12.** (b)	**13.** (c)	**14.** (a)	**15.** (d)	**16.** (b)	**17.** (d)	**18.** (b)	**19.** (a)	**20.** (b)
21. (d)	**22.** (c)	**23.** (c)	**24.** (c)	**25.** (d)					

More than One Correct Options

1. (b,d)	**2.** (a,d)	**3.** (a,c)	**4.** (a,c,d)	**5.** (a,b)	**6.** (All)	**7.** (a,d)

Assertion and Reason

1. (d)	**2.** (d)	**3.** (c)	**4.** (d)	**5.** (a)	**6.** (c)	**7.** (b)	**8.** (d)	**9.** (c)	**10.** (a)
11. (c)	**12.** (d)	**13.** (d)	**14.** (d)						

Match the Columns

1. (A→ r, B→ t, C→ q, D→ s) **2.** (A→ s, B→ r, C→ q, D→ p) **3.** (A→ q, B→ t, C→ s, D→ t)
4. (A→ t, B→ t, C→ t) **5.** (A→ r, B→ p, C→ q, D→ s)

Entrance Gallery

1. (d)	**2.** (d)	**3.** (c)	**4.** (c)	**5.** (d)	**6.** (c)	**7.** (d)	**8.** (a)	**9.** (c)	**10.** (a)
11. (a)	**12.** (d)	**13.** (a)	**14.** (c)	**15.** (b)	**16.** (d)	**17.** (a)	**18.** (c)	**19.** (a)	**20.** (a)
21. (b)									

Solutions

Level 1 : Objective Problems

2. Temperature shown on faulty thermometer will be

$$t = 5 + \left(\frac{95-5}{100}\right) \times 40 = 41°$$

3. Strain $= \frac{\Delta l}{l} = \alpha \Delta \theta$

$$\text{Stress} = Y \times \text{ strain} = Y \alpha \Delta \theta$$

$\therefore$ Force or tension, T = Stress × area

$$= YA\alpha\Delta\theta$$

$$= \frac{\pi Y \pi d^2 \Delta\theta}{4} \left(A = \frac{\pi d^2}{4}\right)$$

or

$$T = \frac{\pi \times 2 \times 10^{11} \times 10^{-5} \times 10^{-4} \times 25}{4}$$

$$= 3926 \text{ N} \approx 4000 \text{ N}$$

4.

$$T = 2\pi\sqrt{\frac{l}{g}}$$

$$T \propto l$$

$\therefore$

$$\frac{T'}{T} = \left(\frac{l'}{l}\right)^2 = 1 + \frac{1}{2}\frac{\Delta l}{l}$$

$$= 1 + \frac{1}{2}\alpha\Delta\theta$$

$\therefore$

$$\Delta T = T' - T$$

$$= \frac{1}{2}\alpha T \Delta\theta$$

5. Coefficient of volume expansion of container will become $3\left(\frac{\alpha}{2}\right)$, which is greater than coefficient of volume expansion of liquid. Hence, container expands more.

6. $\Delta l_1 = \Delta l_2$

or $\quad l_1\alpha_1\Delta\theta = l_2\alpha_2\Delta\theta$

or $\quad \frac{l_1}{l_2} = \frac{\alpha_2}{\alpha_1}$

7. $L\alpha\Delta\theta + 2L(2\alpha)(\Delta\theta) = (3L)(\alpha_e)\Delta\theta$

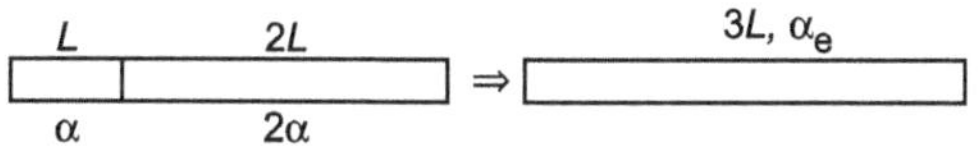

or $\quad \alpha_e = \frac{5}{3}\alpha$

8. Photographic expansion takes place.

9. Original value of circumference, $l = 2\pi R$

$\therefore \quad \Delta l = l\alpha\theta = 2\pi R\alpha\theta$

10. Strain $= \frac{\Delta l}{l} = \frac{l\alpha\Delta\theta}{l} = \alpha\Delta\theta$

$$\text{Stress} = Y \times \text{Strain} = Y\alpha\Delta\theta$$

$$(\text{Stress})_1 = (\text{Stress})_2$$

$\therefore \quad Y_1\alpha_1 = Y_2\alpha_2 \qquad (\Delta\theta \to \text{same})$

or $\quad \frac{Y_1}{Y_2} = \frac{\alpha_2}{\alpha_1} = \frac{3}{2}$

11. On this thermometer, 34°C will read as

$$-20 + \frac{150}{100} \times 34 = 31°$$

12. $\Delta T = -273°C$ (absolute temperature). As no matter can attain this temperature, hence temperature can never be negative on kelvin scale.

13. $\frac{F-32}{9} = \frac{K-273}{5}$

$\Rightarrow \quad \frac{x-32}{9} = \frac{x-273}{5}$

$\Rightarrow \quad x = 574.25$

14. Pyrometer is used to measure very high temperature.

15. Fractional change in period

$$\frac{\Delta T}{T} = \frac{1}{2}\alpha\Delta\theta$$

$$= \frac{1}{2} \times 2 \times 10^{-6} \times 10 = 10^{-5}$$

$$\% \text{ change} = \frac{\Delta T}{T} \times 100 = 10^{-5} \times 100 = 10^{-3} \%$$

16. Suppose, height of liquid in each arm before rising the temperature is l.

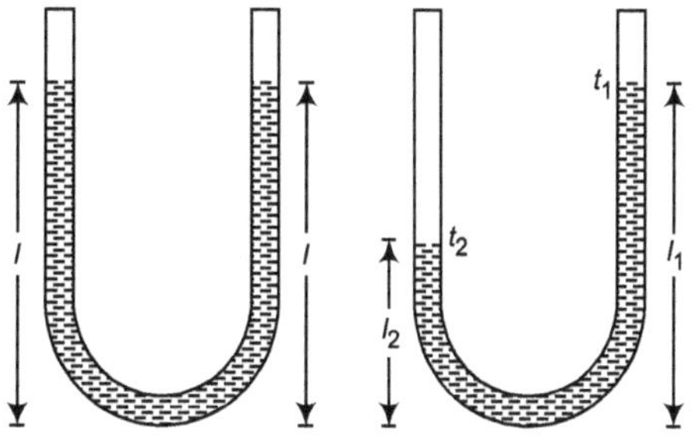

With temperature rise height of liquid in each arm increases i.e. $l_1 > l$ and $l_2 > l$

Also, $\quad l = \frac{l_1}{1+\gamma t_1}$

$$= \frac{l_2}{1+\gamma t_2}$$

$\Rightarrow \quad l_1 + \gamma l_1 t_2 = l_2 + \gamma l_2 t_1$

$\Rightarrow \quad \gamma = \frac{l_1 - l_2}{l_2 t_1 - l_1 t_2}$

17. $V = V_0(1 + \gamma\Delta\theta)$

$L^3 = L_0(1 + \alpha_1\Delta\theta) L_0^2(1 + \alpha_2\Delta\theta)^2$

$\quad = L_0^3(1 + \alpha_1\Delta\theta)(1 + \alpha_2\Delta\theta)^2$

Since, $\quad L_0^3 = V_0$

and $\quad L^3 = V$

Hence, $\quad 1 + \gamma\Delta\theta = (1 + \alpha_1\Delta\theta)(1 + \alpha_2\Delta\theta)^2$

$$\cong (1 + \alpha_1\Delta\theta)(1 + 2\alpha_2\Delta\theta)$$

$$\cong (1 + \alpha_1\Delta\theta + 2\alpha_2\Delta\theta)$$

$\Rightarrow \quad \gamma = \alpha_1 + 2\alpha_2$

18. $\frac{T_C - 0}{100} = \frac{T_F - 32}{180}$

$\therefore \quad -\frac{273}{100} = \frac{T_F - 32}{180}$

Solving, we get $T_F = -460°F$

19. $\frac{\theta}{80-50} = \frac{100}{75-50}$

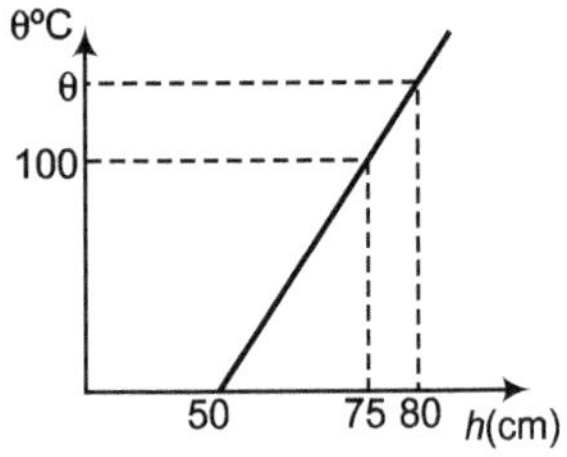

$\therefore \quad \theta = 120°C$

21. $v_{rms} \propto \sqrt{T}$, v_{rms} will become two times when temperature is made four times.

22. $T_1 = 273 + 27 = 300$ K

$T_2 = 273 + 327 = 600$ K

$T_2 = 2T_1$ and $v_{rms} \propto \sqrt{T}$

23. $v_{rms} \propto \sqrt{\frac{3RT}{M}}, v_{av} = \sqrt{\frac{8RT}{\pi M}}$ and $v_{mp} = \sqrt{\frac{2RT}{M}}$

24. $v_{rms} \propto \sqrt{\frac{T}{M}}$

T is doubled and M has become half. Therefore, v_{rms} will become two times.

25. Density of water is 1 g/cm^3, i.e. 1 cm^3 = 1 g of water

In 1 g mole or 18 g water there are total 6.02×10^{23} molecules.

Hence, in 1g of water (or 1cm^3 of water) number of molecules should be $\frac{6.20 \times 10^{23}}{18} \approx \frac{1}{3} \times 10^{23}$

26. $p = p_1 + p_2 + p_3 = \left(\frac{nRT}{V}\right)_1 + \left(\frac{nRT}{V}\right)_2 + \left(\frac{nRT}{V}\right)_3$

$= (n_1 + n_2 + n_3)\frac{RT}{V}$

$= \frac{(0.5 + 0.5 + 0.25)(8.31)(300)}{5 \times 10^{-3}}$

$= 6.23 \times 10^5$ N/m^2

27. Heat given = Heat taken

$\therefore \quad n_1 C_{V_1} \Delta T = n_2 C_{V_2} \Delta T$

$\therefore \quad \left(\frac{1}{2}\right)(3R)(T - 300) = \left(\frac{1}{2}\right)\left(\frac{5}{2}R\right)(300 - T)$

$6T - 1800 = 1550 - 5T$

$\therefore \quad T = 304.54$ K $= 31.5°C$

28. Speed of sound in a gas, $v = \sqrt{\frac{\gamma RT}{M}}$

or $\quad v \propto \sqrt{\frac{\gamma}{M}}$

$\frac{v_{N_2}}{v_{He}} = \sqrt{\frac{\gamma_{N_2}}{\gamma_{He}} \times \frac{M_{He}}{M_{N_2}}} = \sqrt{\frac{(7/5)}{(5/3)} \times \frac{4}{28}}$

$= \frac{\sqrt{3}}{5}$

29. $v_{rms} = \sqrt{\frac{(3v)^2 + (4v)^2 + (5v)^2}{3}} = \sqrt{\frac{50}{3}}\, v$

30. $pV = nRT = (nN)\frac{R}{N}T = (nN)\, kT$

or $\frac{nN}{V} = \frac{p}{kT}$. But, $\frac{nN}{V}$ are the number of molecules per unit volume.

31. $v_{rms} \propto \sqrt{T}$

rms speed will remain half if temperature become $\frac{1}{4}$ th or

$\left(\frac{273}{4} - 273\right)°C$ or $-204.75°C$

32. $v_{rms} = \sqrt{\frac{(1)^2 + (2)^2 + (3)^2 + (4)^2}{4}} = \sqrt{\frac{15}{2}}$ km/s

33. In $pV = nRT$, $pV = RT$ if $n = 1$

34. V-T graph is a straight line passing through origin. Hence, process is isobaric.

$V = \left(\frac{nR}{p}\right)T$

$\text{Slope} = \frac{nR}{p}$

Slope of 2 > slope of 1

$\therefore \quad p_2 < p_1$

35. $v_{rms} \propto \sqrt{T}$. If temperature is doubled, rms speed will become two times.

38. $V \propto T \Rightarrow \frac{V_1}{V_2} = \frac{T_1}{T_2} \Rightarrow \frac{V}{3V} = \frac{(273 + 27)}{T_2}$

$\Rightarrow \quad T_2 = 900$ K $= 627°C$

39. For a given pressure, volume will be more if temperature is more (Charles' law)

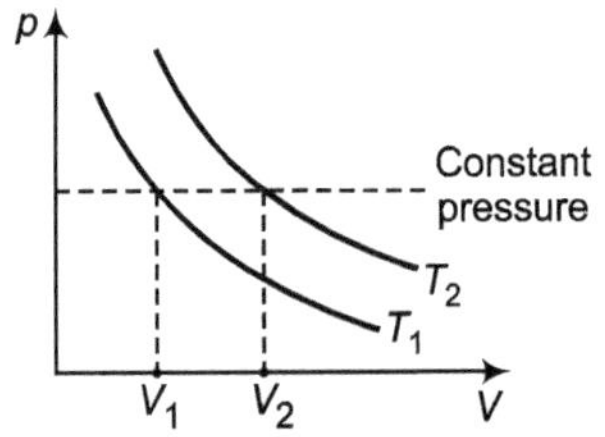

From the graph, it is clear that $V_2 > V_1 \Rightarrow T_2 > T_1$

40. $\rho = \frac{Mp}{RT}$ or $p = \left(\frac{MT}{M}\right)\rho$

Temperature is directly proportional to the slope of p-ρ graph. So, $T_1 > T_2$.

41. As, $\theta_2 > \theta_1 \Rightarrow \tan\theta_2 > \tan\theta_1$

$\Rightarrow \quad \left(\frac{T}{p}\right)_2 > \left(\frac{T}{p}\right)_1$

$\frac{T}{p} \propto V \Rightarrow V_2 > V_1$

42. $pV = \mu RT = \frac{m}{M}RT \Rightarrow \frac{pV}{T} \propto \frac{1}{M}$ $\quad$ ($\because M$ = Molecular mass)

From graph $\quad \left(\frac{pV}{T}\right)_A < \left(\frac{pV}{T}\right)_B < \left(\frac{pV}{T}\right)_C$

$\Rightarrow \quad M_A > M_B > M_C$

43. $p = \frac{\mu RT}{V} = \frac{mRT}{MV}$ $\left(\mu = \frac{m}{M}\right)$

So, at constant volume pressure *versus* temperature garph is a straight line passing through origin with slope $\frac{mp}{MV}$. As the mass is doubled and volume is halved slope becomes four times. Therefore, pressure *versus* temperature graph will be shown by the line *B*.

44. For an ideal gas pV = constant at constant temperature, i.e. pV doesn't vary with V.

45. $v_{rms} = \sqrt{\frac{3RT}{M}} \Rightarrow v_{rms}^2 \propto T$

46. If T is in °C, then $V_T = V_0\left(1 + \frac{T}{273}\right)$

47. $T_1 = 273 + 27 = 300$ K

$$T_2 = 273 + 327 = 600 \text{ K}$$
$$T_2 = 2T_1$$

and $$v_{rms} \propto \sqrt{T}$$

48. $v_{rms} \propto \frac{1}{\sqrt{M}}$ as $v_{rms} \propto \sqrt{\frac{3RT}{M}}$

49. $v_{rms} = \sqrt{\frac{3RT}{M}}$ and $v_{sound} = \sqrt{\frac{\gamma RT}{M}}$

$$\frac{v_{sound}}{v_{rms}} = \sqrt{\frac{\gamma}{3}}$$

Degree of freedom is 6.

$\therefore$ $$\gamma = 1 + \frac{2}{f} = 1 + \frac{2}{6} = \frac{4}{3}$$

$\therefore$ $$v_{sound} = \sqrt{\frac{4/3}{3}}\, v_{rms}$$
$$= \frac{2}{3} v_{rms} = \frac{2c}{3}$$

51. In monoatomic gases, only translational kinetic energy is present, which is proportional to temperature.

53. Average translational kinetic energy per molecule is $\frac{3}{2}kT$ which depends only on temperature not on molecular mass.

55. C_V of the mixture is

$$\frac{n_1 C_{V1} + n_2 C_{V2}}{n_1 + n_2} = \frac{2 \times \frac{3}{2}R + 1 \times \frac{5}{2}R}{3} = \frac{11}{6}R$$

$\therefore$ C_p of mixture is $\frac{11}{6}R + R = \frac{17}{6}R$

$\therefore$ $$\gamma = \frac{C_p}{C_V} = \frac{17}{11} = 1.54$$

56. $C_p - C_V = R$

Dividing the equation by C_V, we get

$$\frac{C_p}{C_V} - 1 = \frac{R}{C_V} \quad \text{or} \quad \gamma - 1 = \frac{R}{C_V}$$

or $$\gamma = 1 + \frac{R}{C_V}$$

$$\frac{1}{1 - \frac{R}{C_p}} = \frac{C_p}{C_p - R} = \frac{C_p}{C_V} = \gamma$$

57. Translational degree of freedom for any type of gas is three.

$\therefore$ Total translational kinetic energy $= 3\left(\frac{1}{2}nRT\right) = \frac{3}{2}pV$

58. Translational degree of freedom for any gas is three.

59. Both O_2 and N_2 are diatomic gases. Rotational degree of freedom in both cases is two. Therefore, average rotational kinetic energy per molecule for each of them is $2\left(\frac{1}{2}kT\right)$ or the ratio is 1 : 1.

60. $U = \frac{f}{2}nRT$. Here, f is degree of freedom and n the number of moles.

$\therefore$ $$1200\,R = \frac{f}{2}(2)(R)(400)$$

or $$f = 3$$

i.e. the given gas is monoatomic. For which, $C_p = \frac{5}{2}R$.

61. $C_p - C_V = R$

Dividing by C_V, we get

$$\gamma - 1 = \frac{R}{C_V}$$

$\therefore$ $$C_V = \frac{R}{\gamma - 1}$$

Further $$U = \frac{f}{2}RT$$

$\therefore$ $$C_V = \frac{dU}{dT} = \frac{f}{2}R$$

62. Mean kinetic energy of gas molecules

$$E = \frac{f}{2}kT$$
$$= \frac{f}{2}k(t + 273)$$
$$= \left(\frac{f}{2}k\right)t + \frac{f}{2} \times 273\,k$$

Comparing it with standard equation of straight line $y = mx + c$. We get $m = \frac{f}{2}k$ and $c = \frac{f}{2}273\,k$. So, the graph between E and t will be straight line with positive intercept on E-axis and positive slope with t-axis.

65. $U = 2\left\{\frac{f_1}{2}RT\right\} + 4\left\{\frac{f_2}{2}RT\right\}$

For O_2, degree of freedom $f_1 = 5$

And for Ar, degree of freedom $f_2 = 3$

$\therefore$ $$U = 2\left\{\frac{5}{2}RT\right\} + 4\left\{\frac{3}{2}RT\right\} = 11\,RT$$

66. $C_p - C_V = R$

Dividing with C_p, we get

$$1 - \frac{1}{\gamma} = \frac{R}{C_p}$$

or $$\frac{\gamma - 1}{\gamma} = \frac{R}{C_p}$$

or $$C_p = \frac{\gamma R}{\gamma - 1}$$

67. $E=\frac{3}{2}kT \Rightarrow E \propto T$

$\therefore \quad \frac{E_2}{E_1}=\frac{T_2}{T_1}=\frac{600}{300}=2$

$\Rightarrow \quad E_2=2E_1$

68. $K_T=\frac{3}{2}RT$

69. $\gamma=1+\frac{2}{f} \quad \therefore f=\frac{2}{\gamma-1}$

70. $\Delta U=nC_V\,\Delta T=n\frac{R}{\gamma-1}(T_f-T_i)$

$$=\frac{p_fV_f-p_iV_i}{\gamma-1}=\frac{pV}{\gamma-1}$$

71. $pV=nRT=n\cdot N\cdot\left(\frac{R}{N}\right)\cdot T=nNkT$

$\therefore \quad \left(\frac{nN}{V}\right)=\frac{p}{kT}$

Here, $\frac{nN}{V}$ are the number of molecules per m^3.

Hence, number of molecules per cm^3 will be

$$\frac{p}{kT}\times10^{-6}=\frac{4\times10^{-10}\times10^{-6}}{1.38\times10^{-23}\times300}\approx1.0\times10^5$$

72. Total internal energy of an ideal diatomic gas will be

$$u=5\left(\frac{1}{2}nRT\right)=\frac{5}{2}pV$$

$\therefore \quad \frac{u}{V}=\frac{5}{2}p$

73. $Q=\frac{mgh}{4.18}\text{cal}=\frac{5\times10\times30}{4.18}=358.8\,\text{cal}$

74. $\frac{V}{T}=\frac{nR}{p}$

V and T for both cases are same. Hence,

$$\frac{n_1}{p_1}=\frac{n_2}{p_2} \quad \text{or} \quad \frac{m_1}{p_1M_1}=\frac{m_2}{p_2M_2}$$

or $\quad m_2=\frac{p_2M_2}{p_1M_1}\cdot m_1$

$$=\frac{(3)(2)}{(5)(28)}\cdot20=0.86\,\text{kg}$$

75. $v=\sqrt{\frac{\gamma p}{\rho}}$

$\therefore \quad \gamma=\frac{\rho v^2}{p}=\frac{1.3\times(330)^2}{1.01\times10^5}=1.4$

i.e. gas is diatomic or degree of freedom is 5.

76. p^2V = constant

$$p=\frac{nRT}{V}$$

or $\quad p=\frac{CT}{V} \quad (C=\text{constant})$

$\therefore \quad \frac{T^2}{V}=\text{constant}$

or $\quad T\propto\sqrt{V}$

V is increased to 3 times. Hence, T will increase $\sqrt{3}$ times.

77. Out of three parameters p, V and T only two should be known to us.

78. Benzene contracts in winter. So, 5 L of benzene will weight more in winter than in summer.

79. Water has maximum density at 4°C, so if the water is heated above 4°C or cooled below 4°C density decreases, i.e. volume increases. In other words, it expands so it overflows in both the cases.

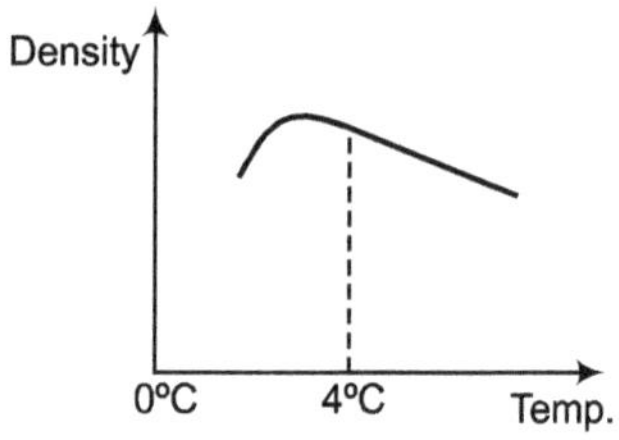

81. According to energy conservation, change in kinetic energy appears in the form of heat (thermal energy).

Thermal energy $=\frac{1}{2}m(v_1^2-v_2^2)$

$$=\frac{1}{2}(100\times10^{-3})(10^2-5^2)$$

$$=3.75\,\text{J}$$

83. At constant pressure, ρT = constant

$\Rightarrow \quad \frac{\rho_1}{\rho_1}=\frac{T_2}{T_1}$

$\Rightarrow \quad \frac{24}{\rho_2}=\frac{(273+127)}{(273+27)}$

$$=\frac{400}{300}$$

$\Rightarrow \quad \rho_2=18\,\text{units}$

85. $pV=\mu RT \Rightarrow p\propto\mu T$

($\because V$ and R = constant)

$\Rightarrow \quad \frac{p_2}{p_1}=\frac{\mu_2}{\mu_1}\times\frac{T_2}{T_1}$

$\Rightarrow \quad \frac{p_{He}}{p}=\frac{1}{1}\times\frac{2T}{T}$

$\Rightarrow \quad p_{He}=2p$

86. $pV=uRT$

$\therefore \quad p\propto\frac{T}{V}$

87. $p=\frac{2}{3}\times(\text{Energy per unit volume})=\frac{2E}{3V}\Rightarrow pV=\frac{2}{3}E$

88. $\rho=\frac{pM}{RT}\Rightarrow\rho\propto\frac{p}{T}$

From graph, $\quad \left(\frac{p}{T}\right)_A=\frac{p_0}{T_0}$

and $\quad \left(\frac{p}{T}\right)_B=\frac{3p_0}{2T_0}$

$\Rightarrow \quad \rho_B=\frac{3}{2}\rho_A$

$$=\frac{3}{2}\rho_0$$

89. $pV = \mu RT = \frac{m}{M}RT$

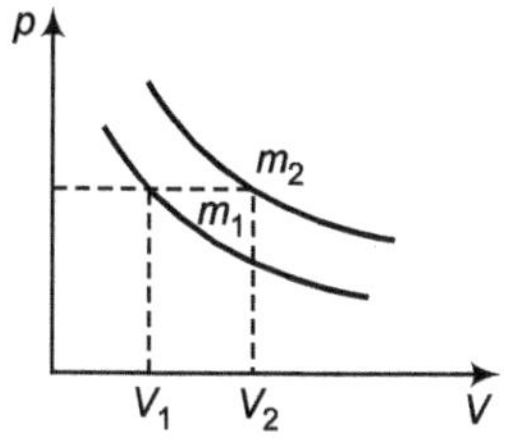

For 1st graph, $p = \frac{m_1}{M}\frac{RT}{V_1}$ ...(i)

For 2nd graph, $p = \frac{m_2}{M}\frac{RT}{V_2}$ ...(ii)

From Eqs. (i) and (ii), we get

$$\frac{m_1}{m_2} = \frac{V_1}{V_2} \Rightarrow m \propto \frac{1}{V}$$

As $V_2 > V_1 \Rightarrow m_1 < m_2$

90. For a gas, $pV = \mu RT = \frac{m}{M}RT$

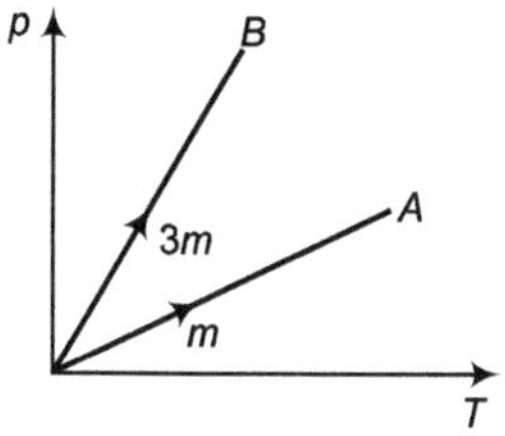

For graph A, $pV = \frac{m}{M}RT$

Slope of graph A, $\left(\frac{p}{V}\right) = \frac{m}{M}\frac{R}{V}$ ...(i)

For graph B, $pV = \frac{3m}{M}RT$

Slope of graph B, $\left(\frac{p}{V}\right) = \frac{3m}{M}\frac{R}{V}$ ...(ii)

$$\frac{\text{Slope of curve } B}{\text{Slope of curve } A} = \frac{\frac{3m}{M}\frac{R}{V}}{\frac{m}{M}\frac{R}{V}} = \frac{3}{1}$$

91. The rms velocity is independent of degree of freedom.

Level 2 : Only One Correct Option

1. $F = \frac{\Delta p}{\Delta t} = (nm)(2v\cos\theta)$

Pressure, $p = \frac{F}{A} = \frac{(nm)(2v\cos\theta)}{A}$

$$= \frac{10^{23} \times 3.32 \times 10^{-27} \times 2 \times 1000 \times \cos 45^\circ}{2 \times 10^{-4}}$$

$$= 2.34 \times 10^3 \text{ N/m}^2$$

2. Pressure on both sides should be same or

$$\frac{n_1 RT}{V_1} = \frac{n_2 RT}{V_2}$$

or $\frac{m}{32(360^\circ - \alpha)} = \frac{m}{28\alpha}$

or $\alpha = 192^\circ$

3. Volume of the gas is constant $\therefore V =$ constant

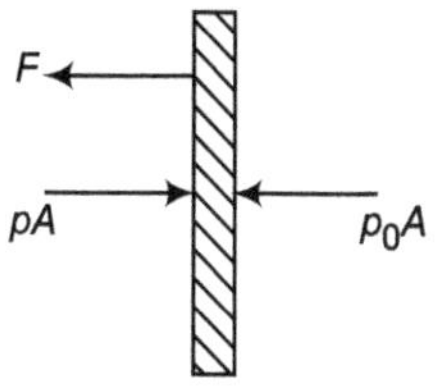

$\therefore p \propto T$ i.e. pressure will be doubled temperature is doubled.

$\therefore$ $p = 2p_0$

Now, let F be the tension in the wire. Then, equilibrium of anyone piston gives

$$F = (p - p_0)A$$
$$= (2p_0 - p_0)A$$
$$= p_0 A$$

4. $v = \sqrt{\frac{\gamma p}{\rho}}$

$\therefore$ $\gamma = \frac{\rho v^2}{p} = \frac{1.3 \times (330)^2}{1.01 \times 10^5}$

$= 1.4$

i.e. gas is diatomic or degree of freedom is 5.

5. $(p_2 - p_1)A = mg$

or $\frac{mg}{A} = \frac{RT_i}{V_1} - \frac{RT_i}{4V_1} = \frac{3RT_i}{4V_1}$ ...(i)

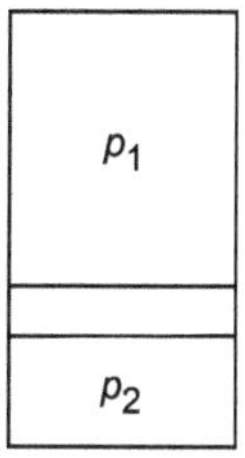

Similarly, in second case

$\frac{mg}{A} = \frac{RT_f}{V_2} - \frac{RT_f}{3V_2} = \frac{2RT_f}{3V_2}$ ...(ii)

Further $5V_1 = 4V_2$

Equating Eqs. (i) and (ii), we get

$$\frac{3T_i}{4V_1} = \frac{2T_f}{3V_2}$$

or $T_f = \frac{9}{8} \times \frac{V_2}{V_1} \times T_i = \frac{9}{8} \times \frac{5}{4} \times 320$

$= 450$ K

6. $V = \left(\frac{nR}{p}\right)T = \left(\frac{mR}{Mp}\right)T$

Slope of V-T graph $\propto \frac{m}{p}$

Since, $\frac{m}{p}$ ratio in both cases is same.

Straight line will be B.

7. From ideal gas equation, $pV = \mu RT$

$\Rightarrow$ Slope of p-T curve, $\frac{p}{T} = \frac{\mu R}{V}$

$\Rightarrow$ Slope $\propto \frac{1}{V}$

It means line of smaller slope represent greater volume of gas. In the given problems, 1 and 2 are on the same line so they will represent same volume, i.e. $V_1 = V_2$

Similarly, points 3 and 4 are on the same line so they will represent same volume, i.e., $V_3 = V_4$

Also slope of line 1-2 is less than 3-4.

Hence, $(V_1 = V_2) > (V_3 = V_4)$

8. **Process 1-2** $V =$ constant, $p \propto T$

T is increasing. Therefore, pressure will be increase.

Process 2-3 $p=$ constant, $V \propto T$

V and T both are increasing.

Process 3-4 $T =$ constant, $p \propto \frac{1}{V}$

V is increasing. Therefore, p will decrease.

Process 4-1 Inverses of process 2-3.

9. Slope of line is 1. Therefore, T-V equation can be written as

$$T = V + T_0$$

$$\therefore \quad \frac{pV}{nR} = V + T_0$$

$$\therefore \quad p = (nR) + \frac{nRT_0}{V}$$

$\therefore$ *p versus V* graph is a sort of rectangular hyperbola.

10. $\underline{AB}$ Isobaric process. Both volume and temperature are increasing.

$\underline{BC}$ Isothermal process. Volume is increasing pressure should decrease. $\left(T = \text{constant}, p \propto \frac{1}{V}\right)$

$\underline{CD}$ Isochoric process. Pressure and temperature both are decreasing. ($V =$ constant, $p \propto T$)

$\underline{DA}$ Again isothermal process. Volume is decreasing and pressure is increasing $\left(T = \text{constant}, p \propto \frac{1}{V}\right)$

11. $U \propto \frac{1}{\rho}$ or $T \propto \frac{1}{(1/V)}$

$$\therefore \quad T \propto V$$

Hence, $p =$ constant

Further ρ is increasing. Therefore, V should decrease.

12. $pV = nRT$

$$\therefore \quad n = \frac{pV}{RT} = \frac{(4\times 10^{-10})\,(10^{-6})}{(8.31)\,(300)} = 1.6\times 10^{-19}$$

$$\therefore \quad N = (6.02\times 10^{23})\,n = 9.65\times 10^{4}$$

$\therefore$ Order is 10^5.

13. 1.0 cc of water $= 1$ g of water

$$= \frac{1}{18} \text{ g-moles of water}$$

$$\therefore \quad N = \frac{1}{18}\times 6.02\times 10^{23} \approx \frac{10^{23}}{3}$$

14. $pV = nRT$

$$= \left(\frac{n_1}{N}\right) RT \;(n_1 = \text{number of molecules})$$

$$= n_1 kt \left(\frac{R}{N} = k\right)$$

$$\therefore \quad \frac{n_1}{V} = \frac{p}{kT}$$

15. $K_T = n\left(\frac{3}{2}RT\right) = \frac{3}{2}pV$

16. $(n_1 + n_2)_i = (n_1 + n_2)_f$

$$\frac{p_0V}{RT_0} + \frac{p_0V}{RT_0} = \frac{pV}{RT_0} + \frac{pV}{2RT_0}$$

$$\therefore \quad p = \frac{4}{3}p_0$$

17. Diameter of brass plate = 2.6 cm and

diameter of steel plate = (2.6 − 0.01) cm = 2.59 cm

Now, $\quad (d + \Delta d)_b = (d + \Delta d)_s$

or $\quad d_b\,(1 + \alpha\Delta\theta)_b = d_s\,(1 + \alpha\Delta\theta)$

$\therefore \quad 2.6\,(1 + 19\times 10^{-6}\,\Delta\theta) = 2.59\,(1 + 11\times 10^{-6}\,\Delta\theta)$

$\therefore \quad \Delta\theta = -478°\text{C}$ or $\theta_f = -458°\text{C}$

18. Change in weight = Upthrust (F)

where, $\quad F = V_s \rho_l g$

$$F' = V_s \rho'_l g$$

$$\frac{F'}{F} = \frac{V'_s}{V_s}\cdot\frac{\rho'_l}{\rho_l}$$

or $\quad \frac{F'}{F} = (1 + \gamma_s\,\Delta\theta)\frac{\rho'_t}{\rho_l}$

$$\left(\frac{46-30.6}{46-30}\right) = (1 + 3\times\alpha_s\times 15)\left(\frac{1.20}{1.24}\right)$$

or $\quad \alpha_s = 2.3\times 10^{-5}/°\text{C}$

19. $pV = nRT$

$$n = \frac{pV}{RT}$$

or $\quad N = n\times 6.02\times 10^{23} = \frac{pV}{RT}\times 6.02\times 10^{23}$

or $\quad N = \frac{(10^{-14}\times 13.6\times 10^{3}\times 10)\,(1)}{(8.31)\,(273+30)}\times 6.02\times 10^{23}$

$$= 3.2\times 10^{11}$$

20. $\gamma_r = \gamma_a + \gamma_V$, where $\gamma_r =$ coefficient of real expansion,

$\gamma_a =$ coefficient of apparent expansion and

$\gamma_V =$ coefficient of expansion of vessel.

For copper, $\quad \gamma_r = C + 3\alpha_{Cu} = C + 3A$

For silver, $\quad \gamma_r = C + 3\alpha_{Ag} = \frac{C - S + 3A}{3}$

$\Rightarrow \quad C + 3A = S + 3\alpha_{Ag} \Rightarrow \alpha_{Ag} = \frac{C - S + 3A}{3}$

21. $(OR)^2 = (PR)^2 - (PO)^2 = l^2 - \left(\frac{l}{2}\right)^2$

$$= [l\,(1 + \alpha_2 t)]^2 - \left[\frac{1}{2}(1 + \alpha_1 t)\right]^2$$

$$l - \frac{l^2}{4} = l^2\,(1 + \alpha_2^2 t^2 + 2\alpha_2 t) - \frac{l^2}{4}(1 + \alpha_1^2 t^2 + 2\alpha_1 t)$$

Neglecting $\alpha_2^2 t^2$ and $\alpha_1^2 t^2$

$$0 = l^2\,(2\alpha_2 t^2) - \frac{l^2}{4}(2\alpha_1 t)$$

$$\Rightarrow \quad 2\alpha_2 = \frac{2\alpha_1}{4}$$

$$\Rightarrow \quad \alpha_1 = 4\alpha_2$$

22. From ideal gas equation, $pV = RT$ ...(i)

or $p\Delta V = R\Delta T$ ...(ii)

Dividing Eq. (ii) by Eq. (i), we get

$$\frac{\Delta V}{V} = \frac{\Delta V}{T}$$

$$\Rightarrow \quad \frac{\Delta V}{V\,\Delta T} = \frac{1}{T} = \delta \quad \text{(given)}$$

$\therefore \delta = 1/T$. So, the graph between δ and T will be rectangular hyperbola.

23. p-T equation from the graph should be like.

$p = aT + b$ (a and b are positive constants)

$$\therefore \quad \frac{p}{T} = a + \frac{b}{T} \Rightarrow T_2 > T_1$$

$$\therefore \quad \left(\frac{p}{T}\right)_2 < \left(\frac{p}{T}\right)_1 \quad \text{or} \quad \left(\frac{T}{p}\right)_2 > \left(\frac{T}{p}\right)_1$$

or $V_2 > V_1$

as $V \propto \frac{T}{p}$

24. Let us draw different isotherms.

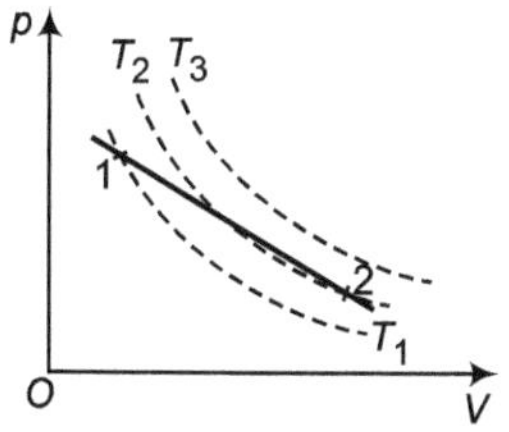

25. Consider the diagram, when the molecules breaks into atoms, the number of moles would become twice.

Now, by ideal gas equation

p = Pressure of gas, n = Number of moles

R = Gas constant, T = Temperature

$pV = nRT$

As volume (V) of the container is constant.

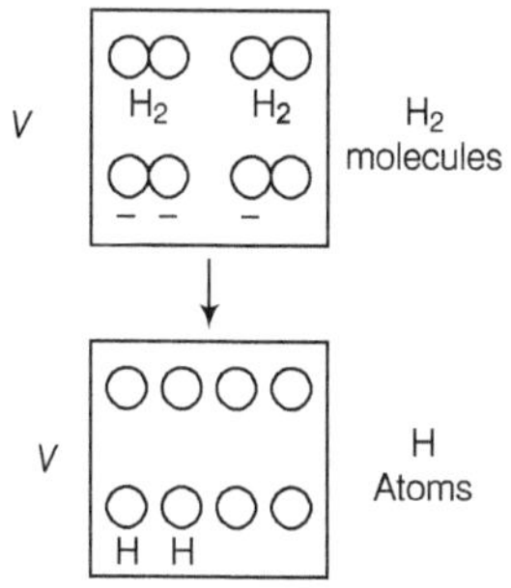

As gases break number of moles become twice of initial, so $n_2 = 2n_1$

So, $p \propto nT$

$$\Rightarrow \quad \frac{p_2}{p_1} = \frac{n_2 T_2}{n_1 T_1} = \frac{(2n_1)(3000)}{n_1 (300)} = 20$$

$$\Rightarrow \quad p_2 = 20p_1$$

Hence, final pressure of the gas would be 20 times the pressure initially.

More than One Correct Options

1. $p \propto \sqrt{\rho}$ as per process. If ρ becomes half, then p will become $\sqrt{2}$ times.

Further $\rho \propto \frac{1}{V}$

$$\therefore \quad \frac{p^2}{(1/V)} = \text{constant}$$

or $(pV)\,p = \text{constant}$

Hence, $Tp = \text{constant}$ (as $pV \propto T$)

$$\therefore \quad p \propto \frac{1}{T}$$

or p - T graph is a rectangular hyperbola.

2. $p \propto \frac{1}{\sqrt{V}}$

If volume becomes 4 times, then p will remain half.

$$\therefore \quad V \propto \frac{T}{p}$$

$$\therefore \quad Vp^2 = \text{constant}$$

$$\therefore \quad \left(\frac{T}{p}\right)p^2 = \text{constant}$$

or $pT = \text{constant}$ or $p \propto \frac{1}{T}$

i.e. p - T graph is a rectangular hyperbola.

If p is halved, then T will become two times.

3. Not required.

4. $\rho \propto \frac{1}{V}$

V is doubled so ρ will remain half.

$$U \propto T \propto pV$$

According to given graph, $p \propto V$

$$\therefore \quad U \propto T \propto p^2 \text{ or } V^2$$

V is doubled, so U and T both will become four times.

$$p \propto V$$

$$\therefore \quad \frac{T}{V} \propto V \quad \left(\text{as } p \propto \frac{T}{V}\right)$$

$$\therefore \quad T \propto V^2$$

or T - V graph is a parabola origin.

5. Not required.

6. $V = \frac{nRT}{p} = \frac{mRT}{Mp} = \left(\frac{mR}{Mp}\right)T$

i.e. V - T graph is a straight line passing through origin of slope,

$$= \left(\frac{mR}{Mp}\right)$$

$$\therefore \quad \text{Slope} \propto \frac{m}{p}$$

Hence, slope depends on both m and p.

7. (a) $p = \frac{nRT}{V} = \frac{(m/V)RT}{M}$

So, given m in the question is the mass of gas per unit volume.

(b) m = total mass of gas

(c) m = mass of one molecule of gas.

Assertion and Reason

1. $K_T = \frac{1}{2}(m)v_{\text{rms}}^2 = \frac{1}{2}m\left(\frac{3RT}{M}\right) = \frac{3}{2}nRT = \frac{3}{2}pV$

$\therefore \quad p = \frac{2}{3}\cdot\frac{K_T}{V} = \frac{2}{3}$

(translational kinetic energy per unit volume).

2. Only that straight line which passes through origin represents isobaric process.
3. In 22.4 L at NTP, number of molecules are 6.02×10^{23}.
4. In isochoric process, ρ and V both are constants. Therefore, ρ-V graph is a dot.
5. At higher temperature, natural length is more. It means rod is compressed.
6. E represents translational kinetic energy per unit volume.
7. Translational degree of freedom of a diatomic gas (O_2) is three and rotational kinetic energy is two. Therefore, according to law of equipartition of energy $\frac{K_T}{K_R} = \frac{3}{2}$.

   Further, intermolecular force between ideal gas molecules is zero. Hence, potential energy is zero.
8. Only translational kinetic energy is $\frac{1}{2}mv_{\text{rms}}^2$.

   Further, translational degree of freedom of any type of gas is three. Therefore, translational kinetic energy of any type of gas (of one mole) is $\frac{3}{2}RT$.

10. At the given conditions, the intermolecular force between the gas molecules is almost zero.
12. Translational motion of the gas chamber increases the total kinetic energy but it does not contribute in internal energy. Therefore, temperature of gas does not increase by translational motion of the gas chamber.
13. It will read less than the actual.
14. Only that straight line passing through origin represents isobaric process.

Match the Columns

1. $T \propto \frac{T}{V}$ or $T \propto \frac{1}{T/p}$

   or $\quad T \propto \sqrt{p}$

   p has become four times. Therefore, T has become two times or V will have become $\frac{1}{2}$ times.

   Since, V is decreasing, work done by the gas is negative.

   Further, T is increasing, hence ΔU will be positive. W is negative and ΔU is positive. Hence, we cannot say about ΔQ.

2. $\frac{C_p}{C_V} = \gamma = \frac{5}{3}$

   Speed of sound in a gas, $v_{\text{sound}} = \sqrt{\frac{\gamma RT}{M}}$

3. Internal energy of n moles of an ideal gas,

   $U = n\left(\frac{f}{2}RT\right) \qquad$ (f = degree of freedom)

   For monoatomic gas, $f = 3$

   $\therefore \quad U = \frac{3nRT}{R}$

   or $U = 3RT$ for $n = 2$ is valid for any gas.

   $$\Delta U = nC_V\,\Delta T$$

4. From the given V-T graph we cannot tell nature of gas

   $$V = \frac{nRT}{p} = \left(\frac{nR}{p}\right)T$$

   If p is constant, V-T graph is a straight line. Slope of this line is $\frac{nR}{p}$.

   $$(\text{Slope})_A > (\text{Slope})_B$$

   or $\frac{n_A}{n_B}$ may be grater than 1.

   or $\frac{p_A}{p_B}$ may be less than 1.

   But, $\left(\frac{n}{p}\right)_A > \left(\frac{n}{p}\right)_B$.

Entrance Gallery

1. We know that

   $$\bar{v}^2 = \frac{3RT}{M} \Rightarrow v_{\text{rms}} = \sqrt{\frac{3RT}{M}}$$

   The root mean square speed of the molecules of a gas is directly proportional to the square root of the absolute temperature of the gas and inversely proportional to the square root of the mass of molecules of the gas.

   So, $\quad v_{\text{rms}} \propto \sqrt{T}$ and $v_{\text{rms}} \propto \frac{1}{\sqrt{m}}$

   or $\quad v = m^{-1/2} \times T^{1/2}$

2. We know that specific heat of a gas (At constant volume)

   $$C_V = f/2 \times R$$

   where, R = number of independent relations

   f = degree of freedom

   $\therefore \quad C_V = \frac{6}{2} \times R = 3R$

3. Total number of degrees of freedom of a rigid diatomic molecule is $3 \times 2 - 1 = 5$.
4. Substances having more specific heat take longer time to get heated to a higher temperature and longer time to get cooled.

   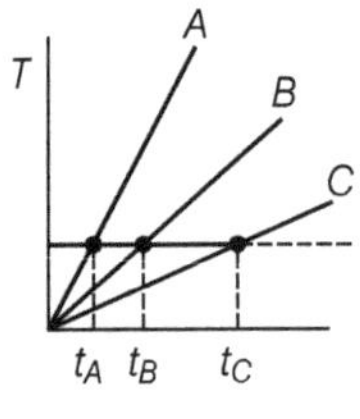

   If we draw a line parallel to the time axis, then it cuts the given graphs at three different points. Corresponding points on the time axes shows that,

   $$t_C > t_B > t_A \Rightarrow c_C > c_B > c_A$$

5. Density of the gas, $\rho = \frac{pM}{RT}$

   $\therefore \quad \rho \propto pM$ or $\frac{\rho_1}{\rho_2} = \left(\frac{p_1}{p_2}\right)\left(\frac{M_1}{M_2}\right) = \left(\frac{2}{3}\right)\left(\frac{4}{3}\right) = \frac{8}{9}$

   $$\rho_1 : \rho_2 = 8 : 9$$

6. The average velocity,

   $$v = \frac{v_1 + v_2 + v_3 + \ldots + v_n}{n} = \frac{1 + 3 + 5 + 7}{4} = 4\,\text{km/s}$$

Root mean square velocity,

$$v_{\text{rms}} = \sqrt{\frac{v_1^2 + v_2^2 + v_3^2 + v_4^2 + \ldots + v_n^2}{n}}$$

$$= \sqrt{\frac{1 + (3)^2 + (5)^2 + (7)^2}{4}} = \sqrt{21} = 4.583\,\text{km/s}$$

Difference between average velocity and root mean square velocity $= 4.583 - 4 = 0.583\,\text{km/s}$

7. Given, ice point $= 10°\text{C}$ and steam point $= 130°\text{C}$

According to the formula,

$$\frac{x - 10}{130 - 10} = \frac{40}{100} \Rightarrow 100x - 1000 = 4800$$

$$100x = 5800 \Rightarrow x = 58°\text{C}$$

8. Cubical expansion, $\Delta V = \gamma V \,\Delta T$

where, γ = coefficient of volumetric expansion

$$\Delta V = 3\alpha V\,\Delta T = 3 \times 23 \times 10^{-6} \times \left(\frac{4}{3}\pi \times 10^3\right) \times 100 = 28.9\,\text{cc}$$

9. Linear expansion $\Delta L = \alpha V\,\Delta t = \dfrac{FL}{AY}$

Stress, $s = \dfrac{F}{A} = Y\alpha\,\Delta t$

where, symbols have their usual meaning.

10. Internal energy of the gas remains constant, hence

$$T_2 = T$$

Using, $p_1V_1 = p_2V_2 \Rightarrow p \cdot \dfrac{V}{2} = p_2 V \Rightarrow p_2 = \dfrac{p}{2}$

11. Conserving total KE, we can write

$$\frac{f}{2}n_1kT_1 + \frac{f}{2}n_2kT_2 + \frac{f}{2}n_3kT_3 = \frac{f}{2}(n_1 + n_2 + n_3)kT$$

$$T = \frac{n_1T_1 + n_2T_2 + n_3T_3}{n_1 + n_2 + n_3}$$

12. $C_V = \dfrac{f}{2}R$ and $C_p = \left(\dfrac{f}{2} + 1\right)R$

$$C_V = \frac{5}{2}R \text{ and } C_p = \frac{7}{2}R$$

$$\therefore \quad \gamma = \frac{C_p}{C_V} = \frac{7}{5}$$

13. Using $pV = nRT$, p = constant $\dfrac{V_1}{V_2} = \dfrac{T_1}{T_2}$

$$\frac{1}{2} = \frac{300}{T_2} \qquad [\because V_2 = 2V_1]$$

$$T_2 = 600\,\text{K} = 600 - 273 = 327\,°\text{C}$$

$$\Delta t = 327 - 27 = 300°\text{C}$$

14. Condition for same reading is $A = B$

$$\frac{A - 42}{110} = \frac{B - 72}{220}$$

$$\Rightarrow \quad \frac{A - 42}{110} = \frac{A - 72}{220}$$

$$2A - 84 = A - 72$$

$$A = 12°\text{C}$$

15. Final pressure is given as $p_f = 2p + \overline{p}$

Saturated vapour pressure will not change, if temperature remains constant, the air pressure becomes twice as volume is halved.

16. A real gas behaves as an ideal gas at low pressure and high temperature.

17. According to Gay Lussac's law, $p \propto T$

$$\therefore \quad \frac{dp}{p} \times 100 = \frac{dT}{T} \times 100 \Rightarrow 1 = \frac{1}{T} \times 100 \Rightarrow T = 100$$

18. Molar specific heat of mixture, $C_V = \dfrac{n_1C_{V_1} + n_2C_{V_2}}{n_1 + n_2}$

$$= \frac{1 \times \frac{3}{2}R + 1 \times \frac{5}{2}R}{1 + 1} = 2R$$

19. We know that $v \propto \sqrt{\dfrac{3RT}{M}}$

$$\Rightarrow \quad T \propto v_{\text{rms}}$$

$$\Rightarrow \quad \frac{T_2}{T_1} = \left[\frac{v_2}{v_1}\right]^2 = \frac{1}{4} \Rightarrow T_2 = \frac{T_1}{4} = \frac{273 + 327}{4}$$

$$= 150 \quad = -123°$$

20. As, $v_{\text{rms}} \propto \sqrt{T}$

$$\Rightarrow \quad \frac{v_1^2}{v_2^2} = \frac{T_1}{T_2} \Rightarrow \frac{v^2}{(2v)^2} = \frac{273}{T_2}$$

$$\Rightarrow \quad T_2 = 1092 \quad = 819°$$

21. The pressure exerted by the gas,

$$p = \frac{1}{3}\rho\bar{c}^2 = \frac{1}{3}\frac{m}{V}\bar{c}^2 = \frac{2}{3}\left(\frac{1}{2}m\bar{c}^2\right)$$

$$\left[\because \frac{1}{2}mc^{-2} = \frac{E}{V} = \text{energy volume}, V = 1\right]$$

$$p = \frac{2}{3}E$$

15

The First Law of Thermodynamics

Chapter Snapshot

- The First Law of Thermodynamics
- Further Explanation of the Three Terms Used in First Law
- Different Thermodynamic Processes
- Efficiency of a Cycle
- Heat Engines
- Refrigerator
- Second Law of Thermodynamics

Thermal Equilibrium

Equilibrium in thermodynamics mean macroscopic variables (like pressure, temperature, volume, etc) of a thermodynamic system (like a gas inside a closed rigid container) do not change in time.

Adiabatic Wall

It is an insulating wall (can be movable) between two thermodynamic systems that does not allow to flow of energy (or heat) from one system to another system.

Diathermic Wall

It is a conducting wall between two thermodynamic systems that allows energy flow (or heat) flow from one system to another system.

Zeroth Law of Thermodynamics

According to this law, if two systems in thermal equilibrium with a third system, then they separately are in thermal equilibrium with each other. Thus, if A and B are separately in equilibrium with C, i.e. if $T_A = T_C$ and $T_B = T_C$. Then, this implies that $T_A = T_B$ i.e. the systems A and B are also in thermal equilibrium.

15.1 The First Law of Thermodynamics

The first law of thermodynamics, is an extension of the principle of conservation of energy. To state energy relationships precisely, we need the concept of a **thermodynamic system**. A thermodynamic system is a system that can interact (and exchange energy) with its surroundings, or environment, at least in one way one of which is heat transfer. In this chapter the thermodynamic system will be an ideal gas contained in a vessel in most of the cases. A process in which there are changes in the state of a thermodynamic system (p, V and T in case of a gas) is called a **thermodynamic process**.

We now come to the first law.

Let a system changes from an initial equilibrium state i to a final equilibrium state f in a definite way, the heat absorbed by the system being Q and the work done by the system being W. Then, we compute the $Q - W$. While Q and W both depend on the thermodynamic path taken between two equilibrium states, their difference $Q - W$ does not depend on it. We do this over and over again, using different paths each time. We find that in every case the quantity $Q - W$ is the same.

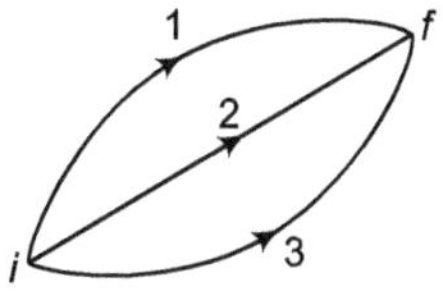

Fig. 15.1

The students may recall from mechanics that when an object is moved from an initial point i to a final point f in a gravitational field in the absence of friction, the work done depends only on the positions of the two points and not at all on the path through which the body is moved. From this we concluded that there is a function of the space coordinates of the body whose final value minus its initial value equals the work done in displacing the body. We called it the potential energy function. In thermodynamics there is a function of the thermodynamic coordinates (p, V and T) whose final value minus its initial value equals the change $Q - W$ in the process. We call this function the **internal energy** function. We have,

$$\Delta U = U_f - U_i = Q - W \qquad \ldots\text{(i)}$$

This equation is known as the **first law of thermodynamics**.

We describe, the energy relations in any thermodynamic process in terms of the quantity of heat Q added to the system and the work done W by the system. Both Q and W may be positive, negative or zero. A positive value of Q represents heat flow into the system, negative Q represents heat flow out of the system. A positive value of W represents work done by the system against its surroundings, such as work done by an expanding gas. A negative value of W represents work done on the system by its surroundings such as work done during compression of a gas.

Table 15.1 Thermodynamic Sign Conventions for Heat and Work

Process	Convention
Heat added to the system	$Q > 0$
Heat removed from the system	$Q < 0$
Work done by the system	$W > 0$
Work done on the system	$W < 0$

Eq. (i) can be written as

$$Q_1 - W_1 = Q_2 - W_2 = \ldots$$

or $$\Delta U_1 = \Delta U_2 = \ldots$$

That is the change in the internal energy of the system between two points is path independent. It depends on thermodynamic coordinates of the two points. For example, in case of an ideal gas it depends only on the initial and final temperatures.

Often the first law must be used in its differential form, which is

$$dU = dQ - dW \qquad \ldots\text{(ii)}$$

This can also be written as

$$dQ = dU + dW \ldots\text{(iii)} \quad \text{or} \quad Q = \Delta U + W \quad \ldots\text{(iv)}$$

The first law can be expressed in other words as under.

Suppose a heat Q is given to a system. This heat is partly used by the system in doing work against its surroundings and partly its internal energy gets increased and from energy conservation principle, $Q = \Delta U + W$. An another analogous example is from our daily life. Consider a person X. Suppose his monthly income is ₹ 50000 (Q).

He spends ₹30000 (W) as his monthly expenditure. Then, obviously the remaining ₹ 20000 goes to his savings (ΔU). In some month it is also possible that he spends more than his income. In that case he will withdraw it from his bank or his savings will get reduced ($\Delta U < 0$). In the similar manner other combinations can be made.

Example 15.1 *When a system goes from state A to state B, it is supplied with 400 J of heat and it does 100 J of work.*

(a) For this transition, what is the system's change in internal energy?

(b) If the system moves from B to A, what is the change in internal energy?

(c) If in moving from A to B along a different path in which $W'_{AB} = 400$ J of work is done on the system, how much heat does it absorb?

Sol. (a) From the first law,

$$\Delta U_{AB} = Q_{AB} - W_{AB} = (400 - 100)\text{ J} = 300\text{ J}$$

(b) Consider a closed path that passes through the state A and B. Internal energy is a state function, so ΔU is zero for a closed path.

Thus, $$\Delta U = \Delta U_{AB} + \Delta U_{BA} = 0$$

or $$\Delta U_{BA} = -\Delta U_{AB} = -300\text{ J}$$

(c) The change in internal energy is the same for any path, so

$$\Delta U_{AB} = \Delta U'_{AB} = Q'_{AB} - W'_{AB}$$

$$300\text{ J} = Q'_{AB} - (-400\text{ J})$$

and the heat exchanged is $Q'_{AB} = -100$ J

The negative sign indicates that the system loses heat in this transition.

Example *The quantities in the following table represent four different paths for the same initial and final state. Find a, b, c, d, e, f and g.*

Table 15.2

Q (J)	W (J)	ΔU (J)
− 80	−120	d
90	c	e
a	40	f
b	−40	g

Ans. $a = 80$ J, $b = 0$, $c = 50$ J, $d = 40$ J
$e = 40$ J, $f = 40$ J, $g = 40$ J

15.2 Further Explanation of the Three Terms Used in First Law

First law of thermodynamics basically revolves round the three terms Q, ΔU and W. If you substitute these three terms correctly with proper signs in the equation $Q = \Delta U + W$, then you are able to solve most of the problems of first law. Let us take each term one by one. Here, we are taking the system an ideal gas.

Heat Transfer (Q or dQ)

For heat transfer apply,

$$Q = nC\Delta T$$

or in differential form,

$$dQ = nCdT$$

where, C is the molar heat capacity of the gas and n is the number of moles of the gas. Always take,

$$\Delta T = T_f - T_i$$

where, T_f is the final temperature and T_i is the initial temperature of the gas. Further, we have discussed in chapter 20, that molar heat capacity of an ideal gas in the process (pV^x = constant) is,

$$C = \frac{R}{\gamma - 1} + \frac{R}{1-x}$$

$$= C_V + \frac{R}{1-x}$$

$$C = C_V = \frac{R}{\gamma - 1} \text{ in isochoric process and}$$

$$C = C_p = C_V + R \text{ in isobaric process}$$

Mostly C_p and C_V are used.

Change in Internal Energy (U or dU)

For change in internal energy of the gas apply,

$$\Delta U = nC_V \Delta T$$

or in differential form,

$$dU = nC_V dT$$

Students are often confused that the result $\Delta U = nC_V \Delta T$ can be applied only in case of an isochoric process (as C_V is here used). However, it is not so. It can be applied in any process, whether it is isobaric, isothermal adiabatic or else.

Work Done (W or dW)

This is the most important of the three.

Work Done During Volume Changes

A gas in a cylinder with a movable piston is a simple example of a thermodynamic system.

Figure shows that a gas confined to a cylinder which has a movable piston at one end. If the gas expands against the piston, it exerts a force through a distance and does work on the piston. If the piston compresses the gas as it is moved inward, work is also done in this case on the gas. The work associated with such volume changes can be determined as follows.

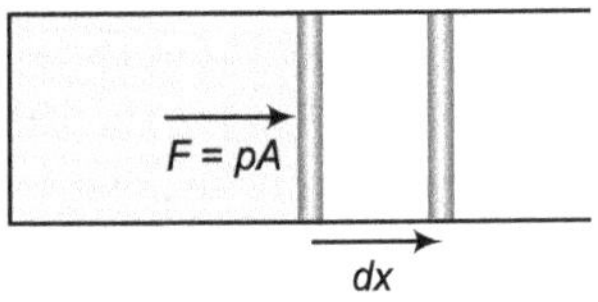

Fig. 15.2

Let the gas pressure on the piston face be p. Then, the force on the piston due to gas is pA, where, A is the area of the face.

When the piston is pushed outward an infinitesimal distance dx, the work done by the gas is

$$dW = F \cdot dx = pA\, dx$$

Which, since the change in volume of the gas is $dV = Adx$, becomes

$$dW = pdV$$

For a finite change in volume from V_i to V_f, this equation is then integrated between V_i to V_f to find the net work done

$$W = \int dW = \int_{V_i}^{V_f} pdV$$

Now there are two methods of finding work done by a gas.

***Method* 1.** *This is used when p-V equation is known to us. Suppose p as a function of V is known to us.*

$$p = f(V)$$

Then, work done can be found by,

$$W = \int_{V_i}^{V_f} f(V)\, dV$$

***Method* 2.** *The work done by a gas is also equal to the area under p-V graph. Following different cases are possible.*

Case I. When volume is constant

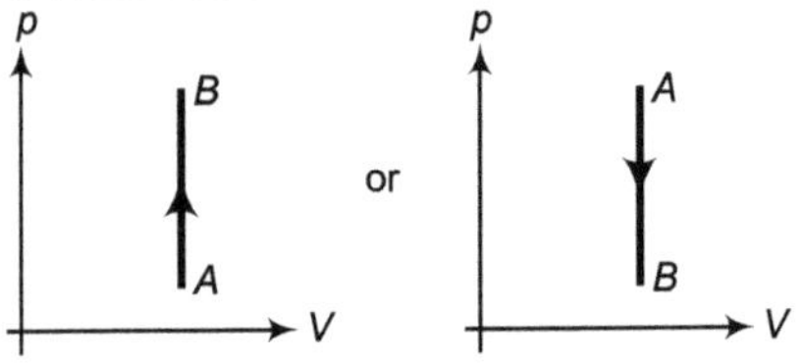

Fig. 15.3

$$V = \text{constant} \quad \Rightarrow \quad \therefore \quad W_{AB} = 0$$

Case II. When volume is increasing

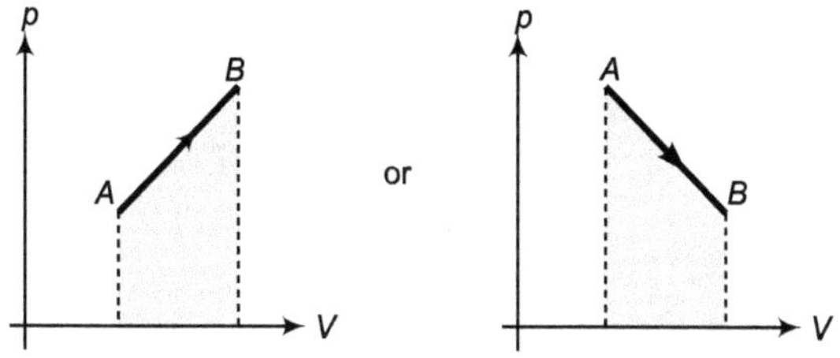

Fig. 15.4

V is increasing

$\therefore \quad W_{AB} > 0$

$W_{AB} = \text{Shaded area}$

Case III. When volume is decreasing

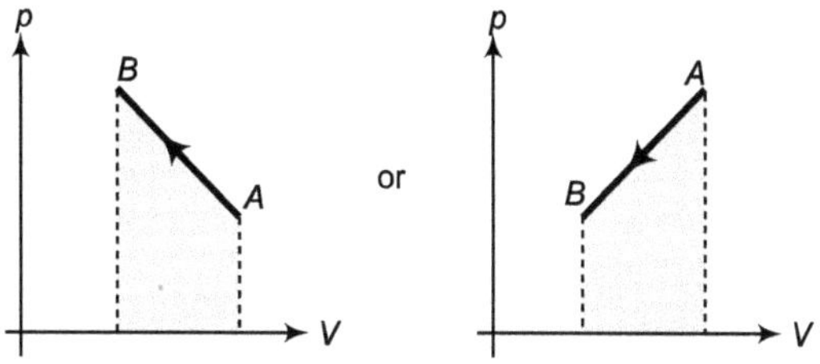

Fig. 15.5

V is decreasing

$\therefore \quad W_{AB} < 0$

$W_{AB} = -\text{Shaded area}$

Case IV. Cyclic process

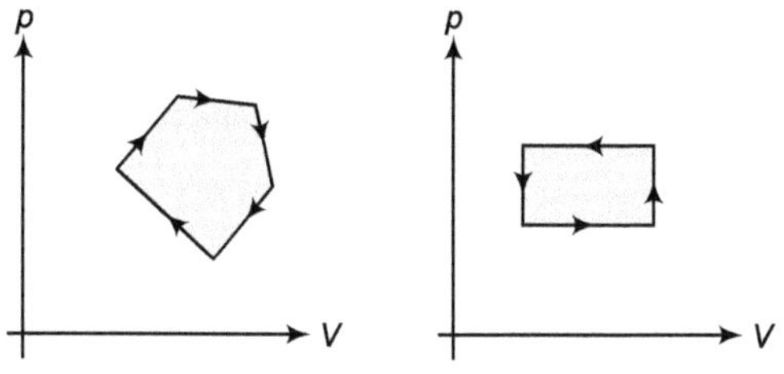

(a) (b)

Fig. 15.6

Cyclic process

$W_{\text{clockwise cycle}} = +\text{Shaded area}$ [in Fig. 15.6(a)]

$W_{\text{anti-clockwise cycle}} = -\text{Shaded area}$ [in Fig. 15.6 (b)]

Case V. Incomplete cycle

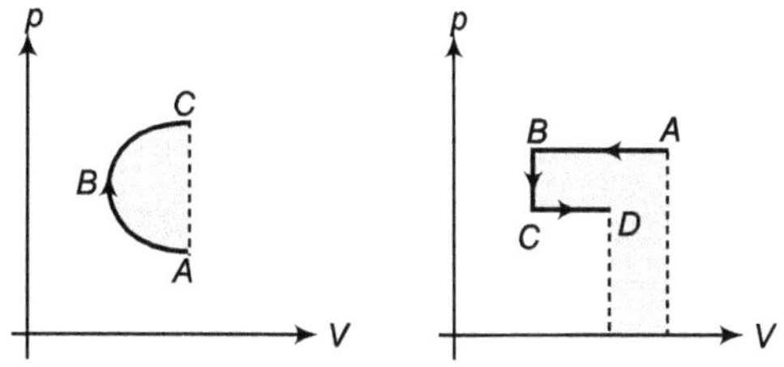

Fig. 15.7

$W_{ABC} = +\text{Shaded area}$

$W_{ABCD} = -\text{Shaded area}$

Extra Knowledge Points

- In the above discussion, we have seen that $W = \int_{V_i}^{V_f} p\, dV$

  From this equation it seems as, if work done can be calculated only when p-V equation is known and the limits V_i and V_f are known to us. But it is not so. We can calculate work done even, if we know the limits of temperature.

  For example, the temperature of n moles of an ideal gas is increased from T_0 to $2T_0$ through a process $p = \frac{\alpha}{T}$ and we are interested in finding the work done by the gas. Then,

  $$pV = nRT \text{(ideal gas equation)} \quad \text{...(i)}$$

  and $$p = \frac{\alpha}{T} \quad \text{...(ii)}$$

  Dividing Eq. (i) by Eq. (ii), we get

  $$V = \frac{nRT^2}{\alpha}$$

  or $$dV = \frac{2nRT}{\alpha} dT$$

  $$\therefore \quad W = \int_{V_i}^{V_f} p\, dV = \int_{T_0}^{2T_0} \left(\frac{\alpha}{T}\right)\left(\frac{2nRT}{\alpha}\right) dT = 2nRT_0$$

  So, we have found the work done without putting the limits of volume.

- Some times the piston (which is massless) is attached to a spring of force constant k and a mass m is placed over the piston. The area of the piston is A. The gas expands. To make the calculation easy we assume that initially the spring was in its natural length. We are required to find the work done by the gas. As, the piston is massless, net force on it at every instant is zero.

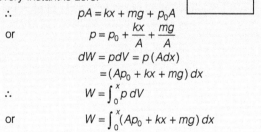

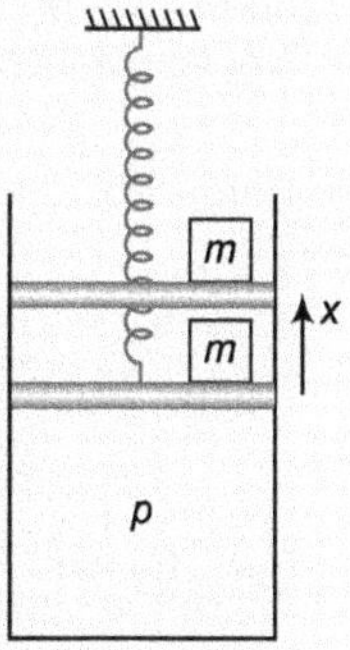

  $$\therefore \quad pA = kx + mg + p_0 A$$

  or $$p = p_0 + \frac{kx}{A} + \frac{mg}{A}$$

  $$dW = p\,dV = p\,(A\,dx)$$

  $$= (Ap_0 + kx + mg)\, dx$$

  $$\therefore \quad W = \int_0^x p\, dV$$

  or $$W = \int_0^x (Ap_0 + kx + mg)\, dx$$

$$= p_0Ax + \frac{1}{2}kx^2 + mgx \quad (\text{as } Ax = \Delta V)$$

or $$W = p_0\Delta V + \frac{1}{2}kx^2 + mgx$$

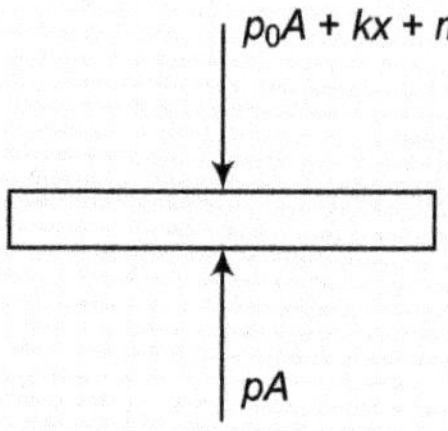

The result can be stated in a different manner as under. The gas does work against the atmospheric pressure p_0 (which is constant), the spring force kx (which varies linearly with x) and the gravity force mg (which is again constant).

$\therefore$ W_1 = Work done against $p_0 = p_0\Delta V$

W_2 = Work done against $kx = \frac{1}{2}kx^2$

and W_3 = Work done against $mg = mgx$

So, $$W_{\text{total}} = W_1 + W_2 + W_3 = p_0\Delta V + \frac{1}{2}kx^2 + mgx$$

From point number (2). We may conclude that work done by a gas is zero, if the other side of the piston is vacuum.

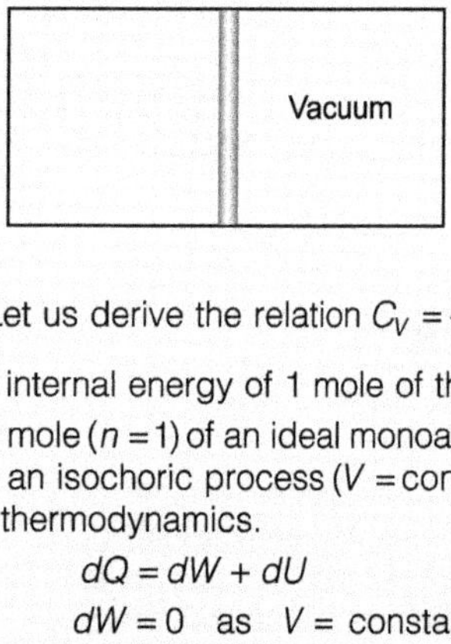

- $\boldsymbol{C_V = \frac{dU}{dT}}$. Let us derive the relation $C_V = \frac{dU}{dT}$

where, U = internal energy of 1 mole of the gas.

Consider 1 mole ($n = 1$) of an ideal monoatomic gas which undergoes an isochoric process (V = constant). From the first law of thermodynamics.

$$dQ = dW + dU \qquad \ldots\text{(i)}$$

Here, $dW = 0$ as V = constant

$$dQ = CdT = C_VdT$$

(In $dQ = nC_V\,dT, n = 1$ and $C = C_V$)

Substituting in Eq. (i), we have

$$C_VdT = dU \quad \text{or} \quad C_V = \frac{dU}{dT}$$

- $\boldsymbol{C_P - C_V = R}$. To prove this relation (also known as Mayer's formula) let us consider 1 mole of an ideal gas which undergoes an isobaric (p = constant) process. From first law of thermodynamics,

$$dQ = dW + dU$$

Here, $dQ = C_pdT \quad (\text{as } n = 1 \text{ and } C = C_p)$

$$dU = C_VdT$$

and $$dW = pdV = pd\left(\frac{RT}{p}\right) \left(\text{as } V = \frac{RT}{p}\right)$$

$$= d(RT) \qquad (\text{as } p = \text{constant})$$

$$= R \cdot dT$$

Substituting these values in Eq. (i),

we have $$C_pdT = R \cdot dT + C_VdT$$

or $$C_p - C_V = R$$

- $C_V = \frac{R}{g - 1}$. We have already derived,

$$C_p - C_V = R$$

Dividing this equation by C_V, we have

$$\frac{C_p}{C_V} - 1 = \frac{R}{C_V}$$

or $$\gamma - 1 = \frac{R}{C_V} \qquad \left(\text{as } \frac{C_p}{C_V} = \gamma\right)$$

$\therefore$ $$C_V = \frac{R}{\gamma - 1}$$

- **Thermodynamic parameters for a mixture of gases**

(i) **Equivalent molar mass** When n_1 moles of a gas with molar mass M_1 are mixed with n_2 moles of a gas with molar mass M_2, then equivalent molar mass of the mixture is given by

$$M = \frac{n_1M_1 + n_2M_2}{n_1 + n_2}$$

(ii) **Internal energy of the mixture** The total energy of the mixture is

$$U = U_1 + U_2$$

(iii) $\boldsymbol{C_V}$ **of the mixture**

$$U = U_1 + U_2$$

$\therefore$ $$dU = dU_1 + dU_2$$

or $$nC_VdT = n_1C_{V_1}dT + n_2C_{V_2}dT \qquad \ldots\text{(ii)}$$

or $$(n_1 + n_2)C_V = n_1C_{V_1} + n_2C_{V_2} \quad (\text{as } n = n_1 + n_2)$$

$\therefore$ $$C_V = \frac{n_1C_{V_1} + n_2C_{V_2}}{n_1 + n_2}$$

(iv) $\boldsymbol{C_p}$ **of the mixture**

$$C_p = C_V + R$$

or $$C_p = \frac{n_1C_{V_1} + n_2C_{V_2}}{n_1 + n_2} + R$$

$$= \frac{n_1(C_{V_1} + R) + n_2(C_{V_2} + R)}{n_1 + n_2}$$

$$= \frac{n_1C_{p_1} + n_2C_{p_2}}{n_1 + n_2}$$

Thus, $$C_p = \frac{n_1C_{p_1} + n_2C_{p_2}}{n_1 + n_2}$$

(v) γ **of the mixture** From Eq. (ii)

$$(n_1 + n_2)C_V = n_1C_{V_1} + n_2C_{V_2}$$

or $$\frac{(n_1 + n_2)R}{\gamma - 1} = \frac{n_1R}{\gamma_1 - 1} + \frac{n_2R}{\gamma_2 - 1}$$

or $$\frac{n_1 + n_2}{\gamma - 1} = \frac{n_1}{\gamma_1 - 1} + \frac{n_2}{\gamma_2 - 1}$$

Thus, γ of the mixture is given by above equation.

Example 15.2 *A certain amount of an ideal gas passes from state A to B first by means of process 1, then by means of process 2. In which of the process is the amount of heat absorbed by the gas greater?*

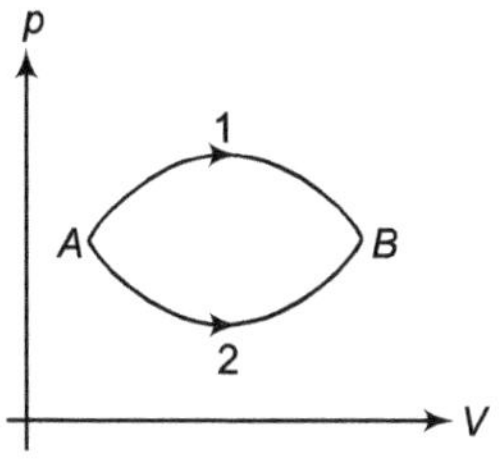

Fig. 15.8

Sol. $Q_1 = W_1 + \Delta U_1$ and $Q_2 = W_2 + \Delta U_2$

U is a state function. Hence, ΔU depends only on the initial and final positions. Therefore,

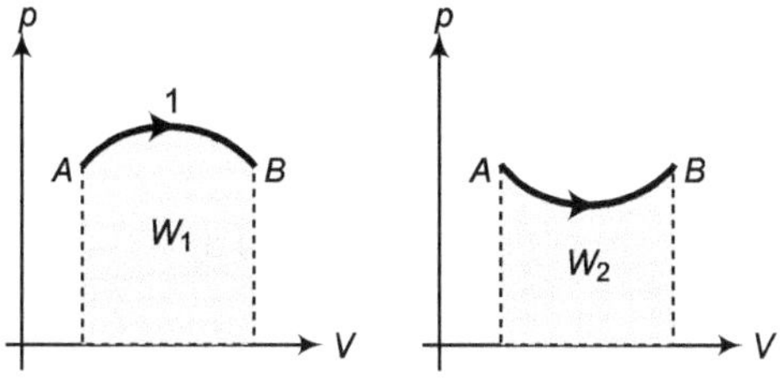

Fig. 15.9

$$\Delta U_1 = \Delta U_2$$

But $W_1 > W_2$

as the area under 1 is greater than area under 2. Hence, $Q_1 > Q_2$

Example 15.3 *Find the ratio of* $\frac{\Delta Q}{\Delta U}$ *and* $\frac{\Delta Q}{\Delta W}$ *in an isobaric process. The ratio of molar heat capacities* $\frac{C_p}{C_V} = \gamma$.

Sol. In an isobaric process, p = constant. Therefore, $C = C_p$.

Now, $$\frac{\Delta Q}{\Delta U} = \frac{nC_p\Delta T}{nC_V\Delta T} = \frac{C_p}{C_V} = \gamma$$

and $$\frac{\Delta Q}{\Delta W} = \frac{\Delta Q}{\Delta Q - \Delta U} = \frac{nC_p\Delta T}{nC_p\Delta T - nC_V\Delta T}$$

$$= \frac{C_p}{C_p - C_V} = \frac{C_p/C_V}{C_p/C_V - 1} = \frac{\gamma}{\gamma - 1}$$

Example 15.4 *Suppose* 1.0 g *of water vaporizes isobarically at atmospheric pressure* $(1.01\times 10^5\ Pa)$. *Its volume in the liquid state is* $V_i = V_{\text{liquid}} = 1.0\ \text{cm}^3$ *and its volume in vapour state is* $V_f = V_{\text{vapour}} = 1671\ \text{cm}^3$. *Find the work done in the expansion and the change in internal energy of the system. Ignore any mixing of the steam and the surrounding air. Take latent heat of vaporisation* $L_v = 2.26\times 10^6\ J/kg$.

Sol. Because, the expansion takes place at constant pressure, the work done is

$$W = \int_{V_i}^{V_f} p_0 dV = p_0 \int_{V_i}^{V_f} dV = p_0 (V_f - V_i)$$

$$= (1.01\times 10^5)(1671\times 10^{-6} - 1.0\times 10^{-6})$$

$$= 169\ \text{J}$$

$$Q = mL_v = (1.0\times 10^{-3})(2.26\times 10^6) = 2260\ \text{J}$$

Hence, from the first law, the change in internal energy

$$\Delta U = Q - W = 2260 - 169$$
$$= 2091\ \text{J}$$

Note Points

- The positive value of ΔU indicates that the internal energy of the system increases. We see that most $\left(\frac{2091\ \text{J}}{2260\ \text{J}} = 93\%\right)$ of the energy transferred to the liquid goes into increasing the internal energy of the system only $\frac{169\ \text{J}}{2260\ \text{J}} = 7\%$ leaves the system by work done by the steam on the surrounding atmosphere.

15.3 Different Thermodynamic Processes

Among the thermodynamic processes we will consider are the following:

(i) An isothermal process during which the system's temperature remains constant.

(ii) An adiabatic process during which no heat is transferred to or from the system.

(iii) An isobaric process during which the pressure of the system is constant.

(iv) An isochoric process during which the system's volume does not change.

Of course there are, many other processes that do not fit into any of these four categories.

We will mostly consider an ideal gas.

Isothermal Process

An isothermal process is a constant temperature process. In an isothermal process

(i) T = constant or $\Delta T = 0$

(ii) $p \propto \frac{1}{V}$ or pV = constant,

i.e. p-V graph is a rectangular hyperbola with $p_iV_i = p_fV_f$.

(iii) As T = constant, hence U = constant for an ideal gas, because U is a function of T only.

(iv) $\Delta U = 0$

$$\therefore \qquad Q = W$$

Work done in isothermal process

$$W = \int_{V_i}^{V_f} p\,dV = \int_{V_i}^{V_f} \left(\frac{nRT}{V}\right) dV \quad \left(\text{as } p = \frac{nRT}{V}\right)$$

$$= nRT \int_{V_i}^{V_f} \frac{dV}{V} \quad (\text{as } T = \text{constant})$$

$$= nRT \ln\left(\frac{V_f}{V_i}\right)$$

$$= nRT \ln\left(\frac{p_i}{p_f}\right) \quad (\text{as } p_iV_i = p_fV_f)$$

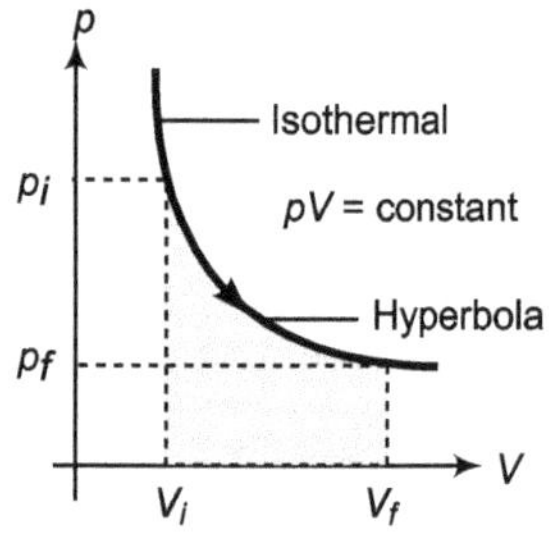

Isothermal expansion of an ideal gas

Fig. 15.10

Thus, in an isothermal process

$$\Delta U = 0 \text{ and } Q = W = nRT \ln\left(\frac{V_f}{V_i}\right) = nRT \ln\left(\frac{p_i}{p_f}\right)$$

Note Point

- For a process to be isothermal, any heat flow into or out of the system must occur **slowly** enough, so that thermal equilibrium is maintained.

Adiabatic Process

An adiabatic process is defined as one with no heat transfer into or out of a system i.e. $Q = 0$. We can prevent heat flow either by surrounding, the system with thermally insulating material or by carrying out the process so quickly that there is not enough time for appreciable heat flow. From the first law we find that for every adiabatic process,

$$W = -\Delta U \quad (\text{as } Q = 0)$$

$$= -nC_V \Delta T$$

$$= -nC_V (T_f - T_i)$$

$$= n\left(\frac{R}{\gamma - 1}\right)(T_i - T_f) \quad \left(\text{as } C_V = \frac{R}{\gamma - 1}\right)$$

$$= \frac{p_iV_i - p_fV_f}{\gamma - 1} \quad (\text{as } nRT = pV)$$

Thus, in an adiabatic process,

$$Q = 0 \quad \text{and} \quad W = -\Delta U = \frac{p_iV_i - p_fV_f}{\gamma - 1}$$

Note Point

- In adiabatic process $\boldsymbol{W = -\Delta U}$. Therefore, if the work done by a gas is positive (*i.e.* volume of the gas is increasing), then ΔU will be negative. That is $\boldsymbol{U}$ and hence $\boldsymbol{T}$ will decrease. The cooling of air can be experienced practically during bursting of a tyre. The process is so fast that it can be assumed as adiabatic, as the gas expands. Therefore, it cools. On the other hand the compression stroke in an internal combustion engine is an approximately adiabatic process. The temperature rises as the air fuel mixture in the cylinder is compressed.

***p-V* relation** In adiabatic process $dQ = 0$

and $\quad dW = -dU$

$\therefore \quad pdV = -C_V dT \quad$ (for $n = 1$)

$\therefore \quad dT = -\dfrac{pdV}{C_V} \quad$ ...(i)

Also for 1 mole of an ideal gas,

$$d(pV) = d(RT)$$

or $\quad pdV + Vdp = RdT$

or $\quad dT = \dfrac{pdV + Vdp}{R} \quad$ ...(ii)

From Eqs. (i) and (ii)

$$C_V Vdp + (C_V + R)\,pdV = 0$$

or $\quad C_V Vdp + C_p\,pdV = 0$

Dividing this equation by pV, we are left with

$$C_V \frac{dp}{p} + C_p \frac{dV}{V} = 0$$

or $\quad \dfrac{dp}{p} + \gamma \dfrac{dV}{V} = 0$

or $\quad \displaystyle\int \frac{dp}{p} + \gamma \int \frac{dV}{V} = 0$

or $\quad \ln(p) + \gamma \ln(V) = \text{constant}$

We can write this in the form of

$$pV^\gamma = \text{constant}$$

This equation is the condition that must be obeyed by an ideal gas in an adiabatic process. For example, if an ideal gas makes an adiabatic transition from a state with pressure and volume p_i and V_i to a state with p_f and V_f, then

$$p_iV_i^\gamma = p_fV_f^\gamma$$

The equation $pV^\gamma = \text{constant}$ can be written in terms of other pairs of thermodynamic variables by combining it with the ideal gas law ($pV = nRT$). In doing, so we will find that,

$$TV^{\gamma - 1} = \text{constant}$$

and $\quad T^\gamma p^{1-\gamma} = \text{constant}$

Slope of p - V Graph

In an adiabatic process (pV^{γ} = constant), the slope of p - V diagram at any point is

$$\frac{dp}{dV} = \frac{d}{dV}\left(\frac{\text{constant}}{V^{\gamma}}\right)$$

$$= -\gamma\left(\frac{p}{V}\right)$$

Thus, $(\text{slope})_{\text{adiabatic}} = -\gamma\left(\frac{p}{V}\right)$

Similarly, in an isothermal process (pV = constant), the slope of p-V diagram at any point is

$$\frac{dp}{dV} = \frac{d}{dV}\left(\frac{\text{constant}}{V}\right) = -\frac{p}{V}$$

or $$(\text{slope})_{\text{isothermal}} = -\frac{p}{V}$$

Because $\gamma > 1$, the isothermal curve is not as steep as that for the adiabatic expansion.

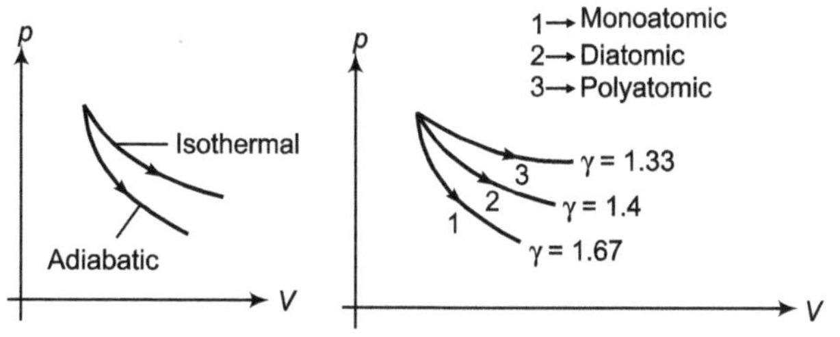

Adiabatic and isothermal expansion of an ideal gas

Adiabatic expansion of mono, dia and polyatomic gases

Fig. 15.11

Isobaric Process

An isobaric process is a constant pressure process. In an isobaric process,

(i) p = constant or $\Delta p = 0$

(ii) $V \propto T$ or $\frac{V}{T}$ = constant

i.e. V-T graph is a straight line passing through origin.

(iii) $Q = nC_p\Delta T$, $\Delta U = nC_V\Delta T$ and therefore

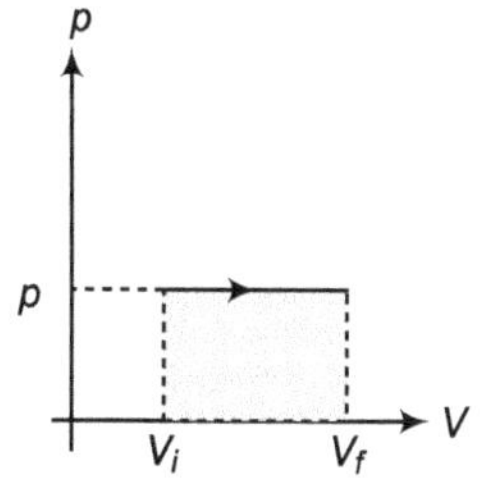

Isobaric expansion of a gas

Fig. 15.12

$$W = Q - \Delta U = n(C_p - C_V)\Delta T = nR\Delta T$$

$$= nR(T_f - T_i)$$

$$= p(V_f - V_i) \qquad (\text{as } nRT = pV)$$

Thus, in an isobaric process

$$Q = nC_p\Delta T,\ \Delta U = nC_V\Delta T$$

and $$W = p(V_f - V_i)$$

Isochoric Process

An isochoric process is a constant volume process. In an isochoric process,

(i) V = constant

or $\Delta V = 0$

(ii) $p \propto T$

or $\frac{p}{T}$ = constant

i.e. p-T graph is a straight line passing through origin.

(iii) As V = constant, hence $W = 0$ and from first law of thermodynamics

$$Q = \Delta U = nC_V\Delta T$$

Thus, in an isochoric process,

$$W = 0$$

and $$Q = \Delta U = nC_V\Delta T$$

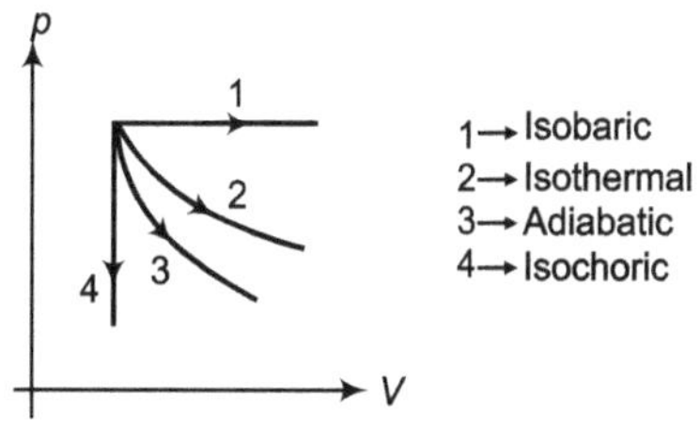

Fig. 15.13

p-V diagram of different processes is shown in one graph.

Table 15.3 shows Q, ΔU and W for different processes discussed above.

Table 15.3

Name of the process	Q	ΔU	W
Isothermal	$Q = W$	0	$nRT \ln\left(\frac{V_f}{V_i}\right) = nRT \ln\left(\frac{p_i}{p_f}\right)$
Adiabatic	0	$nC_V\Delta T$	$\frac{p_iV_i - p_fV_f}{\gamma - 1} = -\Delta U$
Isobaric	$nC_p\,\Delta T$	$nC_V\Delta T$	$p(V_f - V_i)$
Isochoric	$Q = \Delta U = nC_V\Delta T$	$nC_V\Delta T$	0

Extra Knowledge Points

- Bulk modulus of a gas is given by

$$B = -\frac{(dp)}{(dV/V)} = V\left(-\frac{dp}{dV}\right)$$

$\left(-\frac{dp}{dV}\right) = \gamma\left(\frac{p}{V}\right)$ in an adiabatic process.

Hence, adiabatic bulk modulus of an ideal gas is

$$B_s = \gamma p$$

Similarly,

$\left(-\frac{dp}{dV}\right) = \left(\frac{p}{V}\right)$ in an isothermal process.

Hence, isothermal bulk modulus of an ideal gas is

$$B_T = p$$

Example 15.5 *What is the heat input needed to raise the temperature of 2 moles of helium gas from 0°C to 100°C*

(a) At constant volume?

(b) At constant pressure?

(c) What is the work done by the gas in part?

(d) Give your answer in terms of R.

Sol. Helium is monoatomic gas.

Therefore, $C_V = \frac{3R}{2}$

and $C_p = \frac{5R}{2}$

(a) At constant volume,

$$Q = nC_V\Delta T = (2)\left(\frac{3R}{2}\right)(100) = 300R$$

(b) At constant pressure,

$$Q = nC_p\,\Delta T = (2)\left(\frac{5R}{2}\right)(100) = 500R$$

(c) At constant pressure,

$$W = Q - \Delta U = nC_p\Delta T - nC_V\Delta T = nR\Delta T = (2)(R)(100) = 200R$$

Example 15.6 *An ideal monoatomic gas at 300 K expands adiabatically to twice its volume. What is the final temperature?*

Sol. For an ideal monoatomic gas, $\gamma = 5/3$

In an adiabatic process,

$$TV^{\gamma-1} = \text{constant}$$

$\therefore$ $T_fV_f^{\gamma-1} = T_iV_i^{\gamma-1}$

or $T_f = T_i\left(\frac{V_i}{V_f}\right)^{\gamma-1}$

$$= (300)\left(\frac{1}{2}\right)^{\frac{5}{3}-1} = 189\text{ K}$$

Example 15.7 *An ideal gas expands isothermally along AB and does 700 J of work.*

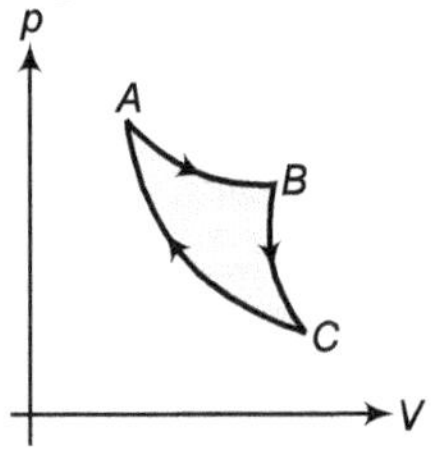

Fig. 15.14

(a) How much heat does the gas exchange along AB?

(b) The gas then expands adiabatically along BC and does 400 J of work. When the gas returns to A along CA, it exhausts 100 J of heat to its surroundings. How much work is done on the gas along this path?

Sol. (a) AB is an isothermal process. Hence,

$$\Delta U_{AB} = 0 \text{ and } Q_{AB} = W_{AB} = 700\text{ J}$$

(b) BC is an adiabatic process. Hence,

$Q_{BC} = 0$

$W_{BC} = 400$ J (given)

$\therefore$ $\Delta U_{BC} = -W_{BC} = -400$ J

ABC is a cyclic process and internal energy is a state function.

Therefore, $(\Delta U)_{\text{whole cycle}} = 0 = \Delta U_{AB} + \Delta U_{BC} + \Delta U_{CA}$ and from first law of thermodynamics,

$$Q_{AB} + Q_{BC} + Q_{CA} = W_{AB} + W_{BC} + W_{CA}$$

Substituting the values, we get

$$700 + 0 - 100 = 700 + 400 + \Delta W_{CA}$$

$\therefore$ $\Delta W_{CA} = -500$ J

Negative sign implies that work is done on the gas.

Table 15.4 shows different values in different processes.

Table 15.4

Process	Q (J)	W (J)	ΔU (J)
AB	700	700	0
BC	0	400	– 400
CA	– 100	– 500	400
For complete cycle	600	600	0

Note Point

- Total work done is 600 J, which implies that area of the closed curve is also 600 J.

15.4 Efficiency of a Cycle

In a cyclic process, $\Delta U = 0$

and $Q_{\text{net}} = W_{\text{net}}$ (from first law of thermodynamics)

First we see what is the meaning of efficiency of a cycle. Suppose, 100 J of heat is supplied to a system (in our case it is an ideal gas) and the system does 60 J of work. Then, efficiency of the cycle is 60%. Thus, efficiency (η) of a cycle can be defined as

$$\eta = \left(\frac{\begin{array}{c}\text{Work done by the working substance} \\ \text{(an ideal gas in our case) during a cycle}\end{array}}{\text{Heat supplied to the gas during the cycle}} \right) \times 100$$

$$= \frac{W_{\text{total}}}{|Q_{+\text{ve}}|} \times 100 = \frac{|Q_{+\text{ve}}| - |Q_{-\text{ve}}|}{|Q_{+\text{ve}}|} \times 100$$

$$= \left\{ 1 - \left| \frac{Q_{-\text{ve}}}{Q_{+\text{ve}}} \right| \right\} \times 100$$

Thus, $$\eta = \frac{W_{\text{total}}}{|Q_{+\text{ve}}|} \times 100 = \left\{ 1 - \left| \frac{Q_{-\text{ve}}}{Q_{+\text{ve}}} \right| \right\} \times 100$$

Note Points

- There cannot be a cycle whose efficiency is 100%. Hence, η is always less than 100%.

  Thus, $W_{\text{total}} \neq Q_{+\text{ve}}$
- It is just like a shopkeeper. He takes some money from you. (Suppose, he takes ₹ 100/- from you). In lieu of this he provides services to you (suppose he provides services of worth ₹ 80/-). Then, the efficiency of the shopkeeper is 80%. There cannot be a shopkeeper whose efficiency is 100%. Otherwise what will he save?

Example 15.8 *In example 15.7 find the efficiency of the given cycle.*

Sol. From table 15.4 we can see that $Q_{+\text{ve}}$ during the cycle is 700 J, while the total work done in the cycle is 600 J.

$$\therefore \quad \eta = \frac{W_{\text{total}}}{|Q_{+\text{ve}}|} \times 100$$

$$= \left(\frac{600}{700} \right) \times 100$$

$$= 85.71\,\%$$

15.5 Heat Engines

A heat engine is a device which converts heat energy into mechanical energy.

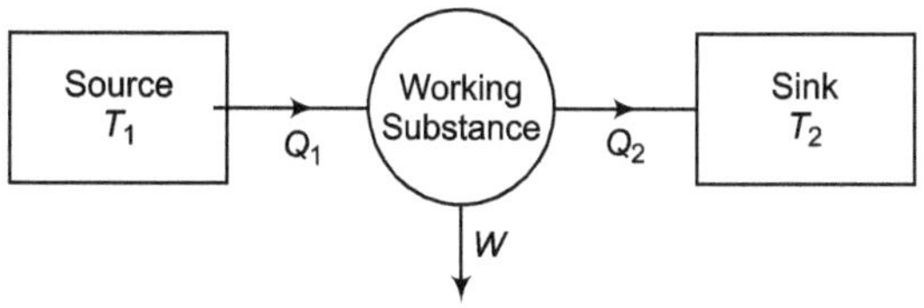

Fig. 15.15

Any heat engine has some working substance (normally a gas in a chamber). In every heat engine the working substance absorbs some heat (Q_1) from a source (at temperature T_1), converts a part of it into work (W) and the rest (Q_2) is rejected to the sink (at temperature T_2).

From conservation of energy,

$$Q_1 = W + Q_2$$

As discussed in article 15.4 thermal efficiency of a heat engine is defined as the ratio of net work done per cycle by the engine to the total amount of heat absorbed per cycle by the working substance from the source. It is denoted by η. Thus,

$$\eta = \frac{W}{Q_1} = \frac{Q_1 - Q_2}{Q_1} = 1 - \frac{Q_2}{Q_1}$$

As some heat is always rejected to the sink $Q_2 \neq 0$. Therefore, η is always less than 1, i.e. thermal efficiency of a heat engine is always less than 100%.

Types of Heat Engines

In practice, heat engine are of two types :

External Combustion Engine

In which heat is produced by burning, the fuel in a chamber outside the main body (working substance) of the engine. Steam engine is an external combustion engine. The thermal efficiency of a steam engine varies from 10 to 20%.

Internal Combustion Engine

In which heat is produced by burning the fuel inside the main body of the engine. Petrol engine and diesel engines are internal combustion engine.

Carnot Engine

Carnot cycle consists of the following four stages :

(i) Isothermal expansion (process AB)

(ii) Adiabatic expansion (process BC)

(iii) Isothermal compression (process CD)

(iv) Adiabatic compression (process DA)

The p-V diagram of the cycle is shown in the figure below:

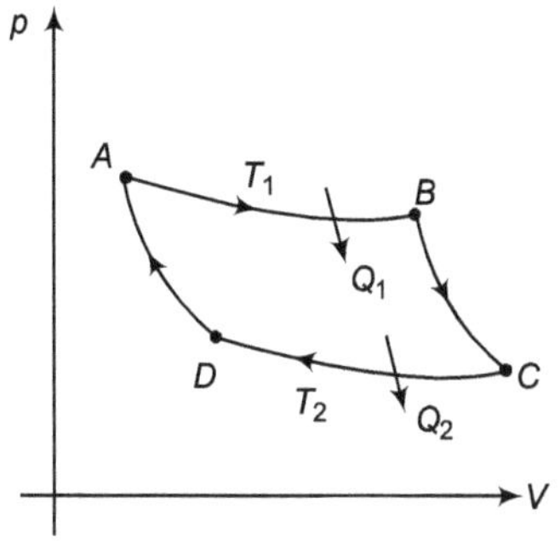

Fig. 15.16

In process AB, heat Q_1 is taken by the working substance at constant temperature T_1 and in process CD, heat Q_2 is liberated from the working substance at constant temperature T_2. The net work done is area of graph $ABCD$.

After doing the calculations for different processes, we can show that

$$\frac{Q_2}{Q_1}=\frac{T_2}{T_1}$$

Therefore, efficiency of the cycle is

$$\eta=1-\frac{T_2}{T_1}$$

Note Point

- That efficiency of Carnot engine is maximum (not 100%) for given temperatures T_1 and T_2. But still Carnot engine is not a practical engine, because many ideal situations have been assumed while designing this engine which can practically not be obtained.

Otto or Petrol Engine

This engine was made by Otto. This is also a four stroke engine. Four stroke means, in a cycle there are four processes. The working substance in it is 2% petrol and 98% air. The four processes are charging stroke, compression stroke, working stroke and exhaust stroke. The efficiency of a petrol engine is about 52%.

Diesel Engine

The diesel engine was made by a German Engineer Diesel. The efficiency of diesel engine is about 64%.

15.6 Refrigerator

Refrigerator is an apparatus which takes heat from a cold body, work is done on it and the work done together with the heat absorbed is rejected to the source.

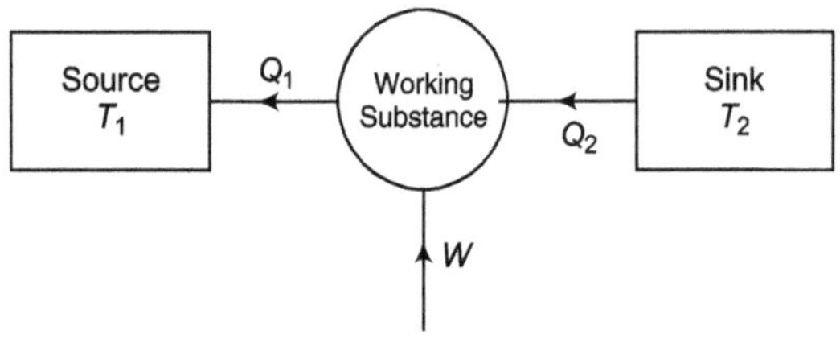

Fig. 15.17

An ideal refrigerator can be regarded as Carnot's ideal heat engine working in the reverse direction.

Coefficient of Performance

Loefficient of performance (β) of a refrigerator is defined as the ratio of quantity of heat removed per cycle (Q_2) to the work done on the working substance per cycle to remove this heat. Thus,

$$\beta=\frac{Q_2}{W}=\frac{Q_2}{Q_1-Q_2}$$

By doing the calculations, we can show that

$$\beta=\frac{T_2}{T_1-T_2}=\frac{1-\eta}{\eta}$$

Here, η is the efficiency of Carnot's cycle.

15.7 Second Law of Thermodynamics

The first law of thermodynamics is the law of conservation of energy. There are many processes that do not violate first law of thermodynamics (energy conservation law) but they are never observed. e.g. a book lying on a table cannot jump by itself. Although, if it jumps, there is no violation of mechanical energy. The second law of thermodynamics is the principle that forbids many phenomena consistent with the first law of thermodynamics.

The second law of thermodynamics gives a fundamental limitation to the efficiency of a heat engine and the coefficient of performance of a refrigerator. It says that efficiency of a heat engine can never be unity (or 100%). This implies that heat released to the cold reservoir can never be made zero.

For a refrigerator, the second law says that the coefficient to performance can never be infinite. This implies that external work can never be zero.

Kelvin-Planck Statement

No process is possible which can completely convert heat absorbed from a reservoir into work.

Clausius Statement

No process is possible which can transfer heat from a cold body to a hot body without doing any external work.

Reversible and Irreversible Processes

Most processes in nature are irreversible. It will violate the second law of thermodynamics if it happens. The free expansion of a gas is irreversible. Cooking gas leaking from a gas cylinder in the kitchen diffuses to the entire process. This is also an irreversible process.

Irreversibility arises mainly due to dissipative forces (like friction, viscosity etc). In the presence of dissipative forces some mechanical energy is lost in the form of heat, sound etc. Since, dissipative effects are present everywhere and can be minimised but not fully eliminated most processes we deal with are irreversible.

A reversible process is an idealised motion. A process is reversible, only if it is quasi-static (system in equilibrium with the surroundings at every stage) and there are no dissipative effects. For example, a quasi-static isothermal expansion of an ideal gas in a cylinder fitted with a frictionless movable piston is a reversible process.

Example 15.9 *The p-V diagram of 0.2 mole of a diatomic ideal gas is shown in figure. Process BC is adiabatic. The value of γ for this gas is 1.4.*

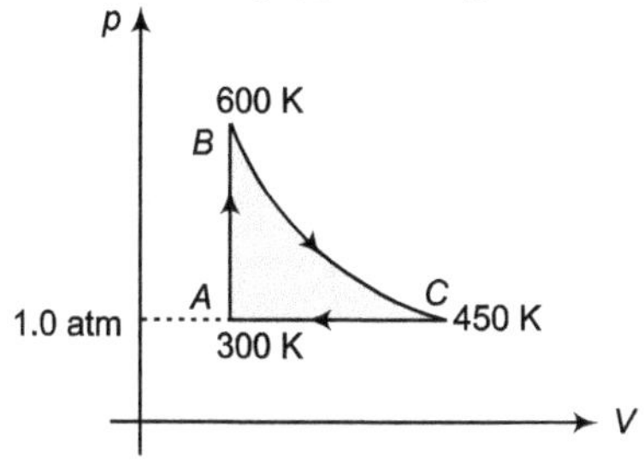

Fig. 15.18

(a) Find the pressure and volume at points A, B and C.
(b) Calculate $\Delta Q, \Delta W$ and ΔU for each of the three processes.
(c) Find the thermal efficiency of the cycle.
(Take, 1 atm $=1.0\times 10^5$ N/m²)

Sol. (a) $p_A = p_C = 1$ atm $= 1.01\times 10^5$ N/m²

Process *AB* is an isochoric process.

$\therefore \quad p \propto T$ or $\frac{p_B}{p_A} = \frac{T_B}{T_A}$

$$\therefore \quad p_B = \left(\frac{T_B}{T_A}\right) p_A = \left(\frac{600}{300}\right) (1 \text{ atm}) = 2 \text{ atm}$$

$$= 2.02\times 10^5 \text{ N/m}^2$$

From ideal gas equation,

$$V = \frac{nRT}{p}$$

$$\therefore \quad V_A = V_B = \frac{nRT_A}{p_A} = \frac{(0.2)\,(8.31)\,(300)}{(1.01\times 10^5)}$$

$$\approx 5.0\times 10^{-3} \text{ m}^3 = 5\text{L}$$

and
$$V_C = \frac{nRT_C}{p_C} = \frac{(0.2)\,(8.31)\,(455)}{(1.01\times 10^5)}$$

$$= 7.6\times 10^{-3} \text{ m}^3 \approx 7.6 \text{ L}$$

State	p	V
A	1 atm	5 L
B	2 atm	5 L
C	1 atm	7.6 L

(b) **Process *AB*** is an isochoric process. Hence,

$$\Delta W_{AB} = 0 \Rightarrow \Delta Q_{AB} = \Delta U_{AB}$$

$$= nC_V\Delta T = n\left(\frac{5}{2}R\right)(T_B - T_A)$$

$$= (0.2)\left(\frac{5}{2}\right)(8.31)\,(600 - 300) \approx 1246 \text{ J}$$

Process *BC* is an adiabatic process. Hence,

$$\Delta Q_{BC} = 0$$

$$\therefore \quad \Delta W_{BC} = -\Delta U_{BC}$$

$$\Delta U_{BC} = nC_V\Delta T = nC_V\,(T_C - T_B)$$

$$= (0.2)\left(\frac{5}{2}R\right)(455 - 600)$$

$$= (0.2)\left(\frac{5}{2}\right)(8.31)(-145) \text{ J} \approx -602 \text{ J}$$

$$\therefore \quad \Delta W_{BC} = -\Delta U_{BC} = 602 \text{ J}$$

Process *CA* is an isobaric process. Hence,

$$\Delta Q_{CA} = nC_p\,\Delta T = n\left(\frac{7}{2}R\right)(T_A - T_C)$$

$$= (0.2)\left(\frac{7}{2}\right)(8.31)\,(300 - 455) \approx -902 \text{ J}$$

$$\Delta U_{CA} = nC_V\Delta T = \frac{\Delta Q_{CA}}{\gamma} \qquad \left(\text{as } \gamma = \frac{C_p}{C_V}\right)$$

$$= -\frac{902}{1.4} \approx -644 \text{ J}$$

$$\therefore \quad \Delta W_{CA} = \Delta Q_{CA} - \Delta U_{CA} = -258 \text{ J}$$

Process	ΔQ (in J)	ΔW (in J)	ΔU (in J)
AB	1246	0	1246
BC	0	602	– 602
CA	– 902	– 258	– 644
Total	344	344	0

(c) Efficiency of the cycle

$$\eta = \frac{W_{\text{total}}}{|Q_{+\text{ve}}|} \times 100 = \frac{344}{1246} \times 100 = 27.6\%$$

Example 15.10 *Carnot engine takes in a 1000 kcal of heat from a reservoir at 827°C and exhausts it to a sink at 27°C. How much work does it perform ? What is the efficiency of the engine?*

Sol. Given, $Q_1 = 10^6$ cal

$$T_1 = (827 + 273) = 1100 \text{ K}$$

and
$$T_2 = (27 + 273) = 300 \text{ K}$$

as,
$$\frac{Q_2}{Q_1} = \frac{T_2}{T_1}$$

$$\therefore \quad Q_2 = \frac{T_2}{T_1}\cdot Q_1 = \left(\frac{300}{1100}\right)(10^6)$$

$$= 2.72\times 10^5 \text{ cal}$$

Efficiency of the cycle,

$$\eta = \left(1 - \frac{T_2}{T_1}\right)\times 100 \quad \text{or} \quad \eta = \left(1 - \frac{300}{1100}\right)\times 100$$

$$= 72.72\%$$

Example 15.11 *Calculate the least amount of work that must be done to freeze one gram of water at 0°C by means of a refrigerator. Temperature of surroundings is 27°C. How much heat in passed on the surroundings in this process. Latent heat of fusion $L = 80$ cal/g.*

Sol. $Q_2 = mL = 1\times 80 = 80$ cal

$$T_2 = 0°\text{C} = 273 \text{ K} \quad \text{and} \quad T_1 = 27°\text{C} = 300 \text{ K}$$

$$\frac{Q_2}{W} = \frac{T_2}{T_1 - T_2}$$

$$\therefore \quad W = \frac{Q_2(T_1 - T_2)}{T_2} = \frac{80\,(300 - 273)}{273} = 7.91 \text{ cal}$$

$$Q_1 = Q_2 + W = (80 + 7.91) = 87.91 \text{ cal}$$

Chapter Summary with Formulae

- Thermodynamics

(i) **Molar heat capacity** C = heat required to raise the temperature of 1 mole of any substance by 1°C or 1 K.

$$= \frac{Q}{n\Delta T} \Rightarrow \therefore \; Q = nC\Delta T$$

Molar heat capacity of solids and liquids is almost constant.

In case of gases C is process dependent. It varies from 0 to ∞.

In isothermal process

$$C = \infty \quad \text{as} \quad \Delta T = 0$$

In adiabatic process

$$C = 0 \quad \text{as} \quad Q = 0$$

C_p (molar heat capacity of isobaric process) and C_V (molar heat capacity of isochoric process) are commonly used. In a general process pV^x = constant, molar heat capacity is given by,

$$C = \frac{R}{\gamma - 1} + \frac{R}{1 - x}$$

(ii) **First law of thermodynamics** It is the law of conservation of energy given by,

$$Q = \Delta U + W$$

(iii) **Detailed discussion of three terms of first law of thermodynamics.**

(a) **Work done** Following methods are generally used to find the work done.

- **Method 1.** $W = \int_{V_i}^{V_f} p dV$ (because $dW = pdV$)

Here, p should be either constant or function of V. If p is constant. It means process is isobaric, $W = p(V_f - V_i) = p\Delta V$

- **Method 2.** Work done can also be obtained by area under p-V diagram with projection on V-axis.

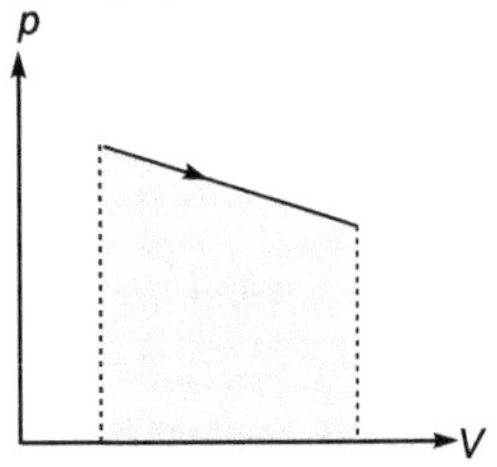

W = +ve as volume is increasing.

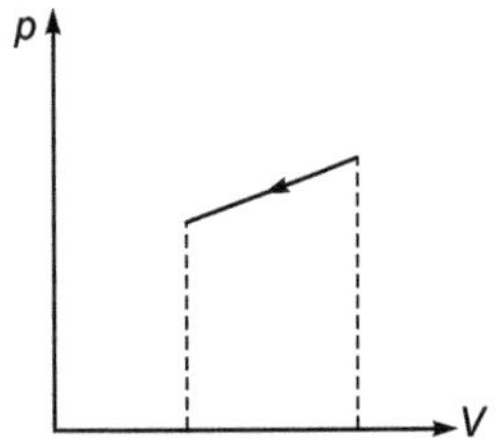

W = –ve as volume is decreasing

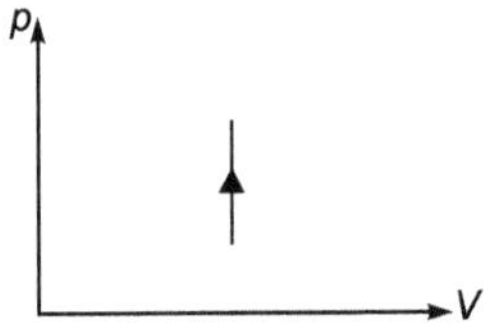

$W = 0$ as volume is constant.

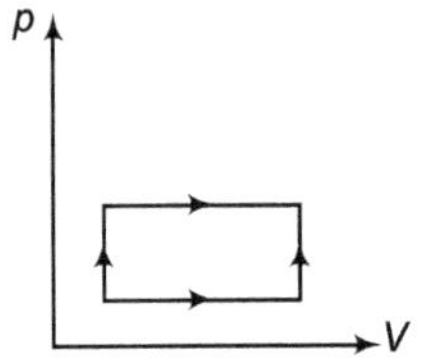

W = +ve as cyclic process is clockwise with p on y-axis.

(b) **Change in internal energy** ΔU

$\Delta U = nC_V\Delta T$ for all processes. For this C_V (or nature of gas), n and ΔT should be known. If either of the three terms is not known, we can calculate ΔU by, $\Delta U = Q - W$

(c) **Heat exchange** Q

$Q = nC\Delta T$. For this n, ΔT and molar heat capacity C should be known. C is a process dependent. So, if either of the three terms (n, ΔT or C) is not known, we can calculate Q by,

$$Q = \Delta U + W$$

(iv) Calorie is the old unit of heat. 1 calorie is the amount of heat required to raise the temperature of 1 g of water from 14.5°C to 15.5°C. 1 cal = 4.186 J

(v) For an ideal gas, the molar specific heat capacities at constant pressure and volume satisfy the relation

$$C_p - C_V = R$$

where, R is the universal gas constant.

(vi) Heat engine is a device in which a system undergoes a cyclic process resulting in conversion of heat into work. If Q_1 is the heat absorbed from the source, Q_2 is the heat released to the sink and the work output in one cycle is W, the efficiency η of the engine is

$$\eta = \frac{W}{Q_1} = 1 - \frac{Q_2}{Q_1}$$

(vii) In a refrigerator or a heat pump, the system extracts heat Q_2 from the cold reservoir and releases Q_1 amount of heat to the hot reservoir, with work done W on the system. The coefficient of performance of a refrigerator is given by

$$\beta = \frac{Q_2}{W} = \frac{Q_2}{Q_1 - Q_2}$$

(viii) Carnot engine is a reversible engine operating between two temperatures T_1 (source) and T_2 (sink). The Carnot cycle consists of two isothermal processes connected by two adiabatic processes. The efficiency of a Carnot engine is given by

$$\eta = 1 - \frac{T_2}{T_1} \quad \text{(Carnot engine)}$$

No engines operating between two temperatures can have efficiency greater than that of the Carnot engine.

Additional Examples

Example 1. *A slab of ice at 273 K and at atmospheric pressure melts*

(*a*) *What is the nature of the work done on the ice water system by the atmosphere ?*

(*b*) *What happens to the internal energy of ice water system ?*

Sol. (*a*) The volume of ice decreases on melting. Hence, the work done by the atmosphere on ice water system is positive in nature.

(*b*) Since, heat is absorbed by the ice during melting, the internal energy of the ice water system increases.

Example 2. *If a door of a working refrigerator is kept open for a long time in a closed room, will it make the room warm or cool ?*

Sol. Suppose that the refrigerator draws some heat from the air in front of it. The compressor has to do some mechanical work (at the expense of electrical energy) to draw heat from the air at lower temperature. The heat drawn from the air together with the work done by the compressor in drawing it, is rejected by the refrigerator with the help of the radiator (provided at the back) to the air.

Example 3. *When a gas is suddenly compressed, its temperature rises. Why ?*

Sol. Sudden compression of a gas is an adiabatic process. The work done in compressing the gas increases the internal energy of the gas. Hence, the temperature of the gas rises.

Example 4. *If an inflated tyre bursts, the air escaping out is cooled, why ?*

Sol. When the tyre bursts, there is an adiabatic expansion of air because the pressure of the air inside is sufficiently greater than the atmospheric pressure. During the expansion, the air does some work against the surroundings, therefore, its internal energy decreases, and as such temperature falls.

Example 5. *Is it possible that there is change in temperature of a body without giving heat to it or taking heat from it ?*

Sol. Yes, for example, during an adiabatic compression temperature increases and in an adiabatic expansion temperature decreases, although no heat is given to or taken from the system in these changes.

Example 6. *Is it possible that there is no change in temperature of a body despite being heated?*

Sol. Yes, for example during a change of state (from solid to liquid or from liquid to gas), the system takes heat, but there is no rise in temperature. Internal energy of the system increases in each case.

Example 7. *First law of thermodynamics does not forbid flow of heat from lower temperature to higher temperature. Comment.*

Sol. First law of thermodynamics simply tells about the conversion of mechanical energy into heat energy and *vice-versa*. It does not put any condition as to why heat cannot flow from lower temperature to higher temperature.

Example 8. *Find the specific heat of the process $p = \frac{a}{T}$ for a monoatomic gas, a being constant.*

Sol. We know that, $dQ = dU + dW$

Specific heat, $$C = \frac{dQ}{dT} = \frac{dU}{dT} + \frac{dW}{dT} \qquad \text{...(i)}$$

Since, $$dU = C_V dT \qquad \text{...(ii)}$$

$$C = C_V + \frac{dW}{dT} = C_V + \frac{pdV}{dT}$$

$$pV = RT$$

∴ For the given process,

$$V = \frac{RT}{p} = \frac{RT^2}{a} \quad \text{or} \quad \frac{dV}{dT} = \frac{2RT}{a}$$

∴ $$C = C_V + p\left(\frac{2RT}{a}\right)$$

$$= C_V + 2R = \frac{3}{2}R + 2R = \frac{7}{2}R$$

Example 9. *A gaseous mixture enclosed in a vessel consists of 1 g mole of a gas A with $\gamma = \left(\frac{5}{3}\right)$ and some amount of gas B with $\gamma = \frac{7}{5}$ at a temperature T. The gases A and B do not react with each other and are assumed to be ideal. Find the number of gram moles of the gas B, if γ for the gaseous mixture is $\left(\frac{19}{13}\right)$.*

Sol. As for an ideal gas, $C_p - C_V = R$ and $\gamma = \left(\frac{C_p}{C_V}\right)$

So, $$C_V = \frac{R}{(\gamma - 1)}$$

∴ $$(C_V)_1 = \frac{R}{\left(\frac{5}{3}\right) - 1} = \frac{3}{2}R,$$

$$(C_V)_2 = \frac{R}{\left(\frac{7}{5}\right) - 1} = \frac{5}{2}R$$

and $$(C_V)_{mix} = \frac{R}{\left(\frac{19}{13}\right) - 1} = \frac{13}{6}R$$

Now, from conservation of energy,

i.e. $$\Delta U = \Delta U_1 + \Delta U_2$$

$$(n_1 + n_2)(C_V)_{mix}\,\Delta T = [n_1(C_V)_1 + n_2(C_V)_2]\,\Delta T$$

i.e. $$(C_V)_{mix} = \frac{n_1 (C_V)_1 + n_2 (C_V)_2}{n_1 + n_2}$$

We have, $$\frac{13}{6}R = \frac{1\times\frac{3}{2}R + n_2\frac{5}{2}R}{1+n_2} = \frac{(3+5n_2)R}{2(1+n_2)}$$

or $13 + 13n_2 = 9 + 15n_2$, i.e. $n_2 = 2$ g-mol

Example 10. *An ideal monoatomic gas at temperature 27°C and pressure 10^6 N/m^2 occupies 10 L volume. 10000 cal of heat is added to the system without changing the volume. Calculate the final temperature of the gas.*
(Given, $R = 8.31$ J/mol-K and $J = 4.18$ J/cal)

Sol. $$\therefore\ n = \frac{pV}{RT} = \frac{10^6 \times 10^{-2}}{8.31 \times 300} = 4.0$$

For monoatomic gas, $C_V = \frac{3}{2}R$

Thus, $$C_V = \frac{3}{2} \times 8.31 \text{ J/mol-K}$$
$$= \frac{3}{2} \times \frac{8.31}{4.18} \approx 3 \text{ cal /(mol-K)}$$

Let ΔT be the rise in temperature, when n mole of the gas is given Q cal of heat at constant volume. Then,

$$Q = nC_V\Delta T$$

or $$\Delta T = \frac{Q}{nC_V}$$

$$= \frac{10000 \text{ cal}}{4.0 \text{ mol} \times 3 \text{ cal (mol - K)}}$$

$$= 833 \text{ K}$$
$$\Delta T = T_f - T_i$$
$$T_f = \Delta T + T_i$$
$$= 833 + 300$$
$$= 1133 \text{ K}$$

Example 11. *An ideal gas is taken through a cyclic thermodynamic process through four steps. The amounts of heat involved in these steps are $Q_1 = 5960\,J, Q_2 = -5585\,J, Q_3 = -2980\,J$ and $Q_4 = 3645\,J$ respectively. The corresponding works involved are $W_1 = 2200\,J, W_2 = -825\,J$, $W_3 = -1100\,J$ and W_4 respectively.*

(a) Find the value of W_4.

(b) What is the efficiency of the cycle?

Sol. (a) According to the given problem

$$\Delta Q = Q_1 + Q_2 + Q_3 + Q_4$$
$$= 5960 - 5585 - 2980 + 3645$$
$$\Delta Q = 9605 - 8565 = 1040 \text{ J}$$
$$\Delta W = W_1 + W_2 + W_3 + W_4$$
$$= 2200 - 825 - 1100 + W_4 = 275 + W_4$$

And as for cyclic process

$$U_F = U_I,\ \Delta U = U_F - U_I = 0$$

So, from first law of thermodynamics, i.e.

$$\Delta Q = \Delta U + \Delta W$$

We have $1040 = (275 + W_4) + 0$, i.e. $W_4 = 756$ J

(b) As, efficiency of a cycle is defined as

$$\eta = \frac{\text{Net work}}{\text{Input heat}} = \frac{\Delta W}{(Q_1 + Q_4)} = \frac{\Delta Q}{(Q_1 + Q_4)}$$

$$\eta = \frac{1040}{9605} = 0.1082 = 10.82\%$$

Example 12. *At 27°C two moles of an ideal monoatomic gas occupy a volume V. The gas expands adiabatically to a volume $2V$. Calculate*

(a) the final temperature of the gas

(b) the change in its internal energy and

(c) the work done by the gas during the process.
($R = 8.31$ J/mol-K)

Sol. (a) In case of adiabatic change

$$TV^{\gamma} = \text{constant}$$

$$\therefore T_1V_1^{\gamma-1} = T_2V_2^{\gamma-1} \text{ with } \gamma = \left(\frac{5}{3}\right)$$

i.e. $300 \times V^{2/3} = T(2V)^{2/3}$ or $T = \frac{300}{(2)^{2/3}} = 189$ K

(b) According to first law of thermodynamics

$$Q = \Delta U + \Delta W$$

And as for adiabatic change $\Delta Q = 0$,

$$\Delta W = -\Delta U$$
$$= 2767.23 \text{ J}$$

Example 13. *One mole of a monoatomic ideal gas is taken through the cycle as shown in figure.*

$A \to B$	*Adiabatic expansion*
$B \to C$	*Cooling at constant volume*
$C \to D$	*Adiabatic compression*
$D \to A$	*Heating at constant volume*

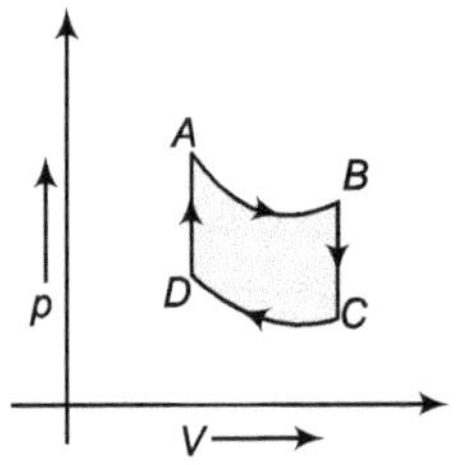

The pressure and temperature at A, B, etc., are denoted by p_A, T_A, p_B, T_B, etc., respectively.

Given, $T_A = 1000$ K, $p_B = \left(\frac{2}{3}\right) p_A$ *and* $p_C = \left(\frac{1}{3}\right) p_A$.

Calculate

(a) the work done by the gas in the process $A \to B$.

(b) the heat lost by the gas in the process $B \to C$.

Given $\left(\frac{2}{3}\right)^{0.4} = 0.85$ *and* $R = 8.31$ *J/mol-K*

Sol. (a) As for adiabatic change $pV^\gamma =$ constant

i.e.

$$p\left(\frac{nRT}{P}\right)^\gamma = \text{constant} \qquad (\text{as } pV = nRT)$$

i.e.

$$\frac{T^\gamma}{p^{\gamma-1}} = \text{constant},$$

So,

$$\left(\frac{T_B}{T_A}\right)^\gamma = \left(\frac{p_B}{p_A}\right)^{\gamma-1} \qquad \left(\text{where } \gamma = \frac{5}{3}\right)$$

i.e.

$$T_B = T_A\left(\frac{2}{3}\right)^{1-\frac{1}{\gamma}}$$

$$= 1000\left(\frac{2}{3}\right)^{2/5} = 850 \text{ K}$$

So,

$$W_{AB} = \frac{nR[T_F - T_I]}{[1-\gamma]}$$

$$= \frac{1 \times 8.31[1000 - 850]}{\left[\left(\frac{5}{3}\right) - 1\right]}$$

i.e.

$$W_{AB} = \left(\frac{3}{2}\right) \times 8.31 \times 150$$

$$= 1869.75 \text{ J}$$

(b) For $B \to C$, $V =$ constant, so $\Delta W = 0$

So, from first law of thermodynamics

$$\Delta Q = \Delta U + \Delta W = nC_V \Delta T + 0$$

or

$$\Delta Q = 1 \times \left(\frac{3}{2}R\right)(T_C - 850) \qquad \left(\text{as } C_V = \frac{3}{2}R\right)$$

Now along path BC, $V =$ constant, $p \propto T$

i.e.

$$\frac{p_C}{p_B} = \frac{T_C}{T_B}$$

$$T_C = \frac{\left(\frac{1}{3}\right)p_A}{\left(\frac{2}{3}\right)p_A} \times T_B$$

$$= \frac{T_B}{2} = \frac{850}{2} = 425 \text{ K} \qquad \text{...(ii)}$$

So,

$$\Delta Q = 1 \times \frac{3}{2} \times 8.31\,(425 - 850)$$

$$= -5297.625 \text{ J}$$

Negative sign means, heat is lost by the system.

Example 14. *The density versus pressure graph of one mole of an ideal monoatomic gas undergoing a cyclic process is shown in figure. The molecular mass of the gas is M.*

(a) Find the work done in each process.

(b) Find heat rejected by gas in one complete cycle.

(c) Find the efficiency of the cycle.

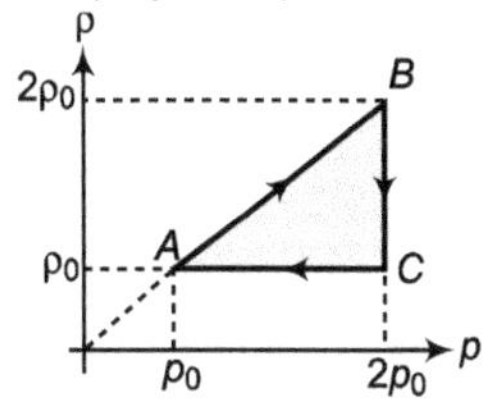

Sol. (a) As $n = 1$, $m = M$

Process *AB* $\rho \propto p$, i.e. it is an isothermal process ($T =$ constant), because $\rho = \frac{pM}{RT}$.

$$\therefore \quad W_{AB} = RT_A \ln\left(\frac{p_A}{p_B}\right) = RT_A \ln\left(\frac{1}{2}\right) = -\frac{p_0 M}{\rho_0}\ln(2)$$

$$\Delta U_{AB} = 0 \quad \text{and} \quad Q_{AB} = W_{ABG} = \frac{\rho_0 M}{\rho_0}\ln(2)$$

Process *BC* is an isobaric process ($p =$ constant)

$$W_{BC} = p_B(V_C - V_B)$$

$$= 2p_0\left(\frac{M}{\rho_C} - \frac{M}{\rho_B}\right) = \frac{2p_0 M}{2\rho_0} = \frac{p_0 M}{\rho_0}$$

$$\Delta U_{BC} = C_V \Delta T$$

$$= \left(\frac{3}{2}R\right)\left[\frac{2p_0 M}{\rho_0 R} - \frac{2p_0 M}{2\rho_0 R}\right] = \frac{3p_0 M}{2\rho_0}$$

$$Q_{BC} = W_{BC} + \Delta U_{BC} = \frac{5p_0 M}{2\rho_0}$$

Process *CA* As $\rho =$ constant $\Rightarrow \therefore V =$ constant.

So, it is an isochoric process.

$$W_{CA} = 0$$

$$\Delta U_{CA} = C_V \Delta T = \left(\frac{3}{2}R\right)(T_A - T_C)$$

$$= \left(\frac{3}{2}R\right)\left[\frac{p_0 M}{\rho_0 R} - \frac{2p_0 M}{\rho_0 R}\right] = -\frac{3p_0 M}{2\rho_0}$$

$$Q_{CA} = \Delta U_{CA} = -\frac{3p_0 M}{2\rho_0}$$

(b) Heat rejected by gas $= |Q_{AB}| + |Q_{CA}|$

$$= \frac{p_0 M}{\rho_0}\left[\frac{3}{2} + \ln(2)\right]$$

(c) Efficiency of the cycle (in fraction)

$$\eta = \frac{\text{Total work done}}{\text{Heat supplied}} = \frac{W_{\text{total}}}{Q_{+\text{ve}}}$$

$$= \frac{\frac{p_0 M}{\rho_0}[1 - \ln(2)]}{\frac{5}{2}\left(\frac{p_0 M}{\rho_0}\right)} = \frac{2}{5}[1 - \ln(2)]$$

NCERT Selected Questions

***Q* 1.** What amount of heat must be supplied to 2.0×10^{-2} kg of nitrogen (at room temperature) to raise its temperature by 45°C at constant pressure? (Molecular mass of $N_2 = 28$, $R = 8.3 \text{ J mol}^{-1} \text{ K}^{-1}$).

Sol. Here, mass of gas, $m = 2 \times 10^{-2} \text{ kg} = 20 \text{ g}$

Rise in temperature, $\Delta T = 45°\text{C}$

Molar mass of $N_2, M = 28$ g

If n = number of moles, then

$$n = \frac{m}{M} = \frac{20}{28} = 0.714$$

If C_p be the molar specific heat of the gas at constant pressure, then

$$C_p = \frac{7}{2}R = \frac{7}{2} \times 8.3 \text{ J mol}^{-1}\text{K}^{-1}$$

Now, $$Q = nC_p\Delta T = 0.714 \times \frac{7}{2} \times 8.3 \times 45 \text{ J}$$

$$= 933.37 \text{ J} \approx 933 \text{ J}$$

***Q* 2.** A cylinder with a movable piston contains 3 moles of hydrogen at standard temperature and pressure. The walls of the cylinder are made of heat insulator, and the piston is insulated by having a pile of sand on it. By what factor does the pressure of the gas increase, if the gas is compressed to half its original volume?

Sol. For diatomic gas or hydrogen,

$$\gamma = \frac{7}{5} = 1.4$$

As no heat is allowed to be exchanged, the process is adiabatic. Now for adiabatic change,

$$p_1V_1^\gamma = p_2V_2^\gamma \quad \text{or} \quad \frac{p_2}{p_1} = \left(\frac{V_1}{V_2}\right)^\gamma$$

$$= (2)^{1.4} = 2.64$$

***Q* 3.** In changing the state of a gas adiabatically from an equilibrium state A to another equilibrium state B, an amount of work equal to 22.3 J is done on the system. If the gas is taken from state A to B *via* a process in which the net heat absorbed by the system is 9.35 cal, how much is the net work done by the system in the latter case?
(Take, 1 cal = 4.19 J).

Sol. In adiabatic process,

ΔU = work done on the system = 2.33 J ...(i)

In second case, when the state A is taken to state B, the heat absorbed by the system ΔQ is

$$\Delta Q = 9.35 \text{ cal}$$

$$= 9.35 \times 4.19 \text{ J} \qquad (\because 1 \text{ cal} = 4.19 \text{ J})$$

$$= 39.18 \text{ J}$$

Let ΔW be the net work done by the system, then using first law of thermodynamics,

$$\Delta Q = \Delta U + \Delta W \Rightarrow \Delta W = \Delta Q - \Delta U$$

$$= 39.18 - 22.3 = 16.88 \text{ J}$$

$$\approx 16.9 \text{ J}$$

***Q* 4.** Two cylinders A and B of equal capacity are connected to each other *via* a stopcock. A contains a gas at standard temperature and pressure. B is completely evacuated. The entire system is thermally insulated. The stopcock is suddenly opened.
Answer the following questions:
(a) What is final pressure of the gas in A and B ?
(b) What is the change in internal energy of the gas?
(c) What is the change in temperature of the gas?

Sol. (a) As, the initial and final temperatures remain the same, then

$$p_1V_1 = p_2V_2$$

$$p_2 = \frac{p_1V_1}{V_2} = \frac{p_1V_1}{2V_1} = \frac{p_1}{2}$$

$$= 0.5 \text{ atm}$$

(b) Zero this is because no work is done on/by the gas, thus there will be no change in the internal energy of the gas under constant temperature conditions.

(c) If the gas is assumed to be ideal, then there is no change in the temperature of the gas as it does no work in expansion, i.e. $\Delta T = 0$.

***Q* 5.** A steam engine delivers 5.4×10^8 J of work per minute and takes 3.6×10^9 J of heat per minute from the boiler. What is the efficiency of the engine? How much heat is wasted per minute?

Sol. Let

Q_1 = heat absorbed per minute

Q_2 = heat rejected per minute

We know that, $$\eta\% = \frac{W}{Q_1} \times 100$$

$$\therefore \quad \eta\% = \frac{5.4 \times 10^8 \text{ J}}{3.6 \times 10^9 \text{ J}} \times 100 = \frac{3}{20} \times 100 = 15\%$$

Also using the relation, $Q_1 = W + Q_2$, we get

$$Q_2 = Q_1 - W = 36 \times 10^8 - 5.4 \times 10^8$$

$$= 30.6 \times 10^8 \text{ J/min} = 3.06 \times 10^9 \text{ J/min}$$

$$\approx 3.1 \times 10^9 \text{ J/min}$$

Q 6. An electric heater supplies heat to a system at a rate of 100 W. If the system performs work at a rate of 75 J/s. At what rate is the internal energy increasing?

Sol. According to the first law of thermodynamics,

$$\Delta U = \Delta Q - \Delta W = 100 - 75 = 25\,\text{J/s}$$
$$= 25\,\text{W}$$

Q 7. A thermodynamic system is taken from an original state to an intermediate state by linear process shown in figure here. Its volume is then reduced to the original value from E to F by an isobaric process. Calculate the total work done by the gas from D to E to F.

Sol.

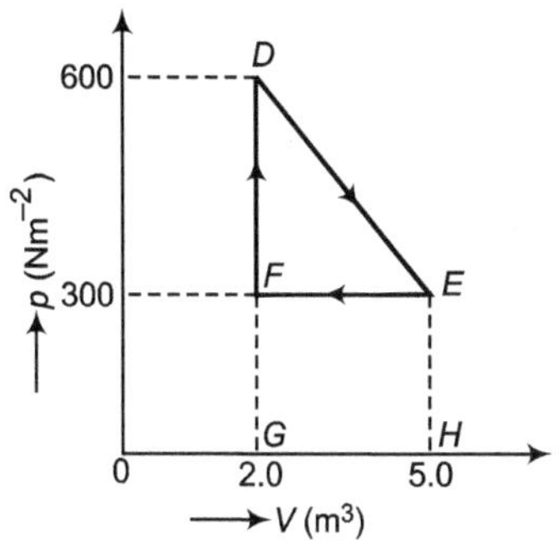

where, W_1 = work done during the process from D to E (expansion)

$= \text{area } DEHGD$

$$\therefore \quad W_1 = \left[\frac{1}{2} \times 3 \times 300 + 3 \times 300\right] \text{J} = 1350\ \text{J} \qquad \text{...(i)}$$

W_2 = work done during the process from E to F (compression)

$$= \text{area } EHGF = -300 \times 3\ \text{J} = -900\ \text{J} \qquad \text{...(ii)}$$

$\therefore$ Total work done

$$W = 1350 - 900 = 450\ \text{J}$$

Q 8. A refrigerator is to remove heat from the eatable kept inside at 9°C. Calculate the coefficient of performance if the room temperature is 36°C.

Sol. $T_1 = 273 + 36 = 309\ \text{K}$

$T_2 = 9°\text{C} = 282\ \text{K}$

Coefficient of performance,

$$\beta = \frac{T_2}{T_1 - T_2},$$

We get,

$$\beta = \frac{282}{309 - 282}$$
$$= \frac{282}{27} = 10.4$$

Objective Problems

[Level 1]

First law of Thermodynamics and Graphs

1. First law of thermodynamics corresponds to
(a) conservation of energy
(b) heat flow from hotter to colder body
(c) law of conservation of angular momentum
(d) Newton's law of cooling

2. A system is taken from state A to state B along two different paths 1 and 2. The heat absorbed and work done by the system along these two paths are Q_1 and Q_2 and W_1 and W_2, respectively. Then
(a) $Q_1 = Q_2$
(b) $W_1 = W_2$
(c) $Q_1 - W_1 = Q_2 - W_2$
(d) $Q_1 + W_1 = Q_2 + W_2$

3. A certain amount of an ideal gas is taken from state A to state B, one time by process I and another time by process II. If the amount of heat absorbed by the gas are Q_1 and Q_2 respectively, then

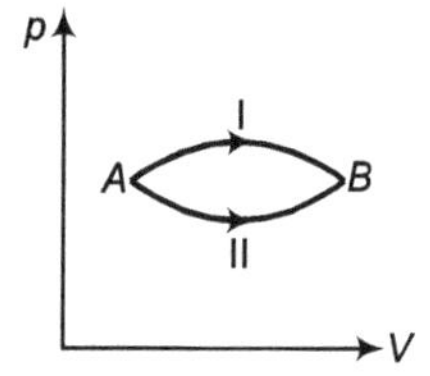

(a) $Q_1 = Q_2$ (b) $Q_1 < Q_2$
(c) $Q_1 > Q_2$ (d) None of these

4. Work done by the gas in the process shown in figure is

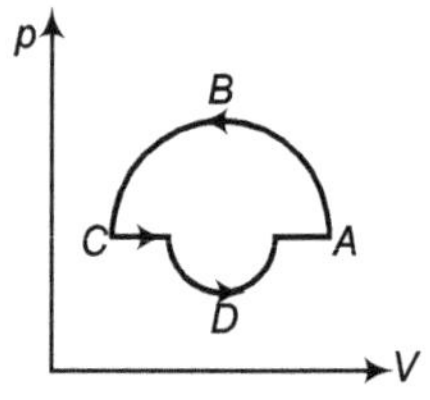

(a) positive
(b) negative
(c) zero
(d) may be positive or negative

5. Identify the incorrect statement related to a cyclic process
(a) the initial and final conditions always coincide
(b) $Q = W$
(c) $W > 0$
(d) None of the above

6. An ideal gas of volume 1.5×10^{-3} m^3 and at pressure 1.0×10^5 Pa is supplied with 70 J of energy. The volume increases to 1.7×10^{-3} m^3, the pressure remaining constant. The internal energy of the gas is
(a) increased by 90 J (b) increased by 70 J
(c) increased by 50 J (d) decreased by 50 J

7. Corresponding to the process shown in figure, what is the heat given to the gas in the process $ABCA$?

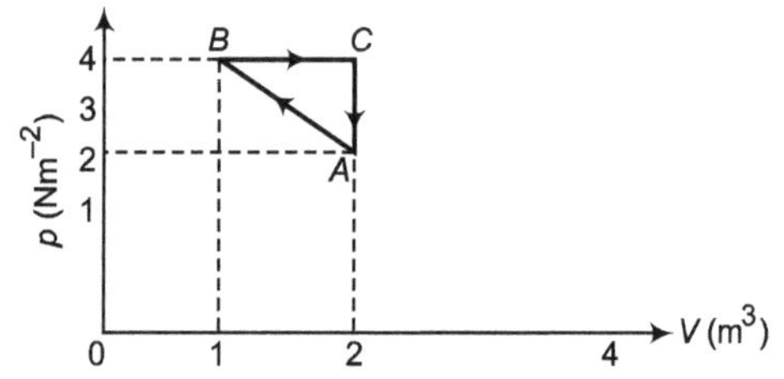

(a) 1 J (b) $\frac{3}{2}$ J
(c) $\frac{1}{2}$ J (d) 0

8. A closed system undergoes a change of state by process $1 \to 2$ for which $Q_{12} = 10$ J and $W_{12} = -5$ J. The system is now returned to its initial state by a different path $2 \to 1$ for which Q_{21} is -3 J. The work done by the gas in the process $2 \to 1$ is
(a) 8 J (b) zero
(c) -2 J (d) $+5$ J

9. A thermodynamical system is changed from state (p_1, V_1) to (p_2, V_2) by two different process. The quantity which will remain same will be?
(a) ΔQ (b) ΔW
(c) $\Delta Q + \Delta W$ (d) $\Delta Q - \Delta W$

10. In thermodynamic process, 200 J of heat is given to a gas and 100 J of work is also done on it. The change in internal energy of the gas is
(a) 100 J (b) 300 J
(c) 419 J (d) 240 J

11. If ΔQ and ΔW represent the heat supplied to the system and the work done on the system respectively, then the first law of thermodynamics can be written as
(a) $\Delta Q = \Delta U + \Delta W$
(b) $\Delta Q = \Delta U - \Delta W$
(c) $\Delta Q = \Delta W - \Delta U$
(d) $\Delta Q = -\Delta W - \Delta U$

where, ΔU is the change in internal energy.

12. In a given process for an ideal gas, $dW = 0$ and $dQ < 0$. Then, for the gas
(a) the temperature will decrease
(b) the volume will increase
(c) the pressure will remain constant
(d) the temperature will increase

13. When 1 g of water at 0°C and $1 \times 10^5 \text{ N/m}^2$ pressure is converted into ice of volume 1.092 cm^3, then work done will be
(a) 0.0091 J (b) 0.0182 J
(c) 0.091 J (d) Data insufficient

14. In an adiabatic expansion of one mole of gas initial and final temperatures are T_1 and T_2 respectively, then the change in internal energy of the gas is
(a) $\frac{R}{\gamma - 1}(T_2 - T_1)$ (b) $\frac{R}{\gamma - 1}(T_1 - T_2)$
(c) $R(T_1 - T_2)$ (d) zero

15. A gas expands under constant pressure p from volume V_1 and V_2. The work done by the gas is
(a) $p(V_2 - V_1)$ (b) $p(V_1 - V_2)$
(c) $p(V_1^\gamma - V_2^\gamma)$ (d) $p\frac{V_1 V_2}{V_2 - V_1}$

16. A gas is compressed at a constant pressure of 50 N/m^2 from a volume of 10 m^3 to a volume of 4 m^3. Energy of 100 J, then added to the gas by heating. Its internal energy is
(a) increased by 400 J (b) increased by 200 J
(c) increased by 100 J (d) decreased by 200 J

17. A thermodynamic system is taken through the cycle *PQRSP* process. The net work done by the system is

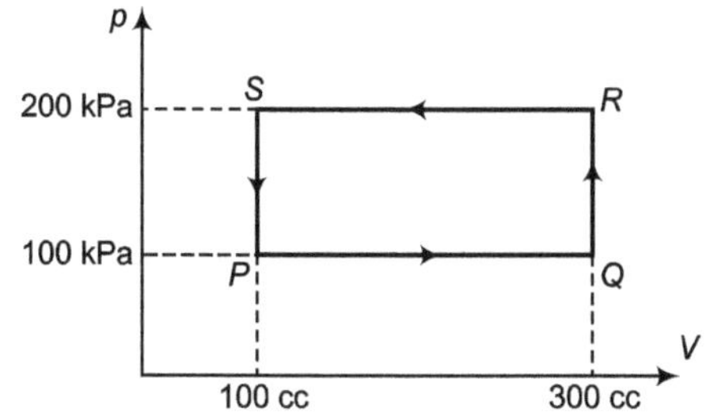

(a) 20 J (b) −20 J (c) 400 J (d) −400 J

18. An ideal gas is taken around *ABCA* as shown in the *p-V* diagram. The work done during the cycle is

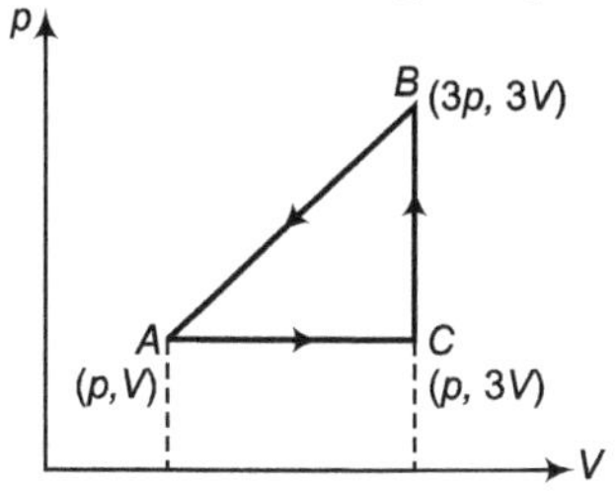

(a) $2pV$ (b) pV (c) $\frac{pV}{2}$ (d) zero

19. The *p-V* diagram of a system undergoing thermodynamic transformation is as shown in figure. The work done on the system in going from $A \rightarrow B \rightarrow C$ is 50 J and 20 cal heat is given to the system. The change in internal energy between *A* and *C* is

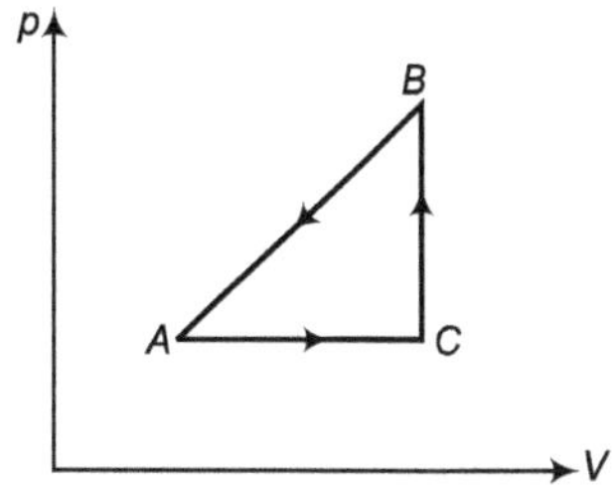

(a) 34 J (b) 70 J (c) 84 J (d) 134 J

20. An ideal gas is taken through the cycle $A \rightarrow B \rightarrow C \rightarrow A$, as shown in the figure. If the net heat supplied to the gas in the cycle is 5 J, then work done by the gas in the process $C \rightarrow A$ is

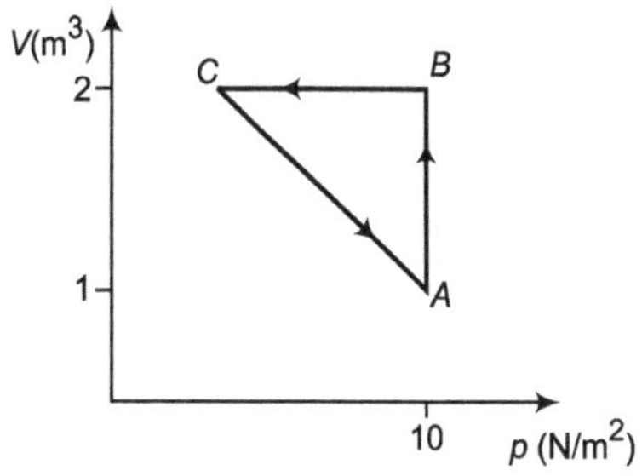

(a) −5 J (b) −10 J (c) −15 J (d) −20 J

21. In the *p-V* diagram, shown in figure the net amount of work done will be

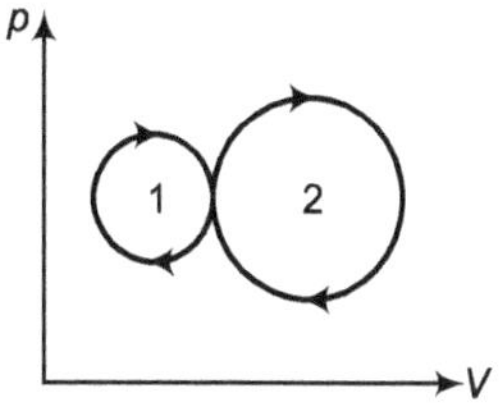

(a) positive (b) negative (c) zero (d) infinity

22. In the cyclic process *ABCDA* shown in the figure, consider the following statements.

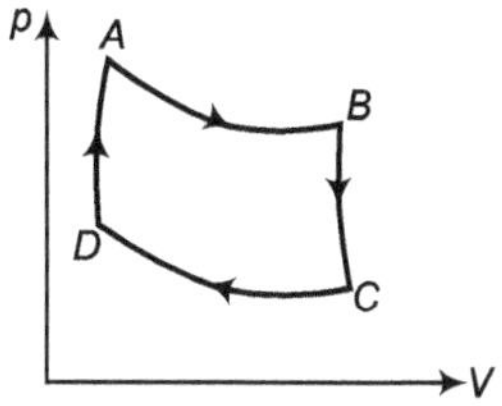

I. Area $ABCD$ = Work done on the gas
II. Area $ABCD$ = Net heat absorbed
III. Change in the internal energy in cycle = 0

Which of these are correct?
(a) Only I (b) Only II (c) II and III (d) I, II and III

23. Carbon monoxide is carried around a closed cycle *abca* in which *bc* is an isothermal process as shown in the figure. The gas absorbs 7000 J of heat as its temperature increases from 300 K to 1000 K in going from *a* to *b*. The quantity of heat rejected by the gas during the process *ca* is approximately

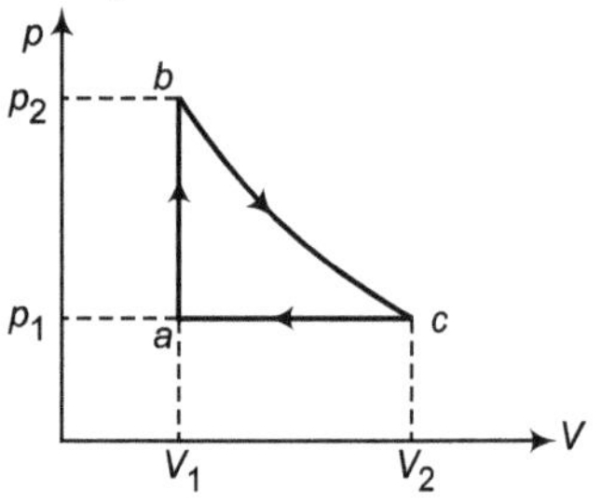

(a) 4200 J (b) 5000 J (c) 9000 J (d) −9800 J

24. When a system is taken from state *i* to a state *f* along path *iaf*, $Q = 50$ J and $W = 20$ J. If $W = -13$ J for the curved return path *fi*, *Q* in this path is

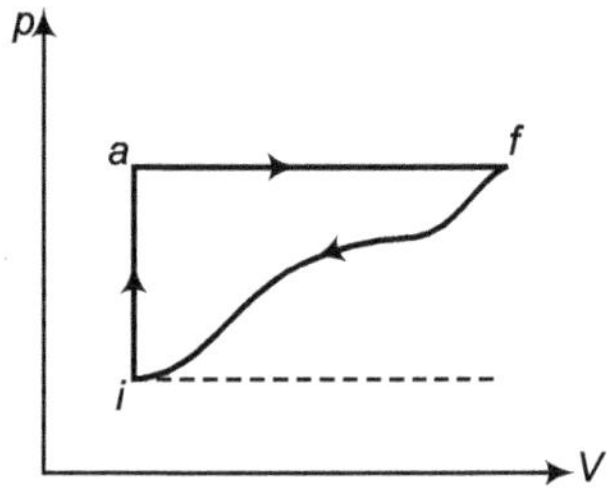

(a) 33 J (b) 23 J (c) −7 J (d) −43 J

25. The *p*-*V* diagram of a system undergoing thermodynamic transformation is as shown in figure. The work done by the system in going from $A \to B \to C$ is 30 J and 40 J heat is given to the system. The change in internal energy between *A* and *C* is

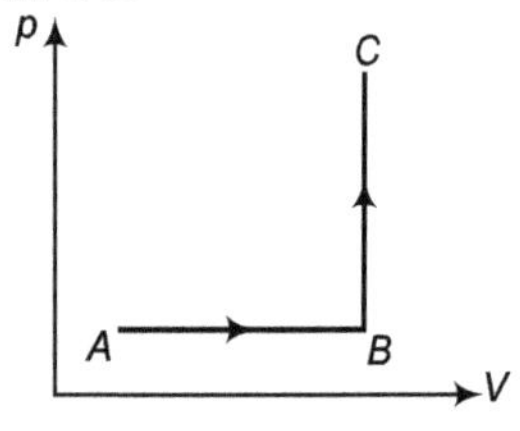

(a) 10 J (b) 70 J (c) 84 J (d) 134 J

26. *p*-*V* diagram of a cyclic process *ABCA* is as shown in figure. Choose the correct statement.

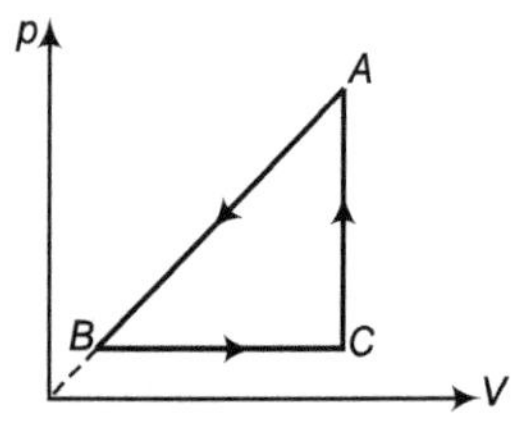

(a) $\Delta Q_{A\to B} =$ constant (b) $\Delta U_{B\to C} =$ positive
(c) $\Delta W_{CAB} =$ negative (d) All of these

27. A sample of an ideal gas is taken through a cycle as shown in figure. It absorbs 50 J of energy during the process *AB*, no heat during *BC*, rejects 70 J during *CA*. 40 J of work is done on the gas during *BC*. Internal energy of gas at *A* is 1500 J, the internal energy at *C* would be

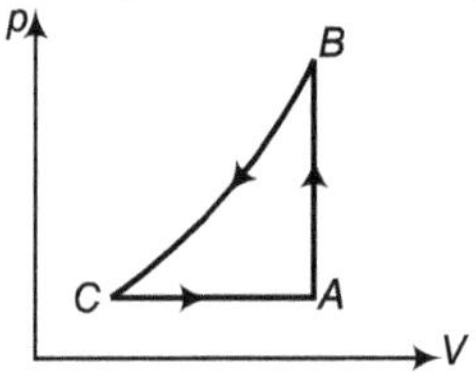

(a) 1590 J (b) 1620 J (c) 1540 J (d) 1570 J

28. An ideal monoatomic gas is taken around the cycle as shown in *p*-*V* diagram. The work done during the cycle is given by

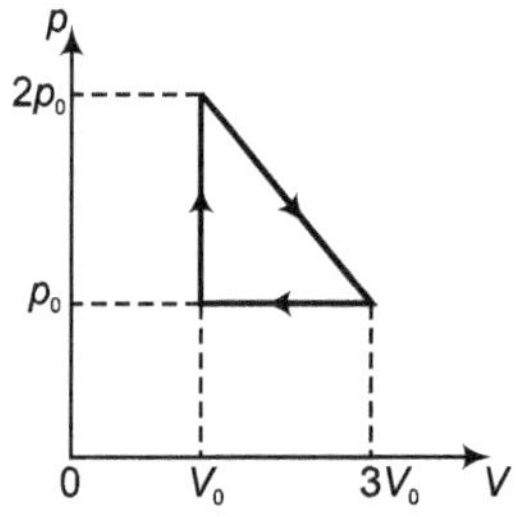

(a) $\frac{1}{2} p_0V_0$ (b) p_0V_0
(c) $2p_0V_0$ (d) None of these

Different known Thermodynamic Processes

29. 70 cal of heat are required to raise the temperature of 2 moles of an ideal gas at constant pressure from 30°C to 35°C. The amount of heat required in calories to raise the temperature of the same gas through, the same range (30°C-35°C) at constant volume is

(a) 30 cal (b) 50 cal (c) 40 cal (d) 90 cal

30. The volume of a gas is reduced adiabatically to 1/4 of its volume at 27°C. If $\gamma = 1.4$, the new temperature will be

(a) $(300)(2)^{0.4}$ K (b) $(300)(4)^{1.4}$ K
(c) $(300)(4)^{0.4}$ K (d) $(300)(2)^{1/4}$ K

31. Which is incorrect?

(a) In an isobaric process, $\Delta p = 0$
(b) In an isochoric process, $\Delta W = 0$
(c) In an isothermal process, $\Delta T = 0$
(d) In an isothermal process, $\Delta Q = 0$

32. The molar heat capacity of a gas at constant volume is C_V. If *n* moles of the gas undergo ΔT change in temperature, its internal energy will change by $nC_V\,\Delta T$

(a) only if the change of temperature occurs at constant volume
(b) only if the change of temperature occurs at constant pressure
(c) in any process which is not adiabatic
(d) in any process

33. During an adiabatic process, the pressure of a gas is found to be proportional to the cube of its absolute temperature. The ratio $\frac{C_p}{C_V} = \gamma$ for the gas is

(a) 2 (b) $\frac{3}{2}$
(c) $\frac{5}{3}$ (d) $\frac{4}{3}$

34. In an adiabatic expansion, a gas does 25 J of work while in an adiabatic compression 100 J of work is done on a gas. The change of internal energy in the two processes respectively are

(a) 25 J and −100 J (b) −25 J and 100 J
(c) −25 J and −100 J (d) 25 J and 100 J

35. For a gas $\gamma = \frac{5}{3}$ and 640 cc of this gas is suddenly compressed to 80 cc. If the initial pressure is p, then the final pressure will be

(a) $8p$ (b) $32p$
(c) $16p$ (d) $64p$

36. When an ideal gas is compressed isothermally, then its pressure increases because

(a) its potential energy increases
(b) its kinetic energy increases and molecules move apart
(c) its number of collisions per unit area with walls of container increases
(d) molecules energy increases

37. In an isothermal expansion of an ideal gas, select wrong statement.

(a) There is no change in the temperature of the gas
(b) There is no change in the internal energy of the gas
(c) the work done by the gas is equal to the heat supplied to the gas
(d) The work done by the gas is equal to the change in its internal energy

38. When an enclosed perfect gas is subjected to an adiabatic process, then

(a) its total internal does not change
(b) its temperature does not change
(c) its pressure varies inversely as a certain power of its volume
(d) the product of its pressure and volume is directly proportional to its absolute temperature

39. One mole of an ideal monoatomic gas is at 360 K and a pressure of 10^5 Pa. It is compressed at constant pressure until its volume is halved. Taking R as 8.3 J mol^{-1}K^{-1} and the initial volume of the gas as 3.0×10^{-2} m^3, the work done on the gas is

(a) −1500 J (b) +1500 J
(c) −3000 J (d) +3000 J

40. A certain mass of an ideal gas is at pressure p_1 and volume V_1. It is compressed isothermally and then allowed to expand adiabatically until its pressure returns to p_1. The gas is then allowed to expand its original volume. Upon which of the following p-V graphs are these processes correctly shown?

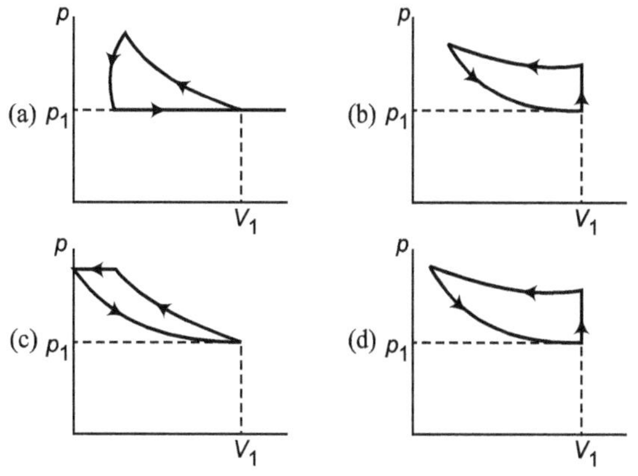

41. A monoatomic ideal gas, initially at temperature T_1, is enclosed in a cylinder fitted with a frictionless piston. The gas is allowed to expand adiabatically to a temperature T_2 by releasing the piston suddenly. If L_1 and L_2 are lengths of the gas column before and after expansion respectively, then $\frac{T_1}{T_2}$ is given by

(a) $\left(\frac{L_1}{L_2}\right)^{2/3}$ (b) $\left(\frac{L_1}{L_2}\right)^{5/3}$ (c) $\left(\frac{L_2}{L_1}\right)^{5/3}$ (d) $\left(\frac{L_2}{L_1}\right)^{2/3}$

42. Starting with the same initial conditions, an ideal gas expands from volume V_1 to V_2 in three different ways. The work done by the gas is W_1, if the process is isothermal, W_2 if isobaric and W_3 if adiabatic, then

(a) $W_2 > W_1 > W_2$ (b) $W_2 > W_3 > W_1$
(c) $W_1 > W_2 > W_3$ (d) $W_1 > W_3 > W_2$

43. A and B are two adiabatic curves for two different gases. Then A and B corresponds to

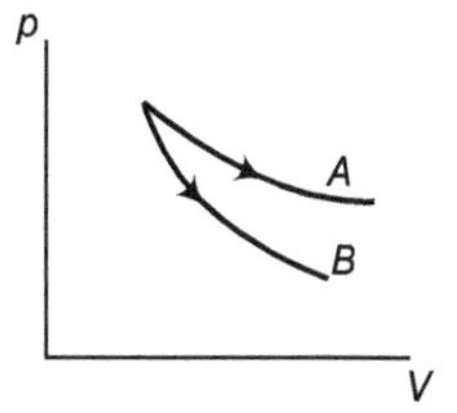

(a) Ar and He respectively (b) He and H_2 respectively
(c) O_2 and H_2 respectively (d) H_2 and He respectively

44. Choose the incorrect statement related to an isobaric process.

(a) $\frac{V}{T}$ = constant
(b) $W = p\,\Delta T$
(c) Only heat given to a system is used up in raising the temperature
(d) None of the above

45. The amount of heat required to raise the temperature of 1 mole of a monoatomic gas from 20°C to 30°C at constant volume is H. Then, the amount of heat required to raise the temperature of 2 mole of a diatomic gas from 20°C to 25°C at constant pressure is

(a) $2H$ (b) $\frac{4}{3}H$

(c) $\frac{5}{3}H$ (d) $\frac{7}{3}H$

46. Specific heat of gas undergoing adiabatic changes is

(a) zero (b) infinite (c) positive (d) negative

47. During an isothermal expansion of an ideal gas

(a) its internal energy decreases
(b) its internal energy does not change
(c) the work done by the gas is equal to the quantity of heat absorbed by it
(d) Both (b) and (c) are correct

48. Two identical samples or a gas are allowed to expand

(i) isothermally
(ii) adiabatically.

Work done is

(a) more in the isothermal process
(b) more in the adiabatic process
(c) zero in both of them
(d) equal in both processes

49. A given system undergoes a change in which the work done by the system equals the decrease in its internal energy. The system must have undergone an

(a) isothermal change (b) adiabatic change
(c) isobaric change (d) isochoric change

50. A cycle tyre bursts suddenly. This represents an

(a) isothermal process (b) isobaric process
(c) isochoric process (d) adiabatic process

51. An ideal gas at a pressure of 1 atm and temperature of 27°C is compressed adiabatically until its pressure becomes 8 times the initial pressure, then the final temperature is ($\gamma = 3/2$)

(a) 627°C (b) 527°C (c) 427°C (d) 327°C

52. In an adiabatic process, the state of a gas is changed from p_1, V_1, T_1 to p_2, V_2, T_2. Which of the following relation is correct?

(a) $T_1V_1^{\gamma-1}$ (b) $p_1V_1^{\gamma-1} = p_2V_2^{\gamma-1}$
(c) $T_1p_1^{\gamma} = T_2p_2^{\gamma}$ (d) $T_1V_1^{\gamma} = T_2V_2^{\gamma}$

53. During an adiabatic process, the pressure of a gas is found to be proportional to the cube of its absolute temperature. The ratio $\frac{C_p}{C_V}$ for the gas is

(a) $\frac{3}{2}$ (b) $\frac{4}{3}$ (c) 2 (d) $\frac{5}{3}$

54. A gas for which $\gamma = 1.5$ is suddenly compressed to $\frac{1}{4}$th of the initial volume. Then, the ratio of the final to the initial pressure is

(a) 1 : 16 (b) 1 : 8
(c) 1 : 4 (d) 8 : 1

55. Which of the following is a slow process?

(a) Isothermal (b) Adiabatic
(c) Isobaric (d) None of these

56. The work done in which of the following processes is zero?

(a) Isothermal process
(b) Adiabatic process
(c) Adiabatic
(d) Isochoric

57. In an isochoric process, if $T_1 = 27°\text{C}$ and $T_2 = 127°\text{C}$, then $\frac{p_1}{p_2}$ will be equal to

(a) $\frac{9}{59}$ (b) $\frac{2}{3}$

(c) $\frac{3}{4}$ (d) None of these

58. A monoatomic gas is supplied the heat Q very slowly keeping the pressure constant. The work done by the gas will be

(a) $\frac{2}{3}Q$ (b) $\frac{3}{5}Q$

(c) $\frac{2}{5}Q$ (d) $\frac{1}{5}Q$

59. The temperature of a hypothetical gas increases to $\sqrt{2}$ times, when compressed adiabatically to half the volume. Its equation can be written as

(a) $pV^{3/2} = \text{constant}$ (b) $pV^{5/2} = \text{constant}$
(c) $pV^{7/3} = \text{constant}$ (d) $pV^{4/3} = \text{constant}$

60. Which of the p-V, diagrams best represents an isothermal process?

(a)
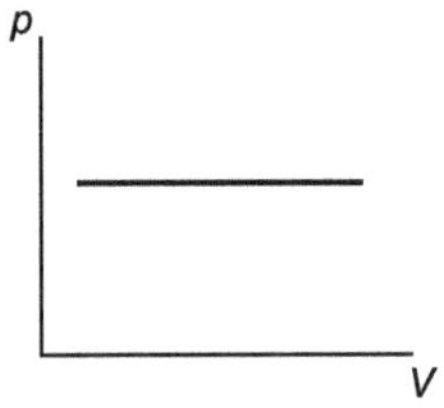

(b)
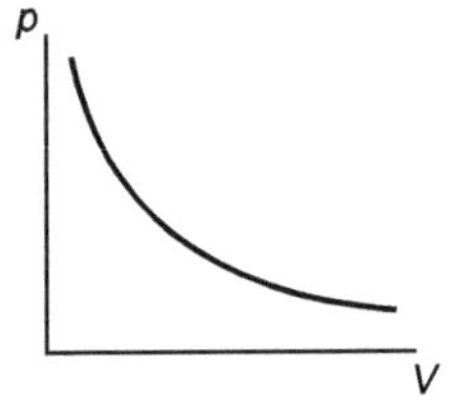

(c)
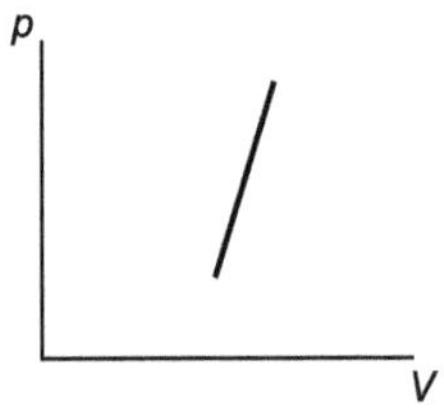

(d)
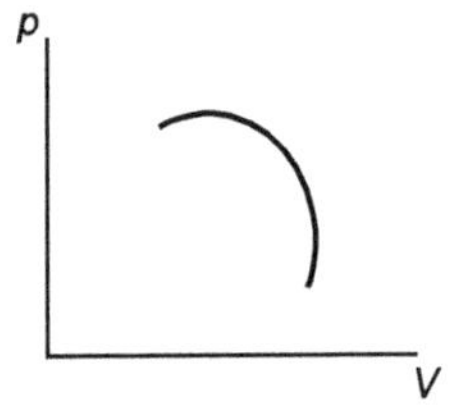

Heat Engines and Refrigerators

61. An ideal heat engine is opening between 227°C and 127°C. It absorbs 10^4 J amount of heat at the higher temperature. The amount of heat converted into work is

(a) 2000 J (b) 4000 J
(c) 8000 J (d) 5600 J

62. A Carnot's engine, with its cold body at 17°C has 50% efficiency. If the temperature of its hot body is now increased by 145°C, then efficiency becomes

(a) 55% (b) 60%
(c) 40% (d) 45%

63. The heat reservoir of an ideal Carnot engine is at 800 K and its sink is at 400 K. The amount of heat taken in it in is to produce useful mechanical work at the rate of 750 J is

(a) 2250 J (b) 1125 J
(c) 1500 J (d) 750 J

64. If a Carnot engine functions at source temperature $= 127°$ C and sink temperature $= 87°$ C, what is its efficiency?

(a) 10% (b) 25% (c) 40% (d) 50%

65. Efficiency of Carnot engine is (input temperature $= T_1$, exhaust temperature $= T_2$)

(a) $\frac{T_1 - T_2}{T_1}$ (b) $\frac{T_2 - T_1}{T_1}$
(c) $\frac{T_1}{T_2}$ (d) $\frac{T_2}{T_1}$

66. A system undergoes a cyclic process in which it absorbs Q_1 heat and gives out Q_2 heat. The efficiency of the process is η and work done is W. Select the correct statement.

(a) $W = \frac{Q_1 - Q_2}{Q_2}$
(b) $\eta = \frac{W}{Q_1}$
(c) Both (a) and (b) are correct
(d) Both (a) and (b) are wrong

67. For which combination of temperatures, the efficiency of Carnot's engine is highest?

(a) 80 K, 60 K (b) 100 K, 80 K
(c) 60 K, 40 K (d) 40 K, 20 K

68. Two Carnot engines A and B are operated in succession. The first one, A receives heat from a source at $T_1 = 800$ K and rejects to sinks at T_2 K. The second engine B receives heat rejected by the first engine and rejects to another sink at $T_3 = 300$ K. If the work outputs of two engines are equal, then the value of T_2 is

(a) 100 K (b) 300 K
(c) 550 K (d) 700 K

69. A Carnot's engine whose low temperature reservoir is at 7°C has an efficiency of 50%. It is desired to increase the efficiency to 70%. By how many degrees should the temperature of the high temperature reservoir be increased

(a) 933 K (b) 432 K
(c) 373 K (d) 267 K

70. Six moles of an ideal gas performs a cycle shown in figure. If the temperatures are $T_A = 600$ K, $T_B = 800$ K, $T_C = 2200$ K and $T_D = 1200$ K, then work done per cycle is approximately

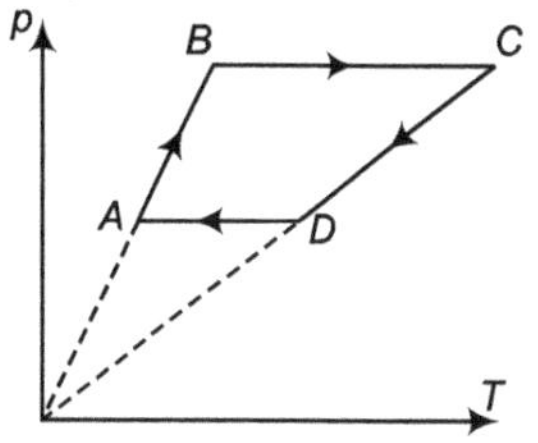

(a) 20 kJ (b) 30 kJ
(c) 40 kJ (d) 60 kJ

71. A Carnot's engine is made to work between 200°C and 0°C first and then between 0° and –200° C. The ratio of efficiencies $\left(\frac{\eta_2}{\eta_1}\right)$ of the engine in the two cases is

(a) 1 : 1.5 (b) 1 : 1
(c) 1 : 2 (d) 1.73 : 1

Miscellaneous Problems

72. The process $\Delta U = 0$, for an ideal gas can be best represented in the form of a graph

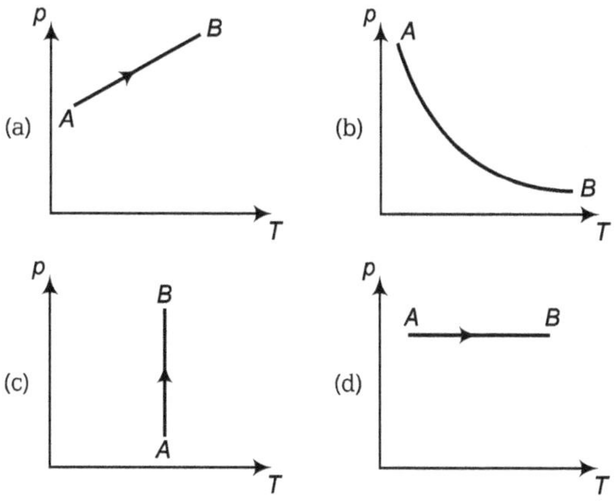

73. A given quantity of monoatomic gas is at pressure p and absolute temperature T. The isothermal bulk modulus of the gas is

(a) $\frac{2}{3}p$ (b) p
(c) $\frac{3}{2}p$ (d) $2p$

74. When an air bubble rises from the bottom to the surface of a lake, its radius becomes double. Find the depth of the lake. Given that the atmospheric pressure is equal to the pressure due to a column of water 10 m high. Assume constant temperature and disregard surface tension

(a) 30 m (b) 40 m
(c) 70 m (d) 80 m

75. Two identical containers joined by a small pipe initially contain the same gas at pressure p_0 and absolute temperature T_0. One container is now maintained at the same temperature while the other is heated to $2T_0$. The common pressure of the gases will be

(a) $\frac{3}{2}p_0$ (b) $\frac{4}{3}p_0$ (c) $\frac{5}{3}p_0$ (d) $2p_0$

76. An ideal monoatomic gas is compressed (no heat being added or removed in the process) so that its volume is halved. The ratio of the new pressure to the original pressure is

(a) $(2)^{3/5}$ (b) $(2)^{4/3}$
(c) $(2)^{3/4}$ (d) $(2)^{5/3}$

77. Two different ideal diatomic gases A and B are initially in the same state. A and B are then expanded to same final volume through adiabatic and isothermal process respectively. If p_A, p_B and T_A, T_B represents the final pressure and temperatures at A and B respectively, then

(a) $p_A < p_B$ and $T_A < T_B$ (b) $p_A > p_B$ and $T_A > T_B$
(c) $p_A > p_B$ and $T_A < T_B$ (d) $p_A < p_B$ and $T_A > T_B$

78. An ideal gas is allowed to expand freely against vacuum in a rigid insulated container. The gas undergoes

(a) an increase in its internal energy
(b) a decrease in its internal energy
(c) neither an increase nor decrease in temperature or internal energy
(d) an increase in temperature

79. Identify the graph (s) which correctly represents an isotherm at two temperatures T_1 and T_2 ($> T_1$)

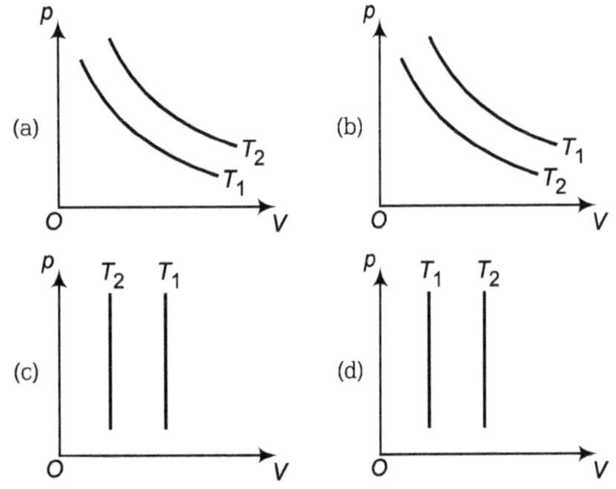

80. Two pistons can move freely inside a horizontal cylinder having two sections of unequal cross-sections. The pistons are joined by an inextensible, light string and some gas is enclosed between the pistons. On heating the system, the piston will

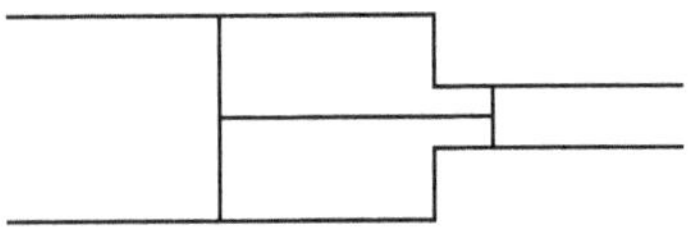

(a) move to the left
(b) move to the right
(c) remain stationary
(d) (a) or (c) depending upon the initial pressure of the gas

81. Two different masses of a gas m and $2m$ are heated separately in vessels of equal volume. The T-p curve for mass $2m$ makes angle an α with T-axis and that for mass m takes angle an β with T-axis, then

(a) $\tan\alpha = \tan\beta$ (b) $\tan\alpha = 2\tan\beta$
(c) $\tan\beta = 2\tan\alpha$ (d) None of these

82. An ideal gas mixture filled inside a balloon expands according to the relation $pV^{2/3}$ = constant. The temperature inside the balloon is

(a) increasing
(b) decreasing
(c) constant
(d) Cannot be defined

83. A container of volume $1\,m^3$ is divided into two equal compartments by a partition. One of these compartments contains an ideal gas at 300 K. The other compartment is vacuum. The whole system is thermally isolated from its surroundings. The partition is removed in the gas container. Its temperature now would be

(a) 300 K (b) greater than 300 K
(c) less than 300 K (d) Data insufficient

84. One mole of helium is adiabatically expanded from its initial state (p_i, V_i, T_i) to its final state (p_f, V_f, T_f). The decrease in the internal energy associated with this expansion is equal to

(a) $C_V(T_i - T_f)$
(b) $C_p(T_i - T_f)$
(c) $\frac{1}{2}(C_p + C_V)(T_i - T_f)$
(d) $(C_p - C_V)(T_i - T_f)$

85. One mole of a perfect gas in a cylinder fitted with a piston has a pressure p, volume V and temperature T. If the temperature is increased by 1 K keeping pressure constant, then increase in volume is

(a) $\frac{2V}{273}$ (b) $\frac{V}{91}$ (c) $\frac{V}{273}$ (d) V

86. Unit mass of a liquid with volume V_1 is completely changed into a gas of volume V_2 at a constant external pressure p and temperature T. If the latent heat of evaporation is L, then the increase in the internal energy of the system is

(a) zero (b) $p(V_2 - V_1)$
(c) $L - p(V_2 - V_1)$ (d) L

87. A gas mixture consists of 2 moles of oxygen and 4 moles argon at temperature T. Neglecting all vibrational modes, the total internal energy of the system is

(a) $4RT$ (b) $15RT$ (c) $9RT$ (d) $11RT$

88. A cylindrical tube of uniform cross-sectional area A is fitted with two air tight frictionless pistons. The pistons are connected to each other by a metallic wire. Initially the pressure of the gas is p_0 and temperature is T_0, atmospheric pressure is also p_0. Now, the temperature of the gas is increased to $2T_0$, the tension in the wire will be

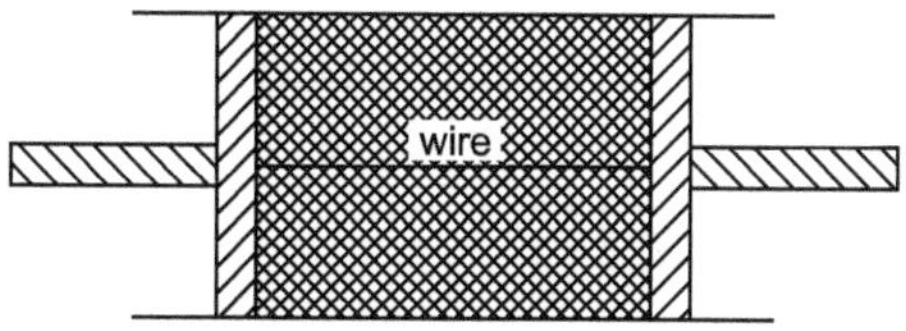

(a) $2p_0A$ (b) p_0A (c) $\frac{p_0A}{2}$ (d) $4p_0A$

89. The molar heat capacity in a process of a diatomic gas if it does a work of $\frac{Q}{4}$, when a heat of Q is supplied to it is

(a) $\frac{2}{5}R$ (b) $\frac{5}{2}R$
(c) $\frac{10}{3}R$ (d) $\frac{6}{7}R$

90. A thermally insulated rigid container contain an ideal gas heated by a filament of resistance 100 Ω through a current of 1 A for 5 min, then change in internal energy is

(a) 0 kJ (b) 10 kJ
(c) 20 kJ (d) 30 kJ

91. An ideal gas expands in such a manner that its pressure and volume can be related by equation pV^2 = constant. During this process, the gas is

(a) heated
(b) cooled
(c) neither heated nor cooled
(d) first heated and then cooled

92. p-V diagram of a diatomic gas is a straight line passing through origin. The molar heat capacity of the gas in the process will be

(a) $4R$ (b) $2.5R$
(c) $3R$ (d) $\frac{4R}{3}$

[Level 2]

1. The equation of a state of a gas is given by $p(V - b) = nRT$. 1 mole of a gas is isothermally expanded from volume V to $2V$, the work done during the process is

(a) $RT \ln\left|\frac{2V - b}{V - b}\right|$ (b) $RT \ln\left|\frac{V - b}{V}\right|$
(c) $RT \ln\left|\frac{V - b}{2V - b}\right|$ (d) $RT \ln\left|\frac{V}{V - b}\right|$

2. A cyclic process for 1 mole of an ideal is shown in the V-T diagram. The work done in AB, BC and CA respectively is

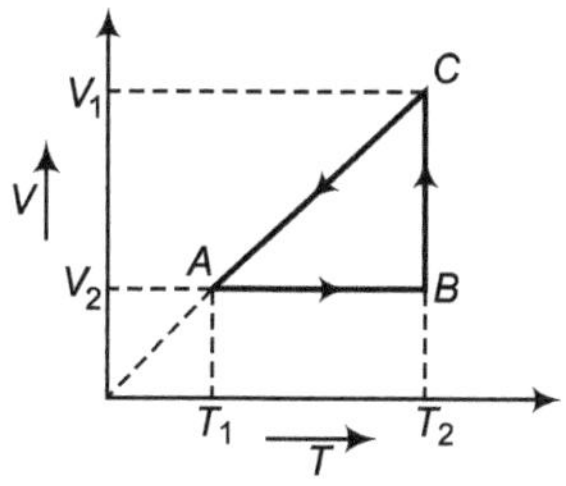

(a) $0, RT_2 \ln\left|\frac{V_1}{V_2}\right|, R(T_2 - T_1)$

(b) $R(T_1 - T_2), 0, RT_1 \ln\left|\frac{V_1}{V_2}\right|$

(c) $0, RT_2 \ln\left|\frac{V_1}{V_2}\right|, R(T_1 - T_2)$

(d) $0, RT_2 \ln\left|\frac{V_2}{V_1}\right|, R(T_2 - T_1)$

3. One mole of a monoatomic gas is carried along process $ABCDEA$ as shown in the diagram. Find the net work done by gas.

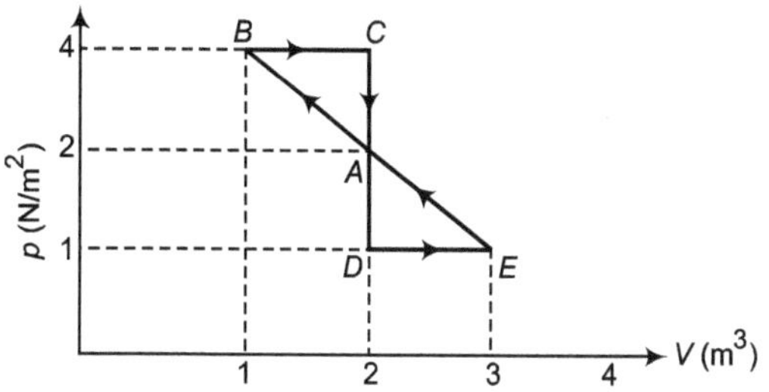

(a) $\frac{3}{2}$ J (b) 1 J
(c) $\frac{1}{2}$ J (d) 0 J

4. One mole of an ideal gas with heat capacity at constant pressure C_p undergoes the process $T = T_0 + \alpha V$ where, T_0 and α are constants. If its volume increases from V_1 to V_2 the amount of heat transferred to the gas is

(a) $C_p\ RT_0 \ln\left|\frac{V_2}{V_1}\right|$

(b) $\alpha C_p(V_2 - V_1) - RT_0 \ln\left|\frac{V_2}{V_1}\right|$

(c) $\alpha C_p(V_2 - V_1) + RT_0 \ln\left|\frac{V_2}{V_1}\right|$

(d) $RT_0 \ln\left|\frac{V_2}{V_1}\right| - \alpha C_p(V_2 - V_1)$

5. Ideal monoatomic gas is taken through a process $dQ = 2dU$. The molar heat capacity for the process is (where, dQ is heat supplied and dU is change in internal energy)

(a) $2.4\ R$
(b) $3R$
(c) R
(d) $2R$

6. Pressure *versus* density graph of an ideal gas is shown in figure

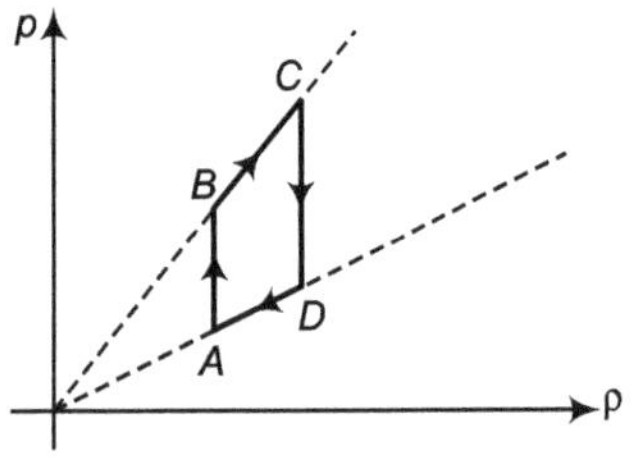

(a) during the process AB work done by the gas is positive
(b) during the process AB work done by the gas is negative
(c) during the process BC internal energy of the gas is increasing
(d) None of the above

7. N moles of a monoatomic gas is carried round the reversible rectangular cycle $ABCDA$ as shown in the diagram. The temperature at A is T_0. The thermodynamic efficiency of the cycle is approximately

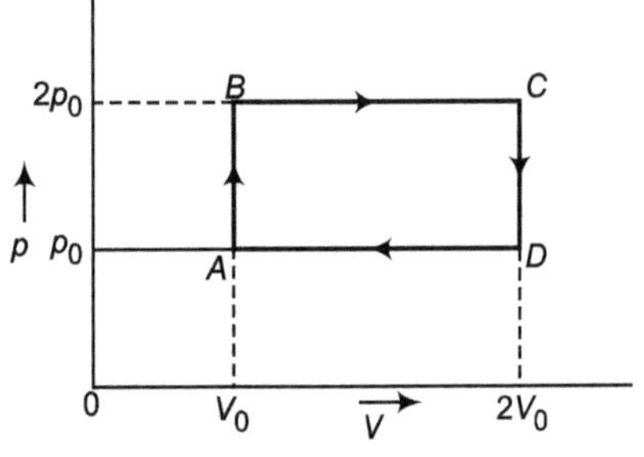

(a) 15% (b) 50%
(c) 20% (d) 25%

8. A closed system undergoes a change of state by process $1 \rightarrow 2$ for which $Q_{12} = 10$ J and $W_{12} = -5$ J. The system is now returned to its initial state by a different path $2 \rightarrow 1$ for which Q_{21} is -3 J. The work done by the gas in the process $2 \rightarrow 1$ is

(a) 8 J (b) zero (c) −2 J (d) +5 J

9. An ideal monoatomic gas undergoes the process AB as shown in the figure. If the heat supplied and the work done in the process are ΔQ and ΔW respectively. The ratio $\Delta Q : \Delta W$ is

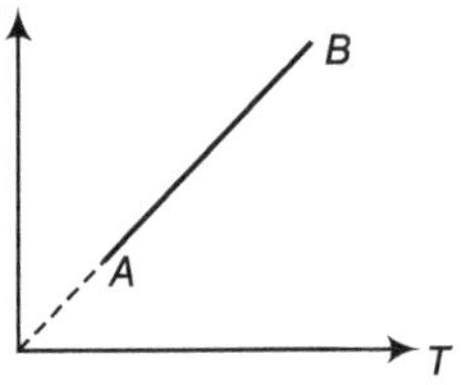

(a) 2.5 (b) 1.67 (c) 1.25 (d) 0.40

10. One mole of a gas expands with temperature T such that its volume, $V = kT^2$, where k is a constant. If the temperature of the gas changes by 60°C, then the work done by the gas is

(a) $120R$ (b) $R \ln 60$
(c) $kR \ln 60$ (d) $60kR$

11. An ideal monoatomic gas undergoes a process in which the gas volume relates to temperature as VT = constant. Then molar specific heat of gas in this process is

(a) $\frac{R}{2}$ (b) R
(c) $\frac{3R}{2}$ (d) None of these

12. In a cyclic process shown in the figure an ideal gas is adiabatically taken from B to A, the work done on the gas during the process $B \rightarrow A$ is 30 J, when the gas is taken from $A \rightarrow B$, the heat absorbed by the gas is 20 J. The change in internal energy of the gas in the process $A \rightarrow B$ is

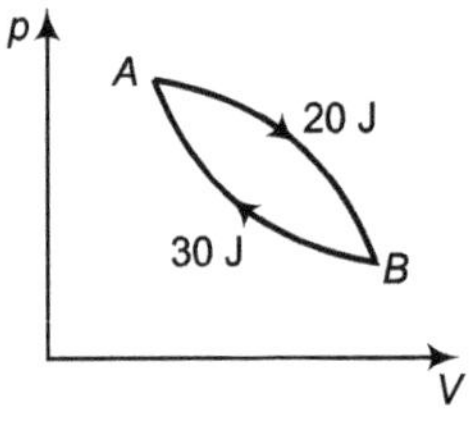

(a) 20 J (b) −30 J (c) 50 J (b) −10 J

13. n moles of an ideal gas undergo a process in which the temperature changes with volume as $T = kV^2$. The work done by the gas as the temperature changes from T_0 to $4T_0$ is

(a) $3nRT_0$ (b) $\left(\frac{5}{2}\right) nRT_0$
(c) $\left(\frac{3}{2}\right) nRT_0$ (d) zero

14. One mole of a monoatomic ideal gas undergoes the process $A \rightarrow B$ in the given p-V diagram. The molar heat capacity for this process is

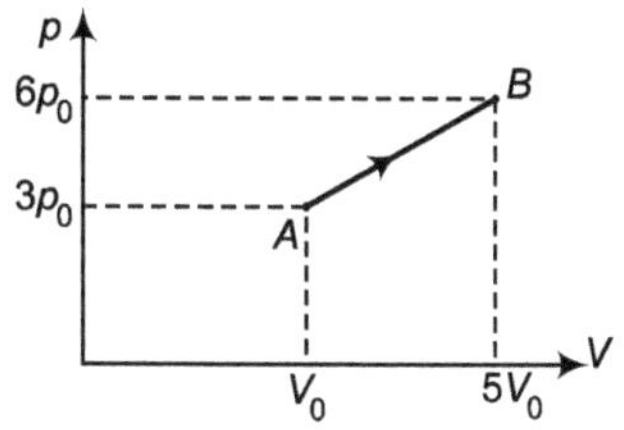

(a) $\frac{3R}{2}$ (b) $\frac{13R}{6}$ (c) $\frac{5R}{2}$ (d) $2R$

15. The p-V diagram of 2 g of helium gas for a certain process $A \rightarrow B$ is shown in the figure. What is the heat given to the gas during the process $A \rightarrow B$?

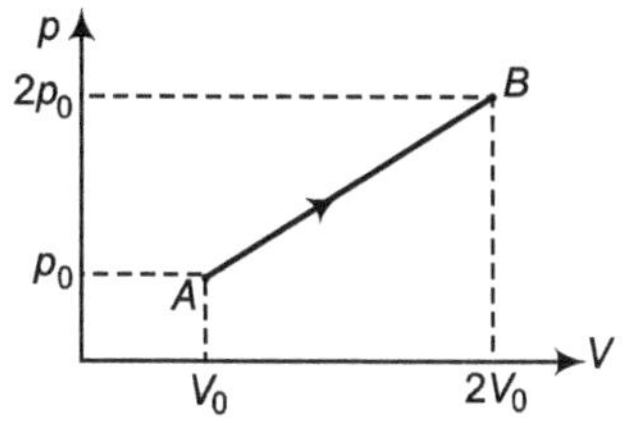

(a) $4p_0V_0$ (b) $6p_0V_0$ (c) $3p_0V_0$ (d) $2p_0V_0$

16. Heat energy absorbed by a system in going through a cyclic process as shown in figure is

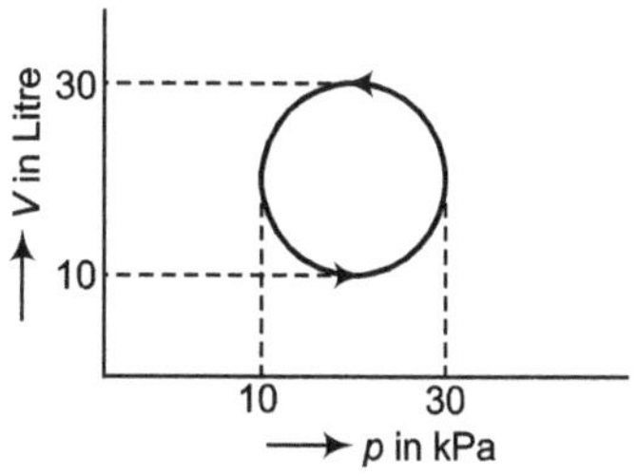

(a) $10^7 \pi$ J (b) $10^4 \pi$ J (c) $10^2 \pi$ J (d) $10^3 \pi$ J

17. A gas undergoes A to B through three different processes 1, 2, and 3 as shown in the figure. The heat supplied to the gas is Q_1, Q_2 and Q_3 respectively, then

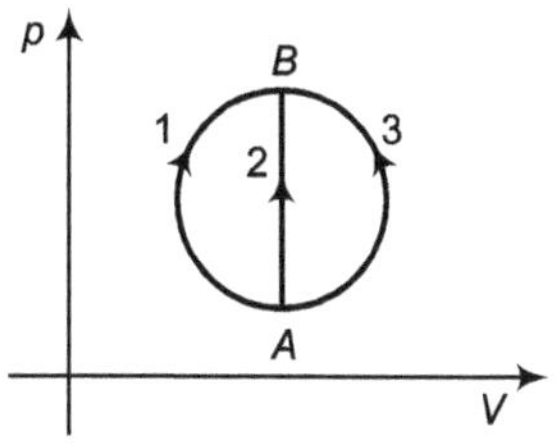

(a) $Q_1 = Q_2 = Q_3$ (b) $Q_1 < Q_2 < Q_3$
(c) $Q_1 > Q_2 > Q_3$ (d) $Q_1 = Q_3 > Q_2$

18. The internal energy of a gas is given by $U = 2pV$. It expands from V_0 to $2V_0$ against a constant pressure p_0. The heat absorbed by the gas in the process is

(a) $2p_0V_0$ (b) $4p_0V_0$
(c) $3p_0V_0$ (d) p_0V_0

19. The figure shown two paths for the change of state of a gas from A to B. The ratio of molar heat capacities in path 1 and path 2 is

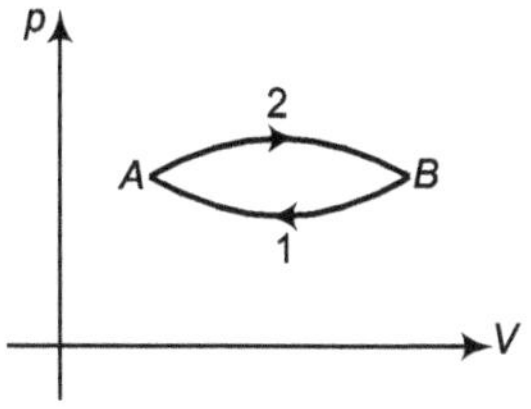

(a) < 1 (b) > 1
(c) 1 (d) data insufficient

20. An ideal monoatomic gas undergoes a process in which its internal energy U and density ρ varies as $U\rho =$ constant. The ratio of change in internal energy and the work done by the gas is

(a) $\frac{3}{2}$ (b) $\frac{-2}{3}$ (c) $\frac{1}{3}$ (d) $\frac{3}{5}$

21. A perfect gas goes from state A to another state B by absorbing 8×10^5 J of heat and doing 6.5×10^5 J of work (by the gas). It is now transferred between the same two states in another process in which it absorbs 10^5 J of heat. Then, in the second process

(a) work done on gas is 0.5×10^5 J
(b) work done by gas in 0.5×10^5 J
(c) work done on gas is 10^5 J
(d) work done by gas is 10^5 J

22. The process $\Delta U = 0$, for an ideal gas can be best represented in the form of a graph

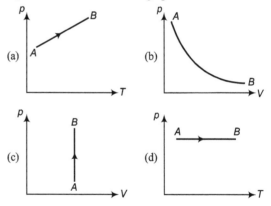

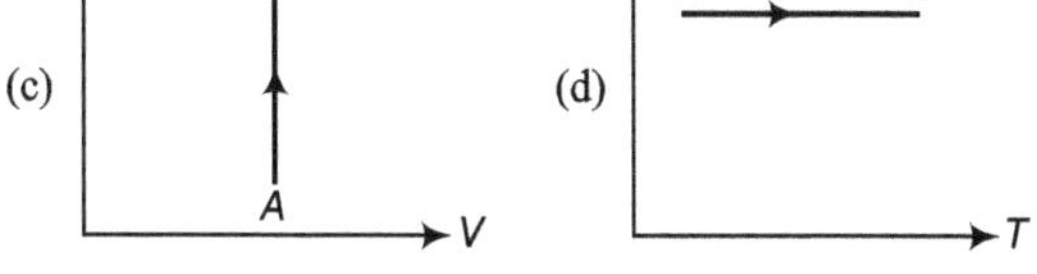

23. Consider two containers A and B containing identical gases at the same pressure, volume and temperature. The gas in container A is compressed to half of its original volume isothermally while the gas in container B is compressed to half of its original value adiabatically. The ratio of final pressure of gas in B to that of gas in A is

(a) $2^{\gamma - 1}$ (b) $\left(\frac{1}{2}\right)^{\gamma - 1}$ (c) $\left(\frac{1}{1-\gamma}\right)^2$ (d) $\left(\frac{1}{\gamma-1}\right)^2$

More than One Correct Options

1. An ideal gas is taken from the state A (pressure p, volume V) to the state B (pressure $\frac{p}{2}$, volume $2V$) along a straight line path in the p-V diagram. Select the correct statement(s) from the following.
(a) The work done by the gas in the process A to B is negative
(b) In the T-V diagram, the path AB becomes a part of a parabola
(c) In the p-T diagram, the path AB becomes a part of a hyperbola
(d) In going from A to B, the temperature T of the gas first increases to a maximum value and then decreases

2. In the process pV^2 = constant, if temperature of gas is increased
(a) change in internal energy of gas is positive
(b) work done by gas is positive
(c) heat is given to the gas
(d) heat is taken out from the gas

3. T-V diagram of two moles of a monoatomic gas is as shown in figure:

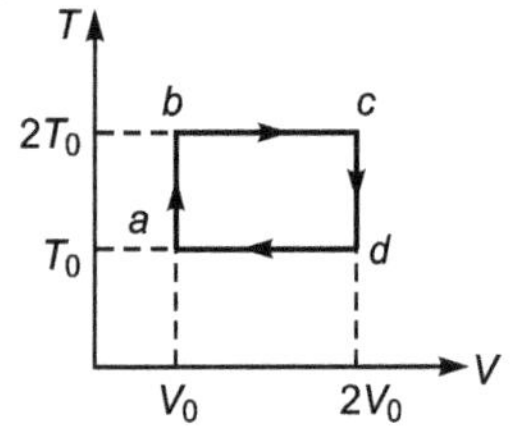

For the process $abcda$, choose the correct options given below
(a) $\Delta U = 0$
(b) work done by gas > 0
(c) heat given to the gas is $4RT_0$
(d) heat given to the gas is $2RT_0$

4. Density (ρ) *versus* internal energy (U) graph of a gas is as shown in figure. Choose the correct options.

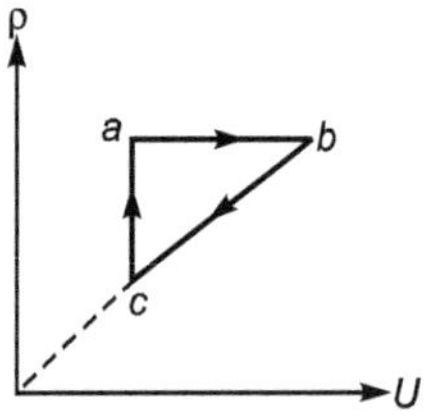

(a) $Q_{bc} = 0$ (b) $W_{bc} = 0$ (c) $W_{ca} < 0$ (d) $Q_{ab} > 0$
Here, W work done by gas and Q is heat given to the gas.

5. Temperature of a monoatomic gas is increased from T_0 to $2T_0$ in three different processes : isochoric, isobaric and adiabatic. Heat given to the gas in these three processes are Q_1, Q_2 and Q_3 respectively. Then, the choose the correct option.
(a) $Q_1 > Q_3$ (b) $Q_2 > Q_1$ (c) $Q_2 > Q_3$ (d) $Q_3 = 0$

6. Three copper blocks of masses M_1, M_2 and M_3 kg respectively are brought into thermal contact till they reach equilibrium. Before contact, they were at T_1, T_2, T_3 ($T_1 > T_2 > T_3$). Assuming there is no heat loss to the surroundings, the equilibrium temperature T is (s is specific heat of copper)
(a) $T = \frac{T_1 + T_2 + T_3}{3}$
(b) $T = \frac{M_1T_1 + M_2T_2 + M_3T_3}{M_1 + M_2 + M_3}$
(c) $T = \frac{M_1T_1 + M_2T_2 + M_3T_3}{3(M_1 + M_2 + M_3)}$
(d) $T = \frac{M_1T_1s + M_2T_2s + M_3T_3s}{M_1 + M_2 + M_3}$

7. Consider a cycle followed by an engine (figure.)

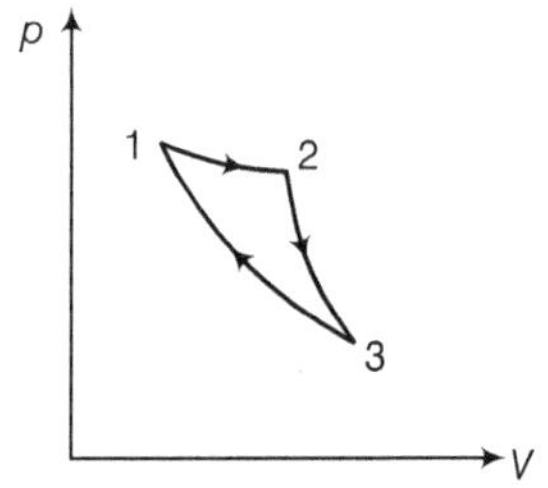

1 to 2 is isothermal
2 to 3 is adiabatic
3 to 1 is adiabatic
Such a process does not exist, because
(a) heat is completely converted to mechanical energy in such a process, which is not possible
(b) mechanical energy is completely converted to heat in this process, which is not possible
(c) curves representing two adiabatic processes don't intersect
(d) curves representing an adiabatic process and an isothermal process don't intersect

8. Consider a heat engine as shown in figure. Q_1 and Q_2 are heat added both to T_1 and heat taken from T_2 in one cycle of engine. W is the mechanical work done on the engine.

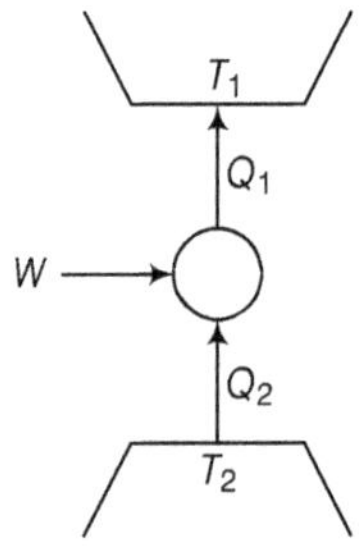

If $W > 0$, then possibilities are
(a) $Q_1 > Q_2 > 0$
(b) $Q_2 > Q_1 > 0$
(c) $Q_2 < Q_1 < 0$
(d) $Q_1 < 0$, $Q_2 > 0$

Comprehension Based Questions

Passage I (Q.1 to 2)

One mole of a monoatomic ideal gas is taken along the cycle ABCA as shown in the diagram.

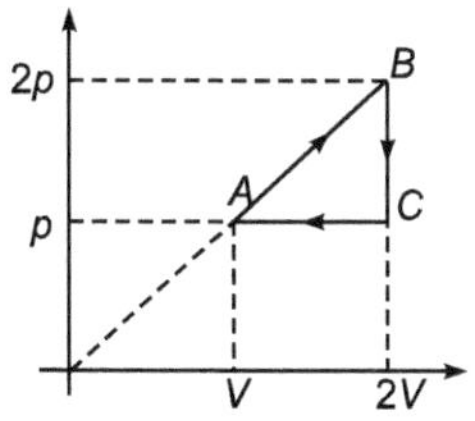

1. The net heat absorbed by the gas in the given cycle is

(a) pV (b) $\frac{pV}{2}$ (c) $2pV$ (d) $4pV$

2. The ratio of specific heat in the process *CA* to the specific heat in the process *BC* is

(a) 2 (b) $\frac{5}{3}$ (c) 4 (d) None of these

Passage II (Q.3 to 5)

One mole of a monoatomic ideal gas is taken through the cycle ABCDA as shown in the figure. $T_A = 1000\,K$ *and* $2p_A = 3p_B = 6p_C$.

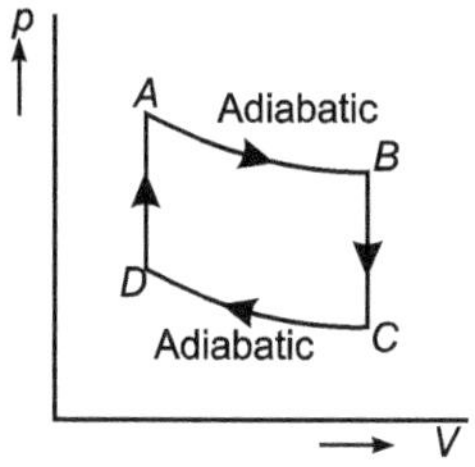

$$\left[\textit{Assume}\left(\frac{2}{3}\right)^{0.4} = 0.85 \textit{ and } R = \frac{25}{3} JK^{-1}\, mol^{-1}\right]$$

3. The temperature at *B* is

(a) 350 K (b) 1175 K (c) 850 K (d) 577 K

4. Work done by the gas in the process $A \to B$ is

(a) 5312.5 J (b) 1875 J (c) zero (d) 8854 J

5. Heat lost by the gas in the process $B \to C$ is

(a) 5312.5 J (b) 1875 J (c) zero (d) 8854 J

Assertion and Reason

Direction (Q. Nos. 1-15) *These questions consists of two statements each printed as assertion and reason. While answering these questions you are required to choose any one of the following five responses.*

(a) If both Assertion and Reason are correct and Reason is the correct explanation of Assertion.
(b) If both Assertion and Reason are true but Reason is not the correct explanation of Assertion.
(c) If Assertion is true but Reason is false.
(d) If Assertion is false but Reason is true.
(e) If both Assertion and Reason are false.

1. Assertion A system of ideal gas is heated by constant power. If temperature *versus* time graph is a straight line, the process must be isochoric.

Reason For isochoric process, $dQ = nC_V\,dT$

$$\Rightarrow \frac{dQ}{dT} = p = nC_V\frac{dT}{dt} \quad \Rightarrow \quad \frac{dT}{dt} = \frac{p}{nC_V} = \text{constant}$$

2. Assertion In isobaric process, $\frac{\Delta Q}{\Delta W}$ for helium gas is $\frac{5}{2}$.

Reason In isobaric process, work done by the gas is $nR\Delta T$.

3. Assertion In adiabatic expansion product of pV always decreases.

Reason In adiabatic expansion work done by the gas is positive.

4. Assertion In isochoric process p-V graph is straight line parallel to p-axis.

Reason In isochoric process density p remains constant.

5. Assertion For given source and sink temperatures efficiency of Carnot's engine is 100%.

Reason For given source and sink temperatures efficiency of Carnot's engine is maximum.

6. Assertion Efficiency of any heat engine cannot be greater than the heat engine of Carnot's engine.

Reason Heat flow never takes place from a body at lower temperature to a body at higher temperature.

7. Assertion If volume of a gas is increasing but temperature of the gas is decreasing, then heat given to the gas may be positive, negative or zero.

Reason Heat given to a gas is a state function. It is not path function.

8. Assertion First law of thermodynamics can be applied only for an ideal or real gas system.

Reason First law of thermodynamics is nothing but law of conservation of energy.

9. Assertion There are two processes : Process 1 is $pV = $ constant and process 2 is $pV^2 = $ constant. In both the processes volume of gas is increased from V_1 to V_2. Initial coordinates (p_1, V_1) of the gas were same. Then, more work is done by the gas in process 1.

Reason In second process pressure drops more rapidly with increase in volume.

10. Assertion Any process taking place in atmosphere is considered as an isobaric process.

Reason Work done by the system in that case is $p_0\,\Delta V$. Where, p_0 is atmospheric pressure.

11. **Assertion** During melting of ice work done by surrounding on (ice + water) system is positive.

Reason Volume of the given system decreases on melting on ice.

12. **Assertion** Molar heat capacity cannot be defined for isothermal process.

Reason In isothermal process (pV) *versus* T graph is a dot.

13. **Assertion** If initial and final volumes are equal, then work done by gas is zero.

Reason In isochoric process initial and final volumes are equal and work done by gas is zero.

14. **Assertion** In adiabatic expansion the product of p and V always decreases.

Reason In adiabatic expansion process work is done by the gas at the cost of internal energy of gas.

15. **Assertion** In isobaric process $Q/\Delta U$ is equal to $\gamma\ (= C_p/C_V)$ of the gas.

Reason For monoatomic gas ratio $Q/\Delta U$ in isobaric process is equal to $\frac{5}{3}$.

Match the Columns

1. In the ρ-T graph shown in figure, match the following.

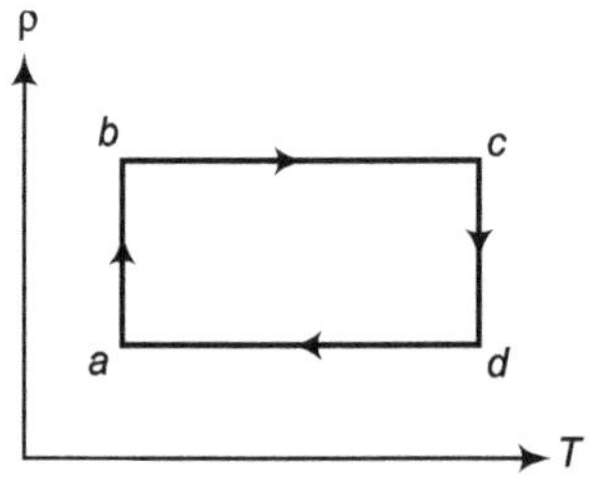

Column I	Column II
(A) Process *a-b*	(p) Isochoric process
(B) Process *b-c*	(q) $\Delta U = 0$
(C) Process *c-d*	(r) p increasing
(D) Process *d-a*	(s) p decreasing

2. Match the following columns.

Column I	Column II
(A) Cyclic process	(p) $\Delta U < 0$
(B) Isobaric process	(q) $\Delta Q = \Delta W$
(C) Isochoric process	(r) $\Delta W = \Delta nR\Delta T$
(D) Adiabatic process	(s) $\Delta Q = \Delta U$

3. For one mole of a monoatomic gas match the following

Column I	Column II
(A) Isothermal bulk modulus	(p) $-\frac{RT}{V^2}$
(B) Adiabatic bulk modulus	(q) $-\frac{5P}{3V}$
(C) Slope of p-V graph in isothermal process	(r) $\frac{T}{V}$
(D) Slope of p-V graph in adiabatic process	(s) $\frac{4T}{3V}$
	(t) None

4. Match the following columns.

Column I	Column II
(A) Adiabatic expansion	(p) No work done
(B) Isobaric expansion	(q) Constant internal energy
(C) Isothermal expansion	(r) Increase in internal energy
(D) Isochoric process	(s) Decrease in internal energy

Entrance Gallery

2014

1. A thermodynamic system is taken from an initial state i with internal energy $U_i = 100$ J to the final state f along two different paths iaf and ibf as schematically shown in the figure. The work done by the system along the paths af, ib and bf are $W_{af} = 200$ J, $W_{ib} = 50$ J and $W_{bf} = 100$ J, respectively. The heat supplied to the system along the path iaf, ib and bf are Q_{iaf}, Q_{ib} and Q_{bf} respectively. If the internal energy of the system in the state b is $U_b = 200$ J and $Q_{iaf} = 500$ J, then ratio Q_{bf}/Q_{ib} is **[JEE Advanced]**

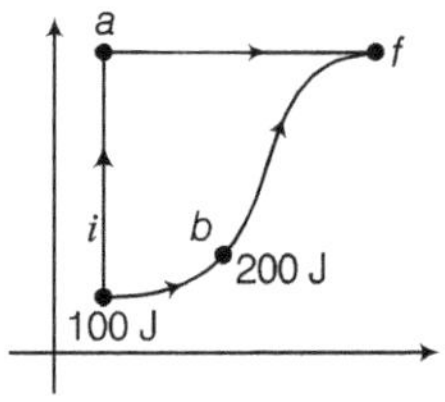

(a) 2 : 1
(b) 1 : 2
(c) 3 : 1
(d) 1 : 3

2. One mole of diatomic ideal gas undergoes a cyclic process ABC as shown in figure. The process BC is adiabatic. The temperature at A, B and C are 400 K, 800 K and 600 K, respectively. Choose the correct statement. **[JEE Main]**

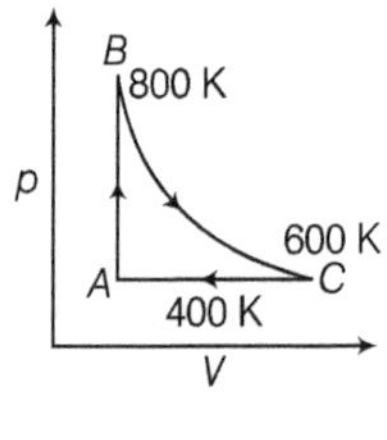

(a) The change in internal energy in whole cyclic process is $250R$
(b) The change in internal energy in the process CA is $700R$
(c) The change in internal energy in the process AB is $-350R$
(d) The change in internal energy in the process BC is $-500R$

3. A Carnot engine operating between temperatures T_1 and T_2 has efficiency 0.2. When T_2 is reduced by 50 K, its efficiency increases to 0.4. Then, T_1 and T_2 are respectively **[Kerala CEE]**

(a) 200 K, 150 K
(b) 250 K, 200 K
(c) 300 K, 250 K
(d) 300 K, 200 K
(e) 300 K, 150 K

4. What is the source temperature of the Carnot engine required to get 70% efficiency?
Given, sink temperature = 27° C. **[Karnataka CET]**

(a) 1000°C (b) 90°C (c) 270°C (d) 727°C

5. A cycle tyre bursts suddenly. What is the type of this process? **[Karnataka CET]**

(a) Isothermal (b) Adiabatic
(c) Isochoric (d) Isobaric

2013

6. One mole of monoatomic ideal gas is taken along two cyclic processes $E \to F \to G \to E$ and $E \to F \to H \to E$ as shown in the p-V diagram.

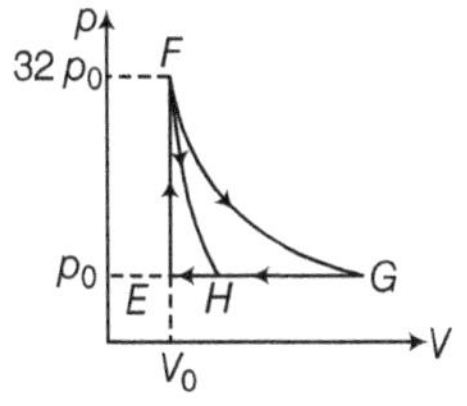

The processes involved are purely isochoric, isobaric, isothermal or adiabatic.

Match the paths in List I with the magnitudes of the work done in List II and select the correct answer using the codes given below the lists. **[JEE Advanced]**

Column I	Column II
P. $G \to E$	1. $160\, p_0V_0 \ln 2$
Q. $G \to H$	2. $36\, p_0V_0$
R. $F \to H$	3. $24\, p_0V_0$
S. $F \to G$	4. $31\, p_0V_0$

Codes

	P	Q	R	S
(a)	4	3	2	1
(b)	4	3	1	2
(c)	3	1	2	4
(d)	1	3	2	4

7. The shown $p\ V$ diagram represents the thermodynamic cycle of an engine, operating with an ideal monoatomic gas. The amount of heat, extracted from the source in a single cycle is **[JEE Main]**

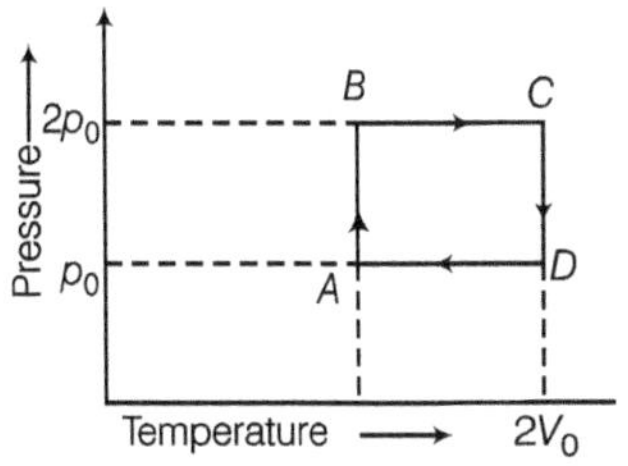

(a) p_0V_0 (b) $\frac{13}{2}p_0V_0$ (c) $\frac{11}{2}p_0V_0$ (d) $4p_0V_0$

8. An ideal gas is taken around $PQRP$ as shown in the figure of $p\ V$ diagram. The work done during a cycle is **[Karnataka CET]**

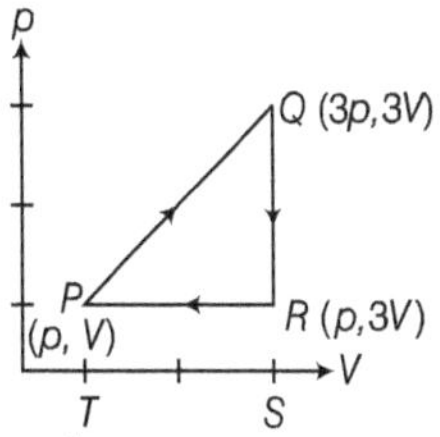

(a) zero (b) $\frac{1}{2}pV$ (c) pV (d) $2pV$

9. For which of the following combination of working temperatures, the efficiency of Carnot engine is highest? **[Karnataka CET]**

(a) 100 K, 80 K (b) 80 K, 60 K
(c) 60 K, 40 K (d) 40 K, 20 K

10. A graph of pressure *versus* volume for an ideal gas for different processes is as shown. In the graph, curve OC represents **[O JEE]**

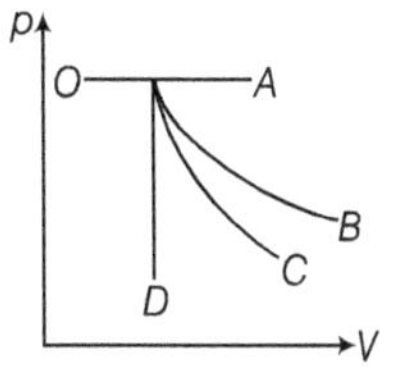

(a) isochoric process (b) isothermal process
(c) isobaric process (d) adiabatic process

11. A gas at state A changes to state B through path I and II shown in figure. The change in internal energy is ΔU_1 and ΔU_2, respectively. Then, **[O JEE]**

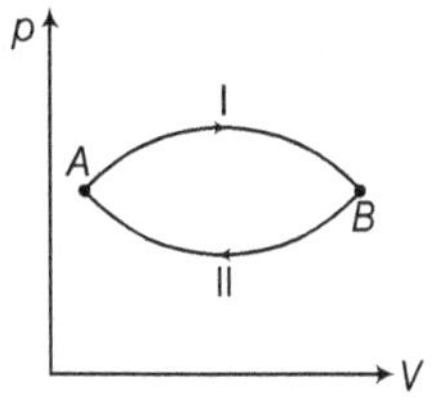

(a) $\Delta U_1 > \Delta U_2$ (b) $\Delta U_1 < \Delta U_2$
(c) $\Delta U_1 = \Delta U_2$ (d) $\Delta U_1 = \Delta U_2 = 0$

2012

12. Two moles of an ideal helium gas are in a rubber balloon at 30° C. The balloon is fully expandable and can be assumed to require no energy in its expansion. The temperature of the gas in the balloon is slowly changed to 35° C. The amount of heat required in raising the temperature is nearly (take, $R = 8.31$ J mol-K) [IIT JEE]

(a) 62 J (b) 104 J (c) 124 J (d) 208 J

13. Helium gas goes through a cycle *ABCDA* (consisting of two isochoric and isobaric lines) as shown in figure. Efficiency of this cycle is nearly (Assume, the gas to be close to ideal gas). [AIEEE]

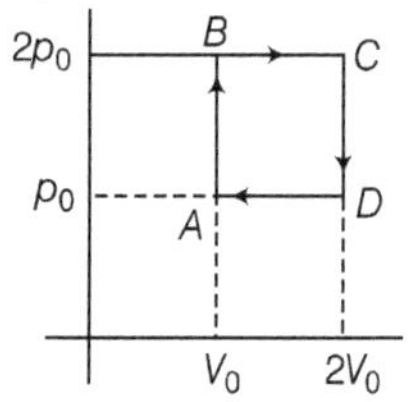

(a) 15.4% (b) 9.1% (c) 10.5% (d) 12.5%

2011

14. 5.6 L of helium gas at STP is adiabatically compressed to 0.7 L. Taking the initial temperature to be T_1, the work done in the process is [IIT JEE]

(a) $-\frac{9}{8}RT_1$ (b) $\frac{3}{2}RT_1$ (c) $\frac{15}{8}RT_1$ (d) $\frac{9}{2}RT_1$

15. A Carnot engine operating between temperatures T_1 and T_2 has efficiency $\frac{1}{6}$. When T_2 is lowered by 62 K, its efficiency increases to $\frac{1}{3}$. Then, T_1 and T_2 are, respectively [AIEEE]

(a) 372 K and 330 K (b) 330 K and 268 K
(c) 310 K and 248 K (d) 372 K and 310 K

16. The specific heat capacity of a metal at low temperature (T) is given as $C_p\ (\text{kJK}^{-1}\text{kg}^{-1}) = 32\left(\frac{T}{400}\right)^3$. A 100 g vessel of this metal is to be cooled from 20 K to 4 K by a special refrigerator operating at room temperature (27°C). The amount of work required to cool the vessel is [AIEEE]

(a) equal to 0.002 kJ
(b) greater than 0.148 kJ
(c) between 0.148 kJ and 0.028 kJ
(d) less than 0.028 kJ

17. A Carnot engine whose efficiency is 40%, receives heat at 500 K. If the efficiency is to be 50%, the source temperature for the same exhaust temperature is [Kerala CEE]

(a) 900 K (b) 600 K (c) 700 K
(d) 800 K (e) 850 K

18. 100 g of water is heated from 30°C to 50°C. Ignoring the slight expansion of the water, the change in its internal energy is (Specific heat of water is 4184 J/kg/K) [Kerala CEE]

(a) 8.4 kJ (b) 84 kJ (c) 2.1 kJ (d) 4.2 kJ
(e) 3.2 kJ

19. The thermodynamic process in which no work is done on or by the gas is [Kerala CEE]

(a) isothermal process (b) adiabatic process
(c) cyclic process (d) isobaric process
(e) isochoric process

20. Two soap bubbles of radii x and y coalesce to constitute a bubble of radius z. Then, z is equal to [WB JEE]

(a) $\sqrt{x^2 + y^2}$ (b) $\sqrt{x + y}$ (c) $x + y$ (d) $\frac{x + y}{2}$

2010

21. One mole of an ideal gas in initial state *A* undergoes a cyclic process *ABCA* as shown in the figure. Its pressure at *A* is p_0. Choose the correct option (s) from the following. [IIT JEE]

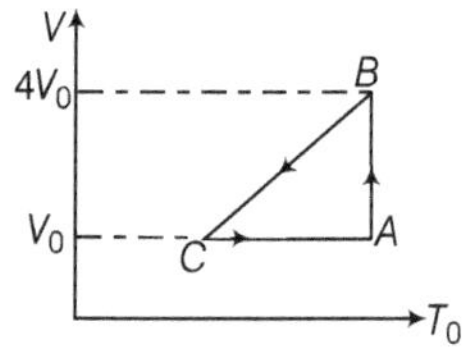

(a) Internal energies at *A* and *B* are the same

(b) Work done by the gas in process *AB* is $p_0V_0 \ln 4$

(c) Pressure at *C* is $\frac{p_0}{4}$

(d) Temperature at *C* is $T_0/4$

22. A diatomic ideal gas is compressed adiabatically to $\frac{1}{32}$ of its initial volume. If the initial temperature of the gas is T_i (in kelvin) and the final temperature is $T_f = aT_i$, then value of a is [IIT JEE]

(a) 4 (b) 6 (c) 5 (d) 9

23. A diatomic ideal gas is used in a car engine as the working substance. If during the adiabatic expansion part of the cycle, volume of the gas increases from V to $32V$. The efficiency of the engine is [AIEEE]

(a) 0.5 (b) 0.75 (c) 0.99 (d) 0.25

24. The efficiency of a Carnot engine working between 800 K and 500 K is [AIEEE]

(a) 0.4 (b) 0.625 (c) 0.375 (d) 0.5

25. The efficiency of Carnot heat engine is 0.5, when the temperature of the source is T_1 and that of sink is T_2. The efficiency of another Carnot heat engine is also 0.5. The temperatures of source and sink of the second engine are respectively [Karnataka CET]

(a) $2T_1, 2T_2$ (b) $2T_1, \frac{T_2}{2}$
(c) $T_1 + 5, T_2 - 5$ (d) $T_1 + 10, T_2 - 10$

Answers

Level 1

Objective Problems

1. (a)	**2.** (c)	**3.** (c)	**4.** (b)	**5.** (c)	**6.** (c)	**7.** (a)	**8.** (a)	**9.** (d)	**10.** (b)
11. (b)	**12.** (a)	**13.** (a)	**14.** (a)	**15.** (a)	**16.** (a)	**17.** (b)	**18.** (a)	**19.** (d)	**20.** (a)
21. (b)	**22.** (c)	**23.** (d)	**24.** (d)	**25.** (a)	**26.** (d)	**27.** (a)	**28.** (b)	**29.** (b)	**30.** (c)
31. (d)	**32.** (d)	**33.** (b)	**34.** (b)	**35.** (b)	**36.** (c)	**37.** (d)	**38.** (c)	**39.** (b)	**40.** (a)
41. (a)	**42.** (a)	**43.** (d)	**44.** (c)	**45.** (c)	**46.** (a)	**47.** (d)	**48.** (a)	**49.** (b)	**50.** (d)
51. (d)	**52.** (a)	**53.** (a)	**54.** (d)	**55.** (a)	**56.** (d)	**57.** (c)	**58.** (c)	**59.** (a)	**60.** (b)
61. (a)	**62.** (b)	**63.** (c)	**64.** (a)	**65.** (a)	**66.** (a)	**67.** (d)	**68.** (c)	**69.** (c)	**70.** (c)
71. (d)	**72.** (b)	**73.** (b)	**74.** (c)	**75.** (b)	**76.** (d)	**77.** (a)	**78.** (c)	**79.** (a)	**80.** (a)
81. (b)	**82.** (a)	**83.** (a)	**84.** (a)	**85.** (c)	**86.** (c)	**87.** (d)	**88.** (b)	**89.** (c)	**90.** (d)
91. (b)	**92.** (c)								

Level 2

Only One Correct Option

1. (a)	**2.** (c)	**3.** (c)	**4.** (c)	**5.** (b)	**6.** (d)	**7.** (a)	**8.** (a)	**9.** (a)	**10.** (a)
11. (b)	**12.** (b)	**13.** (c)	**14.** (b)	**15.** (b)	**16.** (c)	**17.** (c)	**18.** (c)	**19.** (a)	**20.** (a)
21. (a)	**22.** (a)	**23.** (a)							

More than One Correct Options

1. (b,d)	**2.** (a,c)	**3.** (a,b)	**4.** (c,d)	**5.** (all)	**6.** (b)	**7.** (a,c)	**8.** (a, c)

Comprehension Based Questions

1. (b)	**2.** (b)	**3.** (c)	**4.** (b)	**5.** (a)

Assertion and Reason

1. (d)	**2.** (b)	**3.** (b)	**4.** (d)	**5.** (d)	**6.** (c)	**7.** (c)	**8.** (d)	**9.** (b)	**10.** (b)
11. (a)	**12.** (b)	**13.** (d)	**14.** (b)	**15.** (b)					

Match the Columns

1. ($A \rightarrow q, r; B \rightarrow p, r; C \rightarrow q, s; D \rightarrow p, s$) **2.** ($A \rightarrow q, B \rightarrow r, C \rightarrow s, D \rightarrow p$) **3.** ($A \rightarrow t, B \rightarrow t, C \rightarrow p, D \rightarrow q$)
4. ($A \rightarrow s, B \rightarrow r, C \rightarrow q, D \rightarrow p$)

Entrance Gallery

1. (a)	**2.** (d)	**3.** (b)	**4.** (d)	**5.** (b)	**6.** (a)	**7.** (b)	**8.** (d)	**9.** (d)	**10.** (d)
11. (c)	**12.** (d)	**13.** (a)	**14.** (a)	**15.** (d)	**16.** (c)	**17.** (b)	**18.** (a)	**19.** (e)	**20.** (a)
21. (a,b)	**22.** (a)	**23.** (b)	**24.** (c)	**25.** (a)					

Solutions

Level 1 : Objective Problems

2. Internal energy of gas is state function, i.e.

$$\Delta U_1 = \Delta U_2$$

or $$Q_1 - W_1 = Q_2 - W_2$$

3. ΔU in both cases will be same, as potential energy is state function. Area under p-V graph in case 1 is more.

Hence, $W_1 > W_2$. Therefore, $Q_1 > Q_2$.

4. In cyclic process, work done is area between the cycle. It is positive when cycle is clockwise and negative when cycle is anti-clockwise. (With p on y-axis and V on x-axis).

5. In cyclic process, work done may be greater than or less than zero.

6. Work done by gas, $W = p(V_f - V_i) = 20\text{J}$

$$\Delta U = Q - W = 70 - 20 = 50\text{J}$$

7. Cycle is clockwise. Hence, heat will be given to the gas, as work done will be positive.

$$Q = W = \text{area of cycle} = 1\text{J}$$

8. In a cyclic process, $\Delta U = 0$

$\therefore$ $$Q_{\text{total}} = W_{\text{total}}$$

or $$10 - 5 = -3 + W_{21}$$

or $$W_{21} = 8\text{J}$$

9. Change in internal energy does not depend upon path so $\Delta U = \Delta Q - \Delta W$ remain constant.

10. $\Delta Q = \Delta U + \Delta W, \Delta Q = 200\text{J}$ and $\Delta W = -100\text{J}$

$\Rightarrow$ $$\Delta U = \Delta Q - \Delta W = 200 - (-100) = 300\text{J}$$

11. $\Delta Q = \Delta U + \Delta W$

$\because$ Heat is supplied to the system so $\Delta Q \rightarrow$ positive and work is done on the system so $\Delta W \rightarrow$ negative

Hence, $$+\Delta Q = \Delta U - \Delta W$$

12. $dU = dQ - dW \;\Rightarrow\; dU = dQ\,(<0)$ $\quad(\because dW = 0)$

$dU < 0$, so temperature will decrease.

13. Work done $= p(V_2 - V_1)$

14. $\Delta U = C_V\,\Delta T = \dfrac{R}{\gamma - 1}(T_2 - T_1)$

15. Work done $= p\Delta V = p(V_2 - V_1)$

16. From first law of thermodynamics

$$\Delta Q = \Delta U + \Delta W = \Delta U + p\Delta V$$

$\Rightarrow$ $$100 = \Delta U + 50 \times (4 - 10)$$

$\Rightarrow$ $$\Delta U = 400\text{J}$$

17. Work done by the system = Area of shaded portion on p-V diagram $= (300 - 100)10^{-6} \times (200 - 100) \times 10^3 = 20\text{J}$

cycle is anti-clockwise. Hence, work done is negative.

18. Work done = Area enclosed by Δ

$$ABC = \frac{1}{2} AC \times BC$$

$$= \frac{1}{2} \times (3V - V) \times (3p - p) = 2pV$$

19. Heat given $\Delta Q = 20\,\text{cal} = 20 \times 4.2 = 84\text{ J}$

Work done $\Delta W = -50\text{J}$ (as process is anti-clockwise)

By first law of thermodynamics,

$\Rightarrow$ $$\Delta U = \Delta Q - \Delta W = 84 - (-50) = 134\text{J}$$

20. For cyclic process

Total work done $= W_{AB} + W_{BC} + W_{CA}$

$$\Delta W_{AB} = p\,\Delta V = 10(2 - 1) = 10\text{J}$$

and $$\Delta W_{BC} = 0 \qquad (\text{as } V = \text{constant})$$

From first law of thermodynamics, $\Delta Q = \Delta U + \Delta W$

$$\Delta U = 0 \qquad (\text{process } ABCA \text{ is cyclic})$$

$\Rightarrow$ $$\Delta Q = \Delta W_{AB} + \Delta W_{BC} + \Delta W_{CA}$$

$\Rightarrow$ $$5 = 10 + 0 + \Delta W_{CA}$$

$\Rightarrow$ $$\Delta W_{CA} = -5\text{J}$$

21. Cyclic process 1 is clockwise whereas process 2 is anti-clockwise. Since, negative area (2) > positive area (1), hence net work done is negative.

22. Work is done by the gas (as cyclic process is clockwise)

$\therefore$ $$\Delta W = \text{Area } ABCD$$

So, from the first law of thermodynamics ΔQ (net heat absorbed) $= \Delta W =$ Area $ABCD$

As change in internal energy in cycle $\Delta U = 0$.

23. **For path ab** $(\Delta U)_{ab} = 7000\text{J}$

By using $$\Delta U = nC_V\,\Delta T$$

$$7000 = n \times \frac{5}{2} R \times 700 \Rightarrow n = 0.48$$

For path ca

$$(\Delta Q)_{ca} = (\Delta U)_{ca} + (\Delta W)_{ca} \qquad \text{...(i)}$$

$\because$ $$(\Delta U)_{ab} + (\Delta U)_{bc} + (\Delta U)_{ca} = 0$$

$\therefore$ $$7000 + 0 + (\Delta U)_{ca} = 0$$

$\Rightarrow$ $$(\Delta U)_{ca} = -7000\text{J} \qquad \text{...(ii)}$$

Also, $(\Delta W)_{ca} = p_1(V_1 - V_2) = nR(T_1 - T_2)$

$$= 0.48 \times 8.31 \times (300 - 1000) = -2792.16\text{ J} \qquad \text{...(iii)}$$

By solving Eqs. (i), (ii) and (iii), we get

$$(\Delta Q)_{ca} = -7000 - 2792.16$$

$$= -9792.16 \approx -9800\text{ J}$$

24. ΔU, remains same of both path

For path iaf $$\Delta U = \Delta Q - \Delta W$$

$$= 50 - 20 = 30\text{ J}$$

For path fi $\Delta U = -30\text{J}$ and $\Delta W = -13\text{J}$

$\Rightarrow$ $$\Delta Q = -30 - 13 = -43\text{J}$$

25. Heat given, $\Delta Q = 40\text{J}$ and work done $\Delta W = 30\text{J}$

$\Rightarrow$ $$\Delta U = \Delta Q - \Delta W = 40 - 30 = 10\text{J}$$

26. During process A to B, pressure and volume both are decreasing. Therefore, temperature and internal energy of the gas will decrease $(T \propto pV)$ or $\Delta U_{A \rightarrow B}$ = negative. Further $\Delta W_{A \rightarrow B}$ is also negative as the volume of the gas is decreasing. Thus, $\Delta Q_{A \rightarrow B}$ is negative.

In process B to C, pressure of the gas is constant while volume is increasing. Hence, temperature should increase or $\Delta U_{C\to A}$ = positive. During process CAB volume of the gas is decreasing. Hence, work done by the gas is negative.

27. $\Delta W_{AB} = 0$ as V = constant

$\therefore \quad \Delta Q_{AB} = \Delta U_{AB} = 50\text{J}$ (given)

$U_A = 1500\text{J}$

$\therefore \quad U_B = (1500 + 50)\text{J} = 1550\text{J}$

$\Delta W_{BC} = -\Delta U_{BC} = -40\text{J}$ (given)

$\therefore \quad \Delta U_{BC} = 40\text{J}$

$\therefore \quad U_C = (1550 + 40)\text{J} = 1590\text{J}$

28. W = area under p-V diagram.

29. $Q_1 = nC_p\,\Delta T$

$Q_2 = nC_V\,\Delta T$

$\therefore \quad Q_1 - Q_2 = nR\,\Delta T \qquad (\because \Delta T = T_2 - T_1)$

or $\quad Q_2 = Q_1 - nR\,\Delta T = (70) - (2)\left(\frac{8.31}{4.18}\right)(35 - 30)$

$= 50$ cal

30. In adiabatic process, $TV^{\gamma-1}$ = constant

or $\quad T_2V_2^{\gamma-1} = T_1V_1^{\gamma-1}$ or $T_2 = \left(\frac{V_1}{V_2}\right)^{\gamma-1} T_1$

$= (4)^{1.4-1}\,(300) = (300)\,(4)^{0.4}$ K

Given the answer of 31 & 32

33. Given, $p \propto T^3$

In adiabatic process, $p^{1-\gamma}\,T^{\gamma}$ = constant

or $\quad p \propto T^{\frac{\gamma}{\gamma-1}}$

Compare with the given equation, we get

$$\frac{\gamma}{\gamma-1} = 3$$

Solving this equation we get, $\gamma = \frac{3}{2}$

34. In adiabatic process, $\Delta U = -\Delta W_1$

In the first process, $\Delta W_1 = +25\text{J}$

$\therefore \quad \Delta U_1 = -25\text{J}$

In second process, $\Delta W_2 = -100\text{J}$

$\therefore \quad \Delta U_2 = +100\text{J}$

35. Fast processes are adiabatic. Hence,

$$p_1V_1^{\gamma-1} = p_2V_2^{\gamma}$$

$\therefore \quad p_2 = p_1\left(\frac{V_1}{V_2}\right)^{\gamma} = p\left(\frac{640}{8}\right)^{5/3}$

$= p(8)^{5/3} = 32p$

38. In adiabatic process, pV^{γ} = constant

or $\quad p \propto \frac{1}{V^{\gamma}}$

39. Work done by the gas should be negative, but work done on the gas should be positive.

$W = p(V_i - V_f)$ (on the gas)

$= 10^5\,(3.0\times10^{-2} - 1.5\times10^{-2}) = +1500\text{J}$

41. In adiabatic process, $TV^{\gamma-1}$ = constant

$\therefore \quad \frac{T_1}{T_2} = \left(\frac{V_2}{V_1}\right)^{\gamma-1} = \left(\frac{AL_2}{AL_1}\right)^{5/3-1} = \left(\frac{L_2}{L_1}\right)^{2/3}$

42. W = area under p-V graph

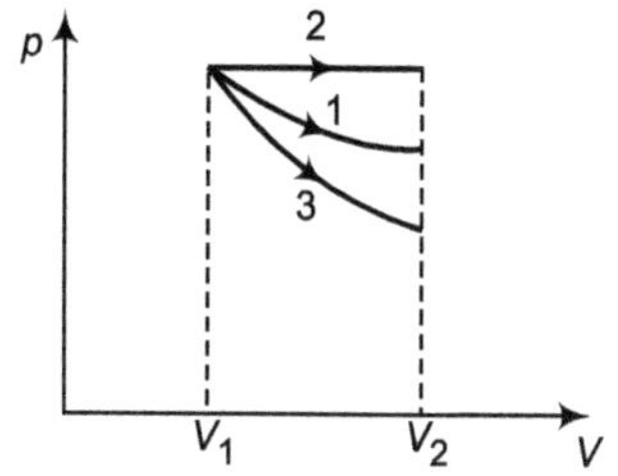

$A_2 > A_1 > A_3$

$\therefore \quad W_2 > W_1 > W_3$

43. In adiabatic process as γ increase slope of p-V graph increases. Slope of B is more. Therefore, for $\gamma_B > \gamma_A$. B should be monoatomic and A diatomic.

44. In isobaric process heat is used raising the temperature of the system.

45. $H = n_1C_V\,\Delta T_1$ and $H' = n_2C_p\,\Delta T_2$

or $\quad \frac{H'}{H} = \frac{n_2}{n_1}\frac{C_p}{C_V}\cdot\frac{\Delta T_2}{\Delta T_1}$

$= \frac{n_2\gamma}{n_1}\frac{\Delta T_2}{\Delta T_1}$

$= \frac{5}{3}\times\frac{5}{10}\times2 = \frac{5}{3}$

$\therefore \quad H' = \frac{5}{3}H$

46. $Q = mS\,\Delta\theta$

$\therefore \quad S = \frac{Q}{m\cdot\Delta\theta}$

In adiabatic process, $Q = 0$

$\therefore \quad S = 0$

47. During isothermal change T = constant $\Rightarrow \Delta U = 0$ also from first law of thermodynamics, $\Delta Q = \Delta W$.

48. Work done = Area under p-V diagram.

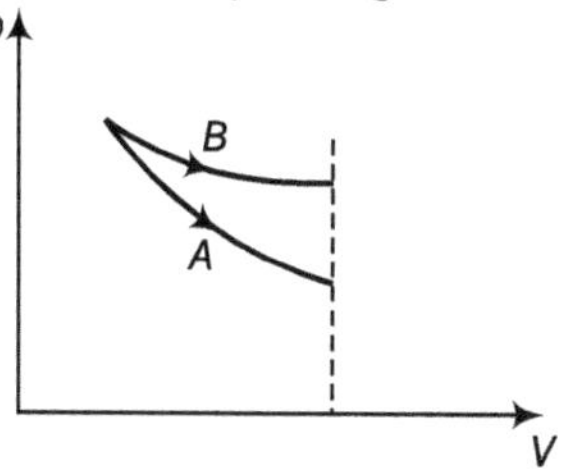

$(\text{area})_{\text{iso}} > (\text{area})_{\text{adi}}$

$\Rightarrow \quad W_{\text{iso}} > W_{\text{adi}}$

49. In adiabatic change, $\Delta Q = 0$

So, $\quad \Delta W = -\Delta U \qquad (\because \Delta Q = \Delta U + \Delta W)$

50. The process is very fast, so the gas fails to gain or loses heat. Hence, this process is adiabatic.

51. Using relation $\frac{T_2}{T_1} = \left(\frac{p_2}{p_1}\right)^{\frac{\gamma-1}{\gamma}} = (8)^{\frac{3/2-1}{3/2}} = 2$

$\Rightarrow \quad T_2 = 2T_1$

$\Rightarrow \quad T_2 = 2(273 + 27)$

$= 600\text{ K} = 327°\text{C}$

53. Given $p \propto T^3$, but we know for an adiabatic process, the pressure $p \propto T^{\gamma/\gamma-1}$.

So, $$\frac{\gamma}{\gamma-1} = 3 \Rightarrow \gamma = \frac{3}{2}$$

$$\Rightarrow \frac{C_p}{C_V} = \frac{3}{2}$$

54. $p_1V_1^\gamma = p_2V_2^\gamma \Rightarrow \frac{p_2}{p_1} = \left[\frac{V_1}{V_2}\right]^\gamma = \left[\frac{4}{1}\right]^{3/2} = \frac{8}{1}$

57. At constant volume, $p \propto T$

$$\Rightarrow \frac{p_1}{p_2} = \frac{T_1}{T_2}$$

$$\Rightarrow \frac{p_1}{p_2} = \frac{300}{400} = \frac{3}{4}$$

58. $\Delta Q = \Delta U + \Delta W$

$$\Rightarrow \Delta W = (\Delta Q)_p - \Delta U = (\Delta Q)_p \left[1 - \frac{(\Delta Q)_V}{(\Delta Q)_p}\right]$$

$$\begin{pmatrix}\text{in isobaric process} \\ \Delta U = (\Delta Q)\end{pmatrix}$$

$$= (\Delta Q)_p \left[1 - \frac{C_V}{C_p}\right] = Q = \left[1 - \frac{3}{5}\right] = \frac{2}{5}Q$$

$\because (\Delta Q)_p = Q$ and $\gamma = \frac{5}{3}$ for monoatomic gas.

59. $TV^{\gamma-1} =$ constant

$$\therefore \frac{T_1}{T_2} = \left(\frac{V_2}{V_1}\right)^{\gamma-1}$$

or $$\left(\frac{1}{2}\right)^{\gamma-1} = \sqrt{\frac{1}{2}}$$

$$\therefore \gamma - 1 = \frac{1}{2} \text{ or } \gamma = \frac{3}{2}$$

$$\therefore pV^{3/2} = \text{constant}$$

60. In isothermal process $p \propto \frac{1}{V}$.

Hence, graph between p and V is a hyperbola.

61. $\frac{W}{Q_1} = \left(1 - \frac{T_2}{T_1}\right)$ or $W = Q_1\left(1 - \frac{T_2}{T_1}\right)$

$$= 10000\left(1 - \frac{400}{500}\right) = 2000 \text{ J}$$

62. $1 - \frac{T_2}{T_1} = 0.5$ or $T_1 = 2T_2$

$$= 2(17 + 273) = 580 \text{ K}$$

Temperature of hot body is increased by 145°C or 145 K.

$$\therefore T_1' = (580 + 145) = 725 \text{ K}$$

and $$T_2 = (17 + 273) = 290 \text{ K}$$

$$\therefore \eta = \left(1 - \frac{290}{725}\right) \times 100 = 60\%$$

63. $\frac{W}{Q} = 1 - \frac{T_2}{T_1} = 1 - \frac{400}{800} = 0.5$

or $$Q = \frac{W}{0.5} = \frac{750}{0.5} = 1500 \text{ J}$$

64. Efficiency of Carnot engine, $\eta = \left(1 - \frac{T_2}{T_1}\right) \times 100$

$$= \left\{1 - \frac{(87 + 273)}{(127 + 273)}\right\} \times 100 = 10\%$$

66. $\eta = \frac{W}{Q_1} = \frac{Q_1 - Q_2}{Q_1}$

67. $\eta = 1 - \frac{T_2}{T_1}$, for η to be maximum ratio $\frac{T_2}{T_1}$ should be minimum.

68. $\eta_A = \frac{T_1 - T_2}{T_1} = \frac{W_A}{Q_1}$

$$\Rightarrow \eta_B = \frac{T_2 - T_3}{T_2} = \frac{W_B}{Q_2}$$

$$\therefore \frac{Q_1}{Q_2} = \frac{T_1}{T_2} \times \frac{T_2 - T_3}{T_1 - T_2} = \frac{T_1}{T_2}$$

$$\therefore W_A = W_B$$

$$\therefore T_2 = \frac{T_1 + T_3}{2}$$

$$= \frac{800 + 300}{2} = 550 \text{ K}$$

69. Initially, $\eta = \frac{T_1 - T_2}{T_1}$

$$\Rightarrow 0.5 = \frac{T_1 - (273 + 7)}{T_1}$$

$$\Rightarrow \frac{1}{2} = \frac{T_1 - 280}{T_1} \Rightarrow T_1 = 560 \text{ K}$$

Finally, $$\eta'_1 = \frac{T'_1 - T_2}{T'_1} \Rightarrow 0.7$$

$$= \frac{T'_1 - (273 + 7)}{T'_1} \Rightarrow T'_1 = 933 \text{ K}$$

$\therefore$ Increase in temperature $= 933 - 560 = 373$ K.

70. Processes A to B and C to D are parts of straight line graphs passing through origin.

$p \propto T$. So, volume remains constant for the graphs AB and CD. So, no work is done during processes for A to B and C to D

i.e. $$W_{AB} = W_{CD} = 0$$

and $$W_{BC} = p_2(V_C - V_B) = nR(T_C - T_B) = 6R(2200 - 800) = 6R \times 1400 \text{ J}$$

Also $$W_{DA} = p_1(V_A - V_D) = nR(T_A - T_B) = 6R(600 - 1200) = -6R \times 600 \text{ J}$$

Hence, work done in complete cycle

$$W = W_{AB} + W_{BC} + W_{CD} + W_{DA}$$
$$= 0 + 6R \times 1400 + 0 - 6R \times 600$$
$$= 6R \times 900 = 6 \times 8.3 \times 800 = 40 \text{ kJ}$$

73. Isothermal bulk modulus, $B = p$

74. $p_1V_1 = p_2V_2$ or $p \propto \frac{1}{V}$

Radius has become two times. Hence, volume will become eight times, i.e. pressure at surface is $\frac{1}{8}$ times the pressure at depth h. On surface pressure is equal to 10 m of water. Hence, pressure at depth h is equal to 80 m of water. Of this 10 m is atmospheric pressure. Therefore, h is 70 m.

75. $(n_1 + n_2)_i = (n_1 + n_2)_f$

or $$\frac{p_0V}{RT_0} + \frac{p_0V}{RT_0} = \frac{pV}{RT_0} + \frac{pV}{2RT_0}$$

or $$\frac{2p_0V}{RT_0} = \frac{3pV}{2RT_0}$$

$\therefore$ $$p = \frac{4}{3}p_0$$

76. Process is adiabatic, as $\Delta Q = 0$

$$p_1V_1^\gamma = p_2V_2^\gamma$$

$\therefore$ $$\frac{p_2}{p_1} = \left(\frac{V_1}{V_2}\right)^\gamma = (2)^{5/3}$$

77. In adiabatic process, pressure will be less. Further temperature is the product of p and V. Hence, $p_A < p_B$ and $T_A < T_B$.

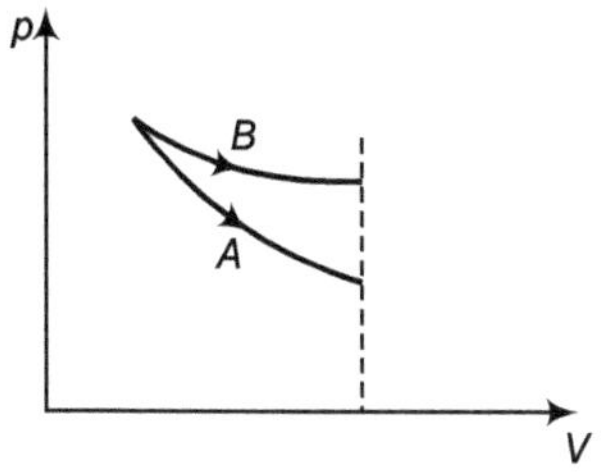

78. Vessel is insulated. Therefore, $Q = 0$

Against vacuum, $W = 0$

$\therefore$ ΔU or $\Delta T = 0$

79. In isothermal process p-V diagram is a rectangular hyperbola. $T_2 > T_1$. Hence, $(pV)_2 > (pV)_1$.

80. On heating volume will increase. Hence, option will move to the left.

81. At constant volume

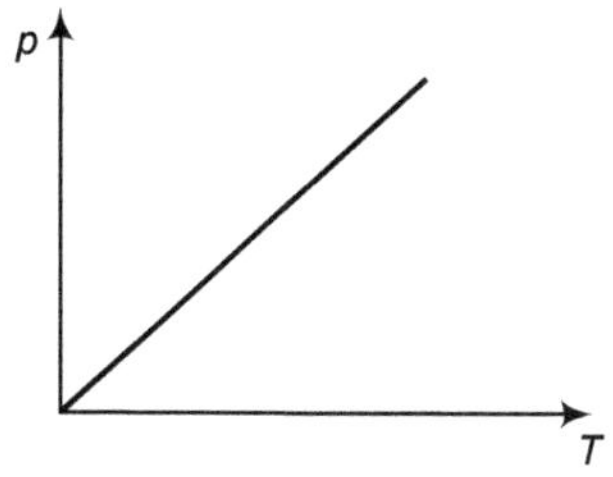

$$p \propto T$$

or $$p = \left(\frac{nR}{V}\right)T = \left(\frac{mR}{MV}\right)T$$

i.e. p-T graph is a straight line passing through origin. Slope of this line is proportional to m, mass of gas.

$\therefore$ $$\tan\alpha = 2\tan\beta$$

82. $pV^{2/3}$ = constant

$\therefore$ $$\left(\frac{T}{V}\right)\cdot V^{2/3} = \text{constant}$$

or $$TV^{-1/3} = \text{constant}$$

or $$T \propto V^{1/3}$$

with increase in V, temperature T will also increase.

83. This is the case of free expansion and in this case $\Delta W = 0$, $\Delta U = 0$ so temperature remains same, i.e. 300 K.

84. $\Delta U = C_V\,\Delta T = C_V(T_f - T_i) = -C_V\,(T_i - T_f)$

$\Rightarrow$ $$|\Delta U| = C_V(T_i - T_f)$$

85. For isobaric process $\frac{V_2}{V_1} = \frac{T_2}{T_1} \Rightarrow V_2 = V \times \frac{274}{273}$

$$\text{Increase} = \frac{274V}{273} - V = \frac{V}{273}$$

86. $\Delta Q = \Delta U + p\Delta V$

$\Rightarrow$ $$mL = \Delta U + p(V_2 - V_1)$$

$\Rightarrow$ $$\Delta U = L - p(V_2 - V_1) \qquad (\because m = 1)$$

87. Oxygen is diatomic gas, hence its energy of two moles

$$= 2 \times \frac{5}{2}RT = 5RT$$

Argon is a monoatomic gas, hence its internal energy of 4 moles $= 4 \times \frac{3}{2}RT = 6RT$

Total internal energy $= (6 + 5)\,RT = 11RT$

88. Volume of the gas V = constant

$\therefore$ $p \propto T$,

i.e. pressure will be doubled, if temperature is doubled

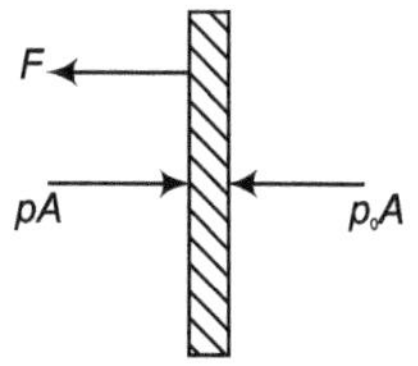

$\therefore$ $$p = 2p_0$$

Now, let F be the tension in the wire. Then equilibrium of any one piston gives.

$$F = (p - p_0)\,A = (2p_0 - p_0)\,A = p_0A$$

89. $dU = C_V\,dT = \left(\frac{5}{2}R\right)dT$ or $dT = \frac{2(dU)}{5R}$

From first law of thermodynamics

$$dU = dQ - dW = Q - \frac{Q}{4} = \frac{3Q}{4}$$

Now, molar heat capacity

$$C = \frac{dQ}{dT} = \frac{Q}{\frac{2(dU)}{5R}} = \frac{5RQ}{2\left(\frac{3Q}{4}\right)} = \frac{10}{3}R$$

90. Volume of the ideal gas is constant so $W = 0$

Using first law of thermodynamics

$$\Delta Q = \Delta U \;\Rightarrow\; \Delta U = i^2rt$$
$$= 1^2 \times 100 \times 5 \times 60 = 30 \times 10^3 = 30\text{ kJ}$$

91. pV^2 = constant

$\therefore$ $$pV \propto \frac{1}{V} \text{ or } T \propto \frac{1}{V}$$

So, when gas is expand them the temperature of the gas is decreases.

92. p-V diagram of the gas is a straight line passing through origin. Hence, $p \propto V$ or pV^{-1} = constant

Molar heat capacity in the process pV^x = constant is

$$C = \frac{R}{\gamma - 1} + \frac{R}{1 - x}; \quad \text{Here } \gamma = 1.4 \qquad \text{(for diatomic gas)}$$

$$\Rightarrow C = \frac{R}{1.4 - 1} + \frac{R}{1 + 1} \Rightarrow C = 3R$$

Level 2 : Only One Correct Option

1. Given $p=\frac{nRT}{V-b}$

$$W=\int_V^{2V} p\,dV = nRT\int_V^{2V}\frac{dV}{V-b} \quad \text{(as } T=\text{constant)}$$

$$= nRT\ln\left(\frac{2V-b}{V-b}\right)$$

$$= RT\ln\left(\frac{2V-b}{V-b}\right) \quad (\text{as } n=1)$$

2. Process AB is isochoric. Hence, $W_{AB}=0$

Process BC is isothermal. Hence,

$$W_{BC}=nRT_2\ln\left(\frac{V_f}{V_i}\right)$$

or $$W_{BC}=RT_2\ln\left(\frac{V_1}{V_2}\right)$$

Process CA is isobaric (as $V\propto T$)

$\therefore$ $$W_{CA}=p(V_f-V_i)=nR(T_f-T_i)$$
$$=R(T_1-T_2)$$

3. Net work done = Area ABC – Area AED

$$=\frac{1}{2}\times2\times1-\frac{1}{2}\times1\times1=\frac{1}{2}\text{J}$$

4. We know that $p=\frac{RT}{V}=\frac{R(T_0+\alpha V)}{V}$

$\therefore$ $$W=\int_{V_1}^{V_2}p\cdot dV=\int_{V_1}^{V_2}\frac{R(T_0+\alpha V)}{V}\cdot dV$$

$$=RT_0\ln\frac{V_2}{V_1}+R\alpha(V_2-V_1)$$

$$\Delta U = nC_V\,\Delta T$$
$$=(C_p-R)[(T_0+\alpha V_2)-(T_0+\alpha V_1)]$$
$$=\alpha(C_p-R)(V_2-V_1)$$

Now, $$Q=W+\Delta U$$
$$=RT_0\ln\frac{V_2}{V_1}+\alpha C_p(V_2-V_1)$$

5. Given $$dQ=2\cdot dU \quad \text{(given)}$$

$\therefore$ $$nC\cdot dT=2(nC_V\,dT)$$

or $$C=2C_V=2\times\frac{3}{2}R=3R$$

6. In process AB, density is constant. Hence, volume is constant. Therefore, work done will be zero.

In process BC, $p\propto\rho$ but $\rho\propto\frac{1}{V}$

$\therefore$ $$p\propto\frac{1}{V}$$

Or the process is isothermal. Hence, internal energy is constant.

7. W_{total} = Area under cycle = p_0V_0

Positive heat $=nC_V(T_B-T_A)+nC_p(T_C-T_B)$

$$=n\left(\frac{3}{2}R\right)(T_B-T_A)+n\left(\frac{5}{2}R\right)(T_C-T_B)$$

$$=\frac{3}{2}(nRT_B-nRT_A)+\frac{5}{2}(nRT_C-nRT_B)$$

$$=\frac{3}{2}(p_BV_B-p_AV_A)+\frac{5}{2}(p_CV_C-p_BV_B)$$

$$=\frac{3}{2}(2p_0V_0-p_0V_0)+\frac{5}{2}(4p_0V_0-2p_0V_0)$$

$$=6.5\,p_0V_0$$

$\therefore$ $$\eta=\frac{W_{\text{total}}}{Q_{+\text{ve}}}\times100$$

$$\approx15\%$$

8. In a cyclic process, $\Delta U=0$

$\therefore$ $$Q_{\text{total}}=W_{\text{total}}$$

or $$10-5=-3+W_{21}$$

or $$W_{21}=8\text{J}$$

9. V-T graph is a straight line passing through origin or $V\propto T$. Hence, the given process is isobaric.

$$\frac{\Delta Q}{\Delta W}=\frac{nC_p\,\Delta T}{nR\Delta T}$$
$$=\frac{C_p}{R}$$
$$=\frac{5}{2}$$
$$=2.5$$

10. $V=kT^2$

$\therefore$ $$2kT\cdot dT$$

and $$p=\frac{RT}{V}=\frac{RT}{kT^2}=\frac{R}{kT}$$

or $$p\,dV=2R\,dT$$
$$W=\int p\,dV=2R\int dT$$
$$=2R(60)$$
$$=120R$$

11. VT = constant, $V(pV)$ = constant (as $T\propto pV$)

$\therefore$ $$pV^2=\text{constant}$$

Molar heat capacity of a gas in the process

$$pV^x=\text{constant}$$
$$C=C_V+\frac{R}{1-x}$$

Here, $$x=2$$
$$C=\frac{3}{2}R+\frac{R}{1-2}$$
$$C=R$$

12. In adiabatic process $\Delta U=-W$

$\therefore$ $$U_A-U_B=-(-30)=30\text{J}$$

or $$U_B-U_A=-30\text{J}$$

13. $T=kV^2$

$\therefore$ $$dT=(2kV)\cdot dV$$

or $$dV=\frac{dT}{2kV}\quad p=\frac{nRT}{V}$$

$$p\,dV=\frac{nRT}{2kV^2}\cdot dT=\frac{nRT}{2(T)}\cdot dT=\frac{nR}{2}\cdot dT$$

$$W=\int p\,dV=\int_{T_0}^{4T_0}\frac{nR}{2}\cdot dT=\frac{3}{2}nRT_0$$

14. $W_{A\to B}$ = Area under p-V diagram $=18p_0V_0$

$$\Delta U_{A\to B}=nC_V\Delta T$$

$$=(1)\left(\frac{3}{2}R\right)\left(\frac{30p_0V_0}{R}-\frac{3p_0V_0}{R}\right)=\frac{81}{2}p_0V_0$$

$$\therefore Q_{A\to B}=W_{A\to B}+\Delta U_{A\to B}=\frac{117}{2}p_0V_0$$

Molar heat capacity $C=\dfrac{Q}{n\Delta T}$

$$=\frac{\frac{117}{2}p_0V_0}{\left(\frac{30p_0V_0}{R}\right)-\left(\frac{3p_0V_0}{R}\right)}=\frac{13}{6}R$$

15. Number of moles $=\dfrac{2}{4}=\dfrac{1}{2}$

$W_{A\to B}$ = Area under p-V graph

$$=\frac{1}{2}\times 3p_0\times V_0=\frac{3}{2}p_0V_0$$

$$\Delta U_{A\to B}=nC_V\Delta T=\frac{1}{2}\left(\frac{3}{2}R\right)\left(\frac{8p_0V_0}{R}-\frac{2p_0V_0}{R}\right)=\frac{9}{2}p_0V_0$$

$$Q_{A\to B}=W_{A\to B}+\Delta U_{A\to B}=6p_0V_0$$

16. In cyclic process $\Delta U=0$

$\therefore$ Heat absorbed = work done

$$= \text{area under } p\text{-}V \text{ graph}$$

$$=(\pi)(R_x)(R_y)$$

$$=\pi(10\times10^3)(10\times10^{-3})$$

$$=(10^2\pi)\,\text{J}$$

17. ΔU = same in all processes.

$$W_1=+\text{ve}, W_2=0$$

and $\quad W_3=-\text{ve}$

$$\therefore \quad Q_1>Q_2>Q_3$$

18. $Q=W+\Delta U$

$$=p_0(2V_0-V_0)+[2p_0\cdot 2V_0-2p_0V_0]$$

$$=p_0V_0+2p_0V_0=3p_0V_0$$

19. $\Delta Q_1<\Delta Q_2$

$$\therefore \quad \frac{\Delta Q_1}{\Delta T}<\frac{\Delta Q_2}{\Delta T} \quad\text{or}\quad C_1<C_2 \quad\text{or}\quad \frac{C_1}{C_2}<1$$

20. $U\rho$ = constant

$$\therefore \quad \frac{T}{V}=\text{constant}$$

$$\therefore \quad p=\text{constant}$$

$$\frac{dU}{dW}=\frac{dU}{dQ-dW}=\frac{C_V}{C_p-C_V}$$

$$=\frac{1}{C_p/C_V-1}=\frac{1}{(5/3)-1}=\frac{3}{2}$$

21. $\Delta U_1=\Delta U_2$ as U is state function.

$$\therefore \quad Q_1-W_1=Q_2-W_2 \quad\text{or}\quad W_2=Q_2+W_1-Q_1$$

$$=(10^5)+(6.5\times10^5)-(8\times10^5)$$

$$=-0.5\times10^5\text{ J}$$

22. $\Delta U=0$

$\therefore T$ = constant or $p\propto\dfrac{1}{V}$

So, the graph is represented by hyperbola

23. Consider the p-V diagram shown for the container A (isothermal) and for container B (adiabatic).

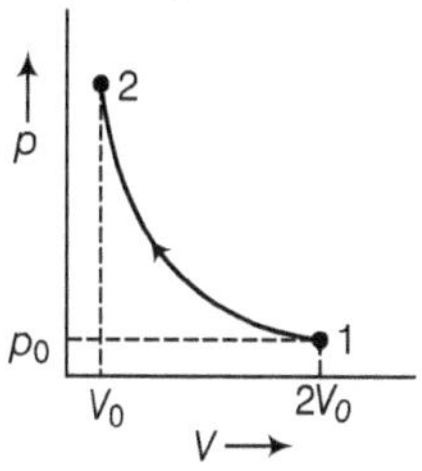

Container A
(Isothermal)

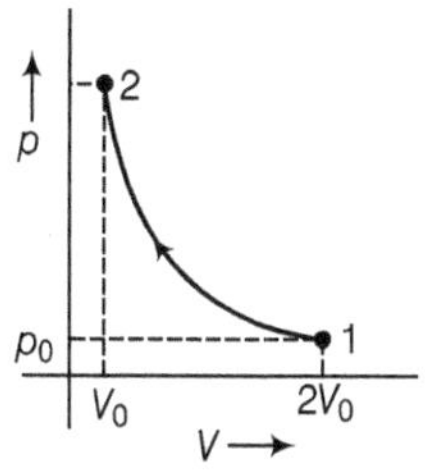

Container B
(Adiabatic)

Both the process involving compression of the gas.

For isothermal process (gas A) (during $1\to2$)

$$p_1V_1=p_2V_2 \;\Rightarrow\; p_0(2V_0)=p_2(V_0)$$

$$\Rightarrow \quad p_2=2p_0$$

For adiabatic process, (gas B) (during $1\to2$)

$$p_1V_1^{\gamma}=p_2V_2^{\gamma}$$

$$\Rightarrow \quad p_0(2V_0)^{\gamma}=p_2(V_0)^{\gamma}$$

$$\Rightarrow \quad p_2=\left(\frac{2V_0}{V_0}\right)^{\gamma}p_0=(2)^{\gamma}p_0$$

Hence, $\dfrac{(p_2)_B}{(p_2)_A}$ = Ratio of final pressure $=\dfrac{(2)^{\gamma}p_0}{2p_0}=2^{\gamma-1}$

where, γ is ratio of specific heat capacities for the gas.

More than One Correct Options

1. Now see the hint of Q-No 14 (d) of subjective questions for JEE mains.

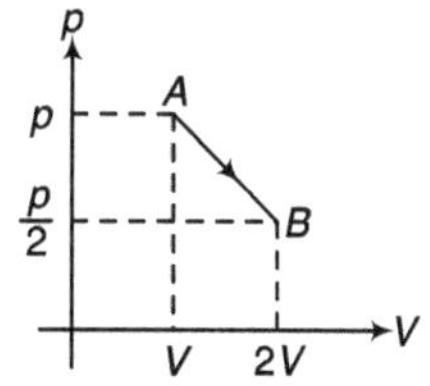

2. Temperature is increases. So, internal energy will also increase.

$$\therefore \quad \Delta U=+\text{ve}$$

Further,

$$pV^2=\text{constant}$$

$$\therefore \quad \left(\frac{T}{V}\right)V^2=\text{constant}$$

$$\text{or} \quad V\propto\frac{1}{T}$$

Temperature is increased. So, volume will decrease and work done will be negative.

In the process pV^x = constant, molar heat capacity is given by

$$C=C_V+\frac{R}{1-x}$$

Here, $x=2$

$$\therefore \quad C=C_V-R$$

C_V of any gas is greater than R.

So, C is positive. Hence, from the equation,

$$Q = nC\Delta T$$

Q is positive, if T is increased.

or, ΔT is positive.

3. Process in a-b, $W = 0$ (as V = constant)

$$Q_1 = \Delta U_1 = nC_V \Delta T$$
$$= (2)\left(\frac{3}{2}R\right)(2T_0 - T_0) = 3RT_0$$

Process in b-c, $\Delta U = 0$ (as T = constant)

$$\therefore \quad Q_2 = W_2 = nRT \ln\left(\frac{V_f}{V_i}\right)$$
$$= (2)(R)(2T_0)\ln(2)$$
$$= 4RT_0 \ln(2)$$

Process in c-d, $W = 0$

$$Q_3 = \Delta U_3 = nC_V \Delta T$$
$$= (2)\left(\frac{3}{2}R\right)(T_0 - 2T_0) = -3RT_0$$

Process in d-a, $\Delta U = 0$

$$Q_4 = W_4 = nRT \ln\left(\frac{V_f}{V_i}\right)$$
$$= (2)(R)(T_0)\ln\left(\frac{1}{2}\right)$$
$$= -2RT_0 \ln(2)$$

Now in complete cycle,

$$\Delta U_{net} = 0$$
$$Q_{net} = W_{net} = 2RT_0 \ln(2) = +\text{ve}$$

4. Process in a-b, ρ = constant

$\therefore \quad V$ = constant

$\therefore \quad W = 0$

$Q = \Delta U$

ΔU is positive, as U is increasing.

Hence, Q is also positive.

Process in b-c, $\quad \rho \propto u$

$$\therefore \quad \frac{1}{V} \propto T$$

ρ is decreasing, so V is increasing. Hence, work done is positive.

Further, $\quad \frac{1}{V} \propto T \; (T \propto pV)$

$\therefore \quad pV^2$ = constant

In the process, pV^x = constant,

Molar heat capacity is given by

$$C = C_V + \frac{R}{1-x}$$

Here $x = 2$

$$\therefore \quad C = C_V - R$$

For any of the gas, $C_V \neq R$.

$$\therefore \quad C \neq 0$$

$$\therefore \quad Q = nC\Delta T \neq 0 \text{ as } \Delta U \neq 0 \text{ and } \Delta T \neq 0$$

Process in c-a,

ρ is increasing. Hence, V is decreasing. So, work done is negative.

5. $Q_1 = nC_V \Delta T$

$$Q_2 = nC_p \Delta T$$
$$Q_3 = 0$$
$$C_P > C_V \quad \therefore \quad Q_2 > Q_1 > Q_3$$

6. Let the equilibrium temperature of the system is T.

Let us assume that $T_1, T_2 < T < T_3$.

According to question, there is no net loss to the surroundings.

Heat lost by M_3 = Heat gained by M_1 + Heat gained by M_2

$\Rightarrow \quad M_3 S(T_3 - T) = M_1 S(T - T_1) + M_2 S(T - T_2)$

(where, S is specific heat of the copper material)

$\Rightarrow \quad T[M_1 + M_2 + M_3] = M_3 T_3 + M_1 T_1 + M_2 T_2$

$$\Rightarrow \quad T = \frac{M_1 T_1 + M_2 T_2 + M_3 T_3}{M_1 + M_2 + M_3}$$

7. (a) The given process is a cyclic process i.e. it returns to the original state 1.

Hence, change in internal energy $dU = 0$

$$\Rightarrow \quad dQ = dU + dW = 0 + dW = dW$$

Hence, total heat supplied is converted to work done by the gas (mechanical energy) which is not possible by second law of thermodynamics.

(c) When the gas expands adiabatically from 2 to 3. It is not possible to return to the same state without being heat supplied, hence the process 3 to 1 cannot be adiabatic.

8. Consider the figure, we can write $Q_1 = W + Q_2$

$\Rightarrow \quad W = Q_1 - Q_2 > 0$ (by question)

$\Rightarrow \quad Q_1 > Q_2 > 0$ (If both Q_1 and Q_2 are positive)

We can also, write $Q_2 < Q_1 < 0$ (If both Q_1 and Q_2 are negative).

Comprehension Based Questions

1. $Q_{net} = W_{net}$ = area under the cycle $= \frac{pV}{2}$

2. $\frac{C_V}{C_p} = \frac{1}{\gamma} = \frac{5}{3}$

3. In adiabatic process,

$$p^{1-\gamma}T^{\gamma} = \text{constant} \quad \text{or} \quad T \propto p^{\frac{\gamma-1}{\gamma}}$$

$$\therefore \quad \frac{T_B}{T_A} = \left(\frac{p_B}{p_A}\right)^{\frac{\gamma-1}{\gamma}}$$

$$\therefore \quad T_B = (1000)\left(\frac{2}{3}\right)^{\frac{5/3-1}{5/3}} = (1000)\left(\frac{2}{3}\right)^{0.4}$$
$$= 850 \text{ K}$$

4. In adiabatic process

$$W_{AB} = -\Delta U_{AB} \quad (\text{as } Q = 0)$$
$$= nC_V(T_A - T_B)$$
$$= (1)\left(\frac{3}{2}R\right)(T_A - T_B)$$
$$= (1)\left(\frac{3}{2} \times \frac{25}{3}\right)(1000 - 850)$$
$$= 1875 \text{ J}$$

5. $W_{BC} = 0$ (as V = constant)

$\therefore \quad T \propto p$ (as V = constant)

$p_C = p_{B/2}$

$\therefore \quad T_C = T_{B/2} = 425\,K$

$Q = \Delta U = nC_V(\Delta T)$

$= (1)\left(\frac{3}{2}R\right)(T_C - T_B)$

$= \frac{3}{2} \times \frac{25}{3} \times (425 - 850)$

$= 5312.5\,J$

Assertion and Reason

1. For isobaric process also T-t graph under given condition may be a straight line. Replace C_V by C_p.
2. $\frac{\Delta Q}{\Delta W} = \frac{nC_p\Delta T}{nR\Delta T} = \frac{C_p}{R} = \frac{5/2R}{R} = \frac{5}{2}$
3. In adiabatic expansion gas is cooled down. Therefore, temperature decreases or product of p and V will decrease. Because, $T \propto pV$
4. In isochoric process p and V both are constants. Therefore, p-V graph is a dot.
5. For given source and sink temperatures efficiency of Carnot engine is maximum but not 100%.
6. For given temperatures T_1 and T_2 efficiency of Carnot engine is maximum.
7. Volume is increasing. Therefore, W is positive. Therefore, dU is negative.

   Now, $Q = W + \Delta U$

   Therefore, Q may be positive, negative or zero.

   Further, heat given to gas in both function.
8. First law can be applied for any type of system.
9. Area under p-V graph in process 1 is more.

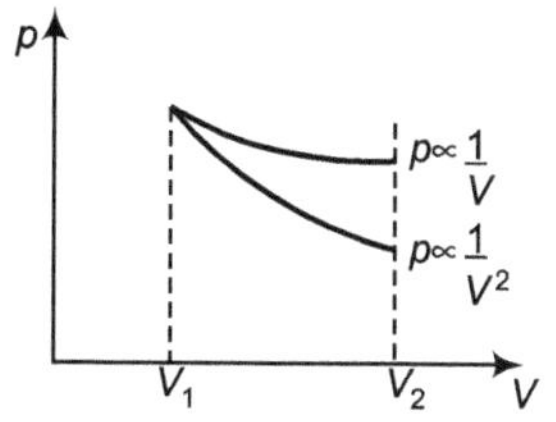

12. $C = \frac{Q}{n\Delta T}$

   In isothermal process, $\Delta T = 0$

   Therefore, C is not defined. Further, in isothermal process pV and T both are constants.

   Therefore, p-V *versus* T graph is a dot.
13. If initial and final volumes are equal, then work done may or may not be zero. For work done by the gas to be zero volume should remain constant throughout the process. For example in isochoric process.
14. In adiabatic expansion,

   $Q = 0$

   $W = +\text{ ve}$

   $\therefore \quad \Delta U = -\text{ ve}$

   Therefore U, T or product of pV will decrease.

   Because, $U \propto T \propto pV$

Match the Columns

1. **Process** a-b $\quad T$ = constant

   $\therefore \quad \Delta T = 0$ or $\Delta U = 0$

   $p \propto \frac{1}{V}$

   p is increasing. Therefore, V is decreasing $V \propto \frac{1}{p}$ or p will increase.

   Process b-c $\quad p$ = constant $\Rightarrow \therefore V$ = constant

   Therefore, $p \propto T$

   Since, temperature is increasing. Hence, pressure should also increase.
2. **In cyclic process** $\quad \Delta U = 0 \Rightarrow \therefore \Delta Q = \Delta W$

   In isobaric process

   $\Delta W = \Delta Q - \Delta U$

   $= nC_p\Delta T - nC_V\Delta T = nR\Delta T$

   In isochoric process $\quad \Delta W = 0, \Rightarrow \therefore \Delta Q = \Delta U$

   In adiabatic expansion $\Delta Q = 0$

   $\therefore \quad \Delta U = -\Delta W$ or $\Delta U < 0$

   As, work done in expansion is positive.
3. $B_T = p = \frac{RT}{V}$ (as $n = 1$)

   $B_S = \gamma p = \frac{5}{3}\frac{RT}{V}$

   Slope of p-V graph is isothermal process $= -\frac{p}{V} = \frac{-RT}{V^2}$

   and Slope of p-V graph in adiabatic process

   $= -\frac{\gamma p}{V} = -\frac{5}{3}\frac{RT}{V^2}$

Entrance Gallery

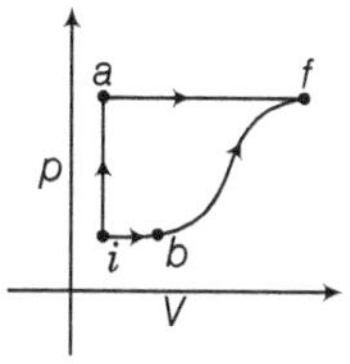

1. Given, $W_{ibf} = 150\,J$

   $W_{iaf} = 200\,J \Rightarrow Q_{iaf} = 500\,J$

   So, $U_{iaf} = 300\,J \Rightarrow U_f = 400\,J$

   $U_{ib} = 100\,J$

   $Q_{ib} = 100 + 50 = 150\,J$

   $Q_{ibf} = 300 + 150 = 450\,J$

   So, the required ratio $\frac{Q_{bf}}{Q_{ib}} = \frac{450 - 150}{150} = \frac{2}{1}$
2. For diatomic gas, $C_V = \frac{5R}{2}$

   $\Delta U = nC_V\Delta T = 1 \times \frac{5R}{2}\Delta T$

   For BC, $\quad \Delta T = -200\,K \Rightarrow \Delta U = -500\,R$
3. By Carnot ideal heat engine,

   $\eta = 1 - \frac{T_2}{T_1}$

   where, $\eta_1 = 0.2, \eta_2 = 0.4$

   For the first condition,

   $\eta_1 = 1 - \frac{T_2}{T_1} \Rightarrow 0.2 = 1 - \frac{T_2}{T_1}$

   or $\quad 0.2 = \frac{T_1 - T_2}{T_1}$

For the second condition

$$\eta_2 = 1 - \frac{T_2 - 50}{T_1} \Rightarrow 0.4 = \frac{T_1 - (T_2 - 50)}{T_1}$$

$$0.4 = \frac{T_1 - T_2 + 50}{T_1} \Rightarrow 0.4 = \frac{T_1 - T_2}{T_1} + \frac{50}{T_1}$$

$$0.4 = 0.2 + \frac{50}{T_1} \quad \text{[from Eq. (i)]}$$

$$0.4 - 0.2 = \frac{50}{T_1}$$

$$T_1 = \frac{50}{0.2} \Rightarrow T_1 = 250$$

Putting the value of T_1 in Eq. (i), we get

$$0.2 = \frac{T_1 - T_2}{T_1} \Rightarrow T_2 = T_1 - 0.2T_1$$

$$T_2 = 250 - 0.2 \times 250$$

$$T_2 = 250 - 50 \Rightarrow T_2 = 200$$

So, $T_1 = 250 \quad T_2 = 200\text{ K}$

4. Given, efficiency, $\eta = 70\%$

Sink temperature, $T_2 = 27 \quad 273 = 300\text{ K}$

Source temperature, $T_1 = ?$

As, we know that efficiency is given by

$$\eta = 1 - \frac{T_2}{T_1} \Rightarrow 70\% = 1 - \frac{300}{T_1}$$

$$\frac{70}{100} = 1 - \frac{300}{T_1} \Rightarrow 0.7 = 1 - \frac{300}{T_1}$$

$$\frac{300}{T_1} = 1 - 0.7$$

$$\Rightarrow \quad T_1 = \frac{3000}{3}$$

$$T_1 = 1000$$

or $T_1 = 1000 - 273 = 727°\text{C}$

5. In a tyre burst, there is too little time for the temperature to be equalized with the surroundings the work done due to the sudden expansion causes the air surrounding the tyre to get cooler. It is adiabatic, because no heat transfer occurs here most of the processes that take very little time is adiabatic.

6. In $F \to G$, work done in isothermal process is

$$nRT \ln\left(\frac{V_f}{V_i}\right) = 32\, p_0 V_0 \ln\left(\frac{32V_0}{V_0}\right)$$

$$= 32 p_0 V_0 \ln 2^5 = 160\, p_0 V_0 \ln 2$$

In $G \to E$, $\Delta W = p_0 \Delta V = p_0(31V_0) = 31\, p_0 V_0$

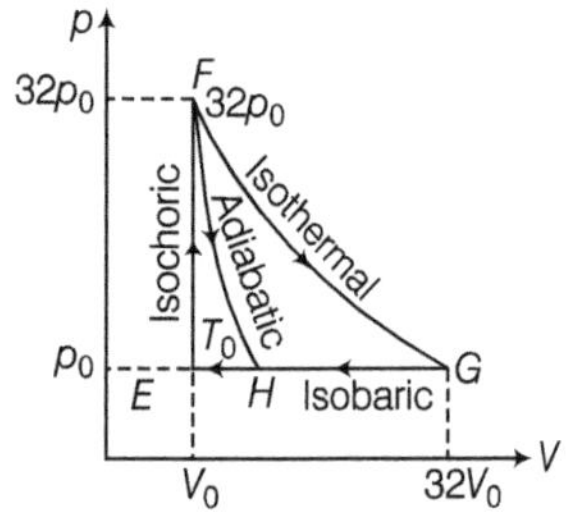

In $G \to H$, work done is less than $31\, p_0 V_0$, i.e. $24\, p_0 V_0$

In $F \to H$, work done is $36\, p_0 V_0$

7. Heat is extracted from the source means heat is given to the system (or gas) or Q is positive. This is positive only along the path ABC.

Heat supplied,

$$\therefore \quad Q_{ABC} = \Delta U_{ABC} + W_{ABC}$$

$$= nC_V(T_f - T_i) + \text{Area under } p\text{-}V \text{ graph}$$

$$\left[\text{for monoatomic gas, } C_V = \frac{3R}{2}\right]$$

$$= n\left(\frac{3}{2}R\right)(T_C - T_A) + 2p_0V_0$$

$$= \frac{3}{2}(nRT_C - nRT_A) + 2p_0V_0$$

$$= \frac{3}{2}(p_CV_C - p_AV_A) + 2p_0V_0$$

$$= \frac{3}{2}(4p_0V_0 - p_0V_0) + 2p_0V_0 = \frac{13}{2}p_0V_0$$

8. The work done during the process from P to Q = area $PQRSTP$ (positive sign is to be taken due to expansion along PQ)

Area of triangle PQR + area of rectangle $PRST$

$$= \frac{1}{2} \times 2V \times 2p + p \times 2V = 4pV$$

Work done during process from R to P = – Area $RSTP$ (negative sign due to compression in atom RP)

$$= -ST \times PT = -2V \times 1p = -2pV$$

Hence, the work done in the complete cycle

$$= 4pV - 2pV = 2pV$$

9. Efficiency of Carnot engine, $\eta = \frac{T_1 - T_2}{T_1}$

The value of efficiency will be higher, if $T_1 - T_2$ is more. As $T_1 - T_2$ is constant in all four cases. Therefore, efficiency will be highest when T_1 is the lowest.

10. (a) Curve OA represents isobaric process (since pressure is constant). Since, the slope of adiabatic process is more steeper than the isothermal process.

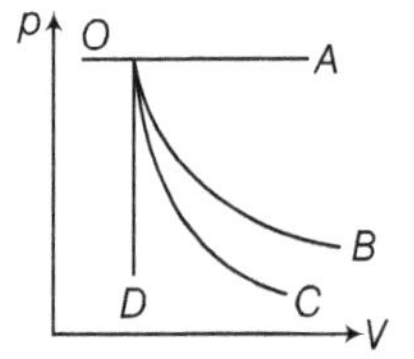

(b) Curve OB represents isothermal process.

(c) Curve OC represents adiabatic process.

(d) Curve OD represents isochoric process (since volume is constant)

11. Change in internal energy of the system does not depend on path followed.

So, $\Delta U_1 = \Delta U_2$

12. The process may be assumed to be isobaric.

$$\therefore \quad Q = nC_p\Delta T$$

$$= (2)\left(\frac{5}{2}R\right)(5) = 5 \times 8.31 \times 5 = 207.75\text{ J} = 208\text{ J}$$

13. Efficiency of a process is defined as the ratio of work done to energy supplied. Here,

$$\eta = \frac{\Delta W}{\Delta Q} = \frac{\text{Area under } p\text{-}V \text{ diagram}}{\Delta Q_{AB} + \Delta Q_{BC}}$$

$$\therefore \quad \eta = \frac{p_0V_0}{nC_V\Delta T_1 + nC_p\Delta T_2}$$

$$= \frac{p_0V_0}{\frac{3}{2}nR(T_B - T_A) + \frac{5}{2}nR(T_C - T_D)}$$

$$= \frac{p_0V_0}{\frac{3}{2}(2p_0V_0 - p_0V_0) + \frac{5}{4}(4p_0V_0 - 2p_0V_0)}$$

$$= \frac{p_0V_0}{\frac{3}{2}p_0V_0 + \frac{5}{2}\cdot 2p_0V_0} = \frac{1}{6.5} = 15.4\%$$

14. At STP,

22.4 L of any gas is 1 mol,

$$\therefore \quad 5.6\ \text{L} = \frac{5.6}{22.4} = \frac{1}{4}\text{mol} = n$$

In adiabatic process,

$TV^{\gamma-1}$ = constant

$$\therefore \quad T_2V_2^{\gamma-1} = T_1V_1^{\gamma-1} \quad \text{or} \quad T_2 = T_1\left(\frac{V_1}{V_2}\right)^{\gamma-1}$$

For monoatomic He gas, $\gamma = \frac{C_p}{C_V} = \frac{5}{3}$

$$\therefore \quad T_2 = T_1\left(\frac{5.6}{0.7}\right)^{\frac{5}{3}-1} = 4T_1$$

Further in adiabatic process,

$$Q = 0 \Rightarrow W + \Delta U = 0$$

or

$$W = -\Delta U = -nC_V\Delta T$$

$$= -n\left(\frac{R}{\gamma - 1}\right)(T_2 - T_1)$$

$$= -\frac{1}{4}\left(\frac{R}{\frac{5}{3}-1}\right)(4T_1 - T_1) = -\frac{9}{8}RT_1$$

15. $\eta_1 = 1 - \frac{T_2}{T_1} \Rightarrow \frac{1}{6} = 1 - \frac{T_2}{T_1}$

$$\Rightarrow \quad \frac{T_2}{T_1} = \frac{5}{6} \quad \dots\text{(i)}$$

$$\eta_2 = 1 - \frac{(T_2 - 62)}{T_1}$$

$$\Rightarrow \quad \frac{1}{3} = 1 - \frac{(T_2 - 62)}{T_1} \quad \dots\text{(ii)}$$

On solving Eqs. (i) and (ii), we get

$$T_1 = 372 \quad \text{and} \quad T_2 = 310$$

16. Heat required to change the temperature of vessel by a small amount dT

$$-dQ = mC_p dT$$

Total heat required, $-Q = m\int_{20}^{4} 32\left(\frac{T}{400}\right)^3 dT$

$$= \frac{100 \times 10^{-3} \times 32}{(400)^3}\left[\frac{T^4}{4}\right]_{20}^{4}$$

$$\Rightarrow \quad Q = 0.001996\ \text{kJ} \approx 0.002\ \text{kJ}$$

Work done required to maintain the temperature of sink to T_2

$$W = Q_1 - Q_2 = \frac{Q_1 - Q_2}{Q_2}Q_2$$

$$= \left(\frac{T_1}{T_2} - 1\right)Q_2 = \left(\frac{T_1 - T_2}{T_2}\right)Q_2$$

For $T_1 = 20$ K,

$$W_1 = \frac{300 - 20}{20} \times 0.001996 = 0.028\ \text{kJ}$$

For $T_2 = 4$ K,

$$W_2 = \frac{300 - 4}{4} \times 0.001996 = 0.148\ \text{kJ}$$

As temperature is changing from 20 K to 4 K, work done required will be more than W_1 but less than W_2.

17. Carnot efficiency,

$$\eta = \frac{T_1 - T_2}{T_1}$$

Case **I**

$$\frac{T_1 - T_2}{T_1} = 0.4$$

$$T_1 - T_2 = 0.4T_1$$

$$\Rightarrow \quad T_2 = 0.6T_1$$

Case **II**

$$\frac{T_1' - T_2}{T_1'} = 0.5$$

$$T_1' = \frac{0.6}{0.5}T_1 = 600$$

18. As work done, $\Delta W = 0$, then there is no expansion of the water.

$$\Delta Q = \Delta U + \Delta W$$

$$\Delta U = mc\Delta T$$

$$= 100 \times 10^{-3} \times 4184 \times (50 - 30) = 8.4\ \text{kJ}$$

19. In case of no work done, $W = 0$, then volume expansion $V = 0$. So, the volume remains zero $V = 0$. This process is called isochoric process.

20. $n = n_1 + n_2$

$$\Rightarrow \quad pV = p_1V_1 + p_2V_2$$

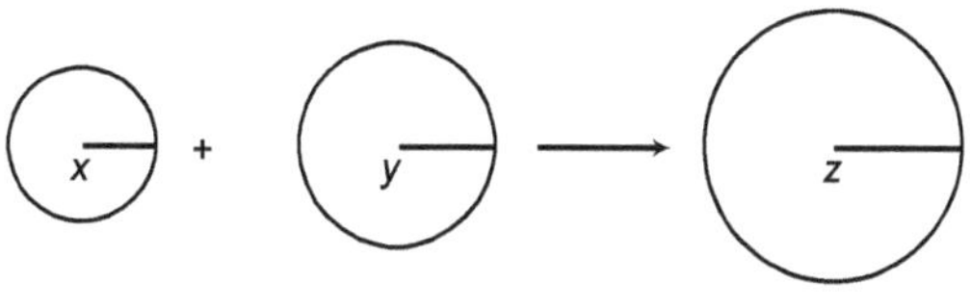

$$p_1 = p_0 + \frac{4T}{x}, \quad p_2 = p_0 + \frac{4T}{y}, \quad p = p_0 + \frac{4T}{z}$$

where, T is surface tension,

If the process takes place is vacuum, then $p_0 = 0$

$$p_1 = \frac{4T}{x}, \quad p_2 = \frac{4T}{y}, \quad p = \frac{4T}{z}$$

The process is isothermal

$$\therefore \quad p_1V_1 + p_2V_2 = pV$$

$$\Rightarrow \quad z = \sqrt{x^2 + y^2}$$

21. $T_A = T_B$

$$\therefore \quad U_A = U_B$$

$$W_{AB} = (1)(R)T_0 \ln\left(\frac{V_f}{V_i}\right)$$

$$= RT_0 \ln\left(\frac{4V_0}{V_0}\right) = p_0V_0 \ln 4$$

Information regarding p and T at C cannot be obtained from the given graph. Unless it is mentioned that line BC passes through origin or not. Hence, the correct options are (a) and (b).

22. In adiabatic process,

$$TV^{\gamma-1} = \text{constant}$$

$\therefore$ $T_i V_i = T_f V_f$ (as $\gamma = 1.4$ for diatomic gas)

or $$T_i V_i^{0.4} = (aT_i)\left(\frac{V_i}{32}\right)^{0.4}$$

or $$a = (32)^{0.4} = 4$$

23. The efficiency of engine is

$$\eta = 1 - \frac{T_2}{T_1}$$

For adiabatic process,

$TV^{\gamma-1} = \text{constant}$

For diatomic gas, $\gamma = \frac{7}{5}$

$$T_1 V_1^{\gamma-1} = T_2 V_2^{\gamma-1}$$

$\Rightarrow$ $$T_1 = T_2\left(\frac{V_2}{V_1}\right)^{\gamma-1}$$

$$T_1 = T_2(32)^{\frac{7}{5}-1}$$

$$= T_2(2^5)^{2/5} = T_2 \times 4$$

$$T_1 = 4T_2$$

$\Rightarrow$ $$\eta = \left(1 - \frac{1}{4}\right)$$

$$= \frac{3}{4} = 0.75$$

24. Efficiency, $\eta = 1 - \frac{T_2}{T_1} = 1 - \frac{500}{800} = \frac{3}{8} = 0.375$

25. Efficiency of Carnot heat engine, $\eta = 1 - \frac{T_2}{T_1}$. Efficiency remains same when both T_1 and T_2 are increased by same factor.

16

Calorimetry and Heat Transfer

16.1 Specific Heat

When heat energy flows into a substance, the temperature of the substance usually rises. An exception occurs during a change in phase, as when water freezes or evaporates. The amount of heat required to produce, the same temperature increase for a given amount of substance varies from one substance to another. The relationship between heat exchanged and the corresponding temperature change is characterised by the specific heat ***c*** of a substance. If the temperature of a substance of mass m changes from T to $T + dT$ when it exchanges an amount of heat dQ with its surroundings, then its specific heat is

$$c = \frac{1}{m} \cdot \frac{dQ}{dT} \qquad \text{...(i)}$$

The SI unit of specific heat ***c*** is J/kg-K. Because heat is so frequently measured in calories, the unit cal/g-°C is also used quite often. The specific heat capacity of water is approximately 1 cal/g-°C.

From Eq. (i), we can define the specific heat of a substance as the amount of energy needed to raise the temperature of unit mass of that substance by 1°C (or 1 K). A closely related quantity is the molar heat capacity C. It is defined as

$$C = \frac{1}{n} \cdot \frac{dQ}{dT} \qquad \text{...(ii)}$$

where, n is the number of moles of the substance. If M is the molecular mass of the substance, then $n = \frac{m}{M}$ where, m is the mass of the substance and

$$C = \frac{M}{m} \cdot \frac{dQ}{dT} \qquad \text{...(iii)}$$

The SI unit of molar heat capacity C is J/mol-K and it can be defined as the amount of energy needed to raise the temperature of one mole of a substance by 1°C (or 1 K) sometimes the product of mc is also written as C, simply the heat capacity, which is defined as the energy needed to raise the temperature of the whole substance by 1°C (or K).

Thus, $$C = mc = \frac{dQ}{dT} \qquad \text{...(iv)}$$

The SI units of molar heat capacity C is J/K.

Chapter Snapshot

- Specific Heat
- Phase Changes and Latent Heat
- Heat Transfer

Note Points

- In general, if c varies with temperature over the interval, then the corresponding expression for Q is

$$Q = m\int_{T_1}^{T_2} c \cdot dT$$

- The specific heat of water is much larger than that of most other substances. Consequently, for the same amount of added heat, the temperature change of a given mass of water is generally less than that for the same mass of another substance. For this reason a large body of water moderates the climate of nearby land. In the winter the water cools off more slowly than the surrounding land and tends to warm the land. In the summer, the opposite effect occurs, as the water heats up more slowly than the land.

Example 16.1 *When 400 J of heat are added to a 0.1 kg sample of metal, its temperature increases by 20°C. What is the specific heat of the metal?*

Sol. Using, $c = \frac{1}{m} \cdot \frac{\Delta Q}{\Delta T}$

We have, $c = \left(\frac{1}{0.1}\right)\left(\frac{400}{20}\right)$

$= 200$ J/kg -°C

16.2 Phase Changes and Latent Heat

Suppose that we slowly heat a cube of ice whose temperature is below 0°C at atmospheric pressure, what changes do we observe in the ice? Initially, we find that its temperature increases according to equation $Q = mc(T_2 - T_1)$. Once 0°C is reached, the additional heat does not increase the temperature of the ice. Instead, the ice melts and temperature remains at 0°C. The temperature of the water, then starts to rise and eventually reaches 100°C, whereupon the water vaporises into steam at this same temperature.

During phase transitions (solid to liquid or liquid to gas) the added heat causes a change in the positions of the molecules relative to one another, without affecting the temperature.

The heat necessary to change a unit mass of a substance from one phase to another is called the **latent heat (L)**. Thus, the amount of heat required for melting and vaporising a substance of mass m are given by

$$Q = mL \quad \ldots(i)$$

For a solid-liquid transition, the latent heat is known as the **latent heat of fusion (L_f)** and for the liquid-gas transition, it is known as the **latent heat of vaporisation (L_v)**.

For water at 1 atm latent heat of fusion is 80.0 cal/g. This simply means 80.0 cal of heat are required to melt 1.0 g of water or 80.0 cal heat is liberated when 1.0 g of water freezes at 0°C. Similarly, latent heat of vaporisation for water at 1 atm is 539 cal/g.

Figure shows how the temperature varies when we add heat continuously to a specimen of ice with an initial temperature below 0°C. Suppose, we have taken 1 g of ice at –20° C specific heat of ice is 0.53 cal/g-°C.

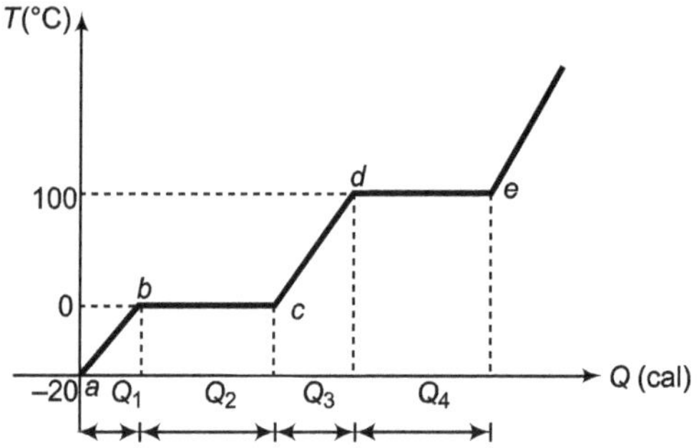

Fig. 16.1

In the figure :

a* to *b Temperature of ice increases until it reaches its melting point 0°C.

$$Q_1 = mc_{\text{ice}}\,[0 - (-20)]$$
$$= (1)\,(0.53)\,(20) = 10.6 \text{ cal}$$

b* to *c Temperature remains constant until all the ice has melted.

$$Q_2 = mL_f = (1)\,(80) = 80 \text{ cal}$$

c* to *d Temperature of water again rises until it reaches its boiling point 100°C.

$$Q_3 = mc_{\text{water}}\,[100 - 0] = (1)\,(1.0)\,(100) = 100 \text{ cal}$$

d* to *e Temperature is again constant until all the water is transformed into the vapour phase.

$$Q_4 = mL_v = (1)\,(539) = 539 \text{ cal}$$

Thus, the net heat required to convert 1 g of ice at – 20° C into steam at 100°C is

$$Q = Q_1 + Q_2 + Q_3 + Q_4 = 729.6 \text{ cal}$$

Example 16.2 *How much heat is required to convert 8.0 g of ice at –15° C to steam at 100°C? (Given, $c_{ice} = 0.53$ cal/g-°C, $L_f = 80$ cal/g and $L_v = 539$ cal/g, and $c_{water} = 1$ cal/g-°C)*

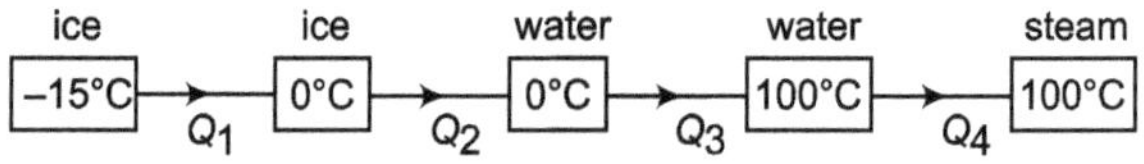

Fig. 16.2

Sol. $Q_1 = mc_{\text{ice}}\,(T_f - T_i) = (8.0)\,(0.53)\,[0 - (-15)] = 63.6$ cal

$Q_2 = mL_f = (8)\,(80) = 640$ cal

$Q_3 = mc_{\text{water}}\,(T_f - T_i) = (8.0)\,(1.0)\,[100 - 0] = 800$ cal

$Q_4 = mL_v = (8.0)\,(539) = 4312$ cal

∴ Net heat required,

$$Q = Q_1 + Q_2 + Q_3 + Q_4 = 5815.6 \text{ cal}$$

Example 16.3 *10 g of water at 70°C is mixed with 5 g of water at 30°C. Find the temperature of the mixture in equilibrium.*

Sol. Let t°C be the temperature of the mixture. From energy conservation,

Heat given by 10 g of water = Heat taken by 5 g of water

or $m_1 c_{water} |\Delta t_1| = m_2 c_{water} |\Delta t_2|$

$\therefore$ $(10)(70 - t) = 5(t - 30)$

$\therefore$ $t = 36.67°C$

Example 16.4 *In a container of negligible mass 30 g of steam at 100°C is added to 200 g of water that has a temperature of 40°C. If no heat is lost to the surroundings, what is the final temperature of the system? Also, find masses of water and steam in equilibrium.*
(Take, $L_v = 539$ cal/g and $c_{water} = 1$ cal/g-°C)

Sol. Let Q be the heat required to convert 200 g of water at 40°C into 100°C, then

$$Q = mc\Delta T = (200)(1.0)(100 - 40) = 12000 \text{ cal}$$

Now, suppose m_0 mass of steam converts into water to liberate this much amount of heat, then

$$m_0 = \frac{Q}{L} = \frac{12000}{539} = 22.26 \text{ g}$$

Since, it is less than 30 g, the temperature of the mixture is 100°C.

Mass of steam in the mixture = 30 − 22.26 = 7.74 g

and mass of water in the mixture

$$= 200 + 22.26 = 222.26 \text{ g}$$

16.3 Heat Transfer

Heat can be transferred from one place to the other by any of three possible ways : conduction, convection and radiation. In the first two processes, a medium is necessary for the heat transfer. Radiation, however, does not have this restriction. This is also the fastest mode of heat transfer, in which heat is transferred from one place to the other in the form of electromagnetic radiation. In competition examinations problems are asked only in first and last. So, we will discuss conduction and radiation in detail.

Conduction

Figure shows a rod whose ends are in thermal contact with a hot reservoir at temperature T_1 and a cold reservoir at temperature T_2.

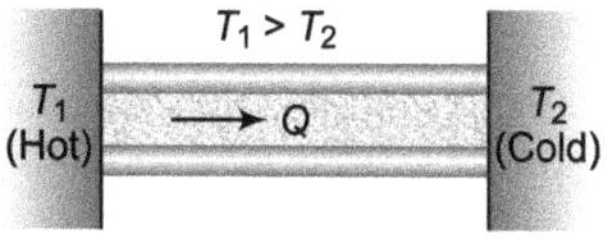

Fig. 16.3

The sides of the rod are covered with insulating medium, so the transport of heat is along the rod, not through the sides. The molecules at the hot reservoir have greater vibrational energy. This energy is transferred by collisions to the atoms at the end face of the rod. These atoms in turn transfer energy to their neighbours further along the rod. Such transfer of heat through a substance in which heat is transported without direct mass transport is called **conduction**.

Most metals use another, more effective mechanism to conduct heat. The free electrons, which move throughout the metal can rapidly carry energy from the hotter to colder regions, so metals are generally good conductors of heat. The presence of free electrons also causes most metals to be good electrical conductors. A metal rod at 5°C feels colder than a piece of wood at 5°C, because heat can flow more easily from your hand into the metal.

Heat transfer occurs only between regions that are at different temperatures and the rate of heat flow is $\frac{dQ}{dt}$. This rate is also called the **heat current,** denoted by H. Experiments show that the heat current is proportional to the cross-section area A of the rod and to the temperature gradient $\frac{dT}{dx}$, which is the rate of change of temperature with distance along the bar. In general,

$$H = \frac{dQ}{dt} = -KA\frac{dT}{dx} \quad \text{...(i)}$$

The negative sign is used to make $\frac{dQ}{dt}$ a positive quantity, since $\frac{dT}{dx}$ is negative. The constant K, called the **thermal conductivity** is a measure of the ability of a material to conduct heat.

A substance with a large thermal conductivity K is a good heat conductor. The value of K depends on the temperature increasing, it is slightly increasing with increasing temperature, but K can be taken to be practically constant throughout a substance, if the temperature difference between its ends is not too great.

Let us apply Eq. (i) to a rod of length L and constant cross-sectional area A in which a steady state has been reached. In a steady state (has been discussed in detail in medical galaxy 16.1), the temperature at each point is constant in time. Hence,

$$-\frac{dT}{dx} = T_1 - T_2$$

Therefore, the heat ΔQ transferred in time Δt is

$$\Delta Q = KA\left(\frac{T_1 - T_2}{l}\right)\Delta t \quad \text{...(ii)}$$

Thermal resistance (R) Eq. (ii) in differential form can be written as

$$\frac{\Delta Q}{\Delta t} = \frac{dQ}{dt} = H = \frac{\Delta T}{l/KA} = \frac{\Delta T}{R} \quad \text{...(iii)}$$

Here, ΔT = temperature difference (TD) and

$R = \frac{l}{KA}$ = thermal resistance of the rod.

Extra Knowledge Points

- Consider a section *ab* of a rod as shown in figure. Suppose Q_1 heat enters into the section at *a* and Q_2 leaves at *b*, then $Q_2 < Q_1$. Part of the energy $Q_1 - Q_2$ is utilized in raising the temperature of section *ab* and the remaining is lost to atmosphere through *ab*.

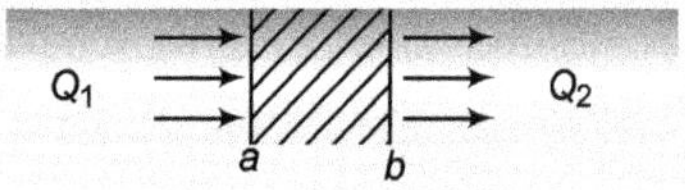

If heat is continuously supplied from the left end of the rod, a stage comes when temperature of the section becomes constant. In that case, $Q_1 = Q_2$, if rod is insulated from the surroundings (or loss through *ab* is zero). This is called the **steady state condition**. Thus, in steady state temperature of different sections of the rod becomes constant (but not same).
Hence, in the figure :

$$T_1 = \text{constant}$$
$$T_2 = \text{constant, etc.}$$
and $$T_1 > T_2 > T_3 > T_4$$

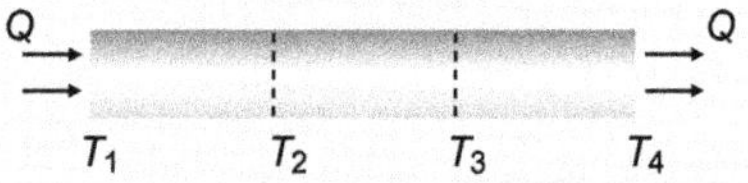

Insulated rod in steady state

Now, a natural question arises, why the temperature of whole rod not becomes equal when heat is being continuously supplied?

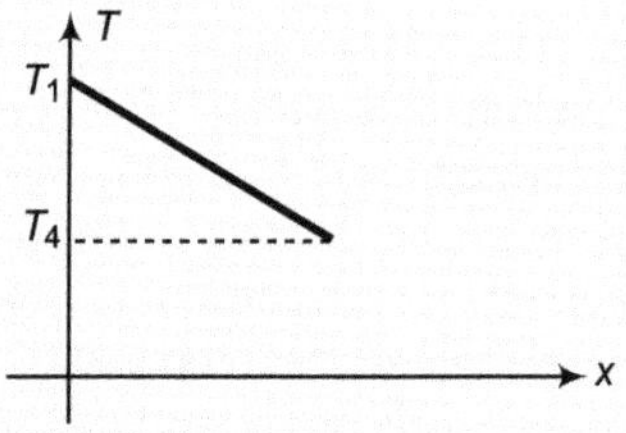

The answer is that there must be a temperature difference in the rod for the heat flow, same as we require a potential difference across a resistance for the current flow through it.

In steady state, the temperature varies linearly with distance along the rod, if it is insulated.

- Comparing equation number (iii), i.e. Heat current,

$$H = \frac{dQ}{dt} = \frac{\Delta T}{R} \quad \left(\text{where, } R = \frac{l}{kA}\right)$$

with the equation, of current flow through a resistance,

$$i = \frac{dq}{dt} = \frac{\Delta V}{R} \quad \left(\text{where, } R = \frac{l}{\sigma A}\right)$$

We find the following similarities in heat flow through a rod and current flow through a resistance.

Table 16.1

Heat flow through a conducting rod	Current flow through a resistance
Heat current, $H = \frac{dQ}{dt}$ = rate of heat flow	Electric current, $i = \frac{dq}{dt}$ = rate of charge flow
$H = \frac{\Delta T}{R} = \frac{TD}{R}$	$i = \frac{\Delta V}{R} = \frac{PD}{R}$
$R = \frac{l}{kA}$	$R = \frac{l}{\sigma A}$
K = thermal conductivity	σ = electrical conductivity

From the above table, it is evident that flow of heat through rods in series and parallel is analogous to the flow of current through resistances in series and parallel. This analogy is of great importance in solving complicated problems of heat conduction.From the above table, it is evident that flow of heat through rods in series and parallel is analogous to the flow of current through resistances in series and parallel. This analogy is of great importance in solving complicated problems of heat conduction.

Convection

Although conduction does occur in liquids and gases also, heat is transported in these media mostly by convection. In this process, the actual motion of the material is responsible for the heat transfer. Familiar examples include hot-air and hot-water home heating systems, the cooling system of an automobile engine and the flow of blood in the body.

You probably have warmed your hands by holding them over an open flame. In this situation, the air directly above the flame is heated and expands. As a result, the density of this air decreases and then air rises. When the movement results from differences in density, as with air around fire, it is referred to as natural convection. Air flow at a beach is an example of natural convection.

When the heated substance is forced to move by a fan or pump, the process is called **forced convection.** If it were not for convection currents, it would be very difficult to boil water. As water is heated in a kettle, the heated water expands and rises to the top, because its density is lowered. At the same time, the denser, cool water at the surface sinks to the bottom of the kettle and is heated. Heating a room by a radiator is an example of forced convection.

It is possible to write an equation for the thermal energy transported by convection and define a coefficient of convection, but the analysis of practical problems is very difficult and will not be treated here. To some approximation, the heat transferred from a body to its surroundings is proportional to the area of the body and to the difference in temperature between the body and the surrounding fluid.

Radiation

The third means of energy transfer is radiation which does not require a medium. The best known example of this process is the radiation from the sun. All objects radiate energy continuously in the form of electromagnetic waves. The rate at which an object radiates energy is proportional to the fourth power of its absolute temperature. This is known as the **Stefan's law** and is expressed in equation form as

$$P = \sigma A e T^4$$

Here, P is the power in watts (J/s) radiated by the object, A is the surface area in m^2, e is a fraction between 0 and 1 called the **emissivity of the object** and σ is a universal constant called Stefan's constant, which has the value

$$\sigma = 5.67 \times 10^{-8}\ W/m^2\text{-}K^4$$

Now, let us define few terms before studying the other topics.

Perfectly Black Body

A body that absorbs all the radiation incident upon it and has an emissivity equal to 1 is called a **perfectly black body**. A black body is also an ideal radiator. It implies that, if a black body and an identical another body are kept at the same temperature, then the black body will radiate maximum power as it is obvious from equation ($P = eA\sigma T^4$) also. Because $e = 1$ for a perfectly black body while for any other body $e < 1$.

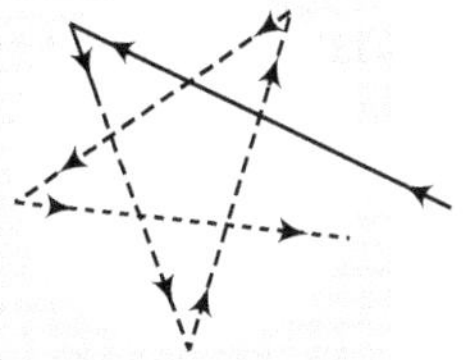

Cavity approximating an ideal black body. Radiation entering the cavity has little chance of leaving before it is completely absorbed.

Fig. 16.4

Materials like black velvet or lamp black come close to being ideal black bodies, but the best practical realisation of an ideal black body is a small hole leading into a cavity, as this absorbs 98% of the radiation incident on them.

Absorptive Power (a)

"It is defined as the ratio of the radiant energy absorbed by it in a given time to the total radiant energy incident on it in the same interval of time."

$$a = \frac{\text{Energy absorbed}}{\text{Energy incident}}$$

As a perfectly black body absorbs all radiations incident on it, the absorptive power of a perfectly black body is maximum and unity.

Spectral Absorptive Power (a_λ)

The absorptive power a refers to radiations of all wavelengths (or the total energy) while the spectral absorptive power is the ratio of radiant energy absorbed by a surface to the radiant energy incident on it for a particular wavelength λ. It may have different values for different wavelengths for a given surface. Let us take an example, suppose $a = 0.6, a_\lambda = 0.4$ for 1000 Å and $a_\lambda = 0.7$ for 2000 Å for a given surface. Then, it means that this surface will absorb only 60% of the total radiant energy incident on it. Similarly, it absorbs 40% of the energy incident on it corresponding to 1000 Å and 70% corresponding to 2000 Å. The spectral absorptive power a_λ is related to absorptive power a through the relation

$$a = \int_0^\infty a_\lambda \, d\lambda$$

Emissive Power (e)

(Don't confuse it with the emissivity e which is different from it, although both have the same symbol e).

For a given surface, it is defined as the radiant energy emitted per second per unit area of the surface. It has the units of W/m^2 or J/s-m^2. For a black body, $e = \sigma T^4$.

Spectral Emissive Power (e_λ)

It is emissive power for a particular wavelength λ. Thus,

$$e = \int_0^\infty e_\lambda \, d\lambda$$

Kirchhoff's Law According to this law, the ratio of emissive power to absorptive power is same for all surfaces at the same temperature.

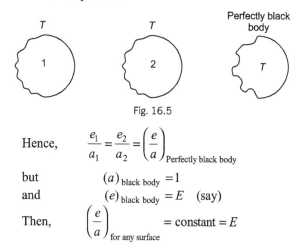

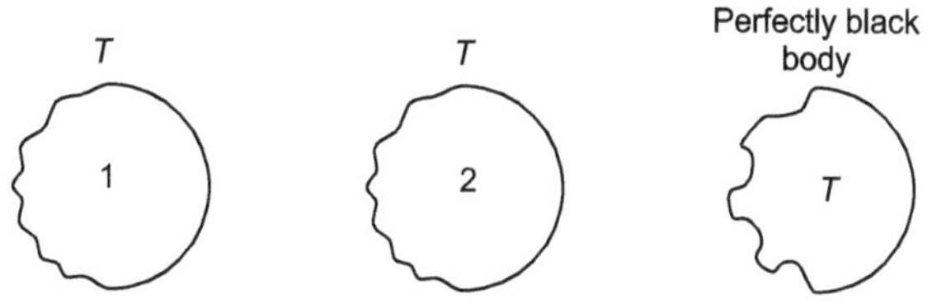

Fig. 16.5

Hence, $\quad \dfrac{e_1}{a_1} = \dfrac{e_2}{a_2} = \left(\dfrac{e}{a}\right)_{\text{Perfectly black body}}$

but $\quad (a)_{\text{black body}} = 1$

and $\quad (e)_{\text{black body}} = E \quad \text{(say)}$

Then, $\quad \left(\dfrac{e}{a}\right)_{\text{for any surface}} = \text{constant} = E$

Similarly, for a particular wavelength λ,

$$\left(\frac{e_\lambda}{a_\lambda}\right)_{\text{for any body}} = E_\lambda$$

Here, E = emissive power of black body at temperature T

$$= \sigma T^4$$

From the above expression, we can see that

$$e_\lambda \propto a_\lambda$$

i.e. good absorbers for a particular wavelength are also good emitters of the same wavelength.

Cooling by Radiation

Consider a hot body at temperature T placed in an environment at a lower temperature T_0. The body emits more radiation than it absorbs and cools down while the surroundings absorb radiation from the body and warm up. The body is losing energy by emitting radiations at a rate,

$$P_1 = eA\sigma T^4$$

and is receiving energy by absorbing radiations at a rate,

$$P_2 = aA\sigma T_0^4$$

Here, a is a pure number between 0 and 1 indicating the relative ability of the surface to absorb radiation from its surroundings. Note that this (a) is different from the absorptive power (a). In thermal equilibrium, both the body and the surroundings have the same temperature (say T_c) and,

$$P_1 = P_2 \quad \text{or} \quad eA\sigma T_c^4 = aA\sigma T_c^4 \quad \text{or} \quad e = a$$

Thus, when $T > T_0$, the net rate of heat transfer from the body to the surroundings is

$$\frac{dQ}{dt} = eA\sigma (T^4 - T_0^4)$$

or $$mc\left(-\frac{dT}{dt}\right) = eA\sigma (T^4 - T_0^4)$$

⇒ Rate of cooling,

$$\left(-\frac{dT}{dt}\right) = \frac{eA\sigma}{mc}(T^4 - T_0^4)$$

or $$-\frac{dT}{dt} \propto (T^4 - T_0^4)$$

Newton's Law of Cooling

According to this law, if the temperature T of the body is not different from that of the surroundings T_0, then rate of cooling $-\frac{dT}{dt}$ is proportional to the temperature difference between them.

To prove it let us assume that

$$T = T_0 + \Delta T$$

So that, $T^4 = (T_0 + \Delta T)^4 = T_0^4\left(1 + \frac{\Delta T}{T_0}\right)^4$

$$\approx T_0^4\left(1 + \frac{4\Delta T}{T_0}\right) \text{ (from binomial expansion)}$$

∴ $$(T^4 - T_0^4) = 4T_0^3 (\Delta T)$$

or $$(T^4 - T_0^4) \propto \Delta T \qquad (\text{as } T_0 = \text{constant})$$

Now, we have already shown that rate of cooling

$$\left(-\frac{dT}{dt}\right) \propto (T^4 - T_0^4)$$

and here we have shown that

$$(T^4 - T_0^4) \propto \Delta T,$$

if the temperature difference is small.

Thus, rate of cooling

$$-\frac{dT}{dt} \propto \Delta T$$

or $$-\frac{d\theta}{dt} \propto \Delta\theta$$

as $$dT = d\theta \quad \text{or} \quad \Delta T = \Delta\theta$$

Variation of Temperature of a Body According to Newton's Law

Suppose a body has a temperature θ_i at time $t = 0$. It is placed in an atmosphere whose temperature is θ_0. We are interested in finding the temperature of the body at time t, assuming Newton's law of cooling to hold good or by assuming that the temperature difference is small. As per this law,

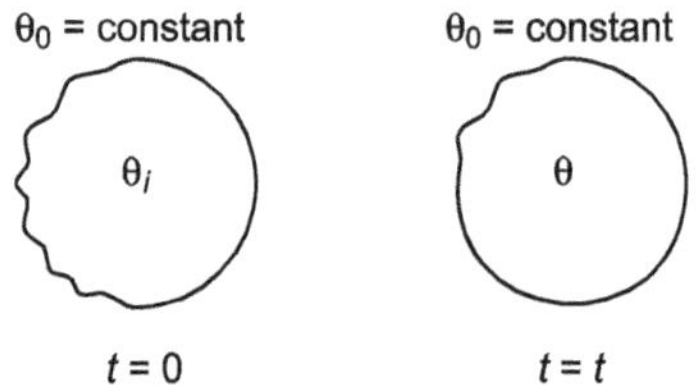

Fig. 16.6

rate of cooling ∝ temperature difference

or $$\left(-\frac{d\theta}{dt}\right) = \left(\frac{eA\sigma}{mc}\right)(4\theta_0^3)(\theta - \theta_0)$$

or $$\left(-\frac{d\theta}{dt}\right) = \alpha(\theta - \theta_0)$$

Here, $$\alpha = \left(\frac{4eA\sigma\theta_0^3}{mc}\right) \text{ is a constant}$$

∴ $$\int_{\theta_i}^{\theta} \frac{d\theta}{\theta - \theta_0} = -\alpha \int_0^t dt$$

∴ $$\theta = \theta_0 + (\theta_i - \theta_0)e^{-\alpha t}$$

From this expression, we see that $\theta = \theta_i$ at $t = 0$ and $\theta = \theta_0$ at $t = \infty$, i.e. temperature of the body varies exponentially with time from θ_i to θ_0 $(<\theta_i)$. The temperature *versus* time graph is as shown in the figure.

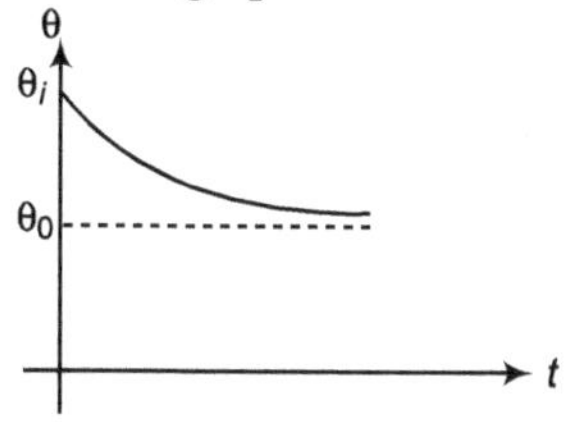

Fig. 16.7

- If the body cools by radiation from θ_1 to θ_2 in time t, then taking the approximation,

$$\left(-\frac{d\theta}{dt}\right) = \frac{\theta_1 - \theta_2}{t} \quad \text{and} \quad \theta = \theta_{av} = \left(\frac{\theta_1 + \theta_2}{2}\right)$$

The equation $\left(-\frac{d\theta}{dt}\right) = \alpha(\theta - \theta_0)$ becomes

$$\frac{\theta_1 - \theta_2}{t} = \alpha\left(\frac{\theta_1 + \theta_2}{2} - \theta_0\right)$$

This form of the law helps in solving numerical problems related to Newton's law of cooling.

Wien's Displacement Law

At ordinary temperatures (below about 600°C), the thermal radiation emitted by a body is not visible, most of it is concentrated in wavelengths much longer than those of visible light.

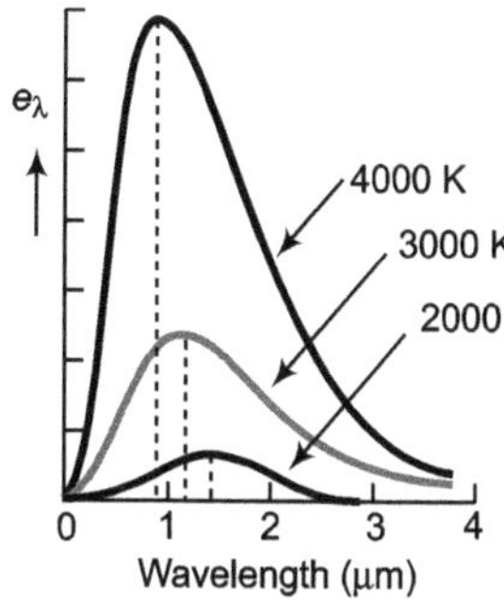

Fig. 16.8
Power of black body radiation *versus* wavelength at three temperatures. Note that the amount of radiation emitted (the area under a curve) increase with increasing temperature

Figure shows how the energy of a black body radiation varies with temperature and wavelength? As the temperature of the black body increases, two distinct behaviours are observed. The first effect is that the peak of the distribution shifts to shorter wavelengths. This shift is found to obey the following relationship called **Wien's displacement law**.

$$\lambda_{max} T = b$$

Here, b is a constant called Wien's constant. The value of this constant in SI unit is 2.898×10^{-3} m-K. Thus,

$$\lambda_{max} \propto \frac{1}{T}$$

Here, λ_{max} is the wavelength corresponding to the maximum spectral emissive power e_λ.

The second effect is that the total amount of energy the black body emits per unit area per unit time $(=\sigma T^4)$ increases with fourth power of absolute temperature T. This is also known as the **emissive power**. We know

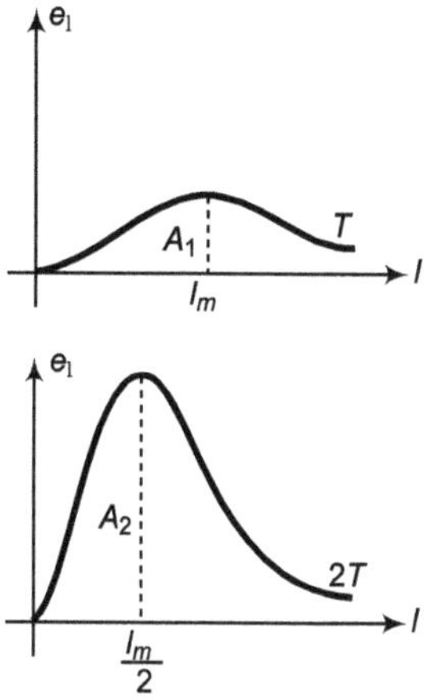

Fig. 16.9

$$e = \int_0^\infty e_\lambda \, d\lambda = \text{Area under } e_\lambda\text{-}\lambda \text{ graph} = \sigma T^4$$

or Area $\propto T^4$ $A_2 = (2)^4 A_1 = 16 A_1$

Thus, if the temperature of the black body is made two fold, λ_{max} remains half while the area becomes 16 times.

Extra Knowledge Points

- In solids (and in mercury) transmission of heat takes place only by conduction, while in liquids and gases it mainly takes place by convection (In liquids and gases it is possible by conduction also).
- Heat conduction in metals. Metals are very good conductor of heat. In fact, the heat conduction in metals mostly takes place by the free electrons present within the metals. These electrons are not bound to any molecule of the metals, but they are free to move within the metal. Thus, these electrons are just like the molecules of a gas. Thus, the transmission of energy (heat) in metals takes place by the free electrons, not by molecules. This is why in metals the transmission of heat takes place very rapidly.
- **Relation between thermal and electrical conductivities of Metals : Wiede mann-Franz Law** All good conductors of heat are also good conductors of electricity. Wiede mann and Franz expressed this fact in the form of an empirical law

which states that, 'the ratio of the thermal and the electrical conductivities are same for all metals at the same temperature.

i.e. $\dfrac{K}{\sigma}$ = constant

- Lorentz extended the law and showed that this ratio is proportional to the absolute temperature,

i.e. $\dfrac{K}{\sigma T}$ = constant

- **Effects and uses of thermal capacity in daily life**
  - (i) **During winter, iron seems colder and in summer seems warmer than wood** During winter the temperature of our body is higher than the room temperature. Hence, when we touch iron, the iron rapidly conducts heat from our hand and gives a cold feeling. Wood on the other hand, is a bad conductor of heat. It conducts heat slowly from our hand and appears less cold. During summers reverse is the process.
  - (ii) **When hot water is poured in a beaker of thick glass, the beaker cracks** When we pour hot water in a glass beaker, the inner surface of the glass expands on heating. But heat from inside does not reach quickly the outer surface of the glass, because glass is a bad conductor of heat. Hence, the outer surface does not expand and the glass cracks.
  - (iii) **In winters woolen cloths or blankets are used** The fibres of woolen clothes have larger interspaces than cotton clothes and air is filled in these spaces. Since, air is a bad conductor of heat, hence, these clothes prevent heat from our body to go outside and our body remains warm. On using woolen clothes for a long time the air is driven out. Hence, old woolen clothes are less warm.
  - (iv) **To prevent ice from melting it is wrapped with blanket or felt** The air filled in the interspaces of these materials, being bad conductors of heat does not allow heat to flow from outside to the ice. Hence, ice does not melt.
- **Solar constant** The amount of heat received from the sun by one square centimetre area of a surface placed normally to the sun rays at mean distance of the earth from the sun is known as solar constant. It is denoted by *S*.

$$S = \left(\frac{r}{R}\right)^2 \cdot \sigma T^4$$

Here, *r* is the radius of the sun and *S* the mean distance of the earth from the centre of the sun. Value of solar constant is 1.937 cal/cm^2- min.

Example 16.5 *A copper rod 2 m long has a circular cross-section of radius 1 cm. One end is kept at 100°C and the other at 0°C, and the surface is insulated so that negligible heat is lost through the surface. Find*

(a) the thermal resistance of the bar.

(b) the thermal current H.

(c) the temperature gradient $\dfrac{dT}{dx}$ and

(d) the temperature 25 cm from the hot end.

Thermal conductivity of copper is 401 W/m-K.

Sol. (a) Thermal resistance, $R = \dfrac{l}{KA} = \dfrac{l}{K(\pi r^2)}$

or $R = \dfrac{(2)}{(401)(\pi)(10^{-2})^2} = 15.9\ \text{K/W}$

(b) Thermal current, $H = \dfrac{\Delta T}{R} = \dfrac{\Delta\theta}{R} = \dfrac{100}{15.9}$

or $H = 6.3\ \text{W}$

(c) Temperature gradient,

$$d = \frac{0-100}{2} = -50\ \text{K/m} = -50°\text{C/m}$$

(d) Let θ_C° be the temperature at 25 cm from the hot end, then

$(\theta - 100)$ = (temperature gradient) × (distance)

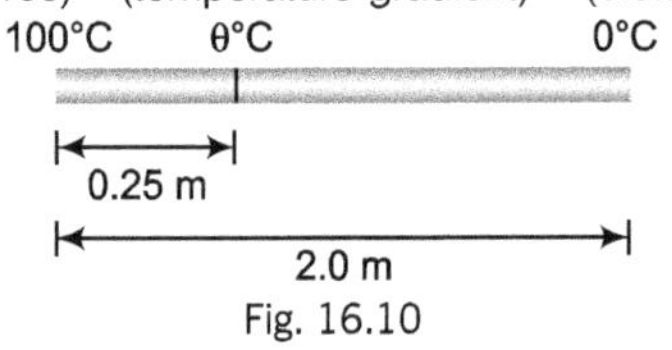

Fig. 16.10

or $\theta - 100 = (-50)(0.25)$ or $\theta = 87.5°\text{C}$

Example 16.6 *Two metal cubes with 3 cm edges of copper and aluminium are arranged as shown in figure. Find*

(a) the total thermal current from one reservoir to the other.

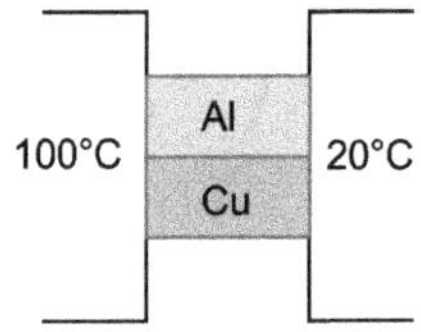

Fig. 16.11

(b) the ratio of the thermal current carried by the copper cube to that carried by the aluminium cube. Thermal conductivity of copper is 401 W/m-K and that of aluminium is 237 W/m-K.

Sol. (a) Thermal resistance of aluminium cube, $R_1 = \dfrac{l}{KA}$

or $R_1 = \dfrac{(3.0 \times 10^{-2})}{(237)(3.0 \times 10^{-2})^2} = 0.14\ \text{K/W}$

and thermal resistance of copper cube, $R_2 = \dfrac{l}{KA}$

or $R_2 = \dfrac{(3.0 \times 10^{-2})}{(401)(3.0 \times 10^{-2})^2} = 0.08\ \text{K/W}$

As these two resistances are in parallel, their equivalent resistance will be,

$$R = \frac{R_1 R_2}{R_1 + R_2} = \frac{(0.14)(0.08)}{(0.14) + (0.08)} = 0.05\ \text{K/W}$$

$\therefore$ Thermal current, $H = \dfrac{\text{Temperature difference}}{\text{Thermal resistance}}$

$$= \frac{(100 - 20)}{0.05} = 1.6 \times 10^3 \text{ W}$$

(b) In parallel thermal current distributes in the inverse ratio of resistance. Hence,

$$\frac{H_{Cu}}{H_{Al}} = \frac{R_{Al}}{R_{Cu}} = \frac{R_1}{R_2} = \frac{0.14}{0.08} = 1.75$$

Example 16.7 *One end of a copper rod of length 1 m and area of cross-section 4.0×10^{-4} m^2 is maintained at 100°C. At the other end of the rod ice is kept at 0°C. Neglecting the loss of heat from the surroundings find the mass of ice melted in 1 h. (Given, $K_{Cu} = 401$ W/m-K and $L_f = 3.35 \times 10^5$ J/kg)*

Sol. Thermal resistance of the rod,

Fig. 16.12

$$R = \frac{l}{KA} = \frac{1.0}{(401)(4 \times 10^{-4})} = 6.23 \text{ K/W}$$

$\therefore$ Heat current, $H = \dfrac{\text{Temperature difference}}{\text{Thermal resistance}}$

$$= \frac{(100 - 0)}{6.23} = 16 \text{ W}$$

Heat transferred in 1 h,

$$Q = Ht \qquad \left(H = \frac{Q}{t}\right)$$

$$= (16)(3600) = 57600 \text{ J}$$

Now, let m mass of ice melts in 1 h, then

$$m = \frac{Q}{L} \qquad (Q = mL)$$

$$= \frac{57600}{3.35 \times 10^5} = 0.172 \text{ kg or } 172 \text{ g}$$

Example 16.8 *A body cools in 10 min from 60°C to 40°C. What will be its temperature after next 10 min? The temperature of the surroundings is 10°C.*

Sol. According to Newton's law of cooling,

$$\left(\frac{\theta_1 - \theta_2}{t}\right) = \alpha\left[\left(\frac{\theta_1 + \theta_2}{2}\right) - \theta_0\right]$$

For the given conditions,

$$\frac{60 - 40}{10} = \alpha\left[\frac{60 + 40}{2} - 10\right] \qquad \text{...(i)}$$

Let θ be the temperature after next 10 min. Then,

$$\frac{40 - \theta}{10} = \alpha\left[\frac{40 + \theta}{2} - 10\right] \qquad \text{...(ii)}$$

Solving Eqs. (i) and (ii), we get

$$\theta = 28°\text{C}$$

Example 16.9 *Two bodies A and B have thermal emissivities of 0.01 and 0.81, respectively. The outer surface areas of the two bodies are same. The two bodies emit total radiant power at the same rate. The wavelength λ_B corresponding to maximum spectral radiancy from B is shifted from the wavelength corresponding to maximum spectral radiancy in the radiation from A by 1.0 μm. If the temperature of A is 5802 K, calculate*

(a) the temperature of B,

(b) wavelength λ_B.

Sol. (a) $P_A = P_B \Rightarrow \therefore e_A \sigma A_A T_A^4 = e_B \sigma A_B T_B^4$

$$\therefore \quad T_B = \left(\frac{e_A}{e_B}\right)^{1/4} T_A \qquad (\text{as } A_A = A_B)$$

Substituting the values

$$T_B = \left(\frac{0.01}{0.81}\right)^{1/4}(5802) = 1934 \text{ K}$$

(b) According to Wien's displacement law,

$$\lambda_A T_A = \lambda_B T_B$$

$$\therefore \quad \lambda_B = \left(\frac{5802}{1934}\right)\lambda_A$$

or $\lambda_B = 3\lambda_A$

Also, $\lambda_B - \lambda_A = 1\,\mu\text{m}$

or $\lambda_B - \left(\frac{1}{3}\right)\lambda_B = 1\,\mu\text{m}$ or $\lambda_B = 1.5\,\mu\text{m}$

Chapter Summary with Formulae

- **Calorimetry**
  (i) $Q = ms\Delta T = c\Delta T$, when temperature changes without change in state.
  (ii) $Q = mL$, when state changes without change in temperature.
- **Heat Transfer**

Heat conduction through a rod

(i) Heat flow in steady state, $Q = \frac{KA(T_1 - T_2)}{L} t$

(ii) Rate of flow of heat = heat current

or $$H = \frac{dQ}{dt} = \frac{\text{TD}}{R}$$

Here, TD = temperature difference = $T_1 - T_2$

and R = thermal resistance = $\frac{L}{KA}$

- **Radiation**

(i) Absorptive power, $a = \frac{\text{Energy absorbed}}{\text{Energy incident}}$

$$a \leq 1$$
$a = 1$ for perfectly black body.

(ii) Spectral absorptive power (a_λ) = absorptive power of wavelength (λ).

$$a_\lambda \leq 1$$
$a_\lambda = 1$ for perfectly black body.

(iii) **Emissive power** (e) Energy radiated per unit surface area per unit time is called emissive power of a body. Its SI units are $Js^{-1}m^{-2}$ or Wm^{-2}.

(iv) **Spectral emissive power** (e_λ) Emissive power of wavelength λ is known as spectral emissive power.

$$e = \int_0^\infty e_\lambda \, d\lambda$$

(v) **Stefan's law** Emissive power of a body is given by

$$e = e_r \sigma T^4$$

Here, e_r = emissivity, emittance, relative emissivity or relative emittance.

$$e_r \leq 1$$
$e_r = 1$ for a perfectly black body.

Sometimes, emissivity is also denoted by *e*. In that case differentiate them by their units. e_r is unitless while *e* has the units Wm^{-2}.

(vi) Total energy radiated by a body

$$E = e_r \sigma T^4 At$$

Here, A = surface area and t = time.

(vii) $a = e_r$ or absorptivity of a body
= its emissivity.

(viii) **Kirchhoff's law** If different bodies (including a perfectly black body) are kept at same temperatures, then

$$e_\lambda \propto a_\lambda \quad \text{or} \quad \frac{e_\lambda}{a_\lambda} = \text{constant} \quad \text{or} \quad \left(\frac{e_\lambda}{a_\lambda}\right)_{\text{Body 1}} = \left(\frac{e_\lambda}{a_\lambda}\right)_{\text{Body 2}}$$
$$= \left(\frac{e_\lambda}{a_\lambda}\right)_{\text{Perfectly black body}}$$
$$= (e_\lambda)_{\text{Perfectly black body}}$$

From this law, following two conclusions can be drawn.

(a) Good absorbers of a particular wavelength λ are also good emitters of same wavelength λ.

(b) At a given temperature, ratio of e_λ and a_λ for any body is constant. This ratio is equal to e_λ of perfectly black body at that temperature.

Additional Examples

Example 1. *Thick bottomed drinking glasses frequently crack, if hot water is poured into them. Why?*

Sol. Glass is a bad conductor of heat. It does not pass down the heat quickly to the lower surface. Different layers of the bottom are at different temperatures and expand differently. This causes breaking of the glass at the bottom.

Example 2. *What kind of thermal conductivity and specific heat requirements would you specify for cooking utensils?*

Sol. A cooking utensil should have

(*i*) high conductivity so that it can conduct heat through itself and transfer it to the contents quickly.

(*ii*) low specific heat so that it immediately attains the temperature of the source.

Example 3. *Why do the metal utensils have wooden handles?*

Sol. Wood is a bad conductor of heat. Wooden handle does not allow heat to be conducted from the hot utensil to the hand. So, we can easily hold the hot utensil with its help.

Example 4. *Can we boil water inside an earth satellite?*

Sol. No. The process of transfer of heat by convection is based on the fact that a liquid becomes lighter on becoming hot and rises up. In condition of weightlessness, this is not possible. So, transfer of heat by convection is not possible in a satellite.

Example 5. *Why do animals curl into a spherical shape, when they feel very cold?*

Sol. The total energy radiated by a body depends on its surface area. Thus, when the animals feel very cold, they curl their bodies into a spherical shape so as to decrease the surface area of their bodies which in turn helps to reduce the amount of heat lost by them.

Example 6. *Why are clear nights colder than cloudy nights?*

Sol. Clouds are opaque to heat radiations. So on a cloudy night, radiations from the earth's surface fail to escape. But on a clear night, the surface of the earth is cooled due to excessive radiation. So, a clear night is colder than a cloudy night.

Example 7. *White clothes are more comfortable in summer while colourful clothes are more comfortable in winter. Why?*

Sol. White clothes absorb very little heat radiation and hence, they are comfortable in summer. Coloured clothes absorb almost whole of the incident radiation and keep the body warm in winter.

Example 8. *What is the effect of pressure on melting point of a solid?*

Sol. The melting point of a solid may increase or decrease depending on the nature of solid. For solids such as ice which contracts on melting, it is lowered while for solids such as sulphur and wax which expand on melting it increases.

Example 9. *Water in a closed tube is heated with one arm placed vertically above an arc lamp. Water will begin to circulate along the tube in a counterclockwise direction. Is this true or false?*

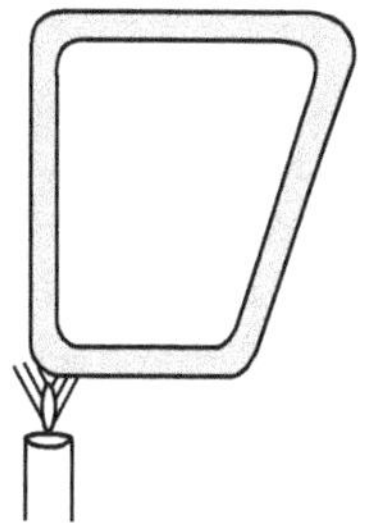

Sol. False. Water will circulate in the clockwise direction. The molecules immediately above the arc receive heat by conduction. They become light and rise up and get replaced by cold molecules from the right side. This will make the water to circulate in clockwise direction.

Example 10. *5 g of water at 30°C and 5 g of ice at −20°C are mixed together in a calorimeter. Find the final temperature of mixture. Water equivalent of calorimeter is negligible, specific heat of ice = 0.5 cal/g°C and latent heat of ice = 80 cal/g.*

Sol. In this case, heat is given by water and taken by ice.

Heat available with water to cool from 30° C to 0° C

$$= ms\Delta\theta = 5 \times 1 \times 30 = 150 \text{ cal}$$

Heat required by 5 g ice to increase its temperature up to 0°C

$$ms\Delta\theta = 5 \times 0.5 \times 20 = 50 \text{ cal}$$

Out of 150 cal heat available, 50 cal is used for increasing temperature of ice from -20° C to 0° C. The remaining heat 100 cal is used for melting the ice.

If mass of ice melted is m g, then

$$m \times 80 = 100 \Rightarrow m = 1.25 \text{ g}$$

Thus, 1.25 g ice out of 5 g melts and mixture of ice and water is at 0° C.

Example 11. *A bullet of mass 10 g moving with a speed of 20 m/s hits an ice block of mass 990 g kept on a frictionless floor and gets stuck in it. How much ice will melt if 50% of the lost kinetic energy goes to ice ? (Temperature of ice block = 0°C).*

Sol. Velocity of bullet + ice block

$$v = \frac{(10\text{ g}) \times (20\text{ m/s})}{1000\text{ g}} = 0.2\text{ m/s}$$

$$\text{Loss of KE} = \frac{1}{2}mv^2 - \frac{1}{2}(m+M)v^2$$

$$= \frac{1}{2}[0.01 \times (20)^2 - 1 \times (0.2)^2]$$

$$= \frac{1}{2}[4 - 0.04] = 1.98\text{ J}$$

$\therefore$ Heat received by ice block $= \frac{1.98}{4.2 \times 2}$ cal $= 0.24$ cal

$\therefore$ Mass of ice melted $= \frac{(0.24\text{ cal})}{(80\text{ cal / g})} = 0.003$ g

Example 12. *The temperature of equal masses of three different liquids A,B and C are 12°C, 19°C and 28°C, respectively. The temperature when A and B are mixed is 16°C and when B and C are mixed it is 23°C. What should be the temperature when A and C are mixed?*

Sol. Let m be the mass of each liquid and S_A, S_B, S_C specific heats of liquids A, B and C respectively. When A and B are mixed. The final temperature is 16° C.

$\therefore$ Heat gained by A = heat lost by B

i.e. $mS_A(16 - 12) = mS_B(19 - 16)$

i.e. $S_B = \frac{4}{3}S_A$ ...(i)

When B and C are mixed. Heat gained by B = heat lost by C

i.e. $mS_B(23 - 19) = mS_C(28 - 23)$

i.e. $S_C = \frac{4}{5}S_B$ ...(ii)

From Eqs. (i) and (ii), we get

$$S_C = \frac{4}{5} \times \frac{4}{3}S_A = \frac{16}{15}S_A$$

When A and C are mixed, let the final temperature be θ

$\therefore$ $mS_A(\theta - 12) = mS_C(28 - \theta)$

i.e. $\theta - 12 = \frac{16}{15}(28 - \theta)$

By solving, we get $\theta = \frac{628}{31} = 20.26°$ C

Example 13. *At 1 atm pressure, 1 g of water having a volume of 1.000 cm^3 becomes 1671 cm^3 of steam when boiled. The heat of vaporisation of water at 1 atm is 539 cal/g. What is the change in internal energy during the process?*

Sol. Heat spent during vaporisation,

$$Q = mL = 1.000 \times 539 = 539\text{ cal}$$

Work done, $W = p(V_v - V_l)$

$$= 1.013 \times 10^5 \times (1671 - 1.000) \times 10^{-6}$$

$$169.2\text{ J} = \frac{169.2}{4.18}\text{ cal} = 40.5\text{ cal}$$

$\therefore$ Change in internal energy, $U = 539$ cal $- 40.5$ cal $= 498.5$ cal

Example 14. *At 1 atm pressure, 1 g of water having a volume of 1 cm^3 becomes 1.091 cm^3 of ice on freezing. The heat of fusion of water at 1 atm is 80.0 cal/g. What is the change in internal energy during the process?*

Sol. Heat given out during freezing,

$$Q = -mL = -1 \times 80 = -80\text{ cal}$$

External work done,

$$W = p(V_{\text{ice}} - V_{\text{water}}) = 1.013 \times 10^5 \times (1.091 - 1.000) \times 10^{-6}$$

$$= 9.22 \times 10^{-3}\text{ J} = \frac{9.22 \times 10^{-3}}{4.18}\text{ cal} = 0.0022\text{ cal}$$

$\therefore$ Change in internal energy $\Delta U = Q - W$

$$= -80 - 0.0022 = -80.0022\text{ cal}$$

Example 15. *A cylinder of radius R made of a material of thermal conductivity K_1 is surrounded by cylindrical shell of inner radius R and outer radius 2R made of a material of thermal conductivity K_2. The two ends of the combined system are maintained at two different temperatures. There is no loss of heat across the cylindrical surface and system is in steady state. What is the effective thermal conductivity of the system?*

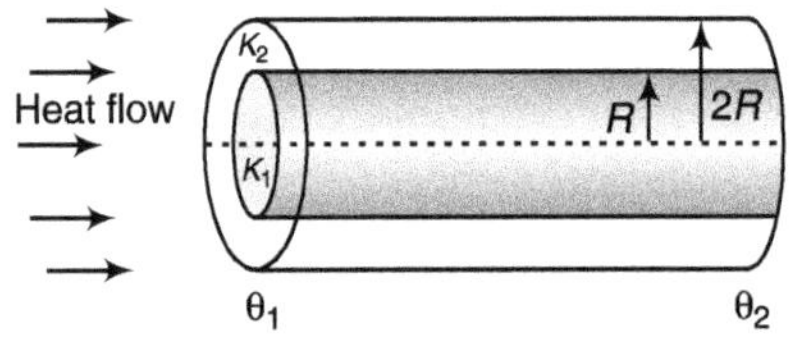

Sol. In this situation, a rod of length L and area of cross-section πR^2 and another rod of same length L and area of cross-section $\pi[(2R)^2 - R^2] = 3\pi R^2$ will conduct heat simultaneously so total heat flowing per second will be

$$\frac{dQ}{dt} = \frac{dQ_1}{dt} + \frac{dQ_2}{dt}$$

$$= \frac{K_1 \pi R^2(\theta_1 - \theta_2)}{L} + \frac{K_2 3\pi R^2(\theta_1 - \theta_2)}{L} \quad ...(i)$$

Now, if the equivalent conductivity is K. Then,

$$\frac{dQ}{dt} = K\frac{4\pi R^2(\theta_1 - \theta_2)}{L} \quad [\text{as, } A = \pi(2R)^2] \quad ...(ii)$$

So from Eqs. (i) and (ii), we have

$$4K = K_1 + 3K_2$$

i.e. $K = \frac{(K_1 + 3K_2)}{4}$

NCERT Selected Questions

Q 1. A geyser heats water flowing at the rate of 3.0 L per min from 27°C to 77°C. If the geyser operates on a gas burner, what is the rate of consumption of fuel, if its heat of consumption is 4.0×10^4 J/g?

Sol. Volume of water heated,

$$V = 3.0 \text{ L min}^{-1} = 3 \times 10^{-3} \text{ m}^3 \text{ min}^{-1}$$

∴ Mass of water heated,

$$m = \rho V = 10^3 \times 3 \times 10^{-3} \text{ kg min}^{-1}$$

$$= 3 \text{ kg min}^{-1}$$

Rise in temperature of water,

$$\Delta\theta = \theta_2 - \theta_1 = 77 - 27 = 50°\text{C}$$

Specific heat of water, $c = 4.2 \text{ Jg}^{-1}\text{C}^{-1}$

$$= 4.2 \times 10^3 \text{ J kg}^{-1}\text{C}^{-1}$$

Heat required by water = $\Delta Q = mc\Delta\theta$

$$= 3 \times 4.2 \times 10^3 \times 50 \text{ J min}$$

$$= 63 \times 10^4 \text{ J min}^{-1} \quad \text{...(i)}$$

Heat of combustion of fuel

$$= 4.0 \times 10^4 \text{ Jg}^{-1}$$

$$= 4.0 \times 10^7 \text{ J kg}^{-1}$$

Let m kg min^{-1} be the rate of combustion of fuel.

∴ Heat supplied by gas burner or combustion of fuel

$$= m \times 4 \times 10^7 \text{ J/min} \quad \text{...(ii)}$$

∴ From Eqs. (i) and (ii), we get

$$63 \times 10^4 = m \times 4 \times 10^7$$

or
$$m = \frac{63 \times 10^4}{4 \times 10^7}$$

$$= 15.75 \times 10^{-3} \text{ kg min}^{-1}$$

or rate of combustion of fuel $\approx 16 \text{ g min}^{-1}$

Q 2. A 10 kW drilling machine is used to drill a bore in a small aluminium block of mass 8.0 kg. How much is the rise in temperature of the block in 2.5 min, assuming 50% of the power is used up in heating the machine itself or lost to the surroundings. Specific heat of aluminium $= 0.91 \text{ Jg}^{-1}\text{K}^{-1}$.

Sol. Power of the machine

$$P = 10 \text{ kW} = 10^4 \text{ W}$$

Time for which the machine is used,

$$t = 2.5 \text{ min}$$

$$= 2.5 \times 60 \text{ s}$$

$$= 150 \text{ s}$$

Specific heat of aluminium, $c = 0.91 \text{ Jg}^{-1}\text{K}^{-1}$

$$= 0.91 \times 10^3 \text{ J kg}^{-1} \text{ K}^{-1}$$

If Q be the energy used up in drilling the bore, then

$$Q = P \times t = 10^4 \times 150$$

$$= 15 \times 10^5 \text{ J}$$

It is given that 50% of energy is lost to the surroundings.

∴ If Q' = Energy transferred to the aluminium block, then

$$Q' = 50\% \text{ of } Q = \frac{50}{100} \times 15 \times 10^5$$

$$= 75 \times 10^4 \text{ J}$$

Let $\Delta\theta$ be the rise in temperature of the block,

∴
$$Q' = mc\Delta\theta \quad \text{or} \quad \Delta\theta = \frac{Q'}{mc}$$

$$= \frac{75 \times 10^4}{8 \times 0.91 \times 10^3} = 103°\text{C}$$

Q 3. A copper block of mass 2.5 kg is heated in a furnace to a temperature of 500°C and then placed on a large ice block. What in the maximum amount of ice that can melt? (Specific heat of copper = $0.39 \text{ Jg}^{-1}\text{K}^{-1}$; heat of fusion of water = 335 Jg^{-1}).

Sol. Specific heat of copper, $c = 0.39 \text{ Jg}^{-1}\text{K}^{-1}$

$$= 0.39 \times 10^3 \text{ J kg}^{-1}\text{K}^{-1}$$

Temperature of furnace, $\Delta\theta = 500°\text{C}$

Latent heat of fusion, $L = 335 \text{ Jg}^{-1}$

$$= 335 \times 10^3 \text{ J kg}^{-1}$$

If Q be the heat absorbed by the copper block, then

$$Q = m_1 c\Delta\theta \quad \text{...(i)}$$

Let m_2 be the mass of ice melted, when copper block is placed on it, then

∴
$$Q = m_2 L \quad \text{...(ii)}$$

∴ From Eqs. (i) and (ii), we get

$$m_1 c\Delta\theta = m_2 L$$

or
$$m_2 = \frac{m_1 c\Delta\theta}{L} = \frac{2.5 \times 0.39 \times 10^3 \times 500}{335 \times 10^3}$$

$$= 1.455 \text{ kg}$$

$$\approx 1.5 \text{ kg}$$

Q 4. A thermocole icebox is a cheap and efficient method for storing small quantities of cooked food in summer in particular. A cubical icebox of side 30 cm has a thickness of 5.0 cm. If 4.0 kg of ice is put in the box, estimate the amount of ice remaining after 6 h. The outside temperature is 45°C and coefficient of thermal conductivity of thermocole is $0.01 \text{ Js}^{-1}\text{M}^{-1}{}^\circ\text{K}^{-1}$.
[Heat of fusion of water = $335 \times 10^3 \text{ J kg}^{-1}$]

Sol. Area of 6 faces of cube = $6 \times 30 \times 30 \text{ cm}^2$

$$= 6 \times 900 \times 10^{-4} \text{ m}^2$$

Distance, $d = 5.0 \text{ m} = 5.0 \times 10^{-2}$ m

Total mass of ice, $M = 4$ kg

Time, $t = 6 \text{ h} = 6 \times 60 \times 60$ s

θ_1 = temperature outside the box = 45°C

θ_2 = temperature inside the box = 0°C

$\therefore \quad \Delta\theta = \theta_1 - \theta_2 = 45 - 0 = 45°C$

Heat of fusion, $L = 335 \times 10^3 \text{ J kg}^{-1}$

Coefficient of thermal conductivity of thermocole,

$$K = 0.01 \text{ Js}^{-1}\text{m}^{-1}\text{K}^{-1}$$

Let m = mass (kg) of ice which melts.

$\therefore$ Heat needed to melt at 0°C, $Q = mL$ ...(i)

Also $$Q = KA\frac{\Delta\theta}{d}t \quad ...(ii)$$

$\therefore$ From Eqs. (i) and (ii), we get

$$m = \frac{KA}{L}\frac{\Delta\theta}{d}t$$

$$= \frac{0.01 \times 6 \times 900 \times 10^{-4} \times 45}{335 \times 10^3 \times 5.0 \times 10^{-2}} \times 6 \times 3600$$

$$= 0.313 \text{ kg}$$

$\therefore$ Mass of ice left in the box $= M - m$

$$= 4.0 - 0.313$$

$$= 3.687 \text{ kg} \approx 3.7 \text{ kg}$$

Q 5. A brass boiler has a base area 0.15 m^2 and thickness 1.0 cm. It boils water at the rate of 6.0 kg/min, when placed on a gas stove. Estimate the temperature of the part of the flame in contact with the boiler. Thermal conductivity of brass = $109 \text{ Js}^{-1}\text{m}^{-1}\text{K}^{-1}$. (Heat of vaporisation of water $= 2256 \times 10^3 \text{ J kg}^{-1}$)

Sol. Let θ_1 = temperature of the part of the boiler in contact with the stove. If Q be the amount of heat flowing per second through the base of the boiler, then

$$Q = \frac{KA(\theta_1 - \theta_2)}{d}$$

or $$Q = \frac{109 \times 0.15 \times (\theta_1 - 100)}{10^{-2}}$$

$$= 1635\,(\theta_1 - 100) \text{ Js}^{-1} \quad ...(i)$$

Rate of boiling of water in the boiler,

$$M = 6.0 \text{ kg min}^{-1}$$

$$= \frac{6.0}{60} = 0.1 \text{ kg - s}^{-1}$$

$\therefore$ Heat received by water per second, $Q = ML$

or $$Q = 0.1 \times 2256 \times 10^3 \text{ Js}^{-1} \quad ...(ii)$$

$\therefore$ From Eqs. (i) and (ii), we get

$$1635\,(\theta_1 - 100) = 2256 \times 10^{-2}$$

or $$\theta_1 - 100 = \frac{2256 \times 100}{1635} = 138$$

$$\theta_1 = 100 + 138$$

$$= 238°C$$

Q 6. Explain why?

(a) A body with large reflectivity is a poor emitter.

(b) A brass tumbler feels much colder than wooden tray on a chilly day.

(c) The earth without its atmosphere would be inhospitably cold.

(d) Heating system based on circulation of steam are more efficient in warming a building than those based on circulation of hot water.

Sol. (a) According to Kirchhoff's law $e \propto a$ that is good absorbers are good emitters and hence poor reflectors or *vice-versa*.

(b) The thermal conductivity of brass is high, i.e. brass is a good conductor of heat. So when a brass tumbler is touched, heat quickly flows from human body to the tumbler. Consequently, the tumbler appears colder.

On the other hand, wood is a bad conductor of heat. So heat does not flow from the human body to the wooden tray, thus it appears relatively hotter.

(c) Gases are generally insulators. The earth's atmosphere acts like an insulating blanket around it and does not allow heat to escape out but reflects it back to the earth. Had this atmosphere been absent, the earth would naturally be colder, as all its heat would have escaped out.

(d) This is because steam at 100°C has more heat content than water at 100°C. (Latent heat = 540 cal/g)

Q 7. A body cools from 80°C to 50°C in 5 min. Calculate the time it takes to cool from 60°C to 30°C. The temperature of the surroundings is 20°C.

Sol. According to Newton's law of cooling, the rate of cooling $\propto$ difference in temperature.

Here, average of 80°C and 50°C = 65°C

Temperature of surroundings = 20°C

$\therefore \quad$ Difference = 65° − 20° = 45°C

Under these conditions, the body cools 30°C in time 5 min.

$$\therefore \quad \frac{\text{Change in temperature}}{\text{Time}} = k\Delta T$$

or $$\frac{30}{5} = k \times 45 \quad ...(i)$$

The average of 60°C and 30°C is 45°C which is 25°C above the room temperature and the body cools by 30°C in a time t (say).

$$\therefore \quad \frac{30}{t} = k \times 25 \quad ...(ii)$$

where, k is same for this situation as for the original.

Eq. (i) divided by Eq. (ii) gives

$$\frac{30/5}{30/t} = \frac{k \times 45}{k \times 25}$$

or $$\frac{t}{5} = \frac{9}{5} \quad \text{or} \quad t = 9 \text{ min}$$

Objective Problems

[Level 1]

Temperature Change and Phase Change

1. 50 g of ice at 0°C is mixed with 50 g of water at 60°C, final temperature of mixture will be

(a) 0°C (b) 40°C
(c) 10°C (d) 15°C

2. A block of ice at −10° C is slowly heated and converted to steam at 100°C. Which of the following curves represent the phenomenon qualitatively?

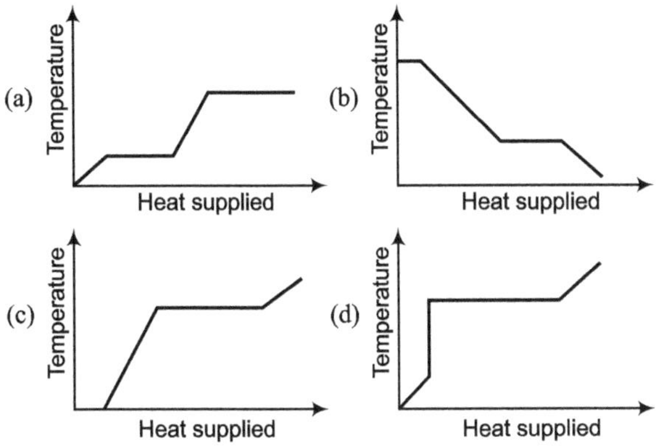

3. An iron ball of mass 0.2 kg is heated to 100°C and put into a block of ice at 0°C. 25 g of ice melts. If the latent heat of fusion of ice is 80 cal/g, then the specific heat of iron in cal/g°C is

(a) 1 (b) 0.1
(c) 0.8 (d) 0.08

4. A metal block is made from a mixture of 2.4 kg of aluminium 1.6 kg of brass and 0.8 kg of copper. The amount of heat required to raise the temperature of this block from 20° C to 80°C is (specific heats of aluminium, brass and copper are 0.216, 0.0917 and 0.0931 cal/kg°C respectively)

(a) 96.2 cal (b) 44.4 cal
(c) 86.2 cal (d) 62.8 cal

5. Steam at 100°C is passed into 1.1 kg of water contained in a calorimeter of water equivalent 0.02 kg at 15°C till the temperature of calorimeter and its contents rises to 80°C. The mass of the steam condensed in kilogram is ($s_w = 10^3$ cal /kg-°C and $L_V = 540 \times 10^3$ cal/kg)

(a) 0.130 (b) 0.065
(c) 0.260 (d) 0.135

6. A block of ice of mass $M = 10$ kg is moved back and forth over the flat horizontal surface of a large block of ice. Both blocks are at 0°C and the force that produces the back and forth motion acts only horizontally. The coefficient of friction between the two surfaces is 0.060. If $m = 15.2$ g of water is produced, the total distance travelled by the upper block relative to the lower is ($L_{ice} = 3.34 \times 10^5$ J/kg)

(a) 432 m (b) 863 m
(c) 368 m (d) 216 m

7. 4 kg of ice at −15°C are added to 5 kg of water at 15°C. The temperature of the resulting mixture equals

(a) −15°C (b) 0°C
(c) 5°C (d) 15°C

8. How much heat energy is gained when 5 kg of water at 20°C is brought to its boiling point (specific heat of water = 4.2 kJ $kg^{-1}\,°C^{-1}$)

(a) 1680 kJ (b) 1700 kJ
(c) 1720 kJ (d) 1740 kJ

9. 80 g of water at 30°C are poured on a large block of ice at 0°C. The mass of ice that melts is

(a) 30 g (b) 80 g
(c) 1600 g (d) 150 g

10. 50 g of copper is heated to increase its temperature by 10°C. If the same quantity of heat is given to 10 g of water, the rise in its temperature is (specific heat of copper = 420 J-$kg^{-1}\,°C^{-1}$)

(a) 5°C (b) 6°C
(c) 7°C (d) 8°C

11. A beaker contains 200 g of water. The heat capacity of the beaker is equal to that of 20 g of water. The initial temperature of water in the beaker is 20°C. If 440 g of hot water at 92°C is poured in it, the final temperature (neglecting radiation loss) will be nearest to

(a) 58°C (b) 68°C
(c) 73°C (d) 78°C

12. A liquid of mass m and specific heat c is heated to a temperature $2T$. Another liquid of mass $\frac{m}{2}$ and specific heat $2c$ is heated to a temperature T. If these two liquids are mixed, the resulting temperature of the mixture is

(a) $\frac{2}{3}T$ (b) $\frac{8}{5}T$ (c) $\frac{3}{5}T$ (d) $\frac{3}{2}T$

13. 100 g of ice at 0°C is mixed with 100 g of water at 100°C. What will be the final temperature of the mixture?
(a) 10°C (b) 20°C
(c) 30°C (d) 0°C

14. A lead bullet of 10 g travelling at 300 m/s strikes against a block of wood and comes to rest. Assuming 50% of heat is absorbed by the bullet, the increase in its temperature is
(a) 100°C (b) 125°C
(c) 150°C (d) 200°C

15. 2 kg of ice at –20°C is mixed with 5 kg of water at 20°C in an insulating vessel having a negligible heat capacity. Calculate the final mass of water remaining in the container. It is given that the specific heats of water and ice are 1 kcal/kg per °C and 0.6 kcal/kg per °C while the latent heat of fusion of ice is 80 kcal/kg
(a) 7 kg (b) 6 kg
(c) 4 kg (d) 2 kg

16. The temperature of equal masses of three different liquids *A, B* and *C* are 12°C, 19°C and 28°C respectively. The temperature when *A* and *B* are mixed is 16°C and when *B* and *C* are mixed is 23°C. The temperature when *A* and *C* are mixed is
(a) 18.2°C (b) 22°C
(c) 20.2°C (d) 25.2°C

17. In an industrial process 10 kg of water per hour is to be heated from 20°C to 80°C. To do this steam at 150°C is passed from a boiler into a copper coil immersed in water. The steam condenses in the coil and is returned to the boiler as water at 90°C, how many kg of steam is required per hour.

(Specific heat of steam = 1 cal per g°C, Latent heat of vaporisation = 540 cal/g)
(a) 1 g
(b) 1 kg
(c) 10 g
(d) 10 kg

18. Which of the substances *A, B* and *C* has the lowest heat capacity, if heat is supplied to all of them at equal rates? The temperature *versus* time graph is shown below:

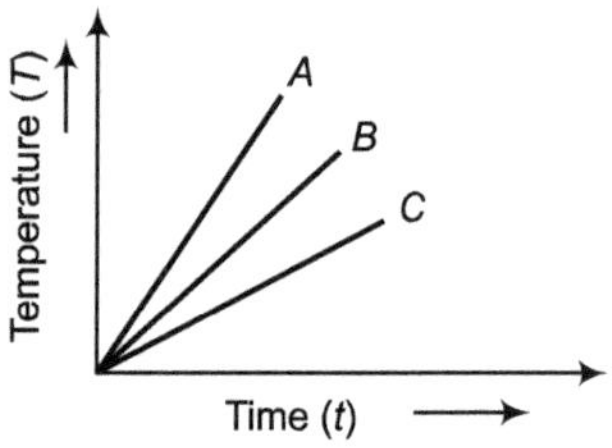

(a) *A*
(b) *B*
(c) *C*
(d) All have equal specific heat

19. The portion *AB* of the indicator diagram representing the state of matter denotes

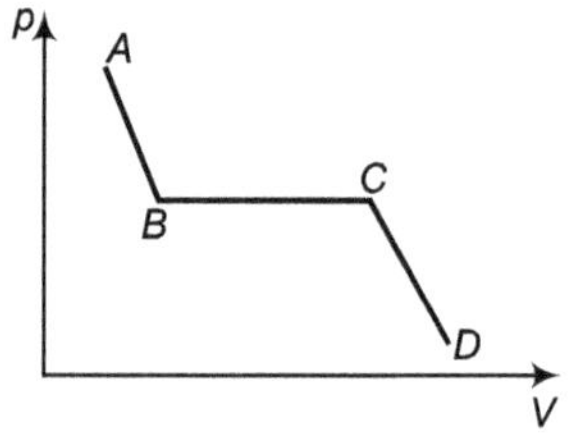

(a) The liquid state of matter
(b) Gaseous state of matter
(c) Change from liquid to gaseous state
(d) Change from gaseous state to liquid state

20. Two substances *A* and *B* of equal mass *m* are heated at uniform rate of 6 cal s^{-1} under similar conditions. A graph between temperature and time is shown in figure. Ratio of heat absorbed $\frac{H_A}{H_B}$ by them for complete fusion is

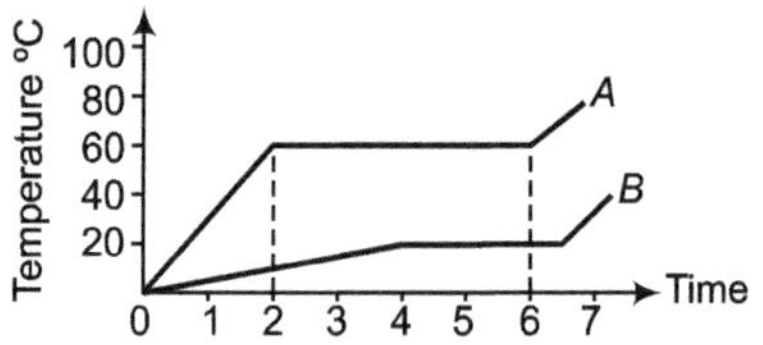

(a) $\frac{9}{4}$ (b) $\frac{4}{9}$
(c) $\frac{8}{5}$ (d) $\frac{5}{8}$

Heat Conduction

21. The dimensional formula for thermal resistance is
(a) $[M^{-1}L^{-2}T^{3}\theta]$ (b) $[ML^{2}T\theta]$
(c) $[M^{-1}L^{2}T^{3}\theta]$ (d) None of these

22. A wall has two layers *A* and *B*, each made of different material. Both the layers have the same thickness. The thermal conductivity for *A* is twice that of *B* and, under steady condition, the temperature difference across the wall is 36°C. The temperature difference across the layer *A* is
(a) 6°C (b) 12°C
(c) 24°C (d) 18°C

23. In a steady state, the temperatures at the end *A* and *B* of 20 cm long rod *AB* are 100°C and 0°C. The temperature of a point 9 cm from *A* is
(a) 45°C (b) 55°C
(c) 5°C (d) 65°C

24. S I unit of thermal conductivity is
(a) J/s-K (b) J/s-m²K (c) J/mK (d) J/s-mK

25. A wall has two layers A and B each made of different materials. The layer A is 10 cm thick and B is 20 cm thick. The thermal conductivity of A is thrice that of B. Under thermal equilibrium temperature difference across the wall is 35°C. The difference of temperature across the layer A is

(a) 20°C (b) 10°C (c) 15°C (d) 5°C

26. In the figure, ABC is a conducting rod whose lateral surfaces are insulated. The length of the section AB is one-half of that of BC, and the respective thermal conductivities of the two sections are as given in the figure. If the ends A and C are maintained at 0°C and 70°C respectively, the temperature of junction B in the steady state is

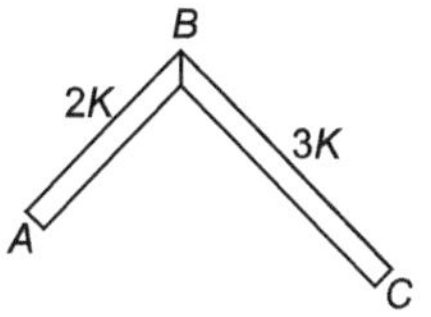

(a) 30°C (b) 40°C (c) 50°C (d) 60°C

27. Equal temperature differences exist between the ends of two metallic rods 1 and 2 of equal lengths. Their thermal conductivities are K_1 and K_2 and area of cross-section are A_1 and A_2, respectively. The condition of equal rates of heat transfer is

(a) $K_1A_2 = K_2A_1$ (b) $K_1A_1 = K_2A_2$
(c) $K_1A_1^2 = K_2A_2^2$ (d) $K_1^2A_2 = K_2^2A_1$

28. Two rods of copper and brass ($K_C > K_B$) of same length and area of cross-section are joined as shown. End A is kept at 100°C and end B at 0°C. The temperature at the junction

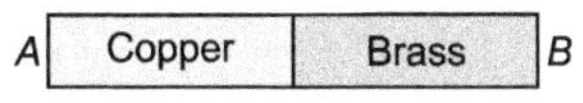

(a) will be more than 50°C
(b) will be less than 50°C
(c) will be 50°C
(d) may be more or less than 50°C depending upon the size of rods

29. Three rods of same dimensions have thermal conductivities $3K, 2K$ and K. They are arranged as shown, with their ends at 100°C, 50°C and 0°C. The temperature of their junction is

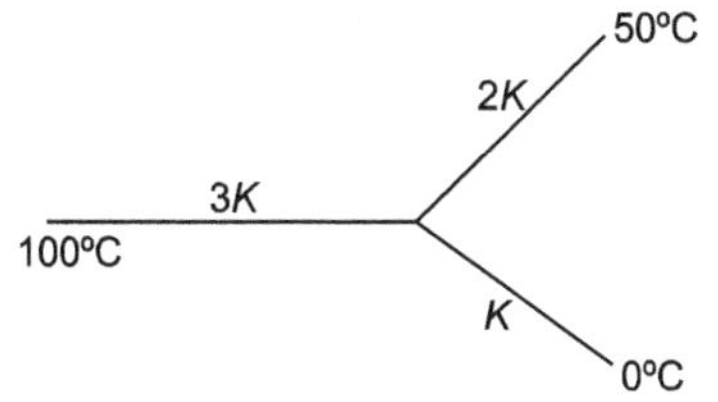

(a) 75°C (b) $\frac{200}{3}$°C (c) 40°C (d) $\frac{100}{3}$°C

30. Three rods made of the same material and having same cross-sectional area but different length 10 cm, 20 cm and 30 cm are joined as shown. The temperature of the junction is

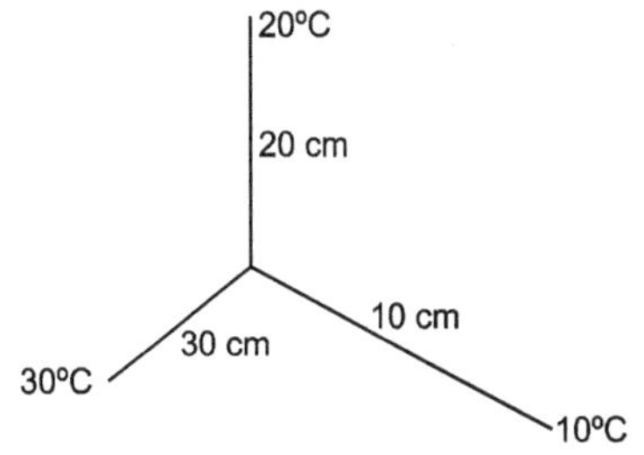

(a) 10.8°C (b) 14.6°C
(c) 16.4°C (d) 18.2°C

31. Two identical square rods of metal are welded end to end as shown in Fig. (i), 20 cal of heat flows through it in 4 min. If the rods are welded as shown in Fig. (ii), the same amount of heat will flow through the rods is

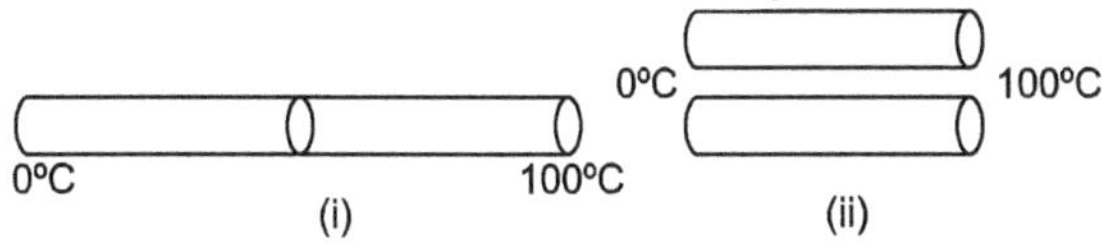

(a) 1 min (b) 2 min
(c) 4 min (d) 16 min

32. The temperature gradient in a rod of 0.5 m long is 80°C/m. If the temperature of hotter end of the rod is 30°C, then the temperature of the cooler end is

(a) 40°C (b) −10°C
(c) 10°C (d) 0°C

33. On heating one end of a rod, the temperature of whole rod will be uniform when

(a) $K = 1$ (b) $K = 0$
(c) $K = 100$ (d) $K = \infty$

34. Two rods of same length and material transfer a given amount of heat in 12 s, when they are joined end to end. But when they are joined lengthwise, then they will transfer same heat in same condition in

(a) 24 s (b) 3 s
(c) 1.5 s (d) 48 s

35. Wires A and B have identical length and have circular cross-sections. The radius of A is twice the radius of B, i.e. $r_A = 2r_B$. For a given temperature difference between the two ends, both wires conduct heat at the same rate. The relation between the thermal conductivities is given by

(a) $K_A = 4K_B$
(b) $K_A = 2K_B$
(c) $K_A = \frac{K_B}{2}$
(d) $K_A = \frac{K_B}{4}$

36. The ends of two rods of different materials with their thermal conductivities, radii of cross-section and lengths all in the ratio 1 : 2 are maintained at the same temperature difference. If the rate of flow of heat in the larger rod is 4 cal/s, then that in the shorter rod in cal/s will be

(a) 1 (b) 2 (c) 8 (d) 16

37. A slab consists of two parallel layers of two different materials of same thickness having thermal conductivities K_1 and K_2. The equivalent conductivity of the combination is

(a) $K_1 + K_2$

(b) $\dfrac{K_1 + K_2}{2}$

(c) $\dfrac{2K_1K_2}{K_1 + K_2}$

(d) $\dfrac{K_1 + K_2}{2K_1K_2}$

38. Three rods of same dimensions are arranged as shown in figure. They have thermal conductivities K_1, K_2 and K_3. The points P and Q are maintained at different temperatures. For the heat flow at the same rate along PRQ and PQ which of the following option is correct?

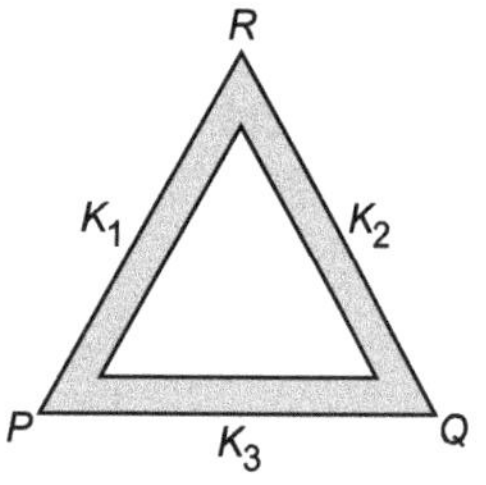

(a) $K_3 = \frac{1}{2}(K_1 + K_2)$ (b) $K_3 = K_1 + K_2$

(c) $K_3 = \dfrac{K_1K_2}{K_1 + K_2}$ (d) $K_3 = 2(K_1 + K_2)$

39. Five rods of same dimensions are arranged as shown in figure. They have thermal conductivities K_1, K_2, K_3, K_4 and K_5. When points A and B are maintained at difference temperatures, no heat flows through the central rod, if

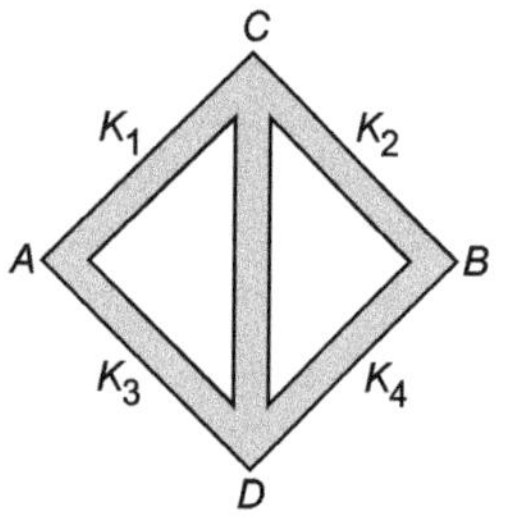

(a) $K_1 = K_4$ and $K_2 = K_3$ (b) $K_1K_4 = K_2K_3$

(c) $K_1K_2 = K_3K_4$ (d) None of these

40. In the figure, the distribution of energy density of the radiation emitted by a black body at a given temperature is shown. The possible temperature of the black body is

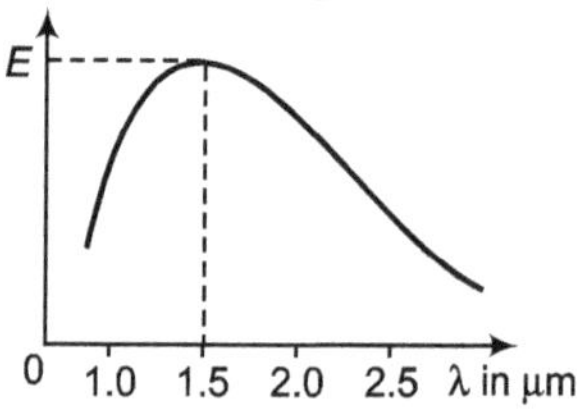

(a) 1500 K (b) 2000 K

(c) 2500 K (d) 3000 K

41. Two similar rods are joined as shown in figure. Then, temperature of junction is (assume no heat loss through lateral surface of rod and temperatures at the ends are shown in steady state)

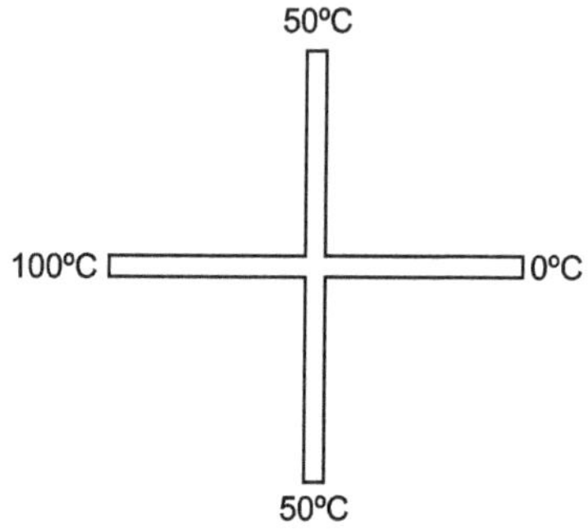

(a) 50°C (b) 75°C (c) 66.6°C (d) 33.3°C

Heat Convection and Radiation

42. If a black body radiates 10 cal/s at 227°C, it will radiate at 727°C

(a) 10 cal/s (b) 80 cal/s

(c) 160 cal/s (d) None of these

43. Two spherical black bodies of radii r_1 and r_2 and with surface temperatures T_1 and T_2, respectively radiate the same power. Then, $\dfrac{r_1}{r_2}$ must be equal to

(a) $\left(\dfrac{T_1}{T_2}\right)^2$ (b) $\left(\dfrac{T_2}{T_1}\right)^2$ (c) $\left(\dfrac{T_1}{T_2}\right)^4$ (d) $\left(\dfrac{T_2}{T_1}\right)^4$

44. The temperature of body is increased from 27°C to 127°C. The radiation emitted by it increases by a factor of

(a) $\dfrac{256}{81}$ (b) $\dfrac{15}{9}$ (c) $\dfrac{4}{5}$ (d) $\dfrac{12}{27}$

45. A sphere has a surface area of 1.0 m^2 and a temperature of 400 K and the power radiated from it is 150 W. Assuming, the sphere is black body radiator, the power in kilowatt radiated when the area expands to 2.0 m^2 and the temperature changes to 800 K.

(a) 6.2 (b) 9.6

(c) 4.8 (d) 16

46. Two spheres of the same material have radii 1 m and 4 m, temperatures 4000 K and 2000 K, respectively. Then, the ratio of energy radiated per second by the first sphere as compared to that by the second is

(a) 4 : 1 (b) 2 : 1
(c) 1 : 1 (d) 1 : 4

47. A spherical black body with radius 12 cm radiates 640 W power at 500 K. If the radius is halved and the temperature doubled, the power radiated in watts would be

(a) 5120 W (b) 640 W
(c) 2560 W (d) 1280 W

48. The ratio of the emissive power to the absorptive power of all substances for a particular wavelength is the same at given temperature. The ratio is known as

(a) the emissive power of a perfectly black body
(b) the emissive power of any type of body
(c) the Stefan's constant
(d) the Wien's constant

49. If wavelength of maximum intensity of radiation emitted by the sun and the moon are 0.5×10^{-6} m and 10^{-4} m respectively, the ratio of their temperatures is

(a) 2000 (b) 1000
(c) 100 (d) 200

50. The spectrum of a black body at two temperatures 27°C and 327°C is shown in the figure. Let A_1 and A_2 be the areas under the two curves, respectively. The value of $\frac{A_2}{A_1}$ is

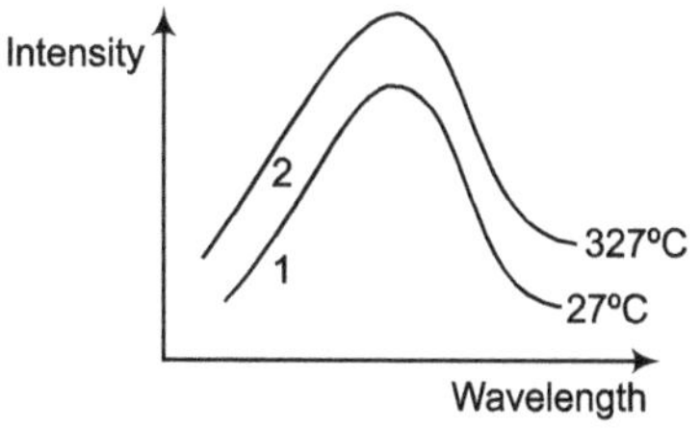

(a) 1 : 16 (b) 4 : 1 (c) 2 : 1 (d) 16 : 1

51. A black body radiates power P and maximum energy is radiated by it around a wavelength λ_0. The temperature of the black body is now changed such that it radiates maximum energy around the wavelength $\frac{3\lambda_0}{4}$. The power radiated by its row is

(a) $\frac{256}{81}P$ (b) $\frac{27}{64}P$ (c) $\frac{64}{27}P$ (d) $\frac{81}{256}P$

52. If a body coated black at 600 K surrounded by atmosphere at 300 K has cooling rate r_0, the same body at 900 K, surrounded by the same atmosphere, will have cooling rate equal to

(a) $\frac{16}{3}r_0$ (b) $\frac{8}{16}r_0$ (c) $16r_0$ (d) $4r_0$

53. For proper ventilation of building, windows must be open near the bottom and top of the walls so as to let pass

(a) more air
(b) cool air near the bottom and hot air near the roof
(c) cool air near the roof and hot air near the bottom
(d) out hot air nears the roof

54. The layers of atmosphere are heated through

(a) convection (b) conduction
(c) radiation (d) None of these

55. In which of the following process, convection does not take place primarily?

(a) sea and land breeze
(b) boiling of water
(c) warming of glass of bulb due to filament
(d) heating air around a furnace

56. If e_λ and a_λ be the emissive power of a body, then according to Kirchhoff's law, which is true (E_λ = emissive power of perfectly black body)

(a) $e_\lambda = a_\lambda = E_\lambda$ (b) $e_\lambda E_\lambda = a_\lambda$
(c) $e_\lambda = a_\lambda E_\lambda$ (d) $e_\lambda a_\lambda E_\lambda$ = constant

57. Distribution of energy in the spectrum of a black body can be correctly represented by

(a) Wien's law (b) Stefan's law
(c) Planck's law (d) Kirchhoff's law

58. There is a black spot on a body. If the body is heated and carried in dark room, then it glows more. This can be explained on the basis of

(a) Newton's law of cooling (b) Wien's law
(c) Kirchhoff's law (d) Stefan's law

59. The maximum wavelength of radiations emitted at 900 K is 4 μm. What will be the maximum wavelength of radiations emitted at 1200 K?

(a) 3 μm (b) 0.3 μm
(c) 1 μm (d) None of these

60. The maximum energy in thermal radiation from a source occurs at wavelength 4000 Å. The effective temperature of the source is approximately

(a) 7225 K (b) 80000 K
(c) 10^4 K (d) 10^6 K

61. The intensity of radiation emitted by the sun has its maximum value at a wavelength of 510 nm and that emitted by the North star has the maximum value at 350 nm. If these stars behave like black bodies, then the ratio of the surface temperature of the sun and North star is

(a) 1.46 (b) 0.69 (c) 1.21 (d) 0.83

62. The area of the hole of heat furnace is 10^{-4} m^2. It radiates 1.58×10^5 cal of heat per hour. If the emissivity of the furnace is 0.80, then its temperature is

(a) 1500 K (b) 2000 K (c) 2500 K (d) 3000 K

63. A spherical black body with a radius of 12 cm radiates 440 W power at 500 K. If the radius were halved and the temperature doubled, the power radiated in watt would be
(a) 440 (b) 1320 (c) 880 (d) 1760

Miscellaneous Problems

64. The calories of heat developed in 200 W heater in 7 min is estimated
(a) 15000 (b) 100
(c) 1000 (d) 20000

65. A sphere, a cube and a thin circular plate are heated to the same temperature. All are made of the same material and have the equal masses. If t_1, t_2 and t_3 are the respective time taken by the sphere, cube and the circular plate in cooling down to a common temperature, then
(a) $t_1 > t_2 > t_3$ (b) $t_1 < t_2 < t_3$
(c) $t_2 > t_1 > t_3$ (d) $t_1 = t_2 = t_3$

66. The sun rays are allowed to fall on a lens of diameter 20 cm. They are then brought to focus on a calorimeter containing 20 g of ice. If the absorption by the lens is negligible, the time required to melt all the ice is

(solar constant = 1.9 cal/min/cm^2 and $L = 80 \text{ cal/g}$)
(a) 6.4 min (b) 3.2 min
(c) 7.2 min (d) 2.7 min

67. If 1 kg water at 100°C is vaporised in open atmosphere. The correct statement is
(a) increase in internal energy is equal to L (L is latent heat of vaporisation for 1 kg)
(b) increase in internal energy is zero
(c) increase in internal energy is less than L
(d) None of the above

68. Thermal capacity of a body depends on
(a) the heat given
(b) the temperature raised
(c) the mass of the body
(d) None of the above

69. The length of the two rods made up of the same metal and having the same area of cross-section are 0.6 m^2 and 0.8 m^2, respectively. The temperature between the ends of first rod is 90°C and 60°C and that for the other rod is 150°C and 110°C. For which rod the rate of conduction will be greater?
(a) First (b) Second
(c) Same for both (d) None of these

70. The thickness of a metallic plate is 0.4 cm. The temperature between its two surfaces is 20°C. The quantity of heat flowing per second is 50 cal from 5 cm^2 area. In CGS system, the coefficient of thermal conductivity will be
(a) 0.4 (b) 0.6
(c) 0.2 (d) 0.5

71. Snow is more heat insulating than ice, because
(a) air is filled in porous of snow
(b) ice is more bad conductor than snow
(c) air is filled in porous of ice
(d) density of ice is more

72. Two thin blankets keep more hotness than one blanket of thickness equal to these two. The reason is
(a) their surface area increases
(b) a layer of air is formed between these two blankets which is bad conductor
(c) these have more wool
(d) they absrob more heat from outside

73. Ice formed over lakes
(a) has very high thermal conductivity and helps in further ice formation
(b) has very low conductivity and retards further formation of ice
(c) permits quick convection and retards further formation of ice
(d) is very good radiator

74. Mud houses are cooler in summer and warmer in winter, because
(a) mud is superconductor of heat
(b) mud is good conductor of heat
(c) mud is bad conductor of heat
(d) None of the above

75. One likes to sit under sunshine in winter seasons, because
(a) the air surrounding the body is hot by which body gets heat
(b) we get energy by the sun
(c) we get heat by conduction by the sun
(d) None of the above

76. Air is bad conductor of heat, still vacuum is preferred between the walls of the thermo flask because
(a) it is difficult to fill the air between the walls of thermo flask
(b) due to more pressure of air, the flask can crack
(c) by convection, heat can flow through air
(d) None of the above

77. Which of the following law states that good absorbers of heat are good emitters?
(a) Stefan's law (b) Kirchhoff's law
(c) Planck's law (d) Wien's law

78. In MKS system, Stefan's constant is denoted by σ. In CGS system, multiplying factor of σ will be
(a) 1 (b) 10^3
(c) 10^5 (d) 10^2

79. A body cools from 50°C to 40°C in 5 min. The surrounding temperature is 20°C. In what further time (in minute) will it cool to 30°C?
(a) 5 (b) $\frac{15}{2}$
(c) $\frac{25}{3}$ (d) 10

80. If a body cools down from 80°C to 60°C in 10 min, when the temperature of the surroundings is 30°C. Then, the temperature of the body after next 10 min will be

(a) 50°C (b) 48°C
(c) 30°C (d) None of these

81. A liquid cools from 50°C to 45°C in 5 min and from 45°C to 41.5°C in the next 5 min. The temperature of the surroundings is

(a) 27°C (b) 40.3°C (c) 23.3°C (d) 33.3°C

82. A substance of mass m kg requires a power input of P watt to remain in the molten state at its melting point. When the power is turned off, the sample completely solidifies in time t s. What is the latent heat of fusion of the substance?

(a) $\frac{Pm}{t}$ (b) $\frac{Pt}{m}$
(c) $\frac{m}{Pt}$ (d) $\frac{t}{Pm}$

[Level 2]

1. Three rods of identical area of cross-section and made from the same metal from the sides of an isosceles ΔABC, right angled at B. The points A and B are maintained at temperatures T and $\sqrt{2}T$, respectively. In the steady state, the temperature of the point C is T_C. Assuming that only heat conduction takes place, $\frac{T_C}{T}$ is equal to

(a) $\frac{1}{(\sqrt{2}+1)}$ (b) $\frac{3}{(\sqrt{2}+1)}$
(c) $\frac{1}{2(2\sqrt{2}-1)}$ (d) $\frac{1}{\sqrt{3}(\sqrt{2}-1)}$

2. Two identical rods are made of different materials whose thermal conductivities are K_1 and K_2. They are placed end to end between two heat reservoirs at temperatures θ_1 and θ_2. The temperature of the junction of the rod is

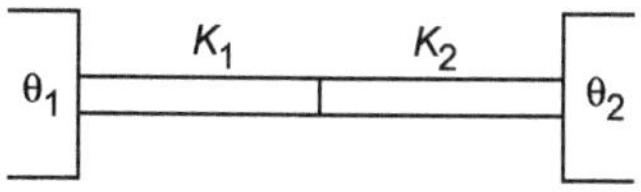

(a) $\frac{\theta_1+\theta_2}{2}$ (b) $\frac{K_1\theta_1+K_2\theta_2}{K_1+K_2}$
(c) $\frac{K_1\theta_2+K_2\theta_1}{K_1+K_2}$ (d) $\frac{K_1\theta_1+K_2\theta_2}{|K_1-K_2|}$

3. Water is being boiled in a flat bottomed kettle placed on a stove. The area of the bottom is 300cm^2 and the thickness is 2 mm. If the amount of steam produced is 1 g min^{-1}, then the difference of the temperature between the inner and the outer surfaces of the bottom is (thermal conductivity of the material of the kettle = 0.5 cal. $\text{cm}^{-1}\,^\circ\text{C}^{-1}\,\text{s}^{-1}$ and latent heat of the steam is equal to 540 cal g^{-1})

(a) 12°C (b) 1.2°C
(c) 0.2°C (d) 0.012°C

4. Three conducting rods of same material and cross-section are shown in figure. Temperatures of A, D and C are maintained at 20°C, 90°C and 0°C respectively. The ratio of lengths BD and BC, if there is no heat flow in AB is

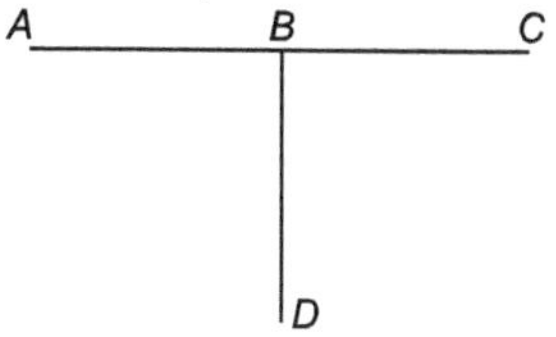

(a) $\frac{2}{7}$ (b) $\frac{7}{2}$ (c) $\frac{9}{2}$ (d) $\frac{2}{9}$

5. Two rods with the same dimensions have thermal conductivities in the ratio 1 : 2. They are arranged between heat reservoirs with the same temperature difference, in two different configurations, A and B. The rates of heat flow in A and B are I_A and I_B, respectively. The ratio $\frac{I_A}{I_B}$ is equal to

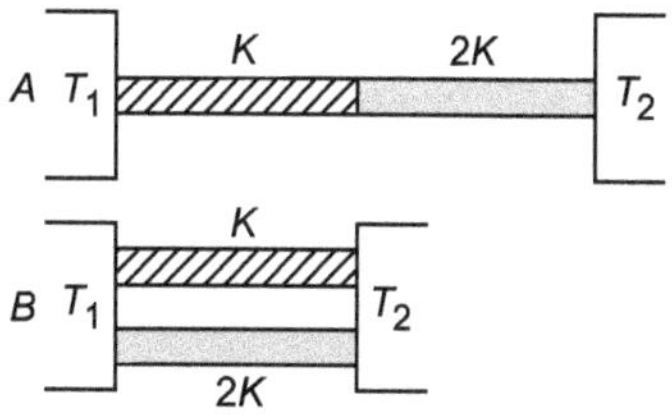

(a) 1 : 2 (b) 1 : 3
(c) 2 : 5 (d) 2 : 9

6. Two identical conducting rods AB and CD are connected to a circular conducting ring at two diametrically opposite points B and C. The radius of the ring is equal to the length of rods AB and CD. The area of cross-section, and thermal conductivity of the rod and ring are equal. Points A and D are maintained at temperatures of 100°C and 0°C. Temperature of point C will be

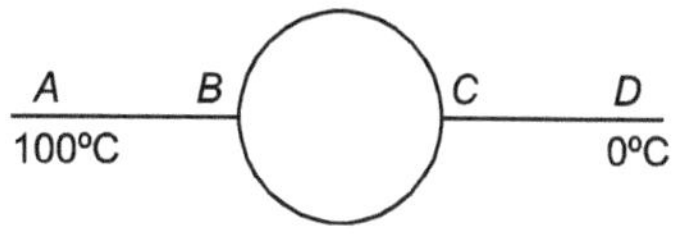

(a) 62°C (b) 37°C (c) 28°C (d) 45°C

7. The temperature change *versus* heat supplied curve is given for 1 kg of a solid block. Then, which of the following statement(s) is/are correct?

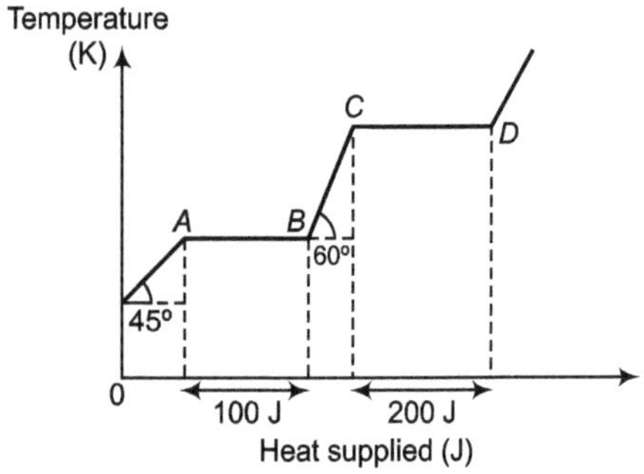

(a) Specific heat capacity of the solid is 1 J/kg-K
(b) Specific heat capacity of liquid phase is $\sqrt{3}$ J/kg-K
(c) Latent heat of vaporisation is 100 J/kg
(d) Latent heat of vaporisation is 200 J/kg

8. The rate of flow of heat through 12 identical conductors made of same material is as shown in the figure. Then, which of the following is correct?

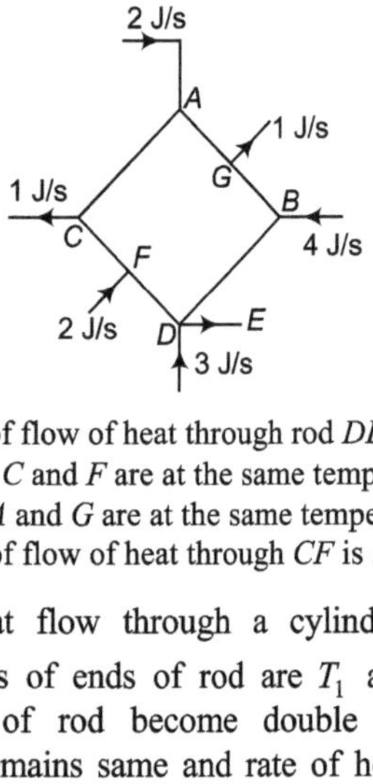

(a) The rate of flow of heat through rod *DE* is 9 J/s
(b) Junctions *C* and *F* are at the same temperature
(c) Junction *A* and *G* are at the same temperature
(d) The rate of flow of heat through *CF* is 5 J/s

9. Rate of heat flow through a cylindrical rod is H_1. Temperatures of ends of rod are T_1 and T_2. If all the dimensions of rod become double and temperature difference remains same and rate of heat flow becomes H_2. Then,

(a) $H_2 = 2H_1$
(b) $H_2 = H_1$
(c) $H_2 = \frac{H_1}{4}$
(d) $H_2 = 4H_1$

10. Equal masses of three liquids *A, B* and *C* have temperatures 10°C, 25°C and 40°C, respectively. If *A* and *B* are mixed, the mixture has a temperature of 15°C. If *B* and *C* are mixed, the mixture has a temperature of 30°C. If *A* and *C* are mixed, the mixture will have a temperature of

(a) 16°C
(b) 20°C
(c) 25°C
(d) 29°C

11. A ring consisting of two parts *ADB* and *ACB* of same conductivity *K* carries an amount of heat *H*. The *ADB* part is now replaced with another metal keeping the temperatures T_1 and T_2 constant. The heat carried increases to 2*H*. What should be the conductivity of the new *ADB* part? (Given, $\frac{ACB}{ADB} = 3$).

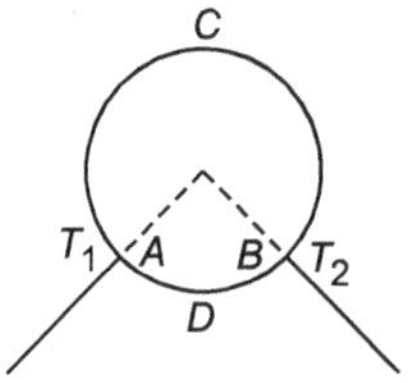

(a) $\frac{7}{3}K$
(b) $2\,K$
(c) $\frac{5}{2}K$
(d) $3\,K$

12. Two identical conducting rods are first connected independently to two vessels, one containing water at 100°C and the other containing ice at 0°C. In the second case, the rods are joined end to end and connected to the same vessels. Let q_1 and q_2 g/s be the rate of melting of ice in two cases, respectively. The ratio of $\frac{q_1}{q_2}$ is

(a) $\frac{1}{2}$
(b) $\frac{2}{1}$
(c) $\frac{4}{1}$
(d) $\frac{1}{4}$

13. The graph, shown in the below diagram, represents the variation of temperature (T) of two bodies, *x* and *y* having same surface area, with time (t) due to the emission of radiation. Find the correct relation between the emissivity (e) and absorptivity (a) of two bodies

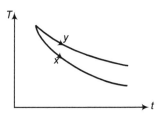

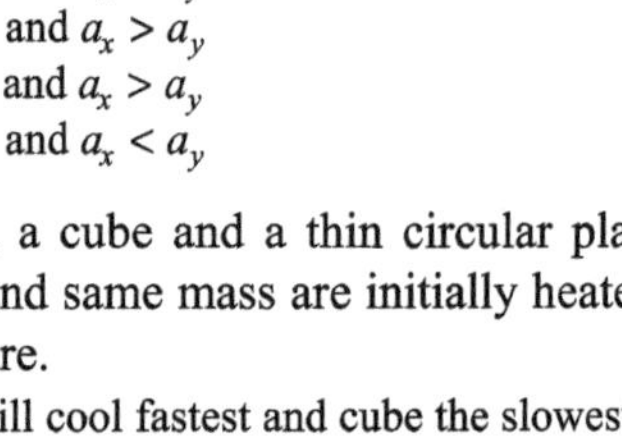

(a) $e_x > e_y$ and $a_x < a_y$
(b) $e_x < e_y$ and $a_x > a_y$
(c) $e_x > e_y$ and $a_x > a_y$
(d) $e_x < e_y$ and $a_x < a_y$

14. A sphere, a cube and a thin circular plate, all of same material and same mass are initially heated to same high temperature.

(a) Plate will cool fastest and cube the slowest
(b) Sphere will cool fastest and cube the slowest
(c) Plate will cool fastest and sphere the slowest
(d) Cube will cool fastest and plate the slowest

15. Three copper blocks of masses M_1, M_2 and M_3 kg respectively, are brought into thermal contact till they reach equilibrium. Before contact, they were at T_1, T_2, T_3 $(T_1 > T_2 > T_3)$. Assuming there is no heat loss to the surroundings, the equilibrium temperature T is (s is specific heat of copper)

(a) $T = \dfrac{T_1 + T_2 + T_3}{3}$

(b) $T = \dfrac{M_1T_1 + M_2T_2 + M_3T_3}{M_1 + M_2 + M_3}$

(c) $T = \dfrac{M_1T_1 + M_2T_2 + M_3T_3}{3(M_1 + M_2 + M_3)}$

(d) $T = \dfrac{M_1T_1s + M_2T_2s + M_3T_3s}{M_1 + M_2 + M_3}$

More than One Correct Options

1. A solid sphere and a hollow sphere of the same material and of equal radii are heated to the same temperature

(a) both will emit equal amount of radiation per unit time in the beginning
(b) both will absorb equal amount of radiation per second from the surroundings in the beginning
(c) the initial rate of cooling will be the same for both the spheres
(d) the two spheres will have equal temperatures at any instant

2. Three identical conducting rods are connected as shown in figure. Given that $\theta_a = 40°$ C, $\theta_c = 30°$ C and $\theta_d = 20°$ C. Choose the correct options.

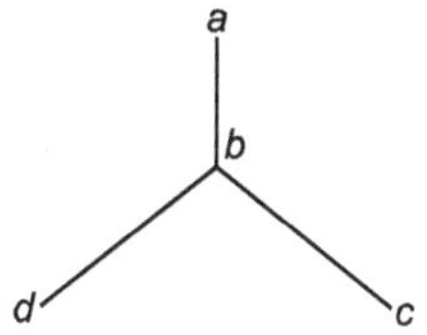

(a) Temperature of junction b is 15°C
(b) Temperature of junction b is 30°C
(c) Heat will flow from c to b
(d) Heat will flow from b to d

3. Two liquids of specific heat ratio 1 : 2 are at temperatures 2θ and θ,

(a) if equal amounts of them are mixed, then temperature of mixture is $1.5\,\theta$
(b) if equal amounts of them are mixed, then temperature of mixture is $\frac{4}{3}\theta$
(c) for their equal amounts, the ratio of heat capacities is 1 : 1
(d)z for their equal amounts, the ratio of their heat capacities is 1 : 2

4. Two conducting rods when connected between two points at constant but different temperatures separately, the rate of heat flow through them is q_1 and q_2.

(a) When they are connected in series, the net rate of heat flow will be $q_1 + q_2$
(b) When they are connected in series, the net rate of heat flow is $\dfrac{q_1q_2}{q_1 + q_2}$
(c) When they are connected in parallel, the net rate of heat flow is $q_1 + q_2$
(d) When they are connected in parallel, the net rate of heat flow is $\dfrac{q_1q_2}{q_1 + q_2}$

5. Choose the correct options.

(a) Good absorbers of a particular wavelength are good emitters of same wavelength. This statement was given by Kirchhoff
(b) At low temperature of a body, the rate of cooling is directly proportional to temperature of the body. This statement was given by the Newton
(c) Emissive power of a perfectly black body is 1
(d) Absorptive power of a perfectly black body is 1

Assertion and Reason

Direction (Q. Nos. 1-17) *These questions consist of two statements each printed as assertion and reason. While answering these questions you are required to choose anyone of the following five responses.*

(a) If both Assertion and Reason are correct and Reason is the correct explanation of Assertion.
(b) If both Assertion and Reason are true but Reason is not the correct explanation of Assertion.
(c) If Assertion is true but Reason is false
(d) If Assertion is false but Reason is true
(e) If both Assertion and Reason are false

1. Assertion A body that is a good radiator is also a good absorber of radiation at a given wavelength.

Reason According to Kirchhoff's law, the absorptivity of a body is equal to its emissivity at a given wavelength.

2. Assertion For higher temperature, the peak emission wavelength of a black body shifts to lower wavelengths.

Reason Peak emission wavelength of a blackbody is proportional to the fourth power of temperature.

3. Assertion Temperatures near the sea coast are moderate.

Reason Water has a high thermal conductivity.

4. Assertion It is hotter over the top of a fire than at the same distance on the sides.

Reason Air surrounding the fire conducts more heat upwards.

5. Assertion Blue star is at high temperature than red star.

Reason Wien's displacement law states that $T \propto \left(\dfrac{1}{\lambda_m}\right)$

6. Assertion Snow is better insulator than ice.

Reason Snow contains air packet and air is good insulator of heat.

7. Assertion All black coloured objects are considered black bodies.

Reason Black colour is a good absorber of heat.

8. Assertion Water vapours at 100°C will burn you more than water at 100°C.

Reason Heat required to convert total mass of any substance from one state to another state is called latent heat.

9. Assertion Heat required to convert 1 g ice at 0°C into vapour at 100°C is 720 cal.

Reason Conversion of solid state directly into vapour state is called vaporisation.

10. Assertion Good conductors of electricity are also good conductors of heat.

Reason In good conductors of electricity, there are a large number of free electrons.

11. Assertion Gravity plays very important role in the process of natural convection.

Reason Convection mainly takes place in liquids and gases.

12. Assertion Absorptive power of any substance is temperature independent. But emissive power depends on the temperature.

Reason Emissive power $\propto T^4$.

13. Assertion If temperature of a body is increased, more number of photons of small wavelengths are radiated.

Reason By increasing the temperature, total energy radiation will increase.

14. Assertion Emissive power of a perfectly black body is one.

Reason Absorptive power of perfectly black body is one.

15. Assertion A normal body cannot radiate energy more than a perfectly black body.

Reason Perfectly black body is always black in colour.

16. Assertion Two conducting rods of same material and same lengths are joined end to end as shown in figure. Heat current H is flowing through them as shown. Temperature difference across rod-1 is more than the temperature difference across rod-2.

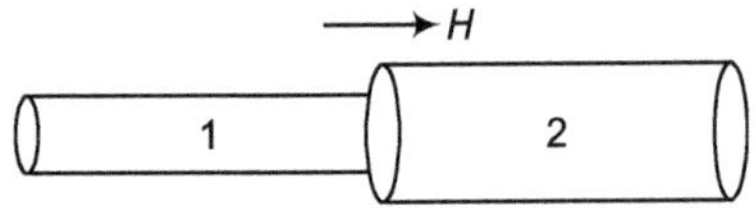

Reason Thermal resistance of rod-1 is less compared to rod-2.

17. Assertion If a body is good absorber of green light, then it will be good reflector of red light.

Reason At a given temperature, the ratio of emissive power to absorptive power is same for all substances.

Match the Columns

1. Match the following columns.

Column I	Column II
(A) Specific heat	(p) $[MLT^{-3}\theta^{-1}]$
(B) Coefficient of thermal conductivity	(q) $[MT^{-3}\theta^{-4}]$
(C) Boltzmann constant	(r) $[L^2T^{-2}\theta^{-1}]$
(D) Stefan's constant	(s) $[ML^2T^{-2}\theta^{-1}]$

2. Three liquids A, B and C having same specific heat and mass m, $2m$ and $3m$ have temperatures 20°C, 40°C and 60°C, respectively. Temperature of the mixture when

Column I	Column II
(A) A and B are mixed	(p) 35°C
(B) A and C are mixed	(q) 52°C
(C) B and C are mixed	(r) 50°C
(D) A, B and C all three are mixed	(s) 45°C
	(t) None

3. Three rods of equal length of same material are joined to form an equilateral ΔABC as shown in figure. Area of cross-sectional of rod AB is S, of rod BC is $2S$ and that of AC is S. Then, match the following columns,

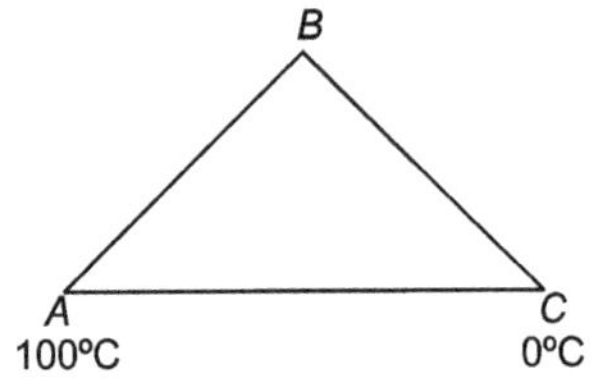

Column I	Column II
(A) Temperature of junction B	(p) Greater than 50°C
(B) Heat current in AB	(q) Less than 50°C
(C) Heat current in BC	(r) Is equal to heat current in BC
	(s) Is $\frac{2}{3}$ times heat current in AC
	(t) None

4. Match the following columns.

Column I	Column II
(A) Thermal resistance	(p) $[MT^{-3}\theta^{-4}]$
(B) Stefan's constant	(q) $[M^{-1}L^{-2}T^3\theta]$
(C) Wien's constant	(r) $[ML^2T^{-3}]$
(D) Heat current	(s) $[L\theta]$

Entrance Gallery

2014

1. Three rods of copper, brass and steel are welded together to from a Y-shaped structure. Area of cross-section of each rod is 4 cm^2. End of copper rod is maintained at 100°C whereas ends of brass and steel are kept at 0°C. Lengths of the copper, brass and steel rods are 46 cm, 13 cm, and 12 cm respectively. The rods are thermally insulated from surroundings except at ends. Thermal conductivities of copper, brass and steel are 0.92, 0.26 and 0.12 in CGS units, respectively. Rate of heat flow through copper rod is **[JEE Main]**

(a) 1.2 cal/s (b) 2.4 cal/s
(c) 5.52 cal/s (d) 6.0 cal/s

2. Consider a black body radiation in a cubical box at absolute temperature T. If the length of each side of the box is doubled and the temperature of the walls of the box and that of the radiation is halved, then the total energy **[WB JEE]**

(a) halves (b) doubles
(c) quadruples (d) remains constant

3. Same quantity of ice is filled in each of the two metal containers P and Q having the same size, shape and wall thickness but made of different materials. The containers are kept in identical surroundings. The ice in P melts completely in time t_1 whereas in Q takes a time t_2. The ratio of thermal conductivities of the materials of P and Q is **[WB JEE]**

(a) $t_2 : t_1$ (b) $t_1 : t_2$
(c) $t_1^2 : t_2^2$ (d) $t_2^2 : t_1^2$

2013

4. Two rectangular blocks, having identical dimensions, can be arranged either in configuration I or in configuration II as shown in the figure. One of the blocks has thermal conductivity K and the other $2K$. The temperature difference between the ends along the X-axis is the same in both the configurations. It takes 9 s to transport a certain amount of heat from the hot end to the cold end in the configuration I. The time to transport the same amount of heat in the configuration II is **[JEE Advanced]**

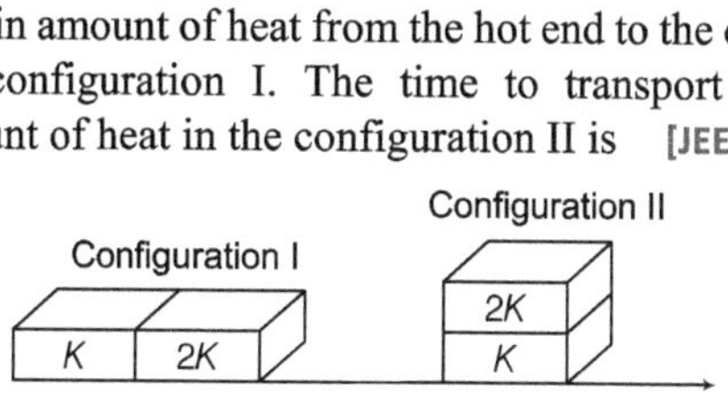

(a) 2.0 s (b) 3.0 s (c) 4.5 s (d) 6.0 s

5. If a piece of metal is heated to temperature θ and then allowed to cool in a room which is at temperature θ_0. The graph between the temperature T of the metal and time t will be closed to **[JEE Main]**

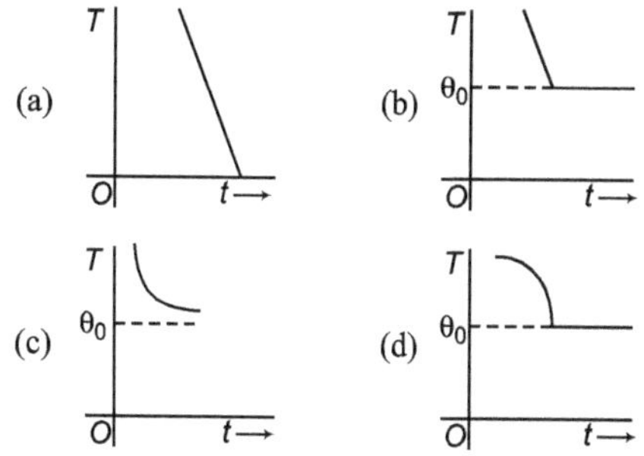

6. The figure shows the temperatures at four faces of a composite slab consisting of four materials S_1, S_2, S_3 and S_4 of identical thickness, through which the heat transfer is steady. Arrange the materials according to their thermal conductivities in decreasing order. **[Karnataka CET]**

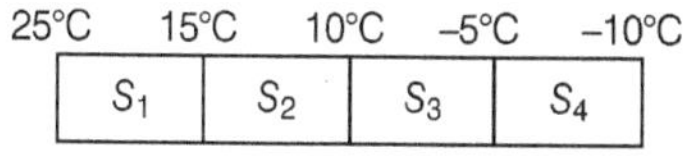

(a) S_2, S_4, S_1, S_3 (b) $S_2 = S_4, S_1, S_3$
(c) $S_1 = S_2, S_3, S_4$ (d) S_1, S_2, S_3, S_4

7. Three rods of same dimensions are arranged as shown in figure, they have thermal conductivities K_1, K_2 and K_3. The points P and Q are maintained at different temperatures for the heat to flow at the same rate along PRQ and PQ. **[Karnataka CET]**

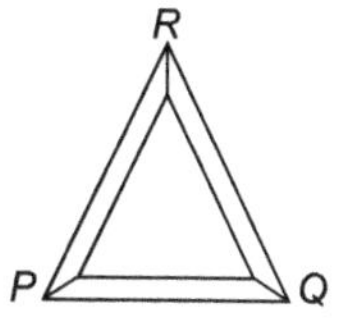

(a) $K_3 = \frac{1}{2}(K_1 + K_2)$ (b) $K_3 = K_1 + K_2$
(c) $K_3 = \frac{K_1 K_2}{K_1 + K_2}$ (d) $K_3 = 2(K_1 + K_2)$

8. A, B and C are three identical conductors but made from different materials. They are kept in contact as shown. Their thermal conductivities are $K, 2K$ and $\frac{K}{2}$. The free end of A is at 100°C and the free end of C is at 0°C. During steady state, the temperature of the junction of A and B is nearly **[Karnataka CET]**

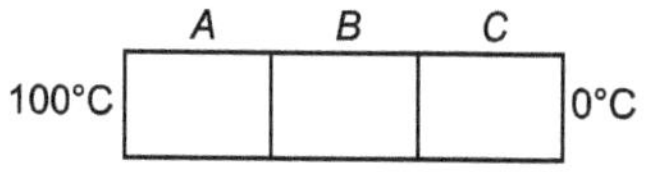

(a) 37°C (b) 71°C (c) 29°C (d) 63°C

9. Two spherical black bodies of radii r_1 and r_2 and with surface temperatures T_1 and T_2 respectively, radiate the same power. Then, the ratio of r_1 and r_2 will be **[Karnataka CET]**

(a) $\left(\frac{T_2}{T_1}\right)^2$ (b) $\left(\frac{T_2}{T_1}\right)^4$
(c) $\left(\frac{T_1}{T_2}\right)^2$ (d) $\left(\frac{T_1}{T_2}\right)^4$

10. A cane is taken out from a refrigerator at 0°C. The atmospheric temperature is 25°C. If t_1 is the time taken to heat from 0°C to 5°C and t_2 is the time taken from 10°C to 15°C, then **[Karnataka CET]**

(a) $t_1 > t_2$ (b) $t_1 < t_2$
(c) $t_1 = t_2$ (d) there is no relation

11. The temperatures of two bodies A and B are respectively, 727°C and 327°C. The ratio $H_A : H_B$ of the rates of heat radiated by them is **[O JEE]**

(a) 727 : 327 (b) 5 : 3
(c) 25 : 9 (d) 625 : 81

12. In a reversible ideal engine, 650 J of heat comes from source of 450 K, heat rejected to sink at 225 K is **[O JEE]**

(a) 325 J (b) 625 J
(c) 700 J (d) 800 J

13. 22320 cal of heat is supplied to 100 g of ice at 0°C. If the latent heat of fusion of ice is 80 cal g^{-1} and latent heat of vaporisation of water is 540 cal g^{-1}, the final obtained amount of water and its temperature respectively, are **[WB JEE]**

(a) 8 g, 100°C (b) 100 g, 90°C
(c) 92 g, 100°C (d) 82 g, 100°C

2012

14. Three very large plates of same area are kept parallel and close to each other. They are considered as ideal black surfaces and have very high thermal conductivity. The first and third plates are maintained at temperatures $2T$ and $3T$, respectively. The temperature of the middle (i.e. second) plate under steady state condition, is **[IIT JEE]**

(a) $\left(\frac{65}{2}\right)^{1/4} T$ (b) $\left(\frac{97}{4}\right)^{1/4} T$
(c) $\left(\frac{97}{2}\right)^{1/4} T$ (d) $(97)^{1/4} T$

15. A liquid in a beaker has temperature $\theta(T)$ at time t and θ_0 is temperature of surroundings, then according to Newton's law of cooling, the correct graph between $\log_e(\theta - \theta_0)$ and t is **[AIEEE]**

(a)
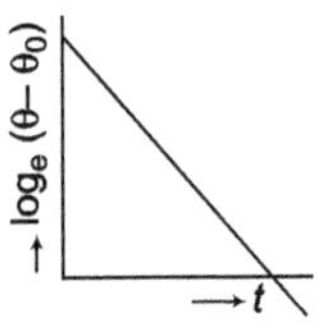

(b)
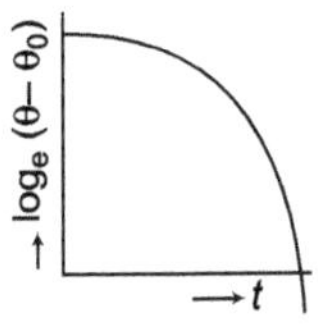

(c)
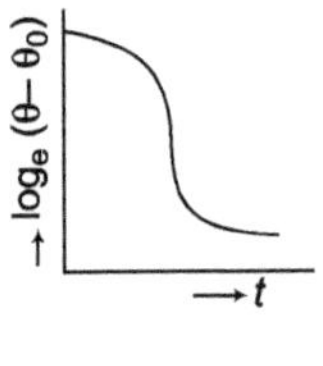

(d)
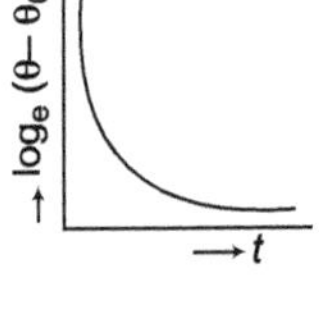

2011

16. A composite block is made of slabs A, B, C, D and E of different thermal conductivities (given in terms of a constant K) and sizes (given in terms of length L) as shown in the figure. All slabs are of same width. Heat Q flows only from left to right through the blocks. Then, in steady state **[IIT JEE]**

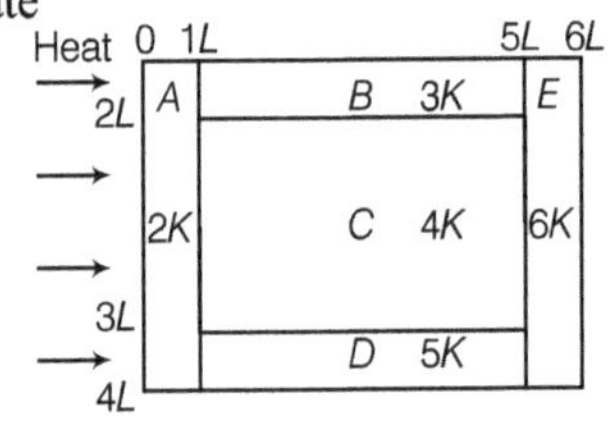

(a) heat flow through A and E slabs are same
(b) heat flow through slab E is maximum
(c) temperature difference across slab E is smallest
(d) heat flow through C = heat flow through B + heat flow through D

17. A lead bullet strikes against a steel plate with a velocity 200 m/s. If the impact is perfectly inelastic and the heat produced is equally shared between the bullet and the target, then the rise in temperature of the bullet is (specific heat capacity of lead = 125 $Jkg^{-1}K^{-1}$) **[Kerala CEE]**

(a) 80°C (b) 60°C
(c) 40°C (d) 120°C
(e) 120°C

18. The quantities of heat required to raise the temperatures of two copper spheres of radii r_1 and r_2 ($r_1 = 1.5 r_2$) through 1 K are in the ratio of **[Karnataka CET]**

(a) 1 (b) $\frac{3}{2}$
(c) $\frac{9}{4}$ (d) $\frac{27}{8}$

19. Three identical rods A, B and C are placed end to end. A temperature difference is maintained between the free ends of A and C. The thermal conductivity of B is thrice that of C and half of that of A.

The effective thermal conductivity of the system will be (K_A is the thermal conductivity of rod A) **[Karnataka CET]**

(a) $\frac{3}{2} K_A$ (b) $2 K_A$ (c) $3 K_A$ (d) $\frac{1}{3} K_A$

20. 1.56×10^5 J of heat is conducted through is 2 m^2 wall of 12 cm thick in one hour. Temperature difference between the two sides of the wall is 20 °C. The thermal conductivity of the material of the wall is (in $Wm^{-1} K^{-1}$) **[WB JEE]**

(a) 0.11 (b) 0.13 (c) 0.15 (d) 1.2

2010

21. A piece of ice (heat capacity $= 2100\ \text{J kg}^{-1}\,^\circ\text{C}^{-1}$ and latent heat $= 3.36\times 10^5\ \text{J kg}^{-1}$) of mass m gram is at -5°C at atmospheric pressure. It is given 420 J of heat, so that the ice starts melting. Finally, when the ice-water mixture is in equilibrium, it is found that 1 g of ice has melted. Assuming there is no other heat exchange in the process, the value of m is **[IIT JEE]**

(a) 8 (b) 6
(c) 4 (d) 8.5

22. Two spherical bodies A (radius 6 cm) and B (radius 18 cm) are at temperatures T_1 and T_2, respectively. The maximum intensity in the emission spectrum of A is at 500 nm and in that of B is at 1500 nm. Considering them to be black bodies, what will be the ratio of the rate of total energy radiated by A to that of B? **[IIT JEE]**

(a) 9 (b) 9.5 (c) 8 (c) 8.5

23. A long metallic bar is carrying heat from one of its ends to the other end under steady-state. The variation of temperature θ along the length x of the bar from its hot end is best described by which of the following figure? **[AIEEE]**

(a)
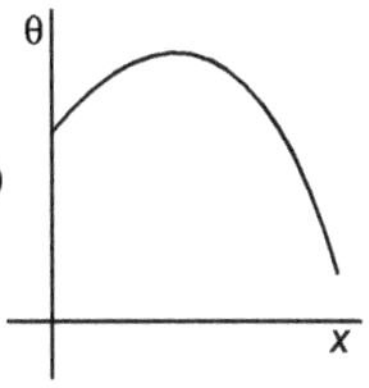

(b)
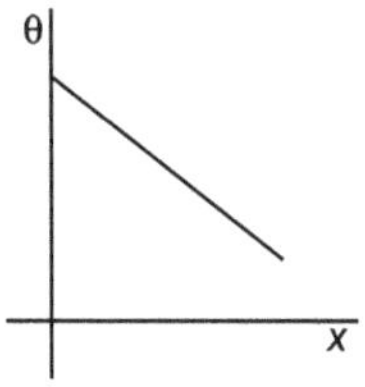

(c)
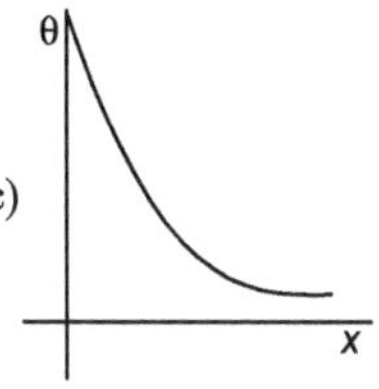

(d)
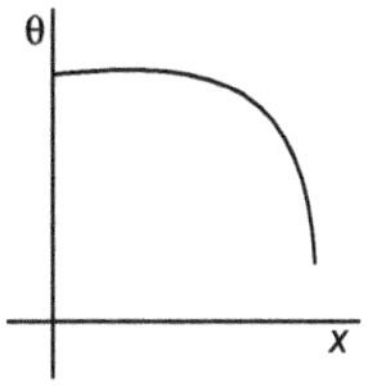

24. Two slabs are of the thicknesses d_1 and d_2. Their thermal conductivities are K_1 and K_2, respectively. They are in series. The free ends of the combination of these two slabs are kept at temperatures θ_1 and θ_2. Assume $\theta_1 > \theta_2$. The temperature θ of their common junction is **[Karnataka CET]**

(a) $\dfrac{K_1\theta_1 + K_2\theta_2}{\theta_1 + \theta_2}$ (b) $\dfrac{K_1\theta_1 d_1 + K_2\theta_2 d_2}{K_1 d_2 + K_2 d_1}$

(c) $\dfrac{K_1\theta_1 d_2 + K_2\theta_2 d_1}{K_1 d_2 + K_2 d_1}$ (d) $\dfrac{K_1\theta_1 + K_2\theta_2}{K_1 + K_2}$

25. For an opaque body, coefficient of transmission is **[MHT CET]**

(a) zero
(b) 1
(c) 0.5
(d) ∞

26. Two spheres of radii 8 cm and 2 cm are cooling. Their temperatures are 127°C and 527°C, respectively. Find the ratio of energy radiated by them in the same time **[MHT CET]**

(a) 0.06 (b) 0.5
(c) 1 (d) 2

Answers

Level 1

Objective Problems

1. (a)	**2.** (a)	**3.** (b)	**4.** (b)	**5.** (a)	**6.** (b)	**7.** (b)	**8.** (a)	**9.** (a)	**10.** (a)
11. (b)	**12.** (d)	**13.** (a)	**14.** (c)	**15.** (b)	**16.** (c)	**17.** (b)	**18.** (a)	**19.** (a)	**20.** (c)
21. (a)	**22.** (b)	**23.** (b)	**24.** (d)	**25.** (d)	**26.** (a)	**27.** (b)	**28.** (a)	**29.** (b)	**30.** (c)
31. (a)	**32.** (b)	**33.** (d)	**34.** (d)	**35.** (d)	**36.** (a)	**37.** (b)	**38.** (c)	**39.** (b)	**40.** (b)
41. (a)	**42.** (c)	**43.** (b)	**44.** (a)	**45.** (c)	**46.** (c)	**47.** (c)	**48.** (a)	**49.** (d)	**50.** (d)
51. (a)	**52.** (a)	**53.** (b)	**54.** (a)	**55.** (c)	**56.** (c)	**57.** (c)	**58.** (c)	**59.** (a)	**60.** (a)
61. (b)	**62.** (c)	**63.** (d)	**64.** (d)	**65.** (a)	**66.** (d)	**67.** (c)	**68.** (c)	**69.** (c)	**70.** (c)
71. (a)	**72.** (b)	**73.** (b)	**74.** (c)	**75.** (a)	**76.** (c)	**77.** (b)	**78.** (b)	**79.** (c)	**80.** (b)
81. (d)	**82.** (b)								

Level 2

Only One Correct Option

1. (b)	**2.** (b)	**3.** (d)	**4.** (b)	**5.** (d)	**6.** (c)	**7.** (a,d)	**8.** (a,b)	**9.** (a)	**10.** (a)
11. (a)	**12.** (c)	**13.** (c)	**14.** (c)	**15.** (b)					

More than One Correct Options

1. (a,b) **2.** (b,d) **3.** (b,d) **4.** (b,c) **5.** (a,d)

Assertion and Reason

1. (a) **2.** (c) **3.** (b) **4.** (c) **5.** (a) **6.** (a) **7.** (d) **8.** (c) **9.** (c) **10.** (a)
11. (b) **12.** (a) **13.** (b) **14.** (d) **15.** (e) **16.** (c) **17.** (d)

Match the Columns

1. (A → r, B → p, C → s, D → q) **2.** (A → t, B → r, C → q, D → t) **3.** (A → p, B → r, C → t)
4. (A → q, B → p, C → s, D → r)

Entrance Gallery

1. (c) **2.** (d) **3.** (a) **4.** (a) **5.** (c) **6.** (b) **7.** (c) **8.** (b) **9.** (a) **10.** (b)
11. (d) **12.** (a) **13.** (a) **14.** (c) **15.** (a) **16.** (a,c,d) **17.** (a) **18.** (c) **19.** (d) **20.** (b)
21. (a) **22.** (a) **23.** (b) **24.** (c) **25.** (a) **26.** (c)

Solutions

Level 1 : Objective Problems

1. Heat liberated when 50 g water at 60°C converts into water at 0°C.

$$Q = ms\Delta\theta = 50 \times 1 \times 60 = 3000 \text{ cal}$$

From this heat mass of ice which can be melted is

$$m = \frac{Q}{L} = \frac{3000}{80} = 37.5 \text{ g}$$

i.e. whole ice is not melted and temperature of mixture will be 0°C.

2. Temperature first increases from −10°C to 0°C. Then, remains constant at 0°C till whole ice is melted. Then, it increases from 0°C to 100°C. Again it remains constant at 100°C till whole water converts into steam.

3. Heat taken = Heat given

or $$Ms\Delta\theta = mL$$

$\therefore$ $$(200)(s)(100) = 25 \times 80$$

$\therefore$ $$s = 0.1 \text{ cal/g°C}$$

4. $Q = m_1 s_1 \Delta\theta + m_2 s_2 \Delta\theta + m_3 s_3 \Delta\theta$

$$= (2.4 \times 0.216 + 1.6 \times 0.0917 + 0.8 \times 0.0931)(60)$$
$$= 44.376 \text{ cal} \approx 44.44 \text{ cal}$$

5. Heat required,

$$Q_1 = (1.1 + 0.02) \times 10^3 \times (80 - 15) = 72800 \text{ cal}$$

Suppose, m kg of steam is condensed and it also reaches to 80°C. Then, heat rejected

$$Q_2 = mL + ms\Delta\theta$$
$$= m(540 \times 10^3 + 20 \times 10^3)$$
$$= 560 \times 10^3 \, m$$

Equating Q_1 and Q_2, we get

$$m = 0.13 \text{ kg}$$

6. $\mu Mgd = mL$

$\therefore$ $$d = \frac{mL}{\mu Mg}$$

$$= \frac{(15.2 \times 10^{-3})(3.34 \times 10^5)}{0.06 \times 10 \times 9.8}$$
$$= 863.4 \text{ m} \approx 863 \text{ m}$$

7. Heat released when 5 kg of water at 15°C converts into 0°C of water is $Q = ms\Delta\theta = (5000)(1)(15) = 75000$ cal. This heat can melt mass of water $m = \frac{Q}{L} = \frac{75000}{80} = 937.5$ g of 0.9375 kg.

Since, whole of ice is not melting, temperature of mixture will be 0°C.

8. $Q = ms\,\Delta\theta = 5 \times (1000 \times 4.2) \times (100 - 20)$

$$= 1680 \times 10^3 \text{ J} = 1680 \text{ kJ}$$

9. If m g ice melts, then

$$\text{Heat lost} = \text{Heat gain}$$
$$80 \times 1 \times (30 - 0) = m \times 80$$

$\Rightarrow$ $$m = 30 \text{ g}$$

10. Same amount of heat is supplied to copper and water.

So, $$m_C c_C \Delta\theta_C = m_W c_W \Delta\theta_W$$

$\Rightarrow$ $$\Delta\theta_\Omega = \frac{\mu_X \chi_X (\Delta\theta)_X}{\mu_\Omega \chi_\Omega}$$

$$= \frac{50 \times 10^{-3} \times 420 \times 10}{10 \times 10^{-3} \times 4200}$$
$$= 5°\text{C}$$

11. Heat lost by hot water = Heat gained by cold water in beaker + Heat absorbed by beaker

$\Rightarrow$ $$440(92° - \theta) = 200 \times (\theta - 20°) + 20 \times (\theta - 20°)$$

$\Rightarrow$ $$\theta = 68°\text{C}$$

12. Temperature of mixture,

$$\theta_{\text{mix}} = \frac{m_1 c_1 \theta_1 + m_2 c_2 \theta_2}{m_1 c_1 + m_2 c_2} = \frac{m \times c \times 2T + \frac{m}{2}(2c)T}{m \times c + \frac{m}{2}(2c)} = \frac{3}{2}T$$

13. $(100 \times 80) + (100 \times 1 \times \theta) = 100 \times 1 \times (100° - \theta)$

or $\theta = 10°C$

14. $ms\Delta\theta = \frac{1}{2}\left(\frac{1}{2}mv^2\right)$

$\therefore \ \Delta\theta = \frac{v^2}{4s} = \frac{(300)^2}{4 \times 150} = 150°C$

15. Initially, ice will absorb heat to raise its temperature 0°C, then its melting takes place.

If m_i = initial mass of ice, m'_i = mass of ice that melts and m_W = initial mass of water

By law of mixture heat gained by ice = heat lost by water

$\Rightarrow \quad m_1 \times c \times (20) + m'_i \times L = m_W c_W \ (20)$

$\Rightarrow \quad 2 \times 0.5 \ (20) + m_i \times 80 = 5 \times 1 \times 20$

$\Rightarrow \quad m'_i = 1 \text{ kg}$

So, final mass of water = initial mass of water + mass of ice that melts $= 5 + 1 = 6 \text{ kg}$

16. Heat gain = Heat lost

$m_A c_A \Delta T_A = m_B c_B \Delta T_B$

$c_A (16 - 12) = c_B (19 - 16) \quad (\because m_A = m_B)$

$\Rightarrow \quad \frac{c_A}{c_B} = \frac{3}{4}$

and $\quad c_B (23 - 19) = c_C (28 - 23)$

$\Rightarrow \quad \frac{c_B}{c_C} = \frac{5}{4}$

$\Rightarrow \quad \frac{c_A}{c_C} = \frac{15}{16} \quad$ ...(i)

Similarly, if θ is the temperature, when A and C are mixed then,

$c_A (\theta - 12°) = c_C (28° - \theta)$

$\Rightarrow \quad \frac{c_A}{c_C} = \frac{28° - \theta}{\theta - 12°} \quad$ ...(ii)

On solving Eqs. (i) and (ii), we have

$\theta = 20.2°C$

17. Suppose, m kg steam is required per hour.

Heat is released by steam in following three steps

(i) When 150°C steam $\xrightarrow[Q_1]{}$ 100 °C steam

$Q_1 = mc_{\text{Steam}} \Delta\theta$

$= m \times 1(150 - 100)$

$= 50m$ cal

(ii) When 100°C steam $\xrightarrow[Q_2]{}$ 100°C water

$Q_2 = mL_V = m \times 540 = 540m \text{cal}$

(iii) When 100°C water $\xrightarrow[Q_2]{}$ 90°C water

$Q_3 = mc_W \Delta\theta$

$= m \times 1 \times (100 - 90)$

$= 10m$

Hence, total heat given by the steam

$Q = Q_1 + Q_2 + Q_3 = 600m \quad$ ...(i)

Heat taken by 10 kg water,

$Q' = mc_W \Delta\theta = 10 \times 10^3 \times 1 \times (80 - 20)$

$= 600 \times 10^3$ cal

Hence, $\quad Q = Q'$

$\Rightarrow \quad 600m = 600 \times 10^3$

$\Rightarrow \quad m = 10^3 \text{ g} = 1 \text{ kg}$

18. Substances having more heat capacity take longer time to get heated to a higher temperature and longer time to get cooled.

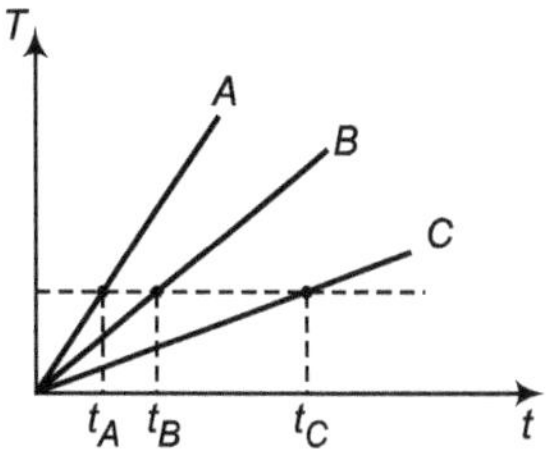

If we draw a line parallel to the time axis, then it cuts the given graphs at three different points. Corresponding points on the time axis shows that

$t_C > t_B > t_A$

$\Rightarrow \quad C_C > C_B > C_A$

19. The volume of matter in portion AB of the curve is almost constant and pressure is decreasing. These are the characteristics of liquid state.

20. From given curve,

Melting point for, $A = 60°C$

And melting point for, $B = 20°C$

Time taken by A for fusion $= (6 - 2) = 4 \text{ min}$

Time taken by B for fusion $= (6.5 - 4) = 2.5 \text{ min}$

Then, $\quad \frac{H_A}{H_B} = \frac{6 \times 4 \times 60}{6 \times 2.5 \times 60} = \frac{8}{5}$

21. Thermal resistance, $R = \frac{\text{Temperature difference}}{\text{Heat current}}$

$\therefore \quad [R] = \frac{[\theta]}{[\text{ML}^2\text{T}^{-2}/\text{T}]}$

$= [\text{M}^{-1}\text{L}^{-2}\text{T}^3\theta]$

22. $K_A = 2K_B \Rightarrow \therefore \ R_A = \frac{R_B}{2} \quad \left(\because R = \frac{l}{KA}\right)$

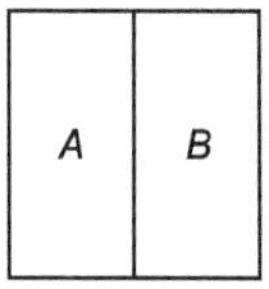

Suppose, $\quad R_A = R$, then $R_B = 2R$

Heat current, $\quad H = \frac{36}{R + 2R} = \frac{36}{3R} = \frac{12}{R}$

$\therefore$ Temperature difference across A

$= H \times R_A = \frac{12}{R} \times R = 12°C$

23. $H = \frac{100 - 0}{R_{AB}} = \text{Heat current} = \frac{100}{R_{AB}}$

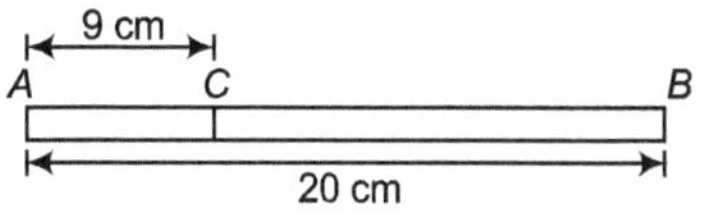

$$100-\theta_C = H\cdot R_{AC} = \frac{100}{R_{AB}}\cdot R_{AC} = \frac{100\times 9}{20} = 45$$

$\therefore \qquad \theta_C = 55\,°C$

24. $Q = \dfrac{KA(\theta_1-\theta_2)t}{l}$

$\therefore \qquad K = \dfrac{Q\times l}{A(\theta_1-\theta_2)t}$

Unit of $K = \dfrac{\text{J-m}}{\text{m}^2\text{K-s}} = \dfrac{\text{J}}{\text{s-mK}}$

25. Thermal resistance of A is $R_A = \dfrac{10}{3KA}$ $\left(\because R = \dfrac{l}{KA}\right)$

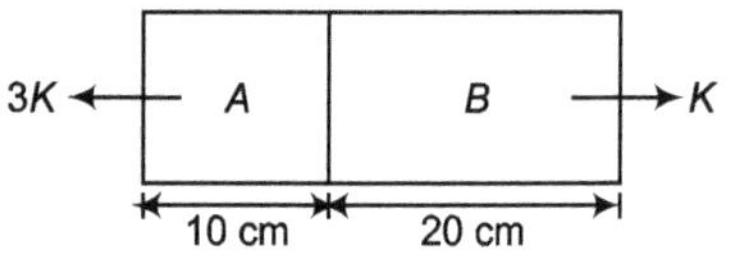

$$R_B = \frac{20}{KA} \Rightarrow \frac{R_A}{R_B} = \frac{1}{6}$$

So let $\qquad R_A = R$, then $R_B = 6R$

Heat current, $H = \dfrac{\text{Temperature difference}}{\text{Total resistance}}$

$$= \frac{35}{7R} = \frac{5}{R}$$

Now, temperature difference across

$$A = H\times R_A = \frac{5}{R}\times R = 5°C$$

26. Heat currents in both the rods are equal.

or $\qquad H_{CB} = H_{BA}$ or $\dfrac{\theta_C-\theta_B}{(2l/3KA)} = \dfrac{\theta_B-\theta_A}{(l/2KA)}$

$\therefore \quad \dfrac{3}{2}(70-\theta_B) = 2(\theta_B-0)$ or $\theta_B = 30°C$

27. $H_1 = H_2$

$\therefore \quad \left(\dfrac{\text{TD}}{R}\right)_1 = \left(\dfrac{\text{TD}}{R}\right)_2$ (TD = Temperature difference)

or $\qquad R_1 = R_2$

$\therefore \qquad \dfrac{l}{K_1A_1} = \dfrac{l}{K_2A_2}$

or $\qquad K_1A_1 = K_2A_2$

28. $K_C > K_B \therefore R_C < R_B$

Temperature difference across copper = $HR_C = (\text{TD})_C$

And temperature difference across brass = $HR_B = (\text{TD})_B$

Heat current H will be same in both as they are in series.

Since, $\qquad R_C < R_B$

$\therefore \qquad (\text{TD})_C < (\text{TD})_B$

Or temperature of junction is more than 50°C.

29. Suppose, θ be the temperature of junction. H_1, H_2 and H_3 are heat currents. Then, $H_1 = H_2 + H_3$

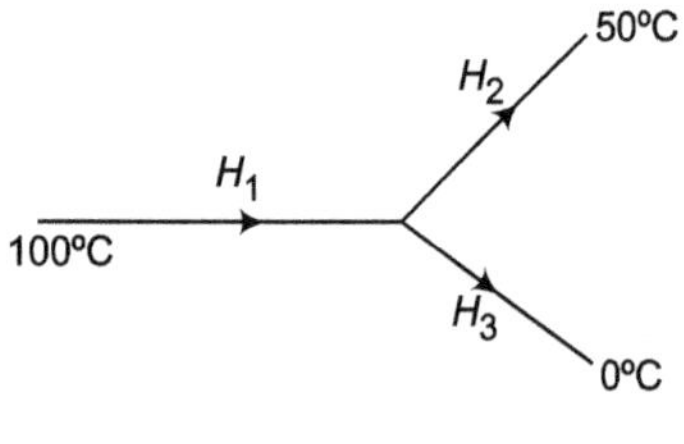

or $\qquad \dfrac{100-\theta}{(l/3KA)} = \dfrac{\theta-50}{(l/2KA)} + \dfrac{\theta-0}{(l/KA)}$

or $\qquad 3(100-\theta) = 2(\theta-50)+\theta$

Solving, we get $\qquad \theta = \dfrac{200}{3}°C$

30. Let θ be the temperature of junction, H_1, H_2 and H_3 the heat currents. Then,

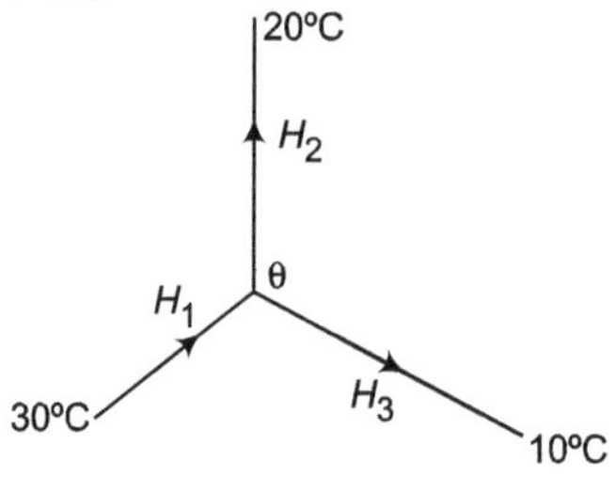

$$H_1 = H_2 + H_3$$

or $\qquad \dfrac{30-\theta}{(30/KA)} = \dfrac{\theta-20}{(20/KA)} + \dfrac{\theta-10}{(10/KA)}$

or $\qquad 2(30-\theta) = 3(\theta-20)+(\theta-10)$

or $\qquad \theta = 16.36°C \approx 16.4°C$

31. $\dfrac{Q}{t} = \dfrac{KA\Delta\theta}{l} = \dfrac{\Delta\theta}{(l/KA)} = \dfrac{\Delta\theta}{R}$ (R = Thermal resistance)

$\Rightarrow \qquad t \propto R \qquad (\because Q$ and $\Delta\theta$ are same)

$\Rightarrow \qquad \dfrac{t_p}{t_s} = \dfrac{R_p}{R_s} = \dfrac{R/2}{2R} = \dfrac{1}{4}$

$\Rightarrow \qquad t_p = \dfrac{t_s}{4} = \dfrac{4}{4} = 1$ min

(Series resistance $R_s = R_1 + R_2$ and parallel resistance $R_p = \dfrac{R_1R_2}{R_1+R_2}$) $\qquad (\therefore R_1 = R_2)$

32. $\dfrac{\theta_1-\theta_2}{l} = 80 \Rightarrow \dfrac{30-\theta_2}{0.5} = 80$

$\Rightarrow \qquad \theta_2 = -10°C$

33. $\dfrac{dQ}{dT} = -KA\dfrac{d\theta}{dx}$, when $K = \infty, \dfrac{d\theta}{dx} = 0$

i.e. θ is independent of x, i.e. constant or uniform.

34. Let the heat transferred be Q

When rods are joined end to end. Heat transferred by each rod.

$$= Q = \frac{K(2A)\Delta\theta}{l}\times 12 \qquad \text{...(i)}$$

When rods are joined lengthwise,

$$Q = \frac{KA\Delta\theta}{2l}t \qquad \text{...(ii)}$$

From Eqs. (i) and (ii), we get $t = 48$ s

35. $\dfrac{Q}{t} = \dfrac{KA\Delta\theta}{l}$

$\Rightarrow \qquad \dfrac{K_A}{K_B} = \dfrac{A_B}{A_A}$

$$= \left(\frac{r_B}{r_A}\right)^2 = \frac{1}{4}$$

$\Rightarrow \qquad K_A = \dfrac{K_B}{4}$

36. $\frac{dQ}{dt} = \frac{K(\pi r^2)\Delta\theta}{l} \Rightarrow \frac{\left(\frac{dQ}{dt}\right)_2}{\left(\frac{dQ}{dt}\right)_1}$

$$= \frac{K_2 \times r_2^2 \times l_1}{K_1 \times r_1^2 \times l_2} = \frac{1}{2} \times \frac{1}{4} \times \frac{2}{1}$$

$$\Rightarrow \left(\frac{dQ}{dt}\right)_2 = \frac{\left(\frac{dQ}{dt}\right)_1}{4} = \frac{4}{4} = 1$$

37. In parallel combination,

$$\frac{1}{R} = \frac{1}{R_1} + \frac{1}{R_2} \quad (R = \text{thermal resistance})$$

or $$\frac{2KA}{l} = \frac{K_1 A}{l} + \frac{K_2 A}{l}$$

$\therefore$ $$K = \frac{K_1 + K_2}{2}$$

38. Thermal resistance across three rods is $R_{PR} + R_{RQ} = R_{PQ}$

$\therefore$ $$\frac{l}{K_1 A} + \frac{l}{K_2 A} = \frac{l}{K_3 A}$$

or $$K_3 = \frac{K_1 K_2}{K_1 + K_2}$$

39. For no current flow between C and D

$$\left(\frac{Q}{t}\right)_{AC} = \left(\frac{Q}{t}\right)_{CB}$$

$$\Rightarrow \frac{K_1 A(\theta_A - \theta_C)}{l} = \frac{K_2 A(\theta_C - \theta_B)}{l}$$

$$\Rightarrow \frac{\theta_A - \theta_C}{\theta_C - \theta_B} = \frac{K_2}{K_1} \quad \text{...(i)}$$

Also, $$\left(\frac{Q}{t}\right)_{AD} = \left(\frac{Q}{t}\right)_{DB}$$

$$\Rightarrow \frac{K_3 A(\theta_A - \theta_D)}{l} = \frac{K_4 A(\theta_D - \theta_B)}{l}$$

$$\Rightarrow \frac{\theta_A - \theta_D}{\theta_D - \theta_B} = \frac{K_4}{K_3} \quad \text{...(ii)}$$

It is given that $\theta_C = \theta_D$, hence from Eqs. (i) and (ii), we get

$$\frac{K_2}{K_1} = \frac{K_4}{K_3}$$

$$\Rightarrow K_1 K_4 = K_2 K_3$$

40. $\lambda_m T = b$, where $b = 2.89 \times 10^{-3}$ mK

$$\Rightarrow T = \frac{b}{\lambda_m} = \frac{2.89 \times 10^{-3}}{1.5 \times 10^{-6}} = 2000\,\text{K}$$

41. Let θ be the temperature of junction.

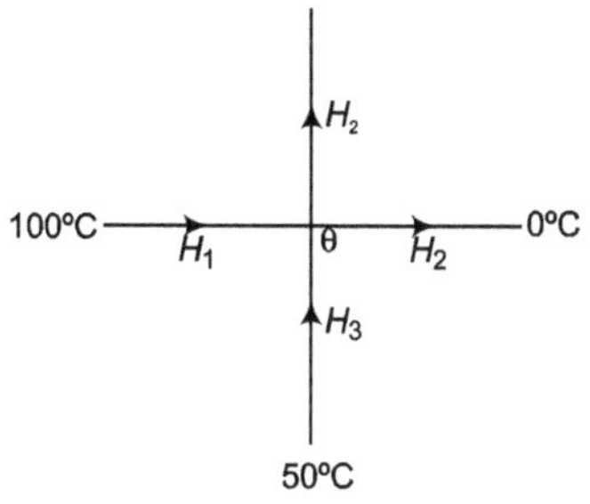

H_1, H_2, H_3 and H_4 be the heat currents. Then

$$H_1 = H_2 + H_3 + H_4$$

or $$\frac{100 - \theta}{R} = \frac{\theta - 50}{R} + \frac{\theta - 50}{R} + \frac{\theta - 0}{R}$$

(R = Thermal resistance of each part of two rods)

Solving these two equations, we get

$$\theta = 50°\text{C}$$

42. $E \propto T^4$,

$$T_1 = 500\,\text{K and } T_2 = 1000\,\text{K}$$

$$T_2 = 2T_1$$

$\therefore$ $$E_2 = (2)^4 E_1 = 16E_1 = 160\,\text{cal/s}$$

43. $4\pi r_1^2 T_1^4 = 4\pi r_2^2 T_2^4$

$\therefore$ $$\frac{r_1}{r_2} = \frac{T_2^2}{T_1^2}$$

44. $E \propto T^4$, $T_1 = 300\,\text{K}$, $T_2 = 400\,\text{K}$

45. $P \propto AT^4$

Area and temperature both are doubled. Hence, power will become $(2)^5$ or 32 times.

$\therefore$ $$P' = 32P = \frac{32 \times 150}{1000}\,\text{kW}$$

$$= 4.8\,\text{kW}$$

46. $\frac{P_1}{P_2} = \frac{A_1 T_1^4}{A_2 T_2^4} = \frac{4\pi (1)^2 (4000)^4}{4\pi (4)^2 (2000)^4} = \frac{1}{1}$

or $$P_1 = P_2$$

47. $\frac{P_1}{P_2} = \frac{A_1 T_1^4}{A_2 T_2^4}$

$$P_2 = \frac{A_2}{A_1}\left(\frac{T_2}{T_1}\right)^4 P_1$$

$$= \left(\frac{1}{2}\right)^2 (2)^4 (640)$$

$$= 2560\,\text{W}$$

48. $\left(\frac{e_\lambda}{a_\lambda}\right)_{\text{any body}} = \left(\frac{e_\lambda}{a_\lambda}\right)_{\text{Perfectly black body}}$

According to Kirchhoff's law,

But $a_\lambda = 1$ for perfectly black body. Hence, $\frac{e_\lambda}{a_\lambda}$ for any body is equal to e_λ for perfectly black body.

49. According to Wien's law, $\lambda_m T$ = constant

$\therefore$ $$(\lambda_m T)_s = (\lambda_m T)_m$$

or $$\frac{T_s}{T_m} = \frac{(\lambda_m)_m}{(\lambda_m)_s}$$

$$= \frac{10^{-4}}{0.5 \times 10^{-6}} = 200$$

50. Area $\propto$ Total energy radiated $\propto T^4$

$$T_1 = 300\,\text{K and } T_2 = 600\,\text{K}$$

Temperature has doubled. Therefore, area will become 16 times.

51. According to Wien's law,

$$\lambda_m T = \text{constant}$$

or $$T \propto \frac{1}{\lambda_m}$$

λ_m has become $\frac{3}{4}$ times.

$\therefore$ Temperature will have become $\frac{4}{3}$ times.

Now, according to Stefan's law, power $P \propto T^4$.
Therefore, new power will be

$$\left(\frac{4}{3}\right)^4 P = \frac{256}{81} P$$

52. Cooling rate $\propto (T^4 - T_0^4)$

$$r = \left\{\frac{(900)^4 - (300)^4}{(600)^4 - (300)^4}\right\} r_0$$
$$= \frac{16}{3} r_0$$

53. Density of hot air is lesser than the density of cold air so hot air rises up.

56. According to Kirchhoff's law, the ratio of emissive power to absorptive power is same for all bodies. Which is equal to the emissive power of a perfectly black body, i.e.

$$\left(\frac{e}{a}\right)_{\text{body}} = E_{\text{Black body}} \text{ for a particular wavelength}$$

$$\left(\frac{e_\lambda}{a_\lambda}\right)_{\text{body}} = (E_\lambda)_{\text{Black body}}$$

$$\Rightarrow \quad e_\lambda = a_\lambda E_\lambda$$

57. Because Planck's law explains the distribution of energy correctly at low temperature as well as at high temperature.

58. According to Kirchhoff's law, a good emitter is also a good absorber.

59. $\frac{\lambda_2}{\lambda_1} = \frac{T_1}{T_2} \Rightarrow \lambda_2 = \frac{T_1}{T_2} \times \lambda_1$

$$= \frac{900}{1200} \times 4$$
$$= 3 \mu\text{m}$$

60. $\lambda_m = \frac{b}{T}$

$$\Rightarrow \quad T = \frac{b}{\lambda_m} = \frac{2.89 \times 10^{-3}}{4000 \times 10^{-10}} = 7225 \text{ K}$$

61. $\frac{T_S}{T_N} = \frac{(\lambda_N)_{\max}}{(\lambda_S)_{\max}} = \frac{350}{510} = 0.69$

62. According to Stefan's law, $E = \sigma\varepsilon A T^4$

$$\Rightarrow \quad \frac{1.58 \times 10^5 \times 4.2}{60 \times 60} = 5.67 \times 10^{-8} \times 10^{-4} \times 0.8 \times T^4$$

$$T = 2500 \text{ K}$$

63. Radiated power by black body,

$$P = \frac{Q}{T} = A\sigma T^4$$

$$\Rightarrow \quad P \propto AT^4 \propto r^2 T^4 \quad \Rightarrow \frac{\Pi_1}{\Pi_2} = \left(\frac{\rho_1}{\rho_2}\right)^2 \left(\frac{T_1}{T_2}\right)^4$$

$$\Rightarrow \quad \frac{440}{P_2} = \left(\frac{12}{6}\right)^2 \left(\frac{500}{1000}\right)^4 \Rightarrow P_2 = 1760 \text{ W}$$

64. $H = \frac{200 \times 7 \times 60}{4.18}$ cal

$= 20095.7 \approx 20000$ cal

65. A body having maximum surface area will cool at fastest rate.

66. Heat required to melt whole ice is

$$Q = mL = 20 \times 80 = 1600 \text{ cal}$$

Energy incident per minute is

$$P = 1.9 \times \frac{\pi}{4} \times d^2$$
$$= 1.9 \times \frac{\pi}{4} \times (20)^2$$
$$= 597 \text{ cal/min}$$

Therefore, time required to melt whole ice is $t = \frac{Q}{P} \approx 2.7$ min.

67. $Q = mL = L$ $\quad (\because m = 1 \text{ kg})$

$$\Delta U = Q - W = L - W$$

i.e. $\quad \Delta U < L$

69. $\frac{Q_1}{t} = \frac{KA(90° - 60°)}{0.6} = 50 KA$

and $\quad \frac{Q_2}{t} = \frac{KA(150 - 110)}{0.8}$

$$= 50 KA$$

70. $\frac{Q}{t} = \frac{KA(\Delta\theta)}{l}$

$$\Rightarrow \quad 50 = \frac{5 \times 20K}{0.4}$$

$$\Rightarrow \quad K = \frac{1}{5} = 0.2$$

74. Mud is bad conductor of heat. So, it prevents the flow of heat between surroundings and inside.

75. Heat flows from hot air to cold body, so person feels comfort.

76. No flow of heat by convection in vacuum.

78. In MKS system, unit of σ is $\frac{\text{J}}{\text{m}^2 \times \text{sec} \times \text{K}^4}$

$$\Rightarrow \quad \frac{\text{J}}{\text{m}^2 \times \text{sec} \times \text{K}^4} = \frac{10^7 \text{ erg}}{10^4 \text{ cm}^2 \times \text{sec} \times \text{K}^4}$$
$$= 10^3 \frac{\text{erg}}{\text{cm}^2 \times \text{sec} \times \text{K}^4}$$

79. Applying, $\frac{\theta_1 - \theta_2}{t} = \alpha\left[\frac{\theta_1 - \theta_2}{2} - \theta_0\right]$ two times. we have

or $\quad \frac{50 - 40}{5} = \alpha\left[\frac{50 + 40}{2} - 20\right] \quad$ ...(i)

and $\quad \frac{40 - 30}{t} = \alpha\left[\frac{40 + 30}{2} - 20\right] \quad$ ...(ii)

Solving these two equations, we get

$$t = \frac{25}{3} \text{ min}$$

80. Applying $\frac{\theta_1 - \theta_2}{t} = \alpha\left[\frac{\theta_1 + \theta_2}{2} - \theta_0\right]$ two times, we get

$$\frac{80 - 60}{10} = \alpha\left[\frac{80 + 60}{2} - 30\right] \quad \text{...(i)}$$

$$\frac{60 - \theta}{10} = \alpha\left[\frac{60 + \theta}{2} - 30\right] \quad \text{...(ii)}$$

Solving these two equations, we get

$$\theta = 48°\text{C}$$

81. Applying, $\frac{\theta_1-\theta_2}{t}=\alpha\left[\frac{\theta_1+\theta_2}{2}-\theta_0\right]$ two times, we get

$$\frac{50-45}{5}=\alpha\left[\frac{50+45}{2}-\theta_0\right] \qquad \text{...(i)}$$

$$\frac{45-41.5}{5}=\alpha\left[\frac{45+41.5}{2}-\theta_0\right] \qquad \text{...(ii)}$$

Solving these two equations, we get

θ_0 = temperature of atmosphere

$= 33.3°C$

82. Heat lost in t second $= mL$ or heat lost per second $=\frac{mL}{t}$. This must be the heat supplied for keeping the substance in molten state per sec.

$\therefore \quad \frac{mL}{t}=P$

or $\quad L=\frac{Pt}{m}$

Level 2 : Only One Correct Option

1. $\because T_B > T_A \Rightarrow$ Heat will flow B to A *via* two paths (i) B to A (ii) along BCA as shown.

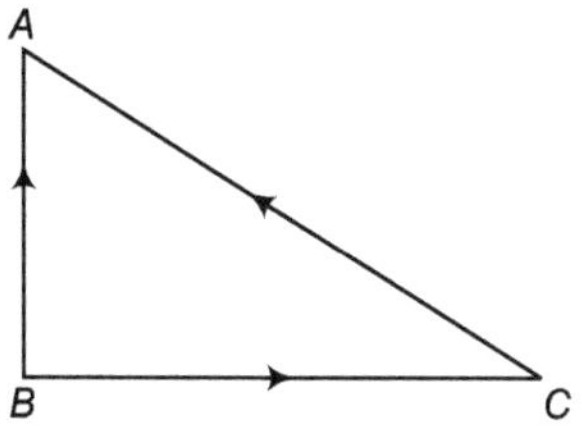

Rate of flow of heat in path BCA will be same

i.e. $\quad \left(\frac{Q}{t}\right)_{BC}=\left(\frac{Q}{t}\right)_{CA}$

$\Rightarrow \quad \frac{K(\sqrt{2}T-T_C)A}{a}=\frac{K(T_C-T)A}{\sqrt{2}a}$

$\Rightarrow \quad \frac{T_C}{T}=\frac{3}{1+\sqrt{2}}$

2. Heat currents in the two rods will be same.

$\therefore \quad H_1=H_2$

or $\quad \frac{\theta_1-\theta}{(l/K_1A)}=\frac{\theta-\theta_2}{(l/K_2A)}$ (θ = temperature of junction)

or $\quad K_1(\theta_1-\theta)=K_2(\theta-\theta_2)$

$\therefore \quad \theta=\frac{K_1\theta_1+K_2\theta_2}{K_1+K_2}$

3. $\frac{d\theta}{dt}=L\cdot\frac{dm}{dt}$ = Heat transfer from bottom $=\frac{\text{TD}}{(l/KA)}$

Heat TD = desired temperature difference.

$\therefore \quad \text{TD}=\frac{L\cdot l\cdot\frac{dm}{dt}}{KA}$

$=\frac{(540)(2\times10^{-1})\left(\frac{1}{60}\right)}{0.5\times300}$

$=0.012°C$

4. There will be no heat flow through AB, if temperature of B is also 20°C and heat current in DB is equal to heat current in BC.

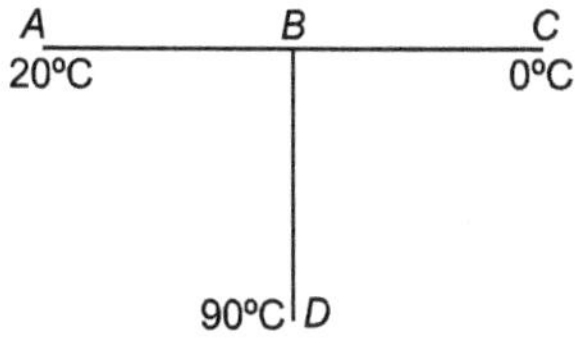

$\therefore \quad H_{DB}=H_{BC} \quad (H=\text{TD}/R)$

or $\quad \frac{70}{(l_{BD}/KA)}=\frac{20}{(l_{BC}/KA)}$

or $\quad \frac{l_{BD}}{l_{BC}}=\frac{7}{2}$

5. Rate of heat flow, $I\propto\frac{\text{TD}}{R}$

$\therefore$ TD = temperature difference is same.

$\therefore \quad I\propto\frac{1}{\text{Thermal resistance}}$

$$\frac{I_A}{I_B}=\frac{R_B}{R_A}=\frac{(R)(2R)/R+2R}{(R+2R)}=\frac{2}{9}$$

6. $R_{AB}=R_{CD}=R$ (say)

Length of semicircle $=\pi l$

$\therefore$ Resistance of semicircle $=\pi R$ as $R\propto l$

$R_{BC}=\frac{\pi R}{2}$ (in parallel)

Heat current, $H=\frac{100-0}{R+R+\frac{\pi}{2}\cdot R}=28$ units

Now, $\quad \theta_C-0°=HR=28°C$

$\therefore \quad \theta_C=28°C$

7. $dQ=ms\cdot dT$

$m=1$ kg

$\therefore \quad s=\frac{dQ}{dT}$ = Slope of Q-T graph

$S_{\text{solid}}=\tan 45°=1$ J/kg-K

$S_{\text{liquid}}=\tan 30°=\frac{1}{\sqrt{3}}$ J/kg-K

Latent heat of vaporisation

= heat required to melt complete 1 kg

= 200 J/kg

8. No heat is flowing through C and F. Therefore, they are at same temperatures.

9. $H=\frac{Q}{t}=\frac{KA(\theta_1-\theta_2)}{l}$

or $\quad H\propto\frac{A}{l}$ or $H_2=2H_1$

10. A and B

$mS_A(15-10)=mS_B(25-15)$

or $\quad \frac{S_A}{S_B}=2 \qquad \text{...(i)}$

B and C

$mS_B(30-25)=mS_C(40-30)$

or $$\frac{S_B}{S_C} = 2 \quad \text{...(ii)}$$

From Eqs. (i) and (ii), we have

$$\frac{S_A}{S_C} = 4$$

Now, let θ be the equilibrium temperature, when A and C are mixed. Then,

$$mS_A(\theta - 10) = mS_C(40 - \theta)$$

or $$4(\theta - 10) = 40 - \theta \quad \left(\text{Using } \frac{S_A}{S_C} = 4\right)$$

or $$\theta = 16°\text{C}$$

11. Heat carried increases to 2 times. Therefore, new net thermal resistance will remain half.

$$\therefore \quad R' = \frac{R}{2}$$

$$\frac{R'_1 R'_2}{R'_1 + R'_2} = \frac{R_1 R_2}{2(R_1 + R_2)}$$

or $$\frac{(3l/KA)(l/K'A)}{(3l/KA) + (l/K'A)} = \frac{(3l/KA)(l/KA)}{2\left\{\frac{3l}{KA} + \frac{l}{KA}\right\}}$$

Solving this we get, $K' = \frac{7}{3}K$

12. Initially, the rods are in parallel

$$\frac{Q}{t} = \frac{(\theta_1 - \theta_2)}{R} \Rightarrow \left(\frac{Q}{t}\right)_1 = \frac{mL}{t}$$

$$= q_1 L = \frac{(100-0)}{R/2} \quad \text{...(i)}$$

Finally, when rods are in series

$$\Rightarrow \left(\frac{Q}{t}\right)_2 = \frac{mL}{t} = q_2 L = \frac{(100-0)}{2R} \quad \text{...(ii)}$$

From Eqs. (i) and (ii), we get

$$\frac{q_1}{q_2} = \frac{4}{1}$$

13. Rate of cooling $\left(-\frac{dT}{dt}\right) \propto$ emissivity (e)

From graph, $\left(-\frac{dT}{dt}\right)_x > \left(-\frac{dT}{dt}\right)_y$

$$\Rightarrow \quad e_x > e_y$$

Further emissivity $e \propto$ absorptive power (a)

$$\Rightarrow \quad a_x > a_y$$

($\because$ good absorbers are good emitters).

14. Consider the diagram where all the three objects are heated to same temperature T. We know that density, $\rho = \frac{\text{mass}}{\text{volume}}$ as ρ is same for all the three objects hence, volume will also be same.

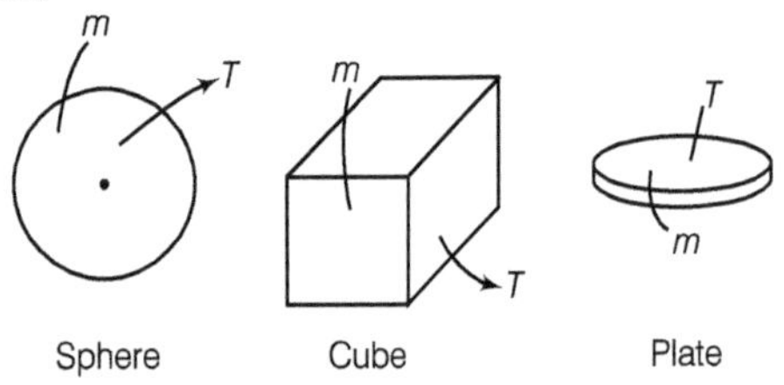

As thickness of the plate is least hence, surface area of the plate is maximum.

We know that, according to Stefan's law of heat loss $H \propto AT^4$

where, A is surface area of for object and T is temperature.

Hence, $$H_{\text{sphere}} : H_{\text{cube}} : H_{\text{plate}} = A_{\text{sphere}} : A_{\text{cube}} : A_{\text{plate}}$$

As A_{plate} is maximum.

Hence, the plate will cool fastest.

As, the sphere is having minimum surface area hence, the sphere cools slowest.

15. Let the equilibrium temperature of the system is T.

Let us assume that $T_1, T_2 < T < T_3$.

According to question, there is no net loss to the surroundings.

Heat lost by M_3 = Heat gained by M_1 + Heat gained by M_2

$$\Rightarrow \quad M_3 s(T_3 - T) = M_1 s(T - T_1) + M_2 s(T - T_2)$$

(where, s is specific heat of the copper material)

$$\Rightarrow \quad T[M_1 + M_2 + M_3] = M_3 T_3 + M_1 T_1 + M_2 T_2$$

$$\Rightarrow \quad T = \frac{M_1 T_1 + M_2 T_2 + M_3 T_3}{M_1 + M_2 + M_3}$$

More than One Correct Options

1. Total radiation per second is given by

$$P = (e_r \sigma T^4 A)$$

Here, $e_r : \sigma : T$ and A all are same.

Hence, P will be same.

Same is the case with absorption per second.

Now, $$P = ms\left(-\frac{dT}{dt}\right)$$

or Rate of cooling $\left(-\frac{dT}{dt}\right) = \frac{P}{ms} \propto \frac{1}{m}$

Mass of hollow sphere is less. So, its initial rate of cooling will be more

2.

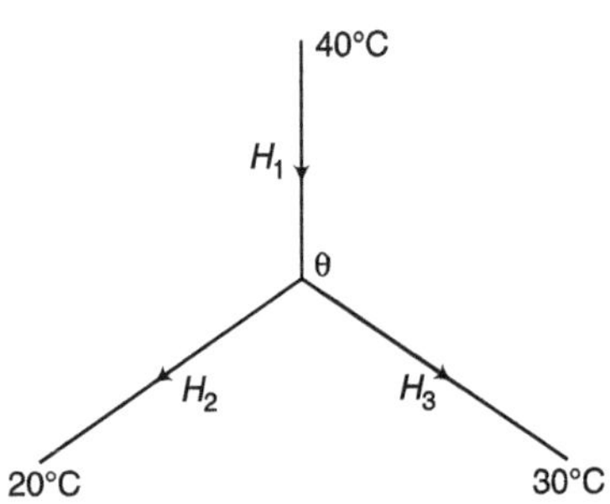

$$\therefore \quad \frac{40 - \theta}{R} = \frac{\theta - 20}{R} + \frac{\theta - 30}{R}$$

where, R = Thermal resistance of each rod.

Solving this equation we get,

$$\theta = 30°\text{C}$$

Heat flows from higher temperature to lower temperature.

3. $ms(2\theta - \theta') = m(2s)(\theta' - \theta)$

Solving this equation, we get

$$\theta' = \frac{4}{3}\theta$$

Further, heat capacity

$$c = ms$$

or $$c \propto s \quad \text{(as } m \text{ is same)}$$

$$\therefore \quad \frac{c_1}{c_2} = \frac{s_1}{s_2} = \frac{1}{2}$$

4. $q_1 = \frac{\text{TD}}{R_1} \Rightarrow \therefore R_1 = \frac{\text{TD}}{q_1}$

Similarly, $$R_2 = \frac{\text{TD}}{q_2}$$

In series,

$$q_s = \frac{\text{TD}}{R_1 + R_2} = \frac{\text{TD}}{\frac{\text{TD}}{q_1} + \frac{\text{TD}}{q_2}}$$

$$= \frac{q_1 q_2}{q_1 + q_2}$$

In parallel, $$q_p = \frac{\text{TD}}{R_{\text{net}}} = \text{TD}\left(\frac{1}{R_{\text{net}}}\right)$$

$$= \text{TD}\left(\frac{1}{R_1} + \frac{1}{R_2}\right)$$

$$= \text{TD}\left(\frac{q_1}{\text{TD}} + \frac{q_2}{\text{TD}}\right) = q_1 + q_2$$

5. *(b)* At low temperature, the rate of cooling is proportional to temperature difference.

Assertion and Reason

8. Water in vapour state has more heat content.

9. $Q = mL_1 + ms\Delta\theta + mL_2$

$$= 1 \times 80 + 1 \times 1 \times 100 + 1 \times 540$$
$$= 720 \text{ cal}$$

11. Due to gravity hot (but lighter) material rises up and cold (but heavier) material comes down.

13. According to Wien's displacement law,

$$\lambda_m \propto \frac{1}{T}$$

14. Emissive power is energy radiated per unit area per unit time. Which is not one.

15. A normal body can also radiate energy more than a perfectly black body, if its temperature is very high.

16. $R = \frac{l}{KA}$ or $R \propto \frac{1}{A} \Rightarrow \therefore R_1 > R_2$

Further, $\text{TD} = H \times R$ or $\text{TD} \propto R$

$$\therefore \quad (\text{TD})_1 > (\text{TD})_2$$

17. Statement of reason is Kirchhoff's law.

Match the Columns

2. A and B

$$(m)(s)(\theta - 20°) = (2m)(s)(40° - \theta)$$

or $$\theta = \frac{100}{3} = 33.3°\text{C}$$

A and *C*

$$(m)(s)(\theta - 20°) = (3m)(s)(60° - \theta) \quad \text{or} \quad \theta = 50°\text{C}$$

B and *C*

$$(2m)(s)(\theta - 40°) = (3m)(s)(60° - \theta) \quad \text{or} \quad \theta = 52°\text{C}$$

A, *B* and *C*

$$(m)(s)(\theta - 20°) + (2m)(s)(40° - \theta) = (3m)(s)(60° - \theta)$$

or $$\theta = 60°\text{ C}$$

3. Thermal resistance $\propto \frac{1}{\text{Area}}$

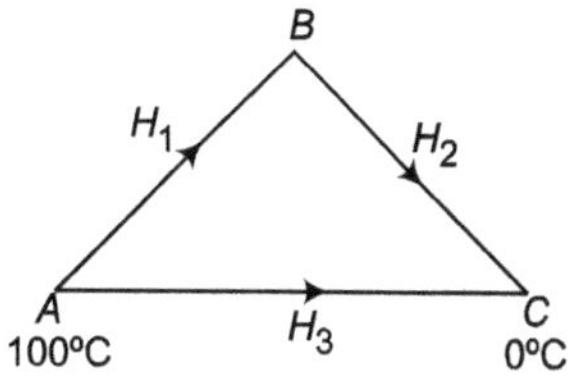

⇓

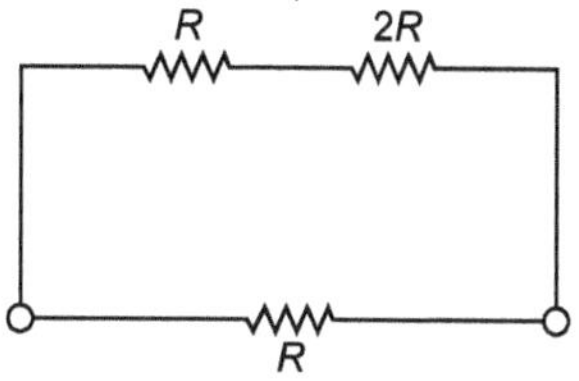

So let, $R_{BC} = R$

Then, $R_{AB} = R_{AC} = 2R$

$$H_1 = H_2$$

$$\therefore \quad \frac{100 - \theta_B}{R} = \frac{\theta_B - 0}{2R}$$

$$\therefore \quad \theta_B = 67.7°\text{C}$$

$$R_{ABC} = 3R \quad \text{and} \quad R_{AC} = 2R$$

$$\therefore \quad H_1 = H_2 = \frac{H_3}{3}$$

Entrance Gallery

1. In thermal conduction, it is found that in steady state the heat current is directly proportional to the area of cross-section A which is proportional to the change in temperature $(T_1 - T_2)$.

Then, $$\frac{\Delta Q}{\Delta t} = \frac{KA(T_1 - T_2)}{x}$$

$$\frac{\Delta Q}{\Delta t} = \text{Heat current}$$

where, K = thermal conductivity

and A = area of cross-section.

According to thermal conductivity, we get

$$\frac{dQ_1}{dt} = \frac{dQ_2}{dt} + \frac{dQ_3}{dt}$$

$$\frac{0.92 \times 4(100 - T)}{46} = \frac{0.26 \times 4(T - 0)}{13} + \frac{0.12 \times 4(T - 0)}{12}$$

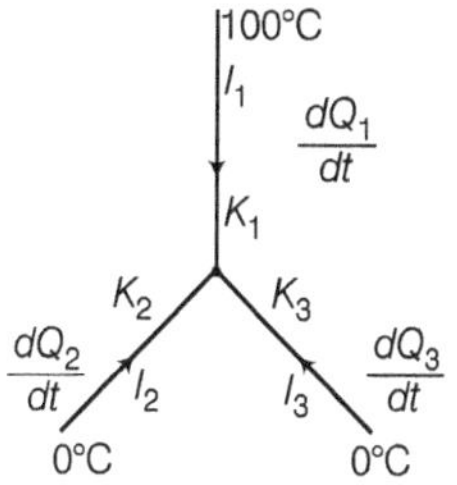

$$\Rightarrow \quad T = 40° \text{C}$$

$$\therefore \quad \frac{dQ_1}{dt} = \frac{0.92 \times 4 (100-40)}{40} = 5.52 \text{ cal/s}$$

2. Assuming the at temperature of the body and cubical box is same initially, i.e. T and finally it becomes $T/2$.

Because, temperature of body and surroundings remain same.

Hence, no net loss of radiation occur through the body. This total energy remains constant.

3. We know that relation between temperature gradient (TG) and thermal conductivity (K)

So, $d\frac{\theta}{dt} = -KA\, d\frac{\theta}{dx} = -KA \times (\text{TG})$

i.e. $t \propto \frac{1}{K}$ $\quad \left(\frac{d\theta}{dt} = \text{constant}\right)$

or $K \propto \frac{1}{t}$

$$K_1 \propto \frac{1}{t_1} \quad \text{...(i)}$$

$$K_2 \propto \frac{1}{t_2} \quad \text{...(ii)}$$

From Eqs. (i) and (ii), we get

$$\frac{K_1}{K_2} = \frac{1/t_1}{1/t_2}$$

$$K_1 : K_2 = t_2 : t_1$$

4. Thermal resistance in configuration I,

$$R_{\text{I}} = R_1 + R_2 = \left(\frac{l}{KA}\right) + \left(\frac{l}{2KA}\right) = \frac{3}{2}\left(\frac{l}{KA}\right)$$

Thermal resistance in configuration II,

$$\frac{1}{R_{\text{II}}} = \frac{1}{R_1} + \frac{1}{R_2} = \frac{KA}{l} + \frac{2KA}{l}$$

or $R_{\text{II}} = \frac{l}{3KA} = \frac{R_{\text{I}}}{4.5}$

Since, thermal resistance R_{II} is 4.5 times less than thermal resistance R_{I}.

$$\therefore \quad t_{\text{II}} = \frac{t_1}{4.5} = \frac{9}{4.5} \text{ s} = 2 \text{ s}$$

5. According to Newton's cooling law, option (c) is correct answer.

6. We know that thermal conductivity, $K \propto \frac{1}{\Delta\theta}$

but $\Delta\theta_1 = -10°\text{C}, \Delta\theta_2 = -5°\text{C}, \Delta\theta_3 = -15°\text{C}$

and $\Delta\theta_4 = -5°\text{C}$

Hence, $S_2 = S_4, S_1, S_3$

7. The relation for the heat flow is given by

$$\frac{Q}{t} = \frac{KA(Q_2 - Q_1)}{l}$$

where, symbols have their usual meaning.

Now equivalent conductivity of PRQ is given by

$$= \frac{l + l}{\frac{l}{K_1} + \frac{l}{K_2}} = \frac{2K_1K_2}{K_1 + K_2}$$

$$\Rightarrow \quad K_3 = \frac{K_1 K_2}{K_1 + K_2}$$

8. Let R be the thermal conductivity of conductor A, then thermal conductivity of conductor $B = \frac{R}{2}$.

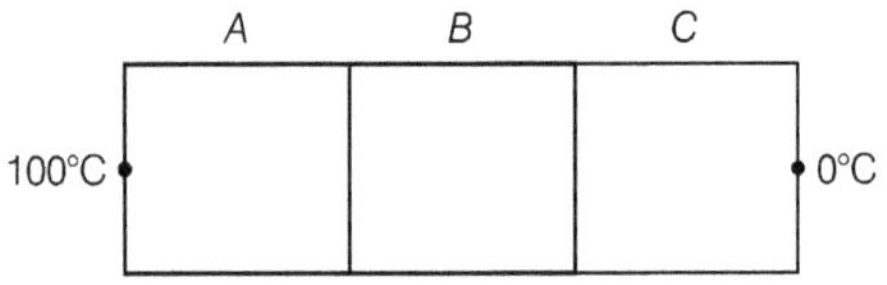

and thermal conductivity of conductor $C = 2R$

$$\therefore \quad \text{Heat current, } H = \frac{100° - 0°}{R + \frac{R}{2} + 2R} = \frac{200}{7R}$$

If T' be the temperature of the junction of A and B, then

$$H = \frac{100 - T'}{R} \text{ or } \frac{200}{7R} = \frac{100 - T'}{R} \text{ or } T' = \frac{500}{7} = 71°\text{C}$$

9. Radius of Ist body $= r_1$, radius of IInd body $= r_2$

Temperature of Ist body $= T_1$

Temperature of IInd body $= T_2$

The emissivity of a black body or radiated power is given by

$E = A\sigma T^4 \times t \propto AT^4$

or $A \propto \frac{E}{T^4}$ (where, A is the surface area of the spherical black body)

As in the condition of question , the power radiated by Ist and IInd body is same.

Hence, $A \propto \frac{1}{T^4}$

So, $\frac{A_1}{A_2} = \left(\frac{T_2}{T_1}\right)^4$

$$\Rightarrow \quad \frac{4\pi r_1^2}{4\pi r_2^2} = \left(\frac{T_2}{T_1}\right)^4$$

Thus, $\frac{r_1}{r_2} = \left(\frac{T_2}{T_1}\right)^2$

10. According to Newton's law of cooling, t_1 will be less than t_2.

11. As we know, $Q \propto T^4$

$$\Rightarrow \quad \frac{H_A}{H_B} = \left[\frac{273 + 727}{273 + 327}\right]^4 = \frac{625}{81}$$

12. We know that for an ideal engine

$$\frac{Q_1}{Q_2} = \frac{T_1}{T_2}$$

where, Q_1 = heat from source,

T_1 = temperature of source

Q_2 = heat rejected by sink,

T_2 = temperature of sink

$$\frac{650}{Q_2} = \frac{450}{225} \Rightarrow Q_2 = \frac{650 \times 225}{450} = 325 \text{J}$$

13. Heat required to convert ice into water at 100°C

$Q = m \times L + ms\Delta T = 18000$ cal

where, L = latent heat, s = specific heat and $m = 100$g

Amount of heat left $= 22320 - 18000 = 4320$ cal

or $m \times L = 4320$ here, $L = 540 \text{ cal g}^{-1}$

$\Rightarrow \quad m = 8$ g steam

14. Let temperature of middle plate in steady state is T_0

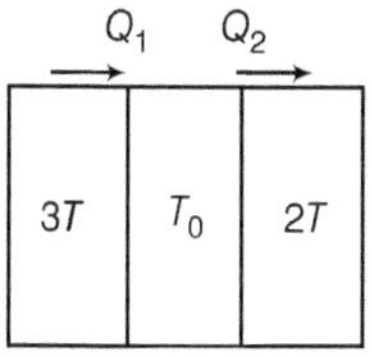

$$Q_1 = Q_2$$

Q = net rate of heat flow

$$\therefore \quad \sigma A(3T)^4 - \sigma A T_0^4 = \sigma A T_0^4 - \sigma A(2T)^4$$

On solving this equation, we get

$$T_0 = \left(\frac{97}{2}\right)^{1/4} T$$

15. According to Newton's law of cooling, "rate of fall in temperature is proportional to the difference in temperature of the body with surroundings", i.e.

$$-\frac{d\theta}{dt} = k(\theta - \theta_0)$$

On integrating above equation, we get

$$\Rightarrow \quad \int \frac{d\theta}{\theta - \theta_0} = \int -k\,dt$$

$$\Rightarrow \quad \ln(\theta - \theta_0) = kt + c$$

Which is a straight line with negative slope.

16. Thermal resistance, $R = \frac{l}{KA}$

where, K = thermal conductivity

A = area of cross-section

$$\therefore \quad R_A = \frac{L}{(2K)(4Lw)} = \frac{1}{8Kw} \qquad \text{(Here, } w = \text{width)}$$

$$R_B = \frac{4L}{(3K)(Lw)} = \frac{4}{3Kw}$$

$$\Rightarrow \quad R_C = \frac{4L}{(4K)(2Lw)} = \frac{1}{2Kw}$$

$$R_D = \frac{4L}{(5K)(Lw)} = \frac{4}{5Kw}$$

$$\Rightarrow \quad R_E = \frac{L}{(6K)(Lw)} = \frac{1}{6Kw}$$

$$R_A : R_B : R_C : R_D : R_E = 15 : 160 : 60 : 96 : 12$$

So, let us write $R_A = 15R$, $R_B = 160R$ etc., and draw a simple electrical circuit as shown in figure below.

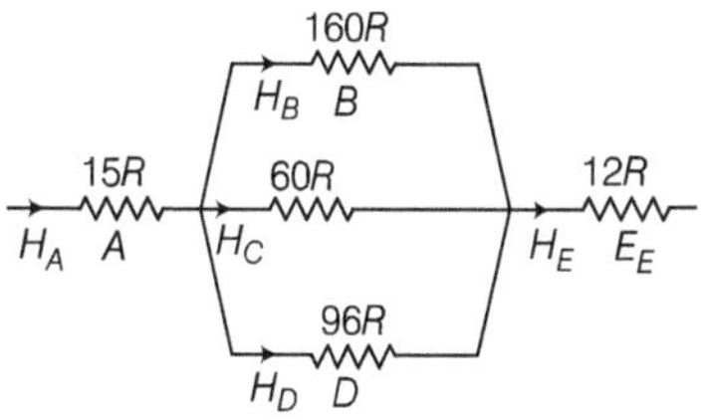

where, H = heat current = rate of heat flow,

$$H_A = H_E = H.$$

$\therefore$ Option (a) is correct.

In parallel circuit current, distributes in inverse ratio of resistance.

$$\therefore \quad H_B : H_C : H_D : \frac{1}{R_B} : \frac{1}{R_C} : \frac{1}{R_D}$$

$$= \frac{1}{160} : \frac{1}{60} : \frac{1}{96} = 9 : 24 : 15$$

$$\therefore \quad H_B = \left(\frac{9}{9 + 24 + 15}\right) H = \frac{3}{16} H$$

$$H_C = \left(\frac{24}{9 + 24 + 15}\right) H = \frac{1}{2} H$$

$$H_D = \left(\frac{15}{9 + 24 + 15}\right) H = \frac{5}{16} H$$

$$H_A = H_B + H_D$$

Temperature difference (let us call it T)

= (heat current) × (thermal resistance)

$$T_A = H_A R_A = (H)(15R) = 15HR$$

$$T_B = H_B R_B = \left(\frac{3}{16} H\right)(160R) = 30HR$$

$$T_C = H_C R_C = \left(\frac{1}{2} H\right)(60R) = 30HR$$

$$T_D = H_D R_D = \left(\frac{5}{16} H\right)(96R) = 30HR$$

$$T_E = H_E R_E = (H)(12R) = 12HR$$

Here, T_E is minimum. Therefore, option (c) is also correct.

17. According to the question, $\frac{1}{2} \times \frac{1}{2} mv^2 = m \times s \times \Delta T$

where, m = mass of bullet

s = specific heat capacity of lead

$$\frac{1}{4} m \times 4 \times 10^4 = 125 \times m \times \Delta T$$

$$\Delta T = \frac{4 \times 10^4}{500} = 80^\circ \text{C}$$

18. Heat required is proportional to square of radius

$$\frac{Q_1}{Q_2} = \frac{r_1^2}{r_2^2} = \frac{(1.5)^2}{(1)^2} = \frac{9}{4}$$

19. According to question, |A|B|C|

$$K_B = \frac{K_A}{2} \quad \Rightarrow \quad K_B = 3K_C$$

$$K_C = \frac{K_A}{6}$$

$$\Rightarrow \quad \frac{l}{K_S} = \frac{l_1}{K_A} + \frac{l_2}{K_B} + \frac{l_3}{K_C}$$

$$\frac{3l}{K_S} = \frac{l}{K_A} + \frac{l}{\frac{K_A}{2}} + \frac{l}{\frac{K_A}{6}}$$

$$\frac{3l}{K_S} = \frac{9l}{K_A}$$

$$\Rightarrow \quad K_S = \frac{K_A}{3}$$

20. As we know that rate of flow of heat is given by

$$\frac{dQ}{dt} = \frac{KA\Delta T}{x}$$

$$\frac{1.56\times10^5}{3600}=\frac{K\times2\times20}{12\times10^{-2}}$$

$$K=\frac{1.56\times10^5\times12\times10^{-2}}{3600\times2\times20}$$

$$=\frac{1.56}{12}=0.13$$

21. Language of question is slightly wrong. As heat capacity and specific heat are two different physical quantities. Unit of heat capacity is $J\,kg^{-1}$, not $J\,kg^{-1}\,{}^\circ C^{-1}$. The heat capacity given in the question is really the specific heat. Now, applying the heat exchange equation.

$$420=(m\times10^{-3})(2100)(5)+(1\times10^{-3})(3.36\times10^5)$$

Solving this equation, we get

$$m=8\text{ g}$$

22. We know that,

$$\frac{\lambda_A}{\lambda_B}=\frac{T_B}{T_A}=\frac{500}{1500}=\frac{1}{3}$$

$E\propto T^4A$ (where, $A=$ surface area $=4\pi R^2$)

$$E\propto T^4R^2$$

$$\frac{E_A}{E_B}=\left(\frac{T_A}{T_B}\right)^4\left(\frac{R_A}{R_B}\right)^2$$

$$=(3)^4\left(\frac{16}{18}\right)^2=9$$

23. We know that, $\frac{dQ}{dt}=KA\frac{d\theta}{dx}$

In steady state, flow of heat

$$d\theta=\frac{dQ}{dt}\cdot\frac{1}{KA}dx$$

$$\Rightarrow \quad \theta_H-\theta=K'x$$

$$\Rightarrow \quad \theta=\theta_H-K'x$$

Equation, $\theta=\theta_H-K'x$ represents a straight line.

24. For first slab,

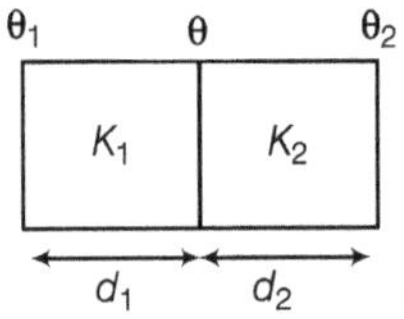

heat current, $H_1=\frac{K_1(\theta_1-\theta)A}{d_1}$

For second slab,

Heat current, $H_2=\frac{K_2(\theta-\theta_2)A}{d_2}$

As slabs are in series

$$H_1=H_2$$

$$\therefore \quad \frac{K_1(\theta_1-\theta)A}{d_1}=\frac{K_2(\theta-\theta_2)A}{d_2}$$

$$\Rightarrow \quad \theta=\frac{K_1\theta_1d_2+K_2\theta_2d_1}{K_2d_1+K_1d_2}$$

25. An opaque body does not transmit any radiation, hence transmission coefficient of an opaque body is zero.

26. Total energy radiated from a body

$$Q=A\varepsilon\sigma T^4t$$

or

$$\frac{Q}{t}\propto AT^4$$

$$\frac{Q}{t}\propto r^2T^4 \qquad (\because A=4\pi r^2)\text{s}$$

$$\frac{Q_1}{Q_2}=\left(\frac{r_1}{r_2}\right)^2\left(\frac{T_1}{T_2}\right)^4$$

$$=\left(\frac{8}{2}\right)^2\left[\frac{273+127}{273+527}\right]^4=1$$

17

Wave Motion

17.1 Introduction

"A wave is any disturbance from a normal or equilibrium condition that propagates without the transport of matter. In general, a wave transports both energy and momentum."

Wave motion appears in almost every branch of physics. We are all familiar with water waves, sound waves and light waves. Waves occur when a system is disturbed from its equilibrium position and this disturbance travels or propagates from one region of the system to other. Energy can be transmitted over considerable distances by wave motion. The waves requiring a medium are called **mechanical waves** and those which do not require a medium are called **non-mechanical waves.** Light waves and all other electromagnetic waves are non-mechanical. The energy in the mechanical waves is the kinetic and potential energy of the matter. In the propagation of mechanical waves, elasticity and inertia of the medium play an important role. This is why mechanical waves sometimes are also referred to as **elastic waves**. Note that the medium itself does not move as a whole along with the wave motion. Apart from mechanical and non-mechanical waves there is also another kind of waves called **matter waves**. These represent wave like properties of particles.

17.2 Transverse and Longitudinal Waves

There are two distinct classes of wave motion :

(i) Transverse (ii) Longitudinal.

In a **transverse wave motion**, the particles of the medium oscillate about their mean or equilibrium position at right angles to the direction of propagation of wave motion itself.

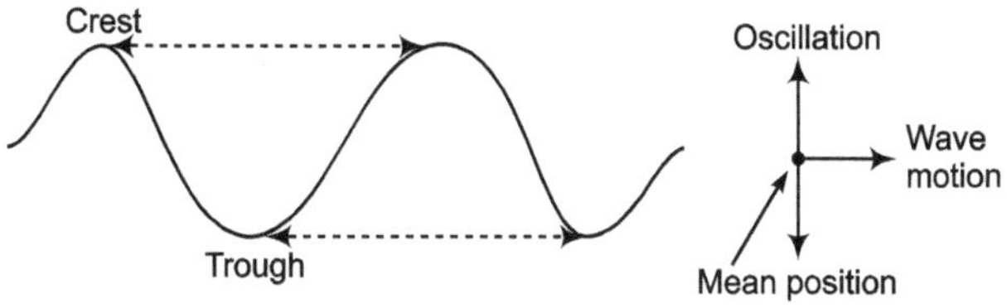

Fig. 17.1

Chapter Snapshot

This form of wave motion travels in the form of **crests** and **troughs**, as for example, waves travelling along a stretched string. This type of waves are possible in media which possess elasticity of shape or rigidity, i.e. in solids. These are also possible on the surface of liquids also, even though they do not possess the property of rigidity. This is because they possess another equally effective property (surface tension) of resisting any vertical displacement of their particles (or keeping their level). Gases, however, possess neither rigidity nor do they resist any vertical displacement of particles (or keep their level). A transverse wave motion is therefore, not possible in a gaseous medium. An electromagnetic wave is necessarily a transverse wave because of the electric and magnetic fields being perpendicular to its direction of propagation. The distance between two successive crests or troughs is known as the **wavelength** (λ) of the wave.

In a **longitudinal wave motion**, the particles of the medium oscillate about their mean or equilibrium position along the direction of propagation of the wave motion itself. This type of wave motion travels in the form of **compressions** and **rarefactions** and is possible in media possessing elasticity of volume, i.e. in solids, liquids and gases.

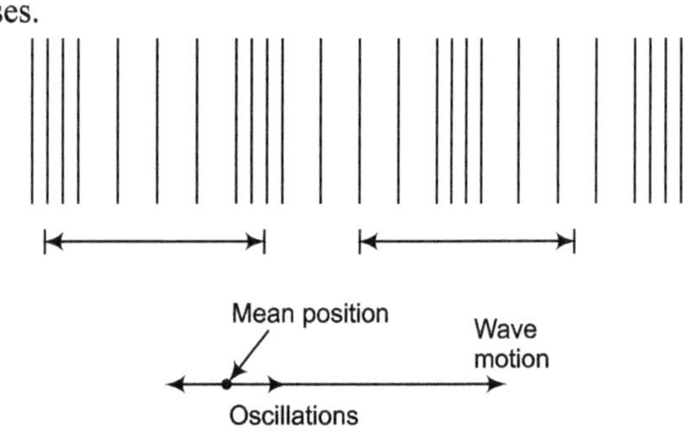

Fig. 17.2

The distance between two successive compressions or rarefactions constitute one wavelength. Sound waves in a gas are longitudinal in nature.

Again waves may be **one dimensional, two dimensional or three dimensional** according as they propagate energy in just one, two or three dimensions.

Transverse waves along a string are one dimensional, ripples on water surface are two dimensional and sound waves proceeding radially from a point source are three dimensional.

17.3 The General Equation of Wave Motion

In a wave motion, some physical quantity (say y) is made to oscillate at one place and these oscillations of y propagate to other places. Thus, y may be

(i) displacement of particles from their mean position in case of transverse wave in a rope or longitudinal sound wave in a gas.

(ii) pressure difference (dp) or density difference ($d\rho$) in case of sound wave or

(iii) electric and magnetic fields in case of electromagnetic waves.

The oscillations of y may or may not be simple harmonic in nature. Now, let us consider a one dimensional wave travelling along x-axis. In this case, y is a function of position (x) and time (t). The reason is that one may be interested in knowing the value of y at a general point x at any time t. Thus, we can say that,

$$y = y(x, t)$$

But only those functions of x and t, represent a wave motion which satisfy the differential equation,

$$\frac{\partial^2 y}{\partial t^2} = k\frac{\partial^2 y}{\partial x^2}$$

Here, k is a constant which is equal to square of the wave speed, or

$$k = v^2$$

Thus, the above equation can be written as,

$$\frac{\partial^2 y}{\partial t^2} = v^2\frac{\partial^2 y}{\partial x^2} \qquad \ldots\text{(i)}$$

The general solution of this equation is of the form

$$y(x, t) = f(ax \pm bt) \qquad \ldots\text{(ii)}$$

Thus, any function of x and t which satisfies Eq. (i) or which can be written as Eq. (ii) represents a wave. The only condition is that it should be finite everywhere and at all times. Further, if these conditions are satisfied, then speed of wave (v) is given by

$$v = \frac{\text{coefficient of } t}{\text{coefficient of } x} = \frac{b}{a}$$

The plus (+) sign between ax and bt implies that the wave is travelling along negative x-direction and minus (–) sign shows that it is travelling along positive x-direction.

Example 17.1 *Which of the following functions represents a wave?*

(a) $(x - vt)^2$ (b) $\ln(x + vt)$

(c) $e^{-(x-vt)^2}$ (d) $\dfrac{1}{x + vt}$

Sol. (c) Although all the four functions are written in the form $f(ax \pm bt)$, only (c) among the four functions is finite everywhere at all times. Hence, only (c) represents a wave.

Example 17.2 *In a wave motion, $y = a\sin(kx - \omega t)$, y can represent*

(a) *electric field* (b) *magnetic field*

(c) *displacement* (d) *pressure*

Sol. (a, b, c, d). In case of sound wave, y can represent pressure and displacement, while in case of an electromagnetic wave it represents electric and magnetic fields.

Note Point

- In general, y is any general physical quantity which is made to oscillate at one place and these oscillations are propagated to other places.

17.4 Plane Progressive Harmonic Waves

When $y(x, t)$ is a sine or cosine function such as, $y(x, t) = A \sin k(x - vt)$ or $y(x, t) = A \cos k(x - vt)$ it is called **plane progressive harmonic wave**. In plane progressive harmonic wave oscillations of y are simple harmonic in nature.

The quantity k has a special meaning. Replacing the value of x by $x + \frac{2\pi}{k}$, we get the same value of y, i.e.

$$y\left(x + \frac{2\pi}{k}, t\right) = A \sin k\left(x + \frac{2\pi}{k} - vt\right)$$
$$= A \sin [k(x - vt) + 2\pi]$$
$$= A \sin k(x - vt) = y(x, t)$$

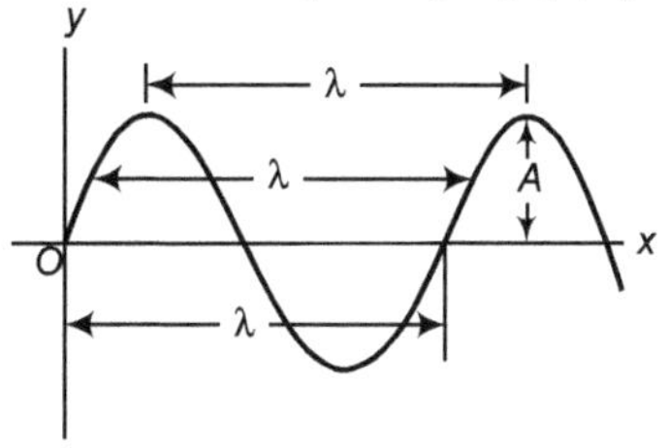

Fig. 17.3

The quantity, $\frac{2\pi}{k} = \lambda$ designated as **wavelength**, is the **space period** of the curve, i.e. the curve repeats itself every length λ. The quantity $k = \frac{2\pi}{\lambda}$ represents the number of wavelengths in the distance 2π and is called the **wave number**.

Therefore, $y(x, t) = A \sin k(x - vt) = A \sin \frac{2\pi}{\lambda}(x - vt)$

represents a plane progressive harmonic wave of wavelength λ propagating towards the positive x-axis with speed v. The above equation can also be written as,

$$y(x, t) = A \sin (kx - \omega t)$$

where, $\omega = kv = \frac{2\pi v}{\lambda}$

gives the **angular frequency** of the wave. Further, where f is the **frequency** with which y oscillates at every point x. We have the important relation,

$$v = \lambda f$$

Also, if T is the **period** of oscillation then,

$$T = \frac{2\pi}{\omega} = \frac{1}{f}$$

We may also write,

$$y = A \sin 2\pi\left(\frac{x}{\lambda} - \frac{t}{T}\right)$$

Thus, the equation of a plane progressive harmonic wave moving along positive x-direction can be written as,

$$y = A \sin k(x - vt) = A \sin (kx - \omega t)$$
$$= A \sin \frac{2\pi}{\lambda}(x - vt) = A \sin 2\pi\left(\frac{x}{\lambda} - \frac{t}{T}\right)$$

Similarly, the expressions

$$y = A \sin k(x + vt) = A \sin (kx + \omega t)$$
$$= A \sin 2\pi\left(\frac{x}{\lambda} + \frac{t}{T}\right)$$

represents a plane progressive harmonic wave travelling in negative x-direction.

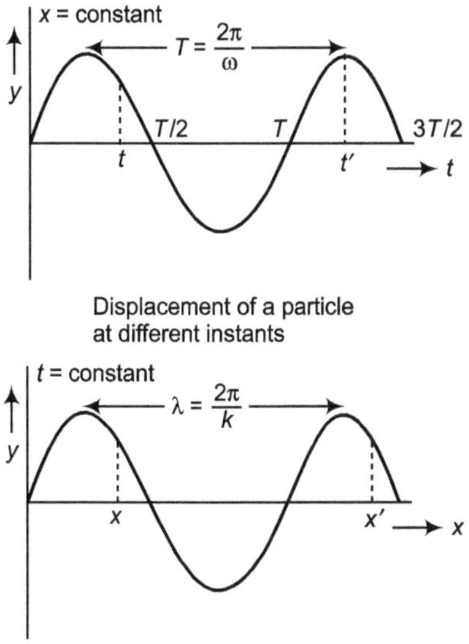

Displacement of a particle at different instants

Position of different particles at same instant in a string

Fig. 17.4

Note that as y propagates in the medium (or space), it repeats itself in space after one period, because

$$\lambda = vT$$

which shows that, the **wavelength is the distance advanced by the wave motion in one period**.

Therefore, in plane progressive harmonic wave we have two periodicities, one in time given by the period T and one in space given by the wavelength λ, with the two related by

$$\lambda = vT$$

Extra Knowledge Points

- As we have read in Art. 17.3, any function of x and t which satisfies equation number (i) of the same article or which can be written in the form of equation number (ii) represents a wave provided it is finite everywhere at all times. What we have read in Art. 17.4 is about plane progressive harmonic wave. If $f(ax \pm bt)$ is a sine or cosine function, it is called plane progressive harmonic wave. The only special characteristic of this wave is that oscillations of y are simple harmonic in nature.
- The general expression of a plane progressive harmonic wave is,
$$y = A \sin (kx \pm \omega t \pm \phi)$$
or
$$y = A \cos (kx \pm \omega t \pm \phi)$$
Here, ϕ represents the initial phase.
- I have seen students often confused whether the equation of a plane progressive wave should be,
$$y = A \sin (kx - \omega t)$$
or
$$y = A \sin (\omega t - kx)$$
Because some books write the first while the others write the second. It hardly matters whether you write the first or the second. Both the equations represent a travelling wave travelling in positive x-direction with speed $v = \frac{\omega}{k}$. The difference between them is that they are out of phase, i.e. phase difference between them is π. It means, if a particle in position $x = 0$ at time $t = 0$ is in its mean position and moving upwards (represented by first wave), then the same particle will be in its mean position but moving downwards (represented by the second wave). Similarly the waves,
$$y = A \sin (kx - \omega t)$$
and
$$y = - A \sin (kx - \omega t)$$
are also out of phase.

Particle velocity (v_p) and acceleration (a_p) in a sinusoidal wave In plane progressive harmonic wave particles of the medium oscillate simple harmonically about their mean position. Therefore, all the formulae what we have read in SHM apply to the particles here also. For example, maximum particle velocity is $\pm A\omega$ at mean position and it is zero at extreme positions etc. Similarly, maximum particle acceleration is $\pm \omega^2 A$ at extreme positions and zero at mean position. However, the wave velocity is different from the particle velocity. This depends on certain characteristics of the medium. Unlike the particle velocity which oscillates simple harmonically (between $+ A\omega$ and $- A\omega$) the wave velocity is constant for given characteristics of the medium.

Suppose the wave function is,
$$y(x, t) = A \sin (kx - \omega t) \quad \text{...(i)}$$
Let us differentiate this function partially with respect to t and x.
$$\frac{\partial y(x,t)}{\partial t} = - A\omega \cos (kx - \omega t) \quad \text{...(ii)}$$
$$\frac{\partial y(x,t)}{\partial x} = Ak \cos (kx - \omega t) \quad \text{...(iii)}$$
Now, these can be written as,
$$\frac{\partial y(x,t)}{\partial t} = - \left(\frac{\omega}{k}\right) \frac{\partial y(x,t)}{\partial x}$$
Here, $\frac{\partial y(x,t)}{\partial t}$ = particle velocity v_p

$\frac{\omega}{k}$ = wave velocity v

and $\frac{\partial y(x,t)}{\partial x}$ = slope of the wave

Thus,
$$v_p = - v \text{ (slope)} \quad \text{...(iv)}$$
i.e. particle velocity at a given position and time is equal to negative of the product of wave velocity with slope of the wave at that point at that instant.

The acceleration of the particle is the second partial derivative of $y(x, t)$ with respect to t,
$$\therefore \quad a_p = \frac{\partial^2 y(x,t)}{\partial t^2} = - \omega^2 A \sin (kx - \omega t)$$
$$= - \omega^2 y(x, t)$$
i.e. the acceleration of the particle equals $- \omega^2$ times its displacement, which is the result we obtained for SHM. Thus,
$$a_p = - \omega^2 \text{ (displacement)} \quad \text{...(v)}$$
which is also the wave equation.

Figure shows the particle velocity (v_p) and acceleration (a_p) given by Eqs. (iv) and (v) for two points 1 and 2 on a string as a sinusoidal wave is travelling in it along positive x-direction.

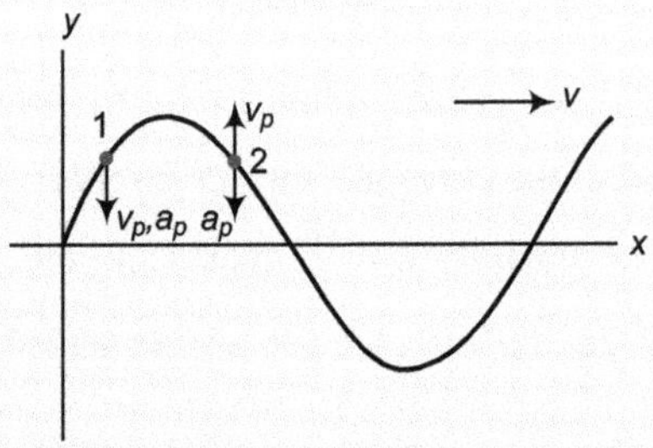

At point 1 Slope of the curve is positive. Hence, from Eq. (iv) particle velocity (v_p) is negative or downwards. Similarly, displacement of the particle is positive, so from Eq. (v) acceleration will be negative or downwards.

At point 2 Slope is negative while displacement is positive. Hence, v_p will be positive (upwards) and a_p is negative (downwards).

- The direction of v_p will change, if the wave travels along negative x-direction.

Example 17.3 *The equation of a wave is,*

$$y(x,t) = 0.05 \sin\left[\frac{\pi}{2}(10x - 40t) - \frac{\pi}{4}\right] m$$

Find

(a) the wavelength, the frequency and the wave velocity,

(b) the particle velocity and acceleration at $x = 0.5$ m and $t = 0.05$ s.

Sol. (a) The equation may be rewritten as,

$$y(x,t) = 0.05 \sin\left(5\pi x - 20\pi t - \frac{\pi}{4}\right) \text{m}$$

Comparing this with equation of plane progressive harmonic wave,

$$y(x,t) = A \sin(kx - \omega t + \phi)$$

we have, wave number $k = \frac{2\pi}{\lambda} = 5\pi$ rad/m

$\therefore \qquad \lambda = 0.4$ m

The angular frequency is,

$$\omega = 2\pi f = 20\pi \text{ rad/s}$$

$\therefore \qquad f = 10$ Hz

The wave velocity is,

$$v = f\lambda = \frac{\omega}{k} = 4 \text{ m/s in positive } x\text{-direction}$$

(b) The particle velocity and acceleration are,

$$\frac{\partial y}{\partial t} = -(20\pi)(0.05)\cos\left(\frac{5\pi}{2} - \pi - \frac{\pi}{4}\right)$$

$$= 2.22 \text{ m/s}$$

$$\frac{\partial^2 y}{\partial t^2} = -(20\pi)^2(0.05)\sin\left(\frac{5\pi}{2} - \pi - \frac{\pi}{4}\right)$$

$$= 140 \text{ m/s}^2$$

Example 17.4 *Figure shows a snapshot of a sinusoidal travelling wave taken at $t = 0.3$ s. The wavelength is 7.5 cm and the amplitude is 2 cm. If the crest P was at $x = 0$ at $t = 0$, write the equation of travelling wave.*

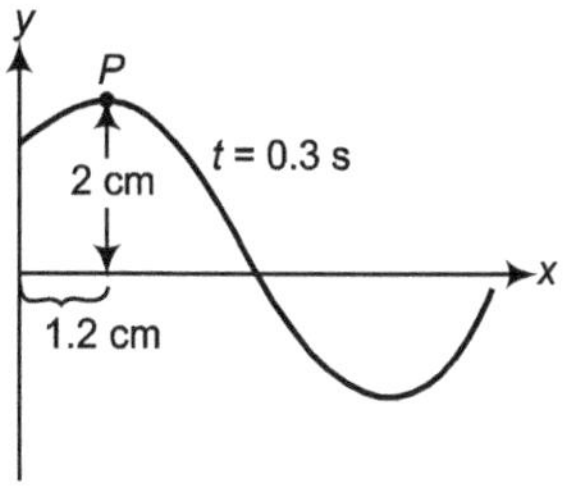

Fig. 17.5

Sol. Given, $A = 2$ cm, $\lambda = 7.5$ cm

$$\therefore \qquad k = \frac{2\pi}{\lambda} = 0.84 \text{ cm}^{-1}$$

The wave has travelled a distance of 1.2 cm in 0.3 s.
Hence, speed of the wave,

$$v = \frac{1.2}{0.3} = 4 \text{ cm/s}$$

$\therefore$ Angular frequency, $\omega = (v)(k) = 3.36$ rad/s

Since, the wave is travelling along positive x-direction and crest (maximum displacement) is at $x = 0$ at $t = 0$, we can write the wave equation as,

$$y(x,t) = A\cos(kx - \omega t)$$

or $\qquad y(x,t) = A\cos(\omega t - kx)$

as $\qquad \cos(-\theta) = \cos\theta$

Therefore, the desired equation is,

$$y(x,t) = (2 \text{ cm})\cos[(0.84 \text{ cm}^{-1})x - (3.36 \text{ rad/s})t] \text{ cm}$$

17.5 Speed of a Transverse Wave on a String

- Derivation of the formula is optional.

One of the key properties of any wave is the wave speed. In this section, we'll see what determines the speed of propagation of transverse waves on a string. The physical quantities that determine the speed of transverse waves on a string are the **tension** in the string and its **mass per unit length** (also called linear mass density). We might guess that increasing the tension showed increase the restoring forces that tend to straighten the string when it is disturbed, thus increasing the wave speed. We might also guess that increasing the mass should make the motion more sluggish and decrease the speed. Both these guesses turn out to be right.

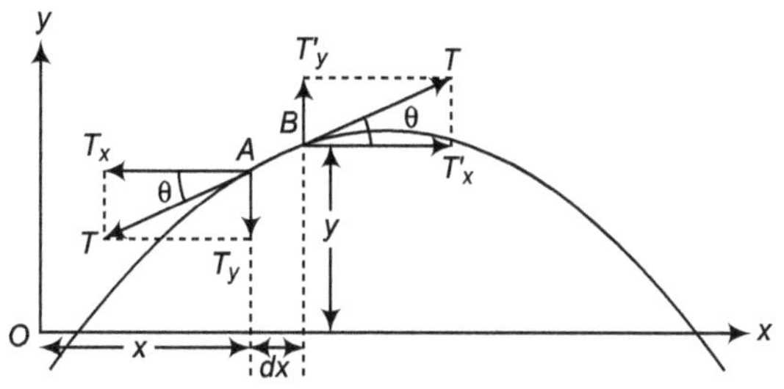

Fig. 17.6

We will develop the exact relationship between wave speed, tension and mass per unit length. Under equilibrium conditions a string subject to a tension T is straight. Suppose that we now displace the string sidewise, or perpendicular to its length, by a small amount as shown in figure. Consider a small section AB of the string of length dx, that has been displaced a distance y from the equilibrium position. On each end a tangential force T is acting. Due to the curvature of the string, the two forces are not directly opposed but make angles θ and θ' with the x-axis. The resultant upward force on the section AB of the string is,

$$F_y = T_y' - T_y \qquad \text{...(i)}$$

Under the acting of this force, the section AB of the string moves up and down.

Rewriting Eq. (i), we have

$$F_y = T(\sin\theta' - \sin\theta)$$

Since, θ and θ' are almost equal, we may write

$$F_y = Td(\sin\theta)$$

If the curvature of the string is not very large, the angles θ and θ' are small, and the sine can be replaced by their tangents. So, the upward force is,

$$F_y = Td\,(\tan\theta) = T\left\{\frac{d}{dx}(\tan\theta)\right\}dx$$

But $\tan\theta$ is the slope of the curve adopted by the string, which is equal to $\frac{dy}{dx}$. Hence,

$$F_y = T\left\{\frac{d}{dx}\left(\frac{dy}{dx}\right)\right\}dx = T\left(\frac{d^2y}{dx^2}\right)dx$$

This force must be equal to the mass of the section AB multiplied by its upward acceleration $\frac{d^2y}{dt^2}$. If μ is the linear density of the string, the mass of the section AB is $\mu\,dx$. We use the relation $F = ma$ and write the equation of motion of this section of the string as,

$$(\mu\,dx)\frac{d^2y}{dt^2} = T\left(\frac{d^2y}{dx^2}\right)dx$$

or

$$\frac{d^2y}{dt^2} = \frac{T}{\mu}\cdot\frac{d^2y}{dx^2}$$

Comparing this with wave equation,

$$\frac{\partial^2 y}{\partial t^2} = v^2\frac{\partial^2 y}{\partial x^2}$$

Thus, wave speed, $v = \sqrt{\frac{T}{\mu}}$

Extra Knowldge Points

- Speed of transverse wave on a string is given by,

$$v = \sqrt{\frac{T}{\mu}}$$

Here, μ = mass per unit length of the string

$$= \frac{m}{l}$$

$$= \frac{mA}{lA} \quad (A = \text{area of cross-section of the string})$$

$$= \left(\frac{m}{V}\right)A \quad (v = \text{volume of string})$$

$$= \rho A \quad (\rho = \text{density of string})$$

Hence, the above expression can also be written as,

$$v = \sqrt{\frac{T}{\rho A}}$$

- Speed of longitudinal wave through a gas (or a liquid) is given by,

$$v = \sqrt{\frac{B}{\rho}}$$

Here, B = bulk modulus of the gas (or liquid)

and ρ = density of the gas (or liquid)

Now, Newton who first deduced this relation for v assumed that during the passage of a sound wave through a gas (or air), the temperature of the gas remains constant i.e. sound wave travels under isothermal conditions and hence took B to be the isothermal elasticity of the gas and which is equal to its pressure p. So, Newton's formula for the velocity of a sound wave (or a longitudinal wave) in a gaseous medium becomes,

$$v = \sqrt{\frac{p}{\rho}}$$

If, however, we calculate the velocity of sound in air at NTP with the help of this formula by substituting

$$p = 1.01\times10^5\,\text{N/m}^2 \quad \text{and} \quad \rho = 1.29\,\text{kg/m}^3$$

then v comes out to be nearly 280 m/s. Actually the velocity of sound in air at NTP as measured by Newton himself, is found to be 332 m/s. Newton could not explain this large discrepancy between his theoretical and experimental results.

Laplace after 140 yr correctly argued that a sound wave passes through a gas (or air) very rapidly. So, adiabatic conditions are developed. So, he took B to be the adiabatic elasticity of the gas, which is equal to γp, where γ is the ratio of C_p (molar heat capacity at constant pressure) and C_V (molar heat capacity at constant volume). Thus, Newton's formula as corrected by Laplace becomes,

$$v = \sqrt{\frac{\gamma p}{\rho}}$$

For air, $\gamma = 1.41$. So, that in air $v = \sqrt{\frac{1.41p}{\rho}}$

which gives 331.6 m/s as the velocity of sound (in air) at NTP which is in agreement with Newton's experimental result.

Note Points

- We will carry out the derivation of formula, $v = \sqrt{\frac{B}{\rho}}$ in the chapter of sound.
- Speed of longitudinal wave in a thin rod or wire is given by,

$$v = \sqrt{\frac{Y}{\rho}}$$

Here, Y is the Young's modulus of elasticity.

Example 17.5 *One end of 12.0 m long rubber tube with a total mass of 0.9 kg is fastened to a fixed support. A cord attached to the other and passes over a pulley and supports an object with a mass of 5.0 kg. The tube is struck a transverse blow at one end. Find the time required for the pulse to reach the other end.* $(g = 9.8\,m/s^2)$

Sol. Tension in the rubber tube AB,

$$T = mg$$

$\Rightarrow \quad T = (5.0)(9.8) = 49\text{ N}$

Mass per unit length of rubber tube,

$$\mu = \frac{0.9}{12} = 0.075\text{ kg/m}$$

$\therefore$ Speed of wave on the tube,

$$v = \sqrt{\frac{T}{\mu}} = \sqrt{\frac{49}{0.075}} = 25.56\text{ m/s}$$

$\therefore$ The required time is,

$$t = \frac{AB}{v} = \frac{12}{25.56} = 0.47\text{ s}$$

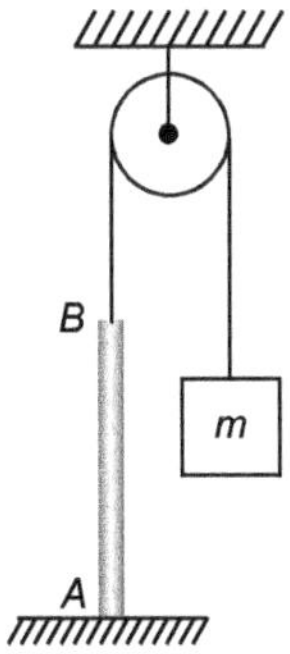

Fig. 17.7

17.6 Energy in Wave Motion

Every wave motion has energy associated with it. In wave motion, energy and momentum are transferred or propagated.

To produce any of the wave motions, we have to apply a force to a portion of the wave medium. The point where the force is applied moves, so we do work on the system. As the wave propagates, each portion of the medium exerts a force and does work on the adjoining portion. In this way a wave can transport energy from one region of space to other.

Regarding the energy in wave motion, we come across three terms namely, energy density (u), power (P) and intensity (I). Now, let us take them one by one.

Energy Density (u)

By the energy density of a plane progressive wave, we mean **the total mechanical energy (kinetic + potential) per unit volume** of the medium through which the wave is passing. Let us proceed to obtain an expression for it.

Energy of a particle in SHM is $\frac{1}{2}m\omega^2 A^2$. In a plane progressive wave, all particles oscillate simple harmonically with same amplitude A and angular frequency ω. Mass per unit volume is called density. Hence, in the above expression mass will be replaced by density for energy per unit volume or energy density.

$$\therefore \qquad u = \frac{1}{2}\rho\omega^2 A^2$$

Power (P)

Power is the instantaneous rate at which energy is transferred along the string (if we consider a transverse wave on a string).

In unit time, the wave will travel a distance v. If S be the area of cross-section of the string, then volume of this length would be sv and energy transmitted per unit time (called power) would be,

$$P = (\text{energy density})\,(\text{volume})$$

$$\therefore \qquad P = \frac{1}{2}\rho\omega^2 A^2 Sv$$

Intensity (I)

Flow of energy per unit area of cross-section of the string in unit time is known as the intensity of the wave. Thus,

$$I = \frac{\text{power}}{\text{area of cross-section}} = \frac{P}{S}$$

or

$$I = \frac{1}{2}\rho\omega^2 A^2 v$$

Note Points

- Although the above relations for power and intensity have been discussed for a transverse wave on a string, they hold good for other waves also.
- **Intensity due to a point source** If a point source emits wave uniformly in all directions, the energy at a distance r from the source is distributed uniformly on a spherical surface of radius r and area $S = 4pr^2$. If P is the power emitted by the source, the power per unit area at a distance r from the source is $\frac{P}{4\pi r^2}$. The average power per unit area that is incident perpendicular to the direction of propagation is called the intensity. Therefore,

$$I = \frac{P}{4\pi r^2} \quad \text{or} \quad I \propto \frac{1}{r^2}$$

Now, as amplitude $A \propto \sqrt{I}$, a spherical harmonic wave emanating from a point source can therefore be written as

$$y(r, t) = \frac{A}{r}\sin(kr - \omega t)$$

Example 17.6 *A stretched string is forced to transmit transverse waves by means of an oscillator coupled to one end. The string has a diameter of 4 mm. The amplitude of the oscillation is 10^{-4} m and the frequency is 10 Hz. Tension in the string is 100 N and mass density of wire is $4.2 \times 10^3\ kg/m^3$. Find*

(a) the equation of the waves along the string,

(b) the energy per unit volume of the wave,

(c) the average energy flow per unit time across any section of the string, and

(d) power required to drive the oscillator.

Sol. (a) Speed of transverse wave on the string is,

$$v = \sqrt{\frac{T}{\rho S}} \qquad (\text{as } \mu = \rho S)$$

Substituting the values, we have

$$v = \sqrt{\frac{100}{(4.2\times 10^3)\left(\frac{\pi}{4}\right)(4.0\times 10^{-3})^2}} = 43.53\text{ m/s}$$

$$\omega = 2\pi f = 20\pi\,\frac{\text{rad}}{\text{s}} = 62.83\,\frac{\text{rad}}{\text{s}}$$

$$k = \frac{\omega}{v} = 1.44\text{ m}^{-1}$$

$\therefore$ Equation of the waves along the string,

$$y(x, t) = A\sin(kx - \omega t)$$

$$= (10^{-4}\text{ m})\sin\left[(1.44\text{ m}^{-1})\,x - \left(62.83\,\frac{\text{rad}}{\text{s}}\right)t\right]$$

(b) Energy per unit volume of the string,

$$u = \text{energy density} = \frac{1}{2}\rho\omega^2 A^2$$

Substituting the values, we have

$$u = \left(\frac{1}{2}\right)(4.2 \times 10^3)(62.83)^2 (10^{-4})^2$$

$$= 8.29 \times 10^{-2} \text{ J/m}^3$$

(c) Average energy flow per unit time,

$$P = \text{power} = \left(\frac{1}{2}\rho\omega^2 A^2\right)(sv) = (u)(sv)$$

Substituting the values, we have

$$P = (8.29 \times 10^{-2})\left(\frac{\pi}{4}\right)(4.0 \times 10^{-3})^2 (43.53)$$

$$= 4.53 \times 10^{-5} \text{ J/s}$$

(d) Power required to drive the oscillator is obviously 4.53×10^{-5} W.

17.7 Principle of Superposition

Two or more waves can travel simultaneously in a medium without affecting the motion of one another. Therefore, the resultant displacement of each particle of the medium at any instant is equal to the vector sum of the displacements produced by the two waves separately. This principle is called 'principle of superposition'. It holds for all types of waves, provided the waves are not of very large amplitude. We can express the superposition principle in the form,

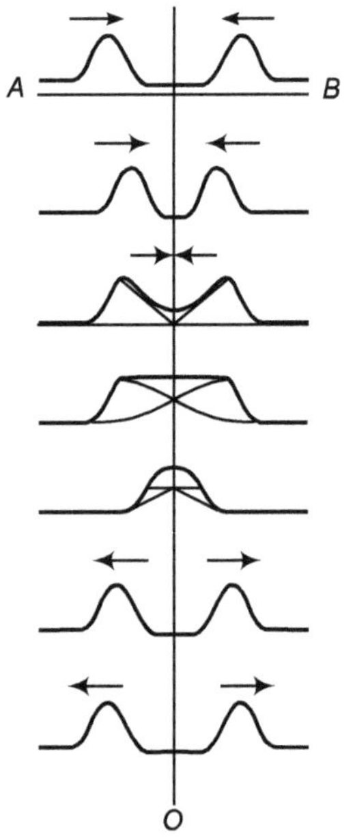

Overlap of two wave pulses travelling in opposite directions. Time increases from top to bottom.

Fig. 17.8

$$y(x, t) = y_1(x, t) + y_2(x, t) + \ldots + y_n(x, t)$$

or $$y(x, t) = \sum_{j=1}^{n} y_j(x, t)$$

Here, y_j are the individual wave functions and their sum, the wave function $y(x, t)$ describes the resultant behaviour of the medium as a function of position and time.

The principle of superposition is of central importance in all types of waves. When a friend talks to you while you are listening to music, you can distinguish the sound of speech and the sound of music from each other. This is precisely because according to principle of superposition two different waves can travel in a medium simultaneously without disturbing the other. Superposition also applies to electromagnetic waves (such as light) and many other types of waves.

As an example, consider a long stretched string AB. At the end A, a pulse is generated which propagates towards B with some speed (say v). At B an identical pulse is generated which travels towards A with the same speed v. The snapshot of the string at different times are shown in figure. From this figure you might have understood how the two waves superimpose and what is the meaning of principle of superposition.

Extra Knowledge Points

- Consider the superposition of two sinusoidal waves of same frequency at a point. Let us assume that the two waves are travelling in the same direction with same velocity. The equation of the two waves reaching at a point can be written as,

$$y_1 = A_1 \sin(kx - \omega t)$$

and $$y_2 = A_2 \sin(kx - \omega t + \phi)$$

The resultant displacement of the point where the waves meet, is

$$y = y_1 + y_2$$

$$= A_1 \sin(kx - \omega t) + A_2 \sin(kx - \omega t + \phi)$$

$$= A_1 \sin(kx - \omega t) + A_2 \sin(kx - \omega t)\cos\phi + A_2 \cos(kx - \omega t)\sin\phi$$

$$= (A_1 + A_2 \cos\phi)\sin(kx - \omega t) + A_2 \sin\phi \cos(kx - \omega t)$$

$$= A\cos\theta \sin(kx - \omega t) + A\sin\theta\cos(kx - \omega t)$$

or $$y = A\sin(kx - \omega t + \theta)$$

Here $$A_1 + A_2\cos\phi = A\cos\theta$$

and $$A_2 \sin\phi = A\sin\theta$$

or $$A^2 = (A_1 + A_2\cos\phi)^2 + (A_2\sin\phi)^2$$

or $$A = \sqrt{A_1^2 + A_2^2 + 2A_1A_2\cos\phi}$$

and $$\tan\theta = \frac{A\sin\theta}{A\cos\theta} = \frac{A_2\sin\phi}{A_1 + A_2\cos\phi}$$

- The above result can be obtained by graphical method as well. Assume a vector $\mathbf{A}_1$ of length A_1 to represent the amplitude of first wave.

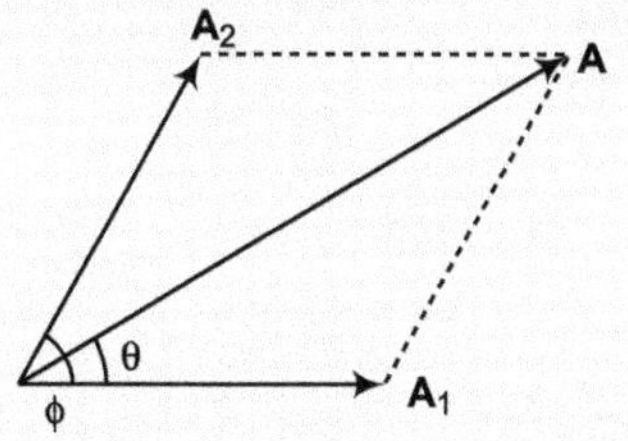

Another vector $\mathbf{A}_2$ of length A_2, making an angle ϕ with $\mathbf{A}_1$ represent the amplitude of second wave. The resultant of $\mathbf{A}_1$ and $\mathbf{A}_2$ represent the amplitude of resulting function y. The angle θ represents the phase difference between the resulting function and the first wave.

Example 17.7 *Two harmonic waves are represented in SI units by,*

$$y_1(x,t) = 0.2 \sin(x - 3.0t)$$

and $$y_2(x,t) = 0.2 \sin(x - 3.0t + \phi)$$

(a) Write the expression for the sum $y = y_1 + y_2$ for $\phi = \frac{\pi}{2}$ rad.

(b) Suppose the phase difference ϕ between the waves is unknown and the amplitude of their sum is 0.32 m, what is ϕ ?

Sol. (a) $y = y_1 + y_2 = 0.2 \sin(x - 3.0t) + 0.2 \sin\left(x - 3.0t + \frac{\pi}{2}\right)$

$= A \sin(x - 3.0t + \theta)$

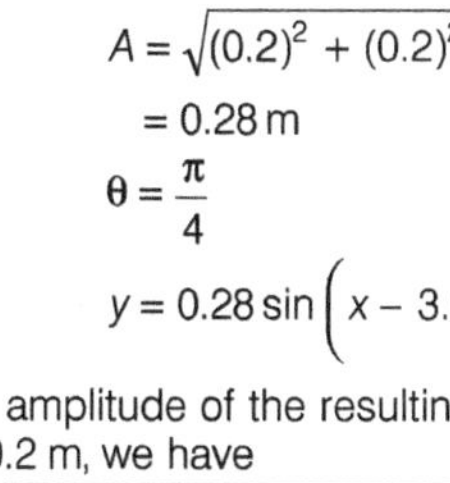

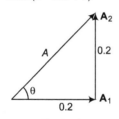

Fig. 17.9

Here, $A = \sqrt{(0.2)^2 + (0.2)^2}$

$= 0.28$ m

and $\theta = \frac{\pi}{4}$

$\therefore$ $y = 0.28 \sin\left(x - 3.0t + \frac{\pi}{4}\right)$

(b) Since, the amplitude of the resulting wave is 0.32 m and $A = 0.2$ m, we have

$$0.32 = \sqrt{(0.2)^2 + (0.2)^2 + (2)(0.2)(0.2)\cos\phi}$$

Solving this, we get

$$\phi = \pm 1.29 \text{ rad}$$

Note Point

- **EXERCISE** What is the resultant wave obtained when $y_1(x,t) = A\sin(kx - \omega t + \phi_1)$ and $y_2 = A\sin(kx - \omega t + \phi_2)$ are added. Use $\phi_1 = \frac{\pi}{6}$ rad and $\phi_2 = \frac{\pi}{2}$ rad.

Ans. $y(x,t) = \sqrt{3}A\sin(kx - \omega t + \pi/3)$

17.8 Interference of Waves

We have seen in Objective Galaxy 17.3 that if two sinusoidal waves of same angular frequency ω meet at a point where phase difference between them is ϕ, the resulting amplitude is given by,

$$A^2 = A_1^2 + A_2^2 + 2A_1A_2\cos\phi \quad \text{...(i)}$$

Further in Art. 17.6, we have also read that intensity of a wave is given by,

$$I = \frac{1}{2}\rho A^2\omega^2 v$$

i.e. $$I \propto A^2$$

So, if ρ, ω and v are same for both interfering waves, Eq. (i) can also be written as,

$$I = I_1 + I_2 + 2\sqrt{I_1 I_2}\cos\phi \quad \text{...(ii)}$$

From Eqs. (i) and (ii), we see that the resulting amplitude A and intensity I depend on the phase difference ϕ between the interfering waves, where, $\cos\phi = +1$

$$A = A_{max} = A_1 + A_2$$

or $$I = I_{max} = (\sqrt{I_1} + \sqrt{I_2})^2$$

and the waves are said to be interfering **constructively.**

Similarly where, $\cos\phi = -1$, $A = A_{min} = A_1 \sim A_2$

or $$I = I_{min} = (\sqrt{I_1} - \sqrt{I_2})^2$$

and the waves are said to be interfering **destructively.**

- Detailed discussion on interference will be done in the chapter of wave optics.

Example 17.8 *Two waves of equal frequencies have their amplitudes in the ratio of 3:5. They are superimposed on each other. Calculate the ratio of maximum and minimum intensities of the resultant wave.*

Sol. Given, $\frac{A_1}{A_2} = \frac{3}{5}$

$\therefore$ $\sqrt{\frac{I_1}{I_2}} = \frac{3}{5}$ (as $I \propto A^2$)

Maximum intensity is obtained, when

$\cos\phi = 1$ and $I_{max} = (\sqrt{I_1} + \sqrt{I_2})^2$

Minimum intensity is found, when

$\cos\phi = -1$ and $I_{min} = (\sqrt{I_1} - \sqrt{I_2})^2$

Hence, $\frac{I_{max}}{I_{min}} = \left(\frac{\sqrt{I_1}+\sqrt{I_2}}{\sqrt{I_1}-\sqrt{I_2}}\right)^2 = \left(\frac{\sqrt{\frac{I_1}{I_2}}+1}{\sqrt{\frac{I_1}{I_2}}-1}\right)^2$

$= \left(\frac{3/5+1}{3/5-1}\right)^2 = \frac{64}{4} = \frac{16}{1}$

Note Point

- **EXERCISE** The ratio of intensities of two waves is 9 : 16. If these two waves interfere, then determine the ratio of the maximum and minimum possible intensities.

Ans. 49 : 1.

17.9 Reflection and Transmission of a Wave

Before starting the reflection and transmission of a wave from a boundary where two media separate each other, let us talk about a denser and a rare medium in reference to its behaviour towards a wave.

A medium is said to be denser (relative to the other), if the speed of wave in this medium is less than the speed of the wave in the other medium. Rather we can say speed of a wave in a denser medium is less than its speed in the rarer medium. Thus, it is the speed of wave which decides whether the medium is denser or rare for that particular wave and

$$v_{denser} < v_{rarer}$$

If a medium is denser for one type of wave, then at the same time the same medium can be rare for the other type of wave. For example, water is denser for electromagnetic (light) waves compared to air, because the speed of electromagnetic waves is less in water than in air. At the same time, for sound wave water is a rare medium because speed of sound wave in water is more. The laws of refraction (or transmission from one medium to the other) and reflection remain the same, i.e.

angle of incidence = angle of reflection (law of reflection)

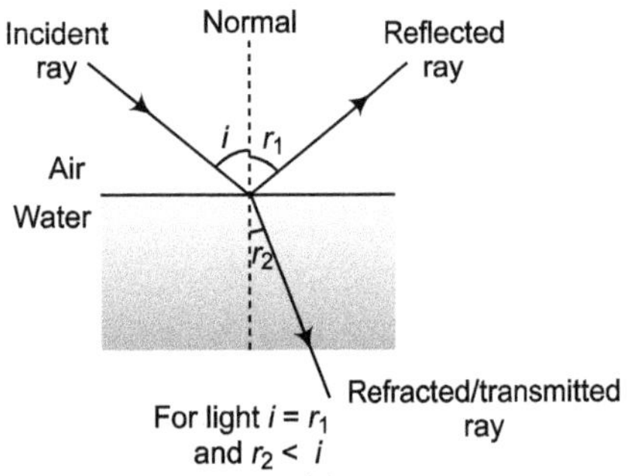

(a)

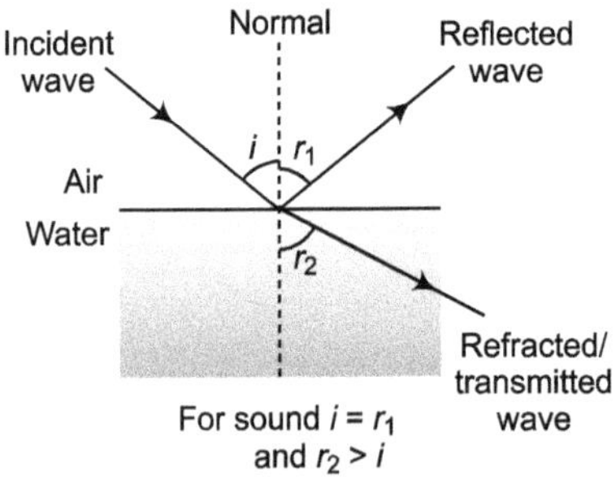

(b)

Fig.. 17.10

and a ray bends towards the normal it travels from a rare medium to a denser medium and *vice-versa* (law of refraction).

Any type of wave is associated with the following physical quantities :

(i) speed of wave (v)

(ii) frequency (f), time period (T) and angular frequency (ω)

(iii) wavelength (λ) and wave number (k)

(iv) amplitude (A) and intensity (I) and

(v) phase (ϕ)

Now, let us see what happens to these physical quantities when they are either reflected or transmitted.

(i) **Speed of wave (v)** Speed of a wave depends on the medium and some of its characteristics. For example, speed of a transverse wave on a stretched string depends on the tension T and its mass per unit length μ. In reflection medium, its characteristics do not change. So, in reflection speed of wave does not change. On the other hand, in transmission medium, the speed of wave change.

(ii) **Frequency (f), time period (T) and angular frequency (ω)** These three are related to each other by the relation

$$\omega = 2\pi f = \frac{2\pi}{T}$$

Or we can say, if any one of them is known, other two can easily be obtained.

Frequency of the wave depends on the source from where wave originates. In reflection and transmission, since source does not change. Hence, none of the three change.

(iii) **Wavelength (λ) and wave number (k)** These two are related to each other by the simple relation,

$$k = \frac{2\pi}{\lambda}, \quad \text{Further,} \quad \lambda = \frac{v}{f}$$

Here, v depends on medium and f on source.

So, if any one of them is changed, the wavelength will change. Therefore, in reflection λ and k do not change as medium and source both remain unchanged, while in transmission, as the medium is changed, so λ and k do change.

(iv) **Amplitude (A) and intensity (I)** Expression for the intensity of a wave is,

$$I = \frac{1}{2}\rho\omega^2 A^2 v$$

$$\therefore \quad I \propto A^2$$

Intensity is the energy transmitted per unit area per unit time. When a wave is incident on a boundary (separating two media) part of it is reflected and part is transmitted. Hence, intensity and amplitude both change in reflection as well as transmission. Unless 100% reflection or 100% transmission is there.

(v) **Phase** In transmission no phase change takes place. While in reflection phase change is zero, if the wave is reflected from a rarer medium and it is π, if it is reflected from a denser medium. For example, in Fig. 17.10(a), the reflected ray suffers a phase difference of π with the incident ray, because for electromagnetic wave water is denser, while in Fig. 17.10(b) the reflected wave is in phase with the incident wave. Because for sound wave water is a rare medium.

The above results in tabular form are given below :

Table 17.1

Wave property	Reflection	Transmission (Refraction)
v	does not change	changes
f, T, ω	do not change	do not change
λ, k	do not change	change
A, I	change	change
ϕ	$\Delta\phi = 0$, from a rarer medium $\Delta\phi = \pi$, from a denser medium	does not change

Example 17.9 *Two strings 1 and 2 are taut between two fixed supports (as shown in figure) such that the tension in both strings is same. Mass per unit length of 2 is more than that of 1. Explain which string is denser for a transverse travelling wave.*

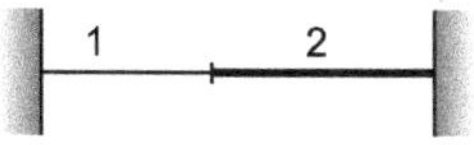

Fig. 17.11

Sol. Speed of a transverse wave on a string,

$$v = \sqrt{\frac{T}{\mu}} \quad \text{or} \quad v \propto \frac{1}{\sqrt{\mu}}$$

Now, $\mu_2 > \mu_1$ (given)

$\therefore \quad v_2 < v_1$

i.e. medium 2 is denser and medium 1 is rarer.

17.10 Standing Waves

We now consider what happens when two harmonic waves of equal frequency and amplitude travel through a medium (say string) in opposite directions. Suppose the two waves are,

$$y_1 = A \sin (kx - \omega t)$$

and $\quad y_2 = A \sin (kx + \omega t)$

By the principle of superposition their sum is

$$y = y_1 + y_2$$

or $\quad y = A[\sin (kx - \omega t) + \sin (kx + \omega t)]$

Using the identity,

$$\sin A + \sin B = 2 \sin \left(\frac{A+B}{2}\right) \cos \left(\frac{A-B}{2}\right),$$

we obtain $y = 2A \sin (kx) \cos (\omega t)$ ...(i)

This expression is different from wave representations that we have encountered up to now. It does not have the form $f(x \pm vt)$ or $f(ax \pm bt)$ and therefore, does not describe a travelling wave. Instead Eq. (i) represents what is known as a **standing wave**.

Eq. (i) can also be written as,

$$y = A(x) \cos \omega t \quad \text{...(ii)}$$

where, $\quad A(x) = 2A \sin kx$ ...(iii)

This equation of a standing wave [Eq. (ii)] is really an equation of simple harmonic motion, whose amplitude [Eq. (iii)] is a function of x.

$$A(x) = 0, \text{ where } \sin kx = 0$$

or $\quad kx = 0, \pi, 2\pi, \ldots, n\pi \quad (n = 0, 1, 2 \ldots)$

Substituting $k = \frac{2\pi}{\lambda}$, we have

$$A(x) = 0 \text{ where, } x = 0, \frac{\lambda}{2}, \lambda, \ldots, \frac{n\lambda}{2}$$

These are the points which never displace from their mean positions. These are known as the nodes of the standing wave.

The distance between two adjacent nodes is $\frac{\lambda}{2}$.

Further, from Eq. (iii), we can see that maximum value of $|A(x)|$ is $2A$, where

or $\quad kx = \frac{\pi}{2}, \frac{3\pi}{2}, \ldots, (2n-1)\frac{\pi}{2} \quad (n = 0, 1, 2, \ldots)$

or $\quad x = \frac{\lambda}{4}, \frac{3\lambda}{4}, \ldots, (2n-1)\frac{\lambda}{4}$

These are the points of maximum displacement called **antinodes**. The distance between two adjacent antinodes is also $\frac{\lambda}{2}$, while that between a node and an antinode is $\frac{\lambda}{4}$.

In Fig. 17.12 (a), (b), (c) and (d), the two travelling waves are their resultant standing waves are shown for four different times over one period $T$of the travelling waves. At $t = 0$, the two travelling waves have the same displacement everywhere and add to produce the standing wave shown. At $t = T/4$, each wave has moved a distance of $\lambda/4$ in opposite directions, so they differ in phase by π rad and completely cancel. At $t = T/2$, they are again in phase but the positions of the crests and the troughs of the standing wave at $t = 0$ and $T/2$ have been interchanged. At $t = \frac{3T}{4}$, the travelling waves completely cancel once more. Finally at $t = T$, or after one period, the standing wave reassumes the shape it had at $t = 0$.

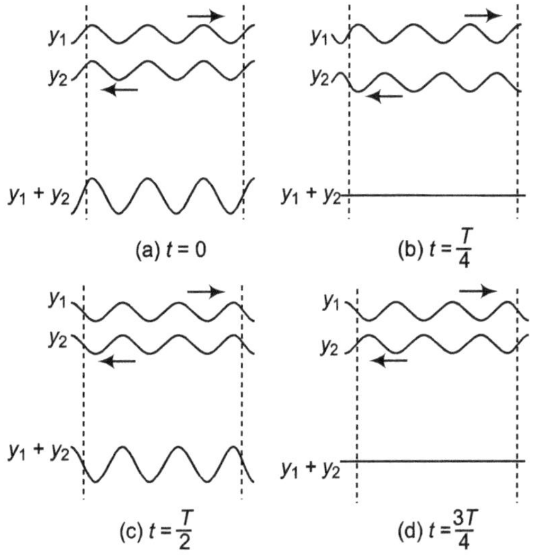

Fig. 17.12

Note that in a travelling wave each particle vibrates with the same amplitude. However, in case of a standing waves it is not the same for different particles but varies with the location x of the particle. Energy is not transported in stationary waves it remains standing, although it alternates between vibrational kinetic energy and the elastic potential energy. We call the motion a wave motion because we can think of it as a superposition of waves travelling in opposite directions. We can equally regard the motion as an oscillation of the string as a whole, each particle oscillating with SHM of angular frequency ω and with an amplitude that depends on its location.

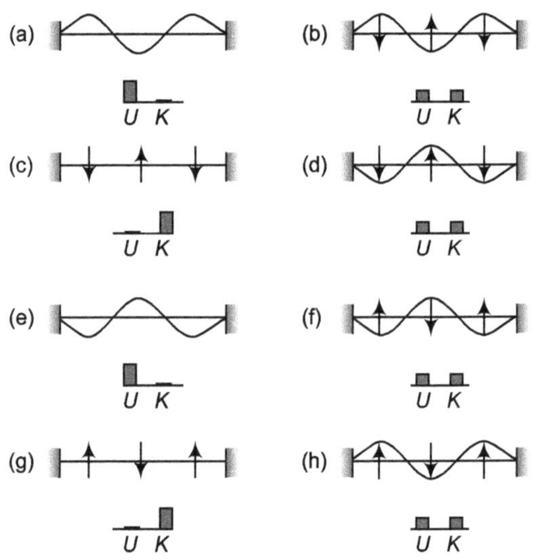

Fig. 17.13

In the figure, we have shown how the energy associated with the oscillation string shifts back and forth between kinetic energy K and potential energy U during one cycle.

Comparison between travelling and stationary waves

Table 17.2

S.No.	Travelling Waves	Stationary Waves
1.	These waves advance in a medium with definite velocity.	These waves remain stationary between two boundaries in the medium.
2.	In these waves, all particles of the medium oscillate with same frequency and amplitude.	In these waves, all particles except nodes oscillate with same frequency but different amplitude is zero at nodes and maximum at antinodes.
3.	At any instant phase of vibration varies continuously from one particle to the other i.e. phase difference between two particles can have any value between 0 and 2π.	At any instant the phase of all particles between two successive nodes is the same, but phase of particles on one side of a node is opposite to the phase of particles on the other side of the node, i.e. phase difference between any two particles can be either zero or π.
4.	In these waves, at no instant all the particles of the medium pass through their mean positions simultaneously.	In these waves all particles of the medium pass through their mean positions simultaneously twice in each time period.
5.	These waves transmit energy in the medium.	These waves do not transmit energy in the medium.

Extra Knowledge Points

- Till now we have come across following three sets of equations :

$$\left.\begin{aligned} y &= A\sin(\omega t \pm \phi) \\ y &= A\cos(\omega t \pm \phi)\end{aligned}\right\}\text{SHM}$$

$$\left.\begin{aligned} y &= A\sin(kx \pm \omega t \pm \phi) \\ y &= A\cos(kx \pm \omega t \pm \phi)\end{aligned}\right\}\text{Travelling wave}$$

$$\left.\begin{aligned} y &= A\sin kx\cos\omega t \text{ or } 2A\sin kx\cos\omega t \\ y &= 2A\sin\omega t\cos kx \text{ or } A\sin\omega t\cos kx \\ y &= A\sin kx\cos\omega t \text{ or } 2A\sin kx\cos\omega t \\ y &= 2A\cos kx\cos\omega t \text{ or } A\cos kx\cos\omega t\end{aligned}\right\}\text{Standing wave}$$

- In standing wave we have given four set of equations . The equation of standing wave basically depends on the component waves. Further, if the maximum amplitude is $2A$, it means amplitude of travelling waves is A and if it is A, then amplitude of travelling waves is $\frac{A}{2}$.
- Standing wave is an example of interference. Nodes means destructive interference and antinodes means constructive interference.
- Two identical waves moving in opposite directions along the string will still produce standing waves even, if their amplitudes are unequal.

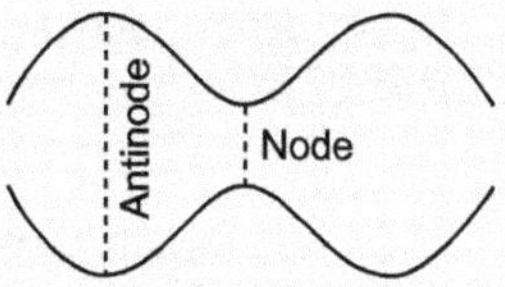

This is the case when an incident travelling wave is only partially reflected from a boundary, the resulting superposition of two waves having different amplitudes and travelling in opposite directions gives a standing wave pattern of waves whose envelope is shown in figure.

The Standing Wave Ratio (SWR) is defined as

$$\frac{A_{max}}{A_{min}} = \frac{A_i + A_r}{A_i - A_r}$$

For 100% reflection, SWR $= \infty$ and for no reflection, SWR $= 1$.

- The intensity of a travelling wave is given by,

$$I = \frac{1}{2}\rho A^2\omega^2 v \quad \text{i.e. } I \propto A^2$$

So, we can write, $\frac{I_1}{I_2} = \left(\frac{A_1}{A_2}\right)^2$ if ρ, ω and v are same for two waves.

For example, when an incident travelling wave is partly reflected and partly transmitted from a boundary, we can write

$$\frac{I_i}{I_r} = \left(\frac{A_i}{A_r}\right)^2$$

As incident and reflected waves are in the same medium hence, they have same values of ρ and v. But we cannot write.

$$\frac{I_i}{I_t} = \left(\frac{A_i}{A_t}\right)^2$$

as they have different value of ρ and v.

17.11 Normal Modes of a String

In an unbounded continuous medium, there is no restriction on the frequencies or wavelengths of the standing waves. However, if the waves are confined in space—for example, when a string is tied at both ends—standing waves can be set-up for a discrete set of frequencies or wavelengths. Consider a string of definite length l, rigidly held at both ends. When we set-up a sinusoidal wave on such a string, it gets reflected from the fixed ends. By the superposition of two identical waves travelling in opposite directions standing waves are established on the string.

The only requirement, we have to satisfy is that the end points be nodes as these points can not oscillate. They are permanently at rest. There may be any number of nodes in between or none at all, so that the wavelength associated with the standing waves can take on many different values. The distance between adjacent nodes is $\lambda/2$, so that in a string of length l there must be exactly an integral number n of half wavelengths, $\lambda/2$. That is,

$$\frac{n\lambda}{2} = l$$

or
$$\lambda = \frac{2l}{n}, \qquad (n = 1, 2, 3\ldots)$$

But $\lambda = \frac{v}{f}$ and $v = \sqrt{\frac{T}{\mu}}$, so that the natural frequencies of oscillation of the system are,

$$f = n\left(\frac{v}{2l}\right)$$

$$= \frac{n}{2l}\sqrt{\frac{T}{\mu}} \qquad (n = 1, 2, 3\ldots)$$

The smallest frequency f_1 corresponds to the largest wavelength $(n = 1)$, $\lambda_1 = 2l$

$$f_1 = \frac{v}{2l}$$

This is called the fundamental frequency. The other standing wave frequencies are,

$$f_2 = \frac{2v}{2l} = 2f_1$$

$$f_3 = \frac{3v}{2l} = 3f_1$$

and so on.

These frequencies are called harmonics. Musicians sometimes call them overtone. Students are advised to remember these frequencies by name.

For example,

f_1 = fundamental tone or first harmonic

$f_2 = 2f_1$ = first overtone or second harmonic

$f_3 = 3f_1$ = second overtone or third harmonic and so on.

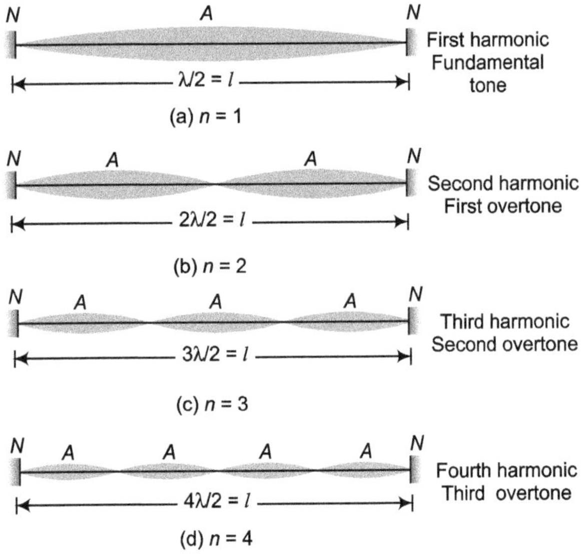

Fig. 17.14

A **normal mode** of an oscillating system is a motion in which all particles of the system move sinusoidally with the same frequency. A harmonic oscillator which has only one oscillating particle, has only one normal mode and one characteristic frequency. By contrast, a string fixed at both ends has infinitely many normal modes because it is made up of a very large (effectively infinite) number of particles. Fig. 17.17 shows the first four normal mode patterns and their associated frequencies and wavelengths. If we could displace a string, so that its shape is the same as one the normal-mode patterns and then release it, it would vibrate with the frequency of that mode.

Example 17.10 *The displacement of a standing wave on a string is given by*

$$y(x, t) = 0.4 = \sin(0.5x)\cos(30t)$$

where, x and y are in centimetres.

(a) Find the frequency, amplitude and wave speed of the component waves.

(b) What is the particle velocity at x = 2.4 cm at t = 0.8 s?

Sol. (a) The given wave can be written as the sum of two component waves as

$$y(x, t) = 0.2\sin(0.5x - 30t) + 0.2\sin(0.5)x + 30t)$$

i.e. the two component waves are

$$y_1(x, t) = 0.2\sin(0.5x + 30t)$$

travelling in positive x-direction and

$$y_2(x, t) = 0.2\sin(0.5x + 30t)$$

travelling in negative x-direction.

Now, $\omega = 30$ rad/s and $k = 0.5\,\text{cm}^{-1}$

$\therefore$ Frequency, $f = \dfrac{\omega}{2\pi} = \dfrac{15}{\pi}$ Hz

amplitude, $A = 0.2$ cm

and wave speed, $v = \dfrac{\omega}{k} = \dfrac{30}{0.5} = 60$ cm/s

(b) Particle velocity

$$v_P(x, t) = \frac{\partial y}{\partial t} = -12\sin(0.5x)\sin(30t)$$

$\therefore$ v_p (x = 2.4 cm, t = 0.8 s)

$$= -12\sin(1.2)\sin(24)$$
$$= 10.12 \text{ cm/s}$$

Example 17.11 *A standing wave is formed by two harmonic waves* $y_1 = A\sin(kx - \omega t)$ *and* $y_2 = A(kx + \omega t)$ *travelling on a string in opposite directions. Mass density of the string is ρ and area of cross-section is s. Find the total mechanical energy between two adjacent nodes on the string.*

Sol. The distance between two adjacent nodes is $\dfrac{\lambda}{2}$ or $\dfrac{\pi}{k}$.

$\therefore$ Volume of string between two nodes will be

V = (area of cross-section) (distance between two nodes)

$$= (S)\left(\frac{\pi}{k}\right)$$

Energy density (energy per unit volume) of a travelling wave is given by

$$u = \frac{1}{2}\rho A^2\omega^2$$

A standing wave is formed by two identical waves travelling in opposite directions. Therefore, the energy stored between two nodes in a standing wave

$E = 2$ [energy stored in a distance of $\dfrac{\pi}{k}$ of a travelling wave]

$$= 2 \text{ (energy density) (volume)}$$

$$= 2\left(\frac{1}{2}\rho A^2\omega^2\right)\left(\frac{\pi S}{k}\right) \Rightarrow E = \frac{\rho A^2\omega^2\,\pi S}{k}$$

Example 17.12 *A string fixed at both ends has consecutive standing wave modes for which the distances between adjacent nodes are* 18 *cm and* 16 *cm respectively.*

(a) What is the minimum possible length of the string?

(b) If the tension is 10 *N and the linear mass density is* 4 *g/m, what is the fundamental frequency?*

Sol. (a)

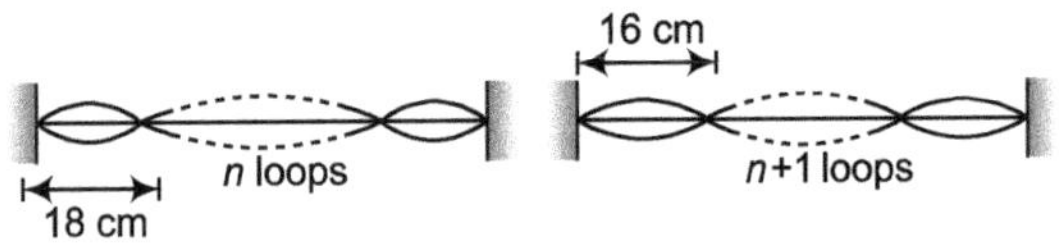

Fig. 17.15

Let l be the length of the string. Then,

$$18n = l \quad \text{...(i)}$$

$$16(n+1) = l \quad \text{...(ii)}$$

From Eqs. (i) and (ii), we get

$$n = 8 \text{ and } l = 144 \text{ cm}$$

Therefore, the minimum possible length of the string can be 144 cm.

(b) For fundamental frequency, $l = \lambda/2$

or $\lambda = 2l = 288\text{ cm} = 2.88\text{ m}$

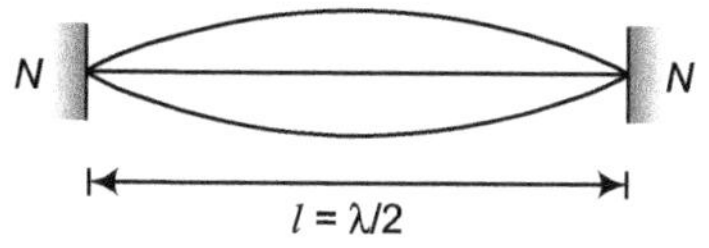

Fig. 17.16

Speed of wave on the string

$$v = \sqrt{\frac{T}{\mu}} = \sqrt{\frac{10}{4 \times 10^{-3}}} = 50\text{ m/s}$$

$\therefore$ Fundamental frequency,

$$f = \frac{v}{\lambda} = \frac{50}{2.88} = 17.36\text{ Hz}$$

17.12 Sound Waves

Of all the mechanical waves that occur in nature, the most important in our everyday lives are longitudinal waves in a medium, usually air, called **sound waves**. The reason is that the human ear is tremendously sensitive and can detect sound waves even of very low intensity. The human ear is sensitive to hear waves in the frequency range from about 20 to 20000 Hz called the **audible range,** but we also use the term sound for similar waves with frequencies above **(ultrasonic)** and below **(infrasonic)** the range of human hearing. Our main concern in this chapter is with sound waves in air, but sound can travel through any gas, liquid or solid.

17.13 Displacement Wave, Pressure Wave and Density Wave

Upto this point we have described mechanical waves primarily in terms of displacement, however, a description of sound waves in terms of pressure fluctuations is often more appropriate, largely because the ear is primarily sensitive to change in pressure.

Harmonic sound waves can be generated by a tuning fork or loudspeaker i.e. vibrating with simple harmonic motion. The vibrating source causes the air molecules next to it to oscillate with simple harmonic motion about their equilibrium position.

These molecules collide with neighbouring molecules, causing them to oscillate, thereby, propagating the sound wave. As we have discussed earlier, a sinusoidal wave equation $y(x, t)$ travelling in positive x-direction can be written as,

$$y(x, t) = A \sin(kx - \omega t)$$

which gives the instantaneous displacement y of a particle in the medium at position x at time t.

Note that in a longitudinal wave the displacements are parallel to the direction of travel of the wave. So, x and y are measured parallel to each other, not perpendicular as in a transverse wave. The amplitude A is the maximum displacement of a particle in a medium from its equilibrium position. These displacements are along the direction of the motion of the wave lead to **variations in the density and pressure** of air.

Hence, a sound wave can also be described in terms of variations of pressure or density at various points. The pressure fluctuations are of the order of 1 Pa $(= 1\text{ N/m}^2)$, whereas atmospheric pressure is about 10^5 Pa. In a sinusoidal sound wave in air, the pressure fluctuates above and below atmospheric pressure p_0 in a sinusoidal variation with the same frequency as the motions of the air particles. Similarly the density of air also vibrates sinusoidally above and below its normal level. So, we can express a sound wave either in terms of $y(x, t)$ or $\Delta p(x, t)$ or $\Delta\rho(x, t)$. All the three equations are related to one another. For example, amplitude of pressure variation $(\Delta p)_m$ is related to amplitude of displacement A by the equation,

$$\Delta p_m = BAk$$

where, B is the bulk modulus of the medium. Moreover pressure or density wave is 90° out of phase with the displacement wave, i.e. when the displacement is zero, the pressure and density changes are either maximum or minimum and when the displacement is a maximum or minimum, the pressure and density changes are zero.

Now, let us find the relation between them.

Relation between Displacement Wave and Pressure Wave

Figure shows a harmonic displacement wave moving through air contained in a long tube of cross-sectional area S. The volume of gas that has a thickness Δx in the horizontal direction is $V_i = S \cdot \Delta x$. The change in volume ΔV is $S \cdot \Delta y$, where Δy is the difference between the value of y at $x + \Delta x$ and the value of y at x. From the definition of bulk modulus, the pressure variation in the gas is

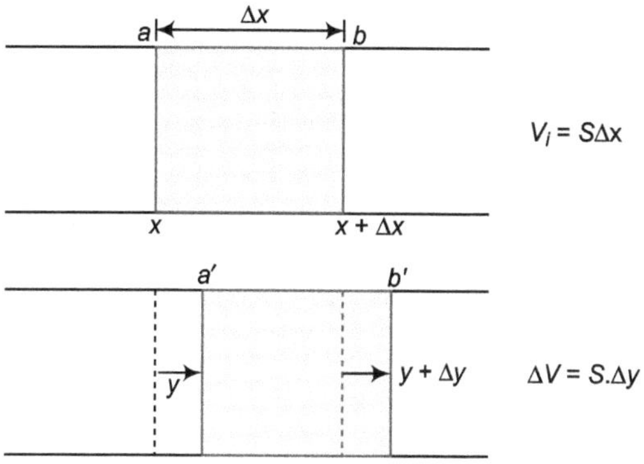

Fig. 17.17

$$\Delta p = -B\frac{\Delta V}{V_i}$$

$$= -B\left(\frac{S\cdot\Delta y}{S\cdot\Delta x}\right) = -B\left(\frac{\Delta y}{\Delta x}\right)$$

As Δx approaches zero, the ratio $\frac{\Delta y}{\Delta x}$ becomes $\frac{\partial y}{\partial x}$ (The partial derivative indicates that we are interested in the variation of y with position at a fixed time).

Therefore, $$\Delta p = -B\frac{\partial y}{\partial x} \quad \ldots(\text{i})$$

So, this is the equation which relates the displacement equation with the pressure equation. Suppose the displacement equation is

$$y = A\cos(kx - \omega t) \quad \ldots(\text{ii})$$

Then, $$\frac{\partial y}{\partial x} = -kA\sin(kx - \omega t) \quad \ldots(\text{iii})$$

From Eqs. (i) and (iii), we find that

$$\Delta p = BAk\sin(kx - \omega t) = (\Delta p)_m \sin(kx - \omega t) \quad \ldots(\text{iv})$$

Here, $$(\Delta p)_m = BAk$$

is the amplitude of pressure variation. From Eqs. (ii) and (iv), we see that pressure equation is 90° out of phase with displacement equation. When the displacement is zero, the pressure variation is either maximum or minimum and *vice-versa*. Fig. 17.18(a) shows displacement from equilibrium of air molecules in a harmonic sound wave *versus* position at some instants. Points x_1 and x_3 are points of zero displacement. Now, refer Fig. 17.18(b), just to the left of x_1, the displacement

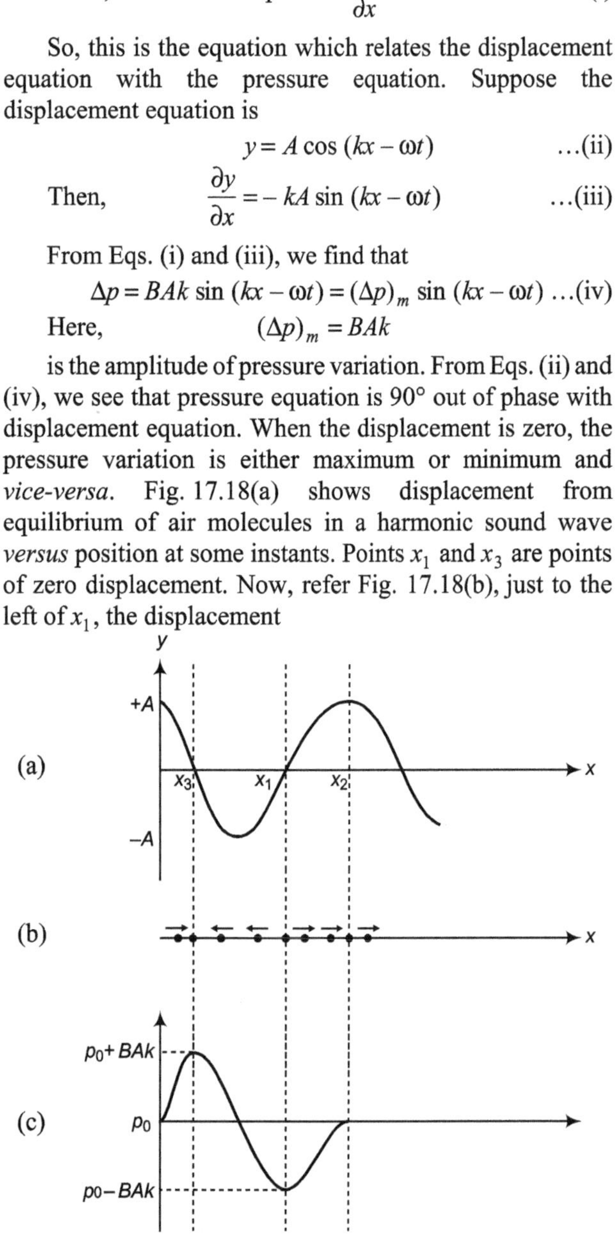

Fig. 17.18

is negative, indicating that the gas molecules are displaced to left, away from point x_1 at this instant. Just to the right of x_1, the displacement is positive, indicating that the molecules are displaced to the right, which is again away from point x_1. So, at point x_1 the pressure of the gas is minimum. So, if p_0 is the atmospheric pressure (normal pressure), the pressure at x_1 will be,

$$p(x_1) = p_0 - (\Delta p)_m$$
$$= p_0 - BAk$$

At point x_3, the pressure (and hence the density also) is maximum because the molecules on both sides of that point are displaced toward point x_3. Hence,

$$p(x_3) = p_0 + (\Delta p)_m = p_0 + BAk$$

At point x_2 the pressure (and hence the density) does not change because the gas molecules on both sides of that point have equal displacements in the same direction or

$$p(x_2) = p_0$$

From Fig. (a) and (c), we see that pressure change and displacement are 90° out of phase.

Relation between Pressure Wave and Density Wave

In this section, we will find the relation between pressure wave and density wave.

According to definition of bulk modulus (B),

$$B = \left(-\frac{dp}{dV/V}\right) \quad \ldots(\text{i})$$

Further, volume $= \frac{\text{mass}}{\text{density}}$ or $V = \frac{m}{\rho}$

or $$dV = -\frac{m}{\rho^2}\cdot d\rho = -\frac{V}{\rho}\cdot d\rho \Rightarrow \frac{dV}{V} = -\frac{d\rho}{\rho}$$

Substituting in Eq. (i), we get

$$d\rho = \frac{\rho\,(dp)}{B} = \frac{dp}{v^2} \qquad \left(\frac{\rho}{B} = \frac{1}{v^2}\right)$$

Or this can be written as,

$$\Delta\rho = \frac{\rho}{B}\cdot\Delta p = \frac{1}{v^2}\cdot\Delta p$$

So, this relation relates the pressure equation with the density equation. For example, if

$$\Delta p = (\Delta p)_m \sin(kx - \omega t)$$

then $$\Delta\rho = (\Delta\rho)_m \sin(kx - \omega t)$$

where, $$(\Delta\rho)_m = \frac{\rho}{B}(\Delta p)_m = \frac{(\Delta p)_m}{v^2}$$

Thus, density equation is in phase with the pressure equation and this is 90° out of phase with the displacement equation.

Example 17.13 *(a) What is the displacement amplitude for a sound wave having a frequency of 100 Hz and a pressure amplitude of 10 Pa ?*

(b) The displacement amplitude of a sound wave of frequency 300 Hz is 10^{-7} m. What is the pressure amplitude of this wave? Speed of sound in air is 340 m/s and density of air is 1.29 kg/m^3.

Sol. (a) $(\Delta p)_m = BAk$

Here, $k = \dfrac{\omega}{v} = \dfrac{2\pi f}{v}$

and $B = \rho v^2$ $\left(\text{as } v = \sqrt{\dfrac{B}{\rho}}\right)$

$\therefore$ $(\Delta p)_m = (\rho v^2)(A)\left(\dfrac{2\pi f}{v}\right)$

$\therefore$ $A = \dfrac{(\Delta p)_m}{2\pi v\rho f}$ ...(i)

Substituting the values, we have

$$A = \frac{(10)}{2 \times 3.14 \times 100 \times 1.29 \times 340}$$

$$= 3.63 \times 10^{5} \text{ m}$$

(b) From Eq. (i),

$$(\Delta p)_m = 2\pi f \rho v A$$

Substituting the values, we have

$$(\Delta p)_m = 2 \times 3.14 \times 300 \times 1.29 \times 340 \times 10^{-7}$$

$$= 8.26 \times 10^{-2} \text{ N/m}^2$$

17.14 Speed of a Longitudinal Wave

- Derivation of the formula is optional.

First we calculate the speed at which a longitudinal pulse propagates through a fluid. We will apply Newton's second law to the motion of an element of the fluid and from this we derive the wave equation.

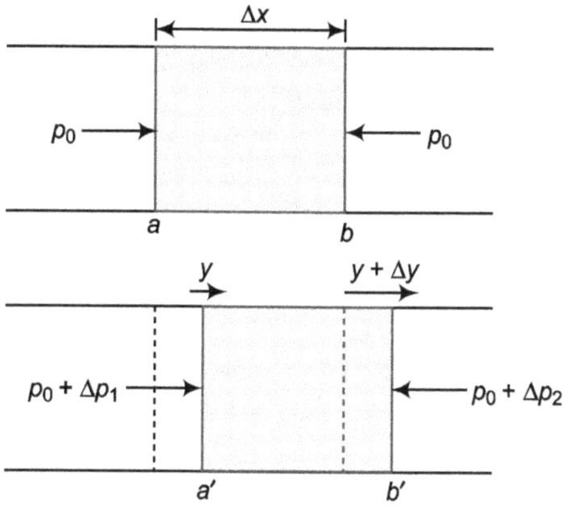

Fig. 17.19

Consider a fluid element ab confined to a tube of cross-sectional area S as shown in figure. The element has a thickness Δx. We assume that the equilibrium pressure of the fluid is p_0. Because of the disturbance, the section a of the element moves a distance y from its mean position and section b moves a distance $y + \Delta y$ to a new position b'. The pressure on the left side of the element becomes $p_0 + \Delta p_1$ and on the right side it becomes $p_0 + \Delta p_2$. If ρ is the equilibrium density, the mass of the element is $\rho S \Delta x$. (When the element moves its mass does not change, even though its volume and density change).

The net force acting on the element is,

$$F = (\Delta p_1 - \Delta p_2)\,S$$

and its acceleration is

$$a = \frac{\partial^2 y}{\partial t^2}$$

Thus, Newton's second law applied to the motion of the element is

$$(\Delta p_1 - \Delta p_2)\,S = \rho S \Delta x \frac{\partial^2 y}{\partial t^2} \quad \text{...(i)}$$

Next we divide both sides by Δx and note that in the limit as $\Delta x \to 0$, we have $(\Delta p_1 - \Delta p_2)/\Delta x \to \partial p/\partial x$, Eq. (i), then takes the form

$$-\frac{\partial p}{\partial x} = \rho \frac{\partial^2 y}{\partial t^2} \quad \text{...(ii)}$$

The excess pressure Δp may be written as

$$\Delta p = -B\frac{\partial y}{\partial x}$$

When this is used in Eq. (ii), we obtain the wave equation

$$\frac{\partial^2 y}{\partial x^2} = \frac{\rho}{B}\cdot\frac{\partial^2 y}{\partial t^2} \quad \text{or} \quad \frac{\partial^2 y}{\partial t^2} = \frac{B}{\rho}\frac{\partial^2 y}{\partial x^2}$$

Comparing this equation with the wave equation

$$\frac{\partial^2 y}{\partial t^2} = v^2\frac{\partial^2 y}{\partial x^2}$$

We have

$$v = \sqrt{\frac{B}{\rho}} \quad \text{(speed of longitudinal wave in a fluid)}$$

This is the speed of longitudinal waves within a **gas** or a **liquid**.

When a longitudinal wave propagates in a solid rod or bar, the rod expands sideways slightly when it is compressed longitudinally and the speed of a longitudinal wave in a rod is given by

$$v = \sqrt{\frac{Y}{\rho}} \quad \text{(speed of a longitudinal wave in a solid rod)}$$

17.15 Sound Waves in Gases

In the preceding section we derived equation

$$v=\sqrt{\frac{B}{\rho}}$$

For the speed of longitudinal waves in a fluid of bulk modulus B and density ρ. We use this to find the speed of sound in an ideal gas. The bulk modulus of the gas, however depends on the process. When a wave travels through a gas, are the compressions and expansions adiabatic, or is there enough heat conduction between adjacent layers of gas to maintain a nearly constant temperature throughout?

Because thermal conductivities of gases are very small, it turns out that for ordinary sound frequencies (20 Hz to 20,000 Hz) propagation of sound is very nearly adiabatic. Thus, in the above equation, we use the **adiabatic bulk modulus** (B_s), which is given by

$$B_s=\gamma p$$

Here, γ is the ratio of molar heat capacity C_p/C_V. Thus,

$$v=\sqrt{\frac{\gamma p}{\rho}}$$

We can get a useful alternative form of the above equation by substituting density ρ of an ideal gas

$$\rho=\frac{pM}{RT}$$

where, R is the gas constant, M is the molecular mass and T is the absolute temperature combining all these equations, we can write

$$v=\sqrt{\frac{\gamma RT}{M}} \quad \text{(speed of sound in an ideal gas)}$$

Effect of temperature, pressure and humidity on the speed of sound in air

(i) **Effect of temperature** From the equation,

$$v=\sqrt{\frac{\gamma RT}{M}}$$

We can see that,

$$v\propto\sqrt{T} \quad \text{or} \quad \frac{v_1}{v_2}=\sqrt{\frac{T_1}{T_2}}$$

At STP, the temperature is 0°C or 273 K. If the speed of sound at 0°C is v_0, its value at t°C will satisfy

$$\frac{v_t}{v_0}=\sqrt{\frac{273+t}{273}}=\left(1+\frac{t}{273}\right)^{\frac{1}{2}}=1+\frac{t}{546}$$

$$\therefore \quad v_t=v_0\left(1+\frac{t}{546}\right)$$

If the speed of sound in air (v_0) at 0°C be taken 332 m/s, then

$$v_t=332\left(1+\frac{t}{546}\right) \quad \text{or} \quad v_t=332+0.61\,t$$

Thus, the velocity of sound in air increases roughly by 0.61 m/s/°C rise in temperature.

(ii) **Effect of pressure** From the formula for the speed of sound in a gas $v=\sqrt{\frac{\gamma p}{\rho}}$, it appears that $v\propto\sqrt{p}$. But actually it is not so.

Because $\frac{p}{\rho}=\frac{RT}{M}=$ constant at constant temperature.

That is, at constant temperature, if p changes then ρ also changes in such a way that p/ρ remains constant. Hence, in the formula $v=\sqrt{\gamma\, p/\rho}$, the value of p/ρ does not change when p changes. From this it is clear that, if the temperature of the gas remains constant, then there is no effect of the pressure change on the speed of sound.

(iii) **Effect of humidity** The density of moist air (i.e. air mixed with water-vapour) is less than the density of dry air. This is because in moist air heavy dust particles settle down due to condensation. Hence, density of air gets decreased thus, increasing the speed of sound. Therefore, assuming the value of γ for moist air same as for dry air (which is actually slightly less than that for dry air) it is clear from the formula $v=\sqrt{\frac{\gamma p}{\rho}}$, that the speed of sound in moist air is slightly greater than in dry air.

Example 17.14 *Calculate the speed of longitudinal waves in the following gases at 0° C and 1 atm (=10^5 Pa)*

(a) oxygen for which the bulk modulus is 1.41×10^5 Pa and density is 1.43 kg/m^3.

(b) helium for which the bulk modulus is 1.7×10^5 Pa and the density is 0.18 kg/m^3.

Sol. (a) $v_{O_2}=\sqrt{\frac{B}{\rho}}=\sqrt{\frac{1.41\times10^5}{1.43}}=314$ m/s

(b) $v_{He}=\sqrt{\frac{B}{\rho}}=\sqrt{\frac{1.7\times10^5}{0.18}}=972$ m/s

Example 17.15 *Find speed of sound in hydrogen gas at 27° C. Ratio C_p/C_V for H_2 is 1.4. Gas constant $R=8.31$ J/mol-K.*

Sol. $v=\sqrt{\frac{\gamma RT}{M}}$

Here, $T=27+273=300$ K, $\gamma=1.4$, $R=8.31$ J/mol-K,

$M=2\times10^{-3}$ kg/mol

$\therefore \quad v = \sqrt{\dfrac{1.4 \times 8.31 \times 300}{2 \times 10^{-3}}} = 1321 \text{ m/s}$

- In the above formula, put $M = 2 \times 10^{-3}$ kg/mol. Don't put it 2 kg.

Example 17.16 *At what temperature will the speed of sound in hydrogen be the same as in oxygen at* $100°C$? *Molar masses of oxygen and hydrogen are in the ratio* $16:1$.

Sol. $v = \sqrt{\dfrac{\gamma RT}{M}} \quad \Rightarrow \quad v_{H_2} = v_{O_2}$

$$\therefore \quad \sqrt{\frac{\gamma_{H_2} RT_{H_2}}{M_{H_2}}} = \sqrt{\frac{\gamma_{O_2} RT_{O_2}}{M_{O_2}}}$$

$$\gamma_{H_2} = \gamma_{O_2} \qquad \text{(as both are diatomic)}$$

$$\therefore \quad T_{H_2} = \left(\frac{M_{H_2}}{M_{O_2}}\right)(T_{O_2}) = \left(\frac{1}{16}\right)(100 + 273)$$

$$= 23.31 \text{ K} \approx -249.7°\text{C}$$

17.16 Sound Intensity

Travelling sound waves, like all other travelling waves, transfer energy from one region of space to another. We define the **intensity** of a wave (denoted by I) to be the time average rate at which energy is transported by the wave, per unit area across a surface perpendicular to the direction of propagation. We have already derived an expression for the intensity of a mechanical wave in Art. 17.6, which is

$$I = \frac{1}{2} \rho A^2 \omega^2 v \qquad \ldots(\text{i})$$

For a sound wave,

$$(\Delta p)_m = BAk = BA\left(\frac{\omega}{v}\right)$$

or

$$\omega = \frac{(\Delta p)_m v}{BA}$$

Substituting this value in Eq. (i), we have

$$I = \frac{1}{2} \rho A^2 \frac{(\Delta p)_m^2 v^2}{B^2 A^2} v \qquad (\rho v^2 = B)$$

or

$$I = \frac{v (\Delta p)_m^2}{2B} \qquad \ldots(\text{ii})$$

Thus, intensity of a sound wave can be calculated by either of the Eqs. (i) or (ii).

Sound Intensity in Decibels

The physiological sensation of loudness is closely related to the intensity of wave producing the sound. At a frequency of 1 kHz people are able to detect sounds with intensities as low as 10^{-12} W/m^2. On the other hand, an intensity of 1 W/m^2 can cause pain, and prolonged exposure to sound at this level will damage a person's ears. Because the range in intensities over which people hear is so large, it is convenient to use a logarithmic scale to specify intensities. This scale is defined as follows :

If the intensity of sound in watts per square metre is I, then the intensity level β in decibels (dB) is given by,

$$\beta = 10 \log \frac{I}{I_0}$$

where, the base of the logarithm is 10 and $I_0 = 10^{-2}$ W/m^2 (roughly the minimum intensity that can be heard).

On the decibel scale, the pain threshold of 1 W/m^2, is then

$$\beta = 10 \log \frac{1}{10^{-12}} = 120 \text{ dB}$$

Table 17.3 gives typical values for the intensity levels of some of the common sounds.

Table 17.3 Sound intensity levels in decibels (Threshold of hearing = 0 dB; threshold of pain = 120 dB)

Source of Sound	dB
Rustling leaves	10
Whisper	20
Quiet room	30
Normal level of speech (inside)	65
Street traffic (inside car)	80
Riveting tool	100
Thunder	110
Indoor rock concert	120

Example 17.17 *A point source of sound emits a constant power with intensity inversely proportional to the square of the distance from the source. By how many decibels does the sound intensity level drop when you move from point* P_1 *to* P_2. *Distance of* P_2 *from the source is two times the distance of source from* P_1.

Sol. We label the two points 1 and 2, and we use the equation $\beta = 10 \log \dfrac{I}{I_0}$ (dB) twice. The difference in sound intensity level $\beta_2 - \beta_1$ is given by

$$\beta_2 - \beta_1 = (10 \text{ dB})\left(\log \frac{I_2}{I_0} - \log \frac{I_1}{I_0}\right)$$

$$= (10 \text{ dB}) [(\log I_2 - \log I_0) - (\log I_1 - \log I_0)]$$

$$= (10 \text{ dB}) \log \frac{I_2}{I_1}$$

Now, $\quad I \propto \dfrac{1}{r^2} \quad \Rightarrow \quad \therefore \quad \dfrac{I_2}{I_1} = \left(\dfrac{r_1}{r_2}\right)^2 = \dfrac{1}{4}$

as $\quad r_2 = 2r_1$

$$\therefore \quad \beta_2 - \beta_1 = (10 \text{ dB}) \log \left(\frac{1}{4}\right) = -6.0 \text{ dB}$$

Example 17.18 *For a person with normal hearing, the faintest sound that can be heard at a frequency of 400 Hz has a pressure amplitude of about* 6.0×10^{-5} *Pa. Calculate the corresponding intensity in* W/m^2. *Take, speed of sound in air as* 344 *m/s and density of air* 1.2 kg/m^3.

Sol. $I = \frac{v\,(\Delta p)_m^2}{2B}$

Substituting $B = \rho v^2$, the above equation reduces to

$$I = \frac{(\Delta p)_m^2}{2\rho v} = \frac{(6.0\times10^{-5})^2}{2\times1.2\times344} = 4.4\times10^{-12}\ \text{W/m}^2$$

17.17 Interference of Sound Waves

The principle of superposition introduced in Art. 17.7 is valid for sound waves as well. When two or more waves meet at some point on a medium, the resultant disturbance is equal to the sum of the disturbances produced by individual waves. Depending on the phase difference, the waves can interfere constructively or destructively leading to a corresponding increase or decrease in the resultant intensity. Before studying the interference of sound waves let us first discuss the two terms :

(i) phase difference and

(ii) coherent sources which will be frequently used later.

Phase Difference $\Delta\phi$

(i) Phase difference between two different points P_1 and P_2 on the path of a travelling wave separated by a distance Δx is

$$\Delta\phi = \left(\frac{2\pi}{\lambda}\right)\Delta x$$

(ii) A phase difference may also arise between two waves generated by the same source when they travel along paths of unequal lengths. Suppose the difference in lengths is Δx, then the phase difference will be same i.e. $\frac{2\pi}{\lambda}\Delta x$.

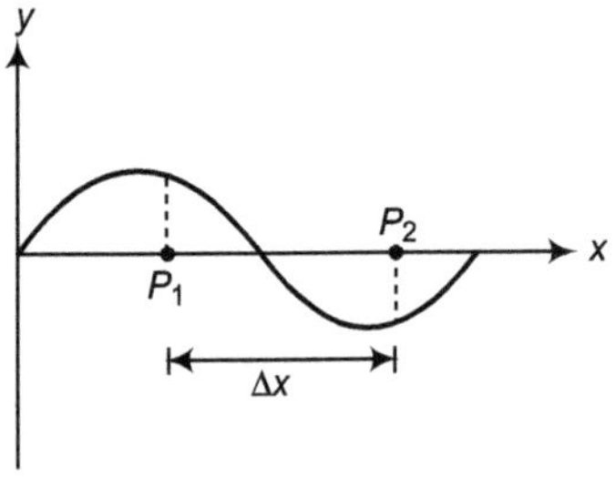

Fig. 17.20

(iii) The phase difference at a point P on the path of a travelling wave at two different times t_1 and $t_2\ (> t_1)$ is

$$\Delta\phi = \frac{2\pi}{T}\Delta t$$

Here, $\Delta t = t_2 - t_1$

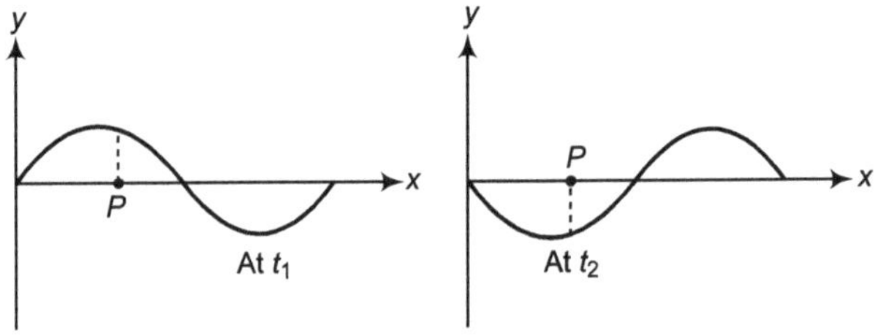

Fig. 17.21

Coherent Sources

Two sources which are in phase or have a constant phase difference are called coherent sources. For the two sources to be coherent their frequencies must be same. But the converse is not always true, i.e. the two different sources having the same frequency are not always coherent. If the phase difference of the sources changes erratically with time, even if they have the same frequency the sources are said to be incoherent. This is what happens with light sources composed of a large number of same kind of atoms, which emit light of the same frequency. Since, there are many atoms involved in each source and they do not oscillate in phase, they are incoherent. Thus, two different light sources can't be coherent.

Now, let us come to the point.

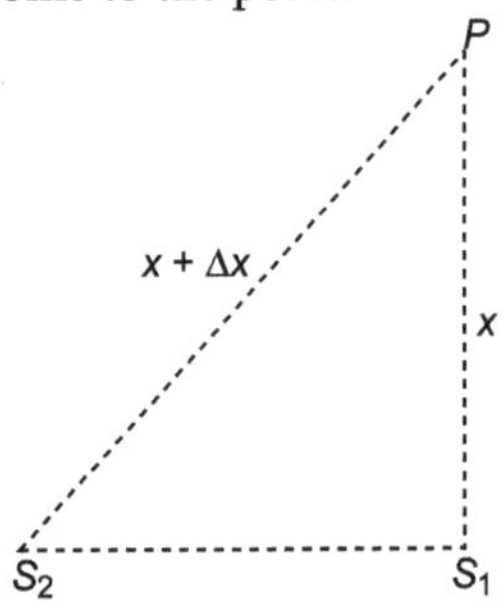

Fig. 17.22

Consider two coherent point sources of sound S_1 and S_2 which oscillate in phase with the same angular frequency ω. A point P is situated at a distance x from S_1 and $x + \Delta x$ from S_2, so that the path difference between the two waves reaching P from S_1 and S_2 is Δx. The displacement equations at P due to the two waves are described by

$$y_1 = A_1 \sin(kx - \omega t)$$

and $$y_2 = A_2 \sin[(kx - \omega t) + \phi]$$

where, $$\phi = \left(\frac{2\pi}{\lambda}\right)\Delta x$$

is the phase difference between the two waves reaching P. These equations are identical to those discussed in Art. 17.8. The resultant wave at P is given by

$$y = A \sin [(kx - \omega t) + \theta]$$

where $$A^2 = A_1^2 + A_2^2 + 2A_1 A_2 \cos \phi \quad \ldots(\text{i})$$

and $$\tan \theta = \frac{A_2 \sin \phi}{A_1 + A_2 \cos \phi} \quad \ldots(\text{ii})$$

Now, as $I \propto A^2$, Eq. (i) can be written as

$$I = I_1 + I_2 + 2\sqrt{I_1 I_2} \cos \phi$$

The resultant amplitude (or intensity) is maximum when $\phi = 2n\pi$ and minimum when $\phi = (2n+1)\pi$, where n is an integer. These are the conditions for constructive and destructive interference.

Thus,

$\phi = 2n\pi$ (condition for constructive interference)

$\phi = (2n+1)\pi$ (condition for destructive interference)

Using $\phi = \frac{2\pi}{\lambda} \Delta x$, the above conditions may be written in terms of the path difference as

$\Delta x = n\lambda$ (constructive interference)

$\Delta x = \left(n + \frac{1}{2}\right)\lambda$ (destructive interference)

At constructive interference

$$A = A_1 + A_2 \quad \text{or} \quad I = (\sqrt{I_1} + \sqrt{I_2})^2$$

and at destructive interference

$$A = A_1 \sim A_2 \quad \text{or} \quad I = (\sqrt{I_1} - \sqrt{I_2})^2$$

If $A_1 = A_2$ then A or $I = 0$ at destructive interference (where $\Delta x = \lambda/2, 3\lambda/2 \ldots$, etc.) or no sound is detected at such a point. If the sources have an initial phase difference ϕ_0 between them, then it is either added or subtracted with ϕ to get the net phase difference at P, depending on the condition whether S_1 leads or lags behind in phase with S_2.

If the sources are incoherent, the phase difference between the sources keep on changing. At any point P, sometimes constructive and sometimes destructive interference takes place. If the intensity due to each source is I, the resultant intensity rapidly and randomly changes between $4I$ and zero, so that the average observable intensity is $2I$. If intensities due to individual sources is I_1 and I_2, the resultant intensity is

$$I = I_1 + I_2 \quad \text{(incoherent sources)}$$

No interference effect is therefore observed. For observable interference, the sources must be coherent. One way to obtain a pair of coherent sources is to obtain two sound waves from the same source by dividing the original wave along two different paths and then combining them. The two waves then differ in phase only because of different paths travelled.

A wave enters from the left and splits equally at point A. At point B, where the waves combine, there is a phase difference because of the path difference. If λ is the wavelength and Δx is the path difference, the phase difference is $\phi = 2\pi\Delta x/\lambda$.

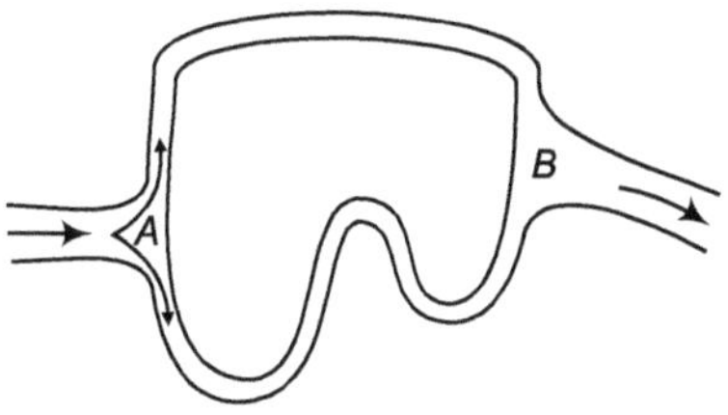

Fig. 17.23

Extra Knowledge Points

- If the two waves emitted from S_1 and S_2 have already a phase difference of π, the conditions of maximas and minimas are interchanged, i.e. path difference

  $\Delta x = \frac{\lambda}{2}, \frac{3\lambda}{2} \ldots$ (for constructive interference)

  and $\Delta x = \lambda, 2\lambda, \ldots$ (for destructive interference)
- Interference of pressure variations can also be obtained in the similar manner.
- Most of the problems of interference can be solved by calculating the path difference Δx and then by putting

  $\Delta x = 0, \lambda, 2\lambda, \ldots$ (for constructive interference)

  $\Delta x = \frac{\lambda}{2}, \frac{3\lambda}{2} \ldots$ (for destructive interference)

  provided the waves emitted from S_1 and S_2 are in phase. If they have a phase difference of π, the conditions are interchanged.

Example 17.19 *Two sound sources S_1 and S_2 emit pure sinusoidal waves in phase. If the speed of sound is 350 m/s, then*

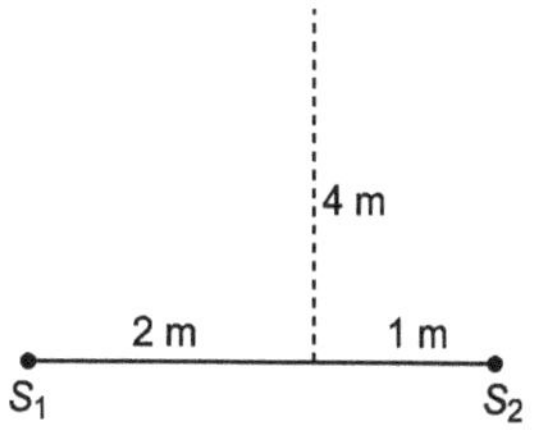

Fig. 17.24

(a) for what frequencies does constructive interference occur at P?

(b) for what frequencies does destructive interference occur at point P?

Sol. Path difference $\Delta x = S_1P - S_2P$

$$= \sqrt{(2)^2 + (4)^2} - \sqrt{(1)^2 + (4)^2}$$

$$= 4.47 - 4.12 = 0.35 \text{ m}$$

(a) Constructive interference occurs when the path difference is an integer number of wavelength.

or $\Delta x = n\lambda = \dfrac{nv}{f}$

$\Rightarrow$ $f = \dfrac{n(v)}{\Delta x}$, where $n = 1, 2, 3, \ldots$

$\therefore$ $f = \dfrac{350}{0.35}, \dfrac{2 \times 350}{0.35}, \dfrac{3 \times 350}{0.35}, \ldots$

$f = 1000\text{ Hz}, 2000\text{ Hz}, 3000\text{ Hz}, \ldots$, etc.

(b) Destructive interference occurs when the path difference is a half-integer number of wavelengths

or $\Delta x = (2n+1)\dfrac{\lambda}{2}$, where $n = 0, 1, 2, \ldots$

$\Rightarrow$ $\Delta x = (2n+1)\dfrac{v}{2f}$

$\therefore$ $f = \dfrac{(2n+1)v}{2\,\Delta x} = \dfrac{350}{2 \times 0.35}, \dfrac{3 \times 350}{2 \times 0.35}, \dfrac{5 \times 350}{2 \times 0.35}, \ldots$

$= 500\text{ Hz}, 1500\text{ Hz}, 2500\text{ Hz}, \ldots$

Example 17.20 *Two coherent narrow slits emitting sound of wavelength λ in the same phase are placed parallel to each other at a small separation of 2λ. The sound is detected by moving a detector on the screen at a distance D ($>> \lambda$) from the slit S_1 as shown in figure. Find the distance y such that the intensity at P is equal to intensity at O.*

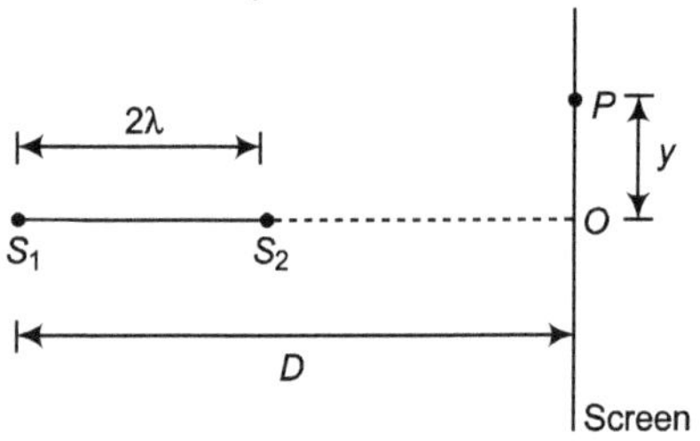

Fig. 17.25

Sol. At point O on the screen the path difference between the sound waves reaching from S_1 and S_2 is 2λ, i.e. constructive interference is obtained at O. At a very large distance from point O on the screen the path difference is zero.

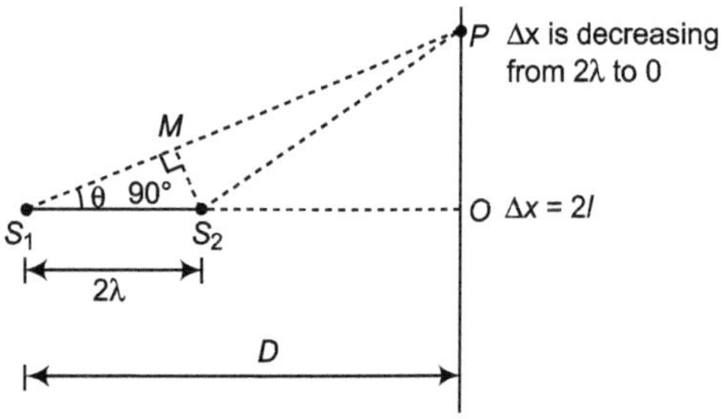

Fig. 17.26

Thus, we can conclude that as we move away from point O on the screen path difference decreases from 2λ to zero. At point O constructive interference is obtained (where $\Delta x = 2\lambda$). So, next constructive interference will be obtained where $\Delta x = \lambda$. Hence,

$$S_1P - S_2P = \lambda$$

$\Rightarrow$ $\sqrt{D^2 + y^2} - \sqrt{y^2 + (D - 2\lambda)^2} = \lambda$

$\therefore$ $\sqrt{D^2 + y^2} - \lambda = \sqrt{y^2 + (D - 2\lambda)^2}$

Squaring on both sides, we get

$$D^2 + y^2 + \lambda^2 - 2\lambda\sqrt{D^2 + y^2} = y^2 + D^2 + 4\lambda^2 - 4\lambda D$$

$\Rightarrow$ $2\sqrt{D^2 + y^2} = 4D - 3\lambda$

as $D >> \lambda$, $4D - 3\lambda \approx 4D$

$\therefore$ $2\sqrt{D^2 + y^2} = 4D$

$\Rightarrow$ $\sqrt{D^2 + y^2} = 2D$

Again squaring on both sides, we get

$$D^2 + y^2 = 4D^2 \quad \text{or} \quad y = \sqrt{3}\,D$$

Alternate method Let $\Delta x = \lambda$ at angle θ as shown. Path difference between the waves is $S_1M = 2\lambda\cos\theta$

$\therefore$ $2\lambda\cos\theta = \lambda$ $\quad (\Delta x = \lambda)$

or $\theta = 60^\circ$

Now, $PO = S_1O\cot 30^\circ$ or $y = \sqrt{3}\,D$

17.18 Standing Longitudinal Waves in Organ Pipes

When longitudinal waves propagate in a fluid in a pipe with finite length, the waves are reflected from the ends in a same way the transverse waves on a string are reflected at its ends. The superposition of the waves travelling in opposite directions forms a longitudinal standing wave.

Transverse waves on a string, including standing waves, are usually described only in terms of the displacement of the string. But longitudinal standing waves in a fluid may be described either in terms of the displacement of the fluid or in terms of the pressure variation in the fluid. To avoid confusion we will use the terms **displacement node** and **displacement antinode** to refer to points where particles of the fluid have zero displacement and maximum displacement, respectively. Similarly, **pressure node** and **pressure antinode** refer to the points where pressure and density variation in the fluid is zero or maximum, respectively. Because the pressure wave is 90° out of phase with the displacement wave. Consequently, the displacement node behaves as a pressure antinode and *vice-versa.*

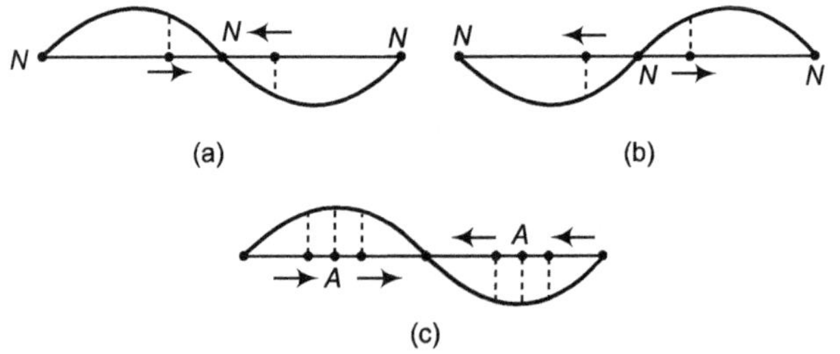

Fig. 17.27

This can be understood physically by realising that two small volume elements of fluid on opposite sides of a displacement node are vibrating in opposite phase. Hence, when they approach each other [see Fig. (a)] the pressure at this node is maximum, and when they recede from each other, [see Fig. (b)] the pressure at this node is a minimum. Similarly, two small elements of fluid which are on opposite sides of a displacement antinode vibrate in phase and therefore give rise to no pressure variations at the antinode [see Fig. (c)].

Pressure Equation of Standing Longitudinal Wave

If two identical (same frequency) longitudinal waves travel in opposite directions, standing waves are produced by their superposition. If the equations of the two waves are written as

$$\Delta p_1 = (\Delta p)_m \sin (kx - \omega t)$$

and $$\Delta p_2 = (\Delta p)_m \sin (kx + \omega t)$$

From the principle of superposition, the resultant wave is

$$\Delta p = \Delta p_1 + \Delta p_2 = 2\,(\Delta p)_m \sin kx \cos \omega t \qquad \ldots\text{(i)}$$

This equation is similar to the equation obtained in chapter 17 for standing waves on a string.

From Eq. (i), we can see that

$\Delta p = 0$ at $x = 0, \lambda/2, \lambda, \ldots$, etc. (pressure nodes)

and $\Delta p =$ maximum at $x = \lambda/4, 3\lambda/4, \ldots$, etc. (pressure antinodes)

The distance between two adjacent nodes or between two adjacent antinodes is $\lambda/2$. Longitudinal standing waves can be produced in air columns trapped in tubes of cylindrical shape. Organ pipes are such vibrating air columns.

Conditions at the Boundary of an Organ Pipe

Let us take an example of a closed organ pipe. In a closed pipe one end is closed and the other is open. When a longitudinal wave encounters the closed end of the pipe it gets reflected from this end. But the reflected wave is 180° out of phase with the incident wave, i.e. a compression is reflected as a compression and a rarefaction is reflected as a rarefaction. This is a necessary condition because the displacement of the small volume elements at the closed end must always be zero. Hence, a closed end is a displacement node.

A sound wave is also reflected from an open end. You may wonder how a sound wave can reflect from an open end, since, there may not appear to be a change in the medium at this point.

At the open end pressure is the same as atmospheric pressure and does not vary. Thus, there is a pressure node (or displacement antinode) at this end. A compression is therefore reflected as a rarefaction and rarefaction as a compression. Now let us see how this reflection takes place. When a rarefaction reaches an open end, the surrounding air rushes towards this region and creates a compression that travels back along the pipe. Similarly, when a compression reaches an open end, the air expands to form a rarefaction. This can be said in a different way as : at the open end of the tube fluid elements are free to move, so there is a displacement antinode. Thus, in a nutshell we can say that closed end of an organ pipe is a displacement node or pressure antinode and open end of the pipe is displacement antinode or pressure node.

Similarly, both ends of an open organ pipe (open at both ends) are displacement antinodes or pressure nodes.

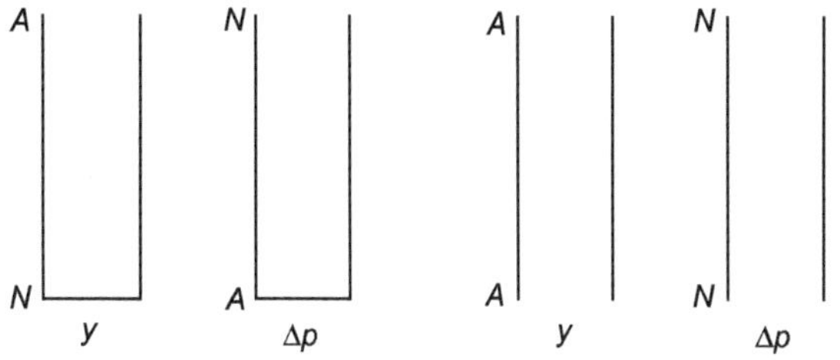

(a) Closed organ pipe (b) Open organ pipe

Fig. 17.28

Note Point

- In practice, the pressure nodes lie slightly beyond the ends of the tube. A compression reaching an open end does not reflect until it passes beyond the end. For a thin walled tube of circular cross section, this **end correction** is approximately $0.6\,R$, where R is the radius of tube. Hence, the effective length of the tube is longer than the true length l. However, we neglect this small correction if the length of the tube is much larger than its diameter.

Standing Waves in a Closed Organ Pipe

Before reading this article, students are advised to read Art. 17.12 for the condition of resonance.

To get resonance in a closed organ pipe sound waves are sent by a source (normally a tuning fork) near the open end. Resonance corresponds to a pressure antinode at the closed end and a pressure node at the open end. The standing wave patterns for the three lowest harmonics in this situation are shown in figure. Since, the node-antinode separation is $\frac{\lambda}{4}$, the resonance condition for the first harmonic is, $l = \frac{\lambda_1}{4}$.

Similarly, the resonance conditions for the higher harmonics are $l = \frac{3\lambda_2}{4}, \frac{5\lambda_3}{4}, \ldots$, etc. The natural frequencies of oscillation of the air in the tube closed at one end and open at the other are, therefore,

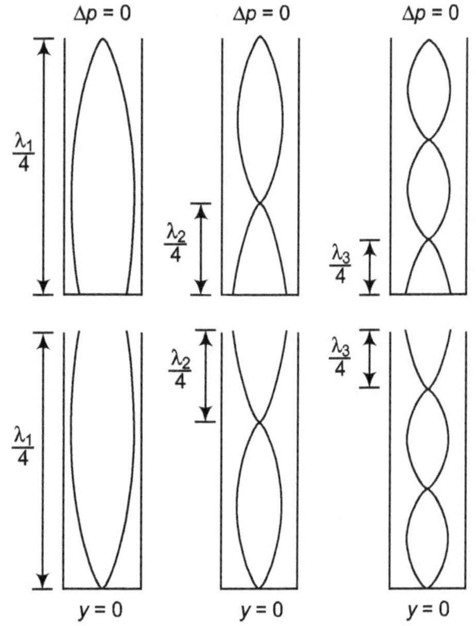

Fig. 17.29

$$f_n = \frac{v}{\lambda_n} = n\frac{v}{4l} = nf_1 \qquad (n = 1, 3, 5, \ldots)$$

Here, v = speed of sound in the tube

$f_1 = \frac{v}{4l}$ (fundamental frequency or the first harmonic)

$f_3 = 3\frac{v}{4l} = 3f_1$ (first overtone or the third harmonic)

and $f_5 = 5\frac{v}{4l} = 5f_1$

(second overtone or the fifth harmonic)

and so on.

Thus, in a pipe closed at one end and open at the other, the natural frequencies of oscillation form a harmonic series that includes only odd integer multiples of the fundamental frequency.

Standing Waves in an Open Organ Pipe

Since, both ends of the tube are open, there are pressure nodes (or displacement antinodes) at both ends. Figure shows the resulting standing waves for the three lowest resonant frequencies since, the distance between pressure nodes is $\lambda/2$, the resonance condition is $l = \frac{n\lambda_n}{2}$

$l = \frac{n\lambda_n}{2}$ where, $n = 1, 2, 3, \ldots$ and l is the length of the tube. The resonant frequencies for a tube open at both ends are then

$$f_n = \frac{v}{\lambda_n} = n\frac{v}{2l} = nf_1 \qquad (n = 1, 2, 3, \ldots)$$

Here, $f_1 = \frac{v}{2l}$ (fundamental frequency or first harmonic)

$f_2 = \frac{v}{l} = 2f_1$ (first overtone or second harmonic)

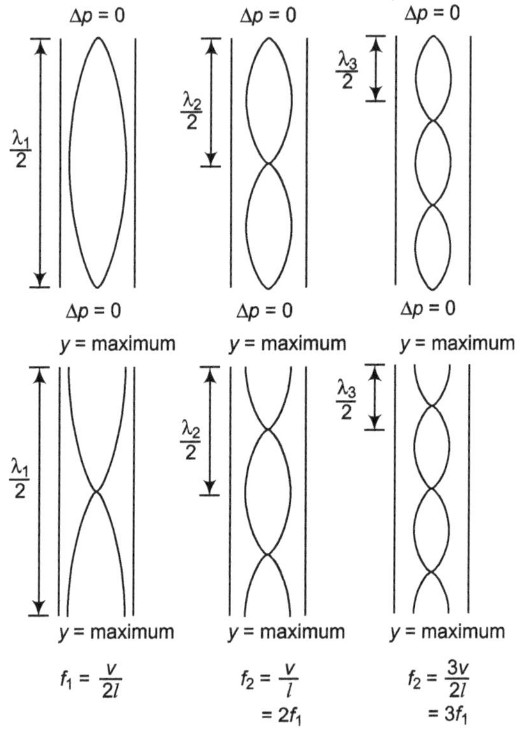

Fig. 17.30

$f_3 = 3\frac{v}{2l} = 3f_1$ (second overtone or third harmonic)

and so on.

Thus, in a pipe open at both ends, the natural frequencies of oscillation form a harmonic series that includes all integral multiples of the fundamental frequency.

Note Point

- It is interesting to investigate what happens to the frequencies of instruments based on air columns and strings as the temperature rises. The sound emitted by a flute, becomes sharp (increases in frequency) because the speed of sound increases inside the flute. The sound produced by a violin becomes flat (decreases in frequency) as the strings expand thermally because the expansion causes their tension to decrease.

Example 17.21 *Third overtone of a closed organ pipe is in unison with fourth harmonic of an open organ pipe. Find the ratio of the lengths of the pipes.*

Sol. Third overtone of closed organ pipe means seventh harmonic. Given,

$$(f_7)_{closed} = (f_4)_{open} \Rightarrow 7\left(\frac{v}{4l_c}\right) = 4\left(\frac{v}{2l_o}\right)$$

$$\therefore \quad \frac{l_c}{l_o} = \frac{7}{8}$$

Example 17.22 *The water level in a vertical glass tube 1.0 m long can be adjusted to any position in the tube. A tuning fork vibrating at 660 Hz is held just over the open top end of the tube. At what positions of the water level will there be resonance. Speed of sound is 330 m/s.*

Sol. Resonance corresponds to a pressure antinode at closed end and pressure node at open end. Further, the distance between a pressure node and a pressure antinode is $\frac{\lambda}{4}$, the condition of resonance would be, length of air column $l = n\frac{\lambda}{4} = n\left(\frac{v}{4f}\right)$

Here, $n = 1, 3, 5, \ldots$

$$l_1 = (1)\left(\frac{330}{4 \times 660}\right) = 0.125\text{ m}$$
$$l_2 = 3l_1 = 0.375\text{ m}$$
$$l_3 = 5l_1 = 0.625\text{ m}$$
$$l_4 = 7l_1 = 0.875\text{ m}$$
$$l_5 = 9l_1 = 1.125\text{ m}$$

Since, $l_5 > 1$ m (the length of tube), the length of air columns can have the values from l_1 to l_4 only. Therefore, level of water at resonance will be

$$(1.0 - 0.125)\text{ m} = 0.875\text{ m}$$
$$(1.0 - 0.375)\text{ m} = 0.625\text{ m}$$
$$(1.0 - 0.625)\text{ m} = 0.375\text{ m}$$
and $$(1.0 - 0.875)\text{ m} = 0.125\text{ m}$$

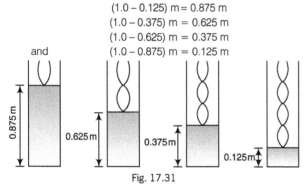

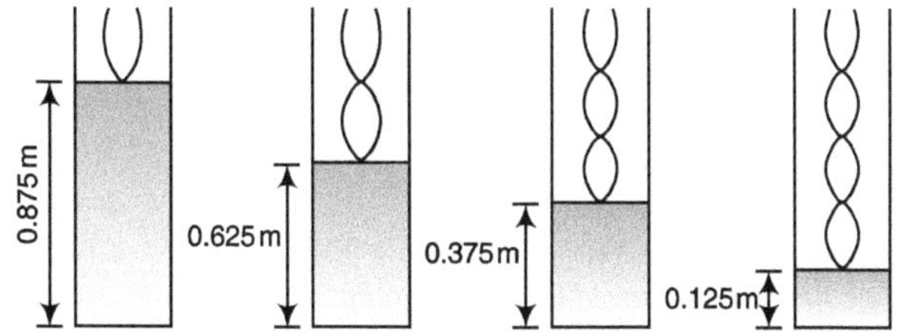

Fig. 17.31

In all the four cases shown in figure, the resonance frequency is 660 Hz but first one is the fundamental tone or first harmonic. Second is first overtone or third harmonic and so on.

Example 17.23 *A tube 1.0 m long is closed at one end. A stretched wire is placed near the open end. The wire is 0.3 m long and has a mass of 0.01 kg. It is held fixed at both ends and vibrates in its fundamental mode. It sets the air column in the tube into vibration at its fundamental frequency by resonance. Find*

(a) the frequency of oscillation of the air column and

(b) the tension in the wire.

Speed of sound in air = 330 m/s.

Sol. (a) Fundamental frequency of closed pipe $= \frac{v}{4l}$

$$= \frac{330}{4 \times 1} = 82.5\text{ Hz}$$

(b) At resonance, given

fundamental frequency of stretched wire (at both ends) = fundamental frequency of air column

$$\therefore \quad \frac{v}{2l} = 82.5\text{ Hz}$$

$$\therefore \quad \frac{\sqrt{\frac{T}{\mu}}}{2l} = 82.5 \quad \text{or} \quad T = \mu\,(2 \times 0.3 \times 82.5)^2$$

$$= \left(\frac{0.01}{0.3}\right)(2 \times 0.3 \times 82.5)^2$$

$$= 81.675\text{ N}$$

Example 17.24 *An open organ pipe has a fundamental frequency of 300 Hz. The first overtone of a closed organ pipe has the same frequency as the first overtone of the open pipe. How long is each pipe? (Speed of sound in air = 330 m/s)*

Sol. Fundamental frequency of an open pipe,

$$f_1 = \frac{v}{2l_o}$$

$$\therefore \quad l_o = \frac{v}{2f_1} = \frac{330}{2 \times 300} = 0.55\text{ m}$$

$$= 55\text{ cm}$$

Given, first overtone of closed pipe = first overtone of open pipe

Hence, $$3\left(\frac{v}{4l_c}\right) = 2\left(\frac{v}{2l_o}\right)$$

$$\therefore \quad l_c = \frac{3}{4}l_o = \left(\frac{3}{4}\right)(0.55)$$

$$= 0.4125\text{ m} = 41.25\text{ cm}$$

Characteristics of Sound

Following are the characteristics :

(i) **Intensity of wave** The amount of energy flowing per unit area and per unit time is called the intensity of wave. It is represented by I. Its units are J/m^2s or W/m^2.

$$I = 2\pi^2 f^2 A^2 \rho v$$

i.e. $$I \propto f^2 \quad \text{and} \quad I \propto A^2$$

If P is the power of an isotropic point source, intensity at a distance r is given by,

$$I = \frac{P}{4\pi r^2}$$

or $$I \propto \frac{1}{r^2} \quad \text{(for a point source)}$$

If P is the power of a line source, then intensity at a distance r is given by,

$$I = \frac{P}{2\pi rl}$$

or $$I \propto \frac{1}{r} \quad \text{(for a line source)}$$

As, $$I \propto A^2$$

Therefore, $$A \propto \frac{1}{r} \quad \text{(for a point source)}$$

and $$A \propto \frac{1}{\sqrt{r}} \quad \text{(for a line source)}$$

(ii) **Intensity level (loudness)** Depends on **intensity** of sound.

$$L = 10 \log_{10}\left(\frac{I}{I_0}\right)$$

Here, I_0 is the intensity of minimum audible sound which is 10^{-12} W/m^2.

- For more details of intensity and loudness read Art. 17.17.

(iii) **Quality** Quality is that characteristics of sound by which we can differentiate between the sounds coming from different sources.

(iv) **Pitch** Pitch is the characteristics of sound that depends on **frequency**. It determines the **shrillness** or **graveness** of sound.

Smaller the frequency smaller is the pitch, greater the frequency greater is the pitch. Frequency of ladies' voice is usually higher than that of gents. Therefore, ladies' voice has higher pitch (sharper) than gents.

Humming of mosquito has high pitch (high frequency) but low intensity (low loudness) while the roar of a lion has high intensity (loudness) but low pitch.

Ultrasonic, Infrasonic and Audible Sound

Sound waves can be classified in three groups according to their range of frequencies.

Infrasonic Waves

Longitudinal waves having frequencies below 20 Hz are called infrasonic waves. They cannot be heard by human beings. They are produced during earthquakes. Infrasonic waves can be heard by snakes.

Audible Waves

Longitudinal waves having frequencies lying between 20–20000 Hz are called audible waves.

Ultrasonic Waves

Longitudinal waves having frequencies above 20000 Hz are called ultrasonic waves. They are produced and heard by bats. They have a large energy content.

Applications of Ultrasonic Waves

Ultrasonic waves have a large range of application. Some of them are as follows :

(i) The fine internal cracks in a metal can be detected by ultrasonic waves.

(ii) Ultrasonic waves can be used for determining the depth of the sea, lakes etc.

(iii) Ultrasonic waves can be used to detect submarines, icebergs etc.

(iv) Ultrasonic waves can be used to clean clothes, fine machinery parts etc.

(v) Ultrasonic waves can be used to kill smaller animals like rats, fish and frogs etc.

Shock Waves

If the speed of the body in air is greater than the speed of the sound, then it is called supersonic speed. Such a body leaves behind a conical region of disturbance which spreads continuously. Such a disturbance is called a '**Shock Waves**'.

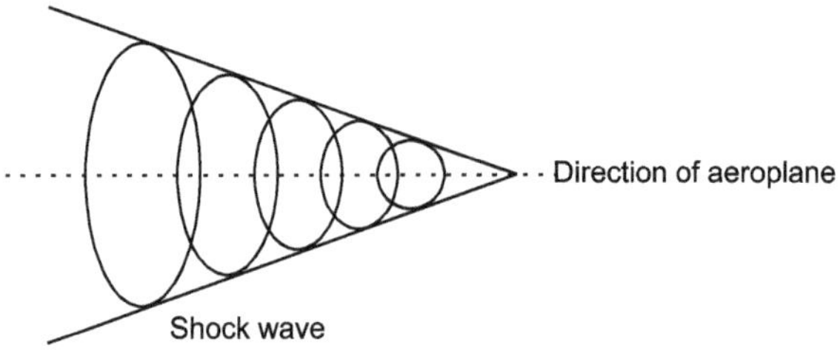

Fig. 17.32

This wave carries huge energy. If it strikes a building, then the building may be damaged.

Musical Scale

The arrangement of notes having definite ratio in respective fundamental frequencies is called a musical scale.

Musical scales are of two types :

(a) Diatomic scale It is known as 'Sargam' in Indian system. It contains eight notes with definite ratios in their frequencies. The note of lowest frequency is called key note and the highest (which is double of first) is called an octave. Harmonium, piano etc. are based on this scale.

(b) Tempered scale It contains 13 notes. The ratio of frequencies of successive notes is $2^{1/12}$.

Musical Interval

The ratio between the frequencies of two notes is called the musical interval.

Following are the names of some musical intervals :

(i) Unison $\frac{n_2}{n_1}=1$ (ii) Octave $\frac{n_2}{n_1}=2$

(iii) Major tone $\frac{n_2}{n_1}=\frac{9}{8}$ (iv) Minor tone $\frac{n_2}{n_1}=\frac{10}{9}$

(v) Semi-tone $\frac{n_2}{n_1}=\frac{16}{15}$ (vi) Fifth tone $\frac{n_2}{n_1}=\frac{3}{2}$

Echo

Multiple reflection of sound is called an echo. If the distance of reflector from the source is d then,

$$2d = vt$$

where, v = speed of sound and t, the time of echo.

$$\therefore \qquad d=\frac{vt}{2}$$

Since, the effect of ordinary sound remains on our ear for 1/10 s, therefore, if the sound returns to the starting point before 1/10 s, then it will not be distinguished from the original sound and no echo will be heard. Therefore, the minimum distance of the reflector is,

$$d=\frac{v\times t}{2}=\left(\frac{330}{2}\right)\left(\frac{1}{10}\right)=16.5 \text{ m}$$

Reverberation Time

A listener receives a series of sound waves due to a large number of reflections from the walls, ceiling and floor of the enclosed space which give him the impression of persistance or prolongation of the sound, which we call reverberation. The sound continues to be heard till the intensity falls below the zero level of intensity or threshold of hearing. The time for which sound continues to be heard after the source has stopped producing sound is called **reverberation time**. The standard reverberation time of a room is defined as the interval of time taken by a sustained note to fall in intensity to 10^{-6} of its original value.

Sabine formula For reverberation time (T) of a hall is,

$$T=\frac{0.16V}{\Sigma aS}$$

where, V is the volume of the hall (in m^3) and $\Sigma aS = a_1S_1 + a_2S_2 + \ldots$ is the absorption of the hall.

Here, $S_1, S_2, \ldots$ are the area of the surfaces (in m^2) which absorb sound and $a_1, a_2, \ldots$ are their respective absorption coefficients.

It has been established that for speech, optimum value of reverberation time is 0.5 s and for music, the optimum time of reverberation time may lie between 1 to 2 s. A room with zero reverberation time is called a **dead room**.

17.19 Beats

When two wavetrains of the same frequency travel along the same line in opposite directions, standing waves are formed in accordance with the principle of superposition. In standing waves amplitude is a function of distance. This illustrates a type of interference that we can call **interference in space.** The same principle of superposition leads us to another type of interference, which we can call **interference in time.** It occurs when two wavetrains of slightly different frequency travel through the same region.

If the waves are in phase at some time (say $t=0$) the interference will be constructive and the resultant amplitude at this moment will be $A_1 + A_2$, where A_1 and A_2 are the amplitudes of individual wavetrains. But at some later time (say $t = t_0$), because the frequencies are different, the waves will be out of phase or the interference will be destructive and the resultant amplitude will be $A_1 - A_2$ (if $A_1 > A_2$). Later, we will see that the time t_0 is $\frac{2}{f_1 - f_2}$.

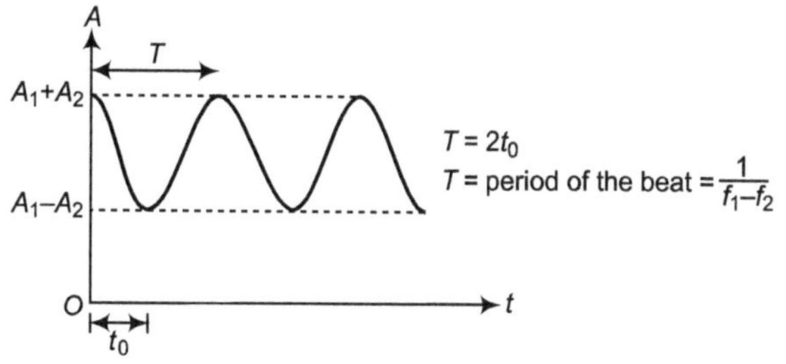

Fig. 17.33

Thus, the resultant amplitude oscillates between $A_1 + A_2$ and $A_1 - A_2$ with a time period $T = 2t_0 = \frac{1}{f_1 - f_2}$ or with a frequency $f = f_1 - f_2$ known as **beat frequency**. Thus,

beat frequency $f = f_1 - f_2$

It is then said that the amplitude is modulated. The situation described arises, when two sound sources of close but different frequencies are vibrating simultaneously at nearby places. A listener observes a fluctuation in the intensity of the sound called **beats**.

Calculation of Beat Frequency

Suppose two waves of frequencies f_1 and f_2 ($< f_1$) are meeting at some point in space. The corresponding periods are T_1 and T_2 ($> T_1$). If the two waves are in phase at

$t = 0$, they will again be in phase when the first wave has gone through exactly one more cycle than the second. This will happen at a time $t = T$, the period of the beat. Let n be the number of cycles of the first wave in time T, then the number of cycles of the second wave in the same time is $(n-1)$. Hence,

$$T = nT_1 \quad \ldots\text{(i)}$$

and
$$T = (n-1)T_1 \quad \ldots\text{(ii)}$$

Eliminating n from these two equations, we have

$$T = \frac{T_1 T_2}{T_2 - T_1} = \frac{1}{\frac{1}{T_1} - \frac{1}{T_2}} = \frac{1}{f_1 - f_2}$$

The reciprocal of the beat period is the beat frequency,

i.e.
$$f = \frac{1}{T} = f_1 - f_2$$

Alternate Method

Let the oscillations at some point in space (say $x = 0$) due to two waves be,

$$y_1 = A_1 \sin 2\pi f_1 t \quad \text{and} \quad y_2 = A_2 \sin 2\pi f_2 t$$

If they are in phase at some time t, then

$$2\pi f_1 t = 2\pi f_2 t$$

$$\Rightarrow \quad f_1 t = f_2 t \quad \ldots\text{(i)}$$

They will be again in phase at time $(t + T)$ if,

$$2\pi f_1 (t+T) = 2\pi f_2 (t+T) + 2\pi$$

$$\Rightarrow \quad f_1 (t+T) = f_2 (t+T) + 1 \quad \ldots\text{(ii)}$$

Solving Eqs. (i) and (ii), we get

$$T = \frac{1}{f_1 - f_2}$$

Note Points

- When $A_1 = A_2 = A$ (say) i.e. when the two amplitudes are equal, we obtain an amplitude which oscillates between zero and $2A$.

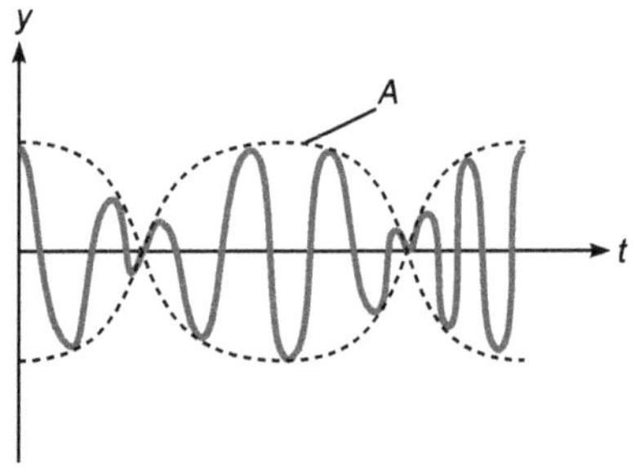

Fig. 17.34

- At frequency difference greater than about 6 or 7 Hz, we no longer hear individual beats. For example, if you listen to a whistle that produces sounds at 2000 Hz and 2100 Hz, you will hear not only these tones but also a much lower 100 Hz tone.
- If the frequency of a tuning fork is f and it produces Δf beats per second with a standard fork of frequency f_0, then

$$f = f_0 \pm \Delta f$$

If on filing the arms of an unknown fork the beat frequency decreases then,

$$f = f_0 - \Delta f$$

This is because filing of an arm of a tuning fork increases its frequency.

Similarly, if on loading/waxing of the unknown fork, the beat frequency decreases then the frequency of the unknown fork is, $f = f_0 + \Delta f$. This is because loading/waxing decreases the frequency of tuning fork.

Similarly, $f = f_0 + \Delta f$, if on filing beat frequency increases and $f = f_0 - \Delta f$ if on loading/waxing beat frequency increases.

Thus,

$f = f_0 - \Delta f$ if on filing beat frequency is decreased

$= f_0 + \Delta f$ if on filing beat frequency is increased

$= f_0 + \Delta f$ if on loading/waxing beat frequency is decreased

$= f_0 - \Delta f$ if on loading/waxing beat frequency is increased

Example 17.25 *The string of a violin emits a note of* 400 *Hz at its correct tension. The string is bit taut and produces* 5 *beats per second with a tuning fork of frequency* 400 *Hz. Find frequency of the note emitted by this taut string.*

Sol. The frequency of vibration of a string increases with increase in the tension. Thus, the note emitted by the string will be a little more than 400 Hz. As it produces 5 beats per second with the 440 Hz tuning fork, the frequency will be 405 Hz.

17.20 The Doppler Effect

If a wave source and a receiver are moving relative to each other, the frequency observed by the receiver (f') is different from the actual source frequency (f). This phenomenon is called the **Doppler effect**, named after the Austrian physicist Christian Johann Doppler (1803–1853), who discovered it in light waves.

Perhaps you might have noticed how the sound of a vehicle's horn changes as the vehicle moves past you. The frequency (pitch) of the sound you hear as the vehicle approaches you is higher than the frequency you hear as it moves away from you.

The Doppler effect, applies to waves in general. Let us apply it to sound waves. We consider the special case in which the source and observer move along the line joining them. We will use the following symbols,

$$v = \text{speed of sound,}$$
$$v_s = \text{speed of source, and}$$
$$v_o = \text{speed of observer.}$$

First we take two special cases. Then, the results obtained there can be generalised.

Source at Rest, Observer Moves

Suppose that the observer O moves towards the source S at speed v_o. The speed of the sound waves relative to O is $v_r = v + v_o$, but wavelength has its normal value $\lambda = \frac{v}{f}$.

Thus, the frequency heard by O is,

$$f' = \frac{v_r}{\lambda} = \left(\frac{v + v_o}{v}\right) f$$

If observer O was moving away from S, the frequency heard by O would be

$$f' = \left(\frac{v - v_o}{v}\right) f$$

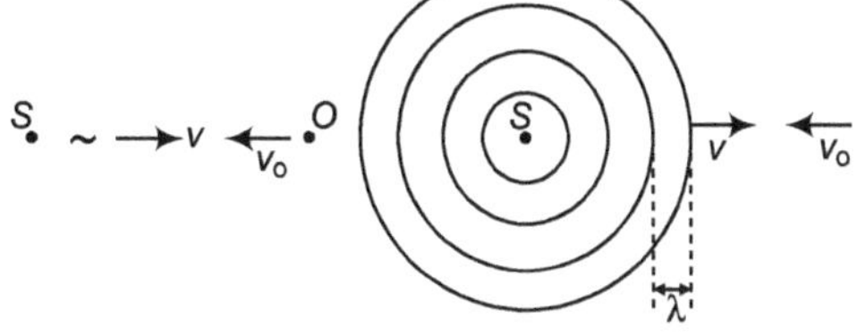

Fig. 17.35

Combining these two expressions, we find

$$f' = \left(\frac{v \pm v_o}{v}\right) f \qquad \ldots\text{(i)}$$

Source Moves, Observer at Rest

Suppose that the source S moves towards O as shown in figure. If S was at rest, the distance between two consecutive wave pulses emitted by sound would be $\lambda = \frac{v}{f} = vT$. (Fig. 17.35) however,

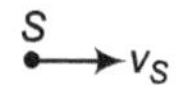

Fig. 17.36

in one time period S moves a distance $v_s T$ before it emits the next pulse. As a result the wavelength is modified. Directly ahead of S the effective wavelength (for both S and O) is

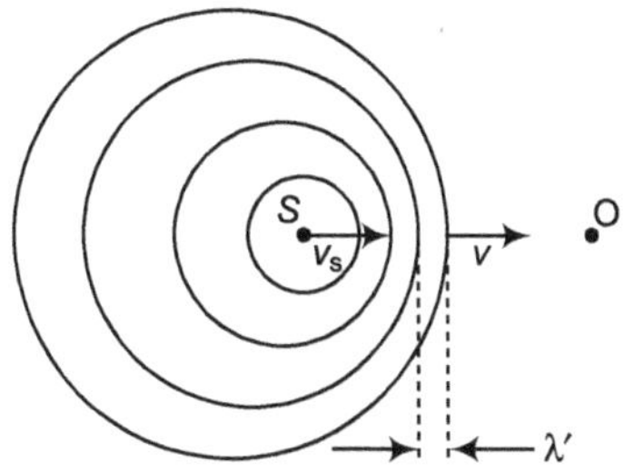

The wavelength in front of the source is less than the normal whereas in the rear it is larger than normal.

Fig. 17.37

$$\lambda' = vT - v_s T = \left(\frac{v - v_s}{f}\right)$$

The speed of sound waves relative to O is simply v. Thus, the frequency observed by O is

$$f' = \frac{v}{\lambda'} = \left(\frac{v}{v - v_s}\right) f$$

If S was moving away from O, the effective wavelength would be $\lambda' = \left(\frac{v + v_s}{f}\right)$ and the apparent frequency would be

$$f' = \left(\frac{v}{v + v_s}\right) f$$

Combining these two results, we have

$$f' = \left(\frac{v}{v \pm v_s}\right) f \qquad \ldots\text{(ii)}$$

All four possibilities can be combined into one equation

$$f' = \left(\frac{v \pm v_o}{v \mp v_s}\right) f \qquad \ldots\text{(iii)}$$

where, the upper signs (+ numerator, – denominator) correspond to the source and observer along the line joining the two in the direction toward the other and the lower signs in the direction away from the other.

Alternate Method

The above formulae can be derived alternately as follows :

Assume that the source and observer are moving along the same line and that the observer O is to the right of the source S. Suppose that at time $t = 0$, when the source and the observer are separated by a distance $SO = l$, the source emits a wave pulse (say p_1) that reaches the observer at a later time t_1. In that time the observer has moved a distance $v_o t_1$ and the total distance travelled by p_1 in the time t_1 has been $l + v_o t_1$. If v is the speed of sound, this distance is also vt_1. Then,

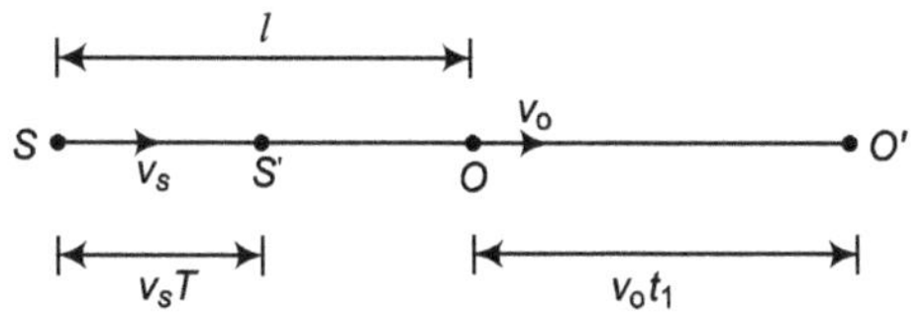

Fig. 17.38

$$vt_1 = l + v_o t_1 \quad \text{or} \quad t_1 = \frac{l}{v - v_o} \qquad \ldots\text{(iv)}$$

At time $t = T$, the source is at S' and the wave pulse (say p_2) emitted at this time will reach the observer at a time say t_2, measured from the same time origin as before. The total distance travelled by p_2 till it is received by the observer (measured from S') is $(l - v_s T) + v_o t_2$. The actual travel time for p_2 is $(t_2 - T)$ and the distance travelled is $v(t_2 - T)$.

Therefore,

$$v(t_2 - T) = (l - v_s T) + v_0 t_2$$

or
$$t_2 = \frac{l + (v - v_s)T}{v - v_o} \qquad \ldots(\text{v})$$

The time interval reckoned by the observer between the two pulses emitted by the source at S and at S' is

$$T' = t_2 - t_1 = \left(\frac{v - v_s}{v - v_o}\right)T$$

This is really the changed time period as observed by the observer. Hence, the new frequency is

$$f' = \frac{1}{T'} = \left(\frac{v - v_o}{v - v_s}\right)\frac{1}{T}$$

or
$$f' = \left(\frac{v - v_o}{v - v_s}\right)f$$

This is a result which we can expect by using equation $f' = \left(\frac{v \pm v_o}{v \mp v_s}\right) f$ in the case when source is moving towards observer and observer is moving away from the source.

Doppler's Effect in Light

Light waves also show Doppler's effect. If a light source is moving away from a stationary observer, then the frequency of light waves appears decreased and wavelength appears increased and *vice-versa.*

If the light source or the observer is moving with a velocity v, such that the distance between them is decreasing, then the apparent frequency of the source will be given by,

$$f' = f\sqrt{\frac{1 + \frac{v}{c}}{1 - \frac{v}{c}}}$$

If the distance between the light source and the observer is increasing, then the apparent frequency of the source is given by,

$$f' = f\sqrt{\frac{1 - \frac{v}{c}}{1 + \frac{v}{c}}}$$

The change in wavelength can be determined by,

$$\Delta\lambda = \frac{v}{c} \cdot \lambda$$

If the light source is moving away from the observer, the shift in the spectrum is towards red and if it is moving towards the observer the shift is towards the violet.

- Doppler's effect in light depends only on the relative motion between the source and the observer while the Doppler's effect in sound also depends upon whether, the source is moving or the observer is moving.

Extra Knowledge Points

- If the medium (air) itself starts moving, appropriate changes must be made in Eq. (iii). If wind blows at a speed v_w from source to observer, take $v \to v + v_w$ (both in numerator and denominator) and if in opposite direction (i.e. from observer to source), take $v \to v - v_w$. Thus, the modified formula is

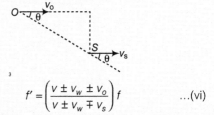

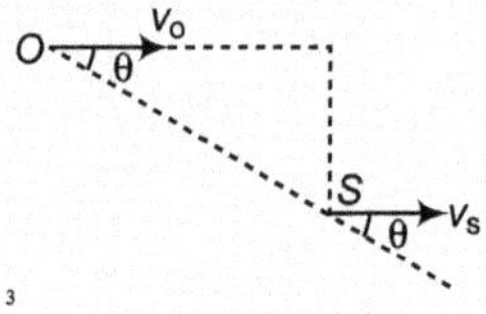

$$f' = \left(\frac{v \pm v_w \pm v_o}{v \pm v_w \mp v_s}\right) f \qquad \ldots(\text{vi})$$

- We have derived Eq.(iii) by assuming that v_o and v_s are along the line joining source and observer. If the motion is along some other direction, the components of velocities along the line joining source and observer are considered.

  For example in the figure shown,

$$f' = \left(\frac{v + v_o \cos\theta}{v + v_o \sin\theta}\right) f$$

- Change in frequency depends on the fact that whether the source is moved towards the observer or the observer is moved towards the source. But when the speed of source and observer are much lesser than that of sound, the change in frequency becomes independent of the fact whether, the source is moving or the observer. This can be shown as under.

  Suppose a source is moving towards a stationary observer with speed u and the speed of sound is v, then

$$f' = \left(\frac{v}{v - u}\right) f = \left(\frac{1}{1 - \frac{u}{v}}\right) f = \left(1 - \frac{u}{v}\right)^{-1} f$$

  Using the binomial expansion, we have

$$\left(1 - \frac{u}{v}\right)^{-1} \approx 1 + \frac{u}{v} + \frac{u^2}{v^2}$$

$$\therefore \qquad f' \approx \left(1 + \frac{u}{v} + \frac{u^2}{v^2}\right) f$$

$$\Rightarrow \qquad f' \approx \left(1 + \frac{u}{v}\right) f$$

  if $\qquad u << v$

  On the other hand, if an observer moves towards a stationary source with same speed u, then

$$f' = \left(\frac{v + u}{v}\right) f = \left(1 + \frac{u}{v}\right) f$$

  which is same as above.

- The Doppler effect is important in light. But the speed of light is so great that only astronomical or atomic sources, which have high velocities compared to speed of light, show pronounced Doppler effect.

  There are differences, however, in the Doppler effect formula for light and for sound. In sound it is not just the relative motion of source and observer that determines the frequency change. Even when the relative motion is the same, we obtain different results, depending on whether the source or the observer is moving. For light however, it is the relative motion between the source and the observer which matters.

Example 17.26 *A siren emitting a sound of frequency 1000 Hz moves away from you toward a cliff at a speed of 10 m/s.*

(a) What is the frequency of the sound you hear coming directly from the siren?

(b) What is the frequency of the sound you hear reflected off the cliff?

(c) What beat frequency would you hear? Take the speed of sound in air is, 330 m/s.

Sol. The situation is as shown in figure.

(a) Frequency of sound reaching directly to us (by S)

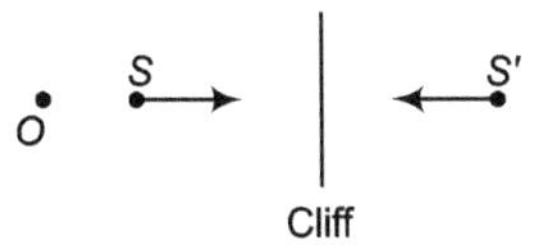

Fig. 17.39

$$f_1 = \left(\frac{v}{v + v_s}\right) f = \left(\frac{330}{330 + 10}\right)(1000)$$

$$= 970.6 \text{ Hz}$$

(b) Frequency of sound which is reflected off from the cliff (from S')

$$f_2 = \left(\frac{v}{v - v_s}\right) f = \left(\frac{330}{330 - 10}\right)(1000)$$

$$= 1031.3 \text{ Hz}$$

(c) Beat frequency $= f_2 - f_1 = 60.7$ Hz

Note Point

- Numerically, the beat frequency comes out to be 60.7 *Hz*. But beats between two tones can be detected by ear upto a frequency of about 7/s. At higher frequencies beats cannot be distinguished in the sound produced. Hence, the correct answer of part (c) should be zero.

Example 17.27 *A whistle of frequency 540 Hz rotates in a circle of radius 2 m at a linear speed of 30 m/s. What is the lowest and highest frequency heard by an observer a long distance away at rest with respect to the centre of circle. Take speed of sound in air is 330 m/s. Can the apparent frequency be ever equal to actual?*

Sol. Apparent frequency will be minimum, when the source is at N and moving away from the observer,

$$f_{min} = \left(\frac{v}{v + v_s}\right) f$$

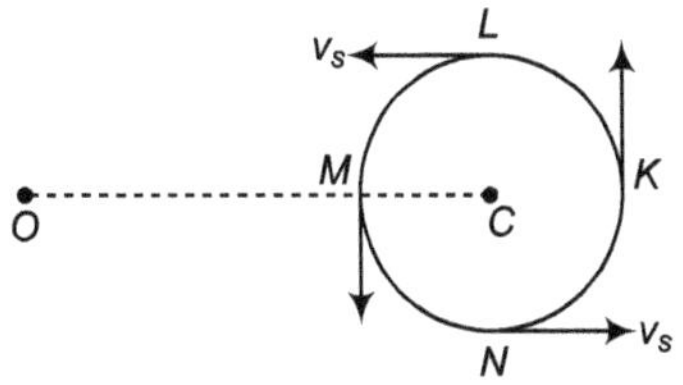

Fig. 17.40

$$= \left(\frac{330}{330 + 30}\right)(540)$$

$$= 495 \text{ Hz}$$

Frequency will be maximum when source is at L and approaching the observer.

$$f_{max} = \left(\frac{v}{v - v_s}\right) f$$

$$= \left(\frac{330}{330 - 30}\right)(540)$$

$$= 594 \text{ Hz}$$

Further when source is at M and K, angle between velocity of source and line joining source and observer is $90°$ or $v_s \cos\theta = v_s \cos 90° = 0$. So, there will be no change in the apparent frequency.

Example 17.28 *A car approaching a crossing at a speed of 20 m/s sounds a horn of frequency 500 Hz, when 80 m from the crossing. Speed of sound in air is 330 m/s. What frequency is heard by an observer 60 m from the crossing on the straight road which crosses car road at right angles?*

Sol. The situation is as shown in figure.

$$\cos\theta = \frac{80}{100} = \frac{4}{5}$$

∴ Apparent frequency is

$$f_{app} = \left(\frac{v}{v - v_s \cos\theta}\right) f$$

$$= \left(\frac{330}{330 - 20 \times \frac{4}{5}}\right) 500$$

$$= 525.5 \text{ Hz}$$

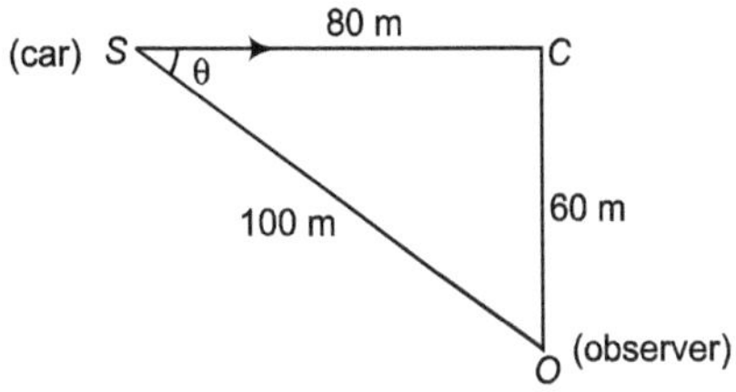

Fig. 17.41

Chapter Summary with Formulae

- In any type of wave, oscillations of a physical quantity y are produced at one place and these oscillations (along with energy and momentum) are transferred to other places also.
- **Classification of Waves** A wave may be classified in following three ways :

First A transverse wave is one in which oscillations of y are perpendicular to wave velocity. Electromagnetic waves are transverse in nature. A longitudinal wave is one in which oscillations of y are parallel to wave velocity. Sound waves are longitudinal in nature.

Second Mechanical waves require medium for their propagation. Sound waves are mechanical in nature. Non-mechanical waves do not require medium for their propagation. Electromagnetic waves are non-mechanical in nature.

Third Transverse string wave is one dimensional. Ripples waves on the surface of water are two dimensional. Sound wave due to a point source is three dimensional.

- **Wave Equation** In any wave equation, value of y is a function of position and time. In case of one dimensional wave position can be represented by one coordinate (say x) only. Hence,

$$y = f(x, t)$$

Only those functions of x and t represent a wave equation which satisfy following condition :

$$\frac{\partial^2 y}{\partial x^2} = (\text{constant}) \frac{\partial^2 y}{\partial t^2}$$

Here, $\text{constant} = \frac{1}{v^2}$

where, v is the wave speed. All functions of x and t of type,

$$y = f(ax \pm bt)$$

satisfy above mentioned condition of wave equation, provided value of y should be finite for any value of t. If $y\ (x, t)$ function is of this type, then following two conclusions can be drawn.

(i) Wave speed $v = \frac{\text{coefficient of } t}{\text{coefficient of } x} = \frac{b}{a}$

(ii) Wave travels along positive x-direction. If ax and bt have opposite signs and it travels along negative x-direction, if they have same signs.

- **Plane Progressive Harmonic Wave** If oscillations of y are simple harmonic in nature, then wave is called plane progressive harmonic wave. General equation of this wave is,

$y = A\sin(\omega t \pm kx \pm \phi)$ or $y = A\cos(\omega t \pm kx \pm \phi)$

In these equations,

(i) A is amplitude of oscillation,

(ii) ω is angular frequency,

$$T = \frac{2\pi}{\omega},\ \omega = 2\pi f \text{ and } f = \frac{1}{T} = \frac{\omega}{2\pi}$$

(iii) k is wave number,

$$k = \frac{2\pi}{\lambda} \qquad (\lambda \rightarrow \text{wavelength})$$

(iv) Wave speed $v = \frac{\omega}{k} = f\lambda$

(v) ϕ is initial phase angle and

(vi) $(\omega t \pm kx \pm \phi)$ is phase angle at time t at coordinate x.

- **Particle Speed (v_p) and Wave Speed (v) in Case of Harmonic Wave**

(i) $y = f(x, t)$, where x and t are two variables.

Then, $$v_p = \frac{\partial y}{\partial t}$$

(ii) In harmonic wave, particles are in SHM. Therefore, all equations of SHM can be applied for particles also.

(iii) Relation between v_p and v

$$v_p = -v \cdot \frac{\partial y}{dx}$$

- **Phase Difference ($\Delta\phi$)**

Case I $\Delta\phi = \omega(t_1 \sim t_2)$ or $\Delta\phi = \frac{2\pi}{T} \cdot \Delta t$

= phase difference of one particle at a time interval of Δt.

Case II $\Delta\phi = k(x_1 \sim x_2)$

$= \frac{2\pi}{\lambda} \cdot \Delta x$

= phase difference at one time between two particles at a path difference of Δx.

- **Wave Speed**

(i) **Speed of transverse wave on a stretched wire**

$$v = \sqrt{\frac{T}{\mu}} = \sqrt{\frac{T}{\rho S}}$$

(ii) **Speed of longitudinal wave**

$$v = \sqrt{\frac{E}{\rho}}$$

(a) In solids, $E = Y$ = Young's modulus of elasticity

$\therefore$ $$v = \sqrt{\frac{Y}{\rho}}$$

(b) In liquids, $E = B$ = Bulk modulus of elasticity

$\therefore$ $$v = \sqrt{\frac{B}{\rho}}$$

(c) In gases, according to Newton,

$E = B_T$ = Isothermal bulk modulus of elasticity $= P$

$\therefore$ $$v = \sqrt{\frac{P}{\rho}}$$

But results did not match with this formula.

Laplace made correction in it. According to him,

$E = B_S$ = Adiabatic bulk modulus of elasticity $= \gamma p$

$\therefore$ $$v = \sqrt{\frac{\gamma p}{\rho}} = \sqrt{\frac{\gamma RT}{M}} = \sqrt{\frac{\gamma kT}{m}}$$

- When two or more waves travel in the same medium, the displacement of any element of the medium is the algebraic sum of the displacements due to each wave. This is known as the principle of superposition of waves.

$$y = \sum_{i=1}^{n} f_1\ (x - vt)$$

- Two sinusoidal waves on the same string exhibit interference, adding or cancelling according to the principle of superposition. If phase difference or phase constant $\phi = 0$ or an integral multiple of 2π, the waves are exactly in phase and the interference is constructive, if $\phi = \pi$, they are exactly out of phase and the interference is destructive.

- A travelling wave, at a rigid boundary or a closed end, is reflected with a phase reversal but the reflection at an open boundary takes place without any phase change.

  For an incident wave
  $$y_i(x, t) = A \sin(kx - \omega t)$$
  the reflected wave at a rigid boundary is
  $$y_r(x, t) = -A \sin(kx - \omega t)$$
  For reflection at an open boundary
  $$y_r(x, t) = A \sin(kx - \omega t)$$

- **Stationary Waves**

  (i) Stationary waves are formed by the superposition of two identical waves travelling in opposite directions.

  (ii) Formation of stationary waves is really the interference of two waves in which coherent (same frequency) sources are required.

  (iii) By the word 'identical waves' we mean that they must have same value of v, ω and k.

  Amplitudes may be different, but same amplitudes are preferred.

  (iv) In stationary waves all particles oscillate with same value of ω but amplitudes varying from $A_1 + A_2$ to $A_1 \sim A_2$. Points where amplitude is maximum (or $A_1 + A_2$) are called antinodes (or points of constructive interference) and points where amplitude is minimum (or $A_1 \sim A_2$) are called nodes (or points of destructive interference).

  (v) If $A_1 = A_2 = A$, then amplitude at antinode is $2A$ and at node is zero. In this case points at node do not oscillate.

  (vi) Points at antinodes have maximum energy of oscillation and points at nodes have minimum energy of oscillation (zero when $A_1 = A_2$).

  (vii) Points lying between two successive nodes are in same phase. They are out of phase with the points lying between two neighbouring successive nodes.

  (viii) Equation of stationary wave is of type,
  $$y = 2A \sin kx \cos \omega t \qquad \ldots(i)$$
  or $$y = A \cos kx \sin \omega t \text{ etc.}$$
  This equation can also be written as,

  $y = A_x \sin \omega t$ or $y = A_x \cos \omega t$

  If $x = 0$ is a node then, $A_x = A_0 \sin kx$

  If $x = 0$ is an antinode then, $A_x = A_0 \cos kx$

  Here, A_0 is maximum amplitude at antinode.

- **Oscillations of Stretched Wire**

  Fundamental tone or first harmonic ($n = 1$)

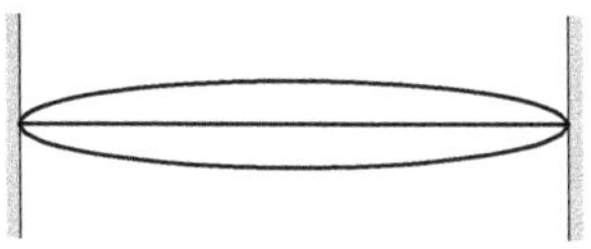

  First overtone or second harmonic ($n = 2$)

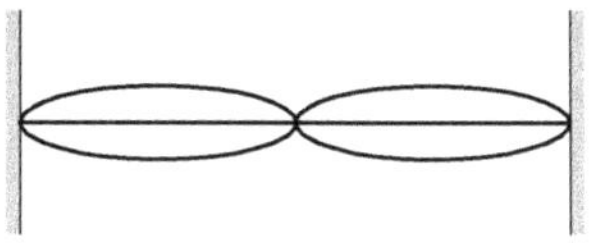

  Second overtone or third harmonic (n = 3)

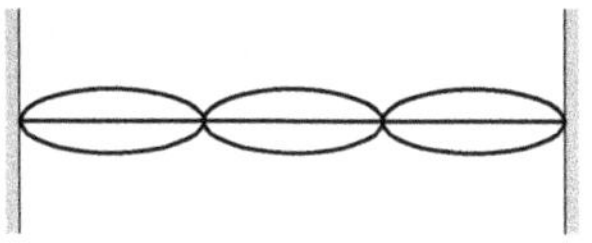

$$f = n\left(\frac{v}{2l}\right) \text{ Here, } n = 1, 2, 3, \ldots..$$

Even and odd both harmonics are obtained.

Here, $$v = \sqrt{\frac{T}{\mu}} \text{ or } \sqrt{\frac{T}{\rho S}}$$

- **Longitudinal Wave**

  (i) There are three equations associated with any longitudinal wave
  $y(x, t)$, $\Delta p(x, t)$ and $\Delta \rho(x, t)$

  (ii) y represents displacement of medium particles from their mean position parallel to direction of wave velocity.

  (iii) From $y(x, t)$ equation, we can make $\Delta p(x, t)$ or $\Delta \rho(x, t)$ equations by using the fundamental relation between them,
  $$\Delta p = -B \cdot \frac{\partial y}{\partial x} \quad \text{and} \quad \Delta \rho = -\rho \cdot \frac{\partial y}{\partial x}$$

  (iv) $\Delta p_0 = BAk$ and $\Delta \rho_0 = \rho Ak$

  (v) $\Delta p(x, t)$ and $\Delta \rho(x, t)$ are in same phase. But $y(x, t)$ equation has a phase difference of $\frac{\pi}{2}$ with rest two equations.

- **Effect of Temperature, Pressure and Relative Humidity in Speed of Sound in Air (or in a Gas)**

  (i) **With temperature** $v \propto \sqrt{T}$

  (ii) **With pressure** Pressure has no effect on speed of sound as long as temperature remains constant.

  (iii) **With relative humidity** With increase in relative humidity in air, density decreases. Hence, speed of sound increases.

- **Sound Level** (L)
  $$L = 10 \log_{10} \frac{I}{I_0} \qquad \text{(in dB)}$$

- Here, I_0 = intensity of minimum audible sound = 10^{-12} Wm^{-2}.

  While comparing loudness of two sounds we may write,
  $$L_2 - L_1 = 10 \log_{10} \frac{I_2}{I_1}$$
  In case of point source,
  $$I \propto \frac{1}{r^2} \text{ or } \frac{I_2}{I_1} = \left(\frac{r_1}{r_2}\right)^2.$$
  In case of line source,
  $$I \propto \frac{1}{r} \text{ or } \frac{I_2}{I_1} = \left(\frac{r_1}{r_2}\right).$$

- **Doppler Effect in Sound**
  $$f' = f\left(\frac{v \pm v_m \pm v_o}{v \pm v_m \pm v_s}\right)$$

- **Beats**
  $$f_b = f_1 - f_2 \qquad (f_1 > f_2)$$

Additional Examples

Example 1. *Name two important properties of a material medium responsible for the propagation of waves through it.*

Sol. Elasticity and inertia.

Example 2. *Why are the longitudinal waves also called pressure waves ?*

Sol. Longitudinal waves travel in a medium as series of alternate compressions and rarefactions i.e. they travel as variations in pressure and hence are called pressure waves.

Example 3. *Can transverse waves be produced in air?*

Sol. No. Transverse waves travel in the form of crests and troughs and so involve change in shape. So, the transverse waves can be produced in a medium which has elasticity of shape or shear modulus of elasticity. As air has no shear modulus of elasticity, hence transverse waves cannot be produced in it.

Example 4. *What is the difference between wave velocity and particle velocity ?*

Sol. The wave velocity (or phase velocity) is constant for a given medium and is given by $v = f\lambda$ while the particle velocity changes harmonically with time. It is maximum at the mean position and zero at the extreme position.

Example 5. *We always see lightning before we hear thundering. Why ?*

Sol. The speed of light (3×10^8 ms^{-1}) is much larger than the speed of sound (330 ms^{-1}). Consequently, the flash of light reaches us much earlier than the sound of thunder.

Example 6. *Two astronauts on the surface of the moon cannot talk to each other. Why?*

Sol. Sound waves being the mechanical waves require material medium for their propagation. As there is no atmosphere on the moon, hence the sound waves cannot propagate on the moon.

Example 7. *How is it possible to detect the approaching of a distant train by placing the ear very close to the railway line ?*

Sol. Sound waves travel much faster in solids than that in air. Moreover, due to high elasticity of solids, sound waves do not die out in solids as soon as in air. This makes possible to detect the sound of a distant approaching train by placing the ear very close to the railway line.

Example 8. *Explosions on other planets are not heard on the earth. Why ?*

Sol. Since, no material medium is present in the space between the planets and the earth, the sound (due to explosions) cannot propagate upto the earth.

Example 9. *Why are the stationary waves called so ?*

Sol. In a stationary wave, the particles of the medium vibrate about their mean positions, but disturbance does not travel in any direction.

Example 10. *When are stationary waves produced ?*

Sol. Stationary waves are produced by the superposition of two identical waves travelling in opposite directions.

Example 11. *Under what condition does a sudden phase reversal of* $180°$ *on reflection takes place ?*

Sol. On reflection from a denser medium, a wave suffers a sudden phase reversal of $180°$.

Example 12. *Does increasing the tension in the string affect, the wavelength of the fundamental note on a guitar string ?*

Sol. No, wavelength depends on the length of the string.

Example 13. *In a vibrating string, what is the separation between a node and a consecutive antinode? What is the phase difference between the two points just on the opposite sides of an antinode ?*

Sol. The separation between a node and a consecutive antinode is $\lambda/4$. Further, the phase difference between the points just on the opposite sides of an antinode is$180°$.

Example 14. *What is the cause of Doppler effect, when the listener is in motion ?*

Sol. It is because of the change in number of sound waves crossing past the ear of the listener due to his motion.

Example 15. *What is the cause of Doppler effect, when the source of sound is in motion?*

Sol. It is because of the apparent change in wavelength of the sound waves due to motion of the source.

Example 16. *Why is the note produced by an open organ pipe sweeter than that produced by the closed organ pipe?*

Sol. The note produced by open organ pipe consists of both odd an even harmonics but the note produced by closed organ pipe onsists of only the odd harmonics. Due to presence of larger number of overtones or harmonics, the note produced by the open organ pipe is sweeter.

Example 17. *Why are there so many holes in a flute?*

Sol. The flute is basically an open organ pipe. The location of the open end can be changed by keeping the one hole open and closing the other holes. Thus, the frequency of the note produced by the flute can be changed.

Example 18. *If density of oil higher than the density of water, is used in a resonance tube, how will the frequency change ?*

Sol. The frequency of vibration depends on the length of the air column. The liquid surface only causes the reflection of water. Hence, frequency does not change if oil of density higher than that of water is used in the resonance tube.

Example 19. *The beats are not heard, if the difference in frequencies of the two sounding notes is more than* 10*. Why ?*

Sol. If the difference in frequencies of the two waves is more than 10, we shall hear more than 10 beats per second. Due to persistence of hearing, our ear is not able to distinguish between two sounds as separate, if the time interval between them is less than (1/10)th of a second. Hence, beats heard will not be distinct, if the number of beats produced per second is more than 10.

Example 20. *Explain why we cannot hear an echo in a smaller room ?*

Sol. For an echo of a simple sound to be heard, the minimum distance between the speaker and the walls should be 17 m. As the length of a room is generally less than 17 m, so we do not hear an echo.

Example 21. *How do we identify our friend from his voice while sitting in a dark room ?*

Sol. On the basis of quality of sound.

Example 22. *What determines the quality of sound ?*

Sol. Quality of sound is determined by the number of harmonic components present in the sound.

Example 23. *Equation of a transverse wave travelling in a rope is given by* $y = 5\sin(4.0t - 0.02x)$ *where, y and x are expressed in cm and time in seconds. Calculate*

(a) the amplitude, frequency, velocity and wavelength of the wave.

(b) the maximum transverse speed and acceleration of a particle in the rope.

Sol. (a) Comparing this with the standard equation of wave motion

$$y = A\sin\left(2\pi ft - \frac{2\pi}{\lambda}x\right)$$

where, A, f and λ are amplitude, frequency and wavelength respectively.

Thus, amplitude $A = 5\,\text{cm}$

$$2\pi f = 4 \Rightarrow \text{Frequency } f = \frac{4}{2\pi} = 0.637\ \text{Hz}$$

Again, $$\frac{2\pi}{\lambda} = 0.02$$

or Wavelength $$\lambda = \frac{2\pi}{0.02} = 100\,\pi\ \text{cm}$$

Velocity of the wave,

$$v = f\lambda = \frac{4}{2\pi}\,\frac{2\pi}{0.02} = 200\ \text{cm/s}$$

(b) Transverse velocity of the particle,

$$u = \frac{dy}{dt} = 5\times 4\cos(4.0\,t - 0.02\,x)$$
$$= 20\cos(4.0\,t - 0.02\,x)$$

Maximum velocity of the particle = 20 cm/s

Particle acceleration

$$a = \frac{d^2y}{dt^2} = -\,20\times 4\sin(4.0\,t - 0.02\,x)$$

Maximum particle acceleration $= 80\ \text{cm/s}^2$.

Example 24. *A wire of uniform cross-section is stretched between two points* 100 *cm apart. The wire is fixed at one end and a weight is hung over a pulley at the other end. A weight of* 9 *kg produces a fundamental frequency of* 750 *Hz.*

(a) What is the velocity of the wave in wire ?

(b) If the weight is reduced to 4 *kg, what is the velocity of wave ? What is the wavelength and frequency ?*

Sol. (a) $L = 100$ cm, $f_1 = 750$ Hz

$$v_1 = 2Lf_1 = 2\times 100\times 750$$
$$= 150000\ \text{cms}^{-1}$$
$$= 1500\ \text{ms}^{-1}$$

(b) $$v_1 = \sqrt{\frac{T_1}{m}}$$

and $$v_2 = \sqrt{\frac{T_2}{m}}$$

$$\frac{v_2}{v_1} = \sqrt{\frac{T_2}{T_1}}$$

$$\Rightarrow \quad \frac{v_2}{1500} = \sqrt{\frac{4}{9}}$$

$$\therefore \quad v_2 = 1000\ \text{ms}^{-1}$$

$$\lambda_2 = \text{wavelength} = 2L$$
$$= 200\ \text{cm} = 2\ \text{m}$$

$$f_2 = \frac{v_2}{\lambda_2} = \frac{1000}{2} = 500\ \text{Hz}$$

Example 25. *Determine the speed of sound waves in water, and find the wavelength of a wave having a frequency of 242 Hz.*
(Take, $B_{water} = 2 \times 10^9$ Pa)

Sol. Speed of sound wave,

$$v = \sqrt{\frac{B}{\rho}} = \sqrt{\frac{(2 \times 10^9)}{10^3}} = 1414 \text{ m/s}$$

Wavelength $\lambda = \frac{v}{f} = 5.84$ m

Example 26. *The faintest sound the human ear can detect at frequency 1 kHz corresponds to an intensity of about 10^{-12} W/m^2. Determine the pressure amplitude and the maximum displacement associated with this sound assuming the density of the air = 1.3 kg/m^3 and velocity of sound in air = 332 m/s.*

Sol. $I = \frac{p^2}{2\rho v}$

$$\Rightarrow \quad p = \sqrt{I \times 2\rho v} = \sqrt{10^{-12} \times 2 \times 1.3 \times 332} = 2.94 \times 10^{-5} \text{ N/m}^2$$

Again, $p = \rho v \omega A$

$$\Rightarrow \quad A = \frac{p}{\rho v \omega} = \frac{2.94 \times 10^{-5}}{1.3 \times 332 \times 2\pi \times 10^3} = 1.1 \times 10^{-11} \text{ m}$$

Example 27. *A flute which we treat as a pipe open at both ends is 60 cm long.*

(a) What is the fundamental frequency when all the holes are covered?

(b) How far from the mouthpiece should a hole be uncovered for the fundamental frequency to be 330 Hz? Take speed of sound in air as 340 m/s.

Sol. (a) Fundamental frequency when the pipe is open at both ends is

$$f_1 = \frac{v}{2l} = \frac{340}{2 \times 0.6} = 283.33 \text{ Hz}$$

(b) Suppose the hole is uncovered at a length l from the mouthpiece, the fundamental frequency will be

$$f_1 = \frac{v}{2l}$$

$$\therefore \quad l = \frac{v}{2f_1} = \frac{340}{2 \times 330} = 0.515 \text{ m} = 51.5 \text{ cm}$$

- Opening holes in the side effectively shortens the length of the resonance column, thus increasing the frequency.

Example 28. *The first overtone of an open organ pipe beats with the first overtone of a closed organ pipe with a beat frequency of 2.2 Hz. The fundamental frequency of the closed organ pipe is 110 Hz. Find the lengths of the pipes. Speed of sound in air $v = 330$ m/s.*

Sol. Let l_1 and l_2 be the lengths of closed and open pipes respectively.

Fundamental frequency of closed organ pipe is given by

$$f_1 = \frac{v}{4l_1}$$

v = speed of sound in air = 330 m/s

Given, $f_1 = 110$ Hz

Therefore, $\frac{v}{4l_1} = 110$ Hz

$$\therefore \quad l_1 = \frac{v}{4 \times 110} = \frac{330}{4 \times 110} \text{ m} = 0.75 \text{ m}$$

First overtone of closed organ pipe will be

$$f_3 = 3f_1 = 3(110) \text{ Hz} = 330 \text{ Hz}$$

This produces a beat frequency of 2.2 Hz with first overtone of open organ pipe.

Therefore, first overtone frequency of open organ pipe is either

(330 + 2.2) Hz = 332.2 Hz

or (330 − 2.2) Hz = 327.8 Hz

If it is 332.2 Hz, then

$$2\left(\frac{v}{2l_2}\right) = 332.2 \text{ Hz}$$

or $$l_2 = \frac{v}{332.2} = \frac{330}{332.2} \text{ m} = 0.99 \text{ m}$$

and if it is 327.8 Hz, then

$$2\left(\frac{v}{2l_2}\right) = 327.8 \text{ Hz}$$

or $$l_2 = \frac{v}{327.8} \text{ m} = \frac{330}{327.8} \text{ m} = 1.0067 \text{ m}$$

Therefore, length of the closed organ pipe is $l_1 = 0.75$ m while length of open pipe is either $l_2 = 0.99$ m or 1.0067 m.

Example 29. *For the speed of a transverse wave on a string, we write the formula*

$$v = \sqrt{\frac{T}{\mu}}$$

and sometimes we also write $v = f\lambda$. Are these two same or different? Explain.

Sol. Speed of a wave depends on some special characteristics of the medium. For example, in case of a string it depends on T and μ. Frequency f depends on the source. Wavelength is self adjusted $\left(=\frac{v}{f}\right)$. Let us take an example.

Suppose tension in a string is 100 N, linear mass density (μ) of this wire is 1 kg/m^3. Then in such a string speed of the transverse wave is,

$$v=\sqrt{\frac{T}{\mu}}=\sqrt{\frac{100}{1}}=10\text{ m/s}$$

This value of 10 m/s is now fixed under the given conditions ($T=100$ N, $\mu=1$ kg/m^3). Now, suppose we keep the frequency of a wave 10 Hz, then wavelength of such a wave will become 1 m $\left(=\frac{v}{f}\right)$. Now, if frequency is reduced to 5 Hz (half), the wavelength will become two times or 2 m, so that product of λ and f remains constant (= 10 m/s).

Example 30. *A wave moves with speed 300 m/s on a wire 2 which is under a tension of 500 N. Find how much the tension must be changed to increase the speed to 312 m/s?*

Sol. Speed of a transverse wave on a wire is,

$$v=\sqrt{\frac{T}{\mu}} \quad \text{...(i)}$$

Differentiating with respect to tension, we have

$$\frac{dv}{dT}=\frac{1}{2\sqrt{\mu T}} \quad \text{...(ii)}$$

Dividing Eq. (ii) by Eq. (i), we have

$$\frac{dv}{v}=\frac{1}{2}\frac{dT}{T} \Rightarrow dT=(2T)\frac{dv}{v}$$

Substituting the proper values, we have

$$dT=\frac{(2)(500)(312-300)}{300}=6.67\text{ N}$$

i.e. tension should be increased by 6.67 N.

Example 31. *For the wave shown in figure, write the equation of this wave if its position is shown at $t=0$. Speed of wave is $v=300$ m/s.*

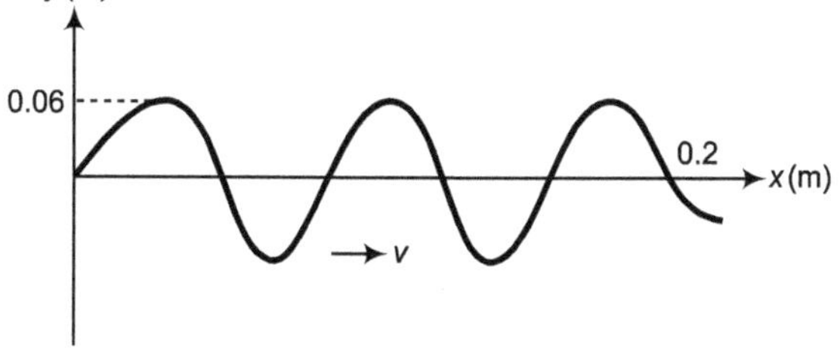

Sol. The amplitude, $A=0.06$ m

$$\frac{5}{2}\lambda=0.2\text{ m}$$

$$\therefore \quad \lambda=0.08\text{ m}$$

$$f=\frac{v}{\lambda}=\frac{300}{0.08}=3750\text{ Hz}$$

$$k=\frac{2\pi}{\lambda}=78.5\text{ m}^{-1}$$

and $\omega=2\pi f=23562$ rad/s

At $t=0$, $x=0$, $\frac{\partial y}{\partial x}=$ positive

and the given curve is a sine curve.

Hence, equation of wave travelling in positive x-direction should have the form,

$$y(x,t)=A\sin(kx-\omega t)$$

Substituting the values, we have

$$y(x,t)=(0.06\text{ m})\sin[(78.5\text{ m}^{-1})x-(23562\text{ s}^{-1})t]\text{ m}$$

Example 32. *A simple harmonic wave of amplitude 8 unit travels along positive x-axis. At any given instant of time, for a particle at a distance of 10 cm from the origin, the displacement is +6 unit, and for a particle at a distance of 25 cm from the origin, the displacement is +4 unit. Calculate the wavelength.*

Sol. $y=A\sin\frac{2\pi}{\lambda}(vt-x)$

or $$\frac{y}{A}=\sin 2\pi\left(\frac{t}{T}-\frac{x}{\lambda}\right)$$

In the first case, $\frac{y_1}{A}=\sin 2\pi\left(\frac{t}{T}-\frac{x_1}{\lambda}\right)$

Here, $y_1=+6$, $A=8$, $x_1=10$ cm

$$\therefore \quad \frac{6}{8}=\sin 2\pi\left(\frac{t}{T}-\frac{10}{\lambda}\right) \quad \text{...(i)}$$

Similarly, in the second case,

$$\frac{4}{8}=\sin 2\pi\left(\frac{t}{T}-\frac{25}{\lambda}\right) \quad \text{...(ii)}$$

From Eq. (i),

$$2\pi\left(\frac{t}{T}-\frac{10}{\lambda}\right)=\sin^{-1}\left(\frac{6}{8}\right)=0.85\text{ rad}$$

or $$\frac{t}{T}-\frac{10}{\lambda}=0.14 \quad \text{...(iii)}$$

Similarly, from Eq. (ii),

$$2\pi\left(\frac{t}{T}-\frac{25}{\lambda}\right)=\sin^{-1}\left(\frac{4}{8}\right)=\frac{\pi}{6}\text{ rad}$$

or $$\frac{t}{T}-\frac{25}{\lambda}=0.08 \quad \text{...(iv)}$$

Subtracting Eq. (iv) from Eq. (iii), we get

$$\frac{15}{\lambda}=0.06 \Rightarrow \therefore \lambda=250\text{ cm}$$

Example 33. *For a wave described by, $y = A\sin(\omega t - kx)$, consider the following points*

(i) $x = 0$, (ii) $x = \frac{\pi}{4k}$,

(iii) $x = \frac{\pi}{2k}$ and (iv) $x = \frac{3\pi}{4k}$.

For a particle at each of these points at $t = 0$, describe whether the particle is moving or not and in what direction and describe whether the particle is speeding up, slowing down or instantaneously not accelerating?

Sol. $y = A\sin(\omega t - kx)$

Particle velocity $v_P(x,t) = \frac{\partial y}{\partial t} = \omega A\cos(\omega t - kx)$ and

particle acceleration $a_P(x,t) = \frac{\partial^2 y}{\partial t^2} = -\omega^2 A\sin(\omega t - kx)$

(i) $t = 0, x = 0$: $v_P = +\omega A$ and $a_P = 0$

i.e. particle is moving upwards but its acceleration is zero.

Note Points

- Direction of velocity can be obtained in a different manner as under,

  At $t = 0$, $y = A\sin(-kx) = -A\sin kx$

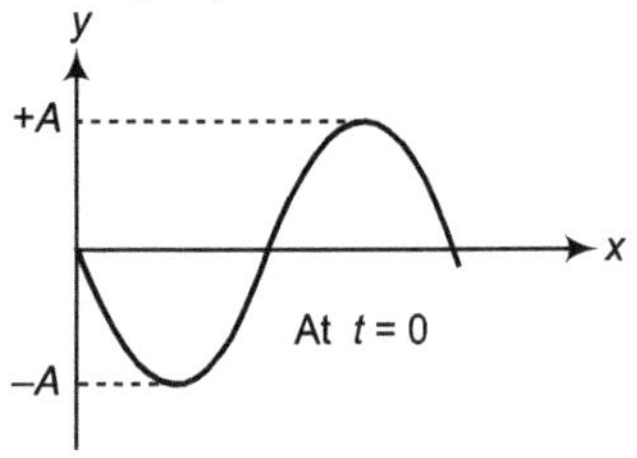

i.e. y-x graph is as shown in figure. At $x = 0$, slope is negative. Therefore, particle velocity is positive ($v_p = -v \times$ slope) as the wave is travelling along positive x-direction.

- $t = 0, x = \frac{\pi}{4k}$: $x = \frac{\pi}{4k} \Rightarrow \therefore kx = \frac{\pi}{4}$

$$v_P = \omega A\cos\left(-\frac{\pi}{4}\right) = +\frac{\omega A}{\sqrt{2}}$$

and $$a_P = -\omega^2 A\sin\left(-\frac{\pi}{4}\right) = +\frac{\omega^2 A}{\sqrt{2}}$$

Velocity of particle is positive, i.e. the particle is moving upwards (along positive y-direction). Further, v_P and a_P are in the same direction (both are positive). Hence, the particle is speeding up.

- $t = 0, x = \frac{\pi}{2k}$: $x = \frac{\pi}{2k} \Rightarrow \therefore kx = \frac{\pi}{2}$

$$v_P = \omega A\cos\left(\frac{-\pi}{2}\right) = 0$$

$$a_P = -\omega^2 A\sin\left(\frac{-\pi}{2}\right) = \omega^2 A$$

i.e. particle is stationary or at its extreme position ($y = -A$). So, it is speeding up at this instant.

- $t = 0, x = \frac{3\pi}{4k}$: $x = \frac{3\pi}{4k}$

$\therefore$ $kx = \frac{3\pi}{4}$

$\therefore$ $v_P = \omega A\cos\left(-\frac{3\pi}{4}\right) = -\frac{\omega A}{\sqrt{2}}$

$a_P = -\omega^2 A\sin\left(-\frac{3\pi}{4}\right) = +\frac{\omega^2 A}{\sqrt{2}}$

Velocity of particle is negative, i.e. the particle is moving downwards. Further v_P and a_P are in opposite directions, i.e. the particle is slowing down.

Example 34. *Consider a wave propagating in the negative x-direction whose frequency is 100 Hz. At $t = 5$ s, the displacement associated with the wave is given by,*

$$y = 0.5\cos(0.1x)$$

where, x and y are measured in centimetres and t in seconds. Obtain the displacement (as a function of x) at $t = 10$ s. What is the wavelength and velocity associated with the wave?

Sol. A wave travelling in negative x-direction can be represented as,

$$y(x,t) = A\cos(kx + \omega t + \phi)$$

At $t = 5$ s,

$$y(x, t = 5) = A\cos(kx + 5\omega + \phi)$$

Comparing this with the given equation,

We have, $A = 0.5$ cm, $k = 0.1\text{ cm}^{-1}$

and $5\omega + \phi = 0$ ...(i)

Now, $\lambda = \frac{2\pi}{k} = \frac{2\pi}{0.1}$

$= 20\pi$ cm

$\omega = 2\pi f = 200\pi \frac{\text{rad}}{\text{s}}$

$\therefore$ $v = \frac{\omega}{k} = \frac{200\pi}{0.1}$

$= 2000\pi \frac{\text{cm}}{\text{s}}$

From Eq. (i), $\phi = -5\omega$

At $t = 10$ s,

$$y(x, t = 10) = 0.5\cos(0.1x + 10\omega - 5\omega)$$
$$= 0.5\cos(0.1x + 5\omega)$$

Substituting $\omega = 200\pi$,

$$y(x, t = 10) = 0.5\cos(0.1x + 1000\pi)$$
$$= 0.5\cos(0.1x)$$

Example 35. *A window whose area is 2 m^2 opens on a street, where the street noise results at the window an intensity level of 60 dB. How much acoustic power enters the window through sound waves ? Now, if a sound absorber is fitted at the window, how much energy from the street will it collect in a day ?*

Sol. Sound level $= 10 \log \frac{I}{I_0} = 60$

or $\frac{I}{I_0} = 10^6$

$$I = 10^{-12} \times 10^6$$
$$= 10^{-6}\ \mu\text{W/m}^2$$

Power entering the room

$$1 \times 10^{-6} \times 2 = 2\,\mu\text{W}$$

Energy collected in a day $= 2 \times 10^{-6} \times 86400$
$= 0.173$ J

Example 36. *Two coherent sound sources are at distances $x_1 = 0.2$ m and $x_2 = 0.48$ m from a point. Calculate the intensity of the resultant wave at that point if the frequency of each wave is $f = 400$ Hz and velocity of wave in the medium is $v = 448$ m/s. The intensity of each wave is $I_0 = 60$ W/m^2.*

Sol. Path difference,

$$p = x_2 - x_1$$
$$= 0.48 - 0.2 = 0.28 \text{ m}$$
$$\phi = \frac{2\pi}{\lambda} p = \left(\frac{2\pi f}{v}\right) p$$
$$= \frac{2\pi (400)(0.28)}{448} = \frac{\pi}{2}$$
$$I = I_1 + I_2 + 2\sqrt{I_1 I_2} \cos \phi$$

or
$$I = I_0 + I_0 + 2I_0 \cos\left(\frac{\pi}{2}\right)$$
$$= 2I_0 = 2(60)$$
$$= 120 \text{ W/m}^2$$

Example 37. *For a certain organ pipe, three successive resonance frequencies are observed at 425, 595 and 765 Hz respectively. Taking the speed of sound in air to be 340 m/s.*

(a) Explain whether the pipe is closed at one end or open at both ends.

(b) Determine the fundamental frequency and length of the pipe.

Sol. (a) The given frequencies are in the ratio 5 : 7 : 9. As the frequencies are odd multiple of 85 Hz the pipe must be closed at one end.

(b) Now, the fundamental frequency is the lowest i.e. 85 Hz.

$$\therefore \quad 85 = \frac{v}{4l}$$
$$\Rightarrow \quad l = \frac{340}{4 \times 85}$$
$$= 1 \text{ m}$$

Example 38. *Two tuning forks P and Q when set vibrating, give 4 beats/s. If a prong of the fork P is filed, the beats are reduced to 2/s. Determine the frequency of P, if that of Q is 250 Hz.*

Sol. There are four beats between P and Q, therefore the possible frequencies of P are 246 or 254 (i.e. 250 ± 4) Hz.

When the prong of P is filed, its frequency becomes greater than the original frequency.

If we assume that the original frequency of P is 254, then on filing its frequency will be greater than 254. The beats between P and Q will be more than 4. But it is given that the beats are reduced to 2, therefore, 254 is not possible.

Therefore, the required frequency must be 246 Hz.

(This is true, because on filing the frequency may increase to 248, giving 2 beats with Q of frequency 250 Hz)

Example 39. *Two tuning forks A and B sounded together give 8 beats/s. With an air resonance tube closed at one end, the two forks give resonances, when the two air columns are 32 cm and 33 cm respectively. Calculate the frequencies of forks.*

Sol. Let the frequency of the first fork be f_1 and that of second be f_2.

We have,

$$f_1 = \frac{v}{4 \times 32}$$

and
$$f_2 = \frac{v}{4 \times 33}$$

We also see that $f_1 > f_2$

$$\therefore \quad f_1 - f_2 = 8 \quad \text{...(i)}$$

and
$$\frac{f_1}{f_2} = \frac{33}{32} \quad \text{...(ii)}$$

Solving Eqs. (i) and (ii), we get

$$f_1 = 264 \text{ Hz}$$

and
$$f_2 = 256 \text{ Hz}$$

Example 40. *A siren emitting a sound of frequency 1000 Hz moves away from you towards a cliff at a speed of 10 m/s.*

(a) What is the frequency of the sound you hear coming directly from the siren ?

(b) What is the frequency of sound you hear reflected off the cliff? Speed of sound in air is 330 m/s?

Sol. (a) Frequency of sound heard directly.

$$f_1 = f_0 \left(\frac{v}{v + v_s}\right)$$
$$v_s = 10 \text{ m/s}$$
$$\therefore \quad f_1 = \left(\frac{330}{330 + 10}\right) \times 1000$$
$$= \frac{33}{34} \times 1000 \text{ Hz}$$

(b) The frequency of the reflected sound is given by

$$f_2 = f_0 \left(\frac{v}{v - v_s} \right)$$

$$f_2 = \left(\frac{330}{330 - 10} \right) \times 1000$$

$$= \frac{33}{32} \times 1000 \text{ Hz}$$

Example 41. *Wavelengths of two notes in air are* $\left(\frac{90}{175}\right)$ *m and* $\left(\frac{90}{173}\right)$ *m. Each note produces 4 beats/s with a third note of a fixed frequency. Calculate the velocity of sound in air.*

Sol. Given, $\lambda_1 = \frac{90}{175}$ m and $\lambda_2 = \frac{90}{173}$ m

Let, f_1 and f_2 be the corresponding frequencies and v be the velocity of sound in air.

$$v = \lambda_1 f_1 \text{ and } v = \lambda_2 f_2$$

$$\therefore \quad f_1 = \frac{v}{\lambda_1} \text{ and } f_2 = \frac{v}{\lambda_2}$$

Since, $\lambda_2 > \lambda_1 \therefore f_1 > f_2$

Let, f be the frequency of the third note.

$$\therefore \quad f_1 - f = 4 \text{ and } f - f_2 = 4$$

$$\therefore \quad f_1 - f_2 = 8$$

$$\therefore \quad \frac{v}{\lambda_1} - \frac{v}{\lambda_2} = 8$$

$$v\left[\frac{175}{90} - \frac{173}{90}\right] = 8$$

$$\therefore \quad v\left[\frac{2}{90}\right] = 8$$

$$\therefore \quad v = 360 \text{ ms}^{-1}$$

Example 42. *Two wires are fixed on a sonometer. Their tensions are in the ratio 8 : 1, their lengths are in the ratio 36:35, the diameters are in the ratio 4:1 and densities are in the ratio 1:2. Find the frequency of the lower pitch, if the note of the higher pitch has a frequency of 360 Hz.*

Sol. Given, $\frac{T_1}{T_2} = \frac{8}{1}$, $\frac{L_1}{L_2} = \frac{36}{35}$

$$\frac{D_1}{D_2} = \frac{4}{1}, \quad \frac{\rho_1}{\rho_2} = \frac{1}{2}$$

Let μ_1 and μ_2 be the linear mass densities.

$$\mu = \rho S = \rho(\pi r^2)$$

$$\therefore \quad \mu_1 = \pi \times \frac{D_1^2}{4} \times \rho_1$$

$$\text{and} \quad \mu_2 = \pi \times \frac{D_2^2}{4} \times \rho_2$$

$$\therefore \quad \frac{\mu_1}{\mu_2} = \left(\frac{D_1}{D_2}\right)^2 \times \frac{\rho_1}{\rho_2}$$

$$= \left(\frac{4}{1}\right)^2 \times \frac{1}{2} = \frac{8}{1}$$

$$f \propto \frac{1}{L}\sqrt{\frac{T}{\mu}}$$

$$\therefore \quad \frac{f_1}{f_2} = \frac{L_2}{L_1} \times \sqrt{\frac{T_1}{T_2} \times \frac{\mu_2}{\mu_1}}$$

$$= \frac{35}{36}\sqrt{\frac{8}{1} \times \frac{1}{8}} = \frac{35}{36}$$

$$f_2 > f_1$$

We have, $f_2 = 360$

$\therefore \quad f_1 = 350$

Example 43. *A column of air and a tuning fork produces 4 beats/s. When sounding together, the tuning fork gives the lower note. The temperature of air is 15°C. When the temperature falls to 10°C, the two produce 3 beats/s. Find the frequency of the tuning fork.*

Sol. The frequency of the air column is given by $f = \frac{v}{\lambda}$ where, v is the velocity of sound in air and λ is the wavelength.

But $v = \sqrt{\frac{\gamma RT}{M}}$, thus it is dependent of temperature.

or $\quad f \propto \sqrt{T}$

Let the frequency of the tuning fork be f.

Thus, the frequency of the air column at $15°\text{C} = f + 4$

and the frequency of the air column at $10°\text{C} = f + 3$

As the frequency decreases with decrease temperature

$$\Rightarrow \quad \frac{f+4}{f+3} = \sqrt{\frac{288}{283}}$$

$$\Rightarrow \quad \frac{f+4}{f+3} = 1.00879$$

$$\Rightarrow \quad f + 4 = 1.00879 f + 3.02638$$

$$\Rightarrow \quad 8.79 \times 10^{-3} f = 0.97362$$

$$f = 110.76 \text{ Hz}$$

NCERT Selected Questions

Q 1. A string of mass 2.50 kg is under a tension of 200 N. The length of the stretched string is 20.0 m. If the transverse jerk is struck at one end of the string, how long does the disturbance take to reach the other end?

Sol. Mass per unit length of the string,

$$\mu = \frac{M}{l} = \frac{2.50}{20.0}$$
$$= 0.125 \text{ kg m}^{-1}$$

Velocity of transverse wave is given by

$$v = \sqrt{\frac{T}{\mu}} = \sqrt{\frac{200}{0.125}}$$
$$= 40 \text{ ms}^{-1}$$

Let t = time taken by the transverse jerk to reach from one end to the other end of the string.

$$\therefore \quad t = \frac{\text{Length of string}}{\text{Velocity of string}}$$
$$= \frac{l}{v} = \frac{20}{40} = 0.5 \text{ s}$$

Q 2. A stone dropped from the top of a tower of height 300 m high splashes into the water of a pond near the base of the tower. When is the splash heard at the top given that the speed of sound in air is 340 ms^{-1}? ($g = 9.8 \text{ ms}^{-2}$)

Sol. Let t_1 = time taken by the stone to fall down,

t_2 = times taken by the sound to travel from the bottom to the top of tower.

If t = time after which splash is heard at the top, then

$$t = t_1 + t_2 \quad \ldots(\text{i})$$

Now, $$t_1 = \sqrt{\frac{2h}{g}}$$

or $$t_1 = \sqrt{\frac{2 \times 300}{9.8}} = 7.82 \text{ s} \quad \ldots(\text{ii})$$

Further, $$t_2 = \frac{\text{Distance}}{\text{Velocity}}$$
$$= \frac{\text{Height of tower}}{\text{Velocity of sound}} = \frac{h}{v}$$

or $$t_2 = \frac{300}{340} = \frac{15}{17} = 0.88 \text{ s} \quad \ldots(\text{iii})$$

$\therefore$ From Eqs. (i), (ii) and (iii), we get

$$t = 7.82 + 0.88 = 8.7 \text{s}$$

Q 3. A steel wire has a length of 12.0 m and a mass of 2.10 kg. What should be the tension in the wire so that speed of a transverse wave on the wire equals, the speed of sound in dry air at $20°\text{C} = 343 \text{ ms}^{-1}$?

Sol. If μ = mass per unit length of the wire, then

$$\mu = \frac{M}{l} = \frac{2.10}{12.0} = 0.175 \text{ kgm}^{-1}$$

Speed of transverse wave in the wire is

$$v = \sqrt{\frac{T}{\mu}} \quad \text{or} \quad v^2 = \frac{T}{\mu}$$

or $$T = \mu v^2 = 0.175 \times (343)^2$$
$$\approx 2.06 \times 10^4 \text{ N}$$

Q 4. Use the formula $v = \sqrt{\frac{\gamma p}{\rho}}$ to explain, why the speed of sound in air

(a) is independent of pressure.
(b) increases with temperature.
(c) increases with humidity.

Sol. (a) $\frac{p}{\rho}$ is also equal to $\frac{RT}{M}$

$$\therefore \quad v = \sqrt{\frac{\gamma p}{\rho}} = \sqrt{\frac{\mu RT}{M}}$$

For given temperature, $\frac{p}{\rho}$ = constant

Hence, v is independent of p.

(b) Effect of temperature

$$v \propto \sqrt{T}$$

i.e. velocity of sound in a gas is directly proportional to the square root of its temperature, hence we conclude that the velocity of sound in air increases with increase in temperature.

(c) Effect of humidity

Let ρ_m = density of moist air.
ρ_d = density of dry air.
v_m = velocity of sound in moist air.
v_d = velocity of sound in dry air.

From formula $v = \sqrt{\frac{\gamma p}{\rho}}$, we get

$$v_m = \sqrt{\frac{\gamma p}{\rho_m}} \quad \ldots(\text{i})$$

and $$v_d = \sqrt{\frac{\gamma p}{\rho_d}} \quad \ldots(\text{ii})$$

$$\therefore \quad \frac{v_m}{v_d} = \sqrt{\frac{\rho_d}{\rho_m}} \quad \ldots(\text{iii})$$

Density of water vapours is less than the density of dry air as the molecular mass of water is less than that of N_2 (28) and O_2(32), so $\rho_m < \rho_d$

or $$\frac{\rho_d}{\rho_m} > 1 \quad \ldots(\text{iv})$$

$\therefore$ From Eqs. (iii) and (iv), we get

$$\frac{v_m}{v_d} > 1 \quad \text{or} \quad v_m > v_d$$

i.e. velocity of sound in air increases with humidity, i.e. velocity of sound in moist air is greater than the velocity of sound in dry air. That is why sound travels faster on a rainy day than on a dry day.

***Q* 5.** You have learnt that a travelling wave in one dimension is represented by a function $y = f(x, t)$, where x and t must appear in the combination $x - vt, x + vt$, i.e. $y = F(x \pm vt)$. Is the converse true? Examine, if the following functions for y can possibly represent a travelling wave

(a) $(x - vt)^2$ (b) $\log\left[\frac{x + vt}{x_0}\right]$

(c) $\frac{1}{(x + vt)}$

Sol. No, the converse is not true.

The basic requirement for a wave function to represent a travelling wave is that for all values of x and t, the wave function must have a finite value.

Out of the given functions for y, none satisfies this condition, so these functions do not represent a travelling wave.

***Q* 6.** A bat emits ultrasonic sound of frequency 1000 kHz in air. If the sound meets a water surface, what is the wavelength of (a) the reflected sound (b) the transmitted sound? Speed of sound in air is 340 ms^{-1} and in water is 1486 ms^{-1}.

Sol. (a) **For reflected sound** After reflection, the ultrasonic sound again travels in air.

$$\therefore \quad \lambda_a = \frac{v_a}{f} = \frac{340}{10^6}$$

$$= 3.4 \times 10^{-4} \text{m}$$

(b) **For transmitted sound** The transmitted sound travels through water. If λ_t be the wavelength of the transmitted sound, then

$$\lambda_t = \frac{\text{Velocity of sound in water}}{\text{Frequency}}$$

$$= \frac{v_w}{f} = \frac{1486}{10^6}$$

$$= 14.86 \times 10^{-4} \text{ m}$$

$$\approx 1.49 \times 10^{-3} \text{ m}$$

Frequency will not change in transmission.

***Q* 7.** A hospital uses an ultrasonic scanner to locate tumours in a tissue. What is the wavelength of sound in the tissue in which the speed of sound is 1.7 kms^{-1}? The operating frequency of the scanner is 4.2 MHz.

Sol. Here, speed of sound

$$v = 1.7 \text{ kms}^{-1}$$

$$= 1.7 \times 10^3 \text{ ms}^{-1}$$

Frequency of the scanner, $f = 4.2 \text{ MHz} = 4.2 \times 10^6$ Hz

$$\lambda = \frac{v}{f}$$

or $$\lambda = \frac{v}{f} = \frac{1.7 \times 10^3}{4.2 \times 10^6} = 4.05 \times 10^{-4} \text{ m}$$

***Q* 8.** A transverse harmonic wave on a string is described by

$$y(x, t) = 3.0 \sin\left(36t + 0.018x + \frac{\pi}{4}\right)$$

where, x and y are in cm and t in second. The positive direction of x is from left to right.

(a) Is this a travelling wave or a stationary wave? If it is travelling, what are the speed and direction of its propagation?

(b) What are its amplitude and frequency?

(c) What is the initial phase at the origin?

(d) What is the least distance between two successive crests in the wave?

Sol. This is a travelling wave

(a) Speed = (Coefficient of t)/(Coefficient of x)

$$v = \left(\frac{36}{0.018}\right) = 2000 \text{ cm/s} = 20 \text{ m/s}$$

(b) Direction of wave velocity is negative x.

$$A = 3 \text{cm}, \; f = \left(\frac{\omega}{2\pi}\right) = \left(\frac{36}{2\pi}\right) \approx 5.73 \text{ Hz}$$

(c) $\phi_i = \frac{\pi}{4}$

(d) Least distance between two crests,

$$\lambda = \frac{v}{f} = \frac{2000}{5.73}$$

$$= 348.9 \text{ cm}$$

***Q* 9.** For the travelling harmonic wave

$$y(x, t) = 2.0 \cos 2\pi\,(10t - 0.0080x + 0.35)$$

where, x and y are in cm and t in second. Calculate the phase difference between oscillatory motion of two points separated by a distance of

(a) 4 m (b) 0.5 m

(c) $\frac{\lambda}{2}$ (d) $\frac{3\lambda}{4}$

Sol. Given equation of a travelling harmonic wave is

$$y(x, t) = 2.0 \cos 2\pi\,(10t - 0.0080x + 0.35) \quad \text{...(i)}$$

The standard equation of a travelling harmonic wave is

$$y(x, t) = A \cos\left[\frac{2\pi}{T} t \pm \frac{2\pi}{\lambda} x \pm \phi_0\right] \quad \text{...(ii)}$$

Comparing Eqs. (i) and (ii), we get

$$\frac{2\pi}{\lambda} = 2\pi \times 0.0080 \text{ cm}^{-1} \quad \text{...(iii)}$$

We know that phase difference $= \frac{2\pi}{\lambda} \times$ path difference ...(iv)

(a) When path difference $= 4 \text{ m} = 400 \text{ cm}$, then from Eq. (iv) phase difference $= \frac{2\pi}{\lambda} \times 400$

$$= 2\pi \times 0.0080 \times 40$$
$$= 6.4\,\pi \text{ rad}$$

(b) When path difference $= 0.5 \text{ m} = 50 \text{ cm}$, then phase difference

$$= 2\pi \times 0.0080 \times 50 = 0.8\,\pi \text{ rad}$$

(c) When path difference $= \frac{\lambda}{2}$, then

$$\text{Phase difference} = \frac{2\pi}{\lambda} \times \frac{\lambda}{2} = \pi \text{ rad}$$

(d) When path difference $= \frac{3\lambda}{4}$, then

$$\text{Phase difference} = \frac{2\pi}{\lambda} \times \frac{3\lambda}{4} = \frac{3\pi}{2} \text{ rad}$$

Q 10. The transverse displacement of a string (clamped at its both ends) is given by $y(x, t) = 0.06 \sin\left(\frac{2\pi x}{3}\right) \cos(120\pi t)$, where x and y are in metres and t in second. The length of the string is 1.5 m and its mass is 3.0×10^{-2} kg.

Answer the following

(a) Does the function represent a travelling or a stationary wave?

(b) Interpret the given wave as a superposition of two waves travelling in opposite directions. What are the wavelength, frequency and speed of each wave?

(c) Determine the tension in the string.

Sol. (a) The given function represents a stationary wave.

(b) When a wave

$$y_1 = A \sin \frac{2\pi}{\lambda}(vt - x)$$

Travelling along positive direction of x-axis is superimposed with the reflected wave

$$y_2 = -A \sin \frac{2\pi}{\lambda}(vt + x)$$

Travelling in the opposite direction, a stationary wave

$$y = y_1 + y_2$$
$$= -2A \sin\left(\frac{2\pi}{\lambda} x\right) \cos\left(\frac{2\pi}{\lambda} vt\right) \quad \text{...(i)}$$

is formed. Comparing with the given equation

$$\frac{2\pi}{\lambda} = \frac{2\pi}{3} \quad \text{or} \quad \lambda = 3 \text{ m}$$

$$\Rightarrow \quad \frac{2\pi}{\lambda} v = 120\pi$$

$$\Rightarrow \quad v = 60\,\lambda = 60 \times 3 = 180 \text{ ms}^{-1}$$

$$\therefore \quad \text{Frequency } f = \frac{v}{\lambda}$$
$$= \frac{180}{3}$$
$$= 60 \text{ Hz}$$

(c) Velocity of transverse wave is given by

$$v = \sqrt{\frac{T}{\mu}}$$

or

$$v^2 = \frac{T}{\mu}$$

$$\therefore \quad T = v^2 \times \mu \quad \text{...(ii)}$$

Here, μ = mass per unit length $= \frac{\text{mass}}{\text{length}}$

$$= \frac{3 \times 10^{-2} \text{ kg}}{1.5 \text{ m}}$$
$$= 2 \times 10^{-2} \text{ kgm}^{-1} \quad \text{...(iii)}$$

$v = 180 \text{ ms}^{-1}$ of each wave

$$\therefore \quad T = (180)^2 \times 2 \times 10^{-2}$$
$$= 648 \text{ N}$$

Q 11. (i) For the wave on a string described in above exercise, do all the points on the string oscillate with the same (a) frequency, (b) phase, (c) amplitude? Explain your answers.

(ii) What is the amplitude of a point 0.375 m away from one end?

Sol. (i) All the points except nodes on the string have the same phase and frequency of oscillation but the amplitude is not same. This is because of the fact that the equation

$y(x, t) = 0.06 \sin\left(\frac{2\pi}{3} x\right) \cos(120\pi t)$ represents a stationary wave in which different points have different amplitudes including zero at node to a certain maximum amplitude at antinode. But time period or frequency of oscillation of different particles have same value.

(ii) The given equation is

$$y(x, t) = 0.06 \sin\left(\frac{2\pi}{3} x\right) \cos(120\pi t)$$

The amplitude of this equation is

$$A = 0.06 \sin\left(\frac{2\pi}{3} x\right)$$

At $x = 0.375$ m

$$A = 0.06 \sin\left(\frac{2\pi}{3} \times 0.375\right)$$
$$= 0.06 \times \sin(0.250\,\pi)$$
$$= 0.042 \text{ m}$$

Q 12. Given below are some functions of x and t to represent the displacement (transverse or longitudinal) of an elastic wave. State which of these represent (i) a travelling wave, (ii) a stationary wave or (iii) None at all

(a) $y = 2\cos(3x)\sin(10t)$

(b) $y = 2\sqrt{x - vt}$

(c) $y = 3\sin(5x - 0.5t) + 4\cos(5x - 0.5t)$

(d) $y = \cos x \sin t + \cos 2x \sin 2t$

Sol. (a) This equation has two harmonic functions of x and t, so it represents a stationary wave.

(b) This function does not represent any wave as it contains no harmonic function.

(c) It represents progressive/travelling harmonic wave as the arguments of cosine and sine functions are same.

(d) This equation is the sum of two functions $\cos x \sin t$ and $\cos 2x \sin 2t$ each representing a stationary wave.

Therefore, it represents superposition of two stationary waves.

Q 13. A wire stretched between two rigid supports vibrates in its fundamental mode with a frequency of 45 Hz. The mass of the wire is 3.5×10^{-2} kg and its linear density is 4.0×10^{-2} kgm^{-1}. What is (a) the speed of a transverse wave in the string and (b) the tension in the string?

Sol. Here, frequency, $f = 45$ Hz

Mass of wire, $M = 3.5 \times 10^{-2}$ kg

μ = mass per unit length = linear density

$= 4.0 \times 10^{-2}$ kgm^{-1}

If L be the length of the wire or string.

Then, from formula, $\mu = \frac{M}{L}$, we get

$$L = \frac{M}{\mu} = \frac{3.5 \times 10^{-2}}{4 \times 10^{-2}} = 0.875 \text{ m}$$

(a) Frequency of fundamental mode of the vibrating string is given by

$$f = \frac{v}{2L}$$

or

$$v = f \times 2L = 45 \times 2 \times 0.875 = 78.75 \text{ ms}^{-1} \approx 79 \text{ ms}^{-1}$$

where, v = speed of transverse wave.

(b) Also, we know that $v = \sqrt{\frac{T}{\mu}}$ = speed of transverse wave.

$$\therefore \quad T = v^2 \times \mu = (78.75)^2 \times 4 \times 10^{-2} \approx 248.06 \text{ N}$$

Q 14. A steel rod 100 cm long is clamped at its middle. The fundamental frequency of longitudinal vibrations of the rod are given to be 2.53 kHz. What is the speed of sound in steel?

Sol. A rod clamped in the middle has antinodes (A) at its ends and node (N) at the point of clamping. In fundamental mode, thus the length of the rod is

$$l = \frac{\lambda}{2} \quad \text{or} \quad \lambda = 2l$$

where l = length or rod

and λ = wavelength of the wave

Here, $l = 100$ cm

$f = 2.53 \text{kHz} = 2.53 \times 10^3$ Hz

$\therefore \quad \lambda = 2 \times 100 = 200$ cm

If v be the speed of sound in steel, then

$$v = f\lambda = 2.53 \times 10^3 \times 200 = 506 \times 10^3 \text{ cms}^{-1}$$

$$\therefore \quad v = 5.06 \text{ kms}^{-1}$$

Q 15. A pipe 20 cm long is closed at one end. Which harmonic mode of the pipe is resonantly excited by a 430 Hz source? Will the same source be in resonance with the pipe, if both ends are open? (speed of sound in air is 340 ms^{-1}).

Sol. Frequency of nth mode of vibration of a closed pipe is given by

$$f_n = (2n-1)\frac{v}{4l}$$

or

$$430 = (2n-1)\frac{340}{4 \times 0.20}$$

or

$$n = 1.01$$

i.e. the organ pipe is in the first harmonic or fundamental mode of vibration.

In case of an open pipe, frequency of nth mode of vibration is given by

$$f_n' = n\frac{v}{2l}$$

where, length l in fundamental mode $= \frac{\lambda}{2}$

or $\lambda = 2l$

$$\Rightarrow \quad 430 = \frac{n \times 340}{2 \times 0.2}$$

$$\Rightarrow \quad n = \frac{430 \times 0.4}{340} = \frac{172}{340} \approx 0.5$$

As, n has to be an integer so, $n = 0.5$ is not valid. Hence, the same source cannot be in resonance with the open pipe.

Q 16. Two sitar strings A and B playing the note Ga are slightly out of tune and produce beats of frequency 6 Hz. The tension in the string A is slightly reduced and the beat frequency is found to reduce to 3 Hz. If the original frequency of A is 324 Hz, what is the frequency of B?

Sol. We know that $v \propto \sqrt{T}$, where f = frequency, T = tension.

The decrease in the tension of a string decreases its frequency. So, let us assume that original frequency f_A of A is more than the frequency f_B of string B.

Thus, $f_A - f_B = \pm 6$ Hz (given)

and $f_A = 324$ Hz

$\therefore$ $324 - f_B = \pm 6$

or $f_B = 324 \pm 6$

$= 318$ Hz or 330 Hz.

On reducing tension of string A, $\Delta f = 3$ Hz.

If $f_B = 330$ Hz and on decreasing tension in A, f_A will be reduced, i.e. number of beats will increase, but this is not so because number of beats becomes 3.

$\therefore$ f_B must be 318 Hz because on reducing the tension in string A, its frequency may be reduced to 321 Hz, thus giving 3 beats with $f_B = 318$ Hz.

Q 17. A train standing at the outer signal of a railway station blows a whistle of frequency 400 Hz in still air.

What is the frequency of the whistle for a platform observed when the train approaches the platform with a speed of $10\ \text{ms}^{-1}$.

The speed of sound in still air can be taken as $340\ \text{ms}^{-1}$.

Sol. When the train approaches the platform, the apparent frequency as heard by the observed on the platform will be

$$f' = \frac{v}{v - v_s} f$$

$$= \frac{340}{340 - 10} \times 400$$

$$= \frac{340}{330} \times 400 = 412.1 \text{ Hz}$$

Objective Problems

[Level 1]

Basic Questions of Wave Motion

1. What does not change when sound enters from one medium to another?

(a) Wavelength (b) Speed
(c) Frequency (d) None of these

2. The displacement y of a particle on a straight line is given by $y = f(x, t)$, as a function of time. Which of the following functions does not represent wave motion?

(a) $y = A \sin (kx - \omega t)$
(b) $y = A \sin^2 (kx - \omega t)$
(c) $y = A \sin (k^2x^2 - \omega^2 t^2)$
(d) $y = A \sin \left(kx + \omega t + \frac{\pi}{10}\right)$

3. In the equation $y = A \sin (kx - \omega t + \phi_0)$, the term phase is defined as

(a) ϕ_0 (b) $\phi_0 - \omega t$
(c) $kx + \phi_0$ (d) $kx - \omega t - \phi_0$

4. 'SONAR' emits which of the following waves?

(a) Radio waves (b) Ultrasonic waves
(c) Light waves (d) Magnetic waves

5. In a wave motion $y = a \sin (kx - \omega t)$, y can represent

(a) electric field (b) magnetic field
(c) displacement (d) pressure

6. Resonance is a special case of

(a) undamped vibration
(b) damped vibration
(c) force vibration
(d) natural vibration

7. The equation of wave is given by $y = A \sin \omega \left[\frac{x}{v} - k\right]$ where, ω is the angular velocity and v is the linear velocity. The dimensional formula of k is

(a) [LT] (b) [T]
(c) $[T^{-1}]$ (d) $[T^2]$

8. Stationary waves are formed when

(a) two waves of equal amplitude and equal frequency travel along the same path in opposite directions
(b) two waves of equal wavelength and equal amplitude travel along the same path with equal speeds in opposite directions
(c) two waves of equal wavelength and equal phase travel along the same path with equal speed
(d) two waves of equal amplitude and equal speed travel along the same path in opposite directions

9. Doppler's effect in sound is due to

(a) motion of source
(b) motion of observer
(c) relative motion of source and observer
(d) None of the above

Plane Progressive Harmonic Waves

10. The distance between two points differing in phase by 60° on a wave having wave velocity 360 ms^{-1} and frequency 500 Hz is

(a) 0.36 m (b) 0.18 m (c) 0.48 m (d) 0.12 m

11. The equations of displacement of two waves are given as

$$y_1 = 10 \sin \left(3\pi t + \frac{\pi}{3}\right)$$

and $$y_2 = 5(\sin 5\pi t + \sqrt{3} \cos 5\pi t)$$

Then, what is the ratio of their amplitudes?

(a) 1 : 2 (b) 2 : 1
(c) 1 : 1 (d) None of these

12. A transverse wave is described by the equation $Y = Y_0 \sin 2\pi \left(ft - \frac{X}{\lambda}\right)$. The maximum particle velocity is equal to four times the wave velocity, if

(a) $\lambda = \frac{\pi Y_0}{4}$ (b) $\lambda = \frac{\pi Y_0}{2}$
(c) $\lambda = \pi Y_0$ (d) $\lambda = 2\pi Y_0$

13. A wave equation which gives the displacement along y-direction is given by $y = 10^{-4} \sin (60t + x)$, where x and y are in metres and t is time in second. This represents a wave

(a) travelling with a velocity of 300 ms^{-1} in the negative x-direction
(b) of wavelength π metre
(c) of frequency $\frac{30}{\pi}$ Hz
(d) of amplitude 10^4 m travelling along the positive x-direction

14. The equation of a transverse wave on a stretched string is given by $y = 0.05 \sin 2\pi \left(\frac{t}{0.002} - \frac{x}{0.1}\right)$, where x and y are expressed in metres and t in second. The speed of the wave is

(a) 100 m/s
(b) 50 m/s
(c) 200 m/s
(d) 400 m/s

15. The phase difference between two points separated by 0.8 m in a wave of frequency 120 Hz is 0.5π. The velocity of wave will be

(a) 720 m/s (b) 384 m/s
(c) 256 m/s (d) 144 m/s

16. If the equation of transverse wave is

$$y = 5 \sin 2\pi \left(\frac{t}{0.04} - \frac{x}{40}\right)$$

where, distance is in cm and time in second, then the wavelength of wave will be

(a) 20 cm (b) 40 cm
(c) 60 cm (d) None of these

17. The frequency of a tuning fork with an amplitude $A = 1\,\text{cm}$ is 250 Hz. The maximum velocity of any particle in air is equal to

(a) 10π m/s (b) 5π m/s
(c) 2π (d) None of these

18. A wave of amplitude $A = 0.2\,\text{m}$, velocity $v = 360\,\text{m/s}$ and wavelength 60 m is travelling along positive x-axis, then the correct expression for the wave is

(a) $y = 0.2 \sin 2\pi \left(6t + \frac{x}{60}\right)$

(b) $y = 0.2 \sin \pi \left(6t + \frac{x}{60}\right)$

(c) $y = 0.2 \sin 2\pi \left(6t - \frac{x}{60}\right)$

(d) $y = 0.2 \sin \pi \left(6t - \frac{x}{60}\right)$

19. A wave of frequency 400 Hz has a wave velocity of 300 m/s. The phase difference between two displacements at a certain point at times $t = 10^{-3}$ s apart is

(a) 72° (b) 102°
(c) 180° (d) 144°

20. What is the phase difference, at a given instant of time, between two particles 25 m apart, when the wave $y(x, t) = 0.03 \sin \pi (2t - 0.01x)$ travels in a medium?

(a) $\frac{\pi}{8}$ (b) $\frac{\pi}{4}$
(c) $\frac{\pi}{2}$ (d) π

21. A progressive wave in a medium is represented by the equation

$$y = 0.1 \sin \left(10\pi t - \frac{5}{11}\pi x\right)$$

where, y and x are in cm and t in second. The maximum speed of a particle of the medium due to the wave is

(a) 1 cm s^{-1}
(b) 10 cm s^{-1}
(c) π cm s^{-1}
(d) 10π cm s^{-1}

22. A wave is represented by the equation

$$y = A \sin \left(10\pi x + 15\pi t + \frac{\pi}{3}\right)$$

where, x is in metre and t is in second. The expression represents

(a) a wave travelling in positive x-direction with a velocity 1.5 m/s
(b) a wave travelling in negative x-direction with a velocity 1.5 m/s
(c) a wave travelling in the negative x-direction having a wavelength 0.4 m
(d) a wave travelling in positive x-direction of wavelength 0.1 m

23. A wave $y(x, t) = 0.03 \sin \pi (2t - 0.01x)$ travels in a medium. Here, x is in metre. The instantaneous phase difference between the two points separated by 25 cm is

(a) $\frac{\pi}{800}$ (b) $\frac{\pi}{400}$
(c) $\frac{\pi}{200}$ (d) $\frac{\pi}{100}$

24. The ratio of the maximum velocity of a particle to the velocity of wave is

(a) 1 (b) ωA
(c) kA (d) No unique value exists

25. Equation of a progressive wave is given by

$$y = 0.2 \cos \pi \left(0.04\, t + 0.02x - \frac{\pi}{6}\right)$$

The distance is expressed in cm and time in second. What will be the minimum distance between two particles having the phase difference of $\frac{\pi}{2}$?

(a) 4 cm (b) 8 cm
(c) 25 cm (d) 12.5 cm

26. The equation of a progressive wave is $y = 8 \sin \left[\pi \left(\frac{t}{10} - \frac{x}{4}\right) + \frac{\pi}{3}\right]$. The wavelength of the wave is

(a) 8 m (b) 4 m
(c) 2 m (d) 10 m

27. A transverse sinusoidal wave of amplitude a, wavelength λ and frequency n is travelling on a stretched string. The maximum speed of particle is $\frac{1}{10}$th of the speed of propagation of the wave. If $a = 10^{-3}$ m and $v = 10\,\text{ms}^{-1}$, then λ and n are given by

(a) $\lambda = 2\pi \times 10^{-2}$ m
(b) $\lambda = 10^{-3}$ m
(c) $n = \frac{10^3}{2\pi}$ Hz
(d) $n = 10^4$ Hz

28. Equation of a progressive wave is given by $y = a \sin \pi \left[\frac{t}{2} - \frac{x}{4}\right]$, where t is in second and x in metre. The distance through which the wave travels in 8 s is (in metre)

(a) 8 (b) 16 (c) 2 (d) 4

29. In a travelling wave

$$y = 0.1 \sin \pi \left(x - 330\,t + \frac{2}{3}\right) \text{ (SI units)}$$

The phase difference between $x_1 = 3$ m and $x_2 = 3.5$ m is

(a) $\frac{\pi}{2}$ (b) π (c) $\frac{3\pi}{2}$ (d) 2π

Stationary Wave, Organ Pipes and Stretched Wire

30. A tube of length 1.05 m is closed at one end. If the velocity of sound in air be 336 m/s, then the fundamental and the next higher overtone in Hz are

(a) 80, 160 (b) 80, 240
(c) 160, 320 (d) 160, 480

31. A string of length 1 m has the mass per unit length $0.1\,\text{g cm}^{-1}$. What would be the fundamental frequency of vibration of this string under tension of 400 N?

(a) 400 Hz (b) 100 Hz
(c) 50 Hz (d) 200 Hz

32. A wave represented by the equation $y = a \cos (kx - \omega t)$ is superposed with another wave to form a stationary wave such that the point $x = 0$ is a node. The equation of the other wave is

(a) $a \sin (kx + \omega t)$ (b) $-a \cos (kx + \omega t)$
(c) $-a \cos (kx - \omega t)$ (d) $-a \sin (kx - \omega t)$

33. Two vibrating strings of same material but lengths l and $2l$ have radii $2r$ and r respectively. They are stretched under the same tension. Both the strings vibrate in their fundamental modes, the one of the length l with frequency n_1 and the other with frequency n_2. The ratio of $\frac{n_1}{n_2}$ is given by

(a) 2 (b) 4
(c) 8 (d) 1

34. A string is tied on a sonometer. Second end is hanging downward through a pulley with tension T. The velocity of the transverse wave produced is proportional to

(a) $T^{-1/2}$ (b) $T^{1/2}$
(c) T (d) T^{-1}

35. Velocity of sound in open ended tube is 330 m/s, the frequency of tuning fork is 1.1 kHz and the length of tube is 30 cm. In which harmonic will it oscillate?

(a) 2nd (b) 3rd (c) 4th (d) 5th

36. Fundamental frequency of a sonometer wire is n, if the tension is made 3 times and length and diameter are also increased 3 times, what is the new frequency?

(a) $\frac{n}{3\sqrt{3}}$ (b) $3n$
(c) $\sqrt{3}\,n$ (d) $\frac{n}{\sqrt{3}}$

37. An organ pipe P_1 is closed at one end and vibrating in its first overtone and another pipe P_2 opened at both ends vibrating in its third overtone are in resonance with a given tuning fork. Then, the ratio of lengths P_1 and P_2 is

(a) $\frac{1}{3}$ (b) $\frac{2}{3}$
(c) $\frac{8}{3}$ (d) $\frac{3}{8}$

38. A standing wave is represented by

$$y = a \sin (100t) \cos (0.01x)$$

where, t is in second and x is in metre. Then, the velocity of the constituent wave is

(a) 10^2 m/s (b) 10 m/s (c) 1 m/s (d) 10^4 m/s

39. A pipe closed at one end produces a fundamental note of 412 Hz. It is cut into two pieces of equal length the fundamental notes produced by the two pieces are

(a) 824 Hz, 1648 Hz (b) 412 Hz, 824 Hz
(c) 206 Hz, 412 Hz (d) 206 Hz, 824 Hz

40. Four wires of identical length, diameters and of the same material are stretched on a sonometer wire. If the ratio of their tensions is 1 : 4 : 9 : 16, then the ratio of their fundamental frequencies are

(a) 16 : 9 : 4 : 1 (b) 4 : 3 : 2 : 1
(c) 1 : 4 : 2 : 16 (d) 1 : 2 : 3 : 4

41. In a resonance column experiment the first resonance is obtained, when the level of the water in tube is 20 cm from the open end. Resonance will also be obtained when the water level is at a distance of

(a) 40 cm from the open end (b) 60 cm from the open end
(c) 80 cm from the open end (d) 90 cm from the open end

42. The 3rd overtone of a closed organ pipe is equal to the 2nd harmonic of an open organ pipe. Then, the ratio of their lengths is equal to

(a) $\frac{7}{4}$ (b) $\frac{3}{5}$
(c) $\frac{3}{2}$ (d) $\frac{7}{6}$

43. The fundamental frequency of a vibrating organ pipe is 100 Hz. Select the wrong alternative.

(a) The first overtone is 200 Hz
(b) The first overtone may be 400 Hz
(c) The first overtone may be 300 Hz
(d) None of the above

44. When two waves meet at a point, their equations are $y_1 = 2a\cos\omega t$ and $y_2 = 3a\cos\left(\omega t + \frac{\pi}{2}\right)$. The amplitude of the resultant wave is

(a) a (b) $5a$
(c) $3.6a$ (d) $2.23a$

45. In a resonance column, two successive resonances are obtained at depths 30 cm and 50 cm respectively. The next resonance will be obtained at a depth

(a) 60 cm (b) 80 cm
(c) 70 cm (d) 90 cm

46. In the equation for a stationary wave given by $y = 5\cos\frac{\pi x}{25}\sin 100\pi t$. Here, x is in cm and t in second. A node will not occur at distance x is equal to

(a) 25 cm (b) 62.5 cm
(c) 12.5 cm (d) 37.5 cm

47. For a string clamped at both ends, which of the following wave equations is valid for a stationary wave set up in it? (Origin is at one end of string)

(a) $y = A\sin kx\sin\omega t$ (b) $y = A\cos kx\sin\omega t$
(c) $y = A\cos kx\cos\omega t$ (d) All of these

48. A tube closed at one end containing air produces, when excited, the fundamental note of frequency 512 kHz. If the tube is opened at both ends, the fundamental frequency, that can be excited, is (in Hz)

(a) 1024 (b) 512
(c) 256 (d) 128

49. Velocity of sound in air is 320 m/s. A pipe closed at one end has length of 1 m. Neglecting end corrections, the air column in the pipe can resonate for sound of frequency

(a) 80 Hz (b) 240 Hz
(c) 400 Hz (d) All of these

50. The vibrations of a string of length 60 cm fixed at both ends are represented by the equation $y = 4\sin\left(\frac{\pi x}{15}\right)\cos(96\pi t)$, where x and y are in cm and t in second. The maximum displacement at $x = 5$ cm is

(a) $2\sqrt{3}$ cm (b) $3\sqrt{2}$ cm
(c) $\sqrt{2}$ cm (d) $\sqrt{3}$ cm

51. If you set up the seventh harmonic on a string fixed at both ends, how many nodes and antinodes are set up in it?

(a) 8, 7 (b) 7, 7
(c) 8, 9 (d) 9, 8

52. A resonating column has resonant frequencies as 100 Hz, 300 Hz, 500 Hz. Then it may

(a) an open pipe
(b) a pipe closed at both ends
(c) pipe closed at one end
(d) Data insufficient

53. A tuning fork of frequency 340 Hz is sounded above a cylindrical tube 1 m long. Water is slowly poured into the tube. If the speed of sound is 340 ms^{-1}, at what levels of water in the tube will the sound of the tuning fork be appreciably intensified?

(a) 25 cm, 75 cm (b) 20 cm, 80 cm
(c) 23 cm, 67 cm (d) 40 cm, 80 cm

54. A standing wave is produced on a string fixed at one end and free at the other. The length of the string

(a) must be an odd integral multiple of $\frac{\lambda}{4}$
(b) must be an integral multiple of $\frac{\lambda}{2}$
(c) must be an integral multiple of λ
(d) may be an integral multiple of $\frac{\lambda}{2}$

55. Two identical strings are stretched at tensions T_A and T_B. A tuning fork is used to set them in vibration. A vibrates in its fundamental mode and B in its second harmonic mode then

(a) $T_A = 2T_B$ (b) $T_A = 4T_B$
(c) $2T_A = T_B$ (d) $4T_A = 4T_B$

56. First overtone frequency of a closed organ pipe is equal to the first overtone frequency of an open organ pipe. Further nth harmonic of closed organ pipe is also equal to the m th harmonic of open pipe, where n and m are

(a) 5, 4 (b) 7, 5 (c) 9, 6 (d) 7, 3

57. The equation of a stationary wave is $y = 0.8\cos\left(\frac{\pi x}{20}\right)\sin 200\pi t$, where x is in cm and t is in second. The separation between consecutive nodes will be

(a) 20 cm (b) 10 cm
(c) 40 cm (d) 30 cm

58. Stationary waves of frequency 300 Hz are formed in a medium in which the velocity of sound is 200 m/s. The distance between a node and the neighbouring antinode is

(a) 1 m (b) 2 m (c) 3 m (d) 4 m

59. When a stationary wave is formed, then its frequency is

(a) same as that of the individual waves
(b) twice as that of the individual waves
(c) half as that of the individual waves
(d) None of the above

60. Consider the three waves z_1, z_2 and z_3 as $z_1 = A\sin(kx - \omega t)$, $z_2 = A\sin(kx + \omega t)$ and $z_3 = A\sin(ky - \omega t)$. Which of the following represents a standing wave?

(a) $z_1 + z_2$ (b) $z_2 + z_3$
(c) $z_3 + z_1$ (d) $z_1 + z_2 + z_3$

61. Two travelling waves $y_1 = A\sin[k(x - ct)]$ and $y_2 = A\sin[k(x + ct)]$ are superimposed on string. The distance between adjacent nodes is

(a) $\frac{c}{\pi}$ (b) $\frac{c}{2\pi}$
(c) $\frac{\pi}{2k}$ (d) $\frac{\pi}{k}$

62. A string of length l fixed at both the ends is vibrating in two segments. The wavelength of the corresponding wave is

(a) $\frac{l}{4}$ (b) $\frac{l}{2}$
(c) l (d) $2l$

63. A 1cm long string vibrates with fundamental frequency of 256 Hz. If the length is reduced to $\frac{1}{4}$ cm keeping the tension unaltered, the new fundamental frequency will be (in Hz)

(a) 64 (b) 256
(c) 512 (d) 1024

64. If you set up the ninth harmonic on a string fixed at both ends, its frequency compared to the seventh harmonic is

(a) higher (b) lower
(c) equal (d) None of these

65. In order to double, the frequency of the fundamental note emitted by a stretched string, the length is reduced to $\frac{3}{4}$th of the original length and the tension is changed. The factor by which the tension is to be changed, is

(a) $\frac{3}{8}$ (b) $\frac{2}{3}$
(c) $\frac{8}{9}$ (d) $\frac{9}{4}$

66. A wave disturbance in a medium is described by $y(x, t) = 0.02\cos\left(50\pi t + \frac{\pi}{2}\right)\cos(10\pi x)$, where x and y are in metres and t in second

(a) a node occurs at $x = 1.5$ m
(b) an antinode occurs at $x = 0.3$ m
(c) the wavelength of the constituent wave is 0.2 m
(d) the speed of the constituent wave is 5.0 m/s

67. A cylindrical resonance tube, open at both ends, has a fundamental frequency f in air. If half of the length is dipped vertically in water. The fundamental of the air column will be

(a) $\frac{3f}{2}$
(b) $2f$
(c) f
(d) $\frac{f}{2}$

68. The length of a sonometer wire AB is 110 cm. Where, should the two bridges be placed from A to divide the wire in 3 segments whose fundamental frequencies are in the ratio of 1 : 2 : 3.

(a) 30 cm and 90 cm (b) 40 cm and 80 cm
(c) 60 cm and 90 cm (d) 30 cm and 60 cm

69. Standing waves are produced in 10 m long stretched string. If the string vibrates in 5 segments and wave velocity is 20 m/s, then its frequency will be

(a) 5 Hz (b) 2 Hz (c) 10 Hz (d) 15 Hz

70. Two transverse sinusoidal waves travel in opposite directions along a string. The speed of transverse waves in the string is 0.5 cm/s. Each has an amplitude of 3.0 cm and wavelength of 6.0 cm. The equation for the resultant wave is

(a) $y = 6\sin\frac{\pi t}{6}\cos\frac{\pi x}{3}$
(b) $y = 6\sin\frac{\pi x}{3}\cos\frac{\pi t}{6}$
(c) Both (a) and (b) are correct
(d) Both (a) and (b) are incorrect

71. A wire of length l having tension T and radius r vibrates with natural frequency f. Another wire of same metal with length $2l$ having tension $2T$ and radius $2r$ will vibrate with natural frequency

(a) f (b) $2f$
(c) $2\sqrt{2}\,f$ (d) $\frac{f}{2\sqrt{2}}$

72. When the string of a sonometer of length L between the bridges vibrates in the second overtone, the amplitude of vibration is maximum at

(a) $\left(\frac{L}{2}\right)$ (b) $\left(\frac{L}{4}\right)$ and $\left(\frac{3L}{4}\right)$
(c) $\left(\frac{L}{6}\right)$ and $\left(\frac{3L}{6}\right)$ (d) $\left(\frac{L}{8}\right)$, $\left(\frac{3L}{8}\right)$ and $\left(\frac{5L}{6}\right)$

73. A stationary wave set up on a string have the equation $y = (2\text{ mm})\sin[(6.25\text{ m}^{-1})x\cos(\omega t)]$. This stationary wave is created by two identical waves, of amplitude A each moving in opposite directions along the string. Then,

(a) $A = 2$ mm
(b) $A = 4$ mm
(c) the smallest length of the string is 50 cm
(d) the smallest length of the string is 2 m

74. For a certain organ pipe, three successive resonance frequencies are observed at 425, 595 and 765 Hz. The speed of sound in air is 340 ms^{-1}. The pipe is a

(a) closed pipe of length 1 m
(b) closed pipe of length 2 m
(c) open pipe of length 1 m
(d) open pipe of length 2 m

75. Two strings A and B have lengths l_A and l_B and carry masses M_A and M_B at their lower ends, the upper ends being supported by rigid supports. If n_A and n_B are their fundamental frequencies of their vibrations, then $n_A : n_B$ is

(a) $\frac{M_A l_A}{M_B l_B}$ (b) $\frac{l_A}{l_B}\sqrt{\frac{M_A}{M_B}}$

(c) $\frac{M_A}{M_B}\sqrt{\frac{l_A}{l_B}}$ (d) None of these

Speed of a Wave, Sound Wave

76. If the frequency of sound produced by a siren increases from 400 Hz to 1200 Hz while the wave amplitude remains constant, the ratio of the intensity of the 1200 Hz to that of the 400 Hz wave will

(a) 1 : 1 (b) 1 : 3
(c) 3 : 1 (d) 9 : 1

77. The pitch of a sound wave is related to its

(a) frequency (b) amplitude
(c) velocity (d) beats

78. The ratio of velocity of the body to the velocity of sound is called

(a) Magic number (b) Laplace number
(c) Natural number (d) Mach number

79. Intensity level of a sound of intensity I is 30 dB. The ratio $\frac{I}{I_0}$ is (where, I_0 is the threshold of hearing)

(a) 30 (b) 300
(c) 1000 (d) 3000

80. The temperature at which the speed of sound in air becomes double of its value at 27°C is

(a) 327°C (b) 127°C
(c) 927°C (d) – 123°C

81. Under the same conditions of pressure and temperature, the velocity of sound in oxygen and hydrogen gases are v_O and v_H then

(a) $v_H = 2v_O$ (b) $v_H = 4v_O$
(c) $v_O = 4v_H$ (d) $v_H = v_O$

82. If the temperature of the gaseous medium drops by 1%, the velocity of sound in that medium

(a) increases by 5%
(b) remains unchanged
(c) decreases by 0.5%
(d) decreases by 2%

83. The velocity of sound through a diatomic gaseous medium of molecular weight M at 0°C

(a) $\sqrt{\frac{R}{M}}$ (b) $\sqrt{\frac{3R}{M}}$

(c) $\sqrt{\frac{382R}{M}}$ (d) $\sqrt{\frac{273\,R}{M}}$

84. The extension in a string obeying Hooke's law is x. The speed of transverse waves in the stretched string is v. If the extension in the string is increased to $1.5\,x$, the speed of transverse waves in it will be

(a) $1.22v$ (b) $2\,v$
(c) $1.5\,v$ (d) v

85. A pulse of sound wave travels a distance l in helium gas in time T at a particular temperature. If at the same temperature a pulse of sound wave is propagated in oxygen gas, it will cover the same distance l in time

(a) $4.36\,T$ (b) $0.23\,T$
(c) $3\,T$ (d) $0.46\,T$

86. A sound is produced between two vertical parallel walls. The echo from one wall is heard after 2 s while from the other 2 s after the first echo. The speed of sound in air is 340 m/s. Choose the correct options.

(a) The distance between two walls is 680 m
(b) The distance between two walls is 1020 m
(c) The next echo will be heard after 8 s from the instant original sound was produced
(d) None of the above

87. Two strings of copper are stretched to the same tension. If their cross-section area are in the ratio 1 : 4, then the respective wave velocities will be in the ratio

(a) 4 : 1 (b) 2 : 1
(c) 1 : 2 (d) 1 : 4

88. A sound has an intensity of 2×10^{-8} Wm^{-2}. Its intensity level (in decibels) is ($\log_{10} 2 = 0.3$)

(a) 23 (b) 4.3
(c) 43 (d) None of these

89. How many times more intense is 90 dB sound than 40 dB sound?

(a) 5 (b) 50
(c) 500 (d) 10^5

90. The ratio of the speed of sound in nitrogen gas to that in helium gas, at 300 K is

(a) $\sqrt{\frac{2}{7}}$ (b) $\sqrt{\frac{1}{7}}$

(c) $\frac{\sqrt{3}}{5}$ (d) $\frac{\sqrt{6}}{5}$

91. Frequency range of the audible sounds is

(a) 0 Hz-30 Hz
(b) 20 Hz-20 kHz
(c) 20 kHz-20,000 kHz
(d) 20 kHz-20 MHz

92. In a medium sound travels 2 km in 3 s and in air, it travels 3 km in 10 s. The ratio of the wavelengths of sound in the two media is

(a) 1 : 8 (b) 1 : 18
(c) 8 : 1 (d) 20 : 9

93. When the temperature of an ideal gas is increased by 600 K, the velocity of sound in the gas become $\sqrt{3}$ times the initial velocity in it. The initial temperature of the gas is

(a) $-73°C$ (b) 27°C
(c) 127°C (d) 327°C

94. The frequency of a sound wave is n and its velocity is v. If the frequency is increased to $4n$, the velocity of the wave will be

(a) v (b) $2v$
(c) $4v$ (d) $\frac{v}{4}$

95. A man standing on a cliff claps his hand hears its echo after 1 s. If sound is reflected from mountain and velocity of sound in air is 340 m/s. Then, the distance between the man and reflection point is

(a) 680 m (b) 340 m
(c) 85 m (d) 170 m

96. Sound velocity is maximum in

(a) H_2 (b) N_2
(c) He (d) O_2

97. Velocity of sound in air

I. increases with temperature.

II. decreases with temperature.

III. increases with pressure.

IV. is independent of pressure.

V. is independent of temperature.

Choose the correct answer.

(a) Only I and II are true
(b) Only I and III are true
(c) Only II and III are true
(d) Only I and IV are true

98. A person speaking normally produces a sound intensity of 40 dB at a distance of 1 m. If the threshold intensity for reasonable audibility is 20 dB, the maximum distance at which person can be heard clearly is

(a) 4 m (b) 5 m
(c) 10 m (d) 20 m

99. The intensity level of a sound wave is 4 dB. If the intensity of the wave is doubled, then the intensity level of the sound would be

(a) 8 (b) 16
(c) 7 (d) 14

100. In a mixture of gases, the average number of degrees of freedom per molecules is 6. The rms speed of the molecule of the gas is c. The velocity of sound in the gas is

(a) $\frac{c}{\sqrt{3}}$ (b) $\frac{c}{\sqrt{2}}$
(c) $\frac{2c}{3}$ (d) $\frac{3c}{4}$

Doppler Effect and Beats

101. Two identical string instruments have frequency of 100 Hz. If tension in one of them increases by 4% and they are sounded together, then the number of beats in one second is

(a) 1 (b) 8
(c) 4 (d) 2

102. If $\lambda_1 = 100\,\text{cm}$, $\lambda_2 = 90\,\text{cm}$ and velocity of sound = 396 m/s. The number of beats (in Hz) are

(a) 4 (b) 2
(c) 3 (d) 44

103. If a particle is travelling with a speed of 0.9 m/s of the speed of sound and is emitting radiations of frequency 1 kHz and moving towards the observer, what is its apparent frequency (in kHz)?

(a) 1.1 (b) 2
(c) 0.1 (d) 10

104. Following two wave trains are approaching each other.

$$y_1 = a \sin 2000\,\pi t$$
$$y_2 = a \sin 2008\,\pi t$$

The number of beats heard per second (in Hz) is

(a) 8 (b) 4
(c) 1 (d) zero

105. A car sounding its horn at 480 Hz moves towards a high wall at a speed of 20 m/s, the frequency of the reflected sound heard by the man sitting in the car will be nearest to (speed of sound = 330 m/s)

(a) 480 Hz (b) 510 Hz
(c) 542 Hz (d) 570 Hz

106. A locomotive engine approaches a railway station and whistles at a frequency of 400 Hz. A stationary observer on the platform observes a change of 40 Hz as the engine passes across him. If the velocity of sound is 340 m/s, the speed of the engine is

(a) 34 m/s (b) 40 m/s
(c) 17 m/s (d) 20 m/s

107. Two sound waves of length 1 m and 1.01 m in a gas produce 10 beats in 3 s. The velocity of sound in gas is

(a) 360 m/s
(b) 300 m/s
(c) 337 m/s
(d) 330 m/s

108. Two sources A and B are sounding notes of frequency 680 Hz. A listener moves from A to B with a constant velocity u. If the speed of sound is 340 ms^{-1}, what must be the value of u so that he hears 10 beats per second?

(a) 2.0 ms^{-1}
(b) 2.5 ms^{-1}
(c) 3.0 ms^{-1}
(d) 3.5 ms^{-1}

109. An accurate and reliable audio oscillator is used to standardise a tuning fork. When the oscillator reading is 514, two beats are heard per second. When the oscillator reading is 510, the beat frequency is 6 Hz. The frequency of the tuning fork (in Hz) is

(a) 506 (b) 510 (c) 516 (d) 512

110. A siren placed at a railway platform is emitting sound of frequency 5 kHz. A passenger sitting in a moving train *A* records a frequency of 5.5 kHz while train approaches the siren. During his return journey in a different train *B*, he records a frequency of 6 kHz while approaching the same siren. The ratio of the velocity of train *B* to that of train *A* is

(a) $\frac{4}{3}$ (b) 2 (c) $\frac{5}{3}$ (d) $\frac{8}{5}$

111. A train moves towards stationary observer with speed 34 m/s. The train sounds a whistle and its frequency registered by the observer is f_1. If the train's speed is reduced to 17 m/s, the frequency registered is f_2. If the speed of sound is 340 m/s, then ratio $\frac{f_1}{f_2}$ is

(a) $\frac{18}{19}$ (b) $\frac{17}{18}$ (c) $\frac{18}{17}$ (d) $\frac{19}{18}$

112. The frequency of a whistle is 300 Hz. It is approaching towards an observer with a speed 1/3 the speed of sound. The frequency of sound as heard by the observer will be

(a) 450 Hz (b) 300 Hz
(c) 400 Hz (d) 425 Hz

113. A tuning fork of frequency 480 Hz produces 10 beats per second, when sounded with a vibrating sonometer string. What must be frequency of the string, if a slight increase in tension produces more beats per second than before?

(a) 460 Hz (b) 470 Hz
(c) 480 Hz (d) 490 Hz

114. When two tuning forks *A* and *B* are sounded together, *x* beat/s are heard. Frequency of *A* is *n*. Now, when one prong of fork *B* is loaded with a little wax, the number of beats decreases. The frequency of fork *B* is

(a) $n + x$ (b) $n - x$
(c) $n + 2x$ (d) $n - 2x$

115. The speed of sound in air is *v*. Both the source and observer are moving towards each other with equal speed *u*. The speed of wind is *w* from source to observer. Then, the ratio $\left(\frac{f}{f_0}\right)$ of the apparent frequency to the actual frequency is given by

(a) $\frac{v+u}{v-u}$ (b) $\frac{v+w+u}{v+w-u}$
(c) $\frac{v+w+u}{v-w-u}$ (d) $\frac{v-w+u}{v-w-u}$

116. When an observer moves toward a stationary source with a certain velocity, he observes an apparent frequency *f*. When it moves away from the source with same velocity, the observed frequency is 0.8 *f*. If the velocity of sound is *v*, then the speed of the observer is

(a) $\frac{v}{4}$ (b) $\frac{v}{8}$
(c) $\frac{v}{9}$ (d) None of these

117. A source of sound with frequency 256 Hz is moving with a velocity *v* towards a wall. When the observer is between source and the wall, he finds that the frequency of two waves received directly from the source is *x* and the frequency of the waves received after reflection from the wall is *y*, then

(a) $x > y$ (b) $x < y$
(c) $x = y$ (d) Nothing can be said

118. A source of sound of frequency 600 Hz is placed inside water. The speed of sound in water is 1500 m/s and in air is 300 m/s. The frequency of sound recorded by an observer who is standing in air, is

(a) 200 Hz (b) 3000 Hz
(c) 120 Hz (d) 600 Hz

119. A tuning fork of frequency 100 when sound together with another tuning fork of unknown frequency produces 2 beats per second. On loading the tuning fork whose frequency is not known and sounded together with the same tuning fork produces one beat, then the frequency of the unknown tuning fork is

(a) 102 (b) 98
(c) 99 (d) 101

120. An unknown tuning fork sounded together with a tuning fork of frequency 256 emits two beats. On loading tuning fork of frequency 256, the number of beats heard is 1 per second. The frequency of unknown tuning fork is

(a) 257 (b) 258
(c) 256 (d) 254

121. The frequency of tuning forks *A* and *B* are respectively 3% more and 2% less than the frequency of tuning fork *C*. When *A* and *B* are simultaneously excited, 5 beats per second are produced. Then the frequency of the tuning fork *A* (in Hz) is

(a) 98 (b) 100
(c) 103 (d) 105

122. Two closed pipes produce 10 beats per second when emitting their fundamental nodes. If their lengths are in ratio of 25 : 26. Then, their fundamental frequency in Hz, are

(a) 270, 280
(b) 260, 270
(c) 260, 250
(d) 260, 280

123. An organ pipe, open from both ends produces 5 beats per second when vibrated with a source of frequency 200 Hz in its fundamental mode. The second harmonic of the same pipe produces 10 beats per second with a source of frequency 420 Hz. The fundamental frequency of pipe is

(a) 195 Hz (b) 205 Hz
(c) 190 Hz (d) 210 Hz

124. The wavelength is 120 cm when the source is stationary. If the source is moving with relative velocity of 60 m/s towards the observer, then the wavelength of the sound wave reaching to the observer will be (velocity of sound = 330 m/s)

(a) 98 cm (b) 140 cm
(c) 120 cm (d) 1440 cm

125. A source of sound emits waves with frequency f Hz and speed v m/s. Two observers move away from this source in opposite directions each with a speed $0.2v$ relative to the source. The ratio of frequencies heard by the two observers will be

(a) 3 : 2 (b) 2 : 3
(c) 1 : 1 (d) 4 : 10

126. A source of sound is travelling towards a stationary observer. The frequency of sound heard by the observer is of three times the original frequency. The velocity of sound is v m/s. The speed of source will be

(a) $\frac{2}{3}v$ (b) v
(c) $\frac{3}{2}v$ (d) $3v$

127. A source of sound is moving with constant velocity of 20 m/s emitting a note of frequency 1000 Hz. The ratio of frequencies observed by a stationary observer while the source is approaching him and after it crosses him will be

(a) 9 : 8 (b) 8 : 9
(c) 1 : 1 (d) 9 : 10
(speed of sound, $v = 340$ m/s)

128. A table is revolving on its axis at 5 revolutions per second. A sound source of frequency 1000 Hz is fixed on the table at 70 cm from the axis. The minimum frequency heard by a listener standing at a distance very far from the table will be (speed of sound = 352 m/s)

(a) 1000 Hz (b) 1066 Hz
(c) 941 Hz (d) 352 Hz

129. A source of sound S of frequency 500 Hz situated between a stationary observer O and a wall, moves towards the wall with a speed of 2 m/s. If the velocity of sound is 332 m/s, then the number of beats per second heard by the observer is (approximately)

(a) 8 Hz
(b) 6 Hz
(c) 4 Hz
(d) 2 Hz

130. A car sounding a horn of frequency 1000 Hz passes an observer. The ratio of frequencies of the horn noted by the observer before and after passing of the car is 11 : 9. If the speed of sound is v, the speed of the car is

(a) $\frac{1}{10}v$ (b) $\frac{1}{2}v$
(c) $\frac{1}{5}v$ (d) v

131. A small source of sound moves on a circle as shown in the figure and an observer is standing on O. Let n_1, n_2 and n_3 be the frequencies heard, when the source is at A, B and C, respectively. Then,

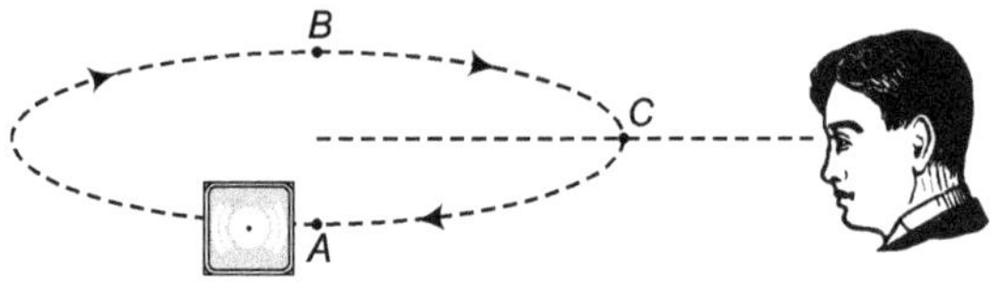

(a) $n_1 > n_2 > n_3$
(b) $n_2 > n_3 > n_1$
(c) $n_1 = n_2 > n_3$
(d) $n_2 > n_1 > n_3$

132. An observer moves towards a stationary source of sound, with a velocity one-fifth of the velocity of sound. What is the percentage increase in the apparent frequency?

(a) 5% (b) 20%
(c) Zero (d) 0.5%

133. Two cars are moving on two perpendicular roads towards a crossing with uniform speeds of 72 km/h and 36 km/h. If first car blows horn of frequency 280 Hz, then the frequency of horn heard by the driver of second car when line joining the cars make 45° angle with the roads, will be

(a) 321 Hz
(b) 298 Hz
(c) 289 Hz
(d) 280 Hz

134. 25 turning forks are arranged in series in the order of decreasing frequency. Any two successive forks produce 3 beats/s. If the frequency of the first turning fork is the octave of the last fork, then the frequency of the 21st fork is

(a) 72 Hz
(b) 288 Hz
(c) 84 Hz
(d) 87 Hz

135. Two whistles A and B have frequencies 660 Hz and 590 Hz, respectively. An observer is standing in the middle of line joining the two sources. Source B and observer are moving towards right with velocity 30 m/s and A is standing to the left side. If the velocity of sound in air is 300 m/s. The number of beats per second listened by the observer are

(a) 8 (b) 6
(c) 4 (d) 2

136. A whistle emitting a sound of frequency 440 Hz is tied to a string of 1.5 m length and rotated with an angular velocity of 20 rad/s in the horizontal plane. Then, the range of frequencies heard by an observer stationed at a large distance from the whistle will be ($v = 330$ m/s)

(a) 400 Hz to 484 Hz
(b) 403.3 Hz to 480 Hz
(c) 400 Hz to 480 Hz
(d) 403.3 Hz to 484 Hz

137. A closed organ pipe and an open organ pipe of same length produce 2 beats, when they are set into vibrations simultaneously in their fundamental mode. The length of open organ pipe is now halved and of closed organ pipe is doubled. The number of beats produced will be

(a) 8 (b) 7
(c) 4 (d) 2

Graphical Problems

138. A railway engine whistling at a constant frequency moves with a constant speed. It goes past a stationary observer standing beside the railway track. The frequency f of the sound by the observer is plotted against time t. Which of the following best represents the resulting curve?

(a)
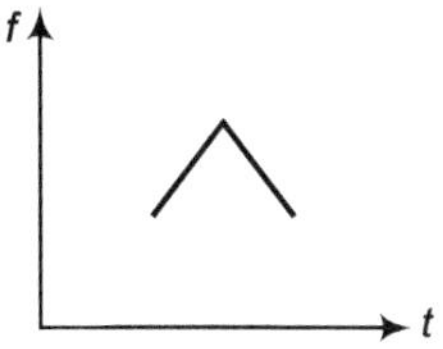

(b)
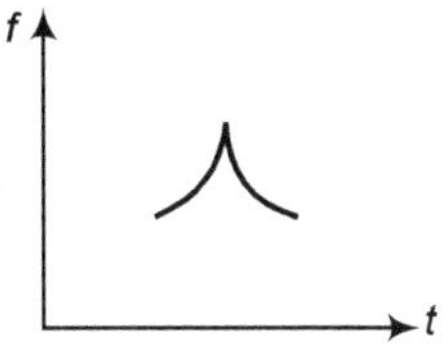

(c)
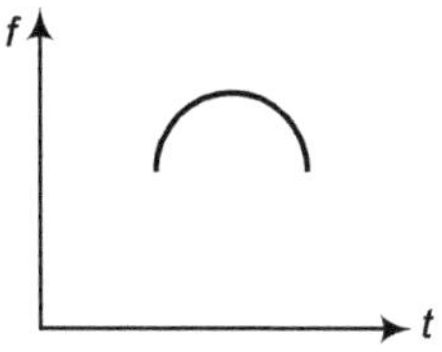

(d)
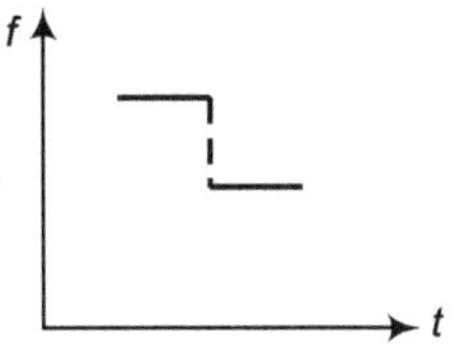

139. The rope shown at an instant is carrying a wave travelling towards right, created by a source vibrating at a frequency n.

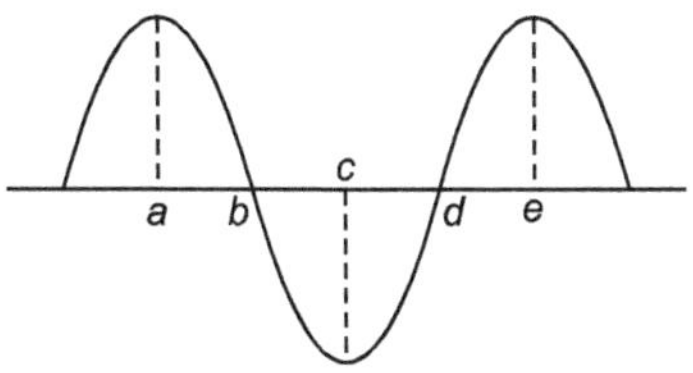

Consider the following statements :
(a) The speed of the wave is $4n \times ab$
(b) The phase difference between b and e is $\frac{3\pi}{2}$
(c) Both (a) and (b) are correct
(d) Both (a) and (b) are wrong

140. The diagram shows the propagation of a wave. Which points are in same phase?

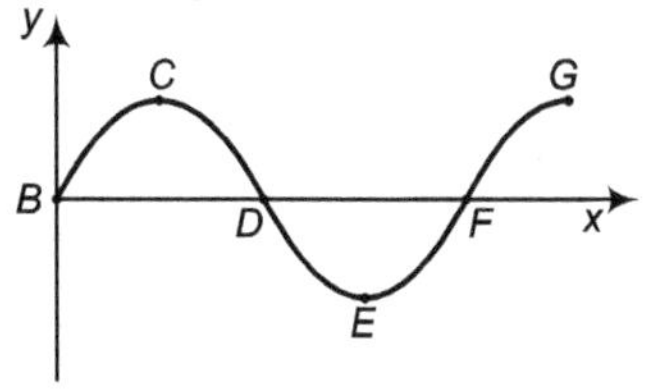

(a) F and G (b) C and E (c) B and G (d) B and F

141. Figure shows the wave $y = A \sin(\omega - kx)$. What is the magnitude of slope of the curve at B?

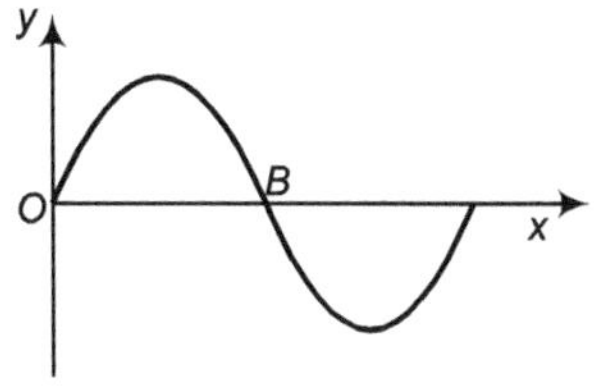

(a) $\frac{\omega}{A}$ (b) $\frac{k}{A}$ (c) kA (d) ωA

142. The correct graph between the frequency n and square root of density (ρ) of a wire, keeping its length, radius and tension constant, is

(a)
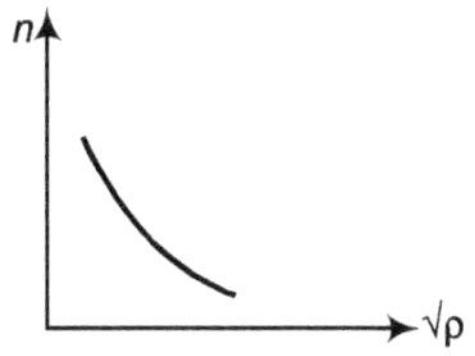

(b)
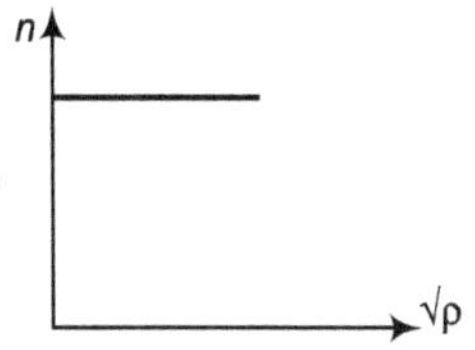

(c)
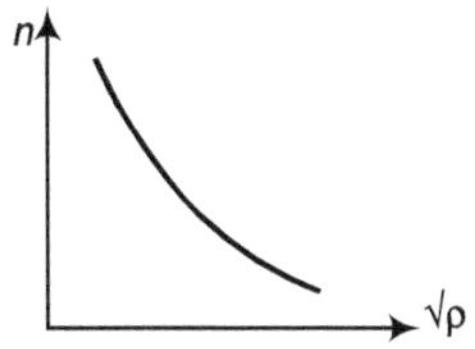

(d)
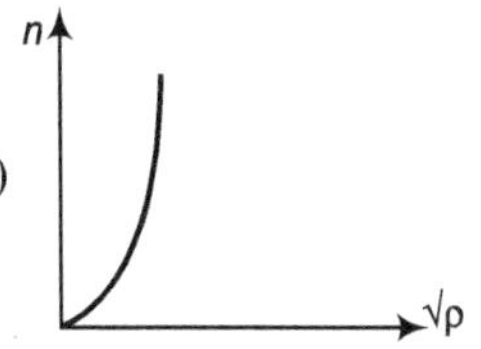

143. A sound source emits sound waves in a uniform medium. If energy density is E and maximum speed of the particles of the medium is v_{max}. The plot between E and v_{max} is best represented by

(a)
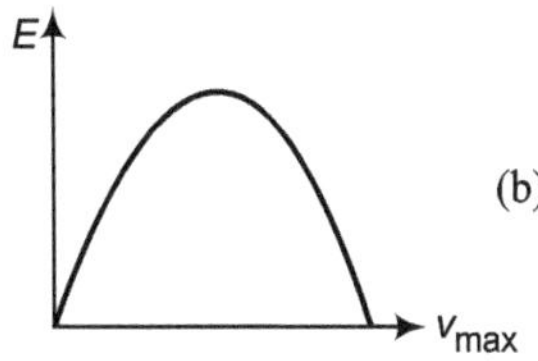

(b)
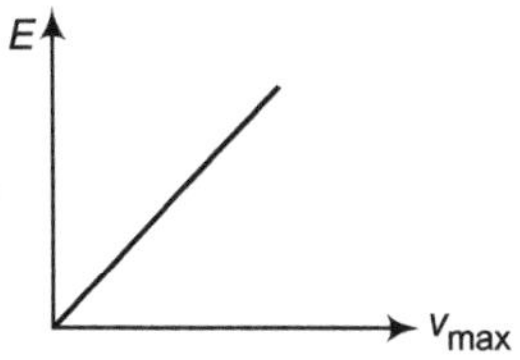

(c)
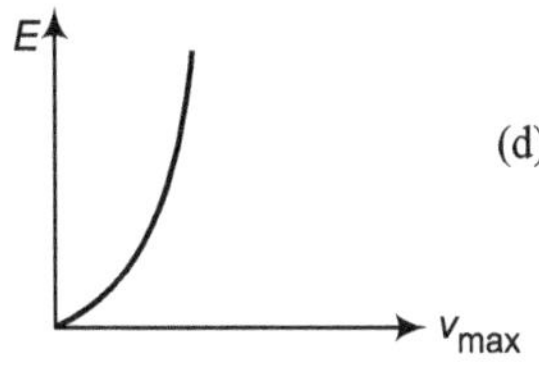

(d)
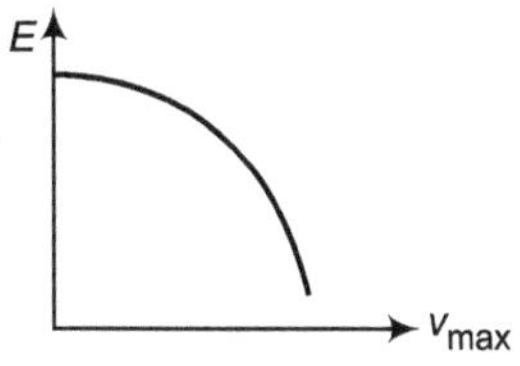

144. If the speed of the wave shown in the figure is 330 m/s in the given medium, then the equation of the wave propagating in the positive x-direction will be (all quantities are in MKS units)

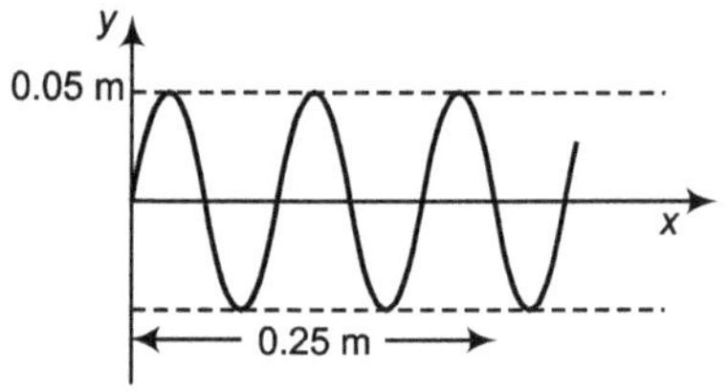

(a) $y = 0.05 \sin 2\pi (4000t - 12.5x)$
(b) $y = 0.05 \sin 2\pi (4000t - 122.5x)$
(c) $y = 0.05 \sin 2\pi (3300t - 10x)$
(d) $y = 0.05 \sin 2\pi (3300x - 10t)$

145. A wave is travelling along a string. At an instant, shape of the string is as shown in figure. At this instant, point A is moving upwards. Which of the following statement(s) is/are correct?

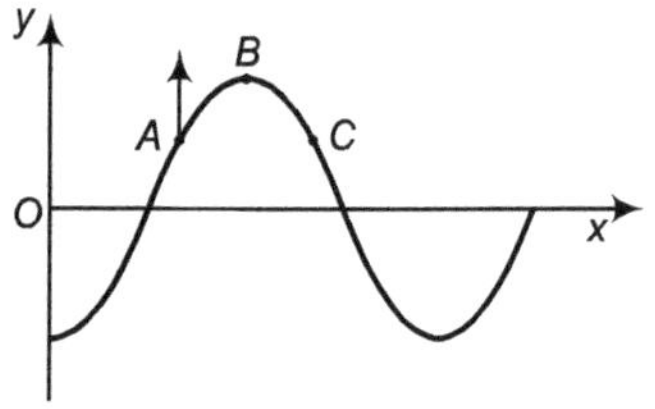

(a) The wave is travelling to the right
(b) Displacement amplitude of the wave is equal to displacement of B at this instant
(c) At this instant velocity of C is also directed upwards
(d) Phase difference between A and C may be equal to $\frac{\pi}{2}$

146. A wave motion has the function $y = a_0 \sin (\omega t - kx)$. The graph in figure shows how the displacement y at a fixed point varies with time t. Which one of the labelled points shows a displacement equal to that at the position $x = \frac{\pi}{2k}$ at time $t = 0$?

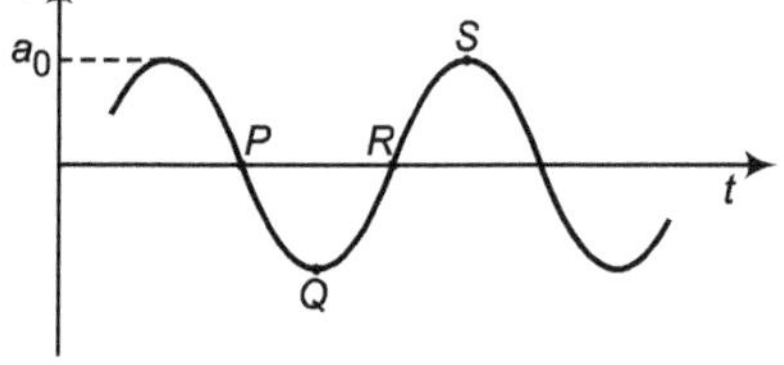

(a) P (b) Q
(c) R (d) S

147. The diagram shows an instantaneous position of a string as a transverse progressive wave travels along it from left to right. Which one of the following correctly shows the direction of velocity of points 1, 2 and 3 on the string?

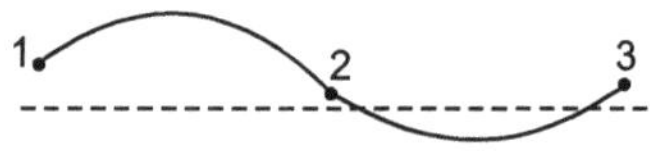

	1	2	3
(a)	↑	↓	↓
(b)	↑	↓	↑
(c)	↓	↓	↓
(d)	↓	↑	↓

148. A train is moving towards a stationary observer. Which of the following curve best represents, the frequency received by observer f as a function of time?

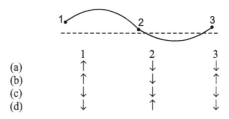

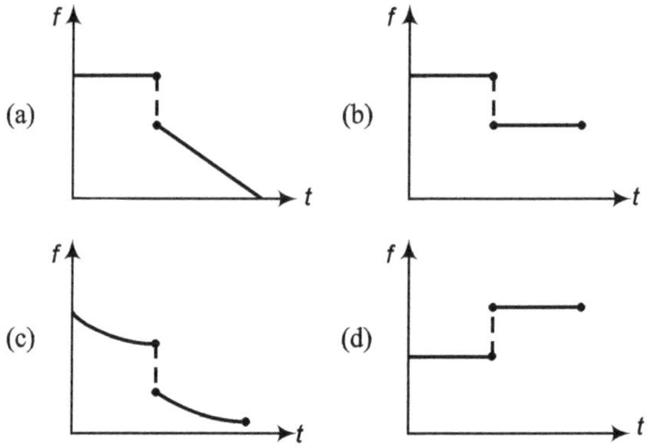

Miscellaneous Problems

149. The intensity of sound gets reduced by 20% on passing through a slab. The reduction in intensity on passing through two consecutive slabs is
(a) 40% (b) 36%
(c) 30% (d) 50%

150. A wave of frequency 100 Hz is sent along a string towards a fixed end. When this wave travels back, after reflection, a node is formed at a distance of 10 cm from the fixed end of the string. The speeds of incident (and reflected) waves are
(a) 48 m/s (b) 20 m/s
(c) 10 m/s (d) 15 m/s

151. The resultant amplitude due to superposition of two harmonic waves expressed by

$$y = a \sin (\omega t - kx)$$

and $y = a \cos (kx - \omega t)$ will be

(a) 0 (b) a (c) $\sqrt{2}\, a$ (d) $2a$

152. In a sinusoidal wave, the minimum time required for a particular point to move from maximum displacement to zero displacement is 0.17 s. The frequency of the wave is
(a) 1.47 Hz (b) 0.36 Hz
(c) 0.73 Hz (d) 2.49 Hz

153. In a medium in which a transverse progressive wave is travelling, the phase difference between two points with a separation of 1.25 cm is $\left(\frac{\pi}{3}\right)$. If the frequency of wave is 1000 Hz, its velocity will be
(a) 75 m/s (b) 125 m/s (c) 100 m/s (d) 50 m/s

154. Two periodic waves of amplitudes a and b pass through a region at the same time in the same direction. If $a > b$, the difference in the maximum and minimum possible amplitudes is

(a) $a + b$ (b) $a - b$ (c) $2a$ (d) $2b$

155. Two waves represented by $y_1 = 10 \sin 200\pi t$ and $y_2 = 20 \sin \left[200\pi\, t + \frac{\pi}{2}\right]$ at a point are superimposed at a particular instant. The amplitude of the resultant wave is

(a) 30 (b) $10\sqrt{5}$
(c) 50 (d) 10

156. When a vibrating tuning fork of frequency 512 Hz is held above the mouth of a resonance tube of adjustable length, the first two successive positions of resonance occur, when the length of the air columns are 15.4 cm and 48.6 cm, respectively. Then, the velocity of sound is

(a) 512 (48.6 – 15.4) cm/s
(b) 1024 (48.6 – 15.4) cm/s
(c) 256 (48.6 – 15.4) cm/s
(d) 2×512 (48.6 + 15.4) cm/s

157. A transverse wave $y = 0.05 \sin (20\pi x - 50\pi t)$ in metre, is propagating along +ve X-axis on a string. A light insect starts crawling on the string with the velocity of 5 cm/s at $t = 0$ along the +ve X-axis from a point, where $x = 5$ cm. After 5s, the difference in the phase of its position is equal to

(a) $150\,\pi$ (b) $250\,\pi$
(c) 10π (d) 5π

158. Two waves represented by the following equations are travelling in the same medium

$$y_1 = 5 \sin 2\pi\,(75t - 0.25x),\ y_2 = 10 \sin 2\pi\,(150t - 0.50x)$$

The intensity ratio $\frac{I_1}{I_2}$ of the two waves is

(a) 1 : 2 (b) 1 : 4
(c) 1 : 8 (d) 1 : 16

159. When two sound waves with a phase difference of $\frac{\pi}{2}$, and each having amplitude A and frequency ω, are superimposed on each other, then the maximum amplitude and frequency of resultant wave is

(a) $\frac{A}{\sqrt{2}}, \frac{\omega}{2}$ (b) $\frac{A}{\sqrt{2}}, \omega$ (c) $\sqrt{2}\,A, \frac{\omega}{2}$ (d) $\sqrt{2}\,A, \omega$

160. Two sources of sound A and B produce the wave of 350 Hz, in the same phase. The particle P is vibrating under the influence of these two waves, if the amplitude at the point P produces by the two waves is 0.3 mm and 0.4 mm, then the resultant amplitude of the point P will be when $AP - BP = 25$ cm and the velocity of sound is 350 m/s

(a) 0.7 mm (b) 0.1 mm (c) 0.2 mm (d) 0.5 mm

161. Two waves are propagating to the point P by two sources A and B of equal frequency. The amplitude of every wave at P is a and the phase of A is ahead by $\frac{\pi}{3}$ than that of B and the distance AP is greater than BP by 50 cm. Then, the resultant amplitude at the point P will be, if the wavelength is 1 m

(a) $2a$ (b) $a\sqrt{3}$ (c) $a\sqrt{2}$ (d) a

162. The amplitude of a wave represented by displacement equation $y = \frac{1}{\sqrt{2}} \sin \omega t \pm \frac{1}{\sqrt{b}} \cos \omega t$ will be

(a) $\frac{a+b}{ab}$ (b) $\frac{\sqrt{a}+\sqrt{b}}{ab}$ (c) $\frac{\sqrt{a}\pm\sqrt{b}}{ab}$ (d) $\sqrt{\frac{a+b}{ab}}$

163. Two sinusoidal waves with same wavelengths and amplitudes travel in opposite directions along a string with a speed $10\ \text{ms}^{-1}$. If the minimum time interval between two instants, when the string is flat is 0.5 s, the wavelength of the waves is

(a) 25 m (b) 20 m (c) 15 m (d) 10 m

164. To increase the frequency from 100 Hz to 400 Hz, the tension in the string has to be changed by

(a) 4 times (b) 16 times (c) 2 times (d) None of these

165. If the temperature increases, then what happens to the frequency of the sound produced by the organ pipe?

(a) Increases (b) Decreases
(c) Unchanged (d) Not definite

166. In a resonance pipe, the first and second resonances are obtained at depths 22.7 cm and 70.2 cm, respectively. What will be the end correction?

(a) 1.05 cm (b) 1115.5 cm (c) 92.5 cm (d) 113.5 cm

167. In the experiment for the determination of the speed of sound in air using the resonance column method, the length of the air column that resonates in the fundamental mode, with a tuning fork is 0.1 m. When this length is changed to 0.35 m, the same tuning fork resonates with the first overtone. Calculate the end correction.

(a) 0.012 m (b) 0.025 m (c) 0.05 m (d) 0.024 m

168. Figure shows an incident pulse P reflected from a rigid support. Which one represents the reflected pulse correctly?

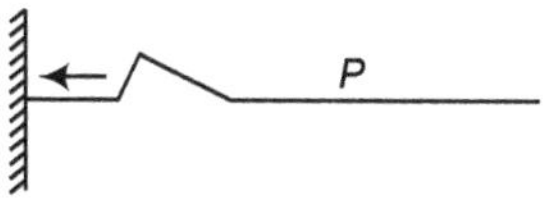

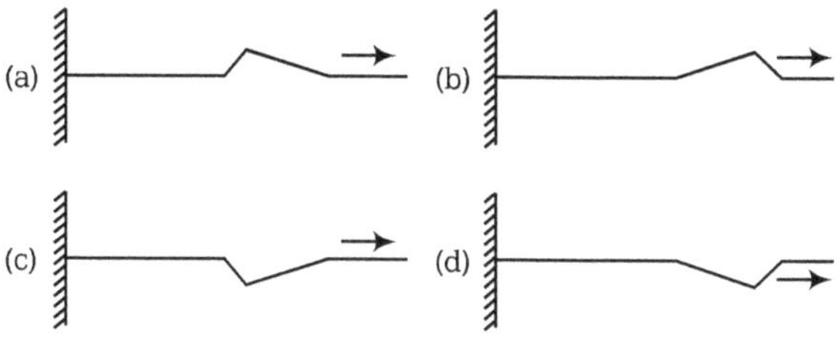

169. In a resonance tube, closed at one end by a smooth moving piston and the other end open, exhibits the first three resonance lengths of air column, L_1, L_2 and L_3 for the same tuning fork. Then, they are related by

(a) $L_3 = 2L_2 = 4L_1$
(b) $L_3 = \frac{5}{3}L_2 = 5L_1$
(c) $(L_3 - L_2) = \left(\frac{L_2 - L_1}{2}\right)$
(d) $(L_3 - L_2) = 2(L_2 - L_1)$

170. Three waves of equal frequencies having amplitudes $10\mu m$, $4\mu m$ and $7\mu m$ arrive at a given point with successive phase difference of $\frac{\pi}{2}$. The amplitude of the resulting wave in μm is given by

(a) 7 (b) 6
(c) 5 (d) 4

171. A glass tube of 1.0 m length is filled with water. The water can be drained out slowly at the bottom of the tube. If a vibrating tuning fork of frequency 500 c/s is brought at the upper end of the tube and the velocity of sound is 330 m/s, then the total number of resonance obtained will be

(a) 4 (b) 3
(c) 2 (d) 1

172. Two identical sonic sources S_1 and S_2 produce plane harmonic progressive waves in the same phase. S_1 is at origin (0, 0) and the wave produced by it is given by $y = a\sin(\omega t - kx)$. If S_2 is d distance from S_1 as shown in the figure, the equation of wave produced by S_2 along the x-axis will be given by

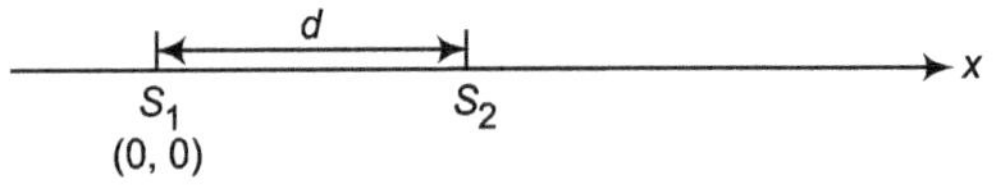

(a) $y = a\sin(\omega t - kx)$
(b) $y = a\sin[\omega t - k(x + d)]$
(c) $y = a\sin[\omega t - k(x - d)]$
(d) Data insufficient

173. The intensity of sound at a point due to a point source is $0.2\,W/m^2$. If the distance of the source is made doubled and the power is also doubled, then the intensity at the point will be

(a) $0.05\,W/m^2$
(b) $0.1\,W/m^2$
(c) $0.2\,W/m^2$
(d) $0.4\,W/m^2$

[Level 2]

Only One Option Correct

1. Three waves

$$y_1 = A\sin(kx - \omega t)$$
$$y_2 = A\sin(kx - \omega t + \phi)$$
$$y_3 = A\sin(kx - \omega t + 2\phi)$$

are superimposed so that $y_1 + y_2 + y_3 = 0$ at all positions. Then, the value of ϕ is

(a) $\frac{2\pi}{3}$ (b) $\frac{\pi}{3}$
(c) $\frac{4\pi}{3}$ (d) $\frac{\pi}{2}$

2. In a large room, a person receives direct sound waves from a source 120 m away from him. He also receives waves from the same source which reaches him after being reflected from the 25 m high ceiling at a point midway between them. The two waves interfere constructively for wavelength (in m) of

(a) 20, 20/3, 20/5, etc.
(b) 10, 5, 2.5, etc.
(c) 10, 20, 30, etc.
(d) 15, 25, 35, etc.

3. A train of sound waves is propagated along a wide pipe and it is reflected from an open end. If the amplitude of the waves is 0.002 cm, the frequency 1000 Hz and the wavelength 40 cm, the amplitude of vibration at a point 10 cm from open end inside the pipe will be

(a) zero (b) 0.002 cm
(c) 0.001 cm (d) 0.005 cm

4. If the speed of longitudinal waves equals 10 times the speed of the transverse waves in a stretched wire of material which has modulus of elasticity E, then the stress in the wire is

(a) $10E$ (b) $100E$
(c) $\frac{E}{10}$ (d) $\frac{E}{100}$

5. A stretched rope having linear mass density $5\times 10^{-2}\,kg/m$ is under a tension of 80 N. The power that has to be supplied to the rope to generate harmonic waves at a frequency of 60 Hz and an amplitude of 6 cm is

(a) 362 W
(b) 251 W
(c) 511 W
(d) 416 W

6. A standing wave $y = A \sin\left(\frac{20\pi x}{3}\right)\cos(1000\pi t)$ is set up a taut string, where x and y are in metre. The distance between two successive points oscillating with the amplitude $\frac{A}{2}$ can be equal to

(a) 10 cm (b) 5 cm
(c) 2.5 cm (d) 4 cm

7. Two identical wires are stretched so as to produce 6 beats/s, when vibrating simultaneously. On changing the tension slightly in one of them, the beat frequency remains unchanged. Denoting by T_1 and T_2, the higher and the lower initial tension in the strings, then it could be said that the following cases may be possible

(a) T_2 was decreased (b) T_2 was increased
(c) T_1 was increased (d) T_1 was decreased

8. Two waves are propagating to the point P along a straight line produces by two sources A and B of simple harmonic and of equal frequency. The amplitude of every wave P is a and the phase of A is ahead by $\frac{\pi}{3}$ than that of B and the distance AP is greater than BP by 50 cm. Then, the resultant amplitude at the point P will be, if the wavelength is 1 m

(a) $2a$ (b) $a\sqrt{3}$ (c) $a\sqrt{2}$ (d) a

9. A sonometer wire when vibrated in full length has frequency n. Now, it is divided by the help of bridges into a number of segments of lengths $l_1, l_2, l_3, \ldots$. When vibrated, these segments have frequencies $n_1, n_2, n_3, \ldots$. Then, the correct relation is

(a) $n = n_1 + n_2 + n_3 + \ldots$
(b) $n^2 = n_1^2 + n_2^2 + n_3^2 + \ldots$
(c) $\frac{1}{n} = \frac{1}{n_1} + \frac{1}{n_2} + \frac{1}{n_3} + \ldots$
(d) $\frac{1}{\sqrt{n}} = \frac{1}{\sqrt{n_1}} + \frac{1}{\sqrt{n_2}} + \frac{1}{\sqrt{n_3}} + \ldots$

10. The frequency of a sonometer wire is f. The frequency becomes $\frac{f}{2}$, when the mass producing the tension is completely immersed in water and on immersing the mass in a certain liquid, frequency becomes $\frac{f}{3}$. The relative density of the liquid is

(a) 1.32 (b) 1.03 (c) 1.41 (d) 1.18

11. A sonometer wire has a length 114 cm between two fixed ends. Where should two bridges be placed so as to divide the wire into three segments whose fundamental frequencies are in the ratio 1 : 3 : 4.

(a) $l_1 = 72$ cm, $l_2 = 24$ cm, $l_3 = 18$ cm
(b) $l_1 = 60$ cm, $l_2 = 40$ cm, $l_3 = 14$ cm
(c) $l_1 = 52$ cm, $l_2 = 30$ cm, $l_3 = 32$ cm
(d) $l_1 = 65$ cm, $l_2 = 30$ cm, $l_3 = 19$ cm

12. Four sources of sound each of sound level 10 dB are sounded together in phase, the resultant intensity level will be ($\log_{10} 2 = 0.3010$)

(a) 40 dB (b) 26 dB
(c) 22 dB (d) 13 dB

13. If at $t = 0$, a travelling wave pulse on a string is described by the function

$$y = \frac{6}{x^2 + 3}$$

What will be the wave function representing the pulse at time t, if the pulse is propagating along positive x-axis with speed 4 m/s?

(a) $y = \frac{6}{(x + 4t)^2 + 3}$ (b) $y = \frac{6}{(x - 4t)^2 + 3}$
(c) $y = \frac{6}{(x - t)^2}$ (d) $y = \frac{6}{(x - t)^2 + 12}$

14. An open pipe is in resonance in its 2nd harmonic with tuning fork of frequency f_1. Now, it is closed at one end. If the frequency of the tuning fork is increased slowly from f_1 then again a resonance is obtained with a frequency f_2. If in this case the pipe vibrates nth harmonic, then

(a) $n = 3, f_2 = \frac{3}{4} f_1$ (b) $n = 3, f_2 = \frac{5}{4} f_1$
(c) $n = 5, f_2 = \frac{5}{4} f_1$ (d) $n = 5, f_2 = \frac{3}{4} f_1$

15. The amplitude of a wave represented by displacement equation $y = \frac{1}{\sqrt{a}} \sin \omega t \pm \frac{1}{\sqrt{b}} \cos \omega t$ will be

(a) $\frac{a + b}{ab}$ (b) $\frac{\sqrt{a} + \sqrt{b}}{ab}$
(c) $\frac{\sqrt{a} \pm \sqrt{b}}{ab}$ (d) $\sqrt{\frac{a + b}{ab}}$

16. A stone hangs from the free end of a sonometer wire whose vibrating length, when tuned to a tuning fork is 40 cm. When the stone hangs wholly immersed in water, the resonant length is reduced to 30 cm in same mode of oscillation. The relative density of the stone is

(a) $\frac{16}{9}$ (b) $\frac{16}{7}$
(c) $\frac{16}{5}$ (d) $\frac{16}{3}$

17. The displacement due to a wave moving in the positive x-direction is given by $y = \frac{1}{(1 + x^2)}$ at time $t = 0$ and by $y = \frac{1}{[1 + (x - 1)^2]}$ at $t = 2$ s, where x and y are in metre. The velocity of the wave in m/s is

(a) 0.5 (b) 1 (c) 2 (d) 4

18. The linear density of a vibrating string is 10^{-4} kg/m. A transverse wave is propagating on the string, which is described by the equation $y = 0.02 \sin (x + 30t)$, where x and y are in metre and the time t is in second. The tension in the string is

(a) 0.09 N (b) 0.36 N
(c) 0.9 N (d) 3.6 N

19. An open pipe is suddenly closed at one end with the result that the frequency of third harmonic of the closed pipe is found to be higher by 100 Hz, then the fundamental frequency of open pipe is

(a) 200 Hz (b) 300 Hz
(c) 240 Hz (d) 480 Hz

20. A string is hanging from a rigid support. A transverse wave pulse is set up at the free end. The velocity v of the pulse related to the distance x covered by it is given as

(a) $v \propto \sqrt{x}$ (b) $v \propto x$
(c) $v \propto \frac{1}{x}$ (d) None of these

21. The loudspeakers L_1 and L_2 driven by a common oscillator and amplifier, are set up as shown in the figure. As the frequency of the oscillator increases from zero, the detector at D recorded as a series of maximum and minimum signals. What is the frequency at which the first maximum is observed? (speed of sound = 330 m/s)

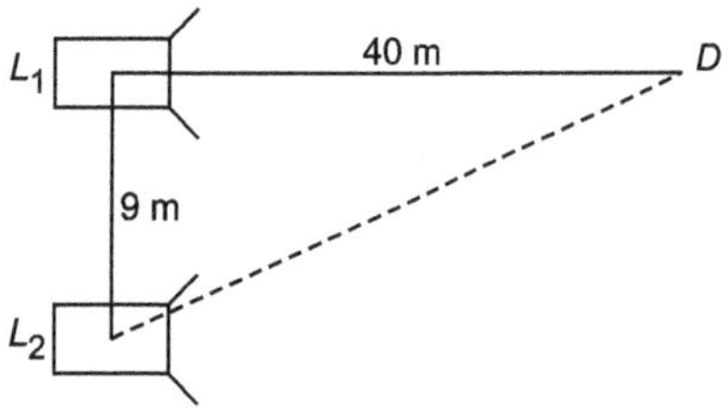

(a) 165 Hz (b) 330 Hz
(c) 495 Hz (d) 660 Hz

22. Two waves are passing through a region in the same direction at the same time. If the equation of these waves are

$$y_1 = a \sin \frac{2\pi}{\lambda}(vt - x)$$
$$y_2 = b \sin \frac{2\pi}{\lambda}[(vt - x) + x_0]$$

then the amplitude of the resultant wave for $x_0 = \frac{\lambda}{2}$ is

(a) $|a - b|$ (b) $(a + b)$
(c) $\sqrt{a^2 + b^2}$ (d) None of these

23. Identify the figure, which correctly represents the given wave function

$$y = 2\sqrt{3} \sin \pi \left(x - 2t + \frac{1}{6}\right)$$

(a)
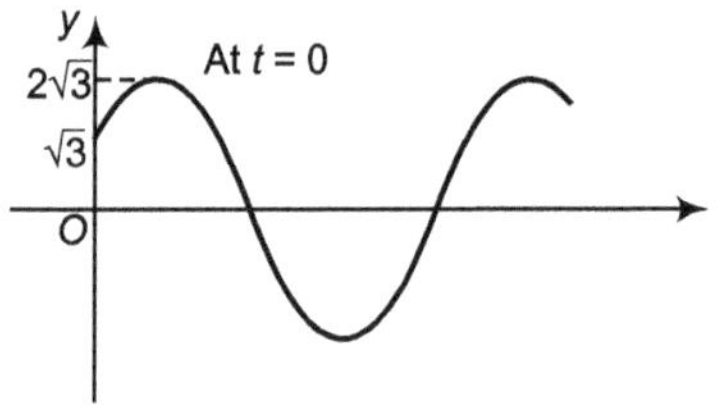

(b)
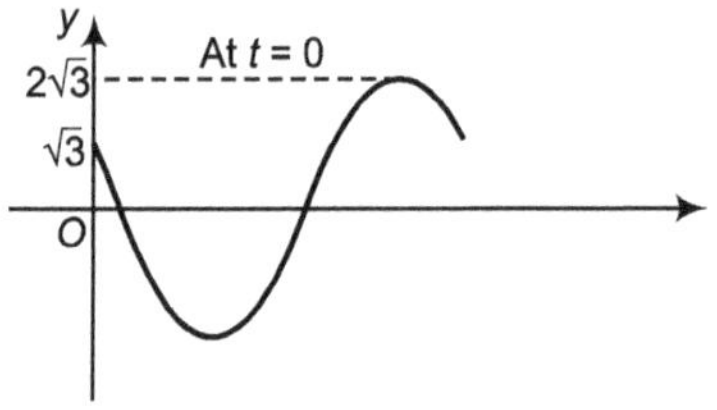

(c) Both graphs are possible
(b) None of the two graphs are possible

24. Two tuning forks P and Q are vibrated together. The numbers of beats produced are represented by the straight line OA in the following graph.

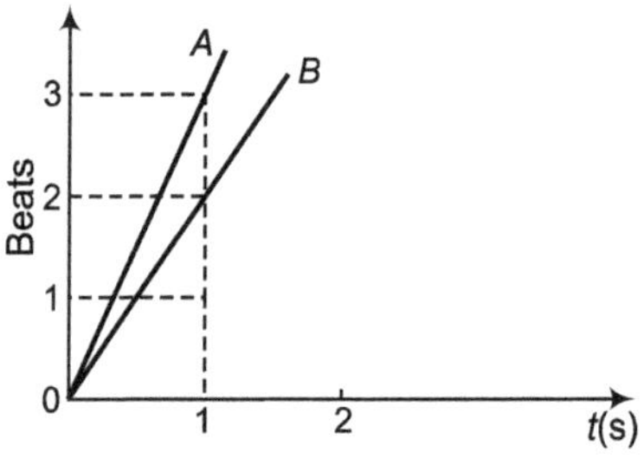

After loading Q with wax again these are vibrated together and the beats produced are represented by the line OB. If the frequency of P is 341 Hz, the frequency of Q will be

(a) 341 Hz (b) 338 Hz
(c) 344 Hz (d) None of these

25. Two sounding bodies producing progressive waves given by and $y_2 = 3 \sin (404\pi t)$, where t is in second, are situated near the ears of a person. The person will hear

(a) 2 beats per second with intensity ratio 4/3 between maxima and minima
(b) 2 beats per second with intensity ratio 49 between maxima and minima
(c) 4 beats per second with intensity ratio 49 between maxima and minima
(d) 4 beats per second with intensity ratio 4/3 between maxima and minima

26. The maximum pressure variation that the human ear can tolerate in loud sound is about 30 N/m^2. The corresponding maximum displacement for a sound wave in air having a frequency of 10^3 Hz is (take velocity of sound in air as 300 m/s and density of air 1.5 kg/m^3)

(a) $\frac{2\pi}{3} \times 10^{-2}$ m (b) $\frac{2 \times 10^{-4}}{\pi}$ m
(c) $\frac{\pi}{3} \times 10^{-2}$ m (d) $\frac{10^{-4}}{3\pi}$ m

27. A standing wave is maintained in a homogeneous string of cross-sectional area s and density ρ. It is formed by the superposition of two waves travelling in opposite directions given by the equation $y_1 = a\sin(\omega t - kx)$ and $y_2 = 2a\sin(\omega t + kx)$. The total mechanical energy confined between the sections corresponding to the adjacent antinodes is

(a) $\frac{3\pi s\rho\omega^2 a^2}{2k}$ (b) $\frac{\pi s\rho\omega^2 a^2}{2k}$

(c) $\frac{5\pi s\rho\omega^2 a^2}{2k}$ (d) $\frac{2\pi s\rho\omega^2 a^2}{k}$

28. A detector is released from rest over a source of sound of frequency $f_0 = 10^3$ Hz. The frequency observed by the detector at time t is plotted in the graph. The speed of sound in air is $(g = 10\ \text{m/s}^2)$

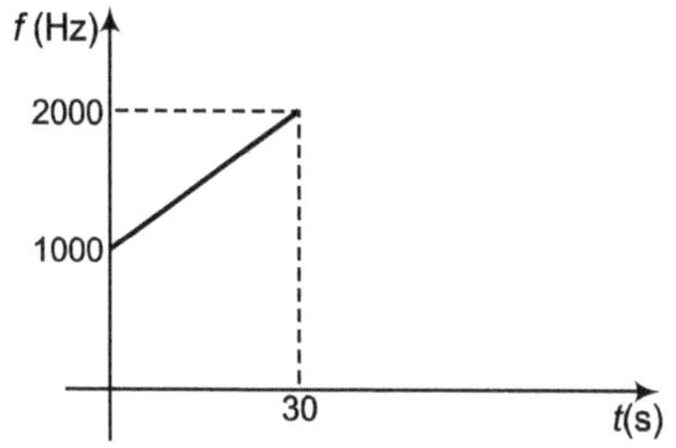

(a) 330 m/s (b) 350 m/s

(c) 300 m/s (d) 310 m/s

29. Two ends of a stretched wire of length L are fixed at $x = 0$ and $x = L$. In one experiment the displacement is $y_1 = A\sin\left(\frac{\pi x}{L}\right)\sin\omega t$ and energy is E_1 and in another experiment its displacement is $y_2 = A\sin\left(\frac{2\pi x}{L}\right)\sin 2\omega t$ and energy is E_1 and in another experiment its displacement is $y_2 = A\sin\left(\frac{2\pi x}{L}\right)\sin 2\omega t$ and energy is E_2, then

(a) $E_2 = E_1$

(b) $E_2 = 2E_1$

(c) $E_2 = 4E_1$

(d) $E_2 = 16E_1$

30. A metal string is fixed between rigid supports. It is initially at negligible tension. Its Young's modulus is Y. Density is ρ and coefficient of thermal expansion is α. If it is now cooled through a temperature $= t$, transverse waves will move along it with speed

(a) $Y\sqrt{\frac{\alpha t}{\rho}}$ (b) $\alpha t\sqrt{\frac{Y}{\rho}}$

(c) $\sqrt{\frac{Y\alpha t}{\rho}}$ (d) $t\sqrt{\frac{Y\alpha}{\rho}}$

31. A closed organ pipe of length L and an open organ pipe contain gases of densities ρ_1 and ρ_2, respectively. The compressibility $\left(= \frac{1}{\text{Bulk modulus}}\right)$ of gases are equal in both the pipes. Both the pipes are vibrating in their first overtone with same frequency. The length of the open organ pipe is

(a) $\frac{L}{3}$ (b) $\frac{4L}{3}$

(c) $\frac{4L}{3}\sqrt{\frac{\rho_1}{\rho_2}}$ (d) $\frac{4L}{3}\sqrt{\frac{\rho_2}{\rho_1}}$

32. The intensity produced by one source is 30 dB. If two such coherent sources are sounded together with zero phase difference, then the intensity level will be

(a) 60 dB (b) 33 dB (c) 40 dB (d) 36 dB

33. A point source of sound S of natural frequency 256 Hz and a receiver R are moving along same line with speed $u = 10$ m/s, towards a reflecting surface which is approaching them with speed u as shown in figure.

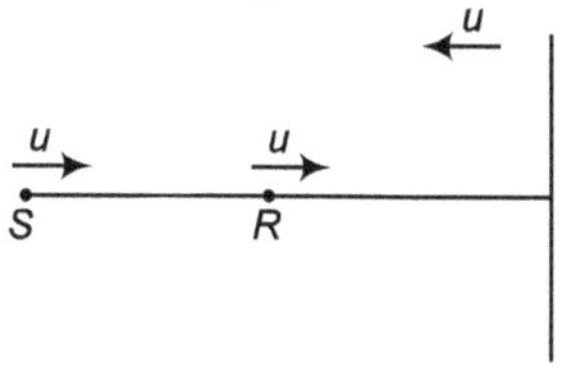

If speed of sound in air is 330 m/s, then wavelength and frequency of reflected wave received by R are respectively

(a) $\frac{330}{289}$ m and 289 Hz (b) $\frac{340}{289}$ m and 289 Hz

(c) $\frac{330}{272}$ m and 289 Hz (d) $\frac{340}{272}$ m and 289 Hz

34. A string of length 0.4 m and mass 10^{-2} kg is tightly clamped at its ends. The tension in the string is 1.6 N. Identical wave pulses are produced at one end at equal intervals of time Δt. The minimum value of Δt which allows constructive interference between successive pulses is

(a) 0.05 s (b) 0.10 s (c) 0.20 s (d) 0.40 s

35. Two speakers connected to the same source of fixed frequency are placed 2 m apart in a box. A sensitive microphone placed at a distance of 4 m from the mid-point along the perpendicular bisector shows maximum response. The box is slowly rotated till the speakers are in line with the microphone. The distance between the mid-point of the speakers and the microphone remains unchanged. Exactly 5 maximum responses (including the initial and last one) are observed in the microphone in doing this. The wavelength of the sound wave is

(a) 0.8 m (b) 0.5 m (c) 0.2 m (d) 1.6 m

More than One Correct Options

1. A transverse wave travelling on a stretched string is represented by the equation $y = \frac{2}{(2x - 6.2t)^2 + 20}$. Then,

(a) velocity of the wave is 3.1 m/s

(b) amplitude of the wave is 0.1 m

(c) frequency of the wave is 20 Hz

(d) wavelength of the wave is 1 m

2. For energy density, power and intensity of any wave choose the correct options.

(a) u = energy density $= \frac{1}{2}\rho\omega^2 A^2$

(b) P = power $= \frac{1}{2}\rho\,\omega^2 A^2 v$

(c) I = intensity $= \frac{1}{2}\rho\,\omega^2 A^2 S v$

(d) $I = \frac{P}{S}$

3. For the transverse wave equation $y = A\sin(\pi x + \pi t)$, choose the correct options at $t = 0$.

(a) Points at $x = 0$ and $x = 1$ are at mean positions

(b) Points at $x = 0.5$ and $x = 1.5$ have maximum accelerations

(c) Points at $x = 0.5$ and $x = 1.5$ are at rest

(d) The given wave is travelling in negative x-direction

4. In the wave equation,

$$y = A\sin\frac{2\pi}{a}(x - bt)$$

(a) speed of wave is a

(b) speed of wave is b

(c) wavelength of wave is a/b

(d) wavelength of wave is a

5. In the wave equation,

$$y = A\sin 2\pi\left(\frac{x}{a} - \frac{t}{b}\right)$$

(a) speed of wave is a/b

(b) speed of wave is b/a

(c) wavelength of wave is a

(d) time period of wave is b

6. Corresponding to y-t graph of a transverse harmonic wave shown in figure

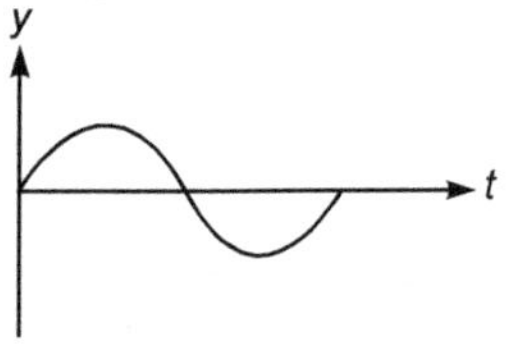

Choose the correct options at same position.

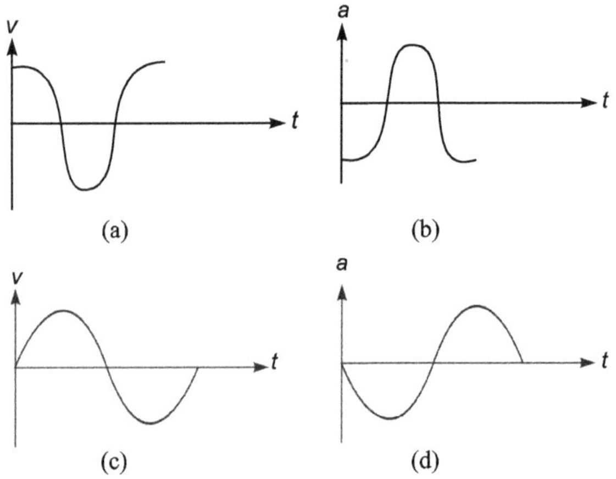

7. If the tension in a stretched string fixed at both ends is increased by 21%, the fundamental frequency is found to change by 15 Hz. Then, the

(a) original frequency is 150 Hz

(b) velocity of propagation of the transverse wave along the string increases by 5%

(c) velocity of propagation of the transverse wave along the string increases by 10%

(d) fundamental wavelength on the string does not change

8. For interference to take place

(a) sources must be coherent

(b) sources must have same amplitude

(c) waves should travel in opposite directions

(d) sources must have same frequency

9. Regarding stationary waves, choose the correct options.

(a) This is an example of interference

(b) Amplitudes of waves may be different

(c) Particles at nodes are always at rest

(d) Energy is conserved

10. When a wave travels from a denser to rarer medium,

(a) speed of wave increases

(b) wavelength of wave decreases

(c) amplitude of wave increases

(d) there is no change in phase angle

11. A wire is stretched and fixed at two ends. Transverse stationary waves are formed in it. It oscillates in its third overtone mode. The equation of stationary wave is

$$Y = A\sin kx\cos\omega t$$

Choose the correct options.

(a) Amplitude of constituent waves is $\frac{A}{2}$

(b) The wire oscillates in three loops

(c) The length of the wire is $\frac{4\pi}{k}$

(d) Speed of stationary wave is $\frac{\omega}{k}$

12. Which of the following equations can form stationary waves?

(i) $Y = A\sin(\omega t - kx)$ (ii) $Y = A\cos(\omega t - kx)$
(iii) $Y = A\sin(\omega t + kx)$ (iv) $Y = A\cos(\omega t + kx)$

(a) (i) and (ii) (b) (i) and (iii)
(c) (iii) and (iv) (d) (ii) and (iv)

13. Two waves

$$y_1 = A\sin(\omega t - kx) \text{ and } y_2 = A\sin(\omega t + kx)$$

superimpose to produce a stationary wave, then

(a) $x = 0$ is a node
(b) $x = 0$ is an antinode
(c) $x = \dfrac{\pi}{k}$ is a node
(d) $\pi = \dfrac{2\pi}{k}$ is an antinode

14. An air column in a pipe, which is closed at one end, is in resonance with a vibrating tuning fork of frequency 264 Hz. If $v = 330$ m/s, the length of the column in cm is (are)

(a) 31.25 (b) 62.50 (c) 93.75 (d) 125

15. Which of the following is/are correct?

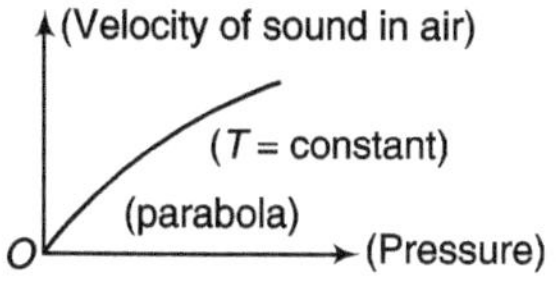

(a)

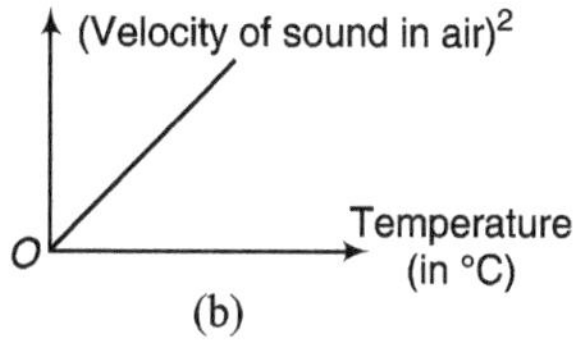

(b)

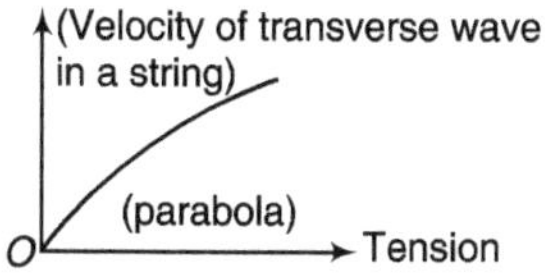

(c)

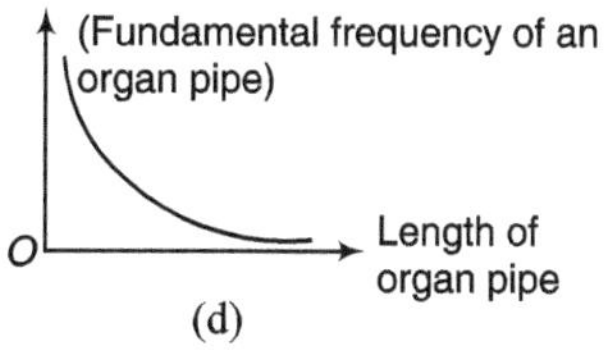

(d)

16. Choose the correct options for longitudinal wave

(a) maximum pressure variation is BAk
(b) maximum density variation is ρAk
(c) pressure equation and density equation are in phase
(d) pressure equation and displacement equation are out of phase

17. Second overtone frequency of a closed pipe and fourth harmonic frequency of an open pipe are same. Then, choose the correct options.

(a) Fundamental frequency of closed pipe is more than the fundamental frequency of open pipe
(b) First overtone frequency of closed pipe is more than the first overtone frequency of open pipe
(c) Fifteenth harmonic frequency of closed pipe is equal to twelfth harmonic frequency of open pipe
(d) Tenth harmonic frequency of closed pipe is equal to eighth harmonic frequency of open pipe

18. For fundamental frequency f of a closed pipe, choose the correct options.

(a) If radius of pipe is increased, f will decrease
(b) If temperature is increased, f will increase
(c) If molecular mass of the gas filled in the pipe is increased, f will decrease.
(d) If pressure of gas (filled in the pipe) is increased without change in temperature, f will remain unchanged

19. A source is approaching towards an observer with constant speed along the line joining them. After crossing the observer, source recedes from observer with same speed. Let f is apparent frequency heard by observer. Then,

(a) f will increase during approaching
(b) f will decrease during receding
(c) f will remain constant during approaching
(d) f will remain constant during receding

20. A transverse harmonic wave on a string is described by

$$y(x,t) = 3.0\sin\left(36t + 0.018x + \frac{\pi}{4}\right)$$

where x and y are in cm and t is in sec. The positive direction of x is from left to right.

(a) the wave is travelling from right to left
(b) the speed of the wave is 20 m/s
(c) frequency of the wave is 5.7 Hz
(d) the least distance between two successive crests in the wave is 2.5 cm

21. The displacement of a string is given by

$$y(x,t) = 0.06\sin\left(\frac{2\pi x}{3}\right)\cos(120\pi t)$$

where x and y are in metre and t in second. The length of the string is 1.5 m and its mass is 3.0×10^{-2} kg.

(a) It represents a progressive wave of frequency 60 Hz
(b) It represents a stationary wave of frequency 60 Hz
(c) It is the result superposition of two waves of wavelength 3 m, frequency 60 Hz each travelling with a speed of 180 m/s in opposite direction
(d) Amplitude of this wave is constant

22. The transverse displacement of a string (clamped at its both ends) is given by

$$y(x,t) = 0.06\sin\left(\frac{2\pi x}{3}\right)\cos(120\pi t).$$

All the points on the string between two consecutive nodes vibrate with

(a) same frequency
(b) same phase
(c) same energy
(d) different amplitude

23. A train, standing in a station yard, blows a whistle of frequency 400 Hz in still air. The wind starts blowing in the direction from the yard to the station with a speed of 10 m/s. Given that the speed of sound in still air is 340 m/s. Then,
(a) the frequency of sound as heard by an observer standing on the platform is 400 Hz
(b) the speed of sound for the observer standing on the platform is 350 m/s
(c) the frequency of sound as heard by the observer standing on the platform will increase
(d) the frequency of sound as heard by the observer standing on the platform will decrease

24. Which of the following statements are true for a stationary waves?
(a) Every particle has a fixed amplitude which is different from the amplitude of its nearest particle
(b) All the particles cross their mean position at the same time
(c) There is no net transfer of energy across any plane
(d) There are some particles which are always at rest

Comprehension Based Questions

Passage I (Q. 1 to 3)

Incident wave $y = A \sin\left(ax + bt + \frac{\pi}{2}\right)$ *is reflected by an obstacle at* $x = 0$, *which reduces intensity of reflected wave by 36%. Due to superposition a resulting wave consists of standing wave and travelling wave given by*

$y = -1.6 \sin ax \sin bt + cA \cos(bt + ax)$
where, A, a, b *and* c *are positive constants.*

1. Amplitude of reflected wave is
(a) $0.6\,A$ (b) $0.8\,A$ (c) $0.4\,A$ (d) $0.2\,A$

2. The value of c is
(a) 0.2 (b) 0.4 (c) 0.6 (d) 0.3

3. The position of second antinode is
(a) $x = \frac{\pi}{3a}$ (b) $x = \frac{3\pi}{a}$ (c) $x = \frac{3\pi}{2a}$ (d) $x = \frac{2\pi}{3a}$

Passage II (Q. 4 to 7)

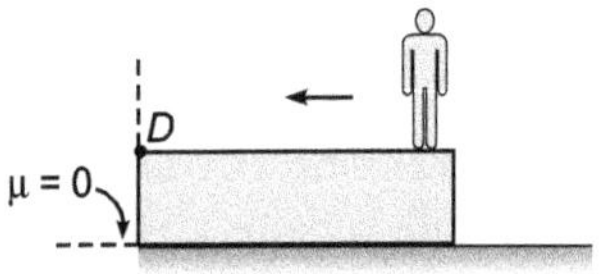

A man of mass 50 kg is running on a plank of mass 150 kg with speed of 8 m/s relative to plank as shown in the figure (both were initially at rest and the velocity of man with respect to ground any how remains constant). Plank is placed on smooth horizontal surface. The man, while running whistles with frequency f_0. *A detector (D) placed on plank detects frequency. The man jumps off with same velocity (w.r.t. to ground) from point D and slides on the smooth horizontal surface [Assume coefficient of friction between man and horizontal is zero]. The speed of sound in still medium is 330 m/s. Answer following questions on the basis of above situations.*

4. The frequency of sound detected by detector D, before man jumps off the plank is
(a) $\frac{332}{324} f_0$ (b) $\frac{330}{322} f_0$
(c) $\frac{328}{336} f_0$ (d) $\frac{330}{338} f_0$

5. The frequency of sound detected by detector D, after man jumps off the plank is
(a) $\frac{332}{324} f_0$ (b) $\frac{330}{322} f_0$
(c) $\frac{328}{336} f_0$ (d) $\frac{330}{338} f_0$

6. Choose the correct plot between the frequency detected by detector *versus* position of the man relative to detector

(a)
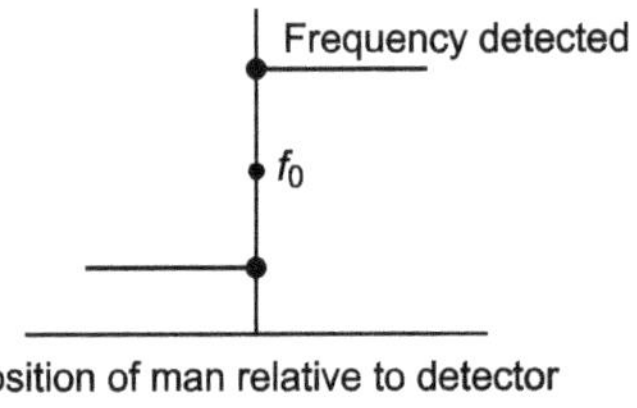

(b)
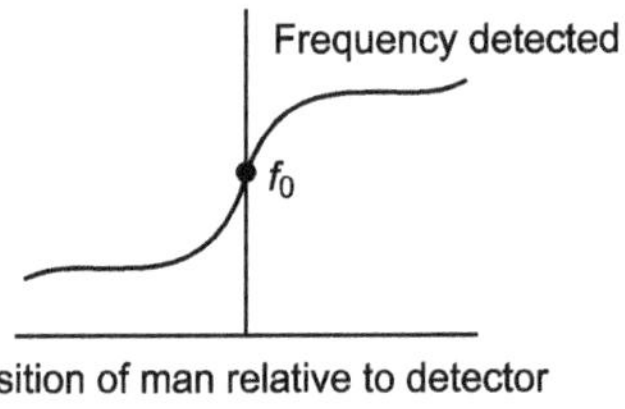

(c)
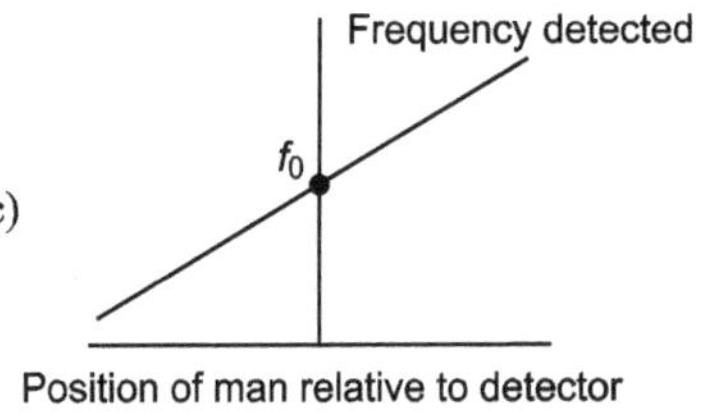

(d)
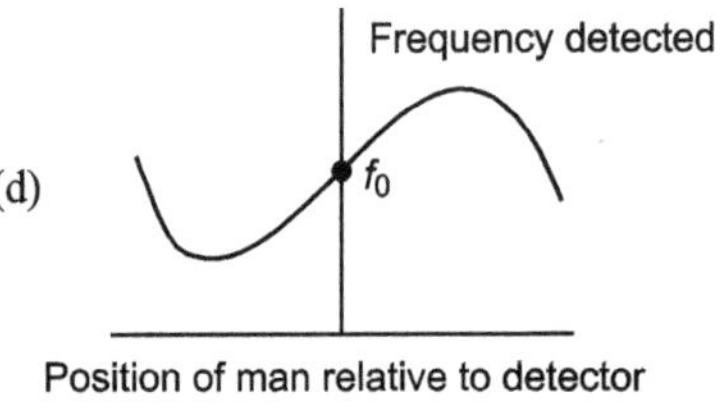

7. Sound waves are travelling along positive x-direction. Displacement of particle at any time t is as shown in figure. Select the wrong statement.

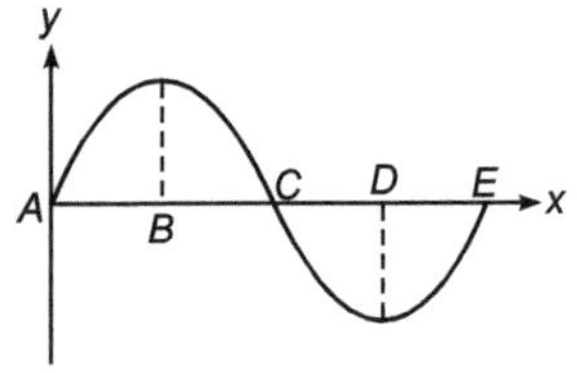

(a) Particle located at E has its velocity in negative x-direction
(b) Particle located at D has zero velocity
(c) Both (a) and (b) are correct
(d) Both (a) and (b) are wrong

Assertion and Reason

Direction (Q. Nos. 1-20) *These questions consist of two statements each printed as Assertion and Reason. While answering these questions you are required to choose any one of the following five responses.*

(a) If both Assertion and Reason are correct and Reason is the correct explanation of Assertion.
(b) If both Assertion and Reason are true but Reason is not the correct explanation of Assertion.
(c) If Assertion is true but Reason is false.
(d) If Assertion is false but Reason is true.
(e) If both Assertion and Reason are false.

1. Assertion When sound wave travels from air to water, it bends towards normal.

Reason When any wave refracts from rarer to denser medium, it bends towards normal.

2. Assertion If temperature is increased, fundamental frequency of an open pipe will increase.

Reason By increasing the temperature speed of sound wave will increase.

3. Assertion In organ pipe experiment, loop size will decrease as the water column length is increased.

Reason Frequency remains unchanged.

4. Assertion In Doppler effect, if the source is moving, change in frequency is observed due to change in wavelength.

Reason By the motion of source wavelength always decreases.

5. Assertion Transverse mechanical wave does not travel in air.

Reason Air has only bulk modulus of elasticity.

6. Assertion A closed pipe and an open pipe cannot be in resonance with the same tuning fork, if the length of the closed pipe is integral multiple of the length of the open pipe.

Reason An odd number is not divisible by an even number.

7. Assertion No Doppler effect will be observed in sound, when source and observer are moving with same speed.

Reason Doppler effect may be due to change in wavelength also.

8. Assertion Ultrasonic is the acoustic analogue of ultraviolet radiation.

Reason Ultraviolet rays do not produce visual sensations while ultrasonic waves are not heard by the human ear.

9. Assertion Longitudinal waves are called pressure waves.

Reason Propagation of longitudinal waves through a medium involves changes in pressure.

10. Assertion If two people talk simultaneously and each creates a sound level of 60 dB at a point P, then total sound level at the point P is 120 dB.

Reason Sound level is defined on a non-linear scale.

11. Assertion Doppler effect does not depend upon the distance between source and observer.

Reason If source and observer are moving with same velocities, then no Doppler effect will be observed.

12. Assertion Speed of sound in O_2 gas at temperature T is v. If temperature is increased to $2T$, the molecular oxygen breaks into atomic oxygen. Then, the speed of sound under these conditions will become $2v$.

Reason With increase in humidity in air, speed of sound increases.

13. Assertion Fundamental frequency of a narrow organ pipe is more than a broad diameter organ pipe of same length.

Reason If we apply Laplace end correction, then fundamental frequency of a closed organ pipe is more than $\frac{v}{4l}$.

14. Assertion If a stretched wire fixed at both ends is vibrating in its second overtone mode, then the total number of nodes and antinodes are four each.

Reason Number of antinodes are equal to number of loops.

15. Assertion Transverse waves cannot travel in gaseous medium.

Reason Gaseous medium does not have modulus of rigidity.

16. Assertion A wire is stretched and then fixed at two ends. Wavelength of second overtone mode is $\frac{2}{3}$ rd of the wavelength of first overtone mode.

Reason Frequency of second overtone mode is $\frac{3}{2}$ times the frequency of first overtone mode.

17. Assertion Sound wave bends towards normal in travelling from air to water.

Reason Whenever a wave travels from a rarer medium to a denser medium, it bends towards normal.

18. Assertion When a wave travels from a denser medium to a rarer medium, its amplitude of oscillation will increase.

Reason In a rarer medium speed of wave is more.

19. Assertion In longitudinal stationary waves, displacement node is pressure antinode and *vice-versa*.

Reason At the point of displacement, node particles are at rest and variation in pressure is maximum.

20. Assertion In the equation $y = A \sin (kx - \omega t)$ particle at $x = 0$ starts from its mean position towards negative y-axis.

Reason The wave corresponding to given equation is travelling towards positive x-direction.

Match the Columns

1. Source has frequency f. Source and observer both have same speed. For the apparent frequency observed by observer, match the following columns.

Column I	Column II
(A) Observer is approaching the source but source is receding from the observer	(p) More than f
(B) Observer and source both approaching towards each other	(q) Less than f
(C) Observer and source both receding from each other	(r) Equal to f
(D) Source is approaching but observer is receding	(s) None

2. In the equation, $y = A \sin 2\pi (ax + bt + \pi/4)$. Match the following columns.

Column I	Column II
(A) Frequency of wave	(p) a
(B) Wavelength of wave	(q) b
(C) Phase difference between two points $\frac{1}{4a}$ distance apart	(r) π
(D) Phase difference of a particle after a time interval of $\frac{1}{8b}$	(s) $\frac{\pi}{2}$
	(t) None

3. A wave is transmitted from a denser to rarer medium. Then, match the following columns.

Column I	Column II
(A) Frequency of wave	(p) Will increase
(B) Speed of wave	(q) Will decrease
(C) Wavelength of wave	(r) Will remain unchanged
(D) Amplitude of wave	(s) May increase or decrease

4. For a closed organ pipe, match the following columns.

Column I	Column II
(A) Third overtone frequency is x times the fundamental frequency. Here, x is equal to	(p) 3
(B) Number of nodes in second overtone	(q) 4
(C) Number of antinodes in second overtone	(r) 5
	(s) None

5. A string is suspended from the ceiling. A wave train is produced at the bottom at regular interval. As the wave moves upwards.

Column I	Column II
(A) Mass per unit length of the string	(p) Increases
(B) Tension in the string	(q) Decreases
(C) Wave speed	(r) Remains same
(D) Wavelength	(s) Missing

6. Regarding speed of sound in a gas match the following columns.

Column I	Column II
(A) Temperature of gas is made 4 times and pressure 2 times	(p) Speed becomes $2\sqrt{2}$ times
(B) Only pressure is made 4 times without change in temperature	(q) Speed becomes 2 times
(C) Only temperature is changed to 4 times	(r) Speed remains unchanged
(D) Molecular mass of the gas is made 4 times	(s) Speed remains half

7. Fundamental frequency of closed pipe is 100 Hz and that of an open pipe is 200 Hz. Match the following columns. ($v_s = 330$ m/s)

Column I	Column II
(A) Length of closed pipe	(p) 0.825 m
(B) Length of open pipe	(q) 1.65 m
(C) Lowest harmonic of closed pipe which is equal to any of the harmonic of open pipe	(r) 5 m
	(s) None

8. Speed of longitudinal wave $v \propto \sqrt{E}$. Here, E is the modulus of elasticity. Match the following columns.

Column I	Column II
(A) In case of solid E is	(p) Bulk modulus of elasticity
(B) In case of liquid E is	(q) Shear modulus of elasticity
(C) In case of gas	(r) Young's modulus of elasticity
	(s) None

9. A string fixed at both ends first oscillates in its fundamental mode, then in second harmonic mode. Then, match the following columns.

Column I	Column II
(A) Frequency	(p) In second case is more
(B) Wavelength	(q) In second case is less
(C) Energy of string	(r) In second case is equal

10. Following is given the equation of a travelling wave (all is SI units)

$$y = (0.02) \sin 2\pi (10t - 5x)$$

Match the following columns.

Column I	Column II
(A) Speed of wave	(p) 10
(B) Frequency of wave	(q) 0.4π
(C) Wavelength of wave	(r) 2
(D) Maximum particle speed	(s) 0.2

11. Following is given the equation of a stationary wave (all in SI units)

$$y = (0.06) \sin (2\pi x) \cos (5\pi t)$$

Match the following columns.

Column I	Column II
(A) Amplitude of constituent wave	(p) 0.06
(B) Position of node at $x = \dots$ m	(q) 0.5
(C) Position of antinode at $x = \dots$ m	(r) 0.25
(D) Amplitude at $x = \frac{3}{4}$ m	(s) 0.03

12. Match the following columns.

Column I	Column II
(A) In refraction	(p) Speed of wave does not change
(B) In reflection	(q) Wavelength is decreased
(C) In refraction in a denser medium	(r) Frequency does not change
(D) In reflection from a denser medium	(s) Phase change of π takes place

Entrance Gallery

2014

1. A pipe of length 85 cm is closed from one end. Find the number of possible natural oscillations of air column in the pipe whose frequencies lie below 1250 Hz. The velocity of sound in air is 340 m/s. **[JEE Main]**

(a) 12
(b) 8
(c) 6
(d) 4

2. A metallic wire of 1 m length has a mass of 10×10^{-3} kg. If a tension of 100 N is applied to a wire, what is the speed of transverse wave? **[Karnataka CET]**

(a) 100 ms^{-1} (b) 10 ms^{-1} (c) 200 ms^{-1} (d) 0.1 ms^{-1}

3. A pipe of 30 cm long and open at both the ends produces harmonics. Which harmonic mode of pipe resonates a 1.1 kHz source? Given, speed of sound in air = 330 ms^{-1}. **[Karnataka CET]**

(a) Fifth harmonic (b) Fourth harmonic
(c) Third harmonic (d) Second harmonic

4. A train is approaching towards a platform with a speed of 10 ms^{-1}, while blowing a whistle of frequency 340 Hz. What is the frequency of whistle by a stationary observer on the whistle heard by a stationary observer on the platform? Given, speed of sound = 340 ms^{-1}. **[Karnataka CET]**

(a) 330 Hz (b) 350 Hz (c) 340 Hz (d) 360 Hz

5. In which of the following phenomena, the heat waves travel along straight lines with the speed of light? **[WB JEE]**

(a) Thermal conduction (b) Forced convection
(c) Natural convection (d) Thermal radiation

6. Two coherent monochromatic beams of intensities $4I$ and I respectively are superimposed. The maximum and minimum intensities in the resulting pattern are **[WB JEE]**

(a) $5I$ and $3I$ (b) $9I$ and $3I$ (c) $4I$ and I (d) $9I$ and I

7. A whistle whose air column is open at both ends, has fundamental frequency of 5100 Hz. If the speed of sound in air is 340 ms^{-1}, the length of the whistle in cm, is **[WB JEE]**

(a) 5/3 (b) 10/3 (c) 5 (d) 20/3

8. Sound waves are passing through two routes-one in straight path and the other along a semicircular path of radius r and are again combined into one pipe and superimposed as shown in the figure. If the velocity of sound waves in the pipe is v, then frequencies of resultant waves of maximum amplitude will be integral multiples of

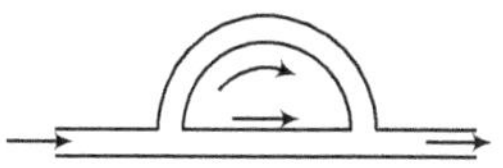

(a) $\frac{v}{r(\pi - 2)}$ (b) $\frac{v}{r(v - 1)}$ (c) $\frac{2v}{r(\pi - 1)}$ (d) $\frac{v}{r(\pi + 1)}$

9. A car is moving with a speed of 72 km-h^{-1} towards a roadside source that emits found at a frequency of 850 Hz. The car driver listens to the sound while approaching, the source and again while moving away from the source after crossing it. If the velocity of sound is 340 ms^{-1}, the difference of the two frequencies, the driver heard is **[WB JEE]**

(a) 50 Hz (b) 85 Hz
(c) 100 Hz (d) 150 Hz

10. Which quantity is transmitted with propagation of longitudinal waves through a medium? **[J&K CET]**

(a) Dispersion (b) Energy
(c) Matter (d) Frequency

11. A string of mass 3 kg is under tension of 400 N. The length of the stretched string is 25 cm. If the transverse jerk is stuck at one end of the string, how long does the disturbance take to reach the other end? **[J&K CET]**

(a) 0.047 s (b) 0.055 s
(c) 0.034 s (d) 0.065 s

12. When the longitudinal wave propagates, what happens in the region of compressions and rarefactions? **[J&K CET]**

(a) Density varies
(b) Density remains constant
(c) There is heat transfers
(d) Boyle's law is obeyed

2013

13. A horizontal stretched string, fixed at two ends, is vibrating in its fifth harmonic according to the equation, $y(x,t) = (0.01\text{ m})[\sin(62.8\text{ m}^{-1})x\cos(628\text{ s}^{-1})t]$.
Assuming, $\pi = 3.14$, the correct statement(s) is (are) **[JEE Advanced]**

(a) The number of nodes is 5
(b) The length of the string is 0.25 m
(c) The maximum displacement of the mid-point of the string from its equilibrium position is 0.01 m
(d) the fundamental frequency is 100 Hz

14. Two vehicles, each moving with speed u on the same horizontal straight road, are approaching each other. Wind blows along the road with velocity. One of these vehicles blows a whistle of frequency f_1. An observer in the other vehicle hears the frequency of the whistle to be f_2. The speed of sound in still air is v. The correct statement(s) is (are) **[JEE Advanced]**

(a) If the wind blows from the observer to the source, $f_2 > f_1$
(b) If the wind blows from the source to the observer, $f_2 > f_1$
(c) If the wind blows from the observer to the source, $f_2 < f_1$
(d) If the wind blows from the source to the observer, $f_2 < f_1$

15. A sonometer wire of length 1.5 m is made of steel. The tension in it produces an elastic strain of 1%. What is the fundamental frequency of steel, if density and elasticity of steel are $7.7\times 10^3\text{ kg/m}^3$ and $2.2\times 10^{11}\text{ N/m}^2$ respectively? **[JEE Main]**

(a) 188.5 Hz (b) 178.2 Hz
(c) 200.5 Hz (d) 770 Hz

16. A bus is moving with a velocity of 5 ms^{-1} towards a huge wall. The driver sounds a horn of frequency 165 Hz. If the speed of sound in air is 335 ms^{-1}, the number of beats heard per second by a passenger on the bus will be **[Karnataka CET]**

(a) 6 (b) 5 (c) 3 (d) 4

17. A sound source is moving towards stationary listener with 1/10th of the speed of the sound. The ratio of apparent to real speed will be **[Karnataka CET]**

(a) $\left(\frac{11}{10}\right)^2$ (b) $\left(\frac{9}{10}\right)^2$
(c) $\frac{10}{9}$ (d) $\frac{11}{10}$

18. A whistle of frequency 500 Hz tied to the end of a string of length of 1.2 m revolves at 400 rev/min. A listener standing some distance away in the plane of rotation of whistle frequencies in the rays (speed of sound $= 340\text{ ms}^{-1}$). **[Karnataka CET]**

(a) 436 to 586 (b) 426 to 574
(c) 436 to 574 (d) 436 to 574

19. A racing car moving towards a cliff, sounds its horn. The driver observes that the sound reflected from the cliff has a pitch one octave higher than the actual sound of the horn. If v is the velocity of sound, then the velocity of the car is **[Karnataka CET]**

(a) $v/2$ (b) $v/\sqrt{2}$
(c) $v/4$ (d) $v/3$

20. A train is approaching with velocity 25 ms^{-1} towards a pedestrian standing on the track, frequency of horn of train is 1 kHz. Frequency heard by the pedestrian is (take, $v = 350\text{ ms}^{-1}$) **[Karnataka CET]**

(a) 1077 Hz (b) 1167 Hz
(c) 985 Hz (d) 954 Hz

21. A string of density 7.5 g cm^{-3} and area of cross-section 0.2 mm^2 is stretched under a tension of 20 N. When it is plucked at the mid-point, the speed of the transverse wave on the wire is **[O JEE]**

(a) 116 ms^{-1} (b) 40 ms^{-1}
(c) 200 ms^{-1} (d) 80 ms^{-1}

22. A man standing between two cliffs, claps his hands and starts hearing a series of echoes at intervals of one second. If the speed of sound in air is 340 ms^{-1}, the distance between the cliffs is **[O JEE]**

(a) 680 m (b) 1700 m (c) 340 m (d) 1620 m

2012

23. A student is performing the experiment of resonance column. The diameter of the column tube is 4 cm. The frequency of the tuning fork is 512 Hz. The air temperature is 38°C in which the speed of sound is 336 m/s. The zero of the meter scale coincides with the top end of the resonance column tube.

When the first resonance occurs, the reading of the water level in the column is **[IIT JEE]**

(a) 14.0 cm (b) 15.2 cm
(c) 16.4 cm (d) 17.6 cm

24. Intensity level of sound whose intensity is 10^{-8} Wm^{-2} is **[Karnataka CET]**

(a) 80 dB (b) 8 dB (c) 4 dB (d) 40 dB

25. A transverse wave is moving in a wire of length of 15 cm long and due to reflection, four anti-nodes are formed. Velocity of wave is 2 m/s, then its frequency is **[OJEE]**

(a) 20 Hz (b) 30 Hz
(c) 40 Hz (d) 45 Hz

26. A train approaching a railway platform with a speed of 20 ms^{-1} starts blowing the whistle, speed of sound in air is 340 ms^{-1}. If the frequency of the emitted sound from the whistle is 640 Hz, the frequency of sound to a person standing on the platform will appear to be **[WB JEE]**

(a) 600 Hz (b) 640 Hz
(c) 680 Hz (d) 720 Hz

2011

27. A police car with a siren of frequency 8 kHz is moving with uniform velocity 36 km/h towards a tall building which reflects the sound waves. The speed of sound in air is 320 m/s. The frequency of the siren heard by the car driver is **[IIT JEE]**

(a) 8.5 kHz (b) 825 kHz
(c) 7.25 kz (d) 7.5 kH

28. A travelling wave represented by $y = a\sin(\omega t - kx)$ is superimposed on another wave represented by, $y = a\sin(\omega t + kx)$. The resultant is **[AIEEE]**

(a) a standing wave having nodes at

$$x = \left(n + \frac{1}{2}\right)\frac{\lambda}{2},\ n = 0, 1, 2$$

(b) a wave travelling along $+x$-direction
(c) a wave travelling along $-x$-direction
(d) a standing wave having nodes at

$$x = \frac{n\lambda}{2},\ n = 0, 1, 2$$

29. The transverse displacement $y(x, t)$ of a wave on a string is given by, $y(x, t) = e^{-(ax^2 + bt^2 + 2\sqrt{ab}xt)}$.

This represents a **[AIEEE]**

(a) wave moving in $-x$-direction with speed $\sqrt{\frac{b}{a}}$
(b) standing wave of frequency $\sqrt{b}$
(c) standing wave of frequency $\frac{1}{\sqrt{b}}$
(d) wave moving in $+x$-direction with speed $\sqrt{\frac{a}{b}}$

30. Tube A has both ends open while tube B has one end closed. Otherwise, they are identical. Their fundamental frequencies are in the ratio **[Kerala CEE]**

(a) 4 : 1 (b) 2 : 1 (c) 1 : 4 (d) 1 : 2
(e) 2 : 3

31. The speed of sound in gas of density ρ at a pressure p is proportional to **[Kerala CEE]**

(a) $\left(\frac{p}{\rho}\right)^2$ (b) $\left(\frac{p}{\rho}\right)^{3/2}$ (c) $\sqrt{\frac{\rho}{p}}$ (d) $\sqrt{\frac{p}{\rho}}$
(e) $\left(\frac{\rho}{p}\right)^2$

32. The equation of a wave is given by $y = 10\sin\left(\frac{2\pi}{45}t + \alpha\right)$.

If the displacement is 5 cm at $t = 0$, then the total phase at $t = 7.5$ s is **[Karnataka CET]**

(a) π (b) $\frac{\pi}{6}$ (c) $\frac{\pi}{2}$ (d) $\frac{\pi}{3}$

33. A plane progressive wave is given by $y = 2\cos 2\pi(330t - x)$. What is period of the wave? **[Karnataka CET]**

(a) $\frac{1}{330}$ s (b) $2\pi \times 330$ s
(c) $(2\pi \times 330)^{-2}$ s (d) $\frac{6.284}{330}$ s

34. Two tuning forks, A and B produce notes of frequencies 258 Hz and 262 Hz. An unknown note sounded with a produces certain beats. When the same note is sounded with B, the beat frequency gets doubled, The unknown frequency is **[Karnataka CET]**

(a) 256 Hz (b) 254 Hz
(c) 300 Hz (d) 280 Hz

35. A wire under tension vibrates with a fundamental frequency of 600 Hz. If the length of the wire is doubled, the radius is halved and the wire is made to vibrate under one-ninth the tension. Then, the fundamental frequency will become **[Karnataka CET]**

(a) 400 Hz (b) 600 Hz (c) 300 Hz (d) 200 Hz

2010

36. A stationary source is emitting a sound at a fixed frequency f_0, which is reflected by two cars approaching the source. The difference between the frequencies of sound reflected from the cars is 1.2% of f_0. What is the difference in the speeds of the cars (in km per hour) to the nearest integer? The cars are moving at constant speeds much smaller than the speed of sound which is 330 ms^{-1}. **[IIT JEE]**

(a) 7.128 km/h (b) 7 km/h
(c) 8.128 km/h (d) 9 km/h

37. A hollow pipe of length 0.8 m is closed at one end. At its open end a 0.5 m long uniform string is vibrating in its second harmonic and it resonates with the fundamental frequency of the pipe. If the tension in the wire is 50 N and the speed of sound is 320 ms^{-1}, the mass of the string is **[IIT JEE]**

(a) 5 g (b) 10 g (c) 20 g (d) 40 g

38. The equation of a wave on a string of linear mass density 0.04 kg m^{-1} is given by

$$y = 0.02 \sin\omega \left[2\pi\left(\frac{t}{0.04\ (\mathrm{s})} - \frac{x}{0.50\ (\mathrm{m})}\right)\right]$$

The tension in the string is **[AIEEE]**

(a) 4.0 N (b) 12.5 N
(c) 0.5 N (d) 6.25 N

39. A closed organ pipe and an open organ pipe of same length produce 2 beats/s while vibrating in their fundamental modes. The length of the open organ pipe is halved and that of closed pipe is doubled. Then, the number of beats produced per second while vibrating in the fundamental mode is **[Karnataka CET]**

(a) 2 (b) 6
(c) 8 (d) 7

40. A uniform wire of length L, diameter D and density ρ is stretched under a tension T. The correct relation between its fundamental frequency f, the length L and the diameter D is **[Karnataka CET]**

(a) $f \propto \frac{1}{LD}$

(b) $f \propto \frac{1}{L\sqrt{D}}$

(c) $f \propto \frac{1}{D^2}$

(d) $f \propto \frac{1}{LD^2}$

41. A bat flies at a steady speed of 4 ms^{-1} emitting a sound of $f = 90 \times 10^3$ Hz. It is flying horizontally towards a vertical wall. The frequency of the reflected sound as detected by the bat will be (take velocity of sound in air as 330 ms^{-1}) **[Karnataka CET]**

(a) 88.1×10^3 Hz
(b) 87.1×10^3 Hz
(c) 92.1×10^3 Hz
(d) 89.1×10^3 Hz

42. Two Cu wires of radii R_1 and R_2, such that $(R_1 > R_2)$. Then, which of the following is true? **[MHT CET]**

(a) Transverse wave travels faster in thicker wire
(b) Transverse wave travels faster in thinner wire
(c) Travels with the same speed in both the wires
(d) Does not travel

43. The equation of a simple harmonic progressive wave is given by, $y = A\sin(100\pi t - 3x)$. Find the distance between 2 particles having a phase difference of $\pi/3$. **[MHT CET]**

(a) $\frac{\pi}{9}$ m (b) $\frac{\pi}{18}$ m
(c) $\frac{\pi}{6}$ m (d) $\frac{\pi}{3}$ m

44. In sine wave, minimum distance between 2 particles having same speed is always **[MHT CET]**

(a) $\frac{\lambda}{2}$ (b) $\frac{\lambda}{4}$
(c) $\frac{\lambda}{3}$ (d) λ

45. In the fundamental mode, time taken by the wave to reach the closed end of the air filled pipe is 0.01 s. The fundamental frequency is **[MHT CET]**

(a) 25 (b) 12.5
(c) 20 (d) 15

46. n_1 is the frequency of the pipe closed at one end and n_2 is the frequency of the pipe open at both ends. If both are joined end to end, find the fundamental frequency of closed pipe, so formed **[MHT CET]**

(a) $\frac{n_1 n_2}{n_2 + 2n_1}$

(b) $\frac{n_1 n_2}{2n_2 + n_1}$

(c) $\frac{n_1 + 2n_2}{n_2 n_1}$

(d) $\frac{2n_1 + n_2}{n_2 n_1}$

Answers

Level 1

Objective Problems

1. (c)	2. (c)	3. (d)	4. (b)	5. (all)	6. (c)	7. (b)	8. (b)	9. (c)	10. (d)
11. (c)	12. (b)	13. (c)	14. (b)	15. (b)	16. (b)	17. (b)	18. (c)	19. (d)	20. (b)
21. (c)	22. (b)	23. (b)	24. (c)	25. (c)	26. (a)	27. (a,c)	28. (b)	29. (a)	30. (b)
31. (b)	32. (b)	33. (d)	34. (b)	35. (a)	36. (a)	37. (d)	38. (d)	39. (a)	40. (d)
41. (b)	42. (a)	43. (b)	44. (c)	45. (c)	46. (a)	47. (a)	48. (a)	49. (d)	50. (a)
51. (a)	52. (c)	53. (a)	54. (a)	55. (b)	56. (c)	57. (a)	58. (a)	59. (a)	60. (a)
61. (d)	62. (c)	63. (d)	64. (a)	65. (d)	66. (all)	67. (c)	68. (c)	69. (a)	70. (c)
71. (d)	72. (c)	73. (c)	74. (a)	75. (d)	76. (d)	77. (a)	78. (d)	79. (c)	80. (c)
81. (b)	82. (c)	83. (c)	84. (a)	85. (c)	86. (b)	87. (b)	88. (c)	89. (d)	90. (c)
91. (b)	92. (d)	93. (b)	94. (a)	95. (d)	96. (a)	97. (d)	98. (c)	99. (c)	100. (c)
101. (d)	102. (d)	103. (d)	104. (b)	105. (c)	106. (c)	107. (c)	108. (b)	109. (c)	110. (b)
111. (d)	112. (a)	113. (d)	114. (a)	115. (b)	116. (c)	117. (c)	118. (d)	119. (a)	120. (d)
121. (c)	122. (c)	123. (b)	124. (a)	125. (c)	126. (a)	127. (a)	128. (c)	129. (b)	130. (a)
131. (b)	132. (b)	133. (b)	134. (c)	135. (c)	136. (d)	137. (b)	138. (d)	139. (c)	140. (d)
141. (c)	142. (c)	143. (c)	144. (c)	145. (b,d)	146. (b)	147. (d)	148. (b)	149. (b)	150. (b)
151. (c)	152. (a)	153. (a)	154. (d)	155. (b)	156. (b)	157. (d)	158. (d)	159. (d)	160. (d)
161. (d)	162. (c)	163. (d)	164. (b)	165. (a)	166. (a)	167. (b)	168. (d)	169. (b)	170. (c)
171. (b)	172. (b)	173. (b)							

Level 2

Objective Problems

1. (a)	2. (a)	3. (a)	4. (d)	5. (c)	6. (a,b)	7. (b,d)	8. (d)	9. (c)	10. (d)
11. (a)	12. (c)	13. (b)	14. (c)	15. (d)	16. (b)	17. (a)	18. (a)	19. (a)	20. (a)
21. (b)	22. (a)	23. (a)	24. (c)	25. (b)	26. (d)	27. (c)	28. (c)	29. (c)	30. (c)
31. (c)	32. (b)	33. (b)	34. (b)	35. (b)					

More than One Correct Options

1. (a,b)	2. (a,d)	3. (all)	4. (b,d)	5. (c,d)	6. (a,d)	7. (a,c,d)	8. (a,d)	9. (a,b,d)	10. (a,c,d)
11. (a,c)	12. (b,d)	13. (b,d)	14. (a,c)	15. (c,d)	16. (a,b,c)	17. (b,c,d)	18. (all)	19. (c,d)	20. (a,b,c)
21. (b,c)	22. (a,b,d)	23. (a,b)	24. (all)						

Comprehension Based Questions

1. (b)	2. (a)	3. (c)	4. (a)	5. (c)	6. (a)	7. (c)

Assertion and Reason

1. (d)	2. (a)	3. (d)	4. (c)	5. (a,b)	6. (a,b)	7. (d)	8. (a,b)	9. (a)	10. (d)
11. (d)	12. (d)	13. (c)	14. (d)	15. (a)	16. (a)	17. (d)	18. (b)	19. (b)	20. (b)

Match the Columns

1. (A $\rightarrow$ r, B $\rightarrow$ p, C $\rightarrow$ q, D $\rightarrow$ r)	2. (A $\rightarrow$ q, B $\rightarrow$ t, C $\rightarrow$ s, D $\rightarrow$ t)	3. (A $\rightarrow$ r, B $\rightarrow$ p, C $\rightarrow$ p, D $\rightarrow$ p)
4. (A $\rightarrow$ s, B $\rightarrow$ p, C $\rightarrow$ p)	5. (A $\rightarrow$ r, B $\rightarrow$ p, C $\rightarrow$ p, D $\rightarrow$ p)	6. (A $\rightarrow$ q, B $\rightarrow$ r, C $\rightarrow$ q, D $\rightarrow$ s)
7. (A $\rightarrow$ p, B $\rightarrow$ p, C $\rightarrow$ s)	8. (A $\rightarrow$ r, B $\rightarrow$ p, C $\rightarrow$ p)	9. (A $\rightarrow$ r, B $\rightarrow$ q, C $\rightarrow$ p)
10. (A $\rightarrow$ r, B $\rightarrow$ p, C $\rightarrow$ s, D $\rightarrow$ q)	11. (A $\rightarrow$ s, B $\rightarrow$ q, C $\rightarrow$ r, D $\rightarrow$ p)	12. (A $\rightarrow$ r, B $\rightarrow$ p, r, C $\rightarrow$ q, r, D $\rightarrow$ p,r,s)

Entrance Gallery

1. (c)	2. (a)	3. (d)	4. (a)	5. (d)	6. (b)	7. (a)	8. (c)	9. (d)	10. (a)
11. (a)	12. (b)	13. (a,b)	14. (b)	15. (b)	16. (b)	17. (c)	18. (a)	19. (d)	20. (a)
21. (a)	22. (c)	23. (b)	24. (d)	25. (b)	26. (c)	27. (a)	28. (a)	29. (a)	30. (b)
31. (d)	32. (c)	33. (a)	34. (b)	35. (d)	36. (a)	37. (c)	38. (d)	39. (d)	40. (a)
41. (c)	42. (b)	43. (a)	44. (a)	45. (a)	46. (a)				

Solutions

Level 1 : Objective Problems

2. All functions of x and t of type $(ax \pm bt)$ represent a wave.
3. Whole angle is called the phase angle.
4. SONAR emits ultrasonic waves.
5. In case of sound wave, y can represent pressure and displacement, while in case of an electromagnetic wave it represents electric and magnetic fields.
(In general y is any general physical quantity which is made to oscillate at one place and these oscillations are propagated to other places also).
7. ωk is dimensionless.

10. $\Delta\phi = \dfrac{2\pi}{\lambda} \cdot \Delta x$

$$\therefore \quad \Delta x = \frac{(\Delta\phi)(\lambda)}{2\pi} = \frac{(\Delta\phi)(v/f)}{2\pi} = \frac{(\pi/3)(360/500)}{2\pi} = 0.12 \text{ m}$$

11. The equation of displacement

$$y_2 = 5(\sin 5\pi t + \sqrt{3}\cos 5\pi t)$$

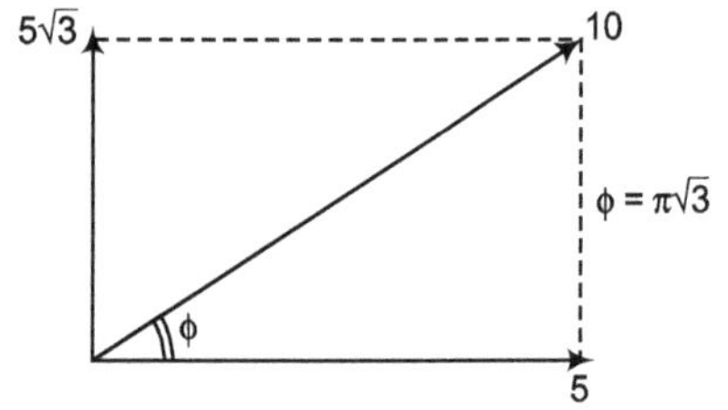

This can also be written as

$$y_2 = 10\sin\left(5\pi t + \frac{\pi}{3}\right)$$

Now, $\quad A_1 = 10 \text{ and } A_2 = 10$

$$\therefore \quad \frac{A_1}{A_2} = \frac{1}{1}$$

12. $(v_P)_{\max} = 4v$

or $\quad Y_0\omega = 4(f\lambda)$

or $\quad Y_0(2\pi f) = 4f\lambda$

$$\therefore \quad \lambda = \frac{\pi Y_0}{2}$$

13. $\omega = 2\pi f$ or $60 = 2\pi f$

$$\therefore \quad f = \frac{30}{\pi} \text{Hz}$$

14. Speed of wave $= \dfrac{\text{coefficient of } t}{\text{coefficient of } x}$

$$= \frac{\dfrac{2\pi}{0.002}}{\dfrac{2\pi}{0.1}} = 50 \text{ m/s}$$

15. $\Delta\phi = \dfrac{2\pi}{\lambda} \cdot \Delta x \Rightarrow \lambda = \left(\dfrac{2\pi}{0.5\pi}\right)(0.8) = 3.2 \text{ m}$

$$\therefore \quad v = f\lambda = 120 \times 3.2 = 384 \text{ m/s}$$

16. $k = \dfrac{2\pi}{40} = \dfrac{2\pi}{\lambda}$

$\therefore \quad \lambda = 40 \text{ cm}$

17. Maximum particle velocity,

$$(v_P)_{\max} = \omega A = 2\pi f A$$
$$= 2\pi(250)(10^{-2}) = 5\pi \text{ m/s}$$

18. $k = \dfrac{2\pi}{\lambda} = \dfrac{2\pi}{60}$

and $\quad \omega = vk = (360)\left(\dfrac{2\pi}{60}\right) = 12\pi$

19. The phase difference

$$\Delta\phi = \left(\frac{2\pi}{T}\right)(\Delta t) = \left(\frac{2\pi}{1/400}\right)(10^{-3}) = 0.8\,\pi = 144^\circ$$

20. The phase difference $\Delta\phi = \dfrac{2\pi}{\lambda} \cdot \Delta x = k \cdot \Delta x = \pi \times 0.01 \times 25 = \dfrac{\pi}{4}$

21. Maximum particle speed is

$$(v_p)_{\max} = \omega A = (10\pi)(0.1) = \pi \text{ cm/s}$$

22. $\omega = 15\pi,\ k = 10\pi$

$$v = \frac{\omega}{k} = 1.5 \text{ m/s}$$
$$\lambda = \frac{2\pi}{k} = \frac{2\pi}{10\pi} = 0.2 \text{ m}$$

Positive sign between kx and ωt means wave is travelling in negative x-direction.

23. $\Delta\phi = \dfrac{2\pi}{\lambda} \cdot \Delta x = k \cdot \Delta x = \pi(0.01)(0.25) = \dfrac{\pi}{400}$

24. $(v_p)_{\max} / v = (\omega A)/(\omega/k) = kA$

25. Comparing with, $y = a\cos(\omega t + kx - \phi)$,

We get $\quad k = \dfrac{2\pi}{\lambda} = 0.02$

$\Rightarrow \quad \lambda = 100 \text{ cm}$

$\Delta\phi = \dfrac{\pi}{2}$. Hence, path difference between them

$$\Delta x = \frac{\lambda}{2\pi} \times \Delta\phi = \frac{\lambda}{2\pi} \times \frac{\pi}{2}$$
$$= \frac{\lambda}{4} = \frac{100}{4} = 25 \text{ cm}$$

26. From the given equation, $k = \dfrac{2\pi}{\lambda} = \dfrac{\pi}{4}$

$\Rightarrow \quad \lambda = 8 \text{ m}$

27. $v_{\max} = a\omega = \dfrac{v}{10} = \dfrac{10}{10} = 1 \text{ m/s}$

$\Rightarrow \quad a\omega = a \times 2\pi n = 1$

$\Rightarrow \quad n = \dfrac{10^3}{2\pi} \qquad (\because a = 10^{-3} \text{ m})$

Since, $\quad v = n\lambda$

$\Rightarrow \quad \lambda = \dfrac{v}{n} = \dfrac{10}{10^3/2\pi}$

$$= 2\pi \times 10^{-2} \text{ m}$$

28. $v = \dfrac{\text{coefficient of } t}{\text{coefficient of } x} = \dfrac{1/2}{1/4} = 2 \text{ m/s}$

Hence, $\quad d = vt = 2 \times 8 = 16 \text{ m}$

29. $\Delta x = k \cdot \Delta x$

$$\Delta\phi = \frac{2\pi}{\lambda} \times \Delta x$$

$$= \pi \times 0.5 = \frac{\pi}{2}$$

30. $f_1 =$ fundamental frequency $= \dfrac{v}{4l} = \dfrac{336}{4 \times 1.05} = 80$ Hz

$f_3 =$ next higher frequency $= 3f_1 = 240$ Hz

31. $\mu = 0.1\ \text{g/cm} = \dfrac{0.1 \times 10^{-3}}{10^{-2}} = 0.01\ \text{kg/m}$

$$v = \sqrt{\frac{T}{\mu}} = \sqrt{\frac{400}{0.01}} = 200\ \text{m/s}$$

Fundamental frequency $= \dfrac{v}{2l} = \dfrac{200}{2 \times 1} = 100$ Hz

32. The given wave is travelling in positive x-direction. The other wave should travel in negative x-direction. Further at $x = 0$, net displacement by both the waves should always be zero for any value of t, as it is node. Only option (b) satisfies both these conditions.

33. $n = \dfrac{v}{2l} = \dfrac{\sqrt{T/\mu}}{2l}$, where $\mu =$ mass per unit length $= \rho S$

or $\quad n = \dfrac{1}{2l}\sqrt{\dfrac{T}{\rho S}}$ but $S = \pi r^2$

$\therefore \quad n \propto \dfrac{1}{rl}$

or $\quad \dfrac{n_1}{n_2} = \dfrac{r_2 l_2}{r_1 l_1} = \dfrac{r \times 2l}{2r \times l} = 1$

34. Velocity of the transverse wave is $v = \sqrt{\dfrac{T}{\mu}}$ or $v \propto T^{1/2}$.

35. Let this be in nth harmonic. Then,

$$n\left(\frac{v}{2l}\right) = 1.1 \times 10^3$$

$$\therefore \quad n = \frac{2.2 \times 10^3 \times l}{v}$$

$$= \frac{2.2 \times 10^3 \times 0.3}{330} = 2$$

36. Fundamental frequency,

$$n = \frac{v}{2l} = \frac{\sqrt{T/\mu}}{2l}$$

$$= \frac{\sqrt{T/\rho S}}{2l} = \frac{\sqrt{T/\rho\pi r^2}}{2l}$$

$$\therefore \quad n \propto \frac{\sqrt{T}}{2l}$$

Given that tension, diameter and length all are made 3 times.

$\therefore$ n will become $\dfrac{1}{3\sqrt{3}}$ times.

37. First overtone of closed pipe $(P_1) =$ third overtone of open pipe (P_2)

$$\therefore \quad 3\left(\frac{v}{4l_1}\right) = 4\left(\frac{v}{2l_2}\right)$$

$$\therefore \quad \frac{l_1}{l_2} = \frac{3}{8}$$

38. Speed of wave, $v = \dfrac{\omega}{k} = \dfrac{100}{0.01} = 10^4$ m/s

39. $f = \dfrac{v}{4l} = 412$ Hz

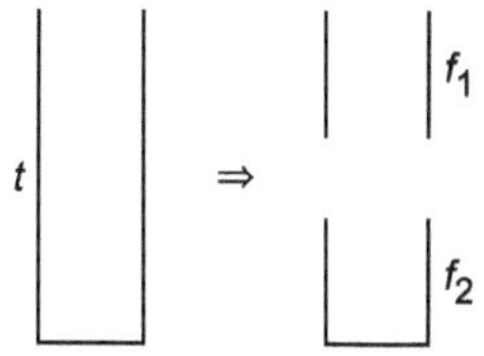

$$f_1 = \frac{v}{2(l/2)} = \frac{v}{l} = 4 \times 412 = 1648\ \text{Hz}$$

$$f_2 = \frac{v}{4(l/2)} = \frac{v}{2l} = 2 \times 412 = 824\ \text{Hz}$$

40. Fundamental frequency $= \dfrac{v}{2l} = \dfrac{\sqrt{T/\mu}}{2l}$ or $f \propto \sqrt{T}$.

So, the rato of their fundamental frequencies are

$$= \sqrt{1:4:9:16} = 1:2:3:4$$

41. $\dfrac{\lambda}{4} = 20$ cm

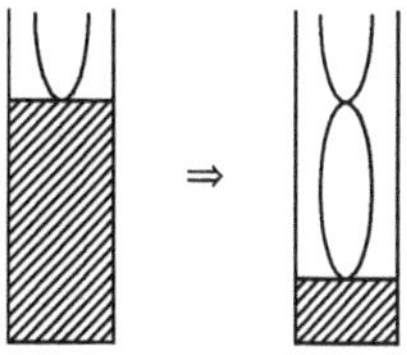

Next is $\dfrac{3\lambda}{4}$ or 60 cm.

42. Third overtone of closed pipe = 2nd harmonic of open pipe

$$\therefore \quad 7\left(\frac{v}{4l_c}\right) = 2\left(\frac{v}{2l_o}\right)$$

$$\therefore \quad \frac{l_c}{l_o} = \frac{7}{4}$$

43. If it is closed pipe, first overtone will be 300 Hz and if it is open pipe, first overtone will be 200 Hz.

44. Phase difference between two equations is $\dfrac{\pi}{2}$. Hence,

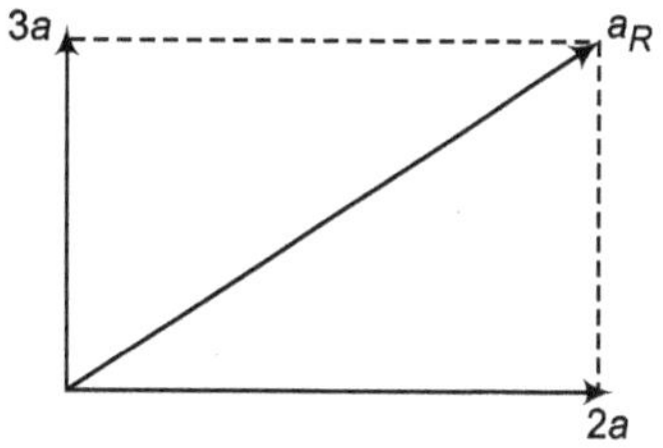

$$a_R = \sqrt{9a^2 + 4a^2} = \sqrt{13}\,a$$

$$= 3.6\,a$$

45. $\dfrac{\lambda}{2} = (50 - 30)\ \text{cm} = 20\ \text{cm}$

Next resonance will be obtained at $50 + \dfrac{\lambda}{2} = 70$ cm

46. Node will occur, where

$$\frac{\pi x}{25} = \frac{\pi}{2}, \frac{3\pi}{2}, \frac{5\pi}{2}, \text{etc.}$$

or $\quad x = 12.5$ cm, 37.5 cm, 62.5 cm, etc.

47. Node should be at $x = 0$.

This is possible when $A_x = A \sin kx$.

48. Fundamental frequency of closed pipe is $\frac{v}{4l}$ and that of open pipe is $\frac{v}{2l}$, two times that of the closed pipe.

Therefore, the fundamental frequency of open pipe at both ends is

$$f = 512 \times 2 = 1024 \text{ Hz}$$

49. Fundamental frequency of closed pipe $f_1 = \frac{v}{4l} = 80$ Hz.

Next is $f_3 = 3f_1 = 240 \text{ Hz} \to$ first overtone

and $f_5 = 5f_1 = 400 \text{ Hz} \to$ second overtone.

50. $A_x = 4\sin\left(\frac{\pi x}{15}\right)$

at $x = 5 \text{ cm}, A_x = 4\sin\frac{\pi}{3} = 2\sqrt{3}$ cm

51. 7th harmonic means 7 loops.

In one loop, there are 2 nodes and 1 antinodes. In 7 loops there will be 8 nodes and 7 antinodes.

52. $f_1 : f_2 : f_3 = 1:3:5$ odd harmonic are obtained in case of a pipe closed at one end.

53. $\lambda = \frac{v}{f} = 1$ m

$$\frac{\lambda}{4} = 0.25 \text{ m} = 25 \text{cm}$$

$$\frac{3\lambda}{4} = 0.75 \text{ m} = 75 \text{cm}$$

$$\frac{5\lambda}{4} = 1.25 \text{ m} = 125 \text{ m}$$

These are the lengths of air column. Hence, length of water column will be 75 cm and 25 cm.

54.

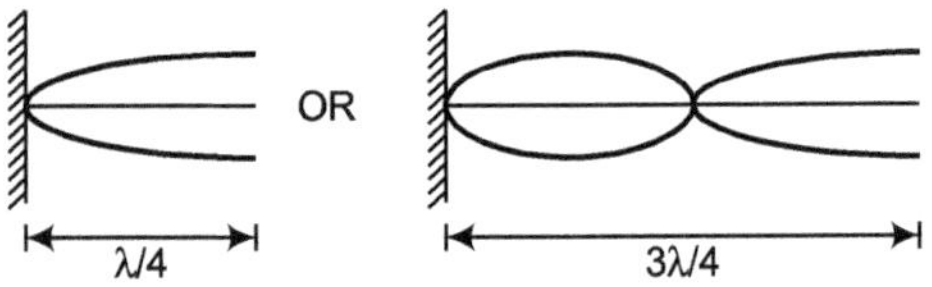

55. $\frac{v_A}{2l} = 2\left(\frac{v_B}{2l}\right) \Rightarrow v_A = 2v_B$ but $v \propto \sqrt{T}$

$\therefore$ $T_A = 4T_B$

56. Given, $3\left(\frac{v}{4l_c}\right) = 2\left(\frac{v}{2l_o}\right)$ or $\frac{l_c}{l_o} = \frac{3}{4}$

Now, $n\left(\frac{v}{4l_c}\right) = m\left(\frac{v}{2l_o}\right)$ or $\frac{n}{m} = 2\frac{l_c}{l_o} = \frac{3}{2}$

Thus ratio $\frac{n}{m}$ should be $\frac{3}{2}$ but n is only odd, while m may be even or odd.

57. $\frac{2\pi}{\lambda} = \frac{\pi}{20} \Rightarrow \lambda = 40$ cm

Separation between two consecutive nodes

$$= \frac{\lambda}{2} = \frac{40}{2} = 20 \text{ cm}$$

58. Required distance $= \frac{\lambda}{4} = \frac{v/n}{4} = \frac{1200}{4 \times 300} = 1$ m

60. Wave $z_1 = A\sin(kx - \omega t)$ is travelling towards positive x-direction.

Wave $z_2 = A\sin(kx + \omega t)$ is travelling towards negative x-direction.

Wave $z_3 = A\sin(ky - \omega t)$ is travelling towards positive y-direction.

Since, waves z_1 and z_2 are travelling along the same line in opposite directions so they will produce stationary wave.

61. The distance between adjacent nodes $x = \frac{\lambda}{2}$

Also, $k = \frac{2\pi}{\lambda}$

Hence, $x = \frac{\pi}{k}$

62.

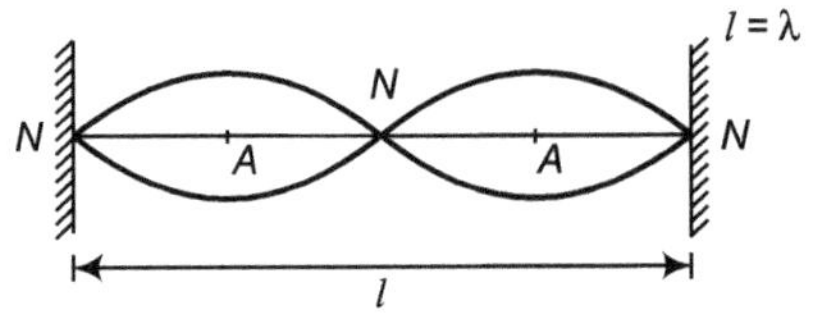

63. $n \propto \frac{1}{l} \Rightarrow \frac{l_1}{l_2} \Rightarrow n_2 = \frac{l_1}{l_2} n_1 = \frac{1 \times 256}{1/4} = 1024$ Hz

65. $n = \frac{1}{2l}\sqrt{\frac{T}{\mu}} \Rightarrow n \propto \frac{\sqrt{T}}{l}$

$$\Rightarrow \quad \frac{T_2}{T_1} = \left(\frac{n_2}{n_1}\right)^2 \left(\frac{l_2}{l_1}\right)^2$$

$$= (2)^2 \left(\frac{3}{4}\right)^2 = \frac{9}{4}$$

66. $y = 0.02\cos(10\pi x)\cos\left(50\pi t + \frac{\pi}{2}\right)$

At node, amplitude $= 0$

$\Rightarrow$ $\cos(10\pi x) = 0$

$\Rightarrow$ $10\pi x = \frac{\pi}{2}, \frac{3\pi}{2}$

$\Rightarrow$ $x = \frac{1}{20} = 0.05$ m, 0.15 m

At antinode, amplitude is maximum

$\Rightarrow$ $\cos(10\pi x) = \pm 1$

$\Rightarrow$ $10\pi x = 0, \pi, 2\pi \ldots$

$\Rightarrow$ $x = 0, 0.1 \text{ m}, 0.2 \text{ m}$

$$v = \frac{50\pi}{10\pi} = 10 \text{ m/s}$$

67. $f = \frac{v}{2l}$ and $f' = \frac{v}{4(l/2)} = \frac{v}{2l} = f$

In half dipping in water, it will become closed pipe.

68. Fundamental frequency $f \propto \frac{1}{l}$.

Given, $f_1 : f_2 : f_3 = 1:2:3$

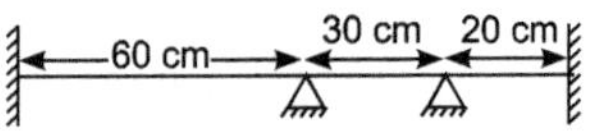

or $\frac{1}{l_1} : \frac{1}{l_2} : \frac{1}{l_3} = 1:2:3$ or $l_1 : l_2 : l_3 = \frac{1}{1} : \frac{1}{2} : \frac{1}{3}$

or $l_1 : l_2 : l_3 = 6:3:2$

$$l_1 = \frac{6}{11} \times 110 = 60\,\text{m}$$

$$l_2 = \frac{3}{11} \times 110 = 30\,\text{cm}$$

and $$l_3 = \frac{2}{11} \times 110 = 20\,\text{cm}$$

69. Five segments means five loops. One loop length is $\frac{\lambda}{2}$

Hence, $$5\frac{\lambda}{2} = 10\,\text{m}$$

$\therefore$ $$\lambda = 4\,\text{m}$$

Now, $$f = \frac{v}{\lambda} = 5\,\text{Hz}$$

70. Standing wave will be formed by their superposition.

$$k = \frac{2\pi}{\lambda} = \frac{2\pi}{6} = \frac{\pi}{3}\,\text{cm}^{-1}$$

$$\omega = v\,k = 0.5 \times \frac{\pi}{3} = \frac{\pi}{6}\,\text{s}^{-1}$$

$$A_{\text{max}} = 2A = 6\,\text{cm}$$

71. $f = \frac{v}{2l} = \frac{\sqrt{T/\mu}}{2l} = \frac{\sqrt{T/\rho S}}{2l} = \frac{\sqrt{T/\pi r^2 \rho}}{2l}$

$\therefore$ $$f \propto \frac{\sqrt{T}}{rl}$$

Now tension, length and radius all are doubled.

Hence, $f' = \frac{f}{2\sqrt{2}}$

72. Second overtone means three loops.

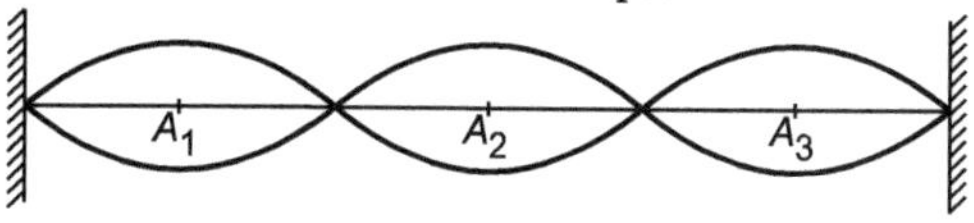

$\therefore$ $$3\frac{\lambda}{2} = L$$

or $$\lambda = \frac{2L}{3}$$

Anitnodes will be obtained at $\frac{\lambda}{4}, \frac{3\lambda}{4}, \frac{5\lambda}{4}$ or $\frac{L}{6}, \frac{3L}{6}$ and $\frac{5L}{6}$

73. $A = \frac{2}{2} = 1\,\text{mm}$

From the given equation we can see that $k = 6.25\,\text{m}^{-1}$, smallest length of string is equivalent to a single loop or length of string equal to $\frac{\lambda}{2}$ or $\frac{\pi}{k}$.

$\therefore$ $$L_{\text{min}} = \frac{\pi}{k} = \frac{\pi}{6.25} = 0.5\,\text{m}$$

74. $f_1 : f_2 : f_3 = 425 : 595 : 765 = 5 : 7 : 9$

Since, these are odd harmonics, pipe is closed.

Further, $5\left(\frac{v}{4l}\right) = 425$

$\therefore$ $$l = \frac{425 \times 4}{5v} = \frac{425 \times 4}{5 \times 340} = 1\,\text{m}$$

75. $n \propto \frac{\sqrt{T}}{l}$ or $n \propto \frac{\sqrt{M}}{l}$

$\therefore$ $$\frac{n_A}{n_B} = \sqrt{\frac{M_A}{M_B}}\frac{l_B}{l_A}$$

76. $I = \frac{1}{2}\rho\omega^2 A^2 v$ or $I \propto \omega^2$ or $I \propto f^2$

$$\frac{I_2}{I_1} = \left(\frac{f_2}{f_1}\right)^2 = \left(\frac{1200}{400}\right)^2 = 9$$

79. Sound level $= 10\log\left(\frac{I}{I_0}\right)$, $30 = 10\log\left(\frac{I}{I_0}\right)$ or $3 = \log\left(\frac{I}{I_0}\right)$

$\therefore$ $$\frac{I}{I_0} = 10^3 = 1000$$

80. Speed of sound $\propto \sqrt{T}$

The temperature $= 27 + 273 = 300\,\text{K}$

Thus, to make speed of sound two times, absolute temperature (in K) has to be raised to 4 times.

81. Both are diatomic. Hence, $\gamma = \frac{C_p}{C_V}$ for both is same $v = \sqrt{\frac{\gamma RT}{M}}$

$\therefore$ $$\frac{v_{\text{H}}}{v_{\text{O}}} = \sqrt{\frac{M_{\text{O}}}{M_{\text{H}}}} = \sqrt{\frac{32}{2}} = 4$$

82. Velocity of sound $v = \sqrt{\frac{\gamma RT}{M}}$ or $v \propto T^{1/2}$.

For small percentage change,

% decrease in $v = \frac{1}{2}$ (% change in T) $= \frac{1}{2}(1\%) = 0.5\%$

83. $v = \sqrt{\frac{\gamma RT}{M}}$. For diatomic gas $\gamma = 1.4$ and $T = 273\,\text{K}$

$\therefore$ $$\gamma T \approx 382 \quad \text{or} \quad v = \sqrt{\frac{382\,R}{M}}$$

84. $v \propto \sqrt{T} \;\Rightarrow\; \therefore \; v' = \sqrt{1.5} \cdot v = 1.22\,v$

85. $$\frac{v_{\text{O}_2}}{v_{\text{He}}} = \frac{\sqrt{\frac{7/5RT}{32}}}{\sqrt{\frac{5/3\,RT}{4}}} = 0.32 \quad \text{or} \quad v_{\text{O}_2} = 0.32\,v_{\text{He}}$$

Time taken $= \frac{1}{0.32}T = 3T$

86. $2d_1 = 340 \times t_1$

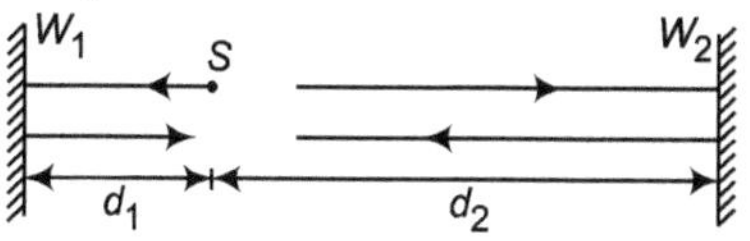

$\therefore$ $$d_1 = 340\,\text{m} \qquad (t_1 = 2\,\text{s})$$

$$2d_2 = 340 \times t_2$$

$\therefore$ $$d_2 = 680\,\text{m}\,(t_2 = t_1 + 2 = 4\,\text{s})$$

$\therefore$ Distance between walls $= d_1 + d_2 = 1020\,\text{m}$

Next echo will be heard at 6 s not at 8 s. Because sound wave reflected from W_2 will be reflected by W_1 in next 2 s.

87. $v = \sqrt{\frac{T}{\mu}} = \sqrt{\frac{T}{\rho s}}$ or $v \propto \frac{1}{\sqrt{s}}$

So, $$\frac{v_1}{v_2} = \sqrt{\frac{S_2}{S_1}} \Rightarrow \sqrt{\frac{4}{1}} = \frac{2}{1}$$

88. Sound level (in dB) $= 10\log_{10}\left(\frac{I}{I_0}\right)$

Here, $$I_0 = 10^{-12}\,\text{W/m}^2$$

$\therefore$ $$\text{Sound level} = 10\log_{10}\frac{2 \times 10^{-8}}{10^{-12}} = 43\,\text{dB}$$

89. $90-40=10\log\frac{I_1}{I_0}-10\log\frac{I_2}{I_0}$

or $50=10\log\left(\frac{I_1}{I_2}\right)$

$\therefore$ $\frac{I_1}{I_2}=10^5$

90. Velocity of sound in gas $v=\sqrt{\frac{\gamma RT}{M}}$

$\Rightarrow$ $v\propto\sqrt{\frac{\gamma}{M}}$

$\Rightarrow$ $\frac{v_{N_2}}{v_{He}}=\sqrt{\frac{\gamma_{N_2}}{\gamma_{He}}\times\frac{M_{He}}{M_{H_2}}}$

$=\sqrt{\frac{\frac{7}{5}\times 4}{\frac{5}{3}\times 28}}=\frac{\sqrt{3}}{5}$

92. $v\propto\lambda\Rightarrow\frac{\lambda_1}{\lambda_2}=\frac{v_1}{v_2}=\frac{2/3}{3/10}=\frac{20}{9}$

93. By using $v=\sqrt{\frac{\gamma RT}{M}}$

$\Rightarrow$ $v\propto\sqrt{T}$

$\frac{v_2}{v_1}=\sqrt{\frac{T_2}{T_1}}$

$=\sqrt{\frac{T+600}{T}}=\sqrt{3}$

$\Rightarrow$ $T=300\,\text{K}=27°\text{C}$

94. Velocity of sound is independent of frequency. Therefore, it is same (v) for frequency n and $4n$.

95. If d is the distance between man and relfecting surface of sound, then for hearing echo

$2d=v\times t$

$\Rightarrow$ $d=\frac{340\times 1}{2}=170\,\text{m}$

96. $v=\sqrt{\frac{\gamma RT}{M}}\Rightarrow v\propto\frac{1}{\sqrt{M}}$. Since, M is minimum for H_2 so sound velocity is maximum in H_2.

97. Speed of sound $v\propto\sqrt{T}$ and it is independent of pressure.

98. $dB=10\log_{10}\left(\frac{I}{I_0}\right)$, where, $I_0=10^{-12}\,\text{Wm}^{-2}$

Since, $40=10\log_{10}\left(\frac{I_1}{I_0}\right)\Rightarrow\frac{I_1}{I_0}=10^4$ ...(i)

Also, $20=10\log_{10}\left(\frac{I_2}{I_0}\right)\Rightarrow\frac{I_2}{I_0}=10^2$ ...(ii)

$\Rightarrow$ $\frac{I_2}{I_1}=10^{-2}=\frac{r_1^2}{r_2^2}$

$\Rightarrow$ $r_2^2=100r_1^2$

$\Rightarrow$ $r_2=10r_1$

$\Rightarrow$ $r_2=10\,\text{m}$ $(\because r_1=1\,\text{m})$

99. $L_2-L_1=10\log\frac{I_2}{I_0}-10\log\frac{I_1}{I_0}=10\log\left(\frac{I_2}{I_1}\right)$

$L_2-4=10\log(2)$ (Given $\frac{I_2}{I_1}=2$)

$\therefore$ $L_2=7\,\text{dB}$

100. $v_{rms}=\sqrt{\frac{3RT}{M}}=c$ ...(i)

$\gamma=1+\frac{2}{f}=\frac{4}{3}$

$v_{sound}=\sqrt{\frac{\gamma RT}{M}}=\sqrt{\frac{4RT}{3M}}$ ...(ii)

From Eqs. (i) and (ii), we have

$v_{sound}=\frac{2}{3}c$

101. $f\propto\sqrt{T}$ or $f\propto T^{1/2}$

% increase in $f=\frac{1}{2}$ (% increase in T) $=\frac{1}{2}(4)=2\%$

$f_1=100\,\text{Hz}$

and $f_2=100+2\%\text{ of }100=102\,\text{Hz}$

Beat frequency, $f=f_2-f_1=2\,\text{Hz}$

102. Number of beats $=f_2-f_1=\frac{396}{0.9}-\frac{396}{1.0}=44\,\text{Hz}$

103. $f'=f\left(\frac{v}{v-v_s}\right)=(1)\left(\frac{v}{v-0.9v}\right)=10\,\text{kHz}$

104. Number of beats $=f_2-f_1=\frac{\omega_2-\omega_1}{2\pi}=\frac{2008\pi-2000\pi}{2\pi}=4\,\text{Hz}$

105. $f'=f\left(\frac{v+v_0}{v-v_s}\right)=480\left(\frac{330+20}{330-20}\right)=542\,\text{Hz}$

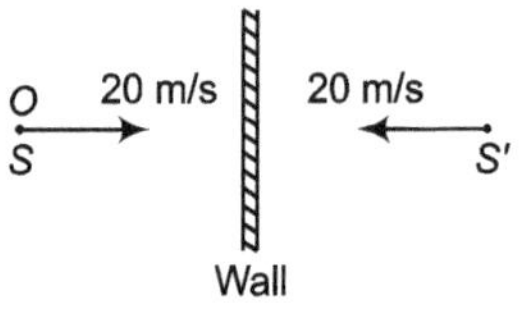

To a man sitting in the car.

106. $f_1-f_2=40$

or $f\left(\frac{340}{340-v_s}\right)-f\left(\frac{340}{340+v_s}\right)=40$

or $400\left(\frac{340}{340-v_s}\right)-400\left(\frac{340}{340+v_s}\right)=40$

Applying Binomial theorem and then solving, we get

$v_s\approx 17\,\text{m/s}$

107. $f_1-f_2=\frac{10}{3}$ or $\frac{v}{1}-\frac{v}{1.01}=\frac{10}{3}$

$\therefore$ $v=336.6\,\text{m/s}$

108. $f_1-f_2=10$

$\therefore$ $680\left(\frac{340+u}{340}\right)-\left(\frac{340-u}{340}\right)(680)=10$

or $u=2.5\,\text{m/s}$

109. With 514 Hz, two beats are heard. Therefore, frequency of tuning fork can be either 512 Hz or 516 Hz. With 510 Hz it gives beats of frequency 6 Hz. Hence, frequency of tuning fork is 516 Hz.

110. $5.5=5\left(\frac{v+v_A}{v}\right)$ or $\frac{v_A}{v}=\frac{1}{10}$ ...(i)

$6=5\left(\frac{v+v_B}{v}\right)$ or $\frac{v_B}{v}=\frac{1}{5}$ ...(ii)

From Eqs. (i) and (ii), we have

$\frac{v_B}{v_A}=2$

111. $f_1 = f\left(\frac{340}{340-34}\right)$

$$f_2 = f\left(\frac{340}{340-17}\right)$$

$$\therefore \quad \frac{f_1}{f_2} = \frac{323}{306} = \frac{19}{18}$$

112. $f = 300\left(\frac{v}{v - v/3}\right) = 450\,\text{Hz}.$

113. Frequency of string may be 470 Hz or 490 Hz.

With increase in tension, frequency of string will increase. If initially it is 470 Hz, beat frequency with given tuning fork will decrease. If it is 490 Hz, beat frequency will increase.

114. Frequency of B can be $n + x$ or $n - x$.

When B is loaded with wax, its frequency will decrease. So, if its original frequency is $(n + x)$, then beat frequency will decrease.

116. Suppose f_0 be the actual frequency. Then,

$$f = f_0\left(\frac{v + v_0}{v}\right)$$

and $$0.8\,f = f_0\left(\frac{v - v_0}{v}\right)$$

Solving these two equations, we get, $v_0 = \frac{v}{9}$

117. Both S and S' are moving towards observer. So both the observed frequencies will be more than the actual, but both will be equal.

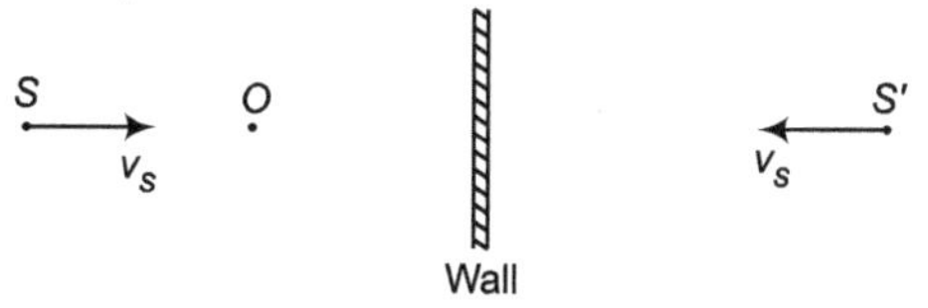

118. Frequency of sound does not change with medium, because it is characteristic of source.

119. Unknown frequency may be 102 Hz or 98 Hz. On loading this tuning fork, its frequency will decrease. Suppose, they become 101 Hz or 97 Hz. With 100 Hz, beat frequency will be 1 Hz or 3 Hz. Given is 1 Hz. Hence, unknown frequency should be 102 Hz.

120. Frequency of unknown tuning fork can be 258 Hz or 254 Hz. On loading the given tuning fork its frequency will decrease. Say it becomes 255 Hz. With 254 Hz, beat frequency is 1 Hz which is given. Hence, frequency of unknown tuning fork is 254 Hz.

121. Let n be the frequency of fork C, then

$$n_A = n + \frac{3n}{100} = \frac{103n}{100}$$

and $$n_B + n - \frac{2n}{100} = \frac{98n}{100}$$

but $$n_A - n_B = 5$$

$$\Rightarrow \quad \frac{5n}{100} = 5$$

$$\Rightarrow \quad n = 100\,\text{Hz}$$

$$\therefore \quad n_A = \frac{(103)(100)}{100}$$

$$= 103\,\text{Hz}$$

122. $n_1 = n_2 = 10$ ...(i)

Using $$n_1 = \frac{v}{4l_1}$$

and $$n_2 = \frac{v}{4l_2}$$

$$\Rightarrow \quad \frac{n_1}{n_2} = \frac{l_2}{l_1} = \frac{26}{25} \qquad \text{...(ii)}$$

After solving these equations

$$n_1 = 260\,\text{Hz}, n_2 = 250\,\text{Hz}$$

123. Initially, number of beats per second = 5

$\therefore$ Frequency of pipe = $200 \pm 5 = 195$ Hz or 205 Hz ...(i)

Frequency of second harmonic of the pipe = $2n$ and number of beats in this case = 10

$$\therefore \quad 2n = 420 \pm 10$$

$$\Rightarrow \quad 410\,\text{Hz} \quad \text{or} \quad 430\,\text{Hz}$$

$$\Rightarrow \quad n = 205\,\text{Hz} \quad \text{or} \quad 215\,\text{Hz} \qquad \text{...(ii)}$$

From Eqs. (i) and (ii), it is clear that $n = 205$ Hz.

124. $n' = n\left(\frac{v}{v - v_s}\right) \Rightarrow \lambda' = \lambda\left(\frac{v - v_s}{v}\right)$

$$\Rightarrow \quad \lambda' = 120\left(\frac{330-60}{330}\right)$$

$$= 98\,\text{cm}$$

125. Both listeners hears the same decreased frequencies.

126. $n' = n\left(\frac{v}{v - v_s}\right) \Rightarrow \frac{n'}{n} = \frac{v}{v - v_s}$

$$\Rightarrow \quad \frac{v}{v - v_s} = 3$$

$$\Rightarrow \quad v_s = \frac{2v}{3}$$

127. When source is approaching the observer, the frequency heard is

$$n_a = \left(\frac{v}{v - v_s}\right) \times n$$

$$= \left(\frac{340}{340-20}\right) \times 1000$$

$$= 1063\,\text{Hz}$$

When source is receding, the frequency heard

$$n_r = \left(\frac{v}{v + v_s}\right) \times n = \left(\frac{340}{340+20}\right) \times 1000 = 944$$

$$\Rightarrow \quad n_a : n_r = 9 : 8$$

Alternately $$\frac{n_a}{n_r} = \frac{v + v_s}{v - v_s}$$

$$= \frac{340+20}{340-20} = \frac{9}{8}$$

128. For source, $v_S = r\omega = 0.70 \times 2\pi \times 5 = 22\,\text{m/s}$

Minimum frequency is heard when the source is receding the man. It is given by $n_{\text{min}} = n\frac{v}{v + v_S}$

$$= 1000 \times \frac{52}{352+22}$$

$$= 941\,\text{Hz}$$

129. For direct sound, source is moving away from the observer so frequency heard in this case

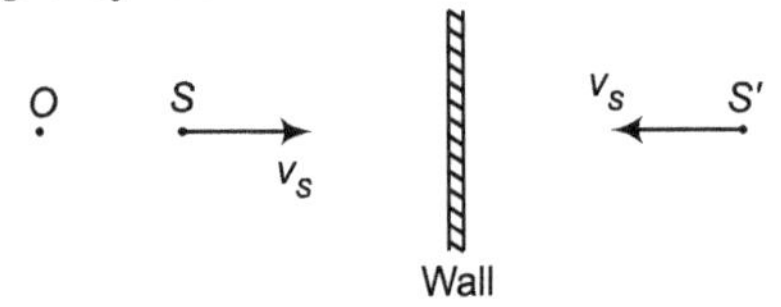

$$n_1 = n\left(\frac{v}{v+v_S}\right)$$
$$= 500\left(\frac{332}{332+2}\right)$$
$$= 500\left(\frac{332}{334}\right)\text{Hz} = 497\text{ Hz}$$

For sound reflected from the wall

$$n_2 = n\left(\frac{v}{v-v_s}\right)$$
$$= 500\left(\frac{332}{332-2}\right) = 503\text{ Hz}$$

$\therefore$ Beat frequency $= n_2 - n_1 = 6$ Hz

130. $n_{\text{Before}} = \dfrac{v}{v-v_C}\cdot n$ and $n_{\text{After}} = \dfrac{v}{v+v_C}\cdot n$

$$\frac{n_{\text{Before}}}{n_{\text{After}}} = \frac{11}{9} = \left(\frac{v+v_C}{v-v_C}\right) \Rightarrow v_C = \frac{v}{10}$$

131. At point A, source is moving away from observer so apparent frequency $n_1 < n$ (actual frequency). At point B, source is coming towards observer so apparent frequency $n_2 > n$ and point C source is moving perpendicular to observer so $n_3 = n$.

Hence, $\quad n_2 > n_3 > n_1$

132. When observer moves towards stationary source, then apparent frequency

$$n' = \left[\frac{v+v_0}{v}\right]n$$
$$= \left[\frac{v+v/5}{v}\right]n$$
$$= \frac{6}{5}n = 1.2n$$

Increment in frequency $= 0.2n$, so percentage change in frequency $= \dfrac{0.2n}{n}\times 100 = 20\%$.

133. Here, $v_A = 72\text{ km/h} = 10\text{ m/s}$

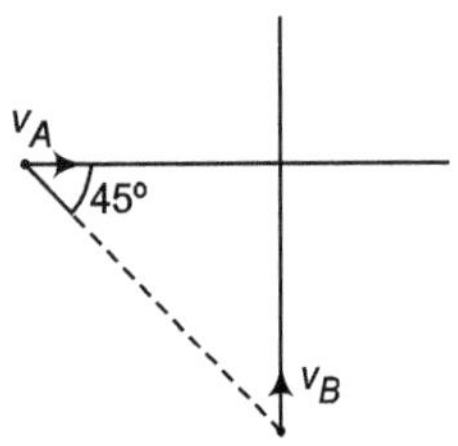

$$v_B = 36\text{ km/h} = 10\text{ m/s}$$
$$n' = n\left(\frac{v+v_B\cos 45^\circ}{v-v_A\cos 45^\circ}\right)$$
$$\Rightarrow \quad n' = 280\left(\frac{340+10/\sqrt{2}}{340-20/\sqrt{2}}\right)$$
$$= 298\text{ Hz}$$

134. According to the question, frequencies of first and last tuning forks are $2n$ and n, respectively.

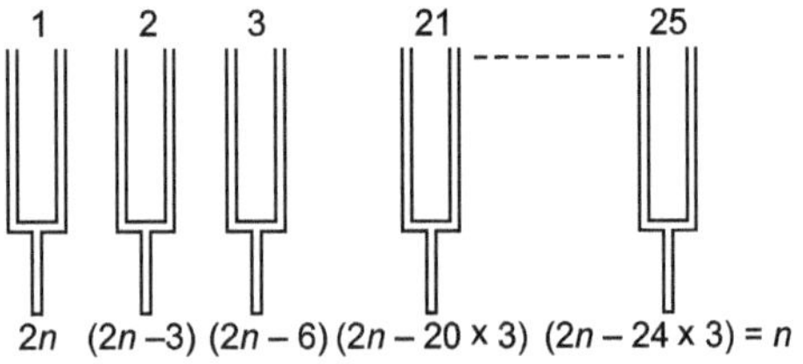

Hence, frequency in given arrangement is as follows

$$\Rightarrow \quad 2n - 24\times 3 = n$$
$$\Rightarrow \quad n = 72\text{ Hz}$$

So, frequency of 21st tuning fork

$$n_{21} = (2\times 72 - 20\times 3) = 84\text{ Hz}$$

135. There is no relative motion between O and B. Hence,

$$f_B' = f_B = 590\text{ Hz}$$

But, $$f_A' = f_A\left(\frac{v-v_0}{v}\right)$$
$$= 660\left(\frac{330-30}{300}\right)$$
$$= 594\text{ Hz}$$

$\therefore$ Beat frequency will be $f_b = f_A' - f_B' = 4$ Hz

136. Velocity of source (or whistle) $v_S = R\omega = 30$ m/s.

Maximum frequency will be heard when whistle is at P and minimum at Q.

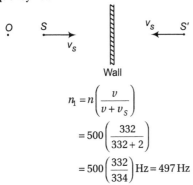

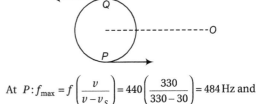

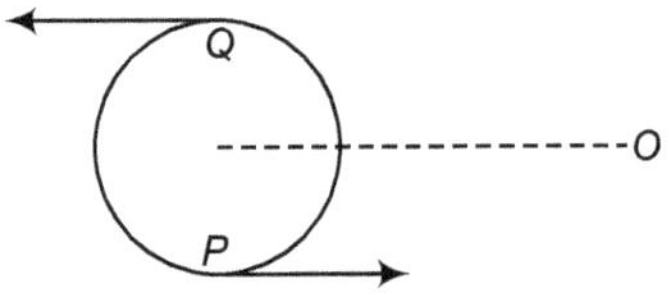

At $P: f_{\max} = f\left(\dfrac{v}{v-v_S}\right) = 440\left(\dfrac{330}{330-30}\right) = 484$ Hz and

At $Q: f_{\min} = f\left(\dfrac{v}{v+v_S}\right) = 440\left(\dfrac{330}{330+30}\right) = 403.3$ Hz

137. $f - \dfrac{f}{2} = 2$ or $f = 4$ Hz

Here, $f = \dfrac{v}{2l}$ = fundamental frequency of open pipe

and $\dfrac{f}{2} = \dfrac{v}{4l}$ = of closed pipe

Now, new beat frequency $= 2f - \dfrac{f}{4} = \dfrac{7f}{4}$

$$= \frac{7}{4}\times 4 = 7\text{ Hz}$$

138. When the engine approaches the observer,

$$f_1 = f\left(\frac{v}{v-v_s}\right) > f \text{ but constant.}$$

When it moves away from the source,

$$f_2 = f\left(\frac{v}{v+v_s}\right) < f \text{ but constant.}$$

139. Speed $= n\lambda = n(4ab) = 4n \times ab$ $\left(\text{As}, ab = \frac{\lambda}{4}\right)$

Path difference between b and e is $\frac{3\lambda}{4}$.

So, the phase difference $= \frac{2\pi}{\lambda} \times$ Path difference

$$= \frac{2\pi}{\lambda} \times \frac{3\lambda}{4} = \frac{3\pi}{2}$$

140. Points B and F are in same phase as they are λ distance apart.

141. The particle velocity is maximum at B and is given by

$$\frac{dy}{dt} = (v_p)_{max} = \omega A$$

Also wave velocity is $\frac{dx}{dt} = v = \frac{\omega}{k}$

So slope, $\frac{dy}{dx} = \frac{(v_p)_{max}}{v} = kA$

142. We know frequency, $n = \frac{p}{2l}\sqrt{\frac{T}{\pi r^2 \rho}} \Rightarrow n \propto \frac{1}{\sqrt{\rho}}$

i.e. graph between n and $\sqrt{\rho}$ will be hyperbola.

143. Energy density $(E) = 2\pi^2 \rho r^2\, \rho n^2 A^2$

$$v_{max} = \omega A = 2\pi n A \Rightarrow E \propto (v_{max})^2$$

i.e. graph between E and v_{max} will be a parabola symmetrical about E-axis.

144. Here, $A = 0.05$ m, $\frac{5\lambda}{2} = 0.25 \Rightarrow \lambda = 0.1$ m

Now, standard equation of wave

$$y = A \sin \frac{2\pi}{\lambda}(vt - x)$$

$$\Rightarrow \quad y = 0.05 \sin 2\pi(3300t - 10x)$$

145. Since, A is moving upwards, therefore, after an elemental time interval the wave will be as shown dotted in below figure. It means the wave is travelling leftward. Therefore, (a) is wrong. Displacement amplitude of the wave means maximum possible displacement of medium particles due to propagation of the wave, which is equal to the displacement at B at the instant shown in figure. Hence, (b) is correct.

From figure, it is clear that C is moving downward at this instant. Hence, (c) is wrong.

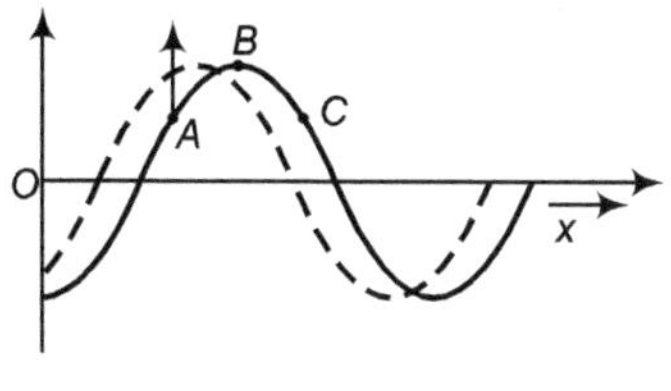

The phase difference between two points will be equal to $\frac{\pi}{2}$, if distance between them is equal to $\frac{\lambda}{4}$. It may be equal to $\frac{\lambda}{4}$. Hence, phase difference between these two points may be equal to $\frac{\pi}{2}$.

146. At $t = 0$ and $x = \frac{\pi}{2k}$.

The displacement, $y = a_0 \sin\left(-k \times \frac{\pi}{2x}\right)$

$$= -a_0 \sin \frac{\pi}{2} = -a_0$$

Point of maximum displacement (a_0) in negative direction is Q.

147. Particle velocity $(v_p) = -v \times$ Slope of the graph at that point.

At point 1 Slope of the curve is positive, hence particle velocity is negative or downward (↓).

At point 2 Slope negative, hence particle velocity is positive or upwards (↑).

At point 3 Again slope of the curve is positive, hence particle velocity is negative or downward (↓).

148. While approaching $f > f_0$ but f = constant.

While receding $f < f_0$ but f = constant.

149. $100 \xrightarrow{\text{First slab}} 80 \xrightarrow{\text{Second slab}} 80\%$ of $80 = 64$

∴ Reduction in intensity on passing through two slabs $= 36\%$

150. $\frac{\lambda}{2} = 10\,\text{cm} = 0.1$ m or $\lambda = 0.2$ m

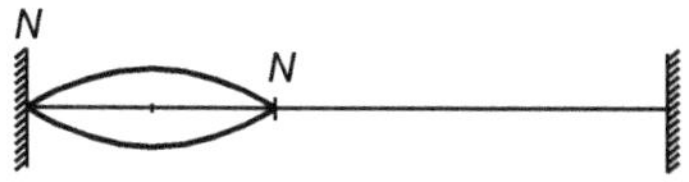

Now, $v = f\lambda = 20$ m/s

151. The two waves have a phase difference of $\frac{\pi}{2}$. Hence,

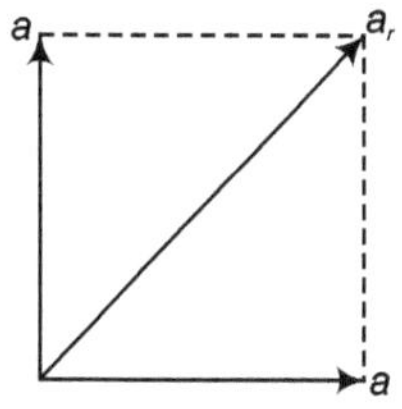

$$a_r = \sqrt{2}\, a$$

152. $\frac{T}{4} = 0.17$ s

∴ $f = \frac{1}{T} = \frac{1}{4 \times 0.17}$ Hz

or $f = 1.47$ Hz

153. $\Delta\phi = \frac{2\pi}{\lambda} \cdot \Delta x$

or $\lambda = \frac{2\pi}{\Delta\phi} \cdot \Delta x$

$$= \frac{2\pi}{(\pi/3)} \cdot 1.25 \times 10^{-2}\ \text{m}$$

$$= 7.5 \times 10^{-2}\ \text{m}$$

$$v = f\lambda$$

$$= 1000 \times 7.5 \times 10^{-2}$$

$$= 75\ \text{m/s}$$

154. $A_{max} = a + b$ and $A_{min} = a - b$

∴ $A_{max} - A_{min} = 2b$

155. Both are 90° out of phase. Hence, $A_R = 10\sqrt{5}$

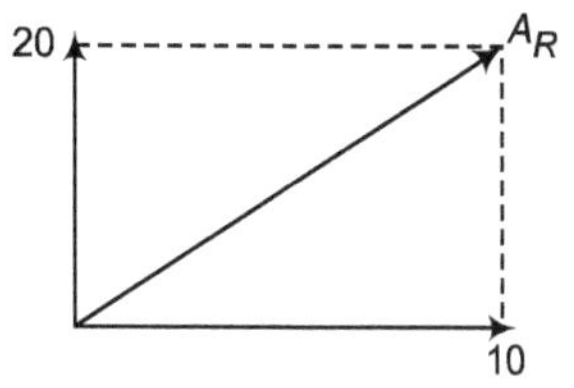

156. Difference in lengths will be equal to one loop or $\frac{\lambda}{2}$.

$\therefore \quad \lambda = 2(48.6 - 15.4)$ cm

Now, $\quad v = f\lambda$

$v = 512 \times 2(48.6 - 15.4)$ cm/s

$v = 1024(48.6 - 15.4)$ cm/s

157. $\Delta\phi = \frac{2\pi}{\lambda} \cdot \Delta x = k(\Delta x) = k(vt)$

Here, vt = distance travelled by insect in given time interval.

or $\quad \Delta\phi = (20\pi)(5 \times 10^{-2} \times 5)$

$= 5\pi$

158. $\frac{I_1}{I_2} = \frac{A_1^2\omega_1^2}{A_2^2\omega_2^2} = \left(\frac{5}{10}\right)^2\left(\frac{2\pi \times 75}{2\pi \times 150}\right)^2 = \frac{1}{16}$

159. $A_{max} = \sqrt{A^2 + A^2} = A\sqrt{2}$, frequency will remain same, i.e. ω.

160. $\lambda = \frac{v}{n} = \frac{350}{350} = 1\text{ m} = 100\text{ cm}$

Also path difference (Δx) between the waves at the point of observation is $AP - BP = 25$ cm. Hence,

$\Rightarrow \quad \Delta\phi = \frac{2\pi}{\lambda}(\Delta x)$

$= \frac{2\pi}{1} \times \left(\frac{25}{100}\right) = \frac{\pi}{2}$

$\Rightarrow \quad A = \sqrt{(a_1)^2 + (a_2)^2}$

$= \sqrt{(0.3)^2 + (0.4)^2}$

$= 0.5$ mm

161. Path difference $(\Delta x) = 50\text{ cm} = \frac{1}{2}$ m

$\therefore$ Phase difference, $\Delta\phi = \frac{2\pi}{\lambda} \times \Delta x$

$\Rightarrow \quad \phi = \frac{2\pi}{1} \times \frac{1}{2} = \pi$

Total phase difference $= \pi - \frac{\pi}{3} = \frac{2\pi}{3}$

$\Rightarrow \quad A = \sqrt{a^2 + a^2 + 2a^2\cos\left(\frac{2\pi}{3}\right)} = a$

162. $y = \frac{1}{\sqrt{a}}\sin\omega t \pm \frac{1}{\sqrt{b}}\sin\left(\omega t + \frac{\pi}{2}\right)$

Here, phase difference $= \frac{\pi}{2}$

$\therefore$ The resultant amplitude.

$= \sqrt{\left(\frac{1}{\sqrt{2}}\right)^2 + \left(\frac{1}{\sqrt{b}}\right)^2}$

$= \sqrt{\frac{1}{a} + \frac{1}{b}}$

$= \sqrt{\frac{a+b}{ab}}$

163. Minimum time interval between two instants when the string is flat $\frac{T}{2} = 0.5\text{ s} \Rightarrow T = 1$ s

Hence, $\quad \lambda = v \times T$

$= 10 \times 1$

$= 10$ m

164. $n \propto \sqrt{T}$

$100 = k\sqrt{T_1}$ ...(i)

$400 = k\sqrt{T_2}$ ...(ii)

From the Eqs. (i) and (ii), we get

Now, $\quad 4 = \sqrt{\frac{T_2}{T_1}}$

$\Rightarrow \quad T_2 = 16T_1$

165. Due to rise in temperature, the speed of sound increases. Since, $n \propto v$, hence n increases.

166. For end correction x, $\frac{l_2 + x}{l_1 + x} = \frac{3\lambda/4}{\lambda/4} = 3$

$x = \frac{l_2 - 3l_1}{2}$

$= \frac{70.2 - 3 \times 22.7}{2}$

$= 1.05$ cm

167. Let x be the end correction, then according to question.

$\frac{v}{4(l_1 + x)} = \frac{3v}{4(l_2 + x)}$

$\Rightarrow \quad x = 2.5$ cm

$= 0.025$ m

168. When pulse is reflected from a rigid support, the pulse is inverted both length and sidewise.

169. $L_1 = \frac{\lambda}{4}, L_2 = \frac{3\lambda}{4}$

and $\quad L_3 = \frac{5\lambda}{4}$

170.

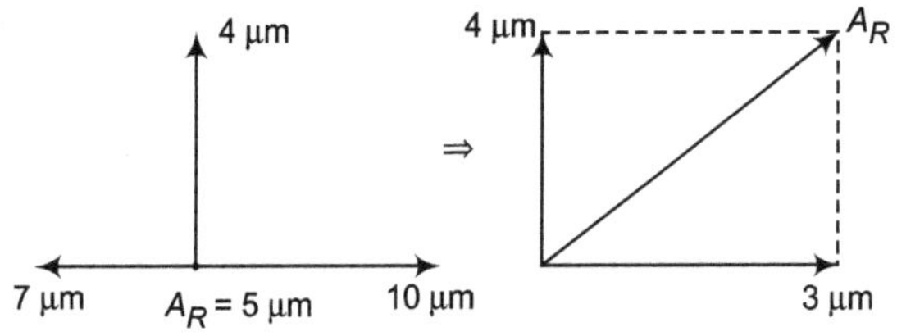

171. $\lambda = \frac{v}{f} = \frac{330}{500} = 0.66$ m

$A_k = \sqrt{4^2 + 3^2}$

$= \sqrt{16 + 9} = \sqrt{25}$

$A_k = 5$

Length of air column at resonance can be $\frac{\lambda}{4}, \frac{3\lambda}{4}, \frac{5\lambda}{4}$, etc.

or 0.165 m, 0.495 m, 0.825 m, 1.155 m, etc.

But, length of tube is only 1 m. Hence, first three resonances will be obtained.

172.

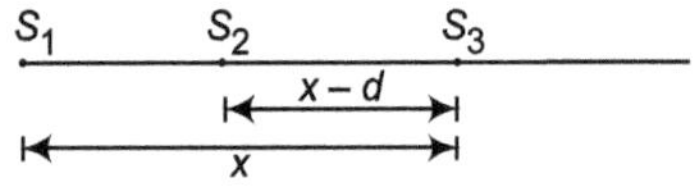

173. Due to a point source,

$I_r = \frac{P}{4\pi r^2}$

$\therefore \quad I_r \propto \frac{P}{r^2}$

If P and r both are doubled, I_r will remain half.

Level 2 : Only One Correct Option

1. $\phi = \frac{2\pi}{3} = 120° \Rightarrow 2\phi = 240°$

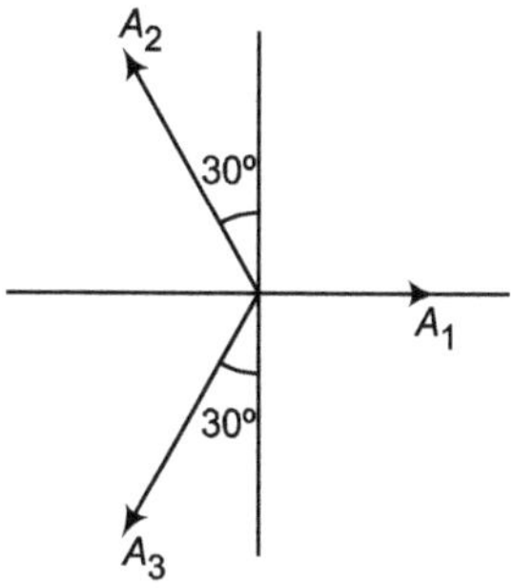

A_1, A_2 and A_3 all are equal to A acting at angles shown in figure. We can see that resultant of these three come out to be zero.

2. Let S be the source of sound and P the person or listener. The waves from S reach point P directly following the path SMP and being reflected from the ceiling at point A following the path SAP. M is mid-point of SP (i.e. $SM = MP$) and $\angle SMA = 90°$

Path difference between waves

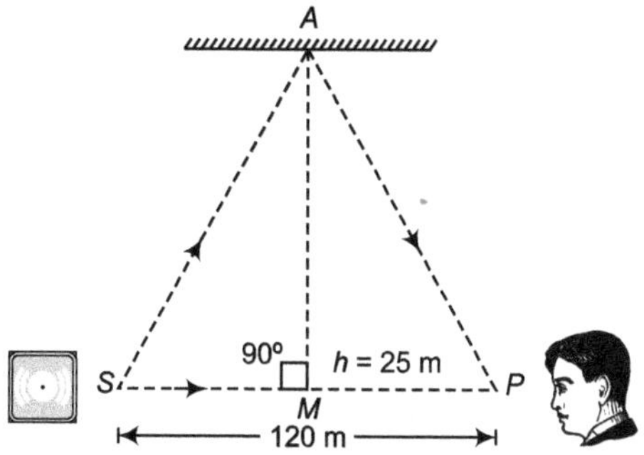

$$\Delta x = SAP - SMP$$

We have $SAP = SA + AP = 2(SA)$

$$= 2\sqrt{(SM)^2 + (MA)^2}$$

$$= 2\sqrt{(60^2 + 25^2)} = 130 \text{ m}$$

$\therefore$ Path difference $= SAP - SMP = 130 - 120 = 10$ m

Path difference due to reflection from ceiling $= \frac{\lambda}{2}$

$\therefore$ If $\Delta x = \frac{\lambda}{2}, \frac{3\lambda}{2}, \frac{5\lambda}{2}$, etc., interference will be constructive

$\therefore \quad \lambda = 20 \text{ m}, \frac{20}{3} \text{ m}, \frac{20}{5}$ m, etc.

3. At 10 cm distance from open end or at distance $\frac{\lambda}{4}$ from the open end, there should be a node.

4. $\sqrt{\frac{E}{\rho}} = 10\sqrt{\frac{T}{\mu}} \Rightarrow 100\left(\frac{T}{\mu}\right) = \frac{E}{\rho}$

$\therefore \quad \frac{T}{S} = \frac{E\mu}{100\rho S}$

But, $\rho S = \mu$

$\therefore \quad \text{Stress} = \frac{T}{S} = \frac{E}{100}$

5. $v = \sqrt{\frac{T}{\mu}} = \sqrt{\frac{80}{5\times10^{-2}}} = 40 \text{ m/s}$

$$P = \frac{1}{2}\rho\omega^2 A^2 Sv$$

$$= \frac{1}{2}\mu(2\pi f)^2 v \qquad (\text{as } \rho S = \mu)$$

$$= \frac{1}{2}\times 5\times10^{-2}(2\pi\times60)^2\times40\times(0.06)^2$$

$$= 512 \text{ W}$$

6. $A_x = A\sin\left(\frac{20\pi x}{3}\right)$

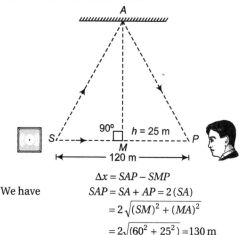

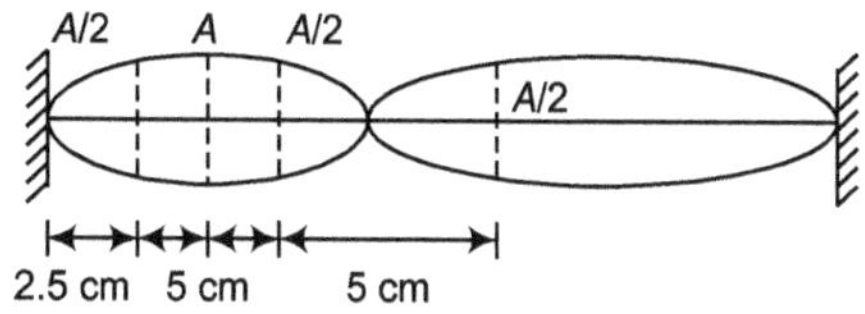

$\Rightarrow x = 0$ is a node

$$A_x = \frac{A}{2}$$

At $\frac{20\pi}{3} = \frac{\pi}{6}$

or $x = \frac{1}{40} \text{ m} = 2.5 \text{cm}$

$$\lambda = \frac{2\pi}{k} = \left(\frac{2\pi}{20\pi/3}\right)$$

$$= 30 \text{cm}$$

$$\frac{\lambda}{4} = 7.5 \text{ cm}$$

7. Since, $T_1 > T_2$

$\therefore \quad f_1 > f_2$

or $\quad f_1 - f_2 = 6 \text{Hz}$

If T_1 is decreased, f_1 will decrease. Therefore, the case may be

$$f_2 - f_1' = 6 \text{Hz}$$

If T_2 is increased, f_2 will increase.

Therefore, the case may be

$$f_2' - f_1 = 6 \text{Hz}$$

8. Path difference $(\Delta x) = 50 \text{cm} = \frac{1}{2}$ m

$\therefore$ Phase difference $\Delta\phi = \frac{2\pi}{\lambda}\times\Delta x$

$\Rightarrow \quad \phi = \frac{2\pi}{1}\times\frac{1}{2} = \pi$

Total phase difference $= \pi - \frac{\pi}{3} = \frac{2\pi}{3}$

$\Rightarrow \quad A = \sqrt{a^2 + a^2 + 2a^2\cos\left(\frac{2\pi}{3}\right)} = a$

9. $n \propto \frac{1}{l}$ or $nl = k$ (constant)

$\therefore \quad l = l_1 + l_2 + l_3 + \ldots$

or $\quad \frac{k}{n} = \frac{k}{n_1} + \frac{k}{n_2} + \frac{k}{n_3} + \ldots$

or $\quad \frac{1}{n} = \frac{1}{n_1} + \frac{1}{n_2} + \frac{1}{n_3} + \ldots$

10. Frequency $\propto \sqrt{T}$

$$\frac{f}{f/2} = \sqrt{\frac{T}{T - F_w}} \quad (F_w = \text{upthrust in water})$$

$\therefore$ $$1 - \frac{F_w}{T} = \frac{1}{4}$$

or $$\frac{F_w}{T} = \frac{3}{4} \quad \ldots\text{(i)}$$

In second case,

$$\frac{f}{f/3} = \sqrt{\frac{T}{T - F_l}} \quad (F_l = \text{upthrust in liquid})$$

$$1 - \frac{F_l}{T} = \frac{1}{9}$$

$\therefore$ $$\frac{F_l}{T} = \frac{8}{9} \quad \ldots\text{(ii)}$$

From Eqs. (i) and (ii), we get

$$\frac{F_l}{F_w} = \frac{8}{9} \times \frac{4}{3} = 1.18$$

Now, upthrust is in the ratio of relative density.

11. $f \propto \frac{1}{l}$

$\therefore$ $$l_1 : l_2 : l_3 = \frac{1}{f_1} : \frac{1}{f_2} : \frac{1}{f_3} = \frac{1}{1} : \frac{1}{3} : \frac{1}{4}$$

or $$l_1 : l_2 : l_3 = 12 : 4 : 3$$

$\therefore$ $$l_1 = \frac{12}{13} \times 114 = 72 \text{ cm}$$

$$l_2 = \frac{4}{19} \times 114 = 24 \text{ cm}$$

and $$l_3 = \frac{3}{19} \times 114 = 18 \text{ cm}$$

12. $A_B = 4A$

$\therefore$ $$I_R = 16I$$

$$10 = 10 \log_{10}\left(\frac{I}{I_0}\right)$$

$$L = 10 \log_{10}\left(\frac{16I}{I_0}\right)$$

$\therefore$ $$L - 10 = 10 \log_{10} (16)$$

$\therefore$ $$L = 22 \text{ dB}$$

13. ax and bt should have opposite signs.
Further, coefficient of $t = (4) \times$ (coefficient of x)

14. $f_1 = 2\left(\frac{v}{2l}\right) = \frac{v}{l}$

$$f_2 = n\left(\frac{v}{4l}\right) \quad (n = 1, 3, 5, 7, \ldots)$$

$$f_2 > f_1$$

or $$f_2 = \frac{n}{4} \cdot f_1$$

15. $y = \frac{1}{\sqrt{a}} \sin \omega t \pm \frac{1}{\sqrt{b}} \sin\left(\omega t + \frac{\pi}{2}\right)$

Here, phase difference $= \frac{\pi}{2}$

$\therefore$ The resultant amplitude

$$= \sqrt{\left(\frac{1}{\sqrt{a}}\right)^2 + \left(\frac{1}{\sqrt{b}}\right)^2} = \sqrt{\frac{1}{a} + \frac{1}{b}} = \sqrt{\frac{a+b}{ab}}$$

16. $f \propto \frac{\sqrt{T}}{l} \Rightarrow f$ remains same. Hence,

$$\frac{\sqrt{T_1}}{l_1} = \frac{\sqrt{T_2}}{l_2}$$

or $$\sqrt{\frac{T_1}{T_2}} = \frac{l_1}{l_2} = \frac{40}{30} = \frac{4}{3}$$

or $$\sqrt{\frac{W}{W - F}} = \frac{4}{3}$$

$\therefore$ $$\sqrt{\frac{Vdg}{Vdg - V \times 1 \times g}} = \frac{4}{3}$$

or $$\sqrt{\frac{d}{d - 1}} = \frac{4}{3}$$

$\therefore$ $$\frac{d}{d - 1} = \frac{16}{9}$$

or $$d = \frac{16}{7}$$

17. In a wave equation, x and t must be related in the form $(x \pm vt)$.

We rewrite the given equations, $y = \frac{1}{1 + (x - vt)^2}$

For $t = 0$, this becomes $y = \frac{1}{(1 + x^2)}$, as given

For $t = 2$, this becomes $y = \frac{1}{[1 + (x - 2v)^2]}$

$$= \frac{1}{[1 + (x - 1)^2]}$$

$\Rightarrow$ $$2v = 1 \quad \text{or} \quad v = 0.5 \text{ m/s}$$

18. $v = \frac{\omega}{k} = \sqrt{\frac{T}{\mu}}$

$\therefore$ $$T = \mu\left(\frac{\omega}{k}\right)^2 = 10^{-4}\left(\frac{30}{1}\right)^2 = 0.09 \text{ N}$$

19. Now given that

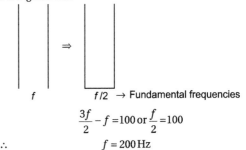

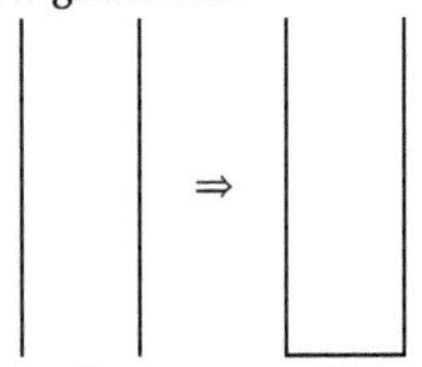

$$\frac{3f}{2} - f = 100 \text{ or } \frac{f}{2} = 100$$

$\therefore$ $$f = 200 \text{ Hz}$$

20. Tension at distance x from free end

$$T_x = W_x = m_x g = (\mu\, x\, g)$$

$$\sqrt{\frac{T_x}{\mu}} = \sqrt{x\, g}$$

or $$v = \sqrt{x\, g}$$

$\therefore$ $$v \propto \sqrt{x}$$

21. $L_2 D = \sqrt{(40)^2 + (9)^2} = 41 \text{ m}$

Path difference,

$$\Delta x = L_2 D - L_1 D = 1 \text{ m} = \lambda$$

$$f = \frac{v}{\lambda} = 330 \text{ Hz}$$

22. They are out of phase at $x_0 = \dfrac{\lambda}{2}$

23. At $x = 0, t = 0, y = \sqrt{3}$

Further, if x is slightly greater than 0, y is slightly greater than $\sqrt{3}$. Therefore, the current graph is (a).

24. $n_Q = 341 \pm 3 = 344$ Hz or 338 Hz

On waxing Q, the number of beats decreases, hence $n_Q = 344$ Hz.

25. Beat frequency, $f_b = f_2 - f_1 = \dfrac{\omega_2 - \omega_1}{2\pi}$

$$= \frac{404\,\pi - 400\pi}{2\pi}$$

$$= 2\text{ Hz}$$

Further $A_1 = 4$ and $A_2 = 3$. Hence,

$$\frac{I_{max}}{I_{min}} = \left(\frac{A_1 + A_2}{A_1 - A_2}\right)^2 = 49$$

26. $(\Delta p)_{max} = BAK$

$$\therefore \quad A = \frac{(\Delta p)_{max}}{BK} \qquad \text{...(i)}$$

Here, $$v = \sqrt{\frac{B}{\rho}} = \frac{\omega}{K}$$

$$\therefore \quad K = \omega\sqrt{\frac{\rho}{B}} = 2\pi f\sqrt{\frac{\rho}{B}}$$

Further $B = \rho v^2$

Substituting in Eq. (i), we get

$$A = \frac{(\Delta p)_{max}}{2\pi f\sqrt{B\rho}} = \frac{(\Delta p)_{max}}{2\pi f\,\rho v} \qquad [\text{as } B = \rho v^2]$$

Substituting the values, we have

$$A = \frac{30}{2\pi \times 10^3 \times 1.5 \times 300}$$

$$= \frac{10^{-4}}{3\pi}\text{ m}$$

27. Distance between two successive antinodes is $\dfrac{\lambda}{2}$ or $\dfrac{\pi}{k}$.

$\therefore$ Volume between two antinodes will be $\dfrac{\pi}{k}s$.

Let u_1 and u_2 be the energy density due to two waves, then $E = (u_1 + u_2)$ volume

$$= \left[\frac{1}{2}\rho\omega^2 a^2 + \frac{1}{2}\rho\omega^2 (2a)^2\right]\frac{\pi}{k}\cdot s$$

$$= \frac{5}{2}\frac{\rho\omega^2 a^2 \pi s}{k}$$

28. $f = f_0\left(\dfrac{v + v_0}{v}\right) = 10^3\left(1 + \dfrac{v_0}{v}\right)$

$$= 10^3 + \frac{10^3}{v}(g\,t)$$

$$= 10^3 + \left(\frac{10^3}{v}\right)t$$

Slope of f-t line should be equal to $\dfrac{10^4}{v}$

$$\therefore \quad \frac{1000}{30} = \frac{10^4}{v}$$

or $$v = 300\text{ m/s}$$

29. E = energy density × volume $= \dfrac{1}{2}\rho\omega^2 A^2\,SL$

or $E \propto \omega^2$, rest all quantities are common for both the waves.

$$\omega_2 = 2\omega \quad \text{and} \quad \omega_1 = \omega$$

$$\therefore \quad E_2 = 4E_1$$

30. $\Delta l = l\alpha t$ or $\dfrac{\Delta l}{l} = \alpha t$, stress $= Y \times$ strain $= Y\alpha t$ (also called thermal stress.)

$\therefore$ Tension, $$T = Y\,s\,\alpha t$$

Now, $$v = \sqrt{\frac{T}{\rho s}} = \sqrt{\frac{Y\alpha s t}{\rho s}} = \sqrt{\frac{Y\alpha t}{\rho}}$$

31. Frequency of first overtone of closed pipe = Frequency of first overtone of open pipe

$$\Rightarrow \quad \frac{3v_1}{4L_1} = \frac{v_2}{L_2}$$

$$\Rightarrow \quad \frac{3}{4L_1}\sqrt{\frac{B}{\rho_1}} = \frac{1}{L_2}\sqrt{\frac{B}{\rho_2}}$$

$$\Rightarrow \quad L_2 = \frac{4L_1}{3}\sqrt{\frac{\rho_1}{\rho_2}} = \frac{4L}{3}\sqrt{\frac{\rho_1}{\rho_2}}$$

33. Assuming the reflecting surface to be at rest.

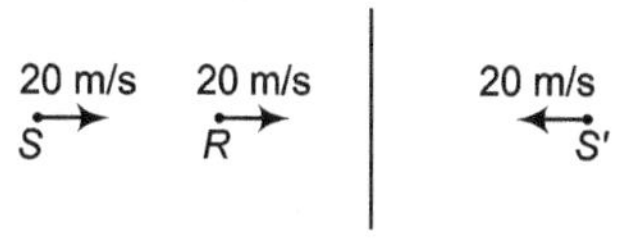

$$f' = 256\left(\frac{330 + 20}{330 - 20}\right)\text{Hz} = 289\text{ Hz}$$

$$\lambda' = \frac{v'}{f'}$$

$$= \frac{(330 + 10)}{289} = \frac{340}{289}\text{ m}$$

34. For string, $\dfrac{\text{Mass}}{\text{Length}} = \mu = \dfrac{10^{-2}}{0.4} = 2.5 \times 10^{-2}$ kg/m

$\therefore$ Velocity, $$v = \sqrt{\frac{T}{\mu}} = \sqrt{\frac{1.6}{2.5 \times 10^{-2}}}$$

$$= 8\text{ m/s}$$

For constructive interference between successive pulses.

$$\Delta t_{min} = \frac{2l}{v} = \frac{2(0.4)}{8} = 0.10\text{ s}$$

35. In the starting $\Delta x = 0$ and maximum intensity is observed.

Next maxima will correspond to $\Delta x = \lambda$, last, i.e. 5th maxima will correspond to

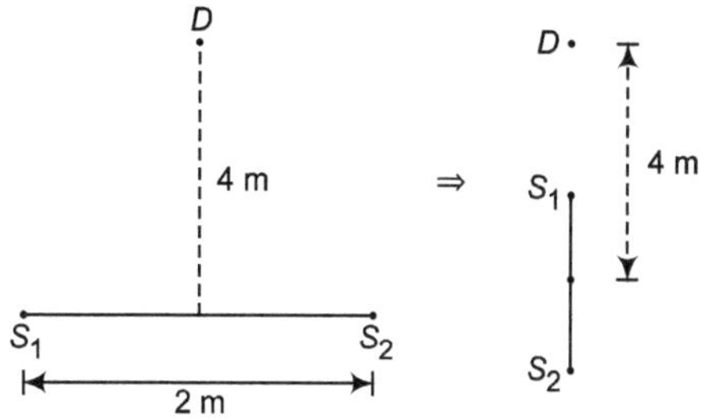

$$\Delta x = 4\lambda. \quad \text{But} \quad \Delta x = S_1S_2 = 2\text{ m}$$

$$\therefore \quad 4\lambda = 2 \quad \text{or} \quad \lambda = 0.5\text{ m}$$

More than One Correct Options

1. $v = \dfrac{\text{Coefficient of } t}{\text{Coefficient of } x}$

$$= \frac{6.2}{2} = 3.1 \text{ m/s}$$

$$A = \frac{2}{20} = 0.1 \text{ m}$$

2. Not required.

3. $y = A \sin(\pi x + \pi t)$

$$v_p = \frac{\partial y}{\partial t} = \pi A \cos(\pi x + \pi t)$$

$$a_p = \frac{\partial^2 y}{\partial t^2} = -\pi^2 A \sin(\pi x + \pi t)$$

Now, substitute $t = 0$ and given value of x.

Since, ωt and kx are of same sign, hence the wave is travelling in negative x-direction.

4. Speed of wave

$$= \frac{\text{Coefficient of } t}{\text{Coefficient of } x} = \frac{b}{1} = b$$

$$\lambda = \frac{2\pi}{k} = \frac{2\pi}{(2\pi/a)} = a$$

5. Speed of wave $= \dfrac{\text{Coefficient of } t}{\text{Coefficient of } x} = \dfrac{1/b}{1/a} = \dfrac{a}{b}$

$$\lambda = \frac{2\pi}{k} = \frac{2\pi}{2\pi/a} = a$$

$$T = \frac{2\pi}{\omega} = \frac{2\pi}{2\pi/b} = b$$

6. $(y - x)$ graph is sine graph. Therefore, $(v - x)$ graph is cosine graph and $(a - x)$ is sine graph as

$$v_p = \frac{\partial y}{\partial t} \quad \text{and} \quad a_p = \frac{\partial^2 y}{\partial t^2}$$

7. $f \propto \sqrt{T}$

$$\frac{f_2}{f_1} = \sqrt{\frac{T_2}{T_1}}$$

$$\therefore \quad \frac{f_1 + 15}{f_1} = \sqrt{\frac{1.21 T_1}{T_1}} = 1.1$$

Solving, we get $f_1 = 150$ Hz

$$v \propto \sqrt{T}$$

$$\therefore \quad \frac{v_2}{v_1} = \sqrt{\frac{T_2}{T_1}}$$

$$\text{or} \quad v_2 = \left(\sqrt{\frac{1.21\, T_1}{T_1}}\right) v_1$$

$$= 1.1 v_1$$

Hence, increase in v is 10%.

$$\frac{\lambda}{2} = l \Rightarrow \therefore \lambda = 2l$$

$\therefore$ Fundamental wavelength $= 2\lambda$ is unchanged.

8. Not required.

9. Not required.

10. $A_t = \left(\dfrac{2v_2}{v_1 + v_2}\right) A_i$

If $v_2 > v_1$ then $A_t > A_i$

11. 

$$l = 4\frac{\lambda}{2} = 2\lambda = 2\left(\frac{2\pi}{k}\right)$$

$$= \frac{4\pi}{k}$$

12. Two waves should travel in opposite directions.

13. $y = y_1 + y_2 = (2A \cos kx) \sin \omega t$

$$= A_x \sin \omega t$$

Here, $A_x = 2A \cos kx$

At $x = 0$, A_x is maximum or $2A$.

So, it is an antinode. Next antinode will occur at.

$$x = \frac{\lambda}{2}, \lambda \ldots \text{etc.}$$

$$\text{or} \quad x = \frac{\pi}{k}, \frac{2\pi}{k} \ldots \text{etc.}$$

14. $f = n\left(\dfrac{v}{4l}\right)$ $(n = 1, 3, 5 \ldots)$

$$\therefore \quad l = \frac{nv}{4f} = n\left(\frac{330}{4 \times 264}\right)$$

$$= (0.3125\, n) \text{ m}$$

$$= (31.25\, n) \text{ cm}$$

Now, keep on substituting $n = 1, 3$, etc.

15. (a) Velocity of sound wave in air is independent of pressure, if T = constant

(b) $v = \sqrt{\dfrac{\gamma RT}{M}} = \sqrt{\dfrac{\gamma R(t + 273)}{M}}$

$$\text{or} \quad v^2 \propto (t + 273)$$

(c) $v = \sqrt{T/\mu}$

$$\therefore \quad v \propto \sqrt{T}$$

$$\text{or} \quad v^2 \propto T$$

(d) $f_0 = \dfrac{v}{2l}$ or $\dfrac{v}{4l}$

$$\Rightarrow \quad f_0 \propto \frac{1}{l}$$

16. Not required.

17. (a) Let f_1 = fundamental frequency of closed pipe and f_2 = fundamental frequency of open pipe.

Given, $5f_1 = 4f_2$

$$\text{or} \quad f_2 = \frac{5}{4} f_1$$

$$\therefore \quad f_2 > f_1$$

(b) First overtone frequency of closed pipe $= 3f_1$ and first overtone frequency of open pipe

$$= 2f_2 = 2\left(\frac{5}{4} f_1\right) = 2.5 f_1$$

(c) Fifteenth harmonic of closed pipe $= 15f_1$

Twelfth harmonic of open pipe $= 12f_2 = 12\left(\dfrac{5}{4} f_1\right)$

(d) Tenth harmonic of closed pipe $= 10 f_1$

Eighth harmonic of open pipe $= 8f_2 = 8\left(\dfrac{5}{4} f_1\right)$

$$= 10 f_1$$

18. $f = \frac{v}{4(l+0.6r)} = \frac{\sqrt{\gamma RT/M}}{4(l+0.6)}$

(d) v does not depend on pressure, if temperature is kept constant.

19. During approach,

$$f = f_0\left(\frac{v}{v - v_s}\right)$$

$> f_0$ but f is constant.

During receding.

$$f = f_0\left(\frac{v}{v - v_s}\right)$$

$< f_0$ but f is again constant.

20. Given equation is given by $y(x,t) = 3.0\sin\left(36t + 0.018x + \frac{\pi}{4}\right)$

Compare the equation with the standard form.

$$y = a\sin(\omega t + kx + \phi)$$

(a) As the equation involves positive sign with x, hence the wave is travelling from right to left. Hence, option (a) is correct.

(b) Given, $\omega = 36 \Rightarrow 2\pi\nu = 36$

$\Rightarrow \quad \nu = \text{frequency} = \frac{36}{2\pi} = \frac{18}{\pi}$

$$k = 0.018 \Rightarrow \frac{2\pi}{\lambda} = 0.018$$

$\Rightarrow \frac{2\pi\nu}{\nu\lambda} = 0.018 \Rightarrow \frac{\omega}{v} = 0.018$ $\quad [\because 2\pi\nu = \omega \text{ and } \nu\lambda = v]$

$\Rightarrow \frac{36}{v} = 0.018 = \frac{18}{1000}$

$\Rightarrow \quad v = 2000 \text{ cm/s} = 20 \text{ m/s}$

(c) $2\pi\nu = 36$

$\Rightarrow \quad \nu = \frac{36}{2\pi}\text{Hz} = \frac{18}{\pi} = 5.7\,\text{Hz}$

(d) $\frac{2\pi}{\lambda} = 0.018 \Rightarrow \lambda = \frac{2\pi}{0.018}\text{cm}$

$= \frac{2000\,\pi}{18}\text{cm}$

$= \frac{20\,\pi}{18}\text{m}$

$= 3.48\,\text{cm}$

Hence, least distance between two successive crests $= \lambda = 3.48\,\text{m}$.

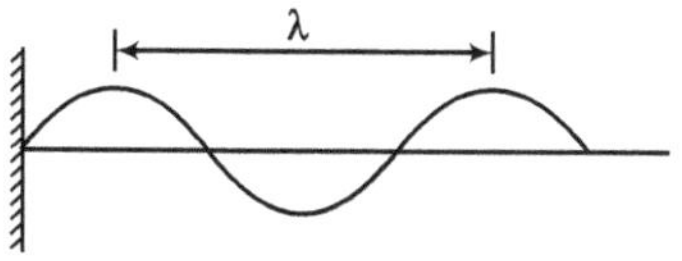

21. Given equation is $y(x,t) = 0.06\sin\left(\frac{2\pi x}{3}\right)\cos(120\pi t)$

(a) Comparing with a standard equation of stationary wave $y(x,t) = a\sin(kx)\cos(\omega t)$

Clearly, the given equation belongs to stationary wave. Hence, option (a) is not correct.

(b) By comparing,

$$\omega = 120\pi$$

$\Rightarrow \quad 2\pi f = 120\,\pi$

$\Rightarrow \quad f = 60\,\text{Hz}$

(c) $k = \frac{2\pi}{3} = \frac{2\pi}{\lambda}$

$\Rightarrow \lambda = \text{wavelength} = 3\,\text{m}$

Frequency $= f = 60$ Hz

Speed $= v = f\lambda = (60\,\text{Hz})(3\,\text{m}) = 180$ m/s

(d) Since in stationary wave, all particles of the medium execute SHM with varying amplitude nodes.

22. Given equation is $y(x,t) = 0.06\sin\left(\frac{2\pi}{3}x\right)\cos(120\pi t)$

Comparing with standard equation of stationary wave $y(x,t) = a\sin(kx)\cos(\omega t)$

It is represented by diagram.

where N denotes nodes and A denotes antinodes.

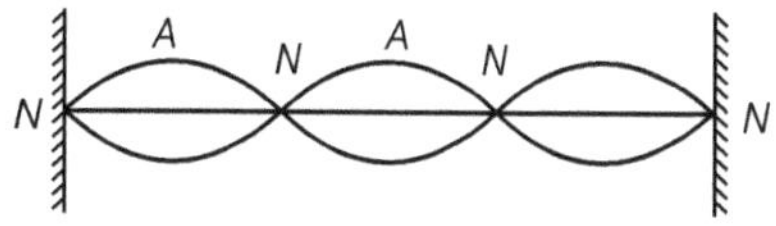

(a) Clearly, frequency is common for all the points.

(b) Consider all the particles between two nodes they are having same phase of $(120\pi t)$ at a given time.

(c) and (d) But are having different amplitudes of $0.06\sin\left(\frac{2\pi}{3}x\right)$ and because of different amplitudes they are having different energies.

23. Given, $\nu_0 = 400\,\text{Hz}$, $v = 340$ m/s

Speed of wind $v_w = 10\,\text{m/s}$

(a) As both source and observer are stationary, hence frequency observed will be same as natural frequency $\nu_0 = 400\,\text{Hz}$

(b) The speed of sound $v = v + v_w$

$$= 340 + 10 = 350 \text{ m/s}$$

(c) and (d) There will be no effect on frequency, because there is no relative motion between source and observer hence (c), (d) are incorrect.

24. Consider the equation of a stationary wave $y = a\sin(kx)\cos\omega t$

(a) Clearly every particle at x will have amplitude $= a\sin kx$ = fixed

(b) For mean position $y = 0$

$\Rightarrow \quad \cos\omega t = 0 \quad \Rightarrow \quad \omega t = (2n-1)\frac{\pi}{2}$

Hence, for a fixed value of n, all particles are having same value of

time, $t = (2n-1)\frac{\pi}{2\omega}$ $\quad [\because \omega = \text{constant}]$

(c) Amplitude of all the particles are $a\sin(kx)$ which is different for different particles at different values of x

(d) The energy is a stationary wave is confined between two nodes

(e) Particles at different nodes are always at rest.

Comprehension Based Questions

1. Reflected and incident rays are in the same medium. Hence,

$$I \propto A^2$$

I_r has become 64% or 0.64 times of I_i

$\therefore \quad A_r = 0.8\,A_i = 0.8\,A$

2. $y = y_i + y_r$

$= A\sin(ax + bt + \pi/2) + 0.8A\sin(ax - bt + \pi/2)$

$= \left[0.8A\sin\left(ax + bt + \frac{\pi}{2}\right) + 0.8A\sin\left(ax - bt + \frac{\pi}{2}\right)\right]$

$+ 0.2A\sin\left(ax + bt + \frac{\pi}{2}\right)$

$= -1.6\sin ax\sin bt + 0.2A\cos(bt + ax)$

$\therefore \quad c = 0.2$

3. $A_x = -1.6\sin ax$

$\therefore \quad x = 0$ is a node

Second antinode is at a distance.

$$x = \frac{\lambda}{4} + \frac{\lambda}{2}$$

$$= \frac{3\lambda}{4} = \frac{3}{4}\left(\frac{2\pi}{k}\right)$$

But $k = a$

$$\therefore \quad x = \frac{3\pi}{2a}$$

4. 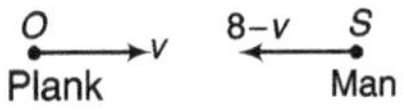

From conservation of linear momentum.

$$50(8 - v) = 150(v)$$

$$\therefore \quad v = 2\text{ m/s}$$

$$8 - v = 6\text{ m/s}$$

$$f_1 = f_0\left(\frac{v + v_0}{v - v_S}\right)$$

$$= f_0\left(\frac{330 + 2}{330 - 6}\right) = \frac{332}{324}f_0$$

5. 

$$f_2 = f_0\left(\frac{v - v_0}{v + v_s}\right)$$

$$= f_0\left(\frac{330 - 2}{330 + 6}\right) = \left(\frac{328}{336}\right)f_0$$

6. $f_1 > f_0$ but $f_1 =$ constant.

Similarly, $f_2 < f_0$ but $f_2 =$ constant

7. $v_p = -v\left(\frac{\partial y}{\partial x}\right)$, where $v = -+$ ve.

At E, $\frac{\partial y}{\partial x}$ or slope is positive.

Hence, v_p is negative.

At D, $\frac{\partial y}{\partial x}$ or slope is zero. Hence, v_p is zero.

Assertion and Reason

1. Speed of sound in water is more, hence water is rarer for sound wave.

2. $f = \frac{v}{2l}$

Here, $\quad v = \sqrt{\frac{\gamma RT}{M}}$

$\therefore \quad f \propto \sqrt{T}$ as $v \propto \sqrt{T}$

3. Number of loops will increase, not the loop size.

5. For the movement of transverse mechanical waves modulus of rigidity is required.

6. $f_c = n_1\left(\frac{v}{4ml}\right)$, where $n_1 = 1, 3, 5\ldots$

$f_o = n_2\left(\frac{v}{2l}\right)$, where $n_2 = 1, 2, 3, \ldots$

$$f_c = f_o \Rightarrow \frac{n_1}{4m} = \frac{n_2}{2}$$

or $\quad \frac{n_1}{mn_2} = 2$ or $\frac{n_1}{n_2} = 2m$

But this condition can't be satisfied with the given numbers.

7. Same speed does not mean relative velocity is zero. Due to motion of source, Doppler effect is observed by change in wavelength.

11. If v_o and v_s are not along the same line, then Doppler effect depends upon the distance between source and observer.

12. $v \propto \sqrt{\frac{T}{M}}$

T has doubles and M has become half. Therefore, v will become two times. Further, with increase in humidity in air density of air decreases. Therefore speed of sound increases.

13. $f = \frac{v}{4(l + 0.6r)}$ or $\frac{v}{2(l + 1.2r)}$

Therefore, if r is less, f will be more.

14. Number of loops = 3

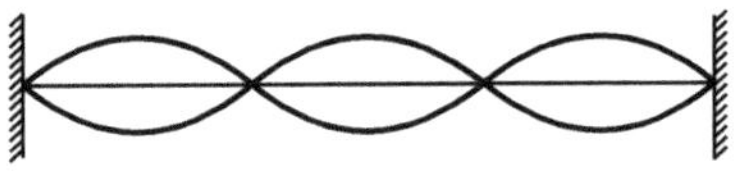

Number of antinodes = 3

Number of nodes = 4

15. Electromagnetic waves are transverse waves. But they can travel in all three media.

16. Frequency of second overtone mode,

$$f_2 = 2\left(\frac{v}{2l}\right)$$

$$f_3 = \frac{3}{2}f_2$$

$$\therefore \quad \lambda_3 = \frac{2}{3}\lambda_2 \text{ as } v = \text{constant}$$

17. For sound wave air is denser and water is rarer.

18. $A_t = \left(\frac{2v_2}{v_1 + v_2}\right)A_i$

If $v_2 > v_1$, then $A_t > A_i$.

20. Putting $x = 0$, we get, $y = -A\sin\omega t$. Further kx and ωt have opposite signs. Therefore, wave is travelling in positive x-direction.

Match the Columns

1. (a) $f' = f\left(\frac{v + v_0}{v + v_s}\right) = f$ (b) $f' = f\left(\frac{v + v_0}{v - v_s}\right) > f$

(c) $f' = f\left(\frac{v - v_0}{v + v_s}\right) < f$ (d) $f' = f\left(\frac{v - v_0}{v - v_s}\right) = f$

2. $k = \frac{2\pi}{\lambda} = 2\pi a$

$$\therefore \quad \lambda = \frac{1}{a}$$

$$\omega = \frac{2\pi}{T} = 2\pi b$$

$$\therefore \quad T = \frac{1}{b} \text{ or } f = b$$

$$\Delta\phi = \frac{2\pi}{\lambda} \times \Delta x = 2\pi a\left(\frac{1}{4a}\right) = \frac{\pi}{2}$$

$$\Delta\phi = \frac{2\pi}{T}\Delta t = 2\pi b\left(\frac{1}{8b}\right) = \frac{\pi}{4}$$

3. Frequency is the property of source. It remains unchanged. In rarer medium, speed of wave is more.

$$\lambda = \frac{v}{f} \text{ or } \lambda \propto v$$

$$A_1 = \left(\frac{2v_2}{v_1 + v_2}\right)A_i$$

Since, $v_2 > v_1 \quad \therefore \quad A_t > A_i$

4. Third overtone frequency of closed pipe means seventh harmonic. Which is 7 times the fundamental frequency.

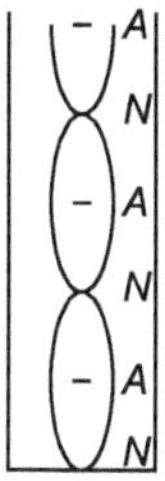

Second overtone of closed pipe.

5. $v = \sqrt{\frac{T}{\mu}}$

Here, μ = mass/length

As, the wave train moves upwards, μ remains unchanged while T increases. Hence, v increases. Further, $\lambda = \frac{v}{f}$ or $\lambda \propto v$

as f is unchanged.

6. $v = \sqrt{\frac{\gamma RT}{M}} = \sqrt{\frac{\gamma p}{\rho}}$

Speed does not change with change in pressure unless temperature is changed.

7. $100 = \frac{v}{4l_c}$

$$\therefore \quad l_c = \frac{v}{400} = \frac{330}{400}$$

$$= 0.825 \text{ m}$$

$$200 = \frac{v}{2l_a}$$

$$\therefore \quad l_0 = \frac{v}{400} = 0.825 \text{ m}$$

Different frequencies of closed pipes will be 100 Hz, 300 Hz, 500 Hz, etc., and different frequencies of open pipe are 200 Hz, 400 Hz, 600 Hz, etc., i.e. none of them matches with each other.

9. $f_2 = 2f_1 \quad \Rightarrow \quad \lambda = \frac{1}{f}$

$$\therefore \quad \lambda_2 = \frac{\lambda_1}{2}$$

$$E \propto f^2 \quad \therefore \quad f_2 > f_1$$

Entrance Gallery

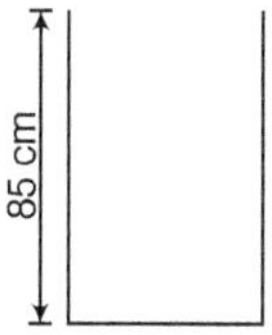

1. For closed organ pipe,

$$\frac{(2n+1)v}{4l} < 1250$$

$$(2n+1) < 1250 \times \frac{4 \times 0.85}{340}$$

$$(2n+1) < 12.50$$

$$n < 5.25$$

So, $n = 0, 1, 2, 3, \cdots, 5$

So, we have 6 possibilities.

2. We know that lLinear mass density is defined as measure of mass per unit of length.

$\therefore$ Linear mass density, $\mu = \frac{\text{Mass}}{\text{Length}}$

$$= \frac{10 \times 10^{-3}}{1}$$

$$= 10 \times 10^{-3} \text{ kg/m}$$

$\therefore$ The speed of transverse wave,

$$v = \sqrt{\frac{T}{\mu}} = \sqrt{\frac{100}{10 \times 10^{-3}}} = \sqrt{10 \times 10^3}$$

(where, μ = volume per unit $\times$ density)

$$= 1 \times 10^2 = 100 \text{ ms}^{-1}$$

3. According to the modes of vibration of air column in open organ pipe,

First mode of vibration,

$$\nu_1 = \frac{v}{2l} \quad \ldots\text{(i)}$$

where, ν_1 = frequency of vibration, v = speed of sound and l = length of pipe.

The frequency of nth mode of vibration,

$$\nu_n = n \times v_1 = n \times \left(\frac{v}{2l}\right)$$

Given, $\nu_n = 1.1 \text{ kHz} = 1100 \text{ Hz} \Rightarrow v = 330 \text{ ms}^{-1}$,

and $\quad l = 30 \text{ cm} = 0.30 \text{ m}$

$$1100 = n \times \left(\frac{330}{2 \times 0.30}\right)$$

$$\Rightarrow \quad n = \frac{1100 \times 2 \times 0.30}{330}$$

$$\Rightarrow \quad n = \frac{660}{330} = 2$$

So, second harmonic mode of pipe resonates.

4. Given, $\nu = 340$ Hz (frequency of sound produced by source) speed of sound, $v = 340 \text{ ms}^{-1}$

Approaching speed of train, $v_s = 10\ \text{ms}^{-1}$, $\nu' = ?$

We know that, when sound source is in motion and observer is stationary, frequency heard

$$\nu' = \nu\left(\frac{v}{v+v_s}\right) = 340\left(\frac{340}{340+10}\right)$$

$$= \frac{340\times340}{350} = \frac{340\times34}{35}$$

$$= \frac{11560}{35}$$

$$= 330.28 \approx 330\ \text{Hz}$$

5. Thermal radiations are the heat waves travel along straight lines with the speed of light.

6. We know that,

Maximum intensity, $I_{max} = (\sqrt{I_1} + \sqrt{I_2})^2$ ...(i)

Minimum intensity, $I_{min} = (\sqrt{I_1} - \sqrt{I_2})^2$ ...(ii)

So, the ratio of the maximum and minimum intensity,

$$\frac{I_{max}}{I_{min}} = \frac{(\sqrt{I_1}+\sqrt{I_2})^2}{(\sqrt{I_1}-\sqrt{I_2})^2}$$

$$\frac{I_{max}}{I_{min}} = \frac{(\sqrt{4I}+\sqrt{I})^2}{(\sqrt{4I}-\sqrt{I})^2} = \frac{9I}{I} = 9:1$$

7. For an open pipe, the frequency,

$$\nu = \frac{v}{2l} \Rightarrow 5100 = \frac{340}{2\times l}$$

$$\Rightarrow \quad l = \frac{340}{5100\times2} \Rightarrow l = \frac{2}{30\times2} = \frac{1}{30}\ \text{m}$$

$$\Rightarrow \quad l = \frac{100}{30} = \frac{10}{3}\ \text{cm}$$

8. Path difference $= (\pi r - 2r) = (\pi - 2)r = n\lambda$

$$v = f\times\lambda \Rightarrow \frac{v}{\lambda} = f = \left[\frac{v}{(\pi-2)r}\right]n$$

So, multiples integral $= \dfrac{v}{(\pi-2)r}$

9. By Doppler's effect,

when observer is moving with velocity v_0, towards a source at rest, then approach frequency

$$N_{\text{Approach}} = N\left(\frac{v+v_o}{v}\right)$$

where, $N = 850\ \text{Hz}, v = 340\ \text{ms}^{-1}, v_o = 72\ \text{kmh}^{-1} = 20\ \text{ms}^{-1}$

$$\therefore \quad N_{\text{Approach}} = 850\left(\frac{340+20}{340}\right)$$

Similarly, when observer is moving away from source, then

$$N_{\text{Separation}} = N\left(\frac{v-v_o}{v}\right) = 850\left(\frac{340-20}{340}\right)$$

The difference of the two frequency,

$$N_{\text{Approach}} - N_{\text{Separation}} = 850\left(\frac{360}{340}\right) - 850\left(\frac{320}{340}\right)$$

$$= \frac{850}{340}\times40 = \frac{850}{8.5} = 100\ \text{Hz}$$

10. Energy transferred with the propagation of longitudinal waves.

11. Speed of transverse wave $= \sqrt{\dfrac{T}{\mu}} = \sqrt{\dfrac{400}{m/l}}$

$$= \sqrt{\frac{\frac{400}{3}}{0.25}} = \sqrt{\frac{400\times25}{3\times100}} = 5.77\ \text{ms}^{-1}$$

So, time taken by disturbance to reach the other end

$$= \frac{0.25}{5.77} = 0.043\ \text{s} \approx 0.047\ \text{s}$$

12. In compression region, density increases and in rarefaction, density decreases. So, density varies due to sound propagation.

13.

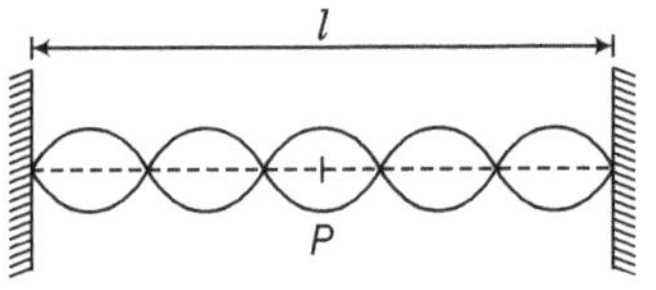

Number of nodes = 6

From the given equation, we can see that

$$k = \frac{2\pi}{\lambda} = 62.8\ \text{m}^{-1} \Rightarrow \lambda = \frac{2\pi}{62.8} = 0.1$$

$$\Rightarrow \quad l = \frac{5\lambda}{2} = 0.25\ \text{m}$$

The mid-point of the string is P, an antinode

$\therefore$ maximum displacement = 0.01 m

$$\omega = 2\pi f = 628\ \text{s}^{-1}$$

$$\Rightarrow \quad \nu = \frac{628}{2\pi} = 100\ \text{Hz}$$

But this is fifth harmonic frequency.

$\therefore$ Fundamental frequency, $\nu_0 = \dfrac{\nu}{5} = 20\ \text{Hz}$

14. When wind blows from S to O

$$f_2 = f_1\left(\frac{v+w+u}{v+w-u}\right)$$

or $\quad f_2 > f_1$

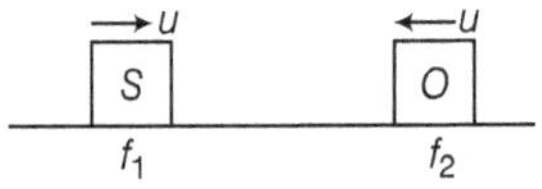

when wind blows from O to S

$$f_2 = f_1\left(\frac{v-w+u}{v-w-u}\right)$$

$$\Rightarrow \quad f_2 > f_1$$

15. Fundamental frequency of sonometer wire,

$$\nu = \frac{v}{2l} = \frac{1}{2l}\sqrt{\frac{T}{\mu}} = \frac{1}{2l}\sqrt{\frac{T}{Ad}} \quad \text{...(i)}$$

Here, μ = mass per unit length of wire.

Also, Young's modulus of elasticity, $Y = \dfrac{Tl}{A\Delta l}$

$$\Rightarrow \quad \frac{T}{A} = \frac{Y\Delta l}{l} = \frac{1}{2l}\sqrt{\frac{Y\Delta l}{ld}}$$

Given, $\quad l = 1.5\ \text{m}, \dfrac{\Delta l}{l} = 0.01, \quad d = 7.7\times10^3\ \text{kg/m}^3$

and $\quad Y = 2.2\times10^{11}\ \text{N/m}^2$

After substituting the values in Eq. (i), we get

$$\nu \approx 178.2\ \text{Hz}$$

16. From the formula,

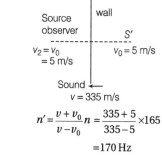

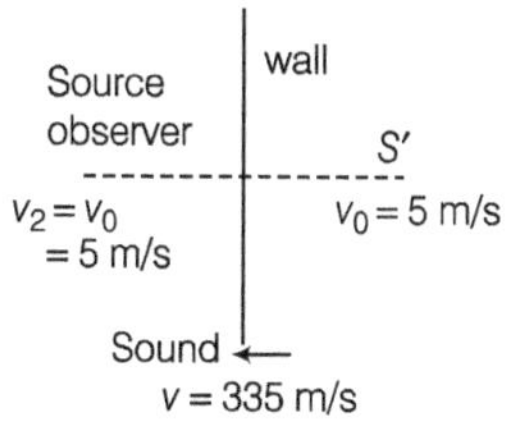

$$\Rightarrow \quad n' = \frac{v + v_0}{v - v_0} n = \frac{335 + 5}{335 - 5} \times 165$$

$$= 170 \text{ Hz}$$

Hence, number of beats heard per second by the passenger is

$$= 170 - 165 = 5 \text{ beats s}^{-1}$$

17. Given, velocity of listener = 0,

velocity of source $= \frac{1}{10} \times v = 0.1v$ (where, v is the velocity of sound)

From the Doppler's effect, the apparent frequency is given by

$$n' = \left(\frac{v}{v - v_s}\right) n = \left(\frac{v}{v - 0.1v}\right) n \Rightarrow \frac{n'}{n} = \frac{10}{9}$$

18. Velocity of the source is given by, $v_s = r\omega = 2\pi f$

Given, $f = 440 \text{ rev min}^{-1} = \frac{400}{60} \text{ rev s}^{-1}$

$$\therefore \quad v_s = 1.2 \times 2 \times 3.14 \times \left(\frac{400}{60}\right) = 50 \text{ ms}^{-1}$$

Minimum frequency of whistle listen by man is given by

$$\nu_{min} = \frac{v}{v + v_s} \nu = \frac{340}{340 + 50} \times 500$$

$$= 435.89 \text{ Hz} \approx 436 \text{ Hz}$$

Maximum frequency of whistle listen by man is given by

$$\nu_{max} = \frac{v}{v - v_s} \nu = \frac{340}{340 - 50} \times 500$$

$$= 586.20 \approx 586 \text{ Hz}$$

19. Let f be the frequency of horn emitted by car.

f' = observed frequency

$\because$ pitch $\propto f$

One octave = Frequency of instrument

$$\therefore \quad f' = \frac{v - v_o}{v + v_s} f$$

where, v_o = velocity of observer, v_s = velocity of source

Both are same and moving with same velocity

$$v_s = v_0 = 4$$

v = velocity of sound

AB = source and sound are in opposite direction, hence, $v + u$ = relative velocity

$$\therefore \quad f' = \frac{v - u}{v + u} f$$

$$\Rightarrow \quad 2f = \left(\frac{v - u}{v + u}\right) d$$

$$2v + 2u = v - u$$

$$\Rightarrow \quad u = v/3$$

20. Since, train (source) is moving towards pedestrian (observer), the perceived frequency will be higher than the original,

$$\nu' = \nu\left(\frac{v + v_0}{v - v_s}\right)$$

Here, $v_0 = 0$ (as observer is stationary)

$v_s = 25 \text{ ms}^{-1}$ (velocity of source)

$v = 350 \text{ ms}^{-1}$ (velocity of sound)

and $\nu = 1$ kHz (original frequency)

Hence, $$\nu' = 1000\left(\frac{350 + 0}{350 - 25}\right)$$

$$= 1000 \times \frac{350}{325}$$

$$= 1077 \text{ Hz}$$

21. The speed of transverse wave,

$$v = \sqrt{\frac{T}{\mu}}$$

Given, $T = 20$ N, $d = 7.5 \text{ g cm}^{-3}$ and $A = 0.2 \text{ mm}^2$

$$v = \sqrt{\frac{T}{Ad}} = \sqrt{\frac{20 \times 10^{-3}}{7.5 \times 0.2 \times (10^{-3})^3}}$$

$$\Rightarrow \quad v \approx 116 \text{ ms}^{-1}$$

22. Let the distance between the two cliffs be d. Since, the man is standing mid-way between the two cliffs, then the distance from either end is $d/2$.

The distance travelled by sound (in producing an echo).

$$2 \times \frac{d}{2} = v \times t$$

$$\Rightarrow \quad d = 340 \times 1 = 340 \text{ m}$$

23. With end correction,

$$\nu = n\left[\frac{v}{4(l + e)}\right], \text{ where, } n = 1, 3, \cdots$$

$$= n\left[\frac{v}{4(l + 0.6\,r)}\right]$$

Because, $e = 0.6r$, where, r is radius of pipe.

For first resonance, $n = 1$

$$\therefore \quad \nu = \frac{v}{4(l + 0.6\,r)} \Rightarrow l = \frac{v}{4f} - 0.6r$$

$$= \left[\left(\frac{336 \times 100}{4 \times 512}\right) - 0.6 \times 2\right] = 15.2 \text{ cm}$$

24. Intensity level of sound $= 10 \log \frac{I}{I_0}$ dB

$$= 10 \log\left(\frac{10^{-8}}{10^{-12}}\right) \text{dB} = 40 \text{ dB} \quad (\because I_0 = 10^{-12} \text{ W/m}^2)$$

25. As, this is a case of vibration of string fixed at one end.

So, frequency of vibration is given by

$$\nu = \left(\frac{2n + 1}{4L}\right) v = \frac{(2 \times 4 + 1)}{4 \times 15} \times 2 \times 100 = \frac{9 \times 2 \times 100}{4 \times 15} = 30 \text{ Hz}$$

26. Given, $v = 340 \text{ ms}^{-1}$, $u_s = 20 \text{ ms}^{-1}$ and $\nu_0 = 640$ Hz

From Doppler's law,

$$\nu = \left(\frac{340}{340 - 20}\right) 640 = 680 \text{ Hz}$$

27. Given, velocity, $v_o = 36\text{ km/h} = 36 \times \frac{5}{18} = 10\text{ m/s}$

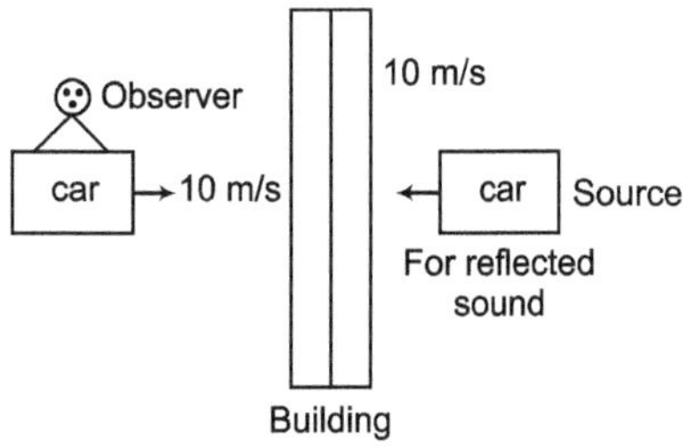

Apparent frequency of sound heard by car driver (observer) reflected from the building will be

$$\nu' = \nu\left(\frac{v + v_o}{v - v_s}\right) = 8\left(\frac{320+10}{320-10}\right)$$

$$\Rightarrow \quad f' = 8.5\text{ kHz}$$

28. $y = y_1 + y_2 = a\sin(\omega t - kx) + a\sin(\omega t + kx)$

$$y = 2a\sin\omega t\cos kx$$

Clearly, it is equation of standing wave for position of nodes $y = 0$.

i.e. $$x = (2n+1)\frac{\lambda}{4}$$

$$= \left(n + \frac{1}{2}\right)\frac{\lambda}{2},\ n = 0, 1, 2, 3$$

29. $y(x, t) = e^{-(ax^2 + bt^2 + 2\sqrt{ab}xt)} = e^{-(\sqrt{a}x + \sqrt{b}t)^2}$

It is a function of type, $y = f(\omega t + kx)$

$\therefore y(x, t)$ represents wave travelling along $-x$ direction.

Speed of wave $= \frac{\omega}{k} = \frac{\sqrt{b}}{\sqrt{a}} = \sqrt{\frac{b}{a}}$

30. As, $\frac{\nu}{\nu_C} = \frac{v/2\lambda}{v/4\lambda} = \frac{2}{1}$

31. We know that, $c = \sqrt{\frac{\gamma p}{\rho}} \Rightarrow c \propto \left(\frac{p}{\rho}\right)^{1/2}$

32. Given, $y = 10\sin\left[\frac{2\pi}{45}t + \alpha\right]$

If $\quad t = 0,\ y = 5\text{ cm} \quad$ (given)

Putting these values in given equation,

$$5 = 10(\sin\alpha)$$

$$\Rightarrow \quad \sin\alpha = \frac{1}{2} \Rightarrow \alpha = \frac{\pi}{6}$$

If $t = 7.5$ s,

Then, total phase $= \frac{2\pi}{45} \times \frac{15}{2} + \frac{\pi}{6} = \frac{\pi}{3} + \frac{\pi}{6} = \frac{\pi}{2}$

33. Given equation is, $y = 2\cos 2\pi(330t - x)$

On comparing with $y = A\cos(\omega t - x)$, we get

$$\omega = 2\pi \times 330$$

$$\therefore \quad T = \frac{1}{330}\text{ s} \qquad \left(\because \omega = \frac{2\pi}{T}\right)$$

34. Given, $\nu_A = 258\text{ Hz}$, $\nu_B = 262\text{ Hz}$

Let ν is the frequency of unknown tuning fork. It produces x beats with 258 and $2x$ with 262.

$$\therefore \quad 262 - (258 - x) = 2x$$

$$\Rightarrow \quad 262 - 258 + x = 2x$$

$$\Rightarrow \quad x = 4$$

35. Fundamental frequency, $\nu = \frac{1}{2rl}\sqrt{\frac{T}{\pi\rho}}$

$$\therefore \quad \frac{\nu_1}{\nu_2} = \frac{l_2}{l_1} \times \frac{r_2}{r_1} \times \sqrt{\frac{T_1}{T_2}} \Rightarrow \frac{600}{f_2} = \frac{2}{1} \times \frac{1}{2} \times \sqrt{\frac{T}{T/9}}$$

36. Firstly, car will be treated as an observer which is approaching the source. Then, it will be treated as a source, which is moving in the direction of sound.

Hence, $$\nu_1 = \nu_o\left(\frac{v + v_1}{v - v_1}\right)$$

$$\nu_2 = \nu_o\left(\frac{v + v_2}{v - v_2}\right)$$

$$\therefore \quad \nu_1 - \nu_2 = \left(\frac{1.2}{100}\right)\nu_o = \nu_o\left[\frac{v + v_1}{v - v_1} - \frac{v + v_2}{v - v_2}\right]$$

or $$\left(\frac{1.2}{100}\right) = \frac{2v(v_1 - v_2)}{(v - v_1)(v - v_2)}$$

As, v_1 and v_2 are very less than v.

We can write, $(v - v_1)$ or $(v - v_2) \approx v$.

$$\therefore \quad \left(\frac{1.2}{100}\right)\nu_o = \frac{2(v_1 - v_2)}{v}\nu_o$$

or $$(v_1 - v_2) = \frac{v \times 1.2}{200}$$

$$= \frac{330 \times 1.2}{200} = 1.98\text{ ms}^{-1}$$

$$= 7.128\text{ kmh}^{-1}$$

37. According to the question, $2\left(\frac{v_1}{2l_1}\right) = \frac{v_2}{4l_2}$

$$\therefore \quad \frac{\sqrt{T/\mu}}{l_1} = \frac{320}{4l_2} \quad \text{(where, } \mu = \text{mass per unit length of wire)}$$

or $$\frac{\sqrt{50/\mu}}{0.5} = \frac{320}{4 \times 0.8}$$

On solving above equation, we get

$$\mu = 0.02\text{ kg/m} = 20\text{ g/m}$$

$\therefore$ Mass of string $= 20\text{ g/m} \times 0.5\text{ m} = 10\text{ g}$

38. Tension, $T = \mu v^2 = \mu\frac{\omega^2}{k^2} = 0.04\frac{(2\pi/0.004)^2}{(2\pi/0.50)^2} = 6.35\text{ N}$

39. Given, $f_o - f_c = 2$ ...(i)

Frequency of fundamental mode for a closed organ pipe,

$$f_c = \frac{v}{4L_c} \Rightarrow L_c = \frac{v}{4f_c}$$

Similarly, frequency of fundamental mode for an open organ pipe,

$$f_o = \frac{v}{2L_o} \Rightarrow L_o = \frac{v}{2f_o}$$

Given, $\quad L_c = L_o \Rightarrow f_o = 2f_c$ ...(ii)

From Eqs. (i) and (ii), we get

$$f_o = 4\text{ Hz} \quad \text{and} \quad f_c = 2\text{ Hz}$$

When the length of the open pipe is halved, its frequency of fundamental mode is

$$f_o' = \frac{v}{2(L_o/2)} = 2f_o = 2 \times 4\text{ Hz}$$

$$= 8\text{ Hz}$$

When the length of the closed pipe is doubled, its frequency of fundamental mode is

$$f_o' = \frac{v}{4(2L_c)}$$

$$= \frac{1}{2} f_c = \frac{1}{2} \times 2 = 1\,\text{Hz}$$

Hence, number of beats produced per second is

$$f_o' - f' = 8 - 1 = 7$$

40. The fundamental frequency is $f = \frac{1}{2L}\sqrt{\frac{T}{\mu}}$

$$f = \frac{1}{2L}\sqrt{\frac{T}{\rho\pi\frac{D^2}{4}}} = \frac{1}{LD}\sqrt{\frac{T}{\pi\rho}} \Rightarrow f \propto \frac{1}{LD}$$

41. Frequency of reflected sound heard by the bat,

$$\nu' = \nu\left[\frac{v-(v_0)}{v-v_s}\right] = \nu\left[\frac{v+v_0}{v-v_s}\right] = \nu\left[\frac{v+v_b}{v-v_b}\right]$$

$$= \nu\left[\frac{330+4}{330-4}\right] \times 90 \times 10^3 = 92.1 \times 10^3\ \text{Hz}$$

42. The velocity of a transverse wave,

$$v = \sqrt{\frac{T}{\rho A}} \Rightarrow v \propto \frac{1}{\sqrt{A}} \quad \text{or} \quad v \propto \frac{1}{R}$$

where, R is the radius of copper wire.

Because, the velocity of wire depends on the radius. So, transverse wave travels faster in thinner wire.

43. Given, $y = A\sin(100\,\pi t - 3x)$

The general equation,

$$y = A\sin(\omega t - kx)$$

$\therefore$ $$k = 3$$

and $$k = \frac{2\pi}{\lambda}$$

or $$\lambda = \frac{2\pi}{k} = \frac{2\pi}{3}$$

Phase difference, $\phi = \frac{\pi}{3} \Rightarrow \frac{2\pi}{\lambda}\cdot x = \frac{\pi}{3}$

or $$x = \frac{\pi}{3} \times \frac{\lambda}{2\pi}$$

$\Rightarrow$ $$x = \frac{\pi}{3} \times \frac{2\pi}{3 \times 2\pi}$$

Distance, $$x = \frac{\pi}{9}\ \text{m}$$

44. Sine wave,

Particle velocity, $v_p = \frac{dy}{dt}$ = slope of wave at that point.

As, slope at A and B is zero. Hence, the velocity at A and B will be same. Distance between A and B is $\frac{\lambda}{2}$.

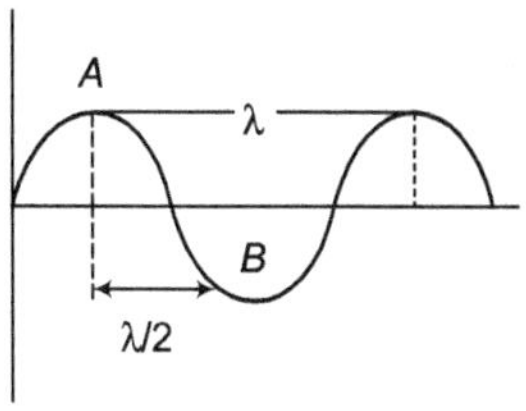

45. In the fundamental mode,

Frequency, $n = \frac{v}{\lambda} \Rightarrow n = \frac{v}{4l} \Rightarrow n = \frac{l}{t \times 4l}$ $\left(\because v = \frac{l}{t}\right)$

$\therefore$ $$n = \frac{1}{0.01 \times 4} = 25\,\text{Hz}$$

46. Frequency of closed pipe, $n_1 = \frac{v}{4l_1} \Rightarrow l_1 = \frac{v}{4n_1}$

Frequency of open pipe, $n_2 = \frac{v}{2l_1} \Rightarrow l_2 = \frac{v}{2n_2}$

When both pipes are joined, then length of closed pipe

$$l = l_1 + l_2$$

$$\frac{v}{4n} = \frac{v}{4n_1} + \frac{v}{2n_2}$$

or $$\frac{1}{2n} = \frac{1}{2n_1} + \frac{1}{n_2}$$

or $$\frac{1}{2n} = \frac{n_2 + 2n_1}{2n_1 n_2}$$

or $$n = \frac{n_1 n_2}{n_2 + 2n_1}$$

JEE Main & Advanced
Solved Paper
2015

Solved Paper 2015

JEE Main

Joint Entrance Examination

Physics

Instructions

- This test consists of 13 questions.
- Each question is allotted 4 marks for correct response.
- Candidates will be awarded marks as stated above for correct response of each question. 1 marks will be deducted for indicating incorrect response of each question. No deduction from the total score will be made if no response is indicated for an item in the answer sheet.
- There is only one correct response for each question. Filling up more than one response in any question will be treated as wrong response and marks for wrong response will be deducted according as per instructions.

---

1. Two stones are thrown up simultaneously from the edge of a cliff 240 m high with initial speed of 10 m/s and 40 m/s respectively. Which of the following graph best represents the time variation of relative position of the second stone with respect to the first? (Assume stones do not rebound after hitting the ground and neglect air resistance, take $g = 10\,\text{m/s}^2$)

(a)

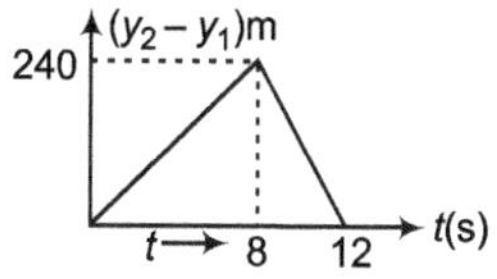

(b)

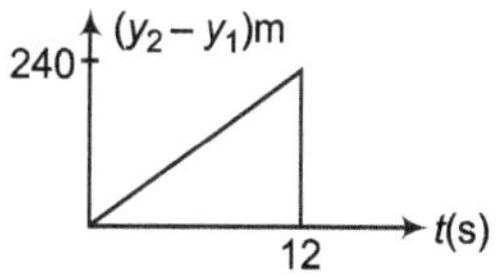

(c)

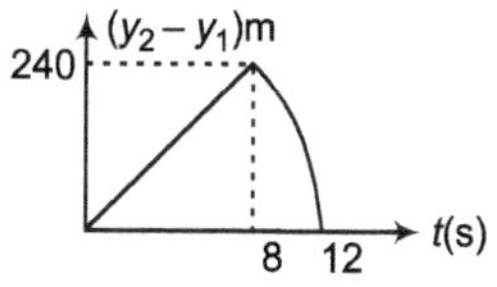

(d)

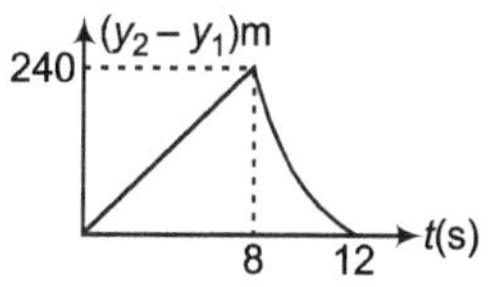

2. The period of oscillation of a simple pendulum is $T = 2\pi\sqrt{\dfrac{L}{g}}$. Measured value of L is 20.0 cm known to 1 mm accuracy and time for 100 oscillations of the pendulum is found to be 90 s using a wrist watch of 1s resolution. The accuracy in the determination of g is

(a) 2%
(b) 3%
(c) 1%
(d) 5%

3. Given in the figure are two blocks A and B of weight 20 N and 100 N respectively. These are being pressed against a wall by a force F as shown in figure. If the coefficient of friction between the blocks is 0.1 and between block B and the wall is 0.15, the frictional force applied by the wall in block B is

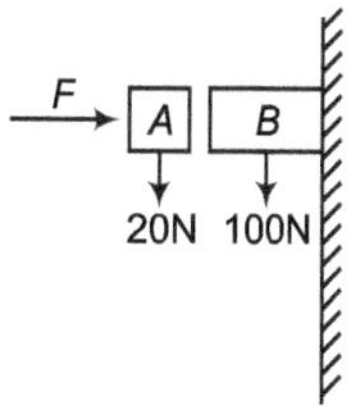

(a) 100 N (b) 80 N
(c) 120 N (d) 150 N

4. A particle of mass m moving in the x-direction with speed $2v$ is hit by another particle of mass $2m$ moving in the y-direction with speed v. If the collision is perfectly inelastic, the percentage loss in the energy during the collision is close to

(a) 44% (b) 50% (c) 56% (d) 62%

5. Distance of the centre of mass of a solid uniform cone from its vertex is z_0. If the radius of its base is R and its height is h, then z_0 is equal to

(a) $\dfrac{h^2}{4R}$ (b) $\dfrac{3h}{4}$
(c) $\dfrac{5h}{8}$ (d) $\dfrac{3h^2}{8R}$

6. From a solid sphere of mass M and radius R, a cube of maximum possible volume is cut. Moment of inertia of cube about an axis passing through its centre and perpendicular to one of its faces is

(a) $\frac{MR^2}{32\sqrt{2}\pi}$ (b) $\frac{MR^2}{16\sqrt{2}\pi}$

(c) $\frac{4MR^2}{9\sqrt{3}\pi}$ (d) $\frac{4MR^2}{3\sqrt{3}\pi}$

7. From a solid sphere of mass M and radius R, a spherical portion of radius $\left(\frac{R}{2}\right)$ is removed as shown in the figure. Taking gravitational potential $V = 0$ at $r = \infty$, the potential at the centre of the cavity thus formed is (G = gravitational constant)

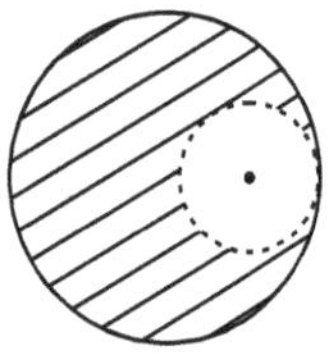

(a) $\frac{-GM}{2R}$ (b) $\frac{-GM}{R}$

(c) $\frac{-2GM}{3R}$ (d) $\frac{-2GM}{R}$

8. A pendulum made of a uniform wire of cross-sectional area A has time period T. When an additional mass M is added to its bob, the time period changes T_M. If the Young's modulus of the material of the wire is Y, then $\frac{1}{Y}$ is equal to (g = gravitational acceleration)

(a) $\left[\left(\frac{T_M}{T}\right)^2 - 1\right]\frac{A}{Mg}$ (b) $\left[\left(\frac{T_M}{T}\right)^2 - 1\right]\frac{Mg}{A}$

(c) $\left[1-\left(\frac{T_M}{T}\right)^2\right]\frac{A}{Mg}$ (d) $\left[1-\left(\frac{T}{T_M}\right)^2\right]\frac{A}{Mg}$

9. Consider a spherical shell of radius R at temperature T. The black body radiation inside it can be considered as an ideal gas of photons with internal energy per unit volume $u = \frac{U}{V} \propto T^4$ and pressure $p = \frac{1}{3}\left(\frac{U}{V}\right)$. If the shell now undergoes an adiabatic expansion, the relation between T and R is

(a) $T \propto e^{-R}$ (b) $T \propto e^{-3R}$

(c) $T \propto \frac{1}{R}$ (d) $T \propto \frac{1}{R^3}$

10. A solid body of constant heat capacity 1 J/°C is being heated by keeping it in contact with reservoirs in two ways

(i) Sequentially keeping in contact with 2 reservoirs such that each reservoir supplies same amount of heat.

(ii) Sequentially keeping in contact with 8 reservoirs such that each reservoir supplies same amount of heat.

In both the cases, body is brought from initial temperature 100°C to final temperature 200°C. Entropy change of the body in the two cases respectively, is

(a) ln2, 4ln2 (b) ln2, ln2

(c) ln2, 2ln2 (d) 2ln2, 8ln2

11. Consider an ideal gas confined in an isolated closed chamber. As the gas undergoes an adiabatic expansion, the average time of collision between molecules increases as V^q, where V is the volume of the gas. The value of q is $\left(\gamma = \frac{C_p}{C_V}\right)$

(a) $\frac{3\gamma + 5}{6}$ (b) $\frac{3\gamma - 5}{6}$ (c) $\frac{\gamma + 1}{2}$ (d) $\frac{\gamma - 1}{2}$

12. For a simple pendulum, a graph is plotted between its Kinetic Energy (KE) and Potential Energy (PE) against its displacement d. Which one of the following represents these correctly? (graphs are schematic and not drawn to scale)

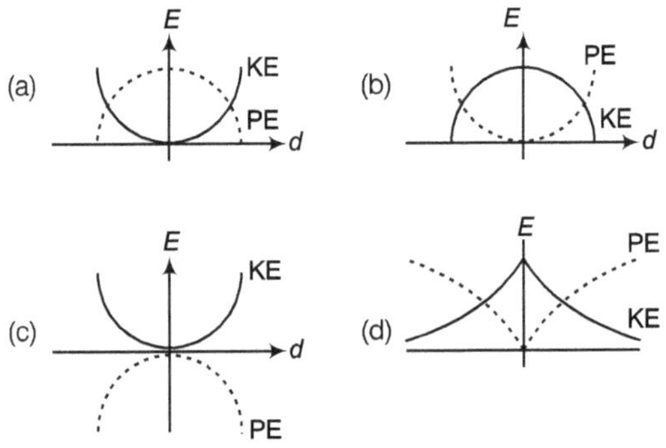

13. A train is moving on a straight track with speed 20 ms^{-1}. It is blowing its whistle at the frequency of 1000 Hz. The percentage change in the frequency heard by a person standing near the track as the train passes him is close to (speed of sound = 320 ms^{-1})

(a) 6% (b) 12% (c) 18% (d) 24%

Answer *with* Explanations

1. (c) ***Central Idea*** *Concept of relative motion can be applied to predict the nature of motion of one particle with respect to the other.*

Consider the stones thrown up simultaneously as shown in the diagram below.

Considering motion of the second particle with respect to the first we have relative acceleration $|\mathbf{a}_{21}| = |\mathbf{a}_2 - \mathbf{a}_1| = g - g = 0$

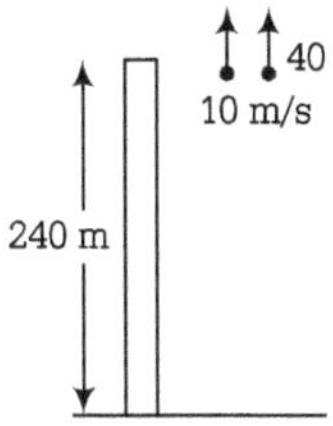

Thus, motion of first particle is straight line with respect to second particle till the first particle strikes ground at a time given by

$$-240 = 10\,t - \frac{1}{2} \times 10 \times t^2$$

or $\quad t^2 - 2t - 48 = 0$

or $\quad t^2 - 8t + 6t - 48 = 0$

or $\quad t = 8, -6 \quad$ (not possible)

Thus, distance covered by second particle with respect to first particle in 8 s is

$$s_{12} = (v_{21})\,t = (40 - 10)(8\text{s})$$
$$= 30 \times 8 = 240 \text{ m}$$

Similarly, time taken by second particle to strike the ground is given by

$$-240 = 40t - \frac{1}{2} \times 10 \times t^2$$

or $\quad -240 = 40t - 5t^2$

or $\quad 5t^2 - 40t - 240 = 0$

or $\quad t^2 - 8t - 48 = 0$

$\quad t^2 - 12t + 4t - 48 = 0$

or $\quad t\,(t - 12) + 4\,(t - 12) = 0$

or $\quad t = 12, -4 \quad$ (not possible)

Thus, after 8 s, magnitude of relative velocity will increase upto 12 s when second particle strikes the ground.

2. (b) ***Central Idea*** *Given time period* $T = 2\pi\sqrt{\frac{L}{g}}$

Thus, changes can be expressed as $\pm \frac{2\Delta T}{T} = \pm \frac{\Delta L}{L} \pm \frac{\Delta g}{g}$

According to the question, we can write

$$\frac{\Delta L}{L} = \frac{0.1\text{cm}}{20.0\text{cm}} = \frac{1}{200}$$

Again time period

$$T = \frac{90}{100}\text{ s}$$

and $\quad \Delta T = \frac{1}{100}\text{s}$

$\Rightarrow \quad \frac{\Delta T}{T} = \frac{1}{90}$

Now,

$\because \quad T = 2\pi\sqrt{\frac{L}{g}}$

$\because \quad g = 4\pi^2 \frac{L}{T^2}$

$\therefore \quad \frac{\Delta g}{g} = \frac{\Delta L}{L} + \frac{2\Delta T}{T}$

or $\quad \frac{\Delta g}{g} \times 100\% = \left(\frac{\Delta L}{L}\right) \times 100\% + \left(\frac{2\Delta T}{T}\right) \times 100\%$

$$= \left(\frac{1}{200} \times 100\right)\% + 2 \times \frac{1}{90} \times 100\%$$

$$\simeq 2.72\% \simeq 3\%$$

Thus, accuracy in the determination of g is approx 3%.

3. (c) ***Central Idea*** *In vertical direction, weights are balanced by frictional forces.*

Consider FBD of block A and B as shown in diagram below.

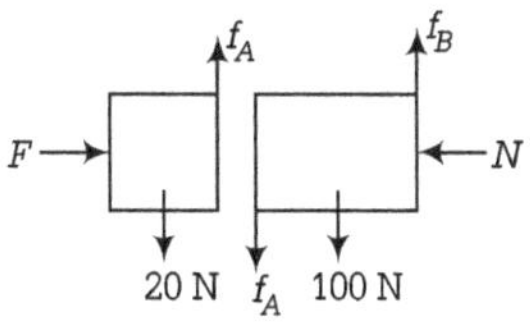

As the blocks are in equilibrium, balance forces are in horizontal and vertical direction.

For the system of blocks $(A + B)$.

$$F = N$$

For block A, $f_A = 20\,\text{N}$ and for block B, $f_B = f_A + 100 = 120$ N

4. (c) ***Central Idea*** *Conservation of linear momentum can be applied but energy is not conserved.*

Consider the movement of two particles as shown below.

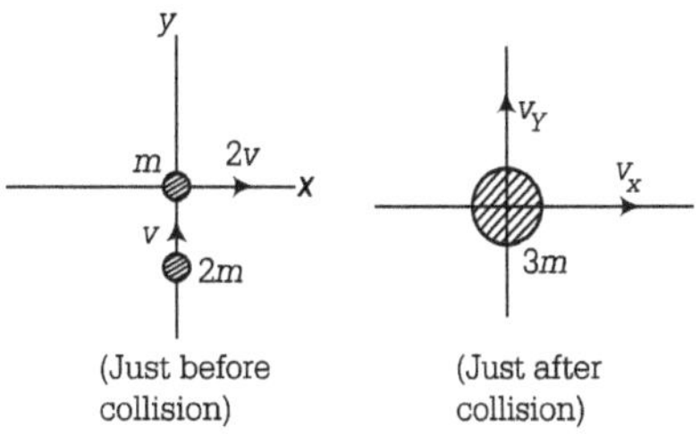

Conserving linear momentum in x-direction

$$(p_i)x = (p_f)x$$

or $\quad 2mv = (2m + m)\,v_x$

or $\quad v_x = \frac{2}{3}v$

Conserving linear momentum in y-direction

$$(p_i)y = (p_f)y$$

or $\quad 2mv = (2m + m)\,v_y$

or $\quad v_y = \frac{2}{3}v$

Initial kinetic energy of the two particles system is

$$E_i = \frac{1}{2} m (2v)^2 + \frac{1}{2} (2m)(v)^2$$
$$= \frac{1}{2} \times 4mv^2 + \frac{1}{2} \times 2mv^2$$
$$= 2mv^2 + mv^2 = 3mv^2$$

Final energy of the combined two particles system is

$$E_f = \frac{1}{2} (3m)(v_x^2 + v_y^2)$$
$$= \frac{1}{2}(3m)\left[\frac{4v^2}{9} + \frac{4v^2}{9}\right]$$
$$= \frac{3m}{2}\left[\frac{8v^2}{9}\right] = \frac{4mv^2}{3}$$

loss in the energy $\Delta E = E_i - E_f$

$$= mv^2 \left[3 - \frac{4}{3}\right] = \frac{5}{3} mv^2$$

Percentage loss in the energy during the collision

$$\frac{\Delta E}{E_i} \times 100 = \frac{\frac{5}{3} mv^2}{3mv^2} \times 100 = \frac{5}{9} \times 100 \simeq 56\%$$

5. *(b)* We know that centre of mass of a uniform solid cone of height h is at height $\frac{h}{4}$ from base, therefore

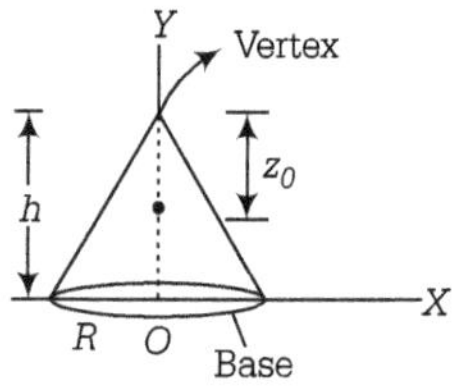

$$h - z_0 = \frac{h}{4}$$

or

$$z_0 = h - \frac{h}{4} = \frac{3h}{4}$$

6. *(c)* ***Central Idea*** *Use geometry of the figure to calculate mass and side length of the cube interms of M and R respectively.*

Consider the cross-sectional view of a diametric plane as shown in the adjacent diagram.

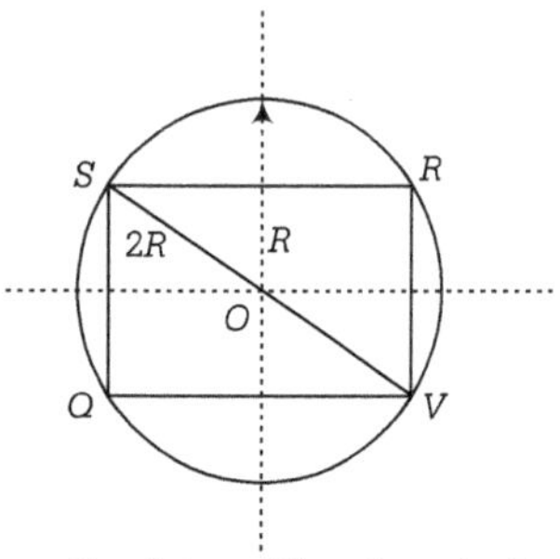

Cross-sectional view of the cube and sphere

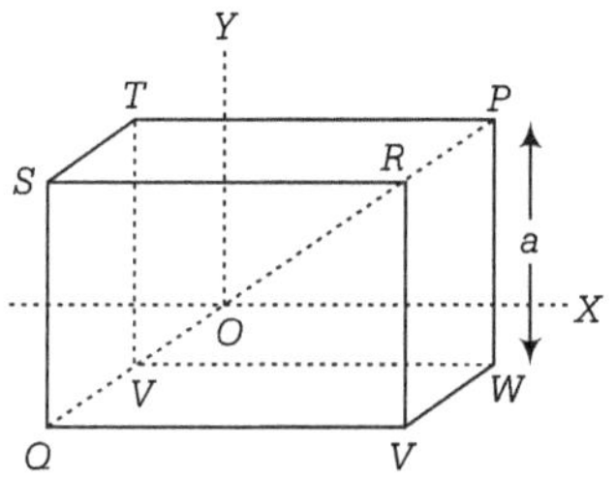

Using geometry of the cube

$$PQ = 2R = (\sqrt{3})\, a$$

or

$$a = \frac{2R}{\sqrt{3}}$$

Volume density of the solid sphere $\rho = \frac{M}{\frac{4}{3}\pi R^3} = \frac{3}{4\pi}\left(\frac{M}{R^3}\right)$

Mass of cube $(m) = (\rho)(a)^3$

$$= \left(\frac{3}{4\pi} \times \frac{M}{R^3}\right)\left[\frac{2R}{\sqrt{3}}\right]^3$$
$$= \frac{3M}{4\pi R^3} \times \frac{8R^3}{3\sqrt{3}} = \frac{2M}{\sqrt{3}\pi}$$

Moment of inertia of the cube about given axis is

$$I_Y = \frac{ma^2}{12}(a^2 + a^2) = \frac{ma^2}{6}$$

$$\Rightarrow \quad I_Y = \frac{ma^2}{6} = \frac{2M}{\sqrt{3}\pi} \times \frac{1}{6} \times \frac{4R^2}{3} = \frac{4MR^2}{9\sqrt{3}\pi}$$

7. *(b)* ***Central Idea*** *Consider cavity as negative mass and apply superposition of gravitational potential.*

Consider the cavity formed in a solid sphere as shown in figure.

$V(\infty) = 0$

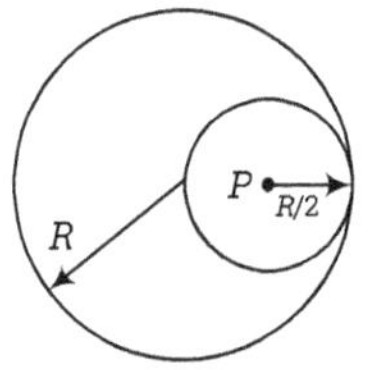

According to the question,we can write potential at an internal point P due to complete solid sphere,

$$V_s = -\frac{GM}{2R^3}\left[3R^2 - \left(\frac{R}{2}\right)^2\right]$$
$$= \frac{-GM}{2R^3}\left[3R^2 - \frac{R^2}{4}\right]$$
$$= \frac{-GM}{2R^3}\left[\frac{11R^2}{4}\right] = \frac{-11\,GM}{8R}$$

Mass of removed part $= \frac{M}{\frac{4}{3} \times \pi R^3} \times \frac{4}{3}\pi\left(\frac{R}{2}\right)^3 = \frac{M}{8}$

Potential at point P due to removed part

$$V_C = \frac{-3}{2} \times \frac{GM/8}{\frac{R}{2}} = \frac{-3GM}{8R}$$

Thus, potential due to remaining part at point P,

$$V_P = V_s - V_c = \frac{-11GM}{8R} - \left(-\frac{3GM}{8R}\right)$$

$$\frac{(-11+3)GM}{8R} = \frac{-GM}{R}$$

8. *(a)* We know that time period, $T = 2\pi\sqrt{\frac{L}{g}}$

When additional mass M is added to its bob

$T_M = 2\pi\sqrt{\frac{L+\Delta L}{g}}$,where ΔL is increase in length.

We know that

$$Y = \frac{Mg/A}{\Delta L/L} = \frac{MgL}{A\Delta L}$$

$$\Rightarrow \quad \Delta L = \frac{MgL}{AY}$$

$$\Rightarrow \therefore \quad T_M = 2\pi\sqrt{\frac{L + \frac{MgL}{AY}}{g}}$$

$$\Rightarrow \quad \left(\frac{T_M}{T}\right)^2 = 1 + \frac{Mg}{AY}$$

or $$\frac{Mg}{AY} = \left(\frac{T_M}{T}\right)^2 - 1$$

or $$\frac{1}{Y} = \frac{A}{Mg}\left[\left(\frac{T_M}{T}\right)^2 - 1\right]$$

9. *(c)* According to question,

$$p = \frac{1}{3}\left(\frac{U}{V}\right)$$

$$\Rightarrow \quad \frac{nRT}{V} = \frac{1}{3}\left(\frac{U}{V}\right) \qquad [\because pV = nRT]$$

or $$\frac{nRT}{V} \propto \frac{1}{3}T^4$$

or $$VT^3 = \text{constant}$$

or $$\frac{4}{3}\pi R^3 T^3 = \text{constant}$$

or $$TR = \text{constant}$$

$$\Rightarrow \quad T \propto \frac{1}{R}$$

10. *(b)* Since, entropy is a state function, therefore change in entropy in both the processes must be same.Therefore, correct option should be (b).

11. *(c)* ***Central Idea*** *For an adiabatic process $TV^{\gamma-1}$ = constant.*

We know that average time of collision between molecules

$$\tau = \frac{1}{n\pi\sqrt{2}\, v_{rms}\, d^2}$$

where, n = number of molecules per unit volume

v_{rms} = rms velocity of molecules

As $$n \propto \frac{1}{V} \text{ and } v_{rms} \propto \sqrt{T}$$

$$\tau \propto \frac{V}{\sqrt{T}}$$

Thus, we can write

$$n = K_1 V^{-1} \text{ and } v_{rms} = K_2\, T^{1/2}$$

where, K_1 and K_2 are constants.

For adiabatic process, $TV^{\gamma-1}$ = constant. Thus, we can write $\tau \propto VT^{-1/2} \propto V\,(V^{1-\gamma})^{-1/2}$ or $\tau \propto V^{\frac{\gamma+1}{2}}$

12. *(b)* During oscillation, motion of a simple pendulum KE is maximum of mean position where PE is minimum. At extreme position, KE is minimum and PE is maximum. Thus, correct graph is depicted in option (b).

13. *(b)* Apparent frequency heard by the person before crossing the train.

$$f_1 = \left(\frac{c}{c - v_s}\right)f_0 = \left(\frac{320}{320-20}\right)1000$$

Similarly, apparent frequency heard, after crossing the trains

$$f_2 = \left(\frac{c}{c + v_s}\right)f_0 = \left(\frac{320}{320+20}\right)1000 \qquad [c = \text{speed of sound}]$$

$$\Delta f = f_1 - f_2 = \left(\frac{2\,cv_s}{c^2 - v_s^2}\right)f_0$$

or $$\frac{\Delta f}{f_0} \times 100 = \left(\frac{2cv_s}{c^2 - v_s^2}\right) \times 100$$

$$= \frac{2 \times 320 \times 20}{300 \times 340} \times 100$$

$$= \frac{2 \times 32 \times 20}{3 \times 34}$$

$$= 12.54\% = 12\%$$

SOLVED PAPER 2015

JEE Advanced

Paper 1

Section 1 (Maximum Marks : 12)

- This section contains THREE questions.
- The answer to each question is a SINGLE DIGIT INTEGER ranging from 0 to 9, both inclusive.
- For each question, darken the bubble corresponding to the correct integer in ORS.
- Marking scheme :
  + 4 If the bubble corresponding to the answer is darkened
  0 in all other cases

1. A bullet is fired vertically upwards with velocity v from the surface of a spherical planet. When it reaches its maximum height, its acceleration due to the planet's gravity is $\frac{1}{4}$ th of its value at the surface of the planet. If the escape velocity from the planet is $v_{sec} = v\sqrt{N}$, then the value of N is (ignore energy loss due to atmosphere)

2. Two identical uniform discs roll without slipping on two different surfaces AB and CD (see figure) starting at A and C with linear speeds v_1 and v_2, respectively, and always remain in contact with the surfaces. If they reach B and D with the same linear speed and $v_1 = 3$ m/s, then v_2 in m/s is $(g = 10 \text{ m/s}^2)$

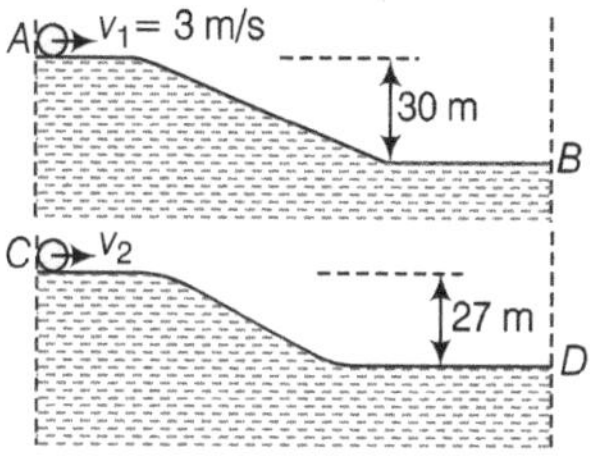

3. Two spherical stars A and B emit blackbody radiation. The radius of A is 400 times that of B and A emits 10^4 times the power emitted from B. The ratio $\left(\frac{\lambda_A}{\lambda_B}\right)$ of their wavelengths λ_A and λ_B at which the peaks occur in their respective radiation curves is

Section 2 (Maximum Marks : 20)

- This section contains FIVE questions.
- Each question has FOUR options (a), (b), (c) and (d), ONE OR MORE THAN ONE of these four option(s) is(are) correct.
- For each question, darken the bubble(s) corresponding to all the correct option(s) in the ORS.
- Marking scheme :
  + 4 If only the bubble(s) corresponding to all the correct option(s) is(are) darkened.
  0 If none of the bubbles is darkened.
  −2 In all other cases.

4. A ring of mass M and radius R is rotating with angular speed ω about a fixed vertical axis passing through its centre O with two point masses each of mass $\frac{M}{8}$ at rest at O. These masses can move radially outwards along two massless rods fixed on the ring as shown in the figure. At some instant, the angular speed of the system is $\frac{8}{9}\omega$ and one of the masses is at a distance of $\frac{3}{5}R$ from O. At this instant, the distance of the other mass from O is

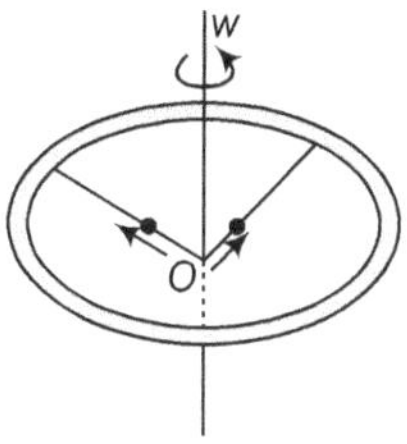

(a) $\frac{2}{3}R$ (b) $\frac{1}{3}R$ (c) $\frac{3}{5}R$ (d) $\frac{4}{5}R$

5. A container of fixed volume has a mixture of one mole of hydrogen and one mole of helium in equilibrium at temperature T. Assuming the gases are ideal, the correct statements is/are

(a) The average energy per mole of the gas mixture is $2RT$

(b) The ratio of speed of sound in the gas mixture to that in helium gas is $\sqrt{\frac{6}{5}}$

(c) The ratio of the rms speed of helium atoms to that of hydrogen molecules is $\frac{1}{2}$

(d) The ratio of the rms speed of helium atoms to that of hydrogen molecules is $\frac{1}{\sqrt{2}}$

6. Consider a Vernier callipers in which each 1 cm on the main scale is divided into 8 equal divisions and a screw gauge with 100 divisions on its circular scale. In the Vernier callipers, 5 divisions of the Vernier scale coincide with 4 divisions on the main scale and in the screw gauge, one complete rotation of the circular scale moves it by two divisions on the linear scale. Then

(a) If the pitch of the screw gauge is twice the least count of the Vernier callipers, the least count of the screw gauge is 0.01 mm

(b) If the pitch of the screw gauge is twice the least count of the Vernier callipers, the least count of the screw gauge is 0.05 mm

(c) If the least count of the linear scale of the screw gauge is twice the least count of the Vernier callipers, the least count of the screw gauge is 0.01 mm

(d) If the least count of the linear scale of the screw gauge is twice the least count of the Vernier callipers, the least count of the screw gauge is 0.005 mm

7. Planck's constant h, speed of light c and gravitational constant G are used to form a unit of length L and a unit of mass M. Then, the correct options is/are

(a) $M \propto \sqrt{c}$

(b) $M \propto \sqrt{G}$

(c) $L \propto \sqrt{h}$

(d) $L \propto \sqrt{G}$

8. Two independent harmonic oscillators of equal masses are oscillating about the origin with angular frequencies ω_1 and ω_2 and have total energies E_1 and E_2, respectively. The variations of their momenta p with positions x are shown in the figures. If $\frac{a}{b} = n^2$ and $\frac{a}{R} = n$, then the correct equation(s) is/are

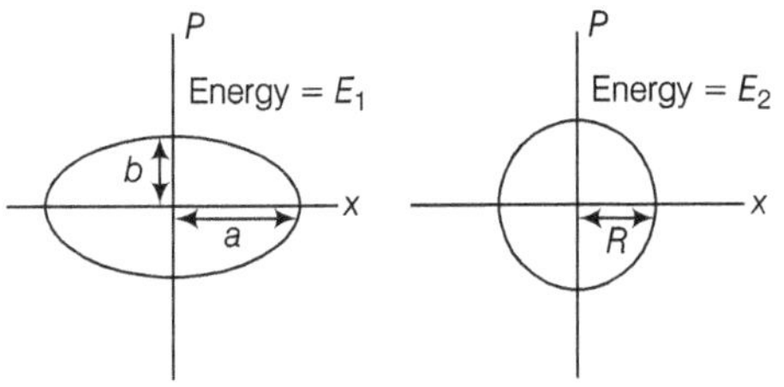

(a) $E_1\omega_1 = E_2\omega_2$

(b) $\frac{\omega_2}{\omega_1} = n^2$

(c) $\omega_1\omega_2 = n^2$

(d) $\frac{E_1}{\omega_1} = \frac{E_2}{\omega_2}$

Section 3 (Maximum Marks : 8)

- This section contains ONE questions.
- Each question contains two columns, Column I and Column II.
- Column I has four entries (A), (B), (C) and (D).
- Column II has five entries (P), (Q), (R), (S) and (T)
- Match the entries in Column I with the entries in Column II.
- One or more entries in Column I may match with one or more entries in Column II.
- The ORS contains a 4 × 5 matrix whose layout will be similar to the one shown below :

(A)	(P)	(Q)	(R)	(S)	(T)
(B)	(P)	(Q)	(R)	(S)	(T)
(C)	(P)	(Q)	(R)	(S)	(T)
(D)	(P)	(Q)	(R)	(S)	(T)

- For each entry in Column I, darken the bubbles of all the matching entries. For example, if entry (A) in Column I matches with entries (Q), (R) and (T), then darken these three bubbles in the ORS. Similarly, for entries (B), (C) and (D).
- Marking schemes:

For each entry in Column I

+ 2 If only the bubble(s) corresponding to all the correct match(es) is (are) darkened.

0 If none of the bubbles is darkened.

– 1 In all other cases.

9. A particle of unit mass is moving along the x-axis under the influence of a force and its total energy is conserved. Four possible forms of the potential energy of the particle are given in Column I (a and U_0 are constants). Match the potential energies in Column I to the corresponding statements in Column II.

	Column I	Column II
A.	$U_1(x) = \frac{U_0}{2}\left[1-\left(\frac{x}{a}\right)^2\right]^2$	P. The force acting on the particle is zero at $x = a$
B.	$U_2(x) = \frac{U_0}{2}\left(\frac{x}{a}\right)^2$	Q. The force acting on the particle is zero at $x = 0$
C.	$U_3(x) = \frac{U_0}{2}\left(\frac{x}{a}\right)^2 \exp\left[-\left(\frac{x}{a}\right)^2\right]$	R. The force acting on the particle is zero at $x = -a$
D.	$U_4(x) = \frac{U_0}{2}\left[\frac{x}{a} - \frac{1}{3}\left(\frac{x}{a}\right)^3\right]$	S. The particle experiences an attractive force towards $x = 0$ in the region $\lvert x\rvert < a$
		T. The particle with total energy $\frac{U_0}{4}$ can oscillate about the point $x = -a$.

Paper 2

Section 1 (Maximum Marks : 16)

- This section contains FOUR questions.
- The answer to each question is a SINGLE DIGIT INTEGER ranging from 0 to 9 both inclusive.
- For each question, darken the bubble corresponding to the correct integer in ORS.
- Marking scheme :
  +4 If the bubble corresponding to the answer is darkened.
  0 in all other cases.

1. Four harmonic waves of equal frequencies and equal intensities I_0 have phase angles $0, \frac{\pi}{3}, \frac{2\pi}{3}$ and π. When they are superposed, the intensity of the resulting wave is nI_0. The value of n is

2. A large spherical mass M is fixed at one position and two identical masses m are kept on a line passing through the centre of M (see figure). The point masses are connected by a rigid massless rod of length l and this assembly is free to move along the line connecting them.

All three masses interact only through their mutual gravitational interaction. When the point mass nearer to M is at a distance $r = 3l$ from M the tension in the rod is zero for $m = k\left(\frac{M}{288}\right)$. The value of k is

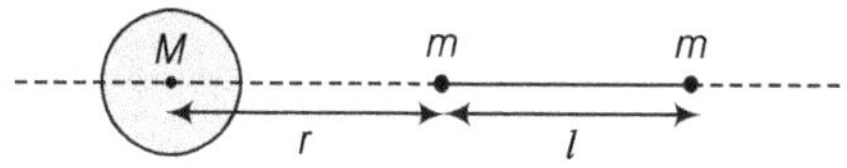

3. The energy of a system as a function of time t is given as $E(t) = A^2 \exp(-\alpha t)$, where $\alpha = 0.2\,s^{-1}$. The measurement of A has an error of 1.25%. If the error in the measurement of time is 1.50%, the percentage error in the value of $E(t)$ at $t = 5$ s is

4. The densities of two solids spheres A and B of the same radii R vary with radial distance r as $\rho_A(r) = k\left(\frac{r}{R}\right)$ and $\rho_B(r) = k\left(\frac{r}{R}\right)^5$, respectively, where k is a constant.

The moments of inertia of the individual spheres about axes passing through their centres are I_A and I_B, respectively.

If $\frac{I_B}{I_A} = \frac{n}{10}$, the value of n is

Section 2 (Maximum Marks : 20)

- This section contains FIVE questions.
- Each question has FOUR options (a), (b), (c) and (d), ONE OR MORE THAN ONE of these four option(s) is(are) correct.
- For each question, darken the bubble(s) corresponding to all the correct option(s) in the ORS.
- Marking scheme :
  + 4 If only the bubble(s) corresponding to all the correct optin(s) is(are) darkened.
  0 If none of the bubbles is darkened.
  – 2 In all other cases.

5. In plotting stress versus strain curves for two materials P and Q, a student by mistake puts strain on the y-axis and stress on the x-axis as shown in the figure. Then the correct statements is/are

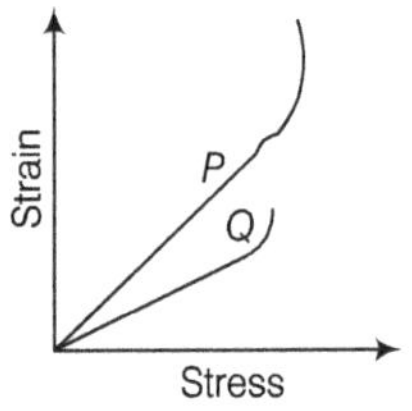

(a) P has more tensile strength than Q
(b) P is more ductile than Q
(c) P is more brittle than Q
(d) The Young's modulus of P is more than that of Q

6. A spherical body of radius R consists of a fluid of constant density and is in equilibrium under its own gravity. If $P(r)$ is the pressure at $r(r < R)$, then the correct options is/are

(a) $P\,(r=0)=0$

(b) $\dfrac{P\left(r=\dfrac{3R}{4}\right)}{P\left(r=\dfrac{2R}{3}\right)}=\dfrac{63}{80}$

(c) $\dfrac{P\left(r=\dfrac{3R}{5}\right)}{P\left(r=\dfrac{2R}{5}\right)}=\dfrac{16}{21}$

(d) $\dfrac{P\left(r=\dfrac{R}{2}\right)}{P\left(r=\dfrac{R}{3}\right)}=\dfrac{20}{27}$

7. An ideal monoatomic gas is confined in a horizontal cylinder by a spring loaded piston (as shown in the figure). Initially the gas is at temperature T_1, pressure P_1 and volume V_1 and the spring is in its relaxed state. The gas is then heated very slowly to temperature T_2, pressure P_2 and volume V_2. During this process the piston moves out by a distance x. Ignoring the friction between the piston and the cylinder, the correct statements is/are

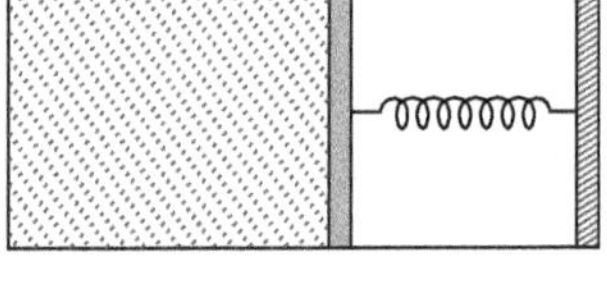

(a) If $V_2 = 2V_1$ and $T_2 = 3T_1$, then the energy stored in the spring is $\frac{1}{4}P_1V_1$

(b) If $V_2 = 2V_1$ and $T_2 = 3T_1$, then the change in internal energy is $3P_1V_1$

(c) If $V_2 = 3V_1$ and $T_2 = 4T_1$, then the work done by the gas is $\frac{7}{3}P_1V_1$

(d) If $V_2 = 3V_1$ and $T_2 = 4T_1$, then the heat supplied to the gas is $\frac{17}{6}P_1V_1$

8. Two spheres P and Q for equal radii have densities ρ_1 and ρ_2, respectively. The spheres are connected by a massless string and placed in liquids L_1 and L_2 of densities σ_1 and σ_2 and viscosities η_1 and η_2, respectively. They float in equilibrium with the sphere P in L_1 and sphere Q in L_2 and the string being taut (see figure). If sphere P alone in L_2 has terminal velocity $\mathbf{v}_P$ and Q alone in L_1 has terminal velocity $\mathbf{v}_Q$, then

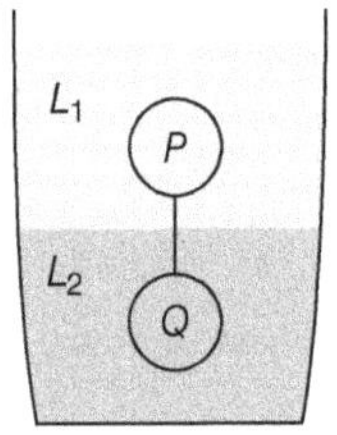

(a) $\dfrac{|\mathbf{v}_P|}{|\mathbf{v}_Q|}=\dfrac{\eta_1}{\eta_2}$

(b) $\dfrac{|\mathbf{v}_P|}{|\mathbf{v}_Q|}=\dfrac{\eta_2}{\eta_1}$

(c) $\mathbf{v}_P \cdot \mathbf{v}_Q > 0$

(d) $\mathbf{v}_P \cdot \mathbf{v}_Q < 0$

9. In terms of potential difference V, electric current I, permittivity ε_0, permeability μ_0 and speed of light c, the dimensionally correct equations is/are

(a) $\mu_0 I^2 = \varepsilon_0 V^2$
(b) $\varepsilon_0 I = \mu_0 V$
(c) $I = \varepsilon_0 c V$
(d) $\mu_0 c I = \varepsilon_0 V$

Answer with Explanations

Paper 1

1. (2) At height h

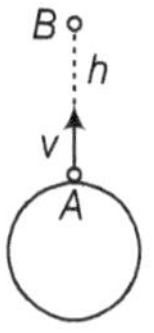

$$g' = \frac{g}{\left(1 + \frac{h}{R}\right)^2} \quad \ldots(i)$$

Given, $g' = \frac{g}{4}$

Substituting in Eq. (i) we get, $h = R$

Now, from A to B,

decrease in kinetic energy = increase in potential energy

$$\Rightarrow \quad \frac{1}{2}mv^2 = \frac{mgh}{1 + \frac{h}{R}} \Rightarrow \frac{v^2}{2} = \frac{gh}{1 + \frac{h}{R}} = \frac{1}{2}gR \qquad (h = R)$$

$$\Rightarrow \quad v^2 = gR \quad \text{or} \quad v = \sqrt{gR}$$

Now, $\quad v_{esc} = \sqrt{2gR} = v\sqrt{2}$

$\Rightarrow \quad N = 2$

2. (7) In case of pure rolling, mechanical energy remains constant (as work-done by friction is zero). Further in case of a disc,

$$\frac{\text{translational kinetic energy}}{\text{rotational kinetic energy}} = \frac{K_T}{K_R} = \frac{\frac{1}{2}mv^2}{\frac{1}{2}I\omega^2}$$

$$= \frac{mv^2}{\left(\frac{1}{2}mR^2\right)\left(\frac{v}{R}\right)^2} = \frac{2}{1}$$

or, $\quad K_T = \frac{2}{3}$ (Total kinetic energy)

or, $\quad$ Total kinetic energy $K = \frac{3}{2}K_T = \frac{3}{2}\left(\frac{1}{2}mv^2\right) = \frac{3}{4}mv^2$

Decrease in potential energy = increase in kinetic energy

or, $\quad mgh = \frac{3}{4}m\,(v_f^2 - v_i^2) \quad$ or $\quad v_f = \sqrt{\frac{4}{3}gh + v_i^2}$

As final velocity in both cases is same.

So, value of $\sqrt{\frac{4}{3}gh + v_i^2}$ should be same in both cases.

$$\therefore \quad \sqrt{\frac{4}{3} \times 10 \times 30 + (3)^2} = \sqrt{\frac{4}{3} \times 10 \times 27 + (v_2)^2}$$

Solving this equation we get,

$$v_2 = 7 \text{ m/s}$$

3. (2) Power, $P = (\sigma T^4 A) = \sigma T^4 (4\pi R^2)$

or, $\quad P \propto T^4 R^2 \quad \ldots(i)$

According to Wien's law,

$$\lambda \propto \frac{1}{T}$$

(λ is the wavelength at which peak occurs)

$\therefore$ Eq. (i) will become,

$$P \propto \frac{R^2}{\lambda^4} \quad \text{or} \quad \lambda \propto \left[\frac{R^2}{P}\right]^{1/4} \Rightarrow \frac{\lambda_A}{\lambda_B} = \left[\frac{R_A}{R_B}\right]^{1/2}\left[\frac{P_B}{P_A}\right]^{1/4}$$

$$= [400]^{1/2}\left[\frac{1}{10^4}\right]^{1/4} = 2$$

4. *(d)* Let the other mass at this instant is at a distance of x from the centre O. Applying law of conservation of angular momentum, we have

$$I_1\omega_1 = I_2\omega_2$$

$$\therefore \quad (MR^2)(\omega) = \left[MR^2 + \frac{M}{8}\left(\frac{3}{5}R\right)^2 + \frac{M}{8}x^2\right]\left(\frac{8}{9}\omega\right)$$

Solving this equation, we get $x = \frac{4}{5}R$.

Note *If we take identical situations with both point masses, then answer will be (c). But in that case, angular momentum is not conserved.*

5. *(a,b,d)*

(a) Total internal energy $U = \frac{f_1}{2}nRT + \frac{f_2}{2}nRT$

$$(U_{\text{ave}})_{\text{per mole}} = \frac{U}{2n} = \frac{1}{4}[5RT + 3RT] = 2RT$$

(b) $$\gamma_{\text{mix}} = \frac{n_1C_{p_1} + n_2C_{p_2}}{n_2C_{v_1} + n_2C_{v_2}} = \frac{(1)\frac{7R}{2} + (1)\frac{5R}{2}}{(1)\frac{5R}{2} + (1)\frac{3R}{2}} = \frac{3}{2}$$

$$M_{\text{mix}} = \frac{n_1M_1 + n_2M_2}{n_1 + n_2} = \frac{M_1 + M_2}{2} = \frac{2 + 4}{2} = 3$$

Speed of sound $V = \sqrt{\frac{\gamma RT}{M}} \Rightarrow V \propto \sqrt{\frac{\gamma}{M}}$

$$\frac{V_{\text{mix}}}{V_{\text{He}}} = \sqrt{\frac{\gamma_{\text{mix}}}{\gamma_{\text{He}}} \times \frac{M_{\text{He}}}{M_{\text{mix}}}} = \sqrt{\frac{3/2}{5/3} \times \frac{4}{3}} = \sqrt{\frac{6}{5}}$$

(d) $V_{\text{rms}} = \sqrt{\frac{3RT}{M}} \Rightarrow V_{\text{rms}} \propto \frac{1}{\sqrt{M}}$,

$$\frac{V_{\text{He}}}{V_{\text{H}}} = \sqrt{\frac{M_{\text{H}}}{M_{\text{He}}}} = \sqrt{\frac{2}{4}} = \frac{1}{\sqrt{2}}$$

6. *(b, c)* **For Vernier callipers**

$$1 \text{ MSD} = \frac{1}{8} \text{ cm}$$

$$5 \text{ VSD} = 4 \text{ MSD}$$

$$\therefore \quad 1 \text{ VSD} = \frac{4}{5} \text{ MSD} = \frac{4}{5} \times \frac{1}{8} = \frac{1}{10} \text{cm}$$

Least count of Vernier callipers = 1 MSD − 1 VSD

$$= \frac{1}{8} \text{ cm} - \frac{1}{10} \text{ cm} = 0.025 \text{ cm}$$

(a) and (b)

Pitch of screw gauge = $2 \times 0.025 = 0.05$ cm

Least count of screw gauge = $\frac{0.05}{100}$ cm = 0.005 mm

(c) and (d)

Least count of linear scale of screw gauge = 0.05

Pitch = $0.05 \times 2 = 0.1$ cm

Least count of screw gauge $= \dfrac{0.1}{100}$ cm = 0.01 mm

7. *(a, c, d)* $M \propto h^a c^b G^c$

$$M^1 \propto (ML^2T^{-1})^a (LT^{-1})^b (M^{-1}L^3T^{-2})^c$$

$$\propto M^{a-c} L^{2a+b+3c} T^{-a-b-2c}$$

$$a - c = 1 \quad \ldots(i)$$

$$2a + b + 3c = 0 \quad \ldots(ii)$$

$$a + b + 2c = 0 \quad \ldots(iii)$$

On solving (i), (ii), (iii), $a = \dfrac{1}{2}, b = +\dfrac{1}{2}, c = -\dfrac{1}{2}$

$\therefore M \propto \sqrt{c}$ only $\rightarrow$ (a) is correct.

In the same way we can find that, $L \propto h^{1/2} c^{-3/2} G^{1/2}$

$L \propto \sqrt{h}, L \propto \sqrt{G} \rightarrow$ (c), (d) are also correct.

8. *(b, d)* **Ist Particle**

$P = 0$ at $x = a \Rightarrow$ 'a' is the amplitude of oscillation 'A_1'.

At $x = 0, P = b$ (at mean position)

$$\Rightarrow \quad mv_{max} = b$$

$$v_{max} = \frac{b}{m}$$

$$E_1 = \frac{1}{2} m v_{max}^2 = \frac{m}{2}\left[\frac{b}{m}\right]^2 = \frac{b^2}{2m}$$

$$A_1 \omega_1 = v_{max} = \frac{b}{m}$$

$$\Rightarrow \quad \omega_1 = \frac{b}{ma} = \frac{1}{mn^2} \quad (A_1 = a, \frac{a}{b} = n^2)$$

IInd Particle

$$P = 0 \text{ at } x = R \Rightarrow A_2 = R$$

At $\quad x = 0, P = R \Rightarrow v_{max} = \dfrac{R}{m}$

$$E_2 = \frac{1}{2} m v_{max}^2 = \frac{m}{2}\left[\frac{R}{m}\right]^2 = \frac{R^2}{2m}$$

$$A_2 \omega_2 = \frac{R}{m} \Rightarrow \omega_2 = \frac{R}{mR} = \frac{1}{m}$$

(b) $\dfrac{\omega_2}{\omega_1} = \dfrac{1/m}{1/mn^2} = n^2$

(c) $\omega_1 \omega_2 = \dfrac{1}{mn^2} \times \dfrac{1}{m} = \dfrac{1}{m^2 n^2}$

(d) $\dfrac{E_1}{\omega_1} = \dfrac{b^2/2m}{1/mn^2} = \dfrac{b^2 n^2}{2} = \dfrac{a^2}{2n^2} = \dfrac{R^2}{2}$

$$\frac{E_2}{\omega_2} = \frac{R^2/2m}{1/m} = \frac{R^2}{2} \Rightarrow \frac{E_1}{\omega_1} = \frac{E_2}{\omega_2}$$

Note *It is not given that the second figure is a circle. But from the figure and as per the requirement of question, we consider it is a circle.*

9. (A)$\rightarrow$(P,Q,R,T); (B)$\rightarrow$(Q, S); (C)$\rightarrow$(P, Q, R, S);(D)$\rightarrow$(P, R, T)

(A) $F_x = \dfrac{-dU}{dx} = -\dfrac{2U_0}{a^3}[x-a][x][x+a]$

$F = 0$ at $x = 0$, $x = a, x = -a$ and $U = 0$ at $x = -a$ and $x = a$

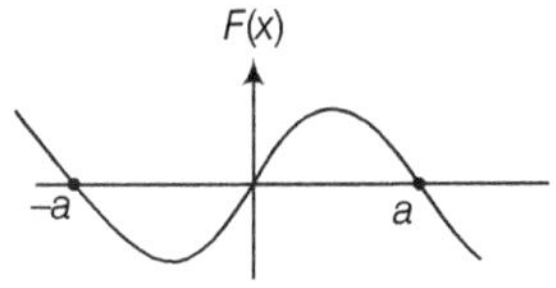

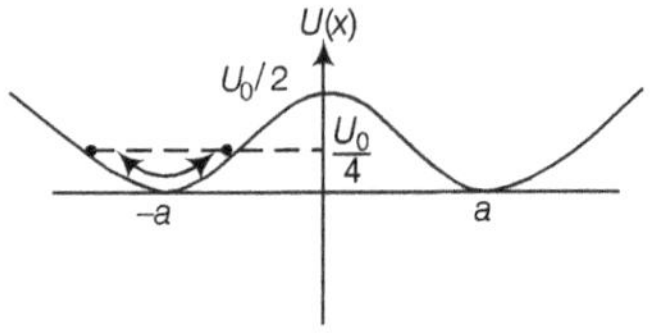

(B) $F_x = -\dfrac{dU}{dx} - U_0\left(\dfrac{x}{a}\right)$

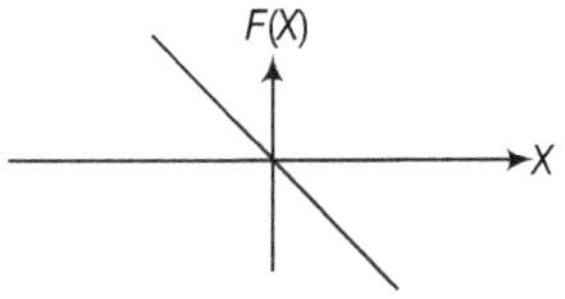

(C) $F_x = -\dfrac{dU}{dx} = U_0 \dfrac{e^{-x^2/x^2}}{a^3}[x][x-a][x+a]$

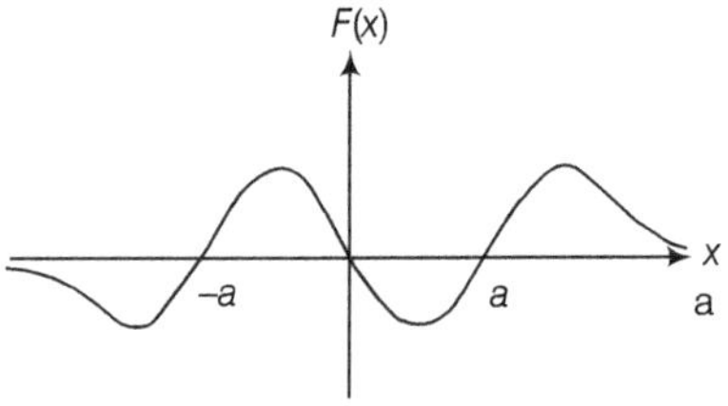

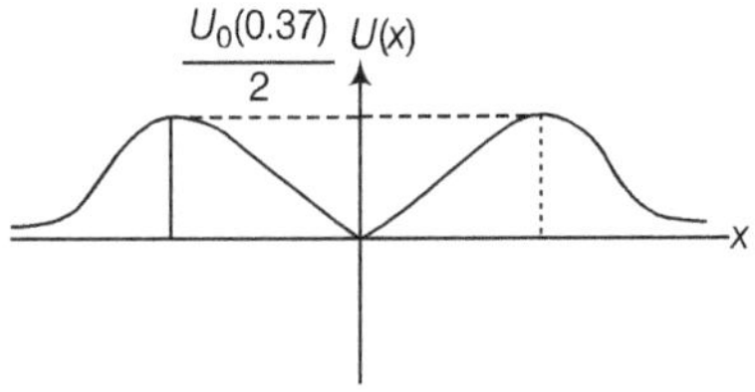

(D) $F_x = -\dfrac{dU}{dx} = -\dfrac{U_0}{2a^3}[(x-a)(x+a)]$

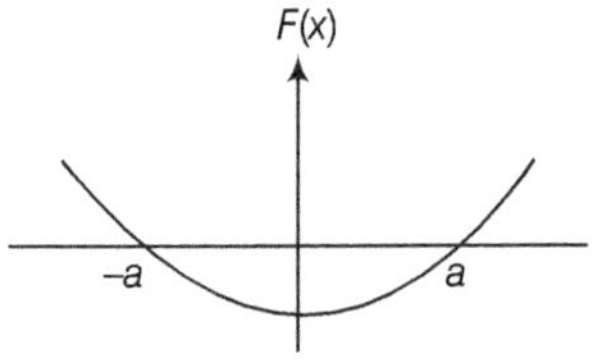

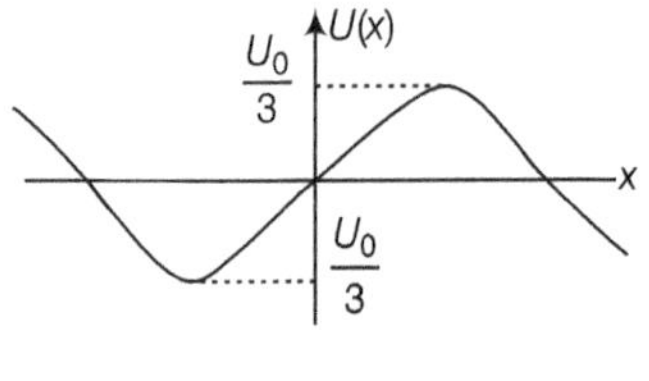

Paper 2

1. **(3)** Let individual amplitudes are A_0 each. Amplitudes can be added by vector method.

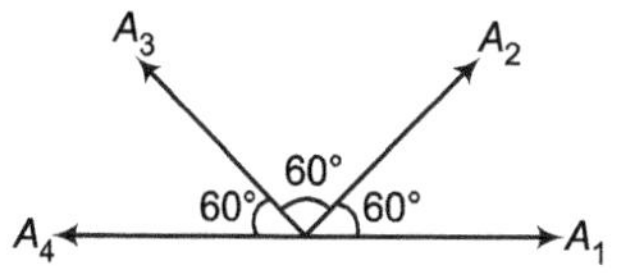

$$A_1 = A_2 = A_3 = A_4 = A_0$$

Resultant of A_1 and A_4 is zero. Resultant of A_2 and A_3 is

$$A = \sqrt{A_0^2 + A_0^2 + 2A_0A_0 \cos 60^\circ} = \sqrt{3}A_0$$

This is also the net resultant.

Now, $I \propto A^2$

$\therefore$ Net intensity will become $3I_0$

$\therefore$ Answer is 3.

2. **(7)** For point mass at distance $r = 3l$

$$\frac{GMm}{(3l)^2} - \frac{Gm^2}{l^2} = ma \quad \text{...(i)}$$

For point mass at distance $r = 4l$

$$\frac{GMm}{(4l)^2} + \frac{Gm^2}{l^2} = ma \quad \text{(ii)}$$

Equating the two equations we have,

$$\frac{GMm}{9l^2} - \frac{Gm^2}{l^2} = \frac{GMm}{16l^2} + \frac{Gm^2}{l^2}$$

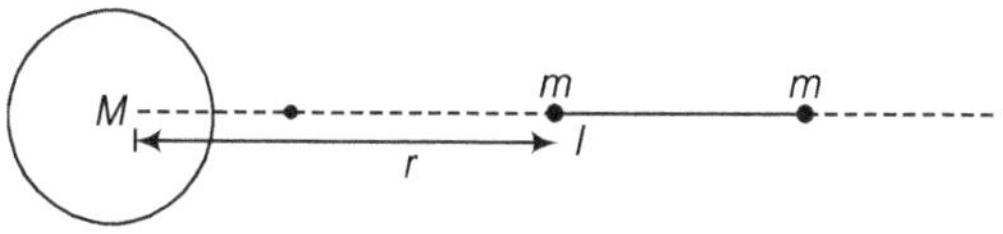

$$\frac{7GMm}{144} = \frac{2Gm^2}{l^2}$$

$$m = \frac{7M}{288}$$

3. **(4)** $$E(t) = A^2 e^{-\alpha t} \quad \text{...(i)}$$

$$\alpha = 0.2 \text{ s}^{-1}$$

$$\left(\frac{dA}{A}\right) \times 100 = 1.25\%$$

$$\left(\frac{dt}{t}\right) \times 100 = 1.50$$

$\Rightarrow$ $(dt \times 100) = 1.5t = 1.5 \times 5 = 7.5$

$\therefore$ $$\left(\frac{dE}{E}\right) \times 100 = \pm 2\left(\frac{dA}{A}\right) \times 100 \pm \alpha\,(dt \times 100)$$

Taking log on both sides of Eq. (i), we get

$$\log E = 2 \log A - \alpha t$$

$$\frac{dE}{E} = \pm 2\frac{dA}{A} \pm \alpha dt$$

$\therefore$ $$\left(\frac{dE}{E}\right) \times 100 = \pm 2\left(\frac{dA}{A}\right) \times 100 \pm \alpha\,(dt \times 100)$$

$$= \pm 2\,(1.25) \pm 0.2\,(7.5)$$

$$= \pm 2.5 \pm 1.5$$

$$= \pm 4\%$$

4. **(6)** Consider a shell of radius r and thickness dr

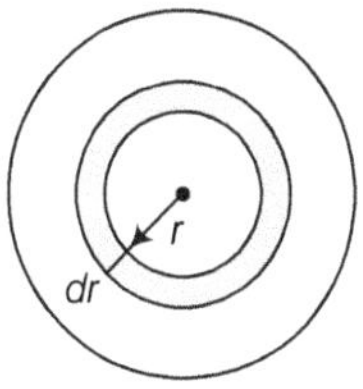

$$dI = (dm)\,r^2$$

$\Rightarrow$ $$dI = \frac{2}{3}(\rho 4\pi r^2 dr) r^2 \Rightarrow I = \int dI$$

$$\frac{I_B}{I_A} = \frac{\int_0^R \frac{2}{3} k \frac{r^5}{R^5} \cdot 4\pi r^2 dr\, r^2}{\int_0^R \frac{2}{3} k \frac{r}{R}\, 4\pi r^2 dr\, r^2} = \frac{6}{10}$$

5. *(a, b)* $Y = \frac{\text{stress}}{\text{strain}}$ or $Y \propto \frac{1}{\text{strain}}$ (for same stress say σ)

$$(\text{strain})_Q < (\text{strain})_P \quad \Rightarrow Y_Q > Y_P$$

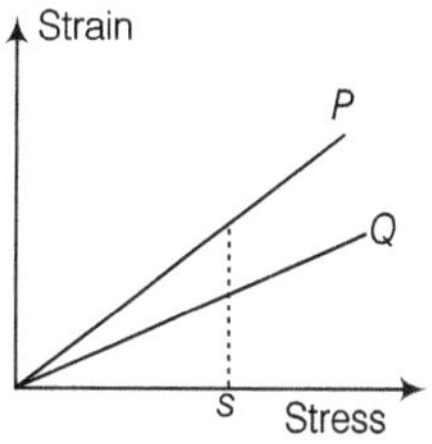

So, P is more ductile than Q. Further, from the given figure we can also see that breaking stress of P is more than Q. So, it has more tensile strength.

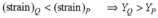

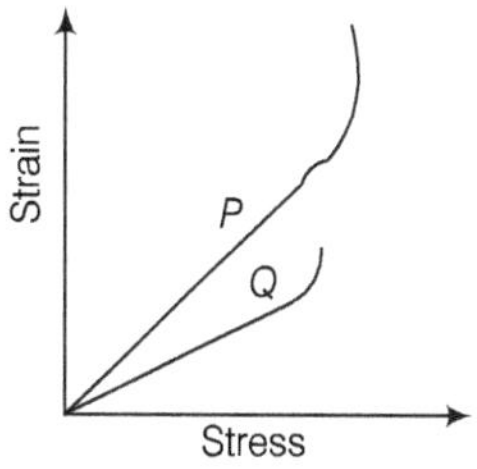

6. *(b, c)* Gravitational field at a distance r due to mass 'm'

$$E = \frac{G\rho \frac{4}{3}\pi r^3}{r^2} = \frac{4G\rho\pi r}{3}$$

Consider a small element of width dr and area ΔA at a distance r

Pressure force on this element outwards = gravitational force on 'dm' from 'm' inwards

$\Rightarrow$ $$(dp)\Delta A = E(dm)$$

$$\Rightarrow \quad -dp\cdot \Delta A = \left(\frac{4}{3}G\pi\rho r\right)(\Delta A\, dr\cdot\rho)$$

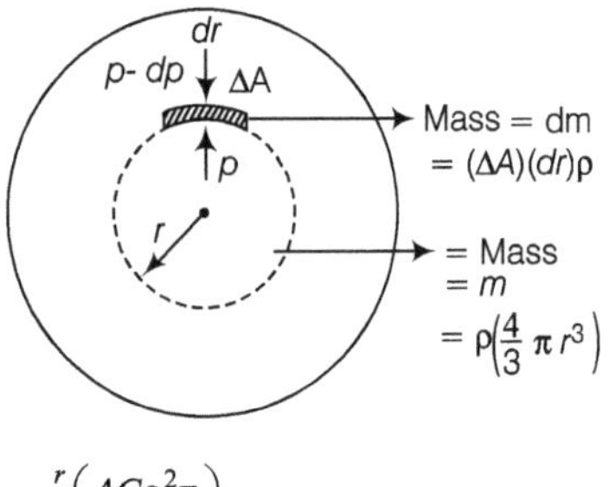

$$-\int_O^P dp = \int_R^r \left(\frac{4G\rho^2\pi}{3}\right) r\,dr$$

$$-p = \frac{4G\rho^2\pi}{3\times 2}[r^2 - R^2] \Rightarrow p = c(R^2 - r^2)$$

$$r = \frac{3R}{4},\ p_1 = c\left(R^2 - \frac{9R^2}{16}\right) = c\left(\frac{7R^2}{16}\right)$$

$$r = \frac{2R}{3},\ p_2 = c\left(R^2 - \frac{4R^2}{9}\right) = c\left(\frac{5R^2}{9}\right)$$

$$\frac{p_1}{p_2} = \frac{63}{80}$$

$$r = \frac{3R}{5},\ p_3 = c\left(R^2 - \frac{9}{25}R^2\right) = c\left(\frac{16R^2}{25}\right)$$

$$r = \frac{2R}{5},\ p_4 = c\left(R^2 - \frac{4R^2}{25}\right) = c\left(\frac{21R^2}{25}\right)$$

$$\frac{p_3}{p_4} = \frac{16}{21}$$

7. *(a, b, c)*

Note *This question can be solved if right hand side hand side chamber is assumed open, so that its pressure remains constant even if the piston shifts towards right.*

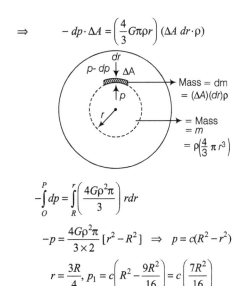

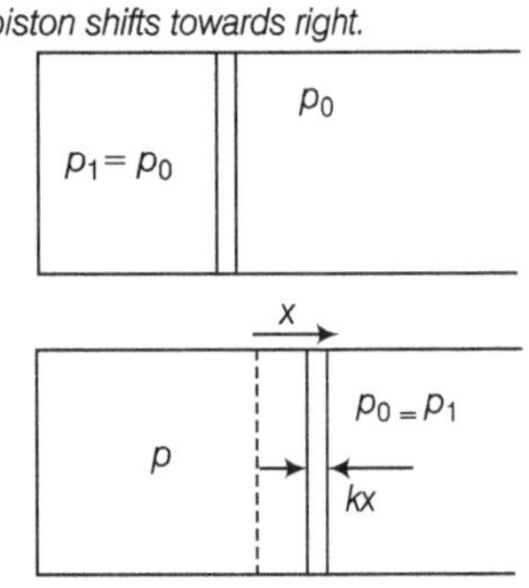

(a) $\quad pV = nRT \quad \Rightarrow \quad p \propto \frac{T}{V}$

Temperature is made three times and volume is doubled

$$\Rightarrow \quad p_2 = \frac{3}{2}p_1$$

Further $\quad x = \frac{\Delta V}{A} = \frac{V_2 - V_1}{A} = \frac{2V_1 - V_1}{A} = \frac{V_1}{A}$

$$p_2 = \frac{3p_1}{2} = p_1 + \frac{kx}{A} \Rightarrow kx = \frac{p_1 A}{2}$$

Energy of spring

$$\frac{1}{2}kx^2 = \frac{p_1 A}{4}x = \frac{p_1 V_1}{4}$$

(b) $\Delta U = nc_V\Delta T = n\left(\frac{3}{2}R\right)\Delta T = \frac{3}{2}(p_2V_2 - p_1V_1)$

$$= \frac{3}{2}\left[\left(\frac{3}{2}p_1\right)(2V_1) - p_1V_1\right] = 3\,p_1V_1$$

(c) $\quad p_2 = \frac{4p_1}{3} \quad \Rightarrow \quad p_2 = \frac{4}{3}p_1 = p_1 + \frac{kx}{A}$

$$\Rightarrow \quad kx = \frac{p_1A}{3} \quad \Rightarrow \quad x = \frac{\Delta V}{A} = \frac{2V_1}{A}$$

$$W_{gas} = (p_0\,\Delta V + W_{spring}) = \left(p_1Ax + \frac{1}{2}kx\cdot x\right)$$

$$= +\left(p_1A\cdot\frac{2V_1}{A} + \frac{1}{2}\cdot\frac{p_1A}{3}\cdot\frac{2V_1}{A}\right)$$

$$= 2p_1V_1 + \frac{p_1V_1}{3} = \frac{7p_1V_1}{3}$$

(d) $\quad \Delta Q = W + \Delta U$

$$= \frac{7p_1V_1}{3} + \frac{3}{2}(p_2V_2 - p_1V_1)$$

$$= \frac{7p_1V_1}{3} + \frac{3}{2}\left(\frac{4}{3}p_1\cdot 3V_1 - p_1V_1\right)$$

$$= \frac{7p_1V_1}{3} + \frac{9}{2}p_1V_1 = \frac{41p_1V_1}{6}$$

Note $\Delta U = \frac{3}{2}(p_2V_2 - p_1V_1)$ *has been obtained in part (b).*

8. *(a, d)* For floating, net weight of system = net upthrust

$\Rightarrow (\rho_1 + \rho_2)Vg = (\sigma_1 + \sigma_2)Vg$

Since string is taut, $\rho_1 < \sigma_1$ and $\rho_2 > \sigma_2$

$$v_P = \frac{2r^2g}{2\eta_2}(\sigma_2 - \rho_1) \quad \text{(upward terminal velocity)}$$

$$v_Q = \frac{2r^2g}{9\eta_1}(\rho_2 - \sigma_1) \quad \text{(downward terminal velocity)}$$

$$\left|\frac{v_P}{v_Q}\right| = \frac{\eta_1}{\eta_2}$$

Further, $\bar{v}_P\cdot\bar{v}_Q$ will be negative as they are opposite to each other.

9. *(a, c)*

(A) Energy of inductor $\frac{1}{2}LI^2 = \frac{1}{2}\frac{\mu_0N^2A}{l}I^2$

Energy of capacitor $= \frac{1}{2}CV^2 = \frac{1}{2}\varepsilon_0\frac{A}{d}V^2$

$\mu_0\frac{A}{l}I^2$ and $\varepsilon_0\frac{A}{d}V^2$ have same dimension.

So, $\mu_0 I^2$ and $\varepsilon_0 V^2$ have same domension.

(C) $Q = CV \quad \Rightarrow \quad \frac{Q}{t} = \frac{CV}{t}$

$$I = \varepsilon_0\frac{A}{l}\frac{V}{t}\quad \frac{A}{lt} \text{ have unit of speed.}$$

So, $\quad I = \varepsilon_0 cV$

JEE Main 2016* (Solved)

1. A student measures the time period of 100 oscillations of a simple pendulum four times. The data set is 90s, 91s, 92s and 95s. If the minimum division in the measuring clock is 1s, then the reported mean time should be

(a) (92 ± 2)s (b) (92 ± 5)s
(c) (92 ± 1.8)s (d) (92 ± 3)s

2. A particle of mass m is moving along the side of a square of side a, with a uniform speed v in the X-Y plane as shown in the figure.

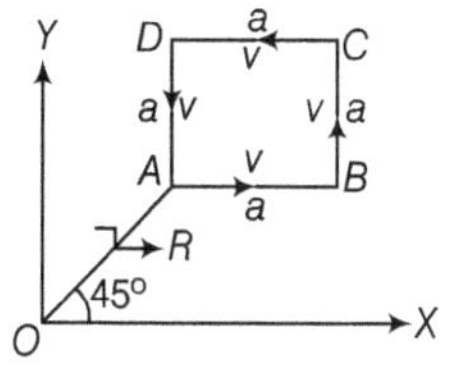

Which of the following statements is false for the angular momentum **L** about the origin?

(a) $\mathbf{L} = \frac{-mv}{\sqrt{2}} R\,\hat{\mathbf{k}}$, when the particle is moving from A to B.

(b) $\mathbf{L} = mv\left(\frac{R}{\sqrt{2}} + a\right)\hat{\mathbf{k}}$, when the particle is moving from B to C.

(c) $\mathbf{L} = mv\left(\frac{R}{\sqrt{2}} - a\right)\hat{\mathbf{k}}$, when the particle is moving from C to D.

(d) $\mathbf{L} = \frac{mv}{\sqrt{2}} R\,\hat{\mathbf{k}}$, when the particle is moving from D to A.

3. A point particle of mass m, moves along the uniformly rough track PQR as shown in the figure. The coefficient of friction, between the particle and the rough track equals μ. The particle is released, from rest, from the point P and it comes to rest at a point R. The energies, lost by the ball, over the parts, PQ and QR, of the track, are equal to each other, and no energy is lost when particle changes direction from PQ to QR. The values of the coefficient of friction μ and the distance x(= QR), are respectively close to

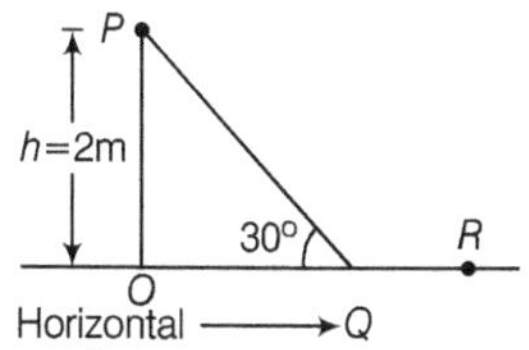

(a) 0.2 and 6.5 m (b) 0.2 and 3.5 m
(c) 0.29 and 3.5 m (d) 0.29 and 6.5 m

4. A person trying to lose weight by burning fat lifts a mass of 10 kg upto a height of 1 m, 1000 times. Assume that the potential energy lost each time he lowers the mass is dissipated. How much fat will he use up considering the work done only when the weight is lifted up? Fat supplies 3.8×10^7 J of energy per kg which is converted to mechanical energy with a 20% of efficiency rate. (Take, $g = 9.8$ ms^{-2})

(a) 2.45×10^{-3} kg (b) 6.45×10^{-3} kg
(c) 9.89×10^{-3} kg (d) 12.89×10^{-3} kg

5. A roller is made by joining together two corners at their vertices O. It is kept on two rails AB and CD which are placed a symmetrically (see the figure), with its axis perpendicular to CD and its centre O at the centre of line joining AB and CD (see the figure). It is given a light path, so that it starts rolling with its centre O moving parallel to CD in the direction shown. As it moves, the roller will tend to

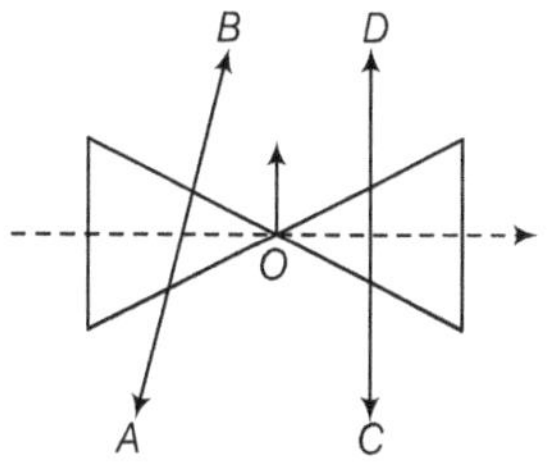

(a) turn left (b) turn right
(c) go straight (d) turn left and right alternately

6. A satellite is revolving in a circular orbit at a height h from the Earth's surface (radius of Earth $R, h << R$). The minimum increase in its orbital velocity required, so that the satellite could escape from the Earth's gravitational field, is close to (Neglect the effect of atmosphere)

(a) $\sqrt{2gR}$ (b) $\sqrt{gR}$ (c) $\sqrt{gR/2}$ (d) $\sqrt{gR}\,(\sqrt{2} - 1)$

7. A pendulum clock loses 12 s a day if the temperature is 40°C and gains 4 s in a day if the temperature is 20°C. The temperature at which the clock will show correct time, and the coefficient of linear expansion (α) of the metal of the pendulum shaft are, respectively.

(a) 25°C, $\alpha = 1.85 \times 10^{-5}$/°C (b) 60°C, $\alpha = 1.85 \times 10^{-4}$/°C
(c) 30°C, $\alpha = 1.85 \times 10^{-3}$/°C (d) 55°C, $\alpha = 1.85 \times 10^{-2}$/°C

8. An ideal gas undergoes a quasistatic, reversible process in which its molar heat capacity C remains constant. If during this process the relation of pressure p and volume V is given by pV^n = constant, then n is given by (Here C_p and C_V are molar specific heat at constant pressure and constant volume, respectively)

(a) $n = \frac{C_p}{C_V}$ (b) $n = \frac{C - C_p}{C - C_V}$
(c) $n = \frac{C_p - C}{C - C_V}$ (d) $n = \frac{C - C_V}{C - C_p}$

*Only Questions Related to Topics Covered in Vol 1

9. n moles of an ideal gas undergoes a process A and B as shown in the figure. The maximum temperature of the gas during the process will be

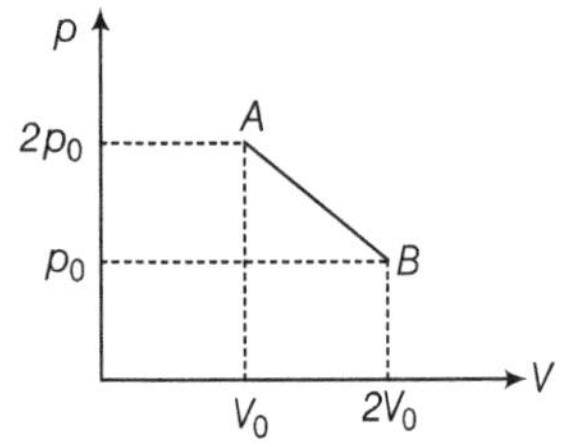

(a) $\frac{9}{4}\frac{p_0V_0}{nR}$ (b) $\frac{3}{2}\frac{p_0V_0}{nR}$
(c) $\frac{9}{2}\frac{p_0V_0}{nR}$ (d) $\frac{9p_0V_0}{nR}$

10. A particle performs simple harmonic motion with amplitude A. Its speed is trebled at the instant that it is at a distance $\frac{2}{3}A$ from equilibrium position. The new amplitude of the motion is

(a) $\frac{A}{3}\sqrt{41}$ (b) $3A$ (c) $A\sqrt{3}$ (d) $\frac{7}{3}A$

11. A uniform string of length 20 m is suspended from a rigid support. A short wave pulse is introduced at its lowest end. It starts moving up the string. The time taken to reach the support is (Take, $g = 10\,\text{ms}^{-2}$)

(a) $2\pi\sqrt{2}$ s (b) 2 s (c) $2\sqrt{2}$ s (d) $\sqrt{2}$ s

12. A screw gauge with a pitch of 0.5 mm and a circular scale with 50 divisions is used to measure the thickness of a thin sheet of aluminium. Before starting the measurement, it is found that when the two jaws of the screw gauge are brought in contact, the 45th division coincides with the main scale line and that the zero of the main scale is barely visible. What is the thickness of the sheet, if the main scale reading is 0.5 mm and the 25th division coincides with the main scale line?

(a) 0.75 mm (b) 0.80 mm (c) 0.70 mm (d) 0.50 mm

13. A pipe open at both ends has a fundamental frequency f in air. The pipe is dipped vertically in water, so that half of it is in water. The fundamental frequency of the air column is now

(a) $\frac{f}{2}$ (b) $\frac{3f}{4}$ (c) $2f$ (d) f

JEE Advanced 2016 (Solved)

Paper I

1. A water cooler of storage capacity 120 litres can cool water at a constant rate of P watts. In a closed circulation system (as shown schematically in the figure), the water from the cooler is used to cool an external device that generates constantly 3 kW of heat (thermal load). The temperature of water fed into the device cannot exceed 30°C and the entire stored 120 litres of water is initially cooled to 10°C. The entire system is thermally insulated. The minimum value of P (in watts) for which the device can be operated for 3 hours is

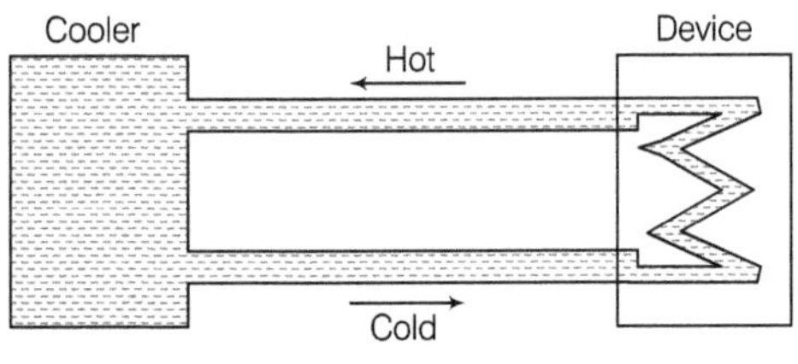

(Specific heat of water is 4.2 kJ $\text{kg}^{-1}\text{K}^{-1}$ and the density of water is $1000\,\text{kg m}^{-3}$) ***(Single Option Correct)***

(a) 1600 (b) 2067
(c) 2533 (d) 3933

2. A uniform wooden stick of mass 1.6 kg of length l rests in an inclined manner on a smooth, vertical wall of height $h(<l)$ such that a small portion of the stick extends beyond the wall. The reaction force of the wall on the stick is perpendicular to the stick. The stick makes an angle of 30° with the wall and the bottom of the stick is on a rough floor. The reaction of the wall on the stick is equal in magnitude to the reaction of the floor on the stick. The ratio $\frac{h}{l}$ and the frictional force f at the bottom of the stick are $(g = 10\,\text{ms}^{-2})$ ***(Single Option Correct)***

(a) $\frac{h}{l} = \frac{\sqrt{3}}{16}, f = \frac{16\sqrt{3}}{3}$ N (b) $\frac{h}{l} = \frac{3}{16}, f = \frac{16\sqrt{3}}{3}$ N

(c) $\frac{h}{l} = \frac{3\sqrt{3}}{16}, f = \frac{8\sqrt{3}}{3}$ N (d) $\frac{h}{l} = \frac{3\sqrt{3}}{16}, f = \frac{16\sqrt{3}}{3}$ N

3. The position vector $\mathbf{r}$ of particle of mass m is given by the following equation $\mathbf{r}(t) = \alpha t^3\hat{\mathbf{i}} + \beta t^2\hat{\mathbf{j}}$

where, $\alpha = \frac{10}{3}\,\text{ms}^{-3}$, $\beta = 5\,\text{ms}^{-2}$ and $m = 0.1$ kg. At $t = 1$s, which of the following statement(s) is (are) true about the particle? ***(One or More Than One Correct Option)***

(a) The velocity $\mathbf{v}$ is given by $\mathbf{v} = (10\hat{\mathbf{i}} + 10\hat{\mathbf{j}})\ \text{ms}^{-1}$
(b) The angular momentum $\mathbf{L}$ with respect to the origin is given by $\mathbf{L} = (5/3)\hat{\mathbf{k}}$ Nms
(c) The force $\mathbf{F}$ is given by $\mathbf{F} = (\hat{\mathbf{i}} + 2\hat{\mathbf{j}})$ N
(d) The torque τ with respect to the origin is given by $\tau = -\dfrac{20}{3}\hat{\mathbf{k}}$ Nm

4. A length-scale (l) depends on the permittivity (ε) of a dielectric material, Boltzmann constant (k_B), the absolute temperature (T), the number per unit volume (n) of certain charged particles, and the charge (q) carried by each of the particles. Which of the following expression(s) for l is (are) dimensionally correct?
(One or More Than One Correct Option)

(a) $l = \sqrt{\left(\dfrac{nq^2}{\varepsilon k_B T}\right)}$ (b) $l = \sqrt{\left(\dfrac{\varepsilon k_B T}{nq^2}\right)}$

(c) $l = \sqrt{\left(\dfrac{q^2}{\varepsilon n^{2/3} k_B T}\right)}$ (d) $l = \sqrt{\left(\dfrac{q^2}{\varepsilon n^{1/3} k_B T}\right)}$

5. Two loudspeakers M and N are located 20 m apart and emit sound at frequencies 118 Hz and 121 Hz, respectively. A car in initially at a point P, 1800 m away from the midpoint Q of the line MN and moves towards Q constantly at 60 km/h along the perpendicular bisector of MN. It crosses Q and eventually reaches a point R, 1800 m away from Q. Let $v(t)$ represent the beat frequency measured by a person sitting in the car at time t. Let ν_P, ν_Q and ν_R be the beat frequencies measured at locations P, Q and R respectively. The speed of sound in air is $330\ \text{ms}^{-1}$. Which of the following statement(s) is (are) true regarding the sound heard by the person?
(One or More Than One Correct Option)

(a) The plot below represents schematically the variation of beat frequency with time

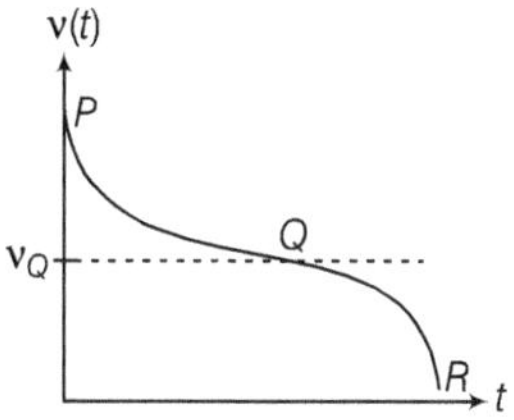

(b) The rate of change in beat frequency is maximum when the car passes through Q
(c) $\nu_P + \nu_R = 2\nu_Q$
(d) The plot below represents schematically the variations of beat frequency with time

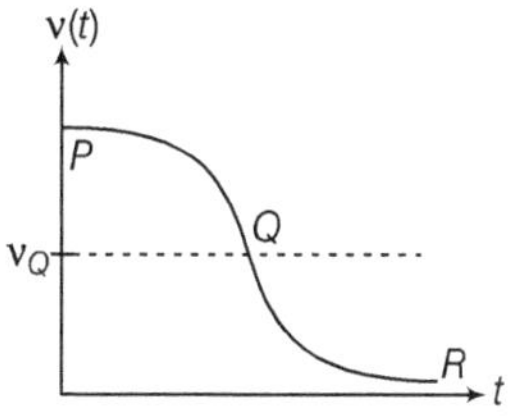

6. Consider two solid spheres P and Q each of density $8\ \text{gm cm}^{-3}$ and diameters 1 cm and 0.5 cm, respectively. Sphere P is dropped into a liquid of density $0.8\ \text{gm cm}^{-3}$ and viscosity $\eta = 3$ poiseulles. Sphere Q is dropped into a liquid of density $1.6\ \text{gm cm}^{-3}$ and viscosity $\eta = 2$ poiseulles. The ratio of the terminal velocities of P and Q is ***(Single Digit Integer)***

7. A metal is heated in a furnace where a sensor is kept above the metal surface to read the power radiated (P) by the metal. The sensor has a scale that displays $\log_2(P/P_0)$, where P_0 is a constant. When the metal surface is at a temperature of 487°C, the sensor shows a value 1. Assume that the emissivity of the metallic surface remains constant. What is the value displayed by the sensor when the temperature of the metal surface is raised to 2767°C? ***(Single Digit Integer)***

Paper II

1. A gas is enclosed in a cylinder with a movable frictionless piston. Its initial thermodynamic state at pressure $P_i = 10^5$ Pa and volume $V_i = 10^{-3}\ \text{m}^3$ changes to a final state at $P_f = (1/32) \times 10^5$ Pa and $V_f = 8 \times 10^{-3}\ \text{m}^3$ in an adiabatic quasi-static process, such that P^3V^5 = constant. Consider another thermodynamic process that brings the system from the same initial state to the same final state in two steps : an isobaric expansion at P_i followed by an isochoric (isovolumetric) process at volume V_f. The amount of heat supplied to the system in the two-step process is approximately ***(Single Option Correct)***
(a) 112 J (b) 294 J (c) 588 J (d) 813 J

2. There are two Vernier calipers both of which have 1 cm divided into 10 equal divisions on the main scale. The Vernier scale of one of the calipers (C_1) has 10 equal divisions that correspond to 9 main scale divisions. The Vernier scale of the other caliper (C_2) has 10 equal divisions that correspond to 11 main scale divisions. The readings of the two calipers are shown in the figure. The measured values (in cm) by calipers C_1 and C_2 respectively, are ***(Single Option Correct)***

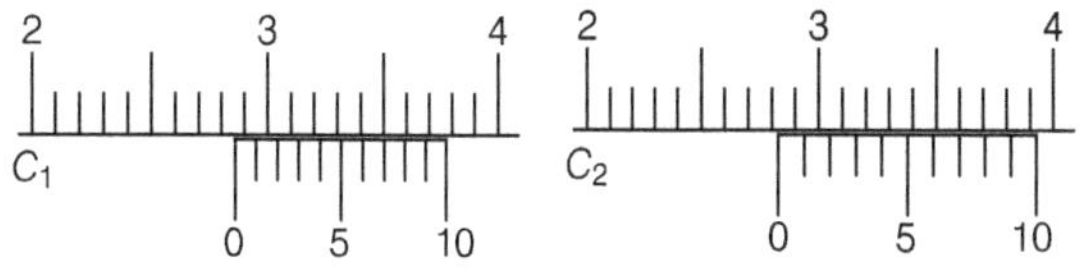

(a) 2.87 and 2.87 (b) 2.87 and 2.83
(c) 2.85 and 2.82 (d) 2.87 and 2.86

3. The ends Q and R of two thin wires, PQ and RS, are soldered (joined) together. Initially, each of the wire has a length of 1 m 10°C. Now, the end P is maintained at 10°C, while the end S is heated and maintained at 400°C. The system is thermally insulated from its surroundings. If the thermal conductivity of wire PQ is twice that of the wire RS and the coefficient of linear thermal expansion of PQ is $1.2 \times 10^{-5}\ K^{-1}$, the change in length of the wire PQ is

(Single Option Correct)

(a) 0.78 mm (b) 0.90 mm (c) 1.56 mm (d) 2.34 mm

4. Two thin circular discs of mass m and $4m$, having radii of a and $2a$, respectively, are rigidly fixed by a massless, rigid rod of length $l = \sqrt{24}\, a$ through their centers. This assembly is laid on a firm and flat surface and set rolling without slipping on the surface so that the angular speed about the axis of the rod is ω. The angular momentum of the entire assembly about the point 'O' is $\mathbf{L}$ (see the figure). Which of the following statement(s) is (are) true?

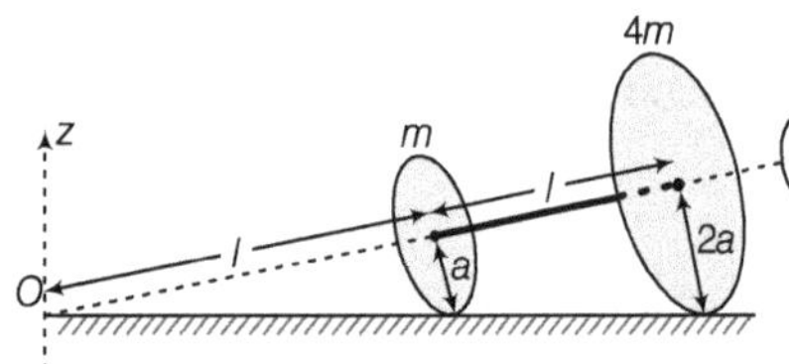

(One or More Than One Option Correct)

(a) The magnitude of the z-component of $\mathbf{L}$ is $55\, ma^2 \omega$
(b) The magnitude of angular momentum of centre of mass of the assembly about the point O is $81\, ma^2 \omega$
(c) The centre of mass of the assembly rotates about the z-axis with an angular speed of $\frac{\omega}{5}$
(d) The magnitude of angular momentum of the assembly about its centre of mass is $17\, ma^2 \frac{\omega}{2}$

5. In an experiment to determine the acceleration due to gravity g, the formula used for the time period of a periodic motion is $T = 2\pi\sqrt{\frac{7(R-r)}{5g}}$. The values of R and r are measured to be (60 ± 1) mm and (10 ± 1) mm respectively. In five successive measurements, the time period is found to be 0.52 s, 0.56s, 0.57 s, 0.54 s and 0.59 s. The least count of the watch used for the measurement of time period is 0.01 s. Which of the following statement(s) is (are) true?

(One or More Than One Option Correct)

(a) The error in the measurement of r is 10%
(b) The error in the measurement of T is 3.57%
(c) The error in the measurement of T is 2%
(d) The error in the determined value of g is 11%

6. A block with mass M is connected by a massless spring with stiffness constant k to a rigid wall and moves without friction on a horizontal surface. The block oscillates with small amplitude A about an equilibrium position x_0. Consider two cases : (i) when the block is at x_0 and (ii) when the block is at $x = x_0 + A$. In both the cases, a particle with mass m ($< M$) is softly placed on the block after which they stick to each other. Which of the following statement(s) is (are) true about the motion after the mass m is placed on the mass M?

(One or More Than One Option Correct)

(a) The amplitude of oscillation in the first case changes by a factor of $\sqrt{\frac{M}{m+M}}$, whereas in the second case it remains unchanged
(b) The final time period of oscillation in both the cases is same
(c) The total energy decreases in both the cases
(d) The instantaneous speed at x_0 of the combined masses decreases in both the cases

Paragraph 1

A frame of the reference that is accelerated with respect to an inertial frame of reference is called a non-inertial frame of reference. A coordinate system fixed on a circular disc rotating about a fixed axis with a constant angular velocity ω is an example of a non-inertial frame of reference. The relationship between the force $\mathbf{F}_{rot}$ experienced by a particle of mass m moving on the rotating disc and the force $\mathbf{F}_{in}$ experienced by the particle in an inertial frame of reference is,

$$\mathbf{F}_{rot} = \mathbf{F}_{in} + 2m(\mathbf{v}_{rot} \times \omega) + m(\omega \times \mathbf{r}) \times \omega,$$

where, $\mathbf{v}_{rot}$ is the velocity of the particle in the rotating frame of reference and $\mathbf{r}$ is the position vector of the particle with respect to the centre of the disc.

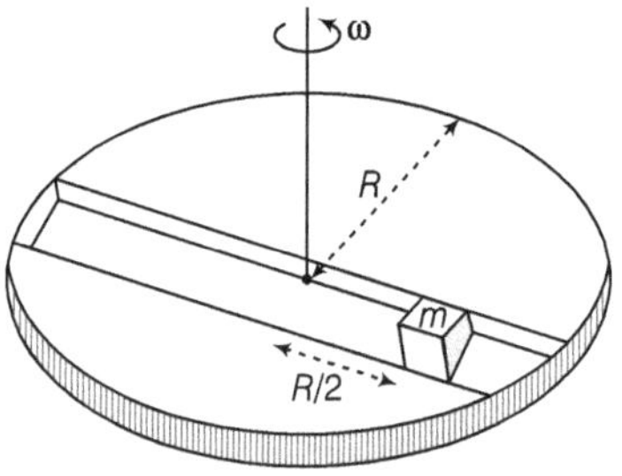

Now consider a smooth slot along a diameter of a disc of radius R rotating counter-clockwise with a constant angular speed ω about its vertical axis through its centre. We assign a coordinate system with the origin at the centre of the disc, the x-axis along the slot, the y-axis perpendicular to the slot and the z-axis along the rotation axis ($\omega = \omega\hat{\mathbf{k}}$). A small block of mass m is gently placed in the slot at $\mathbf{r} = (R/2)\hat{\mathbf{i}}$ at $t = 0$ and is constrained to move only along the slot.

7. The distance r of the block at time t is

(a) $\frac{R}{2}\cos 2\omega t$ (b) $\frac{R}{2}\cos \omega t$
(c) $\frac{R}{4}(e^{\omega t} + e^{-\omega t})$ (d) $\frac{R}{4}(e^{2\omega t} + e^{-2\omega t})$

8. The net reaction of the disc on the block is

(a) $m\omega^2 R \sin\omega t\hat{\mathbf{j}} - mg\hat{\mathbf{k}}$
(b) $\frac{1}{2}m\omega^2 R\,(e^{\omega t} - e^{-\omega t})\hat{\mathbf{j}} + mg\hat{\mathbf{k}}$
(c) $\frac{1}{2}m\omega^2 R\,(e^{2\omega t} - e^{-2\omega t})\hat{\mathbf{j}} + mg\hat{\mathbf{k}}$
(d) $-m\omega^2 R\cos\omega t\,\hat{\mathbf{j}} - mg\hat{\mathbf{k}}$

SOLUTIONS

JEE Main

1. *(a)* Arithmetic mean time of a oscillating simple pendulum

$$= \frac{\Sigma x_i}{N} = \frac{90 + 91 + 92 + 95}{4} = 92 \text{ s}$$

Mean deviation of a simple pendulum

$$= \frac{\Sigma |\bar{x} - x_i|}{N} = \frac{2 + 1 + 3 + 0}{4} = 1.5$$

Given, minimum division in the measuring clock, i.e. simple pendulum = 1 s. Thus, the reported mean time of a oscillating simple pendulum = (92± 2) s

2. *(b, d)* For a particle of mass m is moving along the side of a square of side a. Such that

Angular momentum **L** about the origin = $\mathbf{L} = \mathbf{r} \times \mathbf{p} = rp \sin\theta\, \hat{\mathbf{n}}$

or $\qquad \mathbf{L} = r\,(p)\,\hat{\mathbf{n}}$

When a particle is moving from D to A,

$$\mathbf{L} = \frac{R}{\sqrt{2}}\, mv(-\hat{\mathbf{k}})$$

A particle is moving from A to B,

$$\mathbf{L} = \frac{R}{\sqrt{2}}\, mv(-\hat{\mathbf{k}})$$

and it moves from C to D,

$$\mathbf{L} = \left(\frac{R}{\sqrt{2}} + a\right) mv\,(\hat{\mathbf{k}})$$

For B to C, we have

$$\mathbf{L} = \left(\frac{R}{\sqrt{2}} + a\right) mv\,(\hat{\mathbf{k}})$$

Hence, options (b) and (d) are incorrect.

3. *(c)* Energy lost over path $PQ = \mu\, mg \cos\theta \times 4$

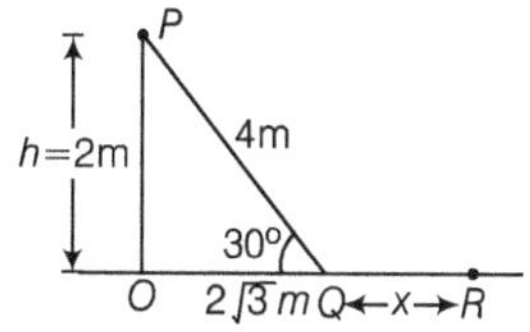

Energy lost over path $QR = \mu\, mgx$

i.e. $\mu\, mg \cos 30° \times 4 = \mu\, mgx$

$$x = 2\sqrt{3} = 3.45 \text{ m}$$

From Q to R energy loss is half of the total energy loss.

i.e. $\qquad \mu mgx = \frac{1}{2} \times mgh \Rightarrow \mu = 0.29$

The values of the coefficient of friction μ and the distance $x(= QR)$ are 0.29 and 3.5.

4. *(d)* Given, potential energy burnt by lifting weight

$$= mgh = 10 \times 9.8 \times 1 \times 1000 = 9.8 \times 10^4 \text{ J}$$

If mass lost by a person be m, then energy dissipated

$$= m \times \frac{2}{10} \times 3.8 \times 10^7 \text{ J}$$

$$\Rightarrow \qquad 9.8 \times 10^4 = m \times \frac{1}{5} \times 3.8 \times 10^7$$

$$\Rightarrow \qquad m = \frac{5}{3.8} \times 10^{-3} \times 9.8$$

$$= 12.89 \times 10^{-3} \text{ kg}$$

5. *(a)* As, the wheel rolls forward the radius of the wheel decreases along AB, hence for the same number of rotations it moves less distance along AB, hence it turns left.

6. *(d)* Given, a satellite is revolving in a circular orbit at a height h from the Earth's surface having radius of Earth R, i.e. $h << R$.

Orbital velocity of a satellite,

$$v = \sqrt{\frac{GM}{R + h}} = \sqrt{\frac{GM}{R}} \qquad (\text{as } h << R)$$

Velocity required to escape,

$$\frac{1}{2} mv'^2 = \frac{GMm}{R + h}$$

$$v' = \sqrt{\frac{2GM}{R + h}} = \sqrt{\frac{2GM}{R}} \qquad (h << R)$$

∴ Minimum increase in its orbital velocity required to escape from the Earth's gravitational field.

$$v' - v = \sqrt{\frac{2GM}{R}} - \sqrt{\frac{GM}{R}}$$

$$= \sqrt{2gR} - \sqrt{gR} = \sqrt{gR}\,(\sqrt{2} - 1) \qquad \left(\because g = \frac{GM}{R^2}\right)$$

7. *(a)* ***Key Idea*** Time period of a pendulum,

$$T = 2\pi \sqrt{\frac{\ }{g}}$$

where is length of pendulum and g is acceleration due to gravity.

Such as change in time period of a pendulum,

$$\frac{\Delta T}{T} = \frac{1}{2} \frac{\Delta}{\ }$$

When clock gains 12 s, we get

$$\frac{12}{T} = \frac{1}{2} \alpha\, (40 - \theta) \qquad \ldots\text{(i)}$$

When clock loses 4 s, we get

$$\frac{4}{T} = \frac{1}{2} \alpha\, (\theta - 20) \qquad \ldots\text{(ii)}$$

Comparing Eqs. (i) and (ii), we get

$$3 = \frac{40 - \theta}{\theta - 20}$$

$$\Rightarrow \qquad 3\theta - 60 = 40 - \theta$$

$$\Rightarrow \qquad 4\theta = 100 \Rightarrow \theta = 25°\text{C}$$

Substituting the value of θ in Eq. (i), we have

$$\frac{12}{T} = \frac{1}{2} \alpha\, (40 - 25)$$

$$\Rightarrow \qquad \frac{12}{24 \times 3600} = \frac{1}{2} \alpha\, (15)$$

$$\alpha = \frac{24}{24 \times 3600 \times 15}$$

$$\alpha = 1.85 \times 10^{-5}/^\circ C$$

Thus, the coefficient of linear expansion in a pendulum clock $= 1.85 \times 10^{-5}/^\circ C$

8. *(b)* For polytropic process, specific heat for an ideal gas,

$$C = \frac{R}{1-n} + C_V$$

$$\therefore \quad \frac{R}{1-n} + C_V = C$$

$$\Rightarrow \quad \frac{R}{1-n} = C - C_V$$

$$\Rightarrow \quad \frac{R}{C - C_V} = 1 - n \qquad \text{(where, } R = C_p - C_V\text{)}$$

$$\Rightarrow \quad \frac{C_p - C_V}{C - C_V} = 1 - n$$

$$\Rightarrow \quad n = 1 - \frac{C_p - C_V}{C - C_V}$$

$$n = \frac{C - C_p}{C - C_V}$$

Thus, number of moles n is given by

$$n = \frac{C - C_p}{C - C_V}$$

9. *(a)* As, T will be maximum temperature where product of pV is maximum.

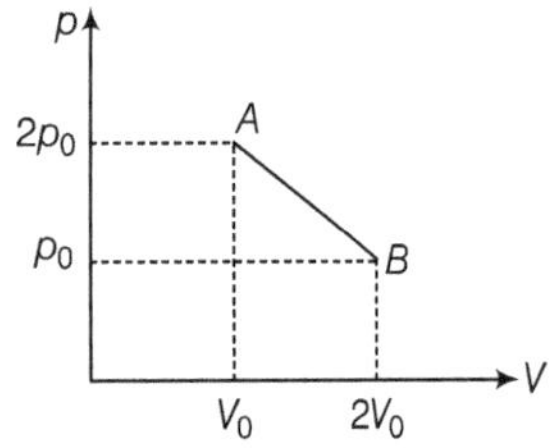

Equation of line AB, we have

$$y - y_1 = \frac{y_2 - y_1}{x_2 - x_1}(x - x_1)$$

$$\Rightarrow \quad p - p_0 = \frac{2p_0 - p_0}{V_0 - 2V_0}(V - 2V_0)$$

$$\Rightarrow \quad p - p_0 = \frac{-p_0}{V_0}(V - 2V_0)$$

$$\Rightarrow \quad p = \frac{-p_0}{V_0}V + 3p_0$$

$$pV = \frac{-p_0}{V_0}V^2 + 3p_0V$$

$$nRT = \frac{-p_0}{V_0}V^2 + 3p_0V$$

$$T = \frac{1}{nR}\left(\frac{-p_0}{V_0}V^2 + 3p_0V\right)$$

For maximum temperature,

$$\frac{\partial T}{\partial V} = 0$$

$$\frac{-p_0}{V_0}(2V) + 3p_0 = 0$$

$$\frac{-p_0}{V_0}(2V) = -3p_0 \Rightarrow V = \frac{3}{2}V_0$$

(condition for maximum temperature)

Thus, the maximum temperature of the gas during the process will be

$$T_{max} = \frac{1}{nR}\left(\frac{-p_0}{V_0} \times \frac{9}{4}V_0^2 + 3p_0 \times \frac{3}{2}V_0\right)$$

$$= \frac{1}{nR}\left(-\frac{9}{4}p_0V_0 + \frac{9}{2}p_0V_0\right) = \frac{9}{4}\frac{p_0V_0}{nR}$$

Alternative Method

Since, initial and final temperature are equal, hence maximum temperature is at the middle of line.

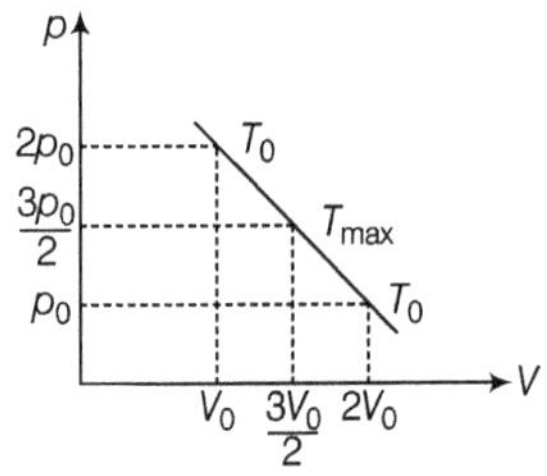

i.e. $\quad pV = nRT \Rightarrow \dfrac{\left(\frac{3}{2}p_0\right)\left(\frac{3V_0}{2}\right)}{nR} = T_{max}$

$$\frac{9}{4}\frac{p_0V_0}{nR} = T_{max}$$

10. *(d)* The velocity of a particle executing SHM at any instant, is defined as the time rate of change of its displacement at that instant.

$$v = \omega\sqrt{A^2 - x^2}$$

where, ω is angular frequency, A is amplitude and x is displacement of a particle.

Suppose that the new amplitude of the motion be A'.

Initial velocity of a particle perform SHM,

$$v^2 = \omega^2\left[A^2 - \left(\frac{2A}{3}\right)^2\right] \qquad \text{... (i)}$$

where, A is initial amplitude and ω is angular frequency.

Final velocity,

$$(3v)^2 = \omega^2\left[A'^2 - \left(\frac{2A}{3}\right)^2\right] \qquad \text{...(ii)}$$

From Eqs. (i) and (ii), we get

$$\frac{1}{9} = \frac{A^2 - \frac{4A^2}{9}}{A'^2 - \frac{4A^2}{9}}$$

$$\Rightarrow \quad A' = \frac{7A}{3}$$

11. *(c)* A uniform string of length 20 m is suspended from a rigid support. Such that the time taken to reach the support,

$$T = \frac{mgx}{}$$

So, velocity at point $P = \sqrt{\frac{mgx}{\frac{}{m/}}}$

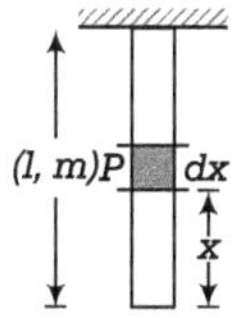

i.e. $v = \sqrt{gx}$

$$\frac{dx}{dt} = \sqrt{gx}$$

$$\Rightarrow \int_0^{20} \frac{dx}{\sqrt{x}} = \int_0^t \sqrt{g}\, dt$$

$$[2\sqrt{x}]_0^{20} = \sqrt{10}\, t$$

$$\Rightarrow 2\sqrt{20} = \sqrt{10}\, t$$

$$t = 2\sqrt{2} \text{ s}$$

12. *(a)* Given that the screw gauge has zero error.

So, least count of a screw gauge $= \frac{0.5}{50}$ mm.

Thickness of the sheet if the main scale reading is 0.5 mm and the 25th division coincides with the main scale line, we have

$$= 0.50 \text{ mm} + (25) \times \frac{0.5}{50} \text{mm}$$

$$= 0.50 \text{ mm} + 0.25 \text{ mm} = 0.75 \text{ mm}$$

13. *(d)* For open ends, fundamental frequency f in air, we have

$$\frac{\lambda}{2} = \quad \Rightarrow \lambda = 2$$

$$v = f\lambda \Rightarrow f = \frac{v}{\lambda} = \frac{v}{2} \qquad \ldots\text{(i)}$$

When a pipe is dipped vertically in water, so that half of it is in water, we have

$$\frac{\lambda}{4} = \frac{}{2}$$

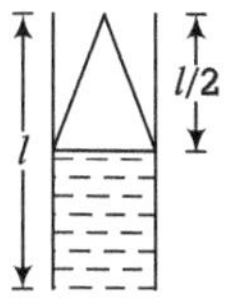

$$\Rightarrow \lambda = 2 \Rightarrow v = f'\lambda$$

$$\Rightarrow f' = \frac{v}{\lambda} = \frac{v}{2} = f \qquad \ldots\text{(ii)}$$

Thus, the fundamental frequency of the air column is now,

$$f = f'$$

JEE Advanced

Paper I

1. *(b)* Heat generated in device in 3 h

$$= \text{time} \times \text{power} = 3 \times 3600 \times 3 \times 10^3 = 324 \times 10^5 \text{ J}$$

Heat used to heat water $= ms\Delta\theta = 120 \times 1 \times 4.2 \times 10^3 \times 20\text{J}$

Heat absorbed by coolant

$$= Pt = 324 \times 10^5 - 120 \times 1 \times 4.2 \times 10^3 \times 20\text{J}$$

$$Pt = (325 - 100.8) \times 10^5 \text{J} = 223.2 \times 10^5 \text{J}$$

$$P = \frac{223.2 \times 10^5}{3600} = 2067 \text{ W}$$

2. *(d)*

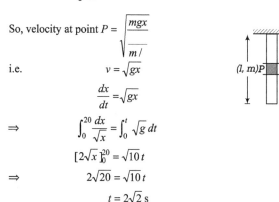

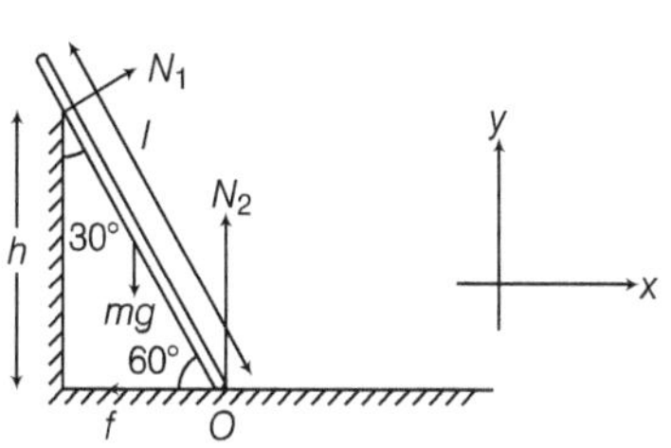

$$\Sigma F_x = 0,$$

$$\therefore \quad N_1 \cos 30° - f = 0 \qquad \ldots\text{(i)}$$

$$\Sigma F y = 0$$

$$\therefore \quad N_1 \sin 30° + N_2 - mg = 0 \qquad \ldots\text{(ii)}$$

$$\Sigma \tau_0 = 0$$

$$mg\frac{l}{2}\cos 60° - N_1 \frac{h}{\cos 30°} = 0 \qquad \ldots\text{(iii)}$$

Also, given $N_1 = N_2$ ...(iv)

Solving Eqs. (i), (ii), (iii) and (iv) we have

$$\frac{h}{l} = \frac{3\sqrt{3}}{16} \text{ and } f = \frac{16\sqrt{3}}{3}$$

3. *(a,b,d)* $\mathbf{r} = \alpha t^3 \hat{\mathbf{i}} + \beta t^2 \hat{\mathbf{j}}$

$$\mathbf{v} = \frac{d\mathbf{r}}{dt} = 3\alpha t^2 \hat{\mathbf{i}} + 2\beta t \hat{\mathbf{j}}$$

$$\mathbf{a} = \frac{d^2\mathbf{r}}{dt^2} = 6\alpha t \hat{\mathbf{i}} + 2\beta \hat{\mathbf{j}}$$

At $t = 1$ s,

(a) $\mathbf{v} = 3 \times \frac{10}{3} \times 1\hat{\mathbf{i}} + 2 \times 5 \times 1\hat{\mathbf{j}} = (10\hat{\mathbf{i}} + 10\hat{\mathbf{j}})\text{m/s}$

(b) $\mathbf{L} = \mathbf{r} \times \mathbf{p} = \left(\frac{10}{3} \times 1\hat{\mathbf{i}} + 5 \times 1\hat{\mathbf{j}}\right) \times 0.1(10\hat{\mathbf{i}} + 10\hat{\mathbf{j}}) = \left(-\frac{5}{3}\hat{\mathbf{k}}\right)\text{N-ms}$

(c) $\mathbf{F} = m\mathbf{a} = m \times \left(6 \times \frac{10}{3} \times 1\hat{\mathbf{i}} + 2 \times 5\hat{\mathbf{j}}\right) = (2\hat{\mathbf{i}} + \hat{\mathbf{j}})\text{N}$

(d) $\tau = r \times \mathbf{F} = \left(\frac{10}{3}\hat{\mathbf{i}} + 5\hat{\mathbf{j}}\right) \times (2\hat{\mathbf{i}} + \hat{\mathbf{j}})$

$$= +\frac{10}{3}\hat{\mathbf{k}} + 10(-\hat{\mathbf{k}}) = \left(-\frac{20}{3}\hat{\mathbf{k}}\right)\text{N-m}$$

4. *(b,d)* $[n]=[L^{-3}]$ $[q]=[AT]$

$[\varepsilon]=[M^{-1}L^{-3}A^2T^4]$ $[T]=[L]$

$[l]=[L]$ $[k_B]=[M^1L^2T^{-2}K^{-1}]$

(a) $\text{RHS} = \sqrt{\dfrac{[L^{-3}A^2T^2]}{[M^{-1}L^{-3}T^4A^2][M^1L^2T^{-2}K^{-1}][K]}}$

$= \sqrt{\dfrac{[L^{-3}A^2T^2]}{[L^{-1}T^2A^2]}} = \sqrt{[L^{-2}]} = [L^{-1}]$ Wrong

(b) $\text{RHS} = \sqrt{\dfrac{[M^{-1}L^{-3}T^4A^2][M^1L^2T^{-2}K^{-1}][K]}{[L^{-3}][A^2T^2]}}$

$= \sqrt{\dfrac{[L^{-1}T^2A^2]}{[L^{-3}T^2A^2]}} = [L]$ Correct

(c) $\text{RHS} = \sqrt{\dfrac{[A^2T^2]}{[M^{-1}L^{-3}T^4A^2][L^{-2}][M^1L^2T^{-2}K^{-1}][K]}}$

$= \sqrt{[L^3]}$ Wrong

(d) $\text{RHS} = \sqrt{\dfrac{[A^2T^2]}{[M^{-1}L^{-3}T^4A^2][L^{-1}][M^1L^2T^{-2}K^{-1}]}}$

$= \sqrt{\dfrac{[A^2T^2]}{[L^{-2}T^2A^2]}} = [L]$ Correct

5. *(b,c,d)*

Speed of car, $V = 60\,\text{km/h} = \dfrac{500}{3}\,\text{m/s}$

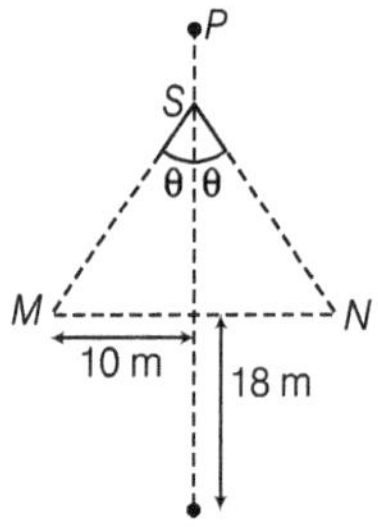

At a point S, between P and Q

$\nu'_M = \nu_M\left(\dfrac{C + V\cos\theta}{C}\right);$

$\nu'_N = \nu_N\left(\dfrac{C + V\cos\theta}{C}\right)$

$\Rightarrow \quad \Delta v = (v_N - v_M)\left(1 + \dfrac{v\cos\theta}{C}\right)$

Similarly, between Q and R

$\Delta\nu = (\nu_N - \nu_M)\left(1 - \dfrac{V\cos\theta}{C}\right)$

$\dfrac{d(\Delta\nu)}{dt} = \pm(\nu_N - \nu_M)\dfrac{V}{C}\sin\theta\dfrac{d\theta}{dt}$

$\theta \approx 0°$ at P and R as they are large distance apart.

$\Rightarrow$ Slope of graph is zero. at Q, $\theta = 90°$

$\sin\theta$ is maximum also value of $\dfrac{d\theta}{dt}$ is maximum

as $\dfrac{d\theta}{dt} = \dfrac{V}{r}$, where V is its velocity and r is the length of the line joining P and S. and r is minimum at Q.

$\Rightarrow$ Slope is maximum at Q.

At P, $\nu_P = \Delta\nu = (\nu_N - \nu_M)\left(1 + \dfrac{V}{C}\right)$ $(\theta \approx 0°)$

At R, $\nu_R = \Delta\nu = (\nu_N - \nu_M)\left(1 - \dfrac{V}{C}\right)$ $(\theta \approx 0°)$

At Q, $\nu_Q = \Delta\nu = (\nu_N - \nu_M)$ $(\theta = 90°)$

From these equations, we can see that $\nu_P + \nu_R = 2\nu_Q$

6. **(3)** Terminal velocity is given by, $v_T = \dfrac{2}{9}\dfrac{r^2}{\eta}(d-\rho)g$

$\dfrac{v_P}{v_Q} = \dfrac{r_P^2}{r_Q^2} \times \dfrac{\eta_Q}{\eta_P} \times \dfrac{(d-\rho_P)}{(d-\rho_Q)}$

$= \left(\dfrac{1}{0.5}\right)^2 \times \left(\dfrac{2}{3}\right) \times \dfrac{(8-0.8)}{(8-1.6)} = 4 \times \dfrac{2}{3} \times \dfrac{7.2}{6.4} = 3$

7. (9) $\log_2 \dfrac{p_1}{p_0} = 1,$

Therefore, $\dfrac{p_1}{p_0} = 2$

According to Stefan's law, $p \propto T^2$

$\Rightarrow \quad \dfrac{p_2}{p_1} = \left(\dfrac{T_2}{T_1}\right)^4 = \left(\dfrac{2767+273}{487+273}\right)^4 = 4^4$

$\dfrac{p_2}{p_1} = \dfrac{p_2}{2p_0} = 4^4 \Rightarrow \dfrac{p_2}{p_0} = 2 \times 4^4$

$\log_2 \dfrac{p_2}{p_0} = \log_2[2 \times 4^4] = \log_2 2 + \log_2 4^4$

$= 1 + \log_2 2^8 = 1 + 8 = 9$

Paper II

1. *(c)* In the first process : $p_iV_i^\gamma = p_fV_f^\gamma$

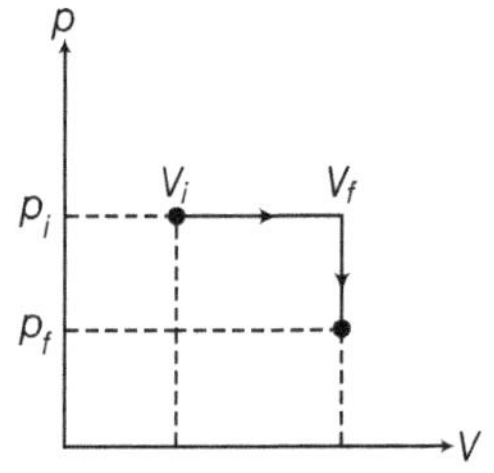

$\Rightarrow \quad \dfrac{p_i}{p_f} = \left(\dfrac{V_f}{V_i}\right)^\gamma \Rightarrow 32 = 8^\gamma$

$\gamma = \dfrac{5}{3}$ ...(i)

For the two step process

$W = p_i(V_f - V_i) = 10^5(7 \times 10^{-3})$

$W = 7 \times 10^2\,\text{J}$

$\Delta U = \dfrac{f}{2}(p_fV_f - p_iV_i)$

$= \dfrac{1}{\gamma - 1}\left(\dfrac{1}{4} \times 10^2 - 10^2\right)$

$\Delta U = -\dfrac{3}{2} \cdot \dfrac{3}{4} \times 10^2$

$= -\dfrac{9}{8} \times 10^2\,\text{J}$

$Q - W = \Delta U$

$\Rightarrow \quad Q = 7 \times 10^2 - \dfrac{9}{8} \times 10^2$

$= \dfrac{47}{8} \times 10^2\,\text{J} = 588\text{ J}$

2. *(b)* For vernier C_1

$$10\text{ VSD} = 9\text{ MSD} = 9\text{ mm}$$
$$1\text{ VSD} = 0.9\text{ mm}$$
$$\Rightarrow \quad LC = 1\text{ MSD} - 1\text{ VSD}$$
$$= 1\text{ mm} - 0.9\text{ mm} = 0.1\text{ mm}$$

Reading of C_1 = MSR + (VSR) (L.C.)
$= 28\text{ mm} + (7)(0.1)$

Reading of $C_1 = 28.7\text{ mm} = 2.87\text{ cm}$

For vernier C_2 : the vernier C_2 is abnormal,

So, we have to find the reading form basics.

The point where both of the marks are matching :

distance measured from main scale = distance measured from vernier scale

$28\text{ mm} + (1\text{ mm})(8) = (28\text{ mm} + x) + (1.1\text{ mm})(7)$

Solving we get, $x = 0.3\text{ mm}$

So, reading of $C_2 = 28\text{ mm} + 0.3\text{ mm} = 2.83\text{ cm}$

3. *(a)*

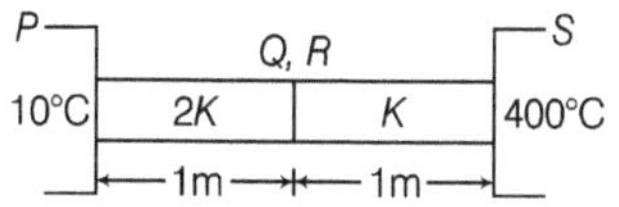

Rate of heat flow from P to Q

$$\frac{dQ}{dt} = \frac{2KA(T-10)}{1}$$

Rate of heat flow from Q to S

$$\frac{dQ}{dt} = \frac{KA(4000-T)}{1}$$

At steady state rate of heat flow is same

$$\therefore \quad \frac{2KA(T-10)}{1} = KA(400-T)$$

or $\quad 2T - 20 = 400 - T$

or $\quad 3T = 420$

$\therefore \quad T = 140°$

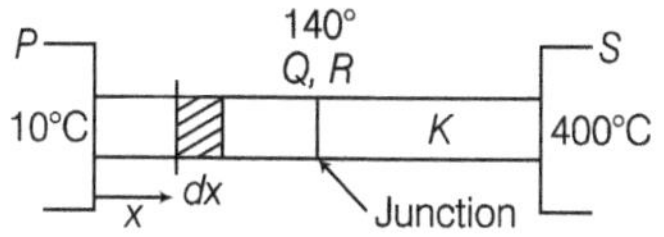

Temperature of junction is 140°C

Temperature at a distance x from end P is

$$T_x = (130x + 10°)$$

Change in length dx is suppose dy

Then, $dy = \alpha dx\,(T_x - 10)$

$$\int_0^{\Delta y} dy = \int_0^1 \alpha dx\,(130x + 10 - 10)$$

$$\Delta y = \left[\frac{\alpha x^2}{2} \times 130\right]_0^1$$

$$\Delta y = 1.2 \times 10^{-5} \times 65$$

$$\Delta y = 78.0 \times 10^{-5}\text{ m} = 0.78\text{ mm}$$

4. *(c,d)*

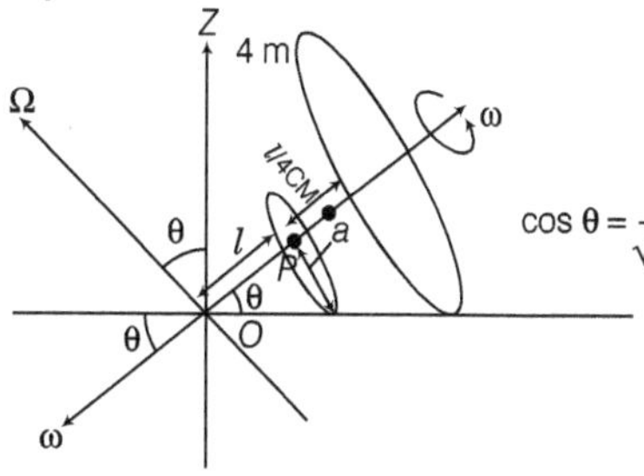

$$\cos\theta = \frac{l}{\sqrt{l^2 + a^2}} = \frac{\sqrt{24}}{5}$$

(a) $\quad L_z = L_{CM-O}\cos\theta - L_{D-CM}\sin\theta$

$$= \frac{81\sqrt{24}}{5}a^2 m\omega \times \frac{\sqrt{24}}{5} - \frac{17\,ma^2\omega}{2} \times \frac{1}{\sqrt{24}}$$

$$= \frac{81 \times 24 ma^2\omega}{25} - \frac{17 ma^2\omega}{2\sqrt{24}}$$

(b) $\quad L_{CM-O} = (5m)\left[\frac{9l}{5}\Omega\right]\frac{9l}{5} = \frac{81 ml^2\Omega}{5}$

$$= \frac{81 ml^2}{5} \times \frac{a\omega}{l}$$

$$L_{CM-O} = \frac{81 mla\omega}{5} = \frac{81\sqrt{24}\,a^2 m\omega}{5}$$

(c) Velocity of point P : $a\omega = l\,\Omega$ then

$\Omega = \frac{a\omega}{l}$ = Angular velocity of C.M. w.r.t point O.

Angular velocity fo C.M. w.r.t. Z-axis = $\Omega\cos\theta$

$$\omega_{CM-z} = \frac{a\omega}{l}\frac{\sqrt{24}}{5} = \frac{a\omega}{\sqrt{24}\,a}\frac{\sqrt{24}}{5}$$

$$\omega_{CM-z} = \frac{\omega}{5}$$

(d) $\quad L_{D-CM} = \frac{ma^2}{2}\omega + \frac{4m(2a)^2}{2}\omega = \frac{17 ma^2\omega}{2}$

5. *(a,b,d)* Mean time period $= \frac{0.52 + 0.56 + 0.57 + 0.54 + 0.59}{5}$

$= 0.556 \cong 0.56$ sec as per significant figures

Error in reading $= |T_{mean} - T_1| = 0.04$

$|T_{mean} - T_2| = 0.00$

$|T_{mean} - T_3| = 0.01$

$|T_{mean} - T_4| = 0.02$

$|T_{mean} - T_5| = 0.03$

Mean error = $0.1/5 = 0.02$

$$\%\text{ error in } T = \frac{\Delta T}{T} \times 100 = \frac{0.02}{0.56} \times 100 = 3.57\%$$

$$\%\text{ error in } r = \frac{0.001 \times 100}{0.010} = 10\%$$

$$\%\text{ error in } R = \frac{0.001 \times 100}{0.60} = 1.67\%$$

$$\%\text{ error in } \frac{\Delta g}{g} \times 100 = \frac{\Delta(R-r)}{R-r} \times 100 + 2 \times \frac{\Delta T}{T}$$

$$= \frac{0.002 \times 100}{0.05} + 2 \times 3.57 = 4\% + 7\% = 11\%$$

6. *(a,b,d)* **Case 1**

Just before m is placed   Just before m is placed

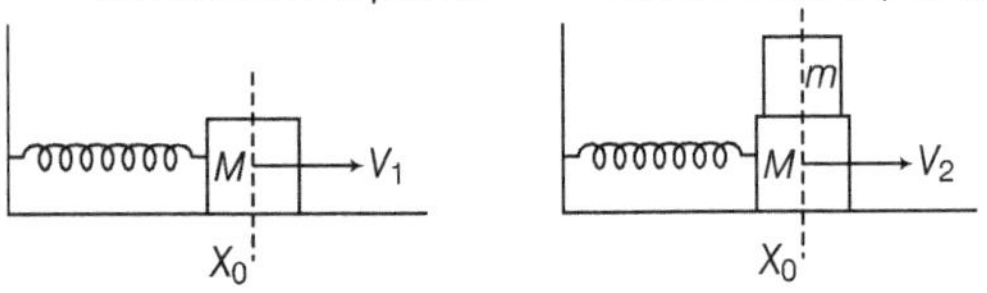

Case 2

Just before m is placed   Just before m is placed

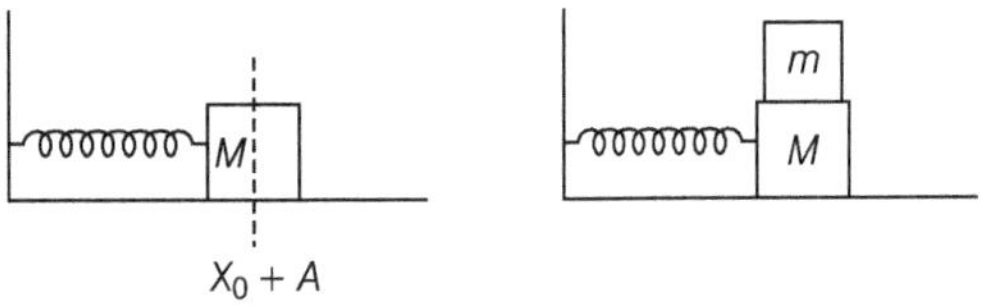

In case 1,

$$Mv_1 = (M + m)v_2$$

$$v_2 = \left(\frac{M}{M + m}\right)v_1$$

$$\sqrt{\frac{k}{M + m}}\,A_2 = \left(\frac{M}{M + m}\right)\sqrt{\frac{k}{M}}\,A_1$$

$$A_2 = \sqrt{\frac{k}{M + m}}\,A_1$$

In case 2, $A_2 = A_1$

$$T = 2\pi\sqrt{\frac{M + m}{k}} \text{ in both cases.}$$

Total energy decreases in first case whereas remain same in 2nd case. Instantaneous speed at x_0 decreases in both cases.

7. *(c)* Force on block along slot $= m\omega^2 r = ma = m\left(\frac{vdv}{dr}\right)$

$$\int_0^v vdv = \int_{R/2}^r \omega^2 rdr \Rightarrow \frac{v^2}{2} = \frac{\omega^2}{2}\left(r^2 - \frac{R^2}{4}\right)$$

$$\Rightarrow \quad v = \omega\sqrt{r^2 - \frac{R^2}{4}} = \frac{dr}{dt} \Rightarrow \int_{R/4}^{r} \frac{dr}{\sqrt{r^2 - \frac{R^2}{4}}} = \int_0^t \omega dt$$

$$\ln\left(\frac{r + \sqrt{r^2 - \frac{R^2}{4}}}{\frac{R}{2}}\right) - \ln\left(\frac{R/2 + \sqrt{\frac{R^2}{4} - \frac{R^2}{4}}}{\frac{R}{4}}\right) = \omega t$$

$$\Rightarrow \quad r + \sqrt{r^2 - \frac{R^2}{4}} = \frac{R}{2}e^{\omega t}$$

$$\Rightarrow \quad r^2 - \frac{R^2}{4} = \frac{R^2}{4}e^{2\omega t} + r^2 - 2r\frac{R}{2}e^{\omega t}$$

$$\Rightarrow \quad r = \frac{\frac{R^2}{4}e^{2\omega t} + \frac{R^2}{4}}{Re^{\omega t}} = \frac{R}{4}(e^{\omega t} + e^{-\omega t})$$

8. *(b)*

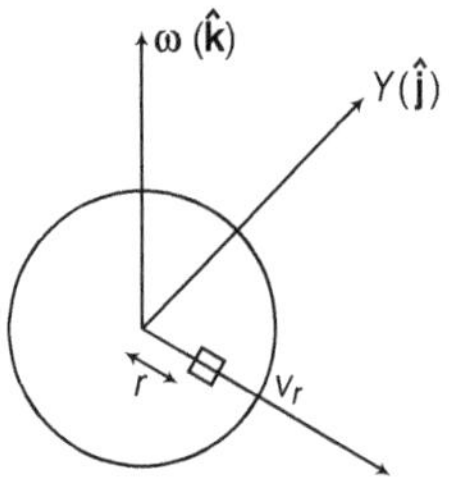

$$\mathbf{F}_{rot} = \mathbf{F}_{in} + 2m(v_{rot}\,\hat{\mathbf{i}}) \times \omega\hat{\mathbf{k}} + m(\omega\hat{\mathbf{k}} \times r\hat{\mathbf{i}}) \times \omega\hat{\mathbf{k}}$$

$$mr\omega^2\hat{\mathbf{i}} = \mathbf{F}_{in} + 2mv_{rot}\,\omega(-\hat{\mathbf{j}}) + m\omega^2 r\hat{\mathbf{i}}$$

$$\mathbf{F}_{in} = 2mv_r\omega\hat{\mathbf{j}}$$

$$r = \frac{R}{4}[e^{\omega t} + e^{-\omega t}]$$

$$\frac{dr}{dt} = v_r = \frac{R}{4}[\omega e^{\omega t} - \omega e^{-\omega t}]$$

$$\mathbf{F}_{in} = 2m\frac{R\omega}{4}[e^{\omega t} - e^{-\omega t}]\omega\hat{\mathbf{j}}$$

$$\mathbf{F}_{in} = \frac{mR\omega^2}{2}[e^{\omega t} - e^{-\omega t}]\hat{\mathbf{j}}$$

Also reaction is due to disc surface then

$$\mathbf{F}_{reaction} = \frac{mR\omega^2}{2}[e^{\omega t} - e^{-\omega t}]\hat{\mathbf{j}} + mg\hat{\mathbf{k}}$$

JEE Main & Adv./BITSAT/KERALA CEE
KCET/AP & TS EAMCET/VIT/MHT CET

SOLVED PAPERS 2017

JEE Main

1. An observer is moving with half the speed of light towards a stationary microwave source emitting waves at frequency 10 GHz. What is the frequency of the microwave measured by the observer? (speed of light $= 3 \times 10^8 \text{ ms}^{-1}$)

(a) 12.1 GHz (b) 17.3 GHz (c) 15.3 GHz (d) 10.1 GHz

2. The following observations were taken for determining surface tension T of water by capillary method. Diameter of capillary, $d = 1.25 \times 10^{-2}$ m rise of water, $h = 1.45 \times 10^{-2}$m. Using $g = 9.80$ m/s^2 and the simplified relation $T = \frac{rhg}{2} \times 10^3$ N/m, the possible error in surface tension is closest to

(a) 1.5% (b) 2.4% (c) 10% (d) 0.15%

3. A body of mass $m = 10^{-2}$kg is moving in a medium and experiences a frictional force $F = -kv^2$. Its initial speed is $v_0 = 10 \text{ ms}^{-1}$. If, after 10 s, its energy is $\frac{1}{8} mv_0^2$, the value of k will be

(a) 10^{-3}kgs^{-1} (b) 10^{-4}kgm^{-1}
(c) $10^{-1}\text{kgm}^{-1}\text{s}^{-1}$ (d) 10^{-3}kgm^{-1}

4. C_p and C_V are specific heats at constant pressure and constant volume, respectively. It is observed that $C_p - C_V = a$ for hydrogen gas $C_p - C_V = b$ for nitrogen gas. The correct relation between a and b is

(a) $a = b$ (b) $a = 14b$ (c) $a = 28b$ (d) $a = \frac{1}{14} b$

5. The moment of inertia of a uniform cylinder of length l and radius R about its perpendicular bisector is I. What is the ratio l/R such that the moment of inertia is minimum?

(a) $\frac{\sqrt{3}}{2}$ (b) 1 (c) $\frac{3}{\sqrt{2}}$ (d) $\sqrt{\frac{3}{2}}$

6. A copper ball of mass 100 g is at a temperature T. It is dropped in a copper calorimeter of mass 100 g, filled with 170 g of water at room temperature. Subsequently, the temperature of the system is found to be 75°C. T is (Given, room temperature = 30°C, specific heat of copper = 0.1 cal/g°C)

(a) 885°C (b) 1250°C (c) 825°C (d) 800°C

7. A slender uniform rod of mass M and length l is pivoted at one end so that it can rotate in a vertical plane (see the figure). There is negligible friction at the pivot. The free end is held vertically above the pivot and then released. The angular acceleration of the rod when it makes an angle θ with the vertical, is

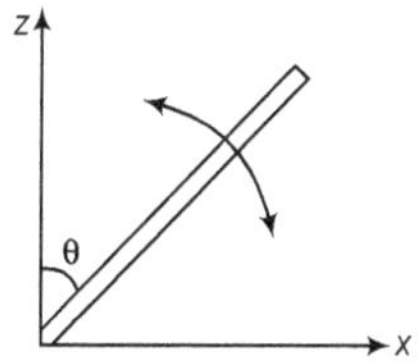

(a) $\frac{2g}{3l} \sin\theta$ (b) $\frac{3g}{2l} \cos\theta$ (c) $\frac{2g}{3l} \cos\theta$ (d) $\frac{3g}{2l} \sin\theta$

8. An external pressure P is applied on a cube at 0°C so that it is equally compressed from all sides. K is the bulk modulus of the material of the cube and α is its coefficient of linear expansion. Suppose we want to bring the cube to its original size by heating. The temperature should be raised by

(a) $\frac{P}{\alpha K}$ (b) $\frac{3\alpha}{PK}$ (c) $3PK\alpha$ (d) $\frac{P}{3\alpha K}$

9. The temperature of an open room of volume 30 m^3 increases from 17°C to 27°C due to the sunshine. The atmospheric pressure in the room remains 1×10^5 Pa. If n_i and n_f are the number of molecules in the room before and after heating, then $n_f - n_i$ will be

(a) 1.38×10^{23} (b) 2.5×10^{25}
(c) -2.5×10^{25} (d) -1.61×10^{23}

10. A time dependent force $F = 6t$ acts on a particle of mass 1 kg. If the particle starts from rest, the work done by the force during the first 1 s will be

(a) 22 J (b) 9 J (c) 18 J (d) 4.5 J

11. The variation of acceleration due to gravity g with distance d from centre of the Earth is best represented by (R = Earth's radius)

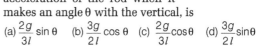

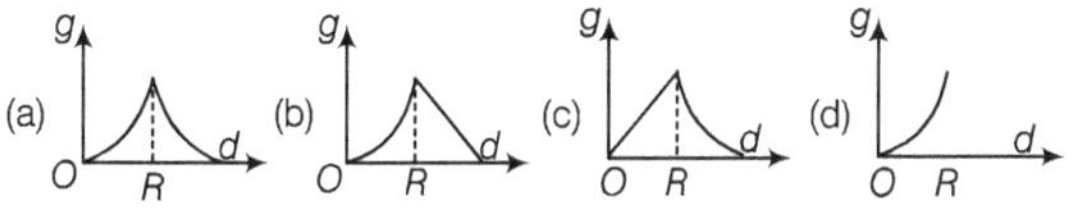

12. A body is thrown vertically upwards. Which one of the following graphs correctly represent the velocity *vs* time?

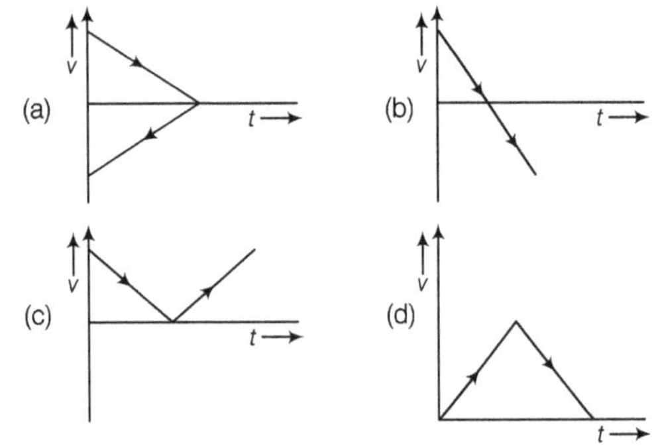

13. A particle is executing simple harmonic motion with a time period T. At time $t = 0$, it is at its position of equilibrium. The kinetic energy-time graph of the particle will look, like

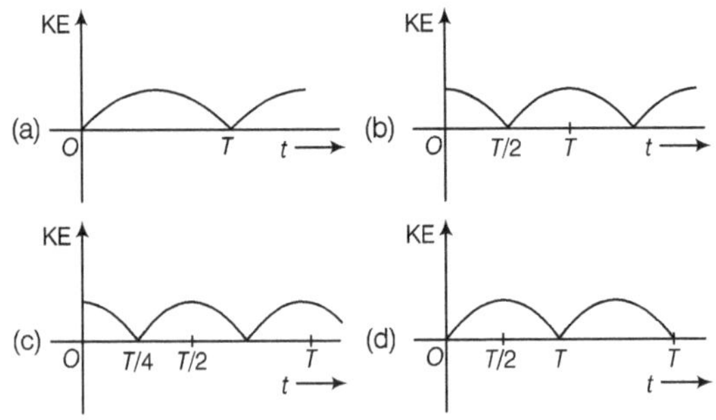

14. A man grows into a giant such that his linear dimensions increase by a factor of 9. Assuming that his density remains same, the stress in the leg will change by a factor of

(a) $\frac{1}{9}$ (b) 81

(c) $\frac{1}{81}$ (d) 9

JEE Advanced

Paper 1

1. A flat plane is moving normal to its plane through a gas under the action of a constant force F. The gas is kept at a very low pressure. The speed of the plate v is much less than the average speed u of the gas molecules. Which of the following options is/are true?

(One or More Than One Correct Option)

(a) At a later time the external force F balances the resistive force

(b) The plate will continue to move with constant non-zero acceleration, at all times

(c) The resistive force experienced by the plate is proportional to v

(d) The pressure difference between the leading and trailing faces of the plate is proportional to uv

2. A block of mass M has a circular cut with a frictionless surface as shown. The block rests on the horizontal frictionless surfaced of a fixed table. Initially the right edge of the block is at $x = 0$, in a coordinate system fixed to the table. A point mass m is released from rest at the topmost point of the path as shown and it slides down. When the mass loses contact with the block, its position is x and the velocity is v. At that instant, which of the following option is/are correct?

(One or More Than One Correct Option)

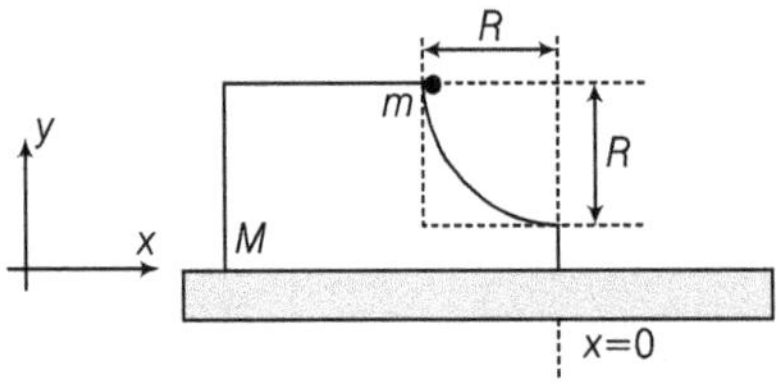

(a) The velocity of the point mass m is $v = \sqrt{\frac{2gR}{1+\frac{m}{M}}}$

(b) The x component of displacement of the centre of mass of the block M is $-\frac{mR}{M+m}$

(c) The position of the point mass is $x = -\sqrt{2}\frac{mR}{M+m}$

(d) The velocity of the block M is $V = -\frac{m}{M}\sqrt{2gR}$

3. A block M hangs vertically at the bottom end of a uniform rope of constant mass per unit length. The top end of the rope is attached to a fixed rigid support at O. A transverse wave pulse (Pulse 1) of wavelength λ_0 is produced at point O on the rope. The pulse takes time T_{OA} to reach point A. If the wave pulse of wavelength λ_0 is produced at point A (Pulse 2) without disturbing the position of M it takes time T_{AO} to reach point O. Which of the following options is/are correct?

(One or More Than One Correct Option)

(a) The time $T_{AO} = T_{OA}$

(b) The wavelength of Pulse 1 becomes longer when it reaches point A

(c) The velocity of any pulse along the rope is independent of its frequency and wavelength

(d) The velocities of the two pulses (Pulse 1 and Pulse 2) are the same at the mid-point of rope

4. A human body has a surface area of approximately $1\ m^2$. The normal body temperature is 10K above the surrounding room temperature T_0. Take the room temperature to be $T_0 = 300$ K. For $T_0 = 300$ K, the value of $\sigma T_0^4 = 460\ Wm^{-2}$ (where σ is the Stefan Boltzmann constant). Which of the following options is/are correct? *(One or More Than One Correct Option)*

(a) If the body temperature rises significantly, then the peak in the spectrum of electromagnetic radiation emitted by the body would shift to longer wavelengths

(b) If the surrounding temperature reduces by a small amount $\Delta T_0 << T_0$, then to maintain the same body temperature the same (living) human being needs to radiate $\Delta W = 4\sigma T_0^3 \Delta T_0$ more energy per unit time

(c) The amount of energy radiated by the body in 1s is close to 60 J

(d) Reducing the exposed surface area of the body (e.g. by curling up) allows humans to maintain the same body temperature while reducing the energy lost by radiation

5. A drop of liquid of radius $R = 10^{-2}$ m having surface tension $S = \dfrac{0.1}{4\pi}$ Nm^{-1} divides itself into K identical drops. In this process the total change in the surface energy $\Delta U = 10^{-3}$ J. If $K = 10^{\alpha}$, then the value of α is *(Single Digit Integer)*

6. A stationary source emits sound of frequency $f_0 = 492$ Hz. The sound is reflected by a large car approaching the source with a speed of 2 ms^{-1}. The reflected signal is received by the source and superposed with the original. What will be the beat frequency of the resulting signal in Hz? (Given that the speed of sound in air is 330 ms^{-1} and the car reflects the sound at the frequency it has received). *(Single Digit Integer)*

Maching Type Questions

Directions (Q.Nos. 7-9) Matching the information given in the three columns of the following table.

An ideal gas is undergoing a cyclic thermodynamic process in different ways as shown in the corresponding *p-V* diagrams in column 3 of the table. Consider only the path from state 1 to state 2. *W* denotes the corresponding work done on the system. The equations and plots in the table have standards notations and used in thermodynamic processes. Here γ is the ratio of heat capacities at constant pressure and constant volume. The number of moles in the gas is *n*.

	Column 1		Column 2		Column 3
(I)	$W_{1\to2} = \dfrac{1}{\gamma - 1}(p_2V_2 - p_1V_1)$	(i)	Isothermal	(P)	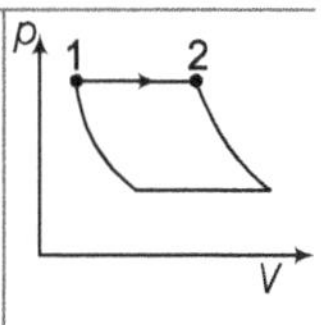
(II)	$W_{1\to2} = -pV_2 + pV_1$	(ii)	Isochoric	(Q)	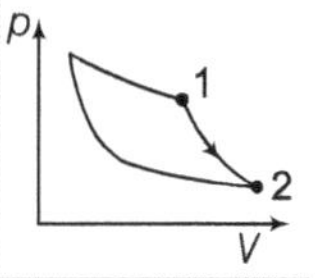
(III)	$W_{1\to2} = 0$	(iii)	Isobaric	(R)	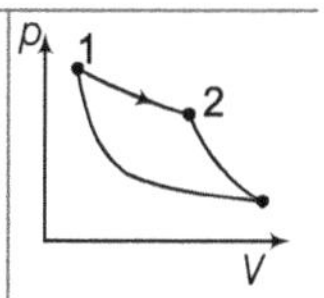
(IV)	$W_{1\to2} = -nRT\ln\left(\dfrac{V_2}{V_1}\right)$	(iv)	Adiabatic	(S)	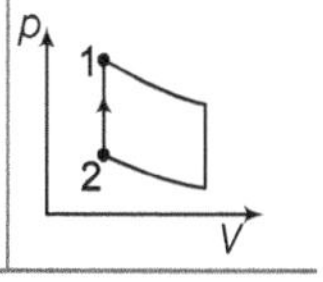

7. Which one of the following options correctly represents a thermodynamic process that is used as a correction in the determination of the speed of sound in an ideal gas?

(a) (IV) (ii) (R) (b) (I) (ii) (Q) (c) (I), (iv) (Q) (d) (III) (iv) (R)

8. Which of the following options is the only correct representation of a process in which $\Delta U = \Delta Q - p\Delta V$?

(a) (II) (iii) (S) (b) (II) (iii) (P) (c) (III) (iii) (P) (d) (II) (iv) (R)

9. Which one of the following options is the correct combination?

(a) (II) (iv) (P) (b) (III) (ii) (S) (c) (II) (iv) (R) (d) (IV) (ii) (S)

Paper 2

1. Consider regular polygons with number of sides $n = 3, 4, 5$ ...... as shown in the figure. The centre of mass of all the polygons is at height h from the ground. They roll on a horizontal surface about the leading vertex without slipping and sliding as depicted. The maximum increase in height of the locus of the centre of mass for each each polygon is Δ. Then, Δ depends on n and h as *(Single Option Correct)*

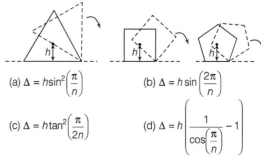

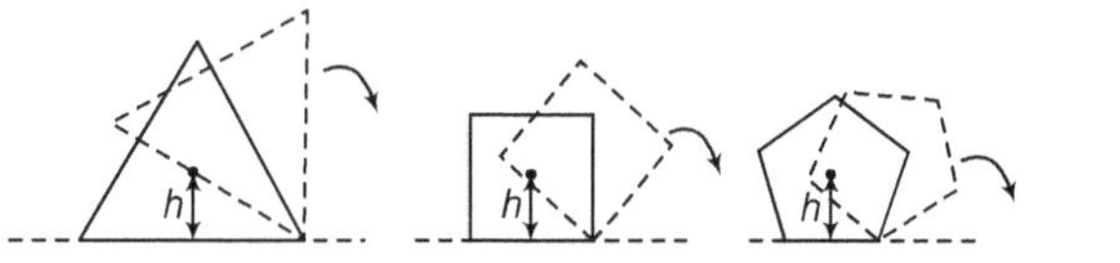

(a) $\Delta = h\sin^2\left(\dfrac{\pi}{n}\right)$ (b) $\Delta = h\sin\left(\dfrac{2\pi}{n}\right)$

(c) $\Delta = h\tan^2\left(\dfrac{\pi}{2n}\right)$ (d) $\Delta = h\left(\dfrac{1}{\cos\left(\dfrac{\pi}{n}\right)} - 1\right)$

2. Consider an expanding sphere of instantaneous radius R whose total mass remains constant. The expansion is such that the instantaneous density ρ remains uniform throughout the volume. The rate of fractional change in density $\left(\dfrac{1}{\rho}\dfrac{d\rho}{dt}\right)$ is constant. The velocity v of any point of the surface of the expanding sphere is proportional to *(Single Option Correct)*

(a) R (b) $\dfrac{1}{R}$ (c) R^3 (d) $R^{\frac{2}{3}}$

3. Three vectors **P**, **Q** and **R** are shown in the figure. Let S be any point on the vector **R**. The distance between the points P and S is $b[\mathbf{R}]$. The general relation among vectors **P**, **Q** and **S** is

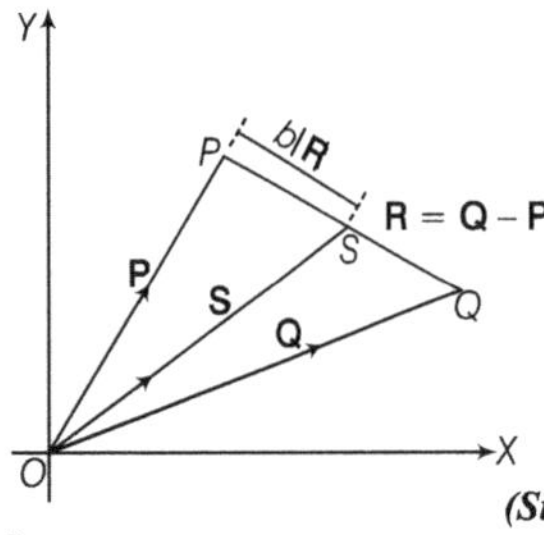

(Single Option Correct)

(a) $\mathbf{S} = (1 - b^2)\mathbf{P} + b\mathbf{Q}$ (b) $\mathbf{S} = (b - 1)\mathbf{P} + b\mathbf{Q}$
(c) $\mathbf{S} = (1 - b)\mathbf{P} + b\mathbf{Q}$ (d) $\mathbf{S} = (1 - b)\mathbf{P} + b^2\mathbf{Q}$

4. A person measures the depth of a well by measuring the time interval between dropping a stone and receiving the sound of impact with the bottom of the well. The error in his measurement of time is $\delta T = 0.01\text{s}$ and he measures the depth of the well to be $L = 20$ m. Take the acceleration due to gravity $g = 10$ ms^{-2} and the velocity of sound is 300 ms^{-1}. Then the fractional error in the measurement, $\frac{\delta L}{L}$, is closest to

(Single Option Correct)

(a) 1% (b) 5%
(c) 3% (d) 0.2%

5. A rocket is launched normal to the surface of the Earth, away from the Sun, along the line joining the Sun and the Earth. The Sun is 3×10^5 times heavier than the Earth and is at a distance 2.5×10^4 times larger than the radius of Earth. The escape velocity from Earth's gravitational field is $v_e = 11.2$ km s^{-1}. The minimum initial velocity (v_s) required for the rocket to be able to leave the Sun-Earth system is closest to

(Ignore the rotation and revolution of the Earth and the presence of any other planet) *(Single Option Correct)*

(a) $v_s = 72$ km s^{-1} (b) $v_s = 22$ km s^{-1}
(c) $v_s = 42$ km s^{-1} (d) $v_s = 62$ km s^{-1}

6. A wheel of radius R and mass M is placed at the bottom of a fixed step of height R as shown in the figure. A constant force is continuously applied on the surface of the wheel so that it just climbs the step without slipping. Consider the torque τ about an axis normal to the plane of the paper passing through the point Q. Which of the following options is/are correct?

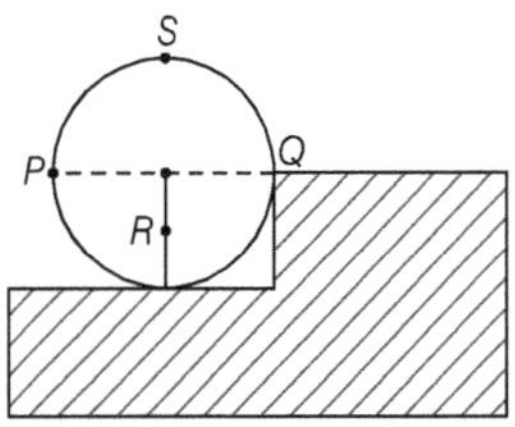

(One or More Than One Correct Option)

(a) If the force is applied normal to the circumference at point P, then τ is zero
(b) If the force is applied tangentially at point S, then $\tau \neq 0$ but the wheel never climbs the step
(c) If the force is applied at point P tangentially, then τ decreases continuously as the wheel climbs
(d) If the force is applied normal to the circumference at point X, then τ is constant

7. A rigid uniform bar AB of length L is slipping from its vertical position on a frictionless floor (as shown in the figure). At some instant of time, the angle made by the bar with the vertical is θ. Which of the following statements about its motion is/are correct?

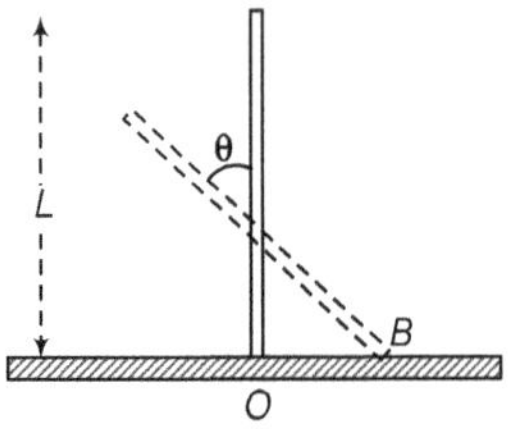

(One or More Than One Correct Option)

(a) Instantaneous torque about the point in contact with the floor is proportional to $\sin\theta$
(b) The trajectory of the point A is parabola
(c) The mid-point of the bar will fall vertically downward
(d) When the bar makes an angle θ with the vertical, the displacement of its mid-point from the initial position is proportional to $(1 - \cos\theta)$

Paragraph Based Questions (Q. No. 8-9)

One twirls a circular ring (of mass M and radius R) near the tip of one's finger as shown in Figure 1. In the process the finger never loses contact with the inner rim of the ring. The finger traces out the surface of a cone, shown by the dotted line. The radius of the path traced out by the point where the ring and the finger is in contact is r. The finger rotates with an angular velocity ω_0. The rotating ring rolls without slipping on the outside of a smaller circle described by the point where the ring and the finger is in contact (Figure 2). The coefficient of friction between the ring and the finger is μ and the acceleration due to gravity is g.

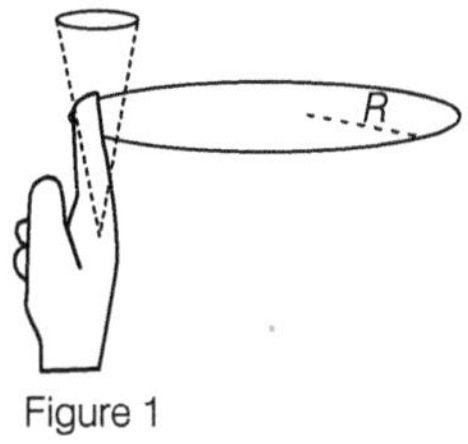

Figure 1

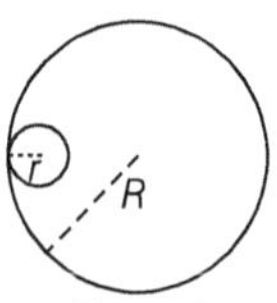

Figure 2

8. The total kinetic energy of the ring is

(a) $M\omega_0^2(R - r)^2$ (b) $\frac{1}{2}M\omega_0^2(R - r)^2$
(c) $M\omega_0^2R^2$ (d) $\frac{3}{2}M\omega_0^2(R - r)^2$

9. The minimum value of ω_0 below which the ring will drop down is

(a) $\sqrt{\frac{g}{2\mu(R - r)}}$ (b) $\sqrt{\frac{3g}{2\mu(R - r)}}$ (c) $\sqrt{\frac{g}{\mu(R - r)}}$ (d) $\sqrt{\frac{2g}{\mu(R - r)}}$

BITSAT

1. If temperature of a black body increases from 300 K to 900 K, then the rate of energy radiation increases by

(a) 81 (b) 3 (c) 9 (d) 2

2. A whistle of frequency 500 Hz tied to the end of a string of length 1.2 m revolves at 400 rev/min. A listener standing some distance away in the plane of rotation of whistle hears frequencies in the range.
(Speed of sound = 340 m/s)

(a) 436 to 574 (b) 426 to 586 (c) 426 to 574 (d) 436 to 586

3. At what angle θ to the horizontal should an object is projected so that the maximum height reached is equal to the horizontal range?

(a) $\tan^{-1}(2)$ (b) $\tan^{-1}(4)$ (c) $\tan^{-1}\left(\frac{2}{3}\right)$ (d) $\tan^{-1}(3)$

4. A body of mass 1 kg is executing simple harmonic motion. Its displacement y (cm) at t seconds is given by $y = 6\sin\left(100t + \frac{\pi}{4}\right)$.
Its maximum kinetic energy is

(a) 6 J (b) 18 J (c) 24 J (d) 36 J

5. Two blocks A and B are placed one over the other on a smooth horizontal surface. The maximum horizontal force that can be applied on lower block B, so that A and B move without separation is 49 N. The coefficient of friction between A and B is

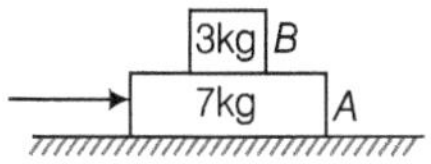

(a) 0.2 (b) 0.3 (c) 0.5 (d) 0.8

6. An aeroplane is flying in a horizontal direction with a velocity u and at a height of 2000 m. When it is vertically below a point A on the ground a food packet is released from it. The packet strikes the ground at point B.
If $AB = 3$ km and $g = 10$ m/s^2, then the value of u is

(a) 54 km/h (b) 540 km/h (c) 150 km/h (d) 300 km/h

7. A mild steel wire of length $2L$ and cross-sectional area A is stretched, well with in the elastic limit, horizontally between two pillars as shown in figure. A mass m is suspended from the mid-point of the wire strain in the wire is

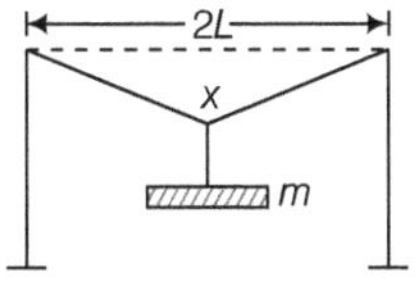

(a) $\frac{x^2}{2L^2}$ (b) $\frac{x}{L}$ (c) $\frac{x^2}{L}$ (d) $\frac{x^2}{2L}$

8. An ice-berg of density 900 kgm^{-3} is floating in water of density 1000 kgm^{-3}. The percentage of volume of ice-berg outside the water is

(a) 20% (b) 35% (c) 10% (d) 11%

9. The horizontal range and maximum height attained by a projectile are R and H respectively. If a constant horizontal acceleration $a = \frac{g}{4}$ is imparted to the projectile due to wind, then its horizontal range and maximum height will be

(a) $(R+H), \frac{H}{2}$ (b) $\left(R + \frac{H}{2}\right), 2H$
(c) $(R+2H), H$ (d) $(R+H), H$

10. A balloon is filled at 27° C and 1 atm pressure by 500 m^3 He. At – 3° C and 0.5 atm pressure, the volume of He will be

(a) 700 m^3 (b) 900 m^3
(c) 1000 m^3 (d) 500 m^3

11. The ratio of intensity at the centre of a bright fringe to the intensity at a point distance one-fourth of the distance between two successive bright fringes will be

(a) 4 (b) 3 (c) 2 (d) 1

12. A rectangular block of mass m and area of cross-section A floats in a liquid of density ρ. If it is given a vertical displacement from equilibrium, it undergoes oscillation with a time period T. Then

(a) $T \propto \sqrt{\rho}$ (b) $T \propto \frac{1}{\sqrt{A}}$ (c) $T \propto \frac{1}{\sqrt{\rho}}$ (d) $T \propto \frac{1}{\sqrt{m}}$

13. A load of mass m falls from a height h on the scale pan hung from a spring as shown. If the spring constant is k and the mass of the scale pan is zero and the mass m does not bounce relative to the pan, then the amplitude of vibration is

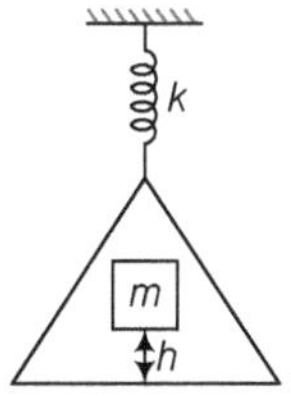

(a) $\frac{mg}{k}$ (b) $\frac{mg}{k}\sqrt{1 + \frac{2hk}{mg}}$
(c) $\frac{mg}{k} + \frac{mg}{k}\sqrt{\frac{1+2hk}{mg}}$ (d) None of these

14. A block A of mass 100 kg rests on another block B of mass 200 kg and is tied to a wall as shown in the figure. The coefficient of friction between A and B is 0.2 and that between B and ground is 0.3. The minimum force required to move the block B is ($g = 10$ ms^{-2})

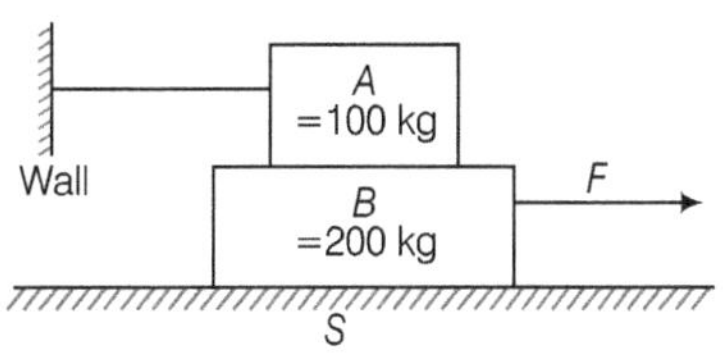

(a) 900 N (b) 200 N
(c) 1100 N (d) 700 N

15. A uniform rod of length l and mass m is free to rotate in a vertical plane about A. The rod initially in horizontal position is released. The initial angular acceleration of the rod is (Moment of inertia of rod about A is $\frac{ml^2}{3}$)

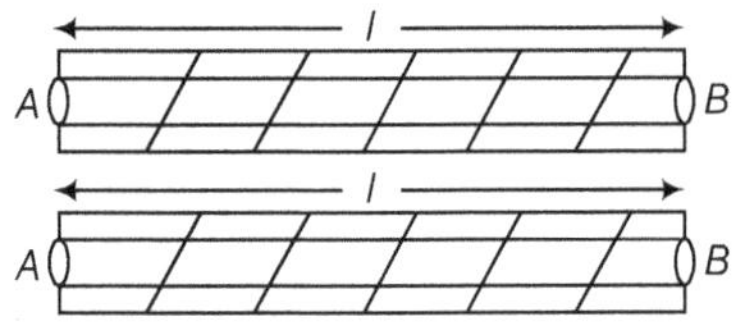

(a) $\frac{3g}{2l}$ (b) $\frac{2l}{3g}$ (c) $\frac{3g}{2l^2}$ (d) $mg\frac{l}{2}$

16. Work done in increasing the size of a soap bubble from radius of (3 to 5) cm is nearly (surface tension of soap solution = 0.03 Nm^{-1})

(a) 0.2 π mJ (b) 2π mJ (c) 0.4 mJ (d) 4π mJ

17. The velocity of a projectile at the initial point A is $(2\hat{i} + 3\hat{j})$ m/s. Its velocity (in m/s) at point B is

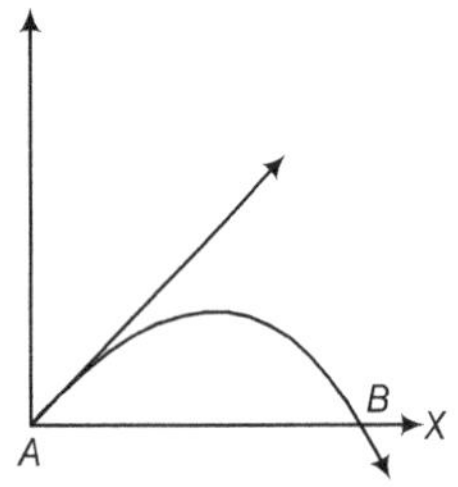

(a) $-2\hat{i} - 3\hat{j}$ (b) $-2\hat{i} + 3\hat{j}$ (c) $2\hat{i} - 3\hat{j}$ (d) $2\hat{i} + 3\hat{j}$

18. Two simple pendulums of lengths 5m and 20m respectively are given small displacement in one direction at the same time they will again be in the same sense when the pendulum of shorter length has completed n oscillations. Then, n is

(a) 5 (b) 1 (c) 2 (d) 3

19. A particle moving along X-axis has acceleration f, at time t given by $f = f_0\left(1 - \frac{t}{T}\right)$, where f_0 and T are constants. The particle at $t = 0$ and the instant when $f = 0$, the particle's velocity v_X is

(a) f_0T (b) $\frac{1}{2}f_0T^2$

(c) f_0T^2 (d) $\frac{1}{2}f_0T$

20. A geostationary satellite orbits around the earth in a circular orbit of radius 36000 km. Then, the time period of a sky satellite orbiting a few 100 km above the earth's surface ($R = 64000$ km) will approximately be

(a) $\frac{1}{2}$ h (b) 1 h (c) 2 h (d) 4 h

21. A transverse wave propagating on a stretched string of linear density 3×10^{-4} kg m^{-1} is represented by the equation

$$y = 0.2 \sin(1.5x + 60t)$$

where, x in metres and t is in seconds. The tension in the string (in Newton) is

(a) 0.24 (b) 0.48 (c) 1.20 (d) 1.80

KERALA CEE

1. A person observes that the full length of a train subtends an angle of 15°. If the distance between the train and the person is 3 km, the length of the train, calculated using parallax method, in meters is

(a) 45 (b) 45 π (c) 250 π
(d) 75 π (e) 450

2. In a measurement, random error

(a) can be decreased by increasing the number of readings and averaging them
(b) can be decreased by changing the person who takes the reading
(c) can be decreased by using new instrument
(d) can be decreased by using a different method in taking the reading
(e) cannot be decreased

3. In order to measure the period of a single pendulum using a stop clock, a student repeated the experiment for 10 times and noted down the time period for each experiment as 5.1, 5.0, 4.9, 4.9, 5.1, 5.0, 4.9, 5.1, 5.0, 4.9 s. The correct way of expressing the result for the period is

(a) 4.99 s (b) 5.0 s (c) 5.00 s (d) 4.9 s (e) 5.1 s

4. The following figure gives the movement of an object. Select the correct statement from the given choices.

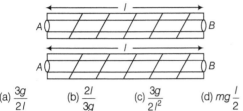

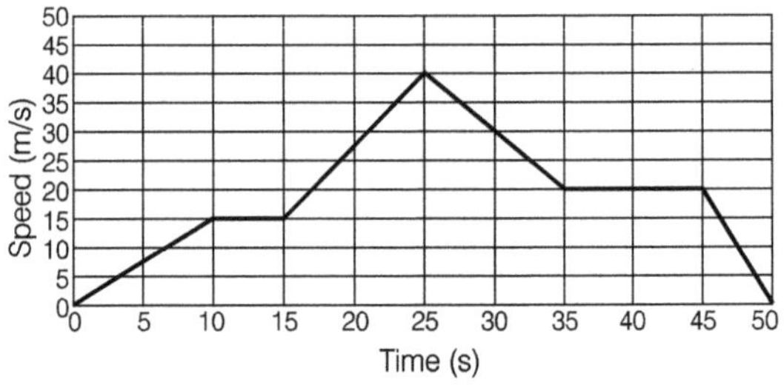

(a) The total distance travelled by the object is 975 m
(b) The maximum acceleration of the object is $2m/s^2$
(c) The maximum deceleration happend between 25th and 85th seconds
(d) The object was at rest between 10th and 15th seconds
(e) At 40th second, the speed of object was decelerating

5. Two object P and Q, travelling in the same direction starts from rest. While the object P starts at time $t = 0$ and object Q starts later at $t = 30$ min. The object P has an acceleration of 40 km/h^2. To catch P at a distance of 20 km, the acceleration of Q should be
(a) 40 km/h^2 (b) 80 km/h^2 (c) 100 km/h^2
(d) 120 km/h^2 (e) 160 km/h^2

6. A train of length L move with a constant speed V_t. A person at the back of the train fires a bullet at time $t = 0$ towards a target which is at a distance of D (at time $t = 0$) from the front of the train (on the same direction of motion). Another person at the front of the train fires another bullet at time $t = T$ towards the same target. Both bullets reach the target at the same time. Assuming the speed of the bullets V_b are same, the length of the train is
(a) $T \times (V_b \times 2V_t)$ (b) $T \times (V_b + V_t)$
(c) $2 \times T \times (V_b + 2V_t)$ (d) $2 \times T \times (V_b - 2V_t)$
(e) $T \times (V_b - 2V_t)$

7. From the ground, a projectile is fired at an angle of 60 degrees to the horizontal with a speed of 20 m/s. Take, acceleration due to gravity as 10 m/s^2. The horizontal range of the projectile is
(a) $10\sqrt{3}$ m (b) 20 m (c) $20\sqrt{3}$ m (d) $40\sqrt{3}$ m
(e) $400\sqrt{3}$ m

8. A person from a truck, moving with a constant speed of 60 km/h, throws a ball upwards with a speed of 60 km/h. Neglecting the effect of Earth and choose the correct answer from the given choice.
(a) The person cannot catch the ball when it comes down since the truck is moving
(b) The person can catch the ball when it comes down, if the truck is stopped immediately after throwing the ball
(c) The person can catch the ball when it comes down, if the truck moves with speed less then 60 km/h but does not stop
(d) The person can catch the ball when it comes down, if the truck moves with speed more than 60 km/h
(e) The person can catch the ball when it comes down, if the truck continues to move with a constant speed of 60 km/h

9. A body of mass $2m$ moving with velocity v makes a head on elastic collision with another body of mass m which is initially at rest. Loss of kinetic energy of the colliding body (mass $2m$) is
(a) 1/9 of its initial kinetic energy
(b) 1/6 of its initial kinetic energy
(c) 1/4 of its initial kinetic energy
(d) 1/2 of its initial kinetic energy
(e) 8/9 of its initial kinetic energy

10. Displacement x (in meters), of body of mass 1 kg as a function of time t, on a horizontal smooth surface is given as $x = 2t^2$. The work done in the first one second by the external force is
(a) 1 J (b) 2 J (c) 4 J (d) 8 J (e) 16 J

11. A massless spring of length l and spring constant k is placed vertically on a table. A ball of mass m is just kept on top of the spring. The maximum velocity of the ball is
(a) $g\sqrt{\frac{m}{k}}$ (b) $g\sqrt{\frac{2m}{k}}$ (c) $2g\sqrt{\frac{m}{k}}$ (d) $\frac{g}{2}\sqrt{\frac{m}{k}}$
(e) $g\sqrt{\frac{m}{2k}}$

12. Under the action of a constant force, a particle is experiencing a constant acceleration. The power is
(a) zero
(b) positive constant
(c) negative constant
(d) increasing uniformly with time
(e) decreasing uniformly with time

13. A comet orbits around the Sun in an elliptical orbit. Which of the following quantities remains constant during the course of its motion?
(a) Linear velocity (b) Angular velocity
(c) Angular momentum (d) Kinetic energy
(e) Potential energy

14. Consider a satellite moving in a circular orbit around Earth. If K and V denote its kinetic energy and potential energy respectively, then (Choose the convention, where $V = 0$ as $r \to \infty$)
(a) $K = V$ (b) $K = 2V$
(c) $V = 2K$ (d) $K = -2V$
(e) $V = -2K$

15. Assuming the mass of Earth to be ten times the mass of Mars, its radius to be twice the radius of Mars and the acceleration due to gravity on the surface of Earth is 10 m/s^2. Then the acceleration due to gravity on the surface of Mars is given by
(a) 0.2 m/s^2 (b) 0.4 m/s^2
(c) 2 m/s^2 (d) 4m/s^2
(e) 5 m/s^2

16. The semi-major axis of the orbit of Saturn is approximately nine times that of Earth. The time period of revolution of Saturn is approximately equal to
(a) 81 years (b) 27 years
(c) 729 years (d) $\sqrt[3]{81}$ years
(e) 9 years

17. A particle of mass 3 kg, attached to a spring with force constant 48 N/m execute simple harmonic motion on a frictionless horizontal surface. The time period of oscillation of the particle, in seconds, is
(a) $\pi/4$ (b) $\pi/2$ (c) 2π (d) 8π (e) $\pi/8$

18. The position and velocity of a particle executing simple harmonic motion at $t = 0$ are given by 3 cm/s and 8 cm/s respectively. If the angular frequency of the particle is 2 rad/s, then the amplitude of oscillation, in centimeters, is
(a) 3 (b) 4 (c) 5 (d) 6 (e) 8

19. A simple harmonic motion is represented by $x(t) = \sin^2\omega t - 2\cos^2\omega t$. The angular frequency of oscillation is given by

(a) ω (b) 2ω (c) 4ω (d) $\omega/2$ (e) $\omega/4$

20. A transverse wave in propagating on a stretched string of mass per unit length 32 g/m. The tension on the string is 80 N. The speed of the wave over the string is

(a) 5/2 m/s (b) $\sqrt{5/2}$ m/s (c) 2/5 m/s (d) $\sqrt{2/5}$ m/s
(e) 50 m/s

21. Consider the propagating sound (with velocity 330 m/s) in a pipe of length 1.5 m with one end closed and the other open. The frequency associated with the fundamental mode is

(a) 11 Hz (b) 55 Hz
(c) 110 Hz (d) 165 Hz
(e) 275 Hz

22. A standing wave propagating with velocity 300 m/s in an open pipe of length 4 m has four nodes. The frequency of the wave is

(a) 75 Hz (b) 100 Hz (c) 150 Hz (d) 300 Hz
(e) 600 Hz

23. Consider the vehicle emitting sound wave of frequency 700 Hz moving towards an observer at a speed 22 m/s. Assuming the observer as well as the medium to be at rest and velocity of sound in the medium to be 330 m/s, the frequency of sound as measured by the observer is

(a) 2525/4 Hz (b) 1960/3 Hz
(c) 2240/3 Hz (d) 750 Hz
(e) 5625/7 Hz

24. The x-t plot shown in the figure below describes the motion of the particle, along x-axis, between two positions A and B. The particle passes through two intermediate points P_1 and P_2 as shown in the figure.

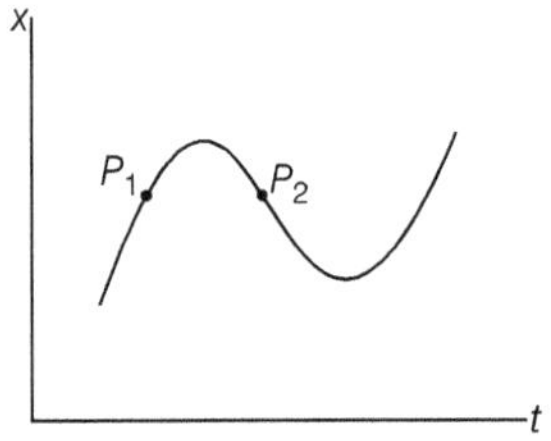

(a) The instantaneous velocity is positive as P_1 and negative at P_2.
(b) The instantaneous velocity is negative at both P_1 and P_2
(c) The instantaneous velocity is negative at P_1 and positive at P_2
(d) The instantaneous velocity is positive at both P_1 and P_2
(e) The instantaneous velocity is always positive

25. A ball falls from a table top with initial horizontal speed V_0. In the absence of air resistance,which of the following statement is correct

(a) The vertical component of the acceleration changes with time
(b) The horizontal component of the velocity does not changes with time
(c) The horizontal component to the acceleration is non zero and finite
(d) The time taken by the ball to touch the ground depends on V_0
(e) The vertical component of the acceleration varies with time

26. A man of mass 60 kg climbed down using an elevator.The elevator had an acceleration $4\ ms^{-2}$. If the acceleration due to gravity is $10\ ms^{-2}$, the man's apparent weight on his way down is

(a) 60 N (b) 240 N (c) 360 N (d) 840 N
(d) 3600 N

27. A uniform rod of length of 1 m and mass of 2 kg is attached to a side support at O as shown in the figure. The rod is at equilibrium due to upward force T acting at P. Assume the acceleration due to gravity as $10\ m/s^2$. The value of T is

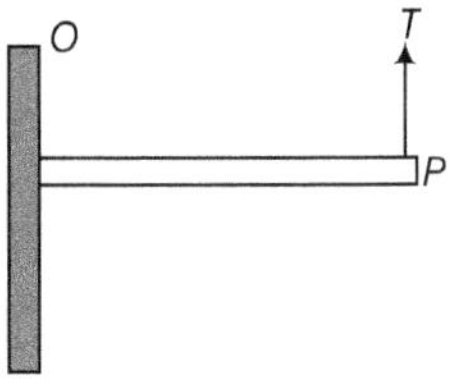

(a) 0 (b) 2 N (c) 5 N (d) 10 N (e) 20 N

28. A capillary tube of radius 0.5 mm is immersed in a beaker of mercury. The level inside he tube is 0.8 cm below the level in beaker and angle of contact is 120°. What is the surface tension of mercury, if the mass density of mercury is $\rho = 13.6 \times 10^3\ kgm^{-3}$ and acceleration due to gravity is $g = 10\ m/s^2$?

(a) 0.225 N/m (b) 0.544 N/m
(c) 0.285 N/m (d) 0.375 N/m
(e) 0.425 N/m

29. Which of the following statement related to stress-strain relation is correct?

(a) Stress is linearly proportional to strain irrespective of the magnitude of the strain
(b) Stress is linearly proportional to strain above
(c) Stress is linearly proportional to strain for stress much smaller than at the yield point
(d) Stress–strain curve is same for all materials
(e) Stress is inversely proportional to strain

30. The lower edge of a square slab of side 50 cm and thickness 20 cm is rigidly fixed to the base of a table. A tangential force of 30 N is applied to the slab. If the shear moduli of the material is $4 \times 10^{10}\ N/m^2$, then displacement of the upper edge, in meters is

(a) 4×10^{-12} (b) 4×10^{-10} (c) 6×10^{-10} (d) 6×10^{-12}
(e) 8×10^{-10}

31. Initially a beaker has 100 g of water at temperature 90°C. Later another 600 g of water at temperature 20°C was poured into the beaker. The temperature, T of the water after mixing is

(a) 20°C (b) 30°C (c) 45°C (d) 55°C (e) 90°C

32. Match the following

I. Isothermal process	1. $\Delta Q = 0$
II. Isobaric process	2. $\Delta V = 0$
III. Isochoric process	3. $\Delta P = 0$
IV. Adiabatic process	4. $\Delta T = 0$

(a) I–4, II–3, III–2, IV–1 (b) I–3, II–2, III–1, IV–4
(c) I–1, II–2, III–3, IV–4 (d) I–4, II–2, II–3, IV–1
(e) I–1, II–4, III–2, IV–3

33. For an ideal gas, the specific heat at constant pressure C_p is greater than the specific heat at constant volume C_v. This is because
(a) There is a finite work done by the gas on its environment when its temperature is increased while the pressure remains constant
(b) There is a finite work done by the gas on its environment when its pressure is increased while the volume remains constant
(c) There is a finite work done by the gas on its environment when its pressure is increased while the temperature remains constant
(d) The pressure of the gas remains constant when its temperature remains constant
(e) The internal energy of the gas at constant pressure is more than at constant volume.

34. Which of the following statement is correct?
(a) Light waves are transverse but sound waves are waves on strings are longitudinal
(b) Sound waves and waves on a string and transverse but light waves are longitudinal
(c) Light waves and waves on a string are transverse but sound waves are longitudinal
(d) Light waves and sound waves are transverse but waves on string are longitudinal
(e) Light waves, sound waves and waves on a string are all longitudinal

35. In Young's double slits experiment, if the separation between the slits is halved, and the distance between the slits and the screen is doubled, then the fringe width compared to the original one will be
(a) Unchanged (b) Halved
(c) Doubled (d) Quadrupled
(e) Fringes will disappear

36. The phase velocity of a wave described by the equations $\psi = \psi_0 \sin(kx + \omega t + \pi/2)$ is
(a) x/t (b) ψ_0/ω (c) ω/k (d) $\pi/2k$ (e) ψ_0

37. Which of the following particle when bombards on ^{65}Cu will turn into ^{66}Cu?
(a) Proton (b) Neutron
(c) Electron (d) Alpha particle
(e) Deutron

38. If the rms value of sinusoidal input to a full wave rectifier is $V_0/\sqrt{2}$, then the rms value of the rectifier's output is
(a) $V_0/\sqrt{2}$ (b) $V_0^2/\sqrt{2}$ (c) $V_0^2/2$ (d) $\sqrt{2}V_0^2$ (e) $2V_0^2$

39. 8 g of Cu^{66} undergoes radioactive decay and after 15 minutes only 1 g remains. The half-life, in minutes, is then
(a) 15 ln (2)/ln (8) (b) 15 ln (8)/ln (2)
(c) 15/8 (d) 8/15
(e) 15 ln (2)

40. For a light nuclei, which of the following relation between the atomic number (Z) and mass number (A) is valid?
(a) $A = Z/2$ (b) $Z = A$ (c) $Z = A/2$ (d) $Z = A^2$ (e) $A = Z^2$

41. A wheel rotating at 12 rev/s is brought to rest in 6 s. The average angular deceleration in rad/s^2 of the wheel during this process is
(a) 4π (b) 4 (c) 72 (d) $1/\pi$ (e) π

42. A torque of 1 N-m is applied to a wheel which is at rest. After 2 second the angular momentum in kg-m^2/s is
(a) 0.5 (b) 1 (c) 2 (d) 4 (e) 3

Karnataka CET

1. A car moving with a velocity of 20 ms^{-1} stopped at a distance of 40 m. If the same car is travelling at double the velocity, the distance travelled by it for same retardation is
(a) 320 m (b) 1280 m
(c) 160 m (d) 640 m

2. A substance of mass 49.53 g occupies 1.5 cm^3 of volume. The density of the substance (in $g\,cm^{-3}$) with correct number of significant figures is
(a) 3.3 (b) 3.300
(c) 3.302 (d) 3.30

3. Two balls are thrown simultaneously in air. The acceleration of the centre of mass of the two balls when in air
(a) is equal to g (acceleration due to gravity)
(b) depends on the speeds of the two balls
(c) depends on the masses of the two balls
(d) depends on the direction of motion of the two balls.

4. A body of the mass 50 kg is suspended using a spring balance inside a lift at rest. If the lift starts falling freely, the reading of the spring balance is
(a) < 50 kg (b) = 50 kg (c) > 50 kg (d) = 0

5. The S.I. unit of specific heat capacity is
(a) J K^{-1} (b) J kg^{-1}
(c) J $mol^{-1}K^{-1}$ (d) J $kg^{-1}K^{-1}$

6. For which combination of working temperatures, the efficiency of Carnot's engine is the least?
(a) 100 K, 80K (b) 40 K, 20 K
(c) 80 K , 60 K (d) 60 K , 40 K

7. According to Huygens' principle, during refraction of light from air to a denser medium
(a) Wavelength decreases but speed increases
(b) Wavelength increases but speed decreases
(c) Wavelength and speed increases
(d) Wavelength and speed decreases

8. The value of acceleration due to gravity at a depth of 1600 km is equal to
(a) $4.9\ ms^{-2}$ (b) $9.8\ ms^{-2}$
(c) $7.35\ ms^{-2}$ (d) $19.6\ ms^{-2}$

9. Two simple pendulums A and B are made to oscillate simultaneously and it is found that A completes 10 oscillations in 20 sec and B completes 8 oscillations in 10 sec. The ratio of the length of A and B is
(a) $\frac{25}{64}$ (b) $\frac{64}{25}$ (c) $\frac{8}{5}$ (d) $\frac{5}{4}$

10. A motor pump lifts 6 tonnes of water from a well of depth 25 m to the first floor of height 35 m from the ground floor in 20 minutes. The power of the pump (in kW) is $[g = 10\ ms^{-2}]$
(a) 3 (b) 12 (c) 1.5 (d) 6

11. The waves set up in a closed pipe are
(a) longitudinal and progressive
(b) transverse and progressive
(c) transverse and stationary
(d) longitudinal and stationary

12. If $\mathbf{A} = 2\hat{\mathbf{i}} + 3\hat{\mathbf{j}} + 8\hat{\mathbf{k}}$ is perpendicular to $\mathbf{B} = 4\hat{\mathbf{j}} - 4\hat{\mathbf{i}} + \alpha\hat{\mathbf{k}}$, then the value of α is
(a) $-\frac{1}{2}$ (b) $\frac{1}{2}$ (c) 1 (d) -1

13. The angle between velocity and acceleration of a particle describing uniform circular motion is
(a) 180° (b) 90° (c) 45° (d) 60°

14. 'Young's modulus' is defined as the ratio of
(a) hydraulic stress and hydraulic strain
(b) shearing stress and shearing strain
(c) tensile stress and longitudinal strain
(d) bulk stress and longitudinal strain

15. A piece of copper is to be shaped into a conducting wire of maximum resistance. The suitable length and diameter are .... and .... respectively.
(a) $2L$ and $d/2$ (b) $L/2$ and $2d$
(c) L and d (d) $2L$ and d

16. During scattering of light, the amount of scattering in inversely proportional to .... of wavelength of light.
(a) square (b) fourth power
(c) half (d) cube

17. 'Hydraulic lift' works on the basis of
(a) Stoke's law (b) Bernoulli's law
(c) Pascal's law (d) Toricelli's law

18. The mean energy of a molecule of an ideal gas is
(a) $\frac{1}{2}KT$ (b) $2KT$ (c) KT (d) $\frac{3}{2}KT$

Andhra Pradesh EAMCET

1. A monoatomic ideal gas goes through a cyclic process as shown in the figure. The efficiency of this process is

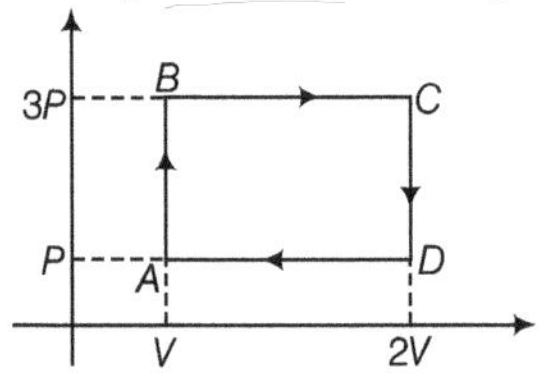

(a) 78% (b) 42% (c) 62% (d) 21%

2. Two situations are shown in fig. (a) and (b).

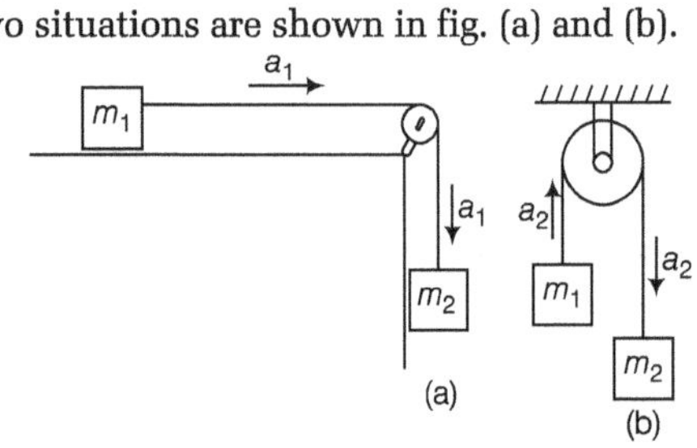

In each case, $m_1 = 3$ kg and $m_2 = 4$ kg. If a_1, a_2 are the respective accelerations of the blocks in these situations, then the values of a_1 and a_2 are respectively $[g = 10\ ms^{-2}]$

(a) $\frac{20}{7}\ ms^{-2}, \frac{10}{7}\ ms^{-2}$ (b) $\frac{10}{7}\ ms^{-2}, \frac{25}{7}\ ms^{-2}$
(c) $\frac{40}{7}\ ms^{-2}, \frac{10}{7}\ ms^{-2}$ (d) $\frac{30}{7}\ ms^{-2}, \frac{5}{7}\ ms^{-2}$

3. Three uniform thin aluminum rods each of length 2 m form an equilateral triangle PQR as shown in the figure. The mid point of the rod PQ is at the origin of the coordinate system. If the temperature of the system of rods increases by 50°C, the increase in y-coordinate of the centre of mass of the system of the rods is .... mm. (Coefficient of volume expansion of aluminium $= 12\sqrt{3} \times 10^{-6}\ K^{-1}$)

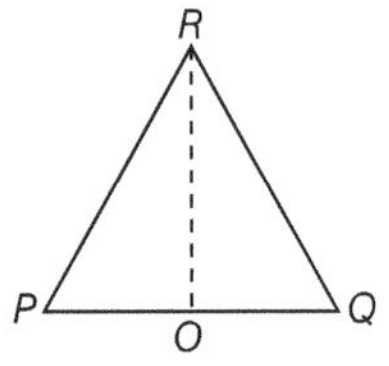

(a) 0.05 (b) 0.8 (c) 0.1 (d) 0.2

4. A body is projected from the top of a tower with a velocity $\overline{u} = 3\hat{\mathbf{i}} + 4\hat{\mathbf{j}} + 5\hat{\mathbf{k}}\ ms^{-1}$, where $\hat{\mathbf{i}}, \hat{\mathbf{j}}$ and $\hat{\mathbf{k}}$ are unit vectors along east, north and vertically upwards respectively. If the height of the tower is 30 m, horizontal range of the body on the ground is $(g = 10\ ms^{-2})$
(a) 15 m (b) 25 m
(c) 9 m (d) 12 m

5. Equation of a projectile is given by $y = Px - Qx^2$, where P and Q are constants. The ratio of maximum height to the range of the projectile is

(a) $\frac{Q^2}{2P}$ (b) $\frac{P^2}{Q}$ (c) $4P$ (d) $\frac{P}{4}$

6. The transverse displacement of a string of a linear density 0.01 kg m^{-1}, clamped at its ends is given by $Y_{(x,t)} = 0.03 \sin\left(\frac{2\pi x}{3}\right) \cos(60\pi t)$, where x and y are in metres and time t is in seconds. Tension in the string is

(a) 9 N (b) 36 N
(c) 162 N (d) 81 N

7. A girl of mass 50 kg swinging on a cradle. If she moves with a velocity of 2 ms^{-1} upwards in a direction making an angle 60° with the vertical, then the power generated is ($g = 9.8$ ms^{-2})

(a) 245 W (b) $490\sqrt{2}$ W (c) $490\sqrt{3}$ W (d) 980 W

8. A wall is made of equally thick layers P and Q of different materials. Thermal conductivity of Q is half of that of the P. In the steady state, If the temperature difference across the wall is 24°C, then the temperature difference across the layer 'P' is ....

(a) 12° C (b) 16°C (c) 4° C (d) 8°C

9. Two bodies of masses 4m and 9m are seperated by a distance 'r'. The gravitational potential at a point on this line joining them where the gravitational field becomes zero is

(a) $\frac{-25\,Gm}{r}$ (b) $\frac{-4Gm}{r}$ (c) $\frac{-9\,Gm}{r}$ (d) $\frac{-13\,Gm}{r}$

10. A bird is tossing (flying to and pro) between two cars moving towards each other on a straight road. One car has speed of 54 kmh^{-1} while the other has the speed of 36 kmh^{-1}. The bird starts moving from first car towards the other and is moving with the speed of 36 kmh^{-1} where the two cars were separated by 36 km. The total distance covered by the bird before the cars meet each other is

(a) 14400 m (b) 1440 m (c) 244 m (d) 24400 m

11. If the average translational kinetic energy of a molecule in a gas is equal to the kinetic energy of an electron accelerating from rest through 10 V, then the temperature of the gas molecule is

(Boltzmann constant = 1.38×10^{-23} JK^{-1})

(a) 7.73×10^3 K (b) 730 K
(c) 73.7 K (d) 77.3×10^3 K

12. A closed organ pipe of length 'L' and an open organ pipe contain gases of densities ρ_1 and ρ_2 respectively. The compressibility of gases are equal in both the pipes. If the frequencies of their first overtones are same, then the length of the open organ pipe is

(a) $\frac{4L}{3}\sqrt{\frac{\rho_2}{\rho_1}}$ (b) $\frac{4L}{3}\sqrt{\frac{\rho_1}{\rho_2}}$ (c) $\frac{4L}{3}$ (d) $\frac{L}{3}$

13. One mole of a gas expands such that its volume 'V' changes with absolute temperature 'T' in accordance with the relation $V = KT^2$ where 'K' is a constant. If the temperature of the gas changes by 60°C, then work done by the gas is (R is universal gas constant).

(a) $KR \ln 60$ (b) $R \ln 60$ (c) $40KR$ (d) $120\,R$

14. Time period of a simple pendulum of length 'L' is T_1. Time period of a uniform rod of same length 'L' suspended from one end and oscillating in a vertical plane is T_2. Amplitude of oscillation is small in both the cases. Then $\frac{T_1}{T_2}$ is

(a) $\sqrt{\frac{2}{3}}$ (b) $\sqrt{\frac{3}{2}}$ (c) $\sqrt{\frac{4}{3}}$ (d) 1

15. A uniform thin rod of 120 cm length and 1600 g mass is bent as shown in the figure. The moment of inertia of the bent rod about an axis passing through the point 'O' and perpendicular to the plane of the paper is .... kg-m^2.

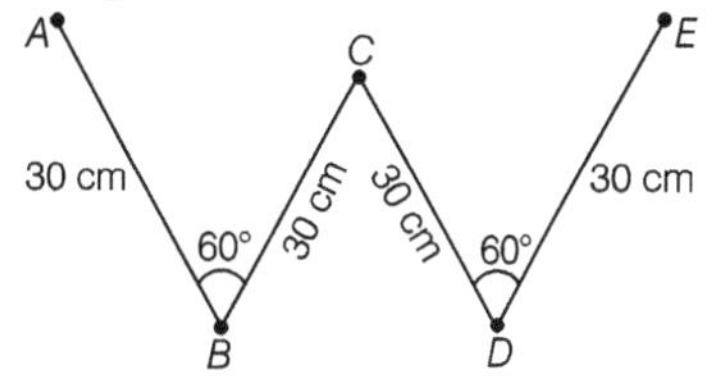

(a) 0.084 (b) 0.360
(c) 0.018 (d) 0.120

16. Two blocks of masses 'M' and 'm' are placed on one another on a smooth horizontal surface as shown in the figure.

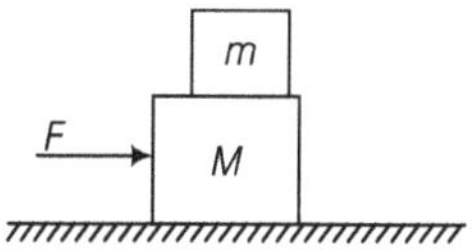

The force 'F' is acting on the mass 'M' horizontally during time interval 't'. Assuming no relative sliding between the blocks, the work done by friction on the blocks is ....

(a) $\frac{Ft}{2(M+m)}$ (b) $\frac{M+m}{mt^2}$
(c) $\frac{mF^2t^2}{2(M+m)^2}$ (d) $\frac{F^2t^2}{(M+m)}$

17. A solid sphere is projected up along an inclined plane of inclination 30° with the horizontal with a speed of 4 ms^{-1}. If it rolls without slipping, the maximum distance traversed by it is .... ($g = 10$ ms^{-2})

(a) 2.24 m (b) 112 m (c) 1.12 m (d) 22.4 m

18. Fully filled open water tank has two holes on either sides of its walls. One is a square hole of side x cm at a depth of 2 m from the top, and the other hole is equilateral triangle of side 4 cm at a depth of 6 m from the top. If the rate of flow of water is same from both the holes, then 'x' is

(a) 1.73 cm (b) 12 cm
(c) 6.92 cm (d) 3.46 cm

19. The ratios of lengths, areas of cross-section and Young's modulii of steel to that of brass wires shown in the figure are a, b and c respectively. The ratio of increase in the lengths of brass to that of steel wires is [Assume that the masses of steel and brass wires are negligible]

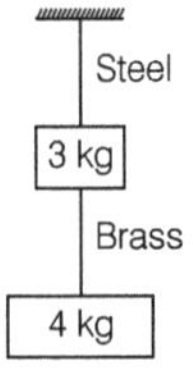

(a) $\frac{4a}{7bc}$ (b) $\frac{7bc}{4a}$ (c) $\frac{4bc}{7a}$ (d) $\frac{7a}{4bc}$

20. A person of 60 kg mass is in a lift which is coming down such that the man exerts a force of 150 N on the floor of the lift. Then the acceleration of the lift is $(g = 10\ \text{ms}^{-2})$

(a) $7.5\ \text{ms}^{-2}$
(b) $40.0\ \text{ms}^{-2}$
(c) $22.5\ \text{ms}^{-2}$
(d) $15.0\ \text{ms}^{-2}$

Telangana State EAMCET

1. A force **F** is applied on a square plate of length L. If the percentage error in the determination of L is 3% and in F is 4%, then permissible error in the calculation of pressure is

(a) 13% (b) 10% (c) 7% (d) 12%

2. A swimmer wants to cross a 200 m wide river which is flowing at a speed of 2 m/s. The velocity of the swimmer with respect to the river is 1 m/s. How far from the point directly opposite to the starting point does the swimmer reach the opposite bank?

(a) 200 m (b) 400 m (c) 600 m (d) 800 m

3. A simple pendulum of length 1 m is freely suspended from the ceiling of an elevator. The time period of small oscillations as the elevator moves up with an acceleration of $2\ \text{m/s}^2$ is (use $g = 10\text{m/s}^2$)

(a) $\frac{\pi}{\sqrt{5}}$ s (b) $\sqrt{\frac{2}{5}}\pi$ s (c) $\frac{\pi}{\sqrt{2}}$ s (d) $\frac{\pi}{\sqrt{3}}$ s

4. The Young's modulus of a material is $2 \times 10^{11}\ \text{N/m}^2$ and its elastic limit is $1 \times 10^8\ \text{N/m}^2$. For a wire of 1 m length of this material, the maximum elongation achievable is

(a) 0.2 mm (b) 0.3 mm (c) 0.4 mm (d) 0.5 mm

5. A wooden box lying at rest on an inclined surface of a wet wood is held at static equilibrium by a constant force **F** applied perpendicular to the incline. If the mass of the box is 1 kg, the angle of inclination is 30° and the coefficient of static friction between the box and the inclined plane is 0.2, the minimum magnitude of **F** is (Use $g = 10\ \text{m/s}^2$)

(a) 0 N, as 30° is less than angle of repose
(b) ≥ 1 N
(c) ≥ 3.3 N
(d) ≥ 16.3 N

6. A meter scale made of steel, reads accurately at 25°C. Suppose in an experiment an accuracy of 0.06 mm in 1 m is required, the range of temperature in which the experiment can be performed with this meter scale is (Coefficient of linear expansion of steel is $11 \times 10^{-6}/°\text{C}$)

(a) 19° C to 31° C (b) 25° C to 32° C
(c) 18° C to 25° C (d) 18° C to 32° C

7. A thermocol box has a total wall area (including the lid) of $1.0\ \text{m}^2$ and wall thickness of 3 cm. It is filled with ice at 0°C. If the average temperature outside the box is 30°C throughout the day, the amount of ice that melts in one day is

[Use $K_{\text{thermocol}} = 0.03$ W/mK,
$L_{\text{fusion (ice)}} = 3.00 \times 10^5$ J/kg]

(a) 1 kg (b) 2.88 kg (c) 25.92 kg (d) 8.64 kg

8. An object is thrown vertically upward with a speed of 30 m/s. The velocity of the object half-a-second before it reaches the maximum height is

(a) 4.9 m/s (b) 9.8 m/s
(c) 19.6 m/s (d) 25.1 m/s

9. Consider the motion of a particle described by $x = a\cos t$, $y = a\sin t$ and $z = t$. The trajectory traced by the particle as a function of time is

(a) helix (b) circular
(c) elliptical (d) straight line

10. Consider a reversible engine of efficiency $\frac{1}{6}$. When the temperature of the sink is reduced by 62°C, its efficiency gets doubled. The temperature of the source and sink respectively are

(a) 372 K and 310 K (b) 273 K and 300 K
(c) 99°C and 10°C (d) 200°C and 37°C

11. An office room contains about 2000 moles of air. The change in the internal energy of this much air when it is cooled from 34°C to 24°C at a constant pressure of 1.0 atm is

[Use $\gamma_{\text{air}} = 1.4$ and universal gas constant $= 8.314$ J/mol-K]

(a) -1.9×10^5 J (b) $+1.9 \times 10^5$ J
(c) -4.2×10^5 J (d) $+0.7 \times 10^5$ J

12. A ball is thrown at a speed of 20 m/s at an angle of 30° with the horizontal. The maximum height reached by the ball is

(Use $g = 10\text{m/s}^2$)

(a) 2 m (b) 3 m (c) 4 m (d) 5 m

13. A horizontal pipeline carrying gasoline has a cross-sectional diameter of 5 mm. If the viscosity and density of the gasoline are 6×10^{-3} Poise and 720 kg/m^3 respectively, the velocity after which the flow becomes turbulent is

(a) > 1.66 m/s (b) > 3.33 m/s
(c) > 1.6×10^{-3} m/s (d) > 0.33 m/s

14. Which of the following principles is being used in Sonar Technology?

(a) Newton's laws of motion
(b) Reflection of electromagnetic waves
(c) Law's of thermodynamics
(d) Reflection of ultrasonic waves

15. A particle of mass M is moving in a horizontal circle of radius R with uniform speed v. When the particle moves from one point to a diametrically opposite point, its

(a) momentum does not change
(b) momentum changes by $2Mv$
(c) kinetic energy changes by $\frac{Mv^2}{4}$
(d) kinetic energy changes by Mv^2

16. A billiard ball of mass M, moving with velocity v_1 collides with another ball of the same mass but at rest. If the collision is elastic, the angle of divergence after the collision is

(a) 0° (b) 30° (c) 90° (d) 45°

17. A planet of mass m moves in a elliptical orbit around an unknown star of mass M such that its maximum and minimum distances from the star are equal to r_1 and r_2 respectively. The angular momentum of the planet relative to the centre of the star is

(a) $m\sqrt{\frac{2GMr_1r_2}{r_1 + r_2}}$ (b) 0
(c) $m\sqrt{\frac{2GM(r_1 + r_2)}{r_1r_2}}$ (d) $\sqrt{\frac{2GMmr_1}{(r_1 + r_2)r_2}}$

18. Consider a frictionless ramp on which a smooth object is made to slide down from an initial height h. The distance d necessary to stop the object on a flat track (of coefficient of friction μ), kept at the ramp end is

(a) h/μ (b) μh (c) $\mu^2 h$ (d) $h^2\mu$

19. A sound wave of frequency ν Hz initially travels a distance of 1 km in air. Then, it gets reflected into a water reservoir of depth 600 m. The frequency of the wave at the bottom of the reservoir is

(V_{air} = 340 m/s V_{water} = 1484 m/s)

(a) > ν Hz
(b) < ν Hz
(c) ν Hz
(d) 0 (the sound wave gets attenuated by water completely)

20. The deceleration of a car traveling on a straight highway is a function of its instantaneous velocity v given by $\omega = a\sqrt{v}$, where a is a constant. If the initial velocity of the car is 60 km/h, the distance of the car will travel and the time it takes before it stops are

(a) $\frac{2}{3}$ m, $\frac{1}{2}$ s (b) $\frac{3}{2a}$ m, $\frac{1}{2a}$ s (c) $\frac{3a}{2}$ m, $\frac{a}{2}$ s (d) $\frac{2}{3a}$ m, $\frac{2}{a}$ s

21. Consider a particle on which constant forces $\mathbf{F}_1 = \hat{\mathbf{i}} + 2\hat{\mathbf{j}} + 3\hat{\mathbf{k}}$ N and $\mathbf{F}_2 = 4\hat{\mathbf{i}} - 5\hat{\mathbf{j}} - 2\hat{\mathbf{k}}$ N act together resulting in a displacement from position $\mathbf{r}_1 = 20\,\hat{\mathbf{i}} + 15\hat{\mathbf{j}}$ cm to $\mathbf{r}_2 = 7\,\hat{\mathbf{k}}$ cm. The total work done on the particle is

(a) – 0.48 J (b) + 0.48 J
(c) – 4.8 J (d) + 4.8 J

VIT

1. A 5 kg stone falls from a height of 100 m and penetrates 2 m in a layer of sand. The time of penetration is

(a) 14.28 sec (b) 0.089 sec (c) 120 kg-m/s (d) 0.89 sec

2. An explosion breaks a rock into three parts in a horizontal plane. Two of them go off at height at right angles to each other. The first part of mass 1 kg moves with a speed of 12 ms^{-1} and the second part of mass 2 kg moves with 8 ms^{-1} speed. If the third part flies of with 4 ms^{-1} speed, then its mass is

(a) 3 kg (b) 5 kg
(c) 7 kg (d) 17 kg

3. A thick rope of rubber of density $1.5 \times 10^3 kg/m^3$ and Young's modulus $Y = 5 \times 10^8 N/m^2$, 8m in length is hung from the ceiling of a room. What is the increase in length due to its own height?

(a) 9.6×10^{-4} metre (b) 10.05×10^{-5} metre
(c) 10.01×10^{-5} metre (d) 11.02×10^{-5} metre

4. A capillary tube of radius 0.25 mm is submerged vertically in water so that 25 mm of its length is outside water. The radius of curvature of the meniscus will be

(ST of water = $75 \times 10^{-3} Nm^{-1}$)

(a) 0.2 mm (b) 0.4 mm
(c) 0.6 mm (d) 0.8 m

5. A block A of mass 100 kg rests on another block B of mass 200 kg and is tied to a wall as shown in the figure.

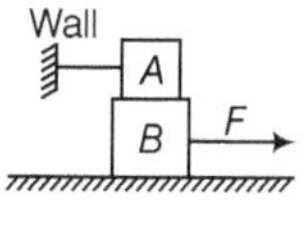

The coefficient of friction between A and B is 0.2 and that between B and the ground is 0.3. The minimum force F required to move the block B is

(a) 900 N (b) 200 N (c) 1100 N (d) 700 N

6. A solid cylindrical rod of radius 3 mm gets depressed under the influence of a load through 8 mm. The depression produced in an identical hollow rod with outer and inner radii of 4 mm and 2 mm respectively, will be

(a) 2.7 mm (b) 1.9 mm (c) 3.2 mm (d) 7.7 mm

7. A small object of uniform density rolls up a curved surface with an initial velocity v'. It reaches upto a maximum height of $\frac{3v^2}{4g}$ with respect to the initial position. The object is

(a) ring (b) solid sphere
(c) hollow sphere (d) disc

8. A block rests on a rough inclined plane making an angle of 30° with the horizontal. Coefficient of static friction between the block and plane is 0.8. If the frictional force on the block is 10 N, find the mass of the block.

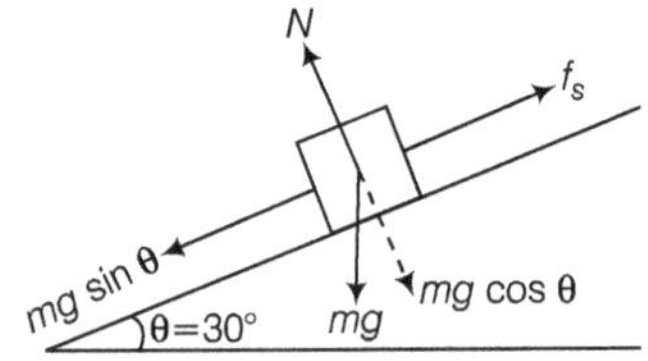

(a) 2.0 kg (b) 4.0 kg (c) 1.6 kg (d) 2.5 kg

9. The coefficient of cubical expansion of mercury is 0.00018 and that of brass 0.00006 per °C . If a barometer having a brass scale were to read 74.5 cm at 30°C. Find the true barometric height at 0°C. The scale is supposed to be correct at 15°C.

(a) 74.122 cm (b) 79.412 cm
(c) 83.421 cm (d) 92.421 cm

MHT CET

1. A wheel of moment of inertia 2 kg m^2 is rotating about an axis passing through centre and perpendicular to its plane at a speed 60 rad/s. Due to friction, it comes to rest in 5 minutes. The angular momentum of the wheel three minutes before it stops rotating is

(a) 24 kg m^2/s (b) 48 kg m^2/s
(c) 72 kg m^2/s (d) 96 kg m^2/s

2. The equation of the progressive wave is $y = 3\sin\left[\pi\left(\frac{t}{3} - \frac{x}{5}\right) + \frac{\pi}{4}\right]$, where x and y are in metre and time in second. Which of the following is correct?

(a) Velocity $v = 1.5$ m/s (b) Amplitude $A = 3$ cm
(c) Frequency $f = 0.2$ Hz (d) Wavelength $\lambda = 10$ m

3. Two spherical black bodies have radii 'r_1' and 'r_2'. Their surface temperature are 'T_1' and 'T_2'. If they radiate same power, then $\frac{r_2}{r_1}$ is

(a) $\frac{T_1}{T_2}$ (b) $\frac{T_2}{T_1}$ (c) $\left(\frac{T_1}{T_2}\right)^2$ (d) $\left(\frac{T_2}{T_1}\right)^2$

4. The closed and open organ pipes have same length. When they are vibrating simultaneously in first overtone, produce three beats. The length of open pipe is made $\frac{1}{3}$rd and closed pipe is made three times the original, the number of beats produced will be

(a) 8 (b) 14
(c) 17 (d) 20

5. A lift of mass 'm' is connected to a rope which is moving upward with maximum acceleration 'a'. For maximum safe stress, the elastic limit of the rope is 'T'. The minimum diameter of the rope is (g = gravitational acceleration).

(a) $\left[\frac{2m(g+a)}{\pi T}\right]^{\frac{1}{2}}$ (b) $\left[\frac{4m(g+a)}{\pi T}\right]^{\frac{1}{2}}$

(c) $\left[\frac{m(g+a)}{\pi T}\right]^{\frac{1}{2}}$ (d) $\left[\frac{m(g+a)}{2\pi T}\right]^{\frac{1}{2}}$

6. A solid sphere of mass 2 kg is rolling on a frictionless horizontal surface with velocity 6m/s. It collides on the free end of an ideal spring whose other end is fixed. The maximum compression produced in the spring will be

(Force constant of the spring = 36 N/m)

(a) $\sqrt{14}$ m (b) $\sqrt{2.8}$ m (c) $\sqrt{1.4}$ m (d) $\sqrt{0.7}$ m

7. A flywheel at rest is to reach an angular velocity of 24rad/s in 8 second with constant angular acceleration. The total angle turned through during this interval is

(a) 24 rad (b) 48 rad (c) 72 rad (d) 96 rad

8. Two uniform wires of the same material are vibrating under the same tension. If the first overtone of the first wire is equal to the second overtone of the second wire and radius of the first wire is twice the radius of the second wire, then the ratio of the lengths of the first wire to second wire is

(a) $\frac{1}{3}$ (b) $\frac{1}{4}$ (c) $\frac{1}{5}$ (d) $\frac{1}{6}$

9. When one end of the capillary is dipped in water, the height of water column is 'h'. The upward force of 105 dyne due to surface tension is balanced by the force due to the weight of water column. The inner circumference of the capillary is

(Surface tension of water $= 7 \times 10^{-2}$ N/m)

(a) 1.5 cm (b) 2 cm (c) 2.5 cm (d) 3 cm

10. For a rigid diatomic molecule, universal gas constant $R = nC_p$, where 'C_p' is the molar specific heat at constant pressure and 'n' is a number. Hence n is equal to

(a) 0.2257 (b) 0.4 (c) 0.2857 (d) 0.3557

11. An ideal gas has pressure 'p', volume 'V' and absolute temperature 'T'. If 'm' is the mass of each molecules and 'K' is the Boltzmann constant, then density of the gas is

(a) $\frac{pm}{MT}$ (b) $\frac{KT}{pm}$ (c) $\frac{Km}{pT}$ (d) $\frac{pK}{Tm}$

12. A big water drop is formed by the combination of 'n' small water drops of equal radii. The ratio of the surface energy of 'n' drops to the surface energy of big drop is

(a) $n^2:1$ (b) $n:1$
(c) $\sqrt{n}:1$ (d) $\sqrt[3]{n}:1$

13. A particle performing SHM starts equilibrium position and its time period is 16 seconds. After 2 seconds its velocity is π m/s. Amplitude of oscillation is $\left(\cos 45^\circ = \frac{1}{\sqrt{2}}\right)$

(a) $2\sqrt{2}$ m (b) $4\sqrt{2}$ m
(c) $6\sqrt{2}$ m (d) $8\sqrt{2}$ m

14. In sonometer experiment, the string of length 'L' under tension vibrates in second overtone between two bridges. The amplitude of vibration is maximum at

(a) $\frac{L}{3}, \frac{2L}{3}, \frac{5L}{6}$ (b) $\frac{L}{8}, \frac{L}{4}, \frac{L}{2}$
(c) $\frac{L}{2}, \frac{L}{4}, \frac{L}{6}$ (d) $\frac{L}{6}, \frac{L}{2}, \frac{5L}{6}$

15. The depth 'd' at which the value of acceleration due to gravity becomes $\frac{1}{n}$ times the value at the earth's surface is (R = radius of earth)

(a) $d = R\left(\frac{n}{n-1}\right)$ (b) $d = R\left(\frac{n-1}{2n}\right)$
(c) $d = R\left(\frac{n-1}{n}\right)$ (d) $d = R^2\left(\frac{n-1}{n}\right)$

16. A particle is performing SHM starting extreme position. Graphical representation shows that, between displacement and acceleration, there is a phase difference of

(a) 0 rad (b) $\frac{\pi}{4}$ rad (c) $\frac{\pi}{2}$ rad (d) π rad

17. The fundamental frequency of an air column in a pipe closed at one end is 100 Hz. If the same pipe is open at both the ends, the frequencies produced in Hz are

(a) 100, 200, 300, 400, ... (b) 100, 300, 500, 700, ...
(c) 200, 300, 400, 500, ... (d) 200, 400, 600, 800

18. For a particle moving in vertical circle, the total energy at different positions along the path

(a) is conserved (b) increases
(c) decreases (d) may increase or decrease

19. A simple pendulum of length 'L' has mass 'M' and it oscillates freely with amplitude 'A'. At extreme position, its potential energy is
(g = acceleration due to gravity)

(a) $\frac{MgA^2}{2L}$ (b) $\frac{MgA}{2L}$ (c) $\frac{MgA^2}{L}$ (d) $\frac{2MgA^2}{L}$

20. A ceiling fan rotates about its own axis with some angular velocity. When the fan is switched off, the angular velocity becomes $\left(\frac{1}{4}\right)$th of the original in time 't' and 'n' revolutions are made in that time. The number of revolutions made by the fan during the time interval between switch off and rest are (Angular retardation is uniform)

(a) $\frac{4n}{15}$ (b) $\frac{8n}{15}$ (c) $\frac{16n}{15}$ (d) $\frac{32n}{15}$

21. A disc of moment of inertia 'I_1' is rotating in horizontal plane about an axis passing through a centre and perpendicular to its plane with constant angular speed 'ω_1'. Another disc of moment of inertia 'I_2' having zero angular speed is placed co-axially on a rotating disc. Now, both the discs are rotating with constant angular speed 'ω_2'. The energy lost by the initial rotating disc is

(a) $\frac{1}{2}\left[\frac{I_1 + I_2}{I_1 I_2}\right]\omega_1^2$ (b) $\frac{1}{2}\left[\frac{I_1 I_2}{I_1 - I_2}\right]\omega_1^2$
(c) $\frac{1}{2}\left[\frac{I_1 - I_2}{I_1 I_2}\right]\omega_1^2$ (d) $\frac{1}{2}\left[\frac{I_1 I_2}{I_1 + I_2}\right]\omega_1^2$

22. A particle performs linear SHM at a particular instant, velocity of the particle is 'u' and acceleration is 'α' while at another instant velocity is 'v' and acceleration is 'β' $(0 < \alpha < \beta)$. The distance between the two position is

(a) $\frac{u^2 - v^2}{\alpha + \beta}$ (b) $\frac{u^2 + v^2}{\alpha + \beta}$
(c) $\frac{u^2 - v^2}{\alpha - \beta}$ (d) $\frac{u^2 + v^2}{\alpha - \beta}$

23. The observer is moving with velocity 'v_0' towards the stationary source of sound and then after crossing moves away from the source with velocity 'v_0'. Assume that the medium through which the sound waves travel is at rest. If v is the velocity of sound and n is the frequency emitted by the source, then the difference between apparent frequencies heard by the observer is

(a) $\frac{2nv_0}{v}$ (b) $\frac{nv_0}{v}$
(c) $\frac{v}{2nv_0}$ (d) $\frac{v}{nv_0}$

24. A metal rod of length 'L' and cross-sectional area 'A' is heated through 'T'°C. What is the force required to prevent the expansion of the rod lengthwise?

(a) $\frac{YA\alpha T}{(1-\alpha T)}$ (b) $\frac{YA\alpha T}{(1+\alpha T)}$
(c) $\frac{(1-\alpha T)}{YA\alpha T}$ (d) $\frac{(1+\alpha T)}{YA\alpha T}$

Answer *with* Explanations

JEE Main

1. (b) As the observer is moving towards the source, so frequency of waves emitted by the source will be given by the formula

$$f_{\text{observed}} = f_{\text{actual}} \cdot \left(\frac{1 + v/c}{1 - v/c}\right)^{1/2}$$

Here, frequency $\frac{v}{c} = \frac{1}{2}$

So, $f_{\text{observed}} = f_{\text{actual}} \left(\frac{3/2}{1/2}\right)^{1/2}$

$\therefore \quad f_{\text{observed}} = 10 \times \sqrt{3} = 17.3 \text{ GHz}$

2. (a) By ascent formula, we have surface tension,

$$T = \frac{rhg}{2} \times 10^3 \frac{\text{N}}{\text{m}}$$

$$= \frac{dhg}{4} \times 10^3 \frac{\text{N}}{\text{m}} \quad \left(\because r = \frac{d}{2}\right)$$

$\Rightarrow \frac{\Delta T}{T} = \frac{\Delta d}{d} + \frac{\Delta h}{h}$ [given, g is constant]

So, percentage

$$= \frac{\Delta T}{T} \times 100 = \left(\frac{\Delta d}{d} + \frac{\Delta h}{h}\right) \times 100$$

$$= \left(\frac{0.01 \times 10^{-2}}{1.25 \times 10^{-2}} + \frac{0.01 \times 10^{-2}}{1.45 \times 10^{-2}}\right) \times 100$$

$= 1.5\%$

$\therefore \quad \frac{\Delta T}{T} \times 100 = 1.5\%$

3. (b) Given, force, $F = -kv^2$

$\therefore$ Acceleration, $a = \frac{-k}{m} v^2$

or $\frac{dv}{dt} = \frac{-k}{m} v^2 \Rightarrow \frac{dv}{v^2} = -\frac{k}{m} \cdot dt$

Now, with limits, we have

$$\int_{10}^{v} \frac{dv}{v^2} = -\frac{k}{m} \int_0^t dt$$

$$\Rightarrow \left(-\frac{1}{v}\right)_{10}^{v} = -\frac{k}{m} t \Rightarrow \frac{1}{v} = 0.1 + \frac{kt}{m}$$

$$\Rightarrow \quad v = \frac{1}{0.1 + \frac{kt}{m}} = \frac{1}{0.1 + 1000k}$$

$$\Rightarrow \quad \frac{1}{2} \times m \times v^2 = \frac{1}{8} m v_0^2$$

$$\Rightarrow \quad v = \frac{v_0}{2} = 5 \Rightarrow \frac{1}{0.1 + 1000\,k} = 5$$

$\Rightarrow \quad 1 = 0.5 + 5000\,k$

$\Rightarrow \quad k = \frac{0.5}{5000} \Rightarrow k = 10^{-4} \text{ kg/m}$

4. (b) By Mayor's relation, for 1 g mole of a gas,

$$C_p - C_V = R$$

So, when n gram moles are given,

$$C_p - C_V = \frac{R}{n}$$

As per given question,

$$a = C_p - C_V = \frac{R}{2}; \text{ for } H_2 \quad \ldots \text{(i)}$$

$$b = C_p - C_V = \frac{R}{28}; \text{ for } N_2 \quad \ldots \text{(ii)}$$

From Eqs. (i) and (ii), we get

$$a = 14b$$

5. (d) MI of a solid cylinder about its perpendicular bisector of length is

$$I = m\left(\frac{l^2}{12} + \frac{R^2}{4}\right)$$

$$\Rightarrow \quad I = \frac{mR^2}{4} + \frac{ml^2}{12} = \frac{m^2}{4\pi\rho l} + \frac{ml^2}{12}$$

$[\because \rho\pi R^2 l = m]$

For I to be maximum,

$$\frac{dI}{dl} = -\frac{m^2}{4\pi\rho}\left(\frac{1}{l^2}\right) + \frac{ml}{6} = 0$$

$$\Rightarrow \quad \frac{m^2}{4\pi\rho} = \frac{ml^3}{6}$$

Now, putting $m = \rho\pi R^2 l$

$$\therefore \quad l^3 = \frac{3}{2\pi\rho} \cdot \rho\pi R^2 l \Rightarrow \frac{l^2}{R^2} = \frac{3}{2}$$

$$\therefore \quad \frac{l}{R} = \sqrt{\frac{3}{2}}$$

6. (a) Heat gained (water + calorimeter) = Heat lost by copper ball

$\Rightarrow \quad m_w s_w \Delta T + m_c s_c \Delta T = m_B s_B \Delta T$

$\Rightarrow \quad 170 \times 1 \times (75 - 30) + 100 \times 0.1 \times (75 - 30)$

$= 100 \times 0.1 \times (T - 75)$

$\therefore \quad T = 885°\text{C}$

Torque applied on a dipole

$\tau = pE \sin\theta$

where θ = angle between axis of dipole and electric field.

For electric field $E_1 = E\hat{i}$

it means field is directed along positive X direction, so angle between dipole and field will remain θ, therefore torque in this direction

$$E_1 = pE_1 \sin\theta$$

In electric field $E_2 = \sqrt{3}\, E\hat{j}$, it means field is directed along positive Y-axis, so angle between dipole and field will be $90 - \theta$

Torque in this direction

$T_2 = pE \sin(90 - \theta) = p\sqrt{3}\, E_1 \cos\theta$

According to question $\tau_2 = -\tau_1$

$\Rightarrow \quad |\tau_2| = |\tau_1|$

$\therefore \quad pE_1 \sin\theta = p\sqrt{3}\, E_1 \cos\theta$

$\tan\theta = \sqrt{3}$

$\Rightarrow \quad \tan\theta = \tan 60° \Rightarrow \theta = 60°$

7. (d) As the rod rotates in vertical plane so a torque is acting on it, which is due to the vertical component of weight of rod.

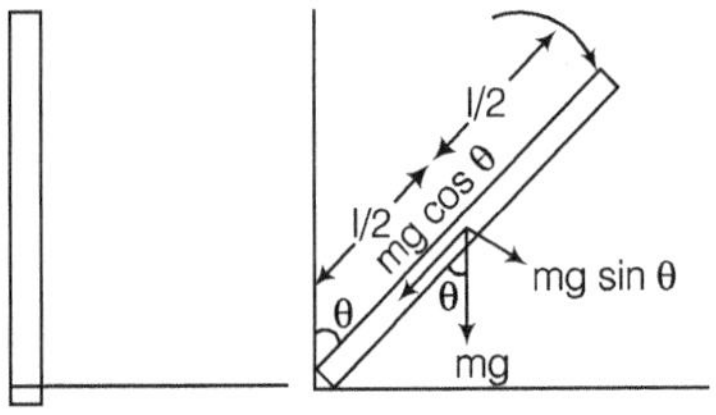

Initial condition — At any time t

Now, Torque τ = force $\times$ perpendicular distance of line of action of force from axis of rotation

$$= mg \sin\theta \times \frac{l}{2}$$

Again, Torque $\tau = I\alpha$

Where, I = moment of inertia $= \frac{ml^2}{3}$

[Force and Torque frequency along axis of rotation passing through in end]

α = angular acceleration

$$\therefore \quad mg \sin\theta \times \frac{l}{2} = \frac{ml^2}{3} \alpha$$

$$\therefore \quad \alpha = \frac{3g \sin\theta}{2l}$$

8. (d) $K = \frac{P}{(-\Delta V / V)}$

$$\Rightarrow \quad \frac{\Delta V}{V} = \frac{P}{K} \Rightarrow -\Delta V = \frac{PV}{K}$$

Change in volume $\Delta V = \gamma V \Delta T$

Where γ = coefficient of volume expansion

Again, $\gamma = 3\alpha$

α is coefficient of linear expansion

$\therefore \quad \Delta V = V(3\alpha)\, \Delta T$

$$\therefore \quad \frac{PV}{K} = V(3\alpha)\,\Delta T \Rightarrow \Delta T = \frac{P}{3\alpha K}$$

9. (c) From $pV = nRT = \frac{N}{N_A} RT$

We have, $n_f - n_i = \frac{pVN_A}{RT_f} - \frac{pVN_A}{RT_i}$

$$\Rightarrow \quad n_f - n_i = \frac{10^5 \times 30}{8.3} \times 6.02 \times 10^{23} \cdot \left(\frac{1}{300} - \frac{1}{290}\right)$$

$$= -2.5 \times 10^{25} \Rightarrow \Delta n = -2.5 \times 10^{25}$$

10. (d) From Newton's second law, $\frac{\Delta p}{\Delta t} = F$

$$\Rightarrow \Delta p = F\Delta t \Rightarrow p = \int dp = \int_0^1 F\,dt$$

$$\Rightarrow \quad p = \int_0^1 6t\,dt = 3 \text{ kg}\left(\frac{\text{m}}{\text{s}}\right)$$

Also, change in kinetic energy

$$\Delta k = \frac{\Delta p^2}{2m} = \frac{3^2}{2\times 1} = 4.5$$

From work-energy theorem, work done = change in kinetic energy.

So, work done = $\Delta k = 4.5$ J

11. (c) Inside the earth surface

$$g = \frac{GM}{R^3} r \text{ i.e. } g \propto r$$

Out the earth surface

$$g = \frac{Gm}{r^2} \text{ i.e. } g \propto \frac{1}{r^2}$$

So, till earth surface 'g' increases linearly with distance r, shown only in graph (c).

12. (b) Initially velocity keeps on decreasing at a constant rate, then it increases in negative direction with same rate.

13. (c) KE is maximum at mean position and minimum at extreme position $\left(\text{at } t = \frac{T}{4}\right)$.

14. (d) Stress $= \frac{\text{Weight}}{\text{Area}}$

Volume will become (9^3) times.

So weight = volume × density × g will also become $(9)^3$ times.

Area of cross section will become $(9)^2$ times.

$$= \frac{9^3 \times W_0}{9^2 \times A_0} = 9\left(\frac{W_0}{A_0}\right)$$

Hence, the stress increases by a factor of 9.

JEE Advanced

Paper 1

1. (a,c,d)

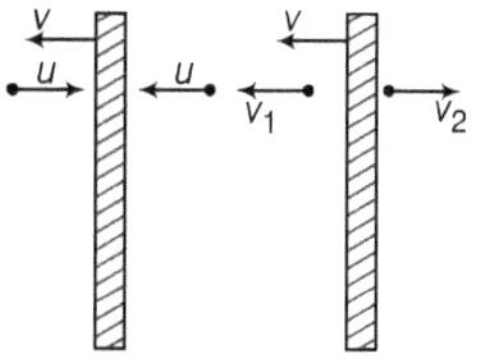

Just before the collision Just after the collision

$$v_1 = u + 2v \Rightarrow \Delta v_1 = (2u + 2v)$$

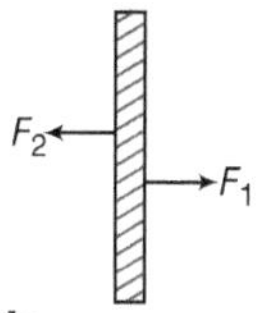

$$F_1 = \frac{dp_1}{dt} = \rho A(u+v)(2u+2v)$$

$$= 2\rho A(u+v)^2 \Rightarrow v_2 = (u - 2v)$$

$$\Delta v_2 = (2u - 2v)$$

$$F_2 = \frac{dp_2}{dt} = \rho A(u - v)(2u - 2v)$$

$$= 2\rho A(u - v)^2$$

$$\Delta F = F_1 - F_2$$

[ΔF is the net force due to the air molecules on the plate]

$$\Delta F = 2\rho A(4uv) = 8\rho Auv$$

$$P = \frac{\Delta F}{A} = 8\rho(uv)$$

$$F_{\text{net}} = (F - \Delta F) = ma$$

[m is mass of the plate]

$$F - (8\rho Au)v = ma$$

2. (a,b) Δx_{cm} of the block and point mass system = 0

$$\therefore \quad m(x + R) + Mx = 0$$

where, x is displacement of the block.

Solving this equation, we get

$$x = -\frac{mR}{M+m}$$

From conservation of momentum and mechanical energy of the combined system

$$0 = mv - MV$$

$$mgR = \frac{1}{2}mv^2 + \frac{1}{2}MV^2$$

Solving these two equations, we get

$$\therefore \quad v = \sqrt{\frac{2gR}{1 + \frac{m}{M}}}$$

3. (a,c,d or a,c) $v = \sqrt{\frac{T}{\mu}}$, so speed at any position will be same for both pulses, therefore time taken by both pulses will be same.

$$\lambda f = v \Rightarrow \lambda = \frac{v}{f} \Rightarrow \lambda \propto v \propto T$$

since when pulse 1 reaches at A tension and hence speed decreases therefore λ decreases.

Note *If we refer velocity by magnitude only, then option (a, c, d) will be correct, else only (a, c) will be correct.*

4. (b,c,d or d) Assumption $e = 1$

[black body radiation]

$$P = \sigma A(T^4 - T_0^4)$$

(c) $P_{\text{rad}} = \sigma AT^4 = \sigma \cdot 1 \cdot (T_0 + 10)^4$

$$= \sigma \cdot T_0^4 \left(1 + \frac{10}{T_0}\right)^4$$

[$T_0 = 300$ K given]

$$= \sigma \cdot (300)^4 \cdot \left(1 + \frac{40}{300}\right)$$

$$\approx 460 \times \frac{17}{15} \approx 520 \text{ J}$$

$$P_{\text{net}} = 520 - 460 \approx 60 \text{ W}$$

$\Rightarrow$ Energy radiated in 1 s = 60 J

(b) $P = \sigma A(T^4 - T_0^4)$

$$dP = \sigma A(0 - 4T_0^3 \cdot dT)$$

and $\quad dT = -\Delta T \Rightarrow dP = 4\sigma A T_0^3 \Delta T$

(d) If surface area decreases, then energy radiation also decreases.

Note *While giving answer (b) and (c) it is assumed that energy radiated refers the net radiation. If energy radiated is taken as only emission, then (b) and (c) will not be included in answer.*

5. (6) From mass conservation,

$$\rho \cdot \frac{4}{3}\pi R^3 = \rho \cdot K \cdot \frac{4}{3}\pi r^3$$

$$\Rightarrow \quad R = K^{1/3} r$$

$$\therefore \quad \Delta U = T\Delta A = T(K \cdot 4\pi r^2 - 4\pi R^2)$$

$$= T(K \cdot 4\pi R^2 K^{-2/3} - 4\pi R^2)$$

$$\Delta U = 4\pi R^2 T[K^{1/3} - 1]$$

Putting the values, we get

$$10^{-3} = \frac{10^{-1}}{4\pi} \times 4\pi \times 10^{-4}[K^{1/3} - 1]$$

$$100 = K^{1/3} - 1$$

$$\Rightarrow \quad K^{1/3} \cong 100 = 10^2$$

Given that $K = 10^{\alpha}$

$$\therefore \quad 10^{\alpha/3} = 10^2$$

$$\Rightarrow \quad \frac{\alpha}{3} = 2 \quad \Rightarrow \alpha = 6$$

6. (6)

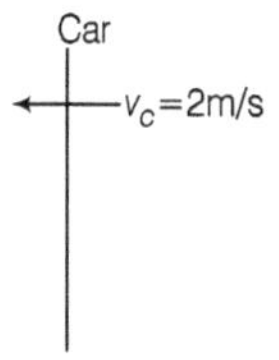

Frequency observed at car

$$f_1 = f_0\left(\frac{v + v_C}{v}\right) \quad (v = \text{speed of sound})$$

Frequency of reflected sound as observed at the source

$$f_2 = f_1\left(\frac{v}{v - v_C}\right) = f_0\left(\frac{v + v_C}{v - v_C}\right)$$

Beat frequency = $f_2 - f_0$

$$= f_0\left[\frac{v+v_C}{v-v_C}-1\right] = f_0\left[\frac{2v_C}{v-v_C}\right]$$

$$= 492 \times \frac{2\times 2}{328} = 6 \text{ Hz}$$

7. (c) $V_{\text{sound}} = \sqrt{\frac{\gamma RT}{M}}$. As the sound wave propagates, the air in a chamber undergoes compression and rarefaction very fastly, hence undergo a adiabatic process. So, curves are steeper than isothermal.

$$\left(\frac{dp}{dV}\right)_{\text{Adi}} = -\gamma\left(\frac{p}{V}\right) \quad \text{...(i)}$$

$$\left(\frac{dp}{dV}\right)_{\text{Iso}} = -\left(\frac{P}{V}\right) \quad \text{...(ii)}$$

Graph 'Q' satisfies Eq. (i)

8. (b) $\Delta U = \Delta Q - p\Delta V$

$$\Delta U + p\Delta V = \Delta Q$$

As $\Delta U \neq 0$, $W \neq 0$, $\Delta Q \neq 0$. The process represents, isobaric process

$$W_{\text{gas}} = -p(\Delta V) = -p(V_2 - V_1)$$
$$= -pV_2 + pV_1$$

Graph 'P' satisfies isobaric process.

9. (b) Work done in isochoric process is zero.

$$W_{12} = 0 \text{ as } \Delta V = 0$$

Graph 'S' represents isochoric process.

Paper 2

1. (d)

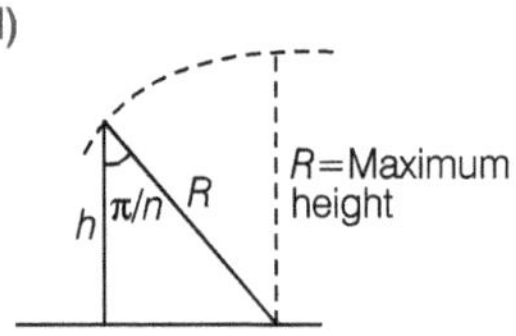

$$\cos\left(\frac{\pi}{n}\right) = \frac{h}{R}$$

$$\Delta = R - h = \frac{h}{\cos\left(\frac{\pi}{n}\right)} - h$$

$$= h\left[\frac{1}{\cos\left(\frac{\pi}{n}\right)} - 1\right]$$

2. (a) $m = \frac{4\pi R^3}{3} \times \rho$

On taking log both sides, we have

$$\ell n(m) = \ell n\left(\frac{4\pi}{3}\right) + \ell n(\rho) + 3\ell n(R)$$

On differentiating with respect to time,

$$0 = 0 + \frac{1}{\rho}\frac{d\rho}{dt} + \frac{3}{R}\frac{dR}{dt}$$

$$\Rightarrow \left(\frac{dR}{dt}\right) = v \propto -R \times \frac{1}{\rho}\left(\frac{d\rho}{dt}\right)$$

$$v \propto R$$

3. (c) $\mathbf{S} = \mathbf{P} + b\mathbf{R} = \mathbf{P} + b(\mathbf{Q} - \mathbf{P})$

$$= \mathbf{P}(1-b) + b\mathbf{Q}$$

4. (a) $t = \sqrt{\frac{L}{5}} + \frac{L}{300}$

$$dt = \frac{1}{\sqrt{5}}\frac{1}{2}L^{-1/2}dL + \left(\frac{1}{300}dL\right)$$

$$dt = \frac{1}{2\sqrt{5}}\frac{1}{\sqrt{20}}dL + \frac{dL}{300} = 0.01$$

$$dL\left(\frac{1}{20} + \frac{1}{300}\right) = 0.01$$

$$dL\left[\frac{15}{300}\right] = 0.01 \Rightarrow dL = \frac{3}{16}$$

$$\frac{dL}{L} \times 100 = \frac{3}{16} \times \frac{1}{20} \times 100 = \frac{15}{16} \simeq 1\%$$

5. (c) Given, $v_e = 11.2 \text{ km/s} = \sqrt{\frac{2GM_e}{R_e}}$

From energy conservation,

$$K_i + U_i = K_f + U_f$$

$$\frac{1}{2}mv_s^2 - \frac{GM_s m}{r} - \frac{GM_e m}{R_e} = 0 + 0$$

Here, r = distane of rocket from sun

$$\Rightarrow v_s = \sqrt{\frac{2GM_e}{R_e} + \frac{2GM_s}{r}}$$

Given, $M_s = 3 \times 10^5 M_e$

and $r = 2.5 \times 10^4 R_e$

$$\Rightarrow v_s = \sqrt{\frac{2GM_e}{R_e} + (2G)\left(\frac{3\times 10^5 M_e}{2.5\times 10^4 R_e}\right)}$$

$$= \sqrt{\frac{2GM_e}{R_e}\left(1 + \frac{3\times 10^5}{2.5\times 10^4}\right)}$$

$$= \sqrt{\frac{2GM_e}{R_e} \times 13} \Rightarrow v_s \simeq 42 \text{ km/s}$$

6. (a,c)

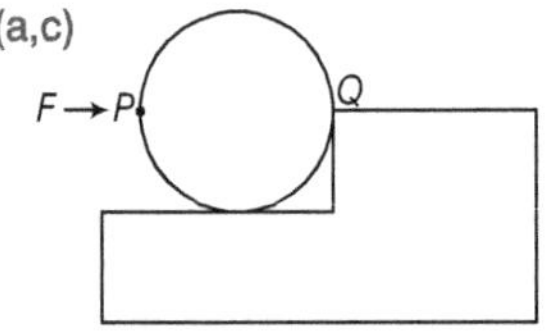

(a) If force is applied normal to surface at P, then line of action of force will pass from Q and thus $\tau = 0$.

(b) Wheel can climb.

(c) $\tau = F(2R\cos\theta) - mgR\cos\theta$

$$\tau \propto \cos\theta$$

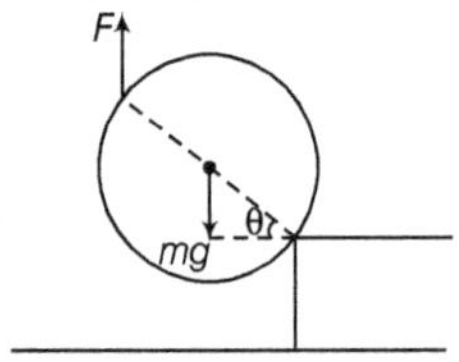

Hence, as θ increases, τ decreases.

So its correct.

(d)

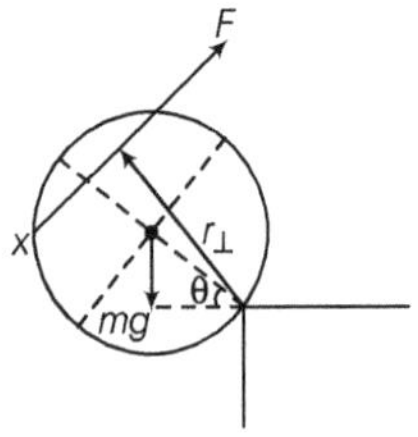

$\tau = Fr_\perp - mg\cos\theta$; τ increases with θ.

7. (a,c,d) When the bar makes an angle θ, the height of its COM (mid-point) is $\frac{L}{2}\cos\theta$.

$\therefore$ Displacement

$$= L - \frac{L}{2}\cos\theta = \frac{L}{2}(1-\cos\theta)$$

Since, force on COM is only along the vertical direction, hence COM is falling vertically downward.

Instantaneous torque about point of contact is

$$\tau = mg \times \frac{L}{2}\sin\theta \text{ or } \tau \propto \sin\theta$$

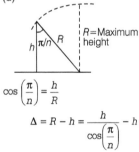

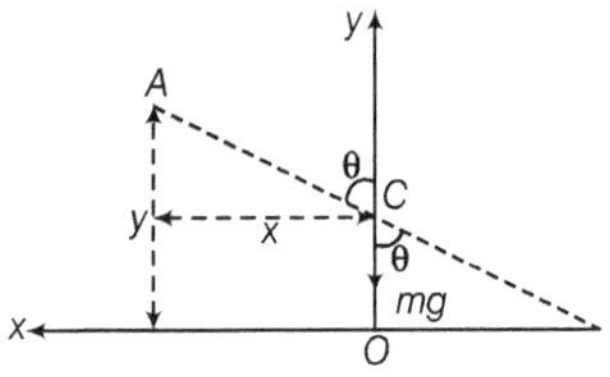

Now, $x = \frac{L}{2}\sin\theta \Rightarrow y = L\cos\theta$

$$\frac{x^2}{(L/2)^2} + \frac{y^2}{L^2} = 1$$

Path of A is an ellipse.

8. Question is not very clear.

9. (c) If height of the cone $h >> r$

Then, $\mu N = mg \Rightarrow \mu m(R-r)\omega_0^2 = mg$

$$\omega_0 = \sqrt{\frac{g}{\mu(R-r)}}$$

BITSAT

1. (a) According to Stefan's law

$$\rho = \frac{E}{t} = \sigma A e T^4$$

So, we can write as

$$\frac{E_2}{E_1} = \left(\frac{T_2}{T_1}\right)^4$$

$$\frac{E_2}{E_1} = \left(\frac{900}{300}\right)^4 \Rightarrow \frac{E_2}{E_1} = (3)^4$$

$$E_2 = 81 E_1 \Rightarrow \frac{E_2}{E_1} = 81$$

2. (d) $v_s = r\omega = 1.2 \times 2\pi f$ $[\because \omega = 2\pi f]$

$$= 1.2 \times 2 \times 3.14 \times \left(\frac{400}{60}\right) = 50 \text{ m/s}$$

$$v_{min} = \frac{v}{v + v_s} \cdot v = \frac{340}{340 + 50} \times 500$$

$$= 436 \text{ Hz}$$

$$v_{max} = \frac{v}{v - v_s} \cdot v = \frac{340}{340 - 50} \times 500$$

$$= 586 \text{ Hz}$$

3. (b) Given, $H = R$

or $$\frac{u^2 \sin^2 \theta}{2g} = \frac{u^2 \sin 2\theta}{g}$$

$$= \frac{2u^2 \sin \theta \cos \theta}{g}$$

or $\tan \theta = 4 \Rightarrow \theta = \tan^{-1}(4)$

4. (b) Given, $a = 6 \text{ cm} = 6 \times 10^{-2}$ m

$\omega = 100$ rad/s, $m = 1$ kg

$$K_{max} = \frac{1}{2} m\omega^2 a^2$$

$$= \frac{1}{2} \times 1 \times (100)^2 \times (6 \times 10^{-2})^2$$

$$= 18 \text{ J}$$

5. (c) Equation of motion of lower block is

$$F = \mu(M_A + M_B) \cdot g$$

$$\mu = \frac{F}{(M_A + M_B)g} = \frac{49}{10 \times 9.8} = \frac{1}{2} = 0.5$$

6. (b) Given $R = 3 \text{ km} = 3000$ m

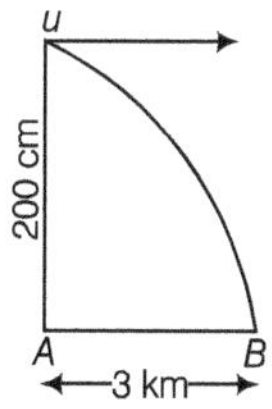

Range, $R = u\sqrt{\frac{2h}{g}} \Rightarrow u = R\sqrt{\frac{g}{2h}}$

or $u = 3000 \times \sqrt{\frac{10}{2 \times 2000}} = 3000 \times \frac{1}{20}$

$$= 150 \text{ m/s} = 150 \times \frac{18}{5} \text{ km/h}$$

$$= 540 \text{ km/h}$$

7. (a) Increase in length

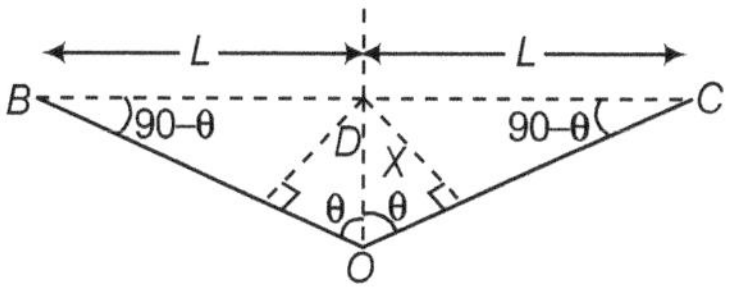

$$\Delta L = BO + OC - (BC) = 2BO - 2BD$$

$$\Rightarrow \Delta L = 2BO - 2L = 2[L^2 + x^2]^{\frac{1}{2}} - 2L$$

or $$\Delta L = 2L\left[1 + \frac{x^2}{L^2}\right]^{\frac{1}{2}} - 2L$$

[$\because$ using Bionomial theorem]

$$\Rightarrow \Delta L \approx 2L\left[1 + \frac{1}{2}\frac{x^2}{L^2} - 1\right] = \frac{x^2}{L}$$

$[\because x << L]$

$$\therefore \text{ Strain} = \frac{\Delta L}{2L} = \frac{\frac{x^2}{L}}{2L} = \frac{x^2}{2L^2}$$

8. (c) Let the volume of ice-berg is V and its density is ρ. If this ice-berg floats in water with volume V_{in} inside it, then

$$V_{in}\sigma g = V\rho g$$

σ = density of water

$$\Rightarrow V_{in} = \left(\frac{\rho}{\sigma}\right) \cdot V$$

$$\Rightarrow V_{out} = V - V_{in} = \left[\frac{\sigma - \rho}{\sigma}\right] V$$

$$= \frac{1000 - 900}{1000} V = \frac{V}{10}$$

$$\Rightarrow V_{out} / V = 0.1 = 10\%$$

9. (d) $T = \frac{2U_y}{g} \Rightarrow H = \frac{U_y^2}{2g}$

and $R = U_x \times T$

When a horizontal acceleration is also given to the projectile U_y, T and H will remains unchanged while range will become

$$R' = U_x \times T + \frac{1}{2} aT^2 = R + \frac{1}{2}\frac{g}{4}\left(\frac{4U_y^2}{g^2}\right)$$

$$= R + H \qquad \left[\because a = \frac{g}{4}\right]$$

10. (b) We have, $\frac{P_1V_1}{T_1} = \frac{P_2V_2}{T_2}$

or $$V_2 = \frac{P_1V_1T_2}{P_2T_1} = \frac{1 \times 500 \times (273 - 3)}{0.5 \times (273 + 27)}$$

$$= \frac{1 \times 500 \times 270}{0.5 \times 300}$$

$$V_2 = 900 \text{ m}^3$$

11. (c) Intensity at centre of bright fringe,

$$I_0 = I + I + 2\sqrt{II} \cdot \cos 0$$

$$I_0 = 2I + 2I \qquad [\because \cos 0° = 1]$$

$$= 4I$$

Similarly Intensity at point $\frac{P}{4}$

(with phase difference $= \frac{2\pi}{4} = \frac{\pi}{2}$)

$$I' = I + I + 2\sqrt{II} \cos \frac{\pi}{2}$$

$$= 2I + 2\sqrt{II} \times 0 = 2I$$

$$\therefore \quad \frac{I_0}{I'} = \frac{4I}{2I} = 2$$

12. (b) Let block is displaced through x m, then weight of displaced water or upthrust (upwards)

$$= -Ax\rho g$$

Where, A is area of a cross-section of the block and ρ is density. This must be equal to force (= ma) applied where m is mass of the block and a is acceleration.

$$\therefore \quad ma = -Ax\rho g$$

or $$a = -\frac{A\rho g}{m} x = -\omega^2 x$$

This is equation of SHM.

The time period of oscillation,

$$T = \frac{2\pi}{\omega} = 2\pi\sqrt{\frac{m}{A\rho g}} \Rightarrow T \propto \frac{1}{\sqrt{A}})$$

13. (b) From the conservation principle,

$$mgh = \frac{1}{2} kX_0^2 - mgX_0$$

where, X_0 is maximum elongation in spring.

$$\Rightarrow \frac{1}{2} kX_0^2 - mgX_0 - mgh = 0$$

$$\Rightarrow X_0^2 - \frac{2mg}{k} X_0 - \frac{2mg}{k} h = 0$$

$$X_0 = \frac{2\frac{mg}{2} \pm \sqrt{\left(\frac{2mg}{k}\right)^2 + 4 \times \frac{2mg}{k} h}}{2}$$

Amplitude = elongation in spring for lowest extreme position – elongation in spring for equilibrium position

$$= X_0 - X_1 = \frac{mg}{k}\sqrt{1 + \frac{2hk}{mg}}$$

$$\left[\because X_1 = \frac{mg}{k}\right]$$

14. (c) Friction force between A and B and between block B and surface S will oppose F

$$\therefore \quad F = F_{AB} + F_{BS}$$

$$= \mu_{AB}M_A g + \mu_{BS}(m_A + m_B)g$$

$$= 0.2 \times 100 \times 10 + 0.3(100 + 200) \times 10$$

$$= 200 + 900 = 1100 \text{ N}$$

15. (a) The moment of inertia of the uniform rod about an axis through one end and perpendicular to its length is

$$= \frac{ml^2}{3}$$

where, m is the mass and l its length.

Torque ($\tau = I \cdot \alpha$) acting on centre of gravity of rod is given by

$$\tau = mg\frac{l}{2}$$

also, $\frac{ml^2}{3} \cdot \alpha = mg\frac{l}{2}$, $\alpha = \frac{3g}{2l}$

16. (d) Work done = Change in surface energy

$$r_2 = 5\text{ cm} \Rightarrow r_1 = 3\text{ cm}$$

$$\Rightarrow \quad W = 2T \times 4\pi(r_2^2 - r_1^2)$$

$$= 2 \times 0.03 \times 4\pi[(5)^2 - (3)^2] \times 10^{-4}$$

$$= 0.4\pi \text{ mJ}$$

17. (c) From the figure, the X-component remains unchanged while the Y-component is reverse. Then, the velocity at point B is $(2\hat{\mathbf{i}} - 3\hat{\mathbf{j}})$ m/s.

18. (c) According to question

$$n_1 T_1 = n_2 T_2$$

$$\Rightarrow \quad n \cdot 2\pi\sqrt{\frac{5}{g}} = (n-1)\, 2\pi\sqrt{\frac{20}{g}}$$

$$\frac{n}{n-1} = 2$$

$$\Rightarrow \quad n = 2$$

18. (d) Acceleration

$$f = \frac{dv}{dt} = f_0\left(1 + \frac{t}{T}\right)$$

or $\quad dv = f_0\left(1 - \dfrac{t}{T}\right) \cdot dt \qquad$ ...(i)

Integrating Eq. (i) on both sides, we get

$$v = f_0 t - \frac{f_0}{T} \cdot \frac{t^2}{2} + C \qquad \text{...(ii)}$$

After applying boundary conditions

$$v = 0 \text{ at } t = 0$$

We get, $\quad C = 0$

$$\Rightarrow \quad v = f_0 t - \frac{f_0}{T} \cdot \frac{t^2}{2} \qquad \text{...(iii)}$$

As $\quad f = f_0\left(1 - \dfrac{t}{T}\right)$

When $f_0 = 0, t = T$

Substituting $t = T$ in Eq. (iii), then velocity

$$v_x = f_0 T - \frac{f_0}{T} \cdot \frac{T^2}{2} = \frac{1}{2} f_0 T$$

20. (c) By Kepler's law of planetary motion,

$$T^2 \propto r^3, \text{ hence } T_1^2 \propto r_1^3$$

Simiarly $\quad T_2^2 \propto r_2^3$

$$\therefore \quad T_1^2 \propto (36000)^3$$

$$T_2^2 \propto (6400 + h)^3$$

Therefore,

$$(T_2)^2 = T_1^2\left(\frac{6400 + h}{36000}\right)^3$$

$$> T_1^2\left[\frac{6400}{3600}\right]^3 > (24)^2\left[\frac{8}{45}\right]^3$$

$$\therefore \quad T_2 > \frac{24 \times 8 \times \sqrt{8}}{45 \times \sqrt{45}} > 1.8 \text{ h}$$

So, $\quad T_2 \cong 2$ h

21. (b) Equation of wave,

$$y = 0.2 \sin[1.5x + 60t]$$

Comparing with standard equation, we get

$$y = A \sin[kx + \omega t]$$

$$k = 1.5 = \frac{2\pi}{\lambda} \text{ and } \quad \omega = 60 = \frac{2\pi}{T}$$

$\therefore$ Velocity of wave, $v = \dfrac{\omega}{k}$

$$= \frac{60}{1.5} = 40 \text{ m/s}$$

Velocity of wave in stretched string,

$$v = \sqrt{\frac{T}{m}},$$

where, m is the linear density, T is tension in the string.

$$\text{So, } T = v^2 \text{ m} = (40)^2 \times 3 \times 10^{-4}$$

$$= 0.48 \text{ N}$$

KERALA CEE

1. (c) Full length of the train subtends 15° angle

$$15° = 15 \times \frac{\pi}{180} \text{ Radian} = \frac{\pi}{12}$$

Radian

Given that, distance between train and person is 3 km.

i.e. $\quad$ 3 km = 3000 meter

Length of the train = Radius × Angle

$$= 3000 \times \frac{\pi}{12} = 250\,\pi$$

2. (a) In a measurement, the random error can be minimixed by increasing the number of readings and averaging them.

3. (b) Mean $= \dfrac{\Sigma x_i f_i}{10}$

$$= \frac{(4.9 \times 4) + (5.0 \times 3) + (5.1 \times 3)}{10} = 4.99 \text{ s}$$

Which can be round off to 5.0 s correct to two significant figures.

4. (a) The total distance travelled by the object will be area of given graph in the figure.

So, the total distance travelled by the object = Area of (1) + Area of (2) + Area of (3) + Area of (4) + Area of (5) + Area of (6) + Area of (7)

$$= \frac{1}{2} \times 10 \times 15 + (20 - 10) \times (15 - 0)$$

$$+ \frac{1}{2m}(25 - 15) \times (40 - 15) + (35 - 25) \times$$

$$(20 - 0) + \frac{1}{2}(35 - 25) \times (40 - 20) +$$

$$(45 - 35) \times (20 - 0) + \frac{1}{2}$$

$$(50 - 45) \times (15 - 0) = 975 \text{ m}$$

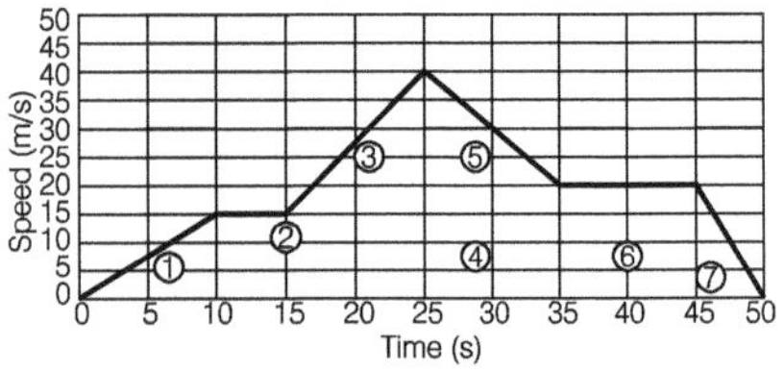

5. (e) According to question,

Object P starts at $t = 0$ and acceleration of $P = 40$ km/h^2

The time taken by object P to cover 20 km will be,

$$s = u + \frac{1}{2}at^2$$

Where, $u = 0$, $s = \dfrac{1}{2}at^2$; $t = \sqrt{\dfrac{2s}{a}}$

Here, $a = 40$ km/h^2 and $s = 20$ km

$$t = \sqrt{\frac{2 \times 20}{40}} = 1 \text{ hour}$$

Time taken by Q will be $\dfrac{1}{2}$ hour, because he started after 30 min.

The acceleration in Q,

$$a = \frac{2s}{t^2} = \frac{2 \times 20}{\left(\frac{1}{2}\right)^2} = 2 \times 4 \times 20$$

$$= 160 \text{ km/h}^2$$

6. (*) (No option matches)

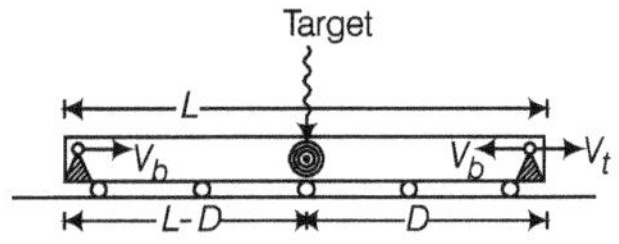

Bullets from both person reaches target at same instant, so we equate time to get,

$$\frac{L - D}{V_b - V_t} = \frac{D}{V_b - V_t} + T$$

$$\frac{L}{V_b - V_t} = \frac{D}{V_b + V_t} + \frac{D}{V_b - V_t} + T$$

$$= \frac{D(V_b - V_t + V_b + V_t)}{V_b^2 - V_t^2} + T$$

$$\frac{L}{V_b - V_t} = \frac{2V_b D}{V_b^2 - V_t^2} + T$$

$$\Rightarrow \quad L = \frac{2V_b D}{V_b + V_t} + T(V_b - V_t)$$

None of the option is matching

7. (c) The horizontal range, $R = \dfrac{u^2 \sin 2\theta}{g}$

Given, $u = 20$ m/s

$$\theta = 60° \Rightarrow g = 10 \text{ m/s}^2$$

$$R = \frac{(20)^2 \sin(2 \times 60°)}{10} = \frac{20 \times 20}{10} \times \frac{\sqrt{3}}{2}$$

$$= 20\sqrt{3} \text{ m}$$

8. (e) When the person throws a ball in upward direction from moving truck, ball also acquire a speed of 60 km/h in horizontal direction. So, if truck continues to move with same speed, ball comes down again in hands of person.

9. (e) Initial K.E of ball of mass $2m = K_1$

$$= \frac{1}{2} \times 2m \times v^2 = mv^2$$

Collision is elastic so both K.E and momentum are conserved. Let velocities of balls are v_1 and v_2 after collision.

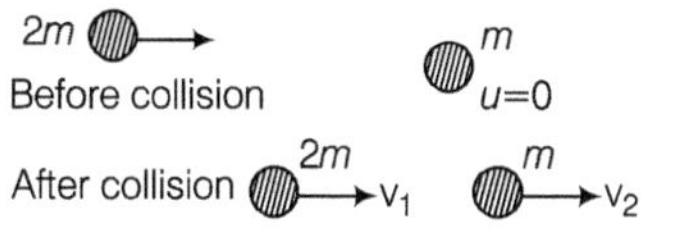

So, KE is conserved

$$\frac{1}{2}(2m)v^2 = \frac{1}{2}(2m)v_1^2 + \frac{1}{2}mv_2^2$$

$$\Rightarrow \quad v^2 = v_1^2 + \frac{1}{2}v_2^2 \quad \ldots\text{(i)}$$

And, momentum is conserved

$(2m)v + m(0) = 2m(v_1) + mv_2$

$$\Rightarrow \quad 2v = 2v_1 + v_2 \quad \ldots\text{(ii)}$$

Now, $v_2 = 2(v - v_1)$

Put this value in Eq. (i), we get

$$v^2 = v_1^2 + \frac{1}{2} \times 4(v - v_1)^2$$

$$\Rightarrow 3v_1^2 - 4vv_1 + v^2 = 0$$

$$\Rightarrow 3\left(\frac{v_1}{v}\right)^2 - 4\left(\frac{v_1}{v}\right) + 1 = 0$$

$$\text{or} \quad \frac{v_1}{v} = -\frac{-(-4) \pm \sqrt{16 - 12}}{2 \times 3}$$

$$\Rightarrow \frac{v_1}{v} = \frac{4 \pm 2}{2 \times 3} \Rightarrow v_1 = v \text{ (Not possible)}$$

$$\text{or } v_1 = \frac{1}{3}v$$

So, final K.E of ball of mass $2m$,

$$k_2 = \frac{1}{2}(2m)(v_1^2) = \frac{1}{2} \times 2m \times \frac{v^2}{9} = \frac{1}{9}(k_1)$$

Hence, loss of K.E. of I[st] ball

$$= K_1 - \frac{1}{g}K_1 = \frac{8}{9}K_1$$

10. (d) Given displacement is,

$$x = 2t^2$$

$$\Rightarrow \quad v = \text{velocity} = \frac{dx}{dt} = 4t$$

$$v_{\text{inital}} = v(t = 0) = 4 \times 0 = 0 \text{ m/s}$$

$$v \quad = v(t = 1) = 4 \times 1 = 4 \text{ m/s}$$

ΔK.E = change in K.E of body

$$= \frac{1}{2}m(v^2 \quad - v^2 \quad)$$

$$= \frac{1}{2} \times 1 \times (16 - 0) = 8 \text{ J}$$

By work-kinetic energy theorem,

Work done = ΔK.E = 8 J

11. (a) If x = displacement of free end of spring, then

$$mg = kx \quad \text{or } x = \frac{mg}{k}$$

Time period of oscillation, $T = 2\pi\sqrt{\frac{m}{k}}$

$$\Rightarrow \omega = \text{Angular frequency} = \frac{2\pi}{T} = \sqrt{\frac{k}{m}}$$

Maximum velocity of oscillating mass

$$= V_{\max} = A\omega = \frac{mg}{k} \times \sqrt{\frac{k}{m}} = g\sqrt{\frac{m}{k}}$$

12. (d) Instantaneous power is

$$P = \ . \ = \ .$$

As force & acceleration are constant.

So, velocity of the particle must be keep on increasing.

Hence, power is increasing uniformly with time. i.e. $P \propto t$.

13. (c) When comet orbits around Sun in an elliptical orbit, it is under action of a central force and its angular momentum remains constant.

14. (e) When a satellite moves in a circular orbit around the earth its

(i) Potential energy,

$$\because \quad V = -\frac{GMm}{r}$$

(ii) Kinetic energy,

$$K = \frac{1}{2}mv^2 = \frac{GMm}{2r} \quad \left[\because v = \sqrt{\frac{GM}{r}}\right]$$

$$\therefore \quad V = -2K$$

15. (d) Given, mass of earth = $10 \times M_m$

Where, M_m = Mass of mars

Radius of earth = $2R_m$

Where, R_m = radius of mass

$$\text{and} \quad g = \frac{GM}{R^2}$$

Let gravity on the surface of mass is g_m

$$\therefore \quad \frac{g_m}{g_E} = \frac{M_m}{M_E} \times \left(\frac{R_E}{R_m}\right)^2$$

$$g_m = g_e \times \frac{M_m}{M_E}\left(\frac{R_E}{R_m}\right)^2$$

$$= 10 \times \frac{M_m}{10M_m}\left(\frac{2R_m}{R_m}\right)^2 = 4 \text{ m/s}^2$$

16. (b) Given, Semi-major axis of the orbit of saturn = $9r_E$

Where, r_E = semi major axis of earth

According to Kepler's law, $T^2 \propto r^3$

Let the time period of revolution of saturn around the sun is T_s

$$\therefore \quad \frac{T_s^2}{T_E^2} = \left(\frac{9r_E}{r_E}\right)^3 \Rightarrow T_s^2 = T_E^2(9)^3$$

$$T_s = \sqrt{T_E^2(9)^3} = 9^{3/2} \times 1 \text{ year}$$

$$\approx 27 \text{ years}$$

17. (b) The time period of mass,

$$T = 2\pi\sqrt{\frac{m}{k}}$$

Given, $m = 3 \text{ kg} \Rightarrow k = 48$ N/m

$$T = 2\pi\sqrt{\frac{3}{48}} = 2\pi\sqrt{\frac{1}{16}} = 2\pi \times \frac{1}{4} = \frac{\pi}{2}$$

18. (c) Given, the position and velocity of the particle executing SHM.

$$y = 3 \text{ cm} \Rightarrow v = 8 \text{ cm/s}$$

Angular frequency, $\omega = 2$ rad's

The velocity,

$$v = \omega\sqrt{a^2 - y^2} \Rightarrow 8 = 2\sqrt{a^2 - (3)^2}$$

$$\Rightarrow 4 = \sqrt{a^2 - (3)^2}$$

$$\Rightarrow 16 = a^2 - 9 \Rightarrow a^2 = 25$$

$$\Rightarrow a = 5 \text{ cm}$$

19. (b) $x = \sin^2 \omega t - 2\cos^2 \omega t$

$$= 1 - 3\cos^2 \omega t = 1 - 3\left(\frac{1 + \cos 2\omega t}{2}\right)$$

$$= -\frac{1}{2} - \frac{3}{2}\cos 2\omega t$$

which is a periodic function with angular frequency of 2ω.

20. (e) Speed of wave in a string,

$$v = \sqrt{\frac{T}{\mu}}$$

Given, $T = 80$ N

$\Rightarrow \mu = 32$ g/m $= 32 \times 10^{-3}$ kg/m

$$v = \sqrt{\frac{80}{32 \times 10^{-3}}} = \sqrt{\frac{8 \times 10^4}{32}}$$

$$= 100 \times \frac{1}{2} = 50 \text{ m/s}$$

21. (b) Given,

Velocity of sound = 330 m/s

Length of closed pipe = 1.5 m

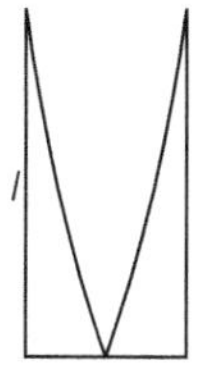

In a closed pipe for fundamental mode

$$\Rightarrow \frac{\lambda}{4} = l$$

$$\Rightarrow \lambda = 4 \times 1.5 = 6 \text{ m} \Rightarrow v = n\lambda$$

$$\Rightarrow n = \frac{v}{\lambda} \Rightarrow = \frac{330}{6} = 55 \text{ Hz}$$

22. (c) Nodes are produced in the open pipe as shown in figure.

So, $2\lambda = 4$

$\Rightarrow \lambda = \frac{4}{2} = 2m$

From wave equation,

$v = n\lambda$

$\Rightarrow n = \frac{v}{\lambda}$

$= \frac{300}{2}$

$= 150$ Hz

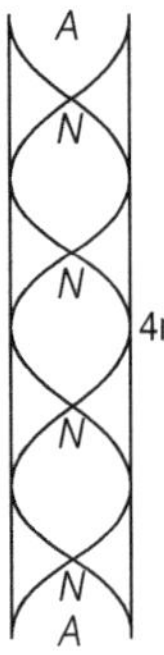

23. (d) Given,

Speed of sound source = 22 m/s

Frequency of emitted sound = 700 Hz

Velocity of sound = 330 m/s

When the sound source is moving the apparent frequency of sound heard by the observer

$$n' = n\left[\frac{v}{v - v_s}\right]$$

$$\Rightarrow n' = 700\left[\frac{330}{300 - 22}\right] = 700\left[\frac{330}{308}\right]$$

$= 749.99 \simeq 750$ Hz

24. (a) According to the figure, the displacement at point P_1 is increasing, so velocity at P_1 will be positive.

At point P_2, displacement is decreasing, so the instantaneous velocity will be negative.

25. (b) When the ball falls from a table top with initial speed v_0, its horizontal component of the velocity will remain unchanged with time because their is no air resistance.

26. (c) The man is climbing down using elevator, so the resultant gravity on the man will be $\because g' = g - a = 10 - 4$

$= 6$ m/s^2

The weight of the person, $W = mg'$

$= 60 \times 6 = 360$ N

27. (d) The one end of the uniform rod is fixed and force T is acting in upward direction

So, at point P, $T = \frac{mg}{2} = \frac{2 \times 10}{2} = 10$ N

28. (b) Given, Radius of capillary tube = 0.5 mm = 0.5×10^{-3} m

Level inside tube = 0.8 cm

$= 0.8 \times 10^{-2}$ m

Angle of contract, $\theta = 120°$

mass density of mercury,

$\rho = 13.6 \times 10^3$ kg/m^3

Acceleration due to gravity, $g = 10$ m/s^2

$$h = \frac{2T\cos\theta}{r\rho g} \Rightarrow T = \frac{hr\rho g}{2\cos\theta}$$

$$= \frac{0.8 \times 10^{-2} \times 0.5 \times 10^{-3} \times 13.6 \times 10^3 \times 10}{2 \times \cos 120}$$

$$= \frac{0.8 \times 10^{-2} \times 0.5 \times 10^{-3} \times 13.6 \times 10^3 \times 10}{2 \times \frac{1}{2}}$$

$= 0.8 \times 0.5 \times 13.6 \times 10^{-1} = 0.544$ N/m

29. (c) Stress is linearly proportional to strain for stress much smaller than at the yield point. Because Hook's law gives (with in elastic limits).

Stress $\propto$ Strain

30. (c)

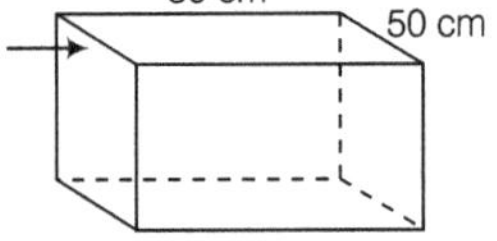

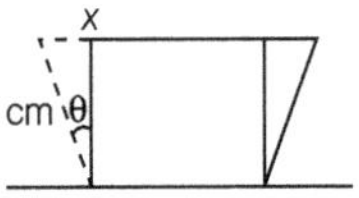

Shear strain, $\varepsilon = \tan\theta = \theta = \frac{x}{l}$

$$\eta = \frac{\sigma}{\varepsilon} \text{ or } x = \frac{F.l}{A.\eta}$$

Given, $F = 30$ N $\Rightarrow A = 50 \times 50 \times 10^{-4}$

$\eta = 4 \times 10^{10}$

$$= \frac{30 \times 20 \times 10^{-2}}{50 \times 50 \times 10^{-4} \times 4 \times 10^{10}}$$

$$= \frac{-6}{24 \times 4 \times 10^{-4} \times 10^{10}} = 6 \times 10^{-10} \text{ m}$$

31. (b) Given,

Mass of water at 90°C = 100 gm

$= 100 \times 10^{-3}$ kg

Mass of water at 20°C = 600 gm

$= 600 \times 10^{-3}$ kg

From calorimetery

$m_1 s_1 t_1 + m_2 s_2 t_2 = (m_1 + m_2) s.T$

$\because s_1 t_1 + s_2 t_2 = st$

[where, T is temperature of mixture].

$100 \times 10^{-3} \times 1 \times 90 + 600 \times 10^{-3} \times 1 \times 20$

$= (100 + 600) \times 10^{-3} \times 1 \times T$

$$T = \frac{100 \times 10^{-3} \times 90 + 600 \times 10^{-3} \times 20}{700 \times 10^{-3}}$$

$$= \frac{(9000 + 12000) \times 10^{-3}}{700 \times 10^{-3}} = \frac{21000}{700}$$

$= 30$°C

32. (a)

I–4	($\because$ In isothermal process, temperature remains constant
II–3	($\because$ In isobaric process, pressure remains constant)
III–2	($\because$ In isochoric process, volume remains constant)
IV–1	($\because$ In adiabatic process, total heat of the system remains constant)

33. (a) For an ideal gas, the specific heat at constant pressure C_p is greater than C_V. This is because some finite work has to be done by the gas on its environment when its temperature is increased while the pressure remains constant.

34. (c) Light and waves on the string transverse but sound waves in air are longitudinal.

35. (d) The fringe width,

$$\beta = \frac{\lambda D}{d}$$

Where, λ = wavelength of light

d = distance of slit

D = distance of screen from slit.

According to question,

$d' = \frac{d}{2} \Rightarrow D = 2D \Rightarrow \beta' = ?$

$$\beta' = \frac{\lambda D'}{d'} = \frac{\lambda 2D}{d/2} = 4\beta$$

So, the fringe width will be quadrupled.

36. (c) Given, $\psi = \psi_0 \sin\left(kx + \omega t + \frac{\pi}{2}\right)$

Phase velocity of wave

$$v = \frac{\omega}{k} = \frac{\text{Angular frequency}}{\text{Propagation constant}}$$

37. (b) When ^{65}Cu is bombard with neutron, the neutron is absorbed in ^{65}Cu and turn into ^{66}Cu.

38. (a) Given,

Input $V_{rms} = \frac{V_0}{\sqrt{2}}$

Where, V_0 = Peak value of voltage

In full wave rectifier, the whole cycle is rectified, so the value of input voltage will be same as output.

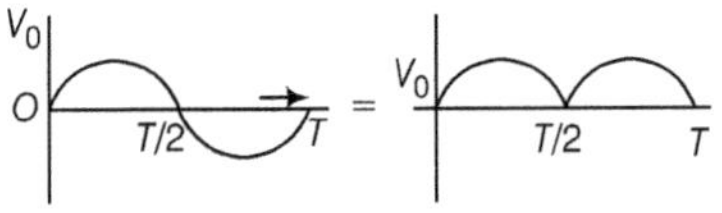

So, the output voltage will be $\frac{V_0}{\sqrt{2}}$,

39. (a) Given, $m_0 = 8\,g \Rightarrow m = 1\,g$

Time of decay = 15 min

$$\ln\frac{M}{M_0} = -\lambda t \Rightarrow \ln\left(\frac{1}{8}\right) = -\lambda(55)$$

$$\therefore \quad \lambda = \frac{\ln 8}{15}$$

So, $T_{1/2} = \frac{\ln 2}{\lambda} = \frac{15 \ln 2}{\ln 8}$

40. (c) Given,

Atomic number = Z

Mass number = A

For lighter nuclei, the relation between Z and A is

$$Z = \frac{A}{2}$$

41. (a) $w_1^0 = 12 \times 2\pi \frac{\text{rad}}{s} = 24\pi \frac{\text{rad}}{s}$

$w_f = 0, \ \Delta t = 6s$

As $w_f = w_i + \alpha \Delta t$

We have, $\alpha = \frac{-w_i}{\Delta t}$

$$= -\frac{24\pi}{6} = -4\pi \frac{\text{rad}}{\text{sec}}$$

42. (c) As; $\tau = I\alpha$

So, $\alpha = \frac{\tau}{I}$

Now $w = \alpha \Delta t$

And $L = Iw = I\alpha\Delta t = I\frac{\tau}{I}\Delta t = \tau \Delta t$

$$= 1(N.m) \times 2(s) = 2 \text{ kgm}^2\text{s}^{-1}$$

KCET

1. (c) Velocity of car $(u) = 20 \text{ ms}^{-1}$

Distance $(s) = 40$ m

From Newton third equation,

$$v^2 - u^2 = 2as \Rightarrow v^2 - u^2 = -2\,as$$

$$(0)^2 - (20)^2 = -2(a) \times 40$$

$$\Rightarrow 400 = 2 \times 40 \times a$$

$$a = \frac{400}{2 \times 40} \Rightarrow a = 5 \text{ ms}^{-2}$$

In the second condition, the velocity becomes twice i.e. $u' = 2u$

Again from Newton's third equation, we get

$$(0)^2 - (2u)^2 = 2 \times (5) \times s$$

$$s = \frac{(2u)^2}{2 \times 5} \Rightarrow s = \frac{4u^2}{2 \times 5}$$

$$s = \frac{4 \times 20 \times 20}{2 \times 5} \Rightarrow s = 160 \text{ m}$$

2. (*) (No option is matched)

Mass of the substance $(m) = 49.53$ g

Volume $(V) = 1.5 \text{ cm}^3$

We know that,

$$\Rightarrow \quad d = \frac{m}{V} \quad \Rightarrow d = \frac{49.53}{1.5}$$

$$\Rightarrow d = 33.02 \quad \Rightarrow d = 33.0$$

3. (a) We know that,

Acceleration of centre of mass

$$\mathbf{a}_{\text{cm}} = \frac{m_1\mathbf{a}_1 + m_2\mathbf{a}_2}{m_1 + m_2}$$

Here, $a_1 = a_2 = g$ (Because balls are thrown in air i.e. under the gravity)

$$\Rightarrow \mathbf{a}_{\text{cm}} = \frac{m_1 g + m_2 g}{m_1 + m_2}$$

$$\Rightarrow \mathbf{a}_{\text{cm}} = \frac{g(m_1 + m_2)}{m_1 + m_2} \Rightarrow \mathbf{a}_{\text{cm}} = g$$

4. (d) Given,

Mass of the body, $m = 50$ kg

When lift is moving down,

Scale reading $= \frac{m(g - a)}{g}$

If lift starts falling freely, then $a = g$

Now, scale reading $= \frac{m(g - g)}{g} = 0$

5. (d) The SI unit of specific heat capacity is J kg^{-1}K^{-1}

6. (a) We know that, $\eta = \left(1 - \frac{T_L}{T_H}\right) \times 100$

According to trial and error method,

$$\Rightarrow \quad \eta = \left(1 - \frac{80}{100}\right) \times 100$$

$$\Rightarrow \quad \eta = \left(\frac{100 - 80}{100}\right) \times 100$$

$$\Rightarrow \quad \eta = \frac{20}{100} \times 100 = 20\%$$

7. (d) When light ray goes from air to denser medium, then its wavelength and speed decreases because every medium have a different velocity of light.

8. (c) Given,

Depth $(d) = 1600$ km

We know that, $g_d = g\left(1 - \frac{d}{R}\right)$

Here, $g = 9.8 \text{ m/s}^2 \Rightarrow R = 6400$ km

$$g_d = g\left(1 - \frac{1600}{6400}\right) = 9.8\left(1 - \frac{1}{4}\right)$$

$$= 9.8\left(\frac{4 - 1}{4}\right) = \frac{9.8 \times 3}{4}$$

$$g_d = 7.35 \text{ ms}^{-2}$$

9. (b) According to questions,

For pendulum A,

$$T_A = 2\pi\sqrt{\frac{l_A}{g}} \quad \Rightarrow \quad \frac{20}{10} = 2\pi\sqrt{\frac{l_A}{g}} \quad \ldots \text{(i)}$$

For pendulum B,

$$T_B = 2\pi\sqrt{\frac{l_B}{g}} \Rightarrow \frac{10}{8} = 2\pi\sqrt{\frac{l_B}{g}} \quad \ldots \text{(ii)}$$

Dividing Eq. (i) by Eq. (ii), we get

$$\frac{l_A}{l_B} = \left(\frac{160}{100}\right)^2 \Rightarrow \frac{l_A}{l_B} = \frac{64}{25}$$

10. (a) Given,

Height of first floor $(h) = 35$ m

Time $(t) = 20$ m

We know that,

Power, $P = \frac{\text{work}}{\text{time}} \Rightarrow P = \frac{mgh}{t}$

$[\because \text{work} = mgh]$

$$= \frac{600 \times 10 \times (35 - 25)}{20} = \frac{600 \times 10 \times 10}{20}$$

$$= \frac{6000}{2} = 3000 \text{ W} = 3 \text{ kW}$$

11. (d) The wave set up in a closed pipe are longitudinal stationary waves.

12. (a) Given, $\mathbf{B} = 2\hat{} + 3\hat{} + 8\hat{}$

$$\mathbf{B} = 4\hat{\mathbf{j}} - 4\hat{\mathbf{i}} + \alpha\hat{\mathbf{k}} = -4\hat{\mathbf{i}} + 4\hat{\mathbf{j}} + \alpha\hat{\mathbf{k}}$$

$\mathbf{B} \perp \mathbf{B}$ Hence, $\mathbf{A} \cdot \mathbf{B} = 0$

$$\Rightarrow (2\hat{\mathbf{i}} + 3\hat{\mathbf{j}} + 8\hat{\mathbf{k}}) \cdot (-4\hat{\mathbf{i}} + 4\hat{\mathbf{j}} + \alpha\hat{\mathbf{k}}) = 0$$

$$\Rightarrow \ -8 + 12 + 8\alpha = 0$$

$$\Rightarrow \ 8\alpha + 4 = 0 \Rightarrow 8\alpha = -4$$

$$\alpha = -\frac{4}{8} = -\frac{1}{2}$$

13. (b) When an object follows a circular path at a constant speed, the motion of object is called uniform circular motion.

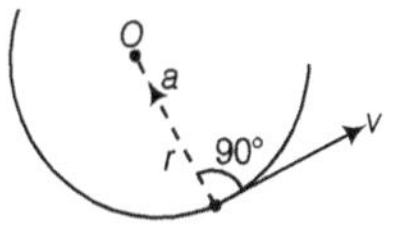

Although the speed does not vary the particle is accelerating because the velocity changes its direction at every point on circular track.

The acceleration is centripetal, which is perpendicular to motion at every point and acts along the radius and directed towards the centre of the curved circular path.

14. (c) Young's modulus

$$= \frac{\text{tensile stress}}{\text{longitudinal strain}}$$

15. (a) We know that,

$$R = \frac{\rho l}{A} = \frac{\rho l}{\pi r^2} = \frac{\rho l}{\pi r^2}$$

If the length of wire is increases, then the radius of wire is decreases.

To get the maximum resistance the suitable length and diameter are $2L$ and $\frac{d}{2}$.

16. (b) During scattering of light, the amount of scattering is inversely proportional to fourth power of wavelength of light i.e. $I \propto \frac{1}{\lambda^4}$

17. (c) Hydraulic lift works on the basis of Pascal's law.

18. (d) Average energy of a molecule of an ideal gas,

$$= \frac{E}{N} = \frac{3}{2}KT$$

AP EAMCET

1. (*) (No options match)

We know that, $W = P\Delta V$

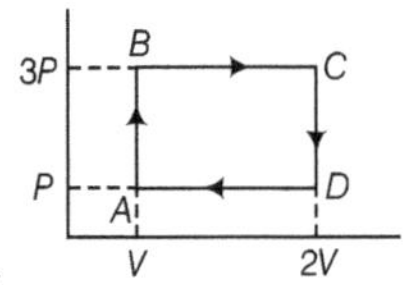

Now,

$$W_{AB} = 0$$
$$W_{BC} = 3P(2V - V) = 3PV$$
$$W_{CD} = 0$$
$$W_{DA} = -PV$$

Total work done,

$$W = W_{AB} + W_{BC} + W_{CD} + W_{DA}$$
$$= 3PV - PV = 2PV$$

Heat given to the system from A to B $= nC_V\Delta T$

$$= n\frac{3}{2}R\Delta T = \frac{3}{2} \times V \times \Delta P$$
$$= \frac{3}{2} \times V \times (3P - P)$$
$$= \frac{3}{2} \times V \times 2P = 3PV$$

Similarly,

Heat given to the system from B to C

$$= nC_P\Delta T = n\left(\frac{5}{2}R\right)\Delta T = \frac{5}{2}(3P)(\Delta V)$$
$$= \frac{5}{2} \times (3P).(2V - V) = \frac{5}{2} \times 3P \times V$$
$$= \frac{15}{2}PV$$

The heat is released from C to D and D to A.

Efficiency (η)

$$= \frac{\text{Work done by gas}}{\text{Heat given to gas}} \times 100$$
$$= \frac{2PV}{3PV + \frac{15}{2}PV} \times 100$$
$$= \frac{2PV}{PV\left(3 + \frac{15}{2}\right)} \times 100 = \frac{2}{\left(\frac{6+15}{2}\right)} \times 100$$
$$= \frac{2 \times 2}{21} \times 100 = \frac{4}{21} \times 100 = 19.04\%$$

2. (c) For fig. (a)

$$(m_1 + m_2)a_1 = m_2 g;\ a_1 = \frac{m_2 g}{m_1 + m_2}$$

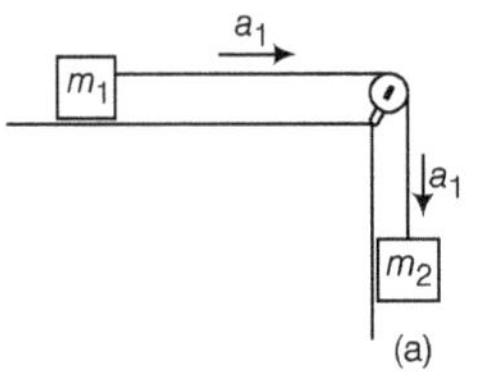

(a)

Given, $m_1 = 3$ kg; $m_2 = 4$ kg

$$g = 10 \text{ m/s}^2$$
$$a_1 = \frac{4 \times 10}{3 + 4} = \frac{40}{7} \text{ m/s}^2$$

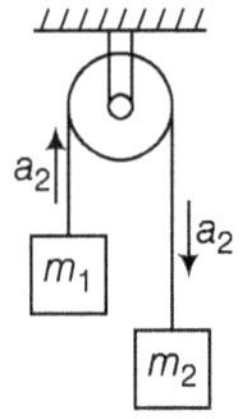

(b)

For fig. (b)

$$(m_1 + m_2)a = (m_2 - m_1)g$$
$$a = \left(\frac{m_2 - m_1}{(m_1 + m_2)}\right)g$$
$$= \left(\frac{4-3}{4+3}\right) \times 10 = \frac{10}{7} \text{ m/s}^2$$

3. (d)

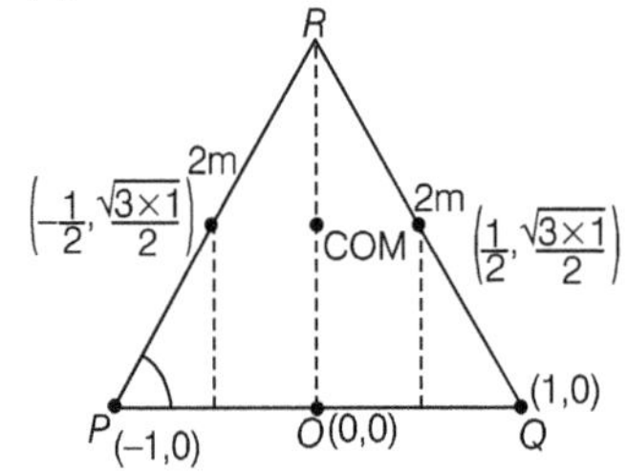

Initially,

$$\mathbf{r}_{COM} = \frac{m_1\mathbf{r}_1 + m_2\mathbf{r}_2 + m_3\mathbf{r}_3}{(m_1 + m_2 + m_3)}$$
$$\mathbf{r}_{COM} = \frac{1}{3}\left(\frac{\sqrt{3}}{2} \times 1 \times 2\right)\hat{\mathbf{j}}$$
$$\Rightarrow \mathbf{r}_{COM} = \frac{1}{\sqrt{3}}\hat{\mathbf{j}}$$

Due to expansion each rod length becomes

$$l_f = l_i(1 + \alpha\Delta T) = l_i\left(1 + \frac{\gamma}{3}\Delta T\right)$$
$$l_f = 2\left(1 + \frac{12\sqrt{3} \times 10^{-6}}{3} \times 50\right)$$
$$l_f = 2(1 + 4 \times 50 \times \sqrt{3} \times 10^{-6})$$
$$= 2(1 + 2\sqrt{3} \times 10^{-4})$$

So, $(\mathbf{r}_{COM})_{\text{final}}$

$$= \frac{1}{3} \times \frac{\sqrt{3}}{2} \times (1 + 2\sqrt{3} \times 10^{-4}) \times 2\hat{\mathbf{j}}$$
$$= \left(\frac{1}{\sqrt{3}} + 2 \times 10^{-4}\right)\hat{\mathbf{j}}$$

Hence, $\Delta y = (r_{COM})_{\text{final}} - (r_{COM})_{\text{initial}}$

$$= 2 \times 10^{-4} \text{ m} = 0.2 \text{ mm}$$

4. (a) Time in which body is dropped to ground is calculated using following data;

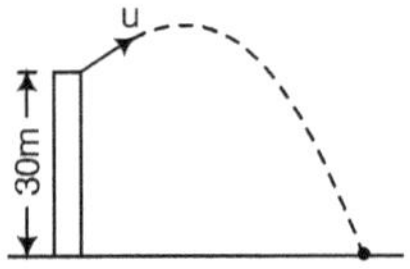

$u = 5$ m/s

($\hat{\mathbf{k}}$ is given vertically upward direction)

$$a = -10 \text{ m/s}^2 \Rightarrow h = -30 \text{ m}$$

Now using the formula

$$h = ut + \frac{1}{2}at^2, \Rightarrow -30 = 5t - \frac{1}{2} \times 10 \times t^2$$
$$\Rightarrow t^2 - t - 6 = 0$$
$$\Rightarrow = t = -2 \text{ s} \quad \text{(Not possible)}$$
or $\quad t = 3$ s

In this time, projectile moving in East and North with speeds 3m/s and 4m/s. Distances covered in these directions are;

In east (x – coordinate).

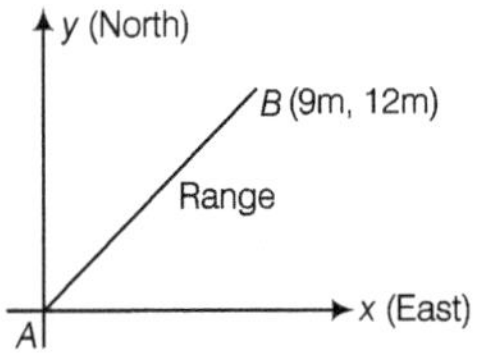

In east (x – coordinate).

$= \text{speed} \times \text{time} = 3 \times 3 = 9$ m

and in North (y – coordinate)

$= 4 \times 3 = 12$ m

So, Projective lando at $(x, y) \equiv (9\text{m}, 12\text{m})$ mark.

So, horizontal range of body on ground is;

$$\text{Range} = \sqrt{9^2 + 12^2} = \sqrt{225} = 15 \text{ m}$$

5. (d) Given, $y = Px - Qx^2$ ...(i)

We know that,

$$y = x\tan\theta - \frac{1}{2}g\frac{x^2}{u^2\cos^2\theta} \quad \text{...(ii)}$$

After comparing Eqs. (i) and (ii), we get

$$P = \tan\theta \quad \text{...(iii)}$$
$$Q = \frac{g}{2u^2\cos^2\theta} \quad \text{...(iv)}$$

We know that

$$H = \frac{u^2 \sin^2 \theta}{2g} \quad \ldots(v)$$

$$R = \frac{u^2 \sin 2\theta}{g} \quad \ldots(vi)$$

Dividing Eq. (v) by Eq. (vi), we get

$$\frac{H}{R} = \frac{u^2 \sin^2 \theta}{2g} \times \frac{g}{u^2 \sin 2\theta}$$

$$\Rightarrow \frac{H}{R} = \frac{\sin^2 \theta}{2 \times 2 \sin\theta \cos\theta}$$

$$\frac{H}{R} = \frac{1}{4} \times \tan\theta$$

From Eq. (iii), we have

$$\frac{H}{R} = \frac{P}{4}$$

6. (d) Transverse displacement

$$y_{(x,t)} = 0.03 \sin\left(\frac{2\pi x}{3}\right) \cos 60\pi t$$

Linear density = 0.01 kg/m

The standard equation of transverse displacement of a wave

$$y = a \sin kx \cos \omega t$$

Velocity, $v = \sqrt{\frac{T}{\mu}} = \frac{\omega}{k}$

$\therefore \sqrt{\frac{T}{\mu}} = \frac{\omega}{k}$ [Here, μ = mass per unit length]

Here, $\omega = 60\pi$, $k = 2\pi/3$

$$\sqrt{\frac{T}{0.01}} = \frac{60\pi}{2\pi/3} = 90$$

$$\frac{T}{0.01} = 90 \times 90$$

$$T = 90 \times 90 \times 0.01 = 81 \text{ N}$$

7. (c) Given,

Mass of the girl (m) = 50 kg

Velocity (v) = 2 m/s

Angle $(\theta) = 60°$

$g = 9.8 \text{ m/s}^2$

Work needed to displace the girl through a small arc of length dS is

$= mg \sin 60° \times ds$

But $ds = vdt$

Now, work done in time dt

$= mg \sin 60° \, vdt$

Power at that instant $= mg \sin 60°. \, v$

$= 50 \times 9.8 \times \sin 60° \times 2$

$= 50 \times 9.8 \times \frac{\sqrt{3}}{2} \times 2$

$= 25 \times 9.8 \times \sqrt{3} \times 2 = 490\sqrt{3}$ W

8. (d) In the steady state, the rate of flow of heat through layer P and Q is same.

$$\frac{k_P \, a(T_1 - T_0)}{x} = \frac{k_Q \, a(T_0 - T_2)}{x}$$

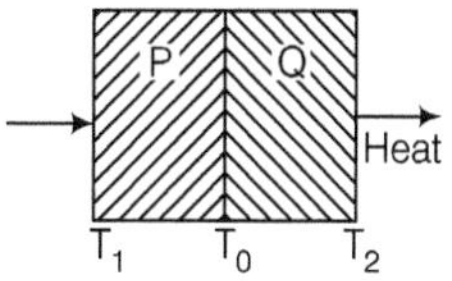

Here,

$$K_Q = \frac{K_P}{2};\ K_p = 2K_Q$$

$$\frac{2k_Q \, a(T_1 - T_0)}{x} = \frac{k_Q \, a(T_0 - T_2)}{x}$$

$$2(T_1 - T_0) = T_0 - T_2$$

$$2T_1 - 2T_0 = T_0 - T_2$$

$$2T_1 - 3T_0 + T_2 = 0 \quad \ldots(i)$$

According to the question,

$$T_1 - T_2 = 24°\text{C} \quad \ldots(ii)$$

Now, from Eq. (i),

$$2T_1 - 3T_0 + T_1 - 24 = 0$$

$$3T_1 - 3T_0 = 24 \Rightarrow T_1 - T_0 = 8°\text{C}$$

9. (a) At the position when the gravitational field is zero. (Distance x is measured from 4m mass)

$$\frac{G(4m)}{x} = \frac{G(9m)}{(r-x)} \Rightarrow \frac{4}{x} = \frac{9}{r-x}$$

$$4(r - x) = 9x \Rightarrow 4r - 4x = 9x$$

$$4r = 9x + 4x \Rightarrow 4r = 13x$$

$$x = \frac{4}{13} r$$

The point P is at a distance $\frac{4}{13} r$ from mass $4m$ and $\left(h - \frac{4}{13} r\right) = \frac{9r}{13}$ from mass m.

$$v = -\frac{G(4m)}{\left(\frac{4}{13} r\right)} - \frac{G(9m)}{\frac{9r}{13}} \Rightarrow v = -\frac{26Gm}{r}$$

Option (a) is nearest but none of the option is matching.

10. (a) The total time duration for the two cars to meet

$$= \frac{\text{Total distance}}{\text{Relative velocity}} = \frac{36}{54 + 36}$$

$$\Rightarrow T = \frac{36}{90} = 0.4 \text{ h}$$

In this duration, the distance travels

$= 36 \times 0.4 = 14.4$ km $= 14400$ m

The total distance covered by the bird before the cars meet each other is 14400 m.

11. (d) Given, $V = 10$ V

$$k = 1.38 \times 10^{-23} \text{ JK}^{-1}$$

According to question, the translation KE of a molecule of gas $= \frac{3}{2} kT$

According to question,

$$\frac{3}{2} kT = eV$$

Temperature, $T = \frac{2eV}{3k}$

$$= \frac{2 \times 1.6 \times 10^{-19} \times 10}{3 \times 1.38 \times 10^{-23}}$$

$$= 7.73 \times 10^4 = 77.3 \times 10^3 \text{ K}$$

12. (b) For closed organ pipe,

$$f = f_2 = 3f_1 = \frac{3 \times v}{4L} = \frac{3}{4L} \times \sqrt{\frac{\gamma p}{\rho_1}}$$

For open organ pipe,

$$f = f_2 = 2f_1 = \frac{2v}{2L'} = \frac{v}{L'} = \frac{1}{L'} \times \sqrt{\frac{\gamma p}{\rho_2}}$$

For same sound frequency,

$$\frac{3}{4L} \times \sqrt{\frac{\gamma P}{\rho_1}} = \frac{1}{L'} \times \sqrt{\frac{\gamma p}{\rho_2}}$$

$$\therefore \quad L' = \frac{4L}{3} \times \sqrt{\frac{\rho_1}{\rho_2}}$$

13. (d) For given gas,

$$pV = RT \text{ and } V = KT^2 \Rightarrow pV = R\left(\frac{V}{K}\right)^{\frac{1}{2}}$$

$pV^{\frac{1}{2}}$ = a constant

$\therefore \gamma = \frac{1}{2}$ for this process.

$\therefore$ Work done in this process,

$$W = \frac{nR(T_i - T_f)}{\gamma - 1} = \frac{R(60)}{\frac{1}{2}} = 120R$$

14. (b) Given,

Length of pendulum = L

Time period $T_1 = 2\pi \sqrt{\frac{L}{g}}$

Length of rod = L

Time period of rod;

$$T_2 = 2\pi \sqrt{\frac{2L}{3g}} \Rightarrow \frac{T_1}{T_2} = \frac{2\pi \sqrt{L/g}}{2\pi \sqrt{2L/3g}} = \sqrt{\frac{3}{2}}$$

15. (a) There are four rods each of mass $m = 400$ g

$$= \frac{400}{1000} = 0.4 \text{ kg}$$

and each having a length of $l = 30$ cm

$$= \frac{30}{100} \text{ m} = 0.3 \text{ m}$$

Now, out of therefore two sets of two rods have same m, I.

So, we can calculate moment of inertia as follows

$$h^2 = l^2 - \left(\frac{l}{2}\right)^2 = \frac{3}{4} l^2$$

M.I system = (MI of I) × 2 + (M.I of II) × 2

$$= \frac{ml^2}{3} \times 2 + 2\left(\frac{ml^2}{12} + mh^2\right)$$

$$= \frac{2}{3}ml^2 + 2\left(\frac{ml^2}{12} + \frac{3}{4}ml^2\right)$$

$$= \frac{2}{3}ml^2 + 2ml^2\left[\frac{1}{12} + \frac{3}{4}\right]$$

$$= \frac{2}{3}ml^2 + 2ml^2 \times \frac{10}{12} = \frac{2}{3}ml^2$$

$$= \frac{7}{3} \times 0.4 \times 0.3 \times 0.3 = 0.084 \text{ kg - m}^2$$

16. (c)

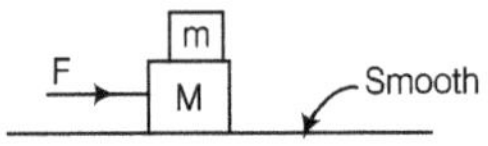

Acceleration of block $= \dfrac{F}{(M+m)}$

So, friction acting over upper block, f

$$ma = \frac{mF}{(M+m)}$$

Distances moved in time $t = \frac{1}{2}at^2$

$$= \frac{1}{2}\left(\frac{F}{M+m}\right)t^2$$

Work done by friction $= f \times s$

$$= \left(\frac{mF}{M+m}\right) \times \frac{1}{2}\frac{Ft^2}{(M+m)} = \frac{mF^2t^2}{2(M+m)^2}$$

17. (a)

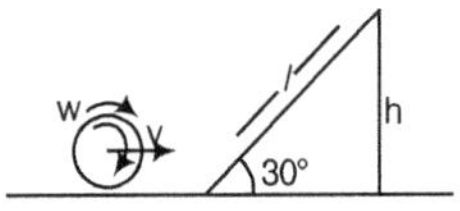

Given, $v = 4\text{ms}^{-1}$ Also $v = r\omega$

Hence,

KE at bottom $= K_{\text{translation}} + K_{\text{rotation}}$

$$= \frac{1}{2}mv^2 + \frac{1}{2}I\omega^2$$

$$= \frac{1}{2}mv^2 + \frac{1}{2}\left[\frac{2}{5}mr^2\frac{v^2}{r^2}\right]$$

$$= \frac{1}{2}mv^2 + \frac{1}{2} \times \frac{2}{5}mv^2$$

$$= \frac{1}{2}\left(\frac{7}{5}mv^2\right)$$

At height h, energy conservation gives

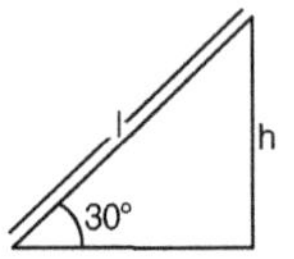

$$\frac{1}{2}\left(\frac{7}{5}\right)mv^2 = mgh$$

$$\Rightarrow h = \frac{7}{10} \times \frac{v^2}{g} = \frac{7}{10} \times \frac{16}{10} = 1.12 \text{ m}$$

As, $l = h / \sin 30°$

$\Rightarrow l = 2 \times h = 2.24$ m

18. (a)

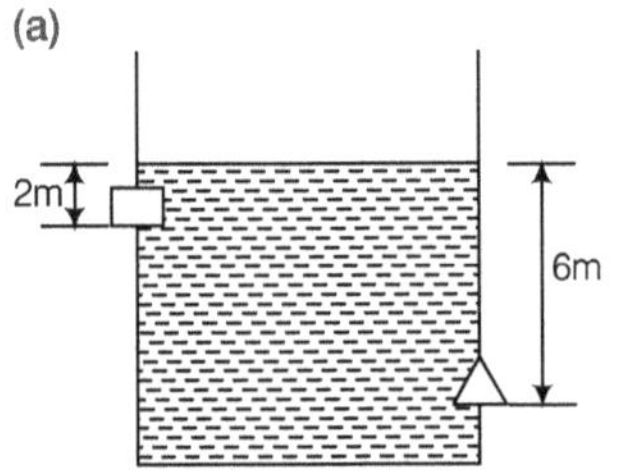

Volume flow rate

$= $ area $\times$ velocity

$= x^2 \times \sqrt{2 \times g \times 2}$ ... square hole

$= \frac{\sqrt{3}}{4} \times 4^2 \times \sqrt{2 \times g \times 6}$ triangle hole

$$\Rightarrow \quad x^2 = \frac{\sqrt{3} \times \sqrt{2g} \times \sqrt{6}}{\sqrt{2g} \times \sqrt{2}}$$

$$x^2 = \sqrt{3} \times \sqrt{3}$$

$$\therefore \quad x = \sqrt{3} \text{ cm} = 1.73 \text{ cm}$$

19. (c) Let the ratio of length be a, ratio of area be b

Ratio of Young's modulus be c

We know that,

$$Y = \frac{F/A}{\Delta L/L} \Rightarrow \Delta L = \frac{FL}{AY}$$

$$\frac{\Delta L_S}{\Delta L_B} = \left(\frac{F_S}{F_B}\right)\left(\frac{L_S}{L_B}\right)\left(\frac{A_B}{A_S}\right)\left(\frac{Y_B}{Y_S}\right)$$

$$= \frac{7}{4} \times (a)\left(\frac{1}{b}\right) \times \left(\frac{1}{c}\right)$$

$$\frac{\Delta L_S}{\Delta L_B} = \frac{7a}{4bc} \Rightarrow \frac{\Delta L_B}{\Delta L_S} = \frac{4bc}{7a}$$

20. (a) Mass of person, $m = 60$ kg

Force on floor of lift, $F = 150$ N

As lift is coming down the resultant acceleration $a' = (g + a)$

Using Newton's force law

$F = ma' \Rightarrow 150 = 60\,(10 + a)$

$$10 + a = \frac{150}{60} = 2.5$$

$$a = -7.5 \text{ m/s}^2 \Rightarrow |a| = 7.5 \text{ m/s}^2$$

TS EAMCET

1. (b) Given,

Percentage error in length = 3%

Percentage error in force = 4%

We know that,

Pressure, $\quad P = \dfrac{F}{A}$

Area of square $= (L)^2$

$$P = \frac{F}{L^2} \Rightarrow \frac{\Delta P}{P} = \frac{\Delta F}{F} + \frac{2\Delta L}{L}$$

$$\left(\frac{\Delta P}{P} \times 100\right) = \left(\frac{\Delta F}{F} \times 100\right) + 2\left(\frac{\Delta L}{L} \times 100\right)$$

$$= 4\% + 2(3\%) = 10\%$$

2. (b) Given,

Width of river $(w) = 200$ m

Velocity of river = 2 m/s

Velocity of man = 1 m/s

We know that,

$$\frac{w}{v_{\text{man}}} = \frac{d}{v_{\text{strem}}} \Rightarrow d = \frac{w \times v_{\text{strem}}}{v}$$

$$d = \frac{200 \times 2}{1} \Rightarrow d = 400 \text{ m}$$

3. (d) Given,

Length of simple pendulum $(L) = 1$ m

Acceleration $(a) = 2 \text{ m/s}^2$

Gravitation acceleration $(g) = 10 \text{ m/s}^2$

We know that,

$$T = 2\pi\sqrt{\frac{l}{g+a}} = 2\pi\sqrt{\frac{1}{10+2}}$$

$$= 2\pi\sqrt{\frac{1}{12}} = \pi\sqrt{\frac{4}{12}} \Rightarrow I = \frac{\pi}{\sqrt{3}} \text{ s}$$

4. (d) Given, Young's modulus of material $= 2 \times 10^{11} \text{ N/m}^2$

Elastic limit or stress $= 1 \times 10^8 \text{ N/m}^2$

Length of the wire = 1 m

We know that,

$$Y = \frac{\text{Stress}}{\text{Strain}} \Rightarrow Y = \frac{\text{Stress}}{\frac{\Delta L}{l}}$$

Here, l = original length of the wire

ΔL = charge in the length of the wire

$$\Delta L = \frac{\text{Stress} \times l}{Y} = \frac{1 \times 10^8 \times 1}{2 \times 10^{11}}$$

$$= 0.5 \times 10^{-3} = 0.5 \text{ mm}$$

5. (d) According to question, we can draw the following diagram.

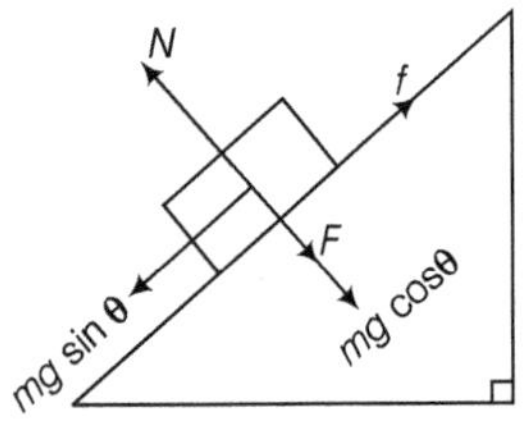

$$mg\sin\theta = \mu\,(F + mg\cos\theta)$$

$$F = \frac{mg\sin\theta}{\mu} - mg\cos\theta$$

$$= mg\left[\frac{\sin\theta}{\mu} - \cos\theta\right]$$

Here, $m = 1$ kg, $g = 10$ m/s^2, $\theta = 30°$, $\mu = 0.2$

$$F = 1 \times 10\left[\frac{\sin 30}{0.2} - \cos 30°\right]$$

$$= 10\left[\frac{1}{2 \times 0.2} - \frac{\sqrt{3}}{2}\right]$$

$$= 10\left[\frac{5}{2} - \frac{\sqrt{3}}{2}\right] = 5\,[5 - \sqrt{3}]$$

$$= 5\,[5 - 1.732] = 16.34 \text{ N}$$

6. (a) Given,

Coefficient of linear expansion of steel, $\alpha = 11 \times 10^{-6}/°C$

We know that,

$$\Delta l = l\alpha\Delta t \Rightarrow \Delta t = \frac{\Delta l}{l\alpha}$$

Here, $\Delta l = 6 \times 10^{-5}$ m, $l = 1$ m

$$\Delta t = \frac{6 \times 10^{-5}}{1 \times 11 \times 10^{-6}} = 5.45° \text{ C}$$

So, the range of temperature in which the experiment can be hence performed him this metre scale will be 19°C to 31°C

Here, option (a) is correct.

7. (d) Given,

Total wall area (including the lid) $(A) = 1.0$ cm^2

Thickness of wall $(l) = 3$ cm $= 3 \times 10^{-2}$ m

Average temperature outside the box $= 30°C$

$$\Delta\theta = 30 - 0 = 30°C$$

$$L_{\text{fusion (ice)}} = 3 \times 10^5 \text{ J/kg}$$

$$K_{\text{thermocol}} = 0.03 \text{ W/m K}$$

We know that,

$$\frac{Q}{t} = \frac{KA}{l}\Delta\theta$$

For one day, $t = 24 \times 60 \times 60$ s

$$\frac{m \times L_{\text{fusion (ice)}}}{t} = \frac{KA}{l}\Delta Q$$

$$\frac{m \times 3 \times 10^5}{24 \times 60 \times 60} = \frac{0.03 \times 1}{3 \times 10^{-2}} \times 30$$

$$m = \frac{0.03 \times 1 \times 30 \times 24 \times 60 \times 60}{3 \times 10^{-2} \times 3 \times 10^5}$$

$$\Rightarrow m = \frac{77760}{9000} = 8.64 \text{ kg}$$

8. (a) According to the question,

Velocity of an object half a second before maximum height
= Velocity of an object half a second after maximum height (return journey)

$$= 0 + gt = 0 + 9.8 \times \frac{1}{2} = 4.9 \text{ m/s}$$

9. (a) Given, $x = a\cos t$, $y = a\sin t$
and $z = t$

Path is circular in xy- plane,

$$\frac{x^2}{a^2} + \frac{y^2}{a^2} = \cos^2\theta + \sin^2\theta$$

$$\Rightarrow \frac{x^2}{a^2} + \frac{y^2}{b^2} = 1$$

In one revolution, the particle move a distance of 1 unit along z-axis.

$$\frac{dz}{dt} = 1$$

Hence, path is helix.

10. (a) Given $\eta_1 = \frac{1}{6}$

According to the question,

$$\eta_1 = \frac{T_1 - T_2}{T_1} \Rightarrow \frac{T_1 - T_2}{T_1} = \frac{1}{6} \quad \text{...(i)}$$

$$\eta_2 = \frac{T_1 - (T_2 - 62)}{T_1}$$

$$\Rightarrow 2 \times \eta_1 = \frac{T_1 - T_2 + 62}{T_1}$$

$$2 \times \frac{1}{6} = \frac{T_1 - T_2 + 62}{T_1}$$

$$\Rightarrow \frac{1}{3} = \frac{T_1 - T_2 + 62}{T_1} \quad \text{...(ii)}$$

From Eq. (i), we get $T_1 - T_2 = \frac{T_1}{6}$

Substituting this value in Eq. (ii), we get

$$\frac{1}{3} = \frac{\frac{T_1}{6} + 62}{T_1} \Rightarrow \frac{1}{3} = \frac{T_1 + 372}{6T_1}$$

$$6T_1 = 3T_1 + 1116 \Rightarrow 3T_1 = 1116$$

$$\Rightarrow T_1 = 372 \text{ K}$$

From Eq. (i), we get

$$\Rightarrow \frac{372 - T_2}{372} = \frac{1}{6} \Rightarrow 6(372 - T_2) = 372$$

$$\Rightarrow 2232 - 6T_2 = 372$$

$$\Rightarrow 6T_2 = 2232 - 372$$

$$\Rightarrow 6T_2 = 1860 \Rightarrow T_2 = \frac{1860}{6}$$

$$\Rightarrow T_2 = 310 \text{ K}$$

11. (c) Given,

Number of moles of air in room $(n) = 2000 = 2 \times 10^3$

Temperature difference $(dT) = 24 - 34 = -10°C$

We know that, $dQ = nC_v dT$

$$= 2 \times 10^3 \times \frac{R}{0.4}[-10]$$

$$= \frac{-2 \times 10^3 \times 8.314 \times 10}{0.4}$$

$$= \frac{2 \times 8.314 \times 10^5}{4} = \frac{16.628}{4} \times 10^5$$

$$= -4.2 \times 10^5 \text{ J}$$

12. (d) Given, Speed of ball $(u) = 20$ m/s

Angle $(\theta) = 30°$. We know that,

$$H = \frac{u^2\sin^2\theta}{2g} \Rightarrow H = \frac{(20)^2(\sin 30)^2}{2 \times 10}$$

$$\Rightarrow H = \frac{20 \times 20 \times \left(\frac{1}{2}\right)^2}{20}$$

$$\Rightarrow H = 20 \times \frac{1}{4} \Rightarrow H = 5 \text{ m}$$

13. (c) Given, Diameter of pipe $(d) = 5$ mm $= 5 \times 10^{-3}$ m

Density of gasoline $(\rho) = 720$ kg/m^3.

Viscocity of gasoline $(\eta) = 6 \times 10^{-3}$ Poise

We know that, $v_c = \frac{\eta}{\rho d}$

$$= \frac{6 \times 10^{-3}}{720 \times 5 \times 10^{-3}} = \frac{6 \times 10^{-3}}{3600 \times 10^{-3}} = \frac{1}{600}$$

$$= \frac{1}{6} \times 10^{-2} = 1.66 \times 10^{-3} \text{ m/s}$$

14. (d) SONAR technology is based on the reflection of ultrasonic waves.

15. (b) Given, Velocity of particle $= v$

Change in momentum = final momentum − initial momentum

$$= Mv - (-Mv) = 2Mv$$

16. (a) When billiard ball collide elastically, they will exchange the kinetic energy, so they may move in straight path or at any angle.

17. (a)

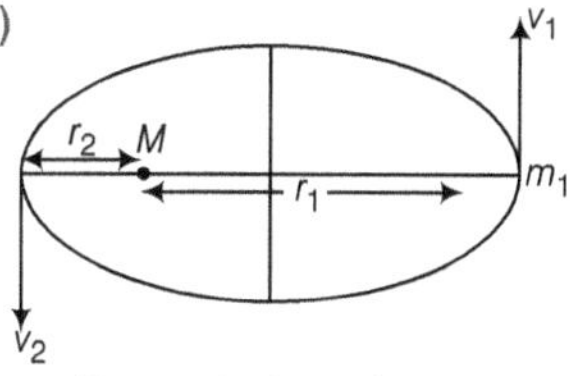

According to the law of conservation of angular momentum,

$$mv_1r_1 = mv_2r_2 \Rightarrow v_2 = \frac{v_1r_1}{r_2} \quad \text{...(i)}$$

From the law of conservation of total mechanical energy.

$$\frac{-GMm}{r_1} + \frac{1}{2}mv_1^2 = -\frac{GMm}{r_2} + \frac{1}{2}mv_2^2 \quad \text{...(ii)}$$

From Eqs. (i) and (ii), we get

$$v_1 = \sqrt{\frac{2GMr_2}{(r_1 + r_2)\,r_1}}$$

Angular momentum,

$$L = mv_1r_1 = m\left(\sqrt{\frac{2GMr_2}{(r_1 + r_2)\,r_1}}\right) \times r_1$$

$$L = m\sqrt{\frac{2GMr_1r_2}{r_1 + r_2}}$$

18. (a) From the conservation of energy,

$$mgh = \mu mgd \Rightarrow d = \frac{h}{\mu}$$

19. (c) The frequency of sound wave remains constant.

20. (d) Deceleration $\omega = -a\sqrt{v}$.

But $\omega = \frac{dv}{dt} \Rightarrow \frac{dv}{dt} = -a\sqrt{v}$

$\frac{-dv}{\sqrt{v}} = a\,.\,dt \Rightarrow \int_{v_i}^{0} \frac{dv}{\sqrt{v}} = \int a\,dt$

$|2\sqrt{v}|_{v_i}^{0} = at \Rightarrow t = \frac{2}{a}\sqrt{v_i s}$

Again, $\frac{-dv}{dt} = a\sqrt{v}$

$\frac{dv}{dx}\cdot\frac{dx}{dt} = -a\sqrt{v} \Rightarrow \frac{dv}{dx}\cdot v = -a\sqrt{v}$

$dv\sqrt{v} = -a\,.\,dx \Rightarrow \int_{v_0}^{0}\sqrt{v}dv = -a\int_0^s ds$

After solving, we get $s = \frac{2}{3a}\cdot v_0^{\frac{3}{2}}$

21. (a) Given, $\mathbf{F}_1 = (\hat{\mathbf{i}} + 2\hat{\mathbf{j}} + 3\hat{\mathbf{k}})$ N

$\mathbf{F} = \hat{\mathbf{i}} - \hat{\mathbf{j}} - \hat{\mathbf{k}}$ N

$\mathbf{r}_1 = 20\hat{\mathbf{i}} + 15\hat{\mathbf{j}}$ cm, $\mathbf{r}_2 = 7\hat{\mathbf{k}}$ cm

Total force on particle, $\mathbf{F} = \mathbf{F}_1 + \mathbf{F}_2$

$= \hat{\mathbf{i}} + 2\hat{\mathbf{j}} + 3\hat{\mathbf{k}} + 4\hat{\mathbf{i}} - 5\hat{\mathbf{j}} - 2\hat{\mathbf{k}}$

$= (5\hat{\mathbf{i}} - 3\hat{\mathbf{j}} + \hat{\mathbf{k}})$ N

Displacement of particle, $\mathbf{s} = \mathbf{r}_2 - \mathbf{r}_1$

$= 7\hat{\mathbf{k}} - 20\hat{\mathbf{i}} - 15\hat{\mathbf{j}}$

$= (-20\hat{\mathbf{i}} - 15\hat{\mathbf{j}} + 7\hat{\mathbf{k}})$ cm

We know that,

Work $(\omega) = \mathbf{F}\cdot\mathbf{s}$

$= (5\hat{\mathbf{i}} - 3\hat{\mathbf{j}} + \hat{\mathbf{k}})(-20\hat{\mathbf{i}} - 15\hat{\mathbf{j}} + 7\hat{\mathbf{k}}) \times 10^{-2}$

$= (-100 + 45 + 7) \times 10^{-2} = -0.48$ J

VIT

1. (b) Initial acceleration $a = g$

Velocity (v) of stone as it hits the sand is given by $v^2 = u^2 + 2as$

$\Rightarrow v^2 = 0 + 2g \times 100 \Rightarrow v = \sqrt{200g}$

When stone traverses in sand initial velocity $u' = u = \sqrt{200g}$, final velocity $u' = 0$

$\therefore v^2 = u^2 + 2as$ gives $0 = 200g + 2a \times 2$

or $a = -50g = 50$ g (retardation)

Now, $v = u + at$ gives

$0 = \sqrt{200g} - 50gt$

$t = \frac{\sqrt{200g}}{50g} = 0\cdot089$ sec

2. (b) We have, $P_1 + P_2 + P_3 = 0$ $[\because P = mu]$

$\therefore \quad 1 \times 12\hat{\mathbf{i}} + 2 \times 8\hat{\mathbf{j}} + P_3 = 0$

$\Rightarrow \quad 12\hat{\mathbf{i}} + 16\hat{\mathbf{j}} + P_3 = 0$

$\Rightarrow \quad P_3 = 12\hat{\mathbf{i}} + 16\hat{\mathbf{j}}$

$|P_3| = \sqrt{(12)^2 + (16)^2}$

$= \sqrt{144 + 256} = 20$ kg m/s

Now, $P_3 = m_3 v_3 \Rightarrow m_s = \frac{P_3}{v_3}$

$= \frac{20}{4} = 5$ kg

3. (a) We have $Y = \frac{F/A}{\Delta L / L'}$

$\therefore$ Increase in length $\Delta L = \frac{FL'}{AY}$

If ρ is density of material, L the length and A the area of rope, longitudinal force,

F = weight of rope $= mg = (LA\rho)g$

As weight acts on centre of gravity, it increases only the length above centre of gravity,

$\therefore L' = \frac{L}{2} \Rightarrow \Delta L = \frac{LA\rho g(L/2)}{AY} = \frac{L^2\rho g}{2Y}$

Substituting given values

$\therefore \quad \Delta L = \frac{(8)^2 \times 1.5 \times 10^3 \times 10}{2 \times 5 \times 10^8}$

$= 9.6 \times 10^{-4}$ m

4. (c) We know that

$h = \frac{2s}{r\rho g} = \frac{2 \times 75 \times 10^{-3}}{0.25 \times 10^{-3} \times 10^3 \times 10}$

$= 60$ mm

Now, $h'r' = hr$

or $r' = \frac{hr}{h'} = \frac{60 \times 0.25}{25} = 0.6$ mm

5. (c) Friction force between blocks A and B and between block B and surface will oppose F

$\therefore F = F_{AB} + F_{BS}$

$= \mu_{AB} m_A g + \mu_{BS}(m_A + m_B)g$

$= 0\cdot2 \times 100 \times 10 + 0.3(100 + 200) \times 10$

$= 200 + 900 = 1100$ N

This is the required minimum force to move the block B.

6. (a) Depression produced in a beam with circular cross section

$\delta = \frac{\omega l^3}{12\pi r^4 Y}$ (for solid rod) and

$\delta = \frac{\omega l^3}{12\pi(r_2^4 - r_1^4)Y}$ (for hollow rod)

$\because \quad \delta_1 \propto \frac{1}{r^4}$ and $\delta_2 \propto \frac{1}{r_2^4 - r_1^4}$

$\therefore \quad \frac{\delta_2}{\delta_1} = \frac{r^4}{r_2^4 - r_1^4}$ or $\frac{\delta_2}{8} = \frac{3^4}{4^4 - 2^4}$

or $\delta = 2.7$ mm

7. (d) $v = \sqrt{\frac{2gh}{1 + \frac{k^2}{r^2}}}$

Given $h = \frac{3v^2}{4g}$

$v^2 = \frac{2gh}{1 + \frac{k^2}{r^2}} = \frac{2g(3v^2)}{4g\left(1 + \frac{k^2}{r^2}\right)}$

$= \frac{6gv^2}{4g\left(1 + \frac{k^2}{r^2}\right)} = \frac{3}{2\left(1 + \frac{k^2}{r^2}\right)}$

or $1 + \frac{k^2}{r^2} = \frac{3}{2}$ or $\frac{k^2}{r^2} = \frac{3}{2} - 1 = 1/2$

$\Rightarrow \quad k^2 = \frac{1}{2}r^2$

(Equation of disc)

Hence the object is disc.

8. (a) A force acting on the block are

(i) weight mg vertically

(ii) normal reaction N

(iii) frictional force f_s

Resolving mg along and normal to plane, normal component is $mg\cos\theta$ and along the plane component is $mg\sin\theta$

For equilibrium, $N = mg\cos\theta$,

$f_s = mg\sin\theta$

Given $f_s = 10$ N

$\Rightarrow \quad m = \frac{f_s}{g\sin\theta} = \frac{10}{10 \times \sin 30°}$

$= \frac{10}{10 \times 0.5} = 2.0$ kg

9. (a) $\alpha_{brass} = \frac{\gamma_{brass}}{3} = \frac{0.00006}{3} = 0.00002$

$= 2 \times 10^{-5}/°C$

The brass scale is true at 15°C, therefore at 30° it graduations will increase in length and so observed reading will be less than actual reading at 30°.

$\therefore$ The change in reading

$\Delta l = l\alpha_{brass}(\Delta T)$

$= 74.5 \times 2 \times 10^{-5} \times (30 - 15) = 0.02235$

$\therefore$ Actual reading at 30°C

$l_{30} = l_{observed} + \Delta l$

$= 74.5 + 0.02235 = 74.522$ cm

Assuming area of cross section to be constant, we have

$V_0 P_0 = V_{30} P_{30}$ or $ah_0\rho_0 = ah_{30}\rho_{30}$

$\therefore$ True height at 0°C

$h_0 = h_{30}\frac{\rho_{30}}{\rho_0} = \frac{h_{30}}{(1 + \gamma_m \Delta T)}$

$= \frac{74.522}{1 + 0.0018 \times 30} = \frac{74.522}{1.0054} = 74.122$ cm

MHT CET

1. (c) Given, $I = 2 \text{ kg m}^2$

$\omega_0 = 60$ rad/s, $\omega = 0$

$t = 5$ min $= 5\times 60 = 300$ s

From the relation,

$$\omega = \omega_0 + \alpha t \Rightarrow \alpha = \frac{\omega - \omega_0}{t}$$

$$\Rightarrow \alpha = \frac{0-60}{300} = \frac{-60}{300} = \frac{-1}{5} \text{ rad/s}^2$$

for $t = 2$ min $= 2\times 60 = 120$ s

$$\omega = \omega_0 + \alpha t = 60 - \frac{1}{5}\times 120 = 60 - 24$$

$$\therefore \quad \omega = 36 \text{ rad/s}$$

As, angular momentum, $L = I\omega$

Substituting the values in the above relation, we get

$$L = 2\times 36 = 72 \text{ kg m}^2\text{/s}.$$

2. (d) Compare the given equation with the standard equation of wave motion,

$$Y = A\sin\left[2\pi\left(\frac{t}{T} - \frac{x}{\lambda}\right) + \frac{\pi}{4}\right]$$

where, A and λ are amplitude and wavelength, respectively.

Amplitude, $A = 3$ m

Wavelength, $\lambda = 10$ m

3. (c) The rate at which an object radiates energy is given as,

$$\frac{Q}{t} = \sigma A T^4$$

$$\Rightarrow \text{Power, } P = \sigma A T^4 \Rightarrow A \propto \frac{1}{T^4}$$

If the body radiates same power, then

$$\Rightarrow \frac{A_2}{A_1} = \frac{T_1^4}{T_2^4} \Rightarrow \frac{4\pi r_2^2}{4\pi r_1^2} = \frac{T_1^4}{T_2^4}$$

$$\therefore \quad \frac{r_2}{r_1} = \left(\frac{T_1}{T_2}\right)^2$$

4. (c) For open pipe first overtone, $\nu_1 = \frac{v}{L}$

For closed pipe first overtone,

$$\nu_1' = \frac{3v}{4L} \Rightarrow \nu_1 - \nu_1' = \frac{v}{L} - \frac{3v}{4L} = 3$$

$$\Rightarrow \frac{v}{3L} = 3 \Rightarrow \frac{v}{L} = 9 \quad \ldots\text{(i)}$$

When length of open pipe is made $\frac{L}{3}$, then fundamental frequency

$$\nu = \frac{v}{2\left(\frac{L}{3}\right)} = \frac{3v}{2L}$$

When length of closed pipe is made 3 times, then fundamental frequency

$$\nu' = \frac{v}{4(3L)} = \frac{v}{12L}$$

Beats produced $= \nu = \nu'$

$$= \frac{3v}{2L} - \frac{v}{12L} = \frac{17}{12}\cdot\frac{v}{L}$$

$$= \frac{17}{12}\times 12 \quad [\because \text{from Eq. (i)}]$$

$$= 17$$

5. (b) The maximum tension in the rope $= m(g+a)$

Stress in the rope, $T = \frac{m(g+a)}{\pi r^2}$

$$\therefore T = \frac{m(g+a)}{\pi r^2} = \frac{m(g+a)}{\pi\left(\frac{d}{2}\right)^2}$$

$$\Rightarrow T = \frac{4m(g+a)}{\pi d^2} \Rightarrow d^2 = \frac{4m(g+a)}{\pi T}$$

$$\therefore \quad d = \left[\frac{4m(g+a)}{\pi T}\right]^{1/2}$$

6. (b) Given, $m = 2$ kg, $v = 6$ m/s

and $k = 36$ N/m

Kinetic energy of rolling solid sphere

$$= \frac{1}{2}mv^2 + \frac{1}{2}I\omega^2 = \frac{1}{2}mv^2 + \frac{1}{2}\times\frac{2}{5}mr^2\omega^2$$

$$= \frac{1}{2}mv^2 + \frac{1}{5}mv^2 = \frac{7}{10}mv^2$$

The potential energy of the spring on maximum compression x

$$= \frac{1}{2}kx^2 \Rightarrow \frac{1}{2}kx^2 = \frac{7}{10}mv^2$$

$$\Rightarrow x^2 = \frac{14}{10}\frac{mv^2}{k}$$

Substituting the given values, we get

$$= \frac{14}{10}\times\frac{2\times(6)^2}{36} = 2.8 \Rightarrow x = \sqrt{2.8} \text{ m}$$

7. (d) Given, $\omega_0 = 0, \omega = 24$ rad/s, $t = 8$ s

From the relation, $\omega = \omega_0 + \alpha t$

$$\Rightarrow \quad \alpha = \frac{\omega - \omega_0}{t}$$

Substituting the given values, we get

$$\alpha = 28/8 = 3 \text{ rad/s}^2$$

Also, $\theta = \omega_0 t + \frac{1}{2}\alpha t^2 = 0 + \frac{1}{2}\times 3\times(8)^2$

$$= \frac{3\times 64}{2} = 96 \text{ rad}$$

8. (a) Fundamental frequency of the first wire is given as,

$$f = \frac{1}{2L_1}\sqrt{\frac{T}{m}} = \frac{1}{2L_1}\sqrt{\frac{T}{\pi r_1^2\rho}} = \frac{1}{2L_1 r_1}\sqrt{\frac{T}{\pi\rho}}$$

The first overtone,

$$f_1 = 2f = \frac{2}{2L_1 r_1}\sqrt{\frac{T}{\pi\rho}} = \frac{1}{L_1 r_1}\sqrt{\frac{T}{\pi\rho}} \quad \ldots\text{(i)}$$

The second overtone of the second wire is given as,

$$f_2 = \frac{3}{2L_2 r_2}\sqrt{\frac{T}{\pi\rho}} \quad \ldots\text{(ii)}$$

Given, $f_1 = f_2$

$$\Rightarrow \frac{1}{L_1 r_1}\sqrt{\frac{T}{\pi\rho}} = \frac{3}{2L_2 r_2}\sqrt{\frac{T}{\pi\rho}}$$

$$\therefore \quad 3L_1 r_1 = 2L_2 r_2$$

$$\Rightarrow \frac{L_1}{L_2} = \frac{2}{3}\cdot\frac{r_2}{r_1} = \frac{2}{3}\cdot\frac{r_2}{2r_2} \quad [\text{given, } r_1 = 2r_2]$$

$$= \frac{1}{3}$$

9. (a) Given, $F = 105$ dyne $= 105\times 10^{-5}$ N

$T = 7\times 10^{-2}$ N/m

As we know, circumference of the capillary × surface tension = upward force $2\pi r T = F$

$$2\pi r = \frac{F}{T} = \frac{105\times 10^{-5}}{7\times 10^{-2}} = 15\times 10^{-3} \text{ m}$$

$$= 1.5\times 10^{-2} \text{ m} = 1.5 \text{ cm}$$

10. (c) As we know, for rigid diatomic molecule.

$$\frac{C_p}{C_V} = \frac{7}{5} \Rightarrow C_V = \frac{5}{7}C_p \quad \ldots\text{(i)}$$

Also, $C_p - C_V = R$

$$\Rightarrow \quad C_p - \frac{5}{7}C_p = R \quad [\because \text{from Eq. (i)}]$$

$$\Rightarrow \frac{2}{7}C_p = R \Rightarrow n = \frac{2}{7} = 0.2857$$

11. (a) For ideal gas, $pV = nRT$

$pV = \frac{m'}{M}RT$ (where, m' is the mass of the gas and M molecular weight)

$$p = \frac{m'T}{V}\frac{RT}{M} \Rightarrow p = \frac{\rho RT}{M}$$

where, $\rho = \frac{m'}{V}$ = density of the gas

$\Rightarrow \rho = \frac{pM}{RT} = \frac{pM}{NkT}$, where N is Avogadro, number

$\Rightarrow \rho = \frac{pm}{KT}$, where $m = \frac{M}{N}$ = mass of each molecule.

12. (d) Volume of big drop = n (Volume of small drop)

$$\Rightarrow \quad \frac{4}{3}\pi R^3 = n\cdot\frac{4}{3}\pi r^3$$

where, R = radius of big water drop and r = radius of small water drop.

$$\Rightarrow \quad R^3 = nr^3 \Rightarrow R = n^{1/3}\cdot r \quad \ldots\text{(i)}$$

Surface energy of n drops, E_2

$$= n\times 4\pi r^2\times T$$

Surface energy of big drop,

$$E_1 = 4\pi R^2 T \Rightarrow \frac{E_2}{E_1} = \frac{nr^2}{R^2} = \frac{nr^2}{(n^{1/3}\cdot r)^2}$$

$$= \frac{nr^2}{n^{2/3}\cdot r^2} = n^{1/3} \quad [\because \text{from Eq. (i)}]$$

$$E_2 : E_1 = \sqrt[3]{n} : 1$$

13. (d) Given, $v = \pi$ m/s, $T = 16$ s

Displacement of the particle $= x = A\sin\omega t$

Velocity of the particle

$$= v = \frac{dx}{dt} = A\omega\cos\omega t \quad \ldots\text{(i)}$$

Angular velocity, $\omega = \frac{2\pi}{T} = \frac{2\pi}{16} = \frac{\pi}{8}$ rad/s

Substituting the values in Eq. (i), we get

$$\pi = A \times \frac{\pi}{8} \times \cos\frac{\pi}{8} \times 2$$

$$\Rightarrow \quad 1 = \frac{A}{8}\cos\frac{\pi}{4} = \frac{A}{8}\cdot\frac{1}{\sqrt{2}} \Rightarrow A = 8\sqrt{2}\text{ m}$$

14. (d) String vibrating in second overtone between two bridges can be represented as,

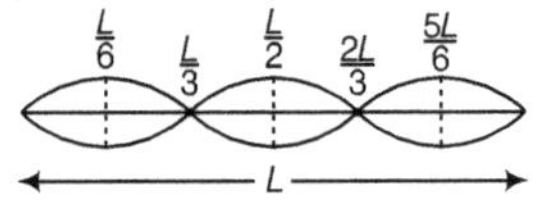

Therefore, the amplitude of vibration is maximum at $\frac{L}{6}, \frac{L}{2}, \frac{L}{6}$

15. (c) Acceleration due to gravity at depth d is given as,

$$g' = g\left(1 - \frac{d}{R}\right) \quad \ldots\text{(i)}$$

Given, $\quad g' = \frac{g}{n}$

Substituting the value of 'g' in Eq. (i), we get

$$\frac{g}{n} = g\left(1 - \frac{d}{R}\right) \Rightarrow \frac{1}{n} = 1 - \frac{d}{R}$$

$$\therefore \quad \frac{d}{R} = 1 - \frac{1}{n} = \frac{n-1}{n} \Rightarrow d = R\left(\frac{n-1}{n}\right)$$

16. (d) The graphical representation showing the relation of displacement and acceleration with time in SHM are shown below,

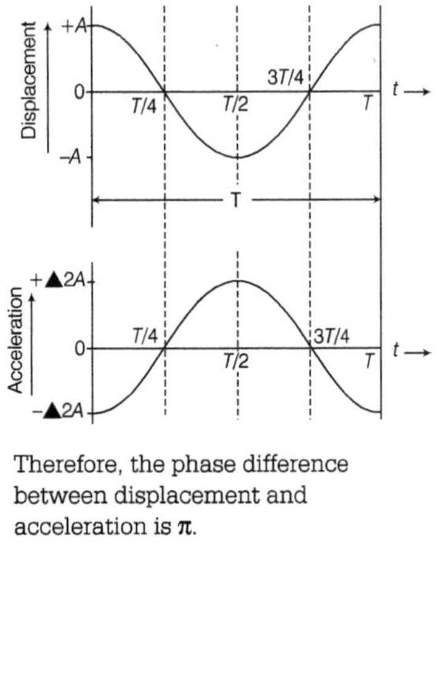

Therefore, the phase difference between displacement and acceleration is π.

17. (d) For a closed pipe fundamental frequency, $\nu_1 = \frac{v}{4L} = 100$ Hz

For an open pipe fundamental frequency, $\nu_1 = \frac{v}{2L} = 200$ Hz

In a pipe open at both the ends, all multiples of the fundamental are produced.

18. (a) When a particle is moving a verticle circle it would move from lowest point to the highest point. Its speed would decrease and becomes minimum at the highest point. Thus, the total mechanical energy remains conserved, kinetic energy changes into potential energy and *vice-versa*.

19. (a) Potential energy of a simple pendulum is given as,

$$= \frac{1}{2}M\omega^2A^2 = \frac{1}{2}M\cdot\frac{g}{L}\cdot A^2 \quad \left(\because \omega = \sqrt{\frac{g}{L}}\right)$$

20. (c) Given, $\omega_0 = \omega$, $\omega = \frac{\omega}{4}$; $\theta = 2\pi n$

From the relation, $\omega^2 = \omega_0^2 - 2\alpha\theta$

Substituting the given values, we get

$$\left(\frac{\omega}{4}\right)^2 = \omega^2 - 2\alpha n(2\pi)$$

$$2\alpha n(2\pi) = \omega^2 - \frac{\omega^2}{16} \Rightarrow 2\pi n = \frac{15}{16}\left(\frac{\omega^2}{2\alpha}\right)$$

Now, when the fan has been switched off,

$$\omega = 0,\ \omega_0 = \omega,\ \theta = 2\pi n'$$

$$\Rightarrow \quad 0 = \omega^2 - 2\alpha n'(2\pi)$$

$$\therefore \quad 2\pi n' = \frac{\omega^2}{2\alpha}$$

$$\Rightarrow \quad n' = \frac{16}{15}n$$

21. (d) Net external torque on the system is zero. Therefore, angular momentum of the system will remain same.

$$\Rightarrow \quad I_1\omega_1 = (I_1 + I_2)\,\omega_2$$

$$\frac{\omega_2}{\omega_1} = \frac{I_1}{I_1 + I_2} \quad \ldots\text{(i)}$$

The energy lost, $E_1 - E_2$

$$= \frac{1}{2}I_1\omega_1^2 - \frac{1}{2}(I_1 + I_2)\,\omega_2^2$$

$$= \frac{1}{2}\omega_1^2\left[I_1 - (I_1 + I_2)\frac{\omega_2^2}{\omega_1^2}\right]$$

$$= \frac{1}{2}\omega_1^2\left[I_1 - (I_1 + I_2)\frac{I_1^2}{(I_1 + I_2)^2}\right]$$

[$\because$ using Eq. (i)]

$$= \frac{1}{2}\omega_1^2\left[\frac{I_1^2 + I_1I_2 - I_1^2}{I_1 + I_2}\right] = \frac{1}{2}\left[\frac{I_1I_2}{I_1 + I_2}\right]\omega_1^2$$

22. (a) Let the distance be p when velocity is u and acceleration α.

Let the distance q when velocity is v and acceleration β.

If ω is the angular frequency, then

$$\alpha = \omega^2 p$$

and $\quad \beta = \omega^2 q$

$$\therefore \quad \alpha + \beta = \omega^2(p + q) \quad \ldots\text{(i)}$$

Also, $\quad u^2 = \omega^2A^2 - \omega^2p^2$

and $\quad v^2 = \omega^2A^2 - \omega^2q^2$

$$\Rightarrow \quad v^2 - u^2 = \omega^2(p^2 - q^2)$$

$$v^2 - u^2 = \omega^2(p - q)(p + q) \quad \ldots\text{(ii)}$$

By Eqs. (i) and (ii), we get

$$v^2 - u^2 = (p - q)(\alpha + \beta)$$

$$\therefore \quad p - q = \frac{v^2 - u^2}{\alpha + \beta}$$

or $\quad q - p = \frac{u^2 - v^2}{\alpha + \beta}$

23. (a) When the observer is moving towards the source and source is stationary, then the apparent frequency is given as,

$$n' = n\left(\frac{v + v_0}{v}\right) \quad \ldots\text{(i)}$$

When the observer is moving away from the source and the source is stationary, then the apparent frequency is given as,

$$n'' = -n\left(\frac{v - v_0}{v}\right) \quad \ldots\text{(ii)}$$

From Eqs. (i) and (ii), we get

$$n' - n'' = \frac{n}{v}(v + v_0 - v + v_0) = \frac{2nv_0}{v}$$

24. (b) Increase in the length of the rod is given as,

$$\Delta L = \alpha L_0 T$$

$$\Rightarrow \quad L = L_0[1 + \alpha T] \quad \ldots\text{(i)}$$

Also, Young's modulus,

$$Y = \frac{FL_0}{A\Delta L} \quad \ldots\text{(ii)}$$

Using relation (i) and substituting in the relation (ii), we get

$$Y = \frac{F}{A}\cdot\frac{L_0(1 + \alpha T)}{\Delta L}$$

or $\quad \Delta L = \frac{FL_0(1 + \alpha T)}{A\cdot Y} \quad \ldots\text{(iii)}$

From Eqs. (i) and (ii), we get

$$\frac{FL_0(1 + \alpha t)}{AY} = \alpha L_0 T$$

$$\Rightarrow \quad F = \frac{YA\alpha T}{(1 + \alpha T)}$$

Engineering Entrances Solved Papers 2018

JEE Main & Advanced/BITSAT/KCET/Andhra Pradesh & Telangana (EAMCET)/VIT/MHT CET

JEE Main

1. Seven identical circular planar discs, each of mass M and radius R are welded symmetrically as shown in the figure. The moment of inertia of the arrangement about an axis normal to the plane and passing through the point P is

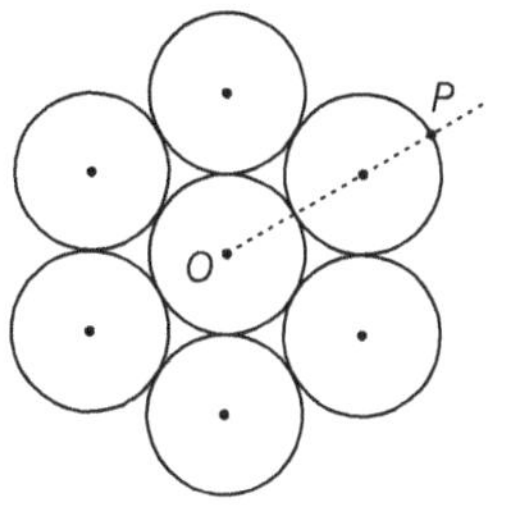

(a) $\frac{181}{2}MR^2$ (b) $\frac{19}{2}MR^2$ (c) $\frac{55}{2}MR^2$ (d) $\frac{73}{2}MR^2$

2. From a uniform circular disc of radius R and mass $9M$, a small disc of radius $\frac{R}{3}$ is removed as shown in the figure. The moment of inertia of the remaining disc about an axis perpendicular to the plane of the disc and passing through centre of disc is

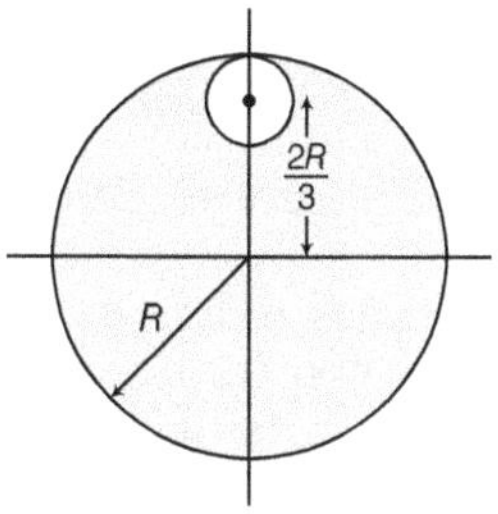

(a) $\frac{37}{9}MR^2$ (b) $4MR^2$ (c) $\frac{40}{9}MR^2$ (d) $10\,MR^2$

3. Two masses $m_1 = 5$ kg and $m_2 = 10$ kg connected by an inextensible string over a frictionless pulley, are moving as shown in the figure. The coefficient of friction of horizontal surface is 0.15. The minimum weight m that should be put on top of m_2 to stop the motion is

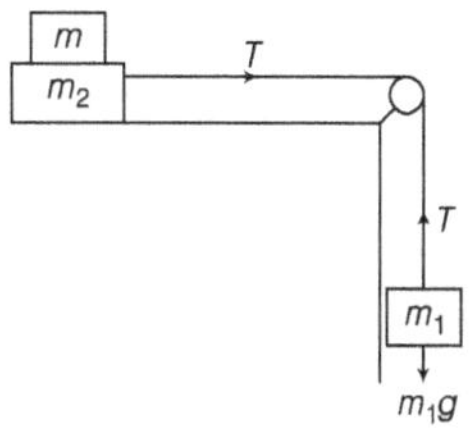

(a) 23.33 kg (b) 18.3 kg (c) 27.3 kg (d) 43.3 kg

4. In a collinear collision, a particle with an initial speed v_0 strikes a stationary particle of the same mass. If the final total kinetic energy is 50% greater than the original kinetic energy, the magnitude of the relative velocity between the two particles after collision, is

(a) $\frac{v_0}{\sqrt{2}}$ (b) $\frac{v_0}{4}$ (c) $\sqrt{2}\,v_0$ (d) $\frac{v_0}{2}$

5. A particle is moving with a uniform speed in a circular orbit of radius R in a central force inversely proportional to the nth power of R. If the period of rotation of the particle is T, then

(a) $T \propto R^{n/2}$ (b) $T \propto R^{3/2}$ for any value of n

(c) $T \propto R^{\frac{n}{2}+1}$ (d) $T \propto R^{\frac{n+1}{2}}$

6. A granite rod of 60 cm length is clamped at its middle point and is set into longitudinal vibrations. The density of granite is 2.7×10^3 kg/m^3 and its Young's modulus is 9.27×10^{10} Pa. What will be the fundamental frequency of the longitudinal vibrations?

(a) 7.5 kHz (b) 5 kHz (c) 2.5 kHz (d) 10 kHz

7. It is found that, if a neutron suffers an elastic collinear collision with deuterium at rest, fractional loss of its energy is P_d; while for its similar collision with carbon nucleus at rest, fractional loss of energy is P_c. The values of P_d and P_c are respectively

(a) (0, 1) (b) (.89, .28) (c) (.28, .89) (d) (0, 0)

8. The density of a material in the shape of a cube is determined by measuring three sides of the cube and its mass. If the relative errors in measuring the mass and length are respectively 1.5% and 1% , the maximum error in determining the density is

(a) 6% (b) 2.5% (c) 3.5% (d) 4.5%

9. Two moles of an ideal monoatomic gas occupies a volume V at 27°C. The gas expands adiabatically to a volume $2V$. Calculate (i) the final temperature of the gas and (ii) change in its internal energy.

(a) (i) 195 K (ii) 2.7 kJ (b) (i) 189 K (ii) 2.7 kJ
(c) (i) 195 K (ii) –2.7 kJ (d) (i) 189 K (ii) –2.7 kJ

10. A solid sphere of radius r made of a soft material of bulk modulus K is surrounded by a liquid in a cylindrical container. A massless piston of area a floats on the surface of the liquid, covering entire cross-section of cylindrical container.

When a mass m is placed on the surface of the piston to compress the liquid, the fractional decrement in the radius of the sphere, $\left(\frac{dr}{r}\right)$ is

(a) $\frac{mg}{Ka}$ (b) $\frac{Ka}{mg}$ (c) $\frac{Ka}{3mg}$ (d) $\frac{mg}{3Ka}$

11. The mass of a hydrogen molecule is 3.32×10^{-27} kg. If 10^{23} hydrogen molecules strike per second, a fixed wall of area 2 cm^2 at an angle of 45° to the normal and rebound elastically with a speed of 10^3 m/s, then the pressure on the wall is nearly

(a) 2.35×10^4 N/m^2 (b) 2.35×10^3 N/m^2
(c) 4.70×10^2 N/m^2 (d) 2.35×10^2 N/m^2

12. All the graphs below are intended to represent the same motion. One of them does it **incorrectly**. Pick it up.

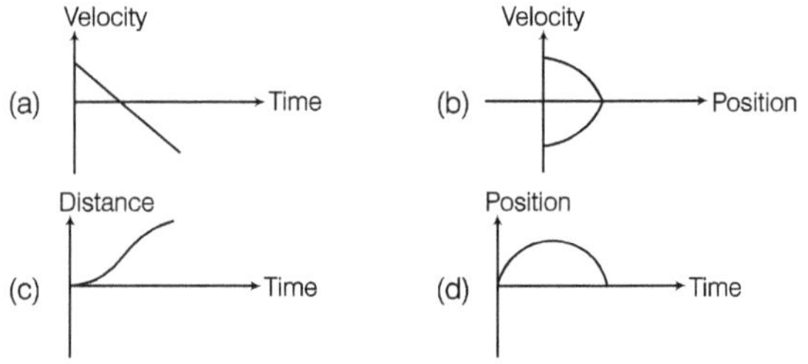

13. A particle is moving in a circular path of radius a under the action of an attractive potential energy $U=-\frac{k}{2r^2}$. Its total energy is

(a) $-\frac{3}{2}\cdot\frac{k}{a^2}$ (b) $-\frac{k}{4a^2}$ (c) $\frac{k}{2a^2}$ (d) zero

14. A silver atom in a solid oscillates in simple harmonic motion in some direction with a frequency of 10^{12} per second. What is the force constant of the bonds connecting one atom with the other? (Take, molecular weight of silver $=108$ and Avogadro number $=6.02\times10^{23}$ g mol^{-1})

(a) 5.5 N/m (b) 6.4 N/m (c) 7.1 N/m (d) 2.2 N/m

JEE Advanced

Paper 1

1. The potential energy of mass m at a distance r from a fixed point O is given by $V(r)=kr^2/2$, where k is a positive constant of appropriate dimensions. This particle is moving in a circular orbit of radius R about the point O.

If v is the speed of the particle and L is the magnitude of its angular momentum about O, which of the following statements is (are) true ?

(One or More than One Option Correct)

(a) $v=\sqrt{\frac{k}{2m}}R$ (b) $v=\sqrt{\frac{k}{m}}R$ (c) $L=\sqrt{mk}R^2$ (d) $L=\sqrt{\frac{mk}{2}}R^2$

2. Consider a body of mass 1.0 kg at rest at the origin at time $t=0$. A force $\mathbf{F}=(\alpha t\,\hat{\mathbf{i}}+\beta\,\hat{\mathbf{j}})$ is applied on the body, where $\alpha=1.0\text{Ns}^{-1}$ and $\beta=1.0\text{N}$. The torque acting on the body about the origin at time $t=1.0$s is τ. Which of the following statements is (are) true ?

(One or More than One Option Correct)

(a) $|\tau|=\frac{1}{3}$N-m

(b) The torque τ is in the direction of the unit vector $+\hat{\mathbf{k}}$

(c) The velocity of the body at $t=1$ s is $\mathbf{v}=\frac{1}{2}(\hat{\mathbf{i}}+2\hat{\mathbf{j}})$ ms^{-1}

(d) The magnitude of displacement of the body at $t=1$ s is $\frac{1}{6}$m

3. A uniform capillary tube of inner radius r is dipped vertically into a beaker filled with water. The water rises to a height h in the capillary tube above the water surface in the beaker. The surface tension of water is σ. The angle of contact between water and the wall of the capillary tube is θ. Ignore the mass of water in the meniscus. Which of the following statements is (are) true ? *(One or More than One Option Correct)*

(a) For a given material of the capillary tube, h decreases with increase in r

(b) For a given material of the capillary tube, h is independent of σ

(c) If this experiment is performed in a lift going up with a constant acceleration, then h decreases

(d) h is proportional to contact angle θ

4. One mole of a monoatomic ideal gas undergoes a cyclic process as shown in the figure (where, V is the volume and T is the temperature). Which of the statements below is (are) true ?

(One or More than One Option Correct)

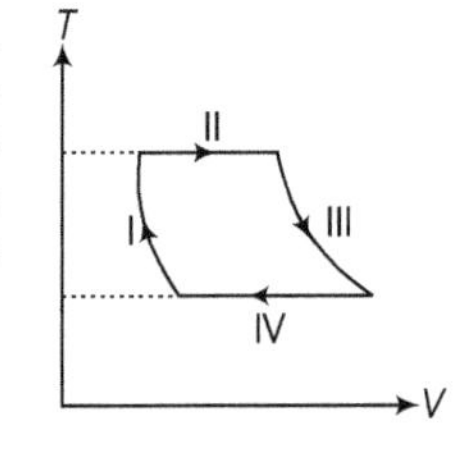

(a) Process I is an isochoric process
(b) In process II, gas absorbs heat
(c) In process IV, gas releases heat
(d) Processes I and III are not isobaric

5. Two vectors **A** and **B** are defined as $\mathbf{A} = a\hat{\mathbf{i}}$ and $\mathbf{B} = a(\cos\omega t\hat{\mathbf{i}} + \sin\omega t\hat{\mathbf{j}})$, where a is a constant and $\omega = \pi/6$ rad s^{-1}. If $|\mathbf{A}+\mathbf{B}| = \sqrt{3}|\mathbf{A}-\mathbf{B}|$ at time $t = \tau$ for the first time, the value of τ, in seconds, is ....

(Numerical Value Based Question)

6. Two men are walking along a horizontal straight line in the same direction. The main in front walks at a speed $1.0\,\text{ms}^{-1}$ and the man behind walks at a speed $2.0\,\text{ms}^{-1}$. A third man is standing at a height 12 m above the same horizontal line such that all three men are in a vertical plane. The two walking men are blowing identical whistles which emit a sound of frequency 1430 Hz. The speed of sound in air 330 ms^{-1}. At the instant, when the moving men are 10 m apart, the stationary man is equidistant from them. The frequency of beats in Hz, heard by the stationary man at this instant, is ....

(Numerical Value Based Question)

7. A ring and a disc are initially at rest, side by side, at the top of an inclined plane which makes an angle 60° with the horizontal. They start to roll without slipping at the same instant of time along the shortest path. If the time difference between their reaching the ground is $(2-\sqrt{3})/\sqrt{10}$ s, then the height of the top of the inclined plane, in metres, is .... (Take, $g = 10\,\text{ms}^{-2}$)

(Numerical Value Based Question)

8. A spring block system is resting on a frictionless floor as shown in the figure. The spring constant is $2.0\,\text{Nm}^{-1}$ and the mass of the block is 2.0 kg. Ignore the mass of the spring. Initially, the spring is in an unstretched condition. Another block of mass 1.0 kg moving with a speed of $2.0\,\text{ms}^{-1}$ collides elastically with the first block. The collision is such that the 2.0 kg block does not hit the wall. The distance, in metres, between the two blocks when the spring returns to its unstretched position for the first time after the collision is ....

(Numerical Value Based Question)

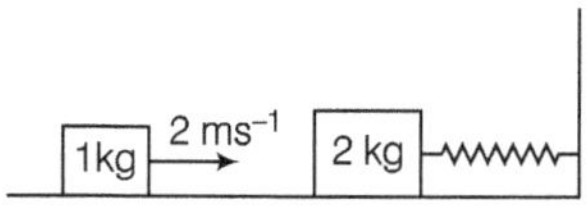

9. In the xy-plane, the region $y > 0$ has a uniform magnetic field $B_1\hat{\mathbf{k}}$ and the region $y < 0$ has another uniform magnetic field $B_2\hat{\mathbf{k}}$. A positively charged particle is projected from the origin along the positive Y-axis with speed $v_0 = \pi\,\text{ms}^{-1}$ at $t = 0$, as shown in figure. Neglect gravity in this problem. Let $t = T$ be the time when the particle crosses the X-axis from below for the first time. If $B_2 = 4B_1$, the average speed of the particle, in ms^{-1}, along the X-axis in the time interval T is............ .

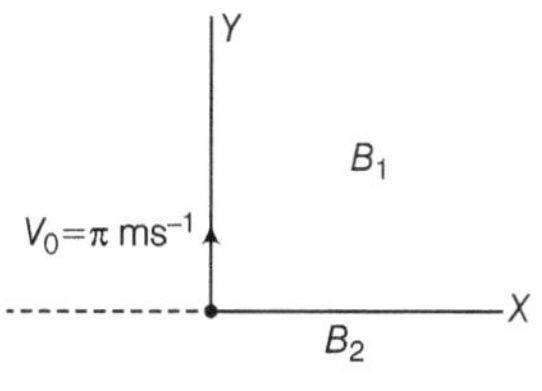

(Numerical Value Based Question)

10. Two conducting cylinders of equal length but different radii are connected in series between two heat baths kept at temperatures $T_1 = 300\text{K}$ and $T_2 = 100\text{K}$, as shown in the figure.

The radius of the bigger cylinder is twice that of the smaller one and the thermal conductivities of the materials of the smaller and the larger cylinders are K_1 and K_2, respectively. If the temperature at the junction of the cylinders in the steady state is 200K, then $K_1/K_2 =$.......... .

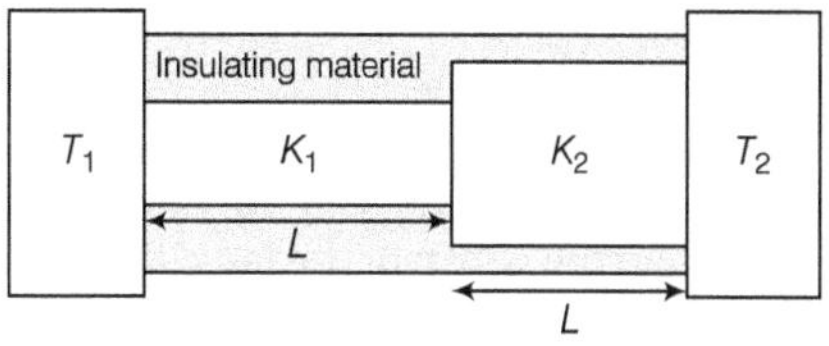

(Numerical Value Based Question)

Paragraph (Q. Nos 11-12)

If the measurement errors in all the independent quantities are known, then it is possible to determine the error in any dependent quantity. This is done by the use of series expansion and truncating the expansion at the first power of the error. For example, consider the relation $z = x/y$. If the errors in x, y and z are Δx, Δy and Δz respectively, then

$$z \pm \Delta z = \frac{x \pm \Delta x}{y \pm \Delta y} = \frac{x}{y}\left(1 \pm \frac{\Delta x}{x}\right)\left(1 \pm \frac{\Delta y}{y}\right)^{-1}$$

The series expansion for $\left(1 \pm \frac{\Delta y}{y}\right)^{-1}$, to first power in $\Delta y/y$, is $1 \mp (\Delta y/y)$. The relative errors in independent variables are always added. So, the error in z will be

$$\Delta z = z\left(\frac{\Delta x}{x} + \frac{\Delta y}{y}\right)$$

The above derivation makes the assumption that $\Delta x/x \ll 1$, $\Delta y/y \ll 1$. Therefore, the higher powers of these quantities are neglected.

(Paragraph Based Question)

11. Consider the ratio $r = \dfrac{(1-a)}{(1+a)}$ to be determined by measuring a dimensionless quantity a. If the error in the measurement of a is $\Delta a\,(\Delta a/a \ll 1)$, then what is the error Δr in determining r?

(a) $\dfrac{\Delta a}{(1+a)^2}$
(b) $\dfrac{2\Delta a}{(1+a)^2}$
(c) $\dfrac{2\Delta a}{(1-a)^2}$
(d) $\dfrac{2a\Delta a}{(1-a^2)}$

12. In an experiment, the initial number of radioactive nuclei is 3000. It is found that 1000 ± 40 nuclei decayed in the first 1.0s. For $|x| \ll 1$, $\ln(1+x) = x$ up to first power in x. The error $\Delta\lambda$, in the determination of the decay constant λ in s^{-1}, is

(a) 0.04
(b) 0.03
(c) 0.02
(d) 0.01

Paper 2

13. A particle of mass m is initially at rest at the origin. It is subjected to a force and starts moving along the X-axis. Its kinetic energy K changes with time as $dK / dt = \gamma t$, where γ is a positive constant of appropriate dimensions. Which of the following statements is (are) true? *(One or More than One Option Correct)*

(a) The force applied on the particle is constant
(b) The speed of the particle is proportional to time
(c) The distance of the particle from the origin increases linearly with time
(d) The force is conservative

14. Consider a thin square plate floating on a viscous liquid in a large tank. The height h of the liquid in the tank is much less than the width of the tank. The floating plate is pulled horizontally with a constant velocity u_0. Which of the following statements is (are) true? *(One or More than One Option Correct)*

(a) The resistive force of liquid on the plate is inversely proportional to h
(b) The resistive force of liquid on the plate is independent of the area of the plate
(c) The tangential (shear) stress on the floor of the tank increases with u_0
(d) The tangential (shear) stress on the plate varies linearly with the viscosity η of the liquid

15. In an experiment to measure the speed of sound by a resonating air column, a tuning fork of frequency 500 Hz is used. The length of the air column is varied by changing the level of water in the resonance tube. Two successive resonances are heard at air columns of length 50.7 cm and 83.9 cm. Which of the following statements is (are) true? *(One or More than One Option Correct)*

(a) The speed of sound determined from this experiment is 332 m s^{-1}
(b) The end correction in this experiment is 0.9 cm
(c) The wavelength of the sound wave is 66.4 cm
(d) The resonance at 50.7 cm corresponds to the fundamental harmonic

16. A solid horizontal surface is covered with a thin layer of oil. A rectangular block of mass $m = 0.4$ kg is at rest on this surface. An impulse of 1.0 N s is applied to the block at time $t = 0$, so that it starts moving along the X-axis with a velocity $v(t) = v_0 e^{-t/\tau}$, where v_0 is a constant and $\tau = 4$ s. The displacement of the block, in metres, at $t = \tau$ is .... (Take, $e^{-1} = 0.37$).

(Numerical Value Based Question)

17. A ball is projected from the ground at an angle of 45° with the horizontal surface. It reaches a maximum height of 120 m and returns to the ground. Upon hitting the ground for the first time, it loses half of its kinetic energy. Immediately after the bounce, the velocity of the ball makes an angle of 30° with the horizontal surface. The maximum height it reaches after the bounce, in metres, is ....

(Numerical Value Based Question)

18. A steel wire of diameter 0.5 mm and Young's modulus 2×10^{11} N m^{-2} carries a load of mass m. The length of the wire with the load is 1.0 m. A vernier scale with 10 divisions is attached to the end of this wire. Next to the steel wire is a reference wire to which a main scale, of least count 1.0 mm, is attached. The 10 divisions of the vernier scale correspond to 9 divisions of the main scale. Initially, the zero of vernier scale coincides with the zero of main scale. If the load on the steel wire is increased by 1.2 kg, the vernier scale division which coincides with a main scale division is .... (Take, $g = 10$ ms^{-2} and $\pi = 3.2$).

(Numerical Value Based Question)

19. One mole of a monoatomic ideal gas undergoes an adiabatic expansion in which its volume becomes eight times its initial value. If the initial temperature of the gas is 100 K and the universal gas constant $R = 8.0$ j mol^{-1} K^{-1}, the decrease in its internal energy in joule, is .... *(Numerical Value Based Question)*

20. A planet of mass M, has two natural satellites with masses m_1 and m_2. The radii of their circular orbits are R_1 and R_2, respectively. Ignore the gravitational force between the satellites. Define v_1, L_1, K_1 and T_1 to be respectively, the orbital speed, angular momentum, kinetic energy and time period of revolution of satellite 1; and v_2, L_2, K_2 and T_2 to be the corresponding quantities of satellite 2. Given, $m_1 / m_2 = 2$ and $R_1 / R_2 = 1/4$, match the ratios in List-I to the numbers in List-II. *(Matching Type Question)*

List-I	List-II
P. v_1/v_2	1. 1/8
Q. L_1/L_2	2. 1
R. K_1/K_2	3. 2
S. T_1/T_2	4. 8

(a) P → 4; Q → 2; R → 1; S → 3
(b) P → 3; Q → 2; R → 4; S → 1
(c) P → 2; Q → 3; R → 1; S → 4
(d) P → 2; Q → 3; R → 4; S → 1

21. One mole of a monoatomic ideal gas undergoes four thermodynamic processes as shown schematically in the pV-diagram below. Among these four processes, one is isobaric, one is isochoric, one is isothermal and one is adiabatic. Match the processes mentioned in List-I with the corresponding statements in List-II. *(Matching Type Question)*

List-I	List-II
P. In process I	1. Work done by the gas is zero
Q. In process II	2. Temperature of the gas remains unchanged
R. In process III	3. No heat is exchanged between the gas and its surroundings
S. In process IV	4. Work done by the gas is $6p_0V_0$

(a) $P \to 4$; $Q \to 3$; $R \to 1$; $S \to 2$
(b) $P \to 1$; $Q \to 3$; $R \to 2$; $S \to 4$
(c) $P \to 3$; $Q \to 4$; $R \to 1$; $S \to 2$
(d) $P \to 3$; $Q \to 4$; $R \to 2$; $S \to 1$

22. In the List-I below, four different paths of a particle are given as functions of time. In these functions, α and β are positive constants of appropriate dimensions and $\alpha \neq \beta$. In each case, the force acting on the particle is either zero or conservative. In List-II, five physical quantities of the particle are mentioned: **p** is the linear momentum, **L** is the angular momentum about the origin, K is the kinetic energy, U is the potential energy and E is the total energy. Match each path in List-I with those quantities in List-II, which are conserved for that path. *(Matching Type Question)*

	List-I	List-II
P.	$\mathbf{r}(t) = \alpha t\,\mathbf{i} + \beta t\,\mathbf{j}$	1. **p**
Q.	$\mathbf{r}(t) = \alpha \cos \omega t\,\mathbf{i} + \beta \sin \omega t\,\mathbf{j}$	2. **L**
R.	$\mathbf{r}(t) = \alpha (\cos \omega t\,\mathbf{i} + \sin \omega t\,\mathbf{j})$	3. K
S.	$\mathbf{r}(t) = at\,\mathbf{i} + \frac{\beta}{2} t^2\,\mathbf{j}$	4. U
		5. E

(a) $P \to 1, 2, 3, 4, 5$; $Q \to 2, 5$; $R \to 2, 3, 4, 5$; $S \to 5$
(b) $P \to 1, 2, 3, 4, 5$; $Q \to 3, 5$; $R \to 2, 3, 4, 5$; $S \to 2, 5$
(c) $P \to 2, 3, 4$; $Q \to 5$; $R \to 1, 2, 4$; $S \to 2, 5$
(d) $P \to 1, 2, 3, 5$; $Q \to 2, 5$; $R \to 2, 3, 4, 5$; $S \to 2, 5$

BITSAT

1. Steam at 100°C is passed into 1.1 kg of water contained in a calorimeter of water equivalent to 0.2 kg at 15°C till the temperature of the calorimeter and its contents rises to 80°C. The mass of steam condensed (in kg) is

(a) 0.130 (b) 0.065 (c) 0.260 (d) 0.135

2. Dimension of which base quantity corresponds to that of $\sqrt{\frac{Gh}{c^3}} = ?$

(a) Time (b) Length (c) Mass (d) Temperature

3. A reservoir is at 827°C and Carnot's engine takes a thousand kilocalories of heat from it and exhausts it to a sink at 27°C. What is the amount of work and the efficiency of the engine?

(a) 2.7×10^5 cal,70.70% (b) 2.72×10^5 cal, 72.72%
(c) 2.70×10^5 cal, 80.70% (d) 3.70×10^5 cal, 70.70%

4. A train moves towards stationary observer with speed 34 m/s. The train blows whistle and its frequency is registered by the observer as f_1. If the train's speed is reduced to 17 m/s, the frequency registered is f_2. If the speed of sound is 340 m/s, then the ratio $\frac{f_1}{f_2}$ is

(a) $\frac{19}{18}$ (b) $\frac{18}{19}$ (c) 2 (d) 1/2

5. An object of mass 5 kg is projected with a velocity $20\,\text{ms}^{-1}$ at an angle 60°, to the horizontal. At the highest point of its path, the projectile explodes and breaks up into two fragments of masses 1 kg and 4 kg. The fragments separate horizontally after the explosion, which releases internal energy such that the kinetic energy of the system at the highest point is doubled. The separation between the two fragments when they reach the ground is

(a) 52.25 m (b) 44.25 m
(c) 65.32 m (d) 78.76 m

6. An automobile moving with a speed of 36 km/h reaches an upward inclined road of angle 30°, its engine becomes switch off. If the coefficient of friction is 0.1, then how much distance will automobile move before coming to rest?

(a) 12.53 m (b) 21.42 m
(c) 15.43 m (d) 8.53 m

7. A block of wood floats in water with $(4/5)$ th of its volume submerged. If the same block just floats in a liquid, the density of liquid in (kg m^{-3}) is

(a) 1250 (b) 600 (c) 400 (d) 800

8. Three rods of identical cross-sectional area and made from the same metal, form the sides of an isosceles triangle ABC right angled at B as shown in figure. The point A and B are maintained at temperature T and $\sqrt{2}\,T$ respectively, in the steady state. Now, assuming that only heat conduction takes place. The temperature of point C will be

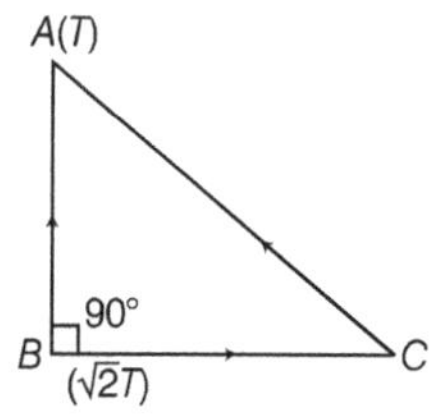

(a) $\frac{T}{\sqrt{2}+1}$ (b) $\frac{T}{\sqrt{2}-1}$ (c) $\frac{3T}{\sqrt{2}+1}$ (d) $\frac{\sqrt{3}T}{(\sqrt{2}+1)}$

9. In Young's double slit experiment, intensity at a point is $\left(\frac{1}{4}\right)$ of the maximum intensity. Angular position of this point is

(a) $\sin^{-1}\left(\frac{\lambda}{d}\right)$ (b) $\sin^{-1}\left(\frac{\lambda}{2d}\right)$ (c) $\sin^{-1}\left(\frac{\lambda}{3d}\right)$ (d) $\sin^{-1}\left(\frac{\lambda}{4d}\right)$

10. The bob of simple pendulum is a spherical hollow ball filled with water. A plugged hole near the bottom of the oscillating bob get suddenly unplugged. During observation, till water is coming out, the time period of oscillation would

(a) first increase and then decrease to the original value
(b) first decrease and then increase to the original value
(c) remain unchanged
(d) increase towards a saturation value

11. Consider the acceleration, velocity and displacement of a tennis ball as its falls to the ground and bounces back. Directions of which of these change in the process?

(a) Velocity only
(b) Displacement and velocity
(c) Acceleration, velocity and displacement
(d) Displacement and acceleration

12. If the time period is doubled, then the angular momentum of the body will (provided the moment of inertia of the body is constant)

(a) remain constant (b) quadruple
(c) become half (d) double

13. Breaking stress of a steel wire is p and the density of steel is ρ. The greatest length of steel wire that can hang vertically without breaking is

(a) $\frac{p}{\rho g}$ (b) $\frac{p}{2\rho g}$
(c) $\frac{2p}{\rho g}$ (d) None of these

14. A stone is projected with velocity $2\sqrt{gh}$, so that it just clears two walls of equal height h, at distance of $2h$ from each other. The time interval of passing between the two walls is

(a) $\sqrt{\frac{h}{g}}$ (b) $\sqrt{\frac{2h}{g}}$
(c) $2\sqrt{\frac{h}{g}}$ (d) $\frac{2h}{g}$

15. An object takes n times as much time to slide down a 45° rough inclined plane as it takes to slide down a perfectly smooth 45° inclined plane. The coefficient of kinetic friction between the rough plane and the object is

(a) $n^2 - 1$ (b) $1 - \frac{1}{n^2}$
(c) $n^2 + 1$ (d) $1 + \frac{1}{n^2}$

16. Helium gas goes through a cycle *ABCDA* (consisting of two isochoric and two isobaric lines) as shown in figure. The efficiency of this cycle is approximately

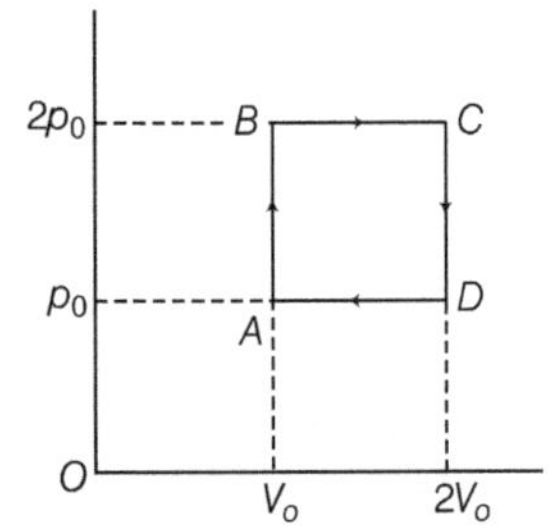

(a) 15.4% (b) 9.1% (c) 10.5% (d) 12.5%

17. An asteroid of mass m is approaching earth, initially at a distance $10\,R_e$ with speed v_i. It hits earth with a speed v_f (R_e and M_e are radius and mass of earth), then

(a) $v_f^2 = v_i^2 + \frac{2Gm}{R_e}\left(1+\frac{1}{10}\right)$ (b) $v_f^2 = v_i^2 + \frac{2GM_e}{R_e}\left(1+\frac{1}{10}\right)$
(c) $v_f^2 = v_i^2 + \frac{2GM_e}{R_e}\left(1-\frac{1}{10}\right)$ (d) $v_f^2 = v_i^2 + \frac{2Gm}{R_e}\left(1-\frac{1}{10}\right)$

18. A pulley of radius 2 m is rotated about its axis by a force $=(20t - 5t^2)$ newton (where t is measured in seconds) applied tangentially. If the moment of inertia of the pulley about its axis of rotation is 10 kg m^2, the number of rotation made by the pulley before its direction of motion is reversed is

(a) more than 3 but less than 6
(b) more than 6 but less than 9
(c) more than 9
(d) less than 3

19. A spherically symmetric gravitational system of particles has mass density

$$\rho = \begin{cases} \rho_0 & \text{for } r \le R \\ 0 & \text{for } r > R \end{cases}$$

where, ρ_0 is a constant. A test mass can undergo circular motion under the influence of the gravitational field of particles. Its speed v as a function of distance r $(0 < r < \infty)$ from the centre of the system is represented by

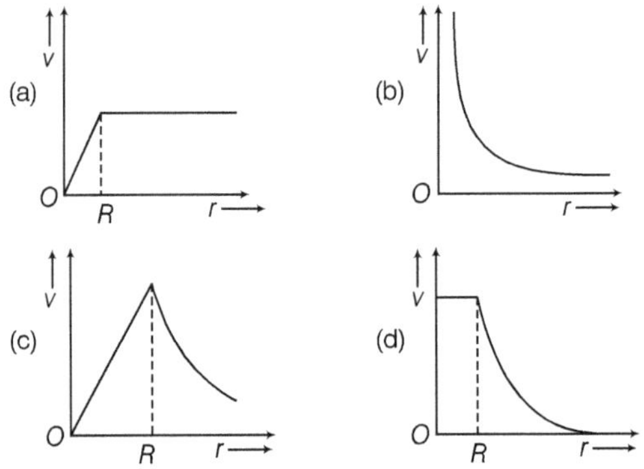

20. A trolley having mass of 200 kg moves with uniform speed of 36 kmh^{-1} on a frictionless track. A child of mass 20 kg runs on the trolley from one end to the other (10 m away) with a speed of 4 ms^{-1} relative to the trolley in a direction opposite to its motion and ultimately jumps out of the trolley. With how much velocity has the trolley moved from the time the child begins to run?

(a) 10.36 ms^{-1} (b) 11.36 ms^{-1} (c) 12.36 ms^{-1} (d) 14.40 ms^{-1}

21. A gas has molar heat capacity $C = 37.55$ J mol^{-1}K^{-1}, in this process pT = constant. The number of degree of freedom of the molecule of gas is

(a) 2 (b) 3
(c) 5 (d) 7

22. If x, v and a denote the displacement, the velocity and the acceleration of a particle executing SHM of time period T. Then, which of the following does not change with time?

(a) $\frac{aT}{x}$ (b) $aT + 2\pi v$ (c) $\frac{aT}{v}$ (d) $a^2T^2 + 4\pi^2v^2$

23. A ball is dropped vertically from a height d above the ground. It hits the ground and bounces up vertically to a height $d/2$. Neglecting subsequent motion and air resistance, its velocity v varies with height h above the ground as

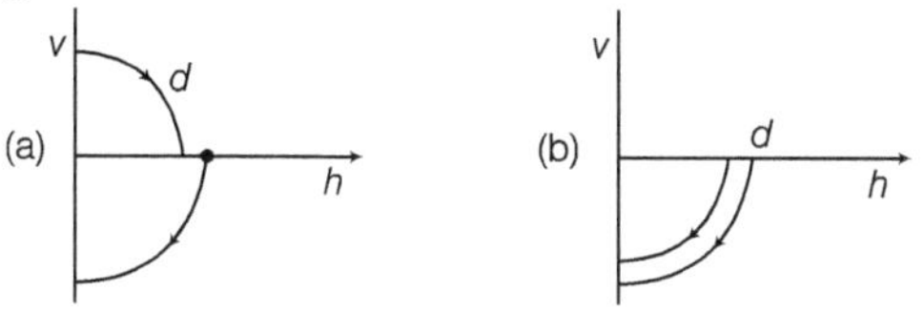

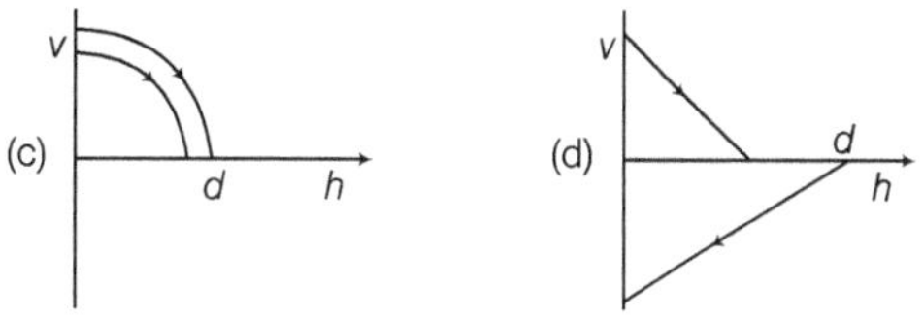

24. If a drop of liquid breaks into smaller droplets, it results in lowering of temperature of the droplets. Let a drop of radius R, break into N small droplets each of radius r, then decrease (drop) in temperature Q' (given, specific heat of liquid drop $= S$ and surface tension $= T$)

(a) $\frac{3T}{\rho S}\left[\frac{1}{r} - \frac{1}{R}\right]$ (b) $-\frac{2T}{\rho S}\left[\frac{1}{r} - \frac{1}{R}\right]$

(c) $\frac{2R}{\rho S}\left[\frac{1}{R} - \frac{1}{r}\right]$ (d) $\frac{3T}{\rho S}\left[\frac{1}{R} - \frac{1}{r}\right]$

KCET

1. The energy equivalent to a substance of mass 1 g is

(a) 18×10^{13} J (b) 9×10^{13} J (c) 18×10^{6} J (d) 9×10^{6} J

2. The radius of the earth is 6400 km. If the height of an antenna is 500 m, then its range is

(a) 800 km (b) 100 km (c) 80 km (d) 10 km

3. A space station is at a height equal to the radius of the earth. If 'v_E' is the escape velocity on the surface of the earth, the same on the space station is ...... times v_E.

(a) $\frac{1}{2}$ (b) $\frac{1}{4}$ (c) $\frac{1}{\sqrt{2}}$ (d) $\frac{1}{\sqrt{3}}$

4. A particle shows distance-time curve as shown in the figure. The maximum instantaneous velocity of the particle is around the point.

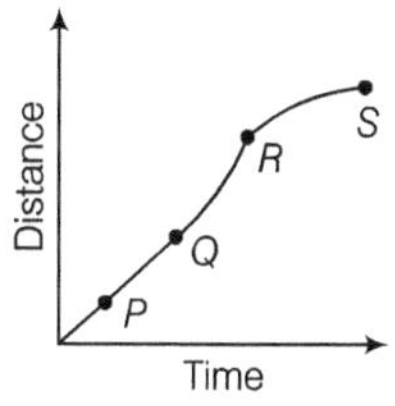

(a) P (b) S (c) R (d) Q

5. Which of the following graphs correctly represents the variation of g on the earth?

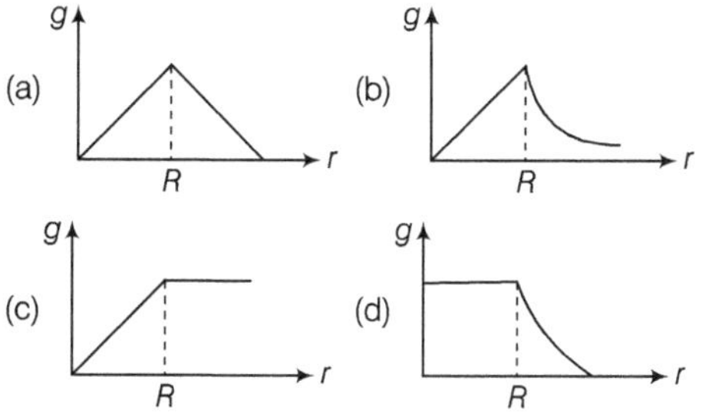

6. A cup of tea cools from 65.5°C to 62.5°C in 1 min in a room at 22.5°C. How long will it take to cool from 46.5°C to 40.5°C in the same room?

(a) 4 min (b) 2 min (c) 1 min (d) 3 min

7. A mass m on the surface of the earth is shifted to a target equal to the radius of the earth. If R is the radius and M is the mass of the earth, then work done in this process is

(a) $\frac{mgR}{2}$ (b) mgR (c) $2\,mgR$ (d) $\frac{mgR}{4}$

8. First overtone frequency of a closed pipe of length l_1 is equal to the second harmonic frequency of an open pipe of length l_2. The ratio $\frac{l_1}{l_2}$ is equal to

(a) $\frac{3}{4}$ (b) $\frac{4}{3}$

(c) $\frac{3}{2}$ (d) $\frac{2}{3}$

9. The resistance $R = \frac{V}{I}$, where $V = (100 \pm 5)$ V and $I = (10 \pm 0.2)$ A. The percentage error in R is

(a) 5.2% (b) 4.8% (c) 7% (d) 3%

10. A block rests on a rough inclined plane making an angle of 30° with the horizontal. The coefficient of static friction between the block and the plane is 0.8. If the frictional force on the block is 10N, the mass of the block is (Take $g = 10\ \text{ms}^{-2}$)

(a) 1 kg (b) 2 kg (c) 3 kg (d) 4 kg

11. Two particles of masses m_1 and m_2 have equal kinetic energies. The ratio of their momenta is

(a) $m_1 : m_2$ (b) $m_2 : m_1$

(c) $\sqrt{m_1} : \sqrt{m_2}$ (d) $m_1^2 : m_2^2$

12. The pressure at the bottom of a liquid tank is not proportional to the
(a) acceleration due to gravity (b) density of the liquid
(c) height of the liquid (d) area of the liquid surface

13. A Carnot engine takes 300 calories of heat from a source at 500 K and rejects 150 calories of heat to the sink. The temperature of the sink is
(a) 125 K (b) 250 K (c) 750 K (d) 1000 K

14. Pressure of an ideal gas is increased by keeping temperature constant. The kinetic energy of molecules
(a) decreases (b) increases
(c) remains same
(d) increases or decreases depending on the nature of gas

15. A man weighing 60 kg is in a lift moving down with an acceleration of 1.8 ms^{-2}. The force exerted by the floor on him is
(a) 588 N (b) 480 N (c) zero (d) 696 N

16. Moment of inertia of a body about two perpendicular axes X and Y in the plane of lamina are 20 kg-m^2 and 25 kg-m^2, respectively. Its moment of inertia about an axis perpendicular to the plane of the lamina and passing through the point of intersection of X and Y-axes is
(a) 5 kg-m^2 (b) 45 kg-m^2
(c) 12.5 kg-m^2 (d) 500 kg-m^2

17. Two wires A and B are stretched by the same load. If the area of cross-section of wire A is double that of B, then the stress on B is
(a) equal to that on A (b) twice that on A
(c) half that on A (d) four times that on A

18. The work done to move a charge on an equipotential surface is
(a) infinity (b) less than 1
(c) greater than 1 (d) zero

AP EAMCET

1. Assertion (A) The velocity of a projectile at a point on its trajectory is equal to the slope at that point.
Reason (R) The velocity vector at a point always along the tangent to the trajectory at that point.
(a) Both A and R are true and R is the correct explanation of A
(b) Both A and R are true but R is not the correct explanation of A
(c) A is true but R is false
(d) A is false but R is true

2. A body is projected from the ground at an angle of $\tan^{-1}\left(\frac{8}{7}\right)$ with the horizontal. The ratio of the maximum height attained by it to its range is
(a) 8 : 7 (b) 4 : 7
(c) 2 : 7 (d) 1 : 7

3. A body is projected with a speed u at an angle θ with the horizontal. The radius of curvature of the trajectory, when it makes an angle $\left(\frac{\theta}{2}\right)$ with the horizontal is (g-acceleration due to gravity)
(a) $\frac{u^2\cos^2\theta\sec^3\left(\frac{\theta}{2}\right)}{\sqrt{3}g}$ (b) $\frac{u^2\cos^2\theta\sec^3\left(\frac{\theta}{2}\right)}{2g}$
(c) $\frac{2u^2\cos^3\theta\sec^2\left(\frac{\theta}{2}\right)}{g}$ (d) $\frac{u^2\cos^2\theta\sec^3\left(\frac{\theta}{2}\right)}{g}$

4. Sand is to be piled up on a horizontal ground in the form of a regular cone of a fixed base of radius R. Coefficient of static friction between the sand layers is μ. Maximum volume of the sand can be piled up in the form of cone without slipping on the ground is
(a) $\frac{\mu R^3}{3\pi}$ (b) $\frac{\mu R^3}{3}$ (c) $\frac{\pi R^3}{3\mu}$ (d) $\frac{\mu\pi R^3}{3}$

5. A block of mass 2 kg is being pushed against a wall by a force $F = 90$ N as shown in the figure. If the coefficient of friction is 0.25, then the magnitude of acceleration of the block is
(Take, $g = 10$ ms^{-2})$\left(\sin 37° = \frac{3}{5}\right)$

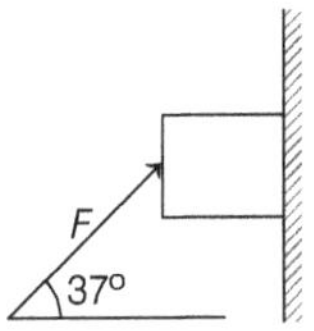

(a) 16 ms^{-2} (b) 8 ms^{-2} (c) 38 ms^{-2} (d) 54 ms^{-2}

6. A body of mass 2 kg thrown vertically from the ground with a velocity of 8 ms^{-1} reaches a maximum height of 3 m. The work done by the air resistance is (acceleration due to gravity = 10 ms^{-2})
(a) 4J (b) 60J
(c) 64J (d) 8J

7. The system of two masses 2 kg and 3 kg as shown in the figure is released from rest. The work done on 3 kg block by the force of gravity during first 2 seconds of its motion is (Take $g = 10$ ms^{-2})

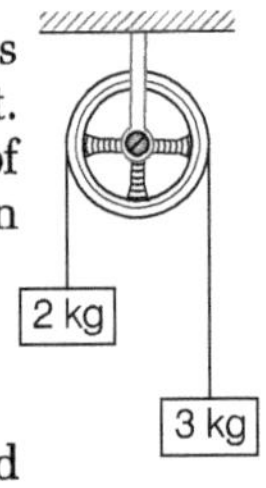

(a) 120 J (b) 80 J
(c) 40 J (d) 30 J

8. A rigid metallic sphere is spinning around its own axis in the absence of external torque. If the temperature is raised, its volume increases by 9%. The change in its angular speed is
(a) increases by 9% (b) decreases by 9%
(c) increases by 6% (d) decreases by 6%

9. Two spheres P and Q, each of mass 200 g are attached to a string of length one metre as shown in the figure. The string and the spheres are then whirled in a horizontal circle about O at a constant angular speed. The ratio of the tension in the string between P and Q to that of between P and O is (P is at mid-point of the line joining O and Q)

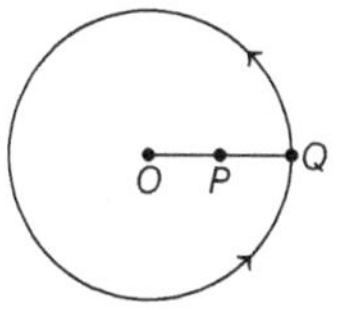

(a) $\frac{1}{2}$ (b) $\frac{2}{3}$ (c) $\frac{3}{2}$ (d) $\frac{2}{1}$

10. The potential energy of a simple harmonic oscillator of mass 2 kg at its mean position is 5 J. If its total energy is 9 J and amplitude is 1 cm, then its time period is

(a) $\frac{\pi}{100}$ s (b) $\frac{\pi}{50}$ s (c) $\frac{\pi}{20}$ s (d) $\frac{\pi}{10}$ s

11. Three masses m, $2m$ and $3m$ are arranged in two triangular configurations as shown in Figs. (a) and (b). Work done by an external agent in changing, the configuration from figure 1 to figure 2 is

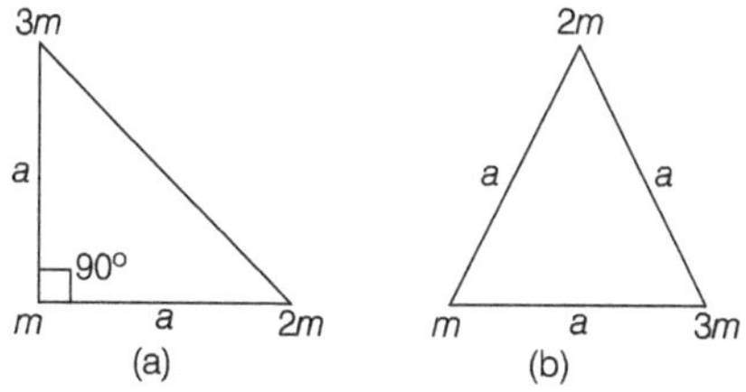

(a) $\frac{6Gm^2}{a}\left[2-\frac{6}{\sqrt{2}}\right]$ (b) 0

(c) $\frac{Gm^2}{a}\left[6+\frac{6}{\sqrt{2}}\right]$ (d) $-\frac{Gm^2}{a}\left[6-\frac{6}{\sqrt{2}}\right]$

12. Two equal and opposite forces each F act on a rod of uniform cross-sectional area a as shown in the figure. Shearing stress on the section AB will be

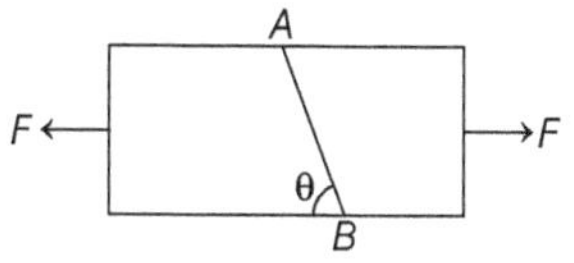

(a) $\frac{F\sin\theta\cos\theta}{a}$ (b) $\frac{F\sin\theta}{a}$

(c) $\frac{F\cos\theta}{a}$ (d) $\frac{F\sin^2\theta}{a}$

13. A body is suspended by a light string. The tensions in the string when the body is in air, when the body is totally immersed in water and when the body is totally immersed in a liquid are respectively 40.2N, 28.4N and 16.6N. The density of the liquid is

(a) 1200 kg - m^{-3} (b) 1600 kg - m^{-3}

(c) 2000 kg - m^{-3} (d) 2400 kg - m^{-3}

14. Steam at 100°C is passed into 1 kg of water contained in a calorimeter at 9°C till the temperature of water and calorimeter is increased to 90°C. The mass of the steam condensed is nearly

(Take, water equivalent of calorimeter = 0.1 kg, specific heat of water = 1 $\text{cal g}^{-1}{}^\circ\text{C}^{-1}$ and latent heat of vaporisation = 540 cal g^{-1})

(a) 81g (b) 162 g (c) 243 g (d) 486 g

15. Three very large plates of same area are kept parallel and close to each other. They are considered as ideal black surfaces and have very high thermal conductivity. First and third plates are maintained at absolute temperatures $2T$ and $3T$ respectively. Temperature of the middle plate in steady state is

(a) $\left(\frac{65}{2}\right)^{\frac{1}{4}}T$ (b) $\left(\frac{97}{4}\right)^{\frac{1}{4}}T$ (c) $\left(\frac{97}{2}\right)^{\frac{1}{4}}T$ (d) $(97)^{\frac{1}{4}}T$

16. A thermally insulated vessel with nitrogen gas at 27°C is moving with a velocity of 100 ms^{-1}. If the vessel is stopped suddenly, then the percentage change in the pressure of the gas is nearly

(assume entire loss in KE of the gas is given as heat to gas and $R = 8.3\ \text{Jmol}^{-1}\text{K}^{-1}$)

(a) 1.1 (b) 0.93 (c) 0.5 (d) 2.25

17. Match the following lists.

	List I		List II
A	Zeroth law of thermodynamics	I	Direction of flow of heat
B	First law of thermodynamics	II	Work done is zero
C	Free expansion of a gas	III	Thermal equilibrium
D	Second law of Thermodynamics	IV	Law of conservation of energy

The correct answer is

	A	B	C	D		A	B	C	D
(a)	II	IV	III	I	(b)	III	IV	II	I
(c)	III	I	II	IV	(d)	I	III	IV	II

18. For a molecule of an ideal gas, the number density is $2\sqrt{2}\times10^8\text{cm}^{-3}$ and the mean free path is $\frac{10^{-2}}{\pi}$ cm. The diameter of the gas molecule is

(a) 5×10^{-4}cm (b) 0.5×10^{-4}cm

(c) 2.5×10^{-4}cm (d) 4×10^{-4}cm

19. A solid ball is suspended from the ceiling of a motor car through a light string. A transverse pulse travels at the speed 60 cm^{-1} on the string, when the car is at rest. When the car accelerates on a horizontal road, then speed of the pulse is 66 cm^{-1}. The acceleration of the car is nearly (Take $g = 10\ \text{ms}^{-2}$)

(a) 4.3 ms^{-2} (b) 2.9 ms^{-2}

(c) 6.8 ms^{-2} (d) 5.5 ms^{-2}

20. A reflector is moving with 20 ms^{-1} towards a stationary source of sound. If the source is producing sound waves of 160 Hz, then the wavelength of the reflected wave is
(Take, speed of sound in air is 340 ms^{-1})

(a) $\frac{17}{8}$m (b) $\frac{17}{11}$m (c) $\frac{17}{9}$m (d) $\frac{17}{16}$m

21. A simple pendulum with a bob of mass 40g and charge $+2\mu C$ makes 20 oscillation in 44 s. A vertical electric field magnitude $4.2 \times 10^4 NC^{-1}$ pointing downward is applied. The time taken by the pendulum to make 15 oscillation in the electric field is (Take, acceleration due to gravity = 10 ms^{-2})

(a) 30 s (b) 60 s (c) 90 s (d) 15 s

TS EAMCET

1. Assertion (A) When we bounce a ball on the ground, it comes to rest after a few bounces, losing all its energy. This is an example of violation of conservation of energy.

Reason (R) Energy can change from one form to another but the total energy is always conserved.

Which of the following is true?

(a) Both (A) and (R) are true and (R) is the correct explanation of (A)
(b) Both (A) and (R) are true, but (R) is not the correct explanation of (A)
(c) (A) is true, but (R) is false
(d) (A) is false, but (R) is true

2. A gas satisfies the relation $pV^{5/3} = k$, where p is pressure, V is volume and k is constant. The dimension of constant k are

(a) $[ML^4T^{-2}]$ (b) $[ML^2T^{-2}]$
(c) $[ML^6T^{-2}]$ (d) $[MLT^{-2}]$

3. A car moves in positive y-direction with velocity v proportional to distance travelled y as $v(y) \propto y^{\beta}$, where β is a positive constant. The car covers a distance L with average velocity $<v>$ proportional to L as $<v> \propto L^{1/3}$. The constant β is given as

(a) $\frac{1}{4}$ (b) $\frac{1}{3}$
(c) $\frac{2}{3}$ (d) $\frac{1}{2}$

4. Consider a particle moving along the positive direction of X-axis. The velocity of the particle is given by $v = \alpha\sqrt{x}$ (α is a positive constant). At time $t = 0$, if the particle is located at $x = 0$, the time dependence of the velocity and the acceleration of the particle are respectively

(a) $\frac{\alpha^2}{2}t$ and $\frac{\alpha^2}{2}$ (b) $\alpha^2 t$ and α^2
(c) $\frac{\alpha}{2}t$ and $\frac{\alpha}{2}$ (d) $\frac{\alpha^2}{4}t$ and $\frac{\alpha^2}{4}$

5. The magnitude of velocity and acceleration and velocity of a particle moving in a plane, whose position vector $\mathbf{r} = 3t^2\hat{\mathbf{i}} + 2t\hat{\mathbf{j}} + \hat{\mathbf{k}}$ at $t = 2$ s are respectively

(a) $\sqrt{148}$, 6 (b) $\sqrt{144}$, 6
(c) $\sqrt{13}$, 3 (d) $\sqrt{14}$, 3

6. Two objects are located at height 10 m above the ground. At some point of time, the objects are thrown with initial velocity $2\sqrt{2}$ m/s at an angle 45° and 135° with the positive X-axis, respectively. Assuming $g = 10$ m/s^2, the velocity vectors will be perpendicular to each other at time is equal to

(a) 0.2 s (b) 0.4 s (c) 0.6 s (d) 0.8 s

7. String AB of unstretched length, L is stretched by applying a force F at the mid-point C such that the segments AC and BC make an angle θ with AB as shown in the figure. The string may be considered as an elastic element with a force to elongation ratio K. The force F is given by

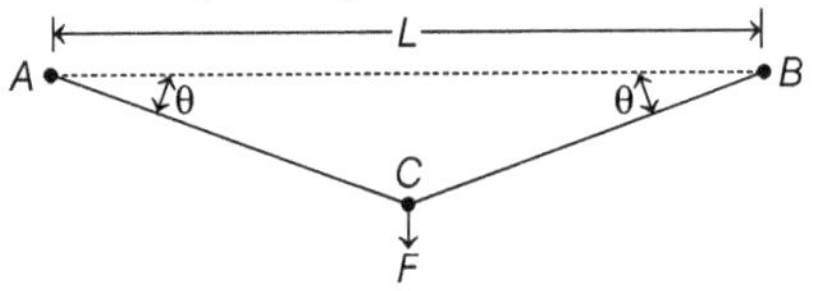

(a) $KL(1 - \tan\theta)\sin\theta$ (b) $2KL(1 - \cos\theta)\tan\theta$
(c) $KL(1 - \cos\theta)\tan\theta$ (d) $2KL(1 - \sin\theta)\tan\theta$

8. A block of mass 5 kg is kept against an accelerating wedge with a wedge angle of 45° to the horizontal. The coefficient of friction between the block and the wedge is $\mu = 0.4$. What is the minimum absolute value of the acceleration of the wedge to keep the block steady? [Assume, $g = 10$ m/s^2]

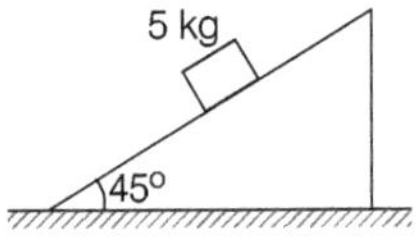

(a) $\frac{60}{7}$ m/s^2 (b) $\frac{30}{7}$ m/s^2 (c) $\frac{30}{\sqrt{7}}$ m/s^2 (d) $\frac{60}{\sqrt{7}}$ m/s^2

9. An object moves along the circle with normal acceleration proportional to t^{α}, where t is the time and α is a positive constant. The power developed by all the forces acting on the object will have time dependence proportional to

(a) $t^{\alpha-1}$ (b) $t^{\alpha/2}$ (c) $t^{\frac{1+\alpha}{2}}$ (d) $t^{2\alpha}$

10. A ball of mass 1 kg moving along x-direction collides elastically with a stationary ball of mass m. The first ball (mass = 1 kg) recoils at right angle to its original direction of motion. If the second ball starts moving at an angle 30° with the X-axis, then the value of m must be

(a) 0.5 kg (b) 1.5 kg (c) 2.5 kg (d) 2 kg

11. A machine gun can fire 200 bullets/min. If 35 g bullets are fired at a speed of 750 m/s, then the average force exerted by the gun on the bullets is

(a) 87.5 N (b) 26.2 N (c) 78.9 N (d) 110.3 N

12. A solid sphere of radius R makes a perfect rolling down on a plane which is inclined to the horizontal axis at an angle θ. If the radius of gyration is K, then its acceleration is

(a) $\dfrac{g\sin\theta}{\left(1+\dfrac{K^2}{R^2}\right)}$ (b) $\dfrac{g\sin\theta}{(R^2+K^2)}$ (c) $\dfrac{g\sin\theta}{2(R^2+K^2)}$ (d) $\dfrac{g\sin\theta}{2\left(1+\dfrac{K^2}{R^2}\right)}$

13. A particle of mass m is attached to four springs with spring constant $k, k, 2k$ and $2k$ as shown in the figure. Four springs are attached to the four corners of a square and a particle is placed at the center. If the particle is pushed slightly towards any sides of the square and released, the period of oscillation will be

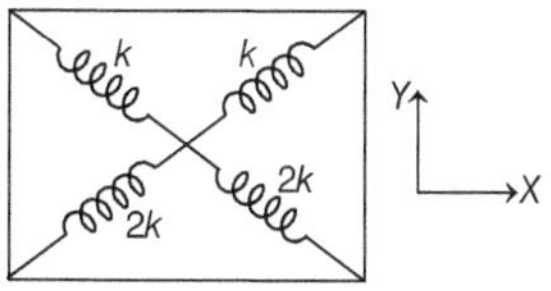

(a) $2\pi\sqrt{\dfrac{m}{3k}}$ (b) $2\pi\sqrt{\dfrac{m}{3\sqrt{2}k}}$ (c) $2\pi\sqrt{\dfrac{m}{6k}}$ (d) $2\pi\sqrt{\dfrac{m}{2k}}$

14. The ratio of the height above the surface of earth to the depth below, the surface of earth, for gravitational accelerations to be the same (assuming small heights) is

(a) 0.25 (b) 0.5 (c) 1.0 (d) 1.25

15. A steel rod has radius 50 mm and length 2 m. It is stretched along its length with a force of 400 kN. This causes an elongation of 0.5 mm. Find the (approximate) Young's modulus of steel from this information.

(a) 2×0^{10} N-m^2 (b) 10^{11} N-m^2
(c) 2×10^{11} N-m^2 (d) 10^{12} N-m^2

16. Consider a vessel filled with a liquid upto height H. The bottom of the vessel lies in the XY-plane passing through the origin. The density of the liquid varies with Z-axis as $\rho(z)=\rho_0\left[2-\left(\dfrac{z}{H}\right)^2\right]$. If p_1 and p_2 are the pressures at the bottom surface and top surface of the liquid, the magnitude of (p_1-p_2) is

(a) $\rho_0 gH$ (b) $\dfrac{8}{5}\rho_0 gH$ (c) $\dfrac{3}{2}\rho_0 gH$ (d) $\dfrac{5}{3}\rho_0 gH$

17. A one mole of ideal gas goes through a process in which pressure p varies with volume V as $p=3-g\left(\dfrac{V}{V_0}\right)^2$, where V_0 is a constant. The maximum attainable temperature by the ideal gas during this process is

(all quantities are is SI units and R is gas constant)

(a) $\dfrac{2V_0}{3R}$ (b) $\dfrac{2V_0}{R}$
(c) $\dfrac{3V_0}{2R}$ (d) None of these

18. The internal energy of the air, in a room of volume V, at temperature T and with outside pressure p increasing linearly with time, varies as

(a) increases linearly (b) increases exponentially
(c) decreases linearly (d) remains constant

19. The efficiency of Carnot engine is η, when its hot and cold reservoirs are maintained at temperature T_1 and T_2, respectively. To increase the efficiency to 1.5 η, the increase in temperature (ΔT) of the hot reservoir by keeping the cold one constant at T_2 is

(a) $\dfrac{T_1T_2}{(1-\eta)(1-1.5\eta)}$ (b) $\dfrac{0.5T_2\eta}{(1-1.5\eta)(1-\eta)}$
(c) $\dfrac{T_1}{(1-\eta)}-\dfrac{T_2}{(1-1.5\eta)}$ (d) $\dfrac{(1-\eta)(1-1.5\eta)}{T_1T_2}$

20. An air bubble rises from the bottom of a water tank of height 5 m. If the initial volume of the bubble is 3 mm^3, then what will be its volume as it reaches the surface? Assume that its temperature does not change.

[$g=9.8$ m/s^2, 1 atm $=10^5$ Pa, density of water $=1$ gm/cc]

(a) 1.5 mm^3 (b) 4.5 mm^3 (c) 9 mm^3 (d) 6 mm^3

21. Two harmonic travelling waves are described by the equation $sy_1=a\sin(kx-\omega t)$ and $y_2=a\sin(-kx+\omega t+\phi)$. The amplitude of the superimposed wave is

(a) $2a\cos\dfrac{\phi}{2}$ (b) $2a\sin\phi$ (c) $2a\cos\phi$ (d) $2a\sin\dfrac{\phi}{2}$

VIT

1. A ball is dropped from the top of a tower of height h. It covers a distance $\dfrac{h}{2}$ in the last second of its motion. Then, for how long does the ball will remain in air?

(a) 2 s (b) $(2+\sqrt{2})$ s
(c) $(2-\sqrt{2})$ s (d) $(2\pm\sqrt{2})$ s

2. A body of mass M is moving with speed v. It collides elastically with a body of mass M (such that $m << M$) which is initially at rest. The speed of lighter body after collision is

(a) 2 v (b) 3 v
(c) v (d) $\dfrac{v}{2}$

3. A wire of area of cross-section 3.0 mm^2 and natural length 25 cm is fixed at one end and a mass of 2.1 kg is hung from the other end. Find the elastic potential energy stored in wire. (Take, $Y = 1.9 \times 10^{11}$ N-m^2 and $g = 10\ ms^{-2}$)

(a) 9.7×10^{-5} J (b) 10.2×10^{-2} J
(c) 3.5×10^{-7} J (d) 2.3×10^{-5} J

4. A stone of relative density K is released from rest on the surface of a lake. If viscous effects are ignored, the stone sinks in water with an acceleration of

(a) $g(1-K)$ (b) $g(1+K)$ (c) $g\left(1-\frac{1}{K}\right)$ (d) $g\left(1+\frac{1}{K}\right)$

5. In an atom, the two electrons move round the nucleus in circular orbit of radii R and $4R$. The ratio of the time taken by them to complete one revolution is

(a) $\frac{1}{4}$ (b) $\frac{4}{1}$ (c) $\frac{8}{1}$ (d) $\frac{1}{8}$

6. Maximum speed at which a car can turn (without skidding) round a curve of 80 m radius on a level road, it the coefficient of friction between the tyres and the road is 0.25, is

(a) 16 m/s (b) 14 m/s (c) 20 m/s (d) 25 m/s

7. A ring, a disc and a solid sphere is rolling down an incline without slipping. The rolling was started from the rest. If the radii of three bodies are same, then which one of the body will attain the maximum speed when reached at the bottom of the incline?

(a) Ring (b) Solid sphere
(c) Disc (d) All have the same speed

8. The moment of inertia of a solid cylinder of mass M with radius R about a line parallel to its geometrical axis and on the surface of cylinder is

(a) $\frac{3}{2}MR^2$ (b) $\frac{MR^2}{2}$ (c) $3MR^2$ (d) $\frac{5}{2}MR^2$

9. A block of mass 1 kg is placed on a wedge shown in figure. Find out minimum coefficient of friction between wedge and block to stop the block on it.

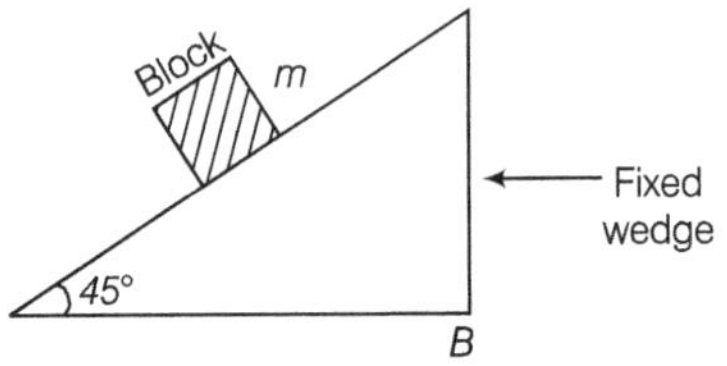

(a) 0.6 (b) 0.9
(c) 1 (d) 0.2

10. A block of copper having mass 2 kg is heated to a temperature of 500 C and then placed in a large block of ice at 0° C. What is the maximum amount of ice that can melt?
The specific heat of copper is 400 J $kg^{-1}C^{-1}$ and the latent heat of fusion of water is 3.5×10^5 J kg^{-1}.

(a) $\frac{4}{3}$ kg (b) $\frac{6}{5}$ kg (c) $\frac{8}{7}$ kg (d) $\frac{10}{9}$ kg

MHT CET

1. The path length of oscillation of simple pendulum of length 1 m is 16 cm. Its maximum velocity is (Take, $g = \pi^2\ m/s^2$)

(a) 2π cm/s (b) 4π cm/s (c) 8π cm/s d. 16π cm/s

2. A vessel completely filled with water has holes A and B at depths h and $3h$ from the top, respectively. Hole A is a square of side L and B is circle of radius r. The water flowing out per second from both the holes is same. Then, L is equal to

(a) $r^{\frac{1}{2}} (\pi)^{\frac{1}{2}} (3)^{\frac{1}{2}}$ (b) $r \cdot (\pi)^{\frac{1}{4}} (3)^{\frac{1}{4}}$ (c) $r \cdot (\pi)^{\frac{1}{2}} (3)^{\frac{1}{4}}$ d. $r^{\frac{1}{2}} (\pi)^{\frac{1}{3}} (3)^{\frac{1}{2}}$

3. A disc has mass M and radius R. How much tangential force should be applied to the rim of the disc, so as to rotate with angular velocity ω in time 't' ?

(a) $\frac{MR\omega}{4t}$ (b) $\frac{MR\omega}{2t}$ (c) $\frac{MR\omega}{t}$ (d) $MR\,\omega t$

4. In a non-uniform circular motion, the ratio of tangential to radial acceleration is (where, r = radius of circle, v = speed of the particle, α = angular acceleration)

(a) $\frac{\alpha^2 r^2}{v}$ (b) $\frac{\alpha^2 r}{v^2}$ (c) $\frac{\alpha r^2}{v^2}$ (d) $\frac{v^2}{r^2\alpha}$

5. A metal wire of density ρ floats on water surface horizontally. If it is not to sink in water, then maximum radius of wire is proportional to
(where, T = surface tension of water,
g = gravitational acceleration)

(a) $\sqrt{\frac{T}{\pi\rho g}}$ (b) $\sqrt{\frac{\pi\rho g}{T}}$ (c) $\frac{T}{\pi\rho g}$ (d) $\frac{\pi\rho g}{T}$

6. A sphere of mass m moving with velocity v collides head-on on another sphere of same mass which is at rest. The ratio of final velocity of second sphere to the initial velocity of the first sphere is (where, e is coefficient of restitution and collision is inelastic)

(a) $\frac{e-1}{2}$ (b) $\frac{e}{2}$ (c) $\frac{e+1}{2}$ (d) e

7. For a particle performing linear SHM, its average speed over one oscillation is (where, a = amplitude of SHM, n = frequency of oscillation)

(a) $2\,an$ (b) $4\,an$ (c) $6\,an$ (d) $8\,an$

8. The molar specific heat of an ideal gas at constant pressure and constant volume is C_p and C_V respectively. If R is the universal gas constant and the ratio of C_p to C_V is γ, then C_V

(a) $\frac{1-\gamma}{1+\gamma}$ (b) $\frac{1+\gamma}{1-\gamma}$ (c) $\frac{\gamma-1}{R}$ (d) $\frac{R}{\gamma-1}$

9. In a capillary tube having area of cross-section A, water rises to a height h. If cross-sectional area is reduced to $\frac{A}{9}$, the rise of water in the capillary tube is

(a) $4h$ (b) $3h$ (c) $2h$ (d) h

10. A string is vibrating in its fifth overtone between two rigid supports 2.4 m apart. The distance between successive node and antinode is

(a) 0.1 m (b) 0.2 m
(c) 0.6 m (d) 0.8 m

11. If $\mathbf{A} = 3\hat{i} - 2\hat{j} + \hat{k}, \mathbf{B} = \hat{i} - 3\hat{j} + 5\hat{k}$ and $\mathbf{C} = 2\hat{i} + \hat{j} - 4\hat{k}$ form a right angled triangle, then out of the following which one is satisfied ?

(a) $\mathbf{A} = \mathbf{B} + \mathbf{C}$ and $\mathbf{A}^2 = \mathbf{B}^2 + \mathbf{C}^2$
(b) $\mathbf{A} = \mathbf{B} + \mathbf{C}$ and $\mathbf{B}^2 = \mathbf{A}^2 + \mathbf{C}^2$
(c) $\mathbf{B} = \mathbf{A} + \mathbf{C}$ and $\mathbf{B}^2 = \mathbf{A}^2 + \mathbf{C}^2$
(d) $\mathbf{B} = \mathbf{A} + \mathbf{C}$ and $\mathbf{A}^2 = \mathbf{B}^2 + \mathbf{C}^2$

12. A square frame $ABCD$ is formed by four identical rods each of mass m and length l. This frame is in xy -plane such that side AB coincides with X-axis and side AD along Y-axis. The moment of inertia of the frame about X-axis is

(a) $\frac{5\,ml^2}{3}$ (b) $\frac{2\,ml^2}{3}$ (c) $\frac{4\,ml^2}{3}$ (d) $\frac{ml^2}{12}$

13. A unit vector is represented as $(0.8\,\hat{\mathbf{i}} + b\hat{\mathbf{j}} + 0.4\hat{\mathbf{k}})$. Hence, the value of b must be

(a) 0.4 (b) $\sqrt{0.6}$ (c) 0.2 (d) $\sqrt{0.2}$

14. A mass is suspended from a vertical spring which is executing SHM of frequency 5 Hz. The spring is unstretched at the highest point of oscillation. Maximum speed of the mass is
(Take, acceleration due to gravity, $g = 10\text{ m/s}^2$)

(a) 2π m/s (b) π m/s (c) $\frac{1}{2\pi}$ m/s (d) $\frac{1}{\pi}$ m/s

15. The moment of inertia of a ring about an axis passing through the centre and perpendicular to its plane is I. It is rotating with angular velocity ω. Another identical ring is gently placed on it, so that their centres coincide. If both the rings are rotating about the same axis, then loss in kinetic energy is

(a) $\frac{I\omega^2}{2}$ (b) $\frac{I\omega^2}{4}$ (c) $\frac{I\omega^2}{6}$ (d) $\frac{I\omega^2}{8}$

16. A bomb at rest explodes into 3 parts of same mass. The momentum of two parts is $-3p\hat{\mathbf{i}}$ and $2p\hat{\mathbf{j}}$, respectively. The magnitude of momentum of the third part is

(a) p (b) $\sqrt{5}\,p$ (c) $\sqrt{11}\,p$ (d) $\sqrt{13}\,p$

17. A mass attached to one end of a string crosses top-most point on a vertical circle with critical speed. Its centripetal acceleration when string becomes horizontal will be
(where, g = gravitational acceleration)

(a) g (b) $3g$ (c) $4g$ (d) $6g$

18. When source of sound moves towards a stationary observer, the wavelength of sound received by him

(a) decreases while frequency increases
(b) remains the same, whereas frequency increases
(c) increases and frequency also increases
(d) decreases while frequency remains the same

19. A body is thrown from the surface of the earth with velocity u m/s. The maximum height in metre above the surface of the earth upto which it will reach is (where, R = radius of earth, g = acceleration due to gravity)

(a) $\frac{u^2R}{2gR - u^2}$ (b) $\frac{2u^2R}{gR - u^2}$ (c) $\frac{u^2R^2}{2gR^2 - u^2}$ (d) $\frac{u^2R}{gR - u^2}$

20. Heat energy is incident on the surface at the rate of 1000 J/min. If coefficient of absorption is 0.8 and coefficient of reflection is 0.1, then heat energy transmitted by the surface in 5 min is

(a) 100 J (b) 500 J (c) 700 J (d) 900 J

21. Two metal wires P and Q of same length and material are stretched by same load. Their masses are in the ratio $m_1 : m_2$. The ratio of elongations of wire P to that of Q is

(a) $m_1^2 : m_2^2$ (b) $m_2^2 : m_1^2$ (c) $m_2 : m_1$ (d) $m_1 : m_2$

22. Let $x = \left[\frac{a^2b^2}{c}\right]$ be the physical quantity. If the percentage error in the measurement of physical quantities a, b and c is 2, 3 and 4 per cent respectively, then percentage error in the measurement of x is

(a) 7% (b) 14% (c) 21% (d) 28%

23. n number of waves are produced on a string in 0.5 s. Now, the tension in the string is doubled (Assume length and radius constant), the number of waves produced in 0.5 s for the same harmonic will be

(a) n (b) $\sqrt{2}\,n$ (c) $\frac{n}{\sqrt{2}}$ (d) $\frac{n}{\sqrt{5}}$

24. The increase in energy of a metal bar of length L and cross-sectional area A when compressed with a load M along its length is
(where, Y = Young's modulus of the material of metal bar)

(a) $\frac{FL}{2\,AY}$ (b) $\frac{F^2L}{2\,AY}$ (c) $\frac{FL}{AY}$ (d) $\frac{F^2L^2}{2\,AY}$

25. A satellite is revolving in a circular orbit at a height h above the surface of the earth of radius R. The speed of the satellite in its orbit is one-fourth the escape velocity from the surface of the earth. The relation between h and R is

(a) $h = 2R$ (b) $h = 3R$ (c) $h = 5R$ (d) $h = 7R$

26. A pipe closed at one end has length 83 cm. The number of possible natural oscillations of air column whose frequencies lie below 1000 Hz are
(take, velocity of sound in air = 332 m/s)

(a) 3 (b) 4 (c) 5 (d) 6

Answer *with* Explanations

JEE Main

1. (a) From theorem of parallel axis,

$$I = I_{CM} + 7M(3R)^2$$

$$= \left[\frac{MR^2}{2} + 6\times\left\{\frac{MR^2}{2} + M(2R)^2\right\}\right]$$

$$+7M(3R)^2 = \frac{181MR^2}{2}$$

2. (b) $I_{remaining} = I_{total} - I_{cavity}$

$$\Rightarrow \quad I = \frac{9MR^2}{2} - \left[\frac{M}{2}\left(\frac{R}{3}\right)^3 + M\left(\frac{2R}{3}\right)^2\right] = 4MR^2$$

3. (a) None of the four options are correct.

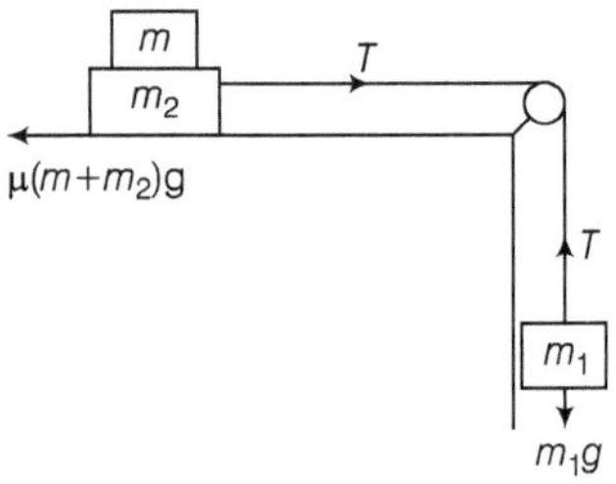

$$\mu(m+m_2)g \ge m_1 g$$

Substituting, $m_1 = 5\,\text{kg}$ and $m_2 = 10\,\text{kg}$,

we get $\mu(10+m)g \ge 5g$

$$\Rightarrow \quad 10 + m \ge \frac{5}{0.15}$$

$$m \ge 23.33\ \text{kg}$$

4. (c) From conservation of linear momentum,

$$mv_0 + 0 = mv_1 + mv_2$$

$$v_0 = v_1 + v_2 \qquad \ldots\text{(i)}$$

Further, $K_f = \frac{3}{2}K$

$$\therefore \quad \frac{3}{2}\left[\frac{1}{2}mv_0^2\right] = \frac{1}{2}mv_1^2 + \frac{1}{2}mv_2^2$$

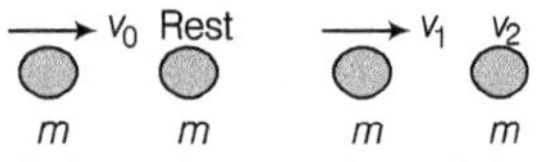

Before collision After collision

$$\frac{3}{2}v_0^2 = v_1^2 + v_2^2 \qquad \ldots\text{(ii)}$$

Solving Eqs. (i) and (ii), we get

$$v_1 = \frac{v_0}{2}(1+\sqrt{2})$$

$$\Rightarrow \quad v_2 = \frac{v_0}{2}(1-\sqrt{2})$$

$$v_{rel} = v_1 - v_2$$

$$\frac{v_0}{2}[1+\sqrt{2}-1+\sqrt{2}]$$

$$= \frac{v_0}{2}\times 2\sqrt{2} = \sqrt{2}v_0$$

5. (d) $F \propto \frac{1}{R^n}$

$$\frac{mv^2}{R} \propto \frac{1}{R^n} \quad \text{or} \quad v \propto \frac{I}{R^{\frac{n-1}{2}}}$$

$$T = \frac{2\pi R}{v} \Rightarrow T \propto \frac{R}{v}$$

$$\Rightarrow \quad T \propto R^{1+\frac{n-1}{2}} \quad \text{or} \quad T \propto R^{\frac{n+1}{2}}$$

6. (b) Wave velocity

$$(v) = \sqrt{\frac{Y}{\rho}} = 5.86 \times 10^3\ \text{m/s}$$

For fundamental mode, $\lambda = 2l = 1.2$ m

∴ Fundamental frequency

$$= \frac{v}{\lambda} = 4.88\ \text{kHz} \approx 5\ \text{kHz}$$

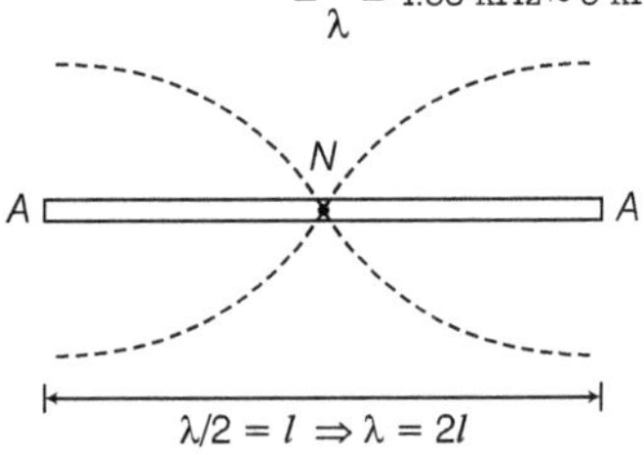

$\lambda/2 = l \Rightarrow \lambda = 2l$

7. (b) **Case I** Just before collision,

Just after collision

From momentum conservation,

$$2v_2 - v_1 = v$$

From the definition of e (= 1 for elastic collision),

$$v_2 + v_1 = v \quad \Rightarrow \quad 3v_2 = 2v$$

$$v_2 = \frac{2v}{3} \quad \Rightarrow \quad v_1 = \frac{v}{3}$$

$$P_d = \frac{\frac{1}{2}mv^2 - \frac{1}{2}mv_1^2}{\frac{1}{2}mv^2} = \frac{1-\frac{1}{9}}{1} = \frac{8}{9} = 0.89$$

Case II Just before collision

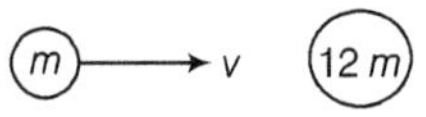

Just after collision,

From momentum conservation,

$$12v_2 - v_1 = v$$

From the definition of e(= 1 for elastic collision),

$$v_2 + v_1 = v$$

$$13v_2 = 2v$$

$$v_2 = \frac{2v}{13} \quad \Rightarrow \quad v_1 = v - \frac{2v}{13} = \frac{11v}{13}$$

$$\Rightarrow \quad P_c = \frac{\frac{1}{2}mv^2 - \frac{1}{2}mv_1^2}{\frac{1}{2}mv^2}$$

$$= \frac{1-\frac{121}{169}}{1} = \frac{48}{169} = 0.28$$

8. (d) $\frac{\Delta m}{m}\times 100 = 1.5 \Rightarrow \frac{\Delta l}{l}\times 100 = 1$

$$d = \frac{m}{l^3} \Rightarrow \frac{\Delta d}{d}\times 100$$

$$= \frac{\Delta m}{m}\times 100 + \frac{3\Delta l}{l}\times 100$$

$$= 1.5 + 3 = 4.5\%$$

9. (d) For adiabatic process,

$$T_1V_1^{\gamma-1} = T_2V_2^{\gamma-1} \qquad \left(\because \gamma = \frac{5}{3}\right)$$

$$\Rightarrow \quad 300(V)^{\frac{2}{3}} = T_2(2V)^{\frac{2}{3}}$$

$$\Rightarrow \quad T_2 = \frac{300}{2^{\frac{2}{3}}} \approx 189\,\text{K}$$

$$\Delta U = \frac{f}{2}nR\Delta T$$

$$= \left(\frac{3}{2}\right)(2)\left(\frac{25}{3}\right)(189-300) = -2.7\ \text{kJ}$$

10. (d) $\Delta P = \frac{F}{A} = \frac{mg}{a}$

Bulk modulus, $K = \frac{-\Delta P}{\Delta V/V}$

Here, $V = \frac{4}{3}\pi r^3$

and ΔV or $dV = (4\pi r^3)dr$

$$\Rightarrow \quad K = -\frac{\frac{mg}{A}}{\frac{4\pi r^2 dr}{\frac{4}{3}\pi r^3}} \quad \text{or} \quad \frac{dr}{r} = -\frac{mg}{3Ka}$$

11. (b) $\text{Pressure} = \frac{\text{Force}}{\text{Area}}$

$$= \frac{\frac{\text{Number of Collisions}}{\text{sec}} \times \frac{\text{Change in momentum}}{\text{collision}}}{\text{Area}}$$

$$= \frac{10^{23}\times 2mv\cos 45^\circ}{2\times 10^{-4}} = 2.35\times 10^3\ \text{N/m}^2$$

12. **(c)** In graph '3' initial slope is zero which is not possible, since initial velocity is not zero. In all other three graphs: initial velocity is not zero.

13. **(d)** Given, $U=-\frac{k}{2r^2}$

$$\Rightarrow \quad F_r=-\frac{dU}{dr}=-\frac{k}{r^3}$$

Since, the particle moves in a circular path of radius a, the required centripetal force is provided by the above force.

Hence, $\frac{mv^2}{a}=\frac{k}{a^3} \Rightarrow mv^2=\frac{k}{a^2}$

Kinetic energy, $K=\frac{1}{2}mv^2=\frac{k}{2a^2}$

Total energy $=K+U=-\frac{k}{2a^2}+\frac{k}{2a^2}=0$

14. **(c)** Given, frequency, $f=10^{12}$ /sec

Angular frequency,

$$\omega=2\pi f=2\pi\times10^{12} \text{ /sec}$$

Force constant, $k=m\omega^2$

$$=\frac{108\times10^{-3}}{6.02\times10^{23}}\times4\pi^2\times10^{24}$$

$$k=7.1 \text{ N/m}$$

JEE Advanced

1. **(b, c)** $V=\frac{Kr^2}{2}$

$$F=-\frac{dV}{dr}=-Kr \text{ (towards centre)}$$

$$\left[\because F=-\frac{dV}{dr}\right]$$

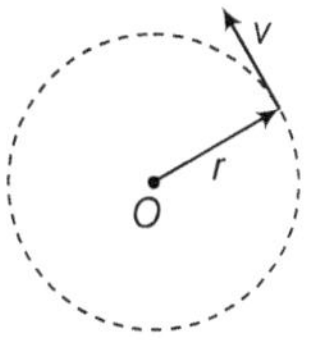

At $r=R$,

$$kR=\frac{mv^2}{R} \text{ (centripetal force)}$$

$$v=\sqrt{\frac{kR^2}{m}}=\sqrt{\frac{k}{m}}R$$

$$L=mvR=\sqrt{\frac{k}{m}}R^2$$

2. **(a, c)** $\mathbf{F}=(\alpha t)\hat{\mathbf{i}}+\beta\hat{\mathbf{j}}$ [at $t=0, v=0$, $\mathbf{r}=0$]

$$\alpha=1, \beta=1$$

$$\mathbf{F}=t\hat{\mathbf{i}}+\hat{\mathbf{j}}$$

$$m\frac{d\mathbf{v}}{dt}=t\hat{\mathbf{i}}+\hat{\mathbf{j}}$$

On integrating,

$$m\mathbf{v}=\frac{t^2}{2}\hat{\mathbf{i}}+t\hat{\mathbf{j}} \text{ [given, } m=1 \text{ kg]}$$

$$\frac{d\mathbf{r}}{dt}=\frac{t^2}{2}\hat{\mathbf{i}}+t\hat{\mathbf{j}} \quad [\mathbf{r}=0 \text{ at } t=0]$$

Again, on integrating,

$$\mathbf{r}=\frac{t^3}{6}\hat{\mathbf{i}}+\frac{t^2}{2}\hat{\mathbf{j}}$$

At $t=1$ s, $\tau=(\mathbf{r}\times\mathbf{F})=\left(\frac{1}{6}\hat{\mathbf{i}}+\frac{1}{2}\hat{\mathbf{j}}\right)\times(\hat{\mathbf{i}}+\hat{\mathbf{j}})$

$$\tau=\frac{1}{3}\hat{\mathbf{k}}$$

$$v=\frac{t^3}{2}\hat{\mathbf{i}}+t\hat{\mathbf{j}}$$

At $t=1$ s, $\mathbf{v}=\left(\frac{1}{2}\hat{\mathbf{i}}+\hat{\mathbf{j}}\right)=\frac{1}{2}(\hat{\mathbf{i}}+2\hat{\mathbf{j}})$ m/s

At $t=1$ s, $\mathbf{r}_1-\mathbf{r}_0=\left[\frac{1}{6}\hat{\mathbf{i}}+\frac{1}{2}\hat{\mathbf{j}}\right]-[0]$

$$\mathbf{s}=\frac{1}{6}\hat{\mathbf{i}}+\frac{1}{2}\hat{\mathbf{j}}$$

$$|\mathbf{s}|=\sqrt{\left(\frac{1}{6}\right)^2+\left(\frac{1}{2}\right)^2} \Rightarrow \frac{\sqrt{10}}{6} \text{ m}$$

3. **(a, c)** $h=\frac{2\sigma\cos\theta}{r\rho g}$

(a) $\rightarrow h\propto\frac{1}{r}$

(b) h depends upon σ.

(c) If lift is going up with constant acceleration.

$$g_{\text{eff}}=(g+a) \Rightarrow h=\frac{2\sigma\cos\theta}{r\rho(g+a)}$$

It means h decreases.

(d) h is proportional to $\cos\theta$.

4. **(b,c,d)** (b) Process II is isothermal expansion,

$$\Delta U=0, W>0$$

$$\Delta Q=W>0$$

(c) Process IV is isothermal compression,

$$\Delta U=0, W<0$$

$$\Delta Q=W<0$$

(d) Processess I and III are not isobaric because in isobaric process $T\propto V$ hence, T-V graph should be a straight line passing through origin.

5. **(2.0)** $\mathbf{A}=a\hat{\mathbf{i}}$ and $\mathbf{B}=a\cos\omega\hat{\mathbf{i}}+a\sin\omega t\hat{\mathbf{j}}$

$$\mathbf{A}+\mathbf{B}=(a+a\cos\omega t)\hat{\mathbf{i}}+a\sin\omega t\hat{\mathbf{j}}$$

$$\mathbf{A}-\mathbf{B}=(a-a\cos\omega t)\hat{\mathbf{i}}+a\sin\omega t\hat{\mathbf{j}}$$

$$\frac{|\mathbf{A}+\mathbf{B}|=\sqrt{3}|\mathbf{A}-\mathbf{B}|}{\sqrt{(a+a\cos\omega t)^2+(a\sin\omega t)^2}=\sqrt{3}\sqrt{(a-a\cos\omega t)^2+(a\sin\omega t)^2}}$$

$$\Rightarrow \quad 2\cos\frac{\omega t}{2}=\pm\sqrt{3}\times2\sin\frac{\omega t}{2}$$

$$\tan\frac{\omega t}{2}=\pm\frac{1}{\sqrt{3}}$$

$$\frac{\omega t}{2}=n\pi\pm\frac{\pi}{6}$$

$$\frac{\pi}{12}t=n\pi\pm\frac{\pi}{6}$$

$$t=(12n\pm2)\text{s}$$

$$=2\text{s}, 10\text{s}, 14\text{s} \text{ and so on.}$$

6. **(5 Hz)**

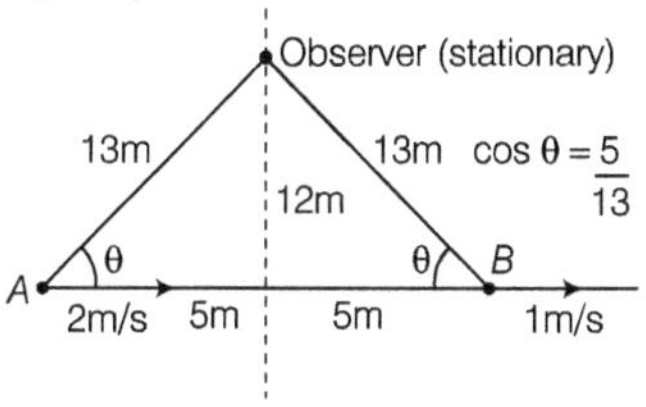

$$f_A=1430\left[\frac{330}{330-2\cos\theta}\right]$$

$$=1430\left[\frac{1}{1-\frac{2\cos\theta}{330}}\right]\approx1430\left[1+\frac{2\cos\theta}{330}\right]$$

[from binomial expansion]

$$f_B=1430\left[\frac{330}{330+1\cos\theta}\right]$$

$$\approx1430\left[1-\frac{\cos\theta}{330}\right]$$

Beat frequency =

$$f_A-f_B=1430\left[\frac{3\cos\theta}{330}\right]=13\cos\theta$$

$$=13\left(\frac{5}{13}\right)=5.00 \text{ Hz}$$

7. **(0.75 m)** $a=\frac{g\sin\theta}{1+\frac{I}{MR^2}}$

$$a_{\text{ring}}=\frac{g\sin\theta}{2} \qquad (\because I=MR^2)$$

$$a_{\text{disc}}=\frac{2g\sin\theta}{3} \qquad \left(\because I=\frac{MR^2}{2}\right)$$

$$s=\frac{h}{\sin\theta}=\frac{1}{2}at^2$$

$$=\frac{1}{2}\left(\frac{g\sin\theta}{2}\right)t_1^2$$

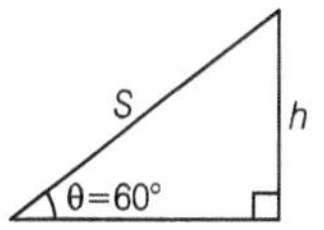

$$\Rightarrow \quad t_1=\sqrt{\frac{4h}{g\sin^2\theta}}=\sqrt{\frac{16h}{3g}}$$

$$s=\frac{h}{\sin\theta}=\frac{1}{2}at^2$$

$$=\frac{1}{2}\left(\frac{2g\sin\theta}{3}\right)t_2^2$$

$$\Rightarrow \quad t_2=\sqrt{\frac{3h}{g\sin^2\theta}}=\sqrt{\frac{4h}{g}}$$

$$t_2 - t_1 = \sqrt{\frac{16h}{3g}} - \sqrt{\frac{4h}{g}} = \frac{2-\sqrt{3}}{\sqrt{10}}$$

$$\sqrt{h}\left[\frac{4}{\sqrt{3}} - 2\right] = 2 - \sqrt{3}$$

Soving this equation we get,

$$h = 0.75 \text{ m.}$$

8. **(2.09 m)** Just before collision,

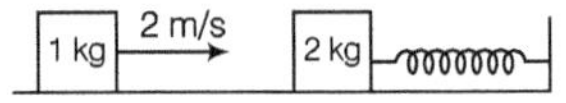

Just after collision,

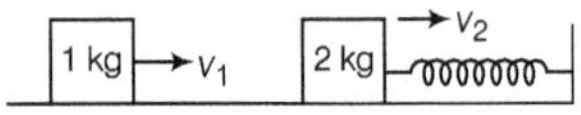

Let velocities of 1 kg and 2 kg blocks just after collision be v_1 and v_2 respectively.

From momentum conservation principle,

$$1 \times 2 = 1v_1 + 2v_2 \quad \text{...(i)}$$

Collision is elastic. Hence $e = 1$ or relative velocity of separation = relation velocity of approach.

$$v_2 - v_1 = 2 \quad \text{...(ii)}$$

From Eqs. (i) and (ii), we get

$$v_2 = \frac{4}{3} \text{ m/s}, v_1 = \frac{-2}{3} \text{ m/s}$$

2 kg block will perform SHM after collision,

$$t = \frac{T}{2} = \pi\sqrt{\frac{m}{k}} = 3.14 \text{ s}$$

$$\text{Distance} = |v_1|t = \frac{2}{3} \times 3.14$$

$$= 2.093 = 2.09 \text{ m}$$

9. **(2 m/s)** If average speed is considered along X-axis,

$$R_1 = \frac{mv_0}{qB_1},$$

$$R_2 = \frac{mv_0}{qB_2} = \frac{mv_0}{4qB_1}$$

$$R_1 > R_2$$

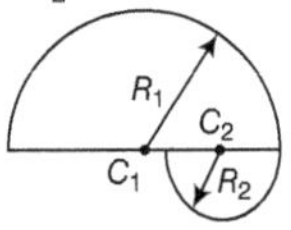

Distance travelled along X-axis,

$$\Delta x = 2(R_1 + R_2) = \frac{5mv_0}{2qB_1}$$

$$\text{Total time} = \frac{T_1}{2} + \frac{T_2}{2} = \frac{\pi m}{qB_1} + \frac{\pi m}{qB_2}$$

$$= \frac{\pi m}{qB_1} + \frac{\pi m}{4qB_1} = \frac{5\pi m}{4qB_1}$$

Magnitude of average speed

$$= \frac{\frac{5mv_0}{2qB_1}}{\frac{5\pi m}{4qB_1}} = 2 \text{ m/s}$$

10. **(d)** Rate of heat flow will be same,

$$\therefore \quad \frac{300-200}{R_1} = \frac{200-100}{R_2}$$

$$\left(\text{as } H = \frac{dQ}{dt} = \frac{T \cdot D}{R}\right)$$

$$\therefore \quad R_1 = R_2 \Rightarrow \frac{L_1}{K_1A_1} = \frac{L_2}{K_2A_2}$$

$$\Rightarrow \quad \frac{K_1}{K_2} = \frac{A_2}{A_1} = 4$$

11. **(b)** $r = \dfrac{1-a}{1+a}$

$$\ln r = \ln(1-a) - \ln(1+a)$$

On differentiating both sides, we get

$$\frac{dr}{r} = -\frac{da}{1-a} - \frac{da}{1+a}$$

or we can write

$$\frac{\Delta r}{r} = -\left[\frac{\Delta a}{1-a} + \frac{\Delta a}{1+a}\right] \quad \frac{\Delta r}{r} = \frac{-2\Delta a}{1-a^2}$$

$$\text{or } \Delta r = -\left(\frac{2\Delta a}{1-a^2}\right)(r) = \frac{-2\Delta a}{(1+a)^2}$$

12. **(c)** $N = N_0 e^{-\lambda t}$

$$\text{In}N = \text{In}N_0 - \lambda t$$

Differentiating w.r.tλ, we get

$$\frac{1}{N} \cdot \frac{dN}{d\lambda} = 0 - t$$

$$\Rightarrow \quad |d\lambda| = \frac{dN}{Nt} = \frac{40}{2000 \times 1} = 0.02$$

13. **(a, b)** $K = \dfrac{1}{2}mv^2 \Rightarrow \dfrac{dK}{dt} = mv\dfrac{dv}{dt}$

$$\text{Given, } \frac{dK}{dt} = \gamma t$$

$$\Rightarrow \quad mv\frac{dv}{dt} = \gamma t$$

$$\Rightarrow \quad \int_0^v v\,dv = \int_0^t \frac{\gamma}{m} t\,dt$$

$$\Rightarrow \quad \frac{v^2}{2} = \frac{\gamma}{m}\frac{t^2}{2}$$

$$\Rightarrow \quad v = \sqrt{\frac{\gamma}{m}}t$$

$$\Rightarrow \quad a = \frac{dv}{dt} = \sqrt{\frac{\gamma}{m}}$$

$$\therefore \quad F = ma = \sqrt{\gamma m} = \text{constant}$$

$$\therefore \quad V = \frac{ds}{dt} = \sqrt{\frac{\gamma}{m}}t$$

$$\Rightarrow \quad s = \sqrt{\frac{\gamma}{m}}\frac{t^2}{2}$$

14. **(a, c, d)**

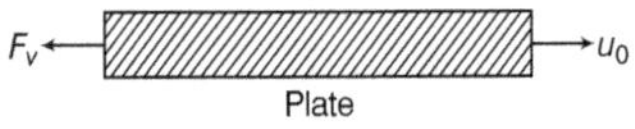

$$F_v = -\eta A\left(\frac{dv}{dy}\right)$$

Since, height h of the liquid in tank is very small.

$$\Rightarrow \quad \frac{dv}{dy} = \frac{\Delta v}{\Delta y} = \left(\frac{u_0}{h}\right)$$

$$F_v = -(\eta)A\left(\frac{u_0}{h}\right)$$

$$F_v \propto \left(\frac{1}{h}\right), F_v \propto u_0, F \propto A, F_v \propto \eta$$

15. **(a, c)** Let n th harmonic is corresponding to 50.7 cm and $(n+1)$ th harmonic is corresponding 83.9 cm.

$\therefore$ Their difference is $\dfrac{\lambda}{2}$.

$$\therefore \quad \frac{\lambda}{2} = (83.9 - 50.7) \text{ cm}$$

$$\text{or} \quad \lambda = 66.4 \text{ cm}$$

$$\therefore \quad \frac{\lambda}{4} = 16.6 \text{ cm}$$

Length corresponding to fundamental mode must be close to $\dfrac{\lambda}{4}$ and 50.7 cm must be an odd multiple of this length. $16.6 \times 3 = 49.8$ cm. Therefore, 50.7 is 3rd harmonic.

If end correction is e, then

$$e + 50.7 = \frac{3\lambda}{4}$$

$$e = 49.8 - 50.7 = -0.9 \text{ cm}$$

$\therefore$ Speed of sound, $v = f\lambda$

$$\Rightarrow v = 500 \times 66.4 \text{ cm/s} = 332 \text{ m/s}$$

16. **(6.30 m)** Linear impulse, $J = mv_0$

$$\therefore \quad v_0 = \frac{J}{m} = 2.5 \text{ m/s}$$

$$\therefore \quad v = v_0 e^{-t/\tau}$$

$$\frac{dx}{dt} = v_0 e^{-t/\tau}$$

$$\int_0^x dx = v_0 \int_0^\tau e^{-t/\tau} dt$$

$$x = v_0 \left[\frac{e^{-t/\tau}}{-\frac{1}{\tau}}\right]_0^\tau$$

$$x = 2.5(-4)(e^{-1} - e^0)$$

$$= 2.5(-4)(0.37 - 1)$$

$$x = 6.30 \text{ m}$$

17. **(30 m)** $\because H = \dfrac{u^2 \sin^2 45°}{2g} = 120 \text{ m}$

$$\Rightarrow \quad \frac{u^2}{4g} = 120 \text{ m}$$

If speed is v after the first collision, then speed should remain $\frac{1}{\sqrt{2}}$ times, as kinetic energy has reduced to half.

$\Rightarrow \quad v = \frac{u}{\sqrt{2}}$

$\therefore \quad h_{max} = \frac{v^2 \sin^2 30°}{2g}$

$= \frac{(u/\sqrt{2})^2 \sin^2 30°}{2g}$

$= \left(\frac{u^2/4g}{4}\right) = \frac{120}{4} = 30 \text{ m}$

18. (3) Given, $d = 0.5$ mm,

$Y = 2 \times 10^{11} \text{ Nm}^{-2}$

$l = 1$ m

$\therefore \quad \Delta l = \frac{Fl}{AY} = \frac{mgl}{\frac{\pi d^2}{4} Y}$

$= \frac{1.2 \times 10 \times 1}{\frac{\pi}{4} \times (5 \times 10^{-4})^2 \times 2 \times 10^{11}} = 0.3$ mm

LC of Vernier $= \left(1 - \frac{9}{10}\right)$ mm $= 0.1$ mm

So, 3rd division of Vernier scale will coincide with main scale.

19. (900) Given, $n = 1, \gamma = \frac{5}{3}$

T-V equation in adiabatic process is

$TV^{\gamma-1} = \text{constant}$

$\therefore \quad T_1 V_1^{\gamma-1} = T_2 V_2^{\gamma-1}$

$\Rightarrow \quad T_2 = T_1 \left(\frac{V_1}{V_2}\right)^{\gamma-1} = 100 \times \left(\frac{1}{8}\right)^{\frac{2}{3}}$

$\Rightarrow \quad T_2 = 25$ K

$C_V = \frac{3}{2} R$ for monoatomic gas

$\therefore \Delta U = nC_V \Delta T = n \times \left(\frac{3R}{2}\right)(T_2 - T_1)$

$= 1 \times \frac{3}{2} \times 8 \times (25 - 100) = -900$ J

$\therefore$ Decrease in internal energy = 900 J

20. (b) $v = \sqrt{\frac{GM}{R}}$

Let $R_1 = R$, then $R_2 = 4R$

If $m_2 = m$, then $m_1 = 2m$

List-I

(P) $\frac{v_1}{v_2} = \sqrt{\frac{R_2}{R_1}} = \sqrt{\frac{4R}{R}} = 2:1$

(Q) $L = mvR$

$\frac{L_1}{L_2} = \frac{R(2m)v_1}{4R(m)v_2} = \frac{1}{2}(2) = 1:1$

(R) $\frac{K_1}{K_2} = \frac{\frac{1}{2}(2m)v_1^2}{\frac{1}{2}(m)v_2^2} = 2(4) = 8:1$

(S) $\frac{T_1}{T_2} = \left(\frac{R_1}{R_2}\right)^{3/2} = \left(\frac{1}{4}\right)^{3/2} = 1:8$

21. (c) $\left(\frac{dp}{dV}\right)_{\text{adiabatic}} = \gamma \left(\frac{dp}{dV}\right)_{\text{isothermal}}$

List-I

(P) Process I $\Rightarrow$ Adiabatic $\Rightarrow Q = 0$

(Q) Process II $\Rightarrow$ Isobaric

$\therefore \quad W = p\Delta V = 3p_0[3V_0 - V_0] = 6p_0V_0$

(R) Process III $\Rightarrow$ Isochoric $\Rightarrow W = 0$

(S) Process (IV) $\Rightarrow$ Isothermal

$\Rightarrow$ Temperature = constant

22. (a) When force $F = 0 \Rightarrow$ potential energy U = constant

$F \neq 0 \Rightarrow$ force is conservative $\Rightarrow$ Total energy E = constant

List-I

(P) $\mathbf{r}(t) = \alpha t\hat{\mathbf{i}} + \beta t\hat{\mathbf{j}}$

$\frac{d\mathbf{r}}{dt} = \mathbf{v} = \alpha\hat{\mathbf{i}} + \beta\hat{\mathbf{j}} = \text{constant}$

$\Rightarrow \quad \mathbf{p} = \text{constant}$

$|\mathbf{v}| = \sqrt{\alpha^2 + \beta^2} = \text{constant}$

$\Rightarrow \quad K = \text{constant}$

$\frac{d\mathbf{v}}{dt} = \mathbf{a} = 0 \Rightarrow F = 0 \Rightarrow U = \text{constant}$

$E = U + K = \text{constant}$

$\mathbf{L} = m(\mathbf{r} \times \mathbf{v}) = 0$

$\mathbf{L} = \text{constant}$

$P \to 1, 2, 3, 4, 5$

(Q) $\mathbf{r}(t) = \alpha \cos \omega t\hat{\mathbf{i}} + \beta \sin \omega t\hat{\mathbf{j}}$

$\frac{d\mathbf{r}}{dt} = \mathbf{v} = \alpha\omega \sin \omega t(-\hat{\mathbf{i}}) + \beta\omega\cos \omega t\hat{\mathbf{j}} \neq \text{constant}$

$\Rightarrow \mathbf{p} \neq \text{constant}$

$|\mathbf{v}| = \omega\sqrt{(\alpha \sin \omega t)^2 + (\beta \cos \omega t)^2} \neq \text{constant} \Rightarrow K \neq \text{constant}$

$\mathbf{a} = \frac{d\mathbf{v}}{dt} = -\omega^2 \mathbf{r} \neq 0$

$\Rightarrow E = \text{constant} = K + U$

But $K \neq$ constant $\Rightarrow U \neq$ constant

$\mathbf{L} = m(\mathbf{r} \times \mathbf{v}) = m\omega\alpha\beta\,(\hat{\mathbf{k}}) = \text{constant}$

$Q \to 2, 5$

(R) $\mathbf{r}(t) = \alpha\,(\cos \omega t\hat{\mathbf{i}} + \sin \omega t\hat{\mathbf{j}})$

$\frac{d\mathbf{r}}{dt} = \mathbf{v} = \alpha\omega\,[\sin \omega\, t(-\hat{\mathbf{i}}) + \cos \omega t\hat{\mathbf{j}}] \neq \text{constant}$

$\Rightarrow \quad \mathbf{p} \neq \text{constant}$

$|\mathbf{v}| = \alpha\omega = \text{constant}$

$\Rightarrow \quad K = \text{constant}$

$\mathbf{a} = \frac{d\mathbf{v}}{dt} = -\omega^2 \mathbf{r} \neq 0$

$= E = \text{constant}, U = \text{constant}$

$\mathbf{L} = m(\mathbf{r} \times \mathbf{v}) = m\omega\alpha^2\,\hat{\mathbf{k}} =$ constant

$R \to 2, 3, 4, 5$

S) $\mathbf{r}(t) = \alpha t\hat{\mathbf{i}} + \frac{\beta}{2}t^2\hat{\mathbf{j}}$

$\frac{d\mathbf{r}}{dt} = \mathbf{v} = \alpha\hat{\mathbf{i}} + \beta t\hat{\mathbf{j}} \neq \text{constant}$

$\Rightarrow \quad \mathbf{p} \neq \text{constant}$

$|\mathbf{v}| = \sqrt{\alpha^2 + (\beta t)^2} \neq$ constant $\Rightarrow K \neq$ constant

$\mathbf{a} = \frac{d\mathbf{v}}{dt} = \beta\hat{\mathbf{j}} \neq 0$

$\Rightarrow \quad E = \text{constant} = K + U$

But $\quad K \neq \text{constant}$

$\therefore \quad U \neq \text{constant}$

$\mathbf{L} = m\,(\mathbf{r} \times \mathbf{v}) = \frac{1}{2}\alpha\beta t^2\,\hat{\mathbf{k}} \neq \text{constant}$

$S \to 5$

BITSAT

1. (a) According to principle of calorimetry, heat gained = heat lost

Heat is lost by steam in two stages

(i) Change of state from steam at 100°C to water at 100°C is $m \times 540$.

(ii) To change water at 100°C to water at 80°C is $m \times 1 \times (100 - 80)$, where m is the mass of the steam condensed.

Total heat lost by steam is

$m \times 540 + m \times 20$

$= m(540 + 20) = 560\text{m}$

Heat gained by calorimeter and its contents is

$(1.1 + 0.2) \times (80 - 15)$

$= 1.12 \times 65$ cal

$\Rightarrow \quad 560\,m = 1.12 \times 65$

$\Rightarrow \quad m = \frac{1.12 \times 65}{560} = 0.130$ kg

2. (b) As, gravitational constant,

$[G] = [M^{-1}L^3T^{-2}]$

Planck's constant, $[h] = [ML^2T^{-1}]$

Hence, $\quad [G]\,[h] = [M^{-1}L^3T^{-2}]\,[ML^2T^{-1}]$

$= [M^0L^5T^{-3}]$

Velocity of light, $[c] = [LT^{-1}]$

Now, $\quad \left[\frac{Gh}{c^3}\right]^{1/2} = \frac{[L^5T^{-3}]^{1/2}}{[L^3T^{-3}]^{1/2}}$

$= [L^2]^{1/2} = [L]$

Hence, [L] = length

3. (b) Given, $Q = 10^6$ cal

$$T_1 = 827°C = (827 + 273) = 1100 \text{ K}$$
$$T_2 = 27°C = (27 + 273) = 300 \text{ K}$$

As, $\dfrac{Q_1}{T_1} = \dfrac{Q_2}{T_2}$

$\therefore \quad Q_2 = \dfrac{T_2}{T_1} Q_1 = \dfrac{300}{1100} \times 10^6$

$= 2.72 \times 10^5$ cal

Efficiency of the engine,

$$\eta = \left(1 - \frac{T_2}{T_1}\right) \times 100$$

$$\eta = \left(1 - \frac{300}{1100}\right) \times 100$$

$= 72.72\%$

4. (a) According to Doppler's effect, the approximate frequency heard by the stationary observer,

$$\nu = \frac{v}{v - v_s} \nu_0$$

Case (i) $v_s = 34$ m/s

where, v = speed of sound in air,

u_s = speed of source

and ν_0 = frequency of the source.

$\therefore \quad \nu_1 = \dfrac{340}{340 - 34} \nu_0$

$= \dfrac{340}{306} \nu_0 \quad \ldots \text{(i)}$

Case (ii) $v_s = 17$ m/s

$\therefore \quad \nu_2 = \dfrac{340}{340 - 17} \nu_0 = \dfrac{340}{323} \nu_0 \quad \ldots \text{(ii)}$

From Eqs. (i) and (ii), we get

$\therefore \quad \dfrac{\nu_1}{\nu_2} = \dfrac{340/306}{340/323} \Rightarrow \dfrac{323}{306} \approx \dfrac{19}{18}$

5. (b) Given, $m = 5$ kg,

$v = 20 \text{ ms}^{-1}, \theta = 60°$

Vertical component of velocity,

$v_y = v\sin 60°$

$= 20 \times \dfrac{\sqrt{3}}{2} = 10\sqrt{3} \text{ ms}^{-1}$

Time taken to reach the highest point = Time taken to reach the ground from highest point.

$$t = \frac{v\sin\theta}{g} = \frac{v_y}{g} = \frac{10\sqrt{3}}{9.8} = 1.77 \text{ s}$$

If the highest point, m splits up into two parts of masses $m_1 = 1$ kg and $m_2 = 4$ kg.

If their velocities v_1 and v_2 respectively, then applying the principle of conservation of linear momentum, we get

$$m_1 v_1 + m_2 v_2 = mv\cos\theta$$

$v_1 + v_2 = 5 \times 20 \times \dfrac{1}{2} \quad [\because \theta = 60°]$

$v_1 + 4v_2 = 5 \times 10 = 50 \quad \ldots \text{(i)}$

Initial KE $= \dfrac{1}{2} m (v\cos\theta)^2$

$= \dfrac{1}{2} \times 5 \times (10)^2 = 250$ J

Final KE $= 2$ (initial KE)

$= 2 \times 250 = 500$ J

$\therefore \quad \dfrac{1}{2} m_1 v_1^2 + \dfrac{1}{2} m_2 v_2^2 = 500$

or $\quad \dfrac{1}{2} \times 1 \times v_1^2 + \dfrac{1}{2} \times 4 \times v_2^2 = 500$

or $\quad v_1^2 + 4v_2^2 = 1000 \quad \ldots \text{(ii)}$

Solving Eqs. (i) and (ii), we get

$v_1 = 30$ m/s, $v_2 = 5$ m/s

Hence, the separation between the two fragments

$= (v_1 - v_2) \times t = (30 - 5) \times 1.77 \text{ m} = 44.25$ m

6. (d) Given, initial speed, $u = 36$ km/h

$= \dfrac{36 \times 1000}{60 \times 60} = 10 \text{ ms}^{-1}$

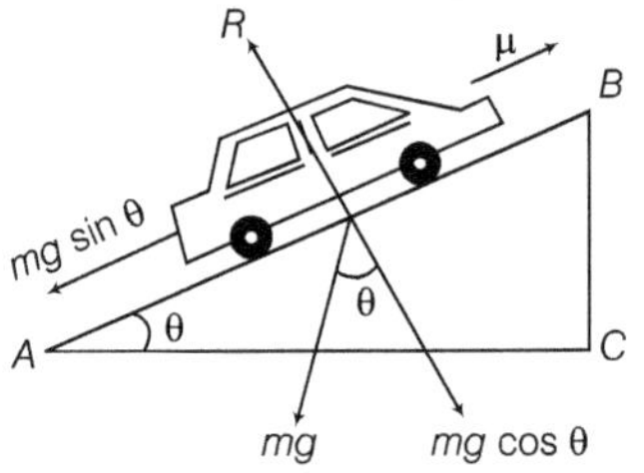

$\theta = 30°, \mu = 0.1, s = ?$

Here, work done in moving up the inclined road

= Kinetic energy of the vehicle

$(mg\sin\theta + F)s = \dfrac{1}{2} mu^2$

$(mg\sin\theta + \mu R) \times s = \dfrac{1}{2} mu^2$

$(mg\sin\theta + \mu\, mg\cos\theta) \times s = \dfrac{1}{2} mu^2$

$$s = \frac{\frac{1}{2} mu^2}{mg(\sin\theta + \mu\cos\theta)} = \frac{u^2}{2g(\sin\theta + \mu\cos\theta)}$$

$$= \frac{10 \times 10}{2 \times 10 \times (\sin 30° + 0.1\cos 30°)} = 8.53 \text{ m}$$

7. (d) Let V be the volume of the block. When block floats in water, then

$$V\rho_{\text{block}}\, g = \left(\frac{4}{5} V\right) \rho_{\text{water}}\, g$$

or $\quad \rho_{\text{block}} = \dfrac{4}{5} \rho_{\text{water}} \quad \ldots \text{(i)}$

When block floats in liquid, then

$V\rho_{\text{block}}\, g = V\rho_{\text{liquid}}\, g$

$\rho_{\text{block}} = \rho_{\text{liquid}}$

$\rho_{\text{liquid}} = \dfrac{4}{5} \rho_{\text{water}} \quad$ [from Eq. (i)]

$= \dfrac{4}{5} \times 10^3 \text{ kg m}^{-3}$

$= 800 \text{ kg m}^{-3}$

8. (c) Let T_0 be the temperature of point C and x be the length of rod AB or BC.

Then, $CA = \sqrt{x^2 + x^2} = \sqrt{2}\, x$

At steady state, the rate of heat flowing from B to C = rate of heat flowing from C to A.

So, $\quad \dfrac{KA(\sqrt{2}\,T - T_0)}{x} = \dfrac{KA(T_0 - T)}{\sqrt{2}\, x}$

$\sqrt{2}(\sqrt{2}\,T - T_0) = T_0 - T$

By solving, $T_0 = \dfrac{3T}{(\sqrt{2} + 1)}$

9. (c) $I = I_{\text{max}} \cos^2\left(\dfrac{\phi}{2}\right)$

$\dfrac{I_{\text{max}}}{4} = I_{\text{max}} \cos^2 \dfrac{\phi}{2}$

$\cos\dfrac{\phi}{2} = \dfrac{1}{2}$

$\dfrac{\phi}{2} = \dfrac{\pi}{3}$

$\Rightarrow \quad \phi = \dfrac{2\pi}{3} = \left(\dfrac{2\pi}{\lambda}\right) \times \Delta x \quad \ldots \text{(i)}$

where, $\Delta x = d\sin\theta$

Putting the value of Δx in Eq. (i), we get

$\sin\theta = \dfrac{\lambda}{3d}$

$\theta = \sin^{-1}\left(\dfrac{\lambda}{3d}\right)$

10. (a) The bob filled completely with water has its centre of mass at its centre. The time period of oscillation is

$$T = 2\pi\sqrt{\frac{l}{g}}$$

As the water starts coming out of the bob, its centre of mass shifts vertically downward as a result effective length of the pendulum increases and hence its period also increases. When the bob is empty, again its centre of mass appears at its centre and as a result, the period of oscillation again reaches to its original value.

11. (b) When a tennis ball falls on the ground and bounces back, its velocity and displacement changes in reverse direction while acceleration remains unchanged.

12. (c) As we know,

$L = I\omega$

$\Rightarrow \quad L = I \times \dfrac{2\pi}{T}$

$$L \propto \frac{1}{T}$$

Given, $T_2 = 2T$

Hence, $\frac{L_1}{L_2} = \frac{2T}{T}$

$\Rightarrow \quad L_2 = \frac{L_1}{2}$

13. (a) Maximum stress

$$= \frac{\text{Maximum weight}}{\text{Cross - sectional area}}$$

Now, maximum weight of steel wire

$= \text{Volume} \times \text{Density} \times g$

$= Al\rho g$

where, l is the maximum length of steel wire that can hang vertically without breaking, ρ is the density of steel and A is the cross-sectional area of steel wire.

$\therefore$ Maximum stress, $p = \frac{Al\rho g}{A} = l\rho g$

$\therefore \quad l = \frac{p}{\rho g}$

14. (c) $2h = ut$

$$u_x = \frac{2h}{\Delta t} \quad [\because u = u_x, t = \Delta t] \ldots \text{(i)}$$

$$h = u_y t - \frac{1}{2} g t^2$$

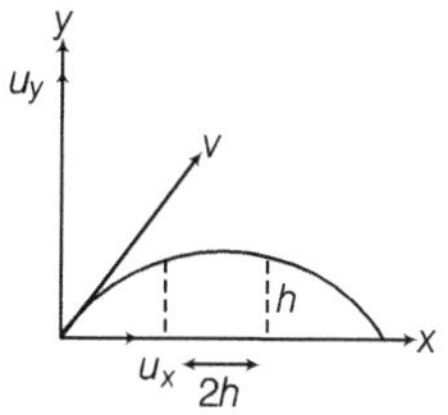

$$gt^2 - 2u_y t + 2h = 0$$

$$t_1 = \frac{2u_y + \sqrt{4u_y^2 - 8gh}}{2g}$$

$$t_2 = \frac{2u_y - \sqrt{4u_y^2 - 8gh}}{2g}$$

$$\Delta t = t_1 - t_2 = \frac{\sqrt{4u_y^2 - 8gh}}{g}$$

$$u_y^2 = \frac{g^2(\Delta t)^2}{4} + 2gh$$

$$u_x^2 + u_y^2 = u^2 = (2\sqrt{gh})^2$$

$$\frac{4h^2}{(\Delta t)^2} + \frac{g^2(\Delta t)^2}{4} + 2gh = 4gh$$

$$\frac{g^2}{4}(\Delta t)^4 - 2gh(\Delta t)^2 + 4h^2 = 0$$

$$(\Delta t)^2 = \frac{2gh \pm \sqrt{4g^2h^2 - 4g^2h^2}}{g^2/2} = \frac{4h}{g}$$

$$\Delta t = 2\sqrt{\frac{h}{g}}$$

15. (b) $s = ut + \frac{1}{2}at^2$, $a = g\sin\theta$, $u = 0$

Then, $s = 0 + \frac{1}{2}at^2$

or $t = \sqrt{\frac{2s}{a}} = \sqrt{\frac{2s}{g\sin\theta}}$

[for smooth plane]

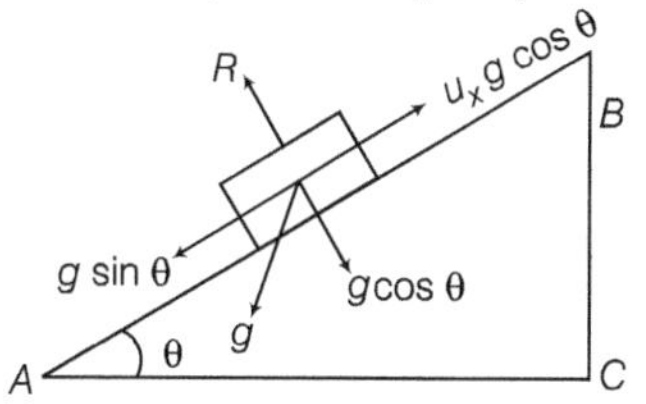

For the rough plane, the effective value of acceleration along the incline is

$$a' = g\sin\theta - \mu_k g\cos\theta$$

and $t' = \sqrt{\frac{2s}{a}} = \sqrt{\frac{2s}{g\sin\theta - \mu_k g\cos\theta}}$

Now, $\frac{t'}{t} = n = \sqrt{\frac{g\sin\theta}{g\sin\theta - \mu_k g\cos\theta}}$

$$= \sqrt{\frac{\sin 45^\circ}{\sin 45^\circ - \mu_k \cos 45^\circ}}$$

$$= \sqrt{\frac{1}{1 - \mu_k}}$$

$\Rightarrow \quad \frac{1}{n^2} = 1 - \mu_k \Rightarrow \mu_k = 1 - \frac{1}{n^2}$

16. (a) Helium is monoatomic gas, for which

$$C_V = \frac{3}{2}R, C_p = \frac{5}{2}R$$

Work done by the gas in one complete cycle W = area $ABCDA = p_0V_0$

From A to B,

Heat given to gas $= nC_V\Delta T$

$$= 1 \times \left(\frac{3}{2}R\right) \times \Delta T = \frac{3}{2}V_0(\Delta p) = \frac{3}{2}V_0p_0$$

From B to C, heat given to gas $= nC_p\Delta T$

$$= 1 \times \left(\frac{5}{2}R\right) \times \Delta T = \frac{5}{2}(2p_0)\Delta V = 5p_0V_0$$

Efficiency of cycle

$$= \frac{\text{Work done by the gas / cycle}}{\text{Total heat given to gas / cycle}}$$

$$= \frac{p_0V_0}{\frac{3}{2}p_0V_0 + 5p_0V_0} = \frac{2}{13}$$

$$\text{Efficiency (\%)} = \frac{2}{13} \times 100 = 15.4\%$$

17. (c) Initial energy of the asteroid is

$$E_i = K_i + U_i = \frac{1}{2}mv_i^2 - \frac{GM_em}{10R_e}$$

Final energy of the asteroid,

$$E_f = \frac{1}{2}mv_f^2 - \frac{GM_em}{R_e}$$

According to law of conservation of energy, $E_i = E_f$

$$\frac{1}{2}mv_i^2 - \frac{GM_em}{10R_e} = \frac{1}{2}mv_f^2 - \frac{GM_em}{R_e}$$

$$v_f^2 - \frac{2GM_e}{R_e} = v_i^2 - \frac{2GM_e}{10R_e}$$

$$\Rightarrow \quad v_f^2 = v_i^2 + \frac{2GM_e}{R_e}\left(1 - \frac{1}{10}\right)$$

18. (a) We have,

Torque $= r \times F = I\alpha$

So, $2(20t - 5t^2) = 10\alpha$

$$\alpha = 4t - t^2$$

But $\alpha = \frac{d\omega}{dt}$, so

$$\frac{d\omega}{dt} = 4t - t^2$$

$$d\omega = (4t - t^2)\,dt$$

Integrating, $\omega = 2t^2 - \frac{t^3}{3}$

ω will be zero at $t = 6$ s.

So, $\omega = \frac{d\theta}{dt} = 2t^2 - \frac{t^3}{3}$

or $d\theta = \left(2t^2 - \frac{t^3}{3}\right)dt$

Again, integrating both sides, we get

$$\theta = \frac{2t^3}{3} - \frac{t^4}{12}$$

Since, $t = 6$ s

so, $\theta = \frac{2 \times 6^3}{3} - \frac{6^4}{12} = 36$

Number of turns, $n = \frac{\theta}{2\pi} = \frac{36}{2\pi} = 5.73$

Hence, option (a) is correct.

19. (c) When $r \le R$, then force on the test mass m at the surface of the sphere $= mg$

Force on the test mass at distance r from the centre of sphere is

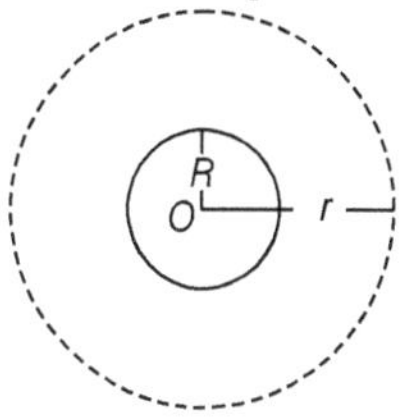

If $r < R$, then

$$F = \frac{GMm}{R^2} \cdot \frac{r}{R} = \frac{GMm}{R^3} \cdot r$$

$$\therefore \quad \frac{mv^2}{r} = \frac{GMm}{R^3} \cdot r$$

$\therefore \quad v \times r$

If $r > R$, then

$$F = \frac{GMm}{r^2}$$

$$\therefore \quad \frac{Mv^2}{r} = \frac{GMm}{r^2}$$

$$\therefore \quad v \times \frac{1}{\sqrt{r}}$$

Hence, option (c) is correct.

20. (a) Since, no external force is acting on the system, we can apply conservation of linear momentum.

Speed of 200 kg trolle

$$= \frac{36 \times 1000}{60 \times 60} = 10 \text{ ms}^{-1}$$

If u be the initial velocity of trolley, v_b be the absolute velocity of the boy after the beginning of journey of the boy, their relative velocity is 4.

So, $\quad v' - v_b = 4$

$\Rightarrow \quad v_b = (v' - 4)$

Now, applying law of conservation of momentum,

Momentum before the boy begins to run

= Momentum after the beginning of boy's running

$\Rightarrow \quad 220 \times 10 = 200\, v' + 20(v' - 4)$

$2200 = 220\, v' - 80$

$\Rightarrow \quad 220\, v' = 2280$

$$v' = \frac{2280}{220} = 10.36 \text{ ms}^{-1}$$

21. (c) Given, $\ C = 37.55 \text{ J mol}^{-1} \text{ K}^{-1}$

Also, $pT = \text{constant}\,(K)$ $\quad$ ... (i)

According to ideal gas equation,

$$pV = RT$$

$\Rightarrow \quad p = \dfrac{RT}{V}$ $\quad$ ... (ii)

Putting the value of p in Eq. (i), we get

$$\frac{RT}{V} \times T = K \ \Rightarrow \ V = \frac{RT^2}{K}$$

On differentiating above equation both sides, we get

$\dfrac{dV}{dT} = \dfrac{2RT}{K}$ $\quad$ ... (iii)

But $\quad \dfrac{T}{K} = \dfrac{1}{p}$ $\quad$ [from Eq. (i)]

Hence, Eq. (iii) becomes

$$\frac{dV}{dT} = \frac{2R}{p}$$

So, $\quad C = C_V + \dfrac{pdV}{dT}$

or $\quad C = C_V + \dfrac{p \times 2R}{p} = C_V + 2R$

or $\quad C_V = C - 2R$ $\quad$ ... (iv)

As $C_V = \dfrac{nR}{2}$, where n = number of degrees of freedom.

Putting the value of C_V in Eq. (iv), we get

$$\frac{nR}{2} = C - 2R$$

$$n = \frac{2(C - 2R)}{R} = \frac{2\,(37.55 - 2 \times 8.3)}{8.3}$$

$$= 5.048 \approx 5$$

22. (a) We know, instantaneous displacement,

$$x = r \sin \omega t$$

$\because$ Instantaneous velocity,

$$v = \frac{dx}{dt} = r\omega \cos \omega t$$

$\because$ Instantaneous acceleration,

$$a = \frac{dv}{dt} = -r\omega^2 \sin \omega t = -\omega^2 x$$

So, $\quad \dfrac{aT}{x} = \dfrac{-\omega^2 x \times T}{x}$

$$= -\omega^2 T = \frac{-4\pi^2}{T^2} \times T$$

$$= \frac{-4\pi^2}{T} = \text{constant}$$

$\Rightarrow$

$$\frac{aT}{V} = \frac{-\omega^2 r \sin \omega t \times T}{\omega r \cos \omega t} = -\omega T \tan \omega T$$

$$= -\frac{2\pi}{T} \times T \tan \omega t = \text{not constant}$$

Similarly, $aT + 2\pi v$ and $a^2T^2 + 4\pi^2 v^2$ is also not constant, i.e. both are function of t.

23. (a) For the uniformly accelerated/decelerated motion,

$$v^2 = u^2 \pm 2gh$$

So, from this equation, we can say v-h graph is parabola.

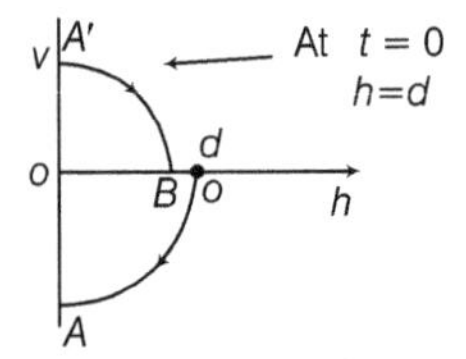

Initially, velocity is downwards (– ve), after collision, it reverses the direction with smaller magnitude and velocity is upwards (+ ve). So, graph (a) satisfies these conditions.

Also,

When $t = 0$, $h = d$

Velocity increases downwards $(0 \to A)$

When $t = 1$, velocity reverses its direction $(A' \to B)$

24. (d) Since, volume remains unchanged, during this phenomenon, so

$$\frac{4}{3}\pi R^3 = N \times \frac{4}{3}\pi r^3$$

$$N = \frac{R^3}{r^3}$$

Now, change in surface area

$$= 4\pi R^2 - N 4\pi r^2$$

$$= 4\pi\,(R^2 - Nr^2)$$

Energy released $(\Delta U) = T \times$ change in surface area

$$= T \times 4\pi\,[R^2 - Nr^2]$$

Here, all this energy released is at the cost of lowering the temperature and mass of the big drop of liquid $= \dfrac{4}{3}\pi R^2 \rho$.

Now, change in temperature,

$$\Delta\theta = \frac{\Delta U}{ms}$$

$$= \frac{T \times 4\pi(R^2 - Nr^2)}{\left(\frac{4}{3}\pi R^3 \rho\right) S}$$

$$= \frac{3T}{\rho S}\left(\frac{1}{R} - \frac{Nr^2}{R^3}\right)$$

$$= \frac{3T}{\rho S}\left(\frac{1}{R} - \frac{R^3 \times r^2}{r^3 \times R^3}\right)$$

$$= \frac{3T}{\rho S}\left(\frac{1}{R} - \frac{1}{r}\right)$$

KCET

1. (b) Given, $m = 1\text{ g} = 10^{-3}$ kg

Energy, $E = mc^2$

$\therefore$ Energy equivalent $= 10^{-3} \times (3 \times 10^8)^2$

$(\because c = 3 \times 10^8 \text{ ms}^{-1})$

$= 9 \times 10^{13}$ J

2. (c) Given, $R = 6400$ km $= 6.4 \times 10^6$ m,

$h = 500$ m

$\therefore$ Range, $\ d = \sqrt{2hR}$

$$= \sqrt{2 \times 500 \times 6.4 \times 10^6}$$

$$= \sqrt{64 \times 10^8} = 8 \times 10^4 = 80 \text{ km}$$

3. (c) Escape velocity at a height h *from the surface of Earth*

$$v'_E = \sqrt{\frac{2GM}{R + h}} \left[\because v_E = \sqrt{\frac{2GM}{R}}\right]$$

Given, $h = R$

$$\Rightarrow \quad v'_E = \sqrt{\frac{2GM}{R + R}} = \frac{1}{\sqrt{2}}\sqrt{\frac{2GM}{R}} = \frac{v_E}{\sqrt{2}}$$

4. (d) Instantaneous velocity is given by,

$$V = \lim_{\Delta t \to 0} \frac{\Delta x}{\Delta t} = \frac{dx}{dt}$$

which is the slope of x-t graph.

Here, $\frac{dx}{dt}$ is maximum for the region QR and minimum for the region RS.

$\therefore$ Instantaneous velocity is maximum around Q.

5. (b) The acceleration due to gravity on the surface of earth, $g = \frac{GM}{R^2}$

At a height h above the surface,

$$g = \frac{GM}{(R+h)^2}$$

At a depth d below the surface,

$$g = \frac{Gm}{(R-d)^2}$$

6. (a) According to Newton's law of cooling,

$$\frac{\theta_1 - \theta_2}{t} = \alpha\left(\frac{\theta_1 + \theta_2}{2}\right) - \theta_0$$

($\because$ α = universal constant)

$$\frac{65.5 - 62.5}{1} = \alpha\left[\frac{(65.5 + 62.5)}{2} - 22.5\right] \quad \text{...(i)}$$

$$\frac{46.5 - 40.5}{t} = \alpha\left[\frac{(46.5 + 40.5)}{2} - 22.5\right] \quad \text{...(ii)}$$

Solving Eqs. (i) and (ii), we get

$$t = 4 \text{ min}$$

7. (b) Work done = PE

$= mgh$

Here, $h = R$

$\therefore \quad W = mgR$

8. (a) First overtone frequency $= \frac{3v}{4l_1}$

Second harmonic frequency $= \frac{2v}{2l_2}$

Given, $\frac{3v}{4l_1} = \frac{2v}{2l_2}$

$\therefore \quad \frac{l_1}{l_2} = \frac{3}{4}$

9. (c) Given, $V = (100 \pm 5)\ \text{V} \Rightarrow \Delta V = 5\ \text{V}$

$I = (10 \pm 0.2)\ \text{A} \Rightarrow \Delta I = 0.2\ \text{A}$

$$R = \frac{V}{I}$$

$$\frac{\Delta R}{R} = \frac{\Delta V}{V} + \frac{\Delta I}{I}$$

$$= \frac{5}{100} + \frac{0.2}{10} = 0.05 + 0.02$$

$= 0.07$

$= 7\%$

10. (b) As the block is at the state of rest,

$\therefore R = mg\cos\theta$, which is balanced.

and f_r = force of friction = 10 N (given)

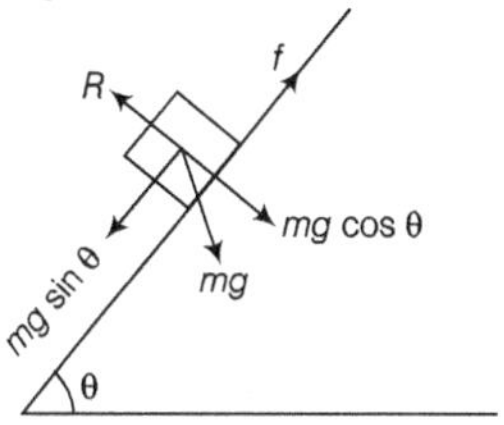

Hence, $f_r = mg\sin\theta$

$10 = m \times 10 \times \sin 30°$

$$m = \frac{10}{10 \times \sin 30°} = 2 \text{ kg}$$

11. (a) $p = 2m\sqrt{\text{KE}}$

p_1, p_2 are momenta of two particles, m_1 and m_2 respectively.

$\therefore \quad \frac{p_1}{p_2} = \frac{2m_1\sqrt{\text{KE}_1}}{2m_2\sqrt{\text{KE}_2}}$

But $\quad \text{KE}_1 = \text{KE}_2$

$\therefore \quad \frac{p_1}{p_2} = \frac{m_1}{m_2}$

12. (d) $p = h\rho g$

$p \propto h$

$p \propto \rho$

$p \propto g$, p does not depend area of the liquid.

13. (b) Given, $Q_1 = 300$ cal, $Q_2 = 150$ cal

$T_1 = 500$ K, $T_2 = ?$

$\therefore \quad \frac{Q_2}{Q_1} = \frac{T_2}{T_1}$

$\frac{150}{300} = \frac{T_2}{500}$

$\Rightarrow \quad T_2 = 250$ K

14. (b) $p = \frac{2}{3}E \Rightarrow p \propto E$

So, as p increases, E also increases.

15. (d) $F = m(g + a) = 60\ (98 + 1.8)$

$= 696$ N

16. (b) $I = I_X + I_Y = 20 + 25 = 45\ \text{kg-m}^2$

17. (b) $\quad \text{Stress} = \frac{\text{Force}}{\text{Area}}$

$$S_A = \frac{F}{A_A}$$

$$S_B = \frac{F}{A_B} = \frac{F}{\frac{A_A}{2}} = 2S_A$$

18. (d) The potential at every point on an equipotential surface is same. Hence, the work done to move a charge on this surface is zero.

AP EAMCET

1. (d) Trajectory of a projectice is of form

$$y = f(x) = x\tan\theta - \frac{gx^2}{2u^2\cos^2\theta}$$

$$\therefore \quad \frac{dy}{dx} = \text{slope} = \frac{dy/dt}{dx/dt} = \frac{v_y}{v_x}$$

So, Assertion (A) is incorrect.

Also, velocity of a projectile is always along tangent to the trajectory (shown)

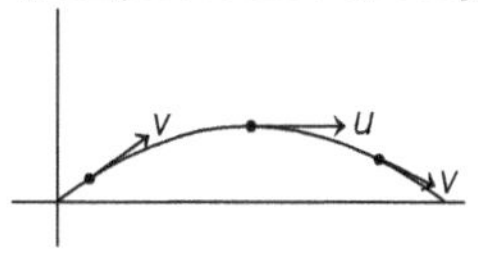

Hence, reason (R) is correct.

2. (c) For a projectile projected at angle θ;

Maximum height, $H_{max} = \frac{u^2\sin^2\theta}{2g}$

and range, $R = \frac{u^2\sin 2\theta}{g}$

$$\therefore \text{Ratio} = \frac{H_{max}}{R} = \frac{\left(\frac{u^2\sin^2\theta}{2g}\right)}{\left(\frac{u^2\sin 2\theta}{g}\right)} = \frac{\tan\theta}{4}$$

Here, $\theta = \tan^{-1}\frac{8}{7} \Rightarrow \tan\theta = \frac{8}{7} = \frac{\frac{8}{7}}{4} = \frac{2}{7}$

3. (d) Let velocity of projectile is v at an angle $\frac{\theta}{2}$ with horizontal

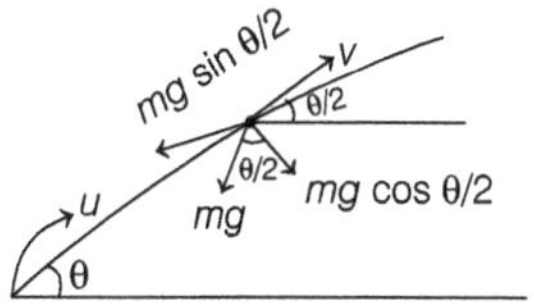

$\therefore \quad v\cos\frac{\theta}{2} = u\cos\theta$

or $\quad v = \frac{u\cos\theta}{\cos\frac{\theta}{2}}$

As horizontal component remains same.

Also, centripetal force is provided by the component of weight.

So, $\quad \frac{mv^2}{r} = mg\cos\frac{\theta}{2}$

Hence, radius of curvature of path,

$$\Rightarrow \quad r = \frac{v^2}{g\cos\frac{\theta}{2}}$$

$$= \frac{\frac{u^2\cos^2\theta}{\left(\cos\frac{\theta}{2}\right)^2}}{g\cos\frac{\theta}{2}} \Rightarrow r = \frac{u^2\cos^2\theta \cdot \sec^3\left(\frac{\theta}{2}\right)}{g}$$

4. (d) Let h = critical height of sand cone of radius R.

Then, for a sand particle to be in equilibrium (it must no slips to the ground)

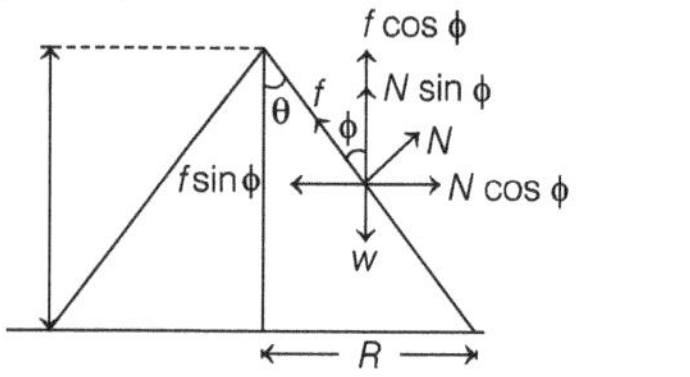

$$f\sin\phi = N\cos\phi$$

where, f = friction,

$$N = \text{normal reaction and } f = \mu N$$

$$\Rightarrow \quad \mu N\sin\phi = N\cos\phi$$

$$\Rightarrow \quad \tan\phi = \frac{1}{\mu} = \frac{R}{h} \text{ (from figure)}$$

So, maximum volume of sand cone that can be formed over level ground is

$$V_{\max} = \frac{1}{3}\pi R^2 h = \frac{1}{3}\pi R^2(\mu R) = \frac{\mu\pi R^3}{3}$$

5. (b) Block's weight (downward), $w = 2 \times 10 = 20$ N

Vertical component of applied force (upwards),

$$F_V = F\sin\theta = 90 \times 3/5 = 54 \text{ N}$$

Maximum frictional force,

$$F_r = \mu F\cos\theta = 0.25 \times 90 \times 4/5 = 18\text{N}$$

∴ Net vertical force,

$$F_{\text{net}} = F_V - (F_r + w) = ma$$

$$\therefore \quad F_{\text{net}} = 54 - 18 - 20 = 16\text{N} = ma$$

$$\Rightarrow \quad a = \frac{16}{2} = 8 \text{ m/s}^2$$

6. (a) Initial energy of particle

$$= \frac{1}{2}mv^2 = \frac{1}{2} \times 2 \times 8^2 = 64 \text{ J}$$

Final energy of particle at maximum height

$$= mgh = 2 \times 10 \times 3 = 60 \text{ J}$$

Work done against air friction

$$= \text{Loss of energy} = 64 - 60 = 4 \text{ J}$$

7. (a) This is the example of Atwood machine, so acceleration is the system

$$a = \left(\frac{M - m}{M + m}\right) g$$

$$a = \frac{1}{5} \times 10 = 2 \text{ m/s}^2$$

Now, $\quad s = ut + \frac{1}{2}at^2$

$$= 0 + \frac{1}{2} \times 2(2)^2$$

$$\Rightarrow \quad s = 4 \text{ m}$$

Work done on block of 3 kg by gravity,

$$W = Mgs = 3 \times 10 \times 4$$

$$W = 120 \text{ J}$$

8. (d) In this case, angular momentum, $J = I\omega$

Taken, $\quad v = 1$

$$v' = 1.09$$

and $\quad \Delta v = 0.09$

$$\Rightarrow \quad \omega \propto \frac{1}{I}$$

But $\quad I \propto r^2$ and $\omega \propto v^{1/3}$

$$\therefore \quad \omega \propto \frac{1}{v^{2/3}} \text{ (or } v^{-2/3}\text{)}$$

∴ Change in angular speed

$$= \frac{\Delta\omega}{\omega} \times 100 = -\frac{2}{3}\frac{\Delta v}{v} \times 100$$

$$= -\frac{2}{3} \times 0.03 \times 100 = -6\%$$

So, angular speed will decrease with 6%.

9. (b) Tension between P and Q is

T_1 = centripetal force on $Q = mr\omega^2$

$$= 200 \times 1 \times \omega^2 \left(\text{g} \cdot \text{m} \cdot \frac{\text{rad}^2}{\text{s}^2}\right)$$

Tension between O and P is

T_2 = centripetal force on Q + centripetal force on P

$$= 200 \times 1 \times \omega^2 + 200 \times \frac{1}{2} \times \omega^2$$

$$= 300 \times 1 \times \omega^2 \left(\text{g} \cdot \text{m} \cdot \frac{\text{rad}^2}{\text{s}^2}\right)$$

Ratio of tension is

$$\frac{T_1}{T_2} = \frac{200 \times 1 \times \omega^2}{300 \times 1 \times \omega^2} = \frac{2}{3}$$

10. (a) Given, total energy = 9J

PE at mean position = 5J

So, maximum KE = 9J − 5J = 4J

Now, in SHM

Maximum (at mean) KE = Maximum PE (at extremes)

$$\therefore \quad \frac{1}{2}ka^2 = 4\text{J}$$

$$\Rightarrow \quad k = \frac{8}{a^2} = \frac{8}{10^{-4}} = 8 \times 10^4 \text{ J/m}^2$$

Now, time period

$$T = 2\pi\sqrt{\frac{m}{k}} = 2\pi \times \sqrt{\frac{2}{8 \times 10^4}}$$

$$\Rightarrow T = \frac{\pi}{100} \text{ s}$$

11. (d) Work done to change configuration from 1 to 2 can be calculated by

$$W_{12} = -(U_f - U_i)$$

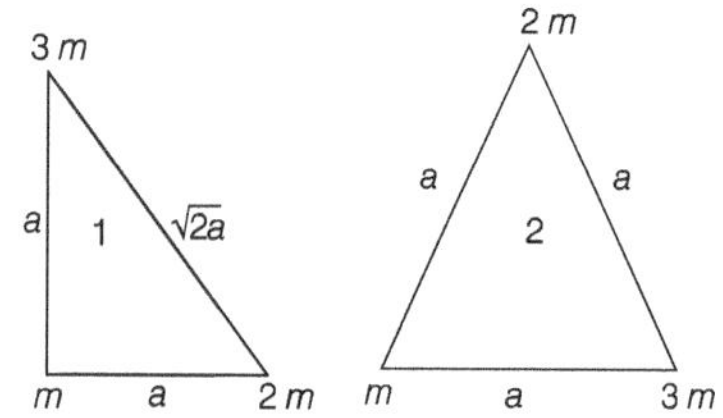

$$= -\left[-\frac{Gm^2}{a}\left(2 + 3 + \frac{6}{\sqrt{2}}\right) - \left(-\frac{Gm^2}{a}(2 + 3 + 6)\right)\right]$$

$$= -\frac{Gm^2}{a}\left(6 - \frac{6}{\sqrt{2}}\right)$$

12. (a)

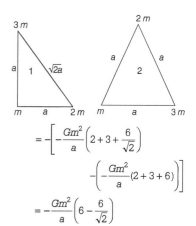

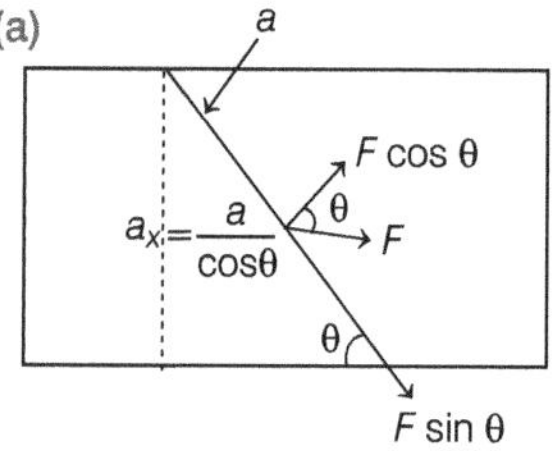

Shearing stress

$$= -\frac{F\sin\theta}{a_x} = -\frac{F}{a}\sin\theta\cos\theta$$

Magnitude of shearing stress

$$= \frac{F}{a}\sin\theta\cos\theta$$

13. (c) Density of water–Density of air ∝ $T_{\text{water}} - T_{\text{air}}$

Density of liquid–Density of water ∝ $T_{\text{liquid}} - T_{\text{water}}$

Now, from given values,

$T_{\text{water}} - T_{\text{air}} = T_{\text{liquid}} - T_{\text{water}}$

Air density = 1 kg / m^3

and water density = 1000 kg / m^3

So, liquid density ≈ 2000 kg / m^3

14. (b) Let mass of the steam condensed is x.

Heat released = Heat gained by water

$\Rightarrow x \times 540 + x \times 1 \times (100 - 90)$

$= 1 \times 1 \times (90 - 9) + 0.1 \times 1 \times (90 - 9)$

$\Rightarrow 540x + 10x = 81 + 8.1$

$$\Rightarrow \quad x = \frac{89.1}{550} = 0.162 \text{ kg}, x = 162 \text{ g}$$

15. (c) Let assume temperature of middle plate is T_0.

As temperatures are constant,

∴ Heat gained by middle surface = Heat lost by middle surface.

So, $\sigma A[(3T)^4 - T_0^4] = \sigma A[T_0^4 - (2T)^4]$

$$\Rightarrow \quad T_0^4 = \frac{97}{2}T^4 \Rightarrow T_0 = \left(\frac{97}{2}\right)^{1/4} T$$

16. (d) Let there are n moles of N_2 gas in the cylinder.

Assuming all of K, E appears in form of heat,

$$n\left(\frac{1}{2}Mv^2\right) = \frac{f}{2}nR\Delta T$$

Here, $M = 28\text{ g} = 28 \times 10^{-3}$ kg, $f = 5$

Increment in pressure due to change of temperature is

Also, $\Delta p = \dfrac{nR\Delta T}{V}$

So, $\dfrac{\Delta p}{p} = \dfrac{\left(\dfrac{nR\Delta T}{V}\right)}{\left(\dfrac{nRT}{V}\right)} = \dfrac{nR\Delta T}{nRT}$

$\Rightarrow \quad \dfrac{\Delta p}{p} = \dfrac{nMv^2}{fnRT} = \dfrac{Mv^2}{fRT}$

So, percentage change in pressure is,

$\therefore \quad \dfrac{\Delta p}{p} \times 100$

$= \dfrac{28 \times 10^{-3} \times 100 \times 100}{5 \times 8.3 \times 300} \times 100$

$= 2.25\%$

17. (b)

(i) Zeroth law of thermodynamics states about thermal equilibrium of different states in contact.

(ii) First law is energy conservation law.

(iii) In free expansion of gases, work done is always zero.

(iv) Second law of thermodynamics discusses about flow of heat in thermal bodies.

18. (a) Mean free path,

$$\lambda = \frac{1}{\sqrt{2}\,\pi n d^2} \Rightarrow d^2 = \frac{1}{\sqrt{2}\pi n \lambda}$$

$$= \frac{1 \times \pi}{\sqrt{2} \times \pi \times 2\sqrt{2} \times 10^8 \times 10^{-2}}$$

$\Rightarrow \quad d^2 = \dfrac{1}{4 \times 10^6} \Rightarrow d = \dfrac{1}{2} \times 10^{-3}$ cm

$\Rightarrow \quad d = 5 \times 10^{-4}$ cm

19. (c) When car is at rest, tension in string is $T = mg$.

$$T = Mg = mv_1^2$$

$$V_1 = \sqrt{\frac{T}{r}} = \sqrt{\frac{Mg}{\mu}} \qquad \ldots\text{(i)}$$

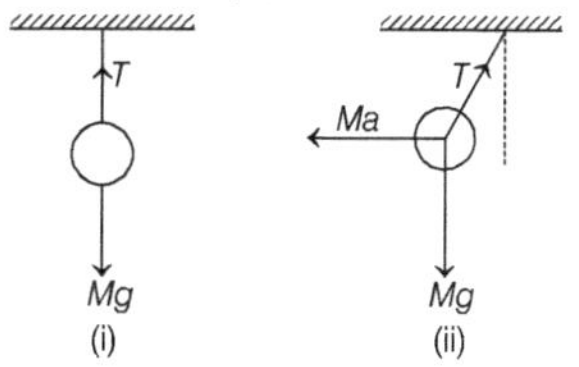

When car is accelerating tension,

$$T = \sqrt{M(a^2 + g^2)^{1/2}}$$

$$\therefore \quad v_2 = \sqrt{\frac{M(a^2 + g^2)^{1/2}}{\mu}}$$

Dividing Eq. (i) by Eq. (ii), we get

$$\frac{v_2}{v_1} = \frac{\sqrt{M(a^2 + g^2)^{1/2}}}{\sqrt{Mg}} = \frac{66}{60}$$

$$\frac{(a^2 + g^2)^{1/2}}{g} = \frac{121}{100}$$

Squaring, we get

$\Rightarrow \quad a^2 + g^2 = \left(\dfrac{121}{100}\right)^2$

$\Rightarrow \quad a^2 = 146.41 - 100$

$\Rightarrow \quad a^2 = 46.41$

So, $\quad a = 6.8\text{ ms}^{-2}$

20. (c) Wavelength of the reflected wave is

$$\lambda' = \left(\frac{v - v_s}{v + v_s}\right)\lambda = \left(\frac{v - v_s}{v + v_s}\right)\frac{v}{\nu}$$

$$= \frac{340 - 20}{340 + 20} \times \frac{340}{160} = \frac{120}{360} \times \frac{340}{160}$$

$\therefore \quad \lambda' = \dfrac{17}{9}$ m

21. (a) Mass, $m = 40\text{ g} = 0.04$ kg,

$$q = 2 \times 10^{-6}\text{ C}$$

Now, $\quad ma = qE$

$\Rightarrow \quad a = \dfrac{q}{m}E$

$= \dfrac{2 \times 10^{-6}}{0.04} \times 4.2 \times 10^4\text{ m/s}^2$

$= 2.1\text{ m/s}^2$ (downward)

So, effective acceleration on bob,

$$a_e = a + g = 12.1\text{ m/s}^2$$

In the absence of electric field,

$$T = 2\pi\sqrt{\frac{l}{g}} = 2\pi\sqrt{\frac{l}{10}}$$

In the presence of electric field,

$$T' = 2\pi\sqrt{\frac{l}{a_e}} = 2\pi\sqrt{\frac{l}{12.1}}$$

$$\frac{T}{T'} = \sqrt{\frac{12.1}{10}} = \frac{11}{10}$$

$\Rightarrow \quad T' = \dfrac{10}{11}T$

Given, $\quad T = \dfrac{44}{20}$

$\Rightarrow \quad T' = \dfrac{10}{11} \times \dfrac{44}{20} = 2\text{ s}$

So, time taken in 15 oscillations

$= 2 \times 15 = 30\text{s}$

TS EAMCET

1. (d)

(A) Energy conservation is a universal law, so there is no violation possibility of this in nature. (False)

(R) This is the statement of energy conservation law. (True)

2. (a) Given, $pV^{5/3} = k$

Using dimensional balance method

Dimension of $\quad p = \text{ML}^{-1}\text{T}^{-2}$

Dimension of $\quad V = \text{L}^3$

$\Rightarrow$ Dimension of $V^{5/3} = \text{L}^5$

As, dimension of LHS = Dimension of RHS

So, dimension of $k = [\text{ML}^{-1}\text{T}^{-2}] \times [\text{L}^5]$

$= [\text{ML}^4\text{T}^{-2}]$

3. (b) Relation between average velocity and instantaneous velocity is

$$\langle v \rangle = \frac{1}{t_2 - t_1}\int_{t_1}^{t_2} v(t)dt$$

In the given problem, $v(t) \propto y^\beta$ (independent of t)

So, $\langle v \rangle \propto v(t)$ or $(v) \propto y^\beta$

But according to question,

$$\langle v \rangle \propto y^{1/3}$$

So, $\quad \beta = \dfrac{1}{3}$

4. (a) Given, $v = \alpha\sqrt{x}$ and at $x = 0, t = 0$

To find velocity and acceleration in terms of time, we need to find x as function of t.

$\because \quad \dfrac{dx}{dt} = \alpha\sqrt{x}$

$\therefore \quad \dfrac{dx}{\sqrt{x}} = \alpha\, dt$

On integrating both sides, we get

$$2\sqrt{x} = \alpha t + C$$

Using, $x = 0$ at $t = 0 \Rightarrow C = 0$

$\therefore \quad \sqrt{x} = \dfrac{\alpha}{2}t$

Hence, velocity, $\quad v = \dfrac{\alpha^2}{2}t \; (\because v = \alpha\sqrt{x})$

and acceleration, $\quad a = \dfrac{dv}{dt} = \dfrac{\alpha^2}{2}$

5. (a) Given, $\mathbf{r} = 3t^2\hat{\mathbf{i}} + 2t\hat{\mathbf{j}} + \hat{\mathbf{k}}$

(i) Velocity vector, $v = \dfrac{d\mathbf{r}}{dt} = 6t\hat{\mathbf{i}} + 2\hat{\mathbf{j}}$

Magnitude of velocity vector,

$$|v|_{t=2} = \sqrt{(12)^2 + (2)^2} = \sqrt{148}$$

(ii) Acceleration vector, $\mathbf{a} = \frac{d\mathbf{v}}{dt} = 6\hat{\mathbf{i}}$

Magnitude of acceleration

$$|\mathbf{a}|_{t=2} = \sqrt{(6^2)} = 6$$

6. (b)

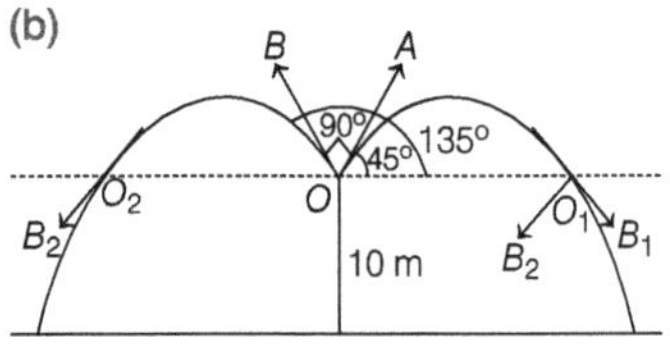

In this projectile motion, both objects are projected at the angle difference of 90°. They will be again have same angle difference, when they are again at same height (of 10 m) from ground. Taking, plane O_2OO_1as reference, time taken to reach at point O_1 by first object is

Time of flight, $T_1 = \frac{2u\sin\theta}{g}$

$$= \frac{2 \times 2\sqrt{2}\sin 45°}{10} = 0.4 \text{ s}$$

Time taken by second object to reach at point O_2 is

$$T_2 = \frac{2 \times 2\sqrt{2}\sin 135°}{10} = 0.4 \text{ s}$$

$\because \quad T_1 = T_2$

So, velocities of both objects will be perpendicular to each other after 0.4 s.

7. (c) Extension x in AC is $L'/2 - L/2$

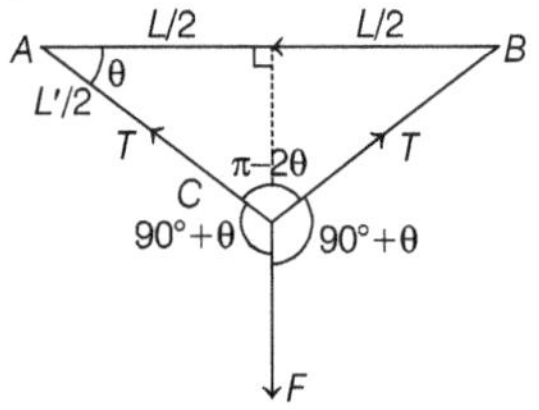

$$= \frac{L}{2\cos\theta} - \frac{L}{2} = \frac{L}{2}\left(\frac{1-\cos\theta}{\cos\theta}\right)$$

Given, $\frac{T}{x} = k$

$$\therefore \quad T = \frac{Lk}{2}\left(\frac{1-\cos\theta}{\cos\theta}\right)$$

Using Lami's theorem,

$$\frac{F}{\sin(\pi - 2\theta)} = \frac{T}{\sin(90° + \theta)}$$

$$\frac{F}{\sin(\pi - 2\theta)} = \frac{\frac{L}{2}k\left(\frac{1-\cos\theta}{\cos\theta}\right)}{\sin(90° + \theta)}$$

$$\Rightarrow \quad F = \frac{kL}{2}\,\frac{(1-\cos\theta)\sin 2\theta}{\cos^2\theta}$$

$$= \frac{kL}{2}\,\frac{(1-\cos\theta)2\sin\theta\cos\theta}{\cos^2\theta}$$

$$F = kL(1-\cos\theta)\cdot\tan\theta$$

8. (b) Let a = acceleration of wedge, then force on block due to acceleration of wedge = ma.

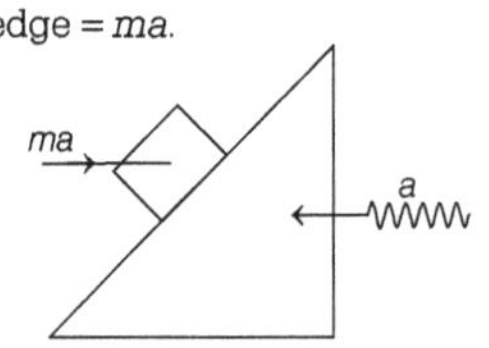

Then, from FBD, we have

For equilibrium of block,

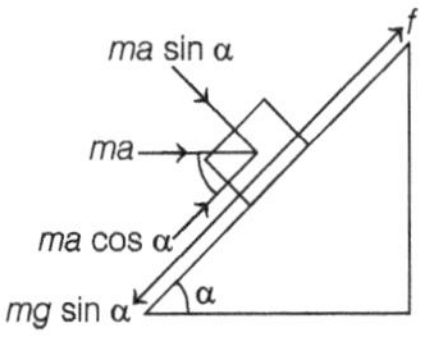

$$mg\sin\alpha = ma\cos\alpha + \mu\,(ma\cos\alpha + ma\sin\alpha)$$

$$\Rightarrow \quad g = a + \mu(g + a) \Rightarrow \frac{0.6}{1.4}g = 9$$

$$\Rightarrow \quad a = \frac{30}{7}\text{ ms}^{-2}$$

9. (a) Given, $a_c \propto t^\alpha$

or $\quad a_c = kt^\alpha$

Now, $\quad a_c = \frac{v_t^2}{r}$

(where, v_t = tangential component of velocity)

$$\Rightarrow \quad v_t = \sqrt{kr\cdot t^\alpha}$$

and $\quad a_t = \frac{dv_t}{dt} = \sqrt{(kr)}\cdot\frac{\alpha}{2}\cdot t^{\frac{\alpha}{2}-1}$

Now, power developed is

$$P = Fv = ma_t v_t$$

$$= m\sqrt{kr}\cdot\frac{\alpha}{2}\cdot t^{\frac{\alpha}{2}-1}$$

$$\Rightarrow \quad P \propto t^{\alpha - 1}$$

10. (d)

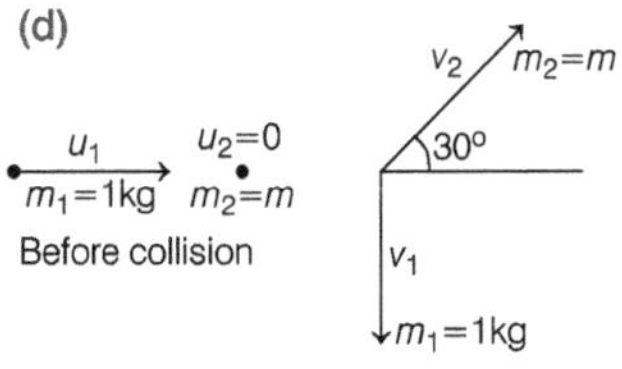

According to the law of conservation of momentum,

$$1 \times u_1 = mv_2\cos 30° \quad (X\text{-axis})$$

$$\Rightarrow \quad u_1 = \frac{\sqrt{3}}{2}mv_2$$

and $\quad 1 \times v_1 = mv_2\sin 30° \quad (Y\text{-axis})$

$$v_1 = \frac{1}{2}mv_2$$

From kinetic energy conservation,

$$\frac{1}{2}u_1^2 = \frac{1}{2}v_1^2 + \frac{1}{2}mv_2^2$$

$$\Rightarrow \quad u_1^2 - v_1^2 = \frac{1}{2}mv_2^2$$

On putting the values of u_1 and v_1, we get

$$\frac{3}{4}m^2v_2^2 - \frac{1}{4}m^2v_2^2 = \frac{1}{2}mv_2^2$$

$$\Rightarrow \quad m = 2\text{ kg}$$

11. (a) Mass of bullet = 35 g = $\frac{35}{1000}$ kg

Number of bullets per second

$$= \frac{200}{60} = \frac{10}{3}s^{-1}$$

Speed of bullet = 750 m/s

$\therefore$ Force exerted by gun on bullets,

$$F = \frac{35}{1000} \times \frac{10}{3} \times 750$$

$$= 35 \times 2.5 = 87.5\text{ N}$$

12. (a) By forces balancing perpendicular to inclined plane,

$$N = mg\cos\theta$$

By forces balancing parallel to inclined plane, $\quad mg\sin\theta - f_r = m\frac{dv}{dt}$

$$\Rightarrow \quad \frac{dv}{dt} = \frac{mg\sin\theta - f_r}{m}$$

$$\Rightarrow v(t) = \int g\sin\theta\,dt - \frac{1}{m}\int f_r\,dt$$

Now, torque, $\quad \tau = I\frac{d\omega}{dt} = \mathbf{r} \times \mathbf{F}_r$

and $\quad I = mK^2$

$$\therefore \quad mK^2\frac{d\omega}{dt} = rF_r \text{ or } \frac{d\omega}{dt} = \frac{rF_r}{mK^2}$$

$$\Rightarrow \quad \omega(t) = \frac{R}{mK^2}\int f_r dt$$

or $\quad \frac{1}{m}\int f_r\,dt = \frac{K^2}{R}\omega(t) = \frac{K^2}{R^2}v(t)$

(using $\omega = v/R$)

$$\Rightarrow \quad v(t) = \int g\sin\theta dt - \frac{K^2}{R^2}v(t)$$

or $\quad v(t) = \frac{1}{\left(1 + \frac{K^2}{R^2}\right)}\int g\sin\theta dt$

So, acceleration,

$$a(t) = \frac{dv(t)}{dt} = \frac{g\sin\theta}{\left(1 + \frac{K^2}{R^2}\right)}$$

13. (b) Along Y-axis, net force on the particle is

$$F = -(kx\sin 45° + kx\sin 45° + 2Kx\sin 135° + 2kx\sin 135°)$$

$$= -\frac{6}{\sqrt{2}}kx$$

$\Rightarrow \quad m\dfrac{d^2x}{dt^2} = -3\sqrt{2}\,kx$

$\Rightarrow \quad \dfrac{d^2x}{dt^2} + \dfrac{3\sqrt{2}\,kx}{m} = 0$

This is the equation of SHM.

Angular frequency, $\omega^2 = 3\sqrt{2}\,\dfrac{k}{m}$

$\Rightarrow \quad f = \left(\dfrac{1}{2}\pi\right)\sqrt{\dfrac{3\sqrt{2}\,k}{m}}$

$\Rightarrow \quad T = 2\pi\sqrt{\dfrac{m}{3\sqrt{2}\,k}}$

14. (b) Gravitation at height h from surface,

$$g_h = g\left(1 - \frac{2h}{R}\right)$$

Gravitation at depth d from surface,

$$g_d = g(1 - d/R)$$

For $\quad g_h = g_d \quad \Rightarrow d = 2h$

So, ratio of height w.r.t. depth is

$$\frac{h}{d} = \frac{1}{2} = 0.5$$

15. (c) Given, $F = 400\text{ kN} = 400 \times 10^3\text{ N}$,

$L = 2\text{ m}$,

$r = 50\text{ mm} = 5 \times 10^{-2}\text{ m}$

$\Delta l = 0.5\text{ mm} = 5 \times 10^{-4}\text{ m}$

Young's modulus, $Y = \dfrac{F \cdot L}{A \cdot \Delta l}$

$$= \frac{4 \times 10^5 \times 2 \times 10^4 \times 10^4}{\frac{22}{7} \times 25 \times 5}$$

$$= \frac{8 \times 7}{22 \times 5 \times 25} \times 10^{13}$$

$$\approx 2 \times 10^{11}\text{ N-m}^2$$

16. (d) Pressure at bottom

$$= g \times \int_{z=0}^{z=H} \frac{1}{\pi r^2} \times \pi r^2 \times dz \times \rho_0\left(2 - \frac{z^2}{H^2}\right)$$

$$= g \times \int_{z=0}^{z=H} \rho_0\left(2 - \frac{z^2}{H^2}\right) dz$$

$$= g \times \left[\rho\left(2z - \frac{z^3}{3H^2}\right)\right]_{z=0}^{z=H}$$

$$= g\rho_0\left(2H - \frac{H^3}{3H^2}\right) = g\rho_0\left(2H - \frac{H}{3}\right)$$

$$= g\rho_0\left(\frac{5H}{3}\right) = \frac{5}{3}\rho_0 gh$$

17. (d) (No option is matching)

Using $pV = RT$ and $p = 3 - g\left(\dfrac{V^2}{V_0^2}\right)$,

we get

$$T = \frac{3V}{R} - \frac{gV^3}{RV_0^2} \quad \text{...(i)}$$

Differentiating w.r.t. volume, we get

$$\frac{dT}{dV} = \frac{3}{R} - \frac{g}{RV_0^2} \times 3V^2$$

For maximum T, $\dfrac{dT}{dV} = 0$

$\Rightarrow \quad \dfrac{3}{R} - \dfrac{g}{RV_0^2} \times 3V^2 = 0$

$\Rightarrow \quad V = \dfrac{V_0}{\sqrt{g}}$

Substituting in Eq. (i), we get

$$T_{\max} = \frac{3V_0}{R\sqrt{g}} - \frac{g(V_0/\sqrt{g})^3}{RV_0^2} = \frac{2V_0}{R\sqrt{g}}.$$

18. (a) Using first law of thermodynamics,

$$dU = dQ - dW$$

Here, $dW = pdV = 0$ (here, V = constant)

$\therefore$ Change in internal energy, $dU = dQ$

$$dU \propto dT \quad (\because dQ = C_V dT)$$

Using, $\quad pV = RT$

$$dT = \frac{V}{R}\,dp$$

$\Rightarrow \quad dT \propto dp$

$\Rightarrow \quad dU \propto dp$

Hence, internal energy will increase with increment in pressure.

19. (b) Efficiency of Carnot cycle is

$$\eta = 1 - \frac{T_2}{T_1}$$

$\Rightarrow \quad 1 - \eta = \dfrac{T_2}{T_1}$

$\Rightarrow \quad \dfrac{T_1}{T_2} = \dfrac{1}{(1 - \eta)} \quad \text{... (i)}$

In second case,

$$1.5\eta = 1 - \frac{T_2}{T_1 + \Delta T}$$

or $\quad 1 - 1.5\eta = \dfrac{T_2}{T_1 + \Delta T}$

$\Rightarrow \quad \dfrac{T_1 + \Delta T}{T_2} = \dfrac{1}{(1 - 1.5\eta)} \quad \text{... (ii)}$

Using Eqs. (i) and (ii), we get

$$\frac{\Delta T}{T_2} = \frac{1}{1 - 1.5\eta} - \frac{1}{1 - \eta}$$

$$= \frac{1 - \eta - 1 + 1.5\eta}{(1 - \eta)(1 - 1.5\eta)}$$

$\Rightarrow \quad \Delta T = \dfrac{0.5\eta T_2}{(1 - \eta)(1 - 1.5\eta)}$

20. (b) Using

$$\frac{(p_a + \rho gh)r_1^3}{T_1} = \frac{p_a r_2^3}{T_2}$$

$\because \quad T_1 = T_2$

$\therefore \quad r_2^3 = \dfrac{1}{p_a}(p_a + \rho gh)r_1^3$

Using all values in SI system,

$$r_2^3 = \frac{1}{10^5}[10^5 + 10^3 \times 9.8 \times 5]r_1^3$$

$$r_2^3 \approx 1.5\,r_1^3$$

$\Rightarrow \quad \dfrac{4}{3}\pi r_2^3 \approx (1.5)\dfrac{4}{3}\pi r_1^3$

$$V_2 = 1.5 \times V_1$$

$\Rightarrow \quad V_2 = 1.5 \times 3 = 4.5\text{ mm}^3$

21. (d) Let $kx - \omega t = \alpha$

$$y_1 = a\sin\alpha,\ y_2 = a\sin(\phi - \alpha)$$

Superimposed wave, $y = y_1 + y_2$

$\Rightarrow \quad y = [a\sin\alpha + a\sin\phi\cos\alpha - a\cos\phi\sin\alpha]$

$\Rightarrow \quad y = [a(1 - \cos\phi)\sin\alpha + (a\cos\phi)\cos\alpha]$

Let $R\cos\theta = a(1 - \cos\phi)$ and $R\sin\theta = a\sin\phi$.

Then, magnitude of R will be the magnitude of resultant wave,

$$R^2 = a^2[\sin^2\phi + (1 - \cos\phi)^2]$$

$$= a^2[1 + 1 - 2\cos\phi]$$

$\Rightarrow \quad R^2 = 4a^2\sin^2\dfrac{\phi}{2}$

$\Rightarrow \quad R = 2a\sin\dfrac{\phi}{2}$

VIT

1. (d) We know that, $D_n = u + \dfrac{a}{2}(2n - 1)$

where, D_n = distance covered in n^{th} second,

u = initial velocity

and $\quad a$ = acceleration of the ball.

Here, $\quad D_n = \dfrac{h}{2}, u = 0, n = t$ s(say)

$\Rightarrow \quad \dfrac{h}{2} = 0 + \dfrac{a}{2}(2 \times t - 1) \quad \text{...(i)}$

Also, distance covered in t time is given as

$$s = ut + \frac{1}{2}at^2$$

$\Rightarrow \quad h = 0 + \dfrac{1}{2}at^2 \quad \text{...(ii)}$

On solving Eqs. (i) and (ii), we get

$$t = (2 \pm \sqrt{2})\text{ s}$$

2. (a) From conservation of linear momentum,

$$Mv + m \times 0 = Mv_1 + mv_2$$

where, v_1 and v_2 are the speeds attained by the bodies of mass M and m, respectively, after collision.

$\Rightarrow \quad M(v - v_1) = mv_2 \quad \text{...(i)}$

Again, from the conservation of kinetic energy (as collision is of elastic nature)

$$\frac{1}{2}Mv^2 + \frac{1}{2}m \times 0 = \frac{1}{2}Mv_1^2 + \frac{1}{2}mv_2^2,$$

$$Mv^2 = Mv_1^2 + mv_2^2$$

or $\quad M(v^2 - v_1^2) = mv_2^2 \quad$ ...(ii)

On solving Eqs. (i) and (ii), we get

$$v_1 = \frac{(M-m)v}{M+m} \text{ and } v_2 = \frac{2Mv}{(M+m)}$$

As $M >> m$

So, $v_1 = v$ and $v_2 = 2v$

3. (a) Given, mass, $m = 2.1$ kg,

$g = 10 \text{ ms}^{-2}$,

Young's modulus, $Y = 1.9 \times 10^{11}$ N-m^2,

natural length of wire,

$l = 25 \text{ cm} = 0.25$ m

Area, $A = 3 \text{ mm}^2 = 3 \times 10^{-6} \text{ m}^2$

Tension in wire, $T = mg$

$= 2.1 \times 10 = 21$ N

$$\text{Strain in wire} = \frac{\text{Stress}}{Y} = \frac{\text{Force / Area}}{Y}$$

$$\Rightarrow \text{Strain} = \frac{\frac{21}{(3\times 10^{-6})}}{1.9 \times 10^{11}} = 3.7 \times 10^{-5}$$

Volume of wire $= l \times A$

$= 0.25 \times 3 \times 10^{-6} = 7.5 \times 10^{-7} \text{ m}^3$

Now, elastic potential energy of wire is given as

$$U = \frac{1}{2} \times \text{stress} \times \text{strain} \times \text{volume}$$

Substituting the values from above, we get

$$U = \frac{1}{2} \times \left(\frac{21}{3\times 10^{-6}}\right) \times (3.7 \times 10^{-5}) \times (7.5 \times 10^{-7})$$

$$U = 9.7 \times 10^{-5} \text{ J}$$

4. (c) When a stone is released on the surface of the lake, then there will be two forces acting on it. The force acting downwards would be equal to $V\sigma g$ due to its weight and secondly the force acting upward would be equal to $v\rho g$ due to the buoyant force. Thus, the net force acting on the stone in downward direction would be equal to

$$= V\sigma g\left(1 - \frac{\rho}{\sigma}\right)$$

$$ma = mg\left(1 - \frac{\rho}{\sigma}\right) = mg\left(1 - \frac{1}{K}\right)$$

where, m is the mass of stone.

Thus, $\quad a = g\left(1 - \frac{1}{K}\right)$

5. (d) As we know, time period of revolution of electron in n[th] orbit,

$$T \propto \frac{n^3}{Z^2}$$

Radius of nth orbit, $R \propto \frac{n^2}{Z}$

For a given atom (Z = constant)

So, $\quad T \propto n^3 \quad$ ...(i)

and $\quad R \propto n^2 \quad$ ...(ii)

From Eqs. (i) and (ii), we get

$$T \propto R^{\frac{3}{2}}$$

$$\frac{T_1}{T_2} = \left(\frac{R_1}{R_2}\right)^{\frac{3}{2}} = \left(\frac{R}{4R}\right)^{\frac{3}{2}} = \frac{1}{8}$$

6. (b) Given, radius (r) = 80 m,

$\mu = 0.25$, maximum speed (v) = ?

On a level road if there will be no skidding, then

Centripetal force $\leq$ Limiting value of frictional force

$$\frac{mv^2}{r} = f = \mu N = \mu\, mg$$

$$\Rightarrow \quad v = \sqrt{\mu \cdot rg}$$

Substituting the given values in the above relation, we get

$$v = \sqrt{0.25 \times 80 \times 9.8}$$

$$= \sqrt{196} = 14 \text{ m/s}$$

7. (b) The speed of the rolling body that reached the bottom of incline is

$$v_{\text{bottom}} = \sqrt{\frac{2gh}{1 + \frac{K^2}{R^2}}}$$

where, k is radius of gyration and h height of incline.

This implies $v_{\text{ring}} = \sqrt{\frac{2gh}{1+1}} = \sqrt{gh}$

$$\left[\because \text{for ring } \frac{K^2}{R^2} = 1\right]$$

$$v_{\text{disc}} = \sqrt{\frac{2gh}{1 + \frac{1}{2}}} \quad \left[\because \text{for disc } \frac{K^2}{R^2} = \frac{1}{2}\right]$$

$$= \sqrt{\frac{4}{3}gh} = \frac{2}{\sqrt{3}}\sqrt{gh}$$

$$v_{\text{sphere}} = \sqrt{\frac{2gh}{1 + \frac{2}{5}}} = \sqrt{\frac{10}{7}gh}$$

$$\left[\because \text{for sphere } \frac{K^2}{R^2} = \frac{2}{5}\right]$$

Clearly, solid sphere has the maximum speed.

8. (a) The moment of inertia of the cylinder about its axis is given as,

$$I_{YY'} = \frac{MR^2}{2}$$

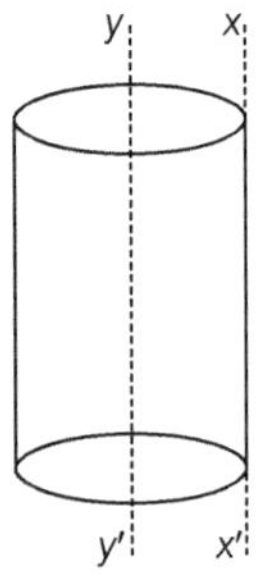

Applying to parallel axes theorem, moment of inertia of solid cylinder about XX' will be

$$I_{XX'} = I_{YY'} + Md^2$$

$$= \frac{MR^2}{2} + MR^2 \quad \text{(where, } d = R\text{)}$$

$$= \frac{3}{2}MR^2$$

9. (c) Free body diagram of the given system shown in the figure below,

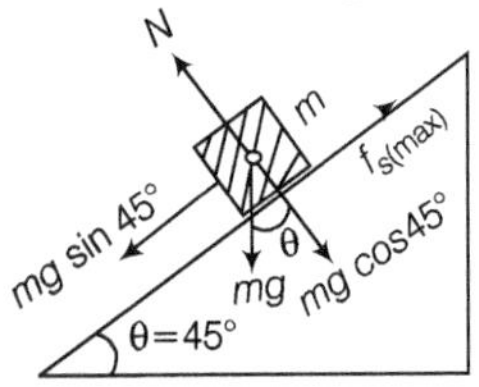

The condition in which the motion of the block will stop on the wedge is given as

$$f_m = mg\sin 45°$$

where, f_m is the limiting frictional force.

As, $\quad f_m = \mu_m N$

$= \mu_m mg\cos 45°$

where, μ_m is the minimum coefficient of friction.

$\Rightarrow \quad \mu_m mg\cos 45° = mg\sin 45°$

or $\quad \mu = 1$

10. (c) Heat energy in copper block

$= ms\Delta\theta$

$= 2 \times 400 \times (500° - 0°)$

$= 4 \times 10^5$ J

Amount of ice that melts will be maximum, if entire heat energy of copper block is used up in melting ice. Now,

Amount of ice melt by 3.5×10^5 J of heat energy = 1 kg

$\therefore$ Amount of ice melt by 4×10^5 J to heat energy

$$= \frac{4 \times 10^5}{3.5 \times 10^5} \text{ kg}^{-1} = \frac{8}{7} \text{ kg}$$

MHT CET

1. (c) Given,

Length of the pendulum $(l) = 1$ m

$\therefore \quad$ Amplitude $(a) = \frac{\text{Path length}}{2}$

$= \frac{16}{2} = 8$ cm

Acceleration due to gravity (g)

$= \pi^2 \text{m/s}^2$

We know that, time period

$(T) = 2\pi\sqrt{\frac{l}{g}} = 2\pi\sqrt{\frac{1}{\pi^2}}$

$T = \frac{2\pi}{\pi}$

$T = 2$ s

$\therefore$ Maximum velocity $(v_{\max}) = a\omega$

$= a \times \frac{2\pi}{T} \quad \left(\because \omega = \frac{2\pi}{T}\right)$

$= \frac{8 \times 2 \times \pi}{2} = 8\pi$ cm/s

2. (c) Given, depth of hole A from the top $= h$

Depth of hole B from the top $= 3h$

According to question, we can draw the following diagram

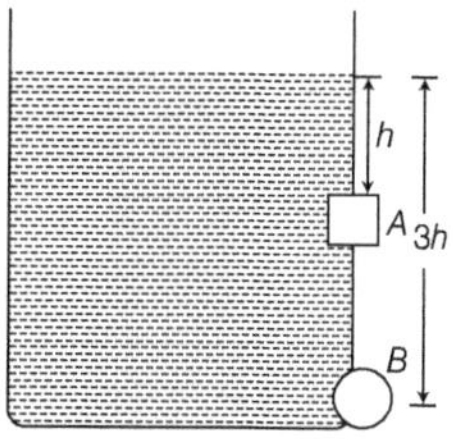

According the continuity equation,

$A_1 v_1 = A_2 v_2$

Velocity of efflux,

$v_1 = \sqrt{2gh}$

(for square hole)

$v_2 = \sqrt{2g(3h)} = \sqrt{6gh}$

(for circular hole)

$L^2\sqrt{2gh} = \pi r^2\sqrt{6gh}$

($\because$ area of square $= L^2$, area of circle $= \pi r^2$)

On squaring both sides, we get

$2L^4 gh = 6\pi^2 r^4 gh$

$2L^4 = 6\pi^2 r^4$

$L^4 = 3\pi^2 r^4$

$L = (3\pi^2)^{1/4} r$

$L = r\pi^{1/2}(3)^{1/4}$

3. (b) Given, mass of disc $= M$

Radius of disc $= R$

We know that,

$\tau = I\alpha$

But $\quad \tau = F \times R$

$\therefore \quad I = \frac{MR^2}{2}$ and $\alpha = \frac{\omega}{t}$

Therefore, $\quad F \times R = \frac{MR^2}{2} \times \frac{\omega}{t}$

$F = \frac{MR}{2} \times \frac{\omega}{t}$

$F = \frac{MR\omega}{2t}$

4. (c) Given, radius of circle $= r$

Speed of particle $= v$

Angular acceleration $= \alpha$

We know that,

tangential acceleration $= \alpha r \qquad$ ...(i)

Radial acceleration $= \frac{v^2}{r} \qquad$ ...(ii)

On dividing Eq. (i) by Eq. (ii), we get

$\frac{\text{Tangential acceleration}}{\text{Radial acceleration}} = \frac{\alpha r}{v^2} \times r$

$= \frac{\alpha r^2}{v^2}$

5. (a) Given, density of metal wire $= \rho$

Surface tension of water $= T$

If l is the length of the wire and f the total force on either side of the wire, then

$f = Tl \qquad$ ...(i)

Also, $\quad f = mg \qquad$ ...(ii)

From Eqs. (i) and (ii), we get

$Tl = mg$

$Tl = v\rho g \qquad \left[\because \text{Density}(\rho) = \frac{m}{v}\right]$

$Tl = \pi r^2 l\rho g$

$r^2 = \frac{T}{\pi\rho g} \Rightarrow r = \sqrt{\frac{T}{\pi\rho g}}$

6. (c) Given, mass of sphere $= m$

According to the question, we can draw the following diagram

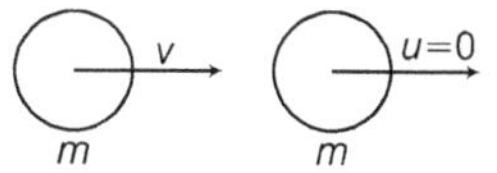

Before collision

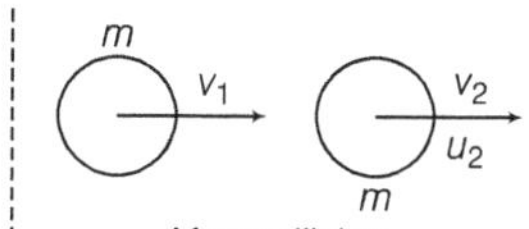

After collision

Applying the conservation law of momentum, we get

$mv + m(0) = mv_1 + mv_2$

$mv = mv_1 + mv_2$

$mv = m(v_1 + v_2)$

$v = v_1 + v_2$

$\Rightarrow \quad v_1 = v - v_2 \qquad$ ...(i)

We know that, coefficient of restitution

$= \frac{\text{Relative velocity of separation}}{\text{Relative velocity of approach}}$

or $\quad e = \frac{v_2 - v_1}{v - 0}$

or $\quad ev = v_2 - v_1$

From Eq. (i), we get

$\Rightarrow \quad ev = v_2 - (v - v_2)$

$\Rightarrow \quad ev = v_2 - v + v_2$

$\Rightarrow \quad ev = 2v_2 - v$

$\Rightarrow \quad ev + v = 2v_2$

$\Rightarrow \quad v(1 + e) = 2v_2$

$\Rightarrow \quad \frac{v_2}{v} = \frac{e + 1}{2}$

7. (b) Given, amplitude of SHM $= a$

Frequency of oscillation $= n$

Distance travelled in one oscillation,

$= 4 \times$ amplitude of SHM

$= 4a$

We know that,

Velocity

$= \frac{\text{Distance travelled in one oscillation}}{\text{Time period}}$

$v = \frac{4a}{T}$

$\therefore \quad v = 4an \qquad \left[\because \text{frequency}, n = \frac{1}{T}\right]$

8. (d) According to Mayer formula,

$C_p - C_V = R \qquad$ ...(i)

where, $C_p =$ specific heat at constant pressure,

$C_V =$ specific heat at constant volume

and $\quad R =$ gas constant.

Now, $\quad \gamma = \frac{C_p}{C_V}$

$\Rightarrow \quad C_p = \gamma C_V \qquad$ ...(ii)

From Eqs. (i) and (ii), we get

$\Rightarrow \quad \gamma C_V - C_V = R$

$\Rightarrow \quad C_V(\gamma - 1) = R$

$\Rightarrow \quad C_V = \frac{R}{(\gamma - 1)}$

9. (b) For a capillary tube,

$rh =$ constant

where, $r =$ radius of capillary tube

and $\quad h =$ height of rised water in capillary tube.

According to the question,

$$r_1 h_1 = r_2 h_2$$

$$\Rightarrow \quad \frac{r_1}{r_2} = \frac{h_2}{h_1} \quad \ldots(\text{i})$$

In the first condition,

$$A_1 = \pi r_1^2 \quad \ldots(\text{ii})$$

In the second condition,

$$A_2 = \pi r_2^2$$

or $$\frac{A}{9} = \pi r_2^2 \quad \left[\text{as, } A_2 = \frac{A}{9}\right] \ldots(\text{iii})$$

On dividing Eq. (ii) by Eq. (iii), we get

$$9 = \frac{r_1^2}{r_2^2}$$

or $$\frac{r_1}{r_2} = \sqrt{9} \Rightarrow \frac{r_1}{r_2} = 3$$

From Eq. (i), we get

$$\frac{h_2}{h_1} = 3 \Rightarrow h_2 = 3h_1$$

$$h_2 = 3h$$

10. (b) Here, fifth overtone = 6th harmonic

∴ String will have 6 loops.

If distance between a mode and an antinode is l, then

$$6 \times 2l = 2.4$$

or $$l = \frac{2.4}{12} = 0.2 \text{ m}$$

11. (b) Given,

$$\mathbf{A} = 3\hat{\mathbf{i}} - 2\hat{\mathbf{j}} + \hat{\mathbf{k}}$$

$$\mathbf{B} = \hat{\mathbf{i}} - 3\hat{\mathbf{j}} + 5\hat{\mathbf{k}}$$

$$\mathbf{C} = 2\hat{\mathbf{i}} + \hat{\mathbf{j}} - 4\hat{\mathbf{k}}$$

Here, $|\mathbf{A}| = \sqrt{(3)^2 + (-2)^2 + (1)^2}$

or $$|\mathbf{A}| = \sqrt{9 + 4 + 1} = \sqrt{14} \quad \ldots(\text{i})$$

$$|\mathbf{B}| = \sqrt{(1)^2 + (-3)^2 + (5)^2}$$

$$|\mathbf{B}| = \sqrt{1 + 9 + 25} = \sqrt{35} \quad \ldots(\text{ii})$$

and $$|\mathbf{C}| = \sqrt{(2)^2 + (1)^2 + (-4)^2}$$

$$|\mathbf{C}| = \sqrt{4 + 1 + 16} = \sqrt{21} \quad \ldots(\text{iii})$$

From Eqs. (i), (ii) and (iii), we get

$$B^2 = A^2 + C^2$$

$$(\sqrt{35})^2 = (\sqrt{14})^2 + (21)^2$$

Now, $$\mathbf{A} \cdot \mathbf{C} = (3\hat{\mathbf{i}} - 2\hat{\mathbf{j}} + \hat{\mathbf{k}}) \cdot (2\hat{\mathbf{i}} + \hat{\mathbf{j}} - 4\hat{\mathbf{k}})$$

$$\mathbf{A} \cdot \mathbf{C} = 6 - 2 - 4 = 0$$

Hence, A and C are perpendicular to each other.

∴ Resultant of A and C is B.

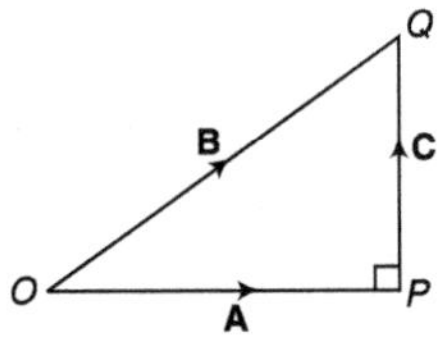

$\mathbf{B} = \mathbf{A} + \mathbf{C}$ [according to triangle law]

12. (a) In the given figure,

$$I = I_{AB} + I_{BC} + I_{CD} + I_{DA}$$

$$= 0 + \frac{ml^2}{3} + ml^2 + \frac{ml^2}{3}$$

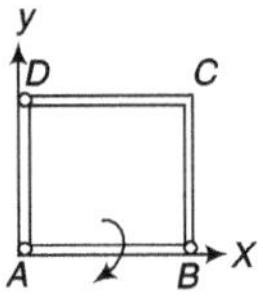

or $$I = \frac{5ml^2}{3}$$

13. (d) Given, unit vector

$$= 0.8\hat{\mathbf{i}} + b\hat{\mathbf{j}} + 0.4\hat{\mathbf{k}}$$

Magnitude of vector

$$(A) = \sqrt{x^2 + y^2 + z^2}$$

Here, $x = 0.8$, $y = b$, $z = 0.4$

Now,

$$1 = \sqrt{(0.8)^2 + (b)^2 + (0.4)^2}$$

$$\Rightarrow \quad 1 = \sqrt{0.64 + b^2 + 0.16}$$

$$\Rightarrow \quad 1 = 0.64 + b^2 + 0.16$$

$$\Rightarrow \quad 1 = 0.80 + b^2$$

$$\Rightarrow \quad b^2 = 1 - 0.80$$

$$\Rightarrow \quad b^2 = 0.2$$

$$\Rightarrow \quad b = \sqrt{0.2}$$

14. (d) Given, frequency of SHM $(n) = 5$ Hz

Acceleration due to gravity $(g) = 10$ m/s^2

We know that,

$$T = 2\pi\sqrt{\frac{m}{k}}$$

But frequency,

$$n = \frac{1}{T} \quad \text{or} \quad n = \frac{1}{2\pi}\sqrt{\frac{k}{m}}$$

or $$5 = \frac{1}{2\pi}\sqrt{\frac{k}{m}}$$

On taking square both sides, we get

$$25 = \frac{1}{4\pi^2}\frac{k}{m}$$

$$k = 100\pi^2 m \quad \ldots(\text{i})$$

But $kA = mg$

$$\Rightarrow \quad A = \frac{mg}{k} \quad \ldots(\text{ii})$$

Now, $v_{\max} = \omega A$

$$= \frac{2\pi}{T} \times \frac{mg}{k} \quad [\text{from Eq. (ii)}]$$

or $$v_{\max} = 2\pi n \times \frac{mg}{k} \quad \left(\because n = \frac{1}{T}\right)$$

$$v_{\max} = 2\pi n \times \frac{mg}{100\pi^2 m} \quad [\text{from Eq. (i)}]$$

$$v_{\max} = \frac{n \times g}{50\pi}$$

$$v_{\max} = \frac{5 \times 10}{50\pi}$$

$$v_{\max} = \frac{1}{\pi} \text{ m/s}$$

15. (b) According to the law of conservation of angular momentum,

$$I\omega = \text{constant}$$

Now, according to the question,

$$I_1\omega_1 = I_2\omega_2 \quad \text{or} \quad I\omega = (2I)\omega_2$$

$$\omega_2 = \frac{\omega}{2}$$

$$\text{New kinetic energy} = \left[\frac{1}{2}I_2\omega_2^2\right]$$

$$= \frac{1}{2}(2I) \times \left(\frac{\omega}{2}\right)^2 = \frac{I\omega^2}{4}$$

Loss in kinetic energy $(K_L) = K_i - K_j$

$$= \frac{1}{2}I\omega^2 - \frac{I\omega^2}{4} = \frac{I\omega^2}{4}$$

16. (d) According to the question, we can draw the following diagram

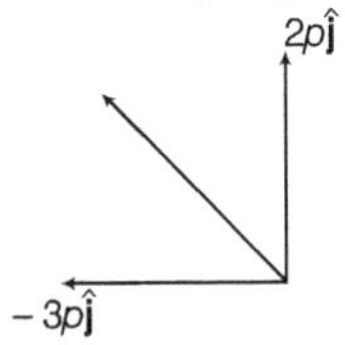

Now, the magnitude of momentum of the third part is given by

$$p_3 = \sqrt{(p_1^2) + (p_2)^2}$$

$$p_3 = \sqrt{(-3p)^2 + (2p)^2}$$

$$p_3 = \sqrt{9p^2 + 4p^2}$$

$$p_3 = \sqrt{13p^2}$$

$$\therefore \quad p_3 = \sqrt{13}\,p$$

17. (b) We know that,

velocity of particle at top-most point on vertical circle,

$$(v_{\text{top}}) = \sqrt{3rg} \quad \ldots(\text{i})$$

But centripetal acceleration $(a_c) = \frac{v^2}{r}$

$$\therefore \quad a_c = \frac{(\sqrt{3rg})^2}{r} \quad [\text{from Eq. (i)}]$$

$$a_c = \frac{3rg}{r}$$

$$a_c = 3g$$

18. (a) When source of sound moving towards to stationary observer.

$$\text{Apparent frequency } (n_a) = n\left[\frac{v}{v - v_s}\right]$$

Hence, the wavelength of sound received by him decreases while frequency increases.

19. (a) Given, velocity of the body = u m/s

According to the question,

$$\frac{-GMm}{R}+\frac{1}{2}mu^2=0+\left(\frac{-GMm}{R+h}\right)$$

$$\frac{GM}{R+h}=\frac{GM}{R}-\frac{u^2}{2}$$

$$\Rightarrow \frac{GM}{R+h}=\frac{2GM-Ru^2}{2R}$$

$$\Rightarrow \frac{R+h}{GM}=\frac{2R}{2GM-Ru^2}$$

$$\Rightarrow h=\frac{2GMR}{2GM-Ru^2}-R$$

$$\Rightarrow h=\frac{2GMR-2GMR+R^2u^2}{2GM-Ru^2}$$

$$\Rightarrow h=\frac{R^2u^2}{2GM-Ru^2} \quad \ldots(i)$$

But $g=\dfrac{GM}{R^2}$

$$\Rightarrow GM=gR^2 \quad \ldots(ii)$$

Substituting the value of Eq. (ii) in Eq. (i), we get

$$\Rightarrow h=\frac{R^2u^2}{2gR^2-Ru^2}$$

$$\Rightarrow h=\frac{Ru^2}{2gR-u^2}$$

20. (b) Given, rate of heat energy

$$(Q_1)=1000 \text{ J/min}$$

We know that,

$$1=a+r+t$$

Here, coefficient of absorption $(a)=0.8$

Coefficient of reflection $(r)=0.1$

Now,

$$1=0.8+0.1+t$$
$$t=1-0.9$$
$$t=0.1$$

The rate of transmitted heat energy in 5 min,

$$Q_t=\theta_i\times t\times(\text{time})$$
$$=1000\times0.1\times5=500 \text{ J}$$

21. (c) We know that,

Young's modulus $(Y)=\dfrac{Fl}{A\Delta l}$

$$\Rightarrow \Delta l=\frac{Fl}{AY}$$

where, A = area of cross-section of wire

Δl = change in the length of the wire,

l = length of the wire and

F = applied force.

According to the question, F, l and Y are same for both wires, i.e.

$$\Delta l\propto\frac{1}{A} \quad \ldots(i)$$

But $m=\rho V$

where, ρ = density, V = volume.

$$m=\rho Al \quad [\because V=Al]$$

$$m\propto A \quad \ldots(ii)$$

From Eqs. (i) and (ii), we get

$$\frac{\Delta l_1}{\Delta l_2}=\frac{A_2}{A_1}=\frac{m_2}{m_1}$$

22. (b) Given,

$$x=\frac{a^2b^2}{c}$$

$$\frac{\Delta x}{x}=2\frac{\Delta a}{a}+2\frac{\Delta b}{b}+\frac{\Delta c}{c}$$

$$\frac{\Delta x}{x}\times100=2\left(\frac{\Delta a}{a}\times100\right)+2\left(\frac{\Delta b}{b}\times b\right)+\left(\frac{\Delta c}{c}\times c\right)$$

Here, percentage error in $a=2$

Percentage error in $b=3$

Percentage error in $c=4$

$$\left(\frac{\Delta x}{x}\times100\right)=2\times2+2\times3+4$$
$$=4+6+4=14\%$$

23. (b) For a stationary wave,

$$n=\frac{1}{2L}\sqrt{\frac{T}{m}} \quad \ldots(i)$$

where, T = tension in string

L = length of string

and m = mass of string.

According to the question,

$$n'=\frac{1}{2L}\sqrt{\frac{T'}{m}}$$

Here, $T'=2T$

$$n'=\frac{1}{2L}\sqrt{\frac{2T}{m}}$$

$$\Rightarrow n'=\sqrt{2}\frac{1}{2L}\sqrt{\frac{T}{m}}$$

From Eq. (i), we get

$$n'=\sqrt{2}n$$

24. (b) Energy $=\dfrac{1}{2}\times$ stress $\times$ strain $\times$ volume

$$=\frac{1}{2}\frac{(\text{stress})^2}{Y}\times\text{volume}$$

$$=\frac{1}{2}\frac{\left(\frac{F}{A}\right)^2}{y}\times L.A$$

$$\left[\text{as stress}=\frac{F}{A},\ \text{volume}=L\times A\right]$$

or $\quad$ energy $(E)=\dfrac{F^2L}{2AY}$

25. (d) According to the question,

$$v_c=\frac{1}{4}v_e$$

$$\sqrt{\frac{GM}{(R+h)}}=\frac{1}{4}\sqrt{\frac{2GM}{R}}$$

On taking square root both sides, we get

$$\Rightarrow \frac{GM}{R+h}=\frac{1}{16}\frac{2GM}{R}$$

$$\Rightarrow \frac{GM}{R+h}=\frac{GM}{8R}$$

$$\Rightarrow R+h=8R$$
$$\Rightarrow h=8R-R$$
$$\Rightarrow h=7R$$

26. (c) Fundamental frequency of one end closed pipe,

$$\nu=\frac{nv}{4L}$$

Here, $n=1$

or $\quad \nu_1=\dfrac{332}{4\times83\times10^{-2}}$

or $\quad \nu_1=100$

As odd harmonics alone are produced in closed organ pipe, therefore possible frequencies are $3\nu_1=300$ Hz, $5\nu_2=500$ Hz, $7\nu_2=700$ Hz, $9\nu_2=900$ Hz

Hence, the number of possible natural oscillation of air column is 5.

Engineering Entrance Questions 2019-21

JEE Main & Adv., BITSAT, AP & TS EAMCET, MHTCET, WBJEE

Units, Dimensions and Error Analysis

1. One main scale division of a vernier callipers is a cm and nth division of the vernier scale coincide with $(n-1)$th division of the main scale. The least count of the callipers (in mm) is **JEE Main 2021**

(a) $\frac{10na}{(n-1)}$ (b) $\frac{10a}{(n-1)}$

(c) $\left(\frac{n-1}{10n}\right)a$ (d) $\frac{10a}{n}$

2. The time period of a simple pendulum is given by $T = 2\pi\sqrt{\frac{l}{g}}$. The measured value of the length of pendulum is 10 cm known to a 1 mm accuracy. The time for 200 oscillations of the pendulum is found to be 100 s using a clock of 1 s resolution. The percentage accuracy in the determination of g using this pendulum is x. The value of x to the nearest integer is **JEE Main 2021**

(a) 2% (b) 3%

(c) 5% (d) 4%

3. A quantity f is given by $f = \sqrt{\frac{hc^5}{G}}$, where c is speed of light, G universal gravitational constant and h is the Planck's constant. Dimensions of f is that of **JEE Main 2020**

(a) area (b) volume

(c) momentum (d) energy

4. If the screw on a screw gauge is given six rotations, it moves by 3 mm on the main scale. If there are 50 divisions on the circular scale, the least count of the screw gauge is **JEE Main 2020**

(a) 0.001 cm (b) 0.01 cm

(c) 0.02 cm (d) 0.0001 cm

5. The force of interaction between two atoms is given by $F = \alpha\beta \exp\left(-\frac{x^2}{\alpha kT}\right)$; where x is the distance, k is the Boltzmann constant and T is temperature and α and β are two constants. The dimensions of β is **JEE Main 2019**

(a) $[MLT^{-2}]$ (b) $[M^0L^2T^{-4}]$

(c) $[M^2LT^{-4}]$ (d) $[M^2L^2T^{-2}]$

6. The least count of the main scale of a screw gauge is 1 mm. The minimum number of divisions on its circular scale required to measure 5 μm diameter of a wire is **JEE Main 2019**

(a) 50 (b) 200 (c) 500 (d) 100

7. The dimensional formula of $\sqrt{\frac{\mu_0}{\varepsilon_0}}$ is **BITSAT 2019**

(a) $[ML^2T^{-3}A^2]$ (b) $[M^0LT^{-1}A^0]$

(c) $[ML^2T^{-3}A^{-2}]$ (d) $[M^{-1}L^{-2}T^3A^2]$

8. A physical quantity obtained from the ratio of the coefficient of thermal conductivity to the universal gravitational constant has a dimensional formula $[M^{2a}L^{4b}T^{2c}K^d]$, then the value of $\frac{a+b}{c+b} - d$ is **AP EAMCET 2019**

(a) $+\frac{3}{2}$ (b) $-\frac{1}{2}$

(c) $-\frac{3}{2}$ (d) $+\frac{1}{2}$

9. A current carrying conductor obeys Ohm's law $(V = RI)$. If the current passing through the conductor is $I = (5 \pm 0.2)$ A and voltage developed is $V = (60 \pm 6)$ V, then find the percentage of error in resistance, R **TS EAMCET 2019**

(a) 18 (b) 6 (c) 14 (d) 2

10. The force F acting on a body of density d are related by the relation $F = \frac{y}{\sqrt{d}}$. The dimensions of y are **MHT CET 2019**

(a) $[L^{-\frac{1}{2}}M^{\frac{3}{2}}T^{-2}]$ (b) $[L^{-1}M^{\frac{1}{2}}T^{-2}]$

(c) $[L^{-1}M^{\frac{3}{2}}T^{-2}]$ (d) $[L^{-\frac{1}{2}}M^{\frac{1}{2}}T^{-2}]$

11. The density of the material of a cube can be estimated by measuring its mass and the length of one of its sides. If the maximum error in the measurement of mass and length are 0.3% and 0.2% respectively, the maximum error in the estimation of the density of the cube is approximately. **WBJEE 2019**

(a) 1.1% (b) 0.5%

(c) 0.9% (d) 0.7%

Vectors

1. If **A, B** and **C** are the unit vectors along the incident ray, reflected ray and outward normal to the reflecting surface, then **Manipal 2021**

(a) $\mathbf{B} = \mathbf{A} - \mathbf{C}$ (b) $\mathbf{B} = \mathbf{A} + (\mathbf{A}\cdot\mathbf{C})\mathbf{C}$

(c) $\mathbf{B} = \mathbf{A} + \mathbf{C}$ (d) $\mathbf{B} = \mathbf{A} - 2(\mathbf{A}\cdot\mathbf{C})\mathbf{C}$

2 An electric dipole of moment $\mathbf{p} = (-\hat{\mathbf{i}} - 3\hat{\mathbf{j}} + 2\hat{\mathbf{k}}) \times 10^{-29}$ C-m is at the origin (0, 0, 0). The electric field due to this dipole at $\mathbf{r} = +\hat{\mathbf{i}} + 3\hat{\mathbf{j}} + 5\hat{\mathbf{k}}$ (note that $\mathbf{r}\cdot\mathbf{p} = 0$) is parallel to **JEE MAIN 2020**

(a) $(+\hat{\mathbf{i}} - 3\hat{\mathbf{j}} - 2\hat{\mathbf{k}})$ (b) $(-\hat{\mathbf{i}} - 3\hat{\mathbf{j}} + 2\hat{\mathbf{k}})$

(c) $(+\hat{\mathbf{i}} + 3\hat{\mathbf{j}} - 2\hat{\mathbf{k}})$ (d) $(-\hat{\mathbf{i}} + 3\hat{\mathbf{j}} - 2\hat{\mathbf{k}})$

3 Two vectors **A** and **B** have equal magnitudes. The magnitude of $(\mathbf{A} + \mathbf{B})$ is n times the magnitude of $(\mathbf{A} - \mathbf{B})$. The angle between **A** and **B** is **JEE MAIN 2019**

(a) $\sin^{-1}\left(\frac{n^2-1}{n^2+1}\right)$ (b) $\sin^{-1}\left(\frac{n-1}{n+1}\right)$

(c) $\cos^{-1}\left(\frac{n^2-1}{n^2+1}\right)$ (d) $\cos^{-1}\left(\frac{n-1}{n+1}\right)$

4 If $\sqrt{A^2+B^2}$ represents the magnitude of resultant of two vectors $(\mathbf{A} + \mathbf{B})$ and $(\mathbf{A} - \mathbf{B})$, then the angle between two vectors is **MHT CET 2019**

(a) $\cos^{-1}\left[-\frac{2(A^2-B^2)}{(A^2+B^2)}\right]$

(b) $\cos^{-1}\left[-\frac{A^2-B^2}{A^2B^2}\right]$

(c) $\cos^{-1}\left[-\frac{(A^2+B^2)}{2(A^2-B^2)}\right]$

(d) $\cos^{-1}\left[-\frac{(A^2-B^2)}{A^2+B^2}\right]$

5 The resultant **R** of **P** and **Q** is perpendicular to **P**. Also $|\mathbf{P}| = |\mathbf{R}|$. The angle between **P** and **Q** is $[\tan 45° = 1]$ **MHT CET 2019**

(a) $\frac{5\pi}{4}$ (b) $\frac{7\pi}{4}$ (c) $\frac{\pi}{4}$ (d) $\frac{3\pi}{4}$

6 **P** and **Q** are two non-zero vectors inclined to each other at an angle 'θ'. '**p**' and '**q**' are unit vectors along **P** and **Q** respectively. The component of **Q** in the direction of **Q** will be **MHT CET 2019**

(a) $\mathbf{P}\cdot\mathbf{Q}$ (b) $\frac{\mathbf{P}\times\mathbf{Q}}{\mathbf{P}}$ (c) $\frac{\mathbf{P}\cdot\mathbf{Q}}{\mathbf{Q}}$ (d) $\mathbf{p}\cdot\mathbf{q}$

7 The vectors $(\mathbf{A} + \mathbf{B})$ and $(\mathbf{A} - \mathbf{B})$ are at right angles to each other. This is possible under the condition **MHT CET 2019**

(a) $|\mathbf{A}| = |\mathbf{B}|$ (b) $\mathbf{A}\cdot\mathbf{B} = 0$

(c) $\mathbf{A}\cdot\mathbf{B} = 1$ (d) $\mathbf{A}\times\mathbf{B} = 0$

Motion in One Dimension

1 The *a versus x* graph for a particle moving with initial velocity 5 m/s is shown in the figure. The velocity of the particle at $x = 35$ m will be **WB JEE 2021**

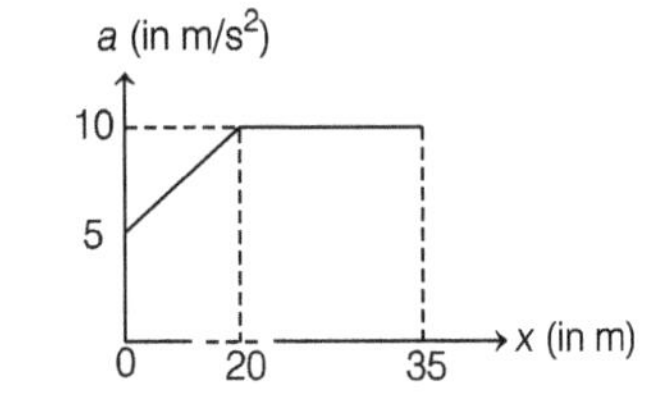

(a) 20.62 m/s (b) 20 m/s
(c) 25 m/s (d) 50 m/s

2 A person standing on an open ground hears the sound of a jet aeroplane, coming from North at an angle 60° with ground level. But he finds the aeroplane right vertically above his position. If v is the speed of sound, then speed of the plane is **JEE MAIN 2019**

(a) $\frac{\sqrt{3}}{2}v$ (b) v (c) $\frac{2v}{\sqrt{3}}$ (d) $\frac{v}{2}$

3 A particle starts from the origin at time $t = 0$ and moves along the positive X-axis. The graph of velocity with respect to time is shown in figure. What is the position of the particle at time $t = 5$s ? **JEE MAIN 2019**

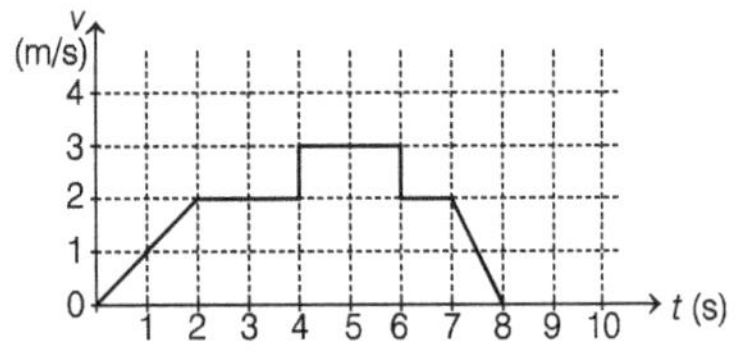

(a) 6 m (b) 3 m (c) 10 m (d) 9 m

4 A particle starts from origin O from rest and moves with a uniform acceleration along the positive X-axis. Identify all figures that correctly represent the motion qualitatively.
(a = acceleration, v = velocity x = displacement, t = time) **JEE MAIN 2019**

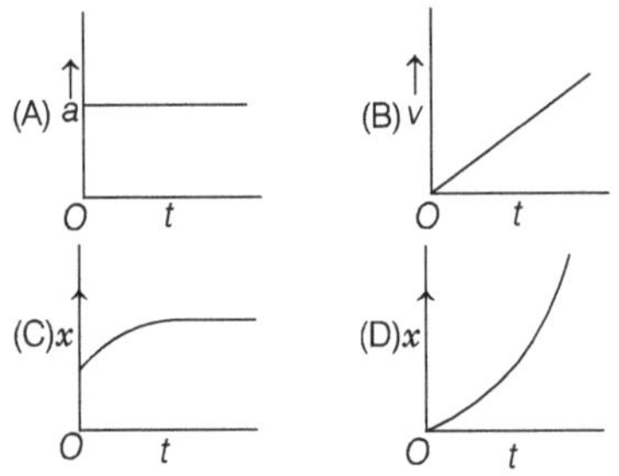

(a) (A) (b) (A), (B), (C)
(c) (B), (C) (d) (A), (B), (D)

5 A particle is moving with a velocity $\mathbf{v} = k(y\hat{\mathbf{i}} + x\hat{\mathbf{j}})$, where k is a constant.
The general equation for its path is **JEE MAIN 2019**
(a) $y = x^2$ + constant
(b) $y^2 = x$ + constant
(c) xy = constant
(d) $y^2 = x^2$ + constant

6 A body starting from rest at $t = 0$ moves along a straight line with a constant acceleration. At $t = 2$ s, the body reverses its direction keeping the acceleration same. The body returns to the initial position at $t = t_0$, then t_0 is **AP EAMCET 2019**
(a) 4 s (b) $(4 + 2\sqrt{2})$ s
(c) $(2 + 2\sqrt{2})$ s (d) $(4 + 4\sqrt{2})$ s

7 A force of $(2.6\hat{\mathbf{i}} + 1.6\hat{\mathbf{j}})$ N acts on a body of mass 2 kg. If the velocity of the body at time, $t = 0$ is $(3.6\hat{\mathbf{i}} - 4.8\hat{\mathbf{j}})$ ms^{-1}, the time at which the body will just have a velocity along X-axis only is **AP EAMCET 2019**
(a) 1 s (b) 2 s (c) 3 s (d) 6 s

8 Ship A is moving Westwards with a speed of 20 km h^{-1} and another ship B which is at 200 km South of A is moving Northwards with a speed of 10 km h^{-1}. The time after which the distance between them is shortest and the shortest distance between them respectively, **AP EAMCET 2019**
(a) 4 h, $80\sqrt{5}$ km
(b) $50\sqrt{2}$ h, $\sqrt{10}$ km
(c) $100\sqrt{2}$ h, $2\sqrt{10}$ km
(d) $80\sqrt{5}$ h, 4 km

9 A body travelling along a straight line path travels first half of the distance with a velocity 7ms^{-1}. During the travel time of the second half of the distance, first half time is travelled with a velocity 14ms^{-1} and the second half time is travelled with a velocity 21 ms^{-1}. Then the average velocity of the body during the journey is **AP EAMCET 2019**
(a) 14 ms^{-1} (b) 10 ms^{-1}
(c) 9 ms^{-1} (d) 12 ms^{-1}

10 Consider a car initially at rest, starts to move along a straight road first with acceleration 5 m/s^2, then with uniform velocity and finally, decelerating at 5 m/s^2, before coming to a stop. Total time taken from start to end is $t = 25$ s. If the average velocity during that time is 72 km/hr, the car moved with uniform velocity for a time of **TS EAMCET 2019**
(a) 15 s (b) 30 s (c) 155 s (d) 2 s

11 A boy runs on a horizontal road with a speed of 4 m/s while it is raining. He sees that the rain is making an angle θ with the vertical while running from West to East. However, when he runs from East to West, the angle is α. The rain is pouring down at an angle 45° with the vertical normal and at a speed of 8 m/s as shown in the figure. The ratio $\frac{\tan\theta}{\tan\alpha}$ is **TS EAMCET 2019**

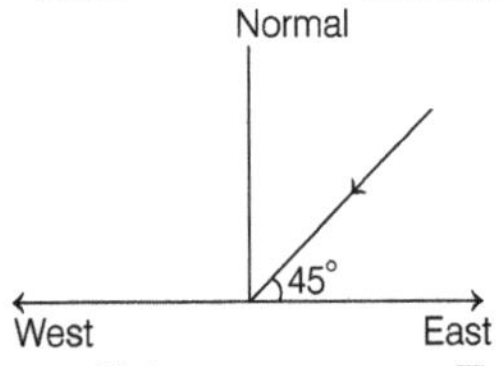

(a) $(1 - \sqrt{2})^2$ (b) $(1 + \sqrt{2})^2$
(c) $(1 + \sqrt{2})$ (d) $(\sqrt{2} - 1)$

Projectile Motion

1 The projectile motion of a particle of mass 5g is shown in the figure.

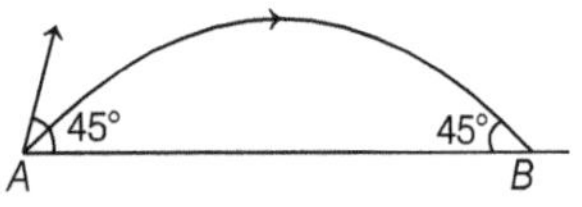

The initial velocity of the particle is $5\sqrt{2}$ ms^{-1} and the air resistance is assumed to be negligible. The magnitude of the change in momentum between the points A and B is $x \times 10^{-2}$ kg-ms^{-1}. The value of x, to the nearest integer is .... **JEE Main 2021**

2 For a projectile motion, the angle between the velocity and acceleration is minimum and acute at **KCET 2021**
(a) only one point
(b) two points
(c) three points
(d) four points

3 A particle starts from the origin at $t = 0$ with an initial velocity of $3.0\hat{\mathbf{i}}$ m/s and moves in the xy-plane with a constant acceleration $(6.0\hat{\mathbf{i}} + 4.0\hat{\mathbf{j}})$m / s^2. The x-coordinate of the particle at the instant when its y-coordinate is 32 m is D metres. The value of D is **AP EAMCET 2021**
(a) 50 (b) 60
(c) 40 (d) 32

4 Two particles are projected from the same point with the same speed u such that they have the same range R, but different maximum heights h_1 and h_2. Which of the following is correct? **JEE MAIN 2019**

(a) $R^2 = 4h_1h_2$ (b) $R^2 = 16h_1h_2$
(c) $R^2 = 2h_1h_2$ (d) $R^2 = h_1h_2$

5 In three dimensional system, the position coordinates of a particle (in motion) are given below
$x = a\cos\omega t$, $y = a\sin\omega t$ $z = a\omega t$.
The velocity of particle will be **JEE MAIN 2019**

(a) $\sqrt{2}\,a\omega$ (b) $2\,a\omega$
(c) $a\omega$ (d) $\sqrt{3}\,a\omega$

6 A body is projected at $t = 0$ with a velocity 10 ms^{-1} at an angle of 60° with the horizontal. The radius of curvature of its trajectory at $t = 1$ s is R. Neglecting air resistance and taking acceleration due to gravity $g = 10\text{ ms}^{-2}$, the value of R is **JEE MAIN 2019**

(a) 10.3 m (b) 2.8 m
(c) 5.1 m (d) 2.5 m

7 A projectile is given an initial velocity of $(\hat{\mathbf{i}} + \sqrt{3}\,\hat{\mathbf{j}})$ m/s, where $\hat{\mathbf{i}}$ is along the ground and $\hat{\mathbf{j}}$ is along the vertical. Then, the equation of the path of projectile is [Take, $g = 10\text{ m/s}^2$] **BITSAT 2019**

(a) $y = \sqrt{3}\,x - 5x^2$
(b) $y = \sqrt{3}\,x + 5x^2$
(c) $x = \sqrt{3}\,y + 5x^2$
(d) $x^2 = y^2 + \sqrt{3}$

8 A projectile is thrown in the upward direction making an angle of 60° with the horizontal with a velocity of 140 ms^{-1}. Then the time after which its velocity makes an angle 45° with the horizontal is (Acceleration due to gravity, $g = 10\text{ ms}^{-2}$) **AP EAMCET 2019**

(a) 0.5124 s (b) 51.24 s
(c) 5.124 s (d) 512.4 s

9 A rough inclined plane BCE of height $\left(\frac{25}{6}\right)$ m is kept on a rectangular wooden block $ABCD$ of height 10 m, as shown in the figure. A small block is allowed to slide down from the top E of the inclined plane. The coefficient of kinetic friction between the block and the inclined plane is $\frac{1}{8}$ and the angle of inclination of the inclined plane is $\sin^{-1}(0.6)$. If the small block finally reaches the ground at a point F, then DF will be (Acceleration due to gravity, g $= 10\text{ ms}^{-2}$) **AP EAMCET 2019**

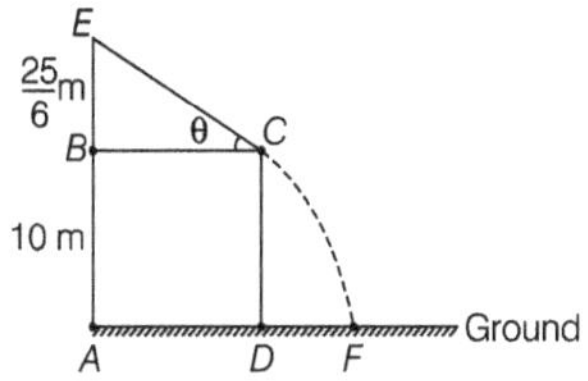

(a) $\frac{5}{3}$ m (b) $\frac{10}{3}$ m (c) $\frac{13}{3}$ m (d) $\frac{20}{3}$ m

10 Two boys conducted experiments on the projectile motion with stopwatch and noted some readings. As one boy throws a stone in air at the same angle with the horizontal, the other boy observes that after 4 s, the stone is moving at an angle 30° to the horizontal and after another 2 s it is travelling horizontally. The magnitude of the initial velocity of the stone is (Acceleration due to gravity, $g = 10\text{ ms}^{-2}$.) **AP EAMCET 2019**

(a) $40\sqrt{3}\text{ ms}^{-1}$ (b) $20\sqrt{3}\text{ ms}^{-1}$
(c) $10\sqrt{3}\text{ ms}^{-1}$ (d) $50\sqrt{3}\text{ ms}^{-1}$

11 A ball is projected vertically up from ground. Boy A standing at the window of first floor of a nearby building observes that the time interval between the ball crossing him while going up and the ball crossing him while going down is 2s.

Another boy B standing on the second floor notices that time interval between the ball passing him twice, during up motion and down motion is 1s. Calculate the difference between the vertical positions of boy B and boy A (Assume, acceleration due to gravity, $g = 10\text{ m/s}^2$) **TS EAMCET 2019**

(a) 8.45 m (b) 3.75 m
(c) 4.25 m (d) 2.50 m

12 Two particles are simultaneously projected in the horizontal direction from a point P at a certain height. The initial velocities of the particles are oppositely directed to each other and have magnitude v each.

The separation between the particles at a time when their position vectors (drawn from the point P) are mutually perpendicular, is **WB JEE 2019**

(a) $\frac{v^2}{2g}$ (b) $\frac{v^2}{g}$
(c) $\frac{4v^2}{g}$ (d) $\frac{2v^2}{g}$

Laws of Motion

1. A body of mass 1 kg rests on a horizontal floor with which it has a coefficient of static friction $\frac{1}{\sqrt{3}}$. It is desired to make the body move by applying the minimum possible force F N. The value of F will be .... (Round off to the nearest integer)
(Take, $g = 10\text{ ms}^{-2}$) **JEE Main 2021**

2. A circular disc with a groove along its diameter is placed horizontally. A block of mass 1 kg is placed as shown. The coefficient of friction between the block and all surfaces of groove in contact is $\mu = \frac{2}{5}$, the disc has an acceleration of 25 m/s^2. Find the acceleration of block with respect to disc. **JEE Main 2021**

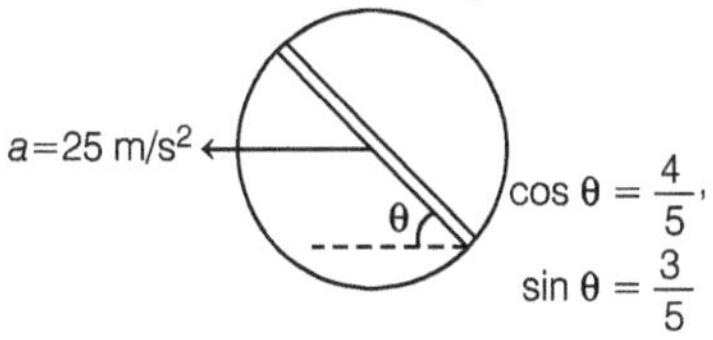

(a) 10 m/s^2 (b) 5 m/s^2
(c) 20 m/s^2 (d) 1 m/s^2

3 A mass of 10 kg is suspended by a rope of length 4 m, from the ceiling. A force F is applied horizontally at the mid-point of the rope such that the top half of the rope makes an angle of 45° with the vertical. Then, F equals (Take, $g = 10\text{ ms}^{-2}$ and the rope to be massless) **JEE Main 2020**

(a) 75 N (b) 70 N
(c) 100 N (d) 90 N

4 Two blocks A and B of masses $m_A = 1$ kg and $m_B = 3$ kg are kept on the table as shown in figure. The coefficient of friction between A and B is 0.2 and between B and the surface of the table is also 0.2. The maximum force F that can be applied on B horizontally, so that the block A does not slide over the block B is
(Take, $g = 10\text{ m/s}^2$) **JEE Main 2020**

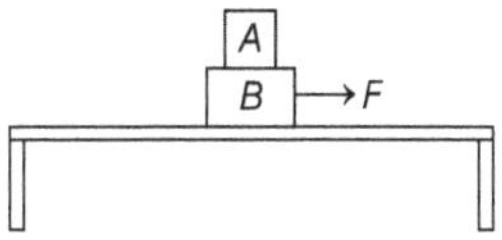

(a) 12 N (b) 16 N
(c) 8 N (d) 40 N

5 A block of mass 5 kg is (i) pushed in case (A) and (ii) pulled in case (B), by a force $F = 20$ N, making an angle of 30° with the horizontal, as shown in the figures. The coefficient of friction between the block, the floor is $\mu = 0.2$. The difference between the accelerations of the block, in case (B) and case (A) will be (Take, $g = 10\,\text{ms}^{-2}$) **JEE Main 2019**

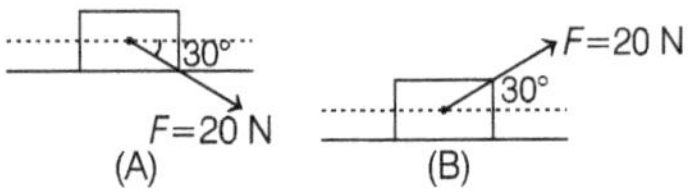

(a) $0.4\ \text{ms}^{-2}$ (b) $3.2\ \text{ms}^{-2}$
(c) $0.8\ \text{ms}^{-2}$ (d) $0\ \text{ms}^{-2}$

6 A block kept on a rough inclined plane, as shown in the figure, remains at rest upto a maximum force 2 N down the inclined plane. The maximum external force up the inclined plane that does not move the block is 10 N. The coefficient of static friction between the block and the plane is (Take, $g = 10\ \text{m/s}^2$) **JEE Main 2019**

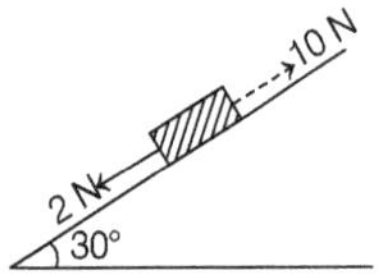

(a) $\frac{2}{3}$ (b) $\frac{\sqrt{3}}{2}$
(c) $\frac{\sqrt{3}}{4}$ (d) $\frac{1}{2}$

7 A body of mass m is moving in a straight line with momentum p. Starting at time $t = 0$, a force $F = at$ acts in the same direction on the moving particle during time interval of T. So that its momentum changes from p to $2p$. The value of T is **BITSAT 2019**

(a) $\sqrt{\frac{2p}{a}}$ (b) $\sqrt{\frac{p}{a}}$ (c) $2\sqrt{\frac{2p}{a}}$ (d) $\frac{2p}{a}$

8 When a maximum force of 3 N is applied on a body kept on rough inclined plane of shown in the figure, then body remains stationary. The maximum external force up the inclined plane that does not move the block is 12 N. The coefficient of static friction between the block and the plane is [Take, $g = 10\ \text{m/s}^2$] **BITSAT 2019**

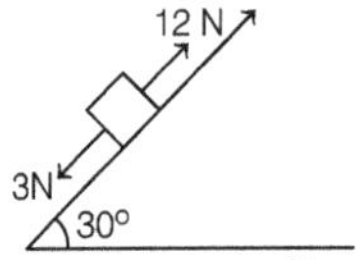

(a) $\frac{\sqrt{3}}{5}$ (b) $\frac{1}{\sqrt{3}}$ (c) $\frac{3\sqrt{3}}{5}$ (d) $\frac{5}{3\sqrt{3}}$

9 The maximun value of the applied force F such that the block as shown in the arrangement does not move is (Acceleration due to gravity, $g = 10\ \text{ms}^{-2}$) **AP EAMCET 2019**

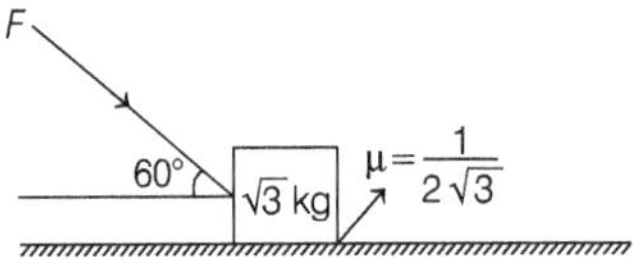

(a) 20 N (b) 15 N
(c) 25 N (d) 10 N

10 The force required to move a body up a rough inclined plane is double the force required to prevent the body from sliding down the plane. If the angle of inclination of the plane is 60°, then the coefficient of friction is **AP EAMCET 2019**

(a) $\frac{1}{3}$ (b) $\frac{1}{\sqrt{2}}$
(c) $\frac{1}{\sqrt{3}}$ (d) $\frac{1}{2}$

11 Three blocks are connected by massless strings on a frictionless inclined plane of 30° as shown in the figure. A force of 104 N is applied upward along the incline to mass m_3 causing an upward motion of the blocks. What is the acceleration of the blocks? (Assume, acceleration due to gravity, $g = 10\ \text{m/s}^2$) **TS EAMCET 2019**

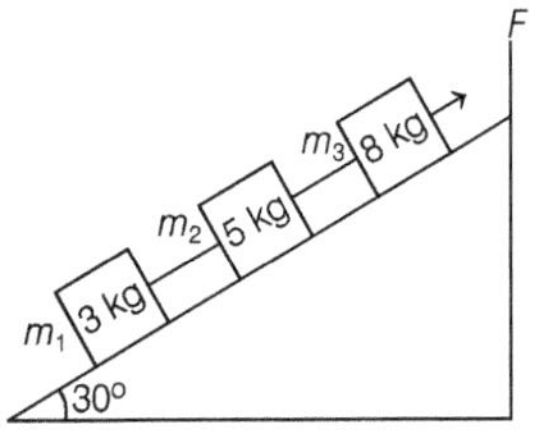

(a) $6.0\ \text{m/s}^2$ (b) $4.5\ \text{m/s}^2$
(c) $3.0\ \text{m/s}^2$ (d) $1.5\ \text{m/s}^2$

12 Two weights of the mass m_1 and $m_2 (> m_1)$ are joined by an inextensible string of negligible mass passing over a fixed frictionless pulley. The magnitude of the acceleration of the loads is **WBJEE 2019**

(a) g (b) $\frac{m_2 - m_1}{m_2} g$
(c) $\frac{m_1}{m_2 + m_1} g$ (d) $\frac{m_2 - m_1}{m_2 + m_1} g$

Work, Energy and Power

1 A ball of mass 4 kg is moving with a velocity of $10\ \text{ms}^{-1}$, collides with a spring of length 8 m and force constant $100\ \text{Nm}^{-1}$. The length of the compressed spring is x m. The value of x, to the nearest integer, is .... **JEE Main 2021**

2 A uniform thin rod of length L, mass m is lying on a smooth horizontal table. A horizontal impulse P is suddenly applied perpendicular to the rod at one end. The total energy of the rod after the impulse is **WB JEE 2021**

(a) $\frac{P^2}{m}$ (b) $\frac{7P^2}{8m}$ (c) $\frac{13P^2}{2m}$ (d) $\frac{2P^2}{m}$

3 Consider a force $\mathbf{F} = -x\hat{\mathbf{i}} + y\hat{\mathbf{j}}$. The work done by this force in moving a particle from point $A(1, 0)$ to $B(0, 1)$ along the line segment is (all quantities are in SI units) **JEE Main 2020**

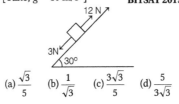

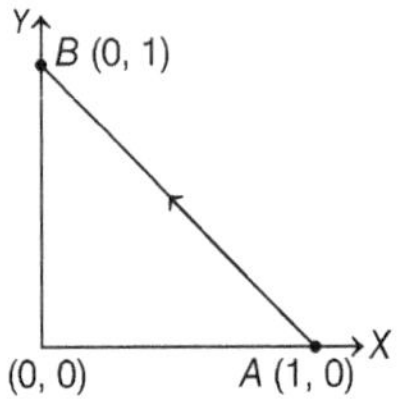

(a) $\frac{3}{2}$ (b) 2 (c) 1 (d) $\frac{1}{2}$

4 An elevator in a building can carry a maximum of 10 persons with the average mass of each person being 68 kg. The mass of the elevator itself is 920 kg and it moves with a constant speed of 3 m/s. The frictional force opposing the motion is 6000 N. If the elevator is moving up with its full capacity, the power delivered by the motor to the elevator ($g = 10\ \text{m/s}^2$) must be at least **JEE Main 2020**

(a) 62360 W (b) 48000 W
(c) 56300 W (d) 66000 W

5 A uniform cable of mass M and length L is placed on a horizontal surface such that its $\left(\frac{1}{n}\right)$th part is hanging below the edge of the surface. To lift the hanging part of the cable upto the surface, the work done should be **JEE Main 2019**

(a) $\frac{2MgL}{n^2}$ (b) $nMgL$
(c) $\frac{MgL}{n^2}$ (d) $\frac{MgL}{2n^2}$

6 A person of mass M is sitting on a swing to length L and swinging with an angular amplitude θ_0. If the person stands up when the swing passes through its lowest point, the work done by him, assuming that his centre of mass moves by a distance $l (l << L)$, is close to **JEE Main 2019**

(a) $Mgl(1-\theta_0^2)$ (b) $Mgl(1+\theta_0^2)$

(c) Mgl (d) $Mgl\left(1+\frac{\theta_0^2}{2}\right)$

7. A mass of 2 kg initially at a height of 1.2 m above an uncompressed spring with spring constant 2×10^4 N/m, is released from rest to fall on the spring. Taking the acceleration due to gravity as 10 m/s^2 and neglecting the air resistance, the compression of the spring in mm is **TS EAMCET 2019**

(a) 20 (b) 40

(c) 50 (d) 60

8. A box of mass 3 kg moves on a horizontal frictionless table and collides with another box of mass 3 kg initially at rest on the edge of the table at height 1 m. The speed of the moving box just before the collision is 4 m/s. The two boxes stick together and fall from the table. The kinetic energy just before the boxes strike the floor is (Assume, acceleration due to gravity, $g = 10$ m/s^2) **TS EAMCET 2019**

(a) 40 J (b) 80 J

(c) 96 J (d) 72 J

9. A body starts from rest, under the action of an engine working at a constant power and moves along a straight line. The displacement s is given as a function of time (t) as **WB JEE 2019**

(a) $s = at + bt^2$, a and b are constants

(b) $s = bt^2$, b is a constant

(c) $s = at^{3/2}$, a is a constant

(d) $s = at$, a is a constant

Circular Motion

1. **Statement I** A cyclist is moving on an unbanked road with a speed of 7 kmh^{-1} and takes a sharp circular turn along a path of radius of 2 m without reducing the speed. The static friction coefficient is 0.2. The cyclist will not slip and pass the curve. (Take, $g = 9.8$ m/s^2)

Statement II If the road is banked at an angle of 45°, cyclist can cross the curve of 2 m radius with the speed of 18.5 kmh^{-1} without slipping.

In the light of the above statements, choose the correct answer from the options given below. **JEE Main 2021**

(a) Statement I is incorrect and statement II is correct.

(b) Statement I is correct and statement II is incorrect.

(c) Both statements I and II are false.

(d) Both statements I and II are true.

2 A particle moves such that its position vector $\mathbf{r}(t) = \cos\omega t\,\hat{\mathbf{i}} + \sin\omega t\,\hat{\mathbf{j}}$, where ω is a constant and t is time. Then, which of the following statements is true for the velocity $\mathbf{v}(t)$ and acceleration $\mathbf{a}(t)$ of the particle? **JEE Main 2020**

(a) $\mathbf{v}$ and $\mathbf{a}$ both are parallel to $\mathbf{r}$.

(b) $\mathbf{v}$ is perpendicular to $\mathbf{r}$ and $\mathbf{a}$ is directed away from the origin.

(c) $\mathbf{v}$ and $\mathbf{a}$ both are perpendicular to $\mathbf{r}$.

(d) $\mathbf{v}$ is perpendicular to $\mathbf{r}$ and $\mathbf{a}$ is directed towards the origin.

3 A particle is moving along a circular path with a constant speed of 10 ms^{-1}. What is the magnitude of the change in velocity of the particle, when it moves through an angle of 60° around the centre of the circle? **JEE Main 2019**

(a) $10\sqrt{2}$ m/s (b) 10 m/s

(c) $10\sqrt{3}$ m/s (d) Zero

4 Two particles A and B are moving on two concentric circles of radii R_1 and R_2 with equal angular speed ω. At $t = 0$, their positions and direction of motion are shown in the figure. The relative velocity $\mathbf{v}_A - \mathbf{v}_B$ at $t = \frac{\pi}{2\omega}$ is given by **JEE Main 2019**

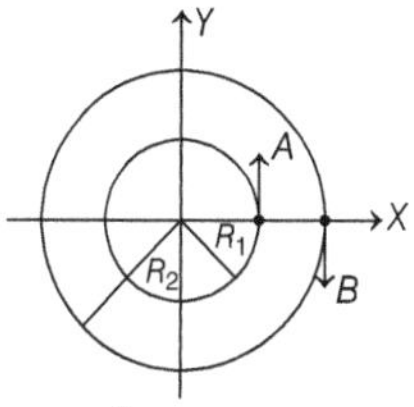

(a) $\omega(R_1 + R_2)\hat{\mathbf{i}}$ (b) $-\omega(R_1 + R_2)\hat{\mathbf{i}}$

(c) $\omega(R_1 - R_2)\hat{\mathbf{i}}$ (d) $\omega(R_2 - R_1)\hat{\mathbf{i}}$

5 If the radius of the circular path and frequency of revolution of a particle of mass m are doubled, then the change in its kinetic energy will be (E_i and E_f are the initial and final kinetic energies of the particle respectively,) **MHT CET2019**

(a) 12 E_f (b) 16 E_i

(c) 8 E_f (d) 15 E_i

COM, Conservation of Linear Momentum, Impulse and Collision

1 The disc of mass M with uniform surface mass density σ is shown in the figure. The centre of mass of the quarter disc (the shaded area) is at the position $\frac{x}{3}\frac{a}{\pi}, \frac{x}{3}\frac{a}{\pi}$ where x is ....

(Round off to the nearest integer)

(Here, a is an area as shown in the figure.) **JEE Main 2021**

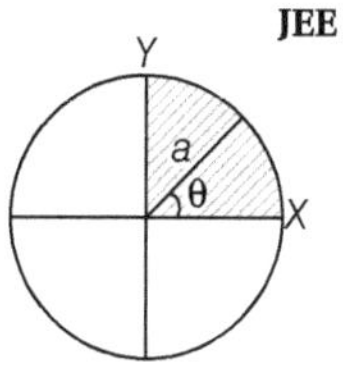

2 A rod of length L has non-uniform linear mass density given by $\rho(x) = a + b\left(\frac{x}{L}\right)^2$, where a and b are constants and $0 \le x \le L$. The value of x for the centre of mass of the rod is at **JEE Main 2020**

(a) $\frac{3}{4}\left(\frac{2a+b}{3a+b}\right)L$ (b) $\frac{4}{3}\left(\frac{a+b}{2a+3b}\right)L$

(c) $\frac{3}{2}\left(\frac{a+b}{2a+b}\right)L$ (d) $\frac{3}{2}\left(\frac{2a+b}{3a+b}\right)L$

3 A particle of mass m is projected with a speed u from the ground at an angle $\theta = \frac{\pi}{3}$ w.r.t. horizontal (X-axis). When it has reached its maximum height, it collides completely inelastically with another particle of the same mass and velocity $u\hat{\mathbf{i}}$. The horizontal distance covered by the combined mass before reaching the ground is **JEE Main 2020**

(a) $\frac{3\sqrt{3}}{8}\frac{u^2}{g}$ (b) $\frac{3\sqrt{2}}{4}\frac{u^2}{g}$

(c) $\frac{5}{8}\frac{u^2}{g}$ (d) $2\sqrt{2}\frac{u^2}{g}$

4 The coordinates of centre of mass of a uniform flag shaped lamina (thin flat plate) of mass 4 kg. (The coordinates of the same are shown in figure) are

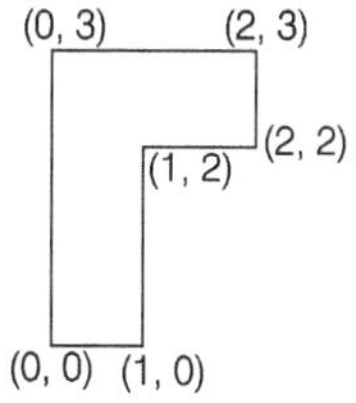

(a) (1.25 m, 1.50 m)

(b) (1 m, 1.75 m)

(c) (0.75 m, 0.75 m)

(d) (0.75 m, 1.75 m)

5 The position vector of the centre of mass $\mathbf{r}_{cm}$ of an asymmetric uniform bar of negligible area of cross-section as shown in figure is **JEE Main 2019**

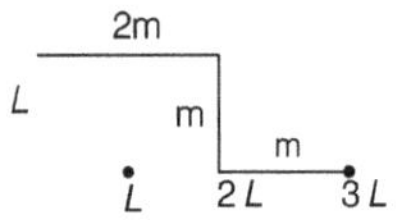

(a) $\mathbf{r} = \frac{13}{8} L\,\hat{\mathbf{x}} + \frac{5}{8} L\,\hat{\mathbf{y}}$

(b) $\mathbf{r} = \frac{11}{8} L\,\hat{\mathbf{x}} + \frac{3}{8} L\,\hat{\mathbf{y}}$

(c) $\mathbf{r} = \frac{3}{8} L\,\hat{\mathbf{x}} + \frac{11}{8} L\,\hat{\mathbf{y}}$

(d) $\mathbf{r} = \frac{5}{8} L\,\hat{\mathbf{x}} + \frac{13}{8} L\,\hat{\mathbf{y}}$

6 A particle X of mass m and initial velocity u collide with another particle Y of mass $\frac{3m}{4}$ which is at rest, The collision is head on and perfectly elastic. The ratio of de-Broglie wavelengths λ_Y and λ_X after the collision is **BITSAT 2019**

(a) 4 : 3 (b) 2 : 32 (c) 3 : 4 (d) 3 : 32

7 Two particles P and Q each of mass 3m lie at rest on the X-axis at points $(-a, 0)$ and $(+a, 0)$, respectively. A third particle R of mass 2 m initially at the origin moves towards the particle Q. If all the collisions of the system of 3 particles are elastic and head on, the total number of collisions in the system is **AP EAMCET 2019**

(a) 2 (b) 3 (c) 4 (d) 5

8 Three identical spheres each of diameter $2\sqrt{3}$ m are kept on a horizontal surface such that each sphere touches the other two spheres. If one of the sphere is removed, then the shift in the position of the centre of mass of the system is **AP EAMCET 2019**

(a) 12 m (b) 1 m (c) 2 m (d) $\frac{3}{2}$ m

9 A ball dropped from a building of height 12 m falls on a slab of 1 m height from the ground and makes a perfect elastic collision. Later the ball falls on a wooden table of height 0.5 m, makes inelastic collision and falls on the ground. If the coefficient of restitution between the ball and the table is 0.5, then the velocity of the ball while touching the ground is about (Acceleration due to gravity, $g = 10\ \text{ms}^{-2}$) **AP EAMCET 2019**

(a) 15.5 ms^{-1} (b) 14.5 ms^{-1}
(c) 9.2 ms^{-1} (d) 8.2 ms^{-1}

10 A circular ring of mass 10 kg rolls along a horizontal floor. The center of mass of the ring has a speed 1.5 m/s. The work required to stop the ring is **TS EAMCET 2019**

(a) 10 J (b) − 6 J (c) 14.5 J (d) − 22.5 J

11 The balls A, B and C of masses 50 g, 100 g and 150 g, respectively are placed at the vertices of an equilateral triangle. The length of each side is 1 m. If A is placed at (0, 0) and B is placed at (1, 0) m, find the coordinates (x, y) for the centre of mass of this system of the balls **TS EAMCET 2019**

(a) $\left(\frac{7}{12}, \sqrt{\frac{3}{4}}\right)$ m (b) $\left(\frac{5}{18}, \sqrt{\frac{1}{4}}\right)$ m

(c) $\left(\frac{7}{12}, \sqrt{\frac{3}{2}}\right)$ m (d) $\left(\frac{5}{18}, \sqrt{\frac{3}{4}}\right)$ m

12 A block of mass 'm' moving on a frictionless surface at speed 'v' collides elastically with a block of same mass, initially at rest. Now the first block moves at an angle 'θ' with its initial direction and has speed 'v_1'. The speed of the second block after collision is **MHT CET 2019**

(a) $\sqrt{v_1^2 - v^2}$ (b) $\sqrt{v^2 - v_1^2}$

(c) $\sqrt{v^2 + v_1^2}$ (d) $\sqrt{v - v_1}$

Rotation

1 A mass M hangs on a massless rod of length l which rotates at a constant angular frequency. The mass M moves with steady speed in a circular path of constant radius. Assume that, the system is in steady circular motion with constant angular velocity ω. The angular momentum of M about point A is L_A which lies in the position z-direction and the angular momentum of M about B is L_B. The correct statement for this system is **JEE Main 2021**

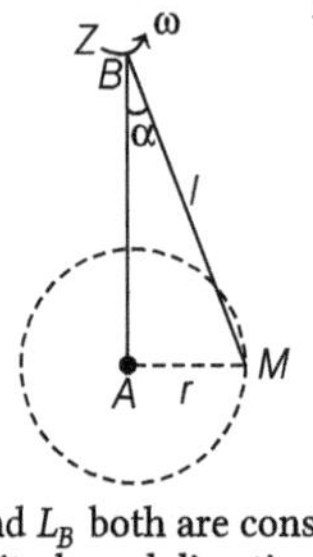

(a) L_A and L_B both are constant in magnitude and direction

(b) L_B is constant in direction with varying magnitude

(c) L_B is constant, both in magnitude and direction

(d) L_A is constant, both in magnitude and direction

2 A copper rod AB of length l is rotated about end A with a constant angular velocity ω. The electric field at a distance x from the axis of rotation is **KCET 2021**

(a) $\frac{m\omega^2 x}{e}$ (b) $\frac{m\omega x}{e}$

(c) $\frac{mx}{\omega^2 l}$ (d) $\frac{me}{\omega^2 x}$

3 Three solid sphere each of mass m and diameter d are stuck together such that the lines connecting the centres form an equilateral triangle of side of length d. The ratio I_0 / I_A of moment of inertia I_0 of the system about an axis passing the centroid and about centre of any of the spheres I_A and perpendicular to the plane of the triangle is **JEE Main 2020**

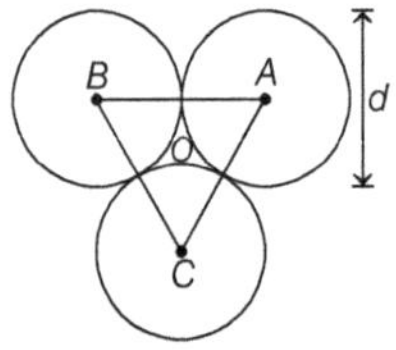

(a) $\frac{15}{13}$ (b) $\frac{13}{15}$

(c) $\frac{13}{23}$ (d) $\frac{23}{13}$

4 A particle of mass m is fixed to one end of a light spring having force constant k and unstretched length l. The other end is fixed. The system is given an angular speed ω about the fixed end of the spring such that it rotates in a circle in gravity free space. Then, the stretch in the spring is **JEE Main 2020**

(a) $\frac{ml\omega^2}{k + m\omega}$ (b) $\frac{ml\omega^2}{k + m\omega^2}$

(c) $\frac{ml\omega^2}{k - m\omega^2}$ (d) $\frac{ml\omega^2}{k - m\omega}$

5 As shown in the figure, a bob of mass m is tied by a massless string whose other end portion is wound on a flywheel (disc) of radius r and mass m. When released from rest the bob starts falling vertically. When it has covered a distance of h, the angular speed of the wheel will be **JEE Main 2020**

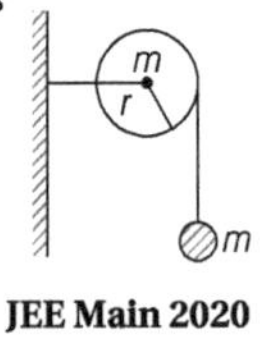

(a) $\frac{1}{r}\sqrt{\frac{2gh}{3}}$ (b) $r\sqrt{\frac{3}{4gh}}$

(c) $\frac{1}{r}\sqrt{\frac{4gh}{3}}$ (d) $r\sqrt{\frac{3}{2gh}}$

6 A rigid massless rod of length $3l$ has two masses attached at each end as shown in the figure. The rod is pivoted at point P on the horizontal axis (see figure). When released from initial horizontal position, its instantaneous angular acceleration will be **JEE Main 2019**

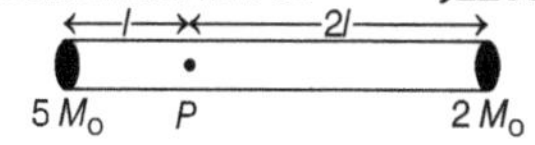

(a) $\frac{g}{13l}$ (b) $\frac{g}{2l}$ (c) $\frac{7g}{3l}$ (d) $\frac{g}{3l}$

7 A circular disc D_1 of mass M and radius R has two identical discs D_2 and D_3 of the same mass M and radius R attached rigidly at its opposite ends (see figure). The moment of inertia of the system about the axis OO' passing through the centre of D_1, as shown in the figure will be **JEE Main 2019**

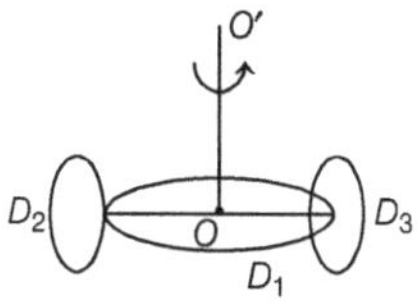

(a) $\frac{2}{3}MR^2$ (b) $\frac{4}{5}MR^2$
(c) $3MR^2$ (d) MR^2

8 A solid sphere of 100 kg and radius 10 m moving in a space becomes a circular disc of radius 20 m in one hour. Then the rate of change of moment of inertia in the process is **AP EAMCET 2019**

(a) $\frac{40}{9}$ kg m^2 s^{-1} (b) $\frac{10}{9}$ kg m^2 s^{-1}
(c) $\frac{50}{9}$ kg m^2 s^{-1} (d) $\frac{25}{9}$ kg m^2 s^{-1}

9 Two identical discs are moving with the same kinetic energy. One rolls and the other slides. The ratio of their speeds is **AP EAMCET 2019**

(a) 1 : 2 (b) 1 : 1
(c) 2 : 3 (d) $\sqrt{2} : \sqrt{3}$

10 A body rotates about a stationary axis. If the angular deceleration is proportional to square root of angular speed, then the mean angular speed of the body, given ω_0 as the initial angular speed, is **TS EAMCET 2019**

(a) $\frac{\omega_0}{\sqrt{2}}$ (b) $\frac{\omega_0}{4}$
(c) $\frac{\omega_0}{2}$ (d) $\frac{\omega_0}{3}$

11 A rod l m long is acted upon by a couple as shown in the figure. The moment of couple is τ Nm. If the force at each end of the rod, then magnitude of each force is ($\sin 30° = \cos 60° = 0.5$) **MHT CET 2019**

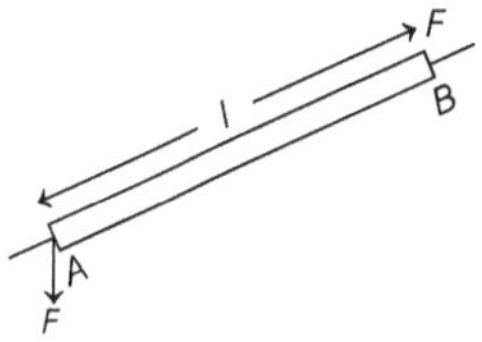

(a) $\frac{\tau}{l}$ (b) $\frac{l}{2\tau}$
(c) $\frac{2\tau}{l}$ (d) $\frac{2l}{\tau}$

12 A solid sphere rolls down from top of inclined plane, 7m high, without slipping. Its linear speed at the foot of plane is ($g = 10$ m/s^2) **MHT CET 2019**

(a) $\sqrt{70}$ m/s
(b) $\sqrt{\frac{140}{3}}$ m/s
(c) $\sqrt{\frac{280}{3}}$ m/s
(d) $\sqrt{100}$ m/s

Gravitation

1 The radius in kilometre to which the present radius of earth ($R = 6400$ km) to be compressed, so that the escape velocity is increased to ten times is ...... km. **JEE Main 2021**

2 A geostationary satellite is orbiting around an arbitrary planet P at a height of $11\,R$ above the surface of P, R being the radius of P. The time period of another satellite (in h) at a height of $2R$ from the surface of P is .... P has the time period of 24 h. **JEE Main 2021**

(a) $6\sqrt{2}$ (b) $\frac{6}{\sqrt{2}}$
(c) 3 (d) 5

3 Planet A has mass M and radius R. Planet B has half the mass and half the radius of planet A. If the escape velocities from the planets A and B are v_A and v_B respectively, then $\frac{v_A}{v_B} = \frac{n}{4}$.

The value of n is **JEE Main 2020**

(a) 1 (b) 2
(c) 3 (d) 4

4 A body A of mass m is moving in a circular orbit of radius R about a planet. Another body B of mass $\frac{m}{2}$ collides with A with a velocity which is half $\left(\frac{\mathbf{v}}{2}\right)$, the instantaneous velocity $\mathbf{v}$ of A. The collision is completely inelastic. Then, the combined body **JEE Main 2020**

(a) escapes from the planet's gravitational field
(b) starts moving in an elliptical orbit around the planet
(c) falls vertically downward towards the planet
(d) continues to move in a circular orbit

5 Consider two solid spheres of radii $R_1 = 1$ m, $R_2 = 2$ m and masses M_1 and M_2, respectively. The gravitational field due to sphere 1 and 2 are shown. The value of $\frac{M_1}{M_2}$ is **JEE Main 2020**

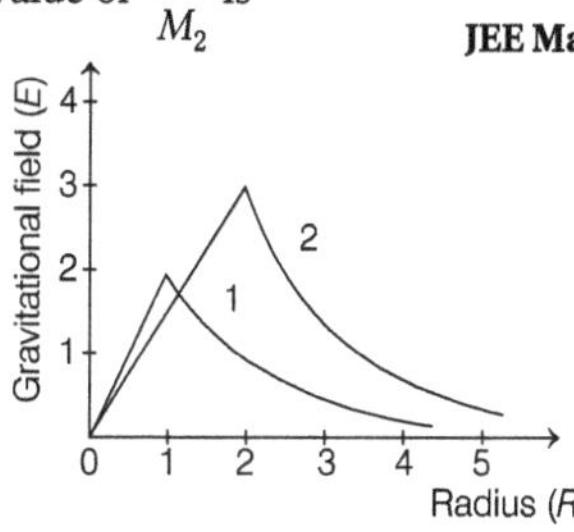

(a) $\frac{1}{3}$ (b) $\frac{1}{2}$ (c) $\frac{1}{6}$ (d) $\frac{2}{3}$

6 A straight rod of length L extends from $x = a$ to $x = L + a$. The gravitational force it exerts on a point mass m at $x = 0$, if the mass per unit length of the rod is $A + Bx^2$, is given by **JEE Main 2019**

(a) $Gm\left[A\left(\frac{1}{a+L} - \frac{1}{a}\right) - BL\right]$
(b) $Gm\left[A\left(\frac{1}{a+L} - \frac{1}{a}\right) + BL\right]$
(c) $Gm\left[A\left(\frac{1}{a} - \frac{1}{a+L}\right) + BL\right]$
(d) $Gm\left[A\left(\frac{1}{a} - \frac{1}{a+L}\right) - BL\right]$

7 Four identical particles of mass M are located at the corners of a square of side a. What should be their speed, if each of them revolves under the influence of other's gravitational field in a circular orbit circumscribing the square? **JEE Main 2019**

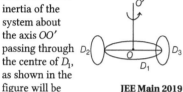

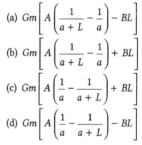

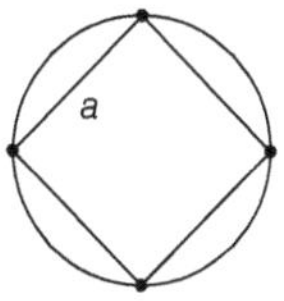

(a) $1.35\sqrt{\frac{GM}{a}}$ (b) $1.16\sqrt{\frac{GM}{a}}$
(c) $1.21\sqrt{\frac{GM}{a}}$ (d) $1.41\sqrt{\frac{GM}{a}}$

8 The value of acceleration due to gravity at earth's surface is 9.8 ms^{-2}. The altitude above its surface at which the acceleration due to gravity decreases to 4.9 ms^{-2}, is close to (Take, radius of earth $= 6.4 \times 10^6$ m) **BITSAT 2019**

(a) 9.0×10^6 m (b) 2.6×10^6 m
(c) 6.4×10^6 m (d) 1.6×10^6 m

9 The escape velocity of the earth surface is v_e. A body is projected with velocity $3v_e$. With what constant velocity will it move in the inter planetary space?

(a) v_e (b) $3v_e$ (c) $\sqrt{2}\, v_e$ (d) $2\sqrt{2}\, v_e$

10 Two bodies each of mass m are hung from a balance whose scale pans differ in a vertical height by h. If the mean density of the earth is ρ, the error in weighing is **AP EAMCET 2019**

(a) $\frac{4\pi\rho\, Gmh}{3}$ (b) $\frac{3\pi\rho\, Gmh}{4}$
(c) $\frac{8\pi\rho\, Gmh}{3}$ (d) $\frac{3\pi\rho\, Gmh}{8}$

11 A hole is drilled half way to the centre of the earth. A body weighs 300 N on the surface of the earth. How much will, it weigh at the bottom of the hole? **MHT CET 2019**

(a) 200 N (b) 250 N (c) 120 N (d) 150 N

12 Two bodies of masses m_1 and m_2 initially at rest at infinite distance apart move towards each other under gravitational force of attraction. Their relative velocity of approach when they are separated by a distance r is (G = universal gravitational constant.) **AP EAMCET 2019**

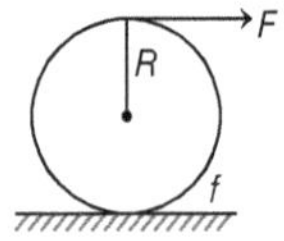

(a) $\left[\frac{2G(m_1 - m_2)}{r}\right]^{1/2}$

(b) $\left[\frac{2G(m_1 + m_2)}{r}\right]^{1/2}$

(c) $\left[\frac{r}{2G(m_1 m_2)}\right]^{1/2}$ (d) $\left[\frac{r}{2G} m_1 m_2\right]^{1/2}$

13 If a satellite has to orbit the earth in a circular path every 6 hrs, at what distance from the surface of the earth should be satellite placed (radius of earth, $R_e = 6400$ km) (Assume, $\frac{GM}{4\pi^2} = 8.0 \times 10^{12}$ N/m^2/kg, where, G and M are gravitational constant and mass of earth and $10^{1/3} = 2.1$). **TS EAMCET 2019**

(a) 15100 km (b) 8720 km
(c) 20600 km (d) 5560 km

14 If the acceleration due to gravity g doubles and the radius of earth becomes half that of the present value, then the value of escape velocity is (Assume, $g = 10$ m/s^2 and radius of earth, $R = 6400$ km) **TS EAMCET 2019**

(a) 12 km/s
(b) $16\sqrt{2}$ km/s
(c) $8\sqrt{2}$ km/s
(d) $4\sqrt{2}$ km/s

Simple Harmonic Motion

1 A block of mass 1 kg attached to a spring is made to oscillate with an initial amplitude of 12 cm. After 2 min, the amplitude decrease to 6 cm. Determine the value of the damping constant for this motion. **JEE Main 2021**

(a) 0.69×10^2 kg s^{-1} (b) 3.3×10^2 kg s^{-1}
(c) 1.16×10^2 kg s^{-1} (d) 5.7×10^{-3} kg s^{-1}

2 Two particles execute simple harmonic motion of the same amplitude and frequency along the same straight line. They pass one another when going in opposite directions and each time their displacement is half the amplitude. The phase difference between them is **BVP 2021**

(a) 45° (b) 60° (c) 90° (d) 120°

3 A cylindrical plastic bottle of negligible mass is filled with 310 mL of water and left floating in a pond with still water. If pressed downward slightly and released, it starts performing simple harmonic motion at angular frequency ω. If the radius of the bottle is 2.5 cm, then ω is close to (Take, density of water $= 10^3$ kg/m^3) **JEE Main 2019**

(a) 2.50 rad s^{-1} (b) 5.00 rad s^{-1}
(c) 1.25 rad s^{-1} (d) 3.75 rad s^{-1}

4 A particle executes simple harmonic motion with an amplitude of 5 cm. When the particle is at 4 cm from the mean position, the magnitude of its velocity in SI units is equal to that of its acceleration. Then, its periodic time (in seconds) is **JEE Main 2019**

(a) $\frac{4\pi}{3}$ (b) $\frac{8\pi}{3}$ (c) $\frac{7}{3}\pi$ (d) $\frac{3}{8}\pi$

5 A simple pendulum of length 1 m is oscillating with an angular frequency 10 rad/s. The support of the pendulum starts oscillating up and down with a small angular frequency of 1 rad/s and an amplitude of 10^{-2} m. The relative change in the angular frequency of the pendulum is best given by **JEE Main 2019**

(a) 1 rad/s (b) 10^{-5} rad/s
(c) 10^{-3} rad/s (d) 10^{-1} rad/s

6 A simple harmonic motion is represented by $y = 5(\sin 3\pi t + \sqrt{3} \cos 3\pi t)$ cm. The amplitude and time period of the motion are **JEE Main 2019**

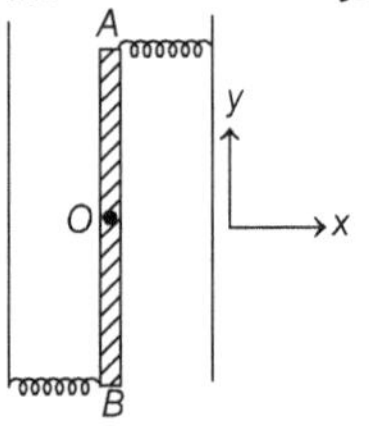

(a) 10 cm, $\frac{3}{2}$ s (b) 5 cm, $\frac{2}{3}$ s
(c) 5 cm, $\frac{3}{2}$ s (d) 10 cm, $\frac{2}{3}$ s

7 Two light identical springs of spring constant k are attached horizontally at the two ends of an uniform horizontal rod AB of length l and mass m. The rod is pivoted at its centre O and can rotate freely in horizontal plane. The other ends of the two springs are fixed to rigid supports as shown in figure.

The rod is gently pushed through a small angle and released. The frequency of resulting oscillation is **JEE Main 2019**

(a) $\frac{1}{2\pi}\sqrt{\frac{2k}{m}}$ (b) $\frac{1}{2\pi}\sqrt{\frac{3k}{m}}$
(c) $\frac{1}{2\pi}\sqrt{\frac{6k}{m}}$ (d) $\frac{1}{2\pi}\sqrt{\frac{k}{m}}$

8 The time period of a bob performing simple harmonic motion in water is 2s. If density of bob is $\frac{4}{3} \times 10^3$ kg/m^3, then time period of bob performing simple harmonic motion in air will be **BITSAT 2019**

(a) 3s (b) 4s (c) 2s (d) 1s

9 A simple pendulum is placed inside a lift, which is moving with a uniform acceleration. If the time periods of the pendulum while the lift is moving upwards and downwards are in the ratio 1:2, then the acceleration of the lift is (Acceleration due to gravity, $g = 10 \text{ ms}^{-2}$) **AP EAMCET 2019**

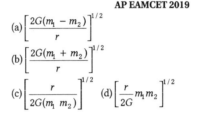

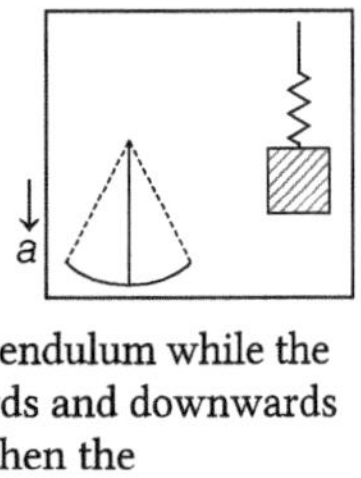

(a) 6 ms^{-2} (b) 0 ms^{-2}
(c) 3 ms^{-2} (d) 2 ms^{-2}

10 A person measures a time period of a simple pendulum inside a stationary lift and finds it to be T. If the lift starts accelerating upwards with an acceleration $\left(\frac{g}{3}\right)$, the time period of the pendulum will be **MHT CET 2019**

(a) $\frac{T}{\sqrt{3}}$ (b) $\sqrt{3}\frac{T}{2}$

(c) $\sqrt{3}\,T$ (d) $\frac{T}{3}$

11 Two pendulums begin to swing simultaneously. The first pendulum makes nine full oscillations when the other makes seven. The ratio of the lengths of the two pendulums is **MHT CET 2019**

(a) $\frac{49}{81}$ (b) $\frac{64}{81}$

(c) $\frac{8}{9}$ (d) $\frac{7}{9}$

Elasticity

1 An object is located at 2 km beneath the surface of the water. If the fractional compression $\frac{\Delta V}{V}$ is 1.36%, the ratio of hydraulic stress to the corresponding hydraulic strain will be (Take, density of water is 1000 kg m^{-3} and $g = 9.8\ ms^{-2}$) **JEE Main 2021**

(a) $1.96\times10^7\ Nm^{-2}$ (b) $1.44\times10^7\ Nm^{-2}$

(c) $2.26\times10^9\ Nm^{-2}$ (d) $1.44\times10^9\ Nm^{-2}$

2 On taking a solid ball from the surface to the bottom of a lake of 200 m depth, the volume of the ball is reduced by 0.1%. Find the bulk modulus of the material of the ball. **AMU 2021**

(a) $2\times10^7\ N/m^2$ (b) $1.6\times10^8\ N/m^2$

(c) $2.0\times10^9\ N/m^2$ (d) $0.6\times10^{10}\ N/m^2$

3 Two steel wires having same length are suspended from a ceiling under the same load. If the ratio of their energy stored per unit volume is 1 : 4, the ratio of their diameters is **JEE Main 2020**

(a) $\sqrt{2}:1$ (b) $1:\sqrt{2}$

(c) 2 : 1 (d) 1 : 2

4 A steel wire having a radius of 2.0 mm, carrying a load of 4 kg, is hanging from a ceiling. Given that $g = 3.1\pi\ ms^{-2}$, what will be the tensile stress that would be developed in the wire? **JEE Main 2020**

(a) $6.2\times10^6\ Nm^{-2}$

(b) $5.2\times10^6\ Nm^{-2}$

(c) $3.1\times10^6\ Nm^{-2}$

(d) $4.8\times10^6\ Nm^{-2}$

5 A boy's catapult is made of rubber cord which is 42 cm long, with 6 mm diameter of cross-section and of negligible mass. The boy keeps a stone weighing 0.02 kg on it and stretches the cord by 20 cm by applying a constant force. When released the stone flies off with a velocity of 20 ms^{-1}. Neglect the change in the area of cross-section of the cord while stretched. The Young's modulus of rubber is closest to **JEE Main 2020**

(a) $10^6\ Nm^{-2}$ (b) $10^4\ Nm^{-2}$

(c) $10^8\ Nm^{-2}$ (d) $10^3\ Nm^{-2}$

6 At 40°C, a brass wire of 1 mm radius is hung from the ceiling. A small mass M is hung from the free end of the wire. When the wire is cooled down from 40°C to 20°C, it regains its original length of 0.2 m. The value of M is close to [Coefficient of linear expansion and Young's modulus of brass are 10^{-5}/°C and $10^{11}\ N/m^2$ respectively, $g = 10\ ms^{-2}$] **JEE Main 2019**

(a) 9 kg (b) 0.5 kg

(c) 1.5 kg (d) 0.9 kg

7 A steel wire of cross-sectional area 4 cm^2 has elastic limit of $2.2\times10^8\ N/m^2$. The maximum upward acceleration that can be given to a 1000 kg elevator supported by this steel wire if the stress is to exceed one-fourth of the elastic limit is [Take, $g = 10\ m/s^2$] **BITSAT 2019**

(a) 10 m/s^2

(b) 9 m/s^2

(c) 15 m/s^2

(d) 12 m/s^2

8 A rubber cord has a cross-sectional area $10^{-6}\ m^2$ and total unstretched length 0.1 m. It is stretched to 0.125 m and then released to project a particle of mass 5.0 g. The velocity of projection is [Given, Young's modulus of rubber, $Y = 5\times10^8\ N/m^2$] **BITSAT 2019**

(a) 45 m/s

(b) 30 m/s

(c) 25 m/s

(d) 15 m/s

9 A one metre steel wire of negligible mass and area of cross-section 0.01 cm^2 is kept on a smooth horizontal table with one end fixed. A ball of mass 1 kg is attached to the other end. The ball and the wire are rotating with an angular velocity of ω. If the elongation of the wire is 2 mm, then ω is (Young's modulus of steel $= 2\times10^{11}\ Nm^{-2}$) **AP EAMCET 2019**

(a) 5 rad s^{-1} (b) 10 rad s^{-1}

(c) 15 rad s^{-1} (d) 20 rad s^{-1}

10 A solid copper cube of 7 cm edge is subjected to a hydraulic pressure of 8000 kPa. The volume contraction of the copper cube is (Bulk modulus of copper = 140 GPa) **AP EAMCET 2019**

(a) $196\times10^{-3}\ cm^3$

(b) $19.6\times10^{-6}\ cm^3$

(c) $19.6\times10^{-3}\ cm^3$

(d) $196\times10^{3}\ cm^3$

11 A uniform rod of length L is rotated in a horizontal plane about a vertical axis through one of its ends. The angular speed of rotation is ω. Find increase in length of the rod, if ρ and Y are the density and Young's modulus of the rod respectively, **TS EAMCET 2019**

(a) $\frac{\rho\omega^3 Y}{4L^2}$ (b) $\frac{4\rho\omega^2 L^3}{3Y}$

(c) $\frac{\rho\omega^2 L^3}{3Y}$ (d) $\frac{\rho\omega^3 L^3}{8Y}$

12 A copper wire of cross-sectional area 0.01 cm^2 is under a tension of 22 N. Find the percentage change in the cross-sectional area. (Young's modulus of copper $= 1.1\times10^{11}\ N/m^2$ and Poisson's ratio = 0.32) **TS EAMCET 2019**

(a) 12.6×10^{-3} (b) 8.6×10^{-3}

(c) 6.4×10^{-3} (d) 2.8×10^{-3}

13 A horizontal aluminium rod of diameter 4 cm projected 6 cm from a wall. An object of mass 400 π kg is suspended from the end of the rod. The shearing modulus of aluminium is $3.0\times10^{10}\ N/m^2$. The vertical deflection of the end of the rod is ($\because g = 10\ m/s^2$) **TS EAMCET 2019**

(a) 0.01 mm (b) 0.02 mm

(c) 0.03 mm (d) 0.04 mm

Fluid Mechanics

1 Consider a water tank as shown in the figure. Its cross-sectional area is 0.4 m^2. The tank has an opening B near the bottom whose cross-sectional area is 1 cm^2. A load of 24 kg is applied on the water at the top when the height of the water level is 40 cm above the bottom, the velocity of water coming out the opening B is v ms^{-1}. The value of v, to the nearest integer, is ....
(Take, $g = 10$ ms^{-2}) **JEE Main 2021**

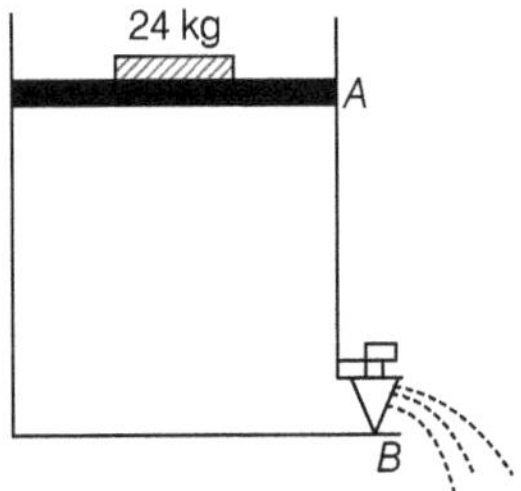

2 The pressure acting on a submarine is 3×10^5 Pa at a certain depth. If the depth is doubled, the percentage increase in the pressure acting on the submarine would be (Assume that, atmospheric pressure $= 1 \times 10^5$ Pa, density of water $= 10^3$ kg m^{-3} and $g = 10$ ms^{-2}) **JEE Main 2021**

(a) $\frac{200}{3}\%$ (b) $\frac{200}{5}\%$ (c) $\frac{5}{200}\%$ (d) $\frac{3}{200}\%$

3 A body of density 1.2×10^3 kg/m^3 is dropped from rest from a height 1 m into a liquid to density 2.4×10^3 kg/m^3. Neglecting all dissipative effects, the maximum depth to which the body sinks before returning to float on the surface is **WB JEE 2021**

(a) 0.1 m (b) 1 m
(c) 0.01 m (d) 2 m

4 Two liquids of densities ρ_1 and $\rho_2 (\rho_2 = 2\rho_1)$ are filled up behind a square wall of side 10 m as shown in figure. Each liquid has a height of 5 m. The ratio of the forces due to these liquids exerted on upper part MN to that at the lower part NO is (assume that the liquids are not mixing) **JEE Main 2020**

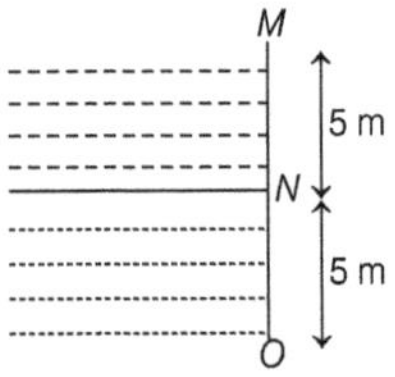

(a) 1/2 (b) 2/3 (c) 1/4 (d) 1/3

5 Water flows in a horizontal tube (see figure). The pressure of water changes by 700 Nm^{-2} between A and B, where the area of cross-section are 40 cm^2 and 20 cm^2, respectively. Find the rate of flow of water through the tube. (Take, density of water = 1000 kgm^{-3}) **JEE Main 2020**

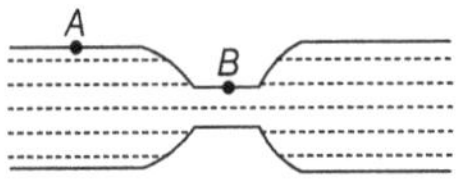

(a) 3020 cm^3/s (b) 2420 cm^3/s
(c) 2720 cm^3/s (d) 1810 cm^3/s

6 A small spherical droplet of density d is floating exactly half immersed in a liquid of density ρ and surface tension T. The radius of the droplet is (take note that the surface tension applies an upward force on the droplet)

(a) $r = \sqrt{\frac{3T}{(2d-\rho)g}}$ (b) $r = \sqrt{\frac{T}{(d-\rho)g}}$

(c) $r = \sqrt{\frac{T}{(d+\rho)g}}$ (d) $r = \sqrt{\frac{2T}{3(d+\rho)g}}$

7 A leak proof cylinder of length 1 m, made of a metal which has very low coefficient of expansion is floating vertically in water at 0 °C such that its height above the water surface is 20 cm. When the temperature of water is increased to 4°C, the height of the cylinder above the water surface becomes 21 cm. The density of water at $T = 4$ °C, relative to the density at $T = 0$ °C is close to **JEE Main 2020**

(a) 1.26 (b) 1.03
(c) 1.01 (d) 1.04

8 A soap bubble, blown by a mechanical pump at the mouth of a tube, increases in volume, with time, at a constant rate. The graph that correctly depicts the time dependence of pressure inside the bubble is given by **JEE Main 2019**

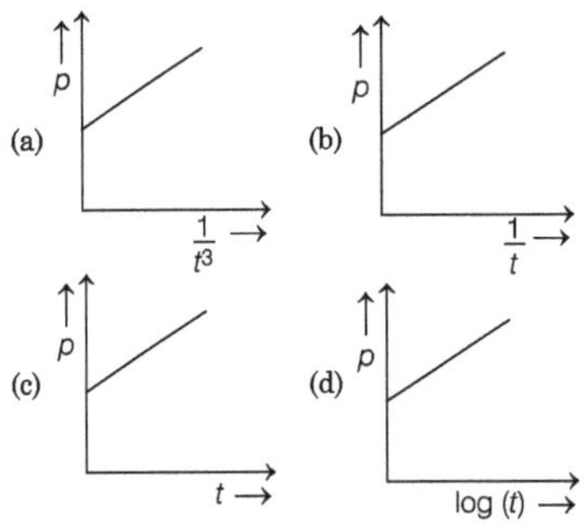

9 A long cylindrical vessel is half-filled with a liquid. When the vessel is rotated about its own vertical axis, the liquid rises up near the wall. If the radius of vessel is 5 cm and its rotational speed is 2 rotations per second, then the difference in the heights between the centre and the sides (in cm) will be **JEE Main 2019**

(a) 0.1 (b) 1.2 (c) 0.4 (d) 2.0

10 Two seperate soap bubbles of radii 8×10^{-3} m and 2×10^{-3} m respectively formed of same liquid (surface tension 6.5×10^{-2} N/m) come together to form a double. The radius of interlace of double bubble is **[BITSAT 2019]**

(a) 6×10^{-3} m (b) 4×10^{-3} m
(c) 15×10^{-3} m (d) 0.66×10^{-3} m

11 A water tank kept on the ground has an orifice of 2 mm diameter on the vertical side. What is the minimum height of the water above the orifice for which the output flow of water is found to be turbulent? (Assume, $g = 10$ m/s^2, $\rho_{water} = 10^3$ kg/m^3, viscosity = 1 centi-poise) **TS EAMCET 2019**

(a) 3 cm (b) 4 cm
(c) 6 cm (d) 2 cm

12 A copper ball of radius 3.0 mm falls in an oil tank of viscosity 1 kg/ms. Then, the terminal velocity of the copper ball will be (Density of oil = 1.5×10^3 kg/m^3, Density of copper= 9×10^3 kg/m^3 and $g = 10$ m/s^2.) **TS EAMCET 2019**

(a) 18×10^{-2} m/s (b) 25×10^{-2} m/s
(c) 15×10^{-2} m/s (d) 20×10^{-2} m/s

13 The radius of a soap bubble is r and the surface tension of the soap solution is S. The electric potential to which the soap bubble be raised by charging it so that the pressure inside the bubble becomes equal to the pressure outside the bubble is (ε_0= permittivity of the free space) **AP EAMCET 2019**

(a) $\sqrt{\frac{Sr}{8\varepsilon_0}}$ (b) $\sqrt{\frac{Sr}{4\varepsilon_0}}$

(c) $\sqrt{\frac{4Sr}{\varepsilon_0}}$ (d) $\sqrt{\frac{8Sr}{\varepsilon_0}}$

14 Two light balls are suspended as shown in figure. When a stream of air passes through the space between them, the distance between the balls will **MHT CET 2019**

(a) remain same
(b) increase
(c) may increase or decrease, depending on speed of air
(d) decrease

Thermometry, Thermal Expansion and Kinetic Theory of Gases

1 The molecules of a given mass of a gas have root mean square speeds of 100 ms^{-1} at 27°C and 1.00 atm pressure. If the root mean square speed of the molecules of the gas at 127°C and 2 atm pressure is found to be $\frac{200}{\sqrt{n}}$ m/s, then find the value of n. **JEE Main 2021**

2 A gas mixture contains monoatomic and diatomic molecules of 2 moles each. The mixture has a total internal energy of (symbols have usual meanings) **KCET 2021**

(a) $3RT$ (b) $5RT$
(c) $8RT$ (d) $9RT$

3 If the degrees of freedom of the molecule of a gas are n, the ratio of its two specific heats $\left(\frac{C_p}{C_V}\right)$ will be **BVP 2021**

(a) $1+\frac{2}{n}$ (b) $1-\frac{2}{n}$
(c) $1+\frac{1}{n}$ (d) $2-\frac{1}{n}$

4 Consider a mixture of n moles of helium gas and $2n$ moles of oxygen gas (molecules taken to be rigid) as an ideal gas. Its C_p / C_V value will be **JEE Main 2020**

(a) 40/27 (b) 23/15
(c) 19/13 (d) 67/45

5 The plot that depicts the behaviour of the mean free time τ (time between two successive collisions) for the molecules of an ideal gas, as a function of temperature (T), qualitatively is (graphs are schematic and not drawn to scale) **JEE Main 2020**

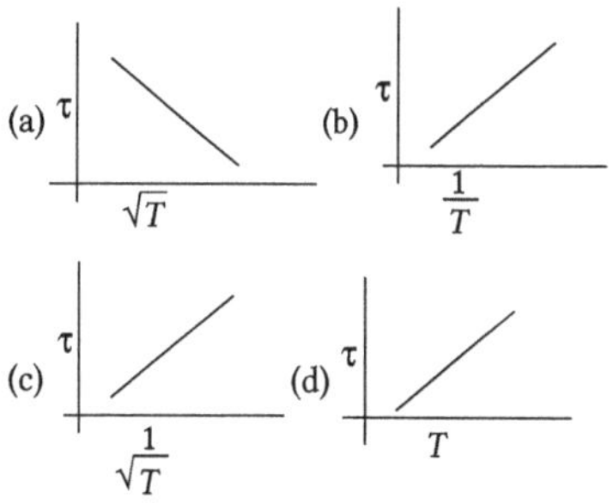

6 Consider two ideal diatomic gases A and B at some temperature T. Molecules of the gas A are rigid and have a mass m. Molecules of the gas B have an additional vibrational mode and have a mass $\frac{m}{4}$. The ratio of the specific heats (C_V^A and C_V^B) of gas A and B respectively is **JEE Main 2020**

(a) 5 : 9
(b) 7 : 9
(c) 3 : 5
(d) 5 : 7

7 A gas mixture consists of 3 moles of oxygen and 5 moles of argon at temperature T. Considering only translational and rotational modes, the total internal energy of the system is **JEE Main 2019**

(a) 12 RT
(b) 15 RT
(c) 20 RT
(d) 4 RT

8 A vertical closed cylinder is separated into two parts by a frictionless piston of mass m and of negligible thickness. The piston is free to move along the length of the cylinder. The length of the cylinder above the piston is l_1 and that below the piston is l_2, such that $l_1 > l_2$. Each part of the cylinder contains n moles of an ideal gas at equal temperature T. If the piston is stationary, its mass m, will be given by (where, R is universal gas constant and g is the acceleration due to gravity) **JEE Main 2019**

(a) $\frac{nRT}{g}\left[\frac{l_1 - l_2}{l_1 l_2}\right]$
(b) $\frac{nRT}{g}\left[\frac{1}{l_2}+\frac{1}{l_1}\right]$
(c) $\frac{RT}{g}\left[\frac{2l_1 + l_2}{l_1 l_2}\right]$
(d) $\frac{RT}{ng}\left[\frac{l_1 - 3l_2}{l_1 l_2}\right]$

9 Pressure versus temperature graph of an ideal gas is shown in the given figure. Density of gas at point A is ρ_0, then density of gas at point B will be **BITSAT 2019**

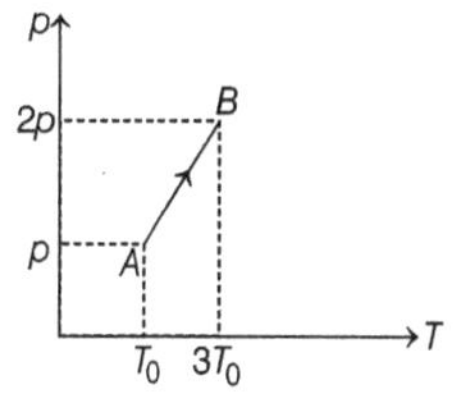

(a) $2\rho_0$ (b) $\frac{2}{3}\rho_0$ (c) $\frac{3}{2}\rho_0$ (d) $3\rho_0$

10 The figure shows the graph of $\frac{pV}{T}$ *versus* p for 2×10^{-4} kg of hydrogen gas at two different temperatures, where p, V and T represents pressure, volume and temperature respectively. Then, the value of $\frac{pV}{T}$, where the curve meet on the vertical axis, is **BITSAT 2019**

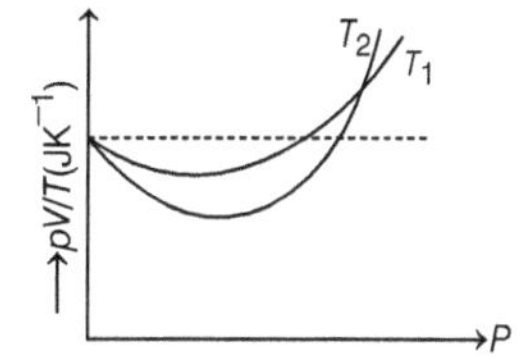

(a) $\frac{R}{5}$ (b) $\frac{2R}{3}$
(c) $\frac{R}{10}$ (d) $\frac{R}{15}$

11 The wavelength of the radiation emitted by a black body is 1 mm and Wien's constant is 3×10^{-3} mK. Then the temperature of the black body will be **TS EAMCET 2019**

(a) 3 K (b) 30 K
(c) 300 K (d) 3000 K

12 An ideal gas in a closed container is heated so that the final rms speed of the gas particles increases by 2 times the initial rms speed. If the initial gas temperature is 27°C, then the final temperature of the ideal gas is : **TS EAMCET 2019**

(a) 1200°C (b) 927°C
(c) 827°C (d) 1473°C

First Law of Thermodynamics

1 For an adiabatic expansion of an ideal gas, the fractional change in its pressure is equal to (where, γ = the ratio of specific heats) **JEE Main 2021**

(a) $-\gamma \frac{dV}{V}$

(b) $-\gamma \frac{V}{dV}$

(c) $-\frac{1}{\gamma}\frac{dV}{V}$

(d) $\frac{dV}{V}$

2 When the temperature of the source of Carnot engine is at 400 K, its efficiency is 25%. The required increase in temperature of the source to increase the efficiency to 50% is **Kerala CEE 2021**

(a) 800 K

(b) 600 K

(c) 100 K

(d) 400 K

(e) 200 K

3 A Carnot engine having an efficiency of $\frac{1}{10}$ is being used as a refrigerator. If the work done on the refrigerator is 10 J, then the amount of heat absorbed from the reservoir at lower temperature is: **JEE Main 2020**

(a) 99 J

(b) 100 J

(c) 90 J

(d) 1 J

4 A thermodynamic cycle $xyzx$ is shown on a V - T diagram.

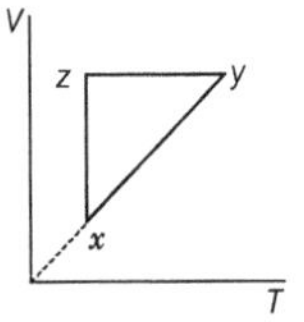

The p - V diagram that best describes this cycle is (diagrams are schematic and not to scale) **JEE Main 2020**

(a)
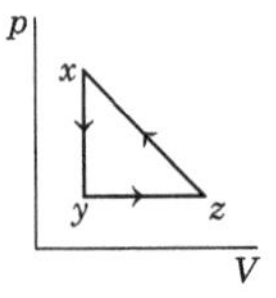

(b)
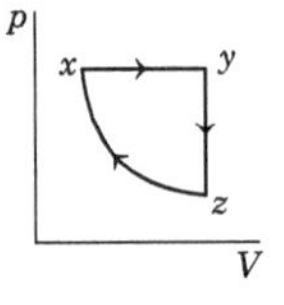

(c)
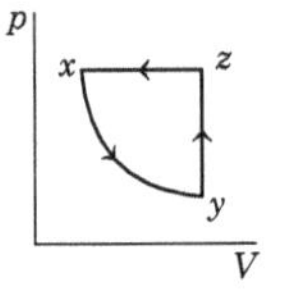

(d)
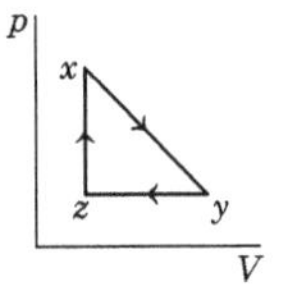

5 A litre of dry air at STP expands adiabatically to a volume of 3 L. If $\gamma = 1.40$, the work done by air is ($3^{1.4} = 4.6555$) [Take, air to be an ideal gas] **JEE Main 2020**

(a) 100.8 J (b) 90.5 J
(c) 48 J (d) 60.7 J

6 Under an adiabatic process, the volume of an ideal gas gets doubled. Consequently, the mean collision time between the gas molecule changes from τ_1 to τ_2. If $\frac{C_p}{C_V} = \gamma$ for this gas, then a good estimate for $\frac{\tau_2}{\tau_1}$ is given by **JEE Main 2020**

(a) $\left(\frac{1}{2}\right)^{\frac{\gamma+1}{2}}$ (b) $\frac{1}{2}$

(c) 2 (d) $\left(\frac{1}{2}\right)^{\gamma}$

7 In a process, temperature and volume of one mole of an ideal monoatomic gas are varied according to the relation $VT = k$, where k is a constant. In this process, the temperature of the gas is increased by ΔT. The amount of heat absorbed by gas is (where, R is gas constant) **JEE Main 2019**

(a) $\frac{1}{2} kR\Delta T$ (b) $\frac{2k}{3} \Delta T$

(c) $\frac{1}{2} R\Delta T$ (d) $\frac{3}{2} R\Delta T$

8 The given diagram shows four processes, i.e. isochoric, isobaric, isothermal and adiabatic. The correct assignment of the processes, in the same order is given by **JEE Main 2019**

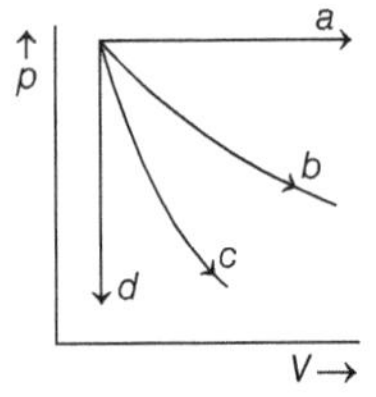

(a) $d\ a\ b\ c$ (b) $a\ d\ b\ c$
(c) $d\ a\ c\ b$ (d) $a\ d\ c\ b$

9 An ideal monoatomic gas at 300 K expands adiabatically to twice its volume. The final temperature of gas is **BITSAT 2019**

(a) $300\sqrt{2}$ (b) $300\sqrt{3}$
(c) $300\left(\frac{1}{2}\right)^{2/3}$ (d) $300(2)^{2/3}$

10 One mole of nitrogen gas being initially at a temperature of $T_0 = 300$ K is adiabatically compressed to increase its pressure 10 times. The final gas temperature after compression is (Assume, nitrogen gas molecules as rigid diatomic and $100^{1/7} = 1.9$) **TS EAMCET 2019**

(a) 120 K (b) 750 K
(c) 650 K (d) 570 K

11 Consider the given diagram. An ideal gas is contained in a chamber (left) of volume V and is at an absolute temperature T. It is allowed to rush freely into the right chamber of volume V which is initially vacuum. The whole system is thermally isolated. What will be the final temperature of the equilibrium has been attained? **WBJEE 2019**

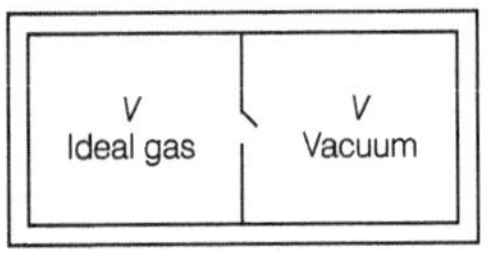

(a) T (b) $\frac{T}{2}$ (c) $2T$ (d) $\frac{T}{4}$

Calorimetry

1 Two identical metal wires of thermal conductivities K_1 and K_2 respectively are connected in series. The effective thermal conductivity of the combination is **JEE Main 2021**

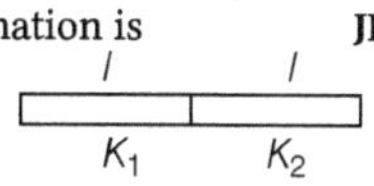

(a) $\frac{2K_1K_2}{K_1 + K_2}$
(b) $\frac{K_1 + K_2}{2K_1K_2}$
(c) $\frac{K_1 + K_2}{K_1 K_2}$
(d) $\frac{K_1 K_2}{K_1 + K_2}$

2 A blackened metal foil is warmed by radiation from a point source whose temperature is T and distance from the foil is d. It is found that the power received by the foil is P. If both the temperature and distance are doubled, the power received by the foil will be **BVP 2021**

(a) P
(b) $2P$
(c) $4P$
(d) $16P$

3 A thermally insulated vessel contains 150 g of water at 0°C. Then, the air from the vessel is pumped out adiabatically. A fraction of water turns into ice and the rest evaporates at 0°C itself. The mass of evaporated water will be closest to (Latent heat of vaporisation of water $= 2.10 \times 10^6 \text{ J kg}^{-1}$ and latent heat of fusion of water $= 3.36 \times 10^5 \text{ J kg}^{-1}$) **JEE Main 2019**

(a) 150 g (b) 20 g
(c) 130 g (d) 35g

4 Two identical beakers A and B contain equal volumes of two different liquids at 60°C each and left to cool down. Liquid in A has density of $8 \times 10^2 \text{ kg / m}^3$ and specific heat of $2000 \text{ J kg}^{-1}\text{K}^{-1}$ while liquid in B has density of 10^3 kg m^{-3} and specific heat of $4000 \text{ J kg}^{-1}\text{K}^{-1}$. Which of the following best describes their temperature *versus* time graph schematically? (Assume the emissivity of both the beakers to be the same) **JEE Main 2019**

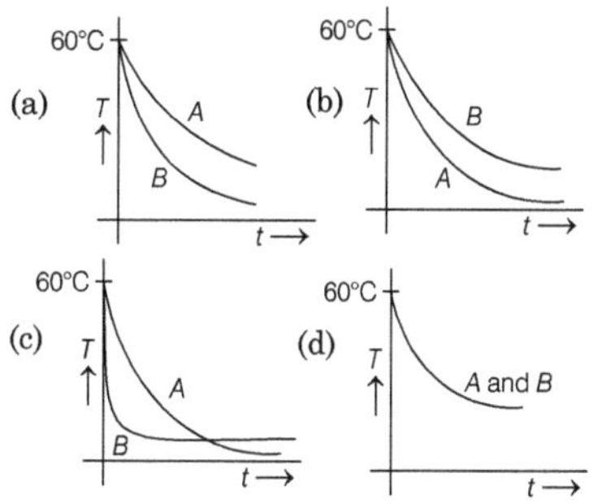

5 Two materials having coefficients of thermal conductivity '$3K$' and 'K' and thickness 'd' and '$3d$' respectively, are joined to form a slab as shown in the figure. The temperatures of the outer surfaces are 'θ_2' and 'θ_1' respectively, ($\theta_2 > \theta_1$). The temperature at the interface is **JEE Main 2019**

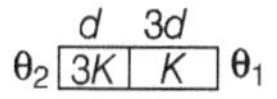

(a) $\frac{\theta_2 + \theta_1}{2}$ (b) $\frac{\theta_1}{3} + \frac{2\theta_2}{3}$
(c) $\frac{\theta_1}{6} + \frac{5\theta_2}{6}$ (d) $\frac{\theta_1}{10} + \frac{9\theta_2}{10}$

6 Temperature difference of 120°C is maintained between two ends of a uniform rod AB of length $2L$. Another bent rod PQ, of same cross-section as AB and length $\frac{3L}{2}$ is connected across AB (see figure). In steady state, temperature difference between P and Q will be close to **JEE Main 2019**

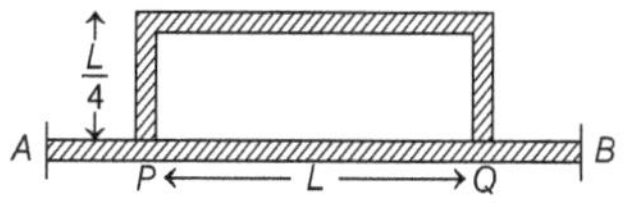

(a) 45°C (b) 35°C (c) 75°C (d) 60°C

7 An unknown metal of mass 192 g heated to a temperature of 100°C was immersed into a brass calorimeter of mass 128 g containing 240 g of water at a temperature of 8.4°C. Calculate the specific heat of the unknown metal, if water temperature stabilises at 21.5°C. (Take, specific heat of brass is $394 \text{ J kg}^{-1}\text{K}^{-1}$) **JEE Main 2019**

(a) $916 \text{ J kg}^{-1} \text{ K}^{-1}$ (b) $654 \text{ J kg}^{-1} \text{ K}^{-1}$
(c) $1232 \text{ J kg}^{-1} \text{ K}^{-1}$ (d) $458 \text{ J kg}^{-1} \text{ K}^{-1}$

8 When 100 g of a liquid A at 100°C is added to 50 g of a liquid B at temperature 75°C, the temperature of the mixture becomes 90°C. The temperature of the mixture, if 100 g of liquid A at 100°C is added to 50 g of liquid B at 50°C will be **JEE Main 2019**

(a) 60°C (b) 80°C (c) 70°C (d) 85°C

9 A metal ball of mass 0.1 kg is heated upto 500°C and dropped into a vessel of heat capacity 800 JK^{-1} and containing 0.5 kg water. The initial temperature of water and vessel is 30°C. What is the approximate percentage increment in the temperature of the water?
[Take, specific heat capacities of water and metal are respectively $4200 \text{ Jkg}^{-1}\text{K}^{-1}$ and $400 \text{ Jkg}^{-1}\text{K}^{-1}$] **JEE Main 2019**

(a) 25% (b) 15% (c) 30% (d) 20%

10 A window used to thermally insulate a room from outside consists of two parallel glass sheets each of area 2.6 m^2 and thickness 1 cm separated by 5 cm thick stagnant air. In the steady state, the room glass interface is at 18°C and the glass-outdoor interface is at −2 °C. If the thermal conductivities of glass and air are respectively 0.8 $\text{Wm}^{-1}\text{K}^{-1}$ and $0.08 \text{ Wm}^{-1} \text{K}^{-1}$, the rate of flow of heat through the window is **AP EAMCET 2019**

(a) 15 W (b) 40 W
(c) 60 W (d) 80 W

11 Two identical blocks of ice move in opposite directions with equal speed and collide with each other. What will be the minimum speed required to make both the blocks melt completely, if the initial temperatures of the blocks were − 8°C each? **WB JEE 2019**
(Specific heat of ice is $2100 \text{ Jkg}^{-1}\text{K}^{-1}$ and latent heat of fusion of ice is $3.36 \times 10^5 \text{ Jkg}^{-1}$)

(a) 840 ms^{-1} (b) 420 ms^{-1}
(c) 8.4 ms^{-1} (d) 84 ms^{-1}

Wave Motion

1 A closed organ pipe of length L and an open organ pipe contain gases of densities ρ_1 and ρ_2, respectively. The compressibility of gases are equal in both the pipes. Both the pipes are vibrating in their first overtone with same frequency.

The length of the open pipe is $\frac{x}{3}L\sqrt{\frac{\rho_1}{\rho_2}}$, where x is .... (Round off to the nearest integer) **JEE Main 2021**

2 A stretched sonometer wire is in unison with a tuning fork. When length of wire is increase by 1%, the number of beats heard per second is 5. Then the frequency of the fork is **Manipal 2021**

(a) 500 Hz (b) 505 Hz
(c) 255 Hz (d) 250 Hz

3 Speed of a transverse wave on a straight wire (mass 6.0 g, length 60 cm and area of cross-section 1.0 mm^2) is 90 ms^{-1}. If the Young's modulus of wire is 16×10^{11} Nm^{-2}, the extension of wire over its natural length is **JEE Main 2020**

(a) 0.01 mm (b) 0.04 mm
(c) 0.03 mm (d) 0.02 mm

4 A transverse wave travels on a taut steel wire with a velocity of v when tension in it is 2.06×10^4 N. When the tension is changed to T, the velocity changed to $v/2$. The value of T is close to **JEE Main 2020**

(a) 10.2×10^2 N (b) 5.15×10^3 N
(c) 2.50×10^4 N (d) 30.5×10^4 N

5 A wire of length $2L$, is made by joining two wires A and B of same length but different radii r and $2r$ and made of the same material. It is vibrating at a frequency such that the joint of the two wires forms a node. If the number of antinodes in wire A is p and that in B is q, then the ratio $p:q$ is **JEE Main 2019**

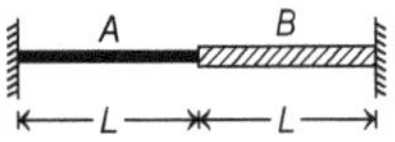

(a) 3 : 5 (b) 4 : 9
(c) 1 : 2 (d) 1 : 4

6 A string is clamped at both the ends and it is vibrating in its 4th harmonic. The equation of the stationary wave is $Y = 0.3\sin(0.157x)\cos(200\pi t)$. The length of the string is (All quantities are in SI units) **JEE Main 2019**

(a) 60 m (b) 40 m
(c) 80 m (d) 20 m

7 A closed organ pipe has a fundamental frequency of 1.5 kHz. The number of overtones that can be distinctly heard by a person with this organ pipe will be (Assume that the highest frequency a person can hear is 20,000 Hz) **JEE Main 2019**

(a) 7 (b) 4 (c) 5 (d) 6

8 Two sound producing sources A and B are moving towards and away from a stationary observer with same speed respectively. If frequency of sound produced by both sources are equal as 400 Hz, then speed of sources (approximately) when observer detects 4 beats per second, is [Given, speed of sound = 340 m/s] **BITSAT 2019**

(a) 1.7 m/s (b) 3.4 m/s
(c) 2.4 m/s (d) 1 m/s

9 If sound travels in air with the speed of 340 m/s, then number of tones present in an open organ pipe of length 2 m and its maximum frequency 1200 Hz, are **BITSAT 2019**

a. 17 b. 11 c. 9 d. 14

10 A siren placed at a railway platform is emitting a sound of frequency 5 kHz. A passenger sitting in a moving train A records the frequency of the siren as 5.5 kHz.

During his return journey by train B he records the frequency of the siren as 6 kHz. The ratio of the speed of train B to that of train A is **AP EAMCET 2019**

(a) $\frac{242}{252}$ (b) 2 (c) $\frac{5}{6}$ (d) $\frac{11}{6}$

11 The speed of a transverse wave travelling in a wire of length 50 cm, cross-sectional area 1 mm^2 and mass 5 g is 80 ms^{-1}. The Young's modulus of the material of the wire is 4×10^{11} Nm^{-2}. The extension in the length of the wire is **AP EAMCET 2019**

(a) 8×10^{-5} m (b) 8×10^{-4} m
(c) 16×10^{-5} m (d) 16×10^{-4} m

12 Standing waves are produced in a string 16 m long. If there are 9 nodes between the two fixed ends of the string and the speed of the wave is 32 m/s, what is the frequency of the wave? **TS EAMCET 2019**

(a) 5 Hz (b) 10 Hz
(c) 30 Hz (d) 20 Hz

13 Two open pipes of different lengths and same diameter in which the air column vibrates with fundamental frequencies n_1, and n_2 respectively. When both pipes are joined to form a single pipe, its fundamental frequency will be **MHT CET 2019**

(a) $\frac{n_1 + n_2}{n_1 n_2}$ (b) $\frac{n_1 n_2}{2n_2 + n_1}$
(c) $\frac{2n_2 + n_1}{n_1 n_2}$ (d) $\frac{n_1 n_2}{n_1 + n_2}$

Solutions with Explanation

Units, Dimensions and Error Analysis

1 (*d*) $na' = (n-1)a$

$\Rightarrow a' = \left(\frac{n-1}{n}\right)a$

$\therefore$ Least Count (LC)

$= 1\text{ MSD} - 1\text{ VSD} = a - a'$

$= a - \left(\frac{n-1}{n}\right)a = \frac{a}{n}\text{ cm} = \frac{10a}{n}\text{ mm}$

2 (*b*) Given, $T = 2\pi\sqrt{\frac{l}{g}} \Rightarrow g = \frac{4\pi^2 l}{T^2}$

$\therefore \frac{\Delta g}{g} = \frac{\Delta l}{l} + \frac{2\Delta T}{T} = \frac{0.1}{10} + 2\left(\frac{1}{0.5 \times 200}\right)$

$\frac{\Delta g}{g} \times 100 = \frac{1}{100} \times 100 + \frac{1}{50} \times 100 = 3\%$

3 (*d*) Dimensions of quantity f are

$$[f] = \frac{[h]^{1/2}\,[c]^{5/2}}{[G]^{1/2}} \qquad \text{...(i)}$$

As, $h = \frac{E}{\nu}$; $\Rightarrow [h] = [ML^2T^{-2}]\,[T] = [ML^2T^{-1}]$

$c = [LT^{-1}]$ and $G = \frac{F \cdot r^2}{m^2}$

$\Rightarrow [G] = \frac{[MLT^{-2}]\,[L^2]}{[M^2]} = [M^{-1}L^3T^{-2}]$

So, dimensions of f using Eq. (i),

$$[f] = \frac{[ML^2T^{-1}]^{\frac{1}{2}}\,[LT^{-1}]^{\frac{5}{2}}}{[M^{-1}L^3T^{-2}]^{\frac{1}{2}}}$$

$$= \left[M^{\frac{1}{2}+\frac{1}{2}}, L^{\frac{5}{2}-\frac{3}{2}+1}, T^{-\frac{1}{2}-\frac{5}{2}+\frac{2}{2}}\right]$$

$= [ML^2T^{-2}]$

Thus, it is the dimensions of energy.

4 (*a*) Pitch of screw of a screw gauge,

$\text{pitch} = \frac{\text{distance moved by screw}}{\text{number of rotations}}$

$= \frac{3\text{ mm}}{6} = 0.5\text{ mm}$

and least count of screw gauge, least count

$= \frac{\text{pitch}}{\text{number of circular scale divisions}}$

$= \frac{0.5\text{ mm}}{50} = 0.01\text{ mm} = 0.001\text{ cm}$

5 (*c*) Force of interaction between two atoms is given as

$$F = \alpha\beta \exp(-x^2/\alpha kT)$$

exponential terms are always dimensionless,

so dimensions of $\left(\frac{-x^2}{\alpha kT}\right) = [M^0L^0T^0]$

$\Rightarrow$ Dimensions of α = Dimensions of (x^2/kT)

Now, substituting the dimensions of individual term in the given equation, we get

$= \frac{[M^0L^2T^0]}{[M^1L^2T^{-2}]}$

{$\because$ Dimensions of kT equivalent to the dimensions of energy $= [M^1L^2T^{-2}]$}

$= [M^{-1}LT^2] \qquad \text{...(i)}$

Now from given equation, we have dimensions of F = dimensions of $\alpha \times$ dimensions of β

$\Rightarrow$ Dimensions of β = Dimensions of $\left(\frac{F}{\alpha}\right)$

$= \frac{[M^1L^1T^{-2}]}{[M^{-1}L^0T^2]}$ [$\because$ using Eq. (i)]

$= [M^2LT^{-4}]$

6 (*b*) In a screw gauge,

Least count

$= \frac{\text{Measure of 1 main scale division (MSD)}}{\text{Number of divisions on circular scale}}$

Here, minimum value to be measured/least count is 5 μm.

$= 5 \times 10^{-6}\text{ m}$

$\therefore$ According to the given values,

$5 \times 10^{-6} = \frac{1 \times 10^{-3}}{N}$

or $N = \frac{10^{-3}}{5 \times 10^{-6}} = \frac{1000}{5} = 200$ divisions

7 (*c*) We know that, velocity of light c is terms of ε_0 and μ_0 is given by

$c = \frac{1}{\sqrt{\mu_0\varepsilon_0}} \Rightarrow c^2 = \frac{1}{\mu_0\,\varepsilon_0}$

$c^2 = \frac{1}{\mu_0^2} \cdot \frac{\mu_0}{\varepsilon_0}$

$\Rightarrow c = \frac{1}{\mu_0}\sqrt{\frac{\mu_0}{\varepsilon_0}} \Rightarrow \sqrt{\frac{\mu_0}{\varepsilon_0}} = c\mu_0 \qquad \text{... (i)}$

Dimensions of magnetic field,

$B = \frac{\tau}{NIA} = \frac{[ML^2T^{-2}]}{[A]\,[L^2]} = [MA^{-1}T^{-2}]$

Again, $B = \mu_0 nI$

$\therefore \mu_0 = \frac{B}{nI} = \frac{[MA^{-1}\,T^{-2}]}{[L^{-1}]\,[A]} = [MLA^{-2}T^{-2}]$

$\mu_0 = [MLA^{-2}T^{-2}] \qquad \text{... (ii)}$

and $c = [LT^{-1}] \qquad \text{... (iii)}$

From Eqs. (i), (ii) and (iii), we get

$\sqrt{\frac{\mu_0}{\varepsilon_0}} = [LT^{-1}]\,[MLA^{-2}T^{-2}] = [ML^2T^{-3}A^{-2}]$

8 (*d*) Dimensional formula of thermal conductivity $[k] = [M^1\,L^1\,T^{-3}\,K^{-1}]$.

Dimensional formula of universal gravitational constant, $[G] = [M^{-1}\,L^3\,T^{-2}]$

Now, $\frac{[k]}{[G]} = [M^2\,L^{-2}\,T^{1}\,K^{-1}]$

Compare above equation with

$[M^{2a}\,L^{4b}\,T^{2c}\,K^d]$

This will give us, $a = 1$, $b = -\frac{1}{2}$, $c = -\frac{1}{2}$ and $d = -1$

Now, $\frac{a+b}{c+b} - d = \frac{1-\frac{1}{2}}{-\frac{1}{2}-\frac{1}{2}} - (-1)$

or $\frac{a+b}{c+b} - d = \frac{1}{2}$

9 (*c*) Given, current passing through the conductor,

$I = (5 \pm 0.2)\text{ A}$, $\Delta I = 0.2\text{ A}$

Voltage developed,

$V = (60 \pm 6)\text{ V} \Rightarrow \Delta V = 6\text{ V}$

By Ohm's law,

$V = RI$, $R = \frac{V}{I}$

$\therefore \frac{\Delta R}{R} = \frac{\Delta V}{V} + \frac{\Delta I}{I} = \frac{6}{60} + \frac{0.2}{5} = 0.1 + 0.04$

$\frac{\Delta R}{R} = 0.14$

$\therefore \frac{\Delta R}{R} \times 100 = 0.14 \times 100 = 14\%$

10 (*a*) The dimensions of force $[F] = [M\,L\,T^{-2}]$ and density $[d] = [M\,L^{-3}\,T^0]$

From the given relation, $F = \frac{y}{\sqrt{d}}$

$\Rightarrow y = F\sqrt{d}$

Substituting the above dimensions, we get

$[y] = [F]\,[d]^{1/2} = [M\,L\,T^{-2}]\,[M\,L^{-3}\,T^0]^{1/2}$

$= [M^{3/2}\,L^{-1/2}\,T^{-2}]$

11 (*c*) Given, Maximum error in the measurement of mass = 0.3%

Maximum error in the measurement of length = 0.2%

We know that,

Density is given as,

Density, $\rho = \frac{\text{mass }(m)}{\text{volume }(V)} = \frac{m}{L^3}$

where, L = side of cube

Error in density is given as,

$\left(\frac{\Delta\rho}{\rho}\right) = \frac{\Delta m}{m} + \frac{3\Delta L}{L}$

or $\left(\frac{\Delta\rho}{\rho}\right) \times 100 = \left(\frac{\Delta m}{m} + \frac{3\Delta L}{L}\right) \times 100$

Substituting the given values, we get

$\left(\frac{\Delta\rho}{\rho}\right)_{max} = (0.3\% + 3(0.2)\%) = 0.3\% + 0.6\%$

$\therefore$ Maximum percentage error measurement of density $\left(\frac{\Delta\rho}{\rho}\right)_{max} = 0.9\%$.

Vectors

1. (*d*) Let **A, B** and **C** be as shown in figure. Let θ be the angle of incidence, which is equal to the angle of reflection. Resolving these vectors in rectangular components, we have

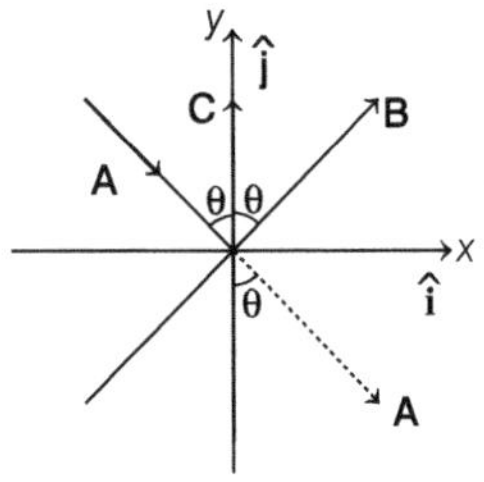

$$\mathbf{A} = \sin\theta\hat{\mathbf{i}} - \cos\theta\hat{\mathbf{j}}$$
$$\mathbf{B} = \sin\theta\hat{\mathbf{i}} + \cos\theta\hat{\mathbf{j}}$$
$$\mathbf{B} - \mathbf{A} = 2\cos\theta\hat{\mathbf{j}}$$
or $$\mathbf{B} = \mathbf{A} + 2\cos\theta\hat{\mathbf{j}}$$

Now,
$\mathbf{A}\cdot\mathbf{C} = (1)(1)\cos(180° - \theta) = -\cos\theta$

$\therefore$ $$\mathbf{B} = \mathbf{A} - 2(\mathbf{A}\cdot\mathbf{C})\hat{\mathbf{j}}$$
or $$\mathbf{B} = \mathbf{A} - 2(\mathbf{A}\cdot\mathbf{C})\mathbf{C} \text{ (as } \hat{\mathbf{j}} = \mathbf{C})$$

2. (*c*) Given, $\mathbf{r}\cdot\mathbf{p} = 0$

So, $\mathbf{r} \perp \mathbf{p}$, i.e. we have following situation,

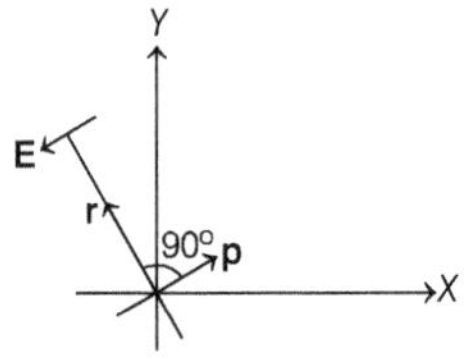

So, we have to find direction of electric field at equatorial line.

As **E** is directed opposite to **p** at all equatorial points, direction of **E** is along $-\mathbf{p}$.

So, $\mathbf{E} = \lambda(-\mathbf{p})$
$$= \lambda[-(-\hat{\mathbf{i}} - 3\hat{\mathbf{j}} + 2\hat{\mathbf{k}})]$$
$$= \lambda(\hat{\mathbf{i}} + 3\hat{\mathbf{j}} - 2\hat{\mathbf{k}})$$

3 (*c*) Given, $|\mathbf{A}| = |\mathbf{B}|$ or $A = B$ ...(i)

Let magnitude of $(\mathbf{A}+\mathbf{B})$ is R and for $(\mathbf{A}-\mathbf{B})$ is R'.

Now, $\mathbf{R} = \mathbf{A} + \mathbf{B}$ and
$$R^2 = A^2 + B^2 + 2AB\cos\theta$$
$$R^2 = 2A^2 + 2A^2\cos\theta \quad \text{...(ii)}$$
[$\because$ using Eq. (i)]

Again, $\mathbf{R}' = \mathbf{A} - \mathbf{B}$
$$\Rightarrow \quad R'^2 = A^2 + B^2 - 2AB\cos\theta$$
$$R'^2 = 2A^2 - 2A^2\cos\theta \quad \text{...(iii)}$$
[$\because$ using Eq. (i)]

Given, $R = nR'$ or $\left(\frac{R}{R'}\right)^2 = n^2$

Dividing Eq. (ii) with Eq. (iii), we get
$$\frac{n^2}{1} = \frac{1+\cos\theta}{1-\cos\theta}$$
or $$\frac{n^2-1}{n^2+1} = \frac{(1+\cos\theta)-(1-\cos\theta)}{(1+\cos\theta)+(1-\cos\theta)}$$
$$\Rightarrow \quad \frac{n^2-1}{n^2+1} = \frac{2\cos\theta}{2} = \cos\theta$$
or $$\theta = \cos^{-1}\left(\frac{n^2-1}{n^2+1}\right)$$

4. (*c*) As we know that the magnitude of the resultant of two vectors **X** and **Y**,
$$R^2 = X^2 + Y^2 + 2XY\cos\theta \quad \text{...(i)}$$
where, θ is the angle between X and Y.

Putting, $X = (\mathbf{A}+\mathbf{B})$
$Y = (\mathbf{A}-\mathbf{B})$

and $R = \sqrt{A^2+B^2}$ in Eq. (i), we get
$$A^2 + B^2 = (\mathbf{A}+\mathbf{B})^2 + (\mathbf{A}-\mathbf{B})^2 + 2(\mathbf{A}+\mathbf{B})(\mathbf{A}-\mathbf{B})\cos\theta$$
$$\Rightarrow \quad A^2 + B^2 = A^2 + B^2 + 2\mathbf{AB} + A^2 + B^2 - 2\mathbf{AB} + 2(A^2 - B^2)\cos\theta$$
$$\Rightarrow \quad \frac{-(A^2+B^2)}{2(A^2-B^2)} = \cos\theta$$
we get, $$\theta = \cos^{-1}\left[-\frac{(A^2+B^2)}{2(A^2-B^2)}\right]$$

5. (*d*) Given that **R** is resultant of **P** and **Q** as shown in the figure,

BC = **P**, **CA** = Q and **BA** = R

Given, BA and BC are perpendicular and equal in magnitude.

So, from property of triangle,
$$\angle ACB = 45°$$

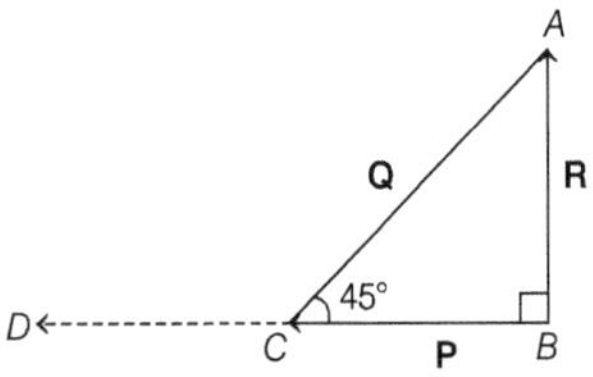

Now, **BC** has to be extended up to D so, that **CD** = **P**

Now, **CD** and **CA** have the initial point C, so the angle between **CD** and **CA**;
$$= 180° - 45° = 135° = \frac{3\pi}{4}$$

So, angle between P and Q is $\frac{3\pi}{4}$.

6 (*a*) Let two vectors **P** and **Q** are represented by graph as below

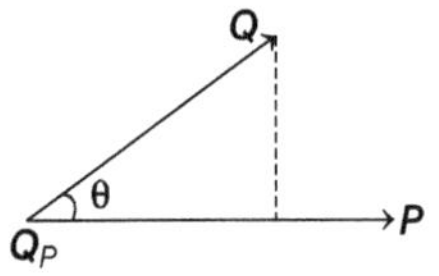

Here, $\mathbf{Q}_p$ is a vector in the direction of **P**.

Then, from the right angle triangle, we get
$$\cos\theta = \frac{Q_P}{Q} \quad \text{...(i)}$$
$$\Rightarrow \quad Q_p = Q\cos\theta$$
Also, $\cos\theta = \frac{\mathbf{P}\cdot\mathbf{Q}}{PQ} \Rightarrow \frac{Q_P}{Q} = \frac{\mathbf{P}\cdot\mathbf{Q}}{PQ}$
$$\Rightarrow \quad Q_P = \frac{\mathbf{P}\cdot\mathbf{Q}}{P} \quad \text{...(ii)}$$

As given that, $\hat{\mathbf{P}}$ is the unit vector along **P**, then
$$\hat{\mathbf{P}} = \frac{\mathbf{P}}{P} \quad \text{...(iii)}$$

Putting the value of **P** from Eq. (iii) to Eq. (ii), we get
$$Q_p = \hat{\mathbf{P}}\cdot\mathbf{Q}$$

7 (*a*) Vectors $(\mathbf{A}+\mathbf{B})$ and $(\mathbf{A}-\mathbf{B})$ are at right angle to each other, therefore
$$(\mathbf{A}+\mathbf{B})\cdot(\mathbf{A}-\mathbf{B}) = 0$$
$$\mathbf{A}\cdot\mathbf{A} + \mathbf{B}\cdot\mathbf{A} - \mathbf{A}\cdot\mathbf{B} - \mathbf{B}\cdot\mathbf{B} = 0$$
$$|\mathbf{A}| + BA\cos\theta - AB\cos\theta - |\mathbf{B}|^2 = 0$$
$$|\mathbf{A}|^2 = |\mathbf{RB}|^2$$
Hence, $|\mathbf{A}| = |\mathbf{B}|$

Motion in One Dimension

1. (*c*) During first half of the motion (i.e. upto $x = 20$ m) acceleration in increasing linearly is given by

$$a = \frac{5}{20}x + 5$$

$$\Rightarrow \quad a = \frac{x}{4} + 5$$

$$\Rightarrow \quad \frac{dv}{dt} = \frac{x}{4} + 5 \qquad \left(\because a = \frac{dv}{dt}\right)$$

$$\Rightarrow \quad \frac{v \cdot dv}{dx} = \frac{x}{4} + 5 \quad \left(\because \frac{dx}{dt} = v \Rightarrow dt = \frac{dx}{v}\right)$$

$$\Rightarrow \quad \int_5^v v \cdot dv = \int_0^{20}\left(\frac{x}{4} + 5\right)\cdot dx$$

$$\Rightarrow \quad \left[\frac{v^2}{2}\right]_5^v = \left[\frac{x^2}{8} + 5x\right]_0^{2}$$

$$\Rightarrow \quad \frac{v^2}{5} - \frac{25}{2} = \frac{400}{8} + 100$$

$$\Rightarrow \quad \frac{v^2}{2} = 150 + \frac{25}{2}$$

$$\Rightarrow \quad v^2 = 325$$

Now, in second half of motion, acceleration is constant, i.e. $a = 10 \text{ m/s}^2$

$$\Rightarrow \quad v^2 - u^2 = 2as$$

$$\Rightarrow \quad v'^2 - v^2 = 2as \text{ (take, } v = v', u = v)$$

$$\Rightarrow \quad v'^2 - 325 = 2 \times 10 \times (35 - 20)$$

$$(\because v^2 = 325)$$

$$\Rightarrow \quad v' = \sqrt{625} = 25 \text{ m/s}$$

2 (*d*) Let P_1 be the position of plane at $t = 0$, when sound waves started towards person A and P_2 is the position of plane observed at time instant t as shown in the figure below.

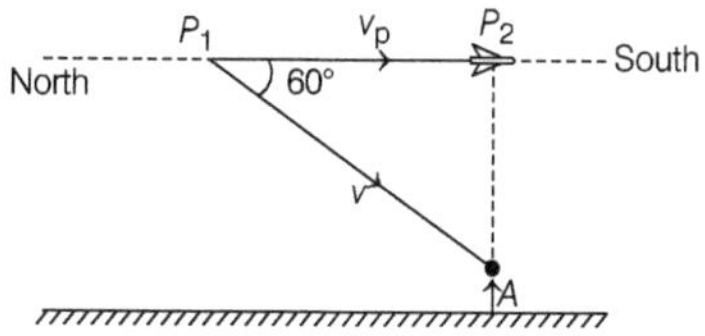

In triangle P_1P_2A,

P_1P_2 = speed of plane × time = $v_P \times t$

P_1A = speed of sound × time = $v \times t$

Now, from ΔP_1P_2A

$$\cos\theta = \frac{\text{base}}{\text{hypotenuse}}$$

$$\cos 60° = \frac{P_1P_2}{P_1A} = \frac{v_P \times t}{v \times t}$$

$$\frac{1}{2} = \frac{v_P}{v}$$

$$\Rightarrow \quad v_P = \frac{v}{2}$$

3. (*d*) **Key Idea** Area under the velocity-time curve represents displacement.

To get exact position at $t = 5$ s, we need to calculate area of the shaded part in the curve as shown below

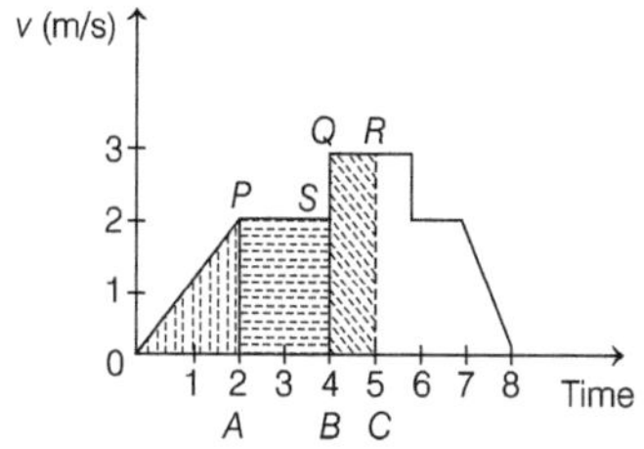

$\therefore$ Displacement of particle = Area of OPA + Area of $PABSP$ + Area of $QBCRQ$

$$= \left(\frac{1}{2} \times 2 \times 2\right) + (2 \times 2) + (3 \times 1)$$

$$= 2 + 4 + 3 = 9 \text{ m}$$

4 (*d*) Since, the particle starts from rest, this means, initial velocity, $u = 0$

Also, it moves with uniform acceleration along positive X-axis. This means, its acceleration (a) is constant.

$\therefore$ Given, a - t graph in (A) is correct.

As we know, for velocity-time graph, slope = acceleration.

Since, the given v-t graph in (B) represents that its slope is constant and non-zero.

$\therefore$ Graph in (B) is also correct.

Also, the displacement of such a particle w.r.t. time is given by

$$x = ut + \frac{1}{2}at^2 = 0 + \frac{1}{2}at^2 \Rightarrow x \propto t^2$$

So, x *versus* t graph would be a parabola with starting from origin.

This is correctly represented in displacement-time graph given in (D).

5. (*d*) Given, velocity of a particle is

$$\mathbf{v} = k(y\hat{\mathbf{i}} + x\hat{\mathbf{j}}) \qquad \text{...(i)}$$

Suppose, it's position is given as

$$\mathbf{r} = x\hat{\mathbf{i}} + y\hat{\mathbf{j}}$$

$$\therefore \quad \mathbf{v} = \frac{d\mathbf{r}}{dt} = \frac{d}{dt}(x\hat{\mathbf{i}} + y\hat{\mathbf{j}})$$

$$= \frac{dx}{dt}\hat{\mathbf{i}} + \frac{dy}{dt}\hat{\mathbf{j}} \qquad \text{...(ii)}$$

Comparing Eqs. (i) and (ii), we get

$$\frac{dx}{dt} = y \qquad \text{...(iii)}$$

$$\text{and} \quad \frac{dy}{dt} = x \qquad \text{...(iv)}$$

Dividing Eq. (iii) and Eq. (iv), we get

$$\frac{\frac{dx}{dt}}{\frac{dy}{dt}} = \frac{y}{x} \Rightarrow x\frac{dx}{dt} = y\frac{dy}{dt}$$

$$\text{or} \quad xdx = ydy$$

Integrating both sides, we get

$$\int xdx = \int ydy$$

$$\text{or} \quad \frac{x^2}{2} + \frac{c_1}{2} = \frac{y^2}{2} + \frac{c_2}{2}$$

where, c_1 and c_2 are the constants of integration.

$$\Rightarrow \quad x^2 + c = y^2$$

[here, c (constant) = $c_1 - c_2$]

$$\text{or} \quad y^2 = x^2 + \text{constant}$$

6 (*b*) According to the question,

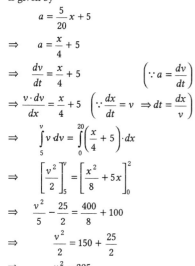

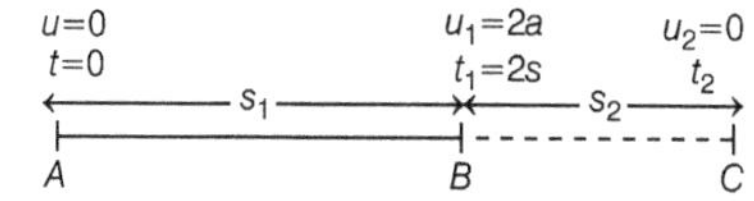

From first equation of the motion,

$$v_1 = u + at_1 \Rightarrow v_1 = 2a$$

For distance BC, from first equation of the motion,

$$v_2 = v_1 - at_2 \Rightarrow 0 = 2a - at_2 \text{ or } t_2 = 2 \text{ s}$$

Hence, total time taken by body to cover the distance $AC, t = 2 + 2 = 4$s

From second equation of motion,

$$s_1 = AB = ut_1 + \frac{1}{2}at_1^2 = 0 + \frac{1}{2}a \times 2^2$$

$$s_1 = 2a$$

$$\because \quad s_1 = s_2 = 2a$$

$$\therefore \quad AC = s_1 + s_2 = 4a$$

Now, body returns from point C to point A.

$$\text{So,} \quad u_1 = 0, AC = ua$$

From second equation of the motion,

$$s = AC = u_1t + \frac{1}{2}at^2 \text{ or } 4a = 0 + \frac{1}{2}at^2$$

$$\Rightarrow t^2 = 8 \Rightarrow t = 2\sqrt{2} \text{ s}$$

Therefore, the total time taken by body,

$$t_0 = t_2 + t = (4 + 2\sqrt{2})\text{s}$$

7 (*d*) Given,

force acts on a body, $\mathbf{F} = (2.6\hat{\mathbf{i}} + 1.6\hat{\mathbf{j}})$ N

mass of a body, $m = 2$ kg

At $t = 0$, velocity of the body,

$$v = (3.6\,\hat{\mathbf{i}} - 4.8\,\hat{\mathbf{j}}) \text{ m/s}$$

We know that,

Force = Mass × Acceleration

$$\therefore \quad \mathbf{F} = m \times \mathbf{a} \text{ or } \mathbf{a} = \frac{\mathbf{F}}{m}$$

$$\text{or} \quad \mathbf{a} = \frac{(2.6\,\hat{\mathbf{i}} + 1.6\,\hat{\mathbf{j}})}{2}$$

$$\text{or} \quad \mathbf{a} = (1.3\hat{\mathbf{i}} + 0.8\,\hat{\mathbf{j}}) \text{ m/s}^2$$

Now, the velocity vector, $\frac{d\mathbf{v}}{dt} = \mathbf{a}$

$$\int d\mathbf{v} = \int \mathbf{a}\, dt$$

$$\text{or} \quad \mathbf{v} = (1.3\hat{\mathbf{i}} + 0.8\hat{\mathbf{j}})t + c$$

$$\text{at } t = 0, c = \mathbf{v} = (3.6\hat{\mathbf{i}} - 4.8\hat{\mathbf{j}}) \text{ ms}^{-1}$$

$$\therefore \mathbf{v} = (3.6 + 1.3\,t)\hat{\mathbf{i}} + (-4.8 + 0.8t)\,\hat{\mathbf{j}} \text{ m/s}$$

$\therefore\ v_y = 0$ (because body will just have a velocity along X-axis.)

$$-4.8 + 0.8t = 0 \Rightarrow 0.8t = 4.8 \Rightarrow t = \frac{4.8}{0.8} = 6\text{s}$$

This is the time at which the body will just have velocity along X-axis

$$t = 6\text{ s}$$

8 (*a*) According to the question,

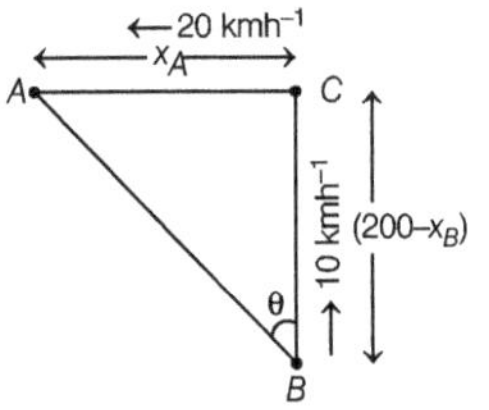

Let ship A travel x_A distance and ship B travel x_B distance, in time t.

So, $u_A = \frac{x_A}{t} \Rightarrow t = \frac{x_A}{20}$

and $u_B = \frac{x_B}{t} \Rightarrow t = \frac{x_B}{10} \Rightarrow \frac{x_A}{20} = \frac{x_B}{10}$

$\Rightarrow x_A = 2\,x_B$

So, $AB = \sqrt{x_A^2 + (200 - x_B)^2}$

$= \sqrt{4x_B^2 + 40000 + x_B^2 - 400x_B}$

$= \sqrt{5x_B^2 - 400x_B + 40000}$

Differentiate distance AB w.r.t. x_B for finding value of x_B,

$$\frac{d(AB)}{dx_B} = \frac{1}{2\sqrt{5x_B^2 - 400x_B + 40000}}$$

$(10x_B - 400) = 0$

or $\quad x_B = 40\text{ m}$

Again differentiating,

So, $\left(\frac{d^2(AB)}{dx_B^2}\right)_{x_B = 40\text{ m}} > 0$

[$\because$ Distance always greater than zero]

Hence x_B at point, $x_B = 40$ m distance, AB will be shortest.

So, $AB = \sqrt{5 \times 40^2 - 400 \times 40 + 40000}$

$AB = \sqrt{32000}$ or $AB = 80\sqrt{5}$ km

The time after which the distance AB is shortest,

$$t = \frac{x_A}{20} = \frac{x_B}{10} = \frac{40}{10} = 4\text{ h}.$$

9 (*b*) According to the question,

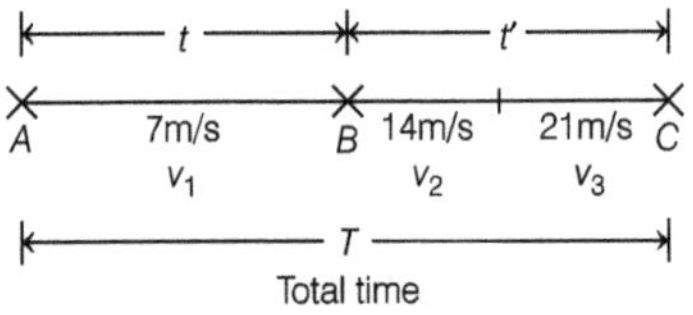

Given, body travels first half of the distance (A to B) with velocity, v_1 =7 m/s

Body travels second half of the distance (B to C) in first half time with velocity, v_2 = 14 m/s and in the second half time with velocity,

$$v_3 = 21\text{ m/s}$$

Let the time taken to travelled from A to B = t s.

Now, distance covered from A to B = d_{AB}

$\therefore$ Distance, d = Velocity × Time

$\therefore \quad d_{AB} = 7t \quad$ ...(i)

Now, distance covered from B to C = d_{BC}

$\therefore$ Average velocity of the body $= \frac{v_2 + v_3}{2}$

or $\quad v' = \frac{14 + 21}{2} = \frac{35}{2}$ m/s

$\therefore \quad d_{BC} = v' \times t' \Rightarrow d_{BC} = \frac{35}{2}t' \quad$ ...(ii)

$\therefore$ Distance travelled by the body from point A to B = distance travelled by the body from point B to C.

$$d_{AB} = d_{BC}$$

From Eqs. (i) and (ii), we get

$7t = \frac{35}{2}t'$ or $t' = \frac{2}{5}t \quad$ ...(iii)

Now, the average velocity from A to C, for finding distance d_{AC},

$v = \frac{v_1 + v_2 + v_3}{3} = \frac{42}{3} = 14$ m/s ...(iv)

$\therefore$ Distance travelled from A to C,

$$d_{AC} = v \times t$$

or $\quad d_{AC} = 14t$ ...(v)[From Eq. (iv)]

Total time taken from A to C,

$$T = t + t'$$

or $\quad T = t + \frac{2t}{5} \quad$ [$\therefore$ From Eq. (iii)]

$T = \frac{7t}{5} \quad$ ...(vi)

Now, average velocity during the whole journey (From A to C),

$v_{avg} = \frac{d_{AC}}{T}$ or $v_{avg} = \frac{14t}{7t} \times 5$

[$\therefore$From Eqs. (v) and (vi)]

or $\quad v_{avg} = 10$ m/s

10 (*a*) Given,Acceleration of car, $a = 5\text{ m/s}^2$, deceleration of car $a = 5\text{ m/s}^2$, total time taken from start of end is, t = 25 s and average velocity of car, v_{avg} = 72 km/hr = 20 m/s $\left(\because 1\frac{\text{km}}{\text{hr}} = \frac{5}{18}\text{ m/s}\right)$

Since, $v_{avg} = \frac{\text{total displacement}}{\text{total time taken}}$

$2t$, total time taken by the car durig acceleration and deceleration

$$v_{avg} = 20 = \frac{d_t + d_{(25-2t)} + d_t}{25}$$

$$= \frac{2d_t + d_{(25-2t)}}{25}$$

Since, $d_t = 0 + \frac{1}{2}at^2 = \frac{1}{2}at^2 = \frac{5}{2}t^2$ and

$d_{(25-2t)} = v_{\text{uni.}}(25 - 2t)$

where, $\quad v_{\text{uni.}} = 5t$

now, $v_{avg} = 20 = \frac{2\left(\frac{5}{2}t^2\right) + 5t\,(25 - 2t)}{25}$

$\therefore \quad 20 \times 25 = 5t^2 + 5t\,(25 - 2t)$

$\Rightarrow \quad 500 = 5t^2 + 125t - 10t^2$

$\Rightarrow \quad t^2 - 25t + 100 = 0$

So, it gives t = 20 and 5 s.

Hence, the time of uniform motion,

$t_{20} = 25 - 2t = 25 - 2 \times 20 = -15$ s

($\because$ Not possible)

or $\quad t_5 = 25 - 10 = 15$ s

11 (*b*) A relative motion between rain and boy is shown in the figure below,

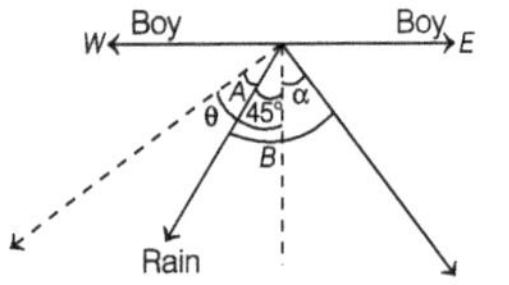

Given, speed of the boy on horizontal road is $v = 4\text{ m/s}$

The boy runs from East to West, the angle is $= \alpha$

The rain is pouring down at an angle = 45°

Now, $\tan A = \frac{4\sin 45°}{8 - 4\cos 45°}$

$$= \frac{4\left(\frac{1}{\sqrt{2}}\right)}{8 - 4\left(\frac{1}{\sqrt{2}}\right)} = \frac{1}{2\sqrt{2} - 1}$$

Similarly,

$$\tan B = \frac{4\sin(90°+45°)}{8 - 4\cos(90°+45°)} = \frac{1}{2\sqrt{2} + 1}$$

As from figure,

$$\tan\theta = \tan(A + 45°) = \frac{\tan A + \tan 45°}{1 - \tan A \tan 45°}$$

$$\Rightarrow \tan\theta = \frac{\frac{1 + 2\sqrt{2} - 1}{2\sqrt{2} - 1}}{\frac{2\sqrt{2} - 1 - 1}{2\sqrt{2} - 1}}$$

$$= \frac{2\sqrt{2}}{2\sqrt{2} - 2}$$

$$= \frac{\sqrt{2}}{\sqrt{2} - 1}$$

Similarly,

$$\tan\alpha = \tan(B - 45°) = \frac{-2\sqrt{2}}{2 + 2\sqrt{2}} = \frac{-\sqrt{2}}{(1 + \sqrt{2})}$$

Hence,

$$\frac{\tan\theta}{\tan\alpha} \simeq \left(\frac{1 + \sqrt{2}}{1 - \sqrt{2}}\right) \times \left(\frac{1 + \sqrt{2}}{1 + \sqrt{2}}\right) = (1 + \sqrt{2})^2$$

So, the correct option is (b).

Projectile Motion

1.

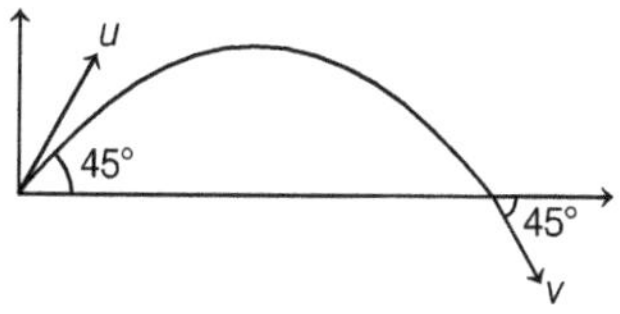

$$|\mathbf{u}| = |\mathbf{v}|$$

$$\mathbf{u} = u\cos 45^\circ \hat{\mathbf{i}} + u\sin 45^\circ \hat{\mathbf{j}} \quad ...(i)$$

$$\mathbf{v} = v\cos 45^\circ \hat{\mathbf{i}} - v\sin 45^\circ \hat{\mathbf{j}} \quad ...(ii)$$

$$|\Delta\mathbf{p}| = |m(\mathbf{v} - \mathbf{u})|$$

$$\Delta\mathbf{p} = 2mu\sin 45^\circ$$

[from Eqs. (i) and (ii)]

$$= 2 \times 5 \times 10^{-3} \times 5\sqrt{2} \times \frac{1}{\sqrt{2}}$$

$$= 50 \times 10^{-3} = 5 \times 10^{-2}$$

$\therefore \quad x = 5$

2. (*a*) In projectile motion, centre of acceleration is directed towards the centre of earth and at peak point, the angle between velocity and acceleration is 90° because at peak point the vertical component of velocity is zero. Thus, it is minimum and acute at only one point (i.e. at highest point).

3. (*b*) Given,

At $t = 0$, initial velocity of particle is $\mathbf{x} = 3\hat{\mathbf{i}}\text{ms}^{-1}$

Acceleration of particle,

$$\mathbf{a} = (6\hat{\mathbf{i}} + 4\hat{\mathbf{j}})\text{ms}^{-2}$$

By $\quad s = ut + \frac{1}{2}at^2$, we have

$$\mathbf{r} = (3\hat{\mathbf{i}})t + \frac{1}{2}(6\hat{\mathbf{i}} + 4\hat{\mathbf{j}})t^2$$

$$\Rightarrow \quad \mathbf{r} = (3t + 3t^2)\hat{\mathbf{i}} + 2t^2\hat{\mathbf{j}}$$

So, at time t,

x-coordinate of particle, $x = 3t + 3t^2$

and y-coordinate of particle, $y = 2t^2$

When $\quad y = 32$ m

$\Rightarrow \quad 2t^2 = 32$

$\Rightarrow \quad t^2 = 16$

or $\quad t = 4$s

Value of x-coordinate at $t = 4$s,

$$x = (3t + 3t^2)_{t=4} = 12 + 48 = 60 \text{ m}$$

4. (*b*)

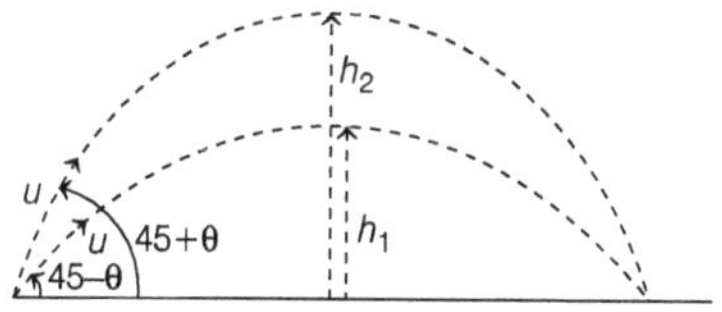

As maximum range occurs at $\theta = 45^\circ$ for a given initial projection speed, we take angles of projection of two particles as

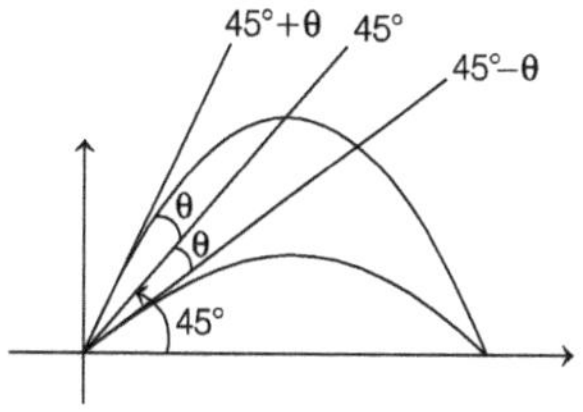

$$\theta_1 = 45^\circ + \theta, \theta_2 = 45^\circ - \theta$$

where, θ is angle of projectiles with 45° line.

So, range of projectiles will be

$$R = R_1 = R_2 = \frac{u^2 \sin 2(\theta_1)}{g}$$

$$\Rightarrow \quad R = \frac{u^2 \sin 2(45^\circ + \theta)}{g}$$

$$\Rightarrow \quad R = \frac{u^2 \sin(90^\circ + 2\theta)}{g}$$

$$\Rightarrow R = \frac{u^2 \cos 2\theta}{g} \quad \Rightarrow R^2 = \frac{u^4 \cos^2 2\theta}{g^2} \quad ...(i)$$

Maximum heights achieved in two cases are

$$h_1 = \frac{u^2 \sin^2 (45^\circ + \theta)}{2g}$$

and $\quad h_2 = \dfrac{u^2 \sin^2 (45^\circ - \theta)}{2g}$

So, $\quad h_1 h_2 = \dfrac{u^4 \sin^2 (45^\circ + \theta) \sin^2 (45^\circ - \theta)}{4g^2}$

Using $2\sin A \cdot \sin B$

$$= \cos(A - B) - \cos(A + B),$$

we have

$$\sin(45^\circ + \theta)\sin(45^\circ - \theta)$$

$$= \frac{1}{2}(\cos 2\theta - \cos 90^\circ)$$

$$\Rightarrow \quad \sin(45^\circ + \theta)\sin(45^\circ - \theta) = \frac{\cos 2\theta}{2}$$

$[\because \cos 90^\circ = 0]$

So, we have

$$h_1 h_2 = \frac{u^4\left(\dfrac{\cos 2\theta}{2}\right)^2}{4g^2}$$

$$\Rightarrow \quad h_1 h_2 = \frac{u^4 \cos^2 2\theta}{16g^2} \quad ...(ii)$$

From Eqs. (i) and (ii), we get

$$\Rightarrow \quad h_1 h_2 = \frac{R^2}{16} \Rightarrow R^2 = 16\, h_1 h_2$$

5. (*a*) Given that the position coordinates of a particle $x = a\cos\omega t$, $y = a\sin\omega t$ and $z = a\omega t$

So, the position vector of the particle is

$$\hat{\mathbf{r}} = x\hat{\mathbf{i}} + y\hat{\mathbf{j}} + z\hat{\mathbf{k}}$$

$$\Rightarrow \quad \hat{\mathbf{r}} = a\cos\omega t\,\hat{\mathbf{i}} + a\sin\omega t\,\hat{\mathbf{j}} + a\,\omega t\,\hat{\mathbf{k}}$$

$$\hat{\mathbf{r}} = a[\cos\omega t\,\hat{\mathbf{i}} + \sin\omega t\,\hat{\mathbf{j}} + \omega t\,\hat{\mathbf{k}}]$$

Therefore, the velocity of the particle is

$$\because \hat{\mathbf{v}} = \frac{d\mathbf{r}}{dt} = \frac{d[a]\,[\cos\omega t\,\hat{\mathbf{i}} + \sin\omega t\hat{\mathbf{j}} + \omega t\,\hat{\mathbf{k}}]}{dt}$$

$$\Rightarrow \hat{\mathbf{v}} = -\,a\omega\sin\omega t\,\hat{\mathbf{i}} + a\omega\cos\omega t\,\hat{\mathbf{j}} + a\omega\hat{\mathbf{k}}$$

The magnitude of velocity is

$$|\mathbf{v}| = \sqrt{v_x^2 + v_y^2 + v_z^2}$$

or

$$|\mathbf{v}| = \sqrt{(-a\omega\sin\omega t)^2 + (a\omega\cos\omega t)^2 + (a\omega)^2}$$

$$= \omega a\sqrt{(-\sin\omega t)^2 + (\cos\omega t)^2 + (1)^2}$$

$$= \sqrt{2}\,\omega a$$

6 (*b*) Components of velocity at an instant of time t of a body projected at an angle θ is

$$v_x = u\cos\theta + g_x t \text{ and } v_y = u\sin\theta + g_y t$$

Here, components of velocity at $t = 1$ s, is

$$v_x = u\cos 60^\circ + 0 \quad [\text{as } g_x = 0]$$

$$= 10 \times \frac{1}{2} = 5 \text{ m/s}$$

and $v_y = u\sin 60^\circ + (-10) \times (1)$

$$= 10 \times \frac{\sqrt{3}}{2} + (-10) \times (1) = 5\sqrt{3} - 10$$

$$\Rightarrow \quad |v_y| = |10 - 5\sqrt{3}| \text{ m/s}$$

Now, angle made by the velocity vector at time of $t = 1$ s

$$|\tan\alpha| = \left|\frac{v_y}{v_x}\right| = \frac{|10 - 5\sqrt{3}|}{5}$$

$$\Rightarrow \quad \tan\alpha = |2 - \sqrt{3}| \text{ or } \alpha = 15^\circ$$

$\therefore$ Radius of curvature of the trajectory of $R = v^2 / g\cos\alpha$

$$= \frac{(5)^2 + (10 - 5\sqrt{3})^2}{10 \times 0.97}$$

$$[\because v^2 = v_x^2 + v_y^2 \text{ and } \cos 15^\circ = 0.97]$$

$$\Rightarrow \quad R = 2.77 \text{ m} \approx 2.8 \text{ m}$$

7 (*a*) Velocity of projectile at time t is given by

$$\mathbf{v} = \mathbf{u} - \mathbf{g}\,t\,\hat{\mathbf{j}}$$

$$\Rightarrow \quad \mathbf{v} = \hat{\mathbf{i}} + \sqrt{3}\,\hat{\mathbf{j}} - 10t\,\hat{\mathbf{j}}$$

$$\therefore \quad \mathbf{r} = \int v\,dt = \int (\hat{\mathbf{i}} + \sqrt{3}\,\hat{\mathbf{j}} - 10t\,\hat{\mathbf{j}})\,dt$$

$$= t\,\hat{\mathbf{i}} + \sqrt{3}\,t\,\hat{\mathbf{j}} - 10\frac{t^2}{2}\hat{\mathbf{j}}$$

$$\mathbf{r} = t\,\hat{\mathbf{i}} + (\sqrt{3}t - 5t^2)\,\hat{\mathbf{j}}$$

or $\quad x\hat{\mathbf{i}} + y\hat{\mathbf{j}} = t\,\hat{\mathbf{i}} + (\sqrt{3}t - 5t^2)\,\hat{\mathbf{j}}$

On comparing, $x = t \quad ...(i)$

and $\quad y = \sqrt{3}\,t - 5t^2 \quad ...(ii)$

From Eqs (i) and (ii), we get

$$y = \sqrt{3}\,x - 5x^2$$

8 (*c*) Given, $\theta = 60^\circ$ and $u = 140 \text{ ms}^{-1}$

This velocity have divided into two components,

$u_x = u\cos 60^\circ = 140\cos 60^\circ$

$u_y = u\sin 60^\circ = 140\sin 60^\circ$

Now, $\tan 45^\circ = \dfrac{v_y}{v_x} \Rightarrow v_x = v_y$

Since, the horizontal component of velocity remains constant, i.e.,

$v_x = u_x = 140\cos 60^\circ$ and $v_y = u_y - gt$

$$140 \cos 60° = 140 \sin 60° - 10t \quad [\because v_y = v_x]$$

$$140 \times \frac{1}{2} = 140 \times \frac{\sqrt{3}}{2} - 10\,t$$

$$70 = 70\sqrt{3} - 10t$$

$$t = \frac{70\,(\sqrt{3}-1)}{10} = 5.124 \text{ s}$$

9 (*d*) According to question, a small block is slide down from top E of inclined plane as shown in figure,

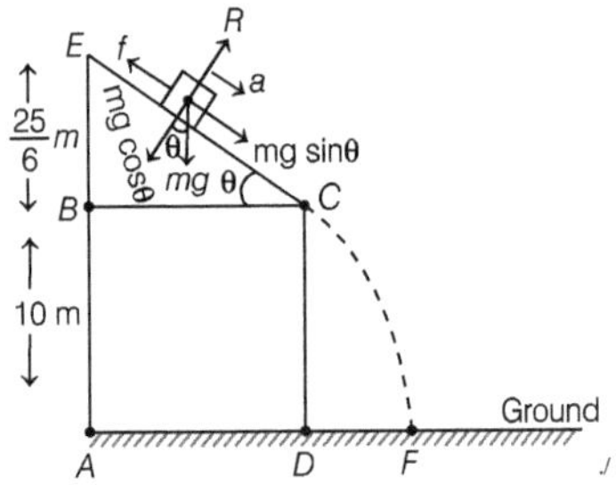

Force equation of a block,

$\Rightarrow \quad mg\sin\theta - f = ma$ ...(i)

$\because$ friction force applied on block, $f = \mu_k R$

or $\quad f = \mu_k R(mg\cos\theta)$ (From figure)

where, μ_k = coefficient of kinetic friction

From Eq. (i), we get

$\Rightarrow \quad mg\sin\theta - \mu_k\, mg\cos\theta = ma$

$\Rightarrow a = 10\sin\theta - 1/8 \times 10\cos\theta$ ... (ii)

$(\because \mu_k = \frac{1}{8}$, given)

$\because$ Given, angle of the inclined plane, $\theta = \sin^{-1}(0.6)$ or $\sin\theta = 0.6$

$\because \quad \cos\theta = \sqrt{1-\sin^2\theta} = \sqrt{1-(0.6)^2}$

$\therefore \quad \cos\theta = 0.8$

Form Eq. (ii),

$a = 10\,(0.6) - \frac{1}{8} \times 10\,(0.8)$ or $a = 5 \text{ ms}^{-2}$

when, block reached to point C, then from third equation of the motion,

$$v^2 = u^2 + 2as$$

where, v = final velocity of the block at point C, u = initial velocity of block at point E

or $\quad v = \sqrt{2as}$ $(\because u = 0)$

From Δ EBC, $EC = \frac{BE}{\sin\theta}$

$$\Rightarrow \quad s = \frac{25/6}{0.6} = \frac{25}{6 \times 0.6} \text{ or } s = \frac{125}{18} \text{ m}$$

Hence, $\quad v = \sqrt{2 \times 5 \times \frac{125}{18}} = \frac{25}{3} \text{ ms}^{-1}$

From second equation of motion of the block at point C.

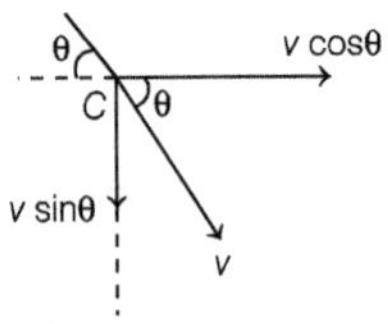

In y-direction,

$$\Rightarrow h = ut + \frac{1}{2}gt^2 \Rightarrow 10 = u\sin\theta t + \frac{1}{2}gt^2$$

$$\Rightarrow 10 = \frac{25}{3} \times 0.6t + \frac{1}{2} \times 10 \times t^2$$

$\Rightarrow \quad 5t + 5t^2 = 10$

$\Rightarrow \quad t^2 + t - 2 = 0$ or $t = 1$ sec

Now, again from second Eq. of motion in x-direction,

$\Rightarrow \quad DF = v\cos\theta . t + 0$ or

$$DF = \frac{25}{3} \times 0.8 = \frac{20}{3} \text{ m}$$

10 (*a*) Given, $g = 10 \text{ m/s}^2$

After 4s, angle between stone and horizontal plane, $\theta = 30°$

After $t = 4$ s, equation of the vertical projectile motion, when $\theta = 30°$

$$\therefore \quad \tan\theta = \frac{v\sin\theta - g(t)}{v\cos\theta} \quad \text{...(i)}$$

$$\tan 30° = \frac{v\sin\theta - g(4)}{v\cos\theta} \quad \text{...(ii)}$$

Total time to reach the stone at horizontal surface, $t = 2 + 4 = 6$ s

After $t = 6$ s, equation of horizontal projectile motion, $\theta = 0°$,

$$\tan 0° = \frac{v\sin\theta - g(6)}{v \times \cos 0} \quad \left[\begin{array}{l}\because \cos 0° = 1 \\ \tan 0° = 0\end{array}\right]$$

or $\quad v\sin\theta - g(6) = 0$

$v\sin\theta = 60$ ...(iii)

$[\because$Given, $g = 10]$

From Eq. (ii), At $t = 4$ s, when particles travelling in horizontal direction,

$\therefore \quad v\cos\theta = 20\sqrt{3}$ ...(iv)

Now, magnitude of initial velocity,

$$v = \sqrt{(v\sin\theta)^2 + (v\cos\theta)^2}$$

[From Eqs. (iii) and (iv)]

or $\quad v = \sqrt{(60)^2 + (20\sqrt{3})^2}$

or $\quad v = 40\sqrt{3}$ m/s

So, the magnitude of initial velocity of stone is $v = 40\sqrt{3}$ m/s.

11 (*b*) The figure given below, shows the vertical motion of a ball,

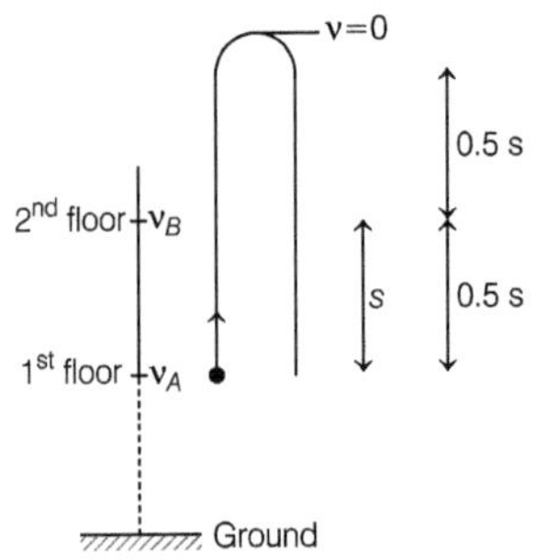

Here, ball crosses the 1st floor in 2s, so it goes up in 1 s and goes down in next one second.

So, $\quad v - v_A = -gt$

$\Rightarrow \quad 0 - v_A = -10(1)$

$\Rightarrow \quad v_A = 10$ m/s

$\because$Boy, B observed the ball crossing him while up and down motion in 1s. So, time taken by the ball to reached the floor B from floor A is 0.5 s.

So, $\quad s = v_A t - \frac{1}{2}gt^2$

$$\Rightarrow \quad = 10 \times 0.5 - \frac{1}{2} \times 10 \times (0.5)^2$$

$= 5 - 1.25$

$= 3.75$ m

12 (*c*) According to the question,

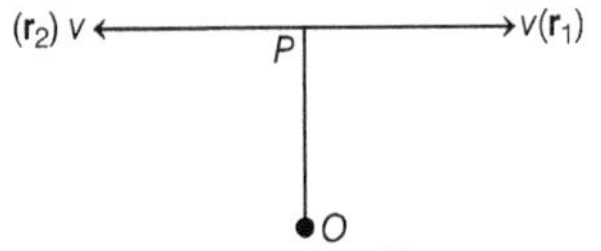

Representation of position vectors of two particles (drawn from the point P)

In two dimension, the position vectors $\mathbf{r}_1$ and $\mathbf{r}_2$ represented as

$$\mathbf{r}_1 = vt\,\hat{\mathbf{i}} - \frac{1}{2}gt^2\hat{\mathbf{j}} \quad \text{...(i)}$$

$$\mathbf{r}_2 = vt(-\hat{\mathbf{i}}) - \frac{1}{2}gt^2(\hat{\mathbf{j}}) \quad \text{...(ii)}$$

$\because$ We know that, when the two vectors are mutually perpendicular, i.e.

$\theta = 90°$

So, $\mathbf{r}_1 \cdot \mathbf{r}_2 = r_1 r_2 \cos 90°$

$\Rightarrow \mathbf{r}_1 \cdot \mathbf{r}_2 = 0$

Substituting the values $\mathbf{r}_1$ and $\mathbf{r}_2$ in the above relation, we get

$$[(vt)\hat{\mathbf{i}} - \frac{1}{2}gt^2\hat{\mathbf{j}}] \cdot [vt(-\hat{\mathbf{i}}) - \frac{1}{2}gt^2(\hat{\mathbf{j}})] = 0$$

$$-v^2t^2 + \frac{1}{4}4g^2t^4 = 0$$

(where, $\hat{\mathbf{i}} \cdot \hat{\mathbf{i}} = \hat{\mathbf{j}} \cdot \hat{\mathbf{j}} = \hat{\mathbf{k}} \cdot \hat{\mathbf{k}} = 1$)

$$v^2t^2 = \frac{1}{4}g^2t^4$$

$$\Rightarrow \quad v^2 = \frac{1}{4}g^2t^2$$

$\therefore$ Magnitude of velocity of the particles, $v = \frac{1}{2}gt$

We know that, separation distance between particles at a time t

$$\Delta x = 2vt$$

$$\Delta x = 2 \times v \times \frac{2v}{g}$$

$$\Rightarrow \quad \Delta x = \frac{4v^2}{g}$$

Laws of Motion

1.

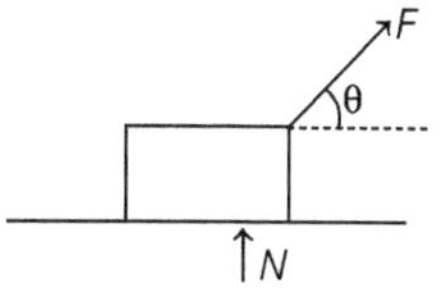

$$F \cos\theta = \mu N$$

$$F \sin\theta + N = mg$$

$$\Rightarrow \quad F = \frac{\mu mg}{\cos\theta + \mu \sin\theta}$$

$$F_{min} = \frac{\mu mg}{\sqrt{1+\mu^2}} = \frac{\frac{1}{\sqrt{3}} \times 10}{\frac{2}{\sqrt{3}}} = 5\text{ N}$$

2. (*a*) The free body diagram of the given system is given as

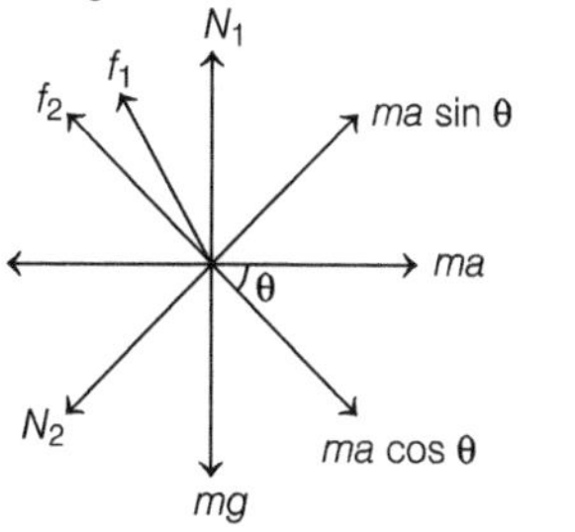

Given, $m = 1$ kg,

$$\mu = \frac{2}{5}, a = 25\text{ ms}^{-2},$$

$$\cos\theta = \frac{4}{5}, \quad \sin\theta = \frac{3}{5}$$

Here, f_1 and f_2 are the two frictional forces corresponding to the two points of contact.

If a' be the acceleration of the block with respect to disc, then from above figure,

$$ma \cos\theta - f_1 - f_2 = ma'$$

$$\Rightarrow \quad ma \cos\theta - \mu N_1 - \mu N_2 = ma'$$

$$\Rightarrow \quad ma \cos\theta - \mu mg - \mu ma \sin\theta = ma'$$

$$\Rightarrow \quad a' = a \cos\theta - \mu g - \mu a \sin\theta$$

$$= 25 \times \frac{4}{5} - \frac{2}{5} \times 10 - \frac{2}{5} \times 25 \times \frac{3}{5}$$

$$= 20 - 4 - 6 = 10\text{ ms}^{-2}$$

3. (*c*) Given situation is as shown below.

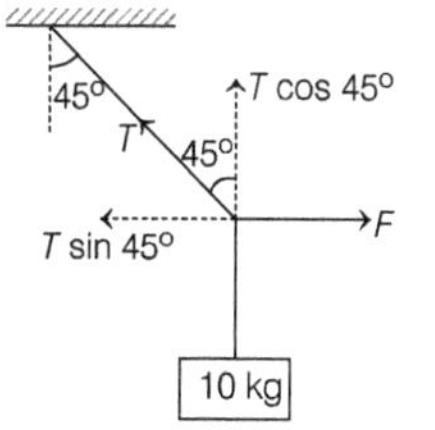

We resolve tension T in string into vertical and horizontal components.

For equilibrium,

$$F = T \sin 45° \qquad \text{...... (i)}$$

and $\quad Mg = T \cos 45° \qquad$ .... (ii)

On dividing Eq. (i) by Eq. (ii), we get

$$\frac{F}{Mg} = \tan 45°$$

or $\quad F = Mg = 10 \times 10 = 100$ N

4. (*b*) Acceleration a of system of blocks A and B is

$$a = \frac{\text{Net force}}{\text{Total mass}} = \frac{F - f_1}{m_A + m_B}$$

where, f_1 = friction between B and the surface $= \mu(m_A + m_B)g$

So, $\quad a = \dfrac{F - \mu(m_A + m_B)g}{(m_A + m_B)} \qquad$ ...(i)

Here, $\mu = 0.2$, $m_A = 1$ kg, $m_B = 3$ kg, $g = 10\text{ ms}^{-2}$

Substituting the above values in Eq. (i), we have

$$a = \frac{F - 0.2(1+3) \times 10}{1+3}$$

$$a = \frac{F - 8}{4} \qquad \text{...(ii)}$$

Due to acceleration of block B, a pseudo force F' acts on A.

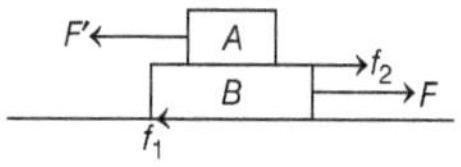

This force F' is given by

$$F' = m_A a$$

where, a is acceleration of A and B caused by net force acting on B.

For A to slide over B; pseudo force on A, i.e. F' must be greater than friction between A and B.

$$\Rightarrow \quad m_A a \geq f_2$$

We consider limiting case,

$$m_A a = f_2 \Rightarrow m_A a = \mu(m_A)g$$

$$\Rightarrow \quad a = \mu g = 0.2 \times 10 = 2\text{ ms}^{-2} \qquad \text{...(iii)}$$

Putting the value of a from Eq. (iii) into Eq. (ii), we get

$$\frac{F - 8}{4} = 2$$

$$\Rightarrow \quad F = 16\text{ N}$$

5. (*c*) **Case I** Block is pushed over surface free body diagram of block is

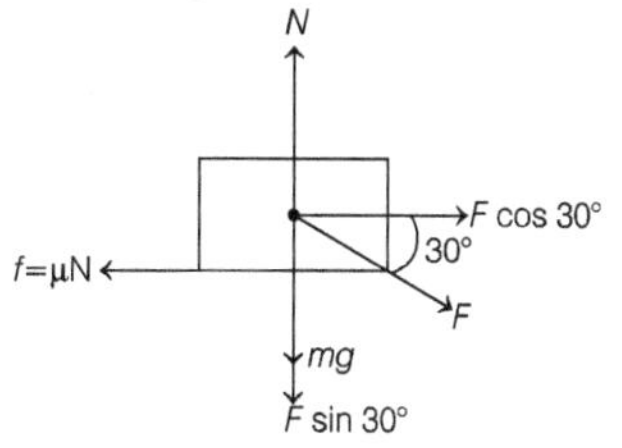

In this case, normal reaction,

$$N = mg + F \sin 30° = 5 \times 10 + 20 \times \frac{1}{2}$$

$= 60$ N $\qquad$ [Given, $m = 5$ kg, $F = 20$ N]

Force of friction, $f = \mu N$

$= 0.2 \times 60 \qquad [\because \mu = 0.2]$

$= 12$ N

So, net force causing acceleration (a_1) is

$$F_{net} = ma_1 = F \cos 30° - f$$

$$\Rightarrow \quad ma_1 = 20 \times \frac{\sqrt{3}}{2} - 12$$

$$\therefore \quad a_1 = \frac{10\sqrt{3} - 12}{5} \approx 1\text{ ms}^{-2}$$

Case II Block is pulled over the surface free body diagram of block is,

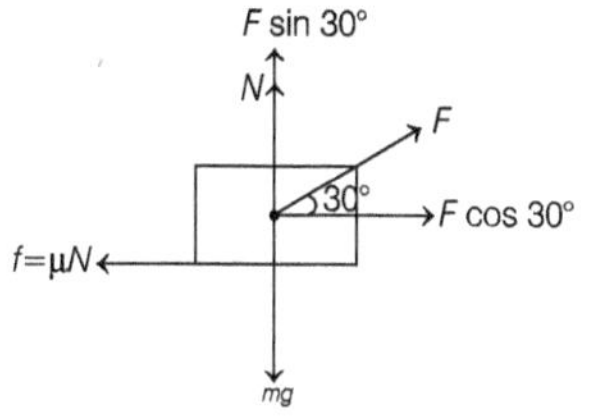

Net force causing acceleration is

$$F_{net} = F \cos 30° - f = F \cos 30° - \mu N$$

$$\Rightarrow F_{net} = F \cos 30° - \mu(mg - F \sin 30°)$$

If acceleration is now a_2, then

$$a_2 = \frac{F_{net}}{m}$$

$$= \frac{F \cos 30° - \mu(mg - F \sin 30°)}{m}$$

$$= \frac{20 \times \frac{\sqrt{3}}{2} - 0.2\left(5 \times 10 - 20 \times \frac{1}{2}\right)}{5}$$

$$= \frac{10\sqrt{3} - 8}{5}$$

$$\Rightarrow a_2 \approx 1.8\text{ ms}^{-2}$$

So, difference $= a_2 - a_1 = 1.8 - 1$
$= 0.8\text{ ms}^{-2}$

6. (*b*) Block does not move upto a maximum applied force of 2N down the inclined plane.

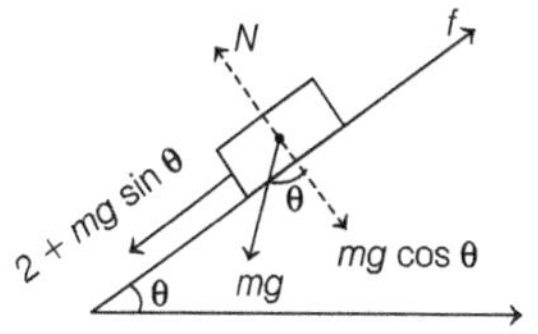

So, equating forces, we have

$$2 + mg \sin\theta = f$$

or $\quad 2 + mg \sin\theta = \mu\, mg \cos\theta \qquad$ ...(i)

Similarly, block also does not move upto a maximum applied force of 10 N up the plane.

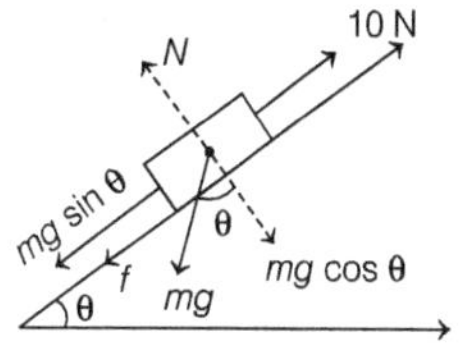

Now, equating forces, we have

$$mg\sin\theta + f = 10\text{ N}$$

or $mg\sin\theta + \mu mg\cos\theta = 10$ ...(ii)

Now, solving Eqs. (i) and (ii), we get

$$mg\sin\theta = 4 \quad \text{...(iii)}$$

and $\mu\, mg\cos\theta = 6$ ...(iv)

Dividing, Eq. (iv) by (iii), we get

$$\mu\cot\theta = \frac{3}{2}$$

$$\Rightarrow \quad \mu = \frac{3\tan\theta}{2} = \frac{3\tan 30^\circ}{2}$$

$$\Rightarrow \quad \mu = \frac{\sqrt{3}}{2}$$

7 (*a*) Given, $F = at$

when, $t = 0$, then linear momentum $= p$

when, $t = T$, then linear momentum $= 2p$

According to Newton's law of motion,

applied force, $F = \frac{dp}{dt}$

or, $dp = Fdt = at\,dt$

or, $\int_p^{2p} dp = a\int_0^T t\,dt$

$$(2p - p) = a\cdot\frac{T^2}{2}$$

$$\Rightarrow \quad \frac{2p}{a} = T^2$$

$$\Rightarrow \quad T = \sqrt{\frac{2p}{a}}$$

8 (*d*) Block does not move upto a maximum applied force of 3 N down the inclined plane.

Equation of motion is $3 + mg\sin 30^\circ = F$

$$3 + mg\sin 30^\circ = \mu\, mg\cos 30^\circ \quad \text{... (i)}$$

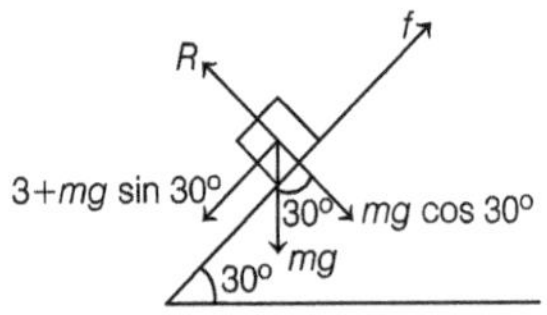

Similarly, block also does not move upto a maximum applied force of 12 N up the plane.

Now, equation of motion is

$$mg\sin 30^\circ + F = 12$$

or, $mg\sin 30^\circ + \mu\, mg\cos 30^\circ = 12$ ... (ii)

Solving Eqs. (i) and (ii), we have

$$mg\sin 30^\circ = \frac{9}{2} \text{ and } \mu\, mg\cos 30^\circ = \frac{15}{2}$$

Hence, on dividing Eq. (i) by Eq. (ii)

$$\therefore \quad \frac{1}{\mu}\tan 30^\circ = \frac{9/2}{15/2} = \frac{3}{2}$$

$$\Rightarrow \quad \mu = \frac{5}{3\sqrt{3}}$$

9 (*a*) Applied force on the block on a horizontal surface, as shown in the figure , we have

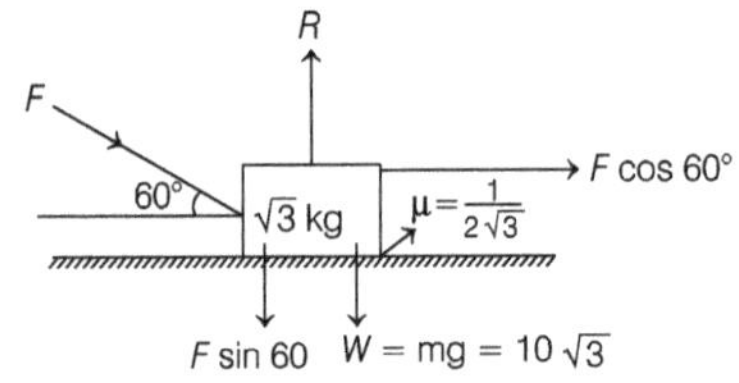

As we know that, force of friction, $f = \mu R$

From the diagram,

$$f = \mu(W + F\sin 60^\circ)$$

$$F\cos 60^\circ = \frac{1}{2\sqrt{3}}(10\sqrt{3} + F\sin 60^\circ)$$

$$\Rightarrow \quad F\times\frac{1}{2} = \frac{1}{2\sqrt{3}}\left(10\sqrt{3} + F\times\frac{\sqrt{3}}{2}\right)$$

$$\left[\because \cos 60^\circ = \frac{1}{2},\ \sin 60^\circ = \frac{\sqrt{3}}{2}\right]$$

$$\Rightarrow \quad \frac{F}{2} = \frac{1}{2\sqrt{3}}\times\sqrt{3}\left(10 + \frac{F}{2}\right)$$

$$\Rightarrow \quad \frac{F}{2} = \frac{1}{2}\frac{(20+F)}{2} \Rightarrow F = \frac{20+F}{2}$$

$$\Rightarrow 2F = 20 + F \Rightarrow 2F - F = 20$$

$$\therefore \quad F = 20\text{ N}$$

10 (*c*) According to the question, angle of inclination $\theta = 60^\circ$.

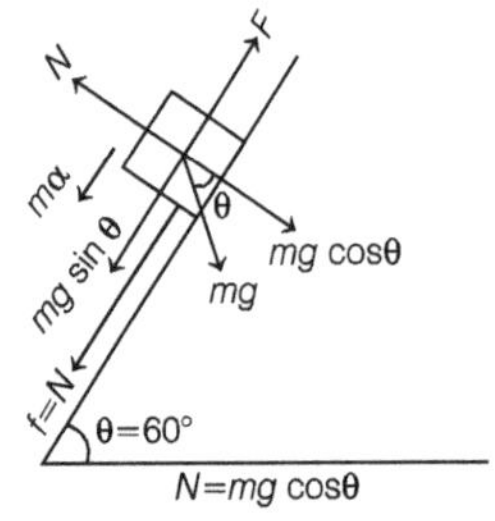

Now, force of friction, $f = \mu N = \mu\, mg\cos\theta$ and net retarding force, $(F_1) = mg\sin\theta + f$

$\therefore$ Net accelerating force down the inclined plane is

$$F = mg\sin\theta - \mu mg\cos\theta \quad \text{...(i)}$$

$\therefore$ External force needed (up the inclined plane) to maintain sliding motion is (net retarding force)

$$F_1 = mg\sin\theta + \mu mg\cos\theta \quad \text{...(ii)}$$

It is given as, $F_1 = 2F$ ...(iii)

From Eqs. (i),(ii) and (iii), we get

$$mg\sin\theta + \mu mg\cos\theta = 2(mg\sin\theta - \mu mg\cos\theta)$$

or $3\mu\, mg\cos\theta = mg\sin\theta$

or $\frac{\sin\theta}{\cos\theta} = 3\mu$

or $\tan\theta = 3\mu$ $(\because \theta = 60^\circ)$

or $\tan 60^\circ = 3\mu$

or $\sqrt{3} = 3\mu$

or $\mu = \frac{1}{\sqrt{3}}$

When inclination of plane is 60° then the coefficient of friction, $\mu = \frac{1}{\sqrt{3}}$.

11 (*d*) The given situation is shown in the figure below,

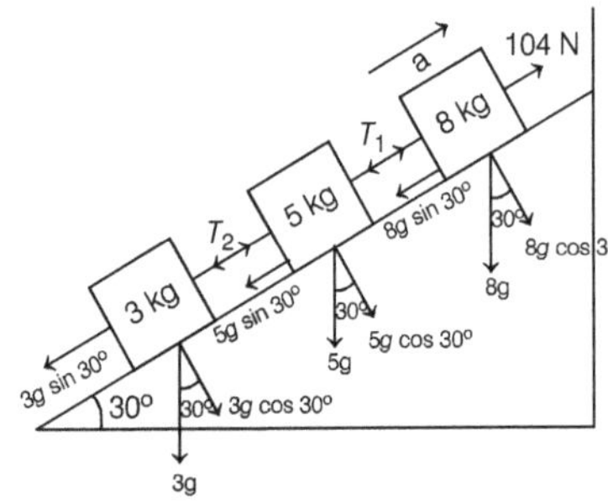

Here, equation of motion for 8 kg block is

$$104 - T_1 - 8g\sin 30^\circ = 8a$$

$$\Rightarrow 104 - T_1 - 8\times 10\times\frac{1}{2} = 8a$$

$$[\because g = 10\text{ m/s}^2]$$

$$64 - T_1 = 8a \quad \text{...(i)}$$

Equation of motion for 5 kg block is

$$T_1 - T_2 - 5g\sin 30^\circ = 5a$$

$$T_1 - T_2 - 25 = 5a \quad \text{...(ii)}$$

Equation of motion for 3 kg block is

$$T_2 - 3g\sin 30^\circ = 3a$$

$$T_2 - 15 = 3a \quad \text{...(iii)}$$

Adding Eqs. (i), (ii) and (iii), we get

$$24 = 16a \Rightarrow a = 1.5\text{ m/s}^2$$

Hence, acceleration a of block is 1.5 m/s^2.

12 (*d*) According to the question, we can draw the following diagram :

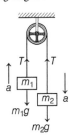

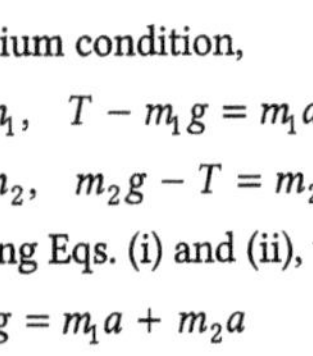

In equilibrium condition,

For mass m_1, $T - m_1 g = m_1 a$ ...(i)

For mass m_2, $m_2 g - T = m_2 a$ ...(ii)

After adding Eqs. (i) and (ii), we get

$$m_2 g - m_1 g = m_1 a + m_2 a$$

$$(m_2 - m_1)g = (m_1 + m_2)a$$

$$a = \frac{(m_2 - m_1)}{(m_1 + m_2)}g$$

Work, Energy and Power

1. Let's say the compression in the spring be x.

So, by work-energy theorem, we have

$\Rightarrow \quad \frac{1}{2}mv^2 = \frac{1}{2}kx^2$

$\Rightarrow \quad x = \sqrt{\frac{m}{k}} \cdot v \Rightarrow x = \sqrt{\frac{4}{100}} \times 10$

$\Rightarrow \quad x = 2\text{ m}$

Final length of spring $= 8 - 2 = 6$ m

2. *(d)*

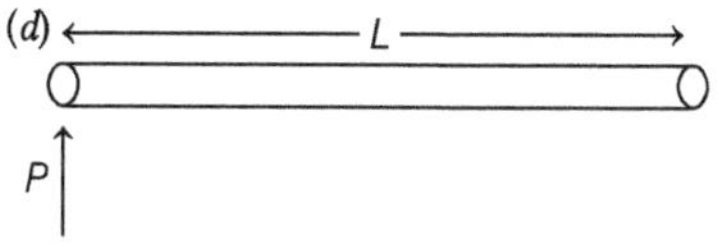

The given impulse acts as both linear and an angular impulse.

Linear impulse $= P = mv_{CM}$ (where, v_{CM} = velocity of centre at mass of rod)

Angular impulse $= P \times \frac{L}{2} = I_{CM}\omega$

where, $I_{CM} = \frac{mL^2}{12}$ and ω is angular velocity at rod about centre of mass.

i.e. $\quad \omega = \frac{PL}{2I_{CM}}$

Kinetic energy

$= \left[\frac{1}{2}mv_{CM}^2 + \frac{1}{2}I_{CM}\omega^2\right]$

$= \left[\frac{1}{2} \times m \times \left(\frac{P}{m}\right)^2 + \frac{1}{2} \times \frac{mL^2}{12} \times \frac{P^2L^2 \times 144}{4 \times m^2 \times L^4}\right]$

$= \frac{P^2}{2m}[1 + 3] = \frac{2P^2}{m}$

3. *(c)* Work done by a variable force on the particle,

$$W = \int F \cdot d\mathbf{r} = \int F \cdot (dx\hat{\mathbf{i}} + dy\hat{\mathbf{j}})$$

$\therefore$ In two dimension, $d\mathbf{r} = dx\hat{\mathbf{i}} + dy\hat{\mathbf{j}}$

and it is given $\mathbf{F} = -x\hat{\mathbf{i}} + y\hat{\mathbf{j}}$.

$\therefore \quad W = \int(-x\hat{\mathbf{i}} + y\hat{\mathbf{j}}) \cdot (dx\hat{\mathbf{i}} + dy\hat{\mathbf{j}})$

$= \int -x\,dx + y\,dy = \int -x\,dx + \int y\,dy$

As particle is displaced from $A(1, 0)$ to $B(0, 1)$, so x varies from 1 to 0 and y varies from 0 to 1.

So, with limits, work will be

$$W = \int_1^0 -x\,dx + \int_0^1 y\,dy = \left[\frac{-x^2}{2}\right]_1^0 + \left[\frac{y^2}{2}\right]_0^1$$

$$= \frac{1}{2}(0 - (-1)^2 + (1)^2 - 0) = 1\text{J}$$

4 *(d)* Mass of elevator, $M = 920$ kg

Mass of all 10 passengers carried by elevator
$= 10 \times m = 10 \times 68 = 680$ kg

Total weight of elevator and passengers
$= (M + 10m)g = 16000$ N
$= (920 + 680) \times 10$

Force of friction = 6000 N

Total force (T) applied by the motor of elevator
$= 16000 + 6000$
$= 22000$ N

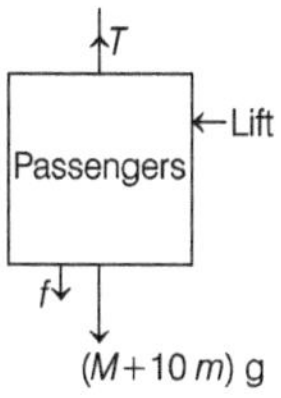

Power delivered by elevator's motor,

$P = F \cdot v = 22000 \times 3 \quad [\because v = 3\text{ms}^{-1}]$
$= 66000$ W

5 *(d)* Given, mass of the cable is M.

So, mass of $\frac{1}{n}$th part of the cable, i.e. hanged part of the cable is

$= M / n \quad \text{...(i)}$

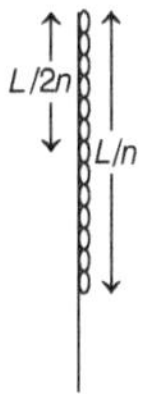

Now, centre of mass of the hanged part will be its middle point.

So, its distance from the top of the table will be $L / 2n$.

$\therefore$ Initial potential energy of the hanged part of cable,

$$U_i = \left(\frac{M}{n}\right)(-g)\left(\frac{L}{2n}\right)$$

$\Rightarrow \quad U_i = -\frac{MgL}{2n^2} \quad \text{...(ii)}$

When whole cable is on the table, its potential energy will be zero.

$\therefore \quad U_f = 0 \quad \text{...(iii)}$

Now, using work-energy theorem,

$W_{net} = \Delta U = U_f - U_i$

$\Rightarrow \quad W_{net} = 0 - \left(-\frac{MgL}{2n^2}\right)$

[using Eqs. (ii) and (iii)]

$\Rightarrow \quad W_{net} = \frac{MgL}{2n^2}$

6 *(b)* Initially, centre of mass is at distance L from the top end of the swing. It shifts to $(L - l)$ distance when the person stands up on the swing.

$\therefore$ Using angular momentum conservation law, if v_0 and v_1 are the velocities before standing and after standing of the person, then

$Mv_0L = Mv_1(L - l)$

$\Rightarrow \quad v_1 = \left(\frac{L}{L-l}\right)v_0 \quad \text{...(i)}$

Now, total work done by (person + gravitation) system will be equal to the change in kinetic energy of the person, i.e.

$W_g + W_p = \text{KE}_1 - \text{KE}_0$

$\Rightarrow \quad -Mgl + W_p = \frac{1}{2}Mv_1^2 - \frac{1}{2}Mv_0^2$

$\Rightarrow \quad W_p = Mgl + \frac{1}{2}M(v_1^2 - v_0^2)$

$= Mgl + \frac{1}{2}M\left[\left(\frac{L}{L-l}\right)^2 v_0^2 - v_0^2\right]$

[from Eq. (i)]

$= Mgl + \frac{1}{2}Mv_0^2\left[\left(1 - \frac{l}{L}\right)^{-2} - 1\right]$

$= Mgl + \frac{1}{2}Mv_0^2\left[\left(1 + \frac{2l}{L}\right) - 1\right]$

[using $(1 + x)^n = 1 + nx$ as higher terms can be neglected, if $n << 1$]

$\Rightarrow \quad W_p = Mgl + \frac{1}{2}Mv_0^2 \times \frac{2l}{L}$

or $\quad W_p = Mgl + Mv_0^2\frac{l}{L} \quad \text{...(ii)}$

Here, $\quad v_0 = WA = \left(\sqrt{\frac{g}{L}}\right)(\theta_0 L)$

$\Rightarrow \quad v_0 = \theta_0\sqrt{gL}$

$\therefore$ Using this value of v_0 in Eq.(ii), we get

$W_p = Mgl + M\theta_0^2\,gL \cdot \frac{l}{L}$

$\Rightarrow \quad W_p = Mgl\,(1 + \theta_0^2)$

7 *(c)* **Key Idea** When a mass m is fall from height h on a uncompressed spring then the maximum compression is given by relation,

$$mgh = \frac{1}{2}kx^2$$

(Principle of energy conservation)

Given, mass of body, $m = 2$ kg, height, $h = 1.2$ m spring constant, $k = 2 \times 10^4$ N/m and acceleration due to gravity, $g = 10\text{ m/s}^2$

Then, $x = \sqrt{\frac{2mgh}{k}}$

$\Rightarrow \quad x = \sqrt{\frac{2 \times 2 \times 1.2 \times 10}{2 \times 10^4}}$

$\Rightarrow \quad = 4.89 \times 10^{-2}$ m
≈ 50 mm

So, the correct option is (c).

8 *(d)* The given situation can be shown as below,

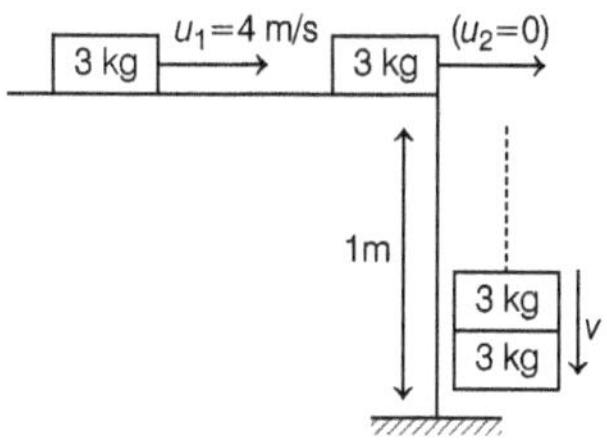

According to law of conservation of momentum,

$m_1u_1 + m_2u_2 = (m_1 + m_2)v$

Given, $m_1 = m_2 = 3$ kg

$u_1 = 4$ m/s and $u_2 = 0$

Putting the given values, we get

$$3 \times 4 + 3 \times 0 = (3 + 3)v$$

$$\Rightarrow \quad v = \frac{12}{6} = 2 \text{ m/s}$$

Applying law of conservation of energy

$$KE_1 + PE_1 = KE_2 + PE_2$$

At the bottom just above the ground, the total energy is KE as height is approximately equals to zero.

$$\frac{1}{2}(m_1 + m_2)v^2 + (m_1 + m_2)gh = KE_2$$

$$\Rightarrow KE_2 = \frac{1}{2} \times 6 \times 4 + 6 \times 10 \times 1$$

$$= 12 + 60 = 72 \text{ J}$$

Hence, the kinetic energy just before the boxes strike the floor is 72 J.

9 *(c)* We know that, $P = \frac{KE}{\Delta t}$

$$\Rightarrow \quad P = \frac{mv^2}{2}$$

$\because \quad P =$ constant,

Hence, velocity of the body $v \propto \sqrt{t}$ ...(i)

As, velocity $v = \frac{ds}{dt}$ ...(ii)

From Eqs. (i) and (ii), we get

So, $\frac{ds}{dt} \propto \sqrt{t}$

Integrating the above equation w.r.t. time (t),

$$\int \frac{ds}{dt} \propto \int \sqrt{t}$$

we get, displacement of the body $s \propto t^{3/2}$

$\because$ Displacement $s = at^{3/2}$,

where a is constant.

Circular Motion

1. Statement I

$$v_{max} = \sqrt{\mu Rg} = \sqrt{(0.2) \times 2 \times 9.8}$$

$$v_{max} = 1.97 \text{ m/s}$$

$$7 \text{ km/h} = 1.944 \text{ m/s}$$

Speed is lower than v_{max}, hence it can take safe turn.

Statement II

$$v_{max} = \sqrt{Rg\left(\frac{\tan\theta + \mu}{1 - \mu \tan\theta}\right)}$$

$$= \sqrt{2 \times 9.8\left(\frac{1 + 0.2}{1 - 0.2}\right)}$$

$$= 5.42 \text{ m/s}$$

$$18.5 \text{ km/h} = 5.14 \text{ m/s}$$

Speed is lower than v_{max}, hence it can take safe turn.

2 *(d)* Position vector of a particle moving around a unit circle $(r = 1)$ in xy-plane is given by

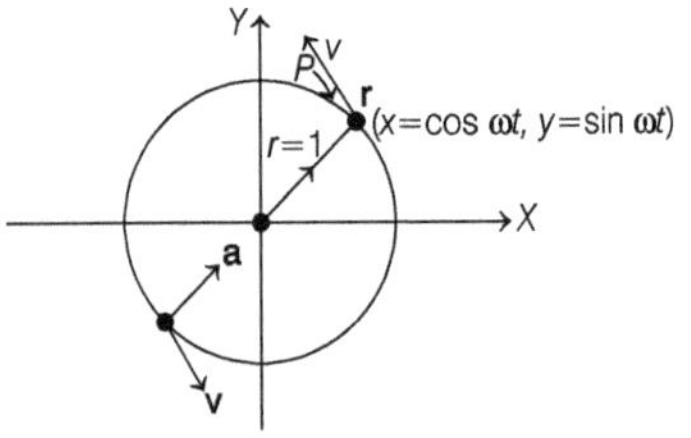

$$\mathbf{r}(t) = \cos\omega t\,\hat{\mathbf{i}} + \sin\omega t\,\hat{\mathbf{j}} \quad \text{...(i)}$$

Velocity,

$$\mathbf{v}(t) = \frac{d\mathbf{r}(t)}{dt} = \omega(-\sin\omega t\,\hat{\mathbf{i}} + \cos\omega t\,\hat{\mathbf{j}})$$

$$\Rightarrow \mathbf{v}(t) = \omega[\cos(\omega t + \pi/2)\,\hat{\mathbf{i}} + \sin(\omega t + \pi/2)\,\hat{\mathbf{j}}] \quad \text{...(ii)}$$

$$\mathbf{a}(t) = \frac{d\mathbf{v}(t)}{dt} = -\omega^2 \cos\omega t\,\hat{\mathbf{i}} - \omega^2 \sin\omega t\,\hat{\mathbf{j}}$$

$$= -\omega^2(\cos\omega t\,\hat{\mathbf{i}} + \omega^2 \sin\omega t\,\hat{\mathbf{j}})$$

$$\mathbf{a}(t) = -\omega^2\,\mathbf{r}(t) \quad \text{...(iii)}$$

From Eqs. (i), (ii) and (iii), it is clear that **v** is perpendicular to **r** and **a** is directed towards the origin.

3 *(b)* Let v_1 be the velocity of the particle moving along the circular path initially, v_1 and v_2 be the velocity when it moves through an angle of 60° as shown below.

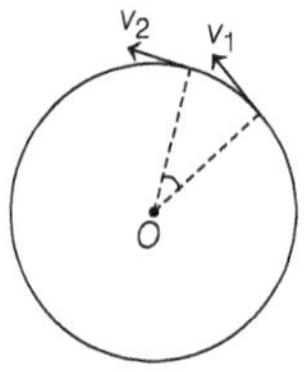

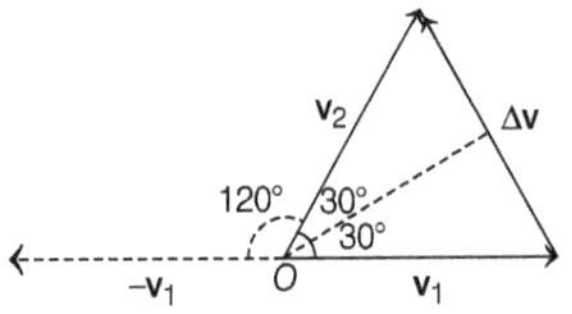

From the figure,

$$\Delta\mathbf{v} = \mathbf{v}_2 - \mathbf{v}_1$$

$$\Rightarrow \quad |\Delta \mathbf{v}| = 2v \sin\frac{\theta}{2} = 2\,\mathbf{v}\sin 30^\circ$$

$$[\because |\mathbf{v}_1| = |\mathbf{v}_2|]$$

$$= 2v \times \frac{1}{2} = v \text{ (Given, } v = 10 \text{ m/s)}$$

$$\Rightarrow \quad |\Delta\mathbf{v}| = 10 \text{ m/s}$$

Alternate method

$$\because \quad \Delta\mathbf{v} = \mathbf{v}_2 - \mathbf{v}_1 = \mathbf{v}_2 + (-\mathbf{v}_1)$$

$$\therefore \quad |\Delta\mathbf{v}|^2 = v_1^2 + v_2^2 + 2v_1v_2 \cos 120^\circ$$

$$= v^2 + v^2 + 2v \times v \times \left(-\frac{1}{2}\right)$$

$$\Rightarrow \quad |\Delta\mathbf{v}| = v = 10 \text{ m/s}.$$

4 *(d)* Angle covered by each particle in time duration 0 to $\frac{\pi}{2\omega}$ is

$$\theta = \omega \times t = \omega \times \frac{\pi}{2\omega} = \frac{\pi}{2} \text{ rad}$$

So, positions of particles at $t = \frac{\pi}{2\omega}$ is as shown below;

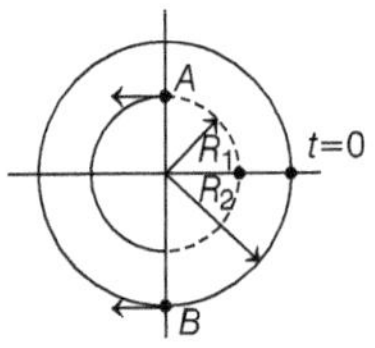

Velocities of particles at $t = \frac{\pi}{2\omega}$ are

$$\mathbf{v}_A = -\omega R_1\hat{\mathbf{i}} \text{ and } \mathbf{v}_B = -\omega R_2\hat{\mathbf{i}}$$

The relative velocity of particles is

$$\mathbf{v}_A - \mathbf{v}_B = -\omega R_1\hat{\mathbf{i}} - (-\omega R_2\hat{\mathbf{i}})$$

5 *(d)* Initial kinetic energy of body,

$$E_i = \frac{1}{2}mv^2 = \frac{1}{2}m\left(\frac{2\pi r_1}{T_1}\right)^2 \quad \left[\because v = \frac{2\pi r}{T}\right]$$

$$E_i = 2\pi^2 m r_1^2 f_1^2 \quad \text{...(i)}$$

Where, f_1 = frequency of revolution of the body. When, $r_2 = 2r_1$ and $f_2 = 2f_1$, then

$$E_f = 2\pi^2 m \cdot r_2^2 \cdot f_2^2$$

$$= 2\pi^2 m\,(2r_1)^2\,(2f_1)^2$$

$$E_f = 32\pi^2 m\,r_1^2\,f_1^2$$

$$= 16 \cdot 2\pi^2 m r_1^2 f_1^2$$

$$E_f = 16\,E_i \quad \text{[using Eq. (i)]}$$

$\therefore$ Change in kinetic energy,

$$\Delta E = E_f - E_i = 16E_i - E_i = 15\,E_i$$

COM, Conservation of Linear Momentum, Impulse and Collision

1. Centre of mass of the quarter disc is at $\frac{4a}{3\pi}, \frac{4a}{3\pi}$

According to the centre of mass of the quarter disc (the shaded area) is at $\frac{x}{3}\cdot\frac{a}{\pi}\cdot\frac{x}{3}\cdot\frac{a}{\pi}$.

So, $x = 4$

2 *(a)* For a continuous mass distribution,

$$x_{CM} = \frac{\int x\,dm}{\int dm}$$

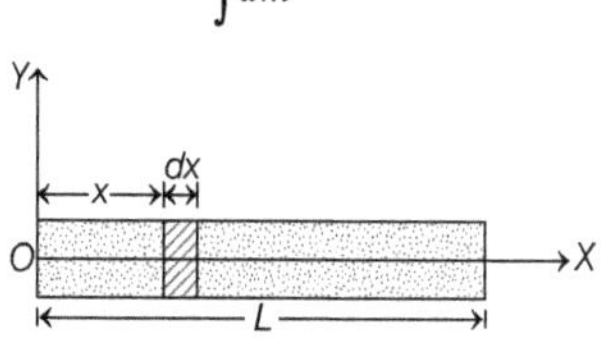

Here, mass of an elemental length dx of rod,

$$dm = \rho dx = \left(a + b\left(\frac{x}{L}\right)^2\right)dx$$

$$\text{So, } x_{CM} = \frac{\int_0^L x\left(a + \frac{bx^2}{L^2}\right)dx}{\int_0^L \left(a + \frac{bx^2}{L^2}\right)dx} = \frac{\int_0^L \left(ax + \frac{bx^3}{L^2}\right)dx}{\int_0^L \left(a + \frac{bx^2}{L^2}\right)dx}$$

$$= \frac{\left[\left(\frac{ax^2}{2} + \frac{b}{L^2}\cdot\frac{x^4}{4}\right)\right]_0^L}{\left[\left(ax + \frac{bx^3}{3L^2}\right)\right]_0^L} = \frac{\frac{aL^2}{2} + \frac{bL^2}{4}}{aL + \frac{bL}{3}} = \left(\frac{\frac{a}{2} + \frac{b}{4}}{a + \frac{b}{3}}\right)L$$

$$\text{So, } x_{CM} = \frac{3}{4}\left(\frac{2a + b}{3a + b}\right)L$$

3 *(a)* Collision is as shown in the figure.

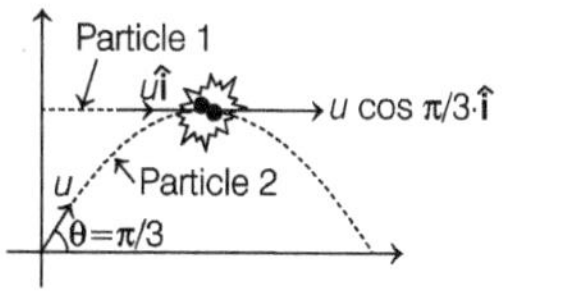

Velocity of the particle projected from origin at its topmost point,

$$\mathbf{u}_2 = u\cos\frac{\pi}{3}\cdot\hat{\mathbf{i}} = \frac{u}{2}\hat{\mathbf{i}}$$

By conservation of momentum [velocity of combined mass after collision (v)], we have

$$mu\hat{\mathbf{i}} + m\frac{u}{2}\hat{\mathbf{i}} = 2m\mathbf{v} \Rightarrow \mathbf{v} = \frac{3}{4}u\hat{\mathbf{i}}$$

Time of fall of combined mass from h_{max},

$$t = \frac{u\sin\theta}{g} = \frac{u\sin\frac{\pi}{3}}{g} = \frac{\sqrt{3}}{2}\frac{u}{g}$$

During this time, combined particle keeps on moving with a horizontal speed of $|\mathbf{v}| = \frac{3}{4}u.$

So, horizontal distance covered by combined mass before reaching the ground,

R = speed × time

$$= \frac{3}{4}u \times \frac{\sqrt{3}}{2}\frac{u}{g} = \frac{3\sqrt{3}}{8}\cdot\frac{u^2}{g}$$

4 *(d)* Given lamina consists of 2 parts I and II as shown in the figure.

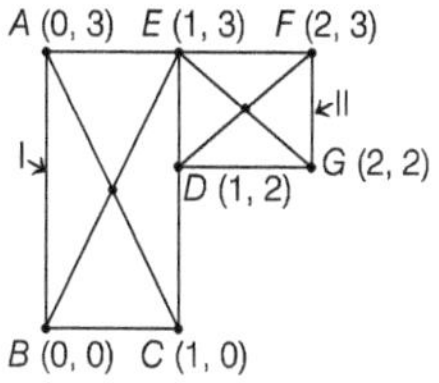

As mass of uniform lamina is 4 kg, mass of part I is $m_1 = 3$ kg and mass of part II is $m_2 = 1$ kg.

These masses can be assumed to be concentrated at geometrical centres of sections I and II.

So, $m_1 = 3$ kg has coordinates $x_1 = 0.5$ m, $y_1 = 1.5$ m

and $m_2 = 1$ kg has coordinates $x_2 = 1.5$ m, $y_2 = 2.5$ m

Now, we use formula of centre of mass (CM) to find X_{CM} and Y_{CM}.

$$\text{So, } X_{CM} = \frac{m_1x_1 + m_2x_2}{m_1 + m_2} = \frac{(3 \times 0.5) + (1 \times 1.5)}{4} = 0.75 \text{ m}$$

$$Y_{CM} = \frac{m_1y_1 + m_2y_2}{m_1 + m_2} = \frac{(3 \times 1.5) + (1 \times 2.5)}{4} = 1.75 \text{ m}$$

5 *(a)* Coordinates of centre of mass (CM) are given by

$$X_{CM} = \frac{m_1x_1 + m_2x_2 + m_3x_3}{m_1 + m_2 + m_3}$$

$$\text{and } Y_{CM} = \frac{m_1y_1 + m_2y_2 + m_3y_3}{m_1 + m_2 + m_3}$$

For given system of rods, masses and coordinates of centre of rods are as shown.

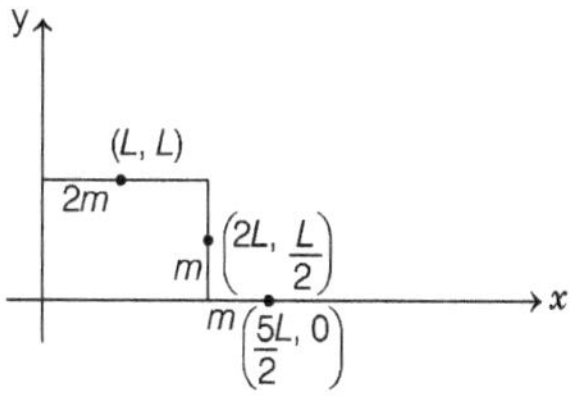

$$\text{So, } X_{CM} = \left(\frac{2mL + m2L + m\frac{5L}{2}}{4m}\right) = \frac{13}{8}L$$

$$\text{and } Y_{CM} = \frac{2mL + m \times \frac{L}{2} + m \times 0}{4m} = \frac{5L}{8}$$

So, position vector of CM is

$$\mathbf{r}_{CM} = X_{CM}\hat{\mathbf{x}} + Y_{CM}\hat{\mathbf{y}} = \frac{13}{8}L\hat{\mathbf{x}} + \frac{5}{8}L\hat{\mathbf{y}}$$

6 *(d)* For perfectly elastic collision,

total momentum before collision

= total momentum after collision

$$\text{i.e. } m\cdot u + \frac{3m}{4} \times 0 = mv_x + \frac{3m}{4}\cdot v_y$$

$$\Rightarrow 4u = 4v_x + 3v_y \quad \text{... (i)}$$

Coefficient of restitution,

$$e = \frac{\text{Velocity of seperation } (v_y - v_x) \text{ after collision}}{\text{Velocity of approach } (u_x - u_y) \text{ before collision}}$$

$$\Rightarrow 1 = \frac{v_y - v_x}{u - 0} \quad \left[\because u_x = v \text{ and } u_y = v\right]$$

$$u = v_y - v_x \quad \text{... (ii)}$$

Solving Eqs. (i) and (ii), we have

$$v_x = \frac{u}{7} \text{ and } u_y = \frac{8u}{7}$$

$$\therefore \frac{\lambda_y}{\lambda_x} = \frac{h/mv_y}{h/\frac{3m}{4}.v_x} = \frac{3}{4}.\frac{v_x}{v_y} = \frac{3}{4}.\frac{u/7}{8u/7} = \frac{3}{32}$$

7 *(a)* According to question, three particles are situated as shown below,

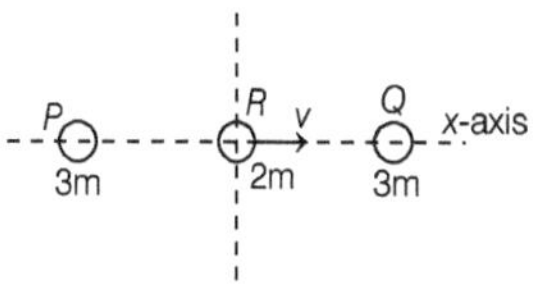

∵ All the collision of the system of 3 particles are elastic and head on.

So, collision of R and Q,

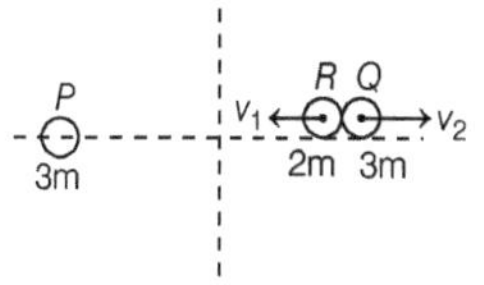

Hence, R and Q will never collide after 1st collision.

∴ Total numbers of collision is 2.

8 *(b)* The centre of mass of the spheres is given by

$$x_{CM} = \frac{m_1x_1 + m_2x_2 + m_3x_3}{m_1 + m_2 + m_3}$$

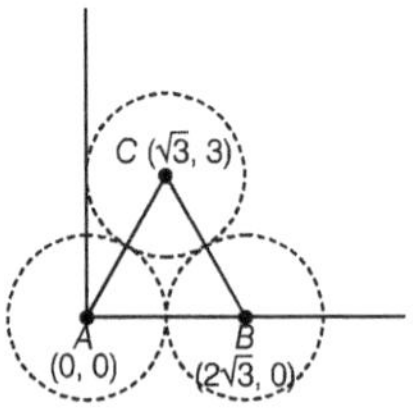

As the spheres are identical,

$$\Rightarrow \quad x_{CM} = \frac{m}{3m}(0 + 2\sqrt{3} + \sqrt{3})$$

$$= \frac{3\sqrt{3}}{3}$$

$$= \sqrt{3}$$

Similarly, $y_{CM} = \frac{m}{3m}(y_1 + y_2 + y_3)$

$$= \frac{(0+0+3)}{3} = 1$$

So, the centre of mass, $C_{CM} = (\sqrt{3}, 1)$

If one sphere is removed (say C), then

$$x'_{CM} = \frac{0 + 2\sqrt{3}}{2} = \sqrt{3}$$

$$y'_{CM} = 0$$

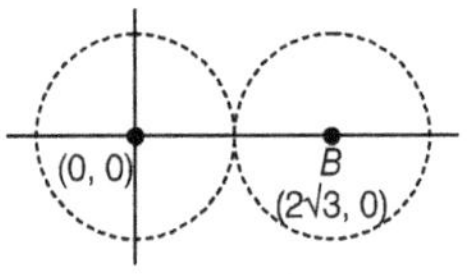

So, $C'_{CM} = (\sqrt{3}, 0)$

Hence, the centre of mass shifted by 1 m in $-y$ direction.

9 *(d)* According to the question,

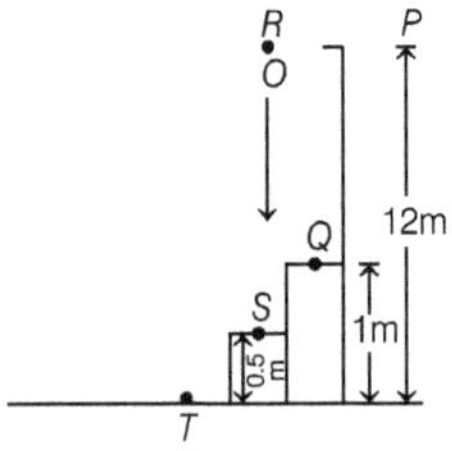

When ball is dropped from point P, and collides at point Q, from third equation of the motion,

$v_Q^2 = u^2 + 2gh = 0 + 2 \times 10 \times (12 - 1) = 220$

$\therefore$ Velocity at a point Q, $V_Q = \sqrt{220}$ m/s

In perfect elastic collision at point Q, there is no loss of kinetic energy.

Hence, the height gain by ball after collision,

$$\Rightarrow \frac{1}{2}mv_Q^2 = mgh'$$

$$\Rightarrow \frac{1}{2}(\sqrt{220})^2 = 10 \times h'$$

$$\Rightarrow \quad h' = 11\text{m}$$

Let velocity of the ball before inelastic collision $= v_s$

From energy conservation law,

$$\Rightarrow \quad \frac{1}{2}mv_s^2 = mg\,(11 + 0.5)$$

$$\Rightarrow \quad v_s^2 = 2 \times 10 \times 11.5$$

$$\Rightarrow \quad v_s = \sqrt{230}\text{ m/s}$$

$\because$ coefficient of restitution,

$$e = 0.5 = \frac{\text{velocity after collision }(v)}{\text{velocity before collision }(v_s)}$$

$$\Rightarrow 0.5 = \frac{v}{\sqrt{230}} \Rightarrow v = 0.5\sqrt{230}\text{ m/s}$$

Again from energy conservation law,

$$\frac{1}{2}mv^2 = mgh''$$

$$\Rightarrow \quad \frac{1}{2}(0.5\sqrt{230})^2 = 10 \times h''$$

$$\Rightarrow h'' = \frac{0.5 \times 230 \times 0.5}{20} = 2.875\text{ m}$$

Hence, the total height from point T,
$h = h'' + 0.5 = 2.875 + 0.5 = 3.375$ m

From energy conservation law,

$$\frac{1}{2}mv_T^2 = mgh$$

Velocity of ball while touching the ground,

$$v_T = \sqrt{2gh}$$

$$= \sqrt{2 \times 10 \times 3.375} = 8.215\text{ m/s}$$

10 *(d)* Given,

mass of a circular ring, $m = 10$ kg

speed of centre of mass of the ring

i.e, linear speed of ring, $u = 1.5$ m/s

Total initial kinetic energy of rolling ring,

$$K_i = K_{\text{rotational}} + K_{\text{linear}} = \frac{1}{2}I\omega^2 + \frac{1}{2}mv^2$$

Here, $I = mR^2$ and $\omega = \frac{v}{R}$

$$= \frac{1}{2}mR^2\left(\frac{v}{R}\right)^2 + \frac{1}{2}mv^2 = \frac{1}{2}mv^2 + \frac{1}{2}mv^2$$

$$= mv^2$$

$$= 10 \times (1.5)^2 = 22.5\text{ J}$$

According to work-energy theorem, work required to stop the ring.

ω = change in kinetic energy

$$= K_f - K_i = 0 - 22.5 = -22.5\text{ J}$$

11 *(*)* Given,

mass of ball A, $m_1 = 50$ g

mass of ball B, $m_2 = 100$ g

mass of ball C, $m_3 = 150$ g

and length of each side of triangle = 1 m

According to the question,

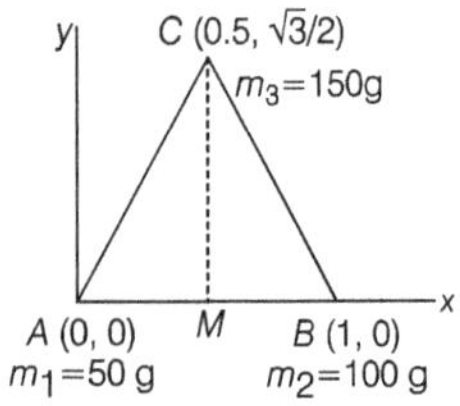

$$\because y_3 = \sqrt{AC^2 - AM^2}$$

$$= \sqrt{1 - (0.5)^2} = \sqrt{0.75} = \frac{\sqrt{3}}{2}$$

$$[\because AM = x_3 = 0.5]$$

Coordinates of centre of mass is (x, y), then

$$x = \frac{m_1x_1 + m_2x_2 + m_3x_3}{m_1 + m_2 + m_3}$$

$$= \frac{50 \times 0 + 100 \times 1 + 150 \times 0.5}{50 + 100 + 150}$$

$$= \frac{175}{300} = \frac{7}{12}$$

and $y = \frac{m_1y_1 + m_2y_2 + m_3y_3}{m_1 + m_2 + m_3}$

$$= \frac{50 \times 0 + 100 \times 0 + 150 \times \frac{\sqrt{3}}{2}}{50 + 100 + 150}$$

$$= \frac{75\sqrt{3}}{300} = \frac{\sqrt{3}}{4}$$

The coordinates (x, y) for the centre of the mass of this system of balls is $\left(\frac{7}{12}\text{m}, \frac{\sqrt{3}}{4}\text{m}\right)$

So, no given option is correct.

12 *(b)* The situation can be shown as,

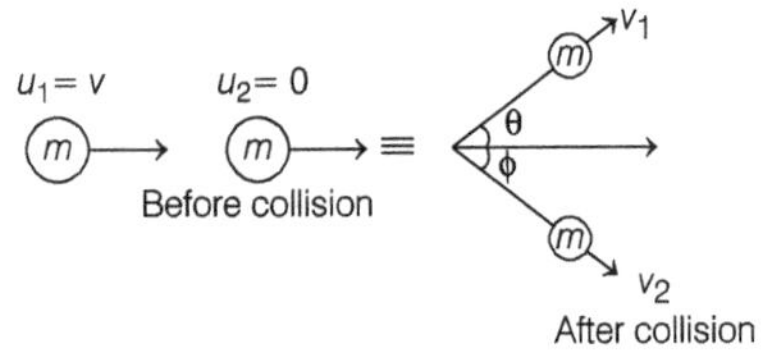

Applying law of conservation of kinetic energy,

KE(before collision) = KE (after collision)

$$\frac{1}{2}mv^2 + \frac{1}{2}m(0)^2$$

$$= \frac{1}{2}mv_1^2 + \frac{1}{2}mv_2^2$$

$$\Rightarrow \quad v^2 = v_1^2 + v_2^2$$

$$\Rightarrow \quad v_2 = \sqrt{v^2 - v_1^2}$$

Thus, the velocity of second block after collision is $\sqrt{v^2 - v_1^2}$

Rotation

1. Angular momentum, $\mathbf{L} = m(\mathbf{r} \times \mathbf{v})$

The direction of $\mathbf{L}$ with respect to A is along the Z-axis and magnitude is mvr. So, L_A is constant, both in magnitude and direction.

But the direction of $\mathbf{L}$ with respect to B is continuously changing with time with constant magnitude.

2. (*a*) In circular motion, net force on the particle is given as

$$F_c = \frac{mv^2}{r} = m\omega^2 r$$

where, ω is angular speed.

When the rod rotates, electrons in it also rotates which produce electric field E at a distance x.

So, force on electron, $F_e = eE$

This force provide the centripetal force,

i.e. $\quad F_e = F_c$

or $\quad E = m\omega^2 x$ or $E = \dfrac{m\omega^2 x}{e}$

3 (*c*) In given arrangement, mass of each sphere $= m$

Radius of each sphere $= \dfrac{d}{2}$

Distance of centre of each sphere from axis of rotation or centroidal axis $= \dfrac{2}{3} \times$ altitude of equilateral triangle of side d

$$= \frac{2}{3} \times \frac{\sqrt{3}}{2} \times d = \frac{d}{\sqrt{3}}$$

Now, moment of inertia of each sphere along an axis through its centre,

$$I_1 = \frac{2}{5} mr^2 = \frac{2}{5} m\left(\frac{d}{2}\right)^2 = \frac{1}{10} md^2$$

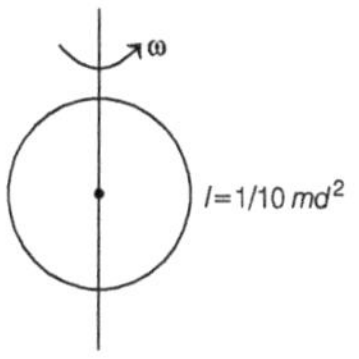

Now, by using parallel axis theorem, moment of inertia of each sphere around axis through O,

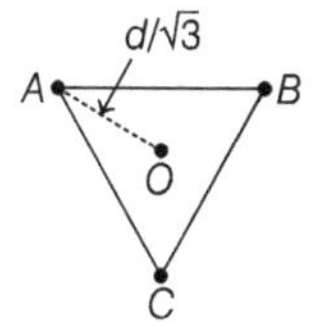

$$I' = I_1 + m(h^2) = I_1 + m\left(\frac{d}{\sqrt{3}}\right)^2$$

$$= \frac{1}{10} md^2 + \frac{md^2}{3} = \frac{13}{30} md^2$$

So, moment of inertia of system of spheres about O,

$$I_0 = 3I' = 3 \times \frac{13}{30} md^2 = \frac{13}{30} md^2 \quad \text{...(i)}$$

Again by using parallel axes theorem, moment of inertia of system about point A,

$$I_A = I_0 + m_{\text{system}} \times h^2$$

$$= \frac{13}{10} md^2 + 3m \times \left(\frac{d}{\sqrt{3}}\right)^2 = \frac{23}{10} md^2 \quad \text{...(ii)}$$

Hence, required ratio from Eqs. (i) and (ii), we get

$$\frac{I_0}{I_A} = \frac{13}{10} md^2 \times \frac{10}{23\, md^2} \Rightarrow \frac{I_0}{I_A} = \frac{13}{23}$$

4 (*c*) Initially, it is given that the unstretched length of the spring is l. When it is given an angular speed ω, then let x = stretched length of spring.

Then, total length of the spring system while rotating will be $(l + x)$ as shown in the figure.

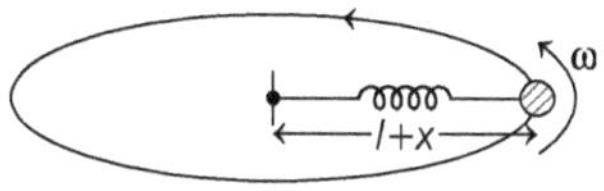

As we know, spring force will give the necessary centripetal force for rotation.

So, $\quad kx = m(l + x)\omega^2$

$\Rightarrow \quad (k - m\omega^2)x = ml\omega^2$

$\Rightarrow \quad x = \dfrac{ml\omega^2}{(k - m\omega^2)}$

5 (*c*) When bob falls, its potential energy is converted into kinetic energy and rotational kinetic energy of disc.

So, decrease in potential energy of bob = kinetic energy of bob + kinetic energy of disc

$$\Rightarrow \quad mgh = \frac{1}{2} mv^2 + \frac{1}{2} I\omega^2 \quad \text{...(i)}$$

If there is no slip, then

$$v = r\omega$$

Also, moment of inertia of disc,

$$I = \frac{mr^2}{2}$$

Substituting these in Eq. (i), we get

$$mgh = \frac{1}{2} m(r\omega)^2 + \frac{1}{2} \times \frac{mr^2}{2} \times \omega^2$$

$\Rightarrow \quad mgh = \dfrac{3}{4} mr^2\omega^2$

$\Rightarrow \quad \omega^2 = \dfrac{4gh}{3r^2}$

or $\quad \omega = \dfrac{1}{r}\sqrt{\dfrac{4gh}{3}}$

6 (*a*)

Key Idea When a rod is pivoted at any point, its angular acceleration is given by $\tau_{net} = I\alpha$.

The given condition can be drawn in the figure below.

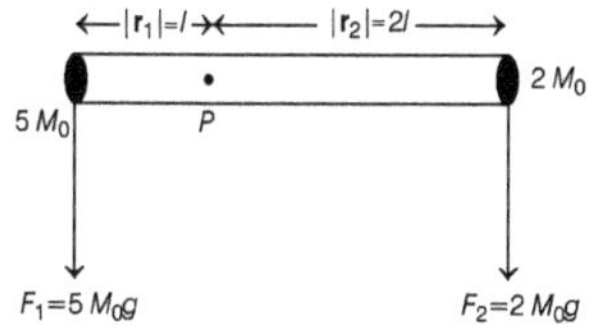

Torque (τ) about $P = \mathbf{r}_1 \times \mathbf{F}_1 + \mathbf{r}_2 \times \mathbf{F}_2$ ... (i)

$\Rightarrow \quad \tau = l \times 5M_0 g$ (outwards) $- 2l \times 2M_0 g$ (inwards)

$\Rightarrow \quad \tau = 5M_0 gl - 4M_0 gl \quad$ (outwards)

$\Rightarrow \quad \tau = M_0 gl$ (outwards)

or $\quad \tau = M_0 gl \quad$ ...(ii)

Now we know that, torque is also given by

$$\tau = I\alpha \quad \text{...(iii)}$$

Here, I = moment of inertia (w.r.t. point P) of rod and α = angular acceleration.

For point P, $I = (5M_0) \times l^2 + (2M_0)(2l)^2$ $\quad [\because I = MR^2]$

$\Rightarrow \quad I = 13M_0 l^2 \quad$ ...(iv)

Putting value of I from Eq. (iv) in Eq. (iii), we get

$$\tau = (13M_0 l^2)\,\alpha \quad \text{...(v)}$$

From Eqs. (ii) and (v), we get

$$M_0 gl = 13M_0 l^2 \alpha$$

$\Rightarrow \quad \alpha = \dfrac{g}{13l}$

7 (*c*) For disc D_1, moment of inertia across axis OO' will be

$$I_1 = \frac{1}{2} MR^2 \quad \text{...(i)}$$

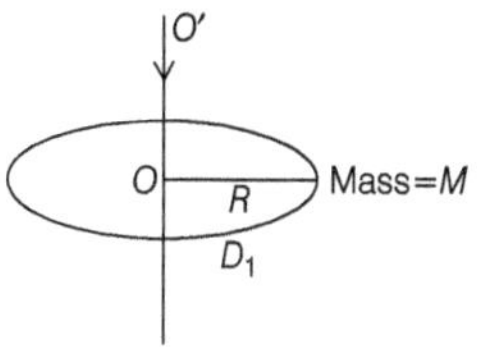

For discs D_2 and D_3, OO' is an axis parallel to the diameter of disc. Using parallel axis theorem,

$$I_2 = I_3 = I_{\text{diameter}} + Md^2 \quad \text{...(ii)}$$

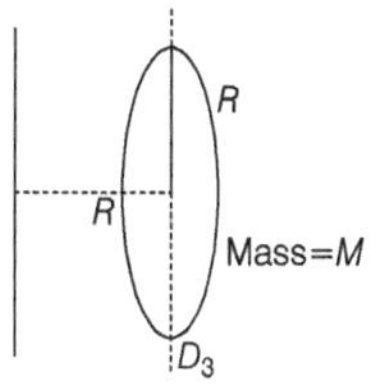

Here, $I_{\text{diameter}} = \dfrac{1}{4} MR^2$

and $\quad d = R$

$\therefore \quad I_2 = I_3 = \dfrac{1}{4} MR^2 + MR^2 = \dfrac{5}{4} MR^2$

Now, total MI of the system

$$I = I_1 + I_2 + I_3 = \frac{1}{2}MR^2 + 2\times\frac{5}{4}MR^2 = 3MR^2$$

8 *(a)* Given, mass of solid sphere, $M_s = 100$ kg

radius of solid sphere, $R_s = 10$ m

radius of circular disc, $R_c = 20$ m

and time = 1 hour = 60 minute = 60 × 60 s

Moment of inertia of the solid sphere,

$$I_s = \frac{2}{5}M_sR_s^2 = \frac{2}{5}\times 100\times(10)^2$$

$$= 4000 \text{ kg-m}^2$$

Similarly, moment of inertia of the disc,

$$I_c = \frac{1}{2}M_cR^2$$

$$= \frac{1}{2}\times 100\times(20)^2 = 20{,}000 \text{ kg- m}$$

2

Rate of change of moment of inertia

$$= \frac{I_c - I_s}{t}$$

$$= \frac{20000 - 4000}{60\times 60}$$

$$= \frac{16000}{60\times 60} = \frac{160}{36}$$

$$= \frac{40}{9}\text{ kg-m}^2\text{s}^{-1}$$

9 *(d)* According to the question,

∴ Kinetic energy for the rolling disc,

or $$\text{KE}_r = \frac{1}{2}mv_1^2 + \frac{1}{2}I\omega^2$$

(∴ Moment of inertia, $I = \frac{mR^2}{2}$)

or $$= \frac{1}{2}mv_1^2 + \frac{1}{2}\frac{mR^2}{2}\left(\frac{v_1}{R}\right)^2$$

or $$\text{KE}_r = \frac{3}{4}mv_1^2 \quad \text{...(i)}$$

Now, KE for the sliding disc,

∴ $$\text{KE}_s = \frac{1}{2}mv_2^2 \quad \text{...(ii)}$$

Given, KE of rolling disc = KE of sliding disc

or, $$\frac{3}{4}mv_1^2 = \frac{1}{2}mv_2^2 \text{ or } \frac{v_1^2}{v_2^2} = \frac{2}{3}$$

or $$\frac{v_1}{v_2} = \sqrt{\frac{2}{3}} \text{ or } v_1 : v_2 = \sqrt{2} : \sqrt{3}$$

10 *(d)* Given,

angular deceleration ∝ $\sqrt{\text{angular speed}}$

i.e., $$\frac{-d\omega}{dt} \propto \sqrt{\omega}$$

or $$\frac{-d\omega}{dt} = k\sqrt{\omega}, \text{ where, } k \text{ is constant}$$

or $$-\int_{\omega_0}^{\omega}\frac{d\omega}{\sqrt{\omega}} = k\int_0^t dt - [2\sqrt{\omega}]_{\omega_0}^{\omega} = k[t]_0^t$$

⇒ $$-2[\sqrt{\omega} - \sqrt{\omega_0}] = kt$$

$$\sqrt{\omega} - \sqrt{\omega_0} = -\frac{kt}{2}$$

⇒ $$\sqrt{\omega} = \sqrt{\omega_0} - \frac{kt}{2} \quad \text{...(i)}$$

When $\omega = 0$, then the total time of rotation,

$$\tau = t = \frac{2\sqrt{\omega_0}}{k}$$

∴ Mean angular speed,

$$<\overline{\omega}> = \frac{\int \omega dt}{\int dt}$$

$$= \frac{\int_0^{2\sqrt{\omega_0}/k}\left(\omega_0 + \frac{k^2t^2}{4} - kt\sqrt{\omega_0}\right)dt}{2\sqrt{\omega_0}/k}$$

$$= \frac{\left[\omega_0 t + \frac{k^2t^3}{12} - \frac{k}{2}\sqrt{\omega_0}t^2\right]_0^{2\sqrt{\omega_0}/k}}{2\sqrt{\omega_0}/k}$$

After solving, we get $= \frac{\omega_0}{3}$

11 *(c)* Length of rod = l

Moment of couple,

$$\tau = Fl\sin\theta$$

∴ $$F = \frac{\tau}{l\sin\theta} = \frac{\tau}{l\sin 30^\circ} = \frac{\tau}{l\times\frac{1}{2}}$$

Force, $F = \frac{2\tau}{l}$

12 *(d)* Given, height of inclined plane, $h = 7$ m

$g = 10 \text{ m/s}^2$

By conservation of energy,

Potential energy lost by the solid sphere in rolling down the inclined plane = Kinetic energy gained by the sphere.

$$mgh = \frac{1}{2}mv^2 + \frac{1}{2}I\omega^2$$

$$= \frac{1}{2}mv^2 + \frac{1}{2}\cdot\frac{2}{5}mR^2\cdot\left(\frac{v}{R}\right)^2$$

$$= \frac{1}{2}mv^2 + \frac{1}{5}mv^2 \quad mgh = \frac{7}{10}mv^2$$

∴ $$v^2 = \frac{10}{7}gh$$

⇒ $$v = \sqrt{\frac{10}{7}gh}$$

$$= \sqrt{\frac{10}{7}\times 10\times 7}$$

$$= \sqrt{100} \text{ m/s}$$

Gravitation

1. Escape velocity of earth, $v_e = \sqrt{\frac{2GM}{R}}$ ...(i)

$\Rightarrow \quad 10v_e = \sqrt{\frac{2GM}{R'}}$ ...(ii)

Dividing Eq. (ii) by Eq. (i), we get

$$10 = \sqrt{\frac{R}{R'}} \Rightarrow R' = \frac{R}{100} = \frac{6400}{100} = 64\,\text{km}$$

2. (*c*) From Kepler's law,

$$T^2 \propto R^3 \Rightarrow T \propto R^{3/2}$$

$$\therefore \quad \frac{T_1}{T_2} = \left(\frac{12R}{3R}\right)^{3/2} = 8$$

$$\Rightarrow \quad T_2 = \frac{T_1}{8} = \frac{24}{8} = 3\,\text{h}$$

3. (*d*) Escape velocity is given by $v_e = \sqrt{\frac{2GM}{R}}$

where, m = mass, R = radius of planet.

So, $v_A = \sqrt{\frac{2\,GM}{R}}$ and $v_B = \sqrt{\frac{2G\left(\frac{M}{2}\right)}{\left(\frac{R}{2}\right)}}$

$$= \sqrt{\frac{2\,GM}{R}} \Rightarrow \frac{v_A}{v_B} = 1 \text{ or } \frac{n}{4} = 1 \Rightarrow n = 4$$

4. (*b*) For a particle or satellite with speed v_p around a planet whose escape speed is v_e and orbital speed of particle v_0, we have following possibilities,

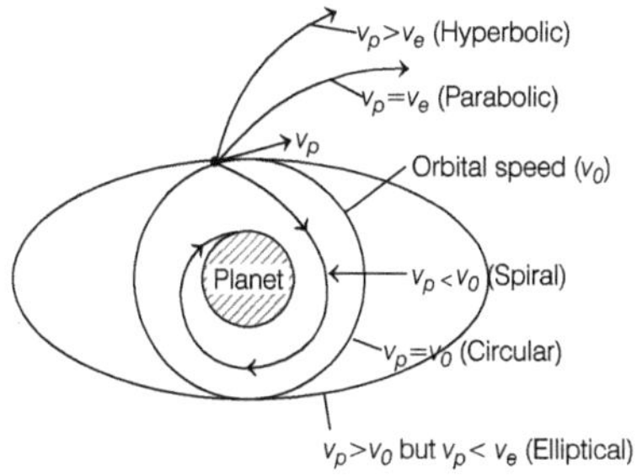

Now, by momentum conservation in given collision, we have, $p_i = p_f$

$$\Rightarrow mv_A + \frac{m}{2}\left(\frac{v_A}{2}\right) = \left(m + \frac{m}{2}\right)v_p \quad \text{...(i)}$$

Body A is orbiting in orbit of radius $R + h$, hence for a small value of h,

$$v_A = \sqrt{\frac{GM}{r}} = \sqrt{\frac{GM}{R}} = \frac{v_e}{\sqrt{2}}$$

So, final speed of combined mass system,

$$v_p = \frac{5}{6}v_0 = \frac{5}{6} \times \frac{1}{\sqrt{2}}v_e$$

As v_p is less than v_0, so it must spiral towards planet. Most nearest option is (b), if we take spiral close to elliptical.

5. (*c*) Gravitational field of a solid sphere is maximum at its surface ($r = R$) and its value at surface,

$$E = \frac{GM}{R^2}$$

From graph given in the question,

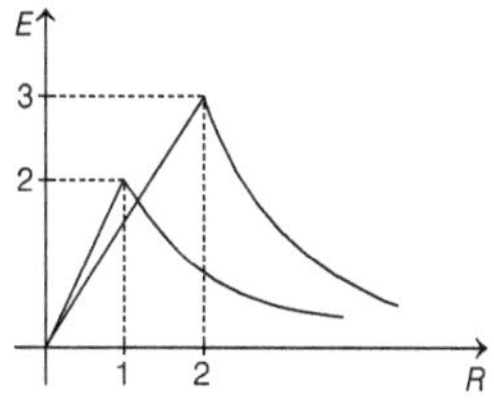

We can observe that

$$E_1 = \frac{GM_1}{(1)^2} = 2 \text{ and } E_2 = \frac{GM_2}{(2)^2} = 3$$

or $\quad GM_1 = 2$ ...(i)

and $\quad GM_2 = 12$ ...(ii)

On dividing Eq. (i) by Eq. (ii), we get

$$\frac{M_1}{M_2} = \frac{2}{12} = \frac{1}{6}$$

6. (*c*) Given, situation is,

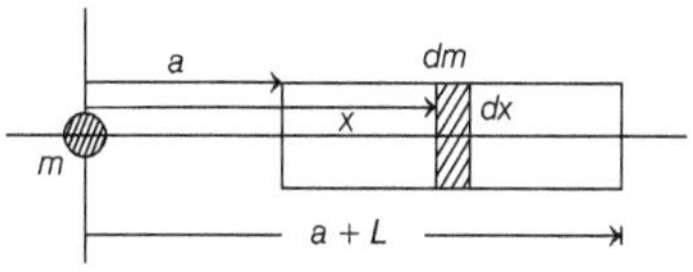

Force of attraction between mass m and an elemental mass dm of rod is

$$dF = \frac{Gmdm}{x^2} = \frac{Gm\,(A + Bx^2)\,dx}{x^2}$$

Total attraction force is sum of all such differential forces produced by elemental parts of rod from $x = a$ to $x = a + L$.

$$\therefore F = \int dF = \int_{x=a}^{x=a+L} \frac{Gm\,(A + Bx^2)}{x^2}dx$$

$$= Gm \int_{x=a}^{x=a+L} \left(\frac{A}{x^2} + B\right)dx$$

$$= Gm\left[-\frac{A}{x} + Bx\right]_{x=a}^{x=a+L}$$

$$= Gm\left(\frac{-A}{a+L} + B(a+L) + \frac{A}{a} - Ba\right)$$

$$= Gm\left(\frac{A}{a} - \frac{A}{a+L} + BL\right)$$

$$= Gm\left[A\left(\frac{1}{a} - \frac{1}{a+L}\right) + BL\right]$$

7 (*b*) **Key Idea** In given configuration of masses, net gravitational force provides the necessary centripetal force for rotation.

Net force on mass M at position B towards centre of circle is

$$F_{BO\text{ net}} = F_{BD} + F_{BA}\sin 45° + F_{BC}\cos 45°$$

$$= \frac{GM^2}{(\sqrt{2}a)^2} + \frac{GM^2}{a^2}\left(\frac{1}{\sqrt{2}}\right) + \frac{GM^2}{a^2}\left(\frac{1}{\sqrt{2}}\right)$$

[where, diagonal length BD is $\sqrt{2}a$]

$$= \frac{GM^2}{a^2}\left(\frac{1}{2} + \sqrt{2}\right)$$

This force will act as centripetal force.

Distance of particle from centre of circle is $\frac{a}{\sqrt{2}}$.

Here, $F_{\text{centripetal}} = \frac{Mv^2}{r} = \frac{Mv^2}{\frac{a}{\sqrt{2}}} = \frac{\sqrt{2}Mv^2}{a}$

$$\left(\because r = \frac{a}{\sqrt{2}}\right)$$

So, for rotation about the centre,

$$F_{\text{centripetal}} = F_{BO(\text{net})}$$

$$\Rightarrow \sqrt{2}\frac{Mv^2}{a} = \frac{GM^2}{a^2}\left(\frac{1}{2} + \sqrt{2}\right)$$

$$\Rightarrow \quad v = 1.16\sqrt{\frac{GM}{a}}$$

8 (*b*) Given that at some height h, acceleration due to gravity,

$$g_h = 4.9\,\text{m/s}^2 \approx \frac{g}{2} \quad \text{... (i)}$$

$\therefore$ The ratio of acceleration due to gravity at earth's surface and at some altitude h is

$$\left(1 + \frac{h}{R_e}\right) = \sqrt{\frac{g}{g_h}} = \sqrt{2} \quad \text{[From Eq. (i)]}$$

$$\therefore \quad \frac{h}{R_e} = \sqrt{2} - 1$$

or $\quad h = 0.414 \times R_e$

$$h = 0.414 \times 6400\,\text{km}$$

($\because$given, radius of earth, $R_e = 6400$ km $= 6.4 \times 10^6$ m)

or $\quad h = 2649.6\,\text{km} = 2.6 \times 10^6\,\text{m}$

9 (*d*) If v be the velocity in the inter planetary space and U_i, U_f initial and final PE

Then, by the law of conservation of energy,

$$U_i + K_i = U_f + K_f$$

$$\frac{-GM_em}{R_e} + \frac{1}{2}m(3v_e)^2 = 0 + \frac{1}{2}mv^2$$

$$\frac{-GM_e}{R_e} + \frac{9}{2}v_e^2 = \frac{1}{2}v^2$$

$$\Rightarrow \quad \frac{-2GM_e}{R_e} + 9V_e^2 = v^2$$

$$\Rightarrow \quad \frac{-2\,gR_e^2}{R_e} + 9v_e^2 = v^2 \quad [\because GM_e = gR_e^2]$$

$$-2gR_e + 9 \times 2 = gR_e = v^2, [\because v_e^2 = 2g\,R_e]$$

$$16\,gR_e = v^2$$

$$8.2\,gR_e = v^2$$

$$\Rightarrow \quad v = \sqrt{8}\,.\,\sqrt{2g\,R_e} = 2\sqrt{2}\,v_e$$

10 (*c*) Gravitational force on a body of mass m at height h due to earth,

$$F = \frac{GM_em}{(R+h)^2}$$

where, M_e is mass of the earth.

$\because$ Density of earth, $\rho = \dfrac{\text{mass of earth } (M_e)}{\text{volume of earth } (V)}$

or $M_e = \rho \cdot V = \rho\left(\dfrac{4}{3}\pi R^3\right)$

$\therefore F = \dfrac{G\left(\frac{4}{3}\pi R^3\right)\rho m}{(R+h)^2}$ or $F = \dfrac{G\left(\frac{4}{3}\pi R\right)\rho m}{\left(1+\frac{h}{R}\right)^2}$

or $\quad F = \dfrac{4}{3}\pi G R\rho m\left(1+\dfrac{h}{R}\right)^{-2}$

By using Binomial expansion,

$$F = \frac{4}{3}\pi G R\rho m\left(1-\frac{2h}{R}\right)$$

[$\because$ $(1+x)^n = 1 + nx + \dfrac{n(n-1)}{2!}.x^2 + \ldots$ and neglecting higher terms.]

Difference in weight

$$= \frac{4}{3}\pi G R\rho m - \frac{4}{3}\pi G\rho m R\left(\frac{2h}{R}\right)$$

Hence, error in weight $= \dfrac{8}{3}\pi\rho G m h$.

11 (*d*) Given, distance of bottom of hole from the surface of earth (d) = half the radius of earth $= \dfrac{R_e}{2}$

If g be the value of gravitational acceleration on the surface of earth, then weight of body

$$mg = 300 \text{ N}$$

If g' be the gravitational acceleration at the bottom of hole, then

$$g' = g\left(1-\frac{d}{R_e}\right) = g\left(1-\frac{\frac{R_e}{2}}{R_e}\right)$$

$$\Rightarrow g' = \frac{g}{2}$$

$\therefore$ Weight of the body on the bottom of hole,

$$mg' = \frac{mg}{2} = \frac{300}{2} = 150 \text{ N}$$

12 (*b*) Initially when the two masses are at an infinite distance from each other, their gravitational potential energy is zero. When they are at a distance r from each, the gravitational PE is

$$\text{PE} = \frac{-Gm_1 m_2}{r^2}$$

The minus sign-indicates that there is a decrease in PE. This gives rise to an increase in kinetic energy. If v_1 and v_2 are their respective velocities when they are a distance r apart then, from the law of conservation of energy, we have

$$\frac{1}{2}m_1 v_1^2 = \frac{Gm_1 m_2}{r} \text{ or } v_1 = \sqrt{\frac{2Gm_2}{r}}$$

and $\quad \dfrac{1}{2}m_2 v_2^2 = \dfrac{Gm_1 m_2}{r}$ or $v_2 = \sqrt{\dfrac{2Gm_1}{r}}$

Therefore, their relative velocity of approach is

$$v_1 + v_2 = \sqrt{\frac{2Gm_2}{r}} + \sqrt{\frac{2Gm_1}{r}}$$

$$= \sqrt{\frac{2G}{r}(m_1+m_2)} = \left(\frac{2G(m_1+m_2)}{r}\right)^{\frac{1}{2}}$$

13 (*b*) Given, time period of a satellite, $T = 6\text{ h} = 2.16\times 10^4$ s and radius of earth, $R = 6.4\times 10^6$ m

As, we know that time period of a satellite

$$T = \frac{2\pi(R+h)}{v_0} \quad \ldots\text{(i)}$$

where, $R+h$ = height of satellite from centre of earth and v_o = orbital velocity

Then, $\quad v_o = \sqrt{\dfrac{GM}{(R+h)}} \quad \ldots\text{(ii)}$

So, from Eqs. (i) and (ii), we get

$$(R+h)^3 = (T)^2\frac{GM}{4\pi^2}$$

$$(\because \frac{GM}{4\pi^2} = 8\times 10^2, \text{N/m}^2 \text{ kg given})$$

$\Rightarrow \quad (R+h)^3 = (2.16\times 10^4)^2 \times 8\times 10^{12}$

$= 373.24\times 10\times 10^{18}$

$\Rightarrow \quad (R+h) = [373.24\times 10\times 10^{18}]^{1/3}$

$\Rightarrow \quad h = 8.72\times 10^6 = 8720$ km

Hence, the correct option is (b).

14 (*a*) Given,

radius of earth, $R_e = 6400$ km $= 6.4\times 10^6$ m and acceleration of gravity, $g = 10$ m/s^2

$\therefore$ Escape velocity of the surface of earth,

$$v_e = \sqrt{2gR_e} = \sqrt{2\times 10\times 6.4\times 10^6}$$

$= 11.3\times 10^3$ m/s

Again, when $g' = 2g$

$$R' = \frac{R}{2}$$

Now, escape velocity is given by

$$\therefore v_e' = \sqrt{2g'R'} = \sqrt{2\times 2g\times\frac{R}{2}}$$

$= \sqrt{2gR} = v_e$

$= 11.3\times 10^3$ m/s

$= 11.3$ km/s ≈ 12 km/s

Simple Harmonic Motion

1. (*) For damped motion, $A = A_0 e^{-rt}$

$$12 = 6e^{-\frac{bt}{2m}}$$

or $\quad \ln 2 = \dfrac{b}{2m}\times 120$

$$b = \frac{0.693\times 2\times 1}{120} = 1.16\times 10^{-2} \text{ kg s}^{-1}$$

$\therefore$ No option is correct.

2. (*d*) Displacement in simple harmonic motion is given by

$$y = a\sin(\omega t + \phi)$$

Given, $\quad y = \dfrac{a}{2}$

$\therefore \quad \dfrac{a}{2} = a\sin(\omega t + \phi)$

or $\quad \sin(\omega t + \phi) = \dfrac{1}{2}$

$= \sin 30^\circ$ or $\sin 150^\circ$

$\therefore \quad \omega t + \phi_1 = 30^\circ$ and $\omega t + \phi_2 = 150^\circ$

$\therefore$ Phase difference

$$(\phi_2 - \phi_1) = 150^\circ - 30^\circ = 120^\circ$$

3. (*) In equilibrium condition, bottle floats in water and its length l inside water is same as the height of water upto which bottle is filled.

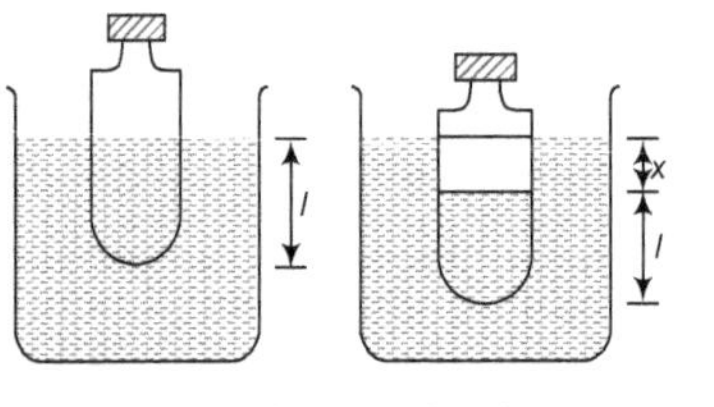

So, l = Volume of water in bottle/Area

$$= \frac{310}{\pi\times(2.5)^2} = 15.8 \text{ cm} = 0.158 \text{ m}$$

When bottle is slightly pushed inside by an amount x then, restoring force acting on the bottle is the upthrust of fluid displaced when bottle goes into liquid by amount x.

So, $\quad$ restoring force is;

$$F = -(\rho A x)\,g \quad \ldots\text{(i)}$$

where ρ = density of water,

A = area of cross-section of bottle and

x = displacement from equilibrium position

But $\quad F = ma \quad \ldots\text{(ii)}$

where, m = mass of water and bottle system $= Al\rho$

From Eqs. (i) and (ii) we have,

$Al\rho\alpha = -\rho A x g$ or $a = -\dfrac{g}{l}x$

As for SHM, $a = -\omega^2 x$

We have $\omega = \sqrt{\dfrac{g}{l}} = \sqrt{\dfrac{10}{0.158}} = \sqrt{63.29}$

≈ 8 rad s^{-1}

$\therefore$ No option is correct.

4. (*b*) In simple harmonic motion, position (x), velocity (v) and acceleration (a) of the particle are given by

$$x = A\sin\omega t$$

$$v = \omega\sqrt{A^2 - x^2}$$

or $\quad v = A\omega\cos\omega t$

and $\quad a = -\omega^2 x \;$ or $a = -\omega^2 A\sin\omega t$

Given, amplitude $A = 5$ cm and displacement $x = 4$ cm.

At this time (when $x = 4$ cm), velocity and acceleration have same magnitude.

$\Rightarrow \quad |v_{x=4}| = |a_{x=4}|$

or $|\omega\sqrt{5^2 - 4^2}| = |-4\omega^2|$

$\Rightarrow \quad 3\omega = +4\omega^2 \Rightarrow \omega = (3/4)$ rad/s

So, time period, $T = \dfrac{2\pi}{\omega}$

$\Rightarrow \quad T = \dfrac{2\pi}{3} \times 4 = \dfrac{8\pi}{3}$ s.

5. *(c)* We know that time period of a pendulum is given by

$$T = 2\pi\sqrt{\frac{l}{g}}$$

So, angular frequency $\omega = \dfrac{2\pi}{T} = \sqrt{\dfrac{g}{l}}$ ..(i)

Now, differentiate both side w.r.t g

$\therefore \quad \dfrac{d\omega}{dg} = \dfrac{1}{2\sqrt{g}\sqrt{l}}$

$$d\omega = \frac{dg}{2\sqrt{g}\sqrt{l}} \quad \text{...(ii)}$$

By dividing Eq. (ii) by Eq. (i), we get

$$\frac{d\omega}{\omega} = \frac{dg}{2g}$$

Or we can write

$$\frac{\Delta\omega}{\omega} = \frac{\Delta g}{2g} \quad \text{...(iii)}$$

As Δg is due to oscillation of support.

$\therefore \quad \Delta g = 2\omega^2 A$

($\omega_1 \rightarrow 1$ rad/s, support)

Putting value of Δg in Eq. (iii), we get

$$\frac{\Delta\omega}{\omega} = \frac{1}{2}\cdot\frac{2\omega_1^2 A}{g} = \frac{\omega_1^2 A}{g};$$

$$(A = 10^{-2}\ \text{m}^2)$$

$\Rightarrow \quad \dfrac{\Delta\omega}{\omega} = \dfrac{1\times10^{-2}}{10} = 10^{-3}$ rad/s

6. *(d)* Equation for SHM is given as

$$y = 5(\sin 3\pi t + \sqrt{3}\cos 3\pi t)$$

$$= 5 \times 2\left(\frac{1}{2}\times\sin 3\pi t + \frac{\sqrt{3}}{2}\cos 3\pi t\right)$$

$$= 5 \times 2\left(\cos\frac{\pi}{3}.\sin 3\pi t + \sin\frac{\pi}{3}.3\pi t\right)$$

$$= 5 \times 2\sin\left(3\pi t + \frac{\pi}{3}\right)$$

[using, $\sin(a+b) = \sin a\cos b + \cos a\sin b$]

or $\quad y = 10\sin\left(3\pi t + \dfrac{\pi}{3}\right)$

Comparing this equation with the general equation of SHM, i.e.

$$y = A\sin\left(\frac{2\pi t}{T} + \phi\right),$$

We get, amplitude, $A = 10$ cm

and $\quad 3\pi = \dfrac{2\pi}{T}$

or $\quad$ Time period, $T = \dfrac{2}{3}$ s

7. *(c)* When a system oscillates, the magnitude of restoring torque of system is given by

$$\tau = C\theta \quad \text{...(i)}$$

where, C = constant that depends on system.

Also, $\quad \tau = I\alpha \quad$ ...(ii)

where, I = moment of inertia

and $\quad \alpha$ = angular acceleration

From Eqs. (i) and (ii),

$$\alpha = \frac{C}{I}\cdot\theta \quad \text{...(iii)}$$

and time period of oscillation of system will be

$$T = 2\pi\sqrt{\frac{I}{C}}$$

In given case, magnitude of torque is

τ = Force $\times$ perpendicular distance

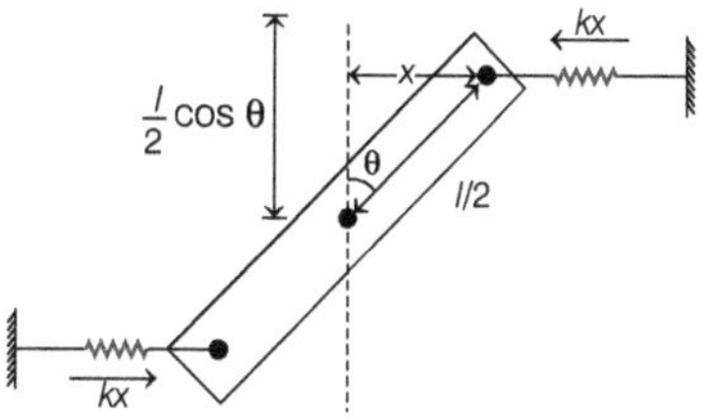

$$\tau = 2kx \times \frac{l}{2}\cos\theta$$

For small deflection, $\tau = \left(\dfrac{kl^2}{2}\right)\theta \quad$ ...(iv)

$\because$ For small deflections, $\sin\theta = \dfrac{x}{(l/2)} \approx \theta$

$\Rightarrow \quad x = \dfrac{l\theta}{2}$

Also, $\cos\theta \approx 1$

comparing Eqs. (iv) and (i), we get

$$C = \frac{kl^2}{2}$$

$$\Rightarrow \alpha = \frac{(kl^2/2)}{\left(\frac{1}{12}ml^2\right)}\cdot\theta \Rightarrow \alpha = \frac{6k}{m}\cdot\theta$$

Hence, time period of oscillation is

$$T = 2\pi\sqrt{\frac{m}{6k}}$$

Frequency of oscillation is given by

$$f = \frac{1}{T} = \frac{1}{2\pi}\sqrt{\frac{6k}{m}}$$

8 *(d)* Given, density of bob,

$$\rho = \frac{4}{3}\times 10^3\ \text{kg/m}^3$$

Density of water, $\sigma = 10^3$ kg/m^3

If g' be gravitational acceleration in water then,

$$g' = g\left(1 - \frac{\sigma}{\rho}\right) = g\left(1 - \frac{10^3}{4/3\times10^3}\right) = \frac{g}{4}$$

As, $\quad T_{\text{air}} = 2\pi\sqrt{\dfrac{l}{g}},$

Similarly, $\quad T_{\text{water}} = 2\pi\sqrt{\dfrac{l}{g'}}$

$$[T_{\text{water}} = 2\text{s}]\left(g' = \frac{g}{4}\right)$$

$$2 = 2\pi\sqrt{\frac{l}{g/4}} = 2\pi\sqrt{\frac{l}{g}}.2$$

$$2 = 2.\,T_{\text{air}} \Rightarrow T_{\text{air}} = 1\ \text{s}$$

9 *(a)* Given, time period of the pendulum while the lift is moving upwards and downward are in the ratio,

$$T_1 : T_2 = 1:2$$

Acceleration due to gravity, $g = 10$ m/s^2

We know that,

If the lift is moving upward, then total time-period,

$$T_1 = 2\pi\sqrt{\frac{l}{g+a}} \quad \text{...(i)}$$

when the lift is moving downwards, then the total time period,

$$T_2 = 2\pi\sqrt{\frac{l}{g-a}} \quad \text{...(ii)}$$

By dividing Eq. (i) to (ii), we get

$\therefore \quad \dfrac{T_1}{T_2} = \sqrt{\dfrac{g-a}{g+a}}$

Now, $\quad \dfrac{1}{2} = \sqrt{\dfrac{g-a}{g+a}}$

Square on the both sides, we get

or $\quad \left(\dfrac{1}{2}\right)^2 = \dfrac{g-a}{g+a}$ or $\dfrac{g-a}{g+a} = \dfrac{1}{4}$

or $\quad 4g - 4a = g + a$ or $3g = 5a$

or $\quad a = \dfrac{3g}{5} \Rightarrow a = \dfrac{30}{5} = 6$ m/s^2

So, the acceleration of the lift is 6 m/s^2.

10 *(b)* Time period of simple pendulum inside a stationary lift,

$$T = 2\pi\sqrt{\frac{l}{g}} \quad \text{...(i)}$$

where, l = length of string

and $\quad g$ = gravitational acceleration

Acceleration of lift in upward direction,

$$a = \frac{g}{3}$$

$\therefore$ Time period of pendulum,

$$T' = 2\pi\sqrt{\frac{l}{g+a}} = 2\pi\sqrt{\frac{l}{g+\frac{g}{3}}}$$

$$= 2\pi\sqrt{\frac{3l}{4g}} = 2\pi\cdot\frac{\sqrt{3}}{2}\cdot\sqrt{\frac{l}{g}}$$

$$= \frac{\sqrt{3}}{2}\cdot 2\pi\sqrt{\frac{l}{g}} = \frac{\sqrt{3}}{2}T \quad \text{[from Eq. (i)]}$$

11 (*a*) As two pendulums begin to swing simultaneously, then

$$n_1 T_1 = n_2 T_2 \quad \text{...(i)}$$

where, n_1 and n_2 are the number of oscillations of first and second pendulum respectively and T_1 and T_2 be their respective time periods. The time period of simple pendulum is given by

$$T = 2\pi\sqrt{\frac{l}{g}}$$

where, l = length of pendulum
and g = acceleration due to gravity

$$\Rightarrow \quad T^2 \propto l \quad \text{...(ii)}$$

So, from Eqs. (i) and (ii), we get

$$\frac{l_1}{l_2} = \frac{T_1^2}{T_2^2} = \frac{n_2^2}{n_1^2}$$

Here, $n_1 = 9, n_2 = 7$

$$\Rightarrow \quad \frac{l_1}{l_2} = \frac{(7)^2}{(9)^2} = \frac{49}{81}$$

Hence, the ratio of pendulum length

$$l_1 : l_2 = 49:81$$

Elasticity

1. (*d*) Pressure at depth, $p = \rho gh$

Bulk modulus, $B = \frac{p}{\Delta V/V} = \frac{\rho gh}{\Delta V/V}$

$$= \frac{10^3 \times 9.8 \times 2 \times 10^3}{1.36 \times 10^{-2}}$$

$$= 1.44 \times 10^9 \text{ Nm}^{-2}$$

2. (*c*) Given, depth of lake, h = 200 m

Decrease in volume, $\frac{\Delta V}{V} = 0.1\%$

$$= \frac{0.1}{100} = \frac{1}{1000}$$

Since, $p = \rho gh$
where, p = pressure,
ρ = density of water = 10^3 kg m^{-3}
and g = acceleration due to gravity = 10 ms^{-2}.

$\therefore \quad p = 10^3 \times 10 \times 200$

$= 2 \times 10^6$ Pa

$\because$ Bulk modulus, $K = \frac{p}{\frac{\Delta V}{V}}$

$$= \frac{2 \times 10^6}{\frac{1}{1000}} = 2 \times 10^9 \text{ Nm}^{-2}$$

3 (*a*) Elastic potential energy stored in a loaded wire,

$$U = \frac{1}{2}(\text{sttress} \times \text{strain} \times \text{volume})$$

$\therefore$ Energy stored per unit volume,

$$u = \frac{U}{\text{volume}} = \frac{1}{2} \times \text{stress} \times \text{strain}$$

$$= \frac{1}{2}\left(\frac{F}{A}\right)^2 \times \frac{1}{Y}$$

Here, both wires are of same material and under same load, so the ratio of stored energies per unit volume, for both the wires will be

$$\frac{u_A}{u_B} = \frac{\frac{1}{2Y}\cdot\frac{F^2}{A_A^2}}{\frac{1}{2Y}\cdot\frac{F^2}{A_B^2}} = \frac{A_B^2}{A_A^2}$$

$$\Rightarrow \quad \frac{u_A}{u_B} = \frac{d_B^4}{d_A^4} \quad \left(\because A = \pi\frac{d^2}{4}\right)$$

Here, $\frac{u_A}{u_B} = \frac{1}{4}$

So, $\frac{d_B^4}{d_A^4} = \frac{1}{4}$

or $\frac{d_B}{d_A} = \frac{1}{\sqrt{2}}$

$\Rightarrow \quad \frac{d_A}{d_B} = \sqrt{2} : 1$

4 (*c*) Given, radius of wire, r = 2 mm

$= 2 \times 10^{-3}$ m

Weight of load, m = 4 kg, $g = 3.1\,\pi$ ms^{-2}

$\therefore$ Tensile stress $= \frac{\text{Force }(F)}{\text{Area }(A)} = \frac{mg}{\pi r^2}$

$$= \frac{4 \times 3.1 \times \pi}{\pi \times (2 \times 10^{-3})^2}$$

$$= 3.1 \times 10^6 \text{ Nm}^{-2}$$

5 (*a*) When rubber cord is stretched, then it stores potential energy and when released, this potential energy is given to the stone as kinetic energy.

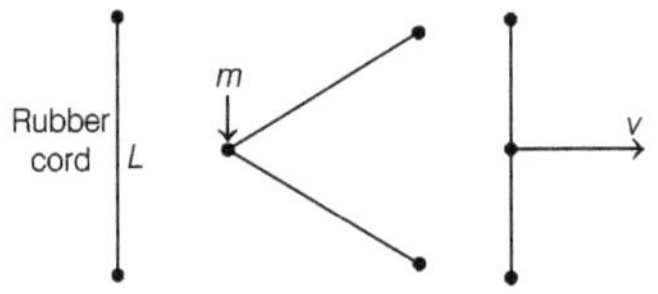

So, potential energy of stretched cord = kinetic energy of stone

$$\Rightarrow \quad \frac{1}{2}Y\left(\frac{\Delta l}{l}\right)^2 A\cdot L = \frac{1}{2}mv^2$$

Here, Δl = 20 cm = 0.2 m, l = 42 cm

= 0.42 m,

v = 20 ms^{-1}, m = 0.02 kg, d = 6 mm

$= 6 \times 10^{-3}$ m

$$\therefore \quad A = \pi r^2 = \pi\left(\frac{d}{2}\right)^2$$

$$= \pi\left(\frac{6 \times 10^{-3}}{2}\right)^2$$

$$= \pi(3 \times 10^{-3})^2$$

$$= 9\pi \times 10^{-6} \text{ m}^2$$

On substituting values, we get

$$Y = \frac{mv^2 L}{A(\Delta L)^2}$$

$$= \frac{0.02 \times (20)^2 \times 0.42}{9\pi \times 10^{-6} \times (0.2)^2}$$

$$\approx 3.0 \times 10^6 \text{ Nm}^{-2}$$

So, the closest value of Young's modulus is 10^6 Nm^{-2}.

6 (*a*) Given, $T_1 = 40°$ C and $T_2 = 20°$ C

$$\Rightarrow \quad \Delta T = T_1 - T_2 = 40 - 20 = 20° \text{ C}$$

Also, Young's modulus,

$$Y = 10^{11} \text{ N/m}^2$$

Coefficient of linear expansion,

$$\alpha = 10^{-5} /°\text{C}$$

Area of the brass wire, $A = \pi \times (10^{-3})^2$ m^2

Now, expansion in the wire due to rise in temperature is

$$\Delta l = l\,\alpha\Delta T$$

$$\Rightarrow \quad \frac{\Delta l}{l} = \alpha\Delta T \quad \text{...(i)}$$

We know that, Young's modulus is defined as

$$Y = \frac{Mgl}{A\Delta l}$$

$$\Rightarrow \quad M = \frac{YA\Delta l}{gl} \quad \text{...(ii)}$$

Using Eq. (i), we get

$$M = \frac{YA}{g} \times \alpha\Delta T$$

$$= \frac{10^{11} \times 22 \times 10^{-6} \times 10^{-5} \times 20}{7 \times 10}$$

$$\Rightarrow \quad M = \frac{22 \times 20}{7 \times 10}$$

$$= \frac{44}{7} = 6.28 \text{ kg}$$

which is closest to 9, so option (a) is nearly correct.

7. (*d*) Given, $Y = 2.2 \times 10^8$ m/s^2

Area of cross-section of steel wire,

$$A = 4 \text{ cm}^2 = 4 \times 10^{-4} \text{m}^2$$

Mass of elevator, $m = 1000$ kg

The maximum tension in the steel wire that can support elevator is given by

$$T = \frac{1}{4} \times \text{stress} \times \text{area of cross-section}$$

$$= \frac{1}{4} \times 2.2 \times 10^8 \times 4 \times 10^{-4}$$

$$= 2.2 \times 10^4 \text{ N/m}^2$$

If f be the maximum upward acceleration of the elevator, then

$$\Rightarrow \quad T = m(g + f)$$

$$2.2 \times 10^4 = 1000\ (10 + f)$$

$$\Rightarrow \quad f = 12 \text{ m/s}^2$$

8 (*c*) Force constant K of rubber is given by

$$K = \frac{YA}{l} = \frac{5 \times 10^8 \times 10^{-6}}{0.1} = 5 \times 10^3 \text{ N/m}$$

Now, from conservation of energy elastic potential energy of cord = kinetic energy of particle

i.e. $\frac{1}{2} K (\Delta l)^2 = \frac{1}{2} mv^2$

$$\Rightarrow \quad v = \sqrt{\frac{K}{m}} \cdot \Delta l$$

$$= \sqrt{\frac{5 \times 10^3}{5 \times 10^{-3}}} (0.125 - 0.1)$$

$$= 10^3 \times 0.025 = 25 \text{ m/s}$$

9 (*d*) Given, elongation of the wire,

$$\Delta l = 2 \text{ mm} = 2 \times 10^{-3} \text{ m}$$

Mass of the ball, $m = 1$ kg

Length of wire, $l = 1$ m

Area of cross-sectional of wire,

$A = 0.01 \text{ cm}^2 = 0.01 \times 10^{-4}$ m

Young's modulus of steel, $Y = 2 \times 10^{11}$ Nm^{-2}

$\because$ Tension force in wire, $T = m\omega^2 l$

$$\because \text{Stress} = \frac{\text{Tension}}{\text{Area}} = \frac{m\omega^2 l}{A}$$

$$\text{Strain} = \frac{\Delta l}{l} = \frac{\text{stress}}{\text{Young's modulus}}$$

$$\text{or } \Delta l = \frac{m\omega^2 l^2}{YA}$$

$$\text{or } \quad \omega = \sqrt{\frac{YA\Delta l}{ml^2}}$$

Putting the given values, we get

$$= \sqrt{\frac{2 \times 10^{11} \times 0.01 \times 10^{-4} \times 2 \times 10^{-3}}{1 \times (1)^2}}$$

$$\omega = 20 \text{ rad s}^{-1}$$

10 (*c*) Given,

edge of solid copper cube, $l = 7$ cm

hydraulic pressure, $p = 8000$ kPa

$= 8000 \times 10^3$ Pa

and Bulk modulus of copper,

$$\beta = 140 \text{ GPa}$$

$$= 140 \times 10^9 \text{ Pa}$$

As we know that,

$$\text{Bulk modulus}, \beta = \frac{p}{\left(\frac{\Delta V}{V}\right)} \text{ or } \beta = \frac{pV}{\Delta V}$$

$$\therefore \Delta V = \frac{pV}{\beta} = \frac{8000 \times 10^3 \times (l)^3}{140 \times 10^9}$$

$$= \frac{8000 \times 10^3 \times (7)^3}{140 \times 10^9}$$

$$= 19.6 \times 10^{-3} \text{ cm}^3$$

11 (*c*) Given,

length of uniform rod = L

angular speed = ω and density of rod = ρ

When rod is rotated in a horizontal plane, centrifugal force is responsible for increasing the length of rod, which is also equal to stress.

If dl be the change in length of small element dx due to rotation, by application of force dF.

Young's modulus,

$$Y = \frac{\text{stress}}{\text{strain}} = \frac{dF/A}{dl/x} \Rightarrow dl = \frac{dF.x}{AY}$$

$$\Rightarrow \quad dF = \frac{AYdl}{x} \quad \text{... (i)}$$

Where, dF = centrifugal force due to mass of the element of length, dx.

$$dF = dm.x\omega^2$$

$$dF = \frac{m}{L}.xdx\omega^2 \quad \text{... (ii)}$$

$$\left[\because \text{Change in mass } dm = \frac{m}{L}dx\right]$$

From Eqs. (i) and (ii), we get

$$\frac{AYdl}{x} = \frac{m}{L}.x\ dx\ \omega^2$$

$$\Rightarrow \quad dl = \frac{mx^2\omega^2}{LAY}.dx$$

$\Rightarrow$ Total change in length,

$$l = \int dl = \int_0^L \frac{mx^2\omega^2}{LAY} dx$$

$$= \frac{m\omega^2 L^2}{3AY} = \frac{m}{AL} \cdot \frac{\omega^2 L^3}{3Y} = \frac{\rho\omega^2 L^3}{3Y}$$

12. (*a*) Given, cross-section area of copper wire, $A = 0.01 \text{ cm}^2 = 10^{-6} \text{ m}^2$ and tension force, $F = 22$ N,

Poisson's ratio, $\sigma = 0.32$, and Young's modulus $Y = 1.1 \times 10^{11}$ N/m^2

Since, Poisson's ratio,

$$\sigma = \frac{\text{Lateral strain}}{\text{Longitudinal strain}}$$

$$\sigma = -\frac{\frac{\Delta D}{D}}{\frac{\Delta L}{L}} \quad \text{...(i)}$$

$\because$ Area, $A = \pi D^2$

$$\Rightarrow \quad \frac{\Delta A}{A} = 2\frac{\Delta D}{D} \quad \text{...(ii)}$$

From Eqs. (i) and (ii), we get

$$\Rightarrow \quad \frac{\Delta A}{A} = 2\sigma\frac{\Delta L}{L} \quad \text{...(iii)}$$

$$\text{Young's modulus, } Y = \frac{FL}{A\Delta L} \quad \text{...(iv)}$$

From Eqs. (iii) and (iv), we get

$$\Rightarrow \quad \frac{\Delta A}{A} = \frac{2\sigma F}{YA}$$

Now, putting the given values,

$$\frac{\Delta A}{A} = \frac{2 \times 0.32 \times 22}{1.1 \times 10^{11} \times 10^{-6}} = 12.8 \times 10^{-5}$$

$$\Rightarrow \quad \frac{\Delta A}{A}\% = 12.8 \times 10^{-3} \simeq 12.6 \times 10^{-3}$$

So, option (a) is correct.

13 (*b*) Modulus of rigidity of a horizontal rod with hanging mass is given as

$$\eta = \frac{F_L}{Ax} = \frac{mgL}{Ax} \quad \text{...(i)}$$

Where, x = verticle deflection of the end of the rod.

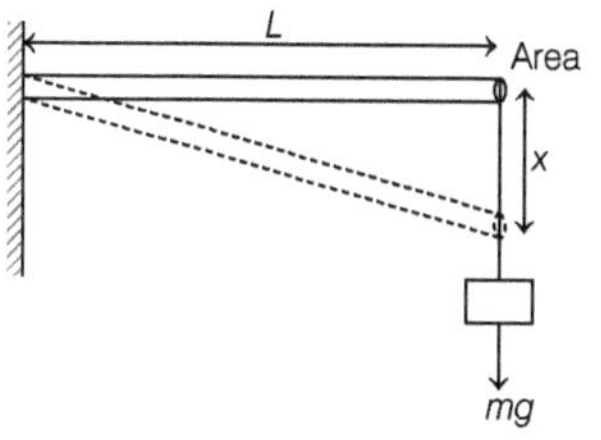

Here, $L = 6 \text{ cm} = 6 \times 10^{-2}$ m, radius $R = 2$ cm $= 2 \times 10^{-2}$ m

mass, $m = 400\ \pi$ kg and $\eta = 3 \times 10^{10}$ N/m^2

Putting these values, in Eq. (i) we get

$$3 \times 10^{10} = \frac{(400\pi \times 10) \times 6 \times 10^{-2}}{\pi\ (2 \times 10^{-2})^2 x}$$

$$\Rightarrow \quad x = \frac{4\pi \times 10^3 \times 6 \times 10^{-2}}{\pi \times 4 \times 10^{-4} \times 3 \times 10^{10}}$$

$$\Rightarrow \quad x = 0.02 \text{ mm}$$

Fluid Mechanics

1

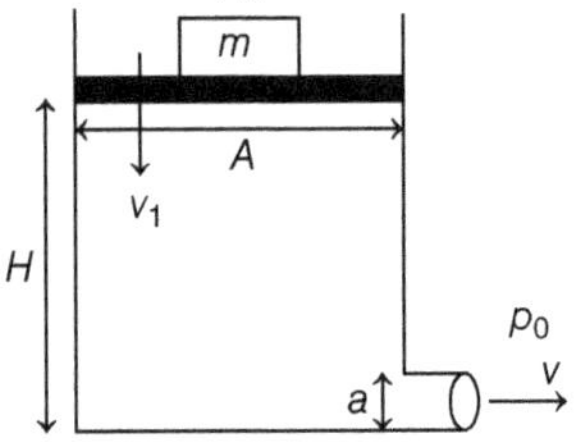

Given, $m = 24$ kg, $A = 0.4\ m^2$

$a = 1\ cm^2$, $H = 40$ cm

Using Bernoulli's equation,

$$\Rightarrow \left(p_0 + \frac{mg}{A}\right) + \rho gH + \frac{1}{2}\rho v_1^2 = p_0 + 0 + \frac{1}{2}\rho v^2 \quad ...(i)$$

Neglecting v_1, we get

$$\Rightarrow v = \sqrt{2gH + \frac{2mg}{A\rho}}$$

$$\Rightarrow v = \sqrt{8 + 1.2}$$

$$\Rightarrow v = 3.033 \text{ m/s}$$

$$\Rightarrow v = 3 \text{ m/s}$$

2 (*a*) Initial pressure, $p_1 = \rho gd + p_0 = 3\times10^5$ Pa

$$\rho gd = 2 \times 10^5 \text{ Pa}$$

Final pressure, $p_2 = 2\rho gd + p_0$

$$= 4 \times 10^5 + 1 \times 10^5 = 5 \times 10^5 \text{ Pa}$$

$$\% \text{ increase in pressure} = \frac{p_2 - p_1}{p_1} \times 100$$

$$= \frac{5\times10^5 - 3\times10^5}{3\times10^5} \times 100 = \frac{200}{3}\%$$

3. (*b*) Given, density of body $= 1.2\times10^3\ kg/m^3 = \rho_b$

Density of liquid $= 2.4\times10^3\ kg/m^3 = \rho_l$

Height of fall = 1m = H

Let the volume of the body be V.

Then, buoyant force acting on the body when it is totally immersed in liquid $= (\rho_l Vg)$

Weight at the body $= (\rho_b Vg)$

Net upward force acting on the body $= (\rho_l - \rho_b)Vg$

Net deceleration produced

$$(a) = \frac{\text{Net upward force}}{\text{Mass of body}}$$

$$(a) = \frac{(\rho_l - \rho_b)Vg}{\rho_b V}$$

$$(a) = \frac{(\rho_l - \rho_b)}{\rho_b} g$$

Let initial velocity of the body be u, then

$$\frac{1}{2}(\rho_b V)u^2 = \rho_b VgH$$

$$\Rightarrow u^2 = 2gH$$

Final velocity of the body will be zero.

$$\Rightarrow v^2 - u^2 = 2as$$

$$\Rightarrow (0)^2 - 2gH = -2\frac{(\rho_l - \rho_b)g}{\rho_b} \times s$$

$$\Rightarrow s = \frac{\rho_b}{\rho_l - \rho_b} \cdot H$$

$$= \left(\frac{1.2\times10^3}{2.4\times10^3 - 1.2\times10^3}\right) \cdot 1$$

$$= \frac{1.2}{1.2} = 1 \text{ m}$$

4 (*c*) Force on a vertical surface of area A and dipped upto height h in a fluid of density ρ is

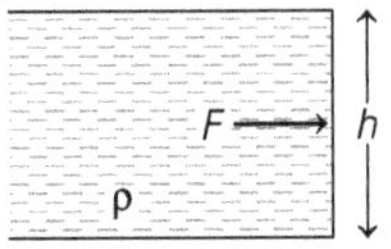

$$\text{Force, } F = \frac{1}{2}(\rho gh \times A)$$

So, in given case,

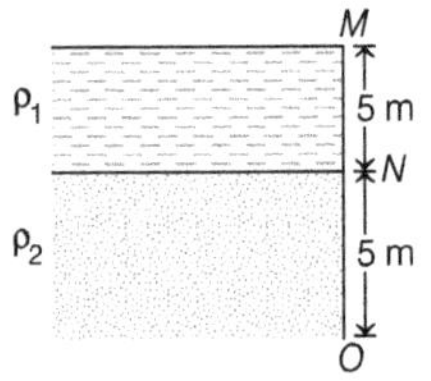

$$\text{Force on portion } MN = \frac{\rho_1 gh}{2} \times A$$

Force on portion NO

= Force due to pressure of liquid in MN portion + Force due to pressure of liquid in NO portion

$$= \rho_1 ghA + \frac{1}{2}(\rho_2 ghA)$$

$$= \rho_1 ghA + \frac{1}{2} \times 2\rho_1 \times ghA \quad (\because \rho_2 = 2\rho_1)$$

$$= 2\rho_1 ghA$$

$$\text{Required ratio, } \frac{F_{MN}}{F_{NO}} = \frac{\frac{1}{2}(\rho_1 ghA)}{2\rho_1 ghA} = \frac{1}{4}$$

5 (*c*) By equation of continuity for sections A and B, we have

$$A_A v_A = A_B v_B$$

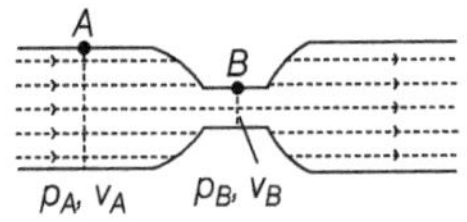

$$\Rightarrow 40\ v_A = 20\ v_B \Rightarrow 2\ v_A = v_B \quad ...(i)$$

Now, using Bernoulli's equation (for horizontal tube), we have

$$p_A + \frac{1}{2}\rho v_A^2 = p_B + \frac{1}{2}\rho v_B^2$$

$$\Rightarrow p_A - p_B = \frac{1}{2}\rho(v_B^2 - v_A^2)$$

Here, $p_A - p_B = 700\ Nm^{-2}$

and $\rho = 1000\ kg\ m^{-3}$

$$\Rightarrow 700 = \frac{1}{2} \times 1000\ (v_B^2 - v_A^2)$$

$$\Rightarrow v_B^2 - v_A^2 = 1.4 \quad ...(ii)$$

From Eqs. (i) and (ii), we have

$$3\ v_A^2 = 1.4 \Rightarrow v_A = \sqrt{0.467}$$

$$= 0.68\ ms^{-1} = 68\ cm\ s^{-1}$$

So, volume flow rate of water $= A_A \cdot v_A$

$$= 40 \times 68 = 2720\ cm^3/s$$

6 (*a*) Weight of the drop is balanced by upthrust or buoyant force (F_B) and surface tension (T).

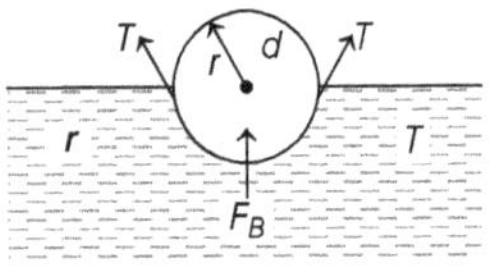

So, weight, $w = F_B + T \times 2\pi r$

$$\Rightarrow d \cdot V \cdot g = \rho \cdot \frac{V}{2} g + T \times 2\pi r$$

$$\left[\because w = dVg \text{ and } F_B = \rho \frac{V}{2} g\right]$$

$$\Rightarrow d \cdot \frac{4}{3}\pi r^3 \cdot g = \rho \cdot \frac{2}{3}\pi r^3 \cdot g + T \times 2\pi r$$

$$\left[\because V = \frac{4}{3}\pi r^3\right]$$

$$\Rightarrow r = \sqrt{\frac{3T}{(2d - \rho)g}}$$

7 (*c*) For a floating body, weight of body = upthrust or buoyant force = weight of water displaced. Initially at 0°C, cylinder floats as shown in the figure.

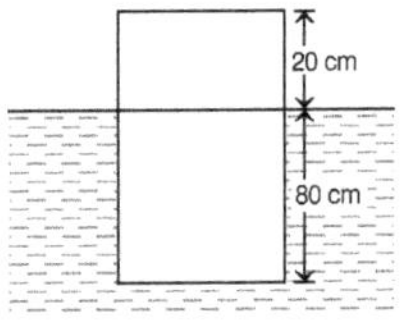

Here, weight of cylinder = weight of water displaced by cylinder of 80 cm length.

$$\Rightarrow mg = \rho_0 \cdot g \cdot 80\ A \quad ...(i)$$

where, ρ_0 = density of water at 0°C and A = area of cross-section of cylinder. When temperature is raised to 4°C, cylinder floats as shown in the figure. In above case, weight of cylinder = weight of water displaced by cylinder of 79 cm length.

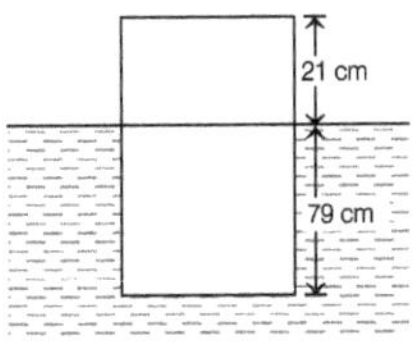

$$\Rightarrow mg = \rho_4 \cdot g \cdot 79\ A \quad ...(ii)$$

From Eqs. (i) and (ii), we have

$$80\ \rho_0 = \rho_4 \cdot 79 \Rightarrow \frac{\rho_4}{\rho_0} = \frac{80}{79} = 1.01$$

8 (*b*) When soap bubble is being inflated and its temperature remains constant

$$pV = \text{constant } (k) \Rightarrow p = \frac{k}{V}$$

Differentiating above equation with time, we get

$\frac{dp}{dt} = k.\frac{d}{dt}\left(\frac{1}{V}\right) \Rightarrow \frac{dp}{dt} = k\left(\frac{-1}{V^2}\right).\frac{dV}{dt}$

It is given that, $\frac{dV}{dt} = c$ (a constant)

So, $\frac{dp}{dt} = \frac{-kc}{V^2}$ ... (i)

Now, from $\frac{dV}{dt} = c$, we get

$dV = cdt$

or $\int dV = \int cdt$ or $V = ct$ ... (ii)

From Eqs. (i) and (ii), we get

$\frac{dp}{dt} = \frac{-kc}{c^2t^2}$ or $\frac{dp}{dt} = -\left(\frac{k}{c}\right)t^{-2}$

$\Rightarrow dp = -\frac{k}{c}.t^{-2}dt$

Integrating both sides, we get

$\int dp = -\frac{k}{c}\int t^{-2}dt \Rightarrow p = -\frac{k}{c}.\left(\frac{t^{-2+1}}{-2+1}\right)$

$= -\frac{k}{c}.\frac{-1}{t} = \frac{k}{ct}$ or $p \propto \frac{1}{t}$

Hence, p *versus* $\frac{1}{t}$ graph is a straight line, which is correctly represented in option (b).

9 *(d)* When liquid filled vessel is rotated the liquid profile becomes a paraboloid due to centripetal force, as shown in the figure below

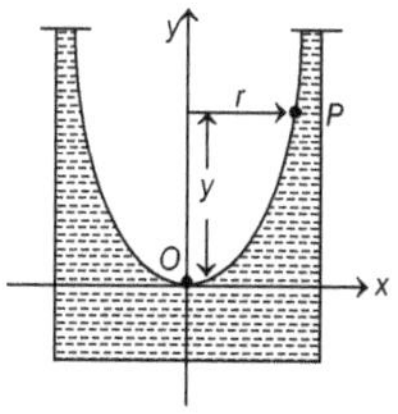

Pressure at any point P due to rotation is

$$p_R = \frac{1}{2}\rho r^2\omega^2$$

Gauge pressure at depth y is $p_G = -\rho gy$

If p_0 is atmospheric pressure, then total pressure at point P is

$$p = p_0 + \frac{1}{2}\rho r^2\omega^2 - \rho gy$$

For any point on surface of rotating fluid,

$$p = p_0$$

Hence, for any surface point;

$$p_0 = p_0 + \frac{1}{2}\rho r^2\omega^2 - \rho gy$$

or $\frac{1}{2}\rho r^2\omega^2 = \rho gy \Rightarrow y = \frac{r^2\omega^2}{2g}$ ... (i)

given case, $\omega = 2\pi s = 2 \times 2\pi = 4\pi$ rad s^{-1}

$r = 5$ cm $= 0.05$ m

and $g = 10$ ms^{-2}

Hence, substituting these values in Eq. (i), we get $y = \frac{\omega^2 r^2}{2g} = \frac{(4\pi)^2(0.05)^2}{2 \times 10} = 0.02$ m

$= 2$ cm

10 *(a)* Excess pressure inside first bubble of radius r_1.

$$p_1 = p_0 + \frac{4T}{r_1}$$

Excess pressure inside second bubble of radius r_2.

$$p_2 = p_0 + \frac{4T}{2}$$

Excess prssure inside double bubble,

$$p = p_2 - p_1 \Rightarrow p = 4T\left(\frac{r_1 - r_2}{r_1 r_2}\right)$$

If R be the radius of double bubble, then

$$p = \frac{4T}{R}$$

or $4T\left(\frac{r_1 - r_2}{r_1 r_2}\right) = \frac{4T}{R}$

$\Rightarrow R = \frac{r_1 r_2}{r_1 - r_2} = \frac{3\times10^{-3}\times2\times10^{-3}}{(3-2)\times10^{-3}}$

$= 6\times10^{-3}$ m

$R = 6\times10^{-3}$ m

11 (*) Here, $D = 2$ mm, $\eta = 1$ centi-poise $= 10^{-3}$ Pa-s and density of the water, $\rho = 10^3$ kg/m^3

For flow to be just turbulent, $R_e = 3000$

$\therefore v = \frac{R_e\eta}{\rho D} = \frac{3000 \times 10^{-3}}{10^3 \times 2 \times 10^{-3}} = 1.5$

We know that the velocity head, $h = \frac{v^2}{2g}$

$\Rightarrow h = \frac{(1.5)^2}{2 \times 10} = 0.1125 \simeq 11$ cm

So, no option is matched.

12 *(c)* Given, $r = 3 \times 10^{-3}$ m,

$\eta = 1$ kg/ms, $\rho = 1.5 \times 10^3$ kg/m^3

$\sigma = 9 \times 10^3$ kg/m^3

$$v_T = \frac{2}{9}\frac{r^2(\sigma - \rho)g}{n}$$

Putting the given values, we get

$= \frac{2}{9} \times \frac{(3 \times 10^{-3})^2(9 \times 10^3 - 1.5 \times 10^3) \times 10}{1}$

$= \frac{2}{9} \times 9 \times 10^{-6} \times 7.5 \times 10^3 \times 10$

$\Rightarrow v_T = 15 \times 10^{-2}$ m/s

Hence, the correct option is (c).

13 *(d)* $\because$ Pressure due to surface tension inside the soap bubble,

$$p_i = \frac{4S}{r} \quad ...(i)$$

Electrostatic pressure outside the soap bubble,

$$p_0 = \frac{\sigma^2}{2\varepsilon_0} \quad ...(ii)$$

$\because$ Electric potential, $V = \frac{kQ}{r} = \frac{k(\sigma A)}{r}$

So, $V = \frac{1}{4\pi\varepsilon_0}\left(\frac{\sigma 4\pi r^2}{r}\right) = \frac{\sigma r}{\varepsilon_0}$ or $\sigma = \frac{\varepsilon_0 V}{r}$

According to the question, $p_i = p_0$

From Eqs. (i) and (ii), we get

$$\frac{4S}{r} = \frac{\sigma^2}{2\varepsilon_0} \Rightarrow \frac{4S}{r} = \frac{\left(\frac{\varepsilon_0 V}{r}\right)^2}{2\varepsilon_0} \Rightarrow V = \sqrt{\frac{8Sr}{\varepsilon_0}}$$

Hence, the pressure outside the bubble is $\sqrt{\frac{8Sr}{\varepsilon_0}}$.

14 *(d)* According to Bernoulli's theorm, when a stream of air passes through the space between the balls, the pressure reduce between them as compared to the atmospheric pressure on either side.

Thermometry, Thermal Expansion and Kinetic Theory of Gases

1. Given, at 27°C, root mean square speed,

$(v_{rms})_1 = 100$ ms^{-1}

$$(v_{rms})_1 = \sqrt{\frac{3p_1}{d_1}} = \sqrt{\frac{3p_1V_1}{M}} \quad ...(i)$$

According to ideal gas equation,

$\Rightarrow \frac{V_1}{V_2} = \frac{p_2T_1}{p_1T_2}$

$= \frac{2 \times 300}{400} = \frac{3}{2}$

$\therefore$ At 127°C, root mean square speed,

$$(v_{rms})_2 = \sqrt{\frac{3p_2}{d_2}} = \sqrt{\frac{3p_2V_2}{M}} \quad ...(ii)$$

From Eqs. (i) and (ii), we get

$\therefore (v_{rms})_2^2 = (v_{rms})_1^2 \times \frac{V_2}{V_1} \times \frac{p_2}{p_1}$

$= (100)^2 \times \frac{2}{3} \times \frac{2}{1}$

or $(v_{rms})_2 = \frac{200}{\sqrt{3}}$ ms^{-1}

$\therefore n = 3$

2. *(c)* Total internal energy of a gas is given as

$$U = \frac{n}{2}fRT$$

where, n = number of moles

and f = degree of freedom.

Given, $n_{diatomic} = n_{monoatomic} = 2$

As, $f_{monoatomic} = 3$

$f_{diatomic} = 5$

$\Rightarrow U_{monoatomic} = \frac{2}{2} \times 3RT = 3RT$

$U_{diatomic} = \frac{2}{2} \times 5RT = 5RT$

$\therefore$ Total internal energy of the mixture of gases,

$U_{total} = U_{monoatomic} + U_{diatomic}$

$= 3RT + 5RT = 8RT$

3. (a) If the degrees of freedom is n, then ratio of specific heat

$$\frac{C_p}{C_V} = \gamma = 1 + \frac{2}{n}$$

4 (c) For a mixture of two gases, ratio of specific heat is given by

$$\gamma_{\text{mixture}} = \frac{C_{p_{\text{mixture}}}}{C_{V_{\text{mixture}}}} = \frac{n_1 C_{p_1} + n_2 C_{p_2}}{n_1 C_{V_1} + n_2 C_{V_2}} \quad \text{....(i)}$$

Here, gas 1 is helium which is monoatomic,

so $n_1 = n, C_{p_1} = \frac{5}{2}R, C_{V_1} = \frac{3}{2}R$

and gas 2 is oxygen which is diatomic,

so $n_2 = 2n$,

$$C_{p_2} = \frac{7}{2}R, \; C_{V_2} = \frac{5}{2}R$$

Hence, from Eq. (i), we get

$$\gamma_{\text{mix}} = \frac{n \times \frac{5R}{2} + 2n \times \frac{7R}{2}}{n \times \frac{3R}{2} + 2n \times \frac{5R}{2}} = \frac{19}{13}$$

5 (c) Mean free time (τ) for a gas molecule is time elapsed between two successive collisions.

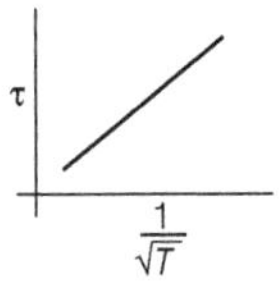

$$\therefore \; \tau = \frac{\text{Mean free path length } (\lambda)}{\text{Average speed } (v_{av})}$$

$$= \frac{\left(\frac{1}{\sqrt{2} \cdot \pi d^2 \cdot n}\right)}{\left(\sqrt{\frac{8}{\pi} \frac{R}{M} \cdot T}\right)}$$

So, for an ideal gas,

$$\tau \propto \frac{1}{\sqrt{T}}$$

Thus, graph of τ *versus* $\frac{1}{\sqrt{T}}$ is a straight line as shown in the figure.

6 (d) For a gas value of specific heat at constant volume, $C_V = \frac{1}{2} fR$

where, f = degrees of freedom
and R = gas constant. For a diatomic gas molecule, degrees of freedom is in general 5 but when vibration occurs, then degrees of freedom increases to 7.

So, for gas A, specific heat is $C_V^A = \frac{1}{2} 5R$

and for gas B, specific heat is $C_V^B = \frac{1}{2} 7R$

$\therefore$ Ratio of specific heats of gas A and B is,

$$\frac{C_V^A}{C_V^B} = \frac{5}{7}$$

7. (b) Internal energy of a gas with f degree of freedom is

$U = \frac{n f RT}{2}$,

Internal energy due to O_2

$$U_1 = \frac{n_1 f_1 RT}{2} = 3 \times \frac{5}{2} RT$$

($\because n_1 = 3$ moles, degree of freedom for a diatomic gas $f_1 = 5$)

Internal energy due to Ar

$$U_2 = \frac{n_2 f_2 RT}{2}$$

$$= 5 \times \frac{3}{2} RT$$

($\because n_2 = 5$ moles, degree of freedom for a monoatomic gas $f_2 = 3$)

$\therefore$ Total internal energy $= U_1 + U_2$

$\Rightarrow \quad U = 15 \, RT$

8. (a) When the piston is stationary, i.e. on equilibrium as shown in the figure below,

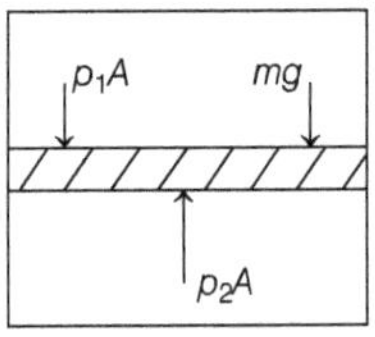

then $\quad p_1 A + mg = p_2 A$

$\Rightarrow \quad mg = p_2 A - p_1 A$

or $\quad mg = \left(\frac{nRTA}{V_2} - \frac{nRTA}{V_1}\right)$

$\{\because pV = nRT$ (ideal gas equation)$\}$

$$= nRT\left(\frac{A}{Al_2} - \frac{A}{Al_1}\right)$$

$$= nRT\left(\frac{l_1 - l_2}{l_1 l_2}\right)$$

or $\quad m = \frac{nRT}{g}\left(\frac{l_1 - l_2}{l_1 l_2}\right)$

9 (b) Given, density of gas at point A,

$\rho_A = \rho_0$

General equation for an ideal gas

$pV = nRT \Rightarrow pV = \frac{m}{M} . RT$ [$m \rightarrow$ mass, $M \rightarrow$ molecular mass]

$$\Rightarrow \quad p = \frac{m}{V} \cdot \frac{R}{M} . T \quad \text{... (i)}$$

Applying Eq. (i) at point A and B, respectively.

$$p = \rho_A \left(\frac{R}{M}\right) T_0 \quad \text{... (ii)}$$

$$2p = \rho_B \left(\frac{R}{M}\right) 3T_0 \quad \text{... (iii)}$$

From Eqs. (ii) and (iii), we get

$$\frac{1}{2} = \frac{\rho_A}{\rho_B} \cdot \frac{1}{3}$$

$$\Rightarrow \quad \rho_B = \frac{2}{3}\rho_A = \frac{2}{3}\rho_0$$

10 (c) Number of moles in 2×10^{-4} kg

$$\mu = \frac{\text{Mass of hydrogen in gram}}{\text{Molecules mass}}$$

$$= \frac{2 \times 10^{-4} \times 10^3}{2}$$

$= 0.1$ mole

$$\therefore \quad \frac{pV}{T} = \mu R = 0.1 \times R$$

$$= \frac{R}{10} \quad (\because pV = \mu RT)$$

11 (a) As we know temperature of a black body is given by,

$\lambda T = b \quad$ (Wien's displacement law)

$$\Rightarrow \quad T = \frac{b}{\lambda} = \frac{3 \times 10^{-3}}{1 \times 10^{-3}} = 3 \text{ K}$$

So, the correct option is (a).

12 (b) As, rms velocity of ideal gas particles is given,

$$v_{\text{rms}} = \sqrt{\frac{3kT}{m}}$$

$$\Rightarrow \quad v_{\text{rms}} \propto \sqrt{T}$$

So, the ratio $\frac{v_{1\,\text{rms}}^2}{v_{2\,\text{rms}}^2} = \frac{T_1}{T_2}$

As given, $v_{2\,\text{rms}} = 2 v_{1\,\text{rms}}$ and

$T_1 = 27\,°\text{C} = 300$ K

$$\Rightarrow \quad \frac{v_{1\,\text{rms}}^2}{4v_{1\,\text{rms}}^2} = \frac{300}{x}$$

$$\Rightarrow \quad x = 300 \times 4 = 1200 \text{ K}$$

Hence, 1200K = (1200 − 273)°C = 927 °C
Hence, the correct option is (b).

First Law of Thermodynamics

1. (*a*) $pV^{\gamma} = \text{constant}$

Differentiating, we get $\dfrac{dp}{dV} = \dfrac{\gamma p}{V}$

$$\frac{dp}{p} = -\frac{\gamma dV}{V}$$

2. (*b*) Given, temperature of source, $T_1 = 400$ K

Initial efficiency, $\eta = 25\%$

Final efficiency, $\eta' = 50\%$

As, efficiency of the Carnot cycle,

$$\eta = 1 - \frac{T_2}{T_1}$$

where, T_2 is the temperature of the sink.

Let $T_2 = T$

$\therefore$ Initial efficiency, $\eta = 1 - \dfrac{T}{T_1} = 25\%$

Substituting the given values, we get

$$\frac{25}{100} = 1 - \frac{T}{400}$$

$$4 \times 25 = 400 - T$$

$\Rightarrow \quad T = 400 - 100 = 300$ K

Final efficiency, $\eta' = 1 - \dfrac{T}{T_1'} = 50\%$

$\Rightarrow \quad 1 - \dfrac{300}{T_1'} = \dfrac{50}{100}$

$\Rightarrow \quad 100(T_1' - 300) = T_1' \times 50$

or $100\, T_1' - 30000 = 50\, T_1'$

or $\quad 50\, T_1' = 30000$

$\Rightarrow \quad T_1' = 600$ K

3 (*c*) Relation of coefficient of performance β of refrigerator and efficiency η of engine,

$$\beta = \frac{1-\eta}{\eta}$$

In given case, $\eta = \dfrac{1}{10}$

So, $\quad \beta = \dfrac{1 - \frac{1}{10}}{\frac{1}{10}} = \dfrac{\frac{9}{10}}{\frac{1}{10}} = 9$

Now, for a Carnot refrigerator, we have

$$\beta = \frac{\text{heat extracted from cold reservoir}}{\text{work done}}$$

$\Rightarrow \quad 9 = \dfrac{Q}{10} \quad \Rightarrow \quad Q = 90\text{J}$

4 (*b*) For the given *V*-*T* graph of thermodynamic cycle *xyzx*,

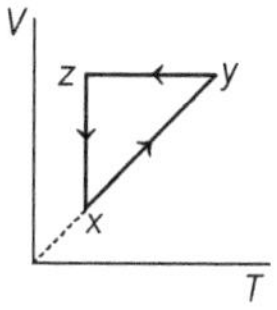

In process *xy*, $V \propto T$

$\Rightarrow$ pressure is constant.

Process *zx* is a isothermal compression process, so pressure increases in this process. Process *yz* is a isochoric process in which temperature decreases, so pressure must decrease in this process. Hence, the *p*-*V* graph is as shown in the figure.

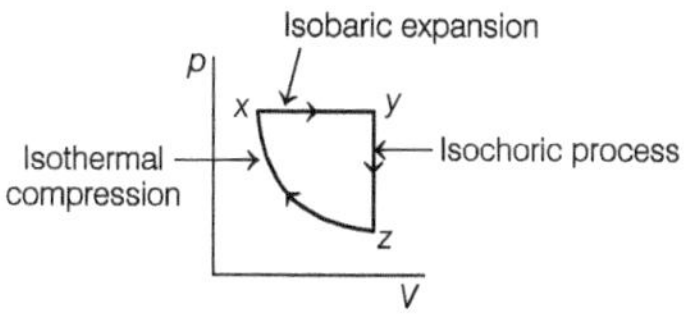

5. (*b*) The given expansion process is adiabatic in nature and we have the following data :

Initial volume, $V_1 = 1\text{ litre} = 10^{-3}\text{ m}^3$

Final volume, $V_2 = 3\text{ litre} = 3 \times 10^{-3}\text{ m}^3$

Initial pressure, $p_1 = 1\text{ atm} = 1.01 \times 10^5$ Pa

Final pressure, $p_2 = ?$

Using $p_1 V_1^{\gamma} = p_2 V_2^{\gamma}$, we have

$$1 \times 1^{\gamma} = p_2 \times (3)^{\gamma}$$

Here, $\gamma = 1.40$ and $(3)^{1.4} = 4.6555$

So, $p_2 = \dfrac{1}{4.6555} \approx 0.22$ atm

$$= 0.22 \times 1.01 \times 10^5 \text{ Pa}$$

Work done in an adiabatic expansion is given by

$$W = \frac{(p_1 V_1 - p_2 V_2)}{\gamma - 1}$$

$$= \frac{(1.01 \times 10^5 \times 10^3 - 0.22 \times 1.01 \times 10^5 \times 3 \times 10^{-3})}{(1.40 - 1)}$$

$= 0.85 \times 1.01 \times 10^5 \times 10^{-3} = 85.85$ J

Closest value of work done, $W \approx 90.5$ J

6 (*a*) Mean free path (λ) of a gas molecule and mean relaxation time (or collision time) τ are related as

$$\tau = \frac{\lambda}{v_{\text{mean}}} = \frac{\dfrac{1}{\sqrt{2}} \dfrac{k_B T}{\pi d^2 p}}{\sqrt{\dfrac{8}{\pi} \cdot \dfrac{k_B}{m} \cdot T}}$$

$$= \frac{1}{4d^2}\left(\frac{k_B m}{\pi}\right)^{1/2} \cdot \left(\frac{T^{1/2}}{p}\right)$$

Using $pV = nRT$, we have

$$\tau = \frac{1}{4d^2} \cdot \left(\frac{k_B m}{\pi}\right)^{1/2} \cdot \frac{VT^{1/2}}{nRT}$$

$$= \left\{\frac{1}{4d^2} \cdot \left(\frac{k_B m}{\pi}\right)^{1/2} \cdot \frac{1}{nR}\right\} \cdot \frac{V}{\sqrt{T}}$$

$\Rightarrow \tau \propto \dfrac{V}{\sqrt{T}}$, as other quantities remains constant during the process.

So, $\quad \dfrac{\tau_1}{\tau_2} = \dfrac{V_1}{V_2} \cdot \left(\dfrac{T_2}{T_1}\right)^{1/2} \quad \ldots \text{(i)}$

As the process is adiabatic,

$$T_1 V_1^{\gamma - 1} = T_2 V_2^{\gamma - 1} \Rightarrow \frac{T_2}{T_1} = \left(\frac{V_1}{V_2}\right)^{\gamma - 1}$$

So, Eq. (i) becomes

$$\frac{\tau_1}{\tau_2} = \frac{V_1}{V_2} \cdot \left(\frac{V_1}{V_2}\right)^{\frac{\gamma-1}{2}} \Rightarrow \frac{\tau_1}{\tau_2} = \left(\frac{V_1}{V_2}\right)^{\frac{\gamma+1}{2}}$$

Now, given $V_2 = 2V_1$, so

$$\frac{\tau_1}{\tau_2} = \left(\frac{1}{2}\right)^{\frac{\gamma+1}{2}}$$

7 (*c*) Given, $VT = k$, (k is constant)

or $\quad T \propto \dfrac{1}{V} \quad \ldots \text{(i)}$

Using ideal gas equation,

$$pV = nRT \Rightarrow pV \propto T$$

$\Rightarrow \quad pV \propto \dfrac{1}{V} \quad$ From eq. ...(i)

or $\quad pV^2 = \text{constant} \quad \ldots \text{(ii)}$

i.e a polytropic process with $x = 2$.

(Polytropic process means, $pV^x = \text{constant}$)

We know that, work done in a polytropic process is given by

$$\Delta W = \frac{p_2 V_2 - p_1 V_1}{1 - x} \text{ (for } x \neq 1) \quad \ldots \text{(iii)}$$

and, $\Delta W = pV \ln\left(\dfrac{V_2}{V_1}\right)$ (for $x = 1$)

Here, $x = 2$,

$\therefore \quad \Delta W = \dfrac{p_2 V_2 - p_1 V_1}{1 - x} = \dfrac{nR(T_2 - T_1)}{1 - x}$

$\Rightarrow \quad \Delta W = \dfrac{nR\Delta T}{1 - 2} = -nR\Delta T \quad \ldots \text{(iv)}$

Now, for monoatomic gas change in internal energy is given by

$$\Delta U = \frac{3}{2} R\Delta T \quad \ldots \text{(v)}$$

Using first law of thermodynamics, heat absorbed

$$\Delta Q = \Delta W + \Delta U = \frac{3}{2} R\Delta T - R\Delta T$$

$\Rightarrow \quad \Delta Q = \dfrac{1}{2} R\Delta T$

8 (*a*) In the given *p*-*V* digram,

For process *a*, pressure is constant.

$\therefore$ *a* is isobaric.

For process *d*, volume is constant.

$\therefore$ *d* is isochoric.

Also, as we know that, slope of adiabatic curve in *p*-*V* diagram is more than that of isothermal curve.

$\therefore$ *b* is isothermal and *c* is adiabatic.

9 (*c*) For an ideal monoatomic gas,

$$\gamma = \frac{5}{3}$$

In an adiabatic process,

$TV^{\gamma - 1} = \text{constant}$

i.e $T_f V_f^{\gamma - 1} = T_i V_i^{\gamma - 1}$

$$\Rightarrow \quad T_f = T_i\left(\frac{V_i}{V_f}\right)^{\gamma - 1} = 300\left(\frac{1}{2}\right)^{\frac{5}{3}-1}$$

$$= 300\left(\frac{1}{2}\right)^{2/3}$$

10 (d) Given, $T_0 = 300$ K

Let, initial pressure of gas, $p_1 = p$

and final pressure of gas, $p_2 = 10\,p$

By adiabatic process relation between p and T,

$$p_1^{1-\gamma}T_1^{\gamma} = p_2^{1-\gamma}\cdot T_2^{\gamma}$$

$$\left(\frac{p_1}{p_2}\right)^{1-\gamma} = \left(\frac{T_2}{T_1}\right)^{\gamma} \Rightarrow \left(\frac{p}{10p}\right)^{1-\gamma} = \left(\frac{T_2}{300}\right)^{\gamma}$$

$$10^{\gamma-1} = \frac{T_2^{\gamma}}{300^{\gamma}} \Rightarrow T_2^{\gamma} = 10^{\gamma-1} \times 300^{\gamma}$$

$$\Rightarrow \quad T_2 = 10^{\frac{\gamma-1}{\gamma}} \times 300 = 10^{1-\frac{1}{\gamma}} \times 300$$

$$= 10^{1-\frac{1}{7/5}} \times 300 \quad \left[\because \text{For diatomic, } \gamma = \frac{7}{5}\right]$$

$$= 10^{2/7} \times 300 = (100)^{1/7} \times 300$$

$$= 1.9 \times 300 \quad [\because 100^{1/7} = 1.9]$$

$$= 570 \text{ K}$$

Therefore, the final gas temperature after compression is 570 K.

11 (a) Since, the whole system is isolated. Also, the right side container is initially vacuum, so the gas could easily rush there without any resistive force. As the right side container has no temperature. Thus, the temperature of the ideal gas would remain same even if it enters right side chamber. Therefore, final temperature attained at the equilibrium will be T.

Calorimetry

1 (a)

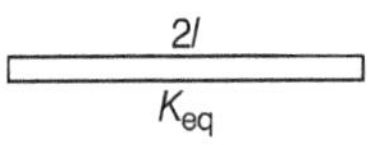

$$R_{\text{eff}} = \frac{l}{K_1A} + \frac{l}{K_2A} = \frac{2l}{K_{\text{eq}}A}$$

$$K_{\text{eq}} = \frac{2K_1K_2}{K_1 + K_2}$$

2 (c) Rate or radiation emitted by the source

$$P \propto T^4$$

(according to Stefan's law)

If d is the distance of the foil from the source,

then $\quad P \propto \frac{1}{d^2} \Rightarrow P \propto \frac{T^4}{d^2}$

$$\therefore \quad \frac{P_1}{P_2} = \frac{T_1^{\,4}}{T_2^{\,4}}\cdot\frac{d_2^{\,2}}{d_1^{\,2}}$$

$$\frac{P}{P_2} = \left(\frac{T}{2T}\right)^4 \times \left(\frac{2d}{d}\right)^2$$

$$= \frac{1}{16} \times \frac{4}{1} = \frac{1}{4}$$

$$\therefore \quad P_2 = 4P$$

3 (b) Let x grams of water is evaporated.

According to the principle of calorimetry,

Heat lost by freezing water (that turns into ice) = Heat gained by evaporated water

Given, mass of water = 150 g

$$\Rightarrow (150 - x) \times 10^{-3} \times 3.36 \times 10^5$$

$$= x \times 10^{-3} \times 2.10 \times 10^6$$

$$\Rightarrow (150 - x) \times 3.36 = 21x$$

$$\Rightarrow x = \frac{150}{7.25} = 20.6 \Rightarrow x \approx 20 \text{ g}$$

4 (b) We know, mass $m = V\cdot\rho$

So, we have

$$\frac{dQ}{dt} = \frac{h}{ms}(T - T_0) = \frac{h(T - T_0)}{V\cdot\rho s}$$

Since, h, $(T - T_0)$ and V are constant for both beaker.

$$\therefore \quad \frac{dQ}{dt} \propto \frac{1}{\rho s}$$

We have given that $\rho_A = 8 \times 10^2$ kgm^{-3},

$\rho_B = 10^3$ kgm^{-3}, $s_A = 2000$ J kg^{-1} K^{-1} and

$s_B = 4000$ J kg^{-1} K^{-1},

$\rho_A s_A = 16 \times 10^5$ and $\rho_B s_B = 4 \times 10^6$

So, $\rho_A < \rho_B$, $s_A < s_B$ and $\rho_A s_A < \rho_B s_B$

$$\Rightarrow \quad \frac{1}{\rho_A s_A} > \frac{1}{\rho_B s_B} \Rightarrow \frac{dQ_A}{dt} > \frac{dQ_B}{dt}$$

So, for container B, rate of cooling is smaller than the container A. Hence, graph of B lies above the graph of A and it is not a straight line (slope of A is greater than B).

5 (d) Let interface temperature in steady state conduction is θ, then assuming no heat loss through sides;

$$\begin{pmatrix}\text{Rate of heat}\\ \text{flow through}\\ \text{first slab}\end{pmatrix} = \begin{pmatrix}\text{Rate of heat}\\ \text{flow through}\\ \text{second slab}\end{pmatrix}$$

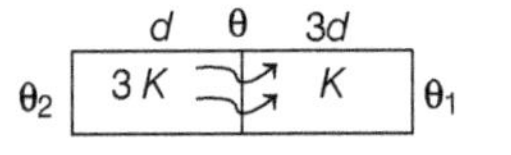

$$\Rightarrow \quad \frac{(3K)\,A(\theta_2 - \theta)}{d} = \frac{KA(\theta - \theta_1)}{3d}$$

$$\Rightarrow \quad 9(\theta_2 - \theta) = \theta - \theta_1$$

$$\Rightarrow \quad 9\theta_2 + \theta_1 = 10\theta$$

$$\Rightarrow \quad \theta = \frac{9\theta_2}{10} + \frac{\theta_1}{10}$$

6 (a) According to the given question, equivalent diagram in

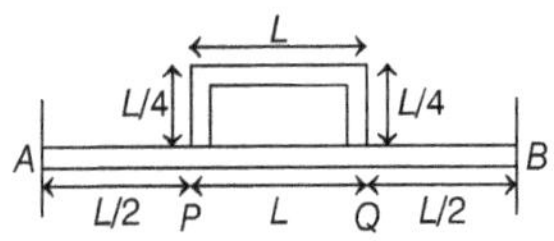

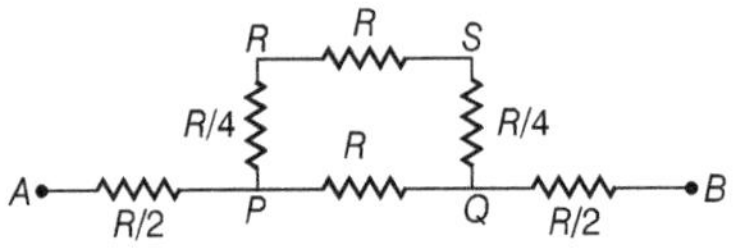

Net resistance of the section $PQRS$ is

$$= \frac{R \times \frac{3R}{2}}{\frac{5R}{2}} = \frac{3R}{5} \quad \text{...(i)}$$

Total resistance of the net network, R_{net}

$$= \frac{R}{2} + \frac{R}{2} + \frac{3R}{5} = \frac{8R}{5}$$

$$\therefore \text{Thermal current, } I = \frac{\Delta T_{AB}}{R_{\text{net}}}$$

$$I = \frac{120 - 0}{\left(\frac{8R}{5}\right)}$$

$$= \frac{120 \times 5}{8R}$$

Thus, the net temperature difference between point P and Q is

$$T_P - T_Q = I \times \frac{3R}{5} \quad \text{[using Eq. (i)]}$$

$$= \frac{120 \times 5}{8R} \times \frac{3R}{5}$$

$$= 45°\text{C}$$

7 (a) Let specific heat of unknown metal is s and heat lost by this metal is ΔQ.

Heat lost and specific heat of a certain material/substance are related as

$$\Delta Q = ms\Delta T \quad \text{... (i)}$$

For unknown metal, $m = 192$ g and

$$\Delta T = (100 - 21.5)°C$$

$$\therefore \quad \Delta Q' = 192(100 - 21.5) \times s \quad \text{...(ii)}$$

Now, this heat is gained by the calorimeter and water inside it.

As, heat gained by calorimeter can be calculated by Eq. (i).

So, for brass specific heat,

$$s = 394 \text{ J kg}^{-1}\text{K}^{-1} \quad \text{(given)}$$

$$= 0.394 \text{ J g}^{-1}\text{K}^{-1}$$

Mass of calorimeter, $m = 128$ g

Change in temperature, $\Delta T = (21.5 - 8.4)°C$

So, using Eq. (i) for calorimeter, heat gained by brass

$$\Delta Q_1 = 128 \times 0.394 \times (21.5 - 8.4) \quad \text{...(iii)}$$

Heat gained by water can be calculated as follows :

mass of water, $m = 240$ g,

specific heat of water, $s = 4.18 \text{ J g}^{-1}\text{K}^{-1}$,

change in temperature,

$$\Delta T = (21.5 - 8.4)°C$$

Using Eq. (i) for water also, we get

heat gained by water,

$$\Delta Q_2 = 240 \times 4.18 \times (21.5 - 8.4) \quad \text{...(iv)}$$

Now, according to the principle of calorimeter,

$$\Delta Q' = \Delta Q_1 + \Delta Q_2$$

Using Eqs. (ii), (iii) and (iv), we get

$$\Rightarrow \quad = 192(100 - 21.5) \times s$$

$$= 128 \times 0.394 \times (21.5 - 8.4) + 240 \times 4.18 \times (21.5 - 8.4)$$

$$\Rightarrow \quad 15072\, s = 660.65 + 13142$$

$$\Rightarrow \quad s = 0.916 \text{ J g}^{-1}\text{K}^{-1}$$

or $\quad s = 916 \text{ J kg}^{-1}\text{K}^{-1}$.

8 (b) As, heat lost by liquid A = heat gained by liquid B

So, $\quad m_A S_A \Delta T_A = m_B S_B \Delta T_B$

$$\Rightarrow 100 \times S_A(100 - 90) = 50 \times S_B(90 - 75)$$

$$\Rightarrow \quad 1000 S_A = 50 \times 15 S_B$$

or $\quad 4S_A = 3S_B \quad \text{...(i)}$

Similarly, in second case,

$$100 \times S_A(100 - T) = 50 \times S_B(T - 50)$$

where, T = Final temperature of the mixture.

$$\Rightarrow \quad 4S_A(100 - T) = 2S_B(T - 50)$$

Using Eq. (i),

$$3S_B(100 - T) = 2S_B(T - 50)$$

or $\quad 300 - 3T = 2T - 100$

or $\quad T = 80°C$

9 (d) Using heat lost or gained without change in state is $\Delta Q = ms\Delta T$,

Let final temperature of ball be T.

Then heat lost by ball is,

$$\Delta Q = 0.1 \times 400(500 - T) \quad \text{...(i)}$$

Heat gained by water,

$$\Delta Q_1 = 0.5 \times 4200(T - 30) \quad \text{...(ii)}$$

and heat gained by vessel is

$$\Delta Q_2 = \text{heat capacity} \times \Delta T$$

$$= 800 \times (T - 30) \quad \text{...(iii)}$$

According to principle of calorimetry, total heat lost = total heat gained

$$\Rightarrow 0.1 \times 400(500 - T)$$

$$= 0.5 \times 4200(T - 30) + 800(T - 30)$$

$$\Rightarrow \quad (500 - T) = \frac{(2100 + 800)(T - 30)}{40}$$

$$\Rightarrow \quad 500 - T = 72.5(T - 30)$$

$$\Rightarrow \quad 500 + 217.5 = 72.5T$$

or $\quad T = 36.39$ K

So, percentage increment in temperature of water

$$= \frac{36.39 - 30}{30} \times 100$$

$$\approx 20\%$$

10 (d) According to the question, the equivalent thermal resistance,

$$R_{eq} = R_1 + R_2 + R_3$$

$$\because \quad R_1 = \frac{l_1}{k_1 A_1}, R_2 = \frac{l_2}{k_2 A_2}, R_3 = \frac{l_3}{k_3 A_3}$$

Where, k_1 and k_3 are thermal conductivity of glass and k_2 is thermal conductivity of air.

or $\quad R_{eq} = \frac{l_1}{k_1 A_1} + \frac{l_2}{k_2 A_2} + \frac{l_3}{k_3 A_3}$

Putting the given values, in question we get

$$\frac{10^{-2}}{0.8 \times 2.6} + \frac{5 \times 10^{-2}}{0.08 \times 2.6} + \frac{10^{-2}}{0.8 \times 2.6}$$

$$= \left(\frac{10^{-2}}{2.6}\right)\left(\frac{52}{0.8}\right) = \frac{1}{4}$$

Hence, flow of heat,

$$H = \frac{\Delta T}{R_{eq}} = \frac{[18 - (-2)]}{\left(\frac{1}{4}\right)}$$

$$\Rightarrow \quad H = 80 \text{ W}$$

11 (a) Maximum loss in KE = Total KE energy before collision of two ice blocks

$$\text{KE}_{\text{max loss}} = \frac{1}{2}mv^2 + \frac{1}{2}mv^2 = mv^2$$

This maximum loss in kinetic energy will be equal to loss in heat to melt two ice block.

$$\Rightarrow \quad mv^2 = (ms\,\Delta\theta + mL)2 \quad \text{...(i)}$$

According to question,

$$L = 3.36 \times 10^5 \text{ J kg}^{-1}$$

$$S = 2100 \text{ J kg}^{-1}\text{K}^{-1} \text{ and}$$

$$\Delta\theta = 8°C$$

From Eq. (i) $\quad v = \sqrt{2(s\,\Delta\theta + L)}$

$$= \sqrt{2(2100 \times 8 + 3.36 \times 10^5)}$$

$$v = 840 \text{ ms}^{-1}$$

Wave Motion

1 Here, $f_c = f_o$

$$\frac{3v_c}{4L} = \frac{2v_o}{2L'}$$

$$L' = \frac{4L}{3}\frac{v_o}{v_c} = \frac{4L}{3}\sqrt{\frac{\rho_1}{\rho_2}}$$

$\therefore \quad x = 4$

2 (*b*) Let frequency of tuning fork be n_1.

$$\therefore \quad \frac{n_2}{n_1} = \frac{100}{101}$$

As, $n_1 - n_2 = 5$

So, $n_1 = 505$ Hz

3 (*c*) Speed of transverse wave over a string,

$$v = \sqrt{\frac{T}{\mu}} \quad \text{...(i)}$$

where, T = tension or force on string

and $\mu = \frac{m}{l}$ = mass per unit length.

Also, Young's modulus of string,

$$Y = \frac{Tl}{A\Delta l} \Rightarrow T = \frac{YA\Delta l}{l} \quad \text{...(ii)}$$

From Eqs. (i) and (ii), we have

$$v^2 = \frac{YA\Delta l}{\mu l}$$

or $$\Delta l = \frac{mv^2}{YA} \quad \text{...(iii)}$$

Here, $m = 6\text{ g} = 6 \times 10^{-3}$ kg, $l = 60$ cm $= 60 \times 10^{-2}$ m,

$A = 1\text{ mm}^2 = 1 \times 10^{-6}\text{ m}^2$,

$Y = 16 \times 10^{11}\text{ Nm}^{-2}$ and $v = 90\text{ ms}^{-1}$

Substituting these given values in Eq. (iii), we get

$$\Delta l = \frac{6 \times 10^{-3} \times (90)^2}{16 \times 10^{11} \times 1 \times 10^{-6}} = 3.03 \times 10^{-5}\text{ m}$$

$$\approx 30 \times 10^{-6}\text{ m} = 0.03\text{ mm}$$

4 (*b*) Transverse wave speed over a string is given by

$$v = \sqrt{\frac{T}{\mu}} \quad \text{... (i)}$$

where, T = tension in string

and μ = mass per unit length of string.

Here, when velocity is v, then tension, $T_1 = 2.06 \times 10^4$ N

Let when velocity is $\frac{v}{2}$, then tension is T, hence from Eq. (i), we get

$$\frac{v}{\frac{v}{2}} = \sqrt{\frac{T_1}{T}}$$

$$\Rightarrow T = \frac{T_1}{4} = \frac{2.06 \times 10^4}{4} \text{ or } T = 5.15 \times 10^3\text{ N}$$

5 (*c*) Let mass per unit length of wires are μ_A and μ_B, respectively.

$\because$ For same material, density is also same.

So, $\mu_A = \frac{\rho\pi r^2 L}{L} = \mu$ and $\mu_B = \frac{\rho 4\pi r^2 L}{L} = 4\mu$

Tension (T) in both connected wires are same.

So, speed of wave in wires are

$$v_A = \sqrt{\frac{T}{\mu_A}} = \sqrt{\frac{T}{\mu}}$$

$$[\because \mu_A = \mu \text{ and } \mu_B = 4\mu]$$

and $$v_B = \sqrt{\frac{T}{\mu_B}} = \sqrt{\frac{T}{4\mu}}$$

So, nth harmonic in such wires system is

$$f_{nth} = \frac{pv}{2L}$$

$$\Rightarrow f_A = \frac{pv_A}{2L} = \frac{p}{2L}\sqrt{\frac{T}{\mu}} \quad (\text{for } p \text{ antinodes})$$

Similarly,

$$f_B = \frac{qv_B}{2L} = \frac{q}{2L}\sqrt{\frac{T}{4\mu}} = \frac{1}{2}\left(\frac{q}{2L}\sqrt{\frac{T}{\mu}}\right)$$

(for q antinodes)

As frequencies f_A and f_B are given equal.

$$\text{So, } f_A = f_B \Rightarrow \frac{p}{2L}\sqrt{\frac{T}{\mu}} = \frac{q}{2}\left[\frac{1}{2L}\sqrt{\frac{T}{\mu}}\right]$$

$$\frac{p}{q} = \frac{1}{2} \Rightarrow p : q = 1 : 2$$

6 (*c*) Given equation of stationary wave is

$Y = 0.3 \sin(0.157x) \cos(200\pi t)$

Comparing it with general equation of stationary wave, i.e. $Y = a \sin kx \cos \omega t$, we get

$$k = \left(\frac{2\pi}{\lambda}\right) = 0.157$$

$$\Rightarrow \lambda = \frac{2\pi}{0.157} = 4\pi^2 \quad \left(\because \frac{1}{2\pi} \approx 0.157\right) \text{...(i)}$$

$$\text{and } \omega = 200\pi = \frac{2\pi}{T} \Rightarrow T = \frac{1}{20}\text{ s}$$

As the possible wavelength associated with nth harmonic of a vibrating string, i.e. fixed at both ends is given as

$$\lambda = \frac{2l}{n} \text{ or } l = n\left(\frac{\lambda}{2}\right)$$

Now, according to question, oscillates in 4th harmonic, so

$$4\left(\frac{\lambda}{2}\right) = l \Rightarrow 2\lambda = l$$

or $l = 2 \times 4\pi^2 = 8\pi^2$ [using Eq. (i)]

Now, $\pi^2 \approx 10 \Rightarrow l \approx 80$ m

7 (*d*) Fundamental frequency of closed organ pipe is given by $f_0 = v / 4L$.

Also, overtone frequencies are given by

$$f = (2n+1)\frac{v}{4L} \quad \text{or} \quad f = (2n+1) f_0$$

Given that, $f_0 = 1500$ Hz and

$f_{max} = 20000$ Hz

This means, $f_{max} > f$

So, $f_{max} > (2n+1) f_0$

$\Rightarrow 20000 > (2n+1) 1500$

$\Rightarrow 2n + 1 < 13.33$

$\Rightarrow 2n < 13.33 - 1$

$\Rightarrow 2n < 12.33$ or $n < 6.16$

or $n = 6$ (integer number)

Hence, total six overtones will be heard.

8 (*a*) Given, $f_A = f_B = 400$ Hz

speed of sound in air, $v = 340$ m/s

Apparent frequency heard by the observer when source A is moving towards him,

$$f_A' = \frac{v}{v - v_S} \times f_A = \frac{340}{340 - v_S} \times 400$$

Apparent frequency heard by the observer when source B is moving away from him,

$$f_B' = \frac{v}{v + v_S} \times f_B = \frac{340}{340 + v_S} \times 400$$

Given, beats $= f_A' - f_B' = 4$

$$\frac{340}{340 - v_S} \times 400 - \frac{340}{340 + v_S} \times 400 = 4$$

$$\Rightarrow \frac{34000 \times 2v_S}{340^2 - v_S^2} = 1$$

$$\Rightarrow \frac{34000 \times 2v_S}{340^2} = 1 \quad [\because v_S^2 << 340^2]$$

$$\Rightarrow v_S = 1.7\text{ m/s}$$

9 (*d*) Given, speed of sound, $v = 340$ m/s

Length of open organ pipe, $l = 2$ m

Frequency, $f = 1200$ Hz

For open organ pipe,

Fundamental frequency,

$$f_0 = \frac{v}{2l} = \frac{340}{2 \times 2} = 85$$

$\therefore$ Number of tones present in the open organ pipe

$$= \frac{f}{f_0} = \frac{1200}{85} = 14.11 \simeq 14$$

10 (*b*) Given, $f_s = 5$ kHz

$f_A = 5.5$ kHz

$f_B = 6$ kHz

Now, let v_s = speed of sound

$\therefore$ For observer A, frequency of siren records by train A is given as,

$$f_A = f_S\left(\frac{v_S + v_A}{v_S}\right)$$

$$\text{or } 5.5 = 5\left(\frac{v_S + v_A}{v_S}\right)$$

$$\text{or} \quad 1.1 = 1 + \frac{v_A}{v_S} \text{ or } v_A = 0.1\,v_S \quad \text{...(i)}$$

$\therefore$ For observer B, frequency of siren records by train B is given as,

$$f_B = f_S\left(\frac{v_S + v_B}{v_S}\right)$$

$$6 = 5\left(\frac{v_S + v_B}{v_S}\right) \text{ or } 6 = 5\left(1 + \frac{v_B}{v_S}\right)$$

$$\Rightarrow \quad \frac{6}{5} = 1 + \frac{v_B}{v_S}$$

$$\Rightarrow \quad \frac{6}{5} - 1 = \frac{v_B}{v_S} \Rightarrow \frac{6-5}{5} = \frac{v_B}{v_S}$$

$$\Rightarrow \quad \frac{1}{5} = \frac{v_B}{v_S} \Rightarrow \frac{v_B}{v_S} = 0.2$$

or $\quad v_B = 0.2\, v_S \quad$ ...(ii)

Now, from Eqs. (i) and (ii), we get

or $\quad \frac{v_B}{v_A} = \frac{0.2 v_S}{0.1 v_S}$ or $\frac{v_B}{v_A} = 2$

So, the ratio of the speed of train *B* to that of train *A* is $v_B : v_A = 2$.

11 *(a)* Given, $l = 50 \text{ cm} = 5 \times 10^{-2}$ m

$A = 1 \text{ mm}^2 = 1 \times 10^{-6} \text{ m}^2$

$m = 5$ g and $v = 80$ m/s

Now, speed of transverse wave given as

$$\therefore \quad v = \sqrt{\frac{T}{\mu}} = \sqrt{\frac{T}{m} \times l} \Rightarrow v = \sqrt{\frac{T}{m} \times l}$$

$$\Rightarrow v^2 = \frac{T}{m} \times l \qquad \Rightarrow l = \frac{v^2 m}{T} \quad \text{...(i)}$$

Now, Young's modulus, $Y = \dfrac{\frac{T}{A}}{\frac{\Delta l}{l}}$

or $\quad \Delta l = \frac{T}{AY} \cdot l \quad$ ...(ii)

From Eqs. (i) and (ii), we get

$$\therefore \quad \Delta l = \frac{v^2 m}{AY}$$

Putting the given values, we get

$$\Delta l = \frac{80^2 \times 5 \times 10^{-3}}{1 \times 10^{-6} \times 4 \times 10^{11}} \text{ m}$$

or $\quad \Delta l = 8 \times 10^{-5}$ m

So, the extension in the length of the wire is $\Delta l = 8 \times 10^{-5}$ m.

12 *(b)* Given, $l = 16$ m, $v = 32$ m/s

and total number of nodes between the two fixed ends of the string = 9

$\therefore$ Total number of segments of vibrations between fixed ends, $p = 9 + 1 = 10$

$\therefore$ frequency of the wave, $f = \frac{p}{2l} \times v$

$$= \frac{10}{2 \times 16} \times 32 = 10 \text{ Hz}$$

13 *(d)* For open pipe with frequency n_1 and length l_1,

$$n_1 = \frac{v}{2l_1} \Rightarrow l_1 = \frac{v}{2n_1} \quad \text{...(i)}$$

Similarly, for open pipe with frequency n_2 and length l_2,

$$n_2 = \frac{v}{2l_2} \Rightarrow l_2 = \frac{v}{2n_2} \quad \text{...(ii)}$$

When both the pipes are joined together then net length of the pipe will be,

$$L = l_1 + l_2 \quad \text{...(iii)}$$

Let, after joining the pipes fundamental frequency will be N_0,

So, $L = \frac{v}{2N_0} = l_1 + l_2$

$$\Rightarrow L = \frac{v}{2n_1} + \frac{v}{2n_2} = \frac{v}{2N_0}$$

Using Eqs. (i) and (ii), we get

$$\Rightarrow \quad N_0 = \frac{n_1 n_2}{n_1 + n_2}$$

Lightning Source UK Ltd.
Milton Keynes UK
UKHW050754050922
408354UK00001B/3